Table of Problem-Solving Strategies

Note for users of the two-volume edition:
Volume 1 (pp. 1–547) includes chapters 1–16.
Volume 2 (pp. 548–1043) includes chapters 17–30.

Table of Math Relationship Boxes

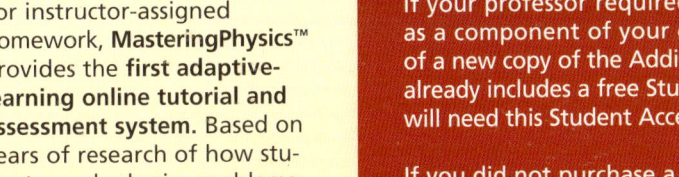

COLLEGE PHYSICS

A STRATEGIC APPROACH

ActivPhysics™ OnLine Activities

www.aw-bc.com/knightjonesfield

COLLEGE PHYSICS

A STRATEGIC APPROACH

RANDALL D. KNIGHT

California Polytechnic State University, San Luis Obispo

BRIAN JONES

Colorado State University

STUART FIELD

Colorado State University

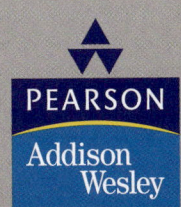

PEARSON

Addison
Wesley

San Francisco   Boston   New York
Cape Town   Hong Kong   London   Madrid   Mexico City
Montreal   Munich   Paris   Singapore   Sydney   Tokyo   Toronto

Vice President and Editorial Director:	Adam Black, Ph.D.
Development Editor:	Alice Houston, Ph.D.
Project Editor:	Martha Steele
Executive Producer:	Claire Masson
Associate Media Producer:	Deb Greco
Editorial Assistant:	Grace Joo
Director of Marketing:	Christy Lawrence
Senior Market Development Manager:	Josh Frost
Market Development Coordinator:	Jessica Lyons
Managing Editor:	Corinne Benson
Production Supervisor:	Shannon Tozier
Production Service and Composition:	WestWords, Inc.
Illustrations:	Precision Graphics
Copyeditor:	Kevin Gleason
Text Design:	Patrick Devine Design and Side by Side Studios
Cover Design:	Yvo Riezebos Design
Manufacturing Manager:	Pam Augspurger
Director, Image Resource Center:	Melinda Patelli
Manager, Rights and Permissions:	Zina Arabia
Photo Research:	Cypress Integrated Systems, Inc.
Printer and Binder:	RR Donnelley, Willard
Cover Printer:	Phoenix Color
Cover Image:	Ken Wilson, Papilio/CORBIS

Library of Congress Cataloging-in-Publication Data

Knight, Randall Dewey.
  College physics : a strategic approach / Randall D. Knight, Brian Jones, Stuart Field. — 1st ed.
    p. cm.
  ISBN 0-8053-0634-X
 1. Physics—Textbooks. I. Jones, Brian. II. Field, Stuart. III. Title.
  QC23.2.K55 2006
  530—dc22

2006032583

ISBN 0-8053-0634-X (Student edition)
ISBN 0-8053-1698-1 (Professional copy)

4 5 6 7 8 9 10—DOW—10 09 08
www.aw-bc.com

Brief Contents

About the Authors

Randy Knight has taught introductory physics for 25 years at Ohio State University and California Polytechnic University, where he is currently Professor of Physics and Director of the Minor in Environmental Studies. Randy received a Ph.D. in physics from the University of California, Berkeley and was a postdoctoral fellow at the Harvard-Smithsonian Center for Astrophysics before joining the faculty at Ohio State University. It was at Ohio State, under the mentorship of Professor Leonard Jossem, that he began to learn about the research in physics education that, many years later, led to *Five Easy Lessons: Strategies for Successful Physics Teaching, Physics for Scientists and Engineers: A Strategic Approach,* and now to this book. Randy's research interests are in the field of lasers and spectroscopy. When he's not in the classroom or in front of a computer, you can find Randy hiking, sea kayaking, playing the piano, or spending time with his wife Sally and their seven cats.

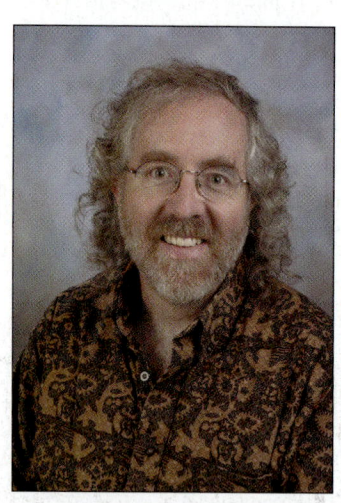

Brian Jones has won several teaching awards at Colorado State University during his 17 years teaching in the Department of Physics. His teaching focus in recent years has been the College Physics class, including writing problems for the MCAT exam and helping students review for this test. Brian is also Director of the *Little Shop of Physics,* the Department's engaging and effective hands-on outreach program, which has merited coverage in publications ranging from the *APS News* to *People* magazine. Brian has been invited to give workshops on techniques of science instruction throughout the United States and internationally, including Belize, Chile, Ethiopia, Azerbaijan and Slovenia. Previously, he taught at Waterford Kamhlaba United World College in Mbabane, Swaziland, and Kenyon College in Gambier, Ohio. Brian and his wife Carol have dozens of fruit trees and bushes in their yard, including an apple tree that was propagated from a tree in Isaac Newton's garden, and they have traveled and camped in most of the United States.

Stuart Field has been interested in science and technology his whole life. While in school he built telescopes, electronic circuits, and computers. After attending Stanford University, he earned a Ph.D. at the University of Chicago, where he studied the properties of materials at ultralow temperatures. After completing a postdoctoral position at the Massachussetts Institute of Technology, he held a faculty position at the University of Michigan. Currently at Colorado State University, Stuart teaches a variety of physics courses, including algebra-based introductory physics, and was an early and enthusiastic adopter of Knight's *Physics for Scientists and Engineers.* Stuart maintains an active research program in the area of superconductivity. His hobbies include woodworking; enjoying Colorado's great outdoors; and ice hockey, where he plays goalie for a local team.

BUILT FROM THE GROUND UP FOR MORE EFFECTIVE LEARNING

- **An NSF-funded educational research program**
- **Input from an unprecedented 4,500 student**
- **Detailed reviews by 250 instructors**

This is just some of the unusual history that shaped *Physics for Scientists and Engineers* by Randy Knight into the most widely adopted new physics text published in more than 30 years. In *College Physics: A Strategic Approach*, Randy Knight is joined by Brian Jones and Stuart Field to apply the best solutions from educational research to the algebra-based introductory physics course and the particular needs of these students.

Built from the ground up on a wealth of research into how students learn physics, and how they can be taught more effectively, *College Physics*:

- **builds** students' problem-solving abilities and confidence
- **integrates** relevant examples and interesting topics
- **promotes** deeper understanding by explicitly addressing students' preconceptions, misconceptions, and common stumbling-blocks
- **provides** a highly readable text and research-based visual pedagogy

The following pages provide an overview of how *College Physics* is designed to achieve these goals based on input from thousands of students and hundreds of instructors to date.

"Visual pedagogy is a critical feature of physics textbooks because so much of physics is expressed in graphics. I am extremely impressed with the care and research-based ideas that have gone into the unique instructional design of the graphics in Knight's textbooks. His physics textbooks are leaders in applying research-based principles of visual pedagogy to support student learning."

—Richard E. Mayer, *Professor of Psychology, University of California, Santa Barbara*

"No text I have seen has done more to lead students to thinking than Knight/Jones/Field."

—Dieter R. Brill, *University of Maryland*

Clear, consistent instruction

Students build confidence and success through a consistent 3-step approach that encourages them to **Prepare** the problem, before trying to **Solve it** and **Assess** their answer.

Topic-specific *Problem-Solving Strategies* follow the same 3-step framework and provide more detailed guidance.

1 **PREPARE** reinforces the value of gathering information, drawing figures, making assumptions, and planning—key steps research shows students otherwise skip.

2 **SOLVE** carefully works through the mathematical steps of the solution, explaining algebraic manipulations and use of key information.

3 **ASSESS** reminds the student to verify whether their answer makes sense—numerically and in context.

(MP) **TACTICS BOX 4.3** **Drawing a free-body diagram** ✏ Exercises 17–22

❶ **Identify all forces acting on the object.** This step was described in Tactics Box 4.2.
❷ **Draw a coordinate system.** Use the axes defined in your visual overview (Tactics Box 2.2). If those axes are tilted, for motion along an incline, then the axes of the free-body diagram should be similarly tilted.
❸ **Represent the object as a dot at the origin of the coordinate axes.** This is the particle model.
❹ **Draw** ~~described~~
❺ **Draw a** ~~diagram~~ ~~check t~~ ~~on your~~

TACTICS BOX 24.1 **Right-hand rule for fields** ✏ Exercises 7–12

❶ Point your *right* thumb in the direction of the current.
❷ Curl your fingers around the wire to indicate a circle.
❸ Your fingers point in the direction of the magnetic field lines around the wire.

Tactics Boxes provide step-by-step procedures that build key skills the student will use over and over—such as drawing free-body diagrams, and using ray tracing.

Math Relationship Boxes ensure the student is confident with the key mathematical relationships most common in this course. Each relationship is consolidated in words, math, and graphics, along with tips on reasoning with limiting cases and scaling. Icons in the text refer back to these boxes to help students see connections in the math they use.

(MP) **PROBLEM-SOLVING STRATEGY 24.1** **Magnetic–field problems**

PREPARE Because current-carrying wires do not lie in the same plane as the fields they produce, you'll need to prepare an especially careful drawing. Generally, you should choose the plane of your drawing so that the magnetic field vectors lie either in the plane of the paper or perpendicular to it.

■ Str... the
■ Us... fiel...
■ Sol... of t...
■ Ma... bes... Th...

If you... syste...

SOLVE noid) each f... solenc... additic...

ASSESS reason...

(MP) **PROBLEM-SOLVING STRATEGY 10.1** **Conservation of energy problems**

PREPARE Choose what to include in your system (see Tactics Box 10... Draw a before-and-after visual overview, as outlined in Tactics Box 9... Note known quantities, and determine what quantity you're trying to fi... If the system is isolated and if there is no friction, your solution will... based on Equation 10.7, otherwise you should use Equation 10.6.
Identify which mechanical energies in the system are changing:

■ If the *speed* of the object is changing, include K_i and K_f in your solutic...
■ If the *height* of the object is changing, include $(U_g)_i$ and $(U_g)_f$.
■ If the *length* of a spring is changing, include $(U_s)_i$ and $(U_s)_f$.
■ If kinetic friction is present, ΔE_{th} will be positive. Some kinetic or pote... tial energy will be transformed into thermal energy.

If an external force acts on the system, you'll need to include the work done by this force in Equation 10.6.

SOLVE Depending on the problem, you'll need to calculate initial and... final values of these energies and insert them into Equation 10.6 or 10... Then you can solve for the unknown energies, and from these any unkno... speeds (from K), positions (from U), or displacements or forces (from W...

ASSESS Check the signs of your energies. Kinetic energy, as we'll see,... always positive. In the systems we'll study in this chapter, thermal ener... can only increase, so that its change is positive. In Chapters 11 and 12 we... study systems for which the thermal energy can decrease.

> "The Prepare, Solve, and Assess steps are right on, not too many steps to confuse the students, but enough to thoroughly discuss the problem. The strategies are well thought-out and expertly executed."
>
> —Toby Moleski, *Muskegon Community College*

◣ **Quadratic relationships** ✏ Exercises 14, 19 **(MP)**

Two quantities are said to have a **quadratic relationship** if y is proportional to the square of x. We write the mathematical relationship as

$$y = Ax^2$$

y is a quadratic function of x

The graph of a quadratic relationship is a parabola.

Doubling x causes y to change by a factor of 4.

SCALING Suppose we double x:

$$x_f = 2x_i$$

This will change y as well:

$$y_f = Ax_f^2 = A(2x_i)^2 = 2^2 Ax_i^2 = 2^2 y_i$$
$$= 4y_i$$

Increasing x by a factor of 2 causes y to increase by a factor of 2^2, or 4. Generally, we can say that

Changing x by a factor C changes y by a factor C^2.

Explicit, guided practice

Worked Examples implement the Strategies and follow the same **PREPARE/SOLVE/ASSESS** framework as part of developing good problem-solving habits. They carefully walk the student through the underlying reasoning and pitfalls to avoid.

EXAMPLE 2.9 Calculating the minimum length of a runway

A fully-loaded 747 with all engines at full thrust accelerates at 2.6 m/s^2. Its minimum takeoff speed is 70 m/s. How much time will the plane take to reach its takeoff speed? What minimum length of runway does the plane require for takeoff?

PREPARE The visual overview of Figure 2.34 summarizes the important details of the problem. We set x_i and t_i equal to zero at the starting point of the motion, when the plane is at rest and the acceleration begins. The final point of the motion is when the plane achieves the necessary takeoff speed of 70m/s. The plane is accelerating to the right, so we will compute the time for the plane to reach a velocity of 70m/s and the position of the plane at this time, giving us the minimum length of the runway.

FIGURE 2.34 Visual overview for an accelerating plane.

SOLVE First we solve for the time required for the plane to reach takeoff speed. We can use the first equation in Table 2.4 to compute this time:

$$(v_x)_f = (v_x)_i + a_x \, \Delta t$$
$$70 \text{ m/s} = 0 \text{ m/s} + (2.6 \text{ m/s}^2) \, \Delta t$$
$$\Delta t = 26.9 \text{ s}$$

We keep an extra significant figure here as we will use this result in the next step of the calculation.

Given the time that the plane takes to reach takeoff speed, we can compute the position of the plane when it reaches this speed using the second equation in Table 2.4:

$$x_f = x_i + (v_x)_i \, \Delta t + \tfrac{1}{2} a_x (\Delta t)^2$$
$$= 0 \text{ m} + (0 \text{ m/s})(26.9 \text{ s}) + \tfrac{1}{2}(2.6 \text{ m/s}^2)(26.9 \text{ s})^2$$
$$= 940 \text{ m}$$

Our final answers are thus that the plane will take 27 s to reach takeoff speed, with a minimum runway length of 940 m.

ASSESS Think about the last time you flew; 27 s seems like a reasonable time for a plane to accelerate on takeoff. Actual runway lengths at major airports are 3000 m or more, a few times greater than the minimum length, because they have to allow for emergency stops during an aborted takeoff. (If we had calculated a distance greater than 3000 m, we would know we had done something wrong!)

EXAMPLE 4.3 Forces on an upward-accelerating elevator

An elevator, suspended by a cable, speeds up as it moves upward from the ground floor. Draw a free-body diagram of the elevator.

PREPARE Figure 4.26 illustrates the steps outlined in Tactics Box 4.3.

FIGURE 4.26 Free-body diagram of an elevator accelerating upward.

Pencil Sketches provide students with an explicit and accessible example of what to draw in solving a problem—a key step research indicates students often skip—often following steps outlined in Tactics Boxes.

Conceptual Examples target students' qualitative reasoning skills. Since no math is involved, they follow a Reason and Assess approach.

CONCEPTUAL EXAMPLE 24.2 Canceling a magnetic field

A current loop carries a counter clockwise current of 3 A. A second current-carrying loop with twice the radius is placed outside the first loop, as shown in Figure 24.25. What current in the outer loop will exactly cancel the field in the center of the first loop?

FIGURE 24.25 Two current loops.

REASON The inner loop produces a magnetic field at its center of magnitude B. By the right-hand rule for fields, the direction of this field is out of the page. To make the total field at the center zero, the outer loop must produce a field at the center that has the same magnitude B but pointing in the opposite direction, into the page. Equation 24.2 shows that the field at the center of a loop is *inversely proportional* to its radius. So if the outer loop also carried 3 A, its field at the center would be only...

CONCEPTUAL EXAMPLE 11.4 Energy transfers and the body

Why—in physics terms—is it more taxing on the body to exercise in very hot weather?

REASON Your body continuously converts chemical energy to thermal energy, as we have seen. In order to maintain a constant body temperature, your body must continuously transfer heat to the environment. This is a simple matter in cool weather when heat is spontaneously transferred to the environment, but when the air temperature is higher than your body temperature, your body cannot cool itself this way and must use other mechanisms to transfer this energy, such as perspiring. These mechanisms require additional energy expenditure.

ASSESS Strenuous exercise in hot weather can easily lead to a rise in body temperature if the body cannot exhaust heat quickly enough.

The **Student Workbook** provides straightforward confidence- and skill-building exercises—bridging the gap between worked examples and end-of-chapter problems. Also included are worksheets—research-based templates that help students structure their approach to solving problems. Workbook activities are referenced throughout the text by ✐.

An active, inductive approach

Drawing from students' majors and the world around them, relevant examples and interesting topics are carefully woven into the text. These provide students with motivation, a means to consolidate their understanding, and a clear context for how physics is valuable to them.

"The writing is clear and engaging. The coverage is well thought-out. *College Physics* is better than any other text I have encountered for algebra-based physics."
—Bruce A. Schumm, *University of California, Santa Cruz*

Worked Examples incorporates scenarios from everyday life and the world around us.

EXAMPLE 10.13 Protecting your head

A bike helmet is basically a shell of hard, crushable foam 3.0 cm thick. In testing, the helmet is strapped onto a 5.0 kg headform that is dropped from a height of 2.0 m onto a hard anvil. What force is encountered by the head in such a fall?

PREPARE A visual overview of the test is shown in Fig-

The foam inside a bike helmet is designed to crush upon

The headform is dropped from a height of 2 m.

The helmet has just touched the anvil. The foam is now pushing up on the headform.

The headform is subject to a large force from the foam. This force does negative work on the headform, slowing it down.

Before:
$y_i = 2.0$ m, $v_i = 0$ m/s

Find: F

$d = 0.030$ m

After:
$y_f = 0$ m, $v_f = 0$ m/s

FIGURE 10.31 The foam in the helmet does negative work on the headform.

EXAMPLE 3.8 Speed of a roller coaster

A classic wooden coaster has cars that go down a big first hill, gaining speed. The cars then ascend a second hill with a slope of 30°. If the cars are going 25 m/s at the bottom and it takes them 2.0 s to climb this hill, how fast are they going at the top?

PREPARE We start with the visual overview in Figure 3.24, which includes a motion diagram, a pictorial representation, and a list of values. Notice how the motion diagram of Figure 3.24 differs from that of the previous example: The velocity decreases as the car moves up the hill, so the acceleration vector is opposite the direction of the velocity vector. The motion is along the *x*-axis, as before, but the acceleration vector points in the negative-*x* direction, so the component a_x is negative.

Known
$x_i = 0$ m
$(v_x)_i = 25$ m/s
$t_i = 0$ s
$t_f = 2.0$ s
$\theta = 30°$

Find
$(v_x)_f$

FIGURE 3.24 The coaster's speed decreases as it goes up the hill.

SOLVE To determine the final speed, we need to know the acceleration. We will assume that there is no friction or air

Free-standing Applications, provided in the margin with photographs and a self-contained caption, connect the physical principles with the real world.

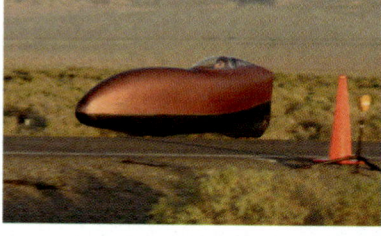

Less of a drag At the high speeds attained by racing cyclists, air drag can become very significant. The world record for the farthest distance traveled in one hour on an ordinary bicycle is 56.38 km, set by Chris Boardman in 1996. But a bicycle with an aerodynamic shell has a much lower drag force, allowing it to attain significantly higher speeds. The bike shown here was pedaled 84.22 km in one hour by Sam Whittingham in 2004, for an amazing average speed of 52.3 mph!

TRY IT YOURSELF

Getting the ketchup out The ketchup stuck at the bottom of the bottle is initially at rest. If you hit the bottom of the bottle, as in (a), the bottle suddenly moves down, taking the ketchup on the bottom of the bottle with it, so that the ketchup just stays stuck to the bottom. But if instead you hit *up* on the bottle, as in (b), you force the bottle rapidly upward. By the first law, the ketchup that was stuck to the bottom stays at rest, so that it separates from the upward-moving bottle: the ketchup has moved forward with respect to the bottle!

TRY IT YOURSELF

Noisy magnets The striped pole structure of refrigerator magnets can be demonstrated by using two *identical* such magnets. Place them together, back to back, then pull them quickly across each other as shown by the arrows. The magnets alternately attract and repel as they bump past each other. If you pull them quickly enough, you will actually hear an audible tone as the rapidly passing magnets are alternately pulled together and pushed apart.

Try It Yourself Activities throughout the text provide students with simple real-world experiments designed to quickly reinforce a key idea through direct experience.

"Extremely readable and instructive."
—Tiffany Landry, *Folsom Lake College*

Optional sections provide **in-depth treatment on key topics for biomedical students**—such as electrical conduction in the nervous system, the workings of an EKG, and how to correctly measure blood pressure.

"College Physics answers better than other texts I've seen the issues particular to teaching physics to life science students."

—Robert W. Lutz, *Drake University*

(a) The electric dipole of the heart.

A cross section of the heart showing muscle tissue.
This line separates cells that have depolarized and those that have not.
Depolarized tissue
Tissue not yet depolarized

The charge separation at the line between the two regions creates an electric dipole.

(b) The field of the heart in the body.

The heart's dipole field extends throughout the torso.

FIGURE 20.35 The beating heart generates a dipole electric field.

■ The electric field is created by charges. Field lines start on a positive charge and end on a negative charge.

We can combine the above information into a technique for drawing electric field lines for an arrangement of charges. For example, Figure 20.34 pictures the electric field of a dipole using electric field lines. You should compare this to Figure 20.29b, which illustrates the field with field vectors.

The Electric Field of the Heart

Nerve and muscle cells have a prominent electrical nature. As we will see in detail in Chapter 23, a cell membrane is an insulator that encloses a conducting fluid and is surrounded by conducting fluid. While resting, the membrane is *polarized* with positive charges on the outside of the cell, negative charges on the inside. When a nerve or a muscle cell is stimulated, the polarity of the membrane switches; we say that the cell *depolarizes*. Later, when the charge balance is restored, we say that the cell *repolarizes*.

All nerve and muscle cells generate an electrical signal when depolarization occurs, but the largest electrical signal in the body comes from the heart. The rhythmic beating of the heart is produced by a highly coordinated wave of depolarization that sweeps across the tissue of the heart. As Figure 20.35a shows, the surface of the heart is positive on one side of the boundary between tissue that is depolarized and tissue that is not yet depolarized, negative on the other. In other words, the heart is a large electric dipole. The orientation and strength of the dipole change during each beat of the heart as the depolarization wave sweeps across it.

The electric dipole of the heart generates a dipole electric field that extends throughout the torso, as shown in Figure 20.35b. As we will see in Chapter 21, an electrocardiogram measures the changing electric field of the heart as it beats. Measurement of the heart's electric field can be used to diagnose the operation of the heart.

Calf muscle

Achilles tend[on]

On each stride, the t[endon] stretches, storing ab[out] of energy.

Fascinating, self-contained *Life-science Applications* are provided throughout the text to illustrate how physics helps students understand the world around them and life-science students with their chosen career.

Magnetotactic Bacteria BIO Several organisms use the earth's magnetic field to navigate. The clearest example of this is *magnetotactic bacteria*. The dark areas in this image are small pieces of iron; each piece is a single domain and hence a very strong magnet. Such a bacterium possesses a very strong magnetic moment: It acts like a bar magnet, lining up with the earth's magnetic field! In the temperate regions where such bacteria live, the earth's field has a large vertical component. The bacteria use their alignment with this vertical field component to navigate up and down.

Spring in your step BIO As you run, [you] lose some of your mechanical energy [each] time your foot strikes the ground; this [energy] is transformed into unrecoverable ther[mal] energy. Luckily, about 35% of the decr[ease in] your mechanical energy when your foo[t lands] is stored as elastic potential energy in t[he] stretchable Achilles tendon of the lowe[r leg.] On each plant of the foot the tendon is [] stretched, storing some energy. The ten[don] springs back as you push off the groun[d] again, helping to propel you forward. T[his] recovered energy reduces the amount [of] internal chemical energy you use, increasing your efficiency.

Dinner at a distance BIO A chameleon's tongue is a powerful tool for catching prey. Certain species can extend the tongue to a distance of over 1 ft in less than 0.1 s! A study of the kinematics of the motion of the chameleon tongue, using techniques like those in this chapter, reveals that the tongue has a period of rapid acceleration followed by a period of constant velocity. This knowledge is a very valuable clue in the analysis of the evolutionary relationships between chameleons and other animals.

Passage Problems

Kangaroo Locomotion BIO

Kangaroos have very stout tendons in their legs that can be used to store energy. When a kangaroo lands on its feet, the tendons stretch, transforming kinetic energy of motion to elastic potential energy. Much of this energy can be transformed back into kinetic energy as the kangaroo takes another hop. The kangaroo's peculiar hopping gait is not very efficient at low speeds but is quite efficient at high speeds.

Figure P11.61 shows the energy cost of human and kangaroo locomotion. The graph shows oxygen uptake (in mL/s) per kg of body mass, allowing a direct comparison between the two species.

For humans, the energy used per second (i.e., power) is proportional to the speed. That is, the human curve nearly passes through the origin, so running twice as fast takes approximately twice as much power. For a hopping kangaroo, the graph of energy use has only a very small slope. In other words, the energy used per second changes very little with speed. Going faster requires very little additional power. Treadmill tests on kangaroos and observations in the wild have shown that they do not become winded at any speed at which they are able to hop. No matter how fast they hop, the necessary power is approximately the same.

61. | A person runs 1 km. How does his speed affect the total energy needed to cover this distance?
   A. A faster speed requires less total energy.
   B. A faster speed requires more total energy.
   C. The total energy is about the same for a fast speed and a slow speed.

62. | A kangaroo hops 1 km. How does its speed affect the total energy needed to cover this distance?
   A. A faster speed requires less total energy.
   B. A faster speed requires more total energy.
   C. The total energy is about the same for a fast speed and a slow speed.

63. | At a speed of 4 m/s,
   A. A running human is more efficient than an equal-mass hopping kangaroo.
   B. A running human is less efficient than an equal-mass hopping kangaroo.
   C. A running human and an equal-mass hopping kangaroo have about the same efficiency.

64. | At approximately what speed would a human use half the power of an equal-mass kangaroo moving at the same speed?
   A. 3 m/s   B. 4 m/s   C. 5 m/s   D. 6 m/s

65. | At approximately what speed would a human use twice the power of a kangaroo of half the mass moving at the same speed?
   A. 3 m/s   B. 5 m/s   C. 7 m/s   D. 9 m/s

Oxygen uptake (mL/kg·s)

Human running

Red kangaroo, hopping

Speed (m/s)

FIGURE P11.61 Oxygen uptake (a measure of energy use per second) for a running human and a hopping kangaroo.

"The authors' use of everyday and especially biological examples is superior to any other text I've seen. Students with biomedical interests will have plenty to sink their teeth into."

—Taner Edis, *Truman State University*

Multiple-choice and Passage Problems are designed to carefully test student understanding by targeting common misconceptions and providing context-rich situations—modeling the latest MCAT and other progressive physics exams.

PROMOTES DEEPER UNDERSTANDING

A structured learning path

This text incorporates many subtle but powerful techniques from cognitive science that improve student learning and retention, including a self-evident and structured learning path and a unique visual pedagogy.

"Clearly written by experienced, knowledgeable teachers who have spent time in the trenches... *College Physics* sets a new standard."

—Marllin L. Simon, *Auburn University*

"The text is exceptionally easy to read... and takes special care to explain the fundamentals and debunk common misconceptions..."

—Michael Pravica, *University of Nevada–Las Vegas*

Looking Ahead ▶▶
The goals of Chapter 3 are to learn more about vectors and to use vectors as a tool to analyze motion in two dimensions. In this chapter, you will learn to:

- ▶ Work with vectors, coordinate systems, and components.
- ▶ Find acceleration vectors on motion diagrams.
- ▶ Use vectors to understand motion on a ramp and relative velocity.
- ▶ Solve problems involving projectiles following parabolic paths.
- ▶ Understand the basic ideas of motion in a circle.

Looking Back ◀◀
This chapter will build naturally on the material from the previous two chapters. Please review:

- ◀ Sections 1.1–1.2 Basic motion concepts.
- ◀ Section 1.5 Vectors.
- ◀ Sections 2.4–2. Acceleration and free fall.

3 VECTORS AND MOTION IN TWO DIMENSIONS

With astonishing precision, an archer fish secures a meal by spitting drops of water at an unsuspecting insect. What factors determine how high and how far the water drops travel?

Looking Ahead ▶▶
The goals of Chapter 3 are to learn more about vectors and to use vectors as a tool to analyze motion in two dimensions. In this chapter, you will learn to:

- ▶ Work with vectors, coordinate systems, and components.
- ▶ Find acceleration vectors on motion diagrams.
- ▶ Use vectors to understand motion on a ramp and relative velocity.
- ▶ Solve problems involving projectiles following parabolic paths.
- ▶ Understand the basic ideas of motion in a circle.

Looking Back ◀◀
This chapter will build naturally on the material from the previous two chapters. Please review:

- ◀ Sections 1.1–1.2 Basic motion concepts.
- ◀ Section 1.5 Vectors.
- ◀ Sections 2.4–2. Acceleration and free fall.

70

Once the drops of water leave the archer fish's mouth and fly through the air, their motion follows a gently curving path. This path, a parabola, is the same type of path followed by a basketball shot toward the basket, a high jumper leaping over a bar, or a skier going off a jump. These are all examples of motion in two dimensions, as the objects are moving horizontally and vertically at the same time.

In order to extend our study of motion to two dimensions, we will need to develop techniques for working with vectors. We will first use vector techniques to analyze motion on a ramp and relative velocity, continue on to look at the parabolic motion of projectiles, then close the chapter by describing the motion of objects moving in circles. Along the way we'll use numerous examples drawn from sports—from skiers gliding down slopes to balls flying through the air.

Stop to Think Questions at the end of a section allow the student to quickly check their understanding. Using powerful ranking-task and graphical techniques, they are designed to efficiently probe key misconceptions and encourage active reading. (Answers are provided at the end of the chapter.)

Each chapter begins with a motivational roadmap—*Looking Ahead* provides learning goals to keep the student focused on the key ideas, and *Looking Back* consolidates connections with earlier material.

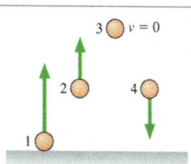

STOP TO THINK 10.4 Rank in order, from largest to smallest, the gravitational potential energies of identical balls 1 to 4.

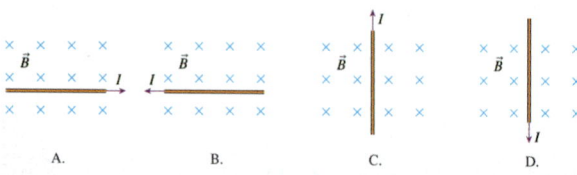

STOP TO THINK 24.5 Four wires carry currents in the directions shown. A uniform magnetic field is directed into the paper as shown. Which wire experiences a force to the left?

NOTE ▶ Since the right-hand rule for forces specifies the orientation of the force on a positive charge, the force on a negative charge will be in the opposite direction. A negatively charged particle will orbit in the opposite sense from that shown in Figure 24.32 for a positive charge. ◀

NOTE Paragraphs throughout guide students away from known preconceptions, around common sticking points, and highlight many math-related issues that cause difficulties.

Proven visual pedagogy

Figures are carefully streamlined in detail and color so students focus on the physics—for instance, the object of interest in mechanics—using an instructional approach well documented to improve learning.

The proven technique of annotation—adding chalkboard-like dialog onto the figure (rather than providing as a lengthy caption beneath)—is used to more effectively guide students in:

- interpreting a graph or figure
- understanding a process
- translating between text, math, graphs, and figures
- grasping a difficult concept through a visual analogy

FIGURE 4.33 External forces and action/reaction pairs.

FIGURE 5.12 The sensation of weight for a man at rest.

FIGURE 2.21 Approximating a velocity-versus-time graph with a series of constant-velocity steps.

"The summaries are very effective in helping students to identify the key ideas and uniting principles of the chapter."

—Kristi D. Concannon, *King's College*

Critically acclaimed **Visual Chapter Summaries** consolidate understanding by providing each concept in words, math, and figures and organizing these into a coherent hierarchy— from General Principles to Applications.

THE BEST-TESTED PROBLEMS

Enhanced by analysis

End-of-chapter problems of superior quality and diversity have been carefully crafted to exercise and test the full range of qualitative and quantitative problem-solving skills. The wealth of 3,000 problems in *College Physics: A Strategic Approach* are the first problems to be pre-tested for reliability, difficulty, and duration on thousands of students nationally before publication in a new textbook. *(See fold-out grid at front of instructor copy for more details).*

Simpler subproblems are provided upon request when a student gets stuck.

Conceptual Questions require careful reasoning and can be used for either group discussions or individual work. The **Multiple-Choice Questions** use carefully chosen distractors to elicit common misconceptions and misunderstandings.

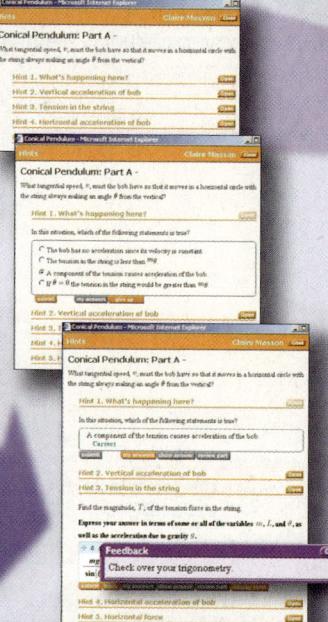

Problems, keyed to sections, allow students to build up their skills, often drawing on real-world or life-science applications to provide motivational examples.

Students select only the specific help they need. Thousands of solution paths are allowed.

More advanced **General Problems** include many context-rich problems which require students to simplify and model more complex real-world situations. Those labeled INT integrate significant material from earlier chapters.

See the **MasteringPhysics™** brochure for more details on how it enables instructors to more quickly build assignments of just the right coverage, difficulty, and duration, and the time-saving and probing diagnostics it provides (including most-common wrong answers by a class at each step, the step-by-step work of each student, even national average comparison data).

Unparalleled teaching diagnostics and individualized tutoring

MasteringPhysics™ is the most widely used and educationally proven physics homework and tutorial system in the world. NSF-sponsored research and independent studies at a wide range of schools show that students who use MasteringPhysics™ significantly improve their scores on final exams and conceptual tests such as the Force Concept Inventory. Drawing from the world's largest metadatabase of the step-by-step work of how real students solve problems, accumulated over eight years of research, MasteringPhysics™ offers unparalleled individualized guidance.

Instantaneous feedback is provided specific to the most common wrong answers.

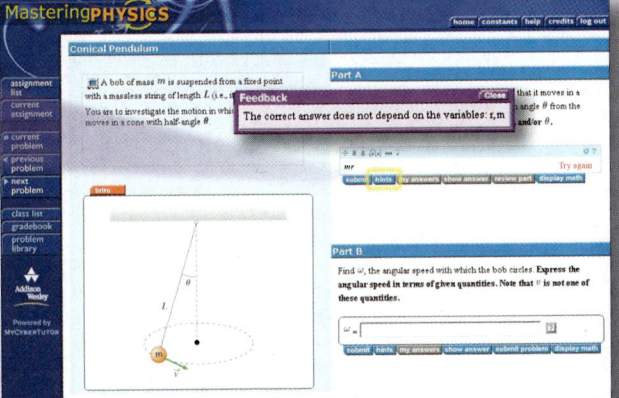

MasteringPhysics™ provides instructors with more than 3,000 problems from which to build wide-ranging weekly assignments that include:

- end-of-chapter textbook problems
- tutorials for practicing all the Problem-Solving Strategies, Tactics Boxes, and Math Relationship Boxes in the textbook
- skill-building tutorials for mastering key concepts, and self-tutoring problems for guided practice
- math-specific skill-building tutorials, ranking task problems, even essay questions

Numerical problems provide randomized numbers and sig-fig feedback, and tutorial problems provide wrong-answer specific feedback and simpler subproblems upon request.

Partial credit is provided for the method, not just the final numerical answer.

A **vector-drawing tool** provides grading and feedback on student-drawn free-body diagrams embedded in the tutorials.

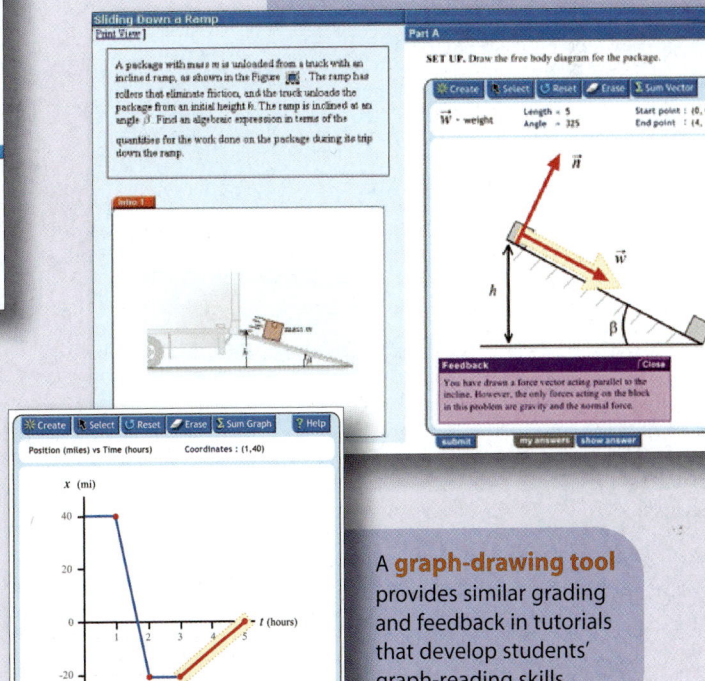

A **graph-drawing tool** provides similar grading and feedback in tutorials that develop students' graph-reading skills.

MasteringPHYSICS

www.masteringphysics.com

Real-World Applications

Applications of biological or medical interest are marked BIO in the list below. MCAT-style Passage Problems are marked BIO below. Other end-of-chapter problems of biological or medical interest are marked BIO in the chapter. "Try It Yourself" experiments are marked **TIY**.

Preface to the Instructor

College Physics: A Strategic Approach is a new algebra-based physics textbook for students majoring in the biological and life sciences, architecture, natural resources, and other disciplines. Students entering this course typically have strong academic abilities and are highly motivated to do well. However, their success and enjoyment of physics is often hampered by a number of obstacles. First, these students do not see the relevance of physics to their major; they tend to view this course as a hurdle to overcome rather than an opportunity to broaden their knowledge of science. Second, they have often learned a rather linear approach to problem solving in other courses and this differs substantially from the more open and wide-ranging approach required in physics. Finally, even if they have taken previous physics courses, students tend to arrive with a wealth of strongly held preconceptions, inconsistent with the way the physical world works.

Objectives

Our primary goal in writing *College Physics: A Strategic Approach* has been to tackle the learning obstacles listed above head on and to thereby provide instructors and students with a more effective teaching and learning resource. Our axioms have been:

- To provide students with a textbook that's a more manageable size, less encyclopedic in its coverage, and better designed for learning.
- To integrate proven techniques from physics education research into the classroom in a way that accommodates a range of teaching and learning styles.
- To help students develop both quantitative reasoning skills and solid conceptual understanding, with special focus on concepts well documented to cause learning difficulties.
- To help students develop problem-solving skills and confidence in a systematic manner using explicit and consistent tactics and strategies.
- To motivate students by integrating real-world examples relevant to their majors—especially from biology, sports, medicine, the animal world—and that build upon their everyday experiences.
- To utilize proven techniques of visual instruction and design from educational research and cognitive psychology that improve student learning and retention and address a range of learner styles.
- To support an active-learning environment.

A more complete explanation of these goals and the rationale behind them can be found in Randy Knight's paperback book, *Five Easy Lessons: Strategies for Successful Physics Teaching* (Addison Wesley, 2002). Please request a copy from your local Addison Wesley sales representative if it would be of interest to you (ISBN 0-8053-8702-1).

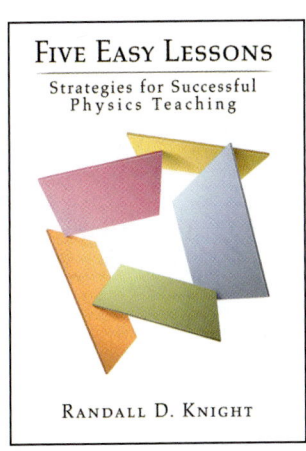

FIVE EASY LESSONS
Strategies for Successful
Physics Teaching

RANDALL D. KNIGHT

Textbook Organization

College Physics: A Strategic Approach is a 30-chapter text intended for use in a two-semester course. The textbook is divided into seven parts: Part I: *Force and Motion,* Part II: *Conservation Laws,* Part III: *Properties of Matter,* Part IV:

- **Complete edition,** with MasteringPhysics™ (ISBN 0-8053-9062-6): chapters 1–30.
- **Volume 1** with MasteringPhysics™ (ISBN 0-8053-1695-7): chapters 1–16.
- **Volume 2** with MasteringPhysics™ (ISBN 0-8053-1696-5): chapters 17–30.
- **Complete edition,** without MasteringPhysics™ (ISBN 0-8053-6582-6): chapters 1–30.
- **Volume 1** without MasteringPhysics™ (ISBN 0-8053-4662-7): chapters 1–16.
- **Volume 2** without MasteringPhysics™ (ISBN 0-8053-7209-1): chapters 17–30.

Oscillations and Waves, Part V: *Optics,* Part VI: *Electricity and Magnetism,* and Part VII: *Modern Physics.*

Part I covers Newton's laws and their applications. We cover the two fundamental conserved quantities, momentum and energy, in Part II for two reasons. First, the way that problems are solved using conservation laws—comparing an *after* situation to a *before* situation—differs fundamentally from the problem-solving strategies used in Newtonian dynamics. Second, the concept of energy has a significance far beyond mechanical (kinetic and potential) energies. In particular, the key idea in thermodynamics is energy, and moving from the study of energy in Part II into thermal physics in Part III allows the uninterrupted development of this important idea.

Optics (Part V) is covered directly after oscillations and waves (Part IV), and *before* electricity and magnetism (Part VI). Further, we treat wave optics before ray optics. Our motivations for this organization are twofold. First, wave optics is largely an extension of the general ideas of waves; in a more traditional organization, students have forgotten much of what they learned about waves by the time they get to wave optics. Second, optics as it is presented in introductory physics makes no use of the properties of electromagnetic fields. The documented difficulties that students have with optics are difficulties with waves, not difficulties with electricity and magnetism. There's little reason other than historical tradition to delay optics. However, the optics chapters can be deferred easily until after Part VI for instructors who prefer that ordering of topics.

The Student Workbook

A key component of *College Physics: A Strategic Approach* is the accompanying *Student Workbook.* The workbook bridges the gap between textbook and homework problems by providing students the opportunity to learn and practice skills prior to using those skills in quantitative end-of-chapter problems, much as a musician practices technique separately from performance pieces. The workbook exercises, which are keyed to each section of the textbook, focus on developing specific skills, ranging from identifying forces and drawing free-body diagrams to interpreting field diagrams.

The workbook exercises, which are generally qualitative and/or graphical, draw heavily upon the physics education research literature. The exercises deal with issues known to cause student difficulties and employ techniques that have proven to be effective at overcoming those difficulties. The workbook exercises can be used in-class as part of an active-learning teaching strategy, in recitation sections, or as assigned homework. More information about effective use of the *Student Workbook* can be found in the *Instructor's Guide.*

Available versions: Volume 1 (ISBN 0-8053-8209-7): chapters 1–16, and Volume 2 (ISBN 0-8053-0626-9): chapters 17–30.

Instructor Supplements

- The **Instructor Guide for** *College Physics: A Strategic Approach* (ISBN 0-8053-0622-6), a comprehensive and highly acclaimed resource written by the textbook authors, provides chapter-by-chapter creative ideas and teaching tips for using *College Physics: A Strategic Approach* in your class. In addition, it contains an extensive review of what has been learned from physics education research, and provides guidelines for using active-learning techniques in your classroom.

- The **Instructor Solutions Manuals, Chapters 1–16** (ISBN 0-8053-0495-9), and **Chapters 17–30** (ISBN 0-8053-0638-2), written by Professor Larry Smith, Snow College; Professor Pawan Kahol, Missouri State University; Professor Marllin Simon, Auburn University; and Walter Polkosnik, University of California-Santa Barbara, provide *complete* solutions to all the end-of-chapter questions and problems. All solutions follow the Prepare/Solve/Assess problem-solving strategy used in the textbook for quantitative problems, and Reason/Assess strategy for qualitative ones. They are available as editable Word documents and as pdf files on the *Media Manager CD-ROMs* for your own use or for posting on your password-protected course website.

- The cross-platform **Media Manager CD-ROMs** (ISBN 0-8053-0306-5) provide invaluable and easy-to-use resources for your class, organized by textbook chapter. The contents include a comprehensive library of more than 220 applets from **ActivPhysics OnLine**™, as well as all figures, tables, summaries, and key equations from the textbook in JPEG format. In addition, all the Problem-Solving Strategies, Math Relationship Boxes, Tactics Boxes, and Key Equations are provided in editable Word format. The **Instructor Guide** and **Instructor Solutions Manuals** are also included as pdfs and editable Word files, along with pdfs of answers to the **Student Workbook** exercises, and **Lecture Outlines (with Classroom Response System "Clicker" Questions)** in PowerPoint® written by the textbook authors.

- The online **Instructor Resource Center** (www.aw-bc.com/irc) provides updates to files on the Media Manager CD-ROMs, and also includes the lecture outlines in Keynote®. To obtain a Login Name and Password, contact your Pearson Addison-Wesley sales representative.

- **MasteringPhysics**™ (www.masteringphysics.com), is the most advanced, educationally effective, and widely used physics homework and tutorial system in the world. It is designed to assign, assess, and track each student's progress using a wide diversity of extensively pre-tested problems. Icons throughout the book indicate that *MasteringPhysics*™ offers specific tutorials for all the textbook's Tactics Boxes, Problem-Solving Strategies, and Math Relationship Boxes, as well as all the End-of-Chapter problems, Test Bank items, and Reading Quizzes. *MasteringPhysics*™ provides instructors with a fast and effective way to assign uncompromising, wide-ranging online homework assignments of just the right difficulty and duration. The powerful post-assignment diagnostics allow instructors to assess the progress of their class as a whole or to quickly identify individual students' areas of difficulty.

- **ActivPhysics OnLine**™ (www.aw-bc.com/knightjonesfield) provides a comprehensive library of more than 420 tried and tested *ActivPhysics* applets. In addition, it provides a suite of highly regarded applet-based tutorials developed by education pioneers Professors Alan Van Heuvelen and Paul D'Alessandris. The *ActivPhysics* icons that appear throughout the book direct students to specific interactive exercises that complement the textbook discussion.

  The online exercises are designed to encourage students to confront misconceptions, reason qualitatively about physical processes, experiment quantitatively, and learn to think critically. They cover all topics from mechanics to electricity and magnetism and from optics to modern physics. The highly acclaimed *ActivPhysics OnLine* companion workbooks (see Student Supplements) help students work through complex concepts and understand them more clearly. More than 220 applets from the *ActivPhysics OnLine* library are also available on the Instructor *Media Manager CD-ROMs*.

- The **Printed Test Bank** (ISBN 0-8053-1697-3) and cross-platform **Computerized Test Bank** (ISBN 0-8053-0618-8), prepared by Sen-Ben Liao, University of California, provides more than 2,000 high-quality problems, with a range of multiple-choice, true/false, short-answer, and regular homework-type questions. The computerized version includes all of the questions from the Printed Test Bank, more than half of which can be assigned with randomized values.
- The **Transparency Acetates** (ISBN 0-8053-1689-2) provide more than 200 key figures from *College Physics: A Strategic Approach* for classroom presentation.

Student Supplements

- The **Student Solutions Manuals, Chapters 1–16** (ISBN 0-8053-0632-3) and **Chapters 17–30** (ISBN 0-8053-0631-5), written by Professor Larry Smith, Snow College; Professor Pawan Kahol, Missouri State University; Marllin Simon, Auburn University; and Walter Polkosnik, University of California-Santa Barbara, provide *detailed* solutions to more than half of the odd-numbered end-of-chapter problems. Following the problem-solving strategy presented in the text, thorough solutions are provided to carefully coach students through both the qualitative (Reason/Assess) and quantitative (Prepare/Solve/Assess) steps in the problem-solving process.

- **MasteringPhysics**™ (www.masteringphysics.com) is the most advanced, widely used, and educationally proven physics tutorial system in the world. It is the result of eight years of detailed studies of how real students work physics problems and of precisely where they need help. Studies show that students who use MasteringPhysics™ significantly improve their scores on final exams and conceptual tests such as the Force Concept Inventory. MasteringPhysics™ achieves this improvement by providing students with instantaneous feedback specific to their wrong answers, simpler subproblems upon request when they get stuck, and partial credit for their method(s) used. This individualized, 24/7 tutor system is recommedned by 9 out of 10 students to their peers as the most effective and time-efficient way to study.

- **ActivPhysics**™ **OnLine** (www.aw-bc.com/knightjonesfield) provides students with a suite of highly regarded applet-based tutorials (see above). The accompanying workbooks help students work though complex concepts and understand them more clearly. The *ActivPhysics* icons throughout the book direct students to specific interactive exercises that complement the textbook discussion. The following workbooks provide a range of tutorial problems designed to use the ActivPhysics OnLine simulations:

- **ActivPhysics OnLine Workbook Volume 1: Mechanics • Thermal Physics • Oscillations & Waves** (ISBN 0-8053-9060-X)

- **ActivPhysics OnLine Workbook Volume 2: Electricity & Magnetism • Optics • Modern Physics** (ISBN 0-8053-9061-8)

- The **Addison-Wesley Tutor Center** (www.aw.com/tutorcenter) provides one-on-one tutoring via telephone, fax, email, or interactive website during evening hours and on weekends. Qualified college instructors answer questions and provide instruction for *Mastering Physics*™ and for the examples, exercises, and problems in *College Physics: A Strategic Approach.*

Acknowledgments

We have relied upon conversations with and, especially, the written publications of many members of the physics education community. Those who may recognize their influence include Arnold Arons, Uri Ganiel, Fred Goldberg, Ibrahim Halloun, Richard Hake, David Hestenes, Leonard Jossem, Jill Larkin, Priscilla

Laws, John Mallinckrodt, Lillian McDermott, Edward "Joe" Redish, Fred Reif, John Rigden, Rachel Scherr, Bruce Sherwood, David Sokoloff, Ronald Thornton, Sheila Tobias, and Alan Van Heuleven.

We are grateful to Marty Gelfand for taking on the significant task of editing and organizing the end-of-chapter problem sets; Larry Smith, Pawan Kahol, Marllin Simon, and Walt Polkosnik for the difficult task of writing the *Instructor Solutions Manuals;* to Jim Andrews for coauthoring the Student Workbook (and literally writing out all its answers); to Wayne Anderson, Jim Andrews, Nancy Beverly, David Cole, Karim Diff, Jim Dove, Marty Gelfand, Kathy Harper, Charlie Hibbard, Robert Lutz, Matt Moelter, Kandiah Manivannan, Ken Robinson, and Cindy Schwarz-Rachmilowitz for their contributions to the end-of-chapter questions and problems; and to Steven Vogel for his careful review of the biological content of many chapters and for helpful suggestions.

We especially want to thank our editor Adam Black, development editor Alice Houston, project editor Martha Steele, production supervisor Shannon Tozier, and all the other staff at Addison Wesley for their enthusiasm and hard work on this project. Jared Sterzer and the team at Westwords, Inc., copy editor Kevin Gleason, and photo researcher Brian Donnelly get much credit for making this complex project all come together. In addition to the reviewers and classroom testers listed below, who gave invaluable feedback, we are particularly grateful to Charlie Hibbard for his close scrutiny of every word, symbol, number, and figure.

Randy Knight: I would like to thank my Cal Poly colleagues, especially Matt Moelter, for many valuable conversations and suggestions. I am endlessly grateful to my wife Sally for her love, encouragement, and patience, and to our many cats for nothing in particular other than being cats.

Brian Jones: I would like to thank my fellow AAPT and PIRA members for their insight and ideas, the creative and wonderful students who are my partners in the Little Shop of Physics, the students in my College Physics classes who help me become a better teacher, and my wife Carol, my best friend and gentlest editor, whose love makes the journey worthwhile.

Stuart Field: I would like to thank my friends and colleagues for their support and insights, both professional and personal, that helped me with the writing of this book. I would especially like to thank my wife Julie for her love, support, and encouragement; and my children Sam and Ellen, who grew up before me (or behind me) as I sat in front of my computer screen.

Reviewers and Classroom Testers

Susmita Acharya, *Cardinal Stritch University*
Ugur Akgun, *University of Iowa*
Ralph Alexander, *University of Missouri-Rolla*
Donald Anderson, *Ivy Tech*
Steve Anderson, *Montana Tech*
James Andrews, *Youngstown State University*
Charles Ardary, *Edmond Community College*
Charles Bacon, *Ferris State University*
John Barry, *Houston Community College*
David H. Berman, *University of Northern Iowa*
Richard Bone, *Florida International University*
Jeff Bodart, *Chipola College*
James Borgardt, *Juniata College*
Daniela Bortoletto, *Purdue University*

Don Bowen, *Stephen F. Austin State University*
Asa Bradley, *Spokane Falls Community College*
Elena Brewer, *SUNY at Buffalo*
Dieter Brill, *University of Maryland*
Hauke Busch, *Augusta State University*
Kapila Castoldi, *Oakland University*
Michael Cherney, *Creighton University*
Lee Chow, *University of Central Florida*
Song Chung, *William Paterson University*
Alice Churukian, *Concordia College*
Kristi Concannon, *Kings College*
Teman Cooke, Georgia *Perimeter College at Lawrenceville*
Jesse Cude, *Hartnell College*
Melissa H. Dancy, *University of North Carolina at Charlotte*

Loretta Dauwe, *University of Michigan-Flint*
Mark Davenport, *San Antonio College*
Carlos Delgado, *Community College of Southern Nevada*
David Donovan, *Northern Michigan University*
Archana Dubey, *University of Central Florida*
Andrew Duffy, *Boston University*
Taner Edis, *Truman State University*
Ralph Edwards, *Lurleen B. Wallace Community College*
Steve Ellis, *University of Kentucky*
Paula Engelhardt, *Tennessee Technical University*
Davene Eryes, *North Seattle Community College*
Gerard Fasel, *Pepperdine University*
Luciano Fleischfresser, *OSSM Autry Tech*
Cynthia Galovich, *University of Northern Colorado*
Bertram Gamory, *Monroe Community College*
Delena Gatch, *Georgia Southern University*
Martin Gelfand, *Colorado State University*
Terry Golding, *University of North Texas*
Robert Gramer, *Lake City Community College*
William Gregg, *Louisiana State University*
Paul Gresser, *University of Maryland*
Robert Hagood, *Washtenaw Community College*
Heath Hatch, *University of Massachusetts*
Carl Hayn, *Santa Clara University*
James Heath, *Austin Community College*
Greg Hood, *Tidewater Community College*
Sebastian Hui, *Florence-Darlington Technical College*
Joey Huston, *Michigan State University*
David Iadevaia, *Pima Community College-East Campus*
Erik Jensen, *Chemeketa Community College*
Todd Kalisik, *Northern Illinois University*
Ju H. Kim, *University of North Dakota*
Armen Kocharian, *California State University Northridge*
J. M. Kowalski, *University of North Texas*
Laird Kramer, *Florida International University*
Chrisopher Kulp, *Eastern Kentucky University*
Richard Kurtz, *Louisiana State University*
Kenneth Lande, *University of Pennsylvania*
Tiffany Landry, *Folsom Lake College*
Todd Leif, *Cloud County Community College*
John Levin, *University of Tennessee-Knoxville*
John Lindberg, *Seattle Pacific University*
Rafael López-Mobilia, *The University of Texas at San Antonio*
Robert W. Lutz, *Drake University*
Lloyd Makorowitz, *SUNY Farmingdale*
Eric Martell, *Millikin University*
Mark Masters, *Indiana University-Purdue*
Denise Meeks, *Pima Community College*
Henry Merrill, *Fox Valley Technical College*

Mike Meyer, *Michigan Technological University*
Tobias Moleski, *Nashville State Tech*
April Moore, *North Harris College*
Gary Morris, *Rice University*
Charley Myles, *Texas Tech University*
Fred Olness, *Southern Methodist University*
Charles Oliver Overstreet, *San Antonio College*
Paige Ouzts, *Lander University*
Russell Palma, *Minnesota State University-Mankato*
Richard Panek, *Florida Gulf Coast University*
Joshua Phiri, *Florence-Darling Technical College*
Iulia Podariu, *University of Nebraska at Omaha*
David Potter, *Austin Community College*
Promod Pratap, *University of North Carolina-Greensboro*
Michael Pravica, *University of Nevada, Las Vegas*
Earl Prohofsky, *Purdue University*
Marilyn Rands, *Lawrence Technological University*
Andrew Rex, *University of Puget Sound*
Andrew Richter, *Valparaiso University*
Phyliss Salmons, *Embry-Riddle Aeronautical University*
Michael Schaab, *Maine Maritime Academy*
Bruce Schumm, *University of California, Santa Cruz*
Mizuho Schwalm, *University of Minnesota Crookston*
Cindy Schwarz, *Vassar College*
Natalia Semushkhina, *Shippensburg University*
Khazhgery (Jerry) Shakov, *Tulane University*
Bart Sheinberg, *Houston Community College*
Marllin Simon, *Auburn University*
Kenneth Smith, *Pennsylvania State University*
Michael Smutko, *Northwestern University*
Jon Son, *Boston University*
Noel Stanton, *Kansas State University*
Donna Stokes, *University of Houston*
Chuck Stone, *North Carolina A&T*
Chun Fu Su, *Mississippi State University*
William Tireman, *Northern Michigan University*
Negussie Tirfessa, *Manchester Community College*
Rajive Tiwari, *Belmont Abbey College*
Herman Trivilino, *College of the Mainland*
Douglas Tussey, *Pennsylvania State University*
James Vesenka, *University of New England*
Christos Valiotis, *Antelope Valley College*
Stamatis Vokos, *Seattle Pacific University*
James Wanliss, *Embry-Riddle Aeronautical University*
Henry Weigel, *Arapahoe Community College*
Courtney Willis, *University of Northern Colorado*
Katherine Wu, *University of Tampa*
Ali Yazdi, *Jefferson State Community College*
David Young, *Louisiana State University*
Hsiao-Ling Zhou, *Georgia State University*
Ulrich Zurcher, *Cleveland State University*

Preface to the Student

The most incomprehensible thing about the universe is that it is comprehensible.
  —Albert Einstein

If you are taking a course for which this book is assigned, you probably aren't a physics major or an engineering major. It's likely that you aren't majoring in a physical science. So why are you taking physics?

It's almost certain that you are taking physics because you are majoring in a discipline that requires it. Someone, somewhere, has decided that it's important for you to take this course. And they are right. There is a lot you can learn from physics, even if you don't plan to be a physicist. We regularly hear from doctors, physical therapists, biologists and others that physics was one of the most interesting and valuable courses they took in college.

So, what can you expect to learn in this course? Let's start by talking about what physics is. Physics is a way of thinking about the physical aspects of nature. Physics is not about "facts." It's far more focused on discovering *relationships* between facts and the *patterns* that exist in nature than on learning facts for their own sake. Our emphasis will be on thinking and reasoning. We are going to look for patterns and relationships in nature, develop the logic that relates different ideas, and search for the reasons *why* things happen as they do.

Once we've figured out a pattern, a set of relationships, we'll look at applications to see where this understanding takes us. Let's look at an example. Part (a) of the figure shows an early mechanical clock. The clock uses a *pendulum,* a mass suspended by a thin rod free to pivot about its end, as its timekeeping element. When you study oscillatory motion, you will learn about the motion of a pendulum. You'll learn that the *period* of its motion, the time for one swing, doesn't depend on the amplitude, the size of the swing. This makes a pendulum the ideal centerpiece of a clock.

But there are other systems that look like pendulums too. The gibbon in part (b) of the figure is moving through the trees by swinging from successive handholds. The gibbon's mass is suspended below a point about which it is free to pivot, so the gibbon's motion can be understood as pendulum motion. You can then use your knowledge of pendulums to describe the motion, explaining, for example, why this gibbon is raising its feet as it swings.

Like any subject, physics is best learned by doing. "Doing physics" in this course means solving problems, applying what you have learned to answer questions at the end of the chapter. When you are given a homework assignment, you may find yourself tempted to simply solve the problems by thumbing through the text looking for a formula that seems like it will work. This isn't how to do physics; if it was, whoever required you to take this course wouldn't bother. The folks who designed your major want you to learn to *reason,* not to "plug and chug." Whatever you end up studying or doing for a career, this ability will serve you well. And that's why someone, somewhere, wants you to take physics.

(a) Pendulum clock

(b) Gibbon locomotion

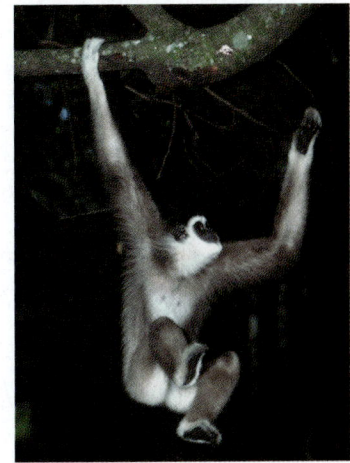

How do you learn to reason in this way? There's no single strategy for studying physics that will work for all students, but we can make some suggestions that will certainly help:

- **Read each chapter *before* it is discussed in class.** Class attendance is much less effective if you have not prepared. When you first read a chapter, focus on learning new vocabulary, definitions, and notation. You won't understand what's being discussed or how the ideas are being used if you don't know what the terms and symbols mean.
- **Participate actively in class.** Take notes, ask and answer questions, take part in discussion groups. There is ample scientific evidence that *active participation* is far more effective for learning science than is passive listening.
- **After class, go back for a careful rereading of the chapter.** In your second reading, pay close attention to the details and the worked examples. Look for the *logic* behind each example, not just at what formula is being used. We have a three-step process by which we solve all of the worked examples in the text. Most chapters have detailed Problem-Solving Strategies to help you see how to apply this procedure to particular topics, and Tactics Boxes that explain specific steps in your analysis.
- **Apply what you have learned to the homework problems at the end of each chapter.** By following the techniques of the worked examples, applying the tactics and problem-solving strategies, you'll learn how to apply the knowledge you are gaining. In short, you'll learn to reason like a physicist.
- **Form a study group with two or three classmates.** There's good evidence that students who study regularly with a group do better than the rugged individualists who try to go it alone.

And we have one final suggestion. As you read the book, take part in class, and work through problems, step back every now and then to appreciate the big picture. You are going to study topics that range from motions in the solar system to the electrical signals in the nervous system that let you order your hand to turn the pages of this book. You will learn quantitative methods to calculate things such as how far a car will move as it brakes to a stop and how to build a solenoid for an MRI machine. It's a remarkable breadth of topics and techniques that is based on a very compact set of organizing principles. It's quite remarkable, really, well worthy of your study.

Now, let's get down to work.

Detailed Contents

Part II **Conservation Laws**

Part III **Properties of Matter**

Part IV **Oscillations and Waves**

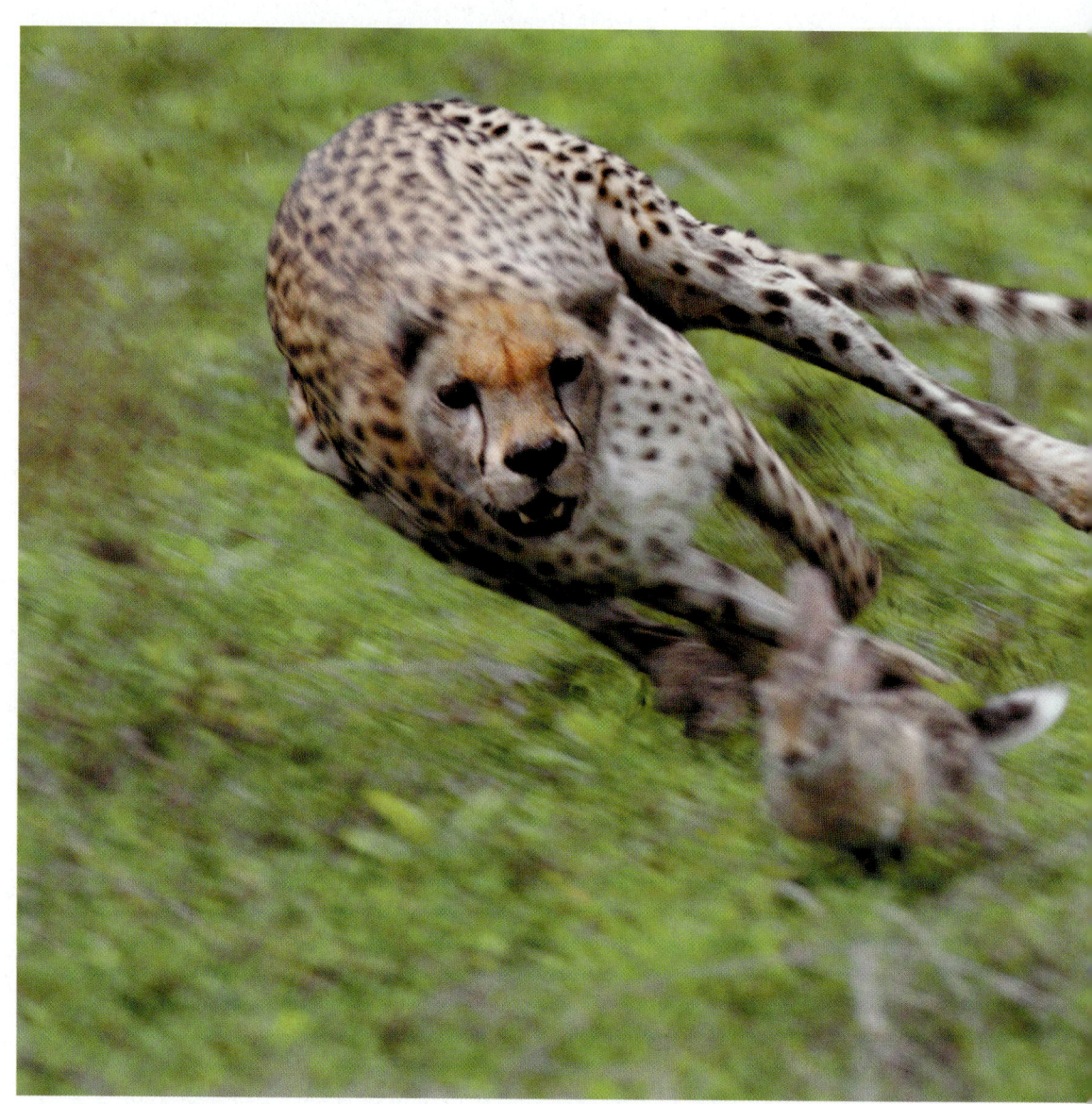

FORCE AND MOTION

Why Things Change

Each of the seven parts of this book opens with an overview that gives you a look ahead, a glimpse of where your journey will take you in the next few chapters. It's easy to lose sight of the big picture while you're busy negotiating the terrain of each chapter. In Part I, the big picture is, in a word, *change*.

Simple observations of the world around you show that most things change. Some changes, such as aging, are biological. Others, such as sugar dissolving in your coffee, are chemical. We will look at changes that involve *motion* of one form or another—running and jumping, throwing balls, lifting weights.

There are two big questions we must tackle to study how things change by moving:

- **How do we describe motion?** How should we measure or characterize the motion if we want to analyze it mathematically?
- **How do we explain motion?** Why do objects have the particular motion they do? Why, when you toss a ball upward, does it go up and then come back down rather than keep going up? Are there "laws of nature" that allow us to predict an object's motion?

Two key concepts that will help answer these questions are *force* (the "cause") and *acceleration* (the "effect"). Our basic tools will be three laws of motion elucidated by Isaac Newton. Newton's laws relate force to acceleration, and we will use them to explain and explore a wide range of problems. As we learn to solve problems, we will learn basic techniques that we can apply in all of the parts of this book.

Simplifying Models

Reality is extremely complicated. We would never be able to develop a science if we had to keep track of every detail of every situation. Suppose we analyze the tossing of a ball. Is it necessary to analyze the way the atoms in the ball are connected? Do we need to analyze what you ate for breakfast and the biochemistry of how that was translated into muscle power? These are interesting questions, of course. But if our task is to understand the motion of the ball, we need to simplify!

We can do a perfectly fine analysis if we treat the ball as a round solid and your hand as another solid that exerts a force on the ball. This is a *model* of the situation. A model is a simplified description of reality—much as a model airplane is a simplified version of a real airplane—that is used to reduce the complexity of a problem to the point where it can be analyzed and understood.

Model building is a major part of the strategy that we will develop for solving problems in all parts of the book. We will introduce different models in different parts. We will pay close attention to where simplifying assumptions are being made, and why. Learning *how* to simplify a situation is the essence of successful modeling—and successful problem solving.

The cheetah is the fastest land animal, able to run at speeds exceeding 60 miles per hour. Nonetheless, the rabbit has an advantage in this chase. It can *change* its motion more quickly, and will likely escape. How can you tell, by looking at the picture, that the cheetah is changing its motion?

1 CONCEPTS OF MOTION AND MATHEMATICAL BACKGROUND

As this snowboarder moves in a graceful arc through the air, the direction of his motion, and the distance between each of his positions and the next, is constantly changing. What language should we use to describe this motion?

Looking Ahead ▶▶

The goal of Chapter 1 is to introduce the fundamental concepts of motion and to review the related basic mathematical principles. In this chapter you will learn to:

▶ Draw and interpret motion diagrams.

▶ Apply the particle model of motion.

▶ Describe motion in terms of distance, displacement, time, and velocity.

▶ Express quantities with appropriate units and the correct number of significant figures.

▶ Describe motion using vectors, and learn how to add vectors.

▶ Understand how different physical quantities can often be expressed using the same mathematical form.

Socrates: *The nature of motion appears to be the question with which we begin.*
Plato, 375 BCE

The universe in which we live is one of change and motion. This snowboarder was clearly in motion in the series of photos that make up the image. In the course of a day you probably walk, run, bicycle, or drive your car, all forms of motion. The clock hands are moving inexorably forward as you read this text. The pages of this book may look quite still, but a microscopic view would reveal jostling atoms and whirling electrons. The stars look as permanent as anything, yet the astronomer's telescope reveals them to be ceaselessly moving within galaxies that rotate and orbit yet other galaxies.

Motion is a theme that will appear in one form or another throughout this entire book. Although we all have intuition about motion, based on our experiences, some of the important aspects of motion turn out to be rather subtle. So rather than jumping immediately into a lot of mathematics and calculations, this first chapter focuses on *visualizing* motion and becoming familiar with the *concepts* needed to describe a moving object. We will use mathematical ideas when needed, because they increase the precision of our thoughts, but we will defer actual calculations until Chapter 2. Our goal is to lay the foundations for understanding motion.

The quest to understand motion dates to antiquity. The ancient Babylonians, Chinese, and Greeks were especially interested in the celestial motions of the night sky. The Greek philosopher and scientist Aristotle wrote extensively about the nature of moving objects. However, our modern understanding of motion did not begin until Galileo (1564–1642) first formulated the concepts of motion in mathematical terms. And it took Newton (1642–1727) to put the concepts of motion on a firm and rigorous footing. This connection between motion and mathematics was the breakthrough that allowed the growth of the science of physics.

One key difference between physics and other sciences is how we set up and solve problems. We'll often use a two-step process to solve motion problems. The first step is to develop a simplified representation of the motion so that key elements stand out. For example, the above photo allows us to observe the position of the snowboarder at several successive times. It is precisely by considering this sort of picture of motion that we will begin our study of this topic. The second step is to analyze the motion with the language of mathematics. The process of putting numbers on nature is often the most challenging aspect of the problems you will solve. In this chapter, we will explore the steps in this process as we introduce the basic concepts of motion.

1.1 Motion: A First Look

As a starting point, let's define **motion** as the change of an object's position or orientation with time. Examples of motion are easy to list. Bicycles, baseballs, cars, airplanes, and rockets are all objects that move. The path along which an object moves, which might be a straight line or might be curved, is called the object's **trajectory.**

Figure 1.1 shows four basic types of motion that we will study in this book. In this chapter, we will start with the first type of motion in the figure, motion along a straight line. In later chapters, we will learn about circular motion, which is the motion of an object along a circular path; projectile motion, the motion of an object through the air; and rotational motion, the spinning of an object about an axis.

Straight-line motion

Circular motion

Projectile motion

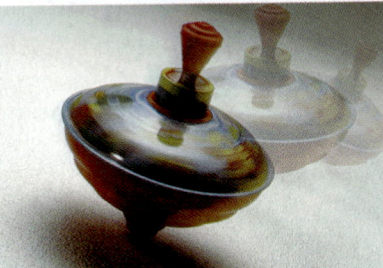
Rotational motion

FIGURE 1.1 Four basic types of motion.

The fundamental question we want to ask is: What *concepts* are needed to give a full and accurate description of motion?

FIGURE 1.2 Several frames from the video of a car.

The same amount of time elapses between each image and the next.

FIGURE 1.3 A motion diagram of the car shows all the frames simultaneously.

Making a Motion Diagram

An easy way to study motion is to record a video of a moving object. A video camera takes images at a fixed rate, typically 30 images every second. Each separate image is called a *frame*. As an example, Figure 1.2 shows a few frames from a video of a car going past. Not surprisingly, the car is in a somewhat different position in each frame.

Suppose we now edit the video, layering the frames on top of each other, and then look at the final result. We end up with the picture in Figure 1.3. This composite image, showing an object's position at several *equally spaced instants of time,* is called a **motion diagram.** As simple as motion diagrams seem, they will turn out to be powerful tools for analyzing motion.

NOTE ▶ It's important to keep the camera in a *fixed position* as the object moves by. Don't "pan" it to track the moving object. ◀

Now let's take our camera out into the world and make a few motion diagrams. The following table illustrates how a motion diagram shows important features of different kinds of motion.

Examples of motion diagrams

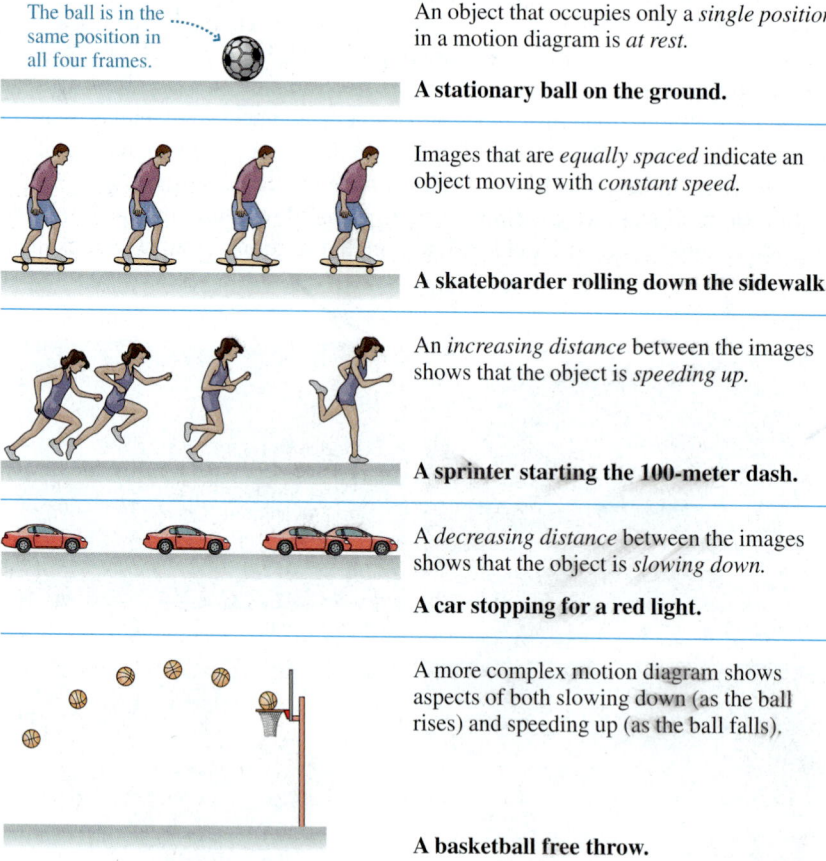

The ball is in the same position in all four frames.

An object that occupies only a *single position* in a motion diagram is *at rest.*

A stationary ball on the ground.

Images that are *equally spaced* indicate an object moving with *constant speed.*

A skateboarder rolling down the sidewalk.

An *increasing distance* between the images shows that the object is *speeding up.*

A sprinter starting the 100-meter dash.

A *decreasing distance* between the images shows that the object is *slowing down.*

A car stopping for a red light.

A more complex motion diagram shows aspects of both slowing down (as the ball rises) and speeding up (as the ball falls).

A basketball free throw.

We have defined several concepts (at rest, constant speed, speeding up, and slowing down) in terms of how the moving object appears in a motion diagram. These are called **operational definitions,** meaning that the concepts are defined in terms of a particular procedure or operation performed by the investigator. For example, we could answer the question "Is the airplane speeding up?" by checking whether or not the images in the plane's motion diagram are getting farther apart. Many of the concepts in physics will be introduced as operational definitions. This reminds us that physics is an experimental science.

STOP TO THINK 1.1   Which car is going faster, A or B? Assume there are equal intervals of time between the frames of both videos.

Car A   Car B

NOTE ▶ Each chapter in this textbook will have several *Stop to Think* questions. These questions are designed to see if you've understood the basic ideas that have been presented. The answers are given at the end of the chapter, but you should make a serious effort to think about these questions before turning to the answers. If you answer correctly, and are sure of your answer rather than just guessing, you can proceed to the next section with confidence. But if you answer incorrectly, it would be wise to reread the preceding sections carefully before proceeding onward. ◀

The Particle Model

For many objects, such as cars and rockets, the motion of the object *as a whole* is not influenced by the "details" of the object's size and shape. To describe the object's motion, all we really need to keep track of is the motion of a single point: You could imagine looking at the motion of a dot painted on the side of the object.

In fact, for the purposes of analyzing the motion, we can often consider the object *as if* it were just a single point, without size or shape. We can also treat the object *as if* all of its mass were concentrated into this single point. An object that can be represented as a mass at a single point in space is called a **particle.** A particle has no size, no shape, and no distinction between top and bottom or between front and back.

If we treat an object as a particle, we can represent the object in each frame of a motion diagram as a simple dot rather than having to draw a full picture. Figure 1.4 shows how much simpler motion diagrams appear when the object is represented as a particle. Note that the dots have been numbered 0, 1, 2, . . . to tell the sequence in which the frames were exposed. These diagrams are more abstract than the pictures, but they are easier to draw and they still convey our full understanding of the object's motion.

Treating an object as a particle is, of course, a simplification of reality. Such a simplification is called a **model.** Models allow us to focus on the important aspects of a phenomenon by excluding those aspects that play only a minor role. The **particle model** of motion is a simplification in which we treat a moving object as if all of its mass were concentrated at a single point. This might seem like an oversimplification, but if all we are concerned with is the motion of an object, it may not be. Using the particle model may allow us to see connections that are very important. Consider the motion of the two objects shown in Figure 1.5. These two very different objects have exactly the same motion diagram! As we will see, all objects falling under the influence of gravity move in exactly the same manner if no other forces act. The simplification of the particle model has revealed something about the physics that underlies both of these situations.

Not all motions can be reduced to the motion of a single point. Consider a rotating gear. The center of the gear doesn't move at all, and each tooth on the gear is moving in a different direction. Rotational motion is qualitatively different from motion along a line, and we'll need to go beyond the particle model later when we study rotational motion.

(a) Motion diagram of a car stopping

(b) Same motion diagram using the particle model

The same amount of time elapses between each frame and the next.

●   ●   ● ●
0   1   2 3

Numbers show the order in which the frames were taken.

A single dot is used to represent the object.

FIGURE 1.4 Simplifying a motion diagram using the particle model.

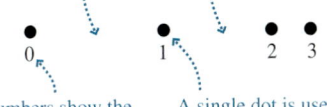

By using the particle model, a falling baseball and a diver are seen to have exactly the same motion diagram.

FIGURE 1.5 The particle model for two falling objects.

STOP TO THINK 1.2   Three motion diagrams are shown. Which is a dust particle settling to the floor at constant speed, which is a ball dropped from the roof of a building, and which is a descending rocket slowing to make a soft landing on Mars?

A.	B.	C.
0 ●	0 ●	0 ●
1 ●		
2 ●	1 ●	
		1 ●
3 ●	2 ●	
		2 ●
	3 ●	3 ●
4 ●		
	4 ●	4 ●
5 ●	5 ●	5 ●

1.2 Position and Time: Putting Numbers on Nature

To develop our understanding of motion further, we need to be able to make quantitative measurements. That is, we need to use numbers. As we look at a motion diagram, it would be useful to know where the object is (its *position*) and when the object was at that position (the *time*). We'll start by considering the motion of an object that can move only along a straight line. Examples of this **one-dimensional** or "1D" motion would be a bicyclist moving along the road, a train moving on a long straight track, or an elevator moving up and down a shaft.

Position and Coordinate Systems

Origin (post office)

Direction
W ←——→ E

Your position

|←— 4 miles —→|

FIGURE 1.6 Describing your position.

Suppose you are driving along a long, straight country road, as in Figure 1.6, and your friend calls and asks where you are. You might reply that you are four miles east of the post office, and your friend would then know just where you were. Your location at a particular instant in time (when your friend phoned) is called your **position.** Notice that to know your position along the road, your friend needed three pieces of information. First, you had to give her a reference point (the post office) from which all distances are to be measured. We call this fixed reference point the **origin.** Second, she needed to know how far you were from that reference point or origin—in this case, four miles. Finally, she needed to know which side of the origin you were on: You could be four miles to the west of it, or four miles to the east.

We will need these same three pieces of information in order to specify any object's position along a line. We first choose our origin, from which we measure the position of the object. The position of the origin is arbitrary, and we are free to place it where we like. Usually, however, there are certain points (such as the well-known post office) that are more convenient choices than others.

In order to specify how far our object is from the origin, we lay down an imaginary axis along the line of the object's motion. Like a ruler, this axis is marked off in equally spaced divisions of distance, perhaps in inches, meters, or miles, depending on the problem at hand. We place the zero mark of this ruler at the origin, allowing us to locate the position of our object by reading the ruler mark where the object is.

Finally, we need to be able to specify which side of the origin our object is on. To do this, we imagine the axis extending from one side of the origin with increasing, positive markings; on the other side, the axis is marked with increasing *negative* numbers. By reporting the position as either a positive or a negative number, we know on what side of the origin the object is.

These elements—an origin and an axis marked in both the positive and negative directions—can be used to unambiguously locate the position of an object. We call this a **coordinate system.** We will use coordinate systems throughout this

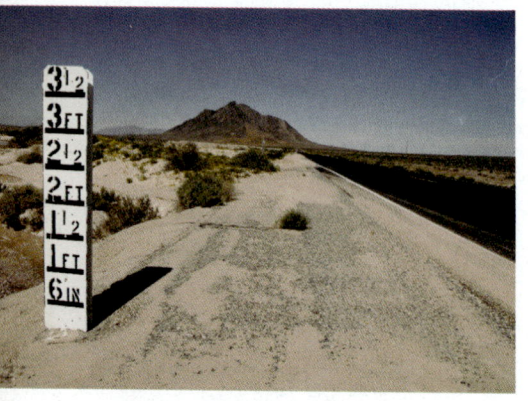

Sometimes measurements have a very natural origin. This snow depth gauge has its origin set at road level.

book, and we will soon develop coordinate systems that can be used to describe the position of objects moving in more complex ways than just along a line. Figure 1.7 shows a coordinate system that can be used to locate various objects along the country road discussed earlier.

Although our coordinate system works well for describing the positions of objects located along the axis, our notation is somewhat cumbersome. We need to keep saying things like "the car is at position $+4$ miles." A better notation, and one that will become particularly important when we study motion in two dimensions, is to use a symbol such as x or y to represent the position along the axis. Then we can say "the cow is at $x = -5$ miles." The symbol that represents a position along an axis is called a **coordinate.** The introduction of symbols to represent positions (and, later, velocities and accelerations) also allows us to work with these quantities mathematically.

Figure 1.8 shows how we would set up a coordinate system for a sprinter running a 50-meter race (we use the standard symbol "m" for meters). For horizontal motion like this we usually use the coordinate x to represent the position.

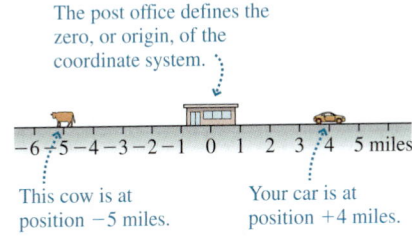

The post office defines the zero, or origin, of the coordinate system.

This cow is at position -5 miles.

Your car is at position $+4$ miles.

FIGURE 1.7 The coordinate system used to describe objects along a country road.

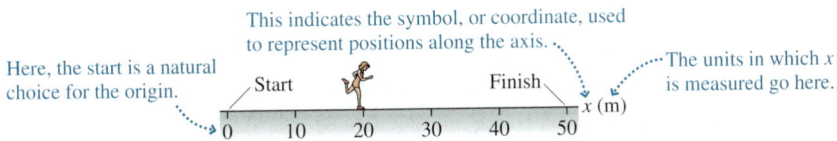

This indicates the symbol, or coordinate, used to represent positions along the axis.

Here, the start is a natural choice for the origin.

The units in which x is measured go here.

FIGURE 1.8 A coordinate system for a 50-meter race.

Motion along a straight line need not be horizontal. As shown in Figure 1.9, a rock falling vertically downward and a skier skiing down a straight slope are also examples of straight-line or one-dimensional motion.

Time

The pictures in Figure 1.9 show the position of an object at just one instant of time. But a full motion diagram represents how an object moves as time progresses. So far, we have labeled the dots in a motion diagram by the numbers 0, 1, 2, ... to indicate the order in which the frames were exposed. But to fully describe the motion, we need to indicate the *time,* as read off a clock or a stopwatch, at which each frame of a video was made. This is important, as we can see from the motion diagram of a stopping car in Figure 1.10. If the frames were taken one second apart, this motion diagram shows a leisurely stop; if 1/10 of a second apart, it represents a screeching halt.

For a complete motion diagram, we thus need to label each frame with its corresponding time (symbol t) as read off a clock. But when should we start the clock? That is, which frame should be labeled $t = 0$? This choice is much like that of choosing the origin $x = 0$ of a coordinate system: You can pick any arbitrary point in the motion and label it "$t = 0$ seconds." This is simply the instant you decide to start your clock or stopwatch, so it is the origin of your time coordinate. A video frame labeled "$t = 4$ seconds" means it was taken 4 seconds after you started your clock. We typically choose $t = 0$ to represent the "beginning" of a problem, but the object may have been moving before then.

To illustrate, Figure 1.11 shows the motion diagram for a car moving at a constant speed, and then braking to a halt. Two possible choices for the frame labeled $t = 0$ seconds are shown; our choice depends on what part of the motion we're interested in. Each successive position of the car is then labeled with the clock reading in seconds (abbreviated by the symbol "s").

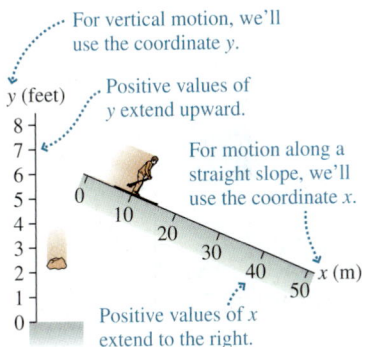

For vertical motion, we'll use the coordinate y.

Positive values of y extend upward.

For motion along a straight slope, we'll use the coordinate x.

Positive values of x extend to the right.

FIGURE 1.9 Examples of one-dimensional motion.

FIGURE 1.10 Is this a leisurely stop or a screeching halt?

If we're interested in the motion of the car from this point, we assign to this point the time $t = 0$ s.

Car starts braking here

$t = 0$ s 1 s   2 s 3 s4 s

If we're only interested in the braking part of the motion, we would assign $t = 0$ s here.

FIGURE 1.11 The motion diagram of a car that travels at constant speed and then brakes to a halt.

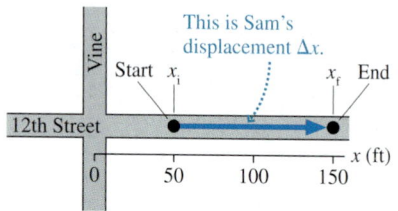

FIGURE 1.12 Sam undergoes a displacement Δx from position x_i to position x_f.

The size and the direction of the displacement both matter. Roy Riegels (pursued above by teammate Benny Lom) found this out in dramatic fashion in the 1928 Rose Bowl when he recovered a fumble and ran 69 yards—toward his own team's end zone. An impressive distance, but in the wrong direction!

FIGURE 1.13 A displacement is a signed quantity. Here Δx is a negative number.

Changes in Position and Displacement

Now that we've seen how to measure position and time, let's return to the problem of motion. To describe motion we'll need to measure the *changes* in position that occur with time. Consider the following:

Sam is standing 50 feet (ft) east of the corner of 12th Street and Vine. He then walks to a second point 150 ft east of Vine. What is Sam's change of position?

Figure 1.12 shows Sam's motion on a map. We've placed a coordinate system on the map, using the coordinate x. We are free to place the origin of our coordinate system wherever we wish, so we have placed it at the intersection. Sam's initial position is then at $x_i = 50$ ft. The positive value for x_i tells us that Sam is east of the origin.

> **NOTE ▶** We will label special values of x or y with subscripts. The value at the start of a problem is usually labeled with a subscript "i," for *initial,* and that at the end is labeled with a subscript "f," for *final.* For cases having several special values, we will usually use subscripts "1," "2," etc. ◀

Sam's final position is $x_f = 150$ ft, indicating that he is 150 feet east of the origin. You can see that Sam has changed position, and a *change* of position is called a **displacement.** His displacement is the distance labeled Δx in Figure 1.12. The Greek letter delta (Δ) is used in math and science to indicate the *change* in a quantity. Thus Δx indicates a change in the position x.

> **NOTE ▶** Δx is a *single* symbol. You cannot cancel out or remove the Δ in algebraic operations. ◀

To get from the 50 ft mark to the 150 ft mark, Sam clearly had to walk 100 ft, so the change in his position—his displacement—is 100 ft. We can think about displacement in a more general way, however. **Displacement is the *difference* between a final position x_f and an initial position x_i.** Thus we can write

$$\Delta x = x_f - x_i = 150 \text{ ft} - 50 \text{ ft} = 100 \text{ ft}$$

> **NOTE ▶** A general principle, used throughout this book, is that the change in any quantity is the final value of the quantity minus its initial value. ◀

Displacement is a *signed quantity.* That is, it can be either positive or negative. If, as shown in Figure 1.13, Sam's final position x_f had been at the origin instead of the 150 ft mark, his displacement would have been

$$\Delta x = x_f - x_i = 0 \text{ ft} - 50 \text{ ft} = -50 \text{ ft}$$

The negative sign tells us that he moved to the *left* along the x-axis, or 50 ft *west.*

Change in Time

A displacement is a change in position. In order to quantify motion, we'll need to also consider changes in *time,* which we call **time intervals.** We've seen how we can label each frame of a motion diagram with a specific time, as determined by our stopwatch. Figure 1.14 on the next page shows the motion diagram of a bicycle moving at a constant speed, with the times of measured points indicated.

The displacement between the initial position x_i and the final position x_f is

$$\Delta x = x_f - x_i = 120 \text{ ft} - 0 \text{ ft} = 120 \text{ ft}$$

Similarly, we define the time interval between these two points to be

$$\Delta t = t_f - t_i = 6 \text{ s} - 0 \text{ s} = 6 \text{ s}$$

A time interval Δt measures the elapsed time as an object moves from an initial position x_i at time t_i to a final position x_f at time t_f. Note that unlike Δx, Δt is always positive because t_f is always greater than t_i.

0 s 1 s 2 s 3 s 4 s 5 s 6 s

0   20   40   60   80 100 120 x (ft)

Initial position x_i Final position x_f

FIGURE 1.14 The motion diagram of a bicycle moving to the right at a constant speed.

STOP TO THINK 1.3   Sarah starts at a positive position along the x-axis. She then undergoes a negative displacement. Her final position

A. Is positive.
B. Is negative.
C. Could be either positive or negative.

1.3 Velocity

We all have an intuitive sense of whether something is moving very fast or just cruising slowly along. To make this intuitive idea more precise, let's start by examining the motion diagrams of some objects moving along a straight line at a *constant* speed, objects that are neither speeding up nor slowing down. This motion at a constant speed is called **uniform motion.** As we saw for the skateboarder in Section 1.1, for an object in uniform motion successive frames of the motion diagram are *equally spaced.* We know now that this means that the object's displacement Δx is the same between successive frames.

To see how an object's displacement between successive frames is related to its speed, consider the motion diagrams of a bicycle and a car, traveling along the same street as shown in Figure 1.15. Clearly the car is moving faster than the bicycle: In any one-second time interval, the car undergoes a displacement $\Delta x = 40$ ft, while the bicycle's displacement is only 20 ft.

The distances traveled in one second by the bicycle and the car are a measure of their speeds. The greater the distance traveled by an object in a given time interval, the greater its speed. This idea leads us to define the speed of an object as

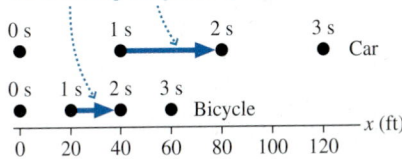

During each second, the car moves twice as far as the bicycle. Hence the car is moving at a greater speed.

0 s   1 s   2 s   3 s   Car

0 s 1 s 2 s 3 s   Bicycle

0   20   40   60   80 100 120 x (ft)

FIGURE 1.15 Motion diagrams for a car and a bicycle.

$$\text{speed} = \frac{\text{distance traveled in a given time interval}}{\text{time interval}} \qquad (1.1)$$

Speed of a moving object

For the bicycle, this gives

$$\text{speed} = \frac{20 \text{ ft}}{1 \text{ s}} = 20 \frac{\text{ft}}{\text{s}}$$

while for the car we have

$$\text{speed} = \frac{40 \text{ ft}}{1 \text{ s}} = 40 \frac{\text{ft}}{\text{s}}$$

The speed of the car is twice that of the bicycle, which seems reasonable.

NOTE ▶ The division gives units that are a fraction: ft/s. This is read as "feet per second," just like the more familiar "miles per hour." ◀

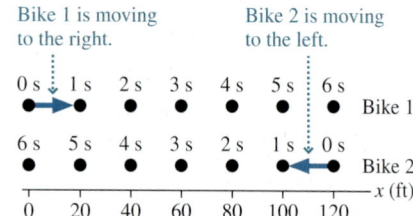

FIGURE 1.16 Two bicycles traveling at the same speed, but with different velocities.

To fully characterize the motion of an object, it is important to specify not only the object's speed but also the *direction* in which it is moving. For example, Figure 1.16 shows the motion diagrams of two bicycles traveling at the same speed of 20 ft/s. The two bicycles have the same speed. but something about their motion is different—the *direction* of their motion.

The problem is that the "distance traveled" in Equation 1.1 doesn't capture any information about the direction of travel. But we've seen that the *displacement* of an object does contain this information. We can then introduce a new quantity, the **velocity,** as

$$\text{velocity} = \frac{\text{displacement}}{\text{time interval}} = \frac{\Delta x}{\Delta t} \qquad (1.2)$$

Velocity of a moving object

The velocity of bicycle 1 in Figure 1.16, computed using the one-second time interval between the $t = 2$ s and $t = 3$ s positions, is

$$v = \frac{\Delta x}{\Delta t} = \frac{x_3 - x_2}{3 \text{ s} - 2 \text{ s}} = \frac{60 \text{ ft} - 40 \text{ ft}}{1 \text{ s}} = +20 \frac{\text{ft}}{\text{s}}$$

while the velocity for bicycle 2, during the same time interval, is

$$v = \frac{\Delta x}{\Delta t} = \frac{x_3 - x_2}{3 \text{ s} - 2 \text{ s}} = \frac{60 \text{ ft} - 80 \text{ ft}}{1 \text{ s}} = -20 \frac{\text{ft}}{\text{s}}$$

NOTE ▶ We have used x_2 for the position at time $t = 2$ seconds, and x_3 for the position at time $t = 3$ seconds. The subscripts 2 and 3 serve the same role as before—identifying particular positions—but in this case the positions are identified by the time at which each position is reached. ◀

The two velocities have opposite signs because the bicycles are traveling in opposite directions. **Speed measures only how fast an object moves, but velocity tells us both an object's speed *and its direction*.** A positive velocity indicates motion to the right or, for vertical motion, upward. Similarly, an object moving to the left, or down, has a negative velocity.

NOTE ▶ Learning to distinguish between speed, which is always a positive number, and velocity, which can be either positive or negative, is one of the most important tasks in the analysis of motion. ◀

The velocity as defined by Equation 1.2 is actually what is called the *average* velocity. On average, over each 1 s interval bicycle 1 moves 20 ft, but we don't know if it sped up or slowed down a little during that 1 s. In Chapter 2, we'll develop the idea of *instantaneous* velocity, the velocity of an object at a particular instant in time. Since our goal in this chapter is to *visualize* motion with motion diagrams, we'll somewhat blur the distinction between average and instantaneous quantities, refining these definitions in Chapter 2, where our goal will be to develop the mathematics of motion.

The "Per" in Meters Per Second

The units for speed and velocity are those of a distance (feet, meters, miles) divided by a unit of time (seconds, hours). Thus we could measure velocity in units of m/s or mph, pronounced "meters *per* second" and "miles *per* hour." The word "per" will often arise in physics when we consider the ratio of two quantities. What do we mean, exactly, by "per"?

If a car moves with a speed of 23 m/s, we mean that it travels 23 meters *for each* 1 second of elapsed time. The word "per" thus associates the number of units in the numerator (23 m) with *one* unit of the denominator (1 s). We'll see many other examples of this idea as the book progresses. You may already know a bit about *density;* you can look up the density of gold and you'll find that it is 19.3 g/cm³ ("grams *per* cubic centimeter"). This means that there are 19.3 grams of gold *for each* 1 cubic centimeter of the metal. Thinking about the word "per" in this way will help you better understand physical quantities whose units are the ratio of two other units.

1.4 A Sense of Scale: Significant Figures, Scientific Notation, and Units

Physics attempts to explain the natural world, from the very small to the exceedingly large. And in order to understand our world, we need to be able to *measure* quantities both minuscule and enormous. A properly reported measurement has three elements. First, we can only measure our quantity with a certain precision. To make this precision clear, we need to make sure that we report our measurement with the correct number of *significant figures.*

Second, writing down the really big and small numbers that often come up in physics can be awkward. To avoid writing all those zeros, scientists use *scientific notation* to express numbers both big and small.

Finally, we need to choose an agreed-upon set of *units* for the quantity. For speed, common units include meters per second and miles per hour. For mass, the kilogram is the most commonly used unit. Later, we'll study more esoteric quantities such as magnetic fields, which have the units of "tesla." Every physical quantity that we can measure has an associated set of units.

Measurements and Significant Figures

When we measure any quantity, such as the length of a bone or the weight of a specimen, we can do so only with a certain *precision.* The digital calipers in Figure 1.17 can make a measurement to within ±0.001 in, so they have a precision of 0.001 in. If you used the tape shown to make a measurement, you probably couldn't do so to better than about ±1 mm, so the precision of the tape measure is about 1 mm. The precision of a measurement can also be affected by the skill or judgment of the person performing the measurement. A stopwatch might have a precision of 0.001 s, but, due to your reaction time, your measurement of the time of a sprinter would be much less precise.

It is important that your measurement be reported in a way that reflects its actual precision. Suppose, for example, that you use a ruler to measure the length of a particular specimen of a newly discovered species of frog. You judge that you can make this measurement with a precision of about 1 mm, or 0.1 cm. In this case, the frog's length should be reported as, say, 6.2 cm. We interpret this to mean that the actual value falls between 6.15 cm and 6.25 and thus rounds to 6.2 cm. Reporting the frog's length as simply 6 cm is saying less than you know; you are withholding information. On the other hand, to report the number as 6.213 cm is wrong. Any person reviewing your work would interpret the number 6.213 cm as meaning that the actual length falls between 6.2125 cm and 6.2135 cm, thus rounding to 6.213 cm. In this case, you are claiming to have knowledge and information that you do not really possess.

This example shows that one way to state your knowledge precisely is through the proper use of **significant figures.** You can think of a significant figure as being a digit that is reliably known. A measurement such as 6.2 cm has *two* significant figures, the 6 and the 2. The next decimal place—the one-hundredths—is

300 million light years
= 2.8×10^{24} m

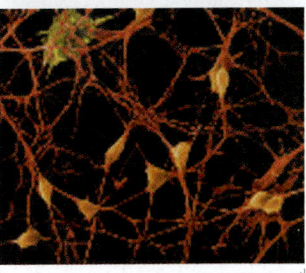

$120 \ \mu m = 1.2 \times 10^{-4}$ m

From galaxies to cells . . . BIO In science, we need to express numbers both very large and very small. The top image is of a computer simulation of the structure of the universe. Bright areas represent regions of clustered galaxies. The bottom image is of cortical nerve cells. Nerve cells relay signals to each other through a complex web of dendrites. These images, though similar in appearance, differ in scale by a factor of about 2×10^{28}!

These calipers have a precision of 0.001 in.

A tape measure has a precision of about 1 mm.

FIGURE 1.17 The precision of a measurement depends on how the measurement is made.

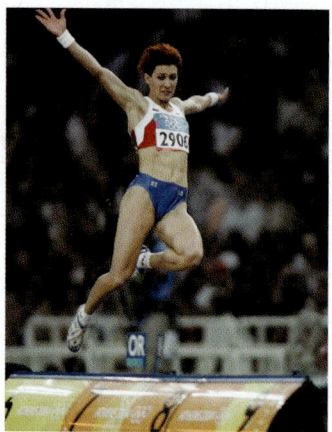

Tatyana Lebedeva won the 2004 Olympic gold medal with a long jump reported as 7.07 m. This number implies that the long jump is measured to an accuracy of about 0.01 m, or 1 cm.

not reliably known, and is thus not a significant figure. Similarly, a time measurement of 34.62 s has four significant figures, implying that the 2 in the hundredths place is reliably known.

When we perform a calculation such as adding or multiplying two or more measured numbers, we can't claim more accuracy for the result than was present in the initial measurements. Calculations with measured numbers follow the "weakest link" rule. The saying, which you probably know, is that "a chain is only as strong as its weakest link." If nine out of ten links in a chain can support a 1000 pound weight, that strength is meaningless if the tenth link can support only 200 pounds. Similarly, nine out of the ten numbers used in a calculation might be known with a precision of 0.01%; but if the tenth number is poorly known, with a precision of only 10%, then the result of the calculation cannot possibly be more precise than 10%. The weak link rules!

Determining the proper number of significant figures is straightforward, but there are a few definite rules to follow. We will often spell out such technical details in what we call a "Tactics Box." A Tactics Box is designed to teach you particular skills and techniques. Each Tactics Box will use the ✐ icon to designate exercises in the *Student Workbook* that you can use to practice these skills.

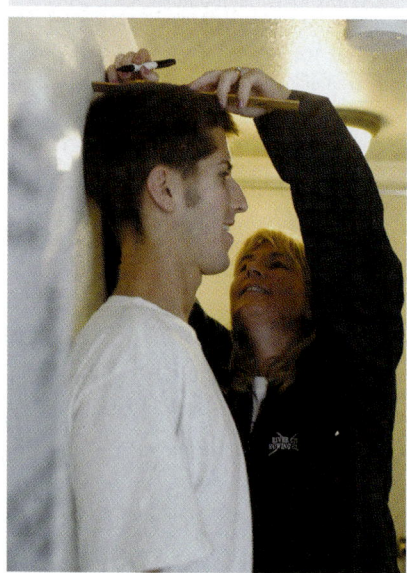

How tall are you really? If you measure your height in the morning, just after you wake up, and then in the evening, after a full day of activity, you'll find that your evening height is *less* by as much as 3/4 inches. Your height decreases over the course of the day as gravity compresses and reshapes your spine. If you give your height as 66 3/16 in, you are claiming more significant figures than are truly warranted; the 3/16 in isn't really reliably known, as your height can vary by more than this. Expressing your height to the nearest inch is plenty!

(MP)   **TACTICS BOX 1.1 Using significant figures**   ✐ Exercise 16

❶ When multiplying or dividing several numbers, or when taking roots, the number of significant figures in the answer should match the number of significant figures of the *least* precisely known number used in the calculation:

Three significant figures

$$3.73 \times 5.7 = 21$$

Two significant figures

Answer should have the *lower* of the two, or two significant figures.

❷ When adding or subtracting several numbers, the number of decimal places in the answer should match the *smallest* number of decimal places of any number used in the calculation.

$$
\begin{array}{r}
18.54 \\
+\,106.6 \\
\hline
=\,125.1
\end{array}
$$

18.54 — Two decimal places
+106.6 — One decimal place
=125.1 — Answer should have the *lower* of the two, or one decimal place.

❸ **Exact numbers** have no uncertainty and, when used in calculations, do not change the number of significant figures of measured numbers. Examples of exact numbers are π and the number 2 in the relation $d = 2r$ between a circle's diameter and radius.

There is one notable exception to these rules:

■ It is acceptable to keep one or two extra digits during *intermediate* steps of a calculation. The goal here is to minimize round-off errors in the calculation. But the *final* answer must be reported with the proper number of significant figures.

EXAMPLE 1.1 Measuring the velocity of a car

To measure the velocity of a car, clocks A and B are set up at two points along the road, as shown in Figure 1.18. Clock A is precise to 0.01 s, while B is precise only to 0.1 s. The distance between these two clocks is carefully measured to be 124.5 m. The two clocks are automatically started when the car passes a trigger in the road; each clock stops automatically when the car passes that clock. After the car has passed both clocks, clock A is found to read $t_A = 1.22$ s, and clock B to read $t_B = 4.5$ s. The time from the less-precise clock B is correctly reported with fewer significant figures than that from A. What is the velocity of the car, and how should it be reported with the correct number of significant figures?

Both clocks start when the car crosses this trigger.

$\Delta x = 124.5$ m

FIGURE 1.18 Measuring the velocity of a car.

SOLVE We've already determined the car's displacement $\Delta x = 124.5$ m as it moves between the two clocks. To calculate its velocity, we need to determine the time interval Δt.

This is

This number has one decimal place.   This number has two decimal places.

$$\Delta t = t_B - t_A = (4.5 \text{ s}) - (1.22 \text{ s}) = 3.3 \text{ s}$$

By rule 2 of Tactics Box 1.1, the result should have *one* decimal place.

We can now calculate the velocity with the displacement and time interval from above:

The displacement has four significant figures.

$$v = \frac{\Delta x}{\Delta t} = \frac{124.5 \text{ m}}{3.3 \text{ s}} = 38 \text{ m/s}$$

The time interval has two significant figures.   By rule 1 of Tactics Box 1.1, the result should have *two* significant figures.

ASSESS Our final value has two significant figures. Suppose you had been hired to measure the speed of a car this way, and you reported 37.72 m/s. It would be reasonable for someone looking at your result to assume that the measurements you used to arrive at this value were correct to four significant figures and thus that you had measured time to the nearest 0.01 second. Our correct result of 38 m/s has all of the accuracy that you can claim, but no more!

Scientific Notation

It's easy to write down measurements of ordinary-sized objects: your height might be 1.72 meters, the weight of an apple 0.34 pounds. But the radius of a hydrogen atom is 0.000 000 000 053 m and the distance to the moon is 384 000 000 m. Writing and keeping track of all those zeros is quite cumbersome.

NOTE ▶ Scientists usually write numbers with many digits by arranging the digits in groups of three, with the groups separated by spaces instead of commas. This makes it easier to read long numbers. ◀

Beyond requiring you to deal with all the zeros, writing quantities this way makes it unclear how many significant figures are involved. In the distance to the moon given above, how many of those digits are significant? Three? Four? All nine?

Writing numbers using scientific notation avoids both these problems. A value in scientific notation is a number with one digit to the left of the decimal point and zero or more to the right of it, multiplied by a power of ten. This solves the problem of all the zeros and makes the number of significant digits immediately apparent. In scientific notation, writing the distance to the sun as 1.50×10^{11} m implies that three digits are significant; writing it as 1.5×10^{11} m implies that only two digits are.

Even for smaller values, scientific notation can clarify the number of significant figures. Suppose a distance is measured as 1200 m. If this distance is known to within 1 m, we could write it as 1.200×10^3 m, showing that all four digits are significant; if it were accurate to only 100 m or so, we would report it as 1.2×10^3 m, indicating two significant figures.

Tactics Box 1.2 shows how to convert a number to scientific notation, and how to correctly indicate the number of significant figures.

Exercises 17, 18

TACTICS BOX 1.2 Using scientific notation

To convert a number into scientific notation:

❶ For a number greater than 10, move the decimal point to the left until only one digit remains to the left of the decimal point. The remaining number is then multiplied by 10 to a power; this power is given by the number of spaces the decimal point was moved. Here we convert the diameter of the earth to scientific notation:

We move the decimal point until there is only one digit to its left, counting the number of steps.

Since we moved the decimal point 6 steps, the power of ten is 6.

$$6\,370\,000 \text{ m} = 6.37 \times 10^6 \text{ m}$$

The number of digits here equals the number of significant figures.

❷ For a number less than 1, move the decimal point to the right until it passes the first digit that isn't a zero. The remaining number is then multiplied by 10 to a negative power; the power is given by the number of spaces the decimal point was moved. For the diameter of a red blood cell we have

We move the decimal point until it passes the first digit that is not a zero, counting the number of steps.

Since we moved the decimal point 6 steps, the power of ten is −6.

$$0.000\,007\,5 \text{ m} = 7.5 \times 10^{-6} \text{ m}$$

The number of digits here equals the number of significant figures.

Proper use of significant figures is part of the "culture" of science. We will frequently emphasize these "cultural issues" because you must learn to speak the same language as the natives if you wish to communicate effectively! Most students "know" the rules of significant figures, having learned them in high school, but many fail to apply them. It is important that you understand the reasons for significant figures and that you get in the habit of using them properly.

Units

As we have seen, in order to measure a quantity we need to give it a numerical value. But a measurement is more than just a number—it requires a *unit* to be given. You can't go to the grocery and ask for "three-and-a-half of flour." You need to use a unit—here, one of weight, such as pounds—in addition to the number.

In your daily life, you generally tend to use the English system of units, in which distances are measured in inches, feet, and miles. These units are well adapted for daily life, but they are rarely used in scientific work. Given that science is an international discipline, it is also important to have a system of units that is recognized around the world. For these reasons, scientists use a system of units called *le Système Internationale d'Unités,* commonly referred to as **SI units.** In casual speaking we often refer to these as *metric units,* because the meter is the basic standard of length.

The importance of units In 1999, the $125 million Mars Climate Orbiter burned up in the Martian atmosphere instead of entering a safe orbit from which it could perform observations. The problem was faulty units! An engineering team had provided critical data on spacecraft performance in English units, but the navigation team assumed these data were in metric units. As a consequence, the navigation team had the spacecraft fly too close to the planet, and it burned up in the atmosphere.

The three basic SI quantities, shown in Table 1.1, are time, length (or distance), and mass. Other quantities needed to understand motion can be expressed as combinations of these basic units. For example, speed or velocity are expressed in meters per second or m/s. This combination is a ratio of the length unit (the meter) to the time unit (the second).

The SI units have a long and interesting history. SI units were originally developed by the French in the late 1700s as a way of standardizing and regularizing numbers for commerce and science. Some of their other innovations of the time did not survive (such as the 10-day week), but their units did.

TABLE 1.1 Common SI units

Quantity	Unit	Abbreviation
time	second	s
length	meter	m
mass	kilogram	kg

Using Prefixes

We will have many occasions to use lengths, times, and masses that are either much less or much greater than the standards of 1 meter, 1 second, and 1 kilogram. We will do so by using *prefixes* to denote various powers of ten. For instance, the prefix "kilo" (abbreviation k) denotes 10^3, or a factor of 1000. Thus 1 km equals 1000 m, 1 MW equals 10^6 watts, and 1 μV equals 10^{-6} V. Table 1.2 lists the common prefixes that will be used frequently throughout this book. Memorize it! Few things in physics are learned by rote memory, but this list is one of them. A more extensive list of prefixes is shown inside the cover of the book.

Although prefixes make it easier to talk about quantities, the proper SI units are meters, seconds, and kilograms. Quantities given with prefixed units must be converted to base SI units before any calculations are done. Thus 23.0 cm must be converted to 0.230 m before starting calculations. The exception is the kilogram, which is already the base SI unit.

TABLE 1.2 Common prefixes

Prefix	Abbreviation	Power of 10
mega-	M	10^6
kilo-	k	10^3
centi-	c	10^{-2}
milli-	m	10^{-3}
micro-	μ	10^{-6}
nano-	n	10^{-9}

Unit Conversions

Although SI units are our standard, we cannot entirely forget that the United States still uses English units. Even after repeated exposure to metric units in classes, most of us "think" in the English units we grew up with. Thus it remains important to be able to convert back and forth between SI units and English units. Table 1.3 shows a few frequently used conversions that will come in handy.

One effective method of performing unit conversions begins by noticing that since, for example, 1 mi = 1.609 km, the ratio of these two distances—*including their units*—is equal to one, so that

$$\frac{1 \text{ mi}}{1.609 \text{ km}} = \frac{1.609 \text{ km}}{1 \text{ mi}} = 1$$

A ratio of values equal to one is called a **conversion factor.** The following Tactics Box shows how to make a unit conversion. It uses the example of converting 60 mi into the equivalent distance in km.

TABLE 1.3 Useful unit conversions

1 inch (in) = 2.54 cm
1 foot (ft) = 0.305 m
1 mile (mi) = 1.609 km
1 mile per hour (mph) = 0.447 m/s
1 m = 39.37 in
1 km = 0.621 mi
1 m/s = 2.24 mph

TACTICS BOX 1.3 Making a unit conversion   Exercise 21

❶ Start with the quantity you wish to convert.

❷ Multiply by the appropriate conversion factor. Because this conversion factor is equal to one, multiplying by it does not change the value of the quantity—only its units.

❺ Remember to convert your final answer to the correct number of significant figures!

$$60 \text{ mi} = 60 \text{ mi} \times \frac{1.609 \text{ km}}{1 \text{ mi}} = 96.54 \text{ km} = 97 \text{ km}$$

❸ You can cancel the original unit (here, miles) because it appears in both the numerator and the denominator.

❹ Calculate the answer; it is in the desired units. Remember, 60 mi and 96.54 km are the same distance; they are simply in different units.

Note that we've rounded the answer to 97 kilometers because the distance we're converting, 60 miles, has only two significant figures.

More complicated conversions can be accomplished with several successive multiplications of conversion factors, as we see in the next example.

EXAMPLE 1.2 Can a bicycle go that fast?

In Section 1.3, we calculated the speed of a bicycle to be 20 ft/s. Is this a reasonable speed for a bicycle?

SOLVE In order to determine whether or not this speed is reasonable, we will convert it to more familiar units. For speed, the unit you are most familiar with is likely miles per hour.

We first collect the necessary unit conversions:

$$1 \text{ mi} = 5280 \text{ ft} \qquad 1 \text{ hour (1 h)} = 60 \text{ min} \qquad 1 \text{ min} = 60 \text{ s}$$

We then multiply our original value by successive factors of 1 in order to convert the units:

We want to cancel feet here in the numerator . . .

. . . so we multiply by $1 = \dfrac{1 \text{ mi}}{5280 \text{ ft}}$ to get the feet in the denominator.

$$20\,\frac{\text{ft}}{\text{s}} = 20\,\frac{\text{ft}}{\text{s}} \times \frac{1 \text{ mi}}{5280 \text{ ft}} \times \frac{60 \text{ s}}{1 \text{ min}} \times \frac{60 \text{ min}}{1 \text{ h}} = 14\,\frac{\text{mi}}{\text{h}} = 14 \text{ mph}$$

The unwanted units cancel in pairs, as indicated by the colors.

Our final result of 14 miles per hour (14 mph) is a very reasonable speed for a bicycle, which gives us confidence in our answer. If we had calculated a speed of 140 miles per hour, we would have suspected that we had made an error, as this is quite a bit faster than the average bicyclist can travel!

How many jellybeans are in the jar? Some reasoning about the size of one bean and the size of the jar can give you a one-significant-figure estimate.

TABLE 1.4 Some approximate lengths

	Length (m)
Circumference of the earth	4×10^7
New York to Los Angeles	5×10^6
Distance you can drive in 1 hour	1×10^5
Altitude of jet planes	1×10^4
Distance across a college campus	1000
Length of a football field	100
Length of a classroom	10
Length of your arm	1
Width of a textbook	0.1
Length of your little fingernail	0.01
Diameter of a pencil lead	1×10^{-3}
Thickness of a sheet of paper	1×10^{-4}
Diameter of a dust particle	1×10^{-5}

Estimation

When scientists and engineers first approach a problem, they may do a quick measurement or calculation to establish the rough physical scale involved. This will help establish the procedures that should be used to make a more accurate measurement—or the estimate may well be all that is needed.

Suppose you see a rock fall off a cliff and would like to know how fast it was going when it hit the ground. By doing a mental comparison with the speeds of familiar objects, such as cars and bicycles, you might judge that the rock was traveling at "about" 20 mph. This is a one-significant-figure estimate. With some luck, you can probably distinguish 20 mph from either 10 mph or 30 mph, but you certainly cannot distinguish 20 mph from 21 mph just from a visual appearance. A one-significant-figure estimate or calculation, such as this estimate of speed, is called an **order-of-magnitude estimate.** An order-of-magnitude estimate is indicated by the symbol $\sim$ which indicates even less precision than the "approximately equal" symbol $\approx$. You would report your estimate of the speed of the falling rock as $v \sim 20$ mph.

A useful skill is to make reliable order-of-magnitude estimates on the basis of known information, simple reasoning, and common sense. This is a skill that is acquired by practice. Several chapters in this book will have homework problems that ask you to make order-of-magnitude estimates. Tables 1.4 and 1.5 have information that will be useful for doing estimates.

Later in the book, we will do some analysis of locomotion and look at the walking and running speeds of different animals. To help put things in perspective, it might be useful to have an estimate of how fast a person walks.

EXAMPLE 1.3 How fast do you walk?

Estimate how fast you walk, in meters per second.

SOLVE In order to compute speed, we will need a distance and a time. If you walked a mile to campus, how long would this take? You'd probably say 30 minutes or so—a half an hour. Let's use this rough number in our estimate.

We have

$$\text{speed} = \frac{\text{distance}}{\text{time}} \sim \frac{1 \text{ mile}}{1/2 \text{ hour}} = 2 \frac{\text{mi}}{\text{h}}$$

But we want this in meters per second. Since our calculation is only an estimate, we use an approximate form of the conversion factor from Table 1.3:

$$1 \frac{\text{mi}}{\text{h}} \approx 0.5 \frac{\text{m}}{\text{s}}$$

This gives an approximate walking speed of about 1 m/s.

ASSESS Is this a reasonable value? Let's do another estimate. Your stride is probably about one yard long—about one meter. And you take about one step per second; next time you are walking, you can count and see. So a walking speed of 1 meter per second sounds pretty reasonable.

This sort of estimation is very valuable. We will see many cases in which we need to know an approximate value for a quantity before we start a problem or after we finish a problem, in order to assess our results.

> **STOP TO THINK 1.4** Rank in order, from the most to the fewest, the number of significant figures in the following numbers. For example, if B has more than C, C has the same number as A, and A has more than D, you would give your answer as B > C = A > D.
>
> A. 0.43   B. 0.0052   C. 0.430   D. 4.321×10^{-10}

1.5 Vectors and Motion: A First Look

Many physical quantities, such as time, mass, and temperature, can be described completely by a number with a unit. For example, the mass of an object might be 6 kg and its temperature 30° C. When a physical quantity is described by a single number (with a unit), we call it a **scalar quantity.** A scalar can be positive, negative, or zero.

Many other quantities, however, have a directional quality and cannot be described by a single number. To describe the motion of a car, for example, you must specify not only how fast it is moving, but also the *direction* in which it is moving. A **vector quantity** is a quantity that has both a *size* (the "How far?" or "How fast?") and a *direction* (the "Which way?"). The size or length of a vector is called its **magnitude.** The magnitude of a vector can be positive or zero, but it cannot be negative.

Some examples of vector and scalar quantities are given at right.

We graphically represent a vector as an *arrow,* as illustrated for the velocity and force vectors in the table at right. The arrow is drawn to point in the direction of the vector quantity, and the *length* of the arrow is proportional to the magnitude of the vector quantity. If we choose to draw an arrow 2 cm long to represent a velocity with magnitude 23 m/s, we will draw an arrow 4 cm long to represent a velocity with magnitude 46 m/s. This graphical notation for representing vectors

TABLE 1.5 Some approximate masses

	Mass (kg)
Large airliner	1×10^5
Small car	1000
Large human	100
Medium-size dog	10
Science textbook	1
Apple	0.1
Pencil	0.01
Raisin	1×10^{-3}
Fly	1×10^{-4}

Vectors and scalars

Scalars

Time, temperature, and weight are all *scalar* quantities. To specify your weight, only one number—150 pounds—need be given. The temperature is reported by a single number, such as 70° F, and the time of day is really just the number of seconds after midnight, although we break it into hours and minutes for convenience.

Vectors

The velocity of the race car is a *vector.* To fully specify a velocity, we need to give its magnitude (e.g., 120 mph) *and* its direction (e.g., west).

The force with which the boy pushes on his friend is another example of a vector. To completely specify this force, we must know not only how hard he pushes (the magnitude) but also in which direction.

is so useful that we often think of the arrow as being the vector itself. Thus we might say, "draw a vector pointing to the right," and you would draw an arrow pointing to the right.

When we want to represent a vector quantity with a *symbol,* we need somehow to indicate that the symbol is for a vector rather than for a scalar. We do this by drawing an arrow over the letter that represents the quantity. Thus $\vec{r}$ and $\vec{A}$ are symbols for vectors, whereas r and A, without the arrows, are symbols for scalars. In handwritten work you *must* draw arrows over all symbols that represent vectors. This may seem strange until you get used to it, but it is very important because we will often use both r and $\vec{r}$, or both A and $\vec{A}$, in the same problem, and they mean different things! Without the arrow, you will be using the same symbol with two different meanings and will likely end up making a mistake. Note that the arrow over the symbol always points to the right, regardless of which direction the actual vector points. Thus we write $\vec{r}$ or $\vec{A}$, never $\overleftarrow{r}$ or $\overleftarrow{A}$.

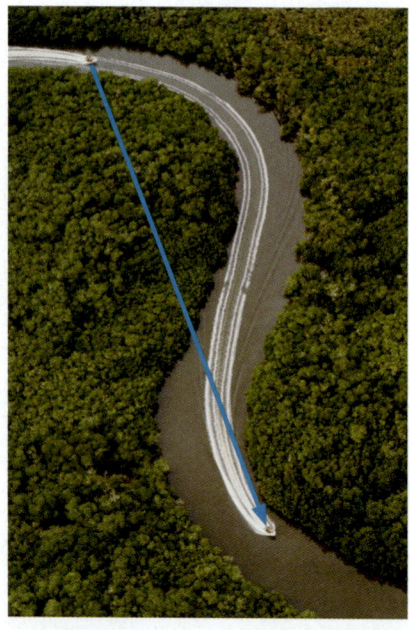

FIGURE 1.19 Two displacement vectors.

Displacement Vectors

For motion along a line, we found in Section 1.2 that the displacement is a quantity that specifies not only how *far* an object moves, but also the *direction*—to the left or to the right—that the object moves. Since displacement is a quantity that has both a magnitude ("How far") and a direction, it can be represented by a vector, the **displacement vector.** Figure 1.19 shows the displacement vector for Sam's trip that we discussed earlier. We've simply drawn an arrow—the vector—from his initial to his final positions and assigned it the symbol $\vec{d}_S$. Because $\vec{d}_S$ has both a magnitude and a direction, it is convenient to write Sam's displacement as $\vec{d}_S = (100 \text{ ft, east})$. The first value in the parentheses is the magnitude of the vector (i.e., the size of the displacement) and the second value specifies its direction.

Also shown in Figure 1.19 is the displacement vector $\vec{d}_J$ for Jane, who started on 12th Street and ended up on Vine. As with Sam, we draw her displacement vector as an arrow from her initial to her final position. In this case, $\vec{d}_J = (100 \text{ ft, } 30° \text{ east of north})$.

Jane's trip illustrates an important point about displacement vectors. Jane started her trip on 12th Street and ended up on Vine, leading to the displacement vector shown. But to get from her initial to her final position, she needn't have walked along the straight-line path denoted by $\vec{d}_J$. If she walked east along 12th Street to the intersection, then headed north on Vine, her displacement would still be the vector shown. **An object's displacement vector is drawn from the object's initial position to its final position, regardless of the actual path followed between these two points.**

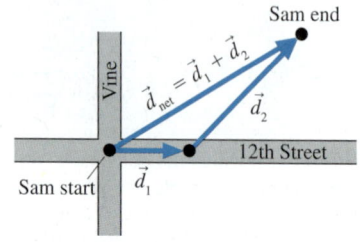

The boat's displacement is the straight-line connection from its initial to its final position.

Vector Addition

Let's consider one more trip for the peripatetic Sam. In Figure 1.20, he starts at the intersection and walks east 50 ft; then he walks 100 ft to the northeast through a vacant lot. His displacement vectors for the two legs of his trip are labeled $\vec{d}_1$ and $\vec{d}_2$ in the figure.

Sam's trip consists of two legs that can be represented by the two vectors $\vec{d}_1$ and $\vec{d}_2$, but we can represent his trip as a whole, from his initial starting position to his overall final position, with the *net* displacement vector labeled $\vec{d}_{net}$. Sam's net displacement is in a sense the *sum* of the two displacements that made it up, so we can write

$$\vec{d}_{net} = \vec{d}_1 + \vec{d}_2$$

Sam's net displacement thus requires the *addition* of two vectors, but vector addition obeys different rules from the addition of two scalar quantities. The directions of the two vectors, as well as their magnitudes, must be taken into account. Sam's trip suggests that we can add vectors together by putting the "tail" of one

FIGURE 1.20 Sam undergoes two displacements.

vector at the tip of the other. This idea, which is reasonable for displacement vectors, in fact is how *any* two vectors are added. Tactics Box 1.4 shows how to add two vectors $\vec{A}$ and $\vec{B}$.

TACTICS BOX 1.4 Vector addition   ✏ Exercise 22

To add $\vec{B}$ to $\vec{A}$:

❶ Draw $\vec{A}$.

❷ Place the tail of $\vec{B}$ at the tip of $\vec{A}$.

❸ Draw an arrow from the tail of $\vec{A}$ to the tip of $\vec{B}$. This is vector $\vec{A} + \vec{B}$.

EXAMPLE 1.4 Finding Sam's net displacement

Find Sam's net displacement in Figure 1.20, writing it in the form $\vec{d}_{net}$ = (magnitude of displacement, direction).

SOLVE We'll solve this graphically, using a ruler and a protractor. As shown in Figure 1.21, we first draw vector $\vec{d}_1$ pointing to the east, or to the right on our paper. We'll choose a scale where 1 cm on our paper represents 25 ft of Sam's neighborhood. Thus the length of $\vec{d}_1$ on the paper is 2 cm, representing the 50 ft magnitude of Sam's first displacement.

We then draw the second vector $\vec{d}_2$ with its tail at the tip of $\vec{d}_1$. Sam walked to the northeast during this leg, so we draw the direction of the vector at 45° to the horizontal; since he walked a distance of 100 ft, we draw the vector with a length of 4 cm.

The net displacement is the vector sum of the two displacements $\vec{d}_1$ and $\vec{d}_2$. It extends from the tail of $\vec{d}_1$ to the tip of $\vec{d}_2$. Using a ruler, we measure its length to be about 5.6 cm, corresponding to 5.6 × 25 ft = 140 ft. We can use a protractor to

find that the angle θ (the Greek letter *theta*) is about 30°. We thus have

$$\vec{d}_{net} = (140 \text{ ft}, 30° \text{ north of east})$$

FIGURE 1.21 Graphical addition of two vectors.

Vectors and Trigonometry

Adding two vectors together using a ruler and protractor isn't very precise or practical. We need a more accurate and efficient method that can be used for adding any two vectors. Trigonometry provides us with just such a method.

Before seeing how trigonometry can be used for vector addition, let's review some of the basic ideas. Suppose we have a right triangle with hypotenuse H, angle θ, side opposite the angle O, and side adjacent to the angle A, as shown in Figure 1.22.

The sine, cosine, and tangent (which we write as "sin," "cos," and "tan") of angle θ are defined as ratios of the sides of the triangle:

$$\sin\theta = \frac{O}{H} \qquad \cos\theta = \frac{A}{H} \qquad \tan\theta = \frac{O}{A} \qquad (1.3)$$

If we know two of the sides of the triangle, we can find the angle θ.

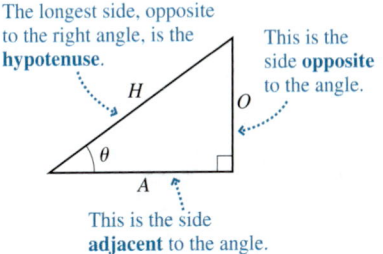

The longest side, opposite to the right angle, is the **hypotenuse**.

This is the side **opposite** to the angle.

This is the side **adjacent** to the angle.

FIGURE 1.22 A right triangle.

Conversely, if we know the angle θ and the length of one side, we can use the sine, cosine, or tangent to find the lengths of the other sides. For example, if you know θ and the length A of the adjacent side, and you need to find the hypotenuse H, you can rearrange the middle Equation 1.3 to give $H = A/\cos\theta$.

We will make regular use of these relationships in the following chapters.

EXAMPLE 1.5 Determining the sides of a triangle

A right triangle has an angle of 30° and a hypotenuse of length 10.0 m, as shown in Figure 1.23. What are the lengths of the other two sides of the triangle?

FIGURE 1.23 A triangle with two unknown sides.

SOLVE This is a problem that uses the trigonometric relationships of Equations (1.3). In Figure 1.23, the hypotenuse is labeled L, the adjacent side B, and the opposite side A. For each problem, the hypotenuse and adjacent and opposite sides will need to be determined; they will not in general be labeled H, O, and A as in Equations (1.3).

Because we know the hypotenuse and an angle, we can compute

$$A = L\sin\theta = (10.0 \text{ m})\sin(30°) = 5.00 \text{ m}$$

$$B = L\cos\theta = (10.0 \text{ m})\cos(30°) = 8.66 \text{ m}$$

ASSESS Since we have found all three sides of a right triangle, we can check our math by seeing if the Pythagorean theorem $L^2 = A^2 + B^2$ holds for our values. We have

$$L^2 = (10.0 \text{ m})^2 = 100 \text{ m}^2$$

and

$$A^2 + B^2 = (5.00 \text{ m})^2 + (8.66 \text{ m})^2 = 100 \text{ m}^2$$

The values agree, giving us confidence that our answer is correct.

As the next example shows, we can use the rules of trigonometry to add vectors together by considering each vector to be the hypotenuse of a right triangle. If A and B are two vertices (corners) of a triangle, then we denote the length of the side between A and B as $\overline{AB}$.

EXAMPLE 1.6 Finding Sam's displacement using trigonometry

Find Sam's net displacement in Figure 1.20, this time using methods of trigonometry.

SOLVE The situation is shown once more in Figure 1.24. We will first find sides $\overline{AB}$ and $\overline{BC}$ of triangle ABC, then use these lengths to find the hypotenuse and angle of triangle DBC. (Here, $\overline{AB}$ means the side between points A and B.)

FIGURE 1.24 Using trigonometry to find Sam's displacement.

The hypotenuse AC has length 100 ft. From Equation 1.3 we have

$$\overline{AB} = \overline{AC}\cos 45° = (100 \text{ ft})\cos 45° = 71 \text{ ft}$$

$$\overline{BC} = \overline{AC}\sin 45° = (100 \text{ ft})\sin 45° = 71 \text{ ft}$$

Now consider triangle DBC. Side $\overline{BC}$, with length 71 ft, is opposite angle θ. The side adjacent to θ is $\overline{DB} = \overline{DA} + \overline{AB} = 50 \text{ ft} + 71 \text{ ft} = 121 \text{ ft}$. Then, using Equation 1.3 once again, we find

$$\tan\theta = \frac{\overline{BC}}{\overline{DB}} = \frac{71 \text{ ft}}{121 \text{ ft}} = 0.59$$

from which $\theta = \tan^{-1}(0.59) = 30°$, where $\tan^{-1}$ is the *arctangent* or *inverse tangent*. Finally, we can find the magnitude of the net displacement from the Pythagorean theorem on triangle DCB:

$$\overline{DC} = \sqrt{(\overline{DB})^2 + (\overline{BC})^2} = \sqrt{(121 \text{ ft})^2 + (71 \text{ ft})^2} = 140 \text{ ft}$$

Thus Sam's net displacement is $\vec{d}_{net} = (140 \text{ ft}, 30°$ north of east). This length, and the angle θ, agree with the values obtained using graphical methods in Example 1.4.

Velocity Vectors

We've seen that a basic quantity that describes the motion of an object is its velocity. Velocity is a vector quantity, since its specification involves not only how fast an object is moving (its speed), but also the direction in which the object is moving. We thus represent the velocity of an object by a **velocity vector** $\vec{v}$ that points in the direction of the object's motion, and whose magnitude is the object's speed. In this section, we'll learn how to draw velocity vectors on a motion diagram.

Figure 1.25a shows (using the particle model) the motion diagram of a car accelerating from rest. We've drawn vectors showing the car's displacement between successive positions of the motion diagram. To draw the car's velocity vectors, we note first that the direction of the displacement vector indicates the direction of motion between successive points in the motion diagram. But the velocity of an object also points in the direction of motion, so **an object's velocity vector points in the same direction as its displacement vector.**

Second, we've already noted that the magnitude of the velocity vector—"how fast"—is the object's speed. From Equation 1.1, the speed of an object is given by

$$\text{speed} = \frac{\text{distance traveled in a given time interval}}{\text{time interval}}$$

Now between any two successive points of the motion diagram, the distance traveled is equal to the magnitude of the displacement, that is, to the length of the displacement vector. Further, in a motion diagram the time interval between successive points is always the same. This means that the speed of an object, and thus the length of the velocity vector, is proportional to the length of the displacement vector between successive points on a motion diagram. Consequently, the vectors connecting each dot of a motion diagram to the next, which we previously labeled as displacement vectors, could equally well be identified as velocity vectors. This is shown for our car in Figure 1.25b. From now on, we'll show and label velocity vectors on motion diagrams rather than displacement vectors.

NOTE ▶ Again, the velocity vectors shown in Figure 1.25 are actually *average* velocity vectors. Because the velocity is increasing, it's actually a bit less than this average at the start of each time interval, and a bit more at the end. In Chapter 2 we'll refine these ideas and develop the idea of instantaneous velocity. ◀

The displacement vectors are lengthening. This means the car is speeding up.

(a)   0 s 1 s 2 s   3 s   4 s

Start

(b)

The longer velocity vectors also indicate that the car is speeding up.

FIGURE 1.25 The motion diagram for a car starting from rest.

EXAMPLE 1.7   Drawing a ball's motion diagram

Jake hits a ball at a 60° angle from the horizontal. It is caught by Jim. Draw a motion diagram of the ball that shows velocity vectors rather than displacement vectors.

SOLVE This example is typical of how many problems in science and engineering are worded. The problem does not give a clear statement of where the motion begins or ends. Are we interested in the motion of the ball just during the time it is in the air between Jake and Jim? What about the motion *as* Jake hits it (ball rapidly speeding up) or *as* Jim catches it (ball rapidly slowing down)? Should we include Jim dropping the ball after he catches it? The point is that *you* will often be called on to make a *reasonable interpretation* of a problem statement. In this problem, the details of hitting and catching the ball are complex. The motion of the ball through the air is easier to describe, and it's a motion you might expect to learn about in a physics class. So our *interpretation* is that the motion diagram should start as the ball leaves Jake's bat (ball already moving) and should end the instant it touches Jim's hand (ball still moving). We will model the ball as a particle.

With this interpretation in mind, Figure 1.26 shows the motion diagram of the ball. Notice how, in contrast to the car of Figure 1.25, the ball is already moving as the motion diagram movie begins. As before, the velocity vectors are found by connecting the dots with arrows. You can see that the velocity vectors get shorter (ball slowing down), get longer (ball speeding up), and change direction. Each $\vec{v}$ is different, so this is *not* constant velocity motion.

FIGURE 1.26 The motion diagram of a ball traveling from Jake to Jim.

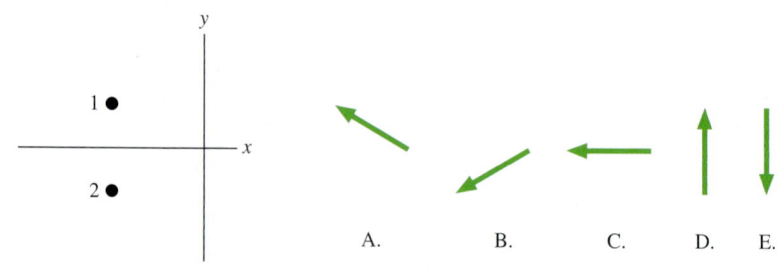

STOP TO THINK 1.5   A particle moves from position 1 to position 2 during the interval Δt. Which vector shows the particle's velocity?

1.6 Making Models: The Power of Physics

You've now seen that we often make drastic simplifications when we analyze a situation in physics. For example, we may represent a speeding car as simply a moving dot. When we analyze a situation or solve a problem, we are making a *model* of a physical situation, an idealized version of the problem that allows us to focus on its most important features.

We will introduce many different models that allow us to focus in this way. Earlier in the chapter we introduced the *particle model,* which allowed us to visualize the motion of an object as a single particle. By ignoring the details of the object, we could concentrate on the object's overall motion. This is a very useful model in many cases, as we have seen.

Another model that we will use regularly is the **atomic model.** Matter is made of atoms, and it is useful to take this into account. In Chapter 8, we will model solids as being composed of particle-like atoms connected by springs. A rubber band modeled in this way is shown in Figure 1.27. This is a simplification, but a simplification grounded in reality: For the purpose of analyzing the elastic behavior of solids, what really matters is that the atoms are connected by bonds—and that these bonds behave quite a bit like little springs.

Applying the concepts that you learn in this course to real problems is also a form of modeling: You are choosing a simplified way of looking at the problem that avoids extraneous detail. It is a skill you will acquire with practice.

FIGURE 1.27 An atomic model of a rubber band.

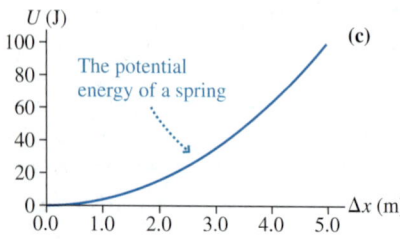

FIGURE 1.28 A mathematical equation models a physical quantity.

Mathematical Forms

In this course, we will also make frequent use of *mathematical relationships* between physical quantities. In fact, we have already introduced some in this chapter. An important part of physics—but not the only part!—is analyzing nature using such mathematical relationships. As we do so, we find that certain ones show up in many different contexts.

As an example, consider the fairly simple mathematical equation

$$y = 4x^2 \tag{1.4}$$

This relationship between y and x is plotted in Figure 1.28a. As x increases, y increases more rapidly, so that the curve gets ever steeper.

In Figure 1.28b, we plot the important physical quantity known as the *kinetic energy K,* which we'll learn about in Chapter 10. The kinetic energy depends on the speed v of an object according to the equation

$$K = \frac{1}{2}mv^2 \tag{1.5}$$

Note how similar the shape of this curve is to that of Figure 1.28a.

Finally, in Figure 1.28c we plot the *potential energy U* of a spring. In Chapter 10 we'll find that this energy depends on the displacement x of the end of the spring according to the equation

$$U = \frac{1}{2}kx^2 \tag{1.6}$$

The shape of this curve is the same as the previous two.

In looking at Figure 1.28, then, it is evident that graphs of all three of these expressions have the same overall appearance. They differ in their variables; kinetic energy depends on v and potential energy on x. They also differ in some of the constants in the equations—the $\frac{1}{2}m$ for kinetic energy, the $\frac{1}{2}k$ for potential energy. But they are all versions of the same mathematical relationship. We say that all three equations have the same mathematical *form*.

The mathematical form that all three of the plots in Figure 1.28 share is that they depend on the *square* of their variable. The variable y in Equation 1.4 depends on the square of x. The kinetic energy depends on the square of v. And the potential energy depends on the square of x. We express this dependence of each equation on the square of its variable using the symbol $\propto$. This symbol is often read as "varies as" or "depends on." We can then write

$$y \propto x^2 \qquad K \propto v^2 \qquad U \propto x^2$$

Writing expressions in this way emphasizes how the quantity depends on its variable. In many problems the form of this variation is more important than other details in the equations, such as m or factors of $1/2$.

Table 1.6 shows four common mathematical forms that we'll encounter in this book, with some examples of each. There will be others as well. Because certain physical principles share the same mathematical form, we can use these similarities to help us learn new concepts. Thus, much of what you learn about gravity in Chapter 6 can later be applied to electricity, because the force of gravity and the electric force share the same mathematical form, $y \propto 1/x^2$.

As we meet each mathematical model for the first time, we will insert a section in the text that gives an overview of the form. When we see the form again, we will refer back to that overview section.

TABLE 1.6 Common mathematical forms

Mathematical form	Physical example
Proportional $y \propto x$	Dependence of acceleration on force
	Force due to a spring
Square $y \propto x^2$	Potential energy of a spring
	Kinetic energy
Inverse $y \propto \dfrac{1}{x}$	Potential energy of two charges
	Magnetic field due to a current
Inverse square $y \propto \dfrac{1}{x^2}$	Gravitational force between two masses
	Electric force between two charges

1.7 Where Do We Go from Here?

This first chapter has been an introduction to some of the fundamental ideas about motion and some of the basic techniques that you will use in the rest of the course. You have seen some examples of how to make *models* of a physical situation, thereby focusing in on the essential elements of the situation. You have learned some practical ideas, such as how to convert quantities from one kind of units to another. The rest of this book—and the rest of your course—will extend these themes. You will learn how to model many kinds of physical systems, and learn the technical skills needed to set up and solve problems using these models.

As we go along, you will learn a set of very practical and useful models that will allow you to discuss and analyze a very wide range of problems. For example,

- We will study the forces in joints, the flow of blood in arteries, how you make the sounds of speech, and how your eyes analyze light from the world around you.
- We will look at steel beams, bones, and spiders' silk and find which is strongest. The answer may surprise you!

Radio signals travel from tower to receiver in the form of invisible *electromagnetic waves*.

This bacterium contains microscopic magnetic particles that allow it to navigate using the earth's magnetic field.

■ We will learn how radio signals are generated, how they travel, how an antenna picks them up, how the speaker in the radio turns the signals into sound, and how this sound is analyzed by your ears.

■ We will look at how electricity is generated and transmitted, whether it's the electricity that comes to your house or the electrical signal that propagates through your nerve cells.

This is all very practical, and will be quite useful in your current field of study. But in the midst of solving such practical problems, you will also learn what physics teaches us about the world, and how truly amazing some of it is. Did you know that:

■ All of the energy that you use on a daily basis comes from nuclear energy?

■ Certain animals, such as mice, can survive falls from any height?

■ There are creatures that use the earth's magnetic field to navigate?

■ It is possible for something to be in two places at once, or for two things to be in the same place at the same time?

■ It is possible to go on a long trip—and return years younger than your twin?

As you work through this course and learn to solve problems, don't lose sight of this big picture! The universe is a remarkable place, and physics is a wonderful tool for showing both its depth and its underlying simplicity.

SUMMARY

The goal of Chapter 1 has been to introduce the fundamental concepts of motion and to review the related basic mathematical principles.

IMPORTANT CONCEPTS

Motion Diagrams

The particle model represents a moving object as if all its mass were concentrated at a single point. Using this model, we can represent motion with a **motion diagram** where dots indicate the object's position at successive times. In a motion diagram, the time interval between successive dots is always the same.

Each dot represents the position of the object. Each position is labeled with the time at which the dot was there.

$t = 0$ s   1 s   2 s   3 s   4 s   5 s   6 s

The time interval between successive positions is the same.

Scalars and Vectors

Scalar quantities have only a magnitude, and can be represented by a single number. Temperature, time, and mass are scalars.

A vector is a quantity described by both a magnitude and a direction. Velocity and displacement are vectors.

Direction

$\vec{A}$

The length of a vector is proportional to its magnitude.

Velocity vectors can be drawn on a motion diagram by connecting successive points with a vector.

Velocity vectors span successive points in a motion diagram.

Start

The velocity vectors are getting longer, so the object is speeding up.

Describing Motion

Position locates an object with respect to a chosen coordinate system. It is described by a **coordinate**.

The *coordinate* is the variable used to describe the position.

x (mi)

This cow is at $x = -5$ miles.   This car is at $x = +4$ miles.

A change in position is called a **displacement**. For motion along a line, a displacement is a signed quantity. The displacement from x_i to x_f is $\Delta x = x_f - x_i$.

Time is measured from a particular instant to which we assign $t = 0$. A **time interval** is the elapsed time between two specific instants t_i and t_f. It is given by $\Delta t = t_f - t_i$.

Velocity is the ratio of the displacement of an object to the time interval during which this displacement occurs:

$$v = \frac{\Delta x}{\Delta t}$$

Units

Every measurement of a quantity must include a unit.

The standard system of units used in science is the SI system. Common SI units include:

- Length: meters (m)
- Time: seconds (s)
- Mass: kilograms (kg)

APPLICATIONS

Working with Numbers

In scientific notation, a number is expressed as a decimal number between 1 and 10 multiplied by a power of ten. In scientific notation, the diameter of the earth is 1.27×10^7 m.

A prefix can be used before a unit to indicate a multiple of 10 or 1/10. Thus we can write the diameter of the earth as 12,700 km, where the k in km denotes 1000.

We can perform a unit conversion to convert the diameter of the earth to a different unit, such as miles. We do so by multiplying by a conversion factor equal to one, such as $1 = 1$ mi/1.61 km.

Significant figures are reliably known digits. The number of significant figures for:

- **Multiplication, division, and powers** is set by the value with the fewest significant figures.

- **Addition and subtraction** is set by the value with the smallest number of decimal places.

An order-of-magnitude-estimate is an estimate that has an accuracy of about one significant figure. Such estimates are usually made using rough numbers from everyday experience.

QUESTIONS

Conceptual Questions

1. a. Write a paragraph describing the *particle model.* What is it, and why is it important?
    b. Give two examples of situations, different from those described in the text, for which the particle model is appropriate.
    c. Give an example of a situation, different from those described in the text, for which it would be inappropriate.
2. A softball player slides into second base. Use the particle model to draw a motion diagram of the player from the time he begins to slide until he reaches the base. Number the dots in order, starting with zero.
3. A car travels to the left at a steady speed for a few seconds, then brakes for a stop sign. Use the particle model to draw a motion diagram of the car for the entire motion described here. Number the dots in order, starting with zero.
4. A ball is dropped from the roof of a tall building and students in a physics class are asked to sketch a motion diagram for this situation. A student submits the diagram shown in Figure Q1.4. Is the diagram correct? Explain.

    FIGURE Q1.4

    ```
    • 0
    • 1
    • 2
    • 3
    • 4
    ```
5. Write a sentence or two describing the difference between position and displacement. Give one example of each.
6. Doug starts at position $x = 10$ m. He then undergoes a displacement $\Delta x = +15$ m. During his motion, is it possible that he passed through the origin, at $x = 0$ m?
7. Write a sentence or two describing the difference between speed and velocity. Give one example of each.
8. The motion of a skateboard along a horizontal axis is observed for 5 s. The initial position of the skateboard is negative with respect to a chosen origin, and its velocity throughout the 5 s is also negative. At the end of the observation time, is the skateboard closer to or further from the origin than initially? Explain.
9. Can the velocity of an object be positive during a time interval in which its position is always negative? Can its velocity be positive during a time interval in which its displacement is negative?
10. Two friends watch a jogger complete a 400 m lap around the track in 100 s. One of the friends states, "The jogger's velocity was 4 m/s during this lap." The second friend objects, saying, "No, the jogger's speed was 4 m/s." Who is correct? Justify your answer.
11. A softball player hits the ball and starts running toward first base. Draw a motion diagram, using the particle model, showing her position and her velocity vectors during the first few seconds of her run.

12. A child is sledding on a smooth, level patch of snow. She encounters a rocky patch and slows to a stop. Draw a motion diagram, using the particle model, showing her position and her velocity vectors.
13. A roof tile falls straight down from a two-story building. It lands in a swimming pool and settles gently to the bottom. Draw a motion diagram, using the particle model, showing the tile's position and its velocity vectors.
14. Your roommate drops a tennis ball from a third story balcony. It hits the sidewalk and bounces as high as the second story. Draw a motion diagram, using the particle model, showing the ball's velocity vectors from the time it is released until it reaches the maximum height on its bounce.
15. A car is driving north at a steady speed. It makes a gradual 90° left turn without losing speed, then continues driving to the west. Draw a motion diagram, using the particle model, showing the car's velocity vectors as seen from a helicopter hovering over the highway.
16. A toy car rolls down a ramp, then across a smooth, horizontal floor. Draw a motion diagram, using the particle model, showing the car's velocity vectors.
17. Estimate the average speed with which you go from home to campus (or another trip you commonly make) via whatever mode of transportation you use most commonly. Give your answer in both mph and m/s. Describe how you arrived at this estimate.
18. Estimate the number of times you sneezed during the past year. Describe how you arrived at this estimate.
19. Density is the ratio of an object's mass to its volume. Would you expect density to be a vector or a scalar quantity? Explain.

Multiple-Choice Questions

20. I A student walks 1.0 mi west and then 1.0 mi north. Afterward, how far is she from her starting point?
    A. 1.0 mi   B. 1.4 mi   C. 1.6 mi   D. 2.0 mi
21. I Which of the following motions is described by the motion diagram of Figure Q1.21?
    A. An ice skater gliding across the ice.
    B. An airplane braking to a stop after landing.
    C. A car pulling away from a stop sign.
    D. A pool ball bouncing off a cushion and reversing direction.

    FIGURE Q1.21
    ```
    0 1  2    3      4        5
    •• •   •     •        •
    ```
22. I A bird flies 3.0 km due west and then 2.0 km due north. What is the magnitude of the bird's displacement?
    A. 2.0 km   B. 3.0 km   C. 3.6 km   D. 5.0 km

23. ▮▮▮ A bird flies 3.0 km due west and then 2.0 km due north. Another bird flies 2.0 km due west and 3.0 km due north. What is the angle between the net displacement vectors for the two birds?
    A. 23°   B. 34°   C. 56°   D. 90°
24. ▮▮ A woman walks briskly at 2.00 m/s. How much time will it take her to walk one mile?
    A. 8.30 min B. 13.4 min C. 21.7 min D. 30.0 min
25. ▮ Compute 3.24 m + 0.532 m to the correct number of significant figures.
    A. 3.7 m   B. 3.77 m   C. 3.772 m   D. 3.7720 m
26. ▮ A rectangle has length 3.24 m and height 0.532 m. To the correct number of significant figures, what is its area?
    A. 1.72 m^2 B. 1.723 m^2
    C. 1.7236 m^2 D. 1.72368 m^2

27. ▮ The earth formed 4.57×10^9 years ago. What is this time in seconds?
    A. 1.67×10^{12} s   B. 4.01×10^{13} s
    C. 2.40×10^{15} s   D. 1.44×10^{17} s
28. ▮ An object's density ρ is defined as the ratio of its mass to its volume: $\rho = M/V$. The earth's mass is 5.94×10^{24} kg, and its volume is 1.08×10^{12} km^3. What is the earth's density?
    A. 5.50×10^3 kg/m^3 B. 5.50×10^6 kg/m^3
    C. 5.50×10^9 kg/m^3 D. 5.50×10^{12} kg/m^3

PROBLEMS

Section 1.1 Motion: A First Look

1. ▮ You've made a video of a car as it skids to a halt to avoid hitting an object in the road. Use the images from the video to draw a motion diagram of the car from the time the skid begins until the car is stopped.
2. ▮ A man rides a bike along a straight road for 5 min, then has a flat tire. He stops for 5 min to repair the flat, but then realizes he cannot fix it. He continues his journey by walking the rest of the way, which takes him another 10 min. Use the particle model to draw a motion diagram of the man for the entire motion described here. Number the dots in order, starting with zero.
3. ▮ A jogger running east at a steady pace suddenly develops a cramp. He is lucky: A westbound bus is sitting at a bus stop just ahead. He gets on the bus and enjoys a quick ride home. Use the particle model to draw a motion diagram of the jogger for the entire motion described here. Number the dots in order, starting with zero.

Section 1.2 Position and Time: Putting Numbers on Nature

4. ▮ Figure P1.4 shows Sue between her home and the cinema. What is Sue's position x if
   a. her home is the origin?   b. the cinema is the origin?

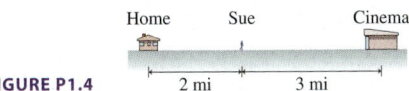

FIGURE P1.4

5. ▮ Keira starts at position $x = 23$ m along a coordinate axis. She then undergoes a displacement of -45 m. What is her final position?
6. ▮ A car travels along a straight east-west road. A coordinate system is established on the road, with x increasing to the east. The car ends up 14 mi west of the intersection with Mulberry Road. If its displacement was -23 mi, how far from and on which side of Mulberry Road did it start?

7. ▮ Lisa is enjoying a bicycle ride on the country road shown earlier in Figure 1.7. Suppose that increasing x means moving east. At noon, she is 2 mi west of the post office. A half hour later, she is 3 mi east of the post office. What is her displacement Δx during that half hour?

Section 1.3 Velocity

8. ▮ A security guard walks 110 m in one trip around the perimeter of the building. It takes him 240 s to make this trip. What is his speed?
9. ▮ List the following items in order of decreasing speed, from greatest to least: (i) A wind-up toy car that moves 0.15 m in 2.5 s. (ii) A soccer ball that rolls 2.3 m in 0.55 s. (iii) A bicycle that travels 0.60 m in 0.075 s. (iv) A cat that runs 8.0 m in 2.0 s.
10. ▮▮ Figure P1.10 shows the motion diagram for a horse galloping in one direction along a straight path. Not every dot is labeled, but the dots are at equally spaced instants of time. What is the horse's velocity
    a. during the first ten seconds of its gallop?
    b. during the interval from 30 s to 40 s?
    c. during the interval from 50 s to 70 s?

    FIGURE P1.10

11. ▮▮ It takes Harry 35 s to walk from $x = -12$ m to $x = -47$ m. What is his velocity?
12. ▮▮ A dog trots from $x = -12$ m to $x = 3$ m in 10 s. What is its velocity?
13. ▮ A ball rolling along a straight line with velocity 0.35 m/s goes from $x = 2.1$ m to $x = 7.3$ m. How much time does this take?

Section 1.4 A Sense of Scale: Significant Figures, Scientific Notation, and Units

14. ▮▮ Convert the following to SI units:
    a. 9.12 μs   b. 3.42 km
    c. 44 cm/ms   d. 80 km/hour

15. ‖ Convert the following to SI units:
    a. 8.0 in   b. 66 ft/s   c. 60 mph
16. | Convert the following to SI units:
    a. 1.0 hour   b. 1.0 day   c. 1.0 year
17. ‖ List the following three speeds in order, from smallest to largest: 1 mm per μs, 1 km per ks, 1 cm per ms.
18. | How many significant figures does each of the following numbers have?
    a. 6.21   b. 62.1   c. 0.620   d. 0.062
19. | How many significant figures does each of the following numbers have?
    a. 0.621   b. 0.006200
    c. 1.0621   d. 6.21×10^3
20. | Compute the following numbers to 3 significant figures.
    a. 33.3×25.4 b. $33.3 - 25.4$
    c. $\sqrt{33.3}$ d. $333.3 \div 25.4$
21. | The Empire State Building has a height of 1250 ft. Express this height in meters, giving your result in scientific notation with three significant figures.
22. ‖ We interpret a measurement of $x_1 = 77$ m to mean a distance somewhere between 76.5 m and 77.5 m, thus rounding to 77 m. Similarly, $x_2 = 813.8$ m implies a distance between 813.75 m and 813.85 m. Use these values to justify the rule for determining significant figures in addition by finding the smallest and largest possible values of $x_1 + x_2$.
23. ‖ We can interpret a distance traveled of 7.43 m to mean a distance between 7.425 m and 7.435 m, thus rounding to 7.43 m. Similarly, a time interval of 0.83 s implies a time interval between 0.825 s and 0.835 s. Use these values to justify the rule for determining significant figures in division by finding the smallest and largest possible values of the speed.
24. ‖ Estimate (don't measure!) the length of a typical car. Give your answer in both feet and meters. Briefly describe how you arrived at this estimate.
25. | Estimate the height of a telephone pole. Give your answer in both feet and meters. Briefly describe how you arrived at this estimate.
26. | Estimate the average speed with which the hair on your
    BIO head grows. Give your answer in both m/s and μm/hr. Briefly describe how you arrived at this estimate.
27. ‖ Estimate the average speed at which your fingernails grow,
    BIO in both m/s and μm/hr. Briefly describe how you arrived at this estimate.

Section 1.5 Vectors and Motion: A First Look

28. | Carol and Robin share a house. To get to work, Carol walks north 2.0 km while Robin drives west 7.5 km. How far apart are their workplaces?
29. | Joe and Max shake hands and say goodbye. Joe walks east 0.55 km to a coffee shop, and Max flags a cab and rides north 3.25 km to a bookstore. How far apart are their destinations?
30. ‖‖ A city has streets laid out in a square grid, with each block 135 m long. If you drive north for three blocks, then west for two blocks, how far are you from your starting point?
31. ‖ A butterfly flies from the top of a tree in the center of a garden to rest on top of a red flower at the garden's edge. The tree is 8.0 m taller than the flower, and the garden is 12 m wide. Determine the magnitude of the butterfly's displacement.

32. ‖ A garden has a circular path of radius 50 m. John starts at the easternmost point on this path, then walks counterclockwise around the path until he is at its southernmost point. What is John's displacement? Use the (magnitude, direction) notation for your answer.
33. ‖ Anna walks 130 m due north, 50 m due east, and then 40 m due south. What is her net displacement? Use the (magnitude, direction) notation for your answer.
34. ‖‖‖ A ball on a porch rolls 60 cm to the porch's edge, drops 40 cm, continues rolling on the grass, and eventually stops 80 cm from the porch's edge. What is the magnitude of the ball's net displacement, in centimeters?

Section 1.6 Making Models: The Power of Physics

35. ‖ Sketch a graph of each function over a range of x values from 0.1 to 10. You may need to use different scales on the y axis.
    a. $y = x$ (proportional)
    b. $y = x^2$ (square)
    c. $y = \dfrac{1}{x}$ (inverse)
    d. $y = \dfrac{1}{x^2}$ (inverse square)

General Problems

Problems 36 through 42 are motion problems similar to those you will learn to solve in Chapter 2. For now, simply *interpret* the problem by drawing a motion diagram showing the object's position and its velocity vectors. **Do *not* solve these problems** or do any mathematics.

36. ‖ A Porsche accelerates from a stoplight at 5.0 m/s^2 for five seconds, then coasts for three more seconds. How far has it traveled?
37. ‖ Billy drops a watermelon from the top of a three-story building, 10 m above the sidewalk. How fast is the watermelon going when it hits?
38. ‖ Sam is recklessly driving 60 mph in a 30 mph speed zone when he suddenly sees the police. He steps on the brakes and slows to 30 mph in three seconds, looking nonchalant as he passes the officer. How far does he travel while braking?
39. ‖ A speed skater moving across frictionless ice at 8.0 m/s hits a 5.0-m-wide patch of rough ice. She slows steadily, then continues on at 6.0 m/s. What is her acceleration on the rough ice?
40. ‖ You would like to stick a wet spit wad on the ceiling, so you toss it straight up with a speed of 10 m/s. How long does it take to reach the ceiling, 3.0 m above?
41. ‖ A ball rolls along a smooth horizontal floor at 10 m/s, then starts up a 20° ramp. How high does it go before rolling back down?
42. ‖ A motorist is traveling at 20 m/s. He is 60 m from a stop light when he sees it turn yellow. His reaction time, before stepping on the brake, is 0.50 s. What steady deceleration while braking will bring him to a stop right at the light?

Problems 43 through 50 show a motion diagram. For each of these problems, write a one or two sentence "story" about a *real object* that has this motion diagram. Your stories should talk about people or objects by name and say what they are doing. Problems 36–42 are examples of motion short stories.

43. ‖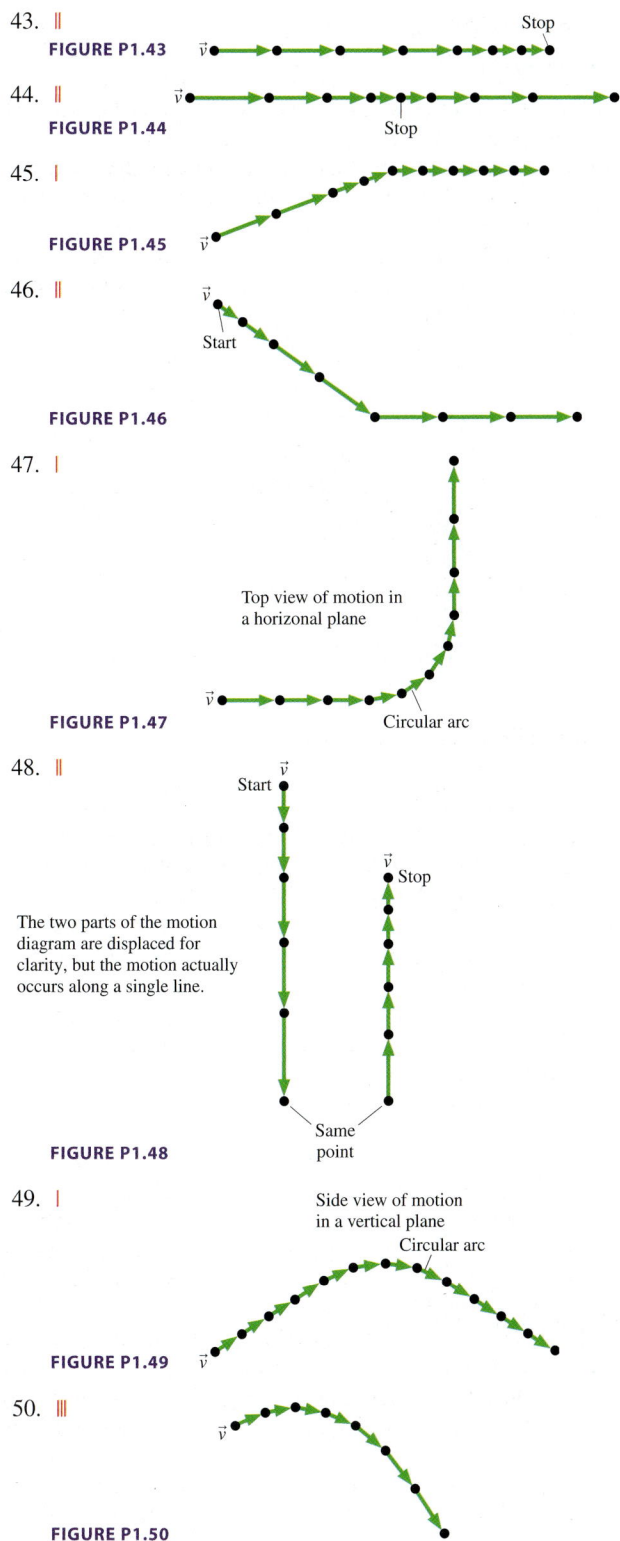
FIGURE P1.43

44. ‖
FIGURE P1.44

45. ⏐
FIGURE P1.45

46. ‖
FIGURE P1.46

47. ⏐
Top view of motion in a horizontal plane
FIGURE P1.47
Circular arc

48. ‖
Start
Stop
The two parts of the motion diagram are displaced for clarity, but the motion actually occurs along a single line.
Same point
FIGURE P1.48

49. ⏐
Side view of motion in a vertical plane
Circular arc
FIGURE P1.49

50. ⦀
FIGURE P1.50

51. ⦀ How many inches does light travel in one nanosecond? The speed of light is 3.0×10^8 m/s.

52. ‖ Joseph watches the roadside mile markers during a long car trip on an interstate highway. He notices that at 10:45 A.M. they are passing a marker labeled 101, and at 11:00 A.M. the car reaches marker 119. What is the car's speed, in mph?

53. ‖ Alberta is going to have dinner at her grandmother's house, but she is running a bit behind schedule. As she gets onto the highway, she knows that she must exit the highway within 45 min if she is not going to arrive late. Her exit is 32 mi away. What is the slowest speed at which she could drive and still arrive in time? Express your answer in miles per hour.

54. ⏐ Dave hits a hockey puck at 10 m/s toward the goal. If the puck travels 15 m to the goal, how long does it take to get there?

55. ⏐ In many living systems, information propagates through
BIO the body in the form of electrical signals called *nerve impulses*. Nerve impulses travel along nerve fibers consisting of many *axons* (fiber-like extensions of nerve cells) aligned in series. For example, when your foot accidentally kicks a rock the feeling of pain is transmitted from the foot to your brain along hundreds of axons that span the distance between your foot and head. Axons come in two varieties: one with a sheath called myelin around it, the other without. Myelinated (sheathed) axons conduct nerve impulses roughly 100 times faster than unmyelinated (unsheathed) axons. Nerve impulses travel at a speed of about 1.0 m/s or 2.2 mph (a walking pace) along unmyelinated axons and about 100 m/s or 220 mph (like a race car) along myelinated axons. Figure P1.55 shows three equal-length nerve fibers consisting of eight axons in a row. Nerve impulses enter at the left side simultaneously and travel to the right.
   a. Draw motion diagrams for the nerve impulses traveling along fibers A, B, and C.
   b. Which nerve impulse arrives at the right side first?
   c. Which will be last?

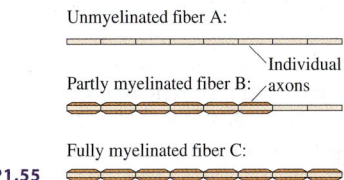
Unmyelinated fiber A:
Individual axons
Partly myelinated fiber B:
Fully myelinated fiber C:
FIGURE P1.55

56. ⦀ The bacterium *Escherichia coli* (or *E. coli*) is a single-celled
BIO organism that lives in the gut of healthy humans and animals. Its body shape can be modeled as a 2-μm-long cylinder with a 1 μm diameter, and it has a mass of 1×10^{-12} g. Its chromosome consists of a single double-stranded chain of DNA 700 times longer than its body length. The bacterium moves at a constant speed of 20 μm/s, though not always in the same direction. Answer the following questions about *E. coli* using SI units (unless specifically requested otherwise) and correct significant figures.
   a. What is its length?
   b. Diameter?
   c. Mass?
   d. What is the length of its DNA, in millimeters?
   e. If the organism were to move along a straight path, how many meters would it travel in one day?

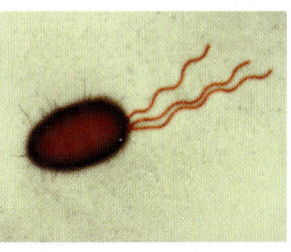

57. ▥ The bacterium *Escher-*
BIO *ichia coli* (or *E. coli*) is a
single-celled organism that
lives in the gut of healthy
humans and animals. When
grown in a uniform medium
rich in salts and amino acids,
it swims along zig-zag paths
at a constant speed. Fig-
ure P1.57 shows the posi-
tions of an *E. coli* as it moves

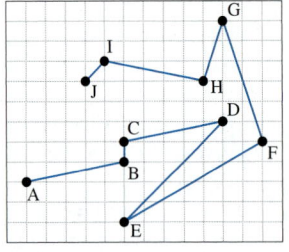

FIGURE P1.57

from point A to point J. Each segment of the motion can be
identified by two letters, such as segment BC. During which
segments, if any, does the bacterium have the same
    a. Displacement?   b. Speed?   c. Velocity?
58. ▯ In 2003, the population of the United States was 291 million people. The per-capita income was $31,459. What was
the total income of everyone in the United States? Express
your answer in scientific notation, with the correct number of
significant figures.
59. ▥ The sun is 30° above the horizon. It makes a 52-m-long
shadow of a tall tree. How high is the tree?
60. ▥ An airplane accelerates down the runway for 500 m, then
lifts into the air at an angle of 20.0 degrees above the ground.
When the plane has traveled 300 m in the air, what is its distance from where it started on the runway?
61. ▥ Starting from its nest, an eagle flies at constant speed for
3.0 min due east, then 4.0 min due north. From there the eagle
flies directly to its nest at the same speed. How long is the
eagle in the air?
62. ▥ John walks 1.00 km north, then turns right and walks
1.00 km east. His speed is 1.50 m/s during the entire stroll.
    a. What is the magnitude of his displacement, from begin-
       ning to end?
    b. If Jane starts at the same time and place as John, but walks
       in a straight line to the endpoint of John's stroll, at what
       speed should she walk to arrive at the endpoint just when
       John does?

Passage Problems

Growth Speed

The images of trees in Figure P1.63 come from a catalog advertis-
ing fast-growing trees. If we mark the position of the top of the tree
in the successive years, as shown in the graph in the figure, we
obtain a motion diagram much like ones we have seen for other
kinds of motion. The motion isn't steady, of course. In some
months the tree grows rapidly; in other months, quite slowly. We
can see, though, that the average speed of growth is fairly constant
for the first few years.

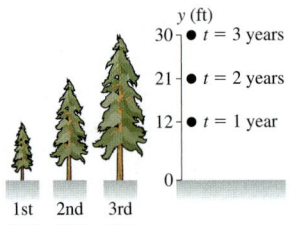

FIGURE P1.63

63. ▯ What is the tree's speed of growth, in feet per year, from
$t = 1$ yr to $t = 3$ yr?
    A. 12 ft/yr   B. 9 ft/yr   C. 6 ft/yr   D. 3 ft/yr
64. ▯ What is this speed in m/s?
    A. 9×10^{-8} m/s   B. 3×10^{-9} m/s
    C. 5×10^{-6} m/s   D. 2×10^{-6} m/s
65. ▮ At the end of year 3, a rope is tied to the very top of the tree
to steady it. This rope is staked into the ground 15 feet away
from the tree. What angle does the rope make with the
ground?
    A. 63°   B. 60°   C. 30°   D. 27°

STOP TO THINK ANSWERS

Stop to Think 1.1: B. The images of B are further apart, so it travels a larger distance than does A during the same intervals of time.

Stop to Think 1.2: A. Dropped ball. **B.** Dust particle. **C.** Descending rocket.

Stop to Think 1.3: C. Depending on her initial positive position,
and how far she moves in the negative direction, she could end up
on either side of the origin.

Stop to Think 1.4: D > C > B = A.

Stop to Think 1.5: E. The velocity vector is found by connecting
one dot on the motion diagram to the next.

2

MOTION IN ONE DIMENSION

A horse can run at 35 mph, much faster than a human. And yet, surprisingly, a man can win a race against a horse if the length of the course is right. When, and why, can a man outrun a horse?

Looking Ahead ▶▶

The goal of Chapter 2 is to describe and analyze linear motion. In this chapter you will learn to:

▶ Represent one-dimensional (straight-line) motion in different ways: using numbers, graphs, pictures, words, and equations.

▶ Use general physics problem-solving strategies and techniques.

▶ Solve problems of motion in one dimension.

Looking Back ◀◀

The material in this chapter builds on what you learned in Chapter 1 about motion and measurement. Please review:

◀ Sections 1.2–1.3 Definitions of displacement and velocity.

◀ Section 1.4 Units and significant figures.

◀ Section 1.5 Velocity vectors and motion diagrams.

A race, whether between runners, bicyclists, or drag racers, exemplifies the idea of motion. Today, we use electronic stopwatches, video recorders, and other sophisticated instruments to analyze motion, but it hasn't always been this way. Galileo, who in the early 1600s was the first scientist to study motion experimentally, used his pulse to measure time! Galileo made a useful distinction between the *cause* of motion and the *description* of motion. The modern name for the mathematical description of motion, without regard to causes, is **kinematics.** The term comes from the Greek word *kinema,* meaning "movement." You know this word through its English variation *cinema*—motion pictures!

This chapter will focus on the kinematics of motion in one dimension—that is, motion along a straight line. The motion of runners, drag racers, and skiers can be considered as motion in one dimension. But even in one dimension, there are many interesting issues to consider. For example, which is faster, a man or a horse? The answer depends on what you mean by "faster." Making such commonplace notions more precise will be one of our goals in this chapter. The other half of Galileo's distinction—the cause of motion—is a topic we will take up in Chapter 4 after first learning to describe motion.

2.1 Describing Motion

In Chapter 1 you learned about quantities that you could use to describe an object's motion. Position and velocity are measured with respect to a coordinate system, a grid or axis that *you* impose on a system. We will use an *x*-axis to analyze both horizontal motion and motion on a ramp; a *y*-axis will be used for vertical motion.

We will adopt the convention that the positive end of an *x*-axis is to the right and the positive end of a *y*-axis is up. This convention is illustrated in Figure 2.1.

Velocity is a vector; it has both a magnitude and a direction. When we draw a velocity vector on a diagram, we will use an arrow labeled with the symbol $\vec{v}$ to represent the magnitude and the direction. For motion in one dimension, vectors are restricted to point only "forward" or "backward" for horizontal motion (or "up" or "down" for vertical motion). This restriction lets us simplify our notation for vectors in one dimension. When we solve problems for motion along an *x*-axis we will represent velocity with the symbol v_x. v_x will be positive or negative, corresponding to motion to the right or the left, as shown in Figure 2.2. For motion along a *y*-axis, we will use the symbol v_y to represent the velocity; the sign conventions are also illustrated in Figure 2.2. We will use the symbol *v*, with no subscript, to represent the speed of an object. Speed is the *magnitude* of the velocity vector and is always positive.

Motion Diagrams and Graphs

We saw in Chapter 1 that a motion diagram is a useful tool for analyzing the one-dimensional motion of an object. Let's continue our discussion of motion by using a motion diagram to study a straightforward situation.

Figure 2.3 is a motion diagram, made at 1 frame per minute, of a student walking to school. We have included velocity vectors connecting successive positions on the motion diagram, as we saw we could do in Chapter 1. The motion diagram shows that she leaves home at a time we choose to call $t = 0$ min, and then makes steady progress for a while. Beginning at $t = 3$ min there is a period in which the distance traveled during each time interval becomes less—perhaps she slowed down to speak with a friend. Then, at $t = 6$ min she begins walking more quickly and the distances traveled within each interval are longer.

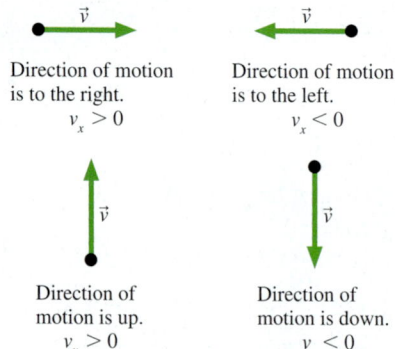

FIGURE 2.1 Sign conventions for position.

FIGURE 2.2 Sign conventions for velocity.

FIGURE 2.3 The motion diagram of a student walking to school and a coordinate axis for making measurements.

Figure 2.3 includes a coordinate axis, and we can see that every dot in a motion diagram occurs at a specific position. Table 2.1 shows the student's positions at different times as measured along this axis. For example, she is at position $x = 120$ m at $t = 2$ min.

The motion diagram of Figure 2.3 is one way to represent the student's motion. Presenting the data as in Table 2.1 is a second way to represent this motion; you can easily see where the student was at any particular time. A third way to repre-

TABLE 2.1 Measured positions of a student walking to school

Time *t* (min)	Position *x* (m)	*t*	*x*
0	0	5	220
1	60	6	240
2	120	7	340
3	180	8	440
4	200	9	540

sent the motion is to make a graph of these measurements. Figure 2.4 is a graph of the position of the student at different times; we say it is a graph of *x* versus *t* for the student. We have merely taken the data from the table and plotted these particular points on the graph.

NOTE ▶ A graph of "*a* versus *b*" means that *a* is graphed on the vertical axis and *b* on the horizontal axis. ◀

We can flesh out the graph of Figure 2.4, though. Common sense tells us a few things. First, the student was *somewhere specific* at all times. That is, there was never a time when she failed to have a well-defined position, nor could she occupy two positions at one time. (As reasonable as this belief appears to be, it will be found to be not entirely accurate when we get to quantum physics!) Second, from the start to the end of her motion, the student moved *continuously* through all intervening points of space. She could not go from $x = 100$ m to $x = 200$ m without passing through every point in between. It is thus quite reasonable to believe that her motion can be shown as a continuous curve that passes through the measured points, as shown in Figure 2.5. A continuous curve that shows an object's position as a function of time is called a **position-versus-time graph** or, sometimes, just a *position graph*.

NOTE ▶ In making the graph continuous, we made some assumptions about her motion. For example, we assumed that the curve is smooth from point to point. This may not be quite true, but in most cases it is a reasonable assumption. ◀

NOTE ▶ A graph is *not* a "picture" of the motion. The student is walking along a straight line, but the graph itself is not a straight line. Further, we've graphed her position on the vertical axis even though her motion is horizontal. A graph is an *abstract representation* of motion. We will place significant emphasis on the process of interpreting graphs, and many of the exercises and problems will give you a chance to practice these skills. ◀

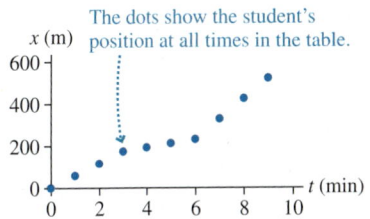

FIGURE 2.4 A graph of the student's motion.

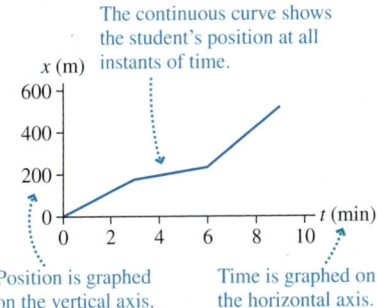

FIGURE 2.5 Extending the graph of Figure 2.4 to a position-versus-time graph.

CONCEPTUAL EXAMPLE 2.1 **Interpreting a car's position-versus-time graph**

The graph in Figure 2.6 represents the motion of a car along a straight road. Describe (in words) the motion of the car.

FIGURE 2.6 Position-versus-time graph for the car.

REASON The vertical axis in Figure 2.6 is labeled as "*x* (km)"; position is measured in kilometers. Our convention for motion along the *x*-axis given in Figure 2.1 tells us that *x* increases as the car moves to the right and *x* decreases as the car moves to the left. The graph thus represents the motion of a car that travels to the left for 30 minutes, stops for 10 minutes, then travels to the right for 40 minutes. It ends up 10 km to the left of where it began. Figure 2.7 gives a full explanation of the reasoning.

1. At $t = 0$ min, the car is 10 km to the right of the origin.

2. The value of *x* decreases for 30 min, indicating that the car is moving to the left.

5. The car reaches the origin at $t = 80$ min.

3. The car stops for 10 min at a position 20 km to the left of the origin.

4. The car starts moving back to the right at $t = 40$ min.

FIGURE 2.7 Looking at the position-versus-time graph in detail.

ASSESS The car travels to the left for 30 minutes and to the right for 40 minutes. Nonetheless, it ends up to the left of where it started. This means that the car was moving faster when it was moving to the left than when it was moving to the right. We can deduce this fact from the graph as well, as we will see in the next section.

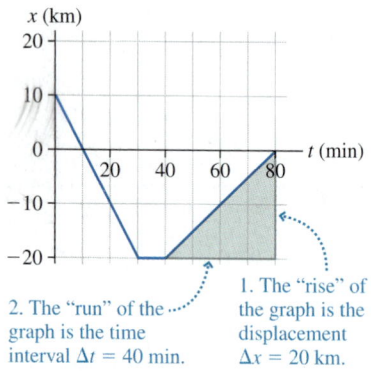

2. The "run" of the graph is the time interval $\Delta t = 40$ min.

1. The "rise" of the graph is the displacement $\Delta x = 20$ km.

3. The slope of the graph, rise over run, is the car's velocity during this phase of motion:

v_x = slope of position-versus-time graph

$= \dfrac{\Delta x}{\Delta t} = \dfrac{20 \text{ km}}{40 \text{ min}} = 0.50$ km/min

FIGURE 2.8 Calculating velocity from a position-versus-time graph.

Velocity Is the Slope of the Position-Versus-Time Graph

Let's look more closely at the graph of Figure 2.6 showing the motion of a car. Suppose we want to compute the velocity of the car during the 40 minutes when it moves to the right. For motion along a line, the definition of velocity from Chapter 1 can be written as

$$v_x = \frac{\Delta x}{\Delta t} \qquad (2.1)$$

We can give a graphical interpretation to Equation 2.1: It is the *slope* of the section of the graph representing this motion. Recall that the slope of a straight-line graph is defined as "rise over run." Because position is graphed on the vertical axis, the "rise" of a position-versus-time graph is the object's displacement Δx. The "run" is the time interval Δt. Consequently, the slope of a straight-line segment is $\Delta x/\Delta t$. This idea is graphically illustrated in Figure 2.8. We can associate the slope of the position-versus-time graph, a *geometrical* quantity, with the *physical* quantity velocity. This will be an important aspect of interpreting position-versus-time graphs, as outlined in Tactics Box 2.1.

 TACTICS BOX 2.1 Interpreting position-versus-time graphs   📝 Exercises 2–4

Information about motion can be obtained from position-versus-time graphs as follows:

❶ Determine the *position* at time t by reading the graph at that instant of time.
❷ Determine the *speed* at time t by finding the magnitude of the slope at that point. Steeper slopes correspond to greater speeds.
❸ Determine the *direction of motion* by noting the sign of the slope. Positive slopes correspond to positive velocities and, hence, to motion to the right (or up). Negative slopes correspond to negative velocities and, hence, to motion to the left (or down).

Time lines BIO This section of the trunk of a pine tree shows the light bands of spring growth and the dark bands of summer and fall growth in successive years. If you focus on the spacing of successive dark bands, you can think of this picture as a motion diagram for the tree, representing its growth in diameter. The years of rapid growth (large distance between dark bands) during wet years and slow growth (small distance between dark bands) during years of drought are readily apparent.

NOTE ▶ The slope is a ratio of intervals, $\Delta x/\Delta t$, not a ratio of coordinates. That is, the slope is *not* simply x/t. ◀

NOTE ▶ We are distinguishing between the actual slope and the *physically meaningful* slope. If you were to use a ruler to measure the rise and the run of the graph, you could compute the actual slope of the line as drawn on the page. That is not the slope we are referring to when we equate the velocity with the slope of the line. Instead, we find the *physically meaningful* slope by measuring the rise and run using the scales along the axes. The "rise" Δx is some number of meters; the "run" Δt is some number of seconds. The physically meaningful rise and run include units, and the ratio of these units gives the units of the slope. ◀

EXAMPLE 2.1 Calculating a car's velocity
Looking again at Figure 2.6, what is the velocity of the car during the period from 0 to 30 min? What is the speed?

SOLVE We find the velocity by computing the slope of the graph:

$$v_x = \frac{\Delta x}{\Delta t} = \frac{-30 \text{ km}}{30 \text{ min}} = -1.0 \text{ km/min}$$

The speed is the magnitude of the velocity:

$$v = |v_x| = 1.0 \text{ km/min}$$

ASSESS The slope of the graph is negative; the value of the distance is decreasing with time, leading to a negative displacement and a negative velocity. The "−" sign in the velocity tells us that the motion is to the left. The speed for this phase of the motion is greater than that of the final phase, as we noted it should be.

Which position-versus-time graph below best describes the motion diagram at left?

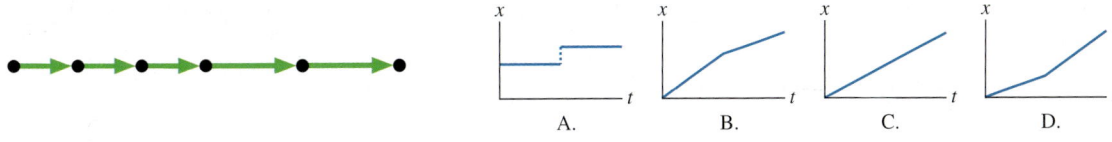

A.   B.   C.   D.

2.2 Uniform Motion

If you drive your car on a straight road at a perfectly steady 60 miles per hour (mph), you will cover 60 mi during the first hour, another 60 mi during the second hour, yet another 60 mi during the third hour, and so on. This is an example of what we call *uniform motion*. In this case, 60 mi is not your position, but rather the *change* in your position during each hour; that is, your displacement Δx. Similarly, 1 hour is a time interval Δt rather than a specific instant of time. This suggests the following definition: **Straight-line motion in which equal displacements occur during any successive equal-time intervals is called uniform motion or constant-velocity motion.**

The qualifier "any" is important. If during each hour you drive 120 mph for 30 min and stop for 30 min, you will cover 60 mi during each successive 1-hour interval. But you would *not* have equal displacements during successive 30-min intervals, so this motion is not uniform. Your constant 60 mph driving is uniform motion because you will find equal displacements no matter how you choose your successive time intervals.

Figure 2.9 shows a motion diagram and a graph for an object in uniform motion. Notice that the position-versus-time graph for uniform motion is a straight line. This follows from the requirement that all values of Δx corresponding to the same value of Δt be equal. In fact, an alternative definition of uniform motion is: **An object's motion is uniform if and only if its position-versus-time graph is a straight line.**

Equations of Uniform Motion

We have seen how to represent motion in four ways:

- Using words
- Using a motion diagram
- Using a table of data of positions at certain times
- Using a graph of position versus time

Now we will add a fifth representation: equations.

Consider an object in uniform motion along the *x*-axis with the linear position-versus-time graph shown in Figure 2.10. Recall from the previous chapter that we denote the object's initial position as x_i at time t_i. The term "initial" refers to the starting point of our analysis or the starting point in a problem. The object may or may not have been in motion prior to t_i. We use the term "final" for the ending point of our analysis or the ending point of a problem, and denote the object's final position x_f at the time t_f. The object's velocity v_x along the *x*-axis can be determined by finding the slope of the graph:

$$v_x = \frac{\text{rise}}{\text{run}} = \frac{\Delta x}{\Delta t} = \frac{x_f - x_i}{t_f - t_i} \tag{2.2}$$

Equation 2.2 can be rearranged to give

$$x_f = x_i + v_x \, \Delta t \tag{2.3}$$

Position equation for an object in uniform motion
(v_x is constant)

A freight train moving steadily on a straight run of track is a practical example of uniform motion.

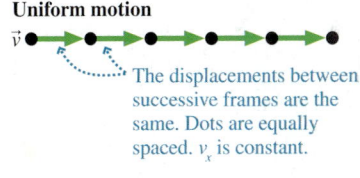

Uniform motion

The displacements between successive frames are the same. Dots are equally spaced. v_x is constant.

The position-versus-time graph is a straight line. The slope of the line is v_x.

Equal displacements

FIGURE 2.9 Motion diagram and position-versus-time graph for uniform motion.

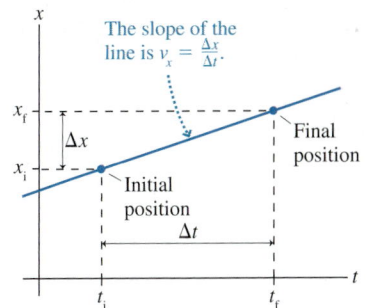

The slope of the line is $v_x = \frac{\Delta x}{\Delta t}$.

Final position

Initial position

FIGURE 2.10 Position-versus-time graph for an object in uniform motion.

where $\Delta t = t_f - t_i$ is the interval of time in which the object moves from position x_i to position x_f. Equation 2.3 applies to any time interval Δt during which the velocity is constant. We can also write this in terms of the object's displacement, $\Delta x = x_f - x_i$:

$$\Delta x = v_x \, \Delta t \qquad (2.4)$$

The velocity of an object in uniform motion tells us the amount by which its position changes during each second. An object with a velocity of 20 m/s *changes* its position by 20 m during every second of motion: by 20 m during the first second of its motion, by another 20 m during the next second, and so on. We say that position is changing at the *rate* of 20 m/s. If the object starts at $x_i = 10$ m, it will be at $x = 30$ m after 1 s of motion and at $x = 50$ m after 2 s of motion. Thinking of velocity like this will help you develop an intuitive understanding of the connection between velocity and position—something we will explore further in the next section.

Physics may seem densely populated with equations, but most equations follow a few basic forms, as we saw in Chapter 1. The mathematical form of Equation 2.3 is a type that we will see again—a *linear relationship*.

NOTE ▶ The important features of a linear relationship are described below. In this text, the first time we use a particular mathematical form we will provide such an overview. In future chapters, when we see other examples of this type of relationship, we will refer back to this overview. ◀

Linear relationships

Exercises 5, 6 (MP)

Two quantities are said to have a **linear relationship** if the graph of y versus x is a straight line. We write the mathematical relationship as

$$y = Ax + B$$

y is a linear function of x

A is the slope of the graph. When x increases by 1, y increases by A.

$y = Ax + B$

The change in y is proportional to the change in x. When x increases by 2, y increases by $2A$.

As we see from the graph, the constant A is the slope of the line; B is the value of y when $x = 0$.

A change in x leads to a **proportional** change in y. If we have initial and final values:

$$y_f = Ax_f + B$$

$$y_i = Ax_i + B$$

We can subtract these equations to find that

$$y_f - y_i = A(x_f - x_i)$$

$$\Delta y = A \, \Delta x$$

Therefore, we can write

$$\Delta y \propto \Delta x$$

SCALING Linear scaling means, for example:

- If you double Δx, you double Δy, as you can see in the graph.
- If you decrease Δx by a factor of 3, you decrease Δy by a factor of 3.

EXAMPLE 2.2   How far does the car travel?

A car takes 2.0 min to travel 1.0 mi. How far does it travel in 10 min?

SOLVE The statement of the problem assumes that the car is in uniform motion—traveling at a constant velocity. The distance will be related to the time by the *linear relationship* of Equation 2.3; this means that the change in the distance is proportional to the change in time. 10 min is a time interval that is 5 times the original 2.0 min time interval, so the car will travel 5 times as far, 5.0 mi.

A Problem-Solving Strategy

Before we introduce any more physics concepts, we'll take a break to look at a strategy for using these concepts to solve practical problems. The first step in solving a seemingly complicated problem is to break it down into a series of smaller steps. In worked examples in the text, we will use a problem-solving strategy that consists of three steps: *prepare, solve* and *assess.* Each of these steps has important elements that you should follow when you solve problems on your own.

 Problem-Solving Strategy

PREPARE The "Prepare" step of a solution is where you identify important elements of the problem and collect information you will need to solve it. It's tempting to jump right to the "solve" step, but a skilled problem solver will spend the most time on this step, the preparation. Preparation includes:

- **Drawing a picture.** In many cases, this is the most important part of a problem. The picture lets you model the problem and identify the important elements. As you add information to your picture, the outline of the solution will take shape. We will give tips for drawing effective pictures for different problems. For the problems in this chapter, a picture could be a motion diagram or a graph—or perhaps both. Later in the chapter you will learn a strategy for drawing a complete *visual overview* of a problem that incorporates these and other elements.

- **Collecting necessary information.** The problem's statement may give you some values of variables. Other important information may be implied, or must be looked up in a table. Gather everything you need to solve the problem, and include it as part of your picture or an accompanying table.

- **Doing preliminary calculations.** In some cases, there are a few calculations, such as unit conversions, that are best done in advance of the main part of the solution.

SOLVE The "Solve" step of a solution is where you actually do the mathematics or reasoning necessary to arrive at the answer needed. This is the part of the problem-solving strategy that you likely think of when you think of "solving problems." But don't make the mistake of starting here! If you just choose an equation and plug in numbers, you will likely go wrong and will waste time trying to figure out why. The "Prepare" step will help you be certain you understand the problem before you start putting numbers in equations.

ASSESS The "Assess" step of your solution is very important. When you have an answer, you should check to see if it makes sense. Ask yourself:

- **Does my solution answer the question that was asked?** Make sure you have addressed all parts of the question and clearly written down your solutions.

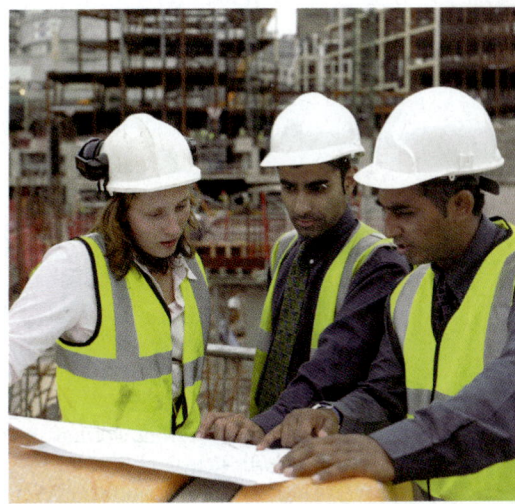

Building a complex structure requires careful planning. The architect's visualization and drawings have to be complete before the detailed procedures of construction get underway. The same is true for solving problems in physics.

Continued

- **Does my answer have the correct units and number of significant figures?**
- **Does the value I computed make physical sense?** In this book all calculations use physically reasonable numbers. You will not be given a problem to solve in which the final velocity of a bicycle is 100 miles per hour! If your final answer seems unreasonable, you should go back and check your work.
- **Can I estimate what the answer should be to check my solution?**
- **Does my final solution make sense in the context of the material I am learning?**

EXAMPLE 2.3   Using the problem-solving strategy to analyze the motion of a baseball

In a game of baseball, a batter hits the ball straight toward the pitcher, who stands 60 ft from the batter. The ball comes off the bat at a speed of 100 ft/s. How far is the ball from the batter 1.0 s after it passes the pitcher?

PREPARE   Start by drawing a picture. In this case, we will draw a position-versus-time graph of the motion, shown in Figure 2.11. We've chosen the origin of our coordinate system to be at the position where the batter hits the ball. As the ball leaves the bat, its position increases at the rate of 100 ft/s. This is uniform motion, so the graph will be a straight line. We now add important information to our picture. One important distance to note on the graph is the position of the pitcher. When the ball is at this position, we can see that the time is a bit over 0.5 s.

We are interested in the position of the ball 1.0 s after this point. A quick look at the graph shows us that this position will be a bit over 150 ft.

SOLVE   As the ball is in uniform motion, we will use Equation 2.3 for our solution. We take x_i to be 60 ft, when the ball passes the pitcher. We are interested in the position x_f 1.0 s later, so Δt is 1.0 s. Thus

$$x_f = x_i + v_x \Delta t = 60 \text{ ft} + (100 \text{ ft/s})(1.0 \text{ s}) = 160 \text{ ft}$$

ASSESS   The final value is close to our estimate from the graph in the Prepare step. It also seems physically reasonable; if you have ever watched a baseball game, you know that 1.0 s after a struck ball passes the pitcher it is well on its way to the outfield.

1. Draw a line for the motion of the ball using two points. At $t = 0$ s, the ball is at 0 ft; at $t = 1.0$ s, the ball will have moved 100 ft. and so will be at 100 ft.

2. The ball passes the pitcher at a position of 60 ft.

3. This is approximately one second after the ball passes the pitcher.

FIGURE 2.11   Graphing the uniform motion of the baseball.

This was a straightforward problem, but you can see that making such a complete picture in the "Prepare" step let us visualize the important points in the motion, aiding our solution. It also allowed us to estimate what the final solution should be. For more involved problems, the picture is much more important.

From Position to Velocity

We've focused thus far on graphs of position versus time. Now let's see what we can learn by graphing velocity versus time. In section 2.1, we considered the motion of a student walking to school; the motion diagram is repeated in Figure 2.12. Rather than looking at her position, let's now focus on her velocity, as represented by the green arrows. A long arrow corresponds to a large velocity, a short arrow a small velocity.

FIGURE 2.12   The motion diagram for the student walking to school, revisited.

The green arrow between $t = 0$ min and $t = 1$ min represents her velocity during this period of time. Looking at the data in Table 2.1, we see that her velocity during this interval is

$$v_x = \frac{\Delta x}{\Delta t} = \frac{60 \text{ m}}{1 \text{ min}} \times \frac{1 \text{ min}}{60 \text{ s}} = 1.0 \text{ m/s}$$

We see from the motion diagram that she moved at this constant velocity until $t = 3$ min, then she suddenly slowed down. We can use the data in Table 2.1 to find that her velocity from $t = 3$ min to $t = 6$ min is 0.33 m/s. She then increased her velocity to 1.7 m/s. Using all this information gives us the **velocity-versus-time graph** shown in Figure 2.13.

There are three clearly defined segments to the motion: an initial segment, a middle segment with a smaller velocity, and a final segment with a larger velocity.

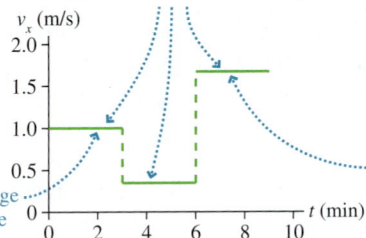

The velocity does not change during each segment, so the graph is simply a straight, level line.

Changing the velocity to another value moves the graph to another straight, level line at a different height on the vertical axis.

FIGURE 2.13 The student's motion is a series of uniform motions at different velocities.

NOTE ▶ This velocity-versus-time graph includes vertical segments in which the velocity changes instantaneously. Such rapid changes are an idealization; it actually takes a small amount of time to change velocity. ◀

Rather than begin with the motion diagram, we could deduce the velocity-versus-time graph from the position-versus-time graph as shown in Figure 2.14. Velocity is the slope of a position-versus-time graph, so we can translate from the position graph to the velocity graph in a straightforward fashion.

During the first segment of the motion, the slope is a constant 60 m/min = 1.0 m/s . . .

The slope decreases during this segment of the motion . . .

. . . so the velocity is a constant 1.0 m/s.

. . . so the velocity is smaller.

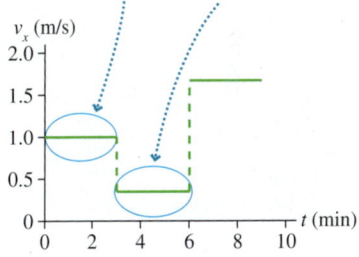

FIGURE 2.14 Deducing the velocity-versus-time graph from the position-versus-time graph.

EXAMPLE 2.4 **Analyzing a car's position graph**

Figure 2.15 gives the position-versus-time graph of a car.

a. Draw the car's velocity-versus-time graph.
b. Describe the car's motion in words.

FIGURE 2.15 The position-versus-time graph of a car.

PREPARE Figure 2.15 is a graphical representation of the motion. The car's position-versus-time graph is a sequence of

three straight lines. Each of these straight lines represents uniform motion at a constant velocity. We can determine the car's velocity during each interval of time by measuring the slope of the line.

SOLVE

a. From $t = 0$ s to $t = 2$ s ($\Delta t = 2$ s) the car's displacement is $\Delta x = -4 \text{ m} - 0 \text{ m} = -4 \text{ m}$. The velocity during this interval is

$$v_x = \frac{\Delta x}{\Delta t} = \frac{-4 \text{ m}}{2 \text{ s}} = -2 \text{ m/s}$$

The car's position does not change from $t = 2$ s to $t = 4$ s ($\Delta x = 0$ m), so $v_x = 0$ m/s. Finally, the displacement between $t = 4$ s and $t = 6$ s ($\Delta t = 2$ s) is $\Delta x = 10$ m. Thus the velocity during this interval is

$$v_x = \frac{10 \text{ m}}{2 \text{ s}} = 5 \text{ m/s}$$

Continued

These velocities are represented graphically in Figure 2.16.

FIGURE 2.16 The velocity-versus-time graph for the car.

b. The velocity-versus-time graph of Figure 2.16 shows the motion in a way that we can describe in a straightforward manner: The car backs up for 2 s at 2 m/s, sits at rest for 2 s, then drives forward at 5 m/s for 2 s.

ASSESS Notice that the velocity graph and the position graph look completely different. The value of the velocity graph at any instant of time equals the *slope* of the position graph. Since the position graph is made up of segments of constant slope, the velocity graph will be made up of segments of constant *value*.

From Velocity to Position

We have seen how to convert from position-versus-time information to velocity-versus-time information. It is possible to go the other way as well: Given information about an object's velocity, we can determine its position.

Suppose a car moves at a constant velocity of 12 m/s for 4.0 s. How far does it travel—that is, what is its displacement during this time interval?

Equation 2.4, $\Delta x = v_x \Delta t$, describes the displacement mathematically; for a graphical interpretation, we consider the graph of velocity versus time in Figure 2.17. In the figure, we've shaded a rectangle whose height is the velocity v_x (12 m/s) and whose base is the time interval Δt (4.0 s). The area of this rectangle is $v_x \Delta t$. Looking at Equation 2.4, we see that this quantity is also equal to the displacement of the car. The area of this rectangle is the area between the axis and the line representing the velocity; we call it the "area under the graph." We see that the **displacement Δx is equal to the area under the velocity graph during interval Δt.**

Whether we use Equation 2.4 or the area under the graph to compute the displacement, we get the same result:

$$\Delta x = v_x \Delta t = (12 \text{ m/s})(4.0 \text{ s}) = 48 \text{ m}$$

Although we've shown that the displacement is the area under the graph only for uniform motion, where the velocity is constant, we'll soon see that this result applies to any one-dimensional motion.

NOTE ▶ Wait a minute! The displacement $\Delta x = x_f - x_i$ is a length. How can a length equal an area? Recall earlier, when we found that the velocity is the slope of the position graph, we made a distinction between the *actual* slope and the *physically meaningful* slope? The same distinction applies here. The velocity graph does indeed bound a certain area on the page. That is the actual area, but it is *not* the area to which we are referring. Once again, we need to measure the quantities we are using, v_x and Δt, by referring to the scales on the axes. Δt is some number of seconds while v_x is some number of meters per second. When these are multiplied together, the *physically meaningful* area has units of meters, appropriate for a displacement. ◀

FIGURE 2.17 Displacement is the area under a velocity-versus-time graph.

STOP TO THINK 2.2   An object moves with the velocity-versus-time graph shown at left below. Which of the position-versus-time graphs matches this motion?

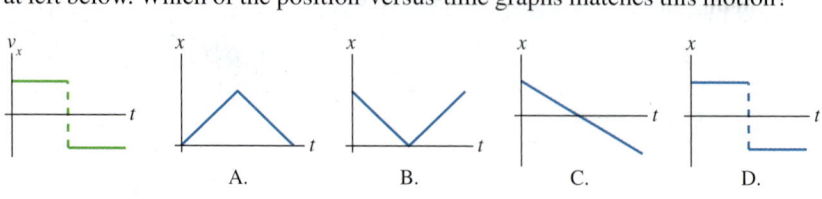

2.3 Motion with Changing Velocity

The objects we've studied so far have moved with a constant, unchanging velocity or, like the car in Example 2.4, changed abruptly from one constant velocity to another. This is not very realistic. Real moving objects speed up and slow down, *changing* their velocity. As an extreme example, think about a drag racer. In a typical race, the car begins at rest but, one second later, is moving at over 25 miles per hour!

For one-dimensional motion, an object changing its velocity is either speeding up or slowing down. When you drive your car, as you speed up or slow down—changing your velocity—a glance at your speedometer tells you how fast you're going *at that instant.* An object's velocity—a speed *and* a direction—at a specific *instant* of time *t* is called the object's **instantaneous velocity.**

But what does it mean to have a velocity "at an instant"? An instantaneous velocity of magnitude 60 mph means that the rate at which your car's position is changing—at that exact instant—is such that it would travel a distance of 60 miles in 1 hour *if* it continued at that rate without change. Said another way, if *just for an instant* your car matches the velocity of another car driving at a steady 60 mph, then your instantaneous velocity is 60 mph. Whether or not your car actually does travel at that velocity for another hour is not relevant. **From now on, the word "velocity" will always mean instantaneous velocity.**

For uniform motion, we found that an object's position-versus-time graph is a straight line and the object's velocity is the slope of that line. In contrast, Figure 2.18 shows that the position-versus-time graph for a drag racer is a *curved* line. The displacement Δx during equal intervals of time gets larger as the car speeds up. Even so, we can still use the slope of the line to measure the car's velocity.

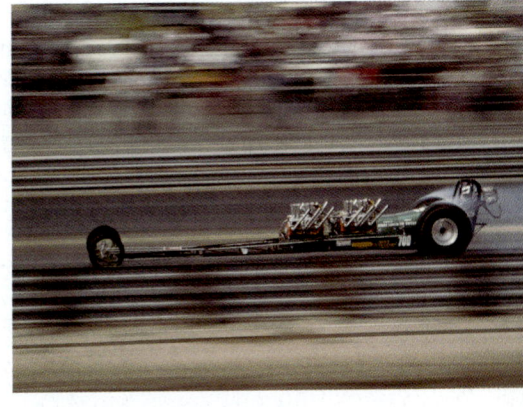

A drag racer moves with rapidly changing velocity.

FIGURE 2.18 Position-versus-time graph for a drag racer.

Finding the instantaneous velocity

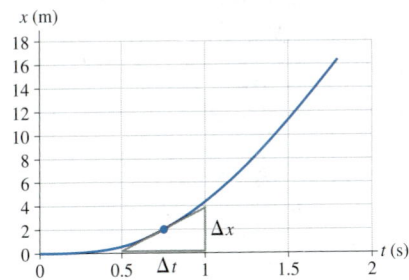

If the velocity changes, the position graph is no longer a straight line: It is clearly curved. How are we to compute the slope to find the velocity? The key is to focus in on a small segment of the motion. Let's look at the motion in a very small time interval right around $t = 0.75$ s. This is highlighted with a circle, and we show a "close-up" in the next graph, at right.

Now that we have magnified a small part of the position graph, we see that the graph in this small part appears to have a constant slope. It is always possible to make the graph appear as a straight line by choosing a small enough time interval. Now we can find the slope of the line by calculating the rise over run, just as we did before:

$$v_x = \frac{1.6 \text{ m}}{0.20 \text{ s}} = 8.0 \text{ m/s}$$

This is the slope of the graph at $t = 0.75$ s, and thus the velocity at this instant of time.

Graphically, the slope of the curve at a particular point is the same as the slope of a straight line drawn *tangent* to the curve at that point. **The slope of the tangent line is the instantaneous velocity at that instant of time.**

Calculating rise over run for the tangent line, we get

$$v_x = \frac{4.0 \text{ m}}{0.50 \text{ s}} = 8.0 \text{ m/s}$$

This is the same value we obtained from considering the close-up view.

CONCEPTUAL EXAMPLE 2.2 **Analyzing an elevator's position graph**

Figure 2.19 shows the position-versus-time graph of an elevator.

a. Sketch an approximate velocity-versus-time graph.
b. At which point or points is the elevator moving the fastest?
c. Is the elevator ever at rest? If so, at which point or points?

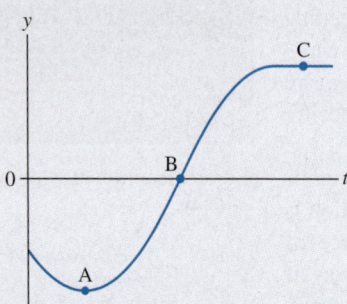

FIGURE 2.19 The position-versus-time graph for an elevator.

REASON a. Notice that the position graph shows y versus t, rather than x versus t, indicating that the motion is vertical rather than horizontal. Our analysis of one-dimensional motion has made no assumptions about the direction of motion, so it applies equally well to both horizontal and vertical motion. Let's start by sketching an approximate velocity-versus-time graph. As we just found, the velocity at a particular instant of time is the slope of a tangent line to the position-versus-time graph at that time. We can move point-by-point along the position-versus-time graph, noting the slope of the tangent at each point. This will give us the velocity at that point.

Initially, to the left of point A, the slope is negative and thus the velocity is negative (i.e., the elevator is moving downward). But the slope decreases as the curve flattens out, and by the time the graph gets to point A, the slope is zero. The slope then increases to a maximum value at point B, decreases back to zero a little before point C, and remains at zero thereafter. This reasoning process is outlined in Figure 2.20a, and Figure 2.20b shows the approximate velocity-versus-time graph that results.

FIGURE 2.20 Finding a velocity graph from a position graph.

The other questions were really answered during the construction of the graph:

b. The elevator moves the fastest at point B where the slope of the position graph is the steepest.
c. A particle at rest has $v_y = 0$. Graphically, this occurs at points where the tangent line to the position-versus-time graph is horizontal and thus has zero slope. Figure 2.20 shows that the slope is zero at points A and C. At point A, the velocity is only instantaneously zero as the particle reverses direction from downward motion (negative velocity) to upward motion (positive velocity). At point C, the elevator has actually stopped and remains at rest.

ASSESS Once again, the shape of the velocity graph bears no resemblance to the shape of the position graph. You must translate between slope information on the position graph and value information on the velocity graph.

For uniform motion we showed that the displacement Δx is the area under the velocity-versus-time graph during time interval Δt. We can generalize this idea to the case of an object whose velocity varies. Figure 2.21a on the next page is the velocity-versus-time graph for an object whose velocity changes with time. Suppose we know the object's position to be x_i at an initial time t_i. Our goal is to find its position x_f at a later time t_f.

Because we know how to handle constant velocities, let's *approximate* the velocity function of Figure 2.21a as a series of constant-velocity steps of width Δt. The velocity during each step is constant (uniform motion), so we can calculate the displacement during each step as the area of the rectangle under the curve. The total displacement of the object between t_i and t_f can be found as the sum of all the individual displacements during each of the constant-velocity steps. We can see in Figure 2.21c that the total displacement is approximately equal to the area under the graph, even in the case where the velocity varies. Although the approximation shown in the figure is rather rough, with only nine steps, we can

FIGURE 2.21 Approximating a velocity-versus-time graph with a series of constant-velocity steps.

imagine that it could be made as accurate as desired by having more and more ever-narrower steps.

Consequently, an object's displacement is related to its velocity by

$$x_f - x_i = \Delta x = \text{area under the velocity graph } v_x \text{ between } t_i \text{ and } t_f \quad (2.5)$$

EXAMPLE 2.5 The displacement during a rapid start

Figure 2.22 shows the velocity-versus-time graph of a car pulling away from a stop. How far does the car move during the first 3.0 s?

PREPARE Figure 2.22 is a graphical representation of the motion. The question "how far" indicates that we need to find a displacement Δx rather than a position x. According to Equation 2.5, the car's displacement $\Delta x = x_f - x_i$ between $t = 0$ s and $t = 3$ s is the area under the curve from $t = 0$ s to $t = 3$ s.

SOLVE The curve in this case is an angled line, so the area is that of a triangle:

$$\Delta x = \text{area of triangle between } t = 0 \text{ s and } t = 3 \text{ s}$$
$$= \tfrac{1}{2} \times \text{base} \times \text{height} = \tfrac{1}{2} \times 3 \text{ s} \times 12 \text{ m/s} = 18 \text{ m}$$

The car moves 18 m during the first 3 seconds as its velocity changes from 0 to 12 m/s.

ASSESS The "area" is a product of s with m/s, so Δx has the proper units of m. Let's check the numbers. The final velocity, 12 m/s, is about 25 mph. Pulling away from a stop, you might expect to reach this speed in about 3 s—at least if you have a reasonably sporty vehicle! If the car had moved at a constant 12 m/s (the final velocity) during these 3 s, the distance would be 36 m. The actual distance traveled during the 3 s is 18 m— half of 36 m. This makes sense, as the velocity was 0 m/s at the start of the problem and increased steadily to 12 m/s.

FIGURE 2.22 Velocity-versus-time graph for the car of Example 2.5.

STOP TO THINK 2.3 Which velocity-versus-time graph goes with the position-versus-time graph on the left?

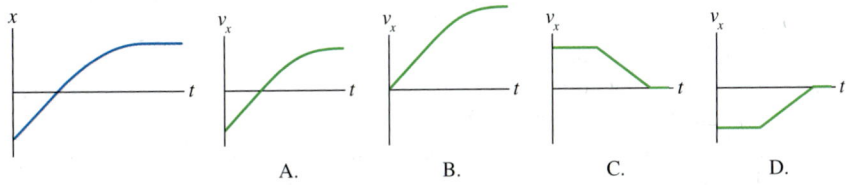

2.4 Acceleration

One of the goals of this chapter is to describe motion. Position, time, and velocity are important concepts, but we need one more motion concept, one that will describe a *change* in the velocity.

As an example, a frequently quoted measurement of car performance is the time it takes the car to change its velocity from 0 to 60 mph. Table 2.2 shows this time for two rather different cars.

TABLE 2.2 Performance data for vehicles

Vehicle	Time to go from 0 to 60 mph
1997 Porsche 911 Turbo S	3.6 s
1973 Volkswagen Super Beetle Convertible	24 s

Let's look at motion diagrams for the Porsche and the Volkswagen in Figure 2.23. We can see two important facts about the motion. First, the lengths of the velocity vectors are increasing, showing that the speeds are increasing. Second, the velocity vectors for the Porsche are increasing in length more rapidly than those of the VW. The quantity we seek is one that measures how rapidly an object's velocity vectors change in length.

FIGURE 2.23 Motion diagrams for the Porsche and Volkswagen.

When we wanted to measure changes in position, the ratio $\Delta x/\Delta t$ was useful. This ratio, which we defined as the velocity, is the *rate of change of position*. Similarly, we can measure how rapidly an object's velocity changes with the ratio $\Delta v_x/\Delta t$. Given our experience with velocity, we can say a couple of things about this new ratio:

- The ratio $\Delta v_x/\Delta t$ is the *rate of change of velocity*.
- The ratio $\Delta v_x/\Delta t$ is the *slope of a velocity-versus-time graph*.

We will define this ratio as the **acceleration,** for which we use the symbol a_x:

$$a_x = \frac{\Delta v_x}{\Delta t} \tag{2.6}$$

Definition of the acceleration as rate of change of velocity

Similarly, $a_y = \Delta v_y/\Delta t$ for vertical motion.

As an example, let's calculate the acceleration for the Porsche and the Volkswagen. For both, the initial velocity $(v_x)_i$ is zero and the final velocity $(v_x)_f$ is 60 mph. Thus the *change* in velocity is $\Delta v_x = 60$ mph. In m/s, our SI unit of velocity, $\Delta v_x = 27$ m/s.

Now we can use Equation 2.6 to compute acceleration. Let's start with the Porsche, which speeds up to 27 m/s in $\Delta t = 3.6$ s:

$$a_{\text{Porsche}\,x} = \frac{\Delta v_x}{\Delta t} = \frac{27 \text{ m/s}}{3.6 \text{ s}} = 7.5 \, \frac{\text{m/s}}{\text{s}}$$

Notice the units; they are velocity over time. In the first second of motion, the Porsche's velocity increases by 7.5 m/s; in the next second, it increases by another 7.5 m/s, and so on. After 1 second, the velocity is 7.5 m/s; after 2 seconds, it is 15 m/s. This increase continues as long as the Porsche has an acceleration of 7.5 (m/s)/s.

The Volkswagen's acceleration is

$$a_{\text{VW}\,x} = \frac{\Delta v_x}{\Delta t} = \frac{27 \text{ m/s}}{24 \text{ s}} = 1.1 \, \frac{\text{m/s}}{\text{s}}$$

In each second, the Volkswagen changes its speed by 1.1 m/s. This is about 1/7 the acceleration of the Porsche! The reasons why the Porsche is capable of greater acceleration have to do with what *causes* the motion. We will explore the reasons for acceleration in Chapter 4. For now, we will simply note that the Porsche is capable of much larger acceleration, something you would have suspected.

> **NOTE** ▶ It is customary to abbreviate the acceleration units (m/s)/s as m/s^2. For example, the Volkswagen has an acceleration of 1.1 m/s^2. We will use this notation, but keep in mind the *meaning* of the notation as "(meters per second) per second." ◀

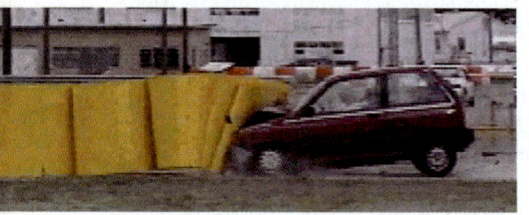

Cushion kinematics When a car hits an obstacle head-on, the damage to the car and its occupants can be reduced by making the acceleration as small as possible. As we can see from Equation 2.6, acceleration can be reduced by making the *time* for a change in velocity as long as possible. This is the purpose of the yellow crash cushion barrels you may have seen in work zones on highways.

Acceleration is a fairly abstract concept. Position and time are our real hands-on measurements of an object, and they are easy to understand. You can "see" where the object is located and the time on the clock. Velocity is a bit more abstract, being a relationship between the change of position and the change of time. Motion diagrams help us visualize velocity as the vector arrows connecting one position of the object to the next. Acceleration is an even more abstract idea about changes in the velocity, but we can use graphs to help us better understand it.

Acceleration Is the Slope of the Velocity-Versus-Time Graph

Let's use the values we have computed for acceleration to make a table of velocities for the Porsche and the Volkswagen. Table 2.3 uses the idea that the VW's velocity increases by 1.1 m/s every second while the Porsche's velocity increases by 7.5 m/s every second. The data in Table 2.3 are the basis for the velocity-versus-time graphs in Figure 2.24.

Look at the graph for the Porsche; it's a straight line. The slope is computed using the rise over the run:

$$\text{slope of velocity-versus-time graph} = \frac{\Delta v_x}{\Delta t}$$

Compare this with Equation 2.6; the equation for the slope is the same as that for the acceleration. **Acceleration is the slope of a velocity-versus-time graph.** The VW has a smaller acceleration, so its graph has a smaller slope. The slope could be negative; this would mean that velocity is *decreasing,* as we will see in the next example.

TABLE 2.3 Velocity data for the Volkswagen and the Porsche

Time (s)	Velocity of VW (m/s)	Velocity of Porsche (m/s)
0	0	0
1	1.1	7.5
2	2.2	15.0
3	3.3	22.5
4	4.4	30.0

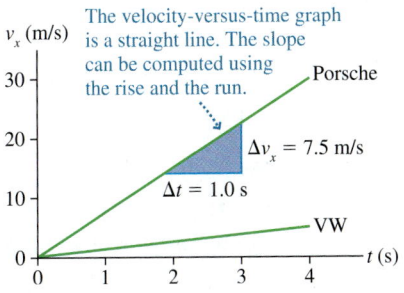

FIGURE 2.24 Velocity-versus-time graphs for the two cars.

CONCEPTUAL EXAMPLE 2.3 **Analyzing a car's velocity graph**

Figure 2.25a is a graph of velocity versus time for a car. Sketch a graph of the car's acceleration versus time.

REASON The graph can be divided into three sections:

- An initial segment, in which the velocity increases at a steady rate.
- A middle segment, in which the velocity is constant.
- A final segment, in which the velocity decreases at a steady rate.

In each section, the acceleration is the slope of the velocity-versus-time graph. Thus the initial segment has constant, posi-

tive acceleration, the middle segment has zero acceleration, and the final segment has a constant, *negative* acceleration. The acceleration graph appears in Figure 2.25b.

ASSESS This process is analogous to finding a velocity graph from the slope of a position graph. In the middle segment, notice that zero acceleration does *not* mean that the velocity is zero. The velocity is constant, which means it is *not changing* and thus the car is not accelerating. The car does accelerate during the initial and final segments. The magnitude of the acceleration is a measure of how quickly the velocity is changing. How about the sign? This is an issue we will address in the next section.

FIGURE 2.25 Finding an acceleration graph from a velocity graph.

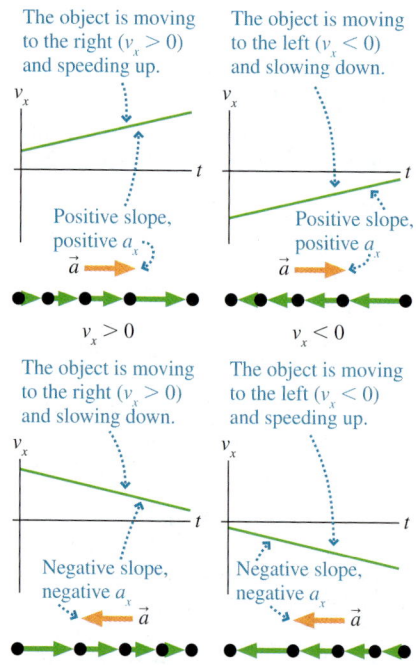

The object is moving to the right ($v_x > 0$) and speeding up.

Positive slope, positive a_x

$\vec{a}$

$v_x > 0$

The object is moving to the left ($v_x < 0$) and slowing down.

Positive slope, positive a_x

$\vec{a}$

$v_x < 0$

The object is moving to the right ($v_x > 0$) and slowing down.

Negative slope, negative a_x

$\vec{a}$

The object is moving to the left ($v_x < 0$) and speeding up.

Negative slope, negative a_x

$\vec{a}$

FIGURE 2.26 Determining the sign of the acceleration.

FIGURE 2.27 The red dots show the position of the top of the Saturn V rocket at equally spaced intervals during liftoff.

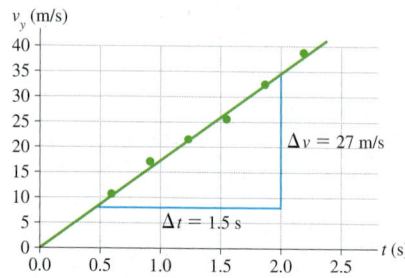

FIGURE 2.28 A graph of the rocket's velocity versus time.

The Sign of the Acceleration

A natural tendency is to think that a positive value of a_x or a_y describes an object that is speeding up while a negative value describes an object that is slowing down (decelerating). Unfortunately, this simple interpretation *does not work.*

Because an object can move right or left (or, equivalently, up and down) while either speeding up or slowing down, there are four situations to consider. Figure 2.26 shows a motion diagram and a velocity graph for each of these situations. As we've seen, an object's acceleration is the slope of its velocity graph, so a positive slope implies a positive acceleration and a negative slope implies a negative acceleration.

Acceleration, like velocity, is really a vector quantity, a concept that we will explore more fully in Chapter 3. Figure 2.26 shows the acceleration vector for the four situations. The acceleration vector points in the same direction as the velocity vector $\vec{v}$ for an object that is speeding up and opposite to $\vec{v}$ for an object that is slowing down.

An object that speeds up as it moves to the right (positive v_x) has a positive acceleration, but an object that speeds up as it moves to the left (negative v_x) has a negative acceleration. Whether or not an object that is slowing down has a negative acceleration depends on whether the object is moving to the right or to the left. This is admittedly a bit more complex than thinking that negative acceleration always means slowing down, but our definition of acceleration as the slope of the velocity graph forces us to pay careful attention to the sign of the acceleration.

STOP TO THINK 2.4   A particle moves with the following velocity-versus-time graph. At which labeled point is the magnitude of the acceleration the greatest?

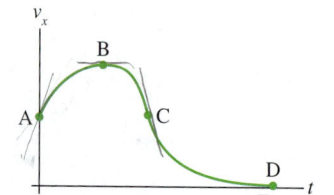

2.5 Motion with Constant Acceleration

For uniform motion—motion with constant velocity—we found in Equation 2.3 a simple relationship between position and time. It's no surprise that there are also simple relationships that connect the various kinematic variables in constant-acceleration motion. We will start with a concrete example, the launch of a Saturn V rocket like the one that carried the Apollo astronauts to the moon in the 1960s and 1970s. Figure 2.27 shows one frame from a video of a rocket lifting off the launch pad. The red dots show the position of the top of the rocket at equally spaced intervals of time. This is a motion diagram for the rocket, and we can see that the velocity is increasing. The graph of velocity versus time in Figure 2.28 shows that the velocity is increasing at a fairly constant rate. We can approximate the rocket's motion as constant acceleration.

We can use the slope of the graph in Figure 2.28 to determine the acceleration of the rocket:

$$a_y = \frac{\Delta v_y}{\Delta t} = \frac{27 \text{ m/s}}{1.5 \text{ s}} = 18 \text{ m/s}^2$$

This acceleration is more than double the acceleration of the Porsche, and it goes on for quite a long time—the first phase of the launch lasts 2 minutes! How fast is the rocket moving at the end of this acceleration, and how far has it traveled? To answer these questions, we first need to work out some basic kinematic formulas for motion with constant acceleration.

Constant Acceleration Equations

Consider an object whose acceleration a_x remains constant during the time interval $\Delta t = t_f - t_i$. At the beginning of this interval, the object has initial velocity $(v_x)_i$ and initial position x_i. Note that t_i is often zero, but it need not be. Figure 2.29a shows the acceleration-versus-time graph. It is a horizontal line between t_i and t_f, indicating a *constant* acceleration.

The object's velocity is changing because the object is accelerating. We can use the acceleration to find $(v_x)_f$ at a later time t_f. We defined acceleration as

$$a_x = \frac{\Delta v_x}{\Delta t} = \frac{(v_x)_f - (v_x)_i}{\Delta t} \tag{2.7}$$

which is rearranged to give

$$(v_x)_f = (v_x)_i + a_x \Delta t \tag{2.8}$$

Velocity equation for an object with constant acceleration

LINEAR
p. 38

NOTE ▶ We have expressed this equation for motion along the x-axis, but it is a general result that will apply to any axis. ◀

The velocity-versus-time graph for this constant-acceleration motion, shown in Figure 2.29b, is a straight line that starts at $(v_x)_i$ and has slope a_x. This is another example of a linear relationship.

We would also like to know the object's position x_f at time t_f. As you learned in the last section, the displacement Δx during the time interval Δt is the area under the velocity-versus-time graph. The shaded area in Figure 2.29b can be subdivided into a rectangle of area $(v_x)_i \Delta t$ and a triangle of area $\frac{1}{2}(a_x \Delta t)(\Delta t) = \frac{1}{2}a_x(\Delta t)^2$. Adding these gives

$$x_f = x_i + (v_x)_i \Delta t + \frac{1}{2}a_x(\Delta t)^2 \tag{2.9}$$

Position equation for an object with constant acceleration

where $\Delta t = t_f - t_i$ is the elapsed time. The fact that the time interval Δt appears in the equation as $(\Delta t)^2$ causes the position-versus-time graph for constant-acceleration motion to have a parabolic shape. For the rocket launch of Figure 2.27, a graph of position of the top of the rocket versus time appears as in Figure 2.30.

Equations 2.8 and 2.9 are two of the basic kinematic equations for motion with constant acceleration. They allow us to predict an object's position and velocity at a future instant of time. We need one more equation to complete our set, a direct relationship between displacement and velocity. To derive this relationship, we first use Equation 2.8 to write $\Delta t = ((v_x)_f - (v_x)_i)/a_x$. We can substitute this into Equation 2.9 to obtain

$$(v_x)_f^2 = (v_x)_i^2 + 2a_x \Delta x \tag{2.10}$$

Relating velocity and displacement for constant-acceleration motion

In Equation 2.10 $\Delta x = x_f - x_i$ is the *displacement* (not the distance!). Notice that Equation 2.10 does not require knowing the time interval Δt. This is an important equation in problems where you're not given information about times.

(a) Acceleration   Constant acceleration a_x

Displacement Δx is the area under the curve. The area can be divided into a rectangle of height $(v_x)_i$ and a triangle of height $a_x \Delta t$.

(b) Velocity

Constant slope $= a_x$

FIGURE 2.29 Acceleration and velocity graphs for motion with constant acceleration.

Solar sailing A rocket achieves a high speed by having a very high acceleration. A different approach is represented by a solar sail. A spacecraft with a solar sail accelerates due to the pressure of sunlight from the sun on a large, mirrored surface. The acceleration is minuscule, but it can continue for a long, long time. After an acceleration period of a few *years*, the spacecraft will reach a respectable speed!

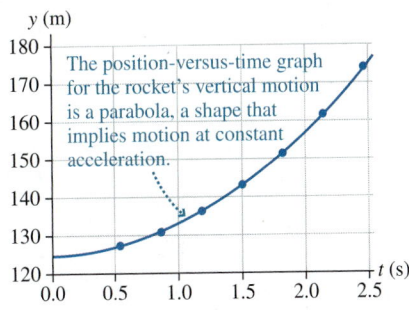

The position-versus-time graph for the rocket's vertical motion is a parabola, a shape that implies motion at constant acceleration.

FIGURE 2.30 Position-versus-time graph for the Saturn V rocket launch.

TABLE 2.4 Kinematic equations for motion with constant acceleration

$$(v_x)_f = (v_x)_i + a_x \Delta t$$
$$x_f = x_i + (v_x)_i \Delta t + \tfrac{1}{2} a_x (\Delta t)^2$$
$$(v_x)_f^2 = (v_x)_i^2 + 2 a_x \Delta x$$

1.1–1.3 Activ Physics

Getting up to speed BIO A bird must have a minimum speed to fly. Generally, the larger the bird, the larger the takeoff speed. Small birds can get moving fast enough to fly with a vigorous jump, but larger birds may need a running start. This swan must accelerate for a long distance in order to achieve the high speed it needs to fly, so it makes a frenzied dash across the frozen surface of a pond. Swans require a long, clear stretch of water or land to become airborne. Airplanes require an even higher takeoff speed and thus an even longer runway, as we will see.

Equations 2.8, 2.9, and 2.10 are the key equations for motion with constant acceleration. These results are summarized in Table 2.4.

Figure 2.31 is a graphical comparison of motion with constant velocity (uniform motion) and motion with constant acceleration. Notice that uniform motion is really a special case of constant-acceleration motion in which the acceleration happens to be zero.

(a) Motion at constant velocity

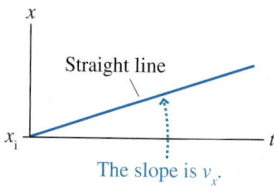

(b) Motion at constant acceleration

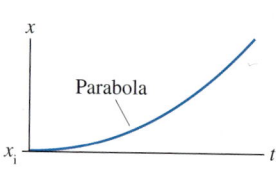

FIGURE 2.31 Motion with constant velocity and constant acceleration. These graphs assume $x_i = 0$, $(v_x)_i > 0$, and (for constant acceleration) $a_x > 0$.

For motion at constant acceleration, a graph of position versus time is a *parabola*. This is a new mathematical form, one that we will see again. If $(v_x)_i = 0$, the second equation in Table 2.4 is simply

$$\Delta x = \tfrac{1}{2} a_x (\Delta t)^2 \tag{2.11}$$

Δx depends on the *square* of Δt; we call this a *quadratic relationship*.

📈 **Quadratic relationships**   ✏️ Exercises 14, 19

Two quantities are said to have a **quadratic relationship** if y is proportional to the square of x. We write the mathematical relationship as

$$y = Ax^2$$

y is a quadratic function of x

The graph of a quadratic relationship is a parabola.

SCALING Suppose we double x:

$$x_f = 2x_i$$

This will change y as well:

$$y_f = Ax_f^2 = A(2x_i)^2 = 2^2 Ax_i^2 = 2^2 y_i$$
$$= 4y_i$$

Increasing x by a factor of 2 causes y to increase by a factor of 2^2, or 4.
Generally, we can say that

Changing x by a factor C changes y by a factor C^2.

Doubling x causes y to change by a factor of 4.

EXAMPLE 2.6 Displacement of a drag racer

A drag racer, starting from rest, travels 6.0 m in 1.0 s. Suppose the car continues this acceleration for an additional 4.0 s. How far from the starting line will the car be?

SOLVE We assume that the acceleration is constant, so the displacement will follow Equation 2.11. This is a *quadratic relationship,* so the displacement will scale as the square of the time. After 1.0 s, the car has traveled 6.0 m; after another 4.0 s, a total of 5.0 s will have elapsed. The time has increased by a factor of 5, so the displacement will increase by a factor of 5^2, or 25. The total displacement is

$$\Delta x = 25(6.0 \text{ m}) = 150 \text{ m}$$

STOP TO THINK 2.5   A car is moving toward the right. The driver puts on the brakes, reducing the speed. The car then continues to the right at a reduced speed. Which of the following velocity-versus-time graphs matches this motion?

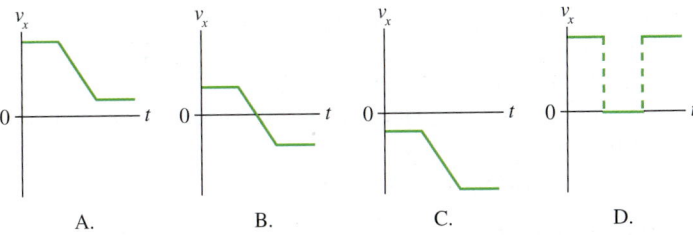

A.   B.   C.   D.

2.6 Solving One-Dimensional Motion Problems

The big challenge when solving a physics problem is to translate the words into symbols that can be manipulated, calculated, and graphed. This translation from words to symbols is the heart of problem solving in physics. Ambiguous words and phrases must be clarified, the imprecise must be made precise, and you must arrive at an understanding of exactly what the question is asking.

In this section we will explore some general problem-solving strategies that we will use throughout the text, applying them to problems of motion along a line.

▶ **Dinner at a distance** BIO A chameleon's tongue is a powerful tool for catching prey. Certain species can extend the tongue to a distance of over 1 ft in less than 0.1 s! A study of the kinematics of the motion of the chameleon tongue, using techniques like those in this chapter, reveals that the tongue has a period of rapid acceleration followed by a period of constant velocity. This knowledge is a very valuable clue in the analysis of the evolutionary relationships between chameleons and other animals.

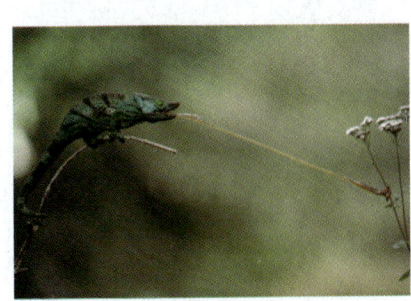

The Pictorial Representation

You'll find that motion problems and other physics problems often have several variables and other pieces of information to keep track of. The best way to tackle such problems is to draw a picture, as we noted when we introduced a general problem-solving strategy. But what kind of picture should we draw?

In this section, we will begin to draw **pictorial representations** as an aid to solving problems. A pictorial representation shows all of the important details that we need to keep track of in a problem. Pictorial representations will be very important in solving motion problems.

TACTICS BOX 2.2 Drawing a pictorial representation

Exercise 21

❶ **Sketch the situation.** Not just any sketch: Show the object at the *beginning* of the motion, at the *end,* and at any point where the character of the motion changes. Very simple drawings are adequate.

❷ **Establish a coordinate system.** Select your axes and origin to match the motion.

❸ **Define symbols.** Use the sketch to define symbols representing quantities such as position, velocity, acceleration, and time. *Every* variable used later in the mathematical solution should be defined on the sketch.

We will generally combine the pictorial representation with a **list of values,** which will include

- *Known information.* Make a table of the quantities whose values you can determine from the problem statement or that can be found quickly with simple geometry or unit conversions.
- *Desired unknowns.* What quantity or quantities will allow you to answer the question?

EXAMPLE 2.7 Drawing a pictorial representation

Complete a pictorial representation and a list of values for the following problem: A rocket sled accelerates at 50 m/s² for 5 s. What are the total distance traveled and the final velocity?

PREPARE Figure 2.32a shows a pictorial representation as drawn by an artist in the style of the figures in this book. This is certainly neater and more artistic than the sketches you will make when solving problems yourself! Figure 2.32b shows a sketch like one you might actually do. It's less formal, but it contains all of the important information you need to solve the problem.

NOTE ▶ Throughout this book we will illustrate select examples with actual hand-drawn figures so that you have them to refer to as you work on your own pictures for homework and practice. ◀

Let's look at how these pictures were constructed. The motion has a clear beginning and end; these are the points sketched. A coordinate system has been chosen with the origin at the starting point. The quantities x, v_x, and t are needed at both points, so these have been defined on the sketch and distinguished by subscripts. The acceleration is associated with an interval between these points. Values for two of these quantities are given in the problem statement. Others, such as $x_i = 0$ m and $t_i = 0$ s, are inferred from our choice of coordinate system. The value $(v_x)_i = 0$ m/s is part of our *interpretation* of the problem. Finally, we identify x_f and $(v_x)_f$ as the quantities that will answer the question. We now understand quite a bit about the problem and would be ready to start a quantitative analysis.

ASSESS We didn't *solve* the problem; that was not our purpose. Constructing a pictorial representation and a list of values is part of a systematic approach to interpreting a problem and getting ready for a mathematical solution.

(a) Artist's version
Pictorial representation

a_x

$x_i, (v_x)_i, t_i$ $x_f, (v_x)_f, t_f$

List of values

Known
$x_i = 0$ m
$(v_x)_i = 0$ m/s
$t_i = 0$ s
$a_x = 50$ m/s²
$t_f = 5$ s

Find
$x_f, (v_x)_f$

(b) Student sketch

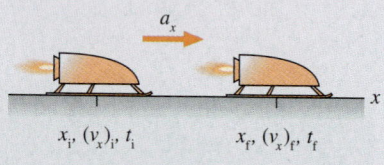

FIGURE 2.32 Constructing a pictorial representation and a list of values.

The Visual Overview

The pictorial representation and the list of values are a very good complement to the motion diagram and other ways of looking at a problem that we have seen. As we translate a problem into a form we can solve, we will combine all of these elements into what we will term a **visual overview.** The visual overview will consist of some or all of the following elements:

- A *motion diagram.* A good strategy for solving a motion problem is to start by drawing a motion diagram.
- A *pictorial representation,* as defined above.
- A *graphical representation.* For motion problems, it is often quite useful to include a graph of position and/or velocity.
- A *list of values.* This list should sum up all of the important values in the problem.

Future chapters will add other elements (such as forces) to this visual overview of the physics.

EXAMPLE 2.8 Kinematics of a rocket launch

A Saturn V rocket is launched straight up with a constant acceleration of 18 m/s². After 150 s, how fast is the rocket moving and how far has it traveled?

PREPARE Figure 2.33 shows a visual overview of the rocket launch that includes a motion diagram, a pictorial representation and a list of values. The visual overview shows the whole problem in a nutshell. The motion diagram illustrates the motion of the rocket, the pictorial representation (produced according to Tactics Box 2.2) shows axes, identifies the important points of the motion and defines variables. Finally, we have included a list of values that gives the known and unknown quantities. In the visual overview we have taken the statement of the problem in words and made it much more precise; it contains everything you need to know about the problem.

SOLVE Our first task is to find the final velocity. We can use the first kinematic equation of Table 2.4:

$$(v_y)_f = (v_y)_i + a_y \, \Delta t = 0 \text{ m/s} + (18 \text{ m/s}^2)(150 \text{ s})$$

$$= 2700 \text{ m/s}$$

The distance traveled is found using the second equation in Table 2.4:

$$y_f = y_i + (v_y)_i \, \Delta t + \tfrac{1}{2} a_y (\Delta t)^2$$

$$= 0 \text{ m} + (0 \text{ m/s})(150 \text{ s}) + \tfrac{1}{2}(18 \text{ m/s}^2)(150 \text{ s})^2$$

$$= 2.0 \times 10^5 \text{ m} = 200 \text{ km}$$

ASSESS The acceleration is very large, and it goes on for a long time, so the large final velocity and large distance traveled seem reasonable.

Motion diagram

1. The motion diagram shows the full range of the motion.

Pictorial representation

$y_f, (v_y)_f, t_f$

$\vec{a}$

$y_i, (v_y)_i, t_i$

2. The pictorial representation identifies the two important points of the motion, the start and the end, and shows that the rocket accelerates between them.

List of values

Known

$y_i = 0$ m
$(v_y)_i = 0$ m/s
$t_i = 0$ s
$a_y = 18$ m/s²
$t_f = 150$ s

Find

$(v_y)_f$ and y_f

3. The list of values makes everything concrete. We define the start of the problem to be at time 0 s, when the rocket has a position of 0 m and a velocity of 0 m/s. The end of the problem is at time 150 s. We are to find the position and velocity at this time.

FIGURE 2.33 Visual overview of the rocket launch.

Problem-Solving Strategy for Motion with Constant Acceleration

1.4–1.6, 1.8, 1.9, 1.11–1.14

Earlier in the chapter, we introduced a general problem-solving strategy. In this section, we will adapt this general strategy to solving problems of motion with constant acceleration. We will introduce such specific problem-solving strategies in future chapters as well.

(MP) **PROBLEM-SOLVING STRATEGY 2.1   Motion with constant acceleration**

PREPARE Draw a visual overview of the problem. This should include a motion diagram, a pictorial representation and a list of values; a graphical representation may be useful for certain problems.

SOLVE The mathematical solution is based on the three equations in Table 2.4.

■ Though the equations are phrased in terms of the variable x, it's customary to use y for motion in the vertical direction.
■ Use the equation that best matches what you know and what you need to find. For example, if you know acceleration and time and are looking for a change in velocity, the first equation is the best one to use.
■ Uniform motion with constant velocity has $a = 0$.

ASSESS Is your result believable? Does it have proper units? Does it make sense?

EXAMPLE 2.9 Calculating the minimum length of a runway

A fully-loaded 747 with all engines at full thrust accelerates at 2.6 m/s^2. Its minimum takeoff speed is 70 m/s. How much time will the plane take to reach its takeoff speed? What minimum length of runway does the plane require for takeoff?

PREPARE The visual overview of Figure 2.34 summarizes the important details of the problem. We set x_i and t_i equal to zero at the starting point of the motion, when the plane is at rest and the acceleration begins. The final point of the motion is when the plane achieves the necessary takeoff speed of 70 m/s. The plane is accelerating to the right, so we will compute the time for the plane to reach a velocity of 70 m/s and the position of the plane at this time, giving us the minimum length of the runway.

Known
$x_i = 0 \text{ m}$ $t_i = 0 \text{ s}$
$(v_x)_i = 0 \text{ m/s}$
$a_x = 2.6 \text{ m/s}^2$
$(v_x)_f = 70 \text{ m/s}$

Find
x_f

FIGURE 2.34 Visual overview for an accelerating plane.

SOLVE First we solve for the time required for the plane to reach takeoff speed. We can use the first equation in Table 2.4 to compute this time:

$$(v_x)_f = (v_x)_i + a_x \, \Delta t$$

$$70 \text{ m/s} = 0 \text{ m/s} + (2.6 \text{ m/s}^2) \, \Delta t$$

$$\Delta t = \frac{70 \text{ m/s}}{2.6 \text{ m/s}^2} = 26.9 \text{ s}$$

We keep an extra significant figure here as we will use this result in the next step of the calculation.

Given the time that the plane takes to reach takeoff speed, we can compute the position of the plane when it reaches this speed using the second equation in Table 2.4:

$$x_f = x_i + (v_x)_i \, \Delta t + \tfrac{1}{2} a_x (\Delta t)^2$$

$$= 0 \text{ m} + (0 \text{ m/s})(26.9 \text{ s}) + \tfrac{1}{2}(2.6 \text{ m/s}^2)(26.9 \text{ s})^2$$

$$= 940 \text{ m}$$

Our final answers are thus that the plane will take 27 s to reach takeoff speed, with a minimum runway length of 940 m.

ASSESS Think about the last time you flew; 27 s seems like a reasonable time for a plane to accelerate on takeoff. Actual runway lengths at major airports are 3000 m or more, a few times greater than the minimum length, because they have to allow for emergency stops during an aborted takeoff. (If we had calculated a distance far greater than 3000 m, we would know we had done something wrong!)

EXAMPLE 2.10 Finding the braking distance

A car is traveling at a speed of 30 m/s, a typical highway speed, on wet pavement. The driver sees an obstacle ahead and decides to stop. From this instant, it takes him 0.75 s to begin applying the brakes. Once the brakes are applied, the car experiences an acceleration of -6.0 m/s². How far does the car travel from the instant the driver notices the obstacle until stopping?

PREPARE This problem is more involved than previous problems we have solved, so we will take more care with the visual overview in Figure 2.35. In addition to a motion diagram and a pictorial representation, we include a graphical representation. Notice that there are two different phases of the motion, a constant-velocity phase before braking begins, and a steady slowing down once the brakes are applied. We will need to do two different calculations, one for each phase. Consequently, we've used numerical subscripts rather than a simple i and f.

SOLVE From t_1 to t_2 the velocity stays constant at 30 m/s. This is uniform motion, so the position at time t_2 is computed using Equation 2.3:

$$x_2 = x_1 + (v_x)_1(t_2 - t_1) = 0 \text{ m} + (30 \text{ m/s})(0.75 \text{ s})$$

$$= 22.5 \text{ m}$$

At t_2, the velocity begins to decrease at a steady -6.0 m/s² until the car comes to rest at t_3. This time interval can be computed using the first equation in Table 2.4, $(v_x)_3 = (v_x)_2 + a_x \Delta t$:

$$\Delta t = t_3 - t_2 = \frac{(v_x)_3 - (v_x)_2}{a_x} = \frac{0 \text{ m/s} - 30 \text{ m/s}}{-6.0 \text{ m/s}^2} = 5.0 \text{ s}$$

The position at time t_3 is computed using the second equation in Table 2.4; we take point 2 as the initial point and point 3 as the final point for this phase of the motion and use $\Delta t = t_3 - t_2$:

$$x_3 = x_2 + (v_x)_2 \Delta t + \tfrac{1}{2}a_x(\Delta t)^2$$

$$= 22.5 \text{ m} + (30 \text{ m/s})(5.0 \text{ s}) + \tfrac{1}{2}(-6.0 \text{ m/s}^2)(5.0 \text{ s})^2$$

$$= 98 \text{ m}$$

x_3 is the position of the car at the end of the problem—and so the car travels 98 m before coming to rest!

ASSESS The numbers for the reaction time and the acceleration on wet pavement are reasonable ones for an alert driver in a car with good tires. The final distance is quite large—it is more than the length of a football field.

$a_x = 0$ a_x

$x_1, (v_x)_1, t_1$ $x_2, (v_x)_2, t_2$ $x_3, (v_x)_3, t_3$

From time t_1 to t_2, the car continues at a constant speed.

From time t_2 to t_3, the car is braking, and the velocity decreases.

v_x (m/s)

$(v_x)_1$

0 t_1 t_2 t_3 t (s)

Known

$t_1 = 0$ s
$x_1 = 0$ m
$(v_x)_1 = 30$ m/s
$t_2 = 0.75$ s
$(v_x)_2 = 30$ m/s
$(v_x)_3 = 0$ m/s

Between t_2 and t_3,
$a_x = -6.0$ m/s²

Find

x_3

FIGURE 2.35 Visual overview for a car braking to a stop.

2.7 Free Fall

1.7, 1.10

A particularly important example of constant acceleration is the motion of an object moving under the influence of gravity only, and no other forces. This motion is called **free fall.** Strictly speaking, free fall occurs only in a vacuum, where there is no air resistance. But if you drop a hammer, air resistance is negligible, so we'll make only a very slight error in treating it *as if* it were in free fall. If you drop a feather, air resistance is *not* negligible, and we can't make this approximation. Motion with air resistance is a problem we will study in Chapter 5. Until then, we will restrict our attention to situations in which air resistance can be ignored, and will make the reasonable assumption that falling objects are in free fall.

"Looks like Mr. Galileo was correct . . ." was the comment made by Apollo 15 astronaut David Scott, who dropped a hammer and a feather on the moon. The objects were dropped from the same height at the same time and hit the ground simultaneously—something that would not happen in the atmosphere of the earth!

As part of his early studies of motion, Galileo did the first careful experiments on free fall and made the somewhat surprising observation that two objects of different weight dropped from the same height will, if air resistance can be neglected, hit the ground at the same time and with the same speed. In fact—as Galileo surmised, and as a famous demonstration on the moon showed—in a vacuum, where there is no air resistance, this holds true for *any* two objects!

Galileo's discovery about free fall means that **any two objects in free fall, regardless of their mass, have the same acceleration.** This is an especially important conclusion. Figure 2.36a shows the motion diagram of an object that was released from rest and falls freely. The motion diagram and graph would be identical for a falling baseball or a falling boulder! Figure 2.36b shows the object's velocity graph. As we can see, the velocity changes at a steady rate. The slope of the velocity-versus-time graph is the free-fall acceleration $a_{\text{free fall}}$.

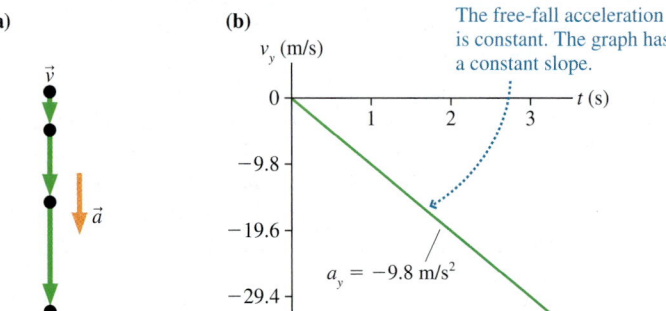

FIGURE 2.36 Motion of an object in free fall.

Careful measurements show that the value of the free-fall acceleration varies slightly at different places on the earth, due to the slightly nonspherical shape of the earth, the fact that different positions on the surface of the earth may be at different heights above sea level, and the fact that the earth is rotating. (We will consider why these factors affect the free-fall acceleration in Chapter 6.) A global average, at sea level, is

$$\vec{a}_{\text{free fall}} = (9.80 \text{ m/s}^2, \text{ vertically downward}) \qquad (2.12)$$

Standard value for the acceleration of an object in free fall

The magnitude of the **free-fall acceleration** has the special symbol g:

$$g = 9.80 \text{ m/s}^2$$

We will generally work with 2 significant figures, and so will use $g = 9.8 \text{ m/s}^2$. Several points about free fall are worthy of note:

■ g, by definition, is *always* positive. **There will never be a problem that uses a negative value for g.**
■ The velocity graph in Figure 2.36b has a negative slope. Even though a falling object speeds up, it has *negative* acceleration. Alternatively, notice that the acceleration vector $\vec{a}_{\text{free fall}}$ points down. Thus g is *not* the object's acceleration, simply the magnitude of the acceleration. The one-dimensional acceleration is

$$a_y = a_{\text{free fall}} = -g$$

It is a_y that is negative, not g.

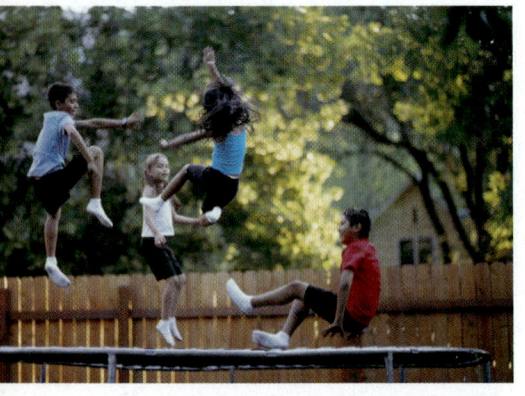

Some of the children are moving up and some are moving down, but all are in free fall—and so are accelerating downward at 9.8 m/s².

- Because free fall is motion with constant acceleration, we can use the kinematic equations of Table 2.4 with the acceleration being due to gravity, $a_y = -g$.
- g is not called "gravity." Gravity is a force, not an acceleration. g is the *free-fall acceleration*.
- $g = 9.80 \text{ m/s}^2$ only on earth. Other planets have different values of g. You will learn in Chapter 6 how to determine g for other planets.
- We will sometimes compute acceleration in units of g. An acceleration of 9.8 m/s^2 is an acceleration of 1 g; an acceleration of 19.6 m/s^2 is 2 g. Generally, we can compute

$$\text{acceleration (in units of } g) = \frac{\text{acceleration (in units of m/s}^2)}{9.8 \text{ m/s}^2} \quad (2.13)$$

This allows us to express accelerations in units that have a definite physical reference.

NOTE ▶ Despite the name, free fall is not restricted to objects that are literally falling. Any object moving under the influence of gravity only, and no other forces, is in free fall. This includes objects falling straight down, objects that have been tossed or shot straight up, objects in projectile motion (such as a passed football) and, as we will see, satellites in orbit. In this chapter we consider only objects that move up and down along a vertical line; projectile motion will be studied in Chapter 3. ◀

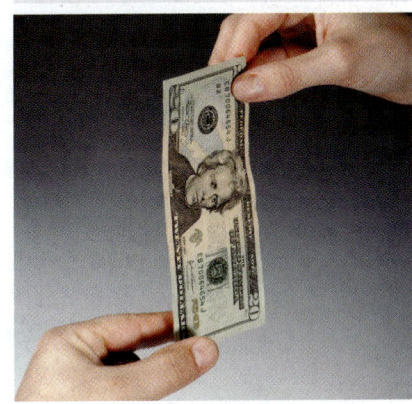

A reaction time challenge Hold a $1 (or larger!) bill by an upper corner. Have a friend prepare to grasp a lower corner, putting her fingers *near but not touching* the bill. Tell her to try to catch the bill when you drop it by simply closing her fingers without moving her hand downward—and that if she can catch it, she can keep it. Don't worry; the bill's free fall will keep your money safe. In the few tenths of a second that it takes your friend to react, the bill will move beyond her grasp.

EXAMPLE 2.11 **Analyzing a rock's fall**

A heavy rock is dropped from rest at the top of a cliff and falls 100 m before hitting the ground. How long does the rock take to fall to the ground, and what is its velocity when it hits?

PREPARE Figure 2.37 shows a visual overview with all necessary data. We have placed the origin at the ground, which makes $y_i = 100$ m.

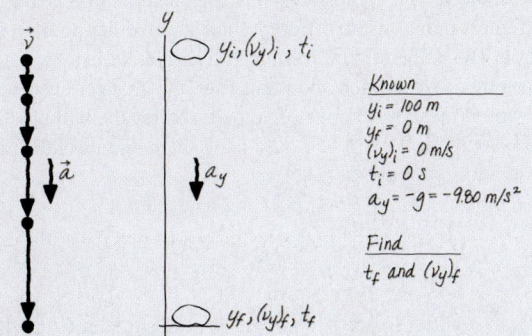

Known
$y_i = 100 \text{ m}$
$y_f = 0 \text{ m}$
$(v_y)_i = 0 \text{ m/s}$
$t_i = 0 \text{ s}$
$a_y = -g = -9.80 \text{ m/s}^2$

Find
t_f and $(v_y)_f$

FIGURE 2.37 Visual overview of a falling rock.

SOLVE Free fall is motion with the specific constant acceleration $a_y = -g$. The first question involves a relation between time and distance, a relation expressed by the second equation in Table 2.4 Using $(v_y)_i = 0$ m/s and $t_i = 0$ s, we find

$$y_f = y_i + (v_y)_i \Delta t + \tfrac{1}{2} a_y \Delta t^2 = y_i - \tfrac{1}{2} g \Delta t^2 = y_i - \tfrac{1}{2} g t_f^2$$

We can now solve for t_f, finding:

$$t_f = \sqrt{\frac{2(y_i - y_f)}{g}} = \sqrt{\frac{2(100 \text{ m} - 0 \text{ m})}{9.80 \text{ m/s}^2}} = 4.52 \text{ s}$$

Now that we know the fall time, we can use the first kinematic equation to find $(v_y)_f$:

$$(v_y)_f = (v_y)_i - g \Delta t = -g t_f = -(9.80 \text{ m/s}^2)(4.52 \text{ s})$$
$$= -44.3 \text{ m/s}.$$

ASSESS Are the answers reasonable? Well, 100 m is about 300 feet, which is about the height of a 30-floor building. How long does it take something to fall 30 floors? Four or five seconds seems pretty reasonable. How fast would it be going at the bottom? Using an approximate version of our conversion factor 1 m/s ≈ 2 mph, we find that 44.3 m/s ≈ 90 mph. That also seems like a pretty reasonable speed for something that has fallen 30 floors. Suppose we had made a mistake. If we misplaced a decimal point we could have calculated a speed of 443 m/s, or about 900 mph! This is clearly *not* reasonable. If we had misplaced the decimal point in the other direction, we would have calculated a speed of 4.3 m/s ≈ 9 mph. This is another unreasonable result, as this is slower than a typical bicycling speed.

CONCEPTUAL EXAMPLE 2.4 Analyzing the motion of a ball tossed upward

Draw a motion diagram and a velocity-versus-time graph for a ball tossed straight up in the air from the point that it leaves the hand until just before it is caught.

REASON You know what the motion of the ball looks like: The ball goes up, and then it comes back down again. This compli-

cates the drawing of a motion diagram a bit, as the ball retraces its route as it falls. A literal motion diagram would show the upward motion and downward motion on top of each other, leading to confusion. We can avoid this difficulty by horizontally separating the upward motion and downward motion diagrams. This will not affect our conclusions because it does not change any of the vectors. The motion diagram and velocity-versus-time graph appear as in Figure 2.38.

FIGURE 2.38 Motion diagram and velocity graph of a ball tossed straight up in the air.

ASSESS The highest point in the ball's motion, where it reverses direction, is called a *turning point*. What are the velocity and the acceleration at this point? We can see from the motion diagram that the velocity vectors are pointing upward but getting shorter as the ball approaches the top. As it starts to fall, the velocity vectors are pointing downward and getting longer. There must be a moment—just an instant as $\vec{v}$ switches from pointing up to pointing down—when the velocity is zero. Indeed, the ball's velocity *is* zero for an instant at the precise top of the motion! We can also see on the velocity graph that there is one instant of time when $v_y = 0$. This is the turning point.

But what about the acceleration at the top? Many people expect the acceleration to be zero at the highest point. But recall that the velocity at the top point is changing—from up to down. If the velocity is changing, there *must* be an acceleration. The slope of the velocity graph at the instant when $v_y = 0$—that is, at the highest point—is no different than at any other point in the motion. The ball is still in free fall with acceleration $a_y = -g$!

Another way to think about this is to note that zero acceleration would mean no change of velocity. When the ball reached zero velocity at the top, it would hang there and not fall if the acceleration were also zero!

EXAMPLE 2.12 Finding the height of a leap

A springbok is an antelope found in southern Africa that gets its name from its remarkable jumping ability. When a springbok is startled, it will leap straight up into the air—a maneuver called a "pronk." If a springbok leaves the ground at 7.0 m/s, a typical value, how high will it go?

PREPARE We will begin with the visual overview shown in Figure 2.39. Even though the springbok is moving upward, this is a free-fall problem because the springbok (after leaving the ground) is moving under the influence of gravity *only*.

FIGURE 2.39 Visual overview of the springbok's leap.

A critical aspect of solving this sort of problem is knowing what to take as the end point of the motion. The clue is this: The very top point of the trajectory is a *turning point,* and the instantaneous velocity at this point is zero. Thus we can characterize the "top" of the trajectory—the highest point of the leap—as the point where $(v_y)_f = 0$ m/s. If we solve for y at this point, we will know the height of the leap. None of this was explicitly stated in the problem. Part of our job in solving a problem is to supply a reasonable *interpretation* of events and thus to gain the information needed to reach a solution.

SOLVE We have information about displacement and velocity, but we don't know anything about the time interval. The third kinematic equation in Table 2.4, a relationship among displacement, velocity, and acceleration, is perfect for situations like this. Because $y_i = 0$, the springbok's displacement is $\Delta y = y_f - y_i = y_f$. Its final velocity is $(v_y)_f = 0$ m/s, so we can compute

$$(v_y)_f^2 = 0 = (v_y)_i^2 - 2g\,\Delta y = (v_y)_i^2 - 2gy_f$$

which gives

$$(v_y)_i^2 = 2gy_f$$

By solving for y_f, we find that the springbok reaches a height of

$$y_f = \frac{(v_y)_i^2}{2g} = \frac{(7.0\ \text{m/s})^2}{2(9.8\ \text{m/s}^2)} = 2.5\ \text{m}$$

ASSESS 2.5 m is a remarkable leap—a bit over 8 ft—but these animals are known for their jumping ability, so this seems reasonable.

We began this chapter with another question about animals and their athletic abilities: comparing the speeds of horses and humans. Who is the winner in a race between a horse and a man? The surprising answer is "It depends." Who the winner will be depends on the length of the race! Figure 2.40 shows position and velocity graphs for an elite male sprinter and a thoroughbred racehorse. The horse's maximum velocity is about twice that of the man, but the man's acceleration from rest is about twice that of the horse. As we can see from the graphs, a man could win a *very* short race. For a longer race, the horse's higher maximum velocity will put it in the lead; the men's world record for the mile is a bit under 4 min, but a horse can easily run this distance in under 2 min.

For a race of many miles, another factor comes into play: energy. A very long race is less about velocity and acceleration than about endurance—the ability to continue expending energy for a long time. In such endurance trials, humans often win. We will explore such energy issues in Chapter 11.

In this chapter, we have explored motion along a line—that is, motion in one dimension. In the next chapter, we will explore motion in two dimensions. We will begin with a more thorough treatment of vectors, and then pick up where we left off, exploring objects moving under the influence of gravity.

FIGURE 2.40 Velocity-versus-time and position-versus-time graphs for a sprint between a man and a horse.

STOP TO THINK 2.6   A volcano ejects a chunk of rock straight up at a velocity of $v_y = 30$ m/s. Ignoring air resistance, what will be the velocity v_y of the rock when it falls back into the volcano's crater?

A. >30 m/s   B. 30 m/s   C. 0 m/s   D. −30 m/s   E. <−30 m/s

SUMMARY

The goal of Chapter 2 has been to describe and analyze linear motion.

GENERAL STRATEGIES

Problem-Solving Strategy

Our general problem-solving strategy has three parts:

PREPARE Set up the problem:

- Draw a picture
- Collect necessary information
- Do preliminary calculations

SOLVE Do the necessary mathematics or reasoning.

ASSESS Check your answer to see if it is complete in all details and makes physical sense.

Visual Overview

A visual overview consists of several pieces that completely specify a problem. This may include any or all of the elements below:

IMPORTANT CONCEPTS

Velocity is the rate of change of position

$$v_x = \frac{\Delta x}{\Delta t}$$

Acceleration is the rate of change of velocity.

$$a_x = \frac{\Delta v_x}{\Delta t}$$

The units of acceleration are m/s^2.

An object is speeding up if v_x and a_x have the same sign, slowing down if they have opposite signs.

A position-versus-time graph plots position on the vertical axis against time on the horizontal axis.

A velocity-versus-time graph plots velocity on the vertical axis against time on the horizontal axis.

APPLICATIONS

Uniform motion
An object in uniform motion has a constant velocity. Its velocity graph is a horizontal line; its position graph is linear.

Kinematic equation for uniform motion:

$$x_f = x_i + v_x \, \Delta t$$

Uniform motion is a special case of constant-acceleration motion, with $a_x = 0$.

Motion with constant acceleration
An object with constant acceleration has a constantly changing velocity. Its velocity graph is linear; its position graph is a parabola.

Kinematic equations for motion with constant acceleration:

$$(v_x)_f = (v_x)_i + a_x \, \Delta t$$

$$x_f = x_i + (v_x)_i \, \Delta t + \tfrac{1}{2} a_x (\Delta t)^2$$

$$(v_x)_f^2 = (v_x)_i^2 + 2 a_x \, \Delta x$$

Free fall
Free fall is a special case of constant-acceleration motion; the acceleration has magnitude $g = 9.80 \text{ m/s}^2$ and is always directed vertically downward whether an object is moving up or down.

QUESTIONS

Conceptual Questions

1. A person gets in an elevator on the ground floor and rides it to the top floor of a building. Sketch a velocity-versus-time graph for this motion.

2. a. Give an example of a vertical motion with a positive velocity and a negative acceleration.
   b. Give an example of a vertical motion with a negative velocity and a negative acceleration.

3. Sketch a velocity-versus-time graph for a rock that is thrown straight upward, from the instant it leaves the hand until the instant it hits the ground.

4. You're driving along the highway at a steady speed of 60 mph when another driver decides to pass you. At the moment when the front of his car is exactly even with the front of your car, and you turn your head to smile at him, do the two cars have equal velocities? Explain.

5. A car is traveling north. Can its acceleration vector ever point south? Explain.

6. Certain animals are capable of running at great speeds; other **BIO** animals are capable of tremendous accelerations. Speculate on which would be more beneficial to a predator—large maximum speed or large acceleration.

7. A ball is thrown straight up into the air. At each of the following instants, is the ball's acceleration a_y equal to g, $-g$, 0, $<g$, or $>g$?
   a. Just after leaving your hand?
   b. At the very top (maximum height)?
   c. Just before hitting the ground?

8. A rock is *thrown* (not dropped) straight down from a bridge into the river below.
   a. Immediately after being released, is the magnitude of the rock's acceleration greater than g, less than g, or equal to g? Explain.
   b. Immediately before hitting the water, is the magnitude of the rock's acceleration greater than g, less than g, or equal to g? Explain.

9. Figure Q2.9 at right shows an object's position-versus-time graph. The letters A to E correspond to various segments of the motion in which the graph has constant slope.
   a. Write a realistic motion short story for an object that would have this position graph.
   b. In which segment(s) is the object at rest?
   c. In which segment(s) is the object moving to the right?
   d. Is the speed of the object during segment C greater than, equal to, or less than its speed during segment E? Explain.

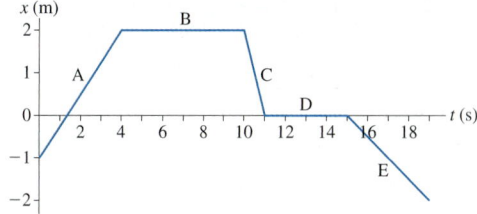

FIGURE Q2.9

10. Figure Q2.10 shows the position graph for an object moving along the horizontal axis.
    a. Write a realistic motion short story for an object that would have this position graph.
    b. Draw the corresponding velocity graph.

FIGURE Q2.10

11. Figure Q2.11 shows the position-versus-time graphs for two objects, A and B, that are moving along the same axis.
    a. At the instant $t = 1$ s, is the speed of A greater than, less than, or equal to the speed of B? Explain.
    b. Do objects A and B ever have the *same* speed? If so, at what time or times? Explain.

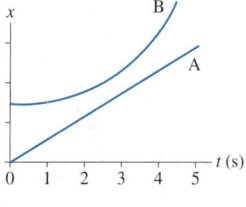

FIGURE Q2.11

12. Do the two position graphs in Figure Q2.12 represent possible motions of an object? If so, describe that motion in words. If not, why not?

(a)

(b)

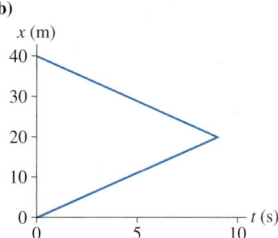

FIGURE Q2.12

13. Figure Q2.13 shows a position-versus-time graph for a moving object. At which lettered point or points
    a. Is the object moving the fastest?
    b. Is the object moving to the left?
    c. Is the object speeding up?
    d. Is the object slowing down?
    e. Is the object turning around?

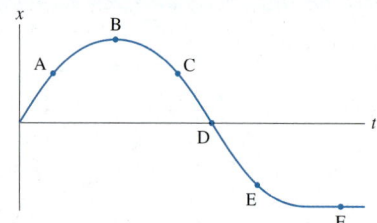

FIGURE Q2.13

14. Figure Q2.14 is the velocity-versus-time graph for an object moving along the x-axis. The graph is divided into six labeled segments.
    a. During which segment(s) is the velocity constant?
    b. During which segment(s) is the object speeding up?
    c. During which segment(s) is the object slowing down?
    d. During which segment(s) is the object standing still?
    e. During which segment(s) is the object moving to the right?

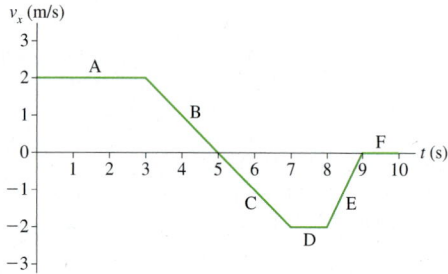

FIGURE Q2.14

15. A car traveling at velocity v takes distance d to stop after the brakes are applied. What is the stopping distance if the car is initially traveling at velocity $2v$? Assume that the acceleration due to the braking is the same in both cases.

Multiple-Choice Questions

16. | Figure Q2.16 shows the position graph of a car traveling on a straight road. At which labeled instant is the speed of the car greatest?

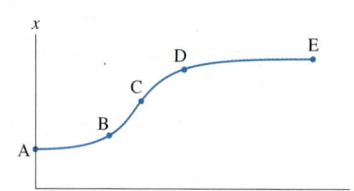

FIGURE Q2.16

17. ‖ Figure Q2.17 shows the position graph of a car traveling on a straight road. The velocity at instant 1 is _____ and the velocity at instant 2 is _____.
    A. positive, negative
    B. positive, positive
    C. negative, negative
    D. negative, zero
    E. positive, zero

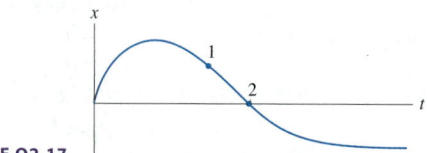

FIGURE Q2.17

18. ‖ Figure Q2.18 shows an object's position-versus-time graph. What is the velocity of the object at $t = 6$ s?
    A. 0.67 m/s   B. 0.83 m/s   C. 3.3 m/s
    D. 4.2 m/s   E. 25 m/s

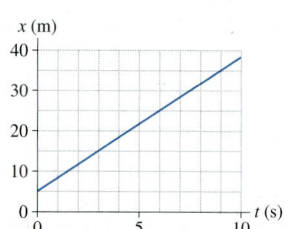

FIGURE Q2.18

19. | The following options describe the motion of four cars A–D. Which car has the largest acceleration?
    A goes from 0 m/s to 10 m/s in 5.0 s
    B goes from 0 m/s to 5.0 m/s in 2.0 s
    C goes from 0 m/s to 20 m/s in 7.0 s
    D goes from 0 m/s to 3.0 m/s in 1.0 s

20. | A car is traveling at $v_x = 20$ m/s. The driver applies the brakes, and the car slows with $a_x = -4.0$ m/s². What is the stopping distance?
    A. 5.0 m   B. 25 m   C. 40 m   D. 50 m

21. ‖ Velocity-versus-time graphs for three drag racers are shown in Figure Q2.21. At $t = 5.0$ s, which car has traveled the furthest?
    A. Andy   B. Betty   C. Carl
    D. All have traveled the same distance

FIGURE Q2.21

22. | Which of the three drag racers in Question 21 had the greatest acceleration at $t = 0$ s?
    A. Andy   B. Betty   C. Carl
    D. All had the same acceleration

23. ‖ Ball 1 is thrown straight up in the air and, at the same instant, ball 2 is released from rest and allowed to fall. Which velocity graph in Figure Q2.23 best represents the motion of the two balls?

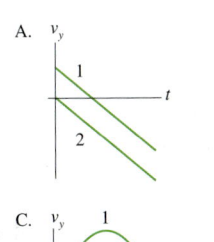

FIGURE Q2.23

Questions 24 and 25 refer to the graph in Figure Q2.24, which is the acceleration graph for an object moving along the x-axis.

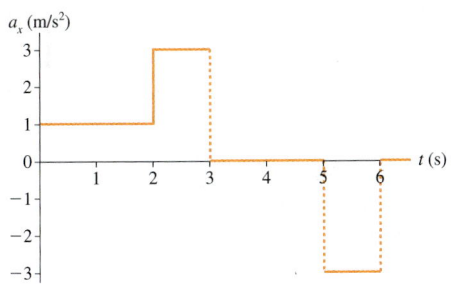

FIGURE Q2.24

24. ‖ If the object's velocity at $t = 0$ s is -1 m/s, what is its velocity at $t = 2$ s?
    A. 0 m/s B. 0.5 m/s C. -1 m/s
    D. 1 m/s E. 2 m/s

25. ‖ Which graph in Figure Q2.25 could represent the velocity of the object?

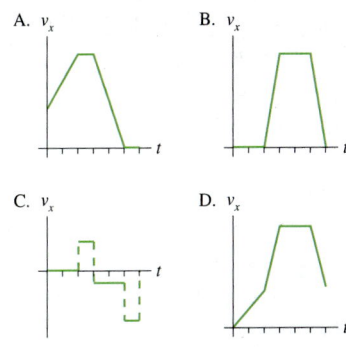

FIGURE Q2.25

26. ‖ Figure Q2.26 shows a motion diagram with the clock reading (in seconds) shown at each position. From $t = 9$ s to $t = 15$ s the object is at the same position. After that, it returns along the same track. The positions of the dots for $t \geq 16$ s are offset for clarity. Which graph best represents the object's *velocity*?

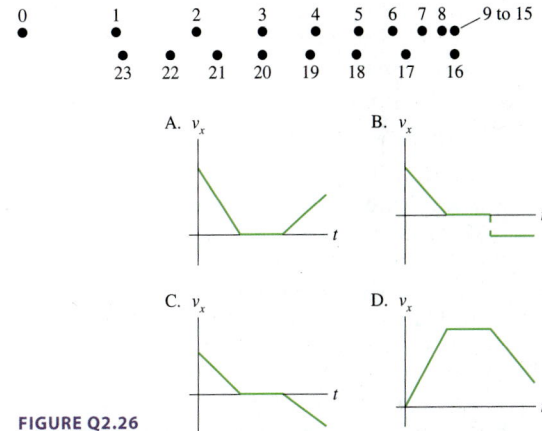

FIGURE Q2.26

PROBLEMS

Section 2.1 Describing Motion

1. ‖ Figure P2.1 shows a motion diagram of a car traveling down a street. The camera took one frame every second. A distance scale is provided.
   a. Measure the x-value of the car at each dot. Place your data in a table, similar to Table 2.1, showing each position and the instant of time at which it occurred.
   b. Make a graph of x versus t, using the data in your table. Because you have data only at certain instants of time, your graph should consist of dots that are not connected together.

FIGURE P2.1

2. ‖ For each motion diagram in Figure P2.2, determine the sign (positive or negative) of the position and the velocity.

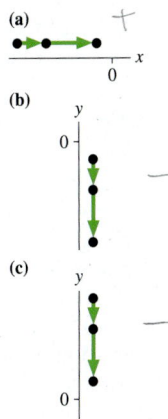

FIGURE P2.2

3. | Write a short description of the motion of a real object for which Figure P2.3 would be a realistic position-versus-time graph.

FIGURE P2.3

4. | Write a short description of the motion of a real object for which Figure P2.4 would be a realistic position-versus-time graph.

FIGURE P2.4

Section 2.2 Uniform Motion

5. ‖ Suppose a car is moving with constant velocity along a straight road. Its position was $x_1 = 15$ m at $t_1 = 0.0$ s and is $x_2 = 30$ m at $t_2 = 3.0$ s.
   a. What was its position at $t = 1.5$ s?
   b. What will its position be at $t = 5.0$ s?

6. | A car starts at the origin and moves with velocity $\vec{v} = (10$ m/s, northeast). How far from the origin will the car be after traveling for 45 s?

7. | Larry leaves home at 9:05 A.M. and runs at constant speed to the lamppost, shown in Figure P2.7. He reaches the lamppost at 9:07 A.M., immediately turns, and runs to the tree, again at constant speed. Larry arrives at the tree at 9:10 A.M. What is Larry's velocity during each of these two intervals?

FIGURE P2.7

8. ‖ Alan leaves Los Angeles at 8:00 A.M. to drive to San Francisco, 400 mi away. He travels at a steady 50 mph. Beth leaves Los Angeles at 9:00 A.M. and drives a steady 60 mph.
   a. Who gets to San Francisco first?
   b. How long does the first to arrive have to wait for the second?

9. ‖ Richard is driving home to visit his parents. 125 mi of the trip are on the interstate highway where the speed limit is 65 mph. Normally Richard drives at the speed limit, but today he is running late and decides to take his chances by driving at 70 mph. How many minutes does he save?

10. ‖ In a 5.00 km race, one runner runs at a steady 12.0 km/h and another runs at 14.5 km/h. How long does the faster runner have to wait at the finish line to see the slower runner cross?

11. ‖ In an 8.00 km race, one runner runs at a steady 11.0 km/h and another runs at 14.0 km/h. How far from the finish line is the slower runner when the faster runner finishes the race?

12. ‖ A bicyclist has the position-versus-time graph shown in Figure P2.12. What is the bicyclist's velocity at $t = 10$ s, at $t = 25$ s, and at $t = 35$ s?

FIGURE P2.12

Section 2.3 Motion with Changing Velocity

13. ‖ Figure P2.13 shows the position graph of a particle.
   a. Draw the particle's velocity graph for the interval 0 s $\leq t \leq 4$ s.
   b. Does this particle have a turning point or points? If so, at what time or times?

FIGURE P2.13

14. ‖ A car starts from $x_i = 10$ m at $t_i = 0$ s and moves with the velocity graph shown in Figure P2.14.
   a. What is the object's position at $t = 2$ s, 3 s, and 4 s?
   b. Does this car ever change direction? If so, at what time?

FIGURE P2.14

Section 2.4 Acceleration

15. ‖ Figure P2.15 shows the velocity graph of a bicycle. Draw the bicycle's acceleration graph for the interval 0 s $\leq t \leq 4$ s. Give both axes an appropriate numerical scale.

FIGURE P2.15

16. | Figure P2.16 shows the velocity graph of a train that starts from the origin at $t = 0$ s.
   a. Draw position and acceleration graphs for the train.
   b. Find the acceleration of the train at $t = 3.0$ s.

FIGURE P2.16

17. | Figure P2.17 shows the velocity graph of a bicycle moving along the x-axis. Its initial position is $x_i = 2.0$ m at $t_i = 0.0$ s. At $t = 2.0$ s, what are the bicycle's (a) position, (b) velocity, and (c) acceleration?

FIGURE P2.17

18. || For each motion diagram shown earlier in Figure P2.2, determine the sign (positive or negative) of the acceleration.

19. || Figure P2.19 is a somewhat simplified velocity graph for Olympic sprinter Carl Lewis starting a 100 m dash. Estimate his acceleration during each of the intervals A, B, and C.

FIGURE P2.19

Section 2.5 Motion with Constant Acceleration

20. | A Thompson's gazelle can reach a speed of 13 m/s in 3.0 s.
   BIO A lion can reach a speed of 9.5 m/s in 1.0 s. A trout can reach a speed of 2.8 m/s in 0.12 s. Which animal has the greatest acceleration?

21. || When striking, the pike, a
   BIO predatory fish, can accelerate from rest to a speed of 4.0 m/s in 0.11 s.
   a. What is the acceleration of the pike during this strike?
   b. How far does the pike move during this strike?

22. ||| a. What constant acceleration, in SI units, must a car have to go from zero to 60 mph in 10 s?
   b. What fraction of g is this?
   c. How far has the car traveled when it reaches 60 mph? Give your answer both in SI units and in feet.

23. ||| A car travels with constant acceleration along a straight road. It was at the origin and at rest at $t_1 = 0$ s; its position is $x_2 = 144$ m at $t_2 = 6.00$ s.
   a. What was the car's position at $t = 3.00$ s?
   b. What will its position be at $t = 9.00$ s?

24. |||| A jet plane is cruising at 250 m/s when suddenly the pilot turns the engines up to full throttle. After traveling an additional 2.0 km, the jet is moving with a speed of 300 m/s.
   a. What is the jet's acceleration, assuming it to be a constant acceleration?
   b. Is your answer reasonable? Explain.

25. ||| A speed skater moving across frictionless ice at 8.0 m/s hits a 5.0-m-wide patch of rough ice. She slows steadily, then continues on at 6.0 m/s. What is her acceleration on the rough ice?

Section 2.6 Solving One-Dimensional Motion Problems

26. ||| A driver has a reaction time of 0.50 s, and the maximum deceleration of her car is 6.0 m/s². She is driving at 20 m/s when suddenly she sees an obstacle in the road 50 m in front of her. Can she stop the car in time to avoid a collision?

27. |||| You're driving down the highway late one night at 20 m/s when a deer steps onto the road 35 m in front of you. Your reaction time before stepping on the brakes is 0.50 s, and the maximum deceleration of your car is 10 m/s².
   a. How much distance is between you and the deer when you come to a stop?
   b. What is the maximum speed you could have and still not hit the deer?

28. | A light-rail train going from one station to the next on a straight section of track accelerates from rest at 1.4 m/s² for 15 s. It then proceeds at constant speed for 1100 m before slowing down at 2.2 m/s² until it stops at the station.
   a. What is the distance between the stations?
   b. How much time does it take the train to go between the stations?

29. || A simple model for a person running the 100 m dash is to assume the sprinter runs with constant acceleration until reaching top speed, then maintains that speed through the finish line. If a sprinter reaches his top speed of 11.2 m/s in 2.14 s, what will be his total time?

Section 2.7 Free Fall

30. ||| Ball bearings can be made by letting spherical drops of molten metal fall inside a tall tower—called a *shot tower*—and solidify as they fall.
   a. If a bearing needs 4.0 s to solidify enough for impact, how high must the tower be?
   b. What is the bearing's impact velocity?

31. | In the chapter, we saw that a person's reaction time is gen-
   BIO erally not quick enough to allow the person to catch a dollar bill dropped between the fingers. If a typical reaction time in this case is 0.25 s, how long would a bill need to be for a person to have a good chance of catching it?

32. || A ball is thrown vertically upward with a speed of 19.6 m/s.
   a. What are the ball's velocity and height after 1.00, 2.00, 3.00, and 4.00 s?
   b. Draw the ball's velocity-versus-time graph. Give both axes an appropriate numerical scale.

33. ‖ A student at the top of a building of height h throws ball A straight upward with speed v_0 and throws ball B straight downward with the same initial speed.

   a. Compare the balls' accelerations, both direction and magnitude, immediately after they leave her hand. Is one acceleration larger than the other? Or are the magnitudes equal?

   b. Compare the final speeds of the balls as they reach the ground. Is one speed larger than the other? Or are they equal?

34. ‖‖‖‖ In an action movie, the villain is rescued from the ocean by grabbing onto the ladder hanging from a helicopter. He is so intent on gripping the ladder that he lets go of his briefcase of counterfeit money when he is 130 m above the water. If the briefcase hits the water 6.0 s later, what was the speed at which the helicopter was ascending?

35. ‖‖‖‖ A rock climber stands on top of a 50-m-high cliff overhanging a pool of water. He throws two stones vertically downward 1.0 s apart and observes that they cause a single splash. The initial speed of the first stone was 2.0 m/s.

   a. How long after the release of the first stone does the second stone hit the water?

   b. What was the initial speed of the second stone?

   c. What is the speed of each stone as they hit the water?

General Problems

36. ‖ Macrophages are cells referred to as "big eaters" or "killing
BIO machines" that destroy any foreign agents (pathogens) or antibodies found in the body. When a pathogen is detected, the macrophage moves toward it and literally devours it almost instantly. Suppose a large number of pathogen molecules are lined up 1.0 μm apart on a straight fiber of tissue. A macrophage initially 0.50 μm from the nearest one moves toward it at a constant speed of 30 nm/s.

   a. How long does it take until the first pathogen molecule is devoured?

   b. How about 100 molecules?

   c. What is the total distance traveled by the macrophage by the time 100 molecules have been eaten?

   d. How many pathogen molecules will one microphage destroy in 10 minutes?

Questions 37 and 38 concern *nerve impulses,* signals that propagate along *nerve fibers* consisting of many *axons* (fiber-like extensions of nerve cells) connected end-to-end. The speed of the signal depends on whether the axon has a sheath called myelin around it or not. Nerve impulses move along axons with myelin at 100 m/s, one hundred times as quickly as on axons without it. Figure P2.37 shows small portions of three nerve fibers consisting of axons of equal size. Two-thirds of the axons in fiber B are myelinated.

Unmyelinated fiber A:

Individual axons

Partly myelinated fiber B:

Fully myelinated fiber C:

FIGURE P2.37

37. ‖ Suppose nerve impulses simultaneously enter the left side
BIO of the nerve fibers sketched in Figure P2.37, then propagate to the right. Draw qualitatively accurate position and velocity graphs for the nerve impulses in all three cases. A nerve fiber is made up of many axons, but show the propagation of the impulses only over the six axons shown here.

38. ‖ Suppose that the nerve fibers in Figure P2.37 connect a
BIO finger to your brain, a distance of 1.2 m.

   a. What are the travel times of a nerve impulse from finger to brain along fibers A and C?

   b. For fiber B, 2/3 of the length is composed of myelinated axons, 1/3 unmyelinated axons. Compute the travel time for a nerve impulse on this fiber.

   c. When you touch a hot stove with the finger, the sensation of pain must reach your brain as a nerve signal along a nerve fiber before your muscles can react. Which of the three fibers gives you the best protection against a burn? Are any of these fibers unsuitable for transmitting urgent sensory information?

39. ‖ A driver travels along a straight east-west road. Starting from rest, he drives east for a while at highway speeds. Spotting some cows in the road, he brakes and comes to a halt. After waiting a few minutes, he backs slowly down the road until reaching the nearest crossroad.

   Draw graphs of the driver's position, velocity, and acceleration as functions of time. There are no numbers in this problem, but your graphs should have the correct shapes. Draw the three graphs one above the other so that the time axes line up.

40. ‖ A truck driver has a shipment of apples to deliver to a destination 440 miles away. The trip usually takes him 8 hours. Today he finds himself daydreaming and realizes 120 miles into his trip that he is running 15 minutes later than his usual pace at this point. At what speed must he drive for the remainder of the trip to complete the trip in the usual amount of time?

41. ‖ When you sneeze, the air in your lungs accelerates from
BIO rest to approximately 150 km/hr in about 0.50 seconds.

   a. What is the acceleration of the air in m/s^2?

   b. What is this acceleration, in units of g?

42. ‖‖‖‖ Two stones are released from the edge of a cliff, one a short time after the other.

   a. As they fall, the first is always going faster than the second. Does the *difference* between their speeds get larger, get smaller, or stay the same? Explain.

   b. Does their separation increase, decrease, or stay the same? Explain.

   c. Will the time interval between the instants at which they hit the ground be smaller than, equal to, or larger than the time interval between the instants of their release? Explain.

43. ‖ Figure P2.43 shows the motion diagram, made at two frames of film per second, of a ball rolling along a track. The track has a 3.0-m-long sticky section.

   a. Use the meter stick to determine the positions of the center of the ball. Place your data in a table, similar to Table 2.1, showing each position and the instant of time at which it occurred.

   b. Make a graph of x versus t for the ball. Because you have data only at certain instants of time, your graph should consist of dots that are not connected together.

c. What is the *change* in the ball's position from $t = 0$ s to $t = 1.0$ s?

d. What is the *change* in the ball's position from $t = 2.0$ s to $t = 4.0$ s?

e. What is the ball's velocity before reaching the sticky section?

f. What is the ball's velocity after passing the sticky section?

g. Determine the ball's acceleration on the sticky section of the track.

FIGURE P2.43

44. ‖ Figure P2.44 shows a velocity-versus-time graph for a particle moving along the *x*-axis. Its initial position is $x_i = 2.0$ m at $t_i = 0$ s.

a. What are the particle's position, velocity, and acceleration at $t = 1.0$ s?

b. What are the particle's position, velocity, and acceleration at $t = 3.0$ s?

FIGURE P2.44

45. ‖ Figure P2.45 shows a velocity graph for a particle having initial position $x_i = 0$ m at $t_i = 0$ s.

a. At what time or times is the particle found at $x = 35$ m? Work with the geometry of the graph, not with kinematic equations.

b. Draw a motion diagram for the particle.

FIGURE P2.45

46. ‖ Julie drives 100 mi to Grandmother's house. On the way to Grandmother's, Julie drives half the *distance* at 40 mph and half the distance at 60 mph. On her return trip, she drives half the *time* at 40 mph and half the time at 60 mph.

a. How long does it take Julie to complete the trip to Grandmother's house?

b. How long does the return trip take?

47. ‖ The takeoff speed for an Airbus A320 jetliner is 80 m/s. Velocity data measured during takeoff are as follows:

t (s)	v_x (m/s)
0	0
10	23
20	46
30	69

a. What is the takeoff speed in miles per hour?

b. What is the jetliner's acceleration during takeoff?

c. At what time do the wheels leave the ground?

d. For safety reasons, in case of an aborted takeoff, the runway must be three times the takeoff distance. Can an A320 take off safely on a 2.5-mi-long runway?

48. ‖‖ Does a real automobile have constant acceleration? Measured data for a Porsche 944 Turbo at maximum acceleration are as follows:

t (s)	v_x (mph)
0	0
2	28
4	46
6	60
8	70
10	78

a. Convert the velocities to m/s, then make a graph of velocity versus time. Based on your graph, is the acceleration constant? Explain.

b. Draw a smooth curve through the points on your graph, then use your graph to *estimate* the car's acceleration at 2.0 s and 8.0 s. Give your answer in SI units. **Hint:** Remember that acceleration is the slope of the velocity graph.

49. ‖ a. How many days will it take a spaceship to accelerate to the speed of light $(3.0 \times 10^8$ m/s) with the acceleration g?

b. How far will it travel during this interval?

c. What fraction of a light year is your answer to part b? A *light year* is the distance light travels in one year.

NOTE ► We know, from Einstein's theory of relativity, that no object can travel at the speed of light. So this problem, while interesting and instructive, is not realistic. ◄

50. ‖‖ You are driving to the grocery store at 20 m/s. You are 110 m from an intersection when the traffic light turns red. Assume that your reaction time is 0.70 s and that your car brakes with constant acceleration.

a. How far are you from the intersection when you begin to apply the brakes?

b. What acceleration will bring you to rest right at the intersection?

c. How long does it take you to stop?

51. ‖ When you blink your eye, the upper lid goes from rest with
BIO your eye open to completely covering your eye in a time of 0.024 s.

a. Estimate the distance that the top lid of your eye moves during a blink.

b. What is the acceleration of your eyelid? Assume it to be constant.

c. What is your upper eyelid's final speed as it hits the bottom eyelid?

52. |||| A bush baby, an African
BIO primate, is capable of leaping vertically to the remarkable height of 2.26 m. To jump this high, the bush baby accelerates over a distance of 0.16 m while rapidly extending its legs. The acceleration during the jump is approximately constant. What is the acceleration in m/s² and in *g*'s?

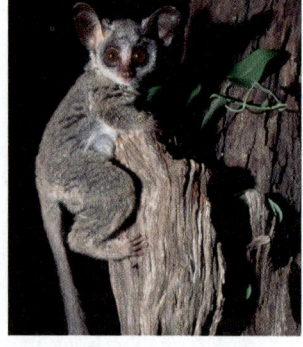

53. |||| When jumping, a flea reaches a takeoff speed of 1.0 m/s
BIO over a distance of 0.50 mm.
    a. What is the flea's acceleration during the jump phase?
    b. How long does the acceleration phase last?
    c. If the flea jumps straight up, how high will it go? (Ignore air resistance for this problem; in reality, air resistance plays a large role, and the flea will not reach this height.)

54. |||| Divers compete by diving into a 3.0-m-deep pool from a platform 10 m above the water. What is the magnitude of the minimum acceleration in the water needed to keep a diver from hitting the bottom of the pool? Assume the acceleration is constant.

55. |||| With a man on first base, Hank hits a ball to the corner of the outfield. The outfielder has trouble retrieving the ball, so Hank decides to try for an inside-the-park home run. As he rounds third base at a speed of 18 ft/s, Hank sees that the player in front of him is two-thirds of the way from third base to home and is limping at 6.0 ft/s. It is 90 ft between bases. If Hank wants to arrive at home plate just behind the slower runner, what should be the magnitude of his acceleration as he slows?

56. |||| A student standing on the ground throws a ball straight up. The ball leaves the student's hand with a speed of 15 m/s when the hand is 2.0 m above the ground. How long is the ball in the air before it hits the ground? (The student moves her hand out of the way.)

57. || A rock is tossed straight up with a speed of 20 m/s. When it returns, it falls into a hole 10 m deep.
    a. What is the rock's velocity as it hits the bottom of the hole?
    b. How long is the rock in the air, from the instant it is released until it hits the bottom of the hole?

58. |||| A 200 kg weather rocket is loaded with 100 kg of fuel and fired straight up. It accelerates upward at 30.0 m/s² for 30.0 s, then runs out of fuel. Ignore any air resistance effects.
    a. What is the rocket's maximum altitude?
    b. How long is the rocket in the air?
    c. Draw a velocity-versus-time graph for the rocket from liftoff until it hits the ground.

59. |||| A juggler throws a ball straight up into the air with a speed of 10 m/s. With what speed would she need to throw a second ball half a second later, starting from the same position as the first, in order to hit the first ball at the top of its trajectory?

60. |||| A lead ball is dropped into a lake from a diving board 5.0 m above the water. After entering the water, it sinks to the bottom with a constant velocity equal to the velocity with which it hit the water. The ball reaches the bottom 3.0 s after it is released. How deep is the lake?

61. ||| A hotel elevator ascends 200 m with a maximum speed of 5.0 m/s. Its acceleration and deceleration both have a magnitude of 1.0 m/s².
    a. How far does the elevator move while accelerating to full speed from rest?
    b. How long does it take to make the complete trip from bottom to top?

62. ||| A car starts from rest at a stop sign. It accelerates at 2.0 m/s² for 6.0 seconds, coasts for 2.0 s, and then slows down at a rate of 1.5 m/s² for the next stop sign. How far apart are the stop signs?

63. || A toy train is pushed forward and released at x_i = 2.0 m with a speed of 2.0 m/s. It rolls at a steady speed for 2.0 s, then one wheel begins to stick. The train comes to a stop 6.0 m from the point at which it was released. What is the train's acceleration after its wheel begins to stick?

64. ||| Bob is driving the getaway car after the big bank robbery. He's going 50 m/s when his headlights suddenly reveal a nail strip that the cops have placed across the road 150 m in front of him. If Bob can stop in time, he can throw the car into reverse and escape. But if he crosses the nail strip, all his tires will go flat and he will be caught. Bob's reaction time before he can hit the brakes is 0.60 s, and his car's maximum deceleration is 10 m/s². Is Bob in jail?

65. || Heather and Jerry are standing on a bridge 50 m above a river. Heather throws a rock straight down with a speed of 20 m/s. Jerry, at exactly the same instant of time, throws a rock straight up with the same speed. Ignore air resistance.
    a. How much time elapses between the first splash and the second splash?
    b. Which rock has the faster speed as it hits the water?

66. ||| A motorist is driving at 20 m/s when she sees that a traffic light 200 m ahead has just turned red. She knows that this light stays red for 15 s, and she wants to reach the light just as it turns green again. It takes her 1.0 s to step on the brakes and begin slowing at a constant deceleration. What is her speed as she reaches the light at the instant it turns green?

67. |||| A "rocket car" is launched along a long straight track at *t* = 0 s. It moves with constant acceleration a_1 = 2.0 m/s². At *t* = 2.0 s, a second car is launched with constant acceleration a_2 = 8.0 m/s².
    a. At what time does the second car catch up with the first one?
    b. How far down the track do they meet?

68. | A Porsche challenges a Honda to a 400 m race. Because the Porsche's acceleration of 3.5 m/s² is larger than the Honda's 3.0 m/s², the Honda gets a 50-m head start. Assume, somewhat unrealistically, that both cars can maintain these accelerations the entire distance. Who wins, and by how much time?

69. |||| The minimum stopping distance for a car traveling at a speed of 30 m/s is 60 m, including the distance traveled during the driver's reaction time of 0.50 s.
    a. What is the minimum stopping distance for the same car traveling at a speed of 40 m/s?
    b. Draw a position-versus-time graph for the motion of the car in part a. Assume the car is at x_i = 0 m when the driver first sees the emergency situation ahead that calls for a rapid halt.

70. |||| A rocket is launched straight up with constant acceleration. Four seconds after liftoff, a bolt falls off the side of the rocket. The bolt hits the ground 6.0 s later. What was the rocket's acceleration?

In Problems 71 through 73 you are given the kinematic equation that is used to solve a problem. For each of these, you are to
a. Write a *realistic* problem for which this is the correct equation. Be sure that the answer your problem requests is consistent with the equation given.
b. Draw the motion diagram and the pictorial representation for your problem.
c. Finish the solution of the problem.

71. ‖ $64 \text{ m} = 0.0 \text{ m} + (32 \text{ m/s})(4.0 \text{ s} - 0.0 \text{ s})$
$+ \frac{1}{2}a_x(4.0 \text{ s} - 0.0 \text{ s})^2$
72. ‖ $0.0 \text{ m/s} = 36 \text{ m/s} - (3.0 \text{ m/s}^2)t$
73. ‖ $(10 \text{ m/s})^2 = (v_y)_i^2 - 2(9.8 \text{ m/s}^2)(10 \text{ m} - 0 \text{ m})$

Passage Problems

Free Fall on Different Worlds

Objects in free fall on the earth have acceleration $a_y = -9.8 \text{ m/s}^2$. On the moon, free-fall acceleration is approximately 1/6 of the acceleration on earth. This changes the scale of problems involving free fall. For instance, suppose you jump straight upward, leaving the ground with velocity v_i and then steadily slowing until reaching zero velocity at your highest point. Because your initial velocity is determined mostly by the strength of your leg muscles, we can assume your initial velocity would be the same on the moon. But considering the final equation in Table 2.4 we can see that, with a smaller free-fall acceleration, your maximum height would be greater. The following questions ask you to think about how certain athletic feats might be performed in this reduced-gravity environment.

74. ‖ If an astronaut can jump straight up to a height of 0.50 m on earth, how high could he jump on the moon?
A. 1.2 m   B. 3.0 m   C. 3.6 m   D. 18 m
75. ‖ On the earth, an astronaut can safely jump to the ground from a height of 1.0 m; her velocity when reaching the ground is low enough to not cause injury. From what height could the astronaut safely jump to the ground on the moon?
A. 2.4 m   B. 6.0 m   C. 7.2 m   D. 36 m
76. ‖ On the earth, an astronaut throws a ball straight upward; it stays in the air for a total time of 3.0 s before reaching the ground again. If a ball were to be thrown upward with the same initial speed on the moon, how much time would pass before it hit the ground?
A. 7.3 s   B. 18 s   C. 44 s   D. 108 s

STOP TO THINK ANSWERS

Stop to Think 2.1: D. The motion consists of two constant-velocity phases; the second one has a larger velocity. The correct graph has two straight-line segments, with the second one having a larger slope.

Stop to Think 2.2: A. The motion consists of two constant-velocity phases; the first one is positive velocity, the second is negative. The correct graph has two straight-line segments, the first with a positive slope and the second with a negative slope.

Stop to Think 2.3: C. Consider the slope of the position-versus-time graph; it starts out positive and constant, then decreases to zero. Thus the velocity graph must start with a constant positive value, then decrease to zero.

Stop to Think 2.4: C. Acceleration is the slope of the velocity-versus-time graph. The largest magnitude of the slope is at point C.

Stop to Think 2.5: A. The car is always moving to the right, so the velocity is positive at all times. There should be two constant-velocity segments, with the second having a smaller velocity.

Stop to Think 2.6: D. The final velocity will have the same *magnitude* as the initial velocity, but the velocity is negative because the rock will be moving downward.

3

VECTORS AND MOTION IN TWO DIMENSIONS

With astonishing precision, an archer fish secures a meal by spitting drops of water at an unsuspecting insect. What factors determine how high and how far the water drops travel?

Looking Ahead ▶▶

The goals of Chapter 3 are to learn more about vectors and to use vectors as a tool to analyze motion in two dimensions. In this chapter, you will learn to:

▶ Work with vectors, coordinate systems, and components.

▶ Find acceleration vectors on motion diagrams.

▶ Use vectors to understand motion on a ramp and relative velocity.

▶ Solve problems involving projectiles following parabolic paths.

▶ Understand the basic ideas of motion in a circle.

Looking Back ◀◀

This chapter will build naturally on the material from the previous two chapters. Please review:

◀ Sections 1.1–1.2 Basic motion concepts.

◀ Section 1.5 Vectors.

◀ Sections 2.4–2.7 Acceleration and free fall.

Once the drops of water leave the archer fish's mouth and fly through the air, their motion follows a gently curving path. This path, a parabola, is the same type of path followed by a basketball shot toward the basket, a high jumper leaping over a bar, or a skier going off a jump. These are all examples of motion in two dimensions, motion in which the objects are moving horizontally and vertically at the same time.

In order to extend our study of motion to two dimensions, we will need to develop techniques for working with vectors. We will first use vector techniques to analyze motion on a ramp and relative velocity, continue on to look at the parabolic motion of projectiles, then close the chapter by describing the motion of objects moving in circles. Along the way we'll use numerous examples drawn from sports—from skiers gliding down slopes to balls flying through the air.

3.1 Using Vectors

In the previous chapter, we solved many problems in which an object moved in a straight-line path. In this chapter, we will look at particles that take curving paths—motion in two dimensions, which we also call motion in a plane. The two most important examples are outlined in the table below. You can see that, in both cases, the changing direction of the motion will be of crucial importance.

Two-dimensional motion

Projectile motion		Circular motion	
	A ball bouncing across a table, captured here in a strobe photograph, is clearly moving in two dimensions. The ball is moving to the right while simultaneously moving up and down. Understanding the curving path of the ball will require us to analyze motion in both dimensions.		This photograph captures the path of a strobe light swung in a circle in a horizontal plane. The flashes are at equally spaced positions, so the speed of the strobe light is constant, but the direction is constantly changing. Understanding the motion will require us to understand this change in direction.

Because the direction of motion will be so important, we need to develop an appropriate mathematical language to describe it—the language of vectors. We introduced the concept of a vector in Chapter 1, and in the next few sections we will develop that concept into a useful and powerful tool. We will practice using vectors by analyzing a problem of motion in one dimension, motion on a ramp, and by studying the interesting notion of relative velocity. We will then be ready to analyze the two-dimensional motion of projectiles and of objects moving in a circle.

Recall from Chapter 1 that a vector is a quantity with both a size, or magnitude, and a direction. Figure 3.1 shows how to represent a particle's velocity as a vector $\vec{v}$. The particle's speed at this point is 5 m/s *and* it is moving in the direction indicated by the arrow. The magnitude of a vector is represented by the letter without an arrow. In this case, the particle's speed—the magnitude of the velocity vector $\vec{v}$—is $v = 5$ m/s. The magnitude of a vector, a *scalar* quantity, cannot be a negative number.

NOTE ▶ Although the vector arrow is drawn across the page, from its tail to its tip, this arrow does *not* indicate that the vector "stretches" across this distance. Instead, the arrow tells us the value of the vector quantity only at the one point where the tail of the vector is placed. ◀

We found in Chapter 1 that the displacement of an object is a vector drawn from its initial position to its position at some later time. Because displacement is an easy concept to think about, we can use it to introduce some of the properties of vectors. However, **all the properties we will discuss in this chapter (addition, subtraction, multiplication, components) apply to all vectors, not just to displacement.**

Suppose that Sam, our old friend from Chapter 1, starts from his front door, walks across the street, and ends up 200 ft to the northeast of where he started. Sam's displacement, which we will label $\vec{d}_S$, is shown in Figure 3.2a on the next page. The displacement vector is a straight-line connection from his initial to his final position, not necessarily his actual path. The dotted line indicates a possible route Sam might have taken, but his displacement is the vector $\vec{d}_S$.

To describe a vector we must specify both its magnitude and its direction. We can write Sam's displacement as

$$\vec{d}_S = (200 \text{ ft, northeast})$$

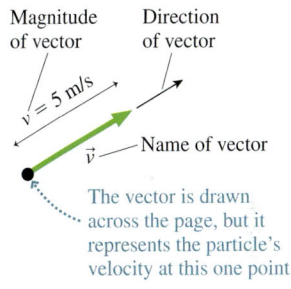

Magnitude of vector · Direction of vector

$v = 5$ m/s

$\vec{v}$ — Name of vector

The vector is drawn across the page, but it represents the particle's velocity at this one point.

FIGURE 3.1 The velocity vector $\vec{v}$ has both a magnitude and a direction.

(a)

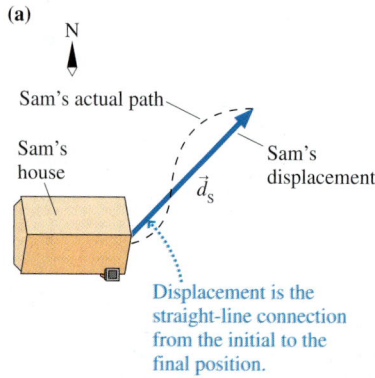

Sam's actual path

Sam's house

Sam's displacement

$\vec{d}_S$

Displacement is the straight-line connection from the initial to the final position.

(b)

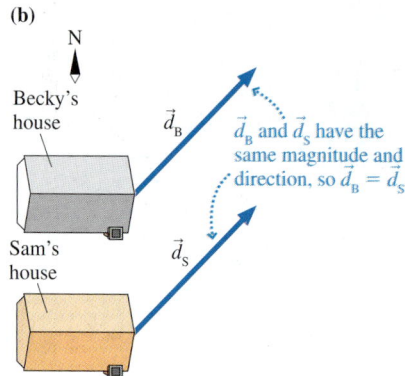

Becky's house

$\vec{d}_B$

$\vec{d}_B$ and $\vec{d}_S$ have the same magnitude and direction, so $\vec{d}_B = \vec{d}_S$.

Sam's house

$\vec{d}_S$

FIGURE 3.2 Displacement vectors.

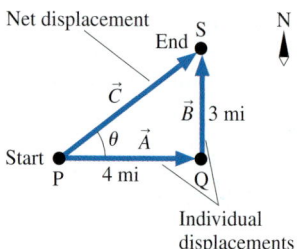

Net displacement

End   S   N

$\vec{C}$

$\vec{B}$ 3 mi

Start θ $\vec{A}$

P   4 mi   Q

Individual displacements

FIGURE 3.3 The net displacement $\vec{C}$ resulting from two displacements $\vec{A}$ and $\vec{B}$.

where the first number specifies the magnitude and the second item gives the direction. The magnitude of Sam's displacement is $d_S = 200$ ft, the distance between his initial and final points.

Sam's next-door neighbor Becky also walks 200 ft to the northeast, starting from her own front door. Becky's displacement $\vec{d}_B = (200$ ft, northeast$)$ has the same magnitude and direction as Sam's displacement $\vec{d}_S$. Because vectors are defined by their magnitude and direction, **two vectors are equal if they have the same magnitude and direction.** This is true regardless of the starting points of the vectors. Thus the two displacements in Figure 3.2b are equal to each other, and we can write $\vec{d}_B = \vec{d}_S$.

Vector Addition

As we saw in Chapter 1, we can combine successive displacements by vector addition. Let's review and extend this concept. Figure 3.3 shows the displacement of a hiker who starts at point P and ends at point S. She first hikes 4 miles to the east, then 3 miles to the north. The first leg of the hike is described by the displacement vector $\vec{A} = (4$ mi, east$)$. The second leg of the hike has displacement $\vec{B} = (3$ mi, north$)$. Now, by definition, a vector from the initial position P to the final position S is also a displacement. This is vector $\vec{C}$ on the figure. $\vec{C}$ is the *net displacement* because it describes the net result of the hiker's first having displacement $\vec{A}$, then displacement $\vec{B}$.

The word *net* implies addition. The net displacement $\vec{C}$ is an initial displacement $\vec{A}$ *plus* a second displacement $\vec{B}$, or

$$\vec{C} = \vec{A} + \vec{B} \tag{3.1}$$

The sum of two vectors is called the **resultant vector.** It's not hard to show that vector addition is commutative: $\vec{A} + \vec{B} = \vec{B} + \vec{A}$. That is, you can add vectors in any order you wish.

Look back at Tactics Box 1.4 on page 21 to see the three-step procedure for adding two vectors. This tip-to-tail method for adding vectors, which is used to find $\vec{C} = \vec{A} + \vec{B}$ in Figure 3.3, is called *graphical addition*. Any two vectors of the same type—two velocity vectors or two force vectors—can be added in exactly the same way.

When two vectors are to be added, it is often convenient to draw them with their tails together, as shown in Figure 3.4a. To evaluate $\vec{D} + \vec{E}$, you could move vector $\vec{E}$ over to where its tail is on the tip of $\vec{D}$, then use the tip-to-tail rule of graphical addition. This gives vector $\vec{F} = \vec{D} + \vec{E}$ in Figure 3.4b. Alternatively, Figure 3.4c shows that the vector sum $\vec{D} + \vec{E}$ can be found as the diagonal of the parallelogram defined by $\vec{D}$ and $\vec{E}$. This method for vector addition, which some of you may have learned, is called the *parallelogram rule* of vector addition.

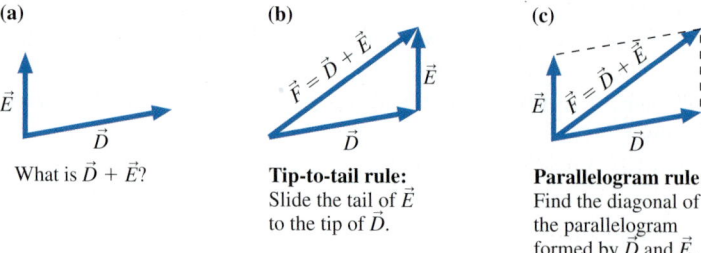

(a)

$\vec{E}$

$\vec{D}$

What is $\vec{D} + \vec{E}$?

(b)

$\vec{F} = \vec{D} + \vec{E}$

$\vec{E}$

$\vec{D}$

Tip-to-tail rule: Slide the tail of $\vec{E}$ to the tip of $\vec{D}$.

(c)

$\vec{F} = \vec{D} + \vec{E}$

$\vec{E}$

$\vec{D}$

Parallelogram rule: Find the diagonal of the parallelogram formed by $\vec{D}$ and $\vec{E}$.

FIGURE 3.4 Two vectors can be added using the tip-to-tail rule or the parallelogram rule.

Vector addition is easily extended to more than two vectors. Figure 3.5 shows a hiker moving from initial position 0 to position 1, then position 2, then position 3, and finally arriving at position 4. These four segments are described by displacement vectors $\vec{d}_1$, $\vec{d}_2$, $\vec{d}_3$, and $\vec{d}_4$. The hiker's *net* displacement, an arrow from position 0 to position 4, is the vector $\vec{d}_{net}$. In this case,

$$\vec{d}_{net} = \vec{d}_1 + \vec{d}_2 + \vec{d}_3 + \vec{d}_4 \qquad (3.2)$$

The vector sum is found by using the tip-to-tail method three times in succession.

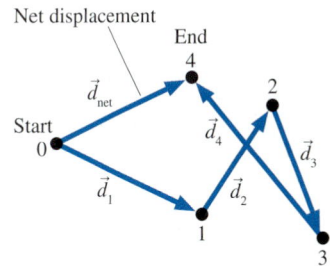

FIGURE 3.5 The net displacement after four individual displacements.

Multiplication by a Scalar

The hiker in Figure 3.3 started with displacement $\vec{A}_1 = (4 \text{ mi, east})$. Suppose a second hiker walks twice as far to the east. The second hiker's displacement will then certainly be $\vec{A}_2 = (8 \text{ mi, east})$. The words "twice as" indicate a multiplication, so we can say

$$\vec{A}_2 = 2\vec{A}_1$$

Multiplying a vector by a positive scalar gives another vector of *different* magnitude but pointing in the *same direction*.

Let the vector $\vec{A}$ be specified as a magnitude A and a direction θ_A. That is, $\vec{A} = (A, \theta_A)$. Now let $\vec{B} = c\vec{A}$, where c is a positive scalar constant. Then

$$\vec{B} = c\vec{A} \text{ means that } (B, \theta_B) = (cA, \theta_A) \qquad (3.3)$$

In other words, the vector is stretched or compressed by the factor c (i.e., vector $\vec{B}$ has magnitude $B = cA$), but $\vec{B}$ points in the same direction as $\vec{A}$. This is illustrated in Figure 3.6.

Suppose we multiply $\vec{A}$ by zero. Using Equation 3.3,

$$0 \cdot \vec{A} = \vec{0} = (0 \text{ m, direction undefined}) \qquad (3.4)$$

The product is a vector having zero length or magnitude. This vector is known as the **zero vector,** denoted $\vec{0}$. The direction of the zero vector is irrelevant; you cannot describe the direction of an arrow of zero length!

What happens if we multiply a vector by a negative number? Equation 3.3 does not apply if $c < 0$ because vector $\vec{B}$ cannot have a negative magnitude. Consider the vector $-\vec{A}$, which is equivalent to multiplying $\vec{A}$ by -1. Because

$$\vec{A} + (-\vec{A}) = \vec{0} \qquad (3.5)$$

the vector $-\vec{A}$ must be such that, when it is added to $\vec{A}$, the resultant is the zero vector $\vec{0}$. In other words, the *tip* of $-\vec{A}$ must return to the *tail* of $\vec{A}$, as shown in Figure 3.7. This will be true only if $-\vec{A}$ is equal in magnitude to $\vec{A}$, but opposite in direction. Thus we can conclude that

$$-\vec{A} = (A, \text{ direction opposite } \vec{A}) \qquad (3.6)$$

That is, multiplying a vector by -1 reverses its direction without changing its length.

As an example, Figure 3.8 shows vectors $\vec{A}$, $2\vec{A}$, and $-3\vec{A}$. Multiplication by 2 doubles the length of the vector but does not change its direction. Multiplication by -3 stretches the length by a factor of 3 *and* reverses the direction.

The length of $\vec{B}$ is "stretched" by the factor c. That is, $B = cA$.

$\vec{B} = c\vec{A}$

$\vec{B}$ points in the same direction as $\vec{A}$.

FIGURE 3.6 Multiplication of a vector by a scalar.

Tail of $-\vec{A}$ at tip of $\vec{A}$

Vector $-\vec{A}$ is equal in magnitude but opposite in direction to $\vec{A}$. Thus $\vec{A} + (-\vec{A}) = \vec{0}$.

Tip of $-\vec{A}$ returns to the starting point. The resultant vector is $\vec{0}$.

FIGURE 3.7 Vector $-\vec{A}$.

Vector Subtraction

How might we *subtract* vector $\vec{B}$ from vector $\vec{A}$ to form the vector $\vec{A} - \vec{B}$? With numbers, subtraction is the same as the addition of a negative number. That is, $5 - 3$ is the same as $5 + (-3)$. Similarly, $\vec{A} - \vec{B} = \vec{A} + (-\vec{B})$. We can use the rules for vector addition and the fact that $-\vec{B}$ is a vector opposite in direction to $\vec{B}$ to form rules for vector subtraction.

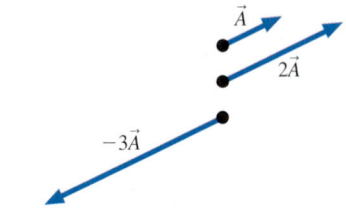

FIGURE 3.8 Vectors $\vec{A}$, $2\vec{A}$, and $-3\vec{A}$.

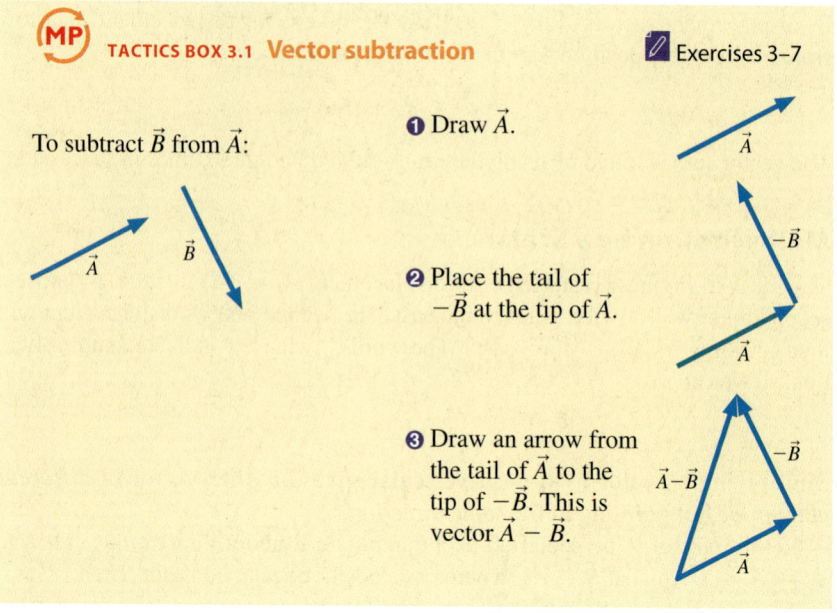

TACTICS BOX 3.1 **Vector subtraction**   Exercises 3–7

To subtract $\vec{B}$ from $\vec{A}$:

❶ Draw $\vec{A}$.

❷ Place the tail of $-\vec{B}$ at the tip of $\vec{A}$.

❸ Draw an arrow from the tail of $\vec{A}$ to the tip of $-\vec{B}$. This is vector $\vec{A} - \vec{B}$.

STOP TO THINK 3.1   Which figure shows $2\vec{P} - \vec{Q}$?

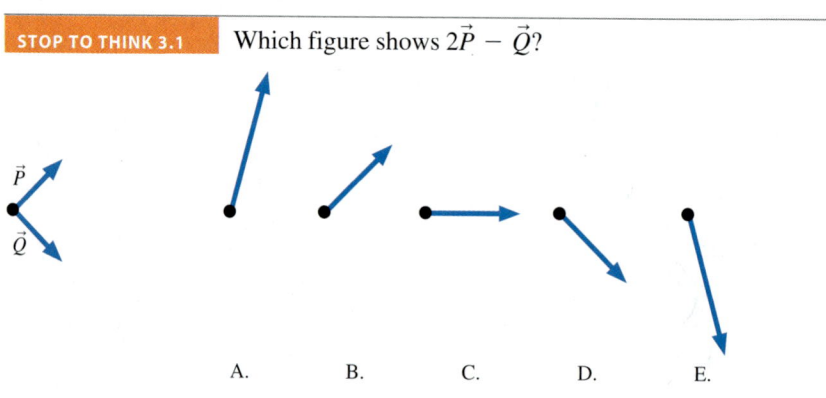

A.   B.   C.   D.   E.

3.2 Using Vectors on Motion Diagrams

In Chapter 2, we defined velocity for one-dimensional motion as an object's displacement—the change in position—divided by the time interval in which the change occurs:

$$v_x = \frac{\Delta x}{\Delta t} = \frac{x_f - x_i}{\Delta t}$$

In two dimensions, an object's displacement is a vector. Suppose an object undergoes displacement $\vec{d}$ during the time interval Δt. Let's define an object's velocity *vector* to be

$$\vec{v} = \frac{\vec{d}}{\Delta t} = \left(\frac{d}{\Delta t}, \text{ same direction as } \vec{d}\right) \qquad (3.7)$$

Definition of velocity in two or more dimensions

Notice that we've multiplied a vector by a scalar: The velocity vector is the displacement vector multiplied by the scalar $1/\Delta t$. Consequently, as we found in Chapter 1, **the velocity vector points in the direction of displacement.** As a result, we can use the dot-to-dot vectors on a motion diagram to visualize the velocity.

NOTE ▶ Strictly speaking, the velocity defined in Equation 3.7 is the *average* velocity for the time interval Δt. This is adequate for using motion diagrams to visualize motion. As we did in Chapter 2, we'll choose Δt to be very small and thus have an *instantaneous* velocity when it's time to start doing calculations. ◀

EXAMPLE 3.1 Finding the velocity of an airplane

A small plane is 100 km due east of Denver. One hour later, it is 200 km due north of Denver. What is the plane's velocity?

PREPARE The initial and final positions of the plane are shown in Figure 3.9; the displacement $\vec{d}$ is the vector that points from the initial to the final position.

FIGURE 3.9 Displacement vector for an airplane.

SOLVE The length of the displacement vector is the hypotenuse of a right triangle:

$$d = \sqrt{(100 \text{ km})^2 + (200 \text{ km})^2} = 224 \text{ km}$$

The direction of the displacement vector is described by the angle θ in Figure 3.9. From trigonometry, this angle is

$$\theta = \tan^{-1}\left(\frac{200 \text{ km}}{100 \text{ km}}\right) = \tan^{-1}(2.00) = 63.4°$$

Thus the plane's displacement vector is

$$\vec{d} = (224 \text{ km}, 63.4° \text{ north of west})$$

Because the plane undergoes this displacement during 1 hour, its velocity is

$$\vec{v} = \left(\frac{d}{\Delta t}, \text{ same direction as } \vec{d}\right) = \left(\frac{224 \text{ km}}{1 \text{ hour}}, 63.4° \text{ north of west}\right)$$

$$= (224 \text{ km/hour}, 63.4° \text{ north of west})$$

ASSESS The plane's *speed* is $v = 224$ km/h. Speed does not have a direction.

We defined an object's acceleration in one dimension as $a_x = \Delta v_x / \Delta t$. In two dimensions, we need to use a vector to describe acceleration. The vector definition of acceleration is a straightforward extension of the one-dimensional version:

$$\vec{a} = \frac{\vec{v}_f - \vec{v}_i}{t_f - t_i} = \frac{\Delta \vec{v}}{\Delta t} \tag{3.8}$$

Definition of acceleration in two or more dimensions

There is an acceleration whenever there is a *change* in velocity. Because velocity is a vector, it can change in two possible ways:

1. The magnitude can change, indicating a change in speed, or
2. The direction of motion can change.

In Chapter 2 we considered the first case, in which an object speeds up or slows down while moving in a straight line. Now we need to examine the second case, in which an object changes its direction of motion.

Suppose an object has an initial velocity $\vec{v}_i$ at time t_i and later, at time t_f has velocity $\vec{v}_f$. The fact that the velocity *changes* tells us the object undergoes an acceleration during the time interval $\Delta t = t_f - t_i$. We see from Equation 3.8 that the acceleration points in the same direction as the vector $\Delta \vec{v}$. This vector is the change in the velocity $\Delta \vec{v} = \vec{v}_f - \vec{v}_i$, so to know which way the acceleration vector points, we have to perform the vector subtraction $\vec{v}_f - \vec{v}_i$. Tactics Box 3.1 showed how to perform vector subtraction. Tactics Box 3.2 shows how to use vector subtraction to find the acceleration vector.

Lunging versus veering BIO The top photo shows a barracuda, a type of fish that catches prey with a rapid linear acceleration, a quick change in speed. The barracuda's body shape is optimized for such a straight-line strike. The butterfly fish in the bottom photo has a very different appearance. It can't rapidly change its speed, but its body shape lets it quickly change its direction. This is a type of acceleration the barracuda isn't good at, so the butterfly fish can avoid being caught.

TACTICS BOX 3.2 Finding the acceleration vector   Exercises 9,10

To find the acceleration between velocity $\vec{v}_i$ and velocity $\vec{v}_f$:

❶ Draw the velocity vector $\vec{v}_f$.

❷ Draw $-\vec{v}_i$ at the tip of $\vec{v}_f$.

❸ Draw $\Delta\vec{v} = \vec{v}_f - \vec{v}_i$
         $= \vec{v}_f + (-\vec{v}_i)$
This is the direction of $\vec{a}$.

❹ Return to the original motion diagram. Draw a vector at the middle point in the direction of $\Delta\vec{v}$; label it $\vec{a}$. This is the average acceleration at the midpoint between $\vec{v}_i$ and $\vec{v}_f$.

Now that we know how to determine acceleration vectors, we can make a complete motion diagram with dots showing the position of the object, average velocity vectors found by connecting dots with an arrow, and acceleration vectors found using Tactics Box 3.2. Note that there is *one* acceleration vector linking each *two* velocity vectors, and $\vec{a}$ is drawn at the dot between the two velocity vectors it links.

EXAMPLE 3.2 Drawing the acceleration for a Mars descent

A spacecraft slows as it safely descends to the surface of Mars. Draw a complete motion diagram for the last few seconds of the descent.

SOLVE Figure 3.10 shows a complete motion diagram. As the spacecraft slows in its descent, the dots get closer together and the velocity vectors get shorter. The inset shows how Tactics Box 3.2 is used to determine the acceleration at one point. All the other acceleration vectors will be similar, because for each pair of velocity vectors the earlier one is longer than the later one.

ASSESS As the spacecraft in Figure 3.10 slows, the acceleration vectors and velocity vectors point in opposite directions, consistent with what we learned about the sign of the acceleration in Chapter 2.

(a) Artist version   (b) Student sketch

The acceleration vector is the same direction as $\Delta\vec{v}$

FIGURE 3.10 Motion diagram for a descending spacecraft.

EXAMPLE 3.3 Drawing the acceleration for a Ferris wheel ride

Anne rides a Ferris wheel at an amusement park. Draw Anne's motion diagram.

SOLVE Figure 3.11 shows 10 points of the motion during one complete revolution of the Ferris wheel. A person riding a Ferris wheel moves in a circle at a constant speed, so we've shown equal distances between successive dots. As before, the velocity vectors are found by connecting each dot to the next. Note that the velocity vectors are *straight lines*, not curves.

We see that all of the velocity vectors have the same length, but each has a different *direction,* and that means Anne is accelerating. This is not a "speeding up" or "slowing down" acceleration, but is, instead, a "change of direction" acceleration. The inset to Figure 3.11 shows how to find the acceleration at one particular position, at the bottom of the circle. Vector $\vec{v}_1$ is the velocity vector that leads into this dot, while $\vec{v}_2$ moves away from it. From the circular geometry of the main figure, the two angles marked α are equal. Thus we see that $\vec{v}_2$ and $-\vec{v}_1$ form an isosceles triangle and vector $\Delta \vec{v} = \vec{v}_2 - \vec{v}_1$ is exactly vertical, toward the center of the circle.

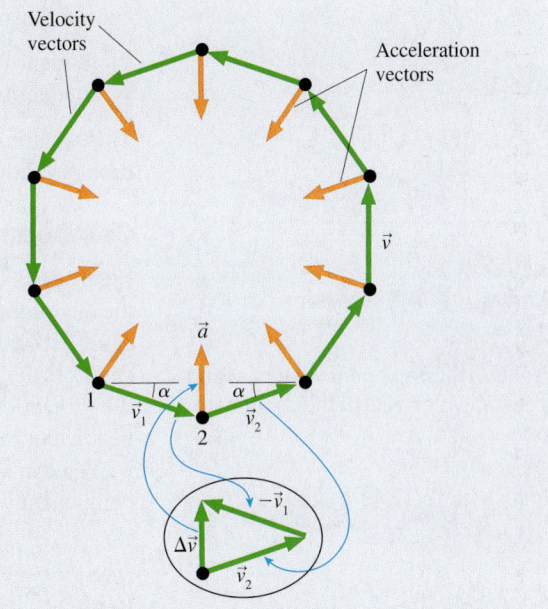

FIGURE 3.11 Motion diagram for Anne on a Ferris wheel.

Interestingly, no matter which dot you select on the motion diagram of Figure 3.11, the velocities change in such a way that the acceleration vector $\vec{a}$ points directly to the center of the circle. An acceleration vector that always points toward the center of a circle is called a *centripetal acceleration.* We will have much more to say about centripetal acceleration later in this chapter.

3.3 Coordinate Systems and Vector Components

In the past two sections, we have seen how to add and subtract vectors graphically, using these operations to deduce important details of motion. But the graphical combination of vectors is not an especially good way to find quantitative results. In this section we will introduce a *coordinate description* of vectors that will be the basis for doing vector calculations.

Coordinate Systems

We introduced the notion of a coordinate system in Chapter 1. The world does not come with a coordinate system attached to it. A coordinate system is an artificially imposed grid that you place on a problem in order to make quantitative measurements. The right choice of coordinate system will make a problem easier to solve.

We will generally use **Cartesian coordinates,** the familiar rectangular grid with perpendicular axes. By convention, the positive y-axis is located 90° *counterclockwise* from the positive x-axis, as illustrated in Figure 3.12. Figure 3.12 also identifies the four **quadrants** of the coordinate system, I through IV. Notice that, by convention, the quadrants are counted counterclockwise from the positive x-axis.

Coordinate axes have a positive end and a negative end, separated by zero at the origin where the two axes cross. When you draw a coordinate system, it is

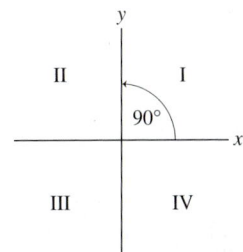

FIGURE 3.12 A Cartesian coordinate system and the quadrants of the xy-plane.

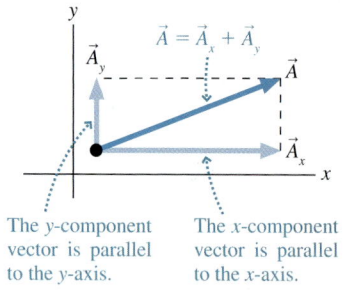

The *y*-component vector is parallel to the *y*-axis.

The *x*-component vector is parallel to the *x*-axis.

FIGURE 3.13 Component vectors $\vec{A}_x$ and $\vec{A}_y$ are drawn parallel to the coordinate axes such that $\vec{A} = \vec{A}_x + \vec{A}_y$.

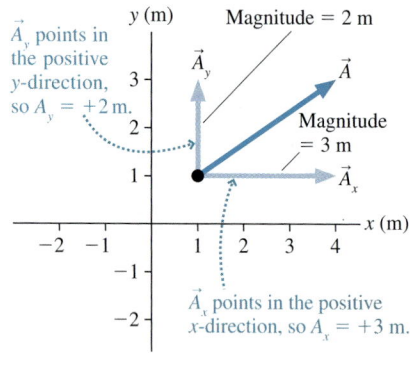

$\vec{A}_y$ points in the positive *y*-direction, so $A_y = +2$ m.

$\vec{A}_x$ points in the positive *x*-direction, so $A_x = +3$ m.

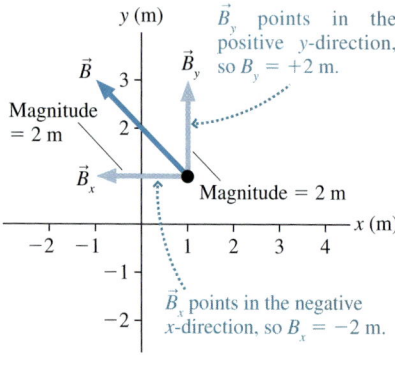

$\vec{B}_y$ points in the positive *y*-direction, so $B_y = +2$ m.

$\vec{B}_x$ points in the negative *x*-direction, so $B_x = -2$ m.

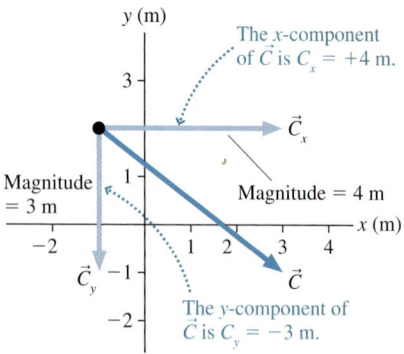

The *x*-component of $\vec{C}$ is $C_x = +4$ m.

The *y*-component of $\vec{C}$ is $C_y = -3$ m.

FIGURE 3.14 Determining the components of a vector.

important to label the axes. This is done by placing *x* and *y* labels at the *positive* ends of the axes, as in Figure 3.12. The purpose of the labels is twofold:

- To identify which axis is which, and
- To identify the positive ends of the axes.

This will be important when you need to determine whether the quantities in a problem should be assigned positive or negative values.

Component Vectors

Figure 3.13 shows a vector $\vec{A}$ and an *xy*-coordinate system that we've chosen. Once the directions of the axes are known, we can define two new vectors *parallel to the axes* that we call the **component vectors** of $\vec{A}$. Vector $\vec{A}_x$, called the *x-component vector,* is the projection of $\vec{A}$ along the *x*-axis. Vector $\vec{A}_y$, the *y-component vector,* is the projection of $\vec{A}$ along the *y*-axis. Notice that the component vectors are perpendicular to each other.

You can see, using the parallelogram rule, that $\vec{A}$ is the vector sum of the two component vectors:

$$\vec{A} = \vec{A}_x + \vec{A}_y \tag{3.9}$$

In essence, we have broken vector $\vec{A}$ into two perpendicular vectors that are parallel to the coordinate axes. We say that we have **decomposed** or **resolved** vector $\vec{A}$ into its component vectors.

NOTE ▶ It is not necessary for the tail of $\vec{A}$ to be at the origin. All we need to know is the *orientation* of the coordinate system so that we can draw $\vec{A}_x$ and $\vec{A}_y$ parallel to the axes. **◀**

Components

You learned in Chapter 2 to give the one-dimensional kinematic variable v_x a positive sign if the velocity vector $\vec{v}$ points toward the positive end of the *x*-axis, a negative sign if $\vec{v}$ points in the negative *x*-direction. The basis of this rule is that v_x is the *x-component* of $\vec{v}$. We need to extend this idea to vectors in general.

Suppose we have a vector $\vec{A}$ that has been decomposed into component vectors $\vec{A}_x$ and $\vec{A}_y$ parallel to the coordinate axes. We can describe each component vector with a single number (a scalar) called the **component.** The *x-component* and *y-component* of vector $\vec{A}$, denoted A_x and A_y, are determined as follows:

(MP) **TACTICS BOX 3.3** **Determining the components of a vector**

✎ Exercises 14–16

① The absolute value $|A_x|$ of the *x*-component A_x is the magnitude of the component vector $\vec{A}_x$.
② The *sign* of A_x is positive if $\vec{A}_x$ points in the positive *x*-direction, negative if $\vec{A}_x$ points in the negative *x*-direction.
③ The *y*-component A_y is determined similarly.

In other words, the component A_x tells us two things: how big $\vec{A}_x$ is and which end of the axis $\vec{A}_x$ points toward. Figure 3.14 shows three examples of determining the components of a vector.

NOTE ▶ $\vec{A}_x$ and $\vec{A}_y$ are *component vectors;* they have a magnitude and a direction. A_x and A_y are simply *components.* The components A_x and A_y are scalars—just numbers (with units) that can be positive or negative. **◀**

Much of physics is expressed in the language of vectors. We will frequently need to decompose a vector into its components or to "reassemble" a vector from its components, moving back and forth between the graphical and the component representations of a vector.

Let's start with the problem of decomposing a vector into its x- and y-components. Figure 3.15a shows a vector $\vec{A}$ at angle θ from the x-axis. It is *essential* to use a picture or diagram such as this to define the angle you are using to describe the vector's direction. $\vec{A}$ points to the right and up, so Tactics Box 3.3 tells us that the components A_x and A_y are both positive. We can use trigonometry to find

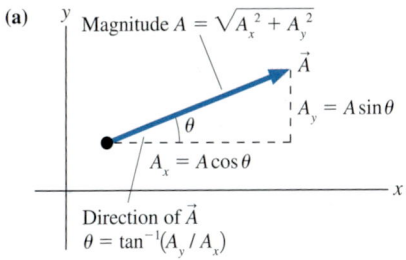

$$A_x = A\cos\theta$$
$$A_y = A\sin\theta \qquad (3.10)$$

where A is the magnitude, or length, of $\vec{A}$. These equations convert the length and angle description of vector $\vec{A}$ into the vector's components, but they are correct *only* if $\vec{A}$ is in the first quadrant, pointing to the right and up.

Figure 3.15b shows vector $\vec{C}$ pointing to the right and down. In this case, where the component vector $\vec{C}_y$ is pointing *down*, in the negative y-direction, the y-component C_y is a *negative* number. The angle ϕ (the Greek letter *phi*) is measured from the y-axis, so the components of $\vec{C}$ are

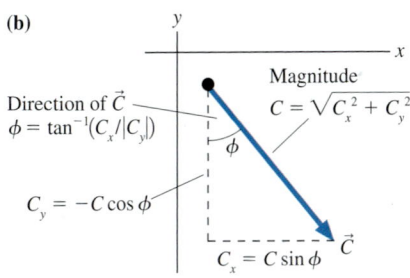

$$C_x = C\sin\phi$$
$$C_y = -C\cos\phi \qquad (3.11)$$

The role of sine and cosine is reversed from that in Equations 3.10 because we are using a different angle.

FIGURE 3.15 Moving between the graphical representation and the component representation.

NOTE ▶ Each decomposition requires that you pay close attention to the direction in which the vector points and the angles that are defined. The minus sign, when needed, must be inserted manually. ◀

We can also go in the opposite direction and determine the length and angle of a vector from its x- and y-components. Because A in Figure 3.15a is the hypotenuse of a right triangle, its length is given by the Pythagorean theorem:

$$A = \sqrt{A_x^2 + A_y^2} \qquad (3.12)$$

Similarly, the tangent of angle θ is the ratio of the far side to the adjacent side, so

$$\theta = \tan^{-1}\left(\frac{A_y}{A_x}\right) \qquad (3.13)$$

Equations 3.12 and 3.13 can be thought of as the "reverse" of Equations 3.10.

Equation 3.12 always works for finding the length or magnitude of a vector because the squares eliminate any concerns over the signs of the components. But finding the angle requires close attention to how the angle is defined and to the signs of the components. Finding the angle of vector $\vec{C}$ in Figure 3.15b requires the length of C_y *without* the minus sign, so vector $\vec{C}$ has direction

$$\phi = \tan^{-1}\left(\frac{C_x}{|C_y|}\right) \qquad (3.14)$$

Notice that the roles of x and y differ from those in Equation 3.13.

EXAMPLE 3.4 Finding the components of an acceleration vector

Find the x- and y-components of the acceleration vector $\vec{a}$ shown in Figure 3.16.

FIGURE 3.16 Acceleration vector $\vec{a}$ of Example 3.4.

PREPARE It's important to *draw* vectors. Figure 3.17 shows the original vector $\vec{a}$ decomposed into component vectors parallel to the axes.

FIGURE 3.17 The components of the acceleration vector.

SOLVE The acceleration vector $\vec{a} = (6.0 \text{ m/s}^2, 30°$ below the negative x-axis) points to the left (negative x-direction) and down (negative y-direction), so the components a_x and a_y are both negative:

$$a_x = -a\cos 30° = -(6.0 \text{ m/s}^2)\cos 30° = -5.2 \text{ m/s}^2$$

$$a_y = -a\sin 30° = -(6.0 \text{ m/s}^2)\sin 30° = -3.0 \text{ m/s}^2$$

ASSESS The units of a_x and a_y are the same as the units of vector $\vec{a}$. Notice that we had to insert the minus signs manually by observing that the vector is in the third quadrant.

STOP TO THINK 3.2 What are the x- and y-components C_x and C_y of vector $\vec{C}$?

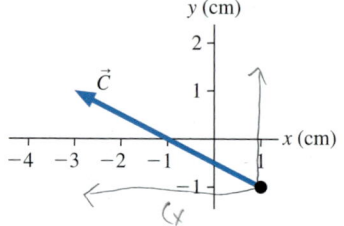

$C_y = \vec{C}\sin\theta$

$C_x = -\vec{C}\cos\theta$

Working with Components

We've seen how to add vectors graphically, but there's an easier way: using components. To illustrate, let's look at the vector sum $\vec{C} = \vec{A} + \vec{B}$ for the vectors shown in Figure 3.18. You can see that the component vectors of $\vec{C}$ are the sum of the component vectors of $\vec{A}$ and $\vec{B}$. The same is true of the components: $C_x = A_x + B_x$ and $C_y = A_y + B_y$.

In general, if $\vec{D} = \vec{A} + \vec{B} + \vec{C} + \cdots$, then the x- and y-components of the resultant vector $\vec{D}$ are

$$D_x = A_x + B_x + C_x + \cdots$$

$$D_y = A_y + B_y + C_y + \cdots \qquad (3.15)$$

This method of vector addition is called *algebraic addition*.

FIGURE 3.18 Using components to add vectors.

EXAMPLE 3.5 Using algebraic addition to find a bird's displacement

A bird flies 100 m due east from a tree, then 200 m northwest (that is, 45° west of north). What is the bird's net displacement?

PREPARE Figure 3.19a shows the displacement vectors $\vec{A} = (100\text{ m, east})$ and $\vec{B} = (200\text{ m, northwest})$ and also the net displacement $\vec{C}$. We draw vectors tip-to-tail if we are going to add them graphically, but it's usually easier to draw them all from the origin if we are going to use algebraic addition. Figure 3.19b redraws the vectors with their tails together.

SOLVE To add the vectors algebraically we must know their components. From the figure these are seen to be

$$A_x = 100\text{ m}$$

$$A_y = 0\text{ m}$$

$$B_x = -(200\text{ m})\cos 45° = -141\text{ m}$$

$$B_y = (200\text{ m})\sin 45° = 141\text{ m}$$

We learned *from the figure* that $\vec{B}$ has a negative x-component. Adding $\vec{A}$ and $\vec{B}$ by components gives

$$C_x = A_x + B_x = 100\text{ m} - 141\text{ m} = -41\text{ m}$$

$$C_y = A_y + B_y = 0\text{ m} + 141\text{ m} = 141\text{ m}$$

The magnitude of the net displacement $\vec{C}$ is

$$C = \sqrt{C_x^2 + C_y^2} = \sqrt{(-41\text{ m})^2 + (141\text{ m})^2} = 147\text{ m}$$

The angle θ, as defined in Figure 3.19, is

$$\theta = \tan^{-1}\left(\frac{C_y}{|C_x|}\right) = \tan^{-1}\left(\frac{141\text{ m}}{41\text{ m}}\right) = 74°$$

Thus the bird's net displacement is $\vec{C} = (147\text{ m}, 74°\text{ north of west})$.

ASSESS The final values of C_x and C_y match what we would expect from the sketch in Figure 3.19. The geometric addition was a valuable check on the answer we found by algebraic addition.

(a)

The bird's net displacement is $\vec{C} = \vec{A} + \vec{B}$.

Angle θ describes the direction of vector $\vec{C}$.

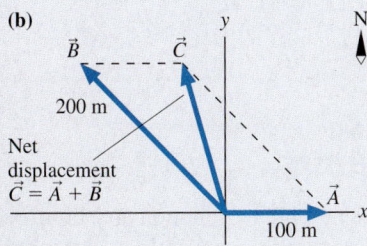

(b)

Net displacement $\vec{C} = \vec{A} + \vec{B}$

FIGURE 3.19 Finding the net displacement.

Vector subtraction and the multiplication of a vector by a scalar are also easily performed using components. To find $\vec{D} = \vec{P} - \vec{Q}$ we would compute

$$D_x = P_x - Q_x$$
$$D_y = P_y - Q_y \tag{3.16}$$

Similarly, $\vec{T} = c\vec{S}$ is

$$T_x = cS_x$$
$$T_y = cS_y \tag{3.17}$$

The next few chapters will make frequent use of *vector equations*. For example, you will learn that the equation to calculate the force on a car skidding to a stop is

$$\vec{F} = \vec{n} + \vec{w} + \vec{f} \tag{3.18}$$

Equation 3.18 is really just a shorthand way of writing the two simultaneous equations:

$$F_x = n_x + w_x + f_x$$
$$F_y = n_y + w_y + f_y \tag{3.19}$$

When you look at a trail map for a hike in a mountainous region, it will give the length of a trail *and* the elevation gain—an important variable! The elevation gain is simply d_y, the vertical component of the displacement for the hike.

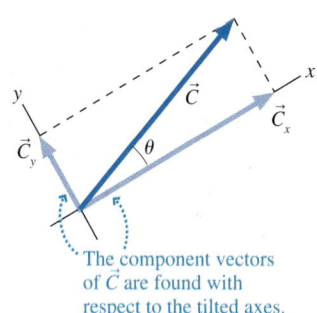

The component vectors of $\vec{C}$ are found with respect to the tilted axes.

FIGURE 3.20 A coordinate system with tilted axes.

In other words, a vector equation is interpreted as meaning: Equate the x-components on both sides of the equals sign, then equate the y-components. Vector notation allows us to write these two equations in a more compact form.

Tilted Axes

While we are used to having the x-axis horizontal, there is no requirement that it has to be that way. In Chapter 1, we saw that for motion on a slope we could put the x-axis along the slope. When we add the y-axis, this gives us a tilted coordinate system such as that shown in Figure 3.20.

Finding components with tilted axes is no harder than what we have done so far. Vector $\vec{C}$ in Figure 3.20 can be decomposed into component vectors $\vec{C}_x$ and $\vec{C}_y$, with $C_x = C\cos\theta$ and $C_y = C\sin\theta$.

STOP TO THINK 3.3 Angle ϕ that specifies the direction of $\vec{C}$ is given by

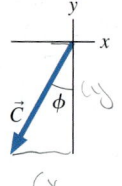

A. $\tan^{-1}(C_x/C_y)$.
B. $\tan^{-1}(C_x/|C_y|)$.
C. $\tan^{-1}(|C_x|/|C_y|)$.
D. $\tan^{-1}(C_y/C_x)$.
E. $\tan^{-1}(C_y/|C_x|)$.
F. $\tan^{-1}(|C_y|/|C_x|)$.

3.4 Motion on a Ramp

In this section, we will examine the problem of motion on a ramp or incline (often called an *inclined plane*). There are three reasons to look at this problem. First, it will provide good practice at using vectors to analyze motion. Second, it is a simple problem for which we can find an exact solution. Third, this seemingly abstract problem has real and important applications. As we will see, both speed skiing and roller coasters can be modeled quite well as the motion of an object sliding down a ramp.

We begin with a constant-velocity example to give us some practice with vectors and components before moving on to the more general case of accelerated motion.

EXAMPLE 3.6 Finding the height gained on a slope

A car drives up a steep 10° slope at a constant speed of 15 m/s. After 10 s, how much height has the car gained?

PREPARE Figure 3.21 is a visual overview, with x- and y-axes defined. The velocity vector $\vec{v}$ points up the slope. We are interested in the vertical motion of the car, so we decompose $\vec{v}$ into component vectors $\vec{v}_x$ and $\vec{v}_y$ as shown.

SOLVE The velocity component we need is v_y; this describes the vertical motion of the car. Using the rules for finding components outlined above, we find

$$v_y = v\sin\theta = (15 \text{ m/s})\sin(10°) = 2.6 \text{ m/s}$$

Because the velocity is constant, the car's vertical displacement (i.e., the height gained) during 10 s is

$$\Delta y = v_y \Delta t = (2.6 \text{ m/s})(10 \text{ s}) = 26 \text{ m}$$

ASSESS The car is traveling at a pretty good clip—15 m/s is a bit more than 30 mph—up a steep slope, so it should climb a respectable height in 10 s. 26 m, or about 80 ft, seems reasonable.

Known
$x_i = y_i = 0$ m
$t_i = 0$ s, $t_f = 10$ s
$v = 15$ m/s
$\theta = 10°$

Find
Δy

FIGURE 3.21 Visual overview of a car moving up a slope.

Accelerated Motion on a Ramp

Figure 3.22a shows a crate sliding down a frictionless (i.e., smooth) ramp tilted at angle θ. The crate accelerates due to the action of gravity, but it is *constrained* to accelerate parallel to the surface. What is the acceleration?

A motion diagram for the crate is drawn in Figure 3.22b. There is an acceleration, because the velocity is changing, with both the acceleration and velocity vectors parallel to the ramp. We can take advantage of the properties of vectors to find the crate's acceleration. To do so, Figure 3.22c sets up a coordinate system with the x-axis along the ramp and the y-axis perpendicular. This choice of coordinate system constrains all motion to be along the x-axis.

If the incline suddenly vanished, the object would have a free-fall acceleration $\vec{a}_{\text{free fall}}$ straight down. This acceleration vector can be decomposed into two component vectors: a vector $\vec{a}_x$ that is *parallel* to the incline and a vector $\vec{a}_y$ that is *perpendicular* to the incline. The vector addition rules studied earlier in this chapter tell us that $\vec{a}_{\text{free fall}} = \vec{a}_x + \vec{a}_y$.

The motion diagram shows that the object's actual acceleration $\vec{a}_x$ is parallel to the incline. The surface of the incline somehow "blocks" the other component of the acceleration $\vec{a}_y$, through a process we will examine in Chapter 5, but $\vec{a}_x$ is unhindered. It is this component of $\vec{a}_{\text{free fall}}$, parallel to the incline, that accelerates the object.

We can use trigonometry to work out the magnitude of this acceleration. Figure 3.22c shows that the three vectors $\vec{a}_{\text{free fall}}$, $\vec{a}_y$, and $\vec{a}_x$ form a right triangle with angle θ as shown. You should satisfy yourself that this is the same as the angle of the incline. By definition, the magnitude of $\vec{a}_{\text{free fall}}$ is g. This vector is the hypotenuse of the right triangle. The vector we are interested in, $\vec{a}_x$, is opposite angle θ. Thus the value of the acceleration along a frictionless slope is

$$a_x = \pm g \sin\theta \qquad (3.20)$$

NOTE ▶ The correct sign depends on the direction in which the ramp is tilted. The acceleration in Figure 3.22 is $+g\sin\theta$, but upcoming examples will show situations in which the acceleration is $-g\sin\theta$. ◀

Let's look at Equation 3.20 to see if it makes sense. A good way to do this is to consider some **limiting cases** in which the angle is at one end of its range. In these cases, the physics is clear and we can check our result. Let's look at two such possibilities:

1. Suppose the plane is perfectly horizontal, with $\theta = 0°$. If you place an object on a horizontal surface, you expect it to stay at rest with no acceleration. Equation 3.20 gives $a_x = 0$ when $\theta = 0°$, in agreement with our expectations.

2. Now suppose you tilt the plane until it becomes vertical, with $\theta = 90°$. You know what happens—the object will be in free fall, parallel to the vertical surface. Equation 3.20 gives $a_x = g$ when $\theta = 90°$, again in agreement with our expectations.

NOTE ▶ Checking your answer by looking at such limiting cases is a very good way to see if your answer makes sense. We will often do this in the "Assess" step of a solution. ◀

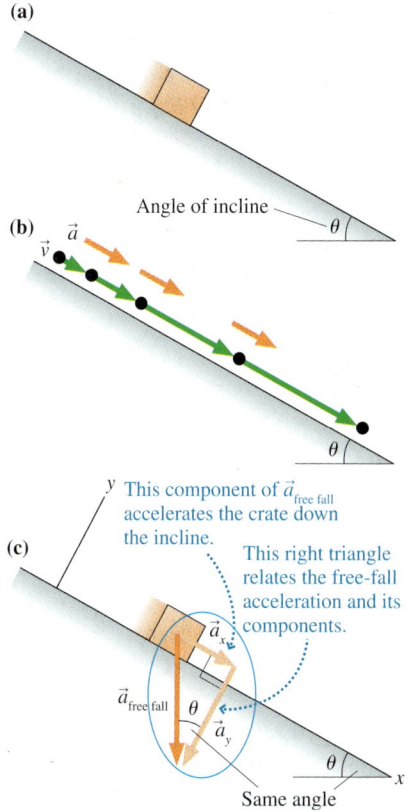

FIGURE 3.22 Acceleration on an inclined plane.

Extreme physics A speed skier, on wide skis with little friction, wearing an aerodynamic helmet and crouched low to minimize air resistance, moves in a straight line down a steep slope—pretty much like an object sliding down a frictionless ramp! There is a maximum speed that a skier could possibly achieve at the end of the slope. Course designers set the starting point to keep this maximum speed within reasonable (for this sport!) limits.

EXAMPLE 3.7 Maximum possible speed for a skier

The Willamette Pass ski area in Oregon was the site of the 1993 U.S. National Speed Skiing Competition. Each competitor's speed was measured at the end of the acceleration portion of the track, which has a reasonably constant slope. During the acceleration phase, the skiers traveled 360 m while dropping a vertical distance of 170 m. What is the fastest speed a skier could achieve at the end of this run? How much time would this fastest run take?

Continued

PREPARE We begin with the visual overview in Figure 3.23. The motion diagram shows the acceleration of the skier and the pictorial representation gives an overview of the problem including the dimensions of the slope. As above, we put the x-axis along the slope.

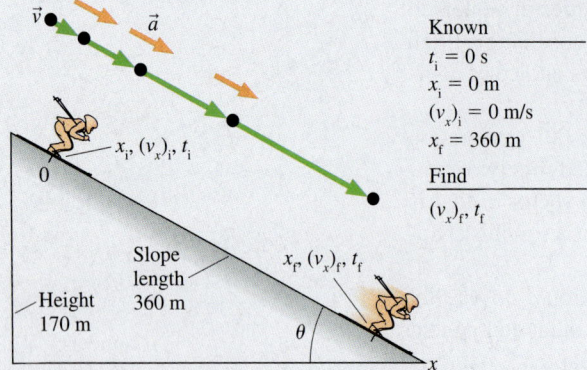

Known
$t_i = 0$ s
$x_i = 0$ m
$(v_x)_i = 0$ m/s
$x_f = 360$ m

Find
$(v_x)_f, t_f$

FIGURE 3.23 Visual overview of a skier accelerating down a slope.

SOLVE The fastest possible run would be one without any friction or air resistance, meaning the acceleration down the slope is given by Equation 3.20. The acceleration is in the positive x-direction, so we use the positive sign. What is the angle in Equation 3.20? Figure 3.23 shows that the 360-m-long slope is the hypotenuse of a triangle of height 170 m, so we use trigonometry to find

$$\sin\theta = \frac{170 \text{ m}}{360 \text{ m}} = 0.472$$

which gives $\theta = \sin^{-1}(0.472) = 28°$. Equation 3.20 then gives

$$a_x = +g\sin\theta = (9.8 \text{ m/s}^2)(\sin 28°) = 4.6 \text{ m/s}^2$$

For linear motion with constant acceleration, we can use the third of the kinematic equations in Table 2.4, $(v_x)_f^2 = (v_x)_i^2 + 2a_x \Delta x$. The initial velocity $(v_x)_i$ is zero, thus:

$$(v_x)_f = \sqrt{2a_x \Delta x} = \sqrt{2(4.6 \text{ m/s}^2)(360 \text{ m})} = 58 \text{ m/s}$$

This is the fastest that any skier could hope to be moving at the end of the run. Any friction or air resistance would decrease this speed. Because the acceleration is constant and the initial velocity $(v_x)_i$ is zero, the time of the fastest-possible run is

$$\Delta t = \frac{(v_x)_f}{a_x} = \frac{58 \text{ m/s}}{4.6 \text{ m/s}^2} = 12 \text{ s}$$

A speed skiing event is a quick affair!

ASSESS The final speed we calculated is 58 m/s, which is about 130 mph. We might expect such a high speed for this sport. In the competition noted, the actual winning speed was 111 mph, not much less than the result we calculated.

Skis on snow have very low friction, but there are other ways to reduce the friction between surfaces. For instance, a roller coaster car rolls along a track on low-friction wheels. No drive force is applied to the cars after they are released at the top of the first hill; the speed changes due to gravity alone. The cars speed up as they go down hills and slow down as they climb.

EXAMPLE 3.8 Speed of a roller coaster

A classic wooden coaster has cars that go down a big first hill, gaining speed. The cars then ascend a second hill with a slope of 30°. If the cars are going 25 m/s at the bottom and it takes them 2.0 s to climb this hill, how fast are they going at the top?

PREPARE We start with the visual overview in Figure 3.24, which includes a motion diagram, a pictorial representation, and a list of values. Notice how the motion diagram of Figure 3.24 differs from that of the previous example: The velocity decreases as the car moves up the hill, so the acceleration vector is opposite the direction of the velocity vector. The motion is along the x-axis, as before, but the acceleration vector points in the negative-x direction, so the component a_x is negative.

Known
$x_i = 0$ m
$(v_x)_i = 25$ m/s
$t_i = 0$ s
$t_f = 2.0$ s
$\theta = 30°$

Find
$(v_x)_f$

FIGURE 3.24 The coaster's speed decreases as it goes up the hill.

SOLVE To determine the final speed, we need to know the acceleration. We will assume that there is no friction or air

（手写）$a_y = -g \sin \theta$

resistance, so the magnitude of the roller coaster's acceleration is given by Equation 3.20 using the minus sign, as noted:

$$a_x = -g\sin\theta = -(9.8 \text{ m/s}^2)\sin 30° = -4.9 \text{ m/s}^2$$

The speed at the top of the hill can then be computed using our kinematic equation for velocity:

$$(v_x)_f = (v_x)_i + a_x\,\Delta t = 25 \text{ m/s} + (-4.9 \text{ m/s}^2)(2.0 \text{ s}) = 15 \text{ m/s}$$

ASSESS The speed is less at the top of the hill than at the bottom, as it should be, but the coaster is still moving at a pretty good clip at the top—almost 35 mph. This seems reasonable. In the motion diagram, notice that we only drew a single acceleration vector—a reasonable shortcut, as we know that the acceleration is constant. One vector can represent the acceleration for the entire motion.

STOP TO THINK 3.4   A block of ice slides down a ramp. For which height and base length is the acceleration the greatest?

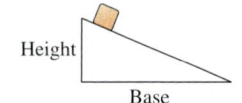

A. Height 4 m, base 12 m
B. Height 3 m, base 6 m
C. Height 2 m, base 5 m
D. Height 1 m, base 3 m

3.5 Relative Motion

You've now dealt many times with problems that say something like "A car travels at 30 m/s" or "A plane travels at 300 m/s." But, as we will see, we may need to be a bit more specific.

In Figure 3.25, Amy, Bill, and Carlos are watching a runner. According to Amy, the runner's velocity is $v_x = 5$ m/s. But to Bill, who's riding alongside, the runner is lifting his legs up and down but going neither forward nor backward relative to Bill. As far as Bill is concerned, the runner's velocity is $v_x = 0$ m/s. Carlos sees the runner receding in his rearview mirror, in the *negative x*-direction, getting 10 m further away from him every second. According to Carlos, the runner's velocity is $v_x = -10$ m/s. Which is the runner's *true* velocity?

Velocity is not a concept that can be true or false. The runner's velocity *relative to Amy* is 5 m/s. That is, his velocity is 5 m/s in a coordinate system attached to Amy and in which Amy is at rest. The runner's velocity relative to Bill is 0 m/s, and the velocity relative to Carlos is −10 m/s. These are all valid descriptions of the runner's motion.

Relative Velocity

Suppose we know that the runner's velocity relative to Amy is 5 m/s; we will call this velocity $(v_x)_{RA}$. The second subscript, "RA" means "**R**unner relative to **A**my." We also know that the velocity of **C**arlos relative to **A**my is 15 m/s; we write this as $(v_x)_{CA} = 15$ m/s. It is equally valid to compute Amy's velocity relative to Carlos. From Carlos's point of view, Amy is moving to the left at 15 m/s; we write **A**my's velocity relative to **C**arlos as $(v_x)_{AC} = -15$ m/s; note that $(v_x)_{AC} = -(v_x)_{CA}$.

Given the runner's velocity relative to Amy and Amy's velocity relative to Carlos, we can compute the runner's velocity relative to Carlos by combining the two velocities we know. The subscripts as we have defined them are our guide for this combination.

$$(v_x)_{RC} = (v_x)_{RA} + (v_x)_{AC} \tag{3.21}$$

The "A" appears on the right of the first expression and on the left of the second; when we combine these velocities, we "cancel" the A to get $(v_x)_{RC}$.

FIGURE 3.25 Amy, Bill, and Carlos each measure the velocity of the runner. The velocities are shown relative to Amy.

Throwing for the gold An athlete throwing the javelin does so while running. It's harder to throw the javelin on the run, but there's a very good reason to do so. The distance of the throw will be determined by the velocity of the javelin with respect to the ground—which is the sum of the velocity of the throw plus the velocity of the athlete. A faster run means a farther throw.

Generally, you can add two relative velocities in this manner, by "canceling" subscripts as in Equation 3.21. In Chapter 27, when we learn about relativity, we will have a more rigorous scheme for computing relative velocities, but this technique will serve our purposes at present.

EXAMPLE 3.9   How fast is a speeding bullet?

The police are chasing a bank robber. While driving at 50 m/s, they fire a bullet to shoot out a tire of his car. The police gun shoots bullets at 300 m/s. What is the bullet's speed as measured by a TV camera crew standing beside the road?

Known	Find
$(v_x)_{bc} = 300$ m/s	$(v_x)_{bg}$
$(v_x)_{cg} = 50$ m/s	

FIGURE 3.26   Relative velocities for Example 3.9.

PREPARE   A visual overview of the situation is given in Figure 3.26. We assume that all motion is along the *x*-axis. The velocity of the **car** is given relative to the **ground**; the velocity of the **bullet** is given relative to the **car**. We wish to find the velocity of the **bullet** relative to the **ground**.

SOLVE   We compute the desired velocity using the technique of Equation 3.21:

$$(v_x)_{bg} = (v_x)_{bc} + (v_x)_{cg} = 300 \text{ m/s} + 50 \text{ m/s} = 350 \text{ m/s}$$

ASSESS   The bullet is fired forward from a speeding vehicle. It moves faster relative to the ground than the speed with which it was shot from the gun, so our result makes sense.

This technique for finding relative velocities also works for two-dimensional situations, as we see in the next example. Relative motion in two dimensions is another good exercise in working with vectors.

EXAMPLE 3.10   Finding the ground speed of an airplane

Cleveland is approximately 300 miles east of Chicago. A plane leaves Chicago flying due east at 500 mph. The pilot forgot to check the weather and doesn't know that the wind is blowing to the south at 100 mph. What is the plane's velocity relative to the ground?

PREPARE   Figure 3.27 is a visual overview of the situation. We are given the speed of the **plane** relative to the **air** ($\vec{v}_{pa}$) and the

Known	Find
$\vec{v}_{pa} = (500 \text{ mph, east})$	$\vec{v}_{pg}$
$\vec{v}_{ag} = (100 \text{ mph, south})$	

FIGURE 3.27   The wind causes a plane flying due east in the air to move to the southeast relative to the ground.

speed of the **air** relative to the **ground** ($\vec{v}_{ag}$); the speed of the **plane** relative to the **ground** will be the vector sum of these velocities:

$$\vec{v}_{pg} = \vec{v}_{pa} + \vec{v}_{ag}$$

This vector sum is shown in Figure 3.27.

SOLVE   The plane's speed relative to the ground is the hypotenuse of the right triangle in Figure 3.27, thus

$$v_{pg} = \sqrt{v_{pa}^2 + v_{ag}^2} = \sqrt{(500 \text{ mph})^2 + (100 \text{ mph})^2} = 510 \text{ mph}$$

The plane's direction can be specified by the angle θ measured from due east:

$$\theta = \tan^{-1}\left(\frac{100 \text{ mph}}{500 \text{ mph}}\right) = \tan^{-1}(0.20) = 11°$$

The velocity of the plane relative to the ground is thus

$$\vec{v}_{pg} = (510 \text{ mph}, 11° \text{ south of east})$$

ASSESS   The good news is that the wind is making the plane move a bit faster relative to the ground; the bad news is that the wind is making the plane move in the wrong direction!

3.6   Motion in Two Dimensions: Projectile Motion

Balls flying through the air, long jumpers, and cars doing stunt jumps are all examples of the two-dimensional motion that we call *projectile motion*. Projectile motion is an extension to two dimensions of the free-fall motion we studied in

Chapter 2. **A projectile is an object that moves in two dimensions under the influence of only gravity and nothing else.** Although real objects are also influenced by air resistance, the effect of air resistance is small for reasonably dense objects moving at modest speeds, so we can ignore it for the cases we consider in this chapter. As long as we can neglect air resistance, any projectile, from a football to the stream of water shot by an archer fish, will follow the same type of path, a trajectory with the mathematical form of a parabola. Because the form of the motion will always be the same, the strategies we develop to solve one projectile problem can be applied to others as well.

Figure 3.28a shows the parabolic arc of a ball tossed into the air; the camera has captured its position at equal intervals of time. In Figure 3.28b we show the motion diagram for this toss, with velocity vectors connecting the points. The acceleration vector points in the same direction as the change in velocity $\Delta\vec{v}$, which we can compute using the techniques of Tactics Box 3.2. You can see that the acceleration vector points straight down; a careful analysis would show that it has magnitude 9.80 m/s^2. Consequently, the acceleration of a projectile is the same as the acceleration of an object falling straight down, namely, the free-fall acceleration

$$\vec{a}_{\text{free fall}} = (9.80 \text{ m/s}^2, \text{ straight down})$$

Because the acceleration is the same for all objects, it is no wonder that the shape of the trajectory—a parabola—is the same as well.

Analyzing Projectile Motion

Suppose you toss a basketball down the court, as shown in Figure 3.29. To study this projectile motion, we've established a coordinate system with the x-axis horizontal and the y-axis vertical. The start of a projectile's motion is called the *launch*, and the angle θ of the initial velocity $\vec{v}_i$ above the horizontal (i.e., above the x-axis) is the **launch angle.** As you learned in Section 3.3, the initial velocity vector $\vec{v}_i$ can be expressed in terms of the x- and y- components $(v_x)_i$ and $(v_y)_i$. You can see from the figure that

$$(v_x)_i = v_i\cos\theta$$
$$(v_y)_i = v_i\sin\theta \qquad (3.22)$$

where v_i is the initial speed.

> **NOTE** ▶ The components $(v_x)_i$ and $(v_y)_i$ are not always positive. A projectile launched at an angle *below* the horizontal (such as a ball thrown downward from the roof of a building) has *negative* values for θ and $(v_y)_i$. However, the *speed* v_i is always positive. ◀

As the projectile moves, the free-fall acceleration will change the vertical component of the velocity, but there will be no change to the horizontal component of the velocity. Therefore, the vertical and horizontal components of the acceleration are

$$a_x = 0 \text{ m/s}^2$$
$$a_y = -g = -9.80 \text{ m/s}^2 \qquad (3.23)$$

The vertical component of acceleration a_y for all projectile motion is just the familiar $-g$ of free fall while the horizontal component a_x is zero.

(a)

(b)

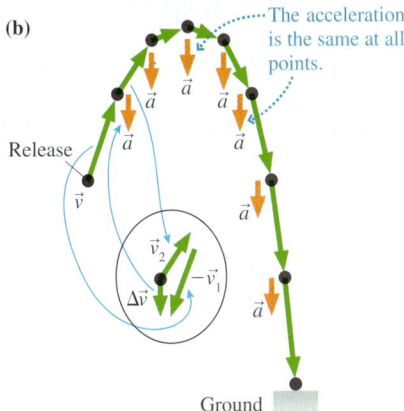

FIGURE 3.28 The motion of a tossed ball. The inset shows how to find the direction of $\Delta\vec{v}$, the change in velocity. This is the direction in which the acceleration $\vec{a}$ points.

FIGURE 3.29 The launch and motion of a projectile.

(a)

(b)

The vertical component of velocity decreases by 9.8 m/s every second.

The horizontal component of velocity is constant throughout the motion.

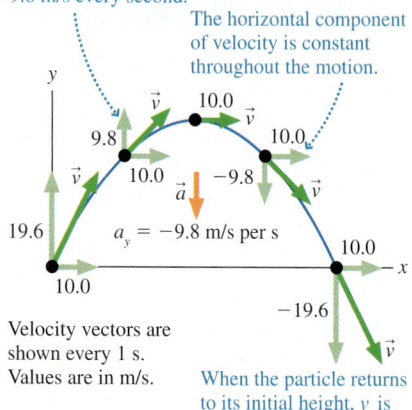

Velocity vectors are shown every 1 s. Values are in m/s.

When the particle returns to its initial height, v_y is opposite its initial value.

FIGURE 3.30 The velocity and acceleration vectors of a projectile.

FIGURE 3.31 A projectile launched horizontally falls in the same time as a projectile that is released from rest.

To see how these accelerations determine the subsequent motion, Figure 3.30 shows a projectile launched at a speed of 22.0 m/s at an angle of 63° to the horizontal. In Figure 3.30a, the initial velocity vector is broken into its horizontal and vertical components. In Figure 3.30b, the velocity and acceleration vectors are shown every subsequent 1.0 s. Because there is no horizontal acceleration ($a_x = 0$), the value of v_x never changes. In contrast, v_y decreases by 9.8 m/s every second. This is what it *means* to accelerate at $a_y = -9.8$ m/s^2 = (-9.8 m/s) per second. Nothing *pushes* the projectile along the curve. Instead, the downward acceleration changes the velocity vector as shown, causing it to increase downward as the motion proceeds. At the end of the motion, when the ball is at the same height as it started, v_y is -19.6 m/s, the negative of its initial value. **The ball finishes its motion moving downward at the same speed as it started moving upward,** just as we saw in the case of one-dimensional free fall in Chapter 2.

You can see from Figure 3.30 that **projectile motion is made up of two independent motions:** uniform motion at constant velocity in the horizontal direction and free-fall motion in the vertical direction. In Chapter 2, we saw kinematic equations for constant-velocity and constant-acceleration motion. We can adapt these general equations to this current case: The horizontal motion is constant-velocity motion at $(v_x)_i$, the vertical motion is constant-acceleration motion with initial velocity $(v_y)_i$ and an acceleration of $a_y = -g$.

$$x_f = x_i + (v_x)_i \, \Delta t \qquad\qquad y_f = y_i + (v_y)_i \, \Delta t - \tfrac{1}{2}g(\Delta t)^2$$

$$(v_x)_f = (v_x)_i = \text{constant} \qquad (v_y)_f = (v_y)_i - g \, \Delta t \tag{3.24}$$

Equations of motion for the parabolic trajectory of a projectile

A close look at these equations reveals a surprising fact: **The horizontal and vertical components of projectile motion are independent of each other.** The initial horizontal velocity has *no* influence over the vertical motion, and vice-versa. This independence of the horizontal and vertical motions is illustrated in Figure 3.31, which shows a strobe photograph of two balls, one shot horizontally and the other released from rest at the same instant. The *vertical* motions of the two balls are identical, and they hit the floor simultaneously. Neither ball has any initial motion in the vertical direction, so both fall distance h in the same amount of time.

Let's extend these ideas to consider a "classic" problem in physics:

A hungry hunter in the jungle wants to shoot down a coconut that is hanging from the branch of a tree. He aims the gun directly at the coconut, but as luck would have it the coconut falls from the branch at the exact instant the hunter pulls the trigger. Does the bullet hit the coconut?

Figure 3.32 shows a useful way to analyze this problem. Figure 3.32a shows the trajectory of a projectile. Without gravity, a projectile would follow a

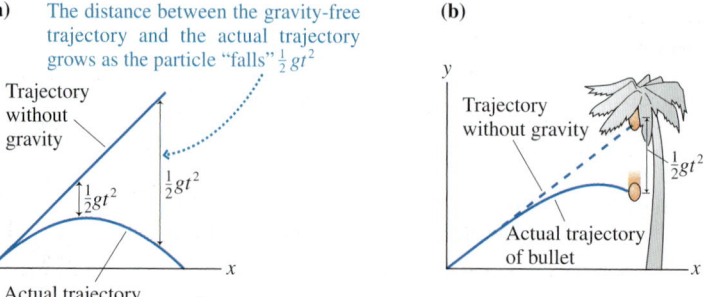

FIGURE 3.32 A projectile follows a parabolic trajectory because it "falls" a distance $\tfrac{1}{2}gt^2$ below a straight-line trajectory.

straight line. Because of gravity, the particle at time t has "fallen" a distance $\frac{1}{2}gt^2$ below this line. The separation grows as $\frac{1}{2}gt^2$, giving the trajectory its parabolic shape. Figure 3.32b applies this reasoning to the bullet and coconut. Although the bullet travels very fast, it follows a slightly curved trajectory, not a straight line. Had the coconut stayed on the tree, the bullet would have curved under its target as gravity causes it to fall a distance $\frac{1}{2}gt^2$ below the straight line. But $\frac{1}{2}gt^2$ is also the distance the coconut falls while the bullet is in flight. Thus, as Figure 3.32b shows, the bullet and the coconut fall the same distance and meet at the same point!

STOP TO THINK 3.5   A 100 g ball rolls off a table and lands 2 m from the base of the table. A 200 g ball rolls off the same table with the same speed. It lands at a distance

A. <1 m.   B. 1 m.   C. Between 1 m and 2 m.
D. 2 m.   E. Between 2 m and 4 m.   F. 4 m.

3.7 Projectile Motion: Solving Problems

Now that we have a good idea of how projectile motion works, we can use that knowledge to solve some true two-dimensional motion problems.

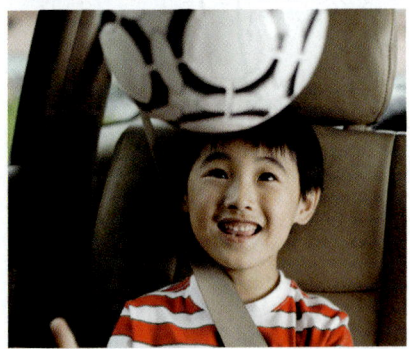

TRY IT YOURSELF

A game of catch in a moving vehicle A great way to illustrate the independence of the horizontal and vertical components of projectile motion is to do a simple experiment. While riding in a car moving at a constant speed, toss a ball or a coin into the air. You can easily catch it! The ball and you and the car continue to move forward at a constant speed during the ball's up and down vertical motion. From the point of a view of a person watching you drive by, the ball's motion would be a parabolic arc. You can also think about this situation in terms of relative motion. How would you do this?

EXAMPLE 3.11 Planning a Hollywood stunt

To get the shots of cars flying through the air in movies, it is sometimes necessary to drive a car off a cliff and film it. Suppose a stunt man drives a car off a 10-m-high cliff at a speed of 20 m/s. How far does the car land from the base of the cliff?

PREPARE We start with a visual overview of the situation in Figure 3.33. Note that we have chosen to put the origin of the coordinate system at the base of the cliff. We assume that the car is moving horizontally as it leaves the cliff. In this case, the x- and y-components of the initial velocity are

$$(v_x)_i = v_i = 20 \text{ m/s}$$
$$(v_y)_i = 0 \text{ m/s}$$

Known
$x_i = 0$ m
$(v_y)_i = 0$ m/s
$t_i = 0$ s
$y_i = 10$ m, $y_f = 0$ m
$(v_x)_i = v_i = 20$ m/s
$a_x = 0$ m/s^2
$a_y = -g$

Find
x_f

FIGURE 3.33 Visual overview for Example 3.11.

SOLVE Each point on the trajectory has x- and y-components of position, velocity, and acceleration but only *one* value of time. The time needed to move horizontally to the final position x_f is

the *same* time needed to fall 10 m vertically. **Although the horizontal and vertical motions are independent, they are connected through the time t.** This is a critical observation for solving projectile motion problems. We will call the time interval between the car leaving the cliff and landing on the ground Δt. In this problem, we'll analyze the vertical motion first. We can solve the vertical motion equations for the time interval Δt. We'll then use that value of Δt in the equation for the horizontal motion.

The vertical motion is just free fall. The initial vertical velocity is zero; the car falls from $y_i = 10$ m to $y_f = 0$ m. We can analyze this motion using the vertical-position equation from Equations 3.24:

$$y_f = y_i + (v_y)_i \Delta t - \tfrac{1}{2}g(\Delta t)^2$$
$$0 \text{ m} = 10 \text{ m} + (0 \text{ m/s})(\Delta t) - \tfrac{1}{2}(9.8 \text{ m/s}^2)(\Delta t)^2$$

Rearranging the terms and then solving for Δt gives

$$-10 \text{ m} = -\tfrac{1}{2}(9.8 \text{ m/s}^2)(\Delta t)^2$$
$$\Delta t = \sqrt{\frac{2(10 \text{ m})}{9.8 \text{ m/s}^2}} = 1.43 \text{ s}$$

Now that we have the time, we can use the horizontal-position equation from Equation 3.24 to find out where the car lands:

$$x_f = x_i + (v_x)_i \Delta t$$
$$x_f = 0 \text{ m} + (20 \text{ m/s})(1.43 \text{ s}) = 29 \text{ m}$$

ASSESS The cliff height is $h \approx 33$ ft and the initial horizontal velocity is $(v_x)_i \approx 40$ mph. At this speed, a car moves over 60 feet per second, so traveling $x_f = 29$ m ≈ 95 ft before hitting the ground seems quite reasonable.

Galileo was the first person to make a serious study of projectile motion, deducing the independence of the horizontal and vertical components. This page from his notes shows his analysis of a projectile launched horizontally. In his day, this topic was cutting-edge science; now it is in Chapter 3 of a 30-chapter book!

The approach of Example 3.11 is a general one. We can condense the relevant details into a problem-solving strategy.

(MP) PROBLEM-SOLVING
STRATEGY 3.1 **Projectile motion problems**

PREPARE There are a number of steps that you should go through in setting up the solution to a projectile motion problem:

- Make simplifying assumptions. Whether the projectile is a car or a basketball, the motion will be the same.
- Draw a visual overview including a pictorial representation showing the beginning and end points of the motion.
- Establish a coordinate system with the x-axis horizontal and the y-axis vertical. In this case, you know that the horizontal acceleration will be zero and the vertical acceleration will be free-fall: $a_x = 0$ and $a_y = -g$.
- Define symbols and write down a list of known values. Identify what the problem is trying to find.

SOLVE There are two sets of kinematic equations for projectile motion, one for the horizontal component and one for the vertical:

Horizontal	Vertical
$x_f = x_i + (v_x)_i \, \Delta t$	$y_f = y_i + (v_y)_i \, \Delta t - \frac{1}{2} g (\Delta t)^2$
$(v_x)_f = (v_x)_i = \text{constant}$	$(v_y)_f = (v_y)_i - g \, \Delta t$

Δt is the same for the horizontal and vertical components of the motion. Find Δt by solving for the vertical or the horizontal component of the motion, then use that value to complete the solution for the other component.

ASSESS Check that your result has the correct units, is reasonable, and answers the question.

3.1–3.7 **Activ Physics** ONLINE

EXAMPLE 3.12 **Checking the feasibility of a Hollywood stunt**

The main characters in the movie *Speed* are on a bus that has been booby-trapped to explode if its speed drops below 50 mph. But there is a problem ahead: A 50 ft section of a freeway overpass is missing. They decide to jump the bus over the gap. The road leading up to the break has an angle of about 5°. A view of the speedometer just before the jump shows that the bus is traveling at 67 mph. The movie bus makes the jump and survives. Is this realistic, or movie fiction?

PREPARE We begin by converting speed and distance to SI units. The initial speed is $v_i = 30$ m/s and the size of the gap is $L = 15$ m. Next, following the problem-solving strategy, we make a sketch, the visual overview shown in Figure 3.34, and a list of values. In choosing our axes, we've placed the origin at the point where the bus starts its jump. The initial velocity vector is tilted 5° above horizontal, so the components of the initial velocity are

$$(v_x)_i = v_i \cos\theta = (30 \text{ m/s})(\cos 5°) = 30 \text{ m/s}$$

$$(v_y)_i = v_i \sin\theta = (30 \text{ m/s})(\sin 5°) = 2.6 \text{ m/s}$$

How do we specify the "end" of the problem? By setting $y_f = 0$ m, we'll solve for the horizontal distance x_f at which the bus returns to its initial height. If x_f exceeds 50 ft, the bus successfully clears the gap. We have optimistically drawn our diagram as if the bus makes the jump, but . . .

FIGURE 3.34 Visual overview of the bus jumping the gap.

SOLVE Problem-Solving Strategy 3.1 suggests using one component of the motion to solve for Δt. We will begin with the vertical motion. The kinematic equation for the vertical position is

$$y_f = y_i + (v_y)_i \Delta t - \tfrac{1}{2}g(\Delta t)^2$$

We know that $y_f = y_i = 0$ m. If we factor out a Δt, the position equation becomes

$$0 = \Delta t\big((v_y)_i - \tfrac{1}{2}g\,\Delta t\big)$$

One solution to this equation is $\Delta t = 0$ s. This is a legitimate solution, but it corresponds to the instant when $y = 0$ at the beginning of the trajectory. We want the second solution, for $y = 0$ at the end of the trajectory, which is when

$$0 = (v_y)_i - \tfrac{1}{2}g\,\Delta t = (2.6 \text{ m/s}) - \tfrac{1}{2}(9.8 \text{ m/s}^2)\,\Delta t$$

Which gives

$$\Delta t = \frac{2 \times (2.6 \text{ m/s})}{9.8 \text{ m/s}^2} = 0.53 \text{ s}$$

During the 0.53 s that the bus is moving vertically it is also moving horizontally. The horizontal distance it travels is $x_f = x_i + (v_x)_i \Delta t$, or

$$x_f = 0 \text{ m} + (30 \text{ m/s})(0.53 \text{ s}) = 16 \text{ m}$$

This is how far the bus has traveled horizontally when it returns to its original height. 16 m is a bit more than the width of the gap, so a bus coming off a 5° ramp at the noted speed would make it—just barely!

ASSESS We can do a quick check on our math by noting that the bus takes off and lands at the same height. This means, as we saw in Figure 3.30b, that the y-velocity at the landing should be the negative of its initial value. We can use the velocity equation for the vertical component of the motion to compute the final value and see that the final velocity value is as we predict:

$$(v_y)_f = (v_y)_i - g\,\Delta t$$

$$= (2.6 \text{ m/s}) - (9.8 \text{ m/s}^2)(0.53 \text{ s}) = -2.6 \text{ m/s}$$

During the filming of the movie, the filmmakers really did jump a bus over a gap in an overpass! The actual jump was a bit more complicated than our example because a real bus, being an extended object rather than a particle, will start rotating as the front end comes off the ramp. The actual stunt jump used an extra ramp to give a boost to the front end of the bus. Nonetheless, our example shows that the filmmakers did their homework and devised a situation in which the physics was correct.

Range and Time of Flight

There are many examples of projectiles that land at the same height as that from which they were launched, such as a football that has been kicked downfield, or a baseball hit to the outfield. In cases like this, we are often interested in how far the projectile went, or how long it was in the air.

Figure 3.35 shows a pictorial representation of a projectile that is launched with initial velocity $\vec{v}_i$ and lands at its starting height. The x- and y-axes are specified, and we have chosen our origin so that $x_i = y_i = 0$. The distance to the landing point, at which $y_f = 0$, is called the **range** of the projectile. The time that the projectile is in the air is the **time of flight.**

The x- and y-components of the projectile's initial velocity are $(v_x)_i = v_i \cos\theta$ and $(v_y)_i = v_i \sin\theta$. We can use the equation for vertical motion to find the time of flight, as we did in Example 3.12:

$$0 = 0 + (v_i \sin\theta)t_f - \tfrac{1}{2}gt_f^2 = (v_i \sin\theta - \tfrac{1}{2}gt_f)t_f$$

Here we've taken advantage of the fact that $t_i = 0$ to write $\Delta t = t_f$. We are looking for a solution for which $t_f \neq 0$, so we set the quantity in brackets equal to zero and solve for the time of flight:

$$\text{Time of flight} = t_f = \frac{2v_i \sin\theta}{g} \qquad (3.25)$$

Let's look at the range now. We can use the time of flight in the equation for horizontal motion to compute the distance to the landing point:

$$x_f = x_i + (v_x)_i t_f = (v_i \cos\theta)t_f$$

$$= (v_i \cos\theta)\frac{2v_i \sin\theta}{g} = \frac{2v_i^2 \sin\theta \cos\theta}{g}$$

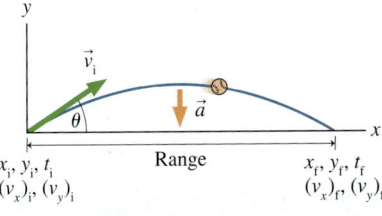

FIGURE 3.35 The parabolic trajectory and range of a projectile.

We can simplify this result by using the trigonometric identity $2\sin\theta\cos\theta = \sin(2\theta)$ to write the distance traveled by the projectile (i.e., its range) when launched at angle θ as

$$\text{Range} = x_f = \frac{v_i^2 \sin(2\theta)}{g} \tag{3.26}$$

NOTE ▶ Equation 3.26 is *not* a general result. It applies *only* in situations in which the projectile lands at the same elevation from which it was launched. ◀

What value of θ will give the largest distance? As you know, the sine function has a maximum value of 1 at an angle of 90°. Because the equation for x_f contains the expression $\sin(2\theta)$, it will be a maximum for $\theta_{max} = 45°$. Thus the maximum possible range, over level ground, is $x_{max} = v_i^2/g$, and this maximum is achieved only for a 45° launch.

Suppose a football player is kicking a ball down the field. If he wants the most distance for the kick, the ball should leave his foot at a 45° angle. But this angle won't give the longest "hang time," the time of flight. You can see from Equation 3.25 that the longest time of flight is achieved by kicking the ball at $\theta = 90°$—straight up! Reducing the angle from 90° will give a greater range, but a lesser hang time. The laws of physics put the range and the hang time in conflict with each other. Both cannot be maximized simultaneously, so the kicker must choose which is more important.

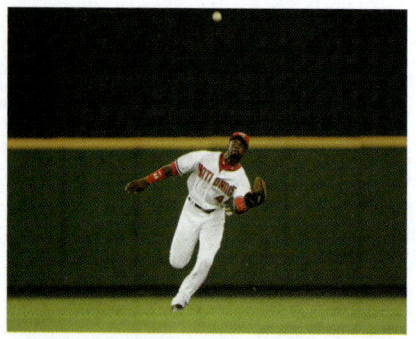

Physics of fielding BIO At the crack of a bat, the fielder begins running. Eyeing the ball, he makes a graceful arc to the exact spot where it will land, catching it on the run. Ball players don't estimate velocity and mentally calculate range, so how they do what they do is a fascinating—and unresolved—physics problem. Perhaps a fielder watches the relative motion of the ball as he runs and makes adjustments in his velocity to keep the ball at a constant angle, leading him to the right spot. If so, he doesn't know where the ball will land, just how to move in order to intercept it.

EXAMPLE 3.13 Finding the range of a kick

A football player kicks a ball straight up in the air, where it stays for a total time of 5.0 s. If he had kicked the ball downfield at the same speed, what is the maximum distance he could have kicked the ball?

PREPARE First, we can use the time of flight for the straight-up kick to calculate how fast the ball was kicked. We will then use that speed to determine the maximum range.

SOLVE Equation 3.25, the time-of-flight equation, with $\theta = 90°$, is

$$\text{Time of flight} = 5.0\text{ s} = \frac{2v_i}{g}$$

Solving for v_i gives

$$v_i = \frac{g(5.0\text{ s})}{2} = 24.5\text{ m/s}$$

The maximum possible range is found by using this speed in Equation 3.26 with $\theta = 45°$:

$$\text{Maximum distance} = \frac{v_i^2}{g} = \frac{(24.5\text{ m/s})^2}{9.8\text{ m/s}^2} = 61\text{ m}$$

ASSESS 61 m is a bit more than 61 yards—an impressive kick, but an achievable one.

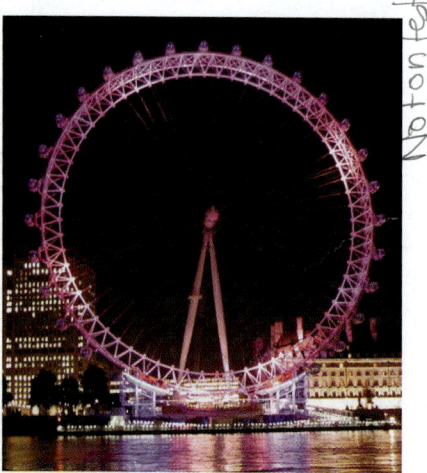

The London Eye Ferris wheel.

3.8 Motion in Two Dimensions: Circular Motion

The 32 cars on the London Eye Ferris wheel move at a constant speed of about 0.5 m/s in a vertical circle of radius 65 m. The cars may move at a constant speed, but they do *not* move with constant velocity. Velocity is a vector that depends on both an object's speed *and* its direction of motion, and the direction of circular motion is constantly changing. This is the hallmark of uniform circular motion: Constant speed, but continuously changing direction. We will introduce some basic ideas about circular motion in this section, then return to treat it in considerably more detail in Chapter 6. For now, we will only consider objects that move around a circular trajectory at constant speed.

Period and Frequency

The time interval it takes an object to go around a circle one time, completing one revolution (abbreviated rev), is called the **period** of the motion. Period is represented by the symbol T.

Rather than specify the time for one revolution, we can specify circular motion by its **frequency,** the number of revolutions per second, for which we use the symbol f. An object with a period of one-half second completes 2 revolutions each second. Similarly, an object can make 10 revolutions in 1 s if its period is one-tenth of a second. This shows that frequency is the inverse of the period:

$$f = \frac{1}{T} \tag{3.27}$$

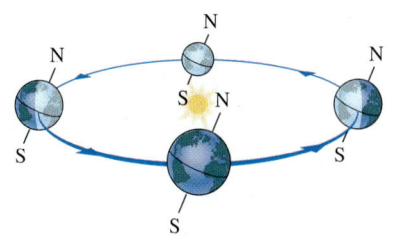

The period of the orbit of the earth around the sun is one year.

Although frequency is often expressed as "revolutions per second," *revolutions* are not true units but merely the counting of events. Thus the SI unit of frequency is simply inverse seconds, or s^{-1}. Frequency may also be given in revolutions per minute (rpm) or another time interval, but these usually need to be converted to s^{-1} before doing calculations.

EXAMPLE 3.14 Period of an audio CD

An audio CD spins at a frequency of up to 540 rpm. At 540 rpm, how much time is required for one revolution of the CD?

SOLVE First, we convert units from rpm to rev/s:

$$f = 540 \, \frac{\text{revolutions}}{\text{minute}} \times \frac{1 \text{ minute}}{60 \text{ seconds}} = 9.0 \, \frac{\text{rev}}{\text{s}} = 9.0 \text{ s}^{-1}$$

We then rearrange Equation 3.27 and compute

$$T = \frac{1}{f} = \frac{1}{9.0 \text{ s}^{-1}} = 0.11 \text{ s}$$

ASSESS 0.11 second is 1/9 of a second, which makes sense, as the frequency is 9 rev/s.

Angular Position

Rather than using xy-coordinates for circular motion, it will be more convenient to describe the position of the particle by its distance r from the center of the circle and its angle θ from the positive x-axis. This is shown in Figure 3.36. The angle θ is the **angular position** of the particle.

We can distinguish a position above the x-axis from a position an equal angle below the x-axis by defining θ to be positive when measured *counterclockwise* from the positive x-axis. An angle measured *clockwise* from the positive x-axis has a negative value. "Clockwise" and "counterclockwise" in circular motion are analogous, respectively, to "left of the origin" and "right of the origin" in linear motion, which we associated with negative and positive values of x.

Rather than measure angles in degrees, mathematicians and scientists usually measure the angle θ in the angular unit of *radians*. In Figure 3.36, we have noted the **arc length** s that the particle has traveled along the edge of the circle of radius r. We define the particle's angle θ in **radians** in terms of this arc length and the radius of the circle:

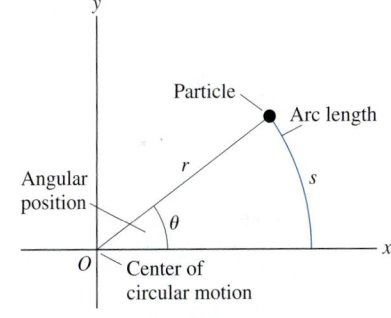

FIGURE 3.36 A particle's position is described by distance r and angle θ.

$$\theta \text{ (radians)} = \frac{s}{r} \tag{3.28}$$

The radian, abbreviated rad, is the SI unit of angle. An angle of 1 rad has an arc length s exactly equal to the radius r. An important consequence of Equation 3.28 is that the arc length spanning the angle θ is

$$s = r\theta \tag{3.29}$$

NOTE ▶ Equation 3.29 is valid only if θ is measured in radians and not degrees. This very simple relationship between angle and arc length is one of the primary motivations for using radians. ◀

The arc length completely around a circle is the circle's circumference $2\pi r$. Thus the angle of a full circle is

$$\theta_{\text{full circle}} = \frac{s}{r} = \frac{2\pi r}{r} = 2\pi \text{ rad}$$

We can use this fact to define conversion factors among revolutions, radians, and degrees:

$$1 \text{ rev} = 360° = 2\pi \text{ rad}$$

$$1 \text{ rad} = 1 \text{ rad} \times \frac{360°}{2\pi \text{ rad}} = 57.3°$$

We will often specify angles in degrees, but keep in mind that the SI unit is the radian. You can visualize angles in radians by remembering that 1 rad is just about 60°.

Acceleration in Circular Motion

It may seem strange to think that an object moving with constant speed can be accelerating, but that's exactly what an object in uniform circular motion is doing. It is accelerating because its velocity is changing as its direction of motion changes. What is the acceleration in this case? We saw in Example 3.3, which is worth reviewing, that for circular motion at a constant speed, **the acceleration vector $\vec{a}$ points toward the center of the circle.** As you can see in Figure 3.37 the velocity is always tangent to the circle, so $\vec{v}$ and $\vec{a}$ are perpendicular to each other at all points on the circle.

An acceleration that always points directly toward the center of a circle is called a *centripetal acceleration*. The word "centripetal" comes from a Greek root meaning "center seeking." Centripetal acceleration is not a new type of acceleration; all we are doing is *naming* an acceleration that corresponds to a particular type of motion. The magnitude of the centripetal acceleration is constant because each successive $\Delta\vec{v}$ in the motion diagram has the same length.

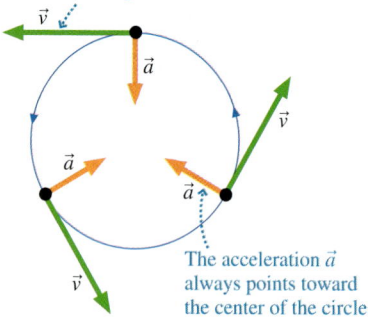

The velocity $\vec{v}$ is always tangent to the circle and perpendicular to $\vec{a}$ at all points.

The acceleration $\vec{a}$ always points toward the center of the circle.

FIGURE 3.37 The velocity and acceleration vectors for circular motion.

CONCEPTUAL EXAMPLE 3.1 Acceleration on a swing

A child is riding a playground swing. At the lowest point of her motion, is she accelerating? If so, what is the direction of her acceleration?

REASON As Figure 3.38 shows, the swing rotates in a circle around a central point where the rope or chain for the swing is attached. Even though speed isn't changing at the lowest point, the *direction* of the velocity vector is changing. Thus there is an acceleration—a centripetal acceleration directed toward the center of the circle. The velocity and acceleration vectors are perpendicular, as was shown in Figure 3.37. The velocity is directed forward, the acceleration is directed upward.

FIGURE 3.38 A child at the lowest point of motion on a swing.

ASSESS In the next chapter, we will find that there must be a force to produce an acceleration. Because the chain or rope is pulling up on the swing, it makes sense to think that the acceleration will be upward as well. But we are getting ahead of ourselves!

4.1

To complete our description of circular motion we need to find a quantitative relationship between the magnitude of the acceleration a and the speed v. Let's return to the case of the Ferris wheel, looking at the two points of the motion shown in Figure 3.39a. During a time Δt in which a car on the Ferris wheel moves from point 1 to point 2, the car moves through an angle θ and travels along an arc of length $r\theta$, as noted in the figure.

The speed of the car is constant, though the direction of motion changes. The speed is the distance divided by the time interval:

$$v = \frac{r\theta}{\Delta t}$$

Rearranging, we find the time Δt to be

$$\Delta t = \frac{r\theta}{v} \qquad (3.30)$$

Figure 3.39b shows velocity vectors for the two points and the difference vector $\Delta \vec{v}$. The angle θ is the same as that in Figure 3.39a. If this angle is small, the magnitude of the difference vector is very nearly equal to the length of a circular arc between the two velocity vectors, so we can approximate

$$\Delta v = v\theta \qquad (3.31)$$

The magnitude of the acceleration is computed using Δv from Equation 3.31 and Δt from Equation 3.30:

$$a = \frac{\Delta v}{\Delta t} = \frac{v\theta}{r\theta/v} = \frac{v^2}{r} \qquad (3.32)$$

Combining this magnitude with the direction we noted above, we can write the centripetal acceleration as

$$\vec{a} = \left(\frac{v^2}{r}, \text{ toward center of circle} \right) \qquad (3.33)$$

Centripetal acceleration of object moving in a circle of radius r at speed v

(a) As the car moves from point 1 to point 2, it goes through a circular arc of angle θ . . .

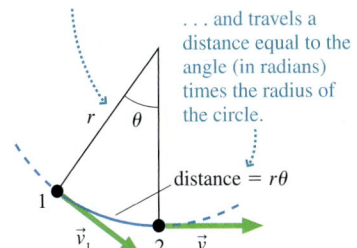

. . . and travels a distance equal to the angle (in radians) times the radius of the circle.

distance $= r\theta$

(b)

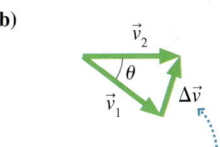

During this motion, the velocity changes direction; the difference vector points toward the center of the circle.

FIGURE 3.39 Changing velocity for an object in circular motion.

EXAMPLE 3.15 Finding the acceleration of a Ferris wheel

A typical carnival Ferris wheel has a radius of 9.0 m and rotates 6.0 times per minute. What magnitude acceleration do the riders experience?

PREPARE The cars on a Ferris wheel move in a circle at constant speed; the acceleration the riders experience is a centripetal acceleration.

SOLVE In order to use Equation 3.33 to compute an acceleration, we need to know the speed v of a rider on the Ferris wheel. The wheel rotates 6.0 times per minute; therefore, the time for one rotation (i.e., the period) is 10 s. During this time, a rider travels one circumference $2\pi r$ of the circle. Thus the speed is

$$v = \frac{\text{distance}}{\text{time interval}} = \frac{2\pi r}{\Delta t} = \frac{2\pi(9.0 \text{ m})}{10 \text{ s}} = 5.7 \text{ m/s}$$

Knowing the speed, we compute the magnitude of the acceleration as

$$a = \frac{v^2}{r} = \frac{(5.7 \text{ m/s})^2}{9.0 \text{ m}} = 3.6 \text{ m/s}^2$$

ASSESS This is about 1/3 of the free-fall acceleration; the acceleration, in units of g, is $0.37g$. This is enough to notice, but not enough to be scary!

What Comes Next: Forces

So far we have been studying motion without saying too much about what actually *causes* motion. Kinematics, the mathematical description of motion, is a good place to start because motion is very visible and very familiar. And in our study of motion we have introduced many of the basic tools, such as vectors, that we will use in the rest of the book.

But kinematics alone has some real limitations. For example, in this chapter we learned about projectile motion. But a real projectile such as a golf ball doesn't actually follow a perfectly parabolic path. Golfers know that the angle for maximum range is quite a bit less than 45°! Air resistance plays a role, and so do the dimples on a golf ball. The dimples give the ball a certain amount of "lift" as the ball spins, thus making it go further than a non-dimpled ball.

To treat more complex problems of this sort, we need to move beyond our treatment of kinematics and begin looking at the cause of motion: forces. By learning about forces, we will be able to explore a much wider range of problems in much more depth. As an example, think about the picture of a roller coaster with an inverted loop. How is it that riders can go through the loop and not fall out of their seats? This is just one of the problems that we will study once we know a bit about forces and the connection between forces and motion.

Amusement park kinematics Acceleration is fun—at least that's what the designer of this roller coaster seems to think! The coaster has ramps that give linear acceleration, parabolic segments in which the coaster follows a projectile path with a free-fall acceleration, and circular arcs in which the centripetal acceleration is larger than g. All of this acceleration means there are forces on the riders—and the coaster must be carefully designed so that these forces are well within safe limits.

STOP TO THINK 3.6   Which of the following particles has the largest centripetal acceleration?

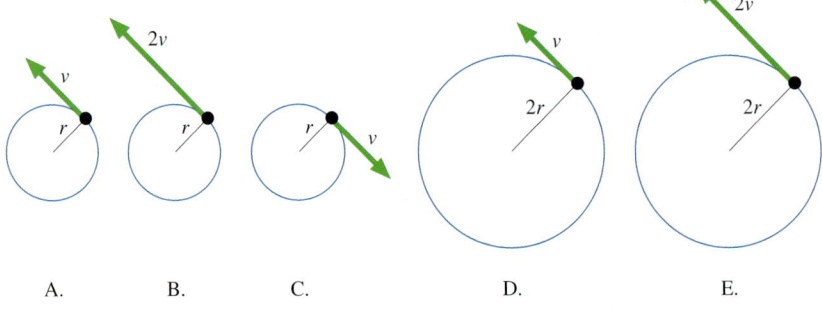

A.   B.   C.   D.   E.

SUMMARY

The goals of Chapter 3 have been to learn more about using vectors and to use vectors as a tool to analyze motion in two dimensions.

GENERAL PRINCIPLES

Projectile Motion

A projectile is an object that moves through the air under the influence of gravity and nothing else.

The path of the motion is a parabola.

The motion consists of two pieces:

1. Vertical motion with free-fall acceleration, $a_y = -g$.
2. Horizontal motion with constant velocity.

Kinematic equations:

$$x_f = x_i + (v_x)_i \, \Delta t$$
$$(v_x)_f = (v_x)_i = \text{constant}$$
$$y_f = y_i + (v_y)_i \, \Delta t - \tfrac{1}{2}g(\Delta t)^2$$
$$(v_y)_f = (v_y)_i - g \, \Delta t$$

Circular Motion

For an object moving in a circle at a constant speed:

- The period, T, is the time for one rotation.
- The velocity is tangent to the circular path.
- The acceleration points toward the center of the circle, and has magnitude

$$a = \frac{v^2}{r}$$

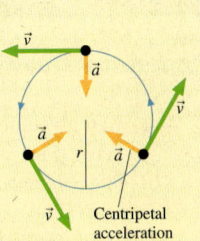

Centripetal acceleration

IMPORTANT CONCEPTS

Vectors and Components

A vector can be decomposed into **component vectors** along the x- and y-axes.

$$\vec{A} = \vec{A}_x + \vec{A}_y$$

The sign of the components depends on the **quadrant:**

$A_x < 0$	$A_x > 0$
$A_y > 0$	$A_y > 0$
$A_x < 0$	$A_x > 0$
$A_y < 0$	$A_y < 0$

Each component vector can be described with a single number specifying the component vector's length and (with a $+$ or $-$ sign) its direction. These two numbers are the x- and y-**components** of the vector.

The magnitude and direction of a vector can be expressed in terms of its components.

Magnitude $A = \sqrt{A_x{}^2 + A_y{}^2}$

$A_y = A \sin \theta$

$A_x = A \cos \theta$

Direction of $\vec{A}$
$\theta = \tan^{-1}(A_y / A_x)$

The Acceleration Vector

We define the acceleration in two or more dimensions as

$$\vec{a} = \frac{\vec{v}_f - \vec{v}_i}{t_f - t_i} = \frac{\Delta \vec{v}}{\Delta t}$$

We find an acceleration vector to complete the motion diagram as follows:

Dots show positions at equal time intervals.

Velocity vectors go dot to dot.

The acceleration vector points in the direction of $\Delta \vec{v}$.

The difference in the velocity vectors is found by adding the negative of $\vec{v}_i$ to $\vec{v}_f$.

APPLICATIONS

Relative motion

Velocities can be expressed relative to an observer. We can add relative velocities to convert to another observer's point of view.

c = car, b = bullet, g = ground

$$(v_x)_{bg} = (v_x)_{bc} + (v_x)_{cg}$$

Motion on a ramp

An object sliding down a ramp will accelerate parallel to the ramp:

$$a_x = \pm g \sin \theta$$

The correct sign depends on the direction in which the ramp is tilted.

Same angle

For instructor-assigned homework, go to www.masteringphysics.com

Problem difficulty is labeled as | (straightforward) to |||| (challenging).

Problems labeled INT integrate significant material from earlier chapters; BIO are of biological or medical interest.

QUESTIONS

Conceptual Questions

1. a. Can a vector have nonzero magnitude if a component is zero? If no, why not? If yes, give an example.
   b. Can a vector have zero magnitude and a nonzero component? If no, why not? If yes, give an example.
2. The zero vector, $\vec{0}$, is defined by the property that $\vec{0} + \vec{A} = \vec{A}$ for any vector $\vec{A}$. What are the components of the zero vector? What is its magnitude?
3. Is it possible to add a scalar to a vector? If so, demonstrate. If not, explain why not.
4. Suppose two vectors have unequal magnitudes. Can their sum be $\vec{0}$? Explain.
5. Suppose $\vec{C} = \vec{A} + \vec{B}$.
   a. Under what circumstances does $C = A + B$?
   b. Could $C = A - B$? If so, how? If not, why not?
6. Suppose $\vec{C} = \vec{A} - \vec{B}$.
   a. Under what circumstances does $C = A - B$?
   b. Could $C = A + B$? If so, how? If not, why not?
7. For a projectile, which of the following quantities are constant during the flight: x, y, v_x, v_y, v, a_x, a_y? Which of the quantities are zero throughout the flight?
8. A baseball player throws a ball at a 40° angle to the ground. The ball lands on the ground some distance away.
   a. Is there any point on the trajectory where $\vec{v}$ and $\vec{a}$ are parallel to each other? If so, where?
   b. Is there any point where $\vec{v}$ and $\vec{a}$ are perpendicular to each other? If so, where?
9. A long jumper trying for the longest possible jump takes off at an angle less than 45°. Why might this be so?
   Hint: Think about the position of the center of the jumper's body on take off and on landing.
   BIO

10. A person trying to throw a ball as far as possible will run forward during the throw. Explain why this increases the distance of the throw.
11. A passenger on a jet airplane claims to be able to walk at a speed in excess of 500 mph. Can this be true? Explain.
12. A roller coaster is designed with a ramp that reverses the direction of the cars. Cars start at the bottom of the ramp, roll up the ramp until stopping, then roll backward down the ramp. In what direction does the acceleration vector of the cars point
   a. as they move up the ramp?
   b. as they move down the ramp?
   c. at the highest point on the ramp?

13. During practice, two basketball players the same distance from the basket each shoot a ball at the same speed at the same time. Assuming their shots are accurate, can they avoid having the balls collide at the basket? If so, how? If not, why not? **Hint:** Are there two different launch angles that will have the same range but different times of flight?
14. A cyclist goes around a level, circular track at constant speed. Do you agree or disagree with the following statement: "Because the cyclist's speed is constant, her acceleration is zero." Explain.
15. You are driving your car in a circular path on level ground at a constant speed of 20 mph. At the instant you are driving north, and turning left, are you accelerating? If so, toward what point of the compass (N, S, E, W) does your acceleration vector point? If not, why not?
16. An airplane has been directed to fly in a clockwise circle, as seen from above, at constant speed until another plane has landed. When the plane is going north, is it accelerating? If so, in what direction does the acceleration vector point? If not, why not?

Multiple-Choice Questions

17. | The gas pedal in a car is sometimes referred to as "the accelerator." Which other controls on the vehicle can be used to produce acceleration?
   A. The brakes.
   B. The steering wheel.
   C. The gear shift.
   D. All of the above.
18. |||| A car travels at constant speed along the curved path shown from above in Figure Q3.18. Five possible vectors are also shown in the figure; the letter E represents the zero vector. Which vector best represents
   a. The car's *velocity* at position 1?
   b. The car's *acceleration* at point 1?
   c. The car's *velocity* at position 2?
   d. The car's *acceleration* at point 2?
   e. The car's *velocity* at position 3?
   f. The car's *acceleration* at point 3?

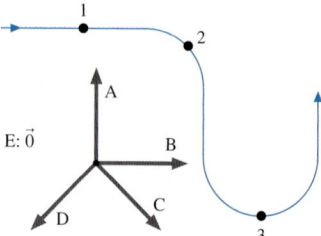

FIGURE Q3.18

19. ‖ A ball is fired from a cannon at point 1 and follows the trajectory shown in Figure Q3.19. Air resistance may be neglected. Five possible vectors are also shown in the figure; the letter E represents the zero vector. Which vector best represents
    a. The ball's *velocity* at position 2?
    b. The ball's *acceleration* at point 2?
    c. The ball's *velocity* at position 3?
    d. The ball's *acceleration* at point 3?

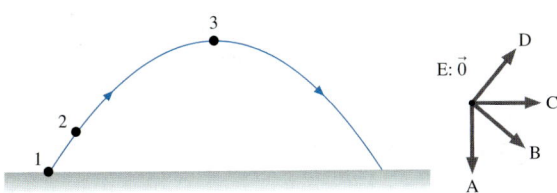

FIGURE Q3.19

20. ‖ A ball thrown at an initial angle of 37.0° and initial velocity of 23.0 m/s reaches a maximum height h, as shown in Figure Q3.20. With what initial speed must a ball be thrown *straight up* to reach the same maximum height h?
    A. 13.8 m/s
    B. 18.4 m/s
    C. 23.0 m/s
    D. 28.8 m/s
    E. 38.2 m/s

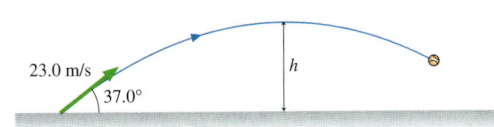

FIGURE Q3.20

21. ‖ A cannon, elevated at 40° is fired at a wall 300 m away on level ground, as shown in Figure Q3.21. The initial speed of the cannonball is 89 m/s.

FIGURE Q3.21

    a. How long does it take for the ball to hit the wall?
       A. 1.3 s   B. 3.3 s   C. 4.4 s
       D. 6.8 s   E. 7.2 s
    b. At what height h does the ball hit the wall?
       A. 39 m   B. 47 m   C. 74 m
       D. 160 m   E. 210 m
22. ‖ A car drives horizontally off a 73-m-high cliff at a speed of 27 m/s. Ignore air resistance.
    a. How long will it take the car to hit the ground?
       A. 2.0 s   B. 3.2 s   C. 3.9 s
       D. 4.9 s   E. 5.0 s
    b. How far from the base of the cliff will the car hit?
       A. 74 m   B. 88 m   C. 100 m
       D. 170 m   E. 280 m
23. ‖ A football is kicked at an angle of 30° with a speed of 20 m/s. To the nearest second, how long will the ball stay in the air?
    A. 1 s   B. 2 s   C. 3 s   D. 4 s
24. ‖ A football is kicked at an angle of 30° with a speed of 20 m/s. To the nearest 5 m, how far will the ball travel?
    A. 15 m   B. 25 m   C. 35 m   D. 45 m
25. ‖ Riders on a Ferris wheel move in a circle with a speed of 4.0 m/s. As they go around, they experience a centripetal acceleration of 2.0 m/s². What is the diameter of this particular Ferris wheel?
    A. 4.0 m   B. 6.0 m   C. 8.0 m
    D. 16 m   E. 24 m

PROBLEMS

Section 3.1 Using Vectors

1. ‖ Trace the vectors in Figure P3.1 onto your paper. Then use graphical methods to draw the vectors (a) $\vec{A} + \vec{B}$ and (b) $\vec{A} - \vec{B}$.

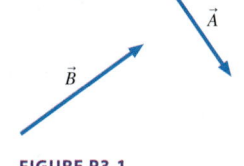

FIGURE P3.1

2. ‖ Trace the vectors in Figure P3.2 onto your paper. Then use graphical methods to draw the vectors (a) $\vec{A} + \vec{B}$ and (b) $\vec{A} - \vec{B}$.

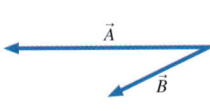

FIGURE P3.2

Section 3.2 Using Vectors on Motion Diagrams

3. ‖ a. Find the average acceleration vector at point 1 of the three-point motion diagram in Figure P3.3.
   b. Is the object's average speed between points 1 and 2 greater than, less than, or equal to its average speed between points 0 and 1? Explain how you can tell.

FIGURE P3.3

4. | a. Find the average acceleration vector at point 1 of the three-point motion diagram in Figure P3.4.
   b. Is the object's average speed between points 1 and 2 greater than, less than, or equal to its average speed between points 0 and 1? Explain how you can tell.

2 ●

1 ●

0 ●

FIGURE P3.4

5. ||| Figure 3.11 showed the motion diagram for Anne as she rode a Ferris wheel that was turning at a constant speed. The inset to the figure showed how to find the acceleration vector at the lowest point in her motion. Use a similar analysis to find Anne's acceleration vector at the 12 o'clock, 4 o'clock, and 8 o'clock positions of the motion diagram. Use a ruler so that your analysis is accurate.

Section 3.3 Coordinate Systems and Vector Components

6. || A position vector in the first quadrant has an x-component of 6.0 m and a magnitude of 10 m. What is the value of its y-component?

7. ||| A velocity vector $40°$ above the positive x-axis has a y-component of 10 m/s. What is the value of its x-component?

8. || Jack and Jill ran up the hill at 3.0 m/s. The horizontal component of Jill's velocity vector was 2.5 m/s.
   a. What was the angle of the hill?
   b. What was the vertical component of Jill's velocity?

9. | A cannon tilted upward at $30°$ fires a cannonball with a speed of 100 m/s. At that instant, what is the component of the cannonball's velocity parallel to the ground?

10. || a. What are the x- and y-components of vector $\vec{E}$ in terms of the angle θ and the magnitude E shown in Figure P3.10?
    b. For the same vector, what are the x- and y-components in terms of the angle ϕ and the magnitude E?

FIGURE P3.10

11. | Draw each of the following vectors, then find its x- and y-components.
    a. $\vec{d} = (100 \text{ m}, 45° \text{ below } +x\text{-axis})$
    b. $\vec{v} = (300 \text{ m/s}, 20° \text{ above } +x\text{-axis})$
    c. $\vec{a} = (5.0 \text{ m/s}^2, -y\text{-direction})$

12. || Draw each of the following vectors, then find its x- and y-components.
    a. $\vec{d} = (2 \text{ km}, 30° \text{ left of } +y\text{-axis})$
    b. $\vec{v} = (5 \text{ cm/s}, -x\text{-direction})$
    c. $\vec{a} = (10 \text{ m/s}^2, 40° \text{ left of } -y\text{-axis})$

13. | Each of the following vectors is given in terms of its x- and y-components. Draw the vector, label an angle that specifies the vector's direction, then find the vector's magnitude and direction.
    a. $v_x = 20 \text{ m/s}, v_y = 40 \text{ m/s}$
    b. $a_x = 2.0 \text{ m/s}^2, a_y = -6.0 \text{ m/s}^2$

14. | Each of the following vectors is given in terms of its x- and y-components. Draw the vector, label an angle that specifies the vector's direction, then find the vector's magnitude and direction.
    a. $v_x = 10 \text{ m/s}, v_y = 30 \text{ m/s}$
    b. $a_x = 20 \text{ m/s}^2, a_y = 10 \text{ m/s}^2$

15. ||| While visiting England, you decide to take a jog and find yourself in the neighborhood shown on the map in Figure P3.15. What is your displacement after running 2.0 km on Strawberry Fields, 1.0 km on Penny Lane, and 4.0 km on Abbey Road?

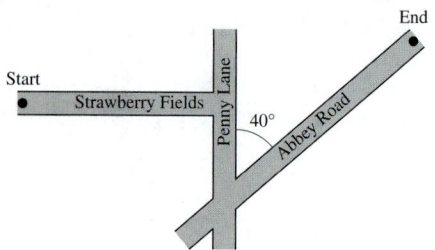

FIGURE P3.15

Section 3.4 Motion on a Ramp

16. ||| Suppose you begin sliding down a $15°$ ski slope. Ignoring friction and air resistance, how fast will you be moving after 10 s?

17. |||| A car traveling at 30 m/s runs out of gas while traveling up a $5.0°$ slope. How far will it coast before starting to roll back down?

18. || Santa loses his footing and slides down a frictionless, snowy roof that is tilted at an angle of $30°$. If Santa slides 10 m before reaching the edge, what is his speed as he leaves the roof?

19. |||| A piano has been pushed to the top of the ramp at the back of a moving van. The workers think it is safe, but as they walk away, it begins to roll down the ramp. If the back of the truck is 1.0 m above the ground and the ramp is inclined at $20°$, how much time do the workers have to get to the piano before it reaches the bottom of the ramp?

20. ||| Starting from rest, several toy cars roll down ramps of differing lengths and angles. Rank them according to their speed at the bottom of the ramp, from slowest to fastest. Car A goes down a 10 m ramp inclined at $15°$, car B goes down a 10 m ramp inclined at $20°$, car C goes down an 8.0 m ramp inclined at $20°$, and car D goes down a 12 m ramp inclined at $12°$.

Section 3.5 Relative Motion

21. | Anita is running to the right at 5 m/s, as shown in Figure P3.21. Balls 1 and 2 are thrown toward her at 10 m/s by friends standing on the ground. According to Anita, what is the speed of each ball?

FIGURE P3.21

22. | Anita is running to the right at 5 m/s, as shown in Figure P3.22. Balls 1 and 2 are thrown toward her by friends

standing on the ground. According to Anita, both balls are approaching her at 10 m/s. According to her friends, with what speeds were the balls thrown?

FIGURE P3.22

23. ▌▌▌ A boat takes 3.0 h to travel 30 km down a river, then 5.0 h to return. How fast is the river flowing?
24. ▌▌▌ Two children who are bored while waiting for their flight at the airport decide to race from one end of the 20-m-long moving sidewalk to the other and back. Phillippe runs on the sidewalk at 2.0 m/s (relative to the sidewalk). Renee runs on the floor at 2.0 m/s. The sidewalk moves at 1.5 m/s relative to the floor. Both make the turn instantly with no loss of speed.
    a. Who wins the race?
    b. By how much time does the winner win?
25. ▌▌▌▌ An assembly line has a staple gun that rolls to the left at 1.0 m/s while parts to be stapled roll past it to the right at 3.0 m/s. The staple gun fires 10 staples per second. How far apart are the staples in the finished part?

Section 3.6 Motion in Two Dimensions: Projectile Motion

Section 3.7 Projectile Motion: Solving Problems

26. ▌▌ An object is launched with an initial velocity of 50.0 m/s at a launch angle of 36.9° above the horizontal.
    a. Make a table showing values of x, y, v_x, v_y, and the speed v every 1 s from $t = 0$ s to $t = 6$ s.
    b. Plot a graph of the object's trajectory during the first 6 s of motion.
27. ▌▌ A ball is thrown horizontally from a 20-m-high building with a speed of 5.0 m/s.
    a. Make a sketch of the ball's trajectory.
    b. Draw a graph of v_x, the horizontal velocity, as a function of time. Include units on both axes.
    c. Draw a graph of v_y, the vertical velocity, as a function of time. Include units on both axes.
    d. How far from the base of the building does the ball hit the ground?
28. ▌ A ball with a horizontal speed of 1.25 m/s rolls off a bench 1.00 m above the floor.
    a. How long will it take the ball to hit the floor?
    b. How far from a point on the floor directly below the edge of the bench will the ball land?
29. ▌▌▌▌ King Arthur's knights fire a cannon from the top of the castle wall. The cannonball is fired at a speed of 50 m/s and an angle of 30°. A cannonball that was accidentally dropped hits the moat below in 1.5 s. How far from the castle wall does the fired cannonball hit the ground?

30. ▌ Two spheres are launched horizontally from a 1.0-m-high table. Sphere A is launched with an initial speed of 5.0 m/s. Sphere B is launched with an initial speed of 2.5 m/s.
    a. What are the times for each sphere to hit the floor?
    b. What are the distances that each travels from the edge of the table?
31. ▌▌ A rifle is aimed horizontally at a target 50 m away. The bullet hits the target 2.0 cm below the aim point.
    a. What was the bullet's flight time?
    b. What was the bullet's speed as it left the barrel?
32. ▌▌ **BIO** A grey kangaroo can leap a distance of 10 m on horizontal ground. What is the minimum takeoff speed for such a leap?
33. ▌▌▌ On the Apollo 14 mission to the moon, astronaut Alan Shepard hit a golf ball with a golf club improvised from a tool. The free-fall acceleration on the moon is 1/6 of its value on earth. Suppose he hit the ball with a speed of 25 m/s at an angle 30° above the horizontal.
    a. How long was the ball in flight?
    b. How far did it travel?
    c. How much farther would it travel on the moon than on earth?

Section 3.8 Motion in Two Dimensions: Circular Motion

34. ▌ An old-fashioned LP record rotates at $33\frac{1}{3}$ rpm.
    a. What is its frequency in rev/s?
    b. What is its period, in seconds?
35. ▌ An audio CD is 12.0 cm in diameter. As it rotates through an angle of 45°, how far does a point at its edge travel?
36. ▌ **BIO** To withstand "g-forces" of up to 10 g's, caused by suddenly pulling out of a steep dive, fighter jet pilots train on a "human centrifuge." 10 g's is an acceleration of 98 m/s². If the length of the centrifuge arm is 12 m, at what speed is the rider moving when she experiences 10 g's?
37. ▌ A particle rotates in a circle with centripetal acceleration $a = 8.0$ m/s². What is a if
    a. The radius is doubled without changing the particle's speed?
    b. The speed is doubled without changing the circle's radius?
38. ▌▌ **BIO** A peregrine falcon in a tight, circular turn can attain a centripetal acceleration 1.5 times the free-fall acceleration. If the falcon is flying at 20 m/s, what is the radius of the turn?
39. ▌ Given the data in the chapter, what is the acceleration (magnitude and direction) of a passenger on the London Eye Ferris wheel?

General Problems

40. ▌ Suppose $\vec{C} = \vec{A} + \vec{B}$ where vector $\vec{A}$ has components $A_x = 5$, $A_y = 2$ and vector $\vec{B}$ has components $B_x = -3$, $B_y = -5$.
    a. What are the x- and y-components of vector $\vec{C}$?
    b. Draw a coordinate system and on it show vectors $\vec{A}$, $\vec{B}$, and $\vec{C}$.
    c. What are the magnitude and direction of vector $\vec{C}$?
41. ▌ Suppose $\vec{D} = \vec{A} - \vec{B}$ where vector $\vec{A}$ has components $A_x = 5$, $A_y = 2$ and vector $\vec{B}$ has components $B_x = -3$, $B_y = -5$.
    a. What are the x- and y-components of vector $\vec{D}$?
    b. Draw a coordinate system and on it show vectors $\vec{A}$, $\vec{B}$, and $\vec{D}$.
    c. What are the magnitude and direction of vector $\vec{D}$?

42. | Suppose $\vec{E} = 2\vec{A} + 3\vec{B}$ where vector $\vec{A}$ has components $A_x = 5$, $A_y = 2$ and vector $\vec{B}$ has components $B_x = -3$, $B_y = -5$.
    a. What are the x- and y-components of vector $\vec{E}$?
    b. Draw a coordinate system and on it show vectors $\vec{A}$, $\vec{B}$, and $\vec{E}$.
    c. What are the magnitude and direction of vector $\vec{E}$?

43. || For the three vectors shown in Figure P3.43, the vector sum $\vec{D} = \vec{A} + \vec{B} + \vec{C}$ has components $D_x = 2$ and $D_y = 0$.
    a. What are the x- and y-components of vector $\vec{B}$?
    b. Write $\vec{B}$ as a magnitude and a direction.

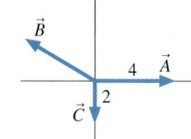

FIGURE P3.43

44. || Let $\vec{A} = (3.0 \text{ m}, 20° \text{ south of east})$, $\vec{B} = (2.0 \text{ m, north})$, and $\vec{C} = (5.0 \text{ m}, 70° \text{ south of west})$.
    a. Draw and label $\vec{A}$, $\vec{B}$, and $\vec{C}$ with their tails at the origin. Use a coordinate system with the x-axis to the east.
    b. Write the x- and y-components of vectors $\vec{A}$, $\vec{B}$, and $\vec{C}$.
    c. Find the magnitude and the direction of $\vec{D} = \vec{A} + \vec{B} + \vec{C}$.

45. ||| While vacationing in the mountains you do some hiking. In the morning, your displacement is $\vec{S}_{\text{morning}} = (2000 \text{ m, east}) + (3000 \text{ m, north}) + (200 \text{ m, vertical})$. After lunch, your displacement is $\vec{S}_{\text{afternoon}} = (1500 \text{ m, west}) + (2000 \text{ m, north}) - (300 \text{ m, vertical})$.
    a. At the end of the hike, how much higher or lower are you compared to your starting point?
    b. What is your total displacement?

46. ||| The minute hand on a watch is 2.0 cm long. What is the displacement vector of the tip of the minute hand
    a. From 8:00 to 8:20 A.M.?
    b. From 8:00 to 9:00 A.M.?

47. ||||| A field mouse trying to escape a hawk runs east for 5.0 m, darts southeast for 3.0 m, then drops 1.0 m down a hole into its burrow. What is the magnitude of the net displacement of the mouse?

48. ||| A pilot in a small plane encounters shifting winds. He flies 26.0 km northeast, then 45.0 km due north. From this point, he flies an additional distance in an unknown direction, only to find himself at a small airstrip that his map shows to be 70.0 km directly north of his starting point. What were the length and direction of the third leg of his trip?

49. || A contestant on "Survivor" is challenged to find buried treasure in a featureless desert. All the contestant knows is that she was driven away blindfolded from the treasure's location for 10 min with an average velocity vector having components 8.6 m/s west and 5.2 m/s north, then given a compass. She can run at a speed of 5.0 m/s. In what direction should she head and for how long should she run in order to retrieve the treasure?

50. ||| The bacterium *Escherichia coli* (or *E. coli*) is a single-
BIO celled organism that lives in the gut of healthy humans and animals. When grown in a uniform medium rich in salts and amino acids, these bacteria swim along zig-zag paths at a constant speed of 20 μm/s. Figure P3.50 shows the trajectory of an *E. coli* as it moves from point A to point E. Each segment of the motion can be identified by two letters, such as segment BC.

a. For each of the four segments in the bacterium's trajectory, calculate the x and y components of its displacement and of its velocity.
b. Calculate both the total distance traveled and the magnitude of the net displacement for the entire motion.
c. What are the magnitude and the direction of the bacterium's average velocity for the entire trip?

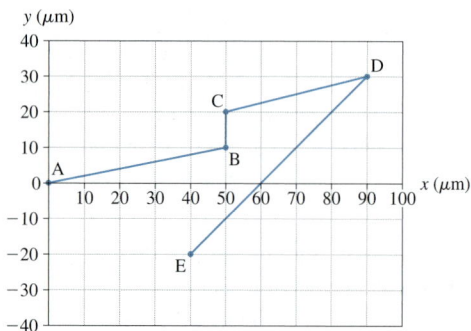

FIGURE P3.50

51. |||| A skier is gliding along at 3.0 m/s on horizontal, frictionless snow. He suddenly starts down a 10° incline. His speed at the bottom is 15 m/s.
    a. What is the length of the incline?
    b. How long does it take him to reach the bottom?

52. ||| A block slides along the frictionless track shown in Figure P3.52 with an initial speed of 5.0 m/s. Assume it turns all the corners smoothly, with no loss of speed.
    a. What is the block's speed as it goes over the top?
    b. What is its speed when it reaches the level track on the right side?
    c. By what percentage does the block's final speed differ from its initial speed? Is this surprising?

FIGURE P3.52

53. | One game at the amusement park has you push a puck up a long, frictionless ramp. You win a stuffed animal if the puck, at its highest point, comes to within 10 cm of the end of the ramp without going off. You give the puck a push, releasing it with a speed of 5.0 m/s when it is 8.5 m from the end of the ramp. The puck's speed after traveling 3.0 m is 4.0 m/s. Are you a winner?

54. |||| When the moving sidewalk at the airport is broken, as it often seems to be, it takes you 50 s to walk from your gate to the baggage claim. When it is working and you stand on the moving sidewalk the entire way, without walking, it takes 75 s to travel the same distance. How long will it take you to travel from the gate to baggage claim if you walk while riding on the moving sidewalk?

55. |||| Ships A and B leave port together. For the next two hours, ship A travels at 20 mph in a direction 30° west of north while ship B travels 20° east of north at 25 mph.
    a. What is the distance between the two ships two hours after they depart?
    b. What is the speed of ship A as seen by ship B?

56. ▮▮ Mary needs to row her boat across a 100-m-wide river that is flowing to the east at a speed of 1.0 m/s. Mary can row the boat with a speed of 2.0 m/s relative to the water.
    a. If Mary rows straight north, how far downstream will she land?
    b. Draw a picture showing Mary's displacement due to rowing, her displacement due to the river's motion, and her net displacement.

57. ▮▮ A flock of ducks is trying to migrate south for the winter, but they keep being blown off course by a wind blowing from the west at 12 m/s. A wise elder duck finally realizes that the solution is to fly at an angle to the wind. If the ducks can fly at 16 m/s relative to the air, in what direction should they head in order to move directly south?

58. ▮▮ A kayaker needs to paddle north across a 100-m-wide harbor. The tide is going out, creating a tidal current that flows to the east at 2.0 m/s. The kayaker can paddle with a speed of 3.0 m/s.
    a. In which direction should he paddle in order to travel straight across the harbor?
    b. How long will it take him to cross?

59. ▮▮ A sailboat is sailing due east at 8.0 mph. The wind appears to blow from the southwest at 12.0 mph.
    a. What are the true wind speed and direction?
    b. What are the true wind speed and direction if the wind appears to blow from the northeast at 12.0 mph?

60. ▮▮ A plane has an airspeed of 200 mph. The pilot wishes to reach a destination 600 mi due east, but a wind is blowing at 50 mph in the direction 30° north of east.
    a. In what direction must the pilot head the plane in order to reach her destination?
    b. How long will the trip take?

61. ▮ A physics student on Planet Exidor throws a ball, and it follows the parabolic trajectory shown in Figure P3.61. The ball's position is shown at 1.0 s intervals until $t = 3.0$ s. At $t = 1.0$ s, the ball's velocity has components $v_x = 2.0$ m/s, $v_y = 2.0$ m/s.

FIGURE P3.61

    a. Determine the x- and y-components of the ball's velocity at $t = 0.0$ s, 2.0 s, and 3.0 s.
    b. What is the value of g on Planet Exidor?
    c. What was the ball's launch angle?

62. ▮ A ball thrown horizontally at 25 m/s travels a horizontal distance of 50 m before hitting the ground. From what height was the ball thrown?

63. ▮▮ BIO In 1780, in what is now referred to as "Brady's Leap," Captain Sam Brady of the U.S. Continental Army escaped certain death from his enemies by running over the edge of the cliff above Ohio's Cuyahoga River, which is confined at that spot to a gorge. He landed safely on the far side of the river. It was reported that he leapt 22 ft across while falling 20 ft.

FIGURE P3.63

    a. Representing the horizontal distance jumped as L and the vertical drop as h, as shown in Figure P3.63, derive an expression for the minimum speed v he would need to make his leap if he ran straight off the cliff.
    b. Evaluate your expression for a 22 ft jump with a 20 ft drop to the other side.
    c. Is it reasonable that a person could make this leap? Use the fact that the world record for the 100 m dash is approximately 10 s to estimate the maximum speed such a runner would have.

64. ▮▮▮ A quarterback passes the football downfield at 20 m/s. It leaves his hand 1.8 m above the ground and is caught by a receiver 30 m away at the same height. What is the maximum height of the ball on its way to the receiver?

65. ▮▮ A spring-loaded gun, fired vertically, shoots a marble 6.0 m straight up in the air. What is the marble's range if it is fired horizontally from 1.5 m above the ground?

66. ▮▮ In a shot-put event, an athlete throws the shot with an initial speed of 12.0 m/s at a 40.0° angle from the horizontal. The shot leaves her hand at a height of 1.80 m above the ground.
    a. How far does the shot travel?
    b. Repeat the calculation of part (a) for angles 42.5°, 45.0°, and 47.5°. Put all your results, including 40.0°, in a table. At what angle of release does she throw the farthest?

67. ▮▮ At what launch angle or angles will a projectile land at half of its maximum possible range on flat ground?

68. ▮▮▮ A tennis player hits a ball 2.0 m above the ground. The ball leaves his racquet with a speed of 20 m/s at an angle 5.0° above the horizontal. The horizontal distance to the net is 7.0 m, and the net is 1.0 m high. Does the ball clear the net? If so, by how much? If not, by how much does it miss?

69. ▮ Your baseball team is up by one run in the bottom of the last inning of the game when a ground ball slips through the infield and comes straight toward you, the right fielder. As you pick up the ball 65 m from home plate, you see a runner rounding third base and heading for home. You throw the ball at an angle of 30° above the horizontal with just the right speed so that the ball is caught by the catcher, standing on home plate, at the same height as you threw it. As you release the ball, the runner is 20 m from home plate and running at 8.0 m/s. Will the ball arrive in time for the catcher to make the tag?

70. ▮▮ A supply plane needs to drop a package of food to scientists working on a glacier in Greenland. The plane flies 100 m above the glacier at a speed of 150 m/s. How far short of the target should it drop the package?

71. ▮▮ A child slides down a frictionless 3.0-m-long playground slide tilted upward at an angle of 40°. At the end of the slide, there is an additional section that curves so that the child is launched off the end of the slide horizontally.
    a. How fast is the child moving at the bottom of the slide?
    b. If the end of the slide is 0.40 m above the ground, how far from the end does she land?

72. ▮▮ A sports car is advertised to be able to "reach 60 mph in 5 seconds flat, corner at 0.85g, and stop from 70 mph in only 168 feet."
    a. In which of those three situations is the magnitude of the car's acceleration the largest? In which is it the smallest?
    b. At 60 mph, what is the smallest turning radius that this car can navigate?

In Problems 73 and 74 you are given the equations that are used to solve a problem. For each of these, you are to
a. Write a realistic problem for which these are the correct equations. Be sure that the answer your problem requests is consistent with the equations given.
b. Finish the solution of the problem.
73. ||| $100 \text{ m} = 0 \text{ m} + (50\cos\theta \text{ m/s})t_1$
$0 \text{ m} = 0 \text{ m} + (50\sin\theta \text{ m/s})t_1 - \frac{1}{2}(9.80 \text{ m/s}^2)t_1^2$
74. ||| $x_1 = 0 \text{ m} + (30 \text{ m/s})t_1$
$0 \text{ m} = 300 \text{ m} - \frac{1}{2}(9.8 \text{ m/s}^2)t_1^2$

Passage Problems

Riding the Water Slide

A rider on a water slide goes through three different kinds of motion, as illustrated in Figure P3.75. Use the data and details from the figure to answer the following questions.

$L = 6.0 \text{ m}$

1. The first section of the motion is a ramp with no friction; riders start at rest and accelerate down the ramp.

2. The second section of the motion is a circular segment that changes the direction of motion; riders go around this circular segment at a constant speed and end up with a velocity that is horizontal.

$\theta = 45°$ $r = 1.5 \text{ m}$ $h = 0.60 \text{ m}$

3. The third section of the motion is a parabolic trajectory through the air at the end of which riders land in the water.

FIGURE P3.75

75. | At the end of the first section of the motion, riders are moving at what approximate speed?
A. 3 m/s   B. 6 m/s   C. 9 m/s   D. 12 m/s
76. | Suppose the acceleration during the second section of the motion is too large to be comfortable for riders. What change could be made to decrease the acceleration during this section?
A. Reduce the radius of the circular segment.
B. Increase the radius of the circular segment.
C. Increase the angle of the ramp.
D. Increase the length of the ramp.
77. || What is the vertical component of the velocity of a rider as he or she hits the water?
A. 2.4 m/s   B. 3.4 m/s   C. 5.2 m/s   D. 9.1 m/s
78. || Suppose the designers of the water slide want to adjust the height h above the water so that riders land twice as far away from the bottom of the slide. What would be the necessary height above the water?
A. 1.2 m   B. 1.8 m   C. 2.4 m   D. 3.0 m
79. || During which section of the motion is the magnitude of the acceleration experienced by a rider the greatest?
A. The first.   B. The second.
C. The third.   D. It is the same in all sections.

Stop to Think 3.1: A. The graphical construction of $2\vec{P} - \vec{Q}$ is shown at right.

Stop to Think 3.2: From the axes on the graph, we can see that the x- and y-components are -4 cm and $+2$ cm, respectively.

Stop to Think 3.3: C. Vector $\vec{C}$ points to the left and down, so both C_x and C_y are negative. C_x is in the numerator because it is the side opposite ϕ.

$-\vec{Q}$ $2\vec{P} - \vec{Q}$ $2\vec{P}$

Stop to Think 3.4: B. The angle of the slope is greatest in this case, leading to the greatest acceleration.

Stop to Think 3.5: D. Mass does not appear in the kinematic equations, so the mass has no effect; the balls will follow the same path.

Stop to Think 3.6: B. The magnitude of the acceleration is v^2/r; this is the case with the largest speed and smallest radius.

4

FORCES AND NEWTON'S LAWS OF MOTION

These ice boats sail across the ice at great speeds. What gets the boats moving in the first place? What keeps them from going even faster?

Looking Ahead ▶▶

The goal of Chapter 4 is to establish a connection between force and motion. In this chapter you will learn to:

▶ Recognize what a force is and is not.
▶ Identify the specific forces acting on an object.
▶ Draw free-body diagrams.
▶ Understand the connection between force and motion.

Looking Back ◀◀

To master the material introduced in this chapter, you must understand how acceleration is determined and how vectors are used. Please review:

◀ Section 1.5 Vectors and motion.
◀ Section 2.4 Acceleration.
◀ Sections 3.2 and 3.3 Vectors, coordinate systems, and vector components.

These ice sailers fly across the frozen lake at some 60 mph. Pretty amazing! We could use kinematics to describe a boat's motion with pictures, graphs, and equations. By defining position, velocity, and acceleration and dressing them in mathematical clothing, kinematics provides a language to describe *how* something moves. But kinematics would tell us nothing about *why* the boat accelerates briskly before reaching a top speed. For the more fundamental task of understanding the *cause* of motion, we turn our attention to **dynamics.** Dynamics joins with kinematics to form **mechanics,** the general science of motion. We study dynamics qualitatively in this chapter, then develop it quantitatively in the next four chapters.

The theory of mechanics originated in the mid-1600s when Sir Isaac Newton formulated his laws of motion. These fundamental principles of mechanics explain how motion occurs as a consequence of forces. Newton's laws are more than 300 years old, but they still form the basis for our contemporary understanding of motion.

A challenge in learning physics is that a textbook is not an experiment. The book can assert that an experiment will have a certain outcome, but you may not be convinced unless you see or do the experiment yourself. Newton's laws are frequently contrary to our intuition, and a lack of familiarity with the evidence for Newton's laws is a source of difficulty for many people. You should have an opportunity through lecture demonstrations and in the laboratory to see for yourself the evidence supporting Newton's laws. Physics is not an arbitrary collection of definitions and formulas, but a consistent theory as to how the universe really works. It is only with experience and evidence that we learn to separate physical fact from fantasy.

4.1 What Causes Motion?

As we remarked in Chapter 1, Aristotle (384–322 BC) and his contemporaries in the world of ancient Greece were very interested in motion. One question they asked was: What is the "natural state" of an object if left to itself? It does not take an expensive research program to see that every moving object on earth, if left to itself, eventually comes to rest. You have certainly seen countless examples of this: You must push a shopping cart to keep it rolling, but when you stop pushing, the cart soon comes to rest; a boulder bounds downhill and then tumbles to a halt. Having observed many such examples himself, Aristotle concluded that the natural state of an earthly object is to be *at rest*. An object at rest requires no explanation; it is doing precisely what comes naturally to it. We'll soon see, however, that this simple viewpoint is *incomplete*.

Aristotle further pondered moving objects. A moving object is *not* in its natural state and thus requires an explanation: Why is this object moving? What keeps it going and prevents it from being in its natural state? When a puck is sliding across the ice, what keeps it going? Why does an arrow fly through the air once it is no longer being pushed by the bowstring? Although these questions seem like reasonable ones to pose, it was Galileo who first showed that the questions being asked were, in fact, the wrong ones.

Galileo reopened the question of the "natural state" of objects. He suggested focusing on the *idealized case* in which resistance to the motion (e.g., friction or air resistance) is zero. He performed many experiments to study motion. Let's imagine a modern experiment of this kind, as shown in Figure 4.1.

The rocks in this rockslide quickly came to rest. Is this the "natural state" of objects?

Interstellar coasting A nearly perfect example of Newton's first law is the pair of Voyager space probes launched in 1977. Both spacecraft long ago ran out of fuel and are now coasting through the frictionless vacuum of space. Although not entirely free of influence from the sun's gravity, they are now so far from the sun and other stars that gravitational influences are very nearly zero. Thus, according to the first law, they will continue at their current speed of about 40,000 miles per hour essentially forever. Billions of years from now, long after our solar system is dead, the Voyagers will still be drifting through the stars.

········ As the friction of the surface is reduced, the sled slides farther before it stops. ····

(a) Smooth snow

(b) Slick ice

(c) Frictionless surface   If friction could be reduced to zero, the sled would *never* stop. ·····

FIGURE 4.1 Sleds sliding on increasingly smooth surfaces.

Tyler slides down a hill on his sled, then out onto a horizontal patch of smooth snow, which is shown in Figure 4.1a. Even if the snow is quite smooth, the friction between the sled and the snow will soon cause the sled to come to rest. What if Tyler slides down the hill onto some very slick ice, as in Figure 4.1b? This gives very low friction, and the sled could slide for quite a distance before stopping. Galileo's genius was to imagine the case where *all* sources of friction, air resistance, and other retarding influences were removed, as for the sled in Figure 4.1c sliding on imaginary *frictionless* ice. We can imagine in that case that the sled, once started in its motion, would continue in its motion *forever,* moving in a straight line with no loss of speed. In other words, **the natural state of an object—its behavior if free of external influences—is** *uniform motion* **with constant velocity!** Further, "at rest" has no special significance in Galileo's view of motion; it is simply uniform motion that happens to have a velocity of zero. This implies that an object at rest, in the absence of external influences, will remain at rest forever.

Galileo's ideas were completely counter to those of the ancient Greeks. We no longer need to explain why a sled continues to slide across the ice; that motion is its "natural" state. What needs explanation, in this new viewpoint, is why objects *don't* continue in uniform motion. Why does a sliding puck eventually slow to a stop? Why does a stone, thrown upward, slow and eventually fall back down? Galileo's new viewpoint was that the stone and the puck are *not* free of "influences": the stone is somehow pulled toward the earth, and some sort of retarding influence acted to slow the sled down. Today, we call such influences that lead to deviations from uniform motion **forces.**

Galileo's experiments were limited to motion along horizontal surfaces. It was left to Newton to generalize Galileo's conclusions, and today we call it Newton's first law of motion.

> **Newton's first law** Consider an object with no force acting on it. If it is at rest, it will remain at rest; if it is moving, it will continue to move in a straight line at a constant speed.

As an important application of Newton's first law, consider the crash test of Figure 4.2. The car contacts the wall and begins to slow. The wall is an external influence—a force—that alters the car's uniform motion. But the wall is a force on the *car,* not on the dummy. In accordance with Newton's first law, the unbelted dummy continues to move straight ahead at his original speed. Only when he collides violently with the dashboard of the stopped car is there a force acting to halt the dummy's uniform motion. If he had been wearing a seatbelt, the influence (i.e., the force) of the seatbelt would have slowed the dummy at the much lower rate with which the car slows down. We'll study the forces of collisions in detail in Chapter 10.

4.2 Force

Newton's first law tells us that an object in motion subject to no forces will continue to move in a straight line forever. But this law does not explain in any detail exactly what a force *is.* Unfortunately, there is no simple one-sentence definition of force. The concept of force is best introduced by looking at examples of some common forces and considering the basic properties shared by all forces. This will be our task in the next two sections. Let's begin by examining the properties that all forces have in common, as outlined in the table on the next page.

TRY IT YOURSELF

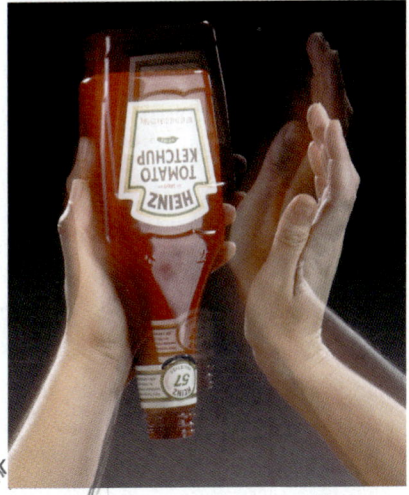

Getting the ketchup out The ketchup stuck at the bottom of the bottle is initially at rest. If you hit the bottom of the bottle, the bottle suddenly moves down, taking the ketchup on the bottom of the bottle with it, so that the ketchup just stays stuck to the bottom. But if instead you hit *up* on the bottle, as shown, you force the bottle rapidly upward. By the first law, the ketchup that was stuck to the bottom stays at rest, so it separates from the upward-moving bottle: the ketchup has moved forward with respect to the bottle!

At the instant of impact, the car and driver are moving at the same speed;

The car slows as it hits, but the driver continues at the same speed . . .

. . . until he hits the now-stationary dashboard. Ouch!

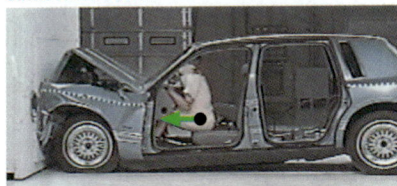

FIGURE 4.2 Newton's first law tells us: "Wear your seatbelts!"

What is a force?

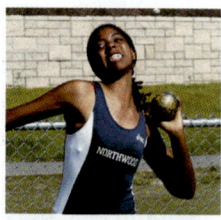

A force is a push or a pull.

Our commonsense idea of a **force** is that it is a *push* or a *pull.* We will refine this idea as we go along, but it is an adequate starting point. Notice our careful choice of words: We refer to "*a* force," rather than simply "force." We want to think of a force as a very specific *action,* so that we can talk about a single force or perhaps about two or three individual forces that we can clearly distinguish. Hence the concrete idea of "a force" acting on an object.

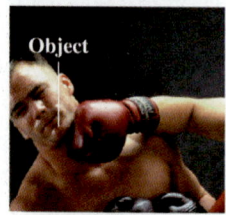

A force acts on an object.

Implicit in our concept of force is that **a force acts on an object.** In other words, pushes and pulls are applied *to* something—an object. From the object's perspective, it has a force *exerted* on it. Forces do not exist in isolation from the object that experiences them.

A force requires an agent.

Every force has an **agent,** something that acts or pushes or pulls. That is, a force has a specific, identifiable *cause.* As you throw a ball, it is your hand, while in contact with the ball, that is the agent or the cause of the force exerted on the ball. *If* a force is being exerted on an object, you must be able to identify a specific cause (i.e., the agent) of that force. Conversely, a force is not exerted on an object *unless* you can identify a specific cause or agent. Note that an agent can be an inert object such as a tabletop or a wall. Such agents are the cause of many common forces.

A force is a vector.

If you push an object, you can push either gently or very hard. Similarly, you can push either left or right, up or down. To quantify a push, we need to specify both a magnitude *and* a direction. It should thus come as no surprise that a force is a vector quantity. The symbol for a force is a vector symbol such as $\vec{F}$, $\vec{w}$, or $\vec{T}$. The size or strength of such a force is its magnitude F (or w or T).

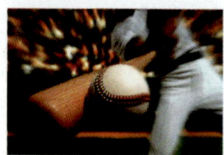

A force can be either a contact force . . .

There are two basic classes of forces, depending on whether the agent touches the object or not. **Contact forces** are forces that act on an object by touching it at a point of contact. The bat must touch the ball to hit it. A string must be tied to an object to pull it. The majority of forces that we will examine are contact forces.

. . . or a long-range force.

Long-range forces are forces that act on an object without physical contact. Magnetism is an example of a long-range force. You have undoubtedly held a magnet over a paper clip and seen the paper clip leap up to the magnet. A coffee cup released from your hand is pulled to the earth by the long-range force of gravity.

Let's summarize these ideas as our definition of force:

- A force is a push or a pull on an object.
- A force is a vector. It has both a magnitude and a direction.
- A force requires an agent. Something does the pushing or pulling. The agent can be an inert object such as a tabletop or a wall.
- A force is either a contact force or a long-range force. Gravity is the only long-range force we will deal with until much later in the book.

There's one more important aspect of forces. If you push against a door (the object) to close it, the door pushes back against your hand (the agent). If a tow rope pulls on a car (the object), the car pulls back on the rope (the agent). In general, if an agent exerts a force on an object, the object exerts a force on the agent. We really need to think of a force as an *interaction* between two objects. Although the interaction perspective is a more exact way to view forces, it adds complications that we would like to avoid for now. Our approach will be to start by focusing on how a single object responds to forces exerted on it. Later in this chapter, we'll return to the larger issue of how two or more objects interact with each other.

Force Vectors

We can use a simple diagram to visualize how forces are exerted on objects. Because we are using the particle model, in which objects are treated as points, the process of drawing a force vector is straightforward. Here is how it goes:

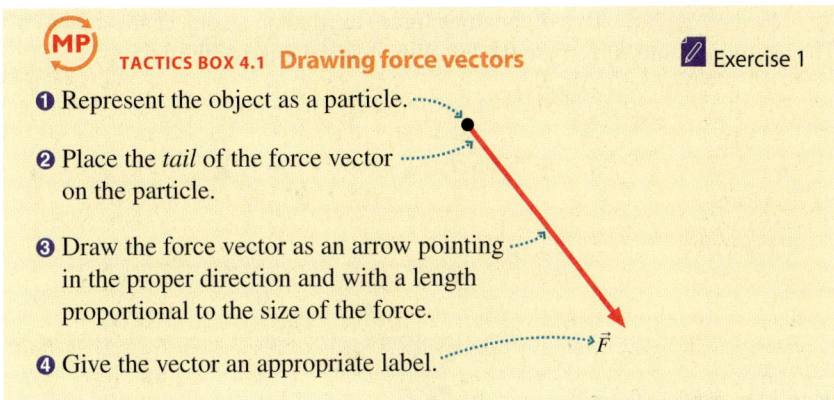

(MP) **TACTICS BOX 4.1 Drawing force vectors**   ✏ Exercise 1

❶ Represent the object as a particle.

❷ Place the *tail* of the force vector on the particle.

❸ Draw the force vector as an arrow pointing in the proper direction and with a length proportional to the size of the force.

❹ Give the vector an appropriate label.

Step 2 may seem contrary to what a "push" should do (it may look as if the force arrow is *pulling* the object rather than *pushing* it), but recall that moving a vector does not change it as long as the length and angle do not change. The vector $\vec{F}$ is the same regardless of whether the tail or the tip is placed on the particle. Our reason for using the tail will become clear when we consider how to combine several forces.

Figure 4.3 shows three examples of force vectors. One is a pull, one a push, and one a long-range force, but in all three the *tail* of the force vector is placed on the particle representing the object.

FIGURE 4.3 Three force vectors.

(a)

Top view of box

Two ropes exerting tension forces on a box

(b)

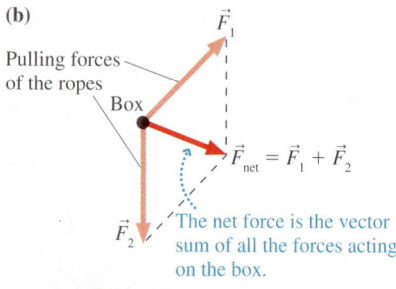

Pulling forces of the ropes

Box

$\vec{F}_1$

$\vec{F}_{net} = \vec{F}_1 + \vec{F}_2$

$\vec{F}_2$

The net force is the vector sum of all the forces acting on the box.

FIGURE 4.4 Two forces applied to a box.

Combining Forces

Figure 4.4a shows a top view of a box being pulled by two ropes, each exerting a force on the box. How will the box respond? Experimentally, we find that when several forces $\vec{F}_1, \vec{F}_2, \vec{F}_3, \ldots$ are exerted on an object, they combine to form a **net force** that is the *vector* sum of all the forces:

$$\vec{F}_{net} = \vec{F}_1 + \vec{F}_2 + \vec{F}_3 \cdots \tag{4.1}$$

Mathematically, this summation is called a *superposition* of forces. The net force is sometimes called the *resultant force*. Figure 4.4b shows the net force on the box.

NOTE ▶ It is important to realize that the net force $\vec{F}_{net}$ is not a new force acting *in addition* to the original forces $\vec{F}_1, \vec{F}_2, \vec{F}_3 \ldots$ Instead, we should think of the original forces being *replaced* by $\vec{F}_{net}$. ◀

STOP TO THINK 4.1   Two of the three forces exerted on an object are shown. The net force points directly to the left. Which is the missing third force?

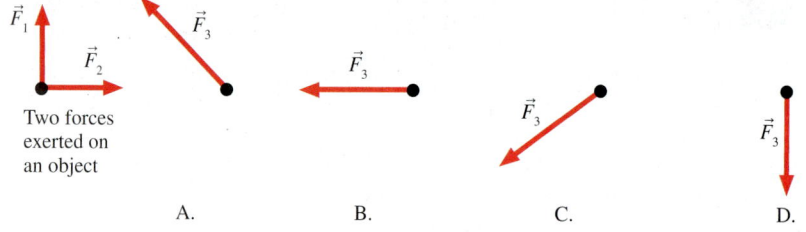

$\vec{F}_1$

$\vec{F}_2$

$\vec{F}_3$

Two forces exerted on an object

$\vec{F}_3$

$\vec{F}_3$

$\vec{F}_3$

A.   B.   C.   D.

4.3 A Short Catalog of Forces

There are many forces we will deal with over and over. This section will introduce you to some of them. Many of these forces have special symbols. As you learn the major forces, be sure to learn the symbol for each.

Weight

A falling rock is pulled toward the earth by the long-range force of gravity. Gravity is what keeps you in your chair, keeps the planets in their orbits around the sun, and shapes the large-scale structure of the universe. We'll have a thorough look at gravity in Chapter 6. For now we'll concentrate on objects on or near the surface of the earth (or other planet).

The gravitational pull of the earth on an object on or near the surface of the earth is called **weight.** The symbol for weight is $\vec{w}$. Weight is the only long-range force we will encounter in the next few chapters. The agent for the weight force is the *entire earth* pulling on an object. The weight force is in some ways the simplest force we'll study. As Figure 4.5 shows, **an object's weight vector always points vertically downward,** no matter how the object is moving.

NOTE ▶ We often refer to "the weight" of an object. This is an informal expression for w, the magnitude of the weight force exerted on the object. Note that **weight is not the same thing as mass.** We will briefly examine mass later in the chapter and explore the connection between weight and mass in Chapter 5. ◀

Free fall   Projectile motion

$\vec{w}$

Rolling   At rest

FIGURE 4.5 Weight points vertically downward.

Spring Force

Springs exert one of the most basic contact forces. A spring can either push (when compressed) or pull (when stretched). Figure 4.6 shows the spring force. In both cases, pushing and pulling, the tail of the force vector is placed on the particle in the force diagram. There is no special symbol for a spring force, so we simply use a subscript label: $\vec{F}_{sp}$.

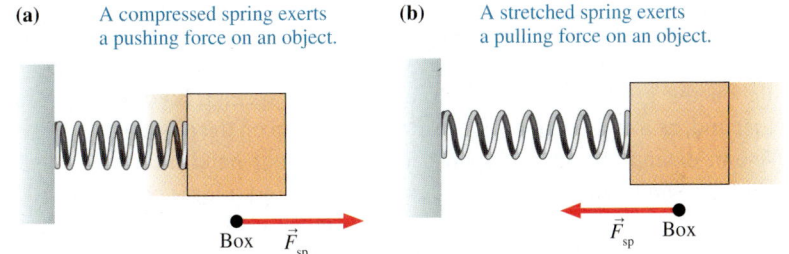

(a) A compressed spring exerts a pushing force on an object.

(b) A stretched spring exerts a pulling force on an object.

Box $\vec{F}_{sp}$ $\vec{F}_{sp}$ Box

FIGURE 4.6 The spring force is parallel to the spring.

Springs come in many forms. When deflected, they push or pull with a spring force.

Although you may think of a spring as a metal coil that can be stretched or compressed, this is only one type of spring. Hold a ruler, or any other thin piece of wood or metal, by the ends and bend it slightly. It flexes. When you let go, it "springs" back to its original shape. This is just as much a spring as is a metal coil.

Tension Force

When a string or rope or wire pulls on an object, it exerts a contact force that we call the **tension force,** represented by a capital $\vec{T}$. **The direction of the tension force is always in the direction of the string or rope,** as you can see in Figure 4.7. When we speak of "the tension" in a string, this is an informal expression for T, the size or magnitude of the tension force. Note that the tension force can only *pull* in the direction of the string; if you try to *push* with a string, it will go slack and be unable to exert a force.

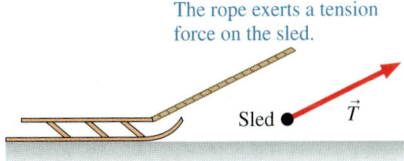

The rope exerts a tension force on the sled.

Sled $\vec{T}$

FIGURE 4.7 Tension is parallel to the rope.

We can think about the tension force using a microscopic picture. If you were to use a very powerful microscope to look inside a rope, you would "see" that it is made of *atoms* joined together by *molecular bonds.* Molecular bonds are not rigid connections between the atoms. They are more accurately thought of as tiny *springs* holding the atoms together, as in Figure 4.8. Pulling on the ends of a string or rope stretches the molecular springs ever so slightly. The tension within a rope and the tension experienced by an object at the end of the rope are really the net spring force being exerted by billions and billions of microscopic springs.

This atomic-level view of tension introduces a new idea: a microscopic **atomic model** for understanding the behavior and properties of **macroscopic** (i.e., containing many atoms) objects. We will frequently use atomic models to obtain a deeper understanding of our observations.

The atomic model of tension also helps to explain one of the basic properties of ropes and strings. When you pull on a rope tied to a heavy box, the rope in turn exerts a tension force on the box. If you pull harder, the tension force on the box becomes greater. How does the box "know" that you are pulling harder on the other end of the rope? According to our atomic model, when you pull harder on the rope its microscopic springs stretch a bit more, increasing the spring force they exert on each other—and on the box they're attached to.

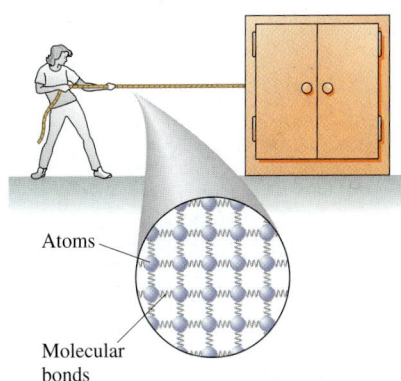

Atoms

Molecular bonds

FIGURE 4.8 An atomic model of tension.

Normal Force

If you sit on a bed, the springs in the mattress compress and, as a consequence of the compression, exert an upward force on you. Stiffer springs would show less

FIGURE 4.9 An atomic model of the force exerted by a table.

The compressed molecular springs push upward on the object.

Atoms

Molecular bonds

The compressed molecular springs in the wall press outward against her hand.

FIGURE 4.10 The wall pushes outward against your hand.

$\vec{n}$

Skier

The surface pushes outward against the bottom of the skis. The force is perpendicular to the surface.

FIGURE 4.11 The normal force is perpendicular to the surface.

compression but would still exert an upward force. The compression of extremely stiff springs might be measurable only by sensitive instruments. Nonetheless, the springs would compress ever so slightly and exert an upward spring force on you.

Figure 4.9 shows a book resting on top of a sturdy table. The table may not visibly flex or sag, but—just as you do to the bed—the book compresses the molecular springs in the table. The size of the compression is very small, but it is not zero. As a consequence, the compressed molecular springs *push upward* on the book. We say that "the table" exerts the upward force, but it is important to understand that the pushing is *really* done by molecular springs. Similarly, an object resting on the ground compresses the molecular springs holding the ground together and, as a consequence, the ground pushes up on the object.

We can extend this idea. Suppose you place your hand on a wall and lean against it, as shown in Figure 4.10. Does the wall exert a force on your hand? As you lean, you compress the molecular springs in the wall and, as a consequence, they push outward *against* your hand. So the answer is "yes," the wall does exert a force on you. It's not hard to see this if you examine your hand as you lean: you can see that your hand is slightly deformed, and becomes more so the harder you lean. This deformation is direct evidence of the force that the wall exerts on your hand. Consider also what would happen if the wall suddenly vanished. Without the wall there to push against you, you would topple forward.

The force the table surface exerts is vertical, while the force the wall exerts is horizontal. In all cases, the force exerted on an object that is pressing against a surface is in a direction *perpendicular* to the surface. Mathematicians refer to a line that is perpendicular to a surface as being *normal* to the surface. In keeping with this terminology, we define the **normal force** as the force exerted by a surface (the agent) against an object that is pressing against the surface. The symbol for the normal force is $\vec{n}$.

We're not using the word *normal* to imply that the force is an "ordinary" force or to distinguish it from an "abnormal force." A surface exerts a force *perpendicular* (i.e., normal) to itself as the molecular springs press *outward*. Figure 4.11 shows an object on an inclined surface, a common situation. Notice how the normal force $\vec{n}$ is perpendicular to the surface.

We have spent a lot of time describing the normal force because many people have a difficult time understanding it. The normal force is a very real force arising from the very real compression of molecular bonds. It is in essence just a spring force, but one exerted by a vast number of microscopic springs acting at once. The normal force is responsible for the "solidness" of solids. It is what prevents you from passing right through the chair you are sitting in and what causes the pain and the lump if you bang your head into a door. Your head can then tell you that the force exerted on it by the door was very real!

Friction

You've certainly observed that a rolling or sliding object, if not pushed or propelled, slows down and eventually stops. You've probably discovered that you can slide better across a sheet of ice than across asphalt. And you also know that most objects stay in place on a table without sliding off even if the table is tilted a bit. The force responsible for these sorts of behavior is **friction.** The symbol for friction is a lower case $\vec{f}$.

Friction, like the normal force, is exerted by a surface. Unlike the normal force, however, **the frictional force is always *parallel* to the surface,** not perpendicular to it. (In many cases, a surface will exert *both* a normal and a frictional force.) On a microscopic level, friction arises as atoms from the object and atoms on the surface run into each other. The rougher the surface is, the more these atoms are forced into close proximity and, as a result, the larger the friction force. We will develop a simple model of friction in the next chapter that will be sufficient for our needs. For now, it is useful to distinguish between two kinds of friction:

- *Kinetic friction,* denoted $\vec{f}_k$, appears as an object slides across a surface. Kinetic friction is a force that always "opposes the motion," meaning that the friction force $\vec{f}_k$ on a sliding object points in a direction opposite to the direction of the object's motion.
- *Static friction,* denoted $\vec{f}_s$, is the force that keeps an object "stuck" on a surface and prevents its motion. Finding the direction of $\vec{f}_s$ is a little trickier than finding it for $\vec{f}_k$. Static friction points opposite the direction in which the object *would* move if there were no friction. That is, it points in the direction necessary to *prevent* motion.

Figure 4.12 shows examples of kinetic and static friction.

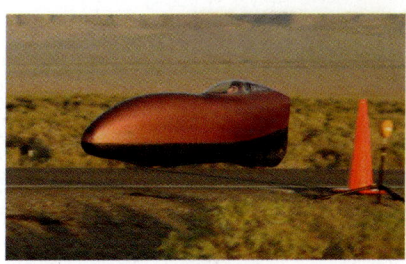

It's a drag At the high speeds attained by racing cyclists, air drag can become very significant. The world record for the longest distance traveled in one hour on an ordinary bicycle is 56.38 km, set by Chris Boardman in 1996. But a bicycle with an aerodynamic shell has a much lower drag force, allowing it to attain significantly higher speeds. The bike shown here was pedaled 84.22 km in one hour by Sam Whittingham in 2004, for an amazing average speed of 52.3 mph!

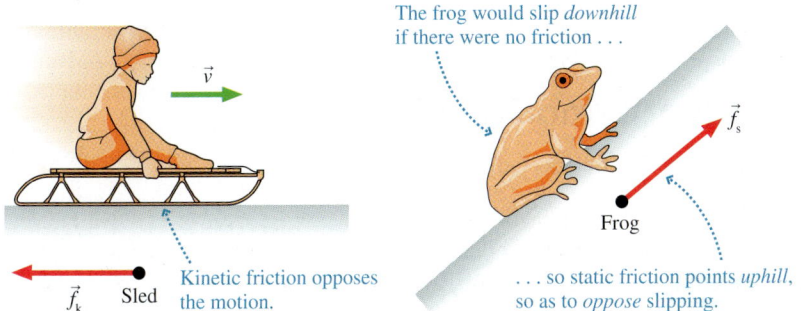

The frog would slip *downhill* if there were no friction . . .

Kinetic friction opposes the motion.

. . . so static friction points *uphill,* so as to *oppose* slipping.

FIGURE 4.12 Kinetic and static friction are parallel to the surface.

Drag

Friction at a surface is one example of a *resistive force,* a force that opposes or resists motion. Resistive forces are also experienced by objects moving through *fluids*—gases (like air) and liquids (like water). This kind of resistive force—the force of a fluid on a moving object—is called **drag** and is symbolized as $\vec{D}$. Like kinetic friction, **drag points opposite to the direction of motion.** Figure 4.13 shows an example of drag.

Drag can be a large force for objects moving at high speeds or in dense fluids. Hold your arm out the window as you ride in a car and feel how hard the air pushes against your arm; note also how the air resistance against your arm increases rapidly as the car's speed increases. Drop a lightweight bead into a beaker of water and watch how slowly it settles to the bottom. The drag force of the water on the bead is very significant.

On the other hand, for objects that are heavy and compact, that move in air, and whose speed is not too great, the drag force of air resistance is fairly small. To keep things as simple as possible, **you can neglect air resistance in all problems unless a problem explicitly asks you to include it.** The error introduced into calculations by this approximation is generally pretty small. Later, when we consider objects moving in liquids, we'll find that drag will be a very significant force that we'll *have* to include.

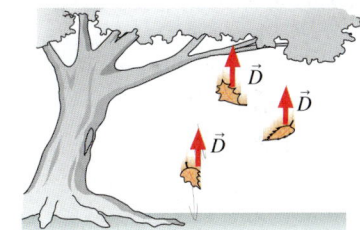

Air resistance is a significant force on falling leaves. It points opposite the direction of motion.

FIGURE 4.13 Air resistance is an example of drag.

Thrust

A jet airplane obviously has a force that propels it forward during takeoff. Likewise for the rocket being launched in Figure 4.14. This force, called **thrust,** occurs when a jet or rocket engine expels gas molecules at high speed. Thrust is a contact force, with the exhaust gas being the agent that pushes on the engine. The process by which thrust is generated is rather subtle, and requires an appreciation of Newton's third law, introduced later in this chapter. For now, we need only

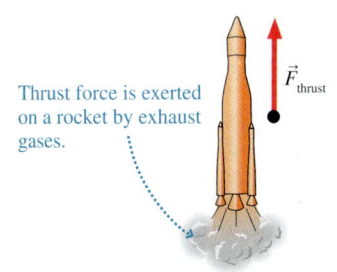

Thrust force is exerted on a rocket by exhaust gases.

FIGURE 4.14 The thrust force on a rocket is opposite to the direction of the expelled gases.

consider that **thrust is a force opposite to the direction in which the exhaust gas is expelled.** There's no special symbol for thrust, so we will call it $\vec{F}_{\text{thrust}}$.

Electric and Magnetic Forces

Electricity and magnetism, like gravity, exert long-range forces. The forces of electricity and magnetism act on charged particles. We will study electric and magnetic forces in detail in Part VI of this book. For now, it is worth noting that the forces holding molecules together—the molecular bonds—are not actually tiny springs. Atoms and molecules are made of charged particles—electrons and protons—and what we call a molecular bond is really an electric force between these particles. So when we say that the normal force and the tension force are due to "molecular springs," or that friction is due to atoms running into each other, what we're really saying is that these forces, at the most fundamental level, are actually electric forces between the charged particles in the atoms.

TABLE 4.1 Common forces and their notation

Force	Notation
General force	$\vec{F}$
Weight	$\vec{w}$
Spring force	$\vec{F}_{\text{sp}}$
Tension	$\vec{T}$
Normal force	$\vec{n}$
Static friction	$\vec{f}_{\text{s}}$
Kinetic friction	$\vec{f}_{\text{k}}$
Drag	$\vec{D}$
Thrust	$\vec{F}_{\text{thrust}}$

4.4 Identifying Forces

Force and motion problems generally have two basic steps:

1. Identify all of the forces acting on an object.
2. Use Newton's laws and kinematics to determine the motion.

Understanding the first step is the primary goal of this chapter. We'll turn our attention to step 2 in the next chapter.

A typical physics problem describes an object that is being pushed and pulled in various directions. Some forces are given explicitly, while others are only implied. In order to proceed, it is necessary to determine all the forces that act on the object. It is also necessary to avoid including forces that do not really exist. Now that you have learned the properties of forces and seen a catalog of typical forces, we can develop a step-by-step method for identifying each force in a problem. A summary of the most common forces we'll come across in the next few chapters is given in Table 4.1.

 TACTICS BOX 4.2 Identifying forces   Exercises 4–8

❶ **Identify "the system" and "the environment."** The system is the object whose motion you wish to study; the environment is everything else.

❷ **Draw a picture of the situation.** Show the object—the system—and everything in the environment that touches the system. Ropes, springs, and surfaces are all parts of the environment.

❸ **Draw a closed curve around the system.** Only the object is inside the curve; everything else is outside.

❹ **Locate every point on the boundary of this curve where the environment touches the system.** These are the points where the environment exerts *contact forces* on the object.

❺ **Name and label each contact force acting on the object.** There is at least one force at each point of contact; there may be more than one. When necessary, use subscripts to distinguish forces of the same type.

❻ **Name and label each long-range force acting on the object.** For now, the only long-range force is weight.

CONCEPTUAL EXAMPLE 4.1 **Identifying forces on a bungee jumper**

A bungee jumper has leapt off a bridge and is nearing the bottom of her fall. What forces are being exerted on the bungee jumper?

REASON

FIGURE 4.15 Forces on a bungee jumper.

CONCEPTUAL EXAMPLE 4.2 **Identifying forces on a skier**

A skier is being towed up a snow-covered hill by a tow rope. What forces are being exerted on the skier?

REASON

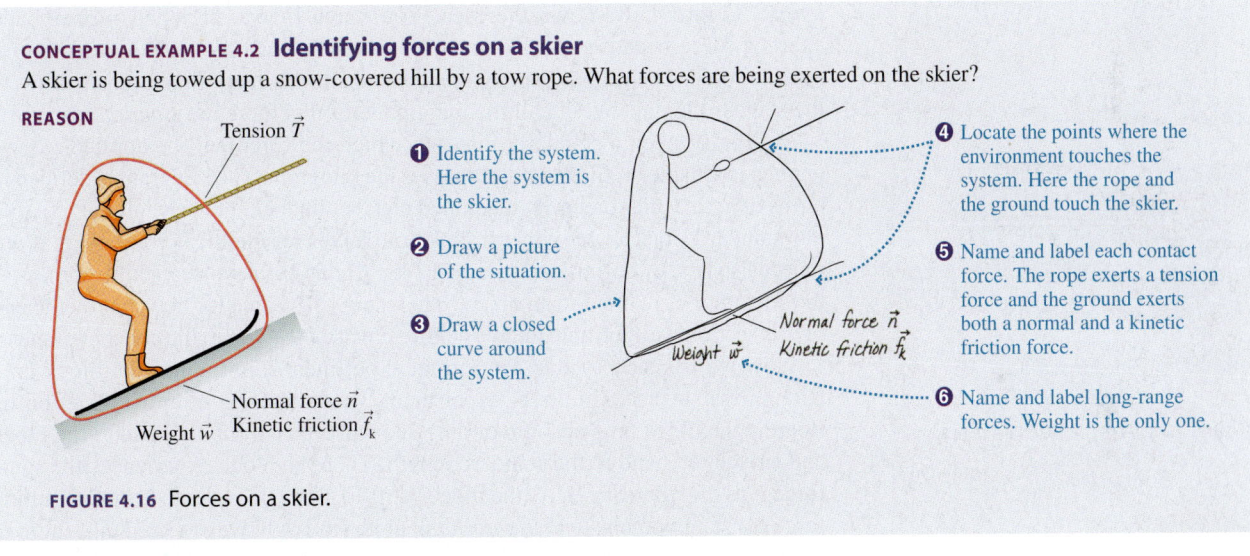

FIGURE 4.16 Forces on a skier.

NOTE ▶ You might have expected two friction forces and two normal forces in Example 4.2, one on each ski. Keep in mind, however, that we're working within the particle model, which represents the skier by a single point. A particle has only one contact with the ground, so there is a single normal force and a single friction force. The particle model is valid if we want to analyze the motion of the skier as a whole, but we would have to go beyond the particle model to find out what happens to each ski. ◀

CONCEPTUAL EXAMPLE 4.3 **Identifying forces on a rocket**

A rocket is being launched to place a new satellite in orbit. Air resistance is not negligible. What forces are being exerted on the rocket?

REASON

FIGURE 4.17 Forces on a rocket.

You've just kicked a rock, and it is now sliding across the ground about 2 meters in front of you. Which of these are forces acting on the rock? List all that apply.

A. Gravity, acting downward.
B. The normal force, acting upward.
C. The force of the kick, acting in the direction of motion.
D. Friction, acting opposite the direction of motion.
E. Air resistance, acting opposite the direction of motion.

4.5 What Do Forces Do?

The fundamental question is: How does an object move when a force is exerted on it? The only way to answer this question is to do experiments. To do experiments, however, we need a way to reproduce the same amount of force again and again, and we need a standard object so our experiments are repeatable.

Let's conduct a "virtual experiment," one you can easily visualize. Imagine using your fingers to stretch a rubber band to a certain length—say 10 centimeters—that you can measure with a ruler. We'll call this the *standard length*. Figure 4.18 shows the idea. You know that a stretched rubber band exerts a force because your fingers *feel* the pull. Furthermore, this is a reproducible force. The rubber band exerts the same force every time you stretch it to the standard length. We'll call the magnitude of this force the *standard force F*.

We'll also need a standard object to which the force will be applied. As we learned in Chapter 1, the SI unit of mass is the kilogram. The kilogram is defined in terms of a particular precisely machined metal cylinder kept in a vault in Paris. For our standard object, we will just make ourselves an identical copy, so that our object also has, by definition, a mass of 1 kg. At this point, you can think of mass as the "quantity of matter" in an object. This idea will suffice for now, but by the end of this section we'll be able to give a more precise meaning to the concept of mass.

Now we're ready to start the virtual experiment. First, place the block on a frictionless surface. (In a real experiment, we can nearly eliminate friction by floating the block on a cushion of air.) Second, attach a rubber band to the block and stretch the band to the standard length. Then the block experiences the same force *F* as did your finger. As the block starts to move, in order to keep the pulling force constant you must *move your hand* in just the right way to keep the length of the rubber band—and thus the force—*constant*. Figure 4.19 shows the experi-

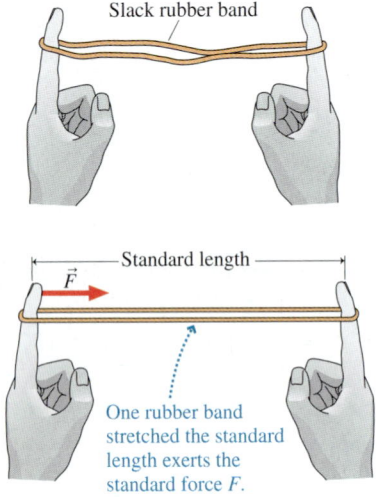

Slack rubber band

$\vec{F}$ ──── Standard length ────

One rubber band stretched the standard length exerts the standard force *F*.

FIGURE 4.18 A reproducible force.

Camera

Maintain constant stretch.

Rubber band ──── Pull

Frictionless surface

$\vec{v}$ ●───●───●───●───●
$\vec{a}$ ──➤ ──➤ ──➤ Motion diagram

FIGURE 4.19 Measuring the motion of a 1 kg block that is pulled with a constant force.

ment being carried out. Once the motion is complete, you can use motion diagrams (made from movie frames from the camera) and kinematics to analyze the block's motion.

The motion diagram in Figure 4.19 shows that the velocity vectors are getting longer, so the velocity is increasing: the block is *accelerating*. In order to study the acceleration more carefully, we make a velocity-versus-time graph, as shown in Figure 4.20. We know that the *slope* of such a graph is the acceleration. We see that the slope of the graph—and thus the acceleration—is the same at all times: it is *constant*. This is the first important finding of this experiment: **an object pulled with a constant force moves with a constant acceleration.** This finding could not have been anticipated in advance. It's conceivable that the object would speed up for a while, then move with a steady speed. Or that it would continue to speed up, but that the *rate* of increase, the acceleration, would steadily decline. These are conceivable motions, but they're not what happens. Instead, the object accelerates *with a constant acceleration* for as long as you pull it with a constant force. Let's use the symbol a_1 for this acceleration of our standard 1 kg block when it's pulled by the standard force.

The next experiment is to see what happens if a different force is used. To get a larger force, we can use more rubber bands. If two rubber bands are each pulling equally hard, like those in Figure 4.21, the net pull is twice that of one rubber band, and the force is $2F$. Three side-by-side rubber bands, each pulled to the standard length, will exert three times the standard force, and so on. So we now have a way of applying forces of different strengths to our block. Figure 4.22 shows what happens in our experiment when we apply larger forces using more bands. When two rubber bands act instead of just one, the slope of the v-versus-t graph is twice as large, hence the acceleration is twice as large. When three bands act, the acceleration is three times greater than with just one. This is our second important finding: **The acceleration is directly proportional to the size of the force.**

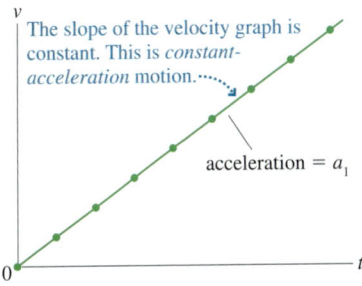

FIGURE 4.20 Velocity-versus-time graph for a 1 kg block pulled with a constant force.

FIGURE 4.21 Doubling the standard force.

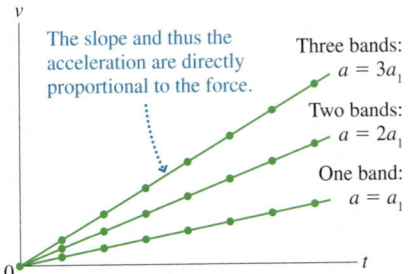

FIGURE 4.22 Velocity-versus-time graphs for a block pulled by one, two, and three rubber bands.

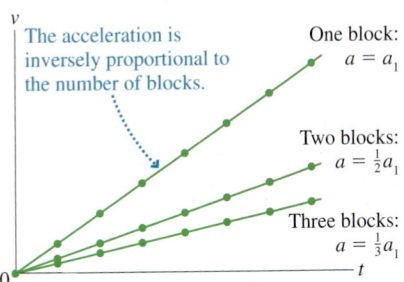

FIGURE 4.23 Velocity-versus-time graphs for one, two, and three blocks pulled with the same force.

The final question for our virtual experiment is: How does the acceleration of an object depend on the "quantity of matter" it contains? To find out, glue two identical 1 kg blocks together, so we have twice the quantity of matter. Now apply the same force—a single rubber band—as you applied to the original block. Figure 4.23 shows that the acceleration is *one-half* as great as that of the original 1 kg block. If we glue three blocks together and pull with one rubber band, we find that the acceleration is only *one-third* of the original acceleration. In general, we find that the acceleration is proportional to the *inverse* of the number of blocks. So our third important result is that **the acceleration is inversely proportional to the number of blocks.**

Inverse proportion ✏ Exercise 12

Two quantities are said to be **inversely proportional** to each other if one quantity is the *inverse* of the other. Mathematically, this means that

$$y = \frac{A}{x}$$

y is inversely proportional to x

If *x* is halved, *y* doubles.

Original point *x, y*

If *x* is doubled, *y* is halved.

Here, *A* is a constant.

If *y* is inversely proportional to *x*, *x* is also inversely proportional to *y*.

SCALING

- If you double *x*, you halve *y*.
- If you triple *x*, *y* is reduced by a factor of 3.
- If you halve *x*, *y* doubles.
- If you reduce *x* by a factor of 3, *y* becomes 3 times as large.

RATIOS For any two values of *x*, say x_1 and x_2, we have

$$y_1 = \frac{A}{x_1} \quad \text{and} \quad y_2 = \frac{A}{x_2}$$

so that

$$\frac{y_1}{y_2} = \frac{x_2}{x_1}$$

That is, the ratio of *y* values is the inverse of the ratio of corresponding values of *x*.

LIMITS

- As *x* gets very large, *y* approaches zero.
- As *x* approaches zero, *y* gets very large.

Mass

So far, our idea of mass is that it is a measure of the "quantity of matter" an object contains. But this definition is a little vague. Exactly what do we mean by the "quantity" of matter in an object? The object's volume? The number of atoms it contains? Or some other measure? Our experiments indicate that an unambiguous way to define the mass of an object is in terms of its *acceleration*. We've just found that the more matter an object has, the more it *resists* accelerating in response to a force. You're familiar with this idea: your car is much harder to get moving than your bicycle. The tendency of an object to resist a *change* in its velocity (i.e., to resist acceleration) is called *inertia*.

To make these ideas more quantitative, we can start with the observation made above that the acceleration is inversely proportional to the number of blocks. But the number of blocks is directly related to the mass: One block has (by definition) a mass of 1 kg, so it seems reasonable that two such blocks, with twice the "quantity of matter," should have a mass of 2 kg, and so on. So we can state more generally that, when acted upon by the standard force, **the acceleration is inversely**

proportional to the mass on which the force acts. We can write this mathematically as $a = A/m$, where A is a constant and m is the mass.

For one block we know that $m = 1$ kg and $a = a_1$. Suppose now we take an unknown object, one not made of a simple combination of 1 kg blocks. We measure its acceleration a. What is its mass m? Because $a = A/m$, we can use the discussion in the math relationship box on inverse proportion to write

$$\frac{a_1}{a} = \frac{m}{1 \text{ kg}}$$

or

$$m = 1 \text{ kg} \times \frac{a_1}{a} \qquad (4.2)$$

This result tells us how to think about mass: The greater an object's mass, the greater its resistance to being accelerated. If an object is more difficult to accelerate than a 1 kg object (i.e., a is smaller than a_1), then its mass is greater than 1 kg. If the object is easier to accelerate, then its mass is less than 1 kg. Any object that undergoes the *same* acceleration as the standard 1 kg object, using the same force, must have a mass of exactly 1 kg. **Mass, then, is the property of an object that determines how it accelerates in response to an applied force.**

TRY IT YOURSELF

Feel the difference Because of its high sugar content, a can of regular soda has a mass about 4% greater than that of a can of diet soda. If you try to judge which can is more massive by simply holding one in each hand, this small difference is almost impossible to detect. If you *move* the cans up and down, however, the difference becomes subtly but noticeably apparent: People evidently are more sensitive to how the mass of each can resists acceleration.

EXAMPLE 4.1 Finding the mass of an unknown block

When a rubber band is stretched to pull on a 1.0 kg block with a constant force, the acceleration of the block is measured to be 3.0 m/s². When a block with an unknown mass is pulled with the same rubber band, using the same force, its acceleration is 5.0 m/s². What is the mass of the unknown block?

PREPARE The acceleration a_1 of the 1 kg block is 3.0 m/s², and the acceleration a of the unknown block is 5.0 m/s².

SOLVE From the definition of mass in terms of acceleration, we find that

$$m = 1.0 \text{ kg} \times \frac{a_1}{a} = 1.0 \text{ kg} \times \frac{3.0 \text{ m/s}^2}{5.0 \text{ m/s}^2} = 0.60 \text{ kg}$$

ASSESS With the same force applied, the unknown block had a *larger* acceleration than the 1.0 kg block. It makes sense, then, that its mass—its resistance to acceleration—is *less* than 1.0 kg.

STOP TO THINK 4.3   Two rubber bands stretched to the standard length cause an object to accelerate at 2 m/s². Suppose another object with twice the mass is pulled by four rubber bands stretched to the standard length. The acceleration of this second object is

A. 1 m/s².   B. 2 m/s².   C. 4 m/s².   D. 8 m/s².   E. 16 m/s².

4.6 Newton's Second Law

We can now summarize the results of our experiments. We've seen that **a force causes an object to accelerate. The acceleration a is directly proportional to the force F and inversely proportional to the mass m.** We can express both these relationships in equation form as

$$a = \frac{F}{m} \qquad (4.3)$$

Note that if we double the size of the force F, the acceleration a will double, as we found experimentally. And if we triple the mass m, the acceleration will be only one-third as great, again in accord with experiment.

Equation 4.3 tells us the magnitude of an object's acceleration in terms of its mass and the force applied. But our experiments also had another important finding: The *direction* of the acceleration was the same as the direction of the force. We can express this fact by writing Equation 4.3 in *vector* form as

$$\vec{a} = \frac{\vec{F}}{m} \qquad (4.4)$$

Recall that any vector (such as $\vec{F}$), when multiplied by an ordinary number (such as $1/m$), gives a vector (such as $\vec{a}$) that points in the *same* direction as the original vector $\vec{F}$. So the relationship between force and acceleration expressed by Equation 4.4 indicates that $\vec{a}$ and $\vec{F}$ point in the same direction.

Finally, our experiment was limited to looking at an object's response to a *single* applied force. Realistically, an object is likely to be subjected to several distinct forces $\vec{F}_1, \vec{F}_2, \vec{F}_3, \ldots$ that may point in different directions. What happens then? Experiments show that the acceleration of the object is determined by the *net force* acting on it. Recall from Figure 4.4 and Equation 4.1 that the net force is the *vector sum* of all forces acting on the object. So if several forces are acting, we use the *net* force in Equation 4.4.

Newton was the first to recognize these connections between force and motion. This relationship is known today as Newton's second law.

Newton's second law An object of mass m subjected to forces $\vec{F}_1, \vec{F}_2, \vec{F}_3, \ldots$ will undergo an acceleration $\vec{a}$ given by

$$\vec{a} = \frac{\vec{F}_{\text{net}}}{m} \qquad (4.5)$$

INVERSE
p. 118

where the net force $\vec{F}_{\text{net}} = \vec{F}_1 + \vec{F}_2 + \vec{F}_3 + \cdots$ is the vector sum of all forces acting on the object. **The acceleration vector $\vec{a}$ points in the same direction as the net force vector $\vec{F}_{\text{net}}$.**

We'll use Newton's second law in Chapter 5 to solve many kinds of motion problems; for the moment, however, the critical idea is that an object accelerates in the direction of the net force acting on it.

The significance of Newton's second law cannot be overstated. There was no reason to suspect that there should be any simple relationship between force and acceleration. Yet a simple but exceedingly powerful equation relates the two. Newton's work, preceded to some extent by Galileo's, marks the beginning of a highly successful period in the history of science during which it was learned that the behavior of physical objects can often be described and predicted by mathematical relationships. While some relationships are found to apply only in special circumstances, others seem to have universal applicability. Those equations that appear to apply at all times and under all conditions have come to be called "laws of nature." Newton's second law is a law of nature; you will meet others as we go through this book.

We can rewrite Newton's second law in the form

$$\vec{F}_{\text{net}} = m\vec{a} \qquad (4.6)$$

which is how you'll see it presented in many textbooks and how, in practice, we'll often use the second law. Equations 4.5 and 4.6 are mathematically equivalent,

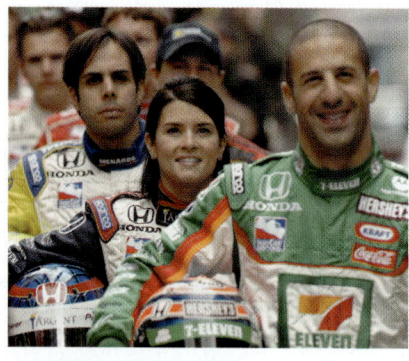

An unfair advantage? Race car driver Danica Patrick was the subject of controversial comments by other drivers who thought her small mass of 45 kg gave her an advantage over heavier drivers; the next-lightest driver's mass was 61 kg. Because every driver's car must have the same mass, Patrick's overall racing mass was lower than any other driver's. Because a car's acceleration is inversely proportional to its mass, her car could be expected to have a slightly greater acceleration.

but Equation 4.5 better describes the central idea of Newtonian mechanics: A force applied to an object causes the object to accelerate.

NOTE ▶ When several forces act on an object, be careful not to think that the strongest force "overcomes" the others to determine the motion on its own. Forces are not in competition with each other! It is $\vec{F}_{net}$, the sum of *all* the forces, that determines the acceleration $\vec{a}$. ◄

CONCEPTUAL EXAMPLE 4.4 Acceleration of a wind-blown basketball

A basketball is released from rest in a stiff breeze directed to the right. In what direction does the ball accelerate?

REASON As shown in Figure 4.24a, two forces are acting on the ball: its weight $\vec{w}$ directed downward, and a wind force $\vec{F}_{wind}$ pushing the ball to the right. Newton's second law tells us that the direction of the acceleration is the same as the direction of the net force $\vec{F}_{net}$. In Figure 4.24b we find $\vec{F}_{net}$ by graphical vector addition of $\vec{w}$ and $\vec{F}_{wind}$. We see that $\vec{F}_{net}$ and therefore $\vec{a}$ point down and to the right.

(a) The force of the wind is to the right.
$\vec{F}_{wind}$
The weight force points down.
$\vec{w}$

(b) The acceleration is in the direction of $\vec{F}_{net}$.
$\vec{a}$
$\vec{F}_{net}$

FIGURE 4.24 A basketball falling in a strong breeze.

Units of Force

Because $\vec{F}_{net} = m\vec{a}$, the unit of force must be mass units multiplied by acceleration units. We've previously specified the SI unit of mass as the kilogram. We can now define the basic unit of force as "the force that causes a 1 kg mass to accelerate at 1 m/s²." From Newton's second law, this force is

$$1 \text{ basic unit of force} = (1 \text{ kg}) \times (1 \text{ m/s}^2) = 1\, \frac{\text{kg} \cdot \text{m}}{\text{s}^2}$$

This basic unit of force is called a *newton:* One **newton** is the force that causes a 1 kg mass to accelerate at 1 m/s². The abbreviation for newton is N. Mathematically, $1 \text{ N} = 1 \text{ kg} \cdot \text{m/s}^2$.

The newton is a *secondary unit,* meaning that it is defined in terms of the *primary units* of kilograms, meters, and seconds. We will introduce other secondary units as needed.

It is important to develop a feeling for what the size of forces should be. Table 4.2 shows some typical forces. As you can see, "typical" forces on "typical" objects are likely to be in the range 0.01–10,000 N. Forces less than 0.01 N are too small to consider unless you are dealing with very small objects. Forces greater than 10,000 N would make sense only if applied to very massive objects.

The unit of force in the English system is the *pound* (abbreviated lb). Although the definition of the pound has varied throughout history, it is now defined in terms of the newton:

$$1 \text{ pound} = 1 \text{ lb} = 4.45 \text{ N}$$

You very likely associate pounds with kilograms rather than with newtons. Everyday language often confuses the ideas of mass and weight, but we're going to need to make a clear distinction between them. We'll have more to say about this in the next chapter.

TABLE 4.2 Approximate magnitude of some typical forces

Force	Approximate magnitude (newtons)
Weight of a U.S. nickel	0.05
Weight of a 1-pound object	5
Weight of a 110-pound person	500
Propulsion force of a car	5,000
Thrust force of a rocket motor	5,000,000

EXAMPLE 4.2 Pulling an airplane

In 2000, a team of 60 British police officers set a world record by pulling a Boeing 747, with a mass of 205,000 kg, a distance of 100 m in 53.3 s. Estimate the force with which each officer pulled on the plane.

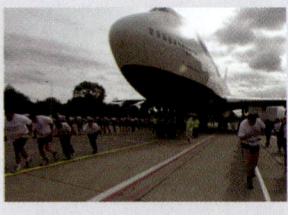

PREPARE If we assume that the plane undergoes a constant acceleration, we can use kinematics to find the magnitude of that acceleration. Then we can use Newton's second law to find the force applied to the airplane. Figure 4.25 shows the visual overview of the airplane.

Known	Find
$x_i = 0$ m, $(v_x)_i = 0$ m/s, $t_i = 0$ s	a_x and F_{net}
$x_f = 100$ m, $t_f = 53.3$ s	

FIGURE 4.25 Visual overview of the airplane accelerating.

SOLVE Because we know the net displacement of the plane, and the time it took to move, we can use the kinematic equation

$$x_f = x_i + (v_x)_i \, \Delta t + \tfrac{1}{2} a_x \, (\Delta t)^2$$

to find the airplane's acceleration a_x. Using the known values $x_i = 0$ m and $(v_x)_i = 0$ m/s, we can solve for the acceleration to get

$$a_x = \frac{2x_f}{(\Delta t)^2} = \frac{2(100 \text{ m})}{(53.3 \text{ s})^2} = 0.0704 \text{ m/s}^2$$

Now we apply Newton's second law. The net force is

$$F_{net} = ma_x = (205{,}000 \text{ kg})(0.0704 \text{ m/s}^2) = 1.44 \times 10^4 \text{ N}$$

This is the force applied by all 60 men. Each man thus applies about 1/60th of this force, or around 240 N.

ASSESS Converting this force to pounds, we have

$$F = 240 \text{ N} \times \frac{1 \text{ lb}}{4.45 \text{ N}} = 54 \text{ lb}$$

Burly policemen can certainly apply a greater force than this. We have neglected the rolling friction of the plane's tires, which apply a force opposite to the plane's motion. Friction lowers the plane's acceleration, leading to an underestimate of the force applied by the men.

STOP TO THINK 4.4 Three forces act on an object. In which direction does the object accelerate?

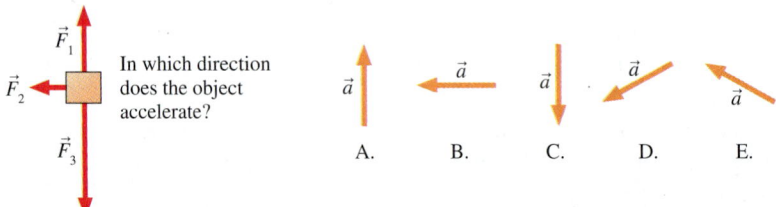

4.7 Free-Body Diagrams

Having discussed at length what is and is not a force, and what forces do to an object, we are ready to assemble our knowledge about force and motion into a single diagram called a **free-body diagram.** A free-body diagram represents the object as a particle and shows *all* of the forces acting on the object. Now that we have forces to consider, we expand our visual overview to include force identification and a free-body diagram. Learning how to draw a correct free-body diagram is a very important skill, one that in the next chapter will become a critical part of our strategy for solving motion problems. For now, let's concentrate on the basic skill of constructing a correct free-body diagram.

TACTICS BOX 4.3 Drawing a free-body diagram   ✐ Exercises 17–22

❶ **Identify all forces acting on the object.** This step was described in Tactics Box 4.2.

❷ **Draw a coordinate system.** Use the axes defined in your pictorial representation (Tactics Box 2.2). If those axes are tilted, for motion along an incline, then the axes of the free-body diagram should be similarly tilted.

❸ **Represent the object as a dot at the origin of the coordinate axes.** This is the particle model.

❹ **Draw vectors representing each of the identified forces.** This was described in Tactics Box 4.1. Be sure to label each force vector.

❺ **Draw and label the *net force* vector** $\vec{F}_{net}$. Draw this vector beside the diagram, not on the particle. Or, if appropriate, write $\vec{F}_{net} = \vec{0}$. Then check that $\vec{F}_{net}$ points in the same direction as the acceleration vector $\vec{a}$ on your motion diagram.

EXAMPLE 4.3 Forces on an upward-accelerating elevator

An elevator, suspended by a cable, speeds up as it moves upward from the ground floor. Draw a free-body diagram of the elevator.

PREPARE Figure 4.26 illustrates the steps outlined in Tactics Box 4.3.

FIGURE 4.26 Free-body diagram of an elevator accelerating upward.

ASSESS The coordinate axes, with a vertical y-axis, are the ones we would use in a pictorial representation of the motion. The elevator is accelerating upward, so $\vec{F}_{net}$ must point upward. For this to be true, the magnitude of $\vec{T}$ must be larger than the magnitude of $\vec{w}$. The diagram has been drawn accordingly.

EXAMPLE 4.4 Forces on a rocket-propelled ice block

Bobby straps a small model rocket to a block of ice and shoots it across the smooth surface of a frozen lake. Friction is negligible. Draw a visual overview—a motion diagram, force identification diagram, and free-body diagram—of the block of ice.

PREPARE Treat the block of ice as a particle. The visual overview consists of a motion diagram to determine $\vec{a}$, a force identification picture, and a free-body diagram. The statement of the situation tells us that friction is negligible. We can draw these three pictures using Problem Solving Strategy 1.1 for the motion diagram, Tactics Box 4.2 to identify the forces, and Tactics Box 4.3 to draw the free-body diagrams. These pictures are shown in Figure 4.27 on the next page.

Continued

FIGURE 4.27 Visual overview for a block of ice shooting across a frictionless frozen lake.

ASSESS The motion diagram tells us that the acceleration is in the positive x-direction. According to the rules of vector addition, this can be true only if the upward-pointing $\vec{n}$ and the downward-pointing $\vec{w}$ are equal in magnitude and thus cancel each other. The vectors have been drawn accordingly, and this leaves the net force vector pointing toward the right, in agreement with $\vec{a}$ from the motion diagram.

EXAMPLE 4.5 Forces on a towed skier

A tow rope pulls a skier up a snow-covered hill at a constant speed. Draw a full visual overview of the skier.

PREPARE This is Example 4.2 again with the additional information that the skier is moving at constant speed. If we were doing a kinematics problem, the pictorial representation would use a tilted coordinate system with the x-axis parallel to the slope, so we use these same tilted coordinate axes for the free-body diagram. The motion diagram, force identification, and free-body diagram are shown in Figure 4.28.

FIGURE 4.28 Visual overview for a skier being towed at a constant speed.

ASSESS We have shown $\vec{T}$ pulling parallel to the slope and $\vec{f}_k$, which opposes the direction of motion, pointing down the slope. The normal force $\vec{n}$ is perpendicular to the surface and thus along the y-axis. Finally, and this is important, the weight $\vec{w}$ is *vertically* downward, *not* along the negative y-axis.

The skier moves in a straight line with constant speed, so $\vec{a} = \vec{0}$. Newton's second law then tells us that $\vec{F}_{net} = m\vec{a} = \vec{0}$. Thus we have drawn the vectors such that the forces add to zero. We'll learn more about how to do this in Chapter 5.

Free-body diagrams will be our major tool for the next several chapters. Careful practice with the workbook exercises and homework in this chapter will pay immediate benefits in the next chapter. Indeed, it is not too much to assert that a problem is half solved, or even more, when you complete the free-body diagram.

An elevator suspended by a cable is moving upward and slowing to a stop. Which free-body diagram is correct?

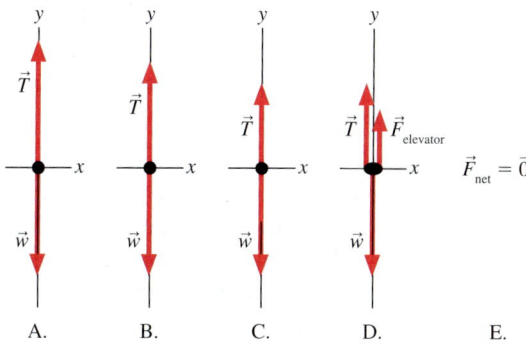

4.8 Newton's Third Law

Thus far, we've focused on the motion of a single particle—which we called the *system*—responding to well-defined forces applied by objects outside the system—that is, from the *environment*. A skier sliding downhill, for instance, is subject to frictional and normal forces from the slope, and the pull of gravity on his body. Once we know these forces, we can use Newton's second law to calculate the acceleration, and hence the overall motion, of the skier.

But motion in the real world often involves two or more objects *interacting* with each other. Consider the two sumo wrestlers in Figure 4.29. The harder one wrestler pushes, the harder the other pushes back. A hammer and a nail, your foot and a soccer ball, and the earth-moon system are other examples of interacting objects.

Newton's second law is not sufficient to explain what happens when two or more objects interact. It does not explain how the force of the hammer on the nail is related to that of the nail on the hammer. In this section we will introduce a new law of physics, Newton's *third* law, that describes how two objects interact with each other.

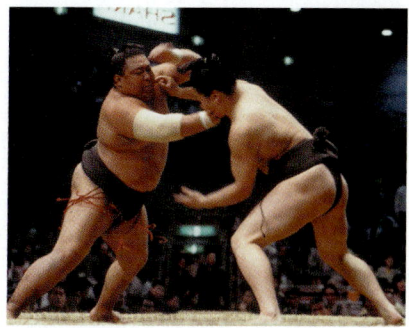

FIGURE 4.29 The two sumo wrestlers are a system of interacting objects.

Interacting Objects

Think about the hammer and nail in Figure 4.30. The hammer certainly exerts a force on the nail as it drives the nail forward. At the same time, the nail exerts a force on the hammer. If you are not sure that it does, imagine hitting the nail with a glass hammer. It's the force of the nail on the hammer that would cause the glass to shatter.

Indeed, if you stop to think about it, any time that object A pushes or pulls on object B, object B pushes or pulls back on object A. As sumo wrestler A pushes on sumo wrestler B, B pushes back on A. (If A pushed forward without B pushing back, A would fall over in the same way you do if someone suddenly opens a door you're leaning against.) Your chair pushes upward on you (a normal force) while, at the same time, you push down on the chair. These are examples of what we call an *interaction*. An **interaction** is the mutual influence of two objects on each other.

These examples illustrate a key aspect of interactions: The forces involved in an interaction between two objects always occur as a *pair*. To be more specific, if object A exerts a force $\vec{F}_{\text{A on B}}$ on object B, then object B exerts a force $\vec{F}_{\text{B on A}}$ on

FIGURE 4.30 The hammer and nail are interacting with each other.

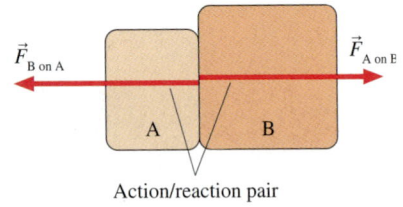

FIGURE 4.31 An action/reaction pair of forces.

object A. This pair of forces, shown in Figure 4.31, is called an **action/reaction pair.** Two objects interact by exerting an action/reaction pair of forces on each other. Notice the very explicit subscripts on the force vectors. The first letter is the *agent*—the source of the force—and the second letter is the *object* on which the force acts. $\vec{F}_{A\,on\,B}$ is thus the force exerted *by* A *on* B. The distinction is important, and this way of labeling forces will be of great help when we discuss how to identify the members of an action/reaction pair.

> **NOTE** ▶ The name "action/reaction pair" is somewhat misleading. The forces occur simultaneously, and we cannot say which is the "action" and which the "reaction." Neither is there any implication about cause and effect; the action does not cause the reaction. **An action/reaction pair of forces exists as a pair, or not at all.** In identifying action/reaction pairs, the labels are the key. Force $\vec{F}_{A\,on\,B}$ is paired with force $\vec{F}_{B\,on\,A}$. ◀

Reasoning with Newton's Third Law

We've discovered that two objects always interact via an action/reaction pair of forces. Newton was the first to recognize how the two members of an action/reaction pair of forces are related to each other. Today we know this as Newton's third law:

> **Newton's third law** Every force occurs as one member of an action/reaction pair of forces.
>
> ■ The two members of an action/reaction pair act on two *different* objects.
> ■ The two members of an action/reaction pair point in *opposite* directions, and are *equal in magnitude*.

Each force in the action/reaction pair acts on a *different* object:

This is a force on the hammer.   This is a force on the nail.

The members of the pair point in *opposite directions*, but are of *equal magnitude*.

FIGURE 4.32 Newton's third law.

Newton's third law is often stated as "For every action there is an equal but opposite reaction." While this is a catchy phrase, it lacks the preciseness of our preferred version. In particular, it fails to capture an essential feature of action/reaction pairs—that they each act on a *different* object. This is shown in Figure 4.32, where a hammer hitting a nail exerts a force $\vec{F}_{hammer\,on\,nail}$ on the nail; by the third law, the nail must exert a force $\vec{F}_{nail\,on\,hammer}$ to complete the action/ reaction pair.

Figure 4.32 also illustrates that these two forces point in *opposite directions*. This feature of the third law is also in accord with our experience. If the hammer hits the nail with a force directed to the right, the force of the nail on the hammer is directed to the left; if the force of my chair on me pushes up, the force of me on the chair pushes down.

Finally, Figure 4.32 shows that, according to Newton's third law, the two members of an action/reaction pair have *equal* magnitudes, so that $F_{hammer\,on\,nail} = F_{nail\,on\,hammer}$. This is something new, and it is by no means obvious. Indeed, this statement causes students the most trouble when applying the third law because it seems so counter to our intuition. Consider, for instance, the collision between a bug and the windshield of a truck. The third law tells us that the magnitude of the force of the windshield on the bug is *equal* to that of the bug on the windshield! How can this be, when the bug is so small compared to the truck? The source of puzzlement in problems like this is that Newton's third law equates the size of the *forces* acting on the two objects, not their *accelerations*. The acceleration of each object depends not only on the force applied to it, but also, according to Newton's second law, its mass. The bug and the truck do in fact feel forces of equal strength from the other, but the bug, with its very small mass, undergoes an extreme acceleration from this force while the acceleration of the heavy truck is negligible. It is important to separate the *effects* of the forces (the accelerations) from the causes (the forces themselves). Because two interacting objects can have very different masses, their accelerations can be very different even though the interaction forces are of the same strength.

Identifying Forces for Interacting Objects

In order to understand the motion of a single object subject to external forces, we took the object to be the *system* and everything else to be the *environment*. Now we're interested in the motion of two or more objects that interact with each other, so we'll expand our system to include all these interacting objects. For example, for a truck pushing a car, we would take the car and truck together as the system, while the road and the earth would make up the environment, as shown in Figure 4.33.

With the system chosen in this way, we can make a distinction between two classes of forces, again shown in Figure 4.33. **External forces** are forces on objects in the system that originate from outside the system; that is, they are forces of the environment on the system. For example, the weight forces of the car and truck are external forces because they are the forces of the earth (part of the environment) on the vehicles. **Internal forces,** on the other hand, are forces *between* objects in the system. The force that the truck exerts on the car is an internal force between the two objects in this system.

Newton's third law tells us that *every* force is a member of an action/reaction pair. For an external force—a force of the environment on the system—the other member of the pair is a force of the *system* on the *environment*. But forces on the *environment* have no effect on the motion of objects in the *system*. Thus, for finding the motion of an object in the system, we don't need to identify action/reaction pairs involving external forces.

Finding both members of an action/reaction pair made up of *internal* forces is crucial, however, because both of these forces act on objects *within* the system and thus affect their motion. Only after identifying such pairs can we use Newton's third law, which relates their directions and magnitudes. The following Tactics Box 4.4 shows you how to correctly identify external forces, and the action/reaction pairs of internal forces.

Revenge of the target We normally think of the damage that the force of a bullet inflicts on its target. But according to Newton's third law, the target exerts an equal force on the bullet. The photo shows the damage sustained by bullets fired at 1600, 1800, and 2000 ft/s, after impacting a test target. The appearance of the bullet before firing is shown at the left.

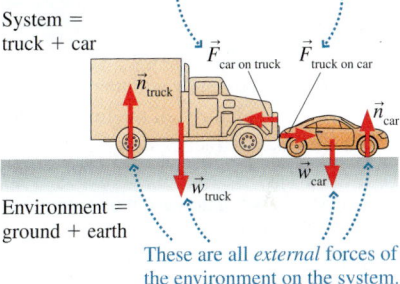

These are *internal* forces between objects in the system. They form an action/reaction pair.

These are all *external* forces of the environment on the system.

FIGURE 4.33 External forces and action/reaction pairs.

(MP) **TACTICS BOX 4.4 Identifying forces for interacting objects**   ✐ Exercises 23–27

❶ **Identify those objects whose motion you wish to study.** These objects make up the system; the environment is everything else.

❷ **Draw each object separately.** Place them in the correct position relative to each other.

❸ **Identify all forces on the system.** For each object in the system, use the techniques of Tactics Box 4.2 to find the forces acting on that object.

❹ ■ **Identify the action/reaction pairs.** For each force acting on an object, decide if it is of the form $\vec{F}_{\text{A on B}}$, where A and B are *both* objects in the system. If so, it is an internal force and forms an action/reaction pair with $\vec{F}_{\text{B on A}}$. Label the forces in a pair using notation like $\vec{F}_{\text{car on truck}}$ and $\vec{F}_{\text{truck on car}}$.

    ■ **Identify the external forces.** External forces are forces of the environment on an object in the system. Name each external force with its appropriate symbols such as $\vec{n}$ or $\vec{w}$. When needed, use subscripted labels such as $\vec{w}_1$ and $\vec{w}_2$ to distinguish between similar forces acting on different objects.

❺ **Draw separate free-body diagrams for each object.** For each object, include the forces acting on it found in Step 3. Connect the force vectors of action/reaction pairs with dotted lines.

CONCEPTUAL EXAMPLE 4.5   **Action/reaction pairs in a balancing act**

An acrobat balances a vase on his head. Define the system, then identify external forces acting on the system and action/reaction pairs of internal forces. Draw free-body diagrams for all objects in the system.

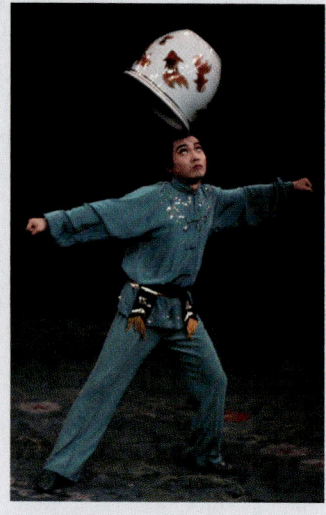

REASON We proceed by following the steps in Tactics Box 4.4, as shown in Figure 4.34a. We're interested in the motions of the acrobat and the vase, so they form the system. In accordance with step 2, we've drawn the objects separately with the vase above the acrobat.

Next we identify all the forces acting on both objects, using the methods of Tactics Box 4.2: We draw a closed curve around each object and decide where that object touches objects outside the curve. The vase makes contact only with the acrobat's head. Because the acrobat is part of the system, this is an internal force that we'll label $\vec{F}_{\text{A on V}}$. The only other force acting on the vase is its weight $\vec{w}_V$.

Now let's look at the forces on the acrobat. He is contacted in only two places, the floor and the top of his head where the vase touches. The floor pushes up on his feet with a normal force $\vec{n}$, an external force. The force on his head is more subtle. You might be tempted to say that this force is "the weight of the vase." But recall that *weight* is a force on an object due to the gravitational pull of the earth. Thus the "weight of the vase" is a force *on the vase* due to gravity; that is, it is the force $\vec{w}_V$ that we've already identified. So the force *on the acrobat* cannot be this weight. In fact, the force on his head is simply that due to the vase itself pushing down on him. It is thus an internal force, and we give it the label $\vec{F}_{\text{V on A}}$. Finally, we must include the long-range force of the acrobat's weight $\vec{w}_A$.

We have identified two internal forces, $\vec{F}_{\text{V on A}}$ and $\vec{F}_{\text{A on V}}$. Because every internal force must be one member of an action/reaction pair, these two must form such a pair. We can check this by noting that the subscripts are reverses of each other: A on V is just V on A flipped.

In Figure 4.34b we show the free-body diagram for the acrobat and vase. Because neither acrobat nor vase is accelerating, we know that the net force on each must be zero by Newton's second law. Thus we draw the vector $\vec{F}_{\text{A on V}}$ to have the same length as the vase's weight $\vec{w}_V$, so that these two vectors cancel. Similarly, the three forces acting on the acrobat must sum to zero. Finally, because $\vec{F}_{\text{V on A}}$ and $\vec{F}_{\text{A on V}}$ form an action/reaction pair, their *magnitudes must be the same,* as we've drawn.

Figure 4.34c shows the drawing as you should draw it. Although much simpler than the artist's version, it still contains all the steps of Tactics Box 4.4.

FIGURE 4.34 Applying Tactics Box 4.4.

ASSESS Using Newton's third law is the key to solving problems of interacting objects. The two forces of the action/reaction pair in Figure 4.34 really connect the two free-body diagrams together. If we know the magnitude of one of these forces in one free-body diagram, we automatically know the magnitude of its paired member in the other free-body diagram.

Propulsion

A sprinter accelerates out of the blocks. Because he's accelerating, there must be a force on him in the forward direction. For a system with an internal source of energy, a force that drives the system is a force of **propulsion.** Propulsion is an important feature not only of walking or running but also of the forward motion of cars, jets, and rockets. Propulsion is somewhat counterintuitive, so it is worth a closer look.

If you tried to walk across a frictionless floor, your foot would slip and slide *backward*. In order for you to walk, the floor needs to have friction so that your foot *sticks* to the floor as you straighten your leg, moving your body forward. The friction that prevents slipping is *static* friction. Static friction, you will recall, acts in the direction that prevents slipping, so the static friction force $\vec{f}_{\text{S on P}}$ (for *Surface* on *Person*) has to point in the *forward* direction to prevent your foot from slipping backward. As shown in Figure 4.35a, it is this forward-directed static friction force that propels you forward! The force of your foot on the floor, $\vec{f}_{\text{P on S}}$, is the other half of the action/reaction pair, and it points in the opposite direction as you push backward against the floor.

Similarly, the car in Figure 4.35b uses static friction to propel itself. The car uses its motor to turn the tires, causing the tires to push backward against the road ($\vec{f}_{\text{tire on road}}$). The road surface responds by pushing the car forward ($\vec{f}_{\text{road on tire}}$). This force of the road on the tire can be seen in photos of drag racers, where the forces are very great (Figure 4.36). Again, the forces involved are *static* friction forces. The tire is rolling, but the bottom of the tire, where it contacts the road, is instantaneously at rest. If it weren't, you would leave one giant skid mark as you drove and would burn off the tread within a few miles.

Rocket motors are somewhat different because they are not pushing *against* anything external. That's why rocket propulsion works in the vacuum of space. Instead, the rocket engine pushes hot, expanding gases out of the back of the rocket, as shown in Figure 4.37. In response, the exhaust gases push the rocket forward with the force we've called *thrust.*

(a) The person pushes backward against the surface. The surface pushes forward on the person.

$\vec{f}_{\text{surface on person}}$
$\vec{f}_{\text{person on surface}}$
Action/reaction pair

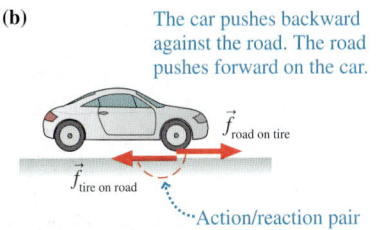

(b) The car pushes backward against the road. The road pushes forward on the car.

$\vec{f}_{\text{road on tire}}$
$\vec{f}_{\text{tire on road}}$
Action/reaction pair

FIGURE 4.35 Examples of propulsion.

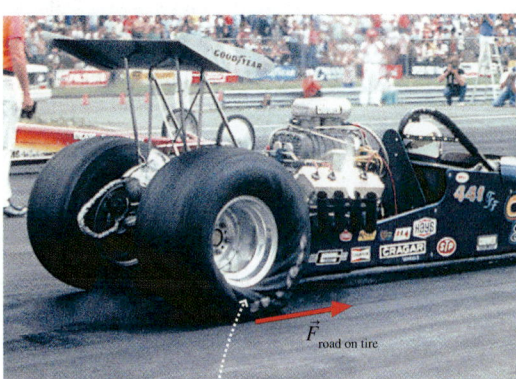

$\vec{F}_{\text{road on tire}}$

You can *see* that the force of the road on the tire points forward by the way it twists the rubber of the tire.

FIGURE 4.36 When the driver hits the gas, the force of the track on the tire is so great that the tire deforms.

The rocket pushes the hot gases backward. The gases push the rocket forward.

$\vec{F}_{\text{gases on rocket}}$
$\vec{F}_{\text{rocket on gases}}$

FIGURE 4.37 Rocket propulsion.

STOP TO THINK 4.6 A small car is pushing a larger truck that has a dead battery. The mass of the truck is larger than the mass of the car. Which of the following statements is true?

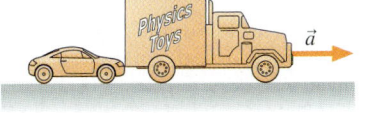

A. The car exerts a force on the truck, but the truck doesn't exert a force on the car.
B. The car exerts a larger force on the truck than the truck exerts on the car.
C. The car exerts the same amount of force on the truck as the truck exerts on the car.
D. The truck exerts a larger force on the car than the car exerts on the truck.
E. The truck exerts a force on the car, but the car doesn't exert a force on the truck.

Now we've assembled all the pieces we need in order to start solving problems in dynamics. We have seen what forces are and how to identify them, and we've learned how forces cause objects to accelerate according to Newton's second law. We've also found how Newton's third law governs the interaction forces between two objects. Our goal in the next several chapters is to apply Newton's laws to a variety of problems involving straight-line and circular motion.

S U M M A R Y

The goal of Chapter 4 has been to establish a connection between force and motion.

GENERAL PRINCIPLES

Newton's First Law

Consider an object with no net force acting on it. If it is at rest, it will remain at rest. If it is in motion, then it will continue to move in a straight line at a constant speed.

The first law tells us that no "cause" is needed for motion. Uniform motion is the "natural state" of an object.

Newton's Second Law

An object with mass m will undergo acceleration

$$\vec{a} = \frac{1}{m}\vec{F}_{net}$$

where $\vec{F}_{net} = \vec{F}_1 + \vec{F}_2 + \vec{F}_3 + \cdots$ is the vector sum of all the individual forces acting on the object.

The second law tells us that a net force causes an object to accelerate. This is the connection between force and motion. The acceleration points in the direction of $\vec{F}_{net}$.

Newton's Third Law

Every force occurs as one member of an **action/reaction** pair of forces. The two members of an action/reaction pair:

- act on two *different* objects.
- point in opposite directions and are equal in magnitude:

$$\vec{F}_{A\,on\,B} = -\vec{F}_{B\,on\,A}$$

IMPORTANT CONCEPTS

Force is a push or pull on an object.

- Force is a vector, with a magnitude and a direction.
- A force requires an agent.
- A force is either a contact force or a long-range force.

The SI unit of force is the **newton** (N). A 1 N force will cause a 1 kg mass to accelerate at 1 m/s².

Net force is the vector sum of all the forces acting on an object.

The net force determines the acceleration according to Newton's second law.

Mass is the property of an object that determines its resistance to acceleration.

If the same force is applied to different objects, their masses in kg are found from

$$m = \frac{a_1}{a} \times 1\ kg$$

where a is an object's acceleration, and a_1 that of a 1 kg object.

APPLICATIONS

Identifying Forces

Forces are identified by locating the points where the environment touches the system. These are points where contact forces are exerted. In addition, objects with mass feel a long-range weight force.

Thrust force $\vec{F}_{thrust}$

Weight $\vec{w}$ Normal force $\vec{n}$

Free-Body Diagrams

A free-body diagram represents the object as a particle at the origin of a coordinate system. Force vectors are drawn with their tails on the particle. The net force vector is drawn beside the diagram.

For instructor-assigned homework, go to www.masteringphysics.com

Problem difficulty is labeled as | (straightforward) to |||| (challenging).

Problems labeled INT integrate significant material from earlier chapters; BIO are of biological or medical interest.

QUESTIONS

Conceptual Questions

1. A hockey puck slides along the surface of the ice. If friction and air resistance are negligible, what force is required to keep the puck moving?

2. If an object is not moving, does that mean that there are no forces acting on it? Explain.

3. An object moves in a straight line at a constant speed. Is it true that there must be no forces of any kind acting on this object? Explain.

4. Write several sentences explaining why you agree or disagree with the statement "Forces cause an object to move."

5. If you know all of the forces acting on a moving object, can you tell in which direction the object is moving? If the answer is Yes, explain how. If the answer is No, give an example.

6. Three arrows are shot horizontally. They have left the bow and are traveling parallel to the ground as shown in Figure Q4.6. Air resistance is negligible. Rank in order, from largest to smallest, the magnitudes of the *horizontal* forces F_1, F_2, and F_3 acting on the arrows. Some may be equal. State your reasoning.

FIGURE Q4.6

7. A carpenter wishes to tighten the heavy head of his hammer onto its light handle. Which method shown in Figure Q4.7 will better tighten the head? Explain.

8. Internal injuries in vehicular BIO accidents may be due to what is called the "third collision." The first collision is the vehicle hitting the external object. The

FIGURE Q4.7

second collision is the person hitting something on the inside of the car, such as the dashboard or windshield. This may cause external lacerations. The third collision, possibly the most damaging to the body, is when organs, such as the heart or brain, hit the ribcage, skull, or other confines of the body, bruising the tissues on the leading edge and tearing the organ from its supporting structures on the trailing edge.

   a. Why is there a third collision? In other words, why are the organs still moving after the second collision?

   b. If the vehicle was traveling at 60 mph before the first collision, would the organs be traveling more than, equal to, or less than 60 mph just before the third collision?

9. a. Give an example of the motion of an object in which the frictional force on the object is directed opposite to the motion.

   b. Give an example of the motion of an object in which the frictional force on the object is in the same direction as the motion.

10. Suppose you are an astronaut in deep space, far from any source of gravity. You have two objects that look identical, but one has a large mass and the other a small mass. How can you tell the difference between the two?

11. Newton's second law says that $\vec{F}_{net} = m\vec{a}$. Is $m\vec{a}$ thus a force? If so, what is its origin?

12. Superman can hover in midair if he wishes. Does this mean that he must exert a force on some other object? Explain.

13. A ball weighs 2.0 N when placed on a scale. It is then thrown straight up. What is its weight at the very top of its motion? Explain.

14. A book sits on a table. List all the forces acting on the book.

15. A person sits on a sloped hillside. Is it ever possible to have the static friction force on this person point down the hill? Explain.

16. Walking without slipping requires a static friction force BIO between your feet (or footwear) and the floor. As described in this chapter, the force on your foot as you push off the floor is forward while the force exerted by your foot on the floor is backward. But what about your *other* foot, the one moved during a stride? What is the direction of the force on that foot as it comes in contact with the floor? Explain.

17. Figure 4.35 shows a case in which the force of the road on the car's tire points forward. In other cases, the force points backward. Give an example of such a case.

18. a. A tightrope walker at the circus steps onto the high wire, causing it to sag slightly. Is the tension in the wire less than, greater than, or equal to the performer's weight? Explain. Include a free-body diagram as part of your explanation.

   b. The leading circus magazine advertises a new wire made of a material called DreamRope. The ad says that a DreamRope wire will remain perfectly straight and horizontal, with absolutely no sag, as the performer walks across. Should you order some? Explain.

19. A very smart three-year-old child is given a wagon for her birthday. She refuses to use it. "After all," she says, "Newton's third law says that no matter how hard I pull, the wagon will exert an equal but opposite force on me. So I will never be able to get it to move forward." What would you say to her in reply?

20. Will hanging a magnet in front of an iron cart, as shown in Figure Q4.20, make it go? Explain why or why not.

FIGURE Q4.20

Multiple-Choice Questions

21. ‖ Figure Q4.21 shows the view looking down onto a frictionless sheet of ice. A puck, tied with a string to point P, slides on the surface of the ice in the circular path shown. If the string suddenly snaps when the puck is in the position shown, which path best represents the puck's subsequent motion?

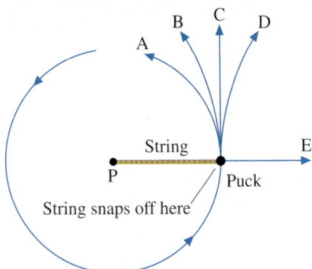

FIGURE Q4.21

22. ‖ A block has acceleration a when pulled by a string. If two identical blocks are glued together and pulled with twice the original force, their acceleration will be
    A. $(1/4)a$ B. $(1/2)a$ C. a D. $2a$ E. $4a$

23. ‖ A person gives a box a shove so that it slides up a ramp, then reverses its motion and slides down. The direction of the force of friction is
    A. Always down the ramp.
    B. Up the ramp and then down the ramp.
    C. Always down the ramp.
    D. Down the ramp and then up the ramp.

24. ‖ Tennis balls experience a large drag force. A tennis ball is hit so that it goes straight up and then comes back down. The direction of the drag force is
    A. Always up.   B. Up and then down.
    C. Always down.   D. Down and then up.

25. ‖ Rachel is pushing a box across the floor while Jon, at the same time, is hoping to stop the box by pushing in the opposite direction. There is friction between the box and floor. If the box is moving at constant speed, then the magnitude of Rachel's pushing force is
    A. Greater than the magnitude of Jon's force.
    B. Equal to the magnitude of Jon's force.
    C. Less than the magnitude of Jon's force.
    D. The problem can't be answered without knowing how large the friction force is.

26. ‖ A person is pushing horizontally on a box with a constant force, causing it to slide across the floor with a constant speed. If the person suddenly stops pushing on the box, the box will
    A. Immediately come to a stop.
    B. Continue moving at a constant speed for a while, then gradually slow down to a stop.
    C. Immediately change to a slower but constant speed.
    D. Immediately begin slowing down and eventually stop.

27. ‖ Figure Q4.27 shows block A sitting on top of block B. A constant force $\vec{F}$ is exerted on block B, causing block B to accelerate to the right. Block A rides on block B without slipping. Which statement is true?
    A. Block B exerts a friction force on block A, directed to the left.
    B. Block B exerts a friction force on block A, directed to the right.
    C. Block B does not exert a friction force on block A.

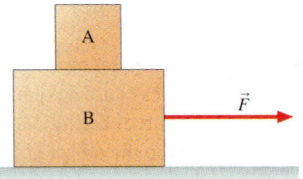

FIGURE Q4.27

28. ‖ Dave pushes his four-year-old son Thomas across the snow on a sled. As Dave pushes, Thomas speeds up. Which statement is true?
    A. The force of Dave on Thomas is larger than the force of Thomas on Dave.
    B. The force of Thomas on Dave is larger than the force of Dave on Thomas.
    C. Both forces have the same magnitude.
    D. It depends on how hard Dave pushes on Thomas.

29. ‖ A truck hits a small car. During this collision,
    A. The truck exerts a larger force on the car than the car on the truck.
    B. The car exerts a larger force on the truck than the truck on the car.
    C. The force of the truck on the car and of the car on the truck have the same magnitude.
    D. The net force is zero.

PROBLEMS

Section 4.1 What Causes Motion?

1. ‖ Whiplash injuries during an automobile accident are
   BIO caused by the inertia of the head. If someone is wearing a seatbelt, her body will tend to move with the car seat. However, her head is free to move until the neck restrains it, causing damage to the neck. Brain damage can also occur.

   Figure P4.1 shows two sequences of head and neck motion for a passenger in an auto accident. One corresponds to a head-on collision, the other to a rear-end collision. Which is which? Explain.

FIGURE P4.1

2. | An automobile has a head-on collision. A passenger in the
BIO car experiences a compression injury to the brain. Is this
injury most likely to be in the front or rear portion of the
brain? Explain.

3. | Passengers in the back seat of an automobile should wear
seatbelts, not only for their own protection, but also for the
protection of the people in the front seats of the car. Explain.

Section 4.2 Force

Problems 4 through 6 show two forces acting on an object at rest.
Redraw the diagram, then add a third force that will allow the
object to remain at rest. Label the new force $\vec{F}_3$.

4. | 5. ‖   6. |

FIGURE P4.4   **FIGURE P4.5**   **FIGURE P4.6**

Section 4.3 A Short Catalog of Forces

Section 4.4 Identifying Forces

7. ‖ A mountain climber is hanging from a rope in the middle
of a crevasse. The rope is vertical. Identify the forces on the
mountain climber.

8. | A circus clown hangs from one end of a large spring. The
other end is anchored to the ceiling. Identify the forces on the
clown.

9. | A baseball player is sliding into second base. Identify the
forces on the baseball player.

10. ‖ A jet plane is speeding down the runway during takeoff.
Air resistance is not negligible. Identify the forces on the jet.

11. | A skier is sliding down a 15° slope. Friction is not negligi-
ble. Identify the forces on the skier.

12. | A tennis ball is flying horizontally across the net. Air resis-
tance is not negligible. Identify the forces on the ball.

Section 4.5 What Do Forces Do?

13. ‖‖‖ Figure P4.13 shows an acceleration-versus-force graph for
three objects pulled by rubber bands. The mass of object 2 is
0.20 kg. What are the masses of objects 1 and 3? Explain your
reasoning.

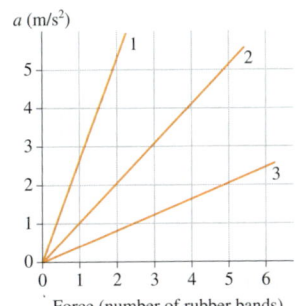

FIGURE P4.13   Force (number of rubber bands)

14. | A constant force applied to object A causes it to accelerate
at 5 m/s². The same force applied to object B causes an accel-

eration of 3 m/s². Applied to object C, it causes an acceleration of 8 m/s².
a. Which object has the largest mass?
b. Which object has the smallest mass?
c. What is the ratio of mass A to mass B (m_A/m_B)?

15. | Two rubber bands pulling on an object cause it to acceler-
ate at 1.2 m/s².
a. What will be the object's acceleration if it is pulled by four
rubber bands?
b. What will be the acceleration of two of these objects glued
together if they are pulled by two rubber bands?

16. | A constant force is applied to an object, causing the object
to accelerate at 10 m/s². What will the acceleration be if
a. The force is halved?
b. The object's mass is halved?
c. The force and the object's mass are both halved?
d. The force is halved and the object's mass is doubled?

17. | A constant force is applied to an object, causing the object
to accelerate at 8.0 m/s². What will the acceleration be if
a. The force is doubled?
b. The object's mass is doubled?
c. The force and the object's mass are both doubled?
d. The force is doubled and the object's mass is halved?

18. ‖‖‖ A man pulling an empty wagon causes it to accelerate at
1.4 m/s². What will the acceleration be if he pulls with the
same force when the wagon contains a child whose mass is
three times that of the wagon?

19. | A car has a maximum acceleration of 5.0 m/s². What will
the maximum acceleration be if the car is towing another car
of the same mass?

Section 4.6 Newton's Second Law

20. | Figure P4.20 shows an acceleration-versus-force graph for
a 500 g object. Redraw this graph and add appropriate acceleration values on the vertical scale.

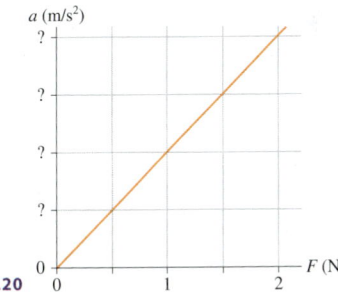

FIGURE P4.20

21. | Figure P4.21 shows an object's acceleration-versus-force
graph. What is the object's mass?

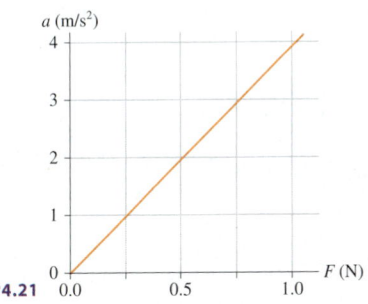

FIGURE P4.21

22. | Two children fight over a 200 g stuffed bear. The 25 kg boy pulls to the right with a 15 N force and the 20 kg girl pulls to the left with a 17 N force. Ignore all other forces on the bear (such as its weight).
    a. At this instant, can you say what the velocity of the bear is? If so, what are the magnitude and direction of the velocity?
    b. At this instant, can you say what the acceleration of the bear is? If so, what are the magnitude and direction of the acceleration?
23. | Based on the information in Table 4.2, estimate in newtons
    a. The weight of a laptop computer.
    b. The propulsion force of a bicycle.
    c. The propulsion force of a sprinter.
24. ‖   Very small forces can have tremendous effects on the motion of very small objects. Consider a single electron, with a mass of 9.1×10^{-31} kg, subject to a single force equal to the weight of a penny, 2.5×10^{-2} N. What is the acceleration of the electron?
25. ‖   The motion of a very massive object is hardly affected by what would seem to be a substantial force. Consider a super-tanker, with a mass of 3.0×10^8 kg. If it is pushed by a rocket motor (see Table 4.2) and is subject to no other forces, what will be the magnitude of its acceleration?

Section 4.7 Free-Body Diagrams

Problems 26 through 28 show a free-body diagram. For each, (a) Redraw the free-body diagram and (b) Write a short description of a real object for which this is the correct free-body diagram. Use Examples 4.3, 4.4, and 4.5 as models of what a description should be like.

FIGURE P4.26   **FIGURE P4.27**

28. |

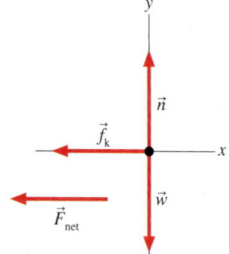

FIGURE P4.28

Problems 29 through 35 describe a situation. For each, identify all forces acting on the object and draw a free-body diagram of the object.
29. | Your car is sitting in the parking lot.
30. | Your car is accelerating from a stop.
31. ‖   Your car is slowing to a stop from a high speed.
32. | An ice hockey puck glides across frictionless ice.
33. | An elevator, hanging from a cable, descends at steady speed.
34. | Your physics textbook is sliding across the table.
35. ‖   You hold a picture motionless against a wall by pressing on it, as shown in Figure P4.35.

FIGURE P4.35

Section 4.8 Newton's Third Law

36. ‖   A weight lifter stands up from a squatting position while holding a heavy barbell across his shoulders. Identify all the third-law pairs of forces, then draw free-body diagrams for the weight lifter and the barbell. Use dotted lines to connect the members of all action/reaction pairs.
37. ‖   A softball player is throwing the ball. Her arm has come forward to where it is beside her head, but she hasn't yet released the ball. Identify all the third-law pairs of forces, then draw free-body diagrams for the ballplayer and the ball. Use dotted lines to connect the members of all action/reaction pairs.
38. | A soccer ball and a bowling ball roll across a hard floor and collide head on. Identify all the third-law pairs of forces at the moment of the collision, then draw free-body diagrams for each ball. Use dotted lines to connect the members of all action/reaction pairs. Friction is negligible.

General Problems

39. | Redraw the motion dia-
INT   gram shown in Figure P4.39, then draw a vector beside it to show the direction of the net force acting on the object. Explain your reasoning.
40. | Redraw the motion dia-
INT   gram shown in Figure P4.40, then draw a vector beside it to show the direction of the net force acting on the object. Explain your reasoning.
41. | Redraw the motion dia-
INT   gram shown in Figure P4.41, then draw a vector beside it to show the direction of the net force acting on the object. Explain your reasoning.

FIGURE P4.39   **FIGURE P4.40**

FIGURE P4.41   **FIGURE P4.42**

42. ‖   Redraw the motion diagram shown in Figure P4.42, then
INT   draw a vector beside it to show the direction of the net force acting on the object. Explain your reasoning.

Problems 43 through 49 show a free-body diagram. For each:
a. Redraw the diagram.
b. Identify the direction of the acceleration vector $\vec{a}$ and show it as a vector next to your diagram. Or, if appropriate, write $\vec{a} = \vec{0}$.
c. If possible, identify the direction of the velocity vector $\vec{v}$ and show it as a labeled vector.
d. Write a short description of a real object for which this is the correct free-body diagram. Use Examples 4.3, 4.4, and 4.5 as models of what a description should be like.

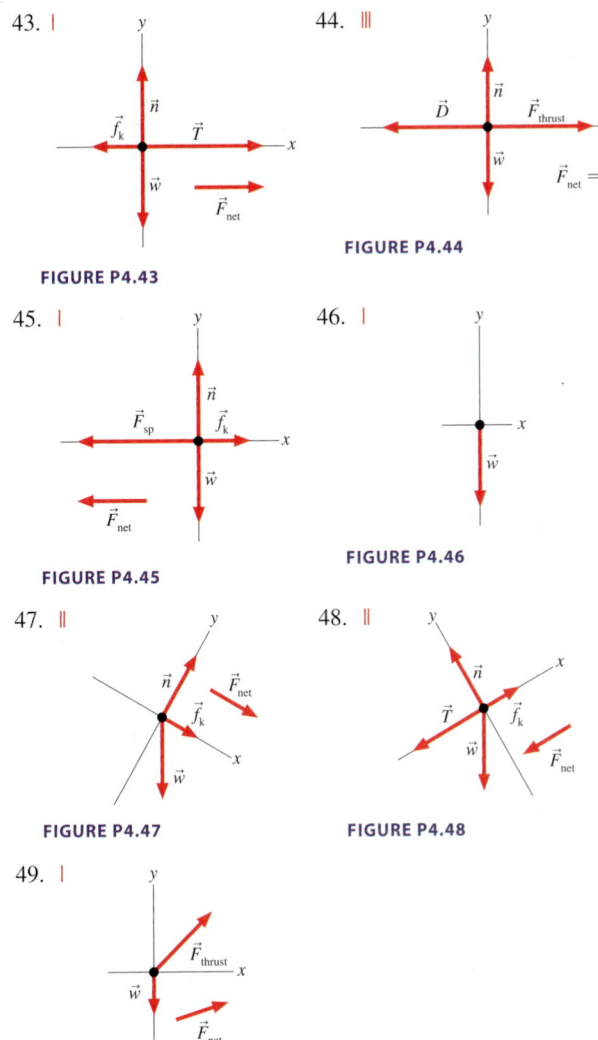

43. FIGURE P4.43

44. FIGURE P4.44

45. FIGURE P4.45

46. FIGURE P4.46

47. FIGURE P4.47

48. FIGURE P4.48

49. FIGURE P4.49

50. ‖ A student draws the flawed free-body diagram shown in Figure P4.50 to represent the forces acting on a car traveling at constant speed on a level road. Identify the errors in the diagram, then draw a correct free-body diagram for this situation.

FIGURE P4.50   FIGURE P4.51

51. ‖ A student draws the flawed free-body diagram shown in Figure P4.51 to represent the forces acting on a golf ball that is traveling upward and to the right a very short time after being hit off the tee. Air resistance is assumed to be relevant. Identify the errors in the diagram, then draw a correct free-body diagram for this situation.

Problems 52 through 63 describe a situation. For each, draw a motion diagram, a force identification diagram, and a free-body diagram.

52. ‖ An elevator, suspended by a single cable, has just left the tenth floor and is speeding up as it descends toward the ground floor.
53. | A rocket is being launched straight up. Air resistance is not negligible.
54. | A jet plane is speeding down the runway during takeoff. Air resistance is not negligible.
55. | You've slammed on the brakes and your car is skidding to a stop while going down a 20° hill.
56. ‖‖ A skier is going down a 20° slope. A *horizontal* headwind is blowing in the skier's face. Friction is small, but not zero.
57. | You've just kicked a rock and it is now sliding down the sidewalk.
58. | A Styrofoam ball has just been shot straight up. Air resistance is not negligible.
59. ‖ A spring-loaded gun shoots a plastic ball. The trigger has just been pulled and the ball is starting to move down the barrel. The barrel is horizontal.
60. ‖ A person on a bridge throws a rock straight down toward the water. The rock has just been released.
61. ‖ A gymnast has just landed on a trampoline. She's still moving downward as the trampoline stretches.
62. ‖ A heavy box is in the back of a truck. The truck is accelerating to the right. Apply your analysis to the box.
63. ‖ A bag of groceries is on the back seat of your car as you stop for a stop light. The bag does not slide. Apply your analysis to the bag.
64. ‖ A rubber ball bounces. We'd like to understand *how* the ball bounces.
   a. A rubber ball has been dropped and is bouncing off the floor. Draw a motion diagram of the ball during the brief time interval that it is in contact with the floor. Show 4 or 5 frames as the ball compresses, then another 4 or 5 frames as it expands. What is the direction of $\vec{a}$ during each of these parts of the motion?
   b. Draw a picture of the ball in contact with the floor and identify all forces acting on the ball.
   c. Draw a free-body diagram of the ball during its contact with the ground. Is there a net force acting on the ball? If so, in which direction?
   d. During contact, is the force of the ground on the ball larger, smaller, or equal to the weight of the ball? Use your answers to parts a–c to explain your reasoning.
65. ‖ If a car stops suddenly, you feel "thrown forward." We'd like to understand what happens to the passengers as a car stops. Imagine yourself sitting on a *very* slippery bench inside a car. This bench has no friction, no seat back, and there's nothing for you to hold on to.
   a. Draw a picture and identify all of the forces acting on you as the car travels in a straight line at a perfectly steady speed on level ground.
   b. Draw your free-body diagram. Is there a net force on you? If so, in which direction?
   c. Repeat parts a and b with the car slowing down.
   d. Describe what happens to you as the car slows down.
   e. Use Newton's laws to explain why you seem to be "thrown forward" as the car stops. Is there really a force pushing you forward?

66. ⫶ The fastest pitched baseball was clocked at 46 m/s. If the
BIO pitcher exerted his force (assumed to be horizontal and constant)
over a distance of 1.0 m, and a baseball has a mass of 145 g,
   a. Draw a free-body diagram of the ball during the pitch.
   b. What force did the pitcher exert on the ball during this
      record-setting pitch?
   c. Estimate the force in (b) as a fraction of the pitcher's
      weight.

67. ⎮ The froghopper, champion leaper of the insect world, can
BIO jump straight up at 4.0 m/s. The jump itself lasts a mere
1.0 ms before the insect is clear of the ground.
   a. Draw a free-body diagram of this mighty leaper while the
      jump is taking place.
   b. While the jump is taking place, is the force that the ground
      exerts on the froghopper greater than, less than, or equal to
      the insect's weight? Explain.

68. ‖ A beach ball is thrown straight up, and some time later it
lands on the sand. Is the magnitude of the net force on the ball
greatest when it is going up or when it is on the way down? Or
is it the same in both cases? Explain. Air resistance should not
be neglected for a large, light object.

Passage Problems

A Simple Solution for a Stuck Car

If your car is stuck in the mud and you don't have a winch to pull
it out, you can use a piece of rope and a tree to do the trick. First,
you tie one end of the rope to your car and the other to a tree, then
pull as hard as you can on the middle of the rope, as shown in
Figure P4.69a. This technique applies a force to the car much larger
than the force that you can apply directly. To see why the car experiences such a large force, look at the forces acting on the center
point of the rope, as shown in Figure 4.69b. The sum of the forces
is zero, thus the tension is much greater than the force you apply. It
is this tension force that acts on the car and, with luck, pulls it free.

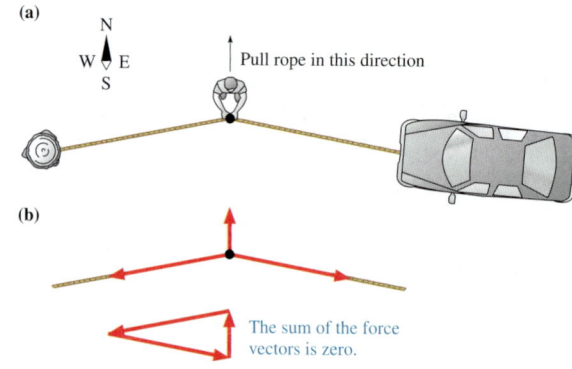

(a)

Pull rope in this direction

(b)

The sum of the force vectors is zero.

FIGURE P4.69

69. ⎮ The sum of the three forces acting on the center point of
the rope is assumed to be zero because
   A. This point has a very small mass.
   B. Tension forces in a rope always cancel.
   C. This point is not accelerating.
   D. The angle of deflection is very small.

70. ⎮ When you are pulling on the rope as shown, what is the
approximate direction of the tension force on the tree?
   A. North   B. South   C. East   D. West

71. ⎮ Assume that you are pulling on the rope but the car is not
moving. What is the approximate direction of the force of the
mud on the car?
   A. North   B. South   C. East   D. West

72. ⎮ Suppose your efforts work, and the car begins to move for-
ward out of the mud. As it does so, the force of the car on the
rope is
   A. Zero.
   B. Less than the force of the rope on the car.
   C. Equal to the force of the rope on the car.
   D. Greater than the force of the rope on the car.

STOP TO THINK ANSWERS

Stop to Think 4.1: C.

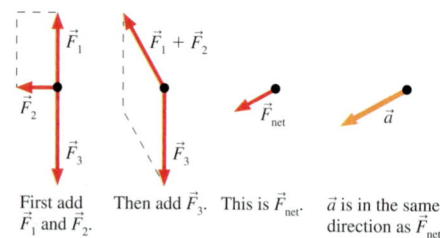

y-component of $\vec{F}_3$
cancels y-component of $\vec{F}_1$.

x-component of $\vec{F}_3$ is to
the left and larger than
the x-component of $\vec{F}_2$.

Stop to Think 4.2: A, B, and D. Friction and the normal force are
the only contact forces. Nothing is touching the rock to provide a
"force of the kick." We've agreed to ignore air resistance unless a
problem specifically calls for it.

Stop to Think 4.3: B. Acceleration is proportional to force, so
doubling the number of rubber bands doubles the acceleration of
the original object from 2 m/s² to 4 m/s². But acceleration is also
inversely proportional to mass. Doubling the mass cuts the acceleration in half, back to 2 m/s².

Stop to Think 4.4: D.

First add $\vec{F}_1$ and $\vec{F}_2$. Then add $\vec{F}_3$. This is $\vec{F}_{net}$. $\vec{a}$ is in the same direction as $\vec{F}_{net}$.

Stop to Think 4.5: C. The acceleration vector points downward as
the elevator slows. $\vec{F}_{net}$ points in the same direction as $\vec{a}$, so $\vec{F}_{net}$
also points down. This will be true if the tension is less than the
weight: $T < w$.

Stop to Think 4.6: C. Newton's third law says that the force of A
on B is *equal* and opposite to the force of B on A. This is always
true. The mass of the objects isn't relevant.

5

APPLYING NEWTON'S LAWS

Before his parachute opens, why does this skydiver fall at a constant speed? And why does he suddenly slow down when his parachute opens?

Looking Ahead ▶▶

The goal of Chapter 5 is to learn how to solve problems about motion in a straight line. In this chapter you will learn to:

▶ Use Newton's first law to solve problems of static and dynamic equilibrium.

▶ Use Newton's second law to solve dynamics problems.

▶ Understand how mass, weight, and apparent weight differ.

▶ Use simple models of friction and drag.

▶ Use Newton's third law to solve problems involving interacting objects.

Looking Back ◀◀

This chapter pulls together many strands of thought from Chapters 1–4. Please review:

◀ Sections 2.4–2.5 Constant acceleration kinematics, including free fall.

◀ Section 3.2–3.3 Vectors and components.

◀ Sections 4.4, 4.7, and 4.8 Identifying forces, drawing free-body diagrams, and identifying action/reaction pairs.

After jumping from the plane, a skydiver accelerates until reaching a *terminal speed* of about 140 mph. But when his parachute is opened, he then slows down to only some 10–20 mph. To understand the skydiver's motion, we need to look closely at the forces exerted on him. We also need to understand how those forces determine his motion.

Chapter 4 introduced Newton's three laws of motion. Now, in Chapter 5, we want to use these laws to solve force and motion problems. As will be the case throughout this textbook, our strategy will be to learn a set of *procedures* that can be applied to a wide variety of problems, not to memorize a set of equations.

This chapter focuses on objects that are at rest or that move in a straight line, such as runners, bicycles, cars, planes, and rockets. Weight, tension, thrust, friction, and drag forces will be essential to our understanding of this motion. Circular motion and rotational motion will be treated in Chapters 6 and 7.

5.1 Equilibrium

The simplest applications of Newton's laws are those for which the net force $\vec{F}_{net}$ on an object is *zero*. According to Newton's first law, an object on which there is no net force moves with constant velocity $(\vec{a} = \vec{0})$. One way an object can have

This human tower is in equilibrium because the net force on each man is zero.

$\vec{a} = \vec{0}$ is to be at rest. An object that remains at rest is said to be in **static equilibrium.** A second way for an object to have $\vec{a} = \vec{0}$ is to move in a straight line at a constant speed. Such an object is in **dynamic equilibrium.** The defining property of both these cases of **equilibrium** is that the net force acting on the object is $\vec{F}_{net} = \vec{0}$.

To use Newton's laws, we have to identify all the forces acting on an object and then evaluate $\vec{F}_{net}$. Recall that $\vec{F}_{net}$ is the vector sum

$$\vec{F}_{net} = \vec{F}_1 + \vec{F}_2 + \vec{F}_3 + \cdots$$

where $\vec{F}_1$, $\vec{F}_2$, and so on are the individual forces, such as tension or friction, acting on the object. We found in Chapter 3 that vector sums can be evaluated in terms of the x- and y-components of the vectors. That is, the x-component of the net force is $(F_{net})_x = F_{1x} + F_{2x} + F_{3x} + \cdots$ If we restrict ourselves to problems where all the forces are in the xy-plane, then the equilibrium requirement $\vec{F}_{net} = \vec{0}$ is a shorthand way of writing two simultaneous equations:

$$(F_{net})_x = F_{1x} + F_{2x} + F_{3x} + \cdots = 0$$
$$(F_{net})_y = F_{1y} + F_{2y} + F_{3y} + \cdots = 0$$

Recall from your math classes that the Greek letter Σ (sigma) stands for "the sum of." It will be convenient to abbreviate the sum of the x-components of all forces as

$$F_{1x} + F_{2x} + F_{3x} + \cdots = \sum F_x$$

With this notation, the requirement that an object be in equilibrium—Newton's first law—can be written as the two equations

$$\sum F_x = 0 \qquad \text{and} \qquad \sum F_y = 0 \qquad (5.1)$$

In equilibrium, the sums of the x- and y-components of the force are zero

Although this may look a bit forbidding, we'll soon see how to use a free-body diagram of the forces to evaluate these sums.

When an object is in equilibrium, we are usually interested in finding the forces that keep it in equilibrium. Newton's first law is the basis for a strategy for solving equilibrium problems.

(MP)   PROBLEM-SOLVING
STRATEGY 5.1   **Equilibrium problems**

PREPARE   First check that the object is in equilibrium: Does $\vec{a} = \vec{0}$?

- An object at rest is in static equilibrium.
- An object moving at a constant velocity is in dynamic equilibrium.

Then identify all forces acting on the object and show them on a free-body diagram. Determine which forces you know and which you need to solve for.

SOLVE   An object in equilibrium must satisfy Newton's first law. In component form, the requirement is

$$\sum F_x = 0 \qquad \text{and} \qquad \sum F_y = 0$$

You can find the force components that go into these sums directly from your free-body diagram. From these two equations, solve for the unknown forces in the problem.

ASSESS   Check that your result has the correct units, is reasonable, and answers the question.

Static Equilibrium

EXAMPLE 5.1 Forces supporting an orangutan

An orangutan weighing 500 N hangs from a vertical vine. What is the tension in the vine?

PREPARE The orangutan is in static equilibrium, so all the forces acting on it must cancel to give zero net force. Figure 5.1 first identifies the forces acting on the orangutan: the upward force of the tension in the vine and the downward, long-range force of gravity. These forces are then shown on a free-body diagram, where it's noted that equilibrium requires $\vec{F}_{net} = \vec{0}$.

Force identification   **Free-body diagram**

Tension $\vec{T}$

Weight $\vec{w}$

Known
$w = 500 \text{ N}$

Find
T

FIGURE 5.1 The forces on an orangutan.

SOLVE Neither force has an x-component, so we need to examine only the y-components of the forces. In this case, the y-component of Newton's first law is

$$\sum F_y = T_y + w_y = 0$$

You might have been tempted to write $T_y - w_y$ because the weight force points down. But remember that T_y and w_y are *components* of vectors, and can thus be positive (for a vector such as $\vec{T}$ that points up) or negative (for a vector such as $\vec{w}$ that points down). The fact that $\vec{w}$ points down is taken into account when we *evaluate* the components; that is, when we write them in terms of the *magnitudes* T and w of the vectors $\vec{T}$ and $\vec{w}$.

Because the tension vector $\vec{T}$ points straight up, in the positive y-direction, its y-component is $T_y = T$. Because the weight vector $\vec{w}$ points straight down, in the negative y-direction, its y-component is $w_y = -w$. This is where the signs enter. With these components, Newton's second law becomes

$$T - w = 0$$

This equation is easily solved for the tension in the vine:

$$T = w = 500 \text{ N}$$

ASSESS It's not surprising that the tension in the vine equals the weight of the orangutan. However, we'll soon see that this is *not* the case if the object is accelerating.

EXAMPLE 5.2 Readying a wrecking ball

A wrecking ball weighing 2500 N hangs from a cable. Prior to swinging, it is pulled back to a 20° angle by a second, horizontal cable. What is the tension in the horizontal cable?

PREPARE The ball hangs in static equilibrium until it is released. In Figure 5.2, we start by identifying all of the forces acting on the ball: a tension force from each cable and the ball's weight. We've used different symbols $\vec{T}_1$ and $\vec{T}_2$ for the two different tension forces. We then construct a free-body diagram for these three forces, noting that $\vec{F}_{net} = \vec{0}$. We're looking for the magnitude T_1 of the tension force $\vec{T}_1$ in the horizontal cable.

Known
$\theta = 20°$
$w = 2500 \text{ N}$

Find
T_1

Tension $\vec{T}_2$
Weight $\vec{w}$
Tension $\vec{T}_1$

FIGURE 5.2 Visual overview of a wrecking ball just before release.

SOLVE The requirement of equilibrium is $\vec{F}_{net} = \vec{0}$. In component form, we have the two equations

$$\sum F_x = T_{1x} + T_{2x} + w_x = 0$$
$$\sum F_y = T_{1y} + T_{2y} + w_y = 0$$

As always, we *add* the force components together. Now we're ready to write the components of each force vector in terms of the magnitudes and directions of those vectors. We learned how to do this in Section 3.3 of Chapter 3. With practice you'll learn to read the components directly off the free-body diagram, but to begin it's worthwhile to organize the components into a table.

Force	Name of x-component	Value of x-component	Name of y-component	Value of y-component
$\vec{T}_1$	T_{1x}	$-T_1$	T_{1y}	0
$\vec{T}_2$	T_{2x}	$T_2 \sin\theta$	T_{2y}	$T_2 \cos\theta$
$\vec{w}$	w_x	0	w_y	$-w$

We see from the free-body diagram that $\vec{T}_1$ points along the negative x-axis, so $T_{1x} = -T_1$ and $T_{1y} = 0$. We need to be careful with our trigonometry as we find the components of $\vec{T}_2$.

Continued

Remembering that the side adjacent to the angle is related to the cosine, we see that the vertical (y) component of $\vec{T}_2$ is $T_2\cos\theta$. Similarly, the horizontal (x) component is $T_2\sin\theta$. The weight vector points straight down, so its y-component is $-w$. Notice that negative signs enter as we evaluate the components of the vectors, *not* when we write Newton's first law. This is a critical aspect of solving force and motion problems. With these components, Newton's first law now becomes

$$-T_1 + T_2\sin\theta + 0 = 0 \quad \text{and} \quad 0 + T_2\cos\theta - w = 0$$

We can rewrite these as

$$T_2\sin\theta = T_1 \quad \text{and} \quad T_2\cos\theta = w$$

These are two simultaneous equations with two unknowns: T_1 and T_2. To eliminate T_2 from the two equations, solve the second equation for T_2, giving $T_2 = w/\cos\theta$. Insert this expression for T_2 into the first equation to get

$$T_1 = \frac{w}{\cos\theta}\sin\theta = \frac{\sin\theta}{\cos\theta}w = w\tan\theta = (2500\text{ N})\tan 20° = 910\text{ N}$$

where we made use of the fact that $\tan\theta = \sin\theta/\cos\theta$.

ASSESS It seems reasonable that to pull the ball back to this modest angle, a force substantially less than the ball's weight will be required.

CONCEPTUAL EXAMPLE 5.1 **Forces in static equilibrium**

A rod is free to slide on a frictionless sheet of ice. One end of the rod is lifted by a string. Once the rod comes to rest, which diagram in Figure 5.3 shows the correct angle of the string?

FIGURE 5.3 Which is the correct angle of the string?

REASON If the rod is to hang motionless, it must be in static equilibrium with $\sum F_x = 0$ and $\sum F_y = 0$. Figure 5.4 shows free-body diagrams for the three string orientations. Remember that tension always acts along the direction of the string and that the weight force always points straight down. The ice pushes up

FIGURE 5.4 Free-body diagrams for three angles of the string.

with a normal force perpendicular to the surface, but frictionless ice cannot exert any horizontal force. If the string is angled, we see that its horizontal component exerts a net force on the rod. Only in case b, where the tension—and the string—are vertical, can the net force be zero.

ASSESS Frictionless surfaces are an idealization, but one that we will often use when friction is very small compared to other forces. This example illustrates that a frictionless surface has no component of force parallel to the surface.

Dynamic Equilibrium

EXAMPLE 5.3 **Tension in towing a car**

A car with a mass of 1500 kg is being towed at a steady speed by a rope held at a 20° angle. A friction force of 320 N opposes the car's motion. What is the tension in the rope?

PREPARE The car is moving in a straight line at a constant speed, so it is in dynamic equilibrium and must have $\vec{F}_{net} = \vec{0}$. Figure 5.5 shows three contact forces acting on the car—the tension force $\vec{T}$, friction $\vec{f}$, and the normal force $\vec{n}$—and the long-range force of gravity $\vec{w}$. These four forces are shown on the free-body diagram.

FIGURE 5.5 Visual overview of a car being towed.

SOLVE This is still an equilibrium problem, even though the car is moving, so our problem-solving procedure is unchanged. With four forces, the requirement of equilibrium is

$$\sum F_x = n_x + T_x + f_x + w_x = 0$$

$$\sum F_y = n_y + T_y + f_y + w_y = 0$$

We can again determine the horizontal and vertical components of the forces by "reading" the free-body diagram. The results are shown in the following table.

Force	Name of x-component	Value of x-component	Name of y-component	Value of y-component
$\vec{n}$	n_x	0	n_y	n
$\vec{T}$	T_x	$T\cos\theta$	T_y	$T\sin\theta$
$\vec{f}$	f_x	$-f$	f_y	0
$\vec{w}$	w_x	0	w_y	$-w$

With these components, Newton's first law reads

$$T\cos\theta - f = 0$$

$$T\sin\theta + n - w = 0$$

The first equation can be used to solve for the tension in the rope:

$$T = \frac{f}{\cos\theta} = \frac{320\ \text{N}}{\cos 20°} = 340\ \text{N}$$

to two significant figures. It turned out that we did not need the y-component equation in this problem. We would need it if we wanted to find the normal force $\vec{n}$.

ASSESS Had we pulled the car with a horizontal rope, the tension would need to exactly balance the friction force of 320 N. Because we are pulling at an angle, however, part of the tension in the rope pulls *up* on the car instead of in the forward direction. Thus we need a little more tension in the rope when it's at an angle.

5.2 Dynamics and Newton's Second Law

Newton's second law is the essential link between force and motion. The essence of Newtonian mechanics can be expressed in two steps:

Activ Physics ONLINE 2.1–2.4

- The forces acting on an object determine its acceleration $\vec{a} = \vec{F}_{net}/m$.
- The object's motion can be found by using $\vec{a}$ in the equations of kinematics.

We want to develop a strategy to solve a variety of problems in mechanics, but first we need to write the second law in terms of its components. To do so, let's first rewrite Newton's second law in the form

$$\vec{F}_{net} = \vec{F}_1 + \vec{F}_2 + \vec{F}_3 + \cdots = m\vec{a}$$

where $\vec{F}_1, \vec{F}_2, \vec{F}_3$, and so on are the forces acting on an object. To write the second law in component form merely requires that we use the x- and y-components of the acceleration. Thus Newton's second law, $\vec{F}_{net} = m\vec{a}$, is

$$\sum F_x = ma_x \quad \text{and} \quad \sum F_y = ma_y \qquad (5.2)$$

Newton's second law in component form

The first equation says that **the component of the acceleration in the x-direction is determined by the sum of the x-components of the forces acting on the object.** A similar statement applies to the y-direction.

There are two basic types of problems in mechanics. In the first, you use information about forces to find an object's acceleration, then use kinematics to determine the object's motion. In the second, you use information about the object's motion to determine its acceleration, then solve for unknown forces. Either way, the two equations of Equation 5.2 are the link between force and motion, and they form the basis of a problem-solving strategy. The primary goal of this chapter is to illustrate the use of this strategy. We'll then follow the strategy with some examples.

(MP) **PROBLEM-SOLVING STRATEGY 5.2** Dynamics problems

PREPARE Sketch a visual overview consisting of

- A list of values that identifies known quantities and what the problem is trying to find.
- A force-identification diagram to help you identify all forces acting on the object.
- A free-body diagram that shows all the forces acting on the object.

If you'll need to use kinematics to find velocities or positions, you'll also need to sketch

- A motion diagram to determine the direction of the acceleration.
- A pictorial representation that establishes a coordinate system, shows important points in the motion, and defines symbols.

It's OK to go back and forth between these steps as you visualize the situation.

SOLVE Write Newton's second law in component form as

$$\sum F_x = ma_x \qquad \text{and} \qquad \sum F_y = ma_y$$

You can find the components of the forces directly from your free-body diagram. Depending on the problem, either

- Solve for the acceleration, then use kinematics to find velocities and positions, or
- Use kinematics to determine the acceleration, then solve for unknown forces.

ASSESS Check that your result has the correct units, is reasonable, and answers the question.

EXAMPLE 5.4 Putting a golf ball

A golfer putts a 46 g ball with a speed of 3.0 m/s. Friction exerts a 0.020 N retarding force on the ball, slowing it down. Will her putt reach the hole, 10 m away?

PREPARE Figure 5.6 is a visual overview of the problem. We've collected the known information, drawn a sketch, and identified what we want to find. The motion diagram shows that the ball is slowing down as it rolls to the right, so the acceleration vector points to the left. Next, we identify the forces acting on the ball and show them on a free-body diagram. Note that the net force points to the left, as it must because the acceleration points to the left.

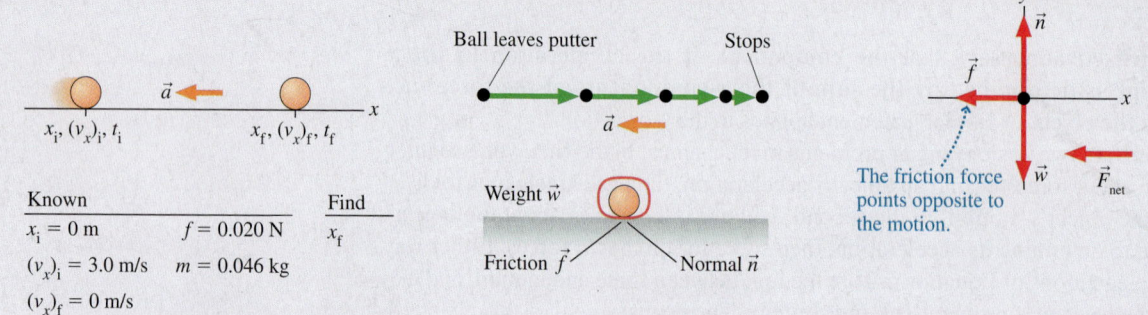

Known
$x_i = 0$ m $f = 0.020$ N
$(v_x)_i = 3.0$ m/s $m = 0.046$ kg
$(v_x)_f = 0$ m/s

Find
x_f

FIGURE 5.6 Visual overview of a golf putt.

SOLVE Newton's second law in component form is

$$\sum F_x = n_x + f_x + w_x = 0 - f + 0 = ma_x$$

$$\sum F_y = n_y + f_y + w_y = n + 0 - w = ma_y = 0$$

We've written the equations as sums, as we did with equilibrium problems, then "read" the values of the force components from the free-body diagram. The components are simple enough in this problem that we don't really need to show them in a table. It is particularly important to notice that we set $a_y = 0$ in the second equation. This is because the ball does not move in the y-direction, so it can't have any acceleration in the y-direction. This will be an important step in many problems.

The first equation is $-f = ma_x$, from which we find

$$a_x = -f/m = -(0.020\ \text{N})/(0.046\ \text{kg}) = -0.43\ \text{m/s}^2$$

(Recall from Chapter 4 that $1\ \text{N} = 1\ \text{kg} \cdot \text{m/s}^2$, so the units above work out correctly.) The negative sign shows that the acceleration is directed to the left, as expected.

Now that we know the acceleration, we can use kinematics to find how far the ball will roll before stopping. We don't have any information about the time it takes for the ball to stop, so we'll use the kinematic equation $(v_x)_f^2 = (v_x)_i^2 + 2a_x(x_f - x_i)$. This gives

$$x_f = x_i + \frac{(v_x)_f^2 - (v_x)_i^2}{2a_x} = 0\ \text{m} + \frac{(0\ \text{m/s})^2 - (3.0\ \text{m/s})^2}{2(-0.43\ \text{m/s}^2)} = 10.5\ \text{m}$$

If her aim is true, the ball will just make it into the hole.

ASSESS The key steps in any dynamics problem are to draw a correct free-body diagram and to use the free-body diagram to write Newton's second law in component form. Once you've mastered these two steps, you're well on your way to solving any dynamics problem!

EXAMPLE 5.5 Finding a rocket cruiser's acceleration

A rocket cruiser with a mass of 2200 kg and weighing 5000 N is flying horizontally over the surface of a distant planet. At its present speed, a 3000 N drag force acts on the cruiser. The cruiser's engines can be tilted so as to provide a thrust angled up or down. The pilot turns the thrust up to 14,000 N while pivoting the engines to continue flying horizontally. What is the cruiser's acceleration?

PREPARE Figure 5.7 is a visual overview in which we've listed known information, identified the forces on the cruiser, and drawn a free-body diagram. (Because kinematics is not needed to find the acceleration, we don't need a pictorial diagram.) As discussed in Chapter 4, the thrust force points *opposite* to the direction of the rocket exhaust, which we've shown at angle θ. The thrust must be directed upward to balance the weight force; otherwise the cruiser would fall. To continue flying horizontally requires the net force to be directed forward.

SOLVE Newton's second law in component form is

$$\sum F_x = (F_{\text{thrust}})_x + D_x + w_x = ma_x$$

$$\sum F_y = (F_{\text{thrust}})_y + D_y + w_y = ma_y$$

From the free-body diagram, we see that $(F_{\text{thrust}})_x = F_{\text{thrust}}\cos\theta$, $(F_{\text{thrust}})_y = F_{\text{thrust}}\sin\theta$, $D_x = -D$, $D_y = 0$, $w_x = 0$, and $w_y = -w$. We know that a_y must be zero because the cruiser is to accelerate *horizontally*. Thus the second law becomes

$$F_{\text{thrust}}\cos\theta - D = ma_x$$

$$F_{\text{thrust}}\sin\theta - w = 0$$

The first of these equations contains a_x, the quantity we want to find, but we can't solve for a_x without knowing what θ is. Fortunately, we can use the second equation to find θ, then use this value of θ in the first equation to find a_x.

The second equation gives

$$\sin\theta = w/F_{\text{thrust}} = (5000\ \text{N})/(14{,}000\ \text{N}) = 0.357$$

$$\theta = \sin^{-1}(0.357) = 20.9°$$

Now we can use this value in the first equation to get

$$a_x = \frac{1}{m}(F_{\text{thrust}}\cos\theta - D)$$

$$= \frac{1}{2200\ \text{kg}}\left[(14{,}000\ \text{N})\cos(20.9°) - 3000\ \text{N}\right] = 4.6\ \text{m/s}^2$$

Known
$m = 2200\ \text{kg}$
$w = 5000\ \text{N}$
$F_{\text{thrust}} = 14{,}000\ \text{N}$
$D = 3000\ \text{N}$

Find
a_x

Thrust $\vec{F}_{\text{thrust}}$ Drag $\vec{D}$

Weight $\vec{w}$

FIGURE 5.7 Visual overview of a rocket cruiser.

Continued

ASSESS An important key to solving this problem was to use the information that the cruiser accelerates only in the horizontal direction. Mathematically, this means that $a_y = 0$. Because the thrust is much greater than the weight, we need only a modest downward component of the thrust to cancel the weight and let the cruiser accelerate horizontally. So our engine tilt seems reasonable.

EXAMPLE 5.6 Towing a car with acceleration

A car with a mass of 1500 kg is being towed by a rope held at a 20° angle. A friction force of 320 N opposes the car's motion. What is the tension in the rope if the car goes from rest to 12 m/s in 10 s?

PREPARE You should recognize that this problem is almost identical to Example 5.3. The difference is that the car is now accelerating, so it is no longer in equilibrium. This means, as shown in Figure 5.8, that the net force is not zero. We've already identified all the forces in Example 5.3.

SOLVE Newton's second law in component form is

$$\sum F_x = n_x + T_x + f_x + w_x = ma_x$$

$$\sum F_y = n_y + T_y + f_y + w_y = ma_y = 0$$

We've again used the fact that $a_y = 0$ for motion that is purely along the x-axis. The components of the forces were worked out in Example 5.3. Using that information, Newton's second law in component form is

$$T\cos\theta - f = ma_x$$

$$T\sin\theta + n - w = 0$$

Because the car speeds up from rest to 12 m/s in 10 s, we can use kinematics to find the acceleration:

$$a_x = \frac{\Delta v_x}{\Delta t} = \frac{(v_x)_f - (v_x)_i}{t_f - t_i} = \frac{(12 \text{ m/s}) - (0 \text{ m/s})}{(10 \text{ s}) - (0 \text{ s})} = 1.2 \text{ m/s}^2$$

We can now use the first Newton's-law equation above to solve for the tension. We have

$$T = \frac{ma_x + f}{\cos\theta} = \frac{(1500 \text{ kg})(1.2 \text{ m/s}^2) + 320 \text{ N}}{\cos 20°} = 2300 \text{ N}$$

ASSESS The tension is substantially more than the 341 N found in Example 5.3. It takes much more force to accelerate the car than to keep it rolling at constant speed.

Known
$x_i = 0 \text{ m}$
$(v_x)_i = 0 \text{ m/s}$
$t_i = 0 \text{ s}, \theta = 20°$
$m = 1500 \text{ kg}$
$f = 320 \text{ N}$
$(v_x)_f = 12 \text{ m/s}$
$t_f = 10 \text{ s}$

Find
T

FIGURE 5.8 Visual overview of a car being towed.

These first examples have shown all the details of our problem-solving strategy. Our purpose has been to demonstrate how the strategy is put into practice. Future examples will be briefer, but the basic *procedure* will remain the same.

STOP TO THINK 5.1 A Martian lander is approaching the surface. It is slowing its descent by firing its rocket motor. Which is the correct free-body diagram for the lander?

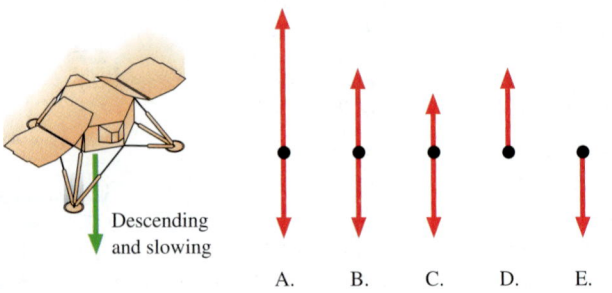

5.3 Mass and Weight

When the doctor asks what you weigh, what does she really mean? We do not make much distinction in our ordinary use of language between the terms *weight* and *mass,* but in physics their distinction is of critical importance.

Mass, you'll recall from Chapter 4, is a quantity that describes an object's inertia, its tendency to resist being accelerated. Loosely speaking, it also describes the amount of matter in an object. Mass, measured in kilograms, is an intrinsic property of an object; it has the same value wherever the object may be and whatever forces might be acting on it.

Weight, on the other hand, is a *force.* Specifically, it is the gravitational force exerted on an object by a planet. Weight is a vector, not a scalar, and the vector's direction is always straight down. Weight is measured in newtons.

Mass and weight are not the same thing, but they are related. We can use Galileo's discovery about free fall to make the connection. Figure 5.9 shows the free-body diagram of an object in free fall. The *only* force acting on this object is its weight $\vec{w}$, the downward pull of gravity. Newton's second law for this object is

$$\vec{F}_{net} = \vec{w} = m\vec{a} \qquad (5.3)$$

Recall Galileo's discovery that *any* object in free fall, regardless of its mass, has the same acceleration:

$$\vec{a}_{free\,fall} = (g, \text{downward}) \qquad (5.4)$$

where $g = 9.80 \text{ m/s}^2$ is the free-fall acceleration at the earth's surface. So a_y in Equation 5.3 is equal to $-g$, and we have $-w = -mg$, or

$$w = mg \qquad (5.5)$$

The magnitude of the weight force, which we call simply "the weight," is directly proportional to the mass, with g as the constant of proportionality. Thus, for example, the weight of a 3.6 kg book is $w = (3.6 \text{ kg})(9.8 \text{ m/s}^2) = 35 \text{ N}$.

> **NOTE** ▶ Although we derived the relationship between mass and weight for an object in free fall, the weight of an object is *independent* of its state of motion. Equation 5.5 holds for an object at rest on a table, sliding horizontally, or moving in any other way. ◀

Because an object's weight depends on g, and the value of g varies from planet to planet, weight is not a fixed, constant property of an object. The value of g at the surface of the moon is about one-sixth its earthly value, so an object on the moon would have only one-sixth its weight on earth. The object's weight on Jupiter would be larger than its weight on earth. Its mass, however, would be the same. The amount of matter has not changed, only the gravitational force exerted on that matter.

So when the doctor asks what you weigh, she really wants to know your *mass.* That's the amount of matter in your body. You can't really "lose weight" by going to the moon, even though you would weigh less there!

Measuring Mass and Weight

A *pan balance,* shown in Figure 5.10, is a device for measuring *mass.* You may have used a pan balance to "weigh" chemicals in a chemistry lab. An unknown mass is placed in one pan, then known masses are added to the other until the pans balance. Gravity pulls down on both sides, effectively *comparing* the masses, and the unknown mass equals the sum of the known masses that balance it. Although a pan balance requires gravity in order to function, it does not depend on the value of g. Consequently, the pan balance would give the same result on another planet.

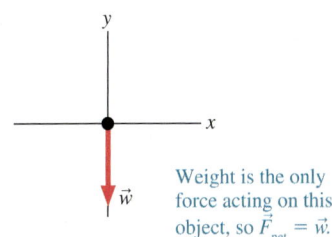

FIGURE 5.9 The free-body diagram of an object in free fall.

On the moon, astronaut John Young jumps 2 feet straight up, despite his spacesuit that weighs 370 pounds on earth. On the moon, where $g = 1.6 \text{ m/s}^2$, he and his suit together weighed only 90 pounds.

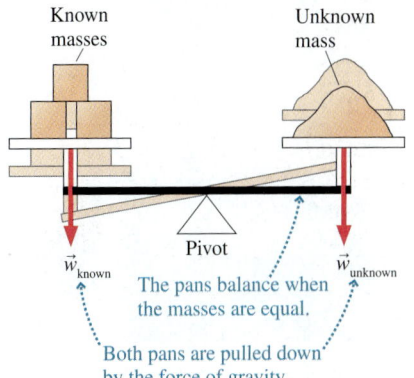

FIGURE 5.10 A pan balance measures mass.

(a)

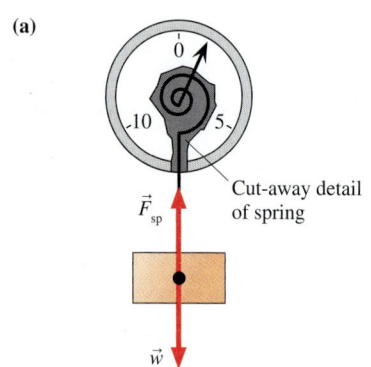

$\vec{F}_{sp}$ Cut-away detail of spring

$\vec{w}$

(b)

$\vec{F}_{sp}$

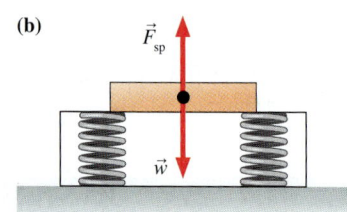

$\vec{w}$

FIGURE 5.11 A spring scale measures weight.

Spring scales, such as the two shown in Figure 5.11, measure weight, not mass. Hanging an item on the scale in Figure 5.11a, which might be used to weigh items in the grocery store, stretches the spring. The spring in the "bathroom scale" in Figure 5.11b is compressed when you stand on it.

A spring scale can be understood on the basis of Newton's first law. The object being weighed is at rest, in static equilibrium, so the net force on it must be zero. The stretched spring in Figure 5.11a *pulls* up on the object with force $\vec{F}_{sp}$; the compressed spring in Figure 5.11b *pushes* up with force $\vec{F}_{sp}$. But in both cases, in order to have $\vec{F}_{net} = \vec{0}$, the upward spring force must exactly balance the downward weight force:

$$F_{sp} = w = mg \qquad (5.6)$$

The *reading* of a spring scale is F_{sp}, the magnitude of the force that the spring is exerting. If the object is in equilibrium, then F_{sp} is exactly equal to the object's weight w. The scale does not "know" the weight of the object. All it can do is to measure how much its spring is stretched or compressed. On a different planet, with a different value for g, the expansion or compression of the spring would be different and the scale's reading would be different.

The unit of force in the English system is the *pound.* We noted in Chapter 4 that the pound is defined as 1 lb = 4.45 N. An object whose weight $w = mg$ is 4.45 N has a mass

$$m = \frac{w}{g} = \frac{4.45\ N}{9.80\ m/s^2} = 0.454\ kg = 454\ g$$

You may have learned in previous science classes that "1 pound = 454 grams" or, equivalently, that "1 kg = 2.2 lb." Strictly speaking, these well-known "conversion factors" are not true. They are comparing a weight (pounds) to a mass (kilograms). The correct statement would be, "A mass of 1 kg has a weight on *earth* of 2.2 pounds." On another planet, the weight of a 1 kg mass would be something other than 2.2 pounds.

EXAMPLE 5.7 Masses of people

What is the mass, in kilograms, of a 90 pound gymnast, a 160 pound professor, and a 240 pound football player?

SOLVE We must convert their weights into newtons; then we can find their masses from $m = w/g$. We have

$$w_{gymnast} = 90\ lb \times \frac{4.45\ N}{1\ lb} = 400\ N \qquad m_{gymnast} = \frac{w_{gymnast}}{g} = \frac{400\ N}{9.80\ m/s^2} = 41\ kg$$

$$w_{prof} = 160\ lb \times \frac{4.45\ N}{1\ lb} = 710\ N \qquad m_{prof} = \frac{w_{prof}}{g} = \frac{710\ N}{9.80\ m/s^2} = 72\ kg$$

$$w_{football} = 240\ lb \times \frac{4.45\ N}{1\ lb} = 1070\ N \qquad m_{football} = \frac{w_{football}}{g} = \frac{1070\ N}{9.80\ m/s^2} = 110\ kg$$

ASSESS It's worth remembering that a *typical* adult has a mass in the range of 60 to 80 kg.

Apparent Weight

The weight of an object is the force of gravity on that object. You may never have thought about it, but gravity is not a force that you can feel or sense directly. Your *sensation* of weight—how heavy you feel—is due to *contact forces* pressing against you. Surfaces touch you and activate nerve endings in your skin. As you read this, your sensation of weight is due to the normal force exerted on you by

This popular amusement park ride shoots you straight up with an acceleration of 4g. Your apparent weight is then five times your true weight.

the chair in which you are sitting. When you stand, you feel the contact force of the floor pushing against your feet. If you hang from a rope, your sensation of weight is due to the tension force pulling up on you.

Figure 5.12a shows how a standing man *feels* the normal force of the ground on his feet. The free-body diagram of Figure 5.12b shows that if he is at rest, with $\vec{a} = \vec{0}$, then the normal force pressing against his feet is exactly equal in magnitude to his weight. So the relationship between your weight and the forces you feel would be simple if objects were always at rest, but what happens if $\vec{a} \neq \vec{0}$?

Recall the sensations you feel while being accelerated. You feel "heavy" when an elevator suddenly accelerates upward or when an airplane accelerates for take-off. This sensation vanishes as soon as the elevator or airplane reaches a steady cruising speed. Your stomach seems to rise a little and you feel lighter than normal as the upward-moving elevator brakes to a halt or a roller coaster goes over the top. Your true weight $w = mg$ has not changed during these events, but your *apparent weight* has.

To investigate this, imagine a man weighing himself by standing on a spring scale in an elevator as it accelerates upward. What does the scale read? How does the scale reading correspond to the man's *sensation* of weight?

As Figure 5.13 shows, the only forces acting on the man are the upward spring force of the scale and the downward weight force. This seems to be the same situation as Figure 5.12, but there's one big difference. The man is accelerating; he's not in equilibrium. Thus, according to Newton's second law, there must be a net force acting on the man in the direction of $\vec{a}$.

For the net force $\vec{F}_{net}$ to point upward, the magnitude of the spring force must be *greater* than the magnitude of the weight force. That is, $F_{sp} > w$. This conclusion has major implications. Looking at the free-body diagram in Figure 5.13, we see that the y-component of Newton's second law is

$$\sum F_y = (F_{sp})_y + w_y = F_{sp} - w = ma_y \tag{5.7}$$

where m is the man's mass.

The scale reading is the value of F_{sp}, the magnitude of the force that the scale exerts on the man. Solving Equation 5.7 for F_{sp} gives

$$F_{sp} = w + ma_y \tag{5.8}$$

If the elevator is either at rest or moving with constant velocity, then $a_y = 0$ and the man is in equilibrium. In that case, $F_{sp} = w$ and the scale correctly reads his weight. But if $a_y \neq 0$, the scale's reading is *not* the man's true weight.

Let's define an object's **apparent weight** w_{app} as the magnitude of the contact force that supports the object. From Equation 5.8, this is

$$w_{app} = w + ma_y = mg + ma_y = m(g + a_y) \tag{5.9}$$

If the elevator is accelerating upward, then $a_y = +a$, and Equation 5.9 reads $w_{app} = w + ma$. Thus $w_{app} > w$ and the man *feels* heavier than normal. If the elevator is accelerating downward, the acceleration vector $\vec{a}$ points downward and $a_y = -a$. Thus $w_{app} < w$ and the man feels lighter. Indeed, the scale reads less than his true weight.

An object doesn't have to be on a scale for its apparent weight to differ from its true weight. An object's apparent weight is the magnitude of the contact force supporting it. It makes no difference whether this is the spring force of the scale or simply the normal force of the floor.

The idea of apparent weight has important applications. Astronauts are nearly crushed by their apparent weight during a rocket launch when a is much greater than g. Much of the thrill of amusement park rides, such as roller coasters, comes from rapid changes in your apparent weight.

FIGURE 5.12 The sensation of weight for a man at rest.

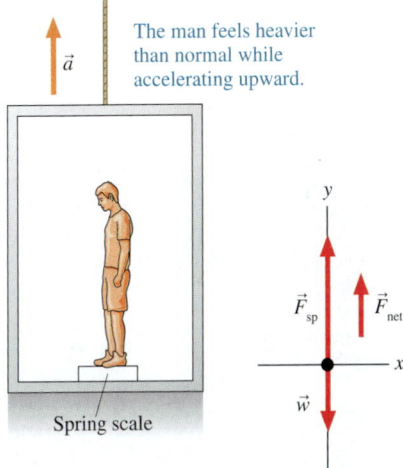

FIGURE 5.13 A man weighing himself in an accelerating elevator.

TRY IT YOURSELF

Physics students can't jump The next time you ride up in an elevator, try jumping in the air just as the elevator starts to rise. You'll feel like you can hardly get off the ground. This is because with $a_y > 0$ your apparent weight is *greater* than your actual weight; for an elevator with a large acceleration it's like trying to jump while carrying an extra 20 pounds. What will happen if you jump as the elevator slows at the top?

EXAMPLE 5.8 Apparent weight in an elevator

Anjay's mass is 70 kg. He's standing on a scale in an elevator. As the elevator stops, the scale reads 750 N. Had the elevator been moving up or down? If the elevator had been moving at 5.0 m/s, how long does it take to stop?

PREPARE The scale reading as he stops is his apparent weight, so $w_{app} = 750$ N. Because we know his mass m, we can then use Equation 5.9 to find the elevator's acceleration a_y. Then we can use kinematics to find the time it takes to stop the elevator.

SOLVE From Equation 5.9 we have $w_{app} = m(g + a_y)$, so that

$$a_y = \frac{w_{app}}{m} - g = \frac{750 \text{ N}}{70 \text{ kg}} - 9.80 \text{ m/s}^2 = 0.91 \text{ m/s}^2$$

This is a *positive* acceleration. If the elevator is stopping with a positive acceleration it must have been moving *down*, with a negative velocity.

To find the stopping time, we can use the kinematic equation

$$(v_y)_f = (v_y)_i + a_y \, \Delta t$$

to get

$$\Delta t = \frac{(v_y)_f - (v_y)_i}{a_y} = \frac{(0 \text{ m/s}) - (-5.0 \text{ m/s})}{0.91 \text{ m/s}^2} = 5.5 \text{ s}$$

Notice that we used −5.0 m/s as the initial velocity because the elevator was moving down before it stopped.

ASSESS Anjay's true weight is $mg = (70 \text{ kg})(9.8 \text{ m/s}^2) = 670$ N. Thus his apparent weight is *greater* than his true weight. You have no doubt experienced this sensation in an elevator that is stopping as it reaches the ground floor. If it had stopped while going up, you'd feel *lighter* than your true weight.

Weightlessness

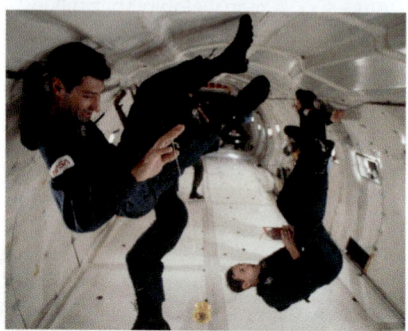

A weightless experience You probably wouldn't want to experience weightlessness in a falling elevator. But, as we learned in Chapter 3, objects undergoing projectile motion are in free fall as well. The special plane shown flies in the same parabolic trajectory as would a projectile with no air resistance. Objects inside, such as these passengers, are then moving along a perfect free fall trajectory. Just as for the man in the elevator, they then float with respect to the plane's interior. Such flights can last up to 30 seconds.

One last issue before leaving this topic: Suppose the elevator cable breaks and the elevator, along with the man and his scale, plunges straight down in free fall! What will the scale read? The acceleration in free fall is $a_y = -g$. When this acceleration is used in Equation 5.9, we find that $w_{app} = 0$! In other words, the man has *no sensation* of weight.

Think about this carefully. Suppose, as the elevator falls, the man inside releases a ball from his hand. In the absence of air resistance, as Galileo discovered, both the man and the ball would fall at the same rate. From the man's perspective, the ball would appear to "float" beside him. Similarly, the scale would float beneath him and not press against his feet. He is what we call *weightless*.

Surprisingly, "weightless" does *not* mean "no weight." An object that is **weightless** has no *apparent* weight. The distinction is significant. The man's weight is still mg, because gravity is still pulling down on him, but he has no *sensation* of weight as he free falls. The term "weightless" is a very poor one, likely to cause confusion because it implies that objects have no weight. As we see, that is not the case.

But isn't this exactly what happens to astronauts orbiting the earth? You've seen films of astronauts and various objects floating inside the Space Shuttle. If an astronaut tries to stand on a scale, it does not exert any force against her feet and reads zero. She is said to be weightless. But if the criterion to be weightless is to be in free fall, and if astronauts orbiting the earth are weightless, does this mean that they are in free fall? This is a very interesting question to which we shall return in Chapter 6.

> **STOP TO THINK 5.2**   An elevator that has descended from the 50th floor is coming to a halt at the 1st floor. As it does, your apparent weight is
>
> A. More than your true weight.   B. Less than your true weight.
> C. Equal to your true weight.   D. Zero.

5.4 Normal Forces

In Chapter 4 we saw that an object at rest on a table is subject to an upward force due to the table. This force is called the *normal force* because it is always directed normal, or perpendicular, to the surface of contact. As we saw, the normal force has its origin in the atomic "springs" that make up the surface. The harder the object bears down on the surface, the more these springs are compressed and the harder they push back. Thus the normal force *adjusts* itself so that the object stays on the surface without penetrating it. This fact is key in solving for the normal force.

EXAMPLE 5.9 Normal force on a pressed book

A 1.2 kg book lies on a table. The book is pressed down from above with a force of 15 N. What is the normal force acting on the book from the table below?

PREPARE The book is not moving and is thus in static equilibrium. We need to identify the forces acting on the book, and prepare a free-body diagram showing these forces. These steps are illustrated in Figure 5.14.

FIGURE 5.14 Finding the normal force of a book pressed from above.

SOLVE Because the book is in static equilibrium, the net force on it must be zero. The only forces acting are in the y-direction, so Newton's first law is

$$\sum F_y = n_y + w_y + F_y = n - w - F = 0$$

We learned in the last section that the weight force is $w = mg$. The weight of the book is thus

$$w = mg = (1.2 \text{ kg})(9.8 \text{ m/s}^2) = 12 \text{ N}$$

With this information, we see that the normal force exerted by the table is

$$n = F + w = 15 \text{ N} + 12 \text{ N} = 27 \text{ N}$$

ASSESS The magnitude of the normal force is *larger* than the weight of the book. From the table's perspective, the extra force from the hand pushes the book further into the atomic springs of the table. These springs then push back harder, giving a larger normal force.

A common situation is that of an object on a ramp or incline. If friction is neglected, there are only two forces acting on the object: gravity and the normal force. However, we need to carefully work out the components of these two forces. Figure 5.15a shows how. Be sure you avoid the two common errors shown in Figure 5.15b.

(a) Analyzing forces on an incline

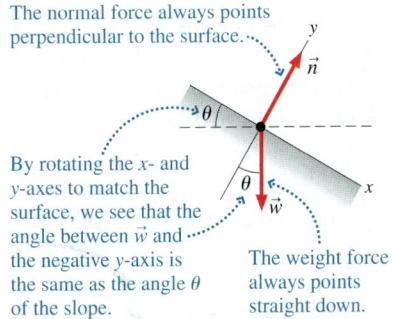

The normal force always points perpendicular to the surface.

By rotating the x- and y-axes to match the surface, we see that the angle between $\vec{w}$ and the negative y-axis is the same as the angle θ of the slope.

The weight force always points straight down.

$\vec{w}$ can be decomposed into x- and y-components.

$w_x = w \sin\theta$

$w_y = -w \cos\theta$

w_y is negative because $\vec{w}$ points in the negative y-direction.

(b) Two common mistakes to avoid

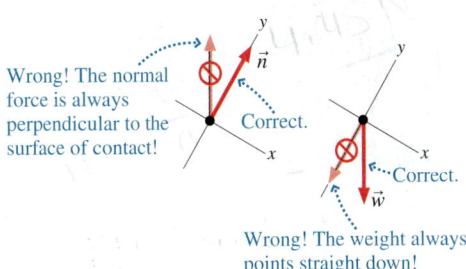

Wrong! The normal force is always perpendicular to the surface of contact!

Correct.

Correct.

Wrong! The weight always points straight down!

FIGURE 5.15 The forces on an object on an incline.

EXAMPLE 5.10 Acceleration of a downhill skier

A skier slides down a steep slope of 27° on ideal, frictionless snow. What is his acceleration?

PREPARE Figure 5.16 on the next page is a visual overview. We choose a coordinate system tilted so that the x-axis points down the slope. This greatly simplifies the analysis, because with this choice $a_y = 0$ (the skier does not move in the y-direction at all). The free-body diagram is based on the information in Figure 5.15.

Continued

FIGURE 5.16 Visual overview of a downhill skier.

SOLVE We can now use Newton's second law in component form to find the skier's acceleration. We have

$$\sum F_x = w_x + n_x = ma_x$$

$$\sum F_y = w_y + n_y = ma_y$$

Because $\vec{n}$ points directly in the positive y-direction, $n_y = n$ and $n_x = 0$. Figure 5.15a showed the important fact that the angle between $\vec{w}$ and the negative y-axis is the *same* as the slope angle θ. With this information, the components of $\vec{w}$ are $w_x = w\sin\theta = mg\sin\theta$ and $w_y = -w\cos\theta = -mg\cos\theta$, where we used the

fact that $w = mg$. With these components in hand, Newton's second law becomes

$$\sum F_x = w_x + n_x = mg\sin\theta = ma_x$$

$$\sum F_y = w_y + n_y = -mg\cos\theta + n = ma_y = 0$$

In the second equation we used the fact that $a_y = 0$. The m cancels in the first Newton's-law equation, leaving us with

$$a_x = g\sin\theta$$

This is the expression for acceleration on a surface that we presented, without proof, in Chapter 3. Now we've justified our earlier assertion. We can use this to calculate the skier's acceleration:

$$a_x = g\sin\theta = (9.8 \text{ m/s}^2)\sin(27°) = 4.4 \text{ m/s}^2$$

ASSESS The skier's acceleration falls between $a_x = 0$, which it would be on a horizontal surface, and $a_x = 9.80 \text{ m/s}^2$, its value for a vertical drop. This seems reasonable. Notice that the mass cancelled out, so we didn't need to know the skier's mass.

5.5 Friction

In everyday life, friction is everywhere. Friction is absolutely essential for many things we do. Without friction you could not walk, drive, or even sit down (you would slide right off the chair!). It is sometimes useful to think about idealized frictionless situations, but it is equally necessary to understand a real world where friction is present. Although friction is a complicated force, many aspects of friction can be described with a simple model.

Static Friction

(a) Force identification

(b) Free-body diagram

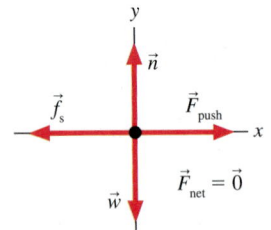

FIGURE 5.17 Static friction keeps an object from slipping.

Chapter 4 defined *static friction* $\vec{f}_s$ as the force on an object that keeps it from slipping. Figure 5.17a shows a woman pushing on a box with horizontal force $\vec{F}_{push}$. If the box remains at rest, "stuck" to the floor, it must be because of a static friction force pushing back to the left. The box is in static equilibrium, so, as shown in 5.17b, the static friction must exactly balance the pushing force:

$$f_s = F_{push}$$

To determine the *direction* of $\vec{f}_s$, decide which way the object would move if there were no friction. The static friction force $\vec{f}_s$ then points in the opposite direction, to prevent the motion. Determining the *magnitude* of $\vec{f}_s$ is a bit trickier. Unlike weight, which has the precise and unambiguous magnitude $w = mg$, the magnitude f_s of the static friction force depends on how hard you push. As shown in Figures 5.18a and 5.18b on the next page, the harder the woman pushes, the harder the friction force from the floor pushes back. If she reduces her pushing force, the friction force will automatically be reduced to match. Static friction acts in *response* to an applied force.

But there's clearly a limit to how big f_s can get. If you push hard enough, the object will slip and start to move. In other words, the static friction force has a *maximum* possible magnitude $f_{s\,max}$, as illustrated in Figure 5.18c. Experiments with friction (first done by Leonardo da Vinci) show that $f_{s\,max}$ is proportional to the magnitude of the normal force between the surface and the object. That is,

$$f_{s\,max} = \mu_s n \qquad (5.10)$$

where μ_s is called the **coefficient of static friction.** The coefficient is a number that depends on the materials of which the object and the surface are made. The higher the coefficient of static friction, the greater the "stickiness" between the object and the surface, and the harder it is to make the object slip. Table 5.1 shows some typical values of coefficients of friction. It is to be emphasized that these are only approximate. The exact value of the coefficient depends on the roughness, cleanliness, and dryness of the surfaces.

NOTE ▶ Equation 5.10 does *not* say $f_s = \mu_s n$. The value of f_s depends on the force or forces that static friction has to balance to keep the object from moving. It can have any value from zero up to, but not exceeding, $\mu_s n$. ◀

It is interesting to note that $f_{s\,max}$ does *not* depend on the contact area between the object and the surface on which it rests.

So our rules for static friction are

- The direction of static friction is such as to oppose motion.
- The magnitude f_s of static friction adjusts itself so that the net force is zero and the object doesn't move.
- The magnitude of static friction cannot exceed the maximum value $f_{s\,max}$ given by Equation 5.10. If the friction force needed to keep the object stationary is larger than $f_{s\,max}$, the object slips and starts to move.

Kinetic Friction

Once the box starts to slide, as in Figure 5.19, the static friction force is replaced by a kinetic (or sliding) friction force $\vec{f_k}$. Kinetic friction is in some ways simpler than static friction: The direction of $\vec{f_k}$ is always opposite to the direction in which an object slides across the surface, and experiments show that kinetic friction, unlike static friction, has a nearly *constant* magnitude, given by

$$f_k = \mu_k n \qquad (5.11)$$

where μ_k is called the **coefficient of kinetic friction.** Equation 5.11 also shows that kinetic friction, like static friction, is proportional to the magnitude of the

(a) Pushing gently: friction pushes back gently.

$\vec{f_s}$ balances $\vec{F}_{push}$ and the box does not move.

(b) Pushing harder: friction pushes back harder.

$\vec{f_s}$ grows as $\vec{F}_{push}$ increases, but they still cancel and the box remains at rest.

(c) Pushing harder still: $\vec{f_s}$ is now pushing back as hard as it can.

Now the magnitude of f_s has reached its maximum value $f_{s\,max}$. If $\vec{F}_{push}$ gets any bigger, the forces will *not* cancel and the box will start to move.

FIGURE 5.18 Static friction acts in *response* to an applied force.

TABLE 5.1 Coefficients of friction

Materials	Static μ_s	Kinetic μ_k	Rolling μ_r
Rubber on concrete	1.00	0.80	0.02
Steel on steel (dry)	0.80	0.60	0.002
Steel on steel (lubricated)	0.10	0.05	
Wood on wood	0.50	0.20	
Wood on snow	0.12	0.06	
Ice on ice	0.10	0.03	

The kinetic friction force is the same, no matter how fast the object slides.

FIGURE 5.19 The kinetic friction force is *opposite* to the direction of motion.

FIGURE 5.20 The bottom of the wheel is stationary.

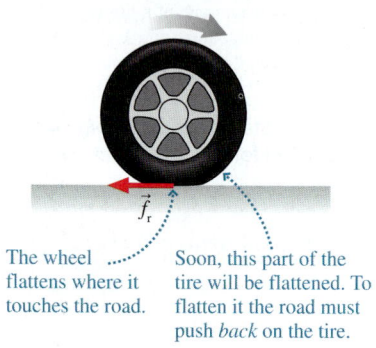

The wheel flattens where it touches the road.

Soon, this part of the tire will be flattened. To flatten it the road must push *back* on the tire.

FIGURE 5.21 Rolling friction is due to deformation of a wheel.

normal force n. Notice that **the magnitude of the kinetic friction force does not depend on how fast the object is sliding.**

Table 5.1 includes typical values of μ_k. You can see that $\mu_k < \mu_s$, which explains why it is easier to keep a box moving than it was to start it moving.

Rolling Friction

If you slam on the brakes hard enough, your car tires slide against the road surface and leave skid marks. This is kinetic friction, because the tire and the road are *sliding* against each other. A wheel *rolling* on a surface also experiences friction, but not kinetic friction: The portion of the wheel that contacts the surface is stationary with respect to the surface, not sliding. The photo in Figure 5.20 was taken with a stationary camera. Note how the part of the wheel touching the ground is not blurred, indicating that this part of the wheel is not moving with respect to the ground.

Textbooks draw wheels as circles, but no wheel is perfectly round. The weight of the wheel, and of any object supported by the wheel, causes the bottom of the wheel to flatten where it touches the surface, as Figure 5.21 shows. As a wheel rolls forward, the leading part of the tire must become deformed. This requires that the road push *backward* on the tire. In this way the road causes a backward force, even without slipping between the tire and the road.

The force of this *rolling friction* can be calculated in terms of a **coefficient of rolling friction** μ_r:

$$f_r = \mu_r n \qquad (5.12)$$

with the *direction* of the force opposing the direction of motion. Thus rolling friction acts very much like kinetic friction, but values of μ_r (see Table 5.1) are much less than values of μ_k. This is why it is easier to roll an object on wheels than to slide it.

STOP TO THINK 5.3 Rank in order, from largest to smallest, the size of the friction forces $\vec{f}_A$ to $\vec{f}_E$ in the 5 different situations (one or more friction forces could be zero). The box and the floor are made of the same materials in all situations.

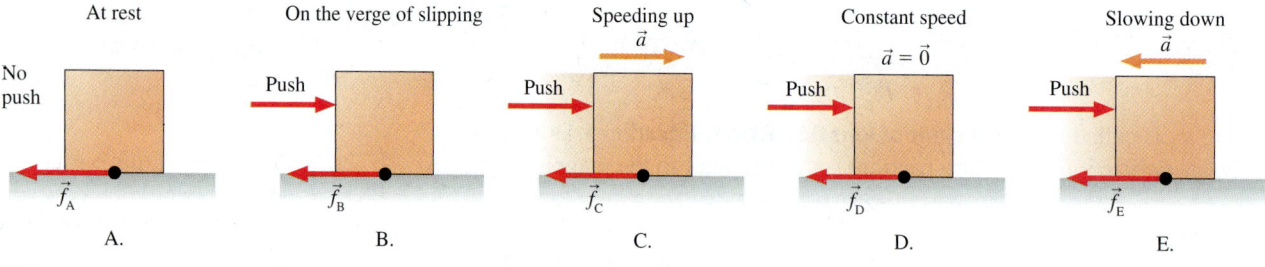

| At rest | On the verge of slipping | Speeding up $\vec{a}$ | Constant speed $\vec{a} = \vec{0}$ | Slowing down $\vec{a}$ |

No push   Push   Push   Push   Push

$\vec{f}_A$ $\vec{f}_B$ $\vec{f}_C$ $\vec{f}_D$ $\vec{f}_E$

A.   B.   C.   D.   E.

Working with Friction Forces

2.5, 2.6   **Activ**Physics

These ideas can be summarized in a *model* of friction:

Static: $\vec{f}_s = $ (magnitude $\leq f_{s\,max} = \mu_s n$, direction as necessary to prevent motion)

Kinetic: $\vec{f}_k = (\mu_k n$, direction opposite the motion)

Rolling: $\vec{f}_r = (\mu_r n$, direction opposite the motion)

(5.13)

LINEAR
p. 38

Here "motion" means "motion relative to the surface." The maximum value of static friction $f_{s\,max} = \mu_s n$ occurs at the point where the object slips and begins to move. Note that only one kind of friction force at a time can act on an object.

NOTE ▶ Equations 5.13 are a "model" of friction, not a "law" of friction. These equations provide a reasonably accurate, but not perfect, description of how friction forces act. For example, we've ignored the surface area of the object because surface area has little effect. Likewise, our model assumes that the kinetic friction force is independent of the object's speed. This is a fairly good, but not perfect, approximation. Equations 5.13 are a simplification of reality that works reasonably well, which is what we mean by a "model." They are not a "law of nature" on a level with Newton's laws. ◀

 TACTICS BOX 5.1 Working with friction forces Exercises 20, 21

❶ If the object is *not moving* relative to the surface it's in contact with, the friction force is **static friction.** Draw a free-body diagram of the object. The *direction* of the friction force is such as to oppose sliding of the object. Then use Problem-Solving Strategy 5.1 or 5.2 to solve for f_s. If f_s is greater than $f_{s\,max} = \mu_s n$, then static friction cannot hold the object in place. The assumption that the object is at rest is not valid, and you need to redo the problem using kinetic friction.

❷ If the object is *sliding* relative to the surface, then **kinetic friction** is acting. From Newton's second law, find the normal force n. Equation 5.13 then gives the magnitude and direction of the friction force.

❸ If the object is *rolling* along the surface, then **rolling friction** is acting. From Newton's second law, find the normal force n. Equation 5.13 then gives the magnitude and direction of the friction force.

Optimized braking If you slam on your brakes, your wheels will lock up and you'll go into a skid. Then it is the *kinetic* friction force between the road and your tires that slows your car to a halt. If, however, you apply the brakes such that you don't quite skid and your tires continue to roll, the force stopping you is the *static* friction force between the road and your tires. This is a better way to brake, because the maximum static friction force is always larger than the kinetic friction force. *Antilock braking systems* (ABS) automatically do this for you when you slam on the brakes, stopping you in the shortest possible distance.

EXAMPLE 5.11 Finding the force to push a box

Carol pushes a 10.0 kg wood box across a wood floor at a steady speed of 2.0 m/s. How much force does Carol exert on the box?

PREPARE Let's assume the box slides to the right. In this case, a kinetic friction force $\vec{f}_k$ opposes the motion by pointing to the left. In Figure 5.22 we identify the forces acting on the box and construct a free-body diagram.

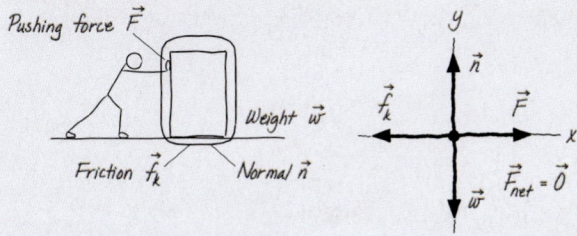

FIGURE 5.22 Forces on a box being pushed across a floor.

SOLVE The box is moving at a constant speed, so it is in dynamic equilibrium with $\vec{F}_{net} = \vec{0}$. This means that the x- and y-components of the net force must each be zero:

$$\sum F_x = n_x + w_x + F_x + (f_k)_x = 0 + 0 + F - f_k = 0$$

$$\sum F_y = n_y + w_y + F_y + (f_k)_y = n - w + 0 + 0 = 0$$

In the first equation, the x-component of $\vec{f}_k$ is equal to $-f_k$ because $\vec{f}_k$ is directed to the left. Similarly, $w_y = -w$ because the weight force points down.

From the first equation, we see that Carol's pushing force is $F = f_k$. To evaluate this, we need f_k. Here we can use our model for kinetic friction:

$$f_k = \mu_k n$$

Because the friction is wood sliding on wood, we can use Table 5.1 to find $\mu_k = 0.20$. Further, we can use the second Newton's-law equation to find that the normal force is $n = w = mg$. Thus

$$F = f_k = \mu_k n = \mu_k mg$$
$$= (0.20)(10.0 \text{ kg})(9.80 \text{ m/s}^2) = 20 \text{ N}$$

This is the force that Carol needs to apply to the box to keep it moving at a steady speed.

ASSESS The speed of 2.0 m/s with which Carol pushes the box does not enter into the answer. This is because our model of kinetic friction does not depend on the speed of the sliding object.

CONCEPTUAL EXAMPLE 5.2 **To push or pull a lawn roller?**

A lawn roller is a heavy cylinder used to flatten a bumpy lawn, as shown in Figure 5.23. Is it easier to push or pull such a roller? Which is more effective for flattening the lawn, pushing or pulling?

FIGURE 5.23 Pushing and pulling a lawn roller.

REASON Figure 5.24 shows free-body diagrams for the two cases. We assume that the roller is pushed at a constant speed so that it is in dynamic equilibrium with $\vec{F}_{net} = \vec{0}$. Because the roller does not move in the y-direction, the y-component of the net force must be zero. According to our model, the magnitude f_r of rolling friction is proportional to the magnitude n of the normal force. If we *push* on the roller, our pushing force $\vec{F}$ will have a downward y-component. To compensate for this, the normal force must increase and, because $f_r = \mu_r n$, the rolling friction will increase as well. This makes the roller harder to move. If we *pull* on the roller, the now upward y-component of

$\vec{F}$ will lead to a *reduced* value of n and hence of f_r. Thus the roller is easier to pull than to push.

FIGURE 5.24 Free-body diagrams for the lawn roller.

However, the purpose of the roller is to flatten the soil. If the normal force $\vec{n}$ of the ground on the roller is larger, by Newton's third law the force of the roller on the ground will be larger as well. So for smoothing your lawn, it's better to push.

ASSESS You've probably experienced this effect while using an upright vacuum cleaner. The vacuum is harder to push on the forward stroke than when drawing it back.

EXAMPLE 5.12 **How to dump a file cabinet**

A 50.0 kg steel file cabinet is in the back of a dump truck. The truck's bed, also made of steel, is slowly tilted. What is the size of the static friction force on the cabinet when the bed is tilted 20°? At what angle will the file cabinet begin to slide?

PREPARE We'll use our model of static friction. The file cabinet will slip when the static friction force reaches its maximum possible value $f_{s\,max}$. Figure 5.25 shows the visual overview when the truck bed is tilted at angle θ. We can make the analysis easier if we tilt the coordinate system to match the bed of the truck. To prevent the file cabinet from slipping, the static friction force must point *up* the slope.

SOLVE Before it slips, the file cabinet is in static equilibrium. Newton's first law reads

$$\sum F_x = n_x + w_x + (f_s)_x = 0$$
$$\sum F_y = n_y + w_y + (f_s)_y = 0$$

From the free-body diagram we see that f_s has only a negative x-component and that n has only a positive y-component. We also have $w_x = w\sin\theta$ and $w_y = -w\cos\theta$. Thus the first law becomes

$$\sum F_x = w\sin\theta - f_s = mg\sin\theta - f_s = 0$$
$$\sum F_y = n - w\cos\theta = n - mg\cos\theta = 0$$

Known	
$\mu_s = 0.80$	$m = 50.0$ kg
$\mu_k = 0.60$	

Find

f_s when $\theta = 20°$
θ at which cabinet slips

Normal $\vec{n}$
Friction $\vec{f}_s$ Weight $\vec{w}$

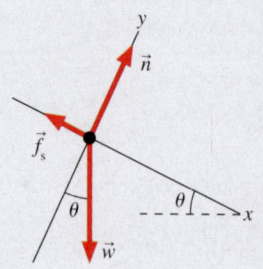

FIGURE 5.25 Visual overview of a file cabinet in a tilted dump truck.

where we've used $w = mg$. The x-component equation allows us to determine the size of the static friction force when $\theta = 20°$:

$$f_s = mg\sin\theta = (50.0 \text{ kg})(9.80 \text{ m/s}^2)\sin 20° = 168 \text{ N}$$

This value does not require knowing μ_s. The coefficient of static friction only enters when we want to find the angle at which the file cabinet slips. Slipping occurs when the static friction reaches its maximum value

$$f_s = f_{s\,max} = \mu_s n$$

From the y-component of Newton's second law we see that $n = mg\cos\theta$. Consequently,

$$f_{s\,max} = \mu_s mg\cos\theta$$

The x-component of the second law gave

$$f_s = mg\sin\theta$$

Setting $f_s = f_{s\,max}$ then gives

$$mg\sin\theta = \mu_s mg\cos\theta$$

The mg in both terms cancels, and we find

$$\frac{\sin\theta}{\cos\theta} = \tan\theta = \mu_s$$

$$\theta = \tan^{-1}\mu_s = \tan^{-1}(0.80) = 39°$$

ASSESS Steel doesn't slide all that well on unlubricated steel, so a fairly large angle is not surprising. The answer seems reasonable. It is worth noting that $n = mg\cos\theta$ in this example. A common error is to use simply $n = mg$. Be sure to evaluate the normal force within the context of each specific problem.

Causes of Friction

It is worth a brief pause to look at the *causes* of friction. All surfaces, even those quite smooth to the touch, are very rough on a microscopic scale. When two objects are placed in contact, they do not make a smooth fit. Instead, as Figure 5.26 shows, the high points on one surface become jammed against the high points on the other surface while the low points are not in contact at all. Only a very small fraction (typically 10^{-4}) of the surface area is in actual contact. The amount of contact depends on how hard the surfaces are pushed together, which is why friction forces are proportional to n.

For an object to slip, you must push it hard enough to force these contact points over each other. Once the two surfaces are sliding against each other, their high points undergo constant collisions, deformations, and even brief bonding that leads to the resistive force of kinetic friction.

5.6 Drag

Fluids—liquids and gases—exert a drag force on objects as they move through the fluid. You experience drag forces every day as you jog, bicycle, ski, or drive your car. The drag force is especially important for the skydiver at the beginning of the chapter and for unicellular animals swimming through water.

The drag force $\vec{D}$

■ Is opposite in direction to the velocity $\vec{v}$.
■ Increases in magnitude as the object's speed increases.

Experimental studies have found that the drag force depends on an object's speed in a complicated way. We'll look separately at drag forces in air and in liquids because different models apply.

Drag in Air

At relatively low speeds, the drag force in air is small and can usually be neglected, but drag plays an important role as speeds increase. Fortunately, we can use a fairly simple *model* of drag if the following three conditions are met:

■ The object's size (diameter) is between a few millimeters and a few meters.
■ The object's speed is less than a few hundred meters per second.
■ The object is moving through the air near the earth's surface.

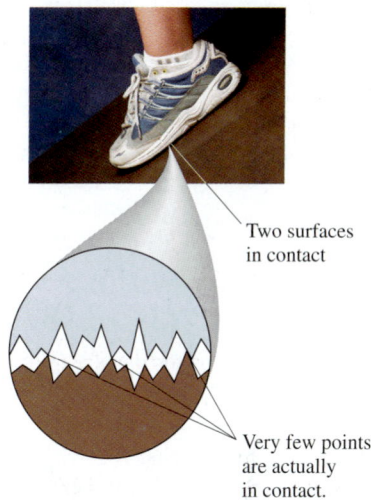

Two surfaces in contact

Very few points are actually in contact.

FIGURE 5.26 A microscopic view of friction.

These conditions are usually satisfied for balls, people, cars, and many other objects of the everyday world. Under these conditions, the drag force can be written

$$\vec{D} = \left(\tfrac{1}{2}C_{\mathrm{D}}\rho A v^2, \text{ direction opposite to the motion}\right) \qquad (5.14)$$

Drag force on an object of cross-section area A moving at speed v

QUADRATIC
p. 50

Here, ρ is the density of air ($\rho = 1.29$ kg/m^3 at sea level), A is the cross-section area of the object (in m^2), and the **drag coefficient** C_{D} depends on the details of the object's shape. However, the value of C_{D} for everyday moving objects is roughly 1/2, so a good approximation to the drag force is

$$D \approx \tfrac{1}{4}\rho A v^2 \qquad (5.15)$$

This is the expression for the magnitude of the drag force that we'll use in this chapter.

The size of the drag force in air is proportional to the *square* of the object's speed: If the speed doubles, the drag increases by a factor of *four*. This model of drag fails for objects that are very small (such as dust particles) or very fast (such as jet planes) or that move in other media (such as water).

Figure 5.27 shows that the area A in Equation 5.14 is the cross section of the object as it "faces into the wind." It's interesting to note that the magnitude of the drag force depends on the object's *size and shape* but not on its *mass*. This has important consequences for the motion of falling objects.

A is the cross-section area of the cyclist as seen from the front. This area is approximated by the rectangle shown, with area $A = h \times w$.

The cross-section area of a sphere is a circle. For this soccer ball, $A = \pi r^2$.

FIGURE 5.27 How to calculate the cross-section area A.

(a) At low speeds D is small and the ball falls with $a \approx g$.

(b) As v increases so does D. The net force and hence a get smaller.

(c) Eventually, v reaches a value such that $D = w$. Then the net force is zero and the ball falls at a constant speed.

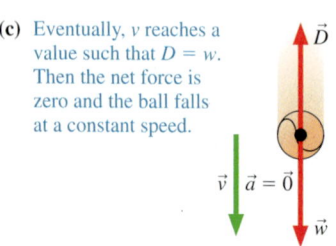

FIGURE 5.28 A falling object eventually reaches terminal speed.

Terminal Speed

Just after an object is released from rest, its speed is low and the drag force is small (as shown in Figure 5.28a). Because the net force is nearly equal to the weight, the object will fall with an acceleration only a little less than g. As it falls further, its speed and hence the drag force increase. Now the net force is smaller, so the acceleration is smaller (as shown in Figure 5.28b). It's still speeding up, but at a lower *rate*. Eventually the speed will increase to a point such that the magnitude of the drag force *equals* the weight (as shown in Figure 5.28c). The net force—and hence the acceleration—at this speed are then *zero,* and the object falls with a *constant* speed. The speed at which the exact balance between the upward drag force and the downward weight force causes an object to fall without acceleration is called the **terminal speed** v_{term}. Once an **object has reached terminal speed, it will continue falling at that speed until it hits the ground.**

It's straightforward to compute the terminal speed. It is the speed, by definition, at which $D = w$ or, equivalently, $\frac{1}{4}\rho A v^2 = mg$. This speed is then

$$v_{\text{term}} \approx \sqrt{\frac{4mg}{\rho A}} \tag{5.16}$$

This equation shows that a more massive object has a larger terminal speed than a less massive object of equal size. A 10-cm-diameter lead ball, with a mass of 6 kg, has a terminal speed of 150 m/s while a 10-cm-diameter Styrofoam ball, with a mass of 50 g, has a terminal speed of only 14 m/s.

EXAMPLE 5.13 Terminal speeds of a skydiver and a mouse

A skydiver and his pet mouse jump from a plane. Estimate their terminal speeds.

PREPARE To use Equation 5.16 we need to estimate the mass m and cross-section area A of both man and mouse. Figure 5.29 shows how. A typical skydiver might be 1.8 m long and 0.40 m wide ($A = 0.72$ m^2) with a mass of 75 kg, while a mouse has a mass of perhaps 20 g (0.020 kg) and is 7 cm long and 3 cm wide ($A = 0.07$ m $\times$ 0.03 m $= 0.0021$ m^2).

FIGURE 5.29 The cross-section areas of a skydiver and a mouse.

SOLVE We can use Equation 5.16 to find that for the skydiver

$$v_{\text{term}} \approx \sqrt{\frac{4mg}{\rho A}} = \sqrt{\frac{4(75 \text{ kg})(9.8 \text{ m/s}^2)}{(1.29 \text{ kg/m}^3)(0.72 \text{ m}^2)}} = 56 \text{ m/s}$$

This is roughly 130 mph. A higher speed can be reached by falling feet first or head first, which reduces the area A. Fortunately the skydiver can open his parachute, greatly increasing A. This brings his terminal speed down to a safe value.

For the mouse we have

$$v_{\text{term}} \approx \sqrt{\frac{4mg}{\rho A}} = \sqrt{\frac{4(0.020 \text{ kg})(9.8 \text{ m/s}^2)}{(1.29 \text{ kg/m}^2)(0.0021 \text{ m}^2)}} = 17 \text{ m/s}$$

The mouse has no parachute—nor does he need one! A mouse's terminal speed is low enough that he can fall from any height, even out of an airplane, and survive. Cats, too, have relatively low terminal speeds. In a study of cats that fell from high rises, over 90% survived—including one that fell 45 stories!

ASSESS The mouse survives the fall not only because of its lower terminal speed. The smaller an animal's body, the proportionally more robust it is. Further, a small animal's low mass and terminal speed mean that it has a very small *kinetic energy*, an idea we'll study in Chapter 10.

Although we've focused our analysis on falling objects, the same ideas apply to objects moving horizontally. If an object is thrown or shot horizontally, $\vec{D}$ causes the object to slow down. An airplane reaches its maximum speed, which is analogous to the terminal speed, when the drag is equal and opposite to the thrust: $D = F_{\text{thrust}}$. The net force is then zero and the plane cannot go any faster. The maximum speed of a passenger jet is about 550 mph.

We will continue to neglect drag unless a problem specifically calls for drag to be considered.

STOP TO THINK 5.4   The terminal speed of a Styrofoam ball is 15 m/s. Suppose a Styrofoam ball is shot straight down with an initial speed of 30 m/s. Which velocity graph is correct?

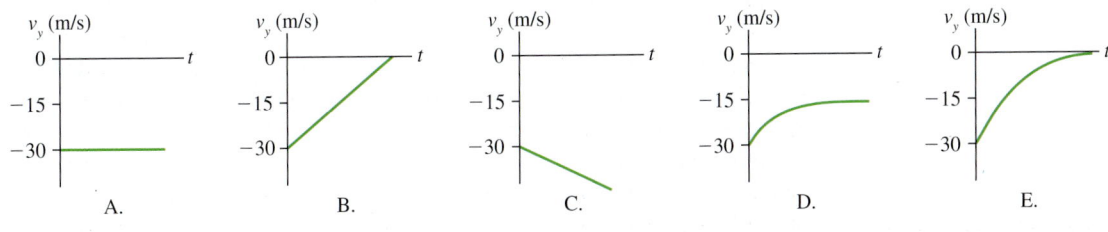

Drag in Liquids

The drag force in a liquid is quite different from that in air. Drag forces in air are largely the result of the object having to push the air out of its way as it moves. For an object moving slowly through a liquid, however, the drag force is mostly due to the *viscosity* of the liquid. We'll study viscosity in Chapter 13. For the moment, you can think of the viscosity of a fluid as a measure of how much resistance to flow the fluid has: Honey, which drizzles slowly out of its container, has a much higher viscosity than water, which flows fairly freely. Drag forces have particularly important effects on the motion of single-celled organisms in water, blood cells in the body, and many aquatic creatures.

The drag force in a liquid depends on the shape of the object, but there is a simple result called **Stokes's law** for the drag on a *sphere*. This law is useful because many biological objects, such as cells, can at least roughly be approximated as spheres. The drag on a sphere moving with speed v is

$$\vec{D} = (6\pi\eta rv, \text{ direction opposite to motion}) \qquad (5.17)$$

Stokes's law for the drag on a sphere of radius r moving at speed v

LINEAR
p. 38

Here, η is the viscosity of the liquid. (Water has a viscosity $\eta = 0.0010 \text{ N} \cdot \text{s/m}^2$ at 20° C.) Notice that the drag force in a liquid depends *linearly* on the speed, whereas drag in air depends on the *square* of the speed.

EXAMPLE 5.14   Drag force on a paramecium

A paramecium is a single-celled animal able to propel itself quite rapidly through water by using its *cilia*, rapidly beating hair-like fibers that ring its body. A typical paramecium has a diameter of 50 μm. What is the drag force on a paramecium swimming at a speed of 0.25 mm/s? If its mass is 1.0×10^{-11} kg, estimate how long it takes for the paramecium to come to a stop once it stops swimming.

PREPARE Although a real paramecium has an elongated shape, we'll model the paramecium as a sphere of radius 25 μm and use Stokes's law to find the drag force on it. Once the paramecium stops swimming, the drag force will be the only force acting, and we can use Newton's second law to find its acceleration and time to come to a stop.

SOLVE The drag force is

$$D = 6\pi\eta rv = (6\pi)(0.0010 \text{ N} \cdot \text{s/m}^2)(25 \times 10^{-6} \text{ m})$$
$$\times (0.25 \times 10^{-3} \text{ m/s}) = 1.2 \times 10^{-10} \text{ N}$$

Once the paramecium stops swimming, its acceleration is $a = F_{\text{net}}/m = -D/m = -(1.2 \times 10^{-10} \text{ N})/(1.0 \times 10^{-11} \text{ kg}) = -12 \text{ m/s}^2$. This acceleration is a little greater than g. We can estimate how long it takes for the paramecium to stop by using the definition $a = \Delta v/\Delta t$. (This is only an estimate, because the drag force and hence the acceleration decrease as the paramecium slows down.) This estimated stopping time is

$$\Delta t = \Delta v/a = \frac{-0.25 \times 10^{-3} \text{ m/s}}{-12 \text{ m/s}^2} = 2 \times 10^{-5} \text{ s} = 20 \text{ } \mu\text{s}$$

This is a *very* short time!

ASSESS In estimating the stopping time, we used the initial speed of the paramecium. Suppose its initial speed v were doubled. Then the drag force, which is proportional to the speed, would also double, and its acceleration a would double as well. But if both v and a double, $\Delta t = \Delta v/a$ would remain the same. Thus the stopping time of an object moving in a liquid is *independent of its initial speed.*

A paramecium, along with bacteria and other small organisms that live in water, lives in a very strange world where inertia—the tendency of a moving body to continue in motion—is negligible. To keep a shopping cart rolling you need to push it, but if you stop pushing it will continue to roll for some distance. A frictionless cart would roll forever at constant speed. In contrast, a paramecium comes to a halt essentially instantaneously, as soon as it stops actively swimming.

The reverse is also true: When it starts swimming again, it almost instantly reaches a constant speed, analogous to a terminal speed, at which the drag force $\vec{D}$ is equal and opposite to the paramecium's propulsion force $\vec{F}_{\text{propulsion}}$. Using Stokes's law for $\vec{D}$, we see that speed of swimming is directly proportional to the propulsion force:

$$v = \frac{F_{\text{propulsion}}}{6\pi\eta r}$$

If a small organism in water wants to go twice as fast, it has to double its propulsion force. (For objects without drag, doubling the force causes the *acceleration*, not the speed, to double.) Life in a world dominated by drag is very different from the world of accelerating objects in which we humans live.

5.7 Interacting Objects

Up to this point we have studied the dynamics of a single object subject to forces from the environment. In Example 5.11, for instance, a sliding box was acted upon by friction, normal, weight, and pushing forces that came from outside the box—from the floor, the earth, and the person pushing. As we've seen, such problems can be solved by an application of Newton's second law after all the forces have been identified.

Activ
Physics   2.7–2.9

But in Chapter 4 we found that real-world motion often involves two or more objects interacting with each other. We further found that forces always come in action/reaction *pairs* that are related by Newton's third law. To remind you, Newton's third law states:

- Every force occurs as one member of an action/reaction pair of forces. The two members of the pair always act on *different* objects.
- The two members of an action/reaction pair point in *opposite* directions and are *equal* in magnitude.

Our goal in this section is to learn how to apply the second *and* third laws to systems of interacting objects.

Acceleration Constraints

Newton's third law is one relationship needed to solve problems of interacting objects. In addition, we frequently have other information about the motion in a problem. For example, consider the two boxes in Figure 5.30. As long as they're touching, box A *has* to have exactly the same acceleration as box B. If they were to accelerate differently, either box B would take off on its own, or it would suddenly slow down and box A would run over it! Our problem implicitly assumes that neither of these is happening. Thus the two accelerations are *constrained* to be equal: $\vec{a}_A = \vec{a}_B$. A well-defined relationship between the accelerations of two or more systems is called an **acceleration constraint.** It is an independent piece of information that can help solve a problem.

In practice, we'll express acceleration constraints in terms of the x- and y-components of $\vec{a}$. Consider the car being towed in Figure 5.31. As long as the cable is under tension, the accelerations are constrained to be equal: $\vec{a}_C = \vec{a}_T$. This is one-dimensional motion, so for problem solving we would use just the x-components a_{Cx} and a_{Tx}. In terms of these components, the acceleration constraint is

$$a_{Cx} = a_{Tx} = a_x$$

Because the accelerations of both systems are equal, we can drop the subscripts C and T and call both of them a_x.

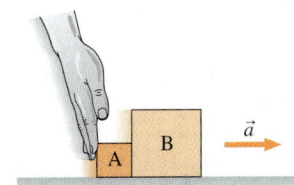
FIGURE 5.30 Two boxes moving together have the same acceleration.

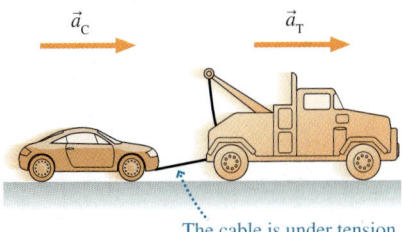
The cable is under tension.

FIGURE 5.31 The car and the truck have the same acceleration.

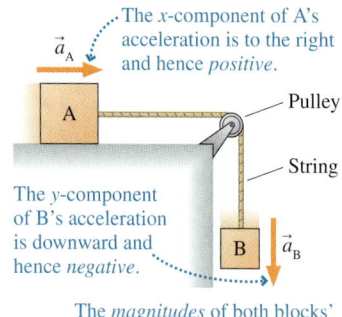

The *x*-component of A's acceleration is to the right and hence *positive.*

$\vec{a}_A$

A

Pulley

String

The *y*-component of B's acceleration is downward and hence *negative.*

B $\vec{a}_B$

The *magnitudes* of both blocks' accelerations are equal.

FIGURE 5.32 The string constrains the two systems to accelerate together.

Don't assume the accelerations of A and B will always have the same sign. Consider blocks A and B in Figure 5.32. The blocks are connected by a string, so they are constrained to move together and their accelerations have equal magnitudes. But A has a positive acceleration (to the right) in the *x*-direction while B has a negative acceleration (downward) in the *y*-direction. Thus the acceleration constraint on the components of the acceleration is

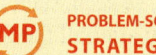

$$a_{Ax} = -a_{By}$$

This relationship does *not* say that a_{Ax} is a negative number. It is simply a relational statement, saying that a_{Ax} is (-1) times whatever a_{By} happens to be. In fact, the acceleration a_{By} in Figure 5.32 is a negative number, so a_{Ax} is actually positive. In many problems, the signs of a_{Ax} and a_{By} may not be known until the problem is solved, but the *relationship* is known from the beginning.

A Revised Strategy for Interacting-Object Problems

Problems of interacting objects can be solved with a few modifications to the basic Problem-Solving Strategy 5.2 we developed earlier in this chapter. A revised problem-solving strategy is shown below.

(MP)   **PROBLEM-SOLVING STRATEGY 5.3**   **Interacting-object problems**

PREPARE   Identify the interacting objects that make up the system. Everything else is part of the environment. Make simplifying assumptions.

Prepare a visual overview: Make a sketch of the situation. Define symbols and identify what the problem is trying to find. Identify all forces acting on each object and all action/reaction pairs. Draw a *separate* free-body diagram for each object and connect the force vectors of action/reaction pairs with dotted lines. Use subscript labels to distinguish forces, such as $\vec{n}$ and $\vec{w}$, that act independently on more than one object. Identify any acceleration constraints and add these to your list of values.

SOLVE   Use Newton's second and third laws:

- Write Newton's second law in component form for each object. Find the force components from the free-body diagrams.
- Equate the magnitudes of action/reaction pairs.
- If relevant, include the acceleration constraints and the friction model.
- Solve for the unknown forces or acceleration.

ASSESS   Check that your result has the correct units, is reasonable, and answers the question.

NOTE ▶ Two steps are especially important when drawing the free-body diagrams. First, draw a *separate* diagram for each object. They need not have the same coordinate system. Second, show only the forces acting *on* that object. The force $\vec{F}_{A\,on\,B}$ goes on the free-body diagram of Object B, but $\vec{F}_{B\,on\,A}$ goes on the diagram of Object A. The two members of an action/reaction pair *always* appear on two different free-body diagrams—*never* on the same diagram. ◀

You might be puzzled that the Solve step calls for the use of the third law to equate just the *magnitudes* of action/reaction forces. What about the "opposite in direction" part of the third law? You have already used it! Your free-body diagrams should show the two members of an action/reaction pair to be opposite in direction, and that information will have been utilized in writing the second-law equations. Because the directional information has already been used, all that is left is the magnitude information.

EXAMPLE 5.15 Pushing two blocks

Figure 5.33 shows a 5.0 kg block A being pushed with a 3.0 N force. In front of this block is a 10 kg block B; the two blocks move together. What force does block A exert on block B?

FIGURE 5.33 Two blocks are pushed by a hand.

PREPARE The visual overview of Figure 5.34 lists known information, what we want to find, and— new to problems of interacting objects—the acceleration constraint. Because the blocks move together in the positive x-direction, they both must have the same x-component of acceleration:

$$a_{Ax} = a_{Bx} = a_x$$

We've identified all the forces acting on blocks A and B, then drawn *separate* free-body diagrams. Because both blocks have a weight force and a normal force, we used subscripts A and B to distinguish between them. A key step is to include the forces of interaction between the two blocks, $\vec{F}_{A\,on\,B}$ and $\vec{F}_{B\,on\,A}$. Notice that $\vec{F}_{A\,on\,B}$ is drawn as acting on box B; it is the force *of* A *on* B. Because action/reaction pairs act in opposite directions, force $\vec{F}_{B\,on\,A}$ pushes *backward* on block A. Force vectors are always drawn on the free-body diagram of the object that *experiences* the force, not the object exerting the force.

SOLVE We begin by writing Newton's second law in component form for each block. Because the motion is only in the x-direction, we need only the x-component of the second law. For block A,

$$\sum F_x = (F_H)_x + (F_{B\,on\,A})_x = m_A a_{Ax}$$

The force components can be "read" from the free-body diagram, where we see $\vec{F}_H$ pointing to the right and $\vec{F}_{B\,on\,A}$ pointing to the left. Thus

$$F_H - F_{B\,on\,A} = m_A a_x$$

We used the acceleration constraint to write $a_{Ax} = a_x$. For B, we have

$$\sum F_x = (F_{A\,on\,B})_x = m_B a_{Bx}$$

or, again using the acceleration constraint,

$$F_{A\,on\,B} = m_B a_x$$

We have an additional piece of information: Newton's third law tells us that $F_{B\,on\,A} = F_{A\,on\,B}$. With this information, the two x-component equations become

$$F_{A\,on\,B} = m_B a_x$$
$$F_H - F_{A\,on\,B} = m_A a_x$$

Our goal is to find $F_{A\,on\,B}$, so we need to eliminate the unknown acceleration a_x. From the first equation, $a_x = F_{A\,on\,B}/m_B$. Substituting this into the second equation gives

$$F_H - F_{A\,on\,B} = \frac{m_A}{m_B} F_{A\,on\,B}$$

This can be solved for the force of block A on block B, giving

$$F_{A\,on\,B} = \frac{F_H}{1 + m_A/m_B} = \frac{3.0\,\text{N}}{1 + (5.0\,\text{kg})/(10\,\text{kg})} = \frac{3.0\,\text{N}}{1.5} = 2.0\,\text{N}$$

ASSESS The critical step in solving this problem was correctly identifying the forces and drawing the two free-body diagrams. The rest of the problem consists of writing the second-law equations directly from the free-body diagram and then using the new information of the acceleration constraint to solve for the desired quantity.

FIGURE 5.34 A visual overview of the two blocks.

Boxes P and Q are sliding to the right across a frictionless table. The hand H is slowing them down. The mass of P is larger than the mass of Q. Rank in order, from largest to smallest, the *horizontal* forces on P, Q, and H.

Frictionless surface

A. $F_{Q\,on\,H} = F_{H\,on\,Q} = F_{P\,on\,Q} = F_{Q\,on\,P}$
B. $F_{Q\,on\,H} = F_{H\,on\,Q} > F_{P\,on\,Q} = F_{Q\,on\,P}$
C. $F_{Q\,on\,H} = F_{H\,on\,Q} < F_{P\,on\,Q} = F_{Q\,on\,P}$
D. $F_{H\,on\,Q} = F_{H\,on\,P} > F_{P\,on\,Q}$

5.8 Ropes and Pulleys

2.10, 2.11

Many objects are connected by strings, ropes, cables, and so on. In single-particle dynamics, we defined *tension* as the force exerted on an object by a rope or string. Now we need to think more carefully about the string itself. Just what do we mean when we talk about the tension "in" a string or rope?

Tension Revisited

Figure 5.35 shows a man pulling on a rope attached to a wall. As we saw in Chapter 4, a rope or string can be thought of as a very stiff spring. When the man pulls, the rope stretches slightly. Like a small stretched spring, each piece of the rope then pulls on its neighboring piece. By Newton's third law, the force of any segment of the rope on its neighbor has the *same* magnitude as the force of the neighbor on that segment. The magnitude of this force—the force holding the rope together—is what we call the tension *in* the rope.

The hand pulls on the rope with a force $\vec{F}_{H\,on\,rope}$. The rope pulls back with a force $\vec{T}_{rope\,on\,H}$. $\vec{F}_{H\,on\,rope}$ and $\vec{T}_{rope\,on\,H}$ form an action/reaction pair, so the tension in the rope is $\vec{T}_{rope\,on\,H} = \vec{F}_{H\,on\,rope}$.

The rope pulls on the wall and the wall pulls back on the rope. These are an action/reaction pair.

Action/reaction pairs

At an imaginary cut anywhere in the rope, each half of the rope pulls on the other half with a force of equal magnitude. The magnitude of this force *is* the tension in the rope.

FIGURE 5.35 Tension forces pull in both directions within a rope.

Is the tension in a rope the *same* at all points, or is there more tension at one end than the other? A rope, like any other object, must obey Newton's second law: $\vec{F}_{net} = m_{rope}\vec{a}$. The stationary rope in Figure 5.35 has no acceleration, so $\vec{F}_{net} = 0$. The external forces acting on the rope are $\vec{F}_{H\,on\,rope}$ and $\vec{F}_{W\,on\,rope}$, so $(F_{net})_x = F_{W\,on\,rope} - F_{H\,on\,rope} = T_{rope\,on\,W} - T_{rope\,on\,H}$. Here we've used the third-law result from Figure 5.35 that the tension at the end of a rope equals the strength of the force pulling on the rope. Thus $(F_{net})_x = 0$ implies $T_{rope\,on\,W} = T_{rope\,on\,H}$. In other words, **the tension in a stationary rope pulls equally hard at both ends of the rope on whatever objects the rope is attached to.**

But the tensions at the two ends of an accelerating rope are *not* the same. Because acceleration requires $(F_{net})_x \neq 0$, the force pulling on one end of the rope must be larger than the force pulling on the other end. That is, there must be *difference* between the tensions at the two ends to supply the net force needed to accelerate the rope itself. However, suppose the rope were massless. If $m_{rope} = 0$, than $(F_{net})_x = m_{rope}a_x = 0$ whether the rope is accelerating or not. It takes no effort to accelerate a massless rope, so the tension *is* the same throughout the rope. While no real rope is massless, this is a very good approximation in the common situation where the mass of a string or rope is much less than the mass of the objects it pulls on. Unless a problem explicitly mentions the mass of a rope or string, we will assume it is "massless" and thus the tension at each end is the same.

This **massless rope (or string) approximation** can greatly simplify situations where objects are connected by ropes or strings. Consider blocks A and B in Figure 5.36. If the string is assumed to be massless, the tension at both ends is the same. But the tension is also the force with which the string pulls on an object, so we see that B is pulled to the left by the string with a tension force of the *same* magnitude with which A is pulled to the right. In fact, as far as the forces on A and B are concerned, we can omit the string entirely and pretend that A and B interact directly, with forces of equal magnitude but opposite. direction. We'll often use this idea in solving problems.

Two blocks connected by a string

We can omit the string if we assume it is massless.

Then it's *as if* A pulls directly on B, and B directly on A.

The forces on each block "from" the other block have the *same* magnitude.

FIGURE 5.36 The blocks act as if they are directly interacting.

CONCEPTUAL EXAMPLE 5.3 Pulling a rope

Figure 5.37a shows a student pulling horizontally with a 100 N force on a rope that is attached to a wall. In Figure 5.37b, two students in a tug-of-war pull on opposite ends of a rope with 100 N each. Is the tension in the second rope larger, smaller, or the same as that in the first?

FIGURE 5.37 Pulling on a rope. Which produces a larger tension?

REASON Surely pulling on a rope from both ends causes more tension than pulling on one end. Right? Before jumping to conclusions, let's analyze the situation carefully. The situation for rope 1 is the same as that analyzed in Figure 5.35, where we learned that the tension in a rope is equal to the magnitude of the force applied to its end. Thus the tension in rope 1 is 100 N, the force with which the student pulls on the rope.

To find the tension in the second rope, suppose we make an imaginary cut through the middle of the rope, as shown in Fig-

ure 5.38. Then we can analyze the forces acting on the segment of the rope to the left of this imaginary cut. The male student pulls on the left end of this segment with a force of magnitude $F_s = 100$ N. The right end of this segment is pulled on by the rest of the rope, to its right, with a tension force of magnitude T. Because this segment of the rope is in equilibrium, we must have $T - F_s = 0$, so the tension is $T = F_s = 100$ N. The tension in the rope has not changed! It is still 100 N.

ASSESS We didn't need to consider the second student. Evidently it makes no difference whether the rope is tied to a wall or is being pulled on by that student. When the rope is tied to the wall, we know that it pulls on the wall with a force equal in magnitude to its tension, or 100 N. By Newton's third law, the wall must pull back on the rope with an equal force, also 100 N. But the second student pulls with a force of 100 N. The forces are the same in both situations! The rope does not care whether it is pulled by a wall or by a hand. It experiences the same forces in both cases, so the rope's tension is the same 100 N in both.

FIGURE 5.38 Analysis of the tension force in rope 2.

Pulleys

Strings and ropes often pass over pulleys. The application might be as simple as lifting a heavy weight or as complex as the internal cable-and-pulley arrangement that precisely moves a robot arm. Figure 5.39 shows a simple situation in which block B drags block A across a frictionless table as it falls. As the string moves, static friction between the string and pulley causes the pulley to turn. If we assume that

- The string *and* the pulley are both massless, and
- There is no friction where the pulley turns on its axle,

The tension in the string is the same on both sides of the pulley.

String

Massless, frictionless pulley

FIGURE 5.39 An ideal pulley changes the direction in which a tension force acts, but not its magnitude.

then no net force is needed to accelerate the string or turn the pulley. In this case, the **tension in a massless string remains constant as it passes over a massless, frictionless pulley.** We'll assume such an ideal pulley for problems in this chapter. Later, when we study rotational motion in Chapter 8, we'll consider the effect of the pulley's mass.

EXAMPLE 5.16 Lifting a stage set

A 200 kg set used in a play is stored in the loft above the stage. The rope holding the set passes up and over a pulley, then is tied backstage. The director tells a 100 kg stagehand to lower the set. When he unties the rope, the set falls and the unfortunate man is hoisted into the loft. What is the stagehand's acceleration?

PREPARE The objects of interest are the stagehand M and the set S, which we will model as particles. Assume a massless rope and a massless, frictionless pulley. Figure 5.40 shows the visual overview. The man's acceleration a_{My} is positive, while the set's acceleration a_{Sy} is negative. These two accelerations have the same magnitude because the two objects are connected by a rope, but they have opposite signs. Thus the acceleration constraint is $a_{Sy} = -a_{My}$. Tension forces $\vec{T}_S$ and $\vec{T}_M$ are due to a massless rope going over an ideal pulley, so their magnitudes are the same and we can use the same symbol T for both magnitudes.

SOLVE From the two free-body diagrams, we can write Newton's second law in component form. For the man we have

$$\sum F_{My} = T_M - w_M = T_M - m_M g = m_M a_{My}$$

while for the set we have

$$\sum F_{Sy} = T_S - w_S = T_S - m_S g = m_S a_{Sy}$$

Only the y-equations are needed. Now we can use the acceleration constraint $a_{Sy} = -a_{My}$ and the fact that the two tension forces have equal magnitude T. Inserting these into the second-law equations gives

$$T - m_M g = m_M a_{My}$$
$$T - m_S g = -m_S a_{My}$$

These are simultaneous equations in the two unknowns T and a_{My}. We can solve for T in the first equation to get

$$T = m_M a_{My} + m_M g$$

Inserting this value of T into the second equation then gives

$$m_M a_{My} + m_M g - m_S g = -m_S a_{My}$$

which we can rewrite as

$$(m_S - m_M)g = (m_S + m_M)a_{My}$$

Finally, we can solve for the hapless stagehand's acceleration:

$$a_{My} = \frac{m_S - m_M}{m_S + m_M}g = \left(\frac{100\ \text{kg}}{300\ \text{kg}}\right) \times 9.80\ \text{m/s}^2 = 3.27\ \text{m/s}^2$$

This is also the acceleration with which the set falls. If the rope's tension was needed, we could now find it from $T = m_M a_{My} + m_M g$.

ASSESS If the stagehand weren't holding on, the set would fall with free-fall acceleration g. The stagehand acts as a *counterweight* to reduce the acceleration.

Since the rope is massless and the pulley ideal, the magnitudes of these two tensions are the same.

Known
$m_M = 100\ \text{kg}$
$m_S = 200\ \text{kg}$

Acceleration constraint
$a_{Sy} = -a_{My}$

Find
a_{My}

FIGURE 5.40 Visual overview for the stagehand and set.

EXAMPLE 5.17 A not-so-clever bank robbery

Bank robbers have pushed a 1000 kg safe to a second-story floor-to-ceiling window. They plan to break the window, then lower the safe 3.0 m to their truck. Not being too clever, they stack up 500 kg of furniture, tie a rope between the safe and the furniture, and place the rope over a pulley. Then they push the safe out the window. What is the safe's speed when it hits the truck? The coefficient of kinetic friction between the furniture and the floor is 0.50.

PREPARE This is essentially the situation that we analyzed in Figure 5.39. The systems of interest are the safe S and the furniture F, which we will model as particles. We will assume a massless rope and a massless, frictionless pulley; the tension is then the same everywhere in the rope. Thus the tension forces $\vec{T}_S$ and $\vec{T}_F$ have the same magnitude, which we'll call T.

The visual overview in Figure 5.41 establishes a coordinate system and defines the symbols that will be needed to calcu-

FIGURE 5.41 Visual overview of the furniture and falling safe.

late the safe's motion. The safe and the furniture are tied together, so their accelerations have the same magnitude. The safe has a y-component of acceleration a_{Sy} that is negative because the safe accelerates in the $-y$-direction. The furniture has an x-component a_F that is positive. Thus the acceleration constraint is

$$a_{Fx} = -a_{Sy}$$

SOLVE We can write Newton's second law directly from the free-body diagrams. For the furniture,

$$\sum F_{Fx} = T_F - f_k = T - f_k = m_F a_{Fx} = -m_F a_{Sy}$$
$$\sum F_{Fy} = n - w_F = n - m_F g = 0$$

And for the safe,

$$\sum F_{Sy} = T_S - w_S = T - m_S g = m_S a_{Sy}$$

Notice how we used the acceleration constraint in the first equation. We also went ahead and made use of the fact that $T_S = T_F = T$. We have one additional piece of information, the model of kinetic friction:

$$f_k = \mu_k n = \mu_k m_F g$$

where we used the y-equation of the furniture to deduce that $n = m_F g$. Substitute this result for f_k into the x-equation of the

furniture, then rewrite the furniture's x-equation and the safe's y-equation:

$$T - \mu_k m_F g = -m_F a_{Sy}$$
$$T - m_S g = m_S a_{Sy}$$

We have succeeded in reducing our knowledge to two simultaneous equations in the two unknowns a_{Sy} and T. Subtract the second equation from the first to eliminate T:

$$(m_S - \mu_k m_F)g = -(m_S + m_F)a_{Sy}$$

Finally, solve for the safe's acceleration:

$$a_{Sy} = -\left(\frac{m_S - \mu_k m_F}{m_S + m_F}\right)g$$
$$= -\frac{1000 \text{ kg} - 0.5(500 \text{ kg})}{1000 \text{ kg} + 500 \text{ kg}} \times 9.80 \text{ m/s}^2 = -4.9 \text{ m/s}^2$$

Now we need to calculate the kinematics of the falling safe. Because the time of the fall is not known or needed, we can use

$$(v_y)_f^2 = (v_y)_i^2 + 2a_{Sy}\Delta y = 0 + 2a_{Sy}(y_f - y_i) = -2a_{Sy}y_i$$
$$(v_y)_f = \sqrt{-2a_{Sy}y_i} = \sqrt{-2(-4.9 \text{ m/s}^2)(3.0 \text{ m})} = 5.4 \text{ m/s}$$

The value of $(v_y)_f$ is negative, but we only needed to find the speed, so we took the absolute value. It seems unlikely that the truck will survive the impact of the 1000 kg safe!

In many situations we are interested in the force exerted on a pulley by a rope that passes through it. Consider the pulley shown in Figure 5.42. Even though the pulley is frictionless, the tension force still pulls on the pulley. Because the rope is under tension on *both* sides of the pulley, the pulley is subject to *two* tension forces, one from each side of the rope. The net force on the pulley due to the rope is then the vector sum of these two tension forces.

The two tension forces have the *same* magnitude.

FIGURE 5.42 Tension forces on a pulley.

EXAMPLE 5.18 Placing a leg in traction

For serious fractures of the leg, the leg may need to have a stretching force applied to it to keep contracting leg muscles from forcing the broken bones together too hard. This is often done using *traction,* an arrangement of a rope, a weight, and pulleys such as that shown in Figure 5.43. For this patient, the horizontal component of the force on the leg—the component that stretches the leg—must have a magnitude of 50 N. In order to apply this force, what must be the weight w?

FIGURE 5.43 A leg in traction.

PREPARE To solve this problem, we need to apply three concepts that we've learned about ropes and pulleys:

1. A rope pulls on what it's attached to with a force equal to the tension in the rope. So the weight is pulled up with a force of magnitude T while being pulled down by its own weight w. But because it's in equilibrium, $w = T$.
2. The tension in a rope is unchanged when it goes around the massless, frictionless pulleys we'll assume here. So T is the same everywhere along the rope.
3. As we just learned, the rope passing through the pulley attached to the foot pulls in two directions, upwards at 70° and downwards at 20°.

We need the patient's leg to be stretched horizontally by a force of magnitude 50 N, so $(F_{\text{PonL}})_x = 50$ N, where P stands for pulley and L for leg. Given that the pulley is in static equilibrium, Newton's third law gives us

$$(F_{\text{PonL}})_x = (F_{\text{LonP}})_x = (F_{\text{RonP}})_x = 50 \text{ N}$$

where R stands for rope. That is, the horizontal component of the sum of the tension forces of the rope on the pulley must also be 50 N. Figure 5.44 shows the forces on the pulley due to the tension in the rope. (This is not a true free-body diagram because we don't show the force on the pulley due to the leg).

FIGURE 5.44 Forces on the pulley due to tension in the rope.

SOLVE From Figure 5.44 we can see that

$$(F_{\text{RonP}})_x = T\cos\theta_1 + T\cos\theta_2$$

From our discussion above, we know that the two tensions are equal to w, and that $(F_{\text{RonP}})_x = 50$ N. Thus we have

$$(F_{\text{RonP}})_x = w(\cos\theta_1 + \cos\theta_2) = 50 \text{ N}$$

which gives

$$w = \frac{50 \text{ N}}{(\cos\theta_1 + \cos\theta_2)} = \frac{50 \text{ N}}{(\cos 70° + \cos 20°)} = 39 \text{ N}$$

ASSESS Although the arrangement of the rope and pulleys looks complicated, the solution is straightforward using our rules for tension in ropes. Thus, for instance, the weight hanging on one end of the rope determines the tension, even though the other end of the rope helps support the unknown weight of the patient's leg.

Newton's three laws form the cornerstone of the science of mechanics. These laws allowed scientists to understand many diverse phenomena, from the motion of a raindrop to the orbits of the planets. These laws were so precise at predicting motion that they went unchallenged for well over two hundred years. At the beginning of the twentieth century, however, it began to be apparent that the laws of mechanics and the laws of electricity and magnetism were somehow inconsistent. Bringing these two apparently disconnected theories into harmony required the genius of a young patent clerk named Albert Einstein. In doing so, he shook the foundations not only of Newtonian mechanics but also of our very notions of space and time as well. We will continue to develop Newtonian mechanics for the next few chapters because of its tremendous importance to the physics of everyday life. But it's worth keeping in the back of your mind that Newton's laws aren't the ultimate statement about motion. Later in this textbook we'll reexamine motion and mechanics from the perspective of Einstein's theory of relativity.

STOP TO THINK 5.6 All three 50 kg blocks are at rest. Is the tension in rope 2 greater than, less than, or equal to the tension in rope 1?

SUMMARY

The goal of Chapter 5 has been to learn how to solve problems about motion in a straight line.

GENERAL STRATEGY

All examples in this chapter follow a three-part strategy. You'll become a better problem solver if you adhere to it as you do the homework problems. The *Dynamics Worksheets* in the *Student Workbook* will help you structure your work in this way.

Equilibrium Problems

Object at rest or moving at constant velocity.

PREPARE Make simplifying assumptions.

- Check that the object is either at rest or moving with constant velocity.
- Identify forces and show them on a free-body diagram.

SOLVE Use Newton's first law in component form:

$$\sum F_x = 0$$
$$\sum F_y = 0$$

"Read" the components from the free-body diagram

ASSESS Is your result reasonable?

Dynamics Problems

Object accelerating.

PREPARE Make simplifying assumptions.
Make a **visual overview:**

- Sketch a pictorial representation
- Identify known quantities and what the problem is trying to find.
- Identify all forces and show them on a free-body diagram.

SOLVE Use Newton's second law in component form:

$$\sum F_x = ma_x \text{ and } \sum F_y = ma_y$$

"Read" the components of the vectors from the free-body diagram. If needed, use kinematics to find positions and velocities.

ASSESS Is your result reasonable?

Interacting Objects

Two or more objects interacting.

PREPARE Identify the system and the environment.
Make a **visual overview:**

- Sketch a pictorial representation.
- Identify any acceleration constraints.
- Identify all forces acting on *each* object.
- Identify action/reaction pairs between objects in the system.
- Draw separate free-body diagrams for each object. Connect action/reaction pairs with dotted lines and show them on the free-body diagram.

SOLVE Write Newton's second law for each object. Use Newton's third law to equate the magnitudes of action/reaction pairs. Include acceleration constraints.

ASSESS Is your result reasonable?

IMPORTANT CONCEPTS

Specific information about three important forces:

Weight $\vec{w} = (mg, \text{ downwards})$

Friction $\vec{f_s} = (0 \text{ to } \mu_s n, \text{ direction as necessary to prevent motion})$
$\vec{f_k} = (\mu_k n, \text{ direction opposite the motion})$
$\vec{f_r} = (\mu_r n, \text{ direction opposite the motion})$

Drag $\vec{D} \approx \left(\frac{1}{4}\rho A v^2, \text{ direction opposite to the motion}\right)$ for motion in air
$\vec{D} = (6\pi\eta r v, \text{ direction opposite the motion})$ for motion in liquid

Newton's laws are vector expressions. You must write them out by components:

$$(F_{net})_x = \sum F_x = ma_x \text{ or } 0$$
$$(F_{net})_y = \sum F_y = ma_y \text{ or } 0$$

APPLICATIONS

Apparent weight is the magnitude of the contact force supporting an object. It is what a scale would read, and it is your sensation of weight.

$$w_{app} = m(g + a_y)$$

Apparent weight equals your true weight $w = mg$ only when $a_y = 0$.

Acceleration constraints
Objects that are constrained to move together must have accelerations of equal magnitude: $a_A = a_B$.

This must be expressed in terms of components, such as

$$a_{Ax} = -a_{By}.$$

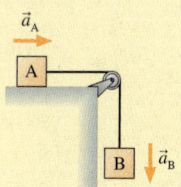

Strings and pulleys

- A string or rope pulls what it's connected to with a force equal to its tension.

- The tension in a rope is equal to the force pulling on the rope.

- The tension in a massless rope is the same at all points in the rope.

- Tension does not change when a rope passes over a massless, frictionless pulley.

(MP) For instructor-assigned homework, go to www.masteringphysics.com

Problem difficulty is labeled as | (straightforward) to |||| (challenging).

Problems labeled ✐ can be done on a Workbook Dynamics Worksheet; INT integrate significant material from earlier chapters; BIO are of biological or medical interest.

QUESTIONS

Conceptual Questions

1. An object is subject to two forces that do not point in opposite directions. Is it possible to choose their magnitudes so that the object is in equilibrium? Explain.

2. Are the objects described here in static equilibrium, dynamic equilibrium, or not in equilibrium at all?
   a. A girder is lifted at constant speed by a crane.
   b. A girder is lowered by a crane. It is slowing down.
   c. You're straining to hold a 200 lb barbell over your head.
   d. A jet plane has reached its cruising speed and altitude.
   e. A rock is falling into the Grand Canyon.
   f. A box in the back of a truck doesn't slide as the truck stops.

3. What forces are acting on you right now? What net force is acting on you right now?

4. Decide whether each of the following is true or false. Give a reason!
   a. The mass of an object depends on its location.
   b. The weight of an object depends on its location.
   c. Mass and weight describe the same thing in different units.

5. An astronaut takes his bathroom scale to the moon and then stands on it. Is the reading of the scale his true weight? Explain.

6. A light block of mass m and a heavy block of mass M are attached to the ends of a rope. A student holds the heavier block and lets the lighter block hang below it, as shown in Figure Q5.6. Then she lets go. Air resistance can be neglected.
   a. What is the tension in the rope while the blocks are falling, before either hits the ground?
   b. Would your answer be different if she had been holding the lighter block initially?

FIGURE Q5.6

7. Four balls are thrown straight up. Figure Q5.7 is a "snapshot" showing their velocities. They have the same size but different mass. Air resistance is negligible. Rank in order, from largest to smallest, the magnitudes of the net forces, F_{net1}, F_{net2}, F_{net3}, F_{net4}, acting on the balls. Some may be equal. Give your answer in the form A > B = C > D, and state your reasoning.

8. Suppose you attempt to pour out 100 g of salt, using a pan balance for measurements, while in an elevator that is accelerating upward. Will the quantity of salt be too much, too little, or the correct amount? Explain.

9. a. Can the normal force on an object be directed horizontally? If not, why not? If so, provide an example.
   b. Can the normal force on an object be directed downward? If not, why not? If so, provide an example.

10. A ball is thrown straight up. Taking the drag force of air into account, does it take longer for the ball to travel to the top of its motion or for it to fall back down again?

11. Three objects move through the air as shown in Figure Q5.11. Rank in order, from largest to smallest, the three drag forces D_1, D_2, and D_3. Some may be equal. Give your answer in the form A > B = C and state your reasoning.

FIGURE Q5.11

12. A skydiver is falling at her terminal speed. Right after she opens her parachute, which has a very large area, what is the direction of the net force on her?

13. Raindrops can fall at different speeds; some fall quite quickly, others quite slowly. Why might this be true?

14. An airplane moves through the air at a constant speed. The jet engine's thrust applies a force in the direction of motion. Reducing thrust will cause the plane to fly at a slower—but still constant—speed. Explain why this is so.

15. Is it possible for an object to travel in air faster than its terminal speed? If not, why not? If so, explain how this might happen.

For Questions 16 to 19, determine the reading of the spring scale.
- All objects are at rest.
- The strings, scales, and pulleys are massless and the pulleys are frictionless.
- The spring scale reads in kg.

16.   17.   18.

Scale

5 kg

FIGURE Q5.16

5 kg

5 kg

FIGURE Q5.17

5 kg

5 kg

FIGURE Q5.18

FIGURE Q5.7

19.

FIGURE Q5.19

For Questions 20 through 23, write the acceleration constraint in terms of *components*. For example, write $a_{1x} = a_{2y}$, if that is the appropriate answer, rather than $a_1 = a_2$.

20.

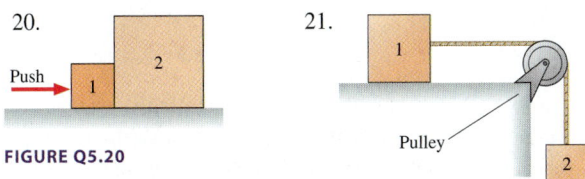

FIGURE Q5.20

21.

FIGURE Q5.21

22.

FIGURE Q5.22

23.

FIGURE Q5.23

Multiple-Choice Questions

24. | A 2.0 kg ball is suspended by two light strings as shown in Figure Q5.24. What is the tension T in the angled string?
   A. 9.5 N   B. 15 N
   C. 20 N   D. 26 N
   E. 30 N

   FIGURE Q5.24

25. | While standing in a low tunnel, you raise your arms and push against the ceiling with a force of 100 N. Your mass is 70 kg.
   a. What force does the ceiling exert on you?
      A. 10 N   B. 100 N   C. 690 N
      D. 790 N   E. 980 N
   b. What force does the floor exert on you?
      A. 10 N   B. 100 N   C. 690 N
      D. 790 N   E. 980 N

26. | A 5.0 kg dog sits on the floor of an elevator that is accelerating *downward* at 1.20 m/s².
   a. What is the magnitude of the normal force of the elevator floor on the dog?
      A. 34.2 N   B. 43.0 N   C. 49.0 N
      D. 55.0 N   E. 74.0 N
   b. What is the magnitude of the force of the dog on the elevator floor?
      A. 4.2 N   B. 49.0 N   C. 55.0 N
      D. 43.0 N   E. 74.0 N

27. ‖ A 3.0 kg puck slides due east on a horizontal frictionless surface at a constant speed of 4.5 m/s. Then a force of magnitude 6.0 N, directed due north, is applied for 1.5 s. Afterward,
   a. What is the northward component of the puck's velocity?
      A. 0.50 m/s   B. 2.0 m/s   C. 3.0 m/s
      D. 4.0 m/s   E. 4.5 m/s
   b. What is the speed of the puck?
      A. 4.9 m/s   B. 5.4 m/s   C. 6.2 m/s
      D. 7.5 m/s   E. 11 m/s

28. ‖ A rocket in space, initially at rest, fires its main engines at a constant thrust. As it burns fuel, the mass of the rocket decreases. Which of the graphs in Figure Q5.28 best represents the velocity of the rocket as a function of time?

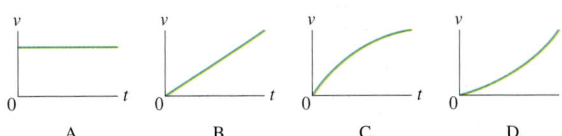

FIGURE Q5.28

29. | Eric has a mass of 60 kg. He is standing on a scale in an elevator that is accelerating downward at 1.7 m/s². What is the approximate reading on the scale?
   A. 0 N   B. 400 N   C. 500 N   D. 600 N

30. | The two blocks in Figure Q5.30 are at rest on frictionless surfaces. What must be the mass of the right block in order that the two blocks remain stationary?

FIGURE Q5.30

   A. 4.9 kg   B. 6.1 kg   C. 7.9 kg
   D. 9.8 kg   E. 12 kg

31. | A football player at practice pushes a 60 kg blocking sled across the field at a constant speed. The coefficient of kinetic friction between the grass and the sled is 0.30. How much force must he apply to the sled?
   A. 18 N   B. 60 N   C. 180 N   D. 600 N

32. | Two football players are pushing a 60 kg blocking sled across the field at a constant speed of 2.0 m/s. The coefficient of kinetic friction between the grass and the sled is 0.30. Once they stop pushing, how far will the sled slide before coming to rest?
   A. 0.20 m   B. 0.68 m   C. 1.0 m   D. 6.6 m

33. ‖ Land Rover ads used to claim that their vehicles could climb a slope of 45°. For this to be possible, what must be the minimum coefficient of static friction between the vehicle's tires and the road?
   A. 0.5   B. 0.7   C. 0.9   D. 1.0

34. ‖ A truck is traveling at 30 m/s on a slippery road. The driver slams on the brakes and the truck starts to skid. If the coefficient of kinetic friction between the tires and the road is 0.20, how far will the truck skid before stopping?
   A. 230 m   B. 300 m   C. 450 m   D. 680 m

PROBLEMS

Section 5.1 Equilibrium

1. | The three ropes in Figure P5.1 are tied to a small, very light ring. Two of the ropes are anchored to walls at right angles, and the third rope pulls as shown. What are T_1 and T_2, the magnitudes of the tension forces in the first two ropes?

FIGURE P5.1   **FIGURE P5.2**

2. || The three ropes in Figure P5.2 are tied to a small, very light ring. Two of these ropes are anchored to walls at right angles with the tensions shown in the figure. What are the magnitude and direction of the tension $\vec{T}_3$ in the third rope?

3. ||| A 20 kg loudspeaker is suspended 2.0 m below the ceiling by two cables that are each 30° from vertical. What is the tension in the cables?

4. | A 1000 kg steel beam is supported by the two ropes shown in Figure P5.4. Each rope can support a maximum sustained tension of 5600 N. Do the ropes break?

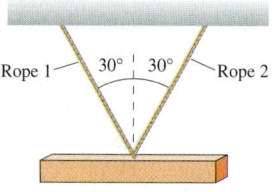

FIGURE P5.4

5. |||| A football coach sits on a sled while two of his players build their strength by dragging the sled across the field with ropes. The friction force on the sled is 1.0 kN and the angle between the two horizontal ropes is 20°. How hard must each player pull to drag the coach at a steady 2.0 m/s?

6. ||| When you bend your knee, BIO the quadriceps muscle is stretched. This increases the tension in the quadriceps tendon attached to your kneecap (patella), which, in turn, increases the tension in the patella tendon that attaches your kneecap to your lower leg bone (tibia). Simultaneously, the end of your upper leg bone (femur) pushes outward on the patella. Figure P5.6 shows how these parts of a knee joint are arranged. What size force does the femur exert on the kneecap if the tendons are oriented as in the figure and the tension in each tendon is 60 N?

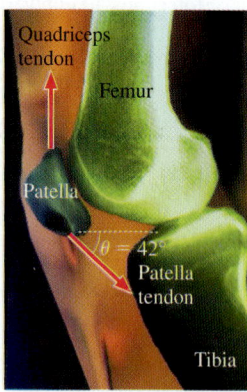

FIGURE P5.6

Section 5.2 Dynamics and Newton's Second Law

7. | A force with x-component F_x acts on a 2.0 kg object as it moves along the x-axis. The object's acceleration graph (a_x versus t) is shown in Figure P5.7. Draw a graph of F_x versus t.

FIGURE P5.7   **FIGURE P5.8**

8. | A force with x-component F_x acts on a 500 g object as it moves along the x-axis. The object's acceleration graph (a_x versus t) is shown in Figure P5.8. Draw a graph of F_x versus t.

9. | A force with x-component F_x acts on a 2.0 kg object as it moves along the x-axis. A graph of F_x versus t is shown in Figure P5.9. Draw an acceleration graph (a_x versus t) for this object.

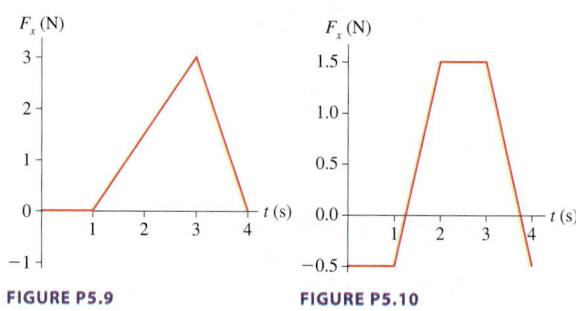

FIGURE P5.9   **FIGURE P5.10**

10. | A force with x-component F_x acts on a 500 g object as it moves along the x-axis. A graph of F_x versus t is shown in Figure P5.10. Draw an acceleration graph (a_x versus t) for this object.

11. || The forces in Figure P5.11 are acting on a 2.0 kg object. Find the values of a_x and a_y, the x- and y-components of the object's acceleration.

FIGURE P5.11   **FIGURE P5.12**

12. | The forces in Figure P5.12 are acting on a 2.0 kg object. Find the values of a_x and a_y, the x- and y-components of the object's acceleration.

13. | A horizontal rope is tied to a 50 kg box on frictionless ice. What is the tension in the rope if
  a. The box is at rest?
  b. The box moves at a steady 5.0 m/s?
  c. The box has $v_x = 5.0$ m/s and $a_x = 5.0$ m/s^2?

14. ‖ A crate pushed along the floor with velocity $\vec{v}_i$ slides a distance d after the pushing force is removed.
    a. If the mass of the crate is doubled but the initial velocity is not changed, what distance does the crate slide before stopping? Explain.
    b. If the initial velocity of the crate is doubled to $2\vec{v}_i$ but the mass is not changed, what distance does the crate slide before stopping? Explain.

15. | In a head-on collision, a car stops in 0.10 s from a speed of 14 m/s. The driver has a mass of 70 kg, and is, fortunately, tightly strapped into his seat. What force is applied to the driver by his seat belt during that fraction of a second?

Section 5.3 Mass and Weight

16. | An astronaut's weight on earth is 800 N. What is his weight on Mars, where $g = 3.76$ m/s²?

17. | A woman has a mass of 55.0 kg.
    a. What is her weight on earth?
    b. What are her mass and her weight on the moon, where $g = 1.62$ m/s²?

18. | A box with a 75 kg passenger inside is launched straight up into the air by a giant rubber band. After the box has left the rubber band but is still moving *upward*,
    a. What is the passenger's true weight?
    b. What is the passenger's apparent weight?

19. | a. How much force does an 80 kg astronaut exert on his chair while sitting at rest on the launch pad?
    b. How much force does the astronaut exert on his chair while accelerating straight up at 10 m/s²?

20. | It takes the elevator in a skyscraper 4.0 s to reach its cruising speed of 10 m/s. A 60 kg passenger gets aboard on the ground floor. What is the passenger's apparent weight
    a. Before the elevator starts moving?
    b. While the elevator is speeding up?
    c. After the elevator reaches its cruising speed?

21. | Zach, whose mass is 80 kg, is in an elevator descending at 10 m/s. The elevator takes 3.0 s to brake to a stop at the first floor.
    a. What is Zach's apparent weight before the elevator starts braking?
    b. What is Zach's apparent weight while the elevator is braking?

22. | Figure P5.22 shows the velocity graph of a 75 kg passenger in an elevator. What is the passenger's apparent weight at $t = 1.0$ s? At 5.0 s? At 9.0 s?

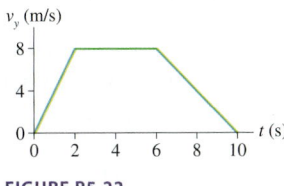

FIGURE P5.22

Section 5.4 Normal Forces

23. | a. A 0.60 kg bullfrog is sitting at rest on a level log. How large is the normal force of the log on the bullfrog?
    b. A second 0.60 kg bullfrog is on a log tilted 30° above horizontal. How large is the normal force of the log on this bullfrog?

24. ‖ A 23 kg child goes down a straight slide inclined 38° above horizontal. The child is acted on by his weight, the normal force from the slide, and kinetic friction.
    a. Draw a free-body diagram of the child.
    b. How large is the normal force of the slide on the child?

Section 5.5 Friction

25. ‖ Bonnie and Clyde are sliding a 300 kg bank safe across the floor to their getaway car. The safe slides with a constant speed if Clyde pushes from behind with 385 N of force while Bonnie pulls forward on a rope with 350 N of force. What is the safe's coefficient of kinetic friction on the bank floor?

26. ‖ A 4000 kg truck is parked on a 15° slope. How big is the friction force on the truck?

27. ‖ A 1000 kg car traveling at a speed of 40 m/s skids to a halt on wet concrete where $\mu_k = 0.60$. How long are the skid marks?

28. | A stubborn 120 kg mule sits down and refuses to move. To drag the mule to the barn, the exasperated farmer ties a rope around the mule and pulls with his maximum force of 800 N. The coefficients of friction between the mule and the ground are $\mu_s = 0.80$ and $\mu_k = 0.50$. Is the farmer able to move the mule?

29. ‖ A 10 kg crate is placed on a horizontal conveyor belt. The materials are such that $\mu_s = 0.50$ and $\mu_k = 0.30$.
    a. Draw a free-body diagram showing all the forces on the crate if the conveyer belt runs at constant speed.
    b. Draw a free-body diagram showing all the forces on the crate if the conveyer belt is speeding up.
    c. What is the maximum acceleration the belt can have without the crate slipping?
    d. If acceleration of the belt exceeds the value determined in part c, what is the acceleration of the crate?

Section 5.6 Drag

30. | What is the drag force on a 1.6 m wide, 1.4 m high car traveling at
    a. 10 m/s ($\approx$22 mph)?   b. 30 m/s ($\approx$65 mph)?

31. ‖ A 75 kg skydiver can be modeled as a rectangular "box" with dimensions 20 cm $\times$ 40 cm $\times$ 1.8 m. What is his terminal speed if he falls feet first?

32. | A 22-cm-diameter bowling ball has a terminal speed of 85 m/s. What is the ball's mass?

33. ‖ In Example 5.14, the stopping of a paramecium due to the BIO Stokes's law drag force was treated in an approximate fashion, taking the drag force to be independent of time.
    a. Use the same approximation to determine how *far* the paramecium drifts after it stops actively swimming at 0.25 mm/s.
    b. Is your answer to part a much less than, comparable to, or much greater than the size of the paramecium?

Section 5.7 Interacting Objects

34. ‖ A 1000 kg car pushes a 2000 kg truck that has a dead battery. When the driver steps on the accelerator, the drive wheels of the car push against the ground with a force of 4500 N.
    a. What is the magnitude of the force of the car on the truck?
    b. What is the magnitude of the force of the truck on the car?

35. ‖ Blocks with masses of 1.0 kg, 2.0 kg, and 3.0 kg are lined up in a row on a frictionless table. All three are pushed forward by a 12 N force applied to the 1.0 kg block. How much force does the 2.0 kg block exert on (a) the 3.0 kg block and (b) the 1.0 kg block?

Section 5.8 Ropes and Pulleys

36. ‖ What is the tension in the rope of Figure P5.36?

37. ‖ A 2.0-m-long, 500 g rope pulls a 10 kg block of ice across a horizontal, frictionless surface. The block accelerates at 2.0 m/s². How much force pulls forward on (a) the block of ice, (b) the rope?

38. ‖ Figure P5.38 shows two 1.00 kg blocks connected by a rope. A second rope hangs beneath the lower block. Both ropes have a mass of 250 g. The entire assembly is accelerated upward at 3.00 m/s² by force $\vec{F}$.
    a. What is F?
    b. What is the tension at the top end of rope 1?
    c. What is the tension at the bottom end of rope 1?
    d. What is the tension at the top end of rope 2?

FIGURE P5.36

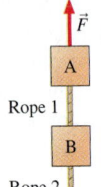

FIGURE P5.38

General Problems

39. ‖‖ A 500 kg piano is being lowered into position by a crane while two people steady it with ropes pulling to the sides. Bob's rope pulls to the left, 15° below horizontal, with 500 N of tension. Ellen's rope pulls toward the right, 25° below horizontal.
    a. What tension must Ellen maintain in her rope to keep the piano descending vertically?
    b. What is the tension in the vertical main cable supporting the piano?

40. ‖‖ In an electricity experiment, a 1.00 g plastic ball is suspended on a 60.0-cm-long string and given an electric charge. A charged rod brought near the ball exerts a horizontal electrical force $\vec{F}_{elec}$ on it, causing the ball to swing out to a 20.0° angle and remain there.
    a. What is the magnitude of $\vec{F}_{elec}$?
    b. What is the tension in the string?

41. ‖ Figure P5.41 shows the velocity graph of a 2.0 kg object as it moves along the x-axis. What is the net force acting on this object at t = 1 s? At 4 s? At 7 s?

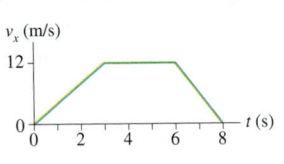

FIGURE P5.41

FIGURE P5.42

42. ‖ Figure P5.42 shows the force acting on a 2.0 kg object as it moves along the x-axis. The object is at rest at the origin at t = 0 s. What are its acceleration and velocity at t = 6.0 s?

43. ‖ A 50 kg box hangs from a rope. What is the tension in the rope if
    a. The box is at rest?
    b. The box has v_y = 5.0 m/s and is speeding up at 5.0 m/s²?

44. ‖ A 50 kg box hangs from a rope. What is the tension in the rope if
    a. The box moves up at a steady 5.0 m/s?
    b. The box has v_y = 5.0 m/s and is slowing down at 5.0 m/s²?

45. ‖ Your forehead can withstand a force of about 6.0 kN force before fracturing, while your cheekbone can only withstand about 1.3 kN.
    a. If a 140 g baseball strikes your head at 30 m/s and stops in 0.0015 s, what is the magnitude of the ball's acceleration?
    b. What is the magnitude of the force that stops the baseball?
    c. What force does the baseball apply to your head? Explain.
    d. Are you in danger of a fracture if the ball hits you in the forehead? In the cheek?

46. ‖ Seat belts and air bags save lives by reducing the forces exerted on the driver and passengers in an automobile collision. Cars are designed with a "crumple zone" in the front of the car. In the event of an impact, the passenger compartment decelerates over a distance of about 1 m as the front of the car crumples. An occupant restrained by seat belts and air bags decelerates with the car. By contrast, an unrestrained occupant keeps moving forward with no loss of speed (Newton's first law!) until hitting the dashboard or windshield, as we saw in Figure 4.2. These are unyielding surfaces, and the unfortunate occupant then decelerates over a distance of only about 5 mm.
    a. A 60 kg person is in a head-on collision. The car's speed at impact is 15 m/s. Estimate the net force on the person if he or she is wearing a seat belt and if the air bag deploys.
    b. Estimate the net force that ultimately stops the person if he or she is not restrained by a seat belt or air bag.
    c. How do these two forces compare to the person's weight?

47. ‖‖ Bob, who has a mass of 75 kg, can throw a 500 g rock with a speed of 30 m/s. The distance through which his hand moves as he accelerates the rock forward from rest until he releases it is 1.0 m.
    a. What constant force must Bob exert on the rock to throw it with this speed?
    b. If Bob is standing on frictionless ice, what is his recoil speed after releasing the rock?

48. ‖‖‖ An 80 kg spacewalking astronaut pushes off a 640 kg satellite, exerting a 100 N force for the 0.50 s it takes him to straighten his arms. How far apart are the astronaut and the satellite after 1.0 min?

49. ‖ What thrust does a 200 g model rocket need in order to have a vertical acceleration of 10.0 m/s²
    a. On earth?
    b. On the moon, where g = 1.62 m/s²?

50. ‖‖ A 20,000 kg rocket has a rocket motor that generates 3.0×10^5 N of thrust.
    a. What is the rocket's initial upward acceleration?
    b. At an altitude of 5.0 km the rocket's acceleration has increased to 6.0 m/s². What mass of fuel has it burned?

51. ‖‖ You've always wondered about the acceleration of the elevators in the 101-story-tall Empire State Building. One day, while visiting New York, you take your bathroom scales into the elevator and stand on them. The scales read 150 lb as the door closes. The reading varies between 120 lb and 170 lb as the elevator travels 101 floors.
    a. What is the magnitude of the acceleration as the elevator starts upward?
    b. What is the magnitude of the acceleration as the elevator brakes to a stop?

52. A 23 kg child goes down a straight slide inclined 38° above horizontal. The child is acted on by his weight, the normal force from the slide, kinetic friction, and a horizontal rope exerting a 30 N force as shown in Figure P5.52. How large is the normal force of the slide on the child?

FIGURE P5.52

53. Researchers often use *force plates* to measure the forces that people exert against the floor during movement. A force plate works like a bathroom scale, but it keeps a record of how the reading changes with time. Figure P5.53 shows the data from a force plate as a woman jumps straight up and then lands.
    a. What was the vertical component of her acceleration during push-off?
    b. What was the vertical component of her acceleration while in the air?
    c. What was the vertical component of her acceleration during the landing?
    d. What was her speed as her feet left the force plate?
    e. How high did she jump?

FIGURE P5.53

54. A 77 kg sprinter is running the 100 m dash. At one instant, early in the race, his acceleration is 4.7 m/s².
    a. What *total* force does the track surface exert on the sprinter? Assume his acceleration is parallel to the ground. Give your answer as a magnitude and an angle with respect to the horizontal.
    b. This force is applied to one foot (the other foot is in the air), which for a fraction of a second is stationary with respect to the track surface. Because the foot is stationary, the net force on it must be zero. Thus the force of the lower leg bone on the foot is equal but opposite to the force of the track on the foot. If the lower leg bone is 60° from horizontal, what are the components of the leg's force on the foot in the directions parallel and perpendicular to the leg? (Force components perpendicular to the leg can cause dislocation of the ankle joint.)

55. Sam, whose mass is 75 kg, takes off across level snow on his jet-powered skis. The skis have a thrust of 200 N and a coefficient of kinetic friction on snow of 0.10. Unfortunately, the skis run out of fuel after only 10 s.
    a. What is Sam's top speed?
    b. How far has Sam traveled when he finally coasts to a stop?

56. A person with compromised pinch strength in his fingers can only exert a normal force of 6.0 N to either side of a pinch-held object, such as the book shown in Figure P5.56. What is the heaviest book he can hold onto vertically before it slips out of his fingers? The coefficient of static friction of the surface between the fingers and the book cover is 0.80.

FIGURE P5.56

57. A 1.0 kg wood block is pressed against a vertical wood wall by a 12 N force as shown in Figure P5.57. If the block is initially at rest, will it move upward, move downward, or stay at rest?

FIGURE P5.57

58. A 50,000 kg locomotive, with steel wheels, is traveling at 10 m/s on steel rails when its engine and brakes both fail. How far will the locomotive roll before it comes to a stop?

59. An Airbus A320 jetliner has a takeoff mass of 75,000 kg. It reaches its takeoff speed of 82 m/s (180 mph) in 35 s. What is the thrust of the engines? You can neglect air resistance but not rolling friction.

60. A 2.0 kg wood block is launched up a wooden ramp that is inclined at a 35° angle. The block's initial speed is 10 m/s.
    a. What vertical height does the block reach above its starting point?
    b. What speed does it have when it slides back down to its starting point?

61. A 2.7 g Ping-Pong ball has a diameter of 4.0 cm.
    a. The ball is shot straight up at twice its terminal speed. What is its initial acceleration?
    b. The ball is shot straight down at twice its terminal speed. What is its initial acceleration?

62. The fastest recorded skydive was by an Air Force officer who jumped from a helium balloon at an elevation of 103,000 ft, three times higher than airliners fly. Because the density of air is so low at these altitudes, he reached a speed of 614 mph at an elevation of 90,000 ft, then gradually slowed as the air became more dense. Assume that he fell in the spread-eagle position and that his low-altitude terminal speed is 125 mph. Use this information to determine the density of air at 90,000 ft.

63. The cable cars in San Francisco are pulled along their tracks by an underground steel cable that moves along at 4.0 m/s. The cable is driven by large motors at a central power station and extends, via an intricate pulley arrangement, for several miles beneath the city streets. The length of a cable stretches by up to 30 m during its lifetime. To keep the tension constant, the cable passes around a 1.5-m-diameter "tensioning pulley" that rolls back and forth on rails, as shown in Figure P5.63. A 2000 kg block is attached to the tensioning pulley's cart, via a rope and pulley, and is suspended in a deep hole. What is the tension in the cable car's cable?

FIGURE P5.63

64. The 100 kg block in Figure P5.64 takes 6.0 s to reach the floor after being released from rest. What is the mass of the block on the left?

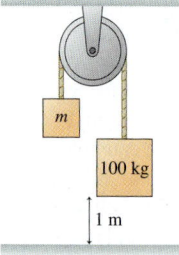

FIGURE P5.64

Problems 65 and 66 show free-body diagrams. For each,

a. Write a realistic dynamics problem for which this is the correct free-body diagram. Your problem should ask a question that can be answered with a value of position or velocity (such as "How far?" or "How fast?"), and should give sufficient information to allow a solution.

b. Solve your problem!

65.

FIGURE P5.65

66.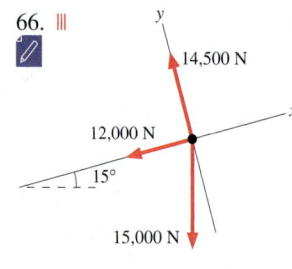

FIGURE P5.66

In Problems 67 through 69 you are given the dynamics equations that are used to solve a problem. For each of these, you are to

a. Write a realistic problem for which these are the correct equations.

b. Draw the free-body diagram and the pictorial representation for your problem.

c. Finish the solution of the problem.

67. $-0.80n = (1500 \text{ kg})a_x$
    $n - (1500 \text{ kg})(9.8 \text{ m/s}^2) = 0$

68. $T - 0.2n - (20 \text{ kg})(9.8 \text{ m/s}^2)\sin 20° = (20 \text{ kg})(2.0 \text{ m/s}^2)$
    $n - (20 \text{ kg})(9.8 \text{ m/s}^2)\cos 20° = 0$

69. $(100 \text{ N})\cos 30° - f_k = (20 \text{ kg})a_x$
    $n + (100 \text{ N})\sin 30° - (20 \text{ kg})(9.8 \text{ m/s}^2) = 0$
    $f_k = 0.20n$

Passage Problems

Sliding on the Ice

In the winter sport of curling, players give a 20 kg stone a push across a sheet of ice. The stone moves approximately 40 m before coming to rest. The final position of the stone, in principle, only depends on the initial speed at which it is launched and the force of friction between the ice and the stone, but team members can use brooms to sweep the ice in front of the stone to adjust its speed and trajectory a bit; they must do this without touching the stone. Judicious sweeping can lengthen the travel of the stone by 3 m.

70. | A curler pushes a stone to a speed of 3.0 m/s over a time of 2.0 s. Ignoring the force of friction, how much force must the curler apply to the stone to bring it up to speed?
    A. 3.0 N   B. 15 N   C. 30 N   D. 150 N

71. | The sweepers in a curling competition adjust the trajectory of the stone by
    A. Decreasing the coefficient of friction between the stone and the ice.
    B. Increasing the coefficient of friction between the stone and the ice.
    C. Changing friction from kinetic to static.
    D. Changing friction from static to kinetic.

72. | Suppose the stone is launched with a speed of 3 m/s and travels 40 m before coming to rest. What is the *approximate* magnitude of the friction force on the stone?
    A. 0 N   B. 2 N   C. 20 N   D. 200 N

73. | Suppose the stone's mass is increased to 40 kg, but it is launched at the same 3 m/s. Which one of the following is true?
    A. The stone would now travel a longer distance before coming to rest.
    B. The stone would now travel a shorter distance before coming to rest.
    C. The coefficient of friction would now be greater.
    D. The force of friction would now be greater.

STOP TO THINK ANSWERS

Stop to Think 5.1: A. The lander is descending and slowing. The acceleration vector points upward, and so $\vec{F}_{net}$ points upward. This can be true only if the thrust has a larger magnitude than the weight.

Stop to Think 5.2: A. You are descending and slowing, so your acceleration vector points upward and there is a net upward force on you. The floor pushes up against your feet harder than gravity pulls down.

Stop to Think 5.3: $f_B > f_C = f_D = f_E > f_A$. Situations C, D, and E are all kinetic friction, which does not depend on either velocity or acceleration. Kinetic friction is smaller than the maximum static friction that is exerted in B. $f_A = 0$ because no friction is needed to keep the object at rest.

Stop to Think 5.4: D. The ball is shot *down* at 30 m/s, so $v_{0y} = -30$ m/s. This exceeds the terminal speed, so the upward drag force is *larger* than the downward weight force. Thus the ball *slows down* even though it is "falling." It will slow until $v_y = -15$ m/s, the terminal velocity, then maintain that velocity.

Stop to Think 5.5: B. $F_{QonH} = F_{HonQ}$ and $F_{PonQ} = F_{QonP}$ because these are action/reaction pairs. Box Q is slowing down and therefore must have a net force to the left. So from Newton's second law we also know that $F_{HonQ} > F_{PonQ}$.

Stop to Think 5.6: Equal to. Each block is hanging in equilibrium, with no net force, so the upward tension force is mg.

6

CIRCULAR MOTION, ORBITS, AND GRAVITY

Motorcyclists in the "Globe of Death" ride their bikes on the inside of a spherical steel frame, seeming to defy gravity as they ride up the sides and upside down over the top. What prevents them from falling?

Looking Ahead ▶▶

The goal of Chapter 6 is to learn about motion in a circle, including orbital motion under the influence of a gravitational force. In this chapter, you will learn to:

▶ Understand the kinematics of circular motion.

▶ Use Newton's laws to analyze the dynamics of circular motion.

▶ Understand the circular orbits of satellites and planets.

▶ Further your understanding of the force of gravity as a long-range force.

Looking Back ◀◀

This chapter uses what you have learned about one-dimensional motion, circular motion, Newton's laws, and the concepts of mass and weight. Please review:

Engines revving, the motorcyclists build up speed until they are riding in excess of 30 mph in small circles inside the spherical enclosure. As their speeds increase, they begin to ride up the sides, culminating in vertical loops that turn them completely upside down. In this chapter we will examine circular motion problems that range from such extreme examples to more everyday situations such as rounding a corner in a car.

You learned in Chapter 3 that circular motion involves an acceleration because the velocity is always changing direction. In this chapter, we will develop a new set of kinematic variables to describe circular motion, then look at the dynamics of uniform circular motion: the forces that produce the motion. One especially important kind of circular motion—the orbital motion of the moon about the earth and the planets about the sun—will lead us to consider the nature of gravity, the force that keeps things in orbit.

This is a very practical chapter, one in which we will learn why you need to turn corners more slowly on slippery roads, why highway curves are banked, and why water stays in a bucket when you swing it over your head.

6.1 Uniform Circular Motion

We began our study of circular motion in Section 3.8. Figure 6.1 is a review of the variables we used to specify circular motion and the relationships among them. We will start this chapter by extending and clarifying these concepts.

(a) Circular motion variables

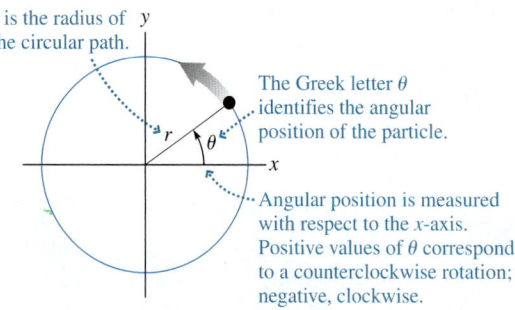

r is the radius of the circular path.

The Greek letter θ identifies the angular position of the particle.

Angular position is measured with respect to the *x*-axis. Positive values of θ correspond to a counterclockwise rotation; negative, clockwise.

T = period = time for one complete orbit

f = frequency = $\frac{1}{T}$

(b) Velocity and acceleration for uniform circular motion

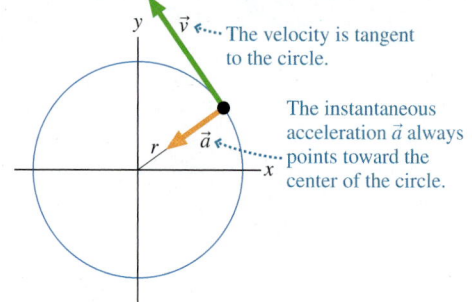

The velocity is tangent to the circle.

The instantaneous acceleration $\vec{a}$ always points toward the center of the circle.

Relationship between velocity and acceleration: $a = -\dfrac{v^2}{r}$

FIGURE 6.1 Reviewing circular motion.

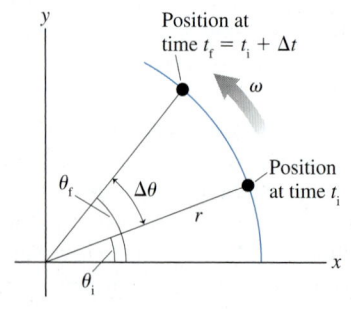

The velocity vectors are all the same length, though the direction changes.

FIGURE 6.2 A particle in uniform circular motion.

If the particle in circular motion is moving at *constant speed* we say that it is in **uniform circular motion.** Figure 6.2 shows a particle in uniform circular motion. The particle might be a satellite moving in an orbit, a ball on the end of a string, or even just a dot painted on the side of a wheel. Regardless of what the particle represents, its velocity vector is always tangent to the circular path. The particle's speed *v* is constant, so the vector's length stays constant as the particle moves around the circle.

Angular Displacement and Angular Velocity

To fully describe uniform circular motion, we will define some new variables. Figure 6.3 shows a particle in uniform circular motion that moves from an initial angular position θ_i at time t_i to a final angular position θ_f at a later time t_f. The change $\Delta\theta = \theta_f - \theta_i$ is called the **angular displacement.** We can measure the particle's circular motion in terms of the rate of change of θ just as we measured a particle's linear motion in terms of the rate of change of its position *x*.

In analogy with linear motion, where $v_x = \Delta x/\Delta t$, we define the **angular velocity** to be

$$\omega = \frac{\text{angular displacement}}{\text{time interval}} = \frac{\Delta\theta}{\Delta t} \qquad (6.1)$$

Angular velocity of a particle in uniform circular motion

The symbol ω is a lowercase Greek omega, *not* an ordinary *w*. The SI unit of angular velocity is rad/s.

NOTE ▶ Recall our discussion of average and instantaneous velocity in Chapter 2. The definition of angular velocity in Equation 6.1 is really the *average* angular velocity. For a very small time interval Δt, the calculation will approximate the *instantaneous* angular velocity. If the angular velocity is constant, as it will be for the problems we consider in this chapter, the average angular velocity and instantaneous angular velocity are the same. ◀

FIGURE 6.3 A particle moves from θ_i to θ_f with angular velocity ω.

Angular velocity is the *rate* at which a particle's angular position is changing as it rotates around the circle. A particle that starts from $\theta = 0$ rad with an angular velocity of 0.5 rad/s will be at angle $\theta = 0.5$ rad after 1 s, at $\theta = 1.0$ rad after 2 s, at $\theta = 1.5$ rad after 3 s, and so on. Its angular position is increasing at the *rate* of 0.5 radians per second. In analogy with uniform linear motion, which you studied in Chapter 2, uniform circular motion is motion in which the angle increases at a *constant* rate: **A particle moves with uniform circular motion if and only if its angular velocity ω is constant.**

NOTE ▶ It's important to keep in mind which quantities are changing and which are constant. For a particle in uniform circular motion, the angular velocity and the speed are constant. This is how we define uniform circular motion. But the velocity is constantly changing because the direction of motion changes. ◀

Angular velocity, like the velocity v_x of one-dimensional motion, can be positive or negative. The signs noted in Figure 6.4 are based on the fact that angles are positive when measured counterclockwise from the positive x-axis.

Circular motion is analogous to linear motion, with angular variables replacing linear variables. Much of what you learned about linear kinematics and dynamics carries over to circular motion. For example, Equation 2.6 gave us a formula for computing a linear displacement during a time interval:

$$x_f - x_i = \Delta x = v_x \, \Delta t$$

You can see from Equation 6.1 that we can write a similar equation for the angular displacement:

$$\theta_f - \theta_i = \Delta\theta = \omega \, \Delta t \qquad (6.2)$$

Angular displacement for uniform circular motion

LINEAR
p. 38

For linear motion, we use the term *speed v* when we are not concerned with the direction of motion, *velocity v_x* when we are. For circular motion, we define the **angular speed** to be the absolute value of the angular velocity, irrespective of its direction of rotation. Although potentially confusing, it is customary to use the symbol ω for angular speed *and* for angular velocity. If the direction of rotation is not important, we will interpret ω to mean angular speed, a positive quantity. In kinematic equations, such as Equation 6.2, ω is the angular velocity, and you need to use a negative value for a clockwise rotation.

The angular speed ω is closely related to the period T and the frequency f of the motion. If a particle in uniform circular motion moves around a circle once, which by definition takes time T, its angular displacement is $\Delta\theta = 360° = 2\pi$ rad. The angular speed is thus

$$\omega = \frac{2\pi \text{ rad}}{T} \qquad (6.3)$$

We can also write the angular speed in terms of the frequency:

$$\omega = (2\pi \text{ rad})f \qquad (6.4)$$

where f must be in rev/s. For example, a particle in circular motion with frequency 10 rev/s would have angular speed $\omega = 20\pi$ rad/s $= 62.8$ rad/s.

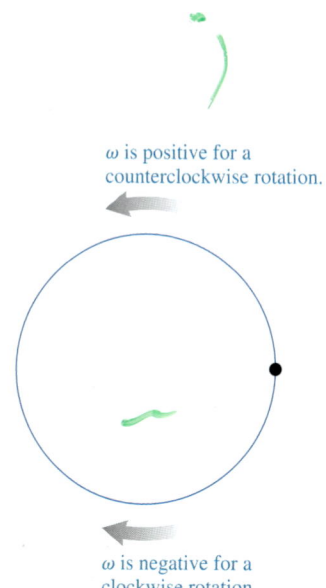

ω is positive for a counterclockwise rotation.

ω is negative for a clockwise rotation.

FIGURE 6.4 Positive and negative angular velocities.

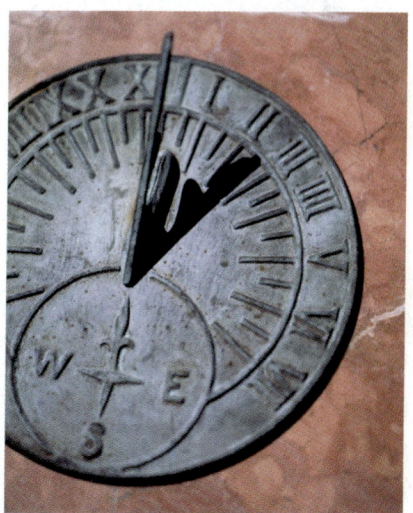

Why do clocks go clockwise? In the northern hemisphere, the rotation of the earth causes the sun to follow a circular arc through the southern sky, rising in the east and setting in the west. For millennia, humans have marked passing time by noting the position of shadows cast by the sun, which sweep in an arc from west to east— eventually leading to the development of the sundial, the first practical timekeeping device. In the northern hemisphere, sundials point north, and the shadow sweeps around the dial from left to right. Early clockmakers used the same convention, which is how it came to be clockwise.

EXAMPLE 6.1 Kinematics at the roulette wheel

A small, steel ball rolls counterclockwise around the inside of a 30.0-cm-diameter roulette wheel. The ball completes exactly 2 rev in 1.20 s.

a. What is the ball's angular velocity?
b. What is the ball's angular position at $t = 2.00$ s? Assume $\theta_i = 0$.

PREPARE Treat the ball as a particle in uniform circular motion.

SOLVE

a. The period of the ball's motion, the time for 1 rev, is $T = 0.600$ s, so the angular speed is

$$\omega = \frac{2\pi \text{ rad}}{T} = \frac{2\pi \text{ rad}}{0.600 \text{ s}} = 10.5 \text{ rad/s}$$

Because the motion is counterclockwise, the angular velocity is positive: $\omega = +10.5$ rad/s.

b. The ball moves with constant angular velocity, so its position is given by Equation 6.2. When we solved problems of this sort for linear motion, we found that a graphical representation was helpful. The same is true here. A graph of θ vs. t is shown in Figure 6.5. The angular position increases steadily with time, starting at $\theta = 0$ rad at $t = 0$ s, and increasing to $\theta = 4\pi$ rad (two full revolutions) at $t = 1.20$ s, and a bit more than three revolutions at $t = 2.00$ s.

The ball's angular position at $t = 2.00$ s is

$$\theta_f = \theta_i + \omega \, \Delta t = 0 \text{ rad} + (10.5 \text{ rad/s})(2.00 \text{ s}) = 21.0 \text{ rad}$$

At 1.20 s, the ball has completed two full revolutions.

At 2.00 s, the ball has completed more than three revolutions.

Because the ball moves at constant angular velocity, a graph of the angular position against time will be a straight line.

FIGURE 6.5 Angular position for the ball on the roulette wheel.

But the ball is back to where it started after each revolution. If what we want is the ball's final angular position, complete revolutions need not be counted. It is common practice to subtract out integer multiples of 2π rad, representing the completed revolutions. Because $21.0/2\pi = 3.34$, we can write

$$\theta_f = 21.0 \text{ rad} = 3.34 \times 2\pi \text{ rad}$$
$$= 3 \times 2\pi \text{ rad} + 0.34 \times 2\pi \text{ rad}$$
$$= 3 \times 2\pi \text{ rad} + 2.1 \text{ rad}$$

In other words, at $t = 2.00$ s, the ball has completed 3 rev and is 2.1 rad = 120° into its fourth revolution. An observer would say that the ball's angular position is $\theta = 120°$.

ASSESS Figure 6.5 shows that the kinematics of uniform circular motion is quite similar to that of uniform motion.

6.2 Speed, Velocity, and Acceleration in Uniform Circular Motion

The previous section described uniform circular motion in terms of angular variables. In Chapter 3, we introduced a description of uniform circular motion in terms of velocity and acceleration vectors. We will now unite these two different descriptions, which will let us consider a much wider range of problems.

Speed

For a particle in uniform circular motion, it's easy to relate the particle's period T to its speed v, as we saw at the end of Chapter 3. For a particle moving with constant speed, speed is simply distance/time. In one period, the particle moves once around a circle of radius r and travels the circumference $2\pi r$. Thus

$$v = \frac{1 \text{ circumference}}{1 \text{ period}} = \frac{2\pi r}{T} \tag{6.5}$$

If we combine this result with Equation 6.3 for the angular speed, we find that speed v and angular speed ω are related by

$$v = \omega r \tag{6.6}$$

Relationship between speed and angular speed

NOTE ▶ In Equation 6.6, ω **must be in units of rad/s.** If you are given a frequency in rev/s or rpm, you should convert it to an angular speed in rad/s. ◀

EXAMPLE 6.2 Finding the speed at two points on a CD

The diameter of an audio compact disc is 12.0 cm. When the disc is spinning at its maximum rate of 540 rpm, what is the speed of a point (a) at the outside edge of the disc, 6.0 cm from the center, and (b) at a distance 3.0 cm from the center?

PREPARE Consider two points A and B on a rotating compact disc in Figure 6.6. During one period, the disc rotates one time, and both points rotate through the same angle, 2π rad. Thus the angular speed, $\omega = 2\pi/T$, is the same for these two points; in fact, it is the same for all points on the disc. But as they go around one time, the two points move different *distances;* the outer point B goes around a larger circle. The two points thus have different *speeds.* We will solve this problem by first finding the angular speed of the disc, then computing the speeds at the two points.

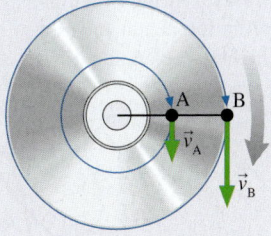

FIGURE 6.6 The rotation of an audio compact disc.

SOLVE We first convert the frequency of the disc from rpm to rev/s:

$$f = \left(540 \, \frac{\text{rev}}{\text{min}}\right) \times \left(\frac{1 \, \text{min}}{60 \, \text{s}}\right) = 9.00 \, \text{rev/s}$$

We can compute the angular speed using Equation 6.4:

$$\omega = (2\pi \, \text{rad})(9.00 \, \text{rev/s}) = 56.5 \, \text{rad/s}$$

The angular speed has the appropriate units. We can now use Equation 6.6 to compute the speeds of points on the disc. At point B, $r = 6.0$ cm $= 0.060$ m so the speed at the outside edge is

$$v_B = \omega r = (56.5 \, \text{rad/s})(0.060 \, \text{m}) = 3.4 \, \text{m/s}$$

At point A, $r = 3.0$ cm $= 0.030$ m, so the speed is

$$v_B = \omega r = (56.5 \, \text{rad/s})(0.030 \, \text{m}) = 1.7 \, \text{m/s}$$

ASSESS The speeds are a few m/s, which seems reasonable. The point farther from the center is moving at a higher speed, as we expected.

CONCEPTUAL EXAMPLE 6.1 Varying the angular speed of a CD

As we saw in Example 6.2, different points on a rotating CD move at different speeds. This creates a technical problem. The data on a CD are read by a laser beam that slowly scans from the inner part of the disc to the outside as the disc is played. The laser beam reflects from "pits" as they go by on the moving disc, so the data *rate* is proportional to the speed at that point on the disc. To have a constant data rate, so the music plays at a constant rate, the position where the data are being read must move at a constant speed. To keep the *speed* constant at the point of the laser beam, the *angular speed* of the disc must change as different parts of the disc are read. As a disc is played, and the laser scans from the inside to the outside, does the angular speed increase or decrease?

REASON We can rewrite Equation 6.6 as follows:

$$\omega = \frac{v}{r}$$

The speed of the disc at the position being read is constant. Thus, as the laser reads points farther from the center, at larger values of r, the angular speed must decrease.

ASSESS The angular speed of an audio compact disc does change as it is read. At the very start, when the laser is near the center of the disc, the frequency is about 540 rpm; by the time the laser is reading data near the outside of the disc, the frequency has dropped to just over 200 rpm.

Acceleration

Although the speed of a particle in uniform circular motion is constant, its velocity is *not* constant because the direction of the motion is always changing. As you learned in Chapter 3, and as Figure 6.7 reminds you, there is an acceleration at every point in the motion, with the acceleration vector $\vec{a}$ pointing toward the center of the circle. We called this the *centripetal acceleration,* and we showed that for uniform circular motion the acceleration was given by $a = v^2/r$. Because $v = \omega r$, we can also write this relationship in terms of the angular speed:

The instantaneous velocity $\vec{v}$ is parallel to the circle at all points.

The instantaneous acceleration $\vec{a}$ is directed toward the center of the circle at all points.

FIGURE 6.7 Velocity and acceleration for uniform circular motion.

$$a = \frac{v^2}{r} = \omega^2 r \qquad (6.7)$$

Centripetal acceleration for uniform circular motion

QUADRATIC
p. 50

Acceleration depends on speed, but also distance from the center of the circle.

CONCEPTUAL EXAMPLE 6.2 Who has the larger acceleration?

Two children are riding in circles on a merry-go-round, as shown in Figure 6.8. Which child experiences the larger acceleration?

FIGURE 6.8 Top view of a merry-go-round.

REASON All points on the merry-go-round move at the same angular speed. The second expression for the acceleration in Equation 6.7 tells us that $a = \omega^2 r$. As the two children are moving with the same angular speed, Emma, with a larger value of r, experiences a larger acceleration.

ASSESS In the previous example, we saw that points farther from the center move at a higher speed. This would imply a higher acceleration as well, so our answer makes sense.

EXAMPLE 6.3 Finding the period of a carnival ride

In the Quasar carnival ride passengers travel in a horizontal 5.0-m-radius circle. For safe operation, the maximum sustained acceleration that riders may experience is 20 m/s², approximately twice the acceleration of gravity. What is the period of the ride when it is being operated at the maximum acceleration?

PREPARE We will assume that the cars on the ride are in uniform circular motion. The visual overview of Figure 6.9 shows a top view of the motion of the ride.

SOLVE The angular speed can be computed from the acceleration by rearranging Equation 6.7:

$$\omega = \sqrt{\frac{a}{r}} = \sqrt{\frac{20 \text{ m/s}^2}{5.0 \text{ m}}} = 2.0 \text{ rad/s}$$

At this angular speed, the period is $T = \dfrac{2\pi}{\omega} = 3.1$ s

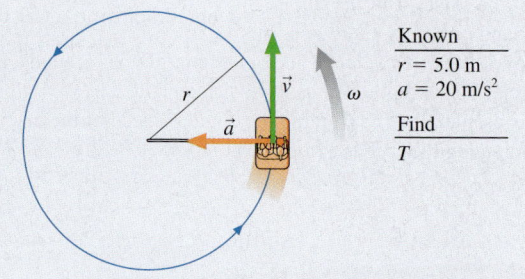

Known
$r = 5.0$ m
$a = 20$ m/s²

Find
T

FIGURE 6.9 Visual overview for the Quasar carnival ride.

ASSESS One rotation in just over three seconds seems reasonable for a pretty zippy carnival ride. The period for this particular ride is actually 3.7 s, so it runs a bit slower than the maximum safe speed.

Hurling the heavy hammer This man is throwing a hammer that weighs over 30 pounds as far as he can by spinning the hammer around in a circle and then letting go. While he holds the handle, the hammer follows a circular path. He must provide a very large force directed toward the center of the circle to produce the centripetal acceleration, as you can see by how he is leaning away from the hammer. When he lets go, there is no longer a force directed toward the center, and the hammer will stop going in a circle and fly across the field.

STOP TO THINK 6.1   Rank in order, from largest to smallest, the centripetal accelerations of particles A to D.

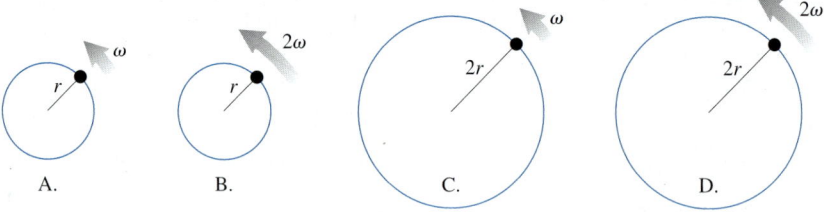

6.3 Dynamics of Uniform Circular Motion

Riders spinning around on a circular carnival ride are accelerating—they are *not* traveling at constant velocity in a straight line, as we have seen. Consequently, according to Newton's first law, the riders *must* have a net force acting on them. In this section, we'll look at the forces necessary for uniform circular motion.

We've already determined the acceleration of a particle in uniform circular motion—the centripetal acceleration of Equation 3.36 and Equation 6.7. Newton's second law tells us exactly what the net force must be to cause this acceleration:

$$\vec{F}_{\text{net}} = m\vec{a} = \left(\frac{mv^2}{r}, \text{toward center of circle}\right) \quad (6.8)$$

Net force producing the centripetal
acceleration of uniform circular motion

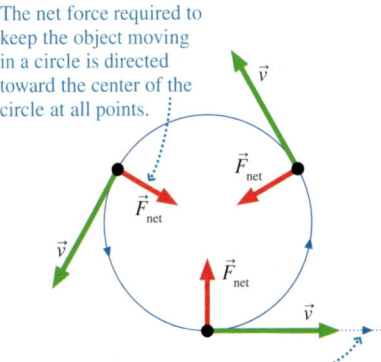

QUADRATIC INVERSE
p.50 p.118

In other words, **a particle of mass m moving at constant speed v around a circle of radius r must always have a net force of magnitude mv^2/r pointing toward the center of the circle**, as in Figure 6.10. Without such a force, the particle would move off in a straight line tangent to the circle.

The force described by Equation 6.8 is not a *new* force. Our rules for identifying forces have not changed. What we are saying is that a particle moves with uniform circular motion *if and only if* a net force always points toward the center of the circle. The force itself must have identifiable agents and must be a combination of our familiar forces, such as tension, friction, or the normal force. Equation 6.8 simply tells us how the net force needs to act—how strong and in which direction—to cause the particle to move with speed v in a circle of radius r.

In each example of circular motion that we will consider in this chapter a physical force or a combination of forces directed toward the center produces the necessary acceleration. In some cases, the circular motion and the force are quite obvious, as in the hammer throw. Other cases are a bit more subtle. For instance, a car going around a corner is following a circular path on a level road; the necessary force is provided by the friction force between the tires and the road.

The net force required to keep the object moving in a circle is directed toward the center of the circle at all points.

In the absence of a net force, the object would continue in a straight line.

FIGURE 6.10 Net force for circular motion.

CONCEPTUAL EXAMPLE 6.3 Will the rope break?

A man has used a rope to hang a swing from a tree branch. The rope is just strong enough to support his weight, but no more. A friend pulls him back on the swing and lets go. Will the rope break?

REASON The swing moves along a circular arc. Although this is not uniform circular motion—it speeds up and then slows down—the speed is nearly constant for a short interval at the bottom of the swing. Thus the motion at the bottom is well approximated as uniform circular motion. There must be a net force toward the center of the circle to keep the swing moving in a circle. You can see from the free-body diagram of Figure 6.11 that the forces acting on it are the man's weight $\vec{w}$, downward, and the tension $\vec{T}$ in the rope, upward. A net force toward the center of the circle requires $T > w$. But the rope can barely support his weight, so $T_{\text{max}} = w$. The rope will break!

FIGURE 6.11 Visual overview for the motion of the swing.

ASSESS Note that the swing is in circular motion even though it does not travel through a complete circle.

Solving Circular Dynamics Problems

We have one basic equation for circular dynamics problems: Equation 6.8. This is just a version of Newton's second law; the right side is a net force, the left a mass times an acceleration, the centripetal acceleration. The techniques for solving circular dynamics problems are thus quite similar to those we have seen for solving other Newton's second law problems.

Activ
Physics 4.2–4.5

(MP) PROBLEM-SOLVING
STRATEGY 6.1   **Circular dynamics problems**

PREPARE Begin your visual overview with a pictorial representation in which you sketch the motion, define symbols, and identify what the problem is trying to find. It is best to draw the free-body diagram with the circle viewed edge-on, the *x*-axis pointing toward the center of the circle, and the *y*-axis perpendicular to the plane of the circle.

SOLVE Newton's second law for uniform circular motion, $\vec{F}_{net} = (mv^2/r$, toward center of circle), is a vector equation. Some forces act in the plane of the circle, some act perpendicular to the circle, and some may have components in both directions. In the coordinate system described above, Newton's second law is

$$\sum F_x = \frac{mv^2}{r} \quad \text{and} \quad \sum F_y = 0$$

That is, the net force toward the center of the circle is mv^2/r, while the net force perpendicular to the circle is zero. The components of the forces are found directly from the free-body diagram. Depending on the problem, either

■ Use the net force to determine the speed *v*, then use circular kinematics to find frequencies or angular velocities.

■ Use circular kinematics to determine the speed *v*, then solve for unknown forces.

ASSESS Make sure your net force points toward the center of the circle. Check that your result has the correct units, is reasonable, and answers the question.

EXAMPLE 6.4 Analyzing the motion of a cart

An energetic father places his 20 kg child on a 5.0 kg cart to which a 2.0-m-long rope is attached. He then holds the end of the rope and spins the cart and child around in a circle, keeping the rope parallel to the ground. If the tension in the rope is 100 N, how much time does it take for the cart to make one rotation?

PREPARE We proceed according to the steps of Problem-Solving Strategy 6.1. Figure 6.12 shows a visual overview of the problem. The main idea of the pictorial representation on the left is to

illustrate the relevant geometry and to define the symbols that will be used. A circular dynamics problem usually does not have starting and ending points like a projectile problem, so numerical subscripts such as x_1 or y_2 are usually not needed. Here we need to define the cart's speed *v* and the radius *r* of the circle.

The object moving in the circle is the cart plus the child, a total mass of 25 kg; the free-body diagram below shows the forces. **For uniform circular motion, we'll draw the free-body diagram looking at the edge of the circle because this is the plane of the forces. Three forces are acting on the cart: the**

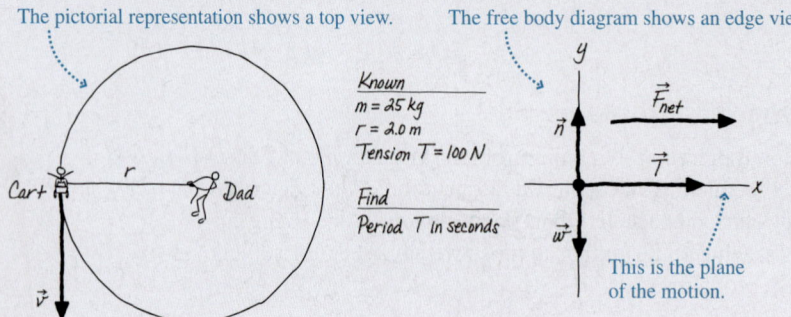

The pictorial representation shows a top view.   The free body diagram shows an edge view.

FIGURE 6.12 A visual overview of the cart spinning in a circle.

weight force $\vec{w}$, the normal force of the ground $\vec{n}$, and the tension force of the rope $\vec{T}$. As the problem-solving strategy suggests, we've shown these forces on a free-body diagram with the x-axis pointing toward the center of the circle and the y-axis perpendicular to the plane of motion.

Notice that there are two quantities for which we use the symbol T: the tension and the period. We will include additional information when necessary to distinguish the two.

SOLVE There is no net force in the y-direction, perpendicular to the circle, so $\vec{w}$ and $\vec{n}$ must be equal and opposite. There is a net force in the x-direction, toward the center of the circle, as there must be to cause the centripetal acceleration of circular motion. Only the tension force has an x-component, so Newton's second law is

$$\sum F_x = T = \frac{mv^2}{r}$$

We know the mass, the radius of the circle, and the tension, so we can solve for v:

$$v = \sqrt{\frac{Tr}{m}} = \sqrt{\frac{(100\ \text{N})(2.0\ \text{m})}{25\ \text{kg}}} = 2.83\ \text{m/s}$$

From this, we can compute the period with a slight rearrangement of Equation 6.5:

$$T = \frac{2\pi r}{v} = \frac{(2\pi)(2.0\ \text{m})}{2.83\ \text{m/s}} = 4.4\ \text{s}$$

ASSESS The speed is about 3 m/s. Because 1 m/s $\approx$ 2mph, this implies that the child is going about 6 mph. A trip around the circle in just over 4 s at a speed of about 6 mph sounds reasonable; it's a fast ride, but not so fast as to be scary!

EXAMPLE 6.5 Finding the maximum speed to turn a corner

What is the maximum speed with which a 1500 kg car can make a turn around a curve of radius 20 m on a level (unbanked) road without sliding? (This radius turn is about what you might expect at a major intersection in a city.)

PREPARE We start with the visual overview in Figure 6.13. The car moves along a circular arc at constant speed—uniform circular motion—for the quarter-circle necessary to complete the turn. The motion before and after the turn is not relevant to the problem. The more interesting issue is *how* a car turns a corner. What force or forces can we identify that cause the direction of the velocity vector to change? Imagine you are driving a car on a completely frictionless road, such as a very icy road. You would not be able to turn a corner. Turning the steering wheel would be of no use; the car would slide straight ahead, in

accordance with both Newton's first law and the experience of anyone who has ever driven on ice! So it must be *friction* that somehow allows the car to turn.

The top view of the tire in Figure 6.13 shows the force on the tire as it turns a corner. If the road surface were frictionless, the tire would slide straight ahead. The force that prevents an object from sliding across a surface is *static friction*. Static friction $\vec{f}_s$ pushes *sideways* on the tire, toward the center of the circle. How do we know the direction is sideways? If $\vec{f}_s$ had a component either parallel to $\vec{v}$ or opposite to $\vec{v}$, it would cause the car to speed up or slow down. Because the car changes direction but not speed, static friction must be perpendicular to $\vec{v}$. Thus $\vec{f}_s$ causes the centripetal acceleration of circular motion around the curve. With this in mind, the free-body diagram, drawn from behind the car, shows the static friction force pointing toward the center of the circle.

Known
$m = 1500\ \text{kg}$
$r = 20\ \text{m}$
$\mu_s = 1.0$
Find
v_{max}

This force prevents the tire from slipping sideways.

$\vec{F}_{net} = \vec{f}_s$

Top view of car   Top view of tire   Rear view of car

FIGURE 6.13 Visual overview of a car turning a corner.

Continued

$$F_{net} = \frac{mv^2}{r} = \frac{1500 \cdot 30^2}{150}$$

SOLVE The only force in the *x*-direction, toward the center of the circle, is static friction. Newton's second law along the *x*-axis is

$$\sum F_x = f_s = \frac{mv^2}{r}$$

The only difference between this example and the previous one is that the tension force toward the center has been replaced by a static friction force toward the center.

Three forces are acting on the car: the weight force $\vec{w}$, the normal force $\vec{n}$, and the force of static friction $\vec{f}_s$. Because there is no vertical motion, the weight force and the normal force must be equal and opposite:

$$n = w = mg$$

The net force toward the center of the circle is the force of static friction. Recall from Equation 5.10 in Chapter 5 that static friction has a maximum possible value:

$$f_{s\,max} = \mu_s n = \mu_s mg$$

Because the static friction force has a maximum value, there will be a maximum speed with which a car can turn without sliding. The maximum speed is reached when the static friction force reaches its maximum $f_{s\,max} = \mu_s mg$. If the car enters the curve at a speed higher than the maximum, static friction will not be large enough to provide the necessary centripetal acceleration and the car will slide.

The maximum speed occurs at the maximum value of the force of static friction:

$$f_{s\,max} = \frac{mv_{max}^2}{r}$$

Using the known value of $f_{s\,max}$, we find

$$\frac{mv_{max}^2}{r} = f_{s\,max} = \mu_s mg$$

Rearrangement of this gives

$$v_{max}^2 = \mu_s gr$$

For rubber tires on pavement, we can find the relevant value of μ_s in Table 5.2; it is 1.0. We can then complete our calculation:

$$v_{max} = \sqrt{\mu_s gr} = \sqrt{(1.0)(9.8 \text{ m/s}^2)(20 \text{ m})} = 14 \text{ m/s}$$

ASSESS 14 m/s ≈ 30 mph, which seems like a reasonable upper limit for the speed at which a car can turn a corner without sliding. There are a few other things to note about the solution:

- The car's mass canceled out. The maximum speed *does not* depend on the mass of the vehicle, though this may seem surprising.
- The final expression for v_{max} *does* depend on the coefficient of friction and the radius of the turn. v_{max} decreases if μ is less (a slipperier road) or if r is smaller (a tighter turn). Both of these make sense.

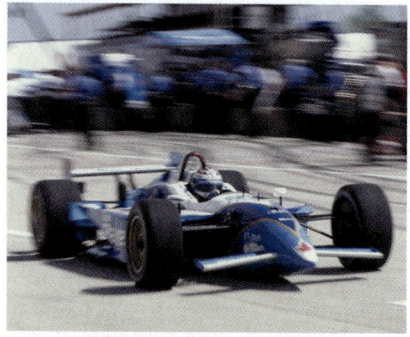

FIGURE 6.14 Wings on an Indy racer.

A banked turn on a racetrack.

Because v_{max} depends on μ_s and because μ_s depends on road conditions, the maximum safe speed through turns can vary dramatically. Wet or icy roads lower the value of μ_s and thus lower the maximum speed of turns. A car that handles normally while driving straight ahead on a wet road can suddenly slide out of control when turning a corner. Icy conditions are even worse. If you lower the value of the coefficient of friction in Example 6.5 from 1.0 (dry pavement) to 0.1 (icy pavement), the maximum speed for the turn goes down to 4.4 m/s—about 10 mph!

Race cars turn corners at much higher speeds than normal passenger vehicles. One design modification of the *cars* to allow this is the addition of wings, as on the car in Figure 6.14. The wings provide an additional force pushing the car *down* onto the pavement by deflecting air upward, causing the air to push down on the car. The extra downward force increases the normal force, thus increasing the maximum static friction force and making faster turns possible.

There are also design modifications of the *track* to allow race cars to corner at high speeds. If the track is banked by raising the outside edge of curved sections, the normal force can provide some of the necessary centripetal force, as we will see in the next example. The curves on race tracks may be quite sharply banked. Curves on ordinary highways are often banked as well, though at more modest angles suiting the lower speeds.

EXAMPLE 6.6 Finding speed on a banked turn

A curve on a racetrack of radius 70 m is banked at a 15° angle. At what speed can a car take this curve without assistance from friction?

PREPARE The purpose of banking becomes clear if you look at the free-body diagram in Figure 6.15. The normal force $\vec{n}$ is perpendicular to the road, so tilting the road causes $\vec{n}$ to have a component toward the center of the circle. The horizontal component n_x is the inward force that causes the centripetal acceleration needed to turn the car. Notice that no friction force is shown because the car is assumed to make the turn without assistance from friction.

Top view

Known
$r = 70$ m
$\theta = 15°$
Find
v

Rear view

FIGURE 6.15 Visual overview for the car on a banked turn.

SOLVE Without friction, $n_x = n\sin\theta$ is the only component of force toward the center of the circle. It is this inward component of the normal force on the car that causes it to turn the corner. Newton's second law is

$$\sum F_x = n\sin\theta = \frac{mv^2}{r}$$

$$\sum F_y = n\cos\theta - w = 0$$

where θ is the angle at which the road is banked and we've assumed that the car is traveling at the correct speed v. From the y-equation,

$$n = \frac{w}{\cos\theta} = \frac{mg}{\cos\theta}$$

Substituting this into the x-equation and solving for v gives

$$\left(\frac{mg}{\cos\theta}\right)\sin\theta = mg\tan\theta = \frac{mv^2}{r}$$

$$v = \sqrt{rg\tan\theta} = 14 \text{ m/s}$$

ASSESS This is ≈ 30 mph, a reasonable speed. Only at this very specific speed can the turn be negotiated without reliance on friction forces.

Maximum Walking Speed

Humans and other two-legged animals have two basic gaits: walking and running. At slow speeds, you walk. When you need to go faster, you run. Why don't you just walk faster? There is an upper limit to the speed of walking, and this limit is set by the physics of circular motion.

Think about the motion of your body as you take a walking stride. You put one foot forward, then push off with your rear foot. Your body pivots over your front foot, and you bring your rear foot forward to take the next stride. As we can see in Figure 6.16a, the pivoting of your body over the forward foot is the arc of a circle. In a walking gait, your body is in circular motion as you pivot on your forward foot.

A force toward the center of the circle is required for this circular motion, as shown in Figure 6.16. Figure 6.16b shows the forces acting on the person's body during the midpoint of the stride: weight, directed down, and the normal force of the ground, directed up. Newton's second law for the x-axis is

$$\sum F_x = w - n = \frac{mv^2}{r}$$

Notice that $n < w$. Your body tries to "lift off" as it pivots over your foot, decreasing the normal force exerted on you by the ground. The normal force becomes smaller as you walk faster, but n cannot be less than zero. Thus the maximum possible walking speed v_{max} occurs when $n = 0$. Setting $n = 0$,

$$w = mg = \frac{mv_{max}^2}{r}$$

Thus

$$v_{max} = \sqrt{gr} \tag{6.9}$$

The maximum possible walking speed is limited by r, the length of the leg, and g, the free-fall acceleration. This formula is a good approximation to the maximum

(a) Walking stride   The speed of the circular motion is the walking speed.

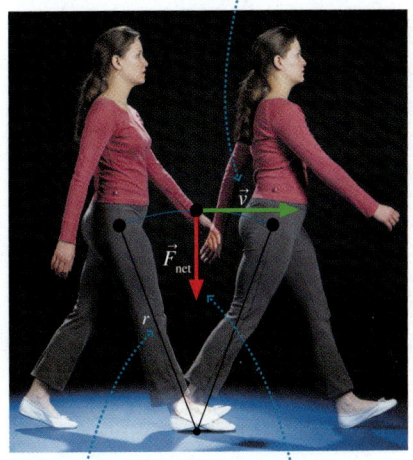

The radius of the circular motion is the length of the leg from the foot to the hip.

The circular motion requires a force directed toward the center of the circle.

(b) Forces in the stride   Side view (same as photo)

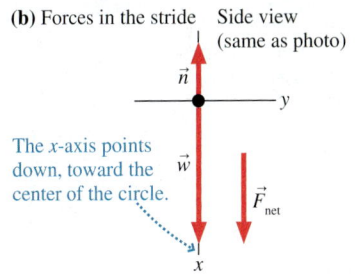

The x-axis points down, toward the center of the circle.

FIGURE 6.16 Analysis of a walking stride.

walking speed for humans and other animals. The maximum walking speed is higher for animals with longer legs. Giraffes, with their very long legs, can walk at high speeds. Animals such as mice with very short legs have such a low maximum walking speed that they rarely use this gait. Mice will generally run to get from one place to another.

For humans, the length of the leg is approximately 0.7 m, so we calculate a maximum speed of

$$v_{\mathrm{max}} \approx 2.6 \text{ m/s} \approx 6 \text{ mph}$$

You *can* walk this fast, though it becomes energetically unfavorable to do so at speeds above 4 mph. Most people will make a transition to a running gait at about this speed. Children, with their shorter legs, must make a transition to a running gait at a much lower speed.

STOP TO THINK 6.2 A block on a string spins in a horizontal circle on a frictionless table. Rank in order, from largest to smallest, the tensions T_A to T_E acting on the blocks A to E.

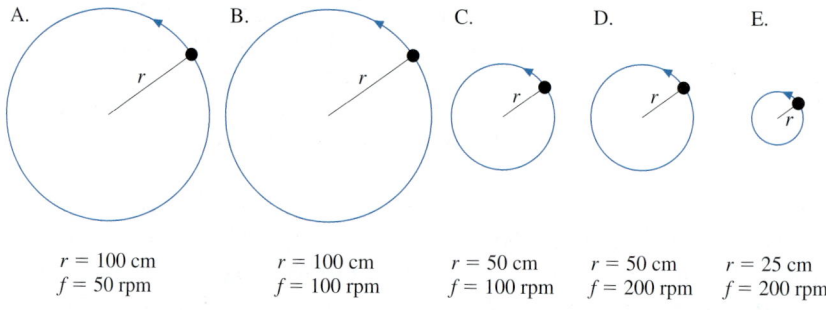

A.	B.	C.	D.	E.
$r = 100$ cm	$r = 100$ cm	$r = 50$ cm	$r = 50$ cm	$r = 25$ cm
$f = 50$ rpm	$f = 100$ rpm	$f = 100$ rpm	$f = 200$ rpm	$f = 200$ rpm

6.4 Apparent Forces in Circular Motion

Figure 6.17 shows a carnival ride that spins the riders around inside a large cylinder. The people are "stuck" to the inside wall of the cylinder! As you probably know from experience, the riders *feel* that they are being pushed outward, into the wall. But our analysis has found that an object in circular motion must have an *inward* force to create the centripetal acceleration. How can we explain this apparent difference?

Centrifugal Force?

If you are a passenger in a car that turns a corner quickly, you may feel "thrown" against the door. But is there really such a force? Figure 6.18 shows a bird's-eye view of you riding in a car as it makes a left turn. You try to continue moving in a straight line, obeying Newton's first law, when—without having been provoked—the door suddenly turns in front of you and runs into you! You do, indeed, then feel the force of the door because it is now the force of the door, pushing *inward* toward the center of the curve, that is causing you to turn the corner. But you were not "thrown" into the door; the door ran into you.

A "force" that *seems* to push an object to the outside of a circle is called a *centrifugal force*. Despite having a name, there really is no such force. What you feel is your body trying to move ahead in a straight line (which would take you away from the center of the circle) as outside forces act to turn you in a circle. The only real forces, those that appear on free-body diagrams, are the ones pushing inward toward the center. **A centrifugal force will never appear on a free-body diagram and never be included in Newton's laws.**

FIGURE 6.17 Inside the Gravitron, a rotating circular room.

You tried to keep moving straight ahead.

$\vec{v}$

The door provides the center-directed force that makes you move in a circle.

$\vec{n}$

Center of curve

FIGURE 6.18 Bird's-eye view of a passenger in a car turning a corner.

With this in mind, let's revisit the rotating carnival ride. A person watching from above would see the riders in the cylinder moving in a circle with the walls providing the inward force that causes their centripetal acceleration. The riders *feel* as if they're being pushed outward because their attempts to move in a straight line are being resisted by the wall of the cylinder, which keeps getting in the way. But feelings aren't forces. The only actual force is the contact force of the cylinder wall pushing *inward*.

Apparent Weight in Circular Motion

Imagine swinging a bucket of water over your head. If you swing the bucket quickly, the water stays in. But you'll get a shower if you swing too slowly. Why does the water stay in the bucket? Or think about a roller coaster that does a loop-the-loop. How does the car stay on the track when it's upside down? You might have said that there was a centrifugal force holding the water in the bucket and the car on the track, but we have seen that there really isn't a centrifugal force. Analyzing these questions will tell us a lot about forces in general and circular motion in particular.

Figure 6.19a shows a roller coaster car going around a vertical loop-the-loop of radius r. If you've ever ridden a roller coaster, you know that your sensation of weight changes as you go over the crests and through the dips. To understand why, let's look at the forces on passengers going through the loop. To simplify our analysis, we will assume that the speed of the car stays constant as it moves through the loop.

Figure 6.19b shows a passenger's free-body diagram at the *bottom* of the loop. The only forces acting on the passenger are her weight $\vec{w}$ and the normal force $\vec{n}$ of the seat pushing up on her. Recall, from Chapter 5, that the passenger's apparent weight, her sensation of weight, is the magnitude of the force supporting her. Here the seat is supporting her with the normal force $\vec{n}$, so her apparent weight is $w_{app} = n$. Based on our understanding of circular motion, we can say that

- She's moving in a circle, so there *must* be a net force toward the center of the circle—above her head—to provide the centripetal acceleration.
- The net force points *upward,* so it must be the case that $n > w$.
- Her apparent weight is $w_{app} = n$, so her apparent weight is larger than her true weight ($w_{app} > w$). Thus she "feels heavy" at the bottom of the circle.

In short, the normal force has to *exceed* the weight force to provide the net force she needs to "turn the corner" at the bottom of the circle. To analyze the situation quantitatively, we note from the force vectors in Figure 6.19b that the magnitude of the net force at the bottom of the loop is

$$(F_{net})_{bottom} = n - w$$

The net force provides the centripetal acceleration, so we can say that

$$(F_{net})_{bottom} = n - w = \frac{mv^2}{r} \qquad (6.10)$$

From Equation 6.10 we find her apparent weight to be:

$$(w_{app})_{bottom} = n = w + \frac{mv^2}{r} \qquad (6.11)$$

The passenger's apparent weight at the bottom is *larger* than her true weight w, which agrees with your experience when you go through a dip or a valley.

Now let's look at the roller coaster car as it crosses the top of the loop. Things are a little trickier here. Whereas the normal force of the track pushes up when the car is at the bottom of the circle, it pushes *down* when the car is at the top and the

FIGURE 6.19 A roller coaster car going around a loop-the-loop.

When "down" is up **BIO** You can tell, even with your eyes closed, what direction is down. This sense is due to small crystals of calcium carbonate, called *otoliths,* in your inner ears. Gravity pulls the otoliths down, so a normal force from a sensitive supporting membrane must push them up. Your brain interprets "down" as the opposite of this normal force. Normally, what your ears tell you is "down" is really down. But at the top of a loop in a roller coaster, the normal force is directed down, so your inner ear tells you that "down" is up! If your ears tell you one thing and your eyes another, it can be quite disorienting.

track is above the car. Figure 6.19c shows the car's free-body diagram at the top of the loop. It's worth thinking carefully about this diagram to make sure you understand what it is showing.

The car is still moving in a circle, so there must be a net force *downward,* toward the center of the circle, to provide the centripetal acceleration. We see from force vectors in the free-body diagram of Figure 6.19c that the net force is now

$$(F_{net})_{top} = n + w$$

Thus we can compute

$$(F_{net})_{top} = n + w = \frac{mv^2}{r} \tag{6.12}$$

The normal force that the track exerts on the car at the top of the circle is

$$n = \frac{mv^2}{r} - w \tag{6.13}$$

The apparent weight of the car (and the passengers in the car) is

$$(w_{app})_{top} = n = \frac{mv^2}{r} - w \tag{6.14}$$

If *v* is sufficiently large, the apparent weight can exceed the true weight.

Let's look at what happens if the car goes slower. Notice from Equation 6.14 that, as *v* decreases, there comes a point when *n* reaches zero. At that point, the track is *not* pushing against the car. Instead, the car is able to complete the circle because the weight force alone provides sufficient centripetal acceleration.

The speed at which $n = 0$ is called the *critical speed* v_c:

$$v_c = \sqrt{\frac{rw}{m}} = \sqrt{\frac{rmg}{m}} = \sqrt{rg} \tag{6.15}$$

The critical speed is the slowest speed at which the car can complete the circle. To understand why, notice that Equation 6.14 gives a negative value for *n* if $v < v_c$. But that is physically impossible. The track can push against the wheels of the car ($n > 0$), but it can't pull on them, so the slowest possible speed is the speed for which $n = 0$ at the top. If $v < v_c$, the car cannot turn the full loop but, instead, comes off the track and becomes a projectile!

Water stays in a bucket swung over your head for the same reason. The bottom of the bucket pushes against the water to provide the inward force that causes circular motion. If you swing the bucket too slowly, the force of the bucket on the water drops to zero. At that point, the water leaves the bucket and becomes a projectile following a parabolic trajectory onto your head!

A fast-spinning world Saturn, a gas giant planet composed largely of fluid matter, is quite a bit larger than the earth. It also rotates much more quickly, completing one rotation in just under 11 hours. The rapid rotation decreases the apparent weight at the equator enough to distort the fluid surface; the planet is noticeably out of round, as the red circle shows. The diameter at the equator is 11% greater than the diameter at the poles.

EXAMPLE 6.7 How slow can you go?

A motorcyclist in the Globe of Death, pictured at the start of the chapter, rides in a 2.2-m-radius vertical loop. To keep control of the bike, the rider wants the normal force on his tires at the top of the loop to equal or exceed his and the bike's combined weight. What is the minimum speed at which the rider can take the loop?

PREPARE The visual overview for this problem is shown in Figure 6.20. At the top of the loop, the normal force of the cage on the tires is a *downward* force. In accordance with Problem-Solving Strategy 6.1, we've chosen the *x*-axis to point toward the center of the circle.

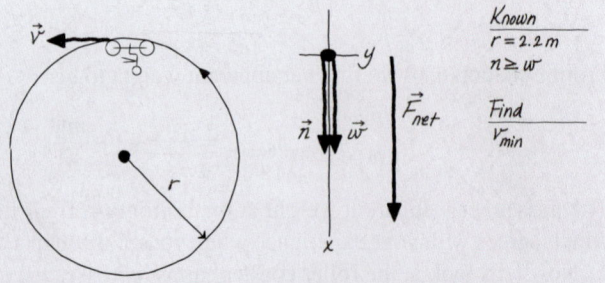

FIGURE 6.20 Riding in a vertical loop around the Globe of Death.

SOLVE We will consider the forces at the top point of the loop. Because the x-axis points downward, Newton's second law is

$$\sum F_x = w + n = \frac{mv^2}{r}$$

The minimum acceptable speed occurs when $n = w$, thus

$$2w = 2mg = \frac{mv_{min}^2}{r}$$

Solving for the speed, we find

$$v_{min} = \sqrt{2gr} = \sqrt{2(9.8 \text{ m/s}^2)(2.2 \text{ m})} = 6.6 \text{ m/s}$$

ASSESS The minimum speed is ≈ 15 mph, which isn't all that fast; the bikes can easily reach this speed. But normally several bikes are in the globe at one time. The big challenge is to keep all of the riders in the cage moving at this speed in synchrony. The period for the circular motion at this speed is $T = 2\pi r/v \approx 2$ s, leaving little room for error!

Centrifuges

The *centrifuge*, an important biological application of circular motion, is used to separate the components of a liquid with different densities. Typically these are different types of cells, or the components of cells, suspended in water. By using very high angular velocities, the centrifuge produces centripetal accelerations that are thousands of times greater than the acceleration due to gravity.

You probably know that a less-dense liquid, such as oil, floats on top of a more-dense liquid, such as water. That is, gravity naturally separates the components of a liquid so that the most dense components are on the bottom and the least dense on top. But suppose the liquid consists of many small objects, such as cells, suspended in water. Although gravity causes the cells to slowly drift toward the bottom of the container, the terminal velocity for very small objects can be minuscule. It could take days or even months for the cells to settle out and separate by density. And the slightest agitation would destroy the fragile separation. It's not practical to wait for biological samples to separate due to gravity alone.

The separation would go faster if the force of gravity could be increased. Although we can't change gravity, we can increase the apparent weight of objects in the sample by spinning them very fast, and that is what the centrifuge in Figure 6.21 does. As the centrifuge effectively increases gravity to thousands of times its normal value, the cells or cell components settle out and separate by density in a matter of minutes or hours.

A centrifuge.

The high angular velocity requires a large normal force, which leads to a large apparent weight.

FIGURE 6.21 The operation of a centrifuge.

EXAMPLE 6.8 Analyzing the ultracentrifuge

An 18-cm-diameter ultracentrifuge produces an extraordinarily large centripetal acceleration of 250,000g. What is the frequency in rpm? What is the apparent weight of a sample with a mass of 0.0030 kg?

PREPARE The acceleration in SI units is

$$a = 250,000(9.80 \text{ m/s}^2) = 2.45 \times 10^6 \text{ m/s}^2$$

The radius is half the diameter, or $r = 9.0 \text{ cm} = 0.090 \text{ m}$.

SOLVE The centripetal acceleration is related to the angular speed by $a = \omega^2 r$. Thus

$$\omega = \sqrt{\frac{a}{r}} = \sqrt{\frac{2.45 \times 10^6 \text{ m/s}^2}{0.090 \text{ m}}} = 5.22 \times 10^3 \text{ rad/s}$$

This corresponds to a frequency

$$f = \frac{\omega}{2\pi} = \frac{5.22 \times 10^3 \text{ rad/s}}{2\pi} = 830 \text{ rev/s}$$

Converting to rpm, we find

$$830 \frac{\text{rev}}{\text{s}} \times \frac{60 \text{ s}}{1 \text{ min}} = 50,000 \text{ rpm}$$

Continued

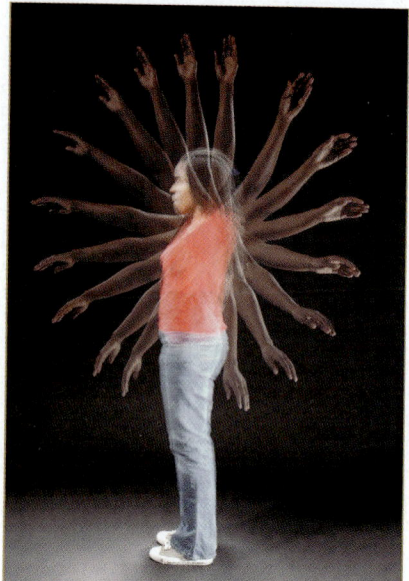

Human centrifuge BIO If you spin your arm rapidly in a vertical circle, the motion will produce an effect like that in a centrifuge. The motion will assist outbound blood flow in your arteries and retard inbound blood flow in your veins. There will be a buildup of fluid in your hand that you will be able to see (and feel!) quite easily.

At this rotation rate, the 0.0030 kg mass has an apparent weight

$$w_{app} = ma = (3.0 \times 10^{-3}\,\text{kg})(2.45 \times 10^6\,\text{m/s}^2) = 7.4 \times 10^3\,\text{N}$$

ASSESS Because the acceleration is 250,000g, the apparent weight is 250,000 times the actual weight. The forces in the ultracentrifuge are very large and can destroy the machine if it is not carefully balanced.

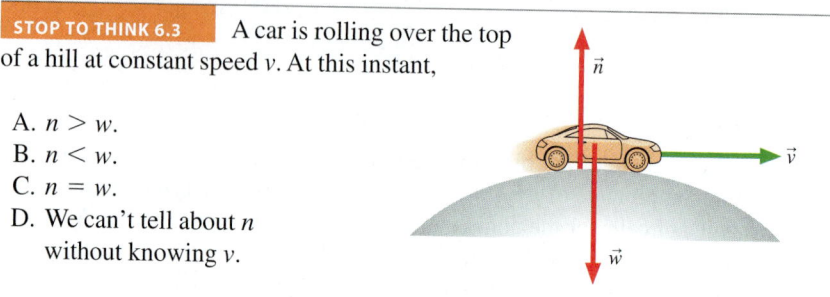

STOP TO THINK 6.3   A car is rolling over the top of a hill at constant speed v. At this instant,

A. $n > w$.
B. $n < w$.
C. $n = w$.
D. We can't tell about n without knowing v.

6.5 Circular Orbits and Weightlessness

The Space Shuttle orbits the earth in a circular path at a speed of over 15,000 miles per hour. What forces act on it? Why does it move in a circle? Before we start considering the physics of orbital motion, let's return, for a moment, to projectile motion. Projectile motion occurs when the only force on an object is gravity. Our analysis of projectiles made an implicit assumption that the earth is flat and that the free-fall acceleration, due to gravity, is everywhere straight down. This is an acceptable approximation for projectiles of limited range, such as baseballs or cannon balls, but there comes a point where we can no longer ignore the curvature of the earth.

Orbital Motion

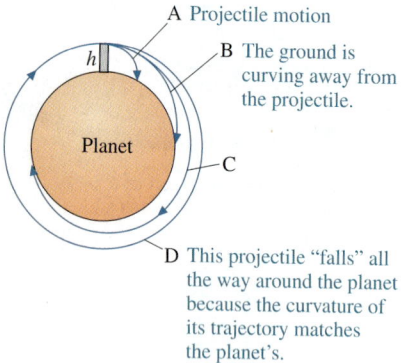

FIGURE 6.22 Projectiles being launched at increasing speeds from height h on a smooth, airless planet.

Figure 6.22 shows a perfectly smooth, spherical, airless planet with one vertical tower of height h. A projectile is launched from this tower with initial speed v_i parallel to the ground. If v_i is very small, as in trajectory A, the "flat-earth approximation" is valid and the problem is identical to Example 3.11 in which a car drove off a cliff. The projectile simply falls to the ground along a parabolic trajectory.

As the initial speed v_i is increased, it seems to the projectile that the ground is curving out from beneath it. It is still falling the entire time, always getting closer to the ground, but the distance that the projectile travels before finally reaching the ground—that is, its range—increases because the projectile must "catch up" with the ground that is curving away from it. Trajectories B and C are of this type.

If the launch speed v_i is sufficiently large, there comes a point at which the curve of the trajectory and the curve of the earth are parallel. In this case, the projectile "falls" but it never gets any closer to the ground! This is the situation for trajectory D. The projectile returns to the point from which it was launched, at the same speed at which it was launched, making a closed trajectory. Such a closed trajectory around a planet or star is called an **orbit.**

The most important point of this qualitative analysis is that, in the absence of air resistance, **an orbiting projectile is in free fall.** This is, admittedly, a strange idea, but one worth careful thought. An orbiting projectile is really no different from a thrown baseball or a car driving off a cliff. The only force acting on it is gravity, but its tangential velocity is so large that the curvature of its trajectory matches the curvature of the earth. When this happens, the projectile "falls" under the influence of gravity but never gets any closer to the surface, which curves away beneath it.

When we first studied free fall in Chapter 2, we said that the free-fall acceleration is always directed vertically downward. As we see in Figure 6.23, this is due to our perspective; "downward" really means "toward the center of the earth." For a projectile in orbit, the direction of the force of gravity changes, always pointing toward the center of the earth.

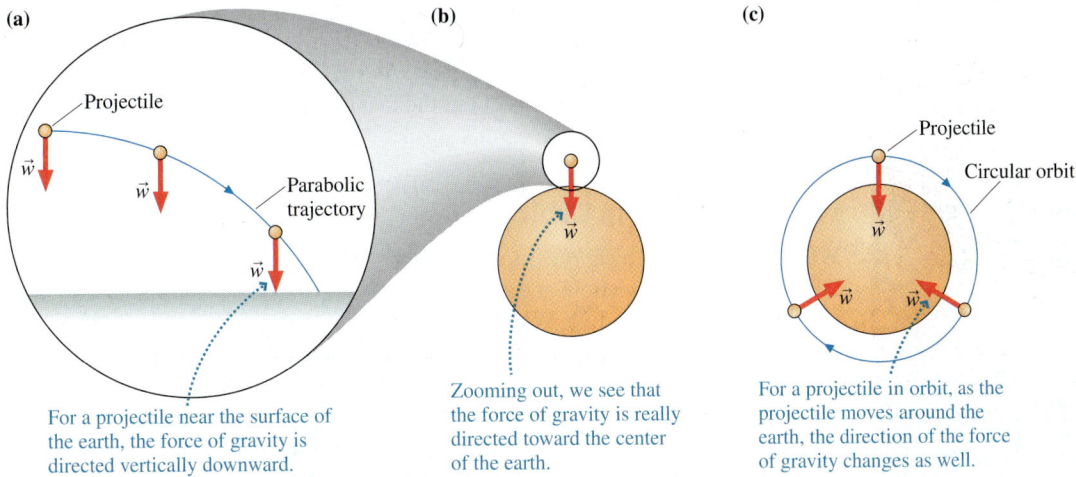

(a)

Projectile

$\vec{w}$

$\vec{w}$

Parabolic trajectory

$\vec{w}$

For a projectile near the surface of the earth, the force of gravity is directed vertically downward.

(b)

$\vec{w}$

Zooming out, we see that the force of gravity is really directed toward the center of the earth.

(c)

Projectile

Circular orbit

$\vec{w}$

$\vec{w}$ $\vec{w}$

For a projectile in orbit, as the projectile moves around the earth, the direction of the force of gravity changes as well.

FIGURE 6.23 The force of gravity is really directed toward the center of the earth.

As you have learned, a force of constant magnitude that always points toward the center of a circle causes the centripetal acceleration of uniform circular motion. Because the only force acting on the orbiting projectile in Figure 6.23 is gravity, and we're assuming the projectile is very near the surface of the earth, we can write

$$a = \frac{F_{net}}{m} = \frac{w}{m} = \frac{mg}{m} = g \tag{6.16}$$

An object moving in a circle of radius r at speed v_{orbit} will have this centripetal acceleration if

$$a = \frac{(v_{orbit})^2}{r} = g \tag{6.17}$$

That is, if an object moves parallel to the surface with the speed

$$v_{orbit} = \sqrt{rg} \tag{6.18}$$

then the free-fall acceleration provides exactly the centripetal acceleration needed for a circular orbit of radius r. An object with any other speed will not follow a circular orbit.

The earth's radius is $r = R_e = 6.37 \times 10^6$ m. The orbital speed of a projectile just skimming the surface of an airless, bald earth is

$$v_{orbit} = \sqrt{R_e g} = \sqrt{(6.37 \times 10^6 \text{ m})(9.80 \text{ m/s}^2)} = 7900 \text{ m/s} \approx 18,000 \text{ mph}$$

We can use v_{orbit} to calculate the period of the satellite's orbit:

$$T = \frac{2\pi r}{v_{orbit}} = 2\pi \sqrt{\frac{r}{g}} \tag{6.19}$$

For this earth-skimming orbit, $T = 5065$ s $= 84.4$ min.

Of course, this orbit is unrealistic; even if there were no trees and mountains, a real projectile moving at this speed would burn up from the friction of air resistance. Suppose, however, that we launched the projectile from a tower of height $h = 200$ mi $\approx 3.2 \times 10^5$ m, above most of the earth's atmosphere. This is approximately the height of low-earth-orbit satellites, such as the Space Shuttle. Note that $h \ll R_e$, so the radius of the orbit $r = R_e + h = 6.69 \times 10^6$ m is only 5% greater than the earth's radius. Many people have a mental image that satellites orbit far above the earth, but in fact most satellites come pretty close to skimming the surface.

At this slightly larger value of r, Equation 6.19 gives $T = 87$ min. The actual period of the Space Shuttle at an altitude of 200 mi is about 91 minutes, so our calculation is very good—but not perfect. As we'll see in the next section, a correct calculation must take into account the fact that gravity gradually gets weaker at higher elevations above the earth's surface.

Weightlessness in Orbit

When we discussed *weightlessness* in Chapter 5, we saw that it occurs during free fall. We asked the question, at the end of Section 5.4, whether astronauts and their spacecraft are in free fall. We can now give an affirmative answer: They are, indeed, in free fall. They are falling continuously around the earth, under the influence of only the gravitational force, but never getting any closer to the ground because the earth's surface curves beneath them. Weightlessness in space is no different from the weightlessness in a free-falling elevator. **Weightlessness does *not* occur from an absence of weight or an absence of gravity.** Instead, the astronaut, the spacecraft, and everything in it are "weightless" (i.e., their *apparent* weight is zero) because they are all falling together.

Zero apparent weight in the Space Shuttle.

The Orbit of the Moon

If a satellite is simply "falling" around the earth, with the gravitational force causing a centripetal acceleration, then what about the moon? Is it obeying the same laws of physics? Or do celestial objects obey laws that we cannot discover by experiments here on earth?

The radius of the moon's orbit around the earth is 3.84×10^8 m. If we use Equation 6.19 to calculate the period of the moon's orbit, the time the moon takes to circle the earth once, we get

$$T = 2\pi \sqrt{\frac{r}{g}} = 2\pi \sqrt{\frac{3.84 \times 10^8 \text{ m}}{9.80 \text{ m/s}^2}} = 655 \text{ min} \approx 11 \text{ hours} \qquad (6.20)$$

This is clearly wrong; the period of the moon's orbit is approximately one month.

Newton believed that the laws of motion he had discovered were *universal* and so should apply to the motion of the moon as well as to the motion of objects in the laboratory. But why should we assume that the free-fall acceleration g is the same at the distance of the moon as it is on or near the earth's surface? If gravity is the force of the earth pulling on an object, it seems plausible that the size of that force, and thus the size of g, should diminish with increasing distance from the earth.

Newton proposed the idea that the earth's force of gravity decreases with the square of the distance from the earth. This is the basis of *Newton's law of gravity,* a topic we will study in the next section. Gravity is less at the distance of the moon—exactly the strength needed to make the moon orbit at the observed rate. The moon, just like the space shuttle, is simply "falling" around the earth!

Rotating space stations BIO The weightlessness astronauts experience in orbit has serious physiological consequences. Astronauts who spend time in weightless environments lose bone and muscle mass and suffer other adverse effects. One solution is to introduce "artificial gravity." On a space station, the easiest way to do this would be to make the station rotate, producing an apparent weight. The designers of this space station model for the movie *2001: A Space Odyssey* made it rotate for just that reason.

6.6 Newton's Law of Gravity

A popular image has Newton thinking of the idea of gravity after an apple fell on his head. This amusing story is at least close to the truth. Newton himself said that the "notion of gravitation" came to him as he "sat in a contemplative mood" and "was occasioned by the fall of an apple." It occurred to him that, perhaps, the apple was attracted to the center of the earth but was prevented from getting there by the earth's surface. And if the apple was so attracted, why not the moon? Newton's genius was his sudden realization that **the force of the earth on the moon (and of the sun on the planets) was identical to the force of the earth on the apple.** In other words, gravitation is a *universal* force between all objects in the universe! This is not shocking today, but no one before Newton had ever thought that the mundane motion of objects on earth had any connection at all with the stately motion of the planets around the sun.

Isaac Newton was born to a poor farming family in 1642, the year of Galileo's death. He entered Trinity College at Cambridge University at age 19 as a "subsizar," a poor student who had to work his way through school. Newton graduated in 1665, at age 23, just as an outbreak of the plague in England forced the universities to close for two years. He returned to his family farm for that period, during which he made important experimental discoveries in optics, laid the foundations for his theories of mechanics and gravitation, and made major progress toward his invention of calculus as a whole new branch of mathematics.

Gravity Obeys an Inverse-Square Law

Newton also recognized that the strength of gravity must decrease with distance, as we saw at the end of the previous section. These two notions about gravity—that it is universal, and that it decreases with distance—form the basis for Newton's law of gravity.

Newton proposed that *every* object in the universe attracts *every other* object with a force that has the following properties:

1. The force is inversely proportional to the square of the distance between the objects.
2. The force is directly proportional to the product of the masses of the two objects.

Figure 6.24 shows two spherical masses m_1 and m_2 separated by distance r. Each mass exerts an attractive force on the other, a force that we call the **gravitational force.** These two forces form an action/reaction pair, so $\vec{F}_{1\,\text{on}\,2}$ is equal in magnitude and opposite in direction to $\vec{F}_{2\,\text{on}\,1}$. The magnitude of the forces is given by Newton's law of gravity.

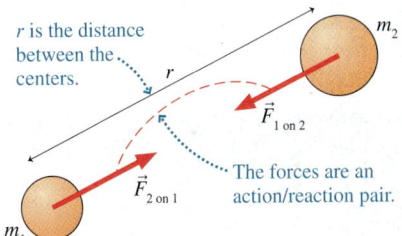

FIGURE 6.24 The gravitational forces on masses m_1 and m_2.

> **Newton's law of gravity** If two objects with masses m_1 and m_2 are a distance r apart, the objects exert attractive forces on each other of magnitude
>
> $$F_{1\,\text{on}\,2} = F_{2\,\text{on}\,1} = \frac{Gm_1m_2}{r^2} \qquad (6.21)$$
>
> The forces are directed along the line joining the two objects.
>   The constant G is called the **gravitational constant.** In the SI system of units, G has the value
>
> $$G = 6.67 \times 10^{-11} \text{ N} \cdot \text{m}^2/\text{kg}^2$$

NOTE ▶ Strictly speaking, Newton's law of gravity applies to *particles* with masses m_1 and m_2. Fortunately, it can be shown that the law also applies to the force between two spherical objects if r is the distance between their centers. ◀

As the distance r between two objects increases, the gravitational force between them decreases. Because the distance appears squared in the denominator, Newton's law of gravity is what we call an **inverse-square** law. Doubling the distance between two masses causes the force between them to decrease by a factor of 4. This mathematical form is one we will see again, so it is worth our while to explore it in more detail.

Inverse-square relationships

 Exercise 23

Two quantities have an **inverse-square relationship** if y is inversely proportional to the *square* of x. We write the mathematical relationship as:

$$y = \frac{A}{x^2}$$

y equals A over (x squared)

$y = \dfrac{A}{x^2}$

When x is 1, y is A.

When x is 2, y is $A/4$. That is, doubling x reduces y by a factor of four (two squared). We don't need to know A to know this.

SCALING Inverse-square scaling means, for example:

- If you double x, you decrease y by a factor of 4, as you can see in the graph.
- If you increase x by a factor of 3, you decrease y by a factor of 9.
- If you decrease x by a factor of 3, you increase y by a factor of 9.

Generally: **An *increase* in x by a factor C results in a *decrease* of y by a factor C^2.**

LIMITS As x becomes large, y becomes very small; as x becomes small, y becomes very large.

CONCEPTUAL EXAMPLE 6.4 Varying gravitational force

The gravitational force between two giant lead spheres is 0.010 N when the spheres are 20 m apart, measured between their centers. What is the distance between their centers when the gravitational force between them is 0.160 N?

REASON We can solve this problem without knowing the masses of the two spheres. The key is to consider the ratios of forces and distances. Gravity is an inverse-square relationship; the force is related to the inverse square of the distance. The force *increases* by a factor $(0.160 \text{ N})/(0.010 \text{ N}) = 16$, so the distance must *decrease* by a factor $\sqrt{16} = 4$. The distance is thus $(20 \text{ m})/4 = 5.0 \text{ m}$.

ASSESS This type of ratio reasoning is a very good way to get a quick handle on the solution to a problem.

EXAMPLE 6.9 Gravitational force between two people

You are seated in your physics class next to another student 0.60 m away. Estimate the magnitude of the gravitational force between you. Assume that you each have a mass of 65 kg.

PREPARE We will model each of you as a sphere; this is not a particularly good model, but it will do for making an estimate. We will take the 0.60 m as the distance between your centers.

SOLVE The gravitational force is given by Equation 6.21:

$$F_{(\text{you}) \text{on} (\text{other student})} = \frac{G m_{\text{you}} m_{\text{other student}}}{r^2}$$

$$= \frac{(6.67 \times 10^{-11} \text{ N} \cdot \text{m}^2/\text{kg}^2)(65 \text{ kg})(65 \text{ kg})}{(0.60 \text{ m})^2}$$

$$= 7.8 \times 10^{-7} \text{ N}$$

ASSESS The force is quite small, roughly the weight of one hair on your head.

There is a gravitational force between all objects in the universe, but the gravitational force between two ordinary-sized objects is very small, and is not something you have ever noticed. Only when one (or both) of the masses is exceptionally large does the force of gravity become important. The downward force of the earth on you—your weight—is large, because the earth has an enormous mass. And the attraction is mutual; by Newton's third law, you exert an upward force on the earth that is equal to your weight. The large mass of the earth makes the effect of this force on the earth negligible, though.

EXAMPLE 6.10 Gravitational force of the earth on a person

What is the magnitude of the gravitational force of the earth on a 60 kg person? The earth has mass 5.98×10^{24} kg and radius 6.37×10^6 m.

PREPARE We'll again model the person as a sphere. The distance r in Newton's law of gravity is the distance between the centers of the two spheres. The size of the person is negligible compared to the size of the earth, so we can use the earth's radius as r.

SOLVE The force of gravity on the person due to the earth can be computed using Equation 6.21:

$$F_{\text{earth on } m} = \frac{GM_e m}{R_e^2} = \frac{(6.67 \times 10^{-11} \text{ N} \cdot \text{kg}^2/\text{m}^2)(5.98 \times 10^{24} \text{ kg})(60 \text{ kg})}{(6.37 \times 10^6 \text{ m})^2}$$

$$= 590 \text{ N}$$

ASSESS This force is exactly the same as we would calculate using the formula for the weight force, $w = mg$. This isn't surprising, though. Chapter 5 introduced the weight of an object as simply the "force of gravity" acting on it. Newton's law of gravity is a more fundamental law for calculating the force of gravity, but it's still the same force that we earlier called "weight."

Variable gravity When we calculated the force of the earth's gravity we assumed that the earth's shape and composition are uniform. Neither is quite true, so there is a very small variation in gravity at the surface of the earth, as shown in this image. Red means slightly greater surface gravity; blue, slightly weaker. These variations are caused by differing distances from the earth's center and by unevenness in density of the earth's crust. Though these variations are important for scientists studying the earth, they are small enough that we can ignore them for the computations we'll do in this textbook.

NOTE ▶ We will use uppercase R and M to represent the large mass and radius of a star or planet, as we did in Example 6.10. ◀

The force of gravitational attraction between the earth and you is responsible for your weight. If you were to venture to another planet, your *mass* would be the same but your *weight* would vary, as we discussed in Chapter 5. We will now explore this concept in more detail.

Gravity on Other Worlds

When astronauts ventured to the moon, television images showed them walking—and even jumping and skipping—with some ease, even though they were wearing life support systems with a mass of over 80 kg. This was a visible reminder that the weight of objects is less on the moon. Let's consider why this is so.

Figure 6.25 shows an astronaut on the moon weighing a rock of mass m. When we compute the weight of an object on the surface of the earth, we use the formula $w = mg$. We can do the same calculation for a mass on the moon, as long as we use the value of g on the moon.

$$w = mg_{\text{moon}} \tag{6.22}$$

This is the "little g" perspective. Falling body experiments on the moon would give the value of g_{moon} as 1.62 m/s^2.

But we can also take a "big G" perspective. The weight of the rock comes from the gravitational attraction of the moon, and we can compute this weight using Equation 6.21. The distance r is the radius of the moon, which we'll call R_{moon}. Thus

$$F_{\text{moon on } m} = \frac{GM_{\text{moon}} m}{R_{\text{moon}}^2} \tag{6.23}$$

Because Equations 6.22 and 6.23 are two names and two expressions for the same force, we can equate the right-hand sides to find that

$$g_{\text{moon}} = \frac{GM_{\text{moon}}}{R_{\text{moon}}^2}$$

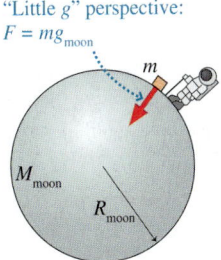

"Little g" perspective:
$F = mg_{\text{moon}}$

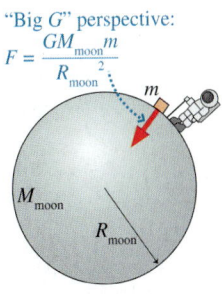

"Big G" perspective:
$F = \dfrac{GM_{\text{moon}} m}{R_{\text{moon}}^2}$

FIGURE 6.25 An astronaut weighing a mass on the moon.

Walking on the moon BIO The low lunar gravity made walking at a reasonable pace difficult for the Apollo astronauts, but the reduced weight made jumping quite easy. Videos from the surface of the moon often show the astronauts getting from place to place by hopping or skipping—not for fun, but for speed and efficiency.

We have done this calculation for an object on the moon, but the result is completely general. At the surface of a planet (or a star), the free-fall acceleration g, a consequence of gravity, can be computed as

$$g_{planet} = \frac{GM_{planet}}{R_{planet}^2} \qquad (6.24)$$

Free-fall acceleration on the surface of a planet

LINEAR
p. 38

INVERSE-
SQUARE
p. 194

If we use values for the mass and the radius of the moon from the table inside the cover of the book, we can compute $g_{moon} = 1.62 \text{ m/s}^2$. This means that an object will weigh less on the moon than it would on the earth, where g is 9.80 m/s^2. A 70 kg astronaut wearing an 80 kg spacesuit will weigh over 330 lb on the earth but only 54 lb on the moon.

The low lunar gravity makes walking very easy, but a walking pace on the moon will be very slow. Earlier in the chapter we found that the maximum walking speed was $v_{max} = \sqrt{gr}$ where r is the length of the leg. For a typical leg length of 0.7 m and the gravity of the moon, the *maximum* walking speed is about 1 m/s, just over 2 mph—a very gentle stroll!

Equation 6.24 gives g at the surface of a planet. More generally, imagine an object at distance $r > R$ from the center of a planet. Its free-fall acceleration at this distance is

$$g = \frac{GM}{r^2} \qquad (6.25)$$

This more general result agrees with Equation 6.24 if $r = R$, but it allows us to determine the "local" free-fall acceleration at distances $r > R$. Equation 6.25 expresses Newton's idea that the size of g should decrease as you get farther from the earth.

As you're flying in a jet airplane at a height of about 10 km, the free-fall acceleration is about 0.3% less than on the ground. At the height of the space shuttle, about 300 km, Equation 6.25 gives $g = 8.9 \text{ m/s}^2$, about 10% less than the free-fall acceleration on the earth's surface. If you use this slightly smaller value of g in Equation 6.19 for the period of a satellite's orbit, you'll get the correct period of about 90 minutes. This value of g, only slightly less than the ground-level value, emphasizes the point that an object in orbit is not "weightless" because there is no gravity in space, but because it is in free fall.

EXAMPLE 6.11 Gravity on Saturn

Saturn, at 5.68×10^{26} kg, has nearly 100 times the mass of the earth. It is also much larger, with a radius of 5.85×10^7 m. What is the value of g on the surface of Saturn?

SOLVE We can use Equation 6.24 to compute the value of g_{Saturn}:

$$g_{Saturn} = \frac{GM_{Saturn}}{R_{Saturn}^2} = \frac{(6.67 \times 10^{-11} \text{ N} \cdot \text{m}^2/\text{kg}^2)(5.68 \times 10^{26} \text{ kg})}{(5.85 \times 10^7 \text{ m})^2} = 11.1 \text{ m/s}^2$$

$$= 11.1 \text{ m/s}^2$$

ASSESS Even though Saturn is much more massive than the earth, its larger radius gives it a surface gravity that is not markedly different from that of the earth. If Saturn had a solid surface, you could walk and move around quite normally.

EXAMPLE 6.12 Finding the speed to orbit Deimos

Mars has two moons, each much smaller than the earth's moon. The smaller of these two bodies, Deimos, has an average radius of only 6.3 km and a mass of 1.8×10^{15} kg. At what speed would a projectile move in a very low orbit around Deimos?

SOLVE The free-fall acceleration at the surface of Deimos is quite small:

$$g_{\text{Deimos}} = \frac{GM_{\text{Deimos}}}{R_{\text{Deimos}}^2} = \frac{(6.67 \times 10^{-11} \text{ N} \cdot \text{m}^2/\text{kg}^2)(1.8 \times 10^{15} \text{ kg})}{(6.3 \times 10^3 \text{ m})^2}$$

$$= 0.0030 \text{ m/s}^2$$

Given this, we can use Equation 6.18 to calculate the orbital speed:

$$v_{\text{orbit}} = \sqrt{gr} = \sqrt{(0.0030 \text{ m/s}^2)(6.3 \times 10^3 \text{ m})} = 4.3 \text{ m/s} \approx 10 \text{ mph}$$

ASSESS This is a quite slow speed. With a good jump, you could easily launch yourself into an orbit around Deimos!

STOP TO THINK 6.4 Rank in order, from largest to smallest, the free-fall accelerations on the surfaces of the following planets.

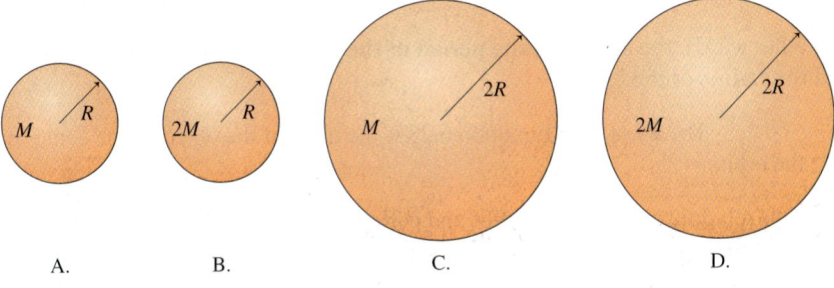

A.　　　B.　　　C.　　　D.

6.7 Gravity and Orbits

The planets of the solar system orbit the sun because the sun's gravitational pull, a force that points toward the center, causes the centripetal acceleration of circular motion. Mercury, the closest planet, experiences the largest acceleration, while Pluto, the most distant, has the smallest.

Figure 6.26 shows a massive body of mass M, such as the earth or the sun, with a lighter body of mass m orbiting it. The lighter body is called a **satellite,** even though it may be a planet orbiting the sun. Newton's second law for the satellite is

$$F_{M \text{ on } m} = \frac{GMm}{r^2} = ma = \frac{mv^2}{r} \quad (6.26)$$

Solving for v, we find that the speed of a satellite in a circular orbit is

$$v = \sqrt{\frac{GM}{r}} \quad (6.27)$$

Speed of a satellite in a circular orbit of radius r
about a star or planet of mass M

Activ Physics ONLINE 4.6

The satellite must have speed $\sqrt{GM/r}$ to maintain a circular orbit of radius r.

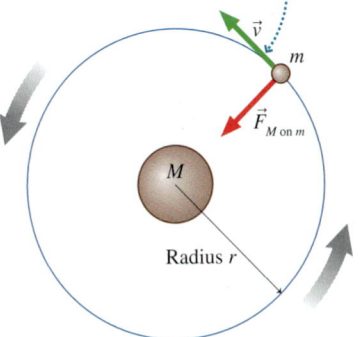

FIGURE 6.26 The orbital motion of a satellite is due to the force of gravity.

A satellite must have this specific speed in order to maintain a circular orbit of radius r about the larger mass M. If the velocity differs from this value, the orbit will become elliptical rather than circular. Notice that the orbital speed does not depend on the satellite's mass m. This is consistent with our previous discoveries that free-fall motion and projectile motion due to gravity are independent of the mass.

For a planet orbiting the sun, the period T is the time to complete one full orbit. The relationship among speed, radius, and period is the same as for any circular motion, $v = 2\pi r/T$. Combining this with the value of v for a circular orbit from Equation 6.27 gives

$$\sqrt{\frac{GM}{r}} = \frac{2\pi r}{T}$$

If we square both sides and rearrange, we find that the period of a satellite is given by

$$T^2 = \left(\frac{4\pi^2}{GM}\right)r^3 \tag{6.28}$$

Relationship between the orbital period T and radius r for a satellite in a circular orbit around an object of mass M

In other words, **the square of the period of the orbit is proportional to the cube of the radius of the orbit.**

NOTE ▶ The mass M in Equation 6.28 is the mass of the object at the center of the orbit. ◀

This relationship between radius and period had been deduced from naked-eye observations of planetary motions by the 17th century astronomer Johannes Kepler. One of Newton's major scientific accomplishments was to use his law of gravity and his laws of motion to prove what Kepler had deduced from observations. Even today, Newton's law of gravity and equations such as Equation 6.28 are essential tools for the NASA engineers who launch probes to other planets in the solar system.

The table inside the back cover of this book contains astronomical information about the sun and the planets that will be useful for many of the end-of-chapter problems. Note that planets farther from the sun have longer periods, in agreement with Equation 6.28.

EXAMPLE 6.13 Locating a geostationary satellite

Communication satellites appear to "hover" over one point on the earth's equator. A satellite that appears to remain stationary as the earth rotates is said to be in a *geostationary orbit*. What is the radius of the orbit of such a satellite?

PREPARE For the satellite to remain stationary with respect to the earth, the satellite's orbital period must be 24 hours; in seconds this is $T = 8.64 \times 10^4$ s.

SOLVE We solve for the radius of the orbit by rearranging Equation 6.28. The mass at the center of the orbit is the earth.

$$r = \left(\frac{GM_eT^2}{4\pi^2}\right)^{\frac{1}{3}} = \left(\frac{(6.67 \times 10^{-11}\ \text{N} \cdot \text{m}^2/\text{kg}^2)(5.98 \times 10^{24}\ \text{kg})(8.64 \times 10^4\ \text{s})^2}{4\pi^2}\right)^{\frac{1}{3}}$$

$$= 4.22 \times 10^7\ \text{m}$$

ASSESS This is quite a high orbit; the radius is about 7 times the radius of the earth.

Gravity on a Grand Scale

Although relatively weak, gravity is a long-range force. No matter how far apart two objects may be, there is a gravitational attraction between them. Consequently, gravity is the most ubiquitous force in the universe. It not only keeps your feet on the ground, but is also at work at a much larger scale. The Milky Way galaxy, the collection of stars of which our sun is a part, is held together by gravity. But why doesn't the attractive force of gravity simply pull all of the stars together?

The reason is that all of the stars in the galaxy are in orbit around the center of the galaxy. The gravitational attraction keeps the stars moving in orbits around the center of the galaxy rather than falling inward, much as the planets orbit the sun rather than falling into the sun. In the nearly 5 billion years that our solar system has existed, it has orbited the center of the galaxy approximately 20 times.

The galaxy as a whole doesn't rotate at a fixed angular speed, though. All of the stars in the galaxy are different distances from the galaxy's center, and so orbit with different periods. Stars closer to the center complete their orbits in less time, as we would expect from Equation 6.28. As the stars orbit, their relative positions shift. Stars that are relatively near neighbors now could be on opposite sides of the galaxy at some later time.

The rotation of a *rigid body* like a wheel is much simpler. As a wheel rotates, all of the points keep the same relationship to each other; every point on the wheel moves with the same angular velocity. The rotational dynamics of such rigid bodies is a topic we will take up in the next chapter.

A spiral galaxy, similar to our Milky Way galaxy.

STOP TO THINK 6.5   If the mass of the moon were doubled but it stayed in its present orbit, how would its orbital period change?

A. The period would increase.
B. The period would decrease.
C. The period would stay the same.

SUMMARY

The goal of Chapter 6 has been to learn about motion in a circle, including orbital motion under the influence of a graviational force.

GENERAL PRINCIPLES

Uniform Circular Motion

An object moving in a circular path is in uniform circular motion if v is constant.

- The speed is constant, but the direction is constantly changing.

- The **centripetal acceleration** is directed toward the center of the circle.

$$a = \frac{v^2}{r}$$

- The acceleration requires a force directed toward the center of the circle. Newton's second law for circular motion is

$$\vec{F}_{net} = m\vec{a} = \left(\frac{mv^2}{r}, \text{ toward center of circle} \right)$$

Universal Gravitation

Two objects with masses m_1 and m_2 at a distance r apart exert attractive gravitational forces on each other of magnitude:

$$F_{1\,on\,2} = F_{2\,on\,1} = \frac{Gm_1 m_2}{r^2}$$

where the gravitational constant is

$$G = 6.67 \times 10^{-11} \text{ N} \cdot \text{m}^2/\text{kg}^2$$

This is **Newton's Law of Gravity.** Gravity is an inverse square law.

IMPORTANT CONCEPTS

Describing circular motion

Circular motion repeats:

Period: T = time for one complete circle.

Frequency: $f = \frac{1}{T}$

We define new variables for circular motion. By convention, counterclockwise is positive.

Angular displacement: $\Delta\theta = \theta_f - \theta_i$

Angular velocity: $\omega = \frac{\Delta\theta}{\Delta t}$

Uniform circular motion kinematics

For uniform circular motion:

$$\omega = 2\pi f \qquad \theta_f - \theta_i = \Delta\theta = \omega\,\Delta t$$

The velocity, acceleration and circular motion variables are related as follows:

$$v = \frac{2\pi r}{T}$$

$$v = \omega r$$

$$a = \frac{v^2}{r} = \omega^2 r$$

APPLICATIONS

Apparent weight and weightlessness

Circular motion requires a net force pointing to the center. The apparent weight $w_{app} = n$ is usually not the same as the true weight w. n must be > 0 for the object to be in contact with a surface.

In orbital motion, the net force is provided by gravity. An astronaut and his spacecraft are both in free fall, so he feels weightless.

Planetary gravity and orbital motion

For a planet of mass M and radius R, the free-fall acceleration on the surface is

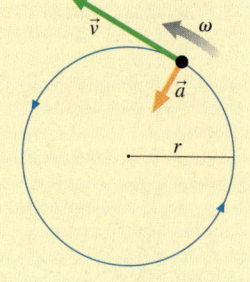

$$g = \frac{GM}{R^2}$$

The speed of a satellite in a low orbit is

$$v = \sqrt{gR}$$

A **satellite** in a circular orbit of radius r around an object of mass M moves at a speed v given by

$$v = \sqrt{\frac{GM}{r}}$$

The period and radius are related as follows:

$$T^2 = \left(\frac{4\pi^2}{GM} \right) r^3$$

For instructor-assigned homework, go to www.masteringphysics.com

Problem difficulty is labeled as I (straightforward) to IIIII (challenging).

Problems labeled 🖊 can be done on a Workbook Dynamics Worksheet; INT integrate significant material from earlier chapters; BIO are of biological or medical interest.

QUESTIONS

Conceptual Questions

1. The batter in a baseball game hits a home run. As he circles the bases, is his angular velocity positive or negative?

2. Viewed from somewhere in space above the north pole, would a point on the earth's equator have a positive or negative angular velocity due to the earth's rotation?

3. A cyclist goes around a level, circular track at constant speed. Do you agree or disagree with the following statement? "Since the cyclist's speed is constant, her acceleration is zero." Explain.

4. In uniform circular motion, which of the following quantities are constant: speed, instantaneous velocity, angular velocity, centripetal acceleration, the magnitude of the net force?

5. A particle moving along a straight line can have nonzero acceleration even when its speed is zero (for instance, a ball in free fall at the top of its path). Can a particle moving in a circle have nonzero *centripetal* acceleration when its speed is zero? If so, give an example. If not, why not?

6. Would having four-wheel drive on a car make it possible to drive faster around corners on an icy road, without slipping, than the same car with two-wheel drive? Explain.

7. BIO Large birds like pheasants often walk short distances. Small birds like chickadees never walk. They either hop or fly. Why might this be?

8. When you drive fast on the highway with muddy tires, you can hear the mud flying off the tires into your wheel wells. Why does the mud fly off?

9. A ball on a string moves in a vertical circle as in Figure Q6.9. When the ball is at its lowest point, is the tension in the string greater than, less than, or equal to the ball's weight? Explain. (You may want to include a free-body diagram as part of your explanation.)

FIGURE Q6.9

10. Give an everyday example of circular motion for which the centripetal acceleration is mostly or completely due to a force of the type specified: (a) Static friction. (b) Tension.

11. Give an everyday example of circular motion for which the centripetal acceleration is mostly or completely due to a force of the type specified: (a) Gravity. (b) Normal force.

12. It's been proposed that future space stations create "artificial gravity" by rotating around an axis. (The space station would have to be much larger than the present space station for this to be feasible.)
    a. How would this work? Explain.
    b. Would the artificial gravity be equally effective throughout the space station? If not, where in the space station would the residents want to live and work?

13. Variation in your apparent weight is desirable when you ride a roller coaster; it makes the ride fun. However, too much variation over a short period of time can be painful. For this reason, the loops of real roller coasters are not simply circles like Figure 6.19a. A typical loop is shown in Figure Q6.13. The radius of the circle that matches the track at the top of the loop is much smaller than that of a matching circle at other places on the track. Explain why this shape gives a more comfortable ride than a circular loop.

FIGURE Q6.13

14. Riding in the back of a pickup truck can be very dangerous. If the truck turns suddenly, the riders can be thrown from the truck bed. Why are the riders ejected from the bed?

15. How far away from the earth does an orbiting spacecraft have to be in order for the astronauts inside to be "weightless"?

16. A small projectile is launched parallel to the ground at height $h = 1$ m with sufficient speed to orbit a completely smooth, airless planet. A bug rides in a small hole inside the projectile. Is the bug weightless? Explain.

17. Why is it impossible for an astronaut inside an orbiting space shuttle to go from one end to the other by walking normally?

18. If every object in the universe feels an attractive gravitational force due to every other object, why don't you feel a pull from someone seated next to you?

19. A mountain climber's weight is less on the top of a tall mountain than at the base, though his mass is the same. Why?

20. Is the earth's gravitational force on the sun larger, smaller, or equal to the sun's gravitational force on the earth? Explain.

21. The mass of Jupiter is $M_{\text{Jupiter}} = 300 M_{\text{earth}}$. Jupiter orbits around the sun with $T_{\text{Jupiter}} = 11.9$ yr in an orbit with $r_{\text{Jupiter}} = 5.2 r_{\text{earth}}$. Suppose the earth could be moved to the distance of Jupiter and placed in a circular orbit around the sun. Would the new period of the earth's orbit be less than, equal to, or greater than T_{Jupiter}? Or is the value impossible to determine, because it would depend on the speed the earth is given? Or is the question meaningless because it is not possible for a planet of earth's mass to orbit at the distance of Jupiter? Explain.

Multiple-Choice Questions

22. | As seen from above, a car rounds the curved path shown in Figure Q6.22 at a constant speed. Which vector best represents the net force acting on the car?

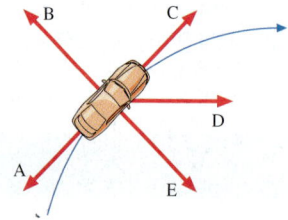

FIGURE Q6.22

23. | Suppose you and a friend, each of mass 60 kg, go to the park and get on a 4.0-m-diameter merry-go-round. You stand on the outside edge of the merry-go-round, while your friend pushes so that it rotates once every 6.0 s. What is the magnitude of the (apparent) outward force that you feel?
    A. 7 N   B. 63 N   C. 130 N   D. 260 N

24. | The cylindrical space station in Figure Q6.24, 200 m in diameter, rotates in order to provide artificial gravity of 1 g for the occupants. How much time does the station take to complete one rotation?
    A. 3 s   B. 20 s   C. 28 s   D. 32 s

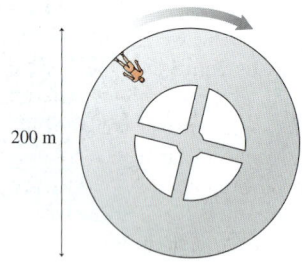

200 m

FIGURE Q6.24

25. ‖ Two cylindrical space stations, the second four times the diameter of the first, rotate so as to provide the same amount of artificial gravity. If the first station makes one rotation in the time T, then the second station makes one rotation in time
    A. $T/4$ B. $2T$ C. $4T$ D. $16T$

26. ‖‖ A newly discovered planet has twice the mass and three times the radius of the earth. What is the free-fall acceleration at its surface, in terms of the free-fall acceleration g at the surface of the earth?
    A. $\frac{2}{9}g$ B. $\frac{2}{3}g$ C. $\frac{3}{4}g$ D. $\frac{4}{3}g$

27. ‖ Suppose one night the radius of the earth doubled but its mass stayed the same. What would be an approximate new value for the free-fall acceleration at the surface of the earth?
    A. 2.5 m/s²   B. 5.0 m/s²   C. 10 m/s²   D. 20 m/s²

28. | Currently, the moon goes around the earth once every 27.3 days. If the moon could be brought into a new orbit with a smaller radius, its orbital period would be
    A. More than 27.3 days.
    B. 27.3 days.
    C. Less than 27.3 days.

29. ‖ Two planets orbit a star. Planet 1 has orbital radius r_1 and planet 2 has $r_2 = 4r_1$. Planet 1 orbits with period T_1. Planet 2 orbits with period
    A. $T_2 = \frac{1}{2}T_1$ B. $T_2 = 2T_1$ C. $T_2 = 4T_1$ D. $T_2 = 8T_1$

Questions 30 through 32 concern a classic figure-skating jump called the axel. A skater starts the jump moving forward as shown in Figure Q6.30, leaps into the air, and turns one-and-a-half revolutions before landing. The typical skater is in the air for about 0.5 s, and the skater's hands are located about 0.8 m from the rotation axis.

FIGURE Q6.30

30. | What is the approximate angular speed of the skater during the leap?
    A. 2 rad/s   B. 6 rad/s   C. 9 rad/s   D. 20 rad/s

31. | The skater's arms are fully extended during the jump. What is the approximate centripetal acceleration of the skater's hand?
    A. 10 m/s²   B. 30 m/s²   C. 300 m/s²   D. 450 m/s²

32. | What is the approximate speed of the skater's hand?
    A. 1 m/s   B. 3 m/s   C. 9 m/s   D. 15 m/s

PROBLEMS

Section 6.1 Uniform Circular Motion

1. ‖ What is the angular speed of the tip of the minute hand on a clock, in rad/s?

2. | An old-fashioned vinyl record rotates on a turntable at 45 rpm. What are (a) the angular speed in rad/s and (b) the period of the motion?

3. ‖ The earth's radius is about 4000 miles. Kampala, the capital of Uganda, and Singapore are both nearly on the equator. The distance between them is 5000 miles.
    a. Through what angle do you turn, relative to the earth, if you fly from Kampala to Singapore? Give your answer in both radians and degrees.
    b. The flight from Kampala to Singapore takes 9 hours. What is the plane's angular speed relative to the earth?

4. ‖ A Ferris wheel rotates at an angular velocity of 0.036 rad/s. At $t = 0$ min, your friend Seth is at the very top of the ride. What is Seth's angular position at $t = 3.0$ min, measured counterclockwise from the top? Give your answer as an angle in degrees between 0° and 360°.

5. ‖‖‖ A turntable rotates counterclockwise at 78 rpm. A speck of dust on the turntable is at $\theta = 0.45$ rad at $t = 0$ s. What is the angle of the speck at $t = 8.0$ s? Your answer should be between 0 and 2π rad.

6. ‖ A fast-moving superhero in a comic book runs around a circular, 70-m-diameter track five and a half times (ending up directly opposite her starting point) in 3.0 s. What is her angular speed, in rad/s?

Section 6.2 Speed, Velocity, and Acceleration in Uniform Circular Motion

7. ‖ A 5.0-m-diameter merry-go-round is turning with a 4.0 s period. What is the speed of a child on the rim?

8. ‖ A skater holds her arms outstretched as she spins at 180 rpm. What is the speed of her hands if they are 140 cm apart?

9. | The horse on a carousel is 4.0 m from the central axis.
   a. If the carousel rotates at 0.10 rev/s, how long does it take the horse to go around twice?
   b. How fast is a child on the horse going (in m/s)?

10. ‖ The radius of the earth's very nearly circular orbit around the sun is 1.50×10^{11} m. Find the magnitude of the earth's (a) velocity, (b) angular velocity, and (c) centripetal acceleration as it travels around the sun. Assume a year of 365 days.

11. ‖ Your roommate is working on his bicycle and has the bike upside down. He spins the 60-cm-diameter wheel, and you notice that a pebble stuck in the tread goes by three times every second. What are the pebble's speed and acceleration?

12. | To withstand "g-forces" of up to 10 g's, caused by suddenly pulling out of a steep dive, fighter jet pilots train on a "human centrifuge." 10 g's is an acceleration of 98 m/s². If the length of the centrifuge arm is 12 m, at what speed is the rider moving when she experiences 10 g's?

Section 6.3 Dynamics of Uniform Circular Motion

13. ‖‖‖ Figure P6.13 is a bird's-eye view of particles on a string moving in horizontal circles on a table top. All are moving at the same speed. Rank in order, from largest to smallest, the tensions T_1 to T_4.

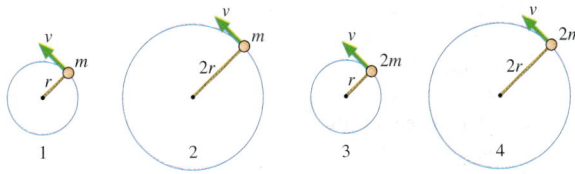

FIGURE P6.13

14. ‖ A 200 g block on a 50-cm-long string swings in a circle on a horizontal, frictionless table at 75 rpm.
   a. What is the speed of the block?
   b. What is the tension in the string?

15. | A 1500 kg car drives around a flat 200-m-diameter circular track at 25 m/s. What are the magnitude and direction of the net force on the car? What causes this force?

16. | A fast pitch softball player does a "windmill" pitch, illustrated in Figure P6.16, moving her hand through a circular arc to pitch a ball at 70 mph. The 0.19 kg ball is 50 cm from the pivot point at her shoulder. At the lowest point of the circle, the ball has reached its maximum speed.
   a. At the bottom of the circle, just before the ball leaves her hand, what is its centripetal acceleration?
   b. What are the magnitude and direction of the force her hand exerts on the ball at this point?

17. | A baseball pitching machine works by rotating a light and stiff rigid rod about a horizontal axis until the ball is moving toward the target. Suppose a 144 g baseball is held 85 cm from the axis of rotation and released at the major league pitching speed of 85 mph.
   a. What is the ball's centripetal acceleration just before it is released?
   b. What is the magnitude of the net force that is acting on the ball just before it is released?

Section 6.4 Apparent Forces in Circular Motion

18. | You hold a bucket in one hand. In the bucket is a 500 g rock. You swing the bucket so the rock moves in a vertical circle 2.2 m in diameter. What is the minimum speed the rock must have at the top of the circle if it is to always stay in contact with the bottom of the bucket?

19. ‖ The passengers in a roller coaster car feel 50% heavier than their true weight as the car goes through a dip with a 30 m radius of curvature. What is the car's speed at the bottom of the dip?

20. ‖‖‖ A roller coaster car crosses the top of a circular loop-the-loop at twice the critical speed. What is the ratio of the car's apparent weight to its true weight?

21. ‖ As a roller coaster car crosses the top of a 40-m-diameter loop-the-loop, its apparent weight is the same as its true weight. What is the car's speed at the top?

22. ‖‖‖ A typical laboratory centrifuge rotates at 4000 rpm. Test
    BIO tubes have to be placed into a centrifuge very carefully
    INT because of the very large accelerations.
   a. What is the acceleration at the end of a test tube that is 10 cm from the axis of rotation?
   b. For comparison, what is the magnitude of the acceleration a test tube would experience if dropped from a height of 1.0 m and stopped in a 1.0-ms-long encounter with a hard floor?

Section 6.5 Circular Orbits and Weightlessness

23. ‖‖‖ A satellite orbiting the moon very near the surface has a period of 110 min. Use this information, together with the radius of the moon from the table on the inside of the back cover, to calculate the free-fall acceleration on the moon's surface.

Section 6.6 Newton's Law of Gravity

24. ‖ The centers of a 10 kg lead ball and a 100 g lead ball are separated by 10 cm.
   a. What gravitational force does each exert on the other?
   b. What is the ratio of this gravitational force to the weight of the 100 g ball?

25. ‖ The gravitational force of a star on an orbiting planet 1 is F_1. Planet 2, which is twice as massive as planet 1 and orbits at twice the distance from the star, experiences gravitational force F_2. What is the ratio F_2/F_1?

26. | The free-fall acceleration at the surface of planet 1 is 20 m/s². The radius and the mass of planet 2 are twice those of planet 1. What is the free-fall acceleration on planet 2?

27. ‖‖‖ What is the ratio of the sun's gravitational force on you to the earth's gravitational force on you?

28. | a. What is the free-fall acceleration at the surface of the sun?
    b. Billions of years from now, the sun will become a red giant star and its radius will swell to the size of the earth's orbit. Assuming its mass doesn't change, what will be the free-fall acceleration on the surface at that time?

29. ‖ a. What is the gravitational force of the sun on the earth?
    b. What is the gravitational force of the moon on the earth?
    c. The moon's force is what percent of the sun's force?

30. | What is the free-fall acceleration at the surface of (a) Mars and (b) Jupiter?

Section 6.7 Gravity and Orbits

31. | Planet X orbits the star Omega with a "year" that is 200 earth days long. Planet Y circles Omega at four times the distance of Planet X. How long is a year on Planet Y?

32. ‖‖ Satellite A orbits a planet with a speed of 10,000 m/s. Satellite B is twice as massive as satellite A and orbits at twice the distance from the center of the planet. What is the speed of satellite B?

33. ‖ The space shuttle is in a 250-mile-high orbit. What are the shuttle's orbital period, in minutes, and its speed?

34. ‖ The *asteroid belt* circles the sun between the orbits of Mars and Jupiter. One asteroid has a period of 5.0 earth years. What are the asteroid's orbital radius and speed?

35. ‖‖ An earth satellite moves in a circular orbit at a speed of 5500 m/s. What is its orbital period?

General Problems

36. | How fast must a plane fly along the earth's equator so that the sun stands still relative to the passengers? In which direction must the plane fly, east to west or west to east? Give your answer in both km/h and mph. The radius of the earth is 6400 km.

37. ‖ The car in Figure P6.37 travels at a constant speed along the road shown. Draw vectors showing its acceleration at the three points A, B, and C, or write $\vec{a} = \vec{0}$. The lengths of your vectors should correspond to the magnitudes of the accelerations.

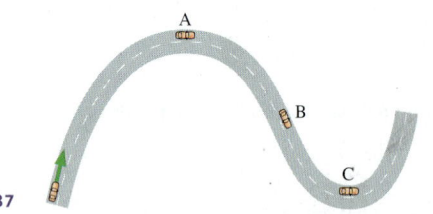

FIGURE P6.37

38. ‖‖ In the Bohr model of the hydrogen atom, an electron (mass $m = 9.1 \times 10^{-31}$ kg) orbits a proton at a distance of 5.3×10^{-11} m. The proton pulls on the electron with an electric force of 8.2×10^{-8} N. How many revolutions per second does the electron make?

39. ‖‖ A 75 kg man weighs himself at the north pole and at the equator. Which scale reading is higher? By how much? Assume the earth is a perfect sphere. Explain why the readings differ.

40. | A 1500 kg car takes a 50-m-radius unbanked curve at 15 m/s. What is the size of the friction force on the car?

41. ‖ A 500 g ball swings in a vertical circle at the end of a 1.5-m-long string. When the ball is at the bottom of the circle, the tension in the string is 15 N. What is the speed of the ball at that point?

42. ‖ Suppose the moon were held in its orbit not by gravity but by a massless cable attached to the center of the earth. What would be the tension in the cable? See the inside of the back cover for astronomical data.

43. ‖‖‖ A 30 g ball rolls around a 40-cm-diameter L-shaped track, shown in Figure P6.43, at 60 rpm. Rolling friction can be neglected.

FIGURE P6.43

    a. How many different contact forces does the track exert on the ball? Name them.
    b. What is the magnitude of the net force of the track on the ball?

44. ‖‖‖ A 5.0 g coin is placed 15 cm from the center of a turntable. The coin has static and kinetic coefficients of friction with the turntable surface of $\mu_s = 0.80$ and $\mu_k = 0.50$. The turntable very slowly speeds up to 60 rpm. Does the coin slide off?

45. ‖‖‖ A *conical pendulum* is formed by attaching a 500 g ball to a 1.0-m-long string, then allowing the mass to move in a horizontal circle of radius 20 cm. Figure P6.45 shows that the string traces out the surface of a cone, hence the name.

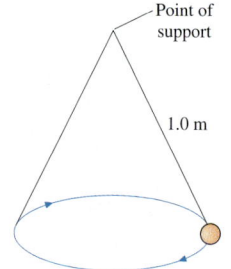

    a. What is the tension in the string?
    b. What is the ball's angular velocity, in rpm?

Hint: Determine the horizontal and vertical components of the forces acting on the ball, and use the fact that the vertical component of acceleration is zero since there is no vertical motion.

FIGURE P6.45

46. ‖‖‖‖ In an old-fashioned amusement park ride, passengers stand inside a 3.0-m-tall, 5.0-m-diameter hollow steel cylinder with their backs against the wall. The cylinder begins to rotate about a vertical axis. Then the floor on which the passengers are standing suddenly drops away! If all goes well, the passengers will "stick" to the wall and not slide. Clothing has a static coefficient of friction against steel in the range 0.60 to 1.0 and a kinetic coefficient in the range 0.40 to 0.70. What is the minimum rotational frequency, in rpm, for which the ride is safe?

47. | A car drives over the top of a hill that has a radius of 50 m. What maximum speed can the car have without flying off the road at the top of the hill?

48. | While at the county fair, you decide to ride the Ferris wheel. Having eaten too many candy apples and elephant ears, you find the motion somewhat unpleasant. To take your mind off your stomach, you wonder about the motion of the ride. You estimate the radius of the big wheel to be 15 m, and you use your watch to find that each loop around takes 25 s.
    a. What are your speed and magnitude of your acceleration?
    b. What is the ratio of your apparent weight to your true weight at the top of the ride?
    c. What is the ratio of your apparent weight to your true weight at the bottom?

49. ‖ While a person is walking, his arms (each with typical length 70 cm measured from the shoulder joint) swing through approximately a 45° angle in 0.5 s. As a reasonable approximation, we can assume that the arm moves with constant speed during each swing.
    a. What is the acceleration of a 1.0 g drop of blood in the fingertips at the bottom of the swing?
    b. Draw a free-body diagram for the drop of blood in part a.
    c. Find the magnitude and direction of the force that the blood vessel must exert on the drop of blood.
    d. What force would the blood vessel exert if the arm were not swinging?

50. ‖ A 100 g ball on a 60-cm-long string is swung in a vertical circle about a point 200 cm above the floor. The string suddenly breaks when it is parallel to the ground and the ball is moving upward. The ball reaches a height 600 cm above the floor. What was the tension in the string an instant before it broke?

51. ‖ The ultracentrifuge is an important tool for separating and analyzing proteins in biological research. Because of the enormous centripetal accelerations that can be achieved, the apparatus (see Figure 6.21) must be carefully balanced so that each sample is matched by another on the opposite side of the rotor shaft. Failure to do so is a costly mistake, as seen in Figure P6.51. Any difference in mass of the opposing samples will cause a net force in the horizontal plane on the shaft of the rotor. Suppose that a scientist makes a slight error in sample preparation, and one sample has a mass 10 mg greater than the opposing sample. If the samples are 10 cm from the axis of the rotor and the ultracentrifuge spins at 70,000 rpm, what is the magnitude of the net force on the rotor due to the unbalanced samples?

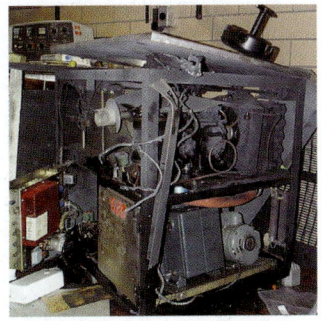

FIGURE P6.51

52. ‖‖ A 1.0-m-diameter lead sphere has a mass of 5900 kg. A dust particle rests on the surface. What is the ratio of the gravitational force of the sphere on the dust particle to the weight of the dust particle?

53. ‖‖ The space shuttle orbits 300 km above the surface of the earth.
    a. What is the force of gravity on a 1.0 kg sphere inside the space shuttle?
    b. The sphere floats around inside the space shuttle, apparently "weightless." How is this possible?

54. ‖‖ A starship is circling a distant planet of radius R. The astronauts find that the free-fall acceleration at their altitude is half the value at the planet's surface. How far above the surface are they orbiting? Your answer will be a multiple of R.

55. ‖‖ A sensitive gravimeter at a mountain observatory finds that the free-fall acceleration is 0.0075 m/s² less than that at sea level. What is the observatory's altitude?

56. ‖ Suppose we could shrink the earth without changing its mass. At what fraction of its current radius would the free-fall acceleration at the surface be three times its present value?

57. ‖ Planet Z is 10,000 km in diameter. The free-fall acceleration on Planet Z is 8.0 m/s².
    a. What is the mass of Planet Z?
    b. What is the free-fall acceleration 10,000 km above Planet Z's north pole?

58. ‖‖ What are the speed and altitude of a geostationary satellite (see Example 6.13) orbiting Mars? Mars rotates on its axis once every 24.8 hours.

59. ‖ a. What is the free-fall acceleration on Mars?
    b. Estimate the maximum speed at which an astronaut can walk on the surface of Mars.

60. ‖‖ How long will it take a rock dropped from 2.0 m above the surface of Mars to reach the ground?

61. ‖‖‖ A 20 kg sphere is at the origin and a 10 kg sphere is at $(x, y) = (20 \text{ cm}, 0 \text{ cm})$. At what point or points could you place a small mass such that the net gravitational force on it due to the spheres is zero?

62. ‖ a. At what height above the earth is the free-fall acceleration 10% of its value at the surface?
    b. What is the speed of a satellite orbiting at that height?

63. ‖ Mars has a small moon, Phobos, that orbits with a period of 7 h 39 min. The radius of Phobos' orbit is 9.4×10^6 m. Use only this information (and the value of G) to calculate the mass of Mars.

64. ‖ You are the science officer on a visit to a distant solar system. Prior to landing on a planet you measure its diameter to be 1.80×10^7 m and its rotation period to be 22.3 h. You have previously determined that the planet orbits 2.20×10^{11} m from its star with a period of 402 earth days. Once on the surface you find that the free-fall acceleration is 12.2 m/s². What are the mass of (a) the planet and (b) the star?

65. ‖ Europa, a satellite of Jupiter, is believed to have a liquid ocean of water (with a possibility of life) beneath its icy surface. In planning a future mission to Europa, what is the fastest that an astronaut with legs of length 0.70 m could walk on the surface of Europa? Europa is 3100 km in diameter and has a mass of 4.8×10^{22} kg.

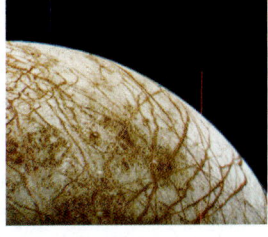

In Problems 66 through 69 you are given the equation (or equations) used to solve a problem. For each of these, you are to

a. Write a realistic problem for which this is the correct equation. The last two questions should involve real planets. Be sure that the answer your problem requests is consistent with the equation given.
b. Finish the solution of the problem.

66. ‖ $60 \text{ N} = (0.30 \text{ kg})\omega^2(0.50 \text{ m})$

67. ‖‖‖ $(1500 \text{ kg})(9.80 \text{ m/s}^2) - 11{,}760 \text{ N} = (1500 \text{ kg})v^2/(200 \text{ m})$

68. ‖ $\dfrac{(6.67 \times 10^{-11} \text{ N} \cdot \text{m}^2/\text{kg}^2)(1.90 \times 10^{27} \text{ kg})}{r^2}$
$= \dfrac{(6.67 \times 10^{-11} \text{ N} \cdot \text{m}^2/\text{kg}^2)(5.98 \times 10^{24} \text{ kg})}{(6.37 \times 10^6 \text{ m})^2}$

69. ‖ $\dfrac{(6.67 \times 10^{-11} \text{ N} \cdot \text{m}^2/\text{kg}^2)(5.98 \times 10^{24} \text{ kg})(1000 \text{ kg})}{r^2}$
$= \dfrac{(1000 \text{ kg})(1997 \text{ m/s})^2}{r}$

Passage Problems

Orbiting the Moon

Suppose a spacecraft orbits the moon in a very low, circular orbit, just a few hundred meters above the lunar surface. The moon has a diameter of 3500 km, and the free-fall acceleration at the surface is 1.6 m/s^2.

70. | The direction of the net force on the craft is
    A. Away from the surface of the moon.
    B. In the direction of motion.
    C. Toward the center of the moon.
    D. Nonexistent, because the net force is zero.
71. | How fast is this spacecraft moving?
    A. 53 m/s   B. 75 m/s   C. 1700 m/s D. 2400 m/s
72. | How much time does it take for the spacecraft to complete one orbit?
    A. 38 min   B. 76 min   C. 110 min D. 220 min

73. | The material that comprises the side of the moon facing the earth is actually slightly more dense than the material on the far side. When the spacecraft is above a more dense area of the surface, the moon's gravitational force on the craft is a bit stronger. In order to stay in a circular orbit of constant height and speed, the spacecraft could fire its rockets while passing over the denser area. The rockets should be fired so as to generate a force on the craft
    A. Away from the surface of the moon.
    B. In the direction of motion.
    C. Toward the center of the moon.
    D. Opposite the direction of motion.

STOP TO THINK ANSWERS

Stop to Think 6.1: D > B > C > A. The centripetal acceleration is $\omega^2 r$. Changing r by a factor of 2 changes the centripetal acceleration by a factor of 2, but changing ω by a factor of 2 changes the centripetal acceleration by a factor of 4.

Stop to Think 6.2: $T_D > T_B = T_E > T_C > T_A.$ The center-directed force is $m\omega^2 r$. Changing r by a factor of 2 changes the tension by a factor of 2, but changing f (and thus ω) by a factor of 2 changes the tension by a factor of 4.

Stop to Think 6.3: B. The car is moving in a circle, so there must be a net force toward the center of the circle. The circle is below the car, so the net force must point downward. This can be true only if $w > n$.

Stop to Think 6.4: B > A > D > C. The free-fall acceleration is proportional to the mass, but inversely proportional to the square of the radius.

Stop to Think 6.5: C. The period of the orbit does not depend on the mass of the orbiting object.

7

ROTATIONAL MOTION

As the earth rotates on its axis, the distant stars appear to move in eternal circles in the sky overhead. In reality, however, the angular velocity of the earth is very slowly decreasing, leading to an increase in the length of the day of $18\ \mu s$ each year. What causes the angular velocity of a rotating object to change?

Looking Ahead ▶▶

The goal of Chapter 7 is to understand the physics of rotating objects. In this chapter you will learn to:

- ▶ Understand the angular and tangential acceleration of a rotating object.
- ▶ Calculate the torque exerted on an object.
- ▶ Determine an object's center of gravity and its moment of inertia.
- ▶ Apply the concepts of torque and moment of inertia to the rotation of an object about a fixed axis.
- ▶ Understand the motion of a rolling object.

Looking Back ◀◀

Rotational motion will revisit many of the major themes introduced in previous chapters, especially the properties of circular motion. Please review:

- ◀ Section 4.6 Newton's second law.
- ◀ Section 6.1 Uniform circular motion.
- ◀ Section 6.2 Speed, velocity, and acceleration in circular motion.
- ◀ Section 6.3 Dynamics of circular motion.

In Chapter 6 we studied circular motion, the motion of a single particle around a circular path. Our goal in this chapter is to extend this treatment to focus on extended rotating objects such as wheels, axles, and spinning tops. Objects such as these, whose shape does not change as they rotate, are called *rigid bodies*. Our own earth is an important example of a rotating rigid body. As we ride along on its surface, we see the stars making circular paths about a pole.

How do forces act to increase the angular velocity of a rotating object? What makes an object easy or hard to get rotating, and what is needed to keep it rotating? How do we describe a rolling object—one that is both rotating and moving as a whole? These are some of the questions we'll address in this chapter.

You will quickly discover that the physics of rotational motion is analogous to the physics of linear motion that you studied in earlier chapters. For example, the new concepts of torque and angular acceleration are the rotational analogs of force and acceleration, and we'll find a new version of Newton's second law that is the rotational equivalent of $\vec{F}_{\text{net}} = m\vec{a}$.

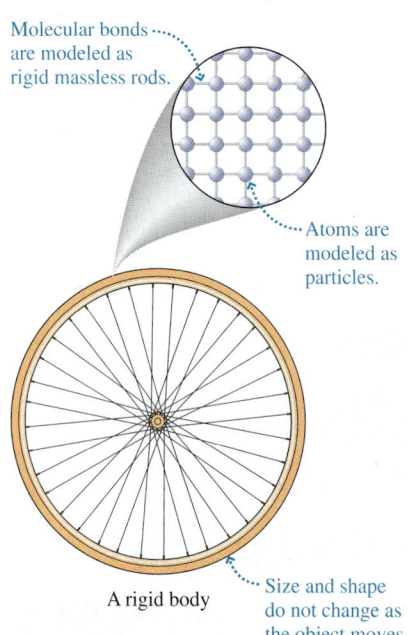

Molecular bonds are modeled as rigid massless rods.

Atoms are modeled as particles.

A rigid body

Size and shape do not change as the object moves.

FIGURE 7.1 The rigid-body model of an extended object.

7.1 The Rotation of a Rigid Body

Thus far, our study of physics has focused almost exclusively on the *particle model* in which an entire object is represented as a single point in space. The particle model is entirely adequate for understanding motion in a wide variety of situations, but there are also cases for which we need to consider the motion of an *extended object*—a system of particles for which the size and shape *do* make a difference and cannot be neglected.

A **rigid body** is an extended object whose size and shape do not change as it moves. For example, a bicycle wheel can be thought of as a rigid body. Figure 7.1 shows a rigid body as a collection of atoms held together by the rigid "massless rods" of molecular bonds.

Real molecular bonds are, of course, not perfectly rigid. That's why an object seemingly as rigid as a bicycle wheel can flex and bend. Thus Figure 7.1 is really a simplified *model* of an extended object, the **rigid-body model.** The rigid-body model is a very good approximation of many real objects of practical interest, such as wheels and axles. Even nonrigid objects can often be modeled as a rigid body during parts of their motion. For example, a diver is well described as a rotating rigid body while she's in the tuck position.

Figure 7.2 illustrates the three basic types of motion of a rigid body: **translational motion, rotational motion,** and **combination motion.** We've already studied translational motion of a rigid body using the particle model. If a rigid body doesn't rotate, this model is often adequate for describing its motion. The rotational motion of a rigid body will be the main focus of this chapter. We'll also discuss an important case of combination motion—that of a *rolling* object—later in this chapter.

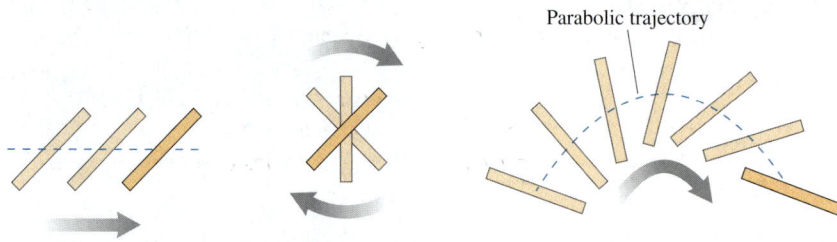

Parabolic trajectory

Translational motion: The object as a whole moves along a trajectory but does not rotate.

Rotational motion: The object rotates about a fixed point. Every point on the object moves in a circle.

Combination motion: An object rotates as it moves along a trajectory.

FIGURE 7.2 Three basic types of motion of a rigid body.

Uniform Rotational Motion of a Rigid Body

Figure 7.3 shows a wheel rotating on an axle. Notice that as the wheel rotates for a time interval Δt, two points on the wheel, marked with dots, turn through the *same angle,* even through their distances r from the axis of rotation may be different. That is, $\Delta\theta_1 = \Delta\theta_2$ during the time interval Δt. As a consequence, the two points have equal angular velocities: $\omega_1 = \omega_2$. In general, **every point on a rotating rigid body has the same angular velocity.** Because of this, we can refer to the angular velocity ω *of the wheel.*

Recall from Chapter 6 that the speed of a particle moving in a circle is $v = \omega r$, so two points of a rotating object will have different *speeds* if they have different distances from the axis of rotation, but *all* points have the *same* angular velocity ω. Thus angular velocity is one of the most important parameters of a rotating object.

Because every point on a rotating object moves in a circle, we can carry forward all the results for circular motion from Chapter 6. Thus the angular displacement of any point on the wheel shown in Figure 7.3 is found from Equation 6.1 as $\Delta\theta = \omega\,\Delta t$; the speed of any particle in the wheel is $v = \omega r$, where r is the particle's distance from the axis; and the particle's centripetal acceleration is $a = \omega^2 r$.

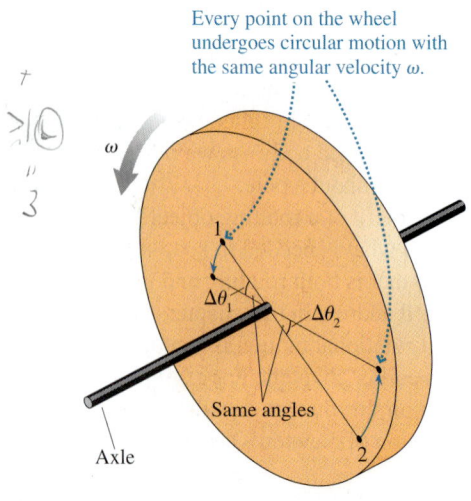

Every point on the wheel undergoes circular motion with the same angular velocity ω.

ω

$\Delta\theta_1$ $\Delta\theta_2$

Same angles

Axle

FIGURE 7.3 All points on a wheel rotate with the same angular velocity.

Angular Acceleration

If you push on the edge of a bicycle wheel, it begins to rotate. If you continue to push, it rotates ever faster. Its angular velocity is *changing*. To understand the dynamics of rotating objects, we'll need to be able to describe this case of changing angular velocity, that is, the case of *nonuniform* circular motion.

Figure 7.4 shows a bicycle wheel whose angular velocity is changing. The dot represents a particular point on the wheel at successive times. At time t_i the angular velocity is ω_i; at a later time $t_f = t_i + \Delta t$ the angular velocity has changed to ω_f. The change in angular velocity during this time interval is

$$\Delta\omega = \omega_f - \omega_i$$

Recall that in Chapter 2 we defined the *linear* acceleration as

$$a_x = \frac{\Delta v_x}{\Delta t} = \frac{(v_x)_f - (v_x)_i}{\Delta t}$$

By analogy, we now define the **angular acceleration** as

$$\alpha = \frac{\text{change in angular velocity}}{\text{time interval}} = \frac{\Delta\omega}{\Delta t} \qquad (7.1)$$

Angular acceleration for a particle in nonuniform circular motion

We use the symbol α (Greek *alpha*) for angular acceleration. Because the units of ω are rad/s, the units of angular acceleration are (rad/s)/s, or rad/s^2. The sign of α depends on the sign of the change in angular velocity—again by analogy with one-dimensional motion. Like ω, the angular acceleration α is the same for every point on a rotating rigid body.

In Chapter 6 we found analogies between linear and angular positions and velocities. Here we've extended those analogies to include linear and angular accelerations. Table 7.1 summarizes all of these analogies between linear and circular motion.

> **NOTE** ▶ Don't confuse the angular acceleration with the centripetal acceleration introduced in Chapter 6. The angular acceleration indicates how rapidly the *angular* velocity is changing. The centripetal acceleration is a vector quantity that points toward the center of a particle's circular path; it is nonzero even if the angular velocity is constant. ◀

In addition, the various equations of one-dimensional kinematics have analogs for rotational or circular motion. Table 7.2 lists the equations for one-dimensional motion and the analogous equations for the kinematics of circular motion.

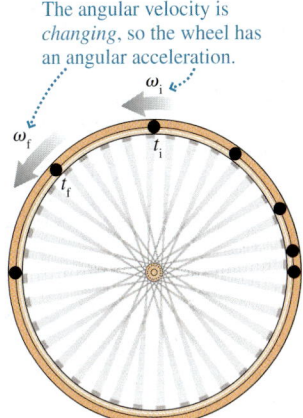

The angular velocity is *changing*, so the wheel has an angular acceleration.

FIGURE 7.4 A rotating wheel with a changing angular velocity.

TABLE 7.1 Linear and circular motion variables

Linear motion	Circular motion
Position x	Angular position θ
Velocity $v_x = \Delta x/\Delta t$	Angular velocity $\omega = \Delta\theta/\Delta t$
Acceleration $a_x = \Delta v_x/\Delta t$	Angular acceleration $\alpha = \Delta\omega/\Delta t$

 7.7

TABLE 7.2 Linear and circular motion equations

Linear motion	Circular motion
Displacement at constant speed: $\Delta x = v\,\Delta t$	Angular displacement at constant angular speed: $\Delta\theta = \omega\,\Delta t$
Displacement at constant acceleration: $\Delta x = v_i\,\Delta t + \frac{1}{2}a\,\Delta t^2$	Angular displacement at constant angular acceleration: $\Delta\theta = \omega_i\,\Delta t + \frac{1}{2}\alpha\,\Delta t^2$
Change in velocity at constant acceleration: $\Delta v = a\,\Delta t$	Change in angular velocity at constant angular acceleration: $\Delta\omega = \alpha\,\Delta t$

EXAMPLE 7.1 Spinning up a computer disk

The disk in a computer disk drive spins up to 5400 rpm in 2.00 s. What is the angular acceleration of the disk? At the end of 2.00 s, how many revolutions has the disk made?

PREPARE The initial angular velocity is $\omega_i = 0$ rad/s. The final angular velocity is 5400 rpm. However, this value is not in the correct SI units of rad/s. The conversion is

$$\omega_f = \frac{5400 \text{ rev}}{\text{min}} \times \frac{1 \text{ min}}{60 \text{ s}} \times \frac{2\pi \text{ rad}}{1 \text{ rev}} = 565 \text{ rad/s}$$

SOLVE From the definition of angular acceleration,

$$\alpha = \frac{\Delta\omega}{\Delta t} = \frac{565 \text{ rad/s} - 0 \text{ rad/s}}{2.00 \text{ s}} = 283 \text{ rad/s}^2$$

We can compute the angular displacement during this acceleration by using the angular displacement equation from Table 7.2:

$$\Delta\theta = \omega_i t + \tfrac{1}{2}\alpha \, \Delta t^2$$

$$= (0 \text{ rad/s})(2.00 \text{ s}) + \tfrac{1}{2}(283 \text{ rad/s}^2)(2.00 \text{ s})^2$$

$$= 566 \text{ rad}$$

Each revolution corresponds to an angular displacement of 2π, so we have

$$\text{Number of revolutions} = \frac{566 \text{ rad}}{2\pi \text{ rad/revolution}}$$

$$= 90.1 \text{ revolutions}$$

The disk completes 90 revolutions during the first two seconds.

ASSESS We solved this problem using the same methods you learned for one-dimensional motion, replacing linear variables and equations with their rotational equivalents.

Tangential Acceleration

(a) Uniform circular motion

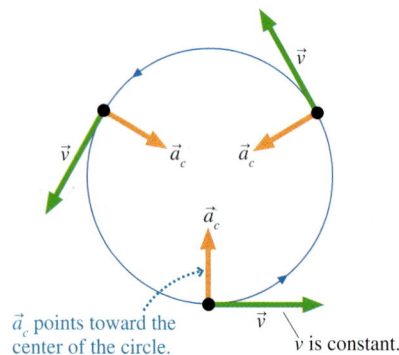

$\vec{a}_c$ points toward the center of the circle.

v is constant.

(b) Nonuniform circular motion

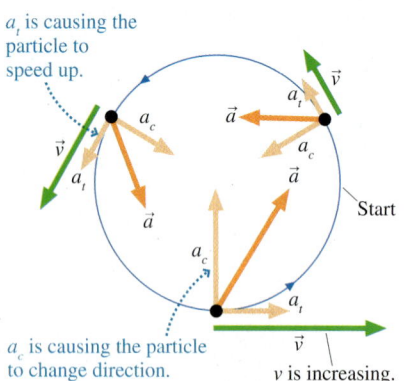

a_t is causing the particle to speed up.

a_c is causing the particle to change direction.

v is increasing.

FIGURE 7.5 Uniform and nonuniform circular motion.

As you learned in Chapter 6, and as Figure 7.5a reminds you, a particle undergoing uniform circular motion has an acceleration directed inward toward the center of the circle. This centripetal acceleration $\vec{a}_c$ is due to the change in the *direction* of the particle's velocity. Recall that the magnitude of the centripetal acceleration is $a_c = v^2/r = \omega^2 r$.

NOTE ▶ Centripetal acceleration will now be denoted a_c to distinguish it from the tangential acceleration a_t, discussed below. ◀

If the particle's circular motion is *nonuniform,* so that the particle's speed is changing, then the particle will have another component to its acceleration. Figure 7.5b shows a particle whose speed is increasing as it moves around its circular path. Because the *magnitude* of the velocity is increasing, this second component of the acceleration is directed *tangentially* to the circle, in the same direction as the velocity. This component of acceleration is called the **tangential acceleration.** As shown in Figure 7.5b, **the full acceleration $\vec{a}$ is then the vector sum of these two components,** the centripetal acceleration $\vec{a}_c$ and the tangential acceleration $\vec{a}_t$.

The tangential acceleration measures the rate at which the particle's speed increases. Thus its magnitude is

$$a_t = \frac{\Delta v}{\Delta t}$$

We can relate the tangential acceleration to the *angular* acceleration by using the relation $v = \omega r$ between the speed of a particle moving in a circle of radius r and its angular velocity ω. We have

$$a_t = \frac{\Delta v}{\Delta t} = \frac{\Delta(\omega r)}{\Delta t} = \frac{\Delta\omega}{\Delta t} r$$

or, because $\alpha = \Delta\omega/\Delta t$ from Equation 7.1,

$$a_t = \alpha r \qquad (7.2)$$

Relationship between tangential and angular acceleration

LINEAR
p. 38

We've seen that all points on a rotating rigid body have the same angular acceleration. From Equation 7.2, however, the tangential acceleration of a point on a rotating object depends on the point's distance r from the axis, so that the tangential acceleration is *not* the same for all points.

> **STOP TO THINK 7.1**   A ball on the end of a string swings in a horizontal circle once every second. State whether the magnitudes of each of the following quantities is zero, constant (but not zero), or changing.
> a. Velocity
> b. Angular velocity
> c. Centripetal acceleration
> d. Angular acceleration
> e. Tangential acceleration

7.2 Torque

Newton's genius, summarized in his second law of motion, was to recognize force as the cause of acceleration. But what about *angular* acceleration? What do Newton's laws have to tell us about rotational motion? To begin our study of rotational motion, we'll need to find a rotational equivalent of force.

Consider the common experience of pushing open a door. Figure 7.6 is a top view of a door that is hinged on the left. Four pushing forces are shown, all of equal strength. Which of these will be most effective at opening the door?

Force $\vec{F}_1$ will open the door, but force $\vec{F}_2$, which pushes straight at the hinge, will not. Force $\vec{F}_3$ will open the door, but not as easily as $\vec{F}_1$. What about $\vec{F}_4$? It is perpendicular to the door, it has the same magnitude as $\vec{F}_1$, but you know from experience that pushing close to the hinge is not as effective as pushing at the outer edge of the door.

The ability of a force to cause a rotation or a twisting motion thus depends on three factors:

1. The magnitude F of the force.
2. The distance r from the pivot to the point at which the force is applied.
3. The angle at which the force is applied.

We can incorporate these three observations into a single quantity called the **torque** τ (Greek tau). Loosely speaking, τ measures the "effectiveness" of a force at causing an object to rotate about a pivot. **Torque is the rotational equivalent of force.** In Figure 7.6, for instance, the torque τ_1 due to $\vec{F}_1$ is greater than τ_4 due to $\vec{F}_4$.

To make these ideas specific, Figure 7.7 shows a force $\vec{F}$ applied at one point of a wrench that can rotate about the nut that it's turning. Figure 7.7 defines the distance r from the pivot to the point at which the force is applied; the **radial line,** the line starting at the pivot and extending through the point at which the force is applied; and the angle ϕ (Greek phi) measured from the radial line to the direction of the force.

We saw in Figure 7.6 that force $\vec{F}_1$, which was directed perpendicular to the door, was quite effective in opening it, but force $\vec{F}_2$, directed toward the hinges, had no effect on its rotation. As shown in Figure 7.8, this suggests breaking the force $\vec{F}$ applied to the wrench into two component vectors: $\vec{F}_\perp$ directed perpendicular to the radial line, and $\vec{F}_\parallel$ directed parallel to it. Because $\vec{F}_\parallel$ points either directly toward or away from the pivot, it has no effect on the wrench's rotation, and thus contributes nothing to the torque. Only $\vec{F}_\perp$ tends to cause rotation of the wrench, so it is this component of the force that determines the torque.

FIGURE 7.6 The four forces are the same strength, but they have different effects on the swinging door.

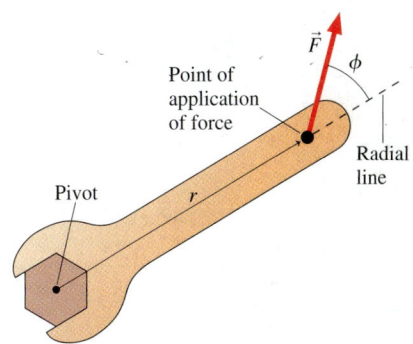

FIGURE 7.7 Force $\vec{F}$ exerts a torque about the pivot point.

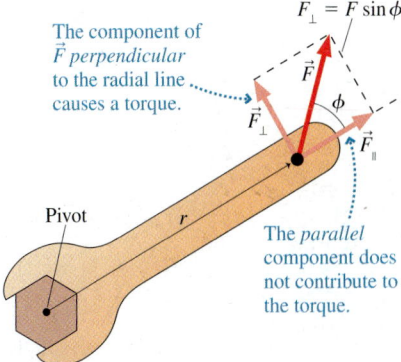

FIGURE 7.8 Torque is due to the component of the force perpendicular to the radial line.

NOTE ▶ The perpendicular component $\vec{F}_\perp$ is pronounced "F perpendicular" and the parallel component $\vec{F}_\parallel$ is pronounced "F parallel." ◀

We've seen that a force applied at a larger distance r from the pivot has a greater effect on rotation, so we expect a larger value of r to give a greater torque. We also saw that only $\vec{F}_\perp$ contributes to the torque. Both these observations are contained in our first expression for torque:

$$\tau = rF_\perp \tag{7.3}$$

Torque due to a force with perpendicular component $F_\perp$
acting at a distance r from the pivot

7.1   *Activ* Physics

From this equation, we see that the SI units of torque are newton-meters, abbreviated N · m.

EXAMPLE 7.2 Torque in opening a door

In trying to open a stuck door, Ryan pushes perpendicular to the door's surface with a force of 240 N at a distance of 0.75 m from the hinges. What torque does Ryan exert on the door?

Top view of door

Radial line

$r = 0.75$ m

Hinge

$\vec{F}$ 240 N

FIGURE 7.9 Ryan's force exerts a torque on the door.

PREPARE In Figure 7.9 the radial line is shown drawn from the pivot—the hinge—through the point at which the force $\vec{F}$ is applied. We see that $\vec{F}$ is perpendicular to the radial line, so $F_\perp = F = 240$ N and $F_\parallel = 0$. The distance from the hinge to the point at which the force is applied is $r = 0.75$ m.

SOLVE We can find the torque on the door from Equation 7.3. We have

$$\tau = rF_\perp = (0.75 \text{ m})(240 \text{ N}) = 180 \text{ N} \cdot \text{m}$$

ASSESS Ryan could slightly increase the torque he exerts by pushing at the very edge of the door in Figure 7.9.

We can see from Figure 7.8 that the component of force $\vec{F}$ perpendicular to the radial line is $F_\perp = F \sin\phi$. Thus a more general expression for the torque exerted by force $\vec{F}$ is

$$\tau = rF \sin\phi \tag{7.4}$$

Torque due to a force F applied at a distance r from the pivot,
at an angle ϕ to the radial line

NOTE ▶ Torque differs from force in a very important way. Torque is calculated or measured *about a particular point*. To say that a torque is 20 N · m is meaningless without specifying the point about which the torque is calculated. Torque can be calculated about any point, but its value depends on the point chosen because this choice determines r and ϕ. In practice, we usually calculate torques about a hinge, pivot, or axle. ◀

Returning to the door of Figure 7.6, you can see that $\vec{F}_1$ is most effective at opening the door because $\vec{F}_1$ exerts the *largest torque* about the pivot point: It has the largest distance r from the pivot *and* acts at $\phi = 90°$, so that $\sin\phi = 1$. $\vec{F}_3$ has equal magnitude, but it is applied at an angle less than 90° and thus exerts less torque. $\vec{F}_2$, pushing straight at the hinge with $\phi = 0°$ and $\sin\phi = 0$, exerts no torque at all. And $\vec{F}_4$, with a smaller value for r, exerts less torque than $\vec{F}_1$.

Figure 7.10 shows an alternative way to calculate torque. The line that is in the direction of the force, and passes through the point at which the force acts,

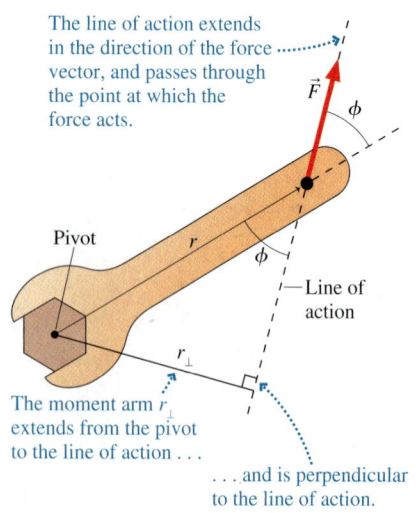

The line of action extends in the direction of the force vector, and passes through the point at which the force acts.

$\vec{F}$

ϕ

Pivot

r

ϕ

Line of action

$r_\perp$

The moment arm $r_\perp$ extends from the pivot to the line of action . . .

. . . and is perpendicular to the line of action.

FIGURE 7.10 You can also calculate torque in terms of the moment arm between the pivot and the line of action.

is called the *line of action.* The perpendicular distance from this line to the pivot is the **moment arm** (or *lever arm*) $r_\perp$. You can see from the figure that $r_\perp = r\sin\phi$. We can then write Equation 7.4 as $\tau = rF\sin\phi = F(r\sin\phi) = Fr_\perp$. Thus an equivalent expression for the torque is

$$\tau = r_\perp F \tag{7.5}$$

Torque due to a force F with moment arm $r_\perp$

Equations 7.3–7.5 are three different ways of thinking about—and calculating—the torque due to a force. Depending on the problem at hand, one might be easier to use than the others. But they are all calculating the *same* torque, and will give the same value for the torque.

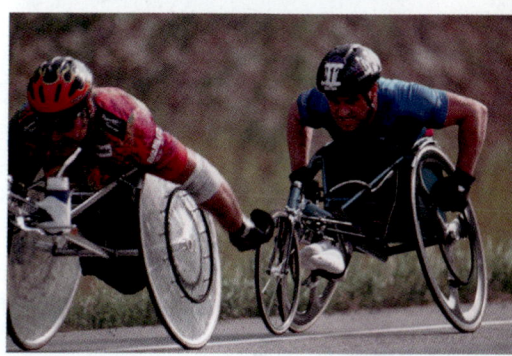

CONCEPTUAL EXAMPLE 7.1 Starting a bike

It is hard to get going if you try to start your bike with the pedal at the highest point. Why is this?

REASON Aided by the weight of the body, the greatest force can be applied to the pedal straight down. But with the pedal at the top, this force is exerted almost directly toward the pivot, causing only a small torque. We could say either that the perpendicular component of the force is small or that the moment arm is small.

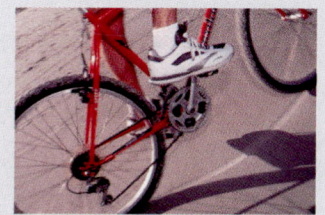

ASSESS If you've ever climbed a steep hill while standing on the pedals, you know that you get the greatest forward motion when one pedal is completely forward with the crank parallel to the ground. This gives the maximum possible torque because the force you apply is entirely perpendicular to the radial line, and the moment arm is as long as it can be.

Torque versus speed To start and stop quickly, the basketball player needs to apply a large torque to her wheel. To make the torque as large as possible, the handrim—the outside wheel that she actually grabs—is almost as big as the wheel itself. The racer needs to move continuously at high speed, so his wheel spins much faster. To allow his hands to keep up, his handrim is much smaller than his chair's wheel, making its linear velocity correspondingly lower. The smaller radius means, however, that the torque he can apply is lower as well.

These equations give only the magnitude of the torque. But torque, like force, has a sign. **A torque that tends to rotate the object in a counterclockwise direction is positive, while a negative torque gives a clockwise rotation.** Figure 7.11 summarizes the signs. Notice that a force pushing straight toward the pivot or pulling straight out from the pivot exerts *no* torque.

NOTE ▶ When calculating a torque, you must supply the appropriate sign by observing the direction in which the torque acts. ◀

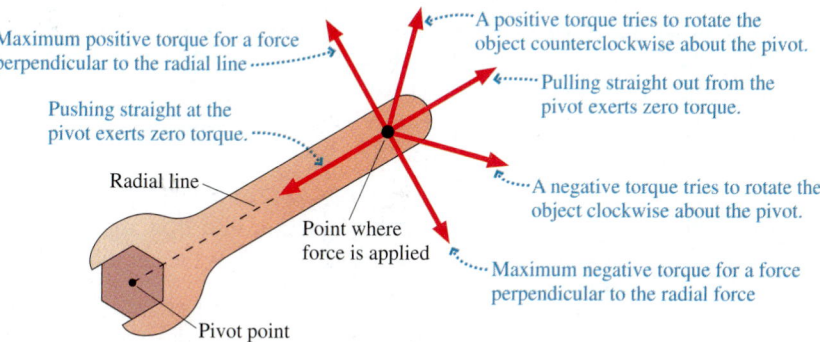

Maximum positive torque for a force perpendicular to the radial line

A positive torque tries to rotate the object counterclockwise about the pivot.

Pushing straight at the pivot exerts zero torque.

Pulling straight out from the pivot exerts zero torque.

Radial line

A negative torque tries to rotate the object clockwise about the pivot.

Point where force is applied

Maximum negative torque for a force perpendicular to the radial force

Pivot point

FIGURE 7.11 Signs and strengths of the torque.

EXAMPLE 7.3 Calculating the torque on a nut

Luis uses a 20-cm-long wrench to turn a nut. The wrench handle is tilted 30° above the horizontal, and Luis pulls straight down on the end with a force of 100 N. How much torque does Luis exert on the nut?

PREPARE Figure 7.12 shows the situation. The two illustrations correspond to two methods of calculating torque, corresponding to Equations 7.3 and 7.5.

(a) **(b)**

FIGURE 7.12 A wrench being used to turn a nut.

SOLVE According to Equations 7.3 and 7.4, the torque can be calculated as $\tau = rF_\perp = rF\sin\phi$. From Figure 7.12a we see that the angle between the force and the radial line is $\phi = 30° + 90° = 120°$.

The torque is then

$$\tau = -rF\sin\phi = -(0.20\text{ m})(100\text{ N})(\sin 120°) = -17\text{ N}\cdot\text{m}$$

We put in the minus sign because the torque is negative—it tries to rotate the nut in a *clockwise* direction.

Alternatively, we can use Equation 7.5 to find the torque. Figure 7.12b shows the moment arm $r_\perp$, the shortest distance from the pivot to the line of action. From the figure we see that

$$r_\perp = r\cos 30° = (0.20\text{ m})(\cos 30°) = 0.17\text{ m}$$

Then the torque is

$$\tau = -r_\perp F = -(0.17\text{ m})(100\text{ N}) = -17\text{ N}\cdot\text{m}$$

Again, we insert the minus sign because the torque acts to give a clockwise rotation.

ASSESS Both methods give the same answer for the torque, as expected. In general, however, you need use only one of Equations 7.3–7.5 to find the torque in any given situation. In using any of these methods to find the torque, remember to include the minus sign if the torque acts to rotate the object in a clockwise direction.

STOP TO THINK 7.2 Rank in order, from largest to smallest, the five torques τ_A to τ_E. The rods all have the same length and are pivoted at the dot.

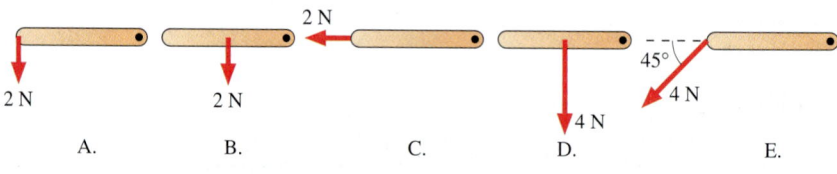

A.　　B.　　C.　　D.　　E.

Net Torque

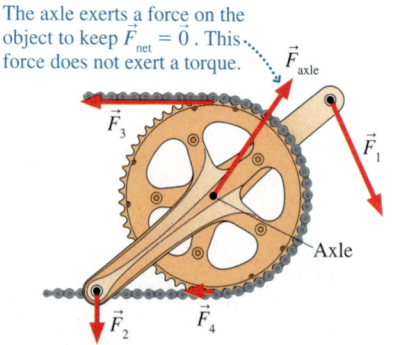

The axle exerts a force on the object to keep $\vec{F}_{net} = \vec{0}$. This force does not exert a torque.

FIGURE 7.13 The forces exert a net torque about the pivot point.

Figure 7.13 shows the forces acting on the crankset of a bicycle. Forces $\vec{F}_1$ and $\vec{F}_2$ are due to the rider pushing on the pedals, and $\vec{F}_3$ and $\vec{F}_4$ are tension forces from the chain. The crankset is free to rotate about a fixed axle, but the axle prevents it from having any translational motion with respect to the bike frame. It does so by exerting force $\vec{F}_{axle}$ on the object to balance the other forces and keep $\vec{F}_{net} = \vec{0}$.

Forces $\vec{F}_1$, $\vec{F}_2$, $\vec{F}_3$, and $\vec{F}_4$ exert torques $\tau_1, \tau_2, \tau_3, \tau_4$ on the crank (measured about the axle), but $\vec{F}_{axle}$ does *not* exert a torque because it is applied at the pivot point—the axle—and so has zero moment arm. Thus the *net* torque about the axle is the sum of the torques due to the *applied* forces:

$$\tau_{net} = \tau_1 + \tau_2 + \tau_3 + \cdots = \sum \tau \qquad (7.6)$$

In practice, usually only a small number of forces exert torques. For example, the only torque we considered in Example 7.3 was due to Luis's pull.

EXAMPLE 7.4 Force in turning a capstan

A capstan is a device used on old sailing ships to raise the anchor. A sailor pushes the long lever, turning the capstan and winding up the anchor rope. If the sailor lifts the anchor at a constant speed, the net torque, as we'll learn later in the chapter, is zero.

Suppose the rope tension due to the weight of the anchor is 1500 N. If the distance from the axis to the point on the lever where the sailor pushes is exactly seven times the radius of the capstan around which the rope is wound, with what force must the sailor push if the net torque on the capstan is to be zero?

PREPARE Shown in Figure 7.14 is a view looking down from above the capstan. The rope pulls with a tension force $\vec{T}$ at distance R from the axis of rotation. The sailor pushes with a force $\vec{F}$ at distance $7R$ from the axis. Both forces are perpendicular to their radial lines, so ϕ in Equation 7.4 is $90°$.

The sailor pushes the capstan in a clockwise direction . . .

. . . while the tension force tries to turn it counterclockwise.

$7R$

R

FIGURE 7.14 Top view of a sailor turning a capstan.

SOLVE The torque due to the tension in the rope is

$$\tau_T = RT\sin 90° = RT$$

We don't know the capstan radius, so we'll just leave it as R for now. This torque is positive because it tries to turn the capstan counterclockwise. The torque due to the sailor is

$$\tau_S = -(7R)F\sin 90° = -7RF$$

We put the minus sign in because this torque acts in the clockwise (negative) direction. The net torque is zero, so we have $\tau_T + \tau_S = 0$, or

$$RT - 7RF = 0$$

Note that the radius R cancels, leaving

$$F = \frac{T}{7} = \frac{1500\text{ N}}{7} = 210\text{ N}$$

ASSESS The force the sailor must exert is one-seventh the force the rope exerts: the long lever helps him lift the heavy anchor. In the HMS *Warrior*, built in 1860, it took 200 men turning the capstan to lift the huge anchor that weighed close to 55 000 N!

Note that forces $\vec{F}$ and $\vec{T}$ point in very different directions. Their torques only depend on their directions with respect to their own radial lines, not on the directions of the forces with respect to each other. The force the sailor needs to apply remains unchanged as he circles the capstan.

STOP TO THINK 7.3 Two forces act on the wheel shown. What third force, acting at point P, will make the net torque on the wheel zero?

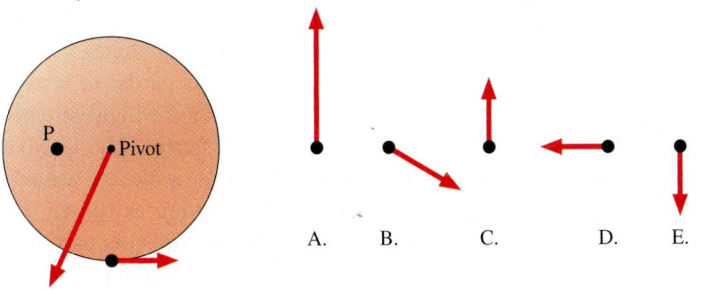

P   Pivot

A.   B.   C.   D.   E.

7.3 Gravitational Torque and the Center of Gravity

As the gymnast in Figure 7.15 pivots around the bar, a torque due to the force of gravity causes her to rotate toward a vertical position. A falling tree or a car hood slamming shut are other examples where gravity exerts a torque on an object. Stationary objects can also experience a torque due to gravity. A diving board experiences a gravitational torque about its fixed end. It doesn't rotate because of a counteracting torque provided by forces from the base at its fixed end.

(a) Gravity exerts a force and a torque on each particle that makes up the gymnast.   Rotation axis

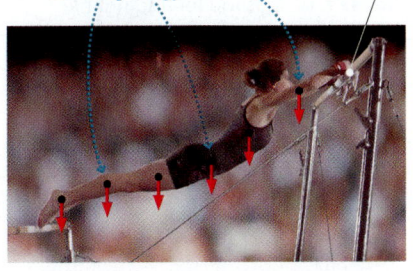

(b) The weight force provides a torque about the rotation axis.

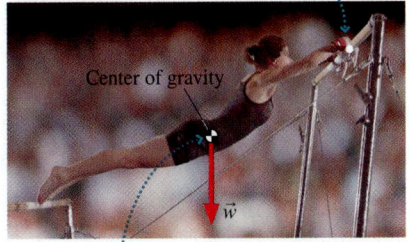

Center of gravity

$\vec{w}$

The gymnast responds *as if* her entire weight acts at her center of gravity.

FIGURE 7.15 The center of gravity is the point where the weight appears to act.

We've learned how to calculate the torque due to a single force acting on an object. But gravity doesn't act at a single point on an object. It pulls downward on *every particle* that makes up the object, as shown for the gymnast in Figure 7.15a, and so each particle experiences a small torque due to the force of gravity that acts upon it. The gravitational torque on the object as a whole is then the *net* torque exerted on all the particles. We won't prove it, but the gravitational torque can be calculated by assuming that the net force of gravity—that is, the object's weight $\vec{w}$—acts at single special point on the object called its **center of gravity** (symbol ☺). Then we can calculate the torque due to gravity by our methods learned earlier for a single force ($\vec{w}$) acting at a single point (the center of gravity). Figure 7.15b shows how we can consider the gymnast's weight as acting at her center of gravity.

Finding the Center of Gravity

To calculate the gravitational torque, we need to know where the object's center of gravity is located. There is a simple experimental method for finding the center of gravity of any object, based on the observation that **any object free to rotate about a pivot will come to rest with its center of gravity directly below the pivot.** To see this, consider the cutout map of the continental United States in Figure 7.16a. If the center of gravity is to the right or left of the blue line, a gravitational torque will cause the map to swing. If the center of gravity lies directly *below* the pivot, however, then the weight force lies along the line of action, and the torque is zero. The map can then remain at rest with no tendency to swing.

We know that the center of gravity lies somewhere along the blue line, but we don't yet know where. To find out, we need to suspend the map from a second pivot, as shown in Figure 7.16b. Then the center of gravity will fall somewhere along the red line shown. Because the center of gravity must lie on both the blue and red lines, it must be at their *intersection*. Interestingly, the geographical center of the continental United States is defined in just this way, as the center of gravity of a map of the contiguous United States. This point is one mile northwest of Lebanon, Kansas.

For a simple symmetrical object, such as a rod, sphere, or cube made from a uniform material, Figure 7.17 shows that **the center of gravity of a symmetrical object lies at its geometrical center.** A particularly simple case of this is a point particle, whose center of gravity lies at the position of the particle.

The center of gravity of an unsymmetrical object is harder to locate and, in some cases, may not even lie inside the object. Consider the high-wire act of Figure 7.18a on the next page, which appears to be a masterful feat of skill and balance. But because the man hangs so far below the motorcycle, the center of gravity of the acrobats and motorcycle actually lies in midair as shown, *below* the wire. This means that the act hangs stably, with no effort on the part of the acrobats. It is actually quite common for an object's center of gravity to lie outside the object. Because of the position of her arms and legs, the center of gravity of the high jumper in Figure 7.18b actually passes *under* the bar as she clears it.

The Center of Mass

In the particle model, all the mass of an object is considered to be concentrated at a single point, and all forces act at this point. The acceleration of this point particle is determined from Newton's second law as $\vec{a} = \vec{F}_{net}/m$. The motion of an extended object is generally more complex. Forces can act at different points on the object, leading to a net torque. And the object can undergo both translational

(a) The center of gravity must lie along this line.

Pivot 1

(b) Pivot 2

The center of gravity must also lie along this line.

So the center of gravity must be here, at the intersection of the two lines.

FIGURE 7.16 Method for finding the center of gravity of an object.

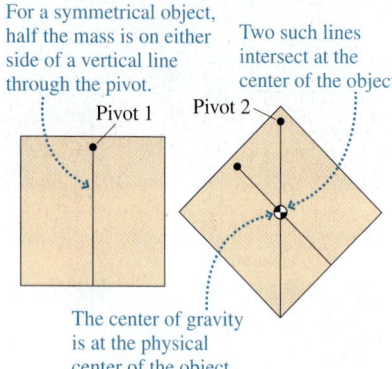

For a symmetrical object, half the mass is on either side of a vertical line through the pivot.

Two such lines intersect at the center of the object.

Pivot 1   Pivot 2

The center of gravity is at the physical center of the object.

FIGURE 7.17 The center of gravity of a symmetrical object lies at its center.

and rotational motion at the same time. This means that different points on the object can have *different* accelerations. To which point on the object does the $\vec{a}$ found from Newton's second law refer?

There is such a special point of an object called its **center of mass.** The concepts of center of mass and center of gravity are closely related. In fact, for everyday objects near the surface of the earth the positions of the center of mass and center of gravity are essentially identical. Their positions differ appreciably only when the size of the object is large enough that $\vec{g}$, the acceleration due to gravity, is different at different points on the object. The distinction between the positions of the center of gravity and center of mass is important only for huge objects such as planets or galaxies, and in this book we'll consider the positions to be the same.

For an extended object, then, Newton's second law gives the acceleration of the center of mass. We can thus write

$$\vec{a}_{cm} = \frac{\vec{F}_{net}}{m} \tag{7.7}$$

where $\vec{a}_{cm}$ is the acceleration of the center of mass.

An important application of this principle is to the case of objects thrown into the air. The net force in this case is just the object's weight $\vec{w} = m\vec{g}$, so that from Equation 7.7 we have $\vec{a}_{cm} = \vec{g}$. This is just the acceleration of a *point particle* during projectile motion, so that **the motion of the center of mass of an extended object undergoing projectile motion is the same as that of a point object undergoing projectile motion.** Figure 7.19 shows a hammer that has been tossed in the air. The hammer is undergoing both translational and rotational motion. Although the overall motion of the hammer is quite complex, the motion of the center of mass is simple: It has the same parabolic path as the ball moving through the air in Figure 3.28.

This is the advantage of the high jump technique illustrated in Figure 7.18b. Once the high jumper leaves the ground, her center of mass (which is the same point as her center of gravity) moves along a fixed parabolic path, no matter how she flexes her body. For a jump just at the limit of her ability, her center of mass will pass *below* the bar. But if she can flex her body as shown, she can just make it over the bar.

Calculating the Position of the Center of Gravity

It is nice to know you can locate an object's center of gravity by suspending it from a pivot, but it is rarely a practical technique. More often, we would like to calculate the center of gravity of an object made up of a combination of particles and other objects whose center of gravity positions are known.

Recall that for the purpose of calculating the torque due to gravity, the center of gravity is the point at which the total weight of the object can be considered as acting. This suggests a method for finding the center of gravity of an object made up of several point particles:

1. Calculate the gravitational torque on the object as the sum of the gravitational torques due to the weight of each of its particles, including the appropriate signs. We can use our already-developed methods for finding torques due to forces acting at known positions.
2. Calculate the torque due to the entire weight $\vec{w}$ acting at the center of gravity, whose position is still unknown. The torques calculated in these two ways must be the same; setting them equal will allow us to find the center-of-gravity position.

(a)

(b)

FIGURE 7.18 An object's center of gravity can lie outside the object.

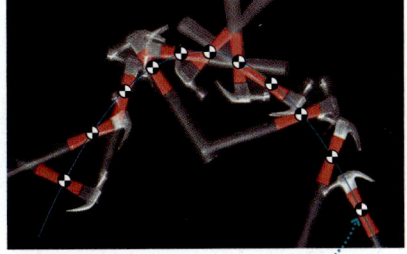

The center of mass of the hammer follows the same trajectory as a point particle undergoing projectile motion.

FIGURE 7.19 A hammer undergoing rotational and translational motion.

(a) Calculating the torque due to the weight of each mass

(b) Calculating the torque due to the total weight acting at the center of gravity

FIGURE 7.20 Finding the center of gravity of a dumbbell.

Holding steady Many moviemaking scenes require handheld shots for which the cameraman must walk along, following the action. The stabilizer shown reduces unwanted camera motion as the cameraman walks. The center of gravity of the camera and its hanging weight arm is located exactly at a pivot that can swing freely in any direction. The frictionless pivot exerts no torque on the camera and arm. Neither does the weight, because it acts at the pivot. With no torque acting on it, the camera has no tendency to rotate. The long arm also increases the system's *moment of inertia,* further decreasing unwanted rotations. More about this later!

To see how this works, consider the unbalanced dumbbell shown in Figure 7.20. Where along the bar should you lift it so that it remains balanced horizontally? From our discussion above, we know that it will remain balanced if the center of gravity is at the pivot—so you need to grasp it at its center of gravity. Let's find the position of the center of gravity by using our method outlined above.

First, we'll calculate the torque due to the weight of each particle; the net torque is the sum of these. Remember that the torque must be calculated about a specific point, so let's calculate the torque around the origin at $x = 0$.

Referring to Figure 7.20a, the torque τ_1 due to the weight of the left particle is

$$\tau_1 = -x_1 w_1 = -x_1 m_1 g$$

We put in the minus sign because the torque is negative—it tries to rotate the mass in a *clockwise* direction around the origin. Similarly, the torque τ_2 due to the right particle is

$$\tau_2 = -x_2 w_2 = -x_2 m_2 g$$

The net torque due to gravity is then

$$\tau_{\text{net}} = \tau_1 + \tau_2 = -(x_1 m_1 + x_2 m_2)g$$

In Figure 7.20b, we calculate the torque by considering the entire weight $\vec{w}$ as acting at the center of gravity. The (as yet unknown) position of the center of gravity is x_{cg}, so we have

$$\tau = -x_{\text{cg}} w = -x_{\text{cg}} m_{\text{total}} g = -x_{\text{cg}} (m_1 + m_2)g$$

This torque is also negative because it tries to rotate the entire dumbbell clockwise.

The torques calculated in these two ways must be the same, so we have

$$-x_{\text{cg}}(m_1 + m_2)g = -(x_1 m_1 + x_2 m_2)g$$

which we can solve to find that the center-of-gravity position is

$$x_{\text{cg}} = \frac{x_1 m_1 + x_2 m_2}{m_1 + m_2} \tag{7.8}$$

The following Tactics Box shows how Equation 7.8 can be generalized to find the center of gravity of *any* number of particles. If the particles don't all lie along the x-axis, then we'll also need to find the y-coordinate of the center of gravity.

(MP) TACTICS BOX 7.1 **Finding the center of gravity** Exercise 12

❶ Choose an origin for your coordinate system. You can choose any convenient point as the origin.
❷ Determine the coordinates (x_1, y_1), (x_2, y_2), (x_3, y_3), . . . for the particles of mass $m_1, m_2, m_3, \ldots$ respectively.
❸ The x-coordinate of the center of gravity is

$$x_{\text{cg}} = \frac{x_1 m_1 + x_2 m_2 + x_3 m_3 + \cdots}{m_1 + m_2 + m_3 + \cdots} \tag{7.9}$$

❹ Similarly, the y-coordinate of the center of gravity is

$$y_{\text{cg}} = \frac{y_1 m_1 + y_2 m_2 + y_3 m_3 + \cdots}{m_1 + m_2 + m_3 + \cdots} \tag{7.10}$$

Because the center of gravity depends on products such as x_1m_1, objects with large masses count more heavily than objects with small masses. Consequently, **the center of gravity tends to lie closer to the heavier objects or particles** that make up the entire object.

EXAMPLE 7.5 Where should the dumbbell be lifted?

A 1.0-m-long dumbbell has a 10 kg mass on the left and a 5.0 kg mass on the right. Find the position of the center of gravity, the point where the dumbbell should be lifted in order to remain balanced.

PREPARE We'll treat the two masses as point particles separated by a massless rod. Then we can use the steps from Tactics Box 7.1 to find the center of gravity. Let's choose the origin to be at the position of the 10 kg mass on the left, making $x_1 = 0$ m and $x_2 = 1.0$ m. Because the dumbbell masses lie on the x-axis, the y-coordinate of the center of gravity must also lie on the x-axis. Thus we only need to solve for the x-coordinate of the center of gravity.

SOLVE The x-coordinate of the center of gravity is found from Equation 7.9:

$$x_{cg} = \frac{x_1m_1 + x_2m_2}{m_1 + m_2} = \frac{(0 \text{ m})(10 \text{ kg}) + (1.0 \text{ m})(5.0 \text{ kg})}{(10 \text{ kg}) + (5.0 \text{ kg})}$$

$$= 0.33 \text{ m}$$

The center of gravity is 0.33 m from the 10 kg mass, or, equivalently, 0.17 m left of the center of the bar.

ASSESS The position of the center of gravity is closer to the greater mass. This is in agreement with our general statement above that the center of gravity tends to lie closer to the heavier particles.

The center of gravity of an extended object can often be found by considering the object as made up of pieces, each of whose mass and center of gravity is known or can be found. Then the coordinates of the entire object's center of gravity are given by Equations 7.9 and 7.10, with (x_1, y_1), (x_2, y_2), (x_3, y_3), . . . the coordinates of the centers of gravity of each piece and m_1, m_2, m_3, . . . their masses.

This method is widely used in biomechanics and kinesiology to calculate the center of gravity of the human body. Figure 7.21 shows how the body can be considered to be made of several segments, each of whose mass and center of gravity has been measured. The numbers shown are appropriate for a man with a total mass of 80 kg. For a given posture the positions of the segments and their centers of gravity can be found, and thus the whole-body center of gravity from Equations 7.9 and 7.10 (and a third equation for the z-coordinate). Example 7.6 explores a simplified version of this method.

FIGURE 7.21 Body segment masses and centers of gravity.

EXAMPLE 7.6 Finding the center of gravity of a gymnast

A gymnast performing on the rings holds himself in the pike position. Figure 7.22 shows how we can consider his body to be made up of two segments whose masses and center-of-gravity positions are shown. The upper segment includes his head, trunk, and arms, while the lower segment consists of his legs. Locate the overall center of gravity of the gymnast.

FIGURE 7.22 Centers of gravity of two segments of a gymnast.

Continued

PREPARE From Figure 7.22 we can find the *x*- and *y*-coordinates of the segment centers of gravity. We have

$$x_{trunk} = 15 \text{ cm}; \quad y_{trunk} = 50 \text{ cm};$$

$$x_{legs} = 30 \text{ cm}; \quad y_{legs} = 20 \text{ cm}$$

SOLVE The *x*- and *y*-coordinates of the center of gravity are given by Equations 7.9 and 7.10:

$$x_{cg} = \frac{x_{trunk} m_{trunk} + x_{legs} m_{legs}}{m_{trunk} + m_{legs}}$$

$$= \frac{(15 \text{ cm})(45 \text{ kg}) + (30 \text{ cm})(30 \text{ kg})}{45 \text{ kg} + 30 \text{ kg}} = 21 \text{ cm}$$

and

$$y_{cg} = \frac{y_{trunk} m_{trunk} + y_{legs} m_{legs}}{m_{trunk} + m_{legs}}$$

$$= \frac{(50 \text{ cm})(45 \text{ kg}) + (20 \text{ cm})(30 \text{ kg})}{45 \text{ kg} + 30 \text{ kg}} = 38 \text{ cm}$$

ASSESS The center of gravity position of the entire body, shown in Figure 7.22, is closer to that of the heavier trunk segment than to that of the lighter legs. It also lies along a line connecting the two segment centers of gravity, just as it would for the center of gravity of two point particles. Note also that the gymnast's hands—the pivot point—must lie directly below his center of gravity. Otherwise he would rotate forward or backward.

STOP TO THINK 7.4 The balls shown below are connected by very lightweight rods pivoted at the point indicated by a dot. The rod lengths are all equal except for A, which is twice as long. Rank in order, from least to greatest, the magnitudes of the gravitational torques about the pivot for arrangements A to E.

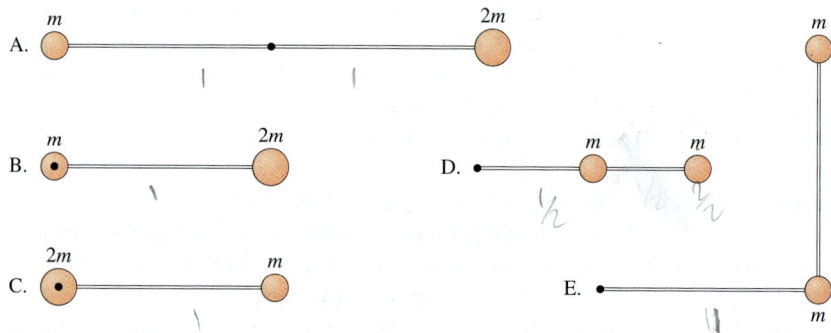

7.4 Rotational Dynamics and Moment of Inertia

We've found that torque is the rotational equivalent of force. Now we need to learn what torque *does*. Let's start by examining a *single particle* subject to a torque. Figure 7.23 shows a particle of mass *m* attached to a lightweight, rigid rod of length *r* that constrains the particle to move in a circle. The particle is subject to two forces. Because it's moving in a circle there must be a centripetal force—here, the tension $\vec{T}$ from the rod—directed toward the center of the circle. As we learned in Chapter 6, this is the force responsible for changing the *direction* of the particle's velocity. The acceleration associated with this change in the particle's velocity is the centripetal acceleration $\vec{a}_c$.

But the particle in Figure 7.23 is also subject to a second force $\vec{F}$, one directed *tangentially* to the circle. Because this force points in the direction of the particle's motion, it is the force responsible for changing the *speed* of the particle. Thus this force causes a tangential acceleration $\vec{a}_t$.

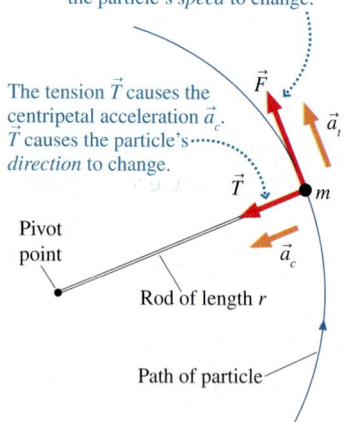

The tangential force $\vec{F}$ causes the tangential acceleration $\vec{a}_t$. $\vec{F}$ causes the particle's *speed* to change.

The tension $\vec{T}$ causes the centripetal acceleration $\vec{a}_c$. $\vec{T}$ causes the particle's *direction* to change.

FIGURE 7.23 A tangential force $\vec{F}$ exerts a torque on the particle and causes an angular acceleration.

To make the connection between torque and angular motion, let's begin by applying Newton's second law in the direction tangent to the circle. We have

$$a_t = \frac{F}{m} \tag{7.11}$$

Now the tangential and angular accelerations are related by $a_t = \alpha r$, so we can write Equation 7.11 as $\alpha r = F/m$, or

$$\alpha = \frac{F}{mr} \tag{7.12}$$

We can write the angular acceleration in terms of the *torque* exerted by F (the tension force exerts no torque because it is directed along the radial line). Because force $\vec{F}$ is perpendicular to the radial line, it exerts torque

$$\tau = rF$$

With this relation between F and τ, we can write Equation 7.12 as

$$\alpha = \frac{\tau}{mr^2} \tag{7.13}$$

In Section 7.2 we asked the question, "What do Newton's laws have to tell us about rotational motion?" We now have the first part of an answer: **A torque causes an angular acceleration.** This is the rotational equivalent of our prior discovery, for motion along a line, that a force causes an acceleration. Now all that remains is to expand this idea from a single particle to an extended object.

Newton's Second Law for Rotational Motion

Figure 7.24 shows a rigid body that undergoes rotation about a fixed and unmoving axis. According to the rigid-body model, we can think of the object as consisting of particles with masses $m_1, m_2, m_3, \ldots$ at fixed distances $r_1, r_2, r_3, \ldots$ from the axis. Suppose forces $\vec{F}_1, \vec{F}_2, \vec{F}_3, \ldots$ act on these particles. These forces exert torques around the rotation axis, so the object will undergo an angular acceleration α. Because all the particles that make up the object rotate together, each particle has this *same* angular acceleration α. Rearranging Equation 7.13 slightly, we can write the torques on the particles as

$$\tau_1 = m_1 r_1^2 \alpha \qquad \tau_2 = m_2 r_2^2 \alpha \qquad \tau_3 = m_3 r_3^2 \alpha$$

and so on for every particle in the object. If we add up all these torques, the *net* torque on the object is

$$\tau_{\text{net}} = \tau_1 + \tau_2 + \tau_3 + \cdots = m_1 r_1^2 \alpha + m_2 r_2^2 \alpha + m_3 r_3^2 \alpha + \cdots$$

$$= \alpha(m_1 r_1^2 + m_2 r_2^2 + m_3 r_3^2 + \cdots) = \alpha \sum m_i r_i^2 \tag{7.14}$$

By factoring α out of the sum, we're making explicit use of the fact that every particle in a rotating rigid body has the *same* angular acceleration α.

The quantity $\sum mr^2$ in Equation 7.14, which is the proportionality constant between angular acceleration and net torque, is called the object's **moment of inertia** I:

$$I = m_1 r_1^2 + m_2 r_2^2 + m_3 r_3^2 + \cdots = \sum m_i r_i^2 \tag{7.15}$$

Moment of inertia of a collection of particles

This large granite ball, with a mass of 26,400 kg, floats with nearly zero friction on a thin layer of pressurized water. As the girl pushes on it, she exerts a torque on the ball. Because of the ball's large moment of inertia, the torque causes only a small angular acceleration, so the ball's angular velocity increases only slowly as she pushes.

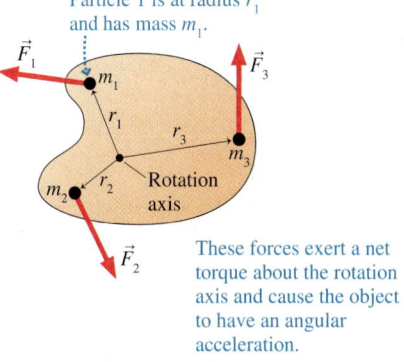

Particle 1 is at radius r_1 and has mass m_1.

These forces exert a net torque about the rotation axis and cause the object to have an angular acceleration.

FIGURE 7.24 The forces on a rigid body exert a torque about the rotation axis.

7.6

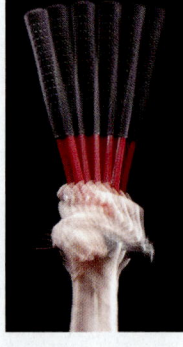

Hammering home inertia Most of the mass of a hammer is in its head, so the hammer's moment of inertia is large when calculated about an axis passing through the end of the handle (far from the head), but small when calculated about an axis passing through the head itself. You can *feel* this difference by attempting to wave a hammer back and forth about the handle end and the head end. It's much harder to do about the handle end because the large moment of inertia keeps the angular acceleration small.

The units of moment of inertia are those of mass times distance squared, or $kg \cdot m^2$. An object's moment of inertia, like torque, *depends on the axis of rotation*. Once the axis is specified, allowing the values of r_1, r_2, r_3, ... to be determined, the moment of inertia *about that axis* can be calculated from Equation 7.15.

NOTE ▶ The word *moment* in *moment of inertia* and *moment arm* has nothing to do with time. These terms stem from the Latin *momentum,* meaning "motion." ◀

Substituting the moment of inertia into Equation 7.14 puts the final piece of the puzzle into place, giving us the fundamental equation for rigid-body dynamics:

> **Newton's second law for rotation** An object that experiences a net torque τ_{net} about the axis of rotation undergoes an angular acceleration
>
> $$\alpha = \frac{\tau_{net}}{I} \qquad (7.16)$$
>
> LINEAR p.38   INVERSE p.118
>
> where I is the moment of inertia of the object *about the rotation axis.*

In practice we often write $\tau_{net} = I\alpha$, but Equation 7.16 better conveys the idea that **a net torque is the cause of angular acceleration.** In the absence of a net torque ($\tau_{net} = 0$), the object either does not rotate ($\omega = 0$) or rotates with *constant angular velocity* ($\omega = $ constant).

Interpreting the Moment of Inertia

Before rushing to calculate moments of inertia, let's get a better understanding of its meaning. First, notice that **moment of inertia is the rotational equivalent of mass.** It plays the same role in Equation 7.16 as does mass m in the now-familiar $\vec{a} = \vec{F}_{net}/m$. Recall that objects with larger mass have a larger *inertia,* meaning that they're harder to accelerate. Similarly, an object with a larger moment of inertia is harder to get rotating: It takes a larger torque to spin up an object with a larger moment of inertia than an object with a smaller moment of inertia. The fact that *moment of inertia* retains the word *inertia* reminds us of this.

But why does the moment of inertia depend on the distances r from the rotation axis? Think about trying to start a merry-go-round from rest, as shown in Figure 7.25. By pushing on the rim of the merry-go-round, you exert a torque on it and its angular velocity begins to increase. If your friends sit at the rim of the merry-go-round, as in Figure 7.25a, their distances r from the axle are large. The moment of inertia is large, according to Equation 7.15, and it will be difficult to get the merry-go-round rotating. If, however, your friends sit near the axle, as in Figure 7.25b, then r and the moment of inertia are small. You'll find it's much easier to get the merry-go-round going.

Thus an object's moment of inertia depends not only on the object's mass but on *how the mass is distributed* around the rotation axis. This is well known to bicycle racers. Every time a cyclist accelerates, she has to "spin up" the wheels and tires. The larger the moment of inertia, the more effort it takes and the slower her acceleration. For this reason, racers use the lightest possible tires, and they put those tires on wheels that have been designed to keep the mass as close as possible to the center without sacrificing the necessary strength and rigidity.

Table 7.3 summarizes the analogies between linear and rotational dynamics.

(a) Mass concentrated around the rim.   **(b)** Mass concentrated at the center.

Larger moment of inertia, harder to get rotating.   Smaller moment of inertia, easier to get rotating.

FIGURE 7.25 Moment of inertia depends both on the mass and how the mass is distributed.

TABLE 7.3 Rotational and linear dynamics

Rotational dynamics		Linear dynamics	
Torque	τ_{net}	Force	$\vec{F}_{net}$
Moment of inertia	I	Mass	m
Angular acceleration	α	Acceleration	$\vec{a}$
Second law	$\alpha = \tau_{net}/I$	Second law	$\vec{a} = \vec{F}_{net}/m$

EXAMPLE 7.7 Calculating the moment of inertia

Your friend is creating an abstract sculpture that consists of three small, heavy spheres attached by very lightweight 10-cm-long rods as shown in Figure 7.26. The spheres have masses $m_1 = 1.0$ kg, $m_2 = 1.5$ kg, and $m_3 = 1.0$ kg. What is the object's moment of inertia if it is rotated about axis a? About axis b?

FIGURE 7.26 Three point particles separated by lightweight rods.

PREPARE We'll use Equation 7.15 for the moment of inertia:

$$I = m_1 r_1^2 + m_2 r_2^2 + m_3 r_3^2$$

In this expression, r_1, r_2, and r_3 are the distances of each particle from the axis of rotation, so they depend on the axis chosen. Particle 1 lies directly on both axes, so $r_1 = 0$ cm in both cases.

Particle 2 lies 10 cm (0.10 m) from both axes. Particle 3 is 10 cm from axis a, but farther from axis b. We can find r_3 for axis b by using the Pythagorean theorem, which gives $r_3 = 14$ cm. These distances are indicated in the figure.

SOLVE For each axis, we can prepare a table of the values of r, m, and mr^2 for each particle, then add the values of mr^2. For axis a we have

Particle	r	m	mr^2
1	0 m	1.0 kg	0 kg·m^2
2	0.10 m	1.5 kg	0.015 kg·m^2
3	0.10 m	1.0 kg	0.010 kg·m^2
			$I_a = 0.025$ kg·m^2

For axis b we have

Particle	r	m	mr^2
1	0 m	1.0 kg	0 kg·m^2
2	0.10 m	1.5 kg	0.015 kg·m^2
3	0.14 m	1.0 kg	0.020 kg·m^2
			$I_b = 0.035$ kg·m^2

ASSESS We've already noted that the moment of inertia of an object is higher when its mass is distributed farther from the axis of rotation. Here, m_3 is farther from axis b than from axis a, leading to a higher moment of inertia about that axis.

The Moments of Inertia of Common Shapes

Newton's second law for rotational motion is easy to write, but we can't make use of it without knowing an object's moment of inertia. Unlike mass, we can't measure moment of inertia by putting an object on a scale. And while we can guess that the center of gravity of a symmetrical object is at the physical center of the object, we can *not* guess the moment of inertia of even a simple object.

For an object consisting of only a few point particles connected by massless rods, we can use Equation 7.15 to directly calculate I. But such an object is pretty unrealistic. All real objects are made up of solid material that is itself comprised of countless atoms. To calculate the moment of inertia of even a simple object requires integral calculus and is beyond the scope of this text. A short list of common moments of inertia is given in Table 7.4 on the next page. We use a capital M for the total mass of an extended object.

TABLE 7.4 Moments of inertia of objects with uniform density and total mass M

Object and axis	Picture	I	Object and axis	Picture	I
Thin rod, about center		$\frac{1}{12}ML^2$	Cylinder or disk, about center		$\frac{1}{2}MR^2$
Thin rod (of any cross section), about end		$\frac{1}{3}ML^2$	Cylindrical hoop, about center		MR^2
Plane or slab, about center		$\frac{1}{12}Ma^2$	Solid sphere, about diameter		$\frac{2}{5}MR^2$
Plane or slab, about edge		$\frac{1}{3}Ma^2$	Spherical shell, about diameter		$\frac{2}{3}MR^2$

Novel golf clubs The latest craze in golf putters is heads with high moment of inertia. When the putter hits the ball, the ball—by Newton's third law—exerts a force on the putter and thus exerts a torque that causes the head of the putter to rotate around the shaft. Any rotation while the putter is still in contact with the ball will affect the ball's direction. If the putter's mass is largely placed rather far from the shaft (the rotation axis), the moment of inertia about the shaft can be greatly increased. The large moment of inertia of the head will keep its angular acceleration small—reducing unwanted rotation and allowing a truer putt.

We can make some general observations about the moments of inertia in Table 7.4. For instance, the cylindrical hoop is composed of particles that are all the same distance R from the axis. Thus each particle of mass m makes the *same* contribution mR^2 to the hoop's moment of inertia. Adding up all these contributions gives

$$I = m_1R^2 + m_2R^2 + m_3R^2 + \cdots = (m_1 + m_2 + m_3 + \cdots)R^2 = MR^2$$

as given in the table. The solid cylinder of the same mass and radius has a *lower* moment of inertia than the hoop because much of the cylinder's mass is nearer its center. In the same way we can see why a slab rotated about its center has a lower moment of inertia than the same slab rotated about its edge: in the latter case, some of the mass is twice as far from the axis as the farthest mass in the former case. Those particles contribute *four times* as much to the moment of inertia, leading to an overall larger moment of inertia for the slab rotated about its edge.

STOP TO THINK 7.5 Four very lightweight disks of equal radii each have three identical heavy marbles glued to them as shown. Rank in order, from largest to smallest, the moments of inertia of the disks about the indicated axis.

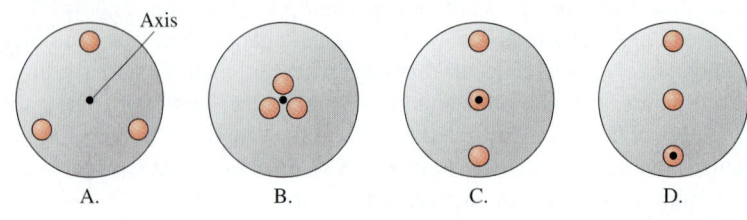

7.5 Using Newton's Second Law for Rotation

In this section we'll look at several examples of rotational dynamics for rigid bodies that rotate about a *fixed axis*. The restriction to a fixed axis avoids complications that arise for an object undergoing a combination of rotational and translational motion. The problem-solving strategy for rotational dynamics is very similar to that of linear dynamics in Chapter 5.

 7.8–7.10

 PROBLEM-SOLVING STRATEGY 7.1   Rotational dynamics problems

PREPARE Model the object as a simple shape. Draw a pictorial representation to clarify the situation, define coordinates and symbols, and list known information.

- Identify the axis about which the object rotates.
- Identify forces and determine their distance from the axis.
- Identify the torques caused by the forces and the signs of the torques.

SOLVE The mathematical representation is based on Newton's second law for rotational motion

$$\tau_{net} = I\alpha \quad \text{or} \quad \alpha = \frac{\tau_{net}}{I}$$

- Find the moment of inertia either by direct calculation using Equation 7.15 or from Table 7.4 for common shapes of objects.
- Use rotational kinematics to find angular positions and velocities.

ASSESS Check that your result has the correct units, is reasonable, and answers the question.

EXAMPLE 7.8 Starting an airplane engine

The engine in a small airplane is specified to have a torque of 500 N · m. This engine drives a 2.0-m-long, 40 kg single-blade propeller. On start-up, how long does it take the propeller to reach 2000 rpm?

PREPARE The propeller can be modeled as a rod that rotates about its center. The engine exerts a torque on the propeller. Figure 7.27 shows the propeller and the rotation axis.

The torque from the engine rotates the propeller.

$M = 40$ kg

$L = 2.0$ m

Axis

FIGURE 7.27 A rotating airplane propeller.

SOLVE The moment of inertia of a rod rotating about its center is found from Table 7.4:

$$I = \tfrac{1}{12}ML^2 = \tfrac{1}{12}(40 \text{ kg})(2.0 \text{ m})^2 = 13.3 \text{ kg} \cdot \text{m}^2$$

The 500 N · m torque of the engine causes an angular acceleration

$$\alpha = \frac{\tau}{I} = \frac{500 \text{ N} \cdot \text{m}}{13.3 \text{ kg} \cdot \text{m}^2} = 37.5 \text{ rad/s}^2$$

The time needed to reach $\omega_f = 2000$ rpm = 33.3 rev/s = 209 rad/s is

$$\Delta t = \frac{\Delta\omega}{\alpha} = \frac{\omega_f - \omega_i}{\alpha} = \frac{209 \text{ rad/s} - 0 \text{ rad/s}}{37.5 \text{ rad/s}^2} = 5.6 \text{ s}$$

ASSESS We've assumed a constant angular acceleration, which is reasonable for the first few seconds while the propeller is still turning slowly. Eventually, air resistance and friction will cause opposing torques and the angular acceleration will decrease. At full speed, the negative torque due to air resistance and friction cancels the torque of the engine. Then $\tau_{net} = 0$ and the propeller turns at *constant* angular velocity with no angular acceleration.

EXAMPLE 7.9 Angular acceleration of a falling pole

A 7.0-m-tall telephone pole with a mass of 260 kg has just been placed in the ground. Before the wires can be connected, the pole is hit by lightning at its base, nearly severing the pole from its base. The pole begins to fall, rotating about the part still connected to the base. Estimate the pole's angular acceleration when it has fallen by 25° from the vertical.

PREPARE The situation is shown in Figure 7.28, where we define our symbols and list the known information. Two forces are acting on the pole: the pole's weight $\vec{w}$, which acts at the center of gravity, and the force of the base on the pole

FIGURE 7.28 A falling telephone pole undergoes an angular acceleration due to a gravitational torque.

(not shown). This second force exerts no torque because it acts at the axis of rotation. The torque on the pole is thus due only to gravity. From the figure we see that this torque tends to rotate the pole in a counterclockwise direction, so the torque is positive.

SOLVE We'll model the pole as a uniform thin rod rotating about one end. Its center of gravity is at its center, a distance $L/2$ from the axis. You can see from the figure that the perpendicular component of $\vec{w}$ is $w_\perp = w\sin\theta$. Thus the torque due to gravity is

$$\tau_{\text{net}} = \left(\frac{L}{2}\right)w_\perp = \left(\frac{L}{2}\right)w\sin\theta = \frac{mgL}{2}\sin\theta$$

From Table 7.4, the moment of inertia of a thin rod rotated about its end is $I = \frac{1}{3}mL^2$. Thus from Newton's second law for rotational motion the angular acceleration is

$$\alpha = \frac{\tau_{\text{net}}}{I} = \frac{\frac{1}{2}mgL\sin\theta}{\frac{1}{3}mL^2} = \frac{3g\sin\theta}{2L}$$

$$= \frac{3(9.8 \text{ m/s}^2)\sin 25°}{2(7.0 \text{ m})} = 0.9 \text{ rad/s}^2$$

ASSESS The answer is given to only one significant figure because the problem asked for an *estimate* of the angular acceleration. This is usually a hint that you should make some simplifying assumptions, as we did here in modeling the pole as a thin rod.

CONCEPTUAL EXAMPLE 7.2 Balancing a meter stick

You've probably tried balancing a rod-shaped object vertically on your fingertip. If the object is very long, like a meter stick or a baseball bat, it's not too hard. But if it's short, like a pencil, it's almost impossible. Why is this?

REASON Suppose you've managed to balance a vertical stick on your fingertip, but then it starts to fall. You'll need to quickly adjust your finger to bring the stick back into balance. As Example 7.9 showed, the angular acceleration α of a thin rod is *inversely proportional* to L. Thus a long object like a

meter stick topples much more slowly than a short one like a pencil. Your reaction time is fast enough to correct for a slowly falling meter stick but not for a rapidly falling pencil.

ASSESS If we double the length of a rod, its mass doubles and its center of gravity is twice as high, so the gravitational torque τ on it is four times as much. But because a rod's moment of inertia is $I = \frac{1}{3}ML^2$, the longer rod's moment of inertia will be *eight* times greater so the angular acceleration will be only half as large.

Constraints Due to Ropes and Pulleys

Many important applications of rotational dynamics involve objects, such as pulleys, that are connected via ropes or belts to other objects. Figure 7.29 shows a rope passing over a pulley and connected to an object in linear motion. If the rope turns on the pulley *without slipping,* then the rope's speed v_{rope} must exactly match the speed of the rim of the pulley, which is $v_{\text{rim}} = \omega R$. If the pulley has an angular acceleration, the rope's acceleration a_{rope} must match the *tangential* acceleration of the rim of the pulley, $a_t = \alpha R$.

The object attached to the other end of the rope has the same speed and acceleration as the rope. Consequently, the object must obey the constraints

$$v_{\text{obj}} = \omega R$$

$$a_{\text{obj}} = \alpha R \qquad\qquad (7.17)$$

Motion constraints for an object connected to a pulley of radius R by a nonslipping rope

These constraints are very similar to the acceleration constraints introduced in Chapter 5 for two objects connected by a string or rope.

NOTE ▶ The constraints are given as magnitudes. Specific problems will require you to specify signs that depend on the direction of motion and on the choice of coordinate system. ◀

Rim speed $= \omega R$.
Rim acceleration $= \alpha R$.

Nonslipping rope

$v_{\text{obj}} = \omega R$
$a_{\text{obj}} = \alpha R$

The motion of the object must match the motion of the rim.

FIGURE 7.29 The rope's motion must match the motion of the rim of the pulley.

EXAMPLE 7.10 Time for a bucket to fall

Josh has just raised a 2.5 kg bucket of water using a well's winch when he accidentally lets go of the handle. The rope is wrapped around a 3.0 kg, 4.0-cm-diameter cylinder, which rotates on an axle through the center. The bucket is released from rest 4.0 m above the water level of the well. How long does it take to reach the water?

PREPARE Assume the rope is massless and does not slip. Figure 7.30a gives a visual overview of the falling bucket. Figure 7.30b shows the free-body diagrams for the cylinder and the bucket. The rope tension exerts an upward force on the bucket and a downward force on the outer edge of the cylinder. The rope is massless, so these two tension forces act as if they are an action/reaction pair with $T_b = T_c$. Because the tensions have equal magnitudes, we'll call both their magnitudes T.

FIGURE 7.30 Visual overview of a falling bucket.

SOLVE Newton's second law applied to the linear motion of the bucket is

$$ma_y = T - mg$$

where, as usual, the y-axis points upward. What about the cylinder? There is a normal force $\vec{n}$ on the cylinder due to the axle and the weight of the cylinder $\vec{w}_c$. However, neither of these forces exerts a torque because each passes through the rotation

axis. The only torque comes from the rope tension. The moment arm for the tension is $r_\perp = R$, and the torque is positive because the rope turns the cylinder counterclockwise. Thus $\tau_{\text{rope}} = TR$ and Newton's second law for the rotational motion is

$$\alpha = \frac{\tau_{\text{net}}}{I} = \frac{TR}{\frac{1}{2}MR^2} = \frac{2T}{MR}$$

The moment of inertia of a cylinder rotating about a center axis was taken from Table 7.4.

The last piece of information we need is the constraint due to the fact that the rope doesn't slip. Equation 7.17 relates only the magnitudes of the linear and angular accelerations, but in this problem α is positive (counterclockwise acceleration), while a_y is negative (downward acceleration). Hence

$$a_y = -\alpha R$$

Using α from the cylinder's equation in the constraint, we find

$$a_y = -\alpha R = -\frac{2T}{MR}R = -\frac{2T}{M}$$

Thus the tension is $T = -\frac{1}{2}Ma_y$. If we use this value of the tension in the bucket's equation, we can solve for the acceleration:

$$ma_y = -\frac{1}{2}Ma_y - mg$$

$$a_y = -\frac{g}{(1 + M/2m)} = -6.1 \text{ m/s}^2$$

The time to fall through $\Delta y = y_f - y_i = -4.0$ m is found from kinematics:

$$\Delta y = \frac{1}{2}a_y(\Delta t)^2$$

$$\Delta t = \sqrt{\frac{2\Delta y}{a_y}} = \sqrt{\frac{2(-4.0 \text{ m})}{-6.1 \text{ m/s}^2}} = 1.1 \text{ s}$$

ASSESS The expression for the acceleration gives $a_y = -g$ if $M = 0$. This makes sense because the bucket would be in free fall if there were no cylinder. When the cylinder has mass, the downward force of gravity on the bucket has to accelerate the bucket *and* spin the cylinder. Consequently, the acceleration is reduced and the bucket takes longer to fall.

FIGURE 7.31 The trajectories of the center of a wheel and of a point on the rim are seen in a time-exposure photograph.

7.6 Rolling Motion

Rolling is a *combination motion* in which an object rotates about an axis that is moving along a straight-line trajectory. For example, Figure 7.31 is a time-exposure photo of a rolling wheel with one light bulb on the axis and a second light bulb at the edge. The axis light moves straight ahead, but the edge light follows a curve called a *cycloid*. Let's see if we can understand this interesting motion. We'll consider only objects that roll without slipping.

Figure 7.32 shows a round object—a wheel or a sphere—that rolls forward exactly one revolution. The point that had been on the bottom follows the cycloid, the curve you saw in Figure 7.31, to the top and back to the bottom. The overall position of the object is measured by the position x_{cm} of its center of mass, located at the object's center. *Because the object doesn't slip,* the center of mass moves forward exactly one circumference: $\Delta x_{cm} = 2\pi R$.

NOTE ▶ We'll use center of mass rather than center of gravity in this section because we're not concerned with gravitational torque. ◀

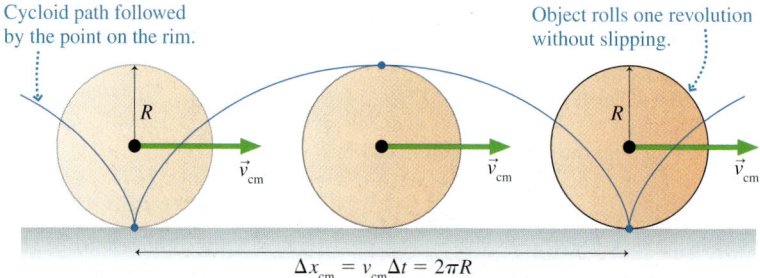

FIGURE 7.32 An object rolling through one revolution.

We can also write the distance traveled in terms of the velocity of the center of mass: $\Delta x_{cm} = v_{cm} \Delta t$. But Δt, the time it takes the object to make one complete revolution, is nothing other than the rotation period T. In other words, $\Delta x_{cm} = v_{cm}T$.

These two expressions for Δx_{cm} come from two perspectives on the motion, one looking at the rotation and the other at the translation of the center of mass. But it's the same distance no matter how you look at it, so these two expressions must be equal. Consequently,

$$\Delta x_{cm} = 2\pi R = v_{cm}T \tag{7.18}$$

If we divide by T, we can write the center-of-mass velocity as

$$v_{cm} = \frac{2\pi}{T}R \tag{7.19}$$

But $2\pi/T$ is the angular velocity ω, as you learned in Chapter 7, leading to

$$v_{cm} = \omega R \tag{7.20}$$

Equation 7.20 is the **rolling constraint,** the basic link between translation and rotation for objects that roll without slipping.

NOTE ▶ The rolling constraint is equivalent to Equation 7.17 for the speed of a rope that doesn't slip as it passes over a pulley. ◀

We can find the velocity for any point on a rolling object by adding the velocity of that point when the object is in pure translation, without rolling, to the velocity of the point when the object is in pure rotation, without translating. Figure 7.33 shows how the velocity vectors at the top, center, and bottom of a rotating wheel are found in this way.

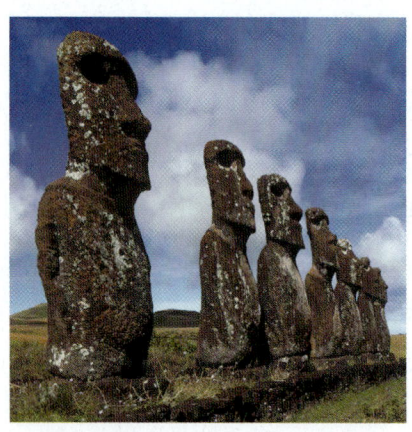

Ancient movers The great stone *moai* of Easter Island were moved as much as 16 km from a quarry to their final positions. Archeologists believe that one possible method of moving these 14 ton statues was to place them on rollers. One disadvantage of this method is that the statues, placed on top of the rollers, move twice as fast as the rollers themselves. Thus rollers are continuously left behind, and have to be carried back to the front and reinserted. Sadly, the indiscriminate cutting of trees for moving *moai* may have hastened the demise of this island civilization.

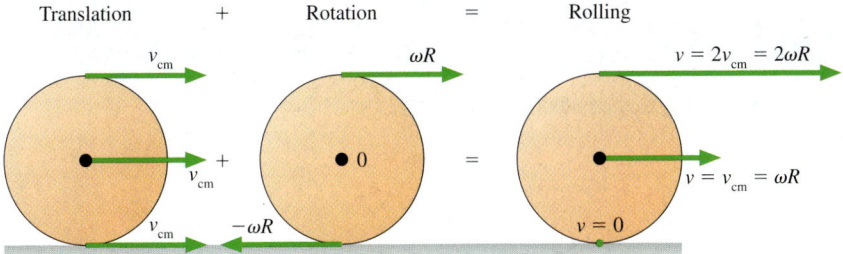

FIGURE 7.33 Rolling is a combination of translation and rotation.

Thus the point at the top of the wheel has a forward speed of v_{cm} due to its translational motion plus a forward speed of $\omega R = v_{cm}$ due to its rotational motion. The speed of a point at the top of a wheel is then $2v_{cm} = 2\omega R$, or *twice* the speed of its center of mass. On the other hand, the point at the bottom of the wheel, where it touches the ground, still has a forward speed of v_{cm} due to its translational motion. But its velocity due to rotation points *backward* with a magnitude of $\omega R = v_{cm}$. Adding these, we find that the velocity of this lowest point is *zero*. In other words, **the point on the bottom of a rolling object is instantaneously at rest.**

Although this seems surprising, it is really what we mean by "rolling without slipping." If the bottom point had a velocity, it would be moving horizontally relative to the surface. In other words, it would be slipping or sliding across the surface. To roll without slipping, the bottom point, the point touching the surface, must be at rest.

EXAMPLE 7.11 Rotating your tires
The diameter of your tires is 0.60 m. You take a 60 mile trip at a speed of 45 mph.

a. During this trip, what was your tires' angular speed?
b. How many times did they revolve?

PREPARE The angular speed is related to the speed of a wheel's center by Equation 7.20, $v_{cm} = \omega R$. Because the center of the wheel turns on an axle fixed to the car, the speed v_{cm} of the wheel's center is the same as that of the car. We prepare by converting the car's speed to SI units to get

$$v_{cm} = (45 \text{ mph}) \times \left(0.447 \frac{\text{m/s}}{\text{mph}}\right) = 20 \text{ m/s}$$

Once we know the angular speed, we can find the number of times the tires turned from the rotational-kinematic equation $\Delta\theta = \omega \Delta t$. We'll need to find the time traveled Δt from $v_{cm} = \Delta x/\Delta t$.

SOLVE a. From Equation 7.20 we have

$$\omega = \frac{v_{cm}}{R} = \frac{20 \text{ m/s}}{0.30 \text{ m}} = 67 \text{ rad/s}$$

b. The time of the trip is

$$\Delta t = \frac{\Delta x}{v_{cm}} = \frac{60 \text{ mi}}{45 \text{ mi/h}} = 1.33 \text{ h} \times \frac{3600 \text{ s}}{1 \text{ h}} = 4800 \text{ s}$$

Thus the total angle through which the tires turn is

$$\Delta\theta = \omega \Delta t = (67 \text{ rad/s})(4800 \text{ s}) = 3.2 \times 10^5 \text{ rad}$$

Because each turn of the wheel is 2π rad, the number of turns is

$$\frac{3.2 \times 10^5 \text{ rad}}{2\pi \text{ rad}} = 51,000 \text{ turns}$$

ASSESS You can see that your tires rotate roughly a thousand times per mile. During the lifetime of a tire, about 50,000 miles, it will rotate about 50 million times!

STOP TO THINK 7.6   A wheel rolls without slipping. Which is the correct velocity vector for point P on the wheel?

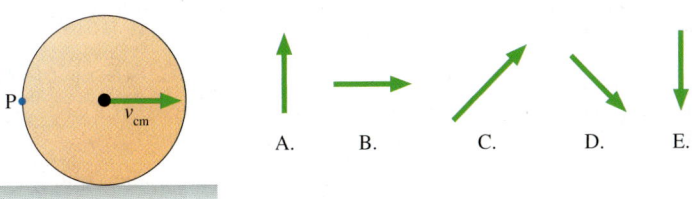

SUMMARY

The goal of Chapter 7 has been to understand the physics of rotating objects.

GENERAL PRINCIPLES

Newton's Second Law for Rotational Motion

If a net torque τ_{net} acts on an object, the object will experience an angular acceleration given by $\alpha = \tau_{net}/I$, where I is the object's moment of inertia about the pivot point.

This law is analogous to Newton's second law for linear motion, $\vec{a} = \vec{F}_{net}/m$.

IMPORTANT CONCEPTS

Torque is the rotational analog of force. Just as a force causes an object to undergo a linear acceleration, torque causes an object to undergo an angular acceleration.

There are two interpretations of torque:

$$1.\ \tau = rF_\perp \qquad 2.\ \tau = r_\perp F$$

The component of $\vec{F}$ perpendicular to the radial line causes a torque.

$F_\perp = F \sin\phi$

Pivot

The moment arm $r_\perp$ extends from the pivot to the line of action.

$r_\perp = r \sin\phi$

Both interpretations give the same expression for the magnitude of the torque: $\tau = rF\sin\phi$

The moment of inertia is the rotational equivalent of mass. The larger the moment of inertia, the more difficult it is to get the object rotating. For an object made up of particles of mass $m_1, m_2, \ldots$ at distances $r_1, r_2, \ldots$ from the axis, the moment of inertia is

$$I = m_1 r_1^2 + m_2 r_2^2 + m_3 r_3^2 + \cdots = \sum mr^2$$

Angular and tangential acceleration

A particle moving in a circle has

- A velocity tangent to the circle
- A centripetal acceleration $\vec{a}_c$ directed towards the center of the circle.

If the particle's speed is increasing, it will also have

- A tangential acceleration $\vec{a}_t$ directed tangent to the circle.
- An angular acceleration α.

The angular and tangential accelerations are related by $a_t = \alpha r$.

For a **rigid body,** the angular velocity and angular acceleration is the same for every point on the object.

Centripetal acceleration   Angular velocity
$\vec{a}_t$ $\vec{a}_c$ r
Tangential acceleration
Velocity

Center of gravity

The **center of gravity** of an object is the point at which gravity can be considered as acting.

Gravity acts on each particle that makes up the object.

The object responds *as if* its entire weight acts at the center of gravity.

$\vec{w}$

The **position of the center of gravity** depends on the distances $x_1, x_2, \ldots$ of each particle of mass $m_1, m_2, \ldots$ from the origin:

$$x_{cg} = \frac{x_1 m_1 + x_2 m_2 + x_3 m_3 + \cdots}{m_1 + m_2 + m_3 + \cdots}$$

APPLICATIONS

Moments of inertia of common shapes

MR^2 $\frac{1}{2}MR^2$

$\frac{2}{5}MR^2$ $\frac{1}{3}ML^2$

$\frac{2}{3}MR^2$ $\frac{1}{12}ML^2$

Rotation about a fixed axis

When a net torque is applied to an object that rotates about a fixed axis, the object will undergo an **angular acceleration** given by

$$\alpha = \frac{\tau_{net}}{I}$$

If a rope unwinds from a pulley of radius R, the angular quantities of the pulley are related to the linear quantities of the rope as

$$a_{obj} = \alpha R \qquad v_{obj} = \omega R$$

Rolling motion

For an object that rolls without slipping

$$v_{cm} = \omega R$$

The velocity of a point at the top of the object is twice that of the center.

$2\vec{v}_{cm}$

$\vec{v}_{cm}$

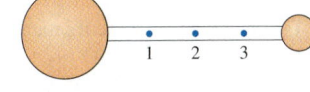

For instructor-assigned homework, go to www.masteringphysics.com

Problem difficulty is labeled as I (straightforward) to IIII (challenging).

Problems labeled INT integrate significant material from earlier chapters; BIO are of biological or medical interest.

QUESTIONS

Conceptual Questions

1. Figure Q7.1 shows four rotating wheels. What are the signs ($+$ or $-$) of the angular velocity and angular acceleration in each case?

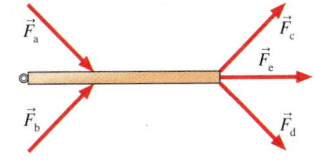

FIGURE Q7.1

2. If you are using a wrench to loosen a very stubborn nut, you can make the job easier by using a "cheater pipe." This is a piece of pipe that slides over the handle of the wrench, as shown in Figure Q7.2, making it effectively much longer. Explain why this would help you loosen the nut.

FIGURE Q7.2

3. Five forces are applied to a door, as seen from above in Figure Q7.3. For each force, is the torque about the hinge positive, negative, or zero?

FIGURE Q7.3

4. A screwdriver with a very thick handle requires less force to operate than one with a very skinny handle. Explain why this is so.

5. If you have ever driven a truck, you likely found that it had a steering wheel with a larger diameter than that of a passenger car. Why is this?

6. A common type of door stop is a wedge made of rubber. Is such a stop more effective when jammed under the door near or far from the hinges? Why?

7. Suppose you are hanging from a tree branch. If you move out along the branch, farther away from the trunk, the branch will be more likely to break. Explain why this is so.

8. A student gives a quick push to a ball at the end of a massless, rigid rod, causing the ball to rotate clockwise in a *horizontal* circle as shown in Figure Q7.8. The rod's pivot is frictionless.
   a. As the student is pushing, is the torque about the pivot positive, negative, or zero?
   b. After the push has ended, what does the ball's angular velocity do? Steadily

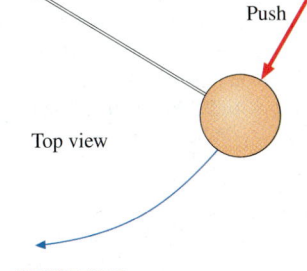

FIGURE Q7.8

increase? Increase for a while, then hold steady? Hold steady? Decrease for a while, then hold steady? Steadily decrease? Explain.
   c. Right after the push has ended, is the torque positive, negative, or zero?

9. Is the center of gravity of the dumbbell shown in Figure Q7.9 at point 1, 2, or 3? Explain.

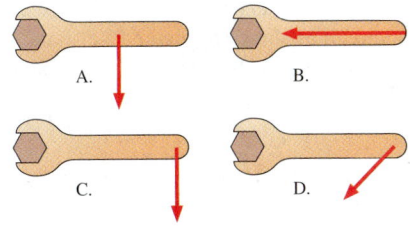

FIGURE Q7.9

10. When you rise from a chair, you have to lean quite far forward (try it!). Why is this?

11. Suppose you have two identical-looking metal spheres of the same size and the same mass. One of them is solid, the other is hollow. How can you tell which is which?

12. The moment of inertia of a uniform rod about an axis through its center is $ML^2/12$. The moment of inertia about an axis at one end is $ML^2/3$. Explain *why* the moment of inertia is larger about the end than about the center.

13. A heavy steel rod, 1.0 m long, and a light pencil, 0.15 m long, are held 15° from the vertical with one end on a table, then released simultaneously. Which will hit the table first? Or will it be a tie? Explain.

14. When they ski, small children fall down more than adults do. This is partially a matter of coordination, but it is partially a matter of physics. Why might you expect shorter people to fall more easily on the ski slope?

15. A car traveling at 60 mph has a pebble stuck in one of its tires. Eventually the pebble works loose, and at the instant of release it is at the top of the tire. Explain why the pebble then slams *hard* into the *front* of the wheel well.

Multiple-Choice Questions

16. I A nut needs to be tightened with a wrench. Which force shown in Figure Q7.16 will apply the greatest torque to the nut?

FIGURE Q7.16

17. I Suppose a bolt on your car engine needs to be tightened to a torque of 20 N · m. You are using a 15-cm-long wrench, and you apply a force at the very end in the direction that produces maximum torque. What force should you apply?
    A. 1300 N   B. 260 N   C. 130 N   D. 26 N

18. ‖ A machine part is made up of two pieces, with centers of gravity shown in Figure Q7.18. Which point could be the center of gravity of the entire part?

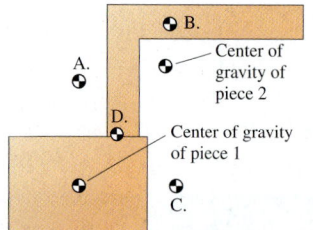

FIGURE Q7.18

19. ‖ A typical compact disk has a mass of 15 g and a diameter of 120 mm. What is its moment of inertia about an axis through its center, perpendicular to the disk?
    A. 2.7×10^{-5} kg · m^2 B. 5.4×10^{-5} kg · m^2
    C. 1.1×10^{-4} kg · m^2 D. 2.2×10^{-4} kg · m^2

20. ‖ Suppose you make a new kind of compact disk that is the same thickness as a current disk but twice the diameter. By what factor will the moment of inertia increase?
    A. 2   B. 4   C. 8   D. 16

21. ‖ Doors 1 and 2 have the same mass, height, and thickness. Door 2 is twice as wide as door 1. Bob pushes straight against the outer edge of door 2 with force F, and Barb pushes straight against the outer edge of door 1 with force $2F$. How do the angular accelerations α_1 and α_2 of the two doors compare?
    A. $\alpha_1 > \alpha_2$ B. $\alpha_1 = \alpha_2$ C. $\alpha_1 < \alpha_2$

22. ‖ A baseball bat has a heavy barrel and a thin handle. If you want to hold a baseball bat on your palm so that it balances vertically, you should
    A. Put the end of the handle in your palm, with the barrel up.
    B. Put the end of the barrel in your palm, with the handle up.
    C. The bat will be equally easy to balance in either configuration.

23. ‖ A car traveling at a steady 30 m/s has 74-cm-diameter tires. What is the approximate acceleration of a piece of the tread on any of the tires?
    A. 24 m/s^2 B. 48 m/s^2 C. 2400 m/s^2 D. 4800 m/s^2

PROBLEMS

Section 7.1 The Rotation of a Rigid Body

1. ‖‖ To throw the discus, the thrower holds the discus with a fully outstretched arm and makes one revolution as rapidly as possible to give maximum speed to the discus at release. The diameter of the circle in which the discus moves is about 1.8 m. If the thrower takes 1.0 s to complete one revolution, starting from rest, what will be the speed of the discus at release?

2. ‖‖ A computer hard disk starts from rest, then speeds up with an angular acceleration of 190 rad/s^2 until it reaches its final angular speed of 7200 rpm. How many revolutions has the disk made 10.0 s after it starts up?

3. ‖ The crankshaft in a race car goes from rest to 3000 rpm in 2.0 s.
    a. What is the crankshaft's angular acceleration?
    b. How many revolutions does it make while reaching 3000 rpm?

Section 7.2 Torque

4. ‖ Reconsider the situation in Example 7.3. If Luis pulls straight down on the end of a wrench that is in the same orientation but is 35 cm long, rather than 20 cm, what force must he apply to exert the same torque?

5. ‖ Balls are attached to light rods and can move in horizontal circles as shown in Figure P7.5. Rank in order, from smallest to largest, the torques τ_1 to τ_4 about the centers of the circles.

6. ‖‖ Six forces, each of magnitude either F or $2F$, are applied to a door as seen from above in Figure P7.6. Rank in order, from smallest to largest, the six torques τ_1 to τ_6 about the hinge.

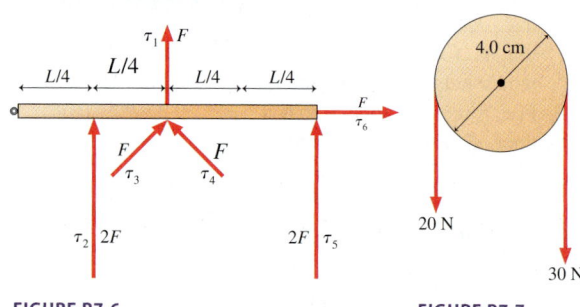

FIGURE P7.6   **FIGURE P7.7**

7. ‖ What is the net torque on the pulley about the axle in Figure P7.7?

8. ‖ The tune-up specifications of a car call for the spark plugs to be tightened to a torque of 38 N · m. You plan to tighten the plugs by pulling on the end of a 25-cm-long wrench. Because of the cramped space under the hood, you'll need to pull at an angle of 120° with respect to the wrench shaft. With what force must you pull?

9. ‖‖ A professor's office door is 0.91 m wide, 2.0 m high, and 4.0 cm thick; has a mass of 25 kg; and pivots on frictionless hinges. A "door closer" is attached to the door and the top of the door frame. When the door is open and at rest, the door closer exerts a torque of 5.2 N · m. What is the least force that you need to apply to the door to hold it open?

10. ‖ In Figure P7.10, force $\vec{F}_2$ acts half as far from the pivot as $\vec{F}_1$. What magnitude of $\vec{F}_2$ causes the net torque on the rod to be zero?

FIGURE P7.5

FIGURE P7.10

11. || Tom and Jerry both push on the 3.0-m-diameter merry-go-round shown in Figure P7.11.

    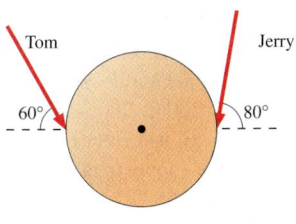
    FIGURE P7.11

    a. If Tom pushes with a force of 40.0 N and Jerry pushes with a force of 35.2 N, what is the net torque on the merry-go-round?
    b. What is the net torque if Jerry reverses the direction he pushes by 180° without changing the magnitude of his force?

12. | What is the net torque of the bar shown in Figure P7.12, about the axis indicated by the dot?

    FIGURE P7.12

    FIGURE P7.13

13. ||| What is the net torque of the bar shown in Figure P7.13, about the axis indicated by the dot?
14. ||| What is the net torque of the bar shown in Figure P7.14, about the axis indicated by the dot?

    FIGURE P7.14

Section 7.3 Gravitational Torque and the Center of Gravity

15. || A 1.7-m-long barbell has a 20 kg weight on its left end and a 35 kg weight on its right end.
    a. If you ignore the weight of the bar itself, how far from the left end of the barbell is the center of gravity?
    b. Where is the center of gravity if the 8.0 kg mass of the barbell itself is taken into account?

16. || Three identical coins lie on three corners of a square 10.0 cm on a side, as shown in Figure P7.16. Determine the x and y coordinates of the center of gravity of the three coins.

    FIGURE P7.16

17. | Hold your arm outstretched so that it is horizontal. Estimate the mass of your arm and the position of its center of gravity. What is the gravitational torque on your arm in this position, computed around the shoulder joint?
    BIO

18. | While sitting in a chair, extend your lower leg so that it is horizontal. Estimate the mass of your leg and the position of its center of gravity. What is the gravitational torque on your leg in this position, computed around the thigh joint?
    BIO

19. ||| A flag pole consists of a 2.0-m-long, 5.0 kg pole with a 3.0 kg decorative ball on its end, as shown in Figure P7.19. What is the gravitational torque on the flagpole, calculated about an axis at the fixed end of the pole?

    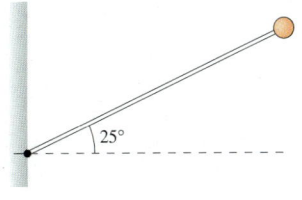
    FIGURE P7.19

20. ||| A 4.00-m-long, 500 kg steel beam extends horizontally from the point where it has been bolted to the framework of a new building under construction. A 70.0 kg construction worker stands at the far end of the beam. What is the magnitude of the torque about the point where the beam is bolted into place?

21. | An athlete at the gym holds a 3.0 kg steel ball in his hand. His arm is 70 cm long and has a mass of 4.0 kg. What is the magnitude of the torque about his shoulder if he holds his arm
    BIO
    a. Straight out to his side, parallel to the floor?
    b. Straight, but 45° below horizontal?

22. || The 2.0-m-long, 15 kg beam in Figure P7.22 is hinged at its left end. It is "falling" (rotating clockwise, under the influence of gravity), and the figure shows its position at three different times. What is the gravitational torque on the beam about an axis through the hinged end when the beam is at the
    a. Upper position?
    b. Middle position?
    c. Lower position?

    FIGURE P7.22

23. |||| Two thin beams are joined end-to-end as shown in Figure P7.23 to make a single object. The left beam is 10.0 kg and 1.00 m long and the right one is 40.0 kg and 2.00 m long.
    a. How far from the left end of the left beam is the center of gravity of the object?
    b. What is the gravitational torque on the object about an axis through its left end?

    FIGURE P7.23

24. ||| Figure P7.24 shows two thin beams joined at right angles. The vertical beam is 15.0 kg and 1.00 m long and the horizontal beam is 25.0 kg and 2.00 m long.

    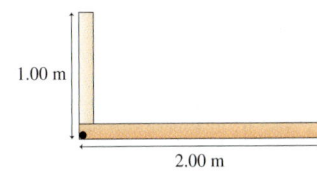
    FIGURE P7.24

    a. Find the center of gravity of the two joined beams. Express your answer in the form (x, y), taking the origin at the corner where the beams join.
    b. Calculate the gravitational torque on the joined beams about an axis through the corner.

Section 7.4 Rotational Dynamics and Moment of Inertia

25. ▐▐▐ A regulation table tennis ball has a mass of 2.7 g and is 40 mm in diameter. What is its moment of inertia about an axis that passes through its center?

26. ▐▐ Three pairs of balls are connected by very light rods as shown in Figure P7.26. Rank in order, from smallest to largest, the moments of inertia I_1, I_2, and I_3 about axes through the centers of the rods.

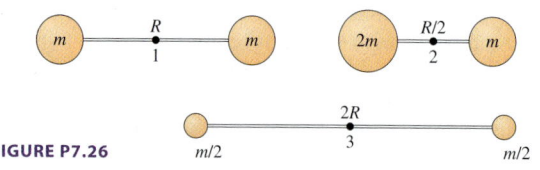

FIGURE P7.26

27. ▐ A playground toy has four seats, each 5.0 kg, attached to very light, 1.5-m-long rods, as seen from above in Figure P7.27. If two children, with masses of 15 kg and 20 kg, sit in seats opposite one another, what is the moment of inertia about the rotation axis?

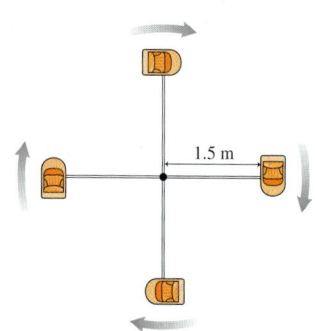

FIGURE P7.27

28. ▐ What is the moment of inertia in Problem 27 if the children sit in adjacent seats, instead of opposite ones?

29. ▐▐▐ You have two steel spheres. Sphere 2 has twice the radius of sphere 1. By what *factor* does the moment of inertia I_2 of sphere 2 exceed the moment of inertia I_1 of sphere 1?

Section 7.5 Using Newton's Second Law for Rotation

30. ▐ The left part of Figure P7.30 shows a bird's-eye view of two identical balls connected by a light rod that rotates about a vertical axis through its center. The right part shows a ball of twice the mass connected to a light rod of half the length, that rotates about its left end. If equal forces are applied as shown in the figure, which of the two rods will have the greater angular acceleration?

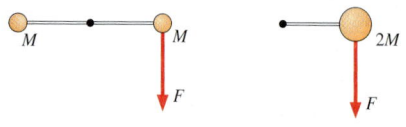

FIGURE P7.30

31. ▐▐ a. What is the moment of inertia of the door in Problem 9?
    b. If you let go of the open door, what is its angular acceleration immediately afterward?

32. ▐ A small grinding wheel has a moment of inertia of 4.0×10^{-5} kg · m². What net torque must be applied to the wheel for its angular acceleration to be 150 rad/s²?

33. ▐▐ While sitting in a swivel chair, you push against the floor with your heel to make the chair spin. The 7.0 N frictional force is applied at a point 40 cm from the chair's rotation axis, in the direction that causes the greatest angular acceleration. If that angular acceleration is 1.8 rad/s², what is the total moment of inertia about the axis of you and the chair?

34. ▐ An object's moment of inertia is 2.0 kg · m². Its angular velocity is increasing at the rate of 4.0 rad/s per second. What is the net torque on the object?

35. ▐▐▐▐ A 200 g, 20-cm-diameter plastic disk is spun on an axle through its center by an electric motor. What torque must the motor supply to take the disk from 0 to 1800 rpm in 4.0 s?

36. ▐▐ The 2.5 kg object shown in Figure P7.36 has a moment of inertia about the rotation axis of 0.085 kg · m². The rotation axis is horizontal. When released, what will be the object's initial angular acceleration?

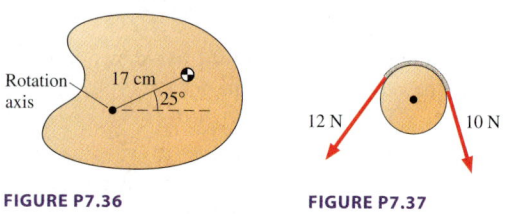

FIGURE P7.36   **FIGURE P7.37**

37. ▐▐ A frictionless pulley, which can be modeled as a 0.80 kg solid cylinder with a 0.30 m radius, has a rope going over it, as shown in Figure P7.37. The tension in the rope is 10 N on one side and 12 N on the other. What is the angular acceleration of the pulley?

38. ▐▐ If you lift the front wheel of a poorly maintained bicycle off the ground and then start it spinning at 0.72 rev/s, friction in the bearings causes the wheel to stop in just 12 s. If the moment of inertia of the wheel about its axle is 0.30 kg · m², what is the magnitude of the frictional torque?

39. ▐▐▐▐ A 5.0-cm-diameter toy top has a moment of inertia of 3.0×10^{-5} kg · m² about its rotation axis. To get the top spinning, its string is pulled with a tension of 0.30 N. How long does it take for the top to complete the first five revolutions? The string is long enough that it is wrapped around the top more than five turns.

40. ▐▐ Reconsider Example 7.9, but imagine that the pole is an electrical pole with a 42 kg electrical transformer attached to its very top. What is the angular acceleration in this case?

41. ▐▐▐ A 1.5 kg block and a 2.5 kg block are attached to opposite ends of a light rope. The rope hangs over a solid, frictionless pulley that is 30 cm in diameter and has a mass of 0.75 kg. What is the acceleration of the lighter block?

Section 7.6 Rolling Motion

42. ▐▐▐ A bicycle with 0.80-m-diameter tires is coasting on a level road at 5.6 m/s. A small blue dot has been painted on the tread of the rear tire.
    a. What is the angular speed of the tires?
    b. What is the speed of the blue dot when it is 0.80 m above the road?
    c. What is the speed of the blue dot when it is 0.40 m above the road?

43. ▐▐▐▐ A 1.2 g pebble is stuck in a tread of a 0.75-m-diameter automobile tire, held in place by static friction that can be at most 3.6 N. The car starts from rest and gradually accelerates on a straight road. How fast is the car moving when the pebble flies out of the tire tread?

General Problems

44. ‖ Figure P7.44 shows the angular-position-versus-time graph
INT for a particle moving in a circle.
  a. Write a description of the particle's motion.
  b. Draw the angular-velocity-versus-time graph.

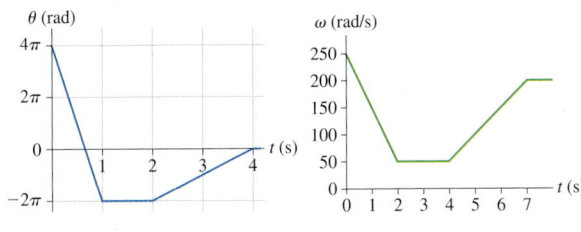

FIGURE P7.44   **FIGURE P7.45**

45. ‖ The graph in Figure P7.45 shows the angular velocity of the
INT crankshaft in a car. Draw a graph of the angular acceleration versus time. Include appropriate numerical scales on both axes.

46. ‖ A computer disk is 8.0 cm in diameter. A reference dot on the edge of the disk is initially located at $\theta = 45°$. The disk accelerates steadily for 0.50 s, reaching 2000 rpm, then coasts at steady angular velocity for another 0.50 s.
  a. What is the tangential acceleration of the reference dot at $t = 0.25$ s?
  b. What is the centripetal acceleration of the reference dot at $t = 0.25$ s?
  c. What is the angular position of the reference dot at $t = 1.0$ s?
  d. What is the speed of the reference dot at $t = 1.0$ s?

47. ‖ The 20-cm-diameter disk in Figure P7.47 can rotate on an axle through its center. What is the net torque about the axle?

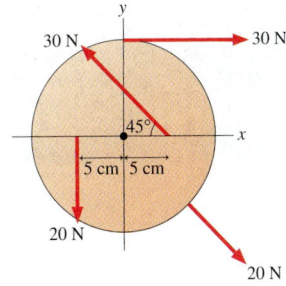

FIGURE P7.47

48. ‖ A combination lock has a 1.0-cm-diameter knob that is
INT part of the dial you turn to unlock the lock. To turn that knob, you grip it between your thumb and forefinger with a force of 0.60 N as you twist your wrist. Suppose the coefficient of static friction between the knob and your fingers is only 0.12 because some oil accidentally got onto the knob. What is the most torque you can exert on the knob without having it slip between your fingers?

49. ‖ A 70 kg man's arm, including the hand, can be modeled as a
BIO 75-cm-long uniform cylinder with a mass of 3.5 kg. In raising both his arms, from hanging down to straight up, by how much does he raise his center of gravity?

50. ‖ A penny has a mass of 2.5 g and is 1.5 mm thick; a nickel has a mass of 5.7 g and is 1.9 mm thick. If you make a stack of coins on a table, starting with five nickels and finishing with four pennies, how far above the tabletop is the center of gravity of the stack?

51. ‖ The machinist's square shown in Figure P7.51 consists of a thin, rectangular blade connected to a rectangular handle.
  a. Determine the x and y coordinates of the center of gravity. Let the lower-left corner be $x = 0, y = 0$.
  b. Sketch how the tool would hang if it were allowed to freely pivot about the point $x = 0, y = 0$.
  c. When hanging from that point, what angle would the long side of the blade make with the vertical?

FIGURE P7.51

52. ‖ The four masses shown in Figure P7.52 are connected by massless, rigid rods.
  a. Find the coordinates of the center of gravity.
  b. Find the moment of inertia about an axis that passes through mass A and is perpendicular to the page.
  c. Find the moment of inertia about a diagonal axis that passes through masses B and D.

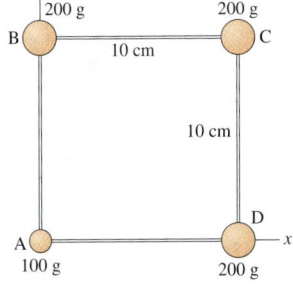

FIGURE P7.52

53. ‖ Three 0.10 kg balls are connected by light rods to form an equilateral triangle with a side length of 0.30 m. What is the moment of inertia of this triangle about an axis perpendicular to its plane and passing through one of the balls?

54. ‖ The three masses shown in Figure P7.54 are connected by massless, rigid rods.
  a. Find the coordinates of the center of gravity.
  b. Find the moment of inertia about an axis that passes through mass A and is perpendicular to the page.
  c. Find the moment of inertia about an axis that passes through masses B and C.

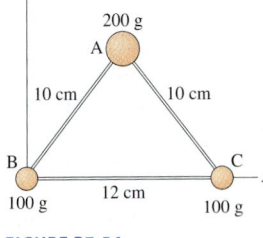

FIGURE P7.54

55. ‖ A closed 55 gallon drum is a 90-cm-long, 58-cm-diameter cylinder with flat ends that is made with 15 kg of sheet metal. If the thickness of the ends is the same as that of the curved sides, what is the moment of inertia about an axis that runs down the center of the drum?
  Hint: The uniform thickness of the sheet metal will allow you to determine the mass of the ends and the mass of the cylinder. What fraction of the surface area of the drum is associated with its ends?

56. ‖ A reasonable estimate of the moment of inertia of an ice
BIO skater spinning with her arms at her sides can be made by
modeling most of her body as a uniform cylinder. Suppose the
skater has a mass of 64 kg. One eighth of that mass is in her
arms, which are 60 cm long and 20 cm from the vertical axis
about which she rotates. The rest of her mass is approximately
in the form of a 20-cm-radius cylinder.
   a. Estimate the skater's moment of inertia to two significant
      figures.
   b. If she were to hold her arms outward, rather than at her
      sides, would her moment of inertia increase, decrease, or
      remain unchanged? Explain.

57. ‖ What is the moment of inertia of a 6.0-cm-diameter, 1.2 kg
cylinder that has a 2.0-cm-diameter cylindrical hole drilled
along its axis?
   Hint: Mathematically, removing mass from an object is
   equivalent to adding "negative mass" at the same location.

58. ‖ Starting from rest, a 12-cm-diameter compact disk takes
3.0 s to reach its operating angular velocity of 2000 rpm.
Assume that the angular acceleration is constant. The disk's
moment of inertia is $2.5 \times 10^{-5} \text{ kg} \cdot \text{m}^2$.
   a. How much torque is applied to the disk?
   b. How many revolutions does it make before reaching full
      speed?

59. ‖ A 60-cm-long, 500 g bar rotates in a horizontal plane on
an axle that passes through the center of the bar, perpendicular to its long axis. Compressed air is fed in through the
axle, passes through a small hole down the length of the bar,
and escapes as air jets from holes at the ends of the bar. The
jets are perpendicular to the bar's axis. Starting from rest,
the bar spins up to an angular velocity of 150 rpm at the end
of 10 s.
   a. How much force does each jet of escaping air exert on the
      bar?
   b. If the axle is moved to one end of the bar while the air jets
      are unchanged, what will be the bar's angular velocity at
      the end of 10 s?

60. ‖ Flywheels are large, massive wheels used to store energy.
They can be spun up slowly, then the wheel's energy can be
released quickly to accomplish a task that demands high
power. An industrial flywheel has a 1.5 m diameter and a
mass of 250 kg. A motor spins up the flywheel with a con-
stant torque of $50 \text{ N} \cdot \text{m}$. How long does it take the flywheel
to reach top angular speed of 1200 rpm?

61. ‖ A 3.0 kg block is attached to a string that is wrapped around
a 2.0 kg, 4.0-cm-diameter *hollow* cylinder that is free to
rotate. (Use Figure 7.30 but treat the cylinder as hollow.) The
block is released 1.0 m above the ground. What is the speed of
the block as it hits the ground?

62. ‖ A 1.5 kg block is connected by a rope
across a 50-cm-diameter, 2.0 kg, frictionless
pulley, as shown in Figure P7.62. A constant
10 N tension is applied to the other end of
the rope. Starting from rest, how long does
it take the block to move 30 cm?

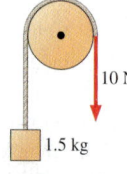

10 N

1.5 kg

FIGURE P7.62

63. ‖ A 1.0 kg ball and a 2.0 kg ball are connected by a 1.0-m-long rigid, massless rod.
The rod is rotating clockwise about its center of gravity at 20 rpm. What torque will
bring the balls to a halt in 5.0 s?

64. ‖ The 2.0 kg, 30-cm-diameter disk in Figure P7.64 is spin-
INT ning at 300 rpm. How much friction force must the brake
apply to the rim to bring the disk to a halt in 3.0 s?

ω

Brake

4.0 kg

1.0 m

2.0 kg

FIGURE P7.64   **FIGURE P7.65**

65. ‖ The two blocks in Figure P7.65 are connected by a massless
rope that passes over a pulley. The pulley is 12 cm in diameter
and has a mass of 2.0 kg. As the pulley turns, friction at the axle
exerts a torque of magnitude $0.50 \text{ N} \cdot \text{m}$. If the blocks are
released from rest, how long does it take the 4.0 kg block to
reach the floor?

66. ‖ The bunchberry flower has the fastest-moving parts ever
BIO seen in a plant. Initially, the stamens are held by the petals in a
bent position, storing elastic energy like a coiled spring. As
the petals release, the tips of the stamens act like medieval
catapults, flipping through a 60° angle in just 0.30 ms to
launch pollen from the anther sacs at their ends. The human
eye just sees a burst of pollen; careful photography (see Figure P7.66a) reveals the details. As shown in Figure P7.66b, we
can model a stamen tip as a 1.0-mm-long, 10 μg rigid rod with
a 10 μg anther sac at one end and a pivot point at the opposite
end. Although oversimplifying, we will assume that the angular acceleration is constant throughout the motion.

(a)   0 ms   0.2 ms   0.4 ms   1.0 ms

(b)   Pollen   Pollen

Anther sac   1.0 mm   60°

                                    Stamen

Initial configuration   Final configuration

FIGURE P7.66

   a. What is the tangential acceleration of the anther sac during
      the motion?
   b. What is the speed of the anther sac as it releases its pollen?
   c. How large is the "straightening torque"? Neglect gravita-
      tional forces in your calculation.
   d. Compute the gravitational torque on the stamen tip
      (including the anther sac) in its initial orientation. Was it
      reasonable to neglect the gravitational torque in part c?

Passage Problems

The Illusion of Flight

The grand jeté is a classic ballet maneuver in which a dancer executes a horizontal leap while moving her arms and legs up and then down. At the center of the leap, the arms and legs are gracefully extended, as we see in Figure P7.67a. The goal of the leap is to create the illusion of flight. As discussed in Section 7.3, the center of mass—and hence the center of gravity—of an extended object follows a parabolic trajectory when undergoing projectile motion. But when you watch a dancer leap through the air, you don't watch her center of gravity, you watch her head. If the translational motion of her head is horizontal—not parabolic—this creates the illusion that she is flying through the air, held up by unseen forces.

FIGURE P7.67

Figure P7.67b illustrates how the dancer creates this illusion. While in the air, she changes the position of her center of gravity relative to her body by moving her arms and legs up, then down. Her center of gravity moves in a parabolic path, but her head moves in a straight line. It's not flight, but it will appear that way, at least for a moment.

67. | To perform this maneuver, the dancer relies on the fact that the position of her center of gravity
   A. Is near the center of the torso.
   B. Is determined by the positions of her arms and legs.
   C. Moves in a horizontal path.
   D. Is outside of her body.

68. | Suppose you wish to make a vertical leap with the goal of getting your head as high as possible above the ground. At the top of your leap, your arms should be
   A. Held at your sides.
   B. Raised above your head.
   C. Outstretched, away from your body.

69. | When the dancer is in the air, is there a gravitational torque on her? Take the dancer's rotation axis to be through her center of gravity.
   A. Yes, there is a gravitational torque.
   B. No, there is not a gravitational torque.
   C. It depends on the positions of her arms and legs.

70. | In addition to changing her center of gravity, a dancer may change her moment of inertia. Consider her moment of inertia about a vertical axis through the center of her body. When she raises her arms and legs, this
   A. Increases her moment of inertia.
   B. Decreases her moment of inertia.
   C. Does not change her moment of inertia.

STOP TO THINK ANSWERS

Stop to Think 7.1: The angular velocity ω is constant. Thus the magnitude of the velocity $v = \omega r$ and the centripetal acceleration $a_c = \omega^2 r$ is constant. This also means that the ball's angular acceleration α and tangential acceleration $a_t = \alpha r$ are both zero.

Stop to Think 7.2: $\tau_E > \tau_A = \tau_D > \tau_B > \tau_C$. The perpendicular component in E is larger than 2 N.

Stop to Think 7.3: A. The force acting at the axis exerts no torque. Thus the third force needs to exert an equal but opposite torque to that exerted by the force acting at the rim. Force A, which has twice the magnitude but acts at half the distance from the axis, does so.

Stop to Think 7.4: $\tau_E = \tau_B > \tau_D > \tau_A = \tau_C$. The torques are $\tau_B = \tau_E = 2mgL$, $\tau_D = \frac{3}{2}mgL$, $\tau_A = \tau_C = mgL$, where L is the length of the rod in B.

Stop to Think 7.5: $I_D > I_A > I_C > I_B$. The moments of inertia are $I_B \approx 0$, $I_C = 2mr^2$, $I_A = 3mr^2$, $I_D = mr^2 + m(2r)^2 = 5mr^2$.

Stop to Think 7.6: C. The velocity of P is the vector sum of v_{cm} directed to the right, and an upward velocity of the same magnitude due to the rotation of the wheel.

8

EQUILIBRIUM AND ELASTICITY

How does a dancer balance so gracefully *en pointe*? And how does her foot withstand the great stresses concentrated on her toes? In this chapter we'll find answers to both these questions.

Looking Ahead ▶▶

The goal of Chapter 8 is to learn about the static equilibrium of extended objects and to understand the basic properties of springs and elastic materials. In this chapter you will learn to:

▶ Understand how torque determines the equilibrium of an extended object.

▶ Use force and torque to analyze extended objects in static equilibrium.

▶ Understand what determines the stability of an object.

▶ Use Hooke's law to calculate the force due to a spring.

▶ Understand the elastic properties of materials and use the concepts of stress, strain, and Young's modulus.

Looking Back ◀◀

This chapter uses what you have learned about equilibrium, torque, and the spring force. Please review:

◀ Section 4.3 Spring forces.

◀ Section 5.1 Equilibrium.

◀ Sections 7.2–7.3 Torque, center of gravity, and gravitational torque.

When you stand on tiptoe, you without thinking about it. with balance, unconsciously adjust the position of your body. You have years of experience adapting to the forces and torques that act on your body as you stand upright in different postures. How any object—a dancer, a chair, or a suspension bridge—remains upright and stable will be a key question we answer in this chapter. We will also look at why some objects are more stable than others, and why it is harder to balance on the toes of one foot than to stand with both feet flat on the floor.

To balance on her toes, the dancer in the photo must adjust the positions of her arms and legs by applying forces with her muscles and tendons. As we will see, these forces may be quite large. How do your bones react to such stresses? We will consider this question by looking at how elastic materials, ranging from spider silk to bones, respond to applied forces.

8.1 Torque and Static Equilibrium

We have now spent several chapters studying motion and its causes. In many disciplines, it is just as important to understand the conditions under which objects do *not* move. In structural engineering, buildings and dams must be designed such that they remain motionless, even when huge forces act on them. In sports science, a correct stationary position is often the starting point for a successful athletic event. And joints in the body must sustain large forces when the body is supporting heavy stationary loads, as in lifting or carrying heavy objects.

Recall from Section 5.1 that an object at rest is in *static equilibrium.* As long as an object can be modeled as a *particle,* the condition necessary for static equilibrium is that the net force $\vec{F}_{net}$ on the particle is zero. Such a situation is shown in Figure 8.1a, where the two forces applied to the particle balance and the particle can remain at rest.

But in Chapter 7 we moved beyond the particle model to study extended objects that can rotate. Consider, for example, the block in Figure 8.1b. In this case the two forces act along the same line, the net force is zero, and the block is in equilibrium. But what about the block in Figure 8.1c? The net force is still zero, but this time the block begins to rotate because the two forces exert a net *torque.* For an extended object, $\vec{F}_{net} = \vec{0}$ is not by itself enough to ensure static equilibrium. There is a second condition for static equilibrium of an extended object: The net torque τ_{net} on the object must also be zero.

If we write the net force in component form, the conditions for static equilibrium of an extended object are

(a) When the net force on a particle is zero, the particle is in equilibrium.

(b) Both the net force and the net torque are zero, so the block is in static equilibrium.

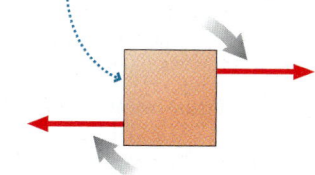

(c) The net force is still zero, but the net torque is *not* zero. The block is not in equilibrium.

FIGURE 8.1 A block with no net force acting on it may still be out of equilibrium.

$$\left.\begin{array}{l} \sum F_x = 0 \\ \sum F_y = 0 \end{array}\right\} \quad \text{No net force}$$

$$\sum \tau = 0 \quad \} \quad \text{No net torque} \tag{8.1}$$

Conditions for static equilibrium of an extended object

If motion is possible in the z-direction, we'd also require that $\sum F_z = 0$. In this chapter, however, we'll only consider motion restricted to the xy-plane.

EXAMPLE 8.1 Finding the force from the biceps tendon

Weightlifting can exert extremely large forces on the body's joints and tendons. In the *strict curl* event, a standing athlete lifts a barbell by moving only his forearms, which pivot at the elbow. The record weight lifted in the strict curl is over 200 pounds (about 900 N). Figure 8.2 shows the arm bones and the main lifting muscle when the forearm is horizontal. The distance from the tendon to the elbow joint is 4.0 cm, and from the barbell to the elbow 35 cm.

a. What is the tension in the tendon connecting the biceps muscle to the bone while holding a 900 N barbell stationary in this position?

b. What is the force exerted by the elbow on the forearm bones?

FIGURE 8.2 An arm holding a barbell.

Continued

PREPARE Figure 8.3 shows a simplified model of the arm and the forces acting on the forearm. $\vec{F}_{\text{tendon}}$ is the tension due to the muscle, $\vec{F}_{\text{barbell}}$ is the downward force of the barbell, and $\vec{F}_{\text{elbow}}$ is the force of the elbow joint on the forearm. As a simplification, we've neglected the weight of the arm itself because it is so much less than the weight of the barbell.

Known
$d_{\text{tendon}} = 4.0 \text{ cm}$
$d_{\text{barbell}} = 35 \text{ cm}$
$F_{\text{barbell}} = 900 \text{ N}$

Find
$F_{\text{tendon}}, F_{\text{elbow}}$

These forces cause torques about the elbow.

FIGURE 8.3 Visual overview of holding a barbell.

SOLVE a. For the forearm to be in static equilibrium, the net force and net torque on it must both be zero. Setting the net force to zero gives

$$\sum F_y = F_{\text{tendon}} - F_{\text{elbow}} - F_{\text{barbell}} = 0$$

Because each arm supports half the weight of the barbell, the magnitude of the weight force is $F_{\text{barbell}} = 450 \text{ N}$. We don't know either of the forces F_{tendon} and F_{elbow}, nor does the force equation give us enough information to find them. But the fact that the torque must be zero gives us that extra information. Choosing the elbow joint to be the axis of rotation, the net torque about this point is

$$\tau_{\text{net}} = F_{\text{elbow}} \times 0 + F_{\text{tendon}} d_{\text{tendon}} - F_{\text{barbell}} d_{\text{barbell}} = 0$$

Note that $\vec{F}_{\text{elbow}}$ makes no contribution to the torque because it acts directly at the pivot, so its moment arm is zero. The tension in the tendon tries to rotate the arm counterclockwise, so it produces a positive torque. Similarly, the torque due to the barbell is negative. We can solve the torque equation for F_{tendon} to find

$$F_{\text{tendon}} = F_{\text{barbell}} \frac{d_{\text{barbell}}}{d_{\text{tendon}}} = (450 \text{ N}) \frac{35 \text{ cm}}{4.0 \text{ cm}} = 3900 \text{ N}$$

b. We now need to make use of the force equation. We have

$$F_{\text{elbow}} = F_{\text{tendon}} - F_{\text{barbell}} = 3900 \text{ N} - 450 \text{ N} = 3450 \text{ N}$$

ASSESS The short distance d_{tendon} from the tendon to the elbow joint means that the force supplied by the biceps has to be very large to counter the torque generated by a force applied at the opposite end of the forearm.

STOP TO THINK 8.1 Which of these objects is in static equilibrium?

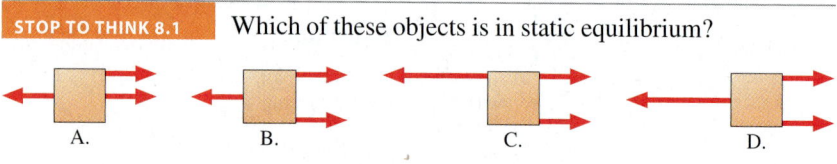

A. B. C. D.

Choosing the Pivot Point

In Example 8.1, we calculated the net torque using the elbow joint as the axis of rotation or pivot point. But we learned in Chapter 7 that the torque depends on which point is chosen as the pivot point. Was there something special about our choice of the elbow joint?

Consider the hammer shown in Figure 8.4, supported on a pegboard by two pegs A and B. Because the hammer is in static equilibrium, the net torque around the pivot at A must be zero: the clockwise torque due to the weight $\vec{w}$ is exactly balanced by the counterclockwise torque due to the force $\vec{n}_B$ of peg B. (Recall that the torque due to $\vec{n}_A$ is zero, because here $\vec{n}_A$ acts at the pivot A.) But if instead we take B as the pivot, the net torque is still zero. The counterclockwise torque due to $\vec{w}$ (with a large force but small moment arm) balances the clockwise torque due to $\vec{n}_A$ (with a small force but large moment arm). Indeed, **for an object in static equilibrium, the net torque about *every* point must be zero.** This means you can pick *any* point you wish as a pivot point for calculating the torque.

Although any choice of a pivot point will work, some choices are better because they simplify the calculations. Often, there is a "natural" axis of rotation in the problem, an axis about which rotation *would* occur if the object were not in static equilibrium. Example 8.1 is of this type, with the elbow joint as a natural axis of rotation.

If no point naturally suggests itself as an axis, look for a point on the object at which several forces act, or at which a force acts whose magnitude you don't know. Such points are good choices because any force acting at that point does not contribute to the torque. For instance, the woman in Figure 8.5 is in equilib-

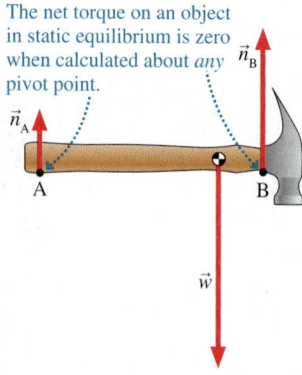

The net torque on an object in static equilibrium is zero when calculated about *any* pivot point.

FIGURE 8.4 A hammer resting on two pegs.

rium as she rests on the rock wall. A good choice of pivot point would be where her foot contacts the wall because this choice eliminates the torque due to the force $\vec{F}$ of the wall on her foot. But don't agonize over the choice of a pivot point! You can still solve the problem no matter which point you choose.

 PROBLEM-SOLVING STRATEGY 8.1   Static equilibrium problems

PREPARE Model the object as a simple shape. Draw a visual overview that shows all forces and distances. List known information.

- Pick an axis or pivot about which the torques will be calculated.
- Determine the torque about this pivot point due to each force acting on the object.
- Determine the sign of each torque about this pivot point.

SOLVE The mathematical steps are based on the fact that an object in static equilibrium has no net force and no net torque.

$$\vec{F}_{net} = \vec{0} \quad \text{and} \quad \tau_{net} = 0$$

- Write equations for $\sum F_x = 0$, $\sum F_y = 0$, $\sum \tau = 0$.
- Solve the resulting equations.

ASSESS Check that your result is reasonable and answers the question.

The torque due to $\vec{F}$ about this point is zero. This makes this point a good choice as the pivot.

FIGURE 8.5 Choosing the pivot for a woman rappelling down a rock wall.

Activ Physics ONLINE   7.2–7.5

EXAMPLE 8.2 Forces on a board on sawhorses

A board weighing 100 N sits across two sawhorses, as shown in Figure 8.6. What are the magnitudes of the normal forces of the sawhorses acting on the board?

FIGURE 8.6 A board sitting on two sawhorses.

PREPARE The board and the forces acting on it are shown in Figure 8.7. $\vec{n}_1$ and $\vec{n}_2$ are the normal forces on the board due to the sawhorses, and $\vec{w}$ is the weight of the board acting at the center of gravity. The distance d_1 to the center of the board is half the board's length, or 1.5 m. Then d_2 is $d_1 - 1.0$ m, or 0.5 m.

The weight acts at the center of gravity.

Known	Find
$d_1 = 1.5$ m	n_1, n_2
$d_2 = 0.5$ m	
$w = 100$ N	

FIGURE 8.7 Visual overview of a board on two sawhorses.

As discussed above, a good choice for the pivot is a point at which an unknown force acts, because that force contributes nothing to the torque. Either the point where $\vec{n}_1$ acts or the point where $\vec{n}_2$ acts will work; let's choose the left end of the board, where $\vec{n}_1$ acts, for this example. With this choice of pivot point, the moment arm for $\vec{w}$, which acts halfway down the board, is $d_1 = 1.5$ m. Because $\vec{w}$ tends to rotate the board clockwise, its torque is negative. The moment arm for $\vec{n}_2$ is the distance $d_1 + d_2 = 2.0$ m, and its torque is positive.

SOLVE The board is in static equilibrium, so the net force $\vec{F}_{net}$ and the net torque τ_{net} must both be zero. The forces have only y-components, so the force equation is

$$F_{net} = n_1 - w + n_2 = 0$$

The torque equation, computed around the left end of the board, is

$$\tau_{net} = -d_1 w + (d_1 + d_2)n_2 = 0$$

We now have two simultaneous equations with the two unknowns n_1 and n_2. To solve these, let's solve for n_2 in the torque equation and then substitute that result into the force equation. From the torque equation,

$$n_2 = \frac{d_1 w}{d_1 + d_2} = \frac{(1.5 \text{ m})(100 \text{ N})}{2.0 \text{ m}} = 75 \text{ N}$$

The force equation is then $n_1 - 100 \text{ N} + 75 \text{ N} = 0$, which we can solve for n_1:

$$n_1 = w - n_2 = 100 \text{ N} - 75 \text{ N} = 25 \text{ N}$$

ASSESS It seems reasonable that $n_2 > n_1$ because more of the board sits over the right sawhorse.

EXAMPLE 8.3   Choosing a different axis

Repeat Example 8.2, but put the pivot point at the center of gravity.

PREPARE   The forces causing torques about the center of gravity are now the normal forces $\vec{n}_1$ and $\vec{n}_2$, with moment arms d_1 and d_2. Force $\vec{n}_2$ creates a positive (counterclockwise) torque and force $\vec{n}_1$ creates a negative (clockwise) torque. The weight $\vec{w}$, acting at the pivot, does not contribute to the torque.

SOLVE   The torque about the center of gravity is

$$\tau_{net} = -d_1 n_1 + d_2 n_2 = 0$$

The force equation still gives $n_1 = w - n_2$. Substituting this into the torque equation we get

$$-d_1(w - n_2) + d_2 n_2 = 0$$

We can solve this for n_2 to get

$$n_2 = \frac{d_1 w}{d_1 + d_2} = \frac{(1.5 \text{ m})(100 \text{ N})}{2.0 \text{ m}} = 75 \text{ N}$$

as before. We can then use the force equation to find that $n_1 = 25$ N.

ASSESS   Even with a very different choice of pivot point from Example 8.2, we get the same result for the forces.

STOP TO THINK 8.2   A beam with a pivot on its left end is suspended from a rope. In which direction is the force of the pivot on the beam?

The scale reads the magnitude of $\vec{F}$.

FIGURE 8.8   Reaction board method of finding a person's center of gravity.

An interesting application of static equilibrium is to find the center of gravity of the human body. Because the human body is highly flexible, the position of the center of gravity is quite variable and depends on just how the body is posed. The horizontal position of the body's center of gravity can be located accurately from simple measurements with a *reaction board* and a scale, as shown in Figure 8.8. The subject lies or stands on the board in the desired posture, and the scale reading F is recorded. The following example shows how a person's center of gravity can be found using this information.

EXAMPLE 8.4   Finding the center of gravity of the human body

A woman weighing 600 N lies on a 2.5-m-long, 60 N reaction board with her feet over the pivot. The scale on the right reads 250 N. What is the distance d from the woman's feet to her center of gravity?

PREPARE   The forces and distances in the problem are shown in Figure 8.9. We'll consider the board and woman as a single object. We've assumed that the board is uniform, so its center of

gravity is at its midpoint. We'll calculate all torques around the pivot at the left end of the board. The torque due to $\vec{F}$ is positive, and those due to $\vec{w}$ and $\vec{w}_b$ are negative. The torque due to $\vec{n}$, which acts at the pivot, is zero.

SOLVE   Because the board and woman are in static equilibrium, the net force and net torque on them must be zero. The force equation reads

$$\sum F_y = n - w_b - w + F = 0$$

and the torque equation gives

$$\sum \tau = -\frac{L}{2} w_b - dw + LF = 0$$

In this case, the force equation isn't needed because we can solve the torque equation for d:

$$d = \frac{LF - \frac{1}{2}Lw_b}{w} = \frac{(2.5 \text{ m})(250 \text{ N}) - \frac{1}{2}(2.5 \text{ m})(60 \text{ N})}{600 \text{ N}}$$

$$= 0.92 \text{ m}$$

ASSESS   If the woman is 5′ 6″ (1.68 m) tall, her center of gravity is $(0.92 \text{ m})/(1.68 \text{ m}) = 55\%$ of her height, or a little more than halfway up her body. This seems reasonable.

The subject's weight acts at her center of gravity.

The board's weight acts at its center of gravity.

Known
$w = 600$ N
$w_b = 60$ N
$L = 2.5$ m
$F = 250$ N

Find
d

FIGURE 8.9   Visual overview of the reaction board and woman.

EXAMPLE 8.5 Will the ladder slip?

A 3.0-m-long ladder leans against a frictionless wall at an angle of 60°. What is the minimum value of μ_s, the coefficient of static friction with the ground, that will prevent the ladder from slipping?

PREPARE The ladder is a rigid rod of length L. To not slip, both the net force and net torque on the ladder must be zero. Figure 8.10 shows the ladder and the forces acting on it. The bottom corner of the ladder is a good choice of a pivot point because two of the forces pass through this point and thus produce no torque about it. With this choice, the weight of the ladder, acting at the center of gravity, exerts torque $d_1 w$ and the force of the wall exerts torque $-d_2 n_2$. The signs are based on the observation that $\vec{w}$ would cause the ladder to rotate counterclockwise, while $\vec{n}_2$ would cause it to rotate clockwise.

Known
$L = 3.0$ m

Find
μ_s

Weight acts at the center of gravity.

Static friction prevents slipping.

$\tau_{net} = 0$ about this point.

FIGURE 8.10 Visual overview of a ladder in static equilibrium.

SOLVE The x- and y-components of $\vec{F}_{net} = \vec{0}$ are

$$\sum F_x = n_2 - f_s = 0$$
$$\sum F_y = n_1 - w = n_1 - Mg = 0$$

The torque about the bottom corner is

$$\tau_{net} = d_1 w - d_2 n_2 = \frac{1}{2}(L\cos 60°)Mg - (L\sin 60°)n_2 = 0$$

Altogether, we have three equations in the three unknowns n_1, n_2, and f_s. If we solve the third for n_2,

$$n_2 = \frac{\frac{1}{2}(L\cos 60°)Mg}{L\sin 60°} = \frac{Mg}{2\tan 60°}$$

we can then substitute this into the first to find

$$f_s = \frac{Mg}{2\tan 60°}$$

Our model of static friction is $f_s \leq f_{s\,max} = \mu_s n_1$. We can find n_1 from the second equation: $n_1 = Mg$. Using this, the model of friction tells us that

$$f_s \leq \mu_s Mg$$

Comparing these two expressions for f_s, we see that μ_s must obey

$$\mu_s \geq \frac{1}{2\tan 60°} = 0.29$$

Thus the minimum value of the coefficient of static friction is 0.29.

ASSESS You know from experience that you can lean a ladder or other object against a wall if the ground is "rough," but it slips if the surface is too smooth. 0.29 is a "medium" value for the coefficient of static friction, which is reasonable.

8.2 Stability and Balance

Figure 8.11 shows several common items on a table. As you know from experience, the tall candlestick can be easily toppled by a bump against the table, while the squat bowl would take a great disturbance to knock it over. To see why this is so, we'll need to understand how the gravitational torque on an object that has been tilted from its equilibrium position acts either to bring the object back upright or to pull it completely over.

Figure 8.12a on the next page shows a soda can and its **base of support**, the area between the points it would pivot on if tilted slightly in either direction. If the can is tilted slightly, as in Figure 8.12b, the gravitational torque tends to rotate it back to its equilibrium position. If the can is tilted further it reaches a special angle, shown in Figure 8.12c, at which the center of gravity is just above the pivot. Then the weight force is directed toward the pivot and so there is no torque on the can; it could fall in either direction if released. Finally, if the can is tilted even further, as in Figure 8.12d, the gravitational torque will rotate the can *away* from its equilibrium position, and the can will topple.

FIGURE 8.11 The tall candlestick is easy to knock over, but the squat bowl is very stable. Why are they different?

(a) Base of support

(b) Torque due to gravity rotates it back.

In this position there is no net torque. A slight nudge will send it either way.

(c) The center of gravity is directly over the pivot.

Original base of support

Torque due to gravity causes it to fall over.

(d) The center of gravity is no longer above the base of support.

FIGURE 8.12 An object will fall over when tilted too far.

For the SUV, the center of gravity height h is 47% of t.

For the car the center of gravity height h is 33% of t.

FIGURE 8.13 Compared to a passenger car, an SUV has a high center of gravity relative to its width.

The vehicle is at the critical angle θ_c when its center of gravity is exactly over the pivot.

Pivot

FIGURE 8.14 The critical angle determines how far an object can tilt before rolling over.

From this discussion, we can see that if a vertical line through an object's center of gravity passes through its base of support, the object will tend to rotate back to its equilibrium position. In this case, we say that the object is **stable.** If such a line lies outside the base of support, the object will fall over; the object is **unstable.** As an example, the photo that opens this chapter shows a dancer balanced on her toes. Her base of support—the tips of her toes—is quite small. To balance, she must get her center of gravity directly over her toes. She does this by positioning her arms and legs. The slightest movement of an arm or leg will cause her to lose her balance because the motion will shift her center of gravity outside the base of support.

One area where stability plays a vital role is in the design of motor vehicles. With the increasing popularity of sport utility vehicles (SUVs), this topic has become one of major concern. Because the center of gravity of an SUV is generally quite high compared to its width, SUVs are more prone to *rollover* accidents than are passenger cars. Figure 8.13 shows an SUV and a passenger car with their centers of gravity marked. The height of the SUV's center of gravity is 47% of its track width whereas the passenger car's center-of-gravity height is only 33% of its track.

Figure 8.14 shows why the height of the center of gravity is important. As a result of an accident, running off the edge of the road, or taking a curve too fast, the vehicle may tilt up onto one set of wheels. Just as for the soda can in Figure 8.12, there is a special *critical angle* θ_c at which the center of gravity is directly over the pivot point. If the vehicle's maximum tilt is less than θ_c it will fall back upright, but if its tilt is greater than θ_c it will continue to tilt further, leading to rollover. The angle θ_c is thus the maximum safe angle to which the vehicle can tilt.

We can see from Figure 8.14 that the critical angle is given by $\tan\theta_c = (t/2)/h$, or

$$\theta_c = \tan^{-1}\left(\frac{t/2}{h}\right) = \tan^{-1}\left(\frac{t}{2h}\right) \tag{8.2}$$

That is, the maximum possible angle of tilt is determined by the *ratio* of track width to center-of-gravity height. Stability does not depend on the absolute height of the center of gravity but, instead, on the center-of-gravity height as a fraction of the track width.

For a passenger car with $h = 0.33t$, the critical angle is $\theta_c = 57°$. But for an SUV with $h = 0.47t$, the critical angle is only $\theta_c = 47°$. Loading an SUV with cargo further raises the center of gravity, especially if the roof rack is used, thus reducing θ_c even more. Various automobile safety groups have determined that a vehicle with $\theta_c > 50°$ is unlikely to roll over in an accident. A rollover becomes increasingly likely when θ_c is reduced below 50°. The same argument that leads to Equation 8.2 for tilted vehicles can be made for any object, leading to the general rule that **a wider base of support and/or a lower center of gravity improve stability.**

CONCEPTUAL EXAMPLE 8.1 How far to walk the plank?

A cat walks along a plank that extends out from a table. If the cat walks too far out on the plank, the plank will begin to tilt. What determines when this happens?

REASON An object is stable if its center of gravity lies over its base of support, and unstable otherwise. Let's take the cat and the plank to be one combined object whose center of gravity lies along a line between the cat's center of gravity and that of the plank.

In Figure 8.15a, when the cat is near the left end of the plank, the combined center of gravity is over the base of support and the plank is stable. As the cat moves to the right, he reaches a point where the combined center of gravity is directly over the edge of the table, as shown in Figure 8.15b. If the cat takes one more step, the cat and plank will become unstable and the plank will begin to tilt.

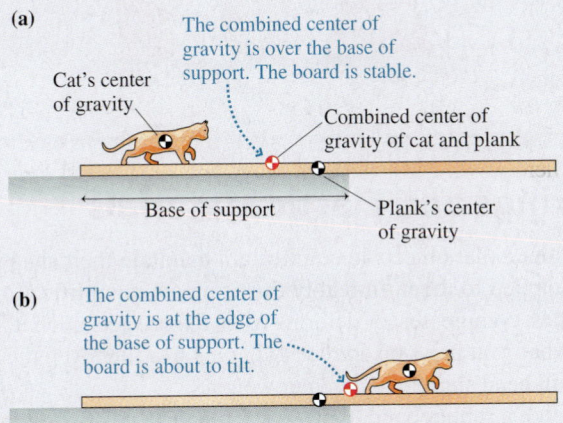

FIGURE 8.15 Changing stability as a cat walks on a plank.

ASSESS Because the plank's center of gravity must be to the left of the edge for it to be stable by itself, the cat can actually walk a short distance out onto the unsupported part of the plank before it starts to tilt. The heavier the plank is, the further the cat can walk.

Stability and Balance of the Human Body

The human body is remarkable for its ability to constantly adjust its stance to remain stable on just two points of support. In walking, running, or even in the simple act of rising from a chair, the position of the body's center of gravity is constantly changing. To maintain stability, we unconsciously adjust the positions of our arms and legs to keep our center of gravity over our base of support.

A simple example of how the body naturally realigns its center of gravity is found in the act of standing up on tiptoes. Figure 8.16a shows the body in its normal standing position. Notice that the center of gravity is well-centered over the base of support (the feet), ensuring stability. If the subject were now to stand on tiptoes *without* otherwise adjusting the body position, her center of gravity would fall behind the base of support, which is now the balls of the feet, and she would fall backwards. To prevent this, as shown in Figure 8.16b, the body naturally leans forward, regaining stability by moving the center of gravity over the balls of the feet.

The simple act of walking can be broken down into a sequence of stable and unstable positions. As you extend your leg forward to initiate a step, a point is reached where your body is no longer stable—your center of gravity has moved forward of the base of support, the rear foot. At this point, your body begins to topple forward—onto the leading foot. You then lean slightly forward, placing your center of gravity over your front foot so that your rear foot can be lifted

Balancing soda can Try to balance a soda can—full or empty—on the narrow bevel at the bottom. It can't be done because, either full or empty, the center of gravity is near the center of the can. If the can is tilted enough to sit on the bevel, the center of gravity lies far outside this small base of support. But if you put about 2 ounces (60 ml) of water in an empty can, the center of gravity will be right over the bevel and the can will balance.

FIGURE 8.16 When you stand on your tiptoes, you must lean forward in order to remain balanced.

Impossible balance Stand facing a wall with your toes touching the base of the wall. Now rise onto your tiptoes. You will not be able to stand without falling backwards. As we see from Figure 8.16b, your body has to lean forward to stand on tiptoes. With the wall in your way, you cannot lean enough to maintain your balance, and you will begin to topple backwards.

without falling. As your rear foot swings forward, the process begins again. This description is a bit oversimplified because it ignores the side-to-side loss of stability as each foot is raised, but it illustrates how physics concepts can be used to understand human motion.

STOP TO THINK 8.3 Rank in order, from least stable to most stable, the three objects shown in the figure. The positions of their centers of gravity are marked. (For the centers of gravity to be positioned like this, the objects must have a nonuniform composition.)

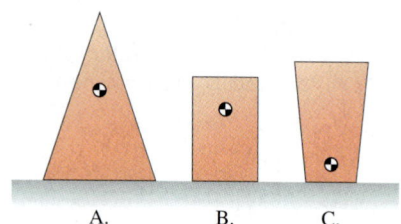

8.3 Springs and Elastic Materials

We have assumed that objects in equilibrium maintain their shape as forces and torques are applied to them. In reality this is an oversimplification. Every solid object stretches, compresses, or deforms when a force acts upon it. This change is easy to see when you press on a green twig on a tree, but even the largest branch on the tree will bend slightly under your weight.

If you stretch a rubber band, there is a force that tries to pull the rubber band back to its equilibrium, or unstretched, length. A force that restores a system to an equilibrium position is called a **restoring force.** Systems that exhibit such restoring forces are called **elastic.** The most basic examples of **elasticity** are things like springs and rubber bands. If you stretch a spring, a tension-like force pulls back. Similarly, a compressed spring tries to re-expand to its equilibrium length. Elasticity and restoring forces are properties of much stiffer systems as well. The steel beams of a bridge bend slightly as you drive your car over it, but they are restored to equilibrium after your car passes by. Your leg bones flex a bit during each step you take. Nearly everything that stretches, compresses, bends, or twists exhibits a restoring force and can be called elastic.

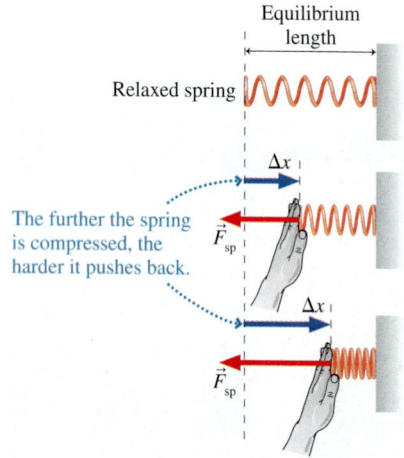

FIGURE 8.17 The spring force depends on how far the spring is compressed.

FIGURE 8.18 Measured data for the restoring force of a real spring.

Springs

The behavior of a simple spring will serve to illustrate the basic ideas of elasticity. When no forces act on a spring to compress or extend it, it will relax to its **equilibrium length.** If we now compress the spring by a displacement Δx, how hard does it push back? Figure 8.17 shows what happens: The further we push the spring, the harder the restoring force of the spring pushes back.

In Figure 8.18, data for the magnitude of the restoring force of a real spring show that **the force of the spring is *proportional* to the displacement of the end of the spring.** That is, compressing or stretching the spring twice as far results in a restoring force that is twice as large. This is a *linear relationship,* and the slope k of the line is the proportionality constant:

$$F_{sp} = k \, \Delta x \tag{8.3}$$

A second important fact about spring forces is illustrated in Figure 8.19. If the spring is compressed, as in Figure 8.19a, Δx is positive and, because $\vec{F}_{sp}$ points to the left, its component $(F_{sp})_x$ is negative. If, however, the spring is stretched, as in Figure 8.19b, Δx is negative and, because $\vec{F}_{sp}$ points to the right, its component

$(F_{sp})_x$ is positive. In general, **the spring force always points in the opposite direction to the displacement.** We can express this fact, along with what we've learned about the magnitude of the spring force, by rewriting Equation 8.3 in terms of the *component* of the spring force:

$$(F_{sp})_x = -k\,\Delta x \qquad (8.4)$$

Hooke's law for the force due to a spring

LINEAR
p. 38

The minus sign in Equation 8.4 reflects the fact that $(F_{sp})_x$ and Δx are always of opposite sign.

The proportionality constant k is called the **spring constant.** The units of the spring constant are N/m. The spring constant k is a property that characterizes a spring, just as mass m characterizes a particle. If k is large, it takes a large pull to cause a significant stretch, and we call the spring a "stiff" spring. If k is small, we can stretch the spring with very little force, and we call it a "soft" spring. Every spring has its own, unique value of k. The spring constant for the spring in Figure 8.18 can be determined from the slope of the straight line to be $k = 3.5$ N/m.

Equation 8.4 for the restoring force of a spring was first suggested by Robert Hooke, a contemporary (and sometimes bitter rival) of Newton. Hooke's law is not a true "law of nature," in the sense that Newton's laws are, but is actually just a *model* of a restoring force. It works extremely well for some springs, as in Figure 8.18, but less well for others. Hooke's law will fail for any spring if it is compressed or stretched too far.

NOTE ▶ Just as we used massless strings, we will adopt the idealization of a *massless spring*. While not a perfect description, it is a good approximation if the mass attached to a spring is much larger than the mass of the spring itself. ◀

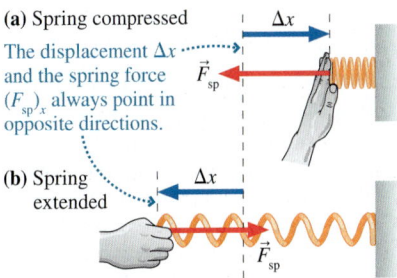

(a) Spring compressed

The displacement Δx and the spring force $(F_{sp})_x$ always point in opposite directions.

(b) Spring extended

FIGURE 8.19 The spring force is always directed opposite to the displacement.

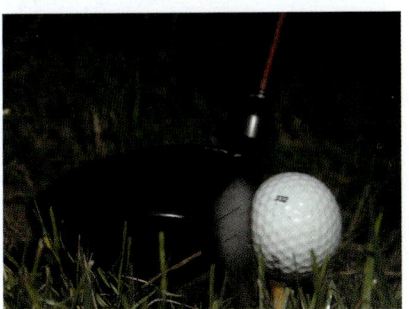

Elasticity in action When a golf ball is struck by a club, it compresses quite a bit during the short time of the collision. The restoring force that pushes the ball back into its original shape helps launch the ball off the face of the club, making for a longer drive. Other deformations happen as well. The club face compresses a small amount, and the flexible shaft of the club may bend by as much as 15°.

EXAMPLE 8.6 Weighing a fish

A scale used to weigh fish consists of a spring connected to the ceiling. The spring's equilibrium length is 30 cm. When a 4.0 kg fish is suspended from the end of the spring, it stretches to a length of 42 cm.

a. What is the spring constant k for this spring?
b. If an 8.0 kg fish is suspended from the spring, what will be the length of the spring?

Known
$L_0 = 0.30$ m
$L = 0.42$ m
$m = 4.0$ kg

Find
k

FIGURE 8.20 Visual overview of a mass suspended from a spring.

PREPARE The visual overview in Figure 8.20 shows the details for the first part of the problem. Because the fish is in static equilibrium, the upward restoring force of the spring must exactly balance the downward weight of the fish.

SOLVE a. The spring stretches by $\Delta x = L - L_0 = 0.42$ m $-$ 0.30 m $= 0.12$ m. The magnitude of the restoring force is equal to the weight of the fish, $w = mg = (4.0\text{ kg})(9.8\text{ m/s}^2) = 39.2$ N. To find k we need only use magnitudes, so we use Equation 8.3 to solve for the spring constant:

$$k = \frac{F_{sp}}{\Delta x} = \frac{39.2\text{ N}}{0.12\text{ m}} = 330\text{ N/m}$$

b. The restoring force (equal to the weight of the fish) is proportional to the stretch of the spring. If we double the mass (and thus the weight) of the fish, the displacement of the end of the spring will double as well, to $\Delta x = 0.24$ m. The new length of the spring will be

$$L = L_0 + \Delta x = 0.30\text{ m} + 0.24\text{ m} = 0.54\text{ m}$$

ASSESS The greater the weight, the more the spring stretches. We could use the displacement as a measure of the weight of an object hung from the end of the spring. This is the principle behind a spring scale.

EXAMPLE 8.7 When does the block slip?

Figure 8.21 shows a spring attached to a 2.0 kg block. The other end of the spring is pulled by a motorized toy train that moves forward at 5.0 cm/s. The spring constant is 50 N/m and the coefficient of static friction between the block and the surface is 0.60. The spring is at its equilibrium length at $t = 0$ s when the train starts to move. When does the block slip?

FIGURE 8.21 A toy train stretches the spring until the block slips.

PREPARE We model the block as a particle and the spring as a massless spring. Figure 8.22 is a free-body diagram for the block. We convert the speed of the train into m/s, $v = 0.050$ m/s.

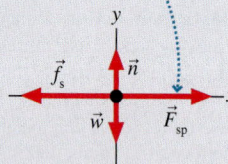

When the spring force exceeds the maximum force of static friction, the block will slip.

FIGURE 8.22 Free-body diagram for the block.

SOLVE Recall that the tension in a massless string pulls equally at *both* ends of the string. The same is true for the spring force: It pulls (or pushes) equally at *both* ends. Imagine holding a rubber band with your left hand and stretching it with your right

hand. Your left hand feels the pulling force, even though it was the right end of the rubber band that moved.

This is the key to solving the problem. As the right end of the spring moves, stretching the spring, the spring pulls backward on the train *and* forward on the block with equal strength. The train is moving to the right, and so the spring force pulls to the left on the train—as we would expect. But the block is at the other end of the spring; the spring force pulls to the right on the block, as shown in Figure 8.22. As the spring stretches, the static friction force on the block increases in magnitude to keep the block at rest. The block is in static equilibrium, so

$$\sum (F_{net})_x = (F_{sp})_x + (f_s)_x = F_{sp} - f_s = 0$$

where F_{sp} is the magnitude of the spring force. This magnitude is $F_{sp} = k\,\Delta x$, where $\Delta x = vt$ is the distance the train has moved. Thus

$$f_s = F_{sp} = k\,\Delta x$$

The block slips when the static friction force reaches its maximum value $f_{s\,max} = \mu_s n = \mu_s mg$. This occurs when the train has moved

$$\Delta x = \frac{f_{s\,max}}{k} = \frac{\mu_s mg}{k} = \frac{(0.60)(2.0\text{ kg})(9.8\text{ m/s}^2)}{50\text{ N/m}} = 0.235\text{ m}$$

The time at which the block slips is

$$t = \frac{\Delta x}{v} = \frac{0.235\text{ m}}{0.050\text{ m/s}} = 4.7\text{ s}$$

ASSESS Note that the distance Δx that the spring can be stretched before the block slips is *inversely proportional* to k. If the spring has a large k, so that it is quite stiff, it need be stretched only slightly before it exerts a large force on the block.

Elasticity: The Springiness of Matter

(a)

1. Your hand pulling the rod . . .

2. . . . stretches the atomic springs . . .

3. . . . which exert a restoring force.

Particle-like atoms
Spring-like bonds
Clamp

In Chapter 4 we noted that we could model most solid materials as being made of particle-like atoms connected by spring-like bonds. We can model a steel rod this way, as illustrated in Figure 8.23a. The spring-like bonds between the atoms in steel are quite stiff, but they can be stretched or compressed, meaning that even a steel rod is elastic. If you pull on the end of a steel rod, as in Figure 8.23a, you will slightly stretch the bonds between the particles that make it up, and the rod itself will stretch. The stretched bonds pull back on your hand with a restoring force that causes the rod to return to its original length when released. In this sense, the entire rod acts like a very stiff spring. As is the case for a spring, a restoring force is also produced by compressing the rod.

In Figure 8.23b, real data for a steel rod show that, just as for a spring, the restoring force is proportional to the change in length. However, the *scale* of the stretch of the rod and the restoring force is much different from that for a spring. If a 1.0 m steel rod with a diameter of 1.0 cm is stretched by 1.0 mm (0.1% of its length), the restoring force is 16,000 N—almost two tons! Steel is elastic, but under normal forces, it experiences only very small changes in dimension. Materials of this sort are called **rigid.**

The behavior of other materials, such as the rubber in a rubber band, is quite different. A rubber band can be stretched quite far—several times its equilibrium length—with a very small force, snapping back to its original

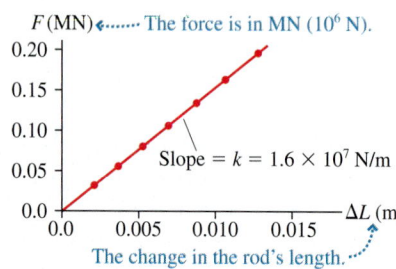

(b) Data for a 1.0-m-long, 1.0-cm-diameter steel rod

F (MN) ← The force is in MN (10^6 N).

Slope = $k = 1.6 \times 10^7$ N/m

ΔL (m)

The change in the rod's length.

FIGURE 8.23 Stretching a steel rod.

shape when released. Materials that show large deformations with small forces are called **pliant.**

The molecular structure of rubber is illustrated in Figure 8.24. The structure is quite different from that of steel. Steel and other rigid materials have atoms in a lattice with strong bonds among them; these spring-like bonds are responsible for the elasticity of the materials. In rubber, the atoms are not in a lattice, but are linked to form long-chain molecules. When the rubber is relaxed, the molecules have no overall ordering, as shown in Figure 8.24a. In Figure 8.24b, we see that as the rubber band is stretched, these long-chain molecules are pulled into alignment with each other. This ordering of the system induces a restoring force, because the molecules "want" to return to their usual random state.

This is a more complicated situation than the spring-like bonds between the atoms of steel, and you can see in Figure 8.24c that the restoring force of a rubber band is not simply proportional to the displacement of the end. Hooke's law typically does not apply to rubber and other pliant materials.

(a) Unstretched rubber band   (b) Stretched rubber band

Long-chain molecules in the rubber are randomly oriented.   When the rubber band is stretched, the molecules straighten out along the stretch direction.

(c) Data for the stretch of a rubber band

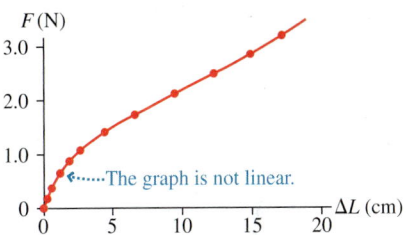

The graph is not linear.

FIGURE 8.24 Stretching a rubber band.

CONCEPTUAL EXAMPLE 8.2 Determining the stretch of a rubber band

A rubber band connected to the ceiling has an equilibrium length of 5.0 cm. A 100 g weight attached to the end stretches the rubber band to a length of 7.0 cm. Suppose an additional 100 g is attached to the end. Will the new length be less than 9.0 cm, exactly 9.0 cm, or more than 9.0 cm? Assume that the rubber band is similar to that of Figure 8.24.

REASON Consider any point on the graph of Figure 8.24c. Because the graph curves downward, doubling the force F (going upward by a factor of two) requires increasing ΔL by more than a factor of 2. If the rubber band stretches by 2.0 cm when supporting 100 g, it will stretch by more than 4.0 cm when supporting 200 g, so the length will be more than 9.0 cm.

▶ **Stretching a molecule** BIO DNA is a long-chain molecule that is normally tightly coiled as in the top figure. Amazingly, it is possible to "grab" the two ends of a DNA molecule and gently stretch it while measuring the restoring force. The computer-generated images show the unstretched molecule (top) beginning to unwind as the force on its ends is increased. The bottom frame shows the molecule just before it ruptures. By measuring the restoring force as the DNA molecule is stretched, scientists can learn, for instance, how various enzymes act to cut and then reseal coils in the DNA structure.

Although Hooke's law does not apply for pliant materials, we can still define an average "spring constant" by finding the average slope of the graph of restoring force versus change in length. Later in the chapter, you will see tables of values for the stiffness and the strength of pliant materials. These numbers are average values defined in just this way.

Elastic behavior occurs in other ways than just stretching or compressing. Consider again a steel rod. If you bend the rod gently, as shown in Figure 8.25, the top surface of the rod stretches slightly while the lower edge compresses. The stretching and compressing of atomic springs in this rigid material produce a restoring force that is trying to return the rod to its original shape. The rod exhibits elasticity in bending as well as in stretching and compressing. Pliant materials exhibit similar restoring forces, but the associated deflections will be much greater.

All elastic systems thus follow the same general rule: if you compress, bend, twist, or otherwise deform them, they develop a restoring force that acts to restore them to their original shape. For rigid materials like steel, these restoring forces all have their origin in the spring-like bonds between atoms. Because this

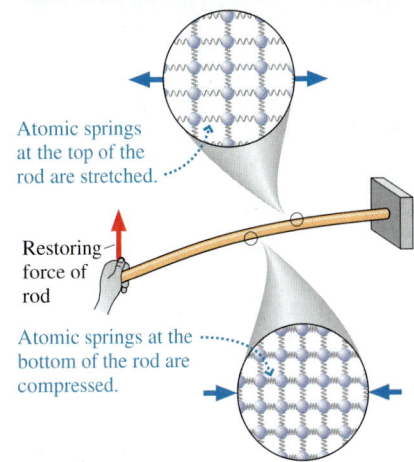

Atomic springs at the top of the rod are stretched.

Restoring force of rod

Atomic springs at the bottom of the rod are compressed.

FIGURE 8.25 When an object is stretched or bent, the spring-like molecular bonds stretch and compress.

restoring force is due to the atomic springs, it is no surprise that (for small deformations) the restoring force is proportional to the deformation; that is, the restoring force obeys Hooke's law. For pliant materials like rubber, the restoring force is typically not linear, but we may use Hooke's law to approximate the behavior of such materials.

STOP TO THINK 8.4   A 1.0 kg weight is suspended from a spring, stretching it by 5.0 cm. How much does the spring stretch if the 1.0 kg weight is replaced by a 3.0 kg weight?

A. 5.0 cm   B. 10.0 cm   C. 15.0 cm   D. 20.0 cm

8.4 Stretching, Compressing, and Bending Materials

An atomic model explains why there are restoring forces accompanying a deformation of rigid and pliant materials. In this section, we will make our models quantitative. We will also take a look at *structures.* Bone is a material with certain properties, but to consider the strength of particular bones in the body, we must consider their geometry. How thick is the bone? Where is the material dense, and where is it porous?

In Figure 8.23a, we saw that a steel rod behaves like a spring with a very large spring constant k. A rod's spring constant depends on several factors, as shown in Figure 8.26. First, we expect that a thick rod, with a large cross-section area A, will be more difficult to stretch than a thinner rod. Second, a rod with a long length L will be easier to stretch by a given amount than a short rod (think of trying to stretch a rope by 1 cm—this would be easy to do for a 10-m-long rope, but it would be pretty hard for a piece of rope only 10 cm long). Finally, the stiffness of the rod will depend on the material that it's made of. Experiments bear out these observations, and it is found that the spring constant of the rod can be written as

$$k = \frac{YA}{L} \tag{8.5}$$

Area A

L

FIGURE 8.26 A stretched rod.

where the constant Y is called **Young's modulus.** Young's modulus characterizes the elasticity of the material from which an object is made, but it does not depend on the object's shape or size. All rods made from steel have the same Young's modulus, regardless of their length or area. Aluminum rods have a different Young's modulus. In this regard, Young's modulus is somewhat analogous to density; all pieces of steel have the same density, regardless of their size and shape, while pieces of aluminum have a different density.

From Equation 8.3, the magnitude of the restoring force for a spring is related to the change in its length as $F_{sp} = k\,\Delta x$. Writing the change in the length of a rod as ΔL, as shown in Figure 8.26, we can use Equation 8.5 to write the restoring force F of a rod as

$$F = \frac{YA}{L}\Delta L \tag{8.6}$$

Equation 8.6 applies both to elongation and to compression.

It's useful to write Equation 8.6 in terms of two new ratios, the *stress* and the *strain,* as follows:

The ratio of force to cross-section area is called **stress**.

$$\frac{F}{A} = Y\left(\frac{\Delta L}{L}\right)$$

The ratio of the change in length to the original length is called **strain**.

(8.7)

The unit of stress is N/m^2. If the stress is due to stretching, we call it a **tensile stress.** The strain is the fractional change in the rod's length. If the rod's length changes by 1%, the strain is 0.01. Because strain is dimensionless, Young's modulus Y has the same units as stress. Table 8.1 gives values of Young's modulus for several rigid materials. Large values of Y characterize materials that are stiff. "Softer" materials have smaller values of Y. Because the values of Young's modulus for rigid materials are very large, it takes a significant stress to produce even a small strain.

TABLE 8.1 Young's modulus for rigid materials

Material	Young's modulus (10^{10} N/m^2)
Cast iron	20
Steel	20
Silicon	13
Copper	11
Aluminum	7
Glass	7
Concrete	3
Wood (Douglas Fir)	1

EXAMPLE 8.8 Finding the stretch of a wire

A *Foucault pendulum* in a physics department (used to prove that the earth rotates) consists of a 120 kg steel ball that swings at the end of a 6.0-m-long steel cable. The cable has a diameter of 2.5 mm. When the ball was first hung from the cable, by how much did the cable stretch?

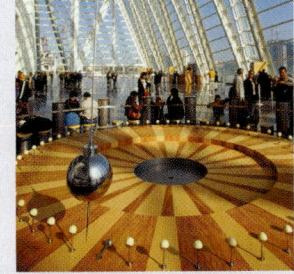

PREPARE The amount by which the cable stretches depends on the elasticity of the steel cable. Young's modulus for steel is given in Table 8.1 as $Y = 20 \times 10^{10}$ N/m^2.

SOLVE Equation 8.7 relates the stretch of the cable ΔL to the restoring force F and to the properties of the cable. Rearranging terms, we find that the cable stretches by

$$\Delta L = L\left(\frac{F}{AY}\right)$$

The cross-section area of the cable is

$$A = \pi r^2 = \pi(0.00125 \text{ m})^2 = 4.91 \times 10^{-6} \text{ m}^2$$

The restoring force of the cable is equal to the ball's weight:

$$F = w = mg = (120 \text{ kg})(9.8 \text{ m/s}^2) = 1180 \text{ N}$$

The change in length is thus

$$\Delta L = \frac{(6.0 \text{ m})(1180 \text{ N})}{(4.9 \times 10^{-6} \text{ m}^2)(20 \times 10^{10} \text{ N/m}^2)}$$

$$= 0.0072 \text{ m} = 7.2 \text{ mm}$$

ASSESS Though the ball is quite heavy—about 270 pounds—the cable stretches just over 0.25 in, 0.12% of the total length of the cable. The strain is only 0.0012. An aluminum or copper cable of similar dimensions would stretch more because they have smaller values of Young's modulus.

Bending Beams

We saw, at the end of the previous section, how bending a rod compresses parts of the rod and stretches others. We can extend our treatment of restoring forces to treat this type of deformation. Figure 8.27 shows a beam of length L, width w, and thickness t fixed at one end and free to move at the other. Deflecting the end of the beam causes a restoring force F at the end of the beam.

The magnitude of the restoring force F depends on the dimensions of the beam, Young's modulus for the material, and the deflection d. For small values of the deflection, it can be shown that the restoring force is given by

$$F = \left(\frac{Ywt^3}{4L^3}\right)d \qquad (8.8)$$

Again, this expression has the same form as Equation 8.3 for a spring, with the quantity in brackets playing the role of the spring constant k. A higher value means a stiffer beam.

If the beam is pushed down, there is an upward restoring force.

FIGURE 8.27 A deflected beam exerts a restoring force.

EXAMPLE 8.9 Imaging biological samples using atomic force microscopy

In atomic force microscopy (AFM), the sharp tip at the end of a flexible microscopic beam is dragged very lightly over a sample. As the tip rides over high spots on the sample, such as the DNA strands shown in the AFM image to the right, the beam flexes very slightly. This bending can be detected using a laser.

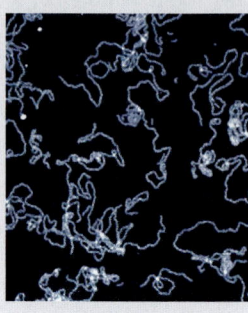

100 μm

The beam shown here is 120 μm long, 35 μm wide, and 1.0 μm thick and is made of silicon. The smallest deflection of the beam that can be detected is 5.0×10^{-11} m. At this deflection, what force is applied to the sample by the AFM tip?

PREPARE We can use Equation 8.8 to find the restoring force of the beam. From Table 8.1, Young's modulus for silicon is 1.3×10^{11} N/m^2.

SOLVE We have

$$F = \left[\frac{(1.3 \times 10^{11} \text{ N/m}^2)(35 \times 10^{-6} \text{ m})(1.0 \times 10^{-6} \text{ m})^3}{4(120 \times 10^{-6} \text{ m})^3} \right] \times$$

$$(5.0 \times 10^{-11} \text{ m}) = 3.3 \times 10^{-11} \text{ N} = 33 \text{ pN}$$

ASSESS In order to image soft biological materials without damage, the force applied must be extremely small. Interestingly, the spring constant of the AFM beam is $k = F/d = 0.7$ N/m, roughly that of an ordinary Slinky. The force is so small because the *deflection* is so small.

When the beam is bent, the restoring force comes mostly from the stretching of the top and the compression of the bottom.

Stretched — d

Compressed

Thus the stiffness of the beam is only slightly lessened by removing much of the steel in the middle of the beam.

FIGURE 8.28 The structure of an I-beam.

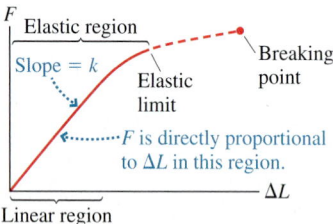

Elastic region

Slope = k

Elastic limit

Breaking point

F is directly proportional to ΔL in this region.

Linear region

FIGURE 8.29 Stretch data for a steel rod.

TABLE 8.2 Tensile strengths of rigid materials

Material	Tensile strength (N/m^2)
Glass	60×10^6
Cast iron	150×10^6
Aluminum	400×10^6
Steel	1000×10^6

Beams used in buildings and bridges usually don't have the simple rectangular cross section shown in Figure 8.27. A common form for buildings and bridges is the "I-beam" with a cross section similar to a capital "I". As we see in Figure 8.28, this form eliminates most of the weight of the beam while preserving most of the restoring force when the beam is bent. This beam will be much stiffer than a square beam of the same mass.

Another shape that is lightweight and resistant to bending is a tube; again, the material is concentrated where the strain is the greatest. Tubes with thin walls are used in many applications where rigidity combined with light weight is of importance. Backpacking tents, hang gliders, and bicycle frames all take advantage of the stiffness of tubes. The long bones in your body also take the form of hollow tubes, as we'll see in the next section.

Beyond the Elastic Limit

In the previous section, we found that if we stretch a rod by a small amount ΔL, it will pull back with a restoring force F, according to Equation 8.6. But if we continue to stretch the rod, this simple linear relationship between ΔL and F will eventually break down. Figure 8.29 is a graph of the rod's restoring force from the start of the stretch until the rod finally breaks.

As you can see, the graph has a *linear region,* the region where F and ΔL are proportional to each other, obeying $F = k \Delta L$—Hooke's law. **As long as the stretch stays within the linear region, a solid rod acts like a spring and obeys Hooke's law.**

How far can you stretch the rod before damaging it? As long as the stretch is less than the **elastic limit,** the rod will return to its initial length L when the force is removed. The elastic limit is the end of the **elastic region.** Stretching the rod beyond the elastic limit will permanently deform it, and the rod won't return to its original length. Finally, at a certain point the rod will reach a breaking point, where it will snap in two. The maximum stress that a material can be subjected to before failing is called the **tensile strength.** Table 8.2 lists values of tensile strength for rigid materials. When we speak of the strength of a material, we are referring to its tensile strength.

EXAMPLE 8.10 Breaking a pendulum cable

After a late night of studying physics, several 80 kg students decide it would be fun to swing on the Foucault pendulum of Example 8.8. What's the maximum number of students that the pendulum cable could support?

SOLVE The tensile strength of steel is given in Table 8.2 as 1000×10^6 N/m², or 1.0×10^9 N/m². This is the largest stress steel can sustain. The stress of the cable is F/A, so we can write

$$F_{max} = A(1.0 \times 10^9 \text{ N/m}^2)$$

From Example 8.8, the diameter of the cable is 2.5 mm, so its radius is 0.00125 m. Thus

$$F_{max} = (\pi(0.00125 \text{ m})^2)(1.0 \times 10^9 \text{ N/m}^2) = 4.9 \times 10^3 \text{ N}$$

This force is the weight of the heaviest mass the cable can support, $w = m_{max}g$. The maximum mass that can be supported is

$$m_{max} = F_{max}/g = 500 \text{ kg}$$

The ball has a mass of 120 kg, leaving 380 kg for the students. Four students have a mass of 320 kg, which is less than this value. But five students, totaling 400 kg, would cause the cable to break.

ASSESS Steel has a very large tensile strength. This very narrow wire can still support 4900 N ≈ 1100 lb.

Spider silk BIO The glands on the abdomen of a spider produce different kinds of silk. The silk that is used in webs can be quite stretchy; that used to subdue prey is generally not. An individual strand of silk may be a mix of fibers of different types, allowing spiders great flexibility in their material.

Biological Materials

Suppose we take equal lengths of spider silk and steel wire, stretch each, and measure the restoring force of each until they break. The graph of stress versus strain might appear as in Figure 8.30.

The spider silk is certainly less stiff: For a given stress, the silk will stretch about 100 times farther than steel. Interestingly, though, spider silk and steel eventually fail at approximately the same stress. In this sense, spider silk is "as strong as steel." Many pliant biological materials share this combination of low stiffness and large tensile strength. These materials can undergo significant deformations without failing. Tendons, the walls of arteries, and the web of a spider are all quite strong but nonetheless capable of significant stretch.

Bone is an interesting example of a rigid biological material. Most bones in your body are made of two different kinds of bony material: dense and rigid cortical (or compact) bone on the outside, and porous, flexible cancellous (or spongy) bone on the inside. Figure 8.31 shows a cross section of a typical bone. Cortical and cancellous bone have very different values of Young's modulus. Young's modulus for cortical bone approaches that of concrete, so it is very rigid with little ability to stretch or compress. In contrast, cancellous bone has a much smaller Young's modulus. Consequently, the elastic properties of bones can be well modeled as those of a hollow cylinder.

The structure of bones in birds actually approximates a hollow cylinder quite well. Figure 8.32 shows that a typical bone is a thin-walled tube of cortical bone with a tenuous structure of cancellous bone inside. Much like an I-beam, most of a cylinder's rigidity comes from the material near its surface. A hollow cylinder retains most of the rigidity of a solid one, but it is much lighter. Bird bones carry this idea to its extreme.

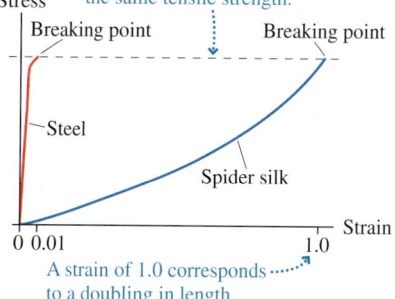

FIGURE 8.30 Stress-versus-strain curves for steel and spider silk.

FIGURE 8.31 Cross section of a long bone.

FIGURE 8.32 Cross section of a bone from a bird.

TABLE 8.3 Young's modulus for biological materials

Material	Young's modulus (10^{10} N/m²)
Tooth enamel	6
Cortical bone	1.6
Cancellous bone	0.02–0.3
Spider silk	0.2
Tendon	0.15
Blood vessel (aorta)	0.00005

TABLE 8.4 Tensile strength of biological materials

Material	Tensile strength (N/m²)
Cancellous bone	5×10^6
Cortical bone	100×10^6
Spider silk	1000×10^6

Table 8.3 gives values of Young's modulus for biological materials. Note the large difference between pliant and rigid materials. Table 8.4 shows the tensile strength for biological materials. Interestingly, spider silk, a pliant material, has a greater tensile strength than bone!

The values in Table 8.4 are for static forces—forces applied for a long time in a testing machine. Bone can withstand significantly higher stresses if the forces are applied for only a very short period of time.

EXAMPLE 8.11 **Finding the compression of a bone**

The femur, the long bone in the thigh, can be modeled as a tube of cortical bone for most of its length. A 70 kg person has a femur with a cross-section area (of the cortical bone) of 4.8×10^{-4} m², a typical value.

a. If this person supports his entire weight on one leg, what fraction of the tensile strength of the bone does this stress represent?
b. By what fraction of its length does the femur shorten?

SOLVE

a. The force compressing the femur is the person's weight, $F = mg = (70 \text{ kg})(9.8 \text{ m/s}^2) = 690$ N. The resulting stress on the femur is

$$\frac{F}{A} = \frac{690 \text{ N}}{4.8 \times 10^{-4} \text{ m}^2} = 1.4 \times 10^6 \text{ N/m}^2$$

A stress of 1.4×10^6 N/m² is 1.4% of the tensile strength of cortical bone given in Table 8.4.

b. We can use Equation 8.7 to compute the strain $\Delta L/L$, with the value of Young's modulus for cortical bone from Table 8.3:

$$\frac{\Delta L}{L} = \left(\frac{1}{Y}\right)\frac{F}{A} = \left(\frac{1}{1.6 \times 10^{10} \text{ N/m}^2}\right)(1.4 \times 10^6 \text{ N/m}^2) = 8.8 \times 10^{-5} \approx 0.0001$$

The femur compression is $\Delta L \approx 0.0001L$, or $\approx 0.01\%$ of its length.

ASSESS Under normal forces, the bones of the body do not compress significantly, and they experience stresses much less than the tensile strength of bone.

The dancer in the chapter opening photo stands *en pointe,* balanced delicately on the tip of her shoe with her entire weight supported on a very small area. The stress on the bones in her toes is very large, but it is still well below the tensile strength of bone.

STOP TO THINK 8.5 A 10 kg mass is hung from a 1-m-long cable, causing the cable to stretch by 2 mm. Suppose a 10 kg mass is hung from a 2 m length of the same cable. By how much does the cable stretch?

A. 0.5 mm   B. 1 mm   C. 2 mm   D. 3 mm   E. 4 mm

The goal of Chapter 8 has been to learn about the static equilibrium of extended objects and to understand the basic properties of springs and elastic materials.

GENERAL PRINCIPLES

Static Equilibrium

An object in **static equilibrium** must have no net force on it and no net torque. Mathematically, we express this as

$$\sum F_x = 0$$
$$\sum F_y = 0$$
$$\sum \tau = 0$$

Since the net torque is zero about *any* point, the pivot point for calculating the torque can be chosen at any convenient location.

Springs and Hooke's Law

When a spring is stretched or compressed, it exerts a force proportional to the change Δx in the spring's length but in the opposite direction. This is known as **Hooke's Law:**

$$(F_{sp})_x = -k \, \Delta x$$

The constant of proportionality k is called the **spring constant.** It is larger for a "stiff" spring.

IMPORTANT CONCEPTS

Stability

An object is **stable** if its center of gravity is over its base of support, and **unstable** if it is not.

If an object is tipped, it will reach the limit of its stability when its center of gravity is over the edge of the base. This defines the **critical angle** θ_c.

Greater stability is possible with a lower center of gravity or a broader base of support.

This object is at its critical angle.

This object has a wider base of support and hence a larger critical angle.

This object has a lower center of gravity, so its critical angle is larger too.

Elastic materials and Young's modulus

A solid rod illustrates how materials respond when stretched or compressed.

Stress is the restoring force of the rod divided by its cross-section area.

$$\left(\frac{F}{A}\right) = Y\left(\frac{\Delta L}{L}\right)$$

Young's modulus.

Strain is the fractional change in the rod's length.

This can be written as

This is the "spring constant" k for the rod.

$$F = \left(\frac{YA}{L}\right)\Delta L$$

showing that a rod obeys Hooke's law, and acts like a very stiff spring.

APPLICATIONS

Forces in the body

Muscles and tendons apply the forces and torques needed to maintain static equilibrium. These forces may be quite large.

The torque from the tendon is due to a *large* force acting with a *short* moment arm.

In equilibrium, the net torque due to these forces must be zero.

Short moment arm

Long moment arm

The torque from the weight is due to a *small* force acting with a *long* moment arm.

The elastic limit and beyond

If a rod or other shape is not stretched too far, when released it will return to its original shape.

If stretched too far, it will permanently deform, and finally break. The stress at which an object breaks is its **tensile stress.**

If not stretched beyond here, the object will return to its original length.

Hooke's law applies

Breaking point

If stretched to this region, the object will be permanently deformed.

For instructor-assigned homework, go to
www.masteringphysics.com

Problem difficulty is labeled as | (straightforward) to |||| (challenging).

Problems labeled INT integrate significant material
from earlier chapters; BIO are of biological or medical
interest.

QUESTIONS

Conceptual Questions

1. An object is acted upon by two (and only two) forces that are of equal magnitude and oppositely directed. Is the object necessarily in static equilibrium?
2. Sketch a force acting at point P in Figure Q8.2 that would make the rod be in static equilibrium. Is there only one such force?
3. Could a ladder on a level floor lean against a wall in static equilibrium if there were no friction forces? Explain.
4. Suppose you are hanging from a tree branch. If you move out the branch, farther away from the trunk, the branch will be more likely to break. Explain why this is so.

FIGURE Q8.2

5. As divers stand on tiptoes on the edge of a diving platform, in preparation for a high dive, as shown in Figure Q8.5, they usually extend their arms in front of them. Why do they do this?
6. Where are the centers of gravity of the two people doing the classic yoga poses shown in Figure Q8.6?

FIGURE Q8.5

(a)

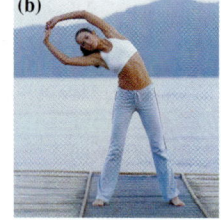

(b)

FIGURE Q8.6

7. A spring exerts a 10 N force after being stretched by 1 cm from its equilibrium length. By how much will the spring force *increase* if the spring is stretched from 4 cm away from equilibrium to 5 cm from equilibrium?
8. The left end of a spring is attached to a wall. When Bob pulls on the right end with a 200 N force, he stretches the spring by 20 cm. The same spring is then used for a tug-of-war between Bob and Carlos. Each pulls on his end of the spring with a 200 N force.
   a. How far does Bob's end of the spring move? Explain.
   b. How far does Carlos's end of the spring move? Explain.

9. A spring is attached to the floor and pulled straight up by a string. The string's tension is measured. The graph in Figure Q8.9 shows the tension in the spring as a function of the spring's length L.
   a. Does this spring obey Hooke's Law? Explain why or why not.
   b. If it does, what is the spring constant?

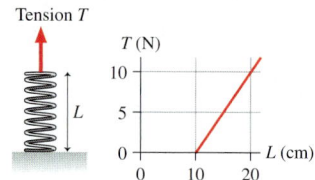

FIGURE Q8.9

10. Take a spring and cut it in half to make two springs. Is the spring constant of these smaller springs larger, smaller, or the same as the spring constant of the original spring? Explain.
11. A wire is stretched right to its breaking point by a 5000 N force. A longer wire made of the same material has the same diameter. Is the force that will stretch it right to its breaking point larger than, smaller than, or equal to 5000 N? Explain.
12. The beam in Figure Q8.12 is being pressed down by a large force on its middle. If the beam is turned 90° about its long axis and then subjected to the same downward force, will its deflection be larger, smaller, or unchanged? Explain.

FIGURE Q8.12

13. Steel nails are rigid and unbending. Steel wool is soft and squishy. How would you account for this difference?

Multiple-Choice Questions

Questions 14 and 15 use the information in the following paragraph and figure.

A student lies on a very light, rigid board with a scale under each end. Her feet are directly over one scale, and her body is positioned as shown in Figure Q8.14. The two scales read the values shown in the figure.

FIGURE Q8.14

14. | What is the student's weight?
    A. 65 lb   B. 75 lb   C. 100 lb   D. 165 lb
15. ||| Approximately how far from her feet is the student's center of gravity?
    A. 0.6 m   B. 0.8 m   C. 1.0 m   D. 1.2 m

Questions 16 through 18 use the information in the following paragraph and figure.

Suppose you stand on one foot while holding your other leg up behind you. Your muscles will have to apply a force to hold your leg in this raised position. We can model this situation as in Figure Q8.16. The leg pivots at the knee joint, and the force to hold the leg up is provided by a tendon attached to the lower leg as shown. Assume that the lower leg and the foot together have a combined mass of 4.0 kg, and that their combined center of gravity is at the center of the lower leg.

The tendon provides the torque to raise the lower leg.

5.0 cm

50 cm

FIGURE Q8.16

16. || How much force must the tendon exert to keep the leg in
BIO this position?
    A. 40 N   B. 200 N   C. 400 N   D. 1000 N
17. || As you hold your leg in this position, the upper leg exerts a
BIO force on the lower leg at the knee joint. What is the direction of this force?
    A. Up   B. Down   C. Right   D. Left
18. || What is the magnitude of the force of the upper leg on the
BIO lower leg at the knee joint?
    A. 40 N   B. 160 N   C. 200 N   D. 240 N

19. ||| A tall ladder is leaning against a wall as shown in Figure Q8.19. There is no friction between the top of the ladder and the wall. The coefficient of static friction between the bottom of the ladder and the ground is small but not zero. A painter climbs up the ladder to reach a high spot on the wall. At which location should the painter be most worried about the ladder slipping?
    A. Near the bottom.
    B. At the middle of the ladder.
    C. Near the top.
    D. The risk of the ladder slipping is the same at all locations.

FIGURE Q8.19

20. ||| A 30.0-cm-long board is placed on a table such that its right end hangs over the edge by 8.0 cm. A second identical board is stacked on top of the first such that its right edge is 8.0 cm past the edge of the first board, as shown in Figure Q8.20. If additional boards are stacked in the same manner, each shifted 8.0 cm to the right, what is the largest number of boards (including the first) that can lie in static equilibrium?

30.0 cm

8.0 cm   8.0 cm

FIGURE Q8.20

    A. 1   B. 2   C. 3   D. 4
    E. As many boards as one likes can be stacked this way without falling over.

PROBLEMS

Section 8.1 Torque and Static Equilibrium

1. | A 64 kg woman stands on a very light, rigid board that rests on a bathroom scale at each end, as shown in Figure P8.1. What is the reading on each of the scales?

1.5 m

2.0 m

FIGURE P8.1

2. || Suppose the woman in Figure P8.1 is 54 kg, and the board she is standing on has a 10 kg mass. What is the reading on each of the scales?

3. || You're carrying a 3.6-m-long, 25 kg pole to a construction site when you decide to stop for a rest. You place one end of the pole on a fence post and hold the other end of the pole 35 cm from its tip. How much force must you exert to keep the pole motionless in a horizontal position?

4. || How much torque must the pin exert to keep the rod in Figure P8.4 from rotating? Calculate this torque about an axis that passes through the point where the pin enters the rod and is perpendicular to the plane of the figure.

80 cm

2.0 kg

Pin

500 g

FIGURE P8.4

5. | Is the object in Figure P8.5 in equilibrium? Explain.

100 N

2.0 m   1.0 m

40 N   Massless   60 N

FIGURE P8.5

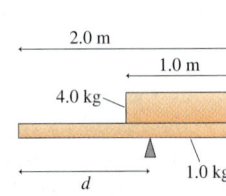

2.0 m

1.0 m

4.0 kg

d

1.0 kg

FIGURE P8.6

6. ||| The two objects in Figure P8.6 are balanced on the pivot. What is distance d?

7. | A 60 kg diver stands at the end of a 30 kg springboard, as shown in Figure P8.7. The board is attached to a hinge at the left end but simply rests on the right support. What is the magnitude of the vertical force exerted by the hinge on the board?

1.5 m

3.0 m

FIGURE P8.7

8. || A uniform beam of length 1.0 m and mass 10 kg is attached to a wall by a cable, as shown in Figure P8.8. The beam is free to pivot at the point where it attaches to the wall. What is the tension in the cable?

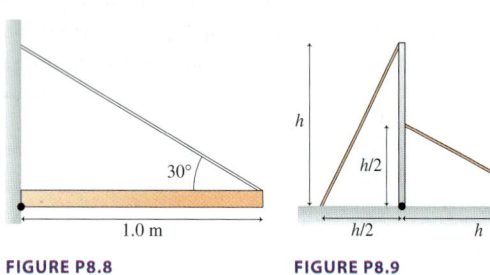

h

$30°$

1.0 m

FIGURE P8.8

h

$h/2$

$h/2$ h

FIGURE P8.9

9. ||| Figure P8.9 shows a vertical pole of height h that can rotate about a hinge at the bottom. The pole is held in position by two wires under tension. What is the ratio of the tension in the left wire to the tension in the right wire?

Section 8.2 Stability and Balance

10. | You want to slowly push a stiff board across a 20 cm gap between two tabletops that are at the same height. If you apply only a horizontal force, how long must the board be so that it doesn't tilt down into the gap before reaching the other side?

11. | A magazine rack has a center of gravity 16 cm above the floor, as shown in Figure P8.11. Through what maximum angle, in degrees, can the rack be tilted without falling over?

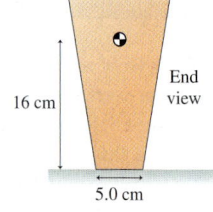

16 cm

End view

5.0 cm

FIGURE P8.11

12. | A car manufacturer claims that you can drive its new vehicle across a hill with a 47° slope before the vehicle starts to tip. If the vehicle is 2.0 m wide, how high is its center of gravity?

13. || A thin 2.00 kg box rests on a 6.00 kg board that hangs over the end of a table, as shown in Figure P8.13. How far can the center of the box be from the end of the table before the board begins to tilt?

30.0 cm   20.0 cm

FIGURE P8.13

14. |||| The object shown in Figure P8.14 is made of a uniform material. What is the greatest that x can be without the object tipping over?

2.0 cm

4.0 cm

x

2.0 cm

FIGURE P8.14

Section 8.3 Springs and Elastic Materials

15. | One end of a spring is attached to a wall. A 25 N pull on the other end causes the spring to stretch by 3.0 cm. What is the spring constant?

16. | Experiments using "optical tweezers" measure the elasticity of individual DNA molecules. For small enough changes in length, the elasticity has the same form as that of a spring. A DNA molecule is anchored at one end, then a force of 1.5 nN $(1.5 \times 10^{-9}$ N) pulls on the other end, causing the molecule to stretch by 5.0 nm $(5.0 \times 10^{-9}$ m). What is the spring constant of that DNA molecule?

BIO

17. |||| A spring has an unstretched length of 10 cm. It exerts a restoring force F when stretched to a length of 11 cm.
   a. For what total stretched length of the spring is its restoring force $3F$?
   b. At what compressed length is the restoring force $2F$?

18. || A 10-cm-long spring is attached to the ceiling. When a 2.0 kg mass is hung from it, the spring stretches to a length of 15 cm.
   a. What is the spring constant?
   b. How long is the spring when a 3.0 kg mass is suspended from it?

19. | A spring stretches 5.0 cm when a 0.20 kg block is hung from it. If a 0.70 kg block replaces the 0.20 kg block, how far does the spring stretch?

20. | A 1.2 kg block is hung from a vertical spring, causing the spring to stretch by 2.4 cm. How much further will the spring stretch if a 0.60 kg block is added to the 1.2 kg block?

21. | A runner wearing spiked shoes pulls a 20 kg sled across frictionless ice using a horizontal spring with spring constant 1.5×10^2 N · m. The spring is stretched 20 cm from its equilibrium length. What is the acceleration of the sled?

22. | You need to make a spring scale to measure the mass of objects hung from it. You want each 1.0 cm length along the scale to correspond to a mass difference of 0.10 kg. What should be the value of the spring constant?

Section 8.4 Stretching, Compressing, and Bending Materials

23. || A force stretches a wire by 1.0 mm.
   a. A second wire of the same material has the same cross section and twice the length. How far will it be stretched by the same force?
   b. A third wire of the same material has the same length and twice the diameter as the first. How far will it be stretched by the same force?

24. |||| What hanging mass will stretch a 2.0-m-long, 0.50-mm-diameter steel wire by 1.0 mm?

25. | How much force does it take to stretch a 10-m-long, 1.0-cm-diameter steel cable by 5.0 mm?

26. | An 80-cm-long, 1.0-mm-diameter steel guitar string must be tightened to a tension of 2.0 kN by turning the tuning screws. By how much is the string stretched?

27. |||| A 2000 N force stretches a wire by 1.0 mm.
   a. A second wire of the same material is twice as long and has twice the diameter. How much force is needed to stretch it by 1.0 mm? Explain.
   b. A third wire of the same material is twice as long as the first and has the same diameter. How far is it stretched by a 4000 N force?

28. | A 1.2-m-long steel rod with a diameter of 0.50 cm hangs vertically from the ceiling. An auto engine weighing 4.7 kN is hung from the rod. By how much does the rod stretch?

29. | A mine shaft has an elevator hung from a single steel-wire cable of diameter 2.5 cm. When the cable is fully extended, the end of the cable is 500 m below the support. How much does the fully extended cable stretch when 3000 kg of ore is loaded into the elevator?

30. || A 3.0-m-tall, 50-cm-diameter concrete column supports a 200,000 kg load. By how much is the column compressed?

31. | A three-legged wooden bar stool made out of solid Douglas fir has legs that are 2.0 cm in diameter. When a 75 kg man sits on the stool, by what percent does the length of the legs decrease? Assume, for simplicity, that the stool's legs are vertical and that each bears the same load.

32. || A force applied to the end of a beam like the one shown in Figure 8.27 deflects the beam by 1.0 cm.
    a. If the same force is applied to the end of a beam that is twice as thick but otherwise identical, by how much will that beam deflect?
    b. If the same force is applied to the end of a beam that is twice as thick, wide, and long, by how much will that beam deflect?

33. | An aluminum beam, like the one shown in Figure 8.27, is deflected by 4.0 mm under a load. If that beam is replaced by another of the same shape and dimensions, but made of cast iron, what will be the deflection?

34. | A long steel bar with a 3.0 cm × 3.0 cm cross-section is imbedded in a concrete wall. 2.0 m of the bar extends horizontally from the wall. If a 0.75 kg brick is placed at the end of the bar, how far is the end of the bar deflected downward?

General Problems

35. ||| A 3.0-m-long rigid beam with a mass of 100 kg is supported at each end, as shown in Figure P8.35. An 80 kg student stands 2.0 m from support 1. How much upward force does each support exert on the beam?

FIGURE P8.35

36. || An 80 kg construction worker sits down 2.0 m from the end of a 1450 kg steel beam to eat his lunch, as shown in Figure P8.36. The cable supporting the beam is rated at 15,000 N. Should the worker be worried?

FIGURE P8.36

37. ||| A forearm can be modeled as a
BIO 1.2 kg, 35-cm-long "beam" that pivots at the elbow and is supported by the biceps, as shown in Figure P8.37. How much force must the biceps exert to hold a 500 g ball with the forearm parallel to the floor?

FIGURE P8.37

38. ||| If the person in Figure P8.37
BIO lowers his forearm to be 15° below horizontal, how much force must the biceps exert to hold the 500 g ball? Note that the "insertion point" where the biceps attaches to the forearm is always 4.0 cm from the elbow joint, and assume that the person's wrist remains unbent as he lowers his forearm.

39. ||| A man is attempting to raise a 7.5-m-long, 28 kg flagpole that has a hinge at the base by pulling on a rope attached to the top of the pole, as shown in Figure P8.39. With what force does the man have to pull on the rope to hold the pole motionless in this position?

FIGURE P8.39

40. |||| Children with masses m_1 and m_2 sit at opposite ends of a seesaw with length L and mass M. What distance D should the pivot point be from the first child if the seesaw is to balance? Your answer will be an expression written in terms of the three masses and the length L.

41. || A 70 kg skier is on flat snow waiting in line for a ski lift.
BIO While goofing around, he bends forward stiffly, as shown in Figure P8.41. To hold this position, tension in his back muscles must support the weight of his upper body. His upper body has 55% of his mass, and the center of gravity of his upper body is 43 cm from his hips, in the middle of his trunk. What torque do his back muscles exert about an axis through his hips?

FIGURE P8.41

FIGURE P8.42

42. || The object shown in Figure P8.42 is made of three identical cubes, each 2.0 cm on a side, glued together.
    a. Taking the pivot to be the origin, what are the x- and y-coordinates of the object's center of gravity?
    b. What is the maximum angle, in degrees, it can be rotated around the pivot and still fall back when released, rather than falling over?

43. ‖ A 40 kg, 5.0-m-long beam is supported by, but not attached to, the two posts in Figure P8.43. A 20 kg boy starts walking along the beam. How close can he get to the right end of the beam without it tipping?

FIGURE P8.43   3.0 m

44. ‖‖ A very light tripod has 70-cm-long legs that can be spread out so that each makes a 20° angle with the vertical. The points where the three legs rest on the ground form an equilateral triangle. If you attach an object to the tripod, such as a camera with a long lens, what is the maximum possible horizontal distance between the center of gravity of the object and the center of the tripod, if the tripod is to remain stable?

45. ‖ A 5.0 kg mass hanging from a spring scale is slowly lowered onto a vertical spring, as shown in Figure P8.45. The scale reads in newtons.
    a. What does the spring scale read just before the mass touches the lower spring?
    b. The scale reads 20 N when the lower spring has been compressed by 2.0 cm. What is the value of the spring constant for the lower spring?
    c. At what compression distance will the scale read zero?

Scale

FIGURE P8.45

46. ‖‖ Two identical, side-by-side springs with spring constant 240 N/m support a 2.00 kg hanging box. By how much is each spring stretched?

47. ‖ Two springs have the same equilibrium length but different spring constants. They are arranged as shown in Figure P8.47, then a block is pushed against them, compressing both by 1.00 cm. With what net force do they push back on the block?

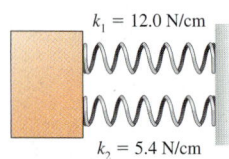

$k_1 = 12.0$ N/cm

$k_2 = 5.4$ N/cm

FIGURE P8.47

2.5 cm

FIGURE P8.48

48. ‖ Two springs have the same spring constant $k = 130$ N/m but different equilibrium lengths, 3.0 cm and 5.0 cm. They are arranged as shown in Figure P8.48, then a block is pushed against them, compressing both to a length of 2.5 cm. With what net force do they push back on the block?

49. ‖‖ Figure P8.49 shows two springs attached to a block that can slide on a friction-less surface. In the block's equilibrium position, the left spring is compressed by 2.0 cm.
    a. By how much is the right spring compressed?
    b. What is the net force on the block if it is moved 15 cm to the right of its equilibrium position?

$k_1 = 10$ N/m $k_2 = 20$ N/m

FIGURE P8.49

50. ‖‖‖ Figure P8.50 shows two springs attached to a box that can slide on a frictionless surface. In the block's equilibrium position, neither spring is stretched. What is the net force on the block if it is moved 15 cm to the right of its equilibrium position? **Hint:** There is zero net force on the point where the two springs meet. This implies a relationship between the amounts the two springs stretch.

$k_1 = 10$ N/m $k_2 = 20$ N/m

FIGURE P8.50

51. ‖ A 60 kg student is standing atop a spring in an elevator that
INT is accelerating upward at 3.0 m/s². The spring constant is 2.5×10^3 N/m. By how much is the spring compressed?

52. ‖ A 25 kg child bounces on a pogo stick. The pogo stick has a
INT spring with spring constant 2.0×10^4 N/m. When the child makes a nice big bounce, she finds that at the bottom of the bounce she is accelerating *upwards* at 9.8 m/s². How much is the spring compressed?

53. ‖‖‖ Two 3.0 kg blocks on a level, frictionless surface are con-
INT nected by a spring with spring constant 1000 N/m, as shown in Figure P8.53. The left block is pushed by a horizontal force $\vec{F}$. At $t = 0$ s, both blocks have velocity 3.2 m/s. For the next second, the spring's compression is a constant 1.5 cm.
    a. What is the velocity of the right block at $t = 1.0$ s?
    b. What is the magnitude of $\vec{F}$ during that 1.0 s interval?

$\vec{F}$

FIGURE P8.53

54. ‖ What is the effective spring constant (that is, the ratio of force to change in length) of a copper cable that is 5.0 mm in diameter and 5.0 m long?

55. ‖ A 3.5 kg potted plant hangs from the end of a 35-cm-long aluminum wall bracket. The bracket is 2.5 cm wide and 5.0 mm thick. Ignoring the weight of the aluminum, how much does the end of the bracket bend downward?

56. ‖ When you walk, your Achilles tendon, which connects your
BIO heel to your calf muscles, repeatedly stretches and contracts, much like a spring. This helps make walking more efficient. Suppose your Achilles tendon is 15 cm long and has a cross-section area of 110 mm², typical values. If you model the Achilles tendon as a spring, what is its spring constant?

57. ‖‖‖ There is a disk of cartilage between each pair of vertebrae
BIO in your spine. Healthy cartilage has a Young's modulus of approximately 1.0×10^6 N/m². Suppose a disk is 0.50 cm thick and 4.0 cm in diameter. If this disk supports half the weight of a 65 kg person, by what fraction of its thickness does the disk compress?

58. ‖‖‖ When you stand on your tiptoes, as the woman in Fig-
BIO ure 8.16 is doing, your feet pivot about your ankle. The forces on your foot are an upward force on your toes from the floor, a downward force on your ankle from the lower leg bone, and an upward force on the heel of your foot from your Achilles tendon. Suppose a 60 kg woman stands on tiptoes with the sole of her foot making a 25° angle with the floor. The distance from her toes to her ankle is 12 cm, the distance from her ankle to the point on the heel where the tendon attaches is

5.0 cm, and each foot supports half her weight. Assume that all forces act vertically.
   a. What is the upward force on the toes of one foot?
   b. What upward force does the Achilles tendon exert on the heel of that foot?
   c. The tension in the Achilles tendon will cause it to stretch. If the Achilles tendon is 15 cm long and has a cross-section area of 110 mm², by how much will it stretch under this force?

59. ‖ In Example 8.2, the tension in the biceps tendon for a per-
BIO son doing a strict curl of a 900 N barbell was found to be 3900 N. What fraction does this represent of the maximum possible tension the biceps tendon can support? The tensile strength of tendon is approximately 100×10^6 N/m², and you can assume a typical cross-section area of 130 mm².

60. ‖ Orb spiders make silk with a typical diameter of 0.15 mm.
BIO a. A typical large orb spider has a mass of 0.50 g. If this spi-
der suspends itself from a single 12-cm-long strand of silk, by how much will the silk stretch?
   b. What is the maximum weight that a single thread of this silk could support?

61. ‖‖ Larger animals have sturdier bones than smaller animals. A
BIO mouse's skeleton is only a few percent of its body weight, compared to 16% for an elephant. To see why this must be so, recall, from Example 8.11, that the stress on the femur for a man standing on one leg is 1.4% of the bone's tensile strength. Suppose we scale this man up by a factor of 10 in all dimensions, keeping the same body proportions. Use the data for Example 8.11 to compute the following.
   a. Both the inside and outside diameter of the femur, the region of cortical bone, will increase by a factor of 10. What will be the new cross-section area?
   b. The man's body will increase by a factor of 10 in each dimension. What will be his new mass?
   c. If the scaled up man now stands on one leg, what fraction of the tensile strength is the stress on the femur?

Passage Problems

The Diving Board

A diving board is a long, flexible board that is mounted as shown in Figure P8.62. The board's left end is fastened to a fixed support by a connection that is free to pivot. The board is supported toward its middle on another fixed support point, as shown. A diver of mass 60 kg stands at the end of the board. In the following problems we will ignore the mass of the board itself.

The board can pivot at this point. The board is supported at this point. Diver: mass 60 kg

0.60 m   1.60 m

FIGURE P8.62

62. ‖ There is a force on the board from the pivot point. What is the direction of this force?
   A. Up   B. Down
   C. To the right, toward the diver
   D. To the left, away from the diver

63. ‖ The support point exerts an upward force on the board. What is the magnitude of this force?
   A. 590 N   B. 1000 N   C. 1600 N   D. 2200 N

64. ‖ Suppose the board deflects by 3.0 cm when the 60 kg diver stands on it. Now he jumps up and comes down on the board; at the lowest point of his motion, the deflection of the board is 6.0 cm. What is the net force on the diver at this point?
   A. 590 N, up   B. 590 N, down
   C. 1180 N, up   D. zero

65. ‖ A new board is installed that is twice as thick as the old board but is made of the same material and is the same length and width. Approximately how much does this board deflect when the 60 kg diver stands on it?
   A. 0.4 cm   B. 0.8 cm   C. 1.5 cm   D. 3.0 cm

Stop to Think 8.1: D. Only object D has both zero net force and zero net torque.

Stop to Think 8.2: B. The tension in the rope and the weight have no horizontal component. To make the net force zero, the force due to the pivot must also have no horizontal component, so we know it points either up or down. Now consider the torque about the point where the rope is attached. The tension provides no torque. The weight exerts a counterclockwise torque. To make the net torque zero, the pivot force must exert a *clockwise* torque, which it can only do if it points *up*.

Stop to Think 8.3: B, A, C. The critical angle θ_c, shown in the figure, measures how far the object can be tipped before falling. B has the smallest critical angle, followed by A, then C.

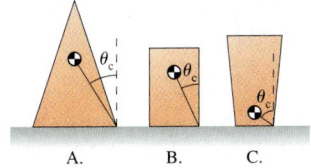

Stop to Think 8.4: C. The restoring force of the spring is proportional to the stretch. Increasing the restoring force by a factor of 3 requires increasing the stretch by a factor of 3.

Stop to Think 8.5: E. The cables have the same diameter, and the force is the same, so the stress is the same in both cases. This means that the strain, $\Delta L/L$, is the same. The 2 m cable will experience twice the change in length of the 1 m cable.

Force and Motion

The goal of Part I has been to discover the connection between force and motion. We started with kinematics, the mathematical description of motion, then we proceeded to dynamics, the explanation of motion in terms of forces. We then used these descriptions to analyze and explain motions ranging from the motion of the moon about the earth to the forces in your elbow when you lift a weight. Newton's three laws of motion formed the basis of all of our explanations.

The table below is a *knowledge structure* for force and motion. The knowledge structure does not represent everything you have learned over the past eight chapters. It's a summary of the "big picture," outlining the basic goals, the general principles, and the primary applications of the part of the book we have just finished. When you are immersed in a chapter, it may be hard to see the connections among all of the different topics. Before we move on to new topics, we will finish each part of the book with a knowledge structure to make these connections clear.

Work through the knowledge structure from top to bottom. First are the goals and general principles. There aren't that many general principles, but we can use them along with the general problem-solving strategy to solve a wide range of problems. Once you recognize a problem as a dynamics problem, you immediately know to start with Newton's laws. You can then determine the category of motion and apply Newton's second law in the appropriate form. The kinematic equations for that category of motion then allow you to reach the solution you seek. These equations and other detailed information from the chapters are summarized in the bottom section.

KNOWLEDGE STRUCTURE I Force and Motion

BASIC GOALS

How can we describe motion?   How does an object respond to a force?
How do systems interact?   What is the nature of the force of gravity?
How can we analyze the motion and deformation of extended objects?

GENERAL PRINCIPLES

Newton's first law   An object will remain at rest or will continue to move with constant velocity (i.e., be in equilibrium), if and only if $\vec{F}_{net} = \vec{0}$.

Newton's second law $\vec{F}_{net} = m\vec{a}$

Newton's third law $\vec{F}_{A\,on\,B} = -\vec{F}_{B\,on\,A}$

Newton's law of gravity $F_{1\,on\,2} = F_{2\,on\,1} = \dfrac{Gm_1m_2}{r^2}$

BASIC PROBLEM-SOLVING STRATEGY

Use Newton's second law for each particle or system. Use Newton's third law to equate the magnitudes of the two members of an action/reaction pair.

Types of forces:
$\vec{w} = (mg,\ \text{downwards})$
$\vec{f}_k = (\mu_k n,\ \text{opposite motion})$
$(F_{sp})_x = -k\,\Delta x$

Linear and projectile motion:

$\sum F_x = ma_x \qquad \sum F_x = 0$
or
$\sum F_y = 0 \qquad \sum F_y = ma_y$

Circular motion:

The force is directed to the center:

$\vec{F}_{net} = \left(\dfrac{mv^2}{r},\ \text{toward center of circle}\right)$

Rigid body motion:

When a torque is exerted on an object with moment of inertia I:

$\tau_{net} = I\alpha$

Equilibrium:

For an object at rest:

$\sum F_x = 0$
$\sum F_y = 0$
$\sum \tau = 0$

Linear and projectile kinematics

Uniform motion: $x_f = x_i + v_x\,\Delta t$
$(a_x = 0, v_x = \text{constant})$

Constant acceleration: $(v_x)_f = (v_x)_i + a_x\,\Delta t$
$(a_x = \text{constant})$

$x_f = x_i + (v_x)_i\,\Delta t + \frac{1}{2}a_x(\Delta t)^2$

$(v_x)_f^2 = (v_x)_i^2 + 2a_x\,\Delta x$

Velocity is the slope of the position-versus-time graph.
Acceleration is the slope of the velocity-versus-time graph.

Projectile motion:
Projectile motion is uniform horizontal motion and constant acceleration vertical motion with $a_y = -g$.

Circular kinematics

Uniform circular motion:

$f = \dfrac{1}{T} \qquad \omega = 2\pi f$

$v = \dfrac{2\pi r}{T} = \omega r \qquad a = \dfrac{v^2}{r} = \omega^2 r$

Rigid bodies

Torque $\tau = rF_\perp = r_\perp F$

Center of gravity $x_{cm} = \dfrac{x_1 m_1 + x_2 m_2 + \cdots}{m_1 + m_2 + \cdots}$

Moment of inertia $I = \sum mr^2$

Dark Matter and the Structure of the Universe

The idea that the earth exerts a gravitational force on us is something we now accept without questioning. But when Isaac Newton developed this idea to show that the gravitational force also holds the moon in its orbit, it was a remarkable, ground-breaking insight. It changed the way that we look at the universe we live in.

Newton's laws of motion and gravity are tools that allow us to continue Newton's quest to better understand our place in the cosmos. But it sometimes seems that the more we learn, the more we realize how little we actually know and understand.

Here's an example. Advances in astronomy over the past 100 years have given us great insight into the structure of the universe. But everything our telescopes can see appears to be only a small fraction of what is out there. As much as 90% of the mass in the universe is *dark matter*— matter that gives off no light or other radiation that we can detect. Everything that we have ever seen through a telescope is merely the tip of the cosmic iceberg.

What is this dark matter? Black holes? Neutrinos? Some form of exotic particle? No one knows. It could be any of these, or all of them—or something entirely different that no one has yet dreamed of. You might wonder how we know that such matter exists if no one has seen it. Even though we can't directly observe dark matter, we see its effects. And you now know enough physics to understand why.

Whatever dark matter is, it has mass, and so it has gravity. This picture of the Andromeda galaxy shows a typical spiral galaxy structure: a dense collection of stars in the center surrounded by a disk of stars and other matter. This is the shape of our own Milky Way galaxy.

The spiral Andromeda galaxy.

This structure is reminiscent of the structure of the solar system, a dense mass (the sun) in the center surrounded by a disk of other matter (the planets, asteroids, and comets.) The sun's gravity keeps the planets in their orbits, but the planets would fall into the sun unless they were in constant motion around it. The same is true of a spiral galaxy; everything in the galaxy orbits its center. Our solar system orbits the center of our galaxy with a period of about 200 million years.

The orbital speed of an object depends on the mass that pulls on it. If you analyze our sun's motion about the center of the Milky Way, or the motion of stars in the Andromeda galaxy about its center, you find that the orbits are much faster than they should be, based on how many stars we see. There must be some other mass present.

There's another problem with the orbital motion of stars around the center of their galaxies. We know that the orbital speeds of planets decrease with distance from the sun; Pluto orbits at a much slower speed than the earth. We might expect something similar for galaxies: Stars farther from the center should orbit at reduced speeds. But they don't. As we measure outward from the center of the galaxy, the orbital speed stays about the same—even as we get to the edge of the visible disk. There must be some other mass—the invisible dark matter—exerting a gravitational force on the stars. This dark matter, which far outweighs the matter we can see, seems to form a halo around the centers of galaxies, providing the gravitational force necessary to produce the observed rotation. Other observations of the motions of galaxies with respect to each other verify this basic idea.

Ultimately, the fate of the universe depends on its mass. The universe is currently expanding. If the universe contains enough mass, the mutual attraction will cause a final collapse—what you might term a "Big Crunch," the antithesis of the Big Bang. If there is not enough mass, the universe will expand forever. Dark matter may well be 90% of the universe. It may hold the key to the universe's future. And we don't know what it is.

This sort of mystery is what drives scientific investigation. It's what drove Newton to wonder about the connection between the fall of an apple and the motion of the moon, and what drove investigators to develop all of the techniques and theories you will learn about in the coming chapters.

PART II

CONSERVATION LAWS

Why Some Things Stay the Same

Part I of this textbook was about *change*. Simple observations show us that most things in the world around us are changing. Even so, there are some things that *don't* change even as everything else is changing around them. Our emphasis in Part II will be on things that stay the same.

Consider, for example, a strong, sealed box in which you have replaced all the air with a mixture of hydrogen and oxygen. The mass of the box plus the gases inside is 600.0 g. Now, suppose you use a spark to ignite the hydrogen and oxygen. As you know, this is an explosive reaction, with the hydrogen and oxygen combining to create water—and quite a bang. But the strong box contains the explosion and all of its products.

What is the mass of the box after the reaction? The gas inside the box is different now, but a careful measurement would reveal that the mass hasn't changed—it's still 600.0 g! We say that the mass is *conserved*. Of course, this is only true if the box has stayed sealed. For conservation of mass to apply, the system must be *closed*.

Conservation Laws

A closed system of interacting particles has another remarkable property. Each system is characterized by a certain number, and no matter how complex the interactions, the value of this number never changes. This number is called the *energy* of the system, and the fact that it never changes is called the *law of conservation of energy*. It is, perhaps, the single most important physical law ever discovered.

The law of conservation of energy is much more general than Newton's laws. Energy can be converted to many different forms, and, in all cases, the total energy stays the same:

- Gasoline, diesel, and jet engines convert the energy of a fuel into the mechanical energy of moving pistons, wheels, and gears.
- A solar cell converts the electromagnetic energy of light into electrical energy.
- An organism converts the chemical energy of food into a variety of other forms of energy, including kinetic energy, sound energy, and thermal energy.

Energy will be *the* most important concept throughout the remainder of this textbook, and much of Part II will focus on understanding what energy is and how it is used.

But energy is not the only conserved quantity. We will begin Part II with the study of two other quantities that are conserved in a closed system: *momentum* and *angular momentum*. Their conservation will help us understand a wide range of physical processes, from the forces when two rams butt heads to the graceful spins of ice skaters.

Conservation laws will give us a new and different *perspective* on motion. Some situations are most easily analyzed from the perspective of Newton's laws, but others make much more sense when analyzed from a conservation-law perspective. An important goal of Part II is to learn which perspective is best for a given problem.

The kestrel is pulling in its wings to begin a steep dive, in which it can achieve a speed of 60 mph. How does the bird achieve such a speed, and why does this speed help the kestrel catch its prey? Such questions are best answered by considering the conservation of energy and momentum.

9

MOMENTUM

Male rams butt heads at high speeds in a ritual to assert their dominance. How can the force of this collision be minimized so as to avoid damage to their brains?

Looking Ahead ▶▶

The goals of Chapter 9 are to introduce the ideas of impulse, momentum, and angular momentum and to learn a new problem-solving strategy based on conservation laws. In this chapter you will learn to:

▶ Understand and use the concepts of impulse and momentum.
▶ Draw a new "before-and-after" visual overview.
▶ Solve problems using the law of conservation of momentum.
▶ Apply these ideas to collisions and explosions.
▶ Understand the concept of angular momentum and solve problems using its conservation law.

Looking Back ◀◀

The law of conservation of momentum is based on Newton's second and third laws, and angular momentum builds on concepts of torque and moment of inertia. Please review:

◀ Section 4.8 Action/reaction force pairs and Newton's third law.
◀ Section 5.2 Newton's second law.
◀ Section 7.2, 7.4 Torque and moment of inertia.

Two rams crashing together is a vivid example of what we'll call a *collision*. During a collision, two objects exert forces on each other that vary in a complex way during the time of the collision. Using Newton's second law to predict the outcome of a collision would thus be a daunting challenge. Nevertheless, some collisions have very simple outcomes. For example, consider a train car rolling along the tracks toward an identical car at rest. The two cars couple together upon impact and then roll down the tracks together. The forces between the train cars during the collision are exceedingly complex. Yet if you were to measure their speeds just before and after the collision, you would find that the two coupled cars have exactly half the speed of the single car before impact. How can such a complex interaction give rise to such a simple outcome?

The opposite of a collision is an interaction that forces two objects apart. These interactions are called *explosions*, even though they may lack a flash or a pop. As an example, imagine you're standing on ice skates, and you throw a ball whose mass is 1% of your mass directly forward. The interaction between the ball and your hand as you propel it forward is quite complex, yet you'll find that no matter how you throw it, your recoil speed is always 1% of the ball's final speed. Another simple outcome.

Our goal in this chapter is to learn how to predict these simple outcomes without having to know all the details of the interaction forces. Two new ideas will make this possible. The first is *momentum,* a concept that unites the masses and velocities of the particles in a system. The second is the idea of a *conservation law,* a statement that will relate the situation "before" an interaction to the situation "after" the interaction. This before-and-after perspective will be a powerful new problem-solving tool.

9.1 Impulse

Suppose that two or more objects have an intense and perhaps complex interaction, such as a collision or an explosion. Our goal is to find a relationship between the velocities of the objects before the interaction and their velocities after the interaction. We'll start by looking at collisions.

A **collision** is a short-duration interaction between two objects. The collision between a tennis ball and a racket, or a baseball and a bat, may seem instantaneous to your eye, but that is a limitation of your perception. A careful look at the tennis ball/racket collision in Figure 9.1 reveals that the right side of the ball is flattened and pressed up against the strings of the racket. It takes time to compress the ball, and more time for the ball to re-expand as it leaves the racket.

The duration of a collision depends on the materials from which the objects are made, but 1 to 10 ms (0.001 to 0.010 s) is typical. This is the time during which the two objects are in contact with each other. The harder the objects, the shorter the contact time. A collision between two steel balls lasts less than 1 ms, while that between a tennis ball and racket might last 10 ms.

Let's begin our discussion by considering a collision that most of us have experienced: kicking a soccer ball. A sequence of high-speed photos of a soccer kick is shown in Figure 9.2. As the foot and the ball just come into contact, as shown in the top frame, the ball is just beginning to compress. By the middle frame of Figure 9.2, the ball has sped up and become greatly compressed. Finally, as shown in the bottom frame, the ball, now moving very fast, is again only slightly compressed.

FIGURE 9.1 A tennis ball collides with a racket. Notice that the right side of the ball is flattened.

  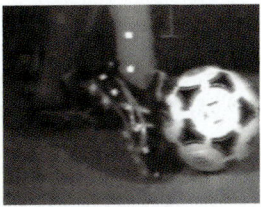

FIGURE 9.2 A sequence of high-speed photos of a soccer ball being kicked.

The amount by which the ball is compressed is a measure of the magnitude of the force the foot exerts on the ball; more compression indicates a greater force. If we were to graph this force versus time, it would look something like Figure 9.3, although the detailed shaped of the curve might be different. The force is zero until the foot first contacts the ball, rises quickly to a maximum value, and then falls back to zero as the ball leaves the foot. Thus there is a well-defined duration Δt of the force. A large force like this exerted during a small interval of time is called an **impulsive force.** The force of a hammer on a nail, or of a bat on a baseball, are other examples of impulsive forces.

The *effect* of an impulsive force depends on the area between the axis and the force-versus-time curve (i.e., the area "under" the force curve). This area, shown in Figure 9.4a on the next page, is called the **impulse** *J* of the force. A harder kick or a kick of longer duration delivers a larger impulse and thus causes the ball to leave the kicker's foot with a higher speed (i.e., it has a larger effect).

FIGURE 9.3 The force on a soccer ball changes rapidly.

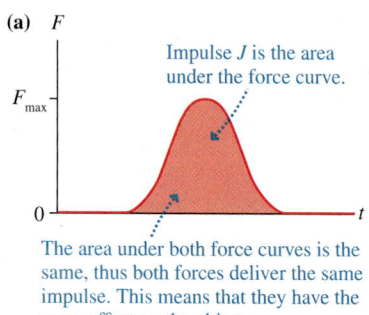

(a) *F*

Impulse *J* is the area under the force curve.

F_{max}

The area under both force curves is the same, thus both forces deliver the same impulse. This means that they have the same effect on the object.

(b) *F*

The area under the rectangle is $F_{avg} \Delta t$, its width Δt multiplied by its height F_{avg}.

F_{avg}

Same duration Δt

FIGURE 9.4 Looking at the impulse graphically.

Impulsive forces can be complex, and the shape of the force-versus-time graph often changes in a complicated way. Consequently, it is often useful to think of the collision in terms of an *average* force F_{avg}. As Figure 9.4b shows, F_{avg} is defined to be the constant force that has the same duration Δt and the same area under the force curve as the real force. You can see from the figure that the area under the force curve can be written simply as $F_{avg} \Delta t$. Thus

$$\text{impulse } J = \text{area under the force curve} = F_{avg} \Delta t \qquad (9.1)$$

Impulse due to a force acting for a duration Δt

From Equation 9.1 we see that impulse has units of N · s, but you should be able to show that N · s are equivalent to kg · m/s. We'll see shortly why the latter are the preferred units for impulse.

So far, we've been assuming the force is directed along a coordinate axis, such as the x-axis. In this case impulse is a *signed* quantity—it can be positive or negative. A positive impulse results from an average force directed in the positive x-direction (that is, F_{avg} is positive), while a negative impulse is due to a force directed in the negative x-direction (F_{avg} is negative). More generally, the impulse is a *vector* quantity pointing in the direction of the average force vector:

$$\vec{J} = \vec{F}_{avg} \Delta t \qquad (9.2)$$

EXAMPLE 9.1 Finding the impulse on a bouncing ball

A rubber ball experiences the force shown in Figure 9.5 as it bounces off the floor.

a. What is the impulse on the ball?
b. What is the average force on the ball?

PREPARE The impulse due to an impulsive force is the area under the force curve. Here the shape of the graph is triangular, so we'll need to use

F

300 N

8.0 ms

FIGURE 9.5 The force of the floor on a bouncing ball.

the fact that the area of a triangle is $\frac{1}{2} \times$ height $\times$ base.

SOLVE a. The impulse is

$$J = \tfrac{1}{2}(300 \text{ N})(0.0080 \text{ s}) = 1.2 \text{ N} \cdot \text{s} = 1.2 \text{ kg} \cdot \text{m/s}$$

b. From Equation 9.1, $J = F_{avg} \Delta t$, we can find the average force that would give this same impulse:

$$F_{avg} = \frac{J}{\Delta t} = \frac{1.2 \text{ N} \cdot \text{s}}{0.0080 \text{ s}} = 150 \text{ N}$$

ASSESS In this particular example, the average value of the force is half the maximum value. This is not surprising for a triangular force because the area of a triangle is *half* the base times the height.

9.2 Momentum and the Impulse-Momentum Theorem

We've noted that the effect of an impulsive force depends on the impulse delivered to the object. The effect also depends on the object's mass. Our experience tells us that giving a kick to a heavy object will change its velocity much less than giving the same kick to a light object. We want now to find a quantitative relationship between impulse, mass, and velocity change.

Consider the puck of mass m in Figure 9.6, sliding with an initial velocity $\vec{v}_i$. It is struck by a hockey stick that delivers an impulse $\vec{J} = \vec{F}_{avg} \Delta t$ to the puck. After the impulse, the puck leaves the stick with a final velocity $\vec{v}_f$. How is this final velocity related to the initial velocity?

From Newton's second law, the average acceleration of the puck during the time the stick is in contact with it is

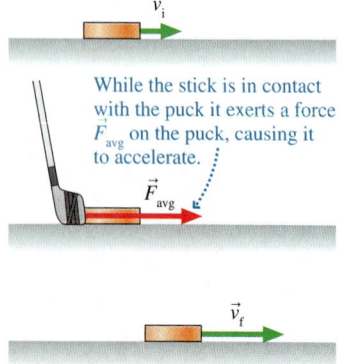

$\vec{v}_i$

While the stick is in contact with the puck it exerts a force $\vec{F}_{avg}$ on the puck, causing it to accelerate.

$\vec{F}_{avg}$

$\vec{v}_f$

FIGURE 9.6 The stick exerts an impulse on the puck, changing its speed.

$$\vec{a}_{avg} = \frac{\vec{F}_{avg}}{m} \qquad (9.3)$$

The average acceleration is related to the change in the velocity by

$$\vec{a}_{\text{avg}} = \frac{\Delta \vec{v}}{\Delta t} = \frac{\vec{v}_{\text{f}} - \vec{v}_{\text{i}}}{\Delta t} \tag{9.4}$$

Combining Equations 9.3 and 9.4, we have

$$\frac{\vec{F}_{\text{avg}}}{m} = \vec{a}_{\text{avg}} = \frac{\vec{v}_{\text{f}} - \vec{v}_{\text{i}}}{\Delta t}$$

or, rearranging,

$$\vec{F}_{\text{avg}} \Delta t = m\vec{v}_{\text{f}} - m\vec{v}_{\text{i}} \tag{9.5}$$

We recognize the left side of this equation as the impulse $\vec{J}$. The right side is the *change* in the quantity $m\vec{v}$. This quantity, the product of the object's mass and velocity, is called the **momentum** of the object. The symbol for momentum is $\vec{p}$:

$$\vec{p} = m\vec{v} \tag{9.6}$$

Momentum of an object of mass m and velocity $\vec{v}$

From Equation 9.6, the units of momentum are those of mass times velocity, or kg · m/s. We noted above that kg · m/s are the preferred units of impulse. Now we see that the reason for that preference is to match the units of momentum.

Figure 9.7 shows that the momentum $\vec{p}$ is a *vector* quantity that points in the same direction as the velocity vector $\vec{v}$. Like any vector, $\vec{p}$ can be decomposed into x- and y-components. Equation 9.6, which is a vector equation, is a shorthand way to write the two equations

$$\begin{aligned} p_x &= mv_x \\ p_y &= mv_y \end{aligned} \tag{9.7}$$

NOTE ▶ One of the most common errors in momentum problems is a failure to use the appropriate signs. The momentum component p_x has the same sign as v_x. Just like velocity, momentum is positive for a particle moving to the right (on the x-axis) or up (on the y-axis), but *negative* for a particle moving to the left or down. ◀

The *magnitude* of an object's momentum is simply the product of the object's mass and speed, or $p = mv$. A heavy, fast-moving object will have a great deal of momentum, while a light, slow-moving object will have very little. Two objects with very different masses can have similar momenta if their speeds are very different as well. Table 9.1 gives some typical values of the momenta (the plural of *momentum*) of various moving objects. You can see that the momenta of a bullet and a fastball are similar. The momentum of a moving car is almost a billion times greater than that of a falling raindrop.

The Impulse-Momentum Theorem

Returning to Equation 9.5, we can now write this equation in terms of impulse and momentum as

$$\vec{J} = \vec{p}_{\text{f}} - \vec{p}_{\text{i}} = \Delta \vec{p} \tag{9.8}$$

Impulse-momentum theorem

where $\vec{p}_{\text{i}} = m\vec{v}_{\text{i}}$ is the object's initial momentum, $\vec{p}_{\text{f}} = m\vec{v}_{\text{f}}$ is its final momentum after the impulse, and $\Delta \vec{p} = \vec{p}_{\text{f}} - \vec{p}_{\text{i}}$ is the *change* in its momentum. This expression is known as the **impulse-momentum** theorem. It states that **an impulse**

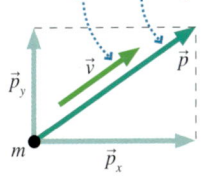

Momentum is a vector that points in the same direction as the object's velocity.

FIGURE 9.7 A particle's momentum vector $\vec{p}$ can be decomposed into x- and y-components.

TABLE 9.1 Some typical momenta (approximate)

Object	Mass (kg)	Speed (m/s)	Momentum (kg · m/s)
Falling raindrop	2×10^{-5}	5	10^{-4}
Bullet	0.004	500	2
Pitched baseball	0.15	40	6
Running person	70	3.0	200
Car on highway	1000	30	3×10^4

Legging it BIO A frog making a jump wants to gain as much momentum as possible before leaving the ground. This means that he wants the greatest impulse $J = F_{avg}\,\Delta t$ delivered to him by the ground. There is a maximum force that muscles can exert, limiting F_{avg}. But the time interval Δt over which the force is exerted can be greatly increased by having long legs. Many animals that are good jumpers have particularly long legs.

delivered to an object causes the object's momentum to change. That is, the *effect* of an impulsive force is to change the object's momentum from $\vec{p}_i$ to

$$\vec{p}_f = \vec{p}_i + \vec{J} \tag{9.9}$$

Equation 9.8 can also be written in terms of its x- and y-components as

$$J_x = \Delta p_x = (p_x)_f - (p_x)_i = m(v_x)_f - m(v_x)_i$$
$$J_y = \Delta p_y = (p_y)_f - (p_y)_i = m(v_y)_f - m(v_y)_i \tag{9.10}$$

The impulse-momentum theorem is illustrated by two examples in Figure 9.8. In the first, the putter strikes the ball, exerting a force on it and delivering an impulse $\vec{J} = \vec{F}_{avg}\,\Delta t$. Notice that the direction of the impulse is the same as that of the force. Because $\vec{p}_i = \vec{0}$ in this situation, we can use the impulse-momentum theorem to find that the ball leaves the putter with momentum $\vec{p}_f = \vec{p}_i + \vec{J} = \vec{J}$.

NOTE ▶ You can think of the putter as changing the ball's momentum by transferring momentum to it as an impulse. Thus we say the putter *delivers* an impulse to the ball, and the ball *receives* an impulse from the putter. ◀

The soccer player in Figure 9.8b presents a more complicated case. Here, the initial momentum of the ball is directed downward to the left. The impulse exerted on it by the player's head, upward to the right, is strong enough to reverse the ball's motion and send it off in a new direction. The graphical addition of vectors in Figure 9.8b again shows that $\vec{p}_f = \vec{p}_i + \vec{J}$.

(a)

(b)

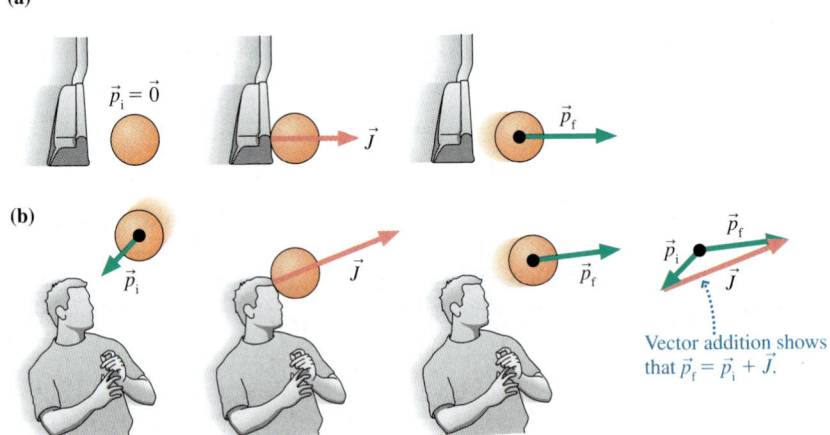

FIGURE 9.8 Impulse causes a *change* in momentum.

EXAMPLE 9.2 Calculating the change in momentum

A ball of mass $m = 0.25$ kg rolling to the right at 1.3 m/s strikes a wall and rebounds to the left at 1.1 m/s. What is the change in the ball's momentum? What is the impulse delivered to it by the wall?

PREPARE A visual overview of the ball bouncing is shown in Figure 9.9. This is a new kind of visual overview, one in which we show the situation "before" and "after" the interaction. We'll have more to say about before-and-after pictures in the next section. The ball is moving along the x-axis, so we'll write the momentum in component form, as in Equation 9.7. The change in momentum is then the difference between the final and initial values of the momentum. By the impulse-momentum theorem, the impulse is equal to this change in momentum.

FIGURE 9.9 Visual overview for a ball bouncing off a wall.

SOLVE The x-component of the initial momentum is

$$(p_x)_i = m(v_x)_i = (0.25 \text{ kg})(1.3 \text{ m/s}) = 0.33 \text{ kg} \cdot \text{m/s}$$

The y-component of the momentum is zero both before and after the bounce. After the ball rebounds, the x-component of its momentum is

$$(p_x)_f = m(v_x)_f = (0.25 \text{ kg})(-1.1 \text{ m/s}) = -0.28 \text{ kg} \cdot \text{m/s}$$

It is particularly important to notice that the x-component of the momentum, like that of the velocity, is negative. This indicates that the ball is moving to the *left*. The change in momentum is

$$\Delta p_x = (p_x)_f - (p_x)_i = (-0.28 \text{ kg} \cdot \text{m/s}) - (0.33 \text{ kg} \cdot \text{m/s})$$
$$= -0.61 \text{ kg} \cdot \text{m/s}$$

The change in the momentum is negative. By the impulse-momentum theorem, the impulse delivered to the ball by the wall is equal to this change, so

$$J_x = \Delta p_x = -0.61 \text{ kg} \cdot \text{m/s}$$

ASSESS The impulse is negative, indicating that the force causing the impulse is pointing to the left, which makes sense.

An interesting application of the impulse-momentum theorem is to the question of how to slow down a fast-moving object in the gentlest possible way. For instance, a car is headed for a collision with a bridge abutment. How can this crash be made survivable? How do the rams in the chapter-opening photo avoid injury when they collide?

In these examples, the object has momentum $\vec{p}_i$ just before impact and zero momentum after (i.e., $\vec{p}_f = \vec{0}$). The impulse-momentum theorem tells us that

$$\vec{J} = \vec{F}_{avg} \, \Delta t = \Delta \vec{p} = \vec{p}_f - \vec{p}_i = -\vec{p}_i$$

or

$$\vec{F}_{avg} = -\frac{\vec{p}_i}{\Delta t} \qquad (9.11)$$

INVERSE
p. 118

That is, the average force needed to stop an object is *inversely proportional* to the duration Δt of the collision. **If the duration of the collision can be increased, the force of the impact will be decreased.** This is the principle used in most impact-lessening techniques.

For example, obstacles such as bridge abutments are made safer by placing a line of water-filled barrels in front of them. Water is heavy but deformable. In case of a collision, the time it takes for the car to plow through these barrels is much longer than the time it would take it to stop if it hit the abutment head on. The force on the car (and on the driver from his or her seat belt) is greatly reduced by the longer-duration collision with the barrels.

The spines of a hedgehog obviously help protect it from predators. But they serve another function as well. If a hedgehog falls from a tree—a not uncommon occurrence—it simply rolls itself into a ball before it lands. Indeed, hedgehogs have been observed to purposely descend to the ground by simply dropping from the tree. Its thick spines then cushion the blow by increasing the time it takes for the animal to come to rest. Along with its small size, this adaptation allows for the hedgehog to easily survive long falls unhurt.

The butting rams shown in the photo at the beginning of this chapter also have adaptations that allow them to collide at high speeds without injury to their brains. The cranium has a double wall to prevent skull injuries, and there is a thick spongy mass that increases the time it takes for the brain to come to rest upon impact, again reducing the magnitude of the force on the brain.

TRY IT YOURSELF

Water balloon catch If you've ever tried to catch a water balloon, you may have learned the hard way not to catch it with your arms rigidly extended. The brief collision time implies a large, balloon-bursting force. A better way to catch a water balloon is to pull your arms in toward your body as you catch it, lengthening the collision time and hence reducing the force on the balloon.

A hedgehog is its own crash cushion!

(a)

(b)

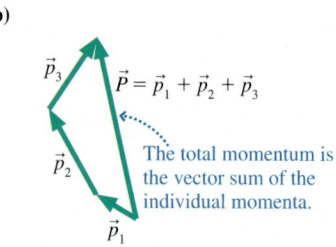

The total momentum is the vector sum of the individual momenta.

FIGURE 9.10 The total momentum of three pool balls.

Total Momentum

If we have more than one object moving—a *system* of particles—then the system as a whole has an overall momentum. The **total momentum** $\vec{P}$ (note the capital P) of a system of particles is the vector sum of the momenta of each of the individual particles:

$$\vec{P} = \vec{p}_1 + \vec{p}_2 + \vec{p}_3 + \cdots$$

Figure 9.10 shows how the momentum vectors of three moving pool balls are graphically added to find the total momentum. The concept of total momentum will be of key importance when we discuss the conservation law for momentum in Section 9.4.

STOP TO THINK 9.1   The cart's change of momentum is

A. -30 kg · m/s.
B. -20 kg · m/s.
C. -10 kg · m/s.
D. 10 kg · m/s.
E. 20 kg · m/s.
F. 30 kg · m/s.

9.3 Solving Impulse and Momentum Problems

Visual overviews have become an important problem-solving tool. The visual overviews and free-body diagrams that you learned to draw in Chapters 1–8 were oriented toward the use of Newton's laws and a subsequent kinematical analysis. Now we are interested in making a connection between "before" and "after."

(MP)   TACTICS BOX 9.1   Drawing a before-and-after visual overview   ✐ Exercises 8–11

❶ **Sketch the situation.** Use two drawings, labeled "Before" and "After," to show the objects *immediately before* they interact and again *immediately after* they interact.

❷ **Establish a coordinate system.** Select your axes to match the motion.

❸ **Define symbols.** Define symbols for the masses and for the velocities before and after the interaction. Position and time are not needed.

❹ **List known information.** Give the values of quantities known from the problem statement or that can be found quickly with simple geometry or unit conversions. Before-and-after pictures are usually simpler than the pictures you used for dynamics problems, so listing known information on the sketch is often adequate.

❺ **Identify the desired unknowns.** What quantity or quantities will allow you to answer the question? These should have been defined as symbols in step 3.

EXAMPLE 9.3 Force in hitting a baseball

A 150 g baseball is thrown with a speed of 20 m/s. It is hit straight back toward the pitcher at a speed of 40 m/s. The impulsive force of the bat on the ball has the shape shown in Figure 9.11. What is the *maximum* force F_{max} that the bat exerts on the ball? What is the *average* force that the bat exerts on the ball?

FIGURE 9.11 The interaction force between the baseball and the bat.

PREPARE Model the interaction as a collision. Figure 9.12 is a before-and-after visual overview in which the steps from Tactics Box 9.1 are explicitly noted. Because F_x is positive (a force to the right), we know the ball was initially moving

toward the left and is hit back toward the right. Thus we converted the statements about *speeds* into information about *velocities*, with $(v_x)_i$ negative.

SOLVE In the last several chapters we've started the mathematical solution with Newton's second law. Now we want to use the impulse-momentum theorem:

$$\Delta p_x = J_x = \text{area under the force curve}$$

We know the velocities before and after the collision, so we can find the change in the ball's momentum:

$$\Delta p_x = m(v_x)_f - m(v_x)_i = (0.15 \text{ kg})(40 \text{ m/s} - (-20 \text{ m/s}))$$
$$= 9.0 \text{ kg} \cdot \text{m/s}$$

The force curve is a triangle with height F_{max} and width 0.60 ms. As in Example 9.1, the area under the curve is

$$J_x = \text{area} = \tfrac{1}{2} \times F_{max} \times (6.0 \times 10^{-4} \text{ s})$$
$$= (F_{max})(3.0 \times 10^{-4})$$

According to the impulse-momentum theorem, $\Delta p_x = J_x$, so we have

$$9.0 \text{ kg} \cdot \text{m/s} = (F_{max})(3.0 \times 10^{-4} \text{ s})$$

Thus the *maximum* force is

$$F_{max} = \frac{9.0 \text{ kg} \cdot \text{m/s}}{3.0 \times 10^{-4} \text{ s}} = 30,000 \text{ N}$$

The *average* force, which depends on the collision duration $\Delta t = 6.0 \times 10^{-4}$ s, has the smaller value

$$F_{avg} = \frac{J_x}{\Delta t} = \frac{\Delta p_x}{\Delta t} = \frac{9.0 \text{ kg} \cdot \text{m/s}}{6.0 \times 10^{-4} \text{ s}} = 15,000 \text{ N}$$

ASSESS F_{max} is a large force, but quite typical of the impulsive forces during collisions. The main thing to focus on is our new perspective: An impulse changes the momentum of an object.

❶ Draw the before-and-after pictures.

❷ Establish a coordinate system.

❸ Define symbols.

❹ List known information.

❺ Identify desired unknowns.

FIGURE 9.12 A before-and-after visual overview.

The Impulse Approximation

Other forces often act on an object during a collision or other brief interaction. In Example 9.3, for instance, the baseball also has a weight force acting on it. Usually these other forces are *much* smaller than the interaction forces. The 1.5 N weight of the ball is vastly less than the 30,000 N force of the bat on the ball. We can reasonably neglect these small forces *during* the brief time of the impulsive force. Doing so is called the **impulse approximation.**

When we use the impulse approximation, $(p_x)_i$ and $(p_x)_f$—and $(v_x)_i$ and $(v_x)_f$—are then the momenta (and velocities) *immediately* before and *immediately* after the collision. For example, the velocities in Example 9.3 are those of the ball just before and after it collides with the bat. We could then do a follow-up problem, including weight and drag, to find the ball's speed a second later as the second baseman catches it.

EXAMPLE 9.4 Height of a bouncing ball

A 100 g rubber ball is thrown straight down onto a hard floor so that it strikes the floor with a speed of 11 m/s. Figure 9.13 shows the force that the floor exerts on the ball. Estimate the height of the ball's bounce.

FIGURE 9.13 The force of the floor on a bouncing rubber ball.

PREPARE The ball experiences an impulsive force while in contact with the floor. Using the impulse approximation, we'll neglect the ball's weight during these 5.0 ms. The ball's rise after the bounce is free-fall motion; that is, motion subject only to the force of gravity. We'll use free-fall kinematics to describe the motion after the bounce.

Figure 9.14 is a visual overview. Here we have a two-part problem, an impulsive collision followed by upward free fall.

FIGURE 9.14 Before-and-after visual overview for a bouncing ball.

The overview thus shows the ball just before the collision, where we label its velocity as v_{1y}; just after the collision, where its velocity is v_{2y}; and at the highest point of its rising free fall, where its velocity is $v_{3y} = 0$.

SOLVE The impulse-momentum theorem tells us that $J_y = \Delta p_y = p_{2y} - p_{1y}$, so that $p_{2y} = p_{1y} + J_y$. The initial momentum, just before the collision, is $p_{1y} = mv_{1y} = (0.10 \text{ kg})(-11 \text{ m/s}) = -1.1 \text{ kg} \cdot \text{m/s}$.

Next, we need to find the impulse J_y, which is the area under the curve in Figure 9.13. Because the force is just given as a smooth curve, we'll have to *estimate* this area. Recall that the area can be written as $F_{avg} \Delta t$. From the curve, we might estimate F_{avg} to be about 400 N, or half the maximum value of the force. With this estimate we have

$$J_y = \text{area under the force curve} \approx (400 \text{ N}) \times (0.0050 \text{ s})$$
$$= 2.0 \text{ N} \cdot \text{s} = 2.0 \text{ kg} \cdot \text{m/s}$$

Thus

$$p_{2y} = p_{1y} + J_y = (-1.1 \text{ kg} \cdot \text{m/s}) + 2.0 \text{ kg} \cdot \text{m/s}$$
$$= 0.9 \text{ kg} \cdot \text{m/s}$$

and the post-collision velocity is

$$v_{2y} = \frac{p_{2y}}{m} = \frac{0.9 \text{ kg} \cdot \text{m/s}}{0.10 \text{ kg}} = 9 \text{ m/s}$$

The rebound speed is less than the impact speed, as expected. Finally, we can use free-fall kinematics to find

$$v_{3y}^2 = 0 = v_{2y}^2 - 2g \Delta y = v_{2y}^2 - 2gy_3$$
$$y_3 = \frac{v_{2y}^2}{2g} = \frac{(9 \text{ m/s})^2}{2(9.8 \text{ m/s}^2)} = 4 \text{ m}$$

We estimate that the ball bounces to a height of 4 m.

ASSESS This is a reasonable height for a rubber ball thrown down quite hard.

STOP TO THINK 9.2   A 10 g rubber ball and a 10 g clay ball are each thrown at a wall with equal speeds. The rubber ball bounces, the clay ball sticks. Which ball receives the greater impulse from the wall?

A. The clay ball receives a larger impulse because it sticks.
B. The rubber ball receives a larger impulse because it bounces.
C. They receive equal impulses because they have equal momenta.
D. Neither receives an impulse because the wall doesn't move.

9.4 Conservation of Momentum

The impulse-momentum theorem was derived from Newton's second law and is really just an alternative way of looking at that law. It is used in the context of single-particle dynamics, much as we used Newton's law in Chapters 4–7.

This chapter opened by noting that very complex interactions, such as two train cars coupling together, sometimes have very simple outcomes. To predict the outcomes, we need to see how Newton's *third* law looks in the language of impulse and momentum. Newton's third law will lead us to one of the most important conservation laws in physics.

Figure 9.15 shows two balls with initial velocities $(v_{1x})_i$ and $(v_{2x})_i$. The balls collide, then bounce apart with final velocities $(v_{1x})_f$ and $(v_{2x})_f$. The forces during the collision, as the balls are interacting, are the action/reaction pair $\vec{F}_{1\,on\,2}$ and $\vec{F}_{2\,on\,1}$. For now, we'll continue to assume that the motion is one dimensional along the x-axis.

> **NOTE ▶** The notation, with all the subscripts, may seem excessive. But there are two balls, and each has an initial and a final velocity, so we need to distinguish between four different velocities. ◀

Figure 9.16 shows the x-components of the two forces $\vec{F}_{1\,on\,2}$ and $\vec{F}_{2\,on\,1}$. Because they make up an action/reaction pair, the two forces always are of equal magnitude but of opposite sign. The impulse J_{2x} delivered to ball 2 is the area under the $\vec{F}_{1\,on\,2}$ curve; this area and hence this impulse are positive, indicating that the impulse on ball 2 is directed to the right. Likewise, the impulse J_{1x} delivered to ball 1 is the area "under" the $\vec{F}_{2\,on\,1}$ curve—but this area is considered to be *negative* because it is below the axis. As a result, the two impulses J_{1x} and J_{2x} are also equal in magnitude but opposite in sign, so that $J_{1x} = -J_{2x}$. For the soccer player in Figure 9.8, for example, this means that the impulse his head delivers to the ball is equal in magnitude but oppositely directed to the impulse the ball delivers to his head.

According to the impulse-momentum theorem, the changes in momenta of the two balls are

$$\Delta p_{1x} = J_{1x} \quad \text{and} \quad \Delta p_{2x} = J_{2x}$$

But because $J_{1x} = -J_{2x}$, we have

$$\Delta p_{1x} = -\Delta p_{2x}$$

or

$$\Delta p_{1x} + \Delta p_{2x} = \Delta(p_{1x} + p_{2x}) = 0$$

If the change in the quantity $p_{1x} + p_{2x}$ is zero, then it must be the case that

$$p_{1x} + p_{2x} = \text{constant} \tag{9.12}$$

If $p_{1x} + p_{2x}$ is a constant, then the sum of the momenta *after* the collision equals the sum of the momenta *before* the collision. That is,

$$(p_{1x})_f + (p_{2x})_f = (p_{1x})_i + (p_{2x})_i \tag{9.13}$$

Furthermore, this equality is independent of the interaction force. We don't need to know anything about $\vec{F}_{1\,on\,2}$ and $\vec{F}_{2\,on\,1}$ to make use of Equation 9.13.

Section 9.2 defined the total momentum $\vec{P}$ of a *system* of particles to be the sum of the momenta of each particle. Because Equation 9.13 is a sum of momenta, we can write it as

$$(P_x)_f = (P_x)_i \tag{9.14}$$

Because it doesn't change during the collision, we say that the x-component of total momentum is *conserved*. Equations 9.13 and 9.14 are our first example of a *conservation law*.

Law of Conservation of Momentum

The same arguments just presented for the two colliding balls can be extended to systems containing any number of objects. Figure 9.17 on the next page

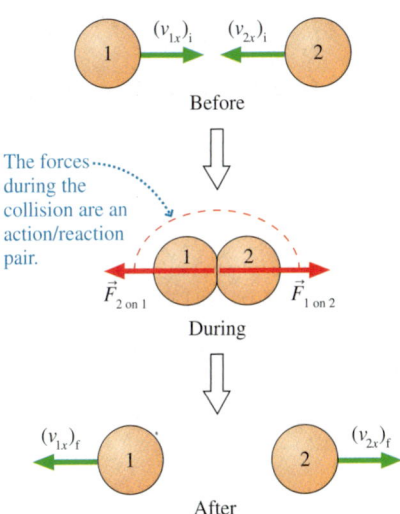

FIGURE 9.15 A collision between two balls.

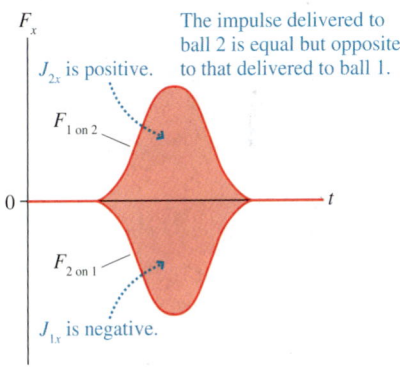

FIGURE 9.16 The impulsive forces acting on balls 1 and 2 during their collision.

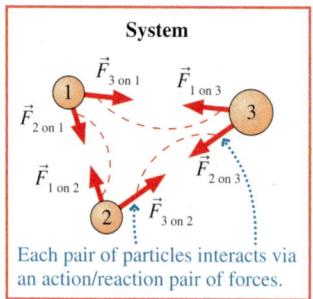

Each pair of particles interacts via an action/reaction pair of forces.

FIGURE 9.17 An isolated system of three particles.

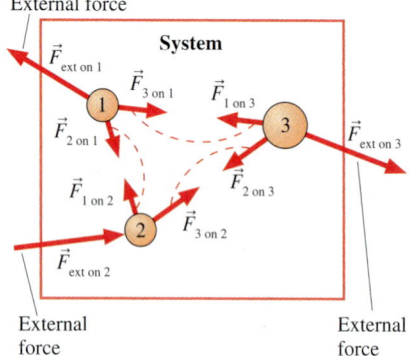

FIGURE 9.18 A system of particles subject to external forces.

shows the idea. Each pair of particles in the system interacts via forces that are an action/reaction pair. Exactly as for the two-particle collision, the change in momentum of particle 2 due to the force from particle 3 is exactly equal in magnitude, but opposite in direction, to the change in particle 3's momentum due to particle 2. The *net* change in the momentum of these two particles due to their interaction forces is thus zero. The same argument holds for every pair, with the net result that the change in the *total* momentum $\vec{P}$ of the system is zero. The total momentum of the system remains constant; that is, it is *conserved*.

Note in Figure 9.17 that the particles interact only with other particles that are part of the system. There are no forces on any of the particles from agents *outside* of the system. Forces that act only between particles within the system are called **internal forces**. A system subject only to internal forces is called an **isolated system** and, as we have just seen, **the total momentum of an isolated system is conserved.**

Most systems are not isolated but are subject to forces from agents outside the system. These forces are called **external forces.** For example, if you toss a ball (the system) into the air, an external force—its weight, the force of the earth's gravity on the ball—acts on the system. How do external forces affect the momentum of a system of particles?

In Figure 9.18 we show the same three-particle system of Figure 9.17, but now with *external* forces acting on the three particles. These external forces *can* change the momentum of the system. During a time interval Δt, for instance, the external force $\vec{F}_{\text{ext on 1}}$ acting on particle 1 changes its momentum, according to the impulse-momentum theorem, by $\Delta \vec{p}_1 = (\vec{F}_{\text{ext on 1}}) \Delta t$. The momenta of the other two particles change similarly. Thus the change in the total momentum is

$$
\begin{aligned}
\Delta \vec{P} &= \Delta \vec{p}_1 + \Delta \vec{p}_2 + \Delta \vec{p}_3 \\
&= (\vec{F}_{\text{ext on 1}} \Delta t) + (\vec{F}_{\text{ext on 2}} \Delta t) + (\vec{F}_{\text{ext on 3}} \Delta t) \\
&= (\vec{F}_{\text{ext on 1}} + \vec{F}_{\text{ext on 2}} + \vec{F}_{\text{ext on 3}}) \Delta t \\
&= \vec{F}_{\text{net}} \Delta t
\end{aligned}
\tag{9.15}
$$

where $\vec{F}_{\text{net}}$ is the net force due to *external forces*.

Equation 9.15 has two very important implications. First, it tells us that we can analyze the motion of the system as a whole without needing to consider the internal forces between the particles that make up the system. In fact, we have been using this idea all along as an *assumption* of the particle model. When we treat cars and rocks and baseballs as particles, we assume that the internal forces between the atoms—the forces that hold the object together—do not affect the motion of the object as a whole. Now we have *justified* that assumption.

The second implication of Equation (9.15), and the key one from the perspective of this chapter, applies to systems for which $\vec{F}_{\text{net}} = \vec{0}$. According to Equation 9.15, if $\vec{F}_{\text{net}} = \vec{0}$, **the *total* momentum $\vec{P}$ of the system does not change.** The total momentum remains constant, *regardless* of whatever interactions are going on *inside* the system.

Earlier, we found that the total momentum is also conserved when the system is isolated, with no external forces acting on it. With no external forces, it is certainly the case that $\vec{F}_{\text{net}} = \vec{0}$. Thus we can extend our definition of an isolated system: **An isolated system is one for which $\vec{F}_{\text{net}} = \vec{0}$.**

The importance of these results is sufficient to elevate them to a law of nature, alongside Newton's laws.

Law of conservation of momentum The total momentum $\vec{P}$ of an isolated system is a constant. Interactions within the system do not change the system's total momentum.

NOTE ▶ It is worth emphasizing the critical role of Newton's third law in the derivation of Equation 9.15. The law of conservation of momentum is a direct consequence of the fact that interactions within an isolated system are action/reaction pairs. ◀

Mathematically, the law of conservation of momentum for an isolated system is

$$\vec{P}_f = \vec{P}_i \qquad (9.16)$$

Law of conservation of momentum for an isolated system

The total momentum after an interaction is equal to the total momentum before the interaction. Because Equation 9.16 is a vector equation, the equality is true for each of the components of the momentum vector. That is,

Final momentum ⎧ Initial momentum

x-component ⋯⋯▶ $(p_{1x})_f + (p_{2x})_f + (p_{3x})_f + \cdots = (p_{1x})_i + (p_{2x})_i + (p_{3x})_i + \cdots$

Particle 1 Particle 2 Particle 3

$$(9.17)$$

y-component ⋯⋯▶ $(p_{1y})_f + (p_{2y})_f + (p_{3y})_f + \cdots = (p_{1y})_i + (p_{2y})_i + (p_{3y})_i + \cdots$

EXAMPLE 9.5 Speed of ice skaters pushing off

Two ice skaters, Sandra and David, stand facing each other on frictionless ice. Sandra has a mass of 45 kg, David a mass of 80 kg. They then push off from each other. After the push, Sandra moves off at a speed of 2.2 m/s. What is David's speed?

PREPARE The two skaters interact with each other, but they form an isolated system because, for each skater, the upward normal force of the ice balances their downward weight force to make $\vec{F}_{net} = \vec{0}$. Thus the total momentum of the system of the two skaters is conserved.

Figure 9.19 shows a before-and-after visual overview for the two skaters. The total momentum before they push off is $\vec{P}_i = \vec{0}$

Before:

$(v_{Dx})_i = 0$ m/s
$m_D = 80$ kg

$(v_{Sx})_i = 0$ m/s
$m_S = 45$ kg

After:

Find: $(v_{Dx})_f$

$(v_{Sx})_f = 2.2$ m/s

FIGURE 9.19 Before-and-after visual overview for two skaters pushing off from each other.

because both skaters are at rest. Consequently, the total momentum will still be $\vec{0}$ *after* they push off.

SOLVE Since the motion is only in the x-direction, we'll only need to consider x-components of momentum. We write Sandra's initial momentum as $(p_{Sx})_i = m_S(v_{Sx})_i$, where m_S is her mass and $(v_{Sx})_i$ her initial velocity. Similarly, we write David's initial momentum as $(p_{Dx})_i = m_D(v_{Dx})_i$. Both these momenta are zero because both skaters are initially at rest.

We can now apply the mathematical statement of momentum conservation, Equation 9.17. Writing the final momentum of Sandra as $m_S(v_{Sx})_f$, and that of David as $m_D(v_{Dx})_f$, we have

$$m_S(v_{Sx})_f + m_D(v_{Dx})_f = m_S(v_{Sx})_i + m_D(v_{Dx})_i = 0$$

The skaters' final momentum . . .   . . . equals their initial momentum . . .   . . . which was zero.

Solving for $(v_{Dx})_f$, we find

$$(v_{Dx})_f = -\frac{m_S}{m_D}(v_{Sx})_f = -\frac{45 \text{ kg}}{80 \text{ kg}} \times 2.2 \text{ m/s} = -1.2 \text{ m/s}$$

David moves backward with a *speed* of 1.2 m/s.

ASSESS The *total* momentum of the system is zero both before and after they push off, but the individual momenta are not. Because $(p_{Sx})_f$ is positive (Sandra moves to the right), $(p_{Dx})_f$ must have the same magnitude but the opposite sign (David moves to the left).

Notice that we didn't need to know any details about the force between David and Sandra in order to find David's final speed. Conservation of momentum *mandates* this result.

A Strategy for Conservation of Momentum Problems

6.3, 6.4, 6.6, 6.7, 6.10

Our derivation of the law of conservation of momentum, and the conditions under which it holds, suggests a problem-solving strategy.

 PROBLEM-SOLVING STRATEGY 9.1   Conservation of momentum problems

PREPARE Clearly define *the system*.

- If possible, choose a system that is isolated ($\vec{F}_{net} = \vec{0}$) or within which the interactions are sufficiently short and intense that you can ignore external forces for the duration of the interaction (the impulse approximation). Momentum is then conserved.
- If it's not possible to choose an isolated system, try to divide the problem into parts such that momentum is conserved during one segment of the motion. Other segments of the motion can be analyzed using Newton's laws or, as you'll learn in Chapter 10, conservation of energy.

Following Tactics Box 9.1, draw a before-and-after visual overview. Define symbols that will be used in the problem, list known values, and identify what you're trying to find.

SOLVE The mathematical representation is based on the law of conservation of momentum: $\vec{P}_f = \vec{P}_i$. In component form, this is

$$(p_{1x})_f + (p_{2x})_f + (p_{3x})_f + \cdots = (p_{1x})_i + (p_{2x})_i + (p_{3x})_i + \cdots$$
$$(p_{1y})_f + (p_{2y})_f + (p_{3y})_f + \cdots = (p_{1y})_i + (p_{2y})_i + (p_{3y})_i + \cdots$$

ASSESS Check that your result has the correct units, is reasonable, and answers the question.

EXAMPLE 9.6 Getaway speed of a cart

Bob is running from the police and thinks he can make a faster getaway by jumping on a stationary cart in front of him. He runs toward the cart, jumps on, and rolls along the horizontal street. Bob has a mass of 75 kg and the cart's mass is 25 kg. If Bob's speed is 4.0 m/s when he jumps onto the cart, what is the cart's speed after Bob jumps on?

PREPARE When Bob lands on and sticks to the cart, a "collision" occurs between Bob and the cart. If we take Bob and the cart together to be the system, the forces involved in this collision—friction forces between Bob's feet and the cart—are internal forces. The net external force on the system is zero, so the total momentum of Bob + cart is conserved: It is the same before and after the collision.

Before: After:

$m_B = 75$ kg
$(v_{Bx})_i = 4.0$ m/s

$(v_{Cx})_i = 0$ m/s

Find: $(v_x)_f$

$(v_{Bx})_f = (v_{Cx})_f = (v_x)_f$

$m_C = 25$ kg

System

FIGURE 9.20 Before-and-after visual overview of Bob and the cart.

The visual overview in Figure 9.20 shows the important point that Bob and the cart move together after he lands on the cart, so $(v_x)_f$ is their common final velocity.

SOLVE To solve for the final velocity of Bob and the cart, we'll use conservation of momentum: $(P_x)_f = (P_x)_i$. Written in terms of the individual momenta, we have

$$(P_x)_i = m_B(v_{Bx})_i + m_C\underbrace{(v_{Cx})_i}_{0 \text{ m/s}} = m_B(v_{Bx})_i$$
$$(P_x)_f = m_B(v_x)_f + m_C(v_x)_f = (m_B + m_C)(v_x)_f$$

In the second equation, we've used the fact that both Bob and the cart travel at the common velocity of $(v_x)_f$. Equating the final and initial total momenta gives

$$(m_B + m_C)(v_x)_f = m_B(v_{Bx})_i$$

Solving this for $(v_x)_f$, we find

$$(v_x)_f = \frac{m_B}{m_B + m_C}(v_{Bx})_i = \frac{75 \text{ kg}}{100 \text{ kg}} \times 4.0 \text{ m/s} = 3.0 \text{ m/s}$$

The cart's speed is 3.0 m/s immediately after Bob jumps on.

ASSESS Bob has *lost* speed because he had to share his initial momentum with the cart. Not a good way to make a getaway!

Notice how easy this was! No forces, no acceleration constraints, no simultaneous equations. Why didn't we think of this before? While conservation laws are indeed powerful, they can only answer certain questions. Had we wanted to know how far Bob slid across the cart before sticking to it, how long the slide took, or what the cart's acceleration was during the collision, we would not have been able to answer such questions on the basis of the conservation law. There is a price to pay for finding a simple connection between before and after, and that price is the loss of information about the details of the interaction. If we are satisfied with knowing only about before and after, then conservation laws are a simple and straightforward way to proceed. But many problems *do* require us to understand the interaction, and for these there is no avoiding Newton's laws and all they entail.

It Depends on the System

The first step in the problem-solving strategy asks you to clearly define *the system*. This is worth emphasizing, because many problem-solving errors arise from trying to apply momentum conservation to an inappropriate system. **The goal is to choose a system whose momentum will be conserved.** Even then, it is the *total* momentum of the system that is conserved, not the momenta of the individual particles within the system.

As an example, consider what happens if you drop a rubber ball and let it bounce off a hard floor. Is momentum conserved during the collision of the ball with the floor? You might be tempted to answer yes because the ball's rebound speed is very nearly equal to its impact speed. But there are two errors in this reasoning.

First, momentum depends on *velocity,* not speed. The ball's velocity and momentum just before the collision are negative. They are positive after the collision. Even if their magnitudes are equal, the ball's momentum after the collision is *not* equal to its momentum before the collision.

But more importantly, we haven't defined the system. The momentum of what? Whether or not momentum is conserved depends on the system. Figure 9.21 shows two different choices of systems. In Figure 9.21a, where the ball itself is chosen as the system, the gravitational force of the earth on the ball is an external force. This force causes the ball to accelerate toward the earth, changing the ball's momentum. The force of the floor on the ball is also an external force. The impulse of $\vec{F}_{\text{floor on ball}}$ changes the ball's momentum from "down" to "up" as the ball bounces. The momentum of this system is most definitely *not* conserved.

Figure 9.21b shows a different choice. Here the system is ball + earth. Now the gravitational forces and the impulsive forces of the collision are interactions *within* the system. This is an isolated system, so the *total* momentum $\vec{P} = \vec{p}_{\text{ball}} + \vec{p}_{\text{earth}}$ is conserved.

In fact, the total momentum is $\vec{P} = \vec{0}$. Before you release the ball, both the ball and the earth are at rest (in the earth's reference frame). The total momentum is zero before you release the ball, so it will *always* be zero. Consider the situation just before the ball hits the floor. If the ball's velocity is v_{By}, it must be the case that

$$m_B v_{By} + m_E v_{Ey} = 0$$

and thus

$$v_{Ey} = -\frac{m_B}{m_E} v_{By}$$

In other words, as the ball is pulled down toward the earth, the ball pulls up on the earth (action/reaction pair of forces) until the entire earth reaches velocity v_{Ey}. The earth's momentum is equal and opposite to the ball's momentum.

(a)
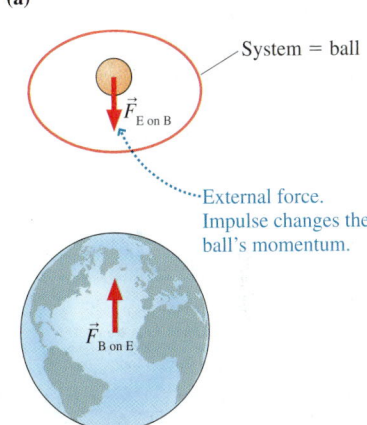
System = ball

External force. Impulse changes the ball's momentum.

(b)
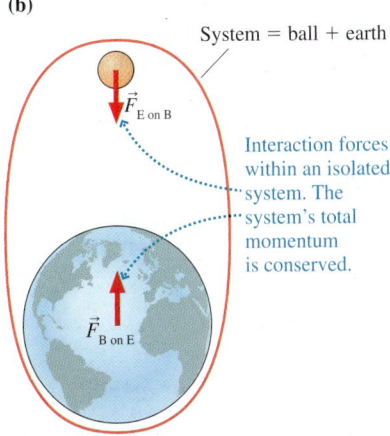
System = ball + earth

Interaction forces within an isolated system. The system's total momentum is conserved.

FIGURE 9.21 Whether or not momentum is conserved as a ball falls to earth depends on your choice of the system.

Why don't we notice the earth "leaping up" toward us each time we drop something? Because of the earth's enormous mass relative to everyday objects. A typical rubber ball has a mass of 60 g and hits the ground with a velocity of about −5 m/s. The earth's upward velocity is thus

$$v_{Ey} \approx -\frac{6 \times 10^{-2}\ \text{kg}}{6 \times 10^{24}\ \text{kg}}(-5\ \text{m/s}) = 5 \times 10^{-26}\ \text{m/s}$$

The earth does, indeed, have a momentum equal and opposite to that of the ball, but the earth is so massive that it needs only an infinitesimal velocity to match the ball's momentum. At this speed, it would take the earth 300 million years to move the diameter of an atom!

Explosions

An **explosion,** where the particles of the system move apart after a brief, intense interaction, is the opposite of a collision. The explosive forces, which could be from an expanding spring or from expanding hot gases, are *internal* forces. If the system is isolated, its total momentum during the explosion will be conserved.

EXAMPLE 9.7 **Recoil speed of a rifle**

A 30 g ball is fired from a 1.2 kg spring-loaded toy rifle with a speed of 15 m/s. What is the recoil speed of the rifle?

PREPARE As the ball moves down the barrel, there are complicated forces exerted on the ball and on the rifle. However, if we take the system to be the ball + rifle, these are *internal* forces that do not change the total momentum.

The *external* forces of the rifle's and ball's weights are balanced by the external force exerted by the person holding the

FIGURE 9.22 Before-and-after visual overview for a toy rifle.

rifle, so $\vec{F}_{net} = \vec{0}$. This is an isolated system and the law of conservation of momentum applies.

Figure 9.22 shows a visual overview before and after the ball is fired. We'll assume the ball is fired in the $+x$-direction.

SOLVE The x-component of the total momentum is $P_x = p_{Bx} + p_{Rx}$. Everything is at rest before the trigger is pulled, so the initial momentum is zero. After the trigger is pulled, the internal force of the spring pushes the ball down the barrel *and* pushes the rifle backwards. Conservation of momentum gives

$$(P_x)_f = m_B(v_{Bx})_f + m_R(v_{Rx})_f = (P_x)_i = 0$$

Solving for the rifle's velocity, we find

$$(v_{Rx})_f = -\frac{m_B}{m_R}(v_{Bx})_f = -\frac{0.030\ \text{kg}}{1.2\ \text{kg}} \times 15\ \text{m/s} = -0.38\ \text{m/s}$$

The minus sign indicates that the rifle's recoil is to the left. The recoil *speed* is 0.38 m/s.

ASSESS Real rifles fire their bullets at much higher velocities, and their recoil is correspondingly higher. Shooters need to brace themselves against the "kick" of the rifle back against their shoulder.

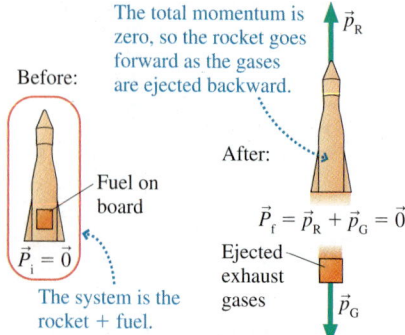

FIGURE 9.23 Rocket propulsion is an example of conservation of momentum.

We would not know where to begin to solve a problem such as this using Newton's laws. But Example 9.7 is a simple problem when approached from the before-and-after perspective of a conservation law. The selection of ball + rifle as "the system" was the critical step. For momentum conservation to be a useful principle, we had to select a system in which the complicated forces due to the spring and to friction were all internal forces. The rifle by itself is *not* an isolated system, so its momentum is *not* conserved.

Much the same reasoning explains how a rocket or jet aircraft accelerates. Figure 9.23 shows a rocket with a parcel of fuel on board. Burning converts the fuel to hot gases that are expelled from the rocket motor. If we choose rocket + gases to be the system, the burning and expulsion are all internal forces. In deep space there are no other forces, so the total momentum of the rocket + gases sys-

▶ **Squid propulsion** BIO Squids use a form of "rocket propulsion" to make quick movements to escape enemies or catch prey. The squid draws in water through a pair of valves in its outer sheath, or mantle, and then quickly expels the water through a funnel, propelling the squid backward. The funnel's direction is adjustable, allowing the squid to move in any backward direction.

tem must be conserved. The rocket gains forward velocity and momentum as the exhaust gases are shot out the back, but the *total* momentum of the system remains zero.

Many people find it hard to understand how a rocket can accelerate in the vacuum of space because there is nothing to "push against." Thinking in terms of momentum, you can see that the rocket does not push against anything *external,* but only against the gases that it pushes out the back. In return, in accordance with Newton's third law, the gases push forward on the rocket. The details of rocket propulsion are more complex than we want to handle, because the mass of the rocket is changing, but you should be able to use the law of conservation of momentum to understand the basic principle by which rocket propulsion occurs.

STOP TO THINK 9.3   An explosion in a rigid pipe shoots out three pieces. A 6 g piece comes out the right end. A 4 g piece comes out the left end with twice the speed of the 6 g piece. From which end, left or right, does the third piece emerge?

9.5 Inelastic Collisions

Collisions can have different possible outcomes. A rubber ball dropped on the floor bounces—it's *elastic*—but a ball of clay sticks to the floor without bouncing; we call such a collision *inelastic*. A golf club hitting a golf ball causes the ball to rebound away from the club (elastic), but a bullet striking a block of wood embeds itself in the block (inelastic).

A collision in which the two objects stick together and move with a common final velocity is called a **perfectly inelastic collision.** The clay hitting the floor and the bullet embedding itself in the wood are examples of perfectly inelastic collisions. Other examples include railroad cars coupling together upon impact and darts hitting a dart board. Figure 9.24 emphasizes the fact that the two objects have a common final velocity after they collide. (We have drawn the combined object moving to the right, but it could have ended up moving to the left, depending on the objects' masses and initial velocities.)

In an *elastic collision,* by contrast, the two objects bounce apart. We've looked at some examples of elastic collisions, but a full analysis requires some ideas about energy. We will return to elastic collisions in Chapter 10.

Two objects approach and collide.

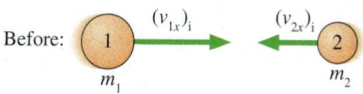

They stick and move together.

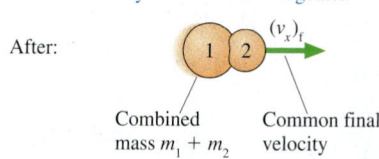

Combined mass $m_1 + m_2$ Common final velocity

FIGURE 9.24 A perfectly inelastic collision.

EXAMPLE 9.8 Speeds in a perfectly inelastic glider collision

In a laboratory experiment, a 200 g air-track glider and a 400 g air-track glider are pushed toward each other from opposite ends of the track. The gliders have Velcro tabs on the front so that they will stick together when they collide. The 200 g glider is pushed with an initial speed of 3.0 m/s. The collision causes it to reverse direction at 0.50 m/s. What was the initial speed of the 400 g glider?

Continued

PREPARE Model the gliders as particles. Define the two gliders as the system. This is an isolated system, so its total momentum is conserved in the collision. The gliders stick together, so this is a perfectly inelastic collision.

Figure 9.25 shows a visual overview. We've chosen to let the 200 g glider (glider 1) start out moving to the right, so $(v_{1x})_i$ is a positive 3.0 m/s. The gliders move to the left after the colli-

Before:

$m_1 = 200$ g $(v_{1x})_i = 3.0$ m/s $(v_{2x})_i$ $m_2 = 400$ g

After:

$(v_x)_f = -0.50$ m/s $m_1 + m_2$

1   2

Find: $(v_{2x})_i$

FIGURE 9.25 Before-and-after visual overview for two gliders colliding on an air track.

sion, so their common final velocity is $(v_x)_f = -0.50$ m/s. You can see that velocity $(v_{2x})_i$ must be negative in order to "turn around" both gliders.

SOLVE The law of conservation of momentum, $(P_x)_f = (P_x)_i$, is

$$(m_1 + m_2)(v_x)_f = m_1(v_{1x})_i + m_2(v_{2x})_i$$

where we made use of the fact that the combined mass $m_1 + m_2$ moves together after the collision. We can easily solve for the initial velocity of the 400 g glider:

$$(v_{2x})_i = \frac{(m_1 + m_2)(v_x)_f - m_1(v_{1x})_i}{m_2}$$

$$= \frac{(0.60 \text{ kg})(-0.50 \text{ m/s}) - (0.20 \text{ kg})(3.0 \text{ m/s})}{0.40 \text{ kg}}$$

$$= -2.3 \text{ m/s}$$

The negative sign, which we anticipated, indicates that the 400 g glider started out moving to the left. The initial *speed* of the glider, which we were asked to find, is 2.3 m/s.

ASSESS The key step in solving inelastic collision problems is that both objects move after the collision with the same velocity. You should thus choose a single symbol (here, $(v_x)_f$) for this common velocity.

STOP TO THINK 9.4 The two particles shown collide and stick together. After the collision, the combined particles

2.5 m/s   1.5 m/s
1   2   1 2
0.40 kg   Before   0.80 kg   After

A. Move to the right as shown. B. Move to the left. C. Are at rest.

9.6 Momentum and Collisions in Two Dimensions

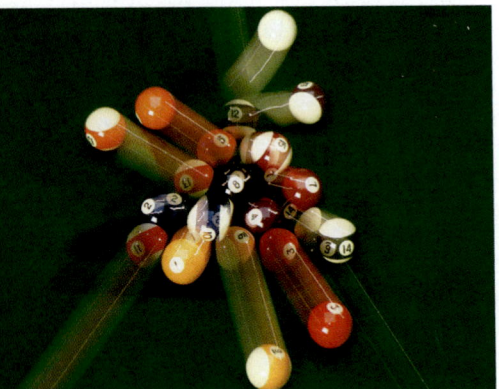

Collisions and explosions often involve motion in two dimensions.

Our examples thus far have been confined to motion along a one-dimensional axis. Many practical examples of momentum conservation involve motion in a plane. The total momentum $\vec{P}$ is a *vector* sum of the momenta $\vec{p} = m\vec{v}$ of the individual particles. Consequently, as we found in Section 9.3, momentum is conserved only if each component of $\vec{P}$ is conserved:

$$(p_{1x})_f + (p_{2x})_f + (p_{3x})_f + \cdots = (p_{1x})_i + (p_{2x})_i + (p_{3x})_i + \cdots$$
$$(p_{1y})_f + (p_{2y})_f + (p_{3y})_f + \cdots = (p_{1y})_i + (p_{2y})_i + (p_{3y})_i + \cdots$$
(9.18)

In this section we'll apply momentum conservation to motion in two dimensions.

EXAMPLE 9.9 **Analyzing a peregrine falcon strike**
Peregrine falcons often grab their prey from above while both falcon and prey are in flight. A falcon, flying at 18 m/s, swoops down at a 45° angle from behind a pigeon flying horizontally at 9.0 m/s. The falcon has a mass of 0.80 kg and the pigeon a mass of 0.36 kg. What are the speed and direction of the falcon (now holding the pigeon) immediately after impact?

PREPARE This is a perfectly inelastic collision because after the collision the falcon and pigeon move at a common velocity. The total momentum of the falcon + pigeon system is conserved. For a two-dimensional collision, this means that the x-component of the total momentum before the collision must equal the x-component of the total momentum after the collision, and similarly for the y-components. Figure 9.26 is a before-and-after visual overview.

FIGURE 9.26 Before-and-after visual overview for a falcon catching a pigeon.

SOLVE We'll start by finding the x- and y-components of the momentum before the collision. For the x-component we have

The x-component of the initial momentum . . .   (Both velocity components are negative since they point to the left.)

$$(P_x)_i = m_F(v_{Fx})_i + m_P(v_{Px})_i = m_F(-v_F \cos\theta) + m_P(-v_P)$$

. . . equals the x-component of the initial momentum of the falcon . . .   . . . plus the x-component of the initial momentum of the pigeon.

$$= (0.80 \text{ kg})(-18 \text{ m/s})(\cos 45°) + (0.36 \text{ kg})(-9.0 \text{ m/s})$$
$$= -13.4 \text{ kg} \cdot \text{m/s}$$

Similarly, for the y-component of the initial momentum we have

$$(P_y)_i = m_F(v_{Fy})_i + m_P(v_{Py})_i = m_F(-v_F \sin\theta) + 0$$
$$= (0.80 \text{ kg})(-18.0 \text{ m/s})(\sin 45°) = -10.2 \text{ kg} \cdot \text{m/s}$$

After the collision, the two birds move with a common velocity $\vec{v}$ that is directed at an angle α from the horizontal. The x-component of the final momentum is then

$$(P_x)_f = (m_F + m_P)(v_x)_f$$

Momentum conservation requires $(P_x)_f = (P_x)_i$, so

$$(v_x)_f = \frac{(P_x)_i}{(m_F + m_P)} = \frac{-13.4 \text{ kg} \cdot \text{m/s}}{(0.80 \text{ kg}) + (0.36 \text{ kg})} = -11.6 \text{ m/s}$$

Similarly, $(P_y)_f = (P_y)_i$ gives

$$(v_y)_f = \frac{(P_y)_i}{(m_F + m_P)} = \frac{-10.2 \text{ kg} \cdot \text{m/s}}{(0.80 \text{ kg}) + (0.36 \text{ kg})} = -8.79 \text{ m/s}$$

From the figure we see that $\tan\alpha = (v_y)_f/(v_x)_f$, so that

$$\alpha = \tan^{-1}\left(\frac{(v_y)_f}{(v_x)_f}\right) = \tan^{-1}\left(\frac{-8.79 \text{ m/s}}{-11.6 \text{ m/s}}\right) = 37°$$

The magnitude of the final velocity (i.e., the speed) can be found from the Pythagorean theorem as

$$v = \sqrt{(v_x)_f^2 + (v_y)_f^2}$$
$$= \sqrt{(-11.6 \text{ m/s})^2 + (-8.79 \text{ m/s})^2} = 15 \text{ m/s}$$

Thus immediately after impact the falcon, with its meal, is moving 37° below horizontal at a speed of 15 m/s.

It's instructive to examine this collision with a picture of the momentum vectors. The vectors $\vec{p}_F$ and $\vec{p}_P$ before the collision, and their sum $\vec{P} = \vec{p}_F + \vec{p}_P$, are shown in Figure 9.27. You can see that the total momentum vector makes a 53° angle with the negative y-axis. The individual momenta change in the collision, *but the total momentum does not.*

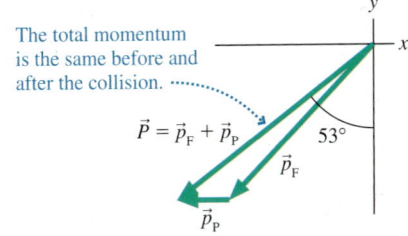

The total momentum is the same before and after the collision.

$\vec{P} = \vec{p}_F + \vec{p}_P$

FIGURE 9.27 The momentum vectors of the falcon strike.

9.7 Angular Momentum

For a single particle, we can think of the law of conservation of momentum as an alternative way of stating Newton's first law. Rather than saying that a particle will continue to move in a straight line at constant velocity unless acted on by a net force, we can say that the momentum of an isolated particle is conserved. Both express the idea that a particle moving in a straight line tends to "keep going" unless something acts on it to change its motion.

Another important motion you've studied is motion in a circle. The momentum $\vec{p}$ is *not* conserved for a particle undergoing circular motion. Momentum is a vector, and the momentum of a particle in circular motion changes as the direction of motion changes.

Nonetheless, a spinning bicycle wheel would keep turning if it were not for friction, and a ball moving in a circle at the end of a string tends to "keep going" in a circular path. The quantity that expresses this idea for circular motion is called *angular momentum.*

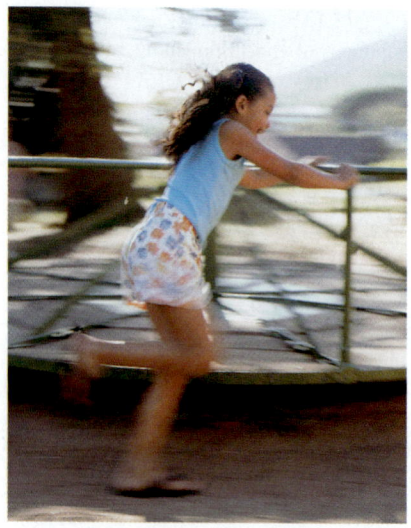

FIGURE 9.28 By applying a torque to the merry-go-round, the girl is increasing its angular momentum.

Let's start by looking at an example from everyday life: pushing a merry-go-round, as in Figure 9.28. If you push tangentially to the rim, you are applying a *torque* to the merry-go-round. As we learned in Chapter 7, the merry-go-round's angular speed will continue to increase for as long as you apply this torque. If you push *harder* (greater torque) or for a *longer time,* the greater the increase in its angular velocity will be. How can we quantify these observations?

Let's apply a constant torque τ_{net} to the merry-go-round for a time Δt. By how much will the merry-go-round's angular speed increase? In Section 7.3 we found that the angular acceleration α is given by the rotational equivalent of Newton's second law, or

$$\alpha = \frac{\tau_{net}}{I} \tag{9.19}$$

where I is the merry-go-round's moment of inertia.

Now the angular acceleration is the rate of change of the angular velocity, so

$$\alpha = \frac{\Delta \omega}{\Delta t} \tag{9.20}$$

Setting Equations 9.19 and 9.20 equal to each other gives

$$\frac{\Delta \omega}{\Delta t} = \frac{\tau_{net}}{I}$$

or, rearranging,

$$\tau_{net}\,\Delta t = I\,\Delta \omega \tag{9.21}$$

If you recall the impulse-momentum theorem for *linear* motion, which is

$$\vec{F}_{net}\,\Delta t = m\,\Delta \vec{v} = \Delta \vec{p} \tag{9.22}$$

you can see that Equation 9.21 is an analogous statement about rotational motion. Because the quantity $I\omega$ is evidently the rotational equivalent of $m\vec{v}$, the linear momentum $\vec{p}$, it seems reasonable to define the **angular momentum** L to be

$$L = I\omega \tag{9.23}$$

Angular momentum of an object with moment of inertia I rotating at angular velocity ω

The SI units of angular momentum are those of moment of inertia times those of angular velocity, or kg · m^2/s.

Just as an object in linear motion can have a large momentum either by having a large mass or a large speed, a rotating object can have a large angular momentum by having a large moment of inertia or a large angular velocity. The merry-go-round in Figure 9.28 has a larger angular momentum if it's spinning fast than if it's spinning slowly. Also, the merry-go-round (large I) has a much larger angular momentum than a toy top (small I) spinning with the same angular velocity.

Table 9.2 summarizes the analogies between linear and rotational quantities that you learned in Chapter 7 and adds the analogy between linear momentum and angular momentum.

TABLE 9.2 Rotational and linear dynamics

Rotational dynamics	Linear dynamics
Torque τ_{net}	Force $\vec{F}_{net}$
Moment of inertia I	Mass m
Angular velocity ω	Velocity $\vec{v}$
Angular momentum $L = I\omega$	Linear momentum $\vec{p} = m\vec{v}$

Conservation of Angular Momentum

Having now defined angular momentum, Equation 9.21 reads

$$\tau_{net}\,\Delta t = \Delta L \tag{9.24}$$

in exact analogy with its linear dynamics equivalent, Equation 9.22. This equation states that the change in the angular momentum of an object is proportional to the net torque applied to the object. If the net torque on an object is *zero,* the rotational analog of an isolated system, then the change in the angular momentum is zero as well. That is, a rotating object will continue to rotate with *constant* angular velocity—to "keep going"—unless acted upon by an external torque. We can state this conclusion as the *law of conservation of angular momentum:*

Activ Physics ONLINE   7.14

> **Law of conservation of angular momentum** The angular momentum of a rotating object subject to no net torque ($\tau_{net} = 0$) is a constant. The final angular momentum L_f is equal to the initial angular momentum L_i.

This law is analogous to that for the conservation of linear momentum; there, linear momentum is conserved if the net *force* is zero. Because the angular momentum is $L = I\omega$, the mathematical statement of the law of conservation of angular momentum is

$$I_i\omega_i = I_f\omega_f \tag{9.25}$$

EXAMPLE 9.10 Period of a merry-go-round

Joey, whose mass is 36 kg, stands at the center of a 200 kg merry-go-round that is rotating once every 2.5 s. While it is rotating, Joey walks out to the edge of the merry-go-round, 2.0 m from its center. What is the rotational period of the merry-go-round when Joey gets to the edge?

PREPARE Take the system to be Joey + merry-go-round and assume frictionless bearings. There is no external torque on this system, so the angular momentum of the system will be conserved. As shown in the visual overview of Figure 9.29, we model the merry-go-round as a uniform disk of radius $R = 2.0$ m. From Table 7.2, the moment of inertia of a disk is $I_{disk} = \frac{1}{2}MR^2$. Modeling Joey as a particle of mass m, his moment of inertia is zero when he is at the center, but it increases to mR^2 when he reaches the edge.

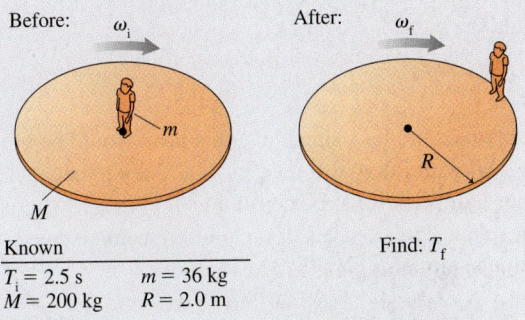

Before: ω_i **After:** ω_f

Known
$T_i = 2.5$ s $m = 36$ kg
$M = 200$ kg $R = 2.0$ m

Find: T_f

FIGURE 9.29 Visual overview of the merry-go-round.

SOLVE The mathematical statement of the law of conservation of momentum is $L_i = L_f$, or, from Equation 9.25, $I_i\omega_i = I_f\omega_f$ which we can rewrite as

$$\omega_f = \frac{I_i}{I_f}\omega_i$$

As Joey moves out to the edge, the moment of inertia of the system increases and, as a result, the angular velocity decreases. Initially, the moment of inertia of the system is just that of the merry-go-round because Joey's contribution is zero. Thus

$$I_i = I_{disk} = \tfrac{1}{2}MR^2 = \tfrac{1}{2}(200 \text{ kg})(2.0 \text{ m})^2 = 400 \text{ kg} \cdot \text{m}^2$$

When Joey reaches the edge, the total moment of inertia becomes

$$I_f = I_{disk} + mR^2 = 400 \text{ kg} \cdot \text{m}^2 + (36 \text{ kg})(2.0 \text{ m})^2$$
$$= 540 \text{ kg} \cdot \text{m}^2$$

The initial angular speed is related to the initial period of rotation T_i by

$$\omega_i = \frac{2\pi}{T_i} = \frac{2\pi}{2.5 \text{ s}} = 2.5 \text{ rad/s}$$

Thus the final angular speed is

$$\omega_f = \frac{I_i}{I_f}\omega_i = \frac{400 \text{ kg} \cdot \text{m}^2}{540 \text{ kg} \cdot \text{m}^2}(2.5 \text{ rad/s}) = 1.9 \text{ rad/s}$$

When Joey reaches the edge, the period of the merry-go-round has increased to

$$T_f = \frac{2\pi}{\omega_f} = \frac{2\pi}{1.9 \text{ rad/s}} = 3.3 \text{ s}$$

ASSESS The merry-go-round rotates *more slowly* after Joey moves out to the edge. If the system's moment of inertia increases, as it does when Joey moves out, the angular velocity must decrease to keep the angular momentum constant.

Large moment of inertia; slow spin.

Small moment of inertia; fast spin.   **FIGURE 9.30** A spinning figure skater.

Conservation of momentum enters into many aspects of sports. Because no external torques act, the angular momentum of a platform diver is conserved while she's in the air. Just as for Joey and the merry-go-round of Example 9.10, she spins slowly when her moment of inertia is large; by decreasing her moment of inertia, she increases her rate of spin. Divers can thus markedly increase their spin rate by changing their body from an extended posture to a tuck position. Figure skaters also increase their spin rate by decreasing their moment of inertia, as shown in Figure 9.30. The following example gives a simplified treatment of this process.

EXAMPLE 9.11 Analyzing a spinning ice skater

An ice skater spins around on the tips of his blades while holding a 5.0 kg weight in each hand. He begins with his arms straight out from his body and his hands 140 cm apart. While spinning at 2.0 rev/s, he pulls the weights in and holds them 50 cm apart against his shoulders. If we neglect the mass of the skater, how fast is he spinning after pulling the weights in?

PREPARE Although the mass of the skater is larger than the mass of the weights, neglecting the skater's mass is not a bad approximation. Angular momentum depends on the radius of the circle. The skater's mass is concentrated in his torso, which has an effective radius (i.e., where most of the mass is concentrated) of only 9 or 10 cm. The weights move in much larger circles and have a disproportionate influence on his motion. The skater's arms exert radial forces on the weights just to keep

them moving in circles, and even larger radial forces as he pulls them in. But there is no torque on the weights, so their total angular momentum is conserved. Figure 9.31 shows a before-and-after visual overview, as seen from above.

SOLVE The two weights have the same mass, move in circles with the same radius, and have the same angular velocity. Thus the total angular momentum is twice that of one weight. The mathematical statement of angular momentum conservation, $I_f\omega_f = I_i\omega_i$, is

$$(2mr_f^2)\omega_f = (2mr_i^2)\omega_i$$

Because the angular velocity is related to the rotation frequency f by $\omega = 2\pi f$, this equation simplifies to

$$f_f = \left(\frac{r_i}{r_f}\right)^2 f_i$$

When he pulls the weights in, his rotation frequency increases to

$$f_f = \left(\frac{0.70 \text{ m}}{0.25 \text{ m}}\right)^2 \times 2.0 \text{ rev/s} = 16 \text{ rev/s}$$

ASSESS Pulling in the weights increases the skater's spin from 2 rev/s to 16 rev/s. This is somewhat high, because we neglected the mass of the skater, but it illustrates how skaters do "spin up" by pulling their mass in toward the rotation axis.

Before:

$f_i = 2$ rev/s

$m = 5$ kg $r_i = 0.70$ m

After:

Find: f_f

$r_f = 0.25$ m

FIGURE 9.31 Top view visual overview of the spinning ice skater.

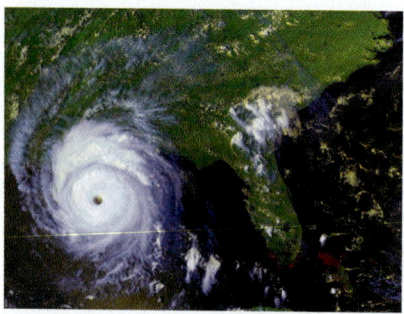

The eye of a hurricane As air masses from the slowing rotating outer zones are drawn toward the low-pressure center, their moment of inertia decreases. Because the angular momentum of these air masses is conserved, their speed must *increase* as they approach the center, leading to the high wind speeds near the center of the storm.

Solving either of these two examples using Newton's laws would be quite difficult. We would have to deal with internal forces, such as Joey's feet against the merry-go-round, and other complications. For problems like these, where we're interested only in the before-and-after aspects of the motion, using a conservation law makes the solution much simpler.

STOP TO THINK 9.5 The left figure shows two boys of equal mass standing halfway to the edge on a turntable that is freely rotating at angular speed ω_i. They then walk to the positions shown in the right figure. The final angular speed ω_f is

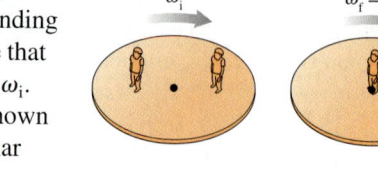

ω_i $\omega_f = ?$

A. Greater than ω_i.   B. Less than ω_i.   C. Equal to ω_i.

SUMMARY

The goals of Chapter 9 have been to introduce the ideas of impulse, momentum, and angular momentum and to learn a new problem-solving strategy based on conservation laws.

GENERAL PRINCIPLES

Law of Conservation of Momentum

The total momentum $\vec{P} = \vec{p}_1 + \vec{p}_2 + \cdots$ of an isolated system is a constant. Thus

$$\vec{P}_f = \vec{P}_i$$

Conservation of Angular Momentum

The angular momentum L of a rotating object subject to zero external torque does not change. Thus

$$L_f = L_i$$

This can be written in terms of the moment of inertia and angular velocity as

$$I_f \omega_f = I_i \omega_i$$

Solving Momentum Conservation Problems

PREPARE Choose an isolated system or a system that is isolated during at least part of the problem. Draw a visual overview of the system before and after the interaction.

SOLVE Write the law of conservation of momentum in terms of vector components

$$(p_{1x})_f + (p_{2x})_f + \cdots = (p_{1x})_i + (p_{2x})_i + \cdots$$
$$(p_{1y})_f + (p_{2y})_f + \cdots = (p_{1y})_i + (p_{2y})_i + \cdots$$

In terms of masses and velocities, this is

$$m_1(v_{1x})_f + m_2(v_{2x})_f + \cdots = m_1(v_{1x})_i + m_2(v_{2x})_i + \cdots$$
$$m_1(v_{1y})_f + m_2(v_{2y})_f + \cdots = m_1(v_{1y})_i + m_2(v_{2y})_i + \cdots$$

ASSESS Is the result reasonable?

IMPORTANT CONCEPTS

Momentum $\vec{p} = m\vec{v}$

Impulse J_x = area under force curve

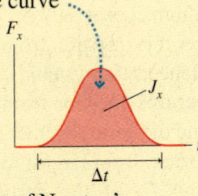

Impulse and momentum are related by the impulse-momentum theorem

$$\Delta p_x = J_x$$

This is an alternative statement of Newton's second law.

Angular momentum $L = I\omega$ is the rotational analog of linear momentum $\vec{p} = m\vec{v}$.

System A group of interacting particles.

Isolated system A system on which the net external force is zero.

Before-and-after visual overview

- Define the system.

- Use two drawings to show the system *before* and *after* the interaction.

- List known information and identify what you are trying to find.

APPLICATIONS

Collisions Two or more particles come together. In a perfectly inelastic collision, they stick together and move with a common final velocity.

Explosions Two or more particles move away from each other.

Two dimensions Both the x- and y-components of the total momentum P must be conserved, giving two simultaneous equations.

For instructor-assigned homework, go to www.masteringphysics.com

Problem difficulty is labeled as I (straightforward) to IIIII (challenging).

Problems labeled ✎ can be done on a Workbook Momentum Worksheet; INT integrate significant material from earlier chapters; BIO are of biological or medical interest.

QUESTIONS

Conceptual Questions

1. Rank in order, from largest to smallest, the momenta p_{1x} through p_{5x} of the objects presented in Figure Q9.1. Explain.

FIGURE Q9.1

2. Explain the concept of *impulse* in nonmathematical language. That is, don't simply put the equation in words to say that "impulse is the area under the force-versus-time curve." Explain it in terms that would make sense to an educated person who had never heard of it.

3. A 0.2 kg plastic cart and a 20 kg lead cart can roll without friction on a horizontal surface. Equal forces are used to push both carts forward for a time of 1 s, starting from rest. After the force is removed at $t = 1$ s, is the momentum of the plastic cart greater than, less than, or equal to the momentum of the lead cart? Explain.

4. Two pucks, of mass m and $4m$, lie on a frictionless table. Equal forces are used to push both pucks forward a distance of 1 m.
   a. Which puck takes longer to travel the distance? Explain.
   b. Which puck has the greater momentum upon completing the distance? Explain.

5. Explain the concept of "isolated system" in nonmathematical language.

6. A golf club continues forward after hitting the golf ball. Is momentum conserved in the collision? Explain, making sure you are careful to identify the "system."

7. Two particles collide, one of which was initially moving and the other initially at rest.
   a. Is it possible for *both* particles to be at rest after the collision? Give an example in which this happens, or explain why it can't happen.
   b. Is it possible for *one* particle to be at rest after the collision? Give an example in which this happens, or explain why it can't happen.

8. Automobiles are designed with "crumple zones" intended to collapse in a collision. Why would a manufacturer design part of a car so that it collapses in a collision?

9. You probably know that it feels better to catch a baseball if you are wearing a padded glove. Explain why this is so, using the ideas of momentum and impulse.

10. In the early days of rocketry, some people claimed that rockets couldn't fly in outer space as there was no air for the rockets to push against. Suppose you were an early investigator in the field of rocketry and met someone who made this argument. How would you convince the person that rockets could travel in space?

11. Two ice skaters, Megan and Jason, push off from each other on frictionless ice. Jason's mass is twice that of Megan.
    a. Which skater, if either, experiences the greater impulse during the push? Explain.
    b. Which skater, if either, has the greater speed after the push-off? Explain.

12. Suppose a rubber ball and a steel ball collide. Which, if either, receives the larger impulse? Explain.

13. While standing still on a basketball court, you throw the ball to a teammate. Why do you not move backward as a result? Is the law of conservation of momentum violated?

14. As you release a ball, it falls, gaining speed and momentum. Is momentum conserved?
    a. Answer this question from the perspective of choosing the ball alone as the system.
    b. Answer this question from the perspective of choosing the ball and the earth together as the system.

15. Rank in order, from largest to smallest, the angular momenta L_1 through L_5 of the particles shown in Figure Q9.15. Explain.

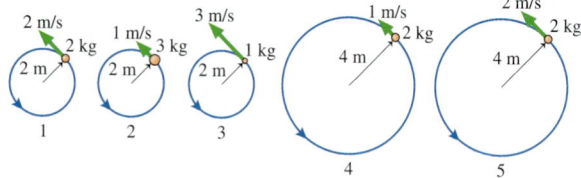

FIGURE Q9.15

16. Figure Q9.16 shows two masses held together by a thread on a rod that is rotating about its center with angular velocity ω. If the thread breaks, the masses will slide out to the ends of the rod. If that happens, will the rod's angular velocity increase, decrease, or remain unchanged? Explain.

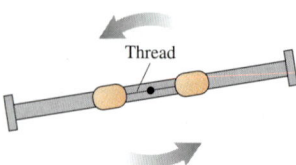

FIGURE Q9.16

17. The disks shown in Figure Q9.17 have equal mass. Is the angular momentum of disk 2, on the right, larger than, smaller than, or equal to the angular momentum of disk 1? Explain.

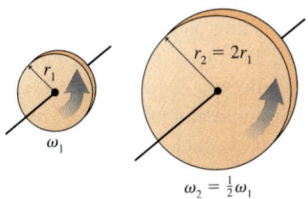

FIGURE Q9.17

18. If the earth warms significantly, the polar ice caps will melt. Water will move from the poles, near the earth's rotation axis, and will spread out around the globe. In principle, this will change the length of the day. Why? Will the length of the day increase or decrease?

Multiple-Choice Questions

19. | Curling is a sport played with 20 kg stones that slide across an ice surface. Suppose a curling stone sliding at 1 m/s strikes another stone and comes to rest in 2 ms. Approximately how much force is there on the stone during the impact?
    A. 200 N   B. 1000 N   C. 2000 N   D. 10,000 N

20. | Two balls are hung from cords. The first ball, of mass 1.0 kg, is pulled to the side and released, reaching a speed of 2.0 m/s at the bottom of its arc. Then, as shown in Figure Q9.20, it hits and sticks to another ball. The speed of the pair just after the collision is 1.2 m/s. What is the mass of the second ball?
    A. 0.67 kg   B. 2.0 kg   C. 1.7 kg   D. 1.0 kg

FIGURE Q9.20

21. ‖ Figure Q9.21 shows two blocks sliding on a frictionless surface. Eventually the smaller block overtakes the larger one, collides with it, and sticks. What is the speed of the two blocks after the collision?
    A. $v_i/2$ B. $4v_i/5$ C. v_i D. $5v_i/4$ E. $2v_i$

FIGURE Q9.21

22. ‖ Two friends are sitting in a stationary canoe. At $t = 3.0$ s the person at the front tosses a sack to the person in the rear, who catches the sack 0.2 s later. Which plot in Figure Q9.22 shows the velocity of the boat as a function of time? Positive velocity is forward, negative velocity is backward. Neglect any drag force on the canoe from the water.

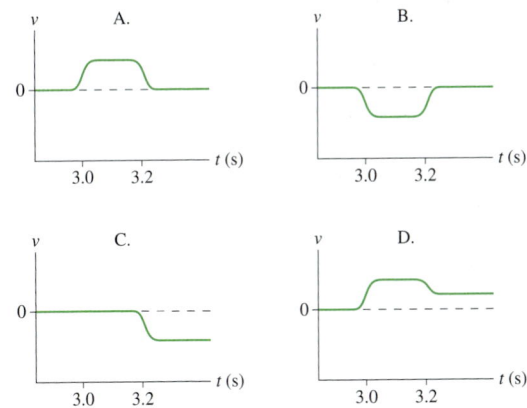

FIGURE Q9.22

23. ‖ A small puck is sliding to the right with momentum $\vec{p}_i$ on a horizontal, frictionless surface, as shown in Figure Q9.23. A force is applied to the puck for a short time and its momentum afterward is $\vec{p}_f$. Which lettered arrow shows the direction of the impulse that was delivered to the puck?

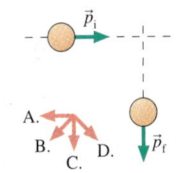

FIGURE Q9.23

24. ‖ A red ball, initially at rest, is simultaneously hit by a blue ball traveling from west to east at 3 m/s and a green ball traveling east to west at 3 m/s. All three balls have equal mass. Afterward, the red ball is traveling south and the green ball is moving to the east. In which direction is the blue ball traveling?
    A. West.
    B. North.
    C. Between north and west.
    D. Between north and east.
    E. Between south and west.

25. ‖ A 24 g, 3-cm-diameter thin, hollow sphere rotates at 30 rpm about a vertical, frictionless axis through its center. A 4 g bug stands at the top of the sphere. He then walks along the surface of the sphere until he reaches its "equator." When he reaches the equator, the sphere is rotating at
    A. 15 rpm   B. 24 rpm   C. 30 rpm
    D. 37 rpm   E. 45 rpm

26. ‖ A 5.0 kg solid cylinder of radius 12 cm rotates with $\omega_i = 3.7$ rad/s about an axis through its center. A torque of 0.040 N · m is applied to the cylinder for 5.0 s. By how much does the cylinder's angular momentum change?
    A. 0.12 kg · m²/s   B. 0.20 kg · m²/s   C. 0.38 kg · m²/s
    D. 0.52 kg · m²/s   E. 0.88 kg · m²/s

PROBLEMS

Section 9.1 Impulse

Section 9.2 Momentum and the Impulse-Momentum Theorem

1. | At what speed do a bicycle and its rider, with a combined mass of 100 kg, have the same momentum as a 1500 kg car traveling at 5.0 m/s?

2. ‖ A 57 g tennis ball is served at 45 m/s. If the ball started from rest, what impulse was applied to the ball by the racket?

3. | A student throws a 120 g snowball at 7.5 m/s at the side of the schoolhouse, where it hits and sticks. What is the magnitude of the average force on the wall if the duration of the collision is 0.15 s?

4. ‖‖ In Figure P9.4, what value of F_{max} gives an impulse of 6.0 N · s?

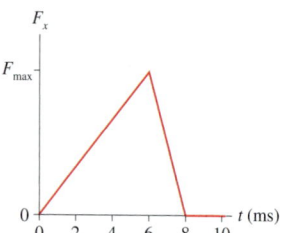

FIGURE P9.4

5. | What is the magnitude of the momentum of
   a. A 1500 kg car traveling at 10 m/s?
   b. A 200 g baseball thrown at 40 m/s?

Section 9.3 Solving Impulse and Momentum Problems

6. | Use the impulse-momentum theorem to find how long a stone falling straight down takes to increase its speed from 5.5 m/s to 10.4 m/s.

7. ‖‖ a. A 2.0 kg object is moving to the right with a speed of 1.0 m/s when it experiences the force shown in Figure P9.7a. What are the object's speed and direction after the force ends?
   b. Answer this question for the force shown in Figure P9.7b.

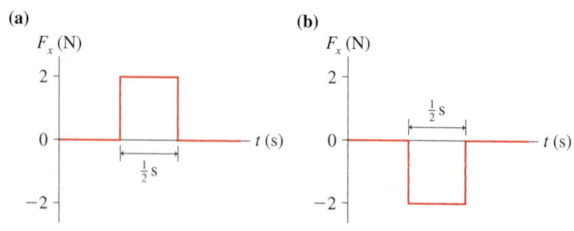

FIGURE P9.7

8. ‖‖ A 60 g tennis ball with an initial speed of 32 m/s hits a wall and rebounds with the same speed. Figure P9.8 shows the force of the wall on the ball during the collision. What is the value of F_{max}, the maximum value of the contact force during the collision?

FIGURE P9.8

9. | A child is sliding on a sled at 1.5 m/s to the right. You stop the sled by pushing on it for 0.50 s in a direction opposite to its motion. If the mass of the child and sled is 35 kg, what average force do you need to apply to stop the sled? Use the concepts of impulse and momentum.

10. ‖ An ice hockey puck slides along the ice at 12 m/s. A hockey stick delivers an impulse of 4.0 kg · m/s, causing the puck to move off in the opposite direction with the same speed. What is the mass of the puck?

11. | As part of a safety investigation, two 1400 kg cars traveling at 20 m/s are crashed into different barriers. Find the average forces exerted on (a) the car that hits a line of water barrels and takes 1.5 s to stop, and (b) the car that hits a concrete barrier and takes 0.1 s to stop.

12. | In a Little League baseball game, the 145 g ball enters the strike zone with a speed of 15.0 m/s. The batter hits the ball, and it leaves his bat with a speed of 20.0 m/s in exactly the opposite direction.
   a. What is the magnitude of the impulse delivered by the bat to the ball?
   b. If the bat is in contact with the ball for 1.5 ms, what is the magnitude of the average force exerted by the bat on the ball?

Section 9.4 Conservation of Momentum

13. ‖‖ A small, 100 g cart is moving at 1.20 m/s on an air track when it collides with a larger, 1.00 kg cart at rest. After the collision, the small cart recoils at 0.850 m/s. What is the speed of the large cart after the collision?

14. ‖ A man standing on very slick ice fires a rifle horizontally. The mass of the man together with the rifle is 70 kg, and the mass of the bullet is 10 g. If the bullet leaves the muzzle at a speed of 500 m/s, what is the final speed of the man?

15. | An 80 kg quarterback jumps straight up in the air right before throwing a 0.43 kg football horizontally at 15 m/s. How fast will he be moving backward just after releasing the ball? Do you think that this would be noticeable?

16. | A strong man is compressing a lightweight spring between two weights. One weight has a mass of 2.3 kg, the other a mass of 5.3 kg. He is holding the weights stationary, but then he loses his grip and the weights fly off in opposite directions. The lighter of the two is shot out at a speed of 6.0 m/s. What is the speed of the heavier weight?

17. | A 10,000 kg railroad car is rolling at 2.00 m/s when a 4000 kg load of gravel is suddenly dropped in. What is the car's speed just after the gravel is loaded?

18. | A 5000 kg open train car is rolling on frictionless rails at 22.0 m/s when it starts pouring rain. A few minutes later, the car's speed is 20.0 m/s. What mass of water has collected in the car?

19. | A 50.0 kg archer, standing on frictionless ice, shoots a 100 g arrow at a speed of 100 m/s. What is the recoil speed of the archer?

20. | A 9.5 kg dog takes a nap in a canoe and wakes up to find the canoe has drifted out onto the lake but now is stationary. He walks along the length of the canoe at 0.50 m/s, relative to the water, and the canoe simultaneously moves in the opposite direction at 0.15 m/s. What is the mass of the canoe?

Section 9.5 Inelastic Collisions

21. ‖ A 300 g bird flying along at 6.0 m/s sees a 10 g insect heading straight toward it with a speed of 30 m/s. The bird opens its mouth wide and enjoys a nice lunch. What is the bird's speed immediately after swallowing?

22. ‖ A 71 kg baseball player jumps straight up to catch a line drive. If the 140 g ball is moving horizontally at 28 m/s, and the catch is made when the ballplayer is at the highest point of his leap, what is his speed immediately after stopping the ball?

23. ‖ A kid at the junior high cafeteria wants to propel an empty milk carton along a lunch table by hitting it with a 3.0 g spit ball. If he wants the speed of the 20 g carton just after the spit ball hits it to be 0.30 m/s, at what speed should his spit ball hit the carton?

24. ‖ The parking brake on a 2000 kg Cadillac has failed, and it is rolling slowly, at 1 mph, toward a group of small children. Seeing the situation, you realize you have just enough time to drive your 1000 kg Volkswagen head-on into the Cadillac and save the children. With what speed should you impact the Cadillac to bring it to a halt?

25. ‖ A 1500 kg car is rolling at 2.0 m/s. You would like to stop the car by firing a 10 kg blob of sticky clay at it. How fast should you fire the clay?

Section 9.6 Momentum and Collisions in Two Dimensions

26. ‖ A 20 g ball of clay traveling east at 3.0 m/s collides with a 30 g ball of clay traveling north at 2.0 m/s. What are the speed and the direction of the resulting 50 g ball of clay?

27. ‖ Two particles collide and bounce apart. Figure P9.27 shows the initial momenta of both and the final momentum of particle 2. What is the final momentum of particle 1? Show your answer by copying the figure and drawing the final momentum vector on the figure.

FIGURE P9.27

28. ‖ A 20 g ball of clay traveling east at 2.0 m/s collides with a 30 g ball of clay traveling 30° south of west at 1.0 m/s. What are the speed and direction of the resulting 50 g blob of clay?

29. ‖ A firecracker in a coconut blows the coconut into three pieces. Two pieces of equal mass fly off south and west, perpendicular to each other, at 20 m/s. The third piece has twice the mass as the other two. What are the speed and direction of the third piece?

Section 9.7 Angular Momentum

30. ‖ What is the angular momentum of the moon around the earth? The moon's mass is 7.4×10^{22} kg and it orbits 3.8×10^{8} m from the earth.

31. ‖ A little girl is going on the merry-go-round for the first time, and wants her 47 kg mother to stand next to her on the ride, 2.6 m from the merry-go-round's center. If her mother's speed is 4.2 m/s when the ride is in motion, what is her angular momentum around the center of the merry-go-round?

32. ‖ What is the angular momentum of the 500 g rotating bar in Figure P9.32?

FIGURE P9.32   **FIGURE P9.33**

33. ‖ What is the angular momentum of the 2.0 kg, 4.0-cm-diameter rotating disk in Figure P9.33?

34. ‖ Divers change their body position in midair while rotating BIO about their center of mass. In one dive, the diver leaves the board with her body nearly straight, then tucks into a somersault position. If the moment of inertia of the diver in a straight position is 14 kg · m² and in a tucked position is 4.0 kg · m², by what factor is her angular velocity when tucked greater than when straight?

35. ‖ Ice skaters often end their performances with spin turns, BIO where they spin very fast about their center of mass with their arms folded in and legs together. Upon ending, their arms extend outward, proclaiming their finish. Not quite as noticeably, one leg goes out as well. Suppose that the moment of inertia of a skater with arms out and one leg extended is 3.2 kg · m² and for arms and legs in is 0.80 kg · m². If she starts out spinning at 5.0 rev/s, what is her angular speed (in rev/s) when her arms and one leg open outward?

General Problems

36. ‖ What is the impulse on a 3.0 kg particle that experiences the force described by the graph in Figure P9.36?

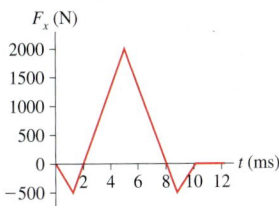

FIGURE P9.36

37. ‖ A 600 g air-track glider collides with a spring at one end of the track. Figure P9.37 shows the glider's velocity and the force exerted on the glider by the spring. How long is the glider in contact with the spring?

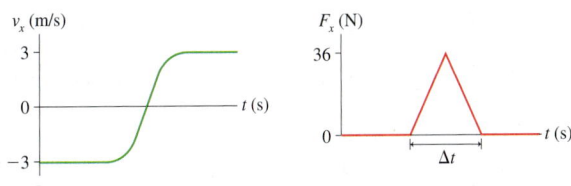

FIGURE P9.37

38. ‖ Far in space, where gravity is negligible, a 425 kg rocket traveling at 75.0 m/s fires its engines. Figure P9.38 shows the thrust force as a function of time. The mass lost by the rocket during these 30.0 s is negligible.
   a. What impulse does the engine impart to the rocket?
   b. At what time does the rocket reach its maximum speed? What is the maximum speed?

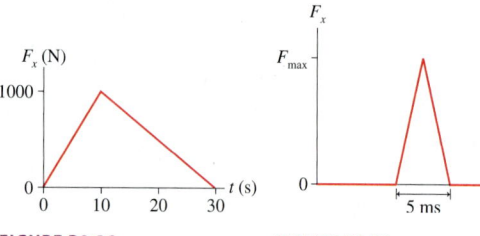

FIGURE P9.38   **FIGURE P9.39**

39. ‖ A 200 g ball is dropped from a height of 2.0 m, bounces on a hard floor, and rebounds to a height of 1.5 m. Figure P9.39 shows the impulse received from the floor. What maximum force does the floor exert on the ball?

40. ‖‖‖ A 200 g ball is dropped from a height of 2.0 m and bounces on a hard floor. The force on the ball from the floor is shown in Figure P9.40. How high does the ball rebound?

FIGURE P9.40   **FIGURE P9.41**

41. ‖‖‖ Figure P9.41 is a graph of the force exerted by the floor on a woman making a vertical jump. At what speed does she leave the ground?
   Hint: The force of the floor is not the only force acting on the woman.

42. ‖ A sled slides along a horizontal surface for which the coefficient of kinetic friction is 0.25. Its velocity at point A is 8.0 m/s and at point B is 5.0 m/s. Use the impulse-momentum theorem to find how long the sled takes to travel from A to B.

43. ‖ A 50.0 g ball is launched from ground level at an angle 30.0° above the horizon. Its initial speed is 25.0 m/s.
   a. What are the values of p_x and p_y an instant after the ball is launched, at the point of maximum altitude, and an instant before the ball hits the ground?
   b. Why is one component of $\vec{p}$ constant? Explain.
   c. For the component of $\vec{p}$ that changes, show that the change in momentum is equal to the weight of the ball multiplied by the time of flight. Explain why this is so.

44. ‖ Squids rely on jet propulsion, a versatile technique to move around in water. A 1.5 kg squid at rest suddenly expels 0.10 kg of water backward to quickly get itself moving forward at 3.0 m/s. If other forces (such as the drag force on the squid) are ignored, what is the speed with which the squid expels the water?

45. ‖‖‖ The flowers of the bunchberry plant open with astonishing force and speed, causing the pollen grains to be ejected out of the flower in a mere 0.30 ms at an acceleration of 2.5×10^4 m/s². If the acceleration is constant, what impulse is delivered to a pollen grain with a mass of 1.0×10^{-7} g?

46. ‖ a. With what speed are pollen grains ejected from a bunchberry flower? See Problem 45 for information.
   b. Suppose that 1000 ejected pollen grains slam into the abdomen of a 5.0 g bee that is hovering just above the flower. If the collision is inelastic, what is the bee's speed immediately afterward? Is the bee likely to notice?

47. ‖‖‖ A tennis player swings her 1000 g racket with a speed of 10 m/s. She hits a 60 g tennis ball that was approaching her at a speed of 20 m/s. The ball rebounds at 40 m/s.
   a. How fast is her racket moving immediately after the impact? You can ignore the interaction of the racket with her hand for the brief duration of the collision.
   b. If the tennis ball and racket are in contact for 10 ms, what is the average force that the racket exerts on the ball?

48. ‖‖‖ A 20 g ball of clay is thrown horizontally at 30 m/s toward a 1.0 kg block sitting at rest on a frictionless surface. The clay hits and sticks to the block.
   a. What is the speed of the block and clay right after the collision?
   b. Use the block's initial and final speeds to calculate the impulse the clay exerts on the block.
   c. Use the clay's initial and final speeds to calculate the impulse the block exerts on the clay.
   d. Does $\vec{J}_{\text{block on clay}} = -\vec{J}_{\text{clay on block}}$?

49. ‖ Dan is gliding on his skateboard at 4.0 m/s. He suddenly jumps backward off the skateboard, kicking the skateboard forward at 8.0 m/s. How fast is Dan going as his feet hit the ground? Dan's mass is 50 kg and the skateboard's mass is 5.0 kg.

50. ‖ James and Sarah stand on a stationary cart with frictionless wheels. The total mass of the cart and riders is 130 kg. At the same instant, James throws a 1.0 kg ball to Sarah at 4.5 m/s, while Sarah throws a 0.50 kg ball to James at 1.0 m/s. James's throw is to the right and Sarah's is to the left.
   a. While the two balls are in the air, what are the speed and direction of the cart and its riders?
   b. After the balls are caught, what are the speed and direction of the cart and riders?

51. ‖ A 10-m-long glider with a mass of 680 kg (including the passengers) is gliding horizontally through the air at 30 m/s when a 60 kg skydiver drops out by releasing his grip on the glider. What is the glider's velocity just after the skydiver lets go?

52. ‖ The cars of a long coal train are filled by pulling them under a hopper, from which coal falls into the cars at a rate of 10,000 kg/s. Ignoring friction due to the rails, what is the average force that the engine must exert on the coal train to keep it moving under the hopper at a speed of 0.50 m/s?

53. ‖ Three identical train cars, coupled together, are rolling east at 2.0 m/s. A fourth car traveling east at 4.0 m/s catches up with the three and couples to make a four-car train. A moment later, the train cars hit a fifth car that was at rest on the tracks, and it couples to make a five-car train. What is the speed of the five-car train?

54. ‖ A 110 kg linebacker running at 2.0 m/s and an 82 kg quarterback running at 3.0 m/s have a head-on collision in midair. The linebacker grabs and holds onto the quarterback. Who ends up moving forward after they hit?

55. | Most geologists believe that the dinosaurs became extinct 65 million years ago when a large comet or asteroid struck the earth, throwing up so much dust that the sun was blocked out for a period of many months. Suppose an asteroid with a diameter of 2.0 km and a mass of 1.0×10^{13} kg hits the earth with an impact speed of 4.0×10^4 m/s.
   a. What is the earth's recoil speed after such a collision? (Use a reference frame in which the earth was initially at rest.)
   b. What percentage is this of the earth's speed around the sun? (Use the astronomical data inside the cover.)

56. | At the center of a 50-m-diameter circular ice rink, a 75 kg skater traveling north at 2.5 m/s collides with and holds onto a 60 kg skater who had been heading west at 3.5 m/s.
   a. How long will it take them to glide to the edge of the rink?
   b. Where will they reach it? Give your answer as an angle north of west.

57. || Two ice skaters, with masses of 50 kg and 75 kg, are at the center of a 60-m-diameter circular rink. The skaters push off against each other and glide to opposite edges of the rink. If the heavier skater reaches the edge in 20 s, how long does the lighter skater take to reach the edge?

58. || One billiard ball is shot east at 2.00 m/s. A second, identical billiard ball is shot west at 1.00 m/s. The balls have a glancing collision, not a head-on collision, deflecting the second ball by 90° and sending it north at 1.41 m/s. What are the speed and direction of the first ball after the collision?

59. || A 10 g bullet is fired into a 10 kg wood block that is at rest on a wood table. The block, with the bullet embedded, slides 5.0 cm across the table. What was the speed of the bullet?

60. || You are part of a search-and-rescue mission that has been called out to look for a lost explorer. You've found the missing explorer, but you're separated from him by a 200-m-high cliff and a 30-m-wide raging river, as shown in Figure P9.60. To save his life, you need to get a 5.0 kg package of emergency supplies across the river. Unfortunately, you can't throw the package hard enough to make it across.

Package
200 m
Explorer
30 m

FIGURE P9.60

Fortunately, you happen to have a 1.0 kg rocket intended for launching flares. Improvising quickly, you attach a sharpened stick to the front of the rocket, so that it will impale itself into the package of supplies, then fire the rocket at ground level toward the supplies. What minimum speed must the rocket have just before impact in order to save the explorer's life?

61. || A 1500 kg weather rocket accelerates upward at 10.0 m/s². It explodes 2.00 s after liftoff and breaks into two fragments, one twice as massive as the other. Photos reveal that the lighter fragment traveled straight up and reached a maximum height of 530 m. What were the speed and direction of the heavier fragment just after the explosion?

62. | Two 500 g blocks of wood are 2.0 m apart on a frictionless table. A 10 g bullet is fired at 400 m/s toward the blocks. It passes all the way through the first block, then embeds itself in the second block. The speed of the first block immediately afterward is 6.0 m/s. What is the speed of the second block after the bullet stops?

63. | A 500 kg cannon fires a 10 kg cannonball with a speed of 200 m/s relative to the muzzle. The cannon is on wheels that roll without friction. When the cannon fires, what is the speed of the cannonball relative to the earth?

64. | Laura, whose mass is 35 kg, jumps horizontally off a 55 kg canoe at 1.5 m/s relative to the canoe. What is the canoe's speed just after she jumps?

65. ||| A spaceship of mass 2.0×10^6 kg is cruising at a speed of 5.0×10^6 m/s when the antimatter reactor fails, blowing the ship into three pieces. One section, having a mass of 5.0×10^5 kg, is blown straight backward with a speed of 2.0×10^6 m/s. A second piece, with mass 8.0×10^5 kg, continues forward at 1.0×10^6 m/s. What are the direction and speed of the third piece?

66. ||| A proton is shot at 5.0×10^7 m/s toward a gold target. The nucleus of a gold atom, with a mass 197 times that of the proton, repels the proton and deflects it straight back with 90% of its initial speed. What is the recoil speed of the gold nucleus?

67. || Figure P9.67 shows a collision between three balls of clay. The three hit simultaneously and stick together. What are the speed and direction of the resulting blob of clay?

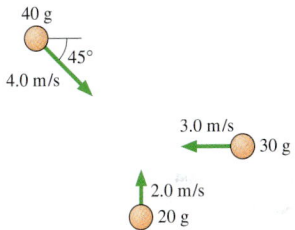
40 g
45°
4.0 m/s
3.0 m/s   30 g
2.0 m/s
20 g

FIGURE P9.67

68. || The carbon isotope ^{14}C is used for carbon dating of archeological artifacts. ^{14}C (mass 2.34×10^{-26} kg) decays by the process known as *beta decay* in which the nucleus emits an electron (the beta particle) and a subatomic particle called a neutrino. In one such decay, the electron and the neutrino are emitted at right angles to each other. The electron (mass 9.11×10^{-31} kg) has a speed of 5.00×10^7 m/s and the neutrino has a momentum of 8.00×10^{-24} kg · m/s. What is the recoil speed of the nucleus?

69. ||| A 1.0-m-long massless rod is pivoted at one end and swings around in a circle on a frictionless table. A block with a hole through the center can slide in and out along the rod. Initially, a small piece of wax holds the block 30 cm from the pivot. The block is spun at 50 rpm, then the temperature of the rod is slowly increased. When the wax melts, the block slides out to the end of the rod. What is the final angular speed? Give your answer in rpm.

70. || A 200 g puck revolves in a circle on a frictionless table at the end of a 50.0-cm-long string. The puck's angular momentum is 3.00 kg · m²/s. What is the tension in the string?

71. ▌ Figure P9.71 shows a 100 g puck revolving in a 20-cm-radius circle on a frictionless table. The string passes through a hole in the center of the table and is tied to two 200 g weights.

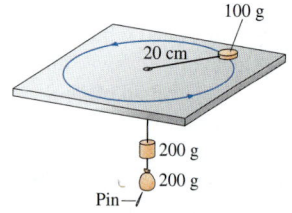

FIGURE P9.71

   a. What speed does the puck need to support the two weights?
   b. The lower weight is a light bag filled with sand. Suppose a hole is poked in the bag and the sand slowly leaks out while the puck is revolving. What will be the puck's speed and the radius of its trajectory after all of the sand is gone?

72. ▌ A 2.0 kg, 20-cm-diameter turntable rotates at 100 rpm on frictionless bearings. Two 500 g blocks fall from above, hit the turntable simultaneously at opposite ends of a diagonal, and stick. What is the turntable's angular speed, in rpm, just after this event?

73. ▌ A 200 g, 40.0-cm-diameter turntable rotates on frictionless bearings at 60.0 rpm. A 20.0 g block sits at the center of the turntable. A compressed spring shoots the block radially outward along a frictionless groove in the surface of the turntable. What is the turntable's angular speed when the block reaches the outer edge?

74. ▌ A 3.0-m-diameter merry-go-round with a mass of 250 kg is spinning at 20 rpm. John runs around the merry-go-round at 5.0 m/s, in the same direction that it is turning, and jumps onto the outer edge. John's mass is 30 kg. What is the merry-go-round's angular speed, in rpm, after John jumps on?

75. ▌ Disk A, with a mass of 2.0 kg and a radius of 40 cm, rotates clockwise about a frictionless vertical axle at 30 rev/s. Disk B, also 2.0 kg but with a radius of 20 cm, rotates counterclockwise about that same axle, but at a greater height than disk A, at 30 rev/s. Disk B slides down the axle until it lands on top of disk A, after which they rotate together. After the collision, what is their common angular speed (in rev/s) and in which direction do they rotate?

Passage Problems

Hitting a Golf Ball

Consider a golf club hitting a golf ball. To a good approximation, we can model this as a collision between the rapidly moving head of the golf club and the stationary golf ball, ignoring the shaft of the club and the golfer.

A golf ball has a mass of 46 g. Suppose a 200 g club head is moving at a speed of 40 m/s just before striking the golf ball. After the collision, the golf ball's speed is 60 m/s.

76. ▌ What is the momentum of the club-ball system right before the collision?
   A. 1.8 kg · m/s   B. 8.0 kg · m/s
   C. 3220 kg · m/s   D. 8000 kg · m/s

77. ▌ Immediately after the collision, the momentum of the club-ball system will be
   A. Less than before the collision.
   B. The same as before the collision.
   C. More than before the collision.

78. ▌ A manufacturer makes a golf ball that compresses more than a traditional golf ball when struck by a club. How will this affect the average force during the collision?
   A. The force will decrease.
   B. The force will not be affected.
   C. The force will increase.

79. ▌ By approximately how much does the club head slow down as a result of hitting the ball?
   A. 4 m/s   B. 6 m/s   C. 14 m/s   D. 26 m/s

STOP TO THINK ANSWERS

Stop to Think 9.1: F. The cart is initially moving in the negative x-direction, so $(p_x)_i = -20$ kg · m/s. After it bounces, $(p_x)_f = 10$ kg · m/s. Thus $\Delta p = (10$ kg · m/s$) - (-20$ kg · m/s$) = 30$ kg · m/s.

Stop to Think 9.2: B. The clay ball goes from $(v_x)_i = v$ to $(v_x)_f = 0$, so $J_{clay} = \Delta p_x = -mv$. The rubber ball rebounds, going from $(v_x)_i = v$ to $(v_x)_f = -v$ (same speed, opposite direction). Thus $J_{rubber} = \Delta p_x = -2mv$. The rubber ball has a larger momentum change, and this requires a larger impulse.

Stop to Think 9.3: Right end. The pieces started at rest, so the total momentum of the system is zero. It's an isolated system, so the total momentum after the explosion is still zero. The 6 g piece has momentum $6v$. The 4 g piece, with velocity $-2v$, has momentum $-8v$. The combined momentum of these two pieces is $-2v$. In order for P to be zero, the third piece must have a *positive* momentum ($+2v$) and thus a positive velocity.

Stop to Think 9.4: B. The momentum of ball 1 is $(0.40$ kg$)(2.5$ m/s$) = 1.0$ kg · m/s, while that of ball 2 is $(0.80$ kg$)(-1.5$ m/s$) = -1.2$ kg · m/s. The total momentum is then 1.0 kg · m/s $- 1.2$ kg · m/s $= -0.2$ kg · m/s. Because it's negative, the total momentum, and hence the final velocity of the balls, is directed to the left.

Stop to Think 9.5: B. Angular momentum $L = I\omega$ is conserved. Both boys have mass m and initially stand distance $R/2$ from the axis. Thus the initial moment of inertia is $I_i = I_{disk} + 2 \times m(R/2)^2 = I_{disk} + \frac{1}{2}mR^2$. The final moment of inertia is $I_f = I_{disk} + 0 + mR^2$, because the boy standing at the axis contributes nothing to the moment of inertia. Because $I_f > I_i$ we must have $\omega_f < \omega_i$.

10

ENERGY AND WORK

Using just a fast run-up and flexible pole, how can a pole vaulter reach an astonishing 6 m (20 ft) off the ground?

Energy. It's a word you hear all the time. We use chemical energy to heat our homes and bodies, electrical energy to run our lights and computers, and solar energy to grow our crops and forests. We're told to use energy wisely and not to waste it. Athletes and weary students consume "energy bars" and "energy drinks."

But just what is energy? The concept of energy has grown and changed with time, and it is not easy to define in a general way just what energy is. Rather than starting with a formal definition, we'll let the concept of energy expand slowly over the course of several chapters. In this chapter we introduce several fundamental forms of energy, including kinetic energy, potential energy, and thermal energy. Our goal is to understand the characteristics of energy, how energy is used, and, especially important, how energy is transformed from one form to another. For example, this pole vaulter, after years of training, has become extraordinarily proficient at transforming his energy of motion into energy associated with height from the ground.

We'll also discover a very powerful conservation law for energy. Some scientists consider the law of conservation of energy to be the most important of all the laws of nature. But all that in due time. First we have to start with the basic ideas.

10.1 A "Natural Money" Called Energy

We will start by discussing what seems to be a completely unrelated topic: money. As you will discover, monetary systems have much in common with energy. Let's begin with a short story.

The Parable of the Lost Penny

John was a hard worker. His only source of income was the paycheck he received each month. Even though most of each paycheck had to be spent on basic necessities, John managed to keep a respectable balance in his checking account. He even saved enough to occasionally buy a few savings bonds, his investment in the future.

John never cared much for pennies, so he kept a jar by the door and dropped all his pennies into it at the end of each day. Eventually, he reasoned, his saved pennies would be worth taking to the bank and converting into crisp new dollar bills.

John found it fascinating to keep track of these various forms of money. He noticed, to his dismay, that the amount of money in his checking account did not spontaneously increase overnight. Furthermore, there seemed to be a definite correlation between the size of his paycheck and the amount of money he had in the bank. So John decided to embark on a systematic study of money.

He began, as would any good scientist, by using his initial observations to formulate a hypothesis, which he called a *model* of the monetary system. He found that he could represent his monetary model with the flowchart in Figure 10.1.

FIGURE 10.1 John's model of the monetary system.

As the chart shows, John divided his money into two basic types, liquid assets and saved assets. The *liquid assets L*, which included his checking account and the cash in his pockets, were moneys available for immediate use. His *saved assets S*, which included his savings bonds as well as the jar of pennies, had the potential to be converted into liquid assets, but they were not available for immediate use.

John decided to call the sum total of assets his *wealth:* $W = L + S$.

John's assets were, more or less, simply definitions. The more interesting question, he thought, was how his wealth depended on his *income I* and *expenditures E*. These represented money transferred *to* him by his employer and money transferred *by* him to stores and bill collectors. After painstakingly collecting and analyzing his data, John finally determined that the relationship between monetary transfers and wealth is

$$\Delta W = I - E$$

John interpreted this equation to mean that the *change* in his wealth, ΔW, was numerically equal to the *net* monetary transfer $I - E$.

During a week-long period when John stayed home sick, isolated from the rest of the world, he had neither income nor expenses. In grand confirmation of his hypothesis, he found that his wealth W_f at the end of the week was identical to his wealth W_i at the week's beginning. That is, $W_f = W_i$. This occurred despite the fact that he had moved pennies from his pocket to the jar and also, by telephone, had sold some bonds and transferred the money to his checking account. In other words, John found that he could make all of the *internal* conversions of assets from one form to another that he wanted, but his total wealth remained constant ($W = $ constant) as long as he was isolated from the world. This seemed such a remarkable rule that John named it the *law of conservation of wealth.*

One day, however, John added up his income and expenditures for the week, and the changes in his various assets, and he was 1¢ off! Inexplicably, some money seemed to have vanished. He was devastated. All those years of careful research, and now it seemed that his monetary hypothesis might not be true. Under some circumstances, yet to be discovered, it looked like $\Delta W \neq I - E$. Off by a measly penny. A wasted scientific life. . . .

But wait! In a flash of inspiration, John realized that perhaps there were other types of assets, yet to be discovered, and that his monetary hypothesis would still be valid if *all* assets were included. Weeks went by as John, in frantic activity, searched fruitlessly for previously *hidden* assets. Then one day, as John lifted the cushion off the sofa to vacuum out the potato chip crumbs—lo and behold, there it was!—the missing penny!

John raced to complete his theory, now including money in the sofa, the washing machine, and behind the radiator as previously unknown forms of assets that were easy to convert from other forms, but often rather difficult to recover.

Other researchers soon discovered other types of assets, such as the remarkable find of the "cash in the mattress." To this day, when *all* known assets are included, monetary scientists have never found a violation of John's simple hypothesis that $\Delta W = I - E$. John was last seen sailing for Stockholm to collect the Nobel Prize for his Theory of Wealth.

10.2 The Basic Energy Model

John, despite his diligent efforts, did not discover a law of nature. The monetary system is a human construction that, by design, obeys John's "laws." Monetary system laws, such as that you cannot print money in your basement, are enforced by society, not by nature. But suppose that physical objects possessed a "natural money" that was governed by a theory, or model, similar to John's. An object might have several forms of natural money that could be converted back and forth, but the total amount of an object's natural money would *change* only if natural money were *transferred* to or from the object. Two key words here, as in John's model, are *transfer* and *change*.

One of the greatest and most significant discoveries of science is that there is such a "natural money" called **energy.** You have heard of some of the many forms of energy, such as solar energy or nuclear energy, but others may be new to you. These forms of energy can differ as much as a checking account differs from loose change in the sofa. Much of our study is going to be focused on the *transformation* of energy from one form to another. Much of modern technology is concerned with transforming energy, such as changing the chemical energy of oil molecules to electrical energy or to the kinetic energy of your car.

As we use energy concepts, we will be "accounting" for energy that is transferred in or out of a system or that is transformed from one form to another within a system. Figure 10.2 shows a simple model of energy that is based on John's model of the monetary system. Many details must be added to this model, but it's a good starting point. The fact that nature "balances the books" for energy is one of the most profound discoveries of science.

A major goal of ours is to discover the conditions under which energy is conserved. Surprisingly, the *law of conservation of energy* was not recognized until the mid-nineteenth century, long after Newton. The reason, similar to John's lost penny, was that it took scientists a long time to realize how many types of energy there are and the various ways that energy can be converted from one form to another. As you'll soon learn, energy ideas go well beyond Newtonian mechanics to include new concepts about heat, about chemical energy, and about the energy of the individual atoms and molecules that comprise an object. All of these forms of energy will ultimately have to be included in our accounting scheme for energy.

There are several kinds of energy within the system. These can be transformed back and forth without loss.

Energy into system

Readily available energy K
Stored energy U
Hard-to-recover energy E_{th}
Total energy $E = K + U + E_{th}$

Energy out of system

FIGURE 10.2 An initial model of energy.

Systems and Energy

In Chapter 9 we introduced the idea of a *system* of interacting objects. A system can be quite simple, such as a saltshaker sliding across the table, or much more complex, such as a city or a human body. But whether simple or complex, every system in nature has associated with it a quantity we call its **total energy** E. Like John's total wealth, which was made up of assets of many kinds, the total energy of a system is made up of many kinds of energies. In the table below, we give a brief overview of some of the more important forms of energy; in the rest of the chapter we'll look at several of these forms of energy in much greater detail.

Some important forms of energy

Kinetic energy K

Kinetic energy is the energy of *motion.* All moving objects have kinetic energy. The heavier an object, and the faster it moves, the more kinetic energy it has. The wrecking ball in this picture is effective in part because of its large kinetic energy.

Gravitational potential energy U_g

Gravitational potential energy is *stored* energy associated with an object's *height above the ground.* As this roller coaster ascends the track, energy is stored as increased gravitational potential energy. As it descends, this stored energy is converted into kinetic energy.

Elastic or spring potential energy U_s

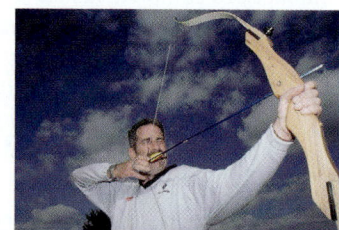

Elastic potential energy is energy stored when a spring or other elastic object, such as this archer's bow, is *stretched.* This energy can later be transformed into the kinetic energy of the arrow. We'll sometimes use the symbol U to represent potential energy when it is not important to distinguish between U_g and U_s.

Thermal energy E_{th}

Hot objects have more *thermal energy* than cold ones because the molecules in a hot object jiggle around more than those in a cold object. Thermal energy is really just the sum of the microscopic kinetic and potential energies of all the molecules in an object. In boiling water, some molecules have enough energy to escape the water as steam.

Chemical energy E_{chem}

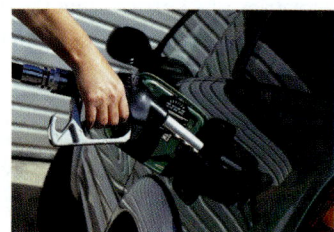

Electric forces cause atoms to bind together to make molecules. Energy can be stored in these bonds, energy that can later be released as the bonds are rearranged during chemical reactions. When we burn fuel to run our car, or eat food to power our bodies, we are using *chemical energy.*

Nuclear energy $E_{nuclear}$

An enormous amount of energy is stored in the *nucleus,* the tiny core of an atom. Certain nuclei can be made to break apart, releasing some of this *nuclear energy,* which is transformed into the kinetic energy of the fragments and then into thermal energy. This is the source of energy of nuclear power plants and nuclear weapons.

A system may have many of these kinds of energy present in it at once. For instance, a moving car has kinetic energy of motion, chemical energy stored in its gasoline, thermal energy in its hot engine, and other forms of energy in its many other parts. The total energy of the system, E, is just the *sum* of the different energies present in the system, so that we have

$$E = K + U_g + U_s + E_{th} + E_{chem} + \cdots \qquad (10.1)$$

The energies shown in this sum are the forms of energy in which we'll be most interested in this and the next chapter. The ellipses (. . .) represent other forms of energy, such as nuclear or electric, that also might be present. We'll treat these and others in later chapters.

Energy Transformations

We've seen that all systems contain energy in many different forms. But if the amounts of each form of energy never changed, the world would be a very dull place. What makes the world interesting is that **energy of one kind can *trans-***

form into energy of another kind. The gravitational potential energy of the roller coaster at the top of the track is rapidly converted into kinetic energy as the coaster descends; the chemical energy of gasoline is converted into the kinetic energy of your moving car. The following table illustrates a few common energy transformations. In this table, we'll use an arrow $\rightarrow$ as a shorthand way of representing an energy transformation.

Some energy transformations

A weightlifter lifts a barbell over her head
The barbell has much more gravitational potential energy when high above her head than when on the floor. To lift the barbell, she is transforming chemical energy in her body into gravitational potential energy of the barbell.

$$E_{chem} \rightarrow U_g$$

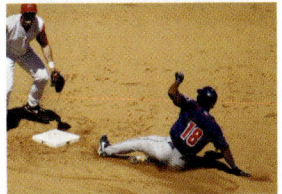

A base runner slides into the base
When running, he has lots of kinetic energy. After sliding, he has none. His kinetic energy is transformed mainly into thermal energy: the ground and his legs are slightly warmer.

$$K \rightarrow E_{th}$$

A burning campfire
The wood contains considerable chemical energy. When the carbon in the wood combines chemically with oxygen in the air, this chemical energy is transformed largely into thermal energy of the hot gases and embers.

$$E_{chem} \rightarrow E_{th}$$

A springboard diver
Here's a two-step energy transformation. The picture shows the diver after his first jump onto the board itself. At the instant shown, the board is flexed to its maximum extent. There is a large amount of elastic potential energy stored in the board. Soon this energy will begin to be transformed into kinetic energy; as he rises into the air and slows, this kinetic energy will be transformed into gravitational potential energy.

$$U_s \rightarrow K \rightarrow U_g$$

Figure 10.3 reinforces the idea that **energy transformations are changes of energy *within* the system from one form to another.** Note that it is easy to convert kinetic, potential, or chemical energies into thermal energy. But converting thermal energy back into these other forms is not so easy. How it can be done, and what possible limitations there might be in doing so, will form a large part of the next chapter.

Energy Transfers: Work and Heat

We've just seen that energy *transformations* occur between forms of energy *within* a system. In our monetary model, these transformations are like John's shifting of money between his own various assets, such as from his savings

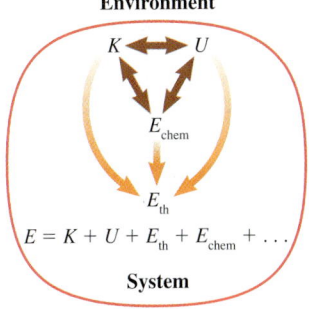

FIGURE 10.3 Energy transformations occur within the system.

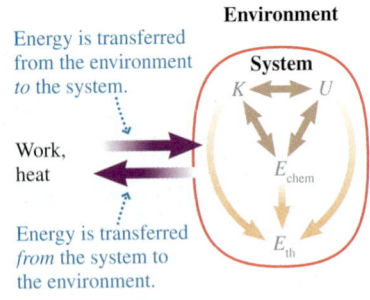

FIGURE 10.4 Work and heat are energy transfers into and out of the system.

One dictionary defines _work_ as:

1. Physical or mental effort; labor.
2. The activity by which one makes a living.
3. A task or duty.
4. Something produced as a result of effort, such as a _work of art_.
5. Plural _works:_ The essential or operating parts of a mechanism.
6. The transfer of energy to a body by application of a force.

account to stocks. But John also interacted with the greater world around him, receiving money as income and outlaying it as expenditures. Every physical system also interacts with the world around it, that is, with its _environment_. In the course of these interactions, the system can exchange energy with the environment. **An exchange of energy between system and environment is called an energy _transfer_.** There are two primary energy transfer processes: **work,** the _mechanical_ transfer of energy to or from a system by pushing or pulling on it, and **heat,** the _nonmechanical_ transfer of energy from the environment to the system (or vice versa) _because of a temperature difference between the two_. Figure 10.4 shows how our energy model is modified to include energy transfers. In this chapter we'll focus mainly on work; the concept of heat will be developed much further in Chapters 11 and 12.

Work is a common word in the English language, with many meanings. When you first think of work, you probably think of the first two definitions in this list. After all, we talk about "working out," or we say, "I just got home from work." But that is _not_ what work means in physics.

In physics we use _work_ in the sense of definition 6: Work is the process of _transferring_ energy from the environment to a system, or from a system to the environment, by the application of mechanical forces—pushes and pulls—to the system. Once the energy has been transferred to the system, it can appear in many forms. Exactly what form it takes depends on the details of the system and how the forces are applied. The table below gives a few examples of energy transfers due to work. We use W as the symbol for work.

Energy transfers: work

Putting a shot

The system: The shot.

The environment: The athlete.

As the athlete pushes on the shot to get it moving, he is doing work on the system. That is, he is transferring energy from himself to the ball. The energy transferred to the system appears as kinetic energy.

The transfer: $W \to K$

Striking a match

The system: The match and matchbox.

The environment: The hand.

As the hand quickly pulls the match across the box, the hand does work on the system, increasing its thermal energy. The matchhead becomes hot enough to ignite.

The transfer: $W \to E_{th}$

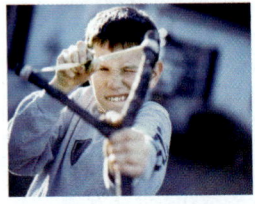

Firing a slingshot

The system: The slingshot.

The environment: The boy.

As the boy pulls back on the elastic bands, he does work on the system, increasing its elastic potential energy.

The transfer: $W \to U_s$

Notice that in each example above, the environment applies a force while the system undergoes a _displacement_. Energy is transferred as work only when the system _moves_ while the force acts. A force applied to a stationary object, such as when you push against a wall, transfers no energy to the object and thus does no work.

NOTE ▶ In the table above, energy is being transferred _from_ the athlete _to_ the shot by the force of his hand. We say he "does work" on the shot, or "work is done" by the force of his hand. ◀

It is also possible to convert work into gravitational potential, electric, or even chemical energy. We'll have much more to say about work in the next section. But the key points to remember are that **work is the transfer of energy to or from a system by the application of forces,** and that **the system must undergo a displacement for this energy to be transferred.**

There is a second, nonmechanical means of transferring energy between a system and its environment, which we discuss here only briefly. As mentioned before, we'll have much more to say about heat in the next two chapters. When a hot object is placed in contact with a cooler one, energy flows naturally from the hot object to the cool one. **The transfer of energy from a hot to a cold object is called *heat*,** and it is given the symbol Q. It is important to note that heat is not an energy *of* a system, as are kinetic energy and chemical energy. Rather, heat is energy *transferred* between two systems.

As the hand in the photo was held against the wall, heat was transferred from the warm hand to the cool wall, warming up the wall. The warm "handprint" can be imaged using a special camera sensitive to the temperature of objects.

STOP TO THINK 10.1 A child slides down a playground slide at constant speed. The energy transformation is

A. $U_g \rightarrow K$ B. $K \rightarrow U_g$ C. $W \rightarrow K$ D. $U_g \rightarrow E_{th}$ E. $K \rightarrow E_{th}$

10.3 The Law of Conservation of Energy

Remember that when John was *isolated* from the rest of the world—having neither income nor expenses—his internal wealth could be converted between its many forms, but his *total* wealth remained constant. A similar but much more fundamental law is found for the "natural money" of energy.

Let's start our study of this law by considering an **isolated system** that is separated from its surrounding environment in such a way that no energy can flow into or out of the system. This means that no work is done on the system, nor is any energy transferred as heat. We've already seen that the total energy of a system is made up of many forms of energy that are continually transforming from one kind to another. It is a deep and remarkable fact of nature that during these transformations, the total energy of an isolated system—the *sum* of all of the individual kinds of energy—remains *constant*. Any increase in, say, the system's kinetic energy must be accompanied by a decrease in its potential or thermal energies so that the total energy remains unchanged, as shown in Figure 10.5. We say that **the total energy of an isolated system is *conserved*,** giving us the following *law of conservation of energy.*

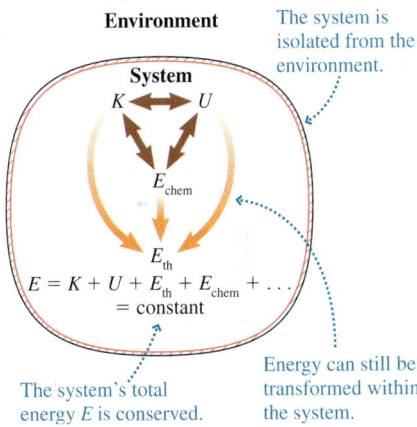

FIGURE 10.5 An isolated system.

Law of conservation of energy for an isolated system The total energy of an isolated system remains constant:

The energies in the system are constantly transforming from one kind to another . . . . . . but their *sum* is a constant: it doesn't change.

$$K + U_g + U_s + E_{th} + E_{chem} + \ldots = E = \text{constant} \quad (10.2)$$

Another way to think of this conservation law is in terms of energy *changes*. Recall that we denote the change in a quantity by the symbol Δ, so we write the change in a system's kinetic energy, for instance, as ΔK. Now suppose that an isolated system has its kinetic energy change by ΔK, its gravitational potential energy by ΔU_g, and so on. Then the sum of these changes is the change in the total energy. But since the total energy is constant, its change is *zero*. We can thus write the law of conservation of energy in an alternate form as

Any increase in one form of energy must be accompanied by a decrease in other forms, so that the total change is zero.

The law of conservation of energy sets a fundamental constraint on those processes that can occur in nature. In any process that occurs within an isolated system, the changes in each form of energy must add up to zero, as required by Equation 10.3.

CONCEPTUAL EXAMPLE 10.1 **Energy changes in a bungee ride**

A popular fair attraction is the trampoline bungee ride. The rider bounces up and down on large bungee cords. During part of her motion she is found to be moving upward with the cords becoming more stretched. Is she speeding up or slowing down during this interval?

REASON We'll take our system to include the rider, the bungee cords, and the *earth*. We'll see later how gravitational potential energy is stored in the *system* consisting of the earth and an object such as the rider. With this choice of system,

to a good approximation the system is isolated, with no energy being transferred into or out of the system. Thus the total energy of the system is constant: $\Delta E = 0$.

Because she's moving upward, her height is increasing—and thus so is her gravitational potential energy. Thus $\Delta U_g > 0$. We also know that the cords are getting more stretched, hence more elastic potential energy is being stored. Thus $\Delta U_s > 0$ as well. Now the law of conservation of energy, Equation 10.3, states that $\Delta E = \Delta K + \Delta U_g + \Delta U_s = 0$, so that $\Delta K = -(\Delta U_g + \Delta U_s)$. Both ΔU_g and ΔU_s are positive, so ΔK must be *negative*. This means that her kinetic energy is *decreasing*. Since kinetic energy is energy of motion, this means that she's slowing down.

ASSESS In that part of her motion where she's moving upward and the cords are stretching, she's approaching the highest point of her motion. It makes sense that she's slowing down here, since at the high point her speed is instantaneously zero.

Systems That Aren't Isolated

When John had income and expenses, his total wealth could change. Indeed, he found that his wealth increased by exactly the amount of his income, and decreased by exactly the amount of his expenditures. Similarly, if a system is *not* isolated, so that it can exchange energy with its environment, the system's energy can change. We have seen that the two primary means of energy exchange are work and heat. If an amount of work W is done on the system, this means that an amount of energy W is transferred from the environment to the system, increasing the system's energy by exactly W. Similarly, if a certain amount of energy is transferred from a hot environment to a cooler system as heat Q, the system's energy will increase by exactly the amount Q. As illustrated in Figure 10.6, **the change in the system's energy is simply the sum of the work done on the system and the heat transferred to the system:**

$$\Delta E = W + Q$$

This gives us a more general statement of conservation of energy:

Energy is transferred to or from the system as work and heat.

Environment

System
Total energy E
$\Delta E = W + Q$

Work W
Heat Q

The *change* in the system's energy equals the amount of work done or heat transferred.

FIGURE 10.6 The law of conservation of energy.

Equation 10.4 is the fullest expression of the law of conservation of energy. It's usually called the **first law of thermodynamics,** but it's really just a restatement of the law of conservation of energy to include the possibility of energy transfers. In this chapter we'll refer to it simply as the law of conservation of energy.

NOTE ▶ It's important to realize that even when the system is not isolated, energy is conserved overall. The energy transferred to the system as, say, work increases the energy of the system. But this energy is *removed* from the environment, so that the total energy of system *plus* environment is still conserved. ◀

Systems and Conservation of Energy

To apply the law of conservation of energy, you need to carefully define which objects make up the system and which belong to the environment. This choice will affect how we analyze the various energy transfers and tranformations that occur. In doing so, we need to make a distinction between two classes of forces. **Internal forces** are forces between objects *within* the system. If a weightlifter and barbell are both part of the system, the forces $\vec{F}_{\text{weightlifter on barbell}}$ and $\vec{F}_{\text{barbell on weightlifter}}$ are both internal forces. Internal forces are responsible for energy transformations within the system. Because they are internal to the system, however, **internal forces cannot do work on the system** and thereby change its energy. **External forces** act on the system, but their agent is part of the *environment.* **External forces *can* do work on the system,** transferring energy in or out of it. Whether a given force is an internal or external force depends on the choice of what's included in the system. The following table shows some choices for a crane accelerating a heavy ball upward.

Airplanes are assisted in takeoff from aircraft carriers by a steam-powered catapult under the flight deck. The force of this catapult does work W on the plane, leading to a large increase ΔK in the plane's kinetic energy.

Different choices of the system

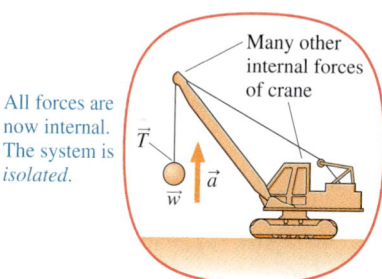

System: The ball only	Ball + earth	Ball + earth + crane
Internal forces: None	$\vec{w}$	$\vec{T}, \vec{w}$, many internal forces of crane
External forces: $\vec{T}, \vec{w}$	$\vec{T}$	None
System energies: K	K, U_g	K, U_g, E_{chem}

| **Energy analysis:** Tension does positive work and the weight does negative work, but since $T > w$ the net work is positive. This work serves to increase the only energy of the system, its kinetic energy. Notice that since the earth is *not* part of the system, the system has no gravitational potential energy. | The weight force is now an *internal* force. That is, it is an interaction force between two objects—the ball and the earth—that are part of the system. The tension force is still an *external* force that does work on the system. This work increases the gravitational potential energy and the kinetic energy of the system. | Now all the forces are internal, and no work is done on the system: The system is *isolated.* With this choice of system, the increased potential and kinetic energy of the ball come from an energy *transformation* from the chemical energy of the crane's fuel. |
| **Energy equation:** $\Delta K = W$ | $\Delta K + \Delta U_g = W$ | $\Delta K + \Delta U_g + \Delta E_{\text{chem}} = 0$ |

There are evidently many possible choices of the system for a given situation. However, certain choices can make problem solving using the law of energy conservation easier. For the crane above, we'd probably choose the second system consisting of the ball and the earth, since it is a good balance between reducing the number of external forces and having only simple system energies such as K

and U_g. The third choice would be hard to work with, since the many complicated internal forces are difficult to calculate. Tactics Box 10.1 gives some suggestions on how to make a good choice for the system.

TACTICS BOX 10.1 Choosing the system for conservation-of-energy problems   Exercise 6

The system should include all of the objects identified as follows:

❶ If the speed of an object or objects is changing, the system should include these moving objects because their kinetic energy is changing.

❷ If the height of an object or objects is changing, the system should include the raised object(s) *plus* the earth. This is because potential energy is stored via the gravitational interaction of the earth and object(s).

❸ If the compression or extension of a spring is changing, the system should include the spring because elastic potential energy is stored in the spring itself.

❹ If kinetic or rolling friction is present, the system should include the moving object and the surface on which it slides or rolls. This is because thermal energy is created in both the moving object and the surface, and we want this thermal energy to all be within the system.

Working with Energy Transformations

The law of conservation of energy applies to every form of energy, from kinetic to chemical to nuclear. For the rest of this chapter, however, we'll narrow our focus a bit and only concern ourselves with the forms of energy typically transformed during the motion of ordinary objects. These energies are the kinetic energy K, the potential energy U (which includes both U_g and U_s), and thermal energy E_{th}. The sum of the kinetic and potential energy, $K + U = K + U_g + U_s$, is called the **mechanical energy** of the system. We'll also limit our analysis to energy transfers in the form of work W. In Chapter 11 we'll expand our scope to include other forms of energy listed in the earlier table, as well as energy transfers as heat Q.

The fact that energy is conserved can be a powerful tool for analyzing the dynamics of moving objects. To see how we can apply the law of conservation of energy to dynamics problems, let's use the fact that the change in any quantity is its final value minus its initial value so that, for example, $\Delta K = K_f - K_i$. Then we can write the law of conservation of energy, Equation 10.4 (with $Q = 0$), as

$$(K_f - K_i) + (U_f - U_i) + \Delta E_{th} = W \tag{10.5}$$

NOTE ▶ We don't rewrite ΔE_{th} as $(E_{th})_f - (E_{th})_i$ because the initial thermal energy of an object is typically unknown. Only the *change* in E_{th} can be measured. ◀

Rearranging, we have

$$K_i + U_i + W = K_f + U_f + \Delta E_{th} \tag{10.6}$$

The initial energy of the system…   …plus the energy transferred to the system as work…   …equals the final system energy, now possibly including extra thermal energy.

If no external forces do work on the system, $W = 0$ in Equation 10.6 and the system is *isolated*. **If no kinetic friction is present, ΔE_{th} will be zero and mechanical energy will be conserved.** Equation 10.6 then becomes the **law of conservation of mechanical energy:**

$$K_i + U_i = K_f + U_f \tag{10.7}$$

Equations 10.6 and 10.7 summarize what we have learned about the conservation of energy, and they will be the basis of our strategy for solving problems using the law of conservation of energy. Much of the rest of this chapter will be concerned with finding quantitative expressions for the different forms of energy in the system and discussing the important question of what to include in the system. We'll use the following Problem-Solving Strategy as we further develop these ideas.

PROBLEM-SOLVING STRATEGY 10.1 Conservation of energy problems

PREPARE   Choose what to include in your system (see Tactics Box 10.1). Draw a before-and-after visual overview, as outlined in Tactics Box 9.1. Note known quantities, and determine what quantity you're trying to find. If the system is isolated and if there is no friction, your solution will be based on Equation 10.7, otherwise you should use Equation 10.6.

Identify which mechanical energies in the system are changing:

- If the *speed* of the object is changing, include K_i and K_f in your solution.
- If the *height* of the object is changing, include $(U_g)_i$ and $(U_g)_f$.
- If the *length* of a spring is changing, include $(U_s)_i$ and $(U_s)_f$.
- If kinetic friction is present, ΔE_{th} will be positive. Some kinetic or potential energy will be transformed into thermal energy.

If an external force acts on the system, you'll need to include the work W done by this force in Equation 10.6.

SOLVE   Depending on the problem, you'll need to calculate initial and/or final values of these energies and insert them into Equation 10.6 or 10.7. Then you can solve for the unknown energies, and from these any unknown speeds (from K), positions (from U), or displacements or forces (from W).

ASSESS   Check the signs of your energies. Kinetic energy, as we'll see, is always positive. In the systems we'll study in this chapter, thermal energy can only increase, so that its change is positive. In Chapters 11 and 12 we'll study systems for which the thermal energy can decrease.

Spring into action BIO A locust can jump as far as one meter, an impressive distance for such a small animal. To make such a jump, its legs must extend much more rapidly than muscles can ordinarily contract. Thus, instead of using its muscles to make the jump directly, the locust uses them to more slowly stretch an internal "spring" near its knee joint. This stores elastic potential energy in the spring. When the muscles relax, the spring is suddenly released, and its energy is rapidly converted into kinetic energy of the insect.

10.4 Work

We've already discussed work as the transfer of energy between a system and its environment by the application of forces on the system. We also noted that in order for energy to be transferred in this way, the system must undergo a displacement—it must *move*—during the time that the force is applied. Let's further investigate the relationship between work, force, and displacement. We'll find that there is a simple expression for work, which we can then use to quantify other kinds of energy as well.

The system's kinetic energy increases and the boarder speeds up.

The force of the wind $\vec{F}$ does work on the system.

FIGURE 10.7 The force of the wind does work on the system, increasing its kinetic energy K.

Consider a system consisting of a windsurfer at rest, as shown on the left in Figure 10.7. Let's assume that there is no friction between his board and the water. Initially the system has no kinetic energy. But if a force from outside the system, such as the force due to the wind, begins to act on the system, the surfer will begin to speed up, and his kinetic energy will increase. In terms of energy transfers, we would say that the energy of the system has increased because of the work done on the system by the force of the wind.

What determines how much work is done by the force of the wind? First, we note that the greater the distance over which the wind pushes the surfer, the faster the surfer goes, and the more his kinetic energy increases. This implies a greater transfer of energy. So **the larger the displacement, the greater the work done.** Second, if the wind pushes with a stronger force, the surfer speeds up more rapidly, and the change in his kinetic energy is greater than with a weaker force. **The stronger the force, the greater the work done.**

This experiment suggests that the amount of energy transferred into a system by a force $\vec{F}$—that is, the amount of work done by $\vec{F}$—depends on both the magnitude F of the force *and* the displacement d of the system. Many experiments of this kind have established that the amount of work done by $\vec{F}$ is *proportional* to both F and d. For the simplest case described above, where the force $\vec{F}$ is constant and points in the direction of the object's displacement, the expression for the work done is found to be

$$W = Fd \qquad (10.8)$$

Work done by a constant force $\vec{F}$
in the direction of a displacement $\vec{d}$

LINEAR
p. 38

The unit of work, that of force multiplied by distance, is N · m. This unit is so important that it has been given its own name, the **joule** (rhymes with *tool*). We define:

$$1 \text{ joule} = 1 \text{ J} = 1 \text{ N} \cdot \text{m}$$

Since work is simply energy being transferred, **the joule is the unit of *all* forms of energy.** Note that work is a *scalar* quantity.

EXAMPLE 10.1 Work done in pushing a crate

Sarah pushes a heavy crate 3.0 m along the floor at a constant speed. She pushes with a constant horizontal force of magnitude 70 N. How much work does Sarah do on the crate?

PREPARE We begin with the visual overview in Figure 10.8. Sarah pushes with a constant force in the direction of the crate's motion, so we can use Equation 10.8 to find the work done.

	Known
	$F = 70$ N
	$d = 3.0$ m
	$v = $ constant
	Find
	W

FIGURE 10.8 Sarah pushing a crate.

SOLVE The work done by Sarah is given by

$$W = Fd = (70 \text{ N})(3.0 \text{ m}) = 210 \text{ J}$$

ASSESS Since the crate moves at a constant speed, it must be in dynamic equilibrium with $\vec{F}_{\text{net}} = \vec{0}$. This means that a friction force (not shown) must act opposite to Sarah's push. If friction is present, Tactics Box 10.1 suggests taking the crate *and* the floor as the system. The work Sarah does represents energy transferred *into* the system. In this case, the work increases the thermal energy in the crate and the part of the floor along which it slid. Contrast this with the windsurfer, where work increased the windsurfer's kinetic energy. Both situations are consistent with the energy model shown in Figure 10.4, which you should review at this point.

Force at an Angle to the Displacement

Pushing a crate in the same direction as the crate's displacement is the most efficient way to transfer energy into the system, and so the largest possible amount of work is done. Less work is done if the force acts at an angle to the displacement. To see this, consider the kite buggy of Figure 10.9a, pulled along a horizontal path by the angled force of the kite string $\vec{F}$. As shown in Figure 10.9b, we can break $\vec{F}$ into a component $F_\perp$ perpendicular to the motion, and a component $F_\parallel$ parallel to the motion. Only the parallel component acts to accelerate the rider and increase his kinetic energy, so only the parallel component does work on the rider. From Figure 10.9b, we see that if the angle between $\vec{F}$ and the displacement is θ, then the parallel component is $F_\parallel = F\cos\theta$. So when the force acts at an angle θ to the direction of the displacement, we have

$$W = F_\parallel d = Fd\cos\theta \qquad (10.9)$$

Work done by a constant force $\vec{F}$ at an angle θ to the displacement $\vec{d}$

Notice that this more general definition of work agrees with Equation 10.8 if $\theta = 0°$.

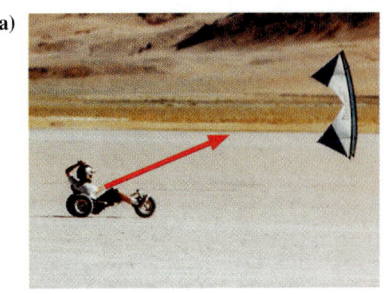

(a)

(b) The rider undergoes a displacement $\vec{d}$.

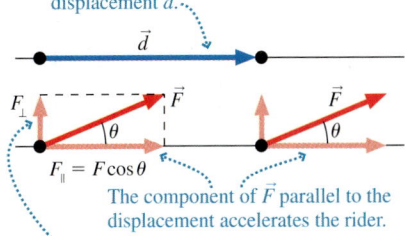

$\vec{d}$

$F_\perp$ $\vec{F}$ θ

$F_\parallel = F\cos\theta$

The component of $\vec{F}$ parallel to the displacement accelerates the rider.

The component of $\vec{F}$ perpendicular to the displacement only pulls up on the rider. It doesn't accelerate him.

FIGURE 10.9 Finding the work done when the force is at an angle to the displacement.

CONCEPTUAL EXAMPLE 10.2 **Work done by a parachute**

A drag racer is slowed by a parachute. What is the sign of the work done?

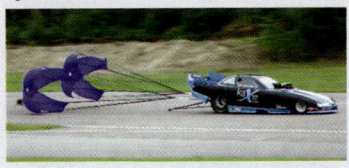

REASON The drag force on the drag racer is shown in Figure 10.10, along with the dragster's displacement as it slows. The force points in the opposite direction to the displacement, so that the angle θ in Equation 10.9 is 180°. Then $\cos\theta = \cos(180°) = -1$. Since F and d in Equation 10.9 are magnitudes, and hence positive, this means that the work $W = Fd\cos\theta = -Fd$ done by the drag force is *negative*.

$\vec{F}$ $\vec{F}$
$\theta = 180°$ $\vec{d}$

FIGURE 10.10 The force acting on a drag racer.

ASSESS Applying Equation 10.4, the law of conservation of energy, to this situation, we have

$$\Delta K = W$$

because the only system energy that changes is the racer's kinetic energy K. Since the kinetic energy is decreasing, its change ΔK is negative. This agrees with the sign of W. This example illustrates the general principle that **negative work represents a transfer of energy out of the system.**

Tactics Box 10.2 shows how to calculate the work done by a force at any angle to the direction of motion. The system illustrated is a block sliding on a frictionless horizontal surface, so that only the kinetic energy is changing. However, the same relationships hold for any object undergoing a displacement.

The quantities F and d are always positive, so **the sign of W is determined entirely by the angle θ between the force and the displacement.** Note that

Activ Physics ONLINE 5.1

Equation 10.9, $W = Fd\cos\theta$, is valid for any angle θ. In three special cases, $\theta = 0°$, $\theta = 90°$, and $\theta = 180°$, however, there are simple versions of Equation 10.9 that you can use. These are noted in Tactics Box 10.2.

MP **TACTICS BOX 10.2 Calculating the work done by a constant force**   ✏ Exercises 9,11,12

Direction of force relative to displacement	Angles and work done	Sign of W	Energy transfer
Before: $\vec{v}_i$ After: $\vec{v}_f$ $\vec{d}$ $\vec{F}$ $\theta = 0°$	$\theta = 0°$ $\cos\theta = 1$ $W = Fd$	+	The force is in the direction of motion. The block has its greatest positive acceleration. K increases the most: **Maximum energy transfer to system.**
$\theta < 90°$ $\vec{F}$ $\vec{d}$	$\theta < 90°$ $W = Fd\cos\theta$	+	The component of force parallel to the displacement is less than F. The block has a smaller positive acceleration. K increases less: **Moderate energy transfer to system.**
$\theta = 90°$ $\vec{F}$ $\vec{d}$	$\theta = 90°$ $\cos\theta = 0$ $W = 0$	0	There is no component of force in direction of motion. The block moves at constant speed. No change in K: **No energy transferred.**
$\theta > 90°$ $\vec{F}$ $\vec{d}$	$\theta > 90°$ $W = Fd\cos\theta$	−	The component of force parallel to the displacement is opposite to the motion. The block slows down, and K decreases: **Moderate energy transfer *out* of system.**
$\theta = 180°$ $\vec{F}$ $\vec{d}$	$\theta = 180°$ $\cos\theta = -1$ $W = -Fd$	−	The force is directly opposite to the motion. The block has it greatest deceleration. K decreases the most. **Maximum energy transfer out of system.**

EXAMPLE 10.2 Work done in pulling a suitcase

A strap inclined upward at a 45° angle pulls a suitcase through the airport. The tension in the strap is 20 N. How much work does the tension do if the suitcase is pulled 100 m at a constant speed?

PREPARE Figure 10.11 shows a visual overview. Since the case moves at a constant speed, there must be a rolling friction force acting to the left. Tactics Box 10.1 suggests in this case that we take as our system the suitcase and the floor upon which it rolls.

SOLVE We can use Equation 10.9 to find that the tension does work

$$W = Td\cos\theta = (20 \text{ N})(100 \text{ m})\cos 45° = 1400 \text{ J}$$

ASSESS Because a person is pulling on the other end of the strap, causing the tension, we would say informally that the person does 1400 J of work on the suitcase. This work represents

FIGURE 10.11 A suitcase pulled by a strap.

energy transferred into the suitcase/floor system. Since the suitcase moves at a constant speed, the system's kinetic energy doesn't change. Thus the work goes entirely into increasing the *thermal* energy E_{th} of the suitcase and the floor.

If several forces act on an object that undergoes a displacement, each does work on the object. The **total (or net) work** W_{total} is the sum of the work done by each force. The total work represents the total energy transfer *to* the system from the environment (if $W_{total} > 0$) or *from* the system to the environment (if $W_{total} < 0$).

Forces That Do No Work

The fact that a force acts on an object doesn't mean that the force will do work on the object. The table below shows three common cases where a force does no work.

Forces that do no work

If the object undergoes no displacement while the force acts, no work is done.

This can sometimes seem counterintuitive. The weightlifter struggles mightily to hold the barbell over his head. But during the time the barbell remains stationary, he does no work on it because its displacement is zero. But why then is it so hard for him to hold it there? We'll see in Chapter 11 that it takes a rapid conversion of his internal chemical energy to keep his arms extended under this great load.

A force perpendicular to the displacement does no work.

The woman exerts only a vertical force on the briefcase she's carrying. This force has no component in the direction of the displacement, so the briefcase moves at a constant velocity and its kinetic energy remains constant. Since the energy of the briefcase doesn't change, it must be that no energy is being transferred to it as work.

(This is the case where $\theta = 90°$ in Tactics Box 10.2.)

If the part of the object on which the force acts undergoes no displacement, no work is done.

Even though the wall pushes on the skater with a normal force $\vec{n}$ and she undergoes a displacement $\vec{d}$, the wall does no work on her, because the point of her body on which $\vec{n}$ acts—her hands—undergoes no displacement. This makes sense: How could energy be transferred as work from an inert, stationary object? So where does her kinetic energy come from? This will be the subject of much of Chapter 11. Can you guess?

STOP TO THINK 10.2 Which force does the most work?

A. The 10 N force.
B. The 8 N force.
C. The 6 N force.
D. They all do the same amount of work.

10 N   8 N   6 N
60°
$\vec{d}$ $\vec{d}$ $\vec{d}$

10.5 Kinetic Energy

We've already qualitatively discussed kinetic energy, an object's energy of motion. Let's now use what we've learned about work, and some simple kinematics, to find a quantitative expression for kinetic energy. Consider the system consisting of a car being pulled by a tow rope as in Figure 10.12. The rope pulls with a constant force $\vec{F}$ while the car undergoes a displacement $\vec{d}$, so that the force does work $W = Fd$ on the car. If we ignore friction and drag, the work done by $\vec{F}$ will

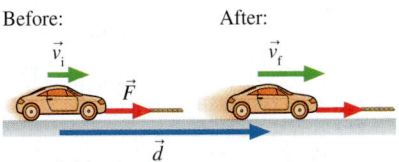

FIGURE 10.12 The work done by the tow rope increases the car's kinetic energy.

be transferred entirely into the car's energy of motion—its kinetic energy. In this case, the law of conservation of energy, Equation 10.6, reads

$$K_i + W = K_f$$

or

$$W = K_f - K_i \qquad (10.10)$$

Using kinematics, we can find another expression for the work done, in terms of the car's initial and final speeds. Recall from Chapter 2 the kinematic equation relating an object's displacement and its change in velocity:

$$v_f^2 = v_i^2 + 2a\Delta x$$

Applied to the motion of our car, $\Delta x = d$ is the car's displacement and, from Newton's second law, the acceleration is $a = F/m$. Thus we can write

$$v_f^2 = v_i^2 + \frac{2Fd}{m} = v_i^2 + \frac{2W}{m}$$

where we have replaced Fd with the work W. If we now solve for the work, we find

$$W = \frac{1}{2}m(v_f^2 - v_i^2) = \frac{1}{2}mv_f^2 - \frac{1}{2}mv_i^2$$

If we compare this result with Equation 10.10, we see that

$$K_f = \frac{1}{2}mv_f^2 \qquad \text{and} \qquad K_i = \frac{1}{2}mv_i^2$$

6.1 **Activ Physics** ONLINE

In general, then, an object moving at a speed v has kinetic energy

$$K = \frac{1}{2}mv^2 \qquad (10.11)$$

QUADRATIC
p.50

Kinetic energy of an object of mass m moving with speed v

From Equation 10.11, the units of kinetic energy are mass times speed squared, or $\text{kg} \cdot (\text{m/s})^2$. But

$$1\,\text{kg} \cdot (\text{m/s})^2 = \underbrace{1\,\text{kg} \cdot (\text{m/s}^2)}_{1\,\text{N}} \cdot \text{m} = 1\,\text{N} \cdot \text{m} = 1\,\text{J}$$

We see that the units of kinetic energy are the same as those of work, as they must be. Table 10.1 gives some approximate kinetic energies. Everyday kinetic energies range from a tiny fraction of a fraction of a joule to nearly a million joules for a speeding car.

TABLE 10.1 Some approximate kinetic energies

Object	Kinetic energy
Walking ant	1×10^{-8} J
Penny dropped 1 m	2.5×10^{-3} J
Person walking	70 J
100 mph fastball	150 J
Bullet	5000 J
Car, 60 mph	5×10^5 J
Supertanker	2×10^{10} J

CONCEPTUAL EXAMPLE 10.3 Kinetic energy changes for a car

Compare the increase in a 1000 kg car's kinetic energy as it speeds up by 5.0 m/s starting from 5.0 m/s, to its increase in kinetic energy as it speeds up by 5.0 m/s starting from 10 m/s.

REASON The change in the car's kinetic energy in going from 5 m/s to 10 m/s is

$$\Delta K_{5\to 10} = \frac{1}{2}mv_f^2 - \frac{1}{2}mv_i^2$$

This gives

$$\Delta K_{5\to 10} = \frac{1}{2}(1000\,\text{kg})(10\,\text{m/s})^2 - \frac{1}{2}(1000\,\text{kg})(5.0\,\text{m/s})^2$$

$$= 3.8 \times 10^4\,\text{J}$$

while

$$\Delta K_{10\to 15} = \frac{1}{2}(1000\,\text{kg})(15\,\text{m/s})^2 - \frac{1}{2}(1000\,\text{kg})(10\,\text{m/s})^2$$

$$= 6.3 \times 10^4\,\text{J}$$

Even though the increase in the car's *speed* was the same in both cases, the increase in kinetic energy is substantially larger in the second case.

ASSESS Kinetic energy depends on the *square* of the speed v. If we plot the kinetic energy of the car as in Figure 10.13, we see that the energy of the car increases rapidly with speed. We can also see graphically why the change in K for a fixed 5 m/s change in v is greater at high speeds than at low speeds. In part this is why it's harder to accelerate your car at high speeds than at low speeds.

FIGURE 10.13 The kinetic energy increases as the *square* of the speed.

EXAMPLE 10.3 Speed of a bobsled after pushing

A two-man bobsled has a mass of 390 kg. Starting from rest, the two racers push the sled for the first 50 m with a net force of 270 N. Neglecting friction, what is the sled's speed at the end of the 50 m?

PREPARE This is the first example where we fully use Problem-Solving Strategy 10.1. We start by identifying the bobsled as the system; the two racers pushing the sled are part of the environment. The racers do work on the system by pushing it with force $\vec{F}$. Because the speed of the sled changes, we'll need to include kinetic energy. Neither U_g nor U_s changes, so we won't need to consider these energies. Figure 10.14 lists the known quantities and the quantity (v_f) that we want to find.

Known
$m = 390$ kg $F = 270$ N
$d = 50$ m $v_i = 0$ m/s

Find: v_f

FIGURE 10.14 The work done by the pushers increases the sled's kinetic energy.

SOLVE With only kinetic energy changing, the conservation of energy equation, Equation 10.6, is

$$K_i + W = K_f$$

Using our expressions for kinetic energy and work, this becomes

$$\frac{1}{2}mv_i^2 + Fd = \frac{1}{2}mv_f^2$$

Because $v_i = 0$, the energy equation reduces to

$$Fd = \frac{1}{2}mv_f^2$$

We can solve for the final speed to get

$$v_f = \sqrt{\frac{2Fd}{m}} = \sqrt{\frac{2(270 \text{ N})(50 \text{ m})}{390 \text{ kg}}} = 8.3 \text{ m/s}$$

ASSESS We solved this problem using the concept of energy conservation. In this case, we could also have solved it using Newton's second law and kinematics. However, we'll soon see that energy conservation can solve problems that would be very difficult for us to solve using Newton's laws alone.

STOP TO THINK 10.3 Rank in order, from greatest to least, the kinetic energies of the sliding pucks.

Rotational Kinetic Energy

We've just found an expression for the kinetic energy of an object moving along a line or some other path. This energy is called **translational kinetic energy.** Consider now an object rotating about a fixed axis, such as the windmill blades in Figure 10.15. Although the blades have no overall translational motion, each

FIGURE 10.15 The large rotating blades of a windmill have a great deal of kinetic energy.

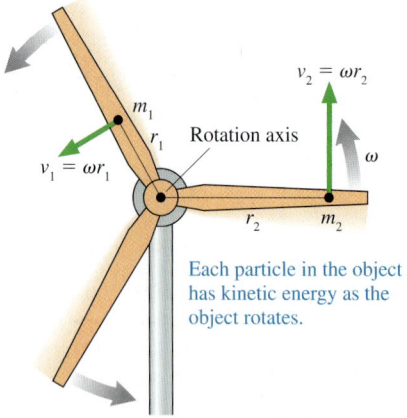

FIGURE 10.16 Rotational kinetic energy is due to the circular motion of the particles.

Rotational recharge The International Space Station (ISS) gets its electrical power from solar panels. But during each 92-min orbit, the ISS is in the earth's shadow for 30 min. The batteries that currently provide power during these blackouts need periodic replacement, which is very expensive in space. A promising new technology would replace the batteries with a *flywheel*—a cylinder rotating at very high angular speed. Energy from the solar cells is used to speed up the flywheel, storing energy as rotational kinetic energy, which can then be converted back into electrical energy when the ISS is in shadow.

particle in the blade is moving and hence has kinetic energy. Adding up the kinetic energy for each particle that makes up the blades, we find that the blades have **rotational kinetic energy,** the kinetic energy due to rotation.

Figure 10.16 shows two of the particles making up a windmill blade that rotates with angular velocity ω. Recall from Section 7.2 that a particle moving with angular velocity ω in a circle of radius r has a speed $v = \omega r$. Thus particle 1, which rotates in a circle of radius r_1, moves with speed $v_1 = r_1\omega$. Particle 2, which rotates in a circle with a larger radius r_2, moves with a larger speed $v_2 = r_2\omega$. The object's rotational kinetic energy is the sum of the kinetic energies of *all* of the particles:

$$K_{\text{rot}} = \frac{1}{2}m_1v_1^2 + \frac{1}{2}m_2v_2^2 + \cdots$$

$$= \frac{1}{2}m_1r_1^2\omega^2 + \frac{1}{2}m_2r_2^2\omega^2 + \cdots = \frac{1}{2}\left(\sum mr^2\right)\omega^2$$

You will recognize the term in parentheses as our old friend, the moment of inertia I. Thus the rotational kinetic energy is

$$K_{\text{rot}} = \frac{1}{2}I\omega^2 \qquad (10.12)$$

QUADRATIC
p. 50

Rotational kinetic energy of object with moment of inertia I and angular velocity ω

NOTE ▶ Rotational kinetic energy is *not* a new form of energy. This is the familiar kinetic energy of motion, only now expressed in a form that is especially convenient for rotational motion. Comparison with the familiar $\frac{1}{2}mv^2$ shows again that the moment of inertia I is the rotational equivalent of mass. ◀

A rolling object, such as a wheel, is undergoing both rotational *and* translational motions. Consequently, its total kinetic energy is the sum of its rotational and translational kinetic energies:

$$K = K_{\text{trans}} + K_{\text{rot}} = \frac{1}{2}mv^2 + \frac{1}{2}I\omega^2$$

Recall from Section 6.3 that v and ω of a rolling object of radius R are related by $\omega = v/R$. Thus we can write the kinetic energy of a rolling object as

$$K_{\text{rolling}} = \frac{1}{2}mv^2 + \frac{1}{2}I\left(\frac{v}{R}\right)^2 = \frac{1}{2}\left(m + \frac{I}{R^2}\right)v^2 \qquad (10.13)$$

This illustrates the important fact that **the kinetic energy of a rolling object is always greater than that of a nonrotating object moving at the same speed.**

EXAMPLE 10.4 Kinetic energy of a bicycle

Bike 1 has a 10.0 kg frame and 1.00 kg wheels, while bike 2 has a 9.00 kg frame and 1.50 kg wheels. Both bikes thus have the same 12.0 kg total mass. What is the kinetic energy of each bike when they are ridden at 12.0 m/s? Model each wheel as a hoop of radius 35.0 cm.

PREPARE Each bike's frame has only translational kinetic energy $K_{\text{frame}} = \frac{1}{2}Mv^2$, where M is the mass of the frame. The kinetic energy of each rolling wheel is given by Equation 10.13. From

Table 7.2, we find that I for a hoop is mR^2, where m is the mass of one wheel.

SOLVE From Equation 10.13 the kinetic energy of each rolling wheel is

$$K_{\text{wheel}} = \frac{1}{2}\left(m + \frac{mR^2}{R^2}\right)v^2 = \frac{1}{2}(2m)v^2 = mv^2$$

Then the total kinetic energy of a bike is

$$K = K_{\text{frame}} + 2K_{\text{wheel}} = \frac{1}{2}Mv^2 + 2mv^2$$

The factor of 2 in the second term occurs because each bike has two wheels. Thus the kinetic energies of the two bikes are

$$K_1 = \frac{1}{2}(10.0 \text{ kg})(12.0 \text{ m/s})^2 + 2(1.00 \text{ kg})(12.0 \text{ m/s})^2$$
$$= 1010 \text{ J}$$

$$K_2 = \frac{1}{2}(9.00 \text{ kg})(12.0 \text{ m/s})^2 + 2(1.50 \text{ kg})(12.0 \text{ m/s})^2$$
$$= 1080 \text{ J}$$

The kinetic energy of bike 2 is about 7% higher than that of bike 1. Note that the radius of the wheels was not needed in this calculation.

ASSESS As the cyclists on these bikes accelerate from rest to 12 m/s, they must convert some of their internal chemical energy into the kinetic energy of the bikes. Racing cyclists want to use as little of their own energy as possible. Although both bikes have the same total mass, the one with the lighter wheels will take less energy to get it moving. Shaving a little extra weight off your wheels is more useful than taking that same weight off your frame.

It's important that racing bike wheels are as light as possible.

10.6 Potential Energy

When two or more objects in a system interact, it is sometimes possible to *store* energy in that system in a way that the energy can be easily recovered. For instance, the earth and a ball interact by the gravitational force between them. If the ball is lifted up into the air, energy is stored in the ball-earth system, energy that can later be recovered as kinetic energy when the ball is released and falls. Similarly, a spring is a system made up of countless atoms that interact via their atomic "springs." If we push a box against a spring, energy is stored that can be recovered when the spring later pushes the box across the table. This sort of stored energy is called **potential energy,** since it has the *potential* to be converted into other forms of energy such as kinetic or thermal energy.

The forces due to gravity and springs are special in that they allow for the storage of energy. Other interaction forces do not. When a crate is pushed across the floor, the crate and the floor interact via the force of friction, and the work done on the system is converted into thermal energy. But this energy is *not* stored up for later recovery—it slowly diffuses into the environment and cannot be recovered.

Interaction forces that can store useful energy are called **conservative forces.** The name comes from the important fact, which we'll soon look at in detail, that when only conservative forces act, the mechanical energy of a system is *conserved.* Gravity and elastic forces are conservative forces, and later we'll see that the electric force is a conservative force as well. Friction, on the other hand, is a **nonconservative force.** When two objects interact via a friction force, energy is not stored. It is usually transformed into thermal energy.

Let's look more closely at the potential energies associated with the two conservative forces—gravity and springs—that we'll study in this chapter.

Gravitational Potential Energy

To find an expression for gravitational potential energy, let's consider the system of the book and the earth shown in Figure 10.17a on the next page. The book is lifted at a constant speed from its initial position at y_i to a final height y_f.

We can analyze this situation using the approach of Problem-Solving Strategy 10.1. The lifting force of the hand is external to the system and so does work W on the system, increasing its energy. The book is lifted at a constant speed, so its kinetic energy doesn't change. Because there's no friction, the book's thermal energy doesn't change either. Thus the work done goes entirely into increasing the gravitational potential energy of the system. The law of conservation of energy, Equation 10.6, then reads

(a)

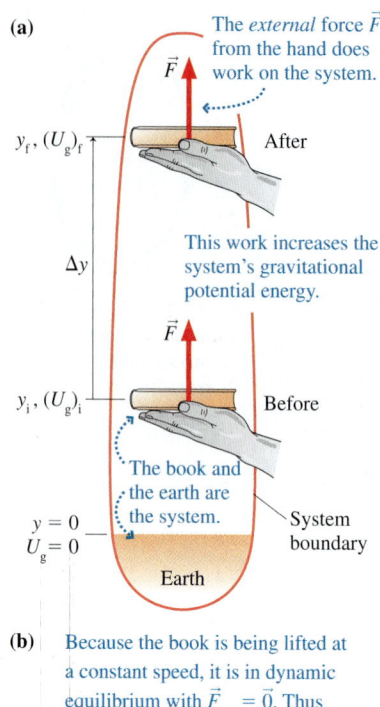

The *external* force $\vec{F}$ from the hand does work on the system.

$y_f, (U_g)_f$

After

This work increases the system's gravitational potential energy.

Δy

$y_i, (U_g)_i$

Before

The book and the earth are the system.

$y = 0$
$U_g = 0$

System boundary

Earth

(b) Because the book is being lifted at a constant speed, it is in dynamic equilibrium with $\vec{F}_{net} = \vec{0}$. Thus $F = w = mg$.

$\vec{F}$

$\vec{w}$

FIGURE 10.17 Lifting a book increases its gravitational potential energy.

The initial gravitational potential energy. . .   . . . .plus the energy put into the system as work. . .

$$(U_g)_i + W = (U_g)_f \qquad (10.14)$$

. . .equals the final gravitational potential energy.

The work done is $W = F\Delta y$, where $\Delta y = y_f - y_i$ is the vertical distance that the book is lifted. From the free-body diagram of Figure 10.17b, we see that $F = mg$. This gives $W = mg\Delta y$, so that

$$(U_g)_i + mg\Delta y = (U_g)_f$$

or

$$(U_g)_f = (U_g)_i + mg\Delta y \qquad (10.15)$$

Since our final height was greater than our initial height, Δy is positive and $(U_g)_f > (U_g)_i$. **The higher the object is lifted, the greater the gravitational potential energy in the object/earth system.**

Equation 10.15 gives the final gravitational potential energy $(U_g)_f$ in terms of its initial value $(U_g)_i$. But what is the value of $(U_g)_i$? We can gain some insight by writing Equation 10.15 in terms of energy *changes*. We have

$$(U_g)_f - (U_g)_i = mg\Delta y$$

or

$$\Delta U_g = mg\Delta y$$

For example, if we lift a 1.5 kg book up by 2.0 m, we increase its gravitational potential energy by $\Delta U_g = (1.5 \text{ kg})(9.8 \text{ m/s}^2)(2.0 \text{ m}) = 29.4$ J. This increase is *independent* of the book's starting height: We would get the *same* increase whether we lifted the book 2.0 m starting at sea level or starting at the top of Mount Everest. If we then dropped the book 2.0 m, we would recover the same 29.4 J as kinetic energy, whether in Miami or on Everest. This illustrates an important general fact about *every* form of potential energy: **Only *changes* in potential energy are significant.**

Because of this fact, we are free to choose a *reference level* where we define U_g to be zero. Our expression for U_g is particularly simple if we choose this reference level to be at $y = 0$. We then have

$$U_g = mgy \qquad (10.16)$$

Gravitational potential energy of an object of mass m at a height y (assuming $U_g = 0$ when the object is at $y = 0$)

LINEAR
p. 38

NOTE ▶ We've emphasized that gravitational potential energy is an energy of the earth-object *system*. In solving problems using the law of conservation of energy, you'll need to include the earth as part of your system. For simplicity, we'll usually speak of "the gravitational potential energy of the ball," but what we really mean is the potential energy of the earth-ball system. ◀

EXAMPLE 10.5 Hitting the bell

At the county fair, Katie tries her hand at the ring-the-bell attraction, as shown in Figure 10.18. She swings the mallet hard enough to give the ball an initial upward speed of 8.0 m/s. Will the ball ring the bell, 3.0 m from the bottom?

PREPARE As discussed above and in Tactics Box 10.1, we'll choose the ball *and* the earth as the system. Figure 10.18 shows the visual overview. If we assume that the track along which the ball moves is frictionless, then only the *mechanical* energy of the system changes. The only force on the ball after it leaves the

We'll calculate how high the ball would go if the bell weren't there. Then we'll see if that height is enough to have reached the bell.

After:
y_f
$v_f = 0$ m/s

Find: y_f

3.0 m

Before:
$v_i = 8.0$ m/s
$y_i = 0$ m

FIGURE 10.18 Before-and-after visual overview of the ring-the-bell attraction.

bottom lever is gravity, but gravity is an *internal* force due to our choice of the ball *plus* the earth as the system. This means that the gravitational interaction is included as gravitational potential energy rather than as external work. Since no *external* forces do work on the earth-ball system, the system is isolated. We can then use the law of conservation of mechanical energy, Equation 10.7.

SOLVE Equation 10.7 tells us that $K_i + (U_g)_i = K_f + (U_g)_f$. We can use our expressions for kinetic and potential energy to write this as

$$\frac{1}{2}mv_i^2 + mgy_i = \frac{1}{2}mv_f^2 + mgy_f$$

Let's ignore the bell for the moment and figure out how far the ball would rise if there were nothing in its way. We know that the ball starts at $y_i = 0$ m and that its speed v_f at the highest point is zero. Thus the energy equation simplifies to

$$mgy_f = \frac{1}{2}mv_i^2$$

This is easily solved for the height y_f:

$$y_f = \frac{v_i^2}{2g} = \frac{(8.0 \text{ m/s})^2}{2(9.8 \text{ m/s}^2)} = 3.3 \text{ m}$$

This is higher than the point where the bell sits, so the ball would actually hit it on the way up.

ASSESS Notice that the mass canceled and wasn't needed, a fact about free fall that you should remember from Chapter 2.

An important conclusion from Equation 10.16 is that **gravitational potential energy depends only on the height of the object above the reference level y = 0, not on the object's horizontal position.** Consider carrying a briefcase while walking on level ground at a constant speed. As shown in the table on page 309, the force of your hand on the briefcase is *vertical* and hence *perpendicular* to the displacement. No work is done on the briefcase and consequently its gravitational potential energy remains constant as long as its height above the ground doesn't change as you walk.

This idea can be applied to more complicated cases, such as the 51 kg hiker in Figure 10.19. His gravitational potential energy depends *only* on his height y above the reference level, so it's the same value $U_g = mgy = 50$ kJ at any point on path A where he is at a height $y = 100$ m above the reference level. If he had instead taken path B, his gravitational potential energy at 100 m would be the same 50 kJ. It doesn't matter *how* he gets to 100 m, his potential energy at that height will be the same. This demonstrates an important aspect of all potential energies: **The potential energy depends only on the *position* of the object and not on the path the object took to get to that position.** This fact will allow us to use the law of conservation of energy to easily solve a variety of problems that would be very difficult to solve using Newton's laws alone, because we won't need to know the details of the path of the object—just its starting and ending points.

The hiker's potential energy at the top is 100 kJ regardless of whether he took path A or path B.
$U_g = 100$ kJ
His potential energy is the same at any point where his elevation is 100 m.
200 m
Path A
100 m
$U_g = 50$ kJ
Path B
0 m
The reference level $y = 0$ m is where $U_g = 0$ J.

FIGURE 10.19 The hiker's gravitational potential energy depends only on his height above the $y = 0$ reference level.

EXAMPLE 10.6 Speed at the bottom of a water slide

Still at the county fair, Katie tries the water slide, whose shape is shown in Figure 10.20. The starting point is 9.0 m above the ground. She pushes off with an initial speed of 2.0 m/s. If the slide is frictionless, how fast will Katie be traveling at the bottom?

PREPARE Figure 10.20 on the next page shows a visual overview of the slide. Because there is no friction, Tactics Box 10.1 suggests that we take as our system Katie (the moving object) and the earth. With this choice of system, the only energies in the system are kinetic and gravitational potential energy. Note that the slope of the slide is not constant, so Katie's

Continued

FIGURE 10.20 Before-and-after visual overview of Katie on the water slide.

acceleration will not be constant either. Thus we can't use constant-acceleration kinematics to find her speed. But we *can* use the law of conservation of energy to easily solve for her speed. Because there is no friction, the mechanical energy is conserved.

SOLVE Conservation of mechanical energy gives

$$K_i + (U_g)_i = K_f + (U_g)_f$$

or

$$\frac{1}{2}mv_i^2 + mgy_i = \frac{1}{2}mv_f^2 + mgy_f$$

Taking $y_f = 0$ m we have

$$\frac{1}{2}mv_i^2 + mgy_i = \frac{1}{2}mv_f^2$$

which we can solve to get

$$v_f = \sqrt{v_i^2 + 2gy_i}$$
$$= \sqrt{(2.0 \text{ m/s})^2 + 2(9.8 \text{ m/s}^2)(9.0 \text{ m})} = 13 \text{ m/s}$$

ASSESS It is important to realize that the *shape* of the slide does not matter because gravitational potential energy depends only on the *height* above a reference level. **In sliding down any (frictionless) slide of the same height, your speed at the bottom would be the same.**

STOP TO THINK 10.4 Rank in order, from largest to smallest, the gravitational potential energies of identical balls 1 to 4.

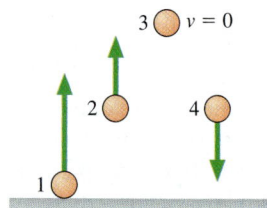

Elastic Potential Energy

Energy can also be stored in a compressed or extended spring as **elastic** (or **spring**) **potential energy** U_s. We can find out how much energy is stored in a spring by using an external force to slowly compress the spring. This external force does work on the spring, transferring energy to the spring. Since only the elastic potential energy of the spring is changing, the law of conservation of energy reads

$$W = \Delta U_s \qquad (10.17)$$

That is, we can find out how much elastic potential energy is stored in the spring by calculating the amount of work needed to compress the spring.

Figure 10.21 shows a spring being compressed by a hand. In Section 8.3 we found that the force that the spring will exert on the hand is equal to $-k\Delta x$, where Δx is the displacement of the end of the spring from its equilibrium position and k is the spring constant. In Figure 10.21 we have set the origin of our coordinate system at the equilibrium position. The displacement from equilibrium Δx is therefore equal to x, and the spring force is then $-kx$. By Newton's third law, this means that the force that the hand exerts on the spring is equal to $+kx$.

As we compress the end of the spring from its equilibrium position to a final displacement x, the force we apply increases from zero to kx. This is not a constant force, so we can't use Equation 10.8, $W = Fd$, to find the work done, because this equation is valid only for a constant force. However, it seems reasonable that we could calculate the work by using the *average* force in Equation 10.8. Because the force varies from $F_i = 0$ to $F_f = kx$, the average force used to compress the spring is

FIGURE 10.21 The force required to compress a spring is not constant.

$$F_{avg} = \frac{1}{2}(F_f + F_i) = \frac{1}{2}(kx + 0) = \frac{1}{2}kx$$

Thus the work done by the hand is

$$W = F_{avg}d = F_{avg}x = \left(\frac{1}{2}kx\right)x = \frac{1}{2}kx^2$$

This work is stored as potential energy in the spring, so we can use Equation 10.17 to find that the elastic potential energy increases by

$$\Delta U_s = \frac{1}{2}kx^2$$

Just as in the case of gravitational potential energy, we have found an expression for the *change* in U_s, not U_s itself. Again, we are free to set $U_s = 0$ at any convenient spring extension. An obvious choice is to set $U_s = 0$ at the point where the spring is in equilibrium, neither compressed nor stretched; that is, at $x = 0$. With this choice we have

$$U_s = \frac{1}{2}kx^2 \qquad (10.18)$$

QUADRATIC
p. 50

Elastic potential energy of a spring displaced a distance x from equilibrium (assuming $U_s = 0$ when the end of the spring is at $x = 0$)

NOTE ▶ Since U_s depends on the *square* of the displacement x, U_s is the same whether x is positive (the spring is compressed as in Figure 10.21) or negative (the spring is stretched). ◀

Calf muscle

Achilles tendon

On each stride, the tendon stretches, storing about 35 J of energy.

Spring in your step BIO As you run, you lose some of your mechanical energy each time your foot strikes the ground; this energy is transformed into unrecoverable thermal energy. Luckily, about 35% of the decrease of your mechanical energy when your foot lands is stored as elastic potential energy in the stretchable Achilles tendon of the lower leg. On each plant of the foot the tendon is stretched, storing some energy. The tendon springs back as you push off the ground again, helping to propel you forward. This recovered energy reduces the amount of internal chemical energy you use, increasing your efficiency.

EXAMPLE 10.7 Speed of a spring-launched ball

A spring-loaded toy gun is used to launch a 10 g plastic ball. The spring, which has a spring constant of 10 N/m, is compressed by 10 cm as the ball is pushed into the barrel. When the trigger is pulled, the spring is released and shoots the ball back out. What is the ball's speed as it leaves the barrel? Assume that friction is negligible.

PREPARE Assume the spring obeys Hooke's law $F = -kx$, and is massless so that it has no kinetic energy of its own. Using Tactics Box 10.1 we choose the system to be the spring and the ball. There's no friction, hence the system's mechanical energy $K + U_s$ is conserved.

Before: $v_i = 0$ m/s

$x_i = -10$ cm $x = 0$ x

After: v_f

$x_f = 0$ cm

Find: v_f

FIGURE 10.22 The before-and-after visual overview of a ball being shot out of a spring-loaded toy gun.

Figure 10.22 shows a before-and-after visual overview. The compressed spring will push on the ball until the spring has returned to its equilibrium length. We have chosen the origin of the coordinate system at the equilibrium position of the free end of the spring, making $x_i = -10$ cm and $x_f = 0$ cm.

SOLVE The energy conservation equation is $K_i + (U_s)_i = K_f + (U_s)_f$. We can use the elastic potential energy of the spring, Equation 10.18, to write this as

$$\frac{1}{2}mv_i^2 + \frac{1}{2}kx_i^2 = \frac{1}{2}mv_f^2 + \frac{1}{2}kx_f^2$$

We know that $x_f = 0$ m and $v_i = 0$ m/s, so this simplifies to

$$\frac{1}{2}mv_f^2 = \frac{1}{2}kx_i^2$$

It is now straightforward to solve for the ball's speed:

$$v_f = \sqrt{\frac{kx_i^2}{m}} = \sqrt{\frac{(10 \text{ N/m})(-0.10 \text{ m})^2}{0.010 \text{ kg}}} = 3.2 \text{ m/s}$$

ASSESS This is *not* a problem that we could have easily solved with Newton's laws. The acceleration is not constant, and we have not learned how to handle the kinematics of nonconstant acceleration. But with conservation of energy—it's easy!

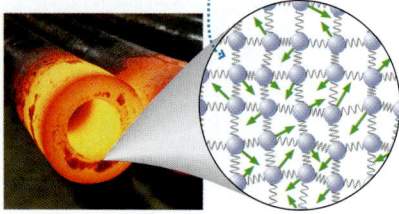

Hot object: Fast-moving molecules have lots of kinetic and elastic potential energy.

Cold object: Slow-moving molecules have little kinetic and elastic potential energy.

FIGURE 10.23 A molecular view of thermal energy.

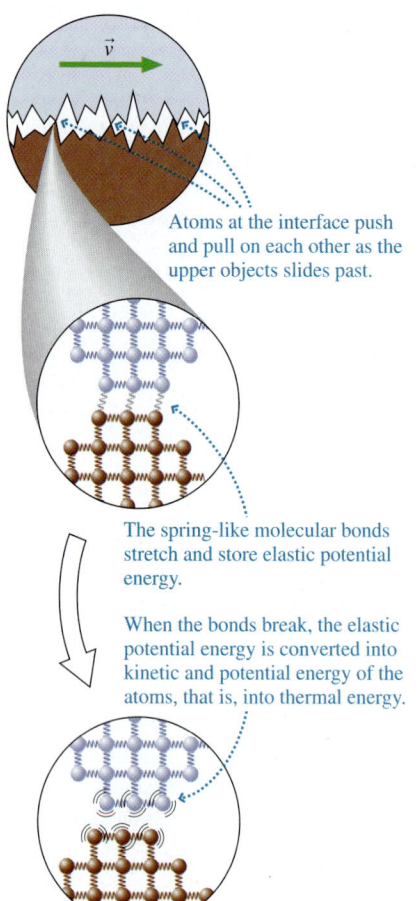

Atoms at the interface push and pull on each other as the upper objects slides past.

The spring-like molecular bonds stretch and store elastic potential energy.

When the bonds break, the elastic potential energy is converted into kinetic and potential energy of the atoms, that is, into thermal energy.

FIGURE 10.24 How friction causes an increase in thermal energy.

> **STOP TO THINK 10.5**   A spring-loaded gun shoots a plastic ball with a speed of 4 m/s. If the spring is compressed twice as far, the ball's speed will be
>
> A. 2 m/s.   B. 4 m/s.   C. 8 m/s.   D. 16 m/s.

10.7 Thermal Energy

We noted earlier that thermal energy is related to the microscopic motion of the molecules of an object. As Figure 10.23 shows, the molecules in a hot object jiggle around their average positions more than the molecules in a cold object. This has two consequences. First, each atom is on average moving faster in the hot object. This means that each atom has a higher *kinetic energy*. Second, each atom in the hot object tends to stray further from its equilibrium position, leading to a greater stretching or compressing of the spring-like molecular bonds. This means that each atom has on average a higher *potential energy*. The potential energy stored in any one bond and the kinetic energy of any one atom are both exceedingly small, but there are incredibly many bonds and atoms. The sum of all these microscopic potential and kinetic energies is what we call **thermal energy.**

Is this microscopic energy worth worrying about? To see, consider a 500 g ($\approx$1 lb) iron ball moving at the respectable speed of $v_{ball} = 20$ m/s ($\approx$45 mph). Its kinetic energy is $K = \frac{1}{2}mv_{ball}^2 = 100$ J.

How fast do the atoms jiggle about their equilibrium positions? This speed depends on the temperature, but at room temperature it's very high—roughly 500 m/s. So each atom is on average traveling in a straight line at 20 m/s, but jiggling about this average motion at a speed of 500 m/s! This a factor of 25 times faster. And since kinetic energy is proportional to the *square* of the speed, the kinetic energy due to the microscopic motion is about 625 times greater than that due to the overall motion. And it turns out that the microscopic potential energy is just as large. Thus the ball that has an ordinary kinetic energy of 100 J has an internal thermal energy of $2 \times 625 \times 100$ J = 125,000 J!

Transforming Mechanical Energy into Thermal Energy

Consider a snowboarder sliding on level snow. After a while, he will glide to a stop because of the friction force of the snow on his board. We can analyze this using the law of conservation of energy. Following Tactics Box 10.1 we'll take the system to be the boarder *plus* the snow. Then there are no forces external to the system that do work on it and, since he's moving horizontally, his potential energy doesn't change. Then the law of conservation of energy is $K_i = K_f + \Delta E_{th}$, or $\Delta E_{th} = K_i - K_f$. He's slowing to a stop, so $K_i > K_f$ and ΔE_{th} is positive. The system's thermal energy *increases* as kinetic energy is transformed into thermal energy.

This increase in thermal energy is a general feature of any system where friction between sliding objects is present: **When two objects slide against each other with friction present, mechanical energy is always transformed into thermal energy.** An atomic-level explanation is shown in Figure 10.24.

The presence of friction has two important consequences for our conservation of energy Problem-Solving Strategy 10.1:

1. As stated in Tactics Box 10.1, we must include in the system not only the moving object but also the surface against which it slides. This is because the thermal energy generated by friction resides in *both* object and surface (as in Figure 10.24), and it is usually impossible to tell what fraction resides

in each. By choosing both object and surface to be in the system, we know that *all* the thermal energy ends up in the system.

2. In addition to the mechanical energy $K + U$ we now must include ΔE_{th} in the conservation of energy equation.

TRY IT YOURSELF

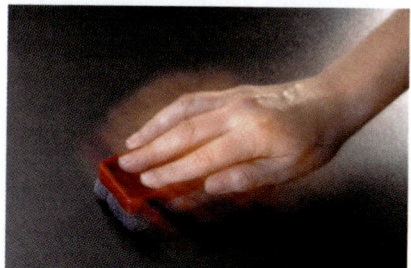

Agitating atoms Vigorously rub a somewhat soft object such as a blackboard eraser on your desktop for about 10 seconds. If you then pass your fingers over the spot where you rubbed, you'll feel a distinct warm area. Congratulations: you've just set some 100,000,000,000,000,000,000,000 atoms into motion!

EXAMPLE 10.8 Thermal energy created sledding down a hill

George jumps onto his sled and starts from rest at the top of a 5.0-m-high hill. His speed at the bottom is 8.0 m/s. The mass of George and the sled is 55 kg. How much thermal energy was produced in this process?

PREPARE Figure 10.25 shows the before-and-after visual overview. The statement of the problem implies that thermal energy will be generated, so following Tactics Box 10.1, we'll take the system to include both George and the sled *and* the slope. Because his height is changing, his gravitational potential energy is changing and we'll need to include the earth in the system as well. No forces act from outside this system, so the work W is zero.

FIGURE 10.25 Visual overview of George sliding down the hill.

SOLVE Here the law of conservation of energy reads

$$K_i + (U_g)_i = K_f + (U_g)_f + \Delta E_{th}$$

In terms of positions and speeds,

$$\frac{1}{2}mv_i^2 + mgy_i = \frac{1}{2}mv_f^2 + mgy_f + \Delta E_{th}$$

Because $v_i = 0$ m/s and $y_f = 0$ m, this simplifies to

$$mgy_i = \frac{1}{2}mv_f^2 + \Delta E_{th}$$

from which we have

$$\Delta E_{th} = mgy_i - \frac{1}{2}mv_f^2$$

$$= (55 \text{ kg})(9.8 \text{ m/s}^2)(5.0 \text{ m}) - \frac{1}{2}(55 \text{ kg})(8.0 \text{ m/s})^2$$

$$= 940 \text{ J}$$

ASSESS The change in E_{th} is positive, as it must be. This extra thermal energy resides in the sled and all along the slope where George slid. You should be able to show that about 35% of George's original gravitational potential energy was transformed into thermal energy as he slid down the hill.

10.8 Further Examples of Conservation of Energy

In this section, we'll tie together what we've learned about using the law of conservation of energy to solve dynamics problems. In each, we use the key idea of setting the "before" energy equal to the "after" energy.

5.2–5.7, 7.11–7.13

EXAMPLE 10.9 Where will the sled stop?

A sledder, starting from rest, slides down a 10-m-high hill. At the bottom of the hill is a long horizontal patch of rough snow. The hill is nearly frictionless, but the coefficient of friction between the sled and the rough snow at the bottom is $\mu_k = 0.30$. How far will the sled slide along the rough patch?

PREPARE A picture of the sledder is shown in Figure 10.26. We'll break this problem into Part A, his motion down the hill; and Part B, his motion along the ice. We know how to use conservation of energy to find his speed at the bottom of the hill. Along the rough patch, however, we'll use kinematics and Newton's laws, as studied in Chapters 2 and 5, to find how far he slides.

FIGURE 10.26 Visual overview of a sledder sliding downhill.

SOLVE We'll solve Part A first and find the sled's speed v_f at the bottom. The hill is frictionless, so mechanical energy is conserved and we have

$$mgy_i + \frac{1}{2}mv_i^2 = mgy_f + \frac{1}{2}mv_f^2$$

Since $v_i = 0$ m/s and $y_f = 0$ m, this reduces to

$$mgy_i = \frac{1}{2}mv_f^2$$

so that

$$v_f = \sqrt{2gy_i} = \sqrt{2(9.8 \text{ m/s}^2)(10 \text{ m})} = 14.0 \text{ m/s}$$

On the rough patch in Part B, where the only horizontal force is the kinetic friction force f_k pointing to the left, the sled's acceleration is

$$a = -\frac{f_k}{m} = -\frac{\mu_k n}{m} = -\frac{\mu_k mg}{m}$$

$$= -\mu_k g = -(0.30)(9.8 \text{ m/s}^2) = -2.94 \text{ m/s}^2$$

The negative acceleration indicates that the sled is slowing down, as expected.

We now use kinematics to find how far the sled slides. We know the acceleration a, as well as the initial and final velocities along the horizontal patch, and we want to know the final position x_2. This suggests using the kinematic equation

$$v_f^2 = v_i^2 + 2a(x_f - x_i)$$

to find the final position. For the motion of Part B, the final velocity $v_f = v_2 = 0$ m/s, the initial velocity is $v_1 = 14.0$ m/s, and the initial position is $x_i = x_1 = 0$ m. We can then solve for the final position in the kinematic equation to get

$$x_2 = -\frac{v_1^2}{2a} = -\frac{(14.0 \text{ m/s})^2}{-5.88 \text{ m/s}^2} = 33 \text{ m}$$

ASSESS When friction is present, mechanical energy is *not* conserved: Some of the mechanical energy of the system is inevitably transformed into thermal energy. Thus we cannot use the law of conservation of mechanical energy for such problems. Instead, we'll need to use Newton's laws and kinematics to find how far objects slide. As in this example, however, there will often be a part of the problem with no friction that we *can* solve using the law of conservation of mechanical energy.

EXAMPLE 10.10 Who wins the great downhill race?

Figure 10.27 shows a contest in which a sphere, a cylinder, and a circular hoop, each with mass M and radius R, are placed at height h on a slope of angle θ. All three are simultaneously released from rest and roll down the ramp without slipping. Which one will win the race to the bottom of the hill?

FIGURE 10.27 Which will win the downhill race?

PREPARE With no sliding friction, the total mechanical energy is conserved. However, the kinetic energy of each object must include a contribution from its rotational kinetic energy.

SOLVE Conservation of energy tells us that the gravitational potential energy $(U_g)_i = Mgh$ at the top will be transformed into an equal amount of kinetic energy K_f at the bottom. Thus

$$(U_g)_i = Mgh = K_f = \frac{1}{2}\left(M + \frac{I}{R^2}\right)v^2$$

where we used Equation 10.13 for the total (translational plus rotational) kinetic energy. The speed at the bottom is then

$$v = \sqrt{\frac{2Mgh}{M + \frac{I}{R^2}}}$$

Table 7.2 gives the moment of inertia for each of the three shapes. We have

Shape	Moment of Inertia I	$v = \sqrt{2Mgh/(M + I/R^2)}$
Sphere	$\frac{2}{5}MR^2$	$v = \sqrt{\frac{10}{7}gh} = 1.19\sqrt{gh}$
Cylinder	$\frac{1}{2}MR^2$	$v = \sqrt{\frac{4}{3}gh} = 1.15\sqrt{gh}$
Hoop	MR^2	$v = \sqrt{gh}$

The sphere has the largest speed at the bottom, a full 19% faster than the hoop. Because the sphere always travels faster than the hoop or the cylinder, it will win the race, followed by the cylinder and the hoop.

ASSESS All the objects have the same kinetic energy at the bottom, because they all started with the same energy, Mgh, at the top. But the object with the smallest I will have the smallest *rotational* kinetic energy at the bottom, and hence the largest *translational* kinetic energy and the largest v. An ordinary sliding object (no rotation) reaches the bottom with speed $v = \sqrt{2gh} = 1.41\sqrt{gh}$. This is significantly faster than any of the rolling objects. The sliding object is faster because *all* its kinetic energy is translational—and it's the translational motion that gets you down the hill.

10.9 Energy in Collisions

In Chapter 9 we studied collisions between two objects. We found that if no external forces are acting on the objects, the total *momentum* of the objects will be conserved. Now we wish to study what happens to *energy* in collisions. The energetics of collisions are important in many applications in biokinetics, such as designing safer automobiles and bicycle helmets.

6.2, 6.5, 6.8, 6.9

Let's first re-examine a perfectly inelastic collision. We studied just such a collision in Example 9.8. Recall that in such a collision the two objects stick together and then move with a common final velocity. What happens to the energy?

EXAMPLE 10.11   Energy transformations in a perfectly inelastic collision

Figure 10.28 shows two air track gliders that are pushed toward each other, collide, and stick together. In Example 9.8, we used conservation of momentum to find the final velocity shown in Figure 10.28 from given initial velocities. Compare the initial and final *mechanical energies* of the system.

FIGURE 10.28 Initial and final velocities in a completely inelastic collision.

PREPARE We'll choose our system to be the two gliders. Because the tracks are horizontal, there is no change in potential energy. Thus the law of conservation of energy, Equation 10.6, reads $K_i = K_f + \Delta E_{th}$. The total energy before the collision must equal the total energy after, but the total *mechanical* energies need not be equal.

SOLVE The initial kinetic energy is

$$K_i = \frac{1}{2}m_1(v_{1x})_i^2 + \frac{1}{2}m_2(v_{2x})_i^2$$

$$= \frac{1}{2}(0.200 \text{ kg})(3.00 \text{ m/s})^2 + \frac{1}{2}(0.400 \text{ kg})(-2.25 \text{ m/s})^2$$

$$= 1.91 \text{ J}$$

Because the gliders stick together and move as a single object with mass $m_1 + m_2$, the final kinetic energy is

$$K_f = \frac{1}{2}(m_1 + m_2)(v_x)_f^2$$

$$= \frac{1}{2}(0.600 \text{ kg})(-0.500 \text{ m/s})^2 = 0.0750 \text{ J}$$

From the conservation of energy equation above, we find that the thermal energy increases by

$$\Delta E_{th} = K_i - K_f = 1.91 \text{ J} - 0.075 \text{ J} = 1.84 \text{ J}$$

This amount of the initial kinetic energy is transformed into thermal energy during the impact of the collision.

ASSESS About 96% of the initial kinetic energy is transformed into thermal energy. This is typical of many real-world collisions.

In a collision between a cue ball and a stationary ball, the mechanical energy of the balls is almost perfectly conserved.

Before: ① →$\vec{v}_{1i}$ ②　　K_i

Energy is stored in compressed molecular bonds, then released as the bonds re-expand.

During: ①②

After: ①→ ②→　$K_f = K_i$
　　　　$\vec{v}_{1f}$　$\vec{v}_{2f}$

FIGURE 10.29 A perfectly elastic collision.

Elastic Collisions

Figure 9.1 showed a collision of a tennis ball with a racket. The ball is compressed and the racket strings stretch as the two collide, then the ball expands and the strings relax as the two are pushed apart. In the language of energy, the kinetic energy of the objects is transformed into the elastic potential energy of the ball and strings, then back into kinetic energy as the two objects spring apart. If *all* of the kinetic energy is stored as elastic potential energy, and then *all* of the elastic potential energy is transformed back into the post-collision kinetic energy of the objects, then mechanical energy is conserved. A collision for which mechanical energy is conserved is called a **perfectly elastic collision.**

Needless to say, most real collisions fall somewhere between perfectly elastic and perfectly inelastic. A rubber ball bouncing on the floor might "lose" 20% of its kinetic energy on each bounce and return to only 80% of the height of the previous bounce. But collisions between two very hard objects, such as two pool balls or two steel balls, come close to being perfectly elastic. And collisions between microscopic particles, such as atoms or electrons, can be perfectly elastic.

Figure 10.29 shows a head-on, perfectly elastic collision of a ball of mass m_1, having initial velocity $(v_{1x})_i$, with a ball of mass m_2 that is initially at rest. The balls' velocities after the collision are $(v_{1x})_f$ and $(v_{2x})_f$. These are velocities, not speeds, and have signs. Ball 1, in particular, might bounce backward and have a negative value for $(v_{1x})_f$.

The collision must obey two conservation laws: conservation of momentum (obeyed in any collision) and conservation of mechanical energy (because the collision is perfectly elastic). Although the energy is transformed into potential energy during the collision, the mechanical energy before and after the collision is purely kinetic energy. Thus

momentum conservation:　$m_1(v_{1x})_i = m_1(v_{1x})_f + m_2(v_{2x})_f$

energy conservation:　$\frac{1}{2}m_1(v_{1x})_i^2 = \frac{1}{2}m_1(v_{1x})_f^2 + \frac{1}{2}m_2(v_{2x})_f^2$

Momentum conservation alone is not sufficient to analyze the collision because there are two unknowns: the two final velocities. That is why we did not consider perfectly elastic collisions in Chapter 9. Energy conservation gives us another condition. The complete solution of these two equations involves straightforward but rather lengthy algebra. We'll just give the solution here, which is:

$$(v_{1x})_f = \frac{m_1 - m_2}{m_1 + m_2}(v_{1x})_i \qquad (v_{2x})_f = \frac{2m_1}{m_1 + m_2}(v_{1x})_i \qquad (10.19)$$

Perfectly elastic collision with object 2 initially at rest

Equations 10.19 allow us to compute the final velocity of each object. Let's look at a common and important example: a perfectly elastic collision between two objects of equal mass.

EXAMPLE 10.12 Velocities in an air hockey collision

On an air hockey table, a moving puck, traveling at 2.3 m/s, makes a head-on collision with a stationary puck. What are the final velocities of each of the pucks?

PREPARE The before-and-after visual overview is shown Figure 10.30. We've sketched in final velocities in the picture, but we don't really know yet which way the pucks will move. Because one puck was initially at rest, we can use Equation 10.19 to find the final velocities of the pucks. The pucks are identical, so we have $m_1 = m_2 = m$.

SOLVE We use Equation 10.19 with $m_1 = m_2 = m$ to get

$$(v_{1x})_f = \frac{m - m}{m + m}(v_{1x})_i = 0 \text{ m/s}$$

$$(v_{2x})_f = \frac{2m}{m + m}(v_{1x})_i = (v_{1x})_i = 2.3 \text{ m/s}$$

Before: $(v_{1x})_i = 2.3$ m/s $(v_{2x})_i = 0$ m/s

$\vec{v}_{1i}$ $\vec{v}_{2i} = \vec{0}$

After: Find: $(v_{1x})_f$ and $(v_{2x})_f$

$\vec{v}_{1f}$ $\vec{v}_{2f}$

FIGURE 10.30 A moving puck collides with a stationary puck.

The incoming puck stops dead, and the initially stationary puck goes off with the same velocity that the incoming one had.

ASSESS You can see that momentum and energy are conserved: the incoming puck's momentum and energy are completely transferred to the outgoing puck. If you've ever played pool, you've probably seen this sort of collision when you hit a ball head-on with the cue ball: the cue ball stops and the other ball picks up the cue ball's velocity.

Other cases where the colliding objects are of unequal mass will be treated in the end-of-chapter problems.

Forces in Collisions

The collision between two pool balls occurs very quickly, and the forces are typically very large and difficult to calculate. Fortunately, by using the concepts of momentum and energy conservation we can often calculate the final velocities of the balls without having to know the forces between them. There are collisions, however, where knowing the forces involved is of critical importance. The following example shows how a helmet helps protect the head from the large forces involved in a bicycle accident.

EXAMPLE 10.13 **Protecting your head**

A bike helmet is basically a shell of hard, crushable foam 3.0 cm thick. In testing, the helmet is strapped onto a 5.0 kg headform that is dropped from a height of 2.0 m onto a hard anvil. What force is encountered by the head in such a fall?

The foam inside a bike helmet is designed to crush upon impact.

PREPARE A visual overview of the test is shown in Figure 10.31. We can use the law of conservation of energy, Equation 10.6, to estimate the force on the headform. We'll choose the headform and the earth to be the system; the foam in the helmet will be part of the environment. We make this choice so that the force on the headform due to the foam is an *external* force that does work W on the headform.

SOLVE The headform starts at initial height $y_i = 2.0$ m above the anvil and ends at rest with the foam fully crushed. Then the law of conservation of energy is

$$K_i + (U_g)_i + W = K_f + (U_g)_f$$

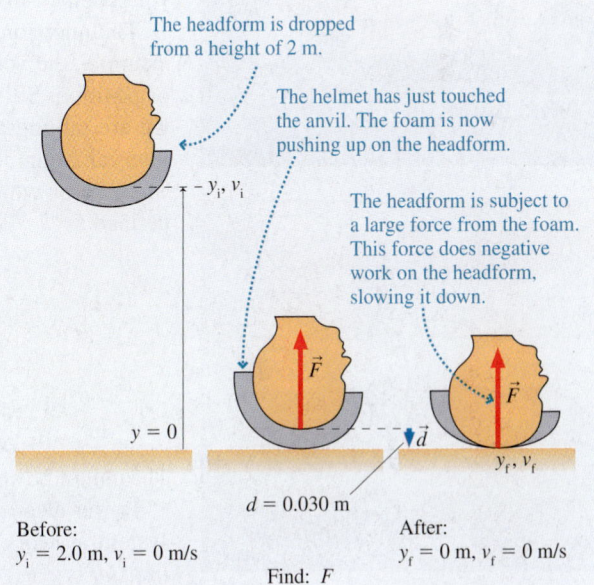

The headform is dropped from a height of 2 m.

The helmet has just touched the anvil. The foam is now pushing up on the headform.

$-y_i, v_i$

The headform is subject to a large force from the foam. This force does negative work on the headform, slowing it down.

$\vec{F}$ $\vec{F}$

$y = 0$ $\vec{d}$ y_f, v_f

$d = 0.030$ m

Before:
$y_i = 2.0$ m, $v_i = 0$ m/s

After:
$y_f = 0$ m, $v_f = 0$ m/s

Find: F

FIGURE 10.31 The foam in the helmet does negative work on the headform.

Continued

In words, this states that the initial energy of the system, plus the energy transferred to the system as work, equals the final energy of the system.

Since the headform starts and ends at rest, both K_i and K_f are zero. Taking our reference height $y = 0$ at the anvil, $(U_g)_f$ is zero as well. Since $(U_g)_i = mgy_i$, conservation of energy gives simply $mgy_i + W = 0$, or

$$mgy_i = -W$$

As the foam is crushed, it pushes up on the headform with force $\vec{F}$, doing work on it. This force is directed *opposite* to the displacement $\vec{d}$ of the headform, so the work done is *negative*—kinetic energy is being *removed* from the headform, slowing it down. The work done is $-Fd$, if we assume the force is relatively constant, so we have

$$mgy_i = -(-Fd)$$

or

$$F = \frac{mgy_i}{d} = \frac{(5.0\ \text{kg})(9.8\ \text{m/s})(2.0\ \text{m})}{0.030\ \text{m}} = 3300\ \text{N}$$

This is the force that acts on the head to bring it to a halt in only 3 cm. More important from the perspective of possible brain injury is the head's *acceleration*

$$a = \frac{F}{m} = \frac{3300\ \text{N}}{5.0\ \text{kg}} = 660\ \text{m/s}^2 = 67g$$

where g is the acceleration due to gravity.

ASSESS The accepted threshold for serious brain injury is around $300g$, so this helmet would protect the rider in all but the most serious accidents. Without the helmet, the rider's head would come to a stop in a much smaller distance and thus be subjected to a much larger acceleration.

It's also interesting to ask where the original energy of the headform went. The work on it was negative, indicating a transfer of energy from the headform to the environment—the foam. As the foam crushes, there is a great deal of internal friction and rubbing between parts of the foam. This causes the foam to get warmer, increasing its thermal energy. This increase must be exactly equal to the energy lost by the headform.

10.10 Power

We've now studied how energy can be transformed from one kind to another and how it can be transferred between the environment and the system as work. In many situations we would like to know *how quickly* the energy is transformed or transferred. Is a transfer of energy very rapid, or does it take place over a long time? In passing a truck, your car needs to transform a certain amount of the chemical energy in its fuel into kinetic energy. It makes a *big* difference whether your engine can do this in 20 s or 60 s!

The question "How quickly?" implies that we are talking about a *rate*. For example, the velocity of an object—how fast it is going—is the *rate of change* of position. So when we raise the issue of how fast the energy is transformed, we are talking about the *rate of transformation* of energy. Suppose in a time interval Δt an amount of energy ΔE is transformed from one form to another. The rate at which this energy is transformed is called the **power** P, and it is defined as

$$P = \frac{\Delta E}{\Delta t} \tag{10.20}$$

Power when amount of energy ΔE is transformed in time interval Δt

The unit of power is the **watt,** which is defined as 1 watt = 1 W = 1 J/s.

Power also measures the rate at which energy is transferred into or out of a system as work W. If work W is done in time interval Δt, the rate of energy *transfer* is

$$P = \frac{W}{\Delta t} \tag{10.21}$$

Power when amount of work W is done in time interval Δt

Both these cars take about the same energy to reach 60 mph, but the race car gets there in a much shorter time, so its *power* is much greater.

A force that is doing work (i.e., transferring energy) at a rate of 3 J/s has an "output power" of 3 W. A system gaining energy at the rate of 3 J/s is said to "consume" 3 W of power. Common prefixes used with power are mW (milliwatts), kW (kilowatts), and MW (megawatts).

We can express Equation 10.21 in a different form. If in the time interval Δt an object undergoes a displacement Δx, the work done by a force acting on the object is $W = F\Delta x$. Then Equation 10.21 can be written

$$P = \frac{W}{\Delta t} = \frac{F\Delta x}{\Delta t} = F\frac{\Delta x}{\Delta t} = Fv$$

The rate at which energy is transferred to an object as work—the power—is the product of the force that does the work, and the velocity of the object:

$$P = Fv \qquad (10.22)$$

Rate of energy transfer due to a force F
acting on an object moving at velocity v

LINEAR
p. 38

The English unit of power is the *horsepower*. The conversion factor to watts is

1 horsepower = 1 hp = 746 W

Many common appliances, such as motors, are rated in hp.

EXAMPLE 10.14 Power to pass a truck

You are behind a 1500 kg truck traveling at 60 mph (27 m/s). To pass it, you speed up to 75 mph (34 m/s) in 6.0 s. What power is required to do this?

PREPARE Your car is undergoing an energy transformation from the chemical energy of your fuel to the kinetic energy of the car. We can calculate the amount of energy transformed by finding the change ΔK in the kinetic energy.

SOLVE We have

$$K_i = \frac{1}{2}mv_i^2 = \frac{1}{2}(1500 \text{ kg})(27 \text{ m/s})^2 = 5.47 \times 10^5 \text{ J}$$

$$K_f = \frac{1}{2}mv_f^2 = \frac{1}{2}(1500 \text{ kg})(34 \text{ m/s})^2 = 8.67 \times 10^5 \text{ J}$$

so that

$$\Delta K = K_f - K_i$$
$$= 8.67 \times 10^5 \text{ J} - 5.47 \times 10^5 \text{ J} = 3.20 \times 10^5 \text{ J}$$

To transform this amount of energy in 6 s, the power required is

$$P = \frac{\Delta K}{\Delta t} = \frac{3.20 \times 10^5 \text{ J}}{6.0 \text{ s}} = 53,000 \text{ W} = 53 \text{ kW}$$

This is about 71 hp. This power is in addition to the power needed to overcome drag and friction and cruise at 60 mph, so the total power required from the engine will be even greater than this.

ASSESS You use a large amount of energy to perform a simple driving maneuver such as this. 3.20×10^5 J is enough energy to lift an 80 kg person 410 m in the air—the height of a tall skyscraper. And 53 kW would lift him there in only 6 s!

STOP TO THINK 10.6 Four students run up the stairs in the time shown. Rank in order, from largest to smallest, their power outputs P_A to P_D.

A. 80 kg 10 m $\Delta t = 10$ s

B. 80 kg $\Delta t = 8$ s

C. 64 kg $\Delta t = 8$ s

D. 80 kg 20 m $\Delta t = 25$ s

SUMMARY

The goal of Chapter 10 has been to learn about energy and how to solve problems using the law of conservation of energy.

GENERAL PRINCIPLES

General Energy Model

Within a system, energy can be **transformed** between various forms.

Energy can be **transferred** into or out of a system in two basic ways:

- **Work:** The transfer of energy by mechanical forces.

- **Heat:** The nonmechanical transfer of energy from a hotter to a colder object.

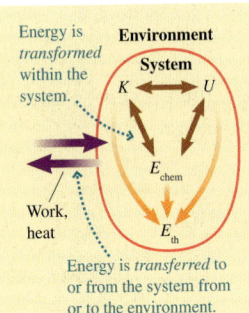

Energy is *transformed* within the system.

Work, heat

Energy is *transferred* to or from the system from or to the environment.

Solving Energy Conservation Problems

PREPARE Choose your system (Tactics Box 10.1). Decide what forms of energy are changing. If there is friction, then thermal energy will be created. Check for external forces that will do work on your system.

SOLVE Use Equation 10.6 to relate the initial energy of your system, plus the work done, to the final energy of the system:

$$K_i + U_i + W = K_f + U_f + \Delta E_{th}$$

ASSESS Kinetic energy is always positive. The *change* in thermal energy should be positive.

Law of Conservation of Energy

Isolated system: No energy is transferred into or out of the system. Each form of energy within the system can change, but the total change in energy is zero. Energy of the system is *conserved:*

The change in the system's energy is zero.

$$\Delta K + \Delta U_g + \Delta U_s + \Delta E_{th} + \Delta E_{chem} + \ldots = 0$$

Nonisolated system: Energy can be exchanged with the environment as work or heat. The energy of the system changes by the amount of work done or heat transferred:

$$\Delta K + \Delta U_g + \Delta U_s + \Delta E_{th} + \Delta E_{chem} + \ldots = W + Q$$

The system's energy changes by the amount of work done and heat transferred.

Systems with mechanical and thermal energy only: The initial mechanical energy, plus the work done, equals the final mechanical energy plus additional thermal energy:

$$K_i + U_i + W = K_f + U_f + \Delta E_{th}$$

In terms of energy changes, this can be written

$$\Delta K + \Delta U_g + \Delta U_s + \Delta E_{th} = W$$

IMPORTANT CONCEPTS

Mechanical energy is the sum of a system's kinetic and potential energies:

Mechanical energy $= K + U = K + U_g + U_s$

Kinetic energy is an energy of motion

$$K = \frac{1}{2}mv^2 + \frac{1}{2}I\omega^2$$

Translational ····· Rotational

Potential energy is energy stored in a system of interacting objects

- **Gravitational potential energy:** $U_g = mgy$

- **Elastic potential energy:** $U_s = \frac{1}{2}kx^2$

Thermal energy is the sum of the microscopic kinetic and potential energy of all the molecules in an object. The hotter an object, the more thermal energy it has. When kinetic (sliding) friction is present, mechanical energy will be transformed into thermal energy.

Work is the process by which energy is transferred to or from a system by the application of mechanical forces.

If a particle moves through a displacement $\vec{d}$ while acted upon by a constant force $\vec{F}$, the force does work

$$W = F_{\parallel}d = Fd\cos\theta$$

Only the component of the force parallel to the

APPLICATIONS

Perfectly elastic collisions
Both mechanical energy and momentum are conserved.

$$(v_{1x})_f = \frac{m_1 - m_2}{m_1 + m_2}(v_{1x})_i$$

$$(v_{2x})_f = \frac{2m_1}{m_1 + m_2}(v_{1x})_i$$

Object 2 initially at rest

Before:
After:

Power is the rate at which energy is transformed . . .

$$P = \frac{\Delta E}{\Delta t}$$

Amount of energy transformed
Time required to transform it

. . . or at which work is done.

$$P = \frac{W}{\Delta t}$$

Amount of work done
Time required to do work

For instructor-assigned homework, go to www.masteringphysics.com

Problem difficulty is labeled as | (straightforward) to ||||| (challenging).

Problems labeled 🖉 can be done on a Workbook Dynamics Worksheet; INT integrate significant material from earlier chapters; BIO are of biological or medical interest.

QUESTIONS

Conceptual Questions

1. The brake shoes of your car are made of a material that can tolerate very high temperatures without being damaged. Why is this so?
2. When you pound a nail with a hammer, the nail gets quite warm. Describe the energy transformations that lead to the addition of thermal energy in the nail.

For Questions 3 through 8, give a specific example of a system with the energy transformation shown. In these questions, W is the work done on the system by the environment, and K and U are the kinetic and potential energies of the system, respectively.

3. $W \rightarrow K$ with $\Delta U = 0$.
4. $W \rightarrow U$ with $\Delta K = 0$.
5. $K \rightarrow U$ with $W = 0$.
6. $K \rightarrow W$ with $\Delta U = 0$.
7. $U \rightarrow K$ with $W = 0$.
8. $U \rightarrow W$ with $\Delta K = 0$.
9. A ball of putty is dropped from a height of 2 m onto a hard floor, where it sticks. What object or objects need to be included within the system if the system is to be isolated during this process?
10. A 0.5 kg mass on a 1-m-long string swings in a circle on a horizontal, frictionless table at a steady speed of 2 m/s. How much work does the tension in the string do on the mass during one revolution? Explain.
11. Particle A has less mass than particle B. Both are pushed forward across a frictionless surface by equal forces for 1 s. Both start from rest.
    a. Compare the amount of work done on each particle. That is, is the work done on A greater than, less than, or equal to the work done on B? Explain.
    b. Compare the impulses delivered to particles A and B. Explain.
    c. Compare the final speeds of particles A and B. Explain.
12. The meaning of the word "work" is quite different in physics from its everyday usage. Give an example of an action a person could do that "feels like work" but that does not involve any work as we've defined it in this chapter.
13. To change a tire, you need to use a jack to raise one corner of your car. While doing so, you happen to notice that pushing the jack handle down 20 cm raises the car only 0.2 cm. Use energy concepts to explain why the handle must be moved so far to raise the car by such a small amount.

Questions 14 through 17 refer to a weightlifter raising a barbell from the floor to above his head. Describe the energy transformations that occur if the system is chosen as specified in the question. Use the notation of Section 10.2 for the various forms of energy and energy transfer.

14. The system is the barbell alone.
15. The system is the weightlifter alone.
16. The system is the barbell plus the earth.
17. The system is the barbell plus the earth plus the weightlifter.

In Questions 18 through 20, imagine yourself doing a chin-up. You start from rest with your arms extended above your head, and end at rest with your elbows bent and your hands still gripping the bar. Describe the energy transformations that occur if the system is chosen as specified in the question. Use the notation of Section 10.2 for the various forms of energy and energy transfer.

18. The system is you alone.
19. The system is you plus the chin-up bar.
20. The system is you plus the chin-up bar plus the earth.
21. One kilogram of matter contains approximately 10^{17} J of nuclear energy. Why don't we need to include this energy when we study ordinary energy transformations?
22. A roller coaster car rolls down a frictionless track, reaching speed v_f at the bottom.
    a. If you want the car to go twice as fast at the bottom, by what factor must you increase the height of the track?
    b. Does your answer to part a depend on whether the track is straight or not? Explain.
23. A spring gun shoots out a plastic ball at speed v_i. The spring is then compressed twice the distance it was on the first shot.
    a. By what factor is the spring's potential energy increased?
    b. By what factor is the ball's speed increased? Explain.
24. Sandy and Chris stand on the edge of a cliff and throw identical mass rocks at the same speed. Sandy throws her rock horizontally while Chris throws his upward at an angle of 45° to the horizontal. Are the rocks moving at the same speed when they hit the ground, or is one moving faster than the other? If one is moving faster, which one? Explain.
25. If you allow a can of chicken broth to join the rolling-object race discussed in Example 10.10, it wins handily. A can of tomato paste, on the other hand, ties with the cylinder. Why? **Hint:** Try to picture how the stuff inside each can moves as the can rolls.
26. A solid cylinder and a cylindrical shell have the same mass, same radius, and turn on frictionless, horizontal axles. (The cylindrical shell has light-weight spokes connecting the shell to the axle.) A rope is wrapped around

FIGURE Q10.26

each cylinder and tied to a block. The blocks have the same mass and are held the same height above the ground as shown in Figure Q10.26. Both blocks are released simultaneously. The ropes do not slip. Which block hits the ground first? Or is it a tie? Explain.

27. You are much more likely to be injured if you fall and your
BIO head strikes the ground than if your head strikes a gymnastics
pad. Use energy and work concepts to explain why this is so.

Multiple-Choice Questions

28. ‖ If you walk up a flight of stairs at constant speed, gaining vertical height h, the work done on you (the system, of mass m) is
    A. $+mgh$, by the normal force of the stairs.
    B. $-mgh$, by the normal force of the stairs.
    C. $+mgh$, by the gravitational force of the earth.
    D. $-mgh$, by the gravitational force of the earth.

29. ‖ You and a friend each carry a 15 kg suitcase up two flights of stairs, walking at a constant speed. Take each suitcase to be the system. Suppose you carry your suitcase up the stairs in 30 s while your friend takes 60 s. Which of the following is true?
    A. You did more work, but both of you expended the same power.
    B. You did more work and expended more power.
    C. Both of you did the same work, but you expended more power.
    D. Both of you did the same work, but you expended less power.

30. ‖ A woman uses a pulley and a rope to raise a 20 kg weight to a height of 2 m. If it takes 4 s to do this, about how much power is she supplying?
    A. 100 W   B. 200 W   C. 300 W   D. 400 W

31. ‖ A hockey puck sliding along frictionless ice with speed v to the right collides with a horizontal spring and compresses it by 2.0 cm before coming to a momentary stop. What will be the spring's maximum compression if the same puck hits it at a speed of $2v$?
    A. 2.0 cm   B. 2.8 cm   C. 4.0 cm
    D. 5.6 cm   E. 8.0 cm

32. ‖ A block slides down a smooth ramp, starting from rest at a height h. When it reaches the bottom it's moving at speed v_i. It then continues to slide up a second smooth ramp. At what height is its speed equal to $v_i/2$?
    A. $h/4$ B. $h/2$ C. $3h/4$ D. $2h$

33. ‖ A wrecking ball is suspended from a 5.0-m-long cable that makes a 30° angle with the vertical. The ball is released and swings down. What is the ball's speed at the lowest point?
    A. 7.7 m/s   B. 4.4 m/s   C. 3.6 m/s   D. 3.1 m/s

PROBLEMS

Section 10.4 Work

1. ‖ During an etiquette class, you walk slowly and steadily at 0.20 m/s for 2.5 m with a 0.75 kg book balanced on top of your head. How much work does your head do on the book?

2. ‖ A 2.0 kg book is lying on a 0.75-m-high table. You pick it up and place it on a bookshelf 2.3 m above the floor.
   a. How much work does gravity do on the book?
   b. How much work does your hand do on the book?

3. ‖ The two ropes seen in Figure P10.3 are used to lower a 255 kg piano exactly 5 m from a second-story window to the ground. How much work is done by each of the three forces?

FIGURE P10.3   **FIGURE P10.4**

4. ‖ The two ropes shown in the bird's-eye view of Figure P10.4 are used to drag a crate exactly 3 m across the floor. How much work is done by each of the ropes on the crate?

5. ‖ a. At the airport, you ride a "moving sidewalk" that carries you horizontally for 25 m at 0.70 m/s. Assuming that you were moving at 0.70 m/s before stepping onto the moving sidewalk and continue at 0.70 m/s afterward, how much work does the moving sidewalk do on you? Your mass is 60 kg.
   b. An escalator carries you from one level to the next in the airport terminal. The upper level is 4.5 m above the lower level, and the length of the escalator is 7.0 m.

How much work does the up escalator do on you when you ride it from the lower level to the upper level?
   c. How much work does the down escalator do on you when you ride it from the upper level to the lower level?

6. ‖ A boy flies a kite with the string at a 30° angle to the horizontal. The tension in the string is 4.5 N. How much work does the string do on the boy if the boy
   a. stands still?
   b. walks a horizontal distance of 11 m away from the kite?
   c. walks a horizontal distance of 11 m toward the kite?

Section 10.5 Kinetic Energy

7. ‖ Which has the larger kinetic energy, a 10 g bullet fired at 500 m/s or a 10 kg bowling ball sliding at 10 m/s?

8. ‖ At what speed does a 1000 kg compact car have the same kinetic energy as a 20,000 kg truck going 25 km/hr?

9. ‖ An oxygen atom is four times as massive as a helium atom. In an experiment, a helium atom and an oxygen atom have the same kinetic energy. What is the ratio v_{He}/v_O of their speeds?

10. ‖ Sam's job at the amusement park is to slow down and bring to a stop the boats in the log ride. If a boat and its riders have a mass of 1200 kg and the boat drifts in at 1.2 m/s, how much work does Sam do to stop it?

11. ‖ A 20 g plastic ball is moving to the left at 30 m/s. How much work must be done on the ball to cause it to move to the right at 30 m/s?

12. ‖ The turntable in a microwave oven has a moment of inertia of 0.040 kg · m² and is rotating once every 4.0 s. What is its kinetic energy?

13. ‖ An energy storage system based on a flywheel (a rotating disk) can store a maximum of 4.0 MJ when the flywheel is rotating at 20,000 revolutions per minute. What is the moment of inertia of the flywheel?

Section 10.6 Potential Energy

14. | The lowest point in Death Valley is 85.0 m below sea level. The summit of nearby Mt. Whitney has an elevation of 4420 m. What is the change in gravitational potential energy of an energetic 65.0 kg hiker who makes it from the floor of Death Valley to the top of Mt. Whitney?

15. | a. What is the kinetic energy of a 1500 kg car traveling at a speed of 30 m/s ($\approx$65 mph)?
    b. From what height should the car be dropped to have this same amount of kinetic energy just before impact?
    c. Does your answer to part b depend on the car's mass?

16. || A boy reaches out of a window and tosses a ball straight up with a speed of 10 m/s. The ball is 20 m above the ground as he releases it. Use conservation of energy to find
    a. The ball's maximum height above the ground.
    b. The ball's speed as it passes the window on its way down.
    c. The speed of impact on the ground.

17. || a. With what minimum speed must you toss a 100 g ball straight up to just barely hit the 10-m-high ceiling of the gymnasium if you release the ball 1.5 m above the floor? Solve this problem using energy.
    b. With what speed does the ball hit the floor?

18. || What minimum speed does a 100 g puck need to make it to the top of a frictionless ramp that is 3.0 m long and inclined at 20°?

19. || A car is parked at the top of a 50-m-high hill. It slips out of gear and rolls down the hill. How fast will it be going at the bottom? (Ignore friction.)

20. || A pendulum is made by tying a 500 g ball to a 75-cm-long string. The pendulum is pulled 30° to one side, then released.
    a. What is the ball's speed at the lowest point of its trajectory?
    b. To what angle does the pendulum swing on the other side?

21. ||| A 1500 kg car is approaching the hill shown in Figure P10.21 at 10 m/s when it suddenly runs out of gas.
    a. Can the car make it to the top of the hill by coasting?
    b. If your answer to (a) is yes, what is the car's speed after coasting down the other side?

FIGURE P10.21

22. | How much energy can be stored in a spring with a spring constant of 500 N/m if its maximum possible stretch is 20 cm?

23. | How far must you stretch a spring with $k = 1000$ N/m to store 200 J of energy?

24. | A student places her 500 g physics book on a frictionless table. She pushes the book against a spring, compressing the spring by 4.00 cm, then releases the book. What is the book's speed as it slides away? The spring constant is 1250 N/m.

25. | A 10 kg runaway grocery cart runs into a spring with spring constant 250 N/m and compresses it by 60 cm. What was the speed of the cart just before it hit the spring?

26. | As a 15,000 kg jet lands on an aircraft carrier, its tail hook snags a cable to slow it down. The cable is attached to a spring with spring constant 60,000 N/m. If the spring stretches 30 m to stop the plane, what was the plane's landing speed?

27. |||| The elastic energy stored in your tendons can contribute up
BIO to 35% of your energy needs when running. Sports scientists have studied the change in length of the knee extensor tendon in sprinters and nonathletes. They find (on average) that the sprinters' tendons stretch 41 mm, while nonathletes' stretch only 33 mm. The spring constant for the tendon is the same for both groups, 33 N/mm. What is the difference in maximum stored energy between the sprinters and the nonathletes?

28. || You're driving at 35 km/hr when the road suddenly descends 15 m into a valley. You take your foot off the accelerator and coast down the hill. Just as you reach the bottom you see the police officer hiding behind the speed limit sign that reads "70 km/hr." Are you going to get a speeding ticket?

29. || Your friend's Frisbee has become stuck 16 m above the ground in a tree. You want to dislodge the Frisbee by throwing a rock at it. The Frisbee is stuck pretty tight, so you figure the rock needs to be traveling at least 5.0 m/s when it hits the Frisbee. If you release the rock 2.0 m above the ground, with what minimum speed must you throw it?

Section 10.7 Thermal Energy

30. | A 1500 kg car traveling at 20 m/s skids to a halt.
    a. What energy transfers and transformations occur during the skid?
    b. What is the change in the thermal energy of the car and the road surface?

31. || A 20 kg child slides down a 3.0-m-high playground slide. She starts from rest, and her speed at the bottom is 2.0 m/s.
    a. What energy transfers and transformations occur during the slide?
    b. What is the change in the thermal energy of the slide and the seat of her pants?

32. ||| A fireman of mass 80 kg slides down a pole. When he reaches the bottom, 4.2 m below his starting point, his speed is 2.2 m/s. By how much has thermal energy increased during his slide?

Section 10.9 Energy in Collisions

33. | A 50 g marble moving at 2.0 m/s strikes a 20 g marble at rest. What is the speed of each marble immediately after the collision? Assume the collision is perfectly elastic.

34. | Ball 1, with a mass of 100 g and traveling at 10 m/s, collides head-on with ball 2, which has a mass of 300 g and is initially at rest. What are the final velocities of each ball if the collision is (a) perfectly elastic? (b) perfectly inelastic?

35. || A proton is traveling to the right at 2.0×10^7 m/s. It has a head-on, perfectly elastic collision with a stationary carbon atom. The mass of the carbon atom is 12 times the mass of the proton. What are the speed and direction of each after the collision?

36. | Two balls undergo a perfectly elastic head-on collision, with one ball initially at rest. If the incoming ball has a speed of 200 m/s, what are the final speed and direction of each ball if
    a. the incoming ball is *much* more massive than the stationary ball?
    b. the stationary ball is *much* more massive than the incoming ball?

37. |||| Derive Equations 10.19 for the final speeds of two objects undergoing a perfectly elastic collision, with one object initially stationary.

Section 10.10 Power

38. | a. How much work does an elevator motor do to lift a 1000 kg elevator a height of 100 m?
    b. How much power must the motor supply to do this in 50 s at constant speed?

39. || a. How much work must you do to push a 10 kg block of steel across a steel table at a steady speed of 1.0 m/s for 3.0 s? The coefficient of kinetic friction for steel on steel is 0.60.
    b. What is your power output while doing so?

40. | Which consumes more energy, a 1.2 kW hair dryer used for 10 min or a 10 W night light left on for 24 hr?

41. ||| A 1000 kg sports car accelerates from 0 to 30 m/s in 10 s. What is the average power of the engine?

42. ||| In just 0.30 s, you compress a spring (spring constant 5000 N/m), which is initially at its equilibrium length, by 4.0 cm. What is your average power output?

43. |||| In the winter sport of curling, players give a 20 kg stone a push across a sheet of ice. A curler accelerates a stone to a speed of 3.0 m/s over a time of 2.0 s.
    a. How much force does the curler exert on the stone?
    b. What average power does the curler use to bring the stone up to speed?

44. |||| A 710 kg car drives at a constant speed of 23 m/s. It is subject to a drag force of 500 N. What power is required from the car's engine to drive the car
    a. on level ground?
    b. up a hill with a slope of 2.0°?

45. || An elevator weighing 2500 N ascends at a constant speed of 8.0 m/s. How much power must the motor supply to do this?

General Problems

46. ||| A 2.3 kg box, starting from rest, is pushed up a ramp by a 10 N force parallel to the ramp. The ramp is 2.0 m long and tilted at 17°. The speed of the box at the top of the ramp is 0.80 m/s. Consider the system to be the box + ramp + earth.
    a. How much work W does the force do on the system?
    b. What is the change ΔK in the kinetic energy of the system?
    c. What is the change ΔU_g in the gravitational potential energy of the system?
    d. What is the change ΔE_{th} in the thermal energy of the system?

47. || A 55 kg skateboarder wants to just make it to the upper edge of a "half-pipe" with a radius of 3.0 m, as shown in Figure P10.47. What speed v_i does he need

FIGURE P10.47

at the bottom if he is to coast all the way up?
    a. First do the calculation treating the skateboarder and board as a point particle, with the entire mass nearly in contact with the half-pipe.
    b. More realistically, the mass of the skateboarder in a deep crouch might be thought of as concentrated 0.75 m from the half-pipe. Assuming he remains in that position all the way up, what v_i is needed to reach the upper edge ?

48. || Fleas have remarkable jumping ability. If a 0.50 mg flea jumps straight up, it will reach a height of 40 cm if there is no air resistance. In reality, air resistance limits the height to 20 cm.
    a. What is the flea's kinetic energy as it leaves the ground?
    b. At its highest point, what fraction of the initial kinetic energy has been converted to potential energy?

49. ||| A marble slides without friction in a *vertical* plane around the inside of a smooth, 20-cm-diameter horizontal pipe. The marble's speed at the bottom is 3.0 m/s; this is fast enough so that the marble makes a complete loop, never losing contact with the pipe. What is its speed at the top?

50. ||| A 20 kg child is on a swing that hangs from 3.0-m-long chains, as shown in Figure P10.50. What is her speed v_i at the bottom of the arc if she swings out to a 45° angle before reversing direction?

FIGURE P10.50   **FIGURE P10.51**

51. | Suppose you lift a 20 kg box by a height of 1.0 m.
    a. How much work do you do in lifting the box?
    Instead of lifting the box straight up, suppose you push it up a 1.0-m-high ramp that makes a 30° degree angle with the horizontal, as shown in Figure P10.51. Being clever, you choose a ramp with no friction.
    b. How much force F is required to push the box straight up the slope at a constant speed?
    c. How long is the ramp?
    d. Use your force and distance results to calculate the work you do in pushing the box up the ramp. How does this compare to your answer to part a?

52. || A cannon tilted up at a 30° angle fires a cannon ball at 80 m/s from atop a 10-m-high fortress wall. What is the ball's impact speed on the ground below? Ignore air resistance.

53. || The sledder shown in Figure P10.53 starts from the top of a frictionless hill and slides down into the valley. What initial speed v_i does the sledder need to just make it over the next hill?

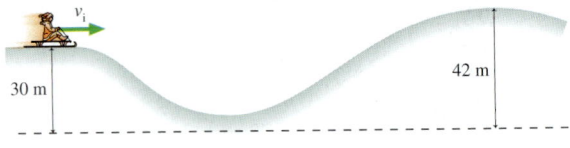

FIGURE P10.53

54. ||| A 100 g granite cube slides down a frictionless 40° incline. At the bottom, just after it exits onto a horizontal table, the granite collides with a 200 g steel cube at rest. How high above the table should the granite cube be released to give the steel cube a speed of 150 cm/s?

55. || A 50 g ice cube can slide without friction up and down a 30° slope. The ice cube is pressed against a spring at the bottom of the slope, compressing the spring 10 cm. The spring constant is 25 N/m. When the ice cube is released, what distance will it travel up the slope before reversing direction?

56. |||| In a physics lab experiment, a spring clamped to the table shoots a 20 g ball horizontally. When the spring is compressed 20 cm, the ball travels horizontally 5.0 m and lands on the floor 1.5 m below the point at which it left the spring. What is the spring constant?

57. | The desperate contestants on a TV survival show are very hungry. The only food they can see is some fruit hanging on a branch high in a tree. Fortunately, they have a spring they can use to launch a rock. The spring constant is 1000 N/m, and they can compress the spring a maximum of 30 cm. All the rocks on the island seem to have a mass of 400 g.
    a. With what speed does the rock leave the spring?
    b. If the fruit hangs 15 m above the ground, will they feast or go hungry?

58. | The maximum energy a bone can absorb without breaking
    BIO is surprisingly small. For a healthy human of mass 60 kg, experimental data show that the leg bones can absorb about 200 J.
    a. From what maximum height could a person jump and land rigidly upright on both feet without breaking his legs? Assume that all the energy is absorbed in the leg bones in a rigid landing.
    b. People jump from much greater heights than this; explain how this is possible.
    Hint: Think about how people land when they jump from greater heights.

59. || In an amusement park water slide, people slide down an essentially frictionless tube. They drop 3.0 m and exit the slide,
    INT moving horizontally, 1.2 m above a swimming pool. What horizontal distance do they travel from the exit point before hitting the water? Does the mass of the person make any difference?

60. || The 5.0-m-long rope in Figure P10.60 hangs vertically from a tree right at the edge of a ravine. A woman wants to use the rope to swing to the other side of the ravine. She runs as fast as she can, grabs the rope, and swings out over the ravine.

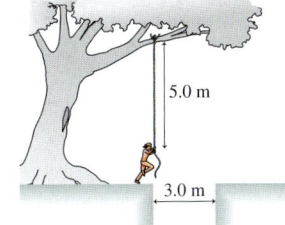

FIGURE P10.60

5.0 m

3.0 m

    a. As she swings, what energy conversion is taking place?
    b. When she's directly over the far edge of the ravine, how much higher is she than when she started?
    c. Given your answers to parts a and b, how fast must she be running when she grabs the rope in order to swing all the way across the ravine?

61. ||| You have been asked to design a "ballistic spring system" to measure the speed of bullets. A bullet of mass m is fired
    INT into a block of mass M. The block, with the embedded bullet, then slides across a frictionless table and collides with a horizontal spring whose spring constant is k. The opposite end of the spring is anchored to a wall. The spring's maximum compression d is measured.
    a. Find an expression for the bullet's initial speed v_B in terms of m, M, k, and d.
    Hint: This is a two-part problem. The bullet's collision with the block is an inelastic collision. What quantity is conserved in an inelastic collision? Subsequently the block hits a spring on a frictionless surface. What quantity is conserved in this collision?
    b. What was the speed of a 5.0 g bullet if the block's mass is 2.0 kg and if the spring, with $k = 50$ N/m, was compressed by 10 cm?
    c. What fraction of the bullet's initial kinetic energy is "lost"? Where did it go?

62. || A new event, shown in Figure P10.62, has been
    INT proposed for the Winter Olympics. An athlete will sprint 100 m, starting from rest, then leap onto a 20 kg bobsled. The person

20°

FIGURE P10.62

and bobsled will then slide down a 50-m-long ice-covered ramp, sloped at 20°, and into a spring with a carefully calibrated spring constant of 2000 N/m. The athlete who compresses the spring the farthest wins the gold medal. Lisa, whose mass is 40 kg, has been training for this event. She can reach a maximum speed of 12 m/s in the 100 m dash.
    a. How far will Lisa compress the spring?
    b. The Olympic committee has very exact specifications about the shape and angle of the ramp. Is this necessary? If the committee asks your opinion, what factors about the ramp will you tell them are important?

63. ||| A 20 g ball is fired horizontally with initial speed v_i toward a 100 g ball that is hanging motionless from a 1.0-m-long string. The balls undergo a head-on, perfectly elastic collision, after which the 100 g ball swings out to a maximum angle $\theta_{max} = 50°$. What was v_i?

64. | A 70 kg human sprinter can accelerate from rest to 10 m/s
    BIO in 3.0 s. During the same interval, a 30 kg greyhound can accelerate from rest to 20 m/s. Compute (a) the change in kinetic energy and (b) the average power output for each.

65. || A 50 g ball of clay traveling at speed v_i hits and sticks to a
    INT 1.0 kg block sitting at rest on a frictionless surface.
    a. What is the speed of the block after the collision?
    b. Show that the mechanical energy is *not* conserved in this collision. What percentage of the ball's initial kinetic energy is "lost"? Where did this kinetic energy go?

66. || A package of mass m is released from rest at a warehouse
    INT loading dock and slides down a 3.0-m-high frictionless chute to a waiting truck. Unfortunately, the truck driver went on a break without having removed the previous package, of mass $2m$, from the bottom of the chute as shown in Figure P10.66.
    a. Suppose the packages stick together. What is their common speed after the collision?
    b. Suppose the collision between the packages is perfectly elastic. To what height does the package of mass m rebound?

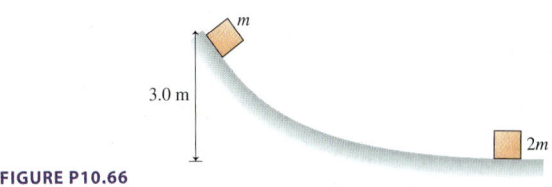

m

3.0 m

$2m$

FIGURE P10.66

67. |||| A 50 kg sprinter, starting from rest, runs 50 m in 7.0 s at
    INT constant acceleration.
    a. What is the magnitude of the horizontal force acting on the sprinter?
    b. What is the sprinter's average power output during the first 2.0 s of his run?
    c. What is the sprinter's average power output during the final 2.0 s?

68. ▌▌▌ Bob can throw a 500 g rock with a speed of 30 m/s. He
INT moves his hand forward 1.0 m while doing so.
    a. How much force, assumed to be constant, does Bob apply
       to the rock?
    b. How much work does Bob do on the rock?

69. ▌▌▌▌ A 2.0 hp electric motor on a water well pumps water from
    10 m below the surface. The density of water is 1.0 kg per L.
    How many liters of water can the motor pump in 1 hr?

70. ▌ The human heart has to pump the average adult's 6.0 L of
BIO blood through the body every minute. The heart must do work
    to overcome frictional forces that resist the blood flow. The
    average blood pressure is 1.3×10^4 N/m².
    a. Compute the work done moving the 6.0 L of blood com-
       pletely through the body, assuming the blood pressure
       always takes its average value.
    b. What power output must the heart have to do this task once
       a minute?
    Hint: When the heart contracts, it applies force to the blood.
    Pressure is just force/area, so we can write work = (pressure)
    (area)(distance). But (area)(distance) is just the blood vol-
    ume passing through the heart.

Passage Problems

Tennis Ball Testing

A tennis ball bouncing on a hard surface compresses and then
rebounds. The details of the rebound are specified in tennis regulations. Tennis balls, to be acceptable for tournament play, must have
a mass of 57.5 g. When dropped from a height of 2.5 m onto a concrete surface, a ball must rebound to a height of 1.4 m. During
impact, the ball compresses by approximately 6 mm.

71. ▌ How fast is the ball moving when it hits the concrete sur-
    face? (Ignore air resistance.)
    A. 5 m/s   B. 7 m/s   C. 25 m/s   D. 50 m/s

72. ▌ If the ball accelerates uniformly when it hits the floor, what is
    its approximate acceleration as it comes to rest before
    rebounding?
    A. 1000 m/s² B. 2000 m/s² C. 3000 m/s² D. 4000 m/s²

73. ▌ The ball's kinetic energy just after the bounce is less than
    just before the bounce. In what form does this lost energy
    end up?
    A. Elastic potential energy
    B. Gravitational potential energy
    C. Thermal energy
    D. Rotational kinetic energy

74. ▌ By what percent does the kinetic energy decrease?
    A. 35%   B. 45%   C. 55%   D. 65%

75. ▌ When a tennis ball bounces from a racket, the ball loses
    approximately 30% of its kinetic energy to thermal energy. A
    ball that hits a racket at a speed of 10 m/s will rebound with
    approximately what speed?
    A. 8.5 m/s   B. 7.0 m/s   C. 4.5 m/s   D. 3.0 m/s

Work and Power in Cycling

When you ride a bicycle at constant speed, almost all of the energy
you expend goes into the work you do against the drag force of the
air. In this problem, assume that *all* of the energy expended goes
into working against drag. As we saw in Section 5.7, the drag force
on an object is approximately proportional to the square of its speed
with respect to the air. For this problem, assume that $F \propto v^2$ exactly
and that the air is motionless with respect to the ground unless
noted otherwise. Suppose a cyclist and her bicycle have a combined mass of 60 kg and she is cycling along at a speed of 5 m/s.

76. ▌ If the drag force on the cyclist is 10 N, how much energy
    does she use in cycling 1 km?
    A. 6 kJ   B. 10 kJ   C. 50 kJ   D. 100 kJ

77. ▌ Under these conditions, how much power does she expend
    as she cycles?
    A. 10 W   B. 50 W   C. 100 W   D. 200 W

78. ▌ If she doubles her speed to 10 m/s, how much energy does
    she use in cycling 1 km?
    A. 20 kJ   B. 40 kJ   C. 400 kJ   D. 400 kJ

79. ▌ How much power does she expend when cycling at that
    speed?
    A. 100 W   B. 200 W   C. 400 W   D. 1000 W

80. ▌ Upon reducing her speed back down to 5 m/s, she hits a
    headwind of 5 m/s. How much power is she expending now?
    A. 100 W   B. 200 W   C. 500 W   D. 1000 W

STOP TO THINK ANSWERS

Stop to Think 10.1: D. Since the child slides at a constant speed,
his kinetic energy doesn't change. But his gravitational potential
energy at the top of the slide decreases as he descends, and is transformed into thermal energy in the slide and his bottom.

Stop to Think 10.2: C. $W = Fd\cos\theta$. The 10 N force at 90° does
no work at all. $\cos 60° = \frac{1}{2}$, so the 8 N force does less work than
the 6 N force.

Stop to Think 10.3: B > D > A = C. $K = (1/2)mv^2$. Using the
given masses and velocities, we find $K_A = 2.0$ J, $K_B = 4.5$ J,
$K_C = 2.0$ J, $K_D = 4.0$ J.

Stop to Think 10.4: $(U_g)_3 > (U_g)_2 = (U_g)_4 > (U_g)_1$. Gravitational potential energy depends only on height, not speed.

Stop to Think 10.5: C. U_s depends on x^2, so doubling the compression increases U_s by a factor of 4. All the potential energy is
converted to kinetic energy, so K increases by a factor of 4. But K
depends on v^2, so v increases by only a factor of $\sqrt{4} = 2$.

Stop to Think 10.6: $P_B > P_A = P_C > P_D$. The power here is the
rate at which each runner's internal chemical energy is converted
into gravitational potential energy. The change in gravitational
potential energy is $mg\Delta y$, so the power is $mg\Delta y/\Delta t$. For runner A,
the ratio $m\Delta y/\Delta t$ equal (80 kg)(10 m)/(10 s) = 80 kg · m/s. For
C, it's the same. For B, it's 100 kg · m/s, while for D the ratio is
64 kg · m/s.

11

USING ENERGY

The strange masks worn by the fighters have a very simple purpose: to measure how much oxygen their bodies are using. How can this tell us how efficiently their bodies are using energy?

Looking Ahead ▶▶

The goal of Chapter 11 is to learn more about energy transformations and transfers, the laws of thermodynamics, and theoretical and practical limitations on energy use. In this chapter, you will learn to:

▶ Use the concept of efficiency.
▶ Discuss how energy is used in the body.
▶ Understand the related concepts of heat, temperature, and thermal energy.
▶ Apply the first and second laws of thermodynamics.
▶ Understand the concept of entropy and the limits it places on energy transformations.

Looking Back ◀◀

This chapter is a companion chapter to Chapter 10. All of the concepts that you learned in that chapter will be used here. Please review especially:

◀ Section 10.2 Basic energy concepts; forms of energy.
◀ Section 10.3 The law of conservation of energy.
◀ Section 10.7 Thermal energy.

Chapter 10 introduced the concept of energy, and in it you learned to solve problems in which energy is transformed from one form to another. For example, you now know how to calculate the energy required to lift a box a certain distance off the floor.

But how much energy would your body actually *use* to complete this task? We will find that the energy used by your body is more than the change in potential energy of the box. Why is this? As you lift the box, where does the energy come from? If you put the box back on the floor, where has the energy gone? Is it "lost"? In this chapter, we will consider just such questions of energy transformations and efficiency. Answering these questions will lead us to learn more about thermal energy, temperature, and heat.

11.1 Transforming Energy

As we saw in Chapter 10, energy can't be created or destroyed; it can only be converted from one form to another. When we say we are *using energy,* we mean that we are transforming it, such as transforming the chemical energy of food into the kinetic energy of your body. Let's revisit the idea of energy transformations, considering some realistic situations that have interesting theoretical and practical limitations.

Energy transformations

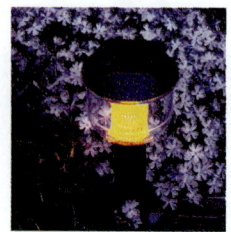

Light energy hitting a solar cell on top of these walkway lights is converted to electric energy and then stored as chemical energy in a battery. At night, the battery's chemical energy is converted to electric energy that is then converted to light energy in a light-emitting diode.

Light energy is absorbed by photosynthetic pigments in soybean plants, which use this energy to create concentrated chemical energy. The soybeans are harvested and their oil is used to make candles. When the candle burns, the stored chemical energy is transformed into light energy and thermal energy.

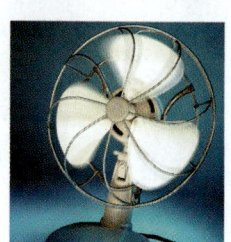

A wind turbine converts the translational kinetic energy of moving air into electric energy. Miles away, this energy can be used by a fan, which transforms the electric energy back into the kinetic energy of moving air.

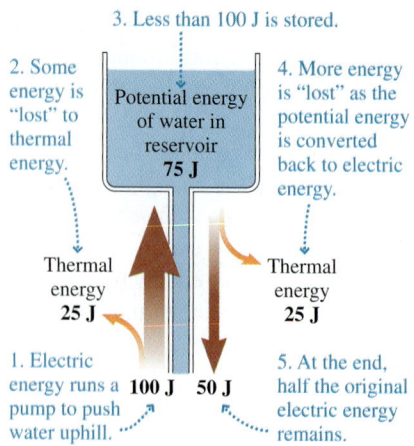

3. Less than 100 J is stored.

2. Some energy is "lost" to thermal energy.

Potential energy of water in reservoir **75 J**

4. More energy is "lost" as the potential energy is converted back to electric energy.

Thermal energy **25 J**

Thermal energy **25 J**

1. Electric energy runs a pump to push water uphill.   **100 J**   **50 J**

5. At the end, half the original electric energy remains.

FIGURE 11.1 Energy transformations for a pumped storage system.

In each of the processes in the table, energy is transformed from one form to another, ending up in the same form as it began. But some energy appears to have been "lost" along the way. The garden lights shine with a glow that is much less bright than the sun that shined on them! No energy is really lost, because energy is conserved; it is merely converted to other forms that are less useful to us.

Let's look at a practical example. Electric power companies sometimes use excess electric energy to pump water uphill. They can then reclaim this energy when demand is greater by letting the water generate electricity as it flows back down. Figure 11.1 shows the energy transformations in this process, assuming 100 J of electric energy at the start.

At the end of the process, only 50 J of electric energy is recovered. No energy is actually lost, but when other forms of energy are transformed to thermal energy, this change is *irreversible:* We cannot easily recover this thermal energy and convert it back to electric energy. **The energy isn't lost, but it is lost to our use.** In order to look at losses in energy transformations in more detail, we will define the notion of *efficiency.*

Efficiency

To get 50 J of energy out of your pumped storage plant, you would actually need to *pay* to put in 100 J of energy, as we saw in Figure 11.1. Because only 50% of the energy is returned as useful energy, we can say that this plant has an efficiency of 50%. Generally, we can define efficiency as

$$e = \frac{\text{what you get}}{\text{what you had to pay}} \qquad (11.1)$$

General definition of efficiency

The larger the energy losses in a system, the lower its efficiency.

The reductions in efficiency can arise from two different sources:

- **Process limitations.** In many cases that we will consider, the energy losses are due to practical details of an energy transformation process. You could, in principle, design a process that entailed smaller losses.
- **Fundamental limitations.** In other cases that we will consider, energy losses are due to physical laws that cannot be circumvented. The best process that could theoretically be designed will have less than 100% efficiency. As we will see, these limitations result from the difficulty of transforming thermal energy into other forms of energy.

As you walk up a flight of stairs, you increase your body's potential energy by about 1800 J. But a measurement of the energy used by your body to climb the stairs would show that your body uses 7200 J to complete this task. The 1800 J increase in potential energy is "what you get"; the 7200 J your body uses is "what you had to pay." We can use Equation 11.1 to compute the efficiency for this action to be

$$e = \frac{1800 \text{ J}}{7200 \text{ J}} = 0.25 = 25\%$$

This low efficiency is due to *process limitations.* The efficiency is limited by the biochemistry of how your food is digested and the biomechanics of how you move. This is not a fundamental limit; the process could be made more efficient. For instance, changing the angle of the stairs can allow a person to climb the same height with a smaller total energy expenditure.

The electric energy you use daily must be generated from other sources, in most cases the chemical energy in coal or other fossil fuels. The chemical energy is converted to thermal energy by burning, and the resulting thermal energy is converted to electric energy. A typical power plant cycle is shown in Figure 11.2. "What you get" is the energy output, the 35 J of electric energy. "What you had to pay" is the energy input, the 100 J of chemical energy, giving an efficiency of

$$e = \frac{35 \text{ J}}{100 \text{ J}} = 0.35 = 35\%$$

This low efficiency turns out to be largely due to a *fundamental limitation.* Thermal energy cannot be transformed to other forms of energy with 100% efficiency. A 35% efficiency is close to the theoretical maximum, and no power plant could be designed that would do better than this maximum. In later sections, we will explore the fundamental properties of thermal energy that make this so.

This idea of efficiency will form a thread that will run through this chapter. We will take some time to explore how to solve problems using this concept.

These cooling towers release thermal energy from a coal-fired power plant.

2. Steam with 100 J of thermal energy enters the turbine.

3. The turbine turns a generator, producing 35 J of electric energy.

1. Burning coal produces 100 J of thermal energy.

4. 65 J of thermal energy is exhausted into the environment.

FIGURE 11.2 Energy transformations in a coal-fired power plant.

PROBLEM-SOLVING STRATEGY 11.1 **Energy efficiency problems**

PREPARE There are two key components to define as we prepare to compute efficiency:

❶ Choose what energy to count as "what you get." This could be the useful energy output of an engine or process or the work that is done in completing a process. For example, when you climb a flight of stairs, "what you get" is your change in potential energy.

❷ "What you had to pay" will generally be the total energy input needed for an engine, task, or process. For example, when you run your air conditioner, "what you had to pay" is the electric energy input.

SOLVE You may need to do additional calculations:

▪ Compute values for "what you get" and "what you had to pay."
▪ Be certain that all energy values are in the same units.

After this, compute the efficiency using $e = \dfrac{\text{what you get}}{\text{what you had to pay}}$.

ASSESS Check your answer to see if it is reasonable, given what you know about typical efficiencies for the process under consideration.

A Case Study in Energy Efficiency

We will now look at a practical example that involves energy transformations and energy transfers by work and heat: the energy transformations and transfers in moving a car along the road. How efficient is your car as you drive down the highway?

Why does your car use energy to move forward at all? In the absence of external forces, your car would continue at a constant speed once it was moving. But if you are driving at highway speed and take your foot off the gas, your car will slow down; this is almost entirely due to one external force, the drag force on the car. The car must apply a forward force to move through the air; it has to do work to move forward.

CONCEPTUAL EXAMPLE 11.1 **Energy transformations and transfers for a car**

A car is traveling along a level road at a steady 65 mph (approximately 30 m/s). What energy transformations and transfers are taking place?

REASON We will take the car as our system. Let's write the conservation of energy equation for a system that isn't isolated, Equation 10.4, including all types of energy which might apply to the car:

$$\Delta K + \Delta U_g + \Delta E_{th} + \Delta E_{chem} = W + Q$$

The car is moving at a constant speed, so the kinetic energy is not changing. The road is level, so there is no change in poten-

tial energy. If the car is running well, its temperature will be constant; its thermal energy isn't changing. Chemical energy does change, decreasing as fuel is burned. The equation thus reduces to

$$\Delta E_{chem} = W + Q$$

Chemical energy in the car's fuel is being used to do work because of the drag force and to transfer energy as heat to the environment. These are the only significant energy transfers taking place at constant speed.

ASSESS The car has kinetic energy, potential energy and thermal energy, but these forms of energy are all constant. Only the chemical energy changes.

We can now use what we know about the drag force to compute the minimum amount of energy necessary to move a car forward. We can take this as "what you get"; it is the minimum amount of energy that could be used to move the car forward. We can then use some real numbers to compute "what you had to pay"—the energy actually used to move the car forward. Once we have these two pieces of information, we can calculate an efficiency.

EXAMPLE 11.1 Minimum power for a car on a freeway

A Chevrolet Corvette has a drag coefficient of 0.30; the area of the front of the car is 1.8 m^2. What minimum power does this car need to move at a steady speed of 30 m/s on level ground?

PREPARE In Conceptual Example 11.1, we saw that the car was doing work because of drag. The car is pushing on the atmosphere with a force F equal in magnitude to the drag force D; this is the force that is doing work on the environment.

SOLVE In Chapter 5, we saw that the drag force on an object moving at a speed v is given by $D = \frac{1}{2}C\rho Av^2$, where C is the drag coefficient, ρ the density of air (approximately 1.2 kg/m^2), A the area of the front of the car, and v the speed. Thus the drag force on the car moving at 30 m/s is

$$D = (\tfrac{1}{2})(0.30)(1.2 \text{ kg/m}^3)(1.8 \text{ m}^2)(30 \text{ m/s})^2 = 290 \text{ N}$$

Now that we know the magnitude of the force we can compute the necessary power by using Equation 10.22:

$$P = Fv = (290 \text{ N})(30 \text{ m/s}) = 8700 \text{ W}$$

ASSESS We can get a better idea of the size of this power by converting to horsepower:

$$P = 8700 \text{ W}\left(\frac{1 \text{ hp}}{746 \text{ W}}\right) = 12 \text{ hp}.$$

A Corvette will likely have a 350 hp engine. The power necessary to drive at highway speeds is just over 3% of the power available from the engine! The extra power is used to provide power for rapid acceleration (as we saw in Example 10.14) and for climbing hills. And, as we'll explore more later, quite a bit of the engine's power is "wasted" as heat energy transferred to the environment.

EXAMPLE 11.2 The efficiency of an automobile

The Corvette of the previous example is rated at 25 miles per gallon of gas at highway speeds. What is the efficiency of a Corvette driven 25 miles (40 km) at 30 m/s? (A gallon of gas contains 1.4 × 10^8 J of chemical energy.)

PREPARE We can use the steps outlined in Problem-Solving Strategy 11.1 for computing efficiency. "What you get" is the minimum energy to move the car forward as in the previous example; "what you had to pay" is the energy in one gallon of gas—the energy the car actually uses to drive this distance.

SOLVE As we saw in the previous example, the power necessary to overcome drag is 8700 W. The minimum energy needed to travel 40 km, which we will label ΔE_{min}, is

$$\Delta E_{min} = P\Delta t = P\left(\frac{\Delta x}{v}\right) = (8700 \text{ W})\left(\frac{40,000 \text{ m}}{30 \text{ m/s}}\right)$$

$$= 11.6 \times 10^6 \text{ J}$$

To drive 25 miles the car will use one gallon of gas, with energy 1.4 × 10^8 J. We can now use Equation 11.1 to find the efficiency:

$$e = \frac{\text{what you get}}{\text{what you had to pay}} = \frac{11.6 \times 10^6 \text{ J}}{1.4 \times 10^8 \text{ J}} = 0.083 = 8.3\%$$

ASSESS Only 8.3% of the chemical energy from the gasoline is used to overcome drag. The rest of the energy—about 92%—is "lost" as heat to the environment.

STOP TO THINK 11.1 Walking or running up a flight of stairs takes the same amount of energy. Which is more efficient?

A. Walking up the stairs.
B. Running up the stairs.
C. The efficiency is the same.

11.2 Energy in the Body: Energy Inputs

In this section, we will look at energy in the body, which will give us the opportunity to explore a number of different energy transformations and transfers in another practical context. Figure 11.3 shows the body considered as the system for energy analysis. The chemical energy in food provides the necessary energy input for your body to function. It is this energy that is used for energy transfers with the environment.

Once food is eaten, it becomes part of the system, the body's supply of chemical energy.

As chemical energy is used by the body, energy is transferred to the environment as work and heat.

Q

W

E_{chem}

FIGURE 11.3 Energy of the body, considered as the system.

Getting Energy from Food

Suppose you do a very simple exercise: lifting a mass and increasing its potential energy. Where did the energy come from to do this? At some point, the energy came from the food you ate, but what were the intermediate steps? The chemical energy in food is made available to the cells in the body by a two-step process. First, the digestive system breaks down food into simpler molecules such as glucose, a simple sugar, or long chains of glucose molecules called glycogen. These molecules are delivered via the bloodstream to cells in the body, where they are metabolized by combining with oxygen to produce simpler molecules, as in Equation 11.2.

Glucose from the digestion of food combines with oxygen that is breathed in to produce…   …carbon dioxide, which is exhaled; water, which can be used by the body; and energy.

$$C_6H_{12}O_6 + 6O_2 \longrightarrow 6CO_2 + 6H_2O + \text{energy} \qquad (11.2)$$

Glucose   Oxygen   Carbon   Water
                         dioxide

This metabolism releases energy, some of which is stored in a molecule called ATP, or adenosine triphosphate. Cells in the body use this ATP to do all the work of life: Muscle cells use it to contract, nerve cells use it to produce electrical signals, and so on.

Oxidation reactions like those in Equation 11.2 "burn" the fuel that you obtain by eating. The oxidation of 1 g of glucose (or any other carbohydrate) will release approximately 17 kJ of energy. Table 11.1 compares the energy content of carbohydrates and other foods to other familiar sources of chemical energy.

It is possible to measure the chemical energy content of food by burning it. Burning food may seem quite different from metabolizing it, but if glucose is burned, the chemical formula for the reaction is exactly that of Equation 11.2; the two reactions are the same. Burning food transforms all of its chemical energy to thermal energy, which can be easily measured. Thermal energy is often measured in units of **calories (cal)** rather than in joules; 1.00 calorie is equivalent to 4.19 joules.

NOTE ▶ The energy content of food is usually quoted in Calories (Cal) (with a capital "c"); one Calorie (also called a "food calorie") is equal to 1000 calories or 1 kcal. If a candy bar contains 230 Cal, this means that, if burned, it would produce 230,000 cal (or 962 kJ) of thermal energy. ◀

TABLE 11.1 Energy in fuels

Fuel	Energy in 1 g of fuel (in kJ)
Hydrogen	121
Gasoline	44
Fat (in food)	38
Coal	27
Carbohydrates (in food)	17
Wood chips	15

Counting calories BIO Most foods burn quite well, as this photo of corn chips illustrates. You could set food on fire to measure its energy content, but this isn't really necessary. The chemical energies of the basic components of food (carbohydrates, proteins, fats) have been carefully measured—by burning—in a device called a *calorimeter.* Foods are analyzed to determine their composition, and their chemical energy can then be calculated.

EXAMPLE 11.3 Energy in food

A 12 oz can of soda contains approximately 40 g (or a bit less than 1/4 cup) of sugar, a simple carbohydrate. How much chemical energy in joules does this sugar contain? How many Calories is this?

SOLVE 1 g of sugar contains 17 kJ of energy; 40 g contains:

$$40 \text{ g} \times \frac{17 \times 10^3 \text{ J}}{1 \text{ g}} = 68 \times 10^4 \text{ J} = 680 \text{ kJ}$$

Converting to Calories, we get:

$$680 \text{ kJ} = 6.8 \times 10^5 \text{ J} = (6.8 \times 10^5 \text{ J}) \frac{1.00 \text{ cal}}{4.19 \text{ J}}$$

$$= 1.6 \times 10^5 \text{ cal} = 160 \text{ Cal}$$

ASSESS 160 Calories is a typical value for the energy content of a 12 oz can of soda (check the nutrition label on one to see!), so this calculation seems reasonable.

TABLE 11.2 Energy content of foods

Food	Energy content in Cal	Energy content in kJ
Fried egg	100	420
Large apple	125	525
Slice of pizza	300	1260
Slice of apple pie	400	1680
Fast-food meal: Burger, fries, drink, large size	1350	5670

The first item on the nutrition label on packaged foods is Calories—a measure of the chemical energy in the food. (In Europe, where SI units are standard, you will also find the energy content listed in kJ.) The energy content of some common foods is given in Table 11.2.

A respiratory apparatus can be used to measure metabolic energy use of animals as well as humans, in this case the energy used by a horse in walking.

11.3 Energy in the Body: Energy Outputs

Your body uses energy in many ways. Even at rest, your body uses energy for tasks such as building and repairing tissue, digesting food, and keeping warm. The number of joules used per second (that is, the power in watts) by different tissues in the resting body is given in Table 11.3.

Even at rest, your body uses energy at the rate of approximately 100 W. This energy comes from chemical energy in your body's stores, and is ultimately converted entirely to thermal energy, which is transferred as heat to the environment. If 100 people are in a lecture hall, they are adding thermal energy to the room at a rate of 10,000 W, and the air conditioning must be designed to take account of this!

Energy Use in Activities

Your body stores very little energy as ATP; as your body uses energy your cells must continually metabolize carbohydrates to produce it. This requires oxygen, as we saw in Equation 11.2. Physiologists can precisely measure the body's energy use by measuring how much oxygen the body is taking up with a respiratory apparatus, as seen in the photograph at the start of the chapter. The spirometer measures the body's *total metabolic energy use*—all of the energy used by the body while performing an activity. This total will include all of the body's basic processes such as breathing plus whatever additional energy is needed to perform the activity.

The metabolic energy used in an activity depends on an individual's size, level of fitness, and other variables. But we can make reasonable estimates for the power used in various activities for a typical individual. Some values are given in Table 11.4.

Your body uses energy above the resting rate to perform different activities. For instance, typing requires 125 W, as seen in Table 11.4. This corresponds to 100 W for basic functions plus 25 W for the work of typing.

Suppose you climb a set of stairs at a constant speed, as in Figure 11.4 on the next page. What is your body's efficiency for this process? To answer this question we will take the system to be your body together with the earth, so that we can consider gravitational potential energy.

TABLE 11.3 Energy usage at rest

Organ	Resting power (W) of 68 kg individual
Liver	26
Brain	19
Heart	7
Kidneys	11
Skeletal muscle	18
Remainder of body	19
Total	**100**

TABLE 11.4 Metabolic power use during activities by a 68 kg (150 lb) individual

Activity	Metabolic power (W) of 68 kg individual
Typing	125
Ballroom dancing	250
Walking at 5 km/hr	380
Cycling at 15 km/hr	480
Swimming at a fast crawl	800
Running at 15 km/hr	1150

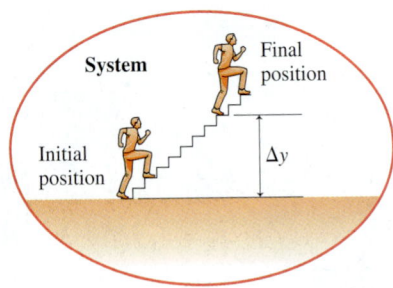

FIGURE 11.4 Pictorial representation of climbing a set of stairs.

Given how we have defined the system, there are no external forces that do work and there is no heat input—the system is isolated. Any energy transformations are internal to the system. We can use Equation 10.3 for the energy changes in the system, using the only forms of energy that change in this situation: gravitational potential, chemical (in your body), and thermal energy; kinetic energy does not change because your speed is constant:

$$\Delta U_g + \Delta E_{th} + \Delta E_{chem} = 0 \qquad (11.3)$$

Thermal energy and gravitational potential energy are increasing, so ΔE_{th} and ΔU_g are positive; chemical energy is being used, so ΔE_{chem} is a negative number. We can get a better feeling about what is happening by rewriting Equation 11.3 as

$$\left|\Delta E_{chem}\right| = \Delta U_g + \Delta E_{th}$$

The *magnitude* of the change in the chemical energy is equal to the sum of the changes in the gravitational potential and thermal energies. Chemical energy from your body is converted into potential energy and thermal energy; in the final position, you are at a greater height, and your body is slightly warmer!

Earlier in the chapter, we noted that the efficiency for stair climbing is about 25%. Let's see where that number comes from.

1. *What you get.* "What you get" is the change in potential energy: You have raised your body to the top of the stairs. If you climb a flight of stairs of vertical height Δy, we can easily compute the increase in potential energy $\Delta U_g = mg\Delta y$. Assuming a mass of 68 kg and a change in height of 2.7 m (about 9 ft, a reasonable value for a flight of stairs) we compute (to two significant figures)

$$\Delta U_g = (68 \text{ kg})(9.8 \text{ m/s}^2)(2.7 \text{ m}) = 1800 \text{ J}$$

2. *What you had to pay.* The cost is the metabolic energy the body used in completing the task. Physiologists can measure directly how much energy $\left|\Delta E_{chem}\right|$ your body uses to perform a task. A typical value for climbing a flight of stairs is

$$\left|E_{chem}\right| = 7200 \text{ J}$$

Given the definition of efficiency of Equation 11.2, we can compute an efficiency for climbing the stairs as

$$e = \frac{\Delta U_g}{\left|\Delta E_{chem}\right|} = \frac{1800 \text{ J}}{7200 \text{ J}} = 0.25 = 25\%$$

For the types of activities we will consider in this chapter, such as running, walking, and cycling, the body's efficiency is typically in the range of 20–30%.

In the above example, 25% of the energy used by your body ends up as increased potential energy; the other 75% ends up as thermal energy. Thermal energy losses arise from several sources, including the metabolism of glucose or glycogen to produce ATP, the use of ATP to make muscle cells contract, and from basic body processes that continue to be performed. Whenever you exercise, most

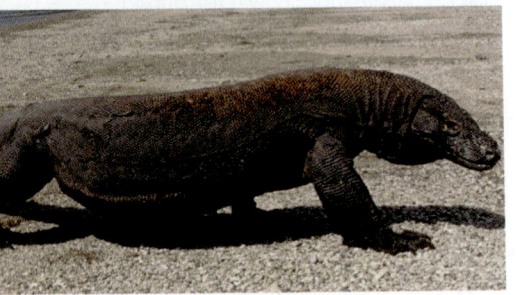

High heating costs? BIO The daily energy use of mammals is much higher than that of reptiles, largely because mammals use energy to maintain a constant body temperature. A 40 kg timber wolf uses approximately 19,000 kJ during the course of a day. A Komodo dragon, a reptilian predator of the same size, uses only 2100 kJ.

of the energy you use is converted to thermal energy; that's why you warm up when you work out!

EXAMPLE 11.4 How many flights?

If you have consumed a 12 oz can of soda, how many flights of stairs could you climb on the chemical energy it contains? Assume a mass of 68 kg and that a flight of stairs has a vertical height of 2.7 m.

PREPARE In Example 11.3, we calculated that the soda contains a chemical energy of 680 kJ. If the body uses this energy to climb stairs, we can assume that 25% of the energy is transformed to increased potential energy:

$$\Delta U_g = (0.25)(680 \times 10^3 \text{ J}) = 1.7 \times 10^5 \text{ J}$$

SOLVE Since $\Delta U_g = mg\Delta y$, the height gained is

$$\Delta y = \frac{\Delta U_g}{mg} = \frac{1.7 \times 10^5 \text{ J}}{(68 \text{ kg})(9.8 \text{ m/s}^2)} = 255 \text{ m}$$

If we assume that each flight has a height of 2.7 m, this gives

$$\text{Number of flights} = \frac{255 \text{ m}}{2.7 \text{ m}} \cong 94 \text{ flights}$$

ASSESS This is almost enough to get to the top of the Empire State Building—all fueled by one can of soda! This is a remarkable result.

The metabolic power values given in Table 11.4 represent the energy *used by the body* while these activities are being performed. Given that the body's efficiency is only 20–30%, the body's actual *useful power output* is quite a bit less than this. The table's value for cycling at 15 km/hr (a bit less than 10 mph) is 480 W. If we assume that the efficiency for cycling is approximately 25%, the actual power going to forward propulsion will only be 120 W. An elite racing cyclist going more than twice as fast as 15 km/hr might be using about 300 W for forward propulsion. This is a surprisingly low figure, as noted in Figure 11.5.

The energy you use per second running is proportional to your speed; running twice as fast takes approximately twice as much power. But running twice as fast takes you twice as far in the same time, so the energy you use to run a certain distance doesn't depend on how fast you run! Running a marathon takes approximately the same amount of energy whether you complete it in 2 hours, 3 hours or 4 hours; it is only the power that varies.

NOTE ▶ It is important to remember the distinction between the metabolic energy used to perform a task and the work done in a physics sense; these values can be quite different. Your muscles use power when applying a force, even when there is no motion. Holding a weight above your head involves no external work, but it clearly takes metabolic power to keep the weight in place! ◀

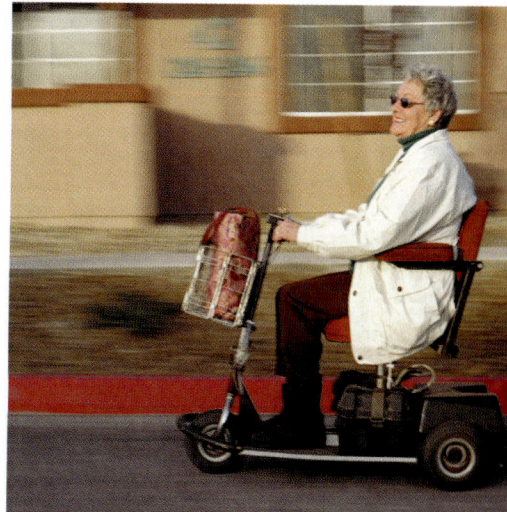

FIGURE 11.5 Amazingly, a racing cyclist moving at 35 km/hr uses about the same power as an electric scooter moving at 5 km/hr.

CONCEPTUAL EXAMPLE 11.2 Energy in weightlifting

A weightlifter lifts a 50 kg bar from the floor to a position over his head and back to the floor again 10 times in succession. At the end of this exercise, what energy transformations have taken place?

REASON We will take the system to be the weightlifter plus the bar. The environment does no work on the system, and we will assume that the time is short enough that no heat is transferred from the system to the environment. The bar has returned to its starting position and is not moving, so there has been no change in potential or kinetic energy. The equation for energy conservation is thus

$$\Delta E_{chem} + \Delta E_{th} = 0$$

Each time the bar is raised or lowered, the muscles use chemical energy. Ultimately, all of this energy is transformed to thermal energy.

ASSESS Most exercises in the gym—lifting weights, running on a treadmill—involve only the transformation of chemical energy to thermal energy.

EXAMPLE 11.5 Energy usage for a cyclist

A cyclist pedals for 20 min at a speed of 15 km/hr. How much metabolic energy is required?

PREPARE Table 11.4 gives a value of 480 W for the power used in cycling at a speed of 15 km/hr. This is the energy that is used per second.

SOLVE The cyclist uses energy at this rate for 20 min, or 1200 s. Power is the rate of using energy, $P = \Delta E/\Delta t$, so the energy required is

$$\Delta E = P\Delta t = (480 \text{ J/s})(1200 \text{ s}) = 580 \text{ kJ}$$

ASSESS How much energy is 580 kJ? A look at Table 11.2 shows that this is slightly more than the amount of energy available in a large apple, and only 10% of the energy available in a large fast-food meal. If you eat such a meal and plan to "work it off" by cycling, you should plan on cycling at a pretty good clip for a bit over 3 hr.

Energy Storage

The body gets energy from food; if this energy is not used, it will be stored. A small amount of energy needed for immediate use is stored as ATP. A larger amount of energy is stored as chemical energy of glycogen and glucose in muscle tissue and the liver. A healthy adult might store 400 g of these carbohydrates, which is a little more carbohydrate than is typically consumed in one day.

If the energy input from food continuously exceeds the energy outputs of the body, this energy will be stored in the form of fat under the skin and around the organs. From an energy point of view, gaining weight is simply explained!

EXAMPLE 11.6 Running out of fuel

The body stores about 400 g of carbohydrates. Approximately how far could a 68 kg runner travel on this stored energy?

PREPARE Table 11.1 gives a value of 17 kJ per g of carbohydrate. The 400 g of carbohydrates in the body contain an energy of

$$E_{chem} = (400 \text{ g})(17 \times 10^3 \text{ J/g}) = 6.8 \times 10^6 \text{ J}$$

SOLVE Table 11.4 gives the power used in running at 15 km/hr as 1150 W. The time that the stored chemical energy will last at this rate is

$$\Delta t = \frac{\Delta E_{chem}}{P} = \frac{6.8 \times 10^6 \text{ J}}{1150 \text{ W}} = 5.91 \times 10^3 \text{ s} = 1.64 \text{ hr}$$

And the distance that can be covered during this time at 15 km/hr is

$$\Delta x = v\Delta t = (15 \text{ km/hr})(1.64 \text{ hr}) = 25 \text{ km}$$

to two significant figures.

ASSESS A marathon is longer than this—just over 42 km. Even with "carbo loading" before the event (eating high-carbohydrate meals), many marathon runners "hit the wall" before the end of the race as they reach the point where they have exhausted their store of carbohydrates. The body has other energy stores (in fats, for instance), but the rate that they can be drawn on is much lower.

Energy and Locomotion

When you walk at a constant speed on level ground, your kinetic energy is constant. Your potential energy is also constant. So why does your body need energy to walk? Where does this energy go?

We use energy to walk because of mechanical inefficiencies in our gait. Figure 11.6 shows how the speed of your foot typically changes during each stride. The kinetic energy of your leg and foot increases, only to go to zero at the end of the stride. The kinetic energy is mostly transformed into thermal energy in your muscles and in your shoe. This thermal energy is lost; it can't be used for making further strides.

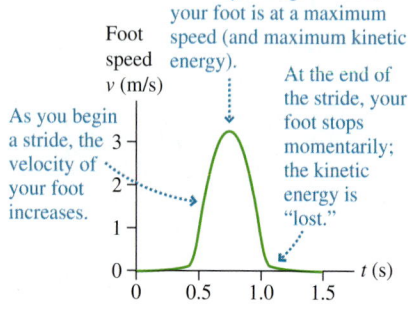

FIGURE 11.6 Human locomotion analysis.

Footwear can be designed to minimize the loss of kinetic energy to thermal energy. A spring in the sole of the shoe can store potential energy, which can be returned to kinetic energy during the next stride. Such a spring will make the collision with the ground more elastic. We saw in Chapter 10 that the tendons in the ankle do store a certain amount of energy during a stride; very stout tendons in the legs of kangaroos store energy even more efficiently. Their peculiar hopping gait is quite efficient at high speeds.

Where do you wear the weights? BIO
If you wear a backpack with a mass equal to 1% of your body mass, your energy expenditure for walking will increase by 1%. But if you wear ankle weights with a combined mass of 1% of your body mass, the increase in energy expenditure is 6%, because you must repeatedly accelerate this extra mass. If you want to "burn more fat," wear the weights on your ankles, not on your back!

STOP TO THINK 11.2   A person is running at a constant speed on level ground. Chemical energy in the runner's body is being transformed to other forms of energy; most of the chemical energy is transformed to

A. Kinetic energy.   B. Potential energy.   C. Thermal energy.

11.4 Thermal Energy and Temperature

We have frequently spoken of the energy that is transformed to thermal energy as being "lost." Regardless of whether "the system" is a car, a power plant, or your body, this thermal energy is exhausted into the environment as heat. But why is thermal energy in your body and other systems not converted to other forms of energy and used for practical purposes? To continue our study of how energy is used, we need to understand a bit more about one particular kind of energy—thermal energy.

What do you mean when you say something is "hot"? Do you mean that it has a high temperature? Or do you mean that it has a lot of thermal energy? Perhaps both of these definitions are the same? Let's give some thought to the definitions of temperature and thermal energy, and the relationship between them.

Measuring Temperature

We are all familiar with the idea of temperature. You hear the word used nearly every day. But just what is temperature a measure of? Velocity is a measure of how fast a system moves. What physical property of the system have you determined if you measure its temperature?

Let's start by looking at how you measure temperature. This is what a *thermometer* does. In a common glass-tube thermometer, for example, a small volume of mercury or alcohol expands or contracts when placed in contact with a "hot" or "cold" object. The object's temperature is determined by the height of the column of liquid.

A thermometer needs a *temperature scale* to be a useful measuring device. In 1742, the Swedish astronomer Anders Celsius sealed mercury into a small capillary tube and observed how it moved up and down the tube as the temperature changed. He selected two temperatures that anyone could reproduce, the freezing and boiling points of pure water, and labeled them 0 and 100. He then marked off the glass tube into one hundred equal intervals between these two reference points. By doing so, he invented the temperature scale that we today call the *Celsius scale*. The units of the Celsius temperature scale are "degrees Celsius," which we abbreviate as °C. The *Fahrenheit scale,* still widely used in the United States, is related to the Celsius scale by

Thermal expansion of the liquid in the thermometer pushes it higher when immersed in hot water than in ice water.

$$T_C = \frac{5}{9}(T_F - 32°) \qquad (11.4)$$

1. The gas is made of a large number N of atoms, each of mass m, all moving randomly.

2. The atoms in the gas are quite far from each other and interact only rarely when they collide.

m

3. The collisions of the atoms with each other (and with walls of the container) are elastic; no energy is lost in these collisions.

FIGURE 11.7 Motion of atoms in an ideal gas.

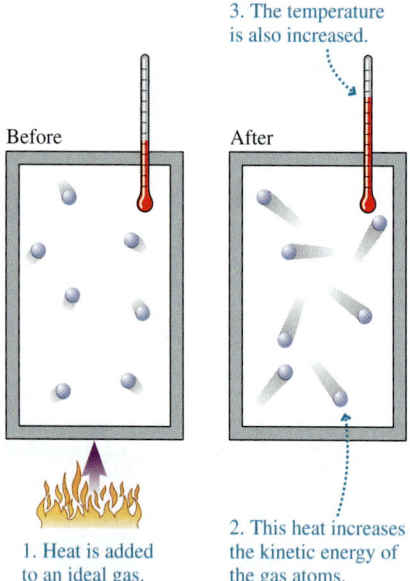

3. The temperature is also increased.

Before   After

1. Heat is added to an ideal gas.

2. This heat increases the kinetic energy of the gas atoms.

FIGURE 11.8 Heating an ideal gas.

8.1–8.3 **Activ Physics** ONLINE

An Atomic View of Thermal Energy and Temperature

When you add heat to an object, you increase its thermal energy; you also increase its temperature. In Chapter 10, we noted that the thermal energy of an object is the sum of the kinetic and potential energy of its atoms or molecules. Now we extend our discussion to include the concept of temperature. Let's start with the simplest possible system, a gas of atoms. In fact, we will make the system even simpler by defining the properties of what we will term an **ideal gas** in Figure 11.7. This is a rather simple model, but it gives a remarkably accurate description of the behavior of many real gases.

Because the ideal-gas atoms do not interact with each other (except for collisions), the system has no potential energy. The only internal energy in the ideal gas is the translational kinetic energy of the individual atoms. Thus, **the thermal energy of an ideal gas is equal to the total kinetic energy of the atoms that make up the gas.**

Adding heat to an ideal gas, as in Figure 11.8, will increase the thermal energy of the gas; the atoms will move faster. Adding heat to the gas will also increase its temperature. So temperature must be related to the kinetic energy of the individual atoms, but how? We can get one hint from the following fact: The temperature of a system doesn't depend on the size of the system. If you have two glasses of water, each at a temperature of 20°C, and you pour them together, you will have a larger volume of water at the same temperature of 20°C. There are more atoms, and therefore there is more *total* thermal energy, but the temperature has not changed. Even though there are more atoms, the *average* energy per atom is the same. It is this average kinetic energy of the atoms that is related to the temperature. **The temperature of an ideal gas is a measure of the average kinetic energy of the atoms that make up the gas.**

Let's consider the average kinetic energy of atoms in the gas. An atom of mass m and velocity v has kinetic energy $K = \frac{1}{2}mv^2$. The average kinetic energy of an atom is computed as the sum of the kinetic energies of all of the atoms divided by the number of atoms:

$$K_{avg} = \frac{\sum \frac{1}{2}mv^2}{N} = \frac{1}{2}m\frac{\sum v^2}{N} = \frac{1}{2}m(v^2)_{avg} \tag{11.5}$$

The quantity $\sum v^2/N$ that appears in Equation 11.5 is the average of the speed squared, not the square of the average speed. If we want to know how fast a typical atom is moving, we should look at the square root of this average:

$$\text{Speed of a typical atom} = \sqrt{(v^2)_{avg}} = v_{rms} \tag{11.6}$$

The quantity that we've defined as v_{rms} is called the **root mean square speed,** often referred to as the *rms speed*. Rewriting Equation 11.5 in terms of v_{rms} gives the average kinetic energy per atom:

$$K_{avg} = \frac{1}{2}mv_{rms}^2 \tag{11.7}$$

If the gas consists of N atoms, the total kinetic energy of the atoms is just NK_{avg}. The total kinetic energy of the atoms in the gas is the thermal energy of the gas, so we find that

$$E_{th} = \frac{1}{2}Nmv_{rms}^2$$

If we can establish a connection between K_{avg} and the temperature T, which we will do in the next section, then we'll know how both the thermal energy E_{th} and the typical atomic speed are related to temperature.

The Meaning of Temperature on the Kelvin Scale

When the Celsius temperature scale was defined, zero degrees was fixed to be the freezing point of water. The Fahrenheit scale has a zero point that was set as the temperature of an ice-salt mix. On both of these scales the choice of zero is arbitrary, and both scales allow temperatures that are negative or "below zero."

If we use the average kinetic energy of atoms as the basis for our definition of temperature, our temperature scale will have a very natural zero—the point at which kinetic energy is zero. Kinetic energy is always positive, so the zero on our temperature scale will be an **absolute zero;** no temperature below this is possible.

This is how zero is defined on the temperature scale called the *Kelvin scale:* **Zero degrees is the point at which the kinetic energy of atoms is zero.** All temperatures on the Kelvin scale are positive, so it is often called an *absolute temperature scale.* The units of the Kelvin temperature scale are "kelvin" (not degrees kelvin!), which is abbreviated K.

The spacing between divisions for the Kelvin scale is the same as that of the Celsius scale; the only difference is the position of the zero point. It is found experimentally that absolute zero—the temperature at which atoms would cease moving—is $-273°C$. Thus the Kelvin scale is obtained from the Celsius scale by shifting the zero point by 273 divisions. For this reason, the conversion between Celsius and Kelvin temperatures is quite straightforward:

$$T = T_C + 273 \qquad (11.8)$$

Thus the freezing point of water at 0°C is $T = 0 + 273 = 273$ K. A 30°C warm summer day is $T = 303$ K on the Kelvin scale. A side-by-side comparison of these scales is given in Figure 11.9.

> **NOTE** ▶ We will use the symbol T for temperature in kelvin. We will use subscripts to denote other scales. In the equations in this chapter and the rest of the text, **T must be interpreted as a temperature in kelvin.** ◀

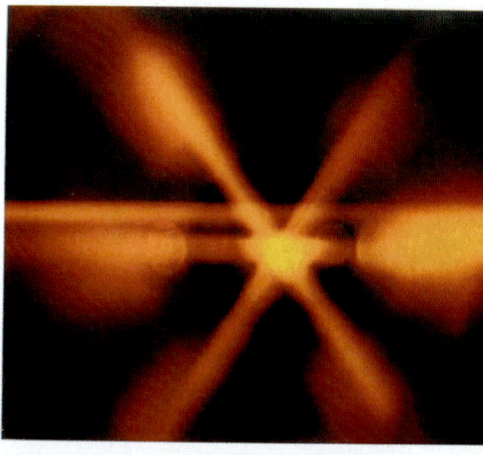

Optical molasses It isn't possible to reach absolute zero, where the atoms would be still, but it is possible to get quite close by slowing the atoms down directly. These crossed laser beams produce what is known as "optical molasses." As we will see in Chapter 28, light is made of photons, which carry energy and momentum. Interactions of the atoms of a diffuse gas with the photons cause the atoms to slow down. Atoms can be slowed to very slow speeds that correspond to a temperature as cold as 5×10^{-10} K!

EXAMPLE 11.7 Temperature scales

The coldest temperature ever measured on earth was $-129°F$, in Antarctica. What is this in °C and K?

SOLVE We use Equation 11.4 to convert the temperature to the Celsius scale:

$$T_C = \frac{5}{9}(-129° - 32°) = -89°C$$

We can then use Equation 11.8 to convert this to kelvin:

$$T = -89 + 273 = 184 \text{ K}$$

ASSESS This is quite cold, but quite a bit warmer than the coldest temperatures achieved in the laboratory.

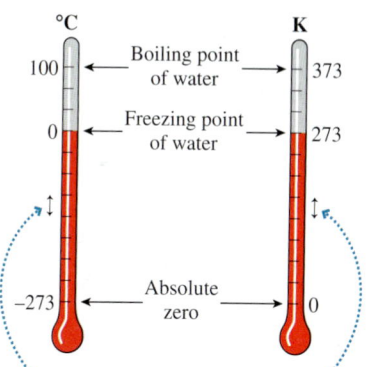

Temperature *differences* are the same on the Celsius and Kelvin scales. The temperature difference between the freezing point and boiling point of water is 100°C or 100 K.

FIGURE 11.9 Celsius and Kelvin temperature scales.

We noted above that the temperature of an ideal gas is a measure of the average kinetic energy of the atoms. It can be shown that temperature on the Kelvin scale is related to the average kinetic energy per atom by

$$T = \frac{2}{3}\frac{K_{avg}}{k_B} \qquad (11.9)$$

LINEAR
p.38

Temperature of an ideal gas in terms of the average kinetic energy per atom

where k_B is a constant known as **Boltzmann's constant.** Its value is

$$k_B = 1.38 \times 10^{-23} \text{ J/K}$$

We can rewrite Equation 11.9 to express the average kinetic energy of the atoms in an ideal gas at a temperature T as

$$K_{avg} = \frac{3}{2}k_B T \qquad (11.10)$$

Interestingly, the average kinetic energy per atom depends *only* on the temperature, not on the atom's mass. For example, the average kinetic energy of an atom (of any gas) at room temperature (20°C, or 293 K) is

$$K_{avg} = \frac{3}{2}(1.38 \times 10^{-23} \text{ J/K})(293 \text{ K}) = 6.07 \times 10^{-21} \text{ J}$$

Combining Equation 11.10 with our definition of the root-mean-square speed of the atoms in a gas in Equation 11.7, we find that

$$v_{rms} = \sqrt{\frac{3k_B T}{m}} \qquad (11.11)$$

The rms speed is proportional to the *square root* of the temperature. This is a new mathematical form that we will see again, so we will take a look at its properties.

Martian airsicles The atmosphere of Mars is mostly carbon dioxide. At night, the temperature may drop so low that the molecules in the atmosphere will slow down enough to stick together—the atmosphere actually freezes. The frost on the surface in this image from the Viking 2 lander is composed partially of frozen carbon dioxide.

Square-root relationships   ✎ Exercise 12 MP

Two quantities are said to have a **square-root relationship** if y is proportional to the square root of x. We write the mathematical relationship as

$$y = A\sqrt{x}$$

y is proportional to the square root of x

The graph of a square-root relationship is a parabola that has been rotated by 90°.

Increasing x by a factor of 4 increases y by a factor of 2.

$y = A\sqrt{x}$

SCALING Suppose we quadruple x, so that $x_f = 4x_i$. Then y will change to

$$y_f = A\sqrt{x_f} = A\sqrt{4x_i} = 2A\sqrt{x_i} = 2y_i$$

Increasing x by a factor of 4 causes y to increase by a factor of $\sqrt{4}$, or 2. Similarly, if we *decrease* x by a factor of 4, y will decrease by a factor of 2. Generally, we can say that:

Changing x by a factor C changes y by a factor $\sqrt{C}$.

We have been treating ideal gases made of atoms, but our results are equally valid for real gases made of atoms or molecules. From Equation 11.11 we see that **a higher temperature corresponds to faster atomic or molecular speeds.** At a given temperature, the speed of atoms or molecules in a gas varies with the atomic or molecular mass. A gas with lighter atoms will have faster atoms, on average, than a gas with heavier atoms.

The masses of atoms and molecules in chemistry are measured in **atomic mass units,** denoted u; this is the atomic mass that you will find in a periodic table. The atomic mass of a hydrogen atom is 1 u, so that of an H_2 molecule is 2 u. However, the equations in this chapter require that the masses of atoms and molecules be in kg. The conversion factor is

$$1 \text{ u} = 1.66 \times 10^{-27} \text{ kg}$$

EXAMPLE 11.8 Speed of nitrogen molecules

Nitrogen gas consists of molecules, N_2. At 20°C, what is the root-mean-square speed of the molecules of nitrogen?

PREPARE In the periodic table, you can find that the mass of a nitrogen atom is 14 u. A N_2 molecule consists of two atoms, so its mass is 28 u. Thus the molecular mass in SI units (i.e., kg) is

$$m = 28 \text{ u} \times \frac{1.66 \times 10^{-27} \text{ kg}}{1 \text{ u}} = 4.65 \times 10^{-26} \text{ kg}$$

Convert the temperature to kelvin:

$$T = 20 + 273 = 293 \text{ K}$$

SOLVE We compute v_{rms} for the nitrogen atoms using Equation 11.11:

$$v_{rms} = \sqrt{\frac{3k_B T}{m}} = \sqrt{\frac{3(1.38 \times 10^{-23} \text{ J/K})(293 \text{ K})}{4.65 \times 10^{-26} \text{ kg}}} = 510 \text{ m/s}$$

ASSESS This is *very* fast, just a bit faster than the speed of sound in air. This result makes sense, however: Sound waves must be carried by moving air molecules, so we would expect the speed of sound to be close to the speed of the molecules.

Computing Thermal Energy

The speeds of gas atoms are quite high. The associated kinetic energy of the atoms in a gas will be quite high as well. The thermal energy of an ideal gas consisting of N atoms can be computed using the above result for the average kinetic energy:

$$E_{th} = NK_{avg} = \frac{3}{2}Nk_B T \qquad (11.12)$$

Thermal energy of an ideal gas of N atoms

LINEAR
p. 38

Thus, **thermal energy is directly proportional to temperature.** Consequently, a change in the thermal energy of an ideal gas is proportional to a change in temperature:

$$\Delta E_{th} = \frac{3}{2}Nk_B \Delta T \qquad (11.13)$$

This relationship between a change in temperature and a change in thermal energy is for an ideal gas, but solids, liquids, and other gases all follow similar rules, as we will see in the next chapter. For now, we will simply note two important conclusions that apply to any substance:

1. The thermal energy of a substance is proportional to the number of atoms. A gas with more atoms has more thermal energy than a gas at the same temperature with fewer atoms.
2. A change in temperature causes a proportional change in the substance's thermal energy. A larger temperature change causes a larger change in thermal energy.

EXAMPLE 11.9 Energy change in heating a gas

100 cm³ of gas at 1 atmosphere pressure and 20°C contains 2.5×10^{21} molecules. (You'll learn how to show this in the next chapter.) By how much does its thermal energy change if it is heated to 30°C?

SOLVE The temerature *change* is $\Delta T = 10$ K. Using Equation 11.13, we find

$$\Delta E_{th} = \frac{3}{2}Nk_B \Delta T = \frac{3}{2}(2.5 \times 10^{21})(1.38 \times 10^{-23} \text{ J/K})(10 \text{ K}) = 0.52 \text{ J}$$

ASSESS Although the number of molecules in a typical gas is vast, a typical change in thermal energy is quite comparable to the values we've been calculating for the kinetic and potential energies of macroscopic objects.

Is it cold in space? The space shuttle orbits in the upper thermosphere, about 300 km above the surface of the earth. There is still a trace of atmosphere left at this altitude, and it has quite a high temperature—over 1000°C. Although the average speed of the air molecules here is quite high, there are so few air molecules present that the thermal energy is extremely low.

Consider gases of the following atoms or molecules, all at the same temperature. Which atoms or molecules have the fastest average speed?

A. H_2 B. He   C. N_2 D. O_2 E. All the same.

11.5 Heat and the First Law of Thermodynamics

In Chapter 10, we saw that a system could exchange energy with the environment through two different means: Work and heat. Work was treated in some detail in Chapter 10; now it is time to look at the transfer of energy by heat. This will begin our study of a topic called *thermodynamics,* the study of thermal energy and heat and their relationships to other forms of energy and energy transfer.

What Is Heat?

Heat is a more elusive concept than work. We use the word "heat" very loosely in the English language, often as synonymous with *hot.* We might say, on a very hot day, that "This heat is oppressive." If your apartment is cold you may say, "Turn up the heat." These expressions date to a time long ago when it was thought that heat was a *substance* with fluid-like properties.

If you place a hot object and a cold object together, they evolve toward a common final temperature. Common sense suggests that "something" flows from the hot object to the cold until equilibrium is achieved. This "heat fluid" was called *caloric.* The notion that objects somehow "contain" heat, and that heat can move around, lingers on in expressions like *heat flow, heat loss,* and *heat capacity*

One of the first to disagree with this notion, in the late 1700s, was Benjamin Thompson, known by his title Count Rumford. While watching the hot metal chips thrown off during the boring of cannons, he began to think about heat. If caloric is a substance, the cannon and borer should eventually run out of caloric and the heat generation ought to decrease with time. But it does not. Rumford noted that the heat generation appears to be "inexhaustible," which is not consistent with the idea of heat as a substance. He concluded that heat is not a substance—it is *motion!*

The British physicist James Joule carried out careful experiments in the 1840s to learn how it is that systems change their temperature. Using experiments like those shown in Figure 11.10, Joule found that you can raise the temperature of a beaker of water by two entirely different means:

1. Heating it with a flame, or
2. Doing work on it with a rapidly spinning paddle wheel.

Surprisingly, the final state of the water is *exactly* the same in both cases. This implies that heat and work are essentially equivalent to each other—**heat and work are simply two different ways of transferring energy to or from a system.**

Thermal energy is transferred between a system and the environment as a consequence of a temperature difference between them. When you place a pan of water on the stove, thermal energy is transferred *from* the hotter flame *to* the cooler water. If you place the water in a freezer, thermal energy is transferred from the warmer water to the colder air in the freezer.

Energy is transferred when the faster molecules in the hotter object collide with the slower molecules in the cooler object. On average, these collisions cause

The flame heats the water.   The spinning paddle does work on the water.

FIGURE 11.10 Joule's experiments to show the equivalence of heat and work.

Energetic cooking We can do a modern version of Joule's experiment in the kitchen. Next time you mix food in a blender, notice how it actually warms the food! As the blades rotate, friction increases the thermal energy of the food in the blender. A blender uses about as much power as a microwave, and most of this energy ends up as thermal energy in the food, so the temperature rises as the food is chopped and blended.

the faster molecules to lose energy and the slower molecules to gain energy. The net result is that energy is transferred from the hotter object to the colder object. The process itself, whereby energy is transferred between the system and the environment via atomic-level collisions, is called a *thermal interaction*. **Heat is the energy transfer during a thermal interaction.**

An Atomic Model of Heat

Let's look at a thermal interaction in some detail, considering what happens at an atomic level. Figure 11.11 shows a rigid, insulated container that is divided into two sections by a very thin membrane. Each side is filled with a gas of the same kind of atoms. The left side, which we'll call system 1, has N_1 atoms at an initial temperature T_{1i}. System 2 on the right has N_2 atoms at an initial temperature T_{2i}. We imagine the membrane to be so thin that atoms can collide at the boundary as if the membrane were not there, yet it is a barrier that prevents atoms from moving from one side to the other.

Suppose that system 1 is initially at a higher temperature: $T_{1i} > T_{2i}$. This means that the atoms in system 1 will have a higher average kinetic energy. Figure 11.12 shows a fast atom and a slow atom approaching the barrier from opposite sides. They undergo a perfectly elastic collision at the barrier. Although no net energy is lost in a perfectly elastic collision, the faster atom loses energy while the slower one gains energy. In other words, there is an energy *transfer* from the faster atom's side to the slower atom's side.

Because the atoms in system 1 are, on average, more energetic than the atoms in system 2, *on average* the collisions transfer energy from system 1 to system 2. This is not true for every collision. Sometimes a fast atom in system 2 collides with a slow atom in system 1, transferring energy from 2 to 1. But the net energy transfer, from all collisions, is from the warmer system 1 to the cooler system 2. This transfer of energy is heat; **thermal energy is transferred from the faster moving atoms on the warmer side to the slower moving atoms on the cooler side.**

This transfer will continue until a stable situation is reached. This is a situation we call **thermal equilibrium.** How do the systems "know" when they've reached thermal equilibrium? Energy transfer continues until the atoms on both sides of the barrier have the *same average kinetic energy*. Once the average kinetic energies are the same, individual collisions will still transfer energy from one side to the other. But since both sides have atoms with the same average kinetic energies, the amount of energy transferred from 1 to 2 will equal that transferred from 2 to 1. Once the average kinetic energies are the same, there will be no more net energy transfer.

An important result from the previous section is that the average kinetic energy of the atoms in a system is directly proportional to the system's temperature. If two systems exchange energy until their atoms have the same average kinetic energy, we can say that

$$T_{1f} = T_{2f} = T_f$$

That is, heat is transferred until the two systems reach a common final temperature; we call this final state **thermal equilibrium.** We considered a rather artificial system in this case, but the result is quite general: **Two systems placed in thermal contact will transfer thermal energy until their final temperatures are the same.** This process is illustrated in Figure 11.13.

Heat is a transfer of energy. The sign that we use for transfers is defined in Figure 11.14 on the next page. In the process of Figure 11.13, Q_1 is negative as system 1 loses energy; Q_2 is positive as system 2 gains energy. Because no energy

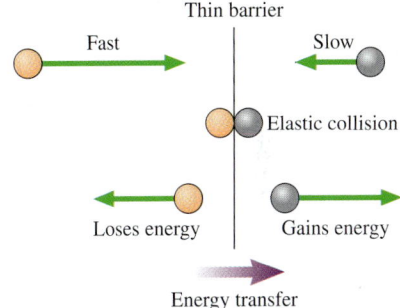

Insulation prevents heat from entering or leaving the container.

A thin barrier prevents atoms from moving from system 1 to 2 but still allows them to collide.

FIGURE 11.11 Two gases are separated by a thin barrier through which they can thermally interact.

FIGURE 11.12 Collisions at a barrier transfer energy from faster molecules to slower molecules.

Collisions transfer energy from the warmer system to the cooler as more energetic atoms lose energy to less energetic atoms. This energy transfer is heat.

Thermal equilibrium occurs when the systems have the same average translational kinetic energy and thus the same temperature.

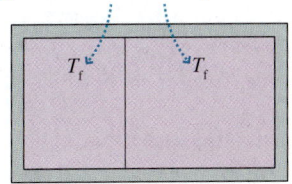

FIGURE 11.13 Two systems in thermal contact exchange thermal energy.

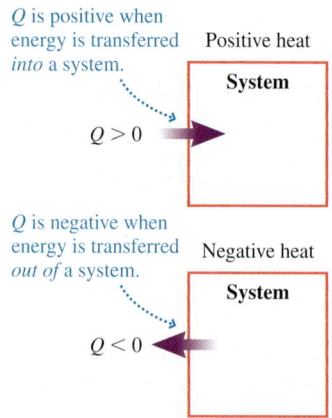

Q is positive when energy is transferred *into* a system.

$Q > 0$

Positive heat

System

Q is negative when energy is transferred *out of* a system.

$Q < 0$

Negative heat

System

FIGURE 11.14 The sign of Q.

escapes from the container, all of the energy that was lost by system 1 was gained by system 2

$$Q_2 = -Q_1$$

as required by energy conservation. That is, the heat energy lost by one system is gained by the other.

The above argument was made for gases, but it is completely general. We could make a similar analysis for different gases, solids, or liquids. In each case, the final result is the same: **Heat is transferred from hot to cold.**

The First Law of Thermodynamics

In Chapter 10, we looked at several versions of the law of conservation of energy. The most general statement was the law of conservation of energy for systems that were not isolated, Equation 10.4:

$$\Delta K + \Delta U_g + \Delta U_s + \Delta E_{th} + \Delta E_{chem} + \cdots = W + Q$$

In Chapter 10, we focused on systems where the potential and kinetic energy could change, such as a sled moving down a hill. Earlier in this chapter, we looked at the body, where the chemical energy changes. Now let's consider systems in which only the thermal energy changes. That is, we will consider systems that aren't moving, that aren't changing chemically, but whose temperatures can change. Such systems are the province of what is called **thermodynamics.** The question of how to keep your house cool in the summer is a question of thermodynamics. If thermal energy is transferred into your house, the temperature will rise; if thermal energy is transferred out, the temperature will drop.

When the law of conservation of energy was introduced, we noted that it was sometimes called the first law of thermodynamics. This is true, but a more common statement of the first law of thermodynamics applies to systems in which only thermal energy changes take place.

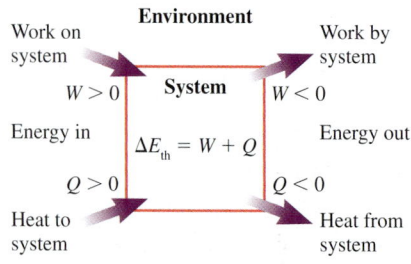

Environment

Work on system
$W > 0$

System

$\Delta E_{th} = W + Q$

Work by system
$W < 0$

Energy in
$Q > 0$

Energy out
$Q < 0$

Heat to system

Heat from system

FIGURE 11.15 Energy transfers in a thermodynamic system.

> **First law of thermodynamics** For systems in which only the thermal energy changes, the change in thermal energy is equal to the energy transferred into or out of the system as work W and/or heat Q:
>
> $$\Delta E_{th} = W + Q \qquad (11.14)$$

Only work and heat, two ways of transferring energy between a system and the environment, cause the system's energy to change. In thermodynamic systems, the only energy change will be a change in thermal energy. Whether this energy increases or decreases depends on the signs of W and Q. In Chapter 10, we noted that when $W > 0$, work is done on the system, increasing its energy, and when $W < 0$ work is done by the system, decreasing its energy. A similar sign convention holds for heat, as we saw in Figure 11.14. The possible energy transfers between a system and the environment are illustrated in Figure 11.15.

CONCEPTUAL EXAMPLE 11.3 Compressing a gas

Suppose a gas is in an insulated container, so that no heat energy can escape. If a piston is used to compress the gas, what happens to the temperature of the gas?

REASON The piston applies a force to the gas, and there is a displacement. This means that work is done on the gas by the piston ($W > 0$). No thermal energy can be exchanged with the environment, meaning $Q = 0$. Since energy is transferred into the system, the thermal energy of the gas must increase. This means that the temperature must increase as well.

ASSESS When you use a bike pump to inflate a tire, you may have noticed that the pump and the tire get warm. This temperature increase is due to the warming of the air by the compression.

EXAMPLE 11.10 Work and heat in an ideal gas

Suppose a container holds 5.0×10^{22} molecules of an ideal gas. 50 J of work are done on the gas by a piston that compresses it. The temperature of the gas changes by 30°C during this process. How much heat is transferred to or from the environment?

SOLVE For an ideal gas, the change in thermal energy is proportional to the change in temperature, as we see in Equation 11.13. We compute the change in thermal energy to be

$$\Delta E_{th} = \frac{3}{2}Nk_B\Delta T$$

$$= \left(\frac{3}{2}\right)(5.0 \times 10^{22})(1.38 \times 10^{-23}\text{ J/K})(30\text{ K}) = 31\text{ J}$$

The first law of thermodynamics, Equation 11.14, tells us that the change in thermal energy ΔE_{th} of the gas is equal to the sum of W and Q. W is known, and we have just calculated ΔE_{th}. Combining, we get

$$Q = \Delta E_{th} - W = 31\text{ J} - 50\text{ J} = -19\text{ J}$$

ASSESS Q is negative, meaning that energy is transferred from the hot gas to the cooler environment in this process.

Energy-Transfer Diagrams

Suppose you drop a hot rock into the ocean. Heat is transferred from the rock to the ocean until the rock and ocean are at the same temperature. Although the ocean warms up ever so slightly, ΔT_{ocean} is so small as to be completely insignificant. For all practical purposes, the ocean is infinite and unchangeable.

An **energy reservoir** is an object or a part of the environment so large that its temperature does not noticeably change when heat is transferred between the system and the reservoir. A reservoir at a higher temperature than the system is called a *hot reservoir*. A vigorously burning flame is a hot reservoir for small objects placed in the flame. A reservoir at a lower temperature than the system is called a *cold reservoir*. The ocean is a cold reservoir for the hot rock. We will use T_H and T_C to designate the temperatures of the hot and cold reservoirs.

Heat energy is transferred between a system and a reservoir if they have different temperatures. We will define

> Q_H = amount of heat transferred to or from a hot reservoir
>
> Q_C = amount of heat transferred to or from a cold reservoir

By definition, Q_H and Q_C are *positive* quantities.

Figure 11.16a shows a heavy copper bar placed between a hot reservoir (at temperature T_H) and a cold reservoir (at temperature T_C). Heat Q_H is transferred from the hot reservoir into the copper and heat Q_C is transferred from the copper to the cold reservoir. Figure 11.16b is an **energy-transfer diagram** for this process. The hot reservoir is always drawn at the top, the cold reservoir at the bottom, and the system—the copper bar in this case—between them. The reservoirs and the system are connected by "pipes" that show the energy transfers. Figure 11.16b shows heat Q_H being transferred into the system and Q_C being transferred out.

Figure 11.17 illustrates an important fact about heat transfers that we have seen: Spontaneous transfers go in one direction only. The fact that heat is not transferred from a colder object to a hotter object is an important result that will have significant practical implications, as we will soon discover.

(a)

Heat energy is transferred from hot to cold.

Copper bar
Q_H
Q_C
Fire
Ice
Hot reservoir at T_H Cold reservoir at T_C

(b)

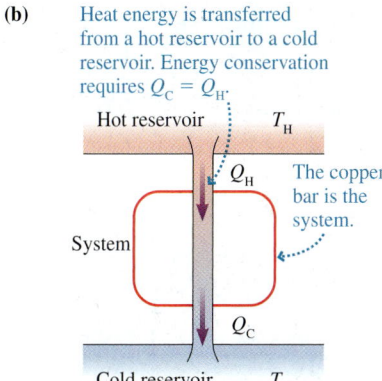

Heat energy is transferred from a hot reservoir to a cold reservoir. Energy conservation requires $Q_C = Q_H$.

Hot reservoir T_H
Q_H
The copper bar is the system.
System
Q_C
Cold reservoir T_C

FIGURE 11.16 Energy-transfer diagrams.

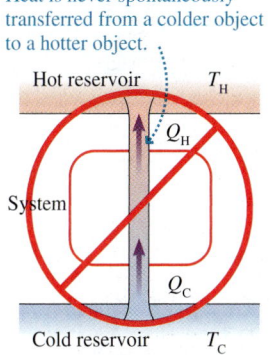

Heat is never spontaneously transferred from a colder object to a hotter object.

Hot reservoir T_H
Q_H
System
Q_C
Cold reservoir T_C

FIGURE 11.17 An impossible energy transfer.

CONCEPTUAL EXAMPLE 11.4   Energy transfers and the body

Why—in physics terms—is it more taxing on the body to exercise in very hot weather?

REASON   Your body continuously converts chemical energy to thermal energy, as we have seen. In order to maintain a constant body temperature, your body must continuously transfer heat to the environment. This is a simple matter in cool weather when heat is spontaneously transferred to the environment, but when the air temperature is higher than your body temperature, your body cannot cool itself this way and must use other mechanisms to transfer this energy, such as perspiring. These mechanisms require additional energy expenditure.

ASSESS   Strenuous exercise in hot weather can easily lead to a rise in body temperature if the body cannot exhaust heat quickly enough.

STOP TO THINK 11.4   The radiator in an automobile is at a higher temperature than the air around it. Considering the radiator as the system, we can say that

A. $Q > 0$ B. $Q = 0$ C. $Q < 0$

11.6 Heat Engines

In the early stages of the industrial revolution, most of the energy needed to run mills and factories came from water power. Water in a high reservoir will naturally flow downhill. A waterwheel can be used to harness this natural flow of water to produce some useful energy. That is, some of the potential energy lost by the water as it flows downhill can be converted into other forms.

It is possible to do something similar with heat. Thermal energy is naturally transferred from a hot reservoir to a cold; it is possible to take some of this energy as it is transferred and convert it to other forms. This is the job of a device known as a **heat engine.**

A simple example of a heat engine is shown in Figure 11.18a. Here a *Peltier device*—a device that produces a voltage if there is a temperature difference between its two sides—is sandwiched between two aluminum vanes that sit in cups of water. The cup of water on the left is hot; the cup on the right is cold. The temperature difference produces a voltage that drives the fan.

Figure 11.18b is a diagram that describes the operation of this device in thermodynamic terms. The Peltier device is the system. Heat energy is transferred through the device from the hot water to the cold water. As the transfer occurs, some of this energy is converted into electric energy to run the fan. The cold-water reservoir is crucial for the operation of the device. The heat energy that is converted to electric energy to run the fan comes from the hot water, but the fan won't run unless some heat energy is transferred into the cold water. Figure 11.19 is a thermodynamic picture of this heat engine—or any heat engine.

This schematic diagram of Figure 11.19 illustrates everything important about the basic physics of a heat engine. It takes in energy as heat from the hot reservoir, turns some of it into useful work, and exhausts the balance as waste heat in the cold reservoir. Any heat engine has exactly the same schematic.

Efficiency of a Heat Engine

We assume, for a heat engine, that the engine's thermal energy doesn't change. This means that there is no net energy transfer into or out of the heat engine. Because energy is conserved, we can say that the useful work extracted is equal to

Falling water turns a waterwheel.

(a)

Peltier device

(b)

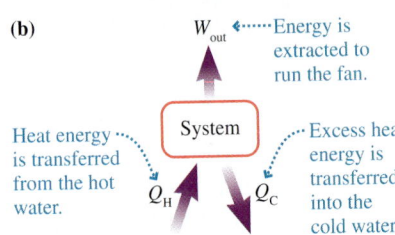

W_{out} ◀······Energy is extracted to run the fan.

Heat energy ···· is transferred from the hot water.

System

Q_H Q_C

···· Excess heat energy is transferred into the cold water.

FIGURE 11.18   A simple heat engine.

the difference between the heat energy transferred from the hot reservoir and the heat exhausted into the cold reservoir:

$$W_{out} = Q_H - Q_C$$

The energy input to the engine is Q_H and the energy output is W_{out}.

> **NOTE** ▶ We earlier defined Q_H and Q_C to be positive quantities. We also define the energy output of a heat engine W_{out} to be a positive quantity. For heat engines, the directions of the transfers will always be clear, and we will take all of these basic quantities to be positive. ◀

We can use the definition of efficiency, from earlier in the chapter, to compute the heat engine's efficiency:

$$e = \frac{\text{what you get}}{\text{what you had to pay}} = \frac{W_{out}}{Q_H} = \frac{Q_H - Q_C}{Q_H} \qquad (11.15)$$

The heat energy that is not converted to work ends up in the cold reservoir as waste heat.

Why should we waste energy this way? Why don't we make a heat engine like the one shown in Figure 11.20 that converts 100% of the heat into useful work? The surprising answer is that we can't. No heat engine can operate without exhausting some fraction of the heat into a cold reservoir. This isn't a limitation on our engineering abilities. As we'll see, it's a fundamental law of nature.

The maximum possible efficiency of a heat engine is fixed by the *second law of thermodynamics,* which we will explore in detail in Section 11.8. We will not do a detailed derivation, but simply note that the second law gives the maximum efficiency of a heat engine as

$$e_{max} = 1 - \frac{T_C}{T_H} \qquad (11.16)$$

Theoretical maximum efficiency of a heat engine

The maximum efficiency of any heat engine is therefore fixed by the ratio of the temperatures of the hot and cold reservoirs. It is possible to increase the efficiency of a heat engine by increasing the temperature of the hot reservoir or decreasing the temperature of the cold reservoir. The efficiency of Equation 11.16 is also called the *Carnot efficiency,* after a particular heat engine that achieves this maximum possible efficiency.

> **NOTE** ▶ The temperatures in Equation 11.16 *must* be in K, not °C. ◀

Because $e_{max} < 1$, the work done is always less than the heat input ($W_{out} < Q_H$). Consequently, there *must* be heat Q_C exhausted to the cold reservoir. That is why the heat engine of Figure 11.20 is impossible. Think of a heat engine as analogous to a water wheel: The engine "siphons off" some of the energy that is spontaneously flowing from hot to cold, but it can't completely shut off the flow.

The fact that heat engine efficiency is limited—and that heat must be exhausted into a cold reservoir—is of tremendous practical importance. Most of the energy that you use daily comes from the conversion of chemical energy into thermal energy and the subsequent conversion of that energy into other forms. Let's look at some common examples of heat engines.

1. Heat energy Q_H is transferred from the hot reservoir to the system.

2. Part of the energy is used to do useful work W_{out}.

3. The remaining energy $Q_C = Q_H - W_{out}$ is exhausted to the cold reservoir as waste heat.

FIGURE 11.19 The operation of a heat engine.

This is impossible! No heat engine can convert 100% of heat into useful work.

FIGURE 11.20 A perfect (and impossible!) heat engine.

Heat engines

Most of the electricity that you use was generated by heat engines. Coal or other fossil fuels are burned to produce the thermal energy of steam. The steam does work by spinning a turbine attached to a generator, which produces electricity. At the end of the proccess, approximately 65% of the thermal energy is deposited in a cold reservoir, usually a lake or a river.

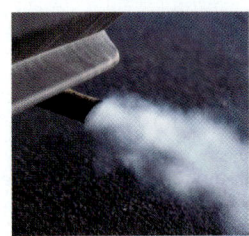

Your car gets the energy it needs to run from the chemical energy in gasoline. The gasoline is burned; the resulting hot gases are the hot reservoir. Some of this thermal energy is converted to kinetic energy of the moving vehicle and electric energy to run systems in the car, but more than 90% may be lost as heat to the environment—the cold reservoir, as we saw earlier.

There are many small, simple heat engines that are part of things you use daily. This fan, which can be put on top of a wood stove, uses the thermal energy of the stove to provide power to drive air around the room. Where are the hot and cold reservoirs in this device?

EXAMPLE 11.11 **The efficiency of a nuclear power plant**

Energy from nuclear reactions in the core or a nuclear reactor produces high-pressure steam at a temperature of 290°C. After the steam is used to spin a turbine, it is condensed (by using cooling water from a nearby river) back to water at 20°C. The excess heat is deposited in the river. The water is then reheated, and the cycle beings again. What is the maximum possible efficiency that this plant could achieve?

PREPARE A nuclear power plant is a heat engine, with energy transfers as illustrated in Figure 11.19. Q_H is the energy in the steam. T_H is the temperature of the steam, 290°C. The steam is cooled and condensed and the heat Q_C exhausted to the river. The river is the cold reservoir, so T_C is 20°C.

In kelvin, these temperatures are:

$$T_H = 290°C = 563 \text{ K}, T_C = 20°C = 293 \text{ K}$$

SOLVE We use Equation 11.16 to compute the maximum possible efficiency:

$$e_{max} = 1 - \frac{T_C}{T_H} = 1 - \frac{293 \text{ K}}{563 \text{ K}} = 0.479 \approx 48\%$$

ASSESS This is the maximum possible efficiency. There are practical limitations as well that limit real power plants, whether nuclear or coal- or gas-fired, to an efficiency $e \approx 0.35$. This means that 65% of the energy from the fuel is exhausted as waste heat into a river or lake, where it may cause problematic warming in the local environment.

Electric energy does work on the Peltier unit . . .

W_{in}

Hot water · Cold water

Q_H Q_C

. . . causing heat to be transferred from the cold water to the hot water.

FIGURE 11.21 A heat pump provides hot and cold water.

STOP TO THINK 11.5 Which of the following changes (there may be more than one) would increase the maximum theoretical efficiency of a heat engine?

A. Increase T_H.   B. Increase T_C.   C. Decrease T_H.   D. Decrease T_C.

11.7 Heat Pumps

As the inside of your refrigerator is colder than the air in your kitchen, heat energy will be transferred from the room to the inside of the refrigerator, warming it. This happens every time you open the door and as heat "leaks" through the walls. But you want the inside of the refrigerator to stay cool. To keep it cool, you need some way to "pump" this heat back out to the warmer room. Transferring heat energy from a cold reservoir to a hot reservoir—the opposite of the natural direction—is the job of a **heat pump.**

The Peltier device that we saw in the last section can be used as a heat pump. If an electric current is put through the device, a temperature difference develops between the two sides. As time goes by, the cold side gets colder and the hot side gets hotter. A practical application of such a heat pump is the water cooler/heater shown in Figure 11.21. The energy that is removed from the cold water ends up as increased thermal energy in the hot water. A single process cools the cold water and heats the hot water.

Heat pumps, like heat engines, are very practical devices. You probably have at least one heat pump in your house and one in your car—a refrigerator and an air conditioner. Each transfers heat energy from someplace cold to someplace warm.

The heat pump in a refrigerator transfers heat from the cold air *inside* the refrigerator to the warmer air in the room. Coils inside the refrigerator that are colder than the inside air take in thermal energy. This energy is transferred to warm coils *outside* the refrigerator that transfer heat to the room. In the antique refrigerator shown here, the cold coils are in the metal trays at the top of the cabinet, the warm coils outside the cabinet in the cylindrical unit on top. When the refrigerator is running, these top coils are quite warm. If you look closely at your refrigerator, you will find warm coils or a warm air exhaust that transfers heat to the room, heat that was removed from the inside.

An air conditioner works similarly, transferring heat from the cool air of a house (or a car) to the warmer air outside. You have probably seen room air conditioners that are single units meant to fit in a window. The side of the air conditioner that faces the room is the cool side; the other side of the air conditioner is warm. Heat is pumped from the cool side to the warm side, cooling the room.

Conversely, you can also use a heat pump to *warm* your house in the winter by moving thermal energy from outside your house to inside. A unit outside the house takes in heat energy that is pumped to a unit inside the house, warming the inside air.

Water doesn't spontaneously flow from a lower to a higher elevation. Nonetheless, you can force water to flow uphill by doing work on it with a pump. Similarly, heat doesn't spontaneously flow from cold to hot, but we can transfer heat energy from a cold reservoir to a hot reservoir by doing work. That's what a heat pump does, as shown in Figure 11.22.

Energy must be conserved, so the heat deposited in the hot side must equal the sum of the heat removed from the cold side and the work input:

$$Q_H = Q_C + W_{in}$$

For heat pumps, rather than compute efficiency, we compute an analogous quantity called the **coefficient of performance COP.**

There are two different ways that one can use a heat pump, as noted above. A refrigerator uses a heat pump for cooling, removing heat from a cold reservoir to keep it cold. You can also use a heat pump for heating, moving heat from a cold reservoir to a hot reservoir, to keep it warm. In both cases, we must do work to make this happen.

If we use a heat pump for cooling, we define the coefficient of performance as

$$\text{COP} = \frac{\text{what you get}}{\text{what you had to pay}} = \frac{\text{energy removed from the cold reservoir}}{\text{work required to perform the transfer}} = \frac{Q_C}{W_{in}}$$

A calculation similar to that for a heat engine can be done to show that the greater the temperature difference between the hot and cold, the more work is required to transfer the energy. The maximum possible coefficient of performance is related to the temperatures of the hot and cold reservoirs as

$$\text{COP}_{max} = \frac{T_C}{T_H - T_C} \tag{11.17}$$

Theoretical maximum coefficient of performance
of a heat pump used for cooling

A refrigerator from 1934.

The amount of heat exhausted to the hot reservoir is larger than the amount of heat extracted from the cold reservoir.

External work is used to remove heat from a cold reservoir and exhaust heat to a hot reservoir.

FIGURE 11.22 The operation of a heat pump.

Hot or cold lunch? Small coolers like this use Peltier devices that run off the 12 V electrical system of your car. They can transfer heat from the interior of the cooler to the outside, keeping food or drinks cool. But Peltier devices, like some other heat pumps, are reversible; switching the direction of current reverses the direction of heat transfer. A flick of a switch will cause heat to be transferred into the interior, keeping your lunch warm!

If we use a heat pump for heating, we define the coefficient of performance as

$$\text{COP} = \frac{\text{what you get}}{\text{what you had to pay}} = \frac{\text{energy added to the hot reservoir}}{\text{work required to perform the transfer}} = \frac{Q_H}{W_{in}}$$

In this case, the maximum possible coefficient of performance is

$$\text{COP}_{max} = \frac{T_H}{T_H - T_C} \qquad (11.18)$$

Theoretical maximum coefficient of performance
of a heat pump used for heating

In both cases, **a larger coefficient of performance means a more efficient heat pump.** Unlike the efficiency of a heat engine, which must be less than 1, the COP of a heat pump can be—and usually is—greater than 1. The following example shows that the COP can be quite high for typical temperatures.

EXAMPLE 11.12 Coefficient of performance of a refrigerator

The inside of your refrigerator is approximately 0°C. Heat from the inside of your refrigerator is deposited into the air in your kitchen, which has a temperature of approximately 20°C. At these operating temperatures, what is the maximum possible coefficient of performance of your refrigerator?

PREPARE The temperatures of the hot side and the cold side must be expressed in kelvin:

$$T_H = 20°C = 293 \text{ K}, T_C = 0°C = 273 \text{ K}$$

SOLVE We use Equation 11.17 to compute the maximum coefficient of performance:

$$\text{COP}_{max} = \frac{T_C}{T_H - T_C} = \frac{273 \text{ K}}{293 \text{ K} - 273 \text{ K}} = 13.6$$

ASSESS A coefficient of performance of 13.6 means that we pump 13.6 J of heat for an energy cost of 1 J. Due to practical limitations, the coefficient of performance of an actual refrigerator is typically ≈5. Other factors affect the overall efficiency of the appliance, including how well insulated it is.

CONCEPTUAL EXAMPLE 11.5 Keeping your cool?

It's a hot day, and your apartment is rather warm. Your roommate suggests cooling off the apartment by keeping the door of the refrigerator open. Will this help the situation?

REASON It's time for a physics lesson for your roommate! What you want to do is remove heat from the room. Your refrigerator, a heat pump, is designed to transfer heat from its inside to the outside. If you leave the door open, all parts of the refrigerator are exposed to the room air. It simply transfers heat from one part of the room to another—it won't make the room cooler.

ASSESS An air conditioner is a heat pump too. It must have a hot side that is outside the house, so that there is a net transfer of heat from inside to outside.

STOP TO THINK 11.6 Which of the following changes would allow your refrigerator to use less energy to run? (There may be more than one correct answer.)

A. Increasing the temperature inside the refrigerator.
B. Increasing the temperature of the kitchen.
C. Decreasing the temperature inside the refrigerator.
D. Decreasing the temperature of the kitchen.

11.8 Entropy and the Second Law of Thermodynamics

Throughout the chapter, we have noticed certain trends and certain limitations in energy transformations and transfers. Heat is transferred spontaneously from hot to cold, not from cold to hot. Once energy is transformed to thermal energy it is (in some sense) "lost." The transfer of heat from hot to cold is an example of an **irreversible** process, a process that can only happen in one direction. Why are some processes irreversible? The spontaneous transfer of heat from cold to hot would not violate any law of physics that we have seen to this point, but it is never observed. There must be another law of physics that prevents it. In this section we will explore the basis for this law, which we will call the **second law of thermodynamics.**

(a) Forward movie

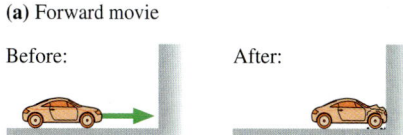

(b) The backward movie is physically impossible.

FIGURE 11.23 Two movies of a collision.

Reversible and Irreversible Processes

Stirring the cream in your coffee mixes the cream and coffee together. No amount of stirring ever unmixes them. If you shake a jar that has red marbles on the top and blue marbles on the bottom, the two colors are quickly mixed together. No amount of shaking ever separates them again. If you watched a movie of someone shaking a jar and saw the red and blue marbles separating, you would be certain that the movie was running backward. In fact, a reasonable definition of an irreversible process is one for which a backward-running movie shows a physically impossible process. In Figure 11.23, we see two possible movies of a particular physical process: the collision of a car with a barrier, one forward, the other backward.

The backward movie of Figure 11.23b is obviously wrong. But what has been violated in the backward movie? To have the crumpled car spring away from the wall would not violate any laws of physics we have so far discussed.

At a microscopic level, collisions are completely reversible. In Figure 11.24 we see two possible movies of a collision between two gas molecules, one forward, the other backward. You cannot tell, just by looking at the two movies, which is really going forward and which is being played backward. Nothing in either collision looks wrong, and no measurements you might make on either would reveal any violations of Newton's laws. **Interactions at the molecular level are reversible processes.**

If microscopic motions are all reversible, how can macroscopic phenomena such as the car crash end up being irreversible? If reversible collisions can cause heat to be transferred from hot to cold, why do they never cause heat to be transferred from cold to hot? There must be something at work that can distinguish the past from the future.

(a) Forward movie

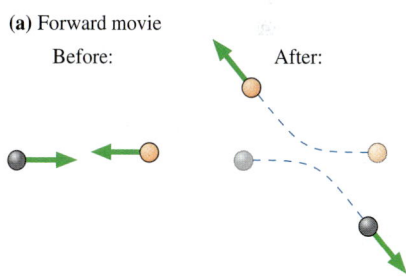

(b) The backward movie is equally plausible.

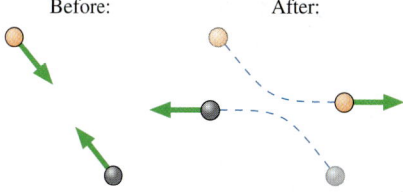

FIGURE 11.24 Molecular collisions are reversible.

Which Way to Equilibrium?

Stated another way, how do two systems initially at different temperatures "know" which way to go to reach equilibrium? Perhaps an analogy will help.

Figure 11.25 shows two boxes, numbered 1 and 2, containing identical balls. Box 1 starts with more balls than box 2, so $N_{1i} > N_{2i}$. Once every second, one ball is chosen at random and moved to the other box. This is a reversible process because a ball can move from box 2 to box 1 just as easily as from box 1 to box 2. What do you expect to see if you return several hours later?

Because balls are chosen at random, and because $N_{1i} > N_{2i}$, it's initially more likely that a ball will move from box 1 to box 2 than from box 2 to box 1. Sometimes a ball will move "backward" from box 2 to box 1, but overall there's a net movement of balls from box 1 to box 2. The system will evolve until $N_1 \approx N_2$. This is a stable situation—equilibrium!—with an equal number of balls moving in both directions.

Balls are chosen at random and moved from one box to the other.

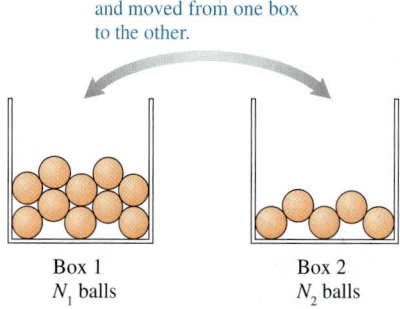

Box 1
N_1 balls

Box 2
N_2 balls

FIGURE 11.25 Moving balls between boxes.

But couldn't it go the other way, with N_1 getting even larger while N_2 decreases? In principle, any arrangement of the balls is possible. But certain arrangements are more likely. Each ball is equally likely to be in either box. With four balls, the odds are 1 in 2^4 or 1 in 16 that, at a randomly chosen instant of time, you would find all the balls in box 1. Were you to do so, you wouldn't find that to be terribly surprising.

With 10 balls, the probability that all of the balls are in the left box is about 1 in 1000. With 100 balls, the probability has dropped to about 1 in 1,000,000,000,000,000,000,000,000,000,000, or 1 in 10^{30}. The larger the number of balls, the more unlikely these extreme arrangements become. For the case of Figure 11.25, though each transfer is reversible, **the statistics of large numbers make it overwhelmingly likely that the system will evolve toward a state in which $N_1 \approx N_2$.** For the 10^{23} or so particles in a realistic system of atoms or molecules, the numbers are so astronomical that we will never see a state that deviates appreciably from the equilibrium case.

In Section 11.5 we considered the exchange of energy between two systems via collisions. Each collision is reversible, just as likely to transfer energy from 1 to 2 as from 2 to 1. But it is overwhelmingly likely that the net result of many, many collisions will be to transfer energy from system 1 to system 2 until the two temperatures are the same—in other words, for heat energy to be transferred from hot to cold.

A system reaches thermal equilibrium not by any plan or by outside intervention, but simply because **equilibrium is the most probable state in which to be.** It is possible that the system will move away from equilibrium, with heat moving from cold to hot, but this is remotely improbable in any realistic system. The consequence of having a vast number of random events is that the system evolves in one direction, toward equilibrium, and not the other. Reversible microscopic events lead to irreversible macroscopic behavior because some macroscopic states are vastly more probable than others.

Order, Disorder, and Entropy

Figure 11.26 shows microscopic views of three containers of gas. The top diagram shows a group of atoms arranged in a very ordered pattern. This is a highly ordered and nonrandom system, with each atom's position precisely specified. Contrast this with the system on the bottom, in which there is no order at all. It is extremely improbable that the atoms in a container would *spontaneously* arrange themselves into the ordered pattern of the top picture; even a small change in the pattern is quite noticeable. By contrast, there are a vast number of arrangements like the one on the bottom that randomly fill the container. A small change in the pattern is hard to detect.

Scientists and engineers use the term **entropy** to quantify the probability that a certain state of a system will occur. The ordered arrangement of the top system, which has a very small probability of spontaneous occurrence, has a very low entropy. The entropy of the randomly filled container is high; it has a large probability of occurrence. The entropy of the middle picture is somewhere in between. It is often said that entropy measures the amount of *disorder* in a system. The entropy in Figure 11.26 increases as you move from the ordered system on the top to the disordered system on the bottom.

Suppose you ordered the atoms in Figure 11.26 as they are in the top diagram, and then you let them go. After you release them, they will move, as gas atoms do, in a random manner. After some time, you would expect their arrangement to appear as in the bottom diagram. In fact, the series of diagrams from top to bottom could be thought of as a series of movie frames showing the evolution of the

The arrangement is unlikely; all of the particles must be precisely arranged. The movement of one particle is easy to notice.

A progression of states of a system of particles. The top state is ordered; the bottom is not.

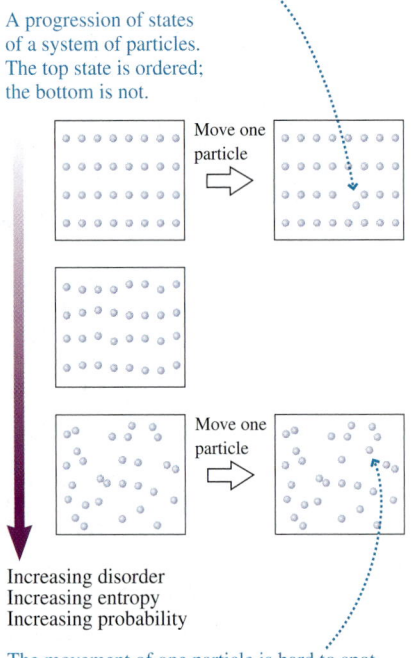

Move one particle

Move one particle

Increasing disorder
Increasing entropy
Increasing probability

The movement of one particle is hard to spot. Many arrangements have a similar appearance, so an arrangement like this is quite likely.

FIGURE 11.26 Ordered and disordered arrangements of atoms in a gas.

positions of the atoms after they are released. The system of atoms would be spontaneously moving toward a state of higher entropy. You would correctly expect this process to be irreversible. That is, the particles will never spontaneously recreate their initial ordered state.

Two thermally interacting systems with different temperatures have a low entropy. These systems are ordered in the sense that the faster atoms are on one side of the barrier, the slower atoms on the other. The most random possible distribution of energy, and hence the least ordered system, corresponds to the situation where the two systems are in thermal equilibrium with equal temperatures. **Entropy increases as two systems with initially different temperatures move toward thermal equilibrium.** Entropy would decrease if heat energy moved from cold to hot, making the hot system hotter and the cold system colder.

The Second Law of Thermodynamics

The fact that macroscopic systems evolve irreversibly toward equilibrium is a statement about nature that is not contained in any of the laws of physics we have encountered. It is a new law of physics, one known as the **second law of thermodynamics.**

The formal statement of the second law of thermodynamics is given in terms of entropy:

> **Second law of thermodynamics** The entropy of an isolated system never decreases. The entropy either increases, until the system reaches equilibrium, or, if the system began in equilibrium, stays the same.

NOTE ▶ The qualifier "isolated" is crucial. We can order the system by reaching in from the outside, perhaps using little tweezers to place atoms in a lattice. Similarly, we can transfer heat from cold to hot by using a refrigerator. The second law is about what a system can or cannot do *spontaneously,* on its own, without outside intervention. ◀

The second law of thermodynamics tells us that an isolated system evolves such that

- Order turns into disorder and randomness.
- Information is lost rather than gained.
- The system "runs down" as other forms of energy are transformed to thermal energy.

An isolated system never spontaneously generates order out of randomness. It is not that the system "knows" about order or randomness, but rather that there are vastly more states corresponding to randomness than there are corresponding to order. As collisions occur at the microscopic level, the laws of probability dictate that the system will, on average, move inexorably toward the most probable and thus most random macroscopic state.

Entropy and Thermal Energy

Suppose we have a very cold, moving baseball, with essentially no thermal energy, and a stationary helium balloon at room temperature, as shown in Figure 11.27. The atoms in the baseball and the atoms in the balloon are all moving, but there is a big difference in their motions. The atoms in the baseball are all moving in the same direction at the same speed, but the atoms in the balloon are moving in random directions. The ordered, organized motion of the baseball has

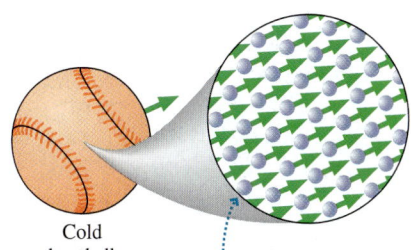

Cold baseball

The molecules in the baseball all move in the same direction at the same speed. This ordered motion is the ball's kinetic energy.

Helium balloon

The molecules in the gas move in different directions at different speeds. This random motion is the thermal energy of the gas.

FIGURE 11.27 Kinetic energy and thermal energy compared.

1. Heat energy Q_H comes into the system from the hot reservoir. Entropy is added to the heat engine.

2. The work out contains no entropy. The only way that entropy can leave is via heat flow to the cold reservoir.

3. Heat energy Q_C goes out of the heat engine into the cold reservoir. Entropy is removed from the heat engine.

FIGURE 11.28 A diagram of a heat engine, with entropy changes noted.

low entropy while the disorganized, random motions of the gas atoms—what we have called thermal energy—has high entropy. You can see that a conversion of macroscopic kinetic energy to thermal energy means an increase in entropy.

Thermal energy has entropy; other forms of energy do not. We saw, in section 11.5, that the conversion of other forms of energy to thermal energy was irreversible. Now, we can explain why: **When another form of energy is converted to thermal energy, there is an increase in entropy. The process will not spontaneously reverse.**

This also explains why it is hard to make a heat engine, a device that turns thermal energy into other forms of energy. We have redrawn our schematic diagram of a heat engine in Figure 11.28, adding arrows representing entropy. Now we can see why heat must be exhausted into the cold reservoir: We need to get rid of entropy! Each unit of entropy to be exhausted requires heat to carry it out, and this limits the efficiency of the engine. Some heat must be "wasted." The best we can do is to exhaust as little heat as we can while getting rid of entropy. Although the full details are beyond the scope of this book, the maximum efficiencies given in equations 11.16, 11.17, and 11.18 can be derived by assuming that we exhaust as little heat as possible to get rid of entropy. Efficiencies or coefficients of performance higher than these would violate the second law of thermodynamics.

Alternative Statements of the Second Law of Thermodynamics

Our formal statement of the second law dealt with entropy. The fact that the entropy of an isolated system always increases has other consequences. Some of these consequences can, in fact, be used as alternative, informal statements of the second law of thermodynamics. We will make a table of these alternative statements as a way of summarizing what we have learned about thermodynamics.

TABLE 11.5 Alternative statements of the second law of thermodynamics

Phenomenon	Related Statement of the Second Law
Heat energy spontaneously flows from hot to cold.	When two systems at different temperatures interact, heat energy is transferred spontaneously from the hotter to the colder system, never from the colder to the hotter.
Entropy considerations limit the possible efficiency of a heat engine.	It is not possible to make a heat engine that converts thermal energy into an equivalent amount of work.
It takes energy to move heat from a cold object to a warm object.	It is not possible to make a heat pump that moves heat from a cold object to a hot object without an external energy input.
The entropy of an isolated system will never spontaneously decrease.	The time direction in which the entropy of an isolated system increases is "the future."

The last of the statements of the second law is perhaps the most compelling. The irreversible evolution of systems from less-likely states to more-likely states is what gives us a direction of time. As we noted, stirring blends your coffee and cream, it never unmixes them. Friction causes an object to stop while increasing its thermal energy; the random atomic motions of thermal energy never spontaneously organize themselves into a macroscopic motion of an entire object. A plant in a sealed, darkened jar dies and decomposes to carbon and various gases; the gases and carbon never spontaneously assemble themselves into a flower. These are all examples of irreversible processes. They each show a clear direction of time, a distinct difference between past and future.

Which of the following processes does not involve a change in entropy?

A. An electric heater raises the temperature of a cup of water by 20°C.
B. A ball rolls up a ramp, decreasing in speed as it rolls higher.
C. A basketball is dropped from 2 m and bounces until it comes to rest.
D. The sun shines on a black surface and warms it.

11.9 Systems, Energy, and Entropy

Over the past two chapters, we have learned a good deal about energy and how it is used. By introducing the concept of entropy, we were able to see how limits on our ability to use energy come about. We will close this chapter, and this part of the book, by considering a few final questions that bring these different pieces together.

The Conservation of Energy and Energy Conservation

We have all heard, for many years, that it is important to "conserve energy." We are asked to turn off lights when we leave rooms, to drive our cars less, to turn down our thermostats. But this brings up an interesting question: If we have a law of conservation of energy, which states that energy can't be created or destroyed, what do we really mean by "conserving energy?" If energy can't be created or destroyed, how can there be an "energy crisis"?

In fact, we aren't, as a society or as a planet, running out of energy. We can't! What we *can* run out of is high-quality sources of energy, forms of energy that can easily be converted into other forms of energy. Oil is a good example. Oil contains chemical energy, in large quantity. It is a liquid, so is easily transported, and it is easily burned in a host of devices to generate heat, electricity, or motion. When we burn oil, we don't use up its energy—we simply convert its chemical energy into thermal energy. As we do this, we decrease the amount of high-quality chemical energy in the world and increase the supply of thermal energy. And, as we have seen, thermal energy is less useful, as it is difficult to convert it to other forms.

Entropy and Life

The second law predicts that systems will run down, that order will evolve toward disorder and randomness, and that complexity will give way to simplicity. But just look around you!

- Plants grow from simple seeds to complex entities.
- Single-celled fertilized eggs grow into complex adult organisms.
- Over the last billion years or so, life has evolved from simple unicellular organisms to very complex forms.

Everywhere we look, it seems, the second law is being violated. How can this be?

There is an important qualification in the second law of thermodynamics: It applies only to *isolated systems,* systems that do not exchange energy with their environment. The situation is entirely different if energy is transferred into or out of the system.

Your body is not an isolated system. Every day, you take in chemical energy in the food you eat. As you use this energy, most of it ends up as thermal energy that you exhaust as heat into the environment—taking away the entropy that natural processes in your body creates. An energy and entropy diagram of this situation is given in Figure 11.29. It is this continual exchange of energy (and entropy) with the environment that makes your life—and all life—possible.

Sealed, but not isolated BIO This glass container is a completely sealed system containing living organisms, shrimp and algae. But the organisms will live and grow for many years. The reason this is possible is that the glass sphere, though sealed, is not an *isolated* system. Energy can be transferred in and out as light and heat. If the container were to be placed in a darkened room, the organisms would quickly perish.

Chemical energy comes into your body in the food you eat.

Energy leaves your body mostly as heat, meaning entropy is transferred as well.

E_{in}

Q_{out}

Entropy out

FIGURE 11.29 Thermodynamic view of the body.

SUMMARY

The goal of Chapter 11 has been to learn more about energy transformations and transfers, the laws of thermodynamics, and theoretical and practical limitations on energy use.

GENERAL PRINCIPLES

Energy and Efficiency

When energy is transformed from one form to another, some may be "lost," usually to thermal energy, due to practical or theoretical constraints. This limits the efficiency of processes. We define **efficiency** as:

$$e = \frac{\text{what you get}}{\text{what you had to pay}}$$

Entropy and Irreversibility

Systems move toward more probable states. These states have higher entropy—more disorder. This change is irreversible. Changing other forms of energy to thermal energy is irreversible.

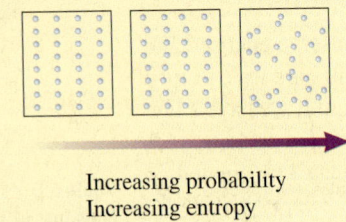

Increasing probability
Increasing entropy

The Laws of Thermodynamics

The **first law of thermodynamics** is the law of conservation of energy for systems in which only thermal energy changes:

$$\Delta E_{th} = W + Q$$

The **second law of thermodynamics** can be stated in a few different ways:

- The entropy of an isolated system always increases.
- Heat energy spontaneously flows only from hot to cold.
- No heat engine can be 100% efficient.
- "The future" is the time direction of entropy increase.

Heat is the transfer of energy via a thermal interaction. Energy will be transferred until thermal equilibrium is reached.

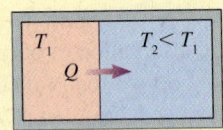

IMPORTANT CONCEPTS

Thermal energy
- For a gas, the thermal energy is the **total kinetic energy** of motion of the atoms.

$$E_{th} = NK_{avg} = \frac{1}{2}Nmv_{rms}^2$$

- Thermal energy is random kinetic energy and thus has entropy.

Temperature
- For a gas, temperature is proportional to the **average kinetic energy** of the motion of the atoms.

$$T = \frac{2}{3}\frac{K_{avg}}{k_B}$$

- Two systems are in **thermal equilibrium** if they are at the same temperature. No heat energy is transferred at thermal equilibrium.

A heat engine converts thermal energy from a hot reservoir into useful work. Some heat is exhausted into a cold reservoir.

$$e_{max} = 1 - \frac{T_C}{T_H}$$

A heat pump uses an energy input to transfer heat from a cold side to a hot side. When used for cooling:

$$COP_{max} = \frac{T_C}{T_H - T_C}$$

APPLICATIONS

Efficiencies

Energy in the body
Cells in the body metabolize chemical energy in food. Efficiency for most actions is about 25%.

Energy used by body at rate of 480 W

Energy for forward propulsion at rate of 120 W

Power plants
A typical power plant converts about 1/3 of the energy input into useful work. The rest is exhausted as waste heat.

Chemical energy in

Waste heat

Useful work out

Temperature scales

Zero on the **Kelvin temperature scale** is the temperature at which the kinetic energy of atoms is zero. This is **absolute zero.** The conversion from °C to K is

$$T = T_C + 273$$

► All temperatures in equations must be in kelvin. ◄

For instructor-assigned homework, go to
www.masteringphysics.com

Problem difficulty is labeled as I (straightforward) to IIII (challenging).

Problems labeled ✏ can be done on a Workbook
Dynamics Worksheet; INT integrate significant material from earlier chapters; BIO are of biological or medical interest.

QUESTIONS

Conceptual Questions

1. Rub your hands together vigorously. What happens? Discuss the energy transfers and transformations that take place.

2. Write a few sentences describing the energy transformations that occur from the time moving water enters a hydroelectric plant until you see some water being pumped out of a nozzle in a public fountain. Use the "Energy transformations" table on page 40 as an example.

3. BIO Describe the energy transfers and transformations that occur from the time you sit down to breakfast until you've completed a fast bicycle ride.

4. BIO According to Table 11.4, cycling at 15 km/hr requires less metabolic energy than running at 15 km/hr. Suggest reasons why this is the case.

5. For most automobiles, the number of miles per gallon decreases as highway speed increases. Fuel economy drops as speeds increase from 55 to 65 mph, then decreases further as speeds increase to 75 mph. Explain why this is the case.

6. When the space shuttle returns to earth, its surfaces get very hot as it passes through the atmosphere at high speed.
   a. Has the space shuttle been heated? If so, what was the source of the heat? If not, why is it hot?
   b. Energy must be conserved. What happens to the space shuttle's initial kinetic energy?

7. One end of a short aluminum rod is in a campfire and the other end is in a block of ice, as shown in Figure Q11.7. If 100 J of energy are transferred from the fire to the rod, and if the temperature at every point in the rod has reached a steady value, how much energy goes from the rod into the ice?

FIGURE Q11.7

8. Two blocks of copper, 1 kg and 3 kg, are at the same temperature. Which block has more thermal energy? If the blocks are placed in thermal contact, will the thermal energy of the blocks change? If so, how?

9. If the temperature T of an ideal gas doubles, by what factor does the average kinetic energy of the atoms change?

10. A bottle of helium gas and a bottle of argon gas contain equal numbers of atoms at the same temperature. Which bottle, if either, has the greater total thermal energy?

For Questions 11 through 16, give a specific example of a process that has the energy changes and transfers described. (For example, if the question states "$\Delta E_{th} > 0$, $W = 0$," you are to describe a process that has an increase in thermal energy and no transfer of energy by work. You could write "Heating a pan of water on the stove.")

11. $\Delta E_{th} < 0$, $W = 0$
12. $\Delta E_{th} > 0$, $Q = 0$
13. $\Delta E_{th} < 0$, $Q = 0$
14. $\Delta E_{th} > 0$, $W \neq 0$, $Q \neq 0$
15. $\Delta E_{th} < 0$, $W \neq 0$, $Q \neq 0$
16. $\Delta E_{th} = 0$, $W \neq 0$, $Q \neq 0$

17. A drop of green ink falls into a beaker of clear water. First *describe* what happens. Then *explain* the outcome in terms of entropy.

18. If you hold a rubber band loosely between two fingers and let it rest against your lips, and then stretch the rubber band, the sensitive skin of your lips will detect a small temperature increase. The rubber band soon returns to its original temperature. What are the signs of W and Q for this process?

19. The exterior unit of a heat pump designed for heating is sometimes buried underground in order to use the earth as a thermal reservoir. Why is it worthwhile to bury the heat exchanger, even if the underground unit costs more to purchase and install than one above ground?

20. Assuming improved materials and better processes, can engineers ever design a heat engine that exceeds the maximum efficiency indicated by Equation 11.16? If not, why not?

Multiple-Choice Questions

21. I A person is walking on level ground at constant speed. What energy transformation is taking place?
    A. Chemical energy is being transformed to thermal energy.
    B. Chemical energy is being transformed to kinetic energy.
    C. Chemical energy is being transformed to kinetic energy and thermal energy.
    D. Chemical energy and thermal energy are being transformed to kinetic energy.

22. II A person walks 1 km, turns around, and runs back to where he started. Compare the energy used and the power during the two segments:
    A. The energy used and the power are the same for both segments.
    B. The energy used while walking is greater, the power while running is greater.
    C. The energy used while running is greater, the power while running is greater.
    D. The energy used is the same for both segments, the power while running is greater.

23. I The temperature of the air in a basketball increases as it is pumped up. This means that:
    A. The total kinetic energy of the air is increasing and the average kinetic energy of each atom is decreasing.
    B. The total kinetic energy of the air is increasing and the average kinetic energy of each atom is increasing.
    C. The total kinetic energy of the air is decreasing and the average kinetic energy of each atom is decreasing.
    D. The total kinetic energy of the air is decreasing and the average kinetic energy of each atom is increasing.

24. | The thermal energy of a container of helium gas is halved. What happens to the temperature, in kelvin?
    A. It decreases to one-fourth its initial value.
    B. It decreases to one-half its initial value.
    C. It stays the same.
    D. It increases to twice its initial value.

25. | An inventor approaches you with a device that he claims will take 100 J of thermal energy input and produce 200 J of electricity. You decide not to invest your money because this device would violate
    A. the first law of thermodynamics.
    B. the second law of thermodynamics.
    C. both the first and second laws of thermodynamics.

26. | While keeping your food cold, your refrigerator transfers energy from the inside to the surroundings. Thus thermal energy goes from a colder object to a warmer one. What can you say about this?
    A. It is a violation of the second law of thermodynamics.
    B. It is not a violation of the second law of thermodynamics because refrigerators can have efficiency of 100%.
    C. It is not a violation of the second law of thermodynamics because the second law doesn't apply to refrigerators.
    D. The second law of thermodynamics applies in this situation, but it is not violated because the energy did not spontaneously go from cold to hot.

PROBLEMS

Section 11.1 Transforming Energy

1. | A 20%-efficient engine accelerates a 1500 kg car from rest to 15 m/s. How much energy is transferred to the engine by burning gasoline?

2. | A 60% efficient device uses chemical energy to generate 600 J of electric energy.
    a. How much chemical energy is used?
    b. A second device uses twice as much chemical energy to generate half as much electric energy? What is its efficiency?

3. | A typical photovoltaic cell delivers 4.0×10^{-3} W of electric energy when illuminated with 1.2×10^{-1} W of light energy. What is the efficiency of the cell?

4. | A 15 W compact fluorescent bulb and a 75 W incandescent bulb each produce 3.0 W of visible light energy. What are the efficiencies of these two types of bulbs for converting electric energy into light?

FIGURE P11.4 Compact fluorescent and incandescent bulbs.

Section 11.2 Energy in the Body: Energy Inputs

5. || A fast-food hamburger (with cheese and bacon) contains 1000 Calories. (a) What is the burger's energy in joules? (b) If all this energy is used to lift a 10 kg mass, how high can it be lifted?

6. || In an average human, basic life processes require energy to be supplied at a steady rate of 100 W. What daily energy intake, in Calories, is required to maintain these basic processes? This is the minimum daily caloric intake needed to avoid starvation.

7. | An "energy bar" contains 6.0 g of fat. How much energy is this in joules? In calories? In Calories?

8. | An "energy bar" contains 22 g of carbohydrates. How much energy is this in joules? In calories? In Calories?

Section 11.3 Energy in the Body: Energy Outputs

9. || An "energy bar" contains 22 g of carbohydrates. If the energy bar was his only fuel, how far could a 68 kg person walk at 5.0 km/hr?

10. || Suppose your body was able to use the chemical energy in gasoline. How far could you pedal a bicycle at 15 km/hr on the energy in 1 gal of gas? (1 gal of gas has a mass of 3.2 kg.)

11. |||| The label on a candy bar says 400 Calories. Assuming a typical efficiency for energy use by the body, if a 60 kg person were to use the energy in this candy bar to climb stairs, how high could she go?

12. || A weightlifter curls a 30 kg bar, raising it each time a distance of 0.60 m. How many times must he repeat this exercise to burn off the energy in one slice of pizza?

13. |||| A weightlifter works out at the gym each day. Part of her routine is to lie on her back and lift a 40 kg barbell straight up from chest height to full arm extension, a distance of 0.50 m.
    a. How much work does the weightlifter do to lift the barbell one time?
    b. If the weightlifter does 20 repetitions a day, what total energy does she expend on lifting? Assume 25% efficiency.
    c. How many 400 Calorie donuts can she eat a day to supply that energy?

Section 11.4 Thermal Energy and Temperature

14. | What are the rms speeds of helium and argon atoms in a gas at 1000°C?

15. | 30 m/s is a typical highway speed for a car. At what temperature do the molecules of nitrogen gas have an rms speed of 30 m/s?

16. || A gas consists of a mixture of neon and argon. The rms speed of the neon atoms is 400 m/s. What is the rms speed of the argon atoms?

17. || At what temperature do hydrogen molecules have the same rms speed as nitrogen molecules at 100°C?

18. || At what temperature, in °C, is the rms speed of oxygen molecules (a) half and (b) twice its value at 0°C?

Section 11.5 Heat and the First Law of Thermodynamics

19. | 500 J of work are done on a system in a process that decreases the system's thermal energy by 200 J. How much energy is transferred to or from the system as heat?

20. | 600 J of heat energy are transferred to a system that does 400 J of work. By how much does the system's thermal energy change?

21. | 300 J of energy are transferred to a system in the form of heat while the thermal energy increases by 150 J. How much work is done on or by the system?
22. | 10 J of heat are removed from a gas sample while it is being compressed by a piston that does 20 J of work. What is the change in the thermal energy of the gas? Does the temperature of the gas increase or decrease?

Section 11.6 Heat Engines

23. ‖‖ A heat engine extracts 55 kJ from the hot reservoir and exhausts 40 kJ into the cold reservoir. What are (a) the work done and (b) the efficiency?
24. | A heat engine does 20 J of work while exhausting 30 J of waste heat. What is the engine's efficiency?
25. | A heat engine does 200 J of work while exhausting 600 J of heat to the cold reservoir. What is the engine's efficiency?
26. | A heat engine with an efficiency of 40% does 100 J of work. How much heat is (a) extracted from the hot reservoir and (b) exhausted into the cold reservoir?
27. ‖ a. At what cold-reservoir temperature (in °C) would an engine operating at maximum theoretical efficiency with a hot-reservoir temperature of 427°C have an efficiency of 60%?
    b. If another engine, operating at maximum theoretical efficiency with a hot-reservoir temperature of 400°C, has the same efficiency, what is its cold-reservoir temperature?
28. ‖ A heat engine operating between energy reservoirs at 20°C and 600°C has 30% of the maximum possible efficiency. How much energy does this engine extract from the hot reservoir to do 1000 J of work?

Section 11.7 Heat Pumps

29. | A refrigerator takes in 20 J of work and exhausts 50 J of heat. What is the refrigerator's coefficient of performance?
30. | Air conditioners are rated by their coefficient of performance at 80°F inside temperature and 95°F outside temperature. An efficient but realistic air conditioner has a coefficient of performance of 3.2. What is the maximum possible coefficient of performance?
31. | 50 J of work are done on a refrigerator with a coefficient of performance of 4.0. How much heat is (a) extracted from the cold reservoir and (b) exhausted to the hot reservoir?
32. ‖ Find the maximum possible coefficient of performance for a heat pump used to heat a house in a northerly climate in winter. The inside is kept at 20°C while the outside is −20°C.

Section 11.8 Entropy and the Second Law of Thermodynamics

33. ‖ Which, if any, of the heat engines in Figure P11.33 at right violate (a) the first law of thermodynamics or (b) the second law of thermodynamics? Explain.
34. ‖ Which, if any, of the refrigerators in Figure P11.34 at right violate (a) the first law of thermodynamics or (b) the second law of thermodynamics? Explain.
35. ‖ Draw all possible distinct arrangements in which three balls (labeled A, B, C) are placed into two different boxes (1 and 2), as in Figure 11.25. If all arrangements are equally likely, what is the probability that all three will be in box 1?

FIGURE P11.33

FIGURE P11.34

General Problems

36. ‖ How many slices of pizza must you eat to walk for 1.0 hr at
BIO a speed of 5.0 km/hr? (Assume your mass is 68 kg.)
37. | A 60 kg person climbs to the top of a 500-m-high hill. If the energy necessary to do this were provided by burning coal, how much coal would be needed?
38. ‖ For how long would a 68 kg person have to swim at a fast
BIO crawl to use all the energy available in a typical fast food meal of burger, fries, and a drink?
39. | a. How much metabolic energy is required for a 68 kg per-
BIO   son to run at a speed of 15 km/hr for 20 min?
    b. How much metabolic energy is required for this person to walk at a speed of 5.0 km/hr for 60 min? Compare your result to your answer to part a.
    c. Compare your results of parts a and b to the result of Example 11.5. Of these three modes of human motion, which is the most efficient?

40. ||| To a good approximation, the only external force that does
BIO work on a cyclist moving on level ground is the force of air
resistance. Suppose a cyclist is traveling at 15 km/hr on level
ground. Assume he is using 480 W of metabolic power.
    a Estimate the amount of power he uses for forward motion.
    b. How much force must he exert to overcome the force of air
       resistance?

41. || The winning time for the 2005 annual race up 86 floors of
BIO the Empire State Building was 10 min and 49 s. The winner's
INT mass was 60 kg.
    a. If each floor was 3.7 m high, what was the winner's
       change in gravitational potential energy?
    b. If the efficiency in climbing stairs is 25%, what total
       energy did the winner expend during the race?
    c. How many food Calories did the winner "burn" in the race?
    d. Of those Calories, how many were converted to thermal
       energy?
    e. What was the winner's metabolic power in watts during
       the race up the stairs?

42. ||| Championship swimmers take about 22 s and about 30 arm
BIO strokes to move through the water in a 50 m freestyle race.
INT a. From Table 12.4, a swimmer's metabolic power is 800 W.
       If the efficiency for swimming is 25%, how much energy
       is expended moving through the water in a 50 m race?
    b. If half the energy is used in arm motion and half in leg
       motion, what is the energy expenditure per arm stroke?
    c. Model the swimmer's hand as a paddle. During one arm
       stroke, the paddle moves halfway around a 90-cm-radius
       circle. If all the swimmer's forward propulsion during an
       arm stroke comes from the hand pushing on the water and
       none from the arm (somewhat of an oversimplification),
       what is the average force of the hand on the water?

43. | Helium remains very nearly an ideal gas to temperatures as
low as 5 K. What are the root-mean-square speed and the
average kinetic energy of helium atoms at this temperature?

44. | The temperature at the center of the sun is $\approx 2 \times 10^7$ K.
What are (a) the average kinetic energy and (b) the rms speed
of a proton in the center of the sun?

45. || From what height must an oxygen molecule fall from rest
in a vacuum so that its kinetic energy at the bottom equals the
INT average kinetic energy of an oxygen molecule at 300 K?

46. | The surface of the sun consists
mostly of hydrogen *atoms* (not molecules) at a temperature of 6000 K. What
are (a) the average translational kinetic
energy per atom and (b) the rms speed
of the atoms?

47. | What is the total translational kinetic
energy at 0°C of 1.0×10^{21} molecules
of (a) hydrogen and (b) oxygen?

48. || A container holding argon atoms changes temperature by
20°C when 30 J of heat are removed. How many atoms are in
the container?

49. | A heat engine with a high-temperature reservoir at 400 K
has an efficiency of 0.20. What is the maximum possible temperature of the cold reservoir?

50. || A heat engine does 10 J of work and exhausts 15 J of waste
heat.
    a. What is the engine's efficiency?
    b. If the cold-reservoir temperature is 20°C, what is the mini-
       mum possible temperature in °C of the hot reservoir?

51. || The heat exhausted to the cold reservoir of an engine oper-
ating at maximum theoretical efficiency is two-thirds the heat
extracted from the hot reservoir. What is the temperature ratio
T_C/T_H?

52. || An engine operating at maximum theoretical efficiency
whose cold-reservoir temperature is 7°C is 40% efficient. By
how much should the temperature of the hot reservoir be
increased to raise the efficiency to 60%?

53. | An ideal heat pump used for heating has a coefficient of
performance of 10 and a cold-reservoir temperature of 0°C.
What is the hot-reservoir temperature in °C?

54. || The coefficient of performance of a refrigerator is 5.0.
    a. If the compressor uses 10 J of energy, how much heat is
       exhausted to the hot reservoir?
    b. If the hot-reservoir temperature is 27°C, what is the lowest
       possible temperature in °C of the cold reservoir?

55. || An engineer claims to have measured the characteristics of
a heat engine that takes in 100 J of thermal energy and produces 50 J of useful work. Is this engine possible? If so, what
is the smallest possible ratio of the temperatures (in kelvin) of
the hot and cold reservoirs?

56. || A 32% efficient electric power plant produces 900 MJ of
electric energy per second and discharges waste heat into 20°C
ocean water. Suppose the waste heat could be used to heat
homes during the winter instead of being discharged into the
ocean. A typical American house requires an average 20 kW
for heating. How many homes could be heated with the waste
heat of this one power plant?

57. ||| A typical coal-fired power plant burns 300 metric tons of
coal *every hour* to generate 2.7×10^6 MJ of electric energy.
1 metric ton = 1000 kg; 1 metric ton of coal has a volume of
1.5 m³. The heat of combustion of coal is 28 MJ/kg. Assume
that *all* heat is transferred from the fuel to the boiler and that
all the work done in spinning the turbine is transformed into
electric energy.
    a. Suppose the coal is piled up in a 10 m × 10 m room. How
       tall must the pile be to operate the plant for one day?
    b. What is the power plant's efficiency?

58. || Each second, a nuclear power plant generates 2000 MJ of
thermal energy from nuclear reactions in the reactor's core.
This energy is used to boil water and produce high-pressure
steam at 300°C. The steam spins a turbine, which produces
700 MJ of electric power, then the steam is condensed and
the water is cooled to 30°C before starting the cycle again.
    a. What is the maximum possible efficiency of the plant?
    b. What is the plant's actual efficiency?

59. | The surface waters of tropical oceans are at a temperature
of 27°C while water at a depth of 1200 m is at 3°C. It has been
suggested these warm and cold waters could be the heat reservoirs for a heat engine, allowing us to do work or generate
electricity from the thermal energy of the ocean. What is the
maximum efficiency possible of such a heat engine?

60. || The light energy that falls on a square meter of ground over
the course of a typical sunny day is about 20 MJ. The average
rate of electric energy consumption in one house is 1.0 kW.
    a. On average, how much energy does one house use during
       each 24 hr day?
    b. If light energy to electric energy conversion using solar
       cells is 5% efficient, how many square miles of land must
       be covered with solar cells to supply the electrical energy
       for 250,000 houses? Assume there is no cloud cover.

Passage Problems

Kangaroo Locomotion BIO

Kangaroos have very stout tendons in their legs that can be used to store energy. When a kangaroo lands on its feet, the tendons stretch, transforming kinetic energy of motion to elastic potential energy. Much of this energy can be transformed back into kinetic energy as the kangaroo takes another hop. The kangaroo's peculiar hopping gait is not very efficient at low speeds but is quite efficient at high speeds.

Figure P11.61 shows the energy cost of human and kangaroo locomotion. The graph shows oxygen uptake (in mL/s) per kg of body mass, allowing a direct comparison between the two species.

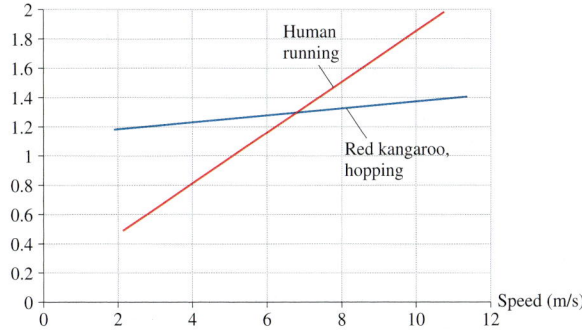

Oxygen uptake (mL/kg·s)

Human running

Red kangaroo, hopping

FIGURE P11.61 Oxygen uptake (a measure of energy use per second) for a running human and a hopping kangaroo.

For humans, the energy used per second (i.e., power) is proportional to the speed. That is, the human curve nearly passes through the origin, so running twice as fast takes approximately twice as much power. For a hopping kangaroo, the graph of energy use has only a very small slope. In other words, the energy used per second changes very little with speed. Going faster requires very little additional power. Treadmill tests on kangaroos and observations in the wild have shown that they do not become winded at any speed at which they are able to hop. No matter how fast they hop, the necessary power is approximately the same.

61. ‖ A person runs 1 km. How does his speed affect the total energy needed to cover this distance?
   A. A faster speed requires less total energy.
   B. A faster speed requires more total energy.
   C. The total energy is about the same for a fast speed and a slow speed.

62. ‖ A kangaroo hops 1 km. How does its speed affect the total energy needed to cover this distance?
   A. A faster speed requires less total energy.
   B. A faster speed requires more total energy.
   C. The total energy is about the same for a fast speed and a slow speed.

63. ‖ At a speed of 4 m/s,
   A. A running human is more efficient than an equal-mass hopping kangaroo.
   B. A running human is less efficient than an equal-mass hopping kangaroo.
   C. A running human and an equal-mass hopping kangaroo have about the same efficiency.

64. ‖ At approximately what speed would a human use half the power of an equal-mass kangaroo moving at the same speed?
   A. 3 m/s   B. 4 m/s   C. 5 m/s   D. 6 m/s

65. ‖ At approximately what speed would a human use twice the power of a kangaroo of half the mass moving at the same speed?
   A. 3 m/s   B. 5 m/s   C. 7 m/s   D. 9 m/s

STOP TO THINK ANSWERS

Stop to Think 11.1: C. In each case, "what you get" is the same because the potential energy change is the same; "What you had to pay" is the same as well.

Stop to Think 11.2: C. As the body uses chemical energy from food, approximately 75% is transformed to thermal energy. Also, kinetic energy of motion of the legs and feet is transformed to thermal energy with each stride. Most of the chemical energy is transformed to thermal energy.

Stop to Think 11.3: A. The temperatures are the same, so all atoms have the same average kinetic energy. Hydrogen molecules, with the smallest mass, with thus have the highest typical speed.

Stop to Think 11.4: C. The radiator is at a higher temperature than the surrounding air. Thermal energy is transferred out of the system to the environment, so $Q < 0$.

Stop to Think 11.5: A, D. The efficiency is fixed by the ratio of T_H to T_C. Increasing this ratio increases efficiency; the heat engine will be more efficient with a hotter hot reservoir or a colder cold reservoir.

Stop to Think 11.6: A, D. The closer the temperatures of the hot and cold reservoirs, the more efficient the heat pump can be. (It is also true that having the two temperatures be closer will cause less thermal energy to "leak" out.) Any change that makes the two temperatures closer will allow the refrigerator to use less energy to run.

Stop to Think 11.7: B. In this case, kinetic energy is transformed to potential energy; there is no entropy change. In the other cases, energy is transformed to thermal energy, meaning entropy increases.

Conservation Laws

In Part II we have discovered that we don't need to know all the details of an interaction to relate the properties of a system "before" the interaction to those "after" the interaction. We also found two important quantities, momentum and energy, that are often conserved. Momentum and energy are characteristics of a system.

Momentum and energy have conditions under which they are conserved. The total momentum $\vec{P}$ and the total energy E are conserved for an *isolated system*. Of course, not all systems are isolated. For both momentum and energy, it was useful to develop a *model* of a system interacting with its environment. Interactions within the system do not change $\vec{P}$ or E. The kinetic, potential, and thermal energy *within* the system can be transformed without changing E. Interactions between the system and the environment *do* change the system's momentum and energy. In particular,

- Impulse is the transfer of momentum to or from the system: $\Delta \vec{p} = \vec{J}$.

- Work is the transfer of energy to or from the system in a mechanical interaction: $\Delta E = W$.
- Heat is the energy transferred to or from the system in a thermal interaction: $\Delta E = Q$.

The laws of conservation of momentum and energy, when coupled with the laws of Newtonian mechanics of Part I, form a powerful set of tools for analyzing motion. But energy is a concept that can be used for much more than the study of motion; it can be used to analyze how your body uses food, how the sun shines, and a host of other problems.

The study of the transformation of energy from one form to another reveals certain limits. Thermal energy is different from other forms of energy. A transformation of energy from another form to thermal energy is *irreversible*. The study of heat and thermal energy thus led us to the discipline of thermodynamics, a subject we will take up further in Part III as we look at properties of matter.

KNOWLEDGE STRUCTURE II Conservation Laws

BASIC GOALS	How is the system "after" an interaction related to the system "before"? What quantities are conserved, and under what conditions? Why are some energy changes more efficient than others?

GENERAL PRINCIPLES **Law of Conservation of Momentum** For an isolated system, $\vec{P}_f = \vec{P}_i$
Law of Conservation of Energy For an isolated system, there is no change in the system's energy:

$$\Delta K + \Delta U_g + \Delta U_s + \Delta E_{th} + \Delta E_{chem} + \cdots = 0$$

Energy can be exchanged with the environment as work or heat:

$$\Delta K + \Delta U_g + \Delta U_s + \Delta E_{th} + \Delta E_{chem} + \cdots = W + Q$$

Laws of Thermodynamics First law: If only thermal energy changes, $\Delta E_{th} = W + Q$
Second law: The entropy of an isolated system always increases.

BASIC PROBLEM-SOLVING STRATEGY Draw a visual overview for the system "before" and "after", then use the conservation of momentum or energy equations to relate the two. If necessary, calculate impulse and/or work.

Momentum and impulse

In a collision, the total momentum

$$\vec{P} = \vec{p}_1 + \vec{p}_2 = m_1\vec{v}_1 + m_2\vec{v}_2$$

before and after is equal.

Before: m_1 ①→ $(v_{1x})_i$ $(v_{2x})_i$ ←② m_2

After: ←① ② → $(v_{1x})_f$ $(v_{2x})_f$

A force can change the momentum of an object. The change is the **impulse:** $\Delta p_x = J_x$

$J_x =$ area under force curve

Basic model of energy

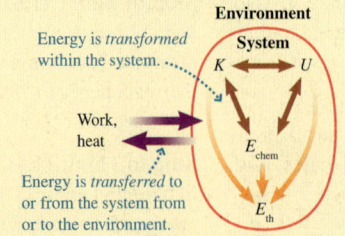

Energy is *transformed* within the system.

Work, heat

Energy is *transferred* to or from the system from or to the environment.

Work $W = F_{\parallel}d$ is done by the component of a force parallel to a displacement.

$F_{\parallel} = F\cos\theta$

Limitations on energy transfers and transformations

Thermal energy is random kinetic energy. Changing other forms of energy to thermal energy is **irreversible.**

When transforming energy from one form to another, some may be "lost" as thermal energy. This limits efficiency:

$$\text{Efficiency: } e = \frac{\text{what you get}}{\text{what you had to pay}}$$

A heat engine can convert thermal energy to other forms. The efficiency must be less than 100%.

Order out of Chaos

The second law of thermodynamics specifies that "the future" is the direction of entropy increase. But, as we have seen, this doesn't mean that systems must invariably become more random. You don't need to look far to find examples of systems that spontaneously evolve to a state of greater order.

A snowflake is a perfect example. As water freezes, the random motion of water molecules is transformed into the orderly arrangement of a crystal. The entropy of the snowflake is less than that of the water vapor from which it formed. Has the second law of thermodynamics been turned on its head?

The entropy of the water molecules in the snowflake certainly decreases, but the water doesn't freeze as an isolated system. For it to freeze, heat energy must be transferred from the water to the surrounding air. The entropy of the air increases by *more* than the entropy of water decreases. Thus the *total* entropy of the water + air system increases when a snowflake is formed, just as the second law predicts. If the system isn't isolated, its entropy can decrease without violating the second law as long as the entropy increases somewhere else.

Systems that become *more* ordered as time passes, and in which the entropy decreases, are called *self-organizing systems.* These systems can't be isolated. It is common in self-organizing systems to find a substantial flow of energy *through* the system. Your body takes in chemical energy from food, makes use of that energy, and then gives waste heat back to the environment. It is this energy flow that allows systems to develop a high degree of order and a very low entropy. The entropy of the environment undergoes a significant *increase* so as to let selected subsystems decrease their entropy and become more ordered.

Self-organizing systems don't violate the second law of thermodynamics, but this fact doesn't really explain their existence. If you toss a coin, no law of physics says that you can't get heads 100 times in a row—but you don't expect this to happen. Can we show that self-organization isn't just possible, but likely?

Let's look at a simple example. Suppose you heat a shallow dish of oil at the bottom, while holding the temperature of the top constant. When the temperature difference between the top and the bottom of the dish is small, heat is transferred from the bottom to the top by conduction. But convection begins when the temperature difference becomes large enough. The pattern of convection needn't be random, though; it can develop in a stable, highly ordered pattern, as we see in the figure. Convection is a much more efficient means of transferring energy than conduction, so the rate of transfer is *increased* as a result of the development of these ordered *convection cells.*

The development of the convection cells is an example of self-organization. The roughly 10^{23} molecules in the fluid had been moving randomly but now have begun behaving in a very orderly fashion. But there is more to the story. The convection cells transfer energy from the hot lower side of the dish to the cold upper side. This hot-to-cold energy transfer increases the entropy of the surrounding environment, as we have seen. In becoming more organized, the system has become more effective at transferring heat, resulting in a greater rate of entropy increase! Order has arisen out of disorder in the system, but the net result is a more rapid increase of the disorder of the universe.

Convection cells are thus a thermodynamically favorable form of order. We should expect this, because convection cells aren't confined to the laboratory. We see them in the sun, where they transfer energy from lower levels to the surface, and in the atmosphere of the earth, where they give rise to some of our most dramatic weather.

Self-organizing systems are a very active field of research in physical and biological sciences. The 1977 Nobel Prize in chemistry was awarded to the Belgian scientist Ilya Prigogine for his studies of *nonequilibrium thermodynamics,* the basic science underlying self-organizing systems. Prigogine and others have shown how energy flow through a system can, when the conditions are right, "bring order out of chaos." And this spontaneous ordering is not just possible—it can be probable. The existence and evolution of self-organizing systems, from thunderstorms to life on earth, might just be nature's preferred way of increasing entropy in the universe.

Convection cells in a shallow dish of oil heated from below (left) and in the sun (right). In both, warmer fluid is rising (lighter color) and cooler fluid is sinking (darker color).

PROPERTIES
OF MATTER

Beyond the Particle Model

The first 11 chapters of this book have made extensive use of the *particle model* in which we represent objects as point masses. The particle model is especially useful for describing how discrete objects move through space and how they interact with each other. Whether a ball is made of metal or wood is irrelevant to calculating its trajectory.

But there are many situations where the distinction between metal and wood is crucial. If you toss a metal ball and a wood ball into a pond, one sinks and the other floats. If you stir a pan on the stove with a metal spoon, it will quickly get too hot to hold unless it has a wooden handle.

Wood and metal have different physical properties. So do air and water. Our goal in Part III is to describe and understand the similarities and differences of different materials. To do so, we must go beyond the particle model and dig deeper into the nature of matter.

Macroscopic Physics

In Part III, we will be concerned with systems that are solids, liquids, or gases. Although it is sometimes useful to think of these systems as collections of particle-like atoms, properties such as pressure, temperature, specific heat, and viscosity are characteristics of the system as a whole, not of individual particles. Solids, liquids, and gases are often called *macroscopic* systems, the prefix *macro* (the opposite of *micro*) meaning "large."

The questions we will be asking are different from those we asked about particles. For example,

- How do the temperature and pressure of a system change if you heat it? Why do some materials respond quickly, others slowly?
- What are the mechanisms by which a system exchanges heat energy with its environment? Why does blowing on a cup of hot coffee cause it to cool off?
- Why are there three phases of matter—solids, liquids, gases? What happens during a phase change?
- Why do some objects float while others, with the same mass, sink? What keeps a massive steel ship afloat?
- What are the laws of motion of a flowing liquid? How do they differ from the laws governing the motion of a particle?

Both Newton's laws and the law of conservation of energy will remain important tools—they are, after all, the basic laws of physics—but we'll have to learn how they apply to macroscopic systems.

It should come as no surprise that an understanding of macroscopic systems and their properties is essential for understanding the world around us. Biological systems, from cells to ecosystems, are macroscopic systems exchanging energy with their environment. On a larger scale, energy transport on earth gives us weather, and the special physical properties of water help explain why the earth is hospitable for life.

Individual bees do not have the ability to regulate their body temperature. But a colony of bees, working together, can very precisely regulate the temperature of their hive. How can the bees use the heat generated by their muscles, the structure of the hive, and the evaporation of water to achieve this control? In Part III, we will learn about the flows of matter and energy that drive such processes.

12

THERMAL PROPERTIES OF MATTER

This image of an elephant shows its surface temperatures. Red represents warm patches (such as the trunk) and blue cool patches (such as the ears). The image shows energy that the elephant *radiates*. Why does an elephant (or any other warm object) radiate energy?

Looking Ahead ▶▶

The goal of Chapter 12 is to use the atomic model of matter to explain and explore many macroscopic phenomena associated with heat, temperature, and the properties of matter. In this chapter, you will learn to:

▶ Use the atomic model of matter to explain the thermal expansion of solids and liquids, changes of phase, pressure in gases, and the transfer of heat.

▶ Apply the ideal-gas law to gases undergoing changes in pressure, volume, and temperature.

▶ Solve problems involving heat transfer between systems.

▶ Understand the heat-transfer processes of conduction, convection, and radiation.

Looking Back ◀◀

This chapter will make extensive use of the concepts of energy and heat from Chapters 10 and 11. Please review:

◀ Section 10.7 The nature of thermal energy.

◀ Section 11.4 The connection between thermal energy and temperature, the nature of thermal energy and temperature for ideal gases.

◀ Section 11.5 Heat and the first law of thermodynamics.

Elephants, humans, and all other animals convert the chemical energy of food to thermal energy. This is the process of metabolism. But if an animal had no means of getting rid of excess thermal energy, its temperature would soon rise to dangerous levels.

Chapter 11 introduced the concepts of thermal energy, temperature, and heat, but many unanswered questions remain. For example:

■ How is heat transferred to or from a system? How can an elephant keep cool even in the hot African savanna?

■ What are the consequences of heating or cooling a system? Changing the temperature is one obvious possibility, but are there others? And what is the connection between the amount of heat transferred to a system and the amount by which its temperature changes?

■ How do the properties of matter depend on temperature? Whether a system is a solid or liquid depends on its temperature. So does the pressure of a gas.

This chapter will cover a very wide range of topics as it addresses such questions, but there is a central organizing theme that will help us see the connections among the diverse phenomena. Namely, we will use the *atomic model of matter* throughout the chapter to predict, explain, and understand the fascinating interplay of matter and heat in our world.

12.1 The Atomic Model of Matter

We have, throughout this book, used a model in which we represent matter as being made of particle-like atoms. This atomic model has been invaluable in explaining friction, elastic forces, and the nature of thermal energy. We now want to use this atomic model to understand the thermal properties of matter.

Each element and most compounds can exist as a solid, liquid, or gas. These three **phases** of matter are familiar from everyday experience. An atomic view of the three phases is shown in Figure 12.1.

- A **gas** is a system in which each particle moves freely through space until, on occasion, it collides with another particle or the wall of its container.
- In a **liquid,** weak bonds permit motion while keeping the particles close together.
- A rigid **solid** has a definite shape, and can be compressed or deformed only slightly, as we saw in Chapter 8. It consists of atoms connected by spring-like molecular bonds.

In a gas, the particles are in random motion and interact only through elastic collisions.

In a liquid, the particles have weak bonds that keep them close together. The particles can slide around each other, so the liquid can flow.

In a rigid solid, the particles are connected by relatively stiff spring-like bonds.

Our atomic model makes some simplifications that are worth noting. The basic particles of the gas in Figure 12.1 are drawn as simple spheres; no mention is made of the nature of the particles. The balloon might contain either helium (in which the basic particles are helium atoms) or air (in which the basic particles are nitrogen and oxygen molecules). A helium atom and a nitrogen molecule are quite different from each other, but many of the properties of the gas as a whole do not depend on the nature of the particles—a gas of helium atoms or oxygen molecules may behave identically. In such cases, we will simply refer to gas *particles,* which may be either atoms or molecules.

The basic particles in the liquid water in Figure 12.1 are water molecules. We ignore the structure of the molecules, so when we speak of the bonds that hold the particles in the liquid state, we are referring to the relatively weak intermolecular bonds among the water molecules, not the strong bonds between the hydrogen and oxygen atoms that form the molecules. The basic particles of the gold bars in Figure 12.1 are gold atoms. The bonds that hold these atoms together are the bonds that give the solid its structure. But there are solids that are composed of molecules that are held together by bonds between the molecules; water ice is one such example. In this case, the particles are the molecules, and the bonds that form the solid are the bonds between the molecules.

FIGURE 12.1 Atomic models of the three phases of matter: solid, liquid, and gas.

Atomic Mass and Atomic Mass Number

Before we see how the atomic model explains the thermal properties of matter, we need to remind you of some "atomic accounting." Recall that atoms of different elements have different masses. The mass of an atom is determined primarily by its most massive constituents, the protons and neutrons in its nucleus. The *sum* of the number of protons and neutrons is called the **atomic mass number** A:

$$A = \text{number of protons} + \text{number of neutrons}$$

TABLE 12.1 Some atomic mass numbers

Element	Symbol	A
Hydrogen	^{1}H	1
Helium	^{4}He	4
Carbon	^{12}C	12
Nitrogen	^{14}N	14
Oxygen	^{16}O	16
Neon	^{20}Ne	20
Aluminum	^{27}Al	27
Argon	^{40}Ar	40
Lead	^{207}Pb	207

A, which by definition is an integer, is written as a leading superscript on the atomic symbol. For example, the primary isotope of carbon, with six protons (which makes it carbon) and six neutrons, has $A = 12$ and is written ^{12}C. The radioactive isotope ^{14}C, used for carbon dating of archeological finds, contains six protons and eight neutrons.

The **atomic mass** scale is established by defining the mass of ^{12}C to be exactly 12 u, where u is the symbol for the *atomic mass unit*. That is, $m(^{12}\text{C}) = 12$ u. In kg, the atomic mass unit is

$$1\text{ u} = 1.66 \times 10^{-27}\text{ kg}$$

Atomic masses are all very nearly equal to the integer atomic mass number A. For example, the mass of ^{1}H, with $A = 1$, is $m = 1.0078$ u. For our present purposes, it will be sufficient to use the integer atomic mass numbers as the values of the atomic mass. That is, we'll use $m(^1\text{H}) = 1$ u, $m(^4\text{He}) = 4$ u, and $m(^{16}\text{O}) = 16$ u. For molecules, the **molecular mass** is the sum of the atomic masses of the atoms forming the molecule. Thus the molecular mass of the diatomic molecule O_2, the constituent of oxygen gas, is $m(O_2) = 2m(^{16}\text{O}) = 32$ u.

> **NOTE** ▶ An element's atomic mass number is *not* the same as its atomic number. The *atomic number*, which gives the element's position in the periodic table, is the number of protons. ◀

Table 12.1 shows the atomic mass numbers of some of the elements that we'll use for examples and homework problems. A complete periodic table, including atomic masses, is found in Appendix B.

The Definition of the Mole

TABLE 12.2 Monatomic and diatomic gases

Monatomic	Diatomic
Helium (He)	Hydrogen (H_2)
Neon (Ne)	Nitrogen (N_2)
Argon (Ar)	Oxygen (O_2)

One way to specify the amount of substance in a system is to give its mass. Another way, one connected to the number of atoms, is to measure the amount of substance in *moles*. By definition, one **mole** of matter is the amount of substance containing as many basic particles as there are atoms in 12 g of ^{12}C. Many decades of ingenious experiments have determined that there are 6.02×10^{23} atoms in 12 g of ^{12}C, so we can say that **1 mole of substance, abbreviated 1 mol, is 6.02×10^{23} basic particles.**

The basic particle depends on the substance, as noted in the previous section. Helium is a **monatomic gas,** meaning that the basic particle is the helium atom. Thus 6.02×10^{23} helium atoms are 1 mol of helium. But oxygen gas is a **diatomic gas** because the basic particle is the two-atom diatomic molecule O_2. 1 mol of oxygen gas contains 6.02×10^{23} *molecules* of O_2 and thus $2 \times 6.02 \times 10^{23}$ oxygen atoms. Table 12.2 lists the monatomic and diatomic gases that we will use for examples and problems.

The number of basic particles per mole of substance is called **Avogadro's number,** N_A. The value of Avogadro's number is thus

$$N_A = 6.02 \times 10^{23}\text{ mol}^{-1}$$

The number n of moles in a substance containing N basic particles is

$$n = \frac{N}{N_A} \tag{12.1}$$

Moles of a substance in terms of the number of basic particles

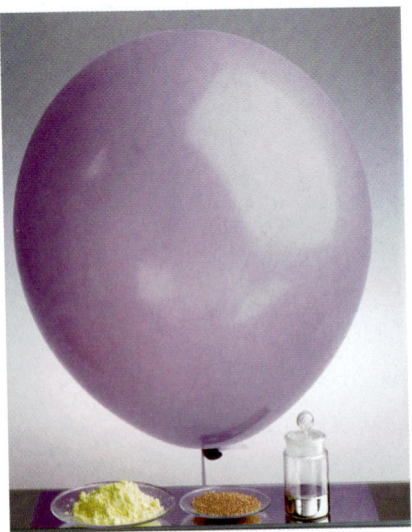

One mole of helium, sulfur, copper, and mercury.

The **molar mass** of a substance M_{mol} is the mass *in grams* of 1 mol of substance. By definition, the molar mass of ^{12}C is $M_{mol}(^{12}\text{C}) = 12$ g/mol. For other substances, the numerical value of the molar mass equals the numerical value of the atomic or molecular mass. That is, the molar mass of He, with $m = 4$ u, is $M_{mol}(\text{He}) = 4$ g/mol and the molar mass of diatomic O_2 is $M_{mol}(O_2) = 32$ g/mol.

You can use the molar mass to determine the number of moles. In one of the few instances where the proper units are *grams* rather than kilograms, the number of moles contained in a system of mass M consisting of atoms or molecules with molar mass M_{mol} is

$$n = \frac{M \text{ (in grams)}}{M_{mol}} \qquad (12.2)$$

Moles of a substance in terms of its mass

EXAMPLE 12.1 Determining quantities of oxygen

A system contains 100 g of oxygen. How many moles does it contain? How many molecules?

SOLVE The diatomic oxygen molecule O_2 has a mass $M_{mol} = 32$ g/mol. From Equation 12.2,

$$n = \frac{100 \text{ g}}{32 \text{ g/mol}} = 3.1 \text{ mol}$$

Each mole contains N_A molecules, so the total number is $N = nN_A = 1.9 \times 10^{24}$ molecules.

Volume

An important property that characterizes a macroscopic system is its volume V, the amount of space the system occupies. The SI unit of volume is m^3. Nonetheless, both cm^3 and, to some extent, liters (L) are widely used metric units of volume. In most cases, you *must* convert these to m^3 before doing calculations.

While it is true that 1 m = 100 cm, it is *not* true that $1 \text{ m}^3 = 100 \text{ cm}^3$. Figure 12.2 shows that the volume conversion factor is $1 \text{ m}^3 = 10^6 \text{ cm}^3$. A liter is 1000 cm³, so $1 \text{ m}^3 = 10^3$ L. A milliliter (1 mL) is the same as 1 cm³.

STOP TO THINK 12.1   Which system contains more atoms: 5 mol of helium $(A = 4)$ or 1 mol of neon $(A = 20)$?

A. Helium.   B. Neon.   C. They have the same number of atoms.

12.2 Thermal Expansion

In our introduction to temperature in Chapter 11, we noted that temperature can be measured with a thermometer. The accompanying figure showed an alcohol thermometer, a glass tube with a bulb of alcohol at the bottom. As the temperature of the alcohol increases, so does its volume; this increase in volume—an example of **thermal expansion**—is responsible for the liquid's motion up the column.

Our atomic model gives a very good way to visualize thermal expansion. Suppose a cube of material, as shown in Figure 12.3a, is initially at temperature T_i. The edge of the cube has length L_i, and the cube's volume is V_i. We then heat the cube, increasing its thermal energy. This increases its temperature to T_f, as shown in Figure 12.3b. The atoms now jiggle around faster, so the bonds are "stretched." This increases the average distance between atoms, and so the cube's volume increases to V_f.

Subdivide the 1 m × 1 m × 1 m cube into little cubes 1 cm on a side. You will get 100 subdivisions along each edge.

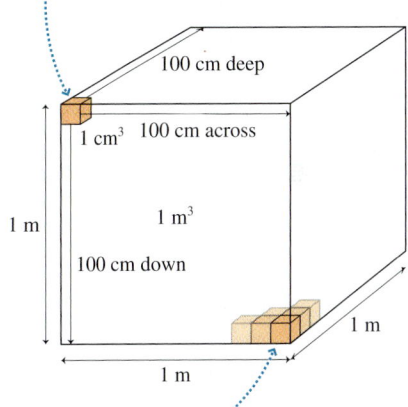

There are 100 × 100 × 100 = 10^6 little 1 cm³ cubes in the big 1 m³ cube.

FIGURE 12.2 There are 10^6 cm³ in 1 m³.

(a)   Temperature T_i

(b)   Temperature T_f

If heat is transferred to an object, its thermal energy increases. The atoms and molecules jiggle around at higher speeds and move farther apart.

FIGURE 12.3 Increasing the temperature of a solid causes it to expand.

Expanding spans A long steel bridge will slightly increase in length on a hot day and decrease on a cold day. Thermal expansion joints let the bridge's length change without causing the roadway to buckle.

For most substances, the change in volume $\Delta V = V_f - V_i$ is linearly related to the change in temperature $\Delta T = T_f - T_i$ by

$$\Delta V = \beta V_i \, \Delta T \qquad (12.3)$$

Volume thermal expansion

LINEAR
p.38

The constant β is known as the **coefficient of volume expansion.** Its value depends on the nature of the bonds in a material; it can vary quite a bit from one substance to another. Because ΔT is measured in K, the units of β are K^{-1}.

As a solid's volume increases, each of its linear dimensions increases as well. We can write a similar expression for this linear thermal expansion. If an object of initial length L_i undergoes a temperature change ΔT, its length changes to L_f. The change in length, $\Delta L = L_f - L_i$, is given by

$$\Delta L = \alpha L_i \, \Delta T \qquad (12.4)$$

Linear thermal expansion

The constant α is the **coefficient of linear expansion.** Note that Equations 12.3 and 12.4 apply equally well to thermal *contractions,* in which case both ΔT and ΔV (or ΔL) are negative.

NOTE ▶ The expressions for thermal expansion are only approximate expressions that apply over a limited range of temperatures. They are what we call **empirical formulas;** they are a good fit to measured data, but they do not represent any underlying fundamental law. There are materials that do *not* closely follow Equations 12.3 or 12.4; water at low temperatures is one example, as we will see. ◀

Values of α and β for some common materials are listed in Table 12.3 The volume expansion of a liquid can be measured but, because a liquid can change shape, we don't assign a coefficient of linear expansion to a liquid. While α and β do vary slightly with temperature, the room-temperature values in Table 12.3 are quite adequate for all problems in this book.

TABLE 12.3 Coefficients of linear and volume thermal expansion at 20°C

Substance	Linear α (K^{-1})	Volume β (K^{-1})
Aluminum	23×10^{-6}	69×10^{-6}
Glass	9×10^{-6}	27×10^{-6}
Iron or steel	12×10^{-6}	36×10^{-6}
Concrete	12×10^{-6}	36×10^{-6}
Ethyl alcohol		1100×10^{-6}
Water		210×10^{-6}
Air (and other gases)		3400×10^{-6}

EXAMPLE 12.2 How much closer to space?

The height of the Space Needle, a steel observation tower in Seattle, is 180 meters on a 0°C winter day. How much taller is it on a hot summer day when the temperature is 30°C?

PREPARE The increase in temperature is

$$\Delta T = T_f - T_i = 30°C - 0°C = 30°C = 30 \text{ K}$$

NOTE ▶ In practice, ΔT is usually measured in °C. But the Kelvin and the Celsius temperature scales have the same step size, so ΔT in K has exactly the same numerical value as ΔT in °C. Thus

- You do not need to convert temperatures from °C to K if you only need a temperature *change* ΔT.
- You do need to convert anytime you need the actual temperature T. ◀

SOLVE The coefficient of linear expansion is given in Table 12.3; we can use this value in Equation 12.4 to compute the increase in height:

$$\Delta L = \alpha L_i \, \Delta T = (12 \times 10^{-6} \text{ K}^{-1})(180 \text{ m})(30 \text{ K}) = 0.065 \text{ m}$$

ASSESS Compared to 180 m, an expansion of 6.5 cm is small—not something you would easily notice—but not negligible. The thermal expansion of structural elements in towers and bridges must be accounted for in the design to avoid damaging stresses. When designers failed to properly account for thermal stresses in marble panels cladding the Amoco Building in Chicago, all 43,000 panels had to be replaced, at great cost.

CONCEPTUAL EXAMPLE 12.1 What happens to the hole?

A metal plate has a circular hole in it. As the plate is heated, does the hole get larger or smaller?

REASON As the plate expands, you might think the hole would shrink (with the metal expanding into the hole). But suppose we took a metal plate and simply drew a circle where a hole could be cut. On heating, the plate and the marked area both expand, as we see in Figure 12.4. We could cut a hole on the marked line before or after heating; the size of the hole would be larger in the latter case. Therefore, the size of the hole must expand as the plate expands.

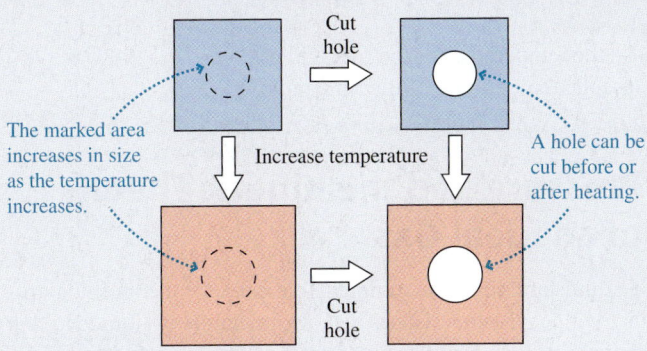

FIGURE 12.4 The thermal expansion of a hole.

ASSESS This somewhat counterintuitive result has many practical applications, as you can see from the "Try It Yourself" exercise on loosening a jar lid.

Special Properties of Water and Ice

Water, the most important molecule for life, differs from other liquids in many important ways. If you cool a sample of water toward its freezing point of 0°C, you would expect its volume to decrease. Figure 12.5a shows a graph of the volume of a mole of water versus temperature. As the water is cooled, you can see that the volume indeed decreases—to a point. Once the water is cooled below 4°C, further cooling results in an *increase* in volume.

FIGURE 12.5 Variation of the volume of one mole of water with temperature.

This anomalous behavior of water is a result of the charge distribution on the water molecules, which determines how nearby molecules interact with each other. As water approaches the freezing point, molecules begin to form clusters that are more strongly bound. Water molecules must actually get a bit farther apart to form such clusters, meaning the volume increases. Freezing the water results in an even greater increase in volume, as illustrated in Figure 12.5b.

TRY IT YOURSELF

Thermal expansion to the rescue If you have a stubborn lid on a glass jar, try this: Put the lid under very hot water for a short time. Heating the lid and the jar makes them both expand, but, as you can see from the data in Table 12.3, the steel lid—and the opening in the lid that fits over the glass jar—expands by more than the glass jar. The jar lid is now looser, and can be more easily removed.

This expansion on freezing has important consequences. In most materials, the solid phase is denser than the liquid phase and so the solid material sinks. Because water *expands* as it freezes, ice is less dense than liquid water and floats. Not only do ice cubes float in your cold drink, a lake freezes by forming ice on the top rather than at the bottom. This layer of ice then insulates the water below from much colder air above. Thus most lakes do not freeze solid, allowing aquatic life to survive even the harshest winters.

STOP TO THINK 12.2   An aluminum ring is tight around a solid iron rod. If we wish to loosen the ring to remove it from the rod, we should

A. Increase the temperature of the ring and rod.
B. Decrease the temperature of the ring and rod.

12.3 Pressure and the Kinetic Theory of an Ideal Gas

Solids and liquids are nearly incompressible because the atomic particles are in close contact with each other. Gases differ from solids or liquids in that gases are highly compressible. That is, a gas can expand or contract not only in response to temperature but also to squeezing the particles closer together or pulling them farther apart. In Chapter 11, we introduced an atomic-level model of an *ideal gas*, reviewed in Figure 12.6. Our goal in this section is to further develop this model of an ideal gas. The explanation of properties of the gas as a whole in terms of randomly moving particles is called **kinetic theory.**

Temperature

We had our first taste of kinetic theory in Chapter 11, when we expressed the temperature of an ideal gas in terms of the motion of individual particles of the gas. To see that the particles are moving with different speeds, Figure 12.7 shows data from an experiment to measure the molecular speeds in nitrogen gas at 20°C. The results are presented as a *histogram,* a bar chart in which the height of the bar indicates what percentage of the molecules have a speed in the range of speeds shown below the bar. For example, 16% of the molecules have speeds in the range from 600 m/s to 700 m/s. The most probable speed, as judged from the tallest bar, is ≈550 m/s. This is really fast, ≈1200 mph!

Even though different molecules have different speeds, we can define an average speed. In Chapter 11, we defined the root-mean-square speed, or rms speed, of ideal gas particles to be

$$v_{rms} = \sqrt{(v^2)_{avg}}$$

where $(v^2)_{avg}$ is the average value of the squares of all the molecular velocities.

In terms of the rms speed, the average kinetic energy of the gas particles is

$$K_{avg} = \tfrac{1}{2}m(v^2)_{avg} = \tfrac{1}{2}mv_{rms}^2 \qquad (12.5)$$

An important result from Chapter 11 was that the average kinetic energy of the gas particles is directly proportional to the absolute temperature T:

$$K_{avg} = \tfrac{3}{2}k_B T \qquad (12.6)$$

where $k_B = 1.38 \times 10^{-23}$ J/K is Boltzmann's constant. Indeed, this equation is really the definition of temperature—temperature measures the average kinetic energy of the particles in a system. So the important concept of temperature is intimately connected to the random motions of the particles in the gas.

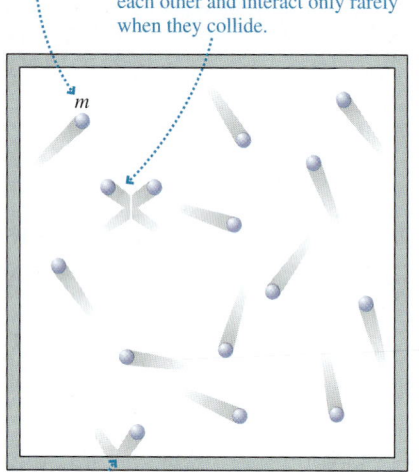

1. The gas is made of a large number N of particles of mass m, each moving randomly.

2. The particles are quite far from each other and interact only rarely when they collide.

3. The collisions of the particles with each other (and with walls of the container) are elastic; no energy is lost in these collisions.

FIGURE 12.6 The ideal-gas model.

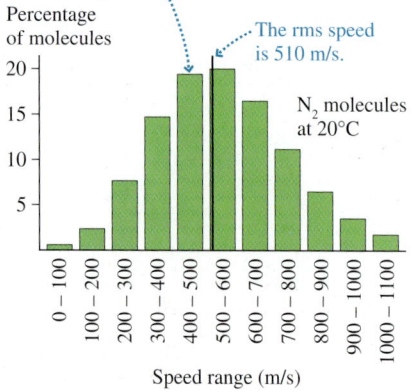

The height of a bar represents the percent of molecules with speed in the range on the horizontal axis.

The rms speed is 510 m/s.

N₂ molecules at 20°C

Speed range (m/s)

FIGURE 12.7 The distribution of molecular speeds in nitrogen gas at 20°C.

Pressure

Everyone has some sense of the concept of *pressure*. If you get a hole in your bicycle tire, the higher-pressure air inside comes squirting out. It's hard to get the lid off a vacuum-sealed jar because of the low pressure inside. But just what is pressure?

Let's take a kinetic theory view of pressure, defining it in terms of the motion of particles of a gas. Suppose we have a sample of gas in a container with rigid walls. As the particles in a gas move around, they sometimes collide with and bounce off the walls. As Figure 12.8 shows, each collision exerts a force on the wall. The force due to any one collision is small, but there are an exceedingly large number of collisions every second. The net force on a surface can be quite substantial.

The force on a surface is proportional to the area. Doubling the surface area will double the number of collisions and thus double the force. Rather than the force itself, a more useful quantity is the force-to-area ratio F/A. This ratio is a property of the gas itself, independent of the surface, and it's what we call the gas **pressure:**

$$p = \frac{F}{A} \qquad (12.7)$$

Definition of pressure in a gas as the force-to-area ratio

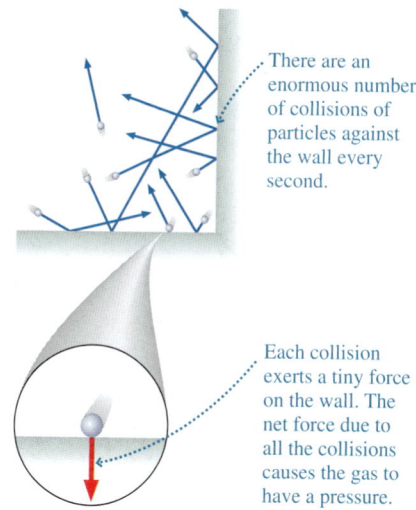

There are an enormous number of collisions of particles against the wall every second.

Each collision exerts a tiny force on the wall. The net force due to all the collisions causes the gas to have a pressure.

FIGURE 12.8 The pressure in a gas is due to the net force of the particles colliding with the walls.

Notice that pressure is a scalar, not a vector. You can see from Equation 12.7 that a gas exerts a force of magnitude

$$F = pA \qquad (12.8)$$

on a surface of area A. The force is *perpendicular* to the surface.

NOTE ▶ Pressure itself is *not* a force, even though we sometimes talk informally about "the force exerted by the pressure." The correct statement is that the *gas* exerts a force on a surface. ◀

From its definition, you can see that pressure has units of N/m². The SI unit of pressure is the **pascal,** defined as

$$1 \text{ pascal} = 1 \text{ Pa} = 1 \frac{N}{m^2}$$

This unit is named for the 17th-century French scientist Blaise Pascal, who was one of the first to study gases. A pascal is a very small pressure, so we usually see pressures in kilopascals, where 1 kPa = 1000 Pa.

The total force on the surface of your body due to the pressure of the atmosphere is over 40,000 pounds. Why doesn't this enormous force crush you? To answer this question, let's look at a surface with pressure on both sides. Figure 12.9a shows a situation where the pressure is the same on both sides of a surface. The forces pushing on both sides may be quite large, but they exactly balance and there's no net force. The pressure of the atmosphere is pushing in on your skin, but it is balanced by an equal pressure inside your body pushing out.

Net forces are exerted only where there is a pressure *difference* Δp, as in Figure 12.9b. The net force on a surface of area A is

$$F_{net} = F_2 - F_1 = p_2 A - p_1 A = A(p_2 - p_1) = A \, \Delta p$$

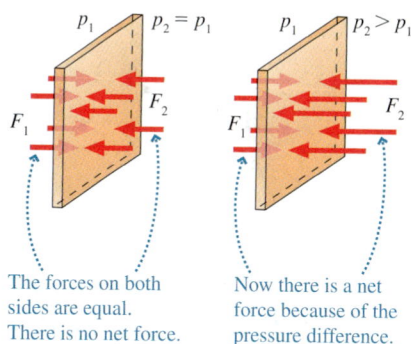

(a) Equal pressures   **(b)** Unequal pressures

p_1 $p_2 = p_1$ p_1 $p_2 > p_1$

F_1 F_2 F_1 F_2

The forces on both sides are equal. There is no net force.

Now there is a net force because of the pressure difference.

FIGURE 12.9 The net force depends on the pressure difference.

▶ Too little pressure too fast **BIO** This photo shows a rockfish, a popular game fish that can be caught at depths up to 1000 ft. The great pressure at these depths is balanced by an equally large pressure inside the fish's swim bladder, a gas-filled organ that we will learn more about in Chapter 13. If the fish is hooked and rapidly raised to the lower pressure of the surface, the pressure inside the swim bladder is suddenly much larger than the pressure outside, causing the swim bladder to expand dramatically, damaging nearby organs. When reeled in from a great depth, these fish seldom survive.

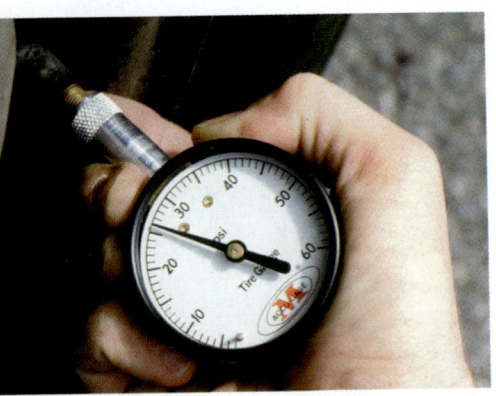

A tire-pressure gauge reads the gauge pressure p_g, not the absolute pressure p. Zero gauge pressure means the pressure inside the tire is the same as the pressure outside the tire—and the tire is flat.

This is the force that holds the lid on a vacuum-sealed jar, where the pressure inside is less than the pressure outside. To remove the lid, you have to exert a force larger than the net force due to the pressure difference.

Decreasing the number of molecules in a container decreases the pressure because there are fewer collisions with the walls. The pressure in a completely empty container would be $p = 0$ Pa. This is called a *perfect vacuum*. A perfect vacuum cannot be achieved because it's impossible to remove every molecule from a region of space. In practice, a **vacuum** is an enclosed space in which $p \ll 1$ atm. Using $p = 0$ Pa is then a good approximation.

Measuring Pressure

The most important pressure for our daily lives is the pressure of the atmosphere. The pressure of the atmosphere varies with altitude and the weather, but the global average pressure at sea level, called the *standard atmosphere,* is

$$1 \text{ standard atmosphere} = 1 \text{ atm} = 101,300 \text{ Pa} = 101.3 \text{ kPa}$$

Just as we measured acceleration in units of g, we will often measure pressure in units of atm.

In the United States, where force is measured in pounds and area in square inches, pressure is often pounds per square inch, or psi. When you measure pressure in the tires on your car or bike, you probably use a gauge that reads in psi. The conversion factor is

$$1 \text{ atm} = 14.7 \text{ psi}$$

Because the effects of pressure depend on pressure differences, many common gauges measure not the actual or *absolute pressure p* but what is called the **gauge pressure.** The gauge pressure, denoted p_g, is the pressure *in excess* of 1 atm. That is,

$$p_g = p - 1 \text{ atm}$$

You need to add 1 atm $= 101.3$ kPa to the reading of a pressure gauge to find the absolute pressure p you need for doing most calculations in this chapter.

EXAMPLE 12.3 Finding the force due to a pressure difference

Patients suffering from decompression sickness and other conditions may be treated in a hyperbaric oxygen chamber filled with oxygen at greater than atmospheric pressure. A cylindrical chamber with flat end plates of diameter 0.75 m is filled with oxygen to a gauge pressure of 27 kPa. What is the net force on the end plate of the cylinder?

PREPARE There is a net force on the end plate because of the pressure *difference* between the inside and outside. 27 kPa is the pressure in excess of 1 atm. If we assume the pressure out-side is 1 atm, then 27 kPa is Δp, the pressure difference across the surface.

SOLVE The end plate has area $A = \pi(0.75 \text{ m}/2)^2 = 0.442 \text{ m}^2$. The pressure difference causes a net force

$$F_{net} = A\,\Delta p = (0.442 \text{ m}^2)(27,000 \text{ Pa}) = 12 \text{ kN}$$

ASSESS This is a large force—more than the weight of one ton. Even modest pressure differences can cause substantial forces on a large enough area. It's remarkable to think that this large force is due to the collisions of individual molecules with the ends of the cylinder.

From Collisions to Pressure and the Ideal-Gas Law

We can use the fact that the pressure in a gas is due to the collision of particles with the walls to make some qualitative predictions. Figure 12.10 presents a few such predictions.

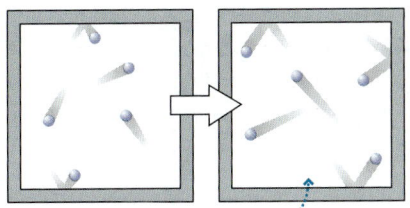

Increasing the temperature of the gas means the particles move at higher speeds. They hit the walls more often and with more force, so there is more pressure.

■ Increasing the temperature leads to an increase in pressure.

Decreasing the volume of the container means more frequent collisions with the walls of the container, and thus more pressure.

■ Decreasing the volume leads to an increase in pressure.

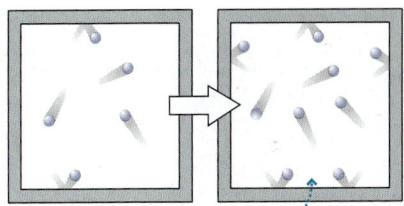

Increasing the number of particles in the container means more frequent collisions with the walls of the container, and thus more pressure.

■ Increasing the number of particles leads to an increase in pressure.

FIGURE 12.10 A kinetic theory model of pressure changes.

We see that temperature, volume, and the number of particles all affect the gas pressure in a container. To analyze pressure, let's start by considering a single collision of a particle with a wall. Figure 12.11 shows a particle moving parallel to one side of the box as it collides with a wall. Collisions in an ideal gas are perfectly elastic, so no energy is lost in the collision. Thus the particle rebounds from the wall with its velocity changed from v to $-v$.

We looked at collisions in Chapter 9. Recall that the average force exerted during a collision of duration Δt is related to the particle's change in momentum by

$$(F_{avg})_x \, \Delta t = \Delta p_x = m(v_x)_f - m(v_x)_i$$

With $(v_x)_i = v$ and $(v_x)_f = -v$ the average force on the particle is

$$(F_{molecule})_{avg\,x} = \frac{m(v_x)_f - m(v_x)_i}{\Delta t} = -m\frac{2v}{\Delta t}$$

The force on the particle is negative because it points to the left. According to Newton's third law, the wall experiences the equal but opposite force

$$(F_{wall})_{avg\,x} = +m\frac{2v}{\Delta t} \qquad (12.9)$$

as a result of this single collision. Not surprisingly, faster particles exert larger forces when they rebound from the wall.

It is the sum of many such collisions that leads to the macroscopic phenomenon of pressure. To find the pressure, we need to multiply the force due to one collision by the total number N_{coll} of collisions that occur during the time interval Δt. If there are N particles moving with speed v in a box of volume V, it can be shown that the number of collisions is

$$N_{coll} = \frac{1}{2}\frac{N}{V}Av\,\Delta t$$

Although we've not proved this result, it makes sense that increasing the number of particles N or increasing the size of the wall A will increase the number of collisions, whereas increasing the volume of the box (which spreads the particles out) will decrease the number of collisions.

The net force on the wall is found by combining this expression for N_{coll} with Equation 12.9. Δt cancels, giving

$$F_{net} = N_{coll}(F_{wall})_{avg\,x} = \left(\frac{1}{2}\frac{N}{V}Av\Delta t\right)\left(\frac{2mv}{\Delta t}\right) = \frac{N}{V}mv^2A$$

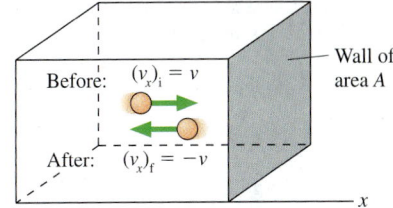

FIGURE 12.11 A particle colliding with the wall exerts a force on it.

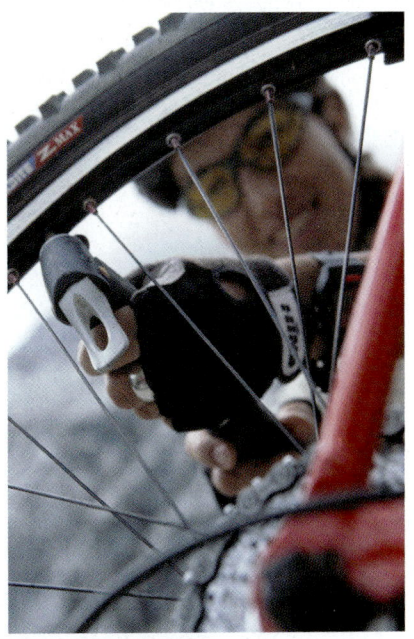

Pumping up a bicycle tire adds more particles to the fixed volume of the tire, increasing the pressure.

This result is based on the assumption that all particles have the same speed and are moving parallel to the x-axis. We can relax this assumption if we replace v by v_{rms} and, because there are three dimensions, divide by 3. Doing so, we find that the force exerted on the wall by a vast number of collisions is

$$F_{net} = \frac{1}{3}\frac{N}{V}mv_{rms}^2 A$$

Dividing F_{net} by A, we finally have an expression for the pressure in the gas:

$$p = \frac{F_{net}}{A} = \frac{1}{3}\frac{N}{V}mv_{rms}^2 = \frac{2}{3}\frac{N}{V}\left(\frac{1}{2}mv_{rms}^2\right)$$

But $\frac{1}{2}mv_{avg}^2$ is the average kinetic energy per particle in the gas, so we're left with

$$p = \frac{2}{3}\frac{N}{V}K_{avg} \tag{12.10}$$

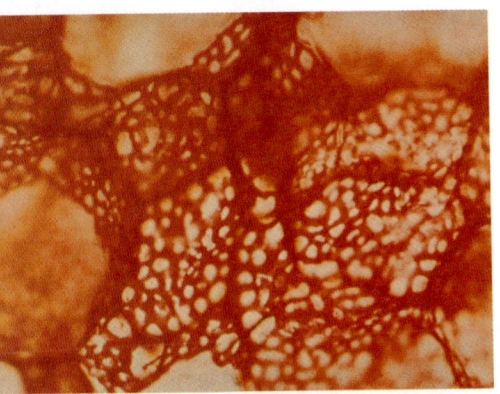

Gas exchange in the lungs BIO When you draw a breath, how does the oxygen get into your bloodstream? Kinetic theory provides the answer. The above picture is a highly magnified view of the alveoli, air sacs in the lungs, that are surrounded by capillaries, fine blood vessels. The thin membranes of the alveoli and the capillaries are permeable—small molecules can move across them. If a permeable membrane separates two regions of space having different concentrations of a molecule, the rapid motion of the molecules causes a net transport of molecules in the direction of lower concentration. This transport—entirely due to the motion of the molecules—is known as **diffusion**. In the lungs, the higher concentration of oxygen in the alveoli drives the diffusion of oxygen into the blood in the capillaries. At the same time, carbon dioxide diffuses from the blood into the alveoli. Diffusion is a rapid and effective means of transport in this case because the membranes are thin. Diffusion won't work to get oxygen from the lungs to other parts of the body, because diffusion is slow over large distances, so the oxygenated blood must be pumped throughout the body.

The pressure of an ideal gas depends on just two things: the number of particles per cubic meter (N/V) and the average kinetic energy of the particles. Surprisingly, it does not depend on the nature of the particles.

As remarkable as this result is, we can take it one more step and "discover" a law you have likely seen in a previous course. If we replace K_{avg} with the temperature expression of Equation 12.6 and take the V to the other side of the equation, we end up with

$$pV = Nk_B T \tag{12.11}$$

Ideal-gas law, version 1

Remarkably, we've derived the **ideal-gas law:** The pressure of a gas multiplied by its volume is equal to the number of gas particles multiplied by Boltzmann's constant multiplied by the temperature. The derivation was longer than we are accustomed to in this text, but it was worthwhile in this particular case. It is not only the final result that is important, but also the process behind it: **By considering the forces of individual particles colliding with the wall of a container (a microscopic picture), we have derived an equation that relates the pressure, volume, and temperature of a gas (a macroscopic picture).** The ideal-gas law is not simply an empirical finding but is a direct consequence of the atomic model of matter.

Equation 12.11 is written in terms of the number N of particles in the gas, whereas the ideal-gas law is stated in chemistry in terms of the number n of moles. But the change is easy to make. The number of particles is $N = nN_A$, so we can rewrite Equation 12.11 as

$$pV = nN_A k_B T = nRT \tag{12.12}$$

Ideal-gas law, version 2

where $R = N_A k_B = 8.31$ J/mol · K is the *gas constant*. The units may seem a little unusual, but the product of Pa and m^3, the units of pV, is equivalent to J.

You may have learned to work gas problems using units of atmospheres and liters. To do so, you had a different numerical value of R that was expressed in those units. In physics we always work gas problem in SI units. Let's review the meaning and the units of the various quantities in the ideal-gas law:

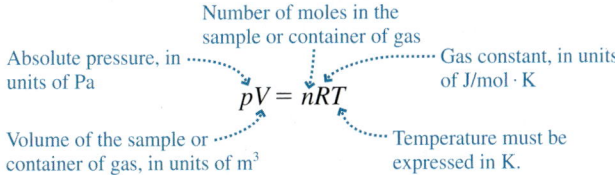

EXAMPLE 12.4 **Finding the volume of a mole**

What volume is occupied by one mole of an ideal gas at a pressure of 1.00 atm and a temperature of 0°C?

PREPARE The first step in ideal-gas law calculations is to convert all quantities to SI units:

$$p = 1.00 \text{ atm} = 101.3 \times 10^3 \text{ Pa}$$

$$T = 0 + 273 = 273 \text{ K}$$

SOLVE We use the ideal-gas law equation to compute

$$V = \frac{nRT}{p} = \frac{(1.00 \text{ mol})(8.31 \text{ J/mol} \cdot \text{K})(273 \text{ K})}{101.3 \times 10^3 \text{ Pa}}$$

$$= 0.0224 \text{ m}^3 = 22.4 \text{ L}$$

ASSESS When we do calculations using gases, it will be useful to keep this volume in mind to see if our answers make physical sense.

NOTE ▶ The conditions of this example, 1 atm pressure and 0°C, are called *standard temperature and pressure*, abbreviated **STP.** They are a common reference point for many gas processes. ◀

Because 1 m³ = 1000 L, this example tells us that the volume of one mole of gas at STP is 22.4 liters, a value you might recall from chemistry.

12.4 Ideal-Gas Processes

The ideal-gas law provides a connection between the pressure, volume, and temperature of a gas. Changing one of these variables causes one or both of the others to change. For example, suppose you measure the pressure in the tires on your car on a cold morning. How much will the tire pressure increase after you've driven at highway speeds for 30 minutes and the tires are much warmer? We will solve a problem like this later in the chapter, but, for now, note these properties of this process:

Actjv ONLINE Physjcs   8.4–8.6, 8.8–8.11

■ The quantity of gas is fixed. No air is added to or removed from the tire. All of the processes we will consider will involve a fixed quantity of gas.
■ There is a well-defined initial state. The initial values of pressure, volume, and temperature will be designated p_i, V_i, T_i.
■ There is a well-defined final state in which pressure, volume, and temperature have values p_f, V_f, T_f.

For gases in sealed containers, the number of moles (and number of molecules) does not change. In that case, the ideal-gas law can be written

$$\frac{pV}{T} = nR = \text{constant}$$

The values of the variables in the initial and final states are then related by

$$\frac{p_f V_f}{T_f} = \frac{p_i V_i}{T_i} \tag{12.13}$$

Initial and final states for an ideal gas in a sealed container

(a)   Each state of an ideal gas is represented as a point on a pV diagram.

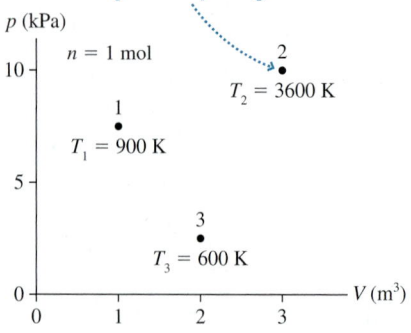

(b)   A process that changes the gas from one state to another is represented by a trajectory on a pV diagram.

FIGURE 12.12 The state of the gas and ideal-gas processes can be shown on a pV diagram.

This before-and-after relationship between the two states, reminiscent of a conservation law, will be valuable for many problems.

NOTE ▶ Because pressure and volume appear on both sides of the equation, this is a rare case of an equation for which we may use any units we wish, not just SI units. However, temperature *must* be in K. The difference is because unit-conversion factors for pressure and volume are multiplicative factors, and the same factor on both sides of the equation cancels. But the conversion from K to °C is an *additive* factor, and additive factors in the denominator don't cancel. ◀

pV Diagrams

It will be useful to represent ideal-gas processes on a graph called a **pV diagram.** The important idea behind a pV diagram is that each point on the graph represents a single, unique state of the gas. This may seem surprising, because a point on a graph specifies only the value of pressure and volume. But knowing p and V, and assuming that n is known for a sealed container, we can find the temperature from the ideal-gas law. Thus each point on a pV diagram actually represents a triplet of values (p, V, T) specifying the state of the gas.

For example, Figure 12.12a is a pV diagram showing three states of a system consisting of 1 mol of gas. The value of p and V can be read from the axes, then the temperature at that point calculated from the ideal-gas law. An ideal-gas process—a process that changes the state of the gas by, for example, heating it or compressing it—can be represented as a "trajectory" in the pV diagram. Figure 12.12b shows one possible process by which the gas of Figure 12.12a is changed from state 1 to state 3.

Constant-Volume Processes

Suppose you have a gas in the closed, rigid container shown in Figure 12.13a. Warming the gas will raise its pressure without changing its volume. This is an example of a **constant-volume process.** $V_f = V_i$ for a constant-volume process.

Because the value of V doesn't change, this process is shown as the vertical line i → f on the pV diagram of Figure 12.13b. **A constant-volume process appears on a pV diagram as a vertical line.**

FIGURE 12.13 A constant-volume process.

EXAMPLE 12.5 Computing tire pressure on a hot day

The pressure in a car tire is 30.0 psi on a cool morning when the air temperature is 0°C. After the day warms up and bright sun shines on the black tire, the temperature of the air inside the tire reaches 30°C. What is the tire pressure at this temperature?

PREPARE A tire is (to a good approximation) a sealed container with constant volume; this is a constant-volume process. The measured tire pressure is a gauge pressure, but the ideal-gas law

requires an absolute pressure. We must correct for this. The initial pressure is

$$p_i = (p_g)_i + 1.00 \text{ atm} = 30.0 \text{ psi} + 14.7 \text{ psi} = 44.7 \text{ psi}$$

Temperatures must be in kelvin, so we convert:

$$T_i = 0°C + 273 = 273 \text{ K}$$

$$T_f = 30°C + 273 = 303 \text{ K}$$

SOLVE The gas is in a sealed container, so we can use the ideal-gas law as given in Equation 12.13 to solve for the final pressure. In this equation, we divide both sides by V_f, and then cancel the ratio of the two volumes, which is equal to 1 for this constant-volume process:

$$p_f = p_i \frac{V_i}{V_f} \frac{T_f}{T_i} = p_i \frac{T_f}{T_i}$$

The units for p_f will be the same as those for p_i, so we can keep the initial pressure in psi. The pressure at the higher temperature is

$$p_f = 44.7 \text{ psi} \times \frac{303 \text{ K}}{273 \text{ K}} = 49.6 \text{ psi}$$

This is an absolute pressure, but the problem asks for the measured pressure in the tire—a gauge pressure. Converting to gauge pressure gives

$$(p_g)_f = p_f - 1.00 \text{ atm} = 49.6 \text{ psi} - 14.7 \text{ psi} = 34.9 \text{ psi}$$

ASSESS This is a significant change in pressure. You may find that you need to add air to your tires in the winter, when the temperature drops.

Constant-Pressure Processes

Many gas processes take place at a constant, unchanging pressure. A constant-pressure process is also called an **isobaric process.** $p_f = p_i$ for an isobaric process.

Figure 12.14a shows one method of changing the state of a gas while keeping the pressure constant. A cylinder of gas has a tight-fitting, massless piston that can slide up and down but seals the container so that no atoms enter or escape. The mass on top applies a constant downward force Mg on the piston. The air also presses down on the piston. The gas pressure is determined by the requirement that the gas must support both the mass on top of the piston and the air pressing inward. This pressure is independent of the temperature of the gas or the height of the piston, so it stays constant as long as M is unchanged. Figure 12.14b shows this trajectory for the isobaric process. Because the pressure does not change **a constant-pressure process appears on a *pV* diagram as a horizontal line.**

(a) The mass maintains a constant pressure in the cylinder.

(b) *p* An isobaric process appears on a *pV* diagram as a horizontal line.

FIGURE 12.14 A constant-pressure (isobaric) process.

EXAMPLE 12.6 A constant-pressure compression
A gas occupying 50.0 cm³ at 50°C is cooled at constant pressure until the temperature is 10°C. What is its final volume?

PREPARE This is a sealed system, so we can use Equation 12.13. The pressure of the gas doesn't change, so this is an isobaric process with $p_i/p_f = 1$.

The temperatures must be in kelvin, so we convert:

$$T_i = 50°C + 273 = 323 \text{ K}$$

and

$$T_f = 10°C + 273 = 283 \text{ K}$$

SOLVE We can use the ideal-gas law for a sealed container to solve for V_f:

$$V_f = V_i \frac{p_i}{p_f} \frac{T_f}{T_i} = 50.0 \text{ cm}^3 \times 1 \times \frac{283 \text{ K}}{323 \text{ K}} = 43.8 \text{ cm}^3$$

ASSESS In this solution and the previous one, we have not converted pressure and volume units as these multiplicative factors will cancel. But we did convert temperature to kelvin as this is an *additive* factor that does *not* cancel.

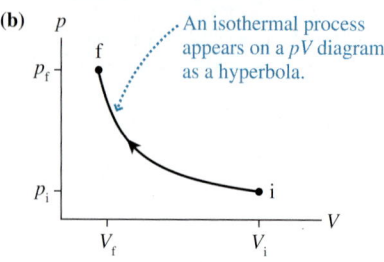

(b)
An isothermal process appears on a pV diagram as a hyperbola.

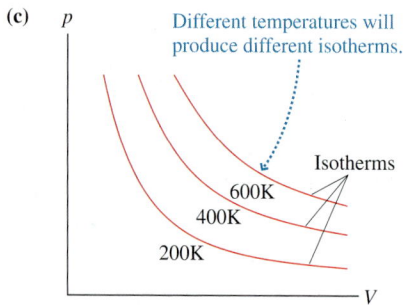

(c)
Different temperatures will produce different isotherms.

Isotherms

FIGURE 12.15 A constant-temperature (isothermal) process.

Constant-Temperature Process

A constant-temperature process is also called an **isothermal process.** For an isothermal process $T_f = T_i$. One possible isothermal process is illustrated in Figure 12.15. A piston is being pushed down to compress a gas, but the gas cylinder is submerged in a large container of liquid that is held at a constant temperature. If the piston is pushed *slowly*, then heat energy transfer through the walls of the cylinder will keep the gas at the same temperature as the surrounding liquid. This would be an *isothermal compression*. The reverse process, with the piston slowly pulled out, would be an *isothermal expansion*.

Representing an isothermal process on the pV diagram is a little more complicated than the two previous processes because both p and V change. As long as T remains fixed, we have the relationship

$$p = \frac{nRT}{V} = \frac{\text{constant}}{V} \qquad (12.14)$$

INVERSE
p. 118

Because there is an inverse relationship between p and V, the graph of an isothermal process is a *hyperbola*.

The process shown as i → f in Figure 12.15b represents the *isothermal compression* shown in Figure 12.15a. An *isothermal expansion* would move in the opposite direction along the hyperbola. The graph of an isothermal process is known as an **isotherm.**

The location of the hyperbola depends on the value of T. If we raise the constant temperature of the process in Figure 12.15a, the isotherm will move farther from the origin of the pV diagram. Figure 12.15c shows three isotherms for this process at three different temperatures. A gas undergoing an isothermal process will move along the isotherm having the appropriate temperature.

EXAMPLE 12.7 Compressing air in the lungs

A snorkeler takes a deep breath at the surface, filling his lungs with 4.0 L of air. He then descends to a depth of 5.0 m, where the pressure is 1.5 atm. At this depth, what is the volume of air in the snorkeler's lungs?

PREPARE The pressure inside the snorkeler's lungs equals the pressure of the surrounding water, since otherwise his lungs would collapse. Further, the air stays at body temperature, making this an isothermal process with $T_f = T_i$.

SOLVE The ideal-gas law for a sealed container (the lungs) gives

$$V_f = V_i \frac{p_i}{p_f} \frac{T_f}{T_i} = 4.0\,\text{L} \times \frac{1.0\,\text{atm}}{1.5\,\text{atm}} = 2.7\,\text{L}$$

ASSESS The air has a smaller volume at the greater pressure, as we would expect.

Work

When the spark plug fires in a cylinder of your car, it ignites the gaseous fuel-air mixture inside. The hot gas expands, pushing the piston out and, through various mechanical linkages, turning the wheels of your car. As the gas expands, it does *work* on the piston. Similarly, the gas in Figure 12.14a does work as it lifts the mass on the piston. pV diagrams are particularly useful for calculating the work done by a gas.

You learned in Chapter 10 that the work done by a constant force F in pushing an object distance d is $W = Fd$. Let's apply this idea to a gas. Figure 12.16a shows a gas cylinder sealed at one end by a movable piston. Force $\vec{F}_{\text{gas}}$ is due to the gas pressure and has magnitude $F_{gas} = pA$. Force $\vec{F}_{ext}$, perhaps a force applied by a piston rod, is equal in magnitude and opposite in direction to $\vec{F}_{gas}$. The gas pressure would blow the piston out if the external force weren't there!

Suppose the gas expands at constant pressure, pushing the piston outward from x_i to x_f, a distance $d = x_f - x_i$, as shown in Figure 12.16b. As it does, the force due to the gas pressure does work

$$W_{gas} = F_{gas}d = (pA)(x_f - x_i) = p(x_f A - x_i A)$$

But $x_i A$ is the cylinder's initial volume V_i (recall that the volume of a cylinder is length times the area of the base), and $x_f A$ is the final volume V_f. Thus the work done is

$$W_{gas} = p(V_f - V_i) = p\,\Delta V \qquad (12.15)$$

Work done by a gas in an isobaric process

where ΔV is the *change* in volume.

Equation 12.15 has a particularly simple interpretation on a pV diagram. As Figure 12.17a shows, $p\,\Delta V$ is the "area under the pV graph" between V_i and V_f. Although we've shown this result only for an isobaric process, it turns out to be true for all ideal-gas processes. That is, as Figure 12.17b shows,

$$W_{gas} = \text{area under the } pV \text{ graph between } V_i \text{ and } V_f$$

The gas pushes on the piston with force $\vec{F}_{gas}$.

(a)

Pressure p

To keep the piston in place, an external force must be equal and opposite to $\vec{F}_{gas}$.

$\vec{F}_{gas}$ $\vec{F}_{ext}$

(b)

As the piston moves a distance d, the gas does work $F_{gas}d$.

Piston area A

d

$0 \qquad x_i \quad x_f \qquad x$

The gas volume changes by $\Delta V = dA$.

FIGURE 12.16 The expanding gas does work on the piston.

(a)

p

For an *isobaric* process, the area under the graph $p\Delta V$ is the work done by the gas.

p

i   f

ΔV

$V_i \qquad V_f \qquad V$

(b)

p

In general, the work done by a gas is the area under the graph.

i

f

$W = \text{area}$

$V_i \qquad V_f \qquad V$

FIGURE 12.17 Calculating the work done in an ideal-gas process.

Thus the work done by a gas can be determined from the geometry of the pV diagram. There are a few things we need to clarify about this determination:

- In order for a gas to do work, the volume must change. Thus no work is done in a constant-volume process.
- The work done by a gas $W_{gas} = p\Delta V$ is positive when the gas expands ($\Delta V > 0$) but negative (meaning work is done on the gas) when the gas is compressed ($\Delta V < 0$).
- The simple relationship of Equation 12.15 applies only to constant-pressure processes. For any other ideal-gas process, you must use the geometry of the pV diagram to calculate the area under the graph.
- To calculate work, pressure must be in Pa and volume in m^3. The product of Pa (i.e., N/m^2) and m^3 is joules, the units of work and energy.
- W_{gas} is *not* the work W that appears in the first law of thermodynamics. The first-law work is work done *on* the system, whereas W_{gas} is work done *by* the system. But $W = -W_{gas}$, so the first law of thermodynamics can be written

$$\Delta E_{th} = Q - W_{gas} \qquad (12.16)$$

EXAMPLE 12.8 Finding the work done by a gas

0.0160 mol of nitrogen follow the ideal-gas process shown in Figure 12.18.

 a. What are the initial and final temperatures of the gas?
 b. How much work is done by the gas?

FIGURE 12.18 pV diagram for Example 12.8.

PREPARE We can use the ideal-gas law to determine the temperatures, but first we must convert the volumes to SI units:

$$V_i = 100 \text{ cm}^3 \times \frac{1 \text{ m}^3}{10^6 \text{ cm}^3} = 1.00 \times 10^{-4} \text{ m}^3$$

$$V_f = 300 \text{ cm}^3 \times \frac{1 \text{ m}^3}{10^6 \text{ cm}^3} = 3.00 \times 10^{-4} \text{ m}^3$$

This is not a constant-pressure process, so we will need to calculate the area under the graph to determine the amount of work done by the gas as it expands.

SOLVE a. The initial and final temperatures are

$$T_i = \frac{p_i V_i}{nR} = \frac{(4.00 \times 10^5 \text{ Pa})(1.00 \times 10^{-4} \text{ m}^3)}{(0.0160 \text{ mol})(8.31 \text{ J/mol} \cdot \text{K})} = 301 \text{ K}$$

$$T_f = \frac{p_f V_f}{nR} = \frac{(2.00 \times 10^5 \text{ Pa})(3.00 \times 10^{-4} \text{ m}^3)}{(0.0160 \text{ mol})(8.31 \text{ J/mol} \cdot \text{K})} = 451 \text{ K}$$

 b. We analyze the area under the graph as a triangle on top of a rectangle, with areas computed as in Figure 12.19. Notice that the areas are in J because they are the product of Pa and m³.

FIGURE 12.19 The work done by the gas is the total area under the graph.

The total work is

$$W_{gas} = \text{area of triangle} + \text{area of rectangle}$$
$$= 20.0 \text{ J} + 40.0 \text{ J} = 60.0 \text{ J}$$

ASSESS The gas did work on the piston as it expanded. In a practical situation, this work would be transferred via gears and levers to a machine that performs some useful function.

Adiabatic Processes

The last ideal-gas process we'll look at is one in which no heat energy is transferred between the gas and the environment and thus $Q = 0$. This is called an **adiabatic process.** There are two ways an adiabatic process can come about: Either the system is thermally insulated, preventing heat transfer, or, more commonly, the process happens too quickly for a transfer to occur. A rapid compression or expansion of a gas doesn't allow time for heat transfer, so a rapid compression is called an *adiabatic compression* and a rapid expansion is an *adiabatic expansion.*

If you've ever pumped up a bicycle tire with a hand pump, you may have noticed that the pump gets quite warm. Why? Pressing the pump handle causes a *rapid* compression of the air inside—an adiabatic process. For an adiabatic process, with $Q = 0$, the first law of thermodynamics, Equation 12.16, is $\Delta E_{th} = -W_{gas}$. You do work *on* the gas to compress it, so the work done *by* the gas is negative ($W_{gas} < 0$). Thus the thermal energy of the gas increases ($\Delta E_{th} > 0$) and the temperature goes up.

Similarly, a gas that does work on a piston by expanding adiabatically ($W_{gas} > 0$) gets colder as its thermal energy decreases. Thus **an adiabatic expansion lowers the temperature of a gas** and **an adiabatic compression raises the temperature of a gas.** You can use an adiabatic process to change the gas temperature without using heat! An adiabatic processes uses work, rather than heat, to change the speeds of the molecules and thus change the temperature.

Warm mountain winds This image shows surface temperatures (in °F) in North America on a winter day in 2003. Notice the bright green area of unseasonably warm temperatures extending north and west from the center of the continent. This warm area was produced by a Chinook wind. (Similar winds in other areas of the world have other names, such as the Santa Ana winds in southern California.) On this day, strong westerly winds were blowing down off the Rocky Mountains. The air was rapidly compressed as it descended to lower regions of higher pressure in a very short time. This rapid compression was an adiabatic process that significantly increased the air temperature.

Adiabatic processes have many important applications. In the atmosphere, large masses of air that move across the planet transfer heat quite slowly, so adiabatic expansions and compressions often lead to changes in temperature, as we see in the discussion of the Chinook wind.

CONCEPTUAL EXAMPLE 12.2 Adiabatic curves on a *pV* diagram

Figure 12.20 shows the *pV* diagram of a gas undergoing an isothermal compression from point 1 to point 2. Roughly how would the *pV* diagram look if the gas were compressed from point 1 to the same final pressure by a rapid adiabatic compression?

FIGURE 12.20 *pV* diagram for an isothermal compression.

REASON An adiabatic compression increases the temperature of the gas as the work done on the gas is transformed into thermal

energy. Consequently, as seen in Figure 12.21, the curve of the adiabatic compression cuts across the isotherms to end on a higher-temperature isotherm when the gas pressure reaches p_2.

FIGURE 12.21 *pV* diagram for an adiabatic compression.

ASSESS The temperature at the final point of an adiabatic compression is higher than at the starting point. The final volume for the adiabatic compression is therefore larger than that for the isothermal compression, in agreement with the diagram as drawn. Similarly, an adiabatic expansion ends on a lower-temperature isotherm.

STOP TO THINK 12.3 What is the ratio T_f/T_i for this process?

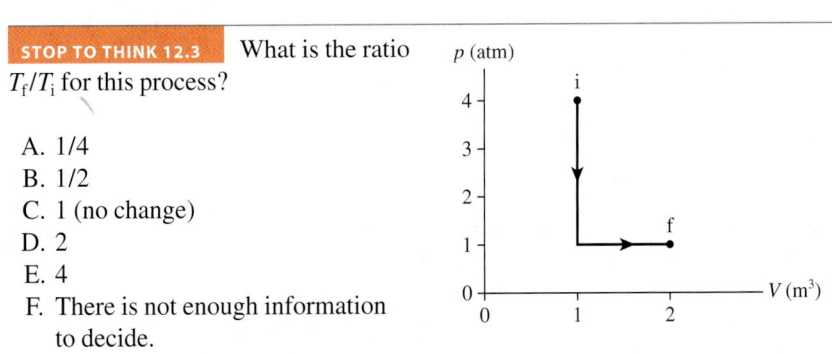

A. 1/4
B. 1/2
C. 1 (no change)
D. 2
E. 4
F. There is not enough information to decide.

12.5 Specific Heat and Heat of Transformation

If you hold a glass of cold water in your hand, the heat from your hand will increase the temperature of the water. Similarly, the heat from your hand will melt an ice cube; this is what we will call a *phase change*. Increasing the temperature and changing the phase are two possible *consequences* of heating a system, and they form an important part of our understanding of the thermal properties of matter.

Specific Heat

Adding 4190 J of heat energy to water raises its temperature by 1 K. If you are fortunate enough to have 1 kg of gold, you need only 129 J of heat to raise its temperature by 1 K. The amount of heat that raises the temperature of 1 kg of a

TABLE 12.4 Specific heats of solids and liquids

Substance	c (J/kg · K)
Solids	
Lead	128
Gold	129
Copper	385
Iron	449
Aluminum	900
Water ice	2090
Mammalian body	3400
Liquids	
Mercury	140
Ethyl alcohol	2400
Water	4190

substance by 1 K is called the **specific heat** of that substance. The symbol for specific heat is c. Water has specific heat $c_{water} = 4190$ J/kg · K, and the specific heat of gold is $c_{gold} = 129$ J/kg · K. Specific heat depends only on the material from which an object is made. Table 12.4 provides some specific heats for common liquids and solids.

NOTE ▶ *Heat* is energy—specifically, the energy transferred between a hotter object and a colder object. *Specific heat,* the property of an object, does not use the word *heat* in the same way. Specific heat is an old idea, dating back to the days when heat was thought to be a substance that was contained in objects. The term has continued in use even though our understanding of heat has changed. ◀

If heat c is required to raise the temperature of 1 kg of substance by 1 K, then heat Mc is needed to raise the temperature of mass M by 1 K and $(Mc) \Delta T$ is needed to raise the temperature of mass M by ΔT. In other words, the heat needed to bring about a temperature change ΔT is

$$Q = Mc \, \Delta T \qquad (12.17)$$

Heat needed to produce a temperature change ΔT for mass M with specific heat c

LINEAR
p. 38

Q can be either positive (temperature goes up) or negative (temperature goes down).

Because $\Delta T = Q/Mc$, it takes more heat energy to change the temperature of a substance with a large specific heat than to change the temperature of a substance with a small specific heat. Water, with a very large specific heat, is slow to warm up and slow to cool down. This is fortunate for us. The large "thermal inertia" of water is essential for the biological processes of life.

Hurricane season During a cold night, the large specific heat of water prevents the temperature of a lake from dropping nearly as much as that of the surrounding air. Early in the morning, water vapor evaporating from the warm lake quickly condenses in the colder air, forming mist. The thermal inertia of the oceans is even greater. Ocean temperatures in the northern hemisphere don't reach their maximum values until the late summer or early fall. This is why hurricanes, which require warm ocean water to form, happen in the fall but not the spring.

EXAMPLE 12.9 Energy to run a fever

A 70 kg student catches the flu, and his body temperature increases from 37.0°C (98.6°F) to 39.0°C (102.2°F). How much energy is required to raise his body's temperature?

SOLVE Raising the temperature of the body uses energy supplied internally from the chemical reactions of the body's metabolism, which transfer heat to the body. The specific heat of the body is given in Table 12.4 as 3400 J/kg · K. We can use Equation 12.17 to find the necessary heat energy:

$$Q = Mc \, \Delta T = (70 \text{ kg})(3400 \text{ J/kg} \cdot \text{K})(2.0 \text{ K}) = 4.8 \times 10^5 \text{ J}$$

ASSESS Raising the temperature of an object as large as the body will take a good deal of energy, so the size of this result seems reasonable. Looking back to Chapter 11, we see that this is approximately the energy in a large apple, or the amount of energy required to walk one mile.

Phase Changes

The temperature inside the freezer compartment of a refrigerator is typically about −20°C. Suppose you were to remove a few ice cubes from the freezer and place them in a sealed container with a thermometer. Then, as Figure 12.22a shows, you put a steady flame under the container. We'll assume that the heating is done slowly so that the inside of the container always has a single, uniform temperature.

Figure 12.22b shows the temperature as a function of time. After steadily rising from the initial $-20°C$, the temperature remains fixed at $0°C$ for an extended period of time. This is the interval of time during which the ice melts. As it's melting, the ice temperature is $0°C$ and the liquid water temperature is $0°C$. Even though the system is being heated, the temperature doesn't begin to rise until all the ice has melted. The energy that is being added to the system is being used to break bonds between the water molecules, changing the form of the water from solid to liquid.

The thermal energy of a solid is the kinetic energy of the vibrating atoms plus the potential energy of stretched and compressed molecular bonds. If you heat a solid, at some point the thermal energy gets so large that the molecular bonds begin to break, allowing the atoms to move around—the solid begins to melt. Continued heating will produce further melting, as the energy is used to break more bonds. The temperature will not rise until all of the bonds are broken. If you wish to return the liquid to a solid, this energy must be removed.

NOTE ▶ In everyday language, the three phases of water are called *ice, water,* and *steam.* That is, the term *water* implies the liquid phase. Scientifically, these are the solid, liquid, and gas phases of the compound called *water.* When we are working with different phases of water, we'll use the term *water* in the scientific sense of a collection of H_2O molecules. We'll say either *liquid* or *liquid water* to denote the liquid phase. ◀

(a)

(b)

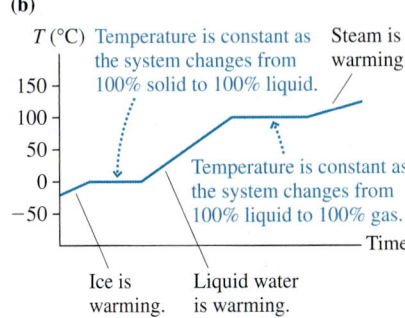

FIGURE 12.22 The temperature as a function of time as water is transformed from solid to liquid to gas.

CONCEPTUAL EXAMPLE 12.3 **Strategy for cooling a drink**

If you have a warm soda that you wish to cool, would it be more effective to add 25 g of liquid water at $0°C$ or 25 g of water ice at $0°C$?

REASON If you add liquid water at $0°C$, heat will be transferred from the soda to the water, raising the temperature of the water and lowering that of the soda. If you add water ice at $0°C$, heat first will be transferred from the soda to the ice to melt it, transforming the $0°C$ ice to $0°C$ liquid water, then will be transferred to the liquid water to raise its temperature. Thus more thermal energy will be removed from the soda, giving it a lower final temperature, if ice is used rather than liquid water.

The temperature at which a solid becomes a liquid or, if the thermal energy is reduced, a liquid becomes a solid is called the **melting point** or the **freezing point.** Melting and freezing are *phase changes.* A system at the melting point is in **phase equilibrium,** meaning that any amount of solid can coexist with any amount of liquid. Raise the temperature ever so slightly and the entire system soon becomes liquid. Lower it slightly and it all becomes solid.

You can see another region of phase equilibrium of water in Figure 12.22b at $100°C$. This is a phase equilibrium between the liquid phase and the gas phase, and any amount of liquid can coexist with any amount of gas at this temperature. As heat is added to the system, the temperature stays the same. The added energy is used to break bonds between the liquid molecules, allowing them to move into the gas phase. The temperature at which a gas becomes a liquid or, if the thermal energy is increased, a liquid becomes a gas is called the **condensation point** or the **boiling point.**

NOTE ▶ Liquid water becomes solid ice at $0°C$, but that doesn't mean the temperature of ice is always $0°C$. Ice reaches the temperature of its surroundings. If the air temperature in a freezer is $-20°C$, then the ice temperature is $-20°C$, Likewise, steam can be heated to temperatures above $100°C$. That doesn't happen when you boil water on the stove because the steam escapes, but steam can be heated far above $100°C$ in a sealed container. ◀

Tin pest Some substances have multiple solid phases in which the atoms are arranged in lattices with different structures. Such substances can undergo a phase change between these solid phases; the atoms rearrange themselves, and the properties of the substance can change. For example, the element tin slowly changes from the usual "white" phase to a brittle "gray" phase at temperatures below $13°C$. The tin sample in this photo was kept below this transition temperature for a year; the resulting "tin pest" is quite obvious. Some historians speculate that this phase change destroyed the tin buttons on the coats of Napoleon's troops during their winter invasion of Russia in 1812, contributing to their defeat as they froze in the bitter cold.

You are cooking pasta on the stove; the water is at a slow boil. Will the pasta cook more quickly if you turn up the burner on the stove so that the water is at a fast boil?

REASON Water boils at 100°C; no matter how vigorously the water is boiling, the temperature is the same. It is the tempera-ture of the water that determines how fast the cooking takes place. Adding heat at a faster rate will make the water boil away more rapidly, but will not change the temperature—and will not alter the cooking time.

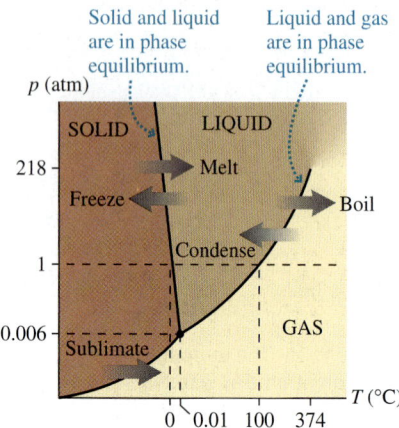

FIGURE 12.23 Phase diagram (not to scale) for water.

A **phase diagram** is used to show how the phases and phase changes of a substance vary with both temperature and pressure. Figure 12.23 shows the phase diagram for water. You can see that the diagram is divided into three regions corresponding to the solid, liquid, and gas phases. The boundary lines separating the regions indicate the phase transitions. The system is in phase equilibrium at a pressure-temperature point that falls on one of these lines.

Phase diagrams contain a great deal of information. Notice on the diagram that the dotted line at $p = 1$ atm crosses the solid-liquid boundary at 0°C and the liquid-gas boundary at 100°C. These well-known melting and boiling point temperatures of water apply only at standard atmospheric pressure. These temperatures only apply for pure water; if the water contains dissolved substances, the melting point is lowered and the boiling point is raised.

The variation of melting and boiling points with pressure has practical consequences. If you boil water in Denver, the "Mile-High City," where the altitude results in atmospheric pressure significantly less than that at sea level, you will find that the boiling point is several degrees below 100°C. It takes a bit longer to cook pasta in Denver than in coastal Miami. A *pressure cooker* works by allowing the pressure inside to exceed 1 atm. This raises the boiling point, so foods in boiling water are at a temperature $> 100°C$ and cook faster.

Heat of Transformation

In Figure 12.22b, the phase changes appeared as horizontal line segments on the graph. During these segments, heat is being transferred to the system but the temperature isn't changing. The thermal energy continues to increase during a phase change, but, as noted, the additional energy goes into breaking molecular bonds rather than speeding up the molecules. **A phase change is characterized by a change in thermal energy without a change in temperature.**

The amount of heat energy that causes 1 kg of a substance to undergo a phase change is called the **heat of transformation** of that substance. For example, laboratory experiments show that 333,000 J of heat are needed to melt 1 kg of ice at 0°C. The symbol for heat of transformation is L. The heat required for the entire system of mass M to undergo a phase change is

$$Q = ML \tag{12.18}$$

Heat of transformation is a generic term that refers to any phase change. Two specific heats of transformation are the **heat of fusion** L_f, the heat of transformation between a solid and a liquid, and the **heat of vaporization** L_v, the heat of transformation between a liquid and a gas. The heat needed for these phase changes is

$$Q = \begin{cases} \pm ML_f & \text{Heat needed to melt/freeze mass } M \\ \pm ML_v & \text{Heat needed to boil/condense mass } M \end{cases} \tag{12.19}$$

where the $\pm$ indicates that heat must be *added* to the system during melting or boiling but *removed* from the system during freezing or condensing. **You must explicitly include the minus sign when it is needed.**

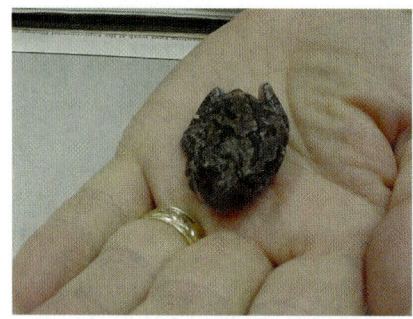

Frozen frogs BIO It seems impossible, but common wood frogs survive the winter with much of their bodies frozen. They use the principle that the freezing point of water decreases when there are substances dissolved in it; this is also the principle behind the antifreeze in your car's radiator. Although the liquid water *between* cells in the frogs' bodies freezes, the water *inside* their cells remains liquid because of high concentrations of dissolved glucose. This prevents the cell damage that accompanies the freezing and subsequent thawing of tissues. When spring arrives, the frogs thaw and appear no worse for their winter freeze.

Table 12.5 lists some heats of transformation. Notice that the heat of vaporization is always much larger than the heat of fusion. Our atomic model explains this. Melting breaks just enough molecular bonds to allow the system to lose rigidity and flow. Even so, the molecules in a liquid remain close together and loosely bonded. Vaporization breaks all bonds completely and sends the molecules flying apart. This process requires a larger increase in the thermal energy and thus a larger quantity of heat.

TABLE 12.5 Melting/boiling temperatures and heats of transformation at standard atmospheric pressure

Substance	T_m (°C)	L_f (J/kg)	T_b (°C)	L_v (J/kg)
Nitrogen (N_2)	-210	0.26×10^5	-196	1.99×10^5
Ethyl alcohol	-114	1.09×10^5	78	8.79×10^5
Mercury	-39	0.11×10^5	357	2.96×10^5
Water	0	3.33×10^5	100	22.6×10^5
Lead	328	0.25×10^5	1750	8.58×10^5

Lava—molten rock—undergoes a phase change from liquid to solid when it contacts liquid water; the transfer of heat to the water causes the water to undergo a phase change from liquid to gas.

EXAMPLE 12.10 Melting a popsicle

A girl eats a 45 g frozen popsicle that was taken out of a $-10°C$ freezer. How much energy does her body use to bring the popsicle up to body temperature?

PREPARE We can assume that the popsicle is pure water. Normal body temperature is 37°C. The specific heats of ice and liquid water are given in Table 12.4; the heat of fusion of water is given in Table 12.5.

SOLVE There are three parts to the problem: The popsicle must be warmed to 0°C, then melted, and then the resulting water must be warmed to body temperature. The heat needed to warm the frozen water by $\Delta T = 10°C = 10$ K to the melting point is

$$Q_1 = Mc_{ice}\Delta T = (0.045 \text{ kg})(2090 \text{ J/kg} \cdot \text{K})(10 \text{ K}) = 940 \text{ J}$$

Note that we use the specific heat of water ice, not liquid water, in this equation. Melting 45 g of ice requires heat

$$Q_2 = ML = (0.045 \text{ kg})(3.33 \times 10^5 \text{ J/kg}) = 15,000 \text{ J}$$

The liquid water must now be warmed to body temperature; this requires heat

$$Q_3 = Mc_{water}\Delta T = (0.045 \text{ kg})(4190 \text{ J/kg} \cdot \text{K})(37 \text{ K})$$
$$= 7000 \text{ J}$$

The total energy is the sum of these three values, $Q_{total} = 23,000$ J.

ASSESS More energy is needed to melt the ice than to warm the water, as we would expect. A commercial popsicle has 40 Calories, which is about 170 kJ. Roughly 15% of the chemical energy in this frozen treat is used to bring it up to body temperature!

Evaporation

Water boils at 100°C. But individual molecules of water can move from the liquid phase to the gas phase at lower temperatures. This process is known as **evaporation.** Water evaporates as sweat from your skin at a temperature well below 100°C. We can use our atomic model to explain this.

In gases, we have seen that there is a variation in particle speeds. The same principle applies to water and other liquids. At any temperature, some molecules will be moving fast enough to go into the gas phase. And they will do so, carrying away thermal energy as they go. The molecules that leave the liquid are the ones that have the highest kinetic energy, so evaporation reduces the average kinetic energy (and thus the temperature) of the liquid left behind.

The evaporation of water, both from sweat and in moisture you exhale, is one of the body's methods of exhausting the excess heat of metabolism to the environment, allowing you to maintain a steady body temperature. The heat to evaporate a mass M of water is $Q = ML_v$, and this amount of heat is removed from your body. However, the heat of vaporization L_v is a little larger than the value for boiling. At a skin temperature of 30°C, the heat of vaporization of water is $L_v = 24 \times 10^5$ J/kg, or 6% larger than the value at 100°C in Table 12.5.

Keeping your cool BIO Humans (and cattle and horses) have sweat glands, so we can perspire to moisten our skin, allowing evaporation to cool our bodies. Animals that do not perspire can also use evaporation to keep cool. Dogs, goats, rabbits, and even birds pant, evaporating water from their respiratory passages. Elephants spray water on their skin; other animals may lick their fur.

EXAMPLE 12.11 **Computing heat loss by perspiration**

The human body can produce approximately 30 g of perspiration per minute. At what rate is it possible to exhaust heat by the evaporation of perspiration?

SOLVE The evaporation of 30 g of perspiration at normal body temperature requires a heat energy

$$Q = (0.030 \text{ kg})(24 \times 10^5 \text{ J/kg}) = 7.2 \times 10^4 \text{ J}$$

This is the heat lost per minute; the rate of heat loss is

$$\frac{Q}{\Delta t} = \frac{7.2 \times 10^4 \text{ J}}{60 \text{ s}} = 1200 \text{ W}$$

ASSESS Given the metabolic power required for different activities, as listed in Chapter 11, this rate of heat removal is quite sufficient to keep the body cool even when exercising in hot weather—as long as the person drinks enough water to keep up this rate of perspiration!

STOP TO THINK 12.4 1 kg of barely molten lead, at 328°C, is poured into a large beaker holding liquid water right at the boiling point, 100°C. The mass of the water that will be boiled away as the lead solidifies is

A. 0 kg B. <1 kg C. 1 kg D. >1 kg

12.6 Calorimetry

At one time or another you've probably put an ice cube into a hot drink to cool it quickly. You were engaged, in a somewhat trial-and-error way, in a practical aspect of heat transfer known as **calorimetry,** the quantitative measurement of the heat transferred between systems or evolved in reactions. You know that heat will be transferred from the hot drink into the cold ice cube, reducing the temperature of the drink, as shown in Figure 12.24.

Let's make this qualitative picture more precise. Figure 12.25 shows two systems thermally interacting with each other but isolated from everything else. Suppose they start at different temperatures T_1 and T_2. As you know, heat energy will be transferred from the hotter to the colder system until they reach a common final temperature T_f. The systems will then be in thermal equilibrium and the temperature will not change further. In the cooling coffee example of Figure 12.24, the coffee is System 1, the ice is System 2, and the insulating barrier is the mug.

The insulation prevents any heat energy from being transferred to or from the environment, so energy conservation tells us that any energy leaving the hotter system must enter the colder system. That is, the systems *exchange* energy with no net loss or gain. The concept is straightforward, but to state the idea mathematically we need to be careful with signs.

Let Q_1 be the energy transferred to system 1 as heat. Q_1 is positive if energy *enters* system 1, negative if energy *leaves* system 1. Similarly, Q_2 is the energy transferred to system 2. The fact that the systems are merely exchanging energy can be written $|Q_1| = |Q_2|$. That is, the energy *lost* by the hotter system is the energy *gained* by the colder system. But Q_1 and Q_2 have opposite signs, so $Q_1 = -Q_2$. No energy is exchanged with the environment, hence it makes more sense to write this relationship as

$$Q_{net} = Q_1 + Q_2 = 0 \qquad (12.20)$$

This idea is easily extended. No matter how many systems interact thermally, Q_{net} must equal zero.

NOTE ▶ The signs are very important in calorimetry problems. ΔT is always $T_f - T_i$, so ΔT and Q are negative for any system whose temperature decreases. The proper sign of Q for any phase change must be supplied *by you*, depending on the direction of the phase change. ◀

FIGURE 12.24 Cooling a hot drink with ice.

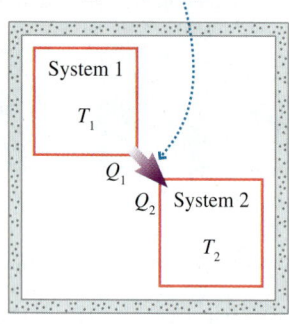

Heat energy is transferred from system 1 to system 2. Energy conservation requires

$$|Q_1| = |Q_2|$$

Opposite signs mean that

$$Q_{net} = Q_1 + Q_2 = 0$$

FIGURE 12.25 Two systems interact thermally. In this particular example, where heat is being transferred from system 1 to system 2, $Q_2 > 0$ and $Q_1 < 0$.

PROBLEM-SOLVING STRATEGY 12.1   **Calorimetry problems**

PREPARE Identify the interacting systems. Assume that they are isolated from the larger environment. List known information and identify what you need to find. Convert all quantities to SI units.

SOLVE The statement of energy conservation is

$$Q_{net} = Q_1 + Q_2 + \cdots = 0$$

- For systems that undergo a temperature change, $Q_{\Delta T} = Mc(T_f - T_i)$. Be sure to have the temperatures T_i and T_f in the correct order.
- For systems that undergo a phase change, $Q_{phase} = \pm ML$. Supply the correct sign by observing whether energy enters or leaves the system during the transition.
- Some systems may undergo a temperature change *and* a phase change. Treat the changes separately. The heat energy is $Q = Q_{\Delta T} + Q_{phase}$.

ASSESS The final temperature should be in the middle of the initial temperatures. A T_f that is higher or lower than all initial temperatures is an indication that something is wrong, usually a sign error.

EXAMPLE 12.12 **Using calorimetry to identify a metal**

200 g of an unknown metal is heated to 200.0°C then dropped into 50.0 g of water at 20.0°C in an insulated container. The water temperature rises within a few seconds to 39.7°C, then changes no further. Identify the metal.

PREPARE The metal and the water interact thermally; there are no phase changes. We know all the initial and final temperatures. We will label the temperatures as follows: The initial temperature of the metal is T_m; the initial temperature of the water is T_w. The common final temperature is T_f. For water, $c_w = 4190 \text{ J/kg} \cdot \text{K}$ is known from Table 12.4. Only the specific heat c_m of the metal is unknown.

SOLVE Energy conservation requires $Q_w + Q_m = 0$. Using $Q = Mc(T_f - T_i)$ for each, we have

$$Q_w + Q_m = M_w c_w(T_f - T_w) + M_m c_m(T_f - T_m) = 0$$

This is solved for the unknown specific heat:

$$c_m = \frac{-M_w c_w(T_f - T_w)}{M_m(T_f - T_m)}$$

$$= \frac{-(0.0500 \text{ kg})(4190 \text{ J/kg} \cdot \text{K})(39.7°C - 20.0°C)}{(0.200 \text{ kg})(39.7°C - 200.0°C)}$$

$$= 129 \text{ J/kg} \cdot \text{K}$$

Referring to Table 12.4, we find we have either 200 g of gold or, if we made an ever-so-slight experimental error, 200 g of lead!

ASSESS The temperature of the unknown metal changed much more than the temperature of the water. That is, the unknown metal had much less thermal inertia, and thus, as we found, its specific heat should be much less than that of water.

EXAMPLE 12.13 **Calorimetry with a phase change**

Your 500 mL soda is at 20°C, room temperature, so you add 100 g of ice from the −20°C freezer. Does all the ice melt? If so, what is the final temperature? If not, what fraction of the ice melts? Assume that you have a well-insulated cup.

PREPARE We need to distinguish between two possible outcomes. If all the ice melts, then $T_f > 0°C$. It's also possible that the soda will cool to 0°C before all the ice has melted, leaving the ice and liquid in equilibrium at 0°C. We need to distinguish between these before knowing how to proceed. All

the initial temperatures, masses, and specific heats are known. The final temperature of the combined soda + ice system is unknown.

SOLVE Let's first calculate the heat needed to melt all the ice and leave it as liquid water at 0°C. To do so, we must warm the ice to 0°C, then change it to water. The heat input for this two-stage process is

$$Q_{melt} = M_{ice}c_{ice}(20 \text{ K}) + M_{ice}L_f = 37,000 \text{ J}$$

where L_f is the heat of fusion of water.

Continued

Q_{melt} is a *positive* quantity because we must *add* heat to melt the ice. Next, let's calculate how much heat energy will leave the soda if it cools all the way to 0°C. The volume is $V = 500$ mL $= 5.00 \times 10^{-4}$ m^3 and its density is that of water, $\rho = 1000$ kg/m^3, thus the mass is $M_{soda} = \rho V = 0.500$ kg. The heat is

$$Q_{cool} = M_{soda}c_{water}(-20 \text{ K}) = -42{,}000 \text{ J}$$

where $\Delta T = -20$ K because the temperature decreases. Because $|Q_{cool}| > Q_{melt}$, the soda has sufficient energy to melt all the ice. Hence the final state will be all liquid at $T_f > 0$. (Had we found $|Q_{cool}| < Q_{melt}$, then the final state would have been an ice-liquid mixture at 0°C.)

Energy conservation requires $Q_{ice} + Q_{soda} = 0$. The heat Q_{ice} consists of three terms: warming the ice to 0°C, melting the ice to water at 0°C, then warming the 0°C water to T_f. The mass will still be M_{ice} in the last of these steps, because it is the "ice system," but we need to use the specific heat of *liquid water*. Thus

$$Q_{ice} + Q_{soda} = \left[M_{ice}c_{ice}(20 \text{ K}) + M_{ice}L_f \right.$$
$$\left. + M_{ice}c_{water}(T_f - 0°C) \right]$$
$$+ M_{soda}c_{water}(T_f - 20°C) = 0$$

We've already done part of the calculation, allowing us to write

$$37{,}000 \text{ J} + M_{ice}c_{water}(T_f - 0°C) + M_{soda}c_{water}(T_f - 20°C) = 0$$

Solving for T_f gives

$$T_f = \frac{20M_{soda}c_{water} - 37{,}000}{M_{ice}c_{water} + M_{soda}c_{water}} = 1.9°C$$

ASSESS A good deal of ice has been put in the soda, so it ends up being cooled nearly to the freezing point, as we might expect.

STOP TO THINK 12.5 1 kg of lead at 100°C is dropped into a container holding 1 kg of water at 0°C. Once the lead and water reach thermal equilibrium, the final temperature is

 A. <50°C.   B. 50°C.   C. >50°C.

12.7 Thermal Properties of Gases

8.7 Activ
Physics
ONLINE

We've examined the thermal properties of solids and liquids—their thermal expansion and their specific heat. We now turn to the thermal properties of gases—what happens when they are heated? Gases are harder to characterize than solids or liquids because the heat required to cause a specified temperature change depends on the *process* by which the gas changes state.

Just as for solids and liquids, heating a gas changes its temperature. But by how much? Figure 12.26 shows two isotherms on the pV diagram for a gas. Processes A and B, which start on the T_i isotherm and end on the T_f isotherm, have the *same* temperature change $\Delta T = T_f - T_i$, so we might expect them both to require the same amount of heat. But it turns out that process A, which takes place at constant volume, requires *less* heat than does process B, which occurs at constant pressure. The reason is that work is done in process B but not in process A.

It is useful to define two different versions of the specific heat of gases, one for constant volume processes and one for constant pressure processes. We will define these as *molar* specific heats because we usually do gas calculations using moles instead of mass. The quantity of heat needed to change the temperature of n moles of gas by ΔT is, for a constant volume process

$$Q = nC_V \Delta T \tag{12.21}$$

and for a constant pressure process

$$Q = nC_P \Delta T \tag{12.22}$$

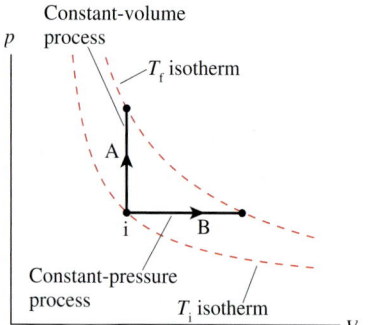

FIGURE 12.26 Processes A and B have the same ΔT and the same ΔE_{th}, but they require different amounts of heat.

C_V is the **molar specific heat at constant volume** and C_P is the **molar specific heat at constant pressure**. Table 12.6 gives the values of C_V and C_P for a few common monatomic and diatomic gases. The units are J/mol · K. The values for air are essentially equal to those for N$_2$.

It's interesting that all the monatomic gases in Table 12.6 have the same value for C_P and for C_V. The same is very nearly true for the diatomic gases. Why should this be? We found in Chapter 11 that the average kinetic energy of the atoms in a gas is $K_{avg} = \frac{3}{2}k_B T$, a fact we used earlier in the kinetic-theory derivation of the pressure of an ideal gas. For an ideal monatomic gas, the thermal energy E_{th} is simply the sum total of the kinetic energies of all the atoms; there is no other way a monatomic gas can store energy. Thus the thermal energy of a gas of N atoms is

$$E_{th} = N\left(\frac{3}{2}k_B T\right) = \frac{3}{2}n(N_A k_B)T = \frac{3}{2}nRT$$

In arriving at this last expression, we first related the number of atoms to the number of moles by $N = nN_A$, then used the definition of the gas constant, $R = N_A k_B$.

If the temperature of a monatomic gas changes by ΔT, its thermal energy changes by

$$\Delta E_{th} = \frac{3}{2}nR\Delta T \qquad (12.23)$$

If we keep the volume of the gas constant so that no work is done, this energy change can come only from heat:

$$Q = \frac{3}{2}nR\Delta T \qquad (12.24)$$

Comparing Equation 12.24 with the definition of molar specific heat in Equation 12.21, we see that the molar specific heat at constant volume must be

$$C_V(\text{monatomic gas}) = \frac{3}{2}R = 12.5 \text{ J/mol} \cdot \text{K} \qquad (12.25)$$

This theoretical prediction of kinetic theory is exactly the measured value of C_V in Table 12.6. **The perfect agreement of theory and experiment is strong evidence that the atomic model of gases that we have been using is correct.**

The molar specific heat at constant pressure differs because in an isobaric (constant-pressure) process, such as process B in Figure 12.26, the gas does work. The expression for ΔE_{th} in Equation 12.23 is still valid, but now, according to the first law of thermodynamics, $\Delta E_{th} = Q - W_{gas}$. Thus

$$Q = \Delta E_{th} + W_{gas} = \frac{3}{2}nR\Delta T + p\Delta V \qquad (12.26)$$

where we used the fact, discovered earlier, that the work done in an isobaric process is $W_{gas} = p\Delta V$.

The ideal-gas law, $pV = nRT$, implies that if p is constant and only V and T change, then $p\Delta V = nR\Delta T$. Using this result in Equation 12.26 gives the heat needed to change the temperature by ΔT in a constant-pressure process:

$$Q = \frac{3}{2}nR\Delta T + nR\Delta T = \frac{5}{2}nR\Delta T$$

A comparison with the definition of molar specific heat shows that

$$C_P(\text{monatomic gas}) = \frac{5}{2}R = 20.8 \text{ J/mol} \cdot \text{K} \qquad (12.27)$$

Checking the data in Table 12.6 shows that the theory-experiment agreement remains perfect.

TABLE 12.6 Molar specific heats of gases (J/mol · K) at 20°C

Gas	C_P	C_V
Monatomic Gases		
He	20.8	12.5
Ne	20.8	12.5
Ar	20.8	12.5
Diatomic Gases		
H_2	28.7	20.4
N_2	29.1	20.8
O_2	29.2	20.9
Triatomic Gases		
Water vapor	33.3	25.0

The thermal energy of a monatomic gas consists exclusively of the translational kinetic energy of the atoms; heating a monatomic gas means that the atoms move faster. Air is primarily made of the diatomic gases N_2 and O_2. The thermal energy of a diatomic gas has more possible forms than for a monatomic gas, as shown in Figure 12.27. Heating a diatomic gas makes the atoms move faster, but it also causes them to rotate more rapidly. For this reason, the specific heat of a diatomic gas is higher than the specific heat of a monatomic gas, as we see in Table 12.6. The table shows that water vapor, a triatomic gas, has an even higher specific heat, as we expect.

The thermal energy of a diatomic gas is the sum of the translational kinetic energy of the molecules . . .

. . . plus the rotational kinetic energy of the molecules.

FIGURE 12.27 Thermal energy of a diatomic gas.

Heat Engines

Chapter 11 introduced the idea of a *heat engine,* a device that takes in heat energy from a hot reservoir, does useful work, and exhausts waste heat to a cold reservoir. Many real heat engines rely on gas processes, and we've now reached the point where we can examine heat engines more closely. A diagram of the (somewhat simplified) operation of a real heat engine, a two-stroke gasoline engine, is shown in Figure 12.28.

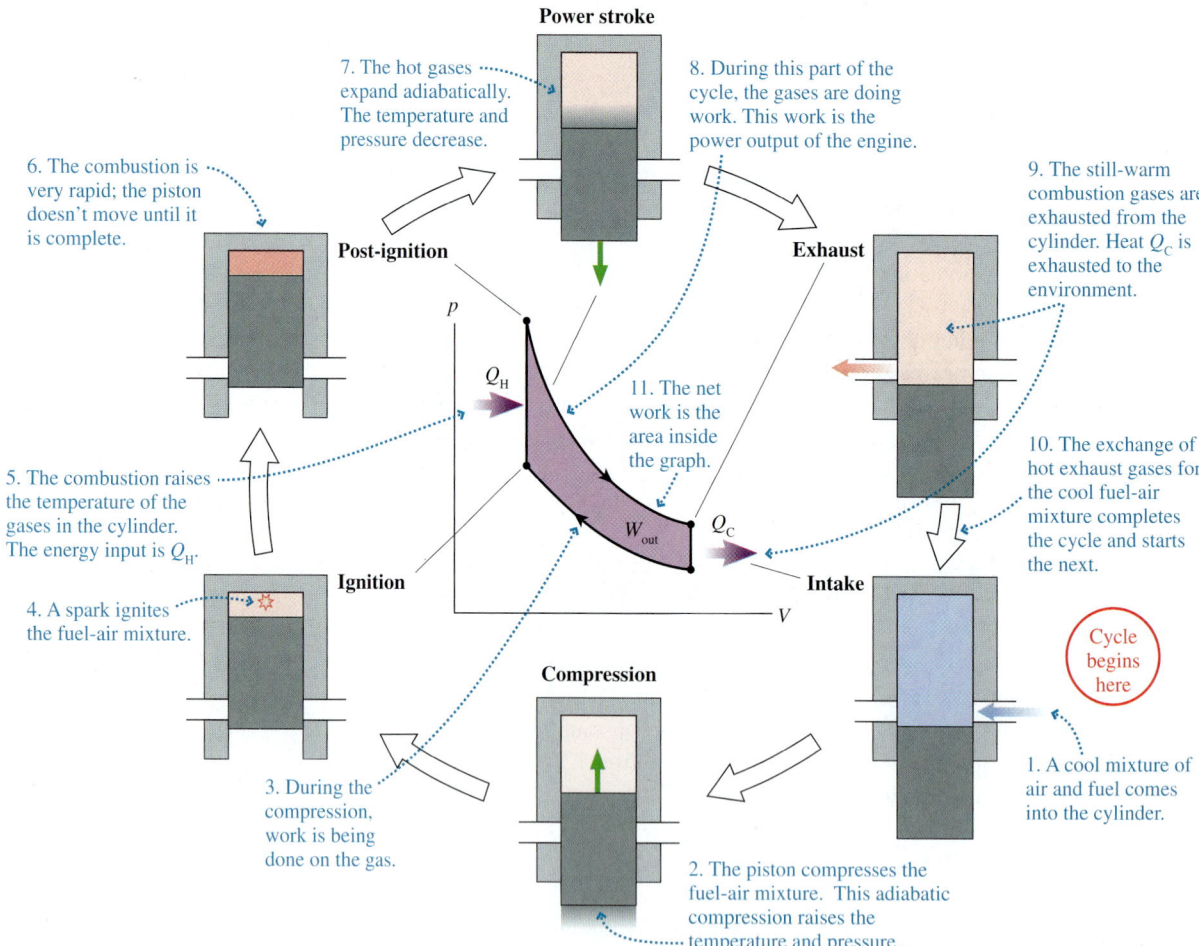

Power stroke

7. The hot gases expand adiabatically. The temperature and pressure decrease.

8. During this part of the cycle, the gases are doing work. This work is the power output of the engine.

9. The still-warm combustion gases are exhausted from the cylinder. Heat Q_C is exhausted to the environment.

6. The combustion is very rapid; the piston doesn't move until it is complete.

Post-ignition

Exhaust

Q_H

11. The net work is the area inside the graph.

10. The exchange of hot exhaust gases for the cool fuel-air mixture completes the cycle and starts the next.

5. The combustion raises the temperature of the gases in the cylinder. The energy input is Q_H.

W_{out} Q_C

Ignition

Intake

4. A spark ignites the fuel-air mixture.

Cycle begins here

Compression

1. A cool mixture of air and fuel comes into the cylinder.

3. During the compression, work is being done on the gas.

2. The piston compresses the fuel-air mixture. This adiabatic compression raises the temperature and pressure.

FIGURE 12.28 A simplified model of a gasoline engine.

NOTE ▶ It is common to use "gas" as shorthand for "gasoline." In this chapter, "gas" always refers to the state of matter. We will use "gasoline" or "fuel" to refer to the chemical energy source that you put in your tank. ◀

In this engine, the gases in the cylinder expand and contract in a repeating cycle as the piston moves back and forth. The cycle begins at the noted point with the introduction of gases (a mixture of air and fuel) into the cylinder. From this point, we can use our knowledge of gas processes to analyze the operation of the engine as it goes through its cycle. Figure 12.28 shows the connection between the operation of the engine and relevant points on the pV diagram of the gases in the engine's cylinder.

There are several important characteristics of this heat engine that apply generally to all heat engines:

- The sequence of processes forms a closed loop in the pV diagram. This must be true for the engine to go through its cycle over and over.
- Heat Q_H is taken in from a hot reservoir (in this case, from the burning fuel) at some point during the cycle. To bring the temperature back down at the end of the cycle, heat Q_C must be exhausted to a cold reservoir. This heat is carried out of the engine by the hot exhaust gases.
- The work done by the engine, W_{out}, is the area *inside* the closed loop on the pV diagram.
- Because the temperature returns to its starting point at the end of each cycle, there's no net change in thermal energy. Thus from energy conservation, $W_{out} = Q_H - Q_C$.

Work is the area inside the closed loop because of our earlier discovery that the work done in a single process is the area under the graph. The work is negative for the compression, because ΔV is negative, and positive for the expansion. Thus the net work is the area under the upper curve *minus* the area under the lower curve, which is just the area inside the closed loop. This is true for all heat engines, not just the example shown here.

 8.12–8.14

EXAMPLE 12.14 **Finding the pressure and temperature in a gasoline engine**

The adiabatic compression stroke in an automobile engine raises the pressure to 17 atm and the temperature to 300°C. The ignition of the fuel-air mixture then provides 800 J of heat. For a typical engine, the cylinder contains 0.021 mol of gas. What are the pressure and temperature immediately after ignition?

PREPARE Ignition happens so fast there is no time for the piston to move, so the gas is heated in a constant-volume process. The gas is mostly air, so we'll use $C_V = 20.8$ J/mol · K as the molar specific heat at constant volume.

SOLVE The temperature increase during a constant-volume heating is

$$\Delta T = \frac{Q}{nC_V} = \frac{800 \text{ J}}{(0.021 \text{ mol})(20.8 \text{ J/mol} \cdot \text{K})} = 1800°C$$

Thus igniting the fuel raises the temperature from $T_i = 300$°C at the end of the compression stroke to $T_f = 2100$°C. The volume doesn't change, so the new pressure is

$$p_f = p_i \frac{V_i}{V_f} \frac{T_f}{T_i} = (17 \text{ atm})(1)\frac{2100 + 273}{300 + 273} = 70 \text{ atm}$$

ASSESS The pressure and temperature are both very large, but that's to be expected following an explosion in a confined volume. When you drive your car, a series of such explosions in your engine's cylinders provide the necessary force to move your car down the road. In a typical car moving at highway speeds, there are over 100 explosions taking place under your hood each second.

12.8 Heat Transfer

You feel warmer when the sun is shining on you, colder when sitting on a cold concrete bench or when a stiff wind is blowing. This is due to the transfer of heat. Although we've talked about heat a lot in these last two chapters, we haven't said much about *how* heat is transferred from a hotter object to a colder object. There are four basic mechanisms, outlined in the table below, by which objects exchange heat with other objects or their surroundings. Evaporation was treated in an earlier section; in this section we will consider the other mechanisms.

Heat transfer mechanisms

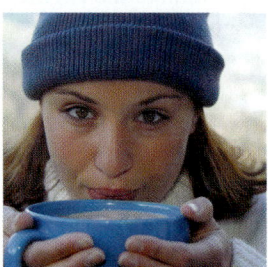

When two objects are in direct physical contact, such as the soldering iron and the circuit board, heat is transferred by *conduction*. **Energy is transferred by direct contact.**

This special photograph shows air currents near a warm glass of water. Air near the glass is warmed and rises, taking thermal energy with it in a process known as *convection*. **Energy is transferred by the motion of molecules with high thermal energy.**

The lamp shines on the lambs huddled below, warming them. The energy is transferred by infrared *radiation,* a form of electromagnetic waves. **Energy is transferred by electromagnetic waves.**

As we saw in a previous section, the *evaporation* of liquid can carry away significant quantities of thermal energy. When you blow on a cup of cocoa, this increases the rate of evaporation, rapidly cooling it. **Energy is transferred by the removal of molecules with high thermal energy.**

Conduction

If you hold a metal spoon in a cup of hot coffee, the handle of the spoon soon gets warm. Thermal energy is transferred along the spoon from the coffee to your hand. The difference in temperature between the two ends drives this heat transfer by a process known as **conduction.** Conduction is the transfer of thermal energy directly through a physical material.

Figure 12.29 shows a copper rod placed between a hot reservoir (a fire) and a cold reservoir (a block of ice). We can use our atomic model to see how thermal energy is transferred along the rod by the interaction between atoms in the rod; fast-moving atoms at the hot end transfer energy to slower-moving atoms at the cold end.

Suppose we set up a series of experiments to measure the heat Q transferred through various rods. We would find the following trends in our data:

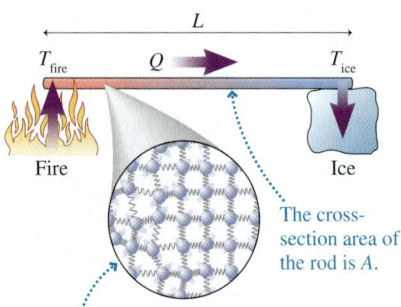

The particles on the left side of the rod are vibrating more vigorously than the particles on the right. The particles on the left transfer energy to the particles on the right via the bonds connecting them.

FIGURE 12.29 Conduction of heat in a solid rod.

- Q increases if the temperature difference ΔT between the hot end and the cold end is increased.
- Q increases if the cross-section area A of the rod is increased.
- Q decreases if the length L of the rod is increased.
- Some materials (such as metals) have a high value of Q; a lot of energy is transferred. Other materials (such as wood) have a low value of Q; very little energy is transferred.

The final observation is one that is familiar to you; if you are stirring a pot of hot soup on the stove, you use a wood or plastic spoon rather than a metal one.

These experimental observations about heat conduction can be summarized in a single formula. If heat Q is transferred in a time interval Δt, the *rate* of heat transfer (joules per second) is $Q/\Delta t$. For a material of cross-section area A and length L, spanning a temperature difference ΔT, the rate of heat transfer is

$$\frac{Q}{\Delta t} = \left(\frac{kA}{L}\right)\Delta T \qquad (12.28)$$

Rate of conduction of heat across a temperature difference

The quantity k, which characterizes whether the material is a good conductor of heat or a poor conductor, is called the **thermal conductivity** of the material. Because the heat transfer rate J/s is a *power,* measured in watts, the units of k are W/m · K. Typical values of k for common materials are given in Table 12.7; a larger number for k means a material is a better conductor of heat.

Note that good conductors of electricity, such as silver and copper, are usually good heat conductors. One exception is diamond: Though diamond is a very poor electrical conductor, the strong bonds among atoms that make diamond such a hard material lead to a rapid transfer of thermal energy.

The weak bonds among molecules make most biological materials poor conductors of heat. Muscle is a better conductor than fat, so sea mammals have thick layers of fat for insulation. Land mammals insulate their bodies with fur, birds with feathers. Both fur and feathers trap a good deal of air, so their conductivity is similar to that of air. Air, like all gases, is a poor conductor of heat, as there are no bonds between adjacent atoms.

TABLE 12.7 Thermal conductivity values (measured at 20°C)

Material	k (W/m · K)
Diamond	1000
Silver	420
Copper	400
Iron	72
Stainless steel	14
Ice	1.7
Concrete	0.8
Plate glass	0.75
Skin	0.50
Muscle	0.46
Fat	0.21
Wood	0.2
Carpet	0.04
Fur, feathers	0.02–0.06
Air (27°C, 100 kPa)	0.026

EXAMPLE 12.15 Heat loss through a window

A 0.90 m by 1.2 m window (approximately 3 ft by 4 ft) in a house is made of a single pane of 3.0-mm-thick (approximately 1/8 in) glass. The temperature inside the house is 21°C; outside, it is a frosty −5°C. What power, in watts, must the furnace provide to compensate for heat loss through the window?

PREPARE The area through which heat is conducted is

$$A = (0.90\ \text{m})(1.2\ \text{m}) = 1.1\ \text{m}^2$$

The length of the conducting medium is the glass thickness, $L = 0.0030$ m. The temperature difference across the glass is

$$T = 21°C - (-5°C) = 26°C = 26\ \text{K}.$$

SOLVE Heat is transferred through the window to the outside. To keep the house at a constant temperature, the furnace must add heat energy at the same rate. The rate of conduction through the window is

$$\frac{Q}{\Delta t} = \left(\frac{(0.75\ \text{W/m · K})(1.1\ \text{m}^2)}{0.0030\ \text{m}}\right)(26\ \text{K}) = 7100\ \text{W}$$

To compensate, the furnace must provide 7100 W of heating power.

ASSESS This is a great deal of power—three or four times the power of common electric space heaters. In modern construction, windows generally consist of two panes of glass with an air space in between. The low thermal conductivity of this air gap dramatically reduces the energy loss by conduction.

Carpet or tile? If you walk around your house barefoot, you will notice that tile floors feel much colder than carpeted floors. The tile floor isn't really colder—the temperatures of all the floors in your house are nearly the same. But tile has a much higher thermal conductivity than carpet, so more heat will flow from your feet into the tile than into carpet. This causes the tile to feel colder. Standing or lying on carpet, with its low thermal conductivity, is much more comfortable than on tile!

Warm water (colored) moves by convection.

A feather coat BIO Penguins are birds and thus have feathers, even though they do not fly. The feathers are more obvious on this penguin chick than on the adults. A penguin's short, dense feathers serve a different role than the flight feathers of other birds: They trap air to provide thermal insulation. The feathers are equipped with muscles at the base of the shafts to flatten them and eliminate these air pockets, otherwise the penguins would be too buoyant to swim underwater.

Convection

We noted that air is a poor conductor of heat. In fact, the data in Table 12.7 show that air has a thermal conductivity that is comparable to that of feathers. So why, on a cold day, will you be more comfortable if you are wearing a down jacket?

Thermal energy is easily transferred through air, water, and other fluids when the air and water can flow. A pan of water on the stove is heated at the bottom. This heated water expands and becomes less dense than the water above it, so it rises to the surface while cooler, denser water sinks to take its place. This transfer of thermal energy by the motion of a fluid is known as **convection.**

Convection is usually the main mechanism for heat transfer in fluid systems. On a small scale, convection mixes the pan of water that you heat on the stove; on a large scale, convection is responsible for making the wind blow and ocean currents circulate. Air is a very poor thermal conductor, but it is very effective at transferring energy by convection. To use air for thermal insulation, it is necessary to trap the air in small pockets to limit convection. And that's exactly what feathers, fur, double-paned windows, and fiberglass insulation do. Convection is much more rapid in water than in air, which is why people can die of hypothermia in 65°F water but can live quite happily in 65°F air.

Forced convection occurs when fluid is pushed by a pump or motor. In most buildings, forced hot-air heating transfers heat from the furnace to the rooms. Forced convection is also important for regulating body temperature. If your body is too hot, blood vessels near the cooler skin dilate to carry more blood. Blood flow transfers heat from the core of your body to the skin and then to your surroundings. Elephants use their ears for forced convection as well. The motion of air across the body from flapping ears is a very effective means of keeping cool.

Radiation

The sunlight that falls on the earth increases the thermal energy of the earth's surface each day. But the earth cools down at night; clearly, there must be some way for the earth to transfer energy back to space.

The earth and all other objects emit energy in the form of **radiation,** electromagnetic waves that are generated by oscillating electric charges in the atoms that form the object, an issue we'll explore further in later chapters on electricity and magnetism. These waves carry energy, transferring energy from the object that emits the radiation to the object that absorbs it. Electromagnetic waves carry energy from the sun; when this sunlight falls on your skin, the energy is absorbed, increasing your skin's thermal energy. Your skin also emits electromagnetic radiation, which reduces your skin's thermal energy. Radiation is a significant part of the energy balance that keeps your body at the proper temperature.

> **NOTE ▶** The word "radiation" comes from "radiate," which means "to beam." You have likely used the word to refer to x-rays and other forms of *ionizing radiation.* This is not the sense we will use in this chapter. Here, we will use radiation to mean light and other forms of electromagnetic waves that can "beam" from an object. ◀

Visible light, the red-to-violet colors of the rainbow, is an electromagnetic wave. Electromagnetic waves of somewhat lower frequency are called *infrared radiation;* we can't see these waves, but we can sometimes sense them as "heat" as they are absorbed on our skin. All objects radiate; objects at room temperature and even somewhat higher temperatures emit virtually all of their radiation as infrared radiation. As the temperature increases, the charged particles oscillate more vigorously and an object becomes "red hot" and eventually "white hot"— some of the radiation is now in the form of visible light. The white light from an incandescent light bulb is radiation emitted by a thin wire filament that has been heated to a very high temperature by an electric current. The radiation that the earth emits, or that you emit, is no different, it is simply lower-frequency infrared radiation that you can't see.

Even though we can't see infrared radiation, some films and detectors are infrared sensitive and can record the infrared radiation from objects. The false-color thermal image of the elephant that opened this chapter shows the infrared emission as the elephant radiates energy into the cooler environment. Warmer patches, such as the trunk, emit more radiation and are easily distinguished from cooler patches that radiate less. Elephants do not sweat, so radiation from their trunk and ears is an important part of their temperature-regulation system.

If an object's temperature is increased, the increasingly vigorous oscillations radiate more energy. The energy radiated by any object thus shows a strong dependence on temperature. If heat energy Q is radiated in a time interval Δt by an object with surface area A and absolute temperature T, the *rate* of heat transfer $Q/\Delta t$ (joules per second) is found to be

$$\frac{Q}{\Delta t} = e\sigma A T^4 \qquad (12.29)$$

Rate of heat transfer by radiation at temperature T (Stefan's Law)

Quantities in this equation are defined as follows:

- e is the **emissivity** of a surface, a measure of the effectiveness of radiation. The value of e ranges from 0 to 1. Human skin is a very effective radiator at body temperature, with $e = 0.97$, regardless of skin color.
- T is the absolute temperature in kelvin.
- A is the surface area in m^2.
- σ is a constant known as the Stefan-Boltzmann constant, with the value $\sigma = 5.67 \times 10^{-8}$ W/m$^2 \cdot$ K^4.

Notice the very strong fourth-power dependence on temperature. Doubling the absolute temperature of an object increases the radiant heat transfer by a factor of 16!

The amount of energy radiated by an animal can be surprisingly large. An adult human with bare skin in a room at a comfortable temperature has a skin temperature of approximately 33°C, or 306 K. A typical value for the skin's surface area is 1.8 m^2. With these values and the emissivity of skin noted above, we can calculate the rate of heat transfer via radiation from the skin:

$$\frac{Q}{\Delta t} = e\sigma A T^4 = (0.97)\left(5.67 \times 10^{-8} \frac{W}{m^2 \cdot K^4}\right)(1.8 \text{ m}^2)(306 \text{ K})^4 = 870 \text{ W}$$

As we learned in Chapter 11, the body at rest generates approximately 100 W of thermal energy. If the body radiated energy at 870 W, it would quickly cool. At this rate of emission, the body temperature would drop by 1°C every seven

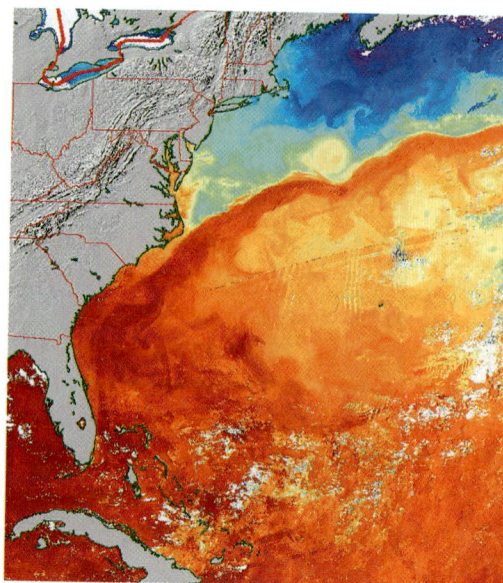

Global heat transfer This satellite image shows radiation emitted by the waters of the ocean off the east coast of the United States. You can clearly see the warm waters of the Gulf Stream, a large-scale convection that transfers heat to northern latitudes. The satellite "sees" the radiation from the earth, so this radiation must readily pass through the atmosphere into space. In fact, the *only* means by which the earth as a whole can reduce its thermal energy is by radiation, as there is no conduction or convection in the vacuum of space. This has consequences for the energy balance of the earth, as we consider in a section on the greenhouse effect at the end of Part III.

The infrared radiation emitted by a dog is captured in this image. His cool nose and paws radiate much less energy than the rest of his body.

minutes! Clearly, there must be some mechanism to balance this emitted radiation: the radiation *absorbed* by the body.

When you sit in the sun, your skin warms due to the radiation you absorb. Even if you are not in the sun, you are absorbing the radiation emitted by the objects surrounding you. Suppose an object at temperature T is surrounded by an environment at temperature T_0. The *net* rate at which the object radiates heat energy—that is, radiation emitted minus radiation absorbed—is

$$\frac{Q_{net}}{\Delta t} = e\sigma A(T^4 - T_0^4) \qquad (12.30)$$

This makes sense. An object should have no net energy transfer by radiation if it's in thermal equilibrium ($T = T_0$) with its surroundings. Note that the emissivity e appears for absorption as well; objects that are good emitters are good absorbers as well.

EXAMPLE 12.16 Determining energy loss by radiation for the body

A person with skin temperature of 33°C is in a room at 24°C. What is the net rate of heat transfer by radiation?

SOLVE Body temperature is $T = 33 + 273 = 306$ K; the temperature of the room is $T_0 = 24 + 273 = 297$ K. The net radiation rate, given by Equation 12.30, is

$$\frac{Q_{net}}{\Delta t} = e\sigma A(T^4 - T_0^4)$$

$$= (0.97)\left(5.67 \times 10^{-8}\ \frac{W}{m^2 \cdot K^4}\right)(1.8\ m^2)\left[(306\ K)^4 - (297\ K)^4\right] = 98\ W$$

ASSESS This is a reasonable value, roughly matching your resting metabolic power. When you are dressed (little convection) and sitting on wood or plastic (little conduction), radiation is your body's primary mechanism for dissipating the excess thermal energy of metabolism.

STOP TO THINK 12.6   Suppose you are an astronaut in space, hard at work in your sealed spacesuit. The only way that you can transfer heat to the environment is by

A. Conduction.   B. Convection.   C. Radiation.   D. Evaporation.

The goal of Chapter 12 has been to use the atomic model of matter to explain and explore many macroscopic phenomena associated with heat, temperature, and the properties of matter.

GENERAL PRINCIPLES

Atomic Model

We model matter as being made of simple basic particles. The relationship of these particles to each other defines the phase.

 Gas   Liquid   Solid

The atomic model explains thermal expansion, specific heat, and heat transfer.

Kinetic Theory

Macroscopic properties of gases can be explained in terms of a microscopic view. The energy of motion of the particles of the gas is related to the temperature of the gas; the collisions of the particles with each other and the walls of the container are related to the pressure.

p, V, T

Ideal-Gas Law

The ideal-gas law can be derived from kinetic theory. It relates volume, pressure, and temperature in gases. The general law is

$$pV = nRT$$

For a gas process in a sealed container:

$$\frac{p_i V_i}{T_i} = \frac{p_f V_f}{T_f}$$

The gauge pressure of a gas is:

$$p_g = p - 1 \text{ atm}$$

IMPORTANT CONCEPTS

Effects of heat transfer

A system that is heated can either change temperature or change phase.

The **specific heat** c of a material is the heat required to raise 1 kg by 1 K.

$$Q = Mc\,\Delta T$$

The **heat of transformation** is the energy necessary to change the phase of 1 kg of a substance. Heat is added to change a solid to a liquid or a liquid to a gas; heat is removed to reverse these changes.

$$Q = \begin{cases} \pm ML_f \text{ (melt/freeze)} \\ \pm ML_v \text{ (boil/condense)} \end{cases}$$

The **molar specific heat** of a gas depends on the process.

$$\begin{cases} \text{For a constant-} \\ \text{volume process:} \quad Q = nC_V\,\Delta T \\ \text{For a constant-} \\ \text{pressure process:} \quad Q = nC_P\,\Delta T \end{cases}$$

Mechanisms of heat transfer

An object can transfer heat to other objects or its environment:

Conduction is the transfer of heat by direct physical contact.

$$\frac{Q}{\Delta t} = \left(\frac{kA}{L}\right)\Delta T$$

Convection is the transfer of heat by the motion of a fluid.

Radiation is the transfer of heat by electromagnetic waves.

$$\frac{Q}{\Delta t} = e\sigma A T^4$$

A *pV* (pressure-volume) diagram is a useful means of looking at a process involving a gas.

A **constant-volume** process has no change in volume.

An **isobaric** process happens at a constant pressure.

An **isothermal** process happens at a constant temperature.

An **adiabatic** process involves no transfer of heat; the temperature changes.

The work done by a gas is the area under the graph.

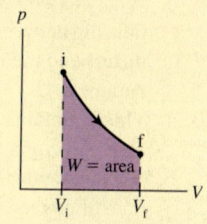

APPLICATIONS

Thermal expansion Objects experience an increase in volume and an increase in length when their temperature changes.

$$\Delta V = \beta V_i\,\Delta T \qquad \Delta L = \alpha L_i\,\Delta T$$

Calorimetry When two or more systems interact thermally, they come to a common final temperature determined by

$$Q_{net} = Q_1 + Q_2 + Q_3 + \cdots = 0$$

The number of **moles** is

$$n = \frac{M \text{ (in grams)}}{M_{mol}}$$

For instructor-assigned homework, go to www.masteringphysics.com

Problem difficulty is labeled as | (straightforward) to |||| (challenging).

Problems labeled INT integrate significant material from earlier chapters; BIO are of biological or medical interest.

QUESTIONS

Conceptual Questions

1. You need to precisely measure the dimensions of a large wood panel for a construction project. Your metal tape measure was left outside for hours in the sun on a hot summer day, and now the tape is so hot it's painful to pick up. How will your measurements differ from those taken by your coworker, whose tape stayed in the shade? Explain.

2. If you buy a sealed bag of potato chips in Miami and drive with it to Denver, where the atmospheric pressure is lower, you will find that the bag gets very "puffy." Explain why.

3. If the pressure of a gas is really due to the *random* collisions of molecules with the walls of the container, why do pressure gauges—even very sensitive ones—give perfectly steady readings? Shouldn't the gauge be continually jiggling and fluctuating? Explain.

4. If you double the average speed of the molecules in a gas, by what factor does the pressure change? Give a simple explanation why the pressure changes by this factor.

5. Two gases have the same number of molecules per cubic meter (N/V) and the same rms speed. The molecules of gas 2 are more massive than the molecules of gas 1.
   a. Do the two gases have the same pressure? If not, which is larger?
   b. Do the two gases have the same temperature? If not, which is larger?

6. According to kinetic theory, the pressure of a gas depends on the number of molecules per cubic meter (N/V) and the rms speed of the gas molecules. Consider a sealed container of gas that is heated at constant volume.
   a. According to the ideal-gas law, does the pressure of the gas increase or stay the same? Explain.
   b. Does the number density of the gas increase or stay the same? Explain.
   c. What can you infer from these observations about a relationship between the gas temperature (a macroscopic parameter) and the rms speed of the molecules (a microscopic parameter)?

7. a. Which contains more particles, a mole of helium gas or a mole of oxygen gas? Explain.
   b. Which contains more particles, a gram of helium gas or a gram of oxygen gas? Explain.

8. You have 100 g of aluminum and 100 g of lead.
   a. Which contains a larger number of moles? Explain.
   b. Which contains more atoms? Explain.

9. Suppose you could suddenly increase the speed of every atom in a gas by a factor of 2.
   a. Does the thermal energy of the gas change? If so, by what factor? If not, why not?
   b. Does the molar specific heat change? If so, by what factor? If not, why not?

10. A gas is in a sealed container. By what factor does the gas pressure change if
    a. The volume is doubled and the temperature is tripled?
    b. The volume is halved and the temperature is tripled?

11. A gas is in a sealed container. By what factor does the gas temperature change if
    a. The volume is doubled and the pressure is tripled?
    b. The volume is halved and the pressure is tripled?

12. A gas is in a rigid container. The gas pressure is tripled and the temperature is doubled. What happens to the number of moles of gas in the container?

13. What is the maximum amount of work that a gas can do during a constant-volume process?

14. Materials A and B have equal densities, but A has a larger specific heat than B. You have 100 g cubes of each material. Cube A, initially at 0°C, is placed in good thermal contact with cube B, initially at 200°C. The cubes are inside a well-insulated container where they don't interact with their surroundings. Is their final temperature greater than, less than, or equal to 100°C? Explain.

15. Two containers hold equal masses of nitrogen gas at equal temperatures. You supply 10 J of heat to container A while not allowing its volume to change, and you supply 10 J of heat to container B while not allowing its pressure to change. Afterward, is temperature T_A greater than, less than, or equal to T_B? Explain.

16. You need to raise the temperature of a gas by 10°C. To use the smallest amount of heat energy, should you heat the gas at constant pressure or at constant volume? Explain.

17. The *rapid* compression of a gas by a fast-moving piston increases the gas temperature. For example, you likely have noticed that pumping up a bicycle tire causes the bottom of the pump to get warm. Consider a gas that is rapidly compressed by a piston.
    a. Does the thermal energy of the gas increase or stay the same? Explain.
    b. Is there a transfer of heat energy to the gas? Explain.
    c. Is work done on the gas? Explain.

18. A student is heating chocolate in a pan on the stove. He uses a cooking thermometer to measure the temperature of the chocolate and sees it varies as shown in Figure Q12.18. Describe what is happening to the chocolate in each of the three portions of the graph.

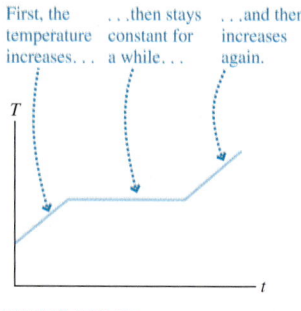

FIGURE Q12.18

19. In early spring, just as the ice on the surface of a lake finishes melting, a scientist measures the water temperature as a function of depth. Describe qualitatively how the temperature should change as the depth increases. Compare this result to the pattern that would be expected in the middle of summer.

20. If you bake a cake at high elevation, where atmospheric pressure is lower than at sea level, you will need to adjust the recipe. You will need to cook the cake for a longer time, and you will need to add less baking powder. (Baking powder is a leavening agent. As it heats, it releases gas bubbles that cause the cake to rise.) Explain why those adjustments are necessary.

21. The specific heat of aluminum is higher than that of iron. 1 kg blocks of iron and aluminum are heated to 100°C, and each is then dropped into its own 1 L beaker of 20°C water. Which beaker will end up with the warmer water? Explain.

22. A student is asked to sketch a pV diagram for a gas that goes through a cycle consisting of (a) an isobaric expansion, (b) a constant-volume reduction in temperature, and (c) an isothermal process that returns the gas to its initial state. The student draws the diagram shown in Figure Q12.22. What, if anything, is wrong with the student's diagram?

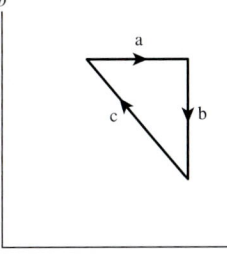

FIGURE Q12.22

23. If you have two spoons of the same size, one silver and one stainless steel, there is a quick test to tell which is which. Hold the end of a spoon in each hand, then lower them both into a cup of very hot water. One spoon will feel hot first. Is that the silver spoon or the stainless steel? Explain.

24. People with body piercings are advised to remove their BIO exposed body jewelry before engaging in cold-weather activities. The skin surrounding the metal in earrings, lip rings, nose rings, or eyebrow rings can get frostbite much more easily than other skin. Explain why.

25. If you live somewhere with cold, clear nights, you may have noticed some mornings when there was frost on open patches of ground but not under trees. This is because the ground under trees does not get as cold as open ground. Explain how tree cover keeps the ground under trees warmer.

Multiple-Choice Questions

26. | A tire is inflated to a gauge pressure of 35 psi. The absolute pressure in the tire is
    A. Less than 35 psi.
    B. Equal to 35 psi
    C. Greater than 35 psi.

27. | The number of atoms in a container is increased by a factor of 2 while the temperature is held constant. The pressure
    A. Decreases by a factor of 4.
    B. Decreases by a factor of 2.
    C. Stays the same.
    D. Increases by a factor of 2.
    E. Increases by a factor of 4.

28. | Suppose you do a calorimetry experiment to measure the specific heat of a penny. You take a number of pennies, measure their mass, heat them to a known temperature, and then drop them into a container of water at a known temperature. You then deduce the specific heat of a penny by measuring the temperature change of the water. Unfortunately, you didn't realize that you dropped one penny on the floor while transferring them to the water. This will
    A. Cause you to underestimate the specific heat.
    B. Cause you to overestimate the specific heat.
    C. Not affect your calculation of specific heat.

29. | A cup of water is heated with a heating coil that delivers 100 W of heat. In one minute, the temperature of the water rises by 20°C. What is the mass of the water?
    A. 72 g   B. 140 g
    C. 720 g   D. 1.4 kg

30. | Three identical beakers each hold 1000 g of water at 20°C. 100 g of liquid water at 0°C is added to the first beaker, 100 g of ice at 0°C is added to the second beaker, and the third beaker gets 100 g of aluminum at 0°C. The contents of which container end up at the lowest final temperature?
    A. The first beaker.
    B. The second beaker.
    C. The third beaker.
    D. All end up at the same temperature.

31. ‖ 100 g of ice at 0°C and 100 g of steam at 100°C interact thermally in a well-insulated container. The final state of the system is
    A. An ice-water mixture at 0°C.
    B. Water at a temperature between 0°C and 50°C.
    C. Water at 50°C.
    D. Water at a temperature between 50°C and 100°C.
    E. A water-steam mixture at 100°C.

32. | Suppose the 600 W of radiation emitted in a microwave oven is absorbed by a very lightweight cup holding 250 g of water. Approximately how long will it take to heat the water from 20°C to 80°C?
    A. 50 s   B. 100 s
    C. 150 s   D. 200 s

33. ‖ 40,000 J of heat is added to 1.0 kg of ice at −10°C. How much ice melts?
    A. 0.012 kg   B. 0.057 kg
    C. 0.12 kg   D. 1.0 kg

34. | Steam at 100°C causes worse burns than liquid water at BIO 100°C. This is because
    A. The steam is hotter than the water.
    B. Heat is transferred to the skin as steam condenses.
    C. Steam has a higher specific heat than water.
    D. Evaporation of liquid water on the skin causes cooling.

PROBLEMS

Section 12.1 The Atomic Model of Matter

1. | Which contains the most moles: 10 g of hydrogen gas, 100 g of carbon, or 50 g of lead?
2. ‖‖ How many grams of water (H_2O) have the same number of oxygen atoms as 1.0 mol of oxygen gas?
3. ‖ How many atoms of hydrogen are in 100 g of hydrogen peroxide (H_2O_2)?
4. | How many cubic millimeters (mm^3) are in 1 L?
5. | A box is 200 cm wide, 40 cm deep, and 3.0 cm high. What is its volume in m^3?

Section 12.2 Thermal Expansion

6. | A straight rod consists of a 1.2-cm-long piece of aluminum attached to a 2.0-cm-long piece of steel. By how much will the length of this rod change if its temperature is increased from 20°C to 40°C?
7. | The length of a steel beam increases by 0.73 mm when its temperature is raised from 22°C to 35°C. What is the length of the beam at 22°C?
8. | Older railroad tracks in the U.S. are made of 12-m-long pieces of steel. When the tracks are laid, gaps are left between the sections to prevent buckling when the steel thermally expands. If a track is laid at 16°C, how large should the gaps be if the track is not to buckle when the temperature is as high as 50°C?
9. | The temperature of an aluminum disk is increased by 120°C. By what percentage does its volume increase?
10. ‖‖ If a container holds 0.865 L of ethyl alcohol at 12.0°C, what is the volume of the alcohol when its temperature is raised to 35.0°C?

Section 12.3 Pressure and the Kinetic Theory of an Ideal Gas

11. | What is the absolute pressure of the air in your car's tires, in psi, when your pressure gauge indicates they are inflated to 35.0 psi? Assume you are at sea level.
12. ‖ BIO Total lung capacity of a typical adult is approximately 5.0 L. Approximately 20% of the air is oxygen. At sea level and at an average body temperature of 37°C, how many moles of oxygen do the lungs contain at the end of an inhalation?
13. ‖‖ BIO Many cultures around the world still use a simple weapon called a blowgun, a tube with a dart that fits tightly inside. A sharp breath into the end of the tube launches the dart. When exhaling forcefully, a healthy person can supply air at a gauge pressure of 6.0 kPa. What force does this pressure exert on a dart in a 1.5-cm-diameter tube?
14. ‖‖ BIO When you stifle a sneeze, you can damage delicate tissues because the pressure of the air that is not allowed to escape may rise by up to 45 kPa. If this extra pressure acts on the inside of your 8.4-mm-diameter eardrum, what is the outward force?
15. ‖‖‖ 7.5 mol of helium are in a 15 L cylinder. The pressure gauge on the cylinder reads 65 psi. What are (a) the temperature of the gas in °C and (b) the average kinetic energy of a helium atom?

16. | 3.0 mol of gas at a temperature of −120°C fills a 2.0 L container. What is the gas pressure?
17. ‖ Determine the number of moles of oxygen gas in the following containers.
    a. Container A holds 0.30 m^3 at 25°C and an absolute pressure of 10 kPa.
    b. Container B holds 0.30 m^3 at 59°F and an absolute pressure of 12 kPa.
    c. Container C holds 0.10 m^3 at 25°C and a gauge pressure of 30 kPa.
18. ‖ Determine the temperature (in °C) of the gases in the following containers.
    a. Container A, with volume 0.10 m^3, holds 2.0 mol of helium at an absolute pressure of 1.0 atm.
    b. Container B, with volume 0.20 m^3, holds 2.0 mol of argon at a gauge pressure of 0.50 atm.
    c. Container C, with volume 100 L holds 2.0 mol of neon at an absolute pressure of 3.0 atm.
19. ‖‖‖ 10 g of liquid water is placed in a flexible bag, the air is excluded, and the bag is sealed. It is then placed in a microwave oven where the water is boiled to make steam at 100°C. What is the volume of the bag after all the water has boiled? Assume that the pressure inside the bag is equal to atmospheric pressure.

Section 12.4 Ideal-Gas Processes

20. ‖ A cylinder contains 3.0 L of oxygen at 300 K and 2.4 atm. The gas is heated, causing a piston in the cylinder to move outward. The heating causes the temperature to rise to 600 K and the volume of the cylinder to increase to 9.0 L. What is the final gas pressure?
21. | A gas with initial conditions p_i, V_i, and T_i expands isothermally until $V_f = 2V_i$. What are (a) T_f and (b) p_f?
22. ‖ 0.10 mol of argon gas is admitted to an evacuated 50 cm^3 container at 20°C. The gas then undergoes heating at constant volume to a temperature of 300°C.
    a. What is the final pressure of the gas?
    b. Show the process on a pV diagram. Include a proper scale on both axes.
23. | 0.10 mol of argon gas is admitted to an evacuated 50 cm^3 container at 20°C. The gas then undergoes an isobaric heating to a temperature of 300°C.
    a. What is the final volume of the gas?
    b. Show the process on a pV diagram. Include a proper scale on both axes.
24. ‖ 0.10 mol of argon gas is admitted to an evacuated 50 cm^3 container at 20°C. The gas then undergoes an isothermal expansion to a volume of 200 cm^3.
    a. What is the final pressure of the gas?
    b. Show the process on a pV diagram. Include a proper scale on both axes.

25. | 0.0040 mol of gas undergoes the process shown in Figure P12.25.
    a. What type of process is this?
    b. What are the initial and final temperatures?

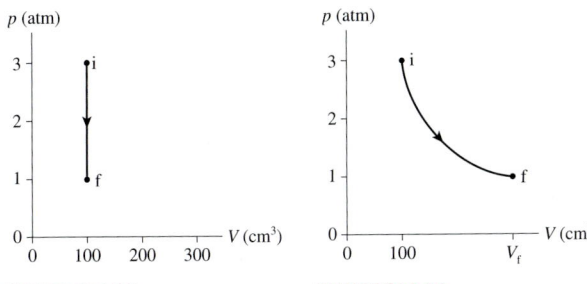

FIGURE P12.25   **FIGURE P12.26**

26. ||| 0.0040 mol of gas follows the hyperbolic trajectory shown in Figure P12.26.
    a. What type of process is this?
    b. What are the initial and final temperatures?
    c. What is the final volume V_f?

27. || A gas with an initial temperature of 900°C undergoes the process shown in Figure P12.27.
    a. What type of process is this?
    b. What is the final temperature?
    c. How many moles of gas are there?

FIGURE P12.27   **FIGURE P12.28**

28. || How much work is done on the gas in the process shown in Figure P12.28?

29. ||| It is possible to make a thermometer by sealing gas in a rigid container and measuring the pressure. Such a constant-volume gas thermometer is placed in an ice-water bath at 0.00°C. After reaching thermal equilibrium, the gas pressure is recorded as 55.9 kPa. The thermometer is then placed in contact with a sample of unknown temperature. After the thermometer reaches a new equilibrium, the gas pressure is 65.1 kPa. What is the temperature of this sample?

Section 12.5 Specific Heat and Heat of Transformation

30. | How much energy must be removed from a 200 g block of ice to cool it from 0°C to −30°C?

31. || How much heat is needed to change 20 g of mercury at 20°C into mercury vapor at the boiling point?

32. | a. 100 J of heat energy are transferred to 20 g of mercury. By how much does the temperature increase?
    b. How much heat is needed to raise the temperature of 20 g of water by the same amount?

33. ||| The maximum amount of water an adult in temperate climates can perspire in one hour is typically 1.8 L. However, after several weeks in a tropical climate the body can adapt, increasing the maximum perspiration rate to 3.5 L/hr. At what rate, in watts, is energy being removed when perspiring that rapidly? Assume all of the perspired water evaporates. At body temperature, the heat of vaporization of water is $L_v = 24 \times 10^5$ J/kg.

34. ||| When air is inhaled, it quickly becomes saturated with water vapor as it passes through the moist upper airways. When breathing dry air, about 25 mg of water are exhaled with each breath. At 12 breaths/min, what is the rate of energy loss due to evaporation? Express your answer in both watts and Calories per day. At body temperature, the heat of vaporization of water is $L_v = 24 \times 10^5$ J/kg.

35. || What minimum heat is needed to bring 100 g of water at 20°C to the boiling point and completely boil it away?

Section 12.6 Calorimetry

36. || 30 g of copper pellets are removed from a 300°C oven and immediately dropped into 100 mL of water at 20°C in an insulated cup. What will the new water temperature be?

37. | A copper block is removed from a 300°C oven and dropped into 1.00 kg of water at 20.0°C. The water quickly reaches 25.5°C and then remains at that temperature. What is the mass of the copper block?

38. ||| A 750 g aluminum pan is removed from the stove and plunged into a sink filled with 10 kg of water at 20.0°C. The water temperature quickly rises to 24.0°C. What was the initial temperature of the pan?

39. ||| A 500 g metal sphere is heated to 300°C, then dropped into a beaker containing 4.08 kg of mercury at 20.0°C. A short time later the mercury temperature stabilizes at 99.0°C. Identify the metal.

40. ||| Marianne really likes coffee, but on summer days she doesn't want to drink a hot beverage. If she is served 200 mL of coffee at 80°C in a well-insulated container, how much ice at 0°C should she add to obtain a final temperature of 30°C?

Section 12.7 Thermal Properties of Gases

41. | A container holds 1.0 g of argon at a pressure of 8.0 atm.
    a. How much heat is required to increase the temperature by 100°C at constant volume?
    b. How much will the temperature increase if this amount of heat energy is transferred to the gas at constant pressure?

42. | A container holds 1.0 g of oxygen at a pressure of 8.0 atm.
    a. How much heat is required to increase the temperature by 100°C at constant pressure?
    b. How much will the temperature increase if this amount of heat energy is transferred to the gas at constant volume?

43. | What is the temperature change of 1.0 mol of a monatomic gas if its thermal energy is increased by 1.0 J?

44. || The temperature of 2.0 g of helium is increased at constant volume by ΔT. What mass of oxygen can have its temperature increased by the same amount at constant volume using the same amount of heat?

45. ▚ How much work is done per cycle by a gas following the pV trajectory of Figure P12.45?

FIGURE P12.45   **FIGURE P12.46**

46. ▚ A gas following the pV trajectory of Figure P12.46 does 60 J of work per cycle. What is p_{max}?

Section 12.8 Heat Transfer

47. ▎ A 1.8-cm-thick wood floor covers a 4.0 m × 5.5 m room. The subfloor on which the flooring sits is at a temperature of 16.2°C, while the air in the room is at 19.6°C. What is the rate of heat conduction through the floor?

48. ▚ A copper-bottomed kettle, its bottom 24 cm in diameter and 3.0 mm thick, sits on a burner. The kettle holds boiling water, and energy flows into the water from the kettle bottom at 800 W. What is the temperature of the bottom surface of the kettle?

49. ▚ What is the rate of energy transfer by radiation from a metal cube 2.0 cm on a side that is at 700°C? Its emissivity is 0.20.

50. ▌ What is the greatest possible rate of energy transfer by radiation for a 5.0-cm-diameter sphere that is at 100°C?

51. ▚ Seals may cool themselves by

BIO using *thermal windows,* patches on their bodies with much higher than average surface temperature. Suppose a seal has a 0.030 m² thermal window at a temperature of 30°C. If the seal's surroundings are a frosty −10°C, what is the net

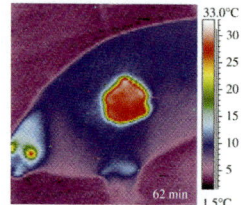

rate of energy loss by radiation? Assume an emissivity equal to that of a human.

52. ▌ The glowing filament in a lamp is radiating energy at a rate of 60 W. At the filament's temperature of 1500°C, the emissivity is 0.23. What is the surface area of the filament?

General Problems

53. ▌ A rigid container holds 2.0 mol of gas at a pressure of 1.0 atm and a temperature of 30°C.
   a. What is the container's volume?
   b. What is the pressure if the temperature is raised to 130°C?

54. ▌ A 15-cm-diameter compressed-air tank is 50 cm tall. The pressure at 20°C is 150 atm.
   a. How many moles of air are in the tank?
   b. What volume would this air occupy at STP?

55. ▚ A 20-cm-diameter cylinder that is 40 cm long contains 50 g of oxygen gas at 20°C.
   a. How many moles of oxygen are in the cylinder?
   b. How many oxygen molecules are in the cylinder?
   c. What is the reading of a pressure gauge attached to the tank?

56. ▚ A 10-cm-diameter cylinder of helium gas is 30 cm long and at 20°C. The pressure gauge reads 120 psi.
   a. How many helium atoms are in the cylinder?
   b. What is the mass of the helium?

57. ▚ An empty room in a house near the Atlantic coast has a volume of 45 m³. What is the total kinetic energy of all of the gas molecules in the room on a 30°C summer day?

58. ▚ Suppose you take a deep breath on a chilly day, inhaling

BIO 4.0 L of air at −10°C. Assume that air pressure is 1.0 atm.
   a. How much heat must your body supply to warm the air to your internal body temperature of 37°C?
   b. How much does the volume of the air increase as it is warmed?

59. ▌ On average, each person in the industrialized world is responsible for the emission of 10,000 kg of carbon dioxide (CO_2) every year. This includes CO_2 that you generate directly, by burning fossil fuels to operate your car or your furnace, as well as CO_2 generated on your behalf by electric generating stations and manufacturing plants. CO_2 is a greenhouse gas that contributes to global warming. If you were to store your yearly CO_2 emissions in a cube at STP, how long would each edge of the cube be?

60. ▚ On a cool morning, when the temperature is 15°C, you measure the pressure in your car tires to be 30 psi. After driving 20 mi on the freeway, the temperature of your tires is 45°C. What pressure will your tire gauge now show?

61. ▚ Suppose you inflate your car tires to 35 psi on a 20°C day. Later, the temperature drops to 0°C. What is the pressure in your tires now?

62. ▎ The volume in a constant-*pressure* gas thermometer is directly proportional to the absolute temperature. A constant-pressure thermometer is calibrated by adjusting its volume to 1000 mL while it is in contact with a reference cell at 0.01°C. The volume increases to 1638 mL when the thermometer is placed in contact with a sample. What is the sample's temperature in °C?

63. ▌ A compressed-air cylinder is known to fail if the pressure exceeds 110 atm. A cylinder that was filled to 25 atm at 20°C is stored in a warehouse. Unfortunately, the warehouse catches fire and the temperature reaches 950°C. Does the cylinder explode?

64. ▌ A rigid sphere has a valve that can be opened or closed. The sphere with the valve open is placed in boiling water in a room where the air pressure is 1.0 atm. After a long period of time has elapsed, the valve is closed. What will be the pressure inside the sphere if it is then placed in (a) a mixture of ice and water and (b) an insulated box filled with dry ice, which is at −78.5°C?

65. ▚ 80 J of work are done on the gas in the process shown in Figure P12.65. What is V_f in cm³?

FIGURE P12.65   **FIGURE P12.66**

66. ‖ How much work is done by the gas in the process shown in Figure P12.66?

67. ‖ A container with 100 cm³ of gas is pressurized at constant volume until the gas pressure triples. It is then expanded at constant temperature until the pressure is one-half of the pressure the gas had before the experiment started.
    a. What is the final volume of the gas?
    b. Show this process on a pV diagram.

68. ‖ 0.10 mol of gas undergoes the process $1 \rightarrow 2$ shown in Figure P12.68.
    a. What are temperatures T_1 and T_2?
    b. What type of process is this?
    c. The gas undergoes constant-volume heating from point 2 until the pressure is restored to the value it had at point 1. What is the final temperature of the gas?

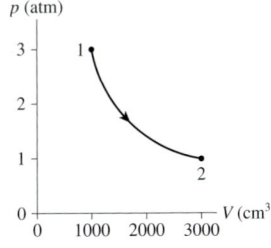

FIGURE P12.68

69. ‖ 10 g of dry ice (solid CO_2) is placed in a 10,000 cm³ container, then all the air is quickly pumped out and the container sealed. The container is warmed to 0°C, a temperature at which CO_2 is a gas.
    a. What is the gas pressure? Give your answer in atm.
    The gas then undergoes an isothermal compression until the pressure is 3.0 atm, immediately followed by an isobaric compression until the volume is 1000 cm³.
    b. What is the final temperature of the gas?
    c. Show the process on a pV diagram.

70. ‖ A 5.0-m-diameter garden pond holds 5.9×10^3 kg of water. Solar energy is incident on the pond at an average rate of 400 W/m². If the water absorbs all the solar energy and does not exchange energy with its surroundings, how many hours will it take to warm from 15°C to 25°C?

71. ‖ A typical nuclear reactor generates 1000 MW (1000 MJ/s) of electrical energy. In doing so, it produces 2000 MW of "waste heat" that must be removed from the reactor. Many reactors are sited next to large bodies of water so that they can use the water for cooling. Consider a reactor where the intake water is at 18°C. State regulations limit the temperature of the output water to 30°C so as not to harm aquatic organisms. How many kilograms of cooling water have to be pumped through the reactor each minute?

72. ‖ INT A 1200 kg car traveling at 60 mph quickly brakes to a halt. The kinetic energy of the car is converted to thermal energy of the disk brakes. The brake disks (one per wheel) are iron disks with a mass of 4.0 kg. Estimate the temperature rise in each disk as the car stops.

73. ‖ INT Suppose you drop a water balloon from a height of 10 m. If the balloon doesn't break on impact, its kinetic energy will be converted to thermal energy. Estimate the temperature rise of the water. Is this likely to be noticeable?

74. ‖ What is the maximum mass of lead you could melt with 1000 J of heat, starting from 20°C?

75. | An experiment measures the temperature of a 200 g substance while steadily supplying heat to it. Figure P12.75 shows the results of the experiment. What are (a) the specific heat of the liquid phase and (b) the heat of vaporization?

FIGURE P12.75

76. ‖‖ 10 g of aluminum at 200°C and 20 g of copper are dropped into 50 cm³ of ethyl alcohol at 15°C. The temperature quickly comes to 25°C. What was the initial temperature of the copper?

77. ‖ A 100 g ice cube at −10°C is placed in an aluminum cup whose initial temperature is 70°C. The system comes to an equilibrium temperature of 20°C. What is the mass of the cup?

78. ‖ A 50.0 g thermometer is used to measure the temperature of 200 g of water. The specific heat of the thermometer, which is mostly glass, is 750 J/kg · K, and it reads 20.0°C while lying on the table. After being completely immersed in the water, the thermometer's reading stabilizes at 71.2°C. What was the actual water temperature before it was measured?

79. ‖‖ Your 300 mL cup of coffee is too hot to drink when served at 90°C. What is the mass of an ice cube, taken from a −20°C freezer, that will cool your coffee to a pleasant 60°C?

80. ‖ A gas is compressed from 600 cm³ to 200 cm³ at a constant pressure of 400 kPa. At the same time, 100 J of heat energy is transferred out of the gas. What is the change in thermal energy of the gas during this process?

81. ‖ An expandable cube, initially 20 cm on each side, contains 3.0 g of helium at 20°C. 1000 J of heat energy are transferred to this gas. What are (a) the final pressure if the process is at constant volume and (b) the final volume if the process is at constant pressure?

82. ‖ 0.10 mol of a monatomic gas follows the process shown in Figure P12.82.
    a. How much heat energy is transferred to or from the gas during process $1 \rightarrow 2$?
    b. How much heat energy is transferred to or from the gas during process $2 \rightarrow 3$?
    c. What is the total change in thermal energy of the gas?

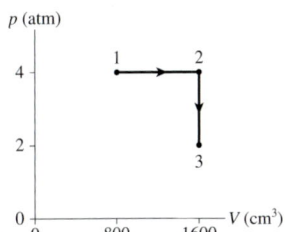

FIGURE P12.82

83. ‖ A monatomic gas follows the process $1 \rightarrow 2 \rightarrow 3$ shown in Figure P12.83. How much heat is needed for (a) process $1 \rightarrow 2$ and (b) process $2 \rightarrow 3$?

FIGURE P12.83

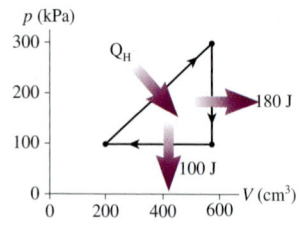

FIGURE P12.84

84. ‖‖‖ What are (a) the heat Q_H extracted from the hot reservoir
INT   and (b) the efficiency for a heat engine described by the pV diagram of Figure P12.84?

85. ‖‖‖ The top layer of your goose down sleeping bag has a thickness of 5.0 cm and a surface area of 1.0 m². When the outside temperature is $-20°C$, you lose 25 Cal/hr by heat conduction through the bag (which remains at a cozy 35°C inside). Assume that you're sleeping on an insulated pad that eliminates heat conduction to the ground beneath you. What is the thermal conductivity of the goose down?

86. ‖ Suppose you go outside in your fiber-filled jacket on a windless but very cold day. The thickness of the jacket is 2.5 cm, and it covers 1.1 m² of your body. The purpose of fiber- or down-filled jackets is to trap a layer of air, and it's really the air layer that provides the insulation. If your skin temperature is 34°C while the air temperature is $-20°C$, at what rate is heat being conducted through the jacket and away from your body?

87. ‖‖‖ Two thin, square copper plates are radiating energy at the same rate. The edge length of plate 2 is four times that of plate 1. What is the ratio of absolute temperatures T_1/T_2 of the plates?

88. ‖ The surface area of an adult human is about 1.8 m². Sup-
BIO pose a person with a skin temperature of 34°C is standing nude in a room where the air is 25°C but the walls are 17°C.
   a. There is a "dead-air" layer next to your skin that acts as insulation. If the dead-air layer is 5.0 mm thick, what is the person's rate of heat loss by conduction?
   b. What is the person's net radiation loss to the walls? The emissivity of skin is 0.97.
   c. Does conduction or radiation contribute more to the total rate of energy loss?
   d. If the person is metabolizing food at a rate of 155 W, does the person feel comfortable, chilly, or too warm?

In Problems 89 through 91 you are given the equation(s) used to solve a problem involving a container of gas. For each of these, you are to
a. Write a realistic problem for which this is the correct equation(s).
b. Draw a pV diagram.
c. Finish the solution of the problem.

89. ‖ $p_f = \dfrac{300 \text{ cm}^3}{100 \text{ cm}^3} \times 1 \times 2 \text{ atm}$

90. ‖ $(T_f + 273) \text{ K} = \dfrac{200 \text{ kPa}}{500 \text{ kPa}} \times 1 \times (400 + 273) \text{ K}$

91. ‖ $V_f = \dfrac{(400 + 273) \text{ K}}{(50 + 273) \text{ K}} \times 1 \times 200 \text{ cm}^3$

92. ‖ You are given an equation below used to solve a problem. You are to
   a. Write a realistic problem for which this is the correct equation.
   b. Finish the solution of the problem.

$$(2.72 \text{ kg})(140 \text{ J/kg} \cdot \text{K})(90°C - 15°C)$$
$$+ (0.50 \text{ kg})(449 \text{ J/kg} \cdot \text{K})(90°C - T_i) = 0$$

Passage Problems

Thermal Properties of the Oceans

Seasonal temperature changes in the ocean only affect the top layer of water, to a depth of 500 m or so. This "mixed" layer is thermally isolated from the cold, deep water below. The average temperature of this top layer of the world's oceans, which has area 3.6×10^8 km², is approximately 17°C.

In addition to seasonal temperature changes, the oceans have experienced an overall warming trend over the last century that is expected to continue as the earth's climate changes. A warmer ocean means a larger volume of water; the oceans will

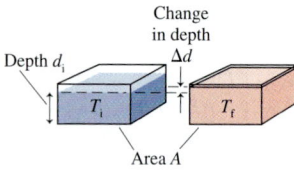

FIGURE P12.93

rise. Suppose the average temperature of the top layer of the world's oceans were to increase from a temperature T_i to a temperature T_f. The area of the oceans will not change, as this is fixed by the size of the ocean basin, so any thermal expansion of the water will cause the water level to rise, as shown in Figure P12.93. The original volume is the product of the original depth and the surface area, $V_i = Ad_i$. The change in volume is given by $\Delta V = A \, \Delta d$.

93. ‖ If the top 500 m of ocean water increased in temperature from 17°C to 18°C, what would be the resulting rise in ocean height?
   A. 0.11 m   B. 0.22 m
   C. 0.44 m   D. 0.88 m

94. ‖ Approximately how much energy would be required to raise the temperature of the top layer of the oceans by 1°C? (1 m³ of water has a mass of 1000 kg.)
   A. 1×10^{24} J   B. 1×10^{21} J
   C. 1×10^{18} J   D. 1×10^{15} J

95. ‖ Water's coefficient of expansion varies with temperature. For water at 2°C, an increase in temperature of 1°C would cause the volume to
   A. Increase.   B. Stay the same.   C. Decrease.

96. ‖ The ocean is mostly heated from the top, by light from the sun. The warmer surface water doesn't mix much with the colder deep ocean water. This lack of mixing can be ascribed to a lack of
   A. Conduction.
   B. Convection.
   C. Radiation.
   D. Evaporation.

Stop to Think 12.1: A. Both helium and neon are monatomic gases, where the basic particles are atoms. 5 mol of helium contain 5 times as many atoms as 1 mol of neon, though both samples have the same mass.

Stop to Think 12.2: A. The thermal expansion coefficients of aluminum are greater than those of iron. Heating the rod and the ring will expand the outer diameter of the rod and the inner diameter of the ring, but the ring's expansion will be greater.

Stop to Think 12.3: B. The product pV/T is constant. During the process, pV decreases by a factor of 2, so T must decrease by a factor of 2 as well.

Stop to Think 12.4: B. To solidify the lead, heat must be removed; this heat boils the water. The heat of vaporization of water is 10 times that of the heat of fusion of lead, so much less than 1 kg of water vaporizes as 1 kg of lead solidifies.

Stop to Think 12.5: A. The lead cools and the water warms as heat is transferred from the lead to the water. The specific heat of water is much larger than that of the lead, so the temperature change of the water is much less than that of the lead.

Stop to Think 12.6: C. With a sealed suit and no matter around you, there is no way to transfer heat to the environment except by radiation.

13

FLUIDS

This woman is floating in the Dead Sea, which has an exceptionally high density because of its salt content of about 35%, ten times that of the open sea. Why does she float so high in the water?

This woman is floating on water, a fluid. The water itself is in motion. Surprisingly, we need no new laws of physics to understand how fluids flow or why some objects float while others sink. The physics of fluids, often called *fluid mechanics,* is an important application of Newton's laws and the law of conservation of energy—physics that you learned in Parts I and II.

We'll begin with *fluid statics,* situations in which the fluid remains at rest. We'll learn why submarines can be crushed by the ocean above, and why hot air balloons float. Then we'll turn to fluids in motion. We'll learn how pressure *differences* are responsible for accelerating fluids. These pressure differences can lead to large forces on objects in the fluids, and they explain how airplanes stay aloft and how blood pressure pushes blood through your circulatory system.

13.1 Fluids and Density

A **fluid** is simply a substance that flows. Because they flow, fluids take the shape of their container rather than retaining a shape of their own. You may think that gases and liquids are quite different, but both are fluids, and their similarities are often more important than their differences.

As you learned in Chapter 12, a gas, shown in Figure 13.1a, is a system in which each molecule moves freely through space until, on occasion, it collides with another molecule or with the wall of the container. The gas you are most familiar with is air, a mixture of mostly nitrogen and oxygen molecules. Gases are *compressible*. That is, the volume of a gas is easily increased or decreased, a consequence of the "empty space" between the molecules in a gas.

Liquids are more complicated than either gases or solids. Liquids, like solids, are nearly *incompressible*. This property tells us that the molecules in a liquid, as in a solid, are about as close together as they can get without coming into contact with each other. At the same time, a liquid flows and deforms to fit the shape of its container. The fluid nature of a liquid tells us that the molecules are free to move around.

Together, these observations suggest the model of a liquid shown in Figure 13.1b. Here you see a system in which the molecules are loosely held together by weak molecular bonds. The bonds are strong enough that the molecules never get far apart but not strong enough to prevent the molecules from sliding around each other.

Density

An important parameter that characterizes a macroscopic system is its *density*. Suppose you have several blocks of copper, each of different size. Each block has a different mass m and a different volume V. Nonetheless, all the blocks are copper, so there should be some quantity that has the *same* value for all the blocks, telling us, "This is copper, not some other material." The most important such parameter is the *ratio* of mass to volume, which we call the **mass density** ρ (lowercase Greek rho):

$$\rho = \frac{m}{V} \tag{13.1}$$

Mass density of an object of mass m and volume V

Conversely, an object of density ρ and volume V has mass

$$m = \rho V \tag{13.2}$$

The SI units of mass density are kg/m^3. Nonetheless, units of g/cm^3 are widely used. You need to convert these to SI units before doing most calculations. You must convert both the grams to kilograms and the cubic centimeters to cubic meters. The net result is the conversion factor

$$1 \text{ g/cm}^3 = 1000 \text{ kg/m}^3$$

The mass density is independent of the object's size. That is, mass and volume are parameters that characterize a *specific piece* of some substance—say copper—whereas the mass density characterizes the substance itself. All pieces of copper have the same mass density, which differs from the mass density of almost any other substance. Thus mass density allows us to talk about the properties of copper in general without having to refer to any specific piece of copper.

The mass density is usually called simply "the density" if there is no danger of confusion. However, we will meet other types of density as we go along, and sometimes it is important to be explicit about which density you are using. Table 13.1 provides a short list of mass densities of various fluids. Notice the enormous difference between the densities of gases and liquids. Gases have lower densities because the molecules in gases are farther apart than in liquids. Also, the density of a liquid varies only slightly with temperature, because its

(a) A gas

Container

Gas molecule moving freely through space

Molecules are far apart. This makes a gas compressible.

Gas molecules occasionally collide with each other . . .

. . . or the wall.

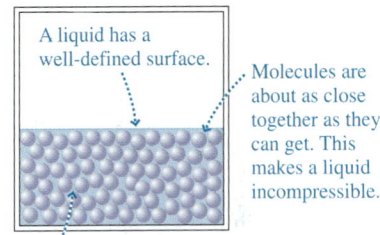

(b) A liquid

A liquid has a well-defined surface.

Molecules are about as close together as they can get. This makes a liquid incompressible.

Molecules make weak bonds with each other that keep them close together. But the molecules can slide around each other, allowing the liquid to flow and conform to the shape of its container.

FIGURE 13.1 Simple atomic-level models of gases and liquids.

TABLE 13.1 Densities of fluids at 1 atm pressure

Substance	ρ (kg/m^3)
Air (0°C)	1.28
Air (20°C)	1.20
Ethyl alcohol	790
Gasoline	680
Glycerin	1260
Helium gas (0°C)	0.179
Mercury	13,600
Oil (typical)	900
Water	1000
Seawater	1030
Dead Sea water	1240
Blood (whole)	1060

molecules are always nearly in contact. The density of a gas, such as air, has a stronger variation with temperature because it's easy to change the already large distance between the molecules.

What does it *mean* to say that the density of gasoline is 680 kg/m³? Recall in Chapter 1 we discussed the meaning of the word "per." We found that it meant "for each," so that 2 miles per hour means you travel 2 miles *for each* hour that passes. In the same way, saying that the density of gasoline is 680 kg per cubic meter means that there are 680 kg of gasoline *for each* one cubic meter of the liquid. If we have two m³ of gasoline, each will have a mass of 680 kg, so the total mass will be 2 × 680 kg = 1360 kg. The product ρV is the number of cubic meters times the mass of each cubic meter; that is, it is the total mass of the object.

EXAMPLE 13.1   Weighing the air in a living room

What is the mass of air in a living room having dimensions 4.0 m × 6.0 m × 2.5 m?

PREPARE Table 13.1 gives air density at a temperature of 20°C, which is about room temperature.

SOLVE The room's volume is

$$V = (4.0\ \text{m}) \times (6.0\ \text{m}) \times (2.5\ \text{m}) = 60\ \text{m}^3$$

The mass of the air is

$$m = \rho V = (1.20\ \text{kg/m}^3)(60\ \text{m}^3) = 72\ \text{kg}$$

ASSESS This is perhaps more mass—about that of an adult person—than you might have expected from a substance that hardly seems to be there. For comparison, a swimming pool this size would contain 60,000 kg of water.

STOP TO THINK 13.1   A piece of glass is broken into two pieces of different size. Rank in order, from largest to smallest, the mass densities of pieces 1, 2, and 3.

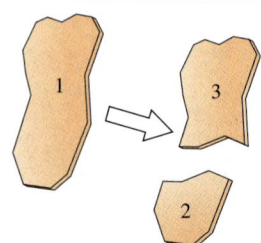

13.2 Pressure

In Chapter 12, you learned how a gas exerts a force on the walls of its container. Liquids also exert forces on the walls of their container, as shown in Figure 13.2, where a force $\vec{F}$ due to the liquid pushes against a small area A of the wall. Just as for a gas, we define the pressure at this point in the fluid to be the ratio of the force to the area on which the force is exerted:

$$p = \frac{F}{A} \tag{13.3}$$

This is the same as Equation 12.7 of Chapter 12. Recall also from Chapter 12 that the SI unit of pressure, the pascal, is defined as

$$1\ \text{pascal} = 1\ \text{Pa} = 1\ \frac{\text{N}}{\text{m}^2}$$

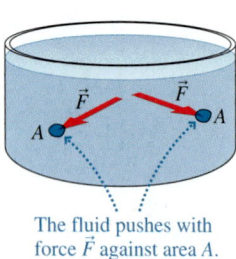

The fluid pushes with force $\vec{F}$ against area A.

FIGURE 13.2 The fluid presses against area A with force $\vec{F}$.

It's important to realize that the force due to a fluid's pressure pushes not only on the walls of its container, but on *all* parts of the fluid itself. If you punch holes in a container of water, the water spurts out from the holes, as in Figure 13.3. It is the force due to the pressure of the water behind each hole that pushes the water forward though the holes.

To measure the pressure at any point within a fluid we can use the simple pressure-measuring device shown in Figure 13.4a. Because the spring constant k and the area A are known, we can determine the pressure by measuring the compression of the spring. Once we've built such a device, we can place it in various liquids and gases to learn about pressure. Figure 13.4b shows what we can learn from a series of simple experiments.

FIGURE 13.3 Water pushes the water *sideways,* out of the holes.

(a)   Piston attached to spring

Vacuum; no fluid force is exerted on the piston from this side.

1. The fluid exerts force $\vec{F}$ on a piston with surface area A.

2. The force compresses the spring. Because the spring constant k is known, we can use the spring's compression to find F.

3. Because A is known, we can find the pressure from $p = F/A$.

(b)   Pressure-measuring device in fluid

1. There is pressure *everywhere* in a fluid, not just at the bottom or at the walls of the container.

2. The pressure at one point in the fluid is the same whether you point the pressure-measuring device up, down, or sideways. The fluid pushes up, down, and sideways with equal strength.

3. In a *liquid*, the pressure increases with depth below the surface. In a *gas*, the pressure is nearly the same at all points (at least in laboratory-size containers).

FIGURE 13.4 Learning about pressure.

The first statement in Figure 13.4b emphasizes again that pressure exists at *all* points within a fluid, not just at the walls of the container. You may recall that tension exists at *all* points in a string, not only at its ends where it is tied to an object. We understood tension as the different parts of the string *pulling* against each other. Pressure is an analogous idea, except that the different parts of a fluid are *pushing* against each other.

Pressure in Liquids

Chapter 12 introduced our model of a gas. The molecules in a gas interact only when they collide with one another. Because of this, a gas almost instantaneously expands to fill the whole volume of a container. The pressure of the gas, which is the same at all points in the container, is then due to the collisions of the molecules with the container's walls.

In contrast to a gas, the molecules of a liquid interact by weak molecular bonds. These bonds tend to keep the molecules close together in a cohesive blob. If you introduce a liquid into a container, the force of gravity pulls the liquid down, causing it to fill the bottom of the container. It is this force of gravity—that is, the weight of the liquid—that is responsible for the pressure in a liquid. Pressure increases with depth in a liquid because the liquid below is being squeezed by all the liquid above, including any other liquid floating on the first liquid, and the weight of the air above the liquid.

We'd like to determine the pressure at a depth d below the surface of the liquid. We will assume that the liquid is at rest; flowing liquids will be considered later in this chapter. The shaded cylinder of liquid in Figure 13.5 extends from the surface to depth d. This cylinder, like the rest of the liquid, is in static equilibrium with $\vec{F}_{net} = \vec{0}$. Several forces act on this cylinder: its weight mg, a downward force p_0A due to the pressure p_0 at the surface of the liquid, an upward force pA due to the

Whatever is above the liquid pushes down on the top of the cylinder.

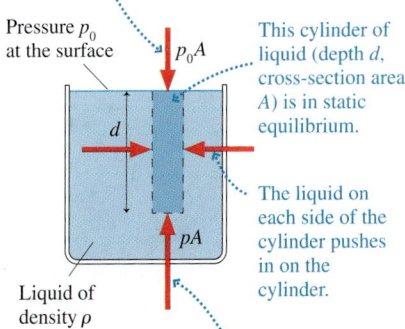

Pressure p_0 at the surface

This cylinder of liquid (depth d, cross-section area A) is in static equilibrium.

Liquid of density ρ

The liquid on each side of the cylinder pushes in on the cylinder.

The liquid beneath the cylinder pushes up on the cylinder. The pressure at depth d is p.

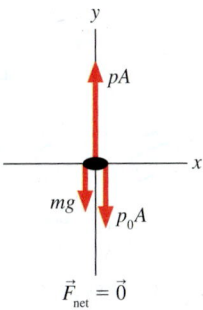

Free-body diagram of the column of liquid. The horizontal forces cancel and are not shown.

FIGURE 13.5 Measuring the pressure at depth d in a liquid.

liquid beneath the cylinder pushing up on the bottom of the cylinder, and inward-directed forces due to the liquid pushing in on the sides of the cylinder. The forces due to the liquid pushing on the cylinder are a consequence of our earlier observation that different parts of a fluid push against each other. Pressure p, which is what we're trying to find, is the pressure at the bottom of the cylinder.

The horizontal forces cancel each other. The upward force balances the two downward forces, so

$$pA = p_0A + mg \qquad (13.4)$$

The liquid is a cylinder of cross-section area A and height d. Its volume is $V = Ad$ and its mass is $m = \rho V = \rho Ad$. Substituting this expression for the mass of the liquid into Equation 13.4, we find that the area A cancels from all terms. The pressure at depth d in a liquid is then

$$p = p_0 + \rho gd \qquad (13.5)$$

Pressure of a liquid with density ρ, at depth d

Because of our assumption that the fluid is at rest, the pressure given by Equation 13.5 is called the **hydrostatic pressure.** The fact that g appears in Equation 13.5 reminds us that the origin of this pressure is the gravitational force on the fluid.

As expected, $p = p_0$ at the surface, where $d = 0$. Pressure p_0 is often due to the air or other gas above the liquid. $p_0 = 1$ atm $= 101.3$ kPa for a liquid that is open to the air at sea level. In other situations, p_0 might be the pressure due to a piston or a closed surface pushing down on the top of the liquid.

NOTE ▶ Equation 13.5 assumes that the fluid is *incompressible*; that is, its density ρ doesn't increase with depth. This is an excellent assumption for liquids, but not a good one for a gas. Equation 13.5 should not be used for calculating the pressure of a gas. Gas pressure is found with the ideal-gas law. ◀

EXAMPLE 13.2 The pressure on a submarine

A submarine cruises at a depth of 300 m. What is the pressure at this depth? Give the answer both in pascals and atmospheres.

SOLVE The density of seawater, from Table 13.1, is $\rho = 1030$ kg/m³. At the surface, $p_0 = 1$ atm $= 101.3$ kPa. The pressure at depth $d = 300$ m is found from Equation 13.5 to be

$p = p_0 + \rho gd$

  $= 1.013 \times 10^5$ Pa $+ (1030 \text{ kg/m}^3)(9.80 \text{ m/s}^2)(300 \text{ m})$

  $= 3.13 \times 10^6$ Pa

Converting the answer to atmospheres gives

$p = 3.13 \times 10^6 \text{ Pa} \times \dfrac{1 \text{ atm}}{1.013 \times 10^5 \text{ Pa}} = 30.9 \text{ atm}$

ASSESS The pressure deep in the ocean is very great. The research submarine *Alvin,* shown in the left photo, can safely dive as deep as 4500 m, where the pressure is over 450 atm! Its viewports are over 3.5 inches thick to withstand this pressure. As shown in the right photo, each viewport is tapered, with its larger face toward the sea. The water pressure then pushes the viewports firmly into their conical seats, helping to seal them tightly.

According to Equation 13.5, the hydrostatic pressure equation $p = p_0 + \rho g d$, the hydrostatic pressure in a liquid depends only on the depth and the pressure at the surface. This observation has some important implications. Figure 13.6a shows two connected tubes. It's certainly true that the larger volume of liquid in the wide tube weighs more than the liquid in the narrow tube. You might think that this extra weight would push the liquid in the narrow tube higher than in the wide tube. But it doesn't. If d_1 were larger than d_2, then, according to the hydrostatic pressure equation, the pressure at the bottom of the narrow tube would be higher than the pressure at the bottom of the wide tube. This *pressure difference* would cause the liquid to *flow* from right to left until the heights were equal.

Thus a first conclusion: **A connected liquid in hydrostatic equilibrium rises to the same height in all open regions of the container.** This is the familiar observation that "water seeks its own level."

Figure 13.6b shows two connected tubes of different shape. The conical tube holds more liquid above the dotted line, so you might think that $p_1 > p_2$. But it isn't. Both points are at the same depth, thus $p_1 = p_2$. You can arrive at the same conclusion by thinking about the pressure at the bottom of the tubes. If p_1 were larger than p_2, the pressure at the bottom of the left tube would be larger than the pressure at the bottom of the right tube. This would cause the liquid to flow until the pressures were equal. Thus a second conclusion: **The pressure is the same at all points on a horizontal line through a connected liquid in hydrostatic equilibrium.**

NOTE ▶ Both of these conclusions are restricted to liquids in hydrostatic equilibrium. The situation is entirely different for flowing fluids, as we'll see later in the chapter. ◀

(a)

Is this possible?

(b)

Is $p_1 > p_2$?

FIGURE 13.6 Some properties of a liquid in hydrostatic equilibrium are not what you might expect.

EXAMPLE 13.3 **Pressure in a closed tube**

Water fills the tube shown in Figure 13.7. What is the pressure at the top of the closed tube?

FIGURE 13.7 A bent tube closed at one end.

PREPARE This is a liquid in hydrostatic equilibrium. The closed tube is not an open region of the container, so the water cannot rise to an equal height. Nevertheless, the pressure is still the

same at all points on a horizontal line. In particular, the pressure at the top of the closed tube equals the pressure in the open tube at the height of the dotted line. Assume $p_0 = 1$ atm.

SOLVE A point 40 cm above the bottom of the open tube is at a depth of 60 cm. The pressure at this depth is

$$p = p_0 + \rho g d$$
$$= 1.013 \times 10^5 \text{ Pa} + (1000 \text{ kg/m}^3)(9.80 \text{ m/s}^2)(0.60 \text{ m})$$
$$= 1.072 \times 10^5 \text{ Pa}$$
$$= 1.06 \text{ atm}$$

The pressure at the top of the closed tube is thus 1.1 atm, to two significant figures.

ASSESS The water in the open tube *pushes* the water in the closed tube up against the top of the tube. Consequently, in accordance with Newton's third law, the top of the tube *presses down on the liquid* with a force of magnitude $F = pA$. This explains why the pressure at the top of the closed tube is greater than atmospheric pressure.

We can draw one more conclusion from the hydrostatic pressure equation $p = p_0 + \rho g d$. If we change the pressure p_0 at the surface to p_1, the pressure at depth d becomes $p' = p_1 + \rho g d$. The *change* in pressure $\Delta p = p_1 - p_0$ is the same at all points in the fluid, independent of the size or shape of the container.

This idea was first recognized by Blaise Pascal (the same Pascal for whom the pressure unit is named), and is called *Pascal's principle:*

> **Pascal's principle** If the pressure at one point in an incompressible fluid is changed, the pressure at every other point in the fluid changes by the same amount.

For example, if we compress the air above the open tube in Example 13.3 to a pressure of 1.5 atm, an increase of 0.5 atm, the pressure at the top of the closed tube will increase to 1.6 atm.

STOP TO THINK 13.2   Water is slowly poured into the container until the water level has risen into tubes 1, 2, and 3. The water doesn't overflow from any of the tubes. How do the water depths in the three columns compare to each other?

A. $d_1 > d_2 > d_3$.
B. $d_1 < d_2 < d_3$.
C. $d_1 = d_2 = d_3$.
D. $d_1 = d_2 > d_3$.
E. $d_1 = d_2 < d_3$.

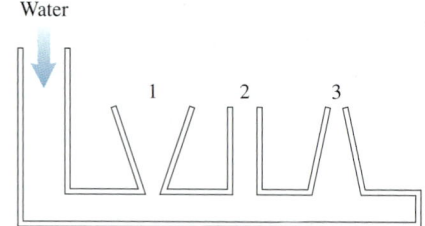

Atmospheric Pressure

We live at the bottom of a "sea" of air that extends up many kilometers. As Figure 13.8 shows, there is no well-defined top to the atmosphere; it just gets less and less dense with increasing height until reaching zero in the vacuum of space. Nonetheless, 99% of the air in the atmosphere is below about 30 km.

If we recall that a gas like air is quite compressible, we can see why the atmosphere becomes less dense with increasing altitude. In a liquid, pressure increases with depth because of the weight of the liquid above. The same holds true for the air in the atmosphere, but because the air is compressible, the weight of the air above compresses the air below, increasing its density. At high altitudes there is very little air above to push down, so the density is less.

We learned in Chapter 12 that the global average sea-level pressure, the *standard atmosphere,* is 1 atm = 101,300 Pa. The standard atmosphere, usually referred to simply as "atmospheres," is a commonly used unit of pressure. But it is not an SI unit, so you must convert atmospheres to pascals before doing most calculations with pressure.

NOTE ▶ Unless you happen to live right at sea level, the atmospheric pressure around you is not exactly 1 atm. Pressure experiments use a barometer to determine the actual atmospheric pressure. For simplicity, this textbook will always assume that the pressure of the air is $p_{atmos} = 1$ atm unless stated otherwise. ◀

Atmospheric pressure varies not only with altitude, but also with changes in the weather. Large-scale regions of low-pressure air are created at the equator, where hot air rises and flows to the north and south temperate zones. There the air falls, creating high pressure zones. Local winds and weather are largely determined by presence and movement of air masses of differing pressure. You may have seen weather maps like the one shown in Figure 13.9 on the evening news. The letters H and L denote regions of high and low atmospheric pressure.

FIGURE 13.8 Atmospheric pressure and density.

3. The density and pressure approach zero in outer space.

2. Because of gravity, the density and pressure decrease with increasing height.

1. The air's density and pressure are greatest at the earth's surface.

FIGURE 13.9 High and low pressure zones on a weather map.

Given that the pressure of the air at sea level is 101.3 kPa, you might wonder why the weight of the air doesn't crush your forearm when you rest it on a table. Your forearm has a surface area of $\approx 200 \text{ cm}^2 = 0.02 \text{ m}^2$, so the force of the air pressing against it is $\approx 2000 \text{ N}$ (≈ 450 pounds). How can you even lift your arm?

The reason, as Figure 13.10 shows, is that a fluid exerts pressure forces in *all* directions. There *is* a downward force of $\approx 2000 \text{ N}$ on your forearm, but the air underneath your arm exerts an *upward* force of the same magnitude. The *net* force is just about zero.

The same arguments hold for deep-sea fish. Some fish have been observed at depths exceeding 8000 m. Here, the pressure is $\rho g h \approx 8 \times 10^7 \text{ Pa} \approx 800 \text{ atm}$! But the pressure inside and outside the fish is the *same,* so there is no net force squeezing in on the fish.

Vacuum cleaners, suction cups, and other similar devices are powerful examples of how strong atmospheric pressure forces can be *if* the air is removed from one side of an object so as to produce an unbalanced force. The fact that we are *surrounded* by the fluid allows us to move around in the air or swim underwater, oblivious of these strong forces.

13.3 Measuring and Using Pressure

The pressure in a fluid is measured with a *pressure gauge,* which is often a device very similar to that in Figure 13.4. The fluid pushes against a spring, and the displacement of the spring is indicated on a scale. The familiar tire-pressure gauge shown in Figure 13.11 works in just this way. The pressure p_{tire} exerts a force $p_{\text{tire}}A$ on the front area A of the piston, while atmospheric pressure exerts a force $p_{\text{atmos}}A$ on the back of the piston. Thus the *net* pressure force on the piston is $(p_{\text{tire}} - p_{\text{atmos}})A$. This force compresses the spring until equilibrium is reached. Thus the movement of the scale depends not on the absolute pressure in the tire, but on the *difference* between the tire pressure and atmospheric pressure. This type of gauge measures the *gauge pressure* $p_{\text{g}} = p_{\text{tire}} - p_{\text{atmos}}$, an idea introduced in Chapter 12.

Solving Hydrostatic Problems

We now have enough information to formulate a set of rules for thinking about hydrostatic problems.

TACTICS BOX 13.1 **Hydrostatics** Exercise 5

❶ **Draw a picture.** Show open surfaces, pistons, boundaries, and other features that affect the pressure. Include height and area measurements and fluid densities. Identify the points at which you need to find the pressure.

❷ **Determine the pressure at surfaces.**
- **Surface open to the air:** $p_0 = p_{\text{atmos}}$, usually 1 atm.
- **Surface covered by a gas:** $p_0 = p_{\text{gas}}$.
- **Closed surface:** $p = F/A$ where F is the force the surface, such as a piston, exerts on the fluid.

❸ **Use horizontal lines.** Pressure in a connected fluid is the same at any point along a horizontal line.

❹ **Allow for gauge pressure.** Pressure gauges read $p_{\text{g}} = p - 1 \text{ atm}$.

❺ **Use the hydrostatic pressure equation.** $p = p_0 + \rho g d$.

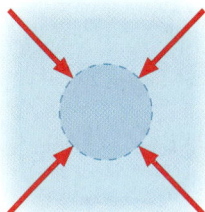

The forces of a fluid push in *all* directions.

FIGURE 13.10 Pressure forces in a fluid.

TRY IT YOURSELF

A crushing experience Heat a very small amount of water in a soda can until the water has boiled for a few seconds. Then, using tongs or gloves, quickly invert the can into a shallow pan of cold water. The can will dramatically implode! Boiling the water fills the can with water vapor at 1 atm, which rapidly condenses back into a liquid when the can is inverted into the cold water. Because liquid water occupies a much smaller volume than its vapor, this condensation creates a partial vacuum in the can. The greater pressure of the atmosphere outside then crushes it.

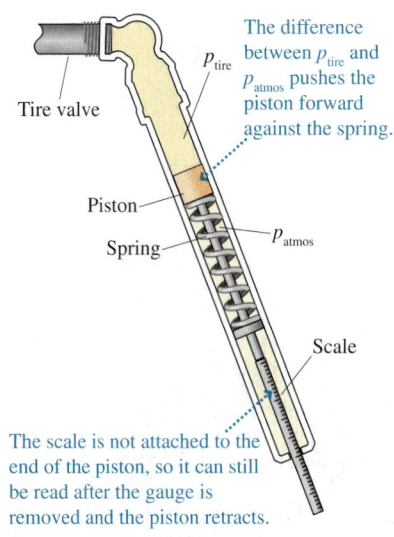

The difference between p_{tire} and p_{atmos} pushes the piston forward against the spring.

Tire valve

p_{tire}

Piston

Spring

p_{atmos}

Scale

The scale is not attached to the end of the piston, so it can still be read after the gauge is removed and the piston retracts.

FIGURE 13.11 A tire gauge measures the difference between the tire's pressure and atmospheric pressure.

❶ Draw a picture.

❷ This is an open surface, so $p_0 = 1$ atm.

Liquid, density ρ

This is a surface covered by a gas, so $p_0 = p_{\text{gas}}$.

Gas at pressure P_{gas}

h

1 2

p_1 p_2

❸ Points 1 and 2 are on a horizontal line, so $p_1 = p_2$.

FIGURE 13.12 A manometer is used to measure gas pressure.

(a) Seal and invert tube.

Liquid, density ρ

(b)

Vacuum (zero pressure)

$p_2 = \rho g h$

$p_1 = p_{\text{atmos}}$

h

1 2

FIGURE 13.13 A barometer.

Manometers and Barometers

Gas pressure is sometimes measured with a device called a *manometer*. A manometer, shown in Figure 13.12, is a U-shaped tube connected to the gas at one end and open to the air at the other end. The tube is filled with a liquid—often mercury—of density ρ. The liquid is in static equilibrium. A scale allows the user to measure the height h of the right side above the left side.

Steps 1–3 from Tactics Box 13.1 lead to the conclusion that the pressures p_1 and p_2 must be equal. Pressure p_1, at the surface on the left, is simply the gas pressure: $p_1 = p_{\text{gas}}$. Pressure p_2 is the hydrostatic pressure at depth $d = h$ in the liquid on the right: $p_2 = 1$ atm $+ \rho g h$. Equating these two pressures gives

$$p_{\text{gas}} = 1 \text{ atm} + \rho g h \tag{13.6}$$

NOTE ▶ The height h has a *positive* value when the liquid is *higher* on the right than on the left ($p_{\text{gas}} > 1$ atm) and a *negative* value when the liquid is *lower* on the right than on the left ($p_{\text{gas}} < 1$ atm). ◀

Another important pressure-measuring instrument is the *barometer*, which is used to measure the atmospheric pressure p_{atmos}. Figure 13.13a shows a glass tube, sealed at the bottom, that has been completely filled with a liquid. If we temporarily seal the top end, we can invert the tube and place it in a beaker of the same liquid. When the temporary seal is removed, some, but not all, of the liquid runs out, leaving a liquid column in the tube that is a height h above the surface of the liquid in the beaker. This device, shown in Figure 13.13b, is a barometer. What does it measure? And why doesn't *all* the liquid in the tube run out?

We can analyze the barometer much as we did the manometer. Points 1 and 2 in Figure 13.13b are on a horizontal line drawn even with the surface of the liquid. The liquid is in hydrostatic equilibrium, so the pressure at these two points must be equal. Liquid runs out of the tube only until a balance is reached between the pressure at the base of the tube and the pressure of the air.

You can think of a barometer as rather like a seesaw. If the pressure of the atmosphere increases, it presses down on the liquid in the beaker. This forces liquid up the tube until the pressures at points 1 and 2 are equal. If the atmospheric pressure falls, liquid has to flow out of the tube to keep the pressures equal at these two points.

The pressure at point 2 is the pressure due to the weight of the liquid in the tube plus the pressure of the gas above the liquid. But in this case there is no gas above the liquid! Because the tube had been completely full of liquid when it was inverted, the space left behind when the liquid ran out is essentially a vacuum. Thus pressure p_2 is simply $p_2 = \rho g h$.

Equating p_1 and p_2, and noting that $p_1 = p_{\text{atmos}}$, gives

$$p_{\text{atmos}} = \rho g h \tag{13.7}$$

Thus we can measure the atmosphere's pressure by measuring the height of the liquid column in a barometer.

Equation 13.7 shows that the liquid height is $h = p_{\text{atmos}}/\rho g$. If a barometer were made using water, with $\rho = 1000$ kg/m^3, the liquid column would be more than 10 m high, which is impractical. Instead, mercury, with its high density of 13,600 kg/m^3, is usually used. The average air pressure at sea level causes a column of mercury in a mercury barometer to stand 760 mm above the surface. We can then use Equation 13.7 to find that the average atmospheric pressure is

$$p_{\text{atmos}} = \rho_{\text{Hg}} g h = (13{,}600 \text{ kg/m}^3)(9.80 \text{ m/s}^2)(0.760 \text{ m})$$

$$= 1.013 \times 10^5 \text{ Pa} = 101.3 \text{ kPa}$$

This is the value given earlier as "one standard atmosphere." Now you can see that 1 atm = 101.3 kPa *because,* on average, a mercury barometer gives a reading of 760 mm.

Pressure Units

In practice, pressure is measured in a number of different units. This plethora of units and abbreviations has arisen historically as scientists and engineers working on different subjects (liquids, high-pressure gases, low-pressure gases, weather, etc.) developed what seemed to them the most convenient units. These units continue in use through tradition, so it is necessary to become familiar with converting back and forth between them. Table 13.2 gives the basic conversions.

TABLE 13.2 Pressure units

Unit	Abbreviation	Conversion to 1 atm	Uses
pascal	Pa	101.3 kPa	SI unit: 1 Pa = 1 N/m^2 use in most calculations
atmosphere	atm	1 atm	general
millimeters of mercury	mm of Hg	760 mm of Hg	gases and barometric pressure
inches of mercury	in	29.92 in	barometric pressure in U.S. weather forecasting
pounds per square inch	psi	14.7 psi	U.S. engineering and industry

Blood Pressure

The last time you had a medical checkup, the doctor may have told you something like, "Your blood pressure is 120 over 80." What does that mean?

About every 0.8 s, assuming a pulse rate of 75 beats per minute, your heart "beats." The heart muscles contract and push blood out into your aorta. This contraction, like squeezing a balloon, raises the pressure in your heart. The pressure increase, in accordance with Pascal's principle, is transmitted through all your arteries.

Figure 13.14 is a pressure graph showing how blood pressure changes during one cycle of the heartbeat. The medical condition of *high blood pressure* usually means that your maximum (*systolic*) blood pressure is higher than necessary for blood circulation. The high pressure causes undue stress and strain on your entire circulatory system, often leading to serious medical problems. Low blood pressure can cause you to faint if you stand up quickly because the pressure isn't adequate to pump the blood up to your brain.

As shown in Figure 13.15, blood pressure is measured with a cuff that goes around your arm. The doctor or nurse pressurizes the cuff, places a stethoscope over the artery in your arm, then slowly releases the pressure while watching a pressure gauge. Initially, the cuff squeezes the artery shut and cuts off the blood flow. When the cuff pressure drops below the systolic pressure, the pressure pulse during each beat of your heart forces the artery open briefly and a squirt of blood goes through. You can feel this, and the doctor or nurse records the pressure when he or she hears the blood start to flow. This is your systolic pressure.

This pulsing of the blood through your artery lasts until the cuff pressure reaches the diastolic pressure. Then the artery remains open continuously and the blood flows smoothly. This transition is easily heard in the stethoscope, and the doctor or nurse records your base or *diastolic* pressure.

Blood pressure is measured in millimeters of mercury. And it is a gauge pressure, the pressure in excess of 1 atm. A fairly typical blood pressure of a healthy young adult is 120/80, meaning that the systolic pressure is $p_g = 120$ mm of Hg (absolute pressure $p = 880$ mm of Hg) and the diastolic pressure is 80 mm of Hg.

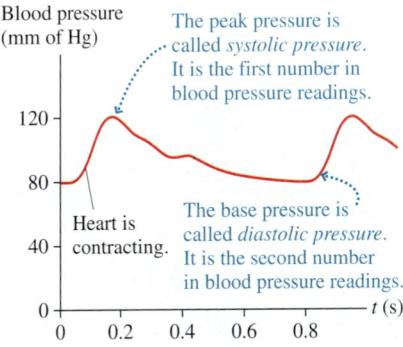

FIGURE 13.14 Blood pressure during one cycle of a heart beat.

FIGURE 13.15 Measuring blood pressure with a manometer.

EXAMPLE 13.4 Measuring blood pressure

During a blood pressure measurement, a mercury-filled manometer is used to measure the pressure in the cuff. When the cuff pressure equals the systolic pressure, the height of the mercury in the open side of the manometer is 130 mm higher than on the side connected to the cuff. What is the patient's systolic pressure?

PREPARE We can use Equation 13.6 to find the pressure in the cuff, which is equal to the patient's blood pressure.

SOLVE Equation 13.6 gives the absolute pressure in the cuff as

$$p_{\text{cuff}} = 1 \text{ atm} + \rho g h$$

We want the blood pressure, which is the cuff pressure converted to gauge pressure. Thus

$$p_{\text{blood}} = p_{\text{cuff}} - 1 \text{ atm} = \rho g h$$

From Table 13.1, the density of mercury is $\rho = 13{,}600 \text{ kg/m}^3$. Using $h = 0.130$ m we have

$$p_{\text{blood}} = \rho g h = (13{,}600 \text{ kg/m}^3)(9.80 \text{ m/s}^2)(0.130 \text{ m})$$
$$= 17.3 \text{ kPa}$$

We can convert our result to millimeters of mercury, the proper unit for blood pressure, by using the conversion factor, 101.3 kPa = 760 mm Hg. Thus

$$p_{\text{blood}} = 17.3 \text{ kPa}\left(\frac{760 \text{ mm Hg}}{101.3 \text{ kPa}}\right) = 130 \text{ mm of Hg}$$

ASSESS The blood pressure in mm of Hg is *equal* to the height difference of the mercury column in the manometer. This is what the units "mm of Hg" *mean:* a gauge pressure of 130 mm of Hg will push the open end of a mercury column 130 mm higher than the pressurized end.

CONCEPTUAL EXAMPLE 13.1 Blood pressure and the height of the arm

In Figure 13.15, the patient's arm is held at about the same height as her heart. Why is this?

REASON The hydrostatic pressure of a fluid varies with height. Although flowing blood is not in hydrostatic equilibrium, it is still true that blood pressure increases with the distance below the heart, and decreases above it. Because the upper arm when held beside the body is at the same height as the heart, the pressure here is the same as the pressure at the heart. If the patient held her arm straight up, the pressure cuff would be a distance $d \approx 25$ cm above her heart and the pressure would be *less* than the pressure at the heart by $\Delta p = \rho_{\text{blood}} g d \approx 20$ mm of Hg.

ASSESS 20 mm of Hg is a substantial fraction of the average blood pressure. Measuring pressure above or below heart level could lead to a misdiagnosis of the patient's condition.

◀ **Pressure at the top** BIO A giraffe's head is some 2.5 m above its heart, compared to a distance of only about 30 cm for humans. To pump blood this extra height requires a blood pressure at the giraffe's heart that is some 170 mm of Hg higher than a human's, making its blood pressure more than twice as high as a human's.

13.4 Buoyancy

A rock, as you know, sinks like a rock. Wood floats on the surface of a lake. A penny with a mass of a few grams sinks, but a massive steel aircraft carrier floats. How can we understand these diverse phenomena?

An air mattress floats effortlessly on the surface of a swimming pool. But if you've ever tried to push an air mattress underwater, you know it is nearly impossible. As you push down, the water pushes up. This upward force of a fluid is called the **buoyant force.**

The basic reason for the buoyant force is easy to understand. Figure 13.16 shows a cylinder submerged in a liquid. The pressure in the liquid increases with depth, so the pressure at the bottom of the cylinder is larger than at the top. Both cylinder ends have equal area, so force $\vec{F}_{up}$ is larger than force $\vec{F}_{down}$. (Remember that pressure forces push in *all* directions.) Consequently, the pressure in the liquid exerts a *net upward force* on the cylinder of magnitude $F_{net} = F_{up} - F_{down}$. This is the buoyant force.

The submerged cylinder illustrates the idea in a simple way, but the result is not limited to cylinders or to liquids. Suppose we isolate a parcel of fluid of arbitrary shape and volume by drawing an imaginary boundary around it, as shown in Figure 13.17a. This parcel is in static equilibrium. Consequently, the parcel's weight force pulling it down must be balanced by an upward force. The upward force, which is exerted on this parcel of fluid by the surrounding fluid, is the buoyant force $\vec{F}_B$. The buoyant force matches the weight of the fluid: $F_B = w$.

Now imagine that we could somehow remove this parcel of fluid and instantaneously replace it with an object having exactly the same shape and size, as shown in Figure 13.17b. Because the buoyant force is exerted by the *surrounding* fluid, and the surrounding fluid hasn't changed, the buoyant force on this new object is *exactly the same* as the buoyant force on the parcel of fluid that we removed.

When an object (or a portion of an object) is immersed in a fluid, it *displaces* fluid that would otherwise fill that region of space. This fluid is called the **displaced fluid.** The displaced fluid's volume is exactly the volume of the portion of the object that is immersed in the fluid. Figure 13.17 leads us to conclude that the magnitude of the upward buoyant force matches the weight of this displaced fluid.

This idea was first recognized by the ancient Greek mathematician and scientist Archimedes, perhaps the greatest scientist of antiquity, and today we know it as *Archimedes' principle.*

> **Archimedes' principle** A fluid exerts an upward buoyant force $\vec{F}_B$ on an object immersed in or floating on the fluid. The magnitude of the buoyant force equals the weight of the fluid displaced by the object.

Suppose the fluid has density ρ_f and the object displaces volume V_f of fluid. The mass of the displaced fluid is $m_f = \rho_f V_f$ and so its weight is $w_f = \rho_f V_f g$. Thus Archimedes' principle in equation form is

$$F_B = \rho_f V_f g \tag{13.8}$$

NOTE ▶ It is important to distinguish the density and volume of the displaced fluid from the density and volume of the object. To do so, we'll use subscript f for the fluid and o for the object. ◀

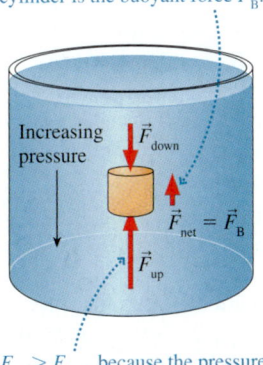

The net force of the fluid on the cylinder is the buoyant force $\vec{F}_B$.

$F_{up} > F_{down}$ because the pressure is greater at the bottom. Hence the fluid exerts a net upward force.

FIGURE 13.16 The buoyant force arises because the fluid pressure at the bottom of the cylinder is larger than at the top.

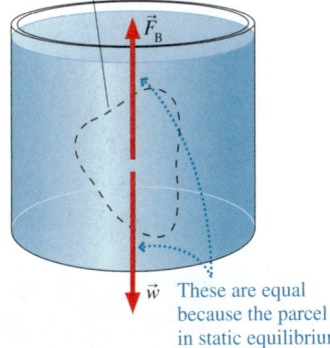

(a) Imaginary boundary around a parcel of fluid

$\vec{w}$ These are equal because the parcel is in static equilibrium.

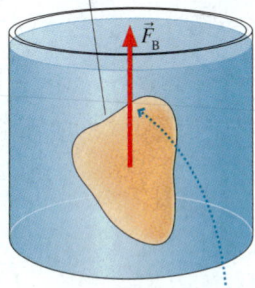

(b) Real object with same size and shape as the parcel of fluid

The buoyant force on the object is the same as on the parcel of fluid because the *surrounding* fluid has not changed.

FIGURE 13.17 The buoyant force on an object is the same as the buoyant force on the fluid it displaces.

EXAMPLE 13.5   Is the crown gold?

Legend has it that Archimedes was asked by King Hiero of Syracuse to determine whether a crown was of pure gold or if it had been adulterated with a lesser metal by an unscrupulous goldsmith. It was this problem that led him to the principle that bears his name. In a modern version of his method, a crown weighing 8.30 N in air is suspended underwater from a string. The tension in the string is measured to be 7.81 N. Is the crown pure gold?

PREPARE To discover whether the crown is pure gold, we need to determine its density ρ_o. Because the crown is submerged, its volume V_o equals the volume V_f of displaced water, which we can find from Equation 13.8.

Figure 13.18 shows the forces acting on the crown. In addition to the familiar tension and weight forces, the water exerts an upward buoyant force on the crown. The size of the buoyant force is given by Archimedes' principle.

Known
$T = 7.81$ N
$w_o = 8.30$ N
$\rho_f = 1000$ kg/m³

Find: ρ_o

FIGURE 13.18 The forces acting on the submerged crown.

SOLVE Because the crown's weight is $w_o = \rho_o V_o g$, its density is

$$\rho_o = \frac{w_o}{V_o g}$$

We know that $V_o = V_f$, and we can rearrange Equation 13.8 to give

$$V_o = V_f = \frac{F_B}{\rho_f g}$$

giving

$$\rho_o = \frac{w_o}{V_o g} = \frac{w_o}{\left(\dfrac{F_B}{\rho_f g}\right)g} = \frac{w_o \rho_f}{F_B}$$

We can find the magnitude F_B of the buoyant force from the fact that the crown is in static equilibrium, so the net force on it must be zero:

$$\sum F_y = F_B + T - w_o = 0$$

so that $F_B = w_o - T$. Putting this all together gives

$$\rho_o = \frac{w_o \rho_f}{F_B} = \frac{w_o \rho_f}{w_o - T}$$

$$= \frac{(8.30 \text{ N})(1000 \text{ kg/m}^3)}{(8.30 \text{ N}) - (7.81 \text{ N})} = 1.69 \times 10^4 \text{ kg/m}^3$$

The crown's density is considerably lower than that of pure gold, which is 1.93×10^4 kg/m³. The crown is not pure gold.

ASSESS Our result shows that the closer the tension T (the *apparent weight*) is to w_o (the actual weight), the larger the density of the object. This makes sense, because the weight of a very dense object is much greater than the buoyant force on it, so that its true and apparent weights are nearly the same.

Float or Sink?

If you *hold* an object underwater and then release it, it either floats to the surface, sinks, or remains "hanging" in the water. How can we predict which it will do? Whether it heads for the surface or the bottom depends on whether the upward buoyant force F_B on the object is larger or smaller than the downward weight force w_o.

The magnitude of the buoyant force is $\rho_f V_f g$. The weight of a uniform object, such as a block of steel, is simply $\rho_o V_o g$. But a compound object, such as a scuba diver, may have pieces of varying density. If we define the **average density** to be $\rho_{avg} = m_o / V_o$, the weight of a compound object can be written $w_o = \rho_{avg} V_o g$.

Comparing $\rho_f V_f g$ to $\rho_{avg} V_o g$, and noting that $V_f = V_o$ for an object that is fully submerged, we see that an object floats or sinks depending on whether the

◄ **Submersible scales**   BIO   In Example 13.5, we saw how the density of an object could be determined by weighing it both underwater and in air. This idea is the basis of an accurate method for determining a subject's percentage of body fat. Fat has a lower density than lean muscle or bone, so a lower overall body density implies a greater proportion of body fat. To determine a subject's density, she is first weighed in air, then lowered completely into water and weighed again. Standard tables accurately relate body density to fat percentage.

fluid density ρ_f is larger or smaller than the object's average density ρ_{avg}. If the densities are equal, the object is in static equilibrium and hangs motionless. This is called **neutral buoyancy.** These conditions are summarized in Tactics Box 13.2.

TACTICS BOX 13.2 Finding whether an object floats or sinks

Exercise 8

❶ Object sinks

❷ Object floats

❸ Object has neutral buoyancy

An object sinks if it weighs more than the fluid it displaces; that is, if its average density is greater than the density of the fluid:

$$\rho_{avg} > \rho_f$$

An object floats on the surface if it weighs less than the fluid it displaces; that is, if its average density is less than the density of the fluid:

$$\rho_{avg} < \rho_f$$

An object hangs motionless in the fluid if it weighs exactly the same as the fluid it displaces. It has neutral buoyancy if its average density equals the density of the fluid:

$$\rho_{avg} = \rho_f$$

As an example, steel is denser than water, so a chunk of steel sinks. Oil is less dense than water, so oil floats on water. Fish use *swim bladders* filled with air and scuba divers use weighted belts to adjust their average density to match the water. Both are examples of neutral buoyancy.

If you release a block of wood underwater, the net upward force causes the block to shoot to the surface. Then what? To understand floating, let's begin with a *uniform* object such as the block shown in Figure 13.19. This object contains nothing tricky, like indentations or voids. Because it's floating, it must be the case that $\rho_o < \rho_f$.

Now that the object is floating, it's in static equilibrium. The upward buoyant force, given by Archimedes' principle, exactly balances the downward weight of the object. That is

$$F_B = \rho_f V_f g = w_o = \rho_o V_o g \qquad (13.9)$$

In this case, the volume of the displaced fluid is *not* the same as the volume of the object. In fact, we can see from Equation 13.9 that the volume of fluid displaced by a floating object of uniform density is

$$V_f = \frac{\rho_o}{\rho_f} V_o \qquad (13.10)$$

which is *less* than V_o because $\rho_o < \rho_f$.

NOTE ▶ Equation 13.10 applies only to *uniform* objects. It does not apply to boats, hollow spheres, or other objects of nonuniform composition. ◀

An object of density ρ_o and volume V_o is floating on a fluid of density ρ_f.

$\vec{F}_B$

Fluid density ρ_f

$\vec{w}_o$

The submerged volume of the object is equal to the volume V_f of displaced fluid.

FIGURE 13.19 A floating object is in static equilibrium.

◄ **Hidden depths** You've probably heard it said that "90% of an iceberg is underwater." Equation 13.10 is the basis for that statement. Most icebergs break off glaciers and are freshwater ice with a density of 917 kg/m³. The density of seawater is 1030 kg/m³. Thus

$$V_f = \frac{917 \text{ kg/m}^3}{1030 \text{ kg/m}^3} V_o = 0.89 V_o$$

V_f, the displaced water, is the volume of the iceberg that is underwater. You can see that, indeed, 89% of the volume of an iceberg is underwater.

CONCEPTUAL EXAMPLE 13.2 Which has the greater buoyant force?

A block of iron sinks to the bottom of a vessel of water while a block of wood of the *same size* floats. On which is the buoyant force greater?

REASON The buoyant force is equal to the volume of water displaced. The iron block is completely submerged, so it displaces a volume of water equal to its own volume. The wood block floats, so it displaces only the fraction of its volume that is under water, which is *less* than its own volume. The buoyant force on the iron block is therefore greater than on the wood one.

ASSESS This result may seem counterintuitive, but remember that the iron block sinks because of its high density, while the wood block floats because of its low density. A smaller buoyant force is sufficient to keep it floating.

EXAMPLE 13.6 Measuring the density of an unknown liquid

You need to determine the density of an unknown liquid. You notice that a block floats in this liquid with 4.6 cm of the side of the block submerged. When the block is placed in water, it also floats but with 5.8 cm submerged. What is the density of the unknown liquid?

PREPARE Assume that the block is an object of uniform composition. Figure 13.20 shows the block and defines the cross-section area A and submerged lengths h_u in the unknown liquid and h_w in water.

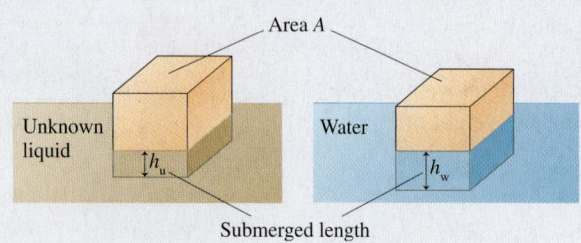

FIGURE 13.20 A block floating in two liquids.

SOLVE The block is floating, so Equation 13.10 applies. The block displaces volume $V_u = Ah_u$ of the unknown liquid. Thus

$$V_u = Ah_u = \frac{\rho_o}{\rho_u} V_o$$

Similarly, the block displaces volume $V_w = Ah_w$ of the water, leading to

$$V_w = Ah_w = \frac{\rho_o}{\rho_w} V_o$$

Because there are two fluids, we've used subscripts w for water and u for the unknown in place of the fluid subscript f. The product $\rho_o V_o$ appears in both equations. In the first, $\rho_o V_o = \rho_u Ah_u$, and in the second $\rho_o V_o = \rho_w Ah_w$. Equating the right-hand sides gives

$$\rho_u Ah_u = \rho_w Ah_w$$

The unknown area A cancels, and the density of the unknown liquid is

$$\rho_u = \frac{h_w}{h_u} \rho_w = \frac{5.8 \text{ cm}}{4.6 \text{ cm}} 1000 \text{ kg/m}^3 = 1300 \text{ kg/m}^3$$

ASSESS Comparison with Table 13.1 shows that the unknown liquid is likely to be glycerin.

Boats and Balloons

A chunk of steel sinks, so how does a steel-hulled boat float? As we've seen, an object floats if the upward buoyant force—the weight of the displaced water—balances the weight of the object. A boat is really a large hollow shell whose weight is determined by the volume of steel in the hull. As Figure 13.21 shows, the volume of water displaced by a shell is *much* larger than the volume of the hull itself. As a boat settles into the water, it sinks until the weight of the displaced water exactly matches the boat's weight. It is then in static equilibrium, so it floats at that level.

The concept of buoyancy and flotation applies to all fluids, not just liquids. An object immersed in a gas such as air feels a buoyant force as well. Because the density of air is so small, this buoyant force is generally negligible. Nonetheless, even though the buoyant force due to air is small, an object will float in air if it weighs less than the air that it displaces. This is why a floating balloon cannot be filled with regular air. If it were, then the weight of the air inside it would equal the weight of the air it displaced, so that it would have no net upward force on it. Adding in the weight of the balloon itself would then lead to a net downward force. For a balloon to float, it must be filled with a gas that has a *lower* density than that of air. The following example illustrates how this works.

The weight of the boat is mostly the weight of the steel in the hull.

The buoyant force is equal to the weight of the water displaced by the hull.

Even though the density of water is much less than that of steel, the volume of water displaced by a hollow shell is very large. The boat will float if the weight of the displaced water matches the weight of the boat.

FIGURE 13.21 How a boat floats.

EXAMPLE 13.7 How big does a balloon need to be?

What diameter must a helium-filled balloon have to float with neutral buoyancy? The mass of the empty balloon is 2.0 g.

PREPARE We'll model the balloon as a sphere. The balloon will float when its weight—the weight of the empty balloon plus that of the helium—equals the weight of the air it displaces. The densities of air and helium are given in Table 13.1.

SOLVE The volume of the balloon is V_{balloon}. Its weight is

$$w_{\text{balloon}} = m_{\text{balloon}}g + \rho_{\text{He}}V_{\text{balloon}}g$$

where m_{balloon} is the mass of the empty balloon. The weight of the displaced air is

$$w_{\text{air}} = \rho_{\text{air}}V_{\text{air}}g = \rho_{\text{air}}V_{\text{balloon}}g$$

where we noted that the volume of displaced air is the volume of the balloon. The balloon will just float when these two forces are equal, or when

$$\rho_{\text{air}}V_{\text{balloon}}g = m_{\text{balloon}}g + \rho_{\text{He}}V_{\text{balloon}}g$$

The g cancels, and we can solve for the volume of the balloon:

$$V_{\text{balloon}} = \frac{m_{\text{balloon}}}{\rho_{\text{air}} - \rho_{\text{He}}}$$

$$= \frac{2.0 \times 10^{-3}\,\text{kg}}{1.3\,\text{kg/m}^3 - 0.18\,\text{kg/m}^3} = 1.8 \times 10^{-3}\,\text{m}^3$$

A sphere has volume $V = (4\pi/3)r^3$. Thus the balloon's radius is

$$r = \left(\frac{3V_{\text{balloon}}}{4\pi}\right)^{\frac{1}{3}} = \left(\frac{3 \times (1.8 \times 10^{-3}\,\text{m}^3)}{4\pi}\right)^{\frac{1}{3}} = 0.075\,\text{m}$$

The diameter of the balloon is twice this, or 15 cm.

ASSESS This example shows why a helium balloon will no longer float once its volume falls below a certain value.

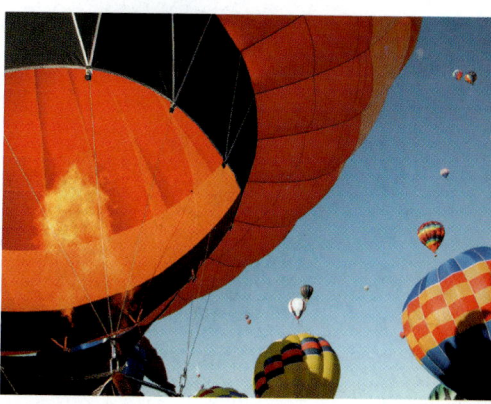

Hot air rising A hot air balloon is filled with a low-density gas: Hot air! You learned in Chapter 12 that gases expand upon heating, thus lowering their density. The air at the top of a hot air balloon is surprisingly hot—about 100°C, the temperature of boiling water. Using the ideal-gas law, you should be able to show that the density of 100°C air is only 79% that of room-temperature air at 20°C. A burst of flame lowers the density of the air and thus the balloon's weight. The balloon rises when its weight becomes less than the weight of the cooler air it has displaced.

STOP TO THINK 13.3 An ice cube is floating in a glass of water that is filled entirely to the brim. When the ice cube melts, the water level will

A. Fall. B. Stay the same. C. Rise, causing the water to spill.

13.5 Fluids in Motion

The wind blowing through your hair, a white-water river, and oil gushing from an oil well are examples of fluids in motion. We've focused thus far on fluid statics, but it's time to turn our attention to fluid *dynamics*.

Fluid flow is a complex subject. Many aspects of fluid flow, especially turbulence and the formation of eddies, are still not well understood and are areas of current research. We will avoid these difficulties by using a simplified *model* of an *ideal* fluid. Our model can be expressed in three assumptions about the fluid:

1. The fluid is *incompressible*. This is a very good assumption for liquids, but it also holds reasonably well for a moving gas, such as air. For instance, even when a 100 mph wind slams into a wall, its density changes by only about one percent.
2. The flow is *steady*. That is, the fluid velocity at each point in the fluid is constant; it does not fluctuate or change with time. Flow under these conditions is called **laminar flow,** and it is distinguished from *turbulent flow*.
3. The fluid is *nonviscous*. Water flows much more easily than cold pancake syrup because the syrup is a very viscous liquid. Viscosity is resistance to flow, and assuming a fluid is nonviscous is analogous to assuming the motion of a particle is frictionless. Gases have very low viscosity, and even many liquids are well approximated as being nonviscous.

Later, in Section 13.7, we'll relax condition 3 and consider the effects of viscosity.

The rising smoke in the photograph of Figure 13.22 begins as laminar flow, recognizable by the smooth contours, but at some point undergoes a transition to turbulent flow. A laminar-to-turbulent transition is not uncommon in fluid flow. Our model of fluids can be applied to the laminar flow, but not to the turbulent flow.

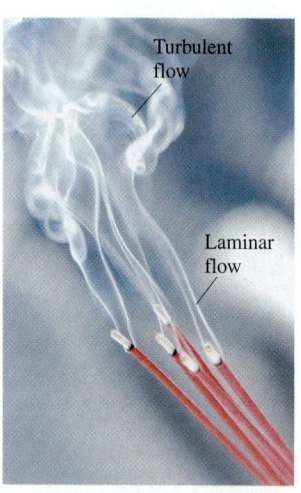

Turbulent flow

Laminar flow

FIGURE 13.22 Rising smoke changes from laminar flow to turbulent flow.

The Equation of Continuity

Consider a fluid flowing through a tube—oil through a pipe or blood through an artery. If the tube's diameter changes, as happens in Figure 13.23, what happens to the speed of the fluid?

When you squeeze a toothpaste tube, the volume of toothpaste that emerges matches the amount by which you reduce the volume of the tube. An *incompressible* fluid flowing through a rigid tube or pipe acts the same way. Fluid is neither created nor destroyed within the tube, and it cannot be stored. If volume V enters the tube during some interval of time Δt, then an equal volume of fluid must leave the tube.

To see the implications of this idea, suppose all molecules of the fluid in Figure 13.23 are moving forward with speed v_1 at a point where the cross-section area is A_1. Later, where the cross-section area is A_2, their speed is v_2. During an interval of time Δt, the first molecules move forward distance $\Delta x_1 = v_1 \Delta t$ and the second move $\Delta x_2 = v_2 \Delta t$. Because the fluid is incompressible, small volumes ΔV_1 and ΔV_2 must be equal. That is,

$$\Delta V_1 = A_1 \Delta x_1 = A_1 v_1 \Delta t = \Delta V_2 = A_2 \Delta x_2 = A_2 v_2 \Delta t \qquad (13.11)$$

Dividing both sides of the equation by Δt gives what is called the **equation of continuity:**

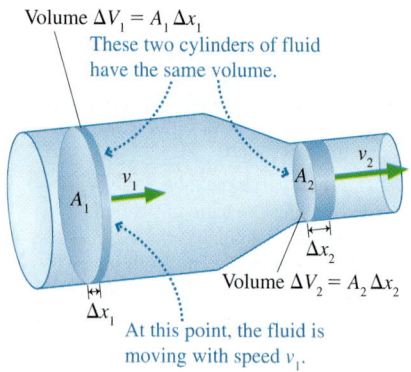

Volume $\Delta V_1 = A_1 \Delta x_1$

These two cylinders of fluid have the same volume.

v_1

A_1

A_2

v_2

Δx_2

Volume $\Delta V_2 = A_2 \Delta x_2$

Δx_1

At this point, the fluid is moving with speed v_1.

FIGURE 13.23 Flow speed changes through a tapered tube.

$$v_1 A_1 = v_2 A_2 \qquad (13.12)$$

The equation of continuity relating the speed v of an incompressible fluid to the cross-section area A of the tube in which it flows

(a) Garden hose

Garden hose with nozzle

(b)

Equations 13.11 and 13.12 say that **the volume of an incompressible fluid entering one part of a tube or pipe must be matched by an equal volume leaving downstream.**

An important consequence of the equation of continuity is that **flow is faster in narrower parts of a tube, slower in wider parts.** You're familiar with this conclusion from many everyday observations. The garden hose shown in Figure 13.24a squirts farther after you put a nozzle on it. This is because the narrower opening of the nozzle gives the water a higher exit speed. Water flowing from the faucet shown in Figure 13.24b picks up speed as it falls. As a result, the flow tube "necks down" to a smaller diameter.

The *rate* at which fluid flows through the tube—volume per second—is $\Delta V/\Delta t$. This is called the **volume flow rate** Q. We can see from Equation 13.11 that

$$Q = \frac{\Delta V}{\Delta t} = vA \qquad (13.13)$$

The SI units of Q are m^3/s, although in practice Q may be measured in cm^3/s, liters per minute, or, in the United States, gallons per minute and cubic feet per minute. Another way to express the meaning of the equation of continuity is to say that **the volume flow rate is constant at all points in a tube.**

NOTE ▶ An ideal fluid has no friction with the walls of its tube, so the speed is the same at all points on a line across the tube (i.e., at the wall and at the center). Real fluids have viscosity so that, as we'll see, the speed of the fluid is different at the tube's wall than at its center. But we can still use the equation of continuity for real fluids if we use the *average* speed of the fluid in Equations 13.12 and 13.13. ◀

FIGURE 13.24 The speed of the fluid changes as the flow diameter changes. This is a consequence of the equation of continuity.

EXAMPLE 13.8 Speed of water through a hose

A garden hose has an inside diameter of 16 mm. The hose can fill a 10 L bucket in 20 s.

a. What is the speed of the water out of the end of the hose?
b. What diameter nozzle would cause the water to exit with a speed 4 times greater than the speed inside the hose?

PREPARE Water is essentially incompressible, so the equation of continuity applies.

SOLVE

a. The volume flow rate is $Q = \Delta V/\Delta t = (10\text{ L})/(20\text{ s}) = 0.50\text{ L/s}$. To convert this to SI units, recall that $1\text{ L} =$

$1000\text{ mL} = 10^3\text{ cm}^3 = 10^{-3}\text{ m}^3$. Thus $Q = 5.0 \times 10^{-4}\text{ m}^3/s$. We can find the speed of the water from Equation 13.13:

$$v = \frac{Q}{A} = \frac{Q}{\pi r^2} = \frac{5.0 \times 10^{-4}\text{ m}^3/s}{\pi (0.0080\text{ m})^2} = 2.5\text{ m/s}$$

b. The quantity $Q = vA$ remains constant as the water flows through the hose and then the nozzle. To increase v by a factor of 4, A must be reduced by a factor of 4. The cross-section area depends on the square of the radius, so the area is reduced by a factor of 4 if the radius is reduced by a factor of 2. Thus the necessary nozzle diameter is 8 mm.

EXAMPLE 13.9 Blood flow in capillaries

The volume flow rate of blood leaving the heart to circulate throughout the body is about 5 L/min for a person at rest. All this blood eventually must pass through the smallest of blood vessels, the capillaries. Microscope measurements show that a

typical capillary is $6\,\mu\text{m}$ in diameter, 1 mm long, and the blood flows through it at an average speed of 1 mm/s.

a. Estimate the total number of capillaries in the body.
b. Estimate the total surface area of all the capillaries.

Continued

The various lengths and areas are defined in Figure 13.25.

Cross-section area A_{cap}

Surface area

$2r \approx 6 \ \mu m$

$L \approx 1 \ mm$

FIGURE 13.25 A capillary (not to scale).

PREPARE We can use the equation of continuity. The volume flow rate through all the capillaries must equal the volume flow rate leaving the heart.

SOLVE

a. We'll start by converting Q to SI units:

$$Q = 5 \frac{L}{min} \times \frac{1 \ m^3}{1000 \ L} \times \frac{1 \ min}{60 \ s} = 8 \times 10^{-5} \ m^3/s$$

to one-significant-figure accuracy. Then the total area of all the capillaries is

$$A_{total} = \frac{Q}{v} = \frac{8 \times 10^{-5} \ m^3/s}{0.001 \ m/s} = 0.08 \ m^2$$

The cross-section area of each capillary is $A_{cap} = \pi r^2$, so the total number of capillaries is approximately

$$N = \frac{A_{total}}{A_{cap}} = \frac{0.08 \ m^2}{\pi (3 \times 10^{-6} \ m)^2} = 3 \times 10^9$$

b. The surface area of one capillary is

$$A = \text{circumference of capillary} \times \text{length}$$

$$= 2\pi r L = 2\pi (3 \times 10^{-6} \ m)(0.001 \ m) = 2 \times 10^{-8} \ m^2$$

so the total surface area of all the capillaries is about

$$A_{surface} = NA = (3 \times 10^9)(2 \times 10^{-8} \ m) = 60 \ m^2$$

ASSESS The number of capillaries is huge, and the total surface area is about the floor area of a good-sized house. As we saw in Chapter 12, oxygen and nutrients move from the blood into cells by the slow process of diffusion. Only by having this large surface area available for diffusion can the required rate of gas and nutrient exchange be attained.

Representing Fluid Flow: Streamlines and Fluid Elements

To represent the flow of fluid is more complicated than representing the motion of a point particle because fluid flow is the collective motion of a vast number of particles. Figure 13.26 is an interesting photograph that gives us an idea of one possible fluid-flow representation. Here smoke is being used to help engineers visualize the air flow around a car in a wind tunnel. The smoothness of the flow tells us this is laminar flow. But notice also how the individual smoke trails retain their identity. They don't cross or get mixed together. Each smoke trail represents a *streamline* in the fluid.

Imagine that we could inject a tiny colored drop of water into a stream of water undergoing laminar flow. Because the flow is steady and the water is incompressible, this colored drop would maintain its identity as it flowed along. The path or trajectory followed by this "particle of fluid" is called a **streamline.** Smoke particles mixed with the air allow you to see the streamlines in the wind-tunnel photograph of Figure 13.26. Figure 13.27 illustrates three important properties of streamlines.

In studying the motion of a fluid, it is often also useful to consider a small *volume* of fluid, a volume containing many particles of fluid. Such a volume is called a **fluid element.** Figure 13.28 on the next page shows some of the important properties of a fluid element. Unlike a particle, a fluid element has an actual shape and volume. Although the shape of a fluid element may change as it moves, the equation of continuity requires that its volume remain constant. The progress of a fluid element as it moves along streamlines and changes shape is another very useful representation of fluid motion.

Streamline

FIGURE 13.26 The laminar air flow around a car in a wind tunnel is made visible with smoke.

1. Streamlines never cross.

$\vec{v}$

2. Fluid particle velocity is tangent to the streamline.

3. The speed is higher where the streamlines are closer together.

◀ **FIGURE 13.27** Particles in a fluid move along streamlines.

STOP TO THINK 13.4

The figure shows volume flow rates (in cm³/s) for all but one tube. What is the volume flow rate through the unmarked tube? Is the flow direction in or out?

Flows in cm³/s

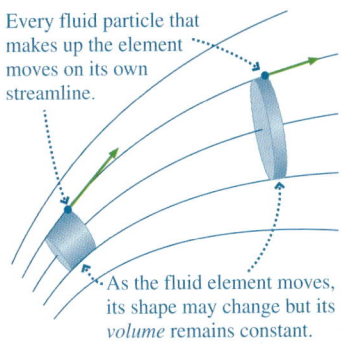

Every fluid particle that makes up the element moves on its own streamline.

As the fluid element moves, its shape may change but its *volume* remains constant.

FIGURE 13.28 Motion of a fluid element.

13.6 Fluid Dynamics

The equation of continuity describes a moving fluid but doesn't tell us anything about *why* the fluid is in motion. To understand the dynamics, consider the ideal fluid moving from left to right through the tube shown in Figure 13.29. The fluid moves at a steady speed v_1 in the wider part of the tube. In accordance with the equation of continuity, its speed is a higher, but steady, v_2 in the narrower part of the tube. If we follow a fluid element through the tube, we see that it undergoes an *acceleration* from v_1 to v_2 in the tapered section of the tube.

In the absence of friction, Newton's first law tells us that a particle will coast forever at a steady speed. Now in a fluid, viscosity is analogous to friction. In a viscous fluid, parts of the fluid moving at different speeds "rub" against each other, tending to slow the fluid down. But an ideal fluid is nonviscous and thus analogous to a particle moving without friction. Thus a fluid element moving through the constant-diameter sections of the tube in Figure 13.29 requires no force to "coast" at steady speed.

On the other hand, a fluid element moving through the neck of the tube accelerates, speeding up from v_1 to v_2. According to Newton's second law, for a particle to accelerate a net force must act on it. Similarly, for this fluid element to speed up there must be a net force acting on it.

What is the origin of this force? There are no external forces, and the horizontal motion rules out gravity. Instead, the fluid element is pushed from both ends by the *surrounding fluid*—that is, by *pressure forces*. The fluid element with cross-section area A in Figure 13.30 has a higher pressure on its left side than on its right. Thus the force $F_L = p_L A$ of the fluid pushing on its left side—a force pushing to the right—is greater than the force $F_R = p_R A$ from the fluid on its right. The net force, which points from the high pressure side of the element to the lower-pressure side, is

$$F_{net} = F_L - F_R = (p_L - p_R)A = A \Delta p$$

In other words, there's a net force on the fluid element, causing it to change speed, if and only if there's a pressure *difference* Δp between the two faces.

Thus, in order to accelerate the fluid elements in Figure 13.29 through the neck, the pressure p_1 in the wider section of the tube must be higher than the pressure p_2 in the narrower section. When the pressure is changing from one point in a fluid to another, we say that there is a **pressure gradient** in that region. We can also say that pressure forces are caused by pressure gradients, so **an ideal fluid accelerates wherever there is a pressure gradient.**

As a result, **the pressure is higher at a point along a streamline where the fluid is moving slower, lower where the fluid is moving faster.** This property of fluids was discovered in the 18th century by the Swiss scientist Daniel Bernoulli and is called the **Bernoulli effect.**

A fluid element coasts at steady speed through the constant-diameter segments of the tube.

As a fluid element flows through the neck, it speeds up. Because it is accelerating, there must be a force acting on it.

FIGURE 13.29 A motion diagram of a fluid element moving through a narrowing tube.

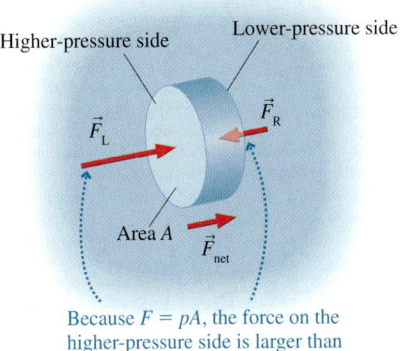

Higher-pressure side   Lower-pressure side

$\vec{F}_L$ $\vec{F}_R$

Area A $\vec{F}_{net}$

Because $F = pA$, the force on the higher-pressure side is larger than that on the lower-pressure side.

FIGURE 13.30 The net force on a fluid element due to pressure points from high to low pressure.

FIGURE 13.31 A Venturi tube measures flow speeds of a fluid.

The centripetal acceleration of a fluid element following a curved streamline.

The pressure is p_{atmos} where the streamlines are undisturbed.

Region of lower pressure. The downward pressure gradient causes the centripetal force.

FIGURE 13.32 Centripetal acceleration of a fluid element is caused by a downward pressure gradient.

NOTE ▶ It is important to realize that it is the change in pressure from high to low that *causes* the fluid to speed up. A high fluid speed doesn't *cause* a low pressure any more than a fast-moving particle causes the force that accelerated it. ◀

This relationship between pressure and fluid speed can be used to measure the speed of a fluid with a device called a *Venturi tube*. Figure 13.31 shows a simple Venturi tube suitable for a flowing liquid. The high pressure at point 1, where the fluid is moving slowly, causes the fluid to rise in the vertical pipe to a total height d_1. Because there's no vertical motion of the fluid, we can use the hydrostatic pressure equation to find that the pressure at point 1 is $p_1 = p_0 + \rho g d_1$. At point 3, on the same streamline, the fluid is moving faster and the pressure is lower, thus the fluid in the vertical pipe rises to a lower height d_3. The pressure *difference* is $\Delta p = \rho g (d_1 - d_3)$, so that the pressure difference across the neck of the pipe can be found by measuring the difference in heights of the fluid. We'll see later in this section how to relate this pressure difference to the increase of speed.

Note that the fluid height d_1 is the same as d_2, and d_3 is the same as d_4. For an ideal fluid, with no viscosity, no pressure difference is needed to keep the fluid moving at a constant speed, as it does in both the large- and small-diameter sections of the tube. In the next section we'll see how this result changes for a viscous fluid.

Lift

For a moving particle, a changing speed is just one way to accelerate. A particle also undergoes an acceleration if its *direction* of motion changes. Similarly, a moving fluid experiences an acceleration if it changes direction, requiring a net force to act on it. Figure 13.32 shows the wind blowing over a hill. Far from the hill, the airflow is undisturbed and the pressure is p_{atmos}. Near the ground, however, the streamlines have to *curve* around the hill, causing them to get bunched closer together.

You learned in Chapter 6 that curved motion requires a force pointing toward the center of curvature. As a fluid element follows the streamlines around the hill, the only force that can give it a centripetal acceleration is a pressure force due to the air above and below it. For the acceleration to point downward, the pressure near the ground, where the streamlines are closer together, must be *less* than the pressure p_{atmos} at slightly higher elevations.

Causing a fluid to follow a curved trajectory has some surprising but important consequences. To see this, try the simple experiment illustrated in Figure 13.33 before reading the next paragraph. Really, do try this!

TRY IT YOURSELF ▶

1. Hold strip at lower edge of bottom lip, just touching lip.

1 inch × 8 inch strip of notebook paper

2. Pucker lips and blow hard straight out over the top of the strip.

A pressure gradient is needed to cause the streamlines to curve along the paper. Thus the pressure on the top edge of the strip is less than atmospheric pressure.

Atmospheric pressure

Because the pressure is higher on this side of the strip, the strip is pushed upward.

FIGURE 13.33 A simple demonstration of pressure forces.

FIGURE 13.34 Why the paper strip rises.

What happened? You probably expected your breath to press the strip of paper down. Instead, the strip *rose*. In fact, the harder you blow, the more nearly the strip becomes parallel to the floor. As shown in Figure 13.34, the pressure near the strip must be less than atmospheric pressure to force the air to follow the curved path. This is just like the air flowing over the hill. Because there's a higher pressure beneath the strip, a net force pushes the strip upward. You can convince yourself that the curvature of the paper is important by placing the strip flat on a table. Blowing across it will not cause it to rise.

The upward motion of the paper strip is a simple example of the idea of *lift*. As you can imagine, an important application of this idea is the upward lift forces that act on airplane wings due to the air flowing past the wings. Figure 13.35 shows how this works. The top of a wing is curved, like the top of a gentle hill, while the bottom is flat. The air curves over the top of the wing, just like it does over the hill in Figure 13.32, which requires a lower pressure near the upper surface of the wing than in the surrounding air. Thus a *low pressure* zone develops just above the wing. The air moving *beneath* the wing is not strongly deflected, so the pressure there is close to p_{atmos}. The atmospheric pressure below pushes up; the low pressure above pushes down, but less strongly. The net result is an upward force—the lift.

A pressure gradient is needed to accelerate the air around the curved upper surface of the wing. Thus the air just above the wing is a zone of low pressure.

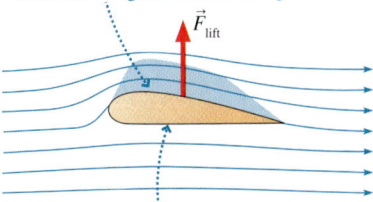

Because the pressure beneath the wing is higher than the pressure above, there's a net upward force on the wing. This is lift.

FIGURE 13.35 Air flow over a wing generates lift by creating unequal pressures above and below.

Smaller, slightly curved hill gives a low pressure.

Tall, strongly curved hill gives an even lower pressure.

The pressure difference between the two ends pushes air through the burrow.

Nature's air conditioning Prairie dogs ventilate their underground burrows with the same aerodynamic forces and pressures that give airplanes lift. The two entrances to their burrows are surrounded by mounds, one higher than the other. When the wind blows across these mounds the pressure is reduced at the top, just as for an airplane's wing. The taller mound, with its greater curvature, has the lower pressure of the two entrances. Air then is pushed through the burrow towards this lower-pressure side.

Bernoulli's Equation

We've seen that a pressure gradient causes a fluid to accelerate horizontally. Not surprisingly, gravity can cause a fluid to speed up or slow down if the fluid changes elevation. These are the key ideas of fluid dynamics. Now we would like to make these ideas quantitative by finding a numerical relationship between pressure, height, and the speed of a fluid. We can do so by applying the statement of conservation of mechanical energy you learned in Chapter 10,

$$\Delta K + \Delta U = W$$

where U is the gravitational potential energy and W is the work done by other forces, in this case pressure forces. Recall that we're still considering ideal fluids, with no friction or viscosity, so there's no dissipation of energy to thermal energy.

Figure 13.36 on the next page shows fluid flowing through a tube. The tube narrows from cross-section area A_1 to area A_2 as it bends uphill. Let's concentrate on the large segment of the fluid that is initially between the two planes, marked a and c, that are perpendicular to the flow. **This moving segment of fluid will be our *system* for the purpose of applying conservation of energy.**

$\vec{F}_2$ due to pressure at top

At time t, the system is the fluid between planes a and c.

A short time Δt later, the system has moved to be between planes b and d.

$\vec{F}_1$ due to pressure at bottom

Volume $A_2 \Delta x_2$

The volumes of the shaded cylinders are equal.

Volume $A_1 \Delta x_1$

FIGURE 13.36 Energy analysis of the flow of an ideal fluid.

A short time interval Δt later, the system has moved a bit further along the tube, so that it is now between planes b and d. Because the tube is not of uniform diameter, the two ends of the fluid system do not move the same distance during Δt: The lower end moves a distance Δx_1 while the upper end moves Δx_2. These two distances are related by the continuity equation, Equation 13.12. We have

$$v_1 A_1 = \frac{\Delta x_1}{\Delta t} A_1 = v_2 A_2 = \frac{\Delta x_2}{\Delta t} A_2$$

or, canceling Δt,

$$A_1 \Delta x_1 = A_2 \Delta x_2 = V$$

Thus V is the volume of either of the small cylinders of fluid shown in Figure 13.36.

As the system moves, positive work is done on it by the force $\vec{F}_1$ due to the pressure p_1 of the fluid to the left of the system, while negative work is done on the system by the force $\vec{F}_2$ due to the pressure p_2 of the fluid to the right of the system. The positive work is

$$W_1 = F_1 \Delta x_1 = (p_1 A_1) \Delta x_1 = p_1 (A_1 \Delta x_1) = p_1 V$$

Similarly the negative work is

$$W_2 = -F_2 \Delta x_2 = -p_2 A_2 \Delta x_2 = -p_2 V$$

Thus the *net* work done on the system is

$$W = W_1 + W_2 = p_1 V - p_2 V = (p_1 - p_2) V$$

By conservation of energy, this work equals the change in the system's energy.

Initially the system had the kinetic and potential energy of all the fluid between points a and c; after time Δt has passed, it has the energy of all the fluid between b and d. The fluid between points b and c is common to both the initial and final systems, so there is *no* change in energy due to this part of the fluid. As the system moves, it *loses* the energy it had in cylinder 1 of length Δx_1, but *gains* energy in cylinder 2 of length Δx_2.

Let's find the kinetic energy contained in each of these cylinders. The mass of each of these cylinders is $m = \rho V$, where ρ is the density of the fluid. The kinetic energies of cylinders 1 and 2 are

$$K_1 = \frac{1}{2} \rho V v_1^2 \qquad \text{and} \qquad K_2 = \frac{1}{2} \rho V v_2^2$$

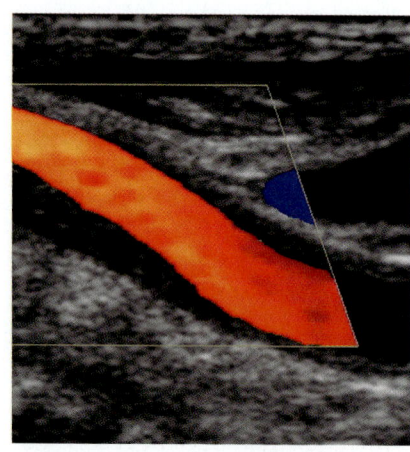

Living under pressure BIO When plaque builds up in major arteries, such as the carotid artery that supplies the head, dangerous drops in blood pressure can result. *Doppler ultrasound*, which you'll study in Chapter 15, uses sound waves to produce images of the interior of the body, and can detect the velocity of flowing blood. The image above shows the blood flow through a carotid artery with significant plaque buildup; yellow indicates a higher blood velocity than red. Once the velocities are known at two points along the flow, Bernoulli's equation can be used to deduce the corresponding pressure drop.

Thus the net *change* in kinetic energy is

$$\Delta K = K_2 - K_1 = \frac{1}{2}\rho V v_2^2 - \frac{1}{2}\rho V v_1^2$$

Similarly, the net change in the gravitational potential energy of our fluid system is

$$\Delta U = U_2 - U_1 = \rho V g y_2 - \rho V g y_1$$

We can use these expressions for ΔK, ΔU, and W to write the energy equation as

$$\underbrace{\frac{1}{2}\rho V v_2^2 - \frac{1}{2}\rho V v_1^2}_{\Delta K} + \underbrace{\rho V g y_2 - \rho V g y_1}_{\Delta U} = \underbrace{(p_1 - p_2)V}_{W}$$

The V cancels, and we can rearrange the remaining terms to get what is called **Bernoulli's equation:**

$$p_2 + \frac{1}{2}\rho v_2^2 + \rho g y_2 = p_1 + \frac{1}{2}\rho v_1^2 + \rho g y_1 \qquad (13.14)$$

Bernoulli's equation relating pressure p, speed v, and height y
at any two points in an ideal fluid

Equation 13.14, a quantitative statement of the ideas we developed earlier in this section, is really nothing more than a statement about work and energy. Using Bernoulli's equation is very much like using the law of conservation of energy. Rather than identifying a "before" and "after," you want to identify two points on a streamline. As the following example shows, Bernoulli's equation is often used in conjunction with the equation of continuity.

EXAMPLE 13.10 Pressure in an irrigation system

Water flows through the pipes shown in Figure 13.37. The water's speed through the lower pipe is 5.0 m/s and a pressure gauge reads 75 kPa. What is the reading of the pressure gauge on the upper pipe?

FIGURE 13.37 The water pipes of an irrigation system.

PREPARE Treat the water as an ideal fluid obeying Bernoulli's equation. Consider a streamline connecting point 1 in the lower pipe with point 2 in the upper pipe.

SOLVE Bernoulli's equation, Equation 13.14, relates the pressure, fluid speed, and heights at points 1 and 2. It is easily solved for the pressure p_2 at point 2:

$$p_2 = p_1 + \frac{1}{2}\rho v_1^2 - \frac{1}{2}\rho v_2^2 + \rho g y_1 - \rho g y_2$$

$$= p_1 + \frac{1}{2}\rho(v_1^2 - v_2^2) + \rho g(y_1 - y_2)$$

All quantities on the right are known except v_2, and that is where the equation of continuity will be useful. The cross-section areas and water speeds at points 1 and 2 are related by

$$v_1 A_1 = v_2 A_2$$

from which we find

$$v_2 = \frac{A_1}{A_2}v_1 = \frac{r_1^2}{r_2^2}v_1 = \frac{(0.030 \text{ m})^2}{(0.020 \text{ m})^2}(5.0 \text{ m/s}) = 11.25 \text{ m/s}$$

The pressure at point 1 is $p_1 = 75 \text{ kPa} + 1 \text{ atm} = 176{,}300 \text{ Pa}$. We can now use the above expression for p_2 to calculate $p_2 = 105{,}900 \text{ Pa}$. This is the absolute pressure; the pressure gauge on the upper pipe will read

$$p_2 = 105{,}900 \text{ Pa} - 1 \text{ atm} = 4.6 \text{ kPa}$$

ASSESS Reducing the pipe size decreases the pressure because it makes $v_2 > v_1$. Gaining elevation also reduces the pressure.

Rank in order, from highest to lowest, the liquid heights h_1 to h_4 in tubes 1 to 4. The air flow is from left to right.

13.7 Viscosity and Poiseuille's Equation

The pressure drops between points 1 and 2.   In this region, the pressure gradient causes the fluid to speed up.   The pressure drops between points 3 and 4.

In these regions, a pressure gradient is needed simply to keep the fluid moving with constant speed.

FIGURE 13.38 A viscous fluid needs a pressure difference to keep it moving. Compare this to Figure 13.31, which showed an ideal fluid.

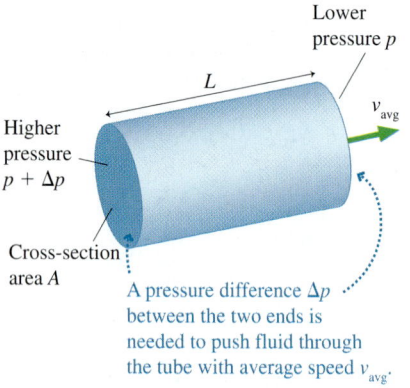

Lower pressure p

L

v_{avg}

Higher pressure $p + \Delta p$

Cross-section area A

A pressure difference Δp between the two ends is needed to push fluid through the tube with average speed v_{avg}.

FIGURE 13.39 The pressure difference needed to keep the fluid flowing is proportional to the fluid's viscosity.

Thus far, the only thing we've needed to know about a fluid is its density. It is the density that determines how the pressure in a static fluid increases with depth, and density appears in Bernoulli's equation for the motion of a fluid. But you know from everyday experience that another property of fluids is often crucial in determining how fluids flow. The densities of honey and water are not too different, but there's a huge difference in the way honey and water pour. The honey is much "thicker." This property of a fluid, which measures its resistance to flow, is called **viscosity.** A very viscous fluid flows slowly when poured and is difficult to force through a pipe or tube. The viscous nature of fluids is of key importance in understanding a wide range of real-world applications, from blood flow to the flight of birds to throwing a curveball.

An ideal fluid, with no viscosity, will "coast" at constant speed through a constant-diameter tube with no change in pressure. That's why the fluid heights are the same in the first and second pressure-measuring columns on the Venturi tube of Figure 13.31. But as Figure 13.38 shows, any real fluid requires a *pressure difference* between the ends of a tube to keep the fluid moving at constant speed. The size of the pressure difference depends on the viscosity of the fluid. Think about how much harder you have to suck on a straw while drinking a thick milkshake than you do to drink water or to pull air through the straw.

Figure 13.39 shows a viscous fluid flowing with a constant average speed v_{avg} through a tube of length L and cross-section area A. Experiments show that the pressure difference needed to keep the fluid moving is proportional to v_{avg} and to L and inversely proportional to A. We can write

$$\Delta p = 8\pi\eta \frac{Lv_{avg}}{A} \qquad (13.15)$$

where $8\pi\eta$ is the constant of proportionality and η (lower-case Greek eta) is called the **coefficient of viscosity** (or just the *viscosity*). (The 8π enters the equation from the technical definition of viscosity, which need not concern us.)

Equation 13.15 makes sense. A more viscous fluid needs a larger pressure difference to push it through the tube; an ideal fluid, with $\eta = 0$, will keep flowing without any pressure difference. We can also see from Equation 13.15 that the units of viscosity are $N \cdot s/m^2$ or, equivalently, $Pa \cdot s$. In practice, a common unit of viscosity is the **poise** (pronounced "pwaz"), defined as 1 poise $= 0.1 \text{ N} \cdot \text{s/m}^2$. However, the poise is not an SI unit and thus is not suitable for most calculations.

Table 13.3 gives values of η for some common fluids. Note that the viscosity of many liquids decreases *very* rapidly with temperature. Cold oil hardly flows at all, but hot oil pours almost like water.

Poiseuille's Equation

Viscosity has a profound effect on how a fluid moves through a tube. Figure 13.40a shows that in an ideal fluid, all fluid particles move with the same speed v, the speed that appears in the equation of continuity. For a viscous fluid, Figure 13.40b shows that the fluid moves fastest in the center of the tube. The speed decreases as you move away from the center of the tube until reaching zero on the walls of the tube. That is, the layer of fluid in contact with the tube doesn't move at all. Whether it be water through pipes or blood through arteries, the fact that the fluid at the outer edges "lingers" and barely moves allows deposits to build on the inside walls of a tube.

Although we can't characterize the flow of a viscous liquid by a single flow speed v, we can still define the *average* flow speed. Suppose a fluid with viscosity η flows through a circular pipe with radius R and cross-section area $A = \pi R^2$. From Equation 13.15, a pressure difference Δp between the ends of the pipe causes the fluid to flow with average speed

$$v_{\text{avg}} = \frac{R^2}{8\eta L} \Delta p \tag{13.16}$$

The average flow speed is directly proportional to the pressure difference; for the fluid to flow twice as fast, you would need to double the pressure difference between the ends of the pipe.

Equation 13.13 defined the volume flow rate $Q = \Delta V/\Delta t$ and found that $Q = vA$ for an ideal fluid. For viscous flow, where v isn't constant throughout the fluid, we simply need to replace v with the average speed v_{avg} found in Equation 13.16. Using $A = \pi R^2$ for a circular tube, we see that a pressure difference Δp causes a volume flow rate

$$Q = v_{\text{avg}}A = \frac{\pi R^4 \, \Delta p}{8\eta L} \tag{13.17}$$

Poiseuille's equation for viscous flow through a tube of radius R and length L

This result is called **Poiseuille's equation** after the French scientist Jean Poiseuille (1797–1869) who first performed this calculation.

One surprising result of Poiseuille's equation is the very strong dependence of the flow on the tube's radius; the volume flow rate is proportional to the *fourth* power of R. If you double the radius of a tube, the flow rate will increase by a factor of $2^4 = 16$. This strong radius dependence comes from two factors. First, the flow rate depends on the area of the tube, which is proportional to R^2; larger tubes carry more fluid. Second, the average speed is larger in a larger tube because the center of the tube, where the flow is fastest, is farther from the "drag" exerted by the wall. As we've seen, the average speed is also proportional to R^2. These two terms combine to give the R^4 dependence.

TABLE 13.3 Viscosities of fluids

Fluid	η (Pa $\cdot$ s)
Air (20°C)	1.8×10^{-5}
Water (20°C)	1.0×10^{-3}
Water (40°C)	0.7×10^{-3}
Water (60°C)	0.5×10^{-3}
Whole blood (37°C)	2.5×10^{-3}
Motor oil (-30°C)	3×10^5
Motor oil (40°C)	0.07
Motor oil (100°C)	0.01
Honey (15°C)	600
Honey (40°C)	20

(a) Ideal fluid

The velocity is the same at all points in the tube.

(b) Viscous fluid

The velocity is maximum at the center of the tube. It decreases away from center.

The velocity is zero on the walls of the tube.

FIGURE 13.40 Viscosity alters the velocities of the fluid particles.

CONCEPTUAL EXAMPLE 13.3 **Blood pressure and cardiovascular disease**

Cardiovascular disease is a narrowing of the arteries due to the buildup of plaque deposits on the interior walls. Magnetic resonance imaging, which you'll learn about in Chapter 24, can create exquisite three-dimensional images of the internal structure of the body. Shown are the carotid arteries that supply blood to the head, with a dangerous narrowing—a *stenosis*—indicated by the arrow.

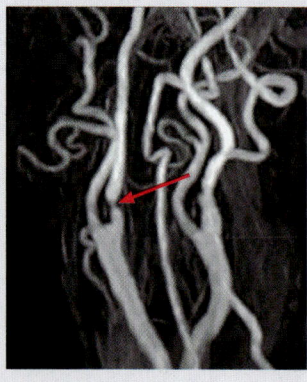

If a section of an artery has narrowed by 8%, not nearly as much as the stenosis shown, by what percent must the blood-pressure difference between the ends of the narrowed section increase to keep blood flowing at the same rate?

REASON According to Poiseuille's equation, the pressure difference Δp must increase to compensate for a decrease in the artery's radius R if the blood flow rate Q is to remain unchanged. If we write Poiseuille's equation as

$$R^4 \, \Delta p = 8\eta L Q / \pi$$

we see that the product $R^4 \, \Delta p$ must remain unchanged if the artery is to deliver the same flow rate. Let the initial artery radius and pressure difference be R_i and Δp_i. Disease decreases the radius by 8%, meaning that $R_f = 0.92 R_i$. The requirement

$$R_i^4 \, \Delta p_i = R_f^4 \, \Delta p_f$$

can be solved for the new pressure difference:

$$\Delta p_f = \frac{R_i^4}{R_f^4} \Delta p_i = \frac{R_i^4}{(0.92 R_i)^4} \Delta p_i = 1.4 \, \Delta p_i$$

That is, the pressure difference must increase by 40% to maintain the flow.

ASSESS Because the flow rate depends on R^4, even a small change in radius requires a large change in Δp to compensate. Either the person's blood pressure must increase, which is dangerous, or he or she will suffer a significant reduction in blood flow. For the stenosis shown in the image above the reduction in radius is much greater than 8%, and the pressure difference will be large and very dangerous.

EXAMPLE 13.11 **Pressure drop along a capillary**

In Example 13.9 we examined blood flow through a capillary. Using the numbers from that example, calculate the pressure "drop" from one end of a capillary to the other.

PREPARE Example 13.9 gives enough information to determine the flow rate through a capillary. We can then use Poiseuille's equation to calculate the pressure difference between the ends.

SOLVE The measured volume flow rate leaving the heart was given as about 5 L/min $= 8 \times 10^{-5}$ m³/s. This flow is divided among all the capillaries, which we found to number $N = 3 \times 10^9$. Thus the flow rate through each capillary is

$$Q_{cap} = \frac{Q_{heart}}{N} = \frac{8 \times 10^{-5} \text{ m}^3/\text{s}}{3 \times 10^9} = 2.7 \times 10^{-14} \text{ m}^3/\text{s}$$

Solving Poiseuille's equation for Δp, we get

$$
\begin{aligned}
\Delta p &= \frac{8\eta L Q_{cap}}{\pi R^4} \\
&= \frac{8(2.5 \times 10^{-3} \text{ Pa} \cdot \text{s})(0.001 \text{ m})(2.7 \times 10^{-14} \text{ m}^3/\text{s})}{\pi (3 \times 10^{-6} \text{ m})^4} \\
&= 2100 \text{ Pa}
\end{aligned}
$$

Converting to mm of mercury, the units of blood pressure, the pressure drop across the capillary is $\Delta p = 16$ mm of mercury.

ASSESS The average blood pressure provided by the heart (the average of systolic and diastolic pressure) is about 100 mm of mercury. A physiology textbook will tell you that the pressure has decreased to 35 mm by the time blood enters the capillaries, and it exits from capillaries into the veins at 17 mm. Thus the pressure drop across the capillaries is 18 mm of mercury. Our calculation, based on the laws of fluid flow and some simple estimates of capillary size, is in almost perfect agreement with measured values.

STOP TO THINK 13.6   A viscous fluid flows through the pipe shown. The three marked segments are of equal length. Across which segment is the pressure difference the greatest?

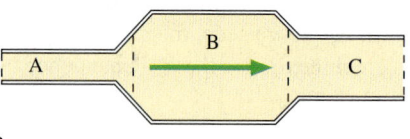

The goal of Chapter 13 has been to understand the static and dynamic properties of fluids.

GENERAL PRINCIPLES

Fluid Statics

Gases

- Freely moving particles
- Compressible
- Pressure mainly due to particle collisions with walls

Liquids

- Loosely bound particles
- Incompressible
- Pressure due to the weight of the liquid
- Hydrostatic pressure at depth d is $p = p_0 + \rho g d$
- The pressure is the same at all points on a horizontal line through a liquid in hydrostatic equilibrium

Fluid Dynamics

Density ρ

Ideal-fluid model

- Incompressible
- Smooth, laminar flow
- Nonviscous

Equation of continuity

Volume flow rate $Q = \dfrac{\Delta V}{\Delta t} = v_1 A_1 = v_2 A_2$

Bernoulli's equation is a statement of energy conservation:

$$p_1 + \frac{1}{2}\rho v_1^2 + \rho g y_1 = p_2 + \frac{1}{2}\rho v_2^2 + \rho g y_2$$

Lift on a curved surface is generated when a fluid moves across the surface. It is due to the pressure gradients that cause the fluid's centripetal acceleration.

Poiseuille's equation governs viscous flow through a tube:

$$Q = v_{\text{avg}} A = \frac{\pi R^4 \, \Delta p}{8 \eta L}$$

IMPORTANT CONCEPTS

Density $\rho = m/V$, where m is mass and V is volume.

Pressure $p = F/A$, where F is force magnitude and A is the area on which the force acts.

- Exists at all points in a fluid
- Pushes equally in all directions
- Gauge pressure $p_g = p - 1$ atm

Viscosity η is the property of a fluid that makes it resist flowing.

Representing fluid flow

Streamlines are the paths of individual fluid particles.

The velocity of a fluid particle is tangent to its streamline.

$\vec{v}$

The speed is higher where the streamlines are closer together.

Fluid elements contain a fixed volume of fluid. Their shape may change as they move.

Every fluid particle that makes up the element moves on its own streamline.

APPLICATIONS

Buoyancy is the upward force of a fluid on an object immersed in the fluid.

Archimedes' principle The magnitude of the buoyant force equals the weight of the fluid displaced by the object.

Sink:	$\rho_{\text{avg}} > \rho_f$	$F_B < w_o$
Float:	$\rho_{\text{avg}} < \rho_f$	$F_B > w_o$
Neutrally buoyant:	$\rho_{\text{avg}} = \rho_f$	$F_B = w_o$

Barometers measure atmospheric pressure. Atmospheric pressure is related to the height of the liquid column by $p_{\text{atmos}} = \rho g h$.

Manometers measure pressure. The pressure at the closed end of the tube is $p = 1$ atm $+ \rho g h$.

These two points are at the same pressure p.

For instructor-assigned homework, go to
www.masteringphysics.com

Problems labeled INT integrate significant material from earlier chapters; BIO are of biological or medical interest.

Problem difficulty is labeled as I (straightforward) to IIIII (challenging).

QUESTIONS

Conceptual Questions

1. Which has the greater density, 1 g of mercury or 1000 g of water?
2. A 1×10^{-3} m³ chunk of material has a mass of 3 kg.
   a. What is the material's density?
   b. Would a 2×10^{-3} m³ chunk of the same material have the same mass? Explain.
   c. Would a 2×10^{-3} m³ chunk of the same material have the same density? Explain.
3. You are given an irregularly shaped chunk of material and asked to find its density. List the *specific* steps that you would follow to do so.
4. Object 1 has an irregular shape. Its density is 4000 kg/m³.
   a. Object 2 has the same shape and dimensions as object 1, but it is twice as massive. What is the density of object 2?
   b. Object 3 has the same mass and the same *shape* as object 1, but its size in all three dimensions is twice that of object 1. What is the density of object 3?
5. BIO When you get a blood transfusion the bag of blood is held above your body, but when you donate blood the collection bag is held below. Why is this?
6. BIO To explore the bottom of a 10-m-deep lake, your friend Tom proposes to get a long garden hose, put one end on land and the other in his mouth for breathing underwater, and descend into the depths. Susan, who overhears the conversation, reacts with horror and warns Tom that he will not be able to inhale when he is at the lake bottom. Why is Susan so worried?
7. Rank in order, from largest to smallest, the pressures at A, B, and C in Figure Q13.7. Explain.
8. Refer to Figure Q13.7. Rank in order, from largest to smallest, the pressures at D, E, and F. Explain.
9. Cylinders A and B contain liquids. The pressure p_A at the bottom of A is higher than the pressure p_B at the bottom of B. Is the ratio p_A/p_B of the absolute pressures larger, smaller, or equal to the ratio of the gauge pressures? Explain.

FIGURE Q13.7

10. In Figure Q13.10, A and B are rectangular tanks full of water. They have equal heights and equal side thicknesses (the dimension into the page), but different widths.

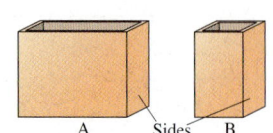

FIGURE Q13.10

   a. Compare the forces the water exerts on the bottoms of the tanks. Is F_A larger, smaller, or equal to F_B? Explain.
   b. Compare the forces the water exerts on the sides of the tanks. Is F_A larger, smaller, or equal to F_B? Explain.

11. Helium-filled weather balloons are spherical when they reach very high altitudes. However, they are only partially inflated with helium before they are released. Explain why this is done.

12. Water expands when heated. Suppose a beaker of water is heated from 10°C to 90°C. Does the pressure at the bottom of the beaker increase, decrease, or stay the same? Explain.
13. In Figure Q13.13, is p_A larger, smaller, or equal to p_B? Explain.

FIGURE Q13.13

380 mm

FIGURE Q13.14

14. At sea level, the height of the mercury column in a sealed glass tube is 380 mm, as shown in Figure Q13.14. What can you say about the contents of the space above the mercury? Be as specific as you can.
15. Rank in order, from largest to smallest, the densities of objects A, B, and C in Figure Q13.15. Explain.

FIGURE Q13.15

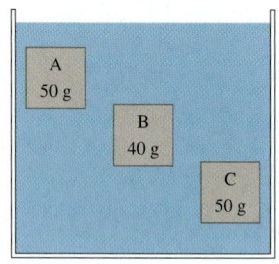

FIGURE Q13.16

16. Objects A, B, and C in Figure Q13.16 have the same volume. Rank in order, from largest to smallest, the sizes of the buoyant forces F_A, F_B, and F_C on A, B, and C. Explain.
17. Refer to Figure Q13.16. Now A, B, and C have the same density, but still have the masses given in the figure. Rank in order, from largest to smallest, the sizes of the buoyant forces on A, B, and C. Explain.
18. When you stand on a bathroom scale, it reads 700 N. Suppose a giant vacuum cleaner sucks half the air out of the room, reducing the pressure to 0.5 atm. Would the scale reading increase, decrease, or stay the same? Explain.

19. Suppose you stand on a bathroom scale that is on the bottom of a swimming pool. The water comes up to your waist. Does the scale read more, less, or the same as your true weight? Explain.

20. When you place an egg in water, it sinks. If you add salt to the water, after some time the egg floats. Explain.

21. Submerged submarines contain tanks filled with water. To rise to the surface, compressed air is used to force the water out of the tanks. Explain why this works.

22. Fish can adjust their buoyancy with an organ called the *swim*
BIO *bladder*. The swim bladder is a flexible gas-filled sac; the fish can increase or decrease the amount of gas in the swim bladder so that it stays neutrally buoyant—neither sinking nor floating. Suppose the fish is neutrally buoyant at some depth and then goes deeper. What needs to happen to the volume of air in the swim bladder? Will the fish need to add or remove gas from the swim bladder to maintain its neutral buoyancy?

23. Figure Q13.23 shows two iden-
tical beakers filled to the same height with water. Beaker B has a plastic sphere floating in it. Which beaker, with all its contents, weighs more? Or are they equal? Explain.

FIGURE Q13.23

24. A tub of water, filled to the brim, sits on a scale. Then a floating block of wood is placed in the tub, pushing some water over the rim. The water that overflows immediately runs off the scale. What happens to the reading of the scale?

25. Ships A and B have the same height and the same mass. Their cross-section profiles are shown in Figure Q13.25. Does one ship ride higher in the water (more height above the water line) than the other? If so, which one? Explain.

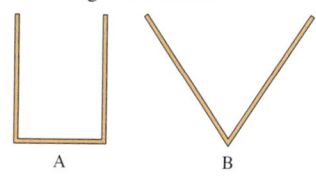

FIGURE Q13.25

26. Gas flows through a pipe, as shown in Figure Q13.26. The pipe's constant outer diameter is shown; you can't see into the pipe to know how the inner diameter changes. Rank in order, from largest to smallest, the gas speeds v_1 to v_3 at points 1, 2, and 3. Explain.

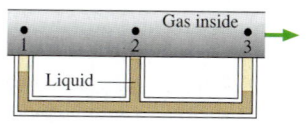

FIGURE Q13.26

27. Liquid flows through a pipe as shown in Figure Q13.27. The pipe's constant outer diameter is shown; you can't see into the pipe to know how the inner diameter changes. Rank in order, from largest to smallest, the flow speeds v_1 to v_3 at points 1, 2, and 3. Explain.

FIGURE Q13.27

28. A liquid with negligible viscosity flows through the pipe shown in Figure Q13.28. This is an overhead view.
   a. Rank in order, from largest to smallest, the flow speeds v_1 to v_4 at points 1 to 4. Explain.
   b. Rank in order, from largest to smallest, the pressures p_1 to p_4 at points 1 to 4. Explain.

FIGURE Q13.28

FIGURE Q13.29

29. Wind blows over a house as shown in Figure Q13.29. A window on the ground floor is open. Is there an air flow through the house? If so, does the air flow in the window and out the chimney, or in the chimney and out the window? Explain.

30. Two pipes have the same inner cross-section area. One has a circular cross section and the other has a rectangular cross section with its height one-tenth its width. Through which pipe, if either, would it be easier to pump a viscous liquid? Explain.

Multiple-Choice Questions

31. | Figure Q13.31 shows a 100 g block of copper ($\rho = 8900 \text{ kg/m}^3$) and a 100 g block of aluminum ($\rho = 2700 \text{ kg/m}^3$) connected by a massless string that runs over two massless, friction-less pulleys. The two blocks exactly balance, since they have the same mass. Now suppose that the whole system is submerged in water. What will happen?

FIGURE Q13.31

   A. The copper block will fall, the aluminum block will rise.
   B. The aluminum block will fall, the copper block will rise.
   C. Nothing will change.
   D. Both blocks will rise.

32. | Masses A and B rest on very light pistons that enclose a fluid, as shown in Figure Q13.32. There is no friction between the pistons and the cylinders they fit inside. Which of the following is true?

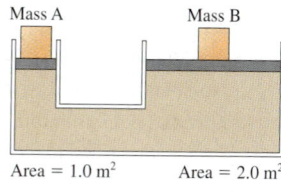

FIGURE Q13.32

   A. Mass A is greater.   B. Mass B is greater.
   C. Mass A and mass B are the same.

33. ‖ If you dive underwater, you notice an uncomfortable pres-
BIO sure on your eardrums due to the increased pressure. The human eardrum has an area of about 70 mm^2 (7×10^{-5} m^2), and it can sustain a force of about 7 N without rupturing. If your body had no means of balancing the extra pressure (which, in reality, it does), what would be the maximum depth you could dive without rupturing your eardrum?
   A. 0.3 m   B. 1 m   C. 3 m   D. 10 m

34. | Suppose you take a basketball and submerge it in a swim-ming pool. To what minimum gauge pressure must you fill the ball for it to stay fully inflated at a depth of 2 m?
   A. 2 kPa   B. 4 kPa   C. 20 kPa   D. 40 kPa

35. ‖ A basketball has a mass of 0.50 kg and a volume of 8.0×10^{-3} m³. What is the magnitude of the net force on a basketball when it is fully submerged in water?
    A. 4.9 N   B. 74 N   C. 78 N   D. 83 N

36. ‖ An object floats in water, with 75% of its volume submerged. What is its approximate density?
    A. 250 kg/m³   B. 750 kg/m³
    C. 1000 kg/m³   D. 1250 kg/m³

37. ‖ A syringe is being used to squirt water as shown in Figure Q13.37. The water is ejected from the nozzle at 10 m/s. At what speed is the plunger of the syringe being depressed?

Radius = 1 cm   Radius = 1 mm   10 m/s

FIGURE Q13.37

    A. 0.01 m/s B. 0.1 m/s   C. 1 m/s   D. 10 m/s

38. ‖ Water flows through a 4.0-cm-diameter horizontal pipe at a speed of 1.3 m/s. The pipe then narrows down to a diameter of 2.0 cm. Ignoring viscosity, what is the pressure difference between the wide and narrow sections of the pipe?
    A. 850 Pa   B. 3400 Pa   C. 9300 Pa
    D. 12,700 Pa   E. 13,500 Pa

39. ‖ A 15-m-long garden hose has an inner diameter of 2.5 cm. One end is connected to a spigot; 20°C water flows from the other end at a rate of 1.2 L/s. What is the gauge pressure at the spigot end of the hose?
    A. 1900 Pa   B. 2700 Pa   C. 4200 Pa
    D. 5800 Pa   E. 7300 Pa

PROBLEMS

Section 13.1 Fluids and Density

1. | A 100 mL beaker holds 120 g of liquid. What is the liquid's density in SI units?

2. | Containers A and B have equal volumes. Container A holds helium gas at 1.0 atm pressure and 0°C. Container B is completely filled with a liquid whose mass is 7000 times the mass of helium gas in container A. Identify the liquid in B.

3. | Air enclosed in a sphere has density $\rho = 1.4$ kg/m³. What will the density be if the radius of the sphere is halved, compressing the air within?

4. | Air enclosed in a cylinder has density $\rho = 1.4$ kg/m³.
    a. What will be the density of the air if the length of the cylinder is doubled while the radius is unchanged?
    b. What will be the density of the air if the radius of the cylinder is halved while the length is unchanged?

5. ‖ a. 50 g of gasoline are mixed with 50 g of water. What is the average density of the mixture?
    b. 50 cm³ of gasoline are mixed with 50 cm³ of water. What is the average density of the mixture?

6. ‖ Ethyl alcohol has been added to 200 mL of water in a container that has a mass of 150 g when empty. The resulting container and liquid mixture has a mass of 512 g. What volume of alcohol was added to the water?

7. | The average density of the body of a fish is 1080 kg/m³. To
BIO keep from sinking, the fish increases its volume by inflating an internal air bladder, known as a swim bladder, with air. By what percent must the fish increase its volume to be neutrally buoyant in fresh water? Use the Table 13.1 value for the density of air.

Section 13.2 Pressure

8. | The deepest point in the ocean is 11 km below sea level, deeper than Mt. Everest is tall. What is the pressure in atmospheres at this depth?

9. ‖ a. What volume of water has the same mass as 8.0 m³ of ethyl alcohol?
    b. If this volume of water completely fills a cubic tank, what is the pressure at the bottom?

10. ‖ A 1.0-m-diameter vat of liquid is 2.0 m deep. The pressure at the bottom of the vat is 1.3 atm. What is the mass of the liquid in the vat?

11. ‖ Find the force on the bottom of a 35-cm-tall, cylindrical beaker with a diameter of 5.0 cm. The beaker is filled to its brim with water.

12. ‖ The gauge pressure at the bottom of a cylinder of liquid is $p_g = 0.40$ atm. The liquid is poured into another cylinder with twice the radius of the first cylinder. What is the gauge pressure at the bottom of the second cylinder?

13. ‖ A research submarine has a 20-cm-diameter window 8.0 cm thick. The manufacturer says the window can withstand forces up to 1.0×10^6 N. What is the submarine's maximum safe depth? The pressure inside the submarine is maintained at 1.0 atm.

14. ‖ The highest that George can suck water up a very long straw is 2.0 m. (This is a typical value.) What is the lowest pressure that he can maintain in his mouth?

15. ‖ The two 60-cm-diameter cylinders in Figure P13.15, closed
BIO at one end, open at the other, are joined to form a single cylinder, then the air inside is removed.
    a. How much force does the atmosphere exert on the flat end of each cylinder?
    b. Suppose one cylinder is bolted to a sturdy ceiling. How many 100 kg football players would need to hang from the lower cylinder to pull the two cylinders apart?

FIGURE P13.15

FIGURE P13.16

Section 13.3 Measuring and Using Pressure

16. | What is the gas pressure inside the box shown in Figure P13.16?

17. | The container shown in Figure P13.17 is filled with oil. It is open to the atmosphere on the left.
    a. What is the pressure at point A?
    b. What is the pressure difference between points A and B? Between points A and C?

18. ‖ Glycerin is poured into an open U-shaped tube until the height in both sides is 20 cm. Ethyl alcohol is then poured into one arm until the height of the alcohol column is 20 cm. The two liquids do not mix. What is the difference in height between the top surface of the glycerin and the top surface of the alcohol?

19. ‖‖ A U-shaped tube, open to the air on both ends, contains mercury. Water is poured into the left arm until the water column is 10.0 cm deep. How far upward from its initial position does the mercury in the right arm rise?

20. | What is the height of a water barometer at atmospheric pressure?

21. | Postural hypotension is the occurrence of low (systolic)
BIO blood pressure when standing up too quickly from a reclined position, causing fainting or lightheadedness. For most people, a systolic pressure less than 90 mm of Hg is considered low. If the blood pressure in your brain is 120 mm when you are lying down, what would it be when you stand up? Assume that your brain is 40 cm from your heart and that $\rho = 1060$ kg/m³ for your blood. Note: Normally, your blood vessels constrict and expand to keep your brain blood pressure stable when you change your posture.

Section 13.4 Buoyancy

22. ‖ A 6.00-cm-diameter sphere with a mass of 89.3 g is neutrally buoyant in a liquid. Identify the liquid.

23. ‖‖‖ A cargo barge is loaded in a saltwater harbor for a trip up a freshwater river. If the rectangular barge is 3.0 m by 20.0 m and sits 0.80 m deep in the harbor, how deep will it sit in the river?

24. | A 10 cm × 10 cm × 10 cm wood block with a density of 700 kg/m³ floats in water.
    a. What is the distance from the top of the block to the water if the water is fresh?
    b. If it's seawater?

25. ‖ What is the tension in the string in Figure P13.25?

100 cm³ of aluminum, density $\rho_{Al} = 2700$ kg/m³

Ethyl alcohol

FIGURE P13.25

26. | A 10 cm × 10 cm × 10 cm block of steel ($\rho_{steel} = 7900$ kg/m³) is suspended from a spring scale. The scale is in newtons.
    a. What is the scale reading if the block is in air?
    b. What is the scale reading after the block has been lowered into a beaker of oil and is completely submerged?

27. ‖‖‖ Styrofoam has a density of 300 kg/m³. What is the maximum mass that can hang without sinking from a 50-cm-diameter Styrofoam sphere in water? Assume the volume of the mass is negligible compared to that of the sphere.

28. ‖ Calculate the buoyant force due to the surrounding air on a man weighing 800 N. Assume his average density is the same as that of water.

Section 13.5 Fluids in Motion

29. ‖ River Pascal with a volume flow rate of 5.0×10^5 L/s joins with River Archimedes, which carries 10.0×10^5 L/s, to form the Bernoulli River. The Bernoulli River is 150 m wide and 10 m deep. What is the speed of the water in the Bernoulli River?

30. ‖ Water flowing through a 2.0-cm-diameter pipe can fill a 300 L bathtub in 5.0 min. What is the speed of the water in the pipe?

31. | A pump is used to empty a 6000 L wading pool. The water exits the 2.5-cm-diameter hose at a speed of 2.1 m/s. How long will it take to empty the pool?

32. ‖ A 1.0-cm-diameter pipe widens to 2.0 cm, then narrows to 0.50 cm. Liquid flows through the first segment at a speed of 4.0 m/s.
    a. What are the speeds in the second and third segments?
    b. What is the volume flow rate through the pipe?

Section 13.6 Fluid Dynamics

33. | What does the top pressure gauge in Figure P13.33 read?

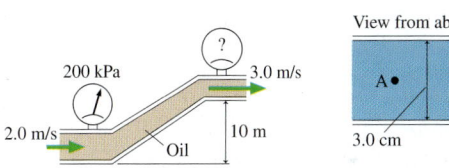

FIGURE P13.33   **FIGURE P13.34**

34. | The 3.0-cm-diameter water line in Figure P13.34 splits into two 1.0-cm-diameter pipes. All pipes are circular and at the same elevation. At point A, the water speed is 2.0 m/s and the gauge pressure is 50 kPa. What is the gauge pressure at point B?

35. ‖‖‖ A rectangular trough, 2.0 m long, 0.60 m wide, and 0.45 m deep, is completely full of water. One end of the trough has a small drain plug right at the bottom edge. When you pull the plug, at what speed does water emerge from the hole?

Section 13.7 Viscosity and Poiseuille's Equation

36. | What pressure difference is required between the ends of a 2.0-m-long, 1.0-mm-diameter horizontal tube for 40°C water to flow through it at an average speed of 4.0 m/s?

37. ‖‖‖ Water flows at 0.25 L/s through a 10-m-long garden hose 2.5 cm in diameter that is lying flat on the ground. The temperature of the water is 20°C. What is the gauge pressure of the water where it enters the hose?

38. ‖ Figure P13.38 shows a water-filled syringe with a 4.0-cm-long needle. What is the gauge pressure of the water at the point P, where the needle meets the wider chamber of the syringe?

FIGURE P13.38

General Problems

39. | The density of gold is 19,300 kg/m^3. 197 g of gold is shaped into a cube. What is the length of each edge?

40. ||| The density of copper is 8920 kg/m^3. How many moles are INT in a 2.0 cm × 2.0 cm × 2.0 cm cube of copper?

41. || The density of aluminum is 2700 kg/m^3. How many atoms INT are in a 2.0 cm × 2.0 cm × 2.0 cm cube of aluminum?

42. || A 50-cm-thick layer of oil floats on a 120-cm-thick layer of water. What is the pressure at the bottom of the water layer?

43. || An oil layer floats on 85 cm of water in a tank. The pressure at the bottom of the tank is 112.0 kPa. How thick is the oil?

44. | The little Dutch boy saved Holland by sticking his finger in the leaking dike. If the water level was 2.5 m above his finger, *estimate* the force of the water on his finger.

45. |||| a. In Figure P13.45, how much force does the fluid exert on the end of the cylinder at A?
    b. How much force does the fluid exert on the end of the cylinder at B?

FIGURE P13.45

46. ||| A friend asks you how much pressure is in your car tires. You know that the tire manufacturer recommends 30 psi, but it's been a while since you've checked. You can't find a tire gauge in the car, but you do find the owner's manual and a ruler. Fortunately, you've just finished taking physics, so you tell your friend, "I don't know, but I can figure it out." From the owner's manual you find that the car's mass is 1500 kg. It seems reasonable to assume that each tire supports one-fourth of the weight. With the ruler you find that the tires are 15 cm wide and the flattened segment of the tire in contact with the road is 13 cm long. What answer will you give your friend?

47. ||| A diver 50 m deep in 10°C fresh water exhales a 1.0-cm-INT diameter bubble. What is the bubble's diameter just as it reaches the surface of the lake, where the water temperature is 20°C?
    Hint: Assume that the air bubble is always in thermal equilibrium with the surrounding water.

48. || A 6.0-cm-tall cylinder floats in water with its axis perpendicular to the surface. The length of the cylinder above water is 2.0 cm. What is the cylinder's mass density?

49. | A sphere completely submerged in water is tethered to the bottom with a string. The tension in the string is one-third the weight of the sphere. What is the density of the sphere?

50. | You need to determine the density of a ceramic statue. If you suspend it from a spring scale, the scale reads 28.4 N. If you then lower the statue into a tub of water so that it is completely submerged, the scale reads 17.0 N. What is the density?

51. | A 5.0 kg rock whose density is 4800 kg/m^3 is suspended by a string such that half of the rock's volume is under water. What is the tension in the string?

52. |||| The bottom of a steel "boat" is a 5.0 m × 10 m × 2.0 cm piece of steel (ρ_{steel} = 7900 kg/m^3). The sides are made of 0.50-cm-thick steel. What minimum height must the sides have for this boat to float in perfectly calm water?

53. ||| A 2.0 mL syringe has an inner diameter of 6.0 mm, a needle BIO inner diameter of 0.25 mm, and a plunger pad diameter (where you place your finger) of 1.2 cm. A nurse uses the syringe to inject medicine into a patient whose blood pressure is 140/100. Assume the liquid is an ideal fluid.
    a. What is the minimum force the nurse needs to apply to the syringe?
    b. The nurse empties the syringe in 2.0 s. What is the flow speed of the medicine through the needle?

54. || A child's water pistol shoots water through a 1.0-mm-INT diameter hole. If the pistol is fired horizontally 70 cm above the ground, a squirt hits the ground 1.2 m away. What is the volume flow rate during the squirt? Ignore air resistance.

55. | A long horizontal tube has a square cross section with sides of width L. A fluid moves through the tube with speed v_1. The tube then changes to a circular cross section with diameter L. What is the fluid's speed in the circular part of the tube?

56. ||| A hurricane wind blows across a 6.00 m × 15.0 m flat roof at a speed of 130 km/hr.
    a. Is the air pressure above the roof higher or lower than the pressure inside the house? Explain.
    b. What is the pressure difference?
    c. How much force is exerted on the roof? If the roof cannot withstand this much force, will it "blow in" or "blow out"?

57. || Water flows from the pipe shown in Figure P13.57 with a speed of 4.0 m/s.
    a. What is the water pressure as it exits into the air?
    b. What is the height h of the standing column of water?

FIGURE P13.57

58. || Air flows through the tube shown in Figure P13.58. Assume that air is an ideal fluid.
    a. What are the air speeds v_1 and v_2 at points 1 and 2?
    b. What is the volume flow rate?

FIGURE P13.58   **FIGURE P13.59**

59. ||| Air flows through the tube shown in Figure P13.59 at a rate of 1200 cm^3/s. Assume that air is an ideal fluid. What is the height h of mercury in the right side of the U-tube?

60. ‖ Water flows at 5.0 L/s through a horizontal pipe that narrows smoothly from 10 cm diameter to 5.0 cm diameter. A pressure gauge in the narrow section reads 50 kPa. What is the reading of a pressure gauge in the wide section?

61. ‖‖ The mercury manometer
INT shown in Figure P13.61 is attached to a gas cell. The mercury height h is 120 mm when the cell is placed in an ice-water mixture. The mercury height drops to 30 mm when the device is carried into an industrial freezer. What is the freezer temperature?

FIGURE P13.61

Hint: The right tube of the manometer is much narrower than the left tube. What reasonable assumption can you make about the gas volume?

62. ‖ Figure P13.62 shows a section of a long tube that narrows near its open end to a diameter of 1.0 mm. Water at 20°C flows out of the open end at 0.020 L/s. What is the gauge pressure at point P, where the diameter is 4.0 mm?

FIGURE P13.62

63. ‖ Smoking tobacco is bad for your circulatory health. In an
BIO attempt to maintain the blood's capacity to deliver oxygen, the body increases its red blood cell production, and this increases the viscosity of the blood. In addition, nicotine from tobacco causes arteries to constrict.

For a nonsmoker, with blood viscosity of $2.5 \times 10^{-3} \, \text{Pa} \cdot \text{s}$, normal blood flow requires a pressure difference of 8.0 mm of Hg between the two ends of an artery. If this person were to smoke regularly, his blood viscosity would increase to $2.7 \times 10^{-3} \, \text{Pa} \cdot \text{s}$, and the arterial diameter would constrict to 90% of its normal value. What pressure difference would be needed to maintain the same blood flow?

64. ‖ A stiff, 10-cm-long tube with an inner diameter of 3.0 mm is attached to a small hole in the side of a tall beaker. The tube sticks out horizontally. The beaker is filled with 20°C water to a level 45 cm above the hole, and it is continually topped off to maintain that level. What is the volume flow rate through the tube?

Passage Problems

Blood Pressure and Blood Flow BIO

The blood pressure at your heart is approximately 100 mm of Hg. As blood is pumped from the left ventricle of your heart, it flows through the aorta, a single large blood vessel with a diameter of about 2.5 cm. The speed of blood flow in the aorta is about 60 cm/s. Any change in pressure as blood flows in the aorta is due to the change in height: the vessel is large enough that viscous drag is not a major factor. As the blood moves through the circulatory system, it flows into successively smaller and smaller blood vessels until it reaches the capillaries. Blood flows in the capillaries at the much lower speed of approximately 0.7 mm/s. The diameter of capillaries and other small blood vessels is so small that viscous drag is a major factor.

65. | There is a limit to how long your neck can be. If your neck were too long, no blood would reach your brain! What is the maximum height a person's brain could be above his heart, given the noted pressure and assuming that there are no valves or supplementary pumping mechanisms in the neck? The density of blood is 1060 kg/m³.
A. 0.97 m   B. 1.3 m   C. 9.7 m   D. 13 m

66. | Because the flow speed in your capillaries is much less than in the aorta, the total cross section area of the capillaries considered together must be much larger than that of the aorta. Given the flow speeds noted, the total area of the capillaries considered together is equivalent to the cross-section area of a single vessel of approximately what diameter?
A. 25 cm   B. 50 cm   C. 75 cm   D. 100 cm

67. | Suppose that in response to some stimulus a small blood vessel narrows to 90% its original diameter. If there is no change in the pressure across the vessel, what is the ratio of the new volume flow rate to the original flow rate?
A. 0.66   B. 0.73   C. 0.81   D. 0.90

68. | Sustained exercise can increase the blood flow rate of the heart by a factor of five with only a modest increase in blood pressure. This is a large change in flow. Although several factors come into play, which of the following physiological changes would most plausibly account for such a large increase in flow with a small change in pressure?
A. A decrease in the viscosity of the blood.
B. Dilation of the smaller blood vessels to larger diameters.
C. Dilation of the aorta to larger diameter.
D. An increase in the oxygen carried by the blood.

STOP TO THINK ANSWERS

Stop to Think 13.1: $\rho_1 = \rho_2 = \rho_3$. Density depends only on what the object is made of, not how big the pieces are.

Stop to Think 13.2: C. These are all open tubes, so the liquid rises to the same height in all three despite their different shapes.

Stop to Think 13.3: B. The weight of the displaced water equals the weight of the ice cube. When the ice cube melts and turns into water, that amount of water will exactly fill the volume that the ice cube is now displacing.

Stop to Think 13.4: 1 cm³/s out. The fluid is incompressible, so the sum of what flows in must match the sum of what flows out.

13 cm³/s is known to be flowing in while 12 cm³/s flows out. An additional 1 cm³/s must flow out to achieve balance.

Stop to Think 13.5: $h_2 > h_4 > h_3 > h_1$. The liquid level is higher where the pressure is lower. The pressure is lower where the flow speed is higher. The flow speed is highest in the narrowest tube, zero in the open air.

Stop to Think 13.6: A. All three segments have the same volume flow rate Q. According to Poiseuille's equation, the segment with the smallest radius R has the greatest pressure difference Δp.

Properties of Matter

The goal of Part III has been to understand macroscopic systems and their properties. We've introduced no new fundamental laws in these chapters. Instead, we've broadened and extended the scope of Newton's law and the law of conservation of energy. In many ways, Part III has focused on *applications* of the general principles introduced in Parts I and II.

That matter exists in three phases—solids, liquids, gases—is perhaps the most basic fact about macroscopic systems. We looked at an atomic-level picture of the three phases of matter, and we also found a connection between *heat* and *phase changes*. Both liquids and gases are *fluids,* and we spent quite a bit of time investigating the properties of fluids, ranging from pressure and the ideal-gas law to buoyancy, fluid flow, and Bernoulli's equation.

Along the way, we introduced *mechanical properties* of matter, such as density and viscosity, and *thermal prop-*

erties, such as thermal-expansion coefficients and specific heats. You should now be able to use these properties to figure out what happens when you heat a system, whether an object will float, and how fast a fluid flows. These are all very practical, useful things to know, with many applications to the world around us.

Last, but not least, we discovered that many *macroscopic* properties are determined by the *microscopic* motions of atoms and molecules. For example, we found that the ideal-gas law is based on the idea that gas pressure is due to the collisions of atoms with the walls of the container. Similarly, regularities in the specific heats of gases were explained on the basis of how atoms and molecules share the thermal energy of a system. Atoms are important, even if we can't see them or follow their individual motions.

KNOWLEDGE STRUCTURE III **Properties of Matter**

BASIC GOAL How can we describe the macroscopic flows of energy and matter in heat transfer and fluid flow?

GENERAL PRINCIPLES **Phases of matter**

Gas: Freely moving particles. Thermal energy is kinetic energy of motion of particles. Pressure due to collisions of particles with walls of container.

Liquid: Particles are loosely bound, but flow is possible. Heat is transferred by convection and conduction. Pressure is due to gravity.

Solid: Particles are joined by spring-like bonds. Increasing temperature causes expansion. Heat is transferred by conduction only.

Heat must be added to change a solid to a liquid and a liquid to a gas. Heat must be removed to reverse these changes.

Ideal-gas processes

We can graph processes on a pV diagram. Work by the gas is the area under the graph.

For a gas process in a sealed container, the ideal gas law relates values of pressure, volume, and temperature:

$$\frac{p_i V_i}{T_i} = \frac{p_f V_f}{T_f}$$

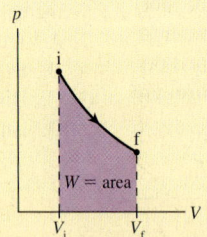

- Constant-volume process $V = \text{constant}$
- Isothermal process $T = \text{constant}$
- Isobaric process $p = \text{constant}$
- Adiabatic process $Q = 0$

Heat and heat transfer

The **specific heat** c of a material is the heat required to raise the temperature of 1 kg by 1 K.

$$Q = Mc\,\Delta T$$

When two ore more systems interact thermally, they come to a common final temperature determined by

$$Q_{net} = Q_1 + Q_2 + Q_3 + \cdots = 0$$

Conduction transfers heat by direct physical contact.
Convection transfers of heat by the motion of a fluid.
Radiation transfers heat by electromagnetic waves.

Fluid statics

Archimedes' principle: The magnitude of the buoyant force equals the weight of the fluid displaced by the object.

An object sinks if it is more dense than the fluid in which it is submerged; if it is less dense, it floats.

Pressure increases with depth in a fluid. The pressure at depth d is

$$p = p_0 + \rho g d$$

Fluid flow Ideal fluid flow is laminar, incompressible, and nonviscous. The **equation of continuity** is

$$v_1 A_1 = v_2 A_2$$

Bernoulli's equation is

$$p_1 + \tfrac{1}{2}\rho v_1^2 + \rho g y_1 =$$
$$p_2 + \tfrac{1}{2}\rho v_2^2 + \rho g y_2$$

Size and Life

Physicists look for simple models and general principles that underlie and explain diverse physical phenomena. In the first 13 chapters of this book, you've seen that just a handful of general principles and laws can be used to solve a wide range of problems. Can this approach have any relevance to a subject like biology? It may seem surprising, but there *are* general "laws of biology" that apply, with quantitative accuracy, to organisms as diverse as elephants and mice.

Let's look at an example. An elephant uses more metabolic power than a mouse. This is not surprising, as an elephant is quite a bit bigger. But recasting the data shows an interesting trend. When we looked at the energy required to raise the temperature of different substances, we considered *specific heat*. The "specific" meant that we considered the heat required for one kilogram. For animals, rather than metabolic rate, we can look at the *specific metabolic rate*, the metabolic power used *per kilogram* of tissue. If we factor out the mass difference between a mouse and an elephant, are their specific metabolic powers the same?

In fact, the specific metabolic rate varies quite a bit among mammals, as the graph of specific metabolic rate versus mass shows. But there is an interesting trend: all of the data points do lie on a single smooth curve. In other words, there really is a biological *law* we can use to predict a mammal's metabolic rate knowing only its mass M. In particular, the specific metabolic rate is proportional to $M^{-0.25}$. Because a 4000 kg elephant is 160,000 times more massive than a 25 g mouse, the mouse's specific metabolic power is $(160,000)^{0.25} = 20$ times that of the elephant. A law that shows how a property scales with the size of a system is called a *scaling law*.

Specific metabolic rate as a function of body mass follows a simple scaling law.

A similar scaling law holds for birds, reptiles—and even bacteria. Why should a single, simple relationship hold true for organisms that range in size from a single cell to a 100 ton blue whale? Interestingly, no one knows for sure. It is a matter of current research to find out just what this and other scaling laws tell us about the nature of life.

Perhaps the metabolic-power scaling law is a result of heat transfer. In Chapter 12, we noted that all metabolic energy used by an animal ends up as heat, which must be transferred to the environment. A 4000 kg elephant has 160,000 times the mass of a 25 g mouse, but it only has about 3,000 times the surface area. The heat transferred to the environment depends on the surface area; the more surface area, the greater the rate of heat transfer. An elephant with a mouse-sized metabolism simply wouldn't be able to dissipate heat fast enough—it would quickly overheat and die.

If heat dissipation were the only factor limiting metabolism, we can show that the specific metabolic rate should scale as $M^{-0.33}$, quite different from the $M^{-0.25}$ scaling observed. Clearly, another factor is at work. Exactly what underlies the $M^{-0.25}$ scaling is still a matter of debate, but some recent analysis suggests the scaling is due to limitations not of heat transfer but of fluid flow. Cells in mice, elephants, and all mammals receive nutrients and oxygen for metabolism from the blood stream. Because the minimum size of a capillary is about the same for all mammals, the structure of the circulatory system must vary from animal to animal. The human aorta has a diameter of about 1 inch; in a mouse, the diameter is approximately $1/20^{th}$ of this. Thus a mouse has fewer levels of branching to smaller and smaller blood vessels as we move from the aorta to the capillaries. The smaller blood vessels in mice mean that viscosity is more of a factor throughout the circulatory system. The circulatory system of a mouse is quite different from that of an elephant.

A model of specific metabolic rate based on blood-flow limitations predicts a $M^{-0.25}$ law, exactly as observed. The model also makes other testable predictions. For example, the model predicts that the smallest possible mammal should have a body mass of about 1 gram—exactly the size of the smallest shrew. Even smaller animals have different types of circulatory systems; in the smallest animals, nutrient transport is by diffusion alone. But the model can be extended to predict that the specific metabolic rate for these animals will follow a scaling law similar to that for mammals, exactly as observed. It is too soon to know if this model will ultimately prove to be correct, but it's indisputable that there are large-scale regularities in biology that follow mathematical relationships based on the laws of physics.

449

OSCILLATIONS AND WAVES

Motion That Repeats Again and Again

Up to this point in the book, we have generally considered processes that have a clear starting and ending point, such as a car accelerating from rest to a final speed, or a solid being heated from an initial to a final temperature. In Part IV, we begin to consider processes that are *periodic*—they repeat. A child on a swing, a boat bobbing on the water, and even the repetitive bass beat of a rock song are *oscillatory motions* that happen over and over without a starting or ending point. The *period*, the time for one cycle of the motion, will be a key parameter for us to consider as we look at oscillatory motion.

Our first goal will be to develop the language and tools needed to describe oscillations, ranging from the swinging of the bob of a pendulum clock to the bouncing of a car on its springs. Once we understand oscillations, we will extend our analysis to consider oscillations that travel—*waves*.

The Wave Model

We've had great success modeling the motion of complex objects as the motion of one or more particles. We were even able to explain the macroscopic properties of matter, such as pressure and temperature, in terms of the motion of the atomic particles that comprise all matter.

Now it's time to explore another way of looking at nature, the *wave model*. Familiar examples of waves include

- Ripples on a pond.
- The sound of thunder.
- The swaying ground of an earthquake.
- A vibrating guitar string.
- The colors of a rainbow.

Despite the great diversity of types and sources of waves, there is a single, elegant physical theory that is capable of describing them all. Our exploration of wave phenomena will call upon water waves, sound waves, and light waves for examples, but our goal will be to emphasize the unity and coherence of the ideas that are common to *all* types of waves. As was the case with the particle model, we will use the wave model to explain a wide range of phenomena.

When Waves Collide

The collision of two particles is a dramatic event. Energy and momentum are transferred as the two particles head off in different directions. Something much gentler happens when two waves come together—the two waves pass through each other unchanged. Where they overlap, we get a *superposition* of the two waves. We will finish our discussion of waves by analyzing the standing waves that result from the superposition of two waves traveling in opposite directions. The physics of standing waves will allow us to understand how your vocal tract can produce such a wide range of sounds, and how your ears are able to analyze them.

Wolves are social animals, and they howl to communicate over distances of several miles with other members of their pack. How are such sounds made? How do they travel through the air? And how are other wolves able to hear these sounds from such a great distance?

451

14

OSCILLATIONS

A gibbon moves in a series of circular arcs as it swings through the forest canopy, pivoting about one handhold and then the next. How does an understanding of oscillatory motion allow us to analyze the swinging motion of a gibbon—and the walking motion of other animals?

Looking Ahead ▶▶

The goal of Chapter 14 is to understand systems that oscillate with simple harmonic motion. In this chapter you will learn to:

- ▶ Describe oscillatory motion and the type of force that leads to it.
- ▶ Use graphical and mathematical representations of oscillatory motion.
- ▶ Understand how conservation of energy applies in simple harmonic motion.
- ▶ Apply concepts of damping and resonance in oscillating systems.

Looking Back ◀◀

This chapter brings together concepts from many previous chapters. Oscillatory motion is produced by a linear restoring force like that of a spring. We will use the concepts of circular motion and conservation of energy to describe and analyze oscillatory motion. Please review:

- ◀ Section 3.8 Period and frequency.
- ◀ Section 6.2 Velocity and acceleration in circular motion.
- ◀ Section 8.3 Springs and restoring forces.
- ◀ Sections 10.2 and 10.3 Energy transformations and the conservation of energy.

A marble rolling back and forth in the bottom of a bowl, a car bouncing up and down on its springs, and a pendulum bob keeping time in a clock are all examples of **oscillatory motion,** a repetitive motion back and forth about an equilibrium position. Swinging motions and vibrations of all kinds are oscillatory motions. All oscillatory motion is **periodic:** the motion repeats, over and over, with a fixed *period,* the time for one cycle of the motion.

Examples of oscillatory motion abound, including pieces of vibrating machinery, ringing bells, and the current in electric circuits used to drive antennas. A vibrating loudspeaker cone pushes air molecules back and forth to send out a sound wave, showing that oscillations are closely related to waves.

Our goal in this chapter is to study the physics of oscillations. Much of our analysis will be focused on the most basic form of oscillatory motion, *simple harmonic motion.* After we explore simple harmonic motion, we will use it to help us understand more complex motions.

14.1 Equilibrium and Oscillation

Consider a marble that is free to roll inside a spherical bowl, as shown in Figure 14.1. The marble has an **equilibrium position** at the bottom of the bowl where it will rest with no net force on it. If you push the marble away from equilibrium, the marble's weight leads to a net force directed back toward the equilibrium position. We call this a **restoring force** because it acts to restore equilibrium. The magnitude of this restoring force increases if the marble is moved farther away from the equilibrium position.

If you pull the marble to the side and release it, it doesn't just roll back to the bottom of the bowl and stay put. It keeps on moving, rolling up and down each side of the bowl, repeatedly moving through its equilibrium position, as we see in Figure 14.2. We call such repetitive motion an **oscillation.** This oscillation is a result of an interplay between the restoring force and the marble's inertia, something we will see in all of the oscillations we consider.

The key characteristic of an oscillation is the *period,* the time for the motion to repeat. We have seen the concepts of frequency and period when we studied circular motion. We will directly connect the ideas of simple harmonic motion and circular motion later in the chapter. For now, we'll review the concepts of frequency and period and see how they apply to oscillatory motion.

Frequency and Period

An electrocardiogram (ECG), such as the one shown in Figure 14.3, is a record of the electrical signals of the heart as it beats. We will explore the ECG in some detail in Chapter 21. Although the shape of a typical ECG is rather complex, notice that it has a *repeating pattern.* For this or any oscillation, the time to complete one full cycle is called the **period** of the oscillation. Period is given the symbol T.

An equivalent piece of information is the number of cycles, or oscillations, completed per second. If the period is $\frac{1}{10}$ s, then the oscillator can complete 10 cycles in one second. Conversely, an oscillation period of 10 s allows only $\frac{1}{10}$ of a cycle to be completed per second. In general, T seconds per cycle implies that $1/T$ cycles will be completed each second. The number of cycles per second is called the **frequency** f of the oscillation. The relationship between frequency and period is therefore

$$f = \frac{1}{T} \quad \text{or} \quad T = \frac{1}{f} \tag{14.1}$$

The units of frequency are **hertz,** abbreviated Hz. By definition,

$$1 \text{ Hz} = 1 \text{ cycle per second} = 1 \text{ s}^{-1}$$

We will frequently deal with very rapid oscillations and make use of the units shown in Table 14.1.

NOTE ▶ Uppercase and lowercase letters *are* important. 1 MHz is 1 megahertz $= 10^6$ Hz, but 1 mHz would be 1 millihertz $= 10^{-3}$ Hz! ◀

EXAMPLE 14.1 Frequency and period of a radio station
An FM radio station broadcasts at a frequency of 100 MHz. What is the period?

SOLVE The frequency f of oscillations in the radio transmitter is 100 MHz = 1.0×10^8 Hz. The period is the inverse of the frequency, hence

$$T = \frac{1}{f} = \frac{1}{1.0 \times 10^8 \text{ Hz}} = 1.0 \times 10^{-8} \text{ s} = 10 \text{ ns}$$

When the ball is displaced from equilibrium . . .

$\vec{n}$ $\vec{F}_{\text{net}}$ $\vec{F}_{\text{net}}$

$\vec{w}$ $\vec{F}_{\text{net}}$

Equilibrium position

. . . a free-body diagram shows a net restoring force, as for a mass on an inclined plane.

The farther away from equilibrium, the greater the net force.

FIGURE 14.1 Equilibrium and restoring forces for a ball in a bowl.

When moved from equilibrium and released, a restoring force pulls the ball back toward equilibrium . . .

$\vec{F}_{\text{net}}$ $\vec{F}_{\text{net}}$

$\vec{v}$

. . . but inertia causes the ball to continue moving to the other side . . .

. . . where the restoring force is directed back toward equilibrium. The ball reverses direction and continues oscillating.

FIGURE 14.2 The motion of a ball rolling in a bowl.

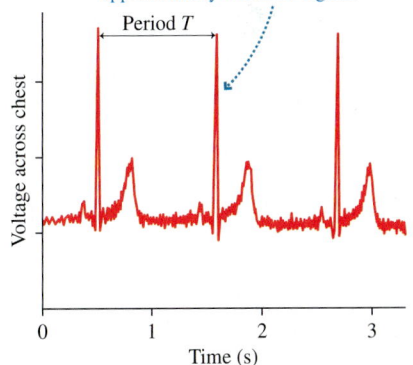

Successive beats of the heart produce approximately the same signal.

Period T

Voltage across chest

Time (s)

FIGURE 14.3 An electrocardiogram.

TABLE 14.1 Common units of frequency

Frequency	Period
10^3 Hz = 1 kilohertz = 1 kHz	1 ms
10^6 Hz = 1 megahertz = 1 MHz	1 μs
10^9 Hz = 1 gigahertz = 1 GHz	1 ns

Oscillatory Motion

Let's return to the marble in the bowl and describe its motion in more detail. We start with a position-versus-time graph for the marble, with positions to the right of equilibrium positive and those to the left negative. Figure 14.4 shows a graph of the motion, with important points noted. This graph has the form of a *cosine function*. A graph or a function that has the form of a sine or cosine function is called **sinusoidal.** A sinusoidal oscillation is called **simple harmonic motion,** often abbreviated SHM.

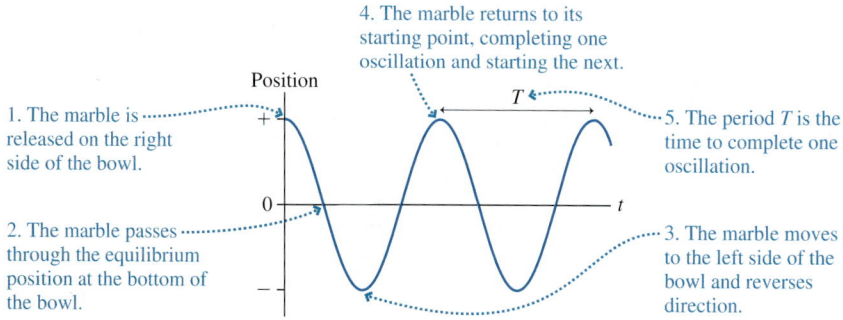

4. The marble returns to its starting point, completing one oscillation and starting the next.

1. The marble is released on the right side of the bowl.

2. The marble passes through the equilibrium position at the bottom of the bowl.

5. The period T is the time to complete one oscillation.

3. The marble moves to the left side of the bowl and reverses direction.

FIGURE 14.4 A position-versus-time graph for a marble rolling in a bowl.

A marble rolling in the bottom of a bowl undergoes simple harmonic motion, as does a car bouncing up and down on its springs. SHM is very common, but we'll find that most cases of SHM can be modeled as one of two simple systems: a mass oscillating on a spring or a pendulum swinging back and forth. The following table shows two examples.

Examples of simple harmonic motion

Oscillating system		Related real-world example	
Mass on a spring	The mass oscillates back and forth due to the restoring force of the spring. The period depends on the mass and the stiffness of the spring.	Vibrations in the ear 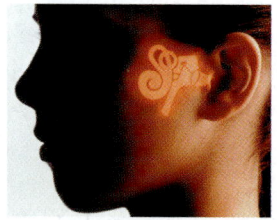	Sound waves entering the ear cause the oscillation of a membrane in the cochlea. The vibration can be modeled as a mass on a spring. The period of oscillation of a segment of the membrane depends on mass (the thickness of the membrane) and stiffness (the rigidity of the membrane.)
Pendulum	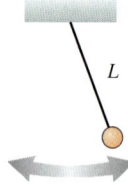 The mass oscillates back and forth due to the restoring gravitational force. The period depends on the length of the pendulum and the acceleration of gravity.	Motion of legs while walking	The motion of a walking animal's legs can be modeled as pendulum motion. The rate at which the legs swing depends on the length of the legs and the acceleration of gravity.

STOP TO THINK 14.1 Two oscillating systems have periods T_1 and T_2, with $T_1 < T_2$. How are the frequencies of the two systems related?

A. $f_1 < f_2$ B. $f_1 = f_2$ C. $f_1 > f_2$

14.2 Linear Restoring Forces and Simple Harmonic Motion

Simple harmonic motion can occur when a system has a restoring force that pushes it back toward equilibrium. We will begin our study with the very simple system of a mass and a spring, analyzing it in detail before moving on to other oscillatory systems.

Figure 14.5 shows a glider that rides with very little friction on an air track. There is a spring connecting the glider to the end of the track. When the spring is neither stretched nor compressed, the net force on the glider is zero. The glider just sits there—this is the equilibrium position.

If the glider is now displaced from this equilibrium position by Δx, the spring exerts a force back toward equilibrium—a restoring force. In Section 8.3, we found that the spring force is given by Hooke's law: $(F_{sp})_x = -k\,\Delta x$. If we set the origin of our coordinate system at the equilibrium position, the displacement from equilibrium Δx is equal to x, thus the spring force can be written as $(F_{sp})_x = -kx$.

The net force on the glider is simply the spring force, so we can write

$$(F_{net})_x = -kx \qquad (14.2)$$

LINEAR
p. 38

This is a **linear restoring force.** That is, **the net force is toward the equilibrium position and is linearly proportional to the distance from equilibrium.**

Motion of a Mass on a Spring

If we pull the air track glider of Figure 14.5 a short distance to the right and release it, it will oscillate back and forth. Figure 14.6 shows actual data from an experiment in which a computer was used to measure the position of a glider 20 times every second. This is a position-versus-time graph that has been rotated 90° from its usual orientation in order for the x-axis to match the motion of the glider.

The object's maximum displacement from equilibrium is called the **amplitude** A of the motion. The object's position oscillates between $x = -A$ and $x = +A$.

> **NOTE ▶** When interpreting a graph, notice that the amplitude is the distance from equilibrium to the maximum, *not* the distance from the minimum to the maximum. ◀

The graph of the position is sinusoidal, so this is simple harmonic motion. The motion is "simple" because the restoring force has a simple, linear form. **Oscillation about an equilibrium position with a linear restoring force is always simple harmonic motion.** There are many other circumstances in which a linear restoring force exists, and we will see simple harmonic motion in these cases as well.

Vertical Mass on a Spring

Figure 14.7 on the following page shows a block of mass m hanging from a spring with spring constant k. Will this mass-spring system have the same simple harmonic motion as the horizontal system we just saw, or will gravity add an additional complication? An important fact to notice is that the equilibrium position of

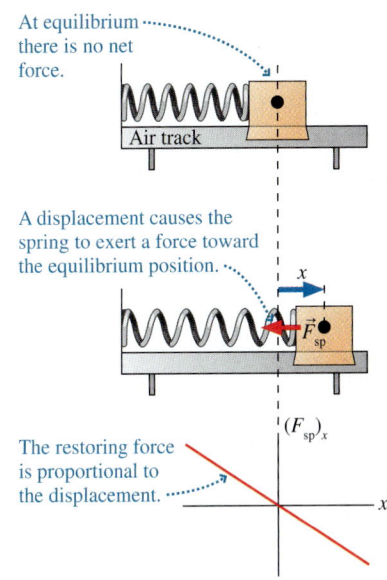

FIGURE 14.5 The restoring force on an air track glider attached to a spring.

FIGURE 14.6 An experiment showing the oscillation of an air-track glider.

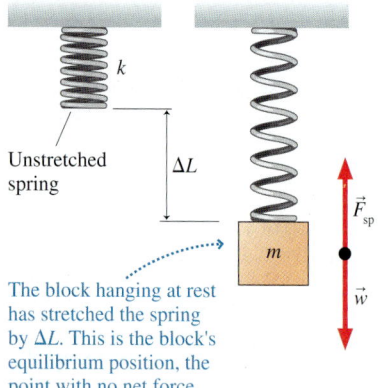

Unstretched spring

ΔL

$\vec{F}_{sp}$

m

$\vec{w}$

The block hanging at rest has stretched the spring by ΔL. This is the block's equilibrium position, the point with no net force.

FIGURE 14.7 The equilibrium position of a mass on a vertical spring.

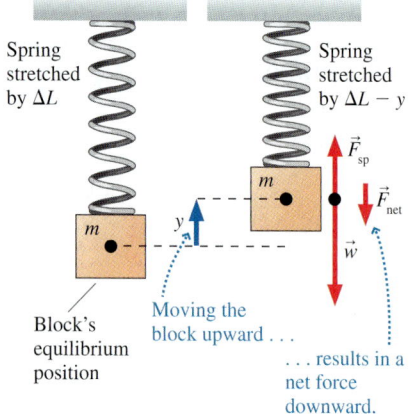

Spring stretched by ΔL

Spring stretched by $\Delta L - y$

$\vec{F}_{sp}$

$\vec{F}_{net}$

m

y

m

$\vec{w}$

Block's equilibrium position

Moving the block upward . . .

. . . results in a net force downward.

FIGURE 14.8 Displacing the block from its equilibrium position produces a restoring force.

Pendulum motion at the playground.

the block is *not* where the spring is at its unstretched length. At the equilibrium position of the block, where it hangs motionless, the spring has stretched by ΔL.

Finding ΔL is a static-equilibrium problem in which the upward spring force balances the downward weight force of the block. The y-component of the spring force is given by Hooke's law,

$$(F_{sp})_y = -k\,\Delta y = +k\,\Delta L \tag{14.3}$$

Equation 14.3 makes a distinction between ΔL, which is simply a *distance* and is a positive number, and the displacement Δy. The block is displaced downward, so $\Delta y = -\Delta L$. Newton's first law for the block in equilibrium is

$$(F_{net})_y = (F_{sp})_y + w_y = k\,\Delta L - mg = 0 \tag{14.4}$$

from which we can find

$$\Delta L = \frac{mg}{k} \tag{14.5}$$

This is the distance the spring stretches when the block is attached to it.

Suppose we now displace the block from this equilibrium position, as shown in Figure 14.8. We've placed the origin of the y-axis at the block's equilibrium position in order to be consistent with our analyses of oscillations throughout this chapter. If the block moves upward, as in the figure, the spring gets shorter compared to its equilibrium length, but the spring is still *stretched* compared to its unstretched length in Figure 14.7. When the block is at position y, the spring is stretched by an amount $\Delta L - y$ and hence exerts an *upward* spring force $F_{sp} = k(\Delta L - y)$. The net force on the block at this point is

$$(F_{net})_y = (F_{sp})_y + w_y = k(\Delta L - y) - mg = (k\,\Delta L - mg) - ky \tag{14.6}$$

But $k\,\Delta L - mg$ is zero, from Equation 14.4, so the net force on the block is simply

$$(F_{net})_y = -ky \tag{14.7}$$

LINEAR
p. 38

Equation 14.7 for a mass hung from a spring has the same form as Equation 14.2 for the horizontal spring, where we found $(F_{net})_x = -kx$. That is, the restoring force for vertical oscillations is identical to the restoring force for horizontal oscillations. **The role of gravity is to determine where the equilibrium position is, but it doesn't affect the restoring force for displacement from the equilibrium position.** Because it has a linear restoring force, **a mass on a vertical spring oscillates with simple harmonic motion.** The motion has the same form as that of the air-track glider.

The Pendulum

The chapter opened with a picture of a gibbon, whose body swings back and forth below a tree branch. The motion of the gibbon is essentially that of a **pendulum**, a mass hung from a pivot point by a light string or rod. A pendulum oscillates about its equilibrium position, but is this simple harmonic motion? To answer this question, we need to examine the restoring force on the pendulum. If the restoring force is linear, the motion will be simple harmonic.

Figure 14.9a shows a mass m attached to a string of length L and free to swing back and forth. The pendulum's position can be described either by the arc of length s or the angle θ, both of which are zero when the pendulum hangs straight down. Because angles are measured counterclockwise, s and θ are positive when the pendulum is to the right of center, negative when it is to the left.

Two forces are acting on the mass: the string tension $\vec{T}$ and the weight $\vec{w}$. The motion is along a circular arc. We choose a coordinate system on the mass with one axis along the radius of the circle and the other tangent to the circle. We divide the forces into tangential components, parallel to the motion (denoted with a subscript "t"), and components directed toward the center of the circle. These are shown on the free-body diagram of Figure 14.9b.

The mass must move along a circular arc, as noted. As Figure 14.9b shows, the net force in this direction is the tangential component of the weight:

$$(F_{net})_t = \sum F_t = w_t = -mg\sin\theta \qquad (14.8)$$

This is the restoring force pulling the mass back toward the equilibrium position.

Suppose we restrict the pendulum's oscillations to *small angles* of less than about 10°, or 0.17 rad. For such small angles, if the angle θ is in radians, it turns out that

$$\sin\theta \approx \theta$$

This result is called the **small-angle approximation.**

> **NOTE ▶** You should check this approximation on your calculator. Put your calculator in radian mode, enter some values that are less than 0.17 rad, take the sine, and see what you get. ◀

Recall from Chapter 6 that the angle is related to the arc length by $\theta = s/L$. Using this and the small-angle approximation, we can write the restoring force as

$$(F_{net})_t = -mg\sin\theta \approx -mg\theta = -mg\frac{s}{L} = -\left(\frac{mg}{L}\right)s \qquad (14.9)$$

The net force is directed toward the equilibrium position, and it is linearly proportional to the displacement s from equilibrium. In other words, **the force on a pendulum is a linear restoring force for small angles, so the pendulum will undergo simple harmonic motion.**

Whenever there is a linear restoring force, there can be simple harmonic motion. There are many examples beyond the few we have considered here, such as sloshing water in a cup or the vibration of a bridge or a building swaying in the wind. We will see these and other examples in the coming sections.

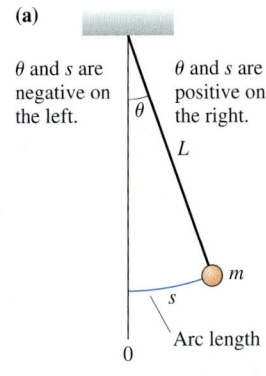

(a) θ and s are negative on the left. θ and s are positive on the right. Arc length

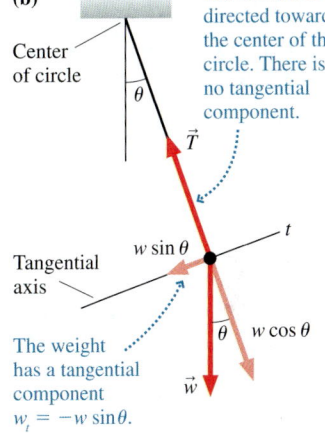

(b) Center of circle. The tension is directed toward the center of the circle. There is no tangential component. Tangential axis. $w\sin\theta$. The weight has a tangential component $w_t = -w\sin\theta$. $w\cos\theta$.

FIGURE 14.9 Describing the motion of and force on a pendulum.

STOP TO THINK 14.2   A ball is hung from a rope, making a pendulum. When it is pulled 5° to the side, the restoring force is 1.0 N. What will be the magnitude of the restoring force if the ball is pulled 10° to the side?

A. 0.5 N
B. 1.0 N
C. 1.5 N
D. 2.0 N

14.3 Describing Simple Harmonic Motion

9.1, 9.2, 9.5

Now that we know what *causes* simple harmonic motion, we can begin to develop a mathematical description. Once we do this for one system, the mass on a spring, we can adapt the description to *any* system. The details will vary, but the basic form of the motion will be exactly the same.

Let's look again at the oscillating air-track glider of Figure 14.5. If we plot data for the position versus time, we get the graph of Figure 14.10a. You can see that the amplitude in this case is $A = 0.17$ m, or 17 cm. You can also measure the period to be $T = 1.60$ s. Thus the oscillation frequency is $f = 1/T = 0.625$ Hz. You can get a sense of the speed of the glider by noting the spacing of the points—where they are close together, the speed is low; where they are far apart, the speed is high. The speed is instantaneously zero at the peaks and valleys of the position-versus-time graph.

Figure 14.10b is a velocity-versus-time graph for the oscillating mass. The velocity graph also has a sinusoidal shape, oscillating between $-v_{max}$ (maximum speed to the left) and v_{max} (maximum speed to the right). As the figure shows,

- The instantaneous velocity is zero at the points where $x = \pm A$. These are the *turning points* of the motion.
- The maximum speed v_{max} is reached as the object passes through the equilibrium position at $x = 0$ m. The *velocity* is positive as the object moves to the right but *negative* as it moves to the left.

If we did a series of experiments on this system, we would find that **the frequency does not depend on the amplitude of the motion;** small oscillations and large oscillations have the same period. This is a key feature of simple harmonic motion that we will explore in more detail later in the chapter. Keep in mind that all these features of the motion of a mass on a spring also apply to other examples of simple harmonic motion.

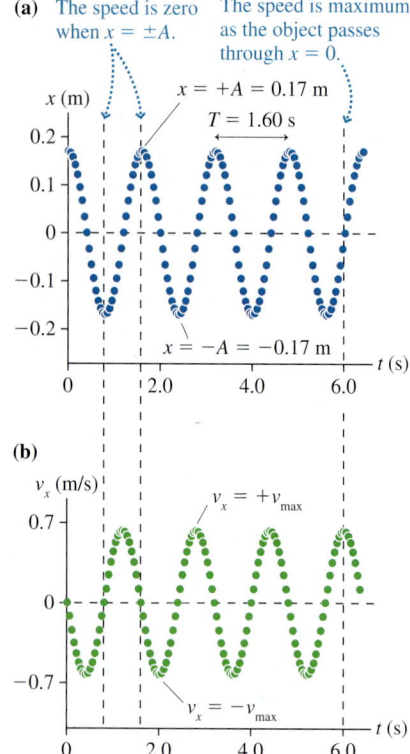

(a) The speed is zero when $x = \pm A$. The speed is maximum as the object passes through $x = 0$.

(b)

FIGURE 14.10 Position and velocity graphs of experimental data for an air-track cart on a spring.

Kinematics of Simple Harmonic Motion

Figure 14.11 redraws the position-versus-time and velocity-versus-time graphs of Figure 14.10 as smooth curves. An acceleration-versus-time graph is introduced as well. The motion divides naturally into phases lasting one-quarter of a cycle:

- At $t = 0$, the mass starts out with $x = A$ and $v_x = 0$. The acceleration a_x is negative. As soon as the mass is released, v_x becomes negative and the mass begins moving back toward equilibrium.
- At $t = \frac{1}{4}T$, the mass passes through its equilibrium position, $x = 0$. The speed is a maximum, with $v_x = -v_{max}$.
- At $t = \frac{1}{2}T$, the mass is at the other end of its motion, with $x = -A$ and $v_x = 0$. The acceleration a_x is positive, so the mass now begins moving in the positive x-direction.
- At $t = \frac{3}{4}T$, the mass again passes through its equilibrium position, $x = 0$, in the opposite direction. The speed is again a maximum, with $v_x = v_{max}$.
- At $t = T$, the mass is back to its starting position, and the cycle begins again.

NOTE ▶ The graphs of Figure 14.11 are for an oscillation in which the object just happened to be at $x = A$ at $t = 0$. You can certainly imagine a different set of *initial conditions,* with the object at $x = -A$, or even somewhere in the middle of an oscillation. But our choice of $t = 0$ is arbitrary. Even if an oscillation begins at a different position, it will certainly pass through the position $x = A$, and we can simply choose to set $t = 0$ at this instant. ◀

The position-versus-time graph in Figure 14.11 is a cosine curve. We can write the object's position as

$$x(t) = A \cos\left(\frac{2\pi t}{T}\right) \qquad (14.10)$$

where the notation $x(t)$ indicates that the position x is a *function* of time t. Because $\cos(2\pi \text{ rad}) = \cos(0 \text{ rad})$, it's easy to see that the position at time $t = T$ is the same as the position at $t = 0$. In other words, this is a cosine function with period T. We can write Equation 14.10 in an alternative form. Because the oscillation frequency is $f = 1/T$, we can write

$$x(t) = A \cos(2\pi f t) \qquad (14.11)$$

NOTE ▶ The argument $2\pi f t$ of the cosine function is in *radians.* That will be true throughout this chapter. Don't forget to set your calculator to radian mode before working oscillation problems. ◀

Just as the position graph was a cosine function, the velocity graph shown in Figure 14.11 is an "upside-down" sine function with the same period T. The velocity v_x, which is a function of time, can be written

$$v_x(t) = -v_{max} \sin\left(\frac{2\pi t}{T}\right) = -v_{max} \sin(2\pi f t) \qquad (14.12)$$

NOTE ▶ v_{max} is the maximum *speed* and thus is inherently a *positive* number. The minus sign in Equation 14.12 is needed to turn the sine function upside down. ◀

We've seen the position and velocity for a mass on a spring; how about the acceleration? As we saw earlier, the restoring force that causes the mass to oscillate with simple harmonic motion, Equation 14.2 is $(F_{net})_x = -kx$. Using Newton's second law, we see that this force causes an acceleration

$$a_x = \frac{(F_{net})_x}{m} = -\frac{k}{m}x \qquad (14.13)$$

Acceleration is proportional to the position x, but with a minus sign. Consequently we expect the acceleration-versus-time graph to be an inverted form of the position-versus-time graph. This is, indeed, what we find; the acceleration-versus-time graph in Figure 14.11 is clearly an "upside-down" cosine function with the same period T. We can write the acceleration as

$$a_x(t) = -a_{max} \cos\left(\frac{2\pi t}{T}\right) = -a_{max} \cos(2\pi f t) \qquad (14.14)$$

In this and coming chapters, we will use sine and cosine functions extensively, so we will summarize some of the key aspects of mathematical relationships using these functions.

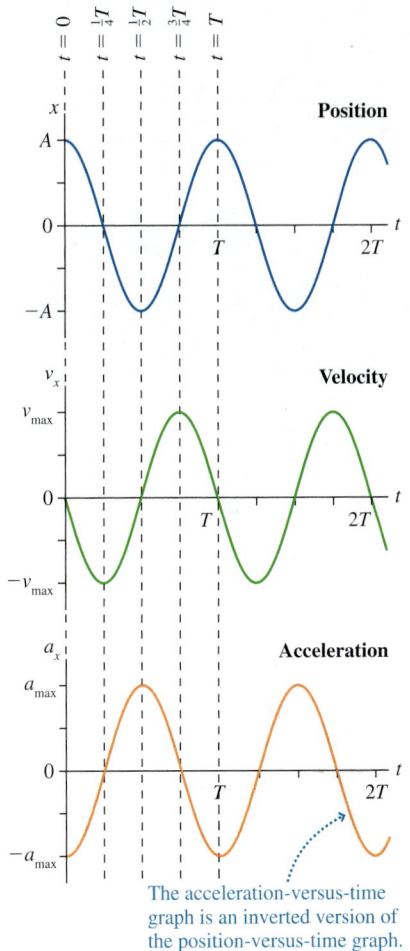

The acceleration-versus-time graph is an inverted version of the position-versus-time graph.

FIGURE 14.11 The position, velocity, and acceleration graphs for simple harmonic motion.

Small bird, fast wings BIO As this rufous hummingbird flaps its wings, the motion of the wing tips is approximately simple harmonic. The frequency of the motion is about 45 Hz, so the period of the wingbeat is only a few hundredths of a second. As we'll see in Equation 14.15, this short period means that the tips of the hummingbird's tiny wings are moving at a maximum speed of over 15 m/s—nearly 35 mph!

 Sinusoidal relationships   Exercise 6

A quantity that oscillates in time and can be written

$$x = A \sin\left(\frac{2\pi t}{T}\right)$$

or

$$x = A \cos\left(\frac{2\pi t}{T}\right)$$

is called a **sinusoidal function** with **period** T. The argument of the functions, $2\pi t/T$, is in radians.

The graphs of both functions have the same shape, but they have different initial values at $t = 0$ s.

The function repeats after a time T.

LIMITS If x is a sinusoidal function, then x is

- *bounded*—it can only take values between A and $-A$.
- *periodic*—it repeats the same sequence of values over and over again. Whatever value x has at time t, it has the same value at $t + T$.

SPECIAL VALUES The function x has special values at certain times:

	$t = 0$	$t = \frac{1}{4}T$	$t = \frac{1}{2}T$	$t = \frac{3}{4}T$	$t = T$
$x = A\sin(2\pi t/T)$	0	A	0	$-A$	0
$x = A\cos(2\pi t/T)$	A	0	$-A$	0	A

EXAMPLE 14.2 Motion of a glider on a spring

An air-track glider oscillates on a spring at a frequency of 0.50 Hz. Suppose the glider is pulled to the right of its equilibrium position by 12 cm and then released. Where will the glider be 1.0 s after its release? What is its velocity at this point?

PREPARE The glider undergoes simple harmonic motion with amplitude 12 cm. The frequency is 0.50 Hz, so the period is $T = 1/f = 2.0$ s. The glider is released at maximum extension from the equilibrium position, meaning that we can take this point to be $t = 0$.

SOLVE 1.0 s is exactly half the period. As the graph of the motion in Figure 14.12 shows, half a cycle brings the glider to its left turning point, 12 cm to the left of the equilibrium position. The velocity at this point is zero.

FIGURE 14.12 Position-versus-time graph for the glider.

We were able to determine the velocity in Example 14.2 because $t = 0$ was a turning point where the instantaneous velocity was zero. To determine the velocity at other points, we need to know v_{max} and use Equation 14.12. How does v_{max} depend on other variables of the motion? Basic reasoning about the motion tells us a few things. A large amplitude implies a high speed because the glider moves a large distance. And a small period implies a high speed as well because the glider must complete its motion in a short time. Later in the chapter we will show that both of these assumptions are correct, and that the maximum speed is

$$v_{max} = 2\pi f A = \frac{2\pi A}{T} \tag{14.15}$$

EXAMPLE 14.3 Analyzing the motion of a spring toy

A classic wooden toy hangs from a spring that lets it bob up and down. If the toy is raised up 20 cm and released, it makes 15 complete oscillations in 10 seconds.

a. What is the period of oscillation?
b. What is the toy's maximum speed?
c. What are the position and velocity at $t = 0.80$ s?

PREPARE The toy is a mass on a vertical spring, and so will undergo simple harmonic motion. The toy begins its motion at its maximum positive displacement.

SOLVE a. The oscillation frequency is

$$f = \frac{15 \text{ oscillations}}{10 \text{ s}} = 1.5 \text{ oscillations/s} = 1.5 \text{ Hz}$$

Thus the period is $T = 1/f = 0.667$ s.

b. The oscillation amplitude is $A = 0.20$ m. The maximum speed, given by Equation 14.15, is

$$v_{max} = \frac{2\pi A}{T} = \frac{2\pi(0.20 \text{ m})}{0.667 \text{ s}} = 1.9 \text{ m/s}$$

c. The object starts at $y = +A$ at $t = 0$. This is exactly the oscillation described by Equations 14.10 and 14.12. The position at $t = 0.80$ s is

$$y = A\cos\left(\frac{2\pi t}{T}\right) = (0.20 \text{ m})\cos\left(\frac{2\pi(0.80 \text{ s})}{0.667 \text{ s}}\right)$$

$$= (0.20 \text{ m})\cos(7.54 \text{ rad}) = 0.062 \text{ m} = 6.2 \text{ cm}$$

The velocity at this instant of time is

$$v_y = -v_{max}\sin\left(\frac{2\pi t}{T}\right) = -(1.9 \text{ m/s})\sin\left(\frac{2\pi(0.80 \text{ s})}{0.667 \text{ s}}\right)$$

$$= -(1.9 \text{ m/s})\sin(7.54 \text{ rad}) = -1.8 \text{ m/s}$$

ASSESS At $t = 0.80$ s, which is slightly more than one period, the object is 6.2 cm above the equilibrium position and moving down at 1.8 m/s. Notice the use of radians in the calculations.

Connection to Uniform Circular Motion

Both circular motion and simple harmonic motion are motions that repeat. Many of the concepts we have used for describing simple harmonic motion were introduced in our study of circular motion. We will use the close connection between the two to extend our knowledge of the kinematics of circular motion to that of simple harmonic motion.

We can demonstrate the relationship between circular and simple harmonic motion with a simple experiment. Suppose a turntable has a small ball glued to the edge. As Figure 14.13a shows, we can make a "shadow movie" of the ball by projecting a light past the ball and onto a screen. The ball's shadow oscillates back and forth as the turntable rotates.

If you place a real object on a real spring directly below the shadow, as shown in Figure 14.13b, and if you adjust the turntable to have the same period as the spring, you will find that the shadow's motion exactly matches the simple harmonic motion of the object on the spring. **Uniform circular motion projected onto one dimension is simple harmonic motion.**

We can show why the projection of circular motion is simple harmonic motion by considering the particle in Figure 14.14 on the next page. As in Chapter 6, we can locate the particle by the angle ϕ measured counterclockwise from the x-axis. Projecting the ball's shadow onto a screen in Figure 14.13 is

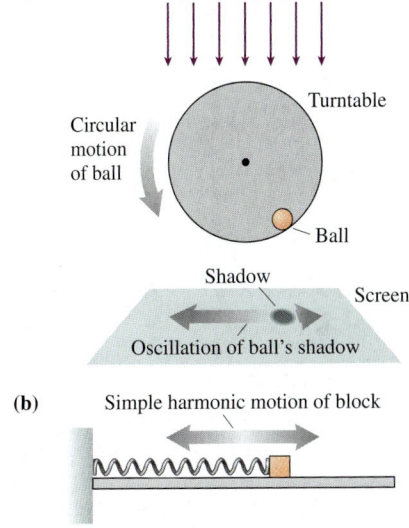

FIGURE 14.13 A projection of the circular motion of a rotating ball matches the SHM of an object on a spring.

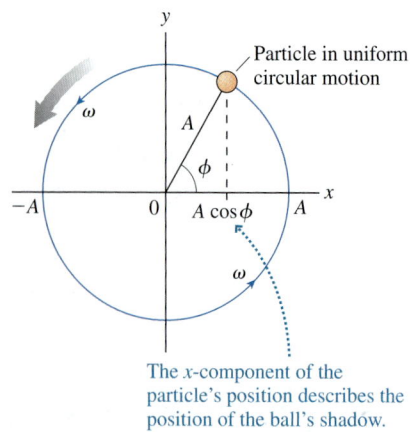

The x-component of the particle's position describes the position of the ball's shadow.

FIGURE 14.14 A particle in uniform circular motion with radius A and angular velocity ω.

TRY IT YOURSELF

SHM in your microwave It's possible to do something like the turntable demonstration right at home. Place a tall (microwave safe) glass or cup on the outside edge of the turntable in your microwave oven. Start the oven. The turntable will rotate, moving the cup in a circle. Stand in front of the oven with your eyes level with the cup. Watch the cup, paying attention to the side-to-side motion. This motion is the horizontal component of the circular motion, and so it is simple harmonic motion.

equivalent to observing just the x-component of the particle's motion. Figure 14.14 shows that the x-component, when the particle is at angle ϕ, is

$$x = A\cos\phi$$

If the particle starts from $\phi_0 = 0$ at $t = 0$, its angle at a later time t is simply

$$\phi = \omega t$$

where ω is the particle's *angular velocity,* as defined in Chapter 6. Recall that the angular velocity is related to the frequency f by

$$\omega = 2\pi f$$

The particle's x-component can therefore be expressed as

$$x(t) = A\cos(2\pi ft) \qquad (14.16)$$

This is identical to Equation 14.11 for the position of a mass on a spring! **The x-component of a particle in uniform circular motion is simple harmonic motion.**

We can use this correspondence to deduce more details. Figure 14.15 is the same motion we looked at in Figure 14.14, but here we've shown the velocity vector (tangent to the circle) and the acceleration vector (a centripetal acceleration pointing to the center of the circle). The magnitude of the velocity vector is the particle's speed. Recall from Chapter 6 that the speed of a particle in circular motion with radius A and frequency f is $v = 2\pi fA$. Thus the x-component of the velocity vector, which is pointing in the negative x-direction, is

$$v_x = -v\sin\phi = -(2\pi f)A\sin(2\pi ft)$$

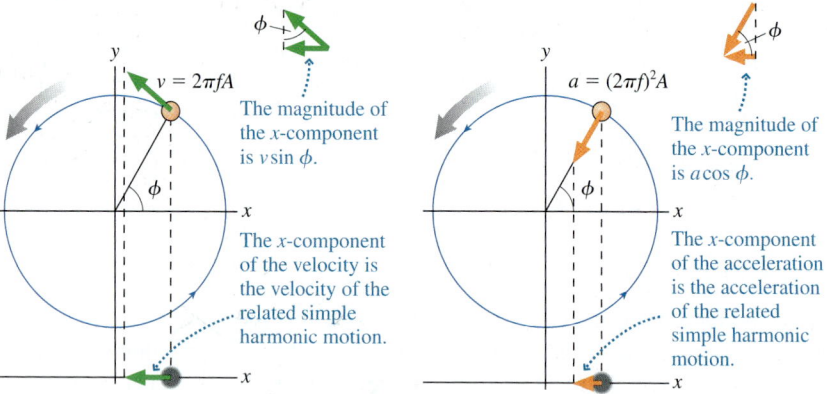

The magnitude of the x-component is $v\sin\phi$.

The x-component of the velocity is the velocity of the related simple harmonic motion.

The magnitude of the x-component is $a\cos\phi$.

The x-component of the acceleration is the acceleration of the related simple harmonic motion.

FIGURE 14.15 Projection of the velocity and acceleration vectors.

According to the correspondence between circular motion and simple harmonic motion, this is the velocity of an object in simple harmonic motion. This is, indeed, exactly Equation 14.12, which we deduced from the graph of the motion, if we define the maximum speed, as we did in Equation 14.15, to be

$$v_{max} = 2\pi fA = \frac{2\pi A}{T}$$

Similarly, the magnitude of the acceleration vector is the centripetal acceleration $a = v^2/A = (2\pi f)^2 A$. The x-component of the acceleration vector, which is the acceleration for simple harmonic motion, is

$$a_x = -a\cos\phi = -(2\pi f)^2 A\cos(2\pi ft)$$

The maximum acceleration is

$$a_{max} = (2\pi f)^2 A \qquad (14.17)$$

We can now summarize our findings for the position, velocity, and acceleration of an object in simple harmonic motion:

$$x(t) = A\cos(2\pi ft)$$
$$v_x(t) = -(2\pi f)A\sin(2\pi ft) \qquad (14.18)$$
$$a_x(t) = -(2\pi f)^2 A\cos(2\pi ft)$$

Position, velocity, and acceleration for an object in simple harmonic motion with frequency f and amplitude A

SINUSOIDAL
p. 460

Any simple harmonic motion will follow these equations. If you know the amplitude and the frequency, the motion is completely specified.

EXAMPLE 14.4 Measuring the sway of a tall building

The John Hancock Center in Chicago is 110 stories high. Strong winds can cause the building to sway, as is the case with all tall buildings. On particularly windy days, the top of the building is known to oscillate with an amplitude of 40 cm ($\approx$16 in) and a period of 7.7 s. What are the maximum speed and acceleration of the top of the building?

PREPARE We will assume that the oscillation of the building is simple harmonic motion with amplitude $A = 0.40$ m. The frequency can be computed from the period:

$$f = \frac{1}{T} = \frac{1}{7.7\text{ s}} = 0.13\text{ Hz}$$

SOLVE We can use Equations 14.15 and 14.17 for the maximum velocity and acceleration to compute:

$$v_{max} = 2\pi fA = 2\pi(0.13\text{ Hz})(0.40\text{ m}) = 0.33\text{ m/s}$$
$$a_{max} = (2\pi f)^2 A = [2\pi(0.13\text{ Hz})]^2(0.40\text{ m}) = 0.27\text{ m/s}^2$$

In terms of the acceleration of gravity, the maximum acceleration is

$$a_{max} = 0.027g$$

ASSESS The acceleration is quite small, as you would expect; if it were large, building occupants would certainly complain! Even if they don't notice the motion directly, office workers on high floors of high buildings may experience a bit of nausea when the oscillations are large because the acceleration affects the equilibrium organ in the inner ear.

STOP TO THINK 14.3   The figure shows four identical oscillators at different points in their motion. Which is moving fastest at the time shown?

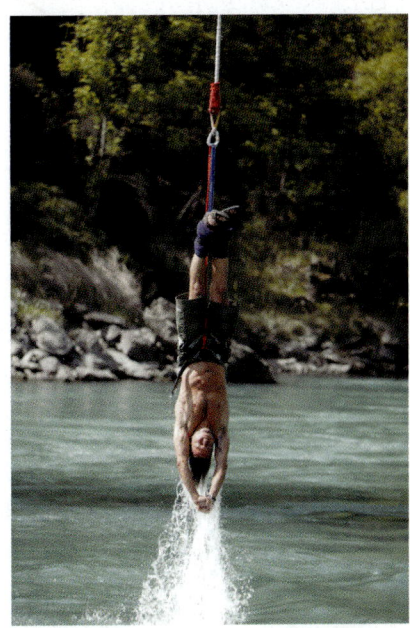

Elastic cords lead to the up-and-down motion of a bungee jump.

14.4 Energy in Simple Harmonic Motion

A bungee jumper falls, increasing in speed, until the elastic cords attached to his ankles start to stretch. The kinetic energy of his motion is transformed into the elastic potential energy of the cords. Once the cords reach their maximum stretch, his velocity reverses. He rises as the elastic potential energy of the cords is transformed back into kinetic energy. If he keeps bouncing (simple harmonic motion), this transformation happens again and again.

This interplay between kinetic and potential energy is very important to understanding simple harmonic motion. The five diagrams in Figure 14.16a show the position and velocity of a mass on a spring at successive points in time. Let's look at the potential and kinetic energy of the object as it moves through these points.

The object begins at rest, with the spring at a maximum extension; the kinetic energy is zero, and the potential energy is a maximum. As the spring contracts, the object speeds up until it reaches the center point of its oscillation, the equilibrium point. At this point, the potential energy is zero and the kinetic energy is a maximum. As the object continues to move, it slows down as it compresses the spring. Eventually, it reaches the turning point, where its instantaneous velocity is zero. At this point, the kinetic energy is zero and the potential energy is again a maximum.

Now we'll specify that the object has mass m, the spring has spring constant k, and the motion takes place on a frictionless surface. You learned in Chapter 10 that the elastic potential energy of a spring stretched by a distance x from its equilibrium position is

$$U = \frac{1}{2}kx^2 \tag{14.19}$$

The potential energy is zero at the equilibrium position and is a maximum when the spring is at its maximum extension or compression. The mechanical energy of the object is the sum of its potential energy and its kinetic energy. Because there is no friction and there are no other forces to do work on the system, the mechanical energy is conserved. We can write

$$E = K + U = \frac{1}{2}mv^2 + \frac{1}{2}kx^2 = \text{constant} \tag{14.20}$$

Figure 14.16b shows a graph of the potential energy, kinetic energy, and total energy for the object as it moves. You can see that, as the object goes through its motion, energy is transformed from potential to kinetic and then back to potential. In particular, you can see that **the object has purely potential energy at $x = \pm A$ and purely kinetic energy as it passes through the equilibrium point at $x = 0$.** At maximum displacement, with $x = \pm A$ and $v_x = 0$ the energy is

$$E(\text{at } x = \pm A) = U_{\text{max}} = \frac{1}{2}kA^2 \tag{14.21}$$

At $x = 0$, where $v_x = \pm v_{\text{max}}$, the energy is

$$E(\text{at } x = 0) = K_{\text{max}} = \frac{1}{2}mv_{\text{max}}^2 \tag{14.22}$$

The system's mechanical energy is conserved, so the energy at maximum displacement and the energy at maximum speed, Equations 14.22 and 14.23, must be equal. That is

$$\frac{1}{2}m(v_{\text{max}})^2 = \frac{1}{2}kA^2 \tag{14.23}$$

By solving Equation 14.23 for the maximum speed we can see that it is related to the amplitude by

$$v_{\text{max}} = \sqrt{\frac{k}{m}}A \tag{14.24}$$

Earlier we found that

$$v_{\text{max}} = 2\pi fA \tag{14.25}$$

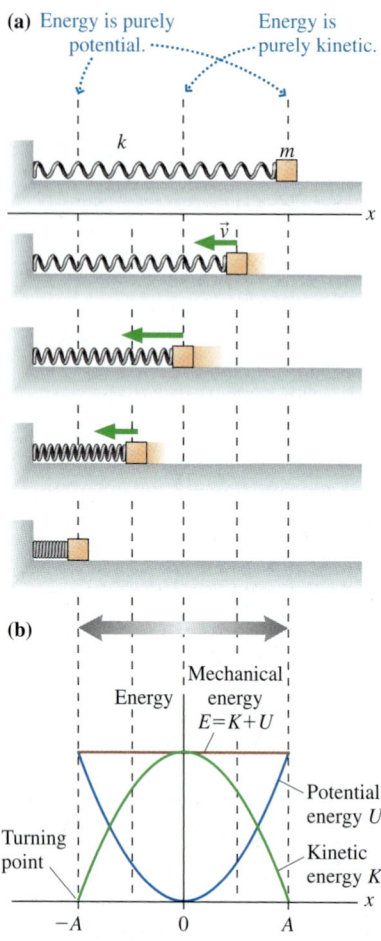

(a) Energy is purely potential. Energy is purely kinetic.

(b)

Energy

Mechanical energy $E = K + U$

Turning point

Potential energy U

Kinetic energy K

$-A$ 0 A x

FIGURE 14.16 Energy transformations for a mass on a spring.

Comparing Equations 14.24 and 14.25, we see that the frequency, and thus the period, of an oscillating mass on a spring is determined by the spring constant k and the object's mass m:

$$f = \frac{1}{2\pi}\sqrt{\frac{k}{m}} \quad \text{and} \quad T = 2\pi\sqrt{\frac{m}{k}} \quad (14.26)$$

SQUARE-ROOT
p. 346

Frequency and period of SHM for
mass m on a spring with spring constant k

We make two observations about this equation:

- **The frequency of simple harmonic motion is determined by the physical properties of the oscillator.** The frequency of a mass on a spring is determined by (1) the mass and (2) the stiffness of the spring, as outlined in Figure 14.17. This dependence of frequency on a force term and an inertia term will also apply to other oscillators.
- **The frequency of simple harmonic motion does not depend on the amplitude A.** A small oscillation and a large oscillation have the same frequency.

Increasing the stiffness of the spring will increase the restoring force. This *increases* the frequency.

High frequency
Stiff spring, low mass

$$f = \frac{1}{2\pi}\sqrt{\frac{k}{m}}$$

Low frequency
Soft spring, high mass

Increasing the mass will increase the inertia of the system. This *decreases* the frequency.

FIGURE 14.17 Frequency dependence on mass and spring stiffness.

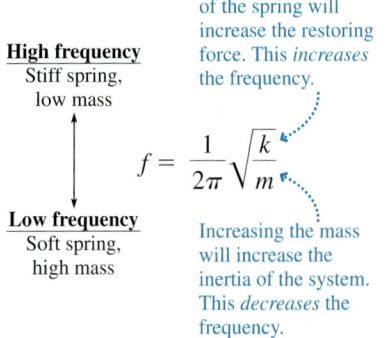

CONCEPTUAL EXAMPLE 14.1 The period of a bobbing bottle

A cylindrical bottle is floating in still water. A push on the top of the bottle causes the bottle to bob up and down. How does the period change as the amplitude of the bobbing slowly decreases?

REASON The motion of the bottle is simple harmonic. The period does not depend on the amplitude, so it will stay constant as the motion dies down.

ASSESS The fact that the period doesn't depend on the amplitude may seem counterintuitive, but it turns out to be a general property of every type of simple harmonic motion. As time goes on, the speed of the bottle decreases, but so does the amplitude—the distance it travels. The bottle is moving more slowly, but it is not going as far. The time for one cycle, the period, stays the same.

Now that we have a complete description of simple harmonic motion in one system, we will summarize the details in a Tactics Box, showing how to use this information to solve oscillation problems.

Measuring mass in space Astronauts on Skylab, a US space station in the 1970's, monitored their weight to determine the effects of prolonged space flight on the body. But the astronauts' weightlessness in space meant they couldn't just hop on a scale! Instead, they became the moving mass in a mass-spring system, holding on tightly to a platform that oscillated back and forth on a spring. By measuring the period of this motion, they could deduce their mass.

 9.3, 9.4, 9.6–9.9

TACTICS BOX 14.1 Identifying and analyzing simple harmonic motion ✏ Exercise 11

① If the net force acting on a particle is a linear restoring force, the motion is simple harmonic motion around the equilibrium position.

② The position, velocity, and acceleration as a function of time are given in Equation 14.18. The equations are given here in terms of x, but they can be written in terms of y, θ, or some other variable if the situation calls for it.

③ The amplitude A is the maximum value of the displacement from equilibrium. The maximum speed and the maximum magnitude of the acceleration are $v_{max} = (2\pi f)A$ and $a_{max} = (2\pi f)^2 A$.

④ The frequency f (and hence the period $T = 1/f$) depends on the physical properties of the particular oscillator, but f does *not* depend on A.

For a mass on a spring, the frequency is given by $f = \frac{1}{2\pi}\sqrt{\frac{k}{m}}$.

⑤ Mechanical energy is conserved. As the oscillation proceeds, energy is transformed from kinetic to potential energy and then back again.

EXAMPLE 14.5 Finding the frequency of an oscillator

A spring has an unstretched length of 10.0 cm. A 25 g mass is hung from the spring, stretching it to a length of 15.0 cm. If the mass is pulled down and released so that it oscillates, what will be the frequency of the oscillation?

FIGURE 14.18 Visual overview of a mass suspended from a spring.

10.0 cm | 15.0 cm

$\Delta L = 5.0$ cm

Adding the mass stretches the spring 5.0 cm.

25 g

$\vec{F}_{sp}$

$\vec{w}$

PREPARE The spring provides a linear restoring force, so the motion will be simple harmonic, as noted in Tactics Box 14.1. The oscillation frequency depends on the spring constant, which we can determine from the stretch of the spring. Figure 14.18 gives a visual overview of the situation.

SOLVE When the mass hangs at rest, after stretching the spring to 15 cm, the net force on it must be zero. Thus the magnitude of the upward spring force equals the downward weight, giving $k\Delta L = mg$. The spring constant is thus

$$k = \frac{mg}{\Delta L} = \frac{(0.025 \text{ kg})(9.8 \text{ m/s}^2)}{0.050 \text{ m}} = 4.9 \text{ N/m}$$

Now that we know the spring constant, we can compute the oscillation frequency:

$$f = \frac{1}{2\pi}\sqrt{\frac{k}{m}} = \frac{1}{2\pi}\sqrt{\frac{4.9 \text{ N/m}}{0.025 \text{ kg}}} = 2.2 \text{ Hz}$$

ASSESS 2.2 Hz is 2.2 oscillations per second. This seems like a reasonable frequency for a mass on a spring. A frequency in the kHz range (thousands of oscillations per second) would have been suspect!

EXAMPLE 14.6 Weighing DNA molecules

It has recently become possible to "weigh" individual DNA molecules by measuring the influence of their mass on a nanoscale oscillator. Figure 14.19 shows a thin rectangular cantilever etched out of silicon (density 2300 kg/m³). If pulled down and released, the end of the cantilever vibrates with simple harmonic motion, moving up and down like a diving board after a jump. When the end of the cantilever is bathed with DNA molecules whose ends have been modified to bind to a surface, one or more molecules may attach to the end of the cantilever. The addition of their mass causes a very slight—but measurable—decrease in the oscillation frequency.

4000 nm

400 nm

Thickness = 100 nm

FIGURE 14.19 A nano-scale cantilever.

A vibrating cantilever of mass M can be modeled as a simple block of mass $\frac{1}{3}M$ attached to a spring. (The factor of $\frac{1}{3}$ arises from the moment of inertia of a bar pivoted at one end: $I = \frac{1}{3}ML^2$.) Neither the mass nor the spring constant can be determined very accurately—perhaps only to two significant figures—but the oscillation frequency can be measured with very high precision simply by counting the oscillations. In one experiment, the cantilever was initially vibrating at exactly 12 MHz. Attachment of a DNA molecule caused the frequency to decrease by 50 Hz. What was the mass of the DNA molecule?

PREPARE We determine the mass of the cantilever as the product of its volume and its density ($M = \rho V$), giving:

$M = (2300 \text{ kg/m}^3)(4000\times10^{-9} \text{ m}\times400\times10^{-9} \text{ m}\times100\times10^{-9} \text{ m})$

$= 3.7 \times 10^{-16} \text{ kg}$

We will model the cantilever as a block of mass $m = \frac{1}{3}M = 1.2 \times 10^{-16}$ kg oscillating on a spring with spring constant k. When the mass increases to $m + m_{DNA}$, the oscillation frequency decreases from $f_0 = 12{,}000{,}000$ Hz to $f_1 = 11{,}999{,}950$ Hz.

SOLVE The oscillation frequency of a mass on a spring is given by Equation 14.26. Addition of mass doesn't change the spring constant, so solving this equation for k allows us to write

$$k = m(2\pi f_0)^2 = (m + m_{DNA})(2\pi f_1)^2$$

The 2π terms cancel, and we can rearrange this equation to give

$$\frac{m + m_{DNA}}{m} = 1 + \frac{m_{DNA}}{m} = \left(\frac{f_0}{f_1}\right)^2 = \left(\frac{12{,}000{,}000 \text{ Hz}}{11{,}999{,}950 \text{ Hz}}\right)^2$$

$$= 1.0000083$$

Subtracting 1 from both sides gives

$$\frac{m_{DNA}}{m} = 0.0000083$$

and thus

$$m_{DNA} = 0.0000083m = (0.0000083)(1.2 \times 10^{-16} \text{ kg})$$

$$= 1.0 \times 10^{-21} \text{ kg} = 1.0 \times 10^{-18} \text{ g}$$

ASSESS This is a reasonable mass for a DNA molecule. It's a remarkable technical achievement to be able to measure a mass this small. With a slight further improvement in sensitivity, scientists will be able to determine the number of base pairs in a strand of DNA simply by weighing it!

EXAMPLE 14.7 Slowing a mass with a spring collision

A 1.5 kg mass slides across a horizontal, frictionless surface at a speed of 2.0 m/s until it collides with and sticks to the free end of a spring with spring constant 50 N/m. The spring's other end is anchored to a wall. How far has the spring compressed when the mass is, at least for an instant, at rest? How much time does it take for the spring to compress to this point?

PREPARE This is a collision problem, but we can solve it using the tools of simple harmonic motion. Figure 14.20 gives a visual overview of the problem. The motion is along the x-axis. We have set $x = 0$ at the uncompressed end of the spring, which is the point of collision. Once the mass hits and sticks, it is going to start oscillating with simple harmonic motion. The position-versus-time graph shows that the motion of the mass until it stops is $\frac{1}{4}$ of a cycle of simple harmonic motion—though the starting point is different from our usual choice. During this quarter cycle of motion, the kinetic energy of the mass is transformed to potential energy of the compressed spring.

SOLVE Conservation of energy tells us that potential energy of the spring when fully compressed is equal to the initial kinetic energy of the mass, so we write

$$\frac{1}{2}m(v_x)_i^2 = \frac{1}{2}kx_f^2$$

The final position of the mass is

$$x_f = (v_x)_i\sqrt{\frac{m}{k}} = (2.0 \text{ m/s})\sqrt{\frac{1.5 \text{ kg}}{50 \text{ N/m}}} = 0.35 \text{ m}$$

This is the compression of the spring.

Because the compression is $\frac{1}{4}$ of a cycle of simple harmonic motion, the time needed to stop the mass is $t_f = \frac{1}{4}T$. The period is computed using Equation 14.26, so we write:

$$t_f = \frac{1}{4}T = \frac{1}{4}2\pi\sqrt{\frac{m}{k}} = \frac{\pi}{2}\sqrt{\frac{1.5 \text{ kg}}{50 \text{ N/m}}} = 0.27 \text{ s}$$

This is the time required for the spring to compress.

ASSESS Interestingly, the time needed to stop the mass does not depend on its initial velocity or on the distance the spring compresses. The motion is simple harmonic, thus the period depends only on the stiffness of the spring and the mass, not on the details of the motion.

Known
$m = 1.5$ kg
$k = 50$ N/m
$t_i = 0$ s
$x_i = 0$ m
$(v_x)_i = 2.0$ m/s
$(v_x)_f = 0$ m/s

Find
x_f, t_f

The spring compresses by x_f as the mass comes to rest.

The mass is instantaneously at rest when the spring reaches maximum compression.

The graph is the start of a sine function, and the motion is $\frac{1}{4}$ of a cycle of SHM.

The mass would, after stopping, reverse direction and continue the cycle of SHM.

FIGURE 14.20 Visual overview for Example 14.7.

Many real-world collisions (such as a pole vaulter landing on a foam pad) involve a moving object that is stopped by an elastic object. In other cases (such as a bouncing rubber ball) the object itself is elastic. The time of the collision is reasonably constant, independent of the speed of the collision, for the reasons noted in the above example.

▶ **Automobile collision times** When a car hits a stationary barrier, it takes approximately 0.1 s to come to rest, regardless of the initial speed, something we can understand with a simple model of the collision. The crumpling of the front of a car during a collision is quite complex, but for many cars the force is approximately proportional to the displacement during the compression of the front of the car. As long as we only consider this initial compression, we can model the body of the car as a mass and the front of the car as a spring. With this model, the collision is similar to that of Example 14.7, and, as in the example, the time for the car to come to rest does not depend on the initial speed.

STOP TO THINK 14.4 Four oscillators have masses and spring constants as shown in the figure. Rank in order, from highest to lowest, the frequencies of the oscillators.

A. k ⎯⎯ $4m$

B. $\frac{1}{2}k$ ⎯⎯ m

C. k ⎯⎯ $2m$

D. $2k$ ⎯⎯ m

9.10–9.12

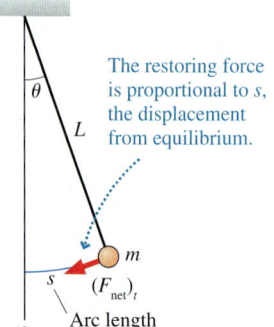

The restoring force is proportional to s, the displacement from equilibrium.

FIGURE 14.21 A simple pendulum.

14.5 Pendulum Motion

As we've already seen, a simple pendulum—a mass at the end of a string or a rod that is free to pivot—is another system that exhibits simple harmonic motion. Everything we have learned about the mass on a spring can be applied to the pendulum as well.

In the first part of the chapter, we looked at the restoring force in the pendulum. For a pendulum of length L displaced by an arc length s, as in Figure 14.21, the tangential restoring force is

$$(F_{net})_t = -\frac{mg}{L}s \qquad (14.27)$$

NOTE ▶ Recall that this equation holds only for small angles. ◀

This linear restoring force has exactly the same form as the net force in a mass-spring system, but with the constants mg/L in place of the constant k. Given this, we can quickly deduce the essential features of pendulum motion by replacing k, wherever it occurs in the oscillating spring equations, with mg/L.

■ The oscillation of a pendulum is simple harmonic motion; the equations of motion can be written for the arc length or the angle:

$$s(t) = A\cos(2\pi ft) \qquad \text{or} \qquad \theta(t) = \theta_{max}\cos(2\pi ft)$$

■ The frequency can be obtained from the equation for the frequency of the mass on a spring by substituting mg/L in place of k:

$$f = \frac{1}{2\pi}\sqrt{\frac{g}{L}} \qquad \text{and} \qquad T = 2\pi\sqrt{\frac{L}{g}} \qquad (14.28)$$

Frequency of a pendulum of length L, with acceleration due to gravity g

■ As for a mass on a spring, the frequency does not depend on the amplitude. Note also that **the frequency, and hence the period, is independent of the mass.** It depends only on the length of the pendulum.

Galileo was the first person to study the pendulum in detail. He realized that the pendulum's fixed frequency would serve as the basis of an accurate clock. Pendulum clocks were the most accurate timepieces available until well into the 20th century.

Pendulum prospecting The period of a pendulum clock does not depend on the amplitude, but it does depend on the strength of gravity. Soon after Christian Huygens built an accurate pendulum clock in 1656 (the photo shows a replica), Jean Richer discovered that the clock ran more slowly near the equator. Richer correctly surmised that this was due to the weaker gravity near the equator because of the greater distance from the center of the earth. In later years, more accurate pendulums were built that could detect much smaller variations in gravity—small enough that they could sense the presence of dense mineral deposits or low-density strata containing petroleum.

EXAMPLE 14.8 Designing a pendulum for a clock

A grandfather clock is designed so that one swing of the pendulum in either direction takes 1.00 s. What is the length of the pendulum?

PREPARE One period of the pendulum is two swings, so the period is $T = 2.00$ s.

SOLVE The period is independent of the mass and depends only on the length. From Equation 14.28,

$$T = \frac{1}{f} = 2\pi\sqrt{\frac{L}{g}}$$

Solving for L, we find

$$L = g\left(\frac{T}{2\pi}\right)^2 = 0.993 \text{ m}$$

ASSESS A pendulum clock with a "tick" or "tock" each second requires a long pendulum of about 1 m—which is why these clocks were originally known as "tall case clocks."

Physical Pendulums and Locomotion

In Chapter 6, we computed maximum walking speed using the ideas of circular motion. We can also model the motion of your legs during walking as pendulum motion. When you walk, you push off with your rear leg and then let it swing forward for the next stride. At normal, comfortable walking speeds, you use very little force to bring your leg forward. Your leg swings forward under the influence of gravity—like a pendulum.

Try this: Stand on one leg, and gently swing your free leg back and forth. There is a certain frequency at which it will naturally swing. This is your leg's pendulum frequency. Now, try swinging your leg at twice this frequency. You can do it, but it is very difficult. The muscles that move your leg back and forth aren't very strong because under normal circumstances they don't need to apply much force.

A pendulum, like your leg, whose mass is distributed along its length is known as a **physical pendulum.** The motion of a physical pendulum is similar to that of a simple pendulum, but its frequency depends on the distribution of mass.

Figure 14.22 shows a simple pendulum and a physical pendulum of the same length. The position of the center of gravity of the physical pendulum is at a distance d from the pivot.

What will be the frequency of a physical pendulum? This is really a rotational motion problem similar to those we considered in Chapter 7. We would expect the frequency to depend on the moment of inertia and the distance to the center of gravity as follows:

- The moment of inertia I is a measure of an object's resistance to rotation. Increasing the moment of inertia while keeping other variables equal should cause the frequency to decrease. In an expression for the frequency of the physical pendulum, we would expect I to appear in the denominator.
- When the pendulum is pushed to the side, a gravitational torque pulls it back. The greater the distance d of the center of gravity from the pivot point, the greater the torque. Increasing this distance while keeping the other variables constant should cause the frequency to increase. In an expression for the frequency of the physical pendulum, we would expect d to appear in the numerator.

A careful analysis of the motion of the physical pendulum produces a result for the frequency that matches these expectations:

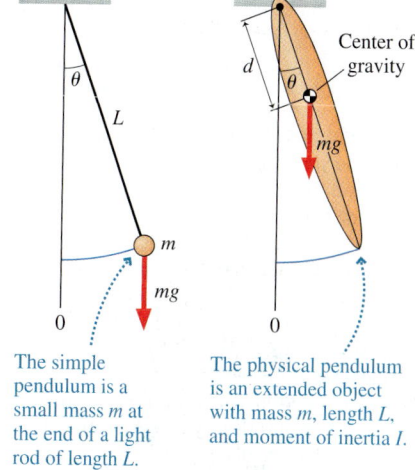

The simple pendulum is a small mass m at the end of a light rod of length L.

The physical pendulum is an extended object with mass m, length L, and moment of inertia I.

FIGURE 14.22 A simple pendulum and a physical pendulum of equal length.

$$f = \frac{1}{2\pi}\sqrt{\frac{mgd}{I}} \qquad (14.29)$$

Frequency of a physical pendulum of mass m,
moment of inertia I, with center of gravity distance d from the pivot

---

EXAMPLE 14.9 Finding the frequency of a swinging leg

A student in a biomechanics lab measures the length of his leg, from hip to heel, to be 0.90 m. What is the frequency of the pendulum motion of the student's leg? What is the period?

PREPARE We can model a human leg reasonably well as a rod of uniform cross section, pivoted at one end (the hip). Recall from Chapter 7 that the moment of inertia of a rod pivoted about its end is $\frac{1}{3}mL^2$. The center of gravity of a uniform leg is at the midpoint, so $d = L/2$.

Continued

How do you hold your arms? You maintain your balance when walking or running by moving your arms back and forth opposite the motion of your legs. You hold your arms so that the natural period of their pendulum motion matches that of your legs. At a normal walking pace, your arms are extended and naturally swing at the same period as your legs. When you run, your gait is more rapid. To decrease the period of the pendulum motion of your arms to match, you bend them at the elbows, shortening their effective length and increasing the natural frequency of oscillation. To test this for yourself, try running fast with your arms fully extended. It's quite awkward!

SOLVE The frequency of a physical pendulum is given by Equation 14.29. Before we put in numbers, we will use symbolic relationships and simplify:

$$f = \frac{1}{2\pi}\sqrt{\frac{mgd}{I}} = \frac{1}{2\pi}\sqrt{\frac{mg(L/2)}{\frac{1}{3}mL^2}} = \frac{1}{2\pi}\sqrt{\frac{3}{2}\frac{g}{L}}$$

The expression for the frequency is similar to that for the simple pendulum, but with an additional numerical factor of 3/2 inside the square root. The numerical value of the frequency is

$$f = \frac{1}{2\pi}\sqrt{\left(\frac{3}{2}\right)\left(\frac{9.8 \text{ m/s}^2}{0.90 \text{ m}}\right)} = 0.64 \text{ Hz}$$

The period is

$$T = \frac{1}{f} = 1.6 \text{ s}$$

ASSESS Notice that we didn't need to know the mass of the leg to find the period. The period of a physical pendulum does not depend on the mass, just as it doesn't for the simple pendulum. The period depends only on the *distribution* of mass. When you walk, swinging your free leg forward to take another stride corresponds to half a period of this pendulum motion. For a period of 1.6 s, this is 0.80 s. For a normal walking pace, one stride in just under one second sounds about right.

As you walk, your legs do swing as physical pendulums as you bring them forward. The frequency is fixed by the length of your legs and their distribution of mass; it doesn't depend on amplitude. Consequently, you don't increase your walking speed by taking more rapid steps—changing the frequency is quite difficult. You simply take longer strides, changing the amplitude, but not the frequency.

Gibbons and other apes move through the forest canopy by a hand-over-hand swinging motion called *brachiation*. In this motion, the body swings under a pivot point where a hand grips a tree branch, a clear example of pendulum motion. A brachiating ape will increase its speed by taking bigger swings; it does this by "pumping" the swinging motion, much as you do when increasing your amplitude on a playground swing. But because the period of the pendulum motion is fixed, a brachiating gibbon can only go so fast. At some point a maximum speed is reached, and gibbons and other apes break into a different gait, launching themselves from branch to branch through the air.

STOP TO THINK 14.5   A pendulum clock is made with a metal rod. It keeps perfect time at a temperature of 20°C. At a higher temperature, the metal rod expands. How will this change the clock's timekeeping?

A. The clock will run fast.
B. The clock will keep perfect time.
C. The clock will run slow.

14.6 Damped Oscillations

A real pendulum clock must have some energy input, otherwise the oscillation of the pendulum would slowly decrease in amplitude due to air resistance. If you strike a bell, the oscillation will soon die away as energy is lost to sound waves in the air and dissipative forces within the metal of the bell.

All real oscillators do run down—some very slowly but others quite quickly—as their mechanical energy is transformed into the thermal energy of the oscillator and its environment. An oscillation that runs down and stops is called a **damped oscillation**.

For a pendulum, the main energy loss is due to air resistance, which we called the *drag force* in Chapter 4. When we learned about the drag force, we noted that it depends on velocity: the faster the motion, the bigger the drag force. For this reason, the decrease in amplitude of an oscillating pendulum will be fastest at the start of the motion. For a pendulum or other oscillator with modest damping, we end up with a graph of motion like that in Figure 14.23a. The maximum displacement, x_{max}, decreases with time. As the oscillation decays, the *rate* of the decay decreases; the difference between successive peaks is less.

If we plot a smooth curve that connects the peaks of successive oscillations (we call such a curve an *envelope*) we get the dotted line shown in Figure 14.23b. It's possible, using calculus, to show that x_{max} decreases with time as

$$x_{max}(t) = Ae^{-t/\tau} \tag{14.30}$$

where $e \approx 2.718$ is the base of the natural logarithm and A is the *initial* amplitude. This steady decrease of x_{max} with time is called an **exponential decay.**

The constant τ (lowercase Greek tau) in Equation 14.30 is called the **time constant.** After one time constant has elapsed—that is, at $t = \tau$—the maximum displacement x_{max} has decreased to

$$x_{max}(\text{at } t = \tau) = Ae^{-1} = A/e \approx 0.37A$$

In other words, the oscillation has decreased after one time constant to about 37% of its initial value. The time constant τ measures the "characteristic time" during which damping causes the amplitude of the oscillation to decay away. An oscillation that decays quickly has a small time constant whereas a "lightly damped" oscillator, which decays very slowly, has a large time constant.

Because exponential decay is something we will see again, we will look at it in more detail.

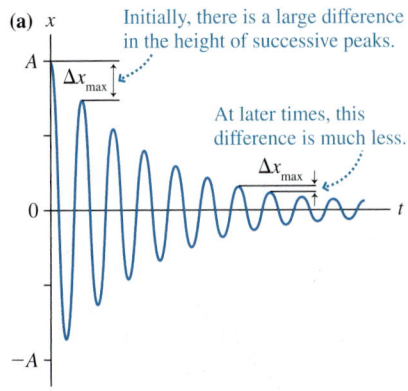

(a) Initially, there is a large difference in the height of successive peaks.

At later times, this difference is much less.

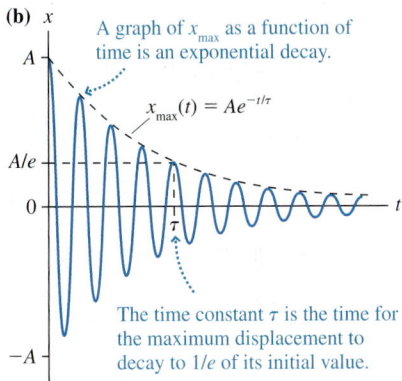

(b) A graph of x_{max} as a function of time is an exponential decay.

$x_{max}(t) = Ae^{-t/\tau}$

The time constant τ is the time for the maximum displacement to decay to $1/e$ of its initial value.

FIGURE 14.23 The motion of a damped oscillator.

 Exponential decay   ✏ Exercises 15, 16   **MP**

Exponential decay occurs when a quantity y is proportional to the number e taken to the power $-t/\tau$. The quantity τ is known as the **time constant.** We write this mathematically as:

$$y = Ae^{-t/\tau}$$

y equals A times (e to the power of minus t over τ)

y starts with initial value A.

$y = Ae^{-t/\tau}$

y has decreased to 37% of its initial value.

Because t appears as the ratio t/τ, the important time intervals are τ, 2τ, and so on.

SCALING Whenever t increases by one time constant, y decreases by a factor of $1/e$. For instance:

- At time $t = 0$, $y = A$.
- Increasing time to $t = \tau$ reduces y to A/e.
- A further increase to $t = 2\tau$ reduces y by another factor of $1/e$ to A/e^2.

Generally, we can say that

At $t = n\tau$, y has the value A/e^n.

LIMITS As t becomes large, y becomes very small and approaches zero.

Damping smoothes the ride A car's wheels are attached to the car's body with springs so that the wheels can move up and down as the car moves over an uneven road. Because the car-spring system is a simple harmonic oscillator, a shock absorber is added to damp unwanted oscillations. The shock absorber is designed to give *critical damping* (see next page) so that, after a bump, the wheel quickly and smoothly returns to its equilibrium position.

Different Amounts of Damping

The damped oscillation shown in Figure 14.23 continues for a long time. The amplitude isn't zero after one time constant, or two, or three. . . . Mathematically, the oscillation never ceases, though the amplitude will eventually be so small as to be undetectable. For practical purposes, we can speak of the time constant τ as the *lifetime* of an oscillation—a measure of about how long it takes to decay. An oscillation with a large time constant will persist for a long time; one with a small time constant will decay quickly.

EXAMPLE 14.10 Finding a clock's decay time

The pendulum in a grandfather clock has a period of 1.00 s. If the clock's driving spring is allowed to run down, damping due to friction will cause the pendulum to slow to a stop. If the time constant for this decay is 300 s, how long will it take for the pendulum's swing to be reduced to half its initial amplitude?

SOLVE Equation 14.30 gives an expression for the decay of the amplitude of a damped harmonic oscillator:

$$x_{max}(t) = Ae^{-t/\tau}$$

As noted in Tactics Box 14.1, we can equally well write this equation in terms of the pendulum's angle in a straightforward manner:

$$\theta_{max}(t) = \theta_i e^{-t/\tau}$$

where θ_i is the initial angle of swing. At some time t, the time we wish to find, the amplitude has decayed to half its initial value. At this time,

$$\theta_{max}(t) = \theta_i e^{-t/\tau} = \frac{1}{2}\theta_i$$

The θ_i cancels, giving

$$e^{-t/\tau} = \frac{1}{2}$$

To solve this for t, first take the natural logarithm of both sides and use the logarithm property $\ln(e^a) = a$. Doing so gives

$$\ln(e^{-t/\tau}) = -\frac{t}{\tau} = \ln\left(\frac{1}{2}\right) = -\ln 2$$

In the last step we used the property $\ln(1/b) = -\ln b$. Now we can solve for t, finding

$$t = \tau \ln 2$$

The time constant was specified as $\tau = 300$ s, so

$$t = (300 \text{ s})(0.693) = 208 \text{ s}$$

It will take 208 s, or about 3.5 min, for the oscillations to decay by half after the spring has run down.

ASSESS The time is less than the time constant, which makes sense. The time constant is the time for the amplitude to decay to 37% of its initial value; we are looking for the time to decay to 50% of its initial value, which should be a shorter time. The time to decay to $\frac{1}{2}$ of the initial amplitude, $t = \tau \ln 2$, could be called the *half-life*. We will see this expression again when we work with radioactivity, another example of an exponential decay.

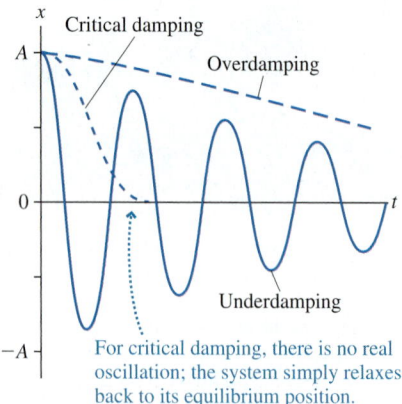

For critical damping, there is no real oscillation; the system simply relaxes back to its equilibrium position.

FIGURE 14.24 Motion of damped oscillator systems.

The time constant is a time that is characteristic of a damped oscillation, but there is another time that characterizes the oscillation—the period T. These two times are independent of each other. The period is fixed by the physical properties of the oscillator; the time constant is fixed by drag or other dissipative forces.

The nature of the decay depends on how the size of the damping time constant τ compares to the period T. Figure 14.24 shows three different levels of damping:

- $\tau \gg T$: If the time constant is much longer than the period, the oscillation will continue for many cycles before dying away, as in the above example of the grandfather clock. In engineering language, this is *underdamping*.
- $\tau \sim T$: If the time constant is just about the same size as the period, the system is *critically damped*. The system doesn't really oscillate. It simply moves smoothly to its equilibrium position in the shortest possible time. A shock absorber in a car is designed to have critical damping.
- $\tau \ll T$: If the time constant is much shorter than the period, the system is so "sticky" that the return to equilibrium is very slow. This is *overdamping*.

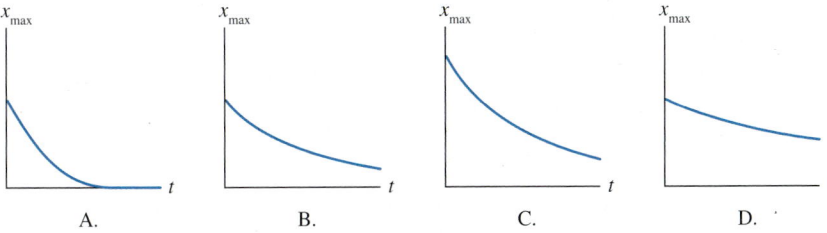

STOP TO THINK 14.6   Rank in order, from largest to smallest, the time constants τ_A to τ_D of the decays shown in the figure. The scales on all of the graphs are the same.

14.7 Driven Oscillations and Resonance

If you jiggle a cup of water, the water sloshes back and forth. This is an example of an oscillator (the water in the cup) subjected to a periodic external force (from your hand). This motion is called a **driven oscillation.**

We can give many examples of driven oscillations. The electromagnetic coil on the back of a loudspeaker cone provides a periodic magnetic force to drive the cone back and forth, causing it to send out sound waves. Earthquakes cause the surface of the earth to move back and forth; this motion causes buildings to oscillate, possibly producing damage or collapse.

Consider an oscillating system that, when left to itself, oscillates at a frequency f_0. We will call this the **natural frequency** of the oscillator. f_0 is simply the frequency of the system if it is displaced from equilibrium and released.

Suppose that this system is now subjected to a *periodic* external force of frequency f_{ext}. This frequency, which is called the **driving frequency,** is completely independent of the oscillator's natural frequency f_0. Somebody or something in the environment selects the frequency f_{ext} of the external force, causing the force to push on the system f_{ext} times every second. The external force causes the oscillation of the system, so it will oscillate at f_{ext}, the driving frequency, not at its natural frequency f_0.

Let's return to the example of the cup of water. If you nudge the cup, you will notice that the water sloshes back and forth at a particular frequency; this is the natural frequency f_0. Now shake the cup at some frequency; this is the driving frequency, f_{ext}. As you shake the cup, the oscillation amplitude of the water depends very sensitively on the frequency f_{ext} of your hand. If the driving frequency is near the natural frequency of the system, the oscillation amplitude may become so large that water splashes out of the cup.

Any driven oscillator will show a similar dependence of amplitude on the driving frequency. Suppose a mass on a spring has a natural frequency $f_0 = 2$ Hz. We can use an external force to push and pull on the mass at frequency f_{ext}, measure the amplitude of the resulting oscillation, and then repeat this over and over for many different driving frequencies. A graph of amplitude versus driving frequency, such as the one in Figure 14.25, is called the oscillator's **response curve.**

At the right and left edges of Figure 14.25, the driving frequency is substantially different from the oscillator's natural frequency. The system oscillates, but its amplitude is very small. The system simply does not respond well to a driving frequency that differs much from f_0. As the driving frequency gets closer and closer to the natural frequency, the amplitude of the oscillation rises dramatically. After all, f_0 is the frequency at which the system "wants" to oscillate, so it is quite happy to respond to a driving frequency near f_0. Hence the amplitude reaches a maximum when the driving frequency matches the system's natural frequency:

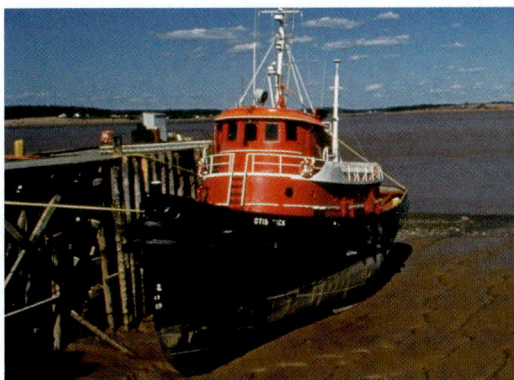

Serious sloshing Water in a cup has a natural frequency at which it will slosh back and forth. The same is true of larger bodies of water. Water in Canada's Bay of Fundy would naturally move into or out of the bay with a period of 12 hours. This is nearly equal to the period of the tidal force of the moon; the two daily high tides are 12.5 hours apart. This *resonance,* a close match between the bay's natural frequency and the moon's driving frequency, produces a huge tidal amplitude. Low tide can be as much as 16 m below high tide, leaving boats high and dry.

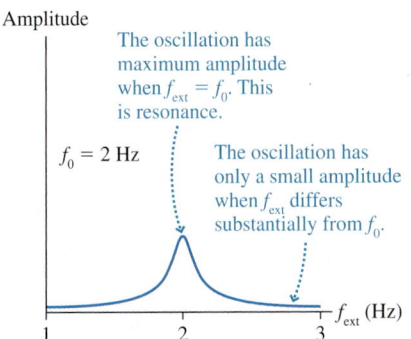

Amplitude

The oscillation has maximum amplitude when $f_{ext} = f_0$. This is resonance.

$f_0 = 2$ Hz

The oscillation has only a small amplitude when f_{ext} differs substantially from f_0.

f_{ext} (Hz)

FIGURE 14.25 The response curve shows the amplitude of a driven oscillator at frequencies near its natural frequency $f_0 = 2$ Hz.

Amplitude

A lightly damped system has a very tall and very narrow response curve.

A system with significant damping has a much smaller response.

$f_0 = 2$ Hz

$\tau = 50T$

$\tau = 20T$

$\tau = 5T$

f_{ext} (Hz)

1   2   3

FIGURE 14.26 The response curve becomes higher and narrower as the damping is reduced.

An optical technique called interferometry reveals the oscillations of a wine glass.

$f_{ext} = f_0$. This large-amplitude response to a driving force whose frequency matches the natural frequency of the system is a phenomenon called **resonance.** Within the context of driven oscillations, the natural frequency f_0 is often called the **resonance frequency.**

The amplitude can become exceedingly large when the frequencies match, especially if there is very little damping. Figure 14.26 shows the response curve of the oscillator of Figure 14.25 with different amounts of damping. Three different graphs are plotted, each with a different time constant for damping. The three graphs have damping that ranges from $\tau = 50T$ (light damping) to $\tau = 5T$ (heavy damping).

A classic example of resonance is the use of sound to break a glass. If you have ever tapped on the edge of a wine glass, you know that it has a certain natural frequency at which it will vibrate. If a sound wave of just the right frequency vibrates the glass, it is possible to increase the amplitude of the oscillation to such a degree that the glass will shatter.

Breaking a glass this way isn't easy. First, you need a glass that has very little damping. As you can see from Figure 14.25, the amplitude of a driven oscillation at resonance is higher if the damping is lower. Second, you need very good pitch control; the sound must be at precisely the right frequency, so that it is an excellent match to the natural frequency of the glass. Third, the sound has to be very, very loud, so that the amplitude of the driving force is quite large. With a very nice crystal wine glass, a signal generator set at just the right frequency, and a very powerful amplifier, it is possible to break a glass via resonance with sound.

> **CONCEPTUAL EXAMPLE 14.2 Fixing an unwanted resonance**
>
> Railroad cars have a natural frequency at which they rock side to side. This can lead to problems on certain stretches of track that have bumps where the rails join. If the joints alternate sides, with a bump on the left rail and then on the right, a train car moving down the track is bumped one way and then the other. For a train moving at just the right speed, the bumps can cause rocking with amplitude large enough to derail the train. One solution is to replace the track with longer rails with greater distance between bumps. How does this solve the problem?
>
> **REASON** The large amplitude of oscillation is produced by a resonance between the frequency at which the train car rocks and the frequency at which the car hits the bumps. Lengthening the distance between bumps will decrease the driving frequency of the bumps, eliminating the resonance.
>
> **ASSESS** Unintended resonances can cause serious difficulties. If the resonance frequency of a building matches the frequency of an earthquake, serious structural damage or failure may occur.

Resonance and Hearing

Resonance in a system means that certain frequencies produce a large response and others do not. This can be very useful. For instance, resonance is responsible for the frequency discrimination of the ear.

As we will see in the next chapter, sound is a vibration in air. Like any vibration, it has a certain frequency. Your ear provides a very sensitive measurement of this frequency—and a resonance in tissues in your ear is responsible. Figure 14.27 on the next page provides an overview of the structures by which sound waves that enter the ear produce vibrations in the cochlea, the coiled, fluid-filled, sound-sensing organ of the inner ear.

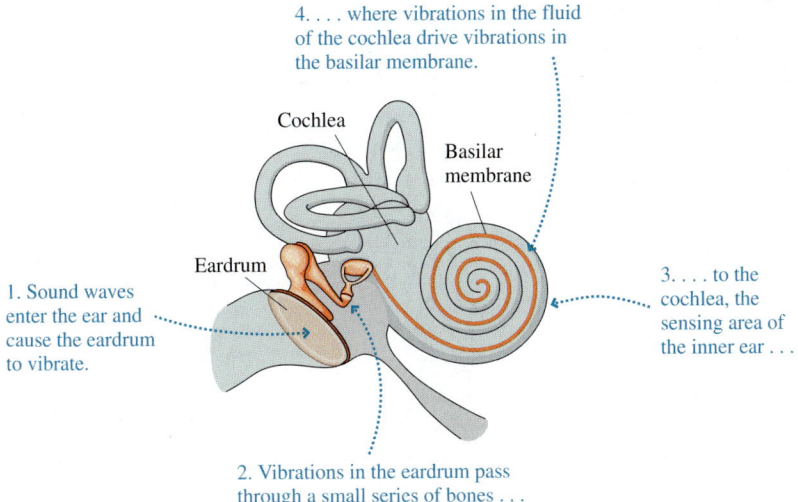

4. . . . where vibrations in the fluid of the cochlea drive vibrations in the basilar membrane.

Cochlea

Basilar membrane

Eardrum

1. Sound waves enter the ear and cause the eardrum to vibrate.

3. . . . to the cochlea, the sensing area of the inner ear . . .

2. Vibrations in the eardrum pass through a small series of bones . . .

FIGURE 14.27 The structures of the ear.

Figure 14.28 shows a very simplified model of the cochlea. As a sound wave travels down the cochlea, it causes a large-amplitude vibration of the basilar membrane at the point where the membrane's natural oscillation frequency matches the sound frequency—a resonance. Lower-frequency sound causes a response farther from the stapes. Sensitive hair cells on the membrane sense the vibration and send nerve signals to your brain. The point on the membrane generating the largest signal tells your brain the frequency of the sound. In this way, your ear and brain can discriminate between two sounds that are very close together in frequency.

We now know a bit about how your ear responds to the vibration of a sound wave—but how does this vibration get from a source to your ear? This is a topic we will consider in the next chapter, when we look at *waves,* oscillations that travel.

▶ **FIGURE 14.28** Resonance plays a role in determining the frequencies of sounds we hear.

To analyze the cochlea, we imagine the spiral structure unrolled, with the basilar membrane separating two fluid-filled chambers.

The stapes, the last of the small bones, transfers vibrations into fluid in the cochlea.

As the distance from the stapes increases, the basilar membrane becomes wider and less stiff, so the resonance frequency of the membrane decreases.

300 Hz   200 Hz   100 Hz   50 Hz

Oscillation amplitude of basilar membrane

0   5   15   25   35
Distance from stapes (mm)

Sounds of different frequencies cause different responses in the basilar membrane.

The goal of Chapter 14 has been to understand systems that oscillate with simple harmonic motion.

GENERAL PRINCIPLES

Restoring Forces

SHM occurs when a **linear restoring force** acts to return a system to an equilibrium position.

Mass on spring

$$(F_{net})_x = -kx$$

Pendulum

$$(F_{net})_t = -\left(\frac{mg}{L}\right)s$$

The frequency of a mass on a spring depends on the mass and the spring constant.

$$f = \frac{1}{2\pi}\sqrt{\frac{k}{m}}$$

The frequency of a pendulum depends on the length and the acceleration due to gravity.

$$f = \frac{1}{2\pi}\sqrt{\frac{g}{L}}$$

Energy

If there is no friction or dissipation, kinetic and potential energy are alternately transformed into each other in SHM, but the total mechanical energy $E = K + U$ is conserved.

$$E = \frac{1}{2}mv_x^2 + \frac{1}{2}kx^2$$
$$= \frac{1}{2}mv_{max}^2$$
$$= \frac{1}{2}kA^2$$

All kinetic

All potential

IMPORTANT CONCEPTS

Oscillation

An **oscillation** is a repetitive motion about an equilibrium position. The **amplitude** A is the maximum displacement from equilibrium. The period T is the time for one cycle. We may also characterize an oscillation by its frequency f.

Simple Harmonic Motion (SHM)

SHM is an oscillation that is described by a sinusoidal function. All systems that undergo SHM can be described by the same functional forms.

Position-versus-time is a cosine function.

$$x(t) = A\cos(2\pi ft)$$

Velocity-versus-time is an inverted sine function.

$$v_x(t) = -v_{max}\sin(2\pi ft)$$
$$v_{max} = 2\pi fA$$

Acceleration-versus-time is an inverted cosine function.

$$a_x(t) = -a_{max}\cos(2\pi ft)$$
$$a_{max} = (2\pi f)^2 A$$

APPLICATIONS

Damping

Simple harmonic motion with damping (due to drag) decreases in amplitude over time. The **time constant** τ determines how quickly the amplitude decays.

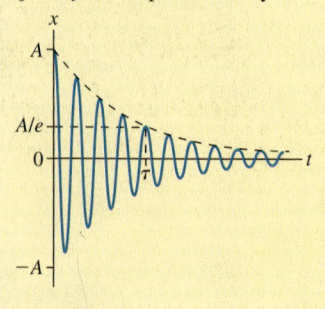

Resonance

A system that oscillates has a **natural frequency** of oscillation f_0. **Resonance** occurs if the system is driven with a frequency f_{ext} that matches this natural frequency. This may produce a large amplitude of oscillation.

Physical pendulum

A **physical pendulum** is a pendulum with mass distributed along its length. The frequency depends on the position of the center of gravity and the moment of inertia.

The motion of legs during walking can be described using a physical pendulum model.

$$f = \frac{1}{2\pi}\sqrt{\frac{mgd}{I}}$$

For instructor-assigned homework, go to www.masteringphysics.com

Problem difficulty is labeled as I (straightforward) to II (challenging).

Problems labeled INT integrate significant material from earlier chapters; BIO are of biological or medical interest.

QUESTIONS

Conceptual Questions

1. Give three examples of *oscillatory* motion. (Note that circular motion is similar to, but not the same as oscillatory motion.)

2. A person's heart rate is given in beats per minute. Is this a period or a frequency?

3. Figure Q14.3 shows the position-versus-time graph of a particle in SHM.
   a. At what time or times is the particle moving to the right at maximum speed?
   b. At what time or times is the particle moving to the left at maximum speed?
   c. At what time or times is the particle instantaneously at rest?

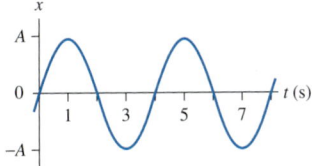

FIGURE Q14.3

4. A block oscillating on a spring has an amplitude of 20 cm. What will be the amplitude if its total mechanical energy is doubled? Explain.

5. A block oscillating on a spring has a maximum speed of 20 cm/s. What will be the block's maximum speed if its total mechanical energy is doubled? Explain.

6. A block oscillating on a spring has 2.0 J of mechanical energy. What will be the energy if its amplitude is doubled? Explain.

7. A block oscillating on a spring has a maximum speed of 30 cm/s. What will be the block's maximum speed if the initial elongation of the spring is doubled?

8. For the graph in Figure Q14.8, determine the frequency f and the oscillation amplitude A.

FIGURE Q14.8

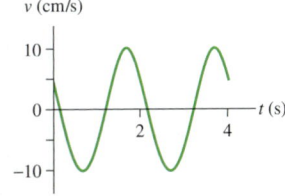

FIGURE Q14.9

9. For the graph in Figure Q14.9, determine the frequency f and the oscillation amplitude A.

10. A block oscillating on a spring has period $T = 2.0$ s.
    a. What is the period if the block's mass is doubled? Explain.
    b. What is the period if the value of the spring constant is quadrupled?
    c. What is the period if the oscillation amplitude is doubled while m and k are unchanged?
    Note: You do not know values for either m or k. Do *not* assume any particular values for them. The required analysis involves thinking about ratios.

11. A pendulum on Planet X, where the value of g is unknown, oscillates with a period of 2.0 s. What is the period of this pendulum if:
    a. Its mass is doubled?
    b. Its length is doubled?
    c. Its oscillation amplitude is doubled?
    Note: You do not know the values of m, L, or g, so do not assume any specific values.

12. As we saw in Chapter 6, the acceleration due to gravity is slightly less in Denver than in Miami. If a pendulum clock keeps perfect time in Miami, will it run fast or slow in Denver? Explain.

13. Sprinters push off from the ball of their foot, then bend their knee to bring their foot up close to the body as they swing their leg forward for the next stride. Why is this an effective strategy for running fast?

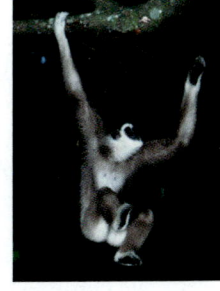

14. Gibbons move through the trees by swinging from successive handholds, as we have seen. To increase their speed, gibbons may bring their legs close to their bodies. How does this help them move more quickly?

15. Describe the difference between τ and T. Don't just *name* them; say what is different about the physical concepts that they represent.

16. What is the difference between the driving frequency and the natural frequency of an oscillator?

17. A person driving a truck on a "washboard" road, one with regularly spaced bumps, notices an interesting effect: When the truck travels at low speed, the amplitude of the vertical motion of the car is small. If the truck's speed is increased, the amplitude of the vertical motion also increases, until it becomes quite unpleasant. But if the speed is increased yet further, the amplitude decreases, and at high speeds the amplitude of the vertical motion is small again. Explain what is happening.

Multiple-Choice Questions

18. | A spring has an unstretched length of 20 cm. A 100 g mass hanging from the spring stretches it to an equilibrium length of 30 cm.
  a. Suppose the mass is pulled down to where the spring's length is 40 cm. When it is released, it begins to oscillate. What is the amplitude of the oscillation?
  A. 5.0 cm   B. 10 cm   C. 20 cm   D. 40 cm
  b. For the data given above, what is the frequency of the oscillation?
  A. 0.10 Hz   B. 0.62 Hz   C. 1.6 Hz   D. 10 Hz
  c. Suppose this experiment were done on the moon, where the acceleration of gravity is approximately 1/6 of that on the earth. How would this change the frequency of the oscillation?
  A. The frequency would decrease.
  B. The frequency would increase.
  C. The frequency would stay the same.

19. ‖ Figure Q14.19 represents the motion of a mass on a spring.

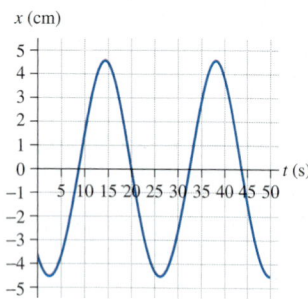

FIGURE Q14.19

  a. What is the period of this oscillation?
  A. 12 s   B. 24 s   C. 36 s
  D. 48 s   E. 50 s
  b. What is the amplitude of the oscillation?
  A. 1.0 cm   B. 2.5 cm   C. 4.5 cm
  D. 5.0 cm   E. 9.0 cm
  c. What is the position of the mass at time $t = 30$ s?
  A. −4.5 cm   B. −2.0 cm   C. 0.0 cm
  D. 4.5 cm   E. 30 cm
  d. When is the first time the velocity of the mass is zero?
  A. 0 s   B. 2 s   C. 8 s
  D. 10 s   E. 13 s
  e. At which of these times does the kinetic energy have its maximum value?
  A. 0 s   B. 8 s   C. 13 s
  D. 26 s   E. 30 s

20. | A ball of mass m oscillates on a spring with spring constant $k = 200$ N/m. The ball's position is $x = (0.350 \text{ m}) \cos 15.0t$, with t measured in seconds.
  a. What is the amplitude of the ball's motion?
  A. 0.175 m   B. 0.350 m   C. 0.700 m
  D. 7.50 m   E. 15.0 m
  b. What is the frequency of the ball's motion?
  A. 0.35 Hz   B. 2.39 Hz   C. 5.44 Hz
  D. 6.28 Hz   E. 15.0 Hz
  c. What is the value of the mass m?
  A. 0.45 kg   B. 0.89 kg   C. 1.54 kg
  D. 3.76 kg   E. 6.33 kg
  d. What is the total mechanical energy of the oscillator?
  A. 1.65 J   B. 3.28 J   C. 6.73 J
  D. 10.1 J   E. 12.2 J
  e. What is the ball's maximum speed?
  A. 0.35 m/s   B. 1.76 m/s   C. 2.60 m/s
  D. 3.88 m/s   E. 5.25 m/s

21. ‖ Figure Q14.21 shows a block on a frictionless surface attached to a spring. The block is pulled out to position $x_i = 20$ cm, then given a "kick" so that it moves to the right with speed v_i. The block then oscillates with an amplitude of 50 cm.

FIGURE Q14.21

  a. What is the oscillation frequency?
  A. 0.025 Hz   B. 0.32 Hz   C. 0.81 Hz
  D. 1.0 Hz   E. 6.4 Hz
  b. What is the total mechanical energy of the oscillator?
  A. 1.7 J   B. 2.5 J   C. 4.6 J
  D. 6.0 J   E. 12 J
  c. What was the initial speed v_i?
  A. 2.0 m/s   B. 2.9 m/s   C. 3.3 m/s
  D. 5.1 m/s   E. 12 m/s

22. | A heavy brass ball is used to make a pendulum with a period of 5.5 s. How long is the cable that connects the pendulum ball to the ceiling?
  A. 4.7 m   B. 6.2 m   C. 7.5 m   D. 8.7 m

23. | Suppose you travel to the moon, and you take with you two timepieces: a pendulum clock and a wristwatch that runs with a wheel and a mainspring. (The wheel and spring works, essentially, like a mass on a spring, but the wheel rotates back and forth rather than moving up and down.) Which will keep good time on the moon?
  A. Only the pendulum clock.
  B. Only the wristwatch.
  C. Both timepieces.
  D. Neither timepiece.

24. ‖ Very loud sounds can damage hearing by injuring the BIO vibration-sensing hair cells on the basilar membrane. Suppose a person has injured hair cells on a segment of the basilar membrane close to the stapes. What type of sound is most likely to have produced this particular pattern of damage?
  A. Loud music with a mix of different frequencies
  B. A very loud, high-frequency sound
  C. A very loud, low-frequency sound

PROBLEMS

Section 14.1 Equilibrium and Oscillation

Section 14.2 Linear Restoring Forces and Simple Harmonic Motion

1. | When a guitar string plays the note "A," the string vibrates at 440 Hz. What is the period of the vibration?
2. || In the aftermath of an intense earthquake, the earth as a whole "rings" with a period of 54 minutes. What is the frequency (in Hz) of this oscillation?
3. | You hear a faucet dripping steadily in your bathroom sink and decide to measure the frequency of the drips. You start a stopwatch on hearing a drip. Calling that the first drip, you count succeeding drips and stop the watch on the fifteenth drip. The watch reads 50 s. What is the frequency?
4. | In taking your pulse, you count 75 heartbeats in 1 min.
   BIO What are the period (in s) and frequency (in Hz) of your heart's oscillations?
5. | Make a table with 3 columns and 8 rows. In row 1, label the columns θ (°), θ (rad), and $\sin\theta$. In the left column, starting in row 2, write 0, 2, 4, 6, 8, 10, and 12.
   a. Convert each of these angles, in degrees, to radians. Put the results in column 2. Show four decimal places.
   b. Calculate the sines. Put the results, showing four decimal places, in column 3.
   c. What is the first angle for which θ and $\sin\theta$ differ by more than 0.0010?
   d. Over what range of angles does the small-angle approximation appear to be valid?

Section 14.3 Describing Simple Harmonic Motion

6. || An air-track glider attached to a spring oscillates between the 10 cm mark and the 60 cm mark on the track. The glider completes 10 oscillations in 33 s. What are the (a) period, (b) frequency, (c) amplitude, and (d) maximum speed of the glider?
7. ||| An air-track glider is attached to a spring. The glider is pulled to the right and released from rest at $t = 0$ s. It then oscillates with a period of 2.0 s and a maximum speed of 40 cm/s.
   a. What is the amplitude of the oscillation?
   b. What is the glider's position at $t = 0.25$ s?
8. | What are the (a) amplitude and (b) frequency of the oscillation shown in Figure P14.8?

FIGURE P14.8

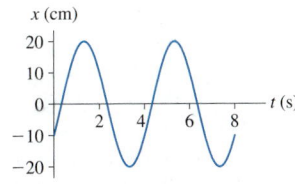

FIGURE P14.9

9. | What are the (a) amplitude and (b) frequency of the oscillation shown in Figure P14.9?
10. | An object in simple harmonic motion has an amplitude of 4.0 cm and a frequency of 2.0 Hz. Draw a position graph showing two cycles of the motion.

11. | An object in simple harmonic motion has an amplitude of 6.0 cm and a frequency of 0.50 Hz. Draw a position graph showing two cycles of the motion.
12. || During an earthquake, the top of a building oscillates with an amplitude of 30 cm at 1.2 Hz. What are the magnitudes of (a) the maximum displacement, (b) the maximum velocity, and (c) the maximum acceleration of the top of the building?
13. | Some passengers on an ocean cruise may suffer from motion sickness as the ship rocks back and forth on the waves. At one position on the ship, passengers experience a vertical motion of amplitude 1 m with a period of 15 s.
   a. To one significant figure, what is the maximum acceleration of the passengers during this motion?
   b. What fraction is this of g?
14. || A child standing at the edge of a 6.0-m-diameter merry-go-
   INT round travels counterclockwise with a period of 10 s. Draw graphs showing just the x-component of (a) the child's position, (b) the child's velocity, and (c) the acceleration of the child, as functions of time for two times around the ride. Be sure to supply appropriate scales on the vertical axes.

Section 14.4 Energy in Simple Harmonic Motion

15. ||| a. When the displacement of a mass on a spring is $\frac{1}{2}A$, what fraction of the mechanical energy is kinetic energy and what fraction is potential energy?
   b. At what displacement, as a fraction of A, is the energy half kinetic and half potential?
16. ||| A 1.0 kg block is attached to a spring with spring constant 16 N/m. While the block is sitting at rest, a student hits it with a hammer and almost instantaneously gives it a speed of 40 cm/s. What are
   a. The amplitude of the subsequent oscillations?
   b. The block's speed at the point where $x = \frac{1}{2}A$?
17. || A block attached to a spring with unknown spring constant oscillates with a period of 2.00 s. What is the period if
   a. The mass is doubled? — 6.005
   b. The mass is halved?
   c. The amplitude is doubled?
   d. The spring constant is doubled?
   Parts a to d are independent questions, each referring to the initial situation.
18. || A 200 g air-track glider is attached to a spring. The glider is pushed in 10.0 cm and released. A student with a stop-watch finds that 10 oscillations take 12.0 s. What is the spring constant?
19. | The position of a 50 g oscillating mass is given by $x(t) = (2.0\ \text{cm})\cos 10t$, where t is in seconds. Determine:
   a. The amplitude.   b. The period.
   c. The spring constant.   d. The maximum speed.
   e. The total energy.   f. The velocity at $t = 0.40$ s.
20. || A 200 g mass attached to a horizontal spring oscillates at a frequency of 2.0 Hz. At one instant, the mass is at $x = 5.0$ cm and has $v_x = -30$ cm/s. Determine:
   a. The period.   b. The amplitude.
   c. The maximum speed.   d. The total energy.

21. ⫼ A 507 g mass oscillates with an amplitude of 10.0 cm on a spring whose spring constant is 20.0 N/m. Determine:
    a. The period.
    b. The maximum speed.
    c. The total energy.

22. ⫼ A 300 g oscillator has a speed of 95.4 cm/s when its displacement is 3.00 cm and 71.4 cm/s when its displacement is 6.00 cm. What is the oscillator's maximum speed?

Section 14.5 Pendulum Motion

23. | A mass on a string of unknown length oscillates as a pendulum with a period of 4.00 s. What is the period if
    a. The mass is doubled?
    b. The string length is doubled?
    c. The string length is halved?
    d. The amplitude is doubled?
    Parts a to d are independent questions, each referring to the initial situation.

24. ‖ A 200 g ball is tied to a string. It is pulled to an angle of 8.00° and released to swing as a pendulum. A student with a stopwatch finds that 10 oscillations take 12.0 s. How long is the string?

25. ⫼ The angle of a pendulum is given by $\theta(t) = (0.10 \text{ rad}) \cos 5t$, where t is in seconds. Determine:
    a. The amplitude.   b. The frequency.
    c. The length of the string.   d. The angle at $t = 2.0$ s.

26. | It is said that Galileo discovered a basic principle of the pendulum—that the period is independent of the amplitude—by using his pulse to time the period of swinging lamps in the cathedral as they swayed in the breeze. Suppose that one oscillation of a swinging lamp takes 5.5 s. How long is the lamp chain?

27. ‖ The acceleration due to gravity on the moon is 1.62 m/s². What is the length of a pendulum whose period on the moon matches the period of a 2.00-m-long pendulum on the earth?

28. | Astronauts on the first trip to Mars take along a pendulum that has a period on earth of 1.50 s. The period on Mars turns out to be 2.45 s. What is the Martian acceleration due to gravity?

29. ‖ A building is being knocked down with a wrecking ball, which is a big metal sphere that swings on a 10-m-long cable. You are (unwisely!) standing directly beneath the point from which the wrecking ball is hung when you notice that the ball has just been released and is swinging directly toward you. How much time do you have to move out of the way?

30. | Interestingly, there have been several studies using cadavers to determine the moment of inertia of human body parts
    BIO by letting them swing as a pendulum about a joint. In one study, the center of gravity of a 5.0 kg lower leg was found to be 18 cm from the knee. When pivoted at the knee and allowed to swing, the oscillation frequency was 1.6 Hz. What was the moment of inertia of the lower leg?

31. ‖ A pendulum clock keeps time by the swinging of a uniform solid rod pivoted at one end. The angular position of the rod is given by $\theta(t) = (0.175 \text{ rad}) \sin(\pi t)$, where t is in seconds.
    a. What is the angular position of the rod at $t = 0.250$ s?
    b. What is the period of oscillation?
    c. How long is the rod?

32. | A thin, circular hoop with a radius of 0.22 m is hanging from its rim on a nail. When pulled to the side and released, the hoop swings back and forth as a physical pendulum. The moment of inertia of a hoop for a rotational axis passing through its edge is $I = 2MR^2$. What is the period of oscillation of the hoop?

Section 14.6 Damped Oscillations

33. ‖ The amplitude of an oscillator decreases to 36.8% of its initial value in 10.0 s. What is the value of the time constant?

34. ‖ A gust of wind hits a flagpole and starts it oscillating. After 10 s, the amplitude of its motion has decreased to 25% of its initial value. What is the value of the time constant?

35. ‖ Calculate and draw an accurate displacement graph from $t = 0$ s to $t = 10$ s of a damped oscillator having a frequency of 1.0 Hz and a time constant of 4.0 s.

36. ‖ A small earthquake starts a lamppost vibrating back and forth. The amplitude of the vibration of the top of the lamppost is 6.5 cm at the moment the quake stops, and 8.0 s later it is 1.8 cm.
    a. What is the time constant for the damping of the oscillation?
    b. What was the amplitude of the oscillation 4.0 s after the quake stopped?

Section 14.7 Driven Oscillations and Resonance

37. | A 25 kg child sits on a 2.0-m-long rope swing. You are going to give the child a small, brief push at regular intervals. If you want to increase the amplitude of her motion as quickly as possible, how much time should you wait between pushes?

38. ‖ Vision is blurred if the head is vibrated at 29 Hz because the
    BIO vibrations are resonant with the natural frequency of the eyeball held by the musculature in its socket. If the mass of the eyeball is 7.5 g, a typical value, what is the effective spring constant of the musculature attached to the eyeball?

General Problems

39. | A spring has an unstretched length of 12 cm. When an 80 g ball is hung from it, the length increases by 4.0 cm. Then the ball is pulled down another 4.0 cm and released.
    a. What is the spring constant of the spring?
    b. What is the period of the oscillation?
    c. Draw a position-versus-time graph showing the motion of the ball for three cycles of the oscillation. Let the equilibrium position of the ball be $y = 0$. Be sure to include appropriate units on the axes so that the period and the amplitude of the motion can be determined from your graph.

40. | A 0.40 kg ball is attached to a spring with spring constant 12 N/m. If the ball is pulled down 0.20 m from the equilibrium position and released, what is its maximum speed while it oscillates?

41. | A spring is hanging from the ceiling. Attaching a 500 g physics book to the spring causes it to stretch 20.0 cm in order to come to equilibrium.
    a. What is the spring constant?
    b. From equilibrium, the book is pulled down 10.0 cm and released. What is the period of oscillation?
    c. What is the book's maximum speed? At what position or positions does it have this speed?

42. ▊ A spring with spring constant 15.0 N/m hangs from the ceiling. A ball is attached to the spring and allowed to come to rest. It is then pulled down 6.00 cm and released. If the ball makes 30 oscillations in 20.0 s, what are its (a) mass and (b) maximum speed?

43. ▊ A spring is hung from the ceiling. When a coffee mug is attached to its end, the spring stretches 2.0 cm before reaching its new equilibrium length. The mug is then pulled down slightly and released. What is the frequency of oscillation?

44. ▊ On your first trip to Planet X you happen to take along a 200 g mass, a 40.0-cm-long spring, a meter stick, and a stopwatch. You're curious about the acceleration due to gravity on Planet X, where ordinary tasks seem easier than on earth, but you can't find this information in your Visitor's Guide. One night you suspend the spring from the ceiling in your room and hang the mass from it. You find that the mass stretches the spring by 31.2 cm. You then pull the mass down 10.0 cm and release it. With the stopwatch you find that 10 oscillations take 14.5 s. Can you now satisfy your curiosity?

45. ▊ An object oscillating on a spring has the position graph shown in Figure P14.45. Draw a position graph if the following changes are made.
    a. The amplitude is halved and the frequency is halved.
    b. The mass is quadrupled.
    Parts a and b are independent questions, each starting from the graph shown.

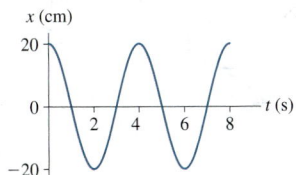

FIGURE P14.45   **FIGURE P14.46**

46. ▊ An object oscillating on a spring has the velocity graph shown in Figure P14.46. Draw a velocity graph if the following changes are made.
    a. The amplitude is doubled and the frequency is halved.
    b. The mass is quadrupled.
    Parts a and b are independent questions, each starting from the graph shown.

47. ▊ The two graphs in Figure P14.47 are for two different vertical mass/spring systems.
    a. What is the frequency of system A? What is the first time at which the mass has maximum speed while traveling in the upward direction?
    b. What is the period of system B? What is the first time at which the mechanical energy is all potential?
    c. If both systems have the same mass, what is the ratio k_A/k_B of their spring constants?

FIGURE P14.47

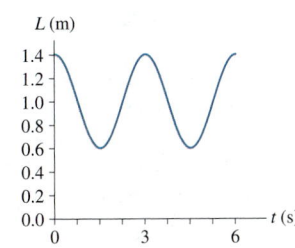

FIGURE P14.48

48. ▊ Astronauts in space cannot weigh themselves by standing on a bathroom scale. Instead, they determine their mass by oscillating on a large spring, as we saw on page 465. Suppose an astronaut attaches one end of a large spring, with spring constant 240 N/m, to her belt and the other end to a hook on the wall of the space capsule. A fellow astronaut then pulls her away from the wall and releases her. The spring's length as a function of time is shown in Figure P14.48. What is her mass?

49. ▊ A 100 g ball attached to a spring with spring constant 2.50 N/m oscillates horizontally on a frictionless table. Its velocity is 20.0 cm/s when $x = -5.00$ cm.
    a. What is the amplitude of oscillation?
    b. What is the speed of the ball when $x = 3.00$ cm?

50. ▊ The ultrasonic transducer used in a medical ultrasound imaging device is a very thin disk ($m = 0.10$ g) driven back and forth in SHM at 1.0 MHz by an electromagnetic coil.
    a. The maximum restoring force that can be applied to the disk without breaking it is 40,000 N. What is the maximum oscillation amplitude that won't rupture the disk?
    b. What is the disk's maximum speed at this amplitude?

51. ▊ A 2 g spider is dangling at the end of a silk thread. You can make the spider bounce up and down on the thread by tapping lightly on his feet with a pencil. You soon discover that you can give the spider the largest amplitude on his little bungee cord if you tap exactly once every 3 seconds. What is the approximate spring constant of the spider's silk thread?

52. ▊ A compact car has a mass of 1200 kg. Assume that the car has one spring on each wheel, that the springs are identical, and that the mass is equally distributed over the four springs.
    a. What is the spring constant of each spring if the empty car bounces up and down 2.0 times each second?
    b. What will be the car's oscillation frequency while carrying four 70 kg passengers?

53. ▊ A 500 g air-track glider attached to a spring with spring constant 10 N/m is sitting at rest on a frictionless air track. A 250 g glider is pushed toward it from the far end of the track at a speed of 120 cm/s. It collides with and sticks to the 500 g glider. What are the amplitude and period of the subsequent oscillations?

54. ▊ A 1.00 kg block is attached to a horizontal spring with spring constant 2500 N/m. The block is at rest on a frictionless surface. A 10.0 g bullet is fired into the block, in the face opposite the spring, and sticks.
    a. What was the bullet's speed if the subsequent oscillations have an amplitude of 10.0 cm?
    b. Could you determine the bullet's speed by measuring the oscillation frequency? If so, how? If not, why not?

55. ▊ Figure P14.55 shows two springs, each with spring constant 20 N/m, connecting a 2.5 kg block to two walls. The block slides on a frictionless surface. If the block is displaced from equilibrium, it will undergo simple harmonic motion. What is the frequency of that motion?

FIGURE P14.55

56. || Bungee Man is a superhero who does super deeds with the help of Super Bungee cords. The Super Bungee cords act like ideal springs no matter how much they are stretched. One day, Bungee Man stopped a school bus that had lost its brakes by hooking one end of a Super Bungee to the rear of the bus as it passed him, planting his feet, and holding on to the other end of the Bungee until the bus came to a halt. (Of course, he then had to quickly release the Bungee before the bus came flying back at him.) The mass of the bus, including passengers, was 12,000 kg, and its speed was 21.2 m/s. The bus came to a stop in 50.0 m.
    a. What was the spring constant of the Super Bungee?
    b. How much time after the Super Bungee was attached did it take the bus to stop?

57. |||| Two 50 g blocks are held 30 cm above a table. As shown in Figure P14.57, one of them is just touching a 30-cm-long spring. The blocks are released at the same time. The block on the left hits the table at exactly the same instant as the block on the right first comes to an instantaneous rest. What is the spring constant?

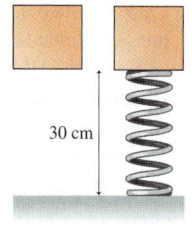

FIGURE P14.57

58. | The earth's acceleration due to gravity varies from 9.78 m/s² at the equator to 9.83 m/s² at the poles. A pendulum whose length is precisely 1.000 m can be used to measure g. Such a device is called a *gravimeter*.
    a. How long do 100 oscillations take at the equator?
    b. How long do 100 oscillations take at the north pole?
    c. Suppose you take your gravimeter to the top of a high mountain peak near the equator. There you find that 100 oscillations take 201 s. What is g on the mountain top?

59. |||| A pendulum clock has a heavy bob supported on a very thin steel rod that is 1.00000 m long at 20°C.
    a. To 6 significant figures, what is the clock's period? Assume that g is 9.80 m/s² exactly.
    b. To 6 significant figures, what is the period if the temperature increases by 10°C?
    c. The clock keeps perfect time at 20°C. At 30°C, after how many hours will the clock be off by 1.0 s?

60. |||| A pendulum consists of a massless, rigid rod with a mass at one end. The other end is pivoted on a frictionless pivot so that it can turn through a complete circle. The pendulum is inverted, so the mass is directly above the pivot point, then released. The speed of the mass as it passes through the lowest point is 5.0 m/s. If the pendulum later undergoes small-amplitude oscillations at the bottom of the arc, what will the frequency be?

61. || Two side-by-side pendulum clocks have heavy bobs at the ends of rigid, very lightweight arms. One pendulum has a 38.8-cm-long rod, the other a 24.8-cm-long rod. Each clock makes one tick for each complete swing of its pendulum.
    a. Determine the frequencies and periods of the two clocks.
    b. Because the two pendulums have different frequencies, their ticks are usually "out of step." However, you notice that they do get back into step (tick at the same instant) at regular intervals. How much time elapses between one such event and the next?
    c. The getting-into-step phenomenon is, itself, periodic. What is the frequency of this phenomenon? Can you see a relationship between its frequency and the frequencies of the two clocks?

62. ||| The 15 g head of a bobble-head doll oscillates in SHM at a frequency of 4.0 Hz.
    a. What is the spring constant of the spring on which the head is mounted?
    b. Suppose the head is pushed 2.0 cm against the spring, then released. What is the head's maximum speed as it oscillates?
    c. The amplitude of the head's oscillations decreases to 0.50 cm in 4.0 s. What is the head's time constant?

63. || An oscillator with a mass of 500 g and a period of 0.50 s has an amplitude that decreases by 2.0% during each complete oscillation. If the initial amplitude is 10 cm, what will be the amplitude after 25 oscillations?

64. || An infant's toy has a 120 g wooden animal hanging from a spring. If pulled down gently, the animal oscillates up and down with a period of 0.50 s. His older sister pulls the spring a bit more than intended. She pulls the animal 30 cm below its equilibrium position, then lets go. The animal flies upward and detaches from the spring right at the animal's equilibrium position. If the animal does not hit anything on the way up, how far above its equilibrium position will it go?

65. |||| A 200 g oscillator in a vacuum chamber has a frequency of 2.0 Hz. When air is admitted, the oscillation decreases to 60% of its initial amplitude in 50 s. How many oscillations will have been completed when the amplitude is 30% of its initial value?

66. || While seated on a tall bench, extend your lower leg a small amount and then let it swing freely about your knee joint, with no muscular engagement. It will oscillate as a damped pendulum. Figure P14.66 is a graph of the lower leg angle versus time in such an experiment.
    a. Is this an overdamped, underdamped, or critically damped oscillation?
    b. Estimate the time constant for this oscillation.

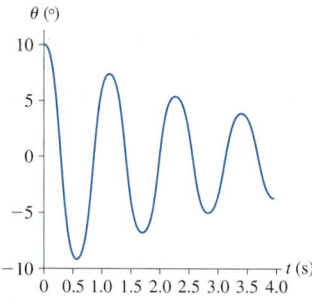

FIGURE P14.66

67. || A 2.0 kg block oscillates up and down on a spring with spring constant 220 N/m. Its initial amplitude is 15 cm. If the time constant for damping of the oscillation is 3.0 s, how much mechanical energy has been dissipated from the block-spring system after 6.0 s?

Passage Problems

Web Spiders and Oscillations BIO

All spiders have special organs that make them exquisitely sensitive to vibrations. Web spiders detect vibrations of their web to determine what has landed in their web, and where.

In fact, spiders carefully adjust the tension of strands to "tune" their web. Suppose an insect lands and is trapped in a web. The silk of the web serves as the spring in a spring-mass system while the body of the insect is the mass. The frequency of oscillation depends on the restoring force of the web and the mass of the insect. Spiders respond more quickly to larger—and therefore more valuable—prey, which they can distinguish by the web's oscillation frequency.

Suppose a 12 mg fly lands in the center of a horizontal spider's web, causing the web to sag by 3.0 mm.

68. | Assuming that the web acts like a spring, what is the spring constant of the web?
  A. 0.039 N/m   B. 0.39 N/m   C. 3.9 N/m   D. 39 N/m

69. | Modeling the motion of the fly on the web as a mass on a spring, at what frequency will the web vibrate when the fly hits it?
  A. 0.91 Hz   B. 2.9 Hz   C. 9.1 Hz   D. 29 Hz

70. | If the web were vertical rather than horizontal, how would he frequency of oscillation be affected?
  A. The frequency would be higher.
  B. The frequency would be lower.
  C. The frequency would be the same.

71. | Spiders are more sensitive to oscillations at higher frequencies. For example, a low-frequency oscillation at 1 Hz can be detected for amplitudes down to 0.1 mm, but a high-frequency oscillation at 1 kHz can be detected for amplitudes as small as 0.1 μm. For these low- and high-frequency oscillations, we can say that
  A. The maximum acceleration of the low-frequency oscillation is greater.
  B. The maximum acceleration of the high-frequency oscillation is greater.
  C. The maximum accelerations of the two oscillations are approximately equal.

STOP TO THINK ANSWERS

Stop to Think 14.1: C. The frequency is inversely proportional to the period, so a shorter period implies a larger frequency.

Stop to Think 14.2: D. The restoring force is proportional to the displacement. If the displacement from equilibrium is doubled, the force is doubled as well.

Stop to Think 14.3: A. The maximum speed occurs when the mass passes through its equilibrium position.

Stop to Think 14.4: $f_D > f_C = f_B > f_A$. The frequency is determined by the ratio of k to m.

Stop to Think 14.5: C. The increase in length will cause the frequency to decrease and thus the period will increase. The time between ticks will increase, so the clock will run slow.

Stop to Think 14.6: $\tau_D > \tau_B = \tau_C > \tau_A$. The time constant is the time to decay to 37% of the initial height. The time constant is independent of the initial height.

15

TRAVELING WAVES AND SOUND

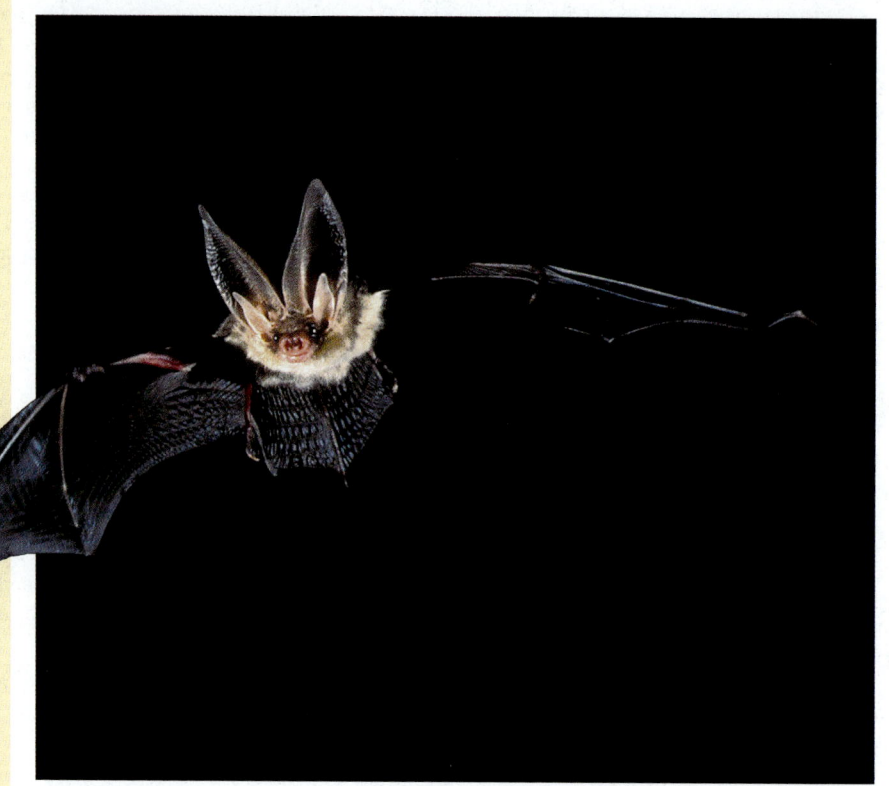

This bat's ears are much more prominent than its eyes. It would appear that hearing is a much more important sense than sight for bats. How does a bat use sound waves to locate prey?

Looking Ahead ▶▶

The goal of Chapter 15 is to learn the basic properties of traveling waves. In this chapter you will learn to:

▶ Understand how a wave travels through a medium.

▶ Recognize the properties of sinusoidal waves.

▶ Discern the important characteristics of sound and light waves.

▶ Apply energy and power concepts to waves.

▶ Understand the decibel scale for loudness of sound.

▶ Use the Doppler effect to find the speed of wave sources and observers.

Looking Back ◀◀

The material in this chapter is closely related to the concept of simple harmonic motion. Please review:

◀ Section 14.2 The properties and mathematical description of simple harmonic motion.

The world is full of waves, although you may not always be aware of them. The "waviness" of a water wave is readily apparent, from the ripples on a pond to ocean waves large enough to surf. It's less apparent that sound and light are also waves because their wave properties are discovered only by careful observations and experiments.

Waves are closely related to oscillations, which we learned about in the previous chapter. If you hold your hand on your throat while you speak, you can feel the vibrations of your vocal cords. This vibration produces a *traveling wave,* a sound wave that moves through the air to the ears of those around you.

In our treatment of waves we will pay special attention to sound waves, but we will consider other traveling waves as well. Light waves travel from the sun to the earth. A sudden fracture in the earth's crust sends out a shock wave that is felt far away as an earthquake. To understand phenomena such as these we need both new models and new mathematics. Our overarching goal in the next two chapters is to understand the properties and characteristics that are common to waves of all types. In other words, we want to find the "essence of waviness" that all waves possess, to have a single model that explains all of the different types of waves we will encounter.

15.1 The Wave Model

The *particle model* that we have been using since Chapter 1 allowed us to simplify the treatment of motion of complex objects by considering them to be particles. Balls, cars, and rockets obviously differ from one another, but the general features of their motions are well described by treating them as particles. As we saw in Chapter 3, a ball or a rock or a car flying through the air will undergo the same motion. The particle model helps us understand this underlying simplicity.

In this chapter we will introduce the basic properties of waves with a **wave model** that emphasizes those aspects of wave behavior common to all waves. Although sound waves, water waves, and radio waves are clearly different, the wave model will allow us to understand many of the important features they share.

The wave model is built around the idea of a **traveling wave,** which is an organized disturbance that travels with a well-defined wave speed. This definition seems straightforward, but several new terms must be understood to gain a complete understanding of the concept of a traveling wave.

Mechanical Waves

Mechanical waves are waves that involve the motion of a substance through which they move, the **medium.** For example, the medium of a water wave is the water, the medium of a sound wave is the air, and the medium of a wave on a stretched string is the string.

As a wave passes through a medium, the atoms that make up the medium are displaced from equilibrium, much like the initial perturbation that begins an oscillation. This is a **disturbance** of the medium. The water ripples of Figure 15.1 are a disturbance of the water's surface.

A wave disturbance is created by a *source*. The source of a wave might be a rock thrown into water, your hand plucking a stretched string, or an oscillating loudspeaker cone pushing on the air. Once created, the disturbance travels outward through the medium at the **wave speed** v. This is the speed with which a ripple moves across the water or a pulse travels down a string.

The disturbance propagates through the medium, and a wave does transfer *energy*, but **the medium as a whole does not travel!** The ripples on the pond (the disturbance) move outward from the splash of the rock, but there is no outward flow of water. Likewise, the particles of a string oscillate up and down but do not move in the direction of a pulse traveling along the string. **A wave transfers energy, but it does not transfer any material or substance outward from the source.**

Electromagnetic and Matter Waves

Mechanical waves require a medium, but there are waves that do not. Two important types of such waves are electromagnetic waves and matter waves.

Electromagnetic waves are waves of an *electromagnetic field*. Electromagnetic waves are very diverse, including visible light, radio waves, microwaves, and x rays. Electromagnetic waves require no material medium and can travel through a vacuum; light can travel through space, though sound cannot. At this point, we have not defined what an "electromagnetic field" is, so we won't worry about the precise nature of what is "waving" in electromagnetic waves. The wave model can describe many of the important aspects of these waves without a detailed description of their exact nature. We'll look more closely at electromagnetic waves in Chapter 25, once we have a full understanding of electric and magnetic fields.

One of the most significant discoveries of the 20th century was that material particles, such as electrons and atoms, have wave-like characteristics. We will learn how a full description of matter at an atomic scale requires an understanding of such **matter waves** in Chapter 28.

The disturbance is the rippling of the water's surface.

The water is the medium.

FIGURE 15.1 Ripples on a pond are a traveling wave.

You may have been at a sporting event in which spectators do "The Wave." The wave moves around the stadium, but the spectators (the medium, in this case) stay right where they are! This is a clear example of the principle that a wave does not transfer any material.

Transverse and Longitudinal Waves

Most waves fall into two general classes, *transverse* and *longitudinal*. For mechanical waves, these terms describe the relationship between the motion of the particles that carry the wave and the motion of the wave itself.

Two types of wave motion

A transverse wave

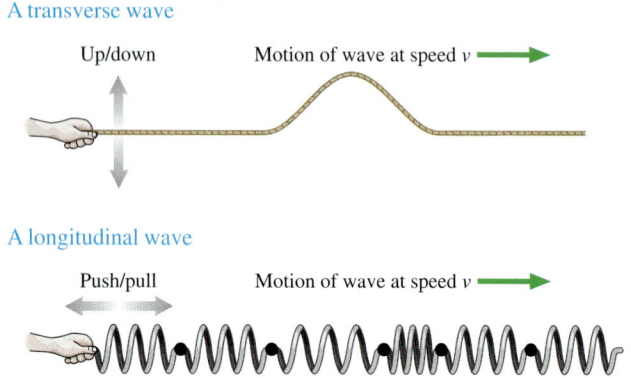

For mechanical waves, a **transverse wave** is a wave in which the particles in the medium move *perpendicular* to the direction in which the wave travels. Shaking the end of a stretched string up and down creates a wave that travels along the string in a horizontal direction while the particles that make up the string oscillate vertically.

A longitudinal wave

In a **longitudinal wave,** the particles in the medium move *parallel* to the direction in which the wave travels. Here we see a chain of masses connected by springs. If you give the first mass in the chain a sharp push, a disturbance travels down the chain by compressing and expanding the springs.

The motion of the water wave.

The motion of the molecules of water carrying the wave.

FIGURE 15.2 A water wave exhibits both transverse and longitudinal characteristics.

Water waves have characteristics of both transverse and longitudinal waves. When a water wave passes, individual water molecules describe a circular path as in Figure 15.2, exhibiting both transverse and longitudinal motion.

The rapid motion of the earth's crust during an earthquake can produce a disturbance that travels through the entire earth. The detection and analysis of these waves is the most important tool we have for analyzing the nature of the earth's interior. The two most important types of earthquake waves are S waves (which are transverse) and P waves (which are longitudinal). The longitudinal P waves are faster, but the transverse S waves are more destructive. Residents of a city a few hundred kilometers from an earthquake will feel the resulting P waves as much as a minute before the S waves, giving them a crucial early warning.

STOP TO THINK 15.1   Spectators at a sporting event do "The Wave," as shown in the photo on the previous page. Is this a transverse or longitudinal wave?

15.2 Traveling Waves

When you drop a pebble in a pond, waves travel outward. But how does this happen? How does a mechanical wave travel through a medium? In answering this question, we must be careful to distinguish the motion of the wave from the motion of the particles that make up the medium. The wave itself is not a particle, so we cannot apply Newton's laws to the wave. However, we can use Newton's laws to examine how the medium responds to a disturbance.

Waves on a String

Figure 15.3 on the next page shows a transverse *wave pulse* traveling to the right along a stretched string. Imagine watching a little dot on the string as a wave pulse passes by. As the pulse approaches from the left, the string near the dot begins to curve. This is the **leading edge** of the pulse. Once the string curves, the tension forces pulling on a small segment of string no longer cancel each other. Instead, as you can see in the first part of the figure, the tension in the string exerts an upward net force on this segment of the string, causing it to accelerate upward.

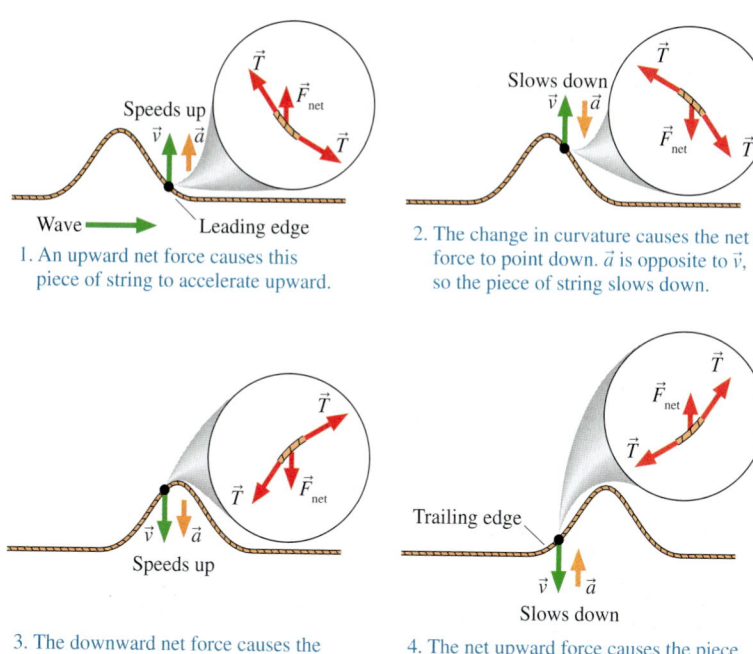

1. An upward net force causes this piece of string to accelerate upward.

2. The change in curvature causes the net force to point down. $\vec{a}$ is opposite to $\vec{v}$, so the piece of string slows down.

3. The downward net force causes the piece of string to speed up in the downward direction.

4. The net upward force causes the piece of string to slow down and stop.

FIGURE 15.3 The motion of a small segment of string as a wave pulse travels to the right.

The string's curvature reverses near the top of the pulse, causing the net force and the acceleration to point downward. Thus this piece of string slows until reaching a turning point at the highest point of the pulse, then it speeds up in the downward direction. This is much like a mass on a spring as the mass approaches the highest point and then reverses direction. Finally, the force and acceleration turn upward on the **trailing edge** of the pulse. The piece of string decelerates and comes to rest when the pulse has completely passed by.

No new physical principles are required to understand how a wave moves. The motion of a pulse along a string is a direct consequence of the tension acting on the segments of the string. An external force may have been required to create the pulse, but **once started the pulse continues to move because of the internal dynamics of the medium.**

Sound Waves

Let's see how a sound wave in air is created using a loudspeaker. When the loudspeaker cone in Figure 15.4a moves forward, it compresses the air in front of it, as shown in Figure 15.4b. The *compression* is the disturbance that travels forward through the air. This is much like the sharp push on the end of the chain of springs on the previous page; thus **a sound wave is a longitudinal wave.** We usually think of sound waves as traveling in air, but sound can travel through any gas, through liquids, and even through solids. A wave similar to that in Figure 15.4b is produced if you hit the end of a metal rod with a hammer.

The motion of a wave on a string is determined by the internal dynamics of the string. Similarly, the motion of a sound wave in air is determined by the physics of gases that we explored in Chapter 12. Once created, the wave in Figure 15.4b will propagate forward; its motion is entirely determined by the properties of the air.

Wave Speed Is a Property of the Medium

The above discussions of waves on a string and sound waves in air show that the motion of these waves is due to the properties of the medium. A careful analysis of these waves would lead to the important and somewhat surprising conclusion

(a) The loudspeaker cone moves in and out in response to electrical signals.

(b) The loudspeaker receives an electrical pulse, and pushes out sharply. This creates a compression of the air.

Compression

v_{sound}

Speaker

Molecules

This compression is the disturbance. It travels at the wave speed v_{sound}.

FIGURE 15.4 A sound wave produced by a loudspeaker.

that **the wave speed does not depend on the shape or size of the pulse, how the pulse was generated, or how far it has traveled.**

What properties of a string determine the speed of waves traveling along the string? The only likely candidates are the string's mass, length, and tension. Because a pulse doesn't travel faster on a longer and thus more massive string, neither the total mass m nor the total length L is important. Instead, the speed depends on the mass-to-length *ratio,* which is called the **linear density** μ of the string:

$$\mu = \frac{m}{L} \qquad (15.1)$$

Linear density characterizes the *type* of string we are using. A fat string has a larger value of μ than a skinny string made of the same material. Similarly, a steel wire has a larger value of μ than a plastic string of the same diameter.

> **NOTE** ▶ We'll assume that strings are *uniform,* meaning that the linear density is the same everywhere along the length of the string. ◀

How does the speed of a wave on a string vary with the tension and the linear density?

- A string with a greater tension will exert larger forces on the segments of the string shown in Figure 15.3. The string will respond more rapidly, so the wave will move at a higher speed. **Wave speed increases with increasing tension.**
- A string with a greater linear density has more inertia. It will respond less rapidly, so the wave will move at a lower speed. **Wave speed decreases with increasing linear density.**

A full analysis of the motion of the string would lead to an expression for the speed of a wave that shows both of these trends:

$$v_{\text{string}} = \sqrt{\frac{T_s}{\mu}} \qquad (15.2)$$

Wave speed on a stretched string with tension T_s and linear density μ

SQUARE-ROOT
p. 346

The subscript s on the symbol T_s for the string's tension will distinguish it from the symbol T for the *period* of oscillation.

Every point on a wave pulse travels with the speed given by Equation 15.2. You can increase the wave speed either by *increasing* the string's tension (make it tighter) or by *decreasing* the string's linear density (make it skinnier). We'll examine the implications for stringed musical instruments in Chapter 16.

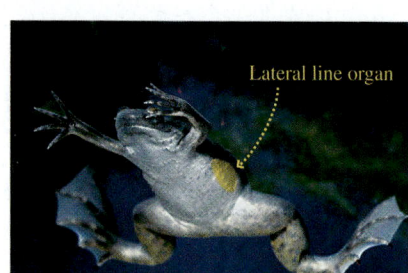

Sensing water waves BIO The African clawed frog has a hunting strategy similar to that of many spiders: It sits and waits for prey to come to it. Like a spider, the frog detects prey animals by the vibrations they cause—not in a web, but in the water. The frog has an array of sensors called the lateral line organ on each side of its body. This organ, highlighted in the above photo, detects oscillations of the water due to passing waves. The frog can determine where the waves come from and what type of animal made them, and thus whether a strike is called for.

Lateral line organ

10.2   Activ
ONLINE
Physics

EXAMPLE 15.1 **When does the spider sense its lunch?**

All spiders are very sensitive to vibrations. An orb spider will sit at the center of its large, circular web and monitor radial threads for vibrations created when an insect lands. Assume that these threads are made of silk, with a linear density of 1.0×10^{-5} kg/m, and are under a tension of 0.40 N. If an insect lands in the web 30 cm from the spider, how long will it take for the spider to find out?

PREPARE When the insect hits the web, a wave pulse will be transmitted along the silk fibers. The speed of the wave depends on the properties of the silk.

SOLVE First, determine the speed of the wave:

$$v = \sqrt{\frac{T_s}{\mu}} = \sqrt{\frac{0.40 \text{ N}}{1.0 \times 10^{-5} \text{ kg/m}}} = 200 \text{ m/s}$$

The time for the wave to travel a distance $d = 30$ cm to reach the spider is

$$\Delta t = \frac{d}{v} = \frac{0.30 \text{ m}}{200 \text{ m/s}} = 1.5 \text{ ms}$$

ASSESS Spider webs are made of very light strings under significant tension, so the wave speed is quite high. The time for signals to travel across the web is very short.

What properties of a gas determine the speed of a sound wave traveling through the gas? It seems plausible that the speed of a sound pulse is related to the speed with which the molecules of the gas move—faster molecules should mean a faster sound wave. In Chapter 11, we found that the typical speed of an atom of mass m, the root-mean-square speed, is

$$v_{\text{rms}} = \sqrt{\frac{3k_{\text{B}}T}{m}}$$

where k_{B} is Boltzmann's constant and T the absolute temperature in kelvin. A thorough analysis finds that the sound speed is slightly less than this rms speed, but has the same dependence on the temperature and the molecular mass:

$$v_{\text{sound}} = \sqrt{\frac{\gamma k_{\text{B}}T}{m}} = \sqrt{\frac{\gamma RT}{M}} \qquad (15.3)$$

SQUARE-ROOT
p. 346

Sound speed in a gas at temperature T

In Equation 15.3 M is the molar mass (kg per mol). γ is a constant that depends on the gas. $\gamma = 1.67$ for monatomic gases such as helium, and $\gamma = 1.40$ for diatomic gases such as nitrogen and oxygen.

 Activ Physics ONLINE 10.3

Certain trends in Equation 15.3 are worth mentioning:

- At a given temperature, the speed of sound in a gas varies significantly with the molecular mass. At room temperature, the speed of sound in helium is much greater than that in air.
- The speed of sound in air (and other gases) varies with temperature. **For calculations in this chapter, use the speed of sound in air at 20°C, 343 m/s, unless otherwise specified.**

Table 15.1 gives the speed of sound in various materials. The greater forces between molecules in liquids and solids result in higher sound speeds than in gases. Generally, sound waves travel faster in liquids than in gases, and faster in solids than in liquids. The speed of sound in a solid depends on its density and stiffness. Light, stiff solids (such as diamond) transmit sound at very high speeds. The sound speed is much less in dense solids such as lead.

TABLE 15.1 The speed of sound

Medium	Speed (m/s)
Air (0°C)	331
Air (20°C)	343
Helium (0°C)	970
Ethyl alcohol	1170
Water	1480
Human tissue (ultrasound)	1540
Lead	1200
Aluminum	5100
Granite	6000
Diamond	12,000

EXAMPLE 15.2 Finding the speed of sound in a cold gas

What is the speed of sound through the gas boiling off a flask of liquid nitrogen? The boiling point of nitrogen is $-196°C$.

PREPARE The speed of sound in a gas depends on the temperature and molar mass of the gas. Because there's no temperature change in a phase transition, the gas boiling off the liquid is at the boiling point temperature, $T = -196°C = 77$ K. The molar mass of molecular nitrogen (N_2) is $M = 0.028$ kg/mol, and $\gamma = 1.40$ for a diatomic gas.

SOLVE The speed of sound in a gas, given by Equation 15.3, is

$$v_{\text{sound}} = \sqrt{\frac{\gamma RT}{M}} = \sqrt{\frac{(1.40)(8.31 \text{ J/mol} \cdot \text{K})(77 \text{ K})}{0.028 \text{ kg/mol}}}$$

$$= 180 \text{ m/s}$$

ASSESS This is much less than the speed of sound in air (which is mostly nitrogen) at 0°C. But that's not unexpected at such a low temperature.

Electromagnetic waves, such as light, travel at a much greater speed than do mechanical waves. As we'll discuss in Chapter 25, all electromagnetic waves travel at the same speed in a vacuum. We call this speed the **speed of light,** which we represent with the symbol c. The value of the speed of light in a vacuum is

$$v_{light} = c = 3.00 \times 10^8 \text{ m/s} \tag{15.4}$$

This is almost one million times the speed of sound in air! At this speed, light could circle the earth 7.5 times in one second.

> **NOTE** ▶ The speed of electromagnetic waves is lower when they travel through a material, but this value for the speed of light in vacuum is also good for electromagnetic waves traveling through air. ◀

TRY IT YOURSELF

Distance to a lightning strike Sound travels approximately 1 km in 3 s, or 1 mi in 5 s. When you see a lightning flash, start counting seconds. When you hear the thunder, stop counting. Divide the result by 5, and you will have the approximate distance to the lightning strike in miles.

EXAMPLE 15.3 How far away was the lightning?

During a thunderstorm, you see a flash from a lightning strike. 8.0 seconds later, you hear the crack of the thunder. How far away did the lightning strike?

PREPARE Two different kinds of waves are involved, with very different wave speeds. The flash of the lightning generates light waves; these will travel from the point of the strike to your position essentially instantaneously. (It takes light about 5 μs to travel 1 mile—not something you will notice.) The strike also generates sound waves that you hear as thunder; these travel much more slowly. The delay between the flash and the thunder is due to the time it takes for the sound wave to travel.

SOLVE We will assume that the speed of sound has its room temperature (20°C) value of 343 m/s. During the time between seeing the flash and hearing the thunder, the sound travels a distance

$$d = v\,\Delta t = (343 \text{ m/s})(8.0 \text{ s}) = 2.7 \times 10^3 \text{ m} = 2.7 \text{ km}$$

ASSESS You can use the time between the flash and the thunder to estimate the distance to a lightning strike, as in the "Try It Yourself" activity.

STOP TO THINK 15.2   Suppose you shake the end of a stretched string to produce a wave. Which of the following actions would increase the speed of the wave down the string? There may be more than one correct answer; if so, give all that are correct.

A. Move your hand up and down more quickly as you generate the wave.
B. Move your hand up and down a greater distance as you generate the wave.
C. Use a heavier string of the same length, under the same tension.
D. Use a lighter string of the same length, under the same tension.
E. Stretch the string tighter to increase the tension.
F. Loosen the string to decrease the tension.

15.3 Graphical and Mathematical Descriptions of Waves

Now that we know a bit about waves and how they travel, it's time to develop our understanding further by describing waves with graphs and equations.

Describing waves and their motion takes a bit more thought than describing particles and their motion. Until now, we have been concerned with quantities, such as position and velocity, that depend only on time. We can write these, as we did in Chapter 14, as $x(t)$ or $v(t)$, indicating that x and v are *functions* of the time variable t. Functions of the single variable t are all right for a particle, because a particle is in only one place at a time, but a wave is not localized. It is spread out through space at each instant of time. To describe a wave mathematically requires a function that specifies what the medium is doing not only at an instant of time (when) but also at a point in space (where); that is, a function of *two* variables.

NOTE ▶ The analysis that follows is for a wave on a string, which is easy to visualize, but the results apply to any traveling wave. ◄

Snapshot and History Graphs

It is helpful to think about waves graphically before we discuss the mathematical details. Consider the wave pulse shown moving along a stretched string in Figure 15.5. (We will consider somewhat artificial triangular and square-shaped wave pulses in this section to clearly show the edges of the pulse.) The graph shows the string's displacement y at a particular instant of time t_1 as a function of position x along the string. This is a "snapshot" of the wave, much like what you might make with a camera whose shutter is opened briefly at t_1. A graph that shows the wave's displacement as a function of position at a single instant of time is called a **snapshot graph.** For a wave on a string, a snapshot graph is literally a picture of the wave at this instant.

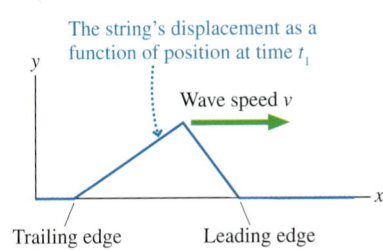

FIGURE 15.5 A snapshot graph of a wave pulse on a string.

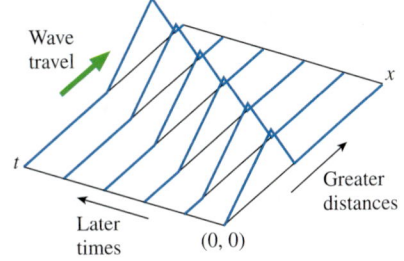

FIGURE 15.6 The traveling wave of Figure 15.5.

As the wave moves, we can take further snapshots. If we place these next to each other, we get a clear picture of the wave's motion, as in Figure 15.6. Note that the wave's shape stays the same as it moves.

We will see this type of picture again, but we will more often simply draw several snapshot graphs one below the other, as in Figure 15.7. This figure shows a sequence of snapshot graphs as the wave of Figure 15.5 continues to move. These are like successive frames from a movie, and reminiscent of the sequences of pictures we saw in Chapter 1. The wave pulse moves forward distance $\Delta x = v \Delta t$ during the time interval Δt. That is, the wave moves with constant speed. Notice also the motion of the point on the string; as we noted, the motion of the medium is distinct from that of the wave.

A snapshot graph tells only half the story. It tells us *where* the wave is and how it varies with position, but only at one instant of time. It gives us no information about how the wave *changes* with time. As a different way of portraying the wave, suppose we follow the dot marked on the string in Figure 15.7 and produce a graph showing how the displacement of this dot changes with time. The result, shown in Figure 15.8, is a displacement-versus-time graph at a single position in space. This is called a **history graph,** as it tells the history of that particular point in the medium.

You might think we have made a mistake; the graph of Figure 15.8 is reversed compared to Figure 15.7. It is not a mistake, but it requires careful thought to see why. As the wave moves toward the dot, the steep leading edge causes the dot to rise quickly. On the displacement-versus-time history graph, *earlier* times

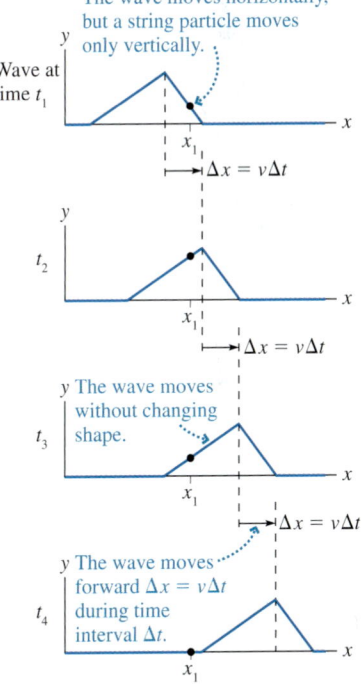

FIGURE 15.7 A sequence of snapshot graphs shows the wave in motion.

FIGURE 15.8 A history graph for the dot on the string in Figure 15.7.

(smaller values of t) are to the *left* and later times (larger t) to the right. Thus the leading edge of the wave is on the *left* side of the Figure 15.8 history graph. As you move to the right on Figure 15.8 you see the slowly falling trailing edge of the wave as it moves past the dot at later times.

Sinusoidal Waves

Waves can come in many different shapes, but for the mathematical description of wave motion we will focus on a particular shape, the **sinusoidal wave.** This is the type of wave produced by a source that oscillates with simple harmonic motion. For example, a loudspeaker cone that oscillates in SHM radiates a sinusoidal sound wave. The sinusoidal electromagnetic waves broadcast by television and FM radio stations are generated by electrons oscillating back and forth in the antenna wire with SHM.

Figure 15.9 shows a sinusoidal wave moving through a medium. The source of the wave, which is undergoing vertical SHM, is located at $x = 0$. Notice how the wave crests move with steady speed toward larger values of x at later times t.

Figure 15.10 shows a history graph for a point in a string as a sinusoidal wave passes by. This graph is identical to the graphs you worked with in Chapter 14, meaning each point in the medium oscillates with simple harmonic motion. The *period* of the wave, shown on the graph, is the time interval for one cycle of the motion. The period is related to the wave *frequency* f by $T = 1/f$, exactly as in SHM.

Because each point on the string oscillates up and down in SHM with period T, we can describe the motion of a point on the string with the familiar expression

$$y(t) = A\cos\left(2\pi\frac{t}{T}\right) \tag{15.5}$$

NOTE ▶ As it was in Chapter 14, the argument of the cosine function is in radians. Make sure that your calculator is in radian mode before starting a calculation with the trigonometric equations in this chapter. ◀

Figure 15.11 shows a snapshot graph for the sinusoidal wave of Figure 15.10, a record of the displacement of the medium at one point in time. Here we see the wave stretched out in space, moving to the right with speed v. The **amplitude** A of the wave is the maximum value of the displacement. The crests of the wave—the high points—have displacement $y_{crest} = A$ and the troughs—the low points—have displacement $y_{trough} = -A$.

An important characteristic of a sinusoidal wave is that it is periodic *in space* as well as in time. As you move from left to right along the "frozen" wave in the snapshot graph of Figure 15.11, the disturbance repeats itself over and over. The distance spanned by one cycle of the motion is called the **wavelength** of the wave. Wavelength is symbolized by λ (lowercase Greek lambda) and, because it is a length, it is measured in units of meters. The wavelength is shown in Figure 15.11 as the distance between two crests, but it could equally well be the distance between two troughs.

NOTE ▶ Wavelength is the spatial analog of period. The period T is the *time* in which the disturbance at a single point in space repeats itself. The wavelength λ is the *distance* in which the disturbance at one instant of time repeats itself. ◀

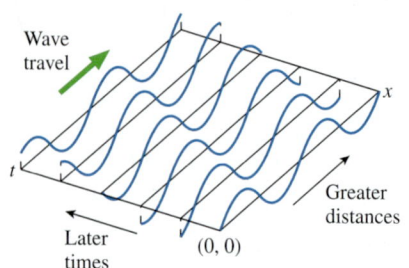

FIGURE 15.9 A sinusoidal wave moving along the x-axis.

FIGURE 15.10 History graph for one point on a string carrying a sinusoidal wave.

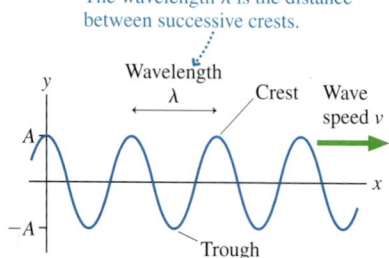

FIGURE 15.11 Snapshot graph of a string carrying a sinusoidal wave.

The snapshot graph shows that the wave, at a frozen instant in time, is also a sinusoidal function of the distance x along the wave, with wavelength λ. At this particular time, the displacement is given by:

$$y(x) = A\cos\left(2\pi\frac{x}{\lambda}\right) \qquad (15.6)$$

Equation 15.5 gives the displacement as a function of time at one point in space, and Equation 15.6 gives the displacement as a function of position at one instant of time. How can we combine these two to form a single expression that is a complete description of the wave? As we'll verify, a wave traveling to the right is described by the equation

$$y(x, t) = A\cos\left(2\pi\left(\frac{x}{\lambda} - \frac{t}{T}\right)\right) \qquad (15.7)$$

SINUSOIDAL
p. 460

Displacement of a traveling wave moving to the right with amplitude A, wavelength λ, and period T

The notation $y(x, t)$ indicates that the displacement y is a function of the *two* variables x and t. We must specify both where (x) and when (t) before we can calculate the displacement of the wave.

We can understand why this expression describes a wave traveling to the right by looking at Figure 15.12. In the figure, we have graphed Equation 15.7 at five instants of time, each separated by one-quarter of the period T, to make five snapshot graphs. One full period has elapsed between the first graph and the last. By following the crest marked with the arrow, you can see that this is a sinusoidal wave moving to the right. As the wave has passed, each point in the medium has undergone exactly one complete oscillation.

For a wave traveling to the left, we have a slightly different form:

$$y(x, t) = A\cos\left(2\pi\left(\frac{x}{\lambda} + \frac{t}{T}\right)\right) \qquad (15.8)$$

Displacement of a traveling wave moving to the left

NOTE ▶ A wave moving to the right (the $+x$-direction) has a $-$ in the expression while a wave moving to the left (the $-x$-direction) has a $+$. Remember that the sign is *opposite* the direction of travel. ◀

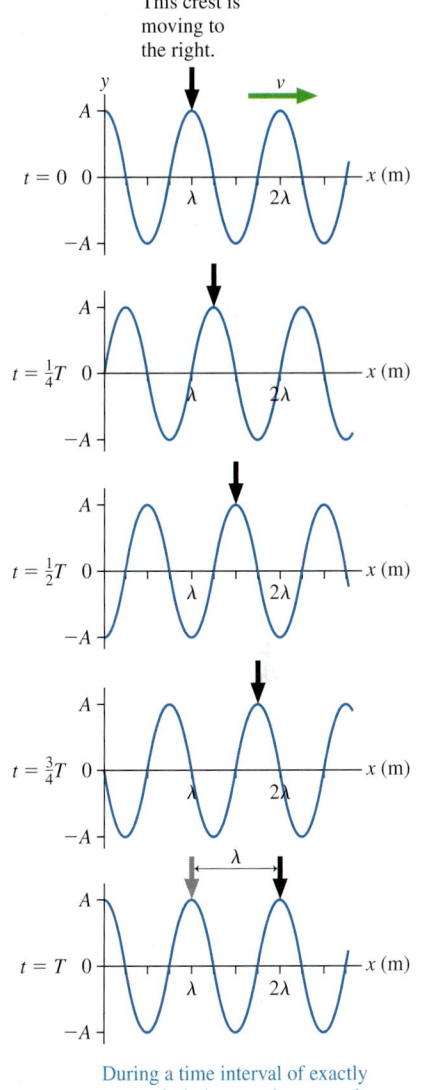

This crest is moving to the right.

During a time interval of exactly one period, the crest has moved forward exactly one wavelength.

FIGURE 15.12 A series of snapshot graphs at time increments of one-quarter of the period T. This is a traveling wave moving to the right.

EXAMPLE 15.4 Determining the rise and fall of a boat

A boat is moving to the right at 5.0 m/s against a sinusoidal ocean wave that is traveling to the left. The waves have 2.0 m between the top of the crests and the bottom of the troughs. The period of the waves is 8.3 s, and their wavelength is 110 m. At one instant, the boat sits on a crest of the wave. 20 s later, what is the vertical displacement of the boat?

PREPARE We begin with a visual overview. Let $t = 0$ be the instant the boat is on the crest, and draw a snapshot graph of the traveling wave at that time. Because the wave is traveling to the left, we will use Equation 15.8 to represent the wave. The boat begins at a crest of the wave, so we see that the boat can start the problem at $x = 0$.

FIGURE 15.13 Visual overview for Example 15.4.

Continued

The distance between the high and low points of the wave is 2.0 m; the amplitude is half this, so $A = 1.0$ m. The wavelength and period are given in the problem.

SOLVE The boat is moving to the right at 5.0 m/s. At $t_f = 20$ s the boat is at position

$$x_f = (5.0 \text{ m/s})(20 \text{ s}) = 100 \text{ m}$$

We need to find the wave's displacement at this position and time. Substituting known values for amplitude, wavelength, and period into Equation 15.8, for a wave traveling to the left, we obtain the following equation for the wave:

$$y(x, t) = (1.0 \text{ m}) \cos\left(2\pi\left(\frac{x}{110 \text{ m}} + \frac{t}{8.3 \text{ s}}\right)\right)$$

At $t_f = 20$ s and $x_f = 100$ m, the boat's displacement on the wave is

$$y_f = y(\text{at } 100 \text{ m}, 20 \text{ s}) = (1.0 \text{ m}) \cos\left(2\pi\left(\frac{100 \text{ m}}{110 \text{ m}} + \frac{20 \text{ s}}{8.3 \text{ s}}\right)\right)$$

$$= -0.42 \text{ m}$$

ASSESS The final displacement is negative—meaning the boat is in a trough of a wave, not underwater. Don't forget that your calculator must be in radian mode when you make your final computation!

The Fundamental Relationship for Sinusoidal Waves

10.1 **Activ** **Physics** ONLINE

In Figure 15.12, one critical observation is that the wave crest marked by the arrow has moved one full wavelength between the first graph and the last. That is, **during a time interval of exactly one period T, each crest of a sinusoidal wave travels forward a distance of exactly one wavelength λ.** Because speed is distance divided by time, the wave speed must be

$$v = \frac{\text{distance}}{\text{time}} = \frac{\lambda}{T} \tag{15.9}$$

Using $f = 1/T$, it is customary to write Equation 15.9 in the form

$$v = \lambda f \tag{15.10}$$

Relationship between velocity, wavelength, and frequency for sinusoidal waves

LINEAR
p. 38

Although Equation 15.10 has no special name, it is *the* fundamental relationship for sinusoidal waves. When using it, keep in mind the *physical* meaning that a wave moves forward a distance of one wavelength during a time interval of one period.

We will frequently see problems in which we are given two of the variables in Equation 15.10 but we need to find the third, as in the following example.

EXAMPLE 15.5 **Writing the equation for a wave**
A sinusoidal wave with an amplitude of 1.5 cm and a frequency of 100 Hz travels at 200 m/s in the positive x-direction. Write the equation for the wave's displacement as it travels.

PREPARE The problem statement gives the following characteristics of the wave:

$$A = 1.5 \text{ cm} = 0.015 \text{ m} \qquad v = 200 \text{ m/s} \qquad f = 100 \text{ Hz}$$

SOLVE To write the equation for the wave, we need the amplitude, the wavelength, and the period. The amplitude was given in the problem statement. We can use the fundamental relationship for sinusoidal waves to find the wavelength:

$$\lambda = \frac{v}{f} = \frac{200 \text{ m/s}}{100 \text{ Hz}} = 2.0 \text{ m}$$

The period can be calculated from the frequency:

$$T = \frac{1}{f} = \frac{1}{100 \text{ Hz}} = 0.010 \text{ s}$$

With these values in hand, we can write the equation for the wave's displacement using Equation 15.7:

$$y(x, t) = (0.015 \text{ m}) \cos\left(2\pi\left(\frac{x}{2.0 \text{ m}} - \frac{t}{0.010 \text{ s}}\right)\right)$$

STOP TO THINK 15.3 Three waves travel to the right with the same speed. Which wave has the highest frequency?

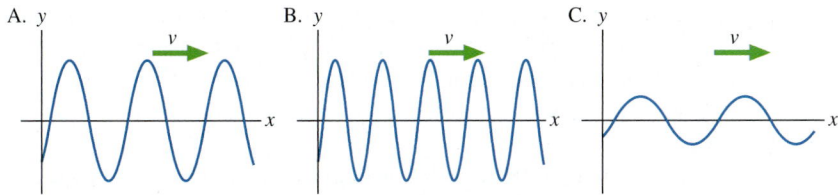

15.4 Sound and Light Waves

Although there are many kinds of waves in nature, two are especially significant for us as humans. These are sound waves and light waves, the basis of hearing and seeing.

Sound Waves

We saw in Figure 15.4 how a loudspeaker creates a sound wave. If the loudspeaker cone moves with simple harmonic motion, it will create a sinusoidal sound wave, as illustrated in Figure 15.14. Each time the cone moves forward, it collides with air molecules and pushes them closer together, creating a region of higher pressure. A half cycle later, as the cone moves backward, the air has room to expand and the pressure decreases. These regions of higher and lower pressure are called **compressions** and **rarefactions.**

As Figure 15.14 suggests, it is often most convenient and informative to think of a sound wave as a pressure wave. As the graph of the pressure shows, the pressure oscillates sinusoidally around the atmospheric pressure p_{atm}. This is a snapshot graph of the wave at one instant of time, and the distance between two adjacent crests (two points of maximum compression) is the wavelength λ.

When the wave reaches your ear, the oscillating pressure causes your eardrums to vibrate. This vibration is transferred through your inner ear to the cochlea, where it is sensed, as we learned in Chapter 14. Humans with normal hearing are able to detect sinusoidal sound waves with frequencies between about 20 Hz and about 20,000 Hz, or 20 kHz. Low frequencies are perceived as a "low pitch" bass note while high frequencies are heard as a "high pitch" treble note.

The loudspeaker cone moves back and forth, creating regions of higher and lower pressure—compressions and rarefactions.

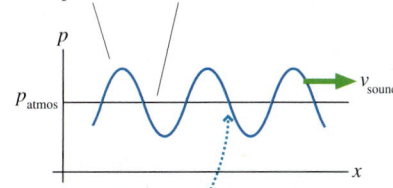

The sound wave is a pressure wave. Compressions are crests; rarefactions are troughs.

FIGURE 15.14 A sound wave is a pressure wave.

EXAMPLE 15.6 Range of wavelengths of sound

What are the wavelengths of sound waves at the limits of human hearing and at the midrange frequency of 500 Hz? Notes sung by human voices are near 500 Hz, as are notes played by striking keys near the center of a piano keyboard.

PREPARE We will do our calculation at room temperature, 20°C, so we will use $v = 343$ m/s for the speed of sound.

SOLVE We can solve for the wavelengths given the fundamental relationship $v = f\lambda$:

$f = 20$ Hz: $\lambda = \dfrac{v}{f} = \dfrac{343 \text{ m/s}}{20 \text{ Hz}} = 17 \text{ m}$

$f = 500$ Hz: $\lambda = \dfrac{v}{f} = \dfrac{343 \text{ m/s}}{500 \text{ Hz}} = 0.69 \text{ m}$

$f = 20$ kHz: $\lambda = \dfrac{v}{f} = \dfrac{343 \text{ m/s}}{20 \times 10^3 \text{ Hz}} = 0.017 \text{ m} = 1.7 \text{ cm}$

ASSESS The wavelength of a 20 kHz note is a small 1.7 cm. At the other extreme, a 20 Hz note has a huge wavelength of 17 m! A wave moves forward one wavelength during a time interval of one period, and a wave traveling at 343 m/s can move 17 m during the $\frac{1}{20}$ s period of a 20 Hz note.

TABLE 15.2 Range of hearing for animals

Animal	Range of hearing (Hz)
Elephant	<5–12,000
Owl	200–12,000
Human	20–20,000
Dog	30–45,000
Mouse	1,000–90,000
Bat	2,000–100,000
Porpoise	75–150,000

Short wavelengths mean sharper images
In the dark depths of the ocean, research submarines use sonar (which stands for "**so**und **n**avigation **a**nd **r**anging") to navigate through and investigate their surroundings. At short distances, high-frequency sound waves can create images of remarkable detail. This sonar image of the Titanic was made from 60 m away with 675 kHz ultrasound; notice the sharp outline, and the small structures visible on the deck.

It is well known that dogs are sensitive to high frequencies that we cannot hear. Other animals also have quite different ranges of hearing than humans; some examples are shown in Table 15.2.

Elephants communicate over great distances with vocalizations at frequencies far too low for us to hear. These low-frequency sounds are transmitted with less energy loss than high-frequency sounds, allowing elephants to hear each other from over 6 km away—important for these social animals that may forage over very large areas.

There are animals that use frequencies well above the range of our hearing as well. High-frequency sounds are useful for *echolocation*—emitting a pulse of sound and listening for its reflection. Bats, which generally feed at night, rely much more on their hearing than their sight. They find and catch insects by echolocation, emitting loud chirps whose reflections are detected by their large, sensitive ears. The frequencies that they use are well above the range of our hearing; we call such sound **ultrasound.** Why do bats and other animals use high frequencies for this purpose?

In Chapter 19, we will look at the *resolution* of optical instruments. The finest detail that your eye—or any optical instrument—can detect is limited by the wavelength of light. Smaller wavelengths allow for the imaging of smaller details.

The same limitations apply to the acoustic image of the world made by bats. In order to sense fine details of their surroundings, bats must use sound of very short wavelength (and thus high frequency). A 50 kHz chirp from a little brown bat has a wavelength of just over half a centimeter, allowing the bat to precisely locate an insect that reflects it. Other animals that use echolocation, such as porpoises, also produce and sense high-frequency sounds.

Sound travels very well though tissues in the body and so can be a useful tool for making images of the body's interior. Again, the fine details necessary for a clinical diagnosis require the short wavelengths of ultrasound. X rays create very good images of bones; ultrasound is used for creating images of soft tissues. It is also useful in cases where the radiation exposure of x rays should be avoided. You have certainly seen ultrasound images taken during pregnancy, where the use of x rays is clearly undesirable.

EXAMPLE 15.7 **Ultrasonic frequencies in medicine**
To make a sufficiently detailed ultrasound image of a fetus in its mother's uterus, a physician has decided that a wavelength of 0.50 mm is needed. What frequency is required?

PREPARE The speed of ultrasound in the body is given in Table 15.1 as 1540 m/s.

SOLVE We can use the familiar relationship among speed, wavelength, and frequency to calculate:

$$f = \frac{v}{\lambda} = \frac{1540 \text{ m/s}}{0.50 \times 10^{-3} \text{ m}} = 3.1 \times 10^6 \text{ Hz} = 3.1 \text{ MHz}$$

Computer processing of an ultrasound image shows fine detail.

ASSESS This is a reasonable result. Clinical ultrasound uses frequencies in the range of 1–20 MHz. Lower frequencies have greater penetration; higher frequencies (and thus shorter wavelengths) show finer detail.

Light and Other Electromagnetic Waves

A light wave is an electromagnetic wave, an oscillation of the electromagnetic field. Other electromagnetic waves, such as radio waves, microwaves, and ultraviolet light, have the same physical characteristics as light waves even though we cannot sense them with our eyes. As we saw earlier, all electromagnetic waves, regardless of wavelength or frequency, travel through vacuum (or air) with the same speed: $c = 3.00 \times 10^8$ m/s.

The wavelengths of light are extremely small. Visible light is an electromagnetic wave with a wavelength (in air) of roughly 400 nm ($=400 \times 10^{-9}$ m) to 700 nm ($=700 \times 10^{-9}$ m). Each wavelength is perceived as a different color. Longer wavelengths, in the 600–700 nm range, are seen as orange or red light; shorter wavelengths, in the 400–500 nm range, are seen as blue or violet light.

Figure 15.15 shows that the visible spectrum is a small slice out of the much broader **electromagnetic spectrum.** We will have much more to say about light and the rest of the electromagnetic spectrum in future chapters.

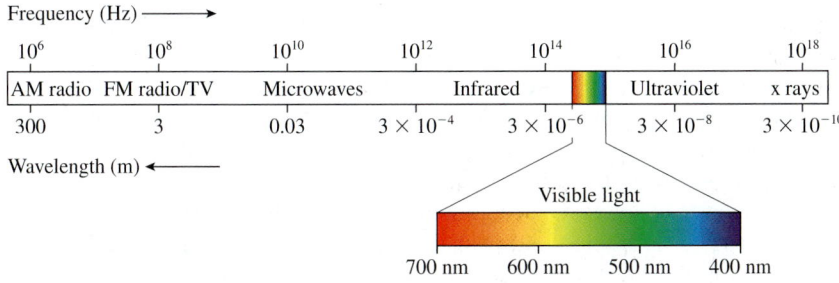

FIGURE 15.15 The electromagnetic spectrum from 10^6 Hz to 10^{18} Hz.

EXAMPLE 15.8 Finding the frequency of microwaves

The wavelength of microwaves in a microwave oven is 12 cm. What is the frequency of the waves?

PREPARE Microwaves are electromagnetic waves, so their speed is the speed of light, $c = 3.00 \times 10^8$ m/s.

SOLVE Using the fundamental relationship for sinusoidal waves, we find

$$f = \frac{3.00 \times 10^8 \text{ m/s}}{0.12 \text{ m}} = 2.5 \times 10^9 \text{ Hz} = 2.5 \text{ GHz}$$

ASSESS Many modern cordless phones work at frequencies of 2.4 GHz, and cell phones at just a bit under 2.0 GHz. These frequencies are not very different from those of the waves that heat your food in a microwave oven, though the power is much less.

NOTE ▶ The wave model we've been developing applies equally well to string, sound, water, and electromagnetic waves. Features such as wavelength, frequency, and speed are characteristics of waves in general. So while we don't yet know much about the details of electromagnetic fields, we can use the wave model to say some significant things about the properties of electromagnetic waves. ◀

15.5 Energy and Intensity

A traveling wave transfers energy from one point to another. The sound wave from a loudspeaker sets your eardrum into motion. Light waves from the sun warm the earth and, if focused with a lens, can start a fire. The *power* of a wave is the rate, in joules per second, at which the wave transfers energy. As you learned

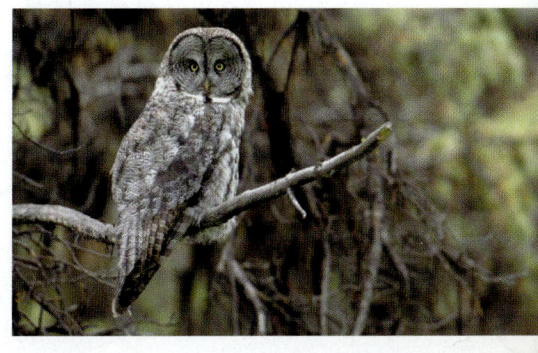

▶ **The better to hear you with** **BIO** The great grey owl has its ears on the front of its face, hidden behind its facial feathers. Its round face works like a radar dish, collecting the energy of sound waves and "funneling" it into the ears. This focusing of the sound allows these owls to sense very quiet sounds. Having ears on the front of the face allows them to precisely determine the source of sounds as well, an asset for a bird of prey. These owls can hear—and locate—mice moving underneath a thick blanket of snow.

(a)

Wave fronts are the crests of the wave. They are spaced one wavelength apart.

The circular wave fronts move outward from the source at speed *v*.

(b)

Very far away from the source, small sections of the wave fronts appear to be straight lines.

FIGURE 15.16 The wave fronts of a circular or spherical wave.

Very far from the source, small segments of spherical wave fronts appear to be planes. The wave is cresting at every point in these planes.

FIGURE 15.17 A plane wave.

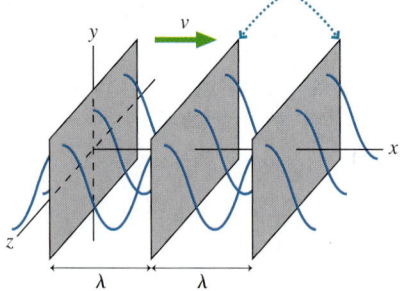

The wave intensity at this surface is $I = P/a$.

Plane waves of power *P*

Area *a*

FIGURE 15.18 Plane waves of power *P* impinge on area *a*.

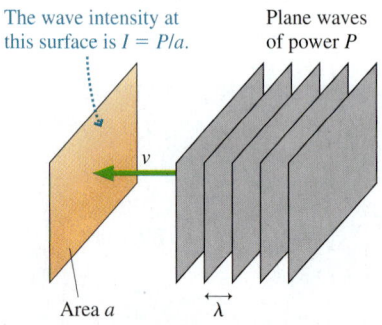

in Chapter 10, power is measured in watts. A person singing or shouting as loud as possible is emitting energy in the form of sound waves at a rate of about 1 W, or 1 J/s. A 25 W light bulb emits about 1 W of visible light, with the other 24 W of power being emitted as heat, or infrared radiation, rather than as visible light. In this section, we will learn how to characterize the power of waves. A first step in doing so is to understand how waves change as they spread out.

Circular, Spherical, and Plane Waves

Suppose you were to take a photograph of ripples spreading on a pond. If you mark the location of the *crests* on the photo, your picture would look like Figure 15.16a. The lines that locate the crests are called **wave fronts,** and they are spaced precisely one wavelength apart. A wave like this is called a **circular wave.** It is a two-dimensional wave that spreads across a surface.

Although the wave fronts are circles, you would hardly notice the curvature if you observed a small section of the wave front very far away from the source. The wave fronts would appear to be parallel lines, still spaced one wavelength apart and traveling at speed *v* as in Figure 15.16b.

Many waves of interest, such as sound waves or light waves, move in three dimensions. For example, loudspeakers and light bulbs emit **spherical waves.** The crests of the wave form a series of concentric spherical shells separated by the wavelength λ. In essence, the waves are three-dimensional ripples. It will still be useful to draw wave-front diagrams such as Figure 15.16a, but now the circles are slices through the spherical shells locating the wave crests.

If you observe a spherical wave very far from its source, the small piece of the wave front that you can see is a little patch on the surface of a very large sphere. If the radius of the sphere is sufficiently large, you will not notice the curvature and this little patch of the wave front appears to be a plane. Figure 15.17 illustrates the idea of a **plane wave.**

Power, Energy, and Intensity

Imagine doing two experiments with a light bulb that emits 2 W of visible light. In the first, you hang the bulb in the center of a room and allow the light to illuminate the walls. In the second experiment, you use mirrors and lenses to "capture" the bulb's light and focus it onto a small spot on one wall. (This is what a slide projector does.) The energy emitted by the bulb is the same in both cases, but, as you know, the light is much brighter when focused onto a small area. We would say that the focused light is more *intense* than the diffuse light that goes in all directions. Similarly, a loudspeaker that beams its sound forward into a small area produces a louder sound in that area than a speaker of equal power that radiates the sound in all directions. Quantities such as brightness and loudness depend not only on the rate of energy transfer, or power, but also on the *area* that receives that power.

Figure 15.18 shows a wave impinging on a surface of area *a*. The surface is perpendicular to the direction in which the wave is traveling. This might be a real, physical surface, such as your eardrum or a solar cell, but it could equally well be a mathematical surface in space that the wave passes right through. If the wave has power *P*, we define the **intensity** *I* of the wave to be

$$I = \frac{P}{a} \tag{15.11}$$

The SI units of intensity are W/m^2. Because intensity is a power-to-area ratio, a wave focused into a small area has a larger intensity than a wave of equal power that is spread out over a large area.

EXAMPLE 15.9 The intensity of a laser beam

A bright, tightly focused laser pointer emits 1.0 mW of light power into a beam that is 1.0 mm in diameter. What is the intensity of the laser beam?

SOLVE The light waves of the laser beam pass through a circle of diameter 1.0 mm. The intensity of the laser beam is

$$I = \frac{P}{a} = \frac{P}{\pi r^2} = \frac{0.0010 \text{ W}}{\pi (0.00050 \text{ m})^2} = 1300 \text{ W/m}^2$$

ASSESS This intensity is roughly equal to the intensity of sunlight at noon on a summer day. The difference between the sun

and a small laser is not their intensities, which are about the same, but their powers. The laser has a small power of 1 mW. It produces a very intense wave only because the area through which the wave passes is very small. In contrast, the sun radiates a total power of about 4×10^{26} W, but this immense power is spread throughout a very large area. Because the beam of the laser is about as intense as the direct, unfiltered light of the sun, you do not want the beam to shine into your eye.

Sound from a loudspeaker and light from a light bulb become less intense as you get farther from the source. This is because spherical waves spread out to fill larger and larger volumes of space. To conserve energy, the wave's amplitude has to decrease with increasing distance r.

If a source of spherical waves radiates uniformly in all directions, then, as Figure 15.19 shows, the power at distance r is spread uniformly over the surface of a sphere of radius r. The surface area of a sphere is $a = 4\pi r^2$, so the intensity of a uniform spherical wave is

$$I = \frac{P_{\text{source}}}{4\pi r^2} \quad (15.12)$$

Intensity of a uniform spherical wave at distance r
from source of power P_{source}

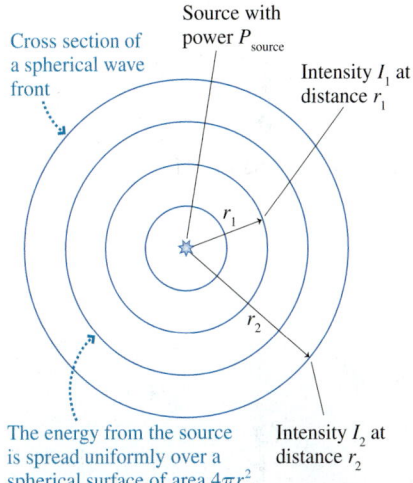

FIGURE 15.19 A source emitting uniform spherical waves.

The inverse-square dependence of r is really just a statement of energy conservation. The source emits energy at the rate P joules per second. The energy is spread over a larger and larger area as the wave moves outward. Consequently, the *energy per area* must decrease in proportion to the surface area of a sphere.

If the intensity at distance r_1 is $I_1 = P_{\text{source}}/4\pi r_1^2$ and the intensity at r_2 is $I_2 = P_{\text{source}}/4\pi r_2^2$, then you can see that the intensity *ratio* is

$$\frac{I_1}{I_2} = \frac{r_2^2}{r_1^2} \quad (15.13)$$

You can use Equation 15.13 to compare the intensities at two distances from a source without needing to know the power of the source.

NOTE ▶ Wave intensities are strongly affected by reflections and absorption. Equations 15.12 and 15.13 apply to situations such as the light from a star or the sound from a firework exploding high in the air. Indoor sound does *not* obey a simple inverse-square law because of the many reflecting surfaces. ◀

STOP TO THINK 15.4   A plane wave, a circular wave, and a spherical wave each have the same intensity. Each of the waves travels the same distance. Afterward, which wave has the largest intensity?

A. The plane wave.   B. The circular wave.   C. The spherical wave.

Hearing hairs BIO This electron microscope picture shows the hair cells in the cochlea of the ear that are responsible for sensing sound. Each cell has a curved row of tiny hairs, which give your ears their remarkable sensitivity. Even very tiny vibrations transmitted into the fluid of the cochlea deflect the hairs, triggering a response in the cells. Motion of the basilar membrane by as little as 0.5 nm, about 5 atomic diameters, can produce an electrical response in the hair cells. Large vibrations can damage these sensitive structures, which is why exposure to loud sounds diminishes hearing.

15.6 Loudness of Sound

Suppose you are listening to one person singing. If nine more people join in, so a total of 10 people are singing in unison, you will certainly perceive the sound of the group as being louder, but it won't be 10 times as loud as the solo singer, even though the intensity is 10 times as great. In fact, you will perceive the sound of 10 people singing together to be approximately twice as loud as the sound of one person.

Now, let's add another 90 people to make a choir of 100 people. This will increase the intensity by another factor of 10, but it will seem to you only twice as loud as the 10-person ensemble, or four times as loud as the original soloist.

Generally, **increasing the sound intensity by a factor of 10 results in an increase in perceived loudness by approximately a factor of 2.** This allows your ears to be sensitive over a very wide range of intensities. The difference in intensity between the quietest sound you can detect and the loudest you can safely hear is a factor of 1,000,000,000,000! A normal conversation has 10,000 times the sound intensity of a whisper, but it sounds only about 16 times as loud.

The loudness of sound is measured by a quantity called the **sound intensity level.** Because of the wide range of intensities we can hear, and the fact that the difference in perceived loudness is much less than the actual difference in intensity, the sound intensity level is measured on a *logarithmic scale*. The next section will explain what this means. The units of sound intensity level (i.e., of loudness) are decibels, a word you have likely heard.

The Decibel Scale

There is a lower limit to the intensity of sound that a human can hear. The exact value varies between individuals and varies with frequency, but an average value for the lowest intensity sound that can be heard in an extremely quiet room is

$$I_0 = 1.0 \times 10^{-12} \text{ W/m}^2$$

This intensity is called the *threshold of hearing*. A sound wave can have less intensity than this, but you won't be able to perceive it.

It's convenient and logical to place the zero of our loudness scale at the threshold of hearing. All other sounds can then be referenced to this intensity. To create a loudness scale, we define the *sound intensity level,* expressed in **decibels** (dB) as

$$\beta = (10 \text{ dB}) \log_{10}\left(\frac{I}{I_0}\right) \qquad (15.14)$$

Sound intensity level in decibels for a sound of intensity *I*

The loudest animal in the world BIO The blue whale is the largest animal in the world, up to 30 m (about 100 ft) long, weighing 150,000 kg or more. It is also the loudest. At close range in the water, the 10-to-30 second calls of the blue whale would be intense enough to damage tissues in your body. Blue whales produce sound with a larynx, much as humans do, but at much lower frequencies of 10–40 Hz that can travel great distances in water. Their loud, low-frequency calls can be heard by other whales hundreds of miles away.

The decibel is named after Alexander Graham Bell, inventor of the telephone. Sound intensity level is actually dimensionless, since it's formed from the ratio of two intensities, so decibels are actually just a *name* to remind us that we're dealing with an intensity *level* rather than a true intensity.

Equation 15.14 takes the base-10 logarithm of the intensity ratio I/I_0. As a reminder, logarithms work like this:

If you express a number   . . . the logarithm
as a power of 10 . . .   is the exponent.

$$\log_{10}(1000) = \log_{10}(10^3) = 3$$

Right at the threshold of hearing, where $I = I_0$, the sound intensity level is

$$\beta = (10 \text{ dB}) \log_{10}\left(\frac{I_0}{I_0}\right) = (10 \text{ dB}) \log_{10}(1) = (10 \text{ dB}) \log_{10}(10^0) = 0 \text{ dB}$$

The threshold of hearing corresponds to 0 dB, as we wanted. 0 dB doesn't mean no sound; it means that, for most people, no sound is heard. Dogs have more sensitive hearing than humans, and most dogs can easily perceive a 0 dB sound.

A comparison between sound intensity in W/m² and the associated sound intensity level in dB is shown in Figure 15.20. The major point to notice is that the sound intensity level increases by 10 dB each time the actual intensity increases by a *factor* of 10. For example, the sound intensity level increases from 70 dB to 80 dB when the sound intensity increases from 10^{-5} W/m² to 10^{-4} W/m². We found earlier that sound is perceived as "twice as loud" when the intensity increases by a factor of 10. In terms of decibels, we can say that the loudness of a sound doubles with each increase in the sound intensity level by 10 dB.

Table 15.3 lists the intensities and sound intensity levels for a number of sounds. Although 130 dB is the threshold of pain, quieter sounds can damage your hearing. A fairly short exposure to 120 dB can cause damage to the hair cells in the ear, but lengthy exposure to sound intensity levels of over 85 dB can produce damage as well.

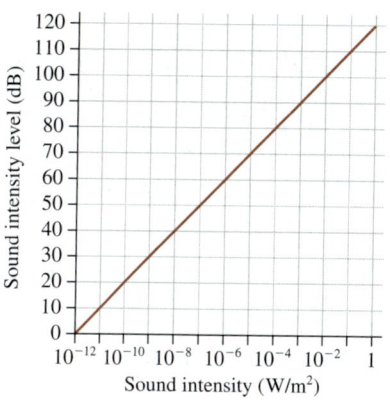

FIGURE 15.20 Sound intensity and sound intensity level scales compared.

TABLE 15.3 Intensity and sound intensity levels of common environmental sounds

Sound	β (dB)	I (W/m²)
Threshold of hearing	0	1.0×10^{-12}
Person breathing, at 3 m	10	1.0×10^{-11}
A whisper, at 1 m	20	1.0×10^{-10}
Classroom during test, no talking	30	1.0×10^{-9}
Residential street, no traffic	40	1.0×10^{-8}
Quiet restaurant	50	1.0×10^{-7}
Normal conversation, at 1 m	60	1.0×10^{-6}
Busy traffic	70	1.0×10^{-5}
Vacuum cleaner, for user	80	1.0×10^{-4}
Niagara Falls, at viewpoint	90	1.0×10^{-3}
Pneumatic hammer, at 2 m	100	0.010
Home stereo at max volume	110	0.10
Rock concert	120	1.0
Threshold of pain	130	10

EXAMPLE 15.10 Finding the loudness of a shout

A person shouting at the top of his lungs emits about 1.0 W of energy as sound waves. What is the sound intensity level 1.0 m from such a person?

PREPARE We will assume that the shouting person emits a spherical sound wave, in which case the intensity decreases according to Equation 15.12.

SOLVE At a distance of 1.0 m, the sound intensity is

$$I = \frac{P}{4\pi r^2} = \frac{1.0 \text{ W}}{4\pi (1.0 \text{ m})^2} = 0.080 \text{ W/m}^2$$

Thus the sound intensity level is

$$\beta = (10 \text{ dB}) \log_{10}\left(\frac{0.080 \text{ W/m}^2}{1.0 \times 10^{-12} \text{ W/m}^2}\right) = 110 \text{ dB}$$

ASSESS This is quite loud (compare with values in Table 15.3), as you might expect. In calculating sound intensity level, be sure to use the $\log_{10}$ button on your calculator, not the natural logarithm button. On many calculators, $\log_{10}$ is labeled LOG and the natural logarithm is labeled LN.

Equation 15.14 allows us to compute the sound intensity level β from the intensity I. We can do the reverse, finding an intensity from the sound intensity level, by taking the inverse of the $\log_{10}$ function. Recall, from the definition of the base-10 logarithm, that $10^{\log(x)} = x$. Applying this to Equation 15.14, we find

$$I = (1.0 \times 10^{-12} \text{ W/m}^2) 10^{(\beta/10 \text{ dB})} \qquad (15.15)$$

EXAMPLE 15.11 **How far away can you hear a conversation?**

The sound intensity level 1.0 m from a person talking in a normal conversational voice is 60 dB. Suppose you are outside, 100 m from the person speaking. If it is a very quiet day with very minimal background noise, will you be able to hear him or her?

PREPARE We know how sound intensity changes with distance, but not sound intensity level, so we need to break this problem into several steps. First, we will convert the sound intensity level at 1.0 m into intensity, using Equation 15.15. Second, we will compute the intensity at a distance of 100 m, using Equation 15.13. Finally, we will convert this result back to a sound intensity level, using Equation 15.14, so that we can judge the loudness.

SOLVE The sound intensity level of a normal conversation at 1.0 m is 60 dB. The intensity is

$$I(1 \text{ m}) = (1.0 \times 10^{-12} \text{ W/m}^2)10^{(60 \text{ dB}/10 \text{ dB})}$$

$$= 1.0 \times 10^{-6} \text{ W/m}^2$$

The intensity at 100 m can be found using Equation 15.13:

$$\frac{I(100 \text{ m})}{I(1 \text{ m})} = \frac{(1 \text{ m})^2}{(100 \text{ m})^2}$$

$$I(100 \text{ m}) = I(1 \text{ m})(1.0 \times 10^{-4}) = 1.0 \times 10^{-10} \text{ W/m}^2$$

This intensity corresponds to a sound intensity level

$$\beta = (10 \text{ dB})\log_{10}\left(\frac{1.0 \times 10^{-10} \text{ W/m}^2}{1.0 \times 10^{-12} \text{ W/m}^2}\right) = 20 \text{ dB}$$

This is above the threshold for hearing—about the level of a whisper—so it could be heard on a very quiet day.

ASSESS The sound is well within what the ear can detect. However, normal background noise is rarely less than 40 dB, which would make the conversation difficult to decipher.

15.7 The Doppler Effect and Shock Waves

In this final section, we will look at sounds from moving objects and for moving observers. You've likely noticed that the pitch of an ambulance's siren drops as it goes past you. A higher pitch suddenly becomes a lower pitch. This change in frequency, which is due to the motion of the ambulance, is called the *Doppler effect*. A more dramatic effect happens when an object moves faster than the speed of sound. The crack of a whip is a *shock wave* produced when the tip moves at such a *supersonic* speed. These examples—and much of this section—concern sound waves, but the phenomena we will explore apply generally to all waves.

Sound Waves from a Moving Source

Figure 15.21a on the next page shows a source of sound waves moving away from Pablo and toward Nancy at a steady speed v_s. The subscript s indicates that this is the speed of the source, not the speed of the waves. The source is emitting sound waves of frequency f_0 as it travels. Part a of the figure is a motion diagram showing the position of the source at times $t = 0$, T, $2T$, and $3T$, where $T = 1/f_0$ is the period of the waves.

Nancy measures the frequency of the wave emitted by the *approaching source* to be f_+. At the same time, Pablo measures the frequency of the wave emitted by the *receding source* to be f_-. Our task is to relate these measured frequencies to the frequency f_0 and speed v_s of the source.

After a wave crest leaves the source, its motion is governed by the properties of the medium. The motion of the source cannot affect a wave that has already been emitted. Thus each circular wave front in Figure 15.21b is centered on the point from which it was emitted. You can see that the wave crests are bunched up in the direction in which the source is moving and are stretched out behind it. The distance between one crest and the next is one wavelength, so the wavelength λ_+ that Nancy measures is *less* than the wavelength $\lambda_0 = v/f_0$ that would be emitted if the source were at rest. Similarly, λ_- behind the source is larger than λ_0.

These crests move through the medium at the wave speed v. Consequently, the frequency $f_+ = v/\lambda_+$ detected by the observer whom the source is approaching is

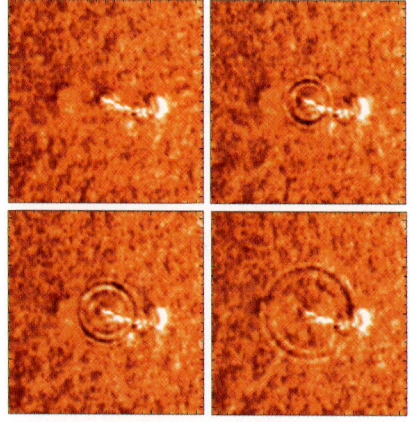

Catching a wave on the sun SOHO (the **So**lar and **H**eliospheric **O**bservatory) is a satellite studying the sun. The instrument on the satellite responsible for images like this measures the Doppler effect of light emitted from the surface of the sun. If the surface is rising toward the satellite, the light is shifted to higher frequencies; these positions are shown darker on the image. If the surface is falling, the light is shifted lower and is shown lighter. This series of images of the sun's surface shows a wave produced by the disruption of a solar flare. The waves reached speeds of 400,000 kilometers per hour and carried an energy 40,000 times that of a large earthquake.

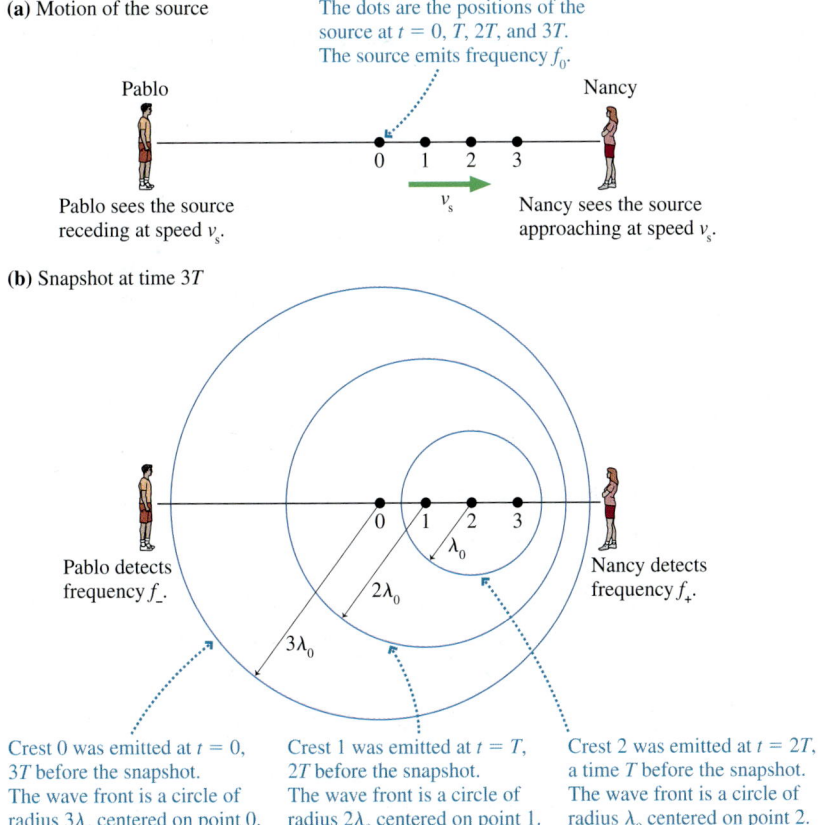

(a) Motion of the source

The dots are the positions of the source at $t = 0$, T, $2T$, and $3T$. The source emits frequency f_0.

Pablo

Nancy

0   1   2   3

v_s

Pablo sees the source receding at speed v_s.

Nancy sees the source approaching at speed v_s.

(b) Snapshot at time $3T$

Pablo detects frequency f_-.

0   1   2   3

λ_0

$2\lambda_0$

$3\lambda_0$

Nancy detects frequency f_+.

Crest 0 was emitted at $t = 0$, $3T$ before the snapshot. The wave front is a circle of radius $3\lambda_0$ centered on point 0.

Crest 1 was emitted at $t = T$, $2T$ before the snapshot. The wave front is a circle of radius $2\lambda_0$ centered on point 1.

Crest 2 was emitted at $t = 2T$, a time T before the snapshot. The wave front is a circle of radius λ_0 centered on point 2.

FIGURE 15.21 A motion diagram showing the wave fronts emitted by a source as it moves to the right at speed v_s.

higher than the frequency f_0 emitted by the source. Similarly, $f_- = v/\lambda_-$ detected behind the source is *lower* than frequency f_0. This change of frequency when a source moves relative to an observer is called the **Doppler effect.** A quantitative analysis based on this information would show that the frequency heard by a stationary observer depends on whether the observer sees the source approaching or receding:

Activ
Physics   10.8, 10.9

$$f_+ = \frac{f_0}{1 - v_s/v}$$

Observed frequency of a wave of speed v emitted from a source approaching at speed v_s

(15.16)

$$f_- = \frac{f_0}{1 + v_s/v}$$

Observed frequency of a wave of speed v emitted from a source receding at speed v_s

As expected, $f_+ > f_0$ (the frequency is higher) for an approaching source, because the denominator is less than 1, and $f_- < f_0$ (the frequency is lower) for a receding source.

EXAMPLE 15.12 How fast are the police driving?

A police siren has a frequency of 550 Hz as the police car approaches you, 450 Hz after it has passed you and is moving away. How fast are the police traveling? Assume that the temperature is 20°C.

PREPARE The siren's frequency is altered by the Doppler effect. The frequency is f_+ as the car approaches and f_- as it moves away. We can write down two equations for these frequencies and solve for the speed of the police car, v_s.

SOLVE Because our goal is to find v_s, we rewrite Equations 15.16 as

$$f_0 = (1 + v_s/v)f_- \quad \text{and} \quad f_0 = (1 - v_s/v)f_+$$

Subtracting the second equation from the first, we get

$$0 = f_- - f_+ + \frac{v_s}{v}(f_- + f_+)$$

Now we can solve for the speed v_s:

$$v_s = \frac{f_+ - f_-}{f_+ + f_-}v = \frac{100 \text{ Hz}}{1000 \text{ Hz}}343 \text{ m/s} = 34 \text{ m/s}$$

ASSESS This is pretty fast (about 75 mph), but is reasonable for a police car speeding with the siren on.

A Stationary Source and a Moving Observer

Suppose the police car in Example 15.12 is at rest while you drive toward it at 34 m/s. You might think that this is equivalent to having the police car move toward you at 34 m/s, but there is an important difference. Mechanical waves move through a medium, and the Doppler effect depends not just on how the source and the observer move with respect to each other but also how they move with respect to the medium. The frequency heard by an observer moving at speed v_o relative to a stationary source emitting frequency f_0 is

$$f_+ = (1 + v_o/v)f_0$$

Doppler effect for an observer approaching a source

$$f_- = (1 - v_o/v)f_0$$ (15.17)

Doppler effect for an observer receding from a source

A quick calculation shows that the frequency of the police siren as you approach it at 34 m/s is 545 Hz, not the 550 Hz you heard as it approached you at 34 m/s.

The Doppler Effect for Light Waves

If a source of light waves is receding from you, the wavelength λ_- that you detect is longer than the wavelength λ_0 emitted by the source. Because the wavelength is shifted toward the red end of the visible spectrum, the longer wavelengths of light, this effect is called the **red shift.** Similarly, the light you detect from a source moving toward you is **blue shifted** to shorter wavelengths.

In the 1920s, an analysis of the spectra of many distant galaxies showed that they *all* had a distinct red shift—all distant galaxies are moving away from us. How can we make sense of this observation? The astronomer Edwin Hubble concluded that the galaxies of the universe are *all* moving apart from each other. Extrapolating backward in time brings us to a point when all the matter of the universe—and even space itself, according to the theory of relativity—began rushing out of a primordial fireball. Many observations and measurements since have given support to the idea that the universe began in a *Big Bang* about 14 billion years ago.

Frequency Shift on Reflection from a Moving Object

A wave striking a barrier or obstacle can *reflect* and travel back toward the source of the wave, a process we will examine more closely in the next chapter. For sound, the reflected wave is called an *echo*. A bat is able to determine the distance to a flying insect by measuring the time between the emission of an ultrasonic

The distant galaxies in this photo all show significant red shifts. The farther away the galaxy, the greater the shift.

chirp and the detection of the echo from the insect. But if the bat is to catch the insect, it's just as important to know where the insect is going—its velocity. The bat can figure this out by noting the *frequency shift* in the reflected wave, another application of the Doppler effect.

Suppose a sound wave of frequency f_0 travels toward a moving object. The object would see the wave's frequency Doppler shifted to a higher frequency f_+, as given by Equation 15.17. The wave reflected back toward the sound source by the moving object is also Doppler shifted to a higher frequency because, to the source, the reflected wave is coming from a moving object. Thus the echo from a moving object is "double Doppler shifted." If the object's speed v_0 is much less than the wave speed v ($v_0 \ll v$), the frequency *shift* for waves reflected from a moving object is

$$\Delta f = \pm 2f_0 \frac{v_o}{v} \qquad (15.18)$$

Frequency shift of waves reflected from an object moving at speed v_o

The $+$ case is for objects moving toward the source of sound, the $-$ for objects moving away. Notice that there is no shift (i.e., the reflected wave has the same frequency as the emitted wave) if the reflecting object is at rest.

The *Doppler blood flow meter* is an important application of the frequency shift of waves reflected from moving objects. If an ultrasound emitter is pressed against the skin, the sound waves reflect off tissues in the body. In most cases, the frequency of the reflected wave is the same as that of the emitted wave. However, some of the sound waves reflect from blood cells moving through arteries toward or away from the emitter. The moving blood cells produce a frequency shift Δf in the reflected wave. By measuring Δf, doctors can determine blood flow speeds in an entirely noninvasive manner.

If we add a frequency-shift measurement to an ultrasound imaging unit, like the one in Example 15.7, we have a device—called *Doppler ultrasound*—that can show not only structure but also motion. *Doppler ultrasound* is a very valuable tool in cardiology because an image can reveal the motion of the heart muscle and the blood in the chambers of the heart in addition to the structure of the heart itself.

A Doppler ultrasound image can show the motion of the tissues of the heart and of blood within the heart.

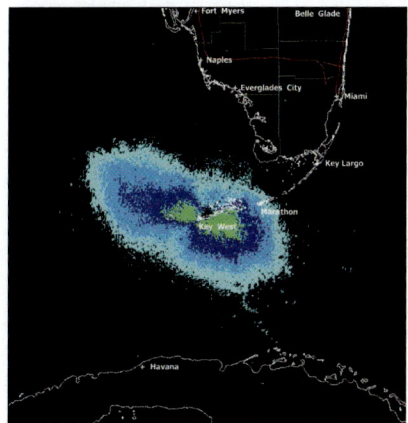

Keeping track of wildlife with radar
Doppler radar is tuned to measure only those reflected radio waves that have a frequency shift, eliminating reflections from stationary objects and showing only objects in motion. Televised Doppler radar images of storms are made using radio waves reflected from moving water droplets. But this Doppler radar image of an area off the tip of Florida was made on a clear night with no rain. The blue and green patch isn't a moving storm, it's a moving flock of birds. Migratory birds move at high altitudes at night, so Doppler radar makes an excellent tool for analyzing their movements.

EXAMPLE 15.13 Ultrasound frequency to measure blood flow
A biomedical engineer is designing a Doppler blood flow meter to measure blood flow in an artery where a typical flow speed is known to be 0.60 m/s. What ultrasound frequency should she use to produce a frequency shift of 1500 Hz when this flow is detected?

SOLVE We can rewrite Equation 15.18 to calculate the frequency of the emitter:

$$f_0 = \left(\frac{\Delta f}{2}\right)\left(\frac{v}{v_{\text{cell}}}\right)$$

The values on the right side are all known. Thus the required ultrasound frequency is

$$f_0 = \left(\frac{1500 \text{ Hz}}{2}\right)\left(\frac{1540 \text{ m/s}}{0.60 \text{ m/s}}\right) = 1.9 \text{ MHz}$$

ASSESS Doppler units to measure blood flow in deep tissues actually work at about 2.0 MHz, so our answer is reasonable.

Frequency shift on reflection is observed for all types of waves, not just sound waves. Radar (which stands for "**ra**dio **d**etection **an**d **r**anging") units work by emitting radio waves and observing the waves that are reflected from objects. The

The source of waves is moving to the right at v_s. The positions at times $t = 0, t = T, t = 2T \ldots$ are marked.

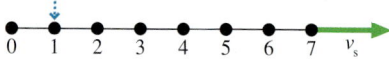

At each point, the source emits a wave that spreads out. A snapshot is taken at $t = 7T$.

Crest 3 was emitted at $t = 3T$, $4T$ before the snapshot. The wave front is a circle of radius 4λ centered on point 3.

The source speed is greater than the wave speed, so during the time $t = 4T$ since crest 3 was emitted, the source has moved farther than crest 3.

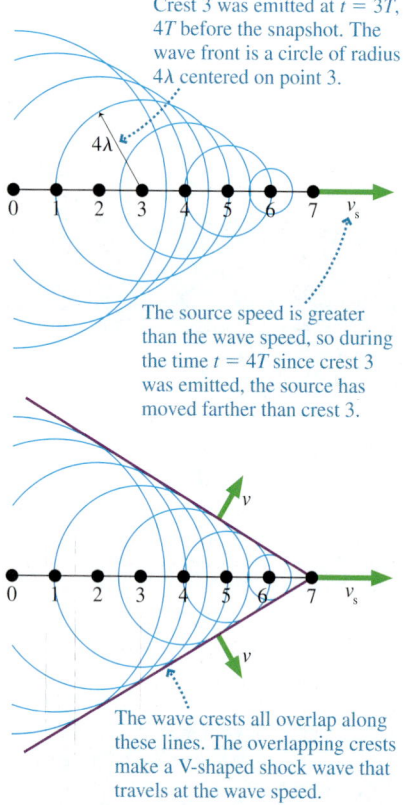

The wave crests all overlap along these lines. The overlapping crests make a V-shaped shock wave that travels at the wave speed.

FIGURE 15.22 Waves emitted by a source traveling faster than the speed of the waves in a medium.

time between the emission of a pulse and its return gives an object's position. The change in frequency of the returned pulse gives the object's speed. This is the principle behind the radar guns used by traffic police, as well as the Doppler radar images you have seen in weather reports.

Shock Waves

When sound waves are emitted by a moving source, the frequency is shifted, as we have seen. Now, let's look at what happens when the source speed v_s exceeds the wave speed v and the source "outruns" the waves it produces.

Earlier in this section, in Figure 15.21, we looked at a motion diagram of waves emitted by a moving source. Figure 15.22 is the same diagram, but in this case the source is moving faster than the waves. This motion causes the waves to overlap. (Compare Figure 15.22 to Figure 15.21.) The amplitudes of the overlapping waves add up to produce a very large amplitude wave—a **shock wave.**

Anything that moves faster than the speed of sound in air will create such a shock wave. The speed of sound was broken on land in 1997 by a British team driving the Thrust SCC in Nevada's Black Rock Desert. Figure 15.23a shows the shock waves being produced during a **supersonic** (faster than the speed of sound) run. You can see distortion of the landscape behind the car where the waves add to produce regions of high pressure. This shock wave travels along with the car. If the car went by you, the passing of the shock wave would produce a **sonic boom,** the distinctive loud sound you may have heard when a supersonic aircraft passes. The crack of a whip is a sonic boom as well, though on a much smaller scale.

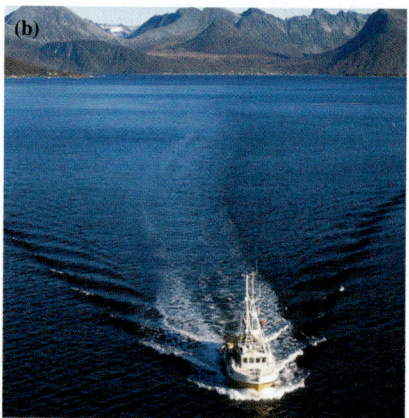

FIGURE 15.23 Extreme and everyday examples of shock waves.

Shock waves are not necessarily an extreme phenomenon. A boat (or even a duck!) can easily travel faster than the speed of the water waves it creates; the resulting wake, with its characteristic "V" shape as shown in Figure 15.23b, is really a shock wave.

STOP TO THINK 15.5   Amy and Zack are both listening to the source of sound waves that is moving to the right. Compare the frequencies each hears.

A. $f_{Amy} > f_{Zack}$
B. $f_{Amy} = f_{Zack}$
C. $f_{Amy} < f_{Zack}$

Amy f_0 Zack

The goal of Chapter 15 has been to learn the basic properties of traveling waves.

GENERAL PRINCIPLES

The Wave Model

This model is based on the idea of a traveling wave, which is an organized disturbance traveling at a well-defined **wave speed** v.

- In transverse waves the particles of the medium move *perpendicular* to the direction in which the wave travels.

- In longitudinal waves the particles of the medium move *parallel* to the direction in which the wave travels.

A wave transfers energy, but there is no material or substance transferred.

Mechanical waves require a material **medium.** The speed of the wave is a property of the medium, not the wave. The speed does not depend on the size or shape of the wave.

- For a **wave on a string,** the string is the medium.

$$v_{string} = \sqrt{\frac{T_s}{\mu}}$$

- A **sound wave** is a wave of compressions and rarefactions of a medium such as air.

In a gas:

$$v_{sound} = \sqrt{\frac{\gamma R T}{M}}$$

Electromagnetic waves are waves of the electromagnetic field. They do not require a medium. All electromagnetic waves travel at the same speed in a vacuum, $c = 3.00 \times 10^8$ m/s.

IMPORTANT CONCEPTS

Graphical representation of waves

A snapshot graph is a picture of a wave at one instant in time. For a periodic wave, the **wavelength** λ is the distance between crests.

Fixed t:

A history graph is a graph of the displacement of one point in a medium versus time. For a periodic wave, the **period** T is the time between crests.

Fixed x:

Mathematical representation of waves

Sinusoidal waves are produced by a source moving with simple harmonic motion. The equation for a sinusoidal wave is a function of position and time.

$$y(x, t) = A \cos\left(2\pi\left(\frac{x}{\lambda} \pm \frac{t}{T}\right)\right)$$

+: wave travels to left
−: wave travels to right

For sinusoidal and other periodic waves:

$$T = \frac{1}{f} \qquad v = f\lambda$$

The intensity of a wave is the ration of the power to the area.

$$I = P/A$$

For a **spherical wave** the power decreases with the surface area of the spherical **wave fronts:**

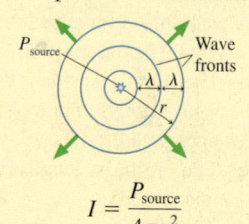

$$I = \frac{P_{source}}{4\pi r^2}$$

APPLICATIONS

The loudness of a sound is given by the sound intensity level. This is a logarithmic function of intensity and is in units of **decibels.**

- The usual **reference level** is the quietest sound that can be heard:

$$I_0 = 1.0 \times 10^{-12} \text{ W/m}^2$$

- The sound intensity level in dB is computed relative to this value:

$$\beta = (10 \text{ dB}) \log_{10}\left(\frac{I}{I_0}\right)$$

- A sound at the reference level corresponds to 0 dB.

The Doppler effect is a shift in frequency when there is relative motion of a wave source (frequency f_0, wave speed v) and an observer.

Moving source, stationary observer:

Receding source:

$$f_- = \frac{f_0}{1 + v_s/v}$$

Approaching source:

$$f_+ = \frac{f_0}{1 - v_s/v}$$

Moving observer, stationary source:

Approaching the source:

$$f_+ = (1 + v_o/v)f_0$$

Moving away from the source:

$$f_- = (1 - v_o/v)f_0$$

Reflection from a moving object:

For $v_o \ll v$, $\Delta f = \pm 2f_0 \dfrac{v_o}{v}$

When an object moves faster than the wave speed in a medium, a shock wave is formed.

For instructor-assigned homework, go to www.masteringphysics.com

Problem difficulty is labeled as | (straightforward) to ||||| (challenging).

Problems labeled INT integrate significant material from earlier chapters; BIO are of biological or medical interest.

QUESTIONS

Conceptual Questions

1. a. In your own words, define what a *transverse wave* is.
    b. Give an example of a wave that, from your own experience, you know is a transverse wave. What observations or evidence tells you this is a transverse wave?
2. a. In your own words, define what a *longitudinal wave* is.
    b. Give an example of a wave that, from your own experience, you know is a longitudinal wave. What observations or evidence tells you this is a longitudinal wave?
3. The wave pulses shown in Figure Q15.3 travel along the same string. Rank in order, from largest to smallest, their wave speeds v_1, v_2, and v_3. Explain.

FIGURE Q15.3

4. Is it ever possible for one sound wave in air to overtake and pass another? Explain.
5. A wave pulse travels along a string at a speed of 200 cm/s. What will be the speed if:
    a. The string's tension is doubled?
    b. The string's mass is quadrupled (but its length is unchanged)?
    c. The string's length is quadrupled (but its mass is unchanged)?
    d. The string's mass and length are both quadrupled?
    Note that parts a–d are independent and refer to changes made to the original string.
6. An ultrasonic range finder sends out a pulse of ultrasound and measures the time between the emission of the pulse and the return of an echo from an object. This time is used to determine the distance to the object. To get good accuracy from the device, a user must enter the air temperature in the room. Why is this?
7. When water freezes, the density decreases and the bonds between molecules become stronger. Do you expect the speed of sound to be greater in liquid water or in water ice?
8. Figure Q15.8 is a history graph showing the displacement as a function of time at one point on a string. Did the displacement at this point reach its maximum of 2 mm *before* or *after* the interval of time when the displacement was a constant 1 mm? Explain how you interpreted the graph to answer this question.

y (mm)

FIGURE Q15.8

9. Figure Q15.9 shows a history graph *and* a snapshot graph for a wave pulse on a string. They describe the same wave from two perspectives.
    a. In which direction is the wave traveling? Explain.
    b. What is the speed of this wave?

Snapshot at $t = 0.01$ s

History at $x = 2$ cm

FIGURE Q15.9

10. Rank in order, from largest to smallest, the wavelengths λ_1 to λ_3 for sound waves having frequencies $f_1 = 100$ Hz, $f_2 = 1000$ Hz, and $f_3 = 10,000$ Hz. Explain.
11. Explain why there is a factor of 2π in Equation 15.5.
12. A laser beam has intensity I_0.
    a. What is the intensity, in terms of I_0, if a lens focuses the laser beam to 1/10 its initial diameter?
    b. What is the intensity, in terms of I_0, if a lens defocuses the laser beam to 10 times its initial diameter?
13. Sound wave A delivers 2 J of energy in 2 s. Sound wave B delivers 10 J of energy in 5 s. Sound wave C delivers 2 mJ of energy in 1 ms. Rank in order, from largest to smallest, the sound powers P_A, P_B, and P_C of these three waves. Explain.
14. When you want to "snap" a towel, the best way to wrap the towel is so that the end that you hold and shake is thick, and the far end is thin. When you shake the thick end, a wave travels down the towel. How does wrapping the towel in a tapered shape help make for a good snap?
15. The volume control on your stereo is likely designed so that each time you turn it by one click, the loudness increases by a certain number of dB. Does each click increase the output power by a fixed amount as well?
16. You are standing at $x = 0$ m, listening to seven identical sound sources described by Figure Q15.16. At $t = 0$ s, all seven are at $x = 343$ m and moving as shown below. The sound from all seven will reach your ear at $t = 1$ s. Rank in order, from highest to lowest, the seven frequencies f_1 to f_7 that you hear at $t = 1$ s. Explain.

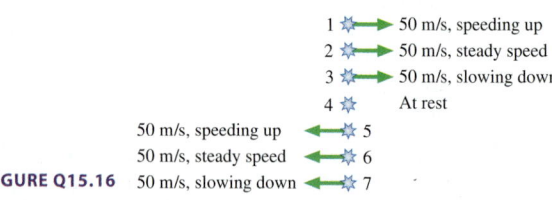

FIGURE Q15.16

17. A bullet can travel at speeds of over 1000 m/s. When a bullet is fired from a rifle, the actual firing makes a distinctive sound, but people at a distance may hear a second, different sound that is even louder. Explain the source of this sound.

Multiple-Choice Questions

18. | Denver, Colorado has an oldies station that calls itself "KOOL 105." This means that they broadcast radio waves at a frequency of 105 MHz. Suppose that they decide to describe their station by its wavelength (in meters), instead of by its frequency. What name would they now use?
    A. KOOL 0.35
    B. KOOL 2.85
    C. KOOL 3.5
    D. KOOL 285

19. | What is the frequency of blue light with a wavelength of 400 nm?
    A. 1.33×10^3 Hz
    B. 7.50×10^{12} Hz
    C. 1.33×10^{14} Hz
    D. 7.50×10^{14} Hz

20. | Ultrasound can be used to deliver energy to tissues for ther-
    BIO apy. It can penetrate tissue to a depth approximately 200 times its wavelength. What is the approximate depth of penetration of ultrasound at a frequency of 5.0 MHz?
    A. 0.29 mm
    B. 1.4 cm
    C. 6.2 cm
    D. 17 cm

21. ‖ A sinusoidal wave traveling on a string has a period of 0.20 s, a wavelength of 32 cm, and an amplitude of 3 cm. The speed of this wave is
    A. 0.60 cm/s
    B. 6.4 cm/s
    C. 15 cm/s
    D. 160 cm/s

22. ‖ Two strings of different linear density are joined together and pulled taut. A sinusoidal wave on these strings is traveling to the right, as shown in Figure Q15.22. When the wave goes across the boundary from string 1 to string 2:

FIGURE Q15.22

    A. The velocity increases.
    B. The velocity stays the same.
    C. The velocity decreases.

23. ‖ You stand at $x = 0$ m, listening to a sound that is emitted at frequency f_0. Figure Q15.23 shows the frequency you hear during a four-second interval. Which of the following describes the motion of the sound source?

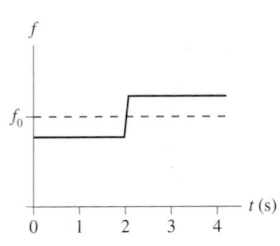

FIGURE Q15.23

    A. It moves from left to right and passes you at $t = 2$ s.
    B. It moves from right to left and passes you at $t = 2$ s.
    C. It moves toward you but doesn't reach you. It then reverses direction at $t = 2$ s.
    D. It moves away from you until $t = 2$ s. It then reverses direction and moves toward you but doesn't reach you.

PROBLEMS

Section 15.1 The Wave Model

Section 15.2 Traveling Waves

1. | The wave speed on a string under tension is 200 m/s. What is the speed if the tension is doubled?
2. | The wave speed on a string is 150 m/s when the tension is 75.0 N. What tension will give a speed of 180 m/s?
3. | A wave travels along a string at a speed of 280 m/s. What will be the speed if the string is replaced by one made of the same material and under the same tension but having twice the radius?
4. | The back wall of an auditorium is 26.0 m from the stage. If you are seated in the middle row, how much time elapses between a sound from the stage reaching your ear directly and the same sound reaching your ear after reflecting from the back wall?
5. ‖‖ A hammer taps on the end of a 4.00-m-long metal bar at room temperature. A microphone at the other end of the bar picks up two pulses of sound, one that travels through the metal and one that travels through the air. The pulses are separated in time by 11.0 ms. What is the speed of sound in this metal?
6. | In an early test of sound propagation through the ocean, an underwater explosion of 1 pound of dynamite in the Bahamas was detected 3200 km away on the coast of Africa. How much time elapsed between the explosion and the detection?

Section 15.3 Graphical and Mathematical Descriptions of Waves

7. | Figure P15.7 is a snapshot graph of a wave at $t = 0$ s. Draw the history graph for this wave at $x = 6$ m, for $t = 0$ s to 6 s.

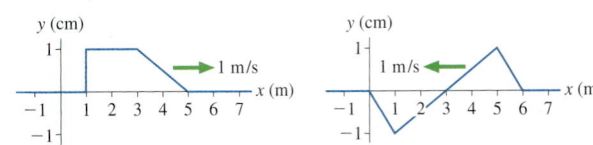

FIGURE P15.7   **FIGURE P15.8**

8. | Figure P15.8 is a snapshot graph of a wave at $t = 2$ s. Draw the history graph for this wave at $x = 0$ m, for $t = 1$ s to 8 s.
9. | Figure P15.9 is a history graph at $x = 0$ m of a wave moving to the right at 1 m/s. Draw a snapshot graph of this wave at $t = 1$ s.

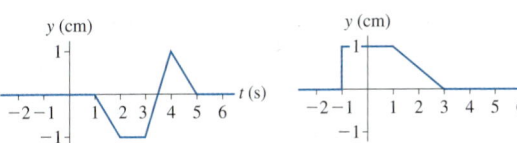

FIGURE P15.9   **FIGURE P15.10**

10. | Figure P15.10 is a history graph at $x = 2$ m of a wave moving to the left at 1 m/s. Draw the snapshot graph of this wave at $t = 0$ s.

11. | A sinusoidal wave has period 0.20 s and wavelength 2.0 m. What is the wave speed?

12. | A sinusoidal wave travels with speed 200 m/s. Its wavelength is 4.0 m. What is its frequency?

13. || The motion detector used in a physics lab sends and receives 40 kHz ultrasonic pulses. A pulse goes out, reflects off the object being measured, and returns to the detector. The lab temperature is 20°C.
    a. What is the wavelength of the waves emitted by the motion detector?
    b. How long does it take for a pulse that reflects off an object 2.5 m away to make a round trip?

14. | The displacement of a wave traveling in the positive x-direction is $y(x, t) = (3.5 \text{ cm}) \cos(2.7x - 124t)$, where x is in m and t is in s. What are the (a) frequency, (b) wavelength, and (c) speed of this wave?

15. || The displacement of a wave traveling in the negative x-direction is $y(x, t) = (5.2 \text{ cm}) \cos(5.5x + 72t)$, where x is in m and t is in s. What are the (a) frequency, (b) wavelength, and (c) speed of this wave?

16. | Figure P15.16 is a snapshot graph at $t = \frac{1}{4}$ s of a traveling wave with displacement $y(x,t) = (2.0 \text{ cm}) \times \cos(2\pi x - 4\pi t)$, where x is in m and t is in s.

FIGURE P15.16

    a. Reproduce this graph on your page, then use a *dotted* line to show the snapshot graph at $t = 0$ s and a *dashed* line to show the snapshot graph at $t = \frac{1}{8}$ s.
    b. What is the speed of this wave?

17. | Figure P15.17 is a snapshot graph of a wave at $t = 0$ s. What are the amplitude, wavelength, and frequency of this wave?

FIGURE P15.17   **FIGURE P15.18**

18. | Figure P15.18 is a history graph at $x = 0$ m of a wave moving to the right at 2 m/s. What are the amplitude, frequency, and wavelength of this wave?

Section 15.4 Sound and Light Waves

19. | A dolphin emits ultrasound at 100 kHz and uses the timing
BIO of reflections to determine the position of objects in the water. What is the wavelength of this ultrasound?

20. || a. What is the wavelength of a 2.0 MHz ultrasound wave traveling through aluminum?
    b. What frequency of electromagnetic wave would have the same wavelength as the ultrasound wave of part a?

21. | a. At 20°C, what is the frequency of a sound wave in air with a wavelength of 20 cm?
    b. What is the frequency of an electromagnetic wave with a wavelength of 20 cm?
    c. What would be the wavelength of a sound wave in water that has the same frequency as the electromagnetic wave of part b?

22. | a. What is the frequency of blue light that has a wavelength of 450 nm?
    b. What is the frequency of red light that has a wavelength of 650 nm?

23. || a. Telephone signals are often transmitted over long distances by microwaves. What is the frequency of microwave radiation with a wavelength of 3.0 cm?
    b. Microwave signals are beamed between two mountaintops 50 km apart. How long does it take a signal to travel from one mountaintop to the other?

24. || a. An FM radio station broadcasts at a frequency of 101.3 MHz. What is the wavelength?
    b. What is the frequency of a sound source that produces the same wavelength in 20°C air?

Section 15.5 Energy and Intensity

25. || Sound is detected when a sound wave causes the eardrum
BIO to vibrate (see Figure 14.27). Typically, the diameter of the
INT eardrum is about 8.4 mm in humans. How much energy is delivered to your eardrum each second when someone whispers (20 dB) a secret in your ear?

26. || A fan at a rock concert is 30 m from the stage, where the
BIO sound intensity level is 110 dB. Her eardrums have a diameter
INT of 8.4 mm. How much energy is transferred to each eardrum in one second?

27. ||| The intensity of electromagnetic waves from the sun is 1.4 kW/m² just above the earth's atmosphere. Eighty percent of this reaches the surface at noon on a clear summer day. Suppose you model your back as a 30 cm × 50 cm rectangle. How many joules of solar energy fall on your back as you work on your tan for 1.0 hr?

28. | The sun emits electromagnetic waves with a power of 4.0×10^{26} W. Determine the intensity of electromagnetic waves from the sun just outside the atmospheres of (a) Venus, (b) Mars, and (c) Saturn. Refer to the table of astronomical data on the inside back cover.

29. | On the other side of the Milky Way, Planet M orbits a star at a distance of 1.31×10^{11} m. Planet N orbits at a distance of 1.49×10^{12} m. Light from the star has an intensity of 1840 W/m² at Planet M. What is the intensity of the light at Planet N?

Section 15.6 Loudness of Sound

30. | What is the sound intensity level of a sound with an intensity of 3.0×10^{-6} W/m²?

31. || What is the sound intensity of a whisper at a distance of 2.0 m, in W/m²? What is the corresponding sound intensity level in dB?

32. | The sound intensity from a jack hammer breaking concrete is 2.0 W/m² at a distance of 2.0 m from the point of impact. This is sufficiently loud to cause permanent hearing damage if the operator doesn't wear ear protection. What are (a) the sound intensity and (b) the sound intensity level for a person watching from 50 m away?

33. || A concert loudspeaker suspended high off the ground emits 35 W of sound power. A small microphone with a 1.0 cm² area is 50 m from the speaker. What are (a) the sound intensity and (b) the sound intensity level at the position of the microphone?

34. | A rock band playing an outdoor concert produces sound at 120 dB 5.0 m away from their single working loudspeaker. What is the sound intensity level 35 m from the speaker?

Section 15.7 The Doppler Effect and Shock Waves

35. | An opera singer in a convertible sings a note at 600 Hz while cruising down the highway at 90 km/hr. What is the frequency heard by
   a. A person standing beside the road in front of the car?
   b. A person standing beside the road behind the car?

36. || An osprey's call is a distinct whistle at 2200 Hz. An osprey
   BIO calls while diving at you, to drive you away from her nest. You hear the call at 2300 Hz. How fast is the osprey approaching?

37. | A whistle you use to call your hunting dog has a frequency of 21 kHz, but your dog is ignoring it. You suspect the whistle may not be working, but you can't hear sounds above 20 kHz. To test it, you ask a friend to blow the whistle, then you hop on your bicycle. In which direction should you ride (toward or away from your friend) and at what minimum speed to know if the whistle is working?

38. | A friend of yours is loudly singing a single note at 400 Hz while driving toward you at 25.0 m/s on a day when the speed of sound is 340 m/s.
   a. What frequency do you hear?
   b. What frequency does your friend hear if you suddenly start singing at 400 Hz?

39. || While anchored in the middle of a lake, you count exactly three waves hitting your boat every 10 s. You raise anchor and start motoring slowly in the same direction the waves are going. When traveling at 1.5 m/s, you notice that exactly two waves are hitting the boat from behind every 10 s. What is the speed of the waves on the lake?

40. || A Doppler blood flow unit emits ultrasound at 5.0 MHz.
   BIO What is the frequency shift of the ultrasound reflected from blood moving in an artery at a speed of 0.20 m/s?

General Problems

41. | You're watching a carpenter pound a nail. He hits the nail twice a second, but you hear the sound of the strike when his hammer is fully raised. What is the minimum distance from you to the carpenter? Assume the air temperature is 20°C.

42. ||| A 2.50 kHz sound wave is transmitted through an aluminum rod.
   a. What is its wavelength in the aluminum?
   b. What is the sound wave's frequency when it passes into the air?
   c. What is its wavelength in air?

43. ||| Oil explorers set off explosives to make loud sounds, then listen for the echoes from underground oil deposits. Geologists suspect that there is oil under 500-m-deep Lake Physics. It's known that Lake Physics is carved out of a granite basin. Explorers detect a weak echo 0.94 s after exploding dynamite at the lake surface. If it's really oil, how deep will they have to drill into the granite to reach it?

44. || A 2.0-m-long string is under 20 N of tension. A pulse travels the length of the string in 50 ms. What is the mass of the string?

45. || Andy (mass 80 kg) uses a 3.0-m-long rope to pull Bob
   INT (mass 60 kg) across the floor ($\mu_k = 0.20$) at a constant speed of 1.0 m/s. Bob signals to Andy to stop by "plucking" the rope, sending a wave pulse forward along the rope. The pulse reaches Andy 150 ms later. What is the mass of the rope?

46. | If a bungee cord is stretched horizontally to a length of 2.5 m, the tension in the cord is 2.1 N. A transverse pulse on the cord makes a round trip in 1.6 s. A pulse makes the round trip in 1.6 s if the cord is stretched to a length of 3.5 m. What is the tension in the 3.5-m-long cord?

47. || String 1 in Figure P15.47 has linear density 2.0 g/m and string 2 has linear density 4.0 g/m. A student sends pulses in both directions by quickly pulling up on the knot, then releasing it. What should the string lengths L_1 and L_2 be if the pulses are to reach the ends of the strings simultaneously?

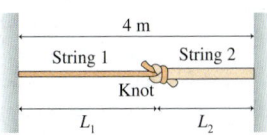

FIGURE P15.47

48. | Ships measure the distance to the ocean bottom with *sonar.* A pulse of sound waves is aimed at the ocean bottom, then sensitive microphones listen for the echo. The graph in Figure P15.48 shows the delay time as a function of the ship's position as it crosses 60 km of ocean. Draw a graph of the ocean bottom. Let the ocean surface define $y = 0$ and ocean bottom have negative values of y. This way your graph will be a picture of the ocean bottom. The speed of sound in ocean water varies slightly with temperature, but you can use 1500 m/s as an average value.

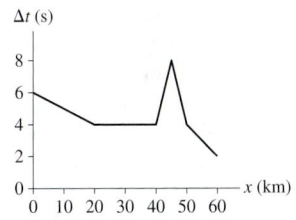

FIGURE P15.48

49. ||| A long string is shaken at its left end at 5.0 Hz, sending a sinusoidal wave down the string at 2.0 m/s. After the wave has traveled 1.0 m, the linear density of the string suddenly decreases by a factor of four. Draw a snapshot graph of the first 3.0 m of the string at an instant when the left end is at its equilibrium position.

50. ||| A coyote can locate a sound source with good accuracy by
   BIO comparing the arrival times of a sound wave at its two ears. Suppose a coyote is listening to a bird whistling at 1000 Hz. The bird is 3.0 m away, directly in front of the coyote's right ear. The coyote's ears are 15 cm apart.
   a. What is the distance between the bird and the coyote's left ear?
   b. What is the difference in the arrival time of the sound at the left ear and the right ear?
   c. What is the ratio of this time difference to the period of the sound wave?
   Hint: You are looking for the difference between two numbers that are nearly the same. What does this near equality imply about the necessary precision during intermediate stages of the calculation?

51. | A 256 Hz sound wave in 20°C air propagates into the water of a swimming pool. What are the water-to-air ratios of the wave's frequency, wave speed, and wavelength?

52. ‖ Earthquakes produce seismic waves that travel through the earth. The speed of longitudinal seismic waves, called P waves, is 8000 m/s. Transverse seismic waves, called S waves, travel at a slower 4500 m/s. A seismograph records the two waves from a distant earthquake. If the S wave arrives 2.0 min after the P wave, how far away was the earthquake? You can assume that the waves travel in straight lines, although actual seismic waves follow more complex routes.

53. ‖‖ One way to monitor global warming is to measure the average temperature of the ocean. Researchers are doing this by measuring the time it takes sound pulses to travel underwater over large distances. At a depth of 1000 m, where ocean temperatures hold steady near 4°C, the average sound speed is 1480 m/s. It's known from laboratory measurements that the sound speed increases 4.0 m/s for every 1.0°C increase in temperature. In one experiment, where sounds generated near California are detected in the South Pacific, the sound waves travel 8000 km. If the smallest time change that can be reliably detected is 1.0 s, what is the smallest change in average temperature that can be measured?

54. ‖ Figure P15.54 shows two snapshot graphs taken 10 ms apart, with the blue curve being the first snapshot. What are the (a) wavelength, (b) speed, (c) frequency, and (d) amplitude of this wave?

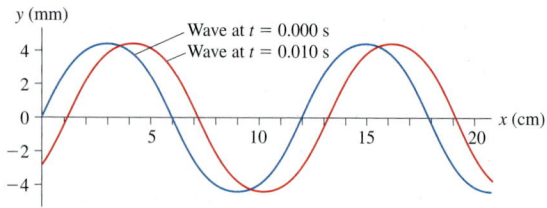

FIGURE P15.54

55. | Low-frequency vertical oscillations are one possible cause
BIO of motion sickness, with 0.3 Hz having the strongest effect. Your boat is bobbing in place at just the right frequency to cause you the maximum discomfort.
   a. How much time elapses between two waves hitting the ship?
   b. If the wave crests appear to be about 30 m apart, what would you estimate to be the speed of the waves?

56. | A wave on a string is described by $y(x, t) = (3.0 \text{ cm}) \times \cos[2\pi(x/(2.4 \text{ m}) + t/(0.20 \text{ s}))]$, where x is in m and t in s.
   a. In what direction is this wave traveling?
   b. What are the wave speed, frequency, and wavelength?
   c. At $t = 0.50$ s, what is the displacement of the string at $x = 0.20$ m?

57. | Write the y-equation for a wave traveling in the negative x-direction with wavelength 50 cm, speed 4.0 m/s, and amplitude 5.0 cm.

58. | Write the y-equation for a wave traveling in the positive x-direction with frequency 200 Hz, speed 400 m/s, and amplitude 0.010 mm.

59. | A wave is described by the expression $y(x, t) = (3.0 \text{ cm}) \times \cos(1.5x - 50t)$, where x is in m and t is in s.
   a. Make an accurate snapshot graph of this wave.
   b. What is the speed of the wave and in what direction is it traveling?

60. ‖‖ A point on a string undergoes simple harmonic motion as a
INT sinusoidal wave passes. When a sinusoidal wave with speed 24 m/s, wavelength 30 cm, and amplitude of 1.0 cm passes, what is the maximum speed of a point on the string?

61. | A simple pendulum is
INT made by attaching a small cup of sand with a hole in the bottom to a 1.2 m long string. The pendulum is mounted on the back of a small motorized car. As the car drives forward, the pendulum swings from side to side and leaves a trail of sand as shown in Figure P15.61. How fast was the car moving?

FIGURE P15.61

62. ‖ a. A typical 100 W lightbulb produces 4.0 W of visible light. (The other 96 W are dissipated as heat and infrared radiation.) What is the light intensity on a wall 2.0 m away from the lightbulb?
   b. A krypton laser produces a cylindrical red laser beam 2.0 mm in diameter with 2.0 W of power. What is the light intensity on a wall 2.0 m away from the laser?

63. ‖ An AM radio station broadcasts with a power of 25 kW at a frequency of 920 kHz. Estimate the intensity of the radio wave at a point 10 km from the broadcast antenna.

64. ‖‖ The earth's average distance from the sun is 1.50×10^{11} m. At this distance, the intensity of radiation from the sun is 1.38 kW/m². The earth's radius is 6.37×10^6 m. What is the total solar power received by the earth? (For comparison, total human power consumption is roughly 10^{13} W.)

65. ‖‖ Lasers can be used to drill or cut material. One such laser generates a series of high-intensity pulses rather than a continuous beam of light. Each pulse contains 500 mJ of energy and lasts 10 ns. The laser fires 10 such pulses per second.
   a. What is the *peak power* of the laser light? The peak power is the power output during one of the 10 ns pulses.
   b. What is the average power output of the laser? The average power is the total energy delivered per second.
   c. A lens focuses the laser beam to a 10-μm-diameter circle on the target. During a pulse, what is the light intensity on the target?
   d. The intensity of sunlight at the surface of the earth at midday is about 1100 W/m². What is the ratio of the laser intensity on the target to the intensity of the midday sun?

66. ‖ The sound intensity 50 m from a wailing tornado siren is 0.10 W/m².
   a. What is the sound intensity level?
   b. In a noisy neighborhood, the weakest sound likely to be heard over background noise is 60 dB. Estimate the maximum distance at which the siren can be heard.

67. ‖ A stereo is claimed to put out 150 W per channel, for a total of 300 W. This is electrical power to the speakers; the actual sound output is much less. Suppose a stereo loudspeaker actually did put out 300 W of sound power, radiated equally in all directions. What would be the sound intensity level 5.0 m from the speaker?

68. ‖ Five trumpet players have a sound intensity level of 75 dB. What would be the sound intensity level of 20 such trumpet players?

69. ‖ A speaker at an open-air concert emits 600 W of sound power, radiated equally in all directions.
    a. What is the intensity of the sound 5.00 m from the speaker?
    b. What sound intensity level would you experience there if you did not have any protection for your ears?
    c. Earplugs you can buy in the drugstore have a noise reduction rating of 23 decibels. If you are wearing those earplugs but your friend Phil is not, how far from the speaker should Phil stand to experience the same loudness as you?

70. | A bat locates insects by emitting ultrasonic "chirps" and
BIO then listening for echoes. The lowest-frequency chirp of a big brown bat is 26 kHz. How fast would the bat have to fly, and in what direction, for you to just barely be able to hear the chirp at 20 kHz?

71. | A physics professor demonstrates the Doppler effect by
INT tying a 600 Hz sound generator to a 1.0-m-long rope and whirling it around her head in a horizontal circle at 100 rpm. What are the highest and lowest frequencies heard by a student in the classroom? Assume the room temperature is 20°C.

72. | Ocean waves with wavelength 1.2 m and period 1.5 s are moving past a pier. A boy runs along the pier, in the direction opposite to the motion of the wave, at 3.5 m/s. How many wave crests pass the boy each second?

73. ‖ A source of sound moves toward you at speed v_s and away from Jane, who is standing on the other side of it. You hear the sound at twice the frequency as Jane. What is the speed of the source? Assume that the speed of sound is 340 m/s.

Passage Problems

Echolocation BIO

As discussed in the chapter, many species of bats find flying insects by emitting pulses of ultrasound and listening for the reflections. This technique is called **echolocation**. Bats possess several adaptations that allow them to echolocate very effectively.

74. | Although we can't hear them, the ultrasonic pulses are very loud. In order not be deafened by the sound they emit, bats can temporarily turn off their hearing. Muscles in the ear cause the bones in their middle ear to separate slightly, so that they don't transmit vibrations to the inner ear. After an ultra-sound pulse ends, a bat can hear an echo from an object 1 m away. Approximately how much time after a pulse is emitted is the bat ready to hear its echo?
    A. 0.5 ms   B. 1 ms   C. 3 ms   D. 6 ms

75. | Bats are sensitive to very small changes in frequency of the reflected waves. What information does this allow them to determine about their prey?
    A. Size   B. Speed   C. Distance   D. Species

76. | Some bats have specially shaped noses that they use to focus the ultrasound pulses in the forward direction. Why is this useful?
    A. They are not distracted by echoes from several directions.
    B. The energy of the pulse is concentrated in a smaller area, so the intensity is larger.
    C. The pulse goes forward only, so it doesn't affect the bat's hearing.

77. | Some bats utilize a sound pulse with a rapidly decreasing frequency. A decreasing-frequency pulse has
    A. Decreasing wavelength.
    B. Decreasing speed.
    C. Increasing wavelength.
    D. Increasing speed.

STOP TO THINK ANSWERS

Stop to Think 15.1: Transverse. The wave moves horizontally through the crowd, but individual spectators move up and down, transverse to the motion of the wave.

Stop to Think 15.2: D, E. Shaking your hand faster or farther will change the shape of the wave, but this will not change the wave speed; the speed is a property of the medium. Changing the linear density of the string or its tension will change the wave speed. To increase the speed, you must decrease the linear density or increase the tension.

Stop to Think 15.3: B. All three waves have the same speed, so the frequency is highest for the wave with the shortest wave-length. (Imagine the three waves moving to the right; the one with the crests closest together has the crests passing by most rapidly.)

Stop to Think 15.4: A. The plane wave does not spread out, so its intensity will be constant. The other two waves spread out, so their intensity will decrease.

Stop to Think 15.5: C. The source is moving toward Zack, so he observes a higher frequency. The source is moving away from Amy, so she observes a lower frequency.

16

SUPERPOSITION AND STANDING WAVES

The didgeridoo is a simple musical instrument played by aboriginal tribes in Australia. It consists of a hollow tube that is open at both ends. How can the player get a wide range of notes out of such a simple device?

A person playing a musical instrument such as a trumpet, flute, or didgeridoo makes sound waves. Listeners hear the traveling sound waves that the instrument emits, but the more important sound waves are the ones that stay inside the instrument's tube. These *standing waves* determine the sound that the instrument produces. Standing waves also exist in laser cavities and microwave ovens, and they are responsible for determining the sound of your voice.

Before we can talk about standing waves, we have to understand how two or more traveling waves interact with each other. What happens when two waves overlap? This combination is called a *superposition*. As with other wave concepts, the principle of superposition will apply to all waves that we study, from water waves to sound waves to light waves.

16.1 The Principle of Superposition

Figure 16.1a shows two baseball players, Alan and Bill, at batting practice. Unfortunately, someone has turned the pitching machines so that pitching machine A throws baseballs toward Bill while machine B throws toward Alan. If two baseballs are launched at the same time, and with the same speed, they collide at the crossing point and bounce away. Two baseballs cannot occupy the same point of space at the same time.

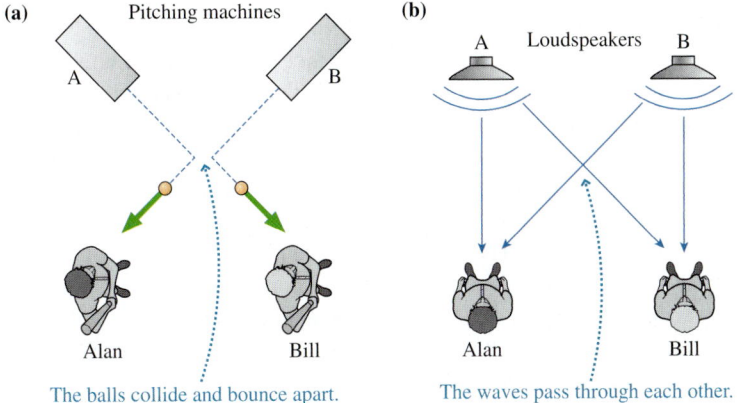

(a) Pitching machines   A   B   Alan   Bill
The balls collide and bounce apart.

(b) Loudspeakers   A   B   Alan   Bill
The waves pass through each other.

FIGURE 16.1 Two baseballs cannot pass through each other. Two waves can.

But unlike baseballs, sound waves *can* pass directly through each other. In Figure 16.1b, Alan and Bill are listening to the stereo system in the locker room after practice. Both hear the music quite well, without distortion or missing sound, so the sound wave that travels from speaker A toward Bill must pass through the wave traveling from speaker B toward Alan, with no effect on either. This is a basic property of waves.

What happens to the medium at a point where two waves are present simultaneously? What is the displacement of the medium at this point? Figure 16.2 shows a sequence of photos of two wave pulses traveling along a stretched string. In the first photo, the waves are approaching each other. In the second, the waves overlap, and the displacement of the string is larger than it was for either of the individual waves. A careful measurement would reveal that the displacement is the sum of the displacements of the two individual waves. In the third frame, the waves have passed through each other and continue on as if nothing had happened.

This result is not limited to stretched strings; the outcome is the same whenever two waves of any type pass through each other. This is known as the *principle of superposition*.

> **Principle of superposition** When two or more waves are *simultaneously* present at a single point in space, the displacement of the medium at that point is the sum of the displacements due to each individual wave.

To use the principle of superposition you must know the displacement that each wave would cause if it traveled through the medium alone. Then you go through the medium *point by point* and add the displacements due to each wave *at that point* to find the net displacement at that point. The outcome will be different at each and every point in the medium because the displacements are different at each point.

Let's illustrate this principle with an idealized example. Figure 16.3 shows five snapshot graphs taken 1 s apart of two waves traveling at the same speed (1 m/s) in opposite directions along a string. The principle of superposition comes into play wherever the waves overlap. The displacement of each wave is shown as a dotted line. The solid line is the sum *at each point* of the two displacements at that point. This is the displacement that you would actually observe as the two waves pass through each other.

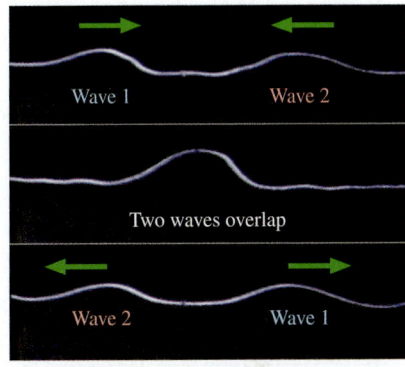

Wave 1   Wave 2

Two waves overlap

Wave 2   Wave 1

FIGURE 16.2 Two wave pulses on a stretched string pass through each other.

1 m/s   1 m/s
$t = 0$ s
x (m)
0 1 2 3 4 5 6 7
Two waves approach each other.

$t = 1$ s
x (m)
0 1 2 3 4 5 6 7

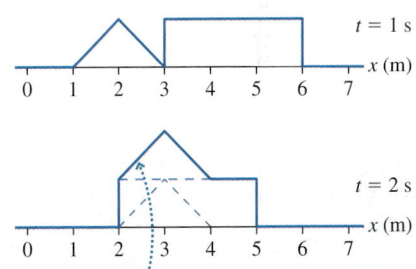

$t = 2$ s
x (m)
0 1 2 3 4 5 6 7
The net displacement is the point-by-point summation of the individual waves.

$t = 3$ s
x (m)
0 1 2 3 4 5 6 7

1 m/s   1 m/s
$t = 4$ s
x (m)
0 1 2 3 4 5 6 7
Both waves emerge unchanged.

FIGURE 16.3 The superposition of two waves on a string as they pass through each other.

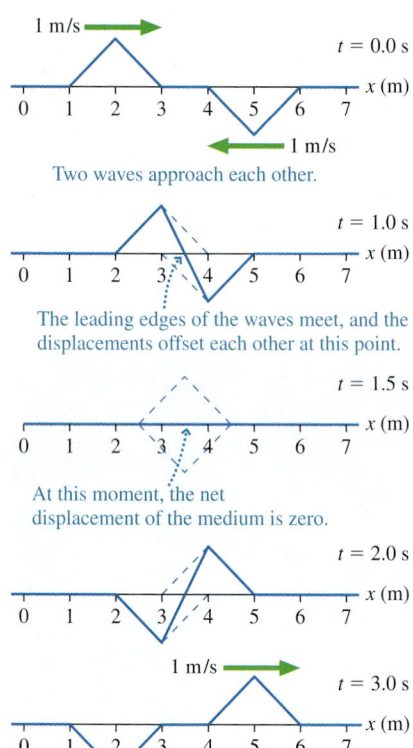

Two waves approach each other.

The leading edges of the waves meet, and the displacements offset each other at this point.

At this moment, the net displacement of the medium is zero.

Both waves emerge unchanged.

FIGURE 16.4 Two waves with opposite displacements produce destructive interference.

Constructive and Destructive Interference

The superposition of two waves is often called **interference.** The displacements of the waves in Figure 16.3 are both positive, so the total displacement of the medium where they overlap is larger than it would be due to either of the waves separately. We call this **constructive interference.**

Figure 16.4 shows another series of snapshot graphs of two counterpropagating waves, but this time one has a negative displacement. The principle of superposition still applies, but now the displacements are opposite each other. The displacement of the medium where the waves overlap is *less* than it would be due to either of the waves separately. We call this **destructive interference.**

In the series of graphs in Figure 16.4 the displacement of the medium at $x = 3.5$ m is always zero. The positive displacement of the wave traveling to the left and the negative displacement of the wave traveling to the right always exactly cancel at this spot. The complete cancellation at one point of two waves traveling in opposite directions is something we will see again. When the displacements of the waves cancel, where does the energy of the wave go? We know that the waves continue on unchanged after their interaction, so no energy is dissipated. Consider the graph at $t = 1.5$ s. There is no net displacement at any point of the medium at this instant, but the *string is moving rapidly.* The energy of the waves hasn't vanished—it is in the form of the kinetic energy of the medium.

STOP TO THINK 16.1   Two pulses on a string approach each other at speeds of 1 m/s. What is the shape of the string at $t = 6$ s?

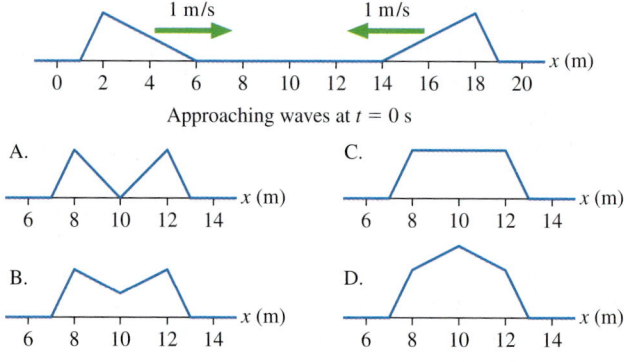

Approaching waves at $t = 0$ s

A. B. C. D.

◄ **Breaking stones with sound** BIO As we saw in Chapter 15, waves carry energy. The energy of high-intensity ultrasonic waves can be used to break up kidney stones so that they can be cleared from the body, a technique known as *shock wave lithotripsy.* This machine uses two generators, each of which produces ultrasonic waves. The two waves, which enter the body at different points, are directed so that they overlap and produce constructive interference at the position of a stone. This allows the individual waves to have lower intensity, minimizing tissue damage as they pass through the body, while still providing a region of high intensity right where it is needed.

16.2 Standing Waves

When you pluck a guitar string or a rubber band stretched between your fingers, you create waves. But how is this possible? There isn't really anywhere for the waves to go, because the string or the rubber band is held between two fixed ends. Figure 16.5 on the next page shows a photograph of waves on a stretched elastic cord. This is a wave, though it may not look like one, because it doesn't "travel" either right or left. Waves that are "trapped" between two boundaries, like those in the photo or on a guitar string, are what we call *standing waves.* **Individual points on the string oscillate up and down, but the wave itself does not travel.** It is called a **standing wave** because the crests and troughs "stand in place" as it oscillates. As we'll see, a standing wave isn't a totally new kind of wave; it is simply the superposition of two traveling waves moving in opposite directions.

Superposition Creates a Standing Wave

Suppose we have a string on which two sinusoidal waves of equal wavelength and amplitude travel in opposite directions, as in Figure 16.6a. When the waves meet, the displacement of the string will be a superposition of these two waves. Figure 16.6b shows nine snapshot graphs, at intervals of $\frac{1}{8}T$, of the two waves as they move through each other. The red and green dots identify particular crests on each of the waves to help you see that the red wave is traveling to the right and the green wave to the left. At *each point,* the net displacement of the medium is found by adding the red displacement and the green displacement. The resulting blue wave is the superposition of the two traveling waves.

FIGURE 16.5 The motion of a standing wave on a string.

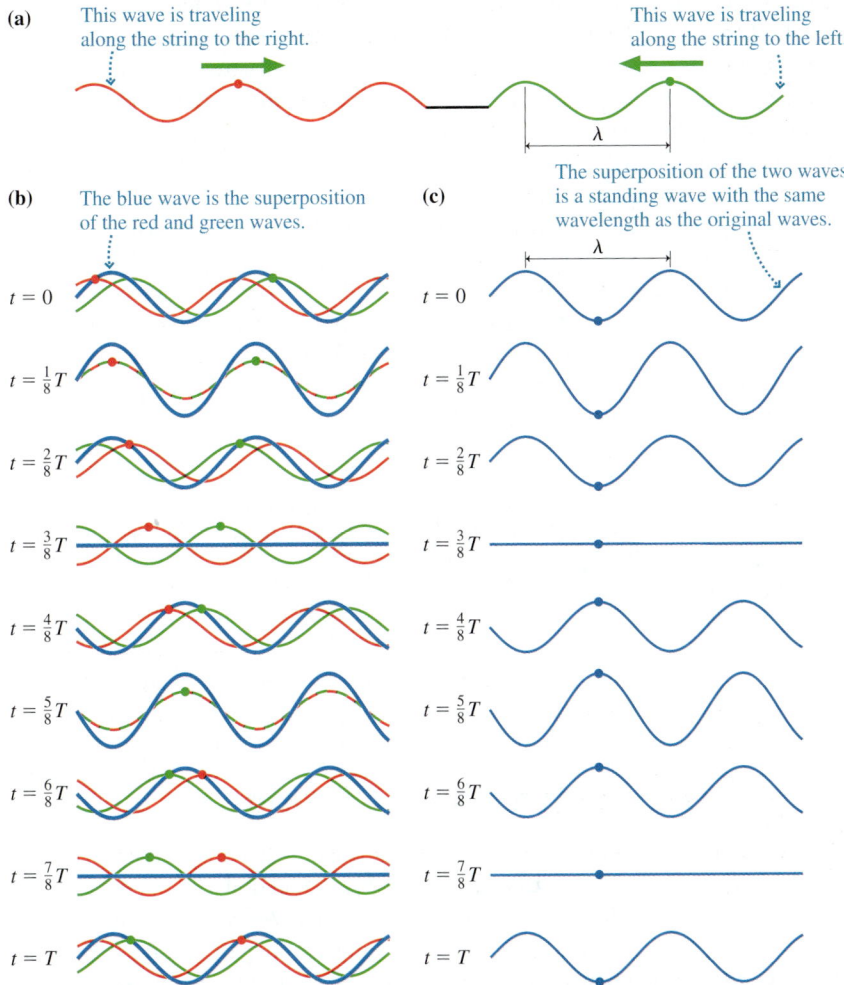

(a) This wave is traveling along the string to the right. This wave is traveling along the string to the left.

λ

(b) The blue wave is the superposition of the red and green waves.

(c) The superposition of the two waves is a standing wave with the same wavelength as the original waves.

λ

$t = 0$

$t = \frac{1}{8}T$

$t = \frac{2}{8}T$

$t = \frac{3}{8}T$

$t = \frac{4}{8}T$

$t = \frac{5}{8}T$

$t = \frac{6}{8}T$

$t = \frac{7}{8}T$

$t = T$

FIGURE 16.6 Two sinusoidal waves traveling in opposite directions.

Figure 16.6b is rather complicated, so Figure 16.6c shows just the blue superposition of the two waves. This is the wave that you would actually observe in the medium. Interestingly, the blue dot shows that the wave in Figure 16.6c is moving neither right nor left. The superposition of the two counterpropagating traveling waves is a standing wave. Notice that the wavelength of the standing wave, the distance between two crests or two troughs, is the same as the wavelengths of the two traveling waves that combine to produce it.

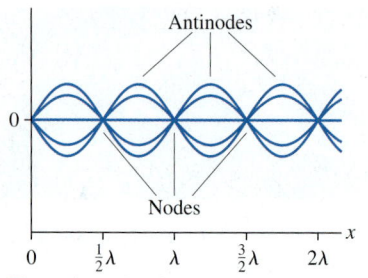

The nodes and antinodes are spaced λ/2 apart.

FIGURE 16.7 Standing waves are often represented as they would be seen in a strobe photograph, which makes the nodes and antinodes clearly visible.

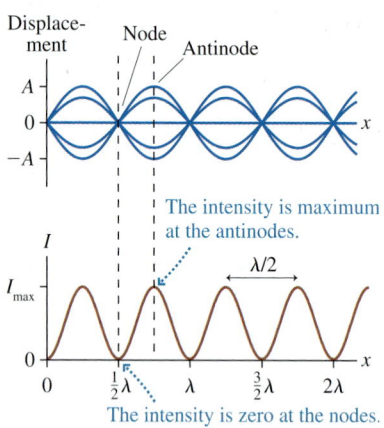

The intensity is maximum at the antinodes.

The intensity is zero at the nodes.

FIGURE 16.8 Intensity of a standing wave.

Nodes and Antinodes

In Figure 16.7 we have superimposed the nine snapshot graphs of Figure 16.6c into a single graphical representation of this standing wave. Compare this to Figure 16.5 and you can see why we called the vibrating string a standing wave. A striking feature of a standing-wave pattern is the existence of points that *never move!* These points, which are spaced λ/2 apart, are called **nodes.** Halfway between the nodes are points where the particles in the medium oscillate with maximum displacement. These points of maximum amplitude are called **antinodes,** and you can see that they are also spaced λ/2 apart. This means that **the wavelength of a standing wave is *twice* the distance between successive nodes or successive antinodes.**

It seems surprising and counterintuitive that some particles in the medium have no motion at all. This happens for the same reason we saw in Figure 16.4: The two waves exactly offset each other at that point. Look carefully at the two traveling waves in Figure 16.6b. You will see that the nodes occur at points where at *every instant* of time the displacements of the two traveling waves have equal magnitudes but *opposite signs.* Thus the superposition of the displacements at these points is always zero—they are points of destructive interference. The antinodes have large displacements. They correspond to points where the two displacements have equal magnitudes and the *same sign* at all times. Constructive interference at these points gives a displacement twice that of each individual wave.

The intensity of a wave is largest at points where it oscillates with maximum amplitude. You can see in Figure 16.8 that the points of maximum intensity along the standing wave occur at the antinodes; the intensity is zero at the nodes. If this is a standing sound wave, the loudness of the sound varies from zero (no sound) at the nodes to a maximum at the antinodes and then back to zero. The key idea is that **the intensity is maximum at points of constructive interference and zero at points of destructive interference.**

EXAMPLE 16.1   Setting up a standing wave

Two children hold an elastic cord at each end. Each child shakes the end of the cord 2.0 times per second, sending waves at 3.0 m/s toward the middle of the cord, where the two waves combine to create a standing wave. What is the distance between adjacent nodes?

SOLVE The distance between adjacent nodes is λ/2. The wavelength, frequency, and speed are related as $v = f\lambda$, as we saw in Chapter 15, so the wavelength is

$$\lambda = \frac{v}{f} = \frac{3.0 \text{ m/s}}{2.0 \text{ Hz}} = 1.5 \text{ m}$$

The distance between adjacent nodes is λ/2 and thus is 0.75 m.

16.3 Standing Waves on a String

10.4–10.6

The oscillation of a guitar string is a standing wave. A standing wave is naturally produced on a string when both ends are fixed (i.e., tied down), as in the case of a guitar string or the string in the photo of Figure 16.5. We also know that a standing wave is produced when there are two counterpropagating traveling waves. But you don't shake both ends of a guitar string to produce the standing wave! How do we actually get two traveling waves on a string with both ends fixed? Before we can answer this question, we need a brief explanation of what happens when a traveling wave encounters a boundary or a discontinuity.

Reflections

We know that light reflects from mirrors; it can also reflect from the surface of a pond or from a pane of glass. As we saw in Chapter 15, sound waves reflect as well; that's how an echo is produced. To understand reflections, we'll look at waves on a string, but the results will be completely general and can be applied to other waves as well.

Suppose we have a string that is attached to a wall or other fixed support, as in Figure 16.9. The wall is what we will call a *boundary*—it's the end of the medium. When the pulse reaches this boundary, it reflects, moving away from the wall. *All* the wave's energy is reflected, hence **the amplitude of a wave reflected from a boundary is unchanged.**

Figure 16.9 shows that the amplitude doesn't change when the pulse reflects, but the pulse is inverted. We can understand why by looking at the forces at the boundary. When the pulse reaches the wall, the piece of string right at the wall feels an upward force from the approaching pulse. But this piece of the string is fixed in place, so there must be an opposite force from the wall on this piece of string. This downward force leads to an inverted reflected pulse.

Waves also reflect from what we will call a *discontinuity,* a point where there is a change in the properties of the medium. Figure 16.10a shows a discontinuity between a string with a larger linear density and one with a smaller linear density. The tension is the same in both strings, so the wave speed is slower on the left, faster on the right. Whenever a wave encounters a discontinuity, some of the wave's energy is *transmitted* forward and some is reflected. Because energy must be conserved, both the transmitted and the reflected pulse have a smaller amplitude than the initial pulse.

In Figure 16.10b, an incident wave encounters a discontinuity at which the wave speed decreases. Once again, some of the wave's energy is transmitted and some is reflected.

In Figure 16.10a, the reflected pulse is right-side up. The string on the right is quite light and provides little resistance, so the junction moves up and down as the pulse passes. This motion of the string is just like the original "snap" of the string that started the pulse, so the reflected pulse has the same orientation as the original pulse. In Figure 16.10b, the string on the right is more massive, so it looks more like the fixed boundary in Figure 16.9 and the reflected pulse is again inverted.

Creating a Standing Wave

Now that we understand reflections, let's create a standing wave. Figure 16.11 shows a string of length L that is tied at $x = 0$ and $x = L$. This string has *two* boundaries where reflections can occur. If you wiggle the string in the middle, sinusoidal waves travel outward in both directions and soon reach the boundaries, where they reflect. The reflections at the ends of the string cause two waves of *equal amplitude and wavelength* to travel in opposite directions along the string. As we've just seen, these are the conditions that cause a standing wave!

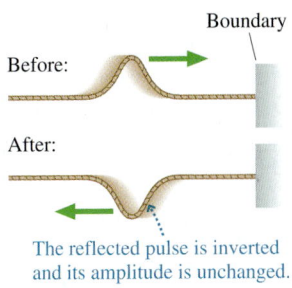

The reflected pulse is inverted and its amplitude is unchanged.

FIGURE 16.9 A wave reflects when it encounters a boundary.

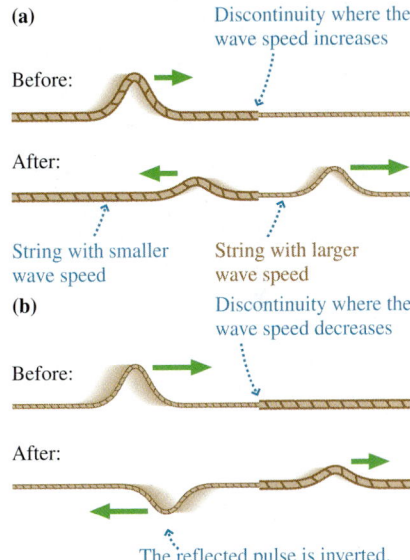

The reflected pulse is inverted.

FIGURE 16.10 The reflection of a wave at a discontinuity.

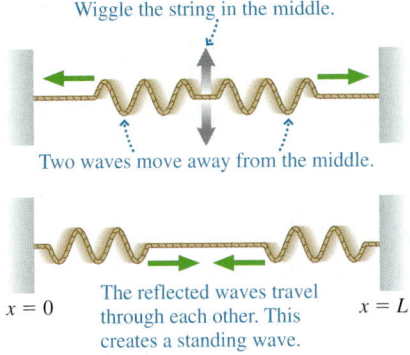

FIGURE 16.11 Reflections at the two boundaries cause a standing wave on the string.

The simplest standing wave has one node at each end, and no more.

The next possible standing wave we can make has an additional node in the center.

The third possible standing wave has two nodes in addition to the nodes at the end.

L

FIGURE 16.12 The first three possible standing waves on a string of length L.

What kind of standing waves might develop on the string? There are two conditions that must be met:

- Because the string is tied down at the ends, the displacements at $x = 0$ and $x = L$ must be zero at all times. Stated another way, we require nodes at both ends of the string.
- We know that standing waves have a spacing of $\lambda/2$ between nodes. This means that the nodes must be equally spaced.

Figure 16.12 sketches the first few possible waves that meet these conditions. These are called the standing wave **modes** of the string. To help quantify the possible waves, we can assign a **mode number** m to each. The first wave in Figure 16.12, with a node at each end, has a mode number $m = 1$. The next wave is $m = 2$, and so on.

NOTE ▶ Figure 16.12 shows only the first three modes, for $m = 1$, $m = 2$, and $m = 3$. But there are many more modes, for all possible values of m. ◀

The distance between adjacent nodes is $\lambda/2$, so the different modes have different wavelengths. For the first mode in Figure 16.12, the distance between nodes is the length of the string, so we can write

$$\lambda_1 = 2L$$

The subscript identifies the mode number; in this case $m = 1$. For $m = 2$, the distance between nodes is $L/2$; this means that $\lambda_2 = L$. Generally, for any mode m the wavelength is given by the following equation:

$$\lambda_m = \frac{2L}{m} \qquad m = 1, 2, 3, 4, \ldots \qquad (16.1)$$

Wavelengths of standing wave modes of a string of length L

LINEAR p. 38   INVERSE p. 118

These are the only possible wavelengths for standing waves on the string. **A standing wave can exist on the string *only* if its wavelength is one of the values given by Equation 16.1.**

NOTE ▶ Other wavelengths, which would be perfectly acceptable wavelengths for a traveling wave, cannot exist as a *standing* wave of length L because they do not meet the constraint of having a node at each end of the string. ◀

If standing waves are possible only for certain wavelengths, then only specific oscillation frequencies are allowed. Because $\lambda f = v$ for a sinusoidal wave, the oscillation frequency corresponding to wavelength λ_m is

$$f_m = \frac{v}{\lambda_m} = \frac{v}{2L/m} = m\left(\frac{v}{2L}\right) \qquad m = 1, 2, 3, 4, \ldots \qquad (16.2)$$

Frequencies of standing wave modes of a string of length L

TRY IT YOURSELF

◀**Through the glass darkly** A piece of window glass is a discontinuity to a light wave, so it both transmits and reflects light. To verify this, look at the windows in a brightly lit room at night. The small percentage of the interior light that reflects from windows is more intense than the light coming in from outside, so reflection dominates and the windows show a mirror-like reflection of the room. Now turn out the lights. With no more reflected interior light you will be able to see the transmitted light from outside.

Figure 16.13 shows the first three modes with their wavelengths and frequencies. You can see that **the mode number m is equal to the number of antinodes of the standing wave.** You can therefore tell a string's mode of oscillation by counting the number of antinodes (*not* the number of nodes).

In Chapter 14, we looked at the concept of *resonance.* A mass on a spring has a certain frequency at which it "wants" to oscillate. If the system is driven at its resonance frequency, it will develop a large amplitude of oscillation. A stretched string will support standing waves, meaning it has a series of frequencies at which it "wants" to oscillate: the frequencies of the different standing wave modes. We can call these **resonant modes,** or more simply, just **resonances.** A small oscillation of a stretched string at a frequency near one of its resonant modes will cause it to develop a standing wave with a large amplitude. In Figure 16.14, a small amplitude oscillation at the $m = 4$ mode frequency of a stretched string produces a large amplitude standing wave.

NOTE ▶ When we draw standing wave modes, as in Figure 16.13, we usually only show the *envelope* of the wave, the greatest extent of the motion of the string. The string's motion is actually continuous and goes through all intermediate positions as well, as we see from the time-exposure photograph of a standing wave in Figure 16.14. ◀

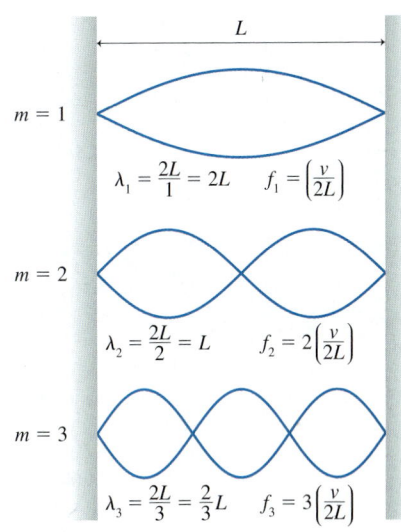

FIGURE 16.13 Possible standing waves of a string fixed at both ends.

The Fundamental and the Higher Harmonics

The sequence of possible frequencies for a standing wave on a string has an interesting pattern that is worth exploring. The first mode has frequency:

$$f_1 = \frac{v}{2L} \qquad (16.3)$$

FIGURE 16.14 A resonant mode of a stretched string.

We call this the **fundamental frequency** of the string. All of the other modes have frequencies that are multiples of this fundamental frequency. We can rewrite Equation 16.2 in terms of the fundamental frequency as

$$f_m = mf_1 \qquad m = 1, 2, 3, 4, \ldots \qquad (16.4)$$

The allowed standing-wave frequencies are all integer multiples of the fundamental frequency. This sequence of possible frequencies is called a set of **harmonics.** The fundamental frequency f_1 is also known as the *first harmonic,* the $m = 2$ wave at frequency f_2 is called the *second harmonic,* the $m = 3$ wave is called the *third harmonic,* and so on. The frequencies above the fundamental frequency, the harmonics with $m = 2, 3, 4, \ldots$, are referred to as the **higher harmonics.**

EXAMPLE 16.2 Identifying harmonics on a string

A 2.50-m-long string vibrates as a 100 Hz standing wave with nodes 1.00 m and 1.50 m from one end of the string and at no points in between these two. Which harmonic is this, and what is the string's fundamental frequency?

PREPARE We begin with the visual overview in Figure 16.15, in which we sketch the appearance of this particular standing wave. We set up an x-axis with one end of the string at $x = 0$ m and the

other end at $x = 2.50$ m. These must be nodes. We also know there are nodes at 1.00 m and 1.50 m, but there are no nodes in between these. This means that the spacing between nodes is 0.50 m. Figure 16.15 is a sketch of a standing wave with a node every 0.50 m.

SOLVE We can simply count the number of antinodes of the standing wave to deduce the mode number; this is mode $m = 5$. The frequencies of the harmonics are given by $f_m = mf_1$, hence the fundamental frequency is

$$f_1 = \frac{f_5}{5} = \frac{100 \text{ Hz}}{5} = 20 \text{ Hz}$$

ASSESS Let's look at this another way to check the value for the mode number. The distance between nodes is 0.50 m. In a 2.5 m length of string, we can fit five such segments, giving us five antinodes, so this is $m = 5$, as we found.

Known
$L = 2.50$ m
$f_m = 100$ Hz

Find
m, f_1

FIGURE 16.15 A visual overview for Example 16.2.

Standing waves on a bridge This photograph shows the Tacoma Narrows suspension bridge on the day in 1940 when it experienced a catastrophic oscillation that led to its collapse. Aerodynamic forces caused the amplitude of a particular resonant mode of the bridge to increase dramatically until the bridge failed. In this photo, the red line shows the original line of the deck of the bridge. You can clearly see the large amplitude of the oscillation and the node at the center of the span.

Stringed Musical Instruments

Think about stringed musical instruments, such as the guitar, the piano, and the violin. These instruments all have strings that are fixed at both ends and tightened to create tension. A disturbance is generated on the string by plucking, striking, or bowing. Regardless of how it is generated, the disturbance creates standing waves on the string. Understanding the sound of a stringed musical instrument means understanding standing waves.

In Chapter 15, we saw that the speed of a wave on a stretched string depended on T_s, the tension in the string and μ, the linear density, as $v = \sqrt{T_s/\mu}$. Combining this with Equation 16.3, we find that the fundamental frequency of a stretched string is

$$f_1 = \frac{v}{2L} = \frac{1}{2L}\sqrt{\frac{T_s}{\mu}} \tag{16.5}$$

When you pluck or bow a string, you initially excite a wide range of frequencies. However, resonance sees to it that the only frequencies to persist are those of the possible standing waves. The musical note you hear when the string is sounded is the fundamental frequency f_1. The higher harmonics determine the *tone quality,* something we will explore later in the chapter.

NOTE ▶ In Equation 16.5, v is the wave speed *on the string,* not the speed of sound in air. ◀

For instruments like the guitar or the violin, the strings are all the same length and under approximately the same tension. Were that not the case, the neck of the instrument would tend to twist toward the side of higher tension. The strings have different frequencies because they differ in linear density. The lower-pitched strings are "fat" while the higher-pitched strings are "skinny." This difference changes the frequency by changing the wave speed. Small adjustments are then made in the tension to bring each string to the exact desired frequency.

CONCEPTUAL EXAMPLE 16.1 Tuning and playing a guitar

A guitar has strings of a fixed length. Plucking a string makes a particular musical note. A player can make other notes by pressing the string against frets, metal bars on the neck of the guitar, as shown in Figure 16.16. The fret becomes the new end of the string, making the effective length shorter.

FIGURE 16.16 Guitar frets.

a. A guitar player plucks a string to play a note. He then presses down on a fret to make the string shorter. Does the new note have a higher or lower frequency?

b. The frequency of one string is too low. (Musically, we say the note is "flat.") How must the tension be adjusted to bring the string to the right frequency?

REASON

a. The fundamental frequency, the note we hear, is $f_1 = (1/2L)\sqrt{T_s/\mu}$. Because f_1 is inversely proportional to L, decreasing the string length increases the frequency.

b. Because f is proportional to the square root of T_s, the player must increase the tension to increase the fundamental frequency.

ASSESS If you watch someone play a guitar, you can see that he or she plays higher notes by moving the fingers to shorten the strings.

EXAMPLE 16.3 Setting the tension in a guitar string

The fifth string on a guitar plays the musical note A, at a frequency of 110 Hz. On a typical guitar, this string is stretched between two fixed points 0.640 m apart, and this length of string has a mass of 2.86 g. What is the tension in the string?

PREPARE Strings sound at their fundamental frequency, so 110 Hz is f_1.

SOLVE The linear density of the string is

$$\mu = (2.86 \times 10^{-3}\,\text{kg})/(0.640\,\text{m}) = 4.47 \times 10^{-3}\,\text{kg/m}$$

We can rearrange Equation 16.5 for the fundamental frequency to solve for the tension in terms of the other variables:

$$\begin{aligned} T_s &= (2Lf_1)^2\mu = (2(0.640\,\text{m})(110\,\text{Hz}))^2(4.47 \times 10^{-3}\,\text{kg/m}) \\ &= 88.6\,\text{N} \end{aligned}$$

ASSESS A guitar has six strings. If each has approximately the same tension, the total force on the neck of the guitar is six times this tension, a bit more than 500 N. This is quite a bit of force.

Standing Electromagnetic Waves

The standing wave descriptions we've found for a vibrating string are actually valid for any transverse wave, including an electromagnetic wave. For example, standing light waves can be established between two parallel mirrors that reflect the light back and forth. The mirrors are boundaries, analogous to the boundaries at the ends of a string. In fact, this is exactly how a laser works. The two facing mirrors in Figure 16.17 form what is called a *laser cavity*.

Because the mirrors act exactly like the points to which a string is tied, the light wave must have a node at the surface of each mirror. (To allow some of the light to escape the laser cavity and form the *laser beam,* one of the mirrors lets some of the light through. This doesn't affect the node.)

FIGURE 16.17 A laser contains a standing light wave between two parallel mirrors.

▶ **Microwave modes** A microwave oven uses a type of electromagnetic wave—microwaves, with a wavelength of about 12 cm—to heat food. The inside walls of a microwave oven are reflective to microwaves, so we have the correct conditions to set up a standing wave. This isn't a good thing! A standing wave has high intensity at the antinodes and low intensity at the nodes, so your oven has hot spots and cold spots, as we see in this thermal image showing the interior of an oven with a thin layer of water that has been "cooked" for a short time. A turntable in a microwave oven keeps the food moving so that no part of your dinner remains at a node or an antinode.

EXAMPLE 16.4 Finding the mode number for a laser

A helium-neon laser (such as the laser in supermarket barcode scanners) emits light of wavelength $\lambda = 633$ nm. A typical cavity for such a laser is 15.0 cm long. What is the mode number of the standing wave in this cavity?

PREPARE Because a light wave is a transverse wave, Equation 16.1 for λ_m applies to a laser as well as a vibrating string.

SOLVE The standing light wave in a laser cavity has a mode number m that is approximately

$$m = \frac{2L}{\lambda} = \frac{2 \times 0.150 \text{ m}}{633 \times 10^{-9} \text{ m}} = 474,000$$

ASSESS The standing light wave inside a laser cavity may have half a million antinodes! This is a consequence of the very small wavelength of light.

STOP TO THINK 16.2   A standing wave on a string is shown. Which of the modes shown below (on the same string) has twice the frequency of the original wave?

Original standing wave

A. B. C. D.

16.4 Standing Sound Waves

Wind instruments like flutes, trumpets, and didgeridoos work very differently from stringed instruments. The player blows air into one end of a tube, producing standing waves of sound that make the notes we hear. In this section, we will look at the properties of such standing sound waves.

Recall that a sound wave is a longitudinal pressure wave. The air molecules oscillate back and forth parallel to the direction in which the wave is traveling, creating *compressions* (regions of higher pressure) and *rarefactions* (regions of lower pressure). Consider a sound wave confined to a long, narrow column of air, such as the air in a tube. A wave traveling down the tube eventually reaches the end, where it encounters the atmospheric pressure of the surrounding environment. This is a discontinuity, much like the small rope meeting the big rope in Figure 16.10b. Part

Fiery interference In this apparatus, a speaker at one end of the metal tube emits a sinusoidal wave. The wave reflects from the other end, which is closed, to make a counter-propagating wave and set up a standing sound wave in the tube. The tube is filled with propane gas that exits through small holes on top. The burning propane allows us to easily discern the nodes and the antinodes of the standing sound wave. An exciting demonstration—but one you shouldn't try yourself!

of the wave's energy is transmitted out into the environment, allowing you to hear the sound, and part is reflected back into the tube. Reflections at both ends of the tube create waves traveling both directions inside the tube, and their superposition, like that of the reflecting waves on a string, is a standing wave.

We start by looking at a sound wave in a tube open at both ends. What kind of standing waves can exist in such a tube? Because the ends of the tube are open to the atmosphere, the pressure at the ends is "fixed" at atmospheric pressure and cannot vary. This is analogous to a stretched string that is fixed at the end. As a result, **the open end of a column of air must be a node of the pressure wave.**

Figure 16.18 shows a column of air open at both ends. We call this an *open-open tube*. Whereas the antinodes of a standing wave on a string are points where the string oscillates with maximum displacement, the antinodes of a standing sound wave are where the pressure has the largest variation, creating, alternately, the maximum compression and the maximum rarefaction. In Figure 16.18, the air molecules squeeze together on the left side of the tube. Then, in Figure 16.18b, half a cycle later, the air molecules squeeze together on the right side. The varying density creates a variation in pressure across the tube, as the graphs show. Figure 16.18c combines the information of Figures 16.18a and 16.18b into a graph of the pressure of the standing sound wave in the tube.

As the standing wave oscillates, the air molecules "slosh" back and forth along the tube with the wave frequency, alternately pushing together (maximum pressure at the antinode) and pulling apart (minimum pressure at the antinode). This makes sense, because sound is a longitudinal wave in which the air molecules oscillate parallel to the tube.

(a) At one instant

(b) Half a cycle later

The shift between compression and rarefaction means a motion of molecules along the tube.

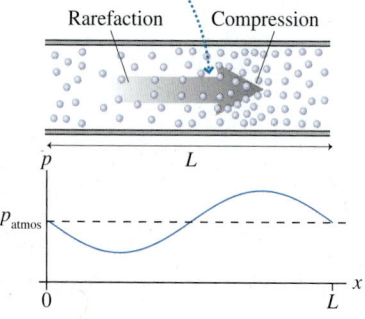

(c) At the ends of the tube, the pressure is equal to atmospheric pressure. These are nodes.

At the antinodes, each cycle sees a change from compression to rarefaction and back to compression.

FIGURE 16.18 The $m = 2$ standing sound wave inside an open-open column of air.

> **NOTE** ▶ The variation in pressure from atmospheric pressure in a real standing sound wave is much smaller than Figure 16.18 implies. When we display graphs of the pressure in sound waves, we won't generally graph the pressure p. Instead, we will graph Δp, the variation from atmospheric pressure. ◀

Many musical instruments, such as a flute, can be modeled as open-open tubes. The flutist blows across one end to create a standing wave inside the tube, and a note of this frequency is emitted from both ends of the flute. The possible standing waves in tubes, like standing waves on strings, are resonances of the system. A gentle puff of air across the mouthpiece of a flute can cause large standing waves at these resonant frequencies.

Other instruments work differently from the flute. A trumpet or a clarinet is a column of air open at the bell end but *closed* by the player's lips at the mouthpiece. To be complete in our treatment of sound waves in tubes, we need to consider tubes that are closed at one or both ends. At a closed end, the air molecules can alternately rush toward the wall, creating a compression, and then rush away from the wall, leaving a rarefaction. Thus **a closed end of an air column is an antinode of pressure.**

Figure 16.19 shows graphs of the first three standing-wave modes of a tube open at both ends (an *open-open tube*), a tube closed at both ends (a *closed-closed tube*), and a tube open at one end but closed at the other (an *open-closed tube*), all with the same length L. These are graphs of the pressure wave, with a node at open ends and an antinode at closed ends. The standing wave in the closed-closed tube looks like the wave in the open-open tube except that the positions of the nodes and antinodes are interchanged. In both cases there are m half-wavelength segments between the ends, thus the wavelengths and frequencies of an open-open tube and a closed-closed tube are the same as those of a string tied at both ends:

$$\begin{cases} \lambda_m = \dfrac{2L}{m} \\[2ex] f_m = m\left(\dfrac{v}{2L}\right) = mf_1 \end{cases} \qquad m = 1, 2, 3, 4, \ldots \qquad (16.6)$$

Wavelengths and frequencies of standing sound wave modes
in an open-open or closed-closed tube

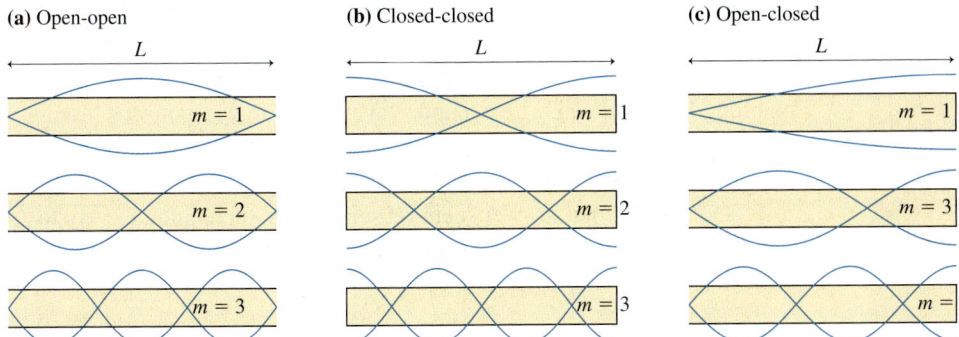

(a) Open-open **(b)** Closed-closed **(c)** Open-closed

$m = 1$ $m = 1$ $m = 1$

$m = 2$ $m = 2$ $m = 3$

$m = 3$ $m = 3$ $m = 5$

FIGURE 16.19 The first three standing sound wave modes in columns of air with different ends. These graphs show the pressure variation in the tube.

The open-closed tube is different, as we can see from Figure 16.19. The $m = 1$ mode has a node at one and an antinode at the other, and so has only one-quarter of a wavelength in a tube of length L. The $m = 1$ wavelength is $\lambda_1 = 4L$, twice the $m = 1$ wavelength of an open-open or a closed-closed tube. Consequently, **the fundamental frequency of an open-closed tube is half that of an open-open or a closed-closed tube of the same length.**

The wavelength of the next mode of the open-closed tube is $4L/3$. Because this is $1/3$ of λ_1, we assign $m = 3$ to this mode. The wavelength of the subsequent mode is $4L/5$, so this is $m = 5$. In other words, an open-closed tube allows only odd-numbered modes. Consequently, the possible wavelengths and frequencies are

$$\begin{cases} \lambda_m = \dfrac{4L}{m} \\[2ex] f_m = m\dfrac{v}{4L} = mf_1 \end{cases} \qquad m = 1, 3, 5, 7, \ldots \qquad (16.7)$$

Wavelengths and frequencies of standing sound wave modes
in an open-closed tube

NOTE ▶ Because sound is a pressure wave, the graphs of Figure 16.19 are *not* "pictures" of the wave as they are for a string wave. The graphs show the pressure variation versus position x. The tube itself is shown merely to indicate the location of the open and closed ends, but the diameter of the tube is *not* related to the amplitude of the wave. ◀

We are now in a position to suggest the following problem-solving strategy, not just for sound waves, but any standing wave.

Standing waves in the sun The sun is a three-dimensional ball of hot gas. Violent convection in the gas as heat emerges from the sun's core can set up standing waves that can be seen in the motion of the sun's surface. This computer model shows one standing-wave mode. The red areas are moving up at this instant while the blue areas are moving down; they'll reverse during the next half cycle. These are the antinodes. The yellow areas in between, which are not moving, are the nodes. This is a more complex standing wave than a one-dimensional gas in a tube, but the basic ideas are the same.

EXAMPLE 16.5   Resonances of the ear canal

The eardrum, which transmits vibrations to the sensory organs of your ear, lies at the end of the ear canal. As Figure 16.20 shows, the ear canal in adults is about 2.5 cm in length. What frequency standing waves can occur within the ear canal that are within the range of human hearing? The speed of sound in the warm air of the ear canal is 350 m/s.

Eardrum

2.5 cm

Ear canal

FIGURE 16.20   The anatomy of the ear.

PREPARE   We proceed according to the steps in Problem-Solving Strategy 16.1. We can treat the ear canal as an open-closed tube: open to the atmosphere at the external end, closed by the eardrum at the other end. The possible standing wave modes appear as in Figure 16.19c. The length of the tube is 2.5 cm.

SOLVE   Equation 16.7 gives the allowed frequencies in an open-closed tube. We are looking for the frequencies in the range 20 Hz–20,000 Hz. The fundamental frequency is

$$f_1 = \frac{v}{4L} = \frac{350 \text{ m/s}}{4(0.025 \text{ m})} = 3500 \text{ Hz}$$

The higher harmonics are odd multiples of this frequency:

$$f_3 = 3(3500 \text{ Hz}) = 10,500 \text{ Hz}$$
$$f_5 = 5(3500 \text{ Hz}) = 17,500 \text{ Hz}$$

These three modes lie within the range of human hearing; higher modes are greater than 20,000 Hz.

ASSESS   These standing wave modes are resonances of the ear canal. There will be an increased amplitude of oscillation within the ear canal in response to sounds at these frequencies.

Curve of equal perceived loudness

The resonances of the ear canal result in an increase in sensitivity at these frequencies.

f_1 f_3

FIGURE 16.21   A curve of equal perceived loudness.

How important are the resonances of the ear canal calculated in Example 16.5? In Chapter 15, you learned about the decibel scale for measuring loudness of sound. In fact, your ears have varying sensitivity to sounds of different frequencies. Figure 16.21 shows a curve of *equal perceived loudness,* the sound intensity level (in dB) required to give the *impression* of equal loudness for sinusoidal waves of the noted frequency. Lower values mean that your ear is more sensitive at that frequency. In general, the curve decreases to about 1000 Hz, the frequency at which your hearing is most acute, then slowly rises at higher frequencies. However, this general trend is punctuated by two dips in the curve, showing two frequencies at which a quieter sound produces the same perceived loudness. As you can see, these two dips correspond to the resonances f_1 and f_3 of the ear canal. The resonances are frequencies at which the air in the ear canal "wants" to oscillate. Incoming sounds at these frequencies produce a larger oscillation, resulting in an increased sensitivity to these frequencies.

Wind Instruments

With a wind instrument, blowing into the mouthpiece creates a standing sound wave inside a tube of air. The player changes the notes by using her fingers to cover holes or open valves, changing the effective length of the tube. The first open hole becomes a node because the tube is open to the atmosphere at that point. The fact that the holes are on the side, rather than literally at the end, makes very little difference. The length of the tube determines the standing wave resonances, and thus the musical note that the instrument produces.

Many wind instruments have a "buzzer" at one end of the tube, such as a vibrating reed on a saxophone or clarinet, or the musician's vibrating lips on a trumpet or trombone. Buzzers like these generate a continuous range of frequencies rather than single notes, which is why they sound like a "squawk" if you play on just the mouthpiece without the rest of the instrument. When connected to the body of the instrument, most of those frequencies cause little response. But the frequencies from the buzzer that match the resonant frequencies of the instrument cause the buildup of large amplitudes at these frequencies—standing-wave resonances. The combination of these frequencies makes the musical note that we hear.

Truly classical music The oldest known musical instruments are bone flutes from burials in central China. The flutes in the photo are up to 9000 years old, and are made from naturally hollow bones from crowned cranes. The positions of the holes determine the frequencies that the flutes can produce. Soon after the first flutes were created, the design was standardized so that different flutes would play the same notes—including the notes in the modern Chinese musical scale.

CONCEPTUAL EXAMPLE 16.2 Comparing the flute and the clarinet

A flute and a clarinet have about the same length, but the lowest note that can be played on the clarinet is much lower than the lowest note that can be played on the flute. Why is this?

REASON A flute is an open-open tube; the frequency of the fundamental mode is $f_1 = v/2L$. A clarinet is open at one end, but the player's lips and the reed close it at the other end. The clarinet is thus an open-closed tube with a fundamental frequency

$f_1 = v/4L$. This is about half the fundamental frequency of the flute, so the lowest note on the clarinet has a much lower pitch. In musical terms, it's about an octave lower than the flute.

ASSESS The possible standing wave modes depend on whether the end of an air column is open or closed. This changes the notes that can be played, as we have seen, and it also affects the way the instrument sounds. A flute and a clarinet sound very different to your ear; we will have more to say about this difference later in the chapter.

EXAMPLE 16.6 The importance of warming up

Players of wind instruments know that it is important to warm up their instruments in order for them to play in tune—that is, at the desired frequency. A "cold" flute plays the note A at 440 Hz when the air temperature is 20°C.

a. How long is the tube? At 20°C, the speed of sound in air is 343 m/s.
b. As the player blows air through the flute, the air inside the instrument warms up. Once the air temperature inside the flute has risen to 32°C, increasing the speed of sound to 350 m/s, what is the frequency?
c. At the higher temperature, how much must the length of the tube be changed to bring the frequency back to 440 Hz?

SOLVE A flute is an open-open tube with fundamental frequency $f_1 = v/2L$.

a. At 20°C, the length corresponding to 440 Hz is

$$L = \frac{v}{2f} = \frac{343 \text{ m/s}}{2(440 \text{ Hz})} = 0.390 \text{ m}$$

b. As the speed of sound increases, the frequency changes to

$$f_1(\text{at } 32°C) = \frac{350 \text{ m/s}}{2(0.390 \text{ m})} = 449 \text{ Hz}$$

c. To bring the flute back into tune, the length must be increased to give a frequency of 440 Hz with a speed of 350 m/s. The new length is

$$L = \frac{v}{2f_1} = \frac{350 \text{ m/s}}{2(440 \text{ Hz})} = 0.398 \text{ m}$$

Thus the flute must be increased in length by 8 mm.

ASSESS The dependence of the frequency on temperature means that wind instruments must have an adjustable joint to change the tube length so that players can keep them in tune as the temperature changes.

STOP TO THINK 16.3   A tube that is open at both ends supports a standing wave with harmonics at 300 Hz and 400 Hz, with no harmonics between. What is the fundamental frequency of this tube?

A. 50 Hz   B. 100 Hz   C. 150 Hz   D. 200 Hz   E. 300 Hz

16.5 Speech and Hearing

When you hear a particular note played on a guitar, it sounds very different from the same note played on a trumpet. And you have perhaps been to a lecture in which the speaker talked at essentially the same pitch the entire time—but you could still understand what was being said. Clearly, there is more to your brain's perception of sound than pitch alone. How do you tell the difference between a guitar and a trumpet? How do you distinguish between an "oo" vowel sound and an "ee" vowel sound at the same pitch?

The Frequency Spectrum

To this point, we have pictured sound waves as sinusoidal waves, with a well-defined frequency. In fact, most of the sounds that you hear are not pure sinusoidal waves. Most sounds are a mix, or superposition, of different frequencies. For example, we have seen how certain standing-wave modes are possible on a stretched string. When you pluck a string on a guitar, you generally don't excite just one standing-wave mode—you simultaneously excite many different modes.

If you play the note "middle C" on a guitar, the fundamental frequency is 262 Hz. There will be a standing wave at this frequency, but there will also be standing waves at the frequencies 524 Hz, 786 Hz, 1048 Hz, . . . all the higher harmonics predicted by Equation 16.4.

Figure 16.22a is a bar chart showing all the frequencies present in the sound of the vibrating guitar string. The height of each bar shows the relative intensity of that harmonic. The fundamental frequency has the highest intensity, but many other harmonics have significant intensities as well. A bar chart showing the relative intensities of the different frequencies is called the **frequency spectrum** of the sound.

When your brain interprets the mix of frequencies from the guitar in Figure 16.22a, it identifies the fundamental frequency as the *pitch*. 262 Hz corresponds to a middle C, so you will identify the pitch as a middle C, even though the sound consists of many different frequencies. Your brain uses the higher harmonics to determine the **tone quality**, which is also called the *timbre*. The tone quality—and therefore the higher harmonics—is what makes a middle C played on a guitar sound quite different from a middle C played on a trumpet. The frequency spectrum of a trumpet would show a very different pattern of the relative intensities of the higher harmonics, and this different mix of higher harmonics gives the trumpet a different sound.

The actual sound wave produced by a guitar playing middle C is shown in Figure 16.22b. The sound wave is periodic, with a period of 3.81 ms that corresponds to the 262 Hz fundamental frequency. But the wave doesn't have a simple sinusoidal shape; it is more complex. **The higher harmonics don't change the period of the sound wave, they only change its shape.** The sound wave of a trumpet playing middle C would also have a 3.81 ms period, but its shape would be entirely different.

(a)

This peak is the fundamental frequency.

These peaks are the higher harmonics.

Relative intensity

262 524 786 1048   1572   2096 f (Hz)

(b)

The sound wave is periodic.

Δp

3.81 ms

t

The vertical axis is the change in pressure from atmospheric pressure due to the sound wave.

FIGURE 16.22 The frequency spectrum and a graph of the sound wave of a guitar playing a note with fundamental frequency 262 Hz.

CONCEPTUAL EXAMPLE 16.3 Playing the didgeridoo

The didgeridoo, a musical instrument developed by aboriginal Australians, is deceptively simple. It consists of a tube (a eucalyptus stem or branch hollowed out by termites) of $1\frac{1}{2}$ m or more in length. The player presses his lips against the end and blows air through his lips as with a trumpet. He may also make sounds with his vocal cords. Skilled players can make a wide variety of sounds. How is this possible with such a simple instrument?

REASON Because the lips seal one end, a didgeridoo has the resonances of an open-closed tube, given by Equation 16.7. The instrument can therefore produce many different frequencies. Changing the vibration of the lips and the sounds from the vocal cords can change the mix of standing-wave modes that are produced, leading to very different sounds.

One instrument, many sounds A synthesizer can be adjusted to sound like a flute, a clarinet, a trumpet, a piano—or any other musical instrument. The keys on a synthesizer determine what fundamental frequency to produce. The other controls adjust the mix of higher harmonics to match the frequency spectrum of various musical instruments, effectively mimicking their sounds.

Vowels and Formants

Try this: Keep your voice at the same pitch, and say the "ee" sound, as in "beet," then the "oo" sound as in "boot". Pay attention to how you reshape your mouth as you move back and forth between the two sounds. The two vowel sounds are at the same pitch, and thus have the same period, but they sound quite different. The difference in sound arises from the difference in the higher harmonics, just as for musical instruments. As you speak, you adjust the properties of your vocal tract to produce different mixes of harmonics that make the "ee," "oo," "ah," and other vowel sounds.

Speech begins with the vibration of your vocal cords, stretched bands of tissue in your throat. The vibration is similar to that of a wave on a stretched string. In ordinary speech, the average fundamental frequency for adult males and females is about 150 Hz and 250 Hz, respectively, but you can change the vibration frequency by changing the tension of your vocal cords. That's how you make your voice higher or lower as you sing. If you have a cold, your vocal cords can become inflamed and swollen, which increases their linear density. This makes the pitch of your voice lower.

As is the case with a guitar string, your vocal cords produce a mix of different frequencies as they vibrate—the fundamental frequency and a rich mixture of higher harmonics. If you were to put a microphone in your throat and measure the sound waves right at your vocal cords, you would measure a frequency spectrum similar to that in Figure 16.23a.

There is more to the story, though. Before reaching the opening of your mouth, sound from your vocal cords must pass through your vocal tract—a series of hollow cavities including your throat, your mouth, and your nose. The vocal tract acts much like a series of tubes, and, as in any tube, certain frequencies will set up standing-wave resonances. The rather broad standing-wave resonances of the vocal tract are called **formants.** Harmonics of your vocal cords at or near the formant frequencies will be amplified; harmonics that are far from a formant will be suppressed. Figure 16.23b shows the formants of an adult male making an "ee" vowel sound. The filtering of the vocal cord harmonics by the formants is clear.

You can change the shape and frequencies of the formants, and thus the sound you make, by changing the shape and length of your vocal tract. You do this by changing your mouth opening and the shape and position of your tongue. The first two formants for an "ee" sound are at roughly 270 Hz and 2300 Hz, but for an "oo" sound they become 300 Hz and 870 Hz. The much lower second formant of the "oo" emphasizes midrange frequencies, making a "calming" sound, while the more strident sound of "ee" comes from enhancing the higher frequencies.

(a) Frequencies from the vocal cords

When you speak, your vocal cords produce a mix of frequencies.

(b) Actual spoken frequencies (Vowel sound "ee")

When you form your vocal tract to make a certain vowel sound it increases the amplitudes of certain frequencies and suppresses others.

The broad peaks at which amplification occurs are due to filtering by the formants.

First formant

Second formant

FIGURE 16.23 The frequency spectrum from the vocal cords, and after passing through the vocal tract.

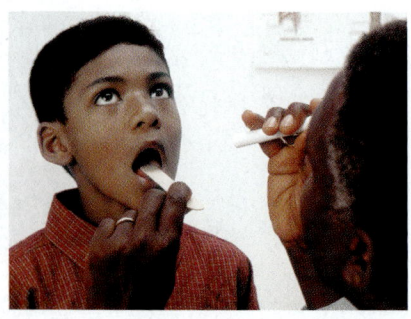

◄**Saying "ah"** BIO Why, during a throat exam, does a doctor ask you to say "ah"? This particular vowel sound is formed by opening the mouth and the back of the throat wide—giving a clear view of the tissues of the throat.

CONCEPTUAL EXAMPLE 16.4 **High-frequency hearing loss**

As you age, your hearing sensitivity will decrease. This decrease is not uniform; for most people, the loss of sensitivity is greater for higher frequencies. The loss of sensitivity at high frequencies may make it difficult to understand what others are saying. Why is this?

REASON It is the high-frequency components of speech that allow us to distinguish different vowel sounds. A decrease in sensitivity to these higher frequencies makes it more difficult to make such distinctions.

ASSESS For high-frequency hearing loss, simple amplification of sound will not help; this will amplify all frequencies equally. A good hearing aid should only amplify the portions of the audio spectrum for which sensitivity is reduced.

STOP TO THINK 16.4 These sound waves represent notes played on different musical instruments. Which has the highest pitch?

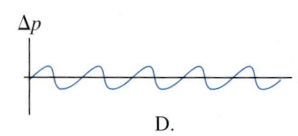

16.6 The Interference of Waves from Two Sources

Perhaps you have seen headphones that offer "active noise reduction." When you turn on the headphones, they produce sound that somehow *cancels* noise from the external environment. How does adding sound to a system make it quieter? This is one of the questions we will answer in this section, in which we look at the interference of waves from two sources.

We began the chapter by noting that waves, unlike particles, can pass through each other. Where they do, the principle of superposition tells us that the displacement of the medium is the sum of the displacements due to each wave acting alone. Consider the two loudspeakers in Figure 16.24, both emitting sound waves with the same frequency. In Figure 16.24a, sound from loudspeaker 2 passes loudspeaker 1, then two overlapped sound waves travel to the right along the *x*-axis. What sound is heard at the point indicated with the dot? And what about at the dot in Figure 16.24b, where the speakers are side by side? These are the two cases we will consider in this section. Although we'll use sound waves for our discussion, the results are completely general and apply to all waves. In Chapter 17, we will use these ideas to study the interference of light waves.

Interference Along a Line

Figure 16.25 on the next page shows two loudspeakers sending traveling waves along a line. The graphs are slightly displaced from each other so that you can see what each wave is doing, but the *physical situation* is one in which the waves are

(a) Two sound waves overlapping along a line

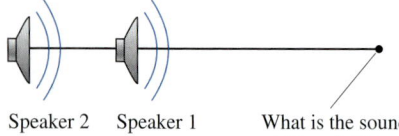

Speaker 2   Speaker 1   What is the sound at this point?

(b) Two overlapping spherical sound waves

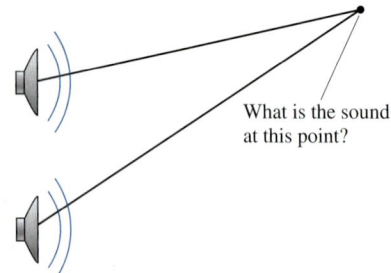

What is the sound at this point?

FIGURE 16.24 Interference of waves from two sources.

traveling *on top of* each other. We assume that the two speakers emit sound waves of identical frequency f, wavelength λ, and amplitude A.

At every point along the line, the net sound pressure wave will be the sum of the pressures from the individual waves. That's the principle of superposition. Because the two speakers are separated by one wavelength, the two waves are aligned crest-to-crest and trough-to-trough. Waves aligned this way are said to be **in phase;** waves that are in phase march along "in step" with each other. If we add the two waves point by point, we see that their superposition is a traveling wave with wavelength λ and twice the amplitude of the individual waves. This is constructive interference.

> **NOTE** ▶ Textbook pictures can be misleading because they're frozen in time. The net sound wave in Figure 16.25 is a *traveling* wave, moving to the right with the speed of sound v. It differs from the two individual waves only by having twice the amplitude. This is not a standing wave with nodes and anti-nodes that remain in one place. ◀

If d_1 and d_2 are the distances from loudspeakers 1 and 2 to a point at which we want to know the sound wave, their difference $\Delta d = d_2 - d_1$ is called the **path-length difference.** It is the *extra* distance traveled by wave 2 on the way to the point where the two waves are combined. In Figure 16.25, we see that constructive interference results from a path-length difference $\Delta d = \lambda$. But increasing Δd by an additional λ would produce exactly the same result, so we will also have constructive interference for $\Delta d = 2\lambda$, $\Delta d = 3\lambda$, and so on. In other words, **two waves will be in phase and will produce constructive interference any time their path-length difference is a whole number of wavelengths.**

In Figure 16.26, the two speakers are separated by half a wavelength. In this case, the crests of one wave align with the troughs of the other and the waves march along "out of step" with each other. We say that the two waves are **out of phase.** When two waves are out of phase, they are equal and opposite at every point. Consequently, the sum of the two waves is zero *at every point.* The super-position of two out-of-phase waves to produce a wave with zero amplitude is destructive interference.

The destructive interference of Figure 16.26 results from a path-length difference $\Delta d = \frac{1}{2}\lambda$. Again, increasing Δd by an additional λ would produce a picture that looks exactly the same, so we will also have destructive interference for $\Delta d = 1\frac{1}{2}\lambda$, $\Delta d = 2\frac{1}{2}\lambda$, and so on. That is, **two waves will be out of phase and will produce destructive interference any time their path-length difference is a whole number of wavelengths plus half a wavelength.**

Summing up, for two identical sources of waves, constructive interference occurs when the path-length difference is

$$\Delta d = m\lambda \qquad m = 0, 1, 2, 3, \ldots \qquad (16.8)$$

and destructive interference occurs when the path-length difference is

$$\Delta d = \left(m + \frac{1}{2}\right)\lambda \qquad m = 0, 1, 2, 3, \ldots \qquad (16.9)$$

> **NOTE** ▶ The path-length difference needed for constructive or destructive interference depends on the wavelength, and hence the frequency. If one particular frequency interferes destructively, another may not. ◀

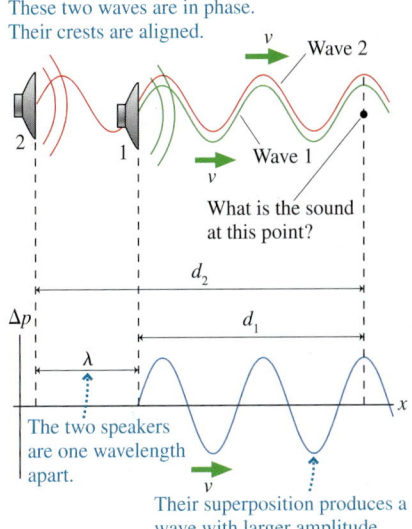

These two waves are in phase. Their crests are aligned.

The two speakers are one wavelength apart.

Their superposition produces a wave with larger amplitude.

FIGURE 16.25 Constructive interference of two waves traveling along the *x*-axis.

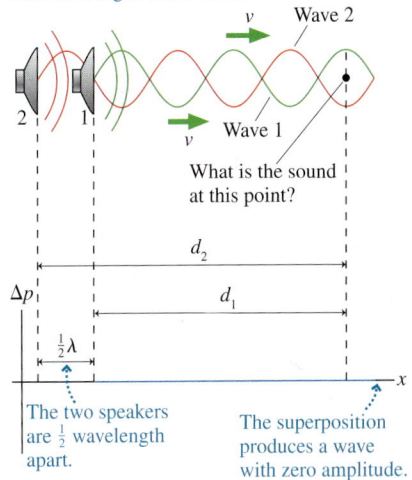

These two waves are out of phase. The crests of one wave are aligned with the troughs of the other.

The two speakers are $\frac{1}{2}$ wavelength apart.

The superposition produces a wave with zero amplitude.

FIGURE 16.26 Destructive interference of two waves traveling along the *x*-axis.

The path-length difference is not necessarily the distance between the speakers, as we see in the next example. It is simply the difference in the distances traveled by the two waves.

EXAMPLE 16.7 Is the sound loud or quiet? Part I

Two loudspeakers 42.0 m apart and facing each other emit identical 115 Hz sinusoidal sound waves. Susan is walking along a line between the speakers. As she walks, she finds herself moving through loud and quiet spots. If Susan stands 19.5 m from one speaker, is she standing at a quiet spot or a loud spot? Assume that the speed of sound is 345 m/s.

PREPARE As Susan walks along the line between the speakers, she moves between points of constructive interference (loud spots) and destructive interference (quiet spots). Is her current position one of constructive or destructive interference? This will depend on the path-length difference. We start with a visual overview of the situation in Figure 16.27.

Known
$f = 115$ Hz
$v = 345$ m/s

Find
$\Delta d = d_2 - d_1$

FIGURE 16.27 Visual overview for Example 16.7

SOLVE At Susan's position, the distances the two waves travel to reach her are

$$d_1 = 19.5 \text{ m} \qquad d_2 = 42.0 \text{ m} - 19.5 \text{ m} = 22.5 \text{ m}$$

At the point where the two waves reach Susan and interfere, their path-length difference is

$$\Delta d = d_2 - d_1 = 3.0 \text{ m}$$

To know if we have constructive or destructive interference, we need to compare this with the wavelength:

$$\lambda = \frac{v}{f} = \frac{345 \text{ m/s}}{115 \text{ Hz}} = 3.0 \text{ m}$$

Because the path-length difference is exactly one wavelength, Susan is standing at a point of constructive interference. That is, she is standing at a loud spot.

ASSESS As you'll recall from earlier in the chapter, two counterpropagating waves create a standing wave. In this case, the conditions for constructive and destructive interference are the same as the conditions for antinodes and nodes, respectively, that we found earlier. Susan is at an antinode of the standing wave. The ideas of interference give us a different perspective on standing waves.

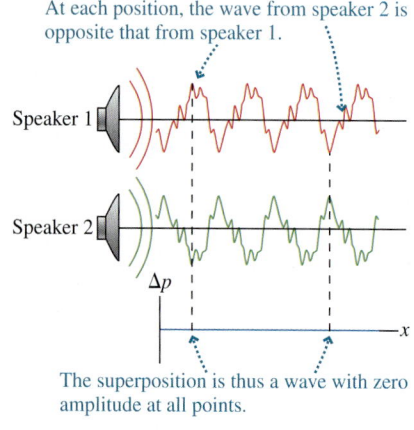

At each position, the wave from speaker 2 is opposite that from speaker 1.

Speaker 1

Speaker 2

Δp

The superposition is thus a wave with zero amplitude at all points.

FIGURE 16.28 Destructive interference of two waves.

The analysis above assumed that the two loudspeakers were emitting identical waves. Another interesting and important case of interference, illustrated in Figure 16.28, occurs when one loudspeaker emits a sound wave that is *the exact inverse* of the wave from the other speaker. If the speakers are side by side, so that $\Delta d = 0$, the superposition of these two waves will result in destructive interference; they will completely cancel. This destructive interference does not require the waves to have any particular frequency or any particular shape. As long as the waves are exact inverses of each other, there will be destructive interference.

Headphones with *active noise reduction* use this technique. A microphone on the outside of the headphones measures ambient sound. A circuit inside the headphones produces an inverted version of the microphone signal and sends it to the headphone speakers. The real ambient sound and the inverted version of the ambient sound coming from the speakers arrive at the ears together and interfere destructively, thus significantly reducing the sound intensity. In this case, *adding* sound results in a *lower* overall intensity inside the headphones!

Interference of Spherical Waves

Interference along a line illustrates the idea of interference, but it's not very realistic. In practice, sound waves from a loudspeaker or light waves from a light bulb spread out as spherical waves. Figure 16.29 on the next page shows a wavefront diagram for a spherical wave. Recall that the wave fronts represent the *crests* of the wave and are spaced by the wavelength λ. Halfway between two wave fronts is a trough of the wave. What happens when two spherical waves overlap? For example, imagine two loudspeakers emitting identical waves radiating sound in all directions. Figure 16.30 shows the wave fronts of the two waves.

This is a static picture, of course, so you have to imagine the wave fronts spreading out as new circular rings are born at the speakers. The waves overlap as they travel, and, as was the case in one dimension, this causes interference.

Consider a particular point like that marked by the red dot in Figure 16.30. The two waves each have a crest at this point, so there is constructive interference here. But at other points, such as that marked by the black dot, a crest overlaps a trough, so this is a point of destructive interference.

Notice—simply by counting the wave fronts—that the red dot is three wavelengths from speaker 2 ($r_2 = 3\lambda$) but only two wavelengths from speaker 1 ($r_1 = 2\lambda$). The path-length difference of the two waves arriving at the red dot is $\Delta r = r_2 - r_1 = \lambda$. That is, the wave from speaker 2 has to travel one full wavelength more than the wave from speaker 1, so the waves are in phase (crest aligned with crest) and interfere constructively. You should convince yourself that Δr is a *whole number of wavelengths* at every point where two wave fronts intersect.

Similarly, the path length difference at the black dot, where the interference is destructive, is $\Delta r = \frac{1}{2}\lambda$. As with interference along a line, destructive interference results when the path-length difference is a whole number of wavelengths plus half a wavelength.

Thus the general rule for determining whether there is constructive or destructive interference at any point is the same for spherical waves as for waves traveling along a line. For identical sources, constructive interference occurs when the path-length difference is

$$\Delta r = m\lambda \qquad m = 0, 1, 2, 3, \ldots \qquad (16.10)$$

Destructive interference occurs when the path-length difference is

$$\Delta r = \left(m + \frac{1}{2}\right)\lambda \qquad m = 0, 1, 2, 3, \ldots \qquad (16.11)$$

The conditions for constructive and destructive interference are the same for spherical waves as for waves along a line. And the treatment we have seen for sound waves can be applied to any wave, as we have noted. For any two wave sources, the following Tactics Box sums up how to determine if the interference at a point is constructive or destructive.

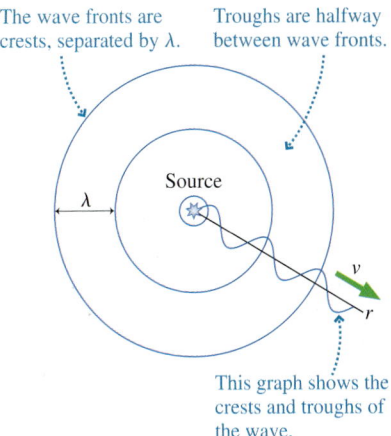

The wave fronts are crests, separated by λ.   Troughs are halfway between wave fronts.

This graph shows the crests and troughs of the wave.

FIGURE 16.29 A spherical wave.

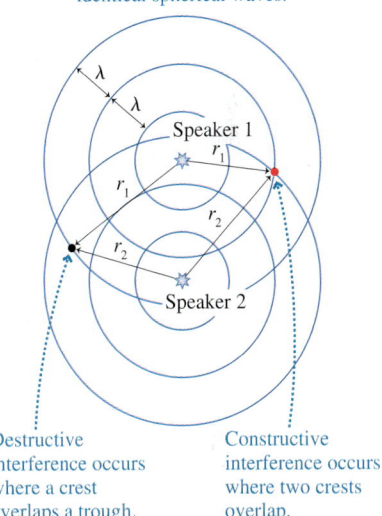

Two sources emit identical spherical waves.

Destructive interference occurs where a crest overlaps a trough.

Constructive interference occurs where two crests overlap.

FIGURE 16.30 The overlapping wave patterns of two sources.

 TACTICS BOX 16.1 Identifying constructive and destructive interference   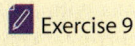 Exercise 9

❶ Identify the path length from each source to the point of interest. Compute the path-length difference $\Delta r = |r_2 - r_1|$.

❷ Find the wavelength, if it is not specified.

❸ If the path-length difference is a whole number of wavelengths ($\lambda, 2\lambda, 3\lambda \ldots$), crests are aligned with crests and there is constructive interference.

❹ If the path-length difference is a whole number of wavelengths plus a half wavelength ($1\frac{1}{2}\lambda, 2\frac{1}{2}\lambda, 3\frac{1}{2}\lambda \ldots$), crests are aligned with troughs and there is destructive interference.

NOTE ▶ Keep in mind that interference is determined by Δr, the path-length *difference*, not by r_1 or r_2. ◀

EXAMPLE 16.8 Is the sound loud or quiet? Part II

Two speakers are 3.0 m apart, and play identical tones of frequency 170 Hz. Sam stands directly in front of one speaker at a distance of 4.0 m. Is this a loud spot or a quiet spot? Assume that the speed of sound in air is 340 m/s.

PREPARE Figure 16.31 shows a visual overview of the situation, showing the positions of and path lengths from each speaker.

FIGURE 16.31 Visual overview for Example 16.8.

SOLVE Following the steps in Tactics Box 16.1, we first compute the path-length difference. r_1, r_2, and the distance between the speakers form a right triangle, so we can use the Pythagorean theorem to find

$$r_2 = \sqrt{(4.0 \text{ m})^2 + (3.0 \text{ m})^2} = 5.0 \text{ m}$$

Thus the path-length difference is

$$\Delta r = r_2 - r_1 = 1.0 \text{ m}.$$

Next, we compute the wavelength:

$$\lambda = \frac{v}{f} = \frac{340 \text{ m/s}}{170 \text{ Hz}} = 2.0 \text{ m}$$

The path-length difference is $\frac{1}{2}\lambda$, so this is a point of destructive interference. Sam is at a quiet spot.

So far, we have only looked at interference at particular points. What can we say about the overall pattern of points at which we have constructive or destructive interference? For instance, the red dot in Figure 16.30 is only one point where $\Delta r = \lambda$; you should be able to locate several more. Taken together, all the points with $\Delta r = \lambda$ form a curved line along which constructive interference is occurring. Another curved line of constructive interference connects all the points at which $\Delta r = 2\lambda$, another connects the $\Delta r = 3\lambda$ points, and so on. These lines, shown as red lines in Figure 16.32 on the next page, are called **antinodal lines.** They are analogous to the antinodes of a standing wave, hence the name. An antinode is a *point* of constructive interference; for spherical waves, oscillation at maximum amplitude occurs along a continuous *line*. To understand this idea better, imagine the static picture of Figure 16.32 evolving with time. As the wave fronts expand, the *intersection point* of two rings moves outward along one of the red lines.

Similarly, we can connect together points where Δr is a multiple of λ plus $\frac{1}{2}\lambda$. The black dot in Figure 16.30 is just one of many points with $\Delta r = \frac{1}{2}\lambda$. Together, these points of destructive interference form a **nodal line** along which the displacement is always zero. The nodal lines in Figure 16.32 are shown in black.

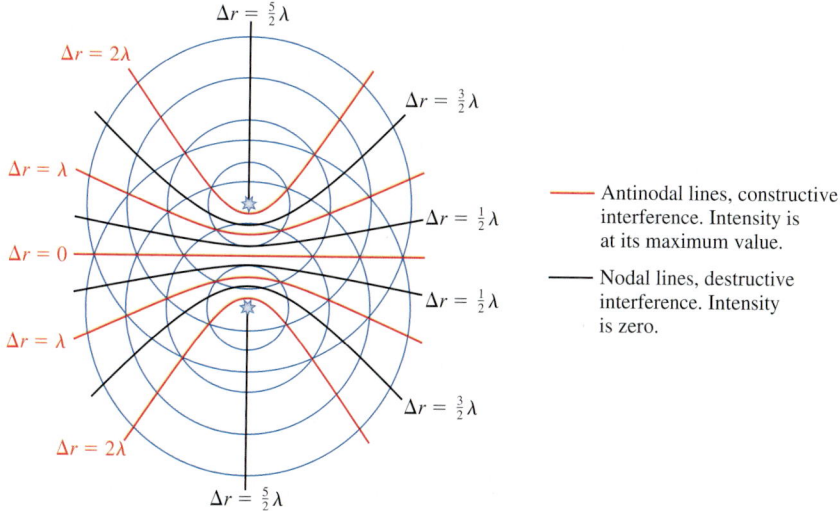

FIGURE 16.32 The points of constructive and destructive interference fall along antinodal and nodal lines.

You are regularly exposed to sound from two separated sources: stereo speakers. When you walk across a room in which a stereo is playing, why don't you hear a pattern of loud and soft sounds as you cross antinodal and nodal lines? First, we don't listen to single frequencies. Music is a complex sound wave, with many frequencies, but only one frequency at a time satisfies the condition for constructive or destructive interference. Most of the sound frequencies are not affected. Second, reflections of sound waves from walls and furniture make the situation much more complex than the idealized two-source picture in Figure 16.32. Sound wave interference can be heard, but it takes careful selection of a pure tone and a room with no hard, reflecting surfaces. But interference that's rather tricky to demonstrate with sound waves is easy to produce with light waves. We'll return to this example in the next chapter.

The two water waves overlap, leading to patterns of constructive and destructive interference.

STOP TO THINK 16.5 These speakers emit identical sound waves with a wavelength of 1.0 m. At the point indicated, is the interference constructive, destructive, or something in between?

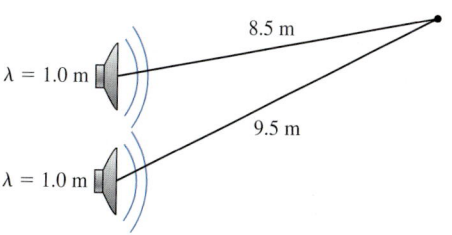

16.7 Beats

Thus far we have looked at the superposition of waves from sources having the same wavelength and frequency. We can also use the principle of superposition to investigate a phenomenon that is easily demonstrated with two sources of slightly *different* frequency and wavelength.

Suppose two sinusoidal waves are traveling toward your ear, as shown in Figure 16.33 on the next page. The two waves have the same amplitude but slightly different frequencies: The red wave has a slightly higher frequency (and thus a slightly shorter wavelength) than the green wave. This slight difference

Act**iv** 10.7
Phys**ics**

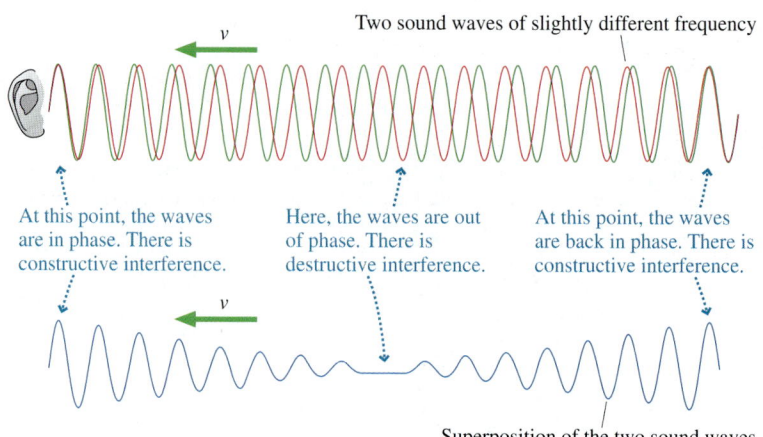

Two sound waves of slightly different frequency

At this point, the waves are in phase. There is constructive interference.

Here, the waves are out of phase. There is destructive interference.

At this point, the waves are back in phase. There is constructive interference.

Superposition of the two sound waves

FIGURE 16.33 The superposition of two sound waves with slightly different frequencies.

causes the waves to combine in a manner that alternates between constructive and destructive interference. Their superposition, drawn in blue below the two waves, is a wave whose amplitude shows a periodic variation. As the waves reach your ear, you will hear a single tone whose intensity is *modulated.* That is, the sound goes up and down in volume, loud, soft, loud, soft, . . . , making a distinctive sound pattern called **beats.**

Suppose the two waves have frequencies f_1 and f_2 that differ only slightly, so that $f_1 \approx f_2$. A complete mathematical analysis would show that the air oscillates against your eardrum at frequency

$$f_{osc} = \frac{1}{2}(f_1 + f_2)$$

This is the *average* of f_1 and f_2, and it differs little from either since the two frequencies are nearly equal. Further, the intensity of the sound is modulated at a frequency called the *beat frequency:*

$$f_{beat} = |f_1 - f_2| \tag{16.12}$$

The beat frequency is simply the *difference* between the two individual frequencies.

Figure 16.34 is a history graph of the wave at the position of your ear. You can see both the sound wave oscillation at frequency f_{osc} and the much slower intensity oscillation at frequency f_{beat}. Frequency f_{osc} determines the pitch you hear while f_{beat} determines the frequency of the loud-soft-loud modulations of the sound intensity.

Musicians can use beats to tune their instruments. If one flute is properly tuned at 440 Hz but the other plays 438 Hz, the flutists will hear two loud-soft-loud beats per second. The second flutist is "flat" and needs to shorten her flute very slightly to bring the frequency up to 440 Hz.

Many measurement devices use beats to determine an unknown frequency by comparing it to a known frequency. For example, Chapter 15 described a Doppler blood-flow meter that used the Doppler shift of ultrasound reflected from moving blood to determine its speed. The meter determines this very small frequency shift by combining the emitted wave and the reflected wave and measuring the resulting beat frequency. The beat frequency is equal to the shift in frequency on reflection, exactly what is needed to determine the blood speed. Another example of using beats to make a measurement is given in the following example.

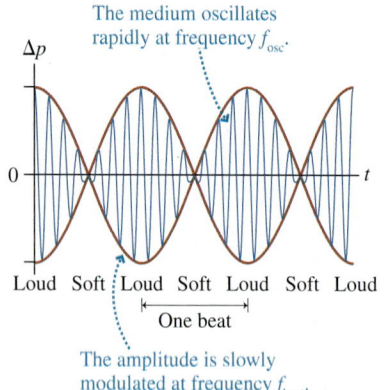

The medium oscillates rapidly at frequency f_{osc}.

Δp

Loud Soft Loud Soft Loud Soft Loud

One beat

The amplitude is slowly modulated at frequency f_{beat}.

FIGURE 16.34 The modulated sound of beats.

EXAMPLE 16.9 Detecting bats using beats

The little brown bat is a common bat species in North America. It emits echolocation pulses at a frequency of 40 kHz, well above the range of human hearing. To
allow observers to "hear" these bats, the bat detector shown in Figure 16.35 combines the bat's sound wave at frequency f_1 with a wave of frequency f_2 from a tunable oscillator. The resulting beat frequency is isolated with a filter, then amplified
and sent to a loudspeaker. To what frequency should the tunable oscillator be set to
produce an audible beat frequency of 3 kHz?

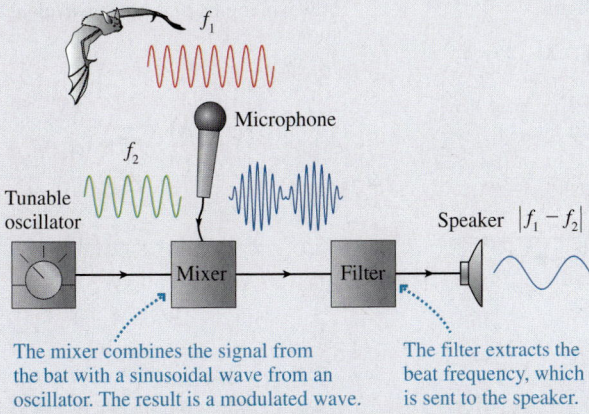

FIGURE 16.35 The operation of a bat detector.

SOLVE The beat frequency is $f_{beat} = |f_1 - f_2|$, so the oscillator frequency and the
bat frequency need to *differ* by 3 kHz. An oscillator frequency of either 37 kHz or
43 kHz will work nicely.

STOP TO THINK 16.6 You hear three beats per second when two sound tones are
generated. The frequency of one tone is known to be 610 Hz. The frequency of
the other is

A. 604 Hz.
B. 607 Hz.
C. 613 Hz.
D. 616 Hz.
E. Either A or D.
F. Either B or C.

The goal of Chapter 16 has been to use the idea of superposition to understand the phenomena of interference and standing waves.

GENERAL PRINCIPLES

Principle of Superposition

The displacement of a medium when more than one wave is present is the sum of the displacements due to each individual wave.

Interference

In general, the superposition of two or more waves into a single wave is called interference.

Constructive interference occurs when crests are aligned with crests and troughs with troughs. We say the waves are in phase. It occurs when the path-length difference $\Delta d = d_2 - d_1$ is a whole number of wavelengths.

Destructive interference occurs when crests are aligned with troughs. We say the waves are out of phase. It occurs when the path-length difference $\Delta d = d_2 - d_1$ is a whole number of wavelengths plus half a wavelength.

IMPORTANT CONCEPTS

Standing Waves

Two identical traveling waves moving in opposite directions create a standing wave.

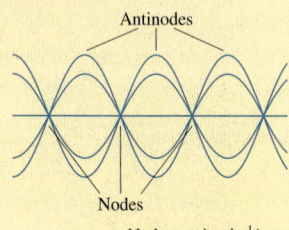

The boundary conditions determine which standing wave frequencies and wavelengths are allowed. The allowed standing waves are **modes** of the system.

A standing wave on a string has a node at each end. Possible modes:

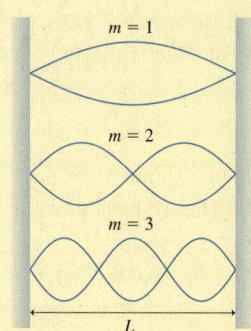

$$\lambda_m = \frac{2L}{m} \qquad f_m = m\left(\frac{v}{2L}\right) = mf_1$$
$$m = 1, 2, 3, \ldots$$

A standing sound wave in a tube can have different boundary conditions: open-open, closed-closed or open-closed.

Open-open

$$f_m = m\left(\frac{v}{2L}\right)$$
$$m = 1, 2, 3, \ldots$$

Closed-closed

$$f_m = m\left(\frac{v}{2L}\right)$$
$$m = 1, 2, 3, \ldots$$

Open-closed

$$f_m = m\left(\frac{v}{4L}\right)$$
$$m = 1, 3, 5, \ldots$$

APPLICATIONS

Beats (loud-soft-loud-soft modulations of intensity) are produced when two waves of slightly different frequency are superimposed.

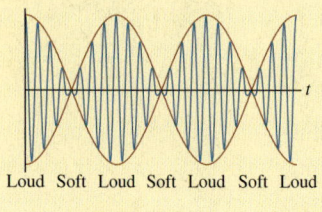

$$f_{\text{beat}} = |f_1 - f_2|$$

Standing waves are multiples of a **fundamental frequency,** the frequency of the lowest mode. The higher modes are the higher **harmonics.**

For sound, the fundamental frequency determines the perceived **pitch;** the higher harmonics determine the **tone quality.**

Our vocal cords create a range of harmonics. The mix of higher harmonics is changed by our vocal tract to create different vowel sounds.

QUESTIONS

Conceptual Questions

1. Light can pass easily through water and through air, but light will reflect from the surface of a lake. What does this tell you about the speed of light in air and in water?

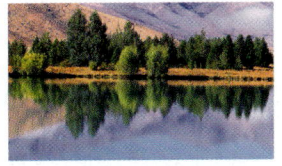

2. Ocean waves are partially reflected from the entrance to a harbor, where the depth of the water is suddenly less. What does this tell you about the speed of waves in water of different depths?

3. A string has an abrupt change in linear density at its midpoint so that the speed of a pulse on the left side is 2/3 of that on the right side.
   a. On which side is the linear density greater? Explain.
   b. From which side would you start a pulse so that its reflection from the midpoint would not be inverted? Explain.

4. A guitarist finds that the frequency of one of her strings is too low by 1.4%. Should she increase or decrease the tension of the string? Explain.

5. Figure Q16.5 shows a standing wave on a string that is oscillating at frequency f_0. How many antinodes will there be if the frequency is doubled to $2f_0$? Explain.

FIGURE Q16.5

6. Figure Q16.6 shows a standing sound wave in a tube of air that is open at both ends.
   a. Which mode (value of m) standing wave is this?
   b. Is the air vibrating horizontally or vertically?

FIGURE Q16.6

7. If you take snapshots of a standing wave on a string at different times, you will find certain times at which the wave is totally flat. What has happened to the energy of the wave at those times?

8. A typical flute is about 66 cm long. A piccolo is a very similar instrument, though it is smaller, with a length of about 32 cm. How does the pitch of a piccolo compare to that of a flute?

9. Some pipes on a pipe organ are open at both ends, others are closed at one end. For pipes that play low-frequency notes, there is an advantage to using pipes that are closed at one end. What is the advantage?

10. An organ pipe is tuned to exactly 384 Hz when the room temperature is 20°C. Later, the room temperature increases to 25°C. What happens to the frequency of the pipe's note? Does it increase, decrease, or stay the same? Explain.

11. A friend's voice sounds different over the telephone than it does in person. This is because telephones do not transmit frequencies over about 3000 Hz. 3000 Hz is well above the normal frequency of speech, so why does eliminating these high frequencies change the sound of a person's voice?

12. Suppose you were to play a trumpet after breathing helium, in which the speed of sound is much greater than in air. Would the pitch of the instrument be higher or lower than normal, or would it be unaffected by being played with helium inside the tube rather than air?

13. If you pour liquid in a tall, narrow glass, you may hear sound with a steadily rising pitch. What is the source of the sound, and why does the pitch rise as the glass fills?

14. When you speak after breathing helium, in which the speed of sound is much greater than in air, your voice sounds quite different. The frequencies emitted by your vocal cords do not change since they are determined by the mass and tension of your vocal cords. So what *does* change when your vocal tract is filled with helium rather than air?

15. Sopranos can sing notes at very high frequencies—over 1000 Hz. When they sing such high notes, it can be difficult to understand the words they are singing. Use the concepts of harmonics and formants to explain this.

16. If a cold gives you a stuffed-up nose, it changes the way your voice sounds, even if your vocal cords are not affected. Explain why this is so.

17. Figure Q16.17 shows wave fronts of a circular wave. Are the displacements at the following pairs of positions *in phase* or *out of phase?* Explain.
    a. A and B   b. C and D   c. E and F

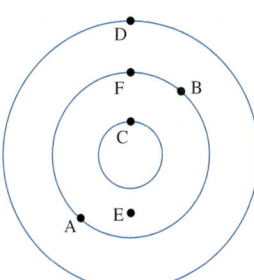

FIGURE Q16.17

Multiple-Choice Questions

Questions 18 through 20 refer to the snapshot graph Figure Q16.18.

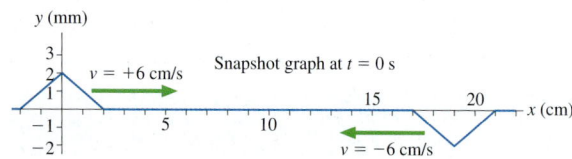

FIGURE Q16.18

18. | At $t = 1$ s, what is the displacement y of the string at $x = 7$ cm?
    A. -1.0 mm   B. 0 mm   C. 0.5 mm
    D. 1.0 mm   E. 2.0 mm
19. | At $x = 3$ cm, what is the earliest time that y will equal 2 mm?
    A. 0.5 s   B. 0.7 s   C. 1.0 s
    D. 1.5 s   E. 2.5 s
20. | The pulses overlap at $t = 1.5$ s. At $x = 10$ cm, what is the value of y at this time?
    A. -2.0 mm   B. -1.0 mm   C. -0.5 mm
    D. 0 mm   E. 1.0 mm
21. ‖ Two sinusoidal waves with the same amplitude A and frequency f travel in opposite directions along a long string. You stand at one point and watch the string. The maximum displacement of the string at that point is
    A. A B. $2A$ C. 0
    D. There is not enough information to decide.

22. | A student in her physics lab measures the standing wave modes of a tube. The lowest frequency that makes a resonance is 20 Hz. As the frequency is increased, the next resonance is at 60 Hz. What will be the next resonance after this?
    A. 80 Hz   B. 100 Hz   C. 120 Hz   D. 180 Hz
23. ‖ Two guitar strings made of the same type of wire have the same length. String 1 has a higher pitch than string 2. Which of the following is true?
    A. The wave speed of string 1 is greater than that of string 2.
    B. The tension in string 2 is greater than that in string 1.
    C. The wavelength of the lowest standing wave mode on string 2 is longer than that on string 1.
    D. The wavelength of the lowest standing wave mode on string 1 is longer than that on string 2.
24. | The frequency of the lowest standing wave mode on a 1.0-m-long string is 20 Hz. What is the wave speed on the string?
    A. 10 m/s   B. 20 m/s   C. 30 m/s   D. 40 m/s
25. | Suppose you pluck a string on a guitar and it produces the note A at a frequency of 440 Hz. Now you press your finger down on the string against one of the frets, making this point the new end of the string. The newly shortened string has 4/5 the length of the full string. When you pluck the string, its frequency will be:
    A. 350 Hz   B. 440 Hz   C. 490 Hz   D. 550 Hz

PROBLEMS

Section 16.1 The Principle of Superposition

1. | Figure P16.1 is a snapshot graph at $t = 0$ s of two waves on a taut string approaching each other at 1 m/s. Draw six snapshot graphs, stacked vertically, showing the string at 1 s intervals from $t = 1$ s to $t = 6$ s.

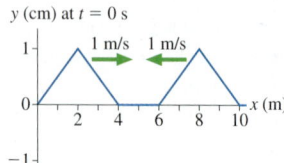

FIGURE P16.1

2. | Figure P16.2 is a snapshot graph at $t = 0$ s of two waves approaching each other at 1 m/s. Draw six snapshot graphs, stacked vertically, showing the string at 1 s intervals from $t = 1$ s to $t = 6$ s.

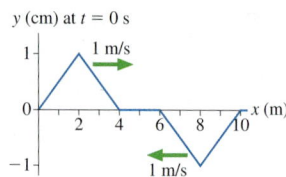

FIGURE P16.2

3. ‖ Figure P16.3 is a snapshot graph at $t = 0$ s of two waves approaching each other at 1 m/s. Draw four snapshot graphs, stacked vertically, showing the string at $t = 2, 4, 6,$ and 8 s.

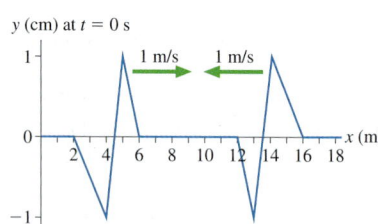

FIGURE P16.3

4. ‖ Figure P16.4 is a snapshot graph at $t = 0$ s of two waves approaching each other at 1 m/s. Draw four snapshot graphs, stacked vertically, showing the string at $t = 2, 4, 6,$ and 8 s.

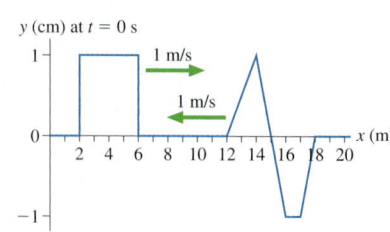

FIGURE P16.4

240 m

5. ‖ Figure P16.5a is a snapshot graph at $t = 0$ s of two waves on a string approaching each other at 1 m/s. At what time was the snapshot graph in Figure P16.5b taken?

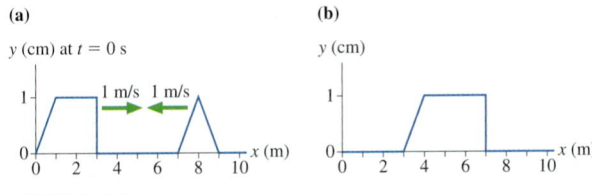

(a) y (cm) at $t = 0$ s

(b) y (cm)

FIGURE P16.5

Section 16.2 Standing Waves

6. ‖ Figure P16.6 is a snapshot graph at $t = 0$ s of two waves moving to the right at 1 m/s. The string is fixed at $x = 8$ m. Draw four snapshot graphs, stacked vertically, showing the string at $t = 2, 4, 6,$ and 8 s.

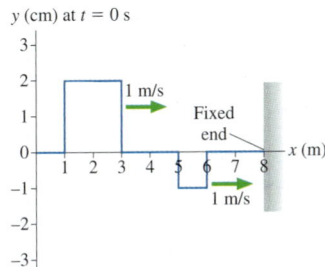

y (cm) at $t = 0$ s

FIGURE P16.6

7. ‖ Figure P16.7 is a snapshot graph at $t = 0$ s of a pulse on a string moving to the right at 1 m/s. The string is fixed at $x = 5$ m. Draw a history graph spanning the time interval $t = 0$ s to $t = 10$ s for the location $x = 3$ m on the string.

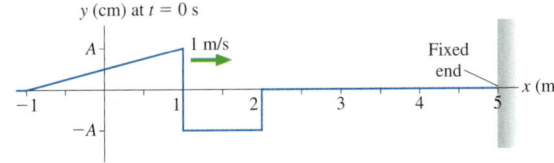

y (cm) at $t = 0$ s

FIGURE P16.7

8. | At $t = 0$ s, a small "upward" (positive y) pulse centered at $x = 6.0$ m is moving to the right on a string with fixed ends at $x = 0.0$ m and $x = 10.0$ m. The wave speed on the string is 4.0 m/s. At what time will the string next have the same appearance that it did at $t = 0$ s?

Section 16.3 Standing Waves on a String

9. ‖ A 2.0-m-long string is fixed at both ends and tightened until the wave speed is 40 m/s. What is the frequency of the standing wave shown in Figure P16.9?

FIGURE P16.9

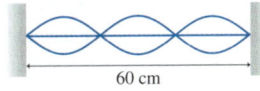

60 cm

FIGURE P16.10

10. | Figure P16.10 (below left) shows a standing wave oscillating at 100 Hz on a string. What is the wave speed?

11. | A bass guitar string is 89 cm long with a fundamental frequency of 30 Hz. What is the wave speed on this string?

12. ‖ The fundamental frequency of a guitar string is 384 Hz. What is the fundamental frequency if the tension in the string is reduced by half?

13. | a. What are the three longest wavelengths for standing waves on a 240-cm-long string that is fixed at both ends?
    b. If the frequency of the second-longest wavelength is 50.0 Hz, what is the frequency of the third-longest wavelength?

14. ‖ Standing waves on a 1.0-m-long string that is fixed at both ends are seen at successive frequencies of 24 Hz and 36 Hz.
    a. What are the fundamental frequency and the wave speed?
    b. Draw the standing-wave pattern when the string oscillates at 36 Hz.

15. | A 121-cm-long, 4.00 g string oscillates in its $m = 3$ mode with a frequency of 180 Hz and a maximum amplitude of 5.00 mm. What are (a) the wavelength and (b) the tension in the string?

16. ‖ A guitar string with a linear density of 2.0 g/m is stretched between supports that are 60 cm apart. The string is observed to form a standing wave with three antinodes when driven at a frequency of 420 Hz. What are (a) the frequency of the fifth harmonic of this string and (b) the tension in the string?

17. ‖ A violin string has a standard length of 32.8 cm. It sounds the musical note A (440 Hz) when played without fingering. How far from the end of the string should you place your finger to play the note C (523 Hz)?

18. ‖ The lowest note on a grand piano has a frequency of 27.5 Hz. The entire string is 2.00 m long and has a mass of 400 g. The vibrating section of the string is 1.90 m long. What tension is needed to tune this string properly?

19. | A carbon dioxide laser emits infrared. A CO_2 laser with a cavity length of 53.00 cm oscillates in the $m = 100,000$ mode. What are the wavelength and frequency of the laser beam?

Section 16.4 Standing Sound Waves

20. | The lowest frequency in the audible range is 20 Hz. (a) What is the length of (a) the shortest open-open tube and (b) the shortest open-closed tube needed to produce this frequency?

21. | The contrabassoon is the wind instrument capable of sounding the lowest pitch in an orchestra. It is folded over several times to fit its impressive 18 ft length into a reasonable size instrument.
    a. If we model the instrument as an open-closed tube, what is its fundamental frequency? The sound speed inside is 350 m/s because the air is warmed by the player's breath.
    b. The actual fundamental frequency of the contrabassoon is 27.5 Hz, which should be different from your answer in part a. This means the model of the instrument as an open-closed tube is a bit too simple. But if you insist on using that model, what is the "effective length" of the instrument?

22. | Figure P16.22 shows a standing sound wave in an 80-cm-long tube. The tube is filled with an unknown gas. What is the speed of sound in this gas?

$f = 500$ Hz

80 cm

FIGURE P16.22

23. | What are the three longest wavelengths for standing sound waves in a 121-cm-long tube that is (a) open at both ends and (b) open at one end, closed at the other?

24. | The lowest pedal note on a large pipe organ has a fundamental frequency of 16 Hz. This extreme bass note is more felt as a rumble than heard with the ears. What is the length of the open-closed pipe that makes that note? Assume a speed of sound of 340 m/s.

25. || The fundamental frequency of an open-open tube is 1500 Hz when the tube is filled with 0°C helium. What is its frequency when filled with 0°C air?

26. ||| A drainage pipe running under a freeway is 30.0 m long. Both ends of the pipe are open, and wind blowing across one end causes the air inside to vibrate.
    a. If the speed of sound on a particular day is 340 m/s, what will be the fundamental frequency of air vibration in this pipe?
    b. What is the frequency of the lowest harmonic that would be audible to the human ear?
    c. What will happen to the frequency in the later afternoon as the air begins to cool?

27. ||| You've decided to install a pipe organ in your living room. Unfortunately, your ceiling is only 3.2 m high. To reach the lowest notes possible, you've decided to use pipes closed at the floor end and open at the top, where they'll be sounded. You need 50 cm clearance at the top end to install the compressed-air box and mount the pipes in the box. If you keep your living room at 20°C, what is the frequency of the lowest note that you will be able to obtain from your organ?

28. ||| A child has an ear canal that is 1.3 cm long. At what sound BIO frequencies in the audible range will the child have increased hearing sensitivity?

29. | Although the vocal tract is quite complicated, we can make BIO a simple model of it as an open-closed tube extending from the opening of the mouth to the diaphragm, the large muscle separating the abdomen and the chest cavity. What is the length of this tube if its fundamental frequency equals a typical speech formant frequency of 200 Hz? Assume a sound speed of 350 m/s. Does this result for the tube length seem reasonable, based on observations on your own body?

Section 16.6 The Interference of Waves from Two Sources

30. |||| Two loudspeakers in a 20°C room emit 686 Hz sound waves along the x-axis. What is the smallest distance between the speakers for which the interference of the sound waves is destructive?

31. || Two loudspeakers emit sound waves along the x-axis. The sound has maximum intensity when the speakers are 20 cm apart. The sound intensity decreases as the distance between the speakers is increased, reaching zero at a separation of 30 cm.
    a. What is the wavelength of the sound?
    b. If the distance between the speakers continues to increase, at what separation will the sound intensity again be a maximum?

32. || Two identical loudspeakers separated by distance d emit 170 Hz sound waves along the x-axis. As you walk along the axis, away from the speakers, you don't hear anything even though both speakers are on. What are three possible values for d? Assume a sound speed of 340 m/s.

33. || Figure P16.33 shows the circular wave fronts emitted by two sources. Make a table with rows labeled P, Q, and R and columns labeled r_1, r_2, Δr, and C/D. Fill in the table for points P, Q, and R, giving the distances as multiples of λ and indicating, with a C or a D, whether the interference at that point is constructive or destructive.

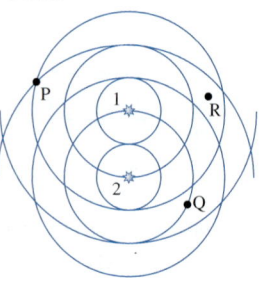

FIGURE P16.33

34. ||| Two identical loudspeakers 2.0 m apart are emitting 1800 Hz sound waves into a room where the speed of sound is 340 m/s. Is the point 4.0 m directly in front of one of the speakers, perpendicular to the plane of the speakers, a point of maximum constructive interference, perfect destructive interference, or something in between?

Section 16.7 Beats

35. | Two strings are adjusted to vibrate at exactly 200 Hz. Then the tension in one string is increased slightly. Afterward, three beats per second are heard when the strings vibrate at the same time. What is the new frequency of the string that was tightened?

36. | A flute player hears four beats per second when she compares her note to a 523 Hz tuning fork (the note C). She can match the frequency of the tuning fork by pulling out the "tuning joint" to lengthen her flute slightly. What was her initial frequency?

General Problems

37. | The fundamental frequency of a standing wave on a 1.0-m-long string is 440 Hz. What would be the wave speed of a pulse moving along this string?

38. | A guitarist plucks the D-string of his instrument, causing it to vibrate at 147 Hz. If he now presses his finger down on the string, creating a new effective end for the string at the string's exact midpoint, what will be the frequency of the string?

39. || An 80-cm-long steel string with a linear density of 1.0 g/m INT is under 200 N tension. It is plucked and vibrates at its fundamental frequency. What is the wavelength of the sound wave that reaches your ear in a 20°C room?

40. | A string, stretched between two fixed posts, forms standing wave resonances at 325 Hz and 390 Hz. What is the greatest possible value of its fundamental frequency?

41. ||| A violinist places her finger so that the vibrating section of INT a 1.0 g/m string has a length of 30 cm, then she draws her bow across it. A listener nearby in a 20°C room hears a note with a wavelength of 40 cm. What is the tension in the string?

42. || A particularly beautiful note reaching your ear from a rare INT Stradivarius violin has a wavelength of 39.1 cm. The room is slightly warm, so the speed of sound is 344 m/s. If the string's linear density is 0.60 g/m and the tension is 150 N, how long is the vibrating section of the violin string?

43. ‖ A heavy piece of hanging sculpture is suspended by a 90-cm-long, 5.0 g steel wire. When the wind blows hard, the wire hums at its fundamental frequency of 80 Hz. What is the mass of the sculpture?

44. ‖ An experiment finds that standing waves on a 0.80-m-long string, fixed at both ends, occur at 24 Hz and 32 Hz, but at no frequencies in between.
    a. What is the fundamental frequency?
    b. What is the wave speed on the string?
    c. Draw the standing wave pattern for the string at 32 Hz.

45. ‖‖‖ Astronauts visiting Planet X have a 2.5-m-long string whose mass is 5.0 g. They tie the string to a support, stretch it horizontally over a pulley 2.0 m away, and hang a 1.0 kg mass on the free end. Then the astronauts begin to excite standing waves on the string. Their data show that standing waves exist at frequencies of 64 Hz and 80 Hz, but at no frequencies in between. What is the value of g, the acceleration due to gravity, on Planet X?

46. ‖‖‖ A 75 g bungee cord has an equilibrium length of 1.2 m. The cord is stretched to a length of 1.8 m, then vibrated at 20 Hz. This produces a standing wave with two antinodes. What is the spring constant of the bungee cord?

47. ‖‖‖ A 2.5-cm-diameter steel cable (with density 7900 kg/m³) that is part of the suspension system for a footbridge stretches 14 m between the tower and the ground). After walking over the bridge, a hiker finds that the cable is vibrating in its fundamental mode with a period of 0.40 s. What is the tension in the cable?

48. ‖ A steel wire is used to stretch a spring, as shown in Figure P16.48. An oscillating magnetic field drives the steel wire back and forth. A standing wave with three antinodes is created when the spring is stretched 8.0 cm. What stretch of the spring produces a standing wave with two antinodes?

FIGURE P16.48

49. ‖ Just as you are about to step into a nice hot bath, a small earthquake rattles your bathroom. Immediately afterward, you notice that the water in the tub is oscillating. The water in the center seems to be motionless while the water at the two ends alternately rises and falls, like a seesaw. You happen to know that your bathtub is 1.4 m long, and you count 10 complete oscillations of the water in 20 s.
    a. What is the wavelength of this standing wave?
    b. What is the speed of the waves that are reflecting back and forth inside the tub to create the standing wave?

50. ‖ A microwave generator can produce microwaves at any frequency between 10 GHz and 20 GHz. As Figure P16.50 shows, the microwaves are aimed, through a small hole, into a "microwave cavity" that consists of a 10-cm-long cylinder with reflective ends.
    a. Which frequencies will create standing waves in the microwave cavity?
    b. For which of these frequencies is the cavity midpoint an antinode?

FIGURE P16.50

51. ‖ An open-open organ pipe is 78.0 cm long. An open-closed pipe has a fundamental frequency equal to the third harmonic of the open-open pipe. How long is the open-closed pipe?

52. ‖ A narrow column of air is found to have standing waves at frequencies of 390 Hz, 520 Hz, and 650 Hz and at no frequencies in between these. The behavior of the tube at frequencies less than 390 Hz or greater than 650 Hz is not known.
    a. Is this an open-open tube or an open-closed tube? Explain.
    b. How long is the tube?

53. ‖ An open-open aluminum pipe has an m = 1 standing wave at 262 Hz. It is also possible to set up a longitudinal standing sound wave in the aluminum itself by striking it on the end. What will be the frequency of the m = 1 mode in the aluminum?

54. ‖ In 1866, the German scientist Adolph Kundt developed a technique for accurately measuring the speed of sound in various gases. A long glass tube, known today as a Kundt's tube, has a vibrating piston at one end and is closed at the other. Very finely ground particles of cork are sprinkled in the bottom of the tube before the piston is inserted. As the vibrating piston is slowly moved forward, there are a few positions that cause the cork particles to collect in small, regularly spaced piles along the bottom. Figure P16.54 shows an experiment in which the tube is filled with pure oxygen and the piston is driven at 400 Hz.

FIGURE P16.54

    a. Do the cork particles collect at standing wave nodes or antinodes? (Hint: consider the appearance of the ends of the tube.)
    b. What is the speed of sound in oxygen?

55. ‖ A 40-cm-long tube has a 40-cm-long insert that can be pulled in and out, as shown in Figure P16.55. A vibrating tuning fork is held next to the tube. As the insert is slowly pulled out, the sound from the tuning fork creates standing waves in the tube when the total length L is 42.5 cm, 56.7 cm, and 70.9 cm. What is the frequency of the tuning fork? The air temperature is 20°C.

FIGURE P16.55

56. ‖‖‖ A 1.0-m-tall vertical tube is filled with 20°C water. A tuning fork vibrating at 580 Hz is held just over the top of the tube as the water is slowly drained from the bottom. At what water heights, measured from the bottom of the tube, will there be a standing wave in the tube?

57. ‖ A 50-cm-long wire with a mass of 1.0 g and a tension of 440 N passes across the open end of an open-closed tube of air. The wire, which is fixed at both ends, is bowed at the center so as to vibrate at its fundamental frequency and generate a sound wave. Then the tube length is adjusted until the fundamental frequency of the tube is heard. What is the length of the tube? Assume the speed of sound is 340 m/s.

58. ⦀ A 25-cm-long wire with a linear density of 20 g/m passes across the open end of an 85-cm-long open-closed tube of air. If the wire, which is fixed at both ends, vibrates at its fundamental frequency, the sound wave it generates excites the second vibrational mode of the tube of air. What is the tension in the wire? Assume the speed of sound is 340 m/s.

59. ⦀ Two loudspeakers located along the *x*-axis as shown in Figure P16.59 produce sounds of equal frequency. Speaker 1 is at the origin, while the location of speaker 2 can be varied by a remote control wielded by the listener. He notices maxima in the sound intensity when speaker 2 is located at $x = 0.75$ m and 1.00 m, but at no points in between. What is the frequency of the sound? Assume the speed of sound is 340 m/s.

FIGURE P16.59

60. ⦀ You are standing 2.50 m directly in front of one of the two loudspeakers shown in Figure P16.60. They are 3.00 m apart and both are playing a 686 Hz tone in phase. As you begin to walk directly away from the speaker, at what distances from the speaker do you hear a *minimum* sound intensity? The room temperature is 20°C.

FIGURE P16.60

61. ⦀ FM station KCOM ("All commercials, all the time") transmits simultaneously, at a frequency of 99.9 MHz, from two broadcast towers placed precisely 31.5 m apart along a north-south line.
   a. What is the wavelength of KCOM's transmissions?
   b. Suppose you stand 90.0 m due east of the point halfway between the two towers with your portable FM radio. Will you receive a strong or weak signal at this position? Why?
   c. You then stand 90.0 m due north of the northern tower with your radio. Will you receive a strong or weak signal at this position? Why?

62. ⦀ Two loudspeakers, 4.0 m apart and facing each other, play identical sounds of the same frequency. You stand halfway between them, where there is a maximum of sound intensity. Moving from this point toward one of the speakers, you encounter a minimum of sound intensity when you have moved 0.25 m.
   a. What is the frequency of the sound?
   b. If the frequency is then increased while you remain 0.25 m from the center, what is the first frequency for which that location will be a maximum of sound intensity?

63. ⦀ Two radio antennas are separated by 2.0 m. Both broadcast identical 750 MHz waves. If you walk around the antennas in a circle of radius 10 m, how many maxima will you detect?

64. ⦀ Two stones, one large and one small, hit water simultaneously and generate ripples. The large stone creates ripples twice as high as the small stone. At the midpoint between the splashes, what is the ratio of the maximum possible ripple amplitude to the minimum possible ripple amplitude?

65. Ⅰ Certain birds produce vocalizations consisting of two distinct frequencies that are not harmonically related—that is, the two frequencies are not harmonics of a common fundamental frequency. These two frequencies must be produced by two different vibrating structures in the bird's vocal tract.
   a. Wood ducks have been observed to make a call with approximately equal intensities at 850 Hz and 1200 Hz. The membranes that produce the vocalizations do not seem to vibrate at frequencies less than 500 Hz. Given this limitation, could the two frequencies noted be higher harmonics of a lower-frequency fundamental?
   b. If we model the duck's vocal tract as an open-closed tube, what length corresponds to a fundamental frequency equal to the lower of the two frequencies in a?

66. Ⅰ Piano tuners tune pianos by listening to the beats between the harmonics of two different strings. When properly tuned, the note A should have the frequency 440 Hz and the note E should be at 659 Hz. The tuner can determine this by listening to the beats between the third harmonic of the A and the second harmonic of the E.
   a. A tuner first tunes the A string very precisely by matching it to a 440 Hz tuning fork. She then strikes the A and E strings simultaneously and listens for beats between the harmonics. What beat frequency indicates that the E string is properly tuned?
   b. The tuner starts with the tension in the E string a little low, then tightens it. What is the frequency of the E string when she hears four beats per second?

67. ⦀ A flutist assembles her flute in a room where the speed of sound is 342 m/s. When she plays the note A, it is in perfect tune with a 440 Hz tuning fork. After a few minutes, the air inside her flute has warmed to where the speed of sound is 346 m/s.
   a. How many beats per second will she hear if she now plays the note A as the tuning fork is sounded?
   b. How far does she need to extend the "tuning joint" of her flute to be in tune with the tuning fork?

68. ⦀ A student waiting at a stoplight notices that her turn signal, which has a period of 0.85 s, makes one blink exactly in sync with the turn signal of the car in front of her. The blinker of the car ahead then starts to get ahead, but 17 s later the two are exactly in sync again. What is the period of the blinker of the other car?

69. ⦀ Two loudspeakers emit 400 Hz notes. One speaker sits on the ground. The other speaker is in the back of a pickup truck. You hear eight beats per second as the truck drives away from you. What is the truck's speed?

70. Ⅰ A Doppler blood flow meter emits ultrasound at a frequency of 5.0 MHz. What is the beat frequency between the emitted waves and the waves reflected from blood cells moving away from the emitter at 0.15 m/s?

71. Ⅰ An ultrasound unit is being used to measure a patient's heartbeat by combining the emitted 2.0 MHz signal with the sound waves reflected from the moving tissue of one point on the heart. The beat frequency between the two signals has a maximum value of 520 Hz. What is the maximum speed of the heart tissue?

Passage Problems

Harmonics and Harmony

You know that certain musical notes sound good together—harmonious—whereas others do not. This harmony is related to the various harmonics of the notes.

The musical notes C (262 Hz) and G (392 Hz) make a pleasant sound when played together; we call this consonance. As Figure P16.72 shows, the harmonics of the two notes are either far from each other or very close to each other (within a few Hz). This is the key to consonance: harmonics that are spaced either far apart or very close. The close harmonics have a beat frequency of a few Hz that is perceived as pleasant. If the harmonics of two notes are close but not too close, the rather high beat frequency between the two is quite unpleasant. This is what we hear as dissonance. Exactly how much a difference is maximally dissonant is a matter of opinion, but harmonic separations of 30 or 40 Hz seem to be quite unpleasant for most people.

FIGURE P16.72

72. | What is the beat frequency between the second harmonic of G and the third harmonic of C?
A. 1 Hz   B. 2 Hz   C. 4 Hz   D. 6 Hz

73. | Would a G-flat (frequency 370 Hz) and a C played together be consonant or dissonant?
A. Consonant
B. Dissonant

74. | An organ pipe open at both ends is tuned so that its fundamental frequency is a G. How long is the pipe?
A. 43 cm   B. 87 cm   C. 130 cm   D. 173 cm

75. | If the C were played on an organ pipe that was open at one end and closed at the other, which of the harmonic frequencies in Figure 16.72 would be present?
A. All of the harmonics in the figure would be present.
B. 262, 786, and 1310 Hz
C. 524, 1048, and 1572 Hz
D. 262, 524, and 1048 Hz

STOP TO THINK ANSWERS

Stop to Think 16.1: C. The figure shows the two waves at $t = 6$ s and their superposition. The superposition is the *point-by-point* addition of the displacements of the two individual waves.

Stop to Think 16.2: C. Standing-wave frequencies are $f_m = mf_1$. The original wave has frequency $f_2 = 2f_1$ because it has two antinodes. The wave with frequency $2f_2 = 4f_1$ is the $m = 4$ mode with four antinodes.

Stop to Think 16.3: B. 300 Hz and 400 Hz are not f_1 and f_2 because 400 Hz $\neq$ 2 × 300 Hz. Instead, both are multiples of the fundamental frequency. Because the difference between them is 100 Hz, we see that $f_3 = 3 \times 100$ Hz and $f_4 = 4 \times 100$ Hz. Thus $f_1 = 100$ Hz.

Stop to Think 16.4: D. Highest pitch, or highest frequency, corresponds to the shortest period. For a complex wave, period is the time required for the entire wave pattern to repeat.

Stop to Think 16.5: Constructive interference. The path-length difference is $\Delta r = 1.0$ m $= \lambda$. Interference is constructive when the path-length difference is a whole number of wavelengths.

Stop to Think 16.6: F. The beat frequency is the difference between the two frequencies.

Oscillations and Waves

As we have studied oscillations and waves, one point we have emphasized is the *unity* of the basic physics. The mathematics of oscillation describes the motion of a mass on a spring or the motion of a gibbon swinging from a branch. The same theory of waves works for string waves and sound waves and light waves. A few basic ideas enable us to understand a wide range of physical phenomena.

The physics of oscillations and waves is not quite as easily summarized as the physics of particles. Newton's laws and the conservation laws are two very general sets of principles about particles, principles that allowed us to develop the powerful problem-solving strategies of Parts I and II. The knowledge structure of oscillations and waves, shown below, rests more heavily on *phenomena* than on general principles. This knowledge structure doesn't

contain a problem-solving strategy or a wide range of general principles, but is instead a logical grouping of the major topics you studied. This is a different way of structuring knowledge, but it still provides you with a mental framework for analyzing and thinking about wave problems.

The physics of oscillations and waves will be with us for the rest of the book. Part V is an exploration of optics, beginning with a detailed study of light as a wave. In Part VI, we will study the nature of electromagnetic waves in more detail. Finally, in Part VII, we will see that matter has a wave nature. There, the wave ideas we have developed in Part IV will lead us into the exciting world of quantum physics, finishing with some quite remarkable insights about the nature of light and matter.

KNOWLEDGE STRUCTURE IV **Oscillations and Waves**

BASIC GOALS   How can we describe oscillatory motion?   What are the distinguishing features of waves?
How does a wave travel through a medium?   What happens when two waves meet?

GENERAL PRINCIPLES   Simple harmonic motion occurs when a linear restoring force acts to return a system to equilibrium.
The period of simple harmonic motion depends on physical properties of the system but not the amplitude.
All sinusoidal waves, whether water waves, sound waves, or light waves, have the same functional form.
When two waves meet, they pass through each other, combining where they overlap by superposition.

Simple harmonic motion

$$x(t) = A\cos(2\pi ft)$$

$$v_x(t) = -v_{max}\sin(2\pi ft)$$
$$v_{max} = 2\pi fA$$

$$a_x(t) = -a_{max}\cos(2\pi ft)$$
$$a_{max} = (2\pi f)^2 A$$

The frequency of a mass on a spring depends on the spring constant and the mass:

$$f = \frac{1}{2\pi}\sqrt{\frac{k}{m}}$$

The frequency of a pendulum depends on the length and the acceleration of gravity:

$$f = \frac{1}{2\pi}\sqrt{\frac{g}{L}}$$

Traveling waves

All traveling waves are described by the same equation:

$$y = A\cos\left(2\pi\left(\frac{x}{\lambda} \pm \frac{t}{T}\right)\right)$$

+: wave travels to left
−: wave travels to right

The wave speed depends on the properties of the medium.

$$v_{string} = \sqrt{\frac{T_s}{\mu}}$$

The speed, wavelength, period and frequency of a wave are related.

$$T = \frac{1}{f} \qquad v = f\lambda$$

Superposition and interference

The superposition of two waves is the sum of the displacements of the two individual waves.

Constructive interference occurs when crests are aligned with crests and troughs with troughs. Constructive interference occurs when the path-length difference is a whole number of wavelengths.

Destructive interference occurs when crests are aligned with troughs. Destructive interference occurs when the path-length difference is a whole number of wavelengths plus half a wavelength.

Standing waves

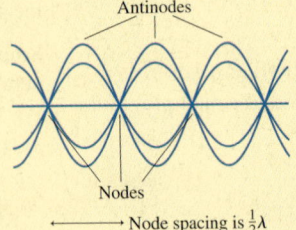

Antinodes

Nodes

Node spacing is $\frac{1}{2}\lambda$

Standing waves are due to the superposition of two traveling waves moving in opposite directions.

The boundary conditions determine which standing wave frequencies and wavelengths are allowed. The allowed standing waves are modes of the system.

For a string of length L, the modes have wavelength and frequency:

$$\lambda_m = \frac{2L}{m} \qquad f_m = m\left(\frac{v_{string}}{2L}\right) = mf_1 \qquad m = 1, 2, 3, 4, \ldots$$

Waves in the Earth and the Ocean

In December 2004, a large earthquake off the coast of Indonesia produced a devastating water wave, called a *tsunami,* that caused tremendous destruction thousands of miles away from the earthquake's epicenter. The tsunami was a dramatic illustration of the energy carried by waves.

It was also a call to action. Many of the communities hardest hit by the tsunami were struck hours after the waves were generated, long after seismic waves from the earthquake that passed through the earth had been detected at distant recording stations, long after the possibility of a tsunami was first discussed. With better detection and more accurate models of how a tsunami is formed and how a tsunami propagates, the affected communities could have received advance warning. The study of physics may seem an abstract undertaking with few practical applications, but on this day a better scientific understanding of these waves could have averted tragedy.

Let's use our knowledge of waves to explore the properties of a tsunami. In Chapter 15, we saw that a vigorous shake of one end of a rope causes a pulse to travel along it, carrying energy as it goes. The earthquake that produced the Indian Ocean tsunami of 2004 caused a sudden upward displacement of the sea floor that produced a corresponding rise in the surface of the ocean. This was

One frame from a computer simulation of the Indian Ocean tsunami three hours after the earthquake that produced it. The disturbance propagating outward from the earthquake is clearly seen, as are wave reflections from the island of Sri Lanka.

the *disturbance* that produced the tsunami, very much like a quick shake on the end of a rope. The resulting wave propagated through the ocean, as we see in the figure.

This simulation of the tsunami looks much like the ripples that spread when you drop a pebble into a pond. But there is a big difference—the scale. The fact that you can see the individual waves on this diagram that spans 5000 km is quite revealing. To show up so clearly, the individual wave pulses must be very wide—up to hundreds of kilometers from front to back.

A tsunami is actually a "shallow water wave," even in the deep ocean, because the depth of the ocean is much less than the width of the wave. Consequently, a tsunami travels differently than normal ocean waves. In Chapter 15 we learned that wave speeds are fixed by the properties of the medium. That is true for normal ocean waves, but the width of the wave causes a tsunami to "feel the bottom." Its wave speed is determined by the depth of the ocean: the greater the depth, the greater the speed. In the deep ocean, a tsunami travels at hundreds of kilometers per hour, much faster than a typical ocean wave. Near shore, as the ocean depth decreases, so does the speed of the wave.

The height of the tsunami in the open ocean was about half a meter. Why should such a small wave—one that ships didn't even notice as it passed—be so fearsome? Again, it's the *width* of the wave that matters. Because a tsunami is the wave motion of a considerable mass of water, great energy is involved. As the front of a tsunami wave nears shore, its speed decreases, and the back of the wave moves faster than the front. Consequently, the width decreases. The water begins to pile up, and the wave dramatically increases in height.

The Indian Ocean tsunami had a height of up to 15 m when it reached shore, with a width of up to several kilometers. This tremendous mass of water was still moving at high speed, giving it a great deal of energy. A tsunami reaching the shore isn't like a typical wave that breaks and crashes. It is a kilometers-wide wall of water that moves onto the shore and just keeps on coming. In many places, the water reached 2 km inland.

The impact of the Indian Ocean tsunami was devastating, but it was the first tsunami for which scientists were able to use satellites and ocean sensors to make planet-wide measurements. An analysis of the data has helped us better understand the physics of these ocean waves. We won't be able to stop future tsunamis, but with a better knowledge of how they are formed and how they travel, we will be better able to warn people to get out of their way.

OPTICS

Light Is a Wave

Isaac Newton is best known for his studies of mechanics and the three laws that bear his name, but he also did important early work on optics. He was the first person to carefully study how a prism breaks white light into colors. Newton was a strong proponent of the "corpuscle" theory of light, arguing that light consists of a stream of particles.

In fact, Newton wasn't quite right. As you will see, the beautiful colors of a peacock's feathers and the shimmery rainbow of a soap bubble both depend on the fact that light is a wave, not a particle. In particular, light is an electromagnetic wave, although these chapters depend on nothing more than the "waviness" of light waves for your understanding. The wave theory we developed in Part IV will be put to good use in Part V as we begin our investigation of light and optics with an analysis of the *wave model* of light.

The Ray Model

Yet Newton was correct in his observation that light seems to travel in straight lines, something we wouldn't expect a wave to do. Consequently, our investigations of how light works will be greatly aided by another model of light, the *ray model,* in which light travels in straight lines, bounces from mirrors, and is bent by lenses.

The ray model will be an excellent tool for analyzing many of the practical applications of optics. When you look in a mirror, you see an *image* of yourself that appears to be behind the mirror. We will use the ray model of light to determine just how it is that mirrors and lenses form images. At the same time, we will need to reconcile the wave and ray models, learning how they are related to each other and when it is appropriate to use each.

Working with Light

The nature of light is quite subtle and elusive. In Parts VI and VII, we will turn to the question of just what light is. As we will see, light has both wave-like *and* particle-like aspects. For now, however, we will set this question aside and work with the wave and ray models to develop a practical understanding of light. This will lead us, in Chapter 19, to an analysis of some common optical instruments. We will explore how a camera captures images and how telescopes and microscopes work.

Ultimately, the fact that you are reading this book is due to the optics of the first optical instrument you ever used, your eye! We will investigate the optics of the eye, learn how the cornea and lens bend light to create an image on your retina, and see how glasses or contact lenses can be used to correct the image should it be out of focus.

These images of animal eyes and light-sensing organs reveal the wide range of structures that can produce a visual sense. There is a certain similarity among eyes as well; did you spot the false eyes, patterns that are designed to mimic the appearance of an eye?

17

WAVE OPTICS

The vivid colors of this hummingbird's feathers have a sheen unlike that of ordinary pigments, and they change subtly depending on the angle at which they're viewed. How does light interact with the feathers to produce this bright display?

You may be surprised to learn that the iridescence of a hummingbird's feathers is closely related to the colors that flash on the surface of a compact disk, as well as to the technology underlying supermarket checkout scanners and the hologram on your credit card. All of these, in one way or another, depend on the interference of light waves. In some, such as the CD, the interference is incidental to the intended use of the product. Others, such as holograms, have been designed to make precise use of the wave-like properties of light.

The study of light is called **optics,** and this is the first of three chapters to explore optics and the nature of light. Light is an elusive topic. You will find, perhaps surprisingly, that there is no simple description of light. Light behaves quite differently in different situations, and we will ultimately need three different *models* of light to capture this behavior. We begin, in this chapter, with situations in which light acts as a wave. The groundwork for *wave optics* has been laid in Chapters 15 and 16, and we will now apply those ideas to light waves.

Although light is an electromagnetic wave, this chapter depends on nothing more than the "waviness" of light waves. You can study this chapter either before or after your study of electricity and magnetism in Part VI.

17.1 What Is Light?

What is light? The first Greek scientists and philosophers did not make a distinction between light and vision. Light, to them, was not something that existed apart from seeing. But gradually there arose a view that light actually "exists," that light is some sort of physical entity that is present regardless of whether or not someone is looking. Newton, in addition to doing pioneering work in mathematics and mechanics in the 1660s, was also one of the early investigators of light. He believed that light consists of very small, light, fast particles, which he called *corpuscles,* traveling in straight lines. Newton was vigorously opposed by Robert Hooke (of Hooke's law) and the Dutch scientist Christian Huygens, who argued that light was some sort of wave. Although the debate was lively, and sometimes acrimonious, Newton eventually prevailed. The belief that light consists of corpuscles was not seriously questioned for more than a hundred years after Newton's death.

The situation changed dramatically in 1801, when the English scientist Thomas Young announced that he had produced *interference* between two waves of light. Young's experiment, which we will analyze in this chapter, was a painstakingly difficult experiment with the technology of his era. Nonetheless, Young's experiment appeared to settle the debate in favor of a wave theory of light because interference is a distinctly wave-like phenomenon. But if light is a wave, why does it sometimes seem to travel in straight lines? And just what kind of wave is it?

Models of Light

As Newton, Young, and others found, the nature of light is elusive. Light is the chameleon of the physical world. Under some circumstances, light acts like particles traveling in straight lines. But change the circumstances, and light shows the same kinds of wave-like behavior as sound waves or water waves. Change the circumstances yet again, and light exhibits behavior that is neither wave-like nor particle-like but has characteristics of both.

Rather than an all-encompassing "theory of light," it will be better to develop several **models of light.** Each model successfully explains the behavior of light within a certain domain—that is, within a certain range of physical situations. Our task will be twofold:

1. To develop clear and distinct models of light.
2. To learn the conditions and circumstances for which each model is valid.

The second task is especially important.

We'll begin with a brief summary of all three models, so that you will have a road map of where we're headed. Each of these models will be developed in the coming chapters.

The wave model: The wave model of light is the most widely applicable model, responsible for the widely known "fact" that light is a wave. It is certainly true that, under many circumstances, light exhibits the same behavior as sound or water waves. Lasers and electro-optical devices, critical technologies of the 21st century, are best understood in terms of the wave model of light. Some aspects of the wave model of light were introduced in Chapters 15 and 16, and the wave model is the primary focus of this chapter. The study of light as a wave is called **wave optics.**

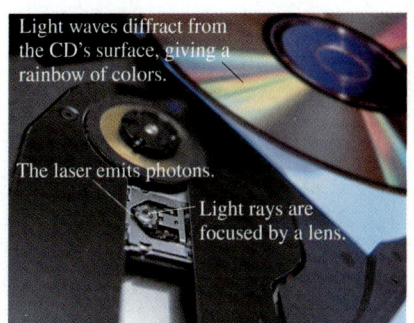

Light waves diffract from the CD's surface, giving a rainbow of colors.

The laser emits photons.

Light rays are focused by a lens.

All three models of light are needed to explain the operation of a CD player, and the rainbow colors of the CD itself.

The ray model: An equally well-known "fact" is that light travels in a straight line. These straight-line paths are called *light rays.* In Newton's view, light rays are the trajectories of particle-like corpuscles of light. The properties of prisms, mirrors, lenses, and optical instruments such as telescopes and microscopes are best understood in terms of light rays. Unfortunately, it's difficult to reconcile the statement "light travels in a straight line" with the statement "light is a wave." For the most part, waves and rays are mutually exclusive models of light. One of our most important tasks will be to learn when each model is appropriate. The ray model of light, the basis of **ray optics,** is the subject of the next chapter.

The photon model: Modern technology is increasingly reliant on quantum physics. In the quantum world, light behaves like neither a wave nor a particle. Instead, light consists of *photons* that have both wave-like and particle-like properties. Photons are the *quanta* of light. Much of the quantum theory of light is beyond the scope of this textbook, but we will take a peek at the important ideas in Chapters 25 and 28 of this text.

The Propagation of Light Waves

Figure 17.1 shows an everyday observation about light. The sunlight passing through the door makes a sharp-edged shadow as it falls upon the floor. This behavior is exactly what you would expect if light consisted of little particles traveling outward from the source in straight lines. Indeed, it was the observation of sharp-edged shadows that led Newton to propose the idea of corpuscles of light.

Is this behavior, where light seems to travel in straight lines, consistent with the motion of a wave? Light waves oscillates so rapidly—at some 10^{14} Hz—that we cannot hope to directly observe the crests and troughs of its electric or magnetic fields. Instead, let's use ordinary water waves to illustrate some basic properties of wave motion common to all waves, including light. Figure 17.2 shows a water wave, as seen looking straight down on the water surface. The wave enters from the top of the picture, then passes though a window-like opening in a barrier. After passing through the opening, the wave *spreads out* to fill the space behind the opening. This spreading of a wave is the phenomenon called **diffraction.** Diffraction is a sure sign that whatever is passing through the opening is a wave.

The straight-line travel of light appears to be incompatible with this spreading of a wave. Notice, however, that the width of the opening in Figure 17.2 is only slightly larger than the wavelength of the water wave. As Figure 17.3 shows, something quite different occurs when we make the opening much wider. Rather than spreading out, now the wave continues to move straight forward, with a well-defined boundary between where the wave is moving and its "shadow," where there is no wave. This is just the behavior observed for light in Figure 17.1.

Whether a wave spreads out or travels straight ahead, with sharp shadows on both sides, evidently depends on the size of the objects with which the wave interacts. The spreading of diffraction becomes noticeable only when an opening or object is "narrow," comparable in size to the wavelength of the wave. Thus we would expect a light wave to spread out in this way only when it passes objects that are comparable in size to the wavelength of light.

But the wavelength of light is *extremely* short. Later in this chapter we'll see how its wavelength can be measured, and we'll find that a typical wavelength of light is only about 0.5 μm. When light waves interact with everyday-sized objects, such as the opening in Figure 17.1, the situation is like that of the water wave in Figure 17.3. The wave travels straight ahead, and we'll be able to use the ray model of light. Only when the size of an object or an opening approaches the wavelength of light does the wavelike spreading become important. For example,

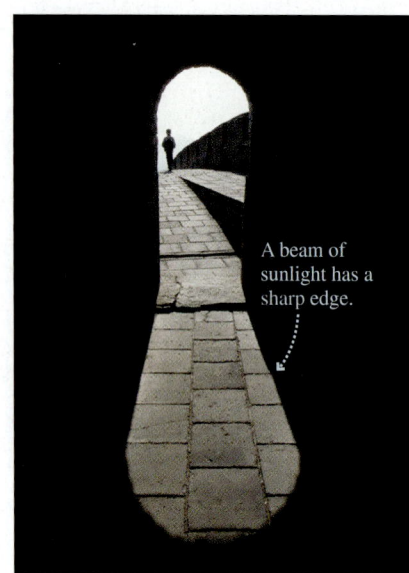

A beam of sunlight has a sharp edge.

FIGURE 17.1 Light passing through an opening makes a sharp-edged shadow.

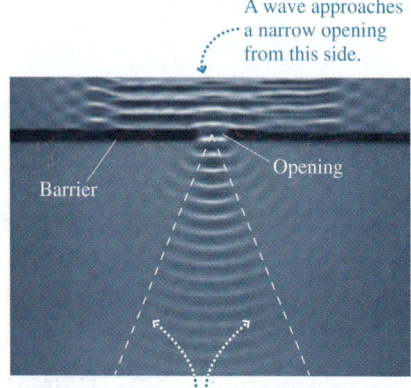

A wave approaches a narrow opening from this side.

Barrier   Opening

The wave spreads out behind the opening.

FIGURE 17.2 A water wave passing through a narrow opening in a barrier.

Barrier   Barrier

"Shadow"   "Shadow"

The wave moves straight forward.

FIGURE 17.3 A water wave passing through an opening that's many wavelengths wide.

Figure 17.4 shows the shadow of a 0.5-mm-diameter pin. For an object this small, the shadow is *not* sharp, but shows a series of bright and dark *fringes* as we move away from the center of the shadow. One of the goals of this chapter is to understand the origin of such fringes, a goal for which we'll need the wave model of light.

Light Is an Electromagnetic Wave

If light is a wave, what is it that is waving? This was the question posed at the start of the 19th century by Young's demonstration of the interference of light. As we briefly noted in Chapter 15, it was ultimately established, through theoretical and experimental efforts by numerous scientists, that light is an *electromagnetic wave,* an oscillation of electric and magnetic fields. We will examine the nature of electromagnetic waves in more detail in Part VI after we introduce the ideas of electric and magnetic fields. For now we can say that light waves are a "self-sustaining oscillation of the electromagnetic field." Being self-sustaining means that electromagnetic waves require *no material medium* in order to travel, hence electromagnetic waves are not mechanical waves. Fortunately, we can learn about the wave properties of light without having to understand electromagnetic fields. In fact, the discovery that light propagates as a wave was made 60 years before it was realized that light is an electromagnetic wave. We, too, will be able to learn much about the wave nature of light without having to know just what it is that is waving.

FIGURE 17.4 Diffraction of light by the point of a pin.

Recall that all electromagnetic waves, including light waves, travel in a vacuum at the same speed, called the *speed of light.* Its value is

$$v_{light} = c = 3.00 \times 10^8 \text{ m/s}$$

where the special symbol c is used to designate the speed of light.

Recall also that the wavelengths of light are extremely small, ranging from about 400 nm for violet light to 700 nm for red light. (We'll see later in this chapter how these wavelengths are determined.) Electromagnetic waves with wavelengths outside this range are not visible to the human eye. A prism is able to spread the different wavelengths apart, from which we learn that "white light" is all the colors, or wavelengths, combined. The spread of colors seen with a prism, or seen in a rainbow, is called the *visible spectrum.*

If the wavelengths of light are incredibly small, the oscillation frequencies are unbelievably large. The frequency for a 600 nm wavelength of light is

$$f = \frac{v}{\lambda} = \frac{3.00 \times 10^8 \text{ m/s}}{600 \times 10^{-9} \text{ m}} = 5.00 \times 10^{14} \text{ Hz}$$

The frequencies of light waves are roughly a factor of a trillion (10^{12}) higher than sound frequencies.

White light passing through a prism is spread out into a band of colors called the *visible spectrum.*

The Index of Refraction

Light waves travel with speed c in a vacuum, but they slow down as they pass through transparent materials such as water or glass or even, to a very slight extent, air. The slowdown is a consequence of interactions between the electromagnetic field of the wave and the electrons in the material. The speed of light in a material is characterized by the material's **index of refraction** n, defined as

$$n = \frac{\text{speed of light in a vacuum}}{\text{speed of light in the material}} = \frac{c}{v} \tag{17.1}$$

where v is the speed of light in the material. The index of refraction of a material

TABLE 17.1 Typical indices of refraction

Material	Index of refraction
Vacuum	1 exactly
Air	1.0003
Water	1.33
Glass	1.50
Diamond	2.42

is always greater than 1 because v is always less than c. A vacuum has $n = 1$ exactly. Table 17.1 shows the index of refraction for several materials. Liquids and solids have larger indices of refraction than gases, simply because they have a much higher density of atoms for the light to interact with.

NOTE ▶ An accurate value for the index of refraction of air is relevant only in very precise measurements. We will assume $n_{air} = 1.00$ in this text. ◀

If the speed of a light wave changes as it enters into a transparent material, such as glass, what happens to the light's frequency and wavelength? Because $v = \lambda f$, either λ or f or both have to change when v changes.

As an analogy, think of a sound wave in the air as it impinges on the surface of a pool of water. As the air oscillates back and forth, it periodically pushes on the surface of the water. These pushes generate the compressions of the sound wave that continues on into the water. Because each push of the air causes one compression of the water, the frequency of the sound wave in the water must be *exactly the same* as the frequency of the sound wave in the air. In other words, **the frequency of a wave does not change as the wave moves from one medium to another.**

The same is true for electromagnetic waves, although the pushes are a bit more complex as the electric and magnetic fields of the wave interact with the atoms at the surface of the material. Nonetheless, the frequency does not change as the wave moves from one material to another.

Figure 17.5 shows a light wave passing through a transparent material with index of refraction n. As the wave travels through vacuum it has wavelength λ_{vac} and frequency f_{vac} such that $\lambda_{vac} f_{vac} = c$. In the material, $\lambda_{mat} f_{mat} = v = c/n$. The frequency does not change as the wave enters ($f_{mat} = f_{vac}$), so the wavelength must change. The wavelength in the material is

A transparent material in which light travels slower, at speed $v = c/n$

Vacuum $n = 1$ Index n $n = 1$

λ_{vac} $\lambda = \lambda_{vac}/n$

The wavelength inside the material decreases, but the frequency doesn't change.

FIGURE 17.5 Light passing through a transparent material with index of refraction n.

$$\lambda_{mat} = \frac{v}{f_{mat}} = \frac{c}{nf_{mat}} = \frac{c}{nf_{vac}} = \frac{\lambda_{vac}}{n} \tag{17.2}$$

The wavelength in the transparent material is less than the wavelength in vacuum. This makes sense. Suppose a marching band is marching at one step per second at a speed of 1 m/s. Suddenly they slow their speed to $\frac{1}{2}$ m/s but maintain their march at one step per second. The only way to go slower while marching at the same pace is to take *smaller steps*. When a light wave enters a material, the only way it can go slower while oscillating at the same frequency is to have a *smaller wavelength*.

EXAMPLE 17.1 Analyzing light traveling through glass

Orange light with a wavelength of 600 nm is incident upon a 1.00-mm-thick glass microscope slide.

a. What is the light speed in the glass?
b. How many wavelengths of the light are inside the slide?

SOLVE

a. From Table 17.1 we see that the index of refraction of glass is $n_{glass} = 1.50$. Thus the speed of light in glass is

$$v_{glass} = \frac{c}{n_{glass}} = \frac{3.00 \times 10^8 \text{ m/s}}{1.50} = 2.00 \times 10^8 \text{ m/s}$$

b. Because $n_{air} = 1.00$, the wavelength of the light is the same in air and vacuum: $\lambda_{vac} = \lambda_{air} = 600$ nm. Thus the wavelength inside the glass is

$$\lambda_{glass} = \frac{\lambda_{vac}}{n_{glass}} = \frac{600 \text{ nm}}{1.50} = 400 \text{ nm} = 4.00 \times 10^{-7} \text{ m}$$

N wavelengths span a distance $d = N\lambda$, so the number of wavelengths in $d = 1.00$ mm is

$$N = \frac{d}{\lambda} = \frac{1.00 \times 10^{-3} \text{ m}}{4.00 \times 10^{-7} \text{ m}} = 2500$$

ASSESS The fact that 2500 wavelengths fit within 1 mm shows how small the wavelengths of light are.

A light wave travels through three transparent materials of equal thickness. Rank in order, from the largest to smallest, the indices of refraction n_1, n_2, and n_3.

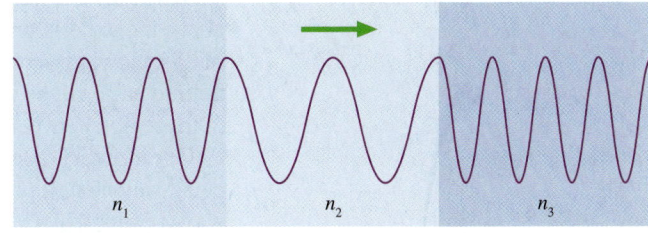

17.2 The Interference of Light

Consider the experiment depicted in Figure 17.6. Light passes through a narrow opening—a *slit*—that is only 0.1 mm wide, about twice the width of a human hair. The photograph shows how the light appears on a viewing screen 2 m behind the slit. If light consists of corpuscles traveling in straight lines, as Newton thought, we should see a narrow strip of light, about 0.1 mm wide, with dark shadows on either side. Instead, we see a band of light extending over about 2.5 cm, a distance much wider than the slit, with dimmer patches of light extending even farther on either side.

If you compare Figure 17.6 to the water wave of Figure 17.2, you see that *the light is spreading out* behind the 0.1-mm-wide slit, just as the water wave did behind the opening in the barrier. The light is exhibiting diffraction, the sure signature of waviness. We will look at diffraction in more detail later in the chapter. For now, we merely need the *observation* that light does, indeed, spread out behind an opening that is sufficiently small.

Young's Double-Slit Experiment

Rather than one small slit, suppose we use two. Figure 17.7a shows an experiment in which a laser beam is aimed at an opaque screen containing two long, narrow slits that are very close together. This pair of slits is called a **double slit,** and in a typical experiment they are ≈0.1 mm wide and spaced ≈0.5 mm apart. We will assume that the laser beam illuminates both slits equally, and any light passing through the slits impinges on a viewing screen. This is the essence of Young's experiment of 1801, although he used sunlight rather than a laser beam.

Viewing screen

The light spreads out behind the slit.

2.5 cm

2 m

0.1-mm-wide slit in an opaque screen

Incident laser beam

FIGURE 17.6 Light, just like a water wave, does spread out behind a slit in a screen *if* the slit is sufficiently small.

(a)

Viewing screen

The drawing is not to scale: the distance to the screen is actually *much* greater than the distance between the slits.

Double slit

Incident laser beam

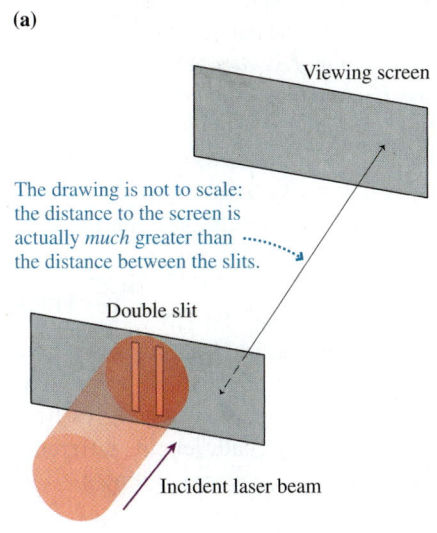

(b)

1. A plane wave is incident on the double slit.

2. Waves spread out behind each slit.

λ

Top view of the double slit

3. The waves interfere in the region where they overlap.

4. Bright fringes occur where the antinodal lines intersect the viewing screen.

$m = 4$
$m = 3$
$m = 2$
$m = 1$
$m = 0$ Central maximum
$m = 1$
$m = 2$
$m = 3$
$m = 4$

FIGURE 17.7 A double-slit interference experiment.

What should we expect to see on the screen? Figure 17.7b is a view from above the experiment, looking down on the top ends of the slits and the top edge of the viewing screen. Because the slits are very narrow, **light spreads out behind each slit** as it did in Figure 17.6, and these two spreading waves overlap in the region between the slits and the screen.

One primary conclusion of Chapter 16 was that two overlapped waves of equal wavelength produce interference. In fact, Figure 17.7b is equivalent to the waves emitted by two loudspeakers, a situation we analyzed in Section 16.6. (It is very useful to compare Figure 17.7b with Figure 16.32.) Nothing in that analysis depended on what type of wave it was, so the conclusions apply equally well to two overlapped light waves. If light really is a wave, we should see interference between the two light waves over the small region, typically 2 or 3 cm wide, where they overlap on the viewing screen.

The photograph in Figure 17.7b shows how the screen looks. As expected, the light is intense at points where an antinodal line intersects the screen. There is no light at all at points where a nodal line intersects the screen. These alternating bright and dark bands of light, due to constructive and destructive interference, are called **interference fringes.** The fringes are numbered $m = 0, 1, 2, 3, \ldots$, going outward from the center. The brightest fringe, at the midpoint of the viewing screen, with $m = 0$, is called the **central maximum.**

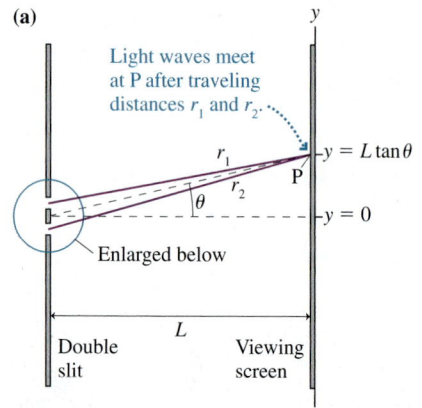

(a)

Light waves meet at P after traveling distances r_1 and r_2.

r_1

r_2

P

θ

$y = L\tan\theta$

$y = 0$

y

Enlarged below

Double slit

L

Viewing screen

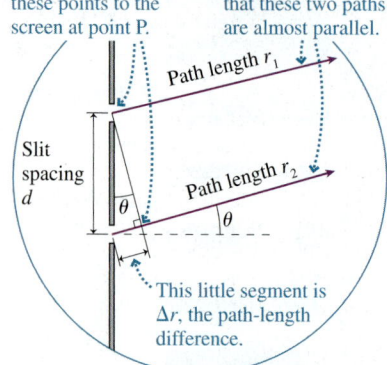

(b)

The two paths have equal lengths from these points to the screen at point P.

The screen is so far away compared to d that these two paths are almost parallel.

Path length r_1

Slit spacing d

θ

Path length r_2

θ

This little segment is Δr, the path-length difference.

FIGURE 17.8 Geometry of the double-slit experiment.

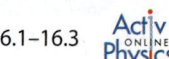

16.1–16.3

STOP TO THINK 17.2   Suppose the viewing screen in Figure 17.7 is moved closer to the double slit. What happens to the interference fringes?

A. They get brighter but otherwise do not change.
B. They get brighter and closer together.
C. They get brighter and farther apart.
D. They get out of focus.
E. They fade out and disappear.

Analyzing Double-Slit Interference

Figure 17.7 showed qualitatively that interference is produced behind a double slit by the overlap of the light waves spreading out behind each opening. Now let's analyze the experiment more carefully. Figure 17.8a shows the geometry of a double-slit experiment in which the spacing between the two slits is d and the distance to the viewing screen is L. We will assume that L is *very* much larger than d.

Our goal, as it was with sound waves in Chapter 16, is to determine if the interference at a particular point is constructive, destructive, or something in between. Let P be a point on the screen that is distance r_1 from one slit and r_2 from the other. We can specify point P either by the distance y from the midpoint ($y = 0$) on the viewing screen or by the angle θ from the midpoint between the slits. Angle θ and distance y are related by

$$y = L\tan\theta \qquad (17.3)$$

Figure 17.7b showed that both slits are illuminated by the *same* wave front from the laser. Consequently, the slits act as sources of identical waves. You learned in Chapter 16 that constructive interference between two waves from identical sources occurs at points for which the path-length difference $\Delta r = r_2 - r_1$ is an integer number of wavelengths:

$$\Delta r = m\lambda \qquad m = 0, 1, 2, 3, \ldots \qquad (17.4)$$

Thus the interference at point P is constructive, producing a bright fringe, if $\Delta r = m\lambda$ at that point.

The midpoint on the viewing screen at $y = 0$ is equally distant from both slits ($\Delta r = 0$). Constructive interference at the midpoint produces the bright fringe identified as the central maximum in Figure 17.7b. The path-length difference increases as you move away from the center of the screen, and the $m = 1$ fringes occur at the positions where $\Delta r = 1\lambda$. That is, one wave has traveled exactly one wavelength farther than the other. In general, **the mth bright fringe occurs where one wave has traveled m wavelengths farther than the other and thus $\Delta r = m\lambda$.**

We need to find the specific positions on the screen where $\Delta r = m\lambda$. Because the screen is very far away in comparison to the spacing between the slits, the two paths to point P are virtually parallel and thus the small triangle in the enlargement in Figure 17.8b is a right triangle whose angle θ is the angle from the slits to point P. The path difference between the two waves is the base of this triangle, so

$$\Delta r = d\sin\theta \tag{17.5}$$

Bright fringes (constructive interference) then occur at angles θ_m such that

$$\Delta r = d\sin\theta_m = m\lambda \qquad m = 0, 1, 2, 3, \ldots \tag{17.6}$$

We have added the subscript m to denote that θ_m is the angle of the mth bright fringe, starting with $m = 0$ at the center.

In practice, the angle θ in a double-slit experiment is almost always a very small angle ($<1°$). Recall, from Chapter 14, the *small-angle approximation* $\sin\theta \approx \theta$, where θ must be in *radians*. We can use this approximation to write Equation 17.6 as

$$\theta_m = m\frac{\lambda}{d} \qquad m = 0, 1, 2, 3, \ldots \tag{17.7}$$

Angles (in radians) of bright fringes for
double-slit interference with slit spacing d

This gives the angular positions *in radians* of the bright fringes in the interference pattern.

It is usually more convenient to measure the *position* of the mth bright fringe, as measured from the center of the viewing screen. Using the small-angle approximation once again, this time in the form $\tan\theta \approx \theta$, we can substitute θ_m from Equation 17.7 for $\tan\theta_m$ in Equation 17.3 to find that the mth bright fringe occurs at position

$$y_m = \frac{m\lambda L}{d} \qquad m = 0, 1, 2, 3, \ldots \tag{17.8}$$

Positions of bright fringes for
double-slit interference at screen distance L

The interference pattern is symmetrical, so there is an mth bright fringe at the same distance on both sides of the center. You can see this in Figure 17.7b. As we've already noted, **the $m = 1$ fringes occur at points on the screen where the light from one slit travels exactly one wavelength farther than the light from the other slit.**

TRY IT YOURSELF

Observing interference It's actually not that hard to observe double-slit interference. Place a piece of aluminum foil on a hard surface and cut two parallel slits, about 1 mm apart, using a razor blade. Now hold the slits up to your eye and look at a distant small bright light, such as a streetlight at night. Because of diffraction, you'll see the light source spread out in a direction *perpendicular* to the long direction of the slits. Superimposed on the diffraction pattern is a fine pattern of interference maxima and minima, as seen in the picture above of Christmas tree lights taken through two slits in foil.

NOTE ▶ Equations 17.7 and 17.8 do *not* apply to the interference of sound waves from two loudspeakers. The approximations we've used (small angles, $L \gg d$) are usually not valid for the much longer wavelengths of sound waves. ◀

Equation 17.8 predicts that **the interference pattern is a series of equally spaced bright lines** on the screen, exactly as shown in Figure 17.7b. How do we know the fringes are equally spaced? The **fringe spacing** between the m fringe and the $m + 1$ fringe is

$$\Delta y = y_{m+1} - y_m = \frac{(m + 1)\lambda L}{d} - \frac{m\lambda L}{d} = \frac{\lambda L}{d} \qquad (17.9)$$

Because Δy is independent of m, *any* two bright fringes have the same spacing.

The dark fringes in the photograph are bands of destructive interference. You learned in Chapter 16 that destructive interference occurs at positions where the path-length difference of the waves is a whole number of wavelengths plus a half wavelength:

$$\Delta r = \left(m + \frac{1}{2}\right)\lambda \qquad m = 0, 1, 2, 3, \ldots \qquad (17.10)$$

We can use Equation 17.6 for Δr and the small-angle approximation to find that the dark fringes are located at positions

$$y'_m = \left(m + \frac{1}{2}\right)\frac{\lambda L}{d} \qquad m = 0, 1, 2, 3, \ldots \qquad (17.11)$$

Positions of dark fringes for double-slit interference

We have used y'_m, with a prime, to distinguish the location of the mth minimum from the mth maximum at y_m. You can see from Equation 17.11 that **the dark fringes are located exactly halfway between the bright fringes.**

Figure 17.9 is a graph of the double-slit intensity versus y. Notice the unusual orientation of the graph, with the intensity increasing toward the left so that the y-axis can match the experimental layout. You can see that the intensity oscillates between dark fringes, where the intensity is zero, and equally spaced bright fringes of maximum intensity. The maxima occur at positions where $y_m = m\lambda L/d$.

The fringe intensity decreases because the light intensity from each slit by itself is not uniform.

Fringe spacing Δy

Slits

Central maximum

The bright fringes are equally spaced.

Increasing light intensity

FIGURE 17.9 Intensity of the interference fringes in the double-slit experiment.

EXAMPLE 17.2 Fringe spacing for a laser beam illuminating a double slit

Light from a helium-neon laser ($\lambda = 633$ nm) illuminates two slits spaced 0.40 mm apart. A viewing screen is 2.0 m behind the slits. What are the distances between the two $m = 2$ bright fringes and between the two $m = 2$ dark fringes?

PREPARE Two closely spaced slits produce a double-slit interference pattern. The interference pattern looks like the photograph of Figure 17.7b. It is symmetrical, with $m = 2$ bright fringes at equal distances on both sides of the central maximum.

SOLVE The $m = 2$ bright fringe is located at position

$$y_m = \frac{m\lambda L}{d} = \frac{2(633 \times 10^{-9}\text{ m})(2.0\text{ m})}{4.0 \times 10^{-4}\text{ m}}$$

$$= 6.3 \times 10^{-3}\text{ m} = 6.3\text{ mm}$$

Each of the $m = 2$ fringes is 6.3 mm from the central maximum, hence the distance between the two $m = 2$ bright fringes is 12.6 mm. The $m = 2$ dark fringe is located at

$$y'_m = \left(m + \frac{1}{2}\right)\frac{\lambda L}{d} = 7.9\text{ mm}$$

Thus the distance between the two $m = 2$ dark fringes is 15.8 mm.

ASSESS Because the fringes are counted outward from the center, the $m = 2$ bright fringe occurs *before* the $m = 2$ dark fringe. Note that this means there are *two* $m = 0$ dark fringes, one on each side of the single $m = 0$ bright fringe.

EXAMPLE 17.3 Measuring the wavelength of light

A double-slit interference pattern is observed on a screen 1.0 m behind two slits spaced 0.30 mm apart. From the center of one particular fringe to the center of the ninth bright fringe from this one is 1.6 cm. What is the wavelength of the light?

PREPARE It is not always obvious which fringe is the central maximum. Slight imperfections in the slits can make the interference fringe pattern less than ideal. However, you do not need to identify the $m = 0$ fringe because you can make use of the fact that the fringe spacing Δy is uniform. The interference pattern looks like the photograph of Figure 17.7b.

SOLVE The fringe spacing is

$$\Delta y = \frac{1.6 \text{ cm}}{9} = 1.8 \times 10^{-3} \text{ m}$$

Using this fringe spacing in Equation 17.9, we find that the wavelength is

$$\lambda = \frac{d}{L} \Delta y = \frac{3.0 \times 10^{-4} \text{ m}}{1.0 \text{ m}} (1.8 \times 10^{-3} \text{ m})$$

$$= 5.4 \times 10^{-7} \text{ m} = 540 \text{ nm}$$

It is customary to express the wavelengths of light in nanometers. Be sure to do this as you solve problems.

ASSESS Young's double-slit experiment not only demonstrated that light is a wave, it provided a means for measuring the wavelength. You saw earlier in this chapter that the wavelengths of visible light span the range 400–700 nm. These lengths are smaller than we can easily comprehend. A wavelength of 540 nm, which is in the middle of the visible spectrum, is only about 1% of the diameter of a human hair.

STOP TO THINK 17.3   Light of wavelength λ_1 illuminates a double slit, and interference fringes are observed on a screen behind the slits. When the wavelength is changed to λ_2, the fringes get closer together. Is λ_2 larger or smaller than λ_1?

A microscopic side-on look at a diffraction grating.

17.3 The Diffraction Grating

Suppose we were to replace the double slit with an opaque screen that has N closely spaced slits. When illuminated from one side, each of these slits becomes the source of a light wave that diffracts, or spreads out, behind the slit. Such a multi-slit device is called a **diffraction grating.** The light intensity pattern on a screen behind a diffraction grating is due to the interference of N overlapped waves.

Figure 17.10 shows a diffraction grating in which N slits are equally spaced a distance d apart. This is a top view of the grating, as we look down on the experiment, and the slits extend above and below the page. Only 10 slits are shown here, but a practical grating will have hundreds or even thousands of slits. Suppose a plane wave of wavelength λ approaches from the left. The crest of a plane wave arrives *simultaneously* at each of the slits, causing the wave emerging from each slit to be *in phase* with the wave emerging from every other slit—that is, all the emerging waves crest and trough simultaneously. Each of these emerging waves spreads out, just like the light wave in Figure 17.6, and after a short distance they all overlap with each other and interfere.

We want to know how the interference pattern will appear on a screen behind the grating. The light wave at the screen is the superposition of N waves, from N slits, as they spread and overlap. As we did with the double slit, we'll assume that the distance L to the screen is very large in comparison with the slit spacing d, hence the path followed by the light from one slit to a point on the screen is *very nearly* parallel to the path followed by the light from neighboring slits. The paths cannot be perfectly parallel, of course, or they would never meet to interfere, but the slight deviation from perfect parallelism is too small to notice. You can see in Figure 17.10 that the wave from one slit travels distance $\Delta r = d\sin\theta$ more than the wave from the slit above it and $\Delta r = d\sin\theta$ less than the wave below it. This is the same reasoning we used in Figure 17.8 to analyze the double-slit experiment.

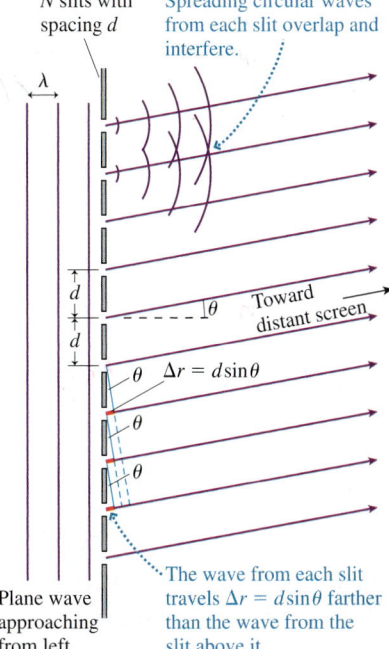

FIGURE 17.10 Top view of a diffraction grating with $N = 10$ slits.

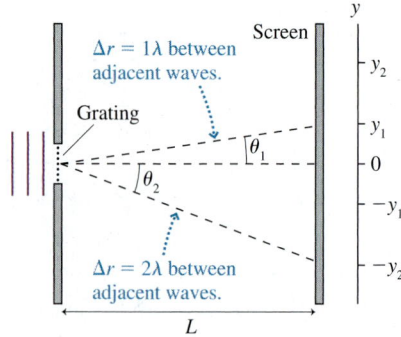

FIGURE 17.11 Angles of constructive interference.

16.4, 16.5

Figure 17.10 was a magnified view of the slits. Figure 17.11 steps back to where we can see the viewing screen. If the angle θ is such that $\Delta r = d\sin\theta = m\lambda$, where m is an integer, then the light wave arriving at the screen from one slit will travel *exactly m* wavelengths more or less than light from the two slits next to it, so these waves will be *exactly in phase* with each other. But each of those waves is in phase with waves from the slits next to them, and so on until we reach the end of the grating. In other words, *N* **light waves, from *N* different slits, will *all* be in phase with each other when they arrive at a point on the screen at angle θ_m such that**

$$d\sin\theta_m = m\lambda \qquad m = 0, 1, 2, 3, \ldots \qquad (17.12)$$

Angles of bright fringes due to a
diffraction grating with slits distance *d* apart

The screen will have bright constructive-interference fringes at the values of θ_m given by Equation 17.12. When this happens, we say that the light is "diffracted at angle θ_m." Because it's usually easier to measure distances rather than angles, the distance y_m from the center to the *m*th maximum is

$$y_m = L\tan\theta_m \qquad (17.13)$$

Positions of bright fringes due to a
diffraction grating distance *L* from screen

The integer *m* is called the **order** of the diffraction. For example, light diffracted at $\theta_2 = 60°$ would be the second-order diffraction. Practical gratings, with very small values for *d*, display only a few orders. Because *d* is usually very small, it is customary to characterize a grating by the number of *lines per millimeter*. Here "line" is synonymous with "slit," so the number of lines per millimeter is simply the inverse of the slit spacing *d* in millimeters.

NOTE ▶ The condition for constructive interference in a grating of *N* slits is identical to Equation 17.6 for just two slits. Equation 17.12 is simply the requirement that the path-length difference between adjacent slits, be they two or *N*, is $m\lambda$. But unlike the angles in double-slit interference, the angles of constructive interference from a diffraction grating are generally *not* small angles. The reason is that the slit spacing *d* in a diffraction grating is usually so small that λ/d is not a small number. Thus you *cannot* use the small-angle approximation to simplify Equations 17.12 and 17.13. ◀

Although the condition for constructive interference is the same for a diffraction grating as it was for the double slit, the intensity of the fringes for a grating differs from that for a double slit in two important ways. First, the bright fringes of a diffraction grating are *narrower* than those of a double slit with the same slit-spacing *d*. Second, the fringes are *brighter*. For a grating with *N* slits, it can be shown that the maximum intensity of a bright fringe is

$$I_{max} = N^2 I_1 \qquad (17.14)$$

Maximum intensity of a bright fringe
for a diffraction grating with *N* slits

QUADRATIC
p. 50

where I_1 is the intensity of the wave from a single slit. For a double slit, with $N = 2$, the intensity of the central maximum is 4 times that of each slit alone. For a practical diffraction grating, with hundreds of slits, the bright fringes are enormously brighter.

Not only do the fringes get brighter as N increases, they also get narrower. If you double the number of slits, for instance, twice as much light will reach the screen. But according to Equation 17.14, the maximum intensity of a bright fringe would increase by a factor of *four*. Since there's only twice as much light to work with, the fringes must be narrower in order to concentrate the light in each fringe enough to make it four times brighter. Figure 17.12 shows the interference pattern behind a diffraction grating both graphically and with a simulation of the viewing screen. A comparison with Figure 17.9 shows that the bright fringes of a diffraction grating are much sharper and more distinct than the fringes of a double slit.

To demonstrate the significance of the number of slits, Figure 17.13 shows the interference pattern behind three diffraction gratings with the same slit spacing d but increasing numbers of slits. The case with $N = 2$ is that of double-slit interference. As the number of slits is increased, the bright fringes become much narrower. For a realistic diffraction grating, with $N > 1000$, the interference pattern consists of a small number of *very* bright and *very* narrow fringes while most of the screen remains dark.

FIGURE 17.12 The intensity pattern behind a diffraction grating.

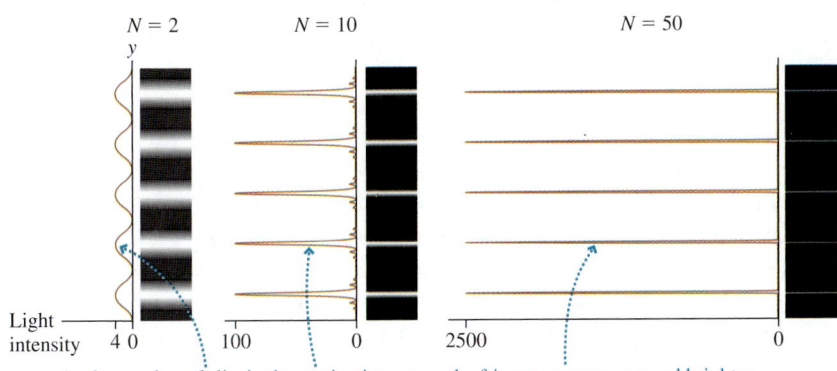

FIGURE 17.13 The intensity on the screen due to three diffraction gratings. Notice that the intensity axes have different scales.

Spectroscopy

As we'll see in Chapter 29, each atomic element in the periodic table, if appropriately excited by light, electricity, or collisions with other atoms, emits light at only certain well-defined wavelengths. By accurately measuring these wavelengths, the various elements in a sample of unknown composition can be deduced. Molecules also emit light that is characteristic of their composition. The science of measuring the wavelengths of atomic and molecular emissions is called **spectroscopy.**

Because their bright fringes are so distinct, diffraction gratings are an ideal tool for spectroscopy. Suppose the light incident on a grating consists of two slightly different wavelengths. According to Equation 17.12, each wavelength will diffract at a slightly different angle and, if N is sufficiently large, we'll see two distinct fringes on the screen. Figure 17.14 illustrates this idea. By contrast, the fringes in a double-slit experiment are so broad that it would not be possible to distinguish the fringes of one wavelength from those of the other.

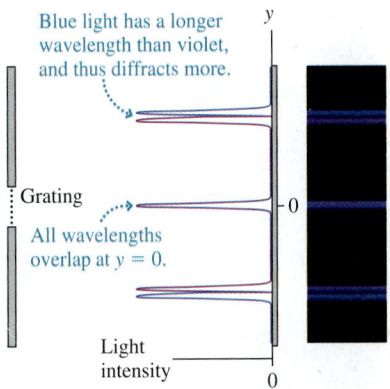

FIGURE 17.14 A diffraction grating can be used to measure the wavelengths of light.

EXAMPLE 17.4 Measuring wavelengths emitted by sodium atoms

Light from a sodium lamp passes through a diffraction grating having 1000 slits per millimeter. The interference pattern is viewed on a screen 1.000 m behind the grating. Two bright yellow fringes are visible 72.88 cm and 73.00 cm from the central maximum. What are the wavelengths of these two fringes?

PREPARE This situation is similar to that in Figure 17.14. The two fringes are very close together, so we expect the wavelengths to be only slightly different. No other yellow fringes are mentioned, so we will assume these two fringes are the first-order diffraction ($m = 1$).

SOLVE The distance y_m of a bright fringe from the central maximum is related to the diffraction angle by $y_m = L\tan\theta_m$. Thus the diffraction angles of these two fringes are

$$\theta_1 = \tan^{-1}\left(\frac{y_1}{L}\right) = \begin{cases} 36.08° & \text{fringe at 72.88 cm} \\ 36.13° & \text{fringe at 73.00 cm} \end{cases}$$

These angles must satisfy the interference condition $d\sin\theta_1 = \lambda$, so the wavelengths are

$$\lambda = d\sin\theta_1$$

What is d? If a 1 mm length of the grating has 1000 slits, then the spacing from one slit to the next must be 1/1000 mm, or $d = 1.00 \times 10^{-6}$ m. Thus the wavelengths creating the two bright fringes are

$$\lambda = d\sin\theta_1 = \begin{cases} 589.0 \text{ nm} & \text{fringe at 72.88 cm} \\ 589.6 \text{ nm} & \text{fringe at 73.00 cm} \end{cases}$$

ASSESS We had data accurate to four significant figures, and all four were necessary to distinguish the two wavelengths.

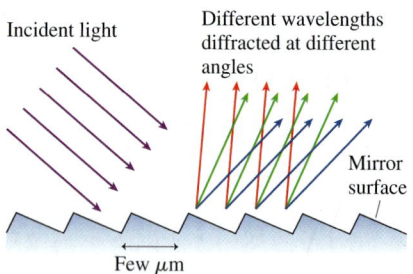

A reflection grating can be made by cutting parallel grooves in a mirror surface. These can be very precise, for scientific use, or mass produced in plastic.

FIGURE 17.15 A reflection grating.

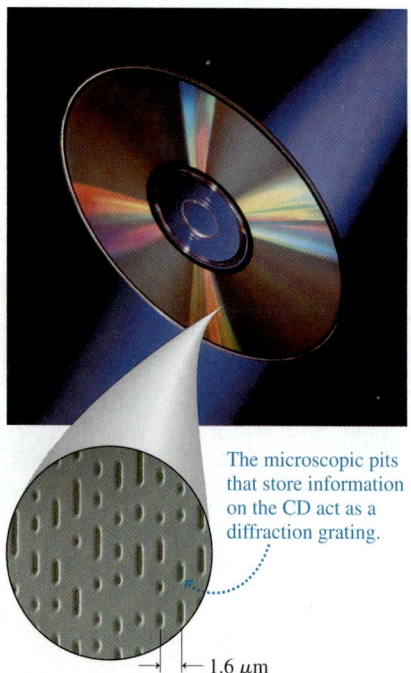

The microscopic pits that store information on the CD act as a diffraction grating.

FIGURE 17.16 A CD's colors are caused by diffraction.

Instruments that measure and analyze spectra, called *spectrophotometers,* are widely used in chemistry, biology, and medicine. Because each molecule has a distinct spectrum—a "fingerprint"—spectroscopy is used to identify specific biomolecules in tissue, drugs in urine, and chlorophyll in seawater.

Reflection Gratings

We have analyzed what is called a *transmission grating,* with many parallel slits. It's difficult to make such a grating with many closely spaced slits. In practice, most diffraction gratings are manufactured as *reflection gratings.* The simplest reflection grating, shown in Figure 17.15, is a mirror with hundreds or thousands of narrow, parallel grooves cut into the surface. The grooves divide the surface into many parallel reflective stripes, each of which, when illuminated, becomes the source of a spreading wave. Thus an incident light wave is divided into N overlapped waves. The interference pattern is exactly the same as the interference pattern of light transmitted through N parallel slits, and so **Equation 17.13 applies to reflection gratings as well as to transmission gratings.**

The rainbow of colors seen on the surface of a CD is an everyday display of interference. The surface of a CD is smooth plastic with a mirror-like reflective coating. As shown in Figure 17.16, millions and millions of microscopic holes, each about 1 μm in diameter, are "burned" into the surface with a laser. The presence or absence of a hole at a particular location on the disk is interpreted as the 0 or 1 of digitally encoded information. But from an optical perspective, the array of holes in a shiny surface is a two-dimensional version of the reflection grating shown in Figure 17.15. Less precise plastic reflection gratings can be manufactured at very low cost simply by stamping holes or grooves into a reflective surface, and these are widely sold as toys and novelty items. Rainbows of color are seen as each wavelength of white light is diffracted at a unique angle.

STOP TO THINK 17.4 White light passes through a diffraction grating and forms rainbow patterns on a screen behind the grating. For each rainbow,

A. The red side is on the right, the violet side on the left.
B. The red side is on the left, the violet side on the right.
C. The red side is closest to the center of the screen, the violet side is farthest from the center.
D. The red side is farthest from the center of the screen, the violet side is closest to the center.

17.4 Thin-Film Interference

In Chapter 16 you learned about the interference of sound waves in one dimension. Depending on whether they are in phase or out of phase, two sound waves of the same frequency, traveling in the same direction, can undergo constructive or destructive interference. Light waves can also interfere in this way. Equal-frequency light waves are produced when *partial reflection* at a boundary splits a light wave into a reflected wave and a transmitted wave. The interference of light waves reflected from the two boundaries of a thin film, such as the thin film of water that makes a soap bubble, is called **thin-film interference.**

Thin-film interference has important applications in the optics industry. Thin-film coatings, less than 1 μm (10^{-6} m) thick, are used for the anti-reflection coatings on the lenses in cameras, microscopes, and other optical equipment. The bright colors of oil slicks and soap bubbles are also due to thin-film interference.

The colors of an oil slick are due to thin-film interference. Different wavelengths of light, and thus different colors, are strongly reflected as the thickness of the oil film slowly changes.

Interference of Reflected Light Waves

As you know, and as we discussed in Chapter 16, a light wave encountering a piece of glass is partially transmitted and partially reflected. In fact, a light wave is partially reflected from *any* boundary between two transparent media with different indices of refraction. Thus light is partially reflected not only from the front surface of a sheet of glass, but from the back surface as well, as it exits from the glass into the air. This leads to the *two* reflections seen in Figure 17.17.

Another important aspect of wave reflections was shown for strings in Figure 16.10b of the last chapter. If a wave moves from a string with a higher wave speed to a string with a lower wave speed, the reflected wave is *inverted* with respect to the incoming wave. It is not inverted if the wave moves from a string with a lower wave speed to a string with a higher wave speed.

The same thing happens for light waves. When a light wave moves from a medium with a higher light speed (lower index of refraction) to a medium with a lower light speed (higher index of refraction), the reflected wave is inverted. This inversion of the wave, called a *phase change,* is equivalent to adding an extra half wavelength $\lambda/2$ to the distance the wave travels. You can see this in Figure 17.18, where a reflected wave with a phase change is compared to a reflection without a phase change. In summary, we can say that **a light wave undergoes a phase change if it reflects from a boundary at which the index of refraction increases.** There's no phase change at a boundary where the index of refraction decreases.

Consider a thin, transparent film with thickness t and index of refraction n coated onto a piece of glass. Figure 17.19 on the next page shows a light wave of wavelength λ approaching the film. Most of the light is transmitted into the film, but, as we've seen, a bit is reflected off the first (air-film) boundary. Further, a bit of the wave that continues into the film is reflected off the second (film-glass) boundary. The two reflected waves, which have exactly the same frequency, travel back out into the air where they overlap and interfere. As we learned in Chapter 16, the two reflected waves will interfere constructively to cause a *strong reflection* if they are *in phase* (i.e., if their crests overlap). If the two reflected waves are *out of phase,* with the crests of one wave overlapping the troughs of the other, they will interfere destructively to cause a *weak reflection* or, if their amplitudes are equal, *no reflection* at all.

We found the interference of two sound waves to be constructive if their path-length difference is $\Delta d = m\lambda$, destructive if $\Delta d = (m + \frac{1}{2})\lambda$, where m is an integer. The same idea holds true for reflected light waves, but we also have to consider the possible phase changes due to reflection. The physical path-length difference is the extra distance traveled by the wave that reflects from the second

FIGURE 17.17 Two reflections are visible in the window, one from each surface.

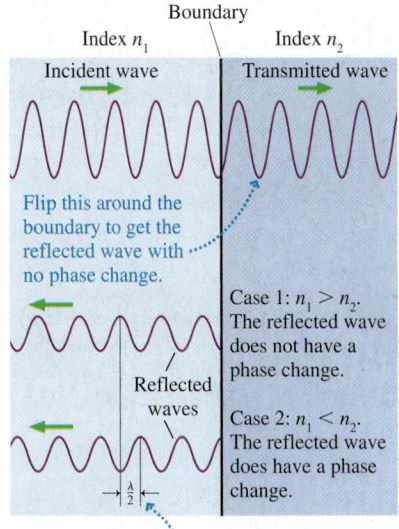

The reflection with the phase change is one half of a wavelength behind, so the effect of the phase change is to increase the path length by $\lambda/2$.

FIGURE 17.18 Reflected waves with and without a phase change.

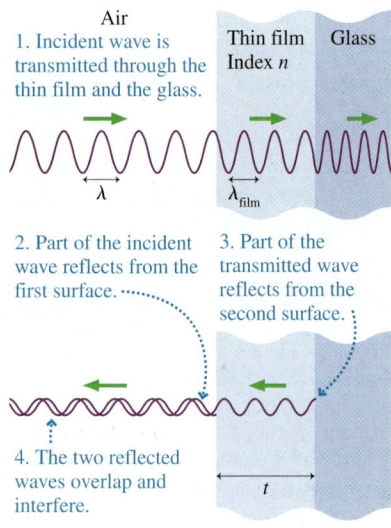

Air

1. Incident wave is transmitted through the thin film and the glass.

Thin film Index n Glass

λ λ_{film}

2. Part of the incident wave reflects from the first surface.

3. Part of the transmitted wave reflects from the second surface.

4. The two reflected waves overlap and interfere.

t

FIGURE 17.19 In thin-film interference, two reflections, one from the coating and one from the glass, overlap and interfere.

×10,000

Iridescent feathers BIO The gorgeous colors of the hummingbird shown at the beginning of this chapter are due not to pigments but to interference. This *iridescence*, present in some bird feathers and insect shells, arises from biological structures whose size is similar to the wavelength of light. The sheen of an insect, for instance, is due to thin-film interference from multiple thin layers in its shell. Peacock feathers are also a layered structure, but each layer itself consists of nearly parallel rods of melanin, as shown in the micrograph, that act as a diffraction grating. Thus a peacock feather combines thin-film interference and grating-like diffraction to produce its characteristic multicolored iridescent hues.

surface. Because this wave travels twice through a film of thickness t, the path-length difference is $\Delta d = 2t$.

We noted above that the phase change when a light wave reflects from a boundary with a higher index of refraction is equivalent to adding an extra half-wavelength to the distance traveled. This leads to two situations:

1. If *neither* or *both* waves have a phase change due to reflection, the net addition to the path-length difference is zero. The *effective path-length difference* is $\Delta d_{\text{eff}} = 2t$.

2. If only *one* wave has a phase change due to reflection, the effective path-length difference is increased by one-half wavelength to $\Delta d_{\text{eff}} = 2t + \frac{1}{2}\lambda$.

The interference of the two reflected waves is then constructive if $\Delta d_{\text{eff}} = m\lambda_{\text{film}}$ and destructive if $\Delta d_{\text{eff}} = \left(m + \frac{1}{2}\right)\lambda_{\text{film}}$. Why λ_{film}? Because the extra distance is traveled inside the film, so we need to compare $2t$ to the wavelength in the film. Further, the film's index of refraction is n, so the wavelength in the film is $\lambda_{\text{film}} = \lambda/n$, where λ is the wavelength of the light in vacuum or air.

With this information, we can write the conditions for constructive and destructive interference of the light waves reflected by a thin film:

$$2t = m\frac{\lambda}{n} \qquad (m = 0, 1, 2, \dots) \qquad (17.15)$$

Constructive interference for 0 or 2 reflective phase changes
Destructive interference for 1 reflective phase change

$$2t = \left(m + \frac{1}{2}\right)\frac{\lambda}{n} \qquad (m = 0, 1, 2, \dots) \qquad (17.16)$$

Destructive interference for 0 or 2 reflective phase changes
Constructive interference for 1 reflective phase change

NOTE ▶ Equations 17.15 and 17.16 give the film thicknesses that yield constructive or destructive interference. At other thicknesses, the waves will interfere neither fully constructively nor fully destructively, and the reflected intensity will fall somewhere in between these two extremes. ◀

These conditions are the basis of a procedure to analyze thin-film interference.

(MP)   **TACTICS BOX 17.1 Analyzing thin-film interference**   ✎ Exercise 10

Follow the light wave as it passes through the film. The wave reflecting from the second boundary travels an extra distance $2t$.

❶ Note the indices of refraction of the three media: the medium before the film, the film itself, and the medium beyond the film. The first and third may be the same. There's a reflective phase change at any boundary where the index of refraction increases.

❷ If *neither* or *both* reflected waves undergo a phase change, the phase changes cancel and the effective path-length difference is $\Delta d = 2t$. Use the version of Equations 17.15 and 17.16 for 0 or 2 reflective changes.

❸ If *only one* wave undergoes a phase change, the effective path-length difference is $\Delta d = 2t + \frac{1}{2}\lambda$. Use the version of Equations 17.15 and 17.16 for 1 reflective change.

EXAMPLE 17.5 Designing an antireflection coating

The glasses on the right have an antireflection coating on them. Those on the left do not.

To keep unwanted light from reflecting from the surface of eyeglasses or other lenses, a thin film of a material with an index of refraction $n = 1.38$ is coated onto the plastic lens ($n = 1.55$). It is desired to have destructive interference for $\lambda = 550$ nm because that is the center of the visible spectrum. What is the thinnest film that will do this?

PREPARE We follow the steps of Tactics Box 17.1. As the light traverses the film, it first reflects at the front surface of the coating. Here, the index of refraction increases from that of air ($n = 1.00$) to that of the film ($n = 1.38$), so there will be a reflective phase change. The light then reflects from the rear surface of the coating. The index again increases from that of

the film ($n = 1.38$) to that of the plastic ($n = 1.55$). With two phase changes, Tactics Box 17.1 tells us that we should use Equation 17.16 for destructive interference.

SOLVE We can solve Equation 17.16 for the thickness t that causes destructive interference:

$$t = \frac{\lambda}{2n}\left(m + \frac{1}{2}\right)$$

The thinnest film is the one for which $m = 0$, giving

$$t = \frac{550 \text{ nm}}{2(1.38)} \times \frac{1}{2} = 100 \text{ nm}$$

ASSESS A 100-nm-thick coating gives perfect destructive interference only at 550 nm. However, the interference is nearly destructive (i.e., crests are nearly aligned with troughs) over a range of wavelengths near 550 nm, hence the net reflection is reduced over the entire range of visible wavelengths. Interestingly, the value of the index of refraction of the plastic lens is unimportant, so long as it is greater than that of the coating. If it were less, there would be a reflective phase change at this surface.

Thin Films of Air

A film need not be of a solid material. A thin layer of air sandwiched between two glass surfaces also exhibits thin-film interference due to the waves that reflect off both air/glass boundaries. Figure 17.20 shows two microscope slides being pressed together. The light and dark "fringes" occur because the slides are not exactly flat and they touch each other only at a few points. Everywhere else there is a thin layer of air between them. At some points, the air layer's thickness is such as to give constructive interference (light fringes), while at other places its thickness results in destructive interference (dark fringes). These fringes can be used to very accurately measure the flatness of two glass plates, as Example 17.6 will show.

FIGURE 17.20 Light and dark fringes caused by thin-film interference due to the air layer between two microscope slides.

EXAMPLE 17.6 Finding the fringe spacing from a wedge-shaped film of air

Two 15-cm-long flat glass plates are separated by a 10-μm-thick spacer at one end, leaving a thin wedge of air between them, as shown in Figure 17.21.

One wave reflects off of this surface . . .
. . . while the other reflects off this surface.

$\updownarrow t$ $\updownarrow T = 10 \mu m$

x

$L = 15$ cm

FIGURE 17.21 Two glass plates with an air wedge between them.

The plates are illuminated by light from a sodium lamp with wavelength $\lambda = 589$ nm. Alternating bright and dark fringes are observed. What is the spacing between two bright fringes?

PREPARE The wave reflected from the lower plate has a reflective phase change, but the top reflection does not because the index of refraction decreases at the glass-air boundary. According to Tactics Box 17.1, we should use Equation 17.16 for constructive interference:

$$2t = \left(m + \frac{1}{2}\right)\frac{\lambda}{n}$$

This is a film of air, so here n is the index of refraction of air. Each integer value of m corresponds to a wedge thickness t for which there is constructive interference and thus a bright fringe.

Continued

SOLVE Let x be the distance from the left end to a bright fringe. From Figure 17.21, by similar triangles we have

$$\frac{t}{x} = \frac{T}{L}$$

or $t = xT/L$. From the condition for constructive interference, we then have

$$2\frac{xT}{L} = \left(m + \frac{1}{2}\right)\frac{\lambda}{n}$$

There will be a bright fringe for any integer value of m, and so the position of the mth fringe, as measured from the left end, is

$$x_m = \frac{\lambda L}{2nT}\left(m + \frac{1}{2}\right)$$

We want to know the spacing between two adjacent fringes, m and $m + 1$, which is

$$\Delta x = x_{m+1} - x_m = \frac{\lambda L}{2nT}\left(m + 1 + \frac{1}{2}\right) - \frac{\lambda L}{2nT}\left(m + \frac{1}{2}\right) = \frac{\lambda L}{2nT}$$

Evaluating, we find

$$\Delta x = \frac{\lambda L}{2nT} = \frac{(5.89 \times 10^{-7}\,\text{m})(0.15\,\text{m})}{2(1.00)(10 \times 10^{-6}\,\text{m})} = 4.4\,\text{mm}$$

ASSESS As the photo shows, if the two plates are very flat, the fringes will appear as straight lines perpendicular to the direction of increasing air thickness. However, if the plates are not quite flat, the fringes will appear curved. The amount of curvature indicates the departure of the plates from perfect flatness.

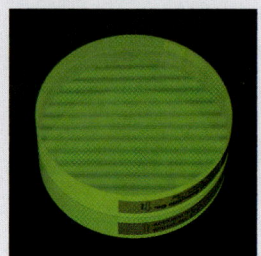

TABLE 17.2 Wavelengths for constructive and destructive interference from a 470-nm-thick soap bubble. Visible wavelengths are shown in **bold**.

Equation	$m = 1$	$m = 2$	$m = 3$
$\lambda_{\text{con}} = \dfrac{2nt}{m + \frac{1}{2}}$	833 nm	**500 nm**	357 nm
$\lambda_{\text{des}} = \dfrac{2nt}{m}$	1250 nm	**625 nm**	**417 nm**

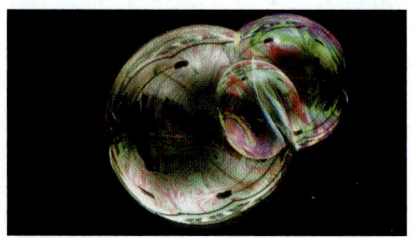

FIGURE 17.22 The colors of a soap bubble.

The Colors of Soap Bubbles and Oil Slicks

So far we have considered thin-film interference only for single wavelengths of light. The bright colors of soap bubbles and oil slicks on water are due to thin-film interference of white light, which, as we've seen, is a mixture of *all* wavelengths.

A soap bubble is a very thin spherical film of soapy water ($n = 1.33$). Consider a soap film with thickness $t = 470$ nm. Light waves reflect from both surfaces of the film, and these reflected waves interfere. The light reflecting from the front (air/water) surface has a reflective phase change, but the back reflection does not. Thus Equation 17.16 describes constructive interference and Equation 17.15 is destructive interference. Table 17.2 shows wavelengths of constructive and destructive interference for three values of m.

Light near the red end (625 nm) and violet end (417) of the spectrum undergoes destructive interference; these colors are *not* reflected by the film. At the same time, light near 500 nm (green) interferes constructively and so is strongly reflected. Consequently, a soap film of this thickness will appear green.

Real soap bubbles and oil slicks have thicknesses that vary from point to point. At some thicknesses, green light is strongly reflected, while at others red or violet light is. Figure 17.22 shows that the colors of a soap bubble are predominately greens or red/violets.

CONCEPTUAL EXAMPLE 17.1 **Colors in a vertical soap film**

Figure 17.23 shows a soap film in a wire loop. The loop is held vertically. Explain the colors seen in the film.

REASON Because of gravity, the film is thicker near the bottom and thinner at the top. It thus has a wedge shape, and the interference pattern consists of lines of alternating constructive and destructive interference, just as for the air wedge of Example 17.6. Because

FIGURE 17.23 A soap film in a wire loop.

this soap film is illuminated by white light, colors form as just discussed for any soap film.

Notice that the very top of the film, which is extremely thin, appears black. This means that it is reflecting no light at all. When the film is very thin—much thinner than the wavelength of light—there is almost no path difference between the two waves reflected off the front and the back of the film. However, the wave reflected off the back undergoes a reflective phase change and is out of phase with the wave reflected off the front. The two waves thus *always* interfere destructively, no matter what their wavelength.

ASSESS This simple experiment shows directly that the two reflected waves have different reflective phase changes.

Reflections from a thin layer of air between two glass plates cause constructive interference for a particular wavelength of light λ. By how much must the thickness of this layer be increased for the interference to be destructive?

A. λ/8   B. λ/4   C. λ/2   D. λ

17.5 Single-Slit Diffraction

We opened this chapter with a photograph of a water wave passing through a hole in a barrier, then spreading out on the other side. You then saw a photograph showing that light, after passing a narrow pin, also spreads out on the other side. This phenomenon is called *diffraction*. We're now ready to look at the details of diffraction.

Figure 17.24 again shows the experimental arrangement for observing the diffraction of light through a narrow slit of width *a*. Diffraction through a tall, narrow slit of width *a* is known as **single-slit diffraction.** A viewing screen is placed a distance *L* behind the slit, and we will assume that $L \gg a$. The light pattern on the viewing screen consists of a *central maximum* flanked by a series of weaker **secondary maxima** and dark fringes. Notice that the central maximum is significantly broader than the secondary maxima. It is also significantly brighter than the secondary maxima, although that is hard to tell here because this photograph has been overexposed to make the secondary maxima show up better.

Huygens' Principle

Our analysis of the superposition of waves from distinct sources, such as two loudspeakers or the two slits in a double-slit experiment, has tacitly assumed that the sources are *point sources,* with no measurable extent. To understand diffraction, we need to think about the propagation of an *extended* wave front. This problem was first considered by the Dutch scientist Christian Huygens, a contemporary of Newton who argued that light is a wave.

Huygens lived before a mathematical theory of waves had been developed, so he developed a geometrical model of wave propagation. His idea, which we now call **Huygens' principle,** has two steps:

1. Each point on a wave front is the source of a spherical *wavelet* that spreads out at the wave speed.
2. At a later time, the shape of the wave front is the line tangent to all the wavelets.

Figure 17.25 illustrates Huygens' principle for a plane wave and a spherical wave. As you can see, the line tangent to the wavelets of a plane wave is a plane that has propagated to the right. The line tangent to the wavelets of a spherical wave is a larger sphere. We've shown only a small number of wavelets, which leaves the new surface looking a bit bumpy, but you can easily imagine that the new surface would be perfectly smooth if we drew a wavelet for *every* point on the wave front.

Huygens' principle is a visual device, not a theory of waves. Nonetheless, the full mathematical theory of waves, as it developed in the 19th century, justifies Huygens' basic idea, although it is beyond the scope of this textbook to prove it.

FIGURE 17.24 A single-slit diffraction experiment.

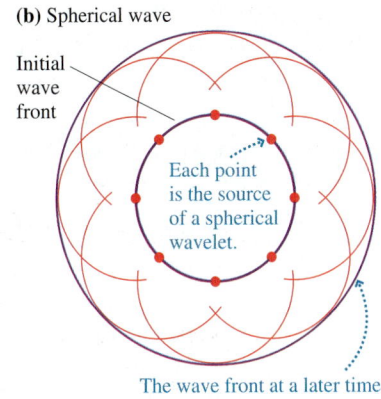

FIGURE 17.25 Huygens' principle applied to the propagation of plane waves and spherical waves.

Analyzing Single-Slit Diffraction

16.6 Activ **Physics**

Figure 17.26a shows a wave front passing through a narrow slit of width a. According to Huygens' principle, each point on the wave front can be thought of as the source of a spherical wavelet. These wavelets overlap and interfere, producing the diffraction pattern seen on the viewing screen. The full mathematical analysis, using *every* point on the wave front, is a fairly difficult problem in calculus. We'll be satisfied with a geometrical analysis based on just a few wavelets.

(a) Greatly magnified view of slit

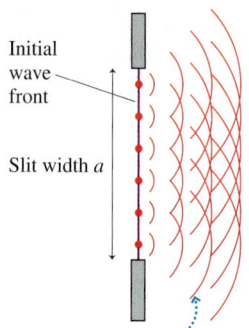

Initial wave front

Slit width a

The wavelets from each point on the initial wave front overlap and interfere, creating a diffraction pattern on the screen.

(b)

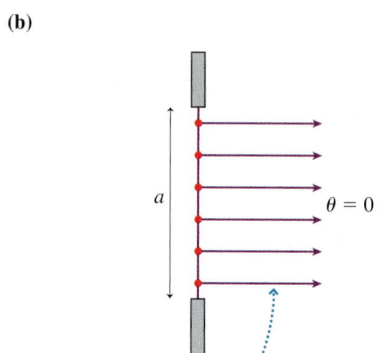

a

$\theta = 0$

The wavelets going straight forward all travel the same distance to the screen. Thus they arrive in phase and interfere constructively to produce the central maximum.

(c)

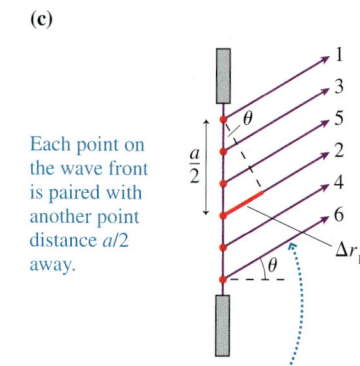

Each point on the wave front is paired with another point distance $a/2$ away.

$\dfrac{a}{2}$

θ

Δr_{12}

These wavelets all meet on the screen at angle θ. Wavelet 2 travels distance $\Delta r_{12} = (a/2)\sin\theta$ farther than wavelet 1.

FIGURE 17.26 Each point on the wave front is a source of spherical wavelets. The superposition of these wavelets produces the diffraction pattern on the screen.

Figure 17.26b shows the paths of several wavelets that travel straight ahead to the central point on the screen. (The screen is *very* far to the right in this magnified view of the slit.) The paths to the screen are very nearly parallel to each other, thus all the wavelets travel the same distance and arrive at the screen *in phase* with each other. The *constructive interference* between these wavelets produces the central maximum of the diffraction pattern at $\theta = 0$.

The situation is different at points away from the center. Wavelets 1 and 2 in Figure 17.26c start from points that are distance $a/2$ apart. Suppose that Δr_{12}, the extra distance traveled by wavelet 2, happens to be $\lambda/2$. In that case, wavelets 1 and 2 arrive out of phase and interfere destructively. But if Δr_{12} is $\lambda/2$, then the difference Δr_{34} between paths 3 and 4 and the difference Δr_{56} between paths 5 and 6 are also $\lambda/2$. Those pairs of wavelets also interfere destructively. The superposition of all the wavelets produces perfect destructive interference.

Figure 17.26c happens to show six wavelets, but our conclusion is valid for any number of wavelets. The key idea is that **every point on the wave front can be paired with another point that is distance $a/2$ away.** If the path-length difference is $\lambda/2$, the wavelets that originate at these two points will arrive at the screen out of phase and interfere destructively. When we sum the displacements of all N wavelets, they will—pair by pair—add to zero. The viewing screen at this position will be dark. This is the main idea of the analysis, one worth thinking about carefully.

You can see from Figure 17.26c that $\Delta r_{12} = (a/2)\sin\theta$. This path-length difference will be $\lambda/2$, the condition for destructive interference, if

$$\Delta r_{12} = \frac{a}{2}\sin\theta_1 = \frac{\lambda}{2} \tag{17.17}$$

or, equivalently, $a\sin\theta_1 = \lambda$.

Water waves can be seen diffracting behind the "slit" between the two breakwaters. The wave pattern can be understood using Huygens' principle.

NOTE ▶ Equation 17.17 cannot be satisfied if the slit width a is less than the wavelength λ. If a wave passes through an opening smaller than the wavelength, the central maximum of the diffraction pattern expands to where it *completely* fills the space behind the opening. There are no minima or dark spots at any angle. This situation is uncommon for light waves, because λ is so small, but quite common in the diffraction of sound and water waves. ◀

We can extend this idea to find other angles of perfect destructive interference. Suppose each wavelet is paired with another wavelet from a point $a/4$ away. If Δr between these wavelets is $\lambda/2$, then all N wavelets will again cancel in pairs to give complete destructive interference. The angle θ_2 at which this occurs is found by replacing $a/2$ in Equation 17.17 with $a/4$, leading to the condition $a\sin\theta_2 = 2\lambda$. This process can be continued, and we find that the general condition for complete destructive interference is

$$a\sin\theta_p = p\lambda \qquad p = 1, 2, 3, \ldots \qquad (17.18)$$

When $\theta_p \ll 1$ rad, which is almost always true for light waves, we can use the small-angle approximation to write

$$\theta_p = p\frac{\lambda}{a} \qquad p = 1, 2, 3, \ldots \qquad (17.19)$$

LINEAR INVERSE
p. 38   p. 118

Angles (in radians) of dark fringes in
single-slit diffraction, slit width a

Equation 17.19 gives the angles *in radians* to the dark minima in the diffraction pattern of a single slit. Notice that $p = 0$ is explicitly *excluded*. $p = 0$ corresponds to the straight-ahead position at $\theta = 0$, but you saw in Figures 17.6 and 17.26b that $\theta = 0$ is the central *maximum,* not a minimum.

NOTE ▶ Equations 17.18 and 17.19 are *mathematically* the same as the condition for the mth *maximum* of the double-slit interference pattern. But the physical meaning here is quite different. Equation 17.19 locates the *minima* (dark fringes) of the single-slit diffraction pattern. ◀

It is possible, although beyond the scope of this textbook, to calculate the entire light intensity pattern. The results of such a calculation are shown graphically in Figure 17.27. You can see the bright central maximum at $\theta = 0$, the weaker secondary maxima, and the dark points of destructive interference at the angles given by Equation 17.19. Compare this graph to the photograph of Figure 17.24 and make sure you see the agreement between the two.

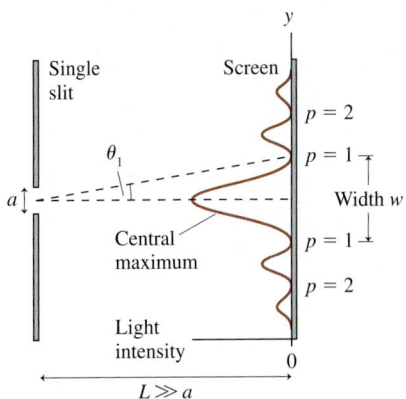

FIGURE 17.27 A graph of the intensity of a single-slit diffraction pattern.

EXAMPLE 17.7 Finding the width of a slit

Light from a helium-neon laser ($\lambda = 633$ nm) passes through a narrow slit and is seen on a screen 2.0 m behind the slit. The first minimum in the diffraction pattern is 1.2 cm from the middle of the central maximum. How wide is the slit?

PREPARE A narrow slit produces a single-slit diffraction pattern. A displacement of only 1.2 cm in a distance of 200 cm means that angle θ_1 is certainly a small angle. The intensity pattern will look like Figure 17.27.

SOLVE We can use the small-angle approximation to find that the angle to the first minimum is

$$\theta_1 = \frac{1.2 \text{ cm}}{200 \text{ cm}} = 0.0060 \text{ rad} = 0.34°$$

The first minimum is at angle $\theta_1 = \lambda/a$, where θ_1 must be in radians, from which we find that the slit width is

$$a = \frac{\lambda}{\theta_1} = \frac{633 \times 10^{-9} \text{ m}}{6.0 \times 10^{-3} \text{ rad}} = 1.1 \times 10^{-4} \text{ m} = 0.11 \text{ mm}$$

ASSESS This is typical of the slit widths used to observe single-slit diffraction. You can see that the small-angle approximation is well satisfied.

The Width of a Single-Slit Diffraction Pattern

We'll find it useful, as we did for the double slit, to measure positions on the screen rather than angles. The position of the pth dark fringe, at angle θ_p, is $y_p = L\tan\theta_p$, where L is the distance from the slit to the viewing screen. Using Equation 17.19 for θ_p and the small-angle approximation $\tan\theta_p \approx \theta_p$, we find that the dark fringes in the single-slit diffraction pattern are located at

$$y_p = \frac{p\lambda L}{a} \qquad p = 1, 2, 3, \ldots \qquad (17.20)$$

LINEAR   INVERSE
p. 38   p. 118

Positions of dark fringes for
single-slit diffraction, screen distance L

$p = 0$ is explicitly excluded because the midpoint on the viewing screen is the central maximum, not a dark fringe.

A diffraction pattern is dominated by the central maximum, which is much brighter than the secondary maxima. The width w of the central maximum, shown in Figure 17.27, is defined as the distance between the two $p = 1$ minima on either side of the central maximum. Because the pattern is symmetrical, the width is simply $w = 2y_1$. This is

$$w = \frac{2\lambda L}{a} \qquad (17.21)$$

Width of the central maximum for single-slit diffraction

The width of the central maximum is *twice* the spacing $\lambda L/a$ between the dark fringes on either side. The farther back you move the screen (increasing L), the wider the pattern of light on it becomes. In other words, the light waves are *spreading out* behind the slit, and they fill a wider and wider region as they travel farther.

An important implication of Equation 17.21, one contrary to common sense, is that a narrower slit (smaller a) causes a *wider* diffraction pattern. **The smaller the opening you squeeze a wave through, the *more* it spreads out on the other side.**

The central maximum of this single-slit diffraction pattern appears white because it is overexposed. The width of the central maximum is clear.

EXAMPLE 17.8 Measuring a hair

Babinet's principle, which we won't prove, states: Except for the intensity of the central maximum, the diffraction pattern produced by an opaque object is the same as that produced by an aperture of the same size and shape in an otherwise opaque screen. Suppose a human hair is illuminated by a helium-neon laser ($\lambda = 633$ nm). A diffraction pattern that looks very much like a single-slit pattern is viewed on a screen 1.2 m behind the hair. The width of the central maximum is 2.5 cm. What is the diameter of the hair?

PREPARE According to Babinet's principle, a thin, opaque rod of diameter a produces the same diffraction pattern as a single, narrow slit of width a except that the intensity in the central

maximum will not be the same. But because everything else is the same, the width of the central maximum in the hair's diffraction pattern is the same as the width of the central maximum for single-slit diffraction.

SOLVE From Equation 17.21, the hair's diameter is

$$a = \frac{2\lambda L}{w} = \frac{2(633 \times 10^{-9}\ \text{m})(1.2\ \text{m})}{0.025\ \text{m}}$$

$$= 6.1 \times 10^{-5}\ \text{m} = 61\ \mu\text{m}$$

ASSESS This idea is used to determine the sizes of cells by measuring the widths of their diffraction patterns.

The figure shows two single-slit diffraction patterns. The distance between the slit and the viewing screen is the same in both cases. Which of the following could be true?

A. The slits are the same for both; $\lambda_1 > \lambda_2$.
B. The slits are the same for both; $\lambda_2 > \lambda_1$.
C. The wavelengths are the same for both; $a_1 > a_2$.
D. The wavelengths are the same for both; $a_2 > a_1$.
E. The slits and the wavelengths are the same for both; $p_1 > p_2$.
F. The slits and the wavelengths are the same for both; $p_2 > p_1$.

17.6 Circular-Aperture Diffraction

Diffraction occurs if a wave passes through an opening of any shape. Diffraction by a single slit establishes the basic ideas of diffraction, but a common situation of practical importance is diffraction of a wave by a **circular aperture.** Circular diffraction is mathematically more complex than diffraction from a slit, and we will present results without derivation.

Activ Physics 16.7

Consider some examples. A loudspeaker cone generates sound by the rapid oscillation of a diaphragm, but the sound wave must pass through the circular aperture defined by the outer edge of the speaker cone before it travels into the room beyond. This is diffraction by a circular aperture. Telescopes and microscopes are the reverse. Light waves from outside need to enter the instrument. To do so, they must pass through a circular lens. In fact, the performance limit of optical instruments is determined by the diffraction of the circular openings through which the waves must pass. This is an issue we'll look at more closely in Chapter 19.

Figure 17.28 shows a circular aperture of diameter D. Light waves passing through this aperture spread out to generate a *circular* diffraction pattern. You should compare this to Figure 17.24 for a single slit to note the similarities and differences. The diffraction pattern still has a *central maximum,* now circular in shape, and it is surrounded by a series of secondary bright fringes. Most of the intensity is contained within the central maximum.

TRY IT YOURSELF

This pinhole acts as a point source of light.

You are observing diffraction due to this pinhole.

 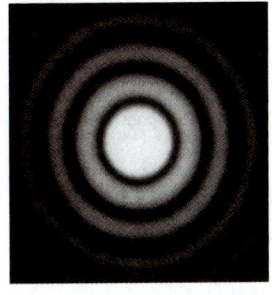

FIGURE 17.28 The diffraction of light by a circular opening.

Angle θ_1 locates the first minimum in the intensity, where there is complete destructive interference. A mathematical analysis of circular diffraction leads to the result

$$\theta_1 = \frac{1.22\lambda}{D} \tag{17.22}$$

where D is the *diameter* of the circular opening. This is very similar to the result for a single slit, but not quite the same. Equation 17.22 has assumed the small-angle approximation, which is almost always valid for the diffraction of light but usually is *not* valid for the diffraction of longer-wavelength sound waves.

Observing diffraction To observe diffraction from a circular aperture, use a pin to prick a *very* small hole in two pieces of aluminum foil. Tape one piece up to a window that faces the sun. Then, holding the other hole close to your eye, observe the sun through the pinhole on the window. You will see clear diffraction fringes around that pinhole. How do the fringes vary if you change the size of the pinhole near your eye? (Don't look directly at the sun except through the two pinholes!)

Within the small-angle approximation, the width of the central maximum is

$$w = 2y_1 = 2L\tan\theta_1 \approx \frac{2.44\lambda L}{D} \qquad (17.23)$$

Width of central maximum for diffraction
from a circular aperture of diameter D

LINEAR
p. 38

INVERSE
p. 118

Again, this is similar to, but not quite identical with, the width of the central maximum in a single-slit diffraction pattern. The diameter of the diffraction pattern increases with distance L, showing that light spreads out behind a circular aperture, but it decreases if the size D of the aperture is increased.

EXAMPLE 17.9 Finding the right viewing distance
Light from a helium-neon laser ($\lambda = 633$ nm) passes through a 0.50-mm-diameter hole. How far away should a viewing screen be placed to observe a diffraction pattern whose central maximum is 3.0 mm in diameter?

SOLVE Equation 17.23 gives us the appropriate screen distance:

$$L = \frac{wD}{2.44\lambda} = \frac{(3.0 \times 10^{-3}\text{ m})(5.0 \times 10^{-4}\text{ m})}{2.44(633 \times 10^{-9}\text{ m})} = 0.97\text{ m}$$

As we've seen, we need to use the wave model of light to understand the passage of light through narrow apertures, where "narrow" means comparable in size to the wavelength of the light. In the next two chapters, we'll consider the interaction of light with objects much larger than the wavelength. There, the ray model of light will be more appropriate for describing how light reflects from mirrors and refracts from lenses. But the wave model will reappear in Chapter 19 when we study telescopes and microscopes. We'll find that the *resolution* of these instruments has a fundamental limit set by the wave nature of light.

The goal of Chapter 17 has been to understand and apply the wave model of light.

GENERAL PRINCIPLES

The wave model

The wave model considers light to be a wave propagating through space. Interference and diffraction are important. The wave model is appropriate when light interacts with objects whose size is comparable to the wavelength of light, or roughly less than about 0.1 mm.

Huygens' principle says that each point on a wave front is the source on a spherical wavelet. The wave front at a later time is tangent to all the wavelets.

IMPORTANT CONCEPTS

The index of refraction of a material determines the speed of light in that material: $v = c/n$. The index of refraction of a material is always greater than 1, so that v is always less than c.

The wavelength λ in a material with index of refraction n is *less* than the wavelength λ_{vac} in vacuum: $\lambda = \lambda_{vac}/n$.

The *frequency* of light does not change as it moves from one material to another.

Diffraction is the spreading of a wave after it passes through an opening.

Constructive and destructive interference are due to the overlap of two or more waves as they spread behind openings.

APPLICATIONS

Diffraction from a single slit

A single slit of width a has a bright **central maximum** of width

$$w = \frac{2\lambda L}{a}$$

that is flanked by weaker **secondary maxima.**

Secondary maxima   Central maximum

Dark fringes

Dark fringes are located at angles such that

$$a\sin\theta_p = p\lambda \qquad p = 1, 2, 3, \ldots$$

If $\lambda/a \ll 1$, then from the small-angle approximation

$$\theta_p = \frac{p\lambda}{a} \qquad y_p = \frac{p\lambda L}{a}$$

Interference from multiple slits

Waves overlap as they spread out behind slits. Constructive interference occurs along antinodal lines. Bright fringes are seen where the antinodal lines intersect the viewing screen.

Double slit with separation d.
Equally spaced bright fringes are located at

$$\theta_m = \frac{m\lambda}{d} \qquad y_m = \frac{m\lambda L}{d} \qquad m = 0, 1, 2, \ldots$$

The **fringe spacing** is $\Delta y = \dfrac{\lambda L}{d}$

Diffraction grating with slit spacing d.
Very bright and narrow fringes are located at angles and positions

$$d\sin\theta_m = m\lambda \qquad y_m = L\tan\theta_m$$

Circular aperture of diameter D.

A bright central maximum of diameter

$$w = \frac{2.44\lambda L}{D}$$

is surrounded by circular secondary maxima. The first dark fringe is located at

$$\theta_1 = \frac{1.22\lambda}{D} \qquad y_1 = \frac{1.22\lambda L}{D}$$

For an aperture of any shape, a smaller opening causes a greater spreading of the wave behind the opening.

Thin-film interference

Interference occurs between the waves reflected from the two surfaces of a thin film. A wave that reflects from a surface at which the index of refraction increases has a phase change.

Interference	0 or 2 phase changes	One phase change
Constructive	$2t = m\dfrac{\lambda}{n}$	$2t = \left(m + \dfrac{1}{2}\right)\dfrac{\lambda}{n}$
Destructive	$2t = \left(m + \dfrac{1}{2}\right)\dfrac{\lambda}{n}$	$2t = m\dfrac{\lambda}{n}$

For instructor-assigned homework, go to
www.masteringphysics.com

Problem difficulty is labeled as I (straightforward) to IIII (challenging).

Problems labeled INT integrate significant material
from earlier chapters; BIO are of biological or medical
interest.

QUESTIONS

Conceptual Questions

1. The frequency of a light wave in air is 5.3×10^{14} Hz. Is the frequency of this wave higher, lower, or the same after it enters a piece of glass?

2. Rank in order the following according to their speeds, from slowest to fastest: (i) 425-nm-wavelength light through a pane of glass, (ii) 500-nm-wavelength light through air, (iii) 540-nm-wavelength light through water, (iv) 670-nm-wavelength light through a diamond, and (v) 670-nm-wavelength light through a vacuum.

3. The wavelength of a light wave is 700 nm in air; this light appears red. If this wave enters a pool of water, its wavelength becomes $\lambda_{air}/n = 530$ nm. If you were swimming underwater, the light would still appear red. Given this, what property of a wave determines its color?

4. A double-slit interference experiment shows fringes on a screen. The entire experiment is then immersed in water. Do the fringes on the screen get closer together, farther apart, remain the same, or disappear entirely? Explain.

5. Consider Figure Q17.5, which shows the viewing screen in a double-slit experiment.

FIGURE Q17.5

   a. What will happen to the fringe spacing if the wavelength of the light is decreased?
   b. What will happen to the fringe spacing if the spacing between the slits is decreased?
   c. What will happen to the fringe spacing if the distance to the screen is decreased?
   d. Suppose the wavelength of the light is 500 nm. How much farther is it from the dot on the screen in the center of fringe E to the left slit than it is from the dot to the right slit?

6. Figure Q17.5 is the interference pattern seen on a viewing screen behind 2 slits. Suppose the 2 slits were replaced by 20 slits having the same spacing d between adjacent slits.
   a. Would the number of fringes on the screen increase, decrease, or stay the same?
   b. Would the fringe spacing increase, decrease, or stay the same?
   c. Would the width of each fringe increase, decrease, or stay the same?
   d. Would the brightness of each fringe increase, decrease, or stay the same?

7. Figure Q17.7 shows the light intensity on a viewing screen behind a single slit of width a. The light's wavelength is λ. Is $\lambda < a$, $\lambda = a$, $\lambda > a$, or is it not possible to tell? Explain.

FIGURE Q17.7

8. Figure Q17.8 shows the light intensity on a viewing screen behind a circular aperture. What happens to the width of the central maximum if
   a. The wavelength is increased?
   b. The diameter of the aperture is increased?
   c. How will the screen appear if the aperture diameter is less than the light wavelength?

FIGURE Q17.8

9. Why does light reflected from peacock feathers change color BIO when you see the feathers at a different angle?

10. White light is incident on a diffraction grating. What color is the central maximum of the interference pattern?

11. A soap bubble usually pops because some part of it becomes too thin due to evaporation or drainage of fluid. The change in thickness also changes the color of light the bubble reflects. Why?

12. An oil film on top of water has one patch that is much thinner than the wavelength of visible light. The index of refraction of the oil is less than that of water. Will the reflection from that extremely thin part of the film be bright or dark? Explain.

13. Should the antireflection coating of a microscope objective lens designed for use with ultraviolet light be thinner, thicker, or the same thickness as the coating on a lens designed for visible light?

14. If the thin wedge of air between the two plates of glass in Figure 17.21 were replaced by water, would the distance between the fringes increase, decrease, or remain the same? Explain.

15. Example 17.5 showed that a thin film whose thickness is one-quarter of the wavelength of light in the film serves as an antireflection coating when coated on glass. In Example 17.5, $n_{film} < n_{glass}$. If a quarter-wave thickness film with $n_{film} > n_{glass}$ were used instead, would the film still serve as an antireflection coating? Explain.

16. You are standing against the wall near a corner of a large building. A friend is standing against the wall that is around the corner from you. You can't see your friend. How is it that you can hear her when she talks to you?

Multiple-Choice Questions

17. | Light of wavelength 500 nm in air enters a glass block with index of refraction $n = 1.5$. When the light enters the block, which of the following properties of the light will not change?
    A. The speed of the light
    B. The frequency of the light
    C. The wavelength of the light

18. | The frequency of a light wave in air is 4.6×10^{14} Hz. What is the wavelength of this wave after it enters a pool of water?
    A. 300 nm   B. 490 nm   C. 650 nm   D. 870 nm

19. | Light passes through a diffraction grating with a slit spacing of 0.001 mm. A viewing screen is 100 cm behind the grating. If the light is blue, with a wavelength of 450 nm, at about what distance from the center of the diffraction pattern will the first-order maximum appear?
    A. 5 cm   B. 25 cm   C. 50 cm   D. 100 cm

20. ‖ Blue light of wavelength 450 nm passes through a diffraction grating with a slit spacing of 0.001 mm and makes a diffraction pattern on the wall. How many bright fringes will be seen?
    A. 1   B. 3   C. 5   D. 7

21. ‖ Yellow light of wavelength 590 nm passes through a diffraction grating and makes a diffraction pattern on a screen 80 cm away. The first bright fringes are 1.9 cm from the center of the pattern. How many lines per mm does this grating have?
    A. 20 lines/mm   B. 40 lines/mm
    C. 80 lines/mm   D. 200 lines/mm

22. | Light passes through a 10-μm-wide slit and is viewed on a screen 1 m behind the slit. If the width of the slit is narrowed, the band of light on the screen will
    A. Become narrower.
    B. Become wider.
    C. Stay about the same.

23. | Blue light of wavelength 450 nm passes through a 0.20-mm-wide slit and illuminates a screen 1.2 m away. How wide is the central maximum of the diffraction pattern?
    A. 1.2 mm   B. 2.0 mm   C. 2.7 mm   D. 5.4 mm

24. | A green laser beam of wavelength 540 nm passes through a pinhole and illuminates a dartboard 3.0 m past the pinhole. The first minimum in the intensity coincides with the ring surrounding the bull's-eye, 12 mm in diameter. What is the diameter of the pinhole?
    A. 0.14 mm   B. 0.33 mm   C. 0.59 mm   D. 1.2 mm

PROBLEMS

Section 17.1 What Is Light?

1. ‖ a. How long does it take light to travel through a 3.0-mm-thick piece of window glass?
    b. Through what thickness of water could light travel in the same amount of time?

2. ‖‖ a. How long (in ns) does it take light to travel 1.0 m in vacuum?
    b. What distance does light travel in water, glass, and diamond during the time that it travels 1.0 m in vacuum?

3. ‖ A 5.0-cm-thick layer of oil ($n = 1.46$) is sandwiched between a 1.0-cm-thick sheet of glass and a 2.0-cm-thick sheet of polystyrene plastic ($n = 1.59$). How long (in ns) does it take light incident perpendicular to the glass to pass through this 8.0-cm-thick sandwich?

4. ‖‖ A light wave has a 670 nm wavelength in air. Its wavelength in a transparent solid is 420 nm.
    a. What is the speed of light in this solid?
    b. What is the light's frequency in the solid?

5. ‖‖ How much time does it take a pulse of light to travel through 150 m of water?

6. ‖ A helium-neon laser beam has a wavelength in air of 633 nm. It takes 1.38 ns for the light to travel through 30 cm of an unknown liquid. What is the wavelength of the laser beam in the liquid?

Section 17.2 The Interference of Light

7. | Two narrow slits 50 μm apart are illuminated with light of wavelength 500 nm. What is the angle of the $m = 2$ bright fringe in radians? In degrees?

8. ‖‖‖ Light from a sodium lamp ($\lambda = 589$ nm) illuminates two narrow slits. The fringe spacing on a screen 150 cm behind the slits is 4.0 mm. What is the spacing (in mm) between the two slits?

9. ‖‖ A double-slit interference pattern is created by two narrow slits spaced 0.20 mm apart. The distance between the first and the fifth minimum on a screen 60 cm behind the slits is 6.0 mm. What is the wavelength of the light used in this experiment?

10. | A double-slit experiment is performed with light of wavelength 600 nm. The bright interference fringes are spaced 1.8 mm apart on the viewing screen. What will the fringe spacing be if the light is changed to a wavelength of 400 nm?

11. ‖ Light from a helium-neon laser ($\lambda = 633$ nm) is used to illuminate two narrow slits. The interference pattern is observed on a screen 3.0 m behind the slits. Twelve bright fringes are seen, spanning a distance of 52 mm. What is the spacing (in mm) between the slits?

12. ‖‖ Two narrow slits are 0.12 mm apart. Light of wavelength 550 nm illuminates the slits, causing an interference pattern on a screen 1.0 m away. Light from each slit travels to the $m = 1$ maximum on the right side of the central maximum. How much farther did the light from the left slit travel than the light from the right slit?

Section 17.3 The Diffraction Grating

13. | A 1.0-cm-wide diffraction grating has 1000 slits. It is illuminated by light of wavelength 550 nm. What are the angles of the first two diffraction orders?

14. | A diffraction grating with 750 slits/mm is illuminated by light that gives a first-order diffraction angle of 34.0°. What is the wavelength of the light?

15. ‖   Light of wavelength 600 nm illuminates a diffraction grating. The second-order maximum is at angle 39.5°. How many lines per millimeter does this grating have?

16. | The central star of a planetary nebula emits ultraviolet light with wavelength 104 nm. This light passes through a diffraction grating with 5000 slits/mm. What is the first-order diffraction angle?

17. ‖   A lab technician uses laser light with a wavelength of 670 nm to test a diffraction grating. When the grating is 40.0 cm from the screen, the first-order maxima appear 6.00 cm from the center of the pattern. How many lines per millimeter does this grating have?

18. ‖   A helium-neon laser ($\lambda = 633$ nm) illuminates a diffraction grating. The distance between the two $m = 1$ bright fringes is 32.0 cm on a screen 2.00 m behind the grating. What is the spacing between slits of the grating?

19. ‖   A diffraction grating with 600 lines/mm is illuminated with light of wavelength 500 nm. A very wide viewing screen is 2.0 m behind the grating.
    a. What is the distance between the two $m = 1$ fringes?
    b. How many bright fringes can be seen on the screen?

20. ‖   A 500 line/mm diffraction grating is illuminated by light of wavelength 510 nm. How many diffraction orders are seen, and what is the angle of each?

Section 17.4 Thin-Film Interference

21. | What is the thinnest film of MgF_2 ($n = 1.39$) on glass that produces a strong reflection for orange light with a wavelength of 600 nm?

22. ‖   A very thin oil film ($n = 1.25$) floats on water ($n = 1.33$). What is the thinnest film that produces a strong reflection for green light with a wavelength of 500 nm?

23. ‖   A film with $n = 1.60$ is deposited on glass. What is the thinnest film that will produce constructive interference in the reflection of light with a wavelength of 550 nm?

24. | Antireflection coatings can be used on the *inner* surfaces
BIO   of eyeglasses to reduce the reflection of stray light into the eye, thus reducing eyestrain.
    a. A 90-nm-thick coating is applied to the lens. What must be the coating's index of refraction to be most effective at 480 nm? Assume that the coating's index of refraction is less than that of the lens.
    b. If the index of refraction of the coating is 1.38, what thickness should the coating be so as to be most effective at 480 nm? The thinnest possible coating is best.

25. | Solar cells are given antireflection coatings to maximize their efficiency. Consider a silicon solar cell ($n = 3.50$) coated with a layer of silicon dioxide ($n = 1.45$). What is the minimum coating thickness that will minimize the reflection at the wavelength of 700 nm, where solar cells are most efficient?

26. ‖   A thin film of MgF_2 ($n = 1.38$) coats a piece of glass. Destructive interference is observed for the reflection of light with wavelengths of 500 nm and 625 nm. What is the thinnest film for which this can occur?

27. ‖   Looking straight downward into a rain puddle whose surface is covered with a thin film of gasoline, you notice a swirling pattern of colors caused by interference inside the gasoline film. The point directly beneath you is colored a beautiful iridescent green. You happen to remember that the index of refraction of gasoline is 1.38 and that the wavelength of green light is about 540 nm. What is the minimum possible thickness of the gasoline layer directly beneath you?

Section 17.5 Single-Slit Diffraction

28. ‖   A helium-neon laser ($\lambda = 633$ nm) illuminates a single slit and is observed on a screen 1.50 m behind the slit. The distance between the first and second minima in the diffraction pattern is 4.75 mm. What is the width (in mm) of the slit?

29. | For a demonstration, a professor uses a razor blade to cut a thin slit in a piece of aluminum foil. When she shines a laser pointer ($\lambda = 680$ nm) through the slit onto a screen 5.5 m away, a diffraction pattern appears. The bright band in the center of the pattern is 8.0 cm wide. What is the width of the slit?

30. ‖   A 0.50-mm-wide slit is illuminated by light of wavelength 500 nm. What is the width of the central maximum on a screen 2.0 m behind the slit?

31. ‖   The second minimum in the diffraction pattern of a 0.10-mm-wide slit occurs at 0.70°. What is the wavelength of the light?

32. | What is the width of a slit for which the first minimum is at 45° when the slit is illuminated by a helium-neon laser ($\lambda = 633$ nm)?
    Hint: The small angle approximation is not valid at 45°.

Section 17.6 Circular-Aperture Diffraction

33. ‖   A 0.50-mm-diameter hole is illuminated by light of wavelength 500 nm. What is the width of the central maximum on a screen 2.0 m behind the slit?

34. ‖   Light from a helium-neon laser ($\lambda = 633$ nm) passes through a circular aperture and is observed on a screen 4.0 m behind the aperture. The width of the central maximum is 2.5 cm. What is the diameter (in mm) of the hole?

35. ‖   You want to photograph a circular diffraction pattern whose central maximum has a diameter of 1.0 cm. You have a helium-neon laser ($\lambda = 633$ nm) and a 0.12-mm-diameter pinhole. How far behind the pinhole should you place the viewing screen?

36. | Infrared light of wavelength 2.5 μm illuminates a 0.20-mm-diameter hole. What is the angle of the first dark fringe in radians? In degrees?

General Problems

37. ‖   An advanced computer sends information to its various parts via infrared light pulses traveling through silicon fibers ($n = 3.50$). To acquire data from memory, the central processing unit sends a light-pulse request to the memory unit. The memory unit processes the request, then sends a data pulse back to the central processing unit. The memory unit takes 0.50 ns to process a request. If the information has to be obtained from memory in 2.00 ns, what is the maximum distance the memory unit can be from the central processing unit?

38. ‖ Figure P17.38 shows the light intensity on a screen behind a double slit. The slit spacing is 0.20 mm and the wavelength of the light is 600 nm. What is the distance from the slits to the screen?

FIGURE P17.38

39. ‖ Figure P17.38 shows the light intensity on a screen behind a double slit. The slit spacing is 0.20 mm and the screen is 2.0 m behind the slits. What is the wavelength of the light?

40. ‖‖ Your friend has been given a laser for her birthday. Unfortunately, she did not receive a manual with it and so she doesn't know the wavelength that it emits. You help her by performing a double-slit experiment, with slits separated by 0.36 mm. You find that the two $m = 1$ bright fringes are 5.5 mm apart on a screen 1.6 m from the slits. What is the wavelength the laser emits?

41. | A double slit is illuminated simultaneously with orange light of wavelength 600 nm and light of an unknown wavelength. The $m = 4$ bright fringe of the unknown wavelength overlaps the $m = 3$ bright orange fringe. What is the unknown wavelength?

42. ‖‖ An interference experiment is done in a tank of water. Light with wavelength 450 nm (in the water) illuminates two slits separated by 0.34 mm. An interference pattern is observed on a screen at the end of the tank, 0.90 m past the slits.
    a. What is the spacing between the $m = 2$ and $m = 3$ bright fringes?
    b. What will the spacing be after the water is drained out of the tank?

43. | A laser beam of wavelength 670 nm shines through a diffraction grating that has 750 lines/mm. Sketch the pattern that appears on a screen 1.0 m behind the grating, noting distances on your drawing and explaining where these numbers come from.

44. | The two most prominent wavelengths in the light emitted by a hydrogen discharge lamp are 656 nm (red) and 486 nm (blue). Light from a hydrogen lamp illuminates a diffraction grating with 500 lines/mm, and the light is observed on a screen 1.50 m behind the grating. What is the distance between the first-order red and blue fringes?

45. | The diffraction grating used for a college physics lab is labeled "10,000 lines/in." You shine a narrow beam of blue light ($\lambda = 450$ nm) through this grating onto a screen 0.50 m behind the grating. At what distance from the center of the interference pattern will you find the second blue interference fringe?

46. ‖ A diffraction grating consists of 100 slits. If the number of slits is increased to 200, with the same spacing, by what factor does the maximum intensity of the bright fringes on the screen increase?

47. | A diffraction grating produces a first-order maximum at an angle of 20.0°. What is the angle of the second-order maximum?

48. | A diffraction grating is illuminated simultaneously with red light of wavelength 660 nm and light of an unknown wavelength. The fifth-order maximum of the unknown wavelength exactly overlaps the third-order maximum of the red light. What is the unknown wavelength?

49. ‖‖ White light (400–700 nm) is incident on a 600 line/mm diffraction grating. What is the width of the first-order rainbow on a screen 2.0 m behind the grating?

50. ‖ For your science fair project you need to design a diffraction grating that will disperse the visible spectrum (400–700 nm) over 30.0° in first order.
    a. How many lines per millimeter does your grating need?
    b. What is the first-order diffraction angle of light from a sodium lamp ($\lambda = 589$ nm)?

51. ‖ Figure P17.51 shows the interference pattern on a screen 1.0 m behind an 800 line/mm diffraction grating. What is the wavelength of the light?

FIGURE P17.51

52. ‖‖ Figure P17.51 shows the interference pattern on a screen 1.0 m behind a diffraction grating. The wavelength of the light is 600 nm. How many lines per millimeter does the grating have?

53. ‖ Because sound is a wave, it is possible to make a diffraction
INT grating for sound from a large board with several parallel slots for the sound to go through. When 10 kHz sound waves pass through such a grating, listeners 10 m from the grating report "loud spots" 1.4 m on both sides of center. What is the spacing between the slots? Use 340 m/s for the speed of sound.

54. | The shiny surface of a CD is imprinted with millions of tiny pits, arranged in a pattern of thousands of essentially concentric circles that act like a reflection grating when light shines on them. You decide to determine the distance between those circles by aiming a laser pointer (with $\lambda = 680$ nm) perpendicular to the disk and measuring the diffraction pattern reflected onto a screen 1.5 m from the disk. The central bright spot you expected to see is blocked by the laser pointer itself. You do find two other bright spots separated by 1.4 m, one on either side of the missing central spot. The rest of the pattern is apparently diffracted at angles too great to show on your screen. What is the distance between the circles on the CD's surface?

55. ‖ If sunlight shines straight onto a peacock feather, the
BIO feather appears bright blue when viewed from 15° on either side of the incident beam of sunlight. The blue color is due to diffraction from the melanin bands in the feather barbules, as was shown in the photograph on p. 564. Blue light with a wavelength of 470 nm is diffracted at 15° by these bands (this is the first-order diffraction) while other wavelengths in the sunlight are diffracted at different angles. What is the spacing of the melanin bands in the feather?

56. ‖ The wings of some beetles
BIO have closely spaced parallel lines of melanin, causing the wing to act as a reflection grating. Suppose sunlight shines straight onto a beetle wing. If the melanin lines on the wing are spaced 2.0 μm apart, what is the first-order diffraction angle for green light ($\lambda = 550$ nm)?

57. ‖ A diffraction grating having 500 lines/mm diffracts visible light at 30°. What is the light's wavelength?

58. ‖ Light emitted by Element X passes through a diffraction grating having 1200 lines/mm. The diffraction pattern is observed on a screen 75.0 cm behind the grating. Bright fringes are *seen* on the screen at distances of 56.2 cm, 65.9 cm, and 93.5 cm from the central maximum. No other fringes are seen.
   a. What is the value of m for each of these diffracted wavelengths? Explain why only one value is possible.
   b. What are the wavelengths of light emitted by Element X?

59. ‖ Helium atoms emit light at several wavelengths. Light from a helium lamp illuminates a diffraction grating and is observed on a screen 50.00 cm behind the grating. The emission at wavelength 501.5 nm creates a first-order bright fringe 21.90 cm from the central maximum. What is the wavelength of the bright fringe that is 31.60 cm from the central maximum?

60. ‖‖ A sheet of glass is coated with a 500-nm-thick layer of oil ($n = 1.42$).
   a. For what *visible* wavelengths of light do the reflected waves interfere constructively?
   b. For what *visible* wavelengths of light do the reflected waves interfere destructively?
   c. What is the color of reflected light? What is the color of transmitted light?

61. ‖‖ A soap bubble is essentially a thin film of water surrounded by air. The colors you see in soap bubbles are produced by interference. What visible wavelengths of light are strongly reflected from a 390-nm-thick soap bubble? What color would such a soap bubble appear to be?

62. | In a single-slit experiment, the slit width is 200 times the wavelength of the light. What is the width of the central maximum on a screen 2.0 m behind the slit?

63. ‖‖ You need to use your cell phone, which broadcasts an 830 MHz signal, but you're in an alley between two massive, radio-wave-absorbing buildings that have only a 15 m space between them. What is the angular width, in degrees, of the electromagnetic wave after it emerges from between the buildings?

64. ‖ Light from a sodium lamp ($\lambda = 589$ nm) illuminates a narrow slit and is observed on a screen 75 cm behind the slit. The distance between the first and third dark fringes is 7.5 mm. What is the width (in mm) of the slit?

65. ‖ INT The opening to a cave is a tall, 30-cm-wide crack. A bat that is preparing to leave the cave emits a 30 kHz ultrasonic chirp. How wide is the "sound beam" 100 m outside the cave opening? Use $v_{sound} = 340$ m/s.

66. ‖‖ For what slit-width-to-wavelength ratio does the first minimum of a single-slit diffraction pattern appear at (a) 30°, (b) 60°, and (c) 90°?
   Hint: The small-angle approximation is not valid.

67. | Figure P17.67 shows the light intensity on a screen behind a single slit. The wavelength of the light is 500 nm and the screen is 1.0 m behind the slit. What is the width (in mm) of the slit?

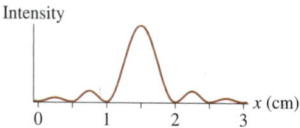

FIGURE P17.67

68. ‖ Figure P17.67 shows the light intensity on a screen behind a single slit. The wavelength of the light is 600 nm and the slit width is 0.15 mm. What is the distance from the slit to the screen?

69. | Figure P17.69 shows the light intensity on a screen 2.5 m behind an aperture. The aperture is illuminated with light of wavelength 600 nm.
   a. Is the aperture a single slit or a double slit? Explain.
   b. If the aperture is a single slit, what is its width? If it is a double slit, what is the spacing between the slits?

FIGURE P17.69   **FIGURE P17.70**

70. | Figure P17.70 shows the light intensity on a screen 2.5 m behind an aperture. The aperture is illuminated with light of wavelength 600 nm.
   a. Is the aperture a single slit or a double slit? Explain.
   b. If the aperture is a single slit, what is its width? If it is a double slit, what is the spacing between the slits?

71. | One day, after pulling down your window shade, you notice that sunlight is passing through a pinhole in the shade and making a small patch of light on the far wall. Having recently studied optics in your physics class, you're not too surprised to see that the patch of light seems to be a circular diffraction pattern. It appears that the central maximum is about 3 cm across, and you estimate that the distance from the window shade to the wall is about 3 m. Knowing that the average wavelength of sunlight is about 500 nm, estimate the diameter of the pinhole.

72. ‖ A radar for tracking aircraft broadcasts a 12 GHz microwave beam from a 2.0-m-diameter circular radar antenna. From a wave perspective, the antenna is a circular aperture through which the microwaves diffract.
   a. What is the diameter of the radar beam at a distance of 30 km?
   b. If the antenna emits 100 kW of power, what is the average microwave intensity at 30 km?

73. ‖ A helium-neon laser ($\lambda = 633$ nm), shown in Figure P17.73, is built with a glass tube of inside diameter 1.0 mm. One mirror is partially transmitting to allow the laser beam out. An electrical discharge in the tube causes it to glow like a neon light. From an optical perspective, the laser beam is a light wave that diffracts out through a 1.0-mm-diameter circular opening.
   a. Explain why a laser beam can't be *perfectly* parallel, with no spreading.
   b. The angle θ_1 to the first minimum is called the *divergence angle* of a laser beam. What is the divergence angle of this laser beam?
   c. What is the diameter (in mm) of the laser beam after it travels 3.0 m?

FIGURE P17.73

   d. What is the diameter of the laser beam after it travels 1.0 km?

74. ‖ Scientists use *laser range-finding* to measure the distance to the moon with great accuracy. A brief laser pulse is fired at the moon, then the time interval is measured until the "echo" is seen by a telescope. A laser beam spreads out as it travels because it diffracts through a circular exit as it leaves the laser. In order for the reflected light to be bright enough to 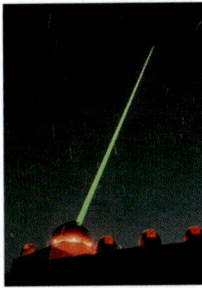 detect, the laser spot on the moon must be no more than 1 km in diameter. Staying within this diameter is accomplished by using a special large-diameter laser. If $\lambda = 532$ nm, what is the minimum diameter of the circular opening from which the laser beam emerges? The earth-moon distance is 384,000 km.

Passage Problems

The Blue Morpho Butterfly BIO

The brilliant blue color of a blue morpho butterfly is, like the colors of peacock feathers, due to interference. Figure P17.75a shows an easy way to demonstrate this: If a drop of the clear solvent acetone is placed on the wing of a blue morpho butterfly, the color changes from a brilliant blue to an equally brilliant green—returning to blue once the acetone evaporates. There would be no change if the color were due to pigment.

Light reflections from different layers have different path lengths.

FIGURE P17.75

A cross-section of a scale from the wing of a blue morpho butterfly reveals the source of the butterfly's color. As Figure P17.75b shows, the scales are covered with structures that look like small Christmas trees. Light striking the wings reflects from different layers of these structures, and the differing path lengths cause the reflected light to interfere constructively or destructively, depending on the wavelength. For light at normal incidence, blue light experiences constructive interference while other colors undergo destructive interference and cancel. Acetone fills the spaces in the scales with a fluid of index of refraction $n = 1.38$; this changes the conditions for constructive interference and results in a change in color.

75. | The coloring of the blue morpho butterfly is protective. As the butterfly flaps its wings, the angle at which light strikes the wings changes. This causes the butterfly's color to change and makes it difficult for a predator to follow. This color change is because:
  A. A diffraction pattern appears only at certain angles.
  B. The index of refraction of the wing tissues changes as the wing flexes.
  C. The motion of the wings causes a Doppler shift in the reflected light.
  D. As the angle changes, the differences in paths among light reflected from different surfaces change, resulting in constructive interference for a different color.

76. | The change in color when acetone is placed on the wing is due to the difference between the indices of refraction of acetone and air. Consider light of some particular color. In acetone,
  A. The frequency of the light is less than in air.
  B. The frequency of the light is greater than in air.
  C. The wavelength of the light is less than in air.
  D. The wavelength of the light is greater than in air.

77. | The scales on the butterfly wings are actually made of a transparent material with index of refraction 1.56. Light reflects from the surface of the scales because
  A. The scales' index of refraction is different from that of air.
  B. The scales' index of refraction is similar to that of glass.
  C. The scales' density is different from that of air.
  D. Different colors of light have different wavelengths.

STOP TO THINK ANSWERS

Stop to Think 17.1: $n_3 > n_1 > n_2$. $\lambda = \lambda_{vac}/n$, so a shorter wavelength corresponds to a larger index of refraction.

Stop to Think 17.2: B. The antinodal lines seen in Figure 17.7b are diverging.

Stop to Think 17.3: Smaller. Shorter-wavelength light doesn't spread as rapidly as longer-wavelength light. The fringe spacing Δy is directly proportional to the wavelength λ.

Stop to Think 17.4: D. Larger wavelengths have larger diffraction angles. Red light has a larger wavelength than violet light, so red light is diffracted farther from the center.

Stop to Think 17.5: B. An extra path difference of $\lambda/2$ must be added to change from constructive to destructive interference. In thin-film interference, one wave passes *twice* through the film. To increase the path length by $\lambda/2$, the thickness need be increased only by one-half this, or $\lambda/4$.

Stop to Think 17.6: B or C. The width of the central maximum, which is proportional to λ/a, has increased. This could occur either because the wavelength has increased or because the slit width has decreased.

18

RAY OPTICS

Why are there two images of this turtle in an aquarium, and why does each image appear to be a different size?

Looking Ahead ▸▸

The goals of Chapter 18 are to understand and apply the ray model of light. In this chapter you will learn to:

- ▸ Use the ray model of light.
- ▸ Understand and apply the laws of reflection and refraction.
- ▸ Understand color and dispersion.
- ▸ Use ray tracing to analyze image formation by lenses.
- ▸ Use ray tracing to analyze image formation by spherical mirrors.

Looking Back ◂◂

The material in this chapter develops the ray model of light, and the reflection and refraction of light. You'll also make extensive use of geometry and trigonometry. Please review:

- ◂ Section 17.1 Models of light and the index of refraction.
- ◂ Appendix A Geometry and Trigonometry.

Humans have always been fascinated by light. Simple mirrors are found in ancient archeological sites from Egypt to China. Our ancestors had learned by 1500 BCE to start fires by focusing sunlight with a simple lens. From there, it's only a small step to drilling holes with a focused laser beam.

Chapter 17 introduced three models of light but emphasized the wave optics of interference and diffraction. This chapter will analyze basic optical systems such as mirrors and lenses in terms of straight-line light trajectories. This is *ray optics,* and it is a subject of immense practical value. The ray model of light will now take center stage, but we'll find that we can't entirely avoid the fact that light is a wave. Wave optics will return in the next chapter, where we'll find that the waviness of light is what ultimately sets the performance limits of optical systems.

18.1 The Ray Model of Light

A flashlight makes a beam of light through the night's darkness. Sunbeams stream into a darkened room through a small hole in the shade, and laser beams are even more well defined. Our everyday experience that light travels in straight lines is the basis of the ray model of light.

The ray model is an oversimplification of reality, but nonetheless is very useful within its range of validity. As we saw in Chapter 17, diffraction and other wave aspects of light are important only for apertures and objects comparable in size to the wavelength of light. Because the wavelength is so small, typically 0.5 μm, the wave nature of light is not apparent when light interacts with ordinary-sized objects. The ray model of light, which ignores diffraction, is valid as long as any apertures through which the light passes (lenses, mirrors, holes, and the like) are larger than about 1 mm.

To begin, let us define a **light ray** as a line in the direction along which light energy is flowing. A light ray is an abstract idea, not a physical entity or a "thing." Any narrow beam of light, such as the laser beam in Figure 18.1, is actually a bundle of many parallel light rays. You can think of a single light ray as the limiting case of a laser beam whose diameter approaches zero. Laser beams are good approximations of light rays, certainly adequate for demonstrating ray behavior, but any real laser beam is a bundle of many parallel rays.

The following table outlines five basic ideas and assumptions of the ray model of light.

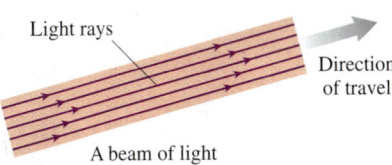

FIGURE 18.1 A laser beam or beam of sunlight is a bundle of parallel light rays.

The ray model of light

Light rays travel in straight lines.

Light travels through a transparent material in straight lines called light rays. The speed of light is $v = c/n$, where n is the index of refraction of the material.

Light rays can cross.

Light rays do not interact with each other. Two rays can cross without either being affected in any way.

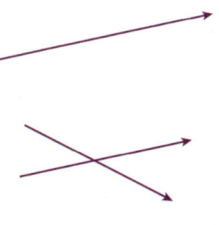

A light ray travels forever unless it interacts with matter.

A light ray continues forever unless it has an interaction with matter that causes the ray to change direction or to be absorbed. Light interacts with matter in four different ways:

■ At an interface between two materials, light can be either *reflected* or *refracted*.

■ Within a material, light can be either *scattered* or *absorbed*.

These interactions are discussed later in the chapter.

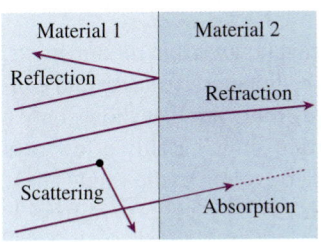

An object is a source of light rays.

An **object** is a source of light rays. Rays originate from *every* point on the object, and each point sends rays in *all* directions. Objects may be self-luminous—they create light rays—or they may be reflective objects that only reflect rays that originate elsewhere.

The eye sees by focusing a bundle of rays.

The eye sees an object when *diverging* bundles of rays from each point on the object enter the pupil and are focused to an image on the retina. Imaging is discussed later in the chapter, and the eye will be treated in much greater detail in Chapter 19.

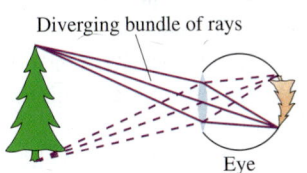

Sources of Light Rays

In the ray model, there are two kinds of objects. **Self-luminous objects** (or *sources*) directly create light rays. Self-luminous objects include light bulbs and the sun. Other objects, such as a piece of paper or a tree, are **reflective objects** that reflect rays originating in self-luminous objects. The table below shows four important kinds of self-luminous sources. Note that although ray and point sources are idealizations, they are very useful in understanding the propagation of rays.

Self-luminous objects

A ray source	A point source	An extended source	A parallel-ray source
Since a light ray is an idealization, there are no true ray sources. Still, the thin beam of a laser is often a good approximation of a single ray.	A point source is also an idealized source of light. It is infinitely small, and emits light rays in every direction. The tiny filaments of these bulbs approximate point sources.	This is the most common light source. **Every point of an extended source acts as a point source.** Light bulbs, flames, and the sun are extended sources.	Certain sources, such as flashlights and movie projectors, produce a bundle of parallel rays. Rays from a very distant object, such as a star, are very nearly parallel.

Reflective objects such as a newspaper, a face, or a mirror, can also be considered as sources of light rays. However, the origin of these rays is not in the object itself. Instead, light rays from a self-luminous object strike a reflective object and "bounce" off of it. These rays can then illuminate other objects, or enter our eyes and form images of the reflective object, just as rays from self-luminous objects do.

Ray Diagrams

Rays originate from *every* point on an object and travel outward in *all* directions, but a diagram trying to show all these rays would be hopelessly messy and confusing. To simplify the picture, we usually use a **ray diagram** that shows only a few rays. For example, Figure 18.2 is a ray diagram showing only a few rays leaving the top and bottom points of the object and traveling to the right. These rays will be sufficient to show us how the object is imaged by lenses or mirrors.

NOTE ▶ Ray diagrams are the basis for a *visual overview* that we'll use throughout this chapter. Be careful not to think that a ray diagram shows all of the rays. The rays shown on the diagram are just a subset of the infinitely many rays leaving the object. ◀

Seeing Objects

How do we *see* an object? The eye works by focusing an image of an object on the retina, a process we'll examine in Chapter 19. For now, we'll ignore the details of image formation and instead make use of a simpler fact from the figure in the ray model table above: **In order for our eye to see an object, rays from that object must enter the eye.** This idea helps explain some subtle points about seeing.

Consider a ray source such as a laser. Can you see a laser beam traveling across the room? Under ordinary circumstances, the answer is no. As we've seen, a laser beam is a good approximation of a single ray. This ray travels in a straight line from the laser to whatever it eventually strikes. As Figure 18.3 shows, no

These are just a few of the infinitely many rays leaving the object.

FIGURE 18.2 A ray diagram simplifies the situation by showing only a few rays.

You can't see a laser beam crossing the room because no light ray enters your eye.

FIGURE 18.3 A laser beam traveling through air is invisible.

light ray enters the eye, so the beam is invisible. The same argument holds for a parallel-ray source.

A point source or extended source behaves differently, as shown in Figure 18.4. Because a point source emits rays in *every* direction, some of these rays will enter the eye no matter where it is located. Thus a point source is visible to everyone looking at it. And, since every point on the surface of an extended source is itself a point source, all parts of an extended source (not blocked by something else) can be viewed by all observers as well.

We can also use our simple model of seeing to explain how we see non-luminous objects. As we've already mentioned, such objects *reflect* rays that strike them. Most ordinary objects—paper, skin, grass—reflect incident light in every direction, a process called **diffuse reflection.** Figure 18.5 illustrates the idea. Single rays are broken into many weaker rays that leave in all directions, a process called **scattering.**

Scattered light is what allows you to read a book by lamplight. Light rays leave the lamp—a self-luminous source—in every direction. As shown in Figure 18.5, every point on the surface of the page is struck by a ray (or rays) from the lamp. Then, because of diffuse reflection, these rays scatter in every direction; some of the scattered rays reach your eye, allowing you to see the page.

It is possible to make a laser beam visible by scattering it from very small particles suspended in air. These particles can be dust, smoke, or water droplets such as fog. Figure 18.6 shows that as the beam strikes such a particle, it scatters rays in every direction. Some of these rays enter the eye, making each particle in the path of the beam visible and outlining the beam's path across the room.

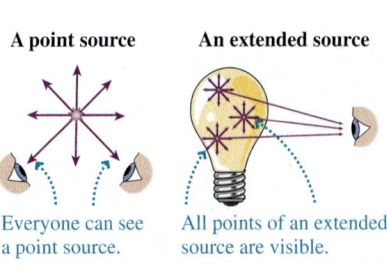

A point source | An extended source

Everyone can see a point source. | All points of an extended source are visible.

FIGURE 18.4 Point and extended sources can be seen by all observers.

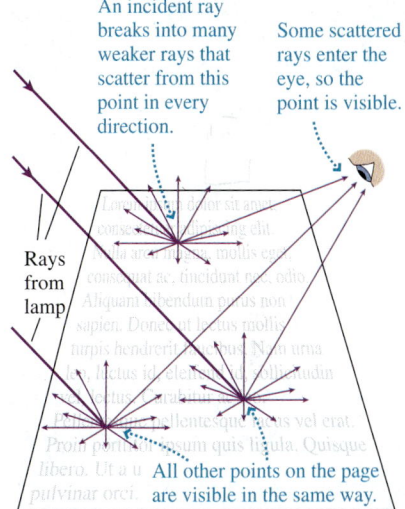

An incident ray breaks into many weaker rays that scatter from this point in every direction.

Some scattered rays enter the eye, so the point is visible.

Rays from lamp

All other points on the page are visible in the same way.

FIGURE 18.5 Reading a book by scattered light.

Rays scatter in every direction from each smoke particle.

Smoke particle

Some of these reach the eye, making the path of the laser beam visible.

FIGURE 18.6 A laser beam is visible if it travels through smoke or dust.

Shadows

Our ray model of light also explains the common phenomenon of *shadows.* Suppose an opaque object (such as a cardboard disk) is placed between a source of light and a screen. The object intercepts some of the rays, leaving a dark area behind it. Other rays travel on to a screen and illuminate it. The simplest shadows are those cast by a point source of light. This process is shown in Figure 18.7a. With a point source, the shadow is completely dark, and the edges of the shadow are sharp.

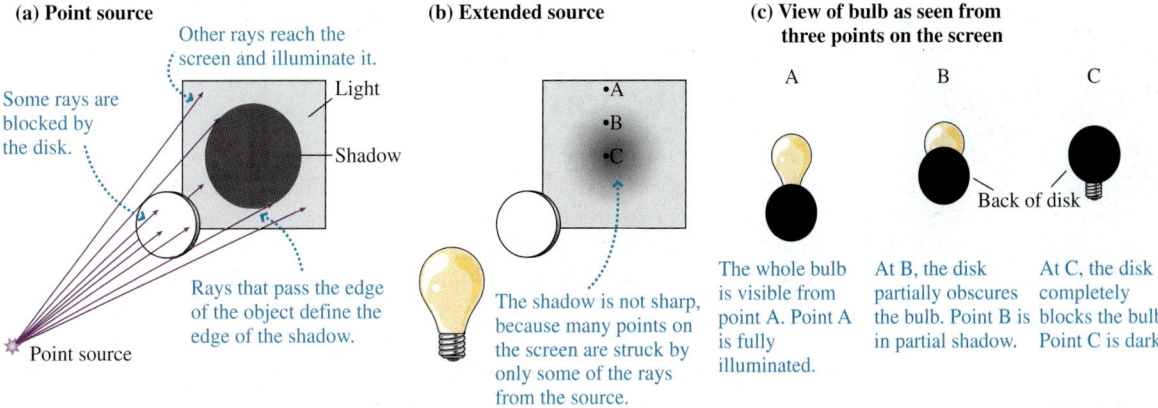

(a) Point source

Other rays reach the screen and illuminate it.

Some rays are blocked by the disk.

Light

Shadow

Rays that pass the edge of the object define the edge of the shadow.

Point source

(b) Extended source

•A
•B
•C

The shadow is not sharp, because many points on the screen are struck by only some of the rays from the source.

(c) View of bulb as seen from three points on the screen

A   B   C

Back of disk

The whole bulb is visible from point A. Point A is fully illuminated.

At B, the disk partially obscures the bulb. Point B is in partial shadow.

At C, the disk completely blocks the bulb. Point C is dark.

FIGURE 18.7 Shadows produced by point and extended sources of light.

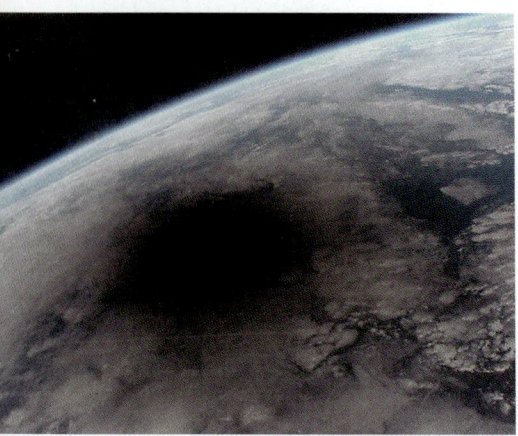

During a solar eclipse, the sun—a small but extended source—casts a shadow of the moon on the earth. In this satellite image, the moon's shadow has a dark center surrounded by a region of increasing brightness, just as in Figure 18.7b.

Shadows cast by extended sources are more complicated. An extended source is a collection of a large number of point sources, each of which casts its own shadow. However, as shown in Figure 18.7b, the patterns of shadow and light from each point overlap and thus the shadow region is no longer sharp. Figure 18.7c shows the view of the *bulb* as seen from three points on the screen. Depending on the size of the source, there is often a true shadow that no light reaches, surrounded by a fuzzy region of increasing brightness.

The Pinhole Camera

One of the simplest devices capable of forming an image is the **pinhole camera,** shown in Figure 18.8, in which a small hole is punched in an otherwise light-proof box. Film, sensitive to light, is placed on the back wall of the box. Figure 18.8b uses the ray model of light passing through a small hole to explain how the pinhole camera works. Each point on an object emits light rays in all directions, but, ideally, only one of these rays passes through the hole and reaches the film. Each point on the object thus illuminates just one point on the film, forming the image. As the figure illustrates, the geometry of the rays causes the image to be upside down.

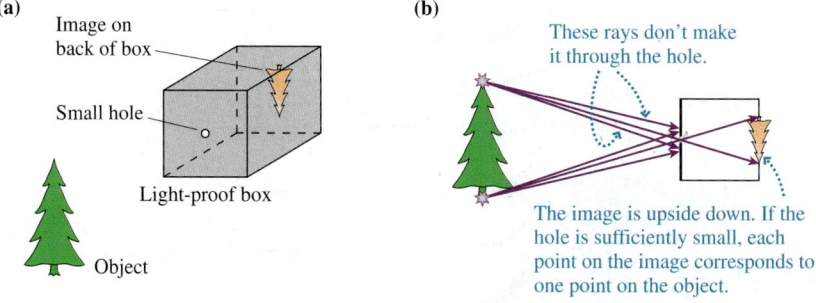

FIGURE 18.8 A pinhole camera.

A pinhole eye BIO The chambered nautilus is the only animal with a true pinhole "camera" as an eye. Light rays passing through the small opening form a crude image on the back surface of the eye, where the rays strike light-sensitive cells. The image may be poor, but it's sufficient to allow the nautilus to catch prey and escape predators.

Actually, as you may have realized, each *point* on the object illuminates a small but finite *patch* on the wall. This is because the finite size of the hole allows several rays from each point on the object to pass through at slightly different angles. As a result, the image is slightly blurred and out of focus. Maximum sharpness is achieved by making the hole smaller and smaller, which makes the image dimmer and dimmer. (Diffraction also becomes an issue if the hole gets too small.) A real pinhole camera has to accept a small amount of blurring as the tradeoff for having an image bright enough to be practical. In Chapter 19 we'll see the dramatic improvement possible by using a *lens* in place of the pinhole.

18.2 Reflection

Reflection of light is a familiar, everyday experience. You see your reflection in the bathroom mirror first thing every morning, reflections in your car's rearview mirror as you drive to school, and the sky reflected in puddles of standing water. Reflection from a flat, smooth surface, such as a mirror or a piece of polished metal, is called **specular reflection,** from *speculum,* the Latin word for "mirror."

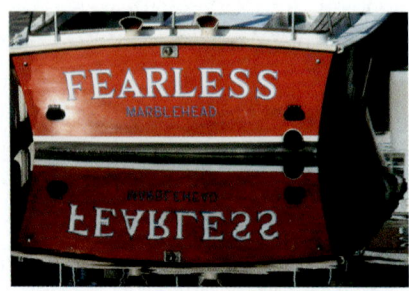

Reflection is an everyday experience.

Figure 18.9a shows a bundle of parallel light rays reflecting from a mirror-like surface. You can see that the incident and reflected rays are both in a plane that is normal, or perpendicular, to the reflective surface. A three-dimensional perspective accurately shows the relation between the light rays and the surface, but figures such as this are hard to draw by hand. Instead, it is customary to represent reflection with the simpler visual overview of Figure 18.9b. In this figure,

- The incident and reflected rays are in the plane of the page. The reflective surface extends into and out of the page.
- A *single* light ray represents the entire bundle of parallel rays. This is oversimplified, but it keeps the figure and the analysis clear.

The angle θ_i between the ray and a line perpendicular to the surface—the *normal* to the surface—is called the **angle of incidence**. Similarly, the **angle of reflection** θ_r is the angle between the reflected ray and the normal to the surface. The **law of reflection,** easily demonstrated with simple experiments, states that

1. The incident ray and the reflected ray are in the same plane normal to the surface, and
2. The angle of reflection equals the angle of incidence: $\theta_r = \theta_i$.

NOTE ▶ Optics calculations *always* use the angle measured from the normal, not the angle between the ray and the surface. ◀

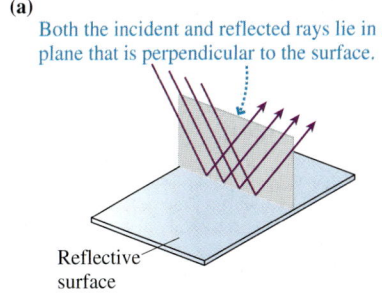

(a) Both the incident and reflected rays lie in a plane that is perpendicular to the surface.

Reflective surface

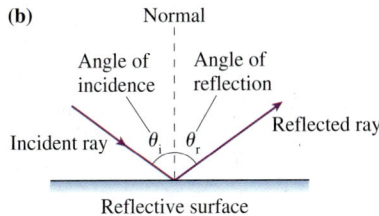

(b)

Normal

Angle of incidence | Angle of reflection

Incident ray θ_i | θ_r Reflected ray

Reflective surface

FIGURE 18.9 Specular reflection of light.

EXAMPLE 18.1 Light reflecting from a mirror

A full-length mirror on a closet door is 2.0 m tall. The bottom touches the floor. A bare light bulb hangs 1.0 m from the closet door, 0.5 m above the top of the mirror. How long is the streak of reflected light across the floor?

PREPARE Treat the light bulb as a point source and use the ray model of light. Figure 18.10 is a visual overview of the light

Bulb   1.0 m

These angles are the same by the law of reflection. θ θ 0.5 m

θ

2.0 m

Mirror

θ

l

FIGURE 18.10 Visual overview of the light rays reflecting from a mirror.

rays. We need to consider only the two rays that strike the edges of the mirror. All other reflected rays will fall between these two.

SOLVE The ray that strikes the bottom of the mirror reflects from it and hits the floor just where the mirror meets the floor. For the top ray, Figure 18.10 has used the law of reflection to set the angle of reflection equal to the angle of incidence; we call both θ. By simple geometry, the other angles shown are also equal to θ. From the small triangle at the upper right,

$$\theta = \tan^{-1}\left(\frac{0.5 \text{ m}}{1.0 \text{ m}}\right) = 26.6°$$

But we also have $\tan\theta = (2.0 \text{ m})/l$, or

$$l = \frac{2.0 \text{ m}}{\tan\theta} = \frac{2.0 \text{ m}}{\tan 26.6°} = 4.0 \text{ m}$$

Since the lower ray struck right at the mirror's base, the total length of the reflected streak is 4.0 m.

Diffuse Reflection

We've already discussed diffuse reflection, the reflection of light rays off of a surface such as paper or cloth that is not shiny like a mirror. If you magnify the surface of a diffuse reflector, you'll find that on the microscopic scale it is quite rough. The law of reflection $\theta_r = \theta_i$ is still obeyed at each point but the irregularities of the surface cause the reflected rays to leave in many random directions. This situation is shown in Figure 18.11. Diffuse reflection is much more common than the mirror-like specular reflection.

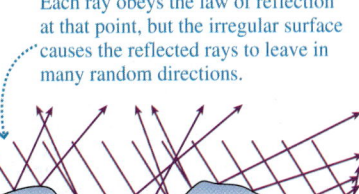

Each ray obeys the law of reflection at that point, but the irregular surface causes the reflected rays to leave in many random directions.

Magnified view of surface

FIGURE 18.11 Diffuse reflection from an irregular surface.

The Plane Mirror

One of the most commonplace observations is that you can see yourself in a mirror. How? Figure 18.12a shows rays from point source P reflecting from a flat mirror, called a **plane mirror.** Consider the particular ray shown in Figure 18.12b. The reflected ray travels along a line that passes through point P′ on the "back side" of the mirror. Because $\theta_r = \theta_i$, simple geometry dictates that P′ is the same distance behind the mirror as P is in front of the mirror. That is, $s' = s$.

(a)

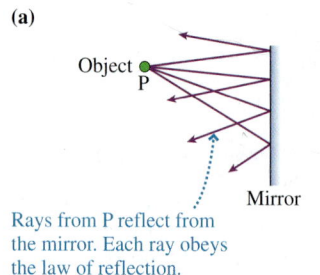

Object
P

Mirror

Rays from P reflect from the mirror. Each ray obeys the law of reflection.

(b)

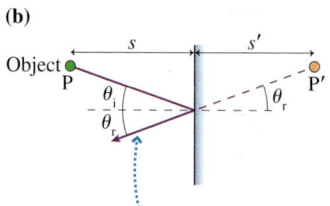

Object
P

This reflected ray appears to have come from point P′.

(c)

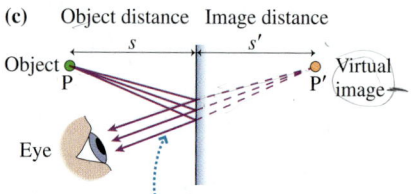

Object distance   Image distance

Object
P

Virtual image

Eye

The reflected rays *all* diverge from P′, which appears to be the source of the reflected rays. Your eye collects the bundle of diverging rays and "sees" the light coming from P′.

FIGURE 18.12 The light rays reflecting from a plane mirror.

The location of point P′ in Figure 18.12b is independent of the value of θ_i. Consequently, *all* reflected rays travel along lines that pass through the *same* point P′. The original light rays diverged from point P, but the reflected rays now *appear* to diverge from point P′. Consequently, as Figure 18.12c shows, **the reflected rays all appear to be coming from point P′.** For a plane mirror, the distance s' to point P′ is equal to the *object distance s*:

$$s' = s \quad \text{(plane mirror)} \tag{18.1}$$

If rays diverge from an *object point* P and interact with a mirror so that the reflected rays diverge from point P′ and *appear* to come from P′, then we call P′ a **virtual image** of point P. The image is "virtual" in the sense that no rays actually leave P′, which is in darkness behind the mirror. But as far as your eye is concerned, the light rays act exactly *as if* the light really originated at P′. So while you may say "I see P in the mirror," what you are actually seeing is the virtual image of P. Distance s' is the *image distance*.

For an extended object, such as the one in Figure 18.13, each point on the object from which rays strike the mirror has a corresponding image point an equal distance on the opposite side of the mirror. The eye captures and focuses diverging bundles of rays from each point of the image in order to see the full image in the mirror. Two facts are worth noting:

1. Rays from each point on the object spread out in all directions and strike *every point* on the mirror. Only a very few of these rays enter your eye, but the other rays are very real and might be seen by other observers.
2. Rays from points P and Q enter your eye after reflecting from *different* areas of the mirror. This is why you can't always see the full image of an object in a very small mirror.

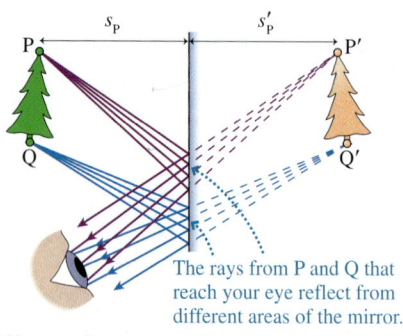

s_P s'_P

P   P′

Q   Q′

The rays from P and Q that reach your eye reflect from different areas of the mirror.

Your eye intercepts only a very small fraction of all the reflected rays.

◄ **FIGURE 18.13** Each point on the extended object has a corresponding image point an equal distance on the opposite side of the mirror.

EXAMPLE 18.2 How high is the mirror?

If your height is h, what is the shortest mirror on the wall in which you can see your full image? Where must the top of the mirror be hung?

PREPARE Use the ray model of light. Figure 18.14 is a visual overview of the light rays. We need to consider only the two rays that leave your head and feet and reflect into your eye.

FIGURE 18.14 Visual overview of light rays from your head and feet reflecting into your eye.

SOLVE Let the distance from your eyes to the top of your head be l_1 and the distance to your feet be l_2. Your height is $h = l_1 + l_2$. A light ray from the top of your head that reflects from the mirror at $\theta_r = \theta_i$ and enters your eye must, by congruent triangles, strike the mirror a distance $\frac{1}{2}l_1$ above your eyes. Similarly, a ray from your foot to your eye strikes the mirror a distance $\frac{1}{2}l_2$ below your eyes. The distance between these two points on the mirror is $\frac{1}{2}l_1 + \frac{1}{2}l_2 = \frac{1}{2}h$. A ray from anywhere else on your body will reach your eye if it strikes the mirror between these two points. Pieces of the mirror outside these two points are irrelevant, not because rays don't strike them but because the reflected rays don't reach your eye. Thus the shortest mirror in which you can see your full reflection is $\frac{1}{2}h$. But this will only work if the top of the mirror is hung midway between your eyes and the top of your head.

ASSESS It is interesting that the answer does not depend on how far you are from the mirror.

TRY IT YOURSELF

Through the looking glass . . . Prop up a small mirror at 90° to a tabletop. Place identical cylindrical objects, such as the candlesticks shown, at equal distances in front of and behind the mirror. As you move your head around, you'll see that the image of the front candlestick always aligns perfectly with the actual candlestick behind the mirror. It is in this sense that the image is behind the mirror: rays from it appear to come from an object behind the mirror.

Left and Right

It's common wisdom that a mirror "reverses left and right." But why, then, does it not also reverse up and down? What's special about left and right?

Hold your hands in front of you so that you can see the back of your right hand but the palm of your left hand. If the lighting were poor so that you could see only the outlines of your hands, as in Figure 18.15, could you tell which is a "right hand" and which is a "left hand"? No! Unlike "up" and "down," the terms "right" and "left" do not have an absolute, unambiguous meaning. Right and left are determined by the orientation of your thumb *relative to* your palm. Without knowing where the palm is, you can't assign a handedness to a hand.

In fact, a mirror does not "reverse right and left" any more than it reverses up and down. Instead, a mirror reverses *front and back*. Hold your right hand out, palm away from you and thumb pointing left. Imagine turning your hand inside-out in the sense that everything on the palm side is pulled through toward you and everything on the back side is pushed through away from you. Your fingers would still point up and your thumb would still point to the left, but you would see your palm rather than the back of your hand. Neither up/down nor left/right have reversed, but your "right hand" has become a "left hand."

Right or left?

FIGURE 18.15 Can you tell which hand this is?

STOP TO THINK 18.1   Two plane mirrors form a right angle. How many images of the ball can you see in the mirrors?

A. 1
B. 2
C. 3
D. 4

Observer

18.3 Refraction

15.1–15.3   Activ Physics

In Figure 18.16, two things happen when a light ray crosses the boundary between the air and the glass:

1. Part of the light *reflects* from the boundary, obeying the law of reflection. This is how you see reflections from pools of water or storefront windows, even though water and glass are transparent.
2. Part of the light continues into the second medium. It is *transmitted* rather than reflected, but the transmitted ray changes direction as it crosses the boundary. The transmission of light from one medium to another, but with a change in direction, is called **refraction.**

The photograph of Figure 18.16 shows the refraction of a light beam as it passes through a glass prism. Notice that the ray direction changes as the light enters and leaves the glass. You can also see two weak reflections leaving the bottom surface of the prism.

Reflection from the boundary between transparent media is usually weak. Typically 95% of the light is transmitted and only 5% is reflected. Our goal in this section is to understand refraction, so we will usually ignore the weak reflection and focus on the transmitted light.

NOTE ▶ A transparent material through which light travels is called the *medium* (plural *media*). ◀

Figure 18.17a shows the refraction of light rays from a parallel beam of light, such as a laser beam, and rays from a point source. It's good to remember that an infinite number of rays are incident on the boundary, but our analysis will be simplified if we focus on a single light ray. Figure 18.17b is a ray diagram showing the refraction of a single ray at a boundary between medium 1 and medium 2. Let the angle between the ray and the normal be θ_1 in medium 1 and θ_2 in medium 2. For the medium in which the ray is approaching the boundary, this is the *angle of incidence* as we've previously defined it. The angle on the transmitted side, *measured from the normal*, is called the **angle of refraction.** Notice that θ_1 is the

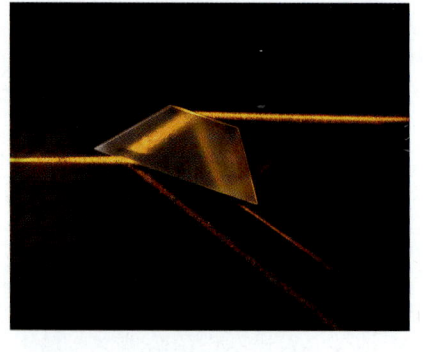

FIGURE 18.16 A light beam refracts through a glass prism.

(a)

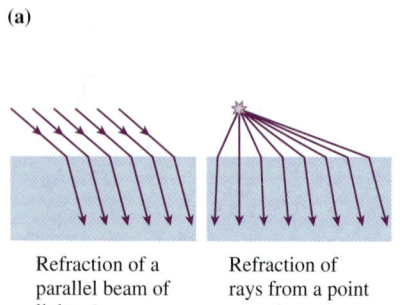

Refraction of a parallel beam of light.

Refraction of rays from a point source.

(b)

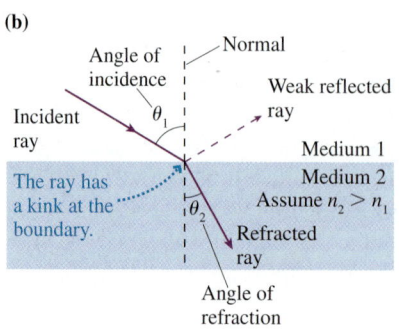

Angle of incidence
Normal
Incident ray
θ_1
Weak reflected ray
Medium 1
The ray has a kink at the boundary.
Medium 2
Assume $n_2 > n_1$
θ_2
Refracted ray
Angle of refraction

(c)

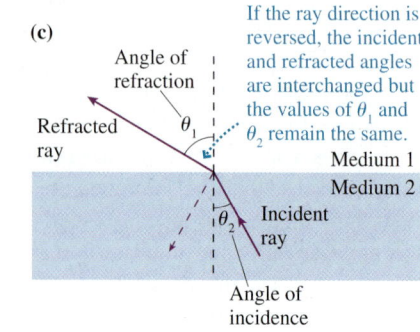

If the ray direction is reversed, the incident and refracted angles are interchanged but the values of θ_1 and θ_2 remain the same.

Angle of refraction
Refracted ray
θ_1
Medium 1
Medium 2
Incident ray
θ_2
Angle of incidence

FIGURE 18.17 Refraction of light rays.

angle of incidence in Figure 18.17b and the angle of refraction in Figure 18.17c, where the ray is traveling in the opposite direction, even though the value of θ_1 has not changed.

The relationship between angles θ_1 and θ_2 is generally believed to have been discovered in 1621 by the Dutch scientist Willebrord Snell, although some historians of science think it may have been known before then. Regardless of its discoverer, we refer to the "law of refraction" as **Snell's law.** If a ray refracts between medium 1 and medium 2, having indices of refraction n_1 and n_2, the ray angles θ_1 and θ_2 in the two media are related by

$$n_1 \sin\theta_1 = n_2 \sin\theta_2 \qquad (18.2)$$

Snell's law for refraction between two media

Notice that Snell's law does not mention which is the incident angle and which the refracted angle.

Table 18.1 lists the indices of refraction for several media. It is interesting to note that the n in the law of refraction, Equation 18.2, is the same index of refraction n we studied in Chapter 17. There we found that the index of refraction determines the speed of a light wave in a medium according to $v = c/n$. Here, it appears to play a different role, determining by how much a light ray is bent when crossing the boundary between two different media. Although we won't do so here, it is possible to use Huygens' principle to show that Snell's law is a *consequence* of the change in the speed of light as it moves across a boundary between media.

Examples of Refraction

Look back at Figure 18.17. As the ray in Figure 18.17b moves from medium 1 to medium 2, where $n_2 > n_1$, it bends closer to the normal. In Figure 18.17c, where the ray moves from medium 2 to medium 1, it bends away from the normal. This is a general conclusion that follows from Snell's law:

■ When a ray is transmitted into a material with a higher index of refraction, it bends to make a smaller angle with the normal.
■ When a ray is transmitted into a material with a lower index of refraction, it bends away from the normal.

This rule becomes a central idea in a procedure for analyzing refraction problems.

TABLE 18.1 Indices of refraction

Medium	n
Vacuum	1.00 exactly
Air (actual)	1.0003
Air (accepted)	1.00
Water	1.33
Ethyl alcohol	1.36
Oil	1.46
Glass (typical)	1.50
Polystyrene plastic	1.59
Zircon	1.96
Diamond	2.42
Silicon (infrared)	3.50

 TACTICS BOX 18.1 Analyzing refraction Exercises 9–13

❶ **Draw a ray diagram.** Represent the light beam with one ray.
❷ **Draw a line normal (perpendicular) to the boundary.** Do this at each point where the ray intersects a boundary.
❸ **Show the ray bending in the correct direction.** The angle is larger on the side with the smaller index of refraction. This is the qualitative application of Snell's law.
❹ **Label angles of incidence and refraction.** Measure all angles from the normal.
❺ **Use Snell's law.** Calculate the unknown angle or unknown index of refraction.

EXAMPLE 18.3 Deflecting a laser beam

A laser beam is aimed at a 1.0-cm-thick sheet of glass at an angle 30° above the glass.

a. What is the laser beam's direction of travel in the glass?
b. What is its direction in the air on the other side of the glass?

PREPARE Represent the laser beam with a single ray and use the ray model of light. Figure 18.18 is a visual overview in which the first four steps of Tactics Box 18.1 have been identified. Notice that the angle of incidence must be measured from the normal, so $\theta_1 = 60°$, not the 30° value given in the problem. The index of refraction of glass was taken from Table 18.1.

① Draw ray diagram.
$n_1 = 1.00$ 30°
② Draw normals to boundary.
$n_2 = 1.50$
$n_1 = 1.00$
③ Show smaller angle in medium with larger n.
④ Label angles, measured from normal.

FIGURE 18.18 The ray diagram of a laser beam passing through a sheet of glass.

SOLVE

a. Snell's law, the final step in the Tactics Box, is $n_1 \sin\theta_1 = n_2 \sin\theta_2$. Using $\theta_1 = 60°$, we find that the direction of travel in the glass is

$$\theta_2 = \sin^{-1}\left(\frac{n_1 \sin\theta_1}{n_2}\right) = \sin^{-1}\left(\frac{\sin 60°}{1.5}\right)$$
$$= \sin^{-1}(0.577) = 35.3°$$

or 35° to two significant figures.

b. Snell's law at the second boundary is $n_2 \sin\theta_3 = n_1 \sin\theta_4$. You can see from Figure 18.18 that the interior angles are equal: $\theta_3 = \theta_2 = 35.3°$. Thus the ray emerges back into the air traveling at angle

$$\theta_4 = \sin^{-1}\left(\frac{n_2 \sin\theta_3}{n_1}\right) = \sin^{-1}(1.5 \sin 35.3°)$$
$$= \sin^{-1}(0.867) = 60°$$

This is the same as θ_1, the original angle of incidence.

ASSESS The laser beam exits the glass still traveling in the same direction as it entered, but its path is *displaced*. This is a general result for light traveling through a medium with parallel sides. As the glass becomes thinner, the displacement becomes less; there is no displacement as the glass thickness becomes zero. This will be an important observation when we later study lenses.

EXAMPLE 18.4 Measuring the index of refraction

Figure 18.19 shows a laser beam deflected by a 30°-60°-90° prism. What is the prism's index of refraction?

30°
Laser beam   22.6°
60°

FIGURE 18.19 A prism deflects a laser beam.

PREPARE Represent the laser beam with a single ray and use the ray model of light. Figure 18.20 uses the steps of Tactics Box 18.1 to draw a ray diagram. The ray is incident perpendicular to the front face of the prism ($\theta_{\text{incident}} = 0°$), thus it is transmitted through the first boundary without deflection. At the second boundary it is especially important to *draw the normal to the surface* at the point of incidence and to *measure angles from the normal*.

$n_2 = 1.00$
Normal
30°
θ_1
60°
$\theta_2 = \theta_1 + \phi$
Angle of refraction
$\theta_1 = 30°$
Angle of incidence
$\phi = 22.6°$
n_1

θ_1 and θ_2 are measured from the normal.

FIGURE 18.20 Visual overview of a laser beam passing through the prism.

SOLVE From the geometry of the triangle you can find that the laser's angle of incidence on the hypotenuse of the prism is $\theta_1 = 30°$, the same as the apex angle of the prism. The ray exits the prism at angle θ_2 such that the deflection is $\phi = \theta_2 - \theta_1 = 22.6°$. Thus $\theta_2 = 52.6°$. Knowing both angles and $n_2 = 1.00$ for air, we can use Snell's law to find n_1:

$$n_1 = \frac{n_2 \sin\theta_2}{\sin\theta_1} = \frac{1.00 \sin 52.6°}{\sin 30°} = 1.59$$

ASSESS Referring to the indices of refraction in Table 18.1, we see that the prism is made of polystyrene plastic.

▶ **Optical image stabilization** When you make a video with a hand-held video camera, the inevitable movement of your hands shows up as unwanted motion in the video itself. High-end video cameras largely eliminate this motion using *optical image stabilization*. By automatically adjusting the angle of an internal liquid-filled prism, rays entering the prism are refracted at the second glass plate so that they always hit the camera's light sensor in the correct spot.

This angle can be rapidly changed so that the prism deflects light rays by the correct amount.

Total Internal Reflection

What would have happened in Example 18.4 if the prism angle had been 45° rather than 30°? The light rays would approach the rear surface of the prism at an angle of incidence $\theta_1 = 45°$. When we try to calculate the angle of refraction at which the ray emerges into the air, we find

$$\sin\theta_2 = \frac{n_1}{n_2}\sin\theta_1 = \frac{1.59}{1.00}\sin 45° = 1.12$$

$$\theta_2 = \sin^{-1}(1.12) = ???$$

Angle θ_2 cannot be computed because the sine of an angle can't be larger than 1. The ray is unable to refract through the boundary. Instead, 100% of the light *reflects* from the boundary back into the prism. This process is called **total internal reflection**, often abbreviated TIR. That it really happens is illustrated in Figure 18.21. Here, three light beams enter a prism from the left. The bottom two refract out through the right side of the prism. The blue beam, which is incident on the prism's top face at a slightly larger angle of incidence, undergoes total internal reflection and then emerges through the right surface.

Figure 18.22 shows several rays leaving a point source in a medium with index of refraction n_1. The medium on the other side of the boundary has $n_2 < n_1$. As we've seen, crossing a boundary into a material with a lower index of refraction causes the ray to bend away from the normal. Two things happen as angle θ_1 increases. First, the refraction angle θ_2 approaches 90°. Second, the fraction of the light energy that is transmitted decreases while the fraction reflected increases.

A **critical angle** θ_c is reached when $\theta_2 = 90°$. Because $\sin 90° = 1$, Snell's law $n_1\sin\theta_c = n_2\sin 90°$ gives

$$\theta_c = \sin^{-1}\left(\frac{n_2}{n_1}\right) \tag{18.3}$$

Critical angle of incidence for total internal reflection

The refracted light vanishes at the critical angle and the reflection becomes 100% for any angle $\theta_1 \geq \theta_c$. The critical angle is well defined because of our assumption that $n_2 < n_1$. **There is no critical angle and no total internal reflection if $n_2 > n_1$.**

As a quick example, the critical angle in a typical piece of glass at the glass-air boundary is

$$\theta_{c\ glass} = \sin^{-1}\left(\frac{1.00}{1.50}\right) = 42°$$

The fact that the critical angle is less than 45° has important applications. For example, Figure 18.23 shows a pair of binoculars. The lenses are much farther apart than your eyes, so the light rays need to be brought together before exiting the eyepieces. Rather than using mirrors, which get dirty, are easily scratched, and require alignment, binoculars use a pair of prisms on each side. Thus the light undergoes two TIRs and emerges from the eyepiece. (The actual prism arrangement in binoculars is a bit more complex, but this illustrates the basic idea.)

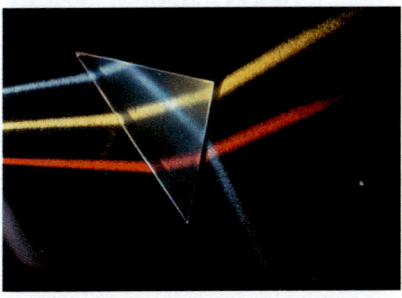

FIGURE 18.21 One of the three beams of light undergoes total internal reflection.

FIGURE 18.22 Refraction and reflection of rays as the angle of incidence increases.

Angles of incidence exceed the critical angle.

FIGURE 18.23 Binoculars and other optical instruments make use of total internal reflection (TIR).

Snell's window We can understand what a diver sees when she's underwater by reversing the direction of all the rays in Figure 18.24 in the adjacent example. The drawing shows that she can see the sun overhead, and clouds at larger angles. She can even see objects sitting at the waterline—but they appear at the edge of a circle as she looks up. Anything outside of this circle is a reflection of something in the water. The photo shows what she sees: A bright circle from the sky above—*Snell's window*—surrounded by the dark reflection of the water below.

EXAMPLE 18.5 Seeing a submerged light

A light bulb is set in the bottom of a 3.0-m-deep swimming pool. What is the diameter of the circle inside which a duck swimming on the surface could see the bulb?

PREPARE Represent the light bulb as a point source and use the ray model of light. Figure 18.24 is a visual overview of the light rays. The light bulb emits rays at all angles, but only some of the rays refract into the air where they can be seen from above. Rays striking the surface at greater than the critical angle undergo TIR and remain within the water. The diameter of the circle of light is the distance D between the two points at which rays strike the surface at the critical angle.

FIGURE 18.24 Visual overview of the rays leaving a light bulb at the bottom of a swimming pool.

SOLVE From trigonometry, the circle diameter is $D = 2h\tan\theta_c$, where h is the depth of the water. The critical angle for a water-air boundary is $\theta_c = \sin^{-1}(1.00/1.33) = 48.7°$. Thus

$$D = 2(3.0 \text{ m})\tan 48.7° = 6.8 \text{ m}$$

ASSESS Light rays emerging at the edge of the circle actually skim the surface of the water. By reversing the direction of the rays, we can understand what a diver sees when she's under water. This idea is explored further in the discussion to the left.

Fiber Optics

The most important modern application of total internal reflection is the transmission of light through optical fibers. Figure 18.25a shows a laser beam shining into the end of a long, narrow-diameter glass tube. The light rays pass easily from the air into the glass, but they then impinge on the inside wall of the glass tube at an angle of incidence θ_1 approaching 90°. This is well above the critical angle, so the laser beam undergoes TIR and remains inside the glass. The laser beam continues to "bounce" its way down the tube as if the light were inside a pipe. Indeed, optical fibers are sometimes called "light pipes." The rays are *below* the critical angle ($\theta_1 \approx 0$) when they finally reach the flat end of the fiber, thus they refract out without difficulty and can be detected.

While a simple glass tube can transmit light, reliance on a glass-air boundary is not sufficiently reliable for commercial use. Any small scratch on the side of the tube alters the rays' angle of incidence and allows leakage of light. Figure 18.25b shows the construction of a practical optical fiber. A small-diameter glass *core* is surrounded by a layer of glass *cladding*. The glasses used for the core and the cladding have $n_{core} > n_{cladding}$, thus light undergoes TIR at the core-cladding boundary and remains confined within the core. This boundary is not exposed to the environment and hence retains its integrity even under adverse conditions.

FIGURE 18.25 Light rays are confined within an optical fiber by total internal reflection.

Optical fibers have found important applications in medical diagnosis and treatment. Thousands of small fibers can be fused together to make a flexible bundle capable of transmitting high-resolution images along its length. Such *endoscopes* are used for minimally invasive inspection of body cavities, joints, and internal organs. One end of the endoscope bundle has a lens that allows an image of what's in front of it to be sent up the bundle where it can be observed on a television monitor by the physician. Optical fibers are also used to send light *down* the bundle to provide illumination, and small instruments such as forceps and clamps can be operated along its length to retrieve samples for biopsy.

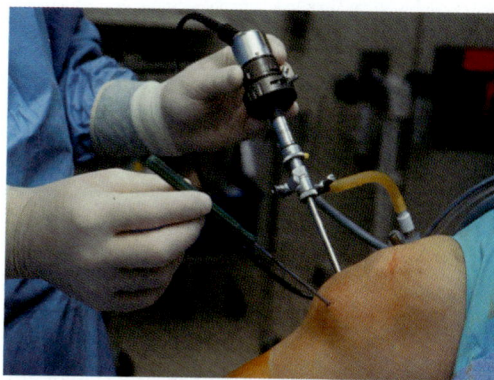

STOP TO THINK 18.2 A light ray travels from medium 1 to medium 3 as shown. For these media,

A. $n_3 > n_1$.
B. $n_3 = n_1$.
C. $n_3 < n_1$.
D. We can't compare n_1 to n_3 without knowing n_2.

Arthroscopic surgery BIO Operations on injured joints can often be performed using an endoscope inserted through a small incision. The endoscope allows the surgeon to observe the procedure, which is performed with instruments inserted through another incision. The recovery time for such surgery is usually much shorter than for conventional operations requiring a full incision to expose the interior of the joint.

18.4 Color and Dispersion

One of the most obvious visual aspects of light is the phenomenon of color. Yet color, for all its vivid sensation, is not inherent in the light itself. Color is a *perception,* not a physical quantity. Color is associated with the wavelength of light, but the fact that we see light with a wavelength of 650 nm as "red" tells us how our visual system responds to electromagnetic waves of this wavelength. There is no "redness" associated with the light wave itself.

Most of the results of optics do not depend on color. Rays of red light pass through glass nearly the same as rays of blue light, so we generally don't need to know the color of light—or, to be more precise, its wavelength—to use the laws of reflection and refraction. Nonetheless, color is an interesting subject, one worthy of a short digression.

Color

It has been known since antiquity that irregularly shaped glass and crystals cause sunlight to be broken into various colors. A common idea was that the glass or crystal somehow altered the properties of the light by *adding* color to the light. Newton suggested a different explanation. He first passed a sunbeam through a prism, producing the familiar rainbow of light. We say that the prism *disperses* the light. Newton's novel idea, shown in Figure 18.26a, was to use a second prism, inverted with respect to the first, to "reassemble" the colors. He found that the light emerging from the second prism was a beam of pure, white light.

But the emerging light beam is white only if *all* the rays are allowed to move between the two prisms. Blocking some of the rays with small obstacles, as in Figure 18.26b, causes the emerging light beam to have color. This suggests that color is associated with the light itself, not with anything that the prism is "doing" to the light. Newton tested this idea by inserting a small aperture between the prisms to pass only the rays of a particular color, such as green. If the prism alters the properties of light, then the second prism should change the green light to other colors. Instead, the light emerging from the second prism is unchanged from the green light entering the prism.

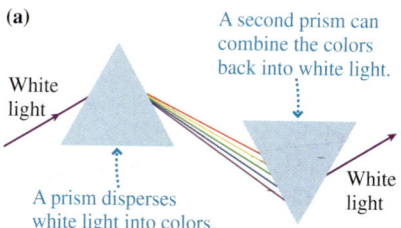

(a)
A second prism can combine the colors back into white light.
White light
A prism disperses white light into colors.
White light

(b)
The second prism does not change pure colors.
White light
Green light
An aperture selects a green ray of light.

FIGURE 18.26 Newton used prisms to study color.

I procured me a triangular glass prism to try therewith the celebrated phenomena of colors.

Isaac Newton

These and similar experiments show that:

1. What we perceive as white light is a mixture of all colors. White light can be dispersed into its various colors and, equally important, mixing all the colors produces white light.
2. The index of refraction of a transparent material differs slightly for different colors of light. Glass has a slightly larger index of refraction for violet light than for green light or red light. Consequently, different colors of light refract at slightly different angles and follow slightly different paths through a piece of glass. A prism does not alter the light or add anything to the light; it simply causes the different colors that are inherent in white light to follow slightly different trajectories.

Dispersion

It was Thomas Young, with his two-slit interference experiment, who showed that what we perceive as different colors are associated with light of different wavelengths. The longest wavelengths are perceived as red light and the shortest wavelengths are perceived as violet light. Table 18.2 is a brief summary of the *visible spectrum* of light. Visible-light wavelengths are used so frequently that it is well worth committing this short table to memory.

Newton's observation that the index of refraction varies slightly with color implies that the index of refraction varies slightly with wavelength. This is known as **dispersion**. Figure 18.27 shows the *dispersion curves* of two common glasses. Notice that *n* is *larger* when the wavelength is *shorter*, thus violet light refracts more than red light.

TABLE 18.2 A brief summary of the visible spectrum of light

Color	Approximate wavelength
Deepest red	700 nm
Red	650 nm
Yellow	600 nm
Green	550 nm
Blue	450 nm
Deepest violet	400 nm

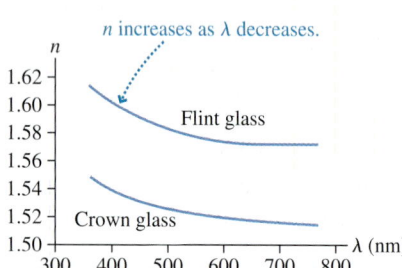

FIGURE 18.27 Dispersion curves show how the index of refraction varies with wavelength.

EXAMPLE 18.6 Dispersing light with a prism

Example 18.4 found that a ray incident on a 30° prism is deflected by 22.6° if the prism's index of refraction is 1.59. Suppose this is the index of refraction of deep violet light, and that deep red light has an index of refraction of 1.54.

a. What is the deflection angle for deep red light?
b. If a beam of white light is dispersed by this prism, how wide is the rainbow spectrum on a screen 2.0 m away?

PREPARE Figure 18.19 in Example 18.4 showed the geometry. A ray of any wavelength is incident on the rear surface of the prism at $\theta_1 = 30°$.

SOLVE

a. If $n_1 = 1.54$ for deep red light, the refraction angle is

$$\theta_2 = \sin^{-1}\left(\frac{n_1 \sin\theta_1}{n_2}\right) = \sin^{-1}\left(\frac{1.54 \sin 30°}{1.00}\right) = 50.4°$$

Example 18.4 showed that the deflection angle is $\phi = \theta_2 - \theta_1$, so deep red light is deflected by $\phi_{red} = 20.4°$. This angle is slightly smaller than the previously observed $\phi_{violet} = 22.6°$.

b. The entire spectrum is spread between $\phi_{red} = 20.4°$ and $\phi_{violet} = 22.6°$. The angular spread is

$$\delta = \phi_{violet} - \phi_{red} = 2.2° = 0.038 \text{ rad}$$

At distance r, the spectrum spans an arc length

$$s = r\delta = (2.0\text{ m})(0.038\text{ rad}) = 0.076\text{ m} = 7.6\text{ cm}$$

ASSESS Notice that we needed three significant figures for ϕ_{red} and ϕ_{violet} in order to determine δ, the *difference* between the two angles, to two significant figures. The angle is so small that there's no appreciable difference between arc length and a straight line. The spectrum will be 7.6 cm wide at a distance of 2.0 m.

Rainbows

One of the most interesting sources of color in nature is the rainbow. The details get somewhat complicated, but Figure 18.28a shows that the basic cause of the rainbow is a combination of refraction, reflection, and dispersion.

FIGURE 18.28 Light seen in a rainbow has undergone refraction + reflection + refraction in a raindrop.

Figure 18.28a might lead you to think that the top edge of a rainbow is violet. In fact, the top edge is red, and violet is on the bottom. The rays leaving the drop in Figure 18.28a are spreading apart, so they can't all reach your eye. As Figure 18.28b shows, a ray of red light reaching your eye comes from a drop *higher* in the sky than a ray of violet light. In other words, the colors you see in a rainbow refract toward your eye from different raindrops, not from the same drop. You have to look higher in the sky to see the red light than to see the violet light.

Colored Filters and Colored Objects

White light passing through a piece of green glass emerges as green light. A possible explanation would be that the green glass *adds* "greenness" to the white light, but Newton found otherwise. Green glass is green because it *removes* any light that is "not green." More precisely, a piece of colored glass *absorbs* all wavelengths except those of one color, and that color is transmitted through the glass without hindrance. We can think of a piece of colored glass or plastic as a *filter* that removes all wavelengths except a chosen few.

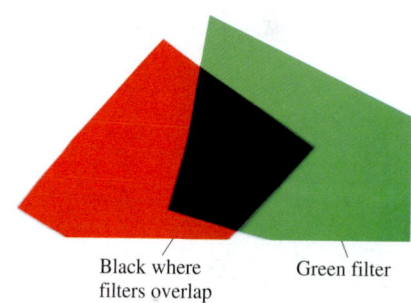

Black where filters overlap   Green filter

No light at all passes through both a green and a red filter.

CONCEPTUAL EXAMPLE 18.1 Filtering light

White light passes through a green filter and is observed on a screen. Describe how the screen will look if a second green filter is placed between the first filter and the screen. Describe how the screen will look if a red filter is placed between the green filter and the screen.

REASON The first filter removes all light except for wavelengths near 550 nm that we perceive as green light. A second green filter doesn't have anything to do. The nongreen wavelengths have already been removed, and the green light emerging from the first filter will pass through the second filter without difficulty. The screen will continue to be green and its intensity will not change. A red filter, by contrast, absorbs all wavelengths except those near 650 nm. The red filter will absorb the green light, and *no* light will reach the screen. The screen will be dark.

Opaque objects appear colored by virtue of *pigments* that absorb light of some wavelengths but *reflect* light at other wavelengths. For example, red paint contains pigments that reflect light at wavelengths near 650 nm while absorbing all other wavelengths. Pigments in paints, inks, and natural objects are responsible for most of the color we observe in the world, from the red of lipstick to the blue of a bluebird's feathers.

As an example, Figure 18.29 shows the absorption curve of *chlorophyll*. Chlorophyll is essential for photosynthesis in green plants. The chemical reactions of photosynthesis are able to use red light and blue/violet light, thus chlorophyll has evolved to absorb red light and blue/violet light from sunlight and put it to use. But green and yellow light are not absorbed. Instead, these wavelengths are mostly *reflected* to give the object a greenish-yellow color. When you look at

FIGURE 18.29 The absorption curve of chlorophyll.

the green leaves on a tree, you're seeing the light that was reflected because it *wasn't* needed for photosynthesis.

STOP TO THINK 18.3 A red apple is viewed through a green filter. The apple appears

A. Red.   B. Green.   C. Yellow.   D. Black.

18.5 Image Formation by Refraction

You've likely made an interesting observation while looking at fish in an aquarium. First you see a fish that appears to be swimming close to the front window of the aquarium. But then, if you look through the side of the aquarium, you find that the fish is actually farther from the front window than you thought. Somehow, as you look through the front window, the fish appears to be closer than it really is. This is a puzzle crying out for an explanation.

To begin, recall how vision functions. A diverging bundle of rays leaves the object, enters the pupil of the eye, and is focused on the retina. By adjusting the eye's lens to achieve a good focus, your brain determines the distance d from which the rays originated. This is where you perceive the object to be. Figure 18.30a shows how you would see a fish out of water at distance d.

Now place the fish back in the aquarium at the same distance d. For simplicity, we'll ignore the glass wall of the aquarium and consider the water-air boundary. (The thin glass of a typical window has only a very small effect on the refraction of the rays and doesn't change the conclusions.) Light rays again leave the fish, but this time they refract at the water-air boundary. Because they're going from a higher to a lower index of refraction, the rays refract *away from* the normal. Figure 18.30b shows the consequences.

A bundle of diverging rays still enters your eye, but now these rays *seem* to be diverging from a closer point, at distance d'. As far as your eye and brain are concerned, it's exactly *as if* the rays really originate at distance d', and this is the location at which you "see" the fish. **The object appears closer than it really is because of the refraction of light at the boundary.**

We found that the rays reflected from a mirror diverge from a point that is not the object point. We called that point a *virtual image*. Similarly, if rays from an object point P refract at a boundary between two media such that the rays then diverge from a point P′ and *appear* to come from P′, we call P′ a virtual image of point P. The virtual image of the fish is what you see.

Let's examine this image formation a bit more carefully. Figure 18.31 shows a boundary between two transparent media having indices of refraction n_1 and n_2. Point P, a source of light rays, is the object. Point P′, from which the rays *appear* to diverge, is the virtual image of P. The figure assumes $n_1 > n_2$, but this assumption isn't necessary. Distance s, measured from the boundary, is called the **object distance.** Our goal is to determine distance s', the **image distance.**

The line through the object and perpendicular to the boundary is called the **optical axis.** Consider a ray that leaves the object at angle θ_1 with respect to the optical axis. θ_1 is also the angle of incidence at the boundary, where the ray refracts into the second medium at angle θ_2. By tracing the refracted ray backward, you can see that θ_2 is also the angle between the refracted ray and the optical axis at point P′.

The distance l is common to both the incident and the refracted rays, and you can see that $l = s\tan\theta_1 = s'\tan\theta_2$. Thus

$$s' = \frac{\tan\theta_1}{\tan\theta_2}s \qquad (18.4)$$

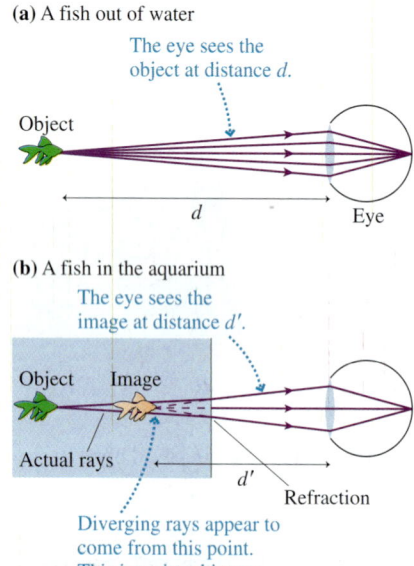

(a) A fish out of water

The eye sees the object at distance d.

Object

d

Eye

(b) A fish in the aquarium

The eye sees the image at distance d'.

Object   Image

Actual rays

d'

Refraction

Diverging rays appear to come from this point. This is a virtual image.

FIGURE 18.30 Refraction of the light rays causes a fish in the aquarium to be seen at distance d'.

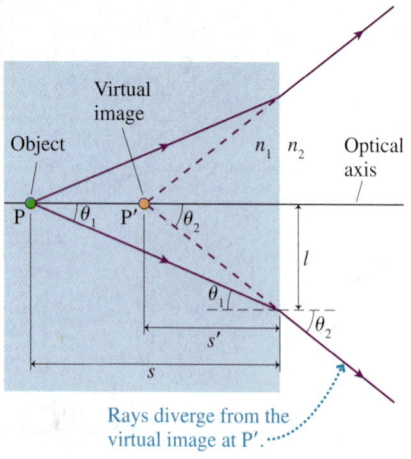

Virtual image

Object

n_1 n_2 Optical axis

P θ_1 P′ θ_2

l

θ_1

θ_2

s'

s

Rays diverge from the virtual image at P′.

FIGURE 18.31 Finding the virtual image P′ of an object at P.

Snell's law relates the sines of angles θ_1 and θ_2. That is,

$$\frac{\sin\theta_1}{\sin\theta_2} = \frac{n_2}{n_1} \tag{18.5}$$

In practice, the angle between any of these rays and the optical axis is very small because the size of the pupil of your eye is very much less than the distance between the object and your eye. (The angles in the figure have been greatly exaggerated.) Rays that are nearly *parallel* to the *axis* are called **paraxial rays.** The small-angle approximation $\sin\theta \approx \tan\theta \approx \theta$, where θ is in radians, can be applied to paraxial rays. Consequently,

$$\frac{\tan\theta_1}{\tan\theta_2} \approx \frac{\sin\theta_1}{\sin\theta_2} = \frac{n_2}{n_1} \tag{18.6}$$

Using this result in Equation 18.4, we find that the image distance is

$$s' = \frac{n_2}{n_1}s \tag{18.7}$$

NOTE ▶ The fact that the result for s' is independent of θ_1 implies that *all* paraxial rays appear to diverge from the same point P′. This property of the diverging rays is essential in order to have a well-defined image. ◀

This section has given us a first look at image formation via refraction. We will extend this idea to image formation with lenses in Section 18.6.

EXAMPLE 18.7   An air bubble in a window

A fish and a sailor look at each other through a 5.0-cm-thick glass porthole in a submarine. There happens to be a small air bubble right in the center of the glass. How far behind the surface of the glass does the air bubble appear to the fish? To the sailor?

PREPARE Represent the air bubble as a point source and use the ray model of light. Paraxial light rays from the bubble refract into the air on one side and into the water on the other. The ray diagram looks like Figure 18.31.

SOLVE The index of refraction of the glass is $n_1 = 1.50$. The bubble is in the center of the window, so the object distance from either side of the window is $s = 2.5$ cm. From the water side, the fish sees the bubble at an image distance

$$s' = \frac{n_2}{n_1}s = \frac{1.33}{1.50}(2.5 \text{ cm}) = 2.2 \text{ cm}$$

The sailor, in air, sees the bubble at an image distance

$$s' = \frac{n_2}{n_1}s = \frac{1.00}{1.50}(2.5 \text{ cm}) = 1.7 \text{ cm}$$

ASSESS The image distance is *less* for the sailor because of the *larger* difference between the two indices of refraction.

18.6 Thin Lenses: Ray Tracing

A pinhole camera forms images on a screen, but the images are faint and not perfectly focused. The ability to create a bright, well-focused image is vastly improved by using a lens. A **lens** is a transparent material that uses refraction of light rays at *curved* surfaces to form an image. In this section we want to establish a pictorial method of understanding image formation. This method is called **ray tracing.** We will defer a mathematical analysis of the image formation by lenses until the next chapter.

Figure 18.32 shows parallel light rays entering two different lenses. The top lens, called a **converging lens,** causes the rays to refract *toward* the optical axis.

Converging lens

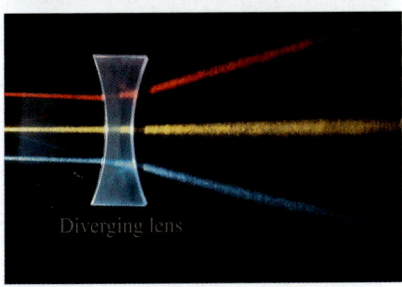
Diverging lens

FIGURE 18.32 Parallel light rays pass through a converging lens and a diverging lens.

The common point through which initially parallel rays pass is called the **focal point** of the lens. The distance of the focal point from the lens is called the **focal length** *f* of the lens. The bottom lens, called a **diverging lens,** refracts parallel rays *away from* the optical axis. This lens also has a focal point, but it is not as obvious in the figure.

NOTE ▶ A converging lens is thicker in the center than at the edges. A diverging lens is thicker at the edges than at the center. ◀

Figure 18.33 clarifies the situation. In the case of a diverging lens, a backward projection of the diverging rays shows that they all *appear* to have started from the same point. This is the focal point of a diverging lens, and its distance from the lens is the focal length of the lens. For both types of lenses, then, we have the definition that **the focal length is the distance from the lens at which rays parallel to the optical axis converge or from which they diverge.**

NOTE ▶ The focal length *f* is a property *of the lens,* independent of how the lens is used. The focal length characterizes a lens in much the same way that a mass *m* characterizes an object or a spring constant *k* characterizes a spring. ◀

Converging Lenses

These basic observations about lenses are enough to understand image formation by a *thin lens,* an idealized lens whose thickness is zero and which lies entirely in a plane called the **lens plane.** Within this *thin-lens approximation,* all refraction occurs as the rays cross the lens plane, and all distances are measured from the lens plane. Fortunately, the thin-lens approximation is quite good for most practical applications of lenses.

NOTE ▶ We'll *draw* lenses as if they have a thickness, because that is how we expect lenses to look, but our analysis will not depend on the shape or thickness of a lens. ◀

Figure 18.34 shows three important situations of light rays passing through a thin, converging lens. Part (a) is familiar from Figure 18.33. If the direction of each of the rays in Figure 18.34a is reversed, Snell's law tells us that each ray will exactly retrace its path and emerge from the lens parallel to the optical axis. This leads to Figure 18.34b, which is the "mirror image" of part (a).

Notice that the lens actually has *two* focal points, located at distances *f* on either side of the lens. The focal point on the side from which the light rays are approaching is the *near focal point.* The focal point opposite the side from which the light rays are approaching is the *far focal point.*

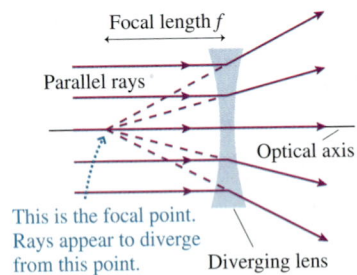

FIGURE 18.33 The focal point and focal length of converging and diverging lenses.

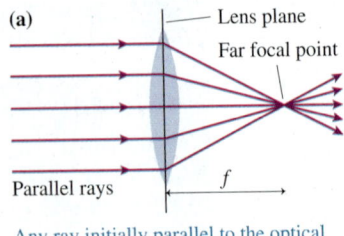

Any ray initially parallel to the optical axis will refract through the focal point on the far side of the lens.

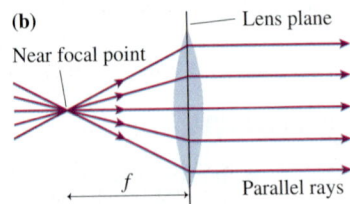

Any ray passing through the near focal point emerges from the lens parallel to the optical axis.

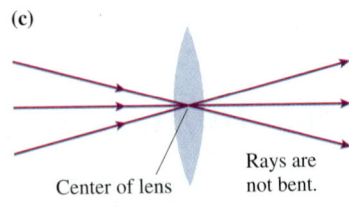

Any ray directed at the center of the lens passes through in a straight line.

FIGURE 18.34 Three important sets of rays passing through a thin, converging lens.

Figure 18.34c shows several rays passing through the *center* of the lens. At the center, the two sides of a lens are very nearly parallel to each other. Earlier, in Example 18.3, we found that a ray passing through a piece of glass with parallel sides is *displaced* but *not bent* and that the displacement becomes zero as the thickness approaches zero. Consequently, a ray through the center of a thin lens, which has zero thickness, is neither bent nor displaced but travels in a straight line.

These three situations form the basis for ray tracing.

Real Images

Figure 18.35 shows a lens and an object whose distance s from the lens is larger than the focal length. Rays from point P on the object are refracted by the lens so as to converge at point P' on the opposite side of the lens, at a distance s' from the lens. If rays diverge from an object point P and interact with a lens such that the refracted rays *converge* at point P', then we call P' a **real image** of point P. Contrast this with our prior definition of a *virtual image* as a point from which rays *diverge*.

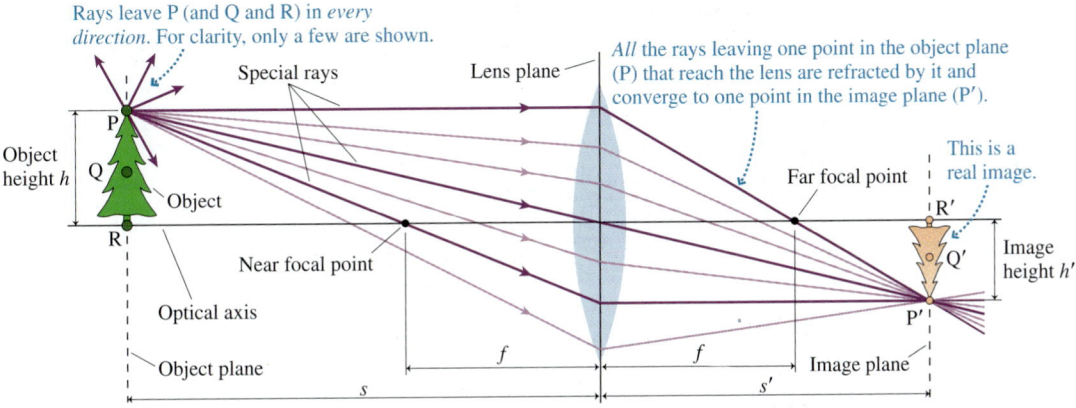

FIGURE 18.35 Rays from an object point P are refracted by the lens and converge to a real image at point P'.

All points on the object that are in the same plane, the **object plane,** converge to image points in the **image plane.** Points Q and R in the object plane of Figure 18.35 have image points Q' and R' in the same plane as point P'. Once we locate *one* point in the image plane, such as point P', we know that the full image lies in the same plane.

There are two important observations to make about Figure 18.35. First, the image is upside down with respect to the object. This is called an **inverted image,** and it is a standard characteristic of real image formation with a converging lens. You have to put slides into a slide tray upside down so that the image seen on the screen is right side up. Second, rays from point P *fill* the entire lens surface, so that all portions of the lens contribute to the image. A larger lens will "collect" more rays and thus make a brighter image. This is the big advantage of a lens over a pinhole camera.

Figure 18.36 is a close-up view of the rays very near the image plane. The rays don't stop at P' unless we place a screen in the image plane. When we do so, we see a sharp, well-focused image on the screen. To focus an image, you must either move the screen to coincide with the image plane or move the lens or object to make the image plane coincide with the screen. For example, the focus knob on a slide projector moves the lens closer to or farther from the slide until the image plane matches the screen position.

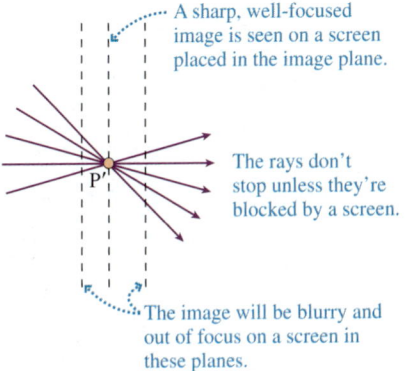

FIGURE 18.36 A close-up look at the rays near the image plane.

NOTE ▶ The ability to view and record *real* images, where the rays actually converge, sets real images apart from *virtual* images. But keep in mind that we need not *see* a real image in order to *have* an image. A real image exists at a point in space where the rays converge even if there's no viewing screen in the image plane. ◀

Figure 18.35 highlights three "special rays" that are based on the three situations of Figure 18.34. Notice that these three rays alone are sufficient to locate the image point P′. That is, we don't need to draw all the rays shown in Figure 18.35. The procedure known as *ray tracing* consists of locating the image by the use of just these three rays.

(MP) TACTICS BOX 18.2 **Ray tracing for a converging lens** Exercise 22

❶ **Draw an optical axis.** Use graph paper or a ruler! Establish an appropriate scale.

❷ **Center the lens on the axis.** Mark and label the focal points at distance *f* on either side.

❸ **Represent the object with an upright arrow at distance *s*.** It's usually best to place the base of the arrow on the axis and to draw the arrow about half the radius of the lens.

❹ **Draw the three "special rays" from the tip of the arrow.** Use a straightedge or a ruler.
  a. A ray parallel to the axis refracts through the far focal point.
  b. A ray that enters the lens along a line through the near focal point emerges parallel to the axis.
  c. A ray through the center of the lens does not bend.

❺ **Extend the rays until they converge.** This is the image point. Draw the rest of the image in the image plane. If the base of the object is on the axis, then the base of the image will also be on the axis.

❻ **Measure the image distance *s′*.** Also, if needed, measure the image height relative to the object height.

EXAMPLE 18.8 **Finding the image of a flower**

A 4.0-cm-diameter flower is 200 cm from the 50-cm-focal-length lens of a camera. How far should the film be placed behind the lens to record a well-focused image? What is the diameter of the image on the film?

PREPARE The flower is in the object plane. Use ray tracing to locate the image.

SOLVE Figure 18.37 shows the ray-tracing diagram and the steps of Tactics Box 18.2. The image has been drawn in the plane where the three special rays converge. You can see *from*

FIGURE 18.37 Ray-tracing diagram for Example 18.8.

the drawing that the image distance is $s' \approx 65$ cm. This is where the film needs to be placed to record a focused image. The heights of the object and image are labeled h and h'. The ray through the center of the lens is a straight line, thus the object and image both subtend the same angle θ. Using similar triangles,

$$\frac{h'}{s'} = \frac{h}{s}$$

Solving for h' gives

$$h' = h\frac{s'}{s} = (4.0 \text{ cm})\frac{65 \text{ cm}}{200 \text{ cm}} = 1.3 \text{ cm}$$

The flower's image has a diameter of 1.3 cm.

ASSESS We've been able to learn a great deal about the image from a simple geometric procedure.

Magnification

The image created by a lens can be either larger or smaller than the object, depending on the location and focal length of the lens. We define the **magnification** m as the ratio of image height h' to object height h:

$$m = \frac{h'}{h} \tag{18.8}$$

While you may think of magnification as meaning "larger," our definition allows the image to be either larger or smaller than the object. If $m > 1$ the image is larger than the object; if $m < 1$ it is smaller. In Example 18.8, for instance, we have $m = (1.3 \text{ cm})/(4.0 \text{ cm}) = 0.33$.

We can also find an expression for the magnification in terms of the image and object *distances*. In Example 18.8, we found that the image heights were related to s and s' by $h'/s' = h/s$. Using Equation 18.8, we can then write the magnification in terms of the image and object distances as

$$m = \frac{h'}{h} = \frac{s'}{s} \tag{18.9}$$

Magnification of a lens or a mirror

Although this equation was developed for the situation depicted in Example 18.8, it applies to images, real or virtual, produced by both lenses and mirrors.

NOTE ▶ In this chapter, we are concerned only with the *graphical* analysis of image formation, using ray tracing. Thus all distances such as s, s', h, and h' are taken to be positive *lengths*. In Chapter 19, we'll develop a mathematical approach to image formation. Using this new approach, we'll find that we need to treat these various lengths as *signed* quantities that can be positive or negative. ◀

STOP TO THINK 18.4 A lens produces a sharply focused, inverted image on a screen. What will you see on the screen if the lens is removed?

A. The image will be inverted and blurry.
B. The image will be upright and sharp.
C. The image will be upright and blurry.
D. The image will be much dimmer but otherwise unchanged.
E. There will be no image at all.

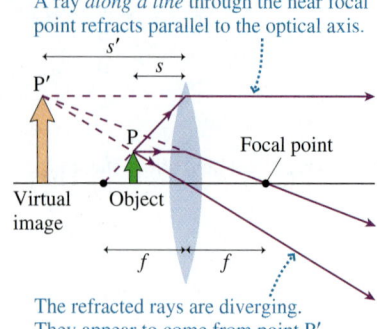

A ray *along a line* through the near focal point refracts parallel to the optical axis.

The refracted rays are diverging. They appear to come from point P'.

FIGURE 18.38 Rays from an object at distance $s < f$ are refracted by the lens and diverge to form a virtual image at point P'.

Virtual Images

The previous section considered a converging lens with the object at distance $s > f$. That is, the object was outside the focal point. What if the object is inside the focal point, at distance $s < f$? Figure 18.38 shows just this situation.

The special rays initially parallel to the axis and through the center of the lens present no difficulties. However, a ray through the near focal point would travel toward the left and would never reach the lens! Referring back to Figure 18.34b, you can see that the rays emerging parallel to the axis entered the lens *along a line* passing through the near focal point. It's the angle of incidence on the lens that is important, not whether the light ray actually passes through the focal point. This was the basis for the wording of step 4b in Tactics Box 18.2 and is the third special ray shown in Figure 18.38.

The three refracted rays don't converge. Instead, all three rays *diverge* from point P'. This is the situation we found for rays reflecting from a mirror and for the rays refracting out of an aquarium. Point P' is a *virtual image* of the object point P. Furthermore, it is an **upright image,** having the same orientation as the object.

The refracted rays, which are all to the right of the lens, *appear* to come from P', but none of the rays were ever at that point. No image would appear on a screen placed in the image plane at P'. So what good is a virtual image?

Your eye collects and focuses bundles of diverging rays. Thus, as Figure 18.39a shows, you can "see" a virtual image by looking *through* the lens. This is exactly what you do with a magnifying glass, producing a scene like the one in Figure 18.39b. In fact, you view a virtual image any time you look *through* the eyepiece of an optical instrument such as a microscope or binoculars.

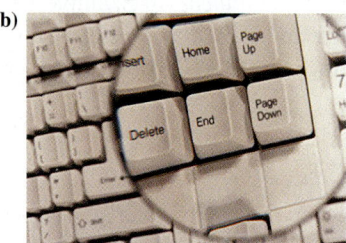

FIGURE 18.39 A converging lens is a magnifying glass when the object distance is $< f$.

NOTE ▶ A lens thicker in the middle than at the edges is classified as a converging lens. The light rays from an object *can* converge to form a real image after passing through such a lens, but only if the object distance is larger than the focal length of the lens: $s > f$. If $s < f$, the rays leaving a converging lens diverge to produce a virtual image. ◀

EXAMPLE 18.9 Magnifying a flower

To see a flower better, you hold a 6.0-cm-focal-length magnifying glass 4.0 cm from the flower. What is the magnification?

PREPARE The flower is in the object plane. Use ray tracing to locate the image. Once the image distance is known, Equation 18.9 can be used to find the magnification.

SOLVE Figure 18.40 shows the ray-tracing diagram. The three special rays diverge from the lens, but we can use a straightedge to extend the rays backward to the point from which they diverge. This point, the image point, is seen to be 12 cm to the left of the lens. From Equation 18.9 the magnification is

$$m = \frac{s'}{s} = \frac{12 \text{ cm}}{4.0 \text{ cm}} = 3.0$$

FIGURE 18.40 Ray-tracing diagram for Example 18.9.

ASSESS The image is three times as large as the object and, as we see from the ray-tracing diagram, upright.

Diverging Lenses

A lens thicker at the edges than in the middle is called a *diverging lens*. Figure 18.41 shows three important sets of rays passing through a diverging lens. These are based on Figures 18.32 and 18.33, where you saw that rays initially parallel to the axis diverge after passing through a diverging lens.

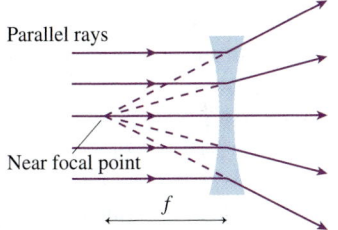

Any ray initially parallel to the optical axis diverges along a line through the near focal point.

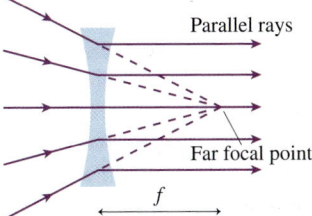

Any ray directed along a line toward the far focal point emerges from the lens parallel to the optical axis.

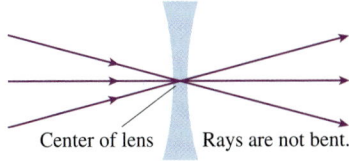

Any ray directed at the center of the lens passes through in a straight line.

FIGURE 18.41 Three important sets of rays passing through a thin, diverging lens.

Ray tracing follows the steps of Tactics Box 18.2 for a converging lens *except* that two of the three special rays in step 4 are different.

Activ
ONLINE   15.9
Physics

(MP) TACTICS BOX 18.3 Ray tracing for a diverging lens   ✐ Exercise 23

❶–❸ **Follow steps 1 through 3 of Tactics Box 18.2.**

❹ **Draw the three "special rays" from the tip of the arrow.** Use a straightedge.
   a. A ray parallel to the axis diverges along a line through the near focal point.
   b. A ray along a line toward the far focal point emerges parallel to the axis.
   c. A ray through the center of the lens does not bend.

❺ **Trace the diverging rays backward.** The point from which they are diverging is the image point, which is always a virtual image.

❻ **Measure the image distance s'.** Also, if needed, measure the image height relative to the object height.

EXAMPLE 18.10 Demagnifying a flower

A diverging lens with a focal length of 50 cm is placed 100 cm from a flower. Where is the image? What is its magnification?

PREPARE The flower is in the object plane. Use ray tracing to locate the image. Then Equation 18.9 can be used to find the magnification.

SOLVE Figure 18.42 shows the ray-tracing diagram. The three special rays (labeled a, b, and c to match the Tactics Box) do not converge. However, they can be traced backward to an intersection ≈33 cm to the left of the lens. Because the rays appear to diverge from the image, this is a virtual image. The magnification is

FIGURE 18.42 Ray-tracing diagram for Example 18.10.

$$m = \frac{s'}{s} = \frac{33 \text{ cm}}{100 \text{ cm}} = 0.33$$

The image, which can be seen by looking *through* the lens, is one-third the size of the object and upright.

ASSESS Ray tracing with a diverging lens is somewhat trickier than for a converging lens, so this example is worth careful study.

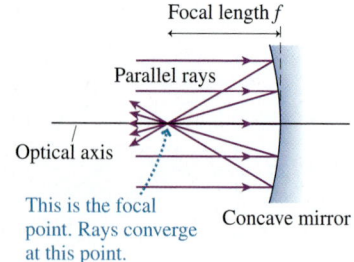

Focal length f

Parallel rays

Optical axis

This is the focal point. Rays converge at this point.

Concave mirror

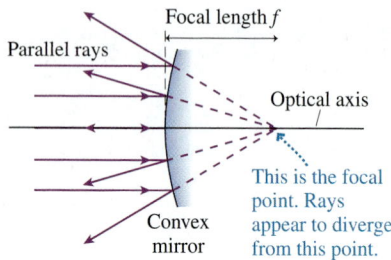

Focal length f

Parallel rays

Optical axis

This is the focal point. Rays appear to diverge from this point.

Convex mirror

FIGURE 18.43 The focal point and focal length of concave and convex mirrors.

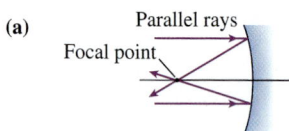

(a) Parallel rays

Focal point

Any ray initially parallel to the optical axis will reflect through the focal point.

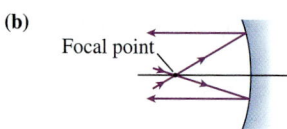

(b) Focal point

Any ray passing through the focal point will, after reflection, emerge parallel to the optical axis.

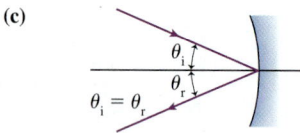

(c)

θ_i

θ_r

$\theta_i = \theta_r$

Any ray directed at the center of the mirror will reflect at an equal angle on the opposite side of the optical axis.

FIGURE 18.44 Three special rays for a concave mirror.

Diverging lenses *always* make virtual images and, for this reason, are rarely used alone. However, they have important applications when used in combination with other lenses. Cameras, eyepieces, and eyeglasses often incorporate diverging lenses.

18.7 Image Formation with Spherical Mirrors

Curved mirrors can also be used to form images. Such mirrors are commonly used in telescopes, security and rearview mirrors, and searchlights. Their images can be analyzed with ray diagrams similar to those used with lenses. We'll only consider the important case of **spherical mirrors,** whose surface is a section of a sphere.

Figure 18.43 shows parallel light rays approaching two spherical mirrors. The upper mirror, where the edges curve toward the light source, is called a **concave mirror.** Parallel rays reflect off the shiny front surface of the mirror and pass through a single point on the optical axis. This is the focal point of the mirror. The lower mirror, where the edges curve away from the light source, is called a **convex mirror.** Parallel rays that reflect off its surface appear to have come from a point behind the mirror. This is the focal point for a convex mirror. For both mirrors, the focal length is the distance from the mirror surface to the focal point.

Concave Mirrors

To understand image formation by a concave mirror, consider the three special rays shown in Figure 18.44. These rays are closely related to those used for ray tracing with lenses. Figure 18.44a shows two incoming rays parallel to the optical axis. As Figure 18.43 showed, these rays reflect off the mirror and pass through the focal point.

Figure 18.44b is the same as Figure 18.44a, but with the directions of the rays reversed. Here we see that rays passing through the focal point emerge parallel to the axis. Finally, Figure 18.44c shows what happens to a ray that is directed toward the center of the mirror. Right at its center, the surface of the mirror is perpendicular to the optical axis. The law of reflection then tells us that the incoming ray will reflect at the same angle, but on the opposite side of the optical axis.

Let's begin by considering the case where the object's distance s from the mirror is larger than the focal length ($s > f$), as shown in Figure 18.45. The three special rays just discussed are enough to locate the position and size of the image. Recall that when ray tracing a thin lens, although we drew the lens as having an actual thickness, the rays refracted at an imaginary plane centered on the lens. Similarly, when ray tracing mirrors, the incoming rays reflect off the **mirror plane** as shown in Figure 18.45, not off the curved surface of the mirror. We see that the image is *real* because rays converge at the image point P′. Further, the image is *inverted.*

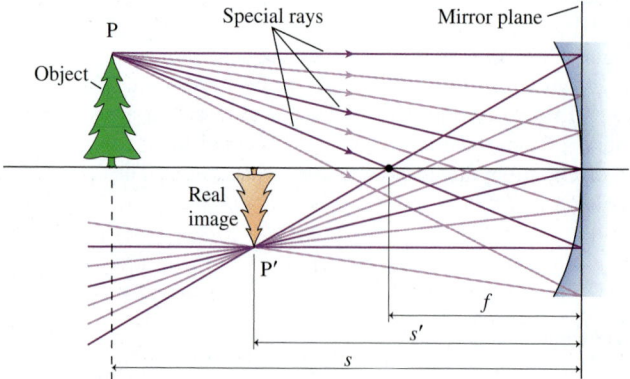

FIGURE 18.45 A real image formed by a concave mirror.

Figure 18.45 suggests the following Tactics Box for using ray tracing with a concave mirror:

(MP)   **TACTICS BOX 18.4 Ray tracing for a concave mirror**   ✎ Exercise 24

❶ **Draw an optical axis.** Use graph paper or a ruler! Establish an appropriate scale.

❷ **Center the mirror on the axis.** Mark and label the focal point at distance f from the mirror's surface.

❸ **Represent the object with an upright arrow at distance s.** It's usually best to place the base of the arrow on the axis and to draw the arrow about half the radius of the mirror.

❹ **Draw the three "special rays" from the tip of the arrow.** Use a straightedge or a ruler. The rays should reflect off the mirror plane.

  a. A ray parallel to the axis reflects through the focal point.

  b. An incoming ray that passes through the focal point emerges parallel to the axis.

  c. A ray that strikes the center of the mirror reflects at an equal angle on the opposite side of the optical axis.

❺ **Extend the rays until they converge.** This is the image point. Draw the rest of the image in the image plane. If the base of the object is on the axis, then the base of the image will also be on the axis.

❻ **Measure the image distance s'.** Also, if needed, measure the image height relative to the object height.

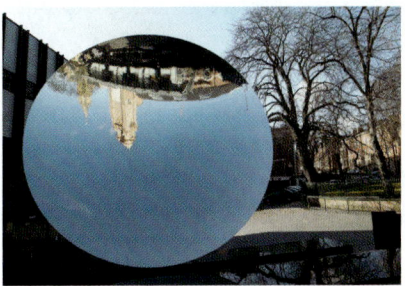

A real, inverted image is formed by a concave mirror.

EXAMPLE 18.11 Analyzing a concave mirror

A 3.0-cm-high object is located 60 cm from a concave mirror. The mirror's focal length is 40 cm. Use ray tracing to find the position, height, and magnification of the image.

PREPARE Figure 18.46 shows the ray-tracing diagram and the steps of Tactics Box 18.4.

SOLVE After preparing a careful drawing, we can use a ruler to find that the image position is $s' \approx 120$ cm and its height is $h' \approx 6$ cm. The magnification is thus

$$m = \frac{s'}{s} \approx \frac{120 \text{ cm}}{60 \text{ cm}} = 2.0$$

ASSESS The image is a *real* image because light rays converge at the image point.

FIGURE 18.46 Ray-tracing diagram for a concave mirror.

FIGURE 18.47 Self-portrait of M. C. Escher.

If the object is inside the focal point ($s < f$), ray tracing can be used to show that the image is a virtual image. This situation is analogous to the formation of a virtual image by a lens when the object is inside the focal point.

Convex Mirrors

A common example of a convex mirror is a silvered ball, such as a tree ornament. You may have noticed that if you look at your reflection in such a ball, your image appears right-side-up but is quite small. Figure 18.47 shows a self-portrait of the Dutch artist M. C. Escher that illustrates these observations. Let's use ray tracing to see why the image appears this way.

Once more, there are three special rays we can use to find the location of the image. These rays are shown in Figure 18.48; they are similar to the special rays we've already studied in other situations.

  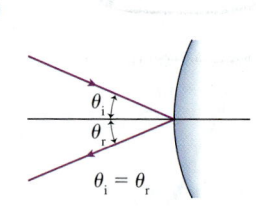

Any ray initially parallel to the optical axis will reflect as though it came from the focal point.

Any ray initially directed toward the focal point will reflect parallel to the optical axis.

Any ray directed at the center of the mirror will reflect at an equal angle on the opposite side of the optical axis.

FIGURE 18.48 Three special rays for a convex mirror.

We can use these three special rays to find the image of an object, as shown in Figure 18.49. We see that the image is virtual—no actual rays converge at the image point P′. Instead, diverging rays *appear* to have come from this point. The image is also upright and much smaller than the object, in accord with our experience and the drawing of Figure 18.47.

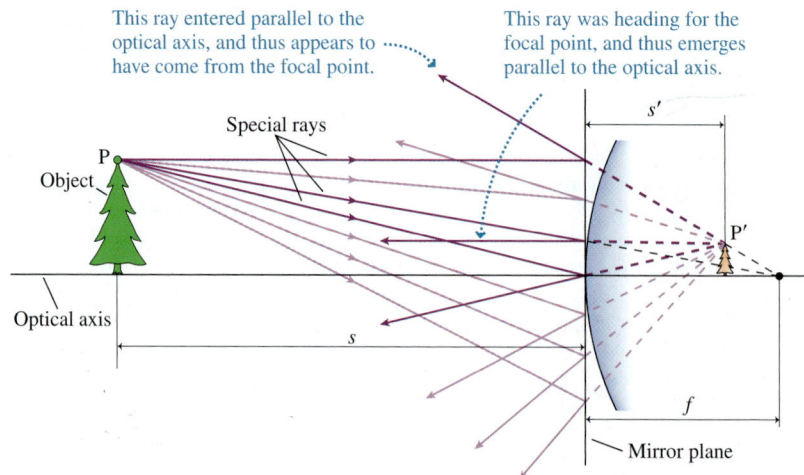

This ray entered parallel to the optical axis, and thus appears to have come from the focal point.

This ray was heading for the focal point, and thus emerges parallel to the optical axis.

FIGURE 18.49 Rays from point P reflect from the mirror and appear to have come from point P′.

These observations form the basis of the following Tactics Box.

TACTICS BOX 18.5 Ray tracing for a convex mirror

 Exercises 25, 26

❶–❸ **Follow steps 1 through 3 of Tactics Box 18.4.**

❹ **Draw the three "special rays" from the tip of the arrow.** Use a straightedge or a ruler.

   a. A ray parallel to the axis reflects as though it came from the focal point.

   b. A ray initially directed toward the focal point reflects parallel to the axis.

   c. A ray that strikes the center of the mirror reflects at an equal angle on the opposite side of the optical axis.

❺ **Extend the emerging rays *behind the mirror* until they converge.** This is the image point. Draw the rest of the image in the image plane. If the base of the object is on the axis, then the base of the image will also be on the axis.

❻ **Measure the image distance s'.** Also, if needed, measure the image height relative to the object height.

Actⓘv ONLINE Physⓘcs   15.5–15.8

The small image in a convex mirror allows a wide-angle view of the store to be visible.

Convex mirrors are used for a variety of safety and monitoring applications, such as passenger-side rearview mirrors and the round mirrors used in stores to keep an eye on the customers. The idea behind such mirrors can be understood from Figure 18.49. When an object is reflected in a convex mirror, the image appears smaller. Because the image is, in a sense, a miniature version of the object, you can *see much more of it* within the edges of the mirror than you could with an equal-sized flat mirror. This wide-angle view is clearly useful for checking traffic behind you, or for checking up on your store.

CONCEPTUAL EXAMPLE 18.2 Driver and passenger mirrors

The side mirror on the driver's side of a car is a plane (flat) mirror, while the mirror on the passenger's side is convex. Why is this?

REASON It is important for the driver to have a wide field of view from either mirror. He sits close to the driver-side mirror, so it appears large and can reflect a fairly wide view of what's behind. The passenger-side mirror is quite far from the driver, so it appears relatively small. If it were flat, it would offer only a narrow view of what's behind. Making it convex, like the security mirror discussed above, provides a wider field of view, but the trade off is a smaller image. That's why the passenger-side mirror usually contains a warning: Objects in mirror are closer than they appear!

In this chapter we've learned how to graphically find the image for a given lens or mirror. Next, we'll want to learn how lenses and mirrors are put to use in a variety of optical instruments such as cameras, microscopes, and telescopes. To do so, we'll need a more precise method for relating image and object distances and heights. This will be our first task in Chapter 19.

STOP TO THINK 18.5 A concave mirror of focal length f forms an image of the moon. Where is the image located?

A. At the mirror's surface.
B. Almost exactly a distance f behind the mirror.
C. Almost exactly a distance f in front of the mirror.
D. At a distance behind the mirror equal to the distance of the moon in front of the mirror.

The goal of Chapter 18 has been to understand and apply the ray model of light.

GENERAL PRINCIPLES

Reflection

Law of reflection: $\theta_r = \theta_i$

Reflection can be **specular** (mirror-like) or **diffuse** (from rough surfaces).

Plane mirrors: A virtual image is formed at P′ with $s = s'$, where s is the **object distance** and s' is the **image distance.**

Refraction

Snell's law of refraction:

$$n_1 \sin\theta_1 = n_2 \sin\theta_2$$

Index of refraction is $n = c/v$. The ray is closer to the normal on the side with the larger index of refraction.

If $n_2 < n_1$, total internal reflection (TIR) occurs when the angle of incidence $\theta_1 > \theta_c = \sin^{-1}(n_2/n_1)$

IMPORTANT CONCEPTS

The ray model of light

Light travels along straight lines, called **light rays,** at speed $v = c/n$.

A light ray continues forever unless an interaction with matter causes it to reflect, refract, scatter, or be absorbed.

Light rays come from self-luminous or reflective **objects.** Each point on the object sends rays in all directions.

Ray diagrams represent all the rays emitted by an object by only a few select rays.

In order for the eye to see an object (or image), rays from the object or image must enter the eye.

Image formation

If rays diverge from P and, after interacting with a lens or mirror, *appear* to diverge from P′ without actually passing through P′, then P′ is a virtual image of P.

These rays *appear* to have come from P′.
Virtual image

If rays diverge from P and interact with a lens so that the refracted rays *converge* at P, then P′ is a real image of P. Rays actually pass through a real image.

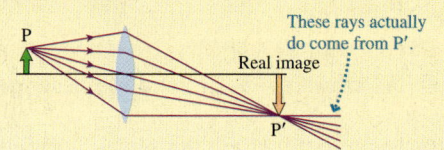
These rays actually do come from P′.
Real image

APPLICATIONS

Ray tracing for lenses

3 special rays in 3 basic situations:

Converging lens
Real image

Converging lens
Virtual image

Diverging lens
Virtual image

Ray tracing for mirrors

3 special rays in 3 basic situations:

Concave mirror
Real image

Concave mirror
Virtual image

Convex mirror
Virtual image

Magnification

The magnification is the ratio of the image height h' to the object height h. It can also be written in terms of the image and object distances:

$$m = \frac{h'}{h} = \frac{s'}{s}$$

Color and dispersion

The eye perceives light of different wavelengths as having different colors.

Dispersion is the dependence of the index of refraction n of a transparent medium on the wavelength of light: Long wavelengths have the lowest n, short wavelengths the highest n.

White light is composed of all wavelengths of light.
A prism breaks white light into its constituent colors. Violet light with its higher n is refracted more than red.
White light
$\lambda = 650$ nm
$\lambda = 550$ nm
$\lambda = 450$ nm

For instructor-assigned homework, go to www.masteringphysics.com

Problem difficulty is labeled as I (straightforward) to IIII (challenging).

Problems labeled INT integrate significant material from earlier chapters; BIO are of biological or medical interest.

QUESTIONS

Conceptual Questions

1. The idea of light rays goes back to the ancient Greeks. However, they believed that "visual rays" were *emitted* by eyes. If you were transported back in time, what arguments would you present to those early scientists to convince them that vision has something to do with rays going into, rather than out of, eyes?

2. Is there any property that distinguishes a light ray emitted by a light bulb and one that has been diffusely reflected by the page of a book? Explain.

3. On a sunny summer day, with the sun overhead, you can stand under a tree and look on the ground at the pattern of light that has passed through gaps between the leaves. You may see illuminated circles of varying brightness. Why are there circles, when the gaps between the leaves have irregular shapes?

4. Suppose you have two pinhole cameras. The first has a small round hole in the front of the camera. The second is identical in every regard, except that it has a square hole of the same area as the round hole in the first camera. Would the pictures taken by these two cameras, under the same conditions, be different in any obvious way? Explain.

5. If you turn on your car headlights during the day, the road ahead of you doesn't appear to get brighter. Why not?

6. Can you see the rays from the sun on a clear day? Why or why not? How about when they stream through a forest on a foggy morning? Why or why not?

7. If you take a walk on a summer night along a dark, unpaved road in the woods, with a flashlight pointing at the ground several yards ahead to guide your steps, any water-filled potholes are noticeable because they appear much darker than the surrounding dry road. Explain why.

8. You are looking at the image of a pencil in a mirror, as shown in Figure Q18.8.
   a. What happens to the image if the top half of the mirror, down to the midpoint, is covered with a piece of cardboard? Explain.
   b. What happens to the image if the bottom half of the mirror is covered with a piece of cardboard? Explain.

FIGURE Q18.8

9. In *The Toilet of Venus* by Velázquez (see Figure Q18.9), we can see the face of Venus in the mirror. Can she see her own face in the mirror, when the mirror is held as shown in the picture? If yes, explain why; if not, what does she see instead?

FIGURE Q18.9

10. In Manet's *A Bar at the Folies-Bergère* (see Figure Q18.10) the reflection of the barmaid is visible in the mirror behind her. Is this the reflection you would expect if the mirror's surface is parallel to the bar? Where would the man seen facing her in the mirror actually be standing?

FIGURE Q18.10

11. Explain briefly why ambulances have the word "AMBU-LANCE" written backward on the front of them.

12. One problem with using optical fibers for communication is that a light ray passing directly down the center of the fiber takes less time to travel from one end to the other than a ray that takes a longer, zigzagging reflected path. Thus light rays that start at the same time but in slightly different directions reach the end of the fiber at different times, so the signals coming out of the fiber are "fuzzier" than the ones that were put in. This problem can be solved by making the refractive index of the glass change gradually from a higher value in the center to a lower value near the edges of the fiber. Explain how this reduces the spread in travel times.

13. A red card is illuminated by red light. What color will it appear to be? What if it's illuminated by blue light?

14. a. Consider *one* point on an object near a lens. What is the minimum number of rays needed to locate its image point?
    b. For each point on the object, how many rays from this point actually strike the lens and refract to the image point?

15. When you look at your reflection in the bowl of a spoon, it is upside down. Why is this?

16. A concave mirror brings the sun's rays to a focus at a distance of 30 cm from the mirror. If the mirror were submerged in a swimming pool, would the sun's rays be focused nearer to, further from, or at the same distance from the mirror?

17. A student draws the ray diagram shown in Figure Q18.17 but forgets to label the object, the image, or the type of lens used. Using the diagram, explain whether the lens is converging or diverging, which arrow represents the object, and which represents the image.

FIGURE Q18.17

Multiple-Choice Questions

Questions 18 through 20 are concerned with the situation sketched in Figure Q18.18, in which a beam of light in the air encounters a transparent block with index of refraction $n = 1.53$. Some of the light is reflected and some is refracted.

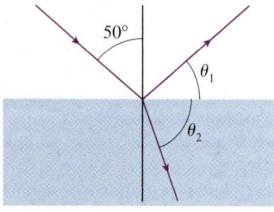

FIGURE Q18.18

18. | What is θ_1?
    A. 40°   B. 45°   C. 50°   D. 90°
19. | What is θ_2?
    A. 20°   B. 30°   C. 50°   D. 60°
20. ‖ Is there an angle of incidence between 0° and 90° such that all of the light will be reflected?
    A. Yes, at an angle greater than 50°.
    B. Yes, at an angle less than 50°.   C. No.

21. | A 2.0-m-tall man is 5.0 m from the converging lens of a camera. His image appears on a piece of film that is 50 mm behind the lens. How tall is his image on the film?
    A. 10 mm   B. 20 mm   C. 25 mm   D. 50 mm

22. ‖ You are 2.4 m from a plane mirror, and you would like to take a picture of yourself in the mirror. You need to manually adjust the focus of the camera by dialing in the distance to what you are photographing. What distance do you dial in?
    A. 1.2 m   B. 2.4 m   C. 3.6 m   D. 4.8 m

23. ‖‖ Figure Q18.23 shows an object and lens positioned to form a well-focused, inverted image on a viewing screen. Then a piece of cardboard is lowered just in front of the lens to cover the *top half* of the lens. What happens to the image on the screen?

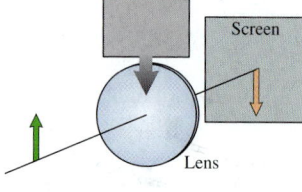

FIGURE Q18.23

    A. Nothing.
    B. The upper half of the image will vanish.
    C. The lower half of the image will vanish.
    D. The image will become fuzzy and out of focus.
    E. The image will become dimmer.

24. | A real image of an object can be formed by
    A. A converging lens.   B. A plane mirror.
    C. A convex mirror.   D. Any of the above.

25. | An object is 40 cm from a converging lens with a focal length of 30 cm. A real image is formed on the other side of the lens, 120 cm from the object. What is the magnification?
    A. 2.0   B. 3.0   C. 4.0
    D. 1.33   E. 0.35

PROBLEMS

Section 18.1 The Ray Model of Light

1. ‖‖ A point source of light illuminates an aperture 2.00 m away. A 12.0-cm-wide bright patch of light appears on a screen 1.00 m behind the aperture. How wide is the aperture?

2. | A student has built a 20-cm-long pinhole camera for a science fair project. She wants to photograph the Washington Monument, which is 167 m (550 ft) tall, and to have the image on the film be 5.0 cm high. How far should she stand from the Washington Monument?

3. | A 5.0-ft-tall girl stands on level ground. The sun is 25° above the horizon. How long is her shadow?

4. | A pinhole camera is made from an 80-cm-long box with a small hole in one end. If the hole is 5.0 m from a 1.8-m-tall person, how tall will the image of the person on the film be?

5. | The human eye has a lot in common with a pinhole camera,
   BIO being essentially a small box with a hole in the front (the pupil) and "film" at the back (the retina). The distance from the pupil to the retina is approximately 24 mm.
   a. Suppose you look at a 180-cm-tall friend who is standing 7.4 m in front of you. Assuming your eye functions like a pinhole camera, what will be the height, in mm, of your friend's image on your retina?
   b. Suppose your friend's image begins to get bigger. How does your brain interpret this information?

Section 18.2 Reflection

6. | The mirror in Figure P18.6 deflects a horizontal laser beam by 60°. What is the angle ϕ?

FIGURE P18.6

7. ‖ Figure P18.7 shows an object O in front of a plane mirror. Use ray tracing to determine from which locations A–D the object's image is visible.

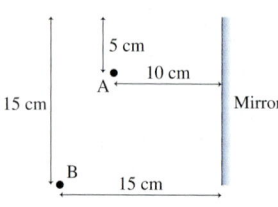

FIGURE P18.7   **FIGURE P18.8**

8. ‖‖ A light ray leaves point A in Figure P18.8, reflects from the mirror, and reaches point B. How far below the top edge does the ray strike the mirror?

9. ‖ It is 165 cm from your eyes to your toes. You're standing 200 cm in front of a tall mirror. How far is it from your eyes to the image of your toes?

10. | A ray of light impinges on a mirror as shown in Figure P18.10. A second mirror is fastened at 90° to the first.
    a. After striking both mirrors, at what angle relative to the incoming ray does the outgoing ray emerge?
    b. What is the answer if the incoming angle is 30°?

FIGURE P18.10   FIGURE P18.11

11. ‖ A red ball is placed at point A in Figure P18.11.
    a. How many images are seen by an observer at point O?
    b. Where is each image located?
    c. Draw a ray diagram showing the formation of each image.

Section 18.3 Refraction

12. ‖ An underwater diver sees the sun 50° above horizontal. How high is the sun above the horizon to a fisherman in a boat above the diver?

13. | A laser beam in air is incident on a liquid at an angle of 37° with respect to the normal. The laser beam's angle in the liquid is 26°. What is the liquid's index of refraction?

14. ‖ A 1.0-cm-thick layer of water stands on a horizontal slab of glass. A light ray in the air is incident on the water 60° from the normal. After entering the glass, what is the ray's angle from the normal?

15. | A 4.0-m-wide swimming pool is filled to the top. The bottom of the pool becomes completely shaded in the afternoon when the sun is 20° above the horizon. How deep is the pool?

16. ‖ A diamond is underwater. A light ray enters one face of the diamond, then travels at an angle of 30° with respect to the normal. What was the ray's angle of incidence on the diamond?

17. | You are playing a video game in which you're chained to one spot and you hear a dragon approaching. A light-activated switch that would open your chains is located in the sealed cavern shown in Figure P18.17. You have a laser cannon that you can fire through the small hole, but from your present position the laser beam will miss the switch. Fortunately, you possess a magical power that will let you momentarily fill with water either the cavern or the room in which you're chained. Can you use your power to release your chains in time to flee from the dragon? If so, what will you do? If not, why not?

FIGURE P18.17

18. ‖ The glass core of an optical fiber has index of refraction 1.60. The index of refraction of the cladding is 1.48. What is the maximum angle a light ray can make with the wall of the core if it is to remain inside the fiber?

19. ‖‖ A thin glass rod is submerged in oil. What is the critical angle for light traveling inside the rod?

Section 18.4 Color and Dispersion

20. ‖ A narrow beam of light with wavelengths from 450 nm to 700 nm is incident perpendicular to one face of a 40.00° prism made of crown glass, for which the index of refraction ranges from $n = 1.533$ to $n = 1.517$ for those wavelengths. What is the angular spread of the beam after passing through the prism?

21. ‖‖ A ray of white light strikes the surface of a 4.0-cm-thick slab of flint glass as shown in Figure P18.21. As the ray enters the glass, it is dispersed into its constituent colors. Estimate how far apart the rays of deepest red and deepest violet light are as they exit the bottom surface. Which exiting ray is closer to point P?

FIGURE P18.21

22. ‖ A ray of red light, for which $n = 1.54$, and a ray of violet light, for which $n = 1.59$, travel through a piece of glass. They meet right at the boundary between the glass and the air, and emerge into the air as one ray with an angle of refraction of 22.5°. What is the angle between the two rays in the glass?

Section 18.5 Image Formation by Refraction

23. | A fish in a flat-sided aquarium sees a can of fish food on the counter. To the fish's eye, the can looks to be 30 cm outside the aquarium. What is the actual distance between the can and the aquarium? (You can ignore the thin glass wall of the aquarium.)

24. ‖‖‖ A biologist keeps a specimen of his favorite beetle embedded in a cube of polystyrene plastic. The hapless bug appears to be 2.0 cm within the plastic. What is the beetle's actual distance beneath the surface?

25. ‖ A swim mask has a pocket of air between your eyes and the flat glass front.
    a. If you look at a fish while swimming underwater with a swim mask on, does the fish appear closer or farther than it really is? Draw a ray diagram to explain.
    b. Does the fish see your face closer or farther than it really is? Draw a ray diagram to explain.

Section 18.6 Thin Lenses: Ray Tracing

26. | An object is 20 cm in front of a converging lens with a focal length of 10 cm. Use ray tracing to determine the location of the image. Is the image upright or inverted? Is it real or virtual?

27. | An object is 30 cm in front of a converging lens with a focal length of 10 cm. Use ray tracing to determine the location of the image. Is the image upright or inverted? Is it real or virtual?

28. | An object is 6.0 cm in front of a converging lens with a focal length of 10 cm. Use ray tracing to determine the location of the image. Is the image upright or inverted? Is it real or virtual?

29. || An object is 20 cm in front of a diverging lens with a focal length of 10 cm. Use ray tracing to determine the location of the image. Is the image upright or inverted? Is it real or virtual?

30. | An object is 15 cm in front of a diverging lens with a focal length of 10 cm. Use ray tracing to determine the location of the image. Is the image upright or inverted? Is it real or virtual?

Section 18.7 Image Formation with Spherical Mirrors

31. | A concave cosmetic mirror has a focal length of 40 cm. A 5-cm-long mascara brush is held upright 20 cm from the mirror. Use ray tracing to determine the location and height of its image. Is the image upright or inverted? Is it real or virtual?

32. | A light bulb is 60 cm from a concave mirror with a focal length of 20 cm. Use ray tracing to determine the location of its image. Is the image upright or inverted? Is it real or virtual?

33. ||| The illumination lights in an
    BIO operating room use a concave mirror to focus an image of a bright lamp onto the surgical site. One such light has a mirror with a focal length of 15.0 cm. Use ray tracing to find the position of its lamp when the patient is positioned 1.0 m from the mirror (you'll need a careful drawing to get a good answer).

34. ||| A dentist uses a curved mirror to view the back side of teeth
    BIO on the upper jaw. Suppose she wants an erect image with a magnification of 2.0 when the mirror is 1.2 cm from a tooth. (Treat this problem as though the object and image lie along a straight line.) Use ray tracing to decide whether a concave or convex mirror is needed, and to estimate its focal length.

35. | A convex mirror, like the passenger-side rearview mirror on a car, has a focal length of 2.0 m. An object is 4.0 m from the mirror. Use ray tracing to determine the location of its image. Is the image upright or inverted? Is it real or virtual?

36. | An object is 6 cm in front of a convex mirror with a focal length of 10 cm. Use ray tracing to determine the location of the image. Is the image upright or inverted? Is it real or virtual?

General Problems

37. | Suppose you point a pinhole camera at a 15-m-tall tree that is 75 m away.
    a. If the film is 22 cm behind the pinhole, what will be the size of the tree's image on the film?
    b. If you would like the image to be larger, should you get closer to the tree or farther from the tree? Explain.
    c. If you had time, you could make the image larger by rebuilding the camera, changing the length or the pinhole size. What one change would give a larger image?

38. | You slowly back away from a plane mirror at a speed of
    INT 0.10 m/s. With what speed does your image appear to be moving away from you?

39. | Starting 3.5 m from a department store mirror, Suzanne
    INT walks toward the mirror at 1.5 m/s for 2.0 s. How far is Suzanne from her image in the mirror after 2.0 s?

40. | You're helping with an experiment in which a vertical cylinder will rotate about its axis by a very small angle. You need to devise a way to measure this angle. You decide to use what is called an *optical lever*. You begin by mounting a small mirror on top of the cylinder. A laser 5.0 m away shoots a laser beam at the mirror. Before the experiment starts, the mirror is adjusted to reflect the laser beam directly back to the laser. Later, you measure that the reflected laser beam, when it returns to the laser, has been deflected sideways by 2.0 mm. How many degrees has the cylinder rotated?

41. |||| At what angle ϕ should the laser beam in Figure P18.41 be aimed at the mirrored ceiling in order to hit the midpoint of the far wall?

FIGURE P18.41

FIGURE P18.42

42. | Figure P18.42 is the top view of a room. As you walk along the wall opposite the mirror (i.e., along the *x*-axis), over what range of *x* can you see the entire green arrow in the mirror?

43. || A laser beam is incident on the left mirror in Figure P18.43. Its initial direction is parallel to a line that bisects the mirrors. What is the angle ϕ of the reflected laser beam?

44. || The place you get your hair cut has two nearly parallel mirrors 5.0 m apart. As you sit in the chair, your head is 2.0 m from the nearer mirror. Looking toward this mirror, you first see your face and then, farther away, the back of your head. (The mirrors need to be slightly nonparallel for you to be able to see the back of your head, but you can treat them as parallel in this problem.) How far away does the back of your head appear to be? Neglect the thickness of your head.

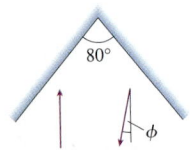

FIGURE P18.43

45. |||| Figure P18.45 shows a light ray incident on a polished metal cylinder. At what angle θ will the ray be reflected?

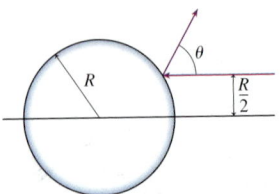

FIGURE P18.45

46. | A ray of light traveling through air encounters a 1.2-cm-thick sheet of glass at a 35° angle of incidence. How far does the light ray travel in the glass before emerging on the far side?

47. ‖ You shine your laser pointer through the flat glass side of a rectangular aquarium at an angle of incidence of 45°. The index of refraction of this type of glass is 1.55.
    a. At what angle from the normal does the beam from the laser pointer enter the water inside the aquarium?
    b. Does your answer to part a depend on the index of refraction of the glass?

48. ‖‖ Figure P18.48 shows a light ray incident on a glass cylinder. At what angle α will the ray be refracted?

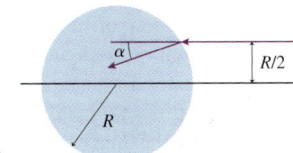

FIGURE P18.48

49. ‖ A 1.0-cm-thick layer of water stands on a horizontal slab of glass. Light from within the glass is incident on the glass-water boundary. What is the maximum angle of incidence for which a light ray can emerge into the air above the water?

50. | If you look at a fish through the corner of a rectangular aquarium you sometimes see two fish, one on each side of the corner, as shown in Figure P18.50. Sketch some of the light rays that reach your eye from the fish to show how this can happen.

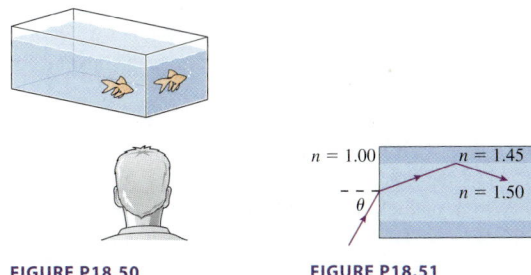

FIGURE P18.50   **FIGURE P18.51**

51. ‖‖ Figure P18.51 shows a cross-section of a fiber optic core and the surrounding cladding. What is the largest entry angle θ for which light will successfully propagate down the fiber if entering from air?

52. ‖‖ It's nighttime, and you've dropped your goggles into a swimming pool that is 3.0 m deep. If you hold a laser pointer 1.0 m above the edge of the pool, you can illuminate the goggles if the laser beam enters the water 2.0 m from the edge. How far are the goggles from the edge of the pool?

53. ‖‖ One of the contests at the school carnival is to throw a spear at an underwater target lying flat on the bottom of a pool. The water is 1.0 m deep. You're standing on a small stool that places your eyes 3.0 m above the bottom of the pool. As you look at the target, your gaze is 30° below horizontal. At what angle below horizontal should you throw the spear in order to hit the target? Your raised arm brings the spear point to the level of your eyes as you throw it, and over this short distance you can assume that the spear travels in a straight line rather than a parabolic trajectory.

54. ‖‖‖ Example 18.3 shows that a laser beam entering a 1.0-cm-thick glass sheet at an incident angle of 60° leaves at the same angle, but with the beam path displaced. Measured perpendicular to the two paths, how far is the beam's final path displaced from its original path?

55. ‖‖‖ What is the angle of incidence in air of a light ray whose angle of refraction in glass is half the angle of incidence?

56. | Figure P18.56 shows a meter stick lying on the bottom of a 100-cm-long tank with its zero mark against the left edge. You look into the tank at a 30° angle, with your line of sight just grazing the upper left edge of the tank. What mark do you see on the meter stick if the tank is (a) empty, (b) half full of water, and (c) completely full of water?

FIGURE P18.56

57. ‖‖ The 80-cm-tall, 65-cm-wide tank shown in Figure P18.57 is completely filled with water. The tank has marks every 10 cm along one wall, and the 0 cm mark is barely submerged. As you stand beside the opposite wall, your eye is level with the top of the water.
    a. Can you see the marks from the top of the tank (the 0 cm mark) going down, or from the bottom of the tank (the 80 cm mark) coming up? Explain.
    b. Which is the lowest or highest mark, depending on your answer to part a, that you can see?

FIGURE P18.57

58. ‖‖‖ What is the smallest angle θ_1 for which a laser beam will undergo total internal reflection on the hypotenuse of the glass prism in Figure P18.58?

FIGURE P18.58   **FIGURE P18.59**

59. ‖ There is just one angle of incidence β onto a prism for which the light inside an isosceles prism travels parallel to the base and emerges at that same angle β, as shown in Figure P18.59.
    a. Find an expression for β in terms of the prism's apex angle α and index of refraction n.
    b. A laboratory measurement finds that $\beta = 52.2°$ for a prism that is shaped as an equilateral triangle. What is the prism's index of refraction?

60. ‖ A 150-cm-tall diver is standing completely submerged on the bottom of a swimming pool full of water. You are sitting on the end of the diving board, almost directly over her. How tall does the diver appear to be?

61. ‖ To a fish, the 4.00-mm-thick aquarium walls appear only 3.50 mm thick. What is the index of refraction of the walls?

62. | A microscope is focused on an amoeba. When a 0.15-mm-thick cover glass ($n = 1.50$) is placed over the amoeba, by how far must the microscope objective be moved to bring the organism back into focus? Must it be raised or lowered?

BIO

63. ‖ A ray diagram can be used to find the location of an object if you are given the location of its image and the focal length of the mirror. Draw a ray diagram to find the height and position of an object that makes a 2.0-cm-high upright virtual image that appears 8.0 cm behind a convex mirror of focal length 20 cm.

64. | A 2.0-cm-tall object is located 8.0 cm in front of a converging lens with a focal length of 10 cm. Use ray tracing to determine the location and height of the image. Is the image upright or inverted? Is it real or virtual?

65. ‖ The image produced by a converging lens is typically a different size from the object itself. However, for a lens with focal length f there is one object distance that will yield an image the same size as the object. What is that object distance?

66. ‖ A near-sighted person might correct his vision by wearing
BIO diverging lenses with focal length $f = 50$ cm. When wearing his glasses, he looks not at actual objects but at the virtual images of those objects formed by his glasses. Suppose he looks at a 12-cm-long pencil held vertically 2.0 m from his glasses. Use ray tracing to determine the location and height of the image.

67. ‖ A 1.0-cm-tall object is 20 cm in front of a converging lens that has a 10 cm focal length. Use ray tracing to find the position and height of the image. To do this accurately, use a ruler or paper with a grid. Determine the image distance and image height by making measurements on your diagram.

68. | A 2.0-cm-tall object is 20 cm in front of a converging lens that has a 60 cm focal length. Use ray tracing to find the position and height of the image. To do this accurately, use a ruler or paper with a grid. Determine the image distance and image height by making measurements on your diagram.

69. ‖ A 1.0-cm-tall object is 7.5 cm in front of a diverging lens that has a 10 cm focal length. Use ray tracing to find the position and height of the image. To do this accurately, use a ruler or paper with a grid. Determine the image distance and image height by making measurements on your diagram.

70. ‖ A 1.5-cm-tall object is 90 cm in front of a diverging lens that has a 45 cm focal length. Use ray tracing to find the position and height of the image. To do this accurately, use a ruler or paper with a grid. Determine the image distance and image height by making measurements on your diagram.

71. | A 1.6-m-tall woman stands 2.0 m in front of a convex fun-house mirror with a focal length of 2/3 m. Use ray tracing to determine the location and height of her image.

Passage Problems

Mirages

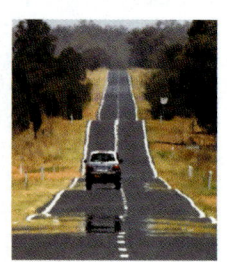

There is an interesting optical effect you have likely noticed while driving along a flat stretch of road on a sunny day. A small, distant dip in the road appears to be filled with water. You may even see the reflection of an oncoming car. But, as you get closer, you find no puddle of water after all; the shimmering surface vanishes, and you see nothing but empty road. It was only a *mirage,* the name for this phenomenon.

The mirage is due to the different index of refraction of hot and cool air. The actual bending of the light rays that produces the mirage is subtle, but we can make a simple model as follows. When air is heated, its density decreases and so does its index of refraction. Consequently, a pocket of hot air in a dip in a road has a lower index of refraction than the cooler air above it. Incident light rays with large angles of incidence (that is, nearly parallel to the road, as shown in Figure P18.72) experience total internal reflection. The mirage that you see is due to this reflection. As you get nearer, the angle goes below the critical angle and there is no more total internal reflection; the "water" disappears!

FIGURE P18.72   Pocket of hot air

72. | The pocket of hot air appears to be a pool of water because
A. Light reflects at the boundary between the hot and cool air.
B. Its density is close to that of water.
C. Light refracts at the boundary between the hot and cool air.
D. The hot air emits blue light that is the same color as the daytime sky.

73. | Which of these changes would allow you to get closer to the mirage before it vanishes?
A. Making the pocket of hot air nearer in temperature to the air above it.
B. Looking for the mirage on a windy day, which mixes the air layers.
C. Increasing the difference in temperature between the pocket of hot air and the air above it.
D. Looking at it from a greater height above the ground.

74. | If you could clearly see the image of an object that was reflected by a mirage, the image would appear
A. Magnified.   B. With up and down reversed.
C. Farther away than the object.
D. With right and left reversed.

STOP TO THINK ANSWERS

Stop to Think 18.1: C. There's one image behind the vertical mirror and a second behind the horizontal mirror. A third image in the corner arises from rays that reflect twice, once off each mirror.

Stop to Think 18.2: A. The ray travels closer to the normal in both media 1 and 3 than in medium 2, so n_1 and n_3 are both larger than n_2. The angle is smaller in medium 3 than in medium 1, so $n_3 > n_1$.

Stop to Think 18.3: D. A green filter lets through only green light, so it blocks the red light from the apple. No light from the apple can pass through the filter, so it appears black.

Stop to Think 18.4: E. The rays from the object are diverging. Without a lens, the rays cannot converge to form any kind of image on the screen.

Stop to Think 18.5: C. For a converging mirror, the focal length f is the distance from the mirror at which incoming parallel rays meet. The moon is so distant that rays from any point on the moon are very nearly parallel. Thus the image of the moon would be very nearly at a distance f in front of the mirror.

19

OPTICAL INSTRUMENTS

This *anableps* is called the "four-eyed fish." How must the top half of its eye differ from the lower half so that it has clear vision both above and below the waterline at the same time?

Looking Ahead ▶

The goal of Chapter 19 is to understand how common optical instruments work. In this chapter you will learn to:

▶ Use the thin-lens equation to find image positions for lenses and mirrors.

▶ Understand how cameras, magnifying glasses, microscopes, and telescopes work.

▶ Analyze the optics of the human eye.

▶ Discuss the practical and fundamental limits to an optical system's resolution.

Looking Back ◀

The material in this chapter draws heavily on what you've learned about ray and wave optics. Please review:

A fish's eye, like ours, uses a lens to focus an image of the outside world on its retina. Although most animals' eyes are not as unusual as those of the *anableps* shown in the photo, nature does use a remarkably wide array of imaging methods for sight, including lenses, mirrors, and even pinholes. Image formation with lenses and mirrors also forms the basis of many modern technologies. If you use a video camera, watch a big-screen TV, or look through a pair of binoculars, you're viewing an image created by lenses and mirrors.

In Chapter 18 you learned how images are formed by single lenses and mirrors. Now we want to apply these ideas to practical instruments such as cameras and microscopes. Chapter 18 emphasized a graphical method—ray tracing—to understand image formation. We'll begin this chapter by developing a more mathematical approach to image formation, then apply this method to several important optical instruments, including the human eye. We'll finish by discussing the important issue of the resolution of optical instruments: How much detail can a microscope or telescope see? We'll find that there's a fundamental limit on the amount of detail observable, a limit set by the wave nature of light.

615

19.1 Finding the Image of a Lens or Mirror

15.10, 15.11 Ray tracing, which you learned in Chapter 18, is an important tool for quickly grasping the overall positions and sizes of an object and its image. For more precise work, however, we would like a mathematical expression that relates the three fundamental quantities of an optical system: the focal length f of the lens or mirror, the object distance s, and the image distance s'. We can find such an expression by considering the converging lens in Figure 19.1. Two of the special rays are shown, one initially parallel to the optical axis that then passes through the far focal point, and the other traveling undeviated through the center of the lens.

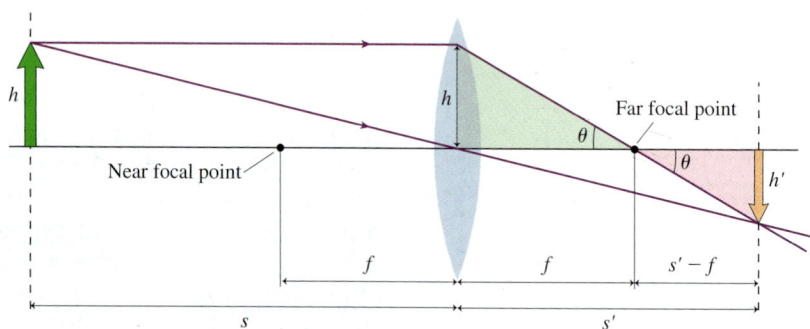

FIGURE 19.1 Deriving the thin-lens equation.

Consider the two right triangles highlighted in green and pink. Because they both have one 90° angle, and a second angle θ that is the same for both, the two triangles are *similar*. This means that they have the same shape, although their sizes may be different. For similar triangles, the ratios of any two similar sides is the same. Thus we have

$$\frac{h'}{h} = \frac{s' - f}{f} \tag{19.1}$$

Further, from Equation 18.9 on page 601, the magnification is

$$m = \frac{h'}{h} = \frac{s'}{s} \tag{19.2}$$

Combining Equations 19.1 and 19.2 gives

$$\frac{s'}{s} = \frac{s' - f}{f}$$

Dividing both sides by s' gives

$$\frac{1}{s} = \frac{s' - f}{s'f} = \frac{1}{f} - \frac{1}{s'}$$

which we can write as

$$\frac{1}{s} + \frac{1}{s'} = \frac{1}{f} \tag{19.3}$$

Thin-lens equation (also works for mirrors)
relating object and image distance to focal length

This equation, the **thin-lens equation,** relates the three important quantities f, s, and s'. In particular, if we know the focal length of a lens and the object's distance from the lens, we can use the thin-lens equation to find the position of the image.

NOTE ▶ Although we derived the thin-lens equation for a converging lens that produced a real image, it works equally well for *any* image—real or virtual—produced by both converging and diverging lenses. And, in spite of its name, the thin-lens equation also describes the images formed by *mirrors*. ◀

It's worth checking that the thin-lens equation describes what we already know about lenses. In Figure 18.34a, we saw that rays initially parallel to the optical axis are focused at the focal point of a converging lens. Initially parallel rays come from an object extremely far away, with $s \rightarrow \infty$. Because $1/\infty = 0$, the thin-lens equation tells us that the image distance is $s' = f$, as we expected. Or suppose an object is located right at the focal point, with $s = f$. Then, according to Equation 19.3, $1/s' = 1/f - 1/s = 0$. This implies that the image distance is infinitely far away ($s' = \infty$), so the rays leave the lens parallel to the optical axis. Indeed, this is what Figure 18.34b showed. Now, it's true that no real object or image can be at infinity. But if either the object or image is more than several focal lengths from the lens ($s \gg f$ or $s' \gg f$), then it's an excellent approximation to consider the distance to be infinite, the rays to be parallel to the axis, and the reciprocal ($1/s$ or $1/s'$) in the thin-lens equation to be zero.

Sign Conventions for Lenses and Mirrors

As we noted in Chapter 18, we take all distances to be positive quantities when performing graphical ray tracing. When using the thin-lens equation, however, the quantities s' and f can be positive *or* negative, depending on the type of lens or mirror, and whether the image is real or virtual. Tactics Box 19.1 summarizes the **sign conventions** that we will use in this text.

(MP) **TACTICS BOX 19.1 Using sign conventions for lenses and mirrors**   ✏ Exercise 1

Quantity	Positive when		Negative when	
Object distance s	Always		We won't treat this case in this book.	
Image distance s'	**Real image**		**Virtual image**	
	Image is on the opposite side of the lens from the object.	Image is in front of the mirror.	Image is on the same side of the lens as the object.	Image is behind the mirror.
Focal length f	Converging lens or concave mirror		Diverging lens or convex mirror	
Image height h', Magnification M	Image is upright		Image is inverted	

NOTE ▶ We won't treat the case of a negative object distance. Such a case occurs for two or more lenses or mirrors in combination, when the image due to the first lens lies behind the second lens. ◀

Note that the image height h' is negative for an inverted image. In Chapter 18 we defined the magnification m as the ratio of image height to object height, with both heights taken as positive numbers. It's useful to now expand that definition to include this new information about the orientation of the image. The revised definition of the **magnification,** which we'll now call M, is

$$M = \frac{h'}{h} = -\frac{s'}{s} \qquad (19.4)$$

With this definition, **the magnification M is positive for an upright image and negative for an inverted one.** In other words, M tells us both the size *and* the orientation of the image.

EXAMPLE 19.1 Analyzing a magnifying lens

A stamp collector uses a magnifying lens that sits 2.0 cm above the stamp. The magnification is 4. What is the focal length of the lens?

FIGURE 19.2 Ray-tracing diagram of a magnifying lens.

PREPARE A magnifying lens is a converging lens with the object distance less than the focal length ($s < f$). Assume it is a thin lens. The user looks *through* the lens and sees a virtual image. Figure 19.2 shows the lens and a ray-tracing diagram.

SOLVE A virtual image is upright, so $M = +4$. The magnification is $M = -s'/s$, thus

$$s' = -4s = -4(2.0 \text{ cm}) = -8.0 \text{ cm}$$

We can use s and s' in the thin-lens equation to find the focal length:

$$\frac{1}{f} = \frac{1}{s} + \frac{1}{s'} = \frac{1}{2.0 \text{ cm}} + \frac{1}{-8.0 \text{ cm}} = 0.375 \text{ cm}^{-1}$$

Thus

$$f = \frac{1}{0.375 \text{ cm}^{-1}} = 2.7 \text{ cm}$$

ASSESS $f > 2$ cm, as expected because the object has to be inside the focal point.

EXAMPLE 19.2 What is the focal length?

An object is 38.0 cm to the left of a lens. Its image is found to be 22.0 cm from the lens on the same side as the object. What is the focal length of the lens? Draw a ray diagram for the lens and object.

PREPARE We know the object distance is $s = 38.0$ cm. According to Tactics Box 19.1, the image distance is a negative number because the image is on the same side of the lens as the object, so $s' = -22.0$ cm.

SOLVE The thin-lens equation, Equation 19.3, is

$$\frac{1}{f} = \frac{1}{s} + \frac{1}{s'} = \frac{1}{38.0 \text{ cm}} + \frac{1}{-22.0 \text{ cm}} = -0.0191 \text{ cm}^{-1}$$

from which we find

$$f = \frac{1}{-0.0191 \text{ cm}^{-1}} = -52.3 \text{ cm}$$

Because f is negative, Tactics Box 19.1 indicates that the lens is a diverging lens. We can use the techniques learned in Chapter 18 to draw the ray diagram shown in Figure 19.3. Notice that the negative value of s' corresponds to a virtual image.

FIGURE 19.3 Ray diagram for a diverging lens.

ASSESS This problem would be difficult to solve graphically because you need to locate the focal points in order to draw a ray diagram. Using the thin-lens equation gives rapid and accurate results. Nonetheless, sketching a ray diagram once f is known is very helpful in fully understanding the situation.

EXAMPLE 19.3 Finding the mirror image of a candle

A concave mirror has a focal length of 2.4 m. Where should a candle be placed so that its image is inverted and twice as large as the object?

PREPARE Even though this is a mirror, we can use the thin-lens equation. We have $f = +2.4$ m because, according to Tactics Box 19.1, a concave mirror has a positive focal length. We also know that the magnification is $M = -2.0$ because an inverted image implies a negative magnification.

SOLVE Equation 19.4, $M = -s'/s$, relates s' and s:

$$s' = -Ms$$

We can insert this expression for s' into the thin-lens equation to get

$$\frac{1}{f} = \frac{1}{s} + \frac{1}{s'} = \frac{1}{s} + \frac{1}{-Ms} = \frac{1}{s} + \frac{1}{-(-2)s} = \frac{1}{s} + \frac{1}{2s} = \frac{3}{2s}$$

from which

$$s = \frac{3}{2}f = \frac{3}{2}(2.4 \text{ m}) = 3.6 \text{ m}$$

The candle should be placed 3.6 m in front of the mirror.

ASSESS The magnification equation tells us that $s' = 2s = 7.2$ m. This is positive, so the image is real and on the same side of the mirror as the object.

Lenses in Combinations

Many optical instruments make use of combinations of lenses. For example, a microscope uses a short-focal-length objective lens in combination with an eyepiece lens. And the objective lens itself is made up of several individual lenses.

 15.12

A simple rule governs how two lenses work in combination: **The image from the first lens acts as the object for the second lens.** If there's a third lens, the image from the second lens acts as its object, and so on.

EXAMPLE 19.4 Finding the image for two lenses in combination

A 5.0-cm-focal-length converging lens is 16.0 cm in front of a 10.0-cm-focal-length diverging lens. A 4.0-cm-tall object is placed 11.0 cm in front of the converging lens. What are the position and size of the final image?

FIGURE 19.4 Two lenses in combination.

PREPARE Let's start with ray tracing. A ray tracing diagram helps us understand the situation and tells us what to expect for an answer. A diagram can often alert you to a calculation error. Figure 19.4a first uses the three special rays of the converging lens to locate its image. We see that the image of the first lens is a real image falling between the two lenses. According to the rule, we then use this image as the object for the second lens. This is done in Figure 19.4b, where we see that the final image is inverted, virtual, and roughly one-fourth the distance between the two lenses. Thus we expect s_2' to be about -4 cm, negative because it's in front of the lens.

Mathematically, we can use the thin-lens equation to find the image location and size due to the first lens, then use this as the object for the second lens in a second use of the lens equation.

SOLVE We first solve for the image due to lens 1. We have

$$\frac{1}{s_1'} = \frac{1}{f_1} - \frac{1}{s_1} = \frac{1}{5.0 \text{ cm}} - \frac{1}{11.0 \text{ cm}}$$

from which $s_1' = 9.2$ cm. Because this is a positive image distance, the image is real and located to the right of the first lens. The magnification of the first lens is $M_1 = -s_1'/s_1 = -(9.2 \text{ cm})/(11.0 \text{ cm}) = -0.84$.

This image is the object for the second lens. Because it is 9.2 cm to the right of the first lens, and the lenses are 16.0 cm apart, it is 16.0 cm − 9.2 cm = 6.8 cm in front of the second lens. Thus $s_2 = 6.8$ cm.

Continued

Applying the thin-lens equation again, we have

$$\frac{1}{s_2'} = \frac{1}{f_2} - \frac{1}{s_2} = \frac{1}{-10.0 \text{ cm}} - \frac{1}{6.8 \text{ cm}}$$

from which we find $s_2' = -4.0$ cm. Because this image distance is negative, the image is virtual and located to the left of lens 2, as shown in Figure 19.4b.

The magnification of the second lens is

$$M_2 = -s_2'/s_2 = -(-4.0 \text{ cm})/(6.8 \text{ cm}) = 0.59.$$

Thus the final image size is

$$h_2' = M_2 h_2 = M_2 h_1' = M_2(M_1 h_1) = M_1 M_2 h_1$$
$$= (-0.84)(0.59)(4.0 \text{ cm}) = -2.0 \text{ cm}$$

Here we used the fact that the object height h_2 for the second lens is equal to the image height h_1' of the first lens.

ASSESS When calculating the final image size, we found that $h_2' = M_1 M_2 h_1$. This shows the important fact that **the total magnification of a combination of lenses is the *product* of the magnifications for each lens alone.** This fact will be important as we later study instruments that use several lenses.

STOP TO THINK 19.1   A candle is placed in front of a converging lens. A well-focused image of the flame is seen on a screen on the opposite side of the lens. If the candle is moved farther away from the lens, how must the screen be adjusted to keep showing a well-focused image?

A. The screen must be moved closer to the lens.
B. The screen must be moved farther away from the lens.
C. The screen does not need to be moved.

Refractive Power

Lenses with shorter focal lengths bend light rays more than lenses with longer focal lengths. The greater the ability of a lens to bend rays, the greater its **refractive power** P. Because a shorter focal length implies a larger refractive power, we define the refractive power of a lens as the *inverse* of its focal length, so that

$$P = \frac{1}{f} \qquad (19.5)$$

Refractive power of a lens with focal length f

INVERSE
p. 118

Because focal lengths are measured in m, in SI units, the units of refractive power are m^{-1}. A common unit for refractive power is the **diopter**, abbreviated D, where

$$1 \text{ D} = 1 \text{ m}^{-1}$$

For instance, a lens with a focal length of 37 cm (0.37 m) has refractive power $P = 1/0.37 \text{ m} = 2.7$ D. Diverging lenses, with their negative focal lengths, have negative refractive powers.

Refractive power is particularly useful for two lenses in contact. In this case it can be shown from the thin-lens equation that the two lenses act as a single lens with refractive power

$$P_{total} = P_1 + P_2 \qquad (19.6)$$

That is, **the total refractive power of two lenses in contact is equal to the *sum* of their individual refractive powers.** Refractive power finds use in optometry, where corrective lenses such as eyeglasses or contact lenses are used in close proximity to the eye. We'll study this application later in the chapter.

EXAMPLE 19.5 Finding the effective focal length of two lenses

Two lenses, one with focal length $f_1 = 36$ cm and the other with $f_2 = -19$ cm, are placed in contact. What are the effective focal length and total refractive power of the combination?

PREPARE The total refractive power of the lens combination is simply the sum of the refractive powers of the two lenses. The effective focal length of the combination is the inverse of this total refractive power.

SOLVE We have $P_1 = 1/(0.36\text{ m}) = 2.8$ D and $P_2 = 1/(-0.19\text{ m}) = -5.3$ D, so that

$$P_{\text{total}} = P_1 + P_2 = 2.8\text{ D} - 5.3\text{ D} = -2.5\text{ D}$$

Then the effective focal length of the combined lenses is

$$f_{\text{eff}} = \frac{1}{P_{\text{total}}} = \frac{1}{-2.5\text{ D}} = -0.40\text{ m}$$

ASSESS The second lens is a diverging lens, with a negative focal length and refractive power. Its refractive power is greater in magnitude than that of the converging, lens. Consequently, it bends rays out from the axis more strongly than the converging lens bends them toward the axis. Thus the combination acts as a diverging lens.

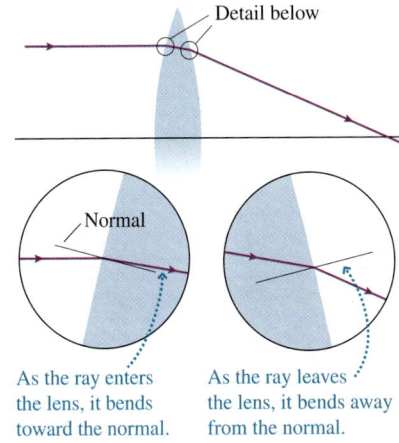

The refractive power of a lens is determined by its shape and its index of refraction. Figure 19.5 shows how, for a converging lens, a ray is bent toward the optical axis at each surface of the lens. According to Snell's law, either a larger angle of incidence or a higher index of refraction will cause the ray to be refracted through a larger angle. The angle of incidence is increased when the lens surfaces are more strongly curved. In general, then, **the more strongly curved a lens's surfaces, or the higher its index of refraction, the higher its refractive power.**

As the ray enters the lens, it bends toward the normal.

As the ray leaves the lens, it bends away from the normal.

FIGURE 19.5 How a lens brings rays to a focus.

STOP TO THINK 19.2 The tube shown contains two lenses in contact. Which one or more of the following statements could be true?

A. Both lenses are converging.
B. One lens is converging, one diverging.
C. Both lenses are diverging.

19.2 The Camera

Although a camera may seem like a very complicated instrument, its basic operation is really fairly simple. As shown in Figure 19.6, a camera consists of a light-tight case with a converging lens. This lens projects an *inverted* real image of objects in front of the camera onto a light-sensitive surface—either film or an electronic detector—where the image is recorded. A shutter (not shown), an opaque barrier, is briefly moved out of the way in order for light to pass through the lens. Let's look at how all this works in a bit more detail.

Figure 19.7 on the next page shows light rays from an object passing through the lens and converging at the image plane B. Here, a single point P on the object focuses to a single point P' in the image plane. If the film is located in this plane, a sharp image will form on it. If, however, the film had been located a bit in *front* of the image plane, at position A, rays from point P would not yet have completely converged and would form a small blurry *circle* on the film instead of a sharp point. Thus the image would appear blurred, as shown. Similarly, if the film

FIGURE 19.6 A camera.

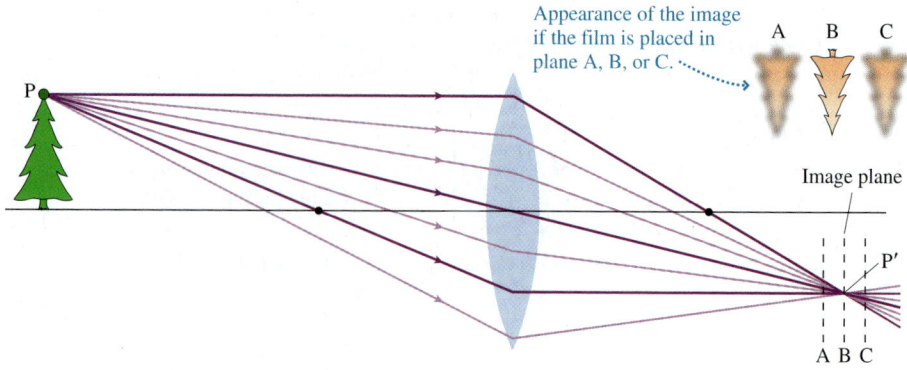

Appearance of the image if the film is placed in plane A, B, or C.

Image plane

FIGURE 19.7 Focusing a camera.

is placed at C, *behind* the image plane, the rays will be diverging from their perfect focus and again form a blurry image.

Thus to get a sharp image, the film must be accurately located in the image plane. Figure 19.7 shows that one way to do this is to move the film until it coincides with the image plane. More commonly, however, a camera is **focused** by moving the *lens* either toward or away from the film plane until the image is sharp. In either case, the lens-film distance is varied.

EXAMPLE 19.6 Focusing a camera

A digital camera whose lens has a focal length of 8.0 mm is used to take a picture of an object 30 cm away. What must be the distance from the lens to the light-sensitive detector in order for the image to be in focus?

PREPARE As shown in Figure 19.7, the image will be in focus when the detector is in the image plane. Thus we need to find the image distance, knowing the object distance $s = 30$ cm and the lens's focal length $f = 8.0$ mm.

SOLVE We can rearrange the thin-lens equation, Equation 19.3, to solve for the image distance s':

$$\frac{1}{s'} = \frac{1}{f} - \frac{1}{s} = \frac{1}{0.0080 \text{ m}} - \frac{1}{0.30 \text{ m}} = 122 \text{ m}^{-1}$$

Thus $s' = 1/122 \text{ m}^{-1} = 0.0082 \text{ m} = 8.2$ mm. The lens-detector distance has to be 8.2 mm.

ASSESS When the object is infinitely far away, the image, by definition, is at the focal length: $s' = f = 8.0$ mm. If the object is brought to 30 cm, the lens has to move forward a distance of only 8.2 mm − 8.0 mm = 0.2 mm to bring the object into focus. In general, camera lenses don't need to move far.

For traditional cameras, the light-sensitive surface is **film.** Light striking microscopic silver halide crystals creates small particles of silver metal on the crystals. The development process converts the entire crystal to dark metallic silver. Thus the film appears *dark* where light has hit it, forming a *negative* image. To get a positive image—your print—the negative image is projected onto paper with its own photosensitive coating. This results in a second negative process, leading to a positive print.

Modern *digital cameras* use an electronic light-sensitive surface called a *charge-coupled device* or **CCD.** A CCD consists of a rectangular array of many millions of small detectors called *pixels*. When light hits one of these pixels, it generates an electric charge proportional to the light intensity. Thus an image is recorded on the CCD in terms of little packets of charge. After the CCD has been

If a digital picture is magnified enough, you can see the individual pixels that make it up.

exposed, the charges are read out, the signal levels are digitized, and the picture stored in the digital memory of the camera.

Figure 19.8 shows a CCD "chip" and, schematically, the magnified appearance of the pixels on its surface. To record color information, different pixels are covered by red, green, or blue filters; a pixel covered by a green filter, for instance, records only the intensity of the green light hitting it. Later, the camera's microprocessor interpolates nearby colors to give each pixel an overall true color. The structure of the retina of the eye is remarkably similar, as we'll see in Chapter 28.

2500 × 2000 pixels

1 pixel

FIGURE 19.8 A CCD chip used in a digital camera.

Controlling the Amount of Light

In taking a photograph it's important to control the amount of light reaching the film or the CCD. Figure 19.9 shows how too little light leads to an image that is overly dark; the picture is **underexposed.** Too much light leads to a bright, **overexposed** image.

There are two ways of controlling the exposure. First, as we've already mentioned, a shutter normally blocks light from reaching the film. When you "press the shutter," you really press a button that causes the shutter to move out of the way, allowing light to reach the film and expose it. The longer the shutter is open, the more light reaches the film. Typical *shutter speeds* are in the range from 1/500 s to 1/30 s (perhaps a better name would be shutter *time,* since speed is not really relevant here). A long shutter speed of 1/60 s = 16.6 ms will let in about 8 times as much light as a short shutter speed of only 1/500 s = 2 ms. However, long shutter speeds can allow the image to become blurred if the object (or the camera) moves during the exposure. Short shutter speeds are better for fast-moving objects, such as those encountered in sports photography.

The second means of controlling the exposure is to effectively change the diameter d of the lens. A small lens will let in less light than a large one. Of course, the actual diameter of the lens cannot be changed. Instead, an *iris diaphragm,* placed behind the lens, is adjusted to change the amount of light entering the camera, as shown in Figure 19.10.

An important measure of the light-gathering ability of a lens is its *f*-**number,** defined as

$$f\text{-number} = \frac{\text{lens focal length}}{\text{lens diameter}} = \frac{f}{d}$$

The *f*-number of a lens directly determines the *brightness* of the image on the film. Somewhat contrary to common sense, the *smaller* the *f*-number, the *brighter* the image. Here's why: For a given focal length, more light enters a lens with a larger diameter than one with a smaller diameter. But since the *f*-number is inversely proportional to d, the larger-diameter, brighter lens has the lower *f*-number. Further, a lens of a given diameter will give a brighter image when the focal length f is *small.* To see why, consider a lens imaging a distant object, so that the image is focused at the far focal point. If f is large, the image is large and its light is spread out and dim. If f is small, the image is small and the light is concentrated and bright.

A lens with its diaphragm fully open will let in the most possible light. But under bright light conditions we may need to close, or *stop down,* the diaphragm to control the exposure. How much must the diaphragm be closed in order to let in half as much light? The amount of light entering the lens is proportional to the *area* of the lens, or to d^2. To let in half as much light, we must reduce d^2 by a factor of two or, equivalently, reduce the diameter of the lens by a factor of $1/\sqrt{2}$. This means that the *f*-number, which is inversely proportional to the diameter, will *increase* by a factor of $\sqrt{2} \approx 1.4$. On most camera lenses, the *f*-**stops** increase in the series 1.4, 2, 2.8, 4, 5.6, 8, 11, 16. Each *f*-number is $\sqrt{2}$ times larger than the previous one, so an increase of one *f*-stop cuts the light intensity in half.

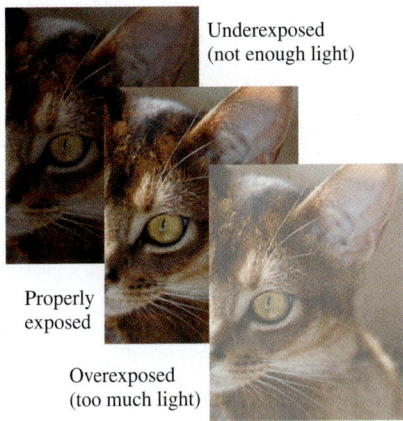

Underexposed (not enough light)

Properly exposed

Overexposed (too much light)

FIGURE 19.9 Photos with increasing amounts of light reaching the film.

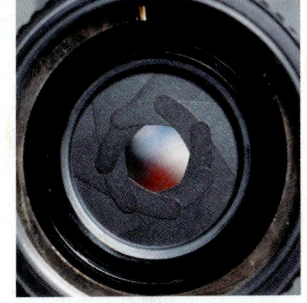

FIGURE 19.10 A camera's iris can change the effective diameter of the lens.

CONCEPTUAL EXAMPLE 19.1 Adjusting a camera lens

A camera takes a perfectly exposed picture when the shutter speed is 1/60 s and the lens diaphragm is set to *f*/11. What *f*-stop should be used to get a correctly exposed picture with a shutter speed of 1/250 s?

REASON We need the same amount of light hitting the film or CCD in both cases. For the second picture the shutter is open only about one-fourth as long, since

$$\frac{1/250 \text{ s}}{1/60 \text{ s}} = \frac{60}{250} = \frac{1}{4.17} \approx \frac{1}{4}$$

To compensate for the reduced time, we need to increase the light intensity by a factor of 4. Because each stop lets in twice as much light as the previous stop, the light intensity will be 4 times as great if we open up the diaphragm by two stops. From the sequence given above, opening the iris by two stops from *f*/11 is *f*/5.6.

ASSESS Camera film and CCD detectors yield a good image over a rather wide range of exposures. Thus the fact that shutter speeds and *f*-numbers change the exposure by factors of two still allows us to get the exposure close enough for a good picture.

STOP TO THINK 19.3 A camera takes a correctly exposed picture with a certain lens. The lens is then replaced by one with the same diameter but twice the focal length. To get the correct exposure when the focal length is doubled, the shutter speed should

A. Be increased.   B. Be decreased.   C. Remain the same.

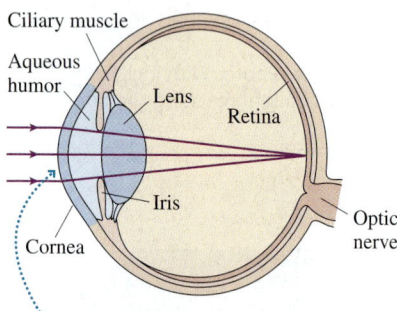

Most of the refraction occurs at the cornea's surface, where Δn is the largest.

FIGURE 19.11 The human eye.

TRY IT YOURSELF

Inverted vision Just like a camera, the lens of the eye produces an *inverted* image on the retina. The brain is wired to "flip" this inverted image and interpret it as being upright. To show this directly, try this simple experiment. Poke a small hole in a card using a pin and, holding the card a few inches away, look through the hole at a light bulb. While doing so, move the head of the pin between the hole and your eye; you'll see an *upside-down* pinhead. The hole acts as a point source that casts an *erect* shadow of the pin on your retina. The brain then inverts this erect shadow, making it appear inverted.

19.3 The Human Eye

The human eye functions much like a camera. Like the camera, it has three main functional groups: an optical system to focus the incoming light, a diaphragm to adjust the amount of light entering the eye, and a light-sensitive surface to detect the resulting image. The parts of the eye making up these three groups are shown in Figure 19.11. The *cornea,* the *aqueous humor,* and the *lens* are together responsible for refracting incoming light rays and producing an image. The adjustable *iris* determines how much light enters the eye, in much the same way as does the diaphragm of a camera. And the *retina* is the light-sensitive surface on which the image is formed. The retina is the biological equivalent of the CCD in a digital camera. Let's look at these three groups in more detail.

Focusing and Accommodation

Like a camera, the eye works by focusing incoming rays onto a light-sensitive surface, here the retina. To do so, light is refracted by, in turn, the cornea, the aqueous humor, and the lens, as shown in Figure 19.11. The indices of refraction of these parts of the eye vary somewhat, but average around 1.4. Perhaps surprisingly, most of the refraction occurs not in the lens but at the surface of the cornea. This is due both to the strong curvature of the cornea and to the large difference between the indices of refraction on either side of the surface. Light refracts less as it passes through the lens because the lens's index of refraction doesn't differ much from the fluid in which it is embedded. If the lens is surgically removed, which it often is for people with cataracts (a clouding of the lens), the cornea alone still provides a marginal level of vision.

The eye must constantly refocus as it views distant objects, then closer ones. It does this so automatically that we're not normally aware of the process. A camera focuses by changing the distance between the lens and the film, but your eye focuses in a different way: by changing the focal length of the lens itself. As shown in Figure 19.12, it does so by using the ciliary muscles to *change the shape* of the lens. This process of changing the lens shape as the eye focuses at different distances is called **accommodation.**

When the eye focuses on distant objects, the ciliary muscles are relaxed and the lens is less curved.

Most of the refraction occurs at the surface of the cornea. The lens is used for fine adjustments.

When the eye focuses on nearby objects, the ciliary muscles are contracted and the lens is more curved.

FIGURE 19.12 Accommodation by the eye.

The most distant point on which the completely relaxed eye can focus is called the eye's **far point.** For normal vision, the far point is at infinity. The closest point on which the eye can focus, with the ciliary muscles fully contracted, is called the **near point.** Objects closer than the near point cannot be brought into sharp focus. The near point is conventionally taken to be 25 cm from the eye, although young people can usually focus on objects closer than this. As people age, however, the near point moves further away because the lens becomes less elastic and more difficult to curve. By the age of 50, the near point has typically receded to some 50 cm; by age 70 it is about 1 m. That's why few people past age 50 can read without the aid of glasses or contact lenses.

Corrective Lenses

Two common conditions of the human eye—**myopia,** or *nearsightedness,* and **hyperopia,** or *farsightedness*—require corrective lenses, either glasses or contact lenses. As Figure 19.13a shows, a normal relaxed eye naturally focuses the rays from a distant object on the retina. In this case the eye can look at distant objects (here, "distant" really means only beyond a few meters) for long periods of time without strain because the muscles need not be contracted.

In a myopic eye the lens is too strongly curved, as shown in Figure 19.13b. When this eye is relaxed, rays from a distant object come to a focus *in front of* the retina and have begun to diverge by the time they strike the retina. Consequently, distant objects appear blurry to a person with myopia. However, if an object is brought close (decreasing s), the image distance s' increases and the light rays come to a focus on the retina. That is, a person with myopia can clearly see close objects with a relaxed eye (so their far point is less than infinity) and, with accommodation, can focus on objects much closer than the normal near point of 25 cm. The term *nearsightedness* refers to the ability to see only things nearby.

In contrast, as Figure 19.13c shows, the lens of a hyperopic eye is too weak, causing rays from a distant object to come to a focus *behind* the retina. Those with hyperopic eyes are usually called *farsighted,* but the term is a bit of a misnomer. With the eye relaxed, distant objects are actually out of focus; only by contracting the ciliary muscles—and introducing eye strain—can distant objects be brought into sharp focus. However, no amount of accommodation is sufficient to focus the rays from a nearby object. Thus a person with hyperopia can, with effort, see distant objects but not close objects. The hyperopic eye's near point is more distant than the normal 25 cm.

Both conditions—myopia and hyperopia—can be corrected with external lenses, either eyeglasses or contact lenses. Let's consider first the case of myopia (nearsightedness). With myopia, the problem is that the optical system of the eye

Seeing underwater BIO When you swim underwater, the difference in refractive indices between the cornea ($n = 1.38$) and water ($n = 1.33$) is too small to allow significant refraction, so the eye cannot focus. If you wear goggles, the surface of the cornea is in contact with air, not water, and the eye can focus normally. Animals that live underwater generally have more sharply curved corneas to compensate for the small difference in refractive indices. The *anableps* fish shown at the beginning of this chapter is particularly unusual in that it lives at the water's surface. To focus simultaneously on objects on both sides of the waterline, it has evolved a very asymmetrical cornea that is more strongly curved below the waterline.

(a) Normal eye

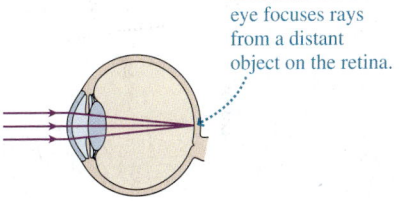

A normal relaxed eye focuses rays from a distant object on the retina.

(b) Myopic eye

A relaxed myopic eye focuses rays from a distant object in front of the retina.

(c) Hyperopic eye

A relaxed hyperopic eye tries to focus rays from a distant object behind the retina.

FIGURE 19.13 Common conditions of the eye.

(a)

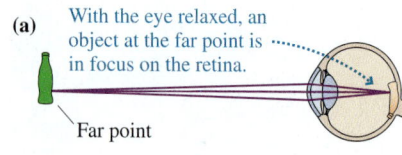

With the eye relaxed, an object at the far point is in focus on the retina.

Far point

(b)

Essentially parallel rays from a distant tree.

A diverging lens is chosen to create a virtual image of the tree at the eye's far point.

(c)

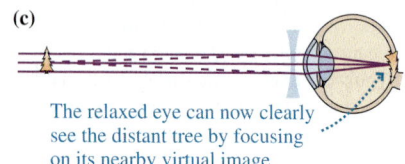

The relaxed eye can now clearly see the distant tree by focusing on its nearby virtual image.

FIGURE 19.14 A diverging lens is used to correct myopia.

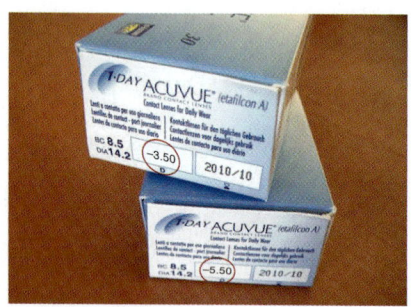

Is the wearer of these lenses nearsighted or farsighted?

is too powerful. As Figure 19.14a shows, this means that the far point of a myopic eye, on which the relaxed eye focuses, is closer than infinity. (Compare this to Figure 19.13b, where we see the parallel rays from a distant object coming to a focus in front of the retina.) How can we help this eye focus the parallel rays from a very distant object?

Because the eye's optics are too powerful, a *diverging* lens, with a negative power, in front of the eye will provide the appropriate correction. Figure 19.14b reminds you that a diverging lens creates a virtual image, a point from which rays *appear* to diverge. For a distant object, the virtual image is at the focal point of the lens. If the focal length of a diverging lens matches the eye's far point, as shown in Figure 19.14c, the lens creates a virtual image of the distant object at the far point of the eye, right where the relaxed eye, as we've seen, naturally focuses. In other words, a diverging lens "brings" distant objects up close where the myopic eye can focus. The correct prescription is a diverging lens whose focal length is the eye's far-point distance.

EXAMPLE 19.7 Correcting nearsightedness

Ryan is nearsighted. The far point of his left eye is 35 cm from his eye. What power lens should be prescribed to correct his vision?

PREPARE We wish to find a corrective lens that will create a virtual image at Ryan's far point of a distant object. We can use the thin-lens equation to do so. For practical purposes, a distant object is at $s = \infty$. We need a lens that will create a virtual image 35 cm *in front* of the lens, so s' is a *negative* number: $s' = -35$ cm.

SOLVE We can solve the thin-lens equation for the required focal length:

$$\frac{1}{f} = \frac{1}{s} + \frac{1}{s'} = \frac{1}{\infty} + \frac{1}{-35 \text{ cm}} = \frac{1}{-35 \text{ cm}}$$

Thus $f = -35$ cm, a diverging lens whose focal length matches Ryan's far-point distance. The power of the lens is

$$P = \frac{1}{f} = \frac{1}{-0.35 \text{ m}} = -2.9 \text{ D}$$

ASSESS Ryan's optometrist would write a prescription saying "−2.9." The lens maker knows the prescription is in diopters, so optometrists don't usually write the D. If you wear glasses or contact lenses, look at your next prescription and you'll see a number like this for each eye, as shown circled in red on the boxes of contact lenses shown at left.

As Figure 19.13c showed, the refractive power of a hyperopic (farsighted) eye is too weak to focus a distant object on the retina. Consequently, a corrective *converging* lens is needed to increase the refractive power of the eye. A person with hyperopia actually has two problems: The eye must strain to see distant objects, and the near point is larger than 25 cm, so the eye cannot focus on near objects at all. Fully correcting both often requires *bifocals,* glasses where the top and bottom halves of the lenses are shaped differently and have different focal lengths. The top half of the lens corrects distant vision while the bottom half corrects near vision. For mild hyperopia, correcting near vision, so that the person can read, also provides reasonably good correction for distant vision, enough to reduce eye strain. We'll consider only the case of mild hyperopia.

Figure 19.15a shows a hyperopic eye that has a near point farther than the normal 25 cm. This means that even with its ciliary mucles fully contracted, the eye cannot focus on objects closer than this. However, we would like this eye to be able to focus on objects as close as the normal near-point distance of 25 cm. A converging lens makes this possible, as shown in Figure 19.15b. Consider an object placed 25 cm from the eye. A properly chosen converging lens creates a *virtual* image of this object at the farther distance of the eye's near point. The eye *can* focus on an image there. In other words, a converging lens "pushes" close objects farther out to where the hyperopic eye can focus. At the same time, this lens will allow distant objects to be focused with less eye strain.

FIGURE 19.15 Correcting the hyperopic eye.

EXAMPLE 19.8 Correcting farsightedness

Angie is farsighted, with a near point 40 cm from her eyes. What power lenses should be prescribed to correct her vision?

PREPARE We need a corrective lens that will create a virtual image at Angie's near point of an object, such as a book, that is 25 cm away, the near point of normal vision. The book is at $s = 25$ cm. We need a lens that will create a virtual image 40 cm *on the same side of the lens* as the object, so $s' = -40$ cm.

SOLVE The thin-lens equation gives us the necessary lens power if we give the object and image distance in meters:

$$P = \frac{1}{f} = \frac{1}{s} + \frac{1}{s'} = \frac{1}{0.25 \text{ m}} + \frac{1}{-0.40 \text{ m}} = 1.5 \text{ m}^{-1} = 1.5 \text{ D}$$

ASSESS Angie's optometrist would write the prescription as "+1.5." This corresponds to a focal length of $f = 1/P = 0.67$ m $= 67$ cm.

STOP TO THINK 19.4 With her right eye, Maria can focus on a vase 0.5 m away, but not on a tree 10 m away. Which of the following could be the eyeglass prescription for her right eye?

A. +3.0 D   B. +10 D   C. −5.0 D   D. −1.5 D

19.4 The Magnifier

You've no doubt used a magnifier, or magnifying glass, to get a better look at a small object such as an insect or a coin. As we saw in Chapter 18, a magnifier is a simple converging lens, but why objects appear larger when viewed through such a lens is actually rather subtle.

Let's begin by considering the simplest way to magnify an object, one that requires no extra optics at all. You simply get closer to the object you're interested in. The closer you get, the bigger the object appears. Obviously the actual size of the object is unchanged as you approach it, so what exactly is getting "bigger"? A half dollar, held at arm's length, will more than cover the distant moon. In what sense is the half dollar "larger" than the moon?

Angular Size and Apparent Size

Consider an object such as the green arrow in Figure 19.16a on the next page. To find the size of its image on the retina, we can trace the two rays shown that go through the center of the eye lens. (Here, we'll use the thin-lens approximation to reduce the eye's entire optical system to one thin lens). As we've learned, such

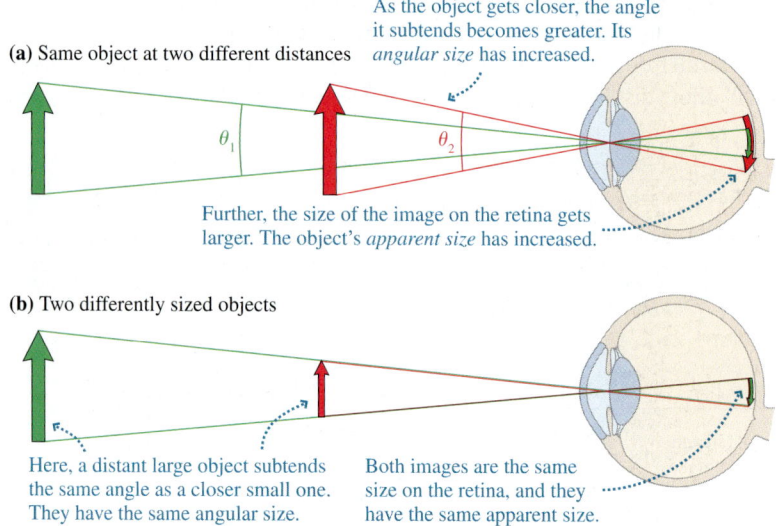

(a) Same object at two different distances

As the object gets closer, the angle it subtends becomes greater. Its *angular size* has increased.

θ_1 θ_2

Further, the size of the image on the retina gets larger. The object's *apparent size* has increased.

(b) Two differently sized objects

Here, a distant large object subtends the same angle as a closer small one. They have the same angular size.

Both images are the same size on the retina, and they have the same apparent size.

FIGURE 19.16 How the apparent size of an object is determined.

rays are undeviated as they pass through the lens, so we can use them to locate the image, shown in green, on the retina. If the arrow is then brought closer to the eye, as shown in red, ray tracing reveals that the size of the arrow's image is *larger*. Our brain interprets a larger image on the retina as representing a larger-appearing *object*. As the object is moved closer, its size doesn't change, but its **apparent size** gets larger.

Angles θ_1 and θ_2 in Figure 19.16a are the angles *subtended* by the green and red arrows. The angle subtended by an object is called its **angular size.** As you can see from the figure, **objects that subtend a larger angle appear larger to the eye.** It's also possible, as Figure 19.16b shows, for objects with different actual sizes to have the same angular size and thus the same apparent size.

Using a Magnifier

There are many ways to use a magnifier. You can hold the lens close to or far from the object, and you can bring your eye right up to the lens or hold it somewhat further away. However, there is a simple case that reflects the way we usually use a magnifier. The lens is held such that the object is at or just inside the lens's focal point. As shown graphically in Figure 19.17, this produces a virtual image that is quite far from the lens. Your eye, looking through the lens, "sees" the virtual image. This is a convenient image location, because your eye's muscles are fully relaxed when looking at a distant image. Thus you can use the magnifier in this way for a long time without eye strain.

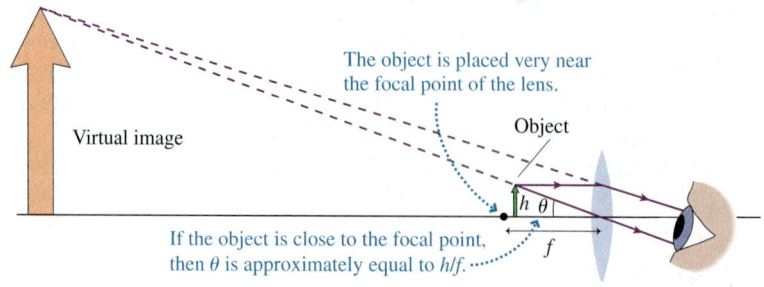

Virtual image

The object is placed very near the focal point of the lens.

Object

If the object is close to the focal point, then θ is approximately equal to h/f.

h θ

f

FIGURE 19.17 The magnifier.

We can analyze the situation using Figure 19.17. Suppose that the object is almost exactly at the focal point, a distance f from the lens. Then, tracing the ray that goes through the lens's center, we can see that the angular size θ of the image

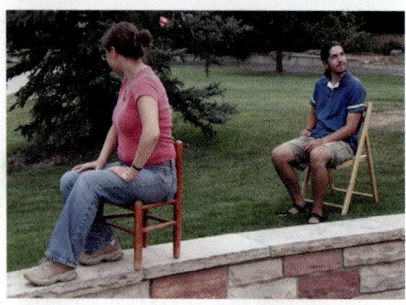

Movie magic? Even if the more distant of two equally sized objects *appears* smaller, we don't usually believe it actually *is* smaller because there are abundant visual clues that tell our brain that it's farther away. If those clues are removed, however, the brain readily accepts the illusion that the farther object is smaller. The technique of *forced perspective* is a special effect used in movies to give this illusion. Here, Camelia, who is actually closer to the camera, looks like a giant compared to Kevin. The lower photo shows how the trick was done.

is such that $\tan\theta = h/f$. If this is a fairly small angle, which it usually is, we can use the small-angle approximation $\tan\theta \approx \theta$ to write

$$\theta \approx \frac{h}{f} \qquad (19.7)$$

Is this an improvement over using no magnifier? With no magnifier, the object has its largest angular size when we bring it as close as possible to the eye. The closest position at which we can focus is the near point of the eye, as shown in Figure 19.18. The object at the near point would have angular size

$$\theta_0 \approx \frac{h}{25\ \text{cm}}$$

where we have taken the conventional value of 25 cm as the near-point distance. Thus the angular size θ when using the magnifier is greater than that without the magnifier by a factor of

$$M = \frac{\theta}{\theta_0} = \frac{h/f}{h/25\ \text{cm}} = \frac{25\ \text{cm}}{f} \qquad (19.8)$$

where, for this calculation, the focal length f must be in cm. M is called the **angular magnification** of the magnifier. With a lens of short focal length it is possible to get magnifications as high as about 20.

FIGURE 19.18 Angular size without a magnifier.

CONCEPTUAL EXAMPLE 19.2 The angular size of a magnified image

An object is placed right at the focal point of a magnifier. How does the apparent size of the image depend on where the *eye* is placed relative to the lens?

Eye close to magnifier   Eye far from magnifier

FIGURE 19.19 Viewing a magnifier with the object at its focus.

REASON When the object is precisely at the focal point of the lens, Equation 19.7 holds exactly. The angular size is equal to *h/f independent* of the position of the eye. Thus the object's apparent size is independent of the eye's position as well. Figure 19.19 shows a calculator at the focal point of a magnifier. The apparent size of the COS button is the same whether the camera taking the picture is close to or far from the lens.

ASSESS When the object is at the magnifier's focus, we've seen that the image is at infinity. The situation is similar to observing any "infinitely" distant object, such as the moon. If you walk closer to or farther from the moon, its apparent size doesn't change at all. The same holds for a virtual *image* at infinity: Its apparent size is independent of the point from which you observe it.

STOP TO THINK 19.5   A student tries to use a *diverging* lens as a magnifier. She observes a coin placed at the focal point of the lens. She sees

A. An upright image, smaller than the object.
B. An upright image, larger than the object.
C. An inverted image, smaller than the object.
D. An inverted image, larger than the object.
E. A blurry image.

Eyepiece

Prism (bends light path so that eyepiece is at a comfortable angle)

Objective lens

Stage (moves up and down to focus sample)

Illuminator

Focus knob

FIGURE 19.20 A microscope.

19.5 The Microscope

To get higher magnifications than are possible using a simple magnifier, a *microscope* is used. A microscope, whose major parts are shown in Figure 19.20, attains a magnification of up to 1000 by a *two-stage* magnification process. A specimen to be observed is placed on the *stage* of the microscope, directly beneath the **objective lens** (or simply the **objective**), a converging lens with a relatively short focal length. The objective creates a magnified real image that is further enlarged by the **eyepiece,** a lens used as an ordinary magnifier. In most modern microscopes a prism bends the path of the rays from the object so that the eyepiece can be held at a comfortable angle. However, we'll consider a simplified version of a microscope without a prism. The light then travels along a straight tube.

Let's examine the magnification process in more detail. In Figure 19.21 we draw a microscope tilted horizontally. The object distance is just slightly larger than the focal length f_o of the objective lens, so the objective forms a highly magnified real image of the object at a distance $s' = L$. This distance, known as the **tube length,** has been standardized for most biological microscopes at $L = 160$ mm. Most microscopes, such as the one shown in Figure 19.20, are focused by moving the sample stage up and down, using the focusing knob, until the object distance is correct for placing the image at L.

Objective lens

Object

f_o f_o

The position of the object is adjusted so that the image is a distance L from the objective lens.

The object is placed just beyond the focal point.

Real image

The eyepiece lens acts as a magnifier with the image as its object.

L = tube length

f_e f_e

FIGURE 19.21 A horizontal view of the optics in a microscope.

Image

Objective

Object

FIGURE 19.22 A microscope objective.

From the magnification equation, Equation 19.4, the magnification of the objective lens is

$$M_o = -\frac{s'}{s} \approx -\frac{L}{f_o} \qquad (19.9)$$

Here we used the fact that the image distance s' is equal to the tube length L and the object distance s is very close to the focal length f_o of the objective. The minus sign tells us that the image is inverted with respect to the object.

Figure 19.22 shows how this works for a real objective lens that was removed from a microscope. The object, a dime, is placed just beyond the focal point of the objective. The objective projects a real image onto the translucent screen, where you can see the enlarged "tails" side of the dime. In an actual microscope, the objective would be closer to the object, and the image would be further away and larger.

EXAMPLE 19.9 Finding the focal length of a microscope objective

A biological microscope objective is labeled "20×." What is its focal length?

PREPARE The "20×" means that the objective has a magnification M_o of −20. We can use Equation 19.9 with L as 160 mm, which we've seen is the standard length for a biological microscope.

SOLVE From Equation 19.9 we have

$$f_o = -\frac{L}{M_o} = -\frac{160 \text{ mm}}{-20} = 8.0 \text{ mm}$$

ASSESS Microscope objectives are specified by their magnification, or "power," not by their focal length. Equation 19.9 relates these two important specifications.

The image created by the objective is a *real* image. The rays converge to a focus in the image plane (which is why we can place a screen there to view the image), then diverge from the image just as they would from an actual object. If you look into a microscope with the eyepieces removed, you can actually see this image seemingly floating just inside the eyepiece tube. This image—now acting as the object for the next lens in a lens combination—is viewed by using the eyepiece lens as a simple magnifier. Recalling from Equation 19.8 that the magnification of a magnifier is

$$M_e = \frac{25 \text{ cm}}{f_e}$$

where f_e is the focal length of the eyepiece, we get the total magnification of a microscope:

$$M = M_o M_e = -\frac{L}{f_o} \frac{25 \text{ cm}}{f_e} \qquad (19.10)$$

Again, the minus sign indicates an inverted image.

Many microscopes have a set of objectives that can be pivoted into place to change the overall magnification. A complete set of objectives might include 5×, 10×, 20×, 40×, and 100×. Eyepieces are also specified by their magnification, and are available with magnifications in the range of 10× to 20×. With these lenses, the lowest magnification available would be 5 × 10 = 50×, while the highest magnification would be 100 × 20 = 2000×.

EXAMPLE 19.10 Magnifying blood cells

A pathologist inspects a sample of 7-μm-diameter human blood cells under a microscope. She selects a 40× objective and a 10× eyepiece. What is the magnification of the microscope? What size object, sitting on her desk, has the same apparent size as a blood cell seen through the microscope?

PREPARE The magnification of the microscope is the product of the magnifications of the objective and the eyepiece.

SOLVE The microscope's magnification is

$$M = M_o M_e = -40 \times 10 = -400$$

where we've used the fact that the objective's magnification is negative. A total magnification of −400 means that the final image is 400 times the size of the object and inverted.

Multiplying the size of a blood cell, 7 μm, by the microscope's magnification gives the size of the object on her desk (as observed from 25 cm) that has the same apparent size. Thus a blood cell appears to be as large as an unmagnified object of diameter

$$d = 400(7 \text{ μm}) \approx 3 \text{ mm}$$

ASSESS 3 mm is a bit bigger than a capital O on this page. Looking at a blood cell through the microscope is similar to looking at the letter O from a distance of 25 cm.

STOP TO THINK 19.6   A biologist rotates the turret of a microscope to replace the 20× objective with a 10× objective. To keep the magnification the same, the focal length of the eyepiece must

A. Be doubled.   B. Be halved.   C. Remain the same.
D. The magnification cannot stay the same if the objective power is changed.

19.6 The Telescope

The microscope magnifies small objects that can be placed near its objective lens. A *telescope* is used to magnify distant objects. The two-lens arrangement, shown in Figure 19.23 on the next page, is similar to that of the microscope, but the objective lens has a long focal length instead of the very short focal length

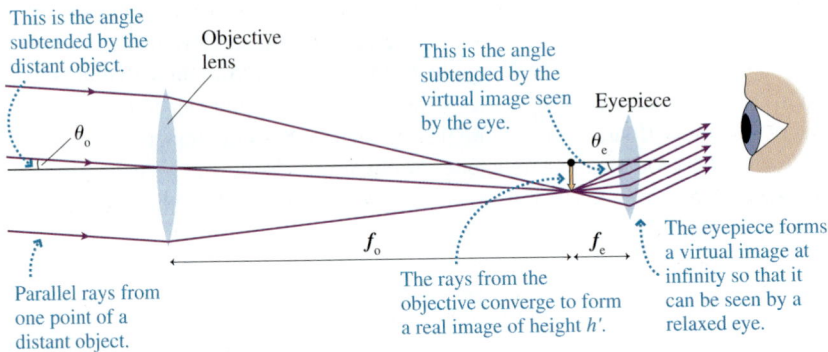

This is the angle subtended by the distant object.

Objective lens

This is the angle subtended by the virtual image seen by the eye.

Eyepiece

θ_o

θ_e

f_o

f_e

The eyepiece forms a virtual image at infinity so that it can be seen by a relaxed eye.

The rays from the objective converge to form a real image of height h'.

Parallel rays from one point of a distant object.

FIGURE 19.23 The telescope.

A clearer view The performance of telescopes on the earth is limited by the atmosphere. Even at night, the atmosphere glows faintly, interfering with the long exposures needed to photograph faint astronomical objects. Further, atmospheric turbulence—visible to the naked eye as the twinkling of stars—obscures the finest details of the object being observed. Because of this, the Hubble telescope orbits the earth high above the atmosphere. With its 2.4-m-diameter mirror, it has produced some of the most spectacular images of astronomical objects, such as this gas cloud surrounding the star V838 Monocerotis.

of a microscope objective. Because the object is very far away ($s \approx \infty$), the converging objective lens forms a real image of the distant object at the lens's focal point. A second lens, the eyepiece, is then used as a simple magnifier to enlarge this real image for final viewing by the eye.

We can use Figure 19.23 to find the magnification of a telescope. The original object subtends an angle θ_o. Because the object is distant, its image is formed in the focal plane of the objective lens, a distance f_o from the objective. From the geometry of Figure 19.23, the height of this image (negative because it's inverted) is

$$h' \approx -f_o\theta_o$$

where we have used the small-angle approximation $\tan\theta \approx \theta$. This image is now the object for the eyepiece lens, which functions as a magnifier. The height of the "object" it views is h', so, from Equation 19.7, the angular size θ_e of the virtual image formed by the eyepiece is

$$\theta_e = \frac{h'}{f_e} = \frac{-f_o\theta_o}{f_e}$$

where f_e is the focal length of the eyepiece. The telescope's angular magnification is the ratio of the angular size seen when looking through the telescope to that seen without the telescope, so we have

$$M = \frac{\theta_e}{\theta_o} = -\frac{f_o}{f_e} \qquad (19.11)$$

The minus sign indicates that you see an upside-down image when looking through a simple telescope. This is not a problem when looking at astronomical objects, but it could be disconcerting to a bird watcher. More sophisticated telescope designs produce an upright image.

To get a large magnification with a telescope, the focal length of the objective should be large and that of the eyepiece small. Contrast this with the magnification of a microscope, Equation 19.10, which is large when the focal lengths of both objective and eyepiece are small.

The telescope shown in Figure 19.23 uses a lens as its objective and hence is known as a *refracting telescope*. It is also possible to make a *reflecting telescope* with a concave mirror instead of a lens, as shown in Figure 19.24. One problem with this arrangement, for small telescopes, is that the image is formed in front of the mirror where it's hard to magnify with an eyepiece without getting one's head in the way. Newton, who built the first such telescope, used a small angled plane mirror, called a *secondary mirror*, to deflect the image to an eyepiece on the side of the telescope.

For large telescopes, such as are used in astronomy, mirrors have several important advantages over lenses. First, objectives of astronomical telescopes

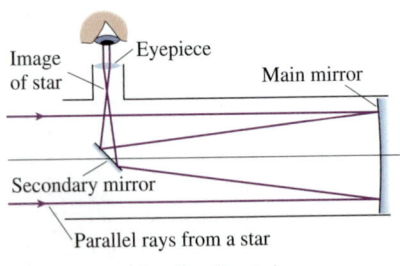

Image of star

Eyepiece

Main mirror

Secondary mirror

Parallel rays from a star

FIGURE 19.24 A reflecting telescope.

must be quite large in order to gather as much light from faint objects as possible. The Subaru Telescope in Hawaii is the world's largest single-mirror telescope; its mirror has a diameter of 8.3 m (27 ft)! A giant lens of this diameter would sag under its own weight. A mirror, on the other hand, can be supported along its entire back surface. Second, mirrors are free from chromatic aberration, the tendency that lenses have of splitting light into its constituent colors, as we'll see in the next section.

19.7 Resolution of Optical Instruments

Suppose you wanted to study the *e. coli* bacterium. It's quite small, about 2 μm long and 0.5 μm wide. You might imagine that you could pair a 150× objective (the highest magnification available) with a 25× eyepiece to get a total magnification of 3750! At that magnification, the *e. coli* would appear about 8 mm across—about the size of Lincoln's head on a penny—with much fine detail revealed. But if you tried this, you'd be disappointed. Although you would see the general shape of a bacterium, you wouldn't be able to make out any real details. Optical instruments are *limited* in the details they can observe. Some limits are practical: Real lenses are imperfect, suffering from **aberrations.** But even a perfect lens has a fundamental limit to the smallest details that can be seen. As we'll see, this limit is set by the diffraction of light, and so is intimately related to the wave nature of light itself. Together, lens aberrations and diffraction set a limit on an optical system's **resolution**—its ability to make out the fine details of an object.

16.8

Aberrations

Consider the simple lens shown in Figure 19.25 imaging an object located at infinity, so that the incoming rays are parallel. An ideal lens would focus all the rays to a single point. However, for a real lens with spherical surfaces, the rays that pass near the lens's center come to a focus a bit further from the lens than those that pass near its edge. There is no single focal point; even at the "best" focus the image is a bit blurred. This inability of a real lens to focus perfectly is called **spherical aberration.**

A careful examination of Figure 19.25 shows that the outer rays are most responsible for the poor focus. Consequently, the effects of spherical aberration can be minimized by using an iris diaphragm to pass only rays near the optical axis. This "stopping down" of a lens improves its imaging characteristics at the expense of its light-gathering capabilities. Part of the function of the iris of the human eye is to improve vision in this way. Our vision is poorer at night with the iris wide open—but our ancestors probably avoided many a predator with this poor but sensitive night vision.

As we learned in Chapter 18, glass has *dispersion.* That is, the index of refraction of glass varies slightly with wavelength. The higher a lens's index of refraction, the more it bends incoming light rays. Because the index of refraction for violet light is larger than for red light, a lens's focal length is slightly shorter for violet light than for red light. Consequently, different colors of light come to a focus at slightly different distances from the lens. If red light is sharply focused on a viewing screen, then blue and violet wavelengths are not well focused. This imaging error, illustrated in Figure 19.26, is called **chromatic aberration.**

The rays don't converge to a straight point.

FIGURE 19.25 Spherical aberration.

Different wavelengths focus at different points.

FIGURE 19.26 Chromatic aberration.

Correcting Aberrations

Single lenses always have aberrations of some kind. For high-quality optics, such as those used in microscopes or telescopes, the aberrations are minimized by using a careful *combinations* of lenses. An important example is the

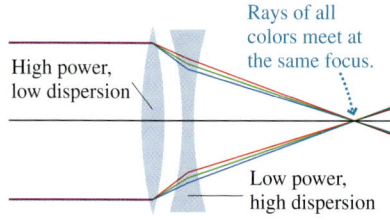

FIGURE 19.27 An achromatic lens.

achromatic doublet (achromatic = "without color"), two lenses used in combination to greatly reduce chromatic aberration. Figure 19.27 shows how this works. A converging lens is paired with a weaker diverging lens; the combination has an overall positive refractive power and so is converging. However, the glasses are chosen so that the weaker diverging lens has a greater dispersion than the stronger converging lens. In this way the colors of white light, separated at first by the converging lens, are brought back together by the diverging lens. Achromatic doublets also minimize spherical aberration. Real microscope objectives are even more complex, but are based on the same principle as the achromatic doublet.

◀ **Eyeglasses in space**   After launch, it was discovered that the mirror of the Hubble space telescope had been ground to the wrong shape, giving it severe spherical aberration. In a later service mission, corrective optics—in essence, very high-tech glasses—were put in place to correct for this spherical aberration. The photos to the left show an image of a galaxy before and after the corrective optics were added.

Resolution and the Wave Nature of Light

Modern lenses can be well-corrected for aberrations, so they might be expected to focus perfectly. According to the ray model of light, a perfect lens should focus parallel rays to a single point in the focal plane. However, we've already hinted that there's a more fundamental limit to the performance of an optical instrument, a limit set by the wave nature of light.

Figure 19.28a shows a plane wave being focused by a lens of diameter D. Only those waves passing *through* the lens can be focused, so the lens acts like a circular aperture of diameter D in an opaque barrier. In other words, the lens both focuses *and diffracts* light waves. Figure 19.28b separates these two effects by modeling a real lens as an "ideal" diffractionless lens behind a circular aperture of diameter D.

(a) A lens acts as a circular aperture.

(b) The aperture and focusing effects can be separated.

(c) The lens produces a diffraction pattern in the focal plane.

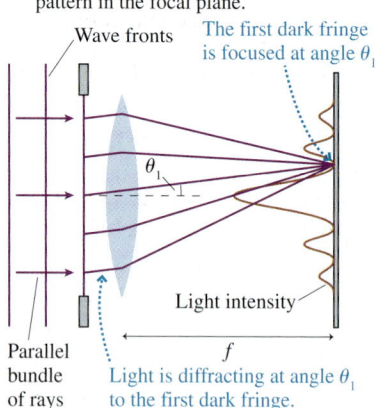

FIGURE 19.28 A lens both focuses and diffracts the light passing through.

You learned in Chapter 17 that a circular aperture produces a diffraction pattern with a bright central maximum surrounded by dimmer circular fringes. A converging lens brings parallel light rays to a focus at distance f. Consequently, as Figure 19.28c shows, a lens behind a circular aperture collects all the light rays diffracting at angle θ and brings these rays together in the focal plane of the lens. Consequently, the image of a parallel bundle of rays is not a perfect point but, instead, a circular diffraction pattern. The width of the diffraction pattern may be very small, but it's not zero and thus it imposes a fundamental limit on how sharp an image can be.

To understand the significance of the diffraction of light by a lens, suppose you point a telescope at two stars near to each other in a galaxy far, far away. If you use the best possible detector, will you be able to distinguish separate images for the two stars, or will they blur into a single blob of light? A similar question could be asked of a microscope. Can two microscopic objects, very close together, be distinguished if sufficient magnification is used? Or is there some size limit at which they will blur together and never be separated? These are important questions about the resolution of optical instruments.

Because of diffraction, the telescope image of a distant star is not a point but a circular diffraction pattern. Our question, then, really amounts to asking how close together two diffraction patterns can be before you can no longer distinguish them. One of the major scientists of the 19th century, Lord Rayleigh, studied this problem and suggested a reasonable rule.

Figure 19.29 shows two objects being imaged by a lens of diameter D. The angular separation between the objects, as seen from the lens, is α. Because of diffraction, each image is a circular diffraction pattern rather than a perfect point. From Chapter 17, recall that the angle to the first dark minimum in a circular diffraction pattern is $\theta_1 = 1.22\lambda/D$. Rayleigh noted that two overlapped diffraction patterns can just barely be resolved if the central maximum of one pattern falls on the first dark minimum of the other pattern. This will be true if the angular separation between the objects is $\alpha = \theta_1$. Thus **Rayleigh's criterion** states that

- The two objects are resolvable if their angular separation is $\alpha > \theta_1$, where $\theta_1 = 1.22\lambda/D$ is the angle of the first dark fringe in the circular diffraction pattern.
- The two objects are not resolvable if $\alpha < \theta_1$ because their diffraction patterns are too overlapped.
- The two objects are marginally resolvable if $\alpha = \theta_1$. The central maximum of one image falls exactly on top of the first dark fringe of the other image. This is the situation shown in Figure 19.29.

Figure 19.30 shows how two closely spaced stars would appear in a telescope. The images are circular diffraction patterns, not points. In the top picture the angular separation α between the stars is greater than θ_1. The stars are easily resolved as two distinct objects. In the middle picture, where $\alpha = \theta_1$, you can see that the two images are just barely resolved. Finally, the two stars in the lower picture are closer together than θ_1, so the two stars appear as one.

For telescopes, the angle $\theta_1 = 1.22\lambda/D$ is called the *angular resolution* of the telescope. The angular resolution is a function of the lens diameter and the wavelength; the magnification is not a factor. Two overlapped, unresolved images will remain overlapped and unresolved no matter what the magnification. For visible light, where λ is pretty much fixed, the only parameter over which the astronomer has any control is the diameter of the lens or mirror of the telescope. The urge to build ever larger telescopes is motivated, in part, by a desire to improve the angular resolution. (Another important motivation is to increase the light-gathering power so as to see objects farther away.)

The Resolution of a Microscope

A microscope differs from a telescope in that it magnifies objects that are very close to the lens, not far away. Nonetheless, the wave nature of light still sets a limit on the ultimate resolution of a microscope. In analogy with a telescope, a

Maximum of image 2 falls on first dark fringe of image 1. The images are marginally resolved.

FIGURE 19.29 Two images that are marginally resolved.

(a) Stars completely resolved

(b) Stars just resolved

(c) Stars not resolved

FIGURE 19.30 The resolution of a telescope.

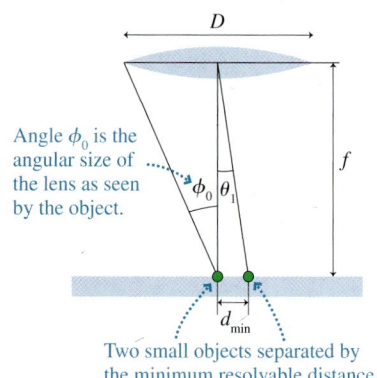

Angle ϕ_0 is the angular size of the lens as seen by the object.

D

f

ϕ_0 θ_1

d_{min}

Two small objects separated by the minimum resolvable distance.

FIGURE 19.31 The resolution of a microscope.

microscope with an objective lens of diameter D should marginally resolve two objects separated by angle $\theta_1 = 1.22\lambda/D$. As shown in Figure 19.31, two objects separated by angle θ_1 are separated by an actual distance

$$d_{min} \approx f\theta_1 = \frac{1.22\lambda f}{D} \tag{19.12}$$

where we've used the small-angle approximation and the fact that $s \approx f$ for a microscope. Because θ_1 is the minimum resolvable angle, d_{min} is the minimum resolvable *distance* between two objects being studied.

Referring again to Figure 19.31, we see that the lens diameter can be written in terms of the angle ϕ_0 as $D = 2f\tan\phi_0$. Consequently, we can write Equation 19.12 as

$$d_{min} = \frac{1.22\lambda f}{2f\tan\phi_0} = \frac{0.61\lambda}{\tan\phi_0} \tag{19.13}$$

This is *nearly* correct for a microscope. Because a microscope focuses rays from a nearby object rather than parallel rays from a distant object, a more thorough analysis finds that the $\tan\phi_0$ in Equation 19.13 should be replaced with $\sin\phi_0$. Thus we have

$$d_{min} = \frac{0.61\lambda}{\sin\phi_0} \tag{19.14}$$

There are two things worth noting about Equation 19.14. First, the angle ϕ_0 is a characteristic of the *lens*. A bigger lens, or one with a smaller focal length, will have a larger value of ϕ_0. Second, λ is the wavelength of light as it travels through the medium between the object and the lens. Often this is air, but biologists sometimes place a transparent fluid, such as oil, in this space. (This is called an *oil immersion microscope*.) Then $\lambda = \lambda_0/n$, where λ_0 is the wavelength in air or vacuum and n is the index of refraction of the fluid. In this case we can write

$$d_{min} = \frac{0.61\lambda_0}{n\sin\phi_0} \tag{19.15}$$

The quantity $n\sin\phi_0$ is called the **numerical aperture** NA of the objective when immersed in a fluid of index n. For a microscope whose objective has numerical aperture NA, the minimum resolvable distance, also called the **resolving power** RP, is

$$\text{RP} = d_{min} = \frac{0.61\lambda_0}{\text{NA}} \tag{19.16}$$

Resolving power of a microscope with numerical aperture NA

Magnification

Tube length

Cover glass thickness range

Cover glass adjustment gauge

Nikon
Plan Apo
60x / 0.95
DIC M
∞/0.11-0.23 WD 0.15

11 14 17 20 23

Numerical aperture

Correction collar

The anatomy of a microscope objective
The two most important specifications of a microscope objective, its magnification and its numerical aperture, are prominently displayed on its barrel. Other important information is shown as well. This objective is designed to project its real image at a tube length of infinity (∞), instead of at the standard distance of 160 mm. Extra lenses in the microscope move this image to just in front of the eyepiece. Many biological studies are conducted through a cover glass. This cover glass can introduce spherical aberration, blurring the image. By turning the correction collar, you can adjust this objective to correct for the exact thickness of the cover glass used.

The smaller the resolving power, the *better* the objective is at seeing small details.

In principle, it would appear from Equation 19.16 that the resolving power of a microscope could be made as small as desired simply by increasing the numerical aperture. But there are rather severe practical limits on how high the numerical aperture can be made. The highest possible numerical aperture for a high-magnification 100× objective used in air is about 0.95. For an oil-immersion objective, the numerical aperture might be as high as 1.3. With such an objective, the resolving power would be

$$\text{RP} \approx 0.5\lambda_0$$

This illustrates the fundamental fact that **the minimum resolving power of a microscope, and thus the size of the smallest detail observable, is about half the wavelength of light.** This is a *fundamental limit* set by the wave nature of light. For $\lambda \approx 400$ nm, at the short-wavelength edge of the visible spectrum, the resolving power is RP ≈ 200 mn.

Figure 19.32 shows an actual micrograph of the *e. coli* bacillus mentioned at the beginning of this section. Their width is about equal to the wavelength of light in the middle of the spectrum, about 500 nm, and the smallest resolved features are about half this. A higher magnification would not reveal any more detail because this micrograph is at the resolution limit set by the diffraction of light.

In contrast, the **electron microscope** micrograph of *e. coli* shows a wealth of detail unobservable in the optical picture. In Chapter 28 we'll find out why the resolving power of an electron microscope is so much smaller than that of an optical microscope.

Optical microscope

Electron microscope

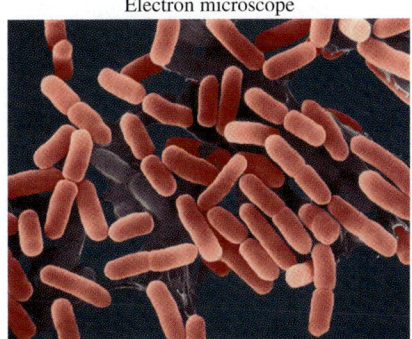

FIGURE 19.32 Optical and electron micrographs of *e. coli*.

EXAMPLE 19.11 Finding the resolving power of a microscope

A microscope objective lens has a diameter of 6.8 mm and a focal length of 4.0 mm. For a sample viewed in air, what is the resolving power of this objective in red light? In blue light?

PREPARE We can use Equation 19.16 to find the resolving power. We'll need the numerical aperture of the objective, given as NA $= n\sin\phi_o$.

SOLVE From the geometry of Figure 19.31,

$$\tan\phi_o = \frac{D/2}{f} = \frac{3.4 \text{ mm}}{4.0 \text{ mm}} = 0.85$$

from which $\phi_o = \tan^{-1}0.85 = 40.4°$ and $\sin\phi_o = \sin 40.4° = 0.65$. Hence the numerical aperture is (since $n = 1$ in air)

$$NA = n\sin\phi_o = 1 \times 0.65 = 0.65$$

Then, from Equation 19.16, the resolving power is

$$RP = \frac{0.61\lambda_0}{0.65} = 0.94\lambda_0$$

Wavelengths of different colors of light were given in Table 18.1. For red light, with $\lambda_0 = 650$ nm, RP $= 610$ nm, while blue light, with $\lambda_0 = 450$ nm, has RP $= 420$ nm.

ASSESS We see that shorter wavelength light yields a higher resolution (smaller RP). Unfortunately, wavelengths much shorter than 400 nm are invisible, and glass lenses are opaque to light of very short wavelength.

STOP TO THINK 19.7   Four lenses are used as microscope objectives, all for light with the same wavelength λ. Rank in order, from largest to smallest, the resolving powers RP_1 to RP_4 of the lenses.

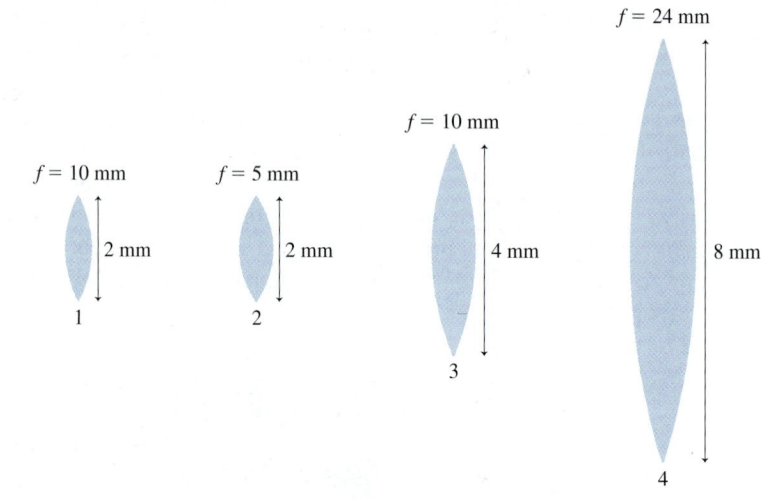

In the last three chapters, we have used the wave and ray models of light to understand a wide variety of optical phenomena and have learned how several important optical instruments function. We have mentioned that light is an "electromagnetic wave," but we don't know yet just what such a wave *is*. In Part VI of this book, to which we now turn, we'll begin our study of electricity and magnetism, a study that will culminate in the discovery of just how light, electricity, and magnetism are linked together.

SUMMARY

The goal of Chapter 19 has been to understand how common optical instruments work.

GENERAL PRINCIPLES

Finding the Image of a Lens or Mirror

For a lens or curved mirror, the object distance s, the image distance s', and the focal length f are related by the thin-lens equation:

$$\frac{1}{s} + \frac{1}{s'} = \frac{1}{f}$$

Sign conventions for the thin-lens equation are:

Quantity	Positive when	Negative when
s	Always	Not treated here
s'	*Real* image, on opposite side of a lens from object, or in front of a mirror.	*Virtual* image, on same side of a lens as the object, or behind a mirror.
f	Converging lens or concave mirror	Diverging lens or convex mirror.
h', M	Image is upright	Image is inverted

The magnification of a lens or mirror is

$$M = -\frac{h'}{h} = -\frac{s'}{s}$$

Resolution of Optical Instruments

The **resolution** of a telescope or microscope is limited by imperfections, or **aberrations,** in the optical elements, and by the more fundamental limits imposed by diffraction.

For a *microscope,* the minimum resolvable distance between two objects is

$$d_{min} = \frac{0.61\lambda_0}{\text{NA}}$$

For a *telescope,* the minimum resolvable angular separation between two objects is

$$\theta_1 = \frac{1.22\lambda}{D}$$

IMPORTANT CONCEPTS

Lenses in combination

When two lenses are used in combination, the image from the first lens serves as the object for the second.

The **refractive power** P of a lens is the inverse of its focal length: $P = 1/f$. When two lenses are in contact, their net refractive power is the sum of their individual refractive powers.

Angular and apparent size

Both objects have the same angular size, and hence the same apparent size.

Same apparent size

Same angular size

APPLICATIONS

The camera and the eye

Both the camera and the eye work by focusing an image on a light-sensitive surface.

Film or CCD

Retina

Light-sensitive surface

Lens

Light-focusing element

Cornea, lens, aqueous humor

The camera focuses by changing the lens-film distance while the eye focuses by changing the focal length of its lens.

The telescope

The telescope magnifies distant objects. The objective lens creates a real image of the distant object. This real image is then further magnified by the eyepiece lens, which acts as a simple magnifier. The angular magnification is $M = -f_o/f_e$.

Rays from distant object

Objective lens

Real image

Eyepiece

f_o

f_e

The magnifier

Without a lens, an object cannot be viewed closer than the eye's near point of ≈ 25 cm. Its angular size θ_0 is $h/25$ cm.

Near point

θ_0

h

25 cm

If the object is now placed at the focal point of a converging lens, its angular size is increased to $\theta = h/f$.

θ

f

The angular magnification is $M = \theta/\theta_0 = 25$ cm$/f$.

The microscope

The microscope magnifies a small, nearby object. The objective lens creates a real image of the object. This real image is then further magnified by the eyepiece lens, which acts as a simple magnifier. The angular magnification is

$$-\frac{L \times 25 \text{ cm}}{f_o f_e}$$

f_o

Objective

Eyepiece

Real image

Tube length L

f_e

For instructor-assigned homework, go to www.masteringphysics.com

Problem difficulty is labeled as | (straightforward) to ||||| (challenging).

Problems labeled INT integrate significant material from earlier chapters; BIO are of biological or medical interest.

QUESTIONS

Conceptual Questions

1. What combination of two thin lenses in contact would have a total refractive power equal to zero? Would this combination of lenses form an image? Explain.

2. An object at distance s from a concave mirror of focal length f produces a real image at distance s' from the mirror. Suppose the mirror is replaced by a new mirror, at the same location, with focal length $\frac{1}{2}f$. Will the new image be real or virtual? Will its distance from the mirror be more or less than s'? Explain.

3. A lens can be used to start a fire by focusing an image of the sun onto a piece of flammable material. All other things being equal, would a lens with a short focal length or a long focal length be better as a fire starter? Explain.

4. A photographer focuses his camera on his subject. The subject then moves closer to the camera. To refocus, should the lens be moved closer to or farther from the film? Explain.

5. Many cameras have a *zoom lens.* This is a lens whose focal length and distance from the film can be varied. If the camera's exposure is correct when the lens has a focal length of 8.0 mm, will it be overexposed, underexposed, or still correct when the focal length is increased to 16.0 mm (assume the lens diameter remains constant)? Explain.

6. A nature photographer taking a close-up shot of an insect replaces the standard lens on his camera with a lens that has a shorter focal length and is positioned farther from the film. Explain why he does this.

7. All humans have what is known as a *blind spot,*
BIO where the optic nerve exits the eye and no light-sensitive cells exist. To locate your blind spot, look at the figure of the cross. Close your left eye and place your index finger on the cross. Slowly move your finger to the left while following it with your right eye. At a certain point the cross will disappear. Is your right eye's blind spot on the right or left side of your retina? Explain.

8. Suppose you wanted special glasses designed to wear underwater, without a face mask. Should the glasses use a converging or diverging lens in order for you to be able to focus under water? Explain.

9. An 8-year-old child and a 75-year-old man both use the same
BIO magnifier to observe a bug. For whom does the magnifier more likely have the higher magnification? Explain.

10. A friend lends you the eyepiece of his microscope to use on your own microscope. He claims that since his eyepiece has the same diameter as yours but twice the focal length, the resolving power of your microscope will be doubled. Is his claim valid? Explain.

11. An astronomer is using a telescope to observe two distant stars. The stars are marginally resolved when she looks at them through a filter that passes green light near 550 nm. Which of the following actions would improve the resolution? Assume that the resolution is not limited by the atmosphere.
   a. Changing the filter to a different wavelength? If so, should she use a shorter or a longer wavelength?
   b. Using a telescope with an objective lens of the same diameter but a different focal length? If so, should she select a shorter or a longer focal length?
   c. Using a telescope with an objective lens of the same focal length but a different diameter? If so, should she select a larger or a smaller diameter?
   d. Using an eyepiece with a different magnification? If so, should she select an eyepiece with more or less magnification?

12. A pair of binoculars has a magnification of 7×. What would be their magnification if you were to look through them the wrong way, that is, through one of their objective lenses instead of the eyepieces?

13. Is the wearer of the glasses in Figure Q19.13 nearsighted or
BIO farsighted? How can you tell?

FIGURE Q19.13

Multiple-Choice Questions

14. | The lens in Figure Q19.14 is used to produce a real image of a candle flame. What is the focal length of the lens?
   A. 9.0 cm   B. 12 cm   C. 24 cm
   D. 36 cm   E. 48 cm

FIGURE Q19.14

15. I A converging lens of focal length 20 cm is used to form a real image 1.0 m away from the lens. How far from the lens is the object?

    A. 20 cm   B. 25 cm   C. 50 cm   D. 100 cm

16. II You look at yourself in a convex mirror. Your image is

    A. Erect.   B. Inverted.
    C. It's impossible to tell without knowing the mirror's focal length.

17. II An object is 50 cm from a diverging lens with a focal length of −20 cm. How far from the lens is the image, and on which side of the lens is it?

    A. 14 cm, on the same side as the object.
    B. 14 cm, on the opposite side from the object.
    C. 30 cm, on the same side as the object.
    D. 33 cm, on the same side as the object.
    E. 33 cm, on the opposite side from the object.

18. II A person looking through a −10 D lens sees an image that appears 8.0 cm from the lens. How far from the lens is the object?

    A. 10 cm   B. 20 cm   C. 25 cm   D. 40 cm

19. II A lens with a focal length of 14.0 cm is placed in contact with one with a focal length of −4.0 cm. The focal length of the combined lens is

    A. −10.0 cm   B. −5.6 cm   C. 10.0 cm   D. 18.0 cm

20. II BIO A nearsighted person has a near point of 20 cm and a far point of 40 cm. What power lens is necessary to correct this person's vision to allow her to see distant objects?

    A. −5.0 D   B. −2.5 D   C. +2.5 D   D. +5.0 D

21. II BIO A nearsighted person has a near point of 20 cm and a far point of 40 cm. When he is wearing glasses to correct his distant vision, what is his near point?

    A. 10 cm   B. 20 cm   C. 40 cm   D. 1.0 m

22. II BIO A 60-year-old man has a near point of 100 cm, making it impossible to read. What power reading glasses would he need to focus on a newspaper held at a comfortable distance of 20 cm?

    A. −6.0 D   B. −4.0 D   C. +4.0 D   D. +6.0 D

PROBLEMS

Section 19.1 Finding the Image of a Lens or Mirror

1. I A 2.0-cm-tall object is 40 cm in front of a converging lens that has a 20 cm focal length. Calculate the image position and height.

2. I A 1.0-cm-tall object is 10 cm in front of a converging lens that has a 30 cm focal length. Calculate the image position and height.

3. I A 2.0-cm-tall object is 15 cm in front of a converging lens that has a 20 cm focal length. Calculate the image position and height.

4. I A 1.0-cm-tall object is 75 cm in front of a converging lens that has a 30 cm focal length. Calculate the image position and height.

5. I A 2.0-cm-tall object is 15 cm in front of a diverging lens that has a −20 cm focal length. Calculate the image position and height.

6. I A 1.0-cm-tall object is 60 cm in front of a diverging lens that has a −30 cm focal length. Calculate the image position and height.

7. I A 3.0-cm-tall object is 15 cm in front of a convex mirror that has a −25 cm focal length. Calculate the image position and height.

8. I A 3.0-cm-tall object is 45 cm in front of a convex mirror that has a −25 cm focal length. Calculate the image position and height.

9. I A 3.0-cm-tall object is 15 cm in front of a concave mirror that has a 25 cm focal length. Calculate the image position and height.

10. I A 3.0-cm-tall object is 45 cm in front of a concave mirror that has a 25 cm focal length. Calculate the image position and height.

11. II Rank in order, from most positive to most negative, the image heights in the following situations:

    A. A 1.5-cm-tall object is 10 cm from a converging lens of focal length 12 cm.
    B. A 1.5-cm-tall object is 16 cm from a converging lens of focal length 12 cm.
    C. A 2.0-cm-tall object is 16 cm from a converging lens of focal length 12 cm.
    D. A 2.0-cm-tall object is 18 cm from a converging lens of focal length 12 cm.

12. II Rank in order, from most positive to most negative, the image heights in the following situations:

    A. A 1.0-cm-tall object is 8.0 cm from a converging lens of focal length 10 cm.
    B. A 1.5-cm-tall object is 7.0 cm from a diverging lens of focal length −10 cm.
    C. A 2.0-cm-tall object is 15 cm from a converging lens of focal length 13 cm.
    D. A 2.5-cm-tall object is 16 cm from a diverging lens of focal length −13 cm.

13. II Rank in order, from closest to farthest, the distances between the object and the image in the following situations:

    4 A. An object is 20 cm from a concave mirror of focal length 18 cm.
    2 B. An object is 20 cm from a concave mirror of focal length 25 cm.
    1 C. An object is 20 cm from a convex mirror of focal length −18 cm.
    3 D. An object is 20 cm from a plane mirror.

CD BA

Section 19.2 The Camera

14. ▮▮▮ In Figure P19.14 the camera lens has a 50 mm focal length. How high is the man's well-focused image on the film?

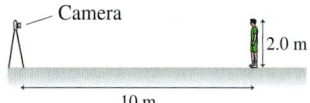

FIGURE P19.14

15. ▮▮ A camera takes a perfectly exposed picture when the lens diaphragm is set to $f/4$ and the shutter speed is 1/250 s. If the diaphragm is changed to $f/11$, what should the new shutter speed be so that the exposure is still correct? (Standard camera shutter speeds include 1/250 s, 1/125 s, 1/60 s, 1/30 s, and 1/15 s.)

16. ▮ A telephoto lens with focal length of 135 mm has f-numbers ranging from $f/2.8$ to $f/22$. What is the diameter of the lens aperture at these two f-numbers?

17. ▮▮ A 50-mm-focal-length camera lens set at its widest aperture of $f/2.0$, together with a 1/125 s shutter speed, gives a perfect exposure. What shutter speed should be used to get the same exposure with a 135 mm telephoto lens set at $f/4$? (Standard camera shutter speeds include 1/250 s, 1/125 s, 1/60 s, 1/30 s, and 1/15 s.)

Section 19.3 The Human Eye

18. ▮ a. Estimate the diameter of your eyeball.

BIO   b. Bring this page up to the closest distance at which the text is sharp—not the closest at which you can still read it, but the closest at which the letters remain sharp. If you wear glasses or contact lenses, leave them on. This distance is called the *near point* of your (possibly corrected) eye. Record it.

   c. Estimate the effective focal length of your eye. The effective focal length includes the focusing due to the lens, the curvature of the cornea, and any corrections you wear. Ignore the effects of the fluid in your eye.

19. ▮▮ A farsighted person has a near point of 50 cm rather than
BIO the normal 25 cm. What strength lens, in diopters, should be prescribed to correct this vision problem?

20. ▮ A nearsighted woman has a far point of 300 cm. What kind
BIO of lens, converging or diverging, should be prescribed for her to see distant objects more clearly? What power should the lens have?

21. ▮▮ The relaxed human eye is about 2 cm from front to back. If
BIO the iris of the human eye can be opened to 7 mm at its widest, what is the f-number of the human eye?

22. ▮ The near point for your myopic uncle is 10 cm. Your own vision is normal; that is, your near point is 25 cm. Suppose you and your uncle hold dimes (which are 1.7 cm in diameter) at your respective near points.
   a. What is the dime's angular size, in radians, according to you?
   b. What is the dime's angular size, in radians, according to your uncle?
   c. Do these calculations suggest any benefit to nearsightedness?

23. ▮▮ For a patient, a doctor prescribes glasses with a converging
BIO lens having a power of 4.0 D.
   a. Is the patient nearsighted or farsighted?
   b. If the patient is nearsighted, what is the location of her eye's far point? If she is farsighted, what is the location of her near point?

24. ▮▮ Rank the following people from the most nearsighted to the
BIO most farsighted, indicating any ties:
   A. Bernie has a prescription of +2.0 D.
   B. Carol needs diverging lenses with a focal length of −0.35 m.
   C. Maria Elena wears converging lenses with a focal length of 0.50 m.
   D. Janet has a prescription of +2.5 D.
   E. Warren's prescription is −3.2 D.

Section 19.4 The Magnifier

25. ▮ A magnifier has a magnification of 5×. How far from the lens should an object be placed so that its (virtual) image is at the near-point distance of 25 cm?

26. ▮▮ A farsighted man has a near point of 40 cm. What power lens should he use as a magnifier to see clearly at a distance of 10 cm without wearing his glasses?

Section 19.5 The Microscope

27. ▮▮ An inexpensive microscope has a tube length of 12.0 cm, and its objective lens is labeled with a magnification of 10×.
   a. Calculate the focal length of the objective lens.
   b. What focal length eyepiece lens should the microscope have to give an overall magnification of 150×?

28. ▮ A standard biological microscope is required to have a magnification of 200×.
   a. When paired with a 10× eyepiece, what power objective is needed to get this magnification?
   b. What is the focal length of the objective?

29. ▮ A forensic scientist is using a standard biological microscope with a 15× objective and a 5× eyepiece to examine a hair from a crime scene. How far from the objective is the hair?

Section 19.6 The Telescope

30. ▮▮▮ A researcher is trying to shoot a tranquilizer dart at a 2.0-m-tall rhino that is 150 m away. Its angular size as seen through the rifle telescope is 9.1°. What is the magnification of the telescope?

31. ▮ The objective lens of the refracting telescope at the Lick Observatory in California has a focal length of 57 ft.
   a. What is the refractive power of this lens?
   b. What focal length (mm) eyepiece would give a magnification of 1000× for this telescope?

32. ▮▮▮ You use your 8× binoculars to focus on a yellow-rumped warbler (length 14 cm) in a tree 18 m away from you. What angle (in degrees) does the image of the warbler subtend on your retina?

33. ‖ Your telescope has a 700-mm-focal-length objective and a 26-mm-focal-length eyepiece. One evening you decide to look at the full moon, which has an angular size of 0.52° when viewed with the naked eye.

a. What angle (in degrees) does the image of the moon subtend when you look at it through your telescope?

b. Suppose you decide to take a photograph of the moon using your telescope. You position film so that it captures the image produced by the objective lens. What is the diameter of that image?

Section 19.7 Resolution of Optical Instruments

34. ‖ Two light bulbs are 1.0 m apart. From what distance can these light bulbs be marginally resolved by a small telescope with a 4.0-cm-diameter objective lens? Assume that the lens is limited only by diffraction and $\lambda = 600$ nm.

35. | A 1.0-cm-diameter microscope objective has a focal length of 2.8 mm. It is used in visible light with a wavelength of 550 nm.

a. What is the objective's resolving power if used in air?

b. What is the resolving power of the objective if it is used in an oil-immersion microscope with $n_{oil} = 1.45$?

36. | A microscope with an objective of focal length 1.6 mm is used to inspect the tiny features of a computer chip. It is desired to resolve two objects only 400 nm apart. What diameter objective is needed if the microscope is used in air with light of wavelength 550 nm?

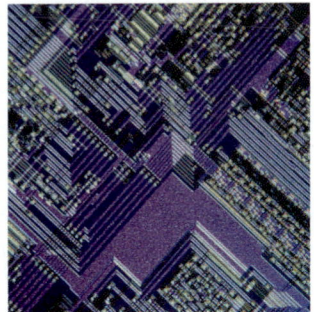

General Problems

37. ‖ A 2.0-cm-diameter spider is 2.0 m from a wall. Determine the focal length and position (measured from the wall) of a lens that will make a half-size image of the spider on the wall.

38. ‖ A 2.0-cm-tall candle flame is 2.0 m from a wall. You happen to have a lens with a focal length of 32 cm. How many places can you put the lens to form a well-focused image of the candle flame on the wall? For each location, what are the height and orientation of the image?

39. ‖ A slide projector needs to create a 98-cm-high image of a 2.0-cm-tall slide. The screen is 300 cm from the slide.

a. What focal length does the lens need? Assume that it is a thin lens.

b. How far should you place the lens from the slide?

40. ‖ Two converging lenses with focal lengths of 40 cm and 20 cm are 10 cm apart. A 2.0-cm-tall object is 15 cm in front of the 40-cm-focal-length lens.

a. Use ray tracing to find the position and height of the image. To do this accurately use a ruler or paper with a grid. Determine the image distance and image height by making measurements on your diagram.

b. Calculate the image position and height. Compare with your ray-tracing answers in part a.

41. | A converging lens with a focal length of 40 cm and a diverging lens with a focal length of −40 cm are 160 cm apart. A 2.0-cm-tall object is 60 cm in front of the converging lens.

a. Use ray tracing to find the position and height of the image. To do this, accurately use a ruler or paper with a grid. Determine the image distance and image height by making measurements on your diagram.

b. Calculate the image position and height. Compare with your ray-tracing answers in part a.

42. | For the combination of two identical lenses shown in Figure P19.42, find the position, size, and orientation of the final image of the 2.0-cm-tall object.

FIGURE P19.42

43. | For the combination of two lenses shown in Figure P19.43, find the position, size, and orientation of the final image of the 1.0-cm-tall object.

FIGURE P19.43

44. ‖ Figure 19.44 shows a meter stick held lengthwise along the optical axis of a concave mirror. How long is the image of the meter stick?

FIGURE P19.44

45. | A lens with a focal length of 25 cm is placed 40 cm in front of a lens with a focal length of 5.0 cm. How far from the second lens is the final image of an object infinitely far from the first lens? Is this image in front of or behind the second lens?

46. | You have a lens with a focal length of 25 cm. You need to combine this lens with another, in contact, to make a combination lens with a focal length of 20 cm. What focal length should you choose for the second lens?

47. ‖ The writing on the passenger-side mirror of your car says "Warning! Objects in mirror are closer than they appear." There is no such warning on the driver's mirror. Consider a typical convex passenger-side mirror with a focal length of −80 cm. A 1.5-m-tall cyclist on a bicycle is 25 m from the mirror. You are 1.0 m from the mirror, and suppose, for simplicity, that the mirror, you, and the cyclist all lie along a line.
    a. How far are you from the image of the cyclist?
    b. How far would you have been from the image if the mirror were flat?
    c. What is the image height?
    d. What would the image height have been if the mirror were flat?
    e. What is the angular size of the image of the cyclist?
    f. What would be the angular size if the mirror were flat?
    g. Why is there a label on the passenger-side mirror?

48. | An object is 4.7 cm from a lens of focal length 5.0 cm.
    a. What is the magnification?
    b. What is the angular magnification of this lens if it is used as a magnifier by a person with a 25 cm near point?

49. | In daylight, the diameter of
BIO the iris is about 3 mm.

Cones   Rods

    a. What is the angle between two distant point objects that can just be resolved? Assume the light's wavelength in air is 600 nm and the index of refraction inside the eye is 1.33.
    b. What is the distance on the retina, 24 mm from the front of the eye, between the images of these two objects?
    c. The distance between *cones,* the color-sensitive cells that cover the retina, is about 2.5 μm. How does this distance compare to the distance between minimally resolved objects you found in part b? Does this comparison make sense from the standpoint of vision?

50. | Your task in physics lab is to make a microscope from two lenses. One lens has a focal length of 10 cm, the other a focal length of 3.0 cm. You plan to use the more powerful lens as the objective, and you want its image to be 16 cm from the lens, as in a standard biological microscope.
    a. How far should the objective lens be from the object to produce a real image 16 cm from the objective?
    b. What will be the magnification of your microscope?

51. | Marooned on a desert island and with a lot of time on your hands, you decide to disassemble your glasses to make a crude telescope with which you can scan the horizon for rescuers. Luckily you're farsighted, and as for most people your two eyes have somewhat different lens prescriptions. Your left eye uses a lens of power +4.5 D and your right eye's lens is +3.0 D.
    a. Which lens should you use for the objective and which for the eyepiece? Explain.
    b. What will be the magnification of your telescope?
    c. Approximately how far apart should the two lenses be when you focus on distant objects?

52. | The 200-inch-diameter objective mirror of the reflecting telescope at the Mt. Palomar Observatory has a focal length of 17 m.
    a. The *f*-number of a mirror is defined exactly the same as the *f*-number of a lens. What is the *f*-number of this mirror?
    b. The *f*-number of the 200-inch telescope is well within the range of *f*-numbers of a cheap camera. So why not just use the camera to take pictures of distant galaxies, instead of constructing this very expensive telescope?

53. ‖ Two stars have an angular separation of 3.3×10^{-6} rad. What diameter telescope objective is necessary to just resolve these two stars, using light with a wavelength of 650 nm?

54. | The planet Neptune is 4.5×10^{12} m from the earth. Its diameter is 4.9×10^{7} m. What diameter telescope objective would be necessary to just barely see Neptune as a disk rather than as a point of light? Assume a wavelength of 550 nm.

55. | What is the angular resolution of the Hubble Space Telescope's 2.4-m-diameter mirror when viewing light with a wavelength of 550 nm? The resolution of a reflecting telescope is calculated exactly the same as for a refracting telescope.

56. ‖ The Hubble Space Telescope has a mirror diameter of 2.4 m. Suppose the telescope is used to photograph stars near the center of our galaxy, 30,000 light years away, using red light with a wavelength of 650 nm.
    a. What is the distance (in km) between two stars that are marginally resolved? The resolution of a reflecting telescope is calculated exactly the same as for a refracting telescope.
    b. For comparison, what is this distance as a multiple of the distance of Jupiter from the sun?

57. | Once dark adapted, the pupil of your eye is approximately
BIO 7 mm in diameter. The headlights of an oncoming car are 120 cm apart. If the lens of your eye is limited only by diffraction, at what distance are the two headlights marginally resolved? Assume the light's wavelength in air is 600 nm and the index of refraction inside the eye is 1.33. (Your eye is not really good enough to resolve headlights at this distance, due both to aberrations in the lens and to the size of the receptors in your retina, but it comes reasonably close.)

58. | The normal human eye has maximum visual acuity with a
BIO pupil size of about 3 mm. For larger pupils, acuity decreases due to increasing aberrations; for smaller pupils, acuity decreases due to increasing effects of diffraction. If your pupil diameter is 2.0 mm, as it would be in fairly bright light, what is the smallest diameter circle that you can barely see as a circle, rather than just a dot, if the circle is at your near point, 25 cm from your eye? Assume the light's wavelength in air is 600 nm and the index of refraction inside the eye is 1.33.

59. | Microtubules are filamentous structures in cells that main-
BIO tain cell shape and facilitate the movement of molecules
within the cell. They are long, hollow cylinders with a diameter of about 25 nm. It is possible to incorporate fluorescent
molecules into microtubules; when illuminated by an ultraviolet light, the fluorescent molecules emit visible light that can
be imaged by the optical system of a microscope. If the emitted light has a wavelength of 500 nm and the NA of the
microscope objective is 1.4, can a biologist looking through
the microscope tell whether she is looking at a single micro-
tubule or at two microtubules lying side by side?

Passage Problems

Surgical Vision Correction BIO

Light that enters your eyes is focused to form an image on your
retina. The optics of your visual system have a total power of
about $+60$ D—about $+20$ D from the lens in your eye and
$+40$ D from the curved shape of your cornea. Surgical procedures to correct vision generally do not work on the lens; they
work to reshape the cornea. In the most common procedures, a
laser is used to remove tissue from the center of the cornea,
reducing its curvature. This change in shape can correct certain
kinds of vision problems.

60. | Flattening the cornea would be a good solution for some-
one who was
A. Nearsighted.
B. Farsighted.
C. Either nearsighted or farsighted.

61. | Suppose a woman has a far point of 50 cm. How much
should the focusing power of her cornea be changed to correct
her vision?
A. -2.0 D   B. -1.0 D   C. $+1.0$ D   D. $+2.0$ D

62. | A *cataract* is a clouding or opacity that develops in the
eye's lens, often in older people. In extreme cases, the lens of
the eye may need to be removed. This would have the effect
of leaving a person
A. Nearsighted.
B. Farsighted.
C. Neither nearsighted nor farsighted.

63. | The length of your eye decreases slightly as you age, mak-
ing the lens a bit closer to the retina. Suppose a man had his
vision surgically corrected at age 30. At age 70, once his eyes
had decreased slightly in length, he would be
A. Nearsighted.
B. Farsighted.
C. Neither nearsighted nor farsighted.

STOP TO THINK ANSWERS

Stop to Think 19.1: A. The thin-lens equation is $1/s + 1/s' = 1/f$.
The focal length of the lens is fixed. Because $1/s$ gets smaller as s
is increased, $1/s'$ must get larger to compensate. Thus s' must
decrease.

Stop to Think 19.2: B and C. The diverging rays could be caused
by two diverging lenses, or a diverging lens combined wth a
weaker converging lens.

Stop to Think 19.3: B. The diameter d of the lens is constant, so
increasing the focal length increases the f-number f/d of the lens.
A lens with a larger f-number needs *more* light for a correct exposure, requiring a *slower* shutter speed.

Stop to Think 19.4: D. From Example 19.7 we know that the
refractive power of the required lens is $P = -1/d_{far}$, where d_{far} is
Maria's far-point distance. We know that d_{far} is farther than 0.5 m
but closer than 10 m. Thus P must be between $-1/10$ m $= -0.1$ D
and $-1/0.5$ m $= -2.0$ D.

Stop to Think 19.5: A. Ray tracing shows why.

Stop to Think 19.6: B. The total magnification is the product of
the objective magnification M_o and the eyepiece magnification M_e.
If M_o is halved, from $20\times$ to $10\times$, M_e must be doubled. Because M_e
is inversely proportional to the eyepiece focal length, the focal
length of the eyepiece must be halved.

Stop to Think 19.7: $RP_1 > RP_4 > RP_2 = RP_3$. The resolving
power is $RP = 0.61\lambda/\sin\phi_0$ for objectives used in air ($n = 1$), so
the resolving power is larger (worse resolution) when the angle ϕ_0
is smaller. From Figure 19.31 you can see ϕ_0 is smaller when the
ratio D/f is smaller. These ratios are $(D/f)_1 = 1/5$, $(D/f)_2 = 2/5$,
$(D/f)_3 = 2/5$, $(D/f)_4 = 1/3$.

Optics

Light is an elusive entity. It is everywhere around us, but exactly what *is* it? One of the more curious aspects of light is that its basic properties depend on the circumstances under which it's studied. Thus it's difficult to develop a single theory of light that applies under all circumstances. Because of this, we have developed two *models* of light in Part V, the wave model and the ray model. We found that each model has its particular realm of applicability.

Many experiments show that light has distinct wave-like properties. Light waves exhibit interference and diffraction, just as water and sound waves do. However, we're usually not aware of the wave aspects of light because the wavelengths of visible light are so small. Wave phenomena become apparent only when light interacts with objects or holes whose size is less than about 0.1 mm.

We can usually ignore the wave nature of light when we consider the propagation of light on larger length scales. In this case, we model light as traveling outward in straight lines, or *rays,* from its source. Light rays change direction at an interface between two media with different indices of refraction (different light speeds). At this interface the rays both reflect, heading back into their original medium, and refract, moving into the new medium but in a new direction. These processes are governed by the laws of reflection and refraction.

Despite light's subtle nature, the practical applications of optics are crucial to many of today's technologies. Cameras, telescopes, and microscopes all employ basic ideas of image formation with lenses and mirrors. We found that the ultimate resolution of an optical instrument is set by the wave nature of light, bringing our study of optics full circle.

KNOWLEDGE STRUCTURE V Optics

BASIC GOALS	What are the consequences of the wave nature of light? In the ray model, how do light rays refract and reflect to form images?
GENERAL PRINCIPLES	Light is understood using two models, the **wave model,** in which light exhibits wave properties such as interference and diffraction, and the **ray model,** in which light travels in straight lines until it reflects or refracts.

Wave model

- Light spreads out when passing through a narrow opening. This is **diffraction.**

The intensity on a screen consists of a bright central maximum and fainter secondary maxima.

Screen

The width of the central maximum is

$$w = \frac{2\lambda L}{a}$$

- Light waves from multiple slits in a screen **interfere** where they overlap. The light intensity is large where the interfering waves are in phase, and small where they are out of phase.

The intensity on a screen consists of equally spaced **interference fringes.**

The fringe spacing is

$$\Delta y = \frac{\lambda L}{d}$$

- Light waves reflected from the two surfaces of a thin transparent film also interfere.

The resolution of optical instruments

Diffraction limits how close together two point objects can be and still be resolved.

For a *telescope,* the minimum resolvable angular separation between two objects is

$$\theta_1 = 1.22\lambda/D$$

For a *microscope,* the minimum resolvable distance between two objects is

$$d_{min} = \frac{0.61\lambda_0}{NA}$$

where the *numerical aperture* NA is a characteristic of the microscope objective.

Ray model

- Light travels out from its source in straight lines, called **rays.**

- Rays reflect off a surface between two media, obeying the **law of reflection,** $\theta_i = \theta_r$.

- Light rays change direction as they cross the surface between two media. The angles of incidence and refraction are related by **Snell's law,**

$$n_1 \sin\theta_1 = n_2 \sin\theta_2.$$

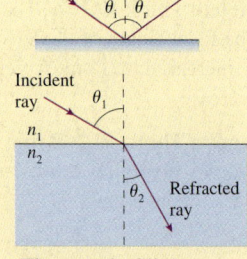

where n is the **index of refraction.** The speed of light in a transparent material is $v = c/n$.

Image formation by lenses and mirrors

A lens or mirror has a characteristic **focal length** f. Rays parallel to the optical axis come to focus a distance f from the lens or mirror.

The **object distance** s, the **image distance** s', and the focal length are related by the *thin-lens equation,* which also works for mirrors:

$$\frac{1}{s} + \frac{1}{s'} = \frac{1}{f}$$

Scanning Confocal Microscopy

Although modern microscopes are marvels of optical engineering, their basic design is not too different from the 1665 compound microscope of Robert Hooke. Recently, advances in optics, lasers, and computer technology have made practical a new kind of optical microscope, the *scanning confocal microscope.* This microscope is capable of taking images of breathtaking clarity.

The figure shows the microscope's basic principle of operation. The left part of the figure shows how the translucent specimen is illuminated by light from a laser. The laser beam is converted to a diverging bundle of rays by suitable optics, reflected off a mirror, then directed through a microscope objective lens to a focus within the sample. The microscope objective focuses the laser beam to a very small ($\approx 0.5~\mu$m) spot. Note that light from the laser passes through other regions of the specimen but, because the rays are not focused in those regions, they are not as intensely illuminated as is the point at the focus. This is the first important aspect of the design: Very intensely illuminate one very small volume of the sample while leaving other regions only weakly illuminated.

As shown in the right half of the figure, light is reflected from all illuminated points in the sample and passes back through the objective lens. The mirror that had reflected the laser light downward is actually a *partially transparent mirror* that reflects 50% of the light and transmits 50%. Thus

half of the light reflected upward from the sample passes through the mirror and is focused on a screen containing a small hole. Because of the hole, only light rays that emanate from the brightly illuminated volume in the sample can completely pass through the hole and reach the light detector behind it. Rays from other points in the sample either miss the hole completely, or they are out of focus when they reach the screen so that only a small fraction of them pass through the hole. This second key design aspect limits the detected light to only those rays that are emitted from the point in the sample at which the laser light was originally focused.

So we see that (a) the point in the sample that is at the focus of the objective is much more intensely illuminated than any other point, so it reflects more rays than any other point, and (b) the hole serves to further limit the detected rays to only those that emanate from the focus. Taken together, these design aspects ensure the detected light comes from a very small, very well-defined volume in the sample.

The microscope as shown would only be useful for examining one small point in the sample. To make an actual *image,* the objective is *scanned* across the sample while the intensity is recorded by a computer. This procedure builds up an image of the sample one *scan line* at a time. The final result is a picture of the sample in the very narrow plane in which the laser beam is focused. Different planes within the sample can be imaged by moving the objective up or down before scanning. It is actually possible to make three-dimensional images of a specimen in this way.

The improvement in contrast and resolution over conventional microscopy can be striking. The images show a section of a human medulla taken using conventional and confocal microscopy. Because light reflected from all parts of the specimen reach the camera in a conventional microscope, that image appears blurred and has low contrast. The confocal microscope image represents a single plane or slice of the sample, and many details become apparent that are invisible in the conventional image.

The screen blocks out light reflected from points other than the focus of the laser beam.

Partially transparent mirror reflects 50%, transmits 50%.

Light detector

Hole in screen

Laser source

Objective lens

The intensity at the focus of the laser beam is very high.

Specimen

$\approx 0.5~\mu$m

The intensity at other points is much weaker.

Bright light reflected from point at focus

Faint light reflected from some other point

A confocal microscope.

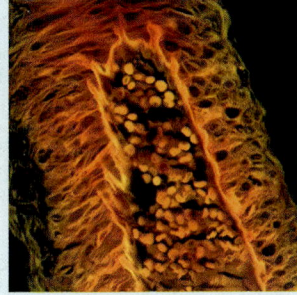

A thick section of fluorescently stained human medulla imaged using standard optical microscopy (left) and scanning confocal microscopy (right).

ELECTRICITY AND MAGNETISM

Charges, Currents, and Fields

The early Greeks discovered that a piece of amber that has been rubbed briskly can attract feathers or small pieces of straw. They also found that certain stones from the region they called *Magnesia* can pick up pieces of iron. These first experiences with the forces of electricity and magnetism began a chain of investigations that has led to today's high-speed computers, lasers, fiber-optic communications, and magnetic-resonance imaging, as well as such mundane modern-day miracles as the light bulb.

The development of a successful electromagnetic theory, which occupied the leading physicists of Europe for most of the nineteenth century, led to sweeping revolutions in both science and technology. The complete formulation of the theory of the electromagnetic field has been called by no less than Einstein "the most important event in physics since Newton's time."

The basic phenomena of electricity and magnetism are not as familiar to most people as those of mechanics. We will deal with this lack of experience by placing a large emphasis on these basic phenomena. We will begin where the Greeks did, by looking at the forces between objects that have been briskly rubbed, exploring the concept of *electric charge*. It is easy to make systematic observations of how charges behave, and we will be led to consider the forces between charges and how charges behave in different materials. *Electric current,* whether it be for lighting a light bulb or changing the state of a computer memory element, is simply a controlled motion of charges through conducting materials. One of our goals will be to understand how charges move through electric circuits.

When we turn to magnetic behavior, we will again start where the Greeks did, noting how magnets stick to some metals. Magnets also affect compass needles. And, as we will see, an electric current can affect a compass needle in exactly the same way as a magnet. This observation shows the close connection between electricity and magnetism, which leads us to the phenomenon of *electromagnetic waves.*

Our theory of electricity and magnetism will introduce the entirely new concept of a *field.* Electricity and magnetism are about the long-range interactions of charges, both static charges and moving charges, and the field concept will help us understand how these interactions take place.

Microscopic Models

The field theory provides a macroscopic perspective on the phenomena of electricity and magnetism, but we can also take a microscopic view. At the microscopic level, we want to know what charges are, how they are related to atoms and molecules, and how they move about through various kinds of materials. Electromagnetic waves are composed of electric and magnetic fields. The interaction of electromagnetic waves with matter can be analyzed in terms of the interactions of these fields with the charges in matter. When you heat food in a microwave oven, you are using the interactions of electric and magnetic fields with charges in a very fundamental way.

Much of what is known about your nervous system comes from the study of an animal that seems quite different from humans—the squid. Nerve fibers conduct electrical signals along their length, allowing the brain to direct the actions of the body. How is an electrical signal generated and transmitted in the nervous system of a human or a squid?

20 ELECTRIC FORCES AND FIELDS

DNA analysis is often done using *gel electrophoresis*. A solution of DNA segments is placed in a well at one end of a plate of gel. Different segments migrate through the gel at different rates, leading to the lines in the photo. What force causes the DNA segments to move through the gel?

Looking Ahead ►►

The goal of Chapter 20 is to develop a basic understanding of electric phenomena in terms of charges, forces, and fields. In this chapter you will learn to:

► Use a charge model to explain basic electric phenomena.

► Use Coulomb's law to calculate the electric force between charges.

► Determine the electric field from charge distributions.

► Calculate forces and torques on charges in electric fields.

Looking Back ◄◄

The mathematical analysis of electric forces and fields makes use of vector addition, as well as the concept of torque. The electric force is in some ways analogous to gravity. Please review:

◄ Sections 3.1–3.3 Vector properties, coordinate systems, and components.

◄ Section 6.6 Newton's law of gravity.

◄ Section 7.2 Torque.

The electric force is one of the fundamental forces of nature. You are familiar with electricity; it is something you use every day to run motors, to power computers and telephones, and to light your home. Electricity is also of great importance in biological systems. Your heart makes an electric signal when it beats, and certain animals use the electric signals their bodies produce to navigate.

What are the properties of electricity and electric forces? How do we generate, transport, and use electricity? How do our bodies use electricity, and how can we use electrical signals to diagnose and treat illness? These are some questions we will begin to explore in this chapter.

Electricity is a big topic, and we cannot hope to answer all these questions at once. We will begin, in this chapter, by investigating some of the basic phenomena of electricity. It's hard to see what rubbing plastic rods with wool has to do with the operation of a computer or an electrocardiogram, but only by starting at the very beginning, with simple observations, will we develop our understanding of how electricity underlies so many facets of our lives.

20.1 Charges and Forces

You can receive a mildly unpleasant shock and produce a little spark if you touch a metal doorknob after scuffing your shoes across a carpet. A plastic comb that you've run through your hair will pick up bits of paper and other small objects. In both of these cases, two objects are *rubbed* together. Why should rubbing an object cause forces and sparks? What kind of forces are these? These are the questions with which we begin our study of electricity.

Our first goal is to develop a model for understanding electric phenomena in terms of *charges* and *forces*. We will later use our contemporary knowledge of atoms to understand electricity on a microscopic level, but the basic concepts of electricity make *no* reference to atoms or electrons. The theory of electricity was well established long before the electron was discovered.

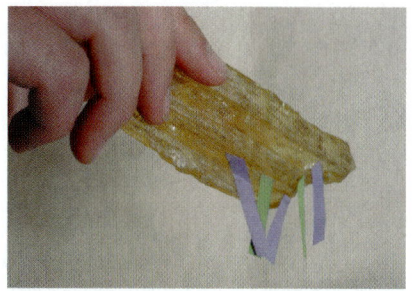

The ancient Greeks first noted the electrical nature of matter by observing amber, a form of fossilized tree resin. When rubbed with fur, amber buttons would attract bits of hair or straw. The Greek word for amber, *elektron*, is the source of our words *electric*, *electricity*, and—of course—*electron*.

Experimenting with Charges

Let us enter a laboratory where we can make observations of electric phenomena. This is a modest laboratory, much like one you would have found in the year 1800. The major tools in the lab are:

- A variety of plastic, glass, and wood rods, each several inches long. These can be held in your hand or suspended by threads from a support.
- A few metal rods with wood handles.
- Pieces of wool and silk.
- Small metal spheres, an inch or two in diameter, on wood stands.

We will manipulate and use these tools with the goal of developing a theory to explain the phenomena we see. The experiments and observations described below are very much like those of early investigators of electric phenomena.

Discovering electricity I

Experiment 1	Experiment 2	Experiment 3

Take a plastic rod that has been undisturbed for a long period of time and hang it by a thread. Pick up another undisturbed plastic rod and bring it close to the hanging rod. Nothing happens to either rod.

Vigorously rub both the hanging plastic rod and the hand-held plastic rod with wool. Now the hanging rod moves away from the handheld rod when you bring the two close together. Rubbing two glass rods with silk produces the same result: The two rods repel each other.

Bring a glass rod that has been rubbed with silk close to a hanging plastic rod that has been rubbed with wool. These two rods *attract* each other.

No forces were observed in Experiment 1. We will say that the original objects are **neutral.** Rubbing the rods (Experiments 2 and 3) somehow causes forces to be exerted between them. We will call the rubbing process **charging** and say that a rubbed rod is *charged*. For now, these are simply descriptive terms. The terms don't tell us anything about the process itself.

Experiment 2 shows that there is a *long-range repulsive force* between two identical objects that have been charged in the *same* way, such as two plastic rods both rubbed with wool. We have seen a long-range force before, gravity, but the gravitational force is always attractive. This is the first time we've observed a repulsive long-range force, so right away we see that new ideas will be needed to understand electricity.

Charges on tape Pull a piece of transparent tape about 6" long off a roll. Now, pull a second piece of tape off the roll, and hold the two pieces near each other. They show a strong repulsive force. What does this tell you about the charges on the two pieces of tape? How could you prepare the two strips of tape so that they attract?

Experiment 3 is a puzzle. Two rods *seem* to have been charged in the same way, by rubbing, but these two rods *attract* each other rather than repel. Why should the outcome of Experiment 3 differ from that of Experiment 2? As you probably know, there are two different types of charge; whether the force is attractive or repulsive depends on the charges on the two objects.

The modern names for the two types of charge are *positive charge* and *negative charge*. You may be surprised to learn that the names were coined by Benjamin Franklin. Franklin was among the first to make quantitative studies of electric phenomena, and he found that charge behaves like positive and negative numbers. Franklin discovered that charge could be transferred from one object to another. If a plastic rod is charged twice, by rubbing, and twice transfers charge to a metal sphere, the charge on the sphere is doubled. Adding charges works like addition: $2 + 2 = 4$. But the sphere is found to be neutral after receiving equal amounts of positive charge and negative charge. This is like $2 + (-2) = 0$. These experiments establish an important property of charge.

Franklin also found that two objects with the same type of charge repel each other, while two objects with different types of charge attract each other. How do we determine which charge to call positive and which to call negative? It's entirely up to us! Franklin established the convention that a glass rod rubbed with silk is *positively* charged. That's it. We can determine the signs of other charges by noting forces. Any other object that repels a charged glass rod is also positively charged. Any charged object that attracts a charged glass rod is negatively charged. Thus the plastic rod rubbed with wool is negative. It was only long afterward, with the discovery of electrons and protons, that electrons were found to be attracted to a charged glass rod while protons were repelled. Thus *by convention* electrons have a negative charge and protons a positive charge.

NOTE ▶ In hindsight, it would have been better had Franklin made the opposite choice. Electrons are the carriers of electric currents in metals, and the convention of assigning a negative charge to electrons will later present us with some sign difficulties that could have been avoided with positive electrons. ◀

Let's refine our model a bit, by returning to the laboratory for a few more experiments.

Discovering electricity II

Experiment 4

Increased distance

Further observations on the force between two charged rods show that:

- The strength of these forces is greater for rods that have been rubbed more vigorously.
- The strength of the forces decreases as the separation between the rods increases.

Experiment 5

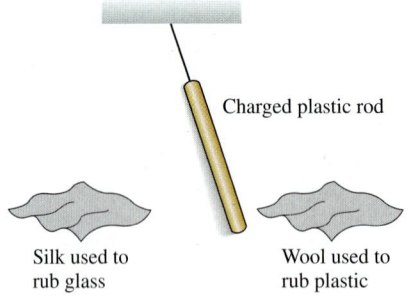

Charged plastic rod

Silk used to rub glass

Wool used to rub plastic

Rub a hanging plastic rod with wool and then hold the *wool* close to the rod. The rod is weakly *attracted* to the wool. The plastic rod is *repelled* by a piece of silk that has been used to rub glass.

Experiment 6

Charged rod

Paper

Hold a charged (i.e., rubbed) plastic rod over small pieces of paper on the table. The pieces of paper leap up and stick to the rod. A charged glass rod does the same. However, a neutral rod has no effect on the pieces of paper.

Experiment 4 shows how the strength of the electric force depends on the properties of the charged objects and the distance between them. When two rods are rubbed more vigorously, meaning they have more charge, the strength of the force between them is increased. Experiment 4 also shows that the force between two charged objects depends on the distance between them. We have seen this variation before, with gravity; as was the case with the gravitational force, the force between two charged objects decreases as the distance between them increases.

Experiment 5 shows us that when a cloth is used to charge a rod, the cloth itself develops a charge. How can we explain this result? As we will see, all matter contains both types of charge. Initially, the cloth is neutral, meaning it has the same amount of positive and negative charge. When rubbed, some positive charge is transferred to the cloth. The rod then has an excess of negative charge.

Experiment 6 shows that there is an attractive force between a charged object and a *neutral* (uncharged) object. This is a bit of a puzzle—especially because the pieces of paper are attracted to a rod with either a positive or negative charge. The paper itself is not charged, but it feels an attractive force to a rod of either sign of charge. We will see that a charged object can *induce* a charge separation on an uncharged object and lead to such an attractive force.

But we are getting ahead of ourselves; for now, let's start to develop a **charge model** to explain our observations.

> **Charge model, part I** The basic postulates of our model are:
>
> 1. Frictional forces, such as rubbing, add something called **charge** to an object or remove it from the object. The process itself is called *charging*. More vigorous rubbing produces a larger quantity of charge.
> 2. There are two kinds of charge, positive and negative.
> 3. Two **like charges** (positive/positive or negative/negative) exert repulsive forces on each other. Two **opposite charges** (positive/negative), exert attractive forces on each other.
> 4. The force between two charges is a long-range force. The magnitude of the force increases as the quantity of charge increases and decreases as the distance between the charges increases.
> 5. *Neutral* objects have an *equal mixture* of positive and negative charge. The rubbing process somehow manages to separate the two.

As we continue, we will develop and refine this model. We still have some very large unexplained puzzles, such as why charged objects exert attractive forces on neutral objects.

Insulators and Conductors

Experiment 5 involved a transfer of charge from one object to another. Let's do some more experiments with charge to look at how charge moves on different materials.

Picking up pollen BIO Rubbing a rod with a cloth gives the rod an electric charge. In a similar fashion, the rapid motion of a bee's wings through the air gives the bee a small positive electric charge. As small pieces of paper are attracted to a charged rod, so are tiny grains of pollen attracted to the charged bee, helping it collect and hold the pollen.

Discovering electricity III

Experiment 7

Charge a plastic rod by rubbing it with wool. Touch a neutral metal sphere with the rubbed area of the rod. The metal sphere then picks up small pieces of paper and repels a charged, hanging plastic rod. The metal sphere appears to have acquired a charge, the same charge as the plastic rod.

Experiment 8

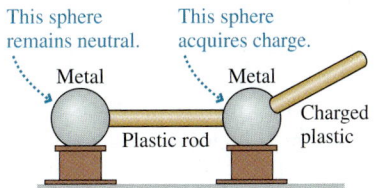

Place two metal spheres close together with a plastic rod connecting them. Charge a second plastic rod, by rubbing, and touch it to one of the metal spheres. Afterward, the metal sphere that was touched picks up small pieces of paper and repels a charged, hanging plastic rod. The other metal sphere does neither.

Experiment 9

Repeat Experiment 8 with a metal rod connecting the two metal spheres. Touch one metal sphere with a charged plastic rod. Afterward, *both* metal spheres pick up small pieces of paper and repel a charged, hanging plastic rod.

A dry day, a plastic slide, and a child with clothes of the right fabric lead to a startling demonstration of electric charges and forces. The rubbing of the child's clothes on the slide has made her build up charge. The body is a good conductor, so the charges spread across her body and her hair. The resulting repulsion produces a dramatic result!

Our final set of experiments has shown that charge can be transferred from one object to another only when the objects *touch*. Contact is required. Removing charge from an object, which you can do by touching it, is called **discharging.** Careful experiments show that discharging an object merely transfers the charge somewhere else; it doesn't eliminate the charge. Further experiments show that, in all cases, **charge is conserved**; it can't be created or destroyed.

Experiment 9, in which a metal rod is used, is in sharp contrast to Experiment 8. Charge somehow *moves through* or along a metal rod, from one sphere to the other, but does not move through or along a plastic or glass rod. Let us define **conductors** as those materials through or along which charge easily moves and **insulators** as those materials on or in which charges remain immobile. Glass and plastic are insulators; metal is a conductor.

This new information allows us to add more postulates to our charge model:

Charge model, part II

6. There are two types of materials. Conductors are materials through or along which charge easily moves. Insulators are materials on or in which charges remain fixed in place.
7. Charge can be transferred from one object to another by contact.
8. Charge is conserved; it cannot be created or destroyed.

NOTE ▶ Both insulators and conductors can be charged. They differ in the *mobility* of the charge. ◀

CONCEPTUAL EXAMPLE 20.1 Transferring charge

In Experiment 9, touching one metal sphere with a charged plastic rod caused a second metal sphere to become charged with the same type of charge as the rod. Use the postulates of the charge model to explain this.

REASON We need the following ideas from the charge model:

1. Charge is transferred upon contact.
2. Metal is a conductor.
3. Like charges repel.

The plastic rod was charged by rubbing with wool, giving it a negative charge. The charge doesn't move around on the rod, because it is an insulator, but some of the charge is transferred to the metal upon contact. Once in the metal, which is a conductor, the charges are free to move around. Furthermore, because like charges repel, these negative charges quickly move as far apart as they possibly can. Some move through the connecting metal rod to the second sphere. Consequently, the second sphere acquires a net negative charge.

Visualizing Charge

Diagrams are going to be an important tool for understanding and explaining charges and the forces on charged objects. A *charge diagram*, which gives a schematic picture of charges on each object and their distribution, is shown in Figure 20.1 for charging a plastic rod by rubbing it with wool.

◀ **How do you charge a powder?** A photocopier makes an image with a mixture of very fine particles called *toner* that are fused to a piece of paper. The toner particles must be evenly charged in order to stick via electric forces to the imaging drum. This charging is accomplished by adding small carrier beads, which are often metal beads coated with plastic. The container with the toner and carrier beads is agitated, and the rubbing between the two gives them opposite charges. This micrograph shows a carrier bead with oppositely charged toner particles held to it by the resulting attraction.

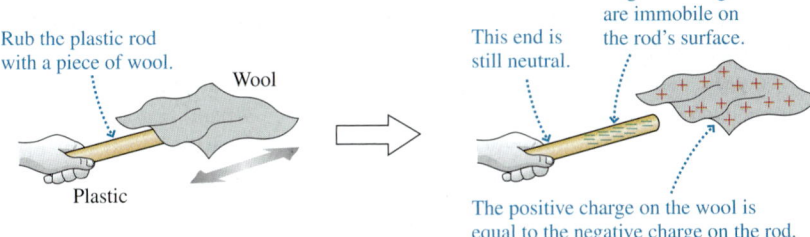

FIGURE 20.1 An insulating rod is charged by rubbing.

The diagram shows certain key features of the charging. The rod and the wool are initially uncharged. Rubbing separates charge, with the rod developing a negative charge and the wool a positive charge. Because charge is conserved, the magnitudes of these two charges must be equal, as reflected in the diagram. And, because the rod is an insulator, the charges on the rod do not move.

Diagrams involving metals are different because **charges in a conductor are free to move.** Suppose we charge a block of metal as in Figure 20.2. Once charge is transferred to the metal, repulsive forces between the negative charges cause them to move apart from each other. The charges get as far apart as possible, so they end up on the surface.

Other than the very brief interval during which the charges are adjusting, the charges on an isolated conductor are in static equilibrium. That is, the charges are at rest and there is no *net* force on any charge. This condition is called **electrostatic equilibrium.**

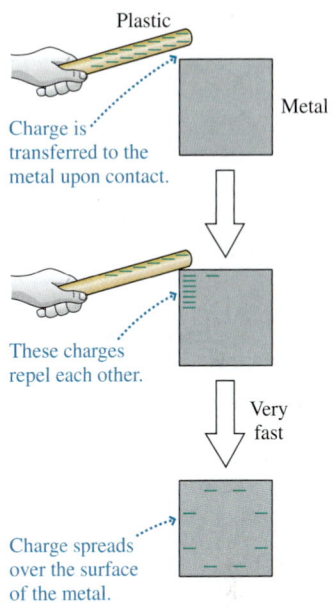

FIGURE 20.2 A conductor is charged by contact with a charged plastic rod.

CONCEPTUAL EXAMPLE 20.2 Drawing a charge diagram for an electroscope

Many electricity demonstrations are carried out with the help of an *electroscope* like the one shown in Figure 20.3. Touching the sphere at the top of an electroscope with a charged plastic rod causes the leaves to fly apart and remain hanging at an angle. Use charge diagrams to explain why.

REASON We will use the charge model and our understanding of insulators and conductors to make a series of charge diagrams in Figure 20.4 that shows the charging of the electroscope.

ASSESS The charges move around, but, because charge is conserved, the total number of negative charges doesn't change from picture to picture.

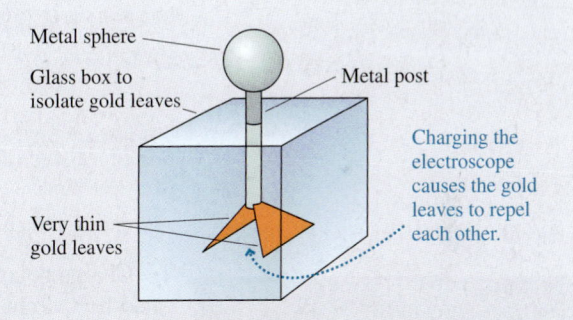

FIGURE 20.3 A charged electroscope.

1. Negative charges (i.e., electrons) are transferred from the rod to the metal sphere upon contact.

2. Metal is a conductor. Therefore charge *very quickly* spreads throughout the entire electroscope. The leaves become negatively charged.

3. Like charges repel. The negatively charged leaves exert repulsive forces on each other, causing them to spread apart.

FIGURE 20.4 Charging an electroscope.

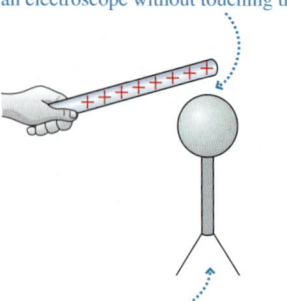

Bring a positively charged glass rod close to an electroscope without touching the sphere.

The electroscope is neutral, yet the leaves repel each other. Why?

FIGURE 20.5 A charged rod held close to an electroscope causes the leaves to repel each other.

We can use our understanding of the electroscope to answer another important question from Experiment 5: How do charged objects of either sign exert an attractive force on a *neutral* object?

Figure 20.5 shows a positively charged rod held close to a *neutral* electroscope *without touching* the metal sphere. The leaves move apart and stay apart as long as you hold the rod near, but they quickly collapse when it is removed. The electroscope is uncharged, because the rod has not touched it. Where does the force come from that causes the leaves to repel?

There is no net charge on the electroscope, but, because the electroscope is metal, the charges that are there are free to move. When a charged object is brought near a conductor, positive and negative charges in the conductor will feel opposite forces and will separate, as in Figure 20.6a. **Charge polarization** is this slight separation of the positive and negative charge in a neutral object. Polarization explains the behavior of the electroscope when the charged rod is brought near. As Figure 20.6b shows, negative charges are attracted to the top of the electroscope, and positive charges are repelled. The positive charges end up on the leaves, where they produce the repulsion between the leaves.

(a)

Negative charges are attracted to the positive rod. The surface of the metal nearest the rod develops a negative charge.

Positive rod

Metal

Positive charges are repelled by the rod. A positive charge is created on the far surface.

The metal's net charge is still zero, but it has been *polarized* by the charged rod.

(b)

The electroscope is polarized by the charged rod. Negative charges move toward the rod.

Although the net charge on the electroscope is still zero, the leaves have excess positive charge and repel each other.

FIGURE 20.6 A charged rod polarizes a metal.

Charge polarization explains not only why the electroscope leaves deflect but also how a charged object exerts an attractive force on a neutral object. Figure 20.7 shows a positively charged rod near a neutral piece of metal. Because the electric force decreases with distance, the attractive force on the negative charges at the top surface is *slightly greater* than the repulsive force on the positive charge on the bottom surface. The net force toward the charged rod is called a **polarization force.** The polarization force arises because the charges in the metal are separated, *not* because the rod and metal are oppositely charged.

Pulling water Turn on your tap so that a thin stream of water flows. Next, run a comb briskly through your hair, and bring the comb close to—but not touching—the stream of water. The deflection of the stream can be quite dramatic! Water from the tap is a reasonably good conductor. The presence of the charged comb separates charges in the water stream, leading to an attractive polarization force.

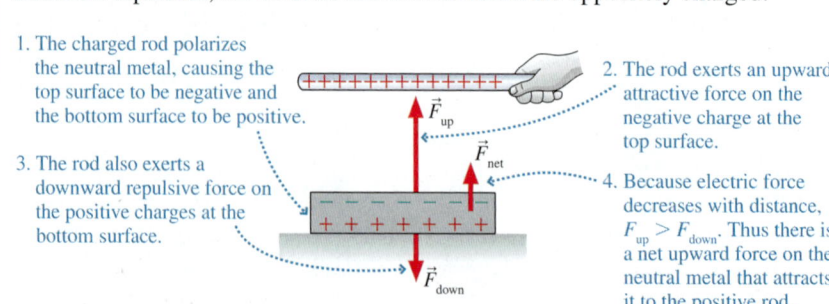

1. The charged rod polarizes the neutral metal, causing the top surface to be negative and the bottom surface to be positive.

2. The rod exerts an upward attractive force on the negative charge at the top surface.

3. The rod also exerts a downward repulsive force on the positive charges at the bottom surface.

4. Because electric force decreases with distance, $F_{up} > F_{down}$. Thus there is a net upward force on the neutral metal that attracts it to the positive rod.

$\vec{F}_{up}$

$\vec{F}_{net}$

$\vec{F}_{down}$

FIGURE 20.7 The polarization force on a neutral piece of metal is due to the slight charge separation.

An electroscope is charged by touching it with a positive glass rod. The electroscope leaves spread apart and the glass rod is removed. Then a negatively charged plastic rod is brought close to the top of the electroscope, but it doesn't touch. What happens to the leaves?

A. The leaves get closer together.
B. The leaves spread farther apart.
C. The leaves do not change their position.

20.2 Charges, Atoms, and Molecules

We have been speaking about giving objects positive or negative charge without explaining what is happening at an atomic level. You already know that the basic constituents of atoms—the nucleus and the electrons surrounding it—are charged. In this section we will connect our observations of the previous section with our understanding of the atomic nature of matter.

Our current model of the atom is that it is made up of a very small and dense positively charged *nucleus,* containing positive *protons* as well as neutral particles called *neutrons,* surrounded by much-less-massive orbiting negatively charged *electrons* that form an **electron cloud** surrounding the nucleus, as illustrated in Figure 20.8. The atom is held together by the attractive electric force between the positive nucleus and the negative electrons.

Experiments show that **charge, like mass, is an inherent property of electrons and protons.** It's no more possible to have an electron without charge than it is to have an electron without mass. As far as we know today, electrons and protons have charges of opposite sign but *exactly* equal magnitude.

An Atomic View of Charging

Electrons and protons are the basic charges in ordinary matter. **There are no other sources of charge.** Consequently, the various observations we made in Section 20.1 need to be explained in terms of electrons and protons.

Because charge is due to electrons and protons, **an object is charged if it has an unequal number of electrons and protons.** An object with a negative charge has more electrons than protons; an object with a positive charge has more protons than electrons. Most macroscopic objects have an *equal number* of protons and electrons. Such an object has no *net* charge; we say it is *electrically neutral.*

> **NOTE** ▶ *Neutral* does *not* mean "no charges" but, instead, means that there is no *net* charge. A typical 1 cm³ solid contains approximately 10^{24} electrons and an equal number of protons. This is a tremendous number of charges, but the equality between positive and negative charges means that most objects are electrically neutral or very close to it. A glass rod loses only approximately 10^{11} electrons as it is charged by rubbing. This corresponds to only 1 electron out of every 10^{13}. ◀

In practice, objects acquire a positive charge not by gaining protons but by losing electrons. Protons are *extremely* tightly bound within the nucleus and cannot be added to or removed from atoms. Electrons, on the other hand, are bound much more loosely than the protons and can be removed with little effort.

The process of removing an electron from the electron cloud of an atom is called **ionization.** An atom that is missing an electron is called a *positive ion.* Some atoms can accommodate an *extra* electron and thus become a *negative ion.* Figure 20.9 shows positive and negative ions.

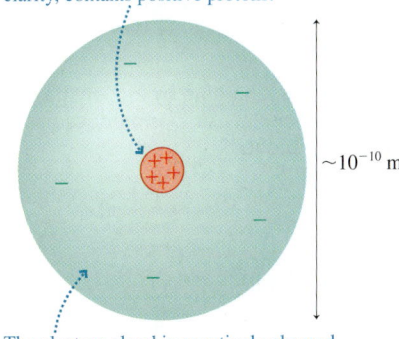

The nucleus, exaggerated for clarity, contains positive protons.

~10^{-10} m

The electron cloud is negatively charged.

FIGURE 20.8 Our modern view of the atom.

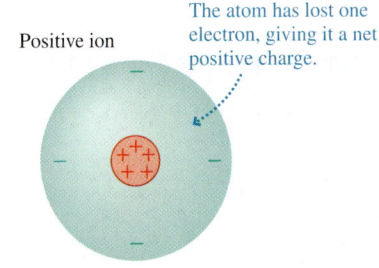

Positive ion

The atom has lost one electron, giving it a net positive charge.

Negative ion

The atom has gained one electron, giving it a net negative charge.

FIGURE 20.9 Positive and negative ions.

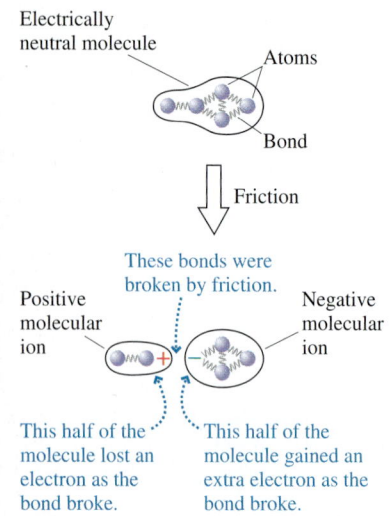

Electrically neutral molecule

Atoms

Bond

Friction

These bonds were broken by friction.

Positive molecular ion

Negative molecular ion

This half of the molecule lost an electron as the bond broke.

This half of the molecule gained an extra electron as the bond broke.

FIGURE 20.10 Charging by friction may result from molecular ions produced as bonds are broken.

TABLE 20.1 Protons and electrons

Particle	Mass (kg)	Charge (C)
Proton	1.67×10^{-27}	$+e = 1.60 \times 10^{-19}$
Electron	9.11×10^{-31}	$-e = -1.60 \times 10^{-19}$

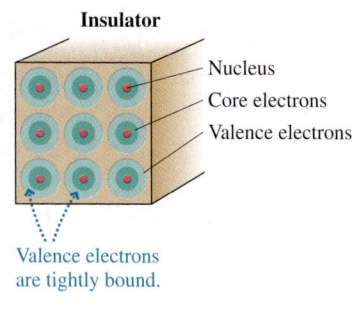

Insulator

Nucleus

Core electrons

Valence electrons

Valence electrons are tightly bound.

Metal

Positive ion cores

Valence electrons form a "sea of electrons."

FIGURE 20.11 A microscopic look at insulators and conductors.

The charging processes we observed in Section 20.1 involved rubbing and friction. The forces of friction often cause molecular bonds at the surface to break as two materials slide past each other. Molecules are electrically neutral, but Figure 20.10 shows that *molecular ions* can be created when one of the bonds in a large molecule is broken. If the positive molecular ions remain on one material and the negative ions on the other, one of the objects being rubbed ends up with a net positive charge and the other with a net negative charge. This is the way in which a plastic rod is charged by rubbing with wool or a comb is charged by passing through your hair. The breaking of such bonds is also responsible for the charge on tape that is pulled from the roll.

Charge Conservation

Charge is represented by the symbol q (or sometimes Q). The SI unit of charge is the **coulomb** (C), named for French scientist Charles Coulomb, one of many scientists investigating electricity in the late 18th century.

Protons and electrons, the charged particles in ordinary matter, have the same amount of charge, but of opposite signs. We use the symbol e for the **fundamental charge,** the magnitude of the charge of an electron or a proton. The fundamental charge e has been measured to have the value

$$e = 1.60 \times 10^{-19} \text{ C}$$

Table 20.1 shows the masses and charges of protons and electrons.

NOTE ▶ The amount of charge produced by rubbing plastic or glass rods is typically in the range 1 nC (10^{-9} C) to 100 nC (10^{-7} C). This corresponds to an excess or deficit of 10^{10} to 10^{12} electrons. ◀

Charge can be transferred from one object to another as electrons and ions move about, but **charge is conserved: The *total* amount of charge remains constant.** For example, charging a plastic rod by rubbing it with wool transfers electrons from the wool to the plastic as the molecular bonds break. The wool is left with a positive charge equal in magnitude but opposite in sign to the negative charge of the rod: $q_{\text{wool}} = -q_{\text{plastic}}$. The *net* charge of the two remains zero.

Insulators and Conductors

Figure 20.11 looks inside an insulator and a metallic conductor. The electrons in the insulator are all tightly bound to the positive nuclei and not free to move around. Charging an insulator by friction leaves patches of molecular ions on the surface, but these patches are immobile.

In metals, the outer atomic electrons (called the *valence electrons* in chemistry) are only weakly bound to the nuclei. As the atoms come together to form a solid, these outer electrons become detached from their parent nuclei and are free to wander about through the entire solid. The solid *as a whole* remains electrically neutral, because we have not added or removed any electrons, but the electrons are now rather like a negatively charged gas or liquid—what physicists like to call a **sea of electrons**—permeating an array of positively charged **ion cores.**

In a metal, the electrons in the sea of electrons are highly mobile, moving quickly and easily in response to electric forces. The motion of charges through a material is what we will later call a **current,** and the charges that physically move are called the **charge carriers.** Although the electrons are highly mobile *within* the metal, they are weakly bound to the ion cores and will not leave the metal.

Electric Dipoles

Figure 20.7 showed how a metal can be polarized by a charged rod. In terms of electrons and protons, we can say that the sea of electrons shifts very slightly toward or away from the charged rod, causing one surface to become positive and the opposite surface to become negative. But why does a charged rod pick up a neutral piece of paper, which is an insulator?

Consider what happens if we bring a positive charge near a neutral atom. As Figure 20.12 shows, the charge polarizes the atom by attracting the electron cloud while repelling the nucleus. The forces cause a slight separation between the center of positive charge and the center of negative charge.

Two equal but opposite charges with a separation between them are called an **electric dipole.** In this case, where the polarization is caused by the external charge, the atom has become an *induced electric dipole.* Because the negative end of the dipole is slightly closer to the positive charge, the attractive force on the negative end slightly exceeds the repulsive force on the positive end, and there is a net force toward the external charge. If a charged rod causes all the atoms in a piece of paper to become induced electric dipoles, the net force is enough to lift the paper to the rod.

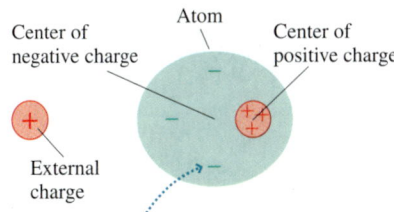

Forces from the external charge cause the atom's negative charge and the positive charge to be slightly offset.

FIGURE 20.12 An induced electric dipole.

Hydrogen Bonding

Some molecules have an asymmetry in their charge distribution that makes them *permanent electric dipoles.* An important example is the water molecule. Bonding between the hydrogen and oxygen atoms results in an unequal sharing of charge that, as shown in Figure 20.13, leaves the hydrogen atoms with a small positive charge and the oxygen atom with a small negative charge.

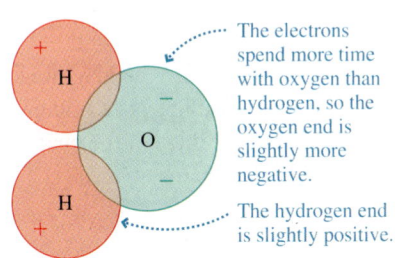

The electrons spend more time with oxygen than hydrogen, so the oxygen end is slightly more negative.

The hydrogen end is slightly positive.

FIGURE 20.13 A water molecule is a permanent electric dipole.

The slightly negative oxygen is attracted to the slightly positive hydrogen of a neighbor.

This water molecule forms hydrogen bonds with four neighbors.

FIGURE 20.14 Hydrogen bonds between water molecules.

When two water molecules are close, the attractive electric force between the positive hydrogen atom of one molecule and the negative oxygen atom of the second molecule can form a weak bond, called a **hydrogen bond,** as illustrated in Figure 20.14. These weak bonds result in a certain "stickiness" between water molecules that is responsible for many of water's special properties, including its expansion on freezing, the wide range of temperatures over which it is liquid, and its high heat of vaporization.

Hydrogen bonds are extremely important in biological systems. As you know, the DNA molecule has the structure of a double helix. Information in DNA is coded in the *nucleotides,* the four molecules guanine, thymine, adenine and cytosine. The nucleotides on one strand of the DNA helix form hydrogen bonds with the nucleotides on the opposite strand.

The nucleotides only bond in certain pairs: Cytosine always forms a bond with guanine, adenine with thymine. This preferential bonding is crucial to

Guanine

Thymine

DNA double helix

Cytosine

Adenine

The base pairs of the two strands can only join in certain combinations.

Guanine

Cytosine

Three hydrogen bonds join these bases.

Thymine

Adenine

Two hydrogen bonds join these bases.

Atoms comprising the bases

H N C O

FIGURE 20.15 Hydrogen bonds in DNA base pairs.

DNA replication. When the two strands of DNA are taken apart, each separate strand of the DNA forms a template on which another complementary strand can form, creating two identical copies of the original DNA molecule.

The preferential bonding of nucleotide base pairs in DNA is explained by hydrogen bonding. In each of the nucleotides, the hydrogen atoms have a small positive charge, oxygen and nitrogen a small negative charge. The positive hydrogen atoms on one nucleotide attract the negative oxygen or nitrogen atoms on another. As the detail in Figure 20.15 shows, the geometry of the nucleotides allows cytosine to form a hydrogen bond only with guanine, adenine only with thymine.

STOP TO THINK 20.2   Rank in order, from most positive to most negative, the charges q_A to q_E of these five systems.

Proton	Electron	17 protons 19 electrons	1,000,000 protons 1,000,000 electrons	Glass ball missing 3 electrons
A.	B.	C.	D.	E.

20.3 Coulomb's Law

11.1–11.3   Activ ONLINE Physics

The last two sections established a *model* of charges and electric forces. This model is very good at explaining electric phenomena and providing a general understanding of electricity. Now we need to become quantitative. Experiment 4 in Section 20.1 found that the electric force increases for objects with more charge and decreases as charged objects are moved farther apart. The force law that describes this behavior is known as *Coulomb's law.*

In the mathematical formulation of Coulomb's law, we will use the magnitude of the charge only, not the sign. We show this by using the absolute value notation we used earlier in the book. $|q|$ therefore represents the magnitude of the charge. It is always a positive number, whether the charge is positive or negative.

> **Coulomb's law**
>
> **Magnitude:** If two charged particles having charges q_1 and q_2 are a distance r apart, the particles exert forces on each other of magnitude
>
> $$F_{1\,\text{on}\,2} = F_{2\,\text{on}\,1} = \frac{K|q_1||q_2|}{r^2} \qquad (20.1)$$
>
> INVERSE-SQUARE p. 194
>
> where the charges are in coulombs (C), and $K = 8.99 \times 10^9 \, \text{N} \cdot \text{m}^2/\text{C}^2$ is called the **electrostatic constant.** These forces are an action/reaction pair, equal in magnitude and opposite in direction.
>
> **Direction:** The forces are directed along the line joining the two particles. The forces are *repulsive* for two like charges and *attractive* for two opposite charges.
>
> It is customary to round K to $9.0 \times 10^9 \, \text{N} \cdot \text{m}^2/\text{C}^2$ for all but extremely precise calculations, and we will do so.

We sometimes speak of the "force between charge q_1 and charge q_2," but keep in mind that we are really dealing with charged *objects* that also have a mass, a size, and other properties. Charge is not some disembodied entity that exists apart from matter. Coulomb's law describes the force between charged *particles*.

NOTE ▶ Coulomb's law applies only to *point charges*. A point charge is an idealized material object with charge and mass but with no size or extension. For practical purposes, two charged objects can be modeled as point charges if they are much smaller than the separation between them. ◀

Coulomb's law looks much like Newton's law of gravity, but there is a key difference: The charge q can be either positive or negative, so the forces can be attractive or repulsive. Consequently, the absolute value signs in Equation 20.1 are especially important. The first part of Coulomb's law gives only the *magnitude* of the force, which is always positive. The direction must be determined from the second part of the law. Figure 20.16 shows the forces between different combinations of positive and negative charges.

Using Coulomb's Law

Coulomb's law is a force law, and forces are vectors. **Electric forces, like other forces, can be superimposed.** If multiple charges 1, 2, 3, . . . are present, the *net* electric force on charge j due to all other charges is therefore the sum of all the individual forces due to each charge. That is,

$$\vec{F}_{net} = \vec{F}_{1\,onj} + \vec{F}_{2\,onj} + \vec{F}_{3\,onj} + \cdots \qquad (20.2)$$

where each of the $\vec{F}_{i\,onj}$ is given by Equation 20.1. These conditions are the basis of a strategy for using Coulomb's law to solve electric force problems.

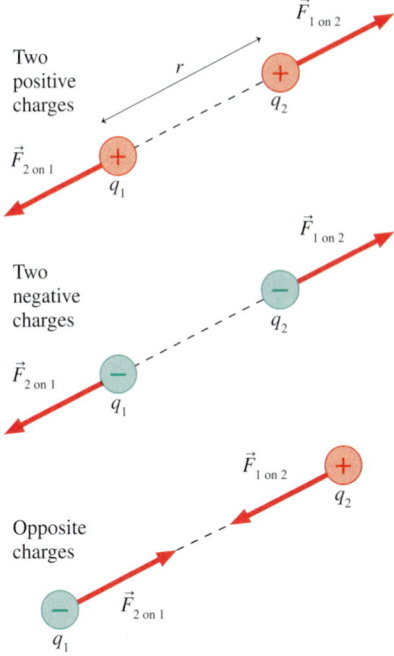

FIGURE 20.16 Attractive and repulsive forces between charges.

PROBLEM-SOLVING STRATEGY 20.1   Electric forces and Coulomb's law

PREPARE Identify point charges or objects that can be modeled as point charges. Create a visual overview in which you establish a coordinate system, show the positions of the charges, show the force vectors on the charges, define distances and angles, and identify what the problem is trying to find.

SOLVE The magnitude of the force between point charges is given by Coulomb's Law

$$F_{1\,on\,2} = F_{2\,on\,1} = \frac{K|q_1||q_2|}{r^2}$$

Use your visual overview as a guide to the use of this law:

- Show the directions of the forces—repulsive for like charges, attractive for opposite charges—on the visual overview.
- When possible, do graphical vector addition on the visual overview. While not exact, it tells you the type of answer you should expect.
- Write each force vector in terms of its x- and y-components, then add the components to find the net force. Use the visual overview to determine which components are positive and which are negative.

ASSESS Check that your result has the correct units, is reasonable, and answers the question.

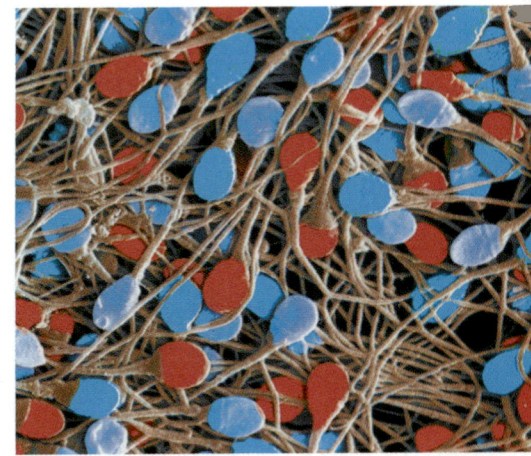

Separating the girls from the boys BIO
Sperm cells can be sorted according to whether they contain an X or a Y chromosome. The cells are put into solution, and the solution is forced through a nozzle, which breaks the solution into droplets. Suppose a droplet contains a sperm cell. An optical test measures which type of chromosome, X or Y, the cell has. The droplet is then given a positive charge if it has an X sperm cell, negative if it contains a Y. The droplets fall between two oppositely charged plates where they are pushed left or right depending on their charge—separating the X from the Y.

EXAMPLE 20.1 Adding electric forces in one dimension

Two $+10$ nC charged particles are 2.0 cm apart on the x-axis. What is the net force on a $+1.0$ nC charge midway between them? What is the net force if the charged particle on the right is replaced by a -10 nC charge?

PREPARE We proceed using the steps of Problem-Solving Strategy 20.1. We model the charged particles as point charges. The visual overview of Figure 20.17 establishes a coordinate system and shows the forces $\vec{F}_{1\,\text{on}\,3}$ and $\vec{F}_{2\,\text{on}\,3}$. Figure 20.17a shows a $+10$ nC charge on the right; Figure 20.17b shows a -10 nC charge.

FIGURE 20.17 A visual overview shows the forces for the two cases.

SOLVE Electric forces are vectors, and the net force on q_3 is the *vector* sum $\vec{F}_{\text{net}} = \vec{F}_{1\,\text{on}\,3} + \vec{F}_{2\,\text{on}\,3}$. Charges q_1 and q_2 each exert a repulsive force on q_3, but they are equal in magnitude and opposite in direction. Consequently, $\vec{F}_{\text{net}} = \vec{0}$. The situation changes if q_2 is negative, as in Figure 20.17b. In this case, the two forces are equal in magnitude but in the *same* direction, so $\vec{F}_{\text{net}} = 2\vec{F}_{1\,\text{on}\,3}$. The magnitude of the force is given by Coulomb's law. The force due to q_1 is

$$F_{1\,\text{on}\,3} = \frac{K|q_1||q_3|}{r_{13}^2}$$

$$= \frac{(9.0 \times 10^9 \text{ N} \cdot \text{m}^2/\text{C}^2)(10 \times 10^{-9} \text{ C})(1.0 \times 10^{-9} \text{ C})}{(0.010 \text{ m})^2}$$

$$= 9.0 \times 10^{-4} \text{ N}$$

There is an equal force due to q_2, so the net force on the 1.0 nC charge is $\vec{F}_{\text{net}} = (1.8 \times 10^{-3} \text{ N}, \text{ to the right})$.

ASSESS This example illustrates the important idea that electric forces are *vectors*. An important part of assessing our answer is to see if it is "reasonable." In the second case, the net force on the charge is approximately 1 mN. Generally, charges of a few nC separated by a few cm experience forces in the range from a fraction of a mN to several mN. With this guideline, the answer appears to be reasonable.

EXAMPLE 20.2 Adding electric forces in two dimensions

Three charged particles with $q_1 = -50$ nC, $q_2 = +50$ nC, and $q_3 = +30$ nC are placed as shown in Figure 20.18. What is the net force on charge q_3 due to the other two charges?

$q_1 = -50$ nC ←———— 10.0 cm ————→ $q_2 = +50$ nC

5.0 cm

$q_3 = +30$ nC

FIGURE 20.18 The arrangement of charges for Example 20.2.

PREPARE We solve for the net force using the steps of Problem Solving Strategy 20.1, beginning with the visual overview shown in Figure 20.19. We have defined a coordinate system, with charge q_3 at the origin. We have drawn the forces on charge q_3, with directions determined by the signs of the charges. We can see from the geometry that the forces $\vec{F}_{1\,\text{on}\,3}$ and $\vec{F}_{2\,\text{on}\,3}$ are at the angles noted in the figure. The vector addition in Figure 20.19b shows the anticipated direction of the net force; this will be a good check on our final result. The distance between charges q_1 and q_3 is the same as that between charges q_2 and q_3; this distance r is $r = \sqrt{(5.0 \text{ cm})^2 + (5.0 \text{ cm})^2} = 7.07$ cm.

FIGURE 20.19 A visual overview of the charges and forces.

SOLVE We are interested in the net force on charge q_3. Let's start by using Coulomb's law to compute the magnitudes of the two forces on charge q_3:

$$F_{1\,\text{on}\,3} = \frac{K|q_1||q_3|}{r^2}$$

$$= \frac{(9.0 \times 10^9 \text{ N} \cdot \text{m}^2/\text{C}^2)(50 \times 10^{-9} \text{ C})(30 \times 10^{-9} \text{ C})}{(0.0707 \text{ m})^2}$$

$$= 2.7 \times 10^{-3} \text{ N}$$

The magnitudes of the charges and the distance are the same for F_{2on3}, so

$$F_{2on3} = \frac{K|q_2||q_3|}{r^2} = 2.7 \times 10^{-3}\,\text{N}$$

The components of these forces are illustrated in Figure 20.20a. Computing values for the components, we find

$$(F_{1on3})_x = -(2.7 \times 10^{-3}\,\text{N})\cos 45° = -1.9 \times 10^{-3}\,\text{N}$$

$$(F_{1on3})_y = (2.7 \times 10^{-3}\,\text{N})\sin 45° = 1.9 \times 10^{-3}\,\text{N}$$

$$(F_{2on3})_x = -(2.7 \times 10^{-3}\,\text{N})\cos 45° = -1.9 \times 10^{-3}\,\text{N}$$

$$(F_{2on3})_y = -(2.7 \times 10^{-3}\,\text{N})\sin 45° = -1.9 \times 10^{-3}\,\text{N}$$

Next, we add components of the net force:

$$F_{3x} = (F_{1on3})_x + (F_{2on3})_x = -1.9 \times 10^{-3}\,\text{N} - 1.9 \times 10^{-3}\,\text{N}$$

$$= 3.8 \times 10^{-3}\,\text{N}$$

$$F_{3y} = (F_{1on3})_y + (F_{2on3})_y$$

$$= +1.9 \times 10^{-3}\,\text{N} - 1.9 \times 10^{-3}\,\text{N} = 0$$

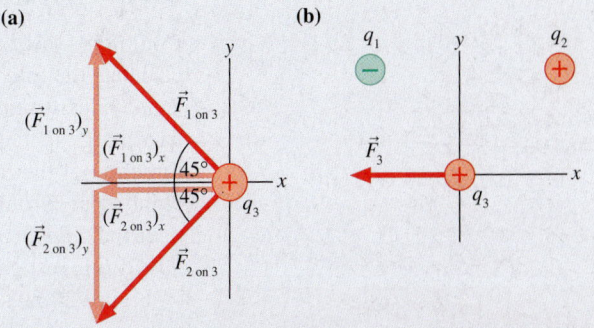

FIGURE 20.20 The net force on q_3 is to the left.

Thus the net force, as shown in Figure 20.20b, is

$$\vec{F}_3 = (3.8 \times 10^{-3}\,\text{N}, -x\text{-direction})$$

ASSESS The net force is directed to the left, as we anticipated. The magnitude of the net force, a few mN, seems reasonable as well.

EXAMPLE 20.3 Comparing electric and gravitational forces

A small plastic sphere is charged to -10 nC. It is held 1.0 cm above a small glass bead at rest on a table. The bead has a mass of 15 mg and a charge of $+10$ nC. Will the glass bead "leap up" to the plastic sphere?

PREPARE We model the plastic sphere and glass bead as point charges. Figure 20.21 establishes a y-axis, identifies the plastic sphere as q_1 and the glass bead as q_2, and shows a free-body diagram. We don't yet know the relative magnitudes of the gravitational and electric forces. In our diagram, we assume that $F_{1on2} < w$, so there is an additional upward normal force. This choice allows us to complete the diagram, but it does not affect our calculations or final answer.

FIGURE 20.21 A visual overview showing the charges and forces.

SOLVE If F_{1on2} is less than the bead's weight $w = m_2 g$, then the bead will remain at rest on the table with $\vec{F}_{1on2} + \vec{w} + \vec{n} = \vec{0}$. But if F_{1on2} is greater than the bead's weight, the glass bead will accelerate upward from the table. Using the values provided,

$$F_{1on2} = \frac{K|q_1||q_2|}{r^2} = 9.0 \times 10^{-3}\,\text{N}$$

$$w = m_2 g = 1.5 \times 10^{-4}\,\text{N}$$

F_{1on2} exceeds the weight by a factor of 60, so the glass bead will leap upward.

ASSESS The final answer is different from what we assumed in the diagram, but this assumption did not affect the final result. The values used in this example are realistic for spheres ≈ 2 mm in diameter. In general, as in this example, electric forces are *significantly* larger than weight forces. Consequently, we can neglect weight forces when working electric-force problems unless the particles are fairly massive.

STOP TO THINK 20.3 Charges 1 and 2 exert repulsive forces on each other. $q_1 = 4q_2$. Which statement is true?

A. $F_{1on2} > F_{2on1}$ B. $F_{1on2} = F_{2on1}$ C. $F_{1on2} < F_{2on1}$

20.4 The Concept of the Electric Field

Coulomb's law is the basic law of electrostatics. We can use Coulomb's law to calculate the force a positive charge exerts on a nearby negative charge. But there is an unanswered question: How does the negative charge "know" that the positive charge is there? Coulomb's law tells us how to calculate the magnitude and direction of the force, but it doesn't tell us how the force is transmitted through empty space from one charge to the other. To answer this question, we will introduce the *field model,* first suggested in the early 19th century by Michael Faraday, a British investigator of electricity and magnetism.

Figure 20.22 shows a photograph of the surface of a shallow pan of oil with tiny grass seeds floating on it. When charged spheres, one positive and one negative, touch the surface of the oil, the grass seeds line up to form a regular pattern. The pattern suggests that some kind of electric influence from the charges *fills the space* around the charges. Perhaps the grass seeds are reacting to this influence, creating the pattern that we see. This alteration of the space around the charges could be the *mechanism* by which the long-range Coulomb's law force is exerted.

This is the essence of the field model. Consider the attractive force between a positive charge A and a negative charge B. Figure 20.23 shows the difference between the force model, which we have been using, and the field model. **In the field model, it is the alteration of space around charge A that is the *agent* that exerts a force on charge B.** This alteration of space is what we call a **field.** The charge makes an alteration *everywhere* in space. Other charges then respond to the alteration at their position.

The field model applies to many branches of physics. The space around a charge is altered to create the **electric field.** The alteration of the space around a mass is called the **gravitational field.** The alteration of space around a magnet is called the **magnetic field,** which we will consider in Chapter 24.

FIGURE 20.22 Visualizing the electric field.

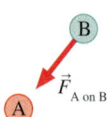

In the force model, A exerts a force directly on B.

In the field model, A alters the space around it. (The wavy lines are poetic license. We don't know what the alteration looks like.)

Particle B then responds to the altered space. The altered space is the agent that exerts the force on B.

FIGURE 20.23 The force and field models for the interaction between two charges.

The Field Model

The field model might seem arbitrary and abstract. We can already calculate forces between charges; why introduce another way of looking at things? We will find that the field model is a very useful tool for visualizing and calculating forces for complex arrangements of charges. However, there's a more fundamental reason for introducing the electric field. When we begin to study fields that change with time, we'll find phenomena that can be understood *only* in terms of fields.

We begin our investigation of electric fields by postulating a **field model** that describes how two charges interact:

1. Some charges, which we will call the **source charges,** alter the space around them by creating an *electric field* $\vec{E}$.

2. A separate charge *in* the electric field then experiences a force $\vec{F}$ exerted *by the field.*

Suppose charge q experiences an electric force $\vec{F}_{onq}$ due to other charges. The strength and direction of this force vary as q is moved from point to point in space. This suggests that "something" is present at each point in space to cause the force that charge q experiences. We define the electric field $\vec{E}$ at the point (x, y, z) as

$$\vec{E} \text{ at } (x, y, z) = \frac{\vec{F}_{onq} \text{ at } (x, y, z)}{q} \qquad (20.3)$$

Electric field at a point defined by the force on charge q

We're *defining* the electric field as a force-to-charge ratio, hence the units of the electric field are newtons per coulomb, or N/C. The magnitude E of the electric field is called the **electric field strength.** Typical orders of magnitude of the electric field strength are given in Table 20.2.

You can think of using charge q as a *probe* to determine if an electric field is present at a point in space. If charge q experiences an electric force at a point in space, as Figure 20.24a shows, we say that there is an electric field at that point causing the force. Further, we *define* the electric field at that point to be the vector given by Equation 20.3. Figure 20.24b shows the electric field at two points, but you can imagine "mapping out" the electric field by moving the charge q all through space.

TABLE 20.2 Typical electric field strengths

Field	Field strength (N/C)
Inside a current-carrying wire	10^{-2}
Near the earth's surface	10^{2}
Near objects charged by rubbing	10^{3} to 10^{6}
Needed to cause a spark in air	10^{6}
Inside a cell membrane	10^{7}
Inside an atom	10^{11}

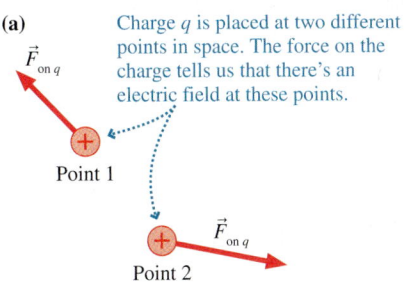

(a) Charge q is placed at two different points in space. The force on the charge tells us that there's an electric field at these points.

Point 1

Point 2

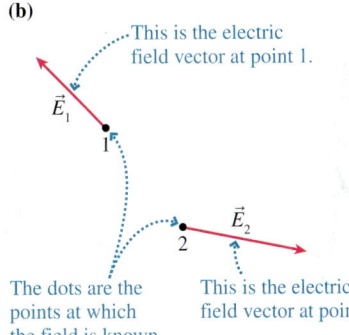

(b) This is the electric field vector at point 1.

The dots are the points at which the field is known.

This is the electric field vector at point 2.

FIGURE 20.24 Charge q is a probe of the electric field.

The basic idea of the field model is that **the field is the agent that exerts an electric force on charge** q. Notice three important things about the field:

1. The electric field, a vector, exists at every point in space. Electric field diagrams will show a sample of the vectors, but there is an electric field vector at every point whether one is shown or not.
2. If the probe charge q is positive, the electric field vector points in the same direction as the force on the charge.
3. Because q appears in Equation 20.3, it may seem that the electric field depends on the size of the charge used to probe the field. It doesn't. We know from Coulomb's law that the force $\vec{F}_{onq}$ is proportional to q. Thus the electric field defined in Equation 20.3 is *independent* of the charge q that probes the field. The electric field depends only on the source charges that create the field.

The Electric Field of a Single Charge

To continue to develop the idea of the electric field, we will determine the electric field of a single point charge q. Figure 20.25a shows a charge q that creates an electric field at all points in space. We can use a second charge, shown as q' in Figure 20.25b, to serve as a probe of the electric field due to charge q.

For the moment, assume both charges are positive. The force on q', which is repulsive and directed straight away from q, is given by Coulomb's law:

$$\vec{F}_{onq'} = \left(\frac{Kqq'}{r^2}, \text{ away from } q \right) \tag{20.4}$$

Equation 20.3 defines the electric field in terms of the force on the probe charge as $\vec{E} = \vec{F}_{onq'}/q'$, so for a positive charge q

$$\vec{E} = \left(\frac{Kq}{r^2}, \text{ away from } q \right) \tag{20.5}$$

The electric field is shown in Figure 20.25c.

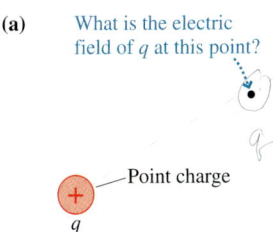

(a) What is the electric field of q at this point?

Point charge

q

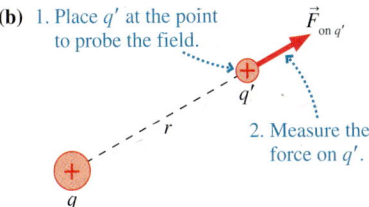

(b) 1. Place q' at the point to probe the field.

$\vec{F}_{onq'}$

q'

r

2. Measure the force on q'.

q

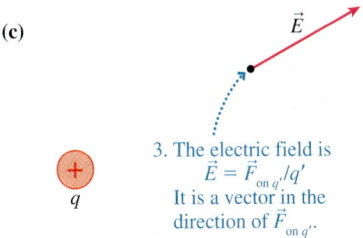

(c)

$\vec{E}$

q

3. The electric field is $\vec{E} = \vec{F}_{onq'}/q'$. It is a vector in the direction of $\vec{F}_{onq'}$.

FIGURE 20.25 Charge q' is used to probe the electric field of point charge q.

If q is negative, the magnitude of the force on the probe charge is the same as in Equation 20.5, but the direction is toward q, so the general expression for the field is

$$\vec{E} = \left(\frac{K|q|}{r^2}, \begin{bmatrix} \text{away from } q \text{ if } q > 0 \\ \text{toward } q \text{ if } q < 0 \end{bmatrix}\right) \quad (20.6)$$

Electric field of point charge q at a distance r

LINEAR
p. 38

INVERSE-SQUARE
p. 194

NOTE ▶ The expression for the electric field is similar to Coulomb's law. To distinguish the two, remember that Coulomb's law has a product of two charges in the numerator. It describes the force between *two* charges. The electric field has a single charge in the numerator. It is the field of a *single* charge. ◀

(a) The electric field of a positive point charge

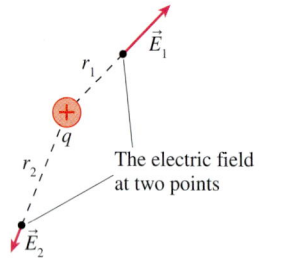

The electric field at two points

EXAMPLE 20.4 Finding the electric field of a proton

The electron in a hydrogen atom orbits the proton at a radius of 0.053 nm. What is the electric field due to the proton at the position of the electron?

SOLVE The proton's charge is $q = e$. At the distance of the electron the magnitude of the field is

$$E = \frac{Ke}{r^2} = \frac{(9.0 \times 10^9 \text{ N} \cdot \text{m}^2/\text{C}^2)(1.60 \times 10^{-19} \text{ C})}{(5.3 \times 10^{-11} \text{ m})^2}$$

$$= 5.1 \times 10^{11} \text{ N/C}$$

Because the proton is positive, the electric field is directed away from the proton:

$$\vec{E} = (5.1 \times 10^{11} \text{ N/C, outward from the proton})$$

ASSESS This is a large field, but Table 20.2 shows that this is the correct magnitude for the field within an atom.

(b) The electric field diagram of a positive point charge

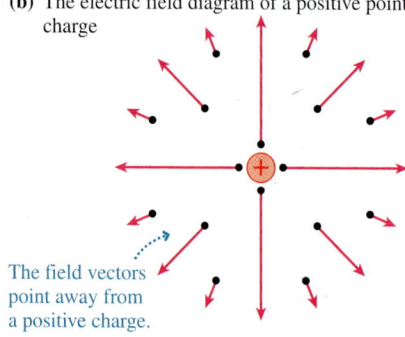

The field vectors point away from a positive charge.

(c) The electric field diagram of a negative point charge

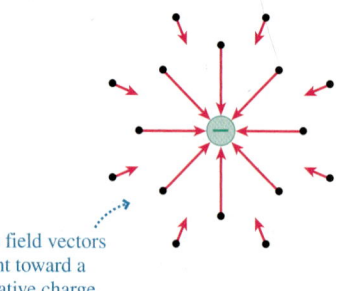

The field vectors point toward a negative charge.

FIGURE 20.26 The electric field near a point charge.

In Figure 20.26a, the field strength E_1 is larger than the field strength E_2 because $r_1 < r_2$. If we calculate the field at a sufficient number of points, we can draw a *field diagram* such as the one shown in Figure 20.26b. Notice that the field vectors all point straight away from charge q. We can draw a field diagram for a negative point charge in a similar fashion, as in Figure 20.26c. In this case, the field vectors point toward the charge, as this would be the direction of the force on a positive probe charge.

In the coming sections, as we use electric field diagrams, keep these points in mind:

1. The diagram is just a representative sample of electric field vectors. The field exists at all the other points. A well-drawn diagram gives a good indication of what the field would be like at a neighboring point.
2. The arrow indicates the direction and the strength of the electric field *at the point to which it is attached*—at the point where the *tail* of the vector is placed. The length of any vector is significant only relative to the lengths of other vectors.
3. Although we have to draw a vector across the page, from one point to another, an electric field vector does not "stretch" from one point to another. Each vector represents the electric field at *one point* in space.

STOP TO THINK 20.4 Rank in order, from largest to smallest, the electric field strengths E_A to E_D at points A to D.

20.5 Applications of the Electric Field

Suppose we want to find the electric field due to more than one source charge. For any number of source charges, the electric field at a point in space is found by looking at the force on a probe charge. Because the net force on the probe charge is the vector sum of the forces due to all of the individual charges, **the electric field due to multiple charges is the vector sum of the electric field due to each of the charges.**

EXAMPLE 20.5 Finding the field near a dipole

A dipole has a positive and negative charge separated by 1.2 cm, as shown in Figure 20.27. What is the electric field along the line connecting the charges at a point 1.2 cm to the right of the positive charge?

FIGURE 20.27 Charges and distances for Example 20.5.

PREPARE We define the x axis to be along the line connecting the two charges, as in Figure 20.28. The dipole has no net charge, but it does have a net electric field. The point at which we calculate the field is 1.2 cm from the positive charge and 2.4 cm from the negative charge. Thus the electric field of the positive charge will be larger, as shown in Figure 20.28. The net electric field of the dipole is the vector sum of these two fields, so the electric field of the dipole at this point is in the positive x-direction.

FIGURE 20.28 Visual overview for Example 20.5.

SOLVE The magnitudes of the fields of the two charges are given by Equation 20.6, so the magnitude of the dipole field is

$$E_{\text{dipole}} = E_+ - E_-$$

$$= \frac{\left(9.0 \times 10^9 \, \dfrac{\text{N} \cdot \text{m}^2}{\text{C}^2}\right)(1.5 \times 10^{-9} \, \text{C})}{(0.012 \, \text{m})^2} - \frac{\left(9.0 \times 10^9 \, \dfrac{\text{N} \cdot \text{m}^2}{\text{C}^2}\right)(1.5 \times 10^{-9} \, \text{C})}{(0.024 \, \text{m})^2}$$

$$= 7.0 \times 10^4 \, \text{N/C}$$

ASSESS This is a reasonable field value for these charges and distances.

(a)

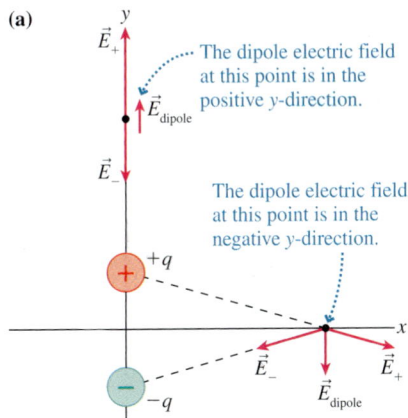

The dipole electric field at this point is in the positive y-direction.

The dipole electric field at this point is in the negative y-direction.

(b)

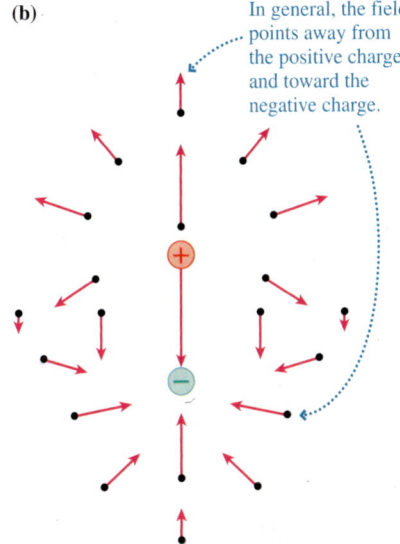

In general, the field points away from the positive charge and toward the negative charge.

The electric dipole is an important charge distribution that we will see many times, so it's worth exploring the full field diagram. Figure 20.29 shows a dipole oriented along the y-axis. We can determine the field at any point by a vector addition of the fields of the two charges, as shown in Figure 20.29a. If we repeat this process at many points, we end up with the field diagram of Figure 20.29b. This is more complex than the field of a single charge, but it accurately shows how two charges alter the space around them.

FIGURE 20.29 The electric field of a dipole.

Uniform Electric Fields

Figure 20.30a shows another important practical situation, one we'll meet many times. Two conducting plates, called **electrodes,** are face-to-face with a narrow gap between them. One electrode has total charge $+Q$ and the other has total charge $-Q$. This arrangement of two electrodes, closely spaced and charged equally but oppositely, is called a **parallel-plate capacitor.** What is the electric field between the two plates? To keep things simple, we will focus on the field in the central region, far from the edges. Figure 20.30b shows a blown-up cross-section view of a region near the center of the plates.

At each point, the electric field is the vector sum of the fields from all of the positive charges and all of the negative charges on the plates. However, the field of a point charge decreases inversely with the square of its distance, so in practice only the nearby charges contribute to the field. As Figure 20.31a shows, the horizontal components of the individual fields cancel while the vertical components add to give an electric field vector pointing from the positive plate toward the negative plate. The exact position of the point we've chosen is not crucial; moving either right or left would produce a similar result.

By mapping the electric field at many points, we find that the field inside a parallel-plate capacitor is the same—in both strength and direction—at every point. This is called a **uniform electric field.** Figure 20.31b shows that a uniform electric field is represented with parallel electric field vectors of equal length. A more detailed analysis finds that the electric field inside a parallel-plate capacitor is

$$\vec{E}_{\text{capacitor}} = \left(\frac{Q}{\epsilon_0 A}, \text{ from positive to negative} \right) \quad (20.7)$$

Electric field in a parallel-plate capacitor
with plate area A and charge Q

LINEAR p. 38 INVERSE p. 118

Equation 20.7 introduces a new constant ϵ_0, pronounced "epsilon zero" or "epsilon naught," called the **permittivity constant.** Its value is related to the electrostatic constant as

$$\epsilon_0 = \frac{1}{4\pi K} = 8.85 \times 10^{-12} \text{ C}^2/\text{N} \cdot \text{m}^2$$

There are a few things to note about the field in a parallel-plate capacitor:

- The field depends on the charge-to-area ratio Q/A, which is often called the *charge density.* If the charges are packed more closely, the field will be larger.
- Our analysis requires that the separation of the plates be small compared to their size. If this is true, the spacing between the plates does not affect the electric field, *and this spacing does not appear in Equation 20.7.*
- Although Figure 20.30 shows circular electrodes, the shape of the electrodes—circular or square or any other shape—is not relevant as long as the electrodes are very close together.

NOTE ▶ The charges on the plates are equal and opposite, $+Q$ and $-Q$, so the net charge is zero. The symbol Q in Equation 20.7 is the *magnitude* of the charge on each plate. ◀

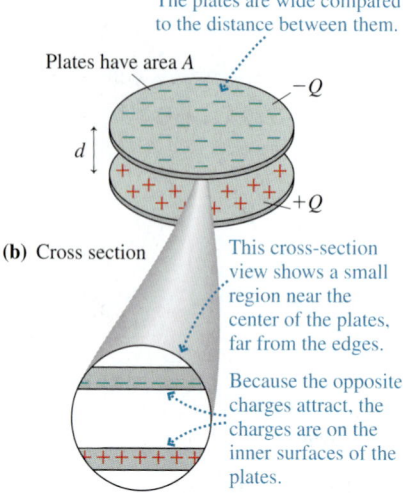

(a) Parallel-plate capacitor

The plates are wide compared to the distance between them.

Plates have area A

$-Q$

d

$+Q$

(b) Cross section

This cross-section view shows a small region near the center of the plates, far from the edges.

Because the opposite charges attract, the charges are on the inner surfaces of the plates.

FIGURE 20.30 A parallel-plate capacitor.

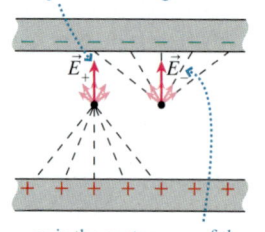

(a) The vector sum of the fields from the positive charges is directed from the positive plate to the negative . . .

$\vec{E}_+$ $\vec{E}$

. . . as is the vector sum of the fields from the negative charges.

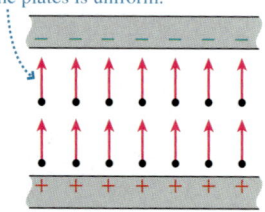

(b) The electric field between the plates is uniform.

FIGURE 20.31 The electric field inside a parallel-plate capacitor.

EXAMPLE 20.6 Finding the field in an air cleaner

Long highway tunnels must have air cleaners to remove dust and soot coming from passing cars and trucks. In one type, known as an *electrostatic precipitator,* air passes between two oppositely charged metal plates, as in Figure 20.32. The large electric field between the plates ionizes dust and soot particles, which then feel a force due to the field. This force causes the charged particles to move toward and stick to one or the other

The two plates have opposite charges.

−540 nC   +540 nC

38.0 cm

Air with suspended dust and smoke particles enters here.

Clean air exits here.

20.6 cm

0.900 cm

FIGURE 20.32 An electrostatic precipitator.

plate, removing them from the air. A typical unit has dimensions and charges as shown in Figure 20.32. What is the electric field between the plates?

PREPARE Because the spacing between the plates is much smaller than their size, this is a parallel-plate capacitor with a uniform electric field between the plates.

SOLVE We find the field using Equation 20.7. The direction is from the positive to the negative plate, which is to the left. The area of the plates is $A = (0.206 \text{ m})(0.380 \text{ m}) = 0.0783 \text{ m}^2$, so the field strength between the plates is

$$E = \frac{Q}{\epsilon_0 A} = \frac{540 \times 10^{-9} \text{ C}}{(8.85 \times 10^{-12} \text{ C}^2/\text{N} \cdot \text{m}^2)(0.0783 \text{ m}^2)}$$

$$= 7.79 \times 10^5 \text{ N/C}$$

The question asked for the electric field, a vector, not just for the field strength. The electric field between the plates is

$$\vec{E} = (7.79 \times 10^5 \text{ N/C, to the left}).$$

ASSESS Table 20.2 says that a field of 10^6 N/C will create a spark in air. The field we calculated between the plates is just a bit less than this, which makes sense. The field should be large, but not large enough to make a spark jump between the plates!

Electric Field Lines

We can't see the electric field, so we use pictorial tools like electric field diagrams to help us visualize the electric field in a region of space. Another way to picture the field is to draw **electric field lines.** These are imaginary lines drawn through a region of space so that

Activ Physics ONLINE   11.4–11.6

- The tangent to a field line at any point is in the direction of the electric field $\vec{E}$ at that point, and
- The field lines are closer together where the electric field strength is larger.

Figure 20.33a shows the relationship between electric field lines and electric field vectors in one region of space. If we know what the field vectors look like, we can extrapolate to the field lines, as in Figure 20.33b and c for the electric field lines near a positive charge and between the plates of a capacitor.

(a) Relationship between field vectors and field lines

The electric field vector is tangent to the electric field line.

Field vector

Field line

The electric field is stronger where the electric field vectors are longer and where the electric field lines are closer together.

(b) Field lines of a positive point charge

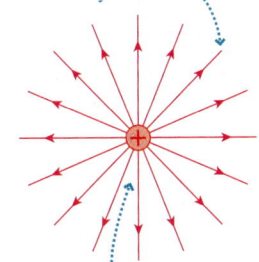

The field is directed away from the positive charge, so the field lines are directed radially outward.

The field lines are closest together near the charge, where the field strength is greatest.

(c) Field lines in an ideal capacitor

The field vectors are directed from the positive to the negative plate, so the field lines are as well.

The field is constant, so the field lines are evenly spaced.

FIGURE 20.33 Field vectors and field lines.

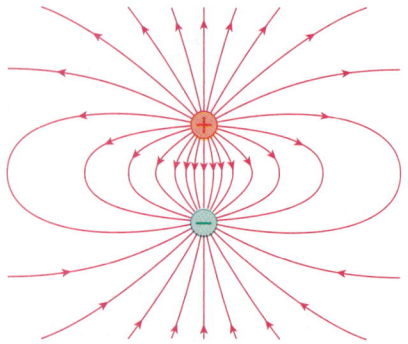

FIGURE 20.34 Electric field lines for a dipole.

(a) The electric dipole of the heart

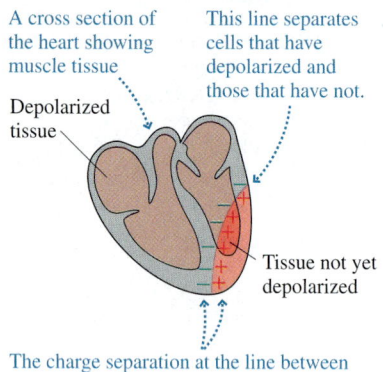

A cross section of the heart showing muscle tissue

This line separates cells that have depolarized and those that have not.

Depolarized tissue

Tissue not yet depolarized

The charge separation at the line between the two regions creates an electric dipole.

(b) The field of the heart in the body

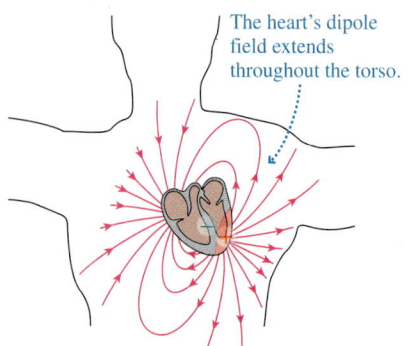

The heart's dipole field extends throughout the torso.

FIGURE 20.35 The beating heart generates a dipole electric field.

If we have an arrangement of charges, we can draw field lines as a guide to what the field looks like. As you generate a field line picture, there are a few rules to keep in mind:

■ Field lines cannot cross. The tangent to the field line is the electric field vector, which indicates the direction of the force on a positive charge. The force must be in a unique, well-defined direction, so two field lines cannot cross.
■ The electric field is created by charges. Field lines start on a positive charge and end on a negative charge.

You can use the above information as the basis of a technique for sketching a field line picture for an arrangement of charges. Draw field lines starting on positive charges and moving toward negative charges. Draw the lines tangent to the field vector at each point. Make the lines close together where the field is strong, far apart where the field is weak. For example, Figure 20.34 pictures the electric field of a dipole using electric field lines. You should compare this to Figure 20.29b, which illustrated the field with field vectors.

The Electric Field of the Heart

Nerve and muscle cells have a prominent electrical nature. As we will see in detail in Chapter 23, a cell membrane is an insulator that encloses a conducting fluid and is surrounded by conducting fluid. While resting, the membrane is *polarized* with positive charges on the outside of the cell, negative charges on the inside. When a nerve or a muscle cell is stimulated, the polarity of the membrane switches; we say that the cell *depolarizes*. Later, when the charge balance is restored, we say that the cell *repolarizes*.

All nerve and muscle cells generate an electrical signal when depolarization occurs, but the largest electrical signal in the body comes from the heart. The rhythmic beating of the heart is produced by a highly coordinated wave of depolarization that sweeps across the tissue of the heart. As Figure 20.35a shows, the surface of the heart is positive on one side of the boundary between tissue that is depolarized and tissue that is not yet depolarized, negative on the other. In other words, the heart is a large electric dipole. The orientation and strength of the dipole change during each beat of the heart as the depolarization wave sweeps across it.

The electric dipole of the heart generates a dipole electric field that extends throughout the torso, as shown in Figure 20.35b. As we will see in Chapter 21, an *electrocardiogram* measures the changing electric field of the heart as it beats. Measurement of the heart's electric field can be used to diagnose the operation of the heart.

STOP TO THINK 20.5 Which of the following is the correct representation of the electric field created by two positive charges?

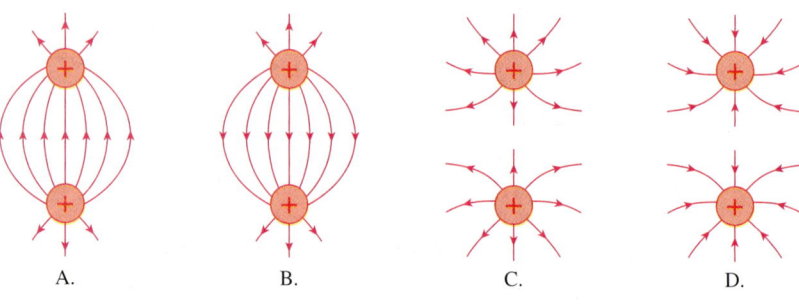

A.   B.   C.   D.

20.6 Conductors in Electric Fields

We started the chapter by looking at experiments with charges and forces. We saw that a charged rod attracts neutral metal objects, and we learned that this attraction is due to the polarization of charges on the conductor. Now that we've introduced electric fields, we can look more closely at this situation. What happens to a conductor in an electric field?

The Electric Field Inside and Outside a Conductor

Consider a conductor in electrostatic equilibrium (recall that this means that none of the charges are moving). Suppose there were an electric field inside the conductor. Electric fields exert forces on charges, so an internal electric field would exert forces on the conductor's charge carriers. The charge carriers in a conductor are free to move—that's what makes it a conductor—and they *would* move in response to these forces. But that would violate the assumption that all the charges are at rest. Thus we're forced to conclude that **the electric field is zero at all points inside a conductor in electrostatic equilibrium.**

Figure 20.36a shows a conductor with excess positive charge. We've already seen that the excess charge lies on the surface because repulsive forces between the charges push them as far apart as possible. We've just seen that the electric field inside the conductor is zero, but the fact that the conductor has a net charge requires an exterior electric field in the space outside the conductor. Figure 20.36b shows that **the electric field right at the surface of a charged conductor is perpendicular to the surface.** To see that this is so, suppose $\vec{E}$ had a component tangent to the surface. This component of $\vec{E}$ would exert a force on charges at the surface and cause them to move along the surface, thus violating the assumption that all charges are at rest. The only exterior electric field consistent with electrostatic equilibrium is one that is perpendicular to the surface.

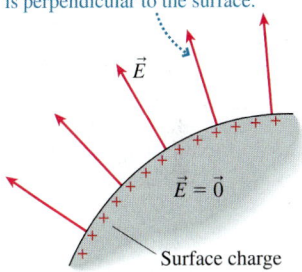

(a) The electric field inside the conductor is zero.

$\vec{E} = \vec{0}$

All excess charge is on the surface.

(b) The electric field at the surface is perpendicular to the surface.

$\vec{E}$

$\vec{E} = \vec{0}$

Surface charge

FIGURE 20.36 The electric field inside and outside a charged conductor.

CONCEPTUAL EXAMPLE 20.3 Drawing electric field lines for a charged sphere and a plate

Figure 20.37 shows a positively charged metal sphere above a conducting plate with a negative charge. Sketch the electric field lines.

REASON Field lines start on positive charges end on negative charges. Thus we draw the field lines from the positive sphere to the negative plate, perpendicular to both surfaces, as shown in Figure 20.38. The single field line that goes upward tells us that there is a field above the sphere, but that it is weak.

FIGURE 20.37 The charged sphere and plate for Conceptual Example 20.3.

These are conductors in electrostatic equilibrium, so the field is perpendicular to the surfaces.

FIGURE 20.38 Drawing field lines from sphere to plate.

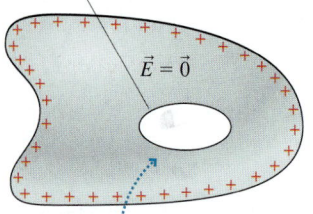

A hollow completely
enclosed by the conductor

$\vec{E} = \vec{0}$

The electric field inside the
enclosed hollow is zero.

FIGURE 20.39 A region of space enclosed by conducting walls is screened from electric fields.

The charges are closer together and the electric field is strongest at the pointed end.

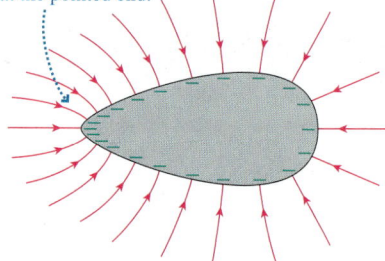

FIGURE 20.40 The electric field is strongest at the pointed end.

TRY IT YOURSELF

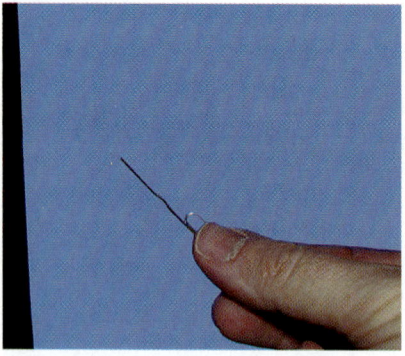

Discharging a television screen A television screen or computer monitor (one that uses an electron beam, but not a plasma or LCD screen) can build up a large static charge. You can discharge the screen by making a "lightning rod" from an unbent paper clip or a pin, as in the photo. Move the point back and forth near to—but not touching—the screen. You should be able to hear the quiet hiss of an electric discharge as the air near the tip is ionized.

Figure 20.39 shows a practical use of these ideas. Here we see a charged conductor with a completely enclosed hollow. The excess charge on the conductor is at the surface and the electric field within the conductor is zero, so there's nothing that could create an electric field within the enclosure. We can conclude that **the electric field within an enclosure is zero.**

A conducting box can be used to exclude electric fields from a region of space; this is called **screening.** Solid metal walls are ideal, but in practice wire screen or wire mesh provides sufficient screening for all but the most sensitive applications.

CONCEPTUAL EXAMPLE 20.4 Analyzing static protection

Computer chips and other electronic components are very sensitive to electric charges and fields. Even a small static charge or field may damage them. Such components are shipped and stored in conducting bags. How do these bags protect the components stored inside?

REASON Such a bag, when sealed, forms a conducting shell around its interior. All excess charge is on the surface of the bag, and the electric field inside is zero. A chip or component inside the bag is protected from damaging charges and fields.

Although any excess charge on a conductor will be found on the surface, it may not be uniformly distributed. Figure 20.40 shows a charged conductor that is more pointed at one end than the other. It turns out that the density of charge is highest—and thus the electric field is strongest—at the pointed end.

The sharper the point, the stronger the field. The electric field near very sharp points may be strong enough to ionize the air around it. Lightning rods on buildings have such a point at the top. If charge begins to accumulate on the building, meaning a lightning strike might be imminent, a large field develops at the tip of the rod. Once the field ionizes the air, excess charge from the building can dissipate into the air, reducing the electric field and thus reducing the probability of a lightning strike. A lightning rod is intended to *prevent* a lightning strike.

A Conductor in an External Electric Field

Suppose we place a piece of metal in the uniform electric field between the plates of a charged capacitor, as shown in Figure 20.41a. What happens next? In equilibrium, the electric field inside the conductor is zero. To reach this equilibrium, the charges in the metal separate as the metal polarizes in response to the field of the capacitor. The charges at the surfaces of the metal create their own electric field, and the net electric field in the metal is then the sum of the electric field of the capacitor plus the electric field of the separated charges. Charge separation proceeds until the interior field is zero, and at that point the conductor is in electrostatic equilibrium. Equilibrium is reached within a few picoseconds, which for nearly all practical purposes is instantaneous.

The induced surface charges on the metal may cancel the field *inside* the piece of metal, but they make the field *outside* the piece of metal quite a bit more complicated, as we can see by comparing Figures 20.41a and 20.41b. Notice, however, that the field remains perpendicular to the conductor at all points. Even though the field is complicated, the fact that it has to be perpendicular to conducting surfaces allows us to roughly sketch the shape of the field.

(a)

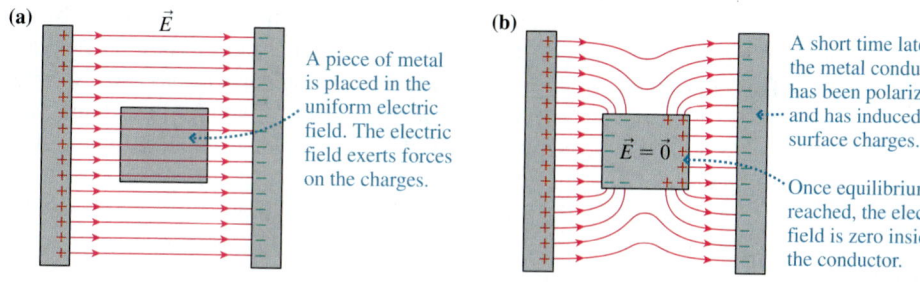

A piece of metal is placed in the uniform electric field. The electric field exerts forces on the charges.

(b)

A short time later, the metal conductor has been polarized and has induced surface charges.

$\vec{E} = \vec{0}$

Once equilibrium is reached, the electric field is zero inside the conductor.

FIGURE 20.41 A conductor in an electric field.

▶ **Electrolocation** BIO Many fish have stacks of specially adapted cells called *electrocytes* that develop electric charges across them. The electrocytes in the tail of this fish (called an elephant nose) form an electric dipole that produces an electric field in the water around it. The elephant nose has sensors along its body that can detect very small changes in this electric field. A nearby conductor—such as another fish—will alter this field. The elephant nose uses these changes in the field to "see" around it. These fish live in very murky water where vision is of little use, so you can see the advantage of having such an alternative form of perception.

20.7 Forces and Torques on Charges in Electric Fields

The electric field was defined in terms of the force on a charge. In practice, we often want to turn the definition around to find the force exerted on a charge in a known electric field. If a charge q is placed at a point in space where the electric field is $\vec{E}$, the charge experiences an electric force

Activ Physics ONLINE 11.9, 11.10

$$\vec{F}_{on\,q} = q\vec{E} \qquad (20.8)$$

Force on a charge due to an electric field

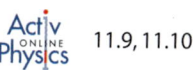

LINEAR p. 38

If q is positive, the force on charge q is in the direction of $\vec{E}$. The force on a negative charge is *opposite* the direction of $\vec{E}$.

EXAMPLE 20.7 Finding the force on an electron in the atmosphere

Under normal circumstances, the earth's electric field outdoors near ground level is uniform, about 100 N/C, directed down. What is the electric force on a free electron in the atmosphere? What acceleration does this force cause?

PREPARE The electric field is uniform, as shown in the field diagram of Figure 20.42. Whatever the position of the electron,

it experiences the same field. Because the electron is negative, the force on it is opposite the field—upward.

SOLVE The magnitude of the force is given by Equation 20.8:

$$F = eE = (1.6 \times 10^{-19}\ \text{C})(100\ \text{N/C}) = 1.6 \times 10^{-17}\ \text{N}$$

Thus the force on the electron is

$$\vec{F} = (1.6 \times 10^{-17}\ \text{N, upward})$$

The electron will accelerate upward, in the direction of the force. The magnitude of the acceleration is

$$a = \frac{F}{m} = \frac{1.6 \times 10^{-17}\ \text{N}}{9.1 \times 10^{-31}\ \text{kg}} = 1.8 \times 10^{13}\ \text{m/s}^2$$

ASSESS This everyday field produces an extremely large acceleration on a free electron. Forces and accelerations at the atomic scale are quite different from what we are used to for macroscopic objects.

The electric field is uniform.

The force is in the direction opposite the field.

$\vec{F}$

$\vec{E}$

FIGURE 20.42 An electron in the earth's electric field.

The electric field between the electrodes exerts a force on the negatively charged DNA fragments.

DNA samples begin here.

$\vec{F}$

$\vec{v}$

Different fragments have different sizes and migrate at different rates.

Gel

FIGURE 20.43 Electrophoresis of DNA samples.

The photo at the start of the chapter showed the colored lines produced by gel electrophoresis of a sample of DNA. The first step of the analysis is to put the sample of DNA into solution. The DNA is then cut into fragments by enzymes. In solution, these fragments have a negative charge. Drops of solution containing the charged DNA fragments are placed in wells at one end of a container of gel. Electrodes at opposite ends of the gel create an electric field that exerts an electric force on the DNA fragments in the solution, as illustrated in Figure 20.43. The electric force makes the fragments move through the gel, but drag forces cause fragments of different size migrate at different rates, with smaller fragments migrating faster than larger ones. After some time, the fragments sort themselves into distinct lines, creating a "genetic fingerprint." Two identical samples of DNA will produce the same set of fragments and thus the same pattern in the gel, but the odds are extremely small that two unrelated DNA samples would produce the same pattern.

EXAMPLE 20.8 Finding the force and torque on a dipole in a uniform field

Two charges, $+q$ and $-q$, are joined by a rigid rod of length L to make a dipole. This dipole is in the uniform electric field shown in Figure 20.44. What is the net force on the dipole? What is the net torque?

PREPARE We begin with the visual overview of Figure 20.45 that shows the forces on the charges.

FIGURE 20.44 A dipole in a uniform field.

(a) Forces

The negative charge feels a force opposite the field; the positive charge, in the direction of the field.

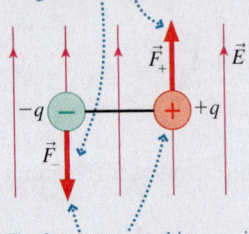

The forces are equal in magnitude but opposite in direction.

(b) Torques

There is a natural pivot point in the center.

Both torques will rotate the dipole counterclockwise.

FIGURE 20.45 Force and torque on a dipole in a uniform field.

SOLVE As we see in Figure 20.45a, the two charges experience forces of equal magnitude $F = qE$ but opposite directions. Thus the net force is zero. The net *torque* is not zero; as shown in Figure 20.45b, the forces cause the dipole to rotate counterclockwise.

To compute torques, we choose a pivot point in the center of the dipole. As we learned in Chapter 7, the torque exerted by a force $\vec{F}$ is $\tau = rF_{\perp}$, where r is the lever arm and $F_{\perp}$ is the component of $\vec{F}$ perpendicular to the lever arm. In this case, both lever arms are $r = L/2$ and forces $\vec{F}_+$ and $\vec{F}_-$ are perpendicular to the lever arms. Thus the net torque is

$$\tau = \tau_+ + \tau_- = F_+\left(\frac{L}{2}\right) + F_-\left(\frac{L}{2}\right) = qEL$$

ASSESS The final result is a force (qE, the magnitude of the electric force) multiplied by a distance (L, the separation of the charges), which is correct for a torque.

It is useful to define the **electric dipole moment,** a vector pointing from the negative to the positive charge of a dipole. As Example 20.9 and Figure 20.46a show, an electric dipole in a uniform electric field experiences a torque that causes it to rotate. When the dipole moment is lined up with the field, as shown in Figure 20.46b, there is no torque. **The equilibrium position of a dipole in an electric field is with the electric dipole moment aligned with the field.**

Earlier, we saw a photo of grass seeds lined up with the electric field from two charged electrodes. Now we can understand why the seeds line up as they do. First, the electric field polarizes the seeds, inducing opposite charges on their ends. The seeds are induced electric dipoles, with dipole moments along the axis of each seed. Second, torques on the dipole moments cause them to line up with the electric field, revealing its structure.

This technique is similar to something else that you may have seen before: the use of iron filings to visualize a *magnetic* field. An iron filing develops a *magnetic* dipole moment that causes it to line up with an applied *magnetic* field. We will begin our study of magnetic fields in Chapter 24 by using iron filings in this fashion to illustrate the structure of the magnetic fields.

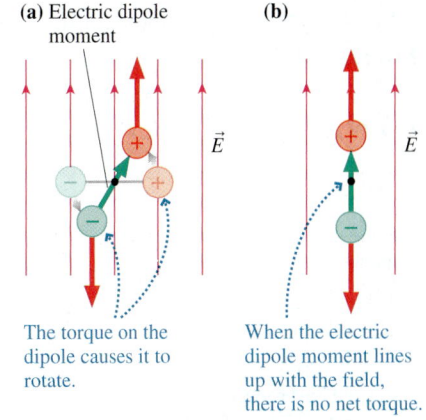

(a) Electric dipole moment   (b)

The torque on the dipole causes it to rotate.

When the electric dipole moment lines up with the field, there is no net torque.

FIGURE 20.46 The torque on a dipole.

| STOP TO THINK 20.6 | Rank in order, from largest to smallest, the forces F_A to F_E a proton would experience if placed at points A to E in this parallel-plate capacitor.

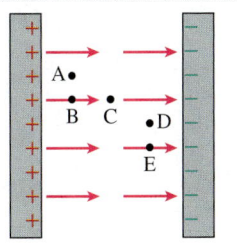

SUMMARY

The goal of Chapter 20 has been to develop a basic understanding of electric phenomena in terms of charges, forces, and fields.

GENERAL PRINCIPLES

Charge

There are two kinds of charges, called positive and negative.

- Atoms have nuclei with positive charge and electrons with negative charge.

- The **fundamental charge** e is the magnitude of the charge on an electron or proton: $e = 1.60 \times 10^{-19}$ C

- Matter with equal amounts of positive and negative charge is neutral.

- Charge is conserved; it can't be created or destroyed.

Coulomb's Law

The forces between two charged particles q_1 and q_2 separated by distance r are

$$F_{1\,\text{on}\,2} = F_{2\,\text{on}\,1} = \frac{K|q_1||q_2|}{r^2}$$

These forces are an action/reaction pair directed along the line joining the particles.

- The forces are repulsive for two like charges, attractive for two opposite charges.

- The net force on a charge is the vector sum of the forces from all other charges.

- The unit of charge is the coulomb (C).

IMPORTANT CONCEPTS

The Electric Field

Charges interact with each other via the electric field $\vec{E}$.

- Charge A alters the space around it by creating an electric field.

- The field is the agent that exerts a force on charge B.

$$\vec{F}_{\text{on B}} = q_B \vec{E}$$

- An electric field is identified and measured in terms of the force on a probe charge q. The unit of the electric field is N/C.

$$\vec{E} = \frac{\vec{F}_{\text{on}\,q}}{q}$$

- The electric field is a vector. The electric field from multiple charges is the vector sum of the fields from the individual charges.

$$\vec{E}_{\text{total}} = \vec{E}_1 + \vec{E}_2 + \cdots$$

Visualizing the electric field

The electric field exists at all points in space.

- An electric field vector shows the field only at one point, the point at the tail of the vector.

- A **field diagram** shows field vectors at several points.

- Electric field lines:

  - are always parallel to the field vectors.

  - are close where the field is strong, far apart where the field is weak.

  - go from positive to negative charges.

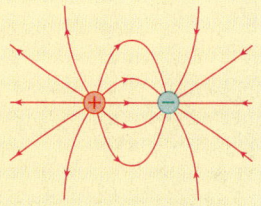

APPLICATIONS

There are two types of material, insulators and conductors.

- Charge remains fixed on an insulator.

- Charge moves easily through conductors.

- Charge is transferred by contact between objects.

A dipole has no net charge, but has a field because the two charges are separated.

A dipole will rotate to align with an electric field.

Electric fields: important cases

The electric field of a **point charge** is

$$\vec{E} = \left(\frac{K|q|}{r^2}, \begin{array}{l} \text{away from } q \text{ if } q > 0 \\ \text{toward } q \text{ if } q < 0 \end{array} \right)$$

The electric field inside a **parallel-plate capacitor** is uniform.

$$\vec{E} = \left(\frac{Q}{\epsilon_0 A}, \text{ from positive to negative} \right)$$

Conductors in electric fields

- The electric field inside a conductor in **electrostatic equilibrium** is zero.

- Any excess charge is on the surface.

- The electric field is perpendicular to the surface.

- The density of charge and the electric field are highest near points.

For instructor-assigned homework, go to www.masteringphysics.com

Problem difficulty is labeled as | (straightforward) to ||||| (challenging).

Problems labeled INT integrate significant material from earlier chapters; BIO are of biological or medical interest.

QUESTIONS

Conceptual Questions

1. What is alike about charges when we say "two like charges"? Do they look, smell, or taste the same?

2. Four lightweight balls A, B, C, and D are suspended by threads. Ball A has been touched by a plastic rod that was rubbed with wool. When the balls are brought close together, without touching, the following observations are made:
   - Balls B, C, and D are attracted to ball A.
   - Balls B and D have no effect on each other.
   - Ball B is attracted to ball C.

   What are the charge states (positive, negative, or neutral) of balls A, B, C, and D? Explain.

3. Plastic and glass rods that have been charged by rubbing with wool and silk, respectively, hang by threads.
   a. An object repels the plastic rod. Can you predict what it will do to the glass rod? If so, what? If not, why not? Explain.
   b. A different object attracts the plastic rod. Can you predict what it will do to the glass rod? If so, what? If not, why not? Explain.

4. a. Can an insulator be charged? If so, how would you charge an insulator? If not, why not?
   b. Can a conductor be charged? If so, how would you charge a conductor? If not, why not?

5. When you take clothes out of the drier right after it stops, the clothes often stick to your hands and arms. Is your body charged? If so, how did it acquire a charge? If not, why does this happen?

6. A lightweight metal ball hangs by a thread. When a charged rod is held near, the ball moves toward the rod, touches the rod, then quickly "flies away" from the rod. Explain this behavior.

7. As shown in Figure Q20.7, metal sphere A has 4 units of negative charge and metal sphere B has 2 units of positive charge. The two spheres are brought into contact. What is the final charge state of each sphere? Explain.

FIGURE Q20.7

8. Figure Q20.8 shows a positively charged rod held near, but not touching, a neutral metal sphere.
   a. Add plusses and minuses to the figure to show the charge distribution on the sphere.
   b. Does the sphere experience a net force? If so, in which direction? Explain.

FIGURE Q20.8

9. A plastic balloon that has been rubbed with wool will stick to a wall.
   a. Can you conclude that the wall is charged? If not, why not? If so, where does the charge come from?
   b. Draw a charge diagram showing how the balloon is held to the wall.

10. Consider two identical insulating spheres, with charge Q uniformly covering the surface of each. If they are placed near each other, but not touching, there will be a repulsive force between them. Now consider two *conducting* spheres of the same diameter, with the same charge Q on each. When they are placed the same distance apart, will the force between the conducting spheres be less than, equal to, or greater than the force between the insulating spheres? Explain, being certain to consider the motion of the charges on the conducting spheres.

11. BIO A honeybee acquires a positive electric charge as it flies through the air. This charge causes pollen grains to be attracted to the bee. Explain, using words and diagrams, how a neutral, conducting pollen grain will be attracted to a positively charged bee.

12. You've been assigned the task of determining the magnitude and direction of the electric field at a point in space. Give a step-by-step procedure of how you will do so. List any objects you will use, any measurements you will make, and any calculations you will need to perform. Make sure that your measurements do not disturb the charges that are creating the field.

13. Each part of Figure Q20.13 shows two points near two charges. Compare the electric field strengths E_1 and E_2 at these two points. Is $E_1 > E_2$, $E_1 = E_2$, or $E_1 < E_2$?

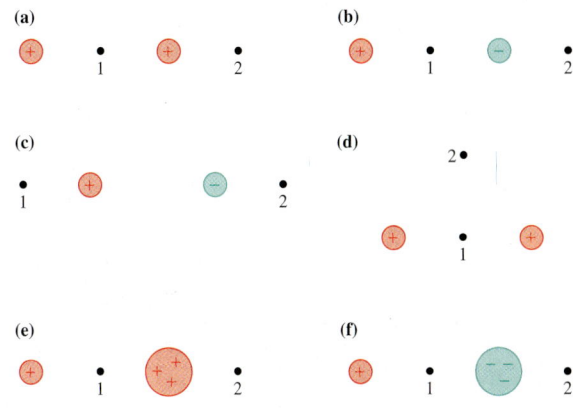

FIGURE Q20.13

14. Iontophoresis is a noninvasive
BIO process that transports drugs through the skin without needles. In the photo, the red electrode is positive and the black electrode is negative. The electric field between the electrodes will drive the negatively charged molecules of an anesthetic through the skin. Should the drug be placed at the red or the black electrode? Explain.

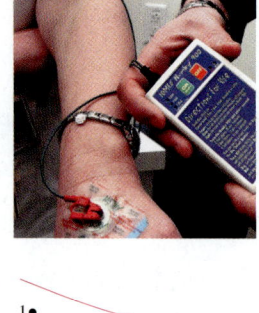

15. Rank in order, from largest to smallest, noting any ties, the electric field strengths E_1 to E_4 at points 1 to 4 in Figure Q20.15.

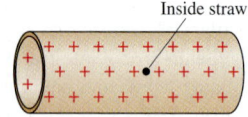

FIGURE Q20.15

16. Explain why electric field lines cannot cross.
17. When DNA breaks into frag-
BIO ments in a cell, electrostatic forces may actually inhibit repair. Explain why this happens.
18. A hollow soda straw is uniformly charged, as shown in Figure Q20.18. What is the electric field at the center (inside) of the straw? Explain.

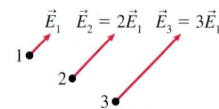

FIGURE Q20.18   **FIGURE Q20.19**

19. A small positive charge q experiences a force of magnitude F_1 when placed at point 1 in Figure Q20.19. In terms of F_1:
    a. What is the magnitude of the force on charge q at point 3?
    b. What is the magnitude of the force on a charge $3q$ at point 1?
    c. What is the magnitude of the force on a charge $2q$ at point 2?
    d. What is the magnitude of the force on a charge $-2q$ at point 2?
20. A typical commercial airplane is struck by lightning about once per year. When this happens, the external metal skin of the airplane might be burned, but the people and equipment inside the aircraft experience no ill effects. Explain why this is so.
21. Microbes such as bacteria have small positive charges when
BIO in solution. Public health agencies are exploring a new way to measure the presence of small numbers of microbes in drinking water by using electric forces to concentrate the microbes. Water is sent between the two oppositely charged electrodes of a parallel-plate capacitor. Any microbes in the water will collect on one of the electrodes.
    a. On which electrode will the microbes collect?
    b. How could the microbes be easily removed from the electrodes for analysis?
22. An electric dipole in a uniform electric field does not experience a net force. Could a dipole in a nonuniform electric field experience a net force? Explain.

Multiple-Choice Questions

23. | Two lightweight, electrically neutral conducting balls hang from threads. Choose the diagram in Figure Q20.23 that shows how the balls hang after:
    a. Both are touched by a negatively charged rod.
    b. Ball 1 is touched by a negatively charged rod and ball 2 is touched by a positively charged rod.
    c. Both are touched by a negatively charged rod but ball 2 picks up more charge than ball 1.
    d. Only ball 1 is touched by a negatively charged rod.
    Note that parts a through d are independent; these are not actions taken in sequence.

FIGURE Q20.23

24. | All the charges in Figure Q20.24 have the same magnitude. In which case does the electric field at the dot have the largest magnitude?

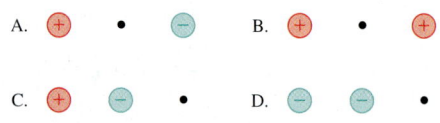

FIGURE Q20.24

25. | All the charges in Figure Q20.25 have the same magnitude. In which case does the electric field at the dot have the largest magnitude?

FIGURE Q20.25

26. | All of the charges in Figure Q20.26 have the same magnitude. In which case does the electric field at the dot have the largest magnitude?

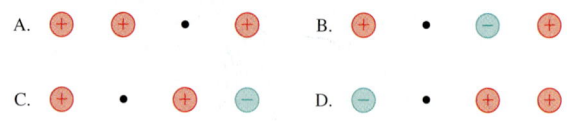

FIGURE Q20.26

27. | A glass bead charged to $+3.5$ nC exerts an 8.0×10^{-4} N repulsive electric force on a plastic bead 2.9 cm away. What is the charge on the plastic bead?
    A. $+2.1$ nC   B. $+7.4$ nC
    C. $+21$ nC   D. $+740$ nC
28. | A $+7.5$ nC point charge and a -2.0 nC point charge are 3.0 cm apart. What is the electric field strength at the midpoint between the two charges?
    A. 3.3×10^3 N/C   B. 5.7×10^3 N/C
    C. 2.2×10^5 N/C   D. 3.8×10^5 N/C

29. ‖ Three point charges are arranged as shown in Figure Q20.29. Which arrow best represents the direction of the electric field vector at the position of the dot?

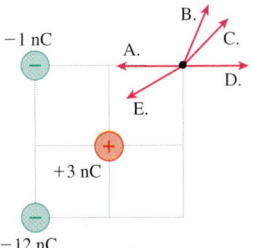

FIGURE Q20.29

30. ‖ A rod has positive charge $+q$ at one end and negative charge $-q$ at the other, forming a dipole. The dipole is placed in a nonuniform electric field represented by the field lines in Figure Q20.30. Which arrow best indicates the direction of the net electric force on the dipole?

FIGURE Q20.30

PROBLEMS

Section 20.1 Charges and Forces

Section 20.2 Charges, Atoms, and Molecules

1. ‖ A glass rod is charged to $+5.0$ nC by rubbing.
   a. Have electrons been removed from the rod or protons added? Explain.
   b. How many electrons have been removed or protons added?
2. ‖ A plastic rod is charged to -20 nC by rubbing.
   a. Have electrons been added to the rod or protons removed? Explain.
   b. How many electrons have been added or protons removed?
3. ‖‖ Suppose you have 1.0 mol of O_2 gas. How many coulombs of positive charge are contained in the atomic nuclei of this gas?
   INT
4. | A plastic rod that has been charged to -15.0 nC touches a metal sphere. Afterward, the rod's charge is -10.0 nC.
   a. What kind of charged particle was transferred between the rod and the sphere, and in which direction? That is, did it move from the rod to the sphere or from the sphere to the rod?
   b. How many charged particles were transferred?
5. ‖ A glass rod that has been charged to $+12.0$ nC touches a metal sphere. Afterward, the rod's charge is $+8.0$ nC.
   a. What kind of charged particle was transferred between the rod and the sphere, and in which direction? That is, did it move from the rod to the sphere or from the sphere to the rod?
   b. How many charged particles were transferred?
6. ‖ Two identical metal spheres A and B are connected by a metal rod. Both are initially neutral. 1.0×10^{12} electrons are added to sphere A, then the connecting rod is removed. Afterward, what are the charge of A and the charge of B?
7. ‖‖ Two identical metal spheres A and B are connected by a plastic rod. Both are initially neutral. 1.0×10^{12} electrons are added to sphere A, then the connecting rod is removed. Afterward, what are the charge of A and the charge of B?
8. ‖ If two identical conducting spheres are in contact, any excess charge will be evenly distributed between the two. Three identical metal spheres are labeled A, B, and C. Initially, A has charge q, B has charge $-q/2$, and C is uncharged.
   a. What is the final charge on each sphere if C is touched to B, removed, and then touched to A?
   b. Starting again from the initial conditions, what is the charge on each sphere if C is touched to A, removed, and then touched to B?

Section 20.3 Coulomb's Law

9. | Two 1.0 kg masses are 1.0 m apart on a frictionless table. Each has $+1.0$ μC of charge.
   INT
   a. What is the magnitude of the electric force on one of the masses?
   b. What is the initial acceleration of each mass if they are released and allowed to move?
10. ‖ Two small plastic spheres each have a mass of 2.0 g and a charge of -50.0 nC. They are placed 2.0 cm apart.
    a. What is the magnitude of the electric force between the spheres?
    b. By what factor is the electric force on a sphere larger than its weight?
11. ‖ A small plastic sphere with a charge of -5.0 nC is near another small plastic sphere with a charge of -12 nC. If the spheres repel one another with a force of magnitude 8.2×10^{-4} N, what is the distance between the spheres?
12. ‖‖‖ Figure P20.12 shows the magnitude of the attractive electric force between two objects as a function of the distance r separating them. One object is charged to $+0.11$ nC. What is the charge on the other object?

FIGURE P20.12

13. ‖‖‖ A small glass bead has been charged to $+20$ nC. A tiny ball bearing 1.0 cm above the bead feels a 0.018 N downward electric force. What is the charge on the ball bearing?
14. | What are the magnitude and direction of the electric force on charge A in Figure P20.14?

FIGURE P20.14

1.0 nC	−1.0 nC	4.0 nC
A +	B −	C +
	1 cm	1 cm

15. ‖ In Figure P20.15, charge q_2 experiences no net electric force. What is q_1?

FIGURE P20.15

q_1	−2.0 nC	q_2
	10 cm	10 cm

16. | Object A, which has been charged to $+10$ nC, is at the origin. Object B, which has been charged to -20 nC, is at $(x, y) = (0.0$ cm, 2.0 cm). What are the magnitude and direction of the electric force on each object?

17. | A small glass bead has been charged to $+20$ nC. What are
INT the magnitude and direction of the acceleration of (a) a proton and (b) an electron that is 1.0 cm from the center of the bead?

Section 20.4 The Concept of the Electric Field

18. ‖‖ What magnitude charge creates a 1.0 N/C electric field at a point 1.0 m away?

19. | What are the strength and direction of the electric field 2.0 cm from a small glass bead that has been charged to $+6.0$ nC?

20. ‖ The electric field 2.0 cm from a small object points toward the object with a strength of 180,000 N/C. What is the object's charge?

21. | What are the strength and direction of the electric field 1.0 mm from (a) a proton and (b) an electron?

22. | A $+10$ nC charge is located at the origin.
   a. What are the strengths of the electric fields at the positions $(x, y) = (5.0$ cm, 0.0 cm), $(-5.0$ cm, 5.0 cm), and $(-5.0$ cm, -5.0 cm)?
   b. Draw a field diagram showing the electric field vectors at these points.

23. ‖‖ A -10 nC charge is located at the origin.
   a. What are the strengths of the electric fields at the positions $(x, y) = (0.0$ cm, 5.0 cm), $(-5.0$ cm, -5.0 cm), and $(-5.0$ cm, 5.0 cm)?
   b. Draw a field diagram showing the electric field vectors at these points.

24. ‖ What are the strength and direction of the electric field at the position indicated by the dot in Figure P20.24? Specify the direction as an angle above or below horizontal.

FIGURE P20.24   **FIGURE P20.25**

25. ‖ What are the strength and direction of the electric field at the position indicated by the dot in Figure P20.25? Specify the direction as an angle above or below horizontal.

Section 20.5 Applications of the Electric Field

26. | What are the strength and direction of an electric field that will balance the weight of a 1.0 g plastic sphere that has been charged to -3.0 nC?

27. ‖ What are the strength and direction of an electric field that will balance the weight of (a) a proton and (b) an electron?

28. ‖ A 0.10 g plastic bead is charged by the addition of 1.0×10^{10} excess electrons. What electric field $\vec{E}$ (strength and direction) will cause the bead to hang suspended in the air?

29. | A parallel-plate capacitor is constructed of two square plates, size $L \times L$, separated by distance d. The plates are given charge $\pm Q$. What is the ratio E_f/E_i of the final electric field strength E_f to the initial electric field strength E_i if:
   a. Q is doubled?   b. L is doubled?   c. d is doubled?

30. | A parallel-plate capacitor is formed from two 4.0 cm $\times$ 4.0 cm electrodes spaced 2.0 mm apart. The electric field strength inside the capacitor is 1.0×10^6 N/C. What is the charge (in nC) on each electrode?

31. ‖ Two closely spaced circular disks form a parallel-plate capacitor. Transferring 1.5×10^9 electrons from one disk to the other causes the electric field strength between them to be 1.0×10^5 N/C. What are the diameters of the disks?

Section 20.6 Conductors in Electric Fields

32. ‖‖ Storm clouds may build up large negative charges near their bottom edges. The earth is a good conductor, so the charge on the cloud attracts an equal and opposite charge on the earth under the cloud. The electric field strength near the earth depends on the shape of the earth's surface, as we can explain with a simple model. The top metal plate in Figure 20.32 has uniformly distributed negative charge. The bottom metal plate, which has a high point, has an equal and opposite charge that is free to move.
   a. Sketch the two plates and the region between them, showing the distribution of positive charge on the bottom plate.
   b. Complete your diagram by sketching electric field lines between the two plates. Be sure to note the direction of the field. Where is the field strongest?
   c. Explain why it is more dangerous to be on top of a hill or mountain during a lightning storm.

FIGURE P20.32   **FIGURE P20.33**

33. ‖‖ A neutral conducting sphere is between two parallel charged plates, as shown in Figure P20.33. Sketch the electric field lines in the region between the plates. Be sure to include the effect of the conducting sphere.

Section 20.7 Forces and Torques on Charges in Electric Fields

34. ‖‖ Two small plastic spheres, one charged to 17 nC and the other to -17 nC, are connected by a 25-mm-long insulating rod. Suppose this dipole is placed in a uniform electric field with strength 7.4×10^5 N/C. What is the maximum possible torque on the dipole?

35. | A protein molecule in an electrophoresis gel has a negative
BIO charge. The exact charge depends on the pH of the solution, but 30 excess electrons is typical. What is the magnitude of the electric force on a protein with this charge in a 1500 N/C electric field?

36. | Large electric fields in cell membranes cause ions to move
BIO through the cell wall, as we will explore in Chapter 23. The field strength in a typical membrane is 1.0×10^7 N/C. What is the magnitude of the force on a calcium ion with charge $+e$?

37. || Molecules of carbon monoxide are permanent electric dipoles due to unequal sharing of electrons between the carbon and oxygen atoms. Figure P20.37 shows the distance and charges. Suppose a carbon monoxide molecule with a horizontal axis is in a vertical electric field of strength 15,000 N/C.

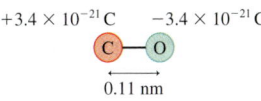

FIGURE P20.37

   a. What is the magnitude of the net force on the molecule?
   b. What is the magnitude of the torque on the molecule?

General Problems

38. ||| A 2.0-mm-diameter copper ball is charged to $+50$ nC. What fraction of its electrons have been removed? The density of copper is 8900 kg/m³.

39. ||| Pennies today are copper-covered zinc, but older pennies are 3.1 g of solid copper. What are the total positive charge and total negative charge in a solid copper penny that is electrically neutral? The density of copper is 8900 kg/m³.

40. |||| Two protons are 2.0 fm apart. (1 fm = 1 femtometer =
INT 1×10^{-15} m.)
   a. What is the magnitude of the electric force on one proton due to the other proton?
   b. What is the magnitude of the gravitational force on one proton due to the other proton?
   c. What is the ratio of the electric force to the gravitational force?

41. |||| The nucleus of a ^{125}Xe atom (an isotope of the element
INT xenon with mass 125 u) is 6.0 fm in diameter. It has 54 protons and charge $q = +54e$. (1 fm = 1 femtometer = 1×10^{-15} m.)
   a. What is the electric force on a proton 2.0 fm from the surface of the nucleus?
   b. What is the proton's acceleration?
   Hint: Treat the spherical nucleus as a point charge.

42. | Two equally charged, 1.00 g spheres are placed with 2.00 cm
INT between their centers. When released, each begins to accelerate at 225 m/s². What is the magnitude of the charge on each sphere?

43. | Objects A and B are both positively charged. Both have a mass of 100 g, but A has twice the charge of B. When A and B are placed with 10 cm between their centers, B experiences an electric force of 0.45 N.
   a. How large is the force on A?
   b. What are the charges q_A and q_B?

44. || An electric dipole is formed from ± 1.0 nC point charges spaced 2.0 mm apart. The dipole is centered at the origin, oriented along the y-axis. What is the electric field strength at the points (a) $(x, y) = (10 \text{ mm}, 0 \text{ mm})$ and (b) $(x, y) = (0 \text{ mm}, 10 \text{ mm})$?

45. |||| What are the strength and direction of the electric field at the position indicated by the dot in Figure P20.45? Specify the direction as an angle above or below horizontal.

FIGURE P20.45   **FIGURE P20.46**

46. || What are the strength and direction of the electric field at the position indicated by the dot in Figure P20.46? Specify the direction as an angle above or below horizontal.

47. ||| What is the force on the 1.0 nC charge in Figure P20.47? Give your answer as a magnitude and a direction.

FIGURE P20.47   **FIGURE P20.48**

48. ||| What is the force on the 1.0 nC charge in Figure P20.48? Give your answer as a magnitude and a direction.

49. ||| What is the magnitude of the force on the 1.0 nC charge in the middle of Figure P20.49 due to the four other charges?

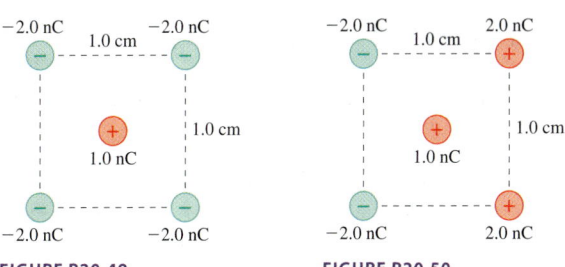

FIGURE P20.49   **FIGURE P20.50**

50. ||| What is the magnitude and direction of the force on the 1.0 nC charge in the middle of Figure P20.50 due to the four other charges?

51. ||| What is the magnitude and direction of the force on the 1.0 nC charge at the bottom of Figure P20.51?

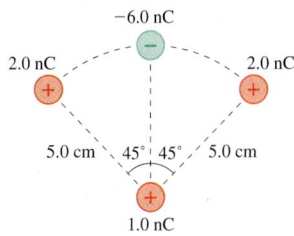

FIGURE P20.51

52. |||| A $+2.0$ nC point charge is at the origin and a -4.0 nC point charge is at $(x, y) = (1.0 \text{ cm}, 0 \text{ cm})$.
   a. At what (x, y) coordinates could you place a proton so that it would experience no net force?
   b. Would the net force be zero for an electron placed at the same position? Explain.

53. ▥ The net force on the 1.0 nC charge in Figure P20.53 is zero. What is q?

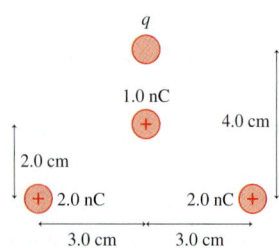

FIGURE P20.53

54. ▥ A positive point charge Q is located at $x = -a$ and a negative point charge $-Q$ is at $x = a$. A positive charge q can be placed anywhere on the y-axis, with coordinates $(0, y)$. Find an expression for $(F_{net})_x$, the x-component of the net force on q.

55. ▥ Figure P20.55 shows four charges at the corners of a square of side L. Assume q and Q are positive.
    a. Draw a diagram showing the three forces on charge q due to the other charges. Give your vectors the correct relative lengths.
    b. Find an expression for the magnitude of the net force on q.

FIGURE P20.55

56. ▥ Suppose the magnitude of the proton charge differs from that of the electron charge by a mere 1 part in 10^9.
    a. What would be the force between two 2.0-mm-diameter copper spheres 1.0 cm apart? Assume that each copper atom has an equal number of electrons and protons. The density of copper is 8900 kg/m^3.
    b. Would this amount of force be detectable? What can you conclude from the fact that no such forces are observed?

57. ▥ INT In a simple model of the hydrogen atom, the electron moves in a circular orbit of radius 0.053 nm around a stationary proton. How many revolutions per second does the electron make?
    Hint: What must be true for a force that causes circular motion?

58. ▥ BIO A 0.10 g honeybee acquires a charge of $+23$ pC while flying.
    a. The electric field near the surface of the earth is typically 100 N/C, directed downward. What is the ratio of the electric force on the bee to the bee's weight?
    b. What electric field strength and direction would allow the bee to hang suspended in the air?

59. ▥ A $+10$ nC charge is located at $(x, y) = (0 \text{ cm}, 10 \text{ cm})$ and a -5.0 nC charge is located $(x, y) = (5.0 \text{ cm}, 0 \text{ cm})$. Where would a -10 nC charge need to be located in order that the electric field at the origin be zero?

60. ▥ INT Two 2.0-cm-diameter disks face each other, 1.0 mm apart. They are charged to ± 10 nC.
    a. What is the electric field strength between the disks?
    b. A proton is shot from the negative disk toward the positive disk. What launch speed must the proton have to just barely reach the positive disk?

61. ▥ INT The electron gun in a television tube uses a uniform electric field to accelerate electrons from rest to 5.0×10^7 m/s in a distance of 1.2 cm. What is the electric field strength?

62. ▥ Air "breaks down" when the electric field strength reaches 3.0×10^6 N/C, causing a spark. A parallel-plate capacitor is made from two 4.0-cm-diameter disks. How many electrons must be transferred from one disk to the other to create a spark between the disks?

63. ▥ INT You have a lightweight spring whose unstretched length is 4.0 cm. You're curious to see if you can use this spring to measure charge. First, you attach one end of the spring to the ceiling and hang a 1.0 g mass from it. This stretches the spring to a length of 5.0 cm. You then attach two small plastic beads to the opposite ends of the spring, lay the spring on a frictionless table, and give each plastic bead the same charge. This stretches the spring to a length of 4.5 cm. What is the magnitude of the charge (in nC) on each bead?

64. ▥ INT Two 3.0 g spheres on 1.0-m-long threads repel each other after being equally charged, as shown in Figure P20.64. What is the charge q?

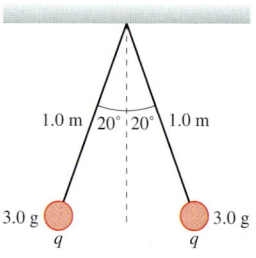

FIGURE P20.64

65. ▥ INT An electric field $\vec{E} = (100{,}000 \text{ N/C, right})$ causes the 5.0 g ball in Figure P20.65 to hang at a 20° angle. What is the charge on the ball?

| **FIGURE P20.65** | **FIGURE P20.66** |

66. Ⅰ An electric field $\vec{E} = (200{,}000 \text{ N/C, right})$ causes the 2.0 g ball in Figure P20.66 to hang at an angle. What is θ?

In Problems 67 through 69 you are given the equation used to solve a problem. For each of these,
a. Write a realistic problem for which this is the correct equation.
b. Finish the solution of the problem.

67. ▥
$$\frac{(9.0 \times 10^9 \text{ N} \cdot \text{m}^2/\text{C}^2) \times N \times (1.60 \times 10^{-19} \text{ C})}{(1.0 \times 10^{-6} \text{ m})^2}$$
$$= 1.5 \times 10^6 \text{ N/C}$$

68. ▥
$$\frac{(9.0 \times 10^9 \text{ N} \cdot \text{m}^2/\text{C}^2)q^2}{(0.015 \text{ m})^2} = 0.020 \text{ N}$$

69. ▥
$$\frac{(9.0 \times 10^9 \text{ N} \cdot \text{m}^2/\text{C}^2)(15 \times 10^{-9} \text{ C})}{r^2} = 54{,}000 \text{ N/C}$$

Passage Problems

Flow Cytometry BIO

Flow cytometry, illustrated in Figure P20.70, is a technique used to sort cells by type. The cells are placed in a conducting saline solution which is then forced from a nozzle. The stream breaks up into small droplets, each containing one cell. A metal collar surrounds the stream right at the point where the droplets separate from the stream. Charging the collar polarizes the conducting liquid, causing the droplets to become charged as they break off from the stream. A laser beam probes the solution just upstream from the charging collar, looking for the presence of certain types of cells. All droplets containing one particular type of cell are given the same charge by the charging collar. Droplets with other desired types of cells receive a different charge, and droplets with no desired cell receive no charge. The charged droplets then pass between two parallel charged electrodes where they receive a horizontal force that directs them into different collection tubes, depending on their charge.

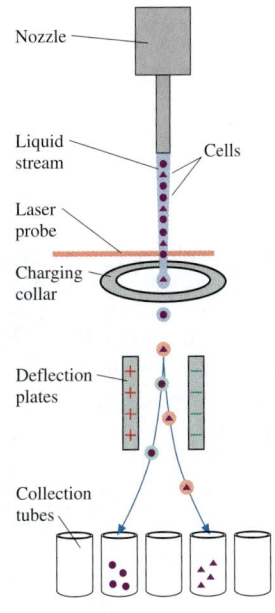

FIGURE P20.70

70. | If the charging collar has a positive charge, the net charge on a droplet separating from the stream will be
A. Positive. B. Negative. C. Neutral.
D. The charge will depend on the type of cell.

71. | Which of the following describes the charges on the droplets that end up in the five tubes, moving from left to right?
A. $+2q, +q, 0, -q, -2q$ B. $+q, +2q, 0, -2q, -q$
C. $-q, -2q, 0, +2q, +q$ D. $-2q, -q, 0, +q, +2q$

72. | Because the droplets are conductors, a droplet's positive and negative charges will separate while the droplet is in the region between the deflection plates. Suppose a neutral droplet passes between the plates. The direction of the droplet's dipole moment will point
A. Up.   B. Down.   C. Left.   D. Right.

73. | Another way to sort the droplets would be to give each droplet the same charge, then vary the electric field between the deflection plates. For the apparatus as sketched, this technique will not work because
A. Several droplets are between the plates at one time, and they would all feel the same force.
B. The cells in the solution have net charges that would affect the droplet charge.
C. A droplet with a net charge would always experience a net force between the plates.
D. The droplets would all repel each other, and this force would dominate the deflecting force.

Stop to Think 20.1: A. The electroscope is originally given a positive charge. The charge spreads out, and the leaves repel each other. When a rod with a negative charge is brought near, some of the positive charge is attracted to the top of the electroscope, away from the leaves. There is less charge on the leaves, and so they move closer together.

Stop to Think 20.2: $q_E(+3e) > q_A(+1e) > q_D(0) > q_B(-1e) > q_C(-2e)$.

Stop to Think 20.3: B. The two forces are an action/reaction pair, opposite in direction but *equal* in magnitude.

Stop to Think 20.4: $E_B > E_A > E_D > E_C$. The field is proportional to the charge, and inversely proportional to the square of the distance.

Stop to Think 20.5: C. Electric field lines *start* on positive charges. Very near to each of the positive charges, the field lines should look like the field lines of a single positive charge.

Stop to Think 20.6: $F_A = F_B = F_C = F_D = F_E$. The field inside a capacitor is the same at all points. Because the field is uniform, the force on the proton will be the same at all points. The electric field exists at all points whether or not a vector is shown at that point.

21

ELECTRIC POTENTIAL

Shown is the electric potential measured on the surface of a patient. This potential is caused by electrical signals originating in the beating heart. Why does the potential have this pattern, and what do these measurements tell us about the heart's condition?

Looking Ahead ▶▶

The goal of Chapter 21 is to calculate and use the electric potential and electric potential energy. In this chapter you will learn to:

▶ Use electric potential energy to analyze the motion of charged particles.

▶ Use the electric potential to find the electric potential energy.

▶ Calculate the electric potential of important charge distributions.

▶ Represent the electric potential graphically.

▶ Understand the relationship between the electric potential and the electric field.

▶ Understand capacitance, capacitors, and dielectrics.

Looking Back ◀◀

This chapter depends on the concepts of work, energy, and conservation of energy. Please review:

◀ Sections 10.4–10.6 Work, kinetic energy, and potential energy.

◀ Section 20.1 Charge polarization.

◀ Sections 20.3–20.4 Coulomb's law and the electric field.

Energy has been a theme throughout most of this textbook. Energy allows things to happen. A system without a source of energy is not terribly interesting; it just sits there. You want your lights to light, your computer to compute, and your stereo to keep your neighbors awake. In other words, you want devices that use electricity to *do* something, and that takes energy. It is time to see how the concept of energy helps us to understand and analyze electric phenomena.

In Chapter 20, we introduced the idea of the *electric field* to understand how one set of charges, the source charges, exerts electric forces on other charges. Now, to understand electric energy, we will introduce a new concept called the *electric potential*. In this chapter, we will study the basic properties of the electric potential and learn how it is connected to electric energy. We'll also explore the relationship between the electric potential and the electric field. These ideas will lead us directly into electric currents and circuits, which are practical applications of the electric potential and electric field.

21.1 Electric Potential Energy and the Electric Potential

Conservation of energy was a powerful tool for understanding the motion of mechanical systems. You will recall that a system's mechanical energy $E_{mech} = K + U$ is conserved for particles that interact with each other via *conservative forces*, where K and U are the kinetic and potential energy. Two important conservative forces that we've already studied are the force of gravity and the elastic force of a spring. Figure 21.1 reminds you what we learned about those conservative forces.

Figure 21.1a shows a system consisting of the earth and a book. These two objects interact via the force of gravity. If a force *external* to the system—here, the force $\vec{F}$ due to the hand—moves the book at steady speed through a displacement, then the force does work on the system. This work represents a *transfer* of energy *from* the environment (here, the hand) *to* the system, so the system's gravitational potential energy U_g increases by the amount of work done: $\Delta U_g = W$.

We can analyze a spring in a similar way, as shown in Figure 21.1b. The external force of the hand compresses the spring, doing work on it. The energy transferred into the system as work appears as elastic potential energy of the spring U_s. We can write this energy transfer as $\Delta U_s = W$. In general, a system's potential energy is found by computing the work an external force must do to change the system's energy.

Let's now apply this idea to the system of charged particles in Figure 21.1c. As in Chapter 20, several of the charges are *source charges*, charges that are fixed in place and don't move. These charges exert an electric force on the charged particle q in the hand, so the hand must exert a counterbalancing external force to move charge q upward at constant speed. The hand does work on the system of charges as it pushes charge q from A to B, increasing the system's energy. The electric force, like the gravitational force, is a conservative force, so this increased energy appears as an increased potential energy, in this case, as **electric potential energy** U_{elec}. The energy transfer is then $\Delta U_{elec} = W$. **Electric potential energy is energy stored in a system of charged particles due to their electrical interactions.**

NOTE ▶ Often we consider a system consisting of several charges, only one of which moves. As this charge moves, it is the electric potential energy of the *system* of charges—moving *and* fixed—that actually changes. However, because we're usually focused on the moving charge, we often speak of the electric potential energy *of the charge*. ◀

The Zero of Electric Potential Energy

For other forms of potential energy, we found that only *changes* in the potential energy were meaningful. The actual value of the potential energy depends on the choice of a particular configuration of the system for which $U = 0$. For the gravitational potential energy of a book, for instance, we might choose to have $U = 0$ when the book is on the ground. The same idea holds for electric potential energy. We can choose a particular position of the moveable charge—whatever is most convenient—to have $U_{elec} = 0$. With this choice, we can then find U_{elec} when the movable charge is at any other position by finding the work done to move it there.

Electric Potential

We introduced the concept of the *electric field* in Chapter 20 to understand forces between charges. In the field model, the electric field provides an intermediary through which two charged particles Q and q exert forces on each other. Charge Q alters the space around it by creating an electric field $\vec{E}$. Charge q then reacts to the field, experiencing force $\vec{F} = q\vec{E}$. The important idea is that the electric field exists whether or not there is a charge q present to experience a force.

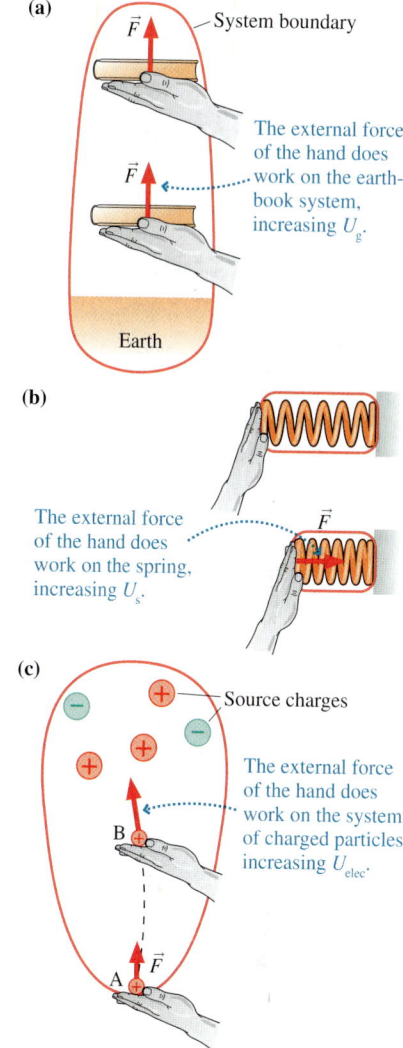

FIGURE 21.1 Gravitational, elastic, and electric potential energy.

Let's try to understand electric potential energy in a similar way. Consider a group of charged particles, all fixed except for one that can move. We imagine that at every point in the space surrounding the fixed charges there is a number that tells us how much potential energy the mobile charge *would* have if it were placed at that point. This works in much the same way that the electric field at a point tells us what the force on a charge *would* be if a charge were placed at that point.

Let's start our discussion by looking in more detail at the situation that was shown in Figure 21.1c. In Figure 21.2a, the moving charge is $q = 10$ nC and, for convenience, we've chosen to let $U_{elec} = 0$ at point A. Suppose that the hand does 3 μJ of work to move this charged particle from point A to point B. Then the charge's electric potential energy at B, as found from the work done in moving it, is $U_{elec} = 3$ μJ. (Recall that the SI unit of energy is the joule, J.)

What if we were to repeat this experiment with a different charge, say $q = 20$ nC. What is this charge's electric potential energy at point B? We *could* repeat the entire experiment, again moving the charge from A to B while measuring the work done, but we don't need to. We've doubled the value of q, from 10 nC to 20 nC. According to Coulomb's law, the electric force on this charged particle is proportional to q, so the force on it also doubles. Because the work done is proportional to the force, the work to move the charge from A to B doubles as well to 6 μJ. Thus $U_{elec} = 6$ μJ at point B. In general, **a charged particle's electric potential energy is proportional to its charge.**

When two quantities are proportional to each other, their *ratio* is constant. In this example, the ratio of the charge's electric potential energy U_{elec} to the charge q is constant. We can see this directly by calculating the ratio for two different charges:

$$\frac{U_{elec}}{q} = \frac{3\ \mu J}{10\ nC} = \frac{6\ \mu J}{20\ nC} = 300\ \frac{J}{C}$$

U/q for U/q for   Both ratios are
$q = 10$ nC $q = 20$ nC   the *same*.

Using this constant ratio, we can write an expression for the electric potential energy that *any* charge q would have at B:

$$U_{elec} = \left(300\frac{J}{C}\right)q$$

This part depends only   This part depends on
on where charge q is.   the movable charge.

It appears there's a special value, 300 J/C, associated with point B. This value is special because it tells us how to find the electric potential energy of *any* charge q that happens to be placed at point B. That is, it tells us the *potential* for creating potential energy (there's a mouthful!) at point B. Thus this value is called the **electric potential,** and it is given the symbol V. At point B, then, the electric potential is $V = 300$ J/C.

This idea can be generalized. Any source charges create an electric potential in the space around them. At a point where the electric potential is V, the electric potential energy of a charged particle q is

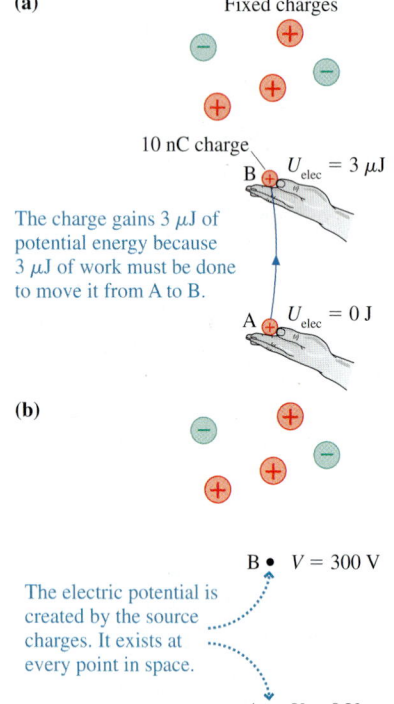

(a) Fixed charges

10 nC charge

B $U_{elec} = 3$ μJ

The charge gains 3 μJ of potential energy because 3 μJ of work must be done to move it from A to B.

A $U_{elec} = 0$ J

(b)

B • $V = 300$ V

The electric potential is created by the source charges. It exists at every point in space.

A • $V = 0$ V

FIGURE 21.2 The change in electric potential energy for a 10 nC charged particle.

You may not realize it, but you use electric potential all the time. An ordinary battery is a source of electric potential, as you can see from the fact that a battery is rated in volts.

$$U_{elec} = qV \qquad (21.1)$$

Relationship between electric potential and electric potential energy

LINEAR
p. 38

Although we've used the work done on a positive charge to justify Equation 21.1, it is also valid if charge q is negative.

The unit of electric potential is the joule per coulomb, which is called the **volt** V:

$$1 \text{ volt} = 1 \text{ V} = 1 \text{ J/C}$$

This unit is named for Alessandro Volta, who invented the electric battery in the year 1800. Microvolts (μV), millivolts (mV), and kilovolts (kV) are commonly used units because the electric potentials used in practical applications differ significantly in magnitude. In Figure 21.2b, the potential due to the source charges at point B is 300 J/C, or 300 V.

Typical electric potentials

Source of potential	Approximate potential
Cells in human body	100 mV
Battery	1–10 V
Household electricity	100 V
Static electricity	10 kV
Transmission lines	1 MV

NOTE ▶ The symbol V is widely used to represent the *volume* of an object, and now we're introducing the same symbol to mean *potential*. To make matters more confusing, V is the abbreviation for *volts*. In printed text, *V* for potential is italicized while V for volts is not, but you can't make such a distinction in handwritten work. This is not a pleasant state of affairs, but these are the commonly accepted symbols. It's incumbent upon you to be especially alert to the *context* in which a symbol is used. ◀

It is important to recognize that the electric potential is not a property of the charge q. It is a property of the source charges, and it depends only on their positions and charge. The electric potential of the source charges exists throughout space whether or not charge q happens to be present. If a charged particle q is brought to a point where the electric potential is V, the charge's electric potential energy is given by $U_{elec} = qV$. Table 21.1 will help you to distinguish between the electric potential and the electric potential energy.

TABLE 21.1 Distinguishing electric potential and potential energy

The *electric potential* is a property of the source charges. The electric potential is present whether or not a charged particle is there to experience it. Potential is measured in J/C, or V.

The *electric potential energy* is the interaction energy of a charged particle with the source charges. Potential energy is measured in J.

EXAMPLE 21.1 Finding the change in potential energy

The electric potentials at points A, B, and C are $V_A = 300$ V, $V_B = 100$ V, and $V_C = -200$ V. A 15 nC charged particle moves from point A to point B to point C. By how much does the particle's electric potential energy change between points A and C?

PREPARE The electric potential V has been created by some configuration of source charges. It exists independently of the 15 nC charge, but we can use $U_{elec} = qV$ to find the potential energy of the charge at any point. The electric force is a conservative force, so that the *change* in potential energy $\Delta U_{elec} = (U_{elec})_C - (U_{elec})_A$ depends only on the potential energy at the end points, not on the path the particle follows from A to C. Knowing the potential at B or even that the particle passed through point B isn't relevant.

SOLVE The electric potential energy of the charged particle at point A is

$$(U_{elec})_A = qV_A = (15 \times 10^{-9} \text{ C})(300 \text{ V}) = 4.5 \text{ } \mu\text{J}$$

Similarly, the electric potential energy of the particle at point C is

$$(U_{elec})_C = qV_C = (15 \times 10^{-9} \text{ C})(-200 \text{ V}) = -3.0 \text{ } \mu\text{J}$$

Thus the change in potential energy is

$$\Delta U_{elec} = (U_{elec})_C - (U_{elec})_A$$
$$= (-3.0 \text{ } \mu\text{J}) - (4.5 \text{ } \mu\text{J}) = -7.5 \text{ } \mu\text{J}$$

ASSESS A positively charged particle loses potential energy as it moves from a higher electric potential at A to a lower electric potential at C.

Visualizing the Electric Potential

Figure 21.2b showed the electric potential at two points in space. But the electric potential exists *everywhere* in space around a set of source charges. Before starting to calculate electric potential, it will be useful to develop some tools to help us *visualize* the electric potential.

Let's start by trying to understand what the potential "looks like" for the three source charges shown in Figure 21.3. We haven't yet found an expression for the electric potential due to point charges, but we can understand the general form of the potential simply by noting that two positive charges repel. We'll have to do work (increasing the potential energy) to bring a positive movable charge closer to a fixed positive charge, thus **the electric potential gets more positive as a positive charge is approached.** Similarly, the electric potential gets more *negative* as you approach a negative charge.

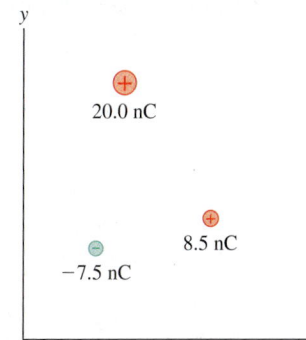

FIGURE 21.3 Three point charges in the *xy*-plane.

Figure 21.4a shows one way of representing the electric potential due to these three charges. Positive values of the potential are plotted above the *xy*-plane, negative values below it. Notice how the potential increases rapidly very near the charges. Because this representation is reminiscent of a mountainous landscape, populated by mountains and valleys, we call this an **elevation graph.**

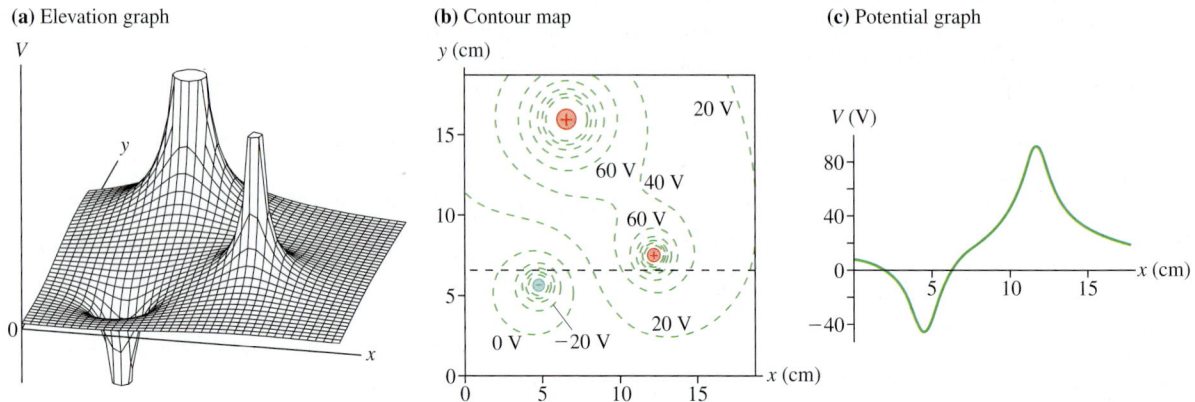

(a) Elevation graph

(b) Contour map

(c) Potential graph

FIGURE 21.4 Three representations of the electric potential of three charges.

For a real landscape, such as the mountain shown in Figure 21.5a, you may be familiar with another method of representing elevations. In the *topographic map* of Figure 21.5b, curves are drawn through points that have the same elevation. In the map shown, the curves—called *contours*—are drawn at 50 ft intervals. An important aspect of topographic maps, one that will be relevant to our study of electric potential and field, is that **the slope of the terrain is steeper where the contours of constant elevation are spaced more closely.**

(a)

Chief Mountain in Glacier National Park, Montana

(b) All points on this contour are at 2000 feet above sea level.

Here you'd have to walk a long way to rise 250 ft. The contours are far apart, so the slope is gradual.

Here you'd have to walk a short distance to rise 250 ft. The contours are close together, so the slope is steep.

2100 ft   2050 ft

Each successive contour is 50 ft higher.

FIGURE 21.5 A topographic map shows elevations above sea level.

A topographic-like map can be drawn for the electric potential as well. If we imagine the heights in the elevation graph of Figure 21.4a as actual heights in a landscape, then the potential can be represented by the **contour map** shown in Figure 21.4b. Here, contours are drawn through points that have the same *potential;* in this map, contours are drawn every 20 V. Note that the contours are closer together near the charges, indicating that the potential is changing rapidly there. Because every point on a contour line is at the same potential, the contours are often called **equipotential lines,** or simply *equipotentials.*

A third representation of potential is shown in Figure 21.4c. Imagine moving parallel to the *x*-axis along the dotted line in Figure 21.4b. The **potential graph** of Figure 21.4c plots the value of the potential at points along this line. You can see that the potential is large and negative as we pass near the negative charge, large and positive near the positive charge.

STOP TO THINK 21.1   A negative charge moves from point 1 to point 2. As it does, its electric potential energy

A. Increases.
B. Decreases.
C. Stays the same.

21.2 Using the Electric Potential

The electric potential is an abstract idea, and it will take some practice to see just what it means and how it is useful. We'll use multiple representations—words, pictures, graphs, and analogies—to explain and describe the electric potential.

The potential energy of a charged particle is determined by the electric potential: $U_{elec} = qV$. Because energy is conserved, charged particles speed up or slow down as they move through a region of changing potential. Figure 21.6 illustrates this idea. Notice that **a positive charge speeds up if it moves into a region of lower potential, slows down if it moves into a region of higher potential.**

It's useful to say that a particle moves through a **potential difference,** which is the difference $\Delta V = V_f - V_i$ between the potential at a starting (initial) point i and an ending (final) point f. The potential difference between two points is often called the **voltage.** In this text, the potential difference between two points is represented by a double-headed green arrow.

A positive charge speeds up as it moves through a negative potential difference (decreasing *V*). Potential energy is transformed into kinetic energy.

A positive charge slows down as it moves through a positive potential difference. Kinetic energy is transformed into potential energy.

FIGURE 21.6 A charged particle speeds up or slows down as it moves through a potential difference.

NOTE ▶ Consider two points with a potential difference between them. If a charge moves from one point to the other, we say that it has moved "through" the potential difference. If the potential difference is between two sides of a circuit element, such as a battery, we speak of the potential difference "across" the battery. ◀

If a particle of charge *q* moves through a potential difference ΔV, the change in its electric potential energy is $\Delta U = q \, \Delta V$. We can write the conservation of energy equation $\Delta E_{mech} = 0$ in terms of the electric potential and kinetic energy *K* as $\Delta K + \Delta U = \Delta K + q \, \Delta V = 0$ or, as is often more practical,

$$K_f + qV_f = K_i + qV_i \tag{21.2}$$

Conservation of energy is the basis of a powerful problem-solving strategy.

EXAMPLE 21.2 Moving through a potential difference

A proton with a speed of 2.00×10^5 m/s enters a region of space in which source charges have created an electric potential. What is the proton's speed after it moves through a potential difference of $+100$ V? What will be the final speed if the proton is replaced by an electron?

PREPARE Energy is conserved. The electric potential determines the potential energy. Figure 21.7 is a before-and-after visual overview of a charged particle moving through a potential difference. A positive charge *slows down* as it moves into a region of higher potential ($K \rightarrow U$). A negative charge *speeds up* ($U \rightarrow K$).

FIGURE 21.7 A charged particle moving through a potential difference.

SOLVE The potential energy of charge q is $U = qV$. Conservation of energy, now expressed in terms of the electric potential V, is $K_f + qV_f = K_i + qV_i$, or

$$K_f = K_i - q\,\Delta V$$

where $\Delta V = V_f - V_i = 100$ V is the potential difference through which the particle moves. In terms of the speeds, energy conservation is

$$\frac{1}{2}mv_f^2 = \frac{1}{2}mv_i^2 - q\,\Delta V$$

We can solve this for the final speed:

$$v_f = \sqrt{v_i^2 - \frac{2q}{m}\Delta V}$$

For a proton, with $q = e$, the final speed is

$$(v_f)_{proton} = \sqrt{(2.00 \times 10^5\ \text{m/s})^2 - \frac{2(1.60 \times 10^{-19}\ \text{C})(100\ \text{V})}{(9.11 \times 10^{-27}\ \text{kg})}}$$

$$= 1.44 \times 10^5\ \text{m/s}$$

An electron, though, with $q = -e$ and its much lower mass, ends up with speed

$$(v_f)_{electron} = \sqrt{(2.00 \times 10^5\ \text{m/s})^2 - \frac{2(-1.60 \times 10^{-19}\ \text{C})(100\ \text{V})}{(9.11 \times 10^{-31}\ \text{kg})}}$$

$$= 5.93 \times 10^6\ \text{m/s}$$

ASSESS The electric potential *already existed* in space due to other charges that are not explicitly given in the problem. The electron and proton have nothing to do with creating the potential. Instead, they *respond* to the potential by having potential energy $U = qV$.

The Electron Volt

The joule is a unit of appropriate size in mechanics and thermodynamics, where we deal with macroscopic objects, but it will be very useful to have an energy unit appropriate to atomic and nuclear events.

Suppose an electron accelerates through a potential difference $\Delta V = 1$ V. The electron might be accelerating from 0 V to 1 V, or from 1000 V to 1001 V.

Regardless of the actual voltages, an electron, being negative, *speeds up* when it moves toward a higher potential, so the 1 V potential difference causes the electron, with $q = -e$, to gain kinetic energy

$$\Delta K = -q\Delta V = e\Delta V = (1.60 \times 10^{-19}\text{ C})(1\text{ V}) = 1.60 \times 10^{-19}\text{ J}$$

Let us define a new unit of energy, called the **electron volt,** as

$$1\text{ electron volt} = 1\text{ eV} = 1.60 \times 10^{-19}\text{ J}$$

With this definition, the kinetic energy gained by the electron in our example is

$$\Delta K = 1\text{ eV}$$

In other words, **1 electron volt is the kinetic energy gained by an electron (or proton) if it accelerates through a potential difference of 1 volt.**

NOTE ▶ The abbreviation eV uses a lower case e but an upper case V. Units of keV (10^3 eV), MeV (10^6 eV), and GeV (10^9 eV) are common. ◀

The electron volt can be a troublesome unit. One difficulty is its unusual name, which looks less like a unit than, say, "meter" or "second." A more significant difficulty is that the name suggests a relationship to volts. But *volts* are units of electric potential, whereas this new unit is a unit of energy! It is crucial to distinguish between the *potential V*, measured in volts, and an *energy* that can be measured either in joules or in electron volts. You can now use electron volts anywhere that you would previously have used joules. Doing so is no different from converting back and forth between pressure units of pascals and atmospheres.

NOTE ▶ The joule remains the SI unit of energy. It will be useful to express energies in eV, but you *must* convert this energy to joules before doing most calculations. ◀

Fighting cancer with high-energy electrons BIO Tightly focused beams of x rays can be used in radiation therapy for cancer patients. The x rays are generated by directing a high-energy beam of electrons at a metal target. The electrons gain their energy by being accelerated in a *linear accelerator* through a potential difference of 20 MV, so their final kinetic energy is 20 MeV.

EXAMPLE 21.3 The speed of a proton

Atomic particles are often characterized by their kinetic energy in MeV. What is the speed of an 8.7 MeV proton?

SOLVE The mass of a proton is $m = 1.67 \times 10^{-27}$ kg. The kinetic energy of this particle is 8.7×10^6 eV. First, convert the energy to joules:

$$K = 8.7 \times 10^6\text{ eV} \times \frac{1.60 \times 10^{-19}\text{ J}}{1.0\text{ eV}} = 1.4 \times 10^{-12}\text{ J}$$

Now we can find the speed from:

$$K = \frac{1}{2}mv^2 = 1.4 \times 10^{-12}\text{ J}$$

which gives

$$v = \sqrt{\frac{2K}{m}} = \sqrt{\frac{2(1.4 \times 10^{-12}\text{ J})}{1.67 \times 10^{-27}\text{ kg}}} = 4.1 \times 10^7\text{ m/s}$$

Because the proton's charge and the electron's charge have the same magnitude, a general rule is that a proton or electron that accelerates (decelerates) through a potential difference of V volts gains (loses) V eV of kinetic energy. In Example 21.2, with a 100 V potential difference, the electron (speeding up) gained 100 eV of kinetic energy while the proton (slowing down) lost 100 eV of kinetic energy. In Example 21.3, the proton had to accelerate through a 8.7×10^6 V = 8.7 MV potential difference to acquire 8.7 MeV of kinetic energy.

STOP TO THINK 21.2   A proton is released from rest at point Q, where the potential is 0 V. Afterward, the proton

A. Remains at rest at Q.
B. Moves toward P with a steady speed.
C. Moves toward P with an increasing speed.
D. Moves toward R with a steady speed.
E. Moves toward R with an increasing speed.

$$-100\text{ V} \qquad 0\text{ V} \qquad +100\text{ V}$$

P•   Q•   R•

21.3 Calculating the Electric Potential

11.11 Activ Physics ONLINE

Now that we've gained some insight into the electric potential, and have several good ways to represent it visually, let's calculate the potential for some important cases. We'll do so using Equation 21.1, the relationship between the potential energy of a charge q at a point in space and the electric potential at that point. Rewriting Equation 21.1 slightly, we have

$$V = \frac{U_{\text{elec}}}{q} \qquad (21.3)$$

Our prescription for finding the potential at a certain point in space, then, is to first calculate the electric potential *energy* of a charge q placed at that point. Then we can use Equation 21.3 to find the electric potential.

The Electric Potential Inside a Parallel-Plate Capacitor

In Chapter 20 we learned that a *uniform* electric field can be created by placing equal but opposite charges on two parallel conducting plates—the **parallel-plate capacitor.** Thus finding the electric potential inside a parallel-plate capacitor is equivalent to finding the potential for the very important case of a uniform electric field.

Figure 21.8 shows a cross-section view of a charged parallel-plate capacitor with separation d between the plates. The charges $\pm Q$ on the plates are the source charges that create both the electric field and an electric potential in the space between the plates. As we found in Chapter 20, the electric field is $\vec{E} = (Q/\epsilon_0 A$, from positive to negative). We'll choose a coordinate system with $x = 0$ at the negative plate and $x = d$ at the positive plate.

We're free to choose the point of zero potential energy anywhere that's convenient, so let $U_{\text{elec}} = 0$ when a mobile charge q is at the negative plate. The charge's potential energy at any other position x is then the amount of work an external force must do to move the charge at steady speed from the negative plate to that position. We'll represent the external force by a hand, although that's not really how charges get moved around.

The electric field in Figure 21.8 points to the left, so the force $\vec{F}_{\text{elec}} = q\vec{E}$ of the field on the charge is also to the left. To move the charge to the right at constant speed ($\vec{F}_{net} = \vec{0}$), the external force $\vec{F}_{\text{hand}}$ must push to the right with a force of the same magnitude: $F_{\text{hand}} = qE$. This force does work on the system as the charge is moved, changing the potential energy of the system.

When the charge has been moved to position x, it has undergone displacement x as well. Because the external force is constant and is parallel to the displacement, the work done is

$$W = \text{force} \times \text{displacement} = F_{\text{hand}}x = qEx$$

Consequently, the electric potential energy when charge q is at position x is

$$U_{\text{elec}} = W = qEx$$

As the final step, we can use Equation 21.3 to find that the electric potential of the parallel-plate capacitor at position x, measured from the positive plate, is

$$V = \frac{U_{\text{elec}}}{q} = Ex = \frac{Q}{\epsilon_0 A}x \qquad (21.4)$$

where, in the last step, we wrote the electric field strength in terms of the amount of charge on the capacitor plates. Keep in mind that this is the potential due to the source charges on the plates. We used the moveable charge q to find the potential, but q does not cause or contribute to the potential.

A first point to notice is that the electric potential increases linearly from the negative plate at $x = 0$, where $V = V_- = 0$, to the positive plate at $x = d$, where

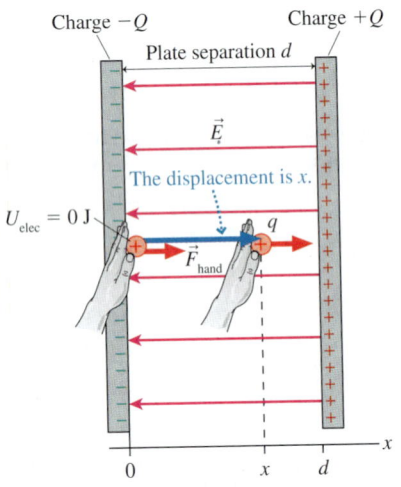

Charge $-Q$ Charge $+Q$

Plate separation d

$\vec{E}$

The displacement is x.

$U_{\text{elec}} = 0 \text{ J}$

$\vec{F}_{\text{hand}}$

q

$0 \qquad x \qquad d$

FIGURE 21.8 Finding the potential of a parallel-plate capacitor.

$V = V_+ = Ed$. Let's define the *potential difference* ΔV_C between the two capacitor plates to be

$$\Delta V_C = V_+ - V_- = Ed \qquad (21.5)$$

NOTE ▶ People who work with circuits would call ΔV_C "the voltage across the capacitor" or simply "the voltage." ◄

In many cases, the capacitor voltage is fixed at some value ΔV_C by connecting its plates to a battery with a known voltage. In this case, the electric field strength inside the capacitor is determined from Equation 21.5 as

$$E = \frac{\Delta V_C}{d} \qquad (21.6)$$

This means that we can establish an electric field of known strength by applying a voltage across a capacitor whose plate spacing is known.

Equation 21.6 implies that the units of electric field are volts per meter, or V/m. We have been using electric field units of newtons per coulomb. In fact, as you can show as a homework problem, these units are equivalent to each other. That is,

$$1 \text{ N/C} = 1 \text{ V/m}$$

NOTE ▶ Volts per meter are the electric field units used by scientists and engineers in practice. We will now adopt them as our standard electric field units. ◄

Returning to the electric potential, we can substitute Equation 21.6 for E into Equation 21.4 for V. In terms of the capacitor voltage ΔV_C, the electric potential inside the capacitor is

$$V = \frac{\Delta V_C}{d}x \qquad (21.7)$$

Let's explore the electric potential inside the capacitor by looking at several different, but related, ways that the potential can be represented. In this example, a battery has established a 1.5 V potential difference across a parallel-plate capacitor with a 3 mm plate spacing.

Graphical representations of the electric potential inside a capacitor

A graph of potential versus x. You can see the potential increasing from 0 V at the negative plate to 1.5 V at the positive plate.

A three-dimensional view showing **equipotential surfaces.** These are mathematical surfaces, not physical surfaces, that have the same value of V at every point. The equipotential surfaces of a capacitor are planes parallel to the capacitor plates. The capacitor plates are also equipotential surfaces.

A two-dimensional contour map. The capacitor plates and the equipotential surfaces are seen edge on, so you need to imagine them extending above and below the plane of the page. The equipotential surfaces seen edge on are the equipotential contours of the contour map.

A three-dimensional elevation graph. The potential is graphed vertically versus the x- and y-coordinates on the other axes. The face-on view of the elevation graph gives you the potential graph.

Potential graph   Equipotential surfaces   Contour map   Elevation graph

EXAMPLE 21.4 A proton in a capacitor

A parallel-plate capacitor is constructed of two disks spaced 2.00 mm apart. It is charged to a potential difference of 500 V. A proton is shot through a small hole in the negative plate with a speed of 2.0×10^5 m/s. Does it reach the other side? If not, what is the furthest distance from the negative plate that the proton reaches?

PREPARE Energy is conserved. The proton's potential energy inside the capacitor can be found from the capacitor's electric potential. Figure 21.9 is a before-and-after visual overview of the proton in the capacitor.

FIGURE 21.9 A before-and-after visual overview of a proton moving in a capacitor.

SOLVE The proton has charge $q = e$, and its potential energy at a point where the capacitor's potential is V is $U_{elec} = eV$. It will gain potential energy $\Delta U_{elec} = e \Delta V_C$ if it moves all the way across the capacitor. The increase in potential energy comes at the expense of kinetic energy, so the proton has sufficient kinetic energy to make it all the way across only if

$$K_i \geq e \Delta V_C$$

We can calculate that $K_i = \frac{1}{2}mv_i^2 = 3.34 \times 10^{-17}$ J and that $e \Delta V_C = 8.00 \times 10^{-17}$ J. The proton does *not* have sufficient kinetic energy to be able to gain 8.00×10^{-17} J of potential energy, so it will not make it across. Instead, the proton will reach a turning point and reverse direction.

The proton starts at the negative plate, where $x_i = 0$ mm. Let the turning point be at x_f. The potential inside the capacitor is given by $V = \Delta V_C x/d$ with $d = 0.0020$ m and $\Delta V_C = 500$ V. Conservation of energy requires $K_f + eV_f = K_i + eV_i$. This is

$$0 + e \Delta V_C \frac{x_f}{d} = \frac{1}{2}mv_i^2 + 0$$

where we used $V_i = 0$ V at the negative plate ($x_i = 0$) and $K_f = 0$ at the turning point. The solution for the turning point is

$$x_f = \frac{mdv_i^2}{2e \Delta V_C} = 0.84 \text{ mm}$$

The proton travels 0.84 mm, less than halfway across, before being turned back.

ASSESS We were able to use the electric potential inside the capacitor to determine the proton's potential energy.

The Potential of a Point Charge

The simplest possible source charge is a single fixed point charge q. To find the electric potential, we'll again start by first finding the electric potential energy when a second charge, which we'll call q', is distance r from charge q. As usual, we'll do this by calculating the work needed to bring q' from a point where $U_{elec} = 0$ to distance r from q. We're free to choose $U_{elec} = 0$ at any point that's convenient. Because the influence of a point charge extends infinitely far, our result for the electric potential will have an especially simple form if we choose $U_{elec} = 0$ (and hence $V = 0$) at a point that is infinitely distant from q.

We can't use the simple expression $W = Fd$ to find the work done in moving q'; this expression is valid only for a *constant* force F and, as we know from Coulomb's law, the force on q' gets larger and larger as it approaches q. To do this calculation properly requires the methods of calculus. However, we can understand *qualitatively* how the potential energy depends on the distance r between the two charges.

Figure 21.10 shows q' at two different distances r from the fixed charge q. When q' is relatively far from q, the electric force on q' is small. Not much external force is needed to push q' closer to q by a small displacement d, so the work done on q' is small and the *change* ΔU_{elec} in the electric potential energy is small as well. This implies, as Figure 21.10 shows, that the graph of U_{elec} versus r is fairly flat when q' is far from q.

On the other hand, the force on q' is quite large when it gets near q, so the work required to move it through the same small displacement is much greater than before. The change in U_{elec} is large, so the graph of U_{elec} versus r is steeper when q' is near q.

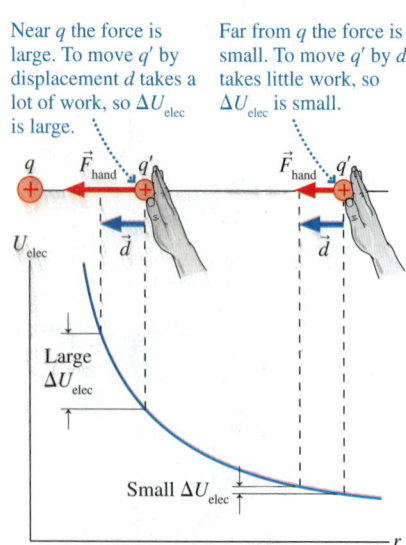

Near q the force is large. To move q' by displacement d takes a lot of work, so ΔU_{elec} is large.

Far from q the force is small. To move q' by d takes little work, so ΔU_{elec} is small.

FIGURE 21.10 The electric potential energy of two point charges.

The general shape of the graph of U_{elec} must be as shown in Figure 21.10. When q' is far from q, the potential energy is small (U_{elec} must go to zero as r goes to infinity). As q' approaches q, the potential energy gets larger and larger. An exact calculation finds the potential energy of two point charges to be

$$U_{elec} = K\frac{qq'}{r} = \frac{1}{4\pi\epsilon_0}\frac{qq'}{r} \qquad (21.8)$$

INVERSE
p. 118

Electric potential energy of two charges
q and q' separated by distance r

NOTE ▶ This expression is very similar to Coulomb's law. The difference is that the electric potential energy depends on the *inverse* of r, that is, on $1/r$, instead of the inverse square dependence of Coulomb's law. Make sure you remember which is which! ◀

Figure 21.10 was a graph of the potential energy for two like charges, where the product qq' is positive. But Equation 21.8 is equally valid for *opposite* charges. In this case, the potential energy of the charges is *negative*. As Figure 21.11 shows, the potential energy of the two charges *decreases* as they get closer together. A particle speeds up as its potential energy decreases ($U \rightarrow K$), so charge q' accelerates toward the fixed charge q. The graph lets us think about the attractive force between opposite charges from the perspective of energy.

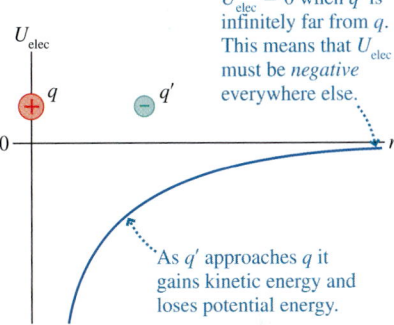

$U_{elec} = 0$ when q' is infinitely far from q. This means that U_{elec} must be *negative* everywhere else.

As q' approaches q it gains kinetic energy and loses potential energy.

FIGURE 21.11 Potential energy diagram for two opposite charges.

EXAMPLE 21.5 Finding the escape velocity

An interaction between two elementary particles causes an electron and a positron (a positively charged electron) to be shot out back-to-back with equal speeds. What minimum speed must each particle have when they are 100 fm apart in order to escape each other? (Recall that 1 fm = 10^{-15} m.)

PREPARE Energy is conserved. The particles end "far apart," which we interpret as sufficiently far to make $(U_{elec})_f \approx 0$ J. Figure 21.12 shows the before-and-after visual overview. The

Before:
v_i v_i
$r_i = 100\,fm$

After:
$v_f = 0$
$v_f = 0$
$r_f \approx \infty$ so $(U_{elec})_f = 0$

FIGURE 21.12 The before-and-after visual overview of an electron and a positron flying apart.

minimum speed to escape is the speed that allows the particles to reach $r_f = \infty$ with $v_f = 0$.

SOLVE Here it is essential to interpret U_{elec} as the potential energy of the electron + positron system. Similarly, K is the *total* kinetic energy of the system. The electron and the positron, with equal masses and equal speeds, have equal kinetic energies. Conservation of energy $K_f + U_f = K_i + U_i$ is

$$0 + 0 = \left(\frac{1}{2}mv_i^2 + \frac{1}{2}mv_i^2\right) + K\frac{q_e q_p}{r_i} = mv_i^2 - \frac{Ke^2}{r_i}$$

Using $r_i = 100$ fm = 1.0×10^{-13} m, we can calculate the minimum initial speed to be

$$v_i = \sqrt{\frac{Ke^2}{mr_i}} = 5.0 \times 10^7 \text{ m/s}$$

ASSESS v_i is a little more than 10% the speed of light, just about the limit of what a "classical" calculation can predict. We would need to use the theory of relativity if v_i were much larger.

Equation 21.8 gives the potential energy of a charge q' when it is a distance r from a point charge q. We know that the electric *potential* is related to the potential energy by $V = U_{elec}/q'$. Thus the electric potential of charge q is

$$V = K\frac{q}{r} = \frac{1}{4\pi\epsilon_0}\frac{q}{r} \qquad (21.9)$$

INVERSE
p. 118

Electric potential at distance r from a point charge q

Notice that only the *source* charge q appears in this expression. This is in line with our picture that a source charge *creates* the electric potential around it.

EXAMPLE 21.6 Calculating the potential of a point charge

What is the electric potential 1.0 cm from a 1.0 nC charge? What is the potential difference between a point 1.0 cm away and a second point 3.0 cm away?

PREPARE We can use Equation 21.9 to find the potential at the two distances from the charge.

SOLVE The potential at $r = 1.0$ cm is

$$V_{1\,cm} = \frac{Kq}{r} = (9.0 \times 10^9 \text{ N} \cdot \text{m}^2/\text{C}^2)\left(\frac{1.0 \times 10^{-9} \text{ C}}{0.010 \text{ m}}\right)$$

$$= 900 \text{ V}$$

We can similarly calculate $V_{3\,cm} = 300$ V. Thus the potential difference between these two points is $\Delta V = V_{1\,cm} - V_{3\,cm} = 600$ V.

ASSESS 1 nC is typical of the electrostatic charge produced by rubbing, and you can see that such a charge creates a fairly large potential nearby. Why are we not shocked and injured when working with the "high voltages" of such charges? As we'll learn in Chapter 26, the sensation of being shocked is a result of current, not potential. Some high-potential sources simply do not have the ability to generate much current.

Figure 21.13 shows four graphical representations of the electric potential of a point charge. These match the four representations of the electric potential inside a capacitor, and a comparison of the two is worthwhile. This figure assumes that q is positive; you may want to think about how the representations would change if q were negative.

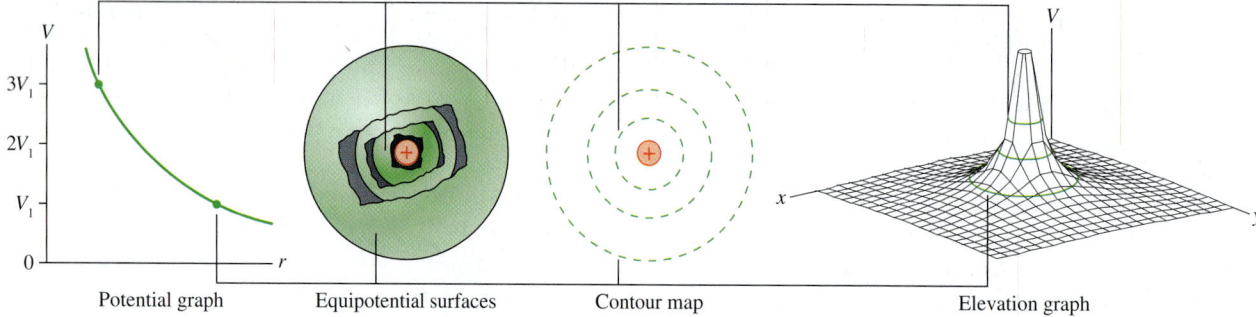

| Potential graph | Equipotential surfaces | Contour map | Elevation graph |

FIGURE 21.13 Four graphical representations of the electric potential of a point charge.

The Electric Potential of a Charged Sphere

In Chapter 20 you learned that the electric field outside a sphere of charge is the same as that of a point charge. Consequently, the electric potential outside a charged sphere is given by Equation 21.9, that is,

$$V = \frac{1}{4\pi\epsilon_0}\frac{Q}{r} \qquad (21.10)$$

INVERSE
p. 118

Electric potential at a distance $r > R$
from the center of a sphere with charge Q

We can cast this result in a more useful form. It is common to charge a metal object, such as a sphere, "to" a certain potential; for instance, this can be done by connecting the sphere to, say, a 30 volt battery. This potential, which we will call V_0, is the potential right on the surface of the sphere. We can see from Equation 21.10 that

$$V_0 = V(\text{at } r = R) = \frac{Q}{4\pi\epsilon_0 R} \qquad (21.11)$$

Consequently, a sphere of radius R that is charged to potential V_0 has total charge

$$Q = 4\pi\epsilon_0 R V_0 \qquad (21.12)$$

If we substitute this expression for Q into Equation 21.10, we can write the potential outside a sphere that is charged to potential V_0 as

$$V = \frac{R}{r} V_0 \qquad (21.13)$$

Equation 21.13 tells us that the potential of a sphere is V_0 on the surface and decreases inversely with the distance. Thus the potential at $r = 3R$ is $\frac{1}{3}V_0$.

EXAMPLE 21.7 Releasing a proton from a charged sphere

A proton is released from rest at the surface of a 1.0-cm-diameter sphere that has been charged to $+1000$ V.

a. What is the charge of the sphere?
b. What is the proton's speed when it is 1.0 cm from the sphere?

PREPARE Energy is conserved. The potential outside the charged sphere is the same as the potential of a point charge at the center. Figure 21.14 is a before-and-after visual overview.

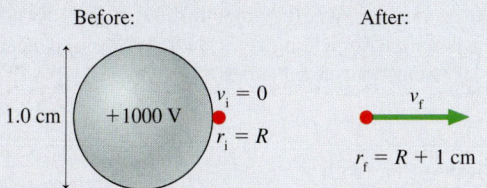

Before: After:

1.0 cm   +1000 V $v_i = 0$ v_f
                    $r_i = R$
                              $r_f = R + 1$ cm

FIGURE 21.14 A before-and-after visual overview of a sphere and a proton.

SOLVE

a. We can use the sphere's potential to find that the charge of the sphere is

$$Q = 4\pi\epsilon_0 R V_0 = 0.56 \times 10^{-9}\,\text{C} = 0.56\,\text{nC}$$

b. A sphere charged to $V_0 = +1000$ V is positively charged. The proton will be repelled by this charge and move away from the sphere. The conservation of energy equation $K_f + eV_f = K_i + eV_i$, with Equation 21.13 for the potential of a sphere, is

$$\frac{1}{2}mv_f^2 + \frac{eR}{r_f}V_0 = \frac{1}{2}mv_i^2 + \frac{eR}{r_i}V_0$$

The proton starts from the surface of the sphere, $r_i = R$, with $v_i = 0$. When the proton is 1.0 cm from the *surface* of the sphere, it has $r_f = 1.0$ cm $+ R = 1.5$ cm. Using these, we can solve for v_f:

$$v_f = \sqrt{\frac{2eV_0}{m}\left(1 - \frac{R}{r_f}\right)} = 3.6 \times 10^5\,\text{m/s}$$

ASSESS This example illustrates how the ideas of electric potential and potential energy work together, yet they are *not* the same thing.

The Electric Potential of Many Charges

Suppose there are many source charges $q_1, q_2, \ldots$. The electric potential V at a point in space is the *sum* of the potentials due to each charge:

$$V = \sum_i \frac{1}{4\pi\epsilon_0}\frac{q_i}{r_i} \qquad (21.14)$$

where r_i is the distance from charge q_i to the point in space where the potential is being calculated. Unlike the electric field of multiple charges, which required vector addition, the electric potential is a simple scalar sum. This makes finding the potential of many charges considerably easier than finding the corresponding electric field. As an example, the contour map and elevation graph in Figure 21.15 on the next page show that the potential of an electric dipole is the sum of the potentials of the positive and negative charges. We'll see later that the electric potential of the heart has the form of an electric dipole.

(a) Contour map   **(b)** Elevation graph

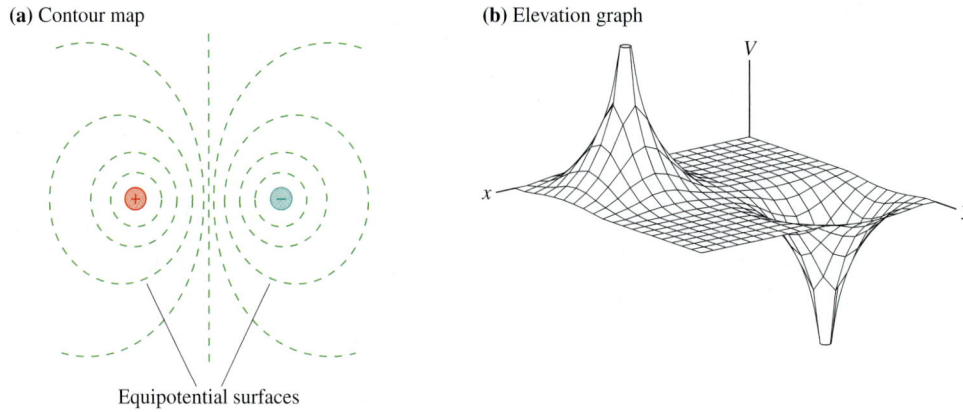

Equipotential surfaces

FIGURE 21.15 The electric potential of an electric dipole.

EXAMPLE 21.8 Finding the potential of two charges

What is the electric potential at the point indicated in Figure 21.16?

FIGURE 21.16 Finding the potential of two charges.

+2.0 nC   3.0 cm   −1.0 nC

5.0 cm   4.0 cm

PREPARE The potential is the sum of the potentials due to each charge.

SOLVE The potential at the indicated point is

$$V = \frac{Kq_1}{r_1} + \frac{Kq_2}{r_2}$$

$$= (9.0 \times 10^9 \text{ N} \cdot \text{m}^2/\text{C}^2)\left(\frac{2.0 \times 10^{-9} \text{ C}}{0.050 \text{ m}} + \frac{-1.0 \times 10^{-9} \text{ C}}{0.040 \text{ m}}\right)$$

$$= 140 \text{ V}$$

ASSESS As noted above, the potential is a *scalar*, so we found the net potential by adding two scalars. We don't need any angles or components to calculate the potential.

STOP TO THINK 21.3   Rank in order, from largest to smallest, the potential differences ΔV_{12}, ΔV_{13}, and ΔV_{23} between points 1 and 2, points 1 and 3, and points 2 and 3.

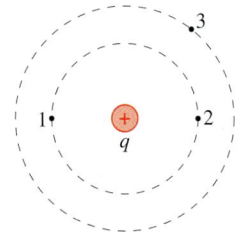

21.4 Sources of Electric Potential

You've probably created a spark and experienced a shock when touching a doorknob after shuffling your feet across a carpet on a dry day. Friction between the carpet and your feet causes a separation of charge, transferring electrons from one object to the other. In the process, a *potential difference* is established between the two objects. The potential difference between you and a doorknob can be many tens of thousands of volts—enough to create a spark as the excess charge on your body moves from higher to lower potential. The main lesson is that **we can create a potential difference by creating a charge separation.**

The **Van de Graaff generator** shown in Figure 21.17a is a mechanical charge separator—essentially a fancy foot shuffler. A moving plastic or leather belt is charged, then the charge is mechanically transported via the conveyor belt to the spherical electrode at the top of the insulating column. The belt is charged by a positively-charged sharp needle that ionizes the air around it. Positive charges in the air are pushed away from the needle and onto the belt.

Lightning is the result of large potential differences built up by charge separation within clouds.

(a)

Hollow metal sphere

2. The plastic or leather belt is the conveyor belt that mechanically transports charge to the top.

Insulating plastic tube

3. A pointed wire draws charge off the belt and charges the sphere.

1. A sharp charged needle charges the belt positively.

Electric motor

(b)

The world's largest Van de Graaff generator, a tandem unit at the Museum of Science in Boston, creates potential differences of up to 2.5 MV.

FIGURE 21.17 A Van de Graaff generator.

A Van de Graaff generator has two noteworthy features:

- Charge is mechanically transported from the negative side to the positive side. This charge separation creates *and sustains* a potential difference between the spherical electrode and its surroundings.
- The electric field of the spherical electrode exerts a downward force on the positive charges moving up the belt. Consequently, *work must be done* to "lift" the positive charges. The work is done by the electric motor that runs the belt.

A classroom demonstration Van de Graaff generator creates a potential difference of several hundred thousand volts between the upper sphere and its surroundings. The maximum potential is reached when the electric field near the sphere becomes large enough to strip electrons off air molecules, making the air conducting. This produces a spark and temporarily discharges the sphere. Sparks from large Van de Graaff generators, like the one shown in Figure 21.17b, can be used as simulated lightning for testing aircraft and power transmission equipment.

For us, the Van de Graaff generator is important because it shows that a potential difference can be created and sustained by a device that separates charge.

Lightning is the result of a charge separation that occurs in clouds. As small ice particles in the clouds collide, charging occurs by frictional rubbing, just as when you scuff your shoes on the carpet. The details are still not well understood, but heavier particles gain a negative charge while lighter particles become positive. Natural convection processes separate the particles, with the lighter ones being lifted toward the top of the cloud while the heavier ones fall toward the bottom. Thus the top of the cloud becomes positively charged and the bottom negatively charged. This natural charge separation creates a huge potential difference—as much as 100 million volts—between the top and bottom of the cloud. Lightning *within* the cloud—the most common type—occurs when this potential difference is so great that the air molecules are torn apart, creating a conducting pathway for the huge rush of charge from top to bottom that we see as lightning. Cloud-to-ground lightning is also a result of charge separation. Here, the negative charge in the bottom of the cloud causes positive charge to accumulate in the ground below. A lightning strike occurs when the potential difference becomes too large for the air to sustain.

Charge separation, and hence potential differences, can also be created by chemical processes. A common and important means of creating a fixed potential difference is the **battery.** We'll study batteries in depth in Chapter 22, but

− terminal

+ terminal

A car battery maintains a fixed potential difference of 12 V between its + and − terminals.

Diffusion tends to equalize the concentrations of Na⁺ and K⁺ inside and outside the cell. It moves Na⁺ in and K⁺ out.

The sodium-potassium pump uses energy from ATP to force Na⁺ out of the cell and K⁺ in, causing a charge separation.

FIGURE 21.18 The membrane potential of a cell is due to a charge separation.

Micro-pipette   Cell body

FIGURE 21.19 Measuring the membrane potential.

A voltmeter always uses two probes to measure a potential difference. Here, we see that the potential difference of a fresh 9 V battery is closer to 9.7 V.

all batteries use chemical reactions to create an internal charge separation. This separation proceeds until a characteristic potential difference—about 1.5 V for a standard alkaline battery—appears between the two terminals of the battery. Different kinds of batteries maintain different potential differences between their terminals.

Chemical means of producing potential differences are also crucial in biological systems. For example, there's a potential difference of about 70 mV between the inside and outside of a cell, with the inside of the cell more negative than the outside. As illustrated in Figure 21.18, this *membrane potential* is caused by an imbalance of potassium (K^+) and sodium (Na^+) ions. The concentration of K^+ is higher inside the cell than outside, while the concentration of Na^+ is higher outside than in. To keep the charge separated in the face of diffusion, a *sodium-potassium exchange pump* continually pumps sodium out of and potassium into the cell. We'll have a careful look at the electrical properties of nerve cells later, in Chapter 23.

Measuring Electric Potential

Measurements of the electric potential play an important role in a broad range of applications. An electrocardiogram measures the potential difference between several locations on the body to diagnose possible heart problems; temperature is often measured using a *thermocouple,* a device that develops a potential difference proportional to temperature; your digital camera can sense when its battery is low by measuring the potential difference between its terminals.

Note that in all these applications, it's the potential *difference* between two points that's measured. The actual value of the potential at a given point depends on where we choose V to be zero, but the difference in potential between two points is independent of this choice. Because of this, a **voltmeter,** the basic instrument for measuring potential differences, always has *two* inputs. Probes are connected from these inputs to the two points between which the potential difference is to be measured. We'll learn more in Chapter 23 about how voltmeters work.

As small as cells are, the membrane potential difference between the inside and outside of a cell can be measured by a (very small) probe connected to a voltmeter. Figure 21.19 is a micrograph of a nerve cell whose membrane potential is being measured. A very small glass pipette, filled with conductive fluid, is actually inserted through the cell's membrane. This pipette is one of the probes. The second probe need not be so small; it is simply immersed in the conducting fluid that surrounds the cell and can be quite far from the cell.

21.5 Connecting Potential and Field

In Chapter 20, we learned how source charges create an electric field around them; in this chapter, we found that source charges also create an electric potential everywhere in the space around them. But these two concepts of field and potential are clearly linked. For instance, we can calculate potential differences by considering the work done on a charge as it is pushed against the electric force due to the field. In this section, we'll build on this idea and will find that **the electric potential and electric field are not two distinct entities but, instead, two different perspectives or two different mathematical representations of how source charges alter the space around them.**

To make the connection between potential and field, Figure 21.20 shows a contour map of the electric potential due to some source charges (which aren't shown here). Suppose a charge q moves a short distance along one of the equipotentials. Because it moves along an equipotential, its potential and hence its potential energy are the *same* at the beginning and end of its displacement. This means that no work is done in moving the charge. As Figure 21.20 shows, the only way that no work can be done is if the electric field is *perpendicular* to the equipotential surface. (You should recall, from Chapter 10, that no work is done by a force perpendicular to a particle's displacement.) This, then, is our first discovery about the connection between field and potential: **The electric field at a point is perpendicular to the equipotential surface at that point.**

We can say more. Suppose we release a positive charge q from rest. Just as a mass falls toward lower potential energy—the ground!—charge q "falls" toward lower electric potential as it transforms potential energy into kinetic energy. But the motion of charge q is governed by the electric force $\vec{F} = q\vec{E}$ and thus by the electric field $\vec{E}$. For q to move toward lower electric potential, the electric field must point toward lower potential. This is our second discovery about the connection between field and potential: **The electric field points "downhill" in the direction of decreasing potential.** The electric field in Figure 21.20 points from the 10 V contour toward the 0 V contour rather than toward the 20 V contour.

Finally, we can learn something about the *magnitude* of $\vec{E}$ by following the charge q as it moves "downhill" in Figure 21.21, parallel to $\vec{E}$, from a higher potential V_+ to a lower potential V_-. Energy is conserved in this motion, so $\Delta K + \Delta U_{elec} = \Delta K + q(V_- - V_+) = 0$. Let's define $\Delta V = V_+ - V_-$, which will always be a positive number. We see that ΔV is related to the charge's gain in kinetic energy ΔK by

$$\Delta V = \frac{\Delta K}{q} \tag{21.15}$$

Charge q gains kinetic energy because it accelerates in the electric field. Suppose the contour lines are distance d apart, and that d is small enough for $\vec{E}$ to be nearly constant over this small distance. In this case, the electric force $F = qE$ is nearly constant and, from Newton's second law, causes the particle to have constant acceleration $a = F/m = qE/m$. One of the important equations of constant-acceleration kinematics related the initial and final velocities of a particle that accelerates over distance d. This was

$$v_f^2 = v_i^2 + 2ad = v_i^2 + \frac{2qEd}{m}$$

We can use this result to find that the change in kinetic energy over distance d is

$$\Delta K = \frac{1}{2}mv_f^2 - \frac{1}{2}mv_i^2 = \frac{1}{2}m(v_f^2 - v_i^2) = qEd$$

With this expression for ΔK, Equation 21.15 becomes

$$\Delta V = Ed$$

Rewriting this, we find our third discovery about the connection between field and potential: **At a point where two equipotential surfaces with potential difference ΔV are distance d apart, the electric field strength is**

$$E = \frac{\Delta V}{d} \tag{21.16}$$

Electric field strength in terms of the potential difference
ΔV between two equipotential surfaces a distance d apart

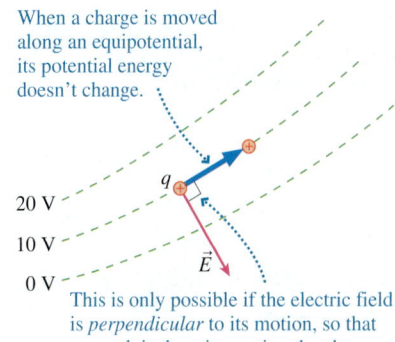

When a charge is moved along an equipotential, its potential energy doesn't change.

20 V

10 V

0 V

$\vec{E}$

This is only possible if the electric field is *perpendicular* to its motion, so that no work is done in moving the charge.

FIGURE 21.20 The electric field is always perpendicular to an equipotential surface and points in the direction of decreasing potential.

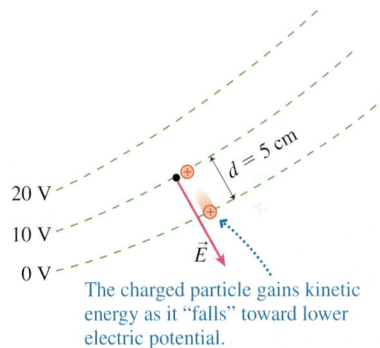

$d = 5$ cm

20 V

10 V

0 V

$\vec{E}$

The charged particle gains kinetic energy as it "falls" toward lower electric potential.

FIGURE 21.21 Finding the strength of the electric field.

Activ
Physics 11.12, 11.13

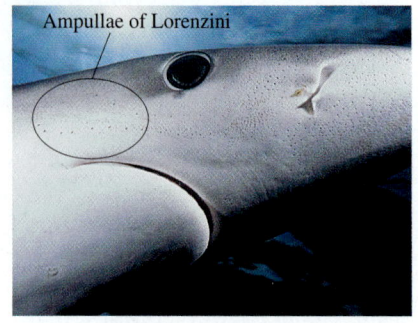

Ampullae of Lorenzini

◀ **Electroreception in sharks** BIO Sharks have a "sixth sense" that aids them in detecting prey: They are able to detect the weak electric potentials created by other fish. A shark has an array of sensor cells, the *ampullae of Lorenzini,* distributed around its snout. These appear as small black spots on the blue shark shown. These cells are highly sensitive to potential differences and can measure voltages as small as a few billionths of a volt! As we saw in Chapter 20, there are electric fields and potentials associated with every beat of the heart; muscle contractions also create potential differences. The shark can detect these small potential differences, even using them to sense animals buried in the sand.

For example, in Figure 21.21 the potential difference is $\Delta V = V_+ - V_- = (10 \text{ V}) - (0 \text{ V}) = 10 \text{ V}$, while $d = 5 \text{ cm} = 0.05 \text{ m}$. Thus the electric field strength between these two equipotential surfaces is

$$E = \frac{\Delta V}{d} = \frac{10 \text{ V}}{0.05 \text{ m}} = 200 \text{ V/m}$$

NOTE ▶ This expression for the electric field strength is similar to the result $E = \Delta V_C/d$ for the electric field strength inside a parallel-plate capacitor. A capacitor has a uniform field, the same at all points, so we can calculate the field using the full potential difference ΔV_C and the full spacing d. In contrast, Equation 21.16 only applies *locally,* at a point where the spacing between two adjacent contour lines is d. As the spacing varies, so too does the field strength. ◀

Figure 21.22 summarizes what we've learned about the connection between potential and field.

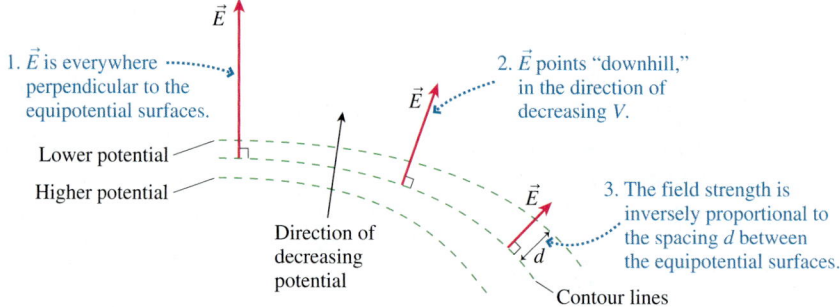

1. $\vec{E}$ is everywhere perpendicular to the equipotential surfaces.

2. $\vec{E}$ points "downhill," in the direction of decreasing V.

Lower potential

Higher potential

Direction of decreasing potential

3. The field strength is inversely proportional to the spacing d between the equipotential surfaces.

Contour lines

FIGURE 21.22 The geometry of potential and field.

Figure 21.23 shows three important arrangements of charges that we've studied in this chapter and Chapter 20. Both electric field lines and equipotentials are shown, and you can see how the connections between field and potential summarized in Figure 21.22 apply in each case.

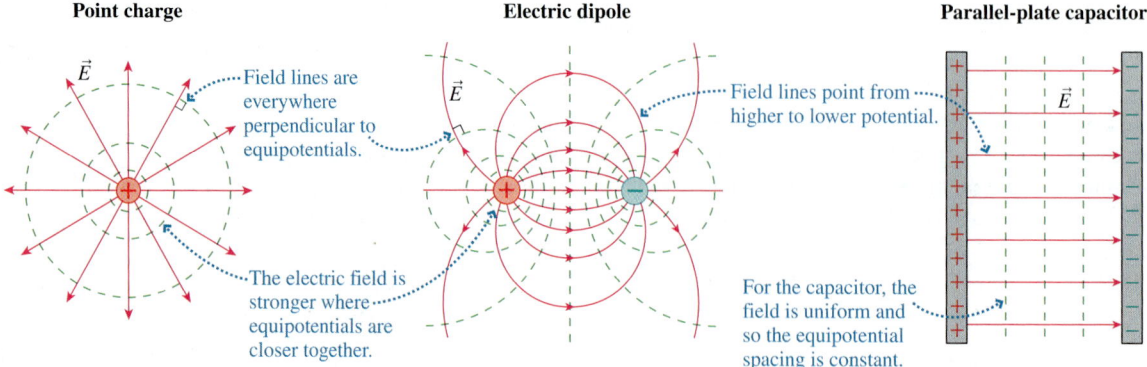

Point charge

Field lines are everywhere perpendicular to equipotentials.

The electric field is stronger where equipotentials are closer together.

Electric dipole

Field lines point from higher to lower potential.

Parallel-plate capacitor

For the capacitor, the field is uniform and so the equipotential spacing is constant.

FIGURE 21.23 Electric field lines and equipotentials for three important cases.

EXAMPLE 21.9 Finding the electric field from equipotential surfaces

In Figure 21.24 a 1 cm × 1 cm grid is superimposed on a contour map of the potential. Estimate the strength and direction of the electric field at points 1, 2, and 3. Show your results graphically by drawing the electric field vectors on the contour map.

FIGURE 21.24 Equipotential lines.

PREPARE The electric field is perpendicular to the equipotential lines, points "downhill," and depends on the spacing between the contour lines. The potential is highest on the bottom and the right. An elevation graph of the potential would look like the lower-right quarter of a bowl or a football stadium.

SOLVE Some distant but unseen source charges have created an electric field and potential. We do not need to see the source charges to relate the field to the potential. Because $E = \Delta V/d$, the electric field is stronger where the equipotential lines are closer together and weaker where they are farther apart.

Figure 21.25 shows how measurements of d from the grid are combined with values of ΔV to determine $\vec{E}$. Point 3 requires an estimate of the spacing between the 0 V and the 100 V surfaces. Notice that we're using the 0 V and 100 V equipotential surfaces to determine $\vec{E}$ at a point on the 50 V equipotential.

FIGURE 21.25 The electric field at points 1 to 3.

ASSESS The *directions* of $\vec{E}$ are found by drawing downhill vectors perpendicular to the equipotentials. The distances between the equipotential surfaces are needed to determine the field strengths.

A Conductor in Electrostatic Equilibrium

In Chapter 20, you learned four important properties about conductors in electrostatic equilibrium:

1. Any excess charge is on the surface.
2. The electric field inside is zero.
3. The exterior electric field is perpendicular to the surface.
4. The field strength is largest at sharp corners.

Now we can add a fifth important property:

5. The entire conductor is at the same potential, and thus the surface is an equipotential surface.

To see why this is so, Figure 21.26 shows two points inside a conductor connected by a line that remains entirely inside the conductor. We can find the potential difference $\Delta V = V_2 - V_1$ between these points by using an external force to push a charge along the line from 1 to 2 and calculating the work done. But because $\vec{E} = \vec{0}$, there is no force on the charge. The work would be zero, and so $\Delta V = 0$. In other words, **any two points inside a conductor in electrostatic equilibrium are at the same potential.**

When a conductor is in electrostatic equilibrium, the *entire conductor* is at the same potential. If we charge a metal electrode, the entire electrode is at a single potential. The fact that the surface is an equipotential surface and that the exterior electric field is perpendicular to the surface can now be seen as a special case of our conclusion from the previous section that electric fields are always perpendicular to equipotentials.

At pointed metal tips, the electric field can be strong enough to ionize air molecules. The field accelerates these ions, which collide, releasing energy in the bluish glow seen here.

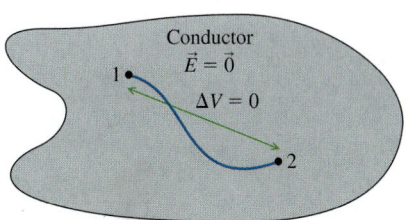

▶ FIGURE 21.26 All points inside a conductor in electrostatic equilibrium are at the same potential.

Figure 21.27 summarizes what we know about conductors in electrostatic equilibrium. These are important and practical conclusions because conductors are the primary components of electrical devices.

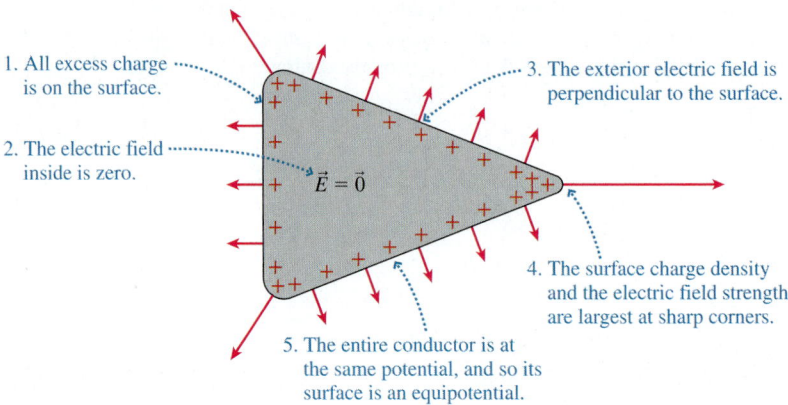

1. All excess charge is on the surface.

2. The electric field inside is zero.

$\vec{E} = \vec{0}$

3. The exterior electric field is perpendicular to the surface.

4. The surface charge density and the electric field strength are largest at sharp corners.

5. The entire conductor is at the same potential, and so its surface is an equipotential.

FIGURE 21.27 Electrical properties of a conductor in electrostatic equilibrium.

STOP TO THINK 21.4 Which set of equipotential surfaces matches this electric field?

$\vec{E}$

| 0 V | 50 V | 0 V | 50 V | 0 V | 50 V |
| A. | | B. | | C. | |

| 50 V | 0 V | 50 V | 0 V | 50 V | 0 V |
| D. | | E. | | F. | |

21.6 The Electrocardiogram

As we saw in Chapter 20, the electrical activity of cardiac muscle cells makes the beating heart an electric dipole. A resting nerve or muscle cell is *polarized,* meaning that the outside is positive and the inside negative. Figure 21.18 showed this situation. Initially, all muscle cells in the heart are polarized. When triggered by an electrical impulse from the heart's sino-auricular node in the right atrium, heart cells begin to *depolarize,* moving ions through the cell wall until the outside becomes negative. This causes the muscle to contract. The depolarization of one cell triggers depolarization in an adjacent cell, causing a "wave" of depolarization to spread across the tissues of the heart.

At any instant during this process, a boundary divides the negative charges of depolarized cells from the positive charges of cells that have not depolarized. As Figure 21.28a shows, this separation of charges creates an electric dipole and produces a dipole electric field and potential. Figure 21.28b shows the equipotential surfaces of the heart's dipole at one instant of time. These equipotential surfaces match those shown earlier in Figure 21.15.

NOTE ▶ The convention is to draw diagrams of the heart as if you were facing the person whose heart is being drawn. The left side of the heart is thus on the right side of the diagram. ◀

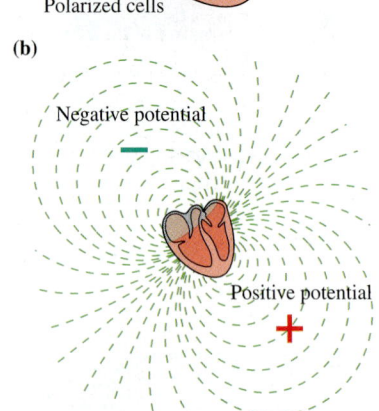

(a)

The boundary between polarized and depolarized cells sweeps rapidly across the atria.

At the boundary there is a charge separation, leading to a net dipole moment.

Depolarized cells

Dipole moment

Polarized cells

(b)

Negative potential

Positive potential

FIGURE 21.28 A contracting heart is an electric dipole.

A measurement of the electric potential of the heart is an invaluable diagnostic tool. Recall, however, that only potential *differences* are meaningful, so we need to measure the potential difference between pairs of electrodes. In practice, as Figure 21.29 shows, the potential difference is measured between several pairs of electrodes (often called leads) attached to the torso. A chart of the potential differences is known as an **electrocardiogram,** abbreviated either ECG or, from its European origin, EKG. A common method of performing an EKG uses 12 leads, and records 12 pairs of potential differences. The image shown at the beginning of this chapter was constructed from information assembled from such measurements.

Figure 21.30 shows a simplified model of electrocardiogram measurement using only two electrodes, one on each arm. As the wave of depolarization moves across the heart muscle during each heart beat, the dipole moment vector of the heart changes its magnitude and direction. As Figure 21.30 shows, both of these affect the potential difference between the electrodes, so each point on the EKG graph corresponds to a particular magnitude and orientation of the dipole moment.

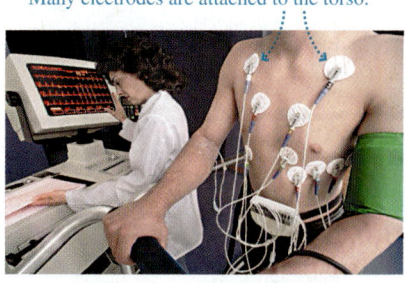
Many electrodes are attached to the torso.

ΔV

Records of the potential difference between various pairs of electrodes allow the doctor to analyze the heart's condition.

FIGURE 21.29 Performing an EKG.

(a) Atrial depolarization — V_2 is positive, V_1 negative.

(b) Septal depolarization — V_2 is negative, V_1 positive.

(c) Ventricular depolarization — V_2 is positive, V_1 negative.

Electrode 1   Electrode 2

(d) $\Delta V = V_2 - V_1$

The record of the potential difference between the two electrodes is the electrocardiogram.

The potential differences at a, b, and c correspond to those measured in the three stages shown to the left.

FIGURE 21.30 The potential difference between the electrodes changes as the heart beats.

21.7 Capacitance and Capacitors

In Section 21.5 we found that potential differences are caused by the separation of charge. One common method of creating a charge separation, shown in Figure 21.31 on the next page, is to move charge Q from one initially uncharged conductor to a second initially uncharged conductor. This results in charge $+Q$ on one conductor and $-Q$ on the other. Two conductors with equal but opposite charge form a **capacitor.** The two conductors that make up a capacitor are its *electrodes* or *plates*. We've already looked at some of the properties of parallel-plate capacitors; now we want to study capacitors that might have any shape. Capacitors can be used to store charge, making them invaluable in all kinds of electronic circuits.

As Figure 21.31 on the next page shows, the electric field strength E and the potential difference ΔV_C increase as the charge on each electrode increases. If we double the amount of charge on each electrode, the work required to move a charge from one electrode to the other doubles. This implies a doubling of the

Capacitors are important elements in electric circuits. They come in a wide variety of sizes and shapes.

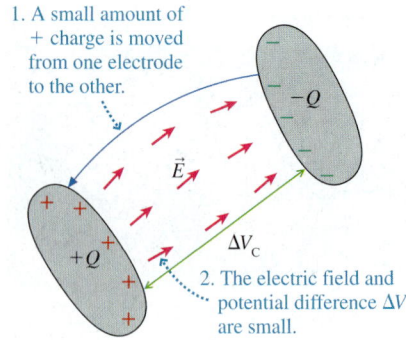

1. A small amount of + charge is moved from one electrode to the other.

2. The electric field and potential difference ΔV are small.

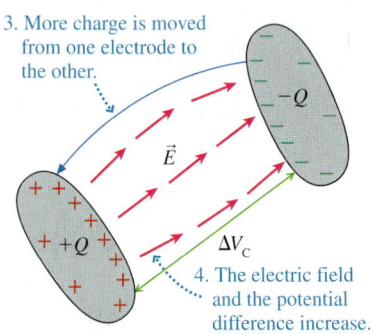

3. More charge is moved from one electrode to the other.

4. The electric field and the potential difference increase.

FIGURE 21.31 Charging a capacitor.

potential difference between the electrodes. Thus **the potential difference between the electrodes is directly proportional to their charge.**

Stated another way, **the charge of a capacitor is directly proportional to the potential difference between its electrodes.** As we'll see, this is actually the most common way that a capacitor is used: A source of potential difference, such as a battery, is connected between the electrodes, causing a charge proportional to the potential difference to be moved from one electrode to the other. We can write the relationship between charge and potential difference as

$$Q = C \Delta V_C \qquad (21.17)$$

Charge on a capacitor with potential difference ΔV_C

LINEAR
p. 38

The constant of proportionality C between Q and ΔV_C is called the **capacitance** of the capacitor. Capacitance depends on the shape, size, and spacing of the two electrodes. A capacitor with a large capacitance holds more charge for a given potential difference than one with a small capacitance.

NOTE ▶ We will only consider situations where the charge on the electrodes is equal in magnitude but opposite in sign. When we say "A capacitor has charge Q," we mean that one electrode has charge $+Q$ and the other charge $-Q$. The potential difference between the electrodes is called the potential difference *of* the capacitor. ◀

The SI unit of capacitance is the **farad,** named in honor of Michael Faraday, the originator of the idea of electric and magnetic fields. One farad is defined as

$$1 \text{ farad} = 1 \text{ F} = 1 \text{ C/V}$$

12.6 **Activ Physics ONLINE**

One farad turns out to be a very large capacitance. Practical capacitors are usually measured in units of microfarads (μF) or picofarads (1 pF $= 10^{-12}$ F).

Charging a Capacitor

To "charge" a capacitor, we need to move charge from one electrode to the other. The simplest way to do this is to use a source of potential difference such as a battery, as shown in Figure 21.32. We learned earlier that a battery uses its internal

(a)

Direction of charge motion

1. Charge flows from this plate, leaving it negative.

ΔV_C ΔV_{bat}

2. The charge then flows through the battery, which acts as a "charge pump."

3. The charge ends up on this plate, making it positively charged.

Charge can move freely through wires.

(b)

The movement of the charge stops when ΔV_C is equal to the battery voltage. The capacitor is then fully charged.

$\Delta V_C = \Delta V_{bat}$ ΔV_{bat}

(c)

If the battery is removed, the capacitor remains charged, with ΔV_C still equal to the battery voltage.

$\Delta V_C = \Delta V_{bat}$

FIGURE 21.32 Charging a capacitor using a battery.

chemistry to maintain a fixed potential difference between its terminals. If we connect a capacitor to a battery, charge flows from the negative electrode of the capacitor, through the battery, and onto the positive electrode. This flow of charge continues until the potential difference between the capacitor's electrodes is the same as the fixed potential difference of the battery. If the battery is then removed, the capacitor remains charged with a potential equal to that of the battery that charged it because there's no conducting path for charge on the positive electrode to move back to the negative electrode. Thus **a capacitor can be used to store charge.**

EXAMPLE 21.10 Charging a capacitor

A 1.3 μF capacitor is connected to a 1.5 V battery. What is the charge on the capacitor?

PREPARE Charge flows through the battery from one capacitor electrode to the other until the potential difference ΔV_C between the electrodes equals that of the battery, or 1.5 V.

SOLVE The charge on the capacitor is given by Equation 21.17:

$$Q = C \, \Delta V_C = (1.3 \times 10^{-6}\,\text{F})(1.5\,\text{V}) = 2.0 \times 10^{-6}\,\text{C}$$

ASSESS This is the charge on the positive plate; the other plate has a charge of -2.0×10^{-6} C. If the battery is removed, the capacitor will retain its charge and maintain a potential difference of 1.5 V.

The Parallel-Plate Capacitor

As we've seen, the *parallel-plate capacitor* is important because it creates a uniform electric field between its flat electrodes. In Chapter 20, we found that the electric field of a parallel-plate capacitor is

$$\vec{E} = \left(\frac{Q}{\epsilon_0 A}, \text{ from positive to negative} \right)$$

where A is the surface area of the electrodes and Q is the charge on the capacitor. We can use this result to find the capacitance of a parallel-plate capacitor.

Earlier in this chapter, we found that the electric field strength of a parallel-plate capacitor is related to the potential difference ΔV_C and the spacing d between the plates by

$$E = \frac{\Delta V_C}{d}$$

Combining these two results, we see that

$$\frac{Q}{\epsilon_0 A} = \frac{\Delta V_C}{d}$$

or, equivalently,

$$Q = \frac{\epsilon_0 A}{d} \Delta V_C \tag{21.18}$$

If we compare Equation 21.18 to Equation 21.17, the definition of capacitance, we see that the capacitance of the parallel-plate capacitor is

$$C = \frac{\epsilon_0 A}{d} \tag{21.19}$$

Capacitance of a parallel-plate capacitor
with plate area A, separation d

Each long structure is one capacitor.

A capacity for memory Your computer's random-access memory, or RAM, uses tiny capacitors to store the digital ones and zeroes that make up your data. A charged capacitor represents a one and an uncharged capacitor a zero. For a billion or more capacitors to fit on a single chip they must be very small. The micrograph is a cross section through the silicon wafer that makes up the memory chip. Each capacitor consists of a very long electrode separated by a thin insulating layer from the common electrode shared by all capacitors. Each capacitor's capacitance is only about 30×10^{-15} F!

− electrode
Insulator
+ electrode
1 μm

LINEAR
p. 38

INVERSE
p. 118

EXAMPLE 21.11 Charging a parallel-plate capacitor

The spacing between the plates of a 1.0 μF parallel-plate capacitor is 0.070 mm.

  a. What is the surface area of the plates?
  b. How much charge is on the plates if this capacitor is attached to a 1.5 V battery?

SOLVE

  a. From the definition of capacitance,

$$A = \frac{dC}{\epsilon_0} = \frac{(0.070 \times 10^{-3}\text{ m})(1.0 \times 10^{-6}\text{ F})}{8.85 \times 10^{-12}\text{ F/m}} = 7.9\text{ m}^2$$

  b. The charge is $Q = C\,\Delta V_C = (1.0 \times 10^{-6}\text{ F})(1.5\text{ V}) = 1.5 \times 10^{-6}\text{ C} = 1.5\ \mu\text{C}$.

ASSESS The surface area needed to construct a 1.0 μF capacitor (a fairly typical value) is enormous and hardly practical. We'll see in the next section that real capacitors can be reduced to a more manageable size by placing an insulator between the capacitor plates.

STOP TO THINK 21.5 If the potential difference across a capacitor is doubled, its capacitance

  A. Doubles.
  B. Halves.
  C. Remains the same.

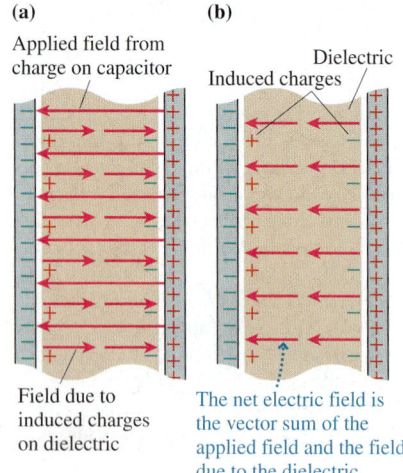

The positive nuclei shift very slightly in the direction of the field.

The negative electrons shift very slightly in the direction opposite to the field.

Because of polarization, this surface has an excess of positive charge . . .

. . . and this surface has an excess of negative charge.

FIGURE 21.33 An insulator in an electric field becomes polarized.

(a)

Applied field from charge on capacitor

Field due to induced charges on dielectric

(b)

Dielectric
Induced charges

The net electric field is the vector sum of the applied field and the field due to the dielectric.

FIGURE 21.34 The electric field inside a dielectric.

21.8 Dielectrics and Capacitors

An insulator consists of vast numbers of atoms. When an insulator is placed in an electric field, each of its atoms polarizes. Recall, from Chapter 20, that *polarization* occurs when an atom's negative electron cloud and positive nucleus shift very slightly in opposite directions in response to an external electric field. The net effect of all these tiny atomic polarizations is shown in Figure 21.33: An *induced* positive charge builds up on one surface of the insulator, and an induced negative charge on the other surface. The insulator has no net charge, but it has become polarized.

An insulator placed between the plates of a capacitor is called a **dielectric.** Figure 21.34a shows a dielectric in the uniform field of a parallel-plate capacitor. The capacitor's electric field polarizes the dielectric, and positive and negative charges build up on opposite surfaces. Notice that this distribution of charge—two equal but opposite layers—is identical to that of a parallel-plate capacitor, so this induced charge will create an electric field identical to that of a parallel-plate capacitor.

However, as Figure 21.34a shows, the field due to the induced charge on the surface on the dielectric is *opposite* to the applied electric field that established the polarization in the first place. The *net* electric field inside the dielectric is the vector sum of these two contributions, the applied field of the capacitor electrodes plus the field due to the polarization of the dielectric. As Figure 21.34b shows, these two fields add to give a net field in the same direction as the applied field, but smaller. Thus **the electric field inside a dielectric is smaller than the applied field.**

Because some atoms are more easily polarized than others, the factor by which the electric field is reduced depends on the dielectric material. This factor is called the **dielectric constant** of the material, and it is given the symbol κ (Greek letter kappa). If E is the strength of the external electric field (i.e., the field inside the capacitor without the dielectric present), the field strength E' with the dielectric present is

$$E' = \frac{E}{\kappa} \qquad (21.20)$$

Table 21.2 gives the dielectric constant for a number of common substances. Note the high dielectric constant of water, which is of great importance in regulating the chemistry of biological processes. As we saw in Section 20.2, a molecule of water is a *permanent* electric dipole because the oxygen atom has a slight negative charge while the two hydrogen atoms have a slight positive charge. Because the charge in a water molecule is *already* separated, the molecules' dipoles easily turn to line up with an applied electric field, leading to water's very high dielectric constant.

If we insert a dielectric between the plates of a capacitor, the electric field between the plates will be reduced by a factor of κ. The electric potential difference between the plates, given by $\Delta V_C = Ed$, is reduced by the same factor, and thus the capacitance $C = Q/\Delta V_C$ is *increased* by a factor of κ. Therefore the capacitance of a dielectric-containing parallel-plate capacitor is

TABLE 21.2 Dielectric constants of some materials at 20°C

Material	Dielectric constant κ
Vacuum	1 (exactly)
Air	1.00054*
Teflon	2.0
Paper	3.0
Pyrex glass	4.8
Cell membrane	9.0
Ethanol	24
Water	80
Strontium titanate	300

*Use 1.00 in all calculations.

$$C = \frac{\kappa\epsilon_0 A}{d} \qquad (21.21)$$

Capacitance of a parallel-plate capacitor
with a dielectric of dielectric constant κ

LINEAR INVERSE
p. 38   p. 118

EXAMPLE 21.12 Finding the dielectric constant

A parallel-plate capacitor is charged using a 100 V battery, then the battery is removed. If a dielectric slab is slid between the plates, filling the space inside, the capacitor voltage drops to 30 V. What is the dielectric constant of the dielectric?

PREPARE The capacitor voltage remains $(\Delta V_C)_1 = 100$ V when it is disconnected from the battery. Placing the dielectric between the plates reduces the voltage to $(\Delta V_C)_2 = 30$ V.

SOLVE The electric field strength between the capacitor plates is $E = \Delta V_C/d$. The plate separation d doesn't change, so

$$\frac{E_1}{E_2} = \frac{(\Delta V_C)_1}{(\Delta V_C)_2} = \frac{100 \text{ V}}{30 \text{ V}} = 3.3$$

But $E_2 = E_1/\kappa$, from Equation 21.20, so

$$\kappa = \frac{E_1}{E_2} = 3.3$$

ASSESS The amount of charge on the capacitor didn't change. The capacitor voltage decreased because the dielectric reduced the electric field strength inside the capacitor.

In Example 21.11, we found that impractically large electrodes would be needed to create a parallel-plate capacitor with $C = 1 \ \mu\text{F}$. Practical capacitors take advantage of dielectrics with $\kappa > 100$. As shown in Figure 21.35, a typical capacitor is a "sandwich" of two modest-sized pieces of aluminum foil separated by a very thin dielectric that increases the capacitance by a large factor. This sandwich is then folded, rolled up, and sealed in a small plastic cylinder. The two wires extending from a capacitor connect to the two electrodes and allow the

FIGURE 21.35 A capacitor disassembled to show its internal rolled-up structure.

capacitor to be charged. Even though the electrodes are no longer planes, the capacitance is reasonably well predicted from the parallel-plate-capacitor equation if A is the area of the foils before being folded and rolled.

STOP TO THINK 21.6   A 100 V battery is connected across the plates of a parallel-plate capacitor. If a Teflon slab is slid between the plates, without disconnecting the battery, the electric field between the plates

A. Increases.   B. Decreases.   C. Remains the same.

21.9 Energy and Capacitors

We've seen that a practical way of charging a capacitor is to attach its plates to a battery. Charge then flows from one plate, through the battery, and onto the other plate, leaving one plate with charge $-Q$ and the other with charge $+Q$. The battery must do *work* to transfer the charge; this work increases the electric potential energy of the charge on the capacitor. **A charged capacitor stores energy as electric potential energy.**

To find out how much energy is stored in a charged capacitor, recall that a charge q moved through a potential difference ΔV gains potential energy $U = q\Delta V$. In charging a capacitor, total charge Q is moved from the negative plate to the positive plate. At first, when the capacitor is uncharged, the potential difference is $\Delta V = 0$. The last little bit of charge to be moved, as the capacitor reaches full charge, has to move through a potential difference $\Delta V \approx \Delta V_C$. *On average,* the potential difference across a capacitor while it's being charged is $\Delta V_{average} = \frac{1}{2}\Delta V_C$. Thus it seems plausible (and can be proved in a more advanced treatment) that the potential energy U_C stored in a charged capacitor is

$$U_C = Q\Delta V_{average} = \frac{1}{2}Q\Delta V_C$$

We can use $Q = C\Delta V_C$ to write this in two different ways:

$$U_C = \frac{1}{2}\frac{Q^2}{C} = \frac{1}{2}C(\Delta V_C)^2 \qquad (21.22)$$

QUADRATIC
p. 50

Electric potential energy of a capacitor
with charge Q and potential difference ΔV_C

The potential energy stored in a capacitor depends on the *square* of the potential difference across it. This result is reminiscent of the potential energy $U_s = \frac{1}{2}k(\Delta x)^2$ stored in a spring, and a charged capacitor really is analogous to a stretched spring. A stretched spring holds the energy until we release it, then that potential energy is transformed into kinetic energy. Likewise, a charged capacitor holds energy until we discharge it.

Taking a picture in a flash When you take a flash picture, the flash is fired using electric potential energy stored in a capacitor. Batteries are unable to deliver the required energy rapidly enough, but capacitors can discharge all their energy in only microseconds. A battery is used to slowly charge up the capacitor, which then rapidly discharges through the flashlamp. This slow recharging process is why you must wait some time between taking flash pictures.

EXAMPLE 21.13   Storing energy in a capacitor
How much energy is stored in a 2.0 μF capacitor that has been charged to 5000 V? What is the average power dissipation if this capacitor is discharged in 10 μs?

SOLVE   The energy stored in the charged capacitor is

$$U_C = \frac{1}{2}C(\Delta V_C)^2 = \frac{1}{2}(2.0 \times 10^{-6}\text{ F})(5000\text{ V})^2 = 25\text{ J}$$

If this energy is released in 10 μs, the average power dissipation is

$$P = \frac{\Delta E}{\Delta t} = \frac{25 \text{ J}}{1.0 \times 10^{-5} \text{ s}} = 2.5 \times 10^6 \text{ W} = 2.5 \text{ MW}$$

ASSESS The stored energy is equivalent to raising a 1 kg mass 2.5 m. This is a rather large amount of energy, which you can see by imagining the damage a 1 kg mass could do after falling 2.5 m. When this energy is released very quickly, which is possible in an electric circuit, it provides an *enormous* amount of power.

The usefulness of a capacitor stems from the fact that it can be charged very slowly, over many seconds, and then can release the energy very quickly. A mechanical analogy would be using a crank to slowly stretch the spring of a catapult, then quickly releasing the energy to launch a massive rock.

An important medical application of the ability of capacitors to rapidly deliver energy is the *defibrillator*. A heart attack or a serious injury can cause the heart to enter a state known as *fibrillation* in which the heart muscles twitch randomly and cannot pump blood. A strong electric shock through the chest can sometimes stop the fibrillation and allow a normal heart rhythm to be restored. A defibrillator has a large capacitor that can store up to 360 J of energy. This energy is released in about 2 ms through two "paddles" pressed against the patient's chest. It takes several seconds to charge the capacitor, which is why, on television medical shows, you hear an emergency room doctor or nurse shout, "Charging!"

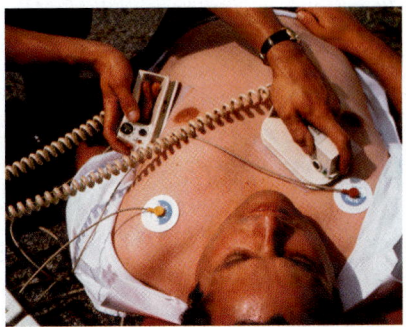

A defibrillator, which can restore a heartbeat, discharges a capacitor through the patient's chest.

The Energy in the Electric Field

We can "see" the potential energy of a stretched spring in the tension of the coils. If a charged capacitor is analogous to a stretched spring, where is the stored energy? It's in the electric field!

Figure 21.36 shows a parallel-plate capacitor in which the plates have area A and are separated by distance d. The potential difference across the capacitor is related to the electric field inside the capacitor by $\Delta V_C = Ed$. The capacitance, which we found in Equation 21.21, is $C = \kappa\epsilon_0 A/d$. Substituting these into Equation 21.22, we find that the energy stored in the capacitor is

$$U_C = \frac{1}{2}C(\Delta V_C)^2 = \frac{1}{2}\frac{\kappa\epsilon_0 A}{d}(Ed)^2 = \frac{\kappa\epsilon_0}{2}(Ad)E^2 \qquad (21.23)$$

The quantity Ad is the volume *inside* the capacitor, the region in which the capacitor's electric field exists. (Recall that an ideal capacitor has $\vec{E} = \vec{0}$ everywhere except between the plates.) Although we talk about "the energy stored in the capacitor," Equation 21.23 suggests that, strictly speaking, **the energy is stored in the capacitor's electric field.**

Because Ad is the volume in which the energy is stored, we can define an **energy density** u_E of the electric field:

$$u_E = \frac{\text{energy stored}}{\text{volume in which it is stored}} = \frac{U_C}{Ad} = \frac{1}{2}\kappa\epsilon_0 E^2 \qquad (21.24)$$

The energy density has units J/m^3. We've derived Equation 21.24 for a parallel-plate capacitor, but it turns out to be the correct expression for any electric field.

From this perspective, charging a capacitor stores energy in the capacitor's electric field as the field grows in strength. Later, when the capacitor is discharged, the energy is released as the field collapses.

Capacitor plate with area A

The capacitor's energy is stored in the electric field in volume Ad between the plates.

FIGURE 21.36 A capacitor's energy is stored in the electric field.

We first introduced the electric field as a way to visualize how a long-range force operates. But if the field can store energy, the field must be real, not merely a pictorial device. We'll explore this idea further in Chapter 25, where we'll find that the energy transported by a light wave—the very real energy of warm sunshine—is the energy of electric and magnetic fields.

EXAMPLE 21.14 Finding the energy density for a defibrillator

A defibrillator unit contains a 150 μF capacitor that is charged to 2000 V. The capacitor plates are separated by a 0.010-mm-thick dielectric with $\kappa = 300$.

a. What is the total area of the capacitor plates?
b. What is the energy density stored in the electric field when the capacitor is charged?

PREPARE Assume the capacitor can be modeled as a parallel-plate capacitor with a dielectric.

SOLVE a. Equation 21.21 for the capacitance of a dielectric-filled parallel-plate capacitor gives the surface area of the electrodes:

$$A = \frac{dC}{\kappa\epsilon_0} = \frac{(1.0 \times 10^{-5}\,\text{m})(150 \times 10^{-6}\,\text{F})}{(300)(8.85 \times 10^{-12}\,\text{C}^2/\text{N} \cdot \text{m}^2)} = 0.56\,\text{m}^2$$

b. The electric field strength is

$$E = \Delta V_C/d = (2000\,\text{V})/(1.0 \times 10^{-5}\,\text{m}) = 2.0 \times 10^8\,\text{V/m}$$

Consequently, the energy density in the electric field is

$$u_E = \frac{1}{2}\kappa\epsilon_0 E^2$$

$$= \frac{1}{2}(300)(8.85 \times 10^{-12}\,\text{C}^2/\text{N} \cdot \text{m}^2)(2.0 \times 10^8\,\text{V/m})^2$$

$$= 5.3 \times 10^7\,\text{J/m}^3$$

ASSESS For comparison, the energy density of gasoline is about 3×10^9 J/m^3, about 60 times higher than this capacitor. Capacitors store less energy than some other devices, but they can deliver this energy *very* rapidly.

STOP TO THINK 21.7 The plates of a parallel-plate capacitor are connected to a battery. If the distance between the plates is halved, the energy of the capacitor

A. Increases by a factor of four.
B. Doubles.
C. Remains the same.
D. Is halved.
E. Decreases by a factor of four.

SUMMARY

The goal of Chapter 21 has been to calculate and use the electric potential and electric potential energy.

GENERAL PRINCIPLES

Electric Potential and Potential Energy

The electric potential V is created by charges and exists at every point surrounding those charges.

Electric potential energy is energy stored in a system of charged particles due to their electrical interactions. A particle with charge q has electric potential energy

$$U_{elec} = qV$$

at a point where source charges have created an electric potential V.

The electric force is a conservative force, so mechanical energy is conserved for a charged particle in an electric potential:

$$K_f + U_f = K_i + U_i$$

These charges create the electric potential.

$V = -100\ V$

$V = 300\ V \qquad V = 200\ V$

Sources of Potential

Potential differences ΔV are created by a *separation of charge*. Two important sources of potential difference are

- A *battery*, which uses chemical means to separate charge and produce a potential difference.
- The opposite charges on the plates of a *capacitor*, which create a potential difference between the plates.

The electric potential of a point charge q is $V = \dfrac{1}{4\pi\epsilon_0}\dfrac{q}{r}$

Connecting potential and field

$\vec{E}$ points "downhill," in the direction of decreasing V.

$\vec{E}$ is everywhere perpendicular to the equipotential surfaces.

Direction of decreasing potential

The field strength is inversely proportional to the distance d between the equipotential surfaces.

IMPORTANT CONCEPTS

For a conductor in electrostatic equilibrium

- The interior electric field is zero.
- The exterior electric field is perpendicular to the surface.
- The surface is an equipotential.
- The interior is at the same potential as the surface.

$\vec{E} = \vec{0}$

The surface is an equipotential.

Graphical representations of the potential

Potential graph **Equipotential surfaces**

Contour map **Elevation graph**

APPLICATIONS

Capacitors and dielectrics

The potential difference ΔV_C between two conductors charged to $\pm Q$ is proportional to the charge:

$$\Delta V_C = Q/C$$

where C is the **capacitance** of the two conductors.

A **parallel-plate capacitor** with plates of area A and separation d has a capacitance

$$C = \kappa\epsilon_0 A/d$$

When a **dielectric** is inserted between the plates of a capacitor, its capacitance is increased by a factor κ, the **dielectric constant** of the material.

The **energy stored in a capacitor** is $U_C = \frac{1}{2}C(\Delta V_C)^2$.

This energy is stored in the electric field, which has energy density $u_E = \frac{1}{2}\kappa\epsilon_0 E^2$.

$-Q$

$+Q$

ΔV_C

Parallel-plate capacitor

For a capacitor charged to ΔV_C the potential a distance x from the negative plate is

$$V = \Delta V_C(x/d)$$

The electric field inside is

$$E = \Delta V_C/d$$

Units

- Electric potential: $1\ V = 1\ J/C$
- Electric field: $1\ V/m = 1\ N/C$
- Energy: 1 electron volt = $1\ eV = 1.60 \times 10^{-19}\ J$ is the kinetic energy gained by an electron upon accelerating through a potential difference of 1 V.

For instructor-assigned homework, go to
www.masteringphysics.com

Problem difficulty is labeled as | (straightforward) to ||||| (challenging).

Problems labeled INT integrate significant material from earlier chapters; BIO are of biological or medical interest.

QUESTIONS

Conceptual Questions

1. By moving a 10 nC charge from point A to point B, you determine that the electric potential at B is 150 V. What would be the potential at B if a 20 nC charge were moved from A to B?

2. Charge q is fired through a small hole in the positive plate of a capacitor, as shown in Figure Q21.2.
   a. If q is a positive charge, does it speed up or slow down inside the capacitor? Answer this question twice: (i) Using the concept of force. (ii) Using the concept of energy.
   b. Repeat part a if q is a negative charge.

FIGURE Q21.2

3. *Why* is the potential energy of two opposite charges a negative number? (Note: Saying that the formula gives a negative number is not an explanation.)

4. An electron ($q = -e$) completes half of a circular orbit of radius r around a nucleus with $Q = +3e$, as shown in Figure Q21.4.
   a. How much work is done on the electron as it moves from i to f? Give either a numerical value or an expression from which you could calculate the value if you knew the radius. Justify your answer.
   b. By how much does the electric potential energy change as the electron moves from i to f?
   c. Is the electron's speed at f greater than, less than, or equal to its speed at i?
   d. Are your answers to parts a and c consistent with each other?

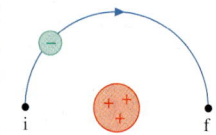

FIGURE Q21.4

5. An electron moves along the trajectory from i to f in Figure Q21.5.
   a. Does the electric potential energy increase, decrease, or stay the same? Explain.
   b. Is the electron's speed at f greater than, less than, or equal to its speed at i? Explain.

FIGURE Q21.5

6. As shown in Figure Q21.6, two protons are launched with the same speed from point 1 inside a parallel-plate capacitor. One proton moves along the path from 1 to 2, the other from 1 to 3. Points 2 and 3 are the same distance from the positive plate.

FIGURE Q21.6

   a. Is $\Delta U_{1 \to 2}$, the change in potential energy along the path $1 \to 2$, larger, smaller, or equal to $\Delta U_{1 \to 3}$? Explain.
   b. Is the proton's speed v_2 at point 2 larger than, smaller than, or equal to v_3? Explain.

7. The graph in Figure Q21.7 shows the electric potential along the x-axis. Draw a graph of the potential energy of a 0.10 C charged particle in this region of space, providing a numerical scale on the energy axis.

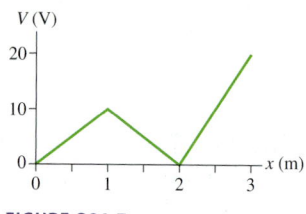

FIGURE Q21.7

8. Figure Q21.8 shows two points inside a capacitor. Let $V = 0$ V at the negative plate.
   a. What is the ratio V_2/V_1 of the electric potential at these two points? Explain.
   b. What is the ratio E_2/E_1 of the electric field strength at these two points? Explain.

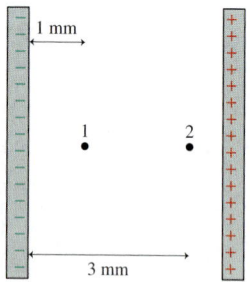

FIGURE Q21.8.

9. A capacitor with plates separated by distance d is charged to a potential difference ΔV_C. All wires and batteries are disconnected, then the two plates are pulled apart (with insulated handles) to a new separation of distance $2d$.
   a. Does the capacitor charge Q change as the separation increases? If so, by what factor? If not, why not?
   b. Does the electric field strength E change as the separation increases? If so, by what factor? If not, why not?
   c. Does the potential difference ΔV_C change as the separation increases? If so, by what factor? If not, why not?

10. Rank in order, from most positive to most negative, the electric potentials V_1 to V_5 at points 1 to 5 in Figure Q21.10. Explain.

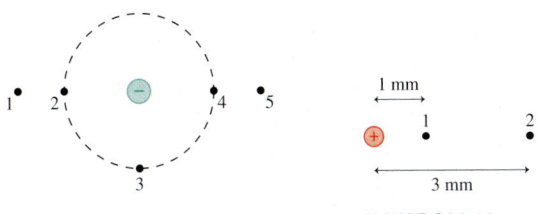

FIGURE Q21.10   **FIGURE Q21.11**

11. Figure Q21.11 shows two points near a positive point charge.
    a. What is the ratio V_1/V_2 of the electric potentials at these two points? Explain.
    b. What is the ratio E_1/E_2 of the electric field strengths at these two points? Explain.

12. Each part of Figure Q21.12 shows three points in the vicinity of two point charges. The charges have equal magnitudes. Rank in order, from largest to smallest, the potentials V_1, V_2, and V_3.

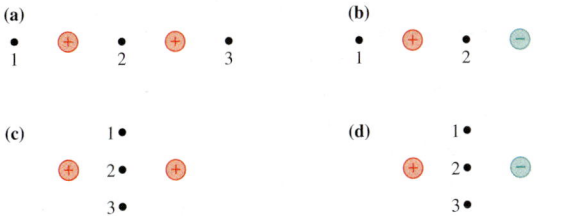

FIGURE Q21.12

13. a. Suppose that $\vec{E} = \vec{0}$ throughout some region of space. Can you conclude that $V = 0$ V in this region? Explain.
    b. Suppose that $V = 0$ V throughout some region of space. Can you conclude that $\vec{E} = \vec{0}$ in this region? Explain.

14. Rank in order, from largest to smallest, the electric field strengths E_1, E_2, E_3, and E_4 at the four labeled points in Figure Q21.14. Explain.

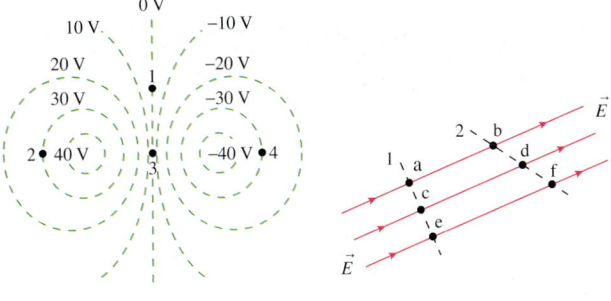

FIGURE Q21.14   **FIGURE Q21.15**

15. Figure Q21.15 shows an electric field diagram. Dotted lines 1 and 2 are two surfaces in space, not physical objects.
    a. Is the electric potential at point a higher, lower, or equal to the electric potential at point b? Explain.
    b. Rank in order, from largest to smallest, the potential differences ΔV_{ab}, ΔV_{cd}, and ΔV_{ef}. Explain.
    c. Is surface 1 an equipotential surface? What about surface 2? Explain why or why not.

16. Figure Q21.16 shows a negatively charged electroscope. The gold leaf stands away from the rigid metal post. Is the electric potential of the leaf higher than, less than, or equal to the potential of the post? Explain.

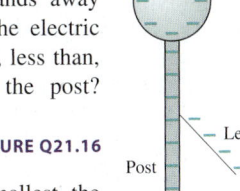

FIGURE Q21.16

17. Rank in order, from largest to smallest, the energies $(U_C)_1$ to $(U_C)_4$ stored in each of the capacitors in Figure Q21.17. Explain.

FIGURE Q21.17

18. A parallel-plate capacitor with plate separation d is connected to a battery that has potential difference ΔV_{bat}. Without breaking any of the connections, insulating handles are used to increase the plate separation to $2d$.
    a. Does the potential difference ΔV_C change as the separation increases? If so, by what factor? If not, why not?
    b. Does the capacitance change? If so, by what factor? If not, why not?
    c. Does the capacitor charge Q change? If so, by what factor? If not, why not?

19. The gap between the capacitor plates shown in Figure Q21.19 is *partially* filled with a dielectric. The capacitor was charged by a 9 V battery, then disconnected from the battery. Rank in order, from smallest to largest, the electric field strengths E_1, E_2, and E_3 at the points labeled in the figure, as well as the field strength E_4 between the plates if the dielectric is removed. Explain.

FIGURE Q21.19

Multiple-Choice Questions

20. | A 1.0 nC positive point charge is located at point A in Figure Q21.20. The electric potential at point B is
    A. 9.0 V
    B. $9.0\sin 30°$ V
    C. $9.0\cos 30°$ V
    D. $9.0\tan 30°$ V

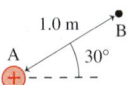

FIGURE Q21.20

21. | For the capacitor shown in Figure Q21.21, the potential difference ΔV_{ab} between points a and b is
    A. 6 V
    B. $6\sin 30°$ V
    C. $6\cos 30°$ V
    D. $6/\sin 30°$ V
    E. $6/\cos 30°$ V

FIGURE Q21.21

22. | The electric potential is 300 V at $x = 0$ cm, is -100 V at $x = 5$ cm, and varies linearly with x. If a positive charge is released from rest at $x = 2.5$ cm, and is subject only to electric forces, the charge will
    A. Move to the right.
    B. Move to the left.
    C. Stay at $x = 2.5$ cm.
    D. Not enough information to tell.

Questions 23 through 27 refer to Figure Q21.23, which shows equipotential lines in a region of space. The equipotential lines are spaced by the same difference in potential, and several of the potentials are given.

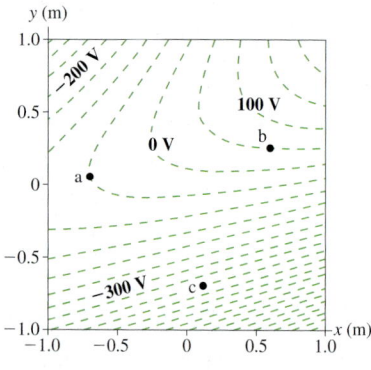

FIGURE Q21.23

23. | What is the potential at point c?
    A. -400 V   B. -350 V   C. -100 V
    D. 350 V   E. 400 V

24. | At which point, a, b, or c, is the magnitude of the electric field the greatest?

25. ‖ What is the approximate magnitude of the electric field at point c?
    A. 100 V/m   B. 300 V/m   C. 800 V/m
    D. 1500 V/m   E. 3000 V/m

26. | The direction of the electric field at point b is closest to which direction?
    A. Right   B. Up   C. Left   D. Down

27. | A $+10$ nC charge is moved from point c to point a. How much work is required in order to do this?
    A. 3.5×10^{-6} J   B. 4.0×10^{-6} J   C. 3.5×10^{-3} J
    D. 4.0×10^{-3} J   E. 3.5 J

28. | A bug zapper consists of two metal plates connected to a high-voltage power supply. The voltage between the plates is set to give an electric field slightly less than 1×10^6 V/m. When a bug flies between the two plates, it increases the field enough to initiate a spark that incinerates the bug. If a bug zapper has a 4000 V power supply, what is the approximate separation between the plates?
    A. 0.05 cm   B. 0.5 cm   C. 5 cm   D. 50 cm

29. ‖ An atom of helium and one of argon are singly ionized—one electron is removed from each. The two ions are then accelerated from rest by the electric field between two plates with a potential difference of 150 V. After accelerating from one plate to the other,
    A. The helium ion has more kinetic energy.
    B. The argon ion has more kinetic energy.
    C. Both ions have the same kinetic energy.
    D. There is not enough information to say which ion has more kinetic energy.

30. | The dipole moment of the heart is shown
BIO   at a particular instant in Figure Q21.30. Which of the following potential differences will have the largest positive value?
    A. $V_1 - V_2$
    B. $V_1 - V_3$
    C. $V_2 - V_1$
    D. $V_3 - V_1$

FIGURE Q21.30

PROBLEMS

Section 21.1 Electric Potential Energy and the Electric Potential

1. | Moving a charge from point A, where the potential is 300 V, to point B, where the potential is 150 V, takes 4.5×10^{-4} J of work. What is the value of the charge?

2. ‖ The graph in Figure P21.2 shows the electric potential energy as a function of separation for two point charges. If one charge is $+0.44$ nC, what is the other charge?

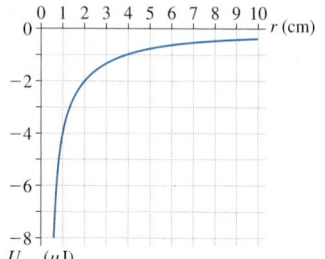

FIGURE P21.2

3. ‖ It takes 3.0 μJ of work to move a 15 nC charge from point A to B. It takes -5.0 μJ of work to move the charge from C to B. What is the potential difference $V_C - V_A$?

Section 21.2 Using the Electric Potential

4. ‖ An electron has been accelerated from rest through a potential difference of 1000 V.
    a. What is its kinetic energy, in electron volts?
    b. What is its kinetic energy, in joules?
    c. What is its speed?

5. | A proton has been accelerated from rest through a potential difference of -1000 V.
    a. What is its kinetic energy, in electron volts?
    b. What is its kinetic energy, in joules?
    c. What is its speed?

6. ‖ What potential difference is needed to accelerate a He$^+$ ion (charge $+e$, mass 4 u) from rest to a speed of 1.0×10^6 m/s?

7. ‖ An electron with an initial speed of 500,000 m/s is brought to rest by an electric field.
   a. Did the electron move into a region of higher potential or lower potential?
   b. What was the potential difference that stopped the electron?
   c. What was the initial kinetic energy of the electron, in electron volts?

8. ‖ A proton with an initial speed of 800,000 m/s is brought to rest by an electric field.
   a. Did the proton move into a region of higher potential or lower potential?
   b. What was the potential difference that stopped the proton?
   c. What was the initial kinetic energy of the proton, in electron volts?

Section 21.3 Calculating the Electric Potential

9. | Starting with the definition of the volt, show that 1 V/m = 1 N/C.

10. ‖ A 2.0 cm × 2.0 cm parallel-plate capacitor has a 2.0 mm spacing. The electric field strength inside the capacitor is 1.0×10^5 V/m.
    a. What is the potential difference across the capacitor?
    b. How much charge is on each plate?

11. | Two 2.00 cm × 2.00 cm plates that form a parallel-plate capacitor are charged to ±0.708 nC. What are the electric field strength inside and the potential difference across the capacitor if the spacing between the plates is (a) 1.00 mm and (b) 2.00 mm?

12. | a. In Figure P21.12, which capacitor plate, left or right, is the positive plate?
    b. What is the electric field strength inside the capacitor?
    c. What is the potential energy of a proton at the midpoint of the capacitor?

FIGURE P21.12

13. | A +25 nC charge is at the origin.
    a. What are the radii of the 1000 V, 2000 V, 3000 V, and 4000 V equipotential surfaces?
    b. Draw a contour map in the xy-plane showing the charge and these four surfaces.

14. | a. What is the electric potential at points A, B, and C in Figure P21.14?
    b. What is the potential energy of an electron at each of these points?
    c. What are the potential differences ΔV_{AB} and ΔV_{BC}?

FIGURE P21.14

15. ‖‖ A 1.0-mm-diameter ball bearing has 2.0×10^9 excess electrons. What is the ball bearing's potential?

16. ‖ What is the electric potential at the point indicated with the dot in Figure P21.16?

FIGURE P21.16   2.0 nC

17. | a. What is the potential of an ordinary AA or AAA battery? (If you're not sure, find one and look at the label.)
    b. An AA battery is connected to a parallel-plate capacitor having 4.0-cm-diameter plates spaced 2 mm apart. How much charge does the battery move from one plate to the other?

Section 21.5 Connecting Potential and Field

18. ‖ a. In Figure P21.18, which point, A or B, has a higher electric potential?
    b. What is the potential difference between A and B?

FIGURE P21.18

19. ‖ What is the potential difference between $x_i = 10$ cm and $x_f = 30$ cm in the uniform electric field $E_x = 1000$ V/m?

20. ‖ The electric potential in a region of uniform electric field is −1000 V at $x = -1.0$ m and +1000 V at $x = +1.0$ m. What is E_x?

21. | What are the magnitude and direction of the electric field at the dot in Figure P21.21?

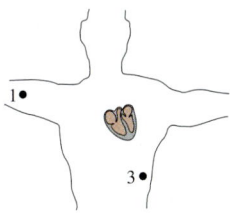

FIGURE P21.21   **FIGURE P21.22**

22. | What are the magnitude and direction of the electric field at the dot in Figure P21.22?

Section 21.6 The Electrocardiogram

23. | One standard location for a pair
    BIO of electrodes during an EKG is shown in Figure P21.23. The potential difference $\Delta V_{31} = V_3 - V_1$ is recorded. For each of the three instants a, b, and c during the heart's cycle shown in Figure 21.30, will ΔV_{31} be positive or negative? Explain.

FIGURE P21.23

24. ‖ Three electrodes, 1–3, are at-
BIO tached to a patient as shown in
Figure P21.24. During ventricular
depolarization (see Figure 21.30),
across which pair of electrodes is
the magnitude of the potential difference likely to be the smallest?
Explain.

FIGURE P21.24

Section 21.7 Capacitance and Capacitors

25. ‖ Two 2.0 cm × 2.0 cm square aluminum electrodes, spaced
0.50 mm apart, are connected to a 100 V battery.
    a. What is the capacitance?
    b. What is the charge on each electrode?
26. ‖ An uncharged capacitor is connected to the terminals of a
3.0 V battery, and 6.0 μC flows to the positive plate. The 3.0 V
battery is then disconnected and replaced with a 5.0 V battery,
with the positive and negative terminals connected in the
same manner as before. How much additional charge flows to
the positive plate?
27. ‖ You need to construct a 100 pF capacitor for a science project. You plan to cut two $L \times L$ metal squares and place spacers between them. The thinnest spacers you have are 0.20 mm
thick. What is the proper value of L?
28. ‖ A switch that connects a battery to a 10 μF capacitor is
closed. Several seconds later you find that the capacitor plates
are charged to $\pm 30 \mu$C. What is the battery voltage?
29. ‖ What is the voltage of a battery that will charge a 2.0 μF
capacitor to $\pm 48 \mu$C?
30. ‖ Two electrodes connected to a 9.0 V battery are charged to
± 45 nC. What is the capacitance of the electrodes?
31. ‖ Initially, the switch in Figure P21.31 is open and the capacitor is uncharged. How much charge flows through the switch
after the switch is closed?

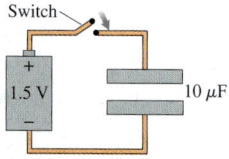

FIGURE P21.31

Section 21.8 Dielectrics and Capacitors

32. ‖ A capacitor with its plates separated by paper stores 4.4 nC
of charge when it is connected to a particular battery. An otherwise identical capacitor, but with its plates separated by
Pyrex glass, is connected to the same battery. How much
charge does that capacitor store?
33. ‖‖ A science-fair radio uses a homemade capacitor made of
two 35 cm × 35 cm sheets of aluminum foil separated by
a 0.25-mm-thick sheet of paper. What is its capacitance?
34. ‖‖ A 25 pF parallel-plate capacitor with an air gap between the
plates is connected to a 100 V battery. A Teflon slab is then
inserted between the plates and completely fills the gap. What
is the change in the charge on the positive plate when the
Teflon is inserted?
35. ‖ Two 2.0-cm-diameter electrodes with a 0.10-mm-thick
sheet of Teflon between them are attached to a 9.0 V battery.
Without disconnecting the battery, the Teflon is removed.
What are the charge, potential difference, and electric field
(a) before and (b) after the Teflon is removed?

36. ‖ A 1.2 nF parallel-plate capacitor has an air gap between its
plates. Its capacitance increases by 3.0 nF when the gap is
filled by a dielectric. What is the dielectric constant of that
dielectric?

Section 21.9 Energy and Capacitors

37. ‖ To what potential should you charge a 1.0 μF capacitor to
store 1.0 J of energy?
38. ‖ A pair of 10 μF capacitors in a high-power laser are
charged to 1.7 kV.
    a. What charge is stored in each capacitor?
    b. How much energy is stored in each capacitor?
39. ‖ Capacitor 2 has half the capacitance and twice the potential
difference as capacitor 1. What is the ratio $(U_C)_1/(U_C)_2$?
40. ‖ Two uncharged metal spheres, spaced 15.0 cm apart, have a
capacitance of 24.0 pF. How much work would it take to
move 12.0 nC of charge from one sphere to the other?
41. ‖‖‖‖ 50 pJ of energy is stored in a 2.0 cm × 2.0 cm × 2.0 cm
region of uniform electric field. What is the electric field
strength?
42. ‖ A 2.0-cm-diameter parallel-plate capacitor with a spacing
of 0.50 mm is charged to 200 V. What are (a) the total energy
stored in the electric field and (b) the energy density?

General Problems

43. ‖ What is the change in electrical potential energy of a 3.0 nC
point charge when it is moved from point A to point B in
Figure P21.43?

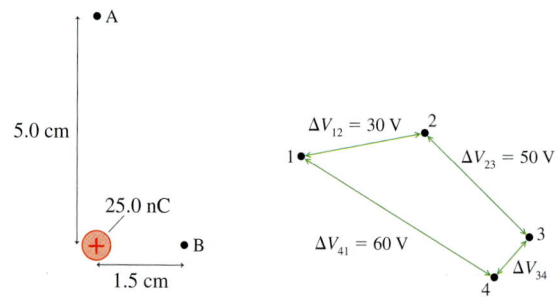

FIGURE P21.43   **FIGURE P21.44**

44. ‖ What is the potential difference ΔV_{43} in Figure P21.44?
45. ‖ What is the electric potential energy of the electron in
Figure P21.45? The protons are fixed and can't move.

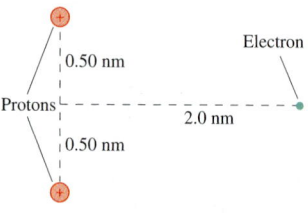

FIGURE P21.45

46. ‖‖ Two point charges 2.0 cm apart have an electric potential
energy -180μJ. The total charge is 30 nC. What are the two
charges?
47. ‖ Two positive point charges are 5.0 cm apart. If the electric
potential energy is 72 μJ, what is the magnitude of the force
between the two charges?

48. ‖‖ A +3.0 nC charge is at $x = 0$ cm and a −1.0 nC charge is at $x = 4$ cm. At what point or points on the x-axis is the electric potential zero?

49. ‖ A −3.0 nC charge is on the x-axis at $x = -9$ cm and a +4.0 nC charge is on the x-axis at $x = 16$ cm. At what point or points on the y-axis is the electric potential zero?

50. ‖ A −2.0 nC charge and a +2.0 nC charge are located on the x-axis at $x = -1.0$ cm and $x = +1.0$ cm, respectively.
    a. At what position or positions on the x-axis is the electric field zero?
    b. At what position or positions on the x-axis is the electric potential zero?
    c. Draw graphs of the electric field strength and the electric potential along the x-axis.

51. ‖‖ A −10.0 nC point charge and a +20.0 nC point charge are 15.0 cm apart on the x-axis.
    a. What is the electric potential at the point on the x-axis where the electric field is zero?
    b. What are the magnitude and direction of the electric field at the point on the x-axis where the electric potential is zero?

52. ‖‖ A 2.0-mm-diameter glass bead is positively charged. The potential difference between a point 2.0 mm from the bead and a point 4.0 mm from the bead is 500 V. What is the charge on the bead?

53. ‖ In a semiclassical model of the hydrogen atom, the electron orbits the proton at a distance of 0.053 nm.
    a. What is the electric potential of the proton at the position of the electron?
    b. What is the electron's potential energy?

54. ‖ What is the electric potential at the point indicated with the dot in Figure P21.54?

FIGURE P21.54   **FIGURE P21.55**

55. ‖ a. What is the electric potential at point A in Figure P21.55?
    b. What is the potential energy of a proton at point A?

56. ‖ A proton's speed as it passes point A is 50,000 m/s. It follows the trajectory shown in Figure P21.56. What is the proton's speed at point B?

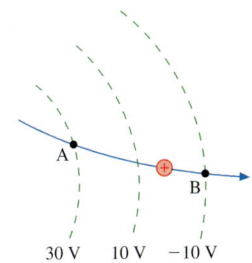

FIGURE P21.56   30 V   10 V −10 V

57. ‖ Electric outlets have a voltage of approximately 120 V between the two parallel slots. Estimate the electric field strength between these two slots.

58. ‖ Estimate the magnitude of the electric field in a cell mem-
    BIO brane with a thickness of 8 nm.

59. ‖ A Na⁺ ion moves from inside a cell, where the electric
    BIO potential is −70 mV, to outside the cell, where the potential is 0 V. What is the change in the ion's electric potential energy as it moves from inside to outside the cell? Does its energy increase or decrease?

60. ‖ Suppose that a molecular ion with charge $-10e$ is embed-
    BIO ded within the 5.0-nm-thick cell membrane of a cell with membrane potential −70 mV. What is the electric force on the molecule?

61. ‖ The electric field strength is 50,000 V/m inside a parallel-plate capacitor with a 2.0 mm spacing. A proton is released from rest at the positive plate. What is the proton's speed when it reaches the negative plate?

62. ‖‖‖ The electric field strength is 20,000 V/m inside a parallel-plate capacitor with a 1.0 mm spacing. An electron is released from rest at the negative plate. What is the electron's speed when it reaches the positive plate?

63. ‖‖ A proton is released from rest at the positive plate of a parallel-plate capacitor. It crosses the capacitor and reaches the negative plate with a speed of 50,000 m/s. What will be the proton's final speed if the experiment is repeated with double the amount of charge on each capacitor plate?

64. ‖‖ A proton is released from rest at the positive plate of a parallel-plate capacitor. It crosses the capacitor and reaches the negative plate with a speed of 50,000 m/s. The experiment is repeated with a He⁺ ion (charge e, mass 4 u). What is the ion's speed at the negative plate?

65. ‖ Two 2.0-cm-diameter disks spaced 2.0 mm apart form a parallel-plate capacitor. The electric field between the disks is 5.0×10^5 V/m.
    a. What is the voltage across the capacitor?
    b. How much charge is on each disk?
    c. An electron is launched from the negative plate. It strikes the positive plate at a speed of 2.0×10^7 m/s. What was the electron's speed as it left the negative plate?

66. ‖ What is the ratio $\Delta V_p/\Delta V_e$ of the potential differences that will accelerate a proton and an electron from rest to (a) the same final speed and (b) the same final kinetic energy?

67. ‖ In a *photoelectric effect* experiment, shining ultraviolet light on a metal surface causes electrons to be ejected from the surface at a speed of 2.65×10^5 m/s. The ejected electrons move toward an electrode located a few mm away. Applying a potential difference between the metal surface and the electrode causes the electrons to speed up or slow down. What potential difference will cause the electrons to come to rest just as they reach the electrode? Give your answer as a positive ΔV if the electrode is at a higher potential than the metal surface, a negative ΔV if the electrode is at a lower potential.

68. ‖ A 2.5-mm-diameter sphere is charged to −4.5 nC. An electron fired directly at the sphere from far away comes to within 0.30 mm of the surface of the target before being reflected.
    a. What was the electron's initial speed?
    b. At what distance from the surface of the sphere is the electron's speed half of its initial value?
    c. What is the acceleration of the electron at its turning point?

69. ‖ A proton is fired from far away toward the nucleus of an iron atom. Iron is element number 26, and the diameter of the nucleus is 9.0 fm. (1 fm = 10^{-15} m.) What initial speed does the proton need to just reach the surface of the nucleus? Assume the nucleus remains at rest.

70. | Two 10.0-cm-diameter electrodes 0.50 cm apart form a parallel-plate capacitor. The electrodes are attached by metal wires to the terminals of a 15 V battery. After a long time, the capacitor is disconnected from the battery but is not discharged. What are the charge on each electrode, the electric field strength inside the capacitor, and the potential difference between the electrodes
    a. Right after the battery is disconnected?
    b. After insulating handles are used to pull the electrodes away from each other until they are 1.0 cm apart?

71. ‖ Two 10.0-cm-diameter electrodes 0.50 cm apart form a parallel-plate capacitor. The electrodes are attached by metal wires to the terminals of a 15 V battery. What are the charge on each electrode, the electric field strength inside the capacitor, and the potential difference between the electrodes
    a. While the capacitor is attached to the battery?
    b. After insulating handles are used to pull the electrodes away from each other until they are 1.0 cm apart? The electrodes remain connected to the battery during this process.

72. | Determine the magnitude and direction of the electric field at points 1 and 2 in Figure P21.72.

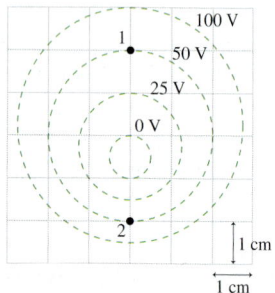

FIGURE P21.72

73. ‖ Figure P21.73 shows a series of equipotential curves.
    a. Is the electric field strength at point A larger than, smaller than, or equal to the field strength at point B? Explain.
    b. Is the electric field strength at point C larger than, smaller than, or equal to the field strength at point D? Explain.
    c. Determine the electric field $\vec{E}$ at point D. Express your answer as a magnitude and direction.

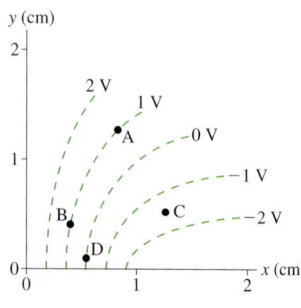

FIGURE P21.73

74. ‖ Figure P21.74 shows the electric potential at points on a 5.0 cm × 5.0 cm grid.
    a. Reproduce this figure on your paper, then draw the 50 V, 75 V, and 100 V equipotential surfaces.
    b. Estimate the electric field (strength and direction) at points A, B, C, and D.
    c. Draw the electric field vectors at points A, B, C, and D on your diagram.

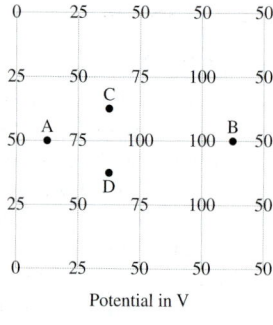

FIGURE P21.74

75. ‖‖ The plates of a 3.0 nF parallel-plate capacitor are each 0.27 m² in area.
    a. How far apart are the plates if there's air between them?
    b. If the plates are separated by a Teflon sheet, how thick is the sheet?

76. ‖ The dielectric in a capacitor serves two purposes. It increases the capacitance, compared to an otherwise identical capacitor with an air gap, and it increases the maximum potential difference the capacitor can support. Recall that air near the sphere of a Van de Graaff generator undergoes breakdown, creating a spark, when the electric field is sufficiently strong. The same is true for any material. That critical field strength, at which breakdown occurs, is 3.0 MV/m for air, but 60 MV/m for Teflon.
    a. A parallel plate capacitor consists of two square plates, 15 cm on a side, spaced 0.50 mm apart with only air between them. What is the maximum energy that can be stored by the capacitor?
    b. What is the maximum energy that can be stored if the plates are separated by a 0.50-mm-thick Teflon sheet?

77. ‖ The flash unit in a camera uses a special circuit to "step up"
INT the 3.0 V from the batteries to 300 V, which charges a capacitor. The capacitor is then discharged through a flashlamp. The discharge takes 10 μs, and the average power dissipated in the flashlamp is 10^5 W. What is the capacitance of the capacitor?

In Problems 78 through 81 you are given the equation(s) used to solve a problem. For each of these,

a. Write a realistic problem for which this is the correct equation(s).
b. Finish the solution of the problem.

78. | $\dfrac{(9.0 \times 10^9 \text{ N} \cdot \text{m}^2/\text{C}^2)q_1 q_2}{0.030 \text{ m}} = 90 \times 10^{-6} \text{ J}$

$q_1 + q_2 = 40 \text{ nC}$

79. ‖ $\frac{1}{2}(1.67 \times 10^{-27} \text{ kg})(2.5 \times 10^6 \text{ m/s})^2 + 0 =$
$\frac{1}{2}(1.67 \times 10^{-27} \text{ kg})v_i^2 +$
$\dfrac{(9.0 \times 10^9 \text{ N} \cdot \text{m}^2/\text{C}^2)(2.0 \times 10^{-9} \text{ C})(1.60 \times 10^{-19} \text{ C})}{0.0010 \text{ m}}$

80. ‖ $\dfrac{(9.0 \times 10^9 \text{ N} \cdot \text{m}^2/\text{C}^2)(3.0 \times 10^{-9} \text{ C})}{0.030 \text{ m}} +$
$\dfrac{(9.0 \times 10^9 \text{ N} \cdot \text{m}^2/\text{C}^2)(3.0 \times 10^{-9} \text{ C})}{(0.030 \text{ m}) + d} = 1200 \text{ V}$

81. ‖ $400 \text{ nC} = (100 \text{ V}) C$
$C = \dfrac{(8.85 \times 10^{-12} \text{ C}^2/\text{N} \cdot \text{m}^2)(0.10 \text{ m} \times 0.10 \text{ m})}{d}$

Passage Problems

A Lightning Strike

Storm clouds build up large negative charges, as described in the chapter. The charges dwell in *charge centers,* regions of concentrated charge. Suppose a cloud has $-25\,C$ in a 1.0-km-diameter spherical charge center located 10 km above the ground, as sketched in Figure P21.84. The negative charge center attracts a similar amount of positive charge that is spread on the ground below the cloud.

FIGURE P21.84

The charge center and the ground function as a charged capacitor, with a potential difference of approximately $4 \times 10^8\,V$. The large electric field between these two "electrodes" may ionize the air, leading to a conducting path between the cloud and the ground. Charges will flow along this conducting path, causing a discharge of the capacitor—a lightning strike.

82. | What is the approximate magnitude of the electric field between the charge center and the ground?
 A. $4 \times 10^4\,V/m$ B. $4 \times 10^5\,V/m$
 C. $4 \times 10^6\,V/m$ D. $4 \times 10^7\,V/m$
83. | Which of the curves sketched in Figure P21.85 best approximates the shape of an equipotential drawn halfway between the charge center and the ground?

FIGURE P21.85

84. | What is the approximate capacitance of the charge center-ground system?
 A. $6 \times 10^{-8}\,F$ B. $2 \times 10^7\,F$
 C. $4 \times 10^6\,F$ D. $8 \times 10^6\,F$
85. | If 12.5 C of charge is transferred from the cloud to the ground in a lightning strike, what fraction of the stored energy is dissipated?
 A. 12% B. 25% C. 50% D. 75%
86. | If the cloud transfers all of its charge to the ground via several rapid lightning flashes lasting a total of 1 s, what is the average power?
 A. 1 GW B. 2 GW C. 5 GW D. 10 GW

STOP TO THINK ANSWERS

Stop to Think 21.1: A. The potential energy of a negative charge is high when near other negative charges, low when near positive charges.

Stop to Think 21.2: C. The proton gains speed by losing potential energy. It loses potential energy by moving in the direction of decreasing electric potential.

Stop to Think 21.3: $\Delta V_{13} = \Delta V_{23} > \Delta V_{12}$. The potential depends only on the *distance* from the charge, not the direction. $\Delta V_{12} = 0$ because these points are at the same distance.

Stop to Think 21.4: C. $\vec{E}$ points "downhill," so V must decrease from right to left. E is larger on the left than on the right, so the contour lines must be closer together on the left.

Stop to Think 21.5: C. Capacitance is a property of the shape and position of the electrodes. It does not depend on the potential difference or charge.

Stop to Think 21.6: C. The electric field is $\Delta V_C/d$. With ΔV_C fixed by the battery, introducing the dielectric does not change E. More charge flows from the battery to compensate for the dielectric.

Stop to Think 21.7: B. The energy is $\frac{1}{2}C(\Delta V_C)^2$. ΔV_C is constant, but C doubles when the distance is halved.

22

CURRENT AND RESISTANCE

This woman is measuring her percentage body fat by gripping a device that passes a small electric current through her body. How does a measurement of the current reveal such details of the body's structure?

Looking Ahead ▶▶

The goal of Chapter 22 is to learn how and why charge moves through a conductor as what we call a current. In this chapter you will learn to:

- ▶ Understand how charge moves through a conductor.
- ▶ Apply the law of conservation of current.
- ▶ Understand the emf of a battery.
- ▶ Use the concepts of resistance and resistivity to predict currents.
- ▶ Analyze the energy transfers as charges move through simple circuits.

Looking Back ◀◀

This chapter depends on the properties of charges, electric fields, and electric potentials, and on the parallel-plate capacitor. Please review:

- ◀ Sections 20.1–20.2 Charges and conductors.
- ◀ Section 20.5 Forces on charges in an electric field.
- ◀ Section 21.3 The definition of potential.

We began our investigation of electric charges in Chapter 20 by looking at the forces between plastic rods rubbed with wool. We continued our investigation by building the concepts of electric field and potential. In this chapter, we will now look at practical applications of these concepts. When you press the power button on your laptop computer, cell phone, or digital camera, the device powers up because the electric potential of a battery causes charges to move through the circuits in what we call a *current*. We can explain the operation of these devices using the physics of charges, electric fields, and electric potentials—the same physics behind the force between the plastic rods.

Our goal in this chapter is to understand how to produce and manipulate electric currents, the controlled flow of charges. We will begin with a description of how charges flow through a wire. This model will help us understand how current depends on the properties of the conductor carrying the current. The measurement of an electric current can reveal something about the physical properties of an object, whether it's the temperature of a device or the percentage of fat in a person's body. The basic properties of current established in this chapter will be necessary groundwork for our further study of electric circuits in the next chapter.

22.1 A Model of Current

Let's start our exploration of current with a very simple experiment. Figure 22.1a shows a charged parallel-plate capacitor. If we connect the two capacitor plates to each other with a metal wire, as shown in Figure 22.1b, the plates quickly become neutral. We say that the capacitor has been *discharged.*

The wire is a conductor, a material through which charge easily moves. In Chapter 20, we defined a *current* as the motion of charges through a material. Later in the chapter we will develop a quantitative expression for current, but for now this definition will suffice. Apparently the excess charge on one capacitor plate is able to move through the wire to the other plate, neutralizing both plates. **The capacitor is discharged by a current in the connecting wire.**

> NOTE ▶ Current is defined as the motion of charges, so we don't say that "current flows." It is *charges* that flow, not current. Current *is* the flow. ◀

If we observe the capacitor discharge, we see other effects. As Figure 22.2 shows, the connecting wire gets warmer. If the wire is very thin in places, such as the thin filament in a lightbulb, the wire gets hot enough to glow. The current-carrying wire also deflects a compass needle. We will explore the connection between currents and magnetism in Chapter 24. For now, we will use "makes the wire warmer" and "deflects a compass needle" as *indicators* that a current is present in a wire.

(a)

A charged parallel-plate capacitor

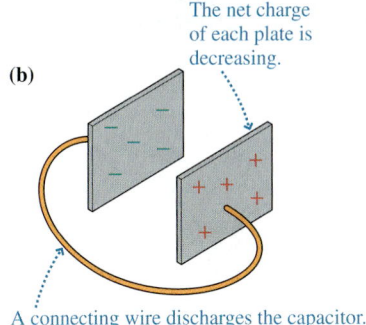

(b)

The net charge of each plate is decreasing.

A connecting wire discharges the capacitor.

FIGURE 22.1 A capacitor is discharged by a metal wire.

The connecting wire gets warm.

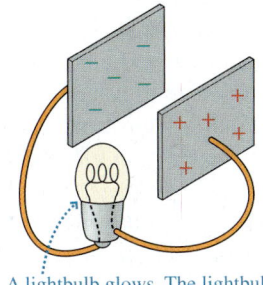

A lightbulb glows. The lightbulb filament is part of the connecting wire.

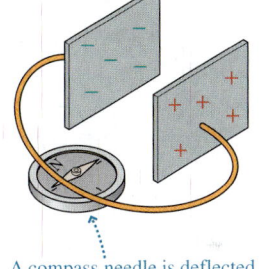

A compass needle is deflected.

FIGURE 22.2 Properties of a current.

Charge Carriers

Opposite charges attract, but the oppositely charged plates of a capacitor don't spontaneously discharge because the charges can't leap from one plate to the other. A connecting wire discharges the capacitor by providing a pathway for charge to move from one side of the capacitor to the other. However, merely observing that a wire discharges a capacitor doesn't answer an important question: Does positive charge move toward the negative plate, or does negative charge move toward the positive plate? *Either* motion would explain the observations we have made.

The charges that move in a current are called the *charge carriers*. The first experiments that could distinguish between positive and negative charge carriers didn't take place until the early 20th century, but the experimental evidence is now clear: **The charge carriers in metals are electrons.** As Figure 22.3 shows, it is the motion of the *conduction electrons,* which are free to move around, that forms a current—a flow of charge—in the metal. An *insulator* does not have such free charges and cannot carry a current. A *semiconductor* is an intermediate case, with relatively few charge carriers, which can be either positive or negative. It will carry a current, but not as easily as a conductor.

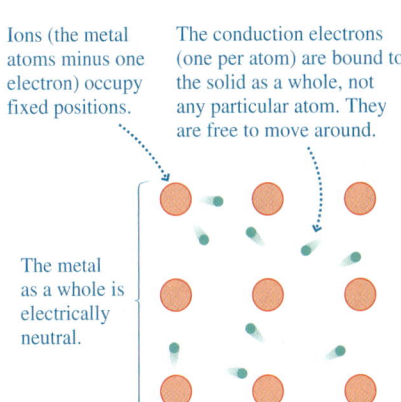

Ions (the metal atoms minus one electron) occupy fixed positions.

The conduction electrons (one per atom) are bound to the solid as a whole, not any particular atom. They are free to move around.

The metal as a whole is electrically neutral.

FIGURE 22.3 Conduction electrons in a metal.

◄ Designer charge carriers Silicon in its natural state is a semiconductor with few charge carriers. Adding other atoms to pure silicon can give it additional charge carriers that are positive or negative, changing the conducting properties of the silicon. An *integrated circuit* is built on a silicon wafer by creating small adjacent regions with different numbers and types of charge carriers, allowing a circuit with thousands or even millions of elements to be built on a single piece of silicon.

NOTE ▶ Electrons are the charge carriers in *metals*. Other materials, such as semiconductors, may have different charge carriers. In ionic solutions, such as seawater, blood, and intercellular fluids, the charge carriers are ions, both positive and negative. ◄

Creating a Current

Suppose you want to slide a book across the table to your friend. You give it a quick push to start it moving, but it begins slowing down because of friction as soon as you take your hand off of it. The book's kinetic energy is transformed into thermal energy, leaving the book and the table slightly warmer. The only way to keep the book moving at a *constant* speed is to continue pushing it.

Something similar happens in a conductor. As we saw in Chapter 20, we can use an electric field to push on the electrons in a conductor. Suppose we take a piece of metal and apply an electric field, as in Figure 22.4. The field exerts a force on the electrons, and they begin to move. But the electrons aren't moving in a vacuum. Collisions between the electrons and the atoms of the metal slow them down, transforming the electrons' kinetic energy into the thermal energy of the metal, making the metal warmer. (Recall that "makes the wire warmer" is one of our indicators of a current.) The motion of the electrons will cease *unless you continue pushing.* To keep the electrons moving, we must maintain an electric field. In a constant field, an electron's average motion will be opposite the field. We call this motion the electron's *drift velocity.* If the field goes to zero, so does the drift velocity.

One of the important conclusions of Chapter 20 was that $\vec{E} = \vec{0}$ inside a conductor in electrostatic equilibrium. But a conductor with electrons moving through it is *not* in electrostatic equilibrium—the charges are still in motion, so the field need not be zero. **A current is a motion of charges sustained by an internal electric field.** Thus the quick answer to "What creates a current?" is "An electric field."

But why is there an electric field in a current-carrying wire? Where does it come from? Let's recall a key piece of information from Chapter 21: Wherever there's a potential difference, an electric field points from higher potential toward lower potential. Figure 22.5 shows how connecting a conducting wire between two points of different potential leads to a current.

Now we can understand how the capacitor in Figure 22.1 gets discharged. Initially, there's a potential difference between the two plates, due to the separation of charges. Connecting a wire between the plates establishes an electric field in the wire, and this electric field causes electrons to flow from the negative plate (which has an excess of electrons) toward the positive plate. In other words, **it is the *potential difference* that creates a current in the wire.** The potential difference across a capacitor depends on the capacitor's charge, so the potential difference—and hence the current—steadily falls as the discharging proceeds. The current ceases when the capacitor is fully discharged because there's no longer a potential difference to maintain the current.

One interesting observation is that the excess electrons on the negative plate need not move all the way to the positive plate. Don't forget that the wire is *already full* of electrons; recall from Chapter 20 the model of the *sea of electrons.* Let's use an analogy with water in a pipe. If a pipe is already full of water, adding a drop to one end quickly pushes a drop out the other end. Likewise with the wire.

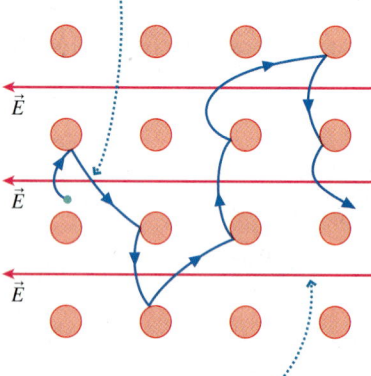

The collisions "reset" the motion of the electron. It then accelerates until the next collision.

$\vec{E}$

$\vec{E}$

$\vec{E}$

The electron has a net displacement opposite the electric field.

FIGURE 22.4 The motion of an electron in a conductor.

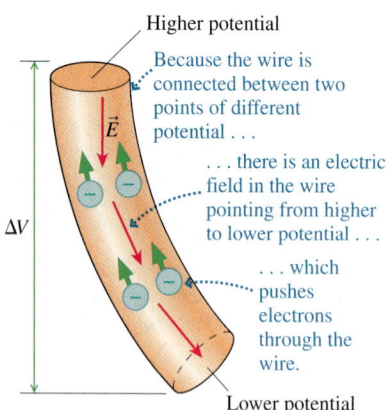

Higher potential

Because the wire is connected between two points of different potential . . .

$\vec{E}$

. . . there is an electric field in the wire pointing from higher to lower potential . . .

ΔV

. . . which pushes electrons through the wire.

Lower potential

FIGURE 22.5 Creating a current.

As soon as some excess electrons move from the negative capacitor plate into the wire, an equal number of electrons are immediately (or very nearly so) pushed out the other end of the wire and onto the positive plate. We don't have to wait for electrons to move all the way through the wire from one plate to the other. Instead, we just need to slightly rearrange the charges on the plates *and* in the wire. No electron moves very far, and the time required is quite short.

CONCEPTUAL EXAMPLE 22.1 Discharging a capacitor

The two plates of a capacitor are charged to ± 10 nC, and are then connected with a wire. In the instant after the wire is connected, the electrons move through the wire with average speed v_i. At a slightly later time, when the discharge is 50% complete, the charge on the plates is ± 5 nC. How does the average speed of the electrons at this time compare to v_i?

REASON The electrons are pushed through the wire by an electric field. The source of the electric field is the potential differ-

ence created by excess charges on the plates of the capacitor. As the capacitor charge decreases, so does the strength of the electric field. With a reduced "push," the electrons move at a speed less than v_i.

ASSESS To verify that this answer makes sense, let's take it to its logical conclusion: Suppose the current continues until the capacitor completely discharges. With no excess charges, there will be no field—and no motion of electrons. The current will cease, exactly as we expect.

Conservation of Current

In Figure 22.6 a lightbulb has been added to the wire connecting two capacitor plates. The bulb glows while the current is discharging the capacitor. How do you think the current at point A compares to the current at point B? Are the currents at these points the same? Or is one larger than the other? Think about this before going on.

You might have predicted that the current at B is less than the current at A because the bulb, in order to glow, must use up some of the current. It's easy to test this prediction; for instance, we could compare the currents at A and B by comparing how far two compass needles at these positions are deflected. All methods give the same result: The current at point B is *exactly equal* to the current at point A. **The current leaving a lightbulb is exactly the same as the current entering the lightbulb.**

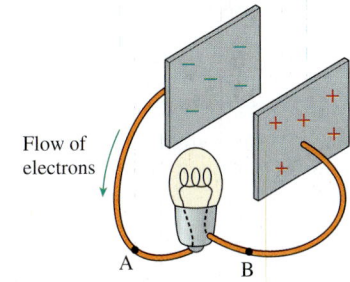

FIGURE 22.6 How does the current at A compare to the current at B?

This is an important observation, one that demands an explanation. After all, "something" makes the bulb glow, so why don't we observe a decrease in the current? Electrons are charged particles. The lightbulb can't destroy electrons without violating both the law of conservation of mass and the law of conservation of charge. Thus the *number* of electrons is not changed by the lightbulb. Further, the lightbulb can't store electrons. Were it to do so, the bulb would become increasingly negative until its repulsive force stopped the flow of new electrons and the bulb would go out. This doesn't happen. Every electron entering the lightbulb must be matched by an electron leaving the bulb, and thus the current at B is the same as at A.

Let's consider again our analogy with water flowing through a pipe. Suppose we put a turbine in the middle of the pipe so that the flow of the water turns the turbine, as in Figure 22.7. Water flows *through* the turbine. It is not consumed by the turbine, and the number of gallons of water per minute leaving the pipe is exactly the same as that entering. Nonetheless, the water must do work to turn the turbine, so the water *loses energy* as it passes through.

Similarly, the lightbulb doesn't "use up" current, but, like the turbine, it *does* use energy. The energy is dissipated by atomic-level friction as the electrons move through the wire, making the wire hotter until, in the case of the lightbulb filament, it glows.

There are many other issues we'll need to examine, but we can draw a first important conclusion:

The amount of water that leaves the turbine is the same as that which entered, and the flow of electrons leaving the bulb is the same as that entering.

FIGURE 22.7 Water in a pipe turns a turbine.

Law of conservation of current The current is the same at all points in a current-carrying wire.

The discharge of a capacitor lights two bulbs. Comparing the current in the two bulbs, we can say that

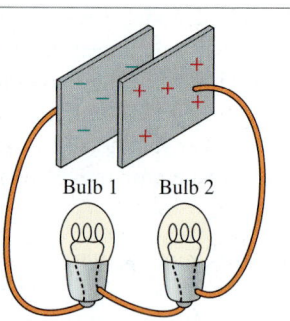

A. The current in bulb 1 is greater than the current in bulb 2.
B. The current in bulb 1 is less than the current in bulb 2.
C. The current in bulb 1 is equal to the current in bulb 2.

22.2 Defining and Describing Current

We have developed the idea of a current as the motion of electrons through metals. But currents were known and studied before it was known what the charge carriers are, and the direction of current was *defined* to be the direction in which positive charges *seem* to move, as illustrated in Figure 22.8. Because the charge carriers turned out to be negative, at least for a metal, the direction of the current in a metal is opposite the direction of motion of the electrons.

The fact that the definition of current is "wrong" in some sense isn't really a problem for the work we will do. A capacitor is discharged regardless of whether positive charges move toward the negative plate or negative charges move toward the positive plate. The primary application of the concept of current is to the analysis of circuits, and in a circuit—a macroscopic device—we simply can't tell the sign of the charges that are moving through the wires. **All of our calculations will be correct and all of our circuits will work perfectly well if we define current to be the flow of positive charge.** The distinction is important only at the microscopic level.

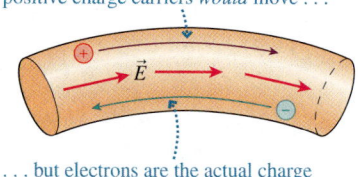

The *current* is defined to point in the direction of $\vec{E}$. It is the direction in which positive charge carriers *would* move . . .

. . . but electrons are the actual charge carriers. They move opposite to $\vec{E}$.

FIGURE 22.8 The current is opposite the direction of motion of the electrons in a metal.

Definition of Current

Because the coulomb is the SI unit of charge, and because currents are charges in motion, we define current as the *rate,* in coulombs per second, at which charge moves through a wire. Figure 22.9 shows a wire in which the electric field is $\vec{E}$. This electric field causes charges to move through the wire. Because we are considering current as the motion of positive charges, the motion is in the direction of the field.

When traffic engineers measure the flow of traffic on a road, they place a wire across the road that detects the passage of cars. The flow of traffic is determined by the number of cars that pass this point in a certain time interval. We use a similar convention for current. As illustrated in Figure 22.9, we can measure the amount of charge ΔQ that passes through a cross section of the wire in a time interval Δt. We then define the current in the wire as

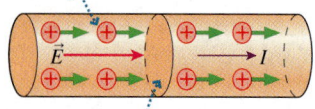

The current I is due to the motion of charges in the electric field.

We imagine an area across the wire through which the charges move. In a time Δt, charge ΔQ moves through this area.

FIGURE 22.9 The current I.

$$I = \frac{\Delta Q}{\Delta t} \qquad (22.1)$$

Definition of current

The current direction in a wire is from higher potential to lower potential or, equivalently, in the direction of the electric field $\vec{E}$. The SI unit for current is the coulomb per second, which is called the **ampere** A:

$$1 \text{ ampere} = 1 \text{ A} = 1 \text{ coulomb per second} = 1 \text{ C/s}$$

The current unit is named for the French scientist André Marie Ampère, who made major contributions to the study of electricity and magnetism in the early 19th century. The *amp* is an informal abbreviation of ampere. Household currents are typically $\approx$1 A. For example, the current through a 100 watt lightbulb is 0.83 A. Currents in electronic circuits, such as those in computers, are much less. They are typically measured in milliamps (1 mA = 10^{-3} A) or microamps (1 μA = 10^{-6} A).

For a *steady current,* which will be our primary focus, the total amount of charge delivered by current I during the time interval Δt is

$$Q = I \, \Delta t \qquad (22.2)$$

EXAMPLE 22.1 **Charge flow in a lightbulb**

A 100 W lightbulb carries a current of 0.83 A. How much charge flows through the bulb in one minute?

SOLVE According to Equation 22.2, the total charge passing through the bulb in 1 min = 60 s is

$$Q = I \, \Delta t = (0.83 \text{ A})(60 \text{ s}) = 50 \text{ C}$$

ASSESS The charge that flows through devices such as light-bulbs or motors is enormous. This is a good check on the con-

cept of conservation of current. If even a minuscule fraction of the charge stayed in the bulb, the bulb would become highly charged. For comparison, a Van de Graaff generator develops a potential of several hundred thousand volts due to an excess charge of just a few μC, a ten-millionth of the charge that flows through the bulb in one minute. Lightbulbs do not develop a noticeable charge, so the current into and out of the bulb must be exactly the same.

Conservation of Current at a Junction

The law of conservation of current tells us that the current I is the same at all points in a current-carrying wire. But what happens when a wire splits, or when two wires come together?

Figure 22.10 shows two wires merging into one and one wire splitting into two. A point where a wire branches is called a **junction.** The presence of a junction doesn't change our basic reasoning. We cannot create or destroy charges in the wire, and neither can we store them in the junction. The rate at which electrons flow into one *or many* wires must, like water in a pipe, be exactly balanced by the rate at which they flow out of others. For a *junction,* the law of conservation of charge requires that

$$\sum I_{\text{in}} = \sum I_{\text{out}} \qquad (22.3)$$

where, as usual, the $\sum$ symbol means "the sum of."

This basic conservation statement, that the sum of the currents into a junction equals the sum of the currents leaving, is called **Kirchhoff's junction law.** The junction law isn't a new law of physics; it is a consequence of the conservation of charge.

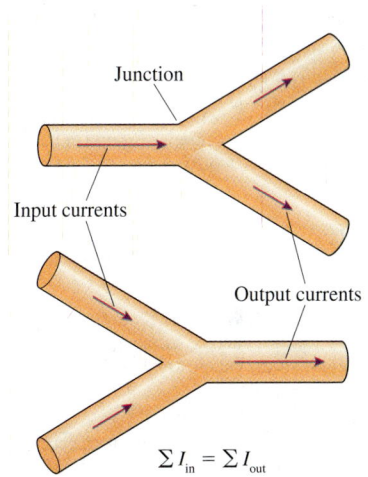

$$\sum I_{\text{in}} = \sum I_{\text{out}}$$

FIGURE 22.10 The sum of the currents into a junction must equal the sum of the currents leaving the junction.

EXAMPLE 22.2 **Currents in a junction**

Four wires have currents as noted in Figure 22.11. What are the direction and the magnitude of the current in the fifth wire?

FIGURE 22.11 The junction of Example 22.2.

SOLVE Two of the wires have currents into the junction:

$$\sum I_{\text{in}} = 3 \text{ A} + 4 \text{ A} = 7 \text{ A}$$

Two of the wires have currents out of the junction:

$$\sum I_{\text{out}} = 6 \text{ A} + 2 \text{ A} = 8 \text{ A}$$

To conserve current, the fifth wire must carry a current of 1 A into the junction.

ASSESS This type of analysis will be very important when we begin to analyze circuits.

The discharge of a capacitor lights three bulbs. Comparing the current in bulbs 1 and 2, we can say that

A. The current in bulb 1 is greater than the current in bulb 2.
B. The current in bulb 1 is less than the current in bulb 2.
C. The current in bulb 1 is equal to the current in bulb 2.

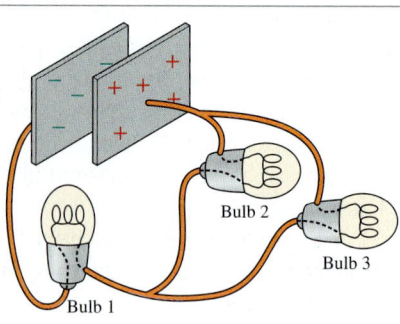

22.3 Batteries and EMF

There are practical devices, such as a camera flash, that use the charge on a capacitor to drive a current. But, as we have seen, a capacitor discharges and the current ceases. A camera flash gives a single, bright flash of light. If you want a light to illuminate your way along a dark path, you need a *continuous* source of light like a flashlight. Continuous light requires the current to be continuous as well.

Figure 22.12 shows a wire connecting the two terminals of a battery, much like the wire that connected the capacitor plates in Figure 22.1. Just like that wire, the wire connecting the battery terminals gets warm, deflects a compass needle, and makes a lightbulb inserted into it glow brightly. These indicators tell us that charges flow through the wire from one terminal to the other. The current in the wire is the same whether it is supplied by a capacitor or a battery. Everything you've learned so far about current applies equally well to the current supplied by a battery, with one important difference.

The difference is the duration of the current. The current that discharges a capacitor is transient; it quickly ceases. By contrast, the wire connecting the battery terminals *continues* to deflect the compass needle and *continues* to light the lightbulb. The capacitor quickly runs out of excess charge, but the battery can keep the charges in motion.

How does a battery produce this sustained motion of charge? A real battery involves a series of chemical reactions, but Figure 22.13 shows a simple model of a battery that illustrates the motion of charges. The inner workings of a battery act like a *charge escalator* between the two terminals. Charges are removed from the negative terminal and "lifted" to the positive terminal. It is the charge escalator that sustains the current in the wire by providing a continually renewed supply of charges at the positive terminal.

Once a charge reaches the positive terminal, it is able to flow through the wire as a current until reaching the negative terminal. The charge escalator then lifts the charge back to the positive terminal where it can start the loop all over again. This flow of charge in a continuous loop is what we call a **complete circuit.**

The charge escalator in the battery must be powered by some external source of energy. It is lifting the electrons "uphill" against the electric field. A battery consists of chemicals, called *electrolytes,* sandwiched between two electrodes made of different materials. The energy to move charges comes from chemical reactions between the electrolytes and the electrodes. These chemical reactions separate charge by moving positive ions to one electrode and negative ions to the other. In other words, chemical reactions, rather than a mechanical conveyor belt, transport charge from one electrode to the other. A dead battery is one in which the supply of chemicals, and thus the supply of chemical energy, has been exhausted.

Battery

A lightbulb lights and a compass needle deflects, just as they do for a wire that discharges a capacitor.

FIGURE 22.12 There is a current in a wire connecting the terminals of a battery.

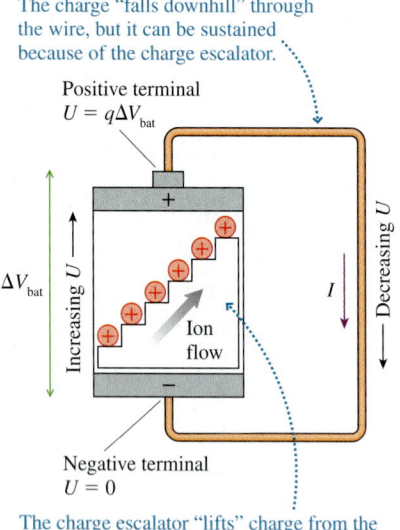

The charge "falls downhill" through the wire, but it can be sustained because of the charge escalator.

Positive terminal
$U = q\Delta V_{bat}$

ΔV_{bat}

Increasing U

Decreasing U

I

Ion flow

Negative terminal
$U = 0$

The charge escalator "lifts" charge from the negative side to the positive side. Charge q gains energy $\Delta U = q\Delta V_{bat}$.

FIGURE 22.13 The charge escalator model of a battery.

By separating charge, the charge escalator establishes the potential difference ΔV_{bat} that is shown between the terminals of the battery in Figure 22.13. The process works like so: Suppose chemical reactions do work W_{chem} to move charge q from the negative to the positive terminal. In an ideal battery, in which there are no internal energy losses, the charge gains electric potential energy $\Delta U = W_{chem}$. This is analogous to a book gaining gravitational potential energy as you do work to lift it from the floor to a shelf.

The quantity W_{chem}/q, which is the work done *per charge* by the charge escalator, is called the **emf** of the battery. It is pronounced as the sequence of three letters "e-m-f." The symbol for emf is $\mathcal{E}$, a script E, and its units are those of the electric potential: joules per coulomb, or volts. The term emf was originally an abbreviation of "electromotive force." That is an outdated term, so today we just call it emf and it's not an abbreviation of anything.

The *rating* of a battery, such as 1.5 V, is the battery's emf. It is determined by the specific chemical reactions employed by the battery. An alkaline battery has an emf of 1.5 V; a rechargeable NiCd battery has an emf of 1.2 V. Larger emfs are created by using several smaller "cells" in a row, much like going to the fourth floor by taking four separate escalators.

By definition, the electric potential is related to the electric potential energy of charge q by $\Delta V = \Delta U/q$. But $\Delta U = W_{chem}$ for the charges in an ideal battery, hence the potential difference between the terminals of an ideal battery is

$$\Delta V_{bat} = \frac{W_{chem}}{q} = \mathcal{E} \qquad (22.4)$$

In other words, the potential difference between the terminals of a battery, often called the **terminal voltage,** is the battery's emf. In practice, inevitable energy losses within the battery cause the terminal voltage of a real battery to be slightly less than the emf. We'll overlook this small difference and assume $\Delta V_{bat} = \mathcal{E}$.

Electric generators, photocells, and power supplies use different means to separate charge, but otherwise they function much like a battery. The common feature of all such devices is that **they use some source of energy to separate charge and, thus, to create a potential difference.** The emf of the device is the work done per charge to separate the charge. In contrast, a capacitor stores separated charges, but a capacitor has no means to *do* the separation. Hence a capacitor has a potential difference, but not an emf.

NOTE ▶ The term *emf,* often capitalized as EMF, is widely used in popular science articles in newspapers and magazines to mean "electromagnetic field." This is *not* how we will use the term *emf.* ◀

CONCEPTUAL EXAMPLE 22.2 Recharging a battery

As a battery creates a current in a circuit, the reactions that run the charge escalator deplete chemicals in the battery. However, you can "recharge" some types of batteries by forcing a current into the positive terminal. Clearly you are not literally *recharging* it, as you are not adding charge to the battery. What are you replenishing in the battery when you recharge it?

REASON When you recharge a battery, you reverse the direction of the current. This reverses the chemical reactions that produce the emf, replenishing the chemicals in the battery and storing energy as chemical energy.

ASSESS Because we are really adding energy to the battery, it would be more appropriate to say that we are "re-energizing" the battery. Note that only certain batteries can be recharged; the chemical reactions must be easily reversible.

Listen to your potential Put on a set of earphones from a portable music player and place the plug on the table. Moisten your fingertips and hold a penny in one hand and a paper clip in the other. This makes a very weak battery; the penny and the clip are the electrodes and your moist skin the electrolyte. Touch the paper clip to the innermost contact on the headphone plug and the penny to the outermost. You will hear a *very* soft click as the potential difference causes a small current in the headphones.

A shocking predator? BIO The torpedo ray captures and eats fish by paralyzing them with electricity. As we will see in Chapter 23, cells in the body use chemical energy to separate charge, just as in a battery. Special cells in the body of the ray called *electrocytes* produce an emf of a bit more than 0.10 V for a short time when stimulated. Such a small emf will not produce a large effect, but the torpedo ray has organs that contain clusters of hundreds of these electrocytes connected in a row. The total emf can be 50 V or more, enough to immobilize nearby prey.

CONCEPTUAL EXAMPLE 22.3 **Potential difference for batteries in series**

Three batteries are connected one after the other as shown in Figure 22.14; we say they are connected in *series*. What's the total potential difference?

REASON We can think of this as three charge escalators, one after the other. Each one lifts charges to a higher potential. Because each battery raises the potential by 1.5 V, the total potential difference of the three batteries in series is 4.5 V.

ASSESS Common AA and AAA batteries are 1.5 V batteries. Many consumer electronics, such as digital cameras, use two or four of these batteries. Wires inside the device connect the batteries in series to produce a total 3.0 V or 6.0 V potential difference.

FIGURE 22.14 Three batteries in series.

Electricity generators, such as coal-burning power plants, burn fuel (a source of chemical energy), transform the resulting thermal energy to the mechanical energy of a spinning turbine, then use a generator, which we'll discuss in Chapter 24, to transform the mechanical energy to electricity. There are unavoidable thermodynamic inefficiencies associated with this process, as we learned in Chapter 11. In contrast, a battery converts chemical energy directly to electric energy. This is more efficient, but batteries have a limited store of chemicals.

A much more elegant solution to the generation of electricity is the *fuel cell*. A particular fuel cell, one that combines hydrogen fuel with oxygen, is illustrated in Figure 22.15. Rather than burning the fuel, as in a power plant, with the resulting thermal losses, the fuel cell's specially designed electrodes allow an electrochemical reaction that transforms the chemical energy of the hydrogen directly to electrical energy. A fuel cell thus works like a battery, but with the chemicals supplied externally. As long as fuel and oxygen are coming in, a fuel cell can produce electricity. A hydrogen-fueled car with a fuel cell and an electric motor under the hood would be very efficient, and it would generate no hazardous exhaust. The only "waste" product is water!

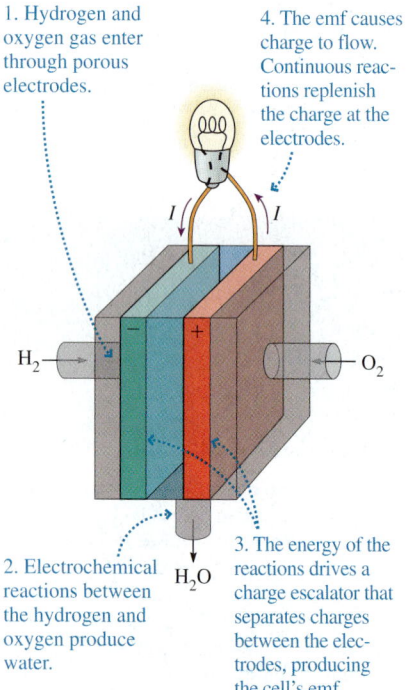

1. Hydrogen and oxygen gas enter through porous electrodes.

4. The emf causes charge to flow. Continuous reactions replenish the charge at the electrodes.

H_2

O_2

2. Electrochemical reactions between the hydrogen and oxygen produce water.

H_2O

3. The energy of the reactions drives a charge escalator that separates charges between the electrodes, producing the cell's emf.

FIGURE 22.15 The electric energy to light a bulb comes from the reaction between hydrogen fuel and oxygen in a fuel cell.

STOP TO THINK 22.3 A battery produces a current in a wire. As the current continues, which of the following quantities (perhaps more than one) decreases?

A. The positive charge in the battery.
B. The emf of the battery.
C. The chemical energy in the battery.

22.4 Connecting Potential and Current

An important conclusion of the charge escalator model is that **a battery is a source of potential difference.** When charges flow through a wire that connects the battery terminals, this current is a *consequence* of the battery's potential difference. You can think of the battery's emf as being the *cause*. Current, heat, light, sound, and so on are all *effects* that happen when the battery is used in certain ways.

Our goal in this section is to find the connection between potential and current. Let's start by connecting a wire between the two terminals of a battery, as shown in Figure 22.16. You learned in Chapter 21 that the potential difference between any two points is independent of the path between them. Consequently, the potential difference between the two ends of the wire, along a path through the wire, is equal to the potential difference between the two terminals of the battery:

$$\Delta V_{\text{wire}} = \Delta V_{\text{bat}} \qquad (22.5)$$

Figure 22.16 shows how the battery causes a current in the wire. Notice that **the current is in the direction of decreasing potential.** We can think of it this way: The charge escalator in the battery raises the charges "uphill," and then they flow "downhill" through the wire.

Resistance

If you connect a wire between the terminals of a battery, the battery's potential difference creates a current in the wire. Suppose we were to do a series of experiments to determine what factors affect this current, as in Figure 22.17. As you would expect, adding a second battery in series increases the current. Careful measurements show that the current in the wire is proportional to ΔV_{wire}. This is easily explained with our model of conduction: A larger potential difference creates a larger electric field that, in turn, pushes charges through the wire faster.

We would also find that increasing the length of the wire decreases the current, while increasing the thickness of the wire increases the current. This seems reasonable, because it should be harder to push charges through a long wire than a short one, and an electric field should be able to push more charges through a fat wire than a skinny one.

Finally, we would find that some materials carry more current than others. A material is a better conductor if the charges are able to move through the wire with fewer collisions. This allows the drift velocity to be higher and leads to a larger current.

For any particular wire, we can define a quantity called the **resistance** that is a measure of how hard it is to push charges through the wire. We use the symbol R for resistance. A large resistance implies that it is hard to move the charges through the wire; in a wire with low resistance, the charges move much more easily. The current in the wire depends on the potential difference ΔV_{wire} between the ends of the wire and the wire's resistance R as follows:

$$I = \frac{\Delta V_{\text{wire}}}{R} \qquad (22.6)$$

Establishing a potential difference ΔV_{wire} between the ends of a wire of resistance R creates an electric field that, in turn, causes a current $I = \Delta V_{\text{wire}}/R$ in the wire. As we would expect, the smaller the resistance, the larger the current.

We can think of Equation 22.6 as the definition of resistance. If a potential difference V_{wire} causes current I in a wire, the wire's resistance is

$$R = \frac{\Delta V_{\text{wire}}}{I} \qquad (22.7)$$

The SI unit of resistance is the **ohm,** defined as

$$1 \text{ ohm} = 1 \ \Omega = 1 \text{ V/A}$$

where Ω is an uppercase Greek omega. The unit takes its name from the German physicist Georg Ohm, whose work in the early 19th century helped clarify the relationship among potential difference, current, and resistance. The ohm is the basic unit of resistance, although kilohms ($1 \text{ k}\Omega = 10^3 \ \Omega$) and megohms ($1 \text{ M}\Omega = 10^6 \ \Omega$) are widely used.

1. The potential difference between these two points is the same whether you go through the battery or the wire.

2. The potential difference between the ends of the wire establishes an electric field inside the wire.

3. The electric field drives a current through the wire, in the direction of decreasing potential.

FIGURE 22.16 The electric field and the current inside the wire.

(a) Adding batteries

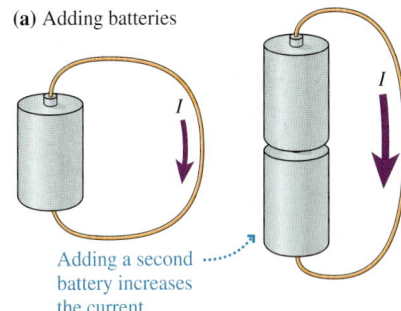

Adding a second battery increases the current.

(b) Changing wire dimensions

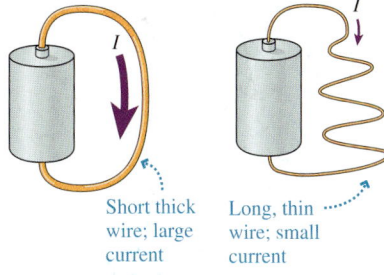

Short thick wire; large current

Long, thin wire; small current

(c) Changing wire material

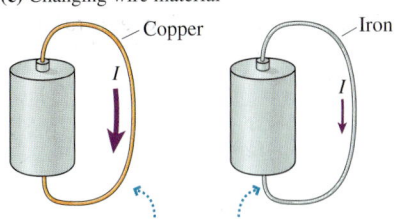

Copper   Iron

A copper wire will carry a larger current than an iron wire of the same dimensions.

FIGURE 22.17 Factors affecting the current in a wire.

EXAMPLE 22.3 Resistance of a lightbulb

The glowing element in an incandescent lightbulb is the *filament,* a long, thin piece of tungsten wire that is heated by the electric current through it. When connected to the 120 V of an electric outlet, a 60 W bulb carries a current of 0.50 A. What is the resistance of the filament in the lamp?

SOLVE We can use Equation 22.7 to compute the resistance:

$$R = \frac{\Delta V_{wire}}{I} = \frac{120 \text{ V}}{0.50 \text{ A}} = 240 \text{ } \Omega$$

ASSESS As we will see below, the resistance of the filament varies with temperature. This value holds for the lightbulb only when the bulb is glowing and the filament is hot.

Resistivity

TABLE 22.1 Resistivity of materials

Material	Resistivity $(\Omega \cdot \text{m})$
Copper	1.7×10^{-8}
Tungsten (20°C)	5.6×10^{-8}
Tungsten (1500°C)	5.0×10^{-7}
Iron	9.7×10^{-8}
Nichrome	1.5×10^{-6}
Seawater	0.22
Blood (average)	1.6
Muscle	13
Fat	25
Pure water	2.4×10^{5}
Cell membrane	3.6×10^{7}

Figure 22.17c noted that the resistance of a wire depends on what it is made of. We define a quantity called **resistivity,** for which we use the symbol ρ (a lowercase Greek rho), to characterize the electrical properties of materials. Materials that are good conductors have low resistivity; materials that are poor conductors (i.e., insulators) have high resistivity. The resistivity ρ has units of $\Omega \cdot \text{m}$. The resistivity of some materials is given in Table 22.1.

The resistivity of materials varies quite a bit. Metals are generally good conductors (and so have very small resistivity), but metals such as copper are much better conductors than metals such as nichrome, an alloy of nickel and chromium that is used to make heating wires. Water is a poor conductor, but the dissolved salts in seawater produce ions that can carry charge, so seawater is a good conductor, with a resistivity one million times less than that of pure water. Glass is an excellent insulator with a resistivity in excess of $10^{14} \text{ } \Omega \cdot \text{m}$, 10^{22} times that of copper.

The resistivity of a material depends on the temperature, as you can see from the two values for tungsten listed in Table 22.1. As the temperature increases, so do the thermal vibrations of the atoms. This makes them "bigger targets" for the moving electrons, causing collisions to be more frequent. Thus the resistivity of a metal increases with increasing temperature.

The resistance of a wire depends both on the resistivity of its material and on the dimensions of the wire. A wire made of a material of resistivity ρ, with length L and cross-section area A, has resistance

$$R = \frac{\rho L}{A} \qquad (22.8)$$

Resistance of a wire in terms of resistivity and dimensions

LINEAR p.38   INVERSE p.118

Resistivity is a property of a material. All copper wires (at the same temperature) have the same resistivity. But resistance is a property of a *specific* wire or conductor because it depends on the conductor's length and diameter as well as on the resistivity of the material from which it is made.

NOTE ▶ It is important to distinguish between resistivity and resistance. *Resistivity* describes just the *material,* not any particular piece of it. *Resistance* characterizes a specific piece of the conductor having a specific geometry. The relationship between resistivity and resistance is analogous to that between density and mass. ◀

▶ **Coils of coils** Most lightbulb filaments are made of tungsten, as it is one of a few materials that can withstand the necessary high temperatures. But a filament must have a large resistance. The resistivity of tungsten is quite low, so a tungsten filament must be very long and very thin. A close view of a typical bulb's filament shows that it is made of very thin wire that is coiled and then coiled again. The double-coil structure is necessary to fit the great length of the filament into the small space of the bulb's globe.

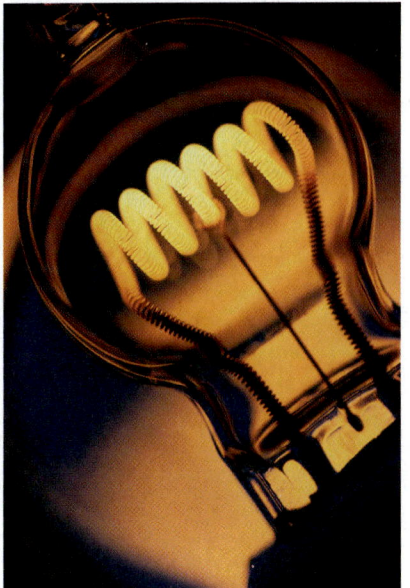

EXAMPLE 22.4 The length of a lightbulb filament

We calculated in Example 22.3 that a 60 W lightbulb has a resistance of 240 Ω. At the operating temperature of the tungsten filament, the resistivity is approximately $5.0 \times 10^{-7}\ \Omega \cdot$ m. If the wire used to make the filament is 0.040 mm in diameter, how long must the filament be?

PREPARE The resistance of a wire depends on its length, its cross-section area, and the material of which it is made.

SOLVE The cross-section area of the wire is $A = \pi r^2 = \pi (2.0 \times 10^{-5}\ \text{m})^2 = 1.26 \times 10^{-9}\ \text{m}^2$. Rearranging Equation 22.8 shows us that the filament must be of length

$$L = \frac{AR}{\rho} = \frac{(1.26 \times 10^{-9}\ \text{m}^2)(240\ \Omega)}{5.0 \times 10^{-7}\ \Omega \cdot \text{m}} = 0.60\ \text{m}$$

ASSESS The resistivity of tungsten is low, so the filament must be quite thin and quite long. This result is quite reasonable.

EXAMPLE 22.5 Making a heater

An amateur astronomer uses a heater to warm her telescope eyepiece so moisture does not collect on it. The heater is a 20-cm-long, 0.50-mm-diameter nichrome wire that wraps around the eyepiece. When the wire is connected to a 1.5 V battery, what is the current in the wire?

PREPARE The current in the wire depends on the emf of the battery and the resistance of the wire. The resistance of the wire depends on the resistivity of nichrome, given in Table 22.1, and the dimensions of the wire. Converted to meters, the relevant dimensions of the wire are $L = 0.20$ m and $r = 2.5 \times 10^{-4}$ m.

SOLVE The wire's resistance is

$$R = \frac{\rho L}{A} = \frac{\rho L}{\pi r^2} = \frac{(1.5 \times 10^{-6}\ \Omega \cdot \text{m})(0.20\ \text{m})}{\pi (2.5 \times 10^{-4}\ \text{m})^2} = 1.5\ \Omega$$

The wire is connected to the battery, so $\Delta V_{\text{wire}} = \Delta V_{\text{bat}} = 1.5$ V. The current in the wire is

$$I = \frac{\Delta V_{\text{wire}}}{R} = \frac{1.5\ \text{V}}{1.5\ \Omega} = 1.0\ \text{A}$$

ASSESS The emf of the battery is small, but so is the resistance of the wire, so this is a reasonable current, enough to warm the wire and the eyepiece.

Electrical Measurements of Physical Properties

Measuring resistance is quite straightforward. Because resistance depends sensitively on the properties of materials, a measurement of resistance can be a simple but effective probe of other quantities of interest.

For example, the resistivity of water is strongly dependent on dissolved substances in the water, so it is easy to make a quick test of water purity by making a measurement of resistivity. In situations in which water must be absolutely pure, such as in the semiconductor industry, the resistivity of the water is continuously monitored. Even a small amount of contaminant can significantly reduce the resistivity, making resistivity a sensitive test of water purity.

CONCEPTUAL EXAMPLE 22.4 Testing drinking water

A house gets its drinking water from a well that has an intermittent problem with salinity. Before the water is pumped into the house, it passes between two electrodes in the circuit shown in Figure 22.18. The current passing through the water is measured with a meter. Which corresponds to increased salinity—an increased current or a decreased current?

REASON Increased salinity causes the water's resistivity to decrease. This decrease causes a decrease in resistance between the electrodes. Current is inversely proportional to resistance, so this leads to an increase in current.

ASSESS Increasing salinity means more ions in solution, and thus more charge carriers, so an increase in current is expected. Electrical systems similar to this can therefore provide a quick check of water purity.

The battery has a fixed emf.

A meter measures the current.

Water flows between two electrodes.

FIGURE 22.18 A water-testing circuit for Conceptual Example 22.4

FIGURE 22.19 An electrical impedance map shows the cross section of a healthy patient's torso. You can clearly see the heart (blue) and lungs (red).

Different tissues in the body have quite different resistivities, as we see in Table 22.1. For example, fat has a higher resistivity than muscle. Consequently, a routine test to estimate the percentage of fat in a person's body is based on a measurement of the body's resistance, as illustrated in the photo at the start of the chapter. A higher resistance of the body means a higher proportion of fat.

More careful measurements of resistance can provide more detailed diagnostic information. Passing a small, safe current between two electrodes on opposite sides of a person's torso permits a measurement of the resistance of the intervening tissue. Figure 22.19 shows an image of a patient's torso generated from measurements of resistance between many pairs of opposing electrodes, created just as the heart contracted, sending blood to the lungs. Blood is a better conductor than the tissues of the heart and lungs, so this motion of blood decreased the resistance of the lungs (shown in red) and increased that of the heart (shown in blue). Such measurements are most useful for studying the heart, in the noninvasive diagnostic procedure called *impedance cardiography.* (Impedance is similar to resistance, but it also applies to AC circuits, which we will explore in Chapter 26.)

Superconductivity

In 1911, the Dutch physicist Kamerlingh Onnes was studying the conductivity of metals at very low temperatures. As we noted above, metals become better conductors (i.e., they have lower resistivity) at lower temperatures. But the resistivity change with temperature is gradual. Onnes, however, found that mercury suddenly and dramatically loses *all* resistance to current when cooled below a temperature of 4.2 K. This complete loss of resistance at low temperatures is called **superconductivity.**

Later experiments established that the resistivity of a superconducting metal is not just small, *it is truly zero.* The electrons are moving in a frictionless environment, and charge will continue to move through a superconductor *without an electric field.* Superconductivity was not understood until the 1950s, when it was explained as being a specific quantum effect.

The discovery of superconductivity was quite unexpected. No less surprising was the discovery of the remarkable magnetic properties of superconductors, illustrated here by the stable levitation of a small, strong magnet above a wafer of high-temperature superconductor.

Superconducting wires can carry enormous currents, and superconducting electromagnets, such as those used in MRI machines, can generate very strong magnetic fields. But applications remained limited for many decades because all known superconductors required temperatures less than 20 K. This situation changed dramatically in 1986 with the discovery of *high-temperature superconductors.* These ceramic-like materials are superconductors at temperatures as "high" as 125 K. Although −150°C may not seem like a high temperature to you, the technology for producing such temperatures is simple and inexpensive. Thus many new superconductor applications are likely to appear in coming years.

A wire connects the positive and negative terminals of a battery. Two identical wires connect the positive and negative terminals of a second, identical battery. Rank in order, from largest to smallest, the currents I_A to I_D at points A to D.

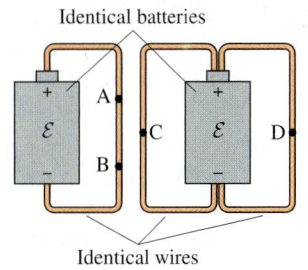

Identical batteries

Identical wires

22.5 Resistors and Ohm's Law

The relationship between the potential difference across a conductor and the current passing through it that we saw in the previous section was first deduced by Georg Ohm, and is known as **Ohm's law:**

$$I = \frac{\Delta V}{R} \qquad (22.9)$$

Ohm's law for a conductor of resistance R

LINEAR INVERSE
p. 38   p. 118

If we know that a wire of resistance R carries a current I, we can compute the potential difference between the ends of the wire as $\Delta V = IR$.

NOTE ▶ We could write the equation for Ohm's law as $\Delta V = IR$, but $I = \Delta V/R$ is a better description of cause and effect because it is the potential difference that causes the current. ◀

Ohmic and Nonohmic Materials

Despite its name, Ohm's law is *not* a law of nature. It is limited to those materials whose resistance R remains constant—or very nearly so—during use. Materials to which Ohm's law applies are called **ohmic.** Figure 22.20a shows that the current through an ohmic material is directly proportional to the potential difference; doubling the potential difference results in a doubling of the current. This is a linear relationship, and the resistance R can be determined from the slope of the graph.

Many materials are ohmic over a reasonable range of operating conditions. The resistance of metals varies slightly with temperature, but a metal wire is ohmic as long as the temperature is reasonably constant, and we can give it a fixed resistance value.

Other materials and devices are **nonohmic,** meaning that the current through the device is *not* directly proportional to the potential difference. An example is a semiconductor device known as a *diode*. The graph in Figure 22.20b shows that a diode does not have a well-defined resistance. Three important examples of nonohmic devices are

1. Batteries, where $\Delta V = \mathcal{E}$ is determined by chemical reactions, independent of I.
2. Semiconductor devices, where the I-versus-ΔV curve is very nonlinear.
3. Capacitors, where, as you'll learn in Chapter 23, the relationship between I and ΔV is very different from that of a resistor.

The main point to remember is that Ohm's law does *not* apply to these nonohmic devices.

(a) Ohmic material: copper wire

The current is directly proportional to the potential difference.

The resistance is $R = \dfrac{1}{\text{slope}}$

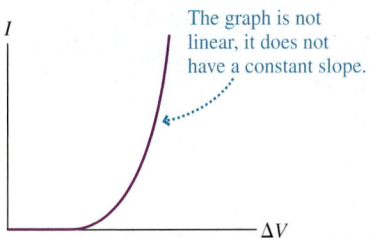

(b) Nonohmic material: diode (semiconductor device)

The graph is not linear, it does not have a constant slope.

FIGURE 22.20 Current-versus-potential-difference graphs for ohmic and nonohmic materials.

Range of Resistance

In circuits, conductors that are designed to have a constant resistance are called **resistors.** A resistor has a specified value of resistance; typical values range from 1 Ω to 1 MΩ. By comparison, the resistance of metal wires is much lower; a 1 foot section of the copper wire in the wall of your house has a resistance of about 0.002 Ω. The resistance of insulators is much greater; a 1 foot section of a thin glass rod would have a resistance of about 1×10^{18} Ω.

We can identify three different classes of ohmic material, based on the size of their resistance:

- *Wires* are made of metals with very small resistivities ρ, thus wires have very small resistances ($R \ll 1\ \Omega$). In most cases, the resistance of wires is too small to make a practical difference, and we can make the approximation that $R = 0\ \Omega$.
- *Insulators* are materials such as glass, plastic, or air. All practical insulators have $R \gg 10^9\ \Omega$, much larger than the range of resistors we will consider. We can generally make the approximation that $R = \infty$ for an insulator.
- *Resistors,* as noted, have resistances in the range of about 1 Ω to 1 MΩ. These circuit elements will be the ones on which we focus our attention.

We will model many different circuit elements as resistors. For example, the filament in a lightbulb functions as a resistor, although its resistance when hot is larger than its room-temperature value. Other examples of resistors that you may work with are illustrated in the table below.

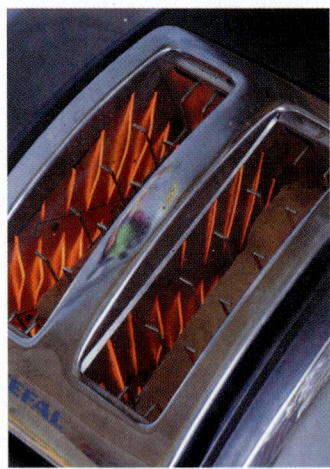

◄ **The resistance goes up when the sun goes down** Many devices use circuit elements whose resistance varies. The sensor on the front of a night light is a *photoresistor*. In the dark, the resistance is large; when light shines on the photoresistor, its resistance decreases. During the day, the resistance is low; at night, the resistance rises. A circuit in the night light monitors the resistance, and switches on the light when the resistance is above a certain value.

Light-sensitive resistor

Real resistors

Turning the dial changes the length of the wire between two contacts, changing the resistance.

Fixed resistors

Thermistor: resistance varies with temperature

Variable resistors: resistance is varied by turning a knob or screw

Photoresistor: resistance varies with light intensity

The heating element in a toaster is a long nichrome wire that functions as a resistor. Current heats the wire until it glows. Other similar resistors include the heating element in a hair dryer, the heating coil on an electric stove, the rear window defroster in a car, and the filament in a lightbulb.

Volume control knobs are usually *variable* resistors. The resistor is a long wire wrapped as a coil. Turning the dial varies the effective length the wire and thus the resistance. This change in resistance affects the operation of a circuit, controlling the volume.

Practical circuits use a wide variety of resistors with different properties. Some are designed to have a constant resistance, others to have a resistance that varies as a dial is turned, still others to have a resistance that varies with light intensity, temperature, or some other physical parameter.

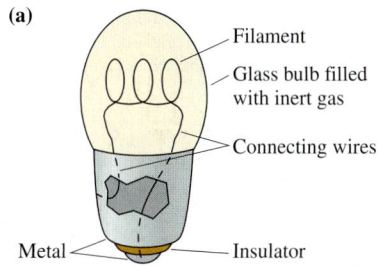

(a)
- Filament
- Glass bulb filled with inert gas
- Connecting wires
- Metal
- Insulator

CONCEPTUAL EXAMPLE 22.5 **The changing current in a toaster**

When you press the lever on a toaster, a switch connects the heating wires to 120 V. The wires are initially cool, but the current in the wires raises the temperature until they are hot enough to glow. As the wire heats up, how does the current in the toaster change?

REASON As the wire heats up, its resistivity increases, as noted above, so the resistance of the wires increases. Because the potential difference stays the same, an increasing resistance causes the current to decrease. The current through a toaster is largest when the toaster is first turned on.

ASSESS This result makes sense. As the wire's temperature increases, the current decreases. This makes the temperature of the wire self-limiting. It will warm up to a certain temperature but no farther.

(b) The filament of the bulb and the connecting wires make a continuous path between the battery terminals.

FIGURE 22.21 The basic circuit of a battery and a bulb.

The Ideal-Wire Model

Figure 22.21a shows the anatomy of a lightbulb. The important point is that a lightbulb, like a wire, has two "ends" and that current passes *through* the bulb. Connections to the filament in the bulb are made at the tip and along the side of the metal cylinder. It is often useful to think of a lightbulb as a resistor that happens to give off light when a current is present. Now, let's look at a circuit using a battery, a lightbulb, and wires to make the connection, as in Figure 22.21b. This is the basic circuit in a flashlight.

A typical flashlight bulb has a resistance of ≈3 Ω, while a wire that one would use to connect such a bulb to a battery has a resistance of ≈0.01 Ω. The resistance of the wires is so much less than that of the bulb that we can, with very little error, adopt the *ideal-wire model* and assume that any connecting wires in a circuit are ideal. An **ideal wire** has $R = 0\ \Omega$, hence the potential difference between the ends of an ideal wire is $\Delta V = 0$ V *even if there is a current in it*.

> **NOTE** ▶ We know that, physically, the potential difference can't be zero. There must be an electric field in the wire for the charges to move, so it must have a potential difference. But in practice this potential difference is so small that we can assume it to be zero with little error. ◄

Figure 22.22 illustrates the ideal-wire model. Here we see wires connected to the two ends of a resistor and a current passing through. Current must be conserved, hence the current I in the resistor is the same as the current in each wire. However, the resistor's resistance is *much* larger than the resistance of the wires: $R_{resist} \gg R_{wire}$. Consequently, the potential difference across the resistor $\Delta V_{resist} = IR_{resist}$ is *much* larger than the potential difference $\Delta V_{wire} = IR_{wire}$ between the ends of each wire.

Figure 22.22b shows the potential along the wire-resistor-wire combination. We saw earlier that current moves in the direction of decreasing potential, so there is a large *voltage drop*—a decrease in potential—across the resistor as we go from left to right, the direction of the current. A very reasonable approximation is to assume that *all* the voltage drop occurs along the resistor, none along the wires. This is the approximation made by the ideal-wire model, which assumes that $R_{wire} = 0\ \Omega$ and $\Delta V_{wire} = 0$ V. With this approximation, shown in Figure 22.22c, the segments of the graph corresponding to the wires are horizontal.

(a) The current is constant along the wire-resistor-wire combination.
Wire   Resistor   Wire
$R_{resist} \gg R_{wire}$

(b) The voltage drop along the wires is much less than across the resistor because the wires have much less resistance.
ΔV_{resist}
Wire   Resistor   Wire

(c) In the ideal wire model, with $R_{wire} = 0\ \Omega$, there is no voltage drop along the wires. All the voltage drop is across the resistor.
Wire   Resistor   Wire

FIGURE 22.22 The potential along a wire-resistor-wire combination.

The electric field inside the resistor is uniform, and points from high to low potential.

The current is in the direction of decreasing potential.

FIGURE 22.23 Electric field inside a resistor.

The linear variation in the potential across the resistor is similar to the linear variation in potential between the plates of a parallel-plate capacitor. In a capacitor, this linear variation in potential corresponds to a uniform electric field; the same will be true here. As we see in Figure 22.23, the electric field in a resistor carrying a current in a circuit is uniform; the strength of the electric field is

$$E = \frac{\Delta V}{L}$$

in analogy to Equation 21.5 for a parallel-plate capacitor. A larger potential difference corresponds to a larger field, as we would expect.

EXAMPLE 22.6   Analyzing a single-resistor circuit

A 15 Ω resistor is connected to the terminals of a 1.5 V battery.

a. Draw a graph showing the potential as a function of distance traveled through the circuit, starting from $V = 0$ V at the negative terminal of the battery.

b. What is the current in the circuit?

PREPARE To help us visualize the change in potential as charges move through the circuit, we begin with the sketch of the circuit in Figure 22.24. The zero point of potential is noted. We have drawn our sketch so that "up" corresponds to higher potential, which will help us make sense of the circuit. Charges are raised to higher potential in the battery, then travel "downhill" from the positive terminal through the resistor and back to the negative terminal. We assume ideal wires.

FIGURE 22.24 Single-resistor circuit for Example 22.6.

SOLVE a. Figure 22.25 is a graphical representation of the potential in the circuit. The distance s is measured from the battery's negative terminal, where $V = 0$ V. As we move around the circuit to the starting point, the potential must return to its original value. Because the wires are ideal, there is no change in potential along the wires. This means that the potential difference across the resistor must be equal to the potential difference across the battery: $\Delta V_R = \mathcal{E} = 1.5$ V.

The potential increases by 1.5 V in the battery.

There's no potential drop along an ideal wire.

The potential decreases by 1.5 V in the resistor.

s measures the distance "around" the loop.

FIGURE 22.25 Potential versus position in the circuit of Example 22.6.

b. Now that we know the potential difference of the resistor, we can compute the current in the resistor by using Ohm's Law:

$$I = \frac{\Delta V_R}{R} = \frac{1.5 \text{ V}}{15 \text{ Ω}} = 0.10 \text{ A}$$

Because current is conserved, this is the current at any point in the circuit. In other words, the battery's charge escalator lifts charge at the rate 0.10 C/s, and charge flows through the wires and the resistor at the rate 0.10 C/s.

ASSESS This is a reasonable value of the current in a battery-powered circuit.

As we noted, there are many devices whose resistance varies as a function of a physical variable that we might like to measure, such as light intensity, temperature, or sound intensity. As the following example shows, we can use resistance measurements to monitor a physical variable.

EXAMPLE 22.7   Using a thermistor

A thermistor is a device whose resistance varies in a well-defined way with temperature. A certain thermistor has a resistance of 2.8 kΩ at 20°C and 0.39 kΩ at 70°C. This thermistor is used in a water bath in a lab to monitor the temperature. The thermistor is connected in a circuit with a 1.5 V battery, and the current measured. What is the change in current in the circuit as the temperature rises from 20°C to 70°C?

SOLVE We can use Ohm's law to find the current in each case:

$$I(20°C) = \frac{\Delta V}{R} = \frac{1.5 \text{ V}}{2.8 \times 10^3 \, \Omega} = 0.54 \text{ mA}$$

$$I(70°C) = \frac{\Delta V}{R} = \frac{1.5 \text{ V}}{0.39 \times 10^3 \, \Omega} = 3.8 \text{ mA}$$

The change in current is thus 3.3 mA.

ASSESS A modest change in temperature leads to a large change in current, making this a very sensitive means of measuring temperature change.

STOP TO THINK 22.5   Two identical batteries are connected in series in a circuit with a single resistor. $V = 0$ V at the negative terminal of the lower battery. Rank in order, from highest to lowest, the potentials V_A to V_E at the labeled points, noting any ties. Assume the wires are ideal.

22.6 Energy and Power

When you flip the switch on a flashlight, a battery is connected to a lightbulb, which then begins to glow. The bulb is radiating energy. Where does this energy come from?

A battery not only supplies a potential difference, it also supplies energy, as shown in the battery and bulb circuit of Figure 22.26. The charge escalator is an energy-transfer process, transferring chemical energy E_{chem} stored in the battery to the electric potential energy U of the charges. That energy is then dissipated as the charges move through the lightbulb, keeping the filament warm and glowing.

Recall that charge q gains potential energy $\Delta U = q \, \Delta V$ as it moves through a potential difference ΔV. The potential difference of a battery is $\Delta V_{bat} = \mathcal{E}$, so the battery supplies energy $\Delta U = q\mathcal{E}$ to charge q as it lifts the charge up the charge escalator from the negative to the positive terminal.

It's more useful to know the *rate* at which the battery supplies energy. You learned in Chapter 10 that the rate at which energy is transformed is *power*, measured in joules per second or *watts*. Suppose an amount of charge Δq moves through the battery in a time Δt. The charge Δq will increase its potential energy by $\Delta U = (\Delta q)\mathcal{E}$. The *rate* at which energy is transferred from the battery to the moving charges is

$$P_{bat} = \text{rate of energy transfer} = \frac{\Delta U}{\Delta t} = \frac{\Delta q}{\Delta t}\mathcal{E} \tag{22.10}$$

But $\Delta q/\Delta t$, the rate at which charge moves through the battery, is the current I. Hence the power supplied by a battery or any emf is

$$P_{emf} = I\mathcal{E} \tag{22.11}$$

Power delivered by an emf

LINEAR
p.38

$I\mathcal{E}$ has units of J/s, or W.

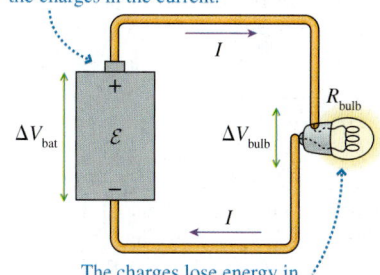

Chemical energy in the battery is transferred to potential energy of the charges in the current.

The charges lose energy in collisions as they pass through the filament of the bulb. This energy is transformed to the thermal energy of the glowing filament.

FIGURE 22.26 Energy transformations in a circuit with a battery and a lightbulb.

Cooking with electricity Before microwave ovens were common, there were devices that used a decidedly lower-tech approach to the rapid cooking of hot dogs. Prongs connected the hot dog to the 120 V of household electricity, making it the resistor in a circuit. The current through the hot dog dissipated energy as thermal energy, cooking the hot dog in about 2 minutes.

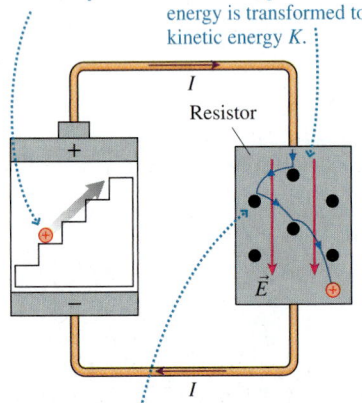

1. Charges gain potential energy U in the battery.

2. As charges accelerate in the electric field in the resistor, potential energy is transformed to kinetic energy K.

I

Resistor

$\vec{E}$

I

3. Collisions with atoms in the resistor transform the kinetic energy of the charges to thermal energy E_{th} of the resistor.

FIGURE 22.27 The power from the battery is dissipated in the resistor.

EXAMPLE 22.8 Power delivered by a car battery
A car battery has $\mathcal{E} = 12$ V. When the car's starter motor is running, the battery current is 320 A. What power does the battery supply?

SOLVE The power is the product of the emf of the battery and the current:

$$P_{bat} = I\mathcal{E} = (320 \text{ A})(12 \text{ V}) = 3.8 \text{ kW}$$

ASSESS This is a lot of power (about 5 hp), but this amount makes sense because turning over a car's engine is hard work. Car batteries are designed to reliably provide such intense bursts of power for starting the engine.

Suppose we consider a circuit consisting of a battery and a single resistor. $P_{bat} = I\mathcal{E}$ is the energy transferred per second from the battery's store of chemicals to the moving charges that make up the current. Figure 22.27 shows the entire sequence of energy transformations, which looks like

$$E_{chem} \rightarrow U \rightarrow K \rightarrow E_{th}$$

The net result is that **the battery's chemical energy is transferred to the thermal energy of the resistor,** raising its temperature.

In the resistor, the amount of charge Δq loses potential energy $\Delta U = (\Delta q)(\Delta V_R)$ as this energy is transformed into kinetic energy and then into the resistor's thermal energy. Thus the rate at which energy is transferred from the current to the resistor is

$$P_R = \frac{\Delta U}{\Delta t} = \frac{\Delta q}{\Delta t} \Delta V_R = I \Delta V_R \tag{22.12}$$

We say that this power—so many joules per second—is *dissipated* by the resistor as charge flows through it.

Our analysis of the single-resistor circuit in Example 22.6 found that $\Delta V_R = \mathcal{E}$. That is, the potential difference across the resistor is exactly the emf supplied by the battery. Because the current is the same in the battery and the resistor, a comparison of Equations 22.11 and 22.12 shows that

$$P_R = P_{bat} \tag{22.13}$$

The power dissipated in the resistor is exactly equal to the power supplied by the battery. The *rate* at which the battery supplies energy is exactly equal to the *rate* at which the resistor dissipates energy. This is, of course, exactly what we would have expected from energy conservation.

EXAMPLE 22.9 Finding the current in a lightbulb
How much current is "drawn" by a 100 W lightbulb connected to a 120 V outlet?

PREPARE We can model the lightbulb as a resistor.

SOLVE Because the lightbulb is operating as intended, it will dissipate 100 W of power. We can rearrange Equation 22.12 to find

$$I = \frac{P_R}{\Delta V_R} = \frac{100 \text{ W}}{120 \text{ V}} = 0.83 \text{ A}$$

ASSESS Most household appliances, such as a 100 W lightbulb or a 1500 W hair dryer, have a power rating. The rating does *not* mean that these appliances *always* dissipate that much power. These appliances are intended for use at a standard household voltage of 120 V, and their rating is the power they will dissipate *if* operated with a potential difference of 120 V. Their power consumption will differ from the rating if they are operated at any other potential difference.

A resistor obeys Ohm's law, $\Delta V_R = IR$. This gives us two alternative ways of writing the power dissipated by a resistor. We can either substitute IR for ΔV_R or substitute $\Delta V_R/R$ for I. Thus

$$P_R = I\,\Delta V_R = I^2 R = \frac{(\Delta V_R)^2}{R} \qquad (22.14)$$

Power dissipated by resistance R with current I and potential difference ΔV_R

QUADRATIC
p. 50

It is worth writing the different forms of this equation to illustrate that the power varies as the square of both the current and the potential difference.

EXAMPLE 22.10 Finding the power of a dim bulb

How much power is dissipated by a 60 W (120 V) lightbulb when operated, using a dimmer switch, at 100 V?

PREPARE The 60 W rating is for operation at 120 V. We can compute the resistance for this case, then compute the power with the dimmer switch.

SOLVE The lightbulb dissipates 60 W at $\Delta V_R = 120$ V. Thus the filament's resistance is

$$R = \frac{(\Delta V_R)^2}{P_R} = \frac{(120 \text{ V})^2}{60 \text{ W}} = 240 \text{ }\Omega$$

The power dissipation when operated at $\Delta V_R = 100$ V is

$$P_R = \frac{(\Delta V_R)^2}{R} = \frac{(100 \text{ V})^2}{240 \text{ }\Omega} = 42 \text{ W}$$

ASSESS Actually, this result is not quite correct. As noted previously, a filament is not a true ohmic material because the resistance changes somewhat with temperature. The filament's resistance at 100 V, where it glows less brightly, is not quite the same as its resistance at 120 V. The voltage in this example is still near 120 V, so the temperature of the filament will decrease only slightly, and our answer should be fairly accurate. However, this calculation would not give a good result for the power dissipation at 20 V, where the filament's temperature would be much less than at 120 V.

EXAMPLE 22.11 Determining the voltage of a stereo

Most stereo speakers are designed to have a resistance of 8.0 Ω. If an 8.0 Ω speaker is connected to a stereo amplifier with a rating of 100 W, what is the maximum possible potential difference the amplifier can apply to the speakers?

PREPARE The rating of an amplifier is the *maximum* power it can deliver. Most of the time it delivers far less, but the maximum might be needed for brief, intense sounds. The maximum potential difference will occur when the amplifier is providing the maximum power, so we will make our computation with this figure. We can model the speaker as a resistor.

SOLVE The maximum potential difference occurs when the power is a maximum. At the maximum power of 100 W,

$$P_R = 100 \text{ W} = \frac{(\Delta V_R)^2}{R} = \frac{(\Delta V_R)^2}{8.0 \text{ }\Omega}$$

$$\Delta V_R = \sqrt{(8.0 \text{ }\Omega)(100 \text{ W})} = 28 \text{ V}$$

This is the maximum potential difference the amplifier might provide.

ASSESS As a check on our result, we note that the resistance of the speaker is less than that of a lightbulb, so a smaller potential difference can provide 100 W of power. The amplifier works like a battery—it is an emf, providing energy to the speaker. Some of the amplifier's energy is transformed to sound energy, but most is simply dissipated as thermal energy because of the resistance of the wires in the speaker.

STOP TO THINK 22.6 Rank in order, from largest to smallest, the powers P_A to P_D dissipated in resistors A to D.

The goal of Chapter 22 has been to learn how and why charge moves through a conductor as a current.

GENERAL PRINCIPLES

Current

The **current** is defined to be the motion of positive charges.

The battery does work to raise the potential of the charges. The potential difference of the battery is its **emf** $\mathcal{E}$.

$$\Delta V_{bat} = \mathcal{E}$$

A **battery** is a source of potential difference. Chemical processes in the battery separate charges. We use a **charge escalator** model to show the lifting of charges to higher potential.

Conservation of current dictates that the current is the same at all points in the circuit.

The battery creates an electric field in the circuit that causes charges to move. Positive charges move in the direction of the electric field, in the direction of decreasing potential.

When charges reach a **resistor**, their motion is impeded by collisions with atoms. The collisions increase the thermal energy of the resistor.

We use the **ideal-wire** model in which we assume that there is no resistance in the wires.

IMPORTANT CONCEPTS

Resistance, resistivity, and Ohm's law

The **resistivity** ρ is a property of a material, a measure of how good a conductor the material is.

- Good conductors have low resistivity.
- Poor conductors have high resistivity.

The **resistance** is a property of a particular wire or conductor. The resistance of a wire depends on its resistivity and dimensions.

$$R = \frac{\rho L}{A}$$

Cross-section area A Length L

Ohm's law describes the relationship between potential difference and current in a resistor.

$$I = \frac{\Delta V}{R}$$

Energy and power

The energy used by a circuit is supplied by the emf of the battery through a series of energy transformations:

$$E_{chem} \rightarrow U_{elec} \rightarrow K \rightarrow E_{th}$$

| Chemical energy in the battery | Potential energy of separated charges | Kinetic energy of moving charges | Thermal energy of atoms in the resistor |

The battery *supplies* power at the rate

$$P = I\mathcal{E}$$

The resistor *dissipates* power at the rate

$$P = I\,\Delta V_R = I^2 R = \frac{(\Delta V_R)^2}{R}$$

APPLICATIONS

Conducting materials

When a potential difference is applied to a wire, if the relationship between potential difference and current is linear, the material is **ohmic**.

The resistance is

$$R = \frac{1}{slope}$$

Resistors are made of ohmic materials and have a well-defined value of resistance.

$$R = \frac{\Delta V}{I}$$

If the variation is not linear, the material is **nonohmic.**

For instructor-assigned homework, go to www.masteringphysics.com

Problems labeled **INT** integrate significant material from earlier chapters; **BIO** are of biological or medical interest.

Problem difficulty is labeled as **I** (straightforward) to **IIII** (challenging).

QUESTIONS

Conceptual Questions

1. Two wires connect a lightbulb to a battery, completing a circuit and causing the bulb to glow. Do the simple observations and measurements that you can make on this circuit prove that something is *flowing* through the wires? If so, state the observations and/or measurements that are relevant and the steps by which you can then infer that something must be flowing. If not, can you offer an alternative hypothesis about why the bulb glows that is at least plausible and that could be tested?

2. Two wires connect a lightbulb to a battery, completing a circuit and causing the bulb to glow. Are the simple observations and measurements you can make on this circuit able to distinguish a current composed of positive charge carriers from a current composed of negative charge carriers? If so, describe how you can tell which it is. If not, why not?

3. Are the charge carriers in a current always electrons? If so, why is this the case? If not, describe a situation in which a current is due to some other charge carrier.

4. What *causes* electrons to move through a wire as a current?

5. A lightbulb is connected to a battery by two copper wires of equal lengths but different thicknesses. A thick wire connects one side of the lightbulb to the positive terminal of the battery and a thin wire connects the other side of the bulb to the negative terminal.
   a. Which wire carries a greater current? Or is the current the same in both? Explain.
   b. If the two wires are switched, will the bulb get brighter, dimmer, or stay the same? Explain.

6. All wires in Figure Q22.6 are made of the same material and have the same diameter. Rank in order, from largest to smallest, the currents I_1 to I_4. Explain.

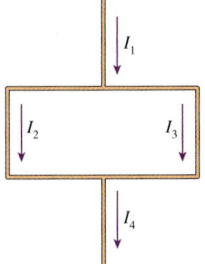

FIGURE Q22.6

7. A wire carries a 4 A current. What is the current in a second wire that delivers twice as much charge in half the time?

8. Metal 1 and metal 2 are each formed into 1-mm-diameter wires. The electric field needed to cause a 1 A current in metal 1 is larger than the electric field needed to cause a 1 A current in metal 2. Which metal has the larger resistivity? Explain.

9. A continuous metal wire connects the two ends of a 3 V battery with a rectangular loop, as shown in Figure Q22.9. The negative terminal of the battery has been chosen as the point where $V = 0$ V.
   a. Redraw this figure on your paper. Then locate and label the approximate points along the wire where $V = 3$ V, $V = 2$ V, and $V = 1$ V.
   b. Estimate the value of ΔV_{14}, the potential difference between points 1 and 4. Explain how you did so.
   c. In moving through the *wire* from point 2 to point 3, does the potential increase, decrease, or not change? If the potential changes, by how much?
   d. In moving through the *battery* from point 2 to point 3, does the potential increase, decrease, or not change? If the potential changes, by how much?
   e. In moving all the way around the loop in a clockwise direction, starting from point 2 and ending at point 2, is the net change in the potential positive, negative, or zero?

FIGURE Q22.9   **FIGURE Q22.10**

10. a. Which direction—clockwise or counterclockwise—does an electron travel through the wire in Figure Q22.10? Explain.
    b. Does an electron's electric potential energy increase, decrease, or stay the same as it moves through the wire? Explain.
    c. If you answered "decrease" in part b, where does the energy go? If you answered "increase" in part b, where does the energy come from?
    d. Which way—up or down—does an electron move through the *battery?* Explain.
    e. Does an electron's electric potential energy increase, decrease, or stay the same as it moves through the battery? Explain.
    f. If you answered "decrease" in part e, where does the energy go? If you answered "increase" in part e, where does the energy come from?

11. The wires in Figure Q22.11 are all made of the same material. Rank in order, from largest to smallest, the resistances R_1 to R_5 of these wires. Explain.

FIGURE Q22.11

12. The two circuits in Figure Q22.12 use identical batteries and wires of equal diameters. Rank in order, from largest to smallest, the currents I_1, I_2, I_3, and I_4 at points 1 to 4.

FIGURE Q22.12

13. The two circuits in Figure Q22.13 use identical batteries and wires of equal diameters. Rank in order, from largest to smallest, the currents I_1 to I_7 at points 1 to 7. Explain.

FIGURE Q22.13

14. Which, if any, of these statements are true? (More than one may be true.) Explain your choice or choices.
    a. A battery supplies the energy to a circuit.
    b. A battery is a source of potential difference. The potential difference between the terminals of the battery is always the same.
    c. A battery is a source of current. The current leaving the battery is always the same.

15. Rank in order, from largest to smallest, the currents I_1 to I_4 through the four resistors in Figure Q22.15. Explain.

FIGURE Q22.15   **FIGURE Q22.16**

16. The circuit in Figure Q22.16 has three batteries of emf $\mathcal{E}$ in series. Assuming the wires are ideal, sketch a graph of the potential as a function of distance traveled around the circuit, starting from $V = 0$ V at the negative terminal of the bottom battery. Note all important points on your graph.

17. Is a superconducting wire an ideal wire? Explain.

18. When lightning strikes the ground, it generates a large electric field along the surface of the ground directed toward the point of the strike. People near a lightning strike are often injured not by the lightning itself but by a large current that flows up one leg and down the other due to this electric field. To minimize this possibility, you are advised to stand with your feet close together if you are trapped outside during a lightning storm. Explain why this is beneficial.
    Hint: The current path through your body, up one leg and down the other, has a certain resistance. The larger the current along this path, the greater the damage.

19. One way to find out if a wire has corroded is to measure its resistance. Explain why the resistance of a wire increases if it becomes corroded.

20. Rank in order, from largest to smallest, the powers P_1 to P_4 dissipated by the four resistors in Figure Q22.20.

FIGURE Q22.20

21. We can model the rear window defroster in a car as a resistor that is connected to the car's 12 V battery. The defroster is made of a material whose resistance increases rapidly as the temperature increases. When the defroster is cold, its resistance is low; when the defroster is warm, its resistance is high. Why is it better to make a defroster with a material like this than with a material whose resistance is independent of temperature? Think about how the resistance, the current, and the power will change as the window warms.

Multiple-Choice Questions

22. | Lightbulbs are typically rated by their power dissipation when operated at a given voltage. Which of the following lightbulbs has the largest current through it when operated at the voltage for which it's rated?
    A. 0.8 W, 1.5 V
    B. 6 W, 3 V
    C. 4 W, 4.5 V
    D. 8 W, 6 V

23. ||| Lightbulbs are typically rated by their power dissipation when operated at a given voltage. Which of the following lightbulbs has the largest resistance when operated at the voltage for which it's rated?
    A. 0.8 W, 1.5 V
    B. 6 W, 3 V
    C. 4 W, 4.5 V
    D. 8 W, 6 V

24. | A copper wire is stretched so that its length increases and its diameter decreases. As a result,
    A. The wire's resistance decreases, but its resistivity stays the same.
    B. The wire's resistivity decreases, but its resistance stays the same.
    C. The wire's resistance increases, but its resistivity stays the same.
    D. The wire's resistivity increases, but its resistance stays the same.

25. | The potential difference across a length of wire is increased. Which of the following does *not* increase as well?
    A. The electric field in the wire.
    B. The power dissipated in the wire.
    C. The resistance of the wire.
    D. The current in the wire.

26. ‖ A stereo amplifier creates a 5.0 V potential difference across a speaker. To double the power dissipation of the speaker, the amplifier's potential difference must be increased to
    A. 7.1 V   B. 10 V   C. 14 V   D. 25 V

27. ‖ If a 1.5 V battery stores 5.0 kJ of energy, for how many minutes could it sustain a current of 1.2 A?
    A. 2.7   B. 6.9   C. 9.0   D. 46

28. | Figure Q22.28 shows a side view of a wire of varying circular cross section. Rank in order the currents flowing in the three sections.

FIGURE Q22.28

    A. $I_1 > I_2 > I_3$
    B. $I_2 > I_3 > I_1$
    C. $I_1 = I_2 = I_3$
    D. $I_1 > I_3 > I_2$

PROBLEMS

Section 22.1 A Model of Current

Section 22.2 Defining and Describing Current

1. ‖ The current in an electric hair dryer is 10 A. How much charge and how many electrons flow through the hair dryer in 5.0 min?

2. ‖ 2.0×10^{13} electrons flow through a transistor in 1.0 ms. What is the current through the transistor?

3. | A wire carries a 1.0 A current for 30 s. How many electrons move past a point in the wire?

4. | A wire carries a 15 μA current. How many electrons pass a given point on the wire in 1.0 s?

5. | The moving belt on a Van de Graaff generator carries charges to the dome on the top at a constant rate. After 30 s, the dome has accumulated a charge of 5.0 μC. What is the current carried by the belt to the dome?

6. | In a typical lightning strike, 2.5 C flows from cloud to ground in 0.20 ms. What is the current during the strike?

7. ‖ A capacitor is charged to 6.0×10^{-4} C, then discharged by connecting a wire between the two plates. 40 μs after the discharge begins, the capacitor still holds 13% of its original charge. What was the average current during the first 40 μs of the discharge?

8. | In an ionic solution, 5.0×10^{15} positive ions with charge $+2e$ pass to the right each second while 6.0×10^{15} negative ions with charge $-e$ pass to the left. What are the magnitude and direction of current in the solution?

9. | The starter motor of a car engine draws a current of 150 A from the battery. The copper wire to the motor is 5.0 mm in diameter and 1.2 m long. The starter motor runs for 0.80 s until the car engine starts. How much charge passes through the starter motor?

10. | A car battery is rated at 90 A · hr, meaning that it can supply a 90 A current for 1 hr before being completely discharged. If you leave your headlights on until the battery is completely dead, how much charge leaves the positive terminal of the battery?

11. ‖ What are the values of currents I_B and I_C in Figure P22.11? The directions are as noted.

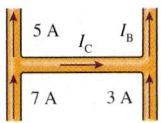

FIGURE P22.11

12. | The currents through several segments of a wire object are shown in Figure P22.12. What are the magnitudes and directions of the currents I_B and I_C in segments B and C?

FIGURE P22.12

Section 22.3 Batteries and EMF

13. | A battery supplies a steady 1.5 A current to a circuit. If the charges moving in the battery are positive ions with charge e, how many ions per second are transported from the negative terminal to the positive terminal?

14. | How much work is done to move 1.0 μC of charge from the negative terminal to the positive terminal of a 1.5 V battery?

15. | What is the emf of a battery that does 0.60 J of work to transfer 0.050 C of charge from the negative to the positive terminal?

16. ‖ A 9.0 V battery supplies a 2.5 mA current to a circuit for 5.0 hr.
    a. How much charge has been transferred from the negative to the positive terminal?
    b. How much work has been done on the charges that passed through the battery?

17. | An individual hydrogen-oxygen fuel cell has an output of 0.75 V. How many cells must be connected to drive a 24.0 V motor?

Section 22.4 Connecting Potential and Current

18. | A wire with resistance R is connected to the terminals of a 6.0 V battery. What is the potential difference ΔV_{ends} between the ends of the wire and the current I through it if the wire has the following resistances? (a) 1.0 Ω (b) 2.0 Ω (c) 3.0 Ω.

19. | Wires 1 and 2 are made of the same metal. Wire 2 has twice the length and twice the diameter of wire 1. What are the ratios (a) ρ_2/ρ_1 of the resistivities and (b) R_2/R_1 of the resistances of the two wires?

20. ‖ A wire has a resistance of 0.010 Ω. What will the wire's resistance be if it is stretched to twice its original length without changing the volume of the wire?

21. ‖ What is the resistance of
    a. A 1.0-m-long copper wire that is 0.50 mm in diameter?
    b. A 10-cm-long piece of iron with a 1.0 mm $\times$ 1.0 mm square cross section?

22. | A 10-m-long wire with a diameter of 0.80 mm has a resistance of 1.1 Ω. Of what material is the wire likely made?

23. | The femoral artery is the large artery that carries blood to
BIO the leg. A person's femoral artery has an inner diameter of
1.0 cm. What is the resistance of a 20-cm-long column of
blood in this artery?

24. ||| A 3.0 V potential difference is applied between the ends of
a 0.80-mm-diameter, 50-cm-long nichrome wire. What is the
current in the wire?

25. || A 1.0-mm-diameter, 20-cm-long copper wire carries a
3.0 A current. What is the potential difference between the
ends of the wire?

26. | a. How long must a 0.60-mm-diameter copper wire be to
carry a 0.50 A current when connected to the terminals
of a 1.5 V flashlight battery?
b. What is the current if the wire is half this length?

27. || The terminals of a 0.70 V watch battery are connected by a
100-m-long copper wire with a diameter of 0.10 mm. What is
the current in the wire?

Section 22.5 Resistors and Ohm's Law

28. | Figure P22.28 is a current-versus-potential-difference
graph for a cylinder. What is the cylinder's resistance?

FIGURE P22.28

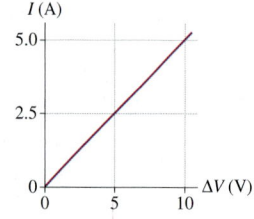
FIGURE P22.29

29. | Figure Q22.29 shows the current-versus-potential-
difference graph for a resistor.
a. What is the resistance of this resistor?
b. Suppose the length of the resistor is doubled while keeping
its cross section the same. (This requires doubling the
amount of material the resistor is made of.) Copy the figure and add to it the current-versus-potential-difference
graph for the longer resistor.

30. | The electric field inside a 30-cm-long copper wire is
0.010 V/m. What is the potential difference between the ends
of the wire?

31. | In Example 22.4 the length of a 60 W, 240 Ω lightbulb fila-
ment was calculated to be 60 cm.
a. If the potential difference across the filament is 120 V,
what is the strength of the electric field inside the filament?
b. Suppose the length of the bulb's filament were doubled
without changing its diameter or the potential difference
across it. What would the electric field strength be in this
case?
c. Remembering that the current in the filament is proportional to the electric field, what is the current in the filament following the doubling of its length?
d. What is the resistance of the filament following the doubling of its length?

Section 22.6 Energy and Power

32. | a. What is the resistance of a 1500 W (120 V) hair dryer?
b. What is the current in the hair dryer when it is used?

33. || You've brought your 1000 W (120 V) hair dryer on vaca-
tion to Europe, where the standard outlet voltages are 220 V.
Assuming the hair dryer can operate safely at the higher voltage, can you actually use it if the outlet is rated at 15 A, or will
it draw more current than this?

34. | A 60-cm-long heating wire is connected to a 120 V outlet.
If the wire dissipates 45 W, what are (a) the current in and
(b) the resistance of the wire?

35. || The total charge a household battery can supply is given in
units of mA · hr. For example, a 9.0 V alkaline battery is rated
450 mA · hr, meaning that such a battery could supply a 1 mA
current for 450 hr, a 2 mA current for 225 hr, etc. How much
energy, in joules, is this battery capable of supplying?

General Problems

36. | A 3.0 V battery powers a flashlight bulb that has a resis-
tance of 6.0 Ω. How much charge moves through the battery
in 10 min?

37. || A sculptor has asked you to help electroplate gold onto a
brass statue. You know that the charge carriers in the ionic
solution are monovalent (charge e) gold ions, and you've calculated that you must deposit 0.50 g of gold to reach the necessary thickness. How much current do you need, in mA, to
plate the statue in 3.0 hr?

38. |||| For a science experiment you need to electroplate a
INT 100-nm-thick zinc coating onto both sides of a very thin,
2.0 cm × 2.0 cm copper sheet. You know that the charge carriers in the ionic solution are divalent (charge $2e$) zinc ions.
The density of zinc is 7140 kg/m^3. If the electroplating apparatus operates at 1.0 mA, how long will it take the zinc to
reach the desired thickness?

39. || The hot dog cooker described in the chapter heats hot dogs
INT by connecting them to 120 V household electricity. A typical
hot dog has a mass of 60 g and a resistance of 150 Ω. How
long will it take for the cooker to raise the temperature of the
hot dog from 20°C to 80°C? The specific heat of a hot dog is
approximately 2500 J/kg · K.

40. | The biochemistry that takes place inside cells depends on
BIO various elements, such as sodium, potassium, and calcium,
that are dissolved in water as ions. These ions enter cells
through narrow pores in the cell membrane known as *ion
channels*. Each ion channel, which is formed from a specialized protein molecule, is selective for one type of ion.
Measurements with microelectrodes have shown that a
0.30-nm-diameter potassium ion (K$^+$) channel carries a current of 1.8 pA. How many potassium ions pass through if
the ion channel opens for 1.0 ms?

41. | High-resolution measurements have shown that an ion
BIO channel (see Problem 40) is a 0.30-nm-diameter cylinder with
length of 5.0 nm. The intracellular fluid filling the ion channel
has resistivity 0.60 Ω · m. What is the resistance of the ion
channel?

42. | When an ion channel opens in a cell wall (see Problem 40),
BIO monovalent (charge e) ions flow through the channel at a rate
of 1.0 × 10^7 ions/s.
a. What is the current through the channel?
b. The potential difference across the ion channel is 70 mV.
What is the power dissipation in the channel?

43. ‖ The total charge a battery can supply is rated in mA · hr. For instance, a battery rated at 1000 mA · hr could supply a 1000 mA current for 1.0 hr, a 500 mA current for 2.0 hr, etc. A rechargeable 1.2 V battery stores 7800 J of energy when fully charged. What is the mA · hr rating of this battery?

44. | The heating element of a simple heater consists of a 2.0-m-long, 0.60-mm-diameter nichrome wire. When plugged into a 120 V outlet, the heater draws 8.0 A of current when hot.
    a. What is the wire's resistance when it is hot?
    b. Use your answer to part a to calculate the resistivity of nichrome. Why is it not the same as the value of ρ given for nichrome in Table 22.1?

45. | Variations in the resistivity of blood can give valuable clues
BIO to changes in the blood's viscosity and other properties. The resistivity is measured by applying a small potential difference and measuring the current. Suppose a medical device attaches electrodes into a 1.5-mm-diameter vein at two points 5.0 cm apart. What is the blood resistivity if a 9.0 V potential difference causes a 230 μA current through the blood in the vein?

46. | The resistance of a very fine tungsten wire with a 10 μm × 10 μm square cross section is 1000 Ω.
    a. How long is the wire?
    b. A 1000 Ω resistor is made by wrapping this wire in a spiral around a 3.0-mm-diameter glass core. How many turns of wire are needed?

47. ‖ Wires aren't really ideal. The voltage drop across a current-carrying wire can be significant unless the resistance of the wire is quite low. Suppose a 50 ft extension cord is being used to provide power to an electric lawn mower. The cord carries a 10 A current. The copper wire in a typical extension cord has a 1.3 mm diameter. What is the voltage drop across a 50 ft length of wire at this current?

48. ‖ When the starter motor on a car is engaged, there is a 300 A current in the wires between the battery and the motor. Suppose the wires are made of copper and have a total length of 1.0 m. What minimum diameter can the wires have if the voltage drop along the wires is to be less than 0.50 V?

49. | The electron beam inside a television picture tube is
INT 0.40 mm in diameter and carries a current of 50 μA. This electron beam impinges on the inside of the picture tube screen.
    a. How many electrons strike the screen each second?
    b. The electrons move with a velocity of 4.0×10^7 m/s. What electric field strength is needed to accelerate electrons from rest to this velocity in a distance of 5.0 mm?
    c. Each electron transfers its kinetic energy to the picture tube screen upon impact. What is the *power* delivered to the screen by the electron beam?
    Hint: What potential difference produced the field that accelerated electrons? This is an emf.

50. | The two segments of the wire in Figure P22.50 have equal diameters and equal lengths but different resistivities ρ_1 and ρ_2. Current I passes through this wire. If the resistivities have the ratio $\rho_2/\rho_1 = 2$, what is the ratio $\Delta V_1/\Delta V_2$ of the potential differences across the two segments of the wire?

FIGURE P22.50

51. ‖ A 15-cm-long nichrome wire is connected across the terminals of a 1.5 V battery. If the current in the wire is 2.0 A, what is the wire's diameter?

52. | A 1.0-m-long copper wire has a diameter of 0.23 mm. How long would a nichrome wire with 0.50 mm diameter need to be in order to have the same resistance as the copper wire?

53. | A wire is 2.3 m long and has a diameter of 0.38 mm. When connected to a 1.2 V battery, there is a current of 0.61 A. What material is the wire likely made of?

54. ‖ A 20-cm-long hollow nichrome tube of inner diameter 2.8 mm, outer diameter 3.0 mm is connected to a 3.0 V battery. What is the current in the tube?

55. ‖ The filament of a 100 W (120 V) lightbulb is a tungsten wire 0.035 mm in diameter. At the filament's operating temperature, the resistivity is 5.0×10^{-7} Ω · m. How long is the filament?

56. ‖ You've made the finals of the Science Olympics! As one of
INT your tasks, you're given 1.0 g of copper and asked to make a wire, using all the metal, with a resistance of 1.0 Ω. Copper has a density of 8900 kg/m³. What length and diameter will you choose for your wire?

57. | Not too long ago houses were protected from excessive currents by fuses rather than circuit breakers. Sometimes a fuse blew out and a replacement wasn't at hand. Because a copper penny happens to have almost the same diameter as a fuse, some people replaced the fuse with a penny. Unfortunately, a penny never blows out, no matter how large the current, and the use of pennies in fuse boxes caused many house fires. Make the appropriate measurements on a penny, then calculate the resistance between the two faces of a solid-copper penny.

58. ‖ An immersion heater used to boil water for a single cup of
INT tea plugs into a 120 V outlet and is rated at 300 W.
    a. What is the resistance of the heater?
    b. Suppose your super-size, super-insulated tea mug contains 400 g of water at a temperature of 18°C. How long will this heater take to bring the water to a boil? You can ignore the energy needed to raise the temperature of the mug and the heater itself.

59. ‖ The graph in Figure P22.59 shows the current through a 1.0 Ω resistor as a function of time.
    a. How much charge flowed through the resistor during the 10 s interval shown?
    b. What was the total energy dissipated by the resistor during this time?

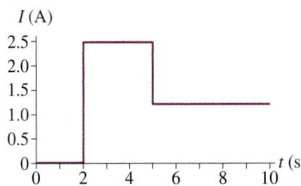

FIGURE P22.59

60. ‖‖ A toy car has a mass of 120 g, including two 1.5 V batteries
INT connected in series. Suppose the motor that drives the car is 80% efficient (i.e., 80% of the electric energy goes to the drive wheels, 20% is dissipated as heat) and that friction and air resistance are negligible. How much charge passes through the batteries during the time it takes the car to accelerate from rest to 1.5 m/s?

61. | If you touch the two terminals of a power supply with your
BIO two fingertips on opposite hands, the potential difference will
produce a current through your torso. The maximum safe current is approximately 5 mA.
   a. If your hands are completely dry, the resistance of your
      body from fingertip to fingertip is approximately 500 kΩ.
      If you accidentally touch both terminals of your 120 V
      household electricity supply with dry fingers, will you
      receive a dangerous shock?
   b. If your hands are moist, your resistance drops to approximately 1 kΩ. If you accidentally touch both terminals of
      your 120 V household supply with moist fingers, will you
      receive a dangerous shock?
   c. For the situation of part a, what electric power is dissipated
      in your body? Would you notice the heating?

62. | The average resistivity of the human body (apart from sur-
BIO face resistance of the skin) is about $5.0 \ \Omega \cdot m$. The conducting
path between the hands can be approximated as a cylinder
1.6 m long and 0.10 m in diameter. The skin resistance can
be made negligible by soaking the hands in salt water.
   a. What is the resistance between the hands if the skin resistance is negligible?
   b. If skin resistance is negligible, what potential difference
      between the hands is needed for a lethal shock current of
      100 mA? Your result shows that even small potential differences can produce dangerous currents when skin is damp.
   c. What power is dissipated in the body by the current in b?

Passage Problems   Lightbulb Failure

You've probably observed that the most common time for an incandescent lightbulb to fail is the moment when it is turned on. Let's
look at the properties of the bulb's filament to see why this happens.

The current in the tungsten filament of a lightbulb heats the filament until it glows. The filament is so hot that some of the atoms
on its surface fly off and end up sticking on a cooler part of the
bulb. Thus the filament gets progressively thinner as the bulb ages.
There will certainly be one spot on the filament that is a bit thinner
than elsewhere. This thin segment will have a higher resistance
than the surrounding filament. More power will be dissipated at
this spot, so it won't only be a thin spot, it also will be a hot spot.

Now, let's look at the resistance of the filament. The graph in
Figure P22.63 shows data for the current in a lightbulb as a function of the potential difference
across it. The graph is not linear, so the filament is not an
ohmic material with a constant
resistance. However, we can
define the resistance at any particular voltage ΔV to be $R = \Delta V/I$. This ratio, and hence the
resistance, increases with ΔV
and thus with temperature.

When the bulb is turned on,

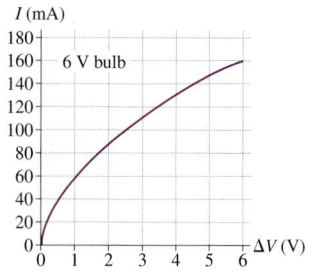

FIGURE P22.63

the filament is cold and its resistance is much lower than during
normal, high-temperature operation. The low resistance causes a
surge of higher-than-normal current lasting a fraction of a second
until the filament heats up. Because power dissipation is I^2R, the
power dissipated during this first fraction of a second is much
larger than the bulb's rated power. This current surge concentrates
the power dissipation at the high-resistance thin spot, perhaps melting it and breaking the filament.

63. | For the bulb in Figure P22.63, what is the approximate
    resistance of the bulb at a potential difference of 6.0 V?
    A. 7.0 Ω   B. 17 Ω   C. 27 Ω   D. 37 Ω
64. | As the bulb ages, the resistance of the filament
    A. Increases.   B. Decreases.   C. Stays the same.
65. | Which of the curves in Figure P22.65 best represents the
    expected variation in current as a function of time in the short
    time interval immediately after the bulb is turned on?
66. | There are devices to put in a light socket that control the
    current through a lightbulb, thereby increasing its lifetime.

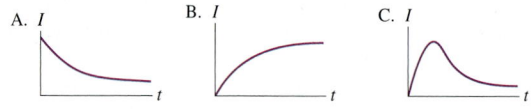

FIGURE P22.65

Which of the following strategies would increase the lifetime
of a bulb without making it dimmer?
   A. Reducing the average current through the bulb.
   B. Limiting the maximum current through the bulb.
   C. Increasing the average current through the bulb.
   D. Limiting the minimum current through the bulb.

STOP TO THINK ANSWERS

Stop to Think 22.1: C. Current is conserved. The connecting wires
and the wires in the lightbulbs form a single current path.

Stop to Think 22.2: A. From Kirchhoff's junction law the current
through bulb 1 is the sum of the currents through bulbs 2 and 3.

Stop to Think 22.3: C. Charge flows out one terminal of the battery
but back into the other; the amount of charge does not change. The
emf is determined by the chemical reactions in the battery and is
constant. But the chemical energy in the battery steadily decreases
as the battery converts it to the potential energy of charges.

Stop to Think 22.4: $I_A = I_B = I_C = I_D$. The current in each wire
is $I = \Delta V/R$. All the wires have the same resistance, because they
are identical, and they all have the same potential difference,
because the battery is a *source of potential*. Therefore all wires
have the same current. Conservation of current requires $I_A = I_B$.

Stop to Think 22.5: $V_B = V_C > V_A > V_D = V_E$. There's no potential difference along ideal wires, so $V_B = V_C$ and $V_D = V_E$. Potential increases in going from the $-$ to the $+$ terminal of a battery, so
$V_A > V_E$ and $V_B > V_A$. These imply $V_C > V_D$, which was expected
because potential decreases as current passes through a resistor.

Stop to Think 22.6: $P_B > P_D > P_A > P_C$. The power dissipated
by a resistor is $P_R = (\Delta V_R)^2/R$. Increasing R decreases P_R; increasing ΔV_R increases P_R. But changing the potential has a larger effect
because P_R depends on the square of ΔV_R.

23

CIRCUITS

The electric eel isn't really an eel; it's a fish. But it is electric, producing pulses of up to 600 V that it uses to stun prey. How does the fish produce such a large potential difference?

Electricity is part of your life. You use a variety of electric devices each day, from lamps to telephones to computers, in which electrical signals process information and deliver energy. But you also use electricity in a more fundamental way. The signals in your body's nervous system are electrical signals, and the same physics that explains the timing of intermittent windshield wipers in your car also explains how signals get from your brain to your muscles. In some animals, such as the electric eel, the cells that produce electrical signals form circuits that create large voltages and currents.

The physics behind the operation of circuits is the physics of charges, forces, and fields—topics we have explored in detail in the previous chapters. This will be a very practical chapter, one in which we apply our knowledge to the analysis of circuits involving batteries, resistors and capacitors. We will look at many practical applications, culminating in a description of electrical conduction in the nervous system, a topic that will bring together many of the concepts we have explored in this part of the book.

23.1 Circuit Elements and Diagrams

In Chapter 22 we analyzed a very simple circuit, a resistor connected to a battery. In this chapter, we will explore more complex circuits involving more and different elements. As was the case with other topics in the book, we will learn a good deal about circuits by making appropriate drawings. We will begin our investigation by developing a system for representing circuits symbolically in a manner that highlights their essential features.

Figure 23.1 shows an electric circuit in which a resistor and a capacitor are connected by wires to a battery. To understand the functioning of this circuit, we do not need to know whether the wires are bent or straight, or whether the battery is to the right or to the left of the resistor. The literal picture of Figure 23.1 provides many irrelevant details. It is customary when describing or analyzing circuits to use a more abstract picture called a **circuit diagram.** A circuit diagram is a *logical* picture of what is connected to what. The actual circuit, once it is built, may *look* quite different from the circuit diagram, but it will have the same logic and connections.

In a circuit diagram we replace pictures of the circuit elements with symbols. Figure 23.2 shows the basic symbols that we will need. Notice that the *longer* line at one end of the battery symbol represents the positive terminal of the battery.

FIGURE 23.1 An electric circuit.

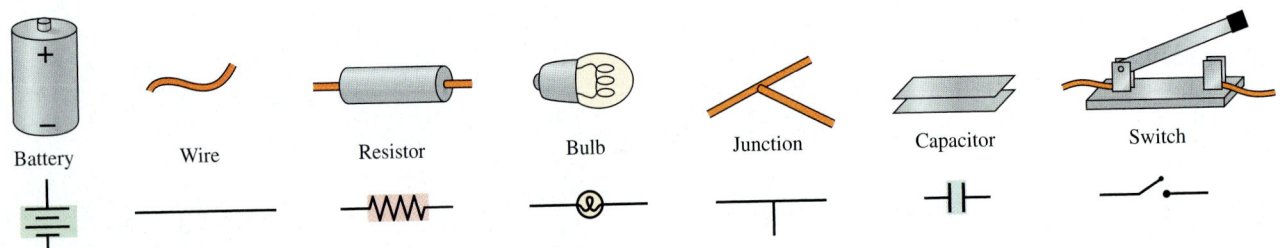

FIGURE 23.2 A library of basic symbols used for electric circuit drawings.

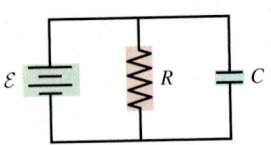

FIGURE 23.3 A circuit diagram for the circuit of Figure 23.1.

Figure 23.3 is a circuit diagram of the circuit shown in Figure 23.1. Notice how circuit elements are labeled. The battery's emf $\mathcal{E}$ is shown beside the battery, and the resistance R of the resistor and capacitance C of the capacitor are written beside them. We would use numerical values for $\mathcal{E}$, R and C if we knew them. The wires, which in practice may bend and curve, are shown as straight-line connections between the circuit elements. The positive potential of the battery is at the top of the diagram; in general, we try to put higher potentials toward the top. You should get into the habit of drawing your own circuit diagrams in a similar fashion.

STOP TO THINK 23.1 Which of these diagrams represent the same circuit?

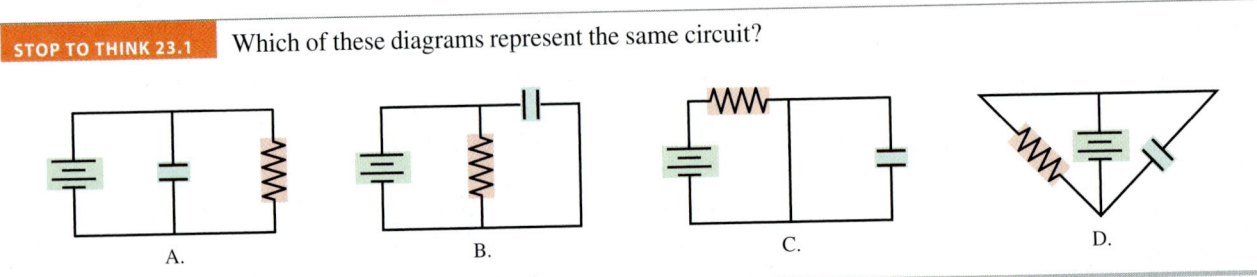

23.2 Kirchhoff's Laws

Once we have a diagram for a circuit, we can analyze it. In this chapter, a full analysis of a circuit means that we find:

1. The potential difference across each circuit component.
2. The current through each circuit component.

Our tools and techniques for analyzing circuits will be based on the physical principles of potential differences and currents.

You learned in Chapter 22 that, as a result of charge and current conservation, the total current into a junction must equal the total current leaving the junction, as in Figure 23.4a This result was called *Kirchhoff's junction law,* which we wrote as follows:

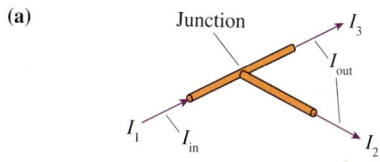

(a)

Junction law: $I_1 = I_2 + I_3$

$$\sum I_{in} = \sum I_{out} \qquad (23.1)$$

Kirchhoff's junction law

Kirchhoff's junction law is one of the two basic tools of circuit analysis. The second tool deals with electric potential. When we learned about gravitational potential energy in Chapter 10, we saw that the gravitational potential energy of an object depends on its position, not on the path it took to get to that position. The same is true of electric potential energy, as we saw in Chapter 21. If a particle moves around a closed loop and returns to its starting point, there will be no net change in electric potential energy, $\Delta U_{elec} = 0$. Because $V = U_{elec}/q$, **the net change in the electric potential around any loop or closed path must be zero** as well.

We can apply this idea to circuits, as shown in Figure 23.4b. If we add all of the potential differences *around* the loop formed by the circuit, the sum of these potential differences must be zero. This result is known as **Kirchhoff's loop law:**

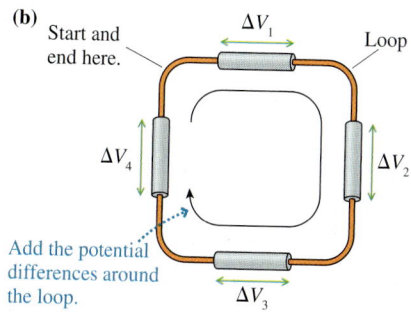

(b) Start and end here.

Add the potential differences around the loop.

Loop law: $\Delta V_1 + \Delta V_2 + \Delta V_3 + \Delta V_4 = 0$

FIGURE 23.4 Kirchhoff's laws apply to junctions and loops in circuits.

$$\Delta V_{loop} = \sum_i \Delta V_i = 0 \qquad (23.2)$$

Kirchhoff's loop law

In Equation 23.2, ΔV_i is the potential difference of the ith component in the loop.

Kirchhoff's loop law can be true only if at least one of the potential differences ΔV_i is negative. To apply the loop law, we need to explicitly identify which potential differences are positive and which are negative.

TACTICS BOX 23.1 Using Kirchhoff's loop law   📝 Exercises 5,6

❶ **Draw a circuit diagram.** Label all known and unknown quantities.
❷ **Assign a direction to the current.** Draw and label a current arrow I to show your choice.
- If you know the actual current direction, choose that direction.
- If you don't know the actual current direction, make an educated guess. All that will happen if you choose wrong is that your value for I will end up negative.

Continued

❸ **"Travel" around the loop.** Start at any point in the circuit, then go all the way around the loop in the direction you assigned to the current in step 2. As you go through each circuit element, ΔV is interpreted to mean

$$\Delta V = V_{downstream} - V_{upstream}$$

■ For a battery with current in the negative-to-positive direction:

$$\Delta V_{bat} = +\mathcal{E}$$

Potential increases

■ For a battery in the positive-to-negative direction (i.e., the current is going into the positive terminal of the battery):

$$\Delta V_{bat} = -\mathcal{E}$$

Potential decreases

■ For a resistor:

$$\Delta V_R = -IR$$

Potential decreases

❹ **Apply the loop law:**

$$\sum \Delta V_i = 0$$

ΔV_{bat} can be positive or negative for a battery, but ΔV_R for a resistor is always negative because the potential in a resistor *decreases* along the direction of the current—charge flows "downhill," as we saw in Chapter 22. Because the potential across a resistor always decreases, we often speak of the *voltage drop* across the resistor.

NOTE ▶ The equation for ΔV_R in Tactics Box 23.1 seems to be the opposite of Ohm's law, but Ohm's law was only concerned with the *magnitude* of the potential difference. Kirchhoff's law requires us to recognize that the electric potential inside a resistor *decreases* in the direction of the current. ◀

The most basic electric circuit is that of a single resistor connected to the two terminals of a battery, as in Figure 23.5. We considered this circuit in Chapter 22, but let's now apply Kirchhoff's laws to its analysis.

This circuit of Figure 23.5 has no junctions, so the current is the same in all parts of the circuit. Kirchhoff's junction law is not needed. Kirchhoff's loop law is the tool we need to analyze this circuit, and Figure 23.6 shows the first three steps of Tactics Box 23.1. Notice that we're assuming the ideal-wire model in which there are no potential differences along the connecting wire. The fourth step is to apply Kirchhoff's loop law, $\sum \Delta V_i = 0$:

$$\Delta V_{loop} = \sum_i \Delta V_i = \Delta V_{bat} + \Delta V_R = 0 \tag{23.3}$$

Let's look at each of the two terms in Equation 23.3:

1. The potential *increases* as we travel through the battery on our clockwise journey around the loop, as we see in the conventions in Tactics Box 23.1. We enter the negative terminal and, farther downstream, exit the positive terminal after having gained potential $\mathcal{E}$. Thus

$$\Delta V_{bat} = +\mathcal{E}$$

2. The *magnitude* of the potential difference across the resistor is $\Delta V = IR$, but Ohm's law does not tell us whether this should be positive or negative—and the difference is crucial. The potential of a resistor

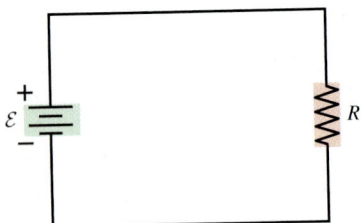

FIGURE 23.5 The basic circuit of a resistor connected to a battery.

FIGURE 23.6 Analysis of the basic circuit using Kirchhoff's loop law.

decreases in the direction of the current, which we've indicated with the $+$ and $-$ signs in Figure 23.6. Thus

$$\Delta V_R = -IR$$

NOTE ▶ Determining which potential differences are positive and which negative is perhaps the most important step in circuit analysis. ◀

With this information about ΔV_{bat} and ΔV_R, the loop equation becomes

$$\mathcal{E} - IR = 0 \qquad (23.4)$$

We can solve the loop equation to find that the current in the circuit is

$$I = \frac{\mathcal{E}}{R} \qquad (23.5)$$

This is exactly the result we saw in the previous chapter. Notice again that the current in the circuit depends on the size of the resistance. The emf of a battery is a fixed quantity; the current that the battery delivers depends jointly on the emf and on the resistance.

EXAMPLE 23.1 Analyzing a circuit with two batteries

What is the current in the circuit of Figure 23.7? What is the potential difference across each resistor?

FIGURE 23.7 Circuit for Example 23.1.

PREPARE We will solve this circuit using Kirchhoff's loop law, as outlined in Tactics Box 23.1. But how do we deal with *two* batteries? What happens when charge flows "backward" through a battery, from positive to negative? Consider the charge-escalator analogy. Left to itself, a charge escalator lifts charge from lower to higher potential. But it *is* possible to run down an up escalator, as many of you have probably done. If two escalators are placed "head to head," whichever is "stronger" will, indeed, force the charge to run down the up

FIGURE 23.8 Analyzing the circuit of Example 23.1.

escalator of the other battery. The current in a battery *can* be from positive to negative if driven in that direction by a larger emf from a second battery. Indeed, this is how batteries are "recharged," as we saw in Chapter 22. The current goes in the direction that the larger emf—the 9.0 V battery—"wants" it to go. We have redrawn the circuit in Figure 23.8, showing the direction of the current and the direction of the potential difference for each circuit element.

SOLVE Kirchhoff's loop law requires us to add the potential differences as we travel around the circuit in the direction of the current. Let's do this starting at the negative terminal of the 9.0 V battery:

$$\sum_i \Delta V_i = +9.0\ \text{V} - I(40\ \Omega) - 6.0\ \text{V} - I(20\ \Omega) = 0$$

The 6.0 V battery has $\Delta V_{bat} = -\mathcal{E}$, in accord with Tactics Box 23.1, because the potential decreases as we travel through this battery in the positive-to-negative direction. We can solve this equation for the current:

$$I = \frac{3.0\ \text{V}}{60\ \Omega} = 0.050\ \text{A} = 50\ \text{mA}$$

Now that the current is known, we can use Ohm's $\Delta V = IR$ law to find the potential difference across each resistor. For the 40 Ω resistor

$$\Delta V_1 = (0.050\ \text{A})(40\ \Omega) = 2.0\ \text{V}$$

and for the 20 Ω resistor

$$\Delta V_2 = (0.050\ \text{A})(20\ \Omega) = 1.0\ \text{V}$$

ASSESS Let's look at the loop law again as a check on our result. As a charge travels around the circuit, the potential increases by 9.0 V in the first battery, then decreases by 2.0 V in the first resistor, decreases by 6.0 V in the second battery, and decreases by 1.0 V in the second resistor. The total decrease is 9.0 V, so the charge returns to its starting potential, as it must.

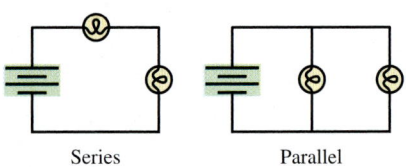

FIGURE 23.9 Series and parallel circuits.

23.3 Series and Parallel Circuits

Example 23.1 involved a circuit with multiple elements—two batteries and two resistors. As we introduce more circuit elements, we have possibilities for different types of connections. Suppose you use a single battery to light two lightbulbs. There are two possible ways that you can connect the circuit, as shown in Figure 23.9. These *series* and *parallel* circuits have very different properties. We will consider these two cases in turn.

We say two bulbs are connected in **series** if they are connected directly to each other with no junction in between. All series circuits share certain characteristics, as we will see.

CONCEPTUAL EXAMPLE 23.1 Brightness of bulbs in series

Figure 23.10 shows two identical lightbulbs connected in series. Which bulb is brighter, A or B? Or are they equally bright?

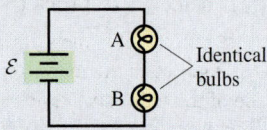

FIGURE 23.10 Two bulbs in series.

REASON Current is conserved, and there are no junctions in the circuit. Thus the current is the same at all points, as we see in Figure 23.11.

We learned in Chapter 22 that the power dissipated by a resistor is $P = I^2R$. If the two bulbs are identical (i.e., the same resistance) and have the same current through them, the power

FIGURE 23.11 The current in the series circuit of Conceptual Example 23.1.

dissipated by each bulb is the same. This means that the brightness of the bulbs must be the same. The voltage across each of the bulbs will be the same as well, as $\Delta V = IR$.

ASSESS It's perhaps tempting to think that bulb A will be brighter than bulb B, thinking that something is "used up" before the current gets to bulb B. It is true that *energy* is being transformed in each bulb, but current must be conserved and so both bulbs dissipate energy at the same rate.

Series Resistors

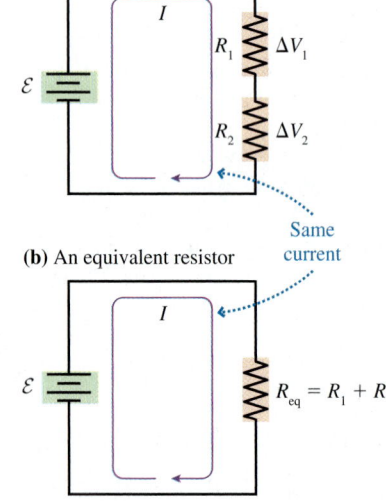

FIGURE 23.12 Replacing two series resistors with an equivalent resistor.

Figure 23.12a shows two resistors in series connected to a battery. Because there are no junctions, the current I must be the same in both resistors.

We can use Kirchhoff's loop law to look at the potential differences. Starting at the battery's negative terminal and following the current clockwise around the circuit, we find:

$$\sum_i \Delta V_i = \mathcal{E} + \Delta V_1 + \Delta V_2 = 0 \tag{23.6}$$

The voltage drops across the two resistors, in the direction of the current, are $\Delta V_1 = -IR_1$ and $\Delta V_2 = -IR_2$, so we can use Equation 23.6 to find the current in the circuit:

$$\mathcal{E} = -\Delta V_1 - \Delta V_2 = IR_1 + IR_2$$
$$I = \frac{\mathcal{E}}{R_1 + R_2} \tag{23.7}$$

Suppose, as in Figure 23.12b, we replace the two resistors with a single resistor having the value $R_{eq} = R_1 + R_2$. The total potential difference across this resistor is still $\mathcal{E}$ because the potential difference is established by the battery. Further, the current in this single-resistor circuit is

$$I = \frac{\mathcal{E}}{R_{eq}} = \frac{\mathcal{E}}{R_1 + R_2}$$

which is the same as it had been in the two-resistor circuit. In other words, this single resistor is *equivalent* to the two series resistors in the sense that the circuit's current and potential difference are unchanged. Nothing anywhere else in the circuit would differ if we took out resistors R_1 and R_2 and replaced them with resistor R_{eq}.

We can extend this analysis to a case with more resistors. If we have N resistors in series, their **equivalent resistance** is the sum of the N individual resistances:

$$R_{eq} = R_1 + R_2 + \cdots + R_N \qquad (23.8)$$

Equivalent resistance of N series resistors

The behavior of the circuit will be unchanged if the N series resistors are replaced by the single resistor R_{eq}.

EXAMPLE 23.2 Analyzing a series resistor circuit

What is the current in the circuit of Figure 23.13?

PREPARE The three resistors are in series, so we can replace them with a single equivalent resistor as shown in Figure 23.14.

SOLVE The equivalent resistance is calculated using Equation 23.8:

$$R_{eq} = 25\ \Omega + 31\ \Omega + 19\ \Omega = 75\ \Omega$$

The current in the equivalent circuit of Figure 23.14 is:

$$I = \frac{\mathcal{E}}{R_{eq}} = \frac{9.0\ \text{V}}{75\ \Omega} = 0.12\ \text{A}$$

This is also the current in the original circuit.

ASSESS The current in the circuit is the same whether there are three resistors or a single equivalent resistor. The equivalent resistance is the sum of the individual resistance values, and so it is always greater than any of the individual values. This is a good check on your work.

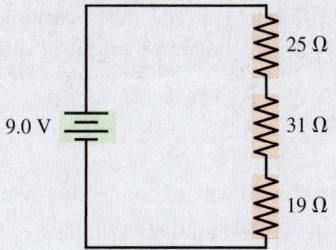

FIGURE 23.13 Circuit for Example 23.2.

FIGURE 23.14 Analyzing a circuit with series resistors.

EXAMPLE 23.3 Potential difference of Christmas-tree minilights

A string of Christmas-tree minilights consists of 50 bulbs wired in series. What is the potential difference across each bulb when the string is plugged into a 120 V outlet?

PREPARE Figure 23.15 shows the minilight circuit, which has 50 bulbs in series. The current in each of the bulbs is the same because they are in series.

FIGURE 23.15 50 bulbs connected in series.

SOLVE Applying Kirchhoff's loop law around the circuit, we find:

$$\mathcal{E} = \Delta V_1 + \Delta V_2 + \cdots + \Delta V_{50}$$

The bulbs are all identical and, because the current in the bulbs is the same, all of the bulbs have the same potential difference. The potential difference across a single bulb is thus

$$\Delta V_1 = \frac{\mathcal{E}}{50} = \frac{120\ \text{V}}{50} = 2.4\ \text{V}$$

ASSESS Minilights are wired in series because the wiring is simpler and the bulbs can be inexpensive low-voltage bulbs. But there is a drawback that is true of all series circuits: If one bulb is removed, there is no longer a complete circuit, and there will be no current. Indeed, if you remove a bulb from a string of minilights, the entire string will go dark.

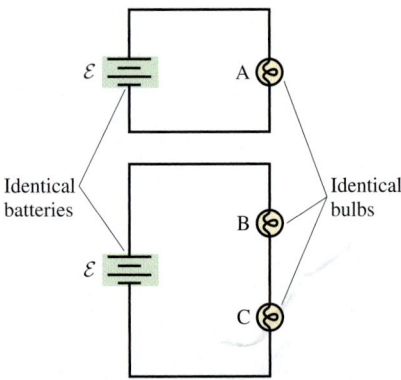

FIGURE 23.16 How does the brightness of bulb B compare to that of bulb A?

Broken filament

Shunt in bulb base

A seasonal series circuit puzzle

Christmas-tree minilights are connected in series. This is easy to verify: When you remove one bulb from a string of lights, the circuit is not complete, and the entire string of lights goes out. But when one bulb *burns* out, meaning its filament has broken, the string of lights stays lit. How is this possible? The secret is a *shunt* in the base of the bulb. Initially, the shunt is a good insulator. But if the filament breaks, the shunt is activated, and its resistance drops. The shunt can now carry the current, so the other bulbs will stay lit.

Let's use our knowledge of series circuits to look at another lightbulb puzzle. Figure 23.16 shows two different circuits, one with one battery and one lightbulb and a second with one battery and two lightbulbs. All of the batteries and bulbs are identical. You now know that B and C, which are connected in series, are equally bright, but how does the brightness of B compare to that of A?

Suppose the resistance of each identical lightbulb is R. In the first circuit, the battery drives current $I_A = \mathcal{E}/R$ through bulb A. In the second circuit, bulbs B and C are in series, with an equivalent resistance $R_{eq} = R_A + R_B = 2R$, but the battery has the same emf $\mathcal{E}$. Thus the current through bulbs B and C is $I_{B+C} = \mathcal{E}/R_{eq} = \mathcal{E}/2R = \frac{1}{2}I_A$. Bulb B has only half the current of bulb A, so B is dimmer.

Many people predict that A and B should be equally bright. It's the same battery, so shouldn't it provide the same current to both circuits? No! **A battery is a source of potential difference, *not* a source of current.** In other words, the battery's emf is the same no matter how the battery is used. When you buy a 1.5 V battery you're buying a device that provides a specified amount of potential difference, not a specified amount of current. The battery does provide the current to the circuit, but the *amount* of current depends on the resistance. Your 1.5 V battery causes 1 A to pass through a 1.5 Ω resistor but only 0.1 A to pass through a 15 Ω resistor.

As an analogy, think about a water faucet. The pressure in the water main underneath the street is a fixed quantity set by the water company. It's like the emf of a battery. But the amount of water coming out of a faucet depends on how far you open it. A faucet opened slightly has a "high resistance," so only a little water flows. A wide-open faucet has a "low resistance," and the water flow is large.

This is a critical idea for understanding circuits. A battery provides a fixed emf (potential difference). It does *not* provide a fixed and unvarying current. **The amount of current depends jointly on the battery's emf *and* the resistance of the circuit attached to the battery.**

Parallel Resistors

In the next example, we consider the second way of connecting two bulbs in a circuit. The two bulbs in Figure 23.17 are connected at *both* ends. We say that they are connected in **parallel.**

CONCEPTUAL EXAMPLE 23.2 **Brightness of bulbs in parallel**

Which lightbulb in the circuit of Figure 23.17 is brighter, A or B? Or are they equally bright?

REASON Both ends of the two lightbulbs are connected together by wires. Because there's no potential difference along ideal wires, the potential at the top of bulb A must be the same as the potential at the top of bulb B. Similarly, the potential at the bottoms of the bulbs must be the same. This means that the potential *difference* ΔV across the two bulbs must be the same, as we see in Figure 23.18. Because the bulbs are identical (i.e., equal resistances), the currents $I = \Delta V/R$ through the two bulbs are equal and thus the bulbs are equally bright.

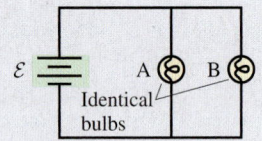

FIGURE 23.17 Two bulbs in parallel.

FIGURE 23.18 The potential differences of the bulbs.

1. The potential at these two points is the same because there is no potential difference across the wire.

$\Delta V_A = \Delta V_B$

3. The potential differences across the two bulbs must be equal.

2. The potential at these two points is the same as well.

ASSESS One might think that A would be brighter than B because current takes the "shortest route." But current is determined by potential difference, and two bulbs connected in parallel have the same potential difference.

Let's look at parallel circuits in more detail. The circuit of Figure 23.19a has a battery and two resistors connected in parallel. If we assume ideal wires, the potential differences across the two resistors are equal. In fact, the potential difference across each resistor is equal to the emf of the battery because both resistors are connected directly to the battery with ideal wires. That is, $\Delta V_1 = \Delta V_2 = \mathcal{E}$.

Now we apply Kirchhoff's junction law. The current I_{bat} from the battery splits into currents I_1 and I_2 at the top junction noted in Figure 23.19b. According to the junction law,

$$I_{bat} = I_1 + I_2 \tag{23.9}$$

We can apply Ohm's law to each resistor to find that the battery current is

$$I_{bat} = \frac{\Delta V_1}{R_1} + \frac{\Delta V_2}{R_2} = \frac{\mathcal{E}}{R_1} + \frac{\mathcal{E}}{R_2} = \mathcal{E}\left[\frac{1}{R_1} + \frac{1}{R_2}\right] \tag{23.10}$$

We found that we could replace a group of series resistors with a single equivalent resistor. Can we do the same for these parallel resistors? To be equivalent, the potential difference across the equivalent resistor must be $\Delta V = \mathcal{E}$, the same as for the two resistors it replaces. Further, so that the battery doesn't know there's been any change, the current through the equivalent resistor must be $I = I_{bat}$. According to Ohm's law, a resistor with this current and potential difference must have resistance

$$R_{eq} = \frac{\Delta V}{I} = \frac{\mathcal{E}}{I_{bat}} = \left[\frac{1}{R_1} + \frac{1}{R_2}\right]^{-1} \tag{23.11}$$

where we used Equation 23.10 for I_{bat}. This is the *equivalent resistance,* so a single resistor R_{eq} acts exactly the same as the two resistors R_1 and R_2.

We can extend this analysis to the case of N resistors in parallel. For this circuit, the equivalent resistance is the inverse of the sum of the inverses of the N individual resistances:

$$R_{eq} = \left(\frac{1}{R_1} + \frac{1}{R_2} + \cdots + \frac{1}{R_N}\right)^{-1} \tag{23.12}$$

Equivalent resistance of N parallel resistors

The behavior of the circuit will be unchanged if the N parallel resistors are replaced by the single resistor R_{eq}.

In Figure 23.19 each of the resistors "sees" the full potential difference of the battery. If one resistor were removed, the conditions of the second resistor would not change. This is an important property of parallel circuits.

Now, let's look at another lightbulb puzzle. Figure 23.20 shows two different circuits, one with one battery and one lightbulb and a second with one battery and two lightbulbs. As before, the batteries and the bulbs are identical. You know that B and C, which are connected in parallel, are equally bright, but how does the brightness of B compare to that of A?

▶ **Parallel circuits for safety** You have certainly seen cars with only one headlight lit. This tells us that automobile headlights are connected in parallel: The currents in the two bulbs are independent, so the loss of one bulb doesn't affect the other. The parallel wiring is very important, as the failure of one headlight will not leave the car without illumination.

(a) Two resistors in parallel

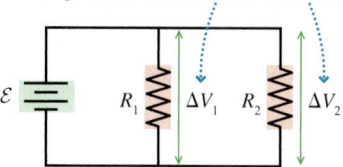
The potential differences are the same.

(b) Applying the junction law

We consider current in and out of this junction.

Same current

(c) An equivalent resistor

FIGURE 23.19 Replacing two parallel resistors with an equivalent resistor.

Activ
Physics 12.2, 12.5

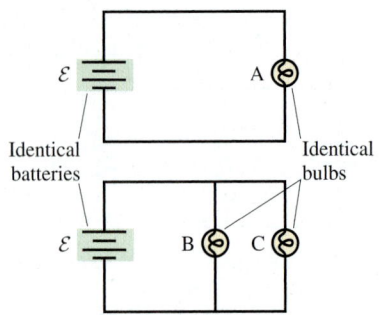

Identical batteries

Identical bulbs

Each of the bulbs A, B and C is connected to the same potential difference, that of the battery, so they each have the *same* brightness. Though all of the bulbs have the same brightness, there is a difference between the circuits. In the second circuit, the battery must power two lightbulbs, and so it must provide twice as much current. Recall that the battery is a source of fixed potential difference; the current depends upon the circuit that is connected to the battery. Adding a second lightbulb doesn't change the potential difference, but it does increase the current from the battery.

◄ **FIGURE 23.20** How does the brightness of bulb B compare to that of bulb A?

EXAMPLE 23.4 Current in a parallel resistor circuit
The three resistors of Figure 23.21 are connected to a 12 V battery. What current is provided by the battery?

FIGURE 23.21 Circuit for Example 23.4.

PREPARE The three resistors are in parallel, so we can reduce them to a single equivalent resistor, as in Figure 23.22.

FIGURE 23.22 Analyzing a circuit with parallel resistors.

SOLVE We can use Equation 23.12 to calculate the equivalent resistance:

$$R_{eq} = \left(\frac{1}{58 \ \Omega} + \frac{1}{70 \ \Omega} + \frac{1}{42 \ \Omega} \right)^{-1} = 18 \ \Omega$$

Once we know the equivalent resistance, we can use Ohm's law to calculate the current leaving the battery:

$$I = \frac{\mathcal{E}}{R_{eq}} = \frac{12 \ V}{18 \ \Omega} = 0.67 \ A$$

Because the battery can't tell the difference between the original three resistors and this single equivalent resistor, the battery in Figure 23.21 provides a current of 0.67 A to the circuit.

ASSESS Note that the resistance of the group of resistors is less than the resistance of any of the resistors in the group. This is always true for parallel resistors, and is a good check on your work.

The value of the total resistance in this example may seem surprising. The equivalent of a parallel combination of 58 Ω, 70 Ω, and 42 Ω is 18 Ω. Shouldn't more resistors imply more resistance? The answer is "yes" for resistors in series, but not for resistors in parallel. Even though a resistor is an obstacle to the flow of charge, parallel resistors provide more pathways for charge to get through. Consequently, **the equivalent of several resistors in parallel is always *less* than any single resistor in the group.** As an analogy, think about driving in heavy traffic. If there is an alternate route or an extra lane for cars to travel, more cars will be able to "flow."

STOP TO THINK 23.2   Rank in order, from brightest to dimmest, the identical bulbs A to D.

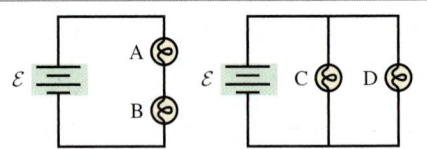

23.4 Measuring Voltage and Current

12.1, 12.4   Activ Physics ONLINE

Suppose you want to use a meter to measure the voltage or the current in a circuit. How would you connect the meter? As we will see, the connection depends on the quantity you wish to measure.

A device that measures the current in a circuit element is called an **ammeter**. Because charge flows *through* circuit elements, an ammeter must be placed *in series* with the circuit element whose current is to be measured.

Figure 23.23a shows a simple one-resistor circuit with a fixed emf $\mathcal{E} = 1.5$ V and an unknown resistance R. Suppose we would like to determine the resistance. To do this, we must know the current in the circuit, which we measure using an ammeter. We insert the ammeter in the circuit as shown in Figure 23.23b. We have to *break the connection* between the battery and the resistor in order to insert the ammeter. The resistor and the ammeter now have the same current, because they are in series, so the reading of the ammeter is the current through the resistor.

Because the ammeter is in series with resistor R, the total resistance seen by the battery is $R_{eq} = R + R_{meter}$. In order to *measure* the current without *changing* the current, the ammeter's resistance must be much less than R. Thus **the resistance of an ideal ammeter is zero.** Real ammeters come quite close to this ideal.

The ammeter in Figure 23.23b reads 0.60 A, meaning that the current in the ammeter—and in the resistor—is $I = 0.60$ A. If the ammeter is ideal, which we will assume, then there is no potential difference across the ammeter ($\Delta V = IR = 0$ if $R = 0\,\Omega$) and thus the potential difference across the resistor is $\Delta V = \mathcal{E}$. The resistance can then be calculated as

$$R = \frac{\mathcal{E}}{I} = \frac{1.5\text{ V}}{0.60\text{ A}} = 2.5\,\Omega$$

A device that measures the potential difference across a circuit element is called a **voltmeter.** Because a potential difference is measured *across* a circuit element, from one side to the other, a voltmeter is placed in *parallel* with the circuit element whose potential difference is to be measured. We want to *measure* the voltage without *changing* the voltage—without affecting the circuit. Because the voltmeter is in parallel with the resistor, the voltmeter's resistance must be very large so that it draws very little current. **An ideal voltmeter has infinite resistance.** Real voltmeters come quite close to this ideal.

Figure 23.24a shows a simple circuit in which a 24 Ω resistor is connected in series with an unknown resistance, with the pair of resistors connected to a 9.0 V battery. To determine the unknown resistance we first characterize the circuit by measuring the potential difference across the known resistor with a voltmeter as shown in Figure 23.24b. The voltmeter is connected in parallel with the resistor; using a voltmeter does *not* require that we break the connections. The resistor and the voltmeter have the same potential difference, because they are in parallel, so the reading of the voltmeter is the voltage across the resistor.

The voltmeter in Figure 23.24b tells us that the potential difference across the 24 Ω resistor is 6.0 V, so the current through the resistor is

$$I = \frac{\Delta V}{R} = \frac{6.0\text{ V}}{24\,\Omega} = 0.25\text{ A} \tag{23.13}$$

The two resistors are in series, so this is also the current in unknown resistor R. We can use Kirchhoff's loop law and the voltmeter reading to find the potential difference across the unknown resistor:

$$\sum_i \Delta V_i = 9.0\text{ V} + \Delta V_R - 6.0\text{ V} = 0 \tag{23.14}$$

from which we find $\Delta V_R = -3.0$ V. We can now use $\Delta V_R = -IR$ to calculate

$$R = \frac{-\Delta V_R}{I} = -\frac{(-3.0\text{ V})}{0.25\text{ A}} = 12\,\Omega \tag{23.15}$$

▶ In this text, we speak of ammeters and voltmeters, but in practice we generally make measurements with a *multimeter*. A dial on the front sets the meter to measure voltage, current, or other electrical quantities. When set to measure current, it works as an ammeter and must be connected in series. When set to measure voltage, it works as a voltmeter and must be connected in parallel. It can also work as an *ohmmeter*, directly measuring the resistance of a resistor not in a circuit.

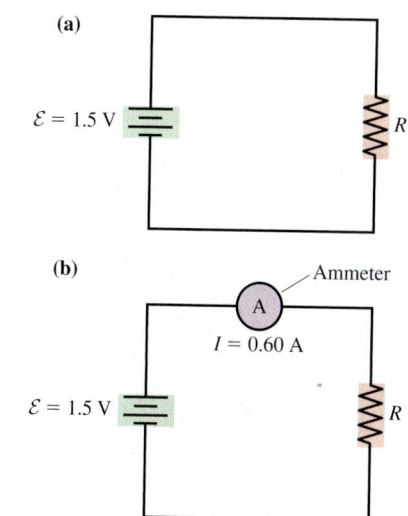

FIGURE 23.23 An ammeter measures the current in a circuit.

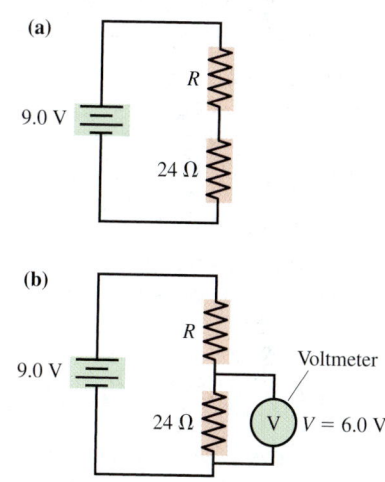

FIGURE 23.24 A voltmeter measures the potential difference across a circuit element.

A circuit for all seasons This device displays wind speed and temperature, but these are computed from basic measurements of voltage and current. The wind turns a propeller attached to a generator; a rapid spin means a high voltage. A circuit in the device contains a *thermistor,* whose resistance varies with temperature; low temperatures mean high resistance and thus a small current.

STOP TO THINK 23.3 Which is the right way to connect the meters to measure the potential difference across and the current through the resistor?

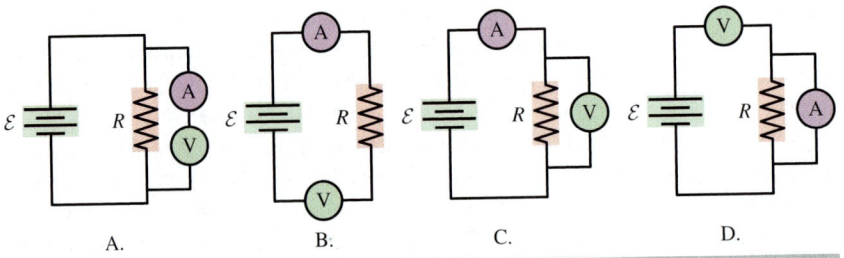

A. B. C. D.

23.5 More Complex Circuits

In this section, we will consider circuits that involve both series and parallel resistors. The analysis of these circuits will take more steps, but the steps are ones with which we are now familiar. We can make the following general statements about series and parallel connections:

- Circuit elements connected in series have the same current.
- Circuit elements connected in parallel have the same potential difference.
- Combinations of resistors can often be reduced to a single equivalent resistance through a step-by-step application of the series and parallel rules.

EXAMPLE 23.5 Combining resistors

What is the equivalent resistance of the group of resistors shown in Figure 23.25?

PREPARE We can analyze this circuit by reducing combinations of series and parallel resistors. We will do this in a series of steps, redrawing the circuit after each step.

FIGURE 23.25 Resistor circuit for Example 23.5.

SOLVE The process of simplifying the circuit is outlined in Figure 23.26. Note that the 10 Ω and 60 Ω resistors are *not* in parallel. They are connected at their top ends but not at their bottom ends. Resistors must be connected at *both* ends to be in parallel. Similarly, the 10 Ω and 45 Ω resistors are *not* in series because of the junction between them.

ASSESS If the original group of four resistors occurred within a larger circuit, they could be replaced with a single 24 Ω resistor without having any effect on the rest of the circuit. The first step in analyzing a complex circuit is often to reduce combinations of resistors in this manner.

FIGURE 23.26 A combination of resistors is reduced to a single equivalent resistor.

Reduce parallel combination:
$R_{eq} = \left(\frac{1}{90\ \Omega} + \frac{1}{45\ \Omega}\right)^{-1} = 30\ \Omega$

Reduce series combination:
$R_{eq} = 30\ \Omega + 10\ \Omega = 40\ \Omega$

Reduce parallel combination:
$R_{eq} = \left(\frac{1}{40\ \Omega} + \frac{1}{60\ \Omega}\right)^{-1} = 24\ \Omega$

Two useful special cases (which will be helpful in the next example) are the equivalent resistances of two identical resistors $R_1 = R_2 = R$ in series and in parallel:

Two identical resistors in series: $R_{eq} = 2R$

Two identical resistors in parallel: $R_{eq} = \frac{R}{2}$

EXAMPLE 23.6 How does the brightness change?

Initially the switch in Figure 23.27 is open. Bulbs A and B are equally bright, and bulb C is not glowing. What happens to the brightness of A and B when the switch is closed? And how does the brightness of C then compare to that of A and B? Assume that all bulbs are identical.

FIGURE 23.27 Lightbulb circuit for Example 23.6.

SOLVE Suppose the resistance of each bulb is R. Initially, before the switch is closed, bulbs A and B are in series; bulb C is not part of the circuit. A and B are identical resistors in series, so their equivalent resistance is $2R$ and the current from the battery is

$$I_{before} = \frac{\mathcal{E}}{R_{eq}} = \frac{\mathcal{E}}{2R} = \frac{1}{2}\frac{\mathcal{E}}{R}$$

This is the initial current in bulbs A and B, so they are equally bright.

Closing the switch places bulbs B and C in parallel with each other. The equivalent resistance of the two identical resistors in parallel is $R_{B+C} = R/2$. This equivalent resistance of B and C is in series with bulb A, hence the total resistance of the circuit is $R_{eq} = R + \frac{1}{2}R = \frac{3}{2}R$ and the current leaving the battery is

$$I_{after} = \frac{\mathcal{E}}{R_{eq}} = \frac{\mathcal{E}}{(3R/2)} = \frac{2}{3}\frac{\mathcal{E}}{R} > I_{before}$$

Closing the switch *decreases* the total circuit resistance and thus *increases* the current leaving the battery.

All the current from the battery passes through bulb A, so A *increases* in brightness when the switch is closed. The current I_{after} then splits at the junction. Bulbs B and C have equal resistance, so the current divides equally. The current in B is $\frac{1}{3}(\mathcal{E}/R)$, which is *less* than I_{before}. Thus B *decreases* in brightness when the switch is closed. With the switch closed, bulbs B and C are in parallel, so bulb C has the same brightness as bulb B.

ASSESS Even though bulb C is connected in parallel with bulb B, adding bulb C changes the brightness of bulb B because this isn't a simple parallel circuit. Bulbs B and C have the same potential difference and will have the same brightness—but this is less than the brightness of bulb B before the switch was closed.

Analyzing Complex Circuits

We can use the information in this chapter to analyze more complex but more realistic circuits. This will give us a chance to bring together the many ideas of this chapter and to see how they are used in practice. The techniques that we use for this analysis are general, so we present them as a Problem-Solving Strategy. You can use the steps outlined to analyze any resistor circuit, as we show in the example following.

This x-ray image of a cell phone shows the complex circuitry inside. Though there are thousands of components, the analysis of such a circuit starts with the same basic rules we are studying in this chapter.

 PROBLEM-SOLVING STRATEGY 23.1 Resistor circuits

PREPARE Draw a circuit diagram. Label all known and unknown quantities.

SOLVE Base your mathematical analysis on Kirchhoff's laws and on the rules for series and parallel resistors:

- Step by step, reduce the circuit to the smallest possible number of equivalent resistors.
- Determine the current through and potential difference across the equivalent resistors.
- Rebuild the circuit, using the facts that the current is the same through all resistors in series and the potential difference is the same across all parallel resistors.

ASSESS Use two important checks as you rebuild the circuit.

- Verify that the sum of the potential differences across series resistors matches ΔV for the equivalent resistor.
- Verify that the sum of the currents through parallel resistors matches I for the equivalent resistor.

 12.3

EXAMPLE 23.7 Analyzing a complex circuit

Find the current through and the potential difference across each of the four resistors in the circuit shown in Figure 23.28.

FIGURE 23.28 Multiple-resistor circuit of Example 23.7.

PREPARE Figure 23.29 shows the circuit diagram. We'll keep redrawing the diagram as we analyze the circuit.

SOLVE First, we break down the circuit, step-by-step, into one with a single resistor. Figure 23.29a does this in three steps, using the rules for series and parallel resistors. The final battery-and-resistor circuit is one we know well how to analyze. The current is

$$I = \frac{\mathcal{E}}{R} = \frac{12 \text{ V}}{400 \text{ }\Omega} = 0.030 \text{ A} = 30 \text{ mA}$$

The potential difference across the 400 Ω equivalent resistor is $\Delta V_{400} = \Delta V_{bat} = \mathcal{E} = 12$ V.

Second, we rebuild the circuit, step-by-step, finding the currents and potential differences at each step. Figure 23.29b repeats the steps of Figure 23.29a exactly, but in reverse order. The 400 Ω resistor came from two 800 Ω resistors in parallel. Because $\Delta V_{400} = 12$ V, it must be true that each $\Delta V_{800} = 12$ V. The current through each 800 Ω is then $I = \Delta V/R = 15$ mA. A check on our work is to note that 15 mA + 15 mA = 30 mA.

The right 800 Ω resistor was formed by 240 Ω and 560 Ω in series. Because $I_{800} = 15$ mA, it must be true that $I_{240} = I_{560} = 15$ mA. The potential difference across each is $\Delta V = IR$, so $\Delta V_{240} = 3.6$ V and $\Delta V_{560} = 8.4$ V. Here the check on our work is to note that 3.6 V + 8.4 V = 12 V = ΔV_{800}, so the potential differences add as they should.

Finally, the 240 Ω resistor came from 600 Ω and 400 Ω in parallel, so they each have the same 3.6 V potential difference as their 240 Ω equivalent. The currents are $I_{600} = 6.0$ mA and $I_{400} = 9.0$ mA. Note that 6.0 mA + 9.0 mA = 15 mA, which is our third check on our work. We now know all currents and potential differences.

ASSESS We *checked our work* at each step of the rebuilding process by verifying that currents summed properly at junctions and that potential differences summed properly along a series of resistances. This "check as you go" procedure is extremely important. It provides you, the problem solver, with a built-in error finder that will immediately inform you if a mistake has been made.

FIGURE 23.29 The step-by-step circuit analysis.

Rank in order, from brightest to dimmest, the identical bulbs A to D.

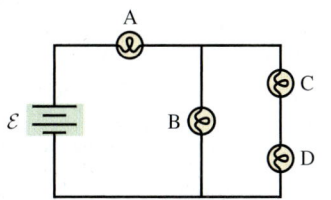

23.6 Household Electricity

Kirchhoff's laws and other principles that we have learned in this chapter apply to any circuit—including the circuits of devices that you plug into outlets in your house. We will use our knowledge of circuits to understand household electricity, focusing on elements of your household electrical supply that are present for your safety.

Getting Grounded

The circuit analysis procedures we have discussed so far deal only with potential *differences.* Although we are free to choose the zero point of potential anywhere that is convenient, our analysis of circuits has not revealed any need to establish a zero point. Potential differences are all we have needed.

Difficulties can begin to arise, however, if you want to connect two *different* circuits together. Perhaps you would like to connect your CD player to your amplifier or your computer monitor to the computer itself. Incompatibilities can arise unless all the circuits to be connected have a *common* reference point for the potential.

You learned previously that the earth itself is a conductor. Suppose we have two circuits. If we connect *one* point of each circuit to the earth by an ideal wire, and we also agree to call the potential of the earth $V_{earth} = 0$ V, then both circuits have a common reference point. A circuit connected to the earth in this way is said to be **grounded.** Figure 23.30 shows a circuit with a ground connection. Under normal circumstances, the ground connection does not carry any current, as it is not part of a complete circuit. In this case, it does not alter the behavior of the circuit.

Grounding serves two functions. First, it provides a common reference potential so that different circuits or instruments can be correctly interconnected. Second, it is an important safety feature. As we will see in the next section, a current to ground can quickly open a circuit breaker if an electric appliance malfunctions.

Electric Outlets Are Grounded Parallel Circuits

The electrical supply to your house is AC, meaning *alternating current.* The voltage isn't constant; it oscillates positive and negative in a sinusoidal manner. We will see why this is important in Chapter 26. For most applications, though, we can model your household electrical supply as a constant 120 V source.

The electrical supply from the utility company is connected to electric outlets in your house via wires that run through the walls, as shown in Figure 23.31 on the next page. One terminal of the electrical supply is grounded with a metal bar sunk into the earth. The side that is grounded is called the **neutral** side; the other side at 120 V is the **hot** side. When you plug something in to an electric outlet, the device completes a circuit between the hot side and the neutral side. The potential difference across the device leads to a current, and the device turns on.

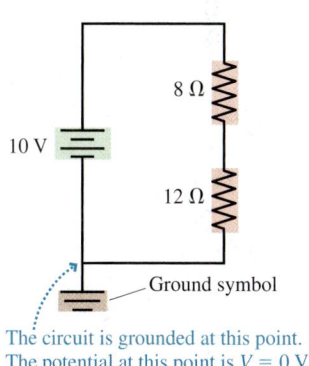

The circuit is grounded at this point. The potential at this point is $V = 0$ V.

FIGURE 23.30 A grounded circuit.

A two-prong "polarized" plug has one large prong and one small one. When plugged in to a standard outlet, the large prong is grounded. A three-prong plug has a round pin that makes a second ground connection.

The circuit breaker opens if the current exceeds a certain value.

Neutral side   Hot side

120 V   15 A

The switch is on the hot side. When the switch is open, the device is neutral.

Ground connection

R_1 R_2 R_3

In a grounded outlet, there is a second ground connection for safety.

When you plug something in, you connect it between the two sides of the outlet. Multiple items on the same circuit are connected in parallel.

FIGURE 23.31 Household electricity.

With a few exceptions, each electric circuit in your house has multiple outlets connected in parallel. Everything that you plug in to an outlet is connected in parallel with other devices on the same circuit. We have seen that adding an additional resistor to a parallel circuit increases the current. The same is true here: When you plug in another device, the total current in the circuit increases. This can create a problem; although the wires in your walls are good conductors, they aren't ideal. The wires have a small resistance and heat up when carrying a current. If there is too much current, the wires could get hot enough to cause a fire.

To keep this from happening, the circuits in your house are protected with circuit breakers. A circuit breaker consists of a switch and an ammeter that measures the current in the circuit. If the ammeter measures too much current in the circuit, it sends a signal to open the switch to disconnect the circuit. You have probably had the experience of having a circuit breaker "trip" if you have too many things plugged in. The circuit breaker has prevented a potentially dangerous situation; the circuit was carrying too much current. To keep the problem from recurring, you must reduce the current in the circuit. Some things need to be turned off, unplugged, or moved to a different circuit.

Grounding of household circuits provides an important reference potential, as noted above, but the main reason for grounding is safety. The two slots in a standard outlet are different sizes; the neutral outlet is a bit larger. Most electrical devices are fitted with plugs that can only be inserted into an outlet in one orientation. A lamp, for instance, will almost certainly have this sort of plug. This is an important safety feature; when you turn the lamp off, the switch disconnects the hot wire, not the neutral wire. The lamp is then neutral, and thus safe, when it is off.

The round hole in a standard electrical outlet is a second ground connection that serves a second safety function. If a device has a metal case, whether it's a microwave oven or an electric drill, the case will likely be connected to this ground connection. Suppose a wire came loose inside the device and contacted the metal case. A person touching the case could get a shock or even be killed. But if the case is grounded, its potential is always 0 V, and it is always safe to touch. In addition, a hot wire touching the grounded case would be a *short circuit,* causing a sudden very large current that would trip the circuit breaker, disconnecting the hot wire and preventing any danger. If you plug something in and the circuit breaker trips, the device likely has an electrical fault and should be discarded or repaired.

A typical circuit breaker panel. Each switch corresponds to a different circuit in the house.

EXAMPLE 23.8 Will the circuit breaker open?

A circuit in a student's room has a 15 A circuit breaker. One evening, the student plugs in a computer (240 W), a lamp (with two 60 W bulbs, for a total of 120 W), and a space heater (1200 W). Will this be enough to trip the circuit breaker?

PREPARE We start by sketching the circuit, as shown in Figure 23.32. Because the three devices are connected in the same circuit, they are in parallel. We can model them as resistors.

FIGURE 23.32 Circuit for Example 23.8.

SOLVE According to Kirchhoff's junction law, the current, which is provided by the 120 V supply and passes through the circuit breaker, is the sum of the currents in the individual devices:

$$I_{total} = I_{computer} + I_{lamp} + I_{heater}$$

The power dissipated by a resistor is $P_R = I\Delta V$, so we can calculate the current in the individual devices as $I = P_R/\Delta V$. All three devices are in parallel with the supply, so $\Delta V = 120$ V. The current is thus

$$I_{total} = \frac{240 \text{ W}}{120 \text{ V}} + \frac{120 \text{ W}}{120 \text{ V}} + \frac{1200 \text{ W}}{120 \text{ V}} = 13 \text{ A}$$

This is not enough to open the circuit breaker.

ASSESS Generally all of the outlets in one room (and perhaps the lights as well) are on the same circuit. You can see that it would be quite easy to plug in enough devices to trip the circuit breaker.

Kilowatt Hours

The energy dissipated (i.e., transformed into thermal energy) by a resistor during time Δt is $E_{th} = P_R \Delta t$. The product of watts and seconds is joules, the SI unit of energy. However, your local electric company prefers to use a different unit, called *kilowatt hours,* to measure the energy you use each month.

A device in your home that consumes P_R kW of electricity for Δt hours has used $P_R \Delta t$ **kilowatt hours** of energy, abbreviated kWh. For example, suppose you run a 1500 W electric water heater for 10 hours. The energy used in kWh is $(1.5 \text{ kW})(10 \text{ hr}) = 15$ kWh.

Despite the rather unusual name, a kilowatt hour is a unit of energy, as it is a power multiplied by a time. The conversion between kWh and J is

$$1.00 \text{ kWh} = (1.00 \times 10^3 \text{ W})(3600 \text{ s}) = 3.60 \times 10^6 \text{ J}$$

Your monthly electric bill specifies the number of kilowatt hours you used last month. This is the amount of energy that the electric company delivered to you, via an electric current, and that you transformed into light and thermal energy inside your home.

The electric meter on the side of your house or apartment records the kilowatt hours of electricity you use each month.

EXAMPLE 23.9 Computing the cost of electric energy

A hair dryer carries a current of 10 A. If electricity costs 10¢ per kWh (an approximate national average), what is the cost to run the dryer for 5.0 min?

SOLVE The power of the hair dryer, in kW, is $P_R = I\Delta V = (10 \text{ A})(120 \text{ V}) = 1200 \text{ W} = 1.2$ kW. The energy consumed is power multiplied by time. With the time in hours, $t = 5.0 \text{ min} = \frac{1}{12}$ hr, the energy is calculated to be

$$E = (1.2 \text{ kW})\left(\frac{1}{12} \text{ hr}\right) = 0.10 \text{ kWh}$$

At 10¢ per kWh, the cost is 1.0¢.

23.7 Capacitors in Parallel and Series

Two conductors separated by an insulating layer make a circuit element called a *capacitor,* a device we have considered in some detail in the past few chapters. Figure 23.33 shows a basic circuit consisting of a battery and a capacitor. When we connect the capacitor to the battery, charge will flow to the capacitor, increasing

FIGURE 23.33 Simple capacitor circuit.

(a) Parallel capacitors

(b) Series capacitors

FIGURE 23.34 Parallel and series capacitors.

(a) Parallel capacitors have the same ΔV_C.

(b) Same ΔV_C as C_1 and C_2

FIGURE 23.35 Replacing two parallel capacitors with an equivalent capacitor.

(a) Series capacitors have the same Q.

(b) Same Q as C_1 and C_2

$\Delta V_C = \Delta V_1 + \Delta V_2$
Same total potential difference as C_1 and C_2

FIGURE 23.36 Replacing two series capacitors with an equivalent capacitor.

its potential difference until $\Delta V_C = \mathcal{E}$. Once the capacitor is fully charged, there will be no further current. We saw in Chapter 21 that the magnitude of the charge on each plate of the capacitor at this point will be $Q = C \Delta V_C = C\mathcal{E}$.

In resistor circuits, we often combine multiple resistors; we can do the same with capacitors. Figure 23.34 illustrates two basic combinations: parallel capacitors and series capacitors.

NOTE ▶ The terms "parallel capacitors" and "parallel-plate capacitor" do not describe the same thing. The former term describes how two or more capacitors are connected to each other, the latter describes how a particular capacitor is constructed. ◀

Parallel or series capacitors can be represented by a single **equivalent capacitance,** just as we have done with resistors. Let's start with the two parallel capacitors C_1 and C_2 of Figure 23.35a. Just as for resistors, **two (or more) capacitors in parallel each have the same potential difference** ΔV_C.

The charges on the two capacitors are $Q_1 = C_1 \Delta V_C$ and $Q_2 = C_2 \Delta V_C$. As the capacitors were being charged, the battery moved total charge $Q = Q_1 + Q_2$ from the negative electrodes to the positive electrodes. Suppose, as in Figure 23.35b, we replace the two capacitors with a single capacitor having charge $Q = Q_1 + Q_2$ and potential difference ΔV_C. This capacitor is equivalent to the original two in the sense that the battery can't tell the difference; it had to establish the same potential difference and move the same amount of charge.

The capacitance of this equivalent capacitor is

$$C_{eq} = \frac{Q}{\Delta V_C} = \frac{Q_1 + Q_2}{\Delta V_C} = \frac{Q_1}{\Delta V_C} + \frac{Q_2}{\Delta V_C} = C_1 + C_2 \tag{23.16}$$

We could easily extend this analysis to more than two capacitors. If N capacitors are in parallel, their equivalent capacitance is the sum of the individual capacitances:

$$C_{eq} = C_1 + C_2 + C_3 + \cdots + C_N \tag{23.17}$$

Equivalent capacitance of N parallel capacitors

Neither the battery nor any other part of a circuit can tell if the parallel capacitors are replaced by a single capacitor having capacitance C_{eq}.

NOTE ▶ The equivalent capacitance of *parallel* capacitors is the sum of the individual values, similar to the calculation for *series* resistors. ◀

Now consider the two series capacitors in Figure 23.36a. The center section, consisting of the bottom plate of C_1, the top plate of C_2, and the connecting wire, is electrically isolated. The battery cannot remove charge from or add charge to this section. If it starts out with no net charge, it must end up with no net charge. As a consequence, the two capacitors in series have equal charges $\pm Q$. The battery transfers Q from the bottom of C_2 to the top of C_1. This transfer polarizes the center section, but it still has $Q_{net} = 0$.

The potential differences across the two capacitors are $\Delta V_1 = Q/C_1$ and $\Delta V_2 = Q/C_2$. The total potential difference across both capacitors is $\Delta V_C = \Delta V_1 + \Delta V_2$. Suppose, as in Figure 23.36b, we replaced the two capacitors with a single capacitor having charge Q and potential difference $\Delta V_C = \Delta V_1 + \Delta V_2$. This capacitor is equivalent to the original two because the battery has to establish the same potential difference and move the same amount of charge in either case.

The inverse of the capacitance of this equivalent capacitor is

$$\frac{1}{C_{eq}} = \frac{\Delta V_C}{Q} = \frac{\Delta V_1 + \Delta V_2}{Q} = \frac{\Delta V_1}{Q} + \frac{\Delta V_2}{Q} = \frac{1}{C_1} + \frac{1}{C_2} \tag{23.18}$$

This analysis hinges on the fact that **series capacitors each have the same charge Q.** We could easily extend this analysis to more than two capacitors. If N capacitors are in series, their equivalent capacitance is the inverse of the sum of the inverses of the individual capacitances:

$$C_{eq} = \left(\frac{1}{C_1} + \frac{1}{C_2} + \frac{1}{C_3} + \cdots + \frac{1}{C_N} \right)^{-1} \qquad (23.19)$$

Equivalent capacitance of N series capacitors

For series capacitors, the equivalent capacitance is less than that of the individual capacitors.

NOTE ▶ The equation for the equivalent capacitance of *series* capacitors has the same form as the equation for the equivalent resistance of *parallel* resistors. ◀

EXAMPLE 23.10 Analyzing a capacitor circuit

a. Find the equivalent capacitance of the combination of capacitors in the circuit of Figure 23.37.
b. What charge flows through the battery as the capacitors are being charged?

FIGURE 23.37 A capacitor circuit.

PREPARE We can use the relationships for parallel and series capacitors to reduce the capacitors to a single equivalent capacitance, much as we did for resistor circuits. We can then com-

pute the charge through the battery using this value of capacitance.

SOLVE

a. Figure 23.38 shows how we find the equivalent capacitance by reducing parallel and series combinations.
b. The battery "sees" a capacitance 2.0 μF. To establish a potential difference of 12 V, the charge that must flow is

$$Q = C_{eq} \Delta V_C = (2.0 \times 10^{-6}\,\text{F})(12\,\text{V}) = 2.4 \times 10^{-5}\,\text{C}$$

ASSESS We solve capacitor circuit problems in a manner very similar to what we followed for resistor circuits.

Reduce parallel combination:

$C_{eq} = 5.0\,\mu\text{F} + 1.0\,\mu\text{F} = 6.0\,\mu\text{F}$

Reduce series combination:

$C_{eq} = \left(\dfrac{1}{3.0\,\mu\text{F}} + \dfrac{1}{6.0\,\mu\text{F}} \right)^{-1} = 2.0\,\mu\text{F}$

FIGURE 23.38 Analyzing a capacitor circuit.

STOP TO THINK 23.5   Rank in order, from largest to smallest, the equivalent capacitance $(C_{eq})_A$ to $(C_{eq})_C$ of circuits A to C.

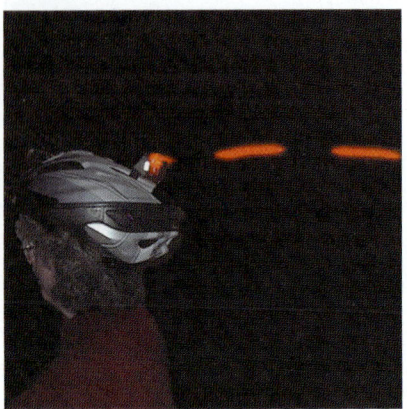

The rear flasher on a bike helmet blinks on and off. The timing is controlled by an *RC* circuit.

23.8 *RC* Circuits

The resistor circuits we have seen have a steady current. If we add a capacitor to a resistor circuit, we can make a circuit in which the current varies with time. Circuits containing resistors and capacitors are known as **RC circuits.** They are very common in electronic equipment. A simple example of an *RC* circuit is the flashing bike-helmet light in the photograph. As we will see, the values of the resistance and capacitance in an *RC* circuit determine the *time* it takes the capacitor to charge or discharge. In the case of the bike light, this time determines the duration of the flash. A large capacitance causes a slow cycle of on-off-on; a smaller capacitance would mean a more rapid flicker.

Figure 23.39a shows an *RC* circuit consisting of a charged capacitor, an open switch, and a resistor. The capacitor has initial charge Q_0 and potential difference $(\Delta V_C)_0 = Q_0/C$. There is no current, so the potential difference across the resistor is zero. Then, at $t = 0$, the switch closes and the capacitor begins to discharge through the resistor, as in Figure 23.39b.

(a) Before the switch closes

The switch will close at $t = 0$.

Charge Q_0
$(\Delta V_C)_0 = Q_0/C$

(b) Immediately after the switch closes

The charge separation on the capacitor produces a potential difference, which causes a current.

Current is the flow of charge, so the current discharges the capacitor.

(c) At a later time

The current has reduced the charge on the capacitor. This reduces the potential difference.

The reduced potential difference leads to a reduced current.

FIGURE 23.39 Discharging an *RC* circuit.

How long does the capacitor take to discharge? How does the current through the resistor vary as a function of time? Let's look at the behavior of the circuit after the switch is closed.

Figure 23.39b shows the circuit *immediately* after the switch closes. The capacitor voltage is still $(\Delta V_C)_0$ because the capacitor hasn't yet had time to lose any charge, but now there's a current I_0 in the circuit that's starting to discharge the capacitor. Applying Kirchhoff's loop law, going around the loop clockwise, we find

$$\sum_i \Delta V_i = \Delta V_C + \Delta V_R = (\Delta V_C)_0 - I_0 R = 0$$

Thus the *initial* current—the initial rate at which the capacitor begins to discharge—is

$$I_0 = \frac{(\Delta V_C)_0}{R} \tag{23.20}$$

As time goes by, the current continues and the charge on the capacitor decreases. Figure 23.39c shows the circuit some time after the switch is closed; as we can see, both the charge on the capacitor (and thus the potential difference) and the current in the circuit have decreased. When the capacitor voltage has decreased to ΔV_C, the current has decreased to

$$I = \frac{\Delta V_C}{R} \tag{23.21}$$

The current discharges the capacitor, which causes ΔV_C to decrease. The capacitor voltage ΔV_C drives the current, so the current I decreases as well. The current I and the voltage ΔV_C both decrease until the capacitor is fully discharged and the current is zero.

If we use a voltmeter and an ammeter to measure the capacitor voltage and the current in the circuit of Figure 23.39 as a function of time, we find the variation shown in the graphs of Figure 23.40. At $t = 0$, when the switch closes, the potential difference across the capacitor is $(\Delta V_C)_0$ and the current suddenly jumps to I_0. The current and the capacitor voltage then "decay" to zero, but *not* linearly.

The graphs in Figure 23.40 have the same shape as that for the decay of the amplitude of a damped simple harmonic oscillator we saw in Chapter 14. Both the voltage and the current are *exponential decays* given by the equations

$$I = I_0 e^{-t/RC}$$

$$\Delta V_C = (\Delta V_C)_0 e^{-t/RC} \qquad (23.22)$$

EXPONENTIAL
p. 471

Current and voltage during a capacitor discharge

In Chapter 14, we saw that we could characterize an exponential decay by a **time constant** τ. The time constant for the decay of current and voltage in an *RC* circuit is

$$\tau = RC \qquad (23.23)$$

If you work with the units, you can show that the product of ohms and farads is seconds, so *RC* really is a time. In terms of this time constant, the current and voltage equations are

$$I = I_0 e^{-t/\tau}$$

$$\Delta V_C = (\Delta V_C)_0 e^{-t/\tau} \qquad (23.24)$$

The current and voltage in the circuit do not drop to zero after one time constant; that's not what the time constant means. Instead, each increase in time by one time constant causes the voltage and current to decrease by a factor of $e^{-1} = 0.37$, as we see in Figure 23.41. The time constant is really a *characteristic time* for the circuit. A long time constant implies a slow decay; a short time constant, a fast decay.

We can understand why the time constant has the form $\tau = RC$. A large value of resistance opposes the flow of charge, so increasing R increases the decay time. A larger capacitance stores more charge, so increasing C also increases the decay time.

After one time constant, the current and voltage in a capacitor circuit have decreased to 37% of their initial values. But when working with capacitor circuits, it is often easier to measure the time for the voltage and current to decrease by 50%—the time for the voltage and current to drop by half. You can use Equation 23.22 to show that the voltage and current drop to 50% of their initial values after a time $t = \tau \ln 2 = 0.69\tau$.

NOTE ▶ There's no specific time at which the capacitor has been fully discharged because ΔV_C approaches zero gradually. After 5τ, the voltage and current have decayed to less than 1% of their initial values. For most purposes, we can say that the capacitor is discharged at this time. ◀

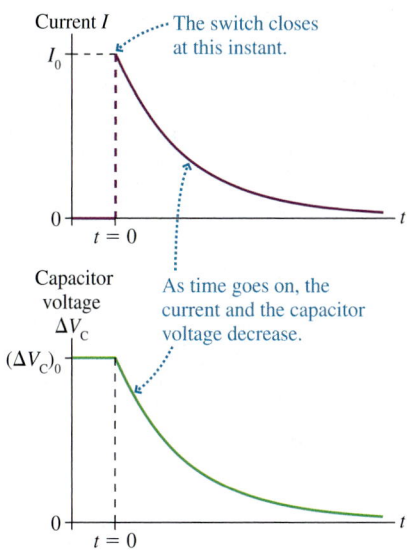

Current *I*   ⋯ The switch closes at this instant.

$t = 0$

Capacitor voltage ΔV_C As time goes on, the current and the capacitor voltage decrease.

$(\Delta V_C)_0$

$t = 0$

FIGURE 23.40 Current and capacitor voltage in an *RC* discharge circuit.

 Act|v Phys|cs   12.8

Capacitor voltage ΔV_C The potential difference has decreased to $e^{-1} = 37\%$ of its initial value at $t = \tau$.

$(\Delta V_C)_0$

The potential difference has decreased to $e^{-2} = 13\%$ of its initial value at $t = 2\tau$.

$0.37(\Delta V_C)_0$
$0.13(\Delta V_C)_0$

$\tau \qquad 2\tau \qquad 3\tau$

FIGURE 23.41 The meaning of the time constant in an *RC* circuit.

EXAMPLE 23.11 Finding the current in an *RC* circuit
The switch in the circuit of Figure 23.42 has been in position a for a long time, so the capacitor is fully charged. The switch is changed to position b at $t = 0$. What is the current in the circuit immediately after the switch is closed? What is the current in the circuit 25 μs later?

9.0 V   1.0 μF   10 Ω

FIGURE 23.42 The *RC* circuit of Example 23.11.

SOLVE The capacitor is connected across the battery terminals, so initially it is charged to $(\Delta V_C)_0 = 9.0$ V. When the switch is closed, the initial current is given by Equation 23.20:

$$I_0 = \frac{(\Delta V_C)_0}{R} = \frac{9.0 \text{ V}}{10 \text{ }\Omega} = 0.90 \text{ A}$$

As charge flows, the capacitor discharges. The time constant for the decay is given by Equation 23.23:

$$\tau = (10 \text{ }\Omega)(1.0 \times 10^{-6} \text{ F}) = 1.0 \times 10^{-5} \text{ s} = 10 \text{ }\mu s$$

The current in the circuit as a function of time is given by Equation 23.22. 25 μs after the switch is closed, the current is

$$I = I_0 e^{-t/\tau} = (0.90 \text{ A})e^{-(25\mu s)/(10\mu s)} = 0.074 \text{ A}$$

ASSESS This result makes sense. 25 μs after the switch has closed is 2.5 time constants, so we expect the current to decrease to a small fraction of the initial current. Notice that we left times in units of μs; this is one of the rare cases where we needn't convert to SI units. Because the exponent is $-t/\tau$, which involves a ratio of two times, we need only be certain that both t and τ are in the same units.

(a)   Switch closes at $t = 0$ s.

(b) Current I

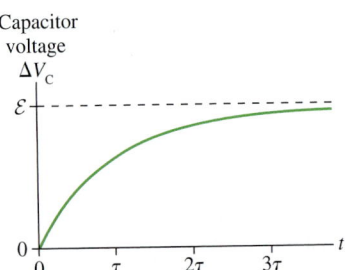

Capacitor voltage ΔV_C

FIGURE 23.43 A circuit for charging a capacitor.

Charging a Capacitor

Figure 23.43a shows a circuit that charges a capacitor. After the switch is closed, the potential difference of the battery causes a current in the circuit, and the capacitor begins to charge. As the capacitor charges, it develops a potential difference that opposes the current, so the current decreases. As the current decreases, so does the rate of charging of the capacitor. The capacitor charges until $\Delta V_C = \mathcal{E}$, when the charging current ceases.

If we measure the current in the circuit and the potential difference across the capacitor as a function of time, we find that they vary according to the graphs in Figure 23.43b. The characteristic time for this charging circuit is the same as for the discharge, the time constant $\tau = RC$.

When the switch is first closed, the potential difference across the uncharged capacitor is zero, so the initial current is

$$I_0 = \frac{\mathcal{E}}{R}$$

The equations that describe the capacitor voltage and the current as a function of time are

$$\Delta V_C = \mathcal{E}(1 - e^{-t/\tau})$$

$$I = I_0 e^{-t/\tau} \tag{23.25}$$

EXPONENTIAL p.471

Current and voltage while charging a capacitor

The time constant τ in an RC circuit can be used to control the behavior of a circuit. For example, a bike flasher uses an RC circuit that alternately charges and discharges, over and over, as a switch opens and closes. A separate circuit turns the light on when the capacitor voltage exceeds some "threshold voltage" and turns the light off when the capacitor voltage goes below this threshold. The time constant of the RC circuit determines how long the capacitor voltage stays above threshold and thus sets the length of the flashes. More complex RC circuits provide timing in computers and other digital electronics. As we will see in the next section, we can also use RC circuits to model the transmission of nerve impulses, and the time constant will be a key factor in determining the speed at which signals can be propagated in the nervous system.

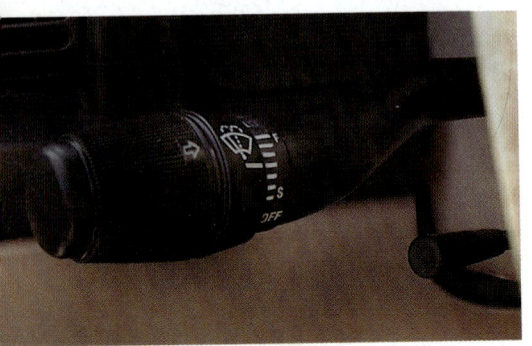

◀ **A rainy-day RC circuit** When you adjust the dial to control the delay of the intermittent windshield wipers in your car, you are adjusting a variable resistor in an RC circuit that triggers the wipers. Increasing the resistance increases the time constant and thus produces a greater delay between swipes of the blades. A light mist calls for a large time constant, and thus a large resistance.

STOP TO THINK 23.6 The time constant for the discharge of this capacitor is

A. 5 s.
B. 4 s.
C. 2 s.
D. 1 s.

2 Ω   2 Ω

1 F

23.9 Electricity in the Nervous System

In the late 1700s, the Italian scientist Galvani discovered that animal tissue has an electrical nature. He found that a frog's leg would twitch when stimulated with electricity, even when no longer attached to the frog. Further investigations by Galvani and others revealed that electrical signals can animate muscle cells, and that a small potential applied to the *axon* of a nerve cell can produce a signal that propagates down its length.

Our goal in this section will be to understand the nature of electrical signals in the nervous system. When your brain orders your hand to move, how does the signal get from your brain to your hand? Answering this question will use our knowledge of fields, potential, resistance, capacitance and circuits, all of the knowledge and techniques that we have learned so far in Part VI.

The Electrical Nature of Nerve Cells

We start our analysis with a very simple *model* of a nerve cell that allows us to describe its electrical properties. The model begins with a *cell membrane,* an insulating layer of lipids approximately 7 nm thick that separates regions of conducting fluid inside and outside the cell.

As we saw in Chapter 21, the cell membrane is not a passive structure. It has channels and pumps that transport ions between the inside and the outside of the cell. In our simple model we will consider the transport of only two positive ions, sodium (Na^+) and potassium (K^+), though other ions are also important to cell function. Ions, rather than electrons, are the charge carriers of the cell. These ions can slowly diffuse across the cell membrane. In addition, sodium and potassium ions are transported via the following structures:

- *Sodium-potassium exchange pumps.* These pump Na^+ ions out of the cell and K^+ ions in. In the cell's resting state, the concentration of sodium ions outside the cell is about ten times the concentration on the inside. Potassium ions are more concentrated on the inside.
- *Sodium and potassium channels.* These channels in the cell membrane are usually closed. When they are open, ions move in the direction of lower concentration. Thus Na^+ ions flow into the cell and K^+ ions flow out.

Our simple model, outlined in Figure 23.44, neglects many of the features of real cells, but it allows us to accurately describe the reaction of nerve cells to a stimulus and the conduction of electrical signals.

The ion exchange pumps act much like the charge escalator of a battery, using chemical energy to separate charge by transporting ions. The transport and subsequent diffusion of charged ions leads to a separation in charge across the cell membrane. Consequently, **a living cell generates an emf.** This emf takes energy to create and maintain. The ion pumps that produce the emf of neural cells account for 25–40% of the energy usage of the brain.

The charge separation produces an electric field inside the cell membrane, and results in a potential difference between the inside and the outside of the cell, as

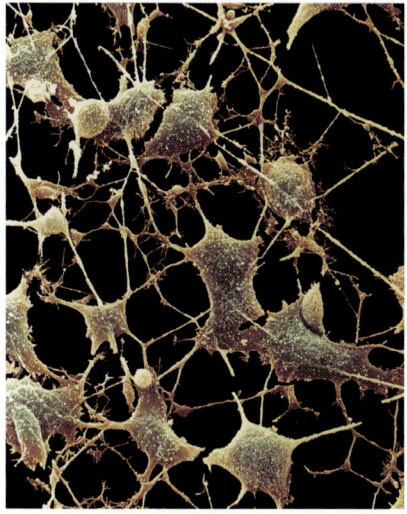

The long fibers connecting these nerve cells are *axons,* the transmission lines for electrical signals between cells.

The pump moves sodium out of the cell and potassium in, so the sodium concentration is higher outside the cell, the potassium concentration is higher inside.

Cell membrane K^+ Exchange pump

Conducting fluids Na^+

K^+

Na^+

Sodium channels   Potassium channels

When a sodium channel is open, the higher sodium concentration outside the cell causes ions to flow into the cell.

When a potassium channel is open, the higher potassium concentration inside the cell causes ions to flow out of the cell.

FIGURE 23.44 A simple model of a nerve cell.

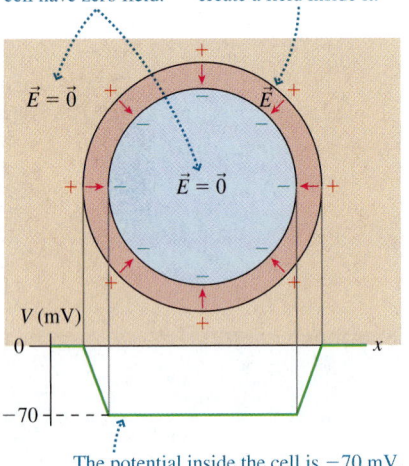

The conducting fluids inside and outside the cell have zero field.

Charges on the inside and outside surfaces of the insulating membrane create a field inside it.

$\vec{E} = \vec{0}$ $\vec{E}$

$\vec{E} = \vec{0}$

V (mV)

0

-70

The potential inside the cell is -70 mV.

FIGURE 23.45 The resting potential of a nerve cell.

shown in Figure 23.45. The potential inside a nerve cell is typically 70 mV less than that outside the cell. This is called the cell's *resting potential*. Because this potential difference is produced by a charge separation across the membrane, we say that the membrane is *polarized*. As the potential difference is entirely across the membrane, we may call this potential difference the *membrane potential*.

EXAMPLE 23.12 Electric field in a cell membrane

The thickness of a typical nerve cell membrane is 7.0 nm. What is the electric field inside the membrane of a resting nerve cell?

PREPARE The potential difference across the membrane of a resting nerve cell is -70 mV. The inner and outer surfaces of the membrane are equipotentials. We learned in Chapter 21 that the electric field is perpendicular to the equipotentials and is related to the potential difference by $E = \Delta V/d$.

SOLVE The magnitude of the potential difference between the inside and the outside of the cell is 70 mV. The field strength is thus

$$E = \frac{\Delta V}{d} = \frac{70 \times 10^{-3} \text{ V}}{7.0 \times 10^{-9} \text{ m}} = 1.0 \times 10^{7} \text{ V/m}$$

The field points from positive to negative, so the electric field is

$$\vec{E} = (1.0 \times 10^{7} \text{ V/m, inward})$$

ASSESS This is a very large electric field; in air it would be large enough to cause a spark! But we expect the fields to be large, to explain the cell's strong electrical character.

EXAMPLE 23.13 Finding the resistance of a cell membrane

Ions can move across the cell membrane, so it is not a perfect insulator; the cell membrane will have a certain resistance. The resistivity of the cell membrane was given in Chapter 22 as $36 \times 10^{6} \; \Omega \cdot \text{m}$. What is the resistance of the 7.0-nm-thick membrane of a spherical cell with diameter 0.050 mm?

PREPARE Current—in this case, the motion of ions—travels *through* the membrane. As we learned in Chapter 22, an object's resistance depends on its resistivity, length, and cross-section area. What this means for a cell membrane is noted in Figure 23.46.

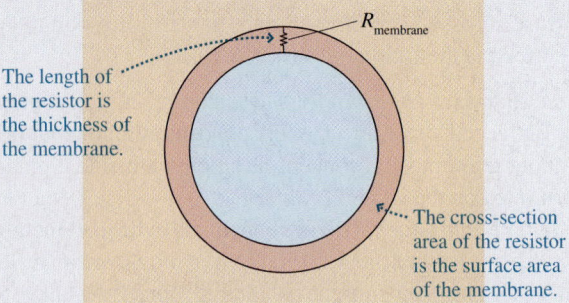

The length of the resistor is the thickness of the membrane.

R_{membrane}

The cross-section area of the resistor is the surface area of the membrane.

FIGURE 23.46 The cell membrane can be modeled as a resistor.

SOLVE The area of the membrane is the surface area of a sphere, $4\pi r^2$. We can calculate the resistance using the equation for the resistance of a conductor of length L and cross-section area A from Chapter 22:

$$R_{\text{membrane}} = \frac{\rho L}{A} = \frac{(3.6 \times 10^{7} \; \Omega \cdot \text{m})(7.0 \times 10^{-9} \text{ m})}{4\pi(2.5 \times 10^{-5} \text{ m})^2}$$

$$= 3.2 \times 10^{7} \; \Omega = 32 \text{ M}\Omega$$

ASSESS The resistance is quite high; the membrane is a good insulator, as we noted.

We can associate a resistance with the cell membrane, but we can associate other electrical quantities as well. The fluids inside and outside of the membrane are good conductors; they are separated by the membrane, which is not. Charges therefore accumulate on the inside and outside surfaces of the membrane. A cell thus looks like two charged conductors separated by an insulator—a capacitor.

EXAMPLE 23.14 Finding the capacitance of a cell membrane

What is the capacitance of the membrane of the spherical cell specified in Example 23.13? The dielectric constant of a cell membrane is approximately 9.0.

PREPARE If we imagine opening up a cell membrane and flattening it out, we would get something that looks like a parallel-plate capacitor with the plates separated by a dielectric. The relevant dimensions are the same as those in Example 23.13.

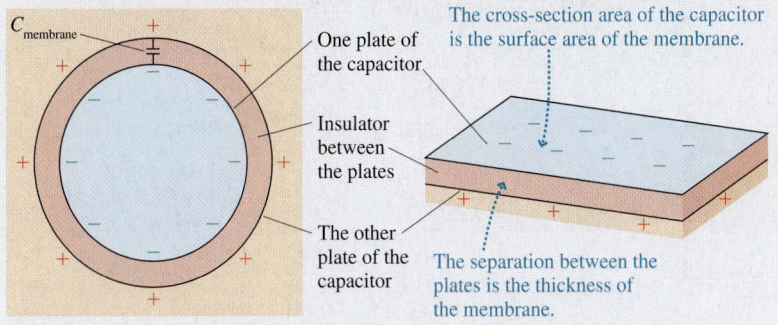

FIGURE 23.47 The cell membrane can also be modeled as a capacitor.

SOLVE The capacitance of the membrane is that of a parallel-plate capacitor filled with a dielectric, which was given in Equation 21.21. Inserting the dimensions from Example 23.13, we find

$$C_{membrane} = \frac{\kappa\epsilon_0 A}{d} = \frac{9.0(8.85 \times 10^{-12}\,\text{C}^2/\text{N}\cdot\text{m}^2)4\pi(2.5 \times 10^{-5}\,\text{m})^2}{7.0 \times 10^{-9}\,\text{m}}$$

$$= 8.9 \times 10^{-11}\,\text{F}$$

ASSESS Though the cell is small, the cell membrane has a reasonably large capacitance of ≈ 90 pF. This makes sense, because the membrane is quite thin.

Because the cell membrane has both resistance and capacitance, it can be modeled as an *RC* circuit, as shown in Figure 23.48. The membrane, like any *RC* circuit, has a time constant. The previous examples calculated the resistance and capacitance of the 7.0-nm-thick membrane of a 0.050-mm-diameter cell. We can use these numbers to compute the membrane's time constant:

$$\tau = RC = (3.2 \times 10^7\,\Omega)(8.9 \times 10^{-11}\,\text{F}) = 2.8 \times 10^{-3}\,\text{s} \approx 3\,\text{ms}$$

Indeed, if we raise the membrane potential of a real nerve cell by 10 mV (large enough to easily measure but not enough to trigger a response in the cell) the potential will decay back to its resting value with a time constant of a few ms.

But the real action happens when some stimulus *is* large enough to trigger a response in the cell. In this case, ion channels open and the potential changes in much less time than the cell's time constant, as we will see in the next section.

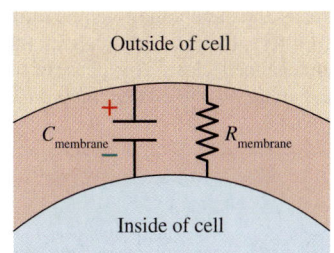

FIGURE 23.48 The cell membrane can be modeled as an *RC* circuit.

The Action Potential

Suppose a nerve cell is sitting quietly at its resting potential. The membrane potential is approximately -70 mV. However, this potential can change drastically in response to a *stimulus. Neurons*—nerve cells—can be stimulated by neurotransmitter chemicals released at synapse junctions. A neuron can also be electrically stimulated by a changing potential, which is why Galvani saw the frog's leg jump. Whatever the stimulus, the result is a rapid change called an *action potential;* this is the "firing" of a nerve cell. There are three phases in the action potential, as outlined below.

The action potential

Depolarization	Repolarization	Reestablishing resting potential

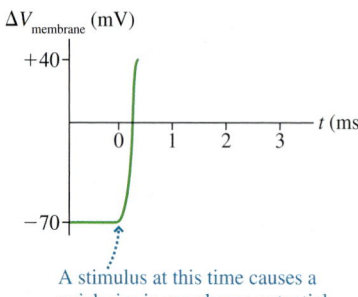

A stimulus at this time causes a quick rise in membrane potential.

The membrane potential drops rapidly, overshooting its initial value.

Diffusion of ions reestablishes the resting potential.

A stimulus causes the cell to "fire"; the first step is the opening of the sodium channels. The concentration of sodium ions is much higher outside the cell, so sodium ions flow rapidly into the cell. In less than 1 ms, this influx of positive ions raises the membrane potential from -70 mV to $+40$ mV, at which point the sodium channels close. This phase of the action potential is called *depolarization.*

The positive membrane potential that causes the sodium channels to close causes the potassium channels to open. The higher potassium concentration inside the cell drives these ions out of the cell, making the membrane potential negative. The negative potential closes the potassium channels, but a delayed response leads to a slight *overshoot* of the resting potential to about -80 mV. This phase of the action potential is called *repolarization.*

The reestablishment of the resting potential after the sodium and potassium channels close is a relatively slow process, because it is controlled by the motion of ions across the membrane. The time constant of the membrane determines how long this takes.

As soon as one action potential is complete, the cell is ready to be triggered again. The action potential is driven by ionic conduction through sodium and potassium channels, so the potential changes are quite rapid. The time for the potential to rise and then to fall is much less than the 3 ms time constant for diffusion of ions across the membrane.

The above discussion concerned nerve cells, but muscle cells undergo a similar cycle of depolarization and repolarization. The resulting potential changes are responsible for the dipole moment of the heart, and so provide the signal that is measured by an electrocardiogram, which we learned about in Chapter 21. The potential differences in the human body are small, as the changes in potential are small. But some fish, such as the electric eel, have electric organs in which the action potentials of thousands of specially adapted cells are added in series, leading to very large potential differences—hundreds of volts in the case of the electric eel.

EXAMPLE 23.15 Counting ions through a channel

Investigators can measure the ion flow through a single ion channel with the *patch clamp* technique, as illustrated in Figure 23.49. A micropipette, a glass tube ≈1 μm in diameter, makes a seal on a patch of cell membrane that includes one sodium channel. This tube is filled with a conducting salt water solution, and a very sensitive ammeter measures the current as sodium ions flow into the cell. A sodium channel passes an average current of 4.0 pA during the 0.40 ms that the channel is open during an action potential. How many sodium ions pass through the channel?

PREPARE Current is the rate of flow of charge. Each ion has charge $q = e$.

SOLVE In Chapter 22, we saw that the charge delivered by a steady current in time Δt is $Q = I\Delta t$. The amount of charge flowing through the channel in $\Delta t = 4.0 \times 10^{-4}$ s is

$$Q = I\Delta t = (4.0 \times 10^{-12} \text{ A})(4.0 \times 10^{-4} \text{ s}) = 1.6 \times 10^{-15} \text{ C}$$

This charge is due to N ions, each with $q = e$, so the number of ions is

$$N = \frac{Q}{e} = \frac{1.6 \times 10^{-15} \text{ C}}{1.6 \times 10^{-19} \text{ C}} = 10,000$$

ASSESS The number of ions flowing through one channel is not large, but a cell has a great many channels. The patch clamp technique and other similar procedures have allowed investigators to elucidate the details of the response of the cell membrane to a stimulus.

FIGURE 23.49 Measuring the current in a single sodium channel.

▶ **Electrical exercising** Perhaps you have seen devices advertised on television that purport to allow you to "effortlessly" exercise the muscles of your body. These devices use electrodes that make good electrical contact with the skin. By applying small potential differences between the electrodes, it is possible to directly stimulate the nerve cells that signal muscles to contract. If you use one of these devices, the stimulation will cause your muscles to contract whether you want them to or not! The benefits of this "exercise" are questionable, and it carries significant risks.

TRY IT YOURSELF

The Propagation of Nerve Impulses

Let's return to the question posed at the start of the section: How is a signal transmitted from the brain to a muscle in the hand? The primary cells of the nervous system, responsible for signal transmission, are known as *neurons*. The transmission of a signal to a muscle is the function of a *motor neuron,* whose structure is sketched in Figure 23.50. The transmission of signals takes place along the *axon*

The *dendrites* are short extensions to the cell that can receive signals from other neurons.

Cell body

The axon may be wrapped in an insulating *myelin sheath*, formed by specialized cells called *Schwann cells.*

A signal is propagated along the axon to where it meets and can stimulate muscle fibers.

Muscle fibers

Dendrites   Axon

The myelin sheath is broken by short, uninsulated segments called *nodes of Ranvier.*

FIGURE 23.50 A motor neuron.

(a) A model of a neuron

FIGURE 23.51 Propagation of a nerve impulse.

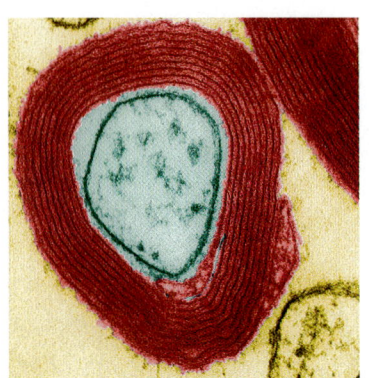

◀ This cross-section view of a myelinated axon shows that the axon (colored aqua) is wrapped several times by a Schwann cell (colored red). There is insulating myelin between each layer.

of the neuron, a long fiber—up to 1 m in length—that connects the cell body to a muscle fiber. This particular neuron has a myelin sheath around the axon, though not all neurons do.

How is a signal transmitted along an axon? The axon is long enough that different points on its membrane may have different potentials. When one point on the axon's membrane is stimulated, the membrane will depolarize at this point. The resulting action potential may trigger depolarization in adjacent parts of the membrane. Stimulating the axon's membrane at one point can trigger a *wave* of action potential—a nerve impulse—that travels along the axon. When this signal reaches a muscle cell, the muscle cell depolarizes and produces a mechanical response.

Let's look at this process in more detail. We will start with a simple model of an axon with no myelin sheath in Figure 23.51a. The sodium channels are normally closed, but if the potential at some point is raised by ≈ 15 mV, from the resting potential of -70 mV to ≈ -55 mV, the sodium channels suddenly open, sodium ions rush into the cell, and an action potential is triggered. This is the key idea: **A small increase in the potential difference across the membrane causes the sodium channels to open, triggering a large action-potential response.**

This process begins at the cell body, in response to signals the neuron receives at its dendrites. If the cell body potential goes up by ≈ 15 mV, an action potential is initiated in the cell body. As the cell body potential quickly rises to a peak of $+40$ mV, it causes the potential on the nearest section of the axon—where the axon attaches to the cell body—to rise by 15 mV. This triggers an action potential in this first section of the axon. The action potential in the first section of the axon triggers an action potential in the next section of the axon, which triggers an action potential in the next section, and so on down the axon until reaching the end.

As Figure 23.51b shows, this causes a wave of action potential to propagate down the axon. The signal moves relatively slowly. At each point on the membrane, channels must open and ions must diffuse through, which takes time. On a typical axon with no myelin sheath, the action potential propagates at a speed of about 1 m/s. If all nerve signals traveled at this speed, a decision to move your hand would take about 1 s to travel from your brain to your hand. Clearly, at least some neurons in the nervous system must transmit signals at a greater speed than this!

One way to make the signals travel more quickly is to increase an axon's diameter. The giant axon in the squid triggers a rapid escape response when the squid is threatened. This axon may have a diameter of 1 mm, a thousand times that of a typical axon, providing for the necessary rapid signal transmission. But your nervous system consists of 300 billion neurons, and they can't all be 1 mm in diameter—there simply isn't enough space in your body. In your nervous system, higher neuron signal speed is achieved in a totally different manner.

Increasing Speed by Insulation

The axons of motor neurons and most other neurons in your body can transmit signals at very high speeds because they are insulated with a myelin sheath. Look back at the diagram of a motor neuron in Figure 23.50. Schwann cells wrap the axon with myelin, insulating it electrically and chemically, with breaks at the nodes of Ranvier. The ion channels are concentrated in these nodes because this is the only place where the extracellular fluid is in contact with the cell membrane. In an insulated axon, a signal propagates by jumping from one node to the next. This process is called *saltatory conduction,* from the Latin *saltare,* "to leap."

Figure 23.52 shows a model for saltatory conduction. The membrane is triggered to depolarize at the left node. An action potential is produced here, but it can't travel down the axon as before because the axon is insulated. Instead, the potential difference between this node and the next causes ions to flow in the body of the axon. As the charges flow down the axon, the potential at the next node rises. When the potential has risen by ≈15 mV, the next node is triggered and an action potential is produced at this node. This process continues, with the depolarization "jumping" from node to node down the axon. This dramatically increases the speed of propagation, as we can show.

How rapidly does a pulse move down a myelinated axon? Figure 23.53 provides a model of the process based on *RC* circuits. The critical time for propagation is the time constant $\tau = RC$ for charging the capacitance of the segments of the axon.

The resistance of an axon between one node and the next is ≈25 MΩ. The myelin insulation increases the separation between the inner conducting fluid and the outer conducting fluid. Because the capacitance of a capacitor depends inversely on the electrode spacing d, the myelin reduces the capacitance of the membrane from the ≈90 pF we calculated earlier to ≈1.6 pF per segment. With these values, the time constant for charging the capacitor in one segment is

$$\tau = R_{axon}C_{membrane} = (25 \times 10^6 \, \Omega)(1.6 \times 10^{-12} \, F) = 40 \, \mu s$$

We've modeled the axon as a series of such segments, and the time constant is a good estimate of how much time it takes for a signal to jump from one node to the next. Because the nodes of Ranvier are spaced about 1 mm apart, the speed at which the nerve impulse travels down the axon is approximately

$$v = \frac{L_{node}}{\tau} = \frac{1.0 \times 10^{-3} \, m}{40 \times 10^{-6} \, s} = 25 \, m/s$$

Although our model of nerve-impulse propagation is very simple, this predicted speed is just about right for saltatory conduction of signals in myelinated axons. This speed is 25 times faster than that in unmyelinated axons; at this speed, your brain can send a signal to your hand in ≈$\frac{1}{25}$ s.

Your electric nature might not be as apparent as that of the electric eel, but the operation of your nervous system is inherently electrical. When you decide to move your hand, the signal from your brain travels to your hand in a process that is governed by the electrical nature of the cells in your body.

FIGURE 23.52 Nerve propagation along a myelinated axon.

(a) A model of a myelinated axon

The fluid in the axon has a certain resistivity. The axon is thin, so the resistance is large.

The interior and exterior of the axon are conducting fluid separated by insulating membrane—a capacitor.

R_{axon} C_{axon}

We model the triggering of an action potential as closing a switch connected to a battery.

(b) Signal propagation in the myelinated axon

1. An action potential is triggered at this node; we close the switch.

2. Once the switch is closed, the action potential emf drives a current down the axon and charges the capacitance of the membrane.

3. When the voltage on the capacitor exceeds a threshold, it triggers an action potential at this node—the next switch is closed.

FIGURE 23.53 A circuit model of nerve-impulse propagation along myelinated axons.

The goal of Chapter 23 has been to learn about the fundamental physical principles that govern electric circuits.

GENERAL PRINCIPLES

Kirchhoff's loop law

For a closed loop:

- Assign a direction to the current.
- Add potential differences around the loop.

$$\sum_i \Delta V_i = 0$$

Kirchhoff's junction law

For a junction:

$$\sum I_{in} = \sum I_{out}$$

Analyzing circuits

PREPARE Draw a circuit diagram.

SOLVE *Break the circuit down:*

- Reduce the circuit to the smallest possible number of equivalent resistors.
- Find the current and potential difference.

Rebuild the circuit:

- Find current and potential difference for each resistor.

ASSESS Verify that

- The sum of potential differences across series resistors matches that for the equivalent resistor.
- The sum of the currents through parallel resistors matches that for the equivalent resistor.

IMPORTANT CONCEPTS

Series elements

A series connection has no junction. The current in each element is the same.

Resistors in series can be reduced to an equivalent:

$$R_{eq} = R_1 + R_2 + R_3 + \cdots$$

Capacitors in series can be reduced to an equivalent:

$$C_{eq} = \left(\frac{1}{C_1} + \frac{1}{C_2} + \frac{1}{C_3} + \cdots\right)^{-1}$$

Parallel elements

Elements connected in parallel are connected by wires at both ends. The potential difference across each element is the same.

Resistors in parallel can be reduced to an equivalent:

$$R_{eq} = \left(\frac{1}{R_1} + \frac{1}{R_2} + \frac{1}{R_3} + \cdots\right)^{-1}$$

Capacitors in parallel can be reduced to an equivalent:

$$C_{eq} = C_1 + C_2 + C_3 + \cdots$$

APPLICATIONS

RC circuits

The discharge of a capacitor through a resistor is an exponential decay:

$$\Delta V_C = (\Delta V_C)_0 e^{-t/\tau}$$

The **time constant** for the decay is:

$$\tau = RC$$

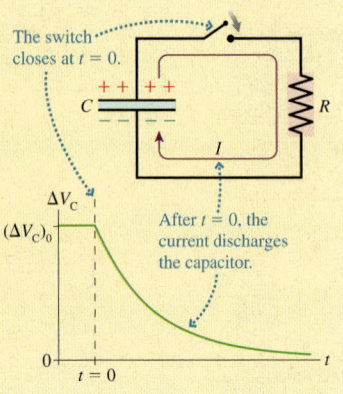

The switch closes at $t = 0$.

After $t = 0$, the current discharges the capacitor.

Electricity in the nervous system

Cells in the nervous system maintain a negative potential inside the cell membrane. When triggered, the membrane depolarizes, and generates an *action potential*.

An action potential travels as a wave along the axon of a neuron. More rapid saltatory conduction can be achieved by insulating the axon with myelin, causing the action potential to jump from node to node.

For instructor-assigned homework, go to
www.masteringphysics.com

Problem difficulty is labeled as I (straightforward) to IIIII (challenging).

Problems labeled INT integrate significant material
from earlier chapters; BIO are of biological or medical
interest.
Problems labeled: can be done on a Workbook

QUESTIONS

Conceptual Questions

1. The tip of a flashlight bulb is touching the top of a 3 V battery as shown in Figure Q23.1. Does the bulb light? Why or why not?

  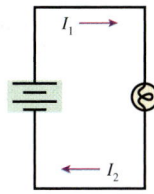

FIGURE Q23.1   **FIGURE Q23.2**

2. A flashlight bulb is connected to a battery and is glowing; the circuit is shown in Figure Q23.2. Is current I_2 greater than, less than, or equal to current I_1? Explain.

3. Current I_{in} flows into three resistors connected together one after the other as shown in Figure Q23.3. The accompanying graph shows the value of the potential as a function of position.
   a. Is I_{out} greater than, less than, or equal to I_{in}? Explain.
   b. Rank in order, from largest to smallest, the three resistances R_1, R_2, and R_3. Explain.

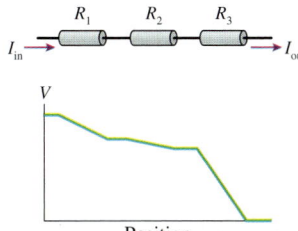

FIGURE Q23.3

4. a. What is the direction of the current through resistor R in Figure Q23.4?
   b. Which end of R is more positive, the right or left? Explain.
   c. If this circuit were analyzed in a clockwise direction, what numerical value would you assign to ΔV_R? Why?
   d. What value would ΔV_R have if the circuit were analyzed in a counterclockwise direction?

FIGURE Q23.4   **FIGURE Q23.5**

5. The wire is broken on the right side of the circuit in Figure Q23.5 at lower left. What is the potential difference ΔV_{12} between points 1 and 2? Explain.

6. The circuit in Figure Q23.6 has two resistors, with $R_1 > R_2$. Which resistor dissipates the larger amount of power? Explain.

FIGURE Q23.6   **FIGURE Q23.7**

7. The circuit in Figure Q23.7 has a battery and two resistors, with $R_1 > R_2$. Which resistor dissipates the larger amount of power? Explain.

8. In the circuit shown in Figure Q23.8, bulbs A and B are glowing. Then the switch is closed. What happens to each bulb? Does it get brighter, stay the same, get dimmer, or go out? Explain.

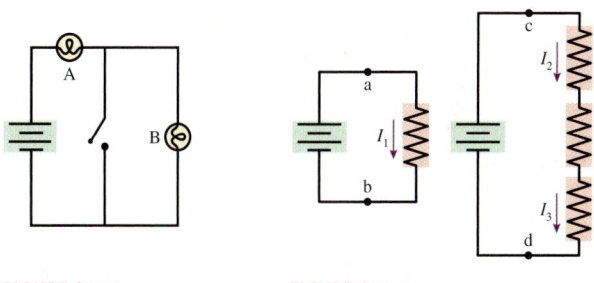

FIGURE Q23.8   **FIGURE Q23.9**

9. Figure Q23.9 shows two circuits. The two batteries are identical and the four resistors all have exactly the same resistance.
   a. Is ΔV_{ab} larger than, smaller than, or equal to ΔV_{cd}? Explain.
   b. Rank in order, from largest to smallest, the currents I_1, I_2, and I_3. Explain.

10. Figure Q23.10 shows two circuits. The two batteries are identical and the four resistors all have exactly the same resistance.
   a. Compare ΔV_{ab}, ΔV_{cd}, and ΔV_{ef}. Are they all the same? If not, rank them in order from largest to smallest. Explain.
   b. Rank in order, from largest to smallest, the five currents I_1 to I_5. Explain.

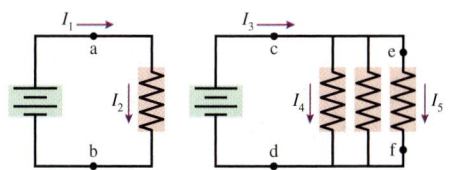

FIGURE Q23.10

11. a. In Figure Q23.11, what fraction of current I goes through the 3 Ω resistor?

    b. If the 9 Ω resistor is replaced with a larger resistor, will the fraction of current going through the 3 Ω resistor increase, decrease, or stay the same?

FIGURE Q23.11   FIGURE Q23.12

12. Two of the three resistors in Figure Q23.12 are unknown but equal. Is the total resistance between points a and b less than, greater than, or equal to 50 Ω? Explain.

13. Two of the three resistors in Figure Q23.13 are unknown but equal. Is the total resistance between points a and b less than, greater than, or equal to 200 Ω? Explain.

FIGURE Q23.13

14. Rank in order, from largest to smallest, the currents I_1, I_2, and I_3 in the circuit diagram in Figure Q23.14.

FIGURE Q23.14

FIGURE Q23.15

15. The three bulbs in Figure Q23.15 are identical. Rank the bulbs from brightest to dimmest. Explain.

16. The four bulbs in Figure Q23.16 are identical. Rank the bulbs from brightest to dimmest. Explain.

FIGURE Q23.16   FIGURE Q23.17

17. Figure Q23.17 shows five identical bulbs connected to a battery. All the bulbs are glowing. Rank the bulbs A to E, from brightest to dimmest. Explain.

18. a. The three bulbs in Figure Q23.18 are identical. Rank the bulbs from brightest to dimmest. Explain.

    b. Suppose a wire is connected between points 1 and 2. What happens to each bulb? Does it get brighter, stay the same, get dimmer, or go out? Explain.

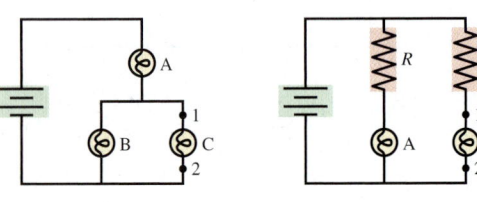

FIGURE Q23.18   FIGURE Q23.19

19. Initially, bulbs A and B in Figure Q23.19 are both glowing. Bulb B is then removed from its socket. Does removing bulb B cause the potential difference ΔV_{12} between points 1 and 2 to increase, decrease, or become zero? Explain.

20. a. Consider the points a and b in Figure Q23.20. Is the potential difference ΔV_{ab} between point a and b zero? If so, why? If not, which point is more positive?

    b. If a wire is connected between points a and b, does it carry a current? If so, in which direction—to the right or to the left? Explain.

FIGURE Q23.20   FIGURE Q23.21

21. Initially the lightbulb in Figure Q23.21 is glowing. It is then removed from its socket.

    a. What happens to the current I when the bulb is removed? Does it increase, stay the same, or decrease? Explain.

    b. What happens to the potential difference ΔV_{12} between points 1 and 2? Does it increase, stay the same, decrease, or become zero? Explain.

22. Rank in order, from largest to smallest, the equivalent capacitances $(C_{eq})_1$ to $(C_{eq})_4$ of the four groups of capacitors shown in Figure Q23.22.

FIGURE Q23.22

23. Figure Q23.23 shows a circuit consisting of a battery, a switch, two identical lightbulbs, and a capacitor that is initially uncharged.

    a. *Immediately* after the switch is closed, are either or both bulbs glowing? Explain.

    b. If both bulbs are glowing, which is brighter? Or are they equally bright? Explain.

    c. For any bulb (A or B or both) that lights up immediately after the switch is closed, does its brightness increase with time, decrease with time, or remain unchanged? Explain.

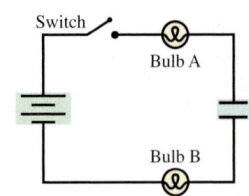

FIGURE Q23.23

24. Figure Q23.24 shows the voltage as a function of time on a capacitor as it is discharged (separately) through three different resistors. Rank in order, from largest to smallest, the values of the resistances R_1 to R_3.

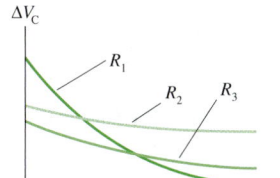

FIGURE Q23.24

25. A charged capacitor could be connected to two identical resistors in either of the two ways shown in Figure Q23.25. Which configuration will discharge of the capacitor in the shortest time once the switch is closed? Explain.

FIGURE Q23.25

26. A flashing light is controlled by the charging and discharging of an *RC* circuit. If the light is flashing too rapidly, describe two changes that you could make to the circuit to reduce the flash rate.

27. Consider the model of nerve conduction in myelinated axons
BIO presented in the chapter. Suppose the distance between the nodes of Ranvier was halved for a particular axon.
   a. How would this affect the resistance and the capacitance of one segment of the axon?
   b. How would this affect the time constant for the charging of one segment?
   c. How would this affect the signal propagation speed for the axon?

28. Adding a myelin sheath to an axon results in faster signal
BIO propagation. It also means that less energy is required for a signal to propagate down the axon. Explain why this is so.

Multiple-Choice Questions

Questions 29 and 30 are about the circuit of Figure Q23.29.

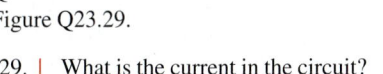

29. | What is the current in the circuit?
   A. 1.0 A
   B. 1.7 A
   C. 2.5 A
   D. 4.2 A

FIGURE Q23.29

30. | Which resistor dissipates the most power?
   A. The 4.0 Ω resistor.
   B. The 6.0 Ω resistor.
   C. Both dissipate the same power.

31. ||| Normally, household lightbulbs are connected in parallel to a power supply. Suppose a 40 W and a 60 W lightbulb are, instead, connected in series, as shown in Figure Q23.31. Which bulb is brighter?
   A. The 60 W bulb.
   B. The 40 W bulb.
   C. The bulbs are equally bright.

FIGURE Q23.31

32. || A metal wire of resistance R is cut into two pieces of equal length. The two pieces are connected together side by side. What is the resistance of the two connected wires?
   A. $R/4$
   B. $R/2$
   C. R
   D. $2R$
   E. $4R$

33. | What is the value of resistor R in Figure Q23.33?
   A. 4.0 Ω
   B. 12 Ω
   C. 36 Ω
   D. 72 Ω
   E. 96 Ω

FIGURE Q23.33

34. | Two capacitors are connected in series. They are then reconnected to be in parallel. The capacitance of the parallel combination
   A. Is less than that of the series combination.
   B. Is more than that of the series combination.
   C. Is the same as that of the series combination.
   D. Could be more or less than that of the series combination depending on the values of the capacitances.

35. | Two 10 pF capacitors are connected in series. The total capacitance of the series combination is
   A. 20 pF
   B. 10 pF
   C. 5.0 pF
   D. 2.5 pF

36. | If the membrane of a cell doubled in thickness but all other
BIO properties stayed the same, how would the resistance and the capacitance of the cell membrane change?
   A. The resistance and the capacitance would increase.
   B. The resistance would increase, the capacitance would decrease.
   C. The resistance would decrease, the capacitance would increase.
   D. The resistance and the capacitance would decrease.

PROBLEMS

Section 23.1 Circuit Elements and Diagrams

1. ‖ Draw a circuit diagram for the circuit of Figure P23.1.

FIGURE P23.1   **FIGURE P23.2**

2. ‖ Draw a circuit diagram for the circuit of Figure P23.2.
3. ‖ Draw a circuit diagram for the circuit of Figure P23.3.

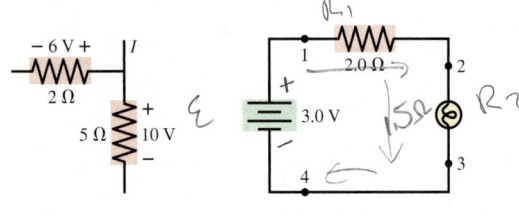

FIGURE P23.3

Section 23.2 Kirchhoff's Laws

4. ‖ In Figure P23.4, what is the current in the wire above the junction? Does charge flow toward or away from the junction?

FIGURE P23.4   **FIGURE P23.5**

5. ‖ The lightbulb in the circuit diagram of Figure P23.5 has a resistance of $1.0 \, \Omega$. Let ΔV_{ab} represent the magnitude of the potential difference between points a and b.
   a. What are the values of ΔV_{12}, ΔV_{23}, and ΔV_{34}?
   b. What are the values if the bulb is removed?
6. | a. What are the magnitude and direction of the current in the $30 \, \Omega$ resistor in Figure P23.6?
   b. Draw a graph of the potential as a function of the distance traveled through the circuit, traveling clockwise from $V = 0$ V at the lower left corner.

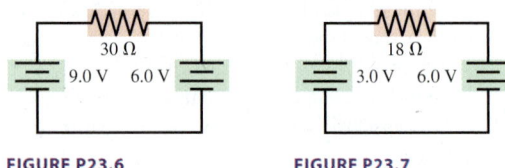

FIGURE P23.6   **FIGURE P23.7**

7. ‖ a. What are the magnitude and direction of the current in the $18 \, \Omega$ resistor in Figure P23.7?
   b. Draw a graph of the potential as a function of the distance traveled through the circuit, traveling clockwise from $V = 0$ V at the lower left corner. See Figure P23.9 for an example of such a graph.

8. | a. What is the potential difference across each resistor in Figure P23.8?
   b. Draw a graph of the potential as a function of the distance traveled through the circuit, traveling clockwise from $V = 0$ V at the lower left corner. See Figure P23.9 for an example of such a graph.

FIGURE P23.8   **FIGURE P23.9**

9. ‖ The current in a circuit with only one battery is 2.0 A. Figure P23.9 shows how the potential changes when going around the circuit in the clockwise direction, starting from the lower left corner. Draw the circuit diagram.

Section 23.3 Series and Parallel Circuits

10. | What is the equivalent resistance of each group of resistors shown in Figure P23.10?

FIGURE P23.10

11. | What is the equivalent resistance of each group of resistors shown in Figure P23.11?

FIGURE P23.11

12. ‖ An 80-cm-long wire is made by welding a 1.0-mm-diameter, 20-cm-long copper wire to a 1.0-mm-diameter, 60-cm-long iron wire. What is the resistance of the composite wire?
13. | You have a collection of $1.0 \, k\Omega$ resistors. How can you connect four of them to produce an equivalent resistance of $0.25 \, k\Omega$?
14. | You have a collection of six $1.0 \, k\Omega$ resistors. What is the smallest resistance you can make by combining them?
15. | You have three $6.0 \, \Omega$ resistors and one $3.0 \, \Omega$ resistor. How can you connect them to produce an equivalent resistance of $5.0 \, \Omega$?
16. ‖ You have four identical resistors. How can you connect them to produce an equivalent resistance which is the same as the resistance of each individual resistor?

Section 23.4 Measuring Voltage and Current

Section 23.5 More Complex Circuits

17. ▮ What is the equivalent resistance between points a and b in Figure P23.17?

FIGURE P23.17

FIGURE P23.18

18. ▮ What is the equivalent resistance between points a and b in Figure P23.18?
19. ▮ Part of a circuit is shown in Figure P23.19.
    a. What is the current through the 3.0 Ω resistor?
    b. What is the value of the current I?

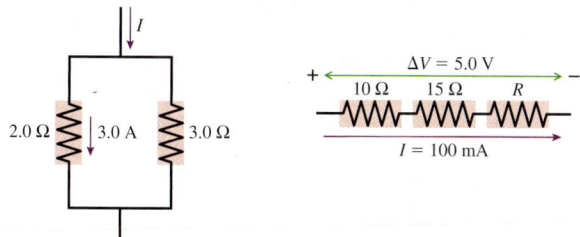

FIGURE P23.19

FIGURE P23.20

20. ▮ What is the value of resistor R in Figure P23.20?
21. ▮ What are the resistance R and the emf of the battery in Figure P23.21?

FIGURE P23.21

FIGURE P23.22

22. ▮▮ The ammeter in Figure P23.22 reads 3.0 A. Find I_1, I_2, and $\mathcal{E}$.
23. ▮ For the circuit shown in Figure P23.23, find the current through and the potential difference across each resistor. Place your results in a table for ease of reading.

FIGURE P23.23

FIGURE P23.24

24. ▮ For the circuit shown in Figure P23.24, find the current through and the potential difference across each resistor. Place your results in a table for ease of reading.

Section 23.6 Household Electricity

25. ▮ A typical American family uses 1000 kWh of electricity a month.
    a. What is the average current in the 120 V power line to the house?
    b. On average, what is the resistance of a household?
26. ▮ The wiring in the wall of your house to and from an outlet
    INT has a total resistance of typically 0.10 Ω. Suppose a device plugged into a 120 V outlet draws 10.0 A of current.
    a. What is the voltage drop along the wire?
    b. How much power is dissipated in the wire?
    c. What is the voltage drop across the device?
    d. At what rate does the device use electric energy?
27. ▮ The following appliances are connected to a single 120 V,
    INT 15 A circuit in a kitchen: a 330 W blender, a 1000 W coffeepot, a 150 W coffee grinder, and a 750 W microwave oven. If these are all turned on at the same time, will they trip the circuit breaker?
28. ▮ A 60 W (120 V) night light is turned on for an average 12 hr a day year round. What is the annual cost of electricity at a billing rate of $0.10/kWh?
29. ▮▮ Suppose you leave a 110 W television and two 100 W light-bulbs on in your house to scare off burglars while you go out dancing. If the cost of electric energy in your town is $0.12/kWh and you stay out for 4.0 hr, how much does this robbery-prevention measure cost?

Section 23.7 Capacitors in Parallel and Series

30. ▮ A 6.0 μF capacitor, a 10 μF capacitor, and a 16 μF capacitor are connected in parallel. What is their equivalent capacitance?
31. ▮ A 6.0 μF capacitor, a 10 μF capacitor, and a 16 μF capacitor are connected in series. What is their equivalent capacitance?
32. ▮ You need a capacitance of 50 μF, but you don't happen to have a 50 μF capacitor. You do have a 30 μF capacitor. What additional capacitor do you need to produce a total capacitance of 50 μF? Should you join the two capacitors in parallel or in series?
33. ▮ You need a capacitance of 50 μF, but you don't happen to have a 50 μF capacitor. You do have a 75 μF capacitor. What additional capacitor do you need to produce a total capacitance of 50 μF? Should you join the two capacitors in parallel or in series?
34. ▮ What is the equivalent capacitance of the three capacitors in Figure P23.34?

FIGURE P23.34

FIGURE P23.35

35. ▮ What is the equivalent capacitance of the three capacitors in Figure P23.35?

Section 23.8 *RC* Circuits

36. | What is the time constant for the discharge of the capacitor in Figure P23.36?

FIGURE P23.36

37. |||| The charging circuit for the flash system of a camera uses a 100 μF capacitor that is charged from a 250 V power supply. What is the most resistance that can be in series with the capacitor if the capacitor is to charge to at least 87% of its final voltage in no more than 8.0 s?

38. | What is the time constant for the discharge of the capacitor in Figure P23.38?

FIGURE P23.38

39. ||| A 10 μF capacitor initially charged to 20 μC is discharged through a 1.0 kΩ resistor. How long does it take to reduce the capacitor's charge to 10 μC?

40. | The switch in Figure P23.40 has been in position a for a long time. It is changed to position b at $t = 0$ s. What are the charge Q on the capacitor and the current I through the resistor (a) immediately after the switch is closed? (b) at $t = 50$ μs? (c) at $t = 200$ μs?

FIGURE P23.40

Section 23.9 Electricity in the Nervous System

41. || Changing the thickness of the myelin sheath surrounding
BIO an axon changes its capacitance and thus the conduction speed. A myelinated nerve fiber has a conduction speed of 55 m/s. If the spacing between nodes is 1.0 mm and the resistance of segments between nodes is 25 MΩ, what is the capacitance of each segment?

42. || To measure signal propagation in a nerve in the arm, the
BIO nerve is triggered near the armpit. The peak of the action potential is measured at the elbow and then, 4.0 ms later, 24 cm away at the wrist.
   a. What is the speed of propagation along this nerve?
   b. A determination of the speed made by measuring the time between the application of a stimulus at the armpit and the peak of an action potential at the elbow or the wrist would be inaccurate. Explain the problem with this approach, and why the noted technique is preferable.

43. | A nerve cell must finish one action potential before it can
BIO fire again. If the action potential has the form shown in the chapter, estimate the maximum possible rate (action potentials per second) at which a nerve cell can fire.

General Problems

44. | How much power is dissipated by each resistor in Fig-
INT ure P23.44?

FIGURE P23.44

45. ||| Two 75 W (120 V) lightbulbs are wired in series, then the
INT combination is connected to a 120 V supply. How much power is dissipated by each bulb?

46. |||| The corroded contacts in a lightbulb socket have 5.0 Ω total
INT resistance. How much actual power is dissipated by a 100 W (120V) lightbulb screwed into this socket?

47. || A real battery is not just an emf. We can
INT model a real 1.5 V battery as a 1.5 V emf in series with a resistor known as the "internal resistance," as shown in Figure P23.47. A typical battery has 1.0 Ω internal resistance due to imperfections that limit current through the battery. When there's no current through the battery, and thus no voltage drop across the internal resistance, the potential difference between its terminals is 1.5 V, the value of the emf. Suppose the terminals of this battery are connected to a 2.0 Ω resistor.

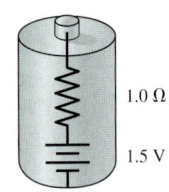

FIGURE P23.47

   a. What is the potential difference between the terminals of the battery?
   b. What fraction of the battery's power is dissipated by the internal resistance?

48. | For the real battery shown in Figure P23.47, calculate the power dissipated by a resistor R connected to the battery when (a) $R = 0.25$ Ω, (b) $R = 0.50$ Ω, (c) $R = 1.0$ Ω, (d) $R = 2.0$ Ω, and (e) $R = 4.0$ Ω. (Your results should suggest that maximum power dissipation is achieved when the external resistance R equals the internal resistance. This is true in general.)

49. ||| When two resistors are connected in parallel across a battery of unknown voltage, one resistor carries a current of 3.2 A while the second carries a current of 1.8 A. What current will be supplied by the same battery if these two resistors are connected to it in series?

50. || The 10 Ω resistor in Figure P23.50 is dissipating 40 W
INT of power. How much power are the other two resistors dissipating?

FIGURE P23.50   **FIGURE P23.51**

51. |||| At this instant, the current in the circuit of Figure P23.51 is
INT 20 mA in the direction shown and the capacitor charge is 200 μC. What is the resistance R?

52. | What is the equivalent resistance between points a and b in Figure P23.52?

FIGURE P23.52

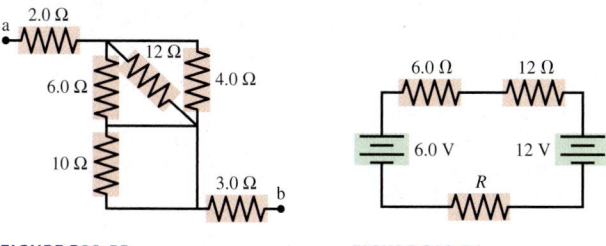

FIGURE P23.53

53. ‖ What is the equivalent resistance between points a and b in Figure P23.53?

54. ‖ You have three 12 Ω resistors. Draw diagrams showing how you could arrange all three so that their equivalent resistance is (a) 4.0 Ω, (b) 8.0 Ω, (c) 18 Ω, and (d) 36 Ω.

55. ‖‖ What is the equivalent resistance between points a and b in Figure P23.55?

FIGURE P23.55

FIGURE P23.56

56. | There is a current of 0.25 A in the circuit of Figure P23.56.
INT a. What is the direction of the current? Explain.
  b. What is the value of the resistance R?
  c. What is the power dissipated by R?
  d. Make a graph of potential versus position, starting from $V = 0$ V in the lower left corner and proceeding clockwise. See Figure P23.9 for an example.

57. ‖ A circuit you're building needs an ammeter that goes from 0 mA to a full-scale reading of 50.0 mA. Unfortunately, the only ammeter in the storeroom goes from 0 μA to a full-scale reading of only 500 μA. Fortunately, you can make this ammeter work by putting it in a measuring circuit, as shown in Figure P23.57. Assume that the ammeter is ideal.

FIGURE P23.57

  a. What value of R must you use so that the meter will go to full scale when the current I is 50.0 mA?
  Hint: When $I = 50.0$ mA, the ammeter is reading its maximum value.
  b. What is the equivalent resistance of your measuring circuit?

58. ‖ A circuit you're building needs a voltmeter that goes from 0 V to a full-scale reading of 5.0 V. Unfortunately, the only meter in the storeroom is an *ammeter* that goes from 0 μA to a full-scale reading of 500 μA. It is possible to use this meter to measure voltages by putting in a measuring circuit as shown in Figure P23.58.

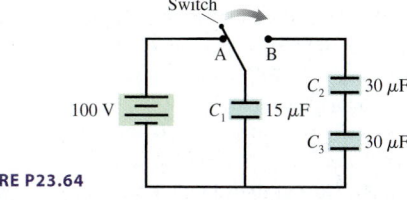

FIGURE P23.58

What value of R must you use so that the meter will go to full scale when the potential difference ΔV is 5.0 V? Assume that the ammeter is ideal.

59. ‖‖‖ For the circuit in Figure P23.59, what are (a) the current
INT through the 2.0 Ω resistor, (b) the power dissipated by the 20 Ω resistor?

FIGURE P23.59

FIGURE P23.60

60. ‖‖‖ For the circuit shown in Figure P23.60, find the current through and the potential difference across each resistor. Place your results in a table for ease of reading.

61. ‖‖‖ For the circuit shown in Figure P23.61, find the current through and the potential difference across each resistor. Place your results in a table for ease of reading.

FIGURE P23.61

62. ‖ You have three 12 μF capacitors. Draw diagrams showing how you could arrange all three so that their equivalent capacitance is (a) 4.0 μF, (b) 8.0 μF, (c) 18 μF, and (d) 36 μF.

63. ‖‖‖ What is the capacitance of the two electrodes in Figure P23.63?
  Hint: Think of this as a combination of capacitors.

Electrodes extend 1.0 cm into page.

FIGURE P23.63

64. ‖ Initially, the switch in Figure P23.64 is in position A and capacitors C_2 and C_3 are uncharged. Then the switch is flipped to position B. Afterward, what are the charge on and the potential difference across each capacitor?

FIGURE P23.64

65. ‖ The capacitor in an *RC* circuit is discharged with a time
INT constant of 10 ms. At what time after the discharge begins is (a) the charge on the capacitor reduced to half its initial value and (b) the energy stored in the capacitor reduced to half its initial value?

66. ‖ What value resistor will discharge a 1.0 μF capacitor to 10% of its initial charge in 2.0 ms?

67. ‖ A capacitor is discharged through a 100 Ω resistor. The discharge current decreases to 25% of its initial value in 2.5 ms. What is the value of the capacitor?

68. ‖‖ A 50 μF capacitor that had been charged to 30 V is discharged through a resistor. Figure P23.68 shows the capacitor voltage as a function of time. What is the value of the resistance?

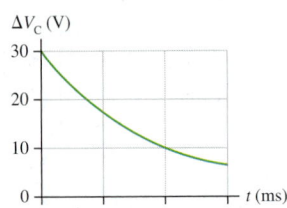

FIGURE P23.68

69. ‖‖‖ The combination of capacitors in Figure P23.69 is charged to 10 V and the switch closes at $t = 0$ s. At what time has the current in the 8.0 Ω resistor decayed to half the value it had immediately after the switch was closed?

FIGURE P23.69

70. ‖‖‖‖ The switch in Figure P23.70 has been closed for a very long time.
    a. What is the charge on the capacitor?
    b. The switch is opened at $t = 0$ s. At what time has the charge on the capacitor decreased to 10% of its initial value?

FIGURE P23.70

71. ‖‖‖‖ In Example 23.14 we estimated the capacitance of the cell
BIO membrane to be 89 pF, and in Example 23.15 we found that approximately 10,000 Na$^+$ ions flow through an ion channel when it opens. Based on this information and what you learned in this chapter about the action potential, estimate the total number of sodium ion channels in a cell.

72. ‖ The giant axon of a squid is 0.5 mm in diameter, 10 cm
BIO long, and not myelinated. Unmyelinated cell membranes
INT behave as capacitors with 1 μF of capacitance per square centimeter of membrane area. When the axon is charged to the -70 mV resting potential, what is the energy stored in this capacitance?

73. ‖ A cell has a 7.0-nm-thick membrane with a total membrane
BIO area of 6.0×10^{-9} m^2.
    a. We can model the cell as a capacitor, as we have seen. What is the magnitude of the charge on each "plate" when the membrane is at its resting potential of -70 mV?
    b. How many sodium ions does this charge correspond to?

Passage Problems

The Defibrillator BIO

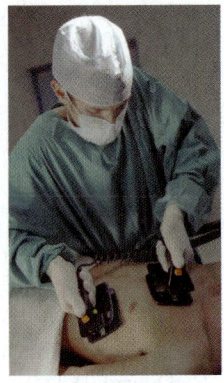

A defibrillator is designed to pass a large current through a patient's torso in order to stop dangerous heart rhythms. Its key part is a capacitor that is charged to a high voltage. The patient's torso plays the role of a resistor in an *RC* circuit. When a switch is closed, the capacitor discharges through the patient's torso. A jolt from a defibrillator is intended to be intense and rapid; the maximum current is very large, so the capacitor discharges quickly. This rapid pulse depolarizes the heart, stopping all electrical activity. This allows the heart's internal nerve circuitry to reestablish a healthy rhythm.

A typical defibrillator has a 32 μF capacitor charged to 5000 V. The electrodes connected to the patient are coated with a conducting gel that reduces the resistance of the skin to where the effective resistance of the patient's torso is 100 Ω.

74. ‖ Which pair of graphs in Figure P23.74 best represents the capacitor voltage and the current through the torso as a function of time after the switch is closed?

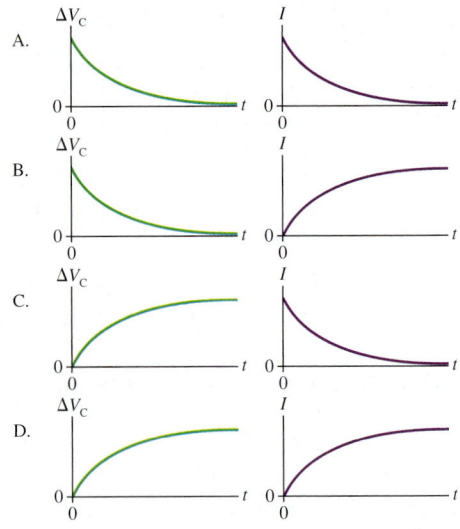

FIGURE P23.74

75. ‖ For the values noted in the passage above, what is the time constant for the discharge of the capacitor?
    A. 3.2 μs B. 160 μs C. 3.2 ms D. 160 ms

76. ‖ If a patient receives a series of jolts, the resistance of the torso may increase. How does such a change affect the initial current and the time constant of subsequent jolts?
    A. The initial current and the time constant both increase.
    B. The initial current decreases, the time constant increases.
    C. The initial current increases, the time constant decreases.
    D. The initial current and the time constant both decrease.

77. ‖ In some cases, the defibrillator may be charged to a lower voltage. How will this affect the time constant of the discharge?
    A. The time constant will increase.
    B. The time constant will not change.
    C. The time constant will decrease.

Electric Fish BIO INT

The voltage produced by a single nerve or muscle cell is quite small, but there are many species of fish that use multiple action potentials in series to produce significant voltages. The electric organs in these fish are composed of specialized disk-shaped cells called *electrocytes*. The cell at rest has the usual potential difference between the inside and the outside, but the net potential difference *across* the cell is zero. An electrocyte is connected to nerve fibers that initially trigger a depolarization in one side of the cell but not the other. For the very short time of this depolarization, there is a net potential difference across the

Electrolyte during depolarization of one side of cell.

Sodium channels open on this side only.

Na^+

Na^+

V

x

ΔV_{cell}

There is a net voltage across the cell.

FIGURE P23.78

cell, as shown in Figure P23.78. Stacks of these cells connected in series can produce a large total voltage. Each stack can produce a small current; for more total current, more stacks are needed, connected in parallel.

78. | In an electric eel, each electrocyte can develop a voltage of 150 mV for a short time. For a total voltage of 450 V, how many electrocytes must be connected in series?
    A. 300   B. 450   C. 1500   D. 3000

79. | An electric eel produces a pulse of current of 0.80 A at a voltage of 500 V. For the short time of the pulse, what is the instantaneous power?
    A. 400 W   B. 500 W   C. 625 W   D. 800 W

80. | Electric eels live in fresh water. The torpedo ray is an electric fish that lives in salt water. The electrocytes in the ray are grouped differently than in the eel; each stack of electrocytes has fewer cells, but there are more stacks in parallel. Which of the following best explains the ray's electrocyte arrangement?
    A. The lower resistivity of salt water requires more current but lower voltage.
    B. The lower resistivity of salt water requires more voltage but lower current.
    C. The higher resistivity of salt water requires more current but lower voltage.
    D. The higher resistivity of salt water requires more voltage but lower current.

81. | The electric catfish is another electric fish that produces a voltage pulse by means of stacks of electrocytes. As the fish grows in length, the magnitude of the voltage pulse the fish produces grows as well. The best explanation for this change is that, as the fish grows,
    A. The voltage produced by each electrocyte increases.
    B. More electrocytes are added to each stack.
    C. More stacks of electrocytes are added in parallel to the existing stacks.
    D. The thickness of the electrocytes increases.

STOP TO THINK ANSWERS

Stop to Think 23.1: A, B, and D. These three are the same circuit because the logic of the connections is the same. In C, the functioning of the circuit is changed by the extra wire connecting the two sides of the capacitor.

Stop to Think 23.2: C = D > A = B. The two bulbs in series are of equal brightness, as are the two bulbs in parallel. But the two bulbs in series have a larger resistance than a single bulb, so there will be less current through the bulbs in series than the bulbs in parallel.

Stop to Think 23.3: C. The voltmeter must be connected in parallel with the resistor, and the ammeter in series.

Stop to Think 23.4: A > B > C = D. All the current from the battery goes through A, so it is brightest. The current divides at the junction, but not equally. Because B is in parallel with C + D, but has half the resistance, twice as much current travels through B as through C + D. So B is dimmer than A but brighter than C and D. C and D are equal because of conservation of current.

Stop to Think 23.5: $(C_{eq})_B > (C_{eq})_A > (C_{eq})_C$. Two capacitors in parallel have a larger capacitance than either alone; two capacitors in series have a smaller capacitance than either alone.

Stop to Think 23.6: B. The two 2 Ω resistors are in series and equivalent to a 4 Ω resistor. Thus $\tau = RC = 4$ s.

24

MAGNETIC FIELDS AND FORCES

This image of a patient's knee was made with magnetic fields, not x rays. How can we use magnetic fields to visualize the inside of the body?

Looking Ahead ▶▶

The goal of Chapter 24 is to learn about magnetic fields, and how magnetic fields exert forces on currents and moving charges. In this chapter, you will learn to:

▶ Understand basic magnetic phenomena.

▶ Draw simple magnetic fields and calculate the magnetic fields due to line and loop currents.

▶ Determine the magnetic force on a moving charged particle and the particle's subsequent motion.

▶ Determine magnetic forces on currents and between currents.

▶ Understand magnetism in materials and the atomic basis of magnetism.

Looking Back ◀◀

This chapter uses what you have learned about circular motion, torque, electric fields and electric dipoles. Please review:

The runner slides into second base and hits the bag hard. He feels a sudden pain in his knee, and he has to be helped off the field. Twenty years ago such a knee injury might have required exploratory surgery, so that the physician could probe for torn tendons or ligaments. Actually repairing the joint would require a second surgery. Today, it's possible to create exquisite pictures of the interior of the knee joint without a single incision being made, by employing the technique of magnetic resonance imaging (MRI). This technique uses a strong magnetic field that penetrates the body and interacts with the microscopic magnets that are part of each atom in our bodies.

Our task in this chapter is to investigate magnets and magnetism. Magnets are all around you. In addition to holding shopping lists and cartoons on refrigerators, magnets allow you to run electric motors, produce a picture on your television screen, store information on computer disks, and listen to music using loudspeakers.

Just what is magnetism? How are magnetic fields created? What are their properties? How are they used? These are some of the questions we will address.

24.1 Magnetism

We began our investigation of electricity in Chapter 20 by looking at the results of simple experiments with charged rods. Let's follow a similar approach with magnetism.

Discovering magnetism

Experiment 1
If a bar magnet is taped to a piece of cork and allowed to float in a dish of water, it always turns to align itself in an approximate north-south direction. The end of a magnet that points north is called the *north-seeking pole,* or simply the **north pole.** The other end is the **south pole.**

North
South
The needle of a compass is a small magnet.

A magnet that is free to pivot like this is called a **compass.**

Experiment 2

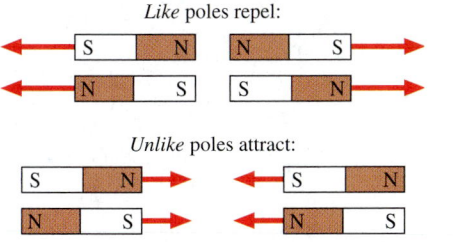
Like poles repel:
Unlike poles attract:

If the north pole of one magnet is brought near the north pole of another magnet, they exert repulsive forces on each other. Two south poles also repel each other, but the north pole of one magnet exerts an attractive force on the south pole of another magnet.

Experiment 3
Since a compass needle is itself a little bar magnet, the north pole of a bar magnet attracts the south pole of a compass needle and repels the north pole.

Compass
Bar magnet

Experiment 4

Cutting a bar magnet in half produces two weaker but still complete magnets, each with a north pole and a south pole. No matter how small the magnets are cut, even down to microscopic sizes, each piece remains a complete magnet with two poles.

Experiment 5
Magnets can pick up some objects, such as paper clips, but not all. If an object is attracted to one pole of a magnet, it is also attracted to the other pole. Most materials, including copper (a penny), aluminum, glass, and plastic, experience no force from a magnet.

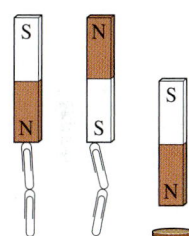

Experiment 6
When a magnet is brought near an electroscope, the leaves of the electroscope remain undeflected. If a charged rod is brought near a magnet, there is a weak *attractive* force on *both* ends of a magnet. However, the force is the same as the force on a metal bar that isn't a magnet, so it is simply a polarization force like the ones we studied in Chapter 21. Other than polarization forces, charged objects have *no effects* on magnets.

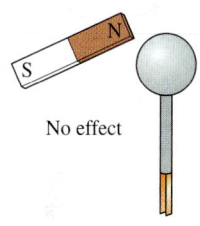
No effect

What do these experiments tell us?

1. Experiment 6 reveals that **magnetism is not the same as electricity.** Magnetic poles and electric charges share some similar behavior, but they are not the same. The magnetic force is a force of nature that we have not previously encountered.

2. **Magnetism is a long-range force.** Paper clips leap up to a magnet. You can feel the pull as you bring a refrigerator magnet close to the refrigerator.

3. **Magnets have two poles,** called north and south poles. The names are merely descriptive; they tell us nothing about how magnetism works. Two like poles exert repulsive forces on each other; two unlike poles exert attractive forces on each other. The behavior is *analogous* to that of electric charges, but, as noted, magnetic poles and electric charges are *not* the same. (In this text we will indicate the north pole of a magnet with red shading and the south pole with white.)

4. The poles of a bar magnet can be identified by using it as a compass. Other magnets, such as flat refrigerator magnets or horseshoe magnets, aren't so easily made into a compass, but their poles can be identified by testing them against a bar magnet. **A pole that repels a known south pole and attracts a known north pole must be a south magnetic pole.**

5. Materials that are attracted to a magnet or that a magnet sticks to are called **magnetic materials.** The most common magnetic material is iron. Others include nickel and cobalt. Magnetic materials are attracted to *both* poles of a magnet. This attraction is analogous to how neutral objects are attracted to both positively and negatively charged rods by the polarization force. The difference is that *all* neutral objects are attracted to a charged rod whereas only a few materials are attracted to a magnet.

Our goal is to develop a theory of magnetism that will enable us to explain these observations.

Magnetic Dipoles and Compasses

It is a strange observation that cutting a magnet in half yields two weaker but still complete magnets, each with a north pole and a south pole. Every magnet that has ever been observed has both a north and a south pole, making every magnet a **magnetic dipole.** A magnetic dipole is analogous to an electric dipole, but the two charges in an electric dipole can be separated and used individually. This appears *not* to be true for a magnetic dipole. The needle of a compass is a small, straight magnet, and so a compass needle is an especially simple magnetic dipole.

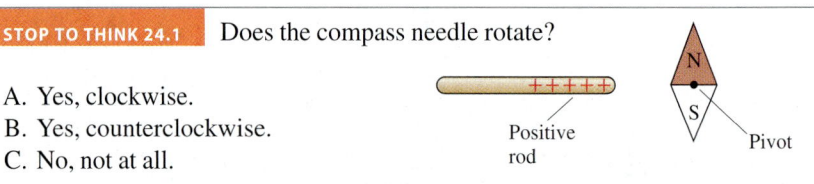

STOP TO THINK 24.1   Does the compass needle rotate?

A. Yes, clockwise.
B. Yes, counterclockwise.
C. No, not at all.

24.2 The Magnetic Field

When we studied the *electric* force between two charges in Chapter 20, we developed a new way to think about forces between charges—the *field model.* In this viewpoint, the space around a charge is not empty: The charge alters the space around it by creating an *electric field.* A second charge brought into this electric field then feels a force due to the *field.*

In Experiment 3 above, we learned that if the north pole of a bar magnet is brought near a compass, the compass needle will turn so that its south pole is toward the magnet. The concept of a field can also be used to describe the force that turns the compass: **Every magnet sets up a *magnetic* field in the space around it.** If another magnet—such as a compass needle—is then brought into this field, the second magnet will feel a force from the *field* of the first magnet. In this section, we'll see how to define the magnetic field, and then we'll study what the magnetic field looks like for some common shapes and arrangements of magnets.

Measuring the Magnetic Field

What does the direction in which a compass needle points tell us about the magnetic field at the position of the compass? We can gain some insight by recalling how an *electric* dipole behaves when placed in an electric field, as shown in Figure 24.1. In Chapter 20 we learned that an electric dipole experiences a *torque* when placed in an electric field, a torque that tends to align the axis of the dipole with the field. This means that the *direction* of the electric field is the same as the direction of the dipole's axis. Furthermore, the torque on the dipole is larger when the electric field is stronger, hence the *magnitude* of the field is proportional to the torque on the dipole.

The magnetic dipole of a compass needle behaves very similarly when it is brought near a magnet: The magnetic field exerts a torque on the compass needle, causing the needle to point in some particular direction. By analogy with how an

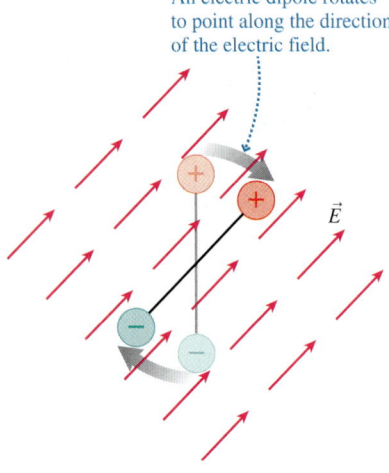

An electric dipole rotates to point along the direction of the electric field.

$\vec{E}$

FIGURE 24.1 An electric dipole in an electric field.

electric dipole behaves in an electric field, then, we are led to the definition of the magnetic field shown in Figure 24.2. Because the magnetic field has both a direction and a magnitude, we need to represent it using a *vector*. We will use the symbol $\vec{B}$ to represent the magnetic field and B to represent the magnitude of the field.

- The *direction* of the magnetic field is the direction that the north pole of a compass needle points.

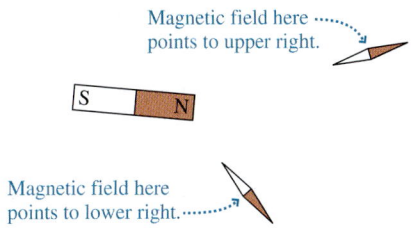

- The *strength* of the magnetic field is proportional to the torque felt by the compass needle if it is turned slightly away from the field direction.

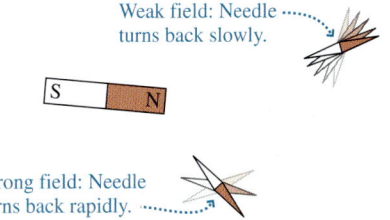

FIGURE 24.2 A compass needle defines the magnetic field.

We can produce a "picture" of the magnetic field by using *iron filings*—very small elongated grains of iron. When these are sprinkled around a magnet, each grain acts as a tiny compass needle, and rotates to point along the field direction. If there are enough grains, iron filings can give a very detailed representation of the magnetic field, as shown in Figure 24.3. Since the north and south poles of iron filings are not marked, we need to use compasses to tell which direction the field points near the magnet's poles. A key observation is that **the magnetic field of a magnet points *away* from the north pole and *toward* the south pole.**

FIGURE 24.3 Revealing the field of a bar magnet using iron filings.

Magnetic Field Vectors and Field Lines

We can draw the field of a magnet such as the one shown in Figure 24.3 in either of two ways. When we want to represent the magnetic field at one particular point, the **magnetic field vector** representation is especially useful. But if we want an overall representation of the field, **magnetic field lines** are often simpler to use. These two representations are similar to the *electric field* vectors and lines used in Chapter 20, and we'll use similar rules to draw them.

Let's start by drawing some magnetic field vectors that represent the magnetic field of our bar magnet. As shown in Figure 24.4, we can imagine placing a number of compasses near the magnet to measure the direction and magnitude of the

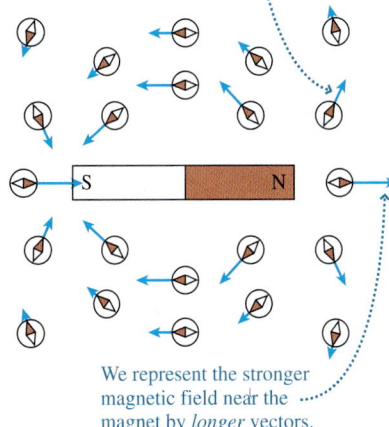

FIGURE 24.4 Mapping out the field of a bar magnet using compasses.

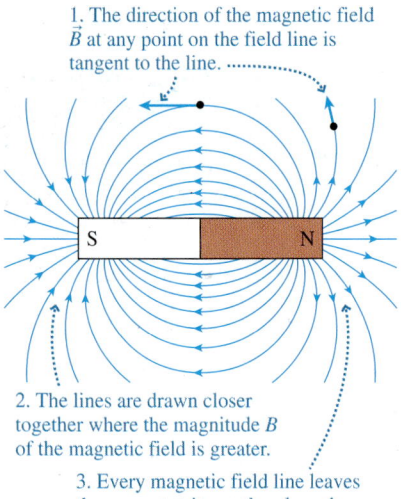

1. The direction of the magnetic field $\vec{B}$ at any point on the field line is tangent to the line. ⋯⋯⋯⋯⋯

2. The lines are drawn closer together where the magnitude B of the magnetic field is greater.

3. Every magnetic field line leaves the magnet at its north pole and enters the magnet at its south pole.

magnetic field. To represent the field at the location of one of the compasses, we then draw a vector with its *tail* at that location. Figure 24.4 shows how to choose the direction and magnitude of this vector.

Although we've drawn magnetic field vectors at only a few points around the magnet, it's important to remember that the magnetic field exists at *every* point around the magnet, including every point in between the places where we've drawn the field vectors.

We can also represent the magnetic field using magnetic field lines. The rules for drawing these lines are similar to those for drawing electric field lines of Chapter 20, and are shown for magnetic fields in Figure 24.5. We also show two magnetic field vectors, so that you can compare the two representations.

Now that we know how to think about magnetic fields, let's look at some magnetic fields from magnets of different arrangements. We'll use the iron filing method to show the lines from real magnets, along with a drawing of the field lines.

◄ **FIGURE 24.5** Drawing the magnetic field lines of a bar magnet.

An atlas of magnetic fields produced by magnets

A single bar magnet	A single bar magnet (closeup)	Two bar magnets, unlike poles facing	Two bar magnets, like poles facing

The magnetic field lines start on the north pole (red) and end on the south pole (white).

As you move away from the magnet, the field lines are farther apart, indicating a weaker field.

Closer to the magnet, we can more clearly see how the field lines always start on north poles and end on south poles.

With more than one magnet, the field lines still start on a north pole and end on a south pole. But they can start on the north pole of one magnet, and end on the south pole of another.

With two like poles placed nearby, the field lines starting on the north poles curve sharply toward their south poles in order to avoid the north pole of the other magnet.

Magnetic Fields Around Us

Magnetic fields exist everywhere around us. We live in a magnetic field that is created in the earth's core, and magnetic fields are used in a wide range of technological applications. An everyday example is the flexible refrigerator magnet. As seen in Figure 24.6, these magnets have an unusual arrangement of long, striped poles. This arrangement forces most of the field to exit the brown side of the magnet, which is why this side sticks better to your refrigerator than the label side (see Conceptual Example 24.5).

Another application that has great technological importance is the use of magnetic materials to store information on computer hard disk drives. As shown in

(a)

The magnetic field extends mostly out of this side of the magnet.

The poles of the magnet are arranged like long U-shaped magnets.

(b)

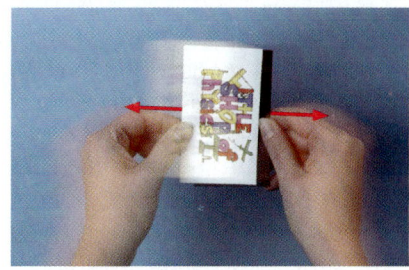

The pole structure can be revealed by a special film that contains fine iron filings.

FIGURE 24.6 The magnetic field of a refrigerator magnet.

Figure 24.7, a hard drive consists of a rapidly rotating disk with a thin magnetic coating on its surface. Information for computers is stored as digital zeros and ones, and on the disk these are stored as tiny magnetic dipoles. Each dipole is less than 100 nm long! The direction of the dipoles can be changed by the *write head*—a tiny switchable magnet that skims over the surface of the disk. The information can then be retrieved by the *read head*—a small probe that is sensitive to the magnetic fields of the tiny dipoles.

Cross-section of the magnetic coating on a hard disk

| S | N S | N N | S N | S |

"0" "1" "0"

Zeros are stored as one longer magnet.

Ones are stored as two short magnets.

This arm rapidly moves the read and write head to the required position over the disk.

FIGURE 24.7 Computer hard disks store information using magnetic fields.

One of the most remarkable facts about magnets is that a bar magnet that is free to pivot—a compass—always swings so that one end, its north pole, points geographically north. But we've also seen that if a magnet is brought near a compass, the compass swings so that its north pole faces the south pole of the magnet. These observations can be reconciled if we assume that the earth itself is a large magnet, as shown in Figure 24.8a, where **the south pole of the earth's magnet is located near the north geographic pole of the earth.** You can see from the figure that the north pole of a compass needle placed, for example, at the equator, will point toward the south pole of the earth's magnet—that is, to the north!

Figure 24.8a shows that the south pole of the earth's magnet is not coincident with the earth's north geographic pole. In fact, the magnetic poles of the earth are in constant motion; as Figure 24.8b shows, the south pole of the earth's magnet has moved some 1100 km through northern Canada over the last century. And on longer time scales, there have been complete reversals of the earth's field, with the north and south magnetic poles switching places. This last occurred some 780,000 years ago. Although these processes are not completely understood, geophysicists generally agree that the earth's magnetic poles arise from currents in its molten iron core.

(a)

The south pole of the earth's magnet is not at the north geographic pole.

Earth's rotational axis

North geographic pole

South magnetic pole

Equator

North magnetic pole

(b)

2001
1994
1984
1972
1962
1948
1904

80° N

70° N

110° W   100° W   90° W

FIGURE 24.8 The earth is a large magnet, whose poles are gradually moving.

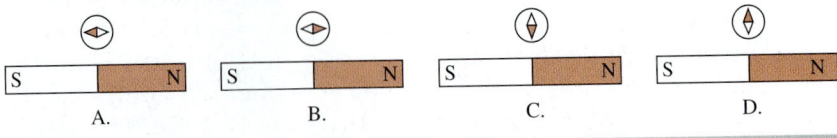

A compass is placed next to a bar magnet as shown. In which direction will it point?

24.3 Electric Currents Also Create Magnetic Fields

As electricity began to be studied seriously in the 18th century, some scientists speculated that there might be a connection between electricity and magnetism. Interestingly, the link between the two was discovered *during a classroom lecture demonstration* in 1819 by the Danish scientist Hans Christian Oersted. Oersted was using a battery to produce a large current in a wire. By chance, a compass was sitting next to the wire, and Oersted noticed that the current caused the compass needle to turn. In other words, the compass responded as if a magnet had been brought near.

The significance of this serendipitous observation was immediately apparent to Oersted. Is the force on a compass due to a current-carrying wire related to the force on a compass due to a magnet? To begin to answer this question, let's look at iron filing patterns produced by currents in different shapes of wires—straight wires and several kinds of loops. Then we can see if the patterns produced by current-carrying wires and those produced by magnets show any similarities.

An atlas of magnetic fields produced by currents in wires

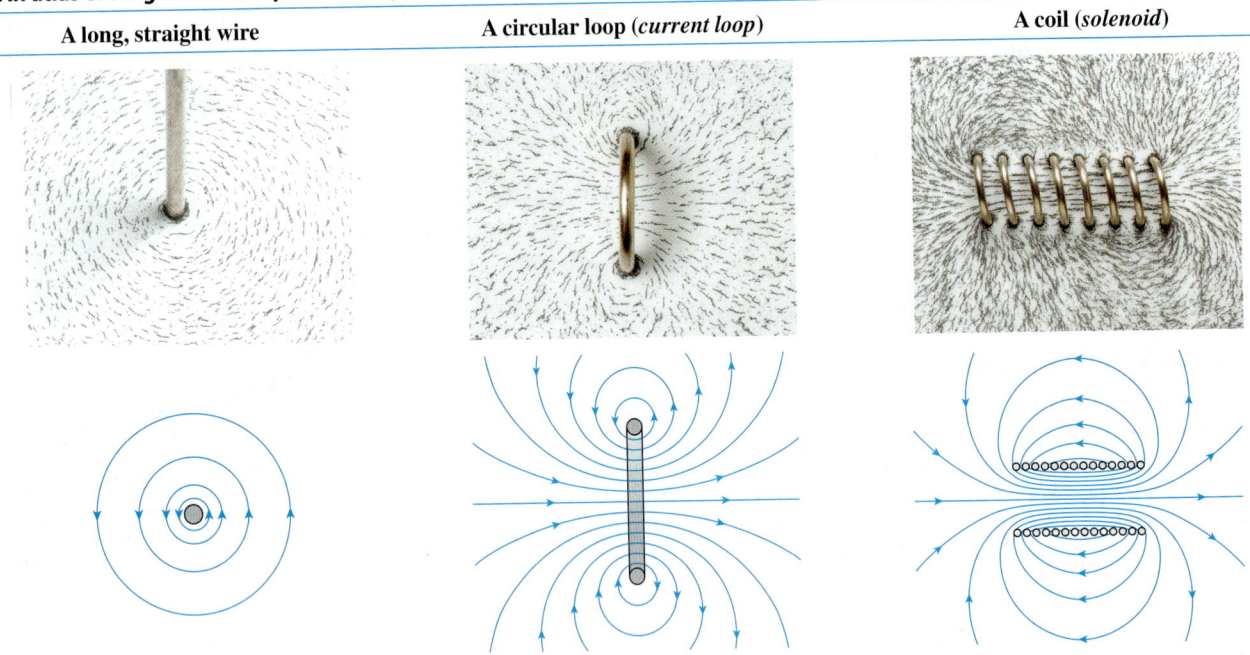

| **A long, straight wire** | **A circular loop (*current loop*)** | **A coil (*solenoid*)** |

The magnetic field lines form *circles* around the wire. The iron filings are less affected by the field as the distance from the wire increases, indicating that the field is getting weaker as the distance increases.

The magnetic field lines curve through the center of the loop, around the outside, and back through the loop's center, forming complete closed curves.

The field lines far from the loop look like the field lines far from a (weak) bar magnet.

A coil or **solenoid** is essentially a series of current loops placed along a common axis. The field outside the coil is very weak compared to the strong and nearly uniform field inside.

Outside the solenoid, note how the field lines are similar to those close to a (weak) bar magnet.

From this atlas we see that there are some striking similarities between the iron filing patterns created by currents in wires and those created by magnets. Both currents and magnets evidently affect iron filings and compasses in very similar ways. It seems reasonable, then, to infer that magnetic fields are created not only by magnets—**magnetic fields are also created by currents.**

Although there are some basic similarities between fields produced by currents and those produced by magnets, there appears to be a key difference as well. Earlier we found that the field lines of magnets start and end on magnetic poles. The field lines due to currents have no start or end: They form complete closed curves. However, if we consider the field lines continuing *inside* a magnet, we find that the magnetic field lines of magnets also form complete closed curves, as shown in Figure 24.9. We see that although outside the magnet the field lines appear to begin and end on poles, the lines actually complete a closed curve *inside* the magnet. In fact, *all* magnetic field lines form complete curves.

In this and the next section, we'll explore in some detail the magnetic fields created by currents; later, we'll look again at fields due to magnets. Ordinary magnets are often called **permanent magnets** to distinguish their unchanging magnetism from that caused by currents that can be switched on and off. We look at magnetism in this order because magnetism from currents is easier to understand, but keep in mind that currents and magnets are both equally important sources of magnetic fields.

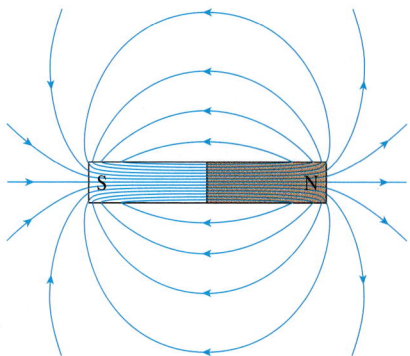

FIGURE 24.9 Field lines form closed curves for magnets, too.

The Magnetic Field of a Straight Current-Carrying Wire

In the atlas above, we see that currents in different shapes of wires produce different magnetic fields. Let's begin with the simplest shape—a long, straight wire. From the atlas picture, we see that the iron filings line up in *circles* around the wire. As Figure 24.10 shows, we can also use our basic instrument, the compass, to explore the magnetic field.

To help remember in which direction compasses will point, we use the *right-hand rule* shown in Tactics Box 24.1. We'll later use this same rule to find the direction of the magnetic field due to several other shapes of current-carrying wire, so we'll call this rule the **right-hand rule for fields.**

(MP) TACTICS BOX 24.1 Right-hand rule for fields   ✐ Exercises 5–10

❶ Point your *right* thumb in the direction of the current.

❷ Curl your fingers around the wire to indicate a circle.

❸ Your fingers point in the direction of the magnetic field lines around the wire.

Magnetism is more demanding than electricity in often requiring a three-dimensional perspective of the sort shown in Tactics Box 24.1. But since two-dimensional figures are easier to draw, we will make as much use of them as we can. Consequently, we will often need to indicate field vectors or currents that are perpendicular to the page. Figure 24.11a on the next page shows the notation we will use. Then Figure 24.11b demonstrates this notation by showing the

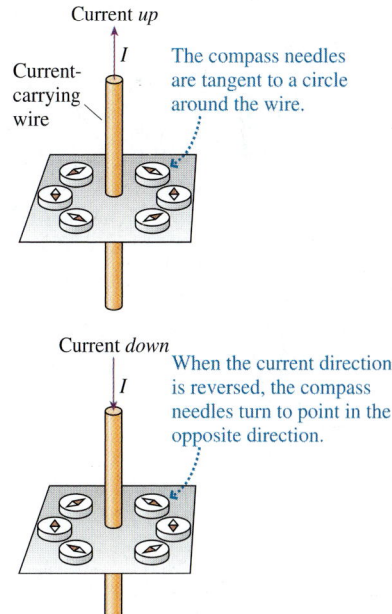

No current

North

South

With no current in the wire, the compass needles point toward the earth's north pole.

Current *up*

Current-carrying wire

I

The compass needles are tangent to a circle around the wire.

Current *down*

I

When the current direction is reversed, the compass needles turn to point in the opposite direction.

FIGURE 24.10 How compasses respond to a current-carrying wire.

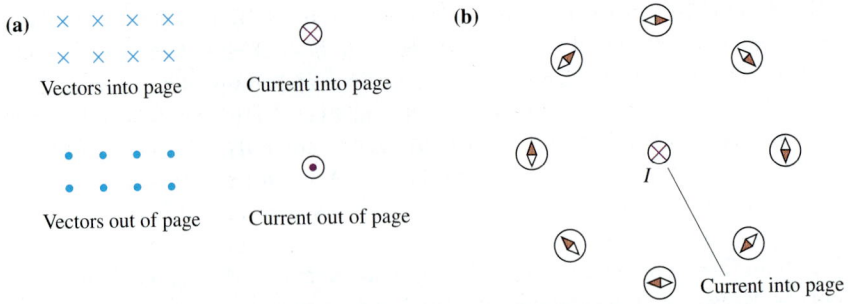

FIGURE 24.11 The notation for vectors and currents that are perpendicular to the page.

compasses around a current that is directed into the page. To use the right-hand rule with this drawing, point your right thumb into the page. Your fingers will curl clockwise, and that is the direction in which the north poles of the compass needles point.

CONCEPTUAL EXAMPLE 24.1   Drawing the magnetic field of a current-carrying wire

Sketch the magnetic field of a long, current-carrying wire, with the current going into the paper. Draw both magnetic field line and magnetic field vector representations.

REASON From the iron filing picture in the atlas, we have seen that the field lines form circles around the wire, and the magnetic field becomes weaker as the distance from the wire is increased. Figure 24.12 shows how we construct both field line and field vector representations of such a field.

ASSESS It is important to understand that the magnetic field exists at *all* points around the wire, out to arbitrarily great distances—although it gets quite weak as we move far from the wire. So in a picture like Figure 24.12, the field line circles should in principle be drawn out to infinity, with the distances between them getting ever greater. Also, Figure 24.12 shows the field lines only in a single plane perpendicular to the wire. A more complex 3D drawing, such as Figure 24.13, is necessary to convey the idea that the field lines also exist in *every* plane along the length of the wire.

❶ ⊗ means current points *into* the page: point your right thumb in this direction.

❷ Your fingers curl clockwise . . .

❸ . . . so the magnetic field lines are clockwise circles around the wire.

Magnetic field vectors are longer where the field is stronger.

Magnetic field vectors point tangent to the field lines.

Field lines are closer together where the field is stronger.

FIGURE 24.12 Drawing the magnetic field of a long, straight, current-carrying wire.

FIGURE 24.13 Field lines exist everywhere along the wire.

The Magnetic Field of a Current Loop

Now that we've seen what the magnetic field looks like for a long, straight wire carrying a current, we can understand what the fields should look like due to other shapes of wires. Here we'll study only two other common shapes: a simple circular loop of wire (a *current loop*) and a tightly wound coil (a *solenoid*).

Let's start with the simple circular current-carrying loop, shown in several views in Figure 24.14. To see what the field due to a current loop looks like, we can imagine bending a straight wire into a loop, as shown in Figure 24.15. As we do so, the field lines near the wire will remain similar to what they looked like when the wire was still straight: circles going around the wire. Further from the wires the field lines are no longer circles, but they still curve through the center of the loop, back around the outside, and then return through the center.

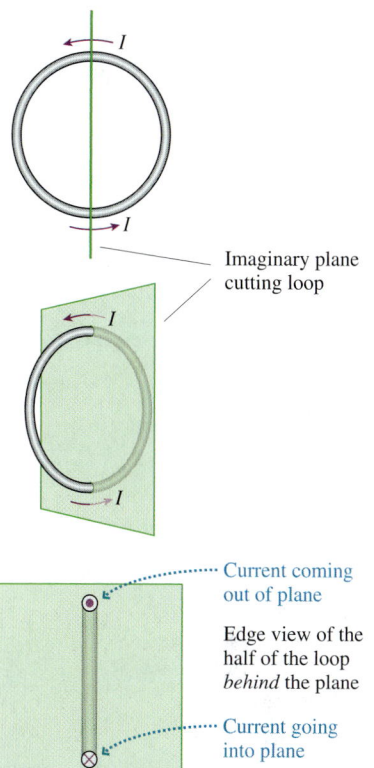

FIGURE 24.14 Three views of a current loop.

Use the right-hand rule for fields to find the direction of the field lines.

Near the wire, the field lines are almost circular.

The field emerges from the center of the loop . . .

. . . and returns around the outside of the loop.

View of loop from left side   Face-on view of loop

FIGURE 24.15 The magnetic field of a current loop.

If we reverse the direction of the current in the loop, all the field lines reverse direction as well. Because a current loop is essentially a straight wire bent into a circle, the same right-hand rule of Tactics Box 24.1, used to find the field direction for a long, straight wire, can also be used to find the field direction for a current loop. As shown in Figure 24.15, you again point your thumb in the direction of the current in the loop and let your fingers curl through the center of the loop. Your fingers are then pointing in the direction in which $\vec{B}$ passes through the *center* of the loop.

The Magnetic Field of a Solenoid

There are many applications of magnetism for which we would like to generate a **uniform magnetic field,** a field that has the same magnitude and the same direction at every point within some region of space. The MRI system used to make the image at the beginning of this chapter needs a very uniform magnetic field to function. Mass spectrometers, used for identifying the chemical composition of unknown substances, also require a uniform field; we'll discuss mass spectrometry in Section 24.5.

In practice, a reasonably uniform magnetic field is generated with a **solenoid.** A solenoid, shown in Figure 24.16, is a coil of wire with the same current I passing through each loop in the coil. Solenoids may have hundreds or thousands of coils, often called *turns,* sometimes wrapped in several layers.

We can see what the magnetic field of a solenoid looks like by examining the iron filing picture in the atlas on page 794. **The field within the solenoid is strong, mainly parallel to the axis, and reasonably uniform, whereas the field outside the solenoid is very weak.** Figure 24.17 on the next page compares the fields of a short and a long solenoid, and shows why the field inside is much stronger than the field outside. Because magnetic field lines form closed curves, all the field lines in the open space outside the solenoid must pack together very tightly as they pass through its interior.

Imaginary plane cutting loop

Current coming out of plane

Edge view of the half of the loop *behind* the plane

Current going into plane

FIGURE 24.16 A solenoid.

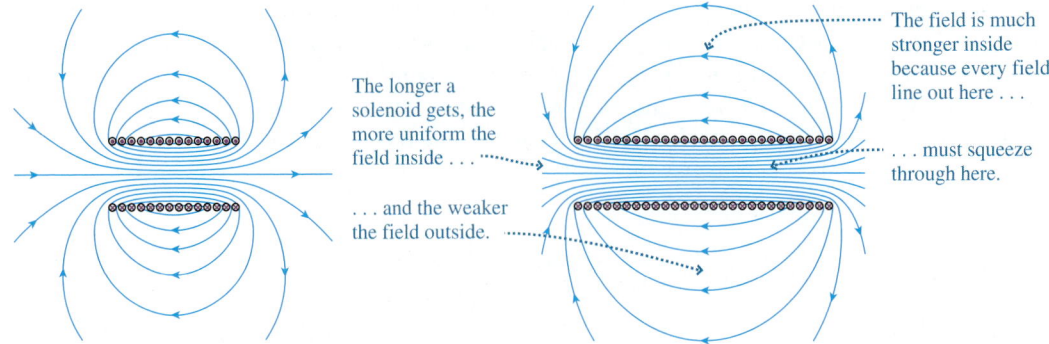

FIGURE 24.17 Comparing the magnetic fields of long and short solenoids.

STOP TO THINK 24.3 A compass is placed a few centimeters above a very long wire with no current. Because of the earth's field, the needle points north. When a large current is turned on in the direction shown, in which direction will the compass point?

24.4 Calculating the Magnetic Field Due to a Current

13.1 **Activ Physics**

The previous section showed qualitatively how the magnetic field looks for several shapes of current-carrying wires. In this section we'll give quantitative expressions for these magnetic fields. We'll also see how to find the magnetic field when there is more than one current-carrying wire present.

Let's start with the simplest case—that of a long, straight, current-carrying wire. We've seen already that the magnetic field lines form circles around the wire, and that the field gets weaker as the distance from the wire increases. Not surprisingly, the magnitude of the field also depends on the *current* through the wire, increasing in proportion to the current. Careful experiments show that the magnitude of the magnetic field a distance r from the wire carrying current I is given by the expression

$$B = \frac{\mu_0 I}{2\pi r} \qquad (24.1)$$

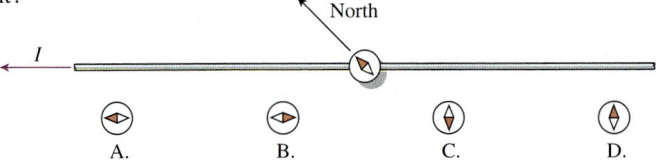

Magnetic field due to a long, straight, current-carrying wire

LINEAR p.38   INVERSE p.118

TABLE 24.1 Typical magnetic field strengths

Field location	Field strength (T)
Surface of the earth	5×10^{-5}
Refrigerator magnet	5×10^{-3}
Laboratory magnet	0.1 to 1
Superconducting magnet	10

Although this equation is exact only for an infinitely long wire, it is quite accurate whenever the length of the wire is significantly greater than the distance r from the wire.

The SI unit of the magnetic field is the **tesla,** abbreviated as T. One tesla is quite a large field. Table 24.1 shows some typical magnetic field strengths. Most magnetic fields are a small fraction of a tesla. We sometimes express magnetic fields in μT (10^{-6} T) or mT (10^{-3} T).

The constant μ_0 in Equation 24.1 is called the **permeability constant.** Its value is

$$\mu_0 = 4\pi \times 10^{-7} \, \text{T} \cdot \text{m/A} = 1.257 \times 10^{-6} \, \text{T} \cdot \text{m/A}$$

Figure 24.18 summarizes what we've learned about the behavior of the magnetic field due to a long, current-carrying wire, and illustrates how the field depends on the distance from the wire, according to Equation 24.1.

Magnetic Fields from More than One Source

When two or more sources of magnetic fields—perhaps a magnet and some current-carrying wires—are brought near each other, how do we find the total magnetic field at any particular point in space? We found for electric fields that the principle of *superposition* applies: The total electric field at any point is the *vector* sum of the individual fields at that point. The same principle is found experimentally to hold for magnetic fields as well. Figure 24.19 illustrates the principle of superposition applied to *magnetic* fields.

We've now learned enough to suggest the following general problem-solving strategy for finding the magnetic field at a point due to known sources of magnetic field. Note that the details of the equations for the field strength due to a current loop and a solenoid, similar to Equation 24.1, are introduced a little later in this section.

 PROBLEM-SOLVING STRATEGY 24.1   Magnetic–field problems

PREPARE Because current-carrying wires do not lie in the same plane as the fields they produce, you'll need to prepare an especially careful drawing. Generally, you should choose the plane of your drawing so that the magnetic field vectors lie either in the plane of the paper or perpendicular to it.

- Straight wires are usually easiest to draw as seen from their ends. Then the field vectors will lie in the plane of the paper.
- Usually, it's best to draw current loops in the plane of the paper. Then the field in the loop's center is perpendicular to the paper.
- Solenoids can be drawn either end-on (field perpendicular to the plane of the paper) or as seen from the side (field in the plane of the paper).
- Many problems have more than one source of magnetic field. It's usually best that your drawing shows the field vectors in the plane of the paper. Then they can be added as vectors, using components if needed.

If you need to add several fields together as vectors, introduce a coordinate system with x- and y-axes.

SOLVE Find the directions of the fields due to each source (wire, loop, solenoid) by using the right-hand rule for fields. Then find the magnitude of each field using the expressions given in this section for a wire, a loop, or a solenoid. Finally, add these fields together using the rules for vector addition.

ASSESS Check if the magnitude and direction of the total field seem reasonable.

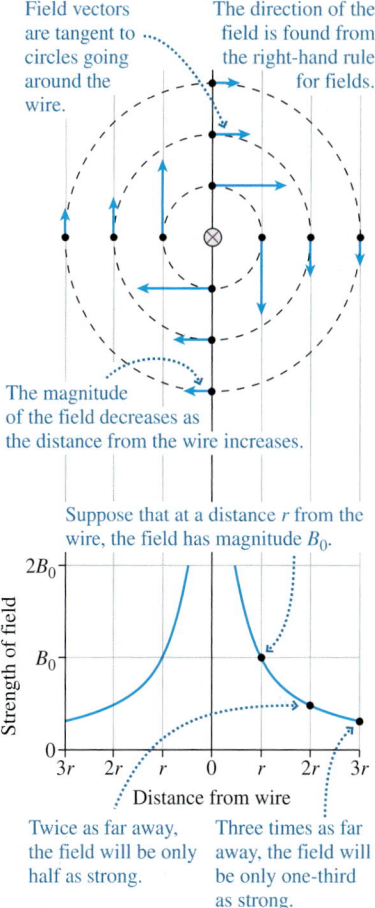

FIGURE 24.18 Summary of the magnetic field due to a long, current-carrying wire.

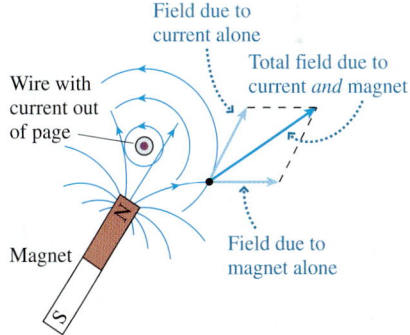

FIGURE 24.19 Adding fields due to more than one source.

EXAMPLE 24.1 Finding the magnetic field of two parallel wires

Two long, straight wires lie parallel to each other as shown in Figure 24.20. They each carry a current of 5.0 A, but in opposite directions. What are the magnitude and direction of the magnetic field at point P?

FIGURE 24.20 Two parallel current-carrying wires.

PREPARE So that we can draw the magnetic field vectors at P in the plane of the paper, let's redraw the wires as seen from their right ends, as in Figure 24.21. In this view, wire 1 is above wire 2. The right-hand rule for fields tells us that the magnetic field at P

FIGURE 24.21 View of the wires from their right ends.

from wire 1 points to the right and that from wire 2 to the left. Because wire 2 is twice as far from P as wire 1, we've drawn its field $\vec{B}_2$ half as large as the field $\vec{B}_1$ from wire 1.

SOLVE From Figure 24.21, we can see that the total field $\vec{B}$—the vector sum of the two fields $\vec{B}_1$ and $\vec{B}_2$—points to the right. To find the magnitude of $\vec{B}$, we'll need the magnitudes of $\vec{B}_1$ and $\vec{B}_2$.

We can use Equation 24.1 to find these magnitudes. From Figure 24.21, r for wire 1 is 8.0 cm (or 0.080 m), while r for wire 2 is 16 cm.

We then have

$$B_1 = \frac{\mu_0 I}{2\pi r} = \frac{(4\pi \times 10^{-7}\ \text{T} \cdot \text{m/A})(5.0\ \text{A})}{2\pi(0.080\ \text{m})} = 1.25 \times 10^{-5}\ \text{T}$$

and

$$B_2 = \frac{\mu_0 I}{2\pi r} = \frac{(4\pi \times 10^{-7}\ \text{T} \cdot \text{m/A})(5.0\ \text{A})}{2\pi(0.16\ \text{m})} = 0.63 \times 10^{-5}\ \text{T}$$

Because $\vec{B}_2$ points to the left, its x-component is negative. Thus the x-component of the total field $\vec{B}$ is

$$B_x = (B_1)_x + (B_2)_x$$
$$= 1.25 \times 10^{-5}\ \text{T} - 0.63 \times 10^{-5}\ \text{T} = 0.62 \times 10^{-5}\ \text{T}$$

ASSESS The positive sign tells us that the total field points to the right, in the same direction as the field due to wire 1. This makes sense, because wire 1 is closer to P than is wire 2.

EXAMPLE 24.2 Finding the magnetic field of a power line

Transmission lines carry electric power across long distances. Consider a transmission line consisting of two parallel wires 3.0 m apart, and 15 m above the ground.

a. If one wire carries a current of 500 A and the other wire has no current, what are the magnitude and direction of the magnetic field at a point on the ground exactly halfway between the wires?

b. What is the field if both wires carry 500 A but in opposite directions?

FIGURE 24.22 Sketch of the power line and its magnetic field.

a. **PREPARE** Since transmission lines are very long wires, and we want to know the field at a point relatively close to the wires, Equation 24.1 for the field due to a long wire is applicable.

This problem requires a *very* careful drawing, shown in Figure 24.22. Again, we choose an end-on view of the wires so that the field vectors will be in the plane of the paper. Review Figure 24.18 for rules on how to draw the field due to a straight current-carrying wire. We choose axes, draw the wires, and label the known distances. Point P, directly below the halfway point, is where we want to find the field.

SOLVE To find the *magnitude* of the field, we use Equation 24.1. We'll need the current $I = 500$ A, and the distance r from the wire to point P. This distance can be calculated from the triangle with sides a and b. We have

$$r = \sqrt{a^2 + b^2} = \sqrt{(1.5 \text{ m})^2 + (15 \text{ m})^2} = 15.1 \text{ m}$$

Then Equation 24.1 gives

$$B = \frac{\mu_0 I}{2\pi r} = \frac{(4\pi \times 10^{-7} \text{ T} \cdot \text{m/A})(500 \text{ A})}{2\pi (15.1 \text{ m})} = 6.6 \times 10^{-6} \text{ T}$$

The direction of $\vec{B}$ can be specified by finding the angle θ. From the triangle with sides a and b we see that

$$\tan\theta = a/b = (1.5 \text{ m})/(15 \text{ m}) = 0.10$$

so that $\theta = 5.7°$.

b. PREPARE With the current in the second wire turned on, in the opposite direction to the first, our sketch now appears as in Figure 24.23. Using the right-hand rule for fields, we see that the second wire creates a field of the same magnitude as the first (so $B_1 = B_2$) but pointing to the left.

SOLVE To find the total field, we add these two fields together as vectors, using components. We have

$$B_x = B_{1x} + B_{2x} = B_1 \cos\theta - B_2 \cos\theta = B_1 \cos\theta - B_1 \cos\theta = 0$$

$$B_y = B_{1y} + B_{2y} = B_1 \sin\theta + B_2 \sin\theta = B_1 \sin\theta + B_1 \sin\theta$$

$$= 2B_1 \sin\theta$$

In these equations we have used the fact that the magnitude of each field has the same value B_1. This is the value found in the

FIGURE 24.23 Fields due to two wires.

first part of this example for one wire. From these components (and from Figure 24.23) we see that the total field points in the vertical or y-direction. Its magnitude is

$$B = 2B_1 \sin\theta = 2(6.6 \times 10^{-6} \text{ T}) \sin(5.7°)$$

$$= 1.3 \times 10^{-6} \text{ T}$$

ASSESS These are rather small fields, only a few percent of the earth's field, even though we are quite close to a wire carrying a sizable current. This example shows that magnetic fields produced by currents are generally much smaller than those produced by magnets, unless very large currents and lots of wires are used.

Current Loops

The magnetic field due to a current loop is more complex than that of a straight wire, as we can see from Figure 24.24, but there is a simple expression for the field at an important place: the *center* of the loop. Because the loop can be thought of as a wire bent into a circle (of radius R), the expression for the field at the center is very similar to that of a wire. The field at the center is given by

Activ Physics $\quad$ 13.2

$$B = \frac{\mu_0 I}{2R} \qquad (24.2)$$

Magnetic field at the center of a current loop of radius R

LINEAR p.38   INVERSE p.118

where I is the current in the loop. We have already seen how to find the *direction* of the field in the center, using our right-hand rule for fields.

If N loops of wire are all tightly wound into a single thin coil, then the field at the center is just N times bigger (since we're superimposing N individual current loops):

$$B = \frac{\mu_0 N I}{2R} \qquad (24.3)$$

Magnetic field at the center of a current loop with N turns

LINEAR p.38   INVERSE p.118

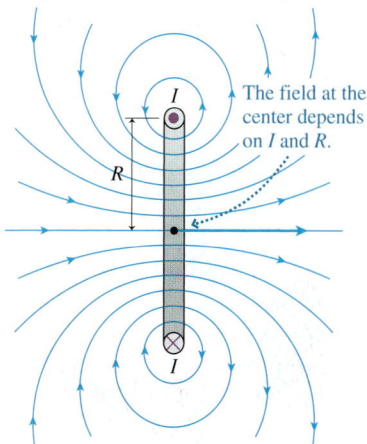

FIGURE 24.24 The magnetic field at the center of a current loop.

CONCEPTUAL EXAMPLE 24.2 Canceling a magnetic field

A current loop carries a counterclockwise current of 3 A. A second current-carrying loop with twice the radius is placed outside the first loop, as shown in Figure 24.25. What current in the outer loop will exactly cancel the field at the common center of the two loops?

FIGURE 24.25 Two current loops.

REASON The inner loop produces a magnetic field at its center of magnitude B. By the right-hand rule for fields, the direction of this field is out of the page. To make the total field at the center zero, the outer loop must produce a field at the center that has the same magnitude B but pointing in the opposite direction, into the page. Equation 24.2 shows that the field at the center of a loop is *inversely proportional* to its radius. If the outer loop also carried 3 A, its field at the center would be only one-half of B. To make it as large as B, we need to *double* the current to 6 A. Finally, since we want the field due to the outer loop to point in the opposite direction to that from the inner loop, the current in the outer loop must be in the opposite direction, or clockwise.

ASSESS We have solved this problem without writing down any equations. We need only understand that the magnetic field at the center of a circular loop is proportional to $1/r$, and proportional to the current. This is why the problem can be solved even though the radius of neither loop is given. See also Problem 24.12 for another way to approach this question.

Solenoids

As we've seen, the field inside a solenoid is fairly uniform, while the field outside is quite small; the greater a solenoid's length in comparison to its diameter, the better these statements hold. Measurements that need a uniform magnetic field are often conducted inside a solenoid, which can be built quite large. The cylinder that surrounds a patient undergoing magnetic resonance imaging (MRI), such as the one shown later, on page 817, contains a large solenoid made of superconducting wire, allowing it to carry the very large currents needed to generate a strong uniform magnetic field. Consider a solenoid of length L having N turns of wire. We expect that the more turns we can pack into a solenoid of a given length—that is, the greater the ratio N/L—the stronger the field inside will be. We further expect that the strength of the field will be proportional to the current I in the turns. We might also expect that the field inside would depend on the radius R of the solenoid, as it did for the current loop. Somewhat surprisingly, this is not the case: The field inside a solenoid does *not* depend on its radius. For this reason, the radius R doesn't appear in the equation for the field inside a solenoid:

The world's most powerful magnet uses both copper and superconducting solenoids to reach fields of 45 T, about 10 million times the field of the earth. It is used for research on new materials.

$$B = \mu_0 I \frac{N}{L} \qquad (24.4)$$

LINEAR
p.38

Magnetic field inside a solenoid of length L with N turns

Activ
ONLINE
Physics   13.3

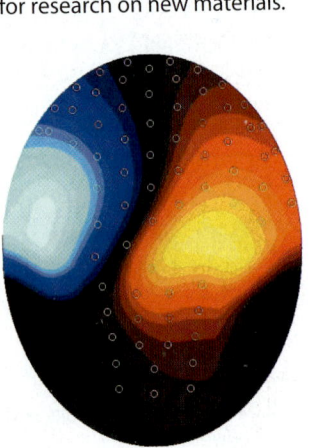

◀ **The magnetocardiogram** BIO When the heart muscle contracts, action potentials create a dipole electric field that can be measured to create an electrocardiogram. These action potentials also cause charges to circulate around the heart, creating a current loop. This current loop creates a small, but measurable, magnetic field. A record of the heart's magnetic field, a *magnetocardiogram*, can provide useful information about the heart in cases where an electrocardiogram is not possible. This image shows the magnetic field of a fetal heartbeat measured at the surface of the mother's abdomen. Here, blue represents a field pointing into the body and red a field out of the body. This is just the field expected from a current loop whose plane lies along the black line between the two colored areas.

A typical MRI solenoid has a length of about 1 m and a diameter of about 1 m. The maximum field inside such a solenoid is about 1 T. How many turns of wire must the solenoid have to produce this field if the largest current the wire can carry is 100 A?

PREPARE This solenoid is not very long compared to its diameter, so using Equation 24.4 will give only an approximate result. This is acceptable, since we have only rough estimates of the field B and the length L.

Equation 24.4 gives the magnetic field B of a solenoid in terms of the current I, the number of turns N, and the length L. Here, however, we want to find the number of turns in terms of the other variables. We'll need $B = 1$ T, $I = 100$ A, and $L = 1$ m.

SOLVE We can solve Equation 24.4 for N to get

$$N = \frac{LB}{\mu_0 I} = \frac{(1\text{ m})(1\text{ T})}{(4\pi \times 10^{-7}\text{ T} \cdot \text{m/A})(100\text{ A})} = 8000 \text{ turns}$$

to one significant figure.

ASSESS The number of turns required is quite large. Since we estimate the diameter of the solenoid to be about 1 m, the length of wire in each turn is about $\pi \times 1$ m, or say 3 m. Thus the total length of wire needed is 8000 turns $\times$ 3 m/turn = 24,000 m = 24 km = 15 miles! If this magnet used ordinary copper wire large enough to carry 100 A, its total resistance R would be about 35 Ω. Recall from Chapter 23 that the power dissipated by a resistor is equal to $I^2 R$. So the power dissipated would be $(100\text{ A})^2(35\text{ }\Omega) = 350{,}000$ W, a huge and impractical value. MRI magnets therefore use *superconducting* wire, which when cooled near absolute zero temperature has *zero* resistance. This allows 24 km of wire to carry 100 A with no power dissipated at all!

24.5 Magnetic Fields Exert Forces on Moving Charges

It's time to switch our attention from what magnetic fields look like and how they are generated to what they actually *do:* exert forces and torques on moving charges and currents. As we saw earlier in the chapter, Oersted discovered that a current passing through a wire causes a magnetic torque to be exerted on a nearby compass needle. Upon hearing of Oersted's discovery, the French scientist André Marie Ampère reasoned that the current was acting like a magnet and, if this were true, that two current-carrying wires should exert magnetic forces on each other, just as two magnets do.

To find out, Ampère (for whom the SI unit of current is named) set up two parallel wires that could carry large currents in either the same direction or in opposite directions. Figure 24.26 shows the outcome of his experiment. Notice that for currents, "likes" attract and "opposites" repel. This is the opposite of what would have happened had the wires been charged and thus exerting electric forces on each other. Ampère's experiment showed that **a magnetic field exerts a force on a current.** Since currents consist of moving charges, Ampère's experiment therefore also implies that **a magnetic field exerts a force on a moving charged particle.** There is a nice symmetry here: Magnetic fields are produced by currents (that is, moving charges) and magnetic fields exert forces on currents (which consist of moving charges.) These forces have a wide range of practical applications. We will begin our treatment of magnetic forces with a look at the force that a magnetic field exerts on a moving charged particle.

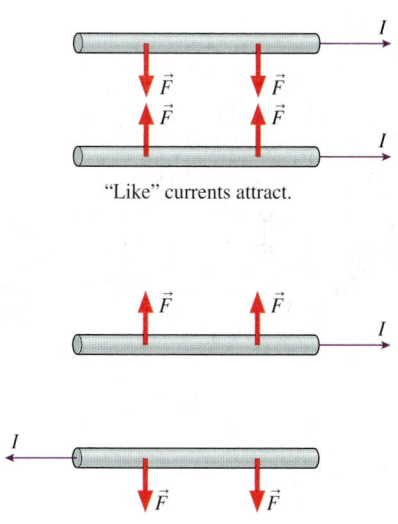

"Like" currents attract.

"Opposite" currents repel.

FIGURE 24.26 Ampère's experiment to observe the forces between parallel current-carrying wires.

Finding the Magnetic Force on a Moving Charged Particle

Let's look at the results of some experiments designed to illustrate the nature of this force. The following series of experiments shows how the magnetic force on a moving charged particle depends on the particle's charge and velocity.

The force on a charged particle moving in a magnetic field

    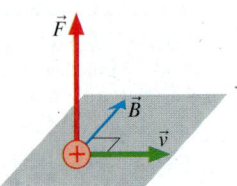

Only a *moving* charged particle experiences a magnetic force. There is no magnetic force on a charge at rest ($v = 0$) in a magnetic field.

There is no force on a charged particle moving *parallel* to a magnetic field. There is a force only if the charged particle is moving at an angle to the field.

A charged particle moving at an angle to the field *does* experience a force. This force is perpendicular to *both* $\vec{v}$ and $\vec{B}$. That is, the force is at right angles to the plane containing $\vec{v}$ and $\vec{B}$.

The force is greater if the particle moves faster or if the magnitude of the charge is increased.

The magnitude of the force depends on the angle between $\vec{v}$ and $\vec{B}$. For given values of v and B, the force is greatest when the angle between $\vec{v}$ and $\vec{B}$ is 90°.

Notice that the magnetic force is quite different from the electric force: A charged particle in an electric field feels a force that is *parallel* to the field, but a charged particle moving in a magnetic field feels a force that is *perpendicular* to the field.

The magnetic force on a moving charged particle is at right angles to the field and the velocity, but this is not sufficient to uniquely determine the direction of the force. We also need the **right-hand rule for forces,** shown in Figure 24.27.

NOTE ▶ The right-hand rule for forces is different from the right-hand rule for fields. ◀

1. There are two possible force vectors that are perpendicular to $\vec{v}$ and $\vec{B}$—up from the plane or down from the plane. We use the right-hand rule for forces to choose the correct one.

2. Spread the fingers of your right hand so that your index finger and thumb point out from your hand as shown. Rotate your hand to point your thumb in the direction of $\vec{v}$, and your index finger in the direction of $\vec{B}$.

3. Now point your middle finger so that it is perpendicular to your palm, as shown. It will point in the direction of $\vec{F}$. In this case, $\vec{F}$ is directed up from the plane.

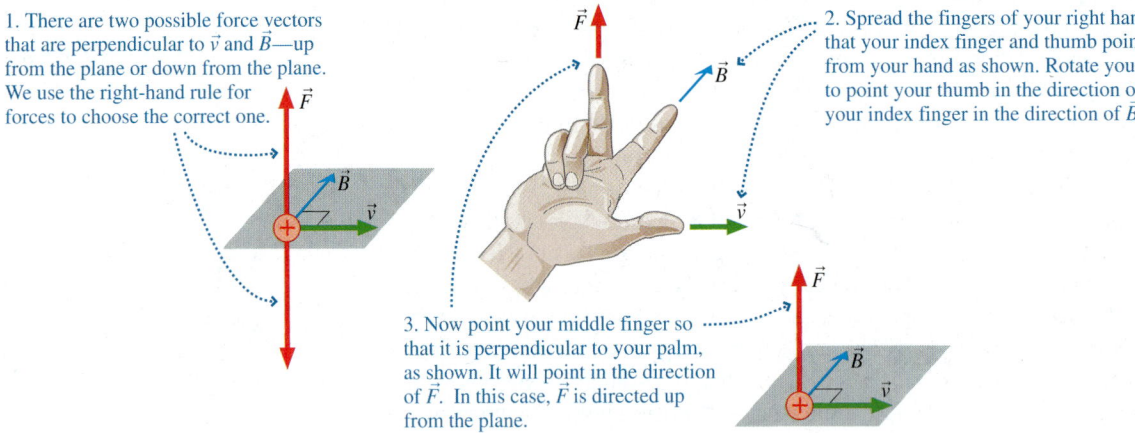

FIGURE 24.27 The right-hand rule for forces.

NOTE ▶ The right-hand rule for forces gives the direction of the force on a *positive* charge. For a negative charge, the force is in the opposite direction. ◀

13.4   ActivPhysics

We can now organize all of the experimental information about the magnetic force on a moving charged particle into a single equation. If a particle of charge q moves with a velocity $\vec{v}$ at an angle α to a magnetic field $\vec{B}$, the force is

$$\vec{F} = (|q|vB\sin\alpha, \text{ direction given by the right-hand rule for forces})$$ (24.5)

LINEAR
p. 38

Force on a charged particle moving in a magnetic field

The velocity and the magnetic field are perpendicular in many practical situations. In this case α is 90°, and we can simplify Equation 24.5 to

$$\vec{F} = (|q|vB, \text{ direction given by the right-hand rule for forces})\quad (24.6)$$

Force on a charged particle moving perpendicular to a magnetic field

TACTICS BOX 24.2 Determining the magnetic force on a moving charged particle   Exercises 16, 17

$$F = |q|vB \sin\alpha$$

❶ Note the direction of $\vec{v}$ and $\vec{B}$, and find the angle α between them.

❷ The force is perpendicular to the plane containing $\vec{v}$ and $\vec{B}$. The direction of $\vec{F}$ is given by the right-hand rule.

❸ The magnitude of the force is given by Equation 24.5.

CONCEPTUAL EXAMPLE 24.3 Determining the force on a moving electron

An electron is moving to the right in a magnetic field that points upward, as in Figure 24.28. What is the direction of the magnetic force?

FIGURE 24.28 An electron moving in a magnetic field.

REASON The direction of the force is specified by the right-hand rule for forces.

- Spread out the thumb and index finger of your right hand as in Figure 24.29.
- Rotate your right hand so that your thumb points in the direction of the velocity (to the right) and your index finger points in the direction of the magnetic field (toward the top of the page) as in Figure 24.29.

- Bend your middle finger so that it's perpendicular to your index finger. Your middle finger, which now points out of the plane of the paper, is in the direction of the force on a positive charge. However, the electron has a negative charge, so the force on the electron is directed *into* the plane of the paper.

FIGURE 24.29 Using the right-hand rule.

ASSESS The force is perpendicular to both the velocity and the magnetic field, as it must be. The force on an electron is into the plane of the paper; the force on a proton would be out of the plane of the paper.

CONCEPTUAL EXAMPLE 24.4 Determining the force on a charged particle moving near a current-carrying wire

A proton is moving to the right above a horizontal wire that carries a current to the right. What is the direction of the magnetic force on the proton?

REASON The current in the wire creates a magnetic field; this magnetic field exerts a force on the moving proton. We follow three steps to solve the problem:

1. Sketch the situation, as shown in Figure 24.30a on the next page.

2. Determine the direction of the field at the position of the proton due to the current in the wire (Figure 24.30b).

3. Determine the direction of the force that this field exerts on the proton (Figure 24.30c). In this case, the force is down.

ASSESS At the start of this section, where we looked at the forces between currents, we saw that "like" currents attract each other. This shows us that our result is sensible. The moving proton could be thought of as making a current to the right that is attracted to the current in the wire.

Continued

(a)

1. The proton moves above the wire. The proton velocity and current in the wire are shown.

(b)

2. The right-hand rule for fields shows that, at the position of the proton, above the wire, the field points out of the plane of the paper.

(c)

3. The right-hand rule for forces shows that the field of the wire exerts a force on the moving proton that points down, toward the wire.

FIGURE 24.30 Determining the direction of the force.

EXAMPLE 24.4 Force on a charged particle in the earth's field

The sun emits streams of charged particles (in what is called the *solar wind*) that move toward the earth at very high speeds. Suppose a proton is moving toward the equator of the earth at a speed of 500 km/s. At this point, the earth's field has a strength of 5.0×10^{-5} T. What are the direction and the magnitude of the force on the proton?

PREPARE Our first step is, as usual, to draw a picture. As we saw in Figure 24.8, the field lines of the earth go from the earth's south pole to the earth's north pole. Figure 24.31 shows the proton entering the field.

We also need to convert the proton's velocity to meters per second:

$$500 \text{ km/s} = 5.0 \times 10^2 \text{ km/s} \times \frac{1 \times 10^3 \text{ m}}{1 \text{ km}} = 5.0 \times 10^5 \text{ m/s}$$

SOLVE We use the steps of Tactics Box 24.2 to determine the force.

❶ $\vec{v}$ and $\vec{B}$ are perpendicular, so $\alpha = 90°$.

❷ The right-hand rule for forces tells us that the force will be into the plane of the paper in Figure 24.31. That is, the force is toward the east.

❸ We compute the magnitude of the force using Equation 24.6:

$$F = |q|vB = (1.6 \times 10^{-19} \text{ C})(5.0 \times 10^5 \text{ m/s})(5.0 \times 10^{-5} \text{ T})$$

$$= 4.0 \times 10^{-18} \text{ N}$$

ASSESS This is a small force, but the proton has an extremely small mass of 1.67×10^{-27} kg. Consequently, this force produces a very large acceleration: 200 million times the acceleration due to gravity!

The direction of the force on the proton is given by the right-hand rule.

Your sketch will look more like this, with $\vec{v}$ and $\vec{B}$ in the plane of the paper. The force is into the plane of the paper.

$\vec{F}$ into page

FIGURE 24.31 A proton in the field of the earth.

Example 24.4 looked at the force on a proton from the sun as it reaches the magnetic field of the earth. How does this force affect the motion of the proton? In the following section, we will consider the general question of how charged particles move in magnetic fields.

Paths of Charged Particles in Magnetic Fields

We just saw that the magnetic force on a moving charged particle is always perpendicular to its velocity. This means that the force changes the *direction* of the velocity but not the *magnitude*. The particle's path will bend, but the particle will not speed up or slow down. Suppose a positively charged particle is moving perpendicular to a uniform magnetic field $\vec{B}$ as shown in Figure 24.32. In Chapter 6, we looked at the motion of objects subject to a force that was always perpendicular to the velocity. The net result was *circular motion at a constant speed.* For a mass moving in a circle at the end of a string, the tension force is always perpendicular to $\vec{v}$. For a satellite moving in a circular orbit, the gravitational force is always perpendicular to $\vec{v}$. Now, for a charged particle moving in a magnetic field, it is the magnetic force that is always perpendicular to $\vec{v}$ and so causes the particle to move in a circle. Thus **a particle moving perpendicular to a uniform magnetic field undergoes uniform circular motion at constant speed.**

NOTE ▶ Because the right-hand rule for forces specifies the direction of the force on a positive charge, the force on a negative charge is in the opposite direction. A negatively charged particle will orbit in the opposite sense from that shown in Figure 24.32 for a positive charge. ◀

Figure 24.33 shows a particle of mass m moving at a speed v in a circle of radius r. We found in Chapter 6 that this motion requires a force directed toward the center of the circle with magnitude

$$F = \frac{mv^2}{r} \tag{24.7}$$

For a charged particle moving in a magnetic field, this force is the force from the interaction of the charged particle with the magnetic field. In Figure 24.32 we assumed that the velocity was perpendicular to the magnetic field, so the angle α was 90°. Thus the magnitude of the force on the charged particle due to the magnetic field is given by Equation 24.6. As this is the force that produces the circular motion, we can equate it to the force in Equation 24.7:

$$F = |q|vB = \frac{mv^2}{r}$$

Solving for the radius of the orbit, we get

$$r = \frac{mv}{|q|B} \tag{24.8}$$

The radius depends on the ratio of the mass of the particle to its charge, a fact we will use later. The radius also depends on the particle's speed and the magnetic field strength: Increasing the speed will increase the radius of the circular motion, while increasing the field will decrease the radius.

A particle moving *perpendicular* to a magnetic field moves in a circle. In the table at the start of this section, we saw that a particle moving *parallel* to a magnetic field experiences no magnetic force, and so continues in a straight line. A more general situation in which a charged particle's velocity $\vec{v}$ is neither parallel to nor perpendicular to the field $\vec{B}$ is shown in Figure 24.34. The net result is a circular motion due to the perpendicular component of the velocity coupled with a constant velocity parallel to the field: The charged particle spirals around the magnetic field lines in a helical trajectory.

High-energy particles and radiation streaming out from the sun in the solar wind create ions and electrons as they strike molecules high in the earth's atmosphere. Some of these charged particles become trapped in the earth's magnetic field. As Figure 24.35 on the next page shows, the electrons spiral in helical trajectories along the earth's magnetic field lines. Some of these particles enter the atmosphere near the north and south poles, ionizing gas and creating the ghostly glow of the **aurora.**

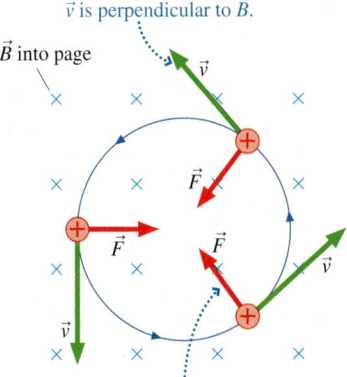

The magnetic force is always perpendicular to $\vec{v}$, causing the particle to move in a circle.

FIGURE 24.32 A charged particle moving perpendicular to a uniform magnetic field.

$\vec{F}$ must point to the center of the circle with magnitude mv^2/r to keep the mass traveling around the circle with constant speed.

FIGURE 24.33 A particle in circular motion.

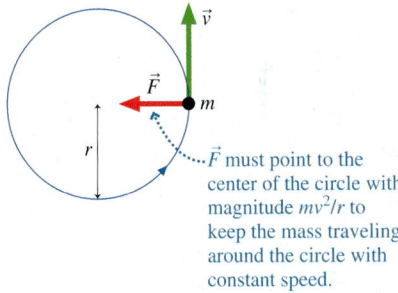

(a) The velocity can be broken into components parallel and perpendicular to the field. The parallel component will continue without change.

(b) A top view shows that the perpendicular component will change, leading to circular motion.

(c) The net result is a helical path that spirals around the field lines.

FIGURE 24.34 A charged particle in a magnetic field follows a helical trajectory.

The earth's magnetic field leads particles into the atmosphere near the poles, causing the aurora.

Where the particles strike the atmosphere, ionized gas creates the glow of the aurora.

FIGURE 24.35 Charged particles in the earth's magnetic field create the aurora.

EXAMPLE 24.5 Force on a charged particle in the earth's field, revisited

In Example 24.4, we considered a proton in the solar wind moving toward the equator of the earth, where the earth's field is 5.0×10^{-5} T, at a speed of 500 km/s (5.0×10^5 m/s). We now know that the proton will move in a circle around the earth's field lines. How long does the proton take to complete one orbit?

PREPARE We begin with a sketch of the situation, noting the proton's orbit, as shown in Figure 24.36.

The proton will orbit the field lines as shown.

$\vec{B}$ out of page

A top view shows the orbit.

FIGURE 24.36 A proton orbits the earth's field lines.

SOLVE Before we use any numbers, we will do some work with symbols. The radius r of the orbit of the proton is given by

Equation 24.8, but we are interested in the time of the orbit. The time t for one orbit is just the distance of one orbit (the circumference $2\pi r$) divided by the speed, as the speed is constant:

$$t = \frac{2\pi r}{v}$$

We substitute Equation 24.8 for the radius of the orbit to get

$$t = \frac{2\pi}{v} r = \frac{2\pi}{v}\left(\frac{mv}{qB}\right) = \frac{2\pi m}{qB}$$

This result is somewhat surprising, because the time for the orbit does not depend on the speed! All protons in the earth's field orbit with the same period, regardless of velocity. Using values for mass, charge, and field, we find the time to be

$$t = \frac{2\pi(1.67 \times 10^{-27} \text{ kg})}{(1.60 \times 10^{-19} \text{ C})(5.0 \times 10^{-5} \text{ T})} = 0.0013 \text{ s}$$

ASSESS The period is just over 1 ms, so the frequency is just under 1000 Hz. This is the natural frequency for the circular motion of protons in the earth's field.

The peaks occur at particular masses, allowing a determination of the particular atom or molecule present.

FIGURE 24.37 Mass spectrum of the atmosphere of Jupiter.

The Mass Spectrometer

In 1995, the Galileo spacecraft dropped a probe into Jupiter's atmosphere. One instrument on board, called a **mass spectrometer,** was used to determine the composition of the atmosphere. The data, shown in Figure 24.37, are called a **mass spectrum,** a record of the masses of atoms and molecules that were encountered.

We will look at a particular kind of mass spectrometer that derives information from the motion of a charged particle in a uniform magnetic field. Such a particle will travel in a circular orbit, with a radius given by Equation 24.8. The radius of the orbit depends on the magnetic field, the velocity of the particle, the charge of the particle, and, most importantly, the mass of the particle.

The operation of such a mass spectrometer is outlined in Figure 24.38. A sample to be analyzed is collected and, if necessary, vaporized to form a gas. The atoms and molecules in this sample are then *ionized:* An electron is removed from each atom or molecule, leaving a positive *ion*. The resulting ions are accelerated through an electric field, and ions of a particular velocity selected. These ions travel into a region of uniform magnetic field, where they follow circular paths. An exit slit allows certain ions to pass through to a detector, where they are counted.

For a fixed field strength, only ions of a certain mass reach the detector. A rearrangement of Equation 24.8 gives the mass of the particles that reach the detector as

$$m = \frac{qBr}{v} \tag{24.9}$$

These ions are counted. If we then change the field strength, the mass of ions reaching the detector also changes, and a new count is made of these ions. If we plot the number of ions recorded as a function of their mass, the net result is a mass spectrum like the one in Figure 24.37. Mass spectrometers find wide application in chemistry and biology, from the detection of trace pollutants in groundwater to the identification of proteins in biological systems.

1. Atoms are ionized and accelerated. Ions of a particular velocity are selected to enter the spectrometer.

2. Ions of different masses follow paths of different radii.

3. Only ions of a particular mass reach the exit slit and continue to the detector.

FIGURE 24.38 A mass spectrometer.

Activ ONLINE Physics   13.7

Electromagnetic Flowmeters

Blood is a fluid that contains many kinds of ions, such as Na^+ and Cl^-. When blood flows through a vessel, these ions move with the blood. An applied magnetic field will produce a force on these moving charges. We can use this principle to make a completely noninvasive device for measuring the blood flow in an artery: an *electromagnetic flowmeter*.

A flowmeter probe clamped to an artery has two active elements: magnets that apply a strong field across the artery and electrodes that contact the artery on opposite sides, as shown in Figure 24.39. The blood flowing in an artery carries a mix of positive and negative ions. Because these ions are in motion, the magnetic field exerts a force on them that produces a measurable voltage. We know from Equation 24.5 that the faster the blood's ions are moving, the greater the forces separating the positive and negative ions. The greater the forces, the greater the degree of separation and the larger the voltage. The measured voltage is therefore directly proportional to the velocity of the blood.

1. The probe's magnets apply a strong field. The electrodes make contact along an axis perpendicular to this field.

2. The magnetic field exerts forces on ions moving with the blood. Positive and negative ions feel forces in opposite directions.

3. An end view shows that these forces create a separation of charge, producing a potential difference between the electrodes.

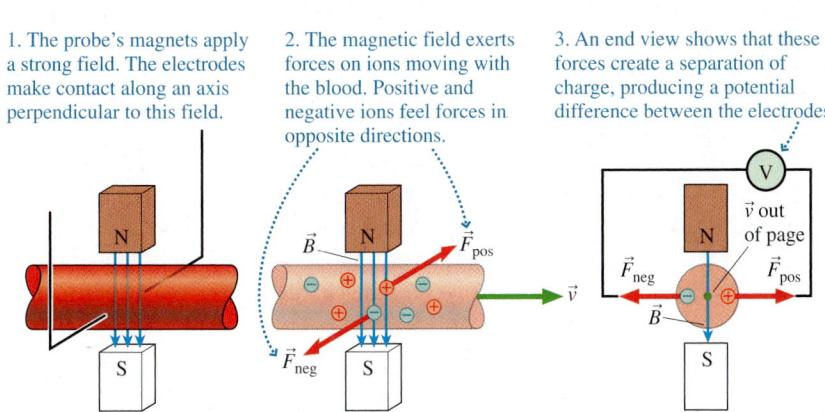

FIGURE 24.39 The operation of an electromagnetic flowmeter.

▶ **Magnets and television screens** Most televisions contain a cathode ray tube (CRT). The image on a CRT television screen is drawn by an electron beam. A magnetic field can exert a force on this beam of electrons. If you place a strong magnet near the TV screen, the electrons will be forced along altered trajectories and will strike different places on the screen than they are supposed to, producing an array of bright colors. (*The magnet can magnetize internal components and permanently alter the image, so do not do this to your television!*)

DON'T TRY IT YOURSELF

These charged particles are traveling in circular orbits with velocities and field directions as noted. Which particles have a negative charge?

A.   B.   C.   D.

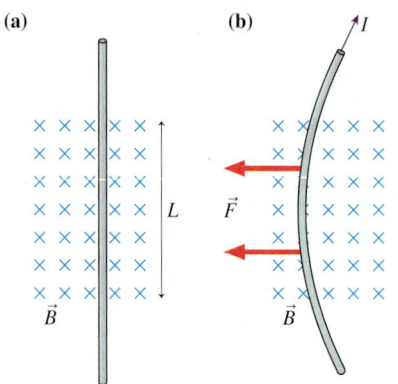

(a) A wire is perpendicular to an externally created magnetic field.

(b) If the wire carries a current, the magnetic field will exert a force on the moving charges, causing a deflection of the wire.

(c) The right-hand rule for forces applies. Your thumb should point in the direction of the current.

FIGURE 24.40 Magnetic force on a current-carrying wire.

1. We consider only the charges in this segment of the wire. As they move to the right, they will all eventually pass the right end.

q = total charge

L

2. The leftmost charge will be the last to exit the wire segment. It will travel a distance L at speed v.

FIGURE 24.41 Charges moving in a wire.

24.6 Magnetic Fields Exert Forces on Currents

We have seen that a magnetic field exerts a force on a current. This force is of great practical importance, being responsible for the operation of loudspeakers, electric motors, and many other devices. We will find an expression for this force by considering a current as a flow of charged particles, then using our results from the previous section.

The Form of the Magnetic Force on a Current

In the table at the start of Section 24.5, we saw that a magnetic field exerts no force on a charged particle moving parallel to a magnetic field. If a current-carrying wire is *parallel* to a magnetic field, we also find that the force on it is zero.

It's more interesting to consider the wire in Figure 24.40a that is *perpendicular* to the magnetic field. (Note that this field is an external magnetic field, created by a permanent magnet or by other currents; it is *not* the field of the current I in the wire.) The direction of the force on the current is found by considering the force on each charge in the current. We model current as the flow of positive charge, so **the right-hand rule for forces applies to currents in the same way it does for moving charges.** With your fingers aligned as usual, point your right thumb in the direction of the current (the direction of the motion of positive charges) and your index finger in the direction of $\vec{B}$. Your middle finger is then pointing in the direction of the force $\vec{F}$ on the wire, as in Figure 24.40c. Consequently, the entire length of wire within the magnetic field experiences a force to the left, perpendicular to both the current direction and the field direction, as shown in Figure 24.40b.

To find the magnitude of the force, we must express the current I in terms of the motion of charges in the wire. Consider a section of wire of length L in which there is a total charge q moving at a speed v, as seen in Figure 24.41.

The current I, by definition, is the rate of flow of charge. That is, the current is the charge q divided by the time Δt it takes the charge to flow out of this section of the wire:

$$I = \frac{q}{\Delta t} \tag{24.10}$$

The time the charges take to move out of the segment of wire is

$$\Delta t = \frac{L}{v} \tag{24.11}$$

Combining Equations 24.10 and 24.11, we find

$$I = \frac{q}{\Delta t} = \frac{q}{L/v} = \frac{qv}{L} \quad \Rightarrow \quad qv = IL \tag{24.12}$$

If we substitute IL for qv in the force equation $F = qvB$, we find that the magnetic force on a length L of a current-carrying wire is

$$F_{\text{wire}} = ILB \qquad (24.13)$$

Magnitude of the force on a current segment of length L
perpendicular to a magnetic field

LINEAR
p. 38

Equation 24.13 is a simple result, but remember the two assumptions behind it: The wire must be perpendicular to the field, and the field must be constant over the length L of the wire. If the wire is at an angle α to the field, the force will depend on this angle:

$$F_{\text{wire}} = ILB \sin \alpha \qquad (24.14)$$

It is sometimes useful to rewrite Equation 24.13 as

$$B = \frac{F_{\text{wire}}}{IL} \qquad (24.15)$$

Given this expression, we can see that the unit for magnetic field, the tesla, can be defined in terms of other units:

$$1\,\text{T} = 1\,\frac{\text{N}}{\text{A} \cdot \text{m}}$$

EXAMPLE 24.6 Finding the magnetic force on a power line

The high-voltage power lines that carry electricity across great distances carry very high currents as well. Suppose we have a DC power line near the equator, where the earth's magnetic field is parallel to the ground. The wire carries a current of 850 A to the east. The length of cable between adjacent towers is 400 m. What is the force of the earth's magnetic field on this length of wire?

PREPARE First, we sketch a top view of the situation as in Figure 24.42. An application of the right-hand rule for forces shows that the force will be directed upward—out of the paper in Figure 24.42.

SOLVE Because the field is perpendicular to the current, the force is given by Equation 24.13. We can use the values for I and L from the statement of the problem. We will use the value $5.0 \times 10^{-5}\,\text{T} = 5.0 \times 10^{-5}\,\text{N/A} \cdot \text{m}$ from Table 24.1 for the magnitude of the earth's magnetic field. Then we calculate

$$F_{\text{wire}} = ILB = (850\,\text{A})(400\,\text{m})\!\left(5.0 \times 10^{-5}\,\frac{\text{N}}{\text{A} \cdot \text{m}}\right) = 17\,\text{N}$$

The field of the earth near the equator is parallel to the ground and points to the north.

FIGURE 24.42 A power line near the equator.

ASSESS Even though the wire carries a large current, the force of the earth's field on the wire is relatively small. This wire can be expected to weigh many thousands of newtons, so the magnetic force on the wire is only a small fraction of the wire's weight.

Forces Between Currents

Because a current produces a magnetic field, and a magnetic field exerts a force on a current, it follows that two current-carrying wires will exert forces on each other, as Ampère discovered. Suppose we have two parallel wires of length L a distance d apart, each carrying a current. Figure 24.43a on the next page shows the currents I_1 and I_2 in the same direction and Figure 24.43b in opposite directions. We will assume that the wires are sufficiently long to allow us to use the earlier result, Equation 24.1, for the magnetic field of a long, straight wire: $B = \mu_0 I/2\pi r$.

Activ
Physics 13.5

Let's look at the situation of Figure 24.43a. There are several steps in our analysis:

(a) Currents in the same direction attract.

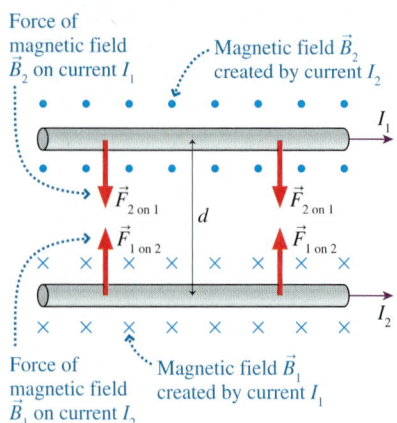

Force of magnetic field $\vec{B}_2$ on current I_1

Magnetic field $\vec{B}_2$ created by current I_2

Force of magnetic field $\vec{B}_1$ on current I_2

Magnetic field $\vec{B}_1$ created by current I_1

(b) Currents in opposite directions repel.

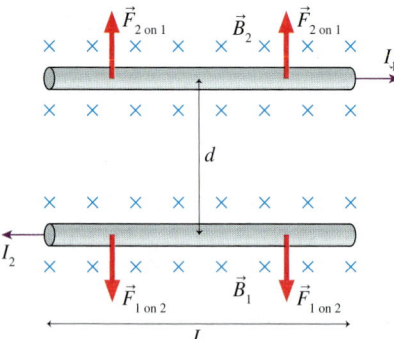

FIGURE 24.43 Forces between currents.

1. The current I_2 in the lower wire creates a magnetic field $\vec{B}_2$ at the position of the upper wire. This field $\vec{B}_2$ points out of the page, perpendicular to the current I_1. **It is this field $\vec{B}_2$, due to the lower wire, that exerts a magnetic force on the upper wire.** At the position of the upper wire, which is a constant distance $r = d$ from the lower wire, the field has the same value at all points along the wire. The field is

$$\vec{B}_2 = \left(\frac{\mu_0 I_2}{2\pi d}, \text{ out of the page}\right) \quad (24.16)$$

2. For the upper wire, the current is to the right and the field $\vec{B}_2$ from the lower wire points out of the page. Using the right-hand rule for forces, you can see that the force on the upper wire is downward, attracting it toward the lower wire.

3. The magnitude of the force on the upper wire is given by Equation 24.13. Using the field from Equation 24.16, we compute

$$F_{\text{parallel wires}} = I_1 L B_2 = I_1 L \frac{\mu_0 I_2}{2\pi d}$$

$$F_{\text{parallel wires}} = \frac{\mu_0 L\, I_1 I_2}{2\pi d} \quad (24.17)$$

Magnetic force between two parallel current-carrying wires

As an exercise, you should convince yourself that the current in the upper wire exerts an upward-directed magnetic force on the lower wire with exactly the same magnitude. (You know that this must be the case: The two forces form a Newton's third law pair.) You should also convince yourself, using the right-hand rule, that the forces are repulsive and tend to push the wires apart if the two currents are in opposite directions, as shown in Figure 24.43b.

Thus two parallel wires exert equal but opposite forces on each other, as required by Newton's third law. **Parallel wires carrying currents in the same direction attract each other; parallel wires carrying currents in opposite directions repel each other.**

EXAMPLE 24.7 Finding the force between wires in jumper cables

You may have used a set of jumper cables to start a car with a dead battery from a running vehicle. Jumper cables are a matched pair of wires, red and black, that are joined together along their length. Suppose we have a set of jumper cables in which the two wires are separated by 1.2 cm along their 3.7 m (12 ft) length. While starting a car, the wires each carry a current of 150 A, in opposite directions. What is the force between the two wires?

PREPARE Our first step is to sketch the situation, noting distances and currents, as shown in Figure 24.44. Because the cur-

rents in the two wires are in opposite directions, the force between the two wires is repulsive.

SOLVE The currents are parallel all along the length of the wires, so we can use Equation 24.17 to compute the force between the wires. The current in each is 150 A, so

$$F = \frac{\mu_0 L\, I_1 I_2}{2\pi d}$$

$$= \frac{(1.26 \times 10^{-6}\,\text{T}\cdot\text{m/A})(3.7\,\text{m})(150\,\text{A})(150\,\text{A})}{2(\pi)(0.012\,\text{m})}$$

$$= 1.4\,\text{T}\cdot\text{m}\cdot\text{A}$$

The units seem a bit odd at first—but recall the definition of a tesla:

$$1\,\text{T} = 1\,\frac{\text{N}}{\text{A}\cdot\text{m}}$$

This gives

$$F = 1.4\,\text{T}\cdot\text{m}\cdot\text{A} = 1.4\left(\frac{\text{N}}{\text{A}\cdot\text{m}}\right)(\text{m}\cdot\text{A}) = 1.4\,\text{N}$$

L = 3.7 m
I_1 = 150 A
d = 1.2 cm
I_2 = 150 A

FIGURE 24.44 Jumper cables carrying opposite currents.

ASSESS These wires are long, close together, and carry very large currents. But the force between them is quite small—much less than the weight of the wires. In practice, the forces between currents are not an important consideration unless there are many coils of wire, leading to a large total force. This is the case in an MRI solenoid, as we will discuss on page 816.

We're now able to summarize the steps in finding the magnetic force on a charge or current due to known magnetic fields:

PROBLEM-SOLVING STRATEGY 24.2 Magnetic–force problems

PREPARE There are two key pieces to identify in magnetic-force problems:
- The source of the magnetic field.
- The charges or currents that feel a force due to this magnetic field.

SOLVE First, determine the magnitude and direction of the magnetic field at the position of the charges or currents that are of interest.
- In some problems the magnetic field is given.
- If the field is due to a current, use the right-hand rule for fields to determine the field direction and the appropriate equation to determine its magnitude (see Problem-Solving Strategy 24.1).

Next, determine the force this field produces. Working with the charges or currents you identified previously,
- Use the right-hand rule for forces to determine the direction of the force on any moving charge or current.
- Use the appropriate equation to determine the magnitude of the force on any moving charge or current.

ASSESS Are the forces you determine perpendicular to velocities of moving charges and to currents? Are the forces perpendicular to the fields?

Forces Between Current Loops

We will now consider the forces between two current loops. Doing so will allow us to begin to make connections with some of the basic phenomena of magnetism that we saw earlier in the chapter.

We've seen that there is an attractive force between two parallel wires that have their currents in the same direction. If these two wires are bent into loops, as in Figure 24.45, then the force between the two loops will also be attractive. The forces will be repulsive if the currents are in opposite directions.

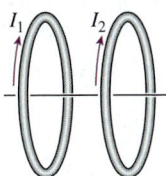

There is an attractive force between parallel loops with currents in the same direction.

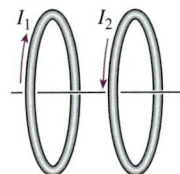

There is a repulsive force between parallel loops with currents in opposite directions.

There is an attractive force between close sections of the loops.

There is a repulsive force between opposite sections, but it is smaller than the attractive force, as their separation is greater.

There is a repulsive force between close sections of the loops.

There is an attractive force between opposite sections, but it is smaller than the repulsive force, as their separation is greater.

FIGURE 24.45 Forces between parallel current loops.

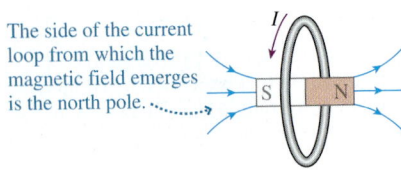

The side of the current loop from which the magnetic field emerges is the north pole.

FIGURE 24.46 We can picture the loop as a small bar magnet.

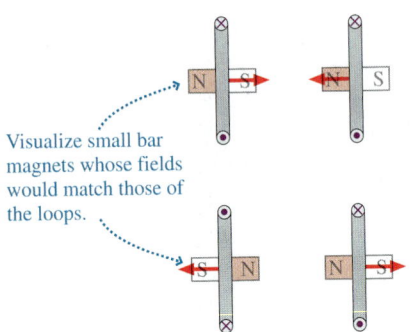

Visualize small bar magnets whose fields would match those of the loops.

FIGURE 24.47 Forces between current loops can be understood in terms of their magnetic poles.

Early in the chapter, we examined the fields from various permanent magnets and current arrangements. Looking back, you can see that a current loop creates a field that is similar to the field of a bar magnet. Both the loop and the bar magnet create what we will call a *dipole field*. This is clearest in the case of the bar magnet, as we have marked the poles. Because their fields are similar, it is often useful to model a current loop as a small bar magnet, as in Figure 24.46. **A current loop, like a bar magnet, is a magnetic dipole.**

We can use this model to understand the forces between current loops. If we label the magnetic poles of the loops as in Figure 24.47, the direction of the forces between current loops becomes immediately obvious. When the currents circulate in the same direction, a north and a south pole face each other and exert attractive forces. When the currents circulate in opposite directions, two like poles face each other and exert repulsive forces. In the case of current loops, **magnetic poles attract or repel because the moving charges in one current exert attractive or repulsive magnetic forces on the moving charges in the other current.**

The dipole is, in some sense, the basic unit of magnetism. We will see this emerge more clearly as the chapter goes on.

STOP TO THINK 24.5   Four wires carry currents in the directions shown. A uniform magnetic field is directed into the paper as shown. Which wire experiences a force to the left?

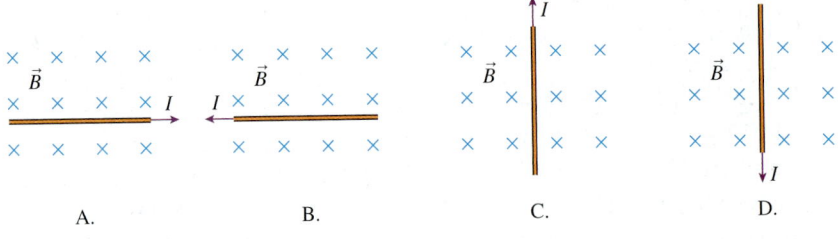

A.   B.   C.   D.

24.7 Magnetic Fields Exert Torques on Dipoles

One of our first observations at the beginning of the chapter was that a compass needle is a small magnet. The fact that it pivots to line up with an external magnetic field means that it experiences a *torque*. In this section, we'll use what we've learned about magnetic forces to understand the torque on a magnetic dipole. We will consider only the case of a current loop (which you'll recall is a magnetic dipole), but the results will be equally applicable to permanent magnets and magnetic dipoles such as compass needles.

A Current Loop in a Uniform Field

Figure 24.48 shows a current loop—a magnetic dipole—in a *uniform* magnetic field. To make our analysis more straightforward, we consider a square current loop, but any other shape would do. The current in each of the four sides of the loop experiences a magnetic force due to the external field $\vec{B}$. Because the field is uniform, the forces on opposite sides of the loop are of equal magnitude. The direction of each force is determined by the right-hand rule for forces. The forces $\vec{F}_{front}$ and $\vec{F}_{back}$ produce no net force or torque. Forces $\vec{F}_{top}$ and $\vec{F}_{bottom}$ also add to give no net force, but because $\vec{F}_{top}$ and $\vec{F}_{bottom}$ don't act along the same line they will rotate the loop by exerting a torque on it.

$\vec{F}_{top}$ and $\vec{F}_{bottom}$ have equal magnitudes and are in opposite directions. They cancel to produce no net force, but they do exert a torque that rotates the loop about the x-axis.

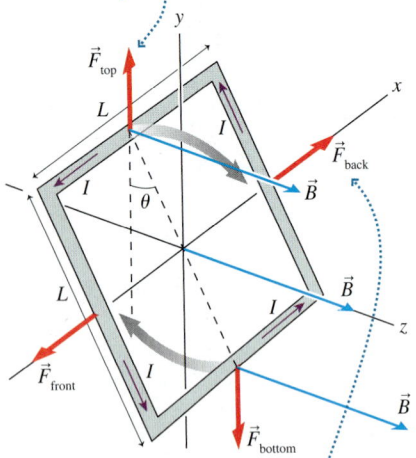

$\vec{F}_{front}$ and $\vec{F}_{back}$ have equal magnitudes and are in opposite directions. They cancel to produce no net force *and* no net torque.

FIGURE 24.48 A loop in a uniform magnetic field experiences a torque.

Although we've shown a current loop, the conclusion is true for any magnetic dipole: **In a uniform field, a dipole experiences a torque but no net force.** For example, a compass needle is not attracted to the earth's poles; it merely feels a torque that lines it up with the earth's field.

We can calculate the torque by looking at a side view of the current loop of Figure 24.48. This is shown in Figure 24.49. In Chapter 7, we calculated the torque by multiplying the force by the moment arm. The loop will rotate about an axis through the center, and we will compute our moment arms from this axis. As you can see from Figure 24.49, the moment arm of $\vec{F}_{top}$ is $\frac{1}{2}L\sin\theta$. $\vec{F}_{top}$ and $\vec{F}_{bottom}$ each produce a torque that tends to rotate the loop. The torque τ_{top} on the top segment of the loop rotates the loop clockwise, as does the torque τ_{bottom} on the bottom segment. The net torque is

$$\tau = \tau_{top} + \tau_{bottom} = F_{top}\left(\tfrac{1}{2}L\right)\sin\theta + F_{bottom}\left(\tfrac{1}{2}L\right)\sin\theta$$

$$= (ILB)\left(\tfrac{1}{2}L\sin\theta\right) + (ILB)\left(\tfrac{1}{2}L\sin\theta\right)$$

$$= (IL^2)B\sin\theta$$

L^2 is the area A of the square loop. Using this, the result can be generalized to any loop of area A:

$$\tau = (IA)B\sin\theta \qquad (24.18)$$

There are two things to note about this torque on a current loop:

1. **The torque depends on properties of the current loop:** Its area (A) and the current (I). The quantity IA, known as the **magnetic dipole moment** (or simply magnetic moment), is a measure of how much torque a dipole will feel in a magnetic field. A permanent magnet can be assigned a magnetic dipole moment as well. It is often useful to think of the magnetic moment as a vector, having a direction. We will represent the magnetic dipole moment as an arrow pointing in the direction of the dipole's magnetic field.

2. **The torque depends on the angle between the loop and the magnetic field.** The torque is maximum when θ is 90°, when the plane of the loop is parallel to the magnetic field. The torque is zero when θ is 0° (or 180°), when the plane of the loop is perpendicular to the magnetic field. As we see in Figure 24.50, a magnetic dipole free to rotate in a field will do so until θ is zero, at which point it will be stable. A magnetic dipole will rotate to line up with a magnetic field just as an electric dipole will rotate to line up with an electric field.

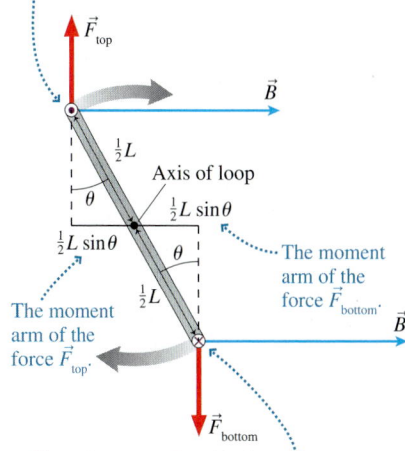

Here, the current is coming out of the paper. The force exerts a torque that will rotate the loop clockwise.

Here, the current is going into the paper. The force exerts a torque that will rotate the loop clockwise.

FIGURE 24.49 Calculating the torque on a current loop in a uniform magnetic field.

13.6

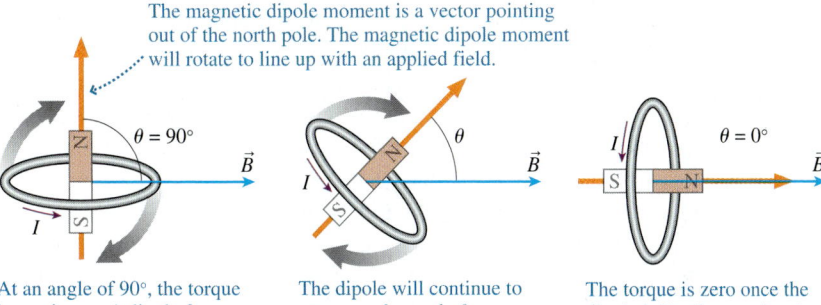

At an angle of 90°, the torque is maximum. A dipole free to rotate will do so in the noted direction.

The dipole will continue to rotate; as the angle θ decreases, the torque decreases as well.

The torque is zero once the dipole is lined up so that the angle θ is zero.

FIGURE 24.50 Torque on a dipole in a magnetic field.

A compass needle, which is a dipole, also rotates until its north pole is in the direction of the magnetic field, as we noted at the very start of the chapter. We are starting to see some connections between the magnetic properties of permanent magnets, such as compass needles, and the magnetic properties of currents.

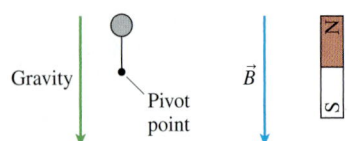

For the pendulum balancing upside down and the magnet aligned exactly opposite the field, there is no torque. But a small displacement from this balance point . . .

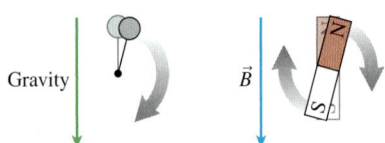

. . . will lead to a torque that will cause a rotation that will continue until. . .

. . . the condition of stable equilibrium is reached. The mass is now below the pivot point; the magnet is lined up with the field.

FIGURE 24.51 Going from unstable to stable equilibrium.

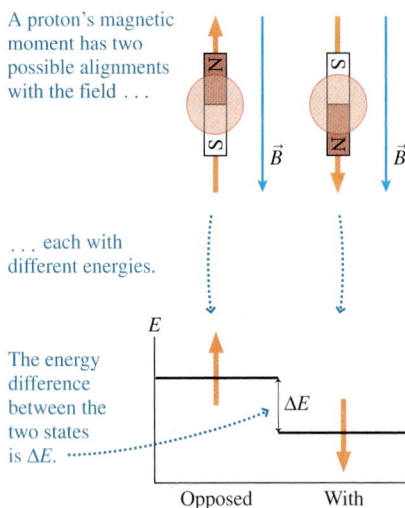

A proton's magnetic moment has two possible alignments with the field . . .

. . . each with different energies.

The energy difference between the two states is ΔE.

FIGURE 24.52 The energy difference between the two possible orientations of a proton's magnetic moment during MRI.

For any dipole in a field, there are actually two angles for which the torque is zero, $\theta = 0°$ and $\theta = 180°$, but there is a difference between these two cases. The $\theta = 0°$ case is *stable:* Once the dipole is in this configuration, it will stay there. The $\theta = 180°$ case is *unstable.* There is no torque if the alignment is perfect, but the slightest rotation will result in a torque that rotates the dipole until it reaches $\theta = 0°$.

We can make a gravitational analogy with this situation in Figure 24.51. For an upside-down pendulum, there will be no torque if the mass is directly above the pivot point. But, as we saw in Chapter 9, this is a position of unstable equilibrium. If displaced even slightly, the mass will rotate until it is below the pivot point. This is the point of lowest potential energy.

We can see, by analogy with the upside-down pendulum, that the unstable alignment of a magnetic dipole has a higher energy. Given a chance, the magnet will rotate "downhill" to the position of lower energy and stable equilibrium. This difference in energy is the key to understanding how the magnetic properties of atoms can be used to image tissues in the body in MRI.

Magnetic Resonance Imaging (MRI)

Magnetic resonance imaging is a modern diagnostic tool that provides detailed images of tissues and structures in the body with no radiation exposure. The key to this imaging technique is the magnetic nature of atoms. **The nuclei of individual atoms have magnetic moments and behave as magnetic dipoles.** Atoms of different elements have different magnetic moments; therefore a magnetic field exerts different torques on different kinds of atoms.

A person receiving an MRI scan is placed in a large solenoid that has a strong magnetic field along its length. Think about one hydrogen atom in this person's body. The nucleus of the hydrogen atom is a single proton. The proton has a magnetic moment, but the proton is a bit different from a simple bar magnet: It is subject to the rules of quantum mechanics, which we will learn about in Chapter 28. A bar magnet can have any angle with the field, but the proton can only line up either *with the field* (the low-energy state) or *opposed to the field* (the high-energy state), as we see in Figure 24.52.

The energy difference ΔE between these two orientations of the proton depends on two key parameters: the magnetic moment of the proton and the strength of the magnetic field.

A tissue sample will have many hydrogen atoms with many protons. Some of the protons will be in the high-energy state and some in the low-energy state. A second magnetic field, called a *probe field,* can be applied to "flip" the dipoles from the low-energy state to the high-energy state. The probe field must be precisely tuned to the energy difference ΔE for this to occur. The probe field is selected to correspond to the magnetic moment of a particular nucleus, in this case hydrogen. If the tuning is correct, dipoles will change state and a signal is measured. A strong signal means that many atoms of this kind are present.

How is this measurement turned into an image? The magnetic field strength of the solenoid and the probe field are varied so that the correct tuning occurs at only one point in the patient's body. The position of the point of correct tuning is swept across a "slice" of tissue. The intensity of the signal is measured at each point, and a cross-sectional image is built up, as shown in Figure 24.53.

Modern MRI scanners use very strong magnetic fields. 1 T fields are common, and some machines have fields that range as high as 4 T! The forces on the current-carrying wires in the solenoid are quite large; the "knocking" that one hears during an MRI scan is due to these forces as the current in the solenoid is changed to vary the magnetic field.

A patient's head in the solenoid of an MRI scanner.

FIGURE 24.53 Cross-sectional image from an MRI scan.

Electric Motors

The torque on a current loop in a magnetic field is the basis for how an electric motor works. The *armature* of a motor is a loop of wire wound on an axle that is free to rotate. This loop is in a strong magnetic field. A current in the loop causes the wires of the loop to feel forces due to this field, as Figure 24.54 shows. The loop becomes a magnetic dipole that will rotate to align itself with the external field. If the current were steady, the armature would simply rotate until its magnetic moment was in its stable position, with the plane of the loop perpendicular to the field. To keep the motor turning, a device called a *commutator* reverses the current direction in the loop every 180°. As the loop reaches its stable configuration, the direction of current in the loop switches, putting the loop back into the unstable configuration, so it will keep rotating to line up in the other direction. This process continues: Each time the loop nears a stable point, the current switches. The loop will keep rotating as long as the current continues. In a real electric motor, the armature is connected to a shaft that can be used to drive gears and wheels.

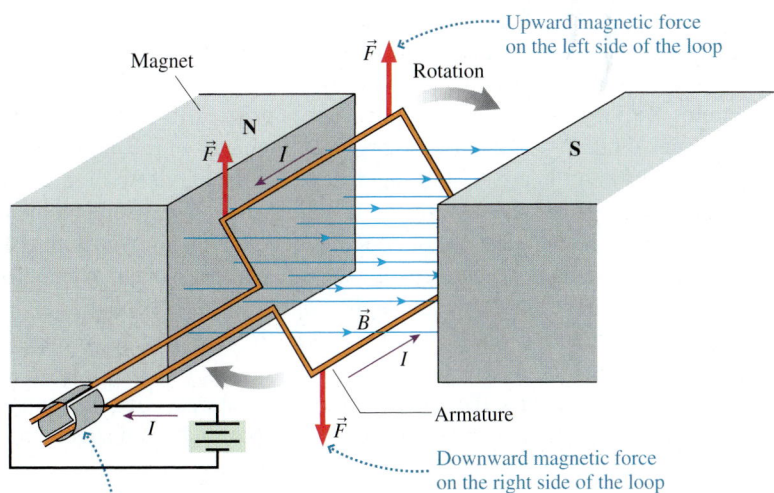

FIGURE 24.54 The operation of a simple motor depends on the torque on a current loop in a magnetic field.

STOP TO THINK 24.6 Which way will this current loop rotate?

A. Clockwise
B. Counterclockwise
C. The loop will not rotate.

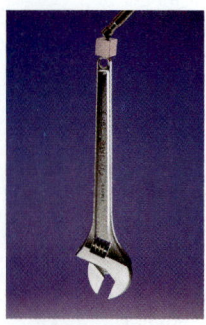

The small magnet supporting the wrench is made of an alloy of neodymium, iron, and boron.

24.8 Magnets and Magnetic Materials

We started the chapter by looking at permanent magnets. We know that permanent magnets produce a magnetic field, but what is the source of this field? There are no electric currents in these magnets. Why can you make a magnet out of certain materials but not others? Why does a magnet stick to the refrigerator? The goal of this section is to answer these questions by developing an atomic-level view of the magnetic properties of matter.

Ferromagnetism

Iron, nickel, and cobalt (and the less common metals gadolinium and dysprosium) are elements that have very strong magnetic behavior: A chunk of iron (or steel, which is mostly iron) will stick to a magnet, and the chunk can be magnetized so that it is itself a magnet. Other metals—such as aluminum and copper—do not exhibit this property.

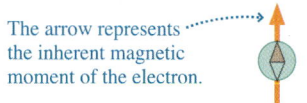

The arrow represents the inherent magnetic moment of the electron.

FIGURE 24.55 Magnetic moment of the electron.

Since magnetic properties of matter were first observed in iron, we call materials that are strongly attracted to magnets and that can be magnetized **ferromagnetic** (from the Latin for iron, *ferrum*). This property of materials is known as *ferromagnetism.*

The key to understanding magnetism at the atomic level was the 1922 discovery that electrons have an *inherent magnetic moment:* An electron itself behaves as a little bar magnet, as we see in Figure 24.55.

We must appeal to the results of quantum physics to find out what happens in an atom with many electrons. The electrons are in atomic *energy levels* or *shells.* Two electrons go into each energy level, one with the magnetic moment pointing up and one with the magnetic moment pointing down. This means that, in a filled shell, the magnetic moments cancel. (These ideas from quantum physics may be familiar to you from chemistry. You will see them in more detail in Chapter 29.) Atoms containing an odd number of electrons must have at least one electron with an unpaired magnetic moment. The magnetic moment of the unpaired electron gives these atoms a net magnetic moment.

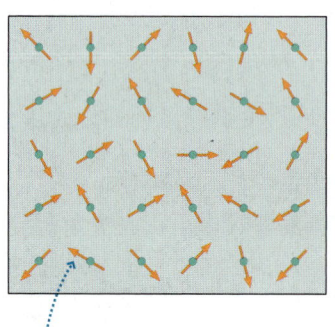

The atomic magnetic moments due to unpaired electrons point in random directions. The sample has no net magnetic moment.

FIGURE 24.56 The random magnetic moments of the atoms in a typical solid.

But atoms with magnetic moments don't necessarily form a solid with magnetic properties. For most elements, the magnetic moments of the atoms are randomly arranged when the atoms join together to form a solid. As Figure 24.56 shows, this random arrangement produces a solid whose net magnetic moment is very close to zero. This agrees with our common experience that most materials are not magnetic; you cannot pick them up with a magnet or make a magnet from them. On the other hand, there are those materials such as iron that do exhibit strong magnetic properties, so we need to discover why these ferromagnetic materials are different.

In ferromagnetic materials, the magnetic moments interact with each other in such a way that they all tend to line up in the *same* direction. Ferromagnetic materials must therefore satisfy *both* of the following conditions:

1. Their atoms have a magnetic moment (due to unpaired electron spins).
2. These atomic magnetic moments line up and reinforce each other.

NOTE ▶ Materials failing one or both of these conditions are not ferromagnetic. Very few materials are ferromagnetic. Only the elements noted earlier and certain alloys incorporating these elements are. ◀

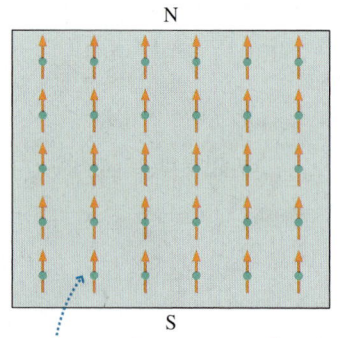

The atomic magnetic moments are aligned. The sample has a north and south magnetic pole.

FIGURE 24.57 In a ferromagnetic solid, the atomic magnetic moments align.

The atomic magnetic moments in a very small piece of iron do not have the random orientations of Figure 24.56. Instead they line up to produce a very strong overall magnetic moment, as in Figure 24.57. This piece of iron has a north and a south magnetic pole, generates a magnetic field, and aligns parallel to an external magnetic field. In other words, it is a magnet—and a very strong one at that.

In a large sample of iron, the magnetic moments will be lined up in local regions called **domains,** each looking like the ordered situation of Figure 24.57, but there will be no long-range ordering. Inside a domain, the atomic magnetic moments will be aligned, but the magnetic moments of individual domains will be randomly oriented. The individual domains are quite small, on the order of 0.1 mm or so, so a piece of iron the size of a nail has thousands of domains. The random orientation of the magnetic moments of the domains, as shown in Figure 24.58, means that there is no overall magnetic moment. So how is it possible to have a magnet that is larger than a single domain? We will get to this next.

Induced Magnetic Moments

Suppose you bring a magnet near a piece of iron, as shown in Figure 24.59. The magnetic field of the magnet penetrates the iron and creates torques on the atomic magnetic moments. The atoms will stay organized in domains, but certain domains will now have more favorable alignments. The result is that domain boundaries move: Domains aligned with the external field become larger at the expense of domains opposed to the field, as atoms in boundary regions realign. After this shift in domain boundaries, the magnetic moments of the domains don't cancel out: Domains that are aligned with the field will predominate. These changes in the size of the domains cause the material to develop a net magnetic moment that is aligned with the external field. The iron will have developed an *induced magnetic moment.*

> **NOTE ▸** Inducing a magnetic moment with a magnetic field is analogous to inducing an electric dipole with an electric field, which we saw in Chapter 21. ◂

Looking at the pole structure of the induced magnetic moment in the iron, we can see that the iron will now be attracted to the magnet. The piece of iron is itself a magnet for now.

The fact that a magnet attracts and picks up ferromagnetic objects was one of the basic observations about magnetism with which we started the chapter. Now we have an explanation of how this works, based on three facts:

1. Electrons are microscopic magnets due to their inherent magnetic moment.
2. A ferromagnetic material in which the magnetic moments are aligned is organized into magnetic domains.
3. The individual domains shift in response to an external magnetic field to produce an induced magnetic moment for the entire object. This induced magnetic moment will be attracted by a magnet that produced the orientation.

When a piece of iron is near a magnet, the iron becomes a magnet as well. But when the applied field is taken away, the domain structure will (generally) return to where it began. That is, the induced magnetic moment will disappear. The iron develops an induced magnetic moment in the presence of an applied field, but it does not become permanently magnetized. In the presence of a *very* strong field, however, a piece of iron can undergo more significant changes to its domain structure, and some domains may permanently change orientation. When the field is removed, the iron may retain some of this magnetic character: The iron will have become permanently magnetized. But pure iron is a rather poor permanent magnetic material; it is very easy to disrupt the ordering of the domains that has created the magnetic moment of the iron. For instance, if you heat (or even just drop!) a piece of magnetized iron, the resulting random atomic motions tend to destroy the alignment of the domains, making it more random. This destroys the magnetic character of the iron.

Alloys of ferromagnetic materials often possess more robust magnetic characters. Alloys of iron and other ferromagnetic elements with rare earth elements can make permanent magnets of incredible strength.

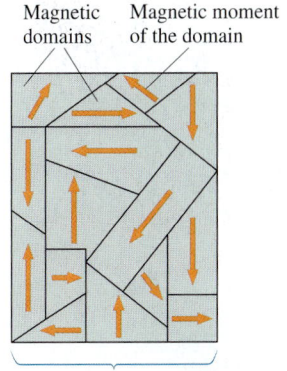

The magnetic moments of the domains tend to cancel one another. The sample as a whole possesses no net magnetic moment.

FIGURE 24.58 Magnetic domains in a ferromagnetic material.

Unmagnetized piece of iron

1. Initially, the magnetic moments of the domains cancel each other; there is no net magnetic moment.

2. Adding the magnetic field from a magnet causes favorably oriented domains (shown in green) to grow at the expense of other domains.

3. The resulting domain structure has a net magnetic moment that is attracted to the magnet. The piece of iron has been magnetized.

FIGURE 24.59 Inducing a magnetic moment in a piece of iron.

Magnetotactic bacteria BIO Several organisms use the earth's magnetic field to navigate. The clearest example of this is *magnetotactic bacteria*. The dark dots in this image are small pieces of iron; each piece is a single domain and hence a very strong magnet. Such a bacterium possesses a very strong magnetic moment: It acts like a bar magnet, lining up with the earth's magnetic field! In the temperate regions where such bacteria live, the earth's field has a large vertical component. The bacteria use their alignment with this vertical field component to navigate up and down.

CONCEPTUAL EXAMPLE 24.5 **Sticking things to the refrigerator**

Everyone has used a magnet to stick papers to the fridge. Why does the magnet stick to the fridge through a layer of paper?

REASON When you bring a magnet near the steel door of the refrigerator, the magnetic field of the magnet induces a magnetic moment in the steel. The direction of this induced moment will be such that it is attracted to the magnet—thus the magnet will stick to the fridge! Magnetic fields go through nonmagnetic materials such as paper, so the magnet will hold the paper to the fridge.

ASSESS The magnetic field strength decreases rapidly with distance, so you won't be able to hold a thick stack of papers to the fridge.

Electromagnets

The magnetic domains in a ferromagnetic material have a strong tendency to line up with an applied magnetic field. This means that it is possible to use a piece of iron or other ferromagnetic material to increase the strength of the field from a current-carrying wire. For example, suppose a solenoid is wound around a piece of iron. When current is passed through the wire, the solenoid's magnetic field lines up the domains in the iron, thus magnetizing it. The resulting **electromagnet** may produce a field that is hundreds of times stronger than the field due to the solenoid itself.

At this point, we've come full circle. We began the chapter by looking at the behavior of bar magnets and compasses. We introduced the idea of a magnetic field. Bar magnets and compasses *produce* a magnetic field, but they also *react to* a magnetic field. We saw that a magnetic field could be produced by moving charges (currents), and that a magnetic field could exert forces on moving charges and currents. Finally, we returned to a microscopic description of magnetism in which we learned that the permanent magnetism of a bar magnet is due to properties of the charges that make it up—the inherent magnetism of electrons. There is clearly a deep connection between electricity and magnetism, as we saw in our final example, the electromagnet. This connection was one of the most important scientific discoveries of the 1800s, and is something we will explore in more detail in the next chapter.

◄ Wrapping a wire around a nail and using a battery to provide a current makes an electromagnet capable of picking up a number of paper clips. The same principle can be applied on a much larger scale.

STOP TO THINK 24.7 A chain of paper clips is hung from a permanent magnet. Which diagram shows the correct induced pole structure of the paper clips?

The goal of Chapter 24 has been to learn about magnetic fields, and how magnetic fields exert forces on currents and moving charges.

GENERAL PRINCIPLES

Sources of Magnetism

At its most fundamental level, magnetism is an interaction between moving charges. Magnetic fields can be created by either:

- electric currents   or by   • permanent magnets

Macroscopic movement of charges as a current

Microscopic magnetism of electrons

The most basic unit of magnetism is the **magnetic dipole,** which consists of a north and south pole.

Three basic kinds of dipoles are:

Current loop   Permanent magnet   Atomic magnet

Consequences of Magnetism

Magnetic fields exert long-range forces on magnetic materials and on moving charges (or currents).

- Unlike poles and like currents attract:

- Like poles and unlike currents repel:

Magnetic fields exert torques on magnetic dipoles, lining them up with the field.

If two or more sources of magnetic field are present, the **principle of superposition** applies to forces and torques.

IMPORTANT CONCEPTS

Magnetic Fields

The direction of the magnetic field

- is the direction in which the north pole of a compass needle points

- due to a current can be found from the **right-hand rule for fields**

The strength of the magnetic field is

- proportional to the torque on a compass needle when turned slightly from the field direction.

- measured in tesla (T): $1 \text{ T} = 1 \dfrac{\text{N}}{\text{A} \cdot \text{m}}$

Magnetic Forces and Torques

The magnitude of the magnetic force on a *moving* charge depends on its charge q, its speed v, and the angle α between the velocity and the field:

$$F = |q|vB\sin\alpha$$

The direction of this force on a positive charge is given by the **right-hand rule for forces.**

The magnitude of the force on a *current-carrying wire* perpendicular to the magnetic field depends on the current and the length of the wire: $F = ILB$

The torque on a *current loop* in a magnetic field depends on the current, the loop's area, and how the loop is oriented in the field: $\tau = (IA)B\sin\theta$

APPLICATIONS

Fields due to common currents

$$B_{\text{wire}} = \frac{\mu_0 I}{2\pi r} \qquad B_{\text{loop}} = \frac{\mu_0 I}{2R}$$

$$B_{\text{solenoid}} = \mu_0 I \frac{N}{L}$$

Charged-particle motion

No force if $\vec{v}$ is parallel to $\vec{B}$.

If $\vec{v}$ is perpendicular to $\vec{B}$ the particle undergoes uniform circular motion with radius $r = mv/|q|B$.

Stability of magnetic dipoles

A magnetic dipole is stable (in a lower energy state) when aligned with the external magnetic field. It is unstable (in a higher energy state) when aligned opposite to the field.

The probe field of an MRI scanner measures the flipping of magnetic dipoles between these two orientations.

QUESTIONS

Conceptual Questions

1. The north pole of a bar magnet is brought near the *center* of another bar magnet as shown in Figure Q24.1. Will the force between the magnets be attractive, repulsive, or zero? Why?

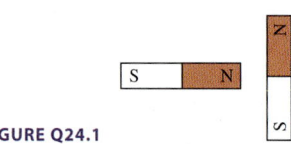

FIGURE Q24.1

2. You have a bar magnet whose poles are not marked. How can you find which pole is north and which is south by using only a piece of string?
3. When you are in the southern hemisphere, does a compass point north or south?
4. Compasses made for use in northern latitudes are weighted so that the south pole of their needle is slightly heavier than the north pole. Compasses for use in southern latitudes are oppositely weighted. Explain why this is done.
5. A *horseshoe* magnet consists of a bar magnet bent into a U-shape as shown in Figure Q24.5. Sketch the magnetic field lines for a horseshoe magnet.

FIGURE Q24.5

6. What is the current direction in the wire of Figure Q24.6? Explain.

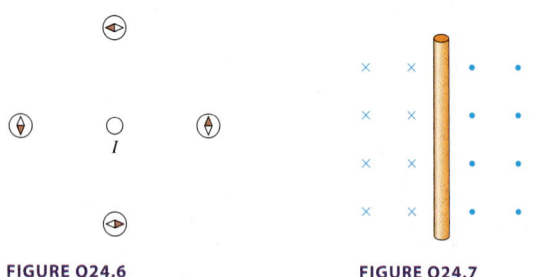

FIGURE Q24.6   **FIGURE Q24.7**

7. What is the current direction in the wire of Figure Q24.7?
8. Given what you know about household wiring, is it possible to use a compass to detect the current in the wires in the walls of your house?

9. Points 1 and 2 in Figure Q24.9 are the same distance from the wires as the point where B = 2.0 mT. What are the strength and direction of $\vec{B}$ at points 1 and 2?

FIGURE Q24.9

10. As shown in Figure Q24.10, a uniform magnetic field points upward, in the plane of the paper. A long wire perpendicular to the paper initially carries no current. When a current is turned on in the wire in the direction shown, the magnetic field at point 1 is found to be zero. Draw the magnetic field vector at point 2 when the current is on.

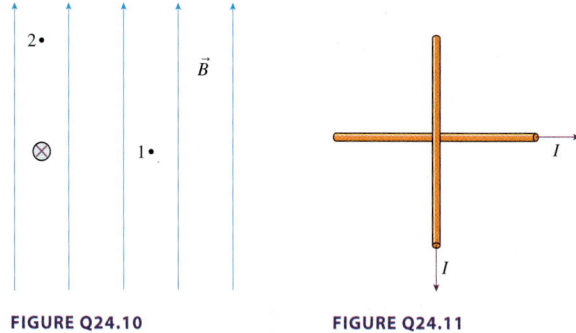

FIGURE Q24.10   **FIGURE Q24.11**

11. Two long wires carry currents in the directions shown in Figure Q24.11. One wire is 10 cm above the other. In which direction is the magnetic field at a point half way between them?
12. If an electron is not moving, is it possible to set it in motion using a magnetic field?
13. Figure Q24.13 shows a short solenoid as seen in cross section. Compasses are placed at points 1 and 2. In which direction will they point when there is a large current in the direction shown? Explain.

FIGURE Q24.13

14. One long solenoid is placed inside another solenoid with twice the diameter but the same length. Each solenoid carries the same current but in opposite directions, as shown in Figure Q24.14. If they also have the same number of turns, in which direction does the magnetic field in the center point? Explain.

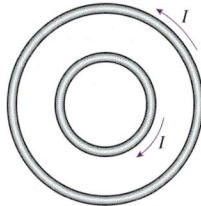

FIGURE Q24.14

15. Can a magnetic field do work on a moving charged particle? If so, explain how. If not, explain why.

16. What is the *initial* direction of deflection for the charged particles entering the magnetic fields shown in Figure Q24.16?

(a)   (b)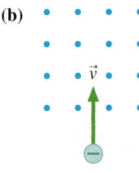

FIGURE Q24.16

17. What is the *initial* direction of deflection for the charged particles entering the magnetic fields shown in Figure Q24.17?

(a)   (b)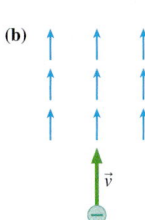

FIGURE Q24.17

18. Determine the magnetic field direction that causes the charged particles shown in Figure Q24.18 to experience the indicated magnetic forces.

(a)   (b)

FIGURE Q24.18 $\vec{F}$ $\vec{F}$ into page

19. Determine the magnetic field direction that causes the charged particles shown in Figure Q24.19 to experience the indicated magnetic forces.

(a)   (b)

FIGURE Q24.19 $\vec{F}$ out of page $\vec{v}$ into page

20. An electron is moving near a long, current-carrying wire as shown in Figure Q24.20. What is the direction of the magnetic force on the electron?

FIGURE Q24.20 $\vec{v}$ I

21. Two positive charges are moving in a uniform magnetic field with velocities as shown in Figure Q24.21. The magnetic force on each charge is also shown. In which direction does the magnetic field point?

FIGURE Q24.21

22. A proton is moving near a long, current-carrying wire. When the proton is at the point shown in Figure Q24.22, in which direction is the force on it?

FIGURE Q24.22   **FIGURE Q24.23**

23. A proton is moving near a long, current-carrying wire. When the proton is at the point shown in Figure Q24.23, in which direction is the force on it?

24. A long wire and a square loop lie in the plane of the paper. Both carry a current in the direction shown in Figure Q24.24. In which direction is the net force on the loop?

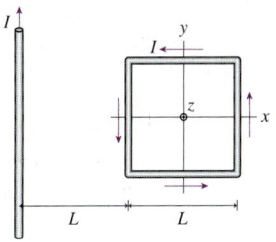

FIGURE Q24.24

25. The computers that control MRI machines cannot have CRT monitors. Explain why this is so.

26. You have a horizontal cathode ray tube (CRT) for which the controls have been adjusted so that the electron beam *should* make a single spot of light exactly in the center of the screen. You observe, however, that the spot is deflected to the right. It is possible that the CRT is broken. But as a clever scientist, you realize that your laboratory might be in either an electric or a magnetic field. Assuming that you do not have a compass, any magnets, or any charged rods, how can you use the CRT itself to determine whether the CRT is broken, is in an electric field, or is in a magnetic field? You cannot remove the CRT from the room.

27. A Slinky is a child's toy that is a long coil spring. Suppose you take a Slinky and let it hang down and stretch out so that the coils do not touch each other, as in Figure Q24.27. Now you connect the Slinky to a power supply and pass a large DC current through it. Think about the current in the coils. Will the Slinky expand or contract?

FIGURE Q24.27

28. A solenoid carries a current that produces a field inside it. A wire carrying a current lies inside the solenoid, at the center, carrying a current along the solenoid's axis. Is there a force on this wire due to the field of the solenoid? Explain.

29. A solenoid carries a current that produces a field inside it. A wire carrying a current lies inside the solenoid, at the center, carrying a current along solenoid's axis. Is there a force on this wire due to the field of the solenoid? Explain.

30. The moon does not have a molten iron core like the earth, but the moon does have a small magnetic field. What might be the source of this field?

Multiple–Choice Questions

31. ‖ A lightweight glass sphere hangs by a thread. The north pole of a bar magnet is brought near the sphere as shown in Figure Q24.31.

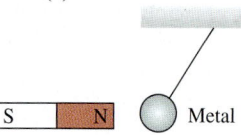

FIGURE Q24.31

    a. Suppose the sphere is electrically neutral. How does it respond?
    A. It is strongly attracted to the magnet.
    B. It is weakly attracted to the magnet.
    C. It does not respond.
    D. It is weakly repelled by the magnet.
    E. It is strongly repelled by the magnet.
    b. How does the sphere respond if it is positively charged? Use the same answer options as in (a).

32. ‖ An unmagnetized metal sphere hangs by a thread. When the north pole of a bar magnet is brought near, the sphere is strongly attracted to the magnet, as shown in Figure Q24.32. Then the magnet is reversed and its south pole is brought near the sphere. How does the sphere respond? Use the same answer options as in Question 31.

Metal

FIGURE Q24.32

33. ‖ If a compass is placed above a current-carrying wire as in Figure Q24.33, the needle will line up with the field of the wire. Which of the views shows the correct orientation of the needle for the noted current direction?

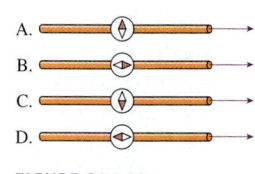

FIGURE Q24.33

34. ‖ Figure Q24.34 shows four particles moving to the right as they enter a region of uniform magnetic field, directed into the paper as noted. All particles move at the same speed and have the same charge. Which particle has the largest mass?

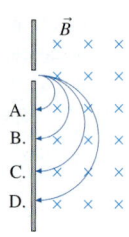

A.
B.
C.
D.

FIGURE Q24.34

35. ‖ Four particles of identical charge and mass enter a region of uniform magnetic field as shown in Figure Q24.35. Which particle has the highest velocity?

FIGURE Q24.35

A. B. C. D.

36. ‖ If all of the particles shown in Figure Q24.35 are electrons, what is the direction of the magnetic field that produced the indicated deflection?
    A. Up (toward the top of the page).
    B. Down (toward the bottom of the page).
    C. Out of the plane of the paper.
    D. Into the plane of the paper.

37. ‖ If two compasses are brought near enough to each other, the magnetic fields of the compasses themselves will be larger than the field of the earth, and the needles will line up with each other. Which of the arrangements of two compasses shown in Figure Q24.37 is a possible stable arrangement?

A.
B.
C.
D.

FIGURE Q24.37

PROBLEMS

Section 24.1 Magnetism

Section 24.2 The Magnetic Field

Section 24.3 Electric Currents Also Create Magnetic Fields

Section 24.4 Calculating the Magnetic Field Due to a Current

1. | What currents are needed to generate the magnetic field strengths of Table 24.1 at a point 1.0 cm from a long, straight wire?

2. | At what distances from a very thin, straight wire carrying a 10 A current would the magnetic field strengths of Table 24.1 be generated?

3. ‖ The magnetic field at the center of a 1.0-cm-diameter loop is 2.5 mT.
    a. What is the current in the loop?
    b. A long, straight wire carries the same current you found in part a. At what distance from the wire is the magnetic field 2.5 mT?

4. ‖‖ For a particular scientific experiment, it is important to be completely isolated from any magnetic field, including the earth's field. A 1.00-m-diameter current loop with 200 turns of wire is set up so that the field at the center is exactly equal to the earth's field in magnitude but opposite in direction. What is the current in the current loop?

5. ‖ What are the magnetic field strength and direction at points 1 to 3 in Figure P24.5?

FIGURE P24.5

6. ‖ Give the magnitude and direction of the magnetic field at points a, b, and c in Figure P24.6?

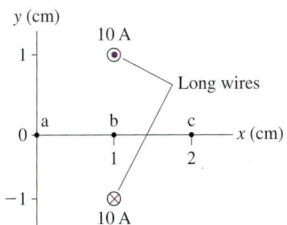

FIGURE P24.6

7. ‖ Although the evidence is weak, there has been concern in
BIO recent years over possible health effects from the magnetic fields generated by transmission lines, the wiring in your house, and electrical appliances.

   a. High-voltage transmission lines, on tall towers, typically carry 200 A at voltages of up to 500,000 V. Although this is much larger than household currents, the lines are roughly 20 m overhead. Estimate the magnetic field strength on the ground underneath such a line.

   b. What percentage of the earth's magnetic field is your answer to part a?

   (Because the percentage is small, and because we live in the earth's field with no harmful effects, it is assumed that any possible health effects are due not to the field strength alone but to the 60 Hz oscillatory nature of fields from power lines.)

   c. A common electrical appliance is an electric blanket. Some consumer groups urge pregnant women not to use electric blankets, just in case there is a health risk. The current through the heater wires is approximately 1 A. Estimate, stating any assumptions you make, the magnetic field strength a fetus might experience. How does this compare to your answer to part a?

   d. It is now standard practice to make electric blankets with minimal external magnetic field. Each wire is paired with another wire that carries current in the opposite direction. How does this reduce the external magnetic field?

8. ‖ The two insulated wires in Figure P24.8 cross at a 30° angle but do not make electrical contact. Each wire carries a 5.0 A current. Points 1 and 2 are each 4.0 cm from the intersection and equally distant from both wires. What are the magnitude and direction of the magnetic fields at points 1 and 2?

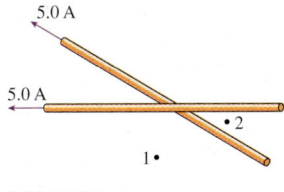

FIGURE P24.8

9. ‖‖ A long wire carrying a 5.0 A current perpendicular to the xy-plane intersects the x-axis at $x = -2.0$ cm. A second, parallel wire carrying a 3.0 A current intersects the x-axis at $x = +2.0$ cm. At what point or points on the x-axis is the magnetic field zero if (a) the two currents are in the same direction and (b) the two currents are in opposite directions?

10. ‖ The element niobium, which is a metal, is a superconductor (i.e., no electrical resistance) at temperatures below 9 K. However, the superconductivity is destroyed if the magnetic field at the surface of the wire of the metal reaches or exceeds 0.10 T. What is the maximum current in a straight, 3.0-mm-diameter superconducting niobium wire?
Hint: You can assume that all the current flows in the center of the wire.

11. ‖ A solenoid used to produce magnetic fields for research purposes is 2.0 m long, with an inner radius of 30 cm and 1000 turns of wire. When running, the solenoid produces a field of 1.0 T in the center. Given this, how large a current does it carry?

12. | Two concentric current loops lie in the same plane. The smaller loop has a radius of 3.0 cm and a current of 12 A. The bigger loop has a current of 20 A. The magnetic field at the center of the loops is found to be zero. What is the radius of the bigger loop?

13. | The magnetic field of the brain has been measured to be
BIO approximately 3.0×10^{-12} T. Although the currents that cause this field are quite complicated, we can get a rough estimate of their size by modeling them as a single circular current loop 16 cm (the width of a typical head) in diameter. What current is needed to produce such a field at the center of the loop?

14. ‖‖ A researcher would like to perform an experiment in zero magnetic field, which means that the field of the earth must be cancelled. Suppose the experiment is done inside a solenoid of diameter 1.0 m, length 4.0 m, with a total of 5000 turns of wire. The solenoid is oriented to produce a field that opposes and exactly cancels the field of the earth. What current is needed in the solenoid's wires?

15. ‖ What is the magnetic field at the center of the loop in Figure P24.15?

FIGURE P24.15

16. ‖‖‖ A straight wire carrying a current of 50 A lies along the axis of a 4.0-cm-diameter solenoid. The solenoid is 50 cm long and has 200 turns carrying a current of 4.0 A.

   a. What is the magnitude of the magnetic field 1.0 cm from the wire?

   b. Estimate the magnitude of the magnetic field 3.0 cm from the wire.

17. ‖ Experimental tests have shown that hammerhead sharks
BIO can detect magnetic fields. In one such test, 100 turns of wire were wrapped around a 7.0-m-diameter cylindrical shark tank. A magnetic field was created inside the tank when this coil of wire carried a current of 1.5 A. Sharks trained by getting a food reward when the field was present would later unambiguously respond when the field was turned on.

   a. What was the magnetic field strength in the center of the tank due to the current in the coil?

   b. Is the strength of the coil's field at the center of the tank larger or smaller than that of the earth?

18. We have seen that the heart produces a magnetic field that
BIO can be used to diagnose problems with the heart. The magnetic field of the heart is a dipole field produced by a loop current in the outer layers of the heart. Suppose the field at the center of the heart is 90 pT (a pT is 10^{-12} T) and that the heart has a diameter of approximately 12 cm. What current circulates around the heart to produce this field?

19. The magnetic field strength at the north pole of a 2.0-cm-diameter, 8.0-cm-long Alnico magnet is 0.10 T. To produce the same field with a solenoid of the same size, carrying a current of 2.0 A, how many turns of wire would you need?

20. You have a 1.0-m-long copper wire. You want to make an N-turn current loop that generates a 1.0 mT magnetic field at the center when the current is 1.0 A. You must use the entire wire. What will be the diameter of your coil?

21. In the Bohr model of the hydrogen atom, the electron moves in a circular orbit of radius 5.3×10^{-11} m with speed 2.2×10^6 m/s. According to this model, what is the magnetic field at the center of a hydrogen atom due to the motion of the electron? **Hint:** Determine the *average* current of the orbiting electron.

Section 24.5 Magnetic Fields Exert Forces on Moving Charges

22. A proton moves with a speed of 1.0×10^7 m/s in the directions shown in Figure P24.22. A 0.50 T magnetic field points in the positive x-direction. For each, what is magnetic force $\vec{F}$ on the proton? Give your answers as a magnitude and a direction.

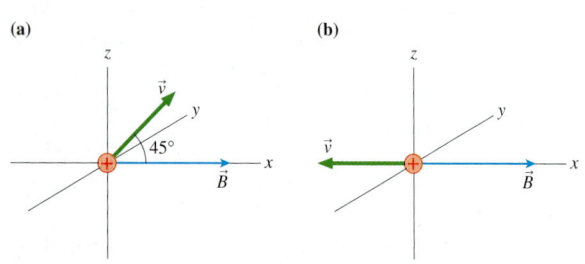

FIGURE P24.22

23. An electron moves with a speed of 1.0×10^7 m/s in the directions shown in Figure P24.23. A 0.50 T magnetic field points in the positive x-direction. For each, what is magnetic force $\vec{F}$ on the electron? Give your answers as a magnitude and a direction.

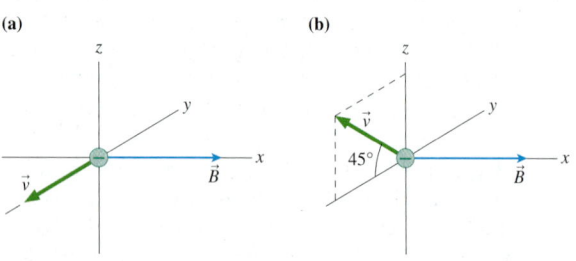

FIGURE P24.23

24. An electromagnetic flowmeter applies a magnetic field of
BIO 0.20 T to blood flowing through a coronary artery at a speed of 15 cm/s. What force is felt by a chlorine ion with a single negative charge?

25. The aurora is caused when electrons and protons, moving in the earth's magnetic field of $\approx 5.0 \times 10^{-5}$ T, collide with molecules of the atmosphere and cause them to glow. What is the radius of the circular orbit for
  a. An electron with speed 1.0×10^6 m/s?
  b. A proton with speed 5.0×10^4 m/s?

26. Problem 24.25 describes two particles that orbit the earth's magnetic field lines. What is the *frequency* of the circular orbit for
  a. An electron with speed 1.0×10^6 m/s?
  b. A proton with speed 5.0×10^4 m/s?

27. The microwaves in a microwave oven are produced in a special tube called a *magnetron*. The electrons orbit in a magnetic field at a frequency of 2.4 GHz, and as they do so they emit 2.4 GHz electromagnetic waves. What is the strength of the magnetic field??

28. A mass spectrometer similar to the one in Figure 24.38 is designed to separate protein fragments. The fragments are ionized by removing a single electron and then enter a 0.80 T uniform magnetic field at a speed of 2.3×10^5 m/s. If a fragment has a mass 85 times the mass of the proton, what will be the distance between the points where the ion enters and exits the magnetic field?

29. In a certain mass spectrometer, particles with a charge of $+e$ are sent into the spectrometer with a velocity of 2.5×10^5 m/s. They are found to move in a circular path with a radius of 0.21 m. If the magnetic field of the spectrometer is 0.050 T, what kind of particles are these likely to be?

30. In Figure P24.30, a proton moves in a 0.40 T magnetic field that is directed along the positive y-axis. The force on the proton is of 6.0×10^{-15} N, directed out of the paper. Draw a possible velocity vector for the proton. What is the x-component of the velocity of the proton?

FIGURE P24.30

31. At $t = 0$, a proton is moving with a speed of 5.5×10^5 m/s at an angle of 30° from the x-axis, as shown in Figure P24.31. A magnetic field of magnitude 1.50 T is pointing in the positive y-direction. What will be the y-coordinate of the proton 10 μs later?

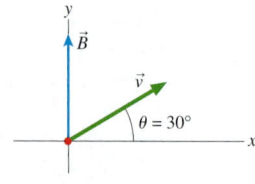

FIGURE P24.31

Section 24.6 Magnetic Fields Exert Forces on Currents

32. What magnetic field strength and direction will levitate the 2.0 g wire in Figure P24.32?

FIGURE P24.32

33. | What is the net force (magnitude and direction) on each wire in Figure P24.33?

FIGURE P24.33

34. | A uniform 2.5 T magnetic field points to the right. A 3.0-m-long wire, carrying 15 A, is placed at an angle of 30° to the field, as shown in Figure 24.34. What is the force (magnitude and direction) on the wire?

FIGURE P24.34

35. | The wire in Figure P24.34 is rotated in the plane of the page until the force on it has a magnitude of 100 N. What is the angle between the wire and the magnetic field?

36. | The four wires in Figure P24.36 are tilted at 20° with respect to the field direction. If each carries 4.5 A and is 0.35 m long, what is the force (direction and magnitude) on each?

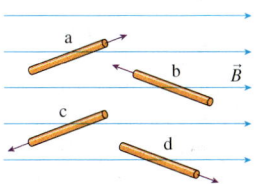

FIGURE P24.36

37. | A loudspeaker cone is moved by the force of a permanent magnet on the current in a coil of wire. The coil in a certain loudspeaker has 20 turns of wire with a total wire length of 1.2 m. The magnet in the speaker applies a magnetic field of 0.15 T that is everywhere perpendicular to this coil of wire, as shown in Figure P24.37. If the wire carries a current of 1.2 A, what is the force on the wire?

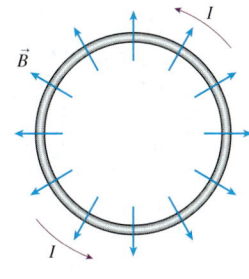

FIGURE P24.37

Section 24.7 Magnetic Fields Exert Torques on Dipoles

38. |||| A current loop in a motor has an area of 0.85 cm². It carries a 240 mA current in a uniform field of 0.62 T. What is the magnitude of the maximum torque on the current loop?

39. || A square current loop 5.0 cm on each side carries a 500 mA current. The loop is in a 1.2 T uniform magnetic field. The axis of the loop, perpendicular to the plane of the loop, is 30° away from the field direction. What is the magnitude of the torque on the current loop?

40. || Figure P24.40 shows two square current loops. The loops are far apart and do not interact with each other.
    a. Use force diagrams to show that both loops are in equilibrium, having a net force of zero and no torque.
    b. One of the loop positions is stable. That is, the forces will return it to equilibrium if it is rotated slightly. The other position is unstable, like an upside-down pendulum: If rotated slightly, it will not return to the position shown. Which is which? Explain.

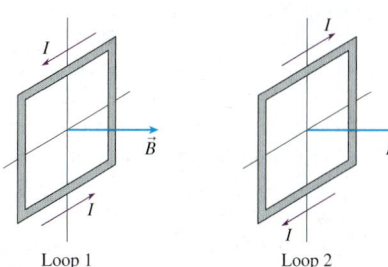

FIGURE P24.40   Loop 1   Loop 2

41. |||| The earth's magnetic field, with a magnetic dipole moment of 8.0×10^{22} A·m², is generated by currents within the molten iron of the earth's outer core. (The inner core is solid iron.) As a simple model, consider the outer core to be a current loop made of a 1000-km-diameter "wire" of molten iron. The loop diameter, measured between the centers of the "wires," is 3000 km. What is the current in the current loop?

42. ||| a. What is the magnitude of the torque on the circular current loop in Figure P24.42?
    b. What is the loop's equilibrium position?

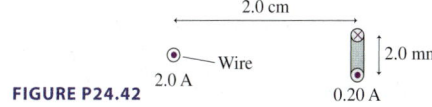

FIGURE P24.42

Section 24.8 Magnets and Magnetic Materials

43. || A computer diskette is a plastic disk with a ferromagnetic coating. A single magnetic domain can have its magnetic moment oriented to point either up or down, and these two orientations can be interpreted as a binary 0 (up) or 1 (down). Each 0 or 1 is called a *bit* of information. A diskette stores roughly 500,000 *bytes* of data on one side, and each byte contains eight bits. Estimate the width of a magnetic domain, and compare your answer to the typical domain size given in the text. List any assumptions you use in your estimate.

44. ||| The ends of two permanent bar magnets can either attract or repel each other. Give a step-by-step description, using both words and pictures, of how these magnetic forces result from the interaction between the electron spins. Consider both the attractive and the repulsive situations.

45. ||| A permanent magnet can pick up a piece of nonmagnetized iron. Give a step-by-step description, using both words and pictures, of how the magnetic force on the iron results from the interaction between the magnetic moments of the iron.

46. || All ferromagnetic materials have a *Curie temperature,* a temperature above which they will cease to be magnetic. Explain in some detail why you might expect this to be so.

General Problems

47. | In Figure P24.47, a compass sits
INT 1 cm above a wire in a circuit containing a 1.0 F capacitor charged to 5.0 V,
a 1.0 Ω resistor, and an open switch.
The compass is lined up with the
earth's magnetic field. The switch is

FIGURE P24.47

then closed, so there is a current in the circuit, and the switch
remains closed until the capacitor has completely discharged.
    a. What is the magnitude of the magnetic field due to the cur-
        rent in the wire right after the switch is closed? How does
        this compare with the magnitude of the field of the earth?
    b. Describe how the compass orientation changes right after
        the switch is closed, and how the compass orientation
        changes as time goes on.
48. ||| The right edge of the circuit in Figure P24.48 extends into a
INT 50 mT uniform magnetic field. What are the magnitude and
direction of the net force on the circuit?

FIGURE P24.48 $B = 50$ mT

49. || The two 10-cm-long parallel wires in Figure P24.49 are
INT separated by 5.0 mm. For what value of the resistor R will the
force between the two wires be 5.4×10^{-5} N?

FIGURE P24.49

50. ||| The capacitor in Figure P24.50 is charged to 50 V. The
INT switch closes at $t = 0$ s. Draw a graph showing the magnetic
field strength as a function of time at the position of the dot.
On your graph indicate the maximum field strength and provide an appropriate numerical scale on the horizontal axis.

FIGURE P24.50   **FIGURE P24.51**

51. || An electron travels with speed 1.0×10^7 m/s between the
INT two parallel charged plates shown in Figure P24.51. The
plates are separated by 1.0 cm and are charged by a 200 V
battery. What magnetic field strength and direction will
allow the electron to pass between the plates without being
deflected?

52. || The two springs in Figure P24.52 each have a spring con-
INT stant of 10 N/m. They are stretched by 1.0 cm when a current
passes through the wire. How big is the current?

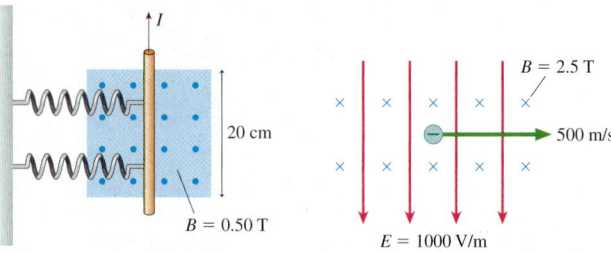

FIGURE P24.52   **FIGURE P24.53**

53. || An antiproton (which has the same properties as a proton
INT except that its charge is $-e$) is moving in the combined elec-
tric and magnetic fields of Figure P24.53.
    a. What are the magnitude and direction of the antiproton's
        acceleration at this instant?
    b. What would be the magnitude and direction of the acceler-
        ation if $\vec{v}$ were reversed?
54. | Typical blood velocities in the coronary arteries range from
BIO 10 to 30 cm/s. An electromagnetic flowmeter applies a mag-
INT netic field of 0.25 T to a coronary artery with a blood velocity
of 15 cm/s. As we saw in Figure 24.39, this field exerts a force
on ions in the blood, which will separate. The ions will separate until they make an electric field that exactly balances the
magnetic force. This electric field produces a voltage that can
be measured.
    a. What force is felt by a singly ionized (positive) sodium
        ion?
    b. Charges in the blood will separate until they produce an
        electric field that cancels this magnetic force. What will be
        the resulting electric field?
    c. What voltage will this electric field produce across an
        artery with a diameter of 3 mm?
55. | A power line consists of two wires, each carrying a current
of 400 A in the same direction. The wires are separated by a
distance of 5.0 m. Which is larger: the force of the earth's
magnetic field on each wire or the magnetic force between the
wires?
56. ||| A 1.0-m-long copper wire of diameter 0.10 cm carries a
INT current of 50.0 A to the east. Suppose we apply to this wire a
magnetic field that produces on it an upward force exactly
equal in magnitude to the wire's weight, causing the wire to
"levitate." What are the field's direction and magnitude?
57. || An insulated copper wire is wrapped
around an iron nail. The resulting coil of
wire consists of 240 turns of wire that cover
1.8 cm of the nail, as shown in Figure P24.57. A current of 0.60 A passes
through the wire. If the ferromagnetic properties of the nail increase the field by a factor
of 100, what is the magnetic field strength
inside the nail?

FIGURE P24.57

58. ||| Figure P24.58 is a cross section through three long wires with linear mass density 50 g/m. They each carry equal currents in the directions shown. The lower two wires are 4.0 cm apart and are attached to a table. What current I will allow the upper wire to "float" so as to form an equilateral triangle with the lower wires?

4.0 cm

FIGURE P24.58

59. ||| A long, straight wire with linear mass density of 50 g/m is suspended by threads, as shown in Figure P24.59. There is a uniform magnetic field pointing vertically downward. A 10 A current in the wire experiences a horizontal magnetic force that deflects it to an equilibrium angle of 10°. What is the strength of the magnetic field $\vec{B}$?

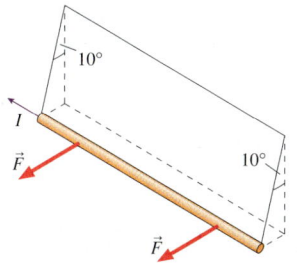

FIGURE P24.59

60. | A mass spectrometer is designed to separate atoms of carbon to determine the fraction of different isotopes. (Isotopes of an element, as we will see in Chapter 30, have the same atomic number but different atomic mass, due to different numbers of neutrons.) There are three main isotopes of carbon, with the following atomic masses:

Atomic masses

^{12}C	1.99×10^{-26} kg
^{13}C	2.16×10^{-26} kg
^{14}C	2.33×10^{-26} kg

The atoms of carbon are singly ionized and enter a mass spectrometer with magnetic field strength $B = 0.200$ T at a speed of 1.50×10^5 m/s. The ions move along a semicircular path and exit through an exit slit. How far from the entrance will the beams of the different isotope ions end up?

61. ||| A solenoid is near a piece of iron, as shown in Figure P24.61. When a current is present in the solenoid, a magnetic field is created. This magnetic field will magnetize the iron, and there will be a net force between the solenoid and the iron.
   a. Make a sketch showing the direction of the magnetic field from the solenoid. On your sketch, label the induced north magnetic pole and the induced south magnetic pole in the iron.
   b. Will the force on the iron be attractive or repulsive?
   c. Suppose this force moves the iron. Which way will the iron move?
   d. As the iron moves, what will happen to the strength of the force?

Current in solenoid is clockwise as viewed from the right end.

Piece of iron is lined up with the axis of the solenoid.

FIGURE P24.61

Passage Problems

The Velocity Selector INT 13.8

In experiments where all the charged particles in a beam are required to have the same velocity (for example, when entering a mass spectrometer), scientists use a *velocity selector*. A velocity selector has a region of uniform electric and magnetic fields that are perpendicular to each other and perpendicular to the motion of the charged particles. Both the electric and magnetic fields exert a force on the charged particles. If a particle has precisely the right velocity, the two forces exactly cancel and the particle is not deflected. Equating the forces due to the electric field and the magnetic field gives the following equation:

$$qE = qvB$$

Solving for the velocity, we get:

$$v = \frac{E}{B}$$

A particle moving at this velocity will pass through the region of uniform fields with no deflection, as shown in Figure P24.62. For higher or lower velocities than this, the particles will feel a net force and will be deflected. A slit at the end of the region allows only the particles with the correct velocity to pass.

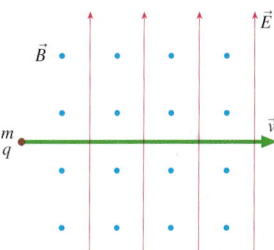

FIGURE P24.62

62. | Assuming the particle in Figure P24.62 is positively charged, what are the directions of the forces due to the electric field and to the magnetic field?
   A. The force due to the electric field is directed up (toward the top of the page); the force due to the magnetic field is directed down (toward the bottom of the page).
   B. The force due to the electric field is directed down (toward the bottom of the page); the force due to the magnetic field is directed up (toward the top of the page).
   C. The force due to the electric field is directed out of the plane of the paper; the force due to the magnetic field is directed into the plane of the paper.
   D. The force due to the electric field is directed into the plane of the paper; the force due to the magnetic field is directed out of the plane of the paper.

63. | How does the kinetic energy of the particle in Figure P24.62 change as it traverses the velocity selector?
   A. The kinetic energy increases.
   B. The kinetic energy does not change.
   C. The kinetic energy decreases.

64. | Suppose a particle with twice the velocity of the particle in Figure P24.62 enters the velocity selector. The path of this particle will curve
   A. Upward (toward the top of the page).
   B. Downward (toward the bottom of the page).
   C. Out of the plane of the paper.
   D. Into the plane of the paper.

65. | Next, a particle with the same mass and velocity as the particle in Figure P24.62 enters the velocity selector. This particle has a charge of $2q$—twice the charge of the particle in Figure P24.62. In this case, we can say that
   A. The force of the electric field on the particle is greater than the force of the magnetic field.
   B. The force of the magnetic field on the particle is greater than the force of the electric field.
   C. The forces of the electric and magnetic fields on the particle are still equal.

STOP TO THINK ANSWERS

Stop to Think 24.1: C. The compass needle will not rotate since there is no force between the stationary charges on the rod and the magnetic poles of the compass needle.

Stop to Think 24.2: A. The compass needle will rotate to line up with the field of the magnet, which goes from the north to the south pole.

Stop to Think 24.3: D. The compass needle will rotate to line up with the field circling the wire. The right-hand rule for fields shows this to be toward the top of the paper in the figure.

Stop to Think 24.4: A, C. The force must be toward the center of the circle. Using the right-hand rule for forces, we see that this will be true for the situations in A and C if the particles are negatively charged.

Stop to Think 24.5: C. The right-hand rule for forces gives the direction of the force. With the field into the paper, the force is to the left if the current is toward the top of the paper.

Stop to Think 24.6: B. Looking at the forces on the top and the bottom of the loop, we can see that the loop will rotate counter-clockwise. Alternatively, we can look at the dipole structure of the loop: With a north pole on the left and a south pole on the right, the loop will rotate counterclockwise.

Stop to Think 24.7: B. All of the dipoles will be aligned the same way as the bar magnet.

25

ELECTROMAGNETIC INDUCTION AND ELECTROMAGNETIC WAVES

The photo of these flowers on the left shows how they appear to our eyes, in visible light. But there's more to the story! The false-color view of the flowers on the right shows their appearance in the ultraviolet, beyond the range of human vision, revealing pigments we can't see. Whose eyes are these pigments intended for?

Looking Ahead ▶▶

The goal of Chapter 25 is to understand the nature of electromagnetic induction and electromagnetic waves. In this chapter, you will learn to:

▶ Find the motional emf in a conductor moving in a magnetic field.

▶ Use Lenz's law and Faraday's law to determine the direction and size of induced currents.

▶ Understand how induced electric and magnetic fields lead to electromagnetic waves.

▶ Apply wave and photon models to the electromagnetic spectrum.

▶ Understand the properties of different types of electromagnetic waves.

Looking Back ◀◀

This chapter will join together ideas about electric and magnetic fields, and will build on your understanding of waves. Please review:

◀ Sections 15.3–15.5 Basic wave concepts, light waves, plane waves, intensity.

◀ Section 20.4 The electric field.

◀ Section 21.4 The electric potential.

◀ Sections 24.2–24.5 Magnetic fields and forces.

In previous chapters, we developed a field model to describe electric and magnetic forces. Electric and magnetic fields may seem like abstractions, but they are very real. Their reality will become more apparent in this chapter as we discover connections between electric and magnetic fields. We will begin by exploring how a changing magnetic field can create an electric field. This phenomenon, *electromagnetic induction,* has a wide range of practical applications.

Studying electromagnetic induction will lead us to *electromagnetic waves.* We have already considered some of the properties of electromagnetic waves in earlier chapters on waves and optics, but we didn't explain exactly what was "waving." That's a task we will take up in this chapter. Electromagnetic waves of different wavelengths have very different properties, as we will see when we explore the nature of different parts of the electromagnetic spectrum, from radio waves to x rays. In doing so, we will introduce the concept of a *photon,* a "particle" of light, our first taste of the modern physics topics to be explored in the final part of this book.

25.1 Induced Currents

In Chapter 24, we learned that a current can create a magnetic field. As soon as this discovery was widely known, investigators began considering a related question: Can a magnetic field create a current?

One of the early investigators was Michael Faraday, who we first encountered in Chapter 20 as the originator of the idea of a field. Faraday was experimenting with two coils of wire wrapped around an iron ring, as shown in Figure 25.1, when he made a remarkable discovery. He had hoped that the magnetic field generated by a current in the coil on the left would create a magnetic field in the iron, and that the magnetic field in the iron might then somehow produce a current in the circuit on the right.

This technique failed to generate a steady current, but Faraday noticed that the needle of the current meter jumped ever so slightly at the instant when he closed the switch in the circuit on the left. After the switch was closed, the needle immediately returned to zero. Faraday's observation suggested to him that a current was generated only if the magnetic field was *changing* as it passed through the coil. Faraday set out to test this hypothesis through a series of experiments.

Closing the switch in the left circuit . . . . . . causes a *momentary* current in the right circuit.

FIGURE 25.1 Faraday's discovery of electromagnetic induction.

Faraday investigates electromagnetic induction

Faraday placed one coil directly above the other, without the iron ring. There was no current in the lower circuit while the switch was in the closed position, but a momentary current appeared whenever the switch was opened or closed.	Faraday pushed a bar magnet into a coil of wire. This action caused a momentary deflection of the needle in the current meter, although *holding* the magnet inside the coil had no effect. A quick withdrawal of the magnet deflected the needle in the other direction.	Must the magnet move? Faraday created a momentary current by rapidly pulling a coil of wire out of a magnetic field, although there was no current if the coil was stationary in the magnetic field. Pushing the coil *into* the magnet caused the needle to deflect in the opposite direction.

Open or close switch.	Push or pull magnet.	Push or pull coil.
Opening or closing the switch creates a momentary current.	Pushing the magnet into the coil or pulling it out creates a momentary current.	Pushing the coil into the magnet or pulling it out creates a momentary current.

All of these experiments served to bolster Faraday's hypothesis: **Faraday found that there is a current in a coil of wire if and only if the magnetic field passing through the coil is *changing*.** This is an informal statement that we'll further develop into a general principle called *Faraday's law*.

It makes no difference what causes the magnetic field to change: current stopping or starting in a nearby circuit, moving a magnet through the coil, or moving the coil in and out of a magnet. The effect is the same in all cases. There is no current if the field through the coil is not changing, so it's not the magnetic field itself that is responsible for the current but, instead, it is the *changing of the magnetic field*.

The current in a circuit due to a changing magnetic field is called an **induced current.** Opening the switch or moving the magnet *induces* a current in a nearby circuit. An induced current is not caused by a battery; it is a completely new way to generate a current. The creation of an electrical current by a changing magnetic field is our first example of **electromagnetic induction.**

▶ **External pacemaker programming** BIO A circuit that creates a changing magnetic field can induce a current in a second circuit with no direct electrical connection to the first. In this photo a coil carrying an alternating current creates a changing magnetic field that induces currents in a sensing circuit in a cardiac pacemaker inside the patient's chest. These currents adjust settings for the pacemaker. The magnetic fields penetrate the clothing and the body, so the pacemaker can be programmed in a doctor's office with no need for surgery.

In the coming sections, we'll explore electromagnetic induction in some detail. There are different ways to induce currents. Some of these can be explained in terms of physics we already understand, others will require new concepts. We'll start our investigation of electromagnetic induction by looking at situations in which the magnetic field is fixed while the circuit moves or changes.

25.2 Motional emf

In 1996, astronauts on the space shuttle deployed a satellite at the end of a 20-km-long conducting tether. A potential difference of up to 3500 V developed between the shuttle and the satellite as the wire between the two swept through the earth's magnetic field. Why would the motion of a wire in a magnetic field produce such a large voltage? Let's explore the mechanism behind this *motional emf*.

To begin, consider a conductor of length *l* that moves with velocity $\vec{v}$ through a uniform magnetic field $\vec{B}$, as shown in Figure 25.2. The charge carriers inside the conductor—assumed to be positive, as in our definition of current—also move with velocity $\vec{v}$, so they each experience a magnetic force. For simplicity, we will assume that $\vec{v}$ is perpendicular to $\vec{B}$, in which case the magnitude of the force is $F_B = qvB$. This force causes the charge carriers to move. For the geometry of Figure 25.2, the right-hand rule tells us that the positive charges move toward the top of the moving conductor, leaving an excess of negative charge at the bottom.

A satellite tethered to the Space Shuttle.

Charge carriers in the conductor experience a force of magnitude $F_B = qvB$. Positive charges are free to move and drift upward.

The resulting charge separation creates an electric field in the conductor. $\vec{E}$ increases as more charge flows.

The charge flow continues until the electric and magnetic forces balance. For a positive charge carrier, the upward magnetic force $\vec{F}_B$ is equal to the downward electric force $\vec{F}_E$.

FIGURE 25.2 The magnetic force on the charge carriers in a moving conductor creates an electric field inside the conductor.

This motion of the charge carriers cannot continue forever. The separation of the charge carriers creates an electric field. The resulting electric force *opposes* the separation of charge, so the charge carriers continue to move only until the electric force has grown to exactly balance the magnetic force:

$$F_E = qE = F_B = qvB$$

When this balance occurs, the charge carriers experience no net force and thus undergo no further motion. The electric field strength at equilibrium is

$$E = vB \qquad (25.1)$$

(a) Chemical reactions separate the charges and cause a potential difference between the ends. This is a chemical emf.

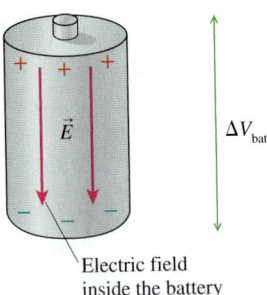

Electric field
inside the battery

(b) Magnetic forces separate the charges and cause a potential difference between the ends. This is a motional emf.

Electric field
inside the moving
conductor

FIGURE 25.3 Two different ways to generate an emf.

Thus, the magnetic force on the charge carriers in a moving conductor creates an electric field $E = vB$ inside the conductor.

The electric field, in turn, creates an electric potential difference between the two ends of the moving conductor. We found in Chapter 21 that the potential difference between two points separated by distance l parallel to an electric field E is $\Delta V = El$. Thus the motion of the wire through a magnetic field *induces* a potential difference

$$\Delta V = vlB \tag{25.2}$$

between the ends of the conductor. The potential difference depends on the strength of the magnetic field and on the wire's speed through the field.

There's an important analogy between this potential difference and the potential difference of a battery. Figure 25.3a reminds you that a battery uses a nonelectric force—which we called the charge escalator—to separate positive and negative charges. We refer to a battery, where the charges are separated by chemical reactions, as a source of *chemical emf*. The moving conductor of Figure 25.3b develops a potential difference because of the work done to separate the charges. The emf of the conductor is due to its motion, rather than to chemical reactions inside, so we can define the **motional emf** of a conductor of length l moving with velocity $\vec{v}$ perpendicular to a magnetic field $\vec{B}$ to be

$$\mathcal{E} = vlB \tag{25.3}$$

EXAMPLE 25.1 Finding the motional emf for an airplane

A Boeing 747 aircraft with a wingspan of 65 m is cruising at 260 m/s over northern Canada, where the magnetic field of the earth (magnitude 5.0×10^{-5} T) is directed straight down. What is the potential difference between the tips of the wings?

PREPARE The wing is a conductor moving through a magnetic field, so there will be a motional emf. We can visualize a top view of this situation exactly as in Figure 25.3b, with the wing as the moving conductor.

SOLVE The magnetic field is perpendicular to the velocity, so we can compute the potential difference using Equation 25.3:

$$\Delta V = vlB = (260 \text{ m/s})(65 \text{ m})(5.0 \times 10^{-5} \text{ T}) = 0.85 \text{ V}$$

ASSESS The earth's magnetic field is small, so the motional emf will be small as well unless the speed and the length are quite large. The tethered satellite generated a much larger voltage due to its much larger speed and the great length of the tether, the moving conductor.

---

1. Positive charge carriers in the wire are pushed upward by the magnetic force.

Positive end of wire

Moving wire

Conducting rail. Fixed to table and doesn't move.

Negative end of wire

2. The charge carriers flow around the conducting loop as an induced current.

FIGURE 25.4 A current is induced in the circuit as the wire moves through a magnetic field.

Induced Current in a Circuit

The moving conductor of Figure 25.3 had an emf, but it couldn't sustain a current because the charges had nowhere to go. We can change this by including the moving conductor in a circuit.

Figure 25.4 shows a conducting wire sliding with speed v along a fixed U-shaped conducting rail. The wire and the rail together form a closed conducting loop—a circuit.

Suppose a magnetic field $\vec{B}$ is perpendicular to the plane of the circuit. Charges in the moving wire will be pushed to the ends of the wire by the magnetic force, just as they were in Figure 25.3, but now the charges can continue to flow around the circuit. The moving wire acts like the battery in a circuit.

The current in the circuit is an induced current, due to magnetic forces on moving charges. In this example, the induced current is counterclockwise. If the total resistance of the circuit is R, the induced current is given by Ohm's law as

$$I = \frac{\mathcal{E}}{R} = \frac{vlB}{R} \tag{25.4}$$

We've assumed that the wire is moving along the rail at constant speed. But we must apply a continuous pulling force $\vec{F}_{pull}$ to make this happen; Figure 25.5 shows why. The moving wire, which now carries induced current I, is in a magnetic field. You learned in Chapter 24 that a magnetic field exerts a force on a current-carrying wire. According to the right-hand rule, the magnetic force $\vec{F}_{mag}$ on the moving wire points to the left. This "magnetic drag" will cause the wire to slow down and stop *unless* we exert an equal but opposite pulling force $\vec{F}_{pull}$ to keep the wire moving.

NOTE ▶ Think about this carefully. As the wire moves to the right, the magnetic force $\vec{F}_B$ pushes the charge carriers *parallel* to the wire. Their motion, as they continue around the circuit, is the induced current I. Now, because we have a current, a second magnetic force $\vec{F}_{mag}$ enters the picture. This force on the current is *perpendicular* to the wire and acts to slow the wire's motion. ◀

The magnitude of the magnetic force on a current-carrying wire was found in Chapter 24 to be $F_{mag} = IlB$. Using that result, along with Equation 25.4 for the induced current, we find that the force required to pull the wire with a constant speed v is

$$F_{pull} = F_{mag} = IlB = \left(\frac{vlB}{R}\right)lB = \frac{vl^2B^2}{R} \qquad (25.5)$$

The induced current flows through the moving wire.

The magnetic force on the current-carrying wire is opposite the motion.

A pulling force to the right must balance the magnetic force to keep the wire moving at constant speed.

FIGURE 25.5 A pulling force is needed to move the wire to the right.

Energy Considerations

Because a force is needed to pull the wire at a constant speed, we must do work to keep the wire moving. Power is the *rate* at which work is done on the wire. You learned in Chapter 10 that the power exerted by a force pushing or pulling an object with velocity v is $P = Fv$. The power provided to the circuit by the force pulling on the wire is

$$P_{input} = F_{pull}v = \frac{v^2l^2B^2}{R} \qquad (25.6)$$

This is the rate at which energy is added to the circuit by the pulling force.

But the circuit dissipates energy by transforming electric energy into the thermal energy of the wires and components, heating them up. You learned in Chapter 22 that the power dissipated by current I as it passes through resistance R is $P = I^2R$. Equation 25.4 for the induced current I gives us the power dissipated by the circuit of Figure 25.4:

$$P_{dissipated} = I^2R = \frac{v^2l^2B^2}{R} \qquad (25.7)$$

You can see that Equations 25.6 and 25.7 are identical. **The rate at which work is done on the circuit exactly balances the rate at which energy is dissipated.** The fact that our final result is consistent with energy conservation is a good check on our work.

▶ **A head for magnetism?** BIO Observations of hammerhead sharks imply that they navigate using the earth's magnetic field, and controlled laboratory experiments verify that they can indeed reliably detect such modest fields. How do they do it? One possibility is that they use their keen electric sense, described in Chapter 21. Sharks can detect small potential differences; perhaps they detect magnetic fields by sensing the motional emf as they move through the water. If so, the width of their oddly shaped heads would be an asset because the magnitude of the potential difference is proportional to the length of the moving conductor.

The spinning propellers at this wind farm turn the axles of generators mounted on top of the towers.

Generators

A device that converts mechanical energy to electric energy is called a **generator.** The example of Figure 25.5 is a simple generator, but it is not very practical. Rather than move a straight wire, it's more practical to rotate a coil of wire, as in Figure 25.6. As the coil rotates, the left edge always moves upward through the magnetic field while the right edge always moves downward. The motion of the wires through the magnetic field induces a current to flow as noted in the figure. The induced current is removed from the rotating loop by *brushes* that press up against rotating *slip rings*. The circuit is completed as shown in the figure.

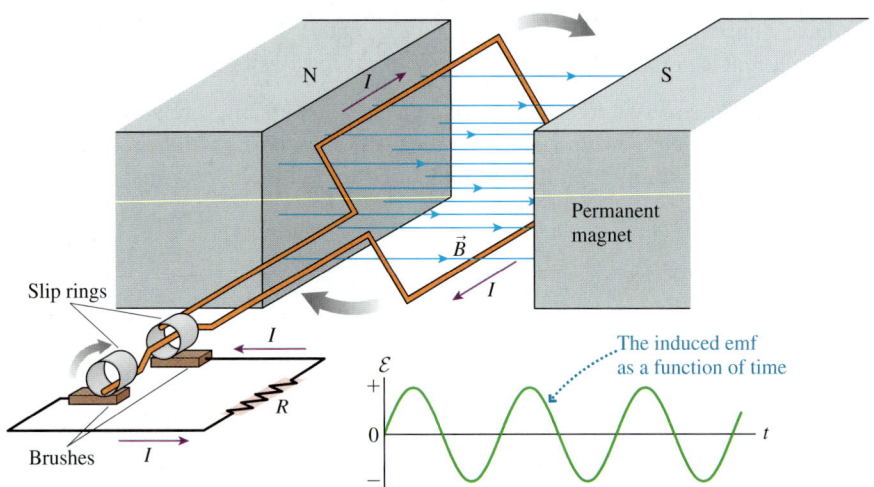

FIGURE 25.6 A generator using a rotating loop of wire.

As the coil in the generator of Figure 25.6 rotates, the sense of the emf changes, giving a sinusoidal variation of emf as a function of time. Electricity is produced using generators of this sort; the electricity in your house has a varying voltage. The alternating sign of the voltage produces an *alternating current,* so we call such electricity AC. Commercial generators use water flowing through a dam, wind turning a propeller, or turbines spun by expanding steam to rotate the generator coils. Work is required to rotate the coil, just as work was required to pull the slide wire, because the magnetic field exerts retarding forces on the currents in the coil. A generator, which turns motion (mechanical energy) into a current (electric energy), is the opposite of a motor, which turns a current into motion.

STOP TO THINK 25.1 A square conductor moves through a uniform magnetic field. Which of the figures shows the correct charge distribution on the conductor?

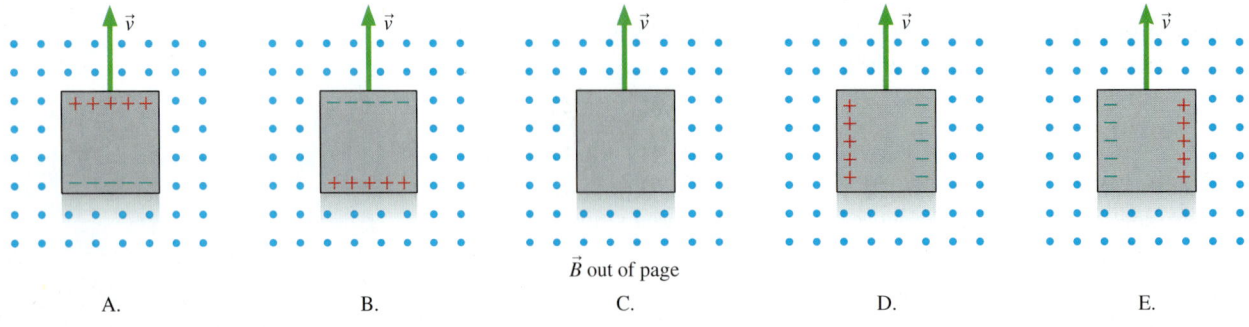

25.3 Magnetic Flux

We've begun our exploration of electromagnetic induction by analyzing a circuit in which one wire moves through a magnetic field. You might be wondering what this has to do with Faraday's discovery. Faraday found that a current is induced when the amount of magnetic field passing through a coil or a loop of wire changes. But that's exactly what happens as the slide wire moves down the rail in Figure 25.4! As the circuit expands, more magnetic field passes through. It's time to define more clearly what we mean by "the amount of field passing through a loop."

Imagine holding a rectangular loop of wire in front of a fan, as shown in Figure 25.7a. The amount of air that flows through the loop depends on the effective area of the loop as seen along the direction of flow. A side view of the situation in 25.7b shows the change in the amount of air flowing through the loop as the angle changes. You can see from Figure 25.7c that the effective area (i.e., as seen facing the fan) is

$$A_{\text{eff}} = ab\cos\theta = A\cos\theta \qquad (25.8)$$

where $A = ab$ is the area of the loop and θ is the tilt angle of the loop. A loop perpendicular to the flow, with $\theta = 0°$, has $A_{\text{eff}} = A$, the full area of the loop. This is the orientation for maximum flow through the loop. No air at all flows through the loop if it is tilted 90°, and you can see that $A_{\text{eff}} = 0$ in this case.

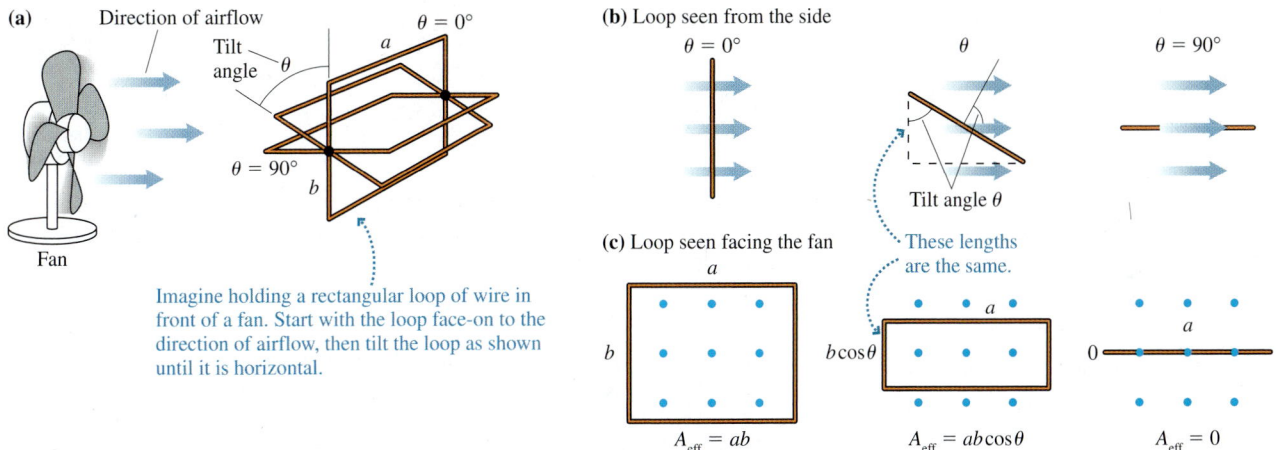

FIGURE 25.7 The amount of air flowing through a loop depends on the effective area of the loop.

We can apply this idea to a magnetic field passing through a loop. Figure 25.8 on the next page shows a loop of area $A = ab$ in a uniform magnetic field. Think of these field vectors as if they were arrows shot through the loop, as air from the fan goes through the loop in Figure 25.7b. The density of arrows (arrows per m²) is proportional to the strength B of the magnetic field; a stronger field would be represented by arrows spaced closer together. The number of arrows passing through a loop of wire depends on two factors:

1. The density of arrows, which is proportional to B, and
2. The effective area $A_{\text{eff}} = A\cos\theta$ of the loop.

The angle θ is the angle between the magnetic field and the axis of the loop. The maximum number of arrows passes through the loop when it is perpendicular to the magnetic field ($\theta = 0°$). No arrows pass through the loop if it is tilted 90°. And a bigger loop with a larger area A will pass more arrows.

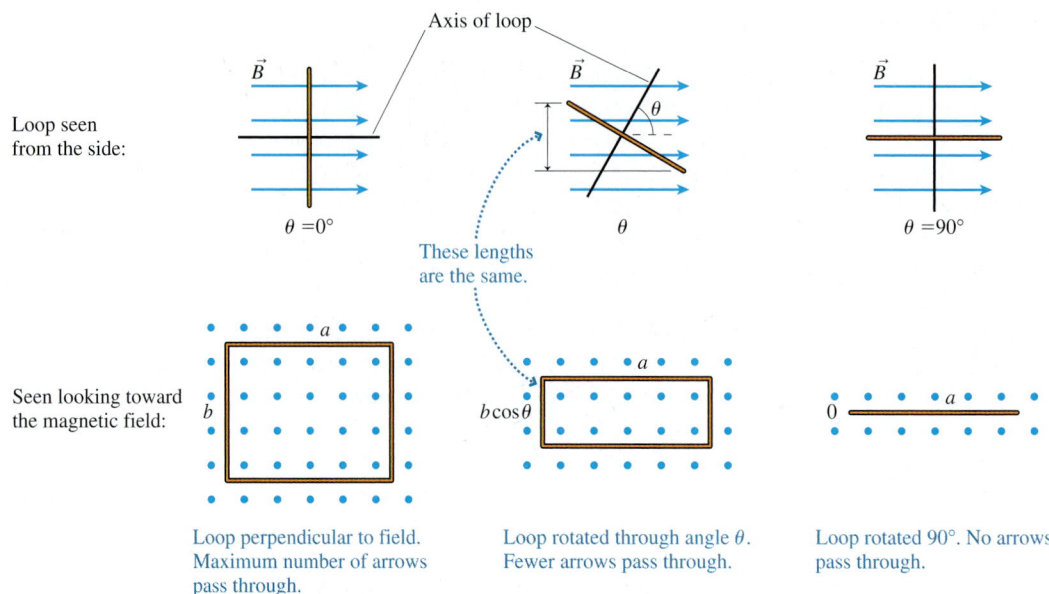

Loop seen from the side:

$\theta = 0°$

θ

$\theta = 90°$

These lengths are the same.

Seen looking toward the magnetic field:

Loop perpendicular to field. Maximum number of arrows pass through.

Loop rotated through angle θ. Fewer arrows pass through.

Loop rotated 90°. No arrows pass through.

FIGURE 25.8 Magnetic field through a loop that is tilted at various angles.

With this in mind, let's define the **magnetic flux** Φ as

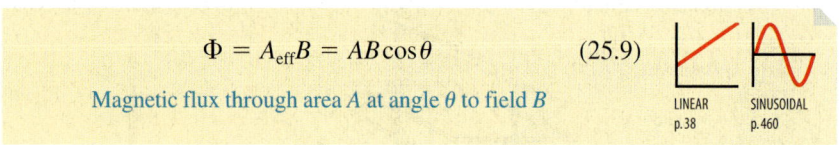

$$\Phi = A_{\text{eff}}B = AB\cos\theta \qquad (25.9)$$

Magnetic flux through area A at angle θ to field B

LINEAR p. 38   SINUSOIDAL p. 460

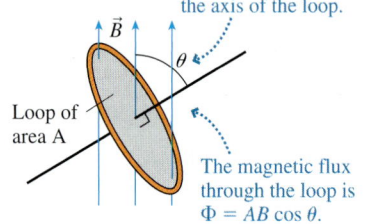

θ is the angle between the magnetic field $\vec{B}$ and the axis of the loop.

Loop of area A

The magnetic flux through the loop is $\Phi = AB\cos\theta$.

FIGURE 25.9 Definition of magnetic flux.

The magnetic flux measures the amount of magnetic field passing through a loop of area A if the loop is tilted at angle θ from the field. The SI unit of magnetic flux is the **weber.** From Equation 25.9 you can see that

$$1 \text{ weber} = 1 \text{ Wb} = 1 \text{ T} \cdot \text{m}^2$$

The relationship of Equation 25.9 is illustrated in Figure 25.9.

EXAMPLE 25.2 Finding the flux for a circular loop rotating in a magnetic field

Figure 25.10 is an edge view of a 10-cm-diameter circular loop rotating in a uniform 0.050 T magnetic field. What is the magnetic flux through the loop when θ is 0°, 30°, 60°, and 90°?

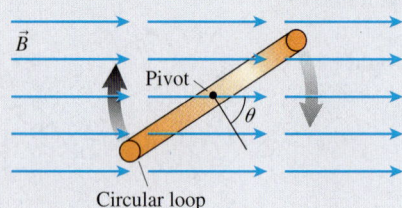

$\vec{B}$

Pivot

θ

Circular loop

FIGURE 25.10 A circular loop in a magnetic field.

PREPARE The angle θ in Figure 25.10 is the angle between the axis of the loop and the magnetic field $\vec{B}$. The area is $A = \pi r^2 = 7.85 \times 10^{-3} \text{ m}^2$.

SOLVE At the noted angles, the magnetic flux is given by Equation 25.9

$$\Phi = AB\cos\theta = \begin{cases} 3.9 \times 10^{-4} \text{ Wb} & \theta = 0° \\ 3.4 \times 10^{-4} \text{ Wb} & \theta = 30° \\ 2.0 \times 10^{-4} \text{ Wb} & \theta = 60° \\ 0 \text{ Wb} & \theta = 90° \end{cases}$$

ASSESS The flux is a maximum when $\theta = 0°$, when the loop is perpendicular to the field, and is zero when $\theta = 90°$, just as we expect.

Lenz's Law

Some of the induction experiments from earlier in the chapter could be explained in terms of motional emf, but others had no motion. What they all have in common, though, is that one way or another the magnetic flux through the coil or loop *changes*. We can summarize all of the discoveries as follows: **Current is induced in a loop of wire when the magnetic flux through the loop changes.**

For example, a momentary current is induced in the loop of Figure 25.11 as the bar magnet is pushed toward the loop because the flux through the loop increases. Pulling the magnet back out of the loop, which decreases the flux, causes the current meter to deflect in the opposite direction. How can we predict the *direction* of the current in the loop?

The German physicist Heinrich Lenz began to study electromagnetic induction after learning of Faraday's discovery. Lenz developed a rule for determining the direction of the induced current. We now call his rule **Lenz's law,** and it can be stated as follows:

> **Lenz's law** There is an induced current in a closed, conducting loop if and only if the magnetic flux through the loop is changing. The direction of the induced current is such that the induced magnetic field opposes the *change* in the flux.

Lenz's law is rather subtle, and it takes some practice to see how to apply it.

> **NOTE ▶** One difficulty with Lenz's law is the term *flux,* from a Latin root meaning "flow." In everyday language, the word *flux* may imply that something is changing. Think of the phrase "The situation is in flux." In physics, *flux* simply means "passes through." A steady magnetic field through a loop creates a steady, *un*changing magnetic flux. ◀

Lenz's law tells us to look for situations where the flux is *changing*. This can happen in three ways.

1. The magnetic field through the loop changes (increases or decreases),
2. The loop changes in area or angle, or
3. The loop moves into or out of a magnetic field.

We can understand Lenz's law this way: If the flux through a loop changes, a current is induced in a loop. That current generates *its own* magnetic field $\vec{B}_{induced}$. **It is this induced field that opposes the flux change.** Let's look at an example to clarify what we mean by this statement.

In Figure 25.11, pushing the bar magnet into the loop causes the magnetic flux through the loop to *increase* in the downward direction. To oppose the *change* in flux, which is what Lenz's law requires, the loop itself needs to generate the *upward*-pointing magnetic field of Figure 25.12. The induced magnetic field at the center of the loop will point upward if the current is counterclockwise, according to the right-hand rule you learned in Chapter 24. Thus pushing the north end of a bar magnet toward the loop induces a counterclockwise current around the loop. The induced current ceases as soon as the magnet stops moving.

Now suppose the bar magnet is pulled back away from the loop, as shown in Figure 25.13a on the next page. There is a downward magnetic flux through the loop, but the flux *decreases* as the magnet moves away. According to Lenz's law,

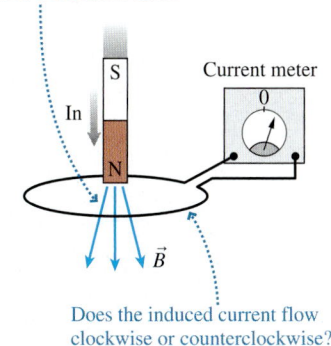

A bar magnet pushed into a loop increases the flux through the loop and induces a current to flow.

Does the induced current flow clockwise or counterclockwise?

FIGURE 25.11 Pushing a bar magnet toward the loop induces a current in the loop.

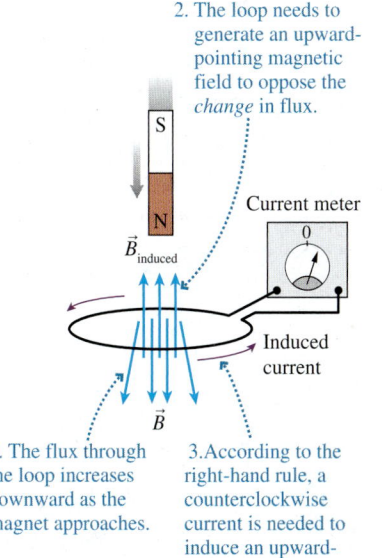

2. The loop needs to generate an upward-pointing magnetic field to oppose the *change* in flux.

1. The flux through the loop increases downward as the magnet approaches.

3. According to the right-hand rule, a counterclockwise current is needed to induce an upward-pointing magnetic field.

FIGURE 25.12 The induced current is counterclockwise.

(a)

The bar magnet is moving away from the loop.

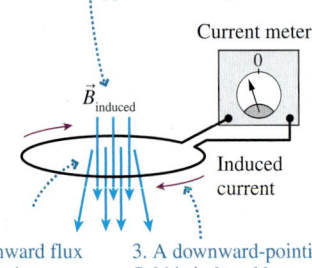

Current meter

Out

$\vec{B}$

(b)

2. A downward-pointing field is needed to oppose the *change*.

Current meter

$\vec{B}_{\text{induced}}$

Induced current

1. Downward flux is decreasing.

3. A downward-pointing field is induced by a clockwise current.

FIGURE 25.13 Pulling the magnet away induces a clockwise current.

the induced magnetic field of the loop will *oppose this decrease*. To do so, the induced field needs to point in the *downward* direction, as shown in Figure 25.13b. Thus as the magnet is withdrawn, the induced current is clockwise, opposite to the induced current of Figure 25.12.

NOTE ▶ Notice that the magnetic field of the bar magnet is pointing downward in both Figures 25.12 and 25.13. It is not the *flux* due to the magnet that the induced current opposes, but the *change* in the flux. This is a subtle but critical distinction. When the field of the magnet points down and is increasing, the induced current opposes the increase by generating an upward field. When the field of the magnet points down but is decreasing, the induced current opposes the decrease by generating a downward field. **◀**

(MP) **TACTICS BOX 25.1** **Using Lenz's law**   🖊 Exercises 5–9

❶ Determine the direction of the applied magnetic field. The field must pass through the loop.
❷ Determine how the flux is changing. Is it increasing, decreasing, or staying the same?
❸ Determine the direction of an induced magnetic field that will oppose the *change* in the flux:
  ■ Increasing flux: The induced magnetic field points opposite the applied magnetic field.
  ■ Decreasing flux: The induced magnetic field points in the same direction as the applied magnetic field.
  ■ Steady flux: There is no induced magnetic field.
❹ Determine the direction of the induced current. Use the right-hand rule to determine the current direction in the loop that generates the induced magnetic field you found in step 3.

EXAMPLE 25.3 Applying Lenz's law 1

The switch in the top circuit of Figure 25.14 has been closed for a long time. What happens in the lower loop when the switch is opened?

PREPARE The current in the upper loop creates a magnetic field. This magnetic field produces a flux through the lower loop. When the current changes in the upper loop, the flux in the lower loop changes.

SOLVE Figure 25.15 shows the four steps of using Lenz's law to find the current in the lower loop. Opening the switch induces a counterclockwise current in the lower loop. This is a momentary current, lasting only until the magnetic field of the upper loop drops to zero.

ASSESS The induced current is in the same direction as the original current. This makes sense, because the induced current is opposing the change, a decrease in the current.

❶ The magnetic field of the upper loop points up, but it is decreasing as the current in the circuit rapidly decreases.

Switch opens.

$\vec{B}$

❷ The flux through the loop is up and decreasing.

Induced current

❸ The induced field needs to point upward to oppose the *change* in flux.

$\vec{B}_{\text{induced}}$

❹ A counterclockwise current induces an upward magnetic field.

FIGURE 25.14 Circuits for Example 25.3.   **FIGURE 25.15** Finding the induced current.

EXAMPLE 25.4 Applying Lenz's law 2

A loop is moved toward a current-carrying wire as shown in Figure 25.16. As the wire is moving, is there a clockwise current around the loop, a counterclockwise current, or no current?

FIGURE 25.16 The moving loop for Example 25.4

PREPARE The current in the wire creates a magnetic field, so there is a magnetic flux through the loop. The magnetic field is stronger closer to the wire, so the flux increases as the loop moves closer.

SOLVE According to the right-hand rule, the wire's magnetic field passes through the loop into the page. The flux through the loop is into the page. As the loop moves toward the wire, the flux through the loop is increasing, as shown in Figure 25.17. To oppose this increase, the induced magnetic field needs to point out of the page. This requires a counterclockwise current.

ASSESS We could have solved this problem using motional emf, but using the concepts of flux change and Lenz's law is a more straightforward way to deduce the current.

FIGURE 25.17 Motion of the wire changes the flux through the loop and induces a current.

STOP TO THINK 25.2 As a coil moves to the right at constant speed, it passes over the north pole of a magnet and then moves beyond it. Which graph best represents the current in the loop for the time of the motion? A counterclockwise current as viewed from above the loop is a positive current, clockwise is a negative current.

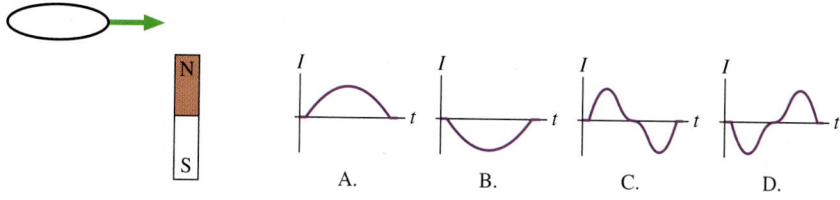

25.4 Faraday's Law

Faraday discovered that a current is induced when the magnetic flux through a conducting loop changes. Lenz's law allows us to find the direction of the induced current. To put electromagnetic induction to practical use, we also need to know the *size* of the induced current.

  13.9, 13.10

Charges don't start moving spontaneously. A current requires an emf to provide the energy. We started our analysis of induced currents with circuits in which there is a *motional emf*. The motional emf can be understood in terms of magnetic forces on moving charges. But we've also seen that a current can be induced by changing the magnetic field through a stationary circuit, a circuit in which there is no motion. There *must* be an emf in this circuit, even though the mechanism for this emf is not yet clear.

▶ **Keep the (flux) change** A credit card has a magnetic strip on the back that has regions of alternating magnetization. "Swiping" the card moves this strip through the circuit of the reader. Different parts of the strip have different fields, so the motion of the strip causes a series of flux changes that induce currents. The pattern of magnetization of the card determines the pattern of the induced currents—and so the card is "read."

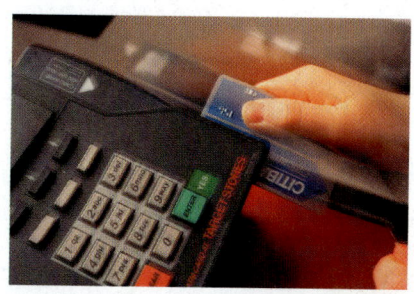

The emf associated with a changing magnetic flux, regardless of what causes the change, is called an **induced emf** $\mathcal{E}$. If this emf is induced in a complete circuit having resistance R, a current

$$I_{\text{induced}} = \frac{\mathcal{E}}{R} \qquad (25.10)$$

is established in the wire as a *consequence* of the induced emf. The direction of the current is given by Lenz's law. The last piece of information we need is the size of the induced emf $\mathcal{E}$.

The research of Faraday and others led to the discovery of the basic law of electromagnetic induction, which we now call **Faraday's law.**

> **Faraday's law** An emf $\mathcal{E}$ is induced in a conducting loop if the magnetic flux through the loop changes. If the flux changes by $\Delta\Phi$ during time interval Δt, the magnitude of the emf is
>
> $$\mathcal{E} = \left|\frac{\Delta\Phi}{\Delta t}\right| \qquad (25.11)$$
>
> and the direction of the emf is such as to drive an induced current in the direction given by Lenz's law.

In other words, the magnitude of the induced emf is the *rate of change* of the magnetic flux through the loop.

A coil of wire consisting of N turns in a changing magnetic field acts like N batteries in series. The induced emf of each of the coils adds, so the induced emf of the entire coil is

$$\mathcal{E}_{\text{coil}} = N\left|\frac{\Delta\Phi_{\text{per coil}}}{\Delta t}\right| \qquad (25.12)$$

LINEAR
p. 38

Faraday's law for an N-turn coil

Faraday's law allows us to find the *magnitude* of induced emfs and currents; Lenz's law allows us to determine the *direction.*

Using Faraday's Law

Most electromagnetic induction problems can be solved with a three-step strategy.

> **(MP)** **PROBLEM-SOLVING STRATEGY 25.1** **Electromagnetic induction**
>
> **PREPARE** Make simplifying assumptions about wires and magnetic fields. Draw a picture or a circuit diagram. Use Lenz's law to determine the direction of the induced current.
>
> **SOLVE** The mathematical representation is based on Faraday's law
>
> $$\mathcal{E} = \left|\frac{\Delta\Phi}{\Delta t}\right|$$
>
> For an N-turn coil, multiply by N. The size of the induced current is $I = \mathcal{E}/R$.
>
> **ASSESS** Check that your result has the correct units, is reasonable, and answers the question.

Let's return to the situation of Figure 25.4, where a wire moves through a magnetic field by sliding on a U-shaped conducting rail. We looked at this problem as an example of motional emf; now, let's look at it using Faraday's law.

EXAMPLE 25.5 Finding the emf using Faraday's law
Figure 25.18 shows a wire of resistance R sliding on a U-shaped conducting rail. Assume that the conducting rail is an ideal wire. Use Faraday's law and the steps of Problem-Solving Strategy 25.1 to derive an expression for the current in the wire.

FIGURE 25.18 A wire sliding on a rail.

PREPARE Figure 25.19 shows the current loop formed by the wire and the rail. Even though the magnetic field is constant, the flux is changing because the loop is increasing in area. The flux is into the loop and increasing. According to Lenz's law, the induced

Magnetic flux $\Phi = AB = xlB$

FIGURE 25.19 Induced current in the sliding wire.

current must be counterclockwise so as to oppose the change, because the induced magnetic field must be out of the loop.

SOLVE The magnetic field $\vec{B}$ is perpendicular to the plane of the loop, so $\theta = 0°$ and the magnetic flux is $\Phi = AB$, where A is the area of the loop. If the sliding wire is distance x from the end, as in Figure 25.19, the area of the loop is $A = xl$ and the flux at that instant of time is

$$\Phi = AB = xlB$$

The flux through the loop increases as the wire moves and x increases. This flux change induces an emf, according to Faraday's law, so we write

$$\mathcal{E} = \left|\frac{\Delta\Phi}{\Delta t}\right| = \left|\frac{\Delta(AB)}{\Delta t}\right| = \left|\frac{\Delta(xlB)}{\Delta t}\right|$$

The only quantity in the final ratio that is changing is the position x, so we can write

$$\mathcal{E} = \left|\frac{\Delta(xlB)}{\Delta t}\right| = lB\left|\frac{\Delta x}{\Delta t}\right|$$

But $|\Delta x/\Delta t|$ is the wire's speed v, so the induced emf is

$$\mathcal{E} = vlB$$

The wire and the loop have a total resistance R, thus the magnitude of the induced current is

$$I = \frac{\mathcal{E}}{R} = \frac{vlB}{R}$$

ASSESS This is exactly the same result we found in Section 25.2, where we analyzed this situation by considering the force on moving charge carriers. This is a good check on our work, and a nice connection between the ideas of motional emf and Faraday's law.

EXAMPLE 25.6 Finding the induced current in a circular loop
A patient having an MRI scan has neglected to remove a copper bracelet. The bracelet is 6.0 cm in diameter and has a resistance of 0.010 Ω. The magnetic field in the MRI solenoid is directed along the person's body from head to foot; her bracelet is perpendicular to $\vec{B}$. As a scan is taken, the magnetic field in the solenoid decreases from 1.00 T to 0.40 T in 1.2 s. What are the magnitude and direction of the current induced in the bracelet?

PREPARE We follow the steps in Problem-Solving Strategy 25.1, beginning with a sketch of the situation. Figure 25.20 shows the bracelet and the applied field looking down along the patient's body. The field is directed down through the loop; as the applied field decreases, the flux into the loop decreases. To oppose the decreasing flux, as required by Lenz's law, the field

from the induced current must be in the direction of the applied field. Thus, from the right-hand rule, the induced current in the bracelet must be clockwise.

FIGURE 25.20 A circular conducting loop in a decreasing magnetic field.

Continued

SOLVE The magnetic field is perpendicular to the plane of the loop, hence $\theta = 0°$ and the magnetic flux is $\Phi = AB = \pi r^2 B$. The area of the loop doesn't change with time, but B does, so $\Delta\Phi = \Delta(AB) = A\Delta B = \pi r^2 \Delta B$. The change in the magnetic field during $\Delta t = 1.2$ s is $\Delta B = 0.40$ T $- 1.00$ T $= -0.60$ T. According to Faraday's law, the magnitude of the induced emf is

$$\mathcal{E} = \left|\frac{\Delta\Phi}{\Delta t}\right| = \pi r^2 \left|\frac{\Delta B}{\Delta t}\right| = \pi(0.030 \text{ m})^2 \left|\frac{-0.60 \text{ T}}{1.2 \text{ s}}\right|$$

$$= \pi(0.030 \text{ m})^2(0.50 \text{ T/s}) = 0.0014 \text{ V}$$

The current induced by this emf is

$$I = \frac{\mathcal{E}}{R} = \frac{0.0014 \text{ V}}{0.010 \text{ } \Omega} = 0.14 \text{ A}$$

The decreasing magnetic field causes a 0.14 A clockwise current during the 1.2 s that the field is decreasing.

ASSESS The emf is quite small, but, because the resistance of the metal bracelet is also very small, the current is respectable. We know that electromagnetic induction produces currents large enough for practical applications, so this result seems plausible. The induced current could easily distort the readings of the MRI machine, and a larger current could cause enough heating to be potentially dangerous. For these and other reasons, operators are careful to have patients remove all metal before an MRI scan.

As these two examples show, there are two fundamentally different ways to change the magnetic flux through a conducting loop:

1. The loop can move or expand or rotate, creating a motional emf.
2. The magnetic field can change.

Faraday's law tells us that the induced emf is simply the rate of change of the magnetic flux through the loop, *regardless* of what causes the flux to change. Motional emf is included within Faraday's law as one way of changing the flux, but any other way of changing the flux will have the same result.

Eddy Currents

Here is a remarkable physics demonstration that you can try: Take a sheet of copper and place it between the pole tips of a strong magnet, as shown in Figure 25.21a. Now, pull hard to move the sheet of copper to the side. Copper is not a magnetic material and thus is not attracted to the magnet, so it comes as quite a surprise to find that it takes a significant effort to pull the metal through the magnetic field.

Let's analyze this situation to discover the origin of the force. Figure 25.21a shows two "loops" lying entirely inside the metal sheet. On the right, the loop is leaving the magnetic field, and the flux through it is decreasing. According to Faraday's law, the flux change will induce a current to flow around this loop, just as in a loop of wire, even though this current does not have a wire to define its path. As a consequence, a clockwise—as given by Lenz's law—"whirlpool" of current begins to circulate in the metal, as shown in Figure 25.21b. Similarly, the loop on the left is entering the field, and the flux through it is increasing. Lenz's law requires this whirlpool of current to circulate the opposite way. These spread-out whirlpools of induced current in a solid conductor are called **eddy currents.**

Figure 25.21b shows the direction of the eddy currents. Notice that both whirlpools are moving in the same direction as they pass through the magnet. The magnet's field exerts a force on this current. By the right-hand rule, this force is to the left, opposite the direction of the pull, and thus it acts as a *braking* force. Because of the braking force, **an external force is required to pull a metal through a magnetic field.** If the pulling force ceases, the magnetic braking force quickly causes the metal to decelerate until it stops. No matter which way the metal is moved, the magnetic forces on the eddy currents act to oppose the motion of the metal. *Magnetic braking* uses the braking force associated with eddy currents to slow trains and transit-system vehicles. It can also provide valuable vibration damping on a much smaller scale, as we see in Figure 25.22.

Eddy currents can also be induced by changing fields. Some modern stoves heat pans by *induction heating*. A coil of wire beneath the stove top is driven by a

(a)

The downward flux through this loop is increasing.

The downward flux through this loop is decreasing.

The loop is being pulled through the magnetic field.

$\vec{F}_{\text{pull}}$

Metal sheet   N   S $\vec{v}$

(b)

The change in flux in these loops induces eddy currents in the metal.

$\vec{F}_{\text{braking}}$

N   S $\vec{v}$

The magnetic field exerts a force on the eddy currents in the direction opposite the motion.

FIGURE 25.21 Eddy currents.

Strong magnets on either side of an aluminum vane induce eddy currents in the vane as it oscillates. The braking force quickly dampens out oscillations.

FIGURE 25.22 Eddy-current damping in a balance.

current that changes direction 120 times a second. The coil's magnetic field passes upward through the metal bottom of a cooking pan. As the field changes direction, it induces an eddy current in the pan that heats the pan to cooking temperature.

A similar technique can induce currents in the tissues of the body. Electrical stimulation of the tissues of the brain is a useful procedure for certain conditions, but there is a significant practical difficulty: The brain is housed in the skull, which is a poor conductor. The solution is to produce electrical stimulation via a changing magnetic field, which can penetrate the skull to reach the tissues of the brain. In a technique known as *transcranial magnetic stimulation* (TMS), a large oscillating magnetic field is applied to the head via a current-carrying coil. Figure 25.23 illustrates how this field produces small eddy currents that stimulate neurons in the tissue of the brain.

TMS produces a short-term inhibitory effect on the neurons in the stimulated region that can produce long-term clinical effects. TMS is being studied for treating migraine headaches and certain types of depression. The technique is also proving to be quite useful in research. By inhibiting the action of certain regions of the brain, researchers can determine the importance of these regions to certain perceptions or tasks.

FIGURE 25.23 Transcranial magnetic stimulation.

25.5 Induced Fields and Electromagnetic Waves

We will start this section with the puzzle shown in Figure 25.24. A long, tightly wound solenoid of radius r_1 passes through the center of a conducting loop having a larger radius r_2. The solenoid carries a current and generates a magnetic field. What happens to the loop if the solenoid current changes?

You learned in Chapter 24 that the magnetic field is strong inside a long solenoid but essentially zero outside. Even so, changing the field inside the solenoid causes the flux through the loop to change, so our theory predicts an induced current in the loop. Indeed, you would find an induced current if you did this experiment.

But the loop is completely outside the solenoid, where the magnetic field is zero. How can the charge carriers in the conducting loop possibly know that the magnetic field inside the solenoid is changing? In other words, how can a current be created?

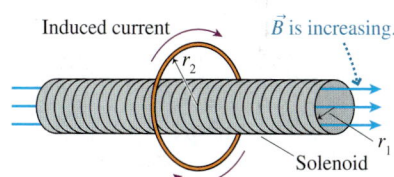

FIGURE 25.24 A changing current in the solenoid induces a current in the loop.

Induced Electric Fields

In order to answer this question, we will first consider another, related question: When a changing flux through a loop induces a current, what actually *causes* the current? What *force* pushes the charges around the loop against the resistive forces of the metal? When we considered currents in Chapter 22, it was an *electric field* that moved charges through a conductor. Somehow, changing a magnetic field must create an electric field.

In fact, a changing magnetic field *does* cause what we call an **induced electric field.** Figure 25.25a shows a conducting loop in an increasing magnetic field. According to Lenz's law, there is an induced current in the counterclockwise direction. Something has to act on the charge carriers to make them move, so we can infer that the current is produced by an induced electric field tangent to the loop at all points. The induced electric field is the *mechanism* we were seeking that creates a current when there is a changing magnetic field inside a stationary loop.

But the induced electric field exists whether there is a conducting loop or not. The space in which the magnetic field is changing is filled with the pinwheel pattern of induced electric fields shown in Figure 25.25b. Charges will move if a conducting path is present, but the induced electric field is there as a direct consequence of the changing magnetic field.

FIGURE 25.25 An induced electric field creates a current in the loop.

This is a rather peculiar electric field. All the electric fields we have examined until now have been created by charges. Electric field vectors pointed away from positive charges and toward negative charges. An electric field created by charges is called a **Coulomb electric field.** The induced electric field of Figure 25.25b is caused not by charges but by a changing magnetic field. It is called a **non-Coulomb electric field.**

Now that we know about induced electric fields, let's revisit the production of an emf by a changing flux described by Faraday's law. We saw two different ways to change the magnetic flux through a conducting loop: expanding, moving, or rotating the loop, or changing the magnetic field. These two different means of changing the flux both induce an emf:

1. If the loop moves, expands or rotates, the motion of charge carriers in the magnetic field produces an electric field that results in a motional emf.
2. If the magnetic field in the loop changes, the changing magnetic field induces an electric field, which results in an emf.

Making this connection between electric and magnetic fields has brought together all of the pieces we have explored so far in this chapter. But there's more to the story of induced fields. Faraday's discovery was the source of inspiration for further investigations that had very far-reaching and practical implications.

Maxwell's Theory

Faraday found that a changing magnetic field creates an induced electric field, a non-Coulomb electric field not tied to charges. The development of a theory of electricity and magnetism continued with work by the Scottish physicist James Clerk Maxwell, who asked the following question: What about a changing *electric* field? Could this induce a *magnetic* field? Maxwell thought the symmetry was compelling. He proposed that a changing electric field creates an **induced magnetic field,** a new kind of magnetic field not tied to the existence of currents or magnets.

If a changing magnetic field can induce an electric field in the absence of any charges, and a changing electric field can induce a magnetic field in the absence of any currents, we have some interesting possibilities. Maxwell soon realized that it would be possible to establish self-sustaining electric and magnetic fields that would be entirely independent of any charges or currents. That is, a changing electric field $\vec{E}$ creates a magnetic field $\vec{B}$, which then changes in just the right way to recreate the electric field, which then changes in just the right way to again recreate the magnetic field, and so on. The fields are continually recreated through electromagnetic induction without any reliance on charges or currents.

Maxwell was able to show that these electric and magnetic fields would be able to sustain themselves, free from charges and currents, if they took the form of an **electromagnetic wave.** The wave would have to have a very specific

◀ **The grandfather of the cell phone** In this photo, Guglielmo Marconi, who is credited with the development of radio communication, or "wireless telegraphy," sits surrounded by high-voltage equipment used in his experiments in the late 1800s. The German scientist Heinrich Hertz discovered that a spark in one circuit emitted a disturbance that affected a second circuit at some distance. Soon after, Marconi developed the technology for producing and detecting these *electromagnetic waves,* transmitting signals, with no wires, over long distances—something we have come to take for granted.

geometry, shown in Figure 25.26a, in which $\vec{E}$ and $\vec{B}$ are perpendicular to each other as well as perpendicular to the direction of travel. Figure 25.26b shows a right-hand rule for determining the relative orientations of the fields and the velocity, reminiscent of the right-hand rule for forces from the previous chapter.

Maxwell's theory predicted that an electromagnetic wave would travel with speed

$$v_{em} = \frac{1}{\sqrt{\epsilon_0 \mu_0}}$$ (25.13)

where ϵ_0 and μ_0 are the permittivity and permeability constants from our expressions for electric and magnetic fields. Maxwell computed that an electromagnetic wave would travel with speed $v_{em} = 3.00 \times 10^8$ m/s.

This is a value you have seen before—it is the speed of light. Making a bold leap of imagination, Maxwell concluded that **light is an electromagnetic wave.** We studied the wave properties of light in Part V, but at that time we were not able to determine just what is "waving." Now we know—light is a wave of electric and magnetic fields.

Maxwell was able to establish the following properties of electromagnetic waves:

■ An electromagnetic wave is a *transverse* wave, as shown in Figure 25.26a.
■ Electromagnetic waves can exist at any frequency, not just the frequencies of visible light.
■ All electromagnetic waves travel in a vacuum with the same speed, a speed that we call the *speed of light,* for which we use the symbol c.
■ At any point on the wave, the electric and magnetic field strengths are related by $E = cB$.

Figure 25.26a shows the values of the electric and magnetic fields at *points* along a single line, the *x*-axis. An $\vec{E}$ vector pointing in the *y*-direction says that *at that point* on the *x*-axis, where the vector's tail is, the electric field points in the *y*-direction and has a certain strength. Nothing is "reaching" to a point in space above the *x*-axis.

Suppose this electromagnetic wave is a *plane wave* traveling in the *x*-direction. Recall from Chapter 15 that the displacement of a plane wave is the same at *all points* in any plane perpendicular to the direction of motion. In this case, the fields are the same in any *yz*-plane. If you were standing on the *x*-axis as the wave moves toward you, the electric and magnetic fields would vary as in the series of pictures of Figure 25.27. The $\vec{E}$ and $\vec{B}$ fields at each point in the *yz*-plane oscillate in time, but they are always synchronized with all the other points in the plane. As the plane wave passed you, you would see a uniform oscillation of the $\vec{E}$ and $\vec{B}$ fields of the wave.

(a) Electromagnetic wave

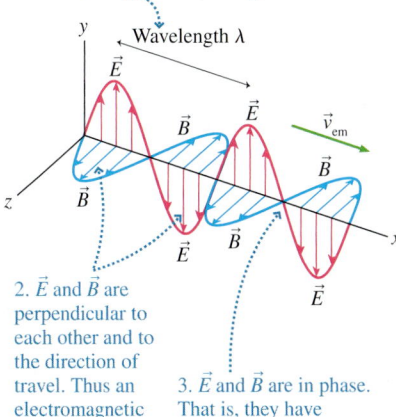

1. The wave is a sinusoidal traveling wave, with frequency f and wavelength λ.

2. $\vec{E}$ and $\vec{B}$ are perpendicular to each other and to the direction of travel. Thus an electromagnetic wave is a transverse wave.

3. $\vec{E}$ and $\vec{B}$ are in phase. That is, they have matching crests, troughs, and zeros.

(b) Right-hand rule for electromagnetic waves

Spread the fingers of your right hand so that your index finger, thumb, and middle finger point out from your hand as shown. Your thumb, index and middle fingers give the directions of $\vec{v}$, $\vec{E}$ and $\vec{B}$, as shown.

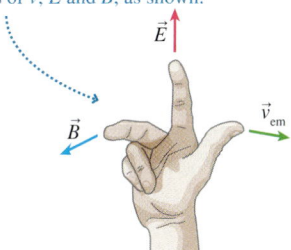

FIGURE 25.26 A sinusoidal electromagnetic wave.

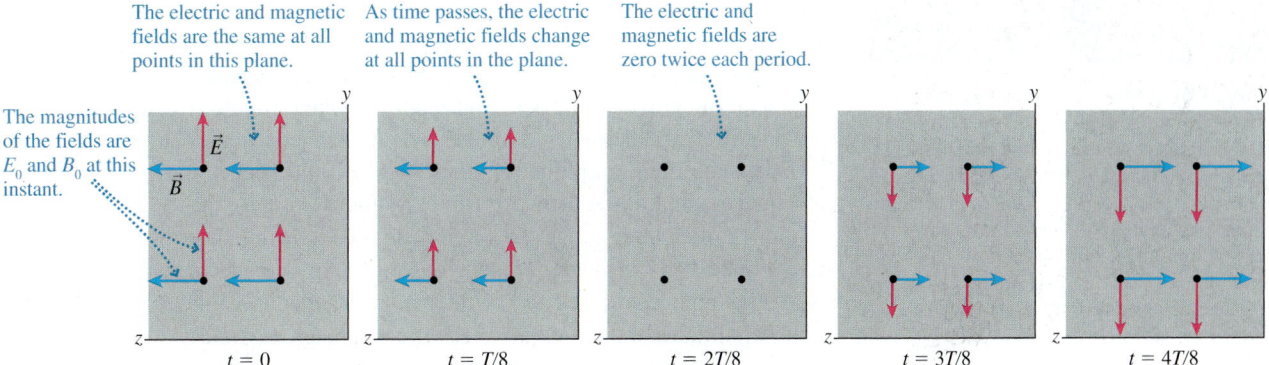

The electric and magnetic fields are the same at all points in this plane.

As time passes, the electric and magnetic fields change at all points in the plane.

The electric and magnetic fields are zero twice each period.

The magnitudes of the fields are E_0 and B_0 at this instant.

$t = 0$ $t = T/8$ $t = 2T/8$ $t = 3T/8$ $t = 4T/8$

FIGURE 25.27 The fields of an electromagnetic plane wave moving toward you, shown every one-eighth period for half a cycle.

We can adapt our equation for traveling waves from Chapter 15 to electromagnetic waves. If a plane electromagnetic wave moves in the x-direction with the electric field along the y-axis, then the magnetic field is along the z-axis. The equations for the electric and magnetic fields of a wave with wavelength λ and period T are

$$E_y = E_0 \sin\left(2\pi\left(\frac{x}{\lambda} - \frac{t}{T}\right)\right) \qquad B_z = B_0 \sin\left(2\pi\left(\frac{x}{\lambda} - \frac{t}{T}\right)\right) \qquad (25.14)$$

E_0 and B_0 are the amplitudes of the oscillating fields. The equation

$$E_0 = cB_0$$

relates the amplitudes of the fields.

STOP TO THINK 25.3   An electromagnetic wave is traveling in the direction shown. What is the direction of the magnetic field at this instant?

A. Toward the top of the page
B. Toward the bottom of the page
C. To the left

D. To the right
E. Into the page
F. Out of the page

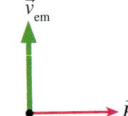

25.6 Properties of Electromagnetic Waves

Electromagnetic waves are oscillations of the electric and magnetic field, but they are still waves. All of the general principles we have learned about waves apply to electromagnetic waves.

In Chapter 15, we learned that we could characterize sinusoidal waves by their speed, wavelength, and frequency, with these related by the fundamental relationship $v = \lambda f$. All electromagnetic waves move at the speed of light, $v_{em} = c$, so this relationship becomes

$$c = \lambda f \qquad (25.15)$$

The *spectrum* of electromagnetic waves ranges from waves of long wavelength and (relatively) low frequency (radio waves and microwaves) to waves of short wavelength and high frequency (visible light, ultraviolet, and x rays), as we saw in Chapter 15. Later in the chapter we will explore the properties of different parts of the electromagnetic spectrum. For now, we will concentrate on physical properties that all electromagnetic waves share.

Energy of Electromagnetic Waves

All waves transfer energy. Ocean waves erode beaches, sound waves set your eardrum to vibrating, and light from the sun warms the earth. In all of these cases, the waves carry energy from the point where they are emitted to another point where their energy is transferred to an object. In water waves, the wave energy is the kinetic and gravitational potential energy of water; for electromagnetic waves, the wave energy is in the form of electric and magnetic fields. The energy of the wave depends on the amplitudes of these fields. If the electric and magnetic fields have greater amplitudes, the wave will carry more energy.

In our earlier study of waves, we defined the *intensity* of a wave (measured in W/m^2) to be $I = P/A$, where P is the power (energy transferred per second) of a wave that impinges on area A. The intensity of an electromagnetic wave depends on the amplitudes of the oscillating electric and magnetic fields:

A field-fired furnace The energy from the sun is carried through space by electromagnetic waves—that is, it is carried by electric and magnetic fields. The mirrors of this solar furnace in the Pyrenees in southern France concentrate the electromagnetic waves from the sun to an intensity 1000 times that of normal sunlight, allowing researchers to test the high-temperature properties of materials at up to 3800°C.

$$I = \frac{P}{A} = \frac{1}{2}c\epsilon_0 E_0^2 = \frac{1}{2}\frac{c}{\mu_0}B_0^2 \qquad (25.16)$$

QUADRATIC
p. 50

Intensity of an electromagnetic wave with field amplitudes E_0 and B_0

The intensity of a plane wave, such as that of a laser beam, does not change with distance. As we saw in Chapter 15, the intensity of a spherical wave, spreading out from a point, must decrease with the square of the distance to conserve energy. If a source with power P_{source} emits waves *uniformly* in all directions, the electromagnetic wave intensity at distance r is

$$I = \frac{P_{source}}{4\pi r^2} \qquad (25.17)$$

The intensities of the electromagnetic waves from antennas, cell phones, and other "point sources" are reasonably well described by Equation 25.17.

EXAMPLE 25.7 Finding the fields of a laser beam

A typical helium-neon laser found in supermarket checkout scanners emits a 1.0-mm-diameter laser beam with a power of 1.0 mW. What are the amplitudes of the oscillating electric and magnetic fields in the laser beam?

PREPARE The laser beam is an electromagnetic plane wave. We will assume that the energy is uniformly distributed over the diameter of the laser beam.

SOLVE 1.0 mW, or 1.0×10^{-3} J/s, is the energy transported per second by the light wave. This energy is carried within a 1.0-mm-diameter beam, so the light intensity is

$$I = \frac{P}{A} = \frac{P}{\pi r^2} = \frac{1.0 \times 10^{-3}\ \text{W}}{\pi(0.00050\ \text{m})^2} = 1270\ \text{W/m}^2$$

We can use Equation 25.16 to relate this intensity to the electric field amplitude:

$$E_0 = \sqrt{\frac{2I}{c\epsilon_0}} = \sqrt{\frac{2(1270\ \text{W/m}^2)}{(3.00 \times 10^8\ \text{m/s})(8.85 \times 10^{-12}\ \text{C}^2/\text{N}\cdot\text{m}^2)}} = 980\ \text{V/m}$$

Because the magnetic and electric field amplitudes are related by $E_0 = cB_0$, we find

$$B_0 = \frac{E_0}{c} = \frac{980\ \text{V/m}}{3.00 \times 10^8\ \text{m/s}} = 3.3 \times 10^{-6}\ \text{T}$$

ASSESS The intensity of the laser beam is approximately equal to that of direct sunlight, which is why you are advised to not look directly into a laser beam! The electric field is large as well, comparable to the electric field near a charged glass or plastic rod. But the magnetic field is quite small, about 1/10 of the earth's field. The interaction of electromagnetic waves with matter is mostly due to the electric field; the magnetic field is small enough that we can generally ignore its effects.

Polarization

The electric field vectors of an electromagnetic wave lie in a plane perpendicular to the direction of propagation. The plane containing the electric field vectors is called the **plane of polarization.** In Figure 25.28a, the wave is traveling along the x-axis, and the plane of polarization is the xy-plane. If the wave were moving toward you, it would appear as in the first diagram in Figure 25.28b. This particular wave is *vertically polarized* ($\vec{E}$ oscillating along the y-axis). The second diagram shows a wave that is *horizontally polarized* ($\vec{E}$ oscillating along the z-axis). The plane of polarization needn't be horizontal or vertical; it can have any orientation, as in the third diagram of Figure 25.28b.

(a)

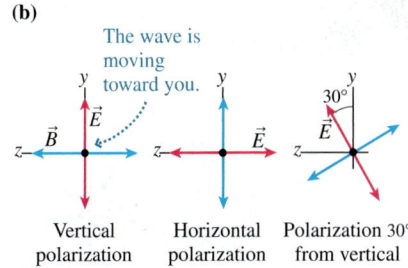

(b)

FIGURE 25.28 The polarization of an electromagnetic wave is determined by the plane in which the electric field oscillates.

16.9 Actjv
ONLINE
Physjcs

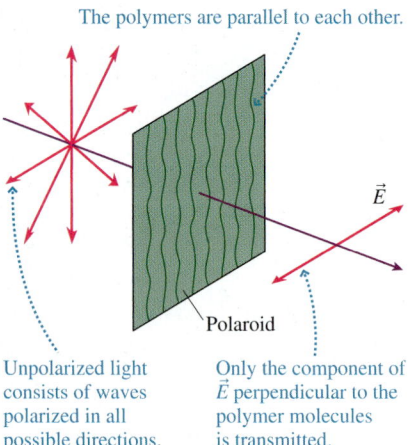

The polymers are parallel to each other.

$\vec{E}$

Polaroid

Unpolarized light consists of waves polarized in all possible directions.

Only the component of $\vec{E}$ perpendicular to the polymer molecules is transmitted.

FIGURE 25.29 A polarizing filter.

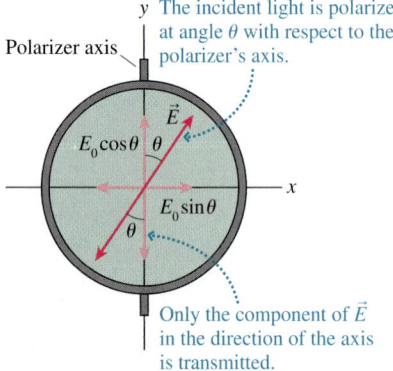

Polarizer axis

y The incident light is polarized at angle θ with respect to the polarizer's axis.

$E_0\cos\theta$ θ

$\vec{E}$

x

$E_0\sin\theta$

θ

Only the component of $\vec{E}$ in the direction of the axis is transmitted.

FIGURE 25.30 An incident electric field can be decomposed into components parallel and perpendicular to a polarizer's axis.

NOTE ▶ This use of the term "polarization" is completely independent of the idea of *charge polarization* that you learned about in Chapter 20. ◀

Most natural sources of electromagnetic radiation are *unpolarized.* Each atom in the sun's hot atmosphere emits light independently of all the other atoms, as does each tiny piece of metal in the incandescent filament of a light bulb. An electromagnetic wave that you see or measure is a superposition of waves from each of these tiny emitters. Although the wave from each individual emitter is polarized, it is polarized in a random direction with respect to the waves from all its neighbors. The resulting wave, a superposition of waves with electric fields in all possible directions, is *unpolarized.*

We can create polarized light by sending unpolarized light through a *polarizing filter.* A typical polarizing filter is made from a material called Polaroid. It is a plastic sheet containing long organic molecules called polymers, as shown in Figure 25.29. The molecules are aligned to form a grid, like the metal bars in a barbecue grill, then treated so they conduct electrons along their length.

As a light wave travels through Polaroid, the component of the electric field oscillating parallel to the polymer grid drives the electrons up and down the molecules. The electrons absorb energy from the light wave, so the parallel component of $\vec{E}$ is absorbed in the filter. But the conduction electrons can't oscillate perpendicular to the molecules, so the component of $\vec{E}$ perpendicular to the polymer grid passes through without absorption. Thus the light wave emerging from a polarizing filter is polarized perpendicular to the polymer grid.

Suppose a *polarized* light wave with electric field amplitude $E_{incident}$ approaches a polarizing filter with a vertical axis (that is, the filter transmits only vertically polarized light). What is the intensity of the light that passes through the filter? Figure 25.30 shows that the oscillating electric field of the polarized light can be decomposed into horizontal and vertical components. The vertical component will pass; the horizontal component will be blocked. As we see in Figure 25.30, the magnitude of electric field of the light transmitted by the filter is

$$E_{transmitted} = E_{incident}\cos\theta \qquad (25.18)$$

Because the intensity depends on the square of the electric field amplitude, the transmitted intensity is related to the incident intensity by what is known as **Malus's law:**

$$I_{transmitted} = I_{incident}\cos^2\theta \qquad (25.19)$$

Malus's law for transmission of polarized light by a polarizing filter

Figure 25.31 shows how Malus's law can be demonstrated with two polarizing filters. The first, called the *polarizer,* is used to produce polarized light of intensity I_0. The second, called the *analyzer,* is rotated by angle θ relative to the

FIGURE 25.31 The intensity of the transmitted light depends on the angle between the polarizing filters.

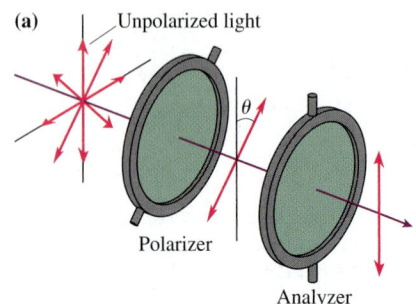

(a) Unpolarized light

θ

Polarizer

Analyzer

(b)

$\theta = 0°$ $\theta = 45°$ $\theta = 90°$

The red lines show the axes of the polarizers.

polarizer. As the photographs of Figure 25.31b show, the transmission of the analyzer is (ideally) 100% when $\theta = 0°$ and steadily decreases to zero when $\theta = 90°$. Two polarizing filters with perpendicular axes, called *crossed polarizers*, block all the light.

Suppose you place an object between two crossed polarizers. Normally, no light would make it through the analyzer, and the object would appear black. But if the object is able to *change* the polarization of the light, some of the light emerging from the object will be able to pass through the analyzer. This can be a valuable analytical technique. Many minerals, crystals, and biological molecules do, indeed, change the polarization of light. As an example, Figure 25.32 shows a micrograph of a very thin section of a meteorite as it appears when viewed between crossed polarizers. Different minerals in the meteorite have different effects on the polarization of the light. The image is not only colorful, it contains valuable clues about the processes that formed the meteorite.

In polarizing sunglasses, the polymer grid is aligned horizontally (when the glasses are in the normal orientation) so that the glasses transmit vertically polarized light. *Glare*—the reflection of the sun and the skylight from lakes and other horizontal surfaces—has a strong horizontal polarization. This light is almost completely blocked by the Polaroid, so the sunglasses "cut glare" without affecting the main scene you wish to see.

FIGURE 25.32 Polarized light micrograph of a thin slice of a meteorite.

The molecules in Polaroid absorb light if its electric field is oriented along the axis of the molecules. Light-sensing *photopigments* in the eye are also long-chain molecules with similar properties; they are much more likely to absorb and respond to light that is polarized along the axis of the molecules. In human eyes, the orientation of the photopigments is random, so we do not have a polarization sense. But honeybees and other insects have light-sensing cells in their eyes in which the axes of the photopigment molecules are aligned. The response of these cells varies with the polarization of the light. Bees and other insects use this polarization sense to navigate, as described in the photo below.

▶ **Making a beeline** BIO   At an angle of 90° to the sun, skylight is partially polarized because of scattering from air molecules. In the left photo, this polarization causes different transmission through polarizers with different axes. Opposite the sun in the sky, the skylight is unpolarized, causing equal transmission for all of the polarizers in the right photo. We can't sense this sky polarization without external filters, but honeybees can. Light sensing cells in honeybee eyes have different polarization axes, so they sense the polarization of the sky in a similar fashion to the photos. A bee can determine the sun's position from a tiny patch of clear sky, so bees can reliably navigate even in dense forest cover.

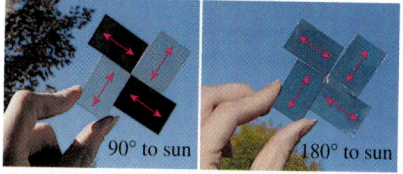

STOP TO THINK 25.4   Unpolarized light of equal intensity is incident on four pairs of polarizing filters. Rank in order, from largest to smallest, the intensities I_A to I_D transmitted through the second polarizer of each pair.

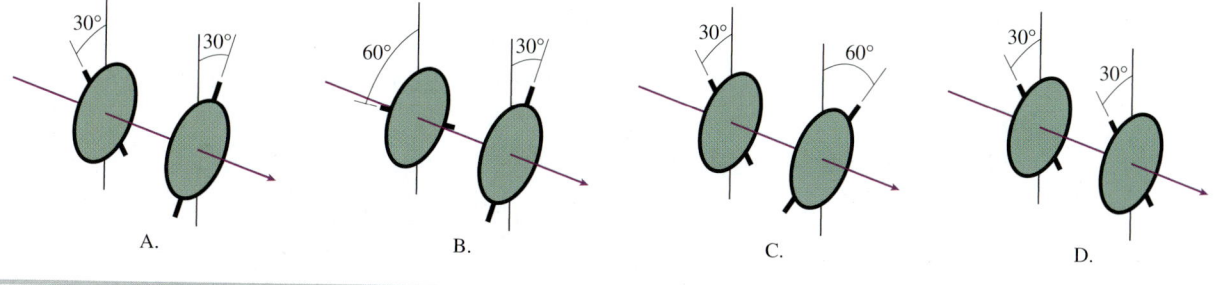

25.7 The Photon Model of Electromagnetic Waves

The photo at very low light levels shows individual points, as if particles are arriving at the detector.

The particle-like behavior is not noticeable at higher light levels.

FIGURE 25.33 Photographs made with an increasing level of light intensity.

Figure 25.33 shows three photographs made with a camera in which the film has been replaced by a special high-sensitivity detector. A correct exposure, at the bottom, shows a perfectly normal photograph of a woman. But with very faint illumination (top), the picture is *not* just a dim version of the properly exposed photo. Instead, it is a collection of dots. A few points on the detector have registered the presence of light, but most have not. As the illumination increases, the density of these dots increases until the dots form a full picture.

This is not what we might expect. If light is a wave, reducing its intensity should cause the picture to grow dimmer and dimmer until disappearing, but the entire picture would remain present. It should be like turning down the volume on your stereo until you can no longer hear the sound. Instead, the left photograph in Figure 25.33 looks as if someone randomly threw "pieces" of light at the detector, causing full exposure at some points but no exposure at others.

If we did not know that light is a wave, we would interpret the results of this experiment as evidence that light is a stream of some type of particle-like object. If these particles arrive frequently enough, they overwhelm the detector and it senses a steady "river" instead of the individual particles in the stream. Only at very low intensities do we become aware of the individual particles.

As we will see in Chapter 28, many experiments convincingly lead to the surprising result that **electromagnetic waves, although they are waves, have a particle-like nature.** These particle-like components of electromagnetic waves are called **photons.**

The **photon model** of electromagnetic waves consists of three basic postulates:

1. Electromagnetic waves consist of discrete, massless units called photons. A photon travels in vacuum at the speed of light, 3.00×10^8 m/s.
2. Each photon has energy

$$E_{\text{photon}} = hf \tag{25.20}$$

where f is the frequency of the wave and h is a *universal constant* called **Planck's constant.** The value of Planck's constant is

$$h = 6.63 \times 10^{-34} \text{ J} \cdot \text{s}$$

In other words, the electromagnetic waves come in discrete "chunks" of energy hf.

3. The superposition of a sufficiently large number of photons has the characteristics of a continuous electromagnetic wave.

EXAMPLE 25.8 Finding the energy of a photon of visible light

550 nm is the average wavelength of visible light.

a. What is the energy of a photon with a wavelength of 550 nm?
b. A 40 W incandescent light bulb emits about 1 J of visible light energy every second. Estimate the number of visible light photons emitted per second.

SOLVE a. The frequency of the photon is

$$f = \frac{c}{\lambda} = \frac{3.00 \times 10^8 \text{ m/s}}{550 \times 10^{-9} \text{ m}} = 5.45 \times 10^{14} \text{ Hz}$$

Equation 25.20 gives us the energy of this photon:

$$E_{\text{photon}} = hf = (6.63 \times 10^{-34} \text{ J} \cdot \text{s})(5.45 \times 10^{14} \text{ Hz}) = 3.61 \times 10^{-19} \text{ J}$$

This is an extremely small energy! In fact, photon energies are so small that they are usually measured in electron volts (eV) rather than joules. Recall that $1\text{ eV} = 1.60 \times 10^{-19}$ J. With this, we find that the photon energy is

$$E_{\text{photon}} = 3.61 \times 10^{-19}\text{ J} \times \frac{1\text{ eV}}{1.60 \times 10^{-19}\text{ J}} = 2.26\text{ eV}$$

b. The photons emitted by a light bulb span a range of energies, because the light spans a range of wavelengths, but the *average* photon energy corresponds to a wavelength near 550 nm. Thus we can estimate the number of photons in 1 J of light as

$$N \approx \frac{1\text{ J}}{3.61 \times 10^{-19}\text{ J/photon}} \approx 3 \times 10^{18}\text{ photons}$$

A typical light bulb emits about 3×10^{18} photons every second.

ASSESS The number of photons emitted per second is staggeringly large. It's not surprising that in our everyday life we would sense only the river and not the individual particles within the flow.

As we saw, a single photon of light at a wavelength of 550 nm has an energy of 2.26 eV. It is worthwhile to see just what 2.26 eV "buys" in interactions with atoms and molecules. Table 25.1 shows some energies required for typical atomic and molecular processes. These values show that 2.26 eV is a significant amount of energy on an atomic scale. It is certainly enough to cause a molecular transformation, and photons with just slightly more energy (shorter wavelength) can break a covalent bond. The photon model of light will be essential as we explore the interaction of electromagnetic waves with matter in coming chapters.

TABLE 25.1 Energies of some atomic and molecular processes

Process	Energy
Breaking a hydrogen bond between two water molecules	0.24 eV
Energy released in metabolizing one molecule of ATP	0.32 eV
Breaking the bond between atoms in a water molecule	4.7 eV
Ionizing a hydrogen atom	13.6 eV

STOP TO THINK 25.5   Two FM radio stations emit radio waves at frequencies of 90.5 MHz and 107.9 MHz. Each station emits the same total power. If you think of the radio waves as photons, which station emits the larger number of photons per second?

A. The 90.5 MHz station.   B. The 107.9 MHz station.
C. Both stations emit the same number of photons per second.

25.8 The Electromagnetic Spectrum

We have now seen two very different ways to look at electromagnetic waves: as oscillating waves of the electric and magnetic field, and as particle-like units of the electromagnetic field called photons. This dual nature of electromagnetic waves is something we will discuss at length in Chapter 28. For now, we will note that each view is appropriate in certain circumstances. For example, we speak of radio *waves* but of x *rays*. The "ray" terminology tells us that x rays are generally better described as photons than as waves.

Figure 25.34 shows the *electromagnetic spectrum* with photon energy (in eV) and wavelength (in m) scales. As you can see, electromagnetic waves span an extraordinarily wide range of wavelengths and energies. Radio waves have wavelengths of many meters but very low photon energies—only a few billionths of an eV. Because the photon energies are so small, radio waves are well described by Maxwell's theory of electromagnetic waves, as we noted above. At the other end of the spectrum, x rays and gamma rays have very short wavelengths and very high photon energies—large enough to ionize atoms and break molecular bonds. Consequently, x rays and gamma rays, although they do have wave-like characteristics, are best described as photons. Visible light is in the middle. As we will see in Chapter 28, we must consider *both* views to fully understand the nature of visible light.

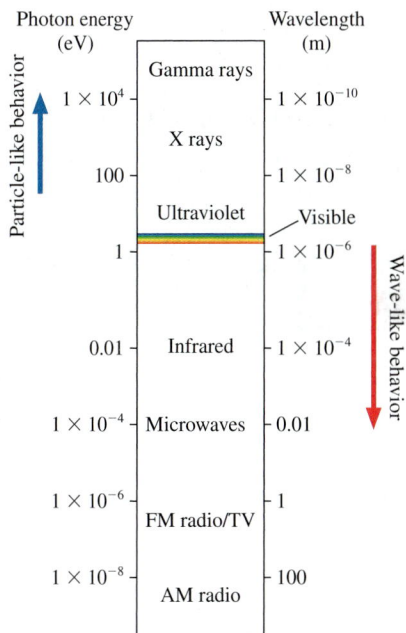

FIGURE 25.34 The electromagnetic spectrum.

CHAPTER 25 · Electromagnetic Induction and Electromagnetic Waves

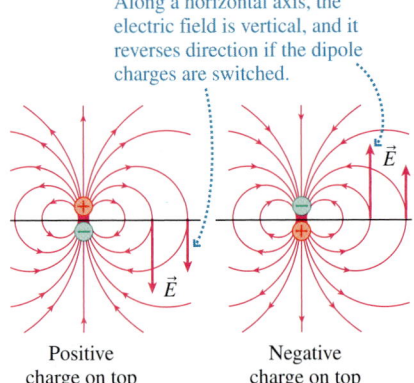

Along a horizontal axis, the electric field is vertical, and it reverses direction if the dipole charges are switched.

$\vec{E}$

Positive charge on top

Negative charge on top

$\vec{E}$

FIGURE 25.35 The electric field of an oscillating dipole.

An oscillating voltage causes the dipole to oscillate.

$\vec{E}$ $\vec{E}$

$\vec{B}$

$\vec{B}$ $\vec{B}$

$\vec{E}$

Antenna wire

The oscillating dipole causes an electromagnetic wave to move away from the antenna at speed $v_{em} = c$.

FIGURE 25.36 An antenna generates a self-sustaining electromagnetic wave.

TRY IT YOURSELF

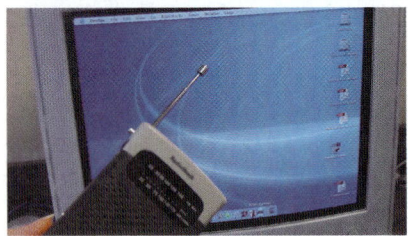

Unwanted transmissions Airplane passengers are asked to turn off electronic devices during takeoff and landing. To see why, hold an AM radio near your computer and adjust the tuning as the computer performs basic operations, such as opening files. You will pick up intense static because the rapid switching of electric currents causes computers—and other electronic devices—to emit radio waves, whether they're designed for communications or not. These electromagnetic waves could interfere with the airplane's electronics.

Radio Waves and Microwaves

An electromagnetic wave is self-sustaining, independent of charges or currents. However, charges and currents are needed at the *source* of an electromagnetic wave. Radio waves and microwaves are generally produced by the motion of charged particles in an antenna.

Figure 25.35 reminds you what the electric field of an electric dipole looks like. If the dipole is vertical, the electric field $\vec{E}$ at points along a horizontal line is also vertical. Reversing the dipole, by switching the charges, reverses $\vec{E}$. If the charges were to *oscillate* back and forth, switching position at frequency f, then $\vec{E}$ would oscillate in a vertical plane. The changing $\vec{E}$ would then create an induced magnetic field $\vec{B}$, which could then create an $\vec{E}$, which could then create a $\vec{B}$, . . . and a vertically polarized electromagnetic wave at frequency f would radiate out into space.

This is exactly what an **antenna** does. Figure 25.36 shows two metal wires attached to the terminals of an oscillating voltage source. The figure shows an instant when the top wire is negative and the bottom is positive, but these will reverse in half a cycle. The wire is basically an oscillating dipole, and it creates an oscillating electric field. The oscillating $\vec{E}$ induces an oscillating $\vec{B}$, and they take off as an electromagnetic wave at speed $v_{em} = c$. The wave does need oscillating charges as a *wave source*, but once created it is self-sustaining and independent of the source.

Radio waves are *detected* by antennas as well. The electric field of a vertically polarized radio wave drives a current up and down a vertical conductor, producing a potential difference that can be amplified. For best reception, the antenna length should be about $\frac{1}{4}$ of a wavelength. A typical cell phone works at 1.9 GHz, with wavelength $\lambda = c/f = 16$ cm. Thus a cell phone antenna should be about 4 cm long, or about $1\frac{1}{2}$ inches. The antenna on your cell phone may seem quite short, but it is the right length to do its job.

AM radio has a lower frequency and thus a longer wavelength—typically 300 m. Having an antenna that is $\frac{1}{4}$ of a wavelength—75 m long!—is simply not practical. Instead, the antenna in an AM radio consists of a coil of wire wrapped around a core of magnetic material. This antenna detects the *magnetic* field of the radio wave. The changing flux of the wave's magnetic field induces an emf in the coil that is detected and amplified by the receiver.

CONCEPTUAL EXAMPLE 25.1 **Orienting a coil antenna**

A vertically polarized AM radio wave is traveling to the right. How should you orient a coil antenna to detect the oscillating magnetic field component of the wave?

REASON You want the oscillating magnetic field of the wave to produce the maximum possible induced emf in the coil, which requires the maximum changing flux. The flux is maximum when the coil is perpendicular to the magnetic field of the electromagnetic wave, as in Figure 25.37. Thus the plane of the coil should match the wave's plane of polarization.

$\vec{E}$ $\vec{E}$ Coil

$\vec{B}$

$\vec{B}$ $\vec{B}$ This orientation produces the maximum magnetic flux through the coil.

$\vec{E}$ c

FIGURE 25.37 A coil antenna.

ASSESS Coil antennas are highly directional. If you turn an AM radio—and thus the antenna—in certain directions, you will no longer have the correct orientation of the magnetic field and the coil, and reception will be quite poor.

The electric fields of radio waves and microwaves interact with matter by exerting a torque on molecules, such as water, that have a permanent electric dipole moment, as shown in 25.38. The molecules acquire kinetic energy from the wave, then their collisions with other molecules transform that energy into thermal energy, increasing the temperature.

This is how a microwave oven heats food. A typical home oven uses microwaves of a frequency of 2.45 GHz and a wavelength of 12.2 cm. Water molecules, with their large dipole moment, rotate in response to the electric field of the microwaves, then transfer this energy to the food via molecular collisions. Physical therapists may use electromagnetic waves for deep heating of tissue. The wavelength is generally longer than that in a microwave oven because the longer wavelengths have greater penetration.

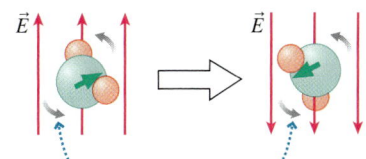

The oscillating electric field of the wave rotates the water molecule by exerting an oscillating torque on its electric dipole moment.

FIGURE 25.38 A radio wave interacts with matter.

Infrared, Visible Light, and Ultraviolet

Radio waves can be produced by oscillating charges in an antenna. At the higher frequencies of infrared, visible light, and ultraviolet, the "antennas" are individual atoms. This portion of the electromagnetic spectrum is *atomic radiation.*

Nearly all the atomic radiation in our environment is *thermal radiation* due to the thermal motion of the atoms in an object. As we saw in Chapter 12, thermal radiation—a form of heat transfer—is described by Stefan's law: If heat energy Q is radiated in a time interval Δt by an object with surface area A and absolute temperature T, the *rate* of heat transfer $Q/\Delta t$ (joules per second) is

$$\frac{Q}{\Delta t} = e\sigma A T^4 \qquad (25.21)$$

The constant e in this equation is the object's emissivity, a measure of its effectiveness at emitting electromagnetic waves, and σ is the Stefan-Boltzmann constant, $\sigma = 5.67 \times 10^{-8}$ W/(m$^2 \cdot$ K^4).

In Chapter 12 we considered the amount of energy radiated and its dependence on temperature. The filament of an incandescent bulb glows simply because it is hot. If you increase the current through a lightbulb, the temperature increases and so does the total energy emitted by the bulb, in accordance with Stefan's law. The three pictures in Figure 25.39 show a glowing lightbulb with the filament at successively higher temperatures. We can clearly see an increase in brightness in the sequence of three photographs.

But it's not just the brightness that varies. The *color* of the emitted radiation changes as well. At low temperatures, the light from the bulb is quite red. (A dim bulb doesn't look this red to your eye because your brain, knowing that the light "should" be white, compensates. But the camera doesn't lie.) Looking at the change in color as the temperature of the bulb rises in Figure 25.39, we see that **the spectrum of thermal radiation changes with temperature.** It's this variation in the spectrum that we want to consider in this chapter.

If we measured the intensity of thermal radiation as a function of wavelength for an object at three temperatures, 3500 K, 4500 K, and 5500 K, the data would appear as in Figure 25.40. Notice two important features of the data:

■ Increasing the temperature increases the intensity at all wavelengths. **Making the object hotter causes it to emit more radiation across the entire spectrum.**

■ Increasing the temperature causes the peak intensity to shift to a shorter wavelength. **The higher the temperature, the shorter the wavelength of the peak of the spectrum.**

Increasing filament temperature →

At lower filament temperatures, the bulb is dim, and the light is noticeably reddish.

When the filament is hotter, the bulb is brighter and the light is whiter.

FIGURE 25.39 The brightness of the bulb varies with the temperature of the filament.

FIGURE 25.40 A thermal emission spectrum depends on the temperature.

The wavelength corresponding to the peak of the intensity graph is given by

$$\lambda_{\text{peak}}(\text{in nm}) = \frac{2.9 \times 10^6 \text{ nm} \cdot \text{K}}{T} \qquad (25.22)$$

INVERSE
p. 118

Wien's law for the peak wavelength of a thermal emission spectrum

where the temperature *must* be in kelvin. The spectrum of a hotter object is a taller graph (more energy radiated) with its peak at a shorter wavelength.

EXAMPLE 25.9 Finding peak wavelengths

What are the wavelengths of peak intensity and the corresponding spectral regions for radiating objects at (a) normal human body temperature of 37°C, (b) the temperature of the filament in an incandescent lamp, 1500°C, and (c) the temperature of the surface of the sun, 5800 K?

PREPARE All of the objects emit thermal radiation.

SOLVE First, we convert temperatures to kelvin. The temperature of the human body is $T = 37 + 273 = 310$ K and the filament temperature is $T = 1500 + 273 = 1773$ K. Equation 25.22 then gives the wavelengths of peak intensity as

a. $\lambda_{\text{peak}}(\text{body}) = \dfrac{2.9 \times 10^6 \text{ nm} \cdot \text{K}}{310 \text{ K}} = 9.4 \times 10^3 \text{ nm} = 9.4 \ \mu\text{m}$

b. $\lambda_{\text{peak}}(\text{filament}) = \dfrac{2.9 \times 10^6 \text{ nm} \cdot \text{K}}{1773 \text{ K}} = 1600 \text{ nm}$

c. $\lambda_{\text{peak}}(\text{sun}) = \dfrac{2.9 \times 10^6 \text{ nm} \cdot \text{K}}{5800 \text{ K}} = 500 \text{ nm}$

ASSESS The peak of the emission curve at body temperature is far into the infrared region of the spectrum, well below the range of sensitivity of human vision. The sun's emission peaks right in the middle of the visible spectrum, which seems reasonable. Interestingly, most of the energy radiated by an incandescent bulb is *not* visible light. The tail of the emission curve extends into the visible region, but the peak of the emission curve—and most of the emitted energy—is in the infrared region of the spectrum. A 100 W bulb emits only a few watts of visible light.

◀ **It's the pits** . . . **BIO** Rattlesnakes can hunt in total darkness. Prey animals are warm, and warm objects emit thermal radiation—which the snakes can sense. Rattlesnakes are in a group of snakes know as *pit vipers*. The name comes from a second set of vision organs that are simply pits with sensitive tissue at the bottom. In the photo, the pits appear as dark spots in front of the eyes. The pits are sensitive to infrared wavelengths of ≈10 μm, near the wavelength of peak emission at mammalian body temperatures. Pit vipers sense the electromagnetic waves *emitted* by warm-blooded animals. They need no light to "see" you. You emit a "glow" they can detect.

Infrared radiation, with its relatively long wavelength and low photon energy, produces effects in tissue similar to those of microwaves—heating—but the penetration is much less than for microwaves. Infrared is absorbed mostly by the top layer of your skin and simply warms you up, as you know from sitting in the sun or under a heat lamp. The wave picture is generally most appropriate for infrared.

In contrast, ultraviolet photons have enough energy to interact with molecules in entirely different ways, ionizing molecules and breaking molecular bonds. The cells in skin are altered by ultraviolet radiation, causing sun tanning and sun burning. DNA molecules can be permanently damaged by ultraviolet radiation. There is a reasonably sharp threshold for such damage at 290 nm (corresponding to 4.3 eV photon energy). At longer wavelengths, damage to cells is slight; at shorter wavelengths, it is extensive. Ultraviolet lamps are very effective at sterilizing surfaces because they disrupt the genetic material of bacteria sufficiently to kill them. These interactions of ultraviolet radiation with matter are best understood from the photon perspective, with the absorption of each photon being associated with a particular molecular event.

Visible light is at a transition point in the electromagnetic spectrum. Your studies of wave optics in Chapter 17 showed you that light has a wave nature. At the same time, the energy of photons of visible light is large enough to cause molecular transitions—which is how your eye detects light. The bending of light by the lens of the eye requires us to think of light as a wave, but the detection of light by the cells in the retina requires us to think of light as photons. When we work with visible light, we will often move back and forth between the wave and photon models.

EXAMPLE 25.10 Finding the photon energy for ultraviolet light

Ultraviolet radiation with a wavelength of 254 nm is used in germicidal lamps. What is the photon energy in eV for such a lamp?

SOLVE The photon energy is $E = hf$:

$$E = hf = \frac{hc}{\lambda} = \frac{(6.63 \times 10^{-34}\,\text{J} \cdot \text{s})(3.00 \times 10^{8}\,\text{m/s})}{254 \times 10^{-9}\,\text{m}}$$

$$= 7.83 \times 10^{-19}\,\text{J}$$

In eV, this is

$$E = 7.83 \times 10^{-19}\,\text{J} \times \frac{1\,\text{eV}}{1.60 \times 10^{-19}\,\text{J}} = 4.89\,\text{eV}$$

ASSESS Table 25.1 shows that this energy is sufficient to break the bonds in a water molecule. It will be enough energy to break other bonds as well, leading to damage on a cellular level.

Color Vision

The cones, the color-sensitive cells in the retina of the eye, each contain one of three slightly different forms of a light-sensitive photopigment. A single photon of light can trigger a reaction in a photopigment molecule, which ultimately leads to a signal being produced by a cell in the retina. The energy of the photon must be matched to the energy of a molecular transition for absorption of the photon energy to take place. Each photopigment has a range of photon energies to which it is sensitive. Our color vision is a result of the differential response of the three types of cones containing these three different pigments, shown in Figure 25.41.

CONCEPTUAL EXAMPLE 25.2 Creating the impression of a color

Computer monitors and color TVs can create millions of different colors by combining light from pixels of only three colors: red, green, and blue. These are called RGB displays. How do they do it?

REASON We've seen that there are three different types of cones in the eye. By using differing amounts of three pure colors, we can independently stimulate each of the cone types and thus mimic the response of the eye to light of almost any color.

ASSESS The fact that there are three primary colors of light—red, green, and blue— is a function of our physiology, not basic physics.

Humans have three color photopigments, mice have two, and chickens four— giving them keener color vision than you. The three color photopigments that bees possess give them excellent color vision, but a bee's color sense is different from a human's. The peak sensitivities of a bee's photopigments are in the yellow, blue, and ultraviolet regions of the spectrum. A bee can't see the red of a rose, but it is quite sensitive to ultraviolet wavelengths well beyond the range of human vision. The flowers in the photo at the start of the chapter look pretty to us, but their coloration is really intended for other eyes. The ring of ultraviolet-absorbing pigments near the center of the flower, which is invisible to humans, helps bees zero in on the pollen.

X Rays and Gamma Rays

At the highest energies of the electromagnetic spectrum we find x rays and gamma rays. There is no sharp dividing line between these two regions of the spectrum; the difference is the source of radiation. High-energy photons emitted by electrons are called x rays. If the source is a nuclear process, we call them gamma rays.

We will look at the emission of x rays in atomic processes and gamma rays in nuclear processes in Part VII. For now, we will focus on the "artificial" production of x rays in an x-ray tube, such as the one shown in Figure 25.42.

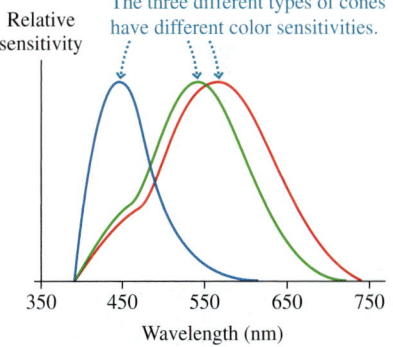

FIGURE 25.41 The sensitivity of different cones in the human eye.

FIGURE 25.42 A simple x-ray tube.

Electrons are emitted from a cathode and accelerated to a kinetic energy of several thousand eV by the electric field between two electrodes connected to a high-voltage power supply. The electrons make a sudden stop when they hit a metal target electrode. The rapid deceleration of an electron can cause the emission of a single photon with a significant fraction of the electron's kinetic energy. These photons, with energies well in excess of 1000 eV, are x rays. The x rays pass through a window in the tube and then may be used to produce an image or to treat a disease.

EXAMPLE 25.11 Determining x-ray energies

An x-ray tube used for medical work has an accelerating voltage of 30 kV. What is the maximum energy of an x-ray photon that can be produced in this tube? What is the wavelength of this x ray?

SOLVE An electron accelerated through a potential difference of 30 kV acquires a kinetic energy of 30 keV. When this electron hits the metal target and stops, energy may be converted to an x ray. The maximum energy that could be converted is 30 keV, so this is the maximum possible energy of an x-ray photon from the tube. In joules, this energy is

$$E = 30 \times 10^3 \text{ eV} \times \frac{1.60 \times 10^{-19} \text{ J}}{1 \text{ eV}} = 4.8 \times 10^{-15} \text{ J}$$

For electromagnetic waves, $c = f\lambda$, so we can calculate

$$\lambda = \frac{c}{f} = \frac{c}{E/h} = \frac{hc}{E} = \frac{(6.63 \times 10^{-34} \text{ J} \cdot \text{s})(3.00 \times 10^8 \text{ m/s})}{4.8 \times 10^{-15} \text{ J}}$$

$$= 4.1 \times 10^{-11} \text{ m} = 0.041 \text{ nm}$$

ASSESS This is a very short wavelength, comparable to the spacing between atoms in a solid.

X rays and gamma rays (and the short-wavelength part of the ultraviolet spectrum) are **ionizing radiation;** the individual photons have sufficient energy to ionize atoms. When such radiation strikes tissue, the resulting ionization can produce cellular damage. When people speak of "radiation" they often mean "ionizing radiation." Ionizing radiation can be harmful to cells, but, as we will see in Chapter 30, it can also be put to good use in radiation therapy to treat cancer. Rapidly dividing cells—such as those in a tumor—are especially sensitive to the damage from ionizing radiation.

X rays and gamma rays are very penetrating, but the absorption of these high-energy photons is greater in materials made of atoms with more electrons. This is why x rays are used in medical and dental imaging. The calcium in bones has many more electrons and thus is much more absorbing than the hydrogen, carbon, and oxygen that make up most of our soft tissue, so we can use x rays to image bones and teeth.

At several points in this chapter we have hinted at places where a full understanding of the phenomena requires some new physics. We have used the photon model of electromagnetic waves, and we have mentioned that nuclear processes can give rise to gamma rays. There are other questions that we did not raise, such as why the electromagnetic spectrum of a hot object has the shape that it does. These puzzles began to arise in the late 1800s and early 1900s, and it soon became clear that the physics of Newton and Maxwell was not sufficient to fully describe the nature of matter and energy. Some new rules, some new models, were needed. After the next chapter, in which we look at AC circuits, we will return to these puzzles as we begin to explore the exciting notions of quantum physics in Part VII.

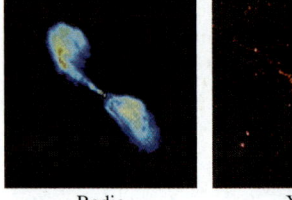

Visible   Infrared

Radio   X ray

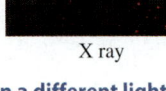

Seeing the universe in a different light
These four images of the Centaurus A galaxy have the same magnification and orientation, but they are records of different types of electromagnetic waves. (All but the visible light image are false-color images.) The visible light images shows a dark dust lane cutting across the galaxy. In the infrared, this dust lane glows quite brightly—telling us that the dust particles are hot. The radio and x ray images show jets of matter streaming out of the galaxy's center, hinting at the presence of a massive black hole. Views of the cosmos beyond the visible range are important tools of modern astronomy.

STOP TO THINK 25.6 A group of four stars, all the same size, have the four different surface temperatures given below. Which of these stars emits the most red light?

A. 3000 K   B. 4000 K   C. 5000 K   D. 6000 K

The goal of Chapter 25 has been to understand the nature of electromagnetic induction and electro-magnetic waves.

GENERAL PRINCIPLES

Electromagnetic Induction

The magnetic flux measures the amount of magnetic field passing through a surface.

$$\Phi = AB\cos\theta$$

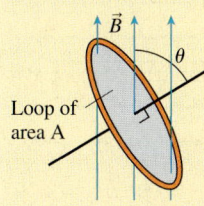

Loop of area A

Lenz's law specifies that there is an induced current in a closed conducting loop if the magnetic flux through the loop is changing. The direction of the induced current is such that the induced magnetic field opposes the *change* in flux.

$\vec{B}_{induced}$ Current meter

$\vec{B}$ Induced current

Faraday's law specifies the magnitude of the induced emf in a closed loop:

$$\mathcal{E} = \left|\frac{\Delta\Phi}{\Delta t}\right|$$

Multiply by N for an N-turn coil.

The size of the induced current is

$$I = \frac{\mathcal{E}}{R}$$

Electromagnetic Waves

An electromagnetic wave is a self-sustaining oscillation of electric and magnetic fields.

- The wave is a transverse wave with $\vec{E}$ perpendicular to $\vec{B}$ and $\vec{v}$.

- The wave propagates with speed

$$v_{em} = c = \frac{1}{\sqrt{\epsilon_0\mu_0}} = 3.00 \times 10^8 \text{ m/s}$$

- The wavelength, frequency, and speed are related by

$$c = f\lambda$$

- The amplitudes of the fields are related by

$$E_0 = cB_0$$

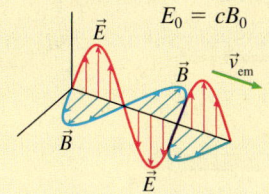

IMPORTANT CONCEPTS

Motional emf

The motion of a conductor through a magnetic field produces a force on the charges. The separation of charges leads to an emf:

$$\mathcal{E} = vlB$$

The photon model

Electromagnetic waves appear to be made of discrete units called photons. The energy of a photon of frequency f is

$$E = hf$$

This photon view becomes increasingly important as the photon energy increases.

The electromagnetic spectrum

Electromagnetic waves come in a wide range of wavelengths and photon energies.

APPLICATIONS

A changing flux in a solid conductor creates eddy currents.

The plane of the electric field of an electromagnetic wave defines its polarization. The intensity of polarized light transmitted through a polarizing filter is given by **Malus's law:**

$$I = I_0\cos^2\theta$$

where θ is the angle between the electric field and the polarizer axis.

Thermal radiation has a peak wavelength that depends on an object's temperature according to Wien's law:

$$\lambda_{peak}(\text{in nm}) = \frac{2.9 \times 10^6 \text{ nm} \cdot \text{K}}{T}$$

QUESTIONS

Conceptual Questions

1. The world's strongest magnet can produce a steady field of 45 T. If a circular wire loop of radius 10 cm were held in this magnetic field, what current would be induced in the loop?

2. The rapid vibration accompanying the swimming motions of BIO mayflies has been measured by gluing a small magnet to a swimming mayfly and recording the emf in a small coil of wire placed nearby. Explain how this technique works.

3. Parts a through f of Figure Q25.3 show one or more metal wires sliding on fixed metal rails in a magnetic field. For each, determine if the induced current is clockwise, counterclockwise, or is zero.

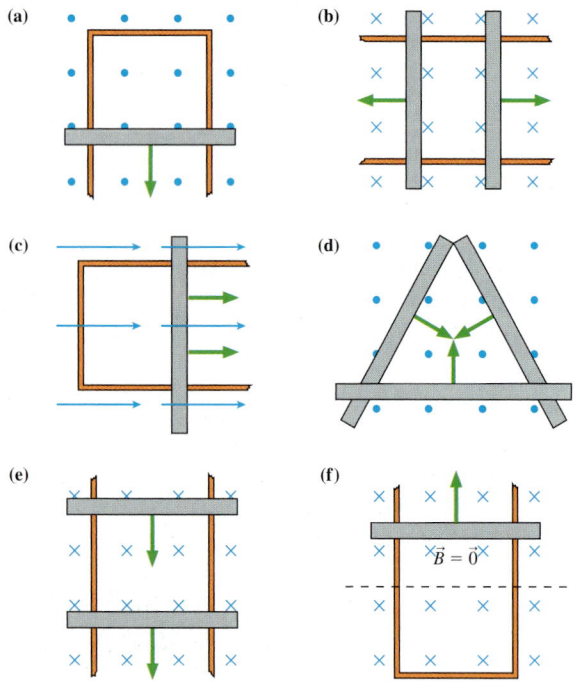

FIGURE Q25.3

4. Figure Q25.4 shows four different loops in a magnetic field. The numbers indicate the lengths of the sides and the strength of the field. Rank in order the magnetic fluxes Φ_1 through Φ_4, from the largest to the smallest. Some may be equal. Explain.

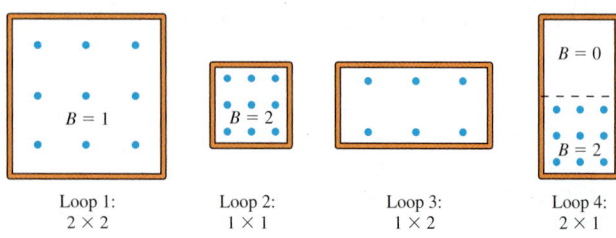

Loop 1:   Loop 2:   Loop 3:   Loop 4:
2 × 2   1 × 1   1 × 2   2 × 1

FIGURE Q25.4

5. Figure Q25.5 shows four different circular loops that are perpendicular to the page. The radius of loops 3 and 4 is twice that of loops 1 and 2. The magnetic field is the same for each. Rank in order the magnetic fluxes Φ_1 through Φ_4, from the largest to the smallest. Some may be equal. Explain.

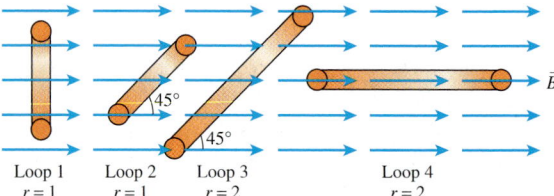

Loop 1   Loop 2   Loop 3   Loop 4
r = 1   r = 1   r = 2   r = 2

FIGURE Q25.5

6. A circular loop rotates at constant speed about an axle through the center of the loop. Figure Q25.6 shows an edge view and defines the angle ϕ, which increases from 0° to 360° as the loop rotates.

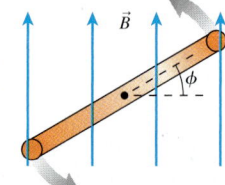

FIGURE Q25.6

   a. At what angle or angles is the magnetic flux a maximum?
   b. At what angle or angles is the magnetic flux a minimum?
   c. At what angle or angles is the magnetic flux *changing* most rapidly?

7. The power lines that run through your neighborhood carry *alternating currents* that reverse direction 120 times per second. As the current changes, so does the magnetic field around a line. Suppose you wanted to put a loop of wire up near the power line to extract power by "tapping" the magnetic field. Sketch a picture of how you would orient the coil of wire next to a power line to develop the maximum emf in the coil. (Note that this is dangerous and illegal, and not something you should try.)

8. The magnetic flux passing through a coil of wire varies as shown in Figure Q25.8. During which time interval(s) will an induced current be present in the coil? Explain.

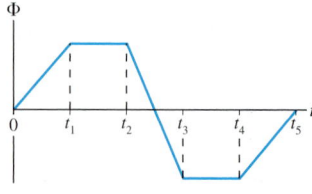

FIGURE Q25.8

9. There is a counterclockwise induced current in the conducting loop shown in Figure Q25.9. Is the magnetic field inside the loop increasing in strength, decreasing in strength, or steady?

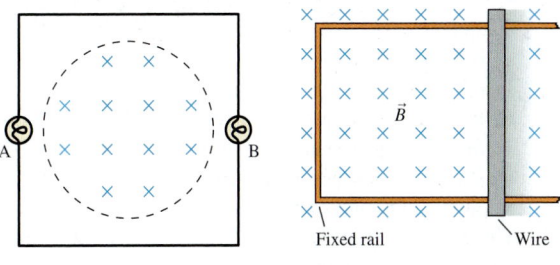

FIGURE Q25.9

10. Cars on the "Tower of Doom" carnival ride are dropped from a great height, giving riders a few seconds of free fall. To stop the cars, strong permanent magnets attached to the bottoms of the cars pass very close to aluminum vanes sticking up from the bottom of the track. Explain how this braking system works.

11. A magnet dropped through a clear plastic tube accelerates as expected in free fall. If dropped through an aluminum tube of exactly the same length and diameter, the magnet falls much more slowly. Explain the behavior of the second magnet.

12. You want to insert a loop of copper wire between the north and south poles of two permanent magnets, as sketched in Figure Q25.12. Is there an attractive magnetic force that tends to *pull* the loop in, like a magnet pulls on a paper clip? Or do you need to *push* the loop in against a repulsive force? Give a step by step analysis to support your answer.

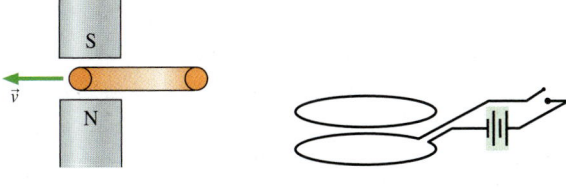

FIGURE Q25.12   **FIGURE Q25.13**

13. Two loops of wire are stacked vertically, one above the other, as shown in Figure Q25.13. Does the upper loop have a clockwise current, a counterclockwise current, or no current at the following times? Explain your reasoning.
    a. Before the switch is closed.
    b. Immediately after the switch is closed.
    c. Long after the switch is closed.
    d. Immediately after the switch is reopened.

14. A loop of wire is horizontal. A bar magnet is pushed toward the loop from below, along the axis of the loop, as shown in Figure Q25.14.
    a. In what direction is the current in the loop? Explain.
    b. Is there a magnetic force on the loop? If so, in which direction? Explain.
    Hint: Recall that a current loop is a magnetic dipole.
    c. Is there a magnetic force on the magnet? If so, in which direction?

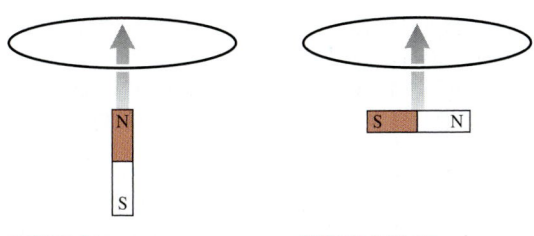

FIGURE Q25.14   **FIGURE Q25.15**

15. A bar magnet is pushed toward a loop of wire, as shown in Figure Q25.15. Is there a current in the loop? If so, in which direction? If not, why not?

16. A conducting loop around a magnetic field contains two light bulbs, as shown in Figure Q25.16. The wires connecting the bulbs are ideal. The magnetic field is increasing rapidly.
    a. Do the bulbs glow? Why or why not?
    b. If they glow, which bulb is brighter? Or are they equally bright? Explain.

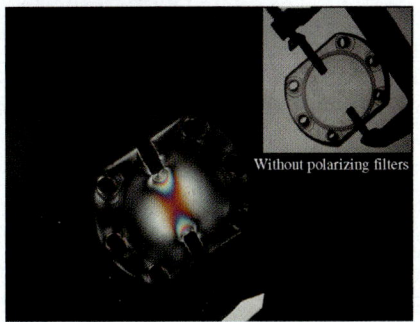

FIGURE Q25.16   **FIGURE Q25.17**

17. A metal wire is resting on a U-shaped conducting rail, as shown in Figure Q25.17. The rail is fixed in position, but the wire is free to move.
    a. If the magnetic field is increasing in strength, what does the wire do? Does it remain in place? Or does it move to the right or left, or up or down, or out of the plane of the page (breaking contact with the rail)? Does it rotate clockwise or counterclockwise? Does it both move and rotate? Explain.
    b. If the magnetic field is decreasing in strength, what does the wire do?

18. Though sunlight is unpolarized, the light that reflects from smooth surfaces may be partially polarized in the direction parallel to the plane of the reflecting surface. How should the long axis of the polarizing molecules in polarized sunglasses be oriented— vertically or horizontally—to reduce the glare from a horizontal surface such as a road or a lake?

19. Two polarizers are oriented with axes at 90°, so no light passes. A piece of plastic is placed between the two. If the plastic is stressed, by being squeezed, light that passes through the first polarizer and the plastic now passes through the second polarizer. The dark and light lines allow the pattern of stress to be determined. What effect does the stressed plastic have on the polarization of light passing through it?

Without polarizing filters

20. Old-fashioned roof-mounted television antennas were designed to pick up signals across a broad frequency range. Explain why these antennas had metal bars of many different lengths.

21. An AM radio detects the oscillating magnetic field of the radio wave with an antenna consisting of a coil of wire wrapped around a ferrite bar, as shown in Figure Q25.21. Ferrite is a magnetic material that "amplifies" the magnetic field of the wave.

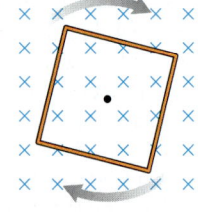

FIGURE Q25.21

Axis of ferrite bar

   a. Explain how the radio antenna detects the magnetic field of the radio wave.
   b. If a radio station is located due north of you, how must the ferrite bar be oriented for best reception? Assume that the station broadcasts with a vertical antenna like the one shown in Figure 25.36.

22. Three laser beams have wavelengths $\lambda_1 = 400$ nm, $\lambda_2 = 600$ nm, and $\lambda_3 = 800$ nm. The power of each laser beam is 1 W.
   a. Rank in order, from largest to smallest, the photon energies E_1, E_2, and E_3 in these three laser beams. Explain.
   b. Rank in order, from largest to smallest, the number of photons per second N_1, N_2, and N_3 delivered by the three laser beams. Explain.

23. The intensity of a beam of light is increased but the light's frequency is unchanged. As a result, which of the following (perhaps more than one) are true? Explain.
   A. The photons travel faster.
   B. Each photon has more energy.
   C. The photons are larger.
   D. There are more photons per second.

24. The frequency of a beam of light is increased but the light's intensity is unchanged. As a result, which of the following (perhaps more than one) are true? Explain.
   A. The photons travel faster.
   B. Each photon has more energy.
   C. There are fewer photons per second.
   D. There are more photons per second.

25. Arc welding uses electric current to make an extremely hot electric arc that can melt metal. The arc emits ultraviolet light that can cause sunburn and eye damage if a welder is not wearing protective gear. Why does the arc give off ultraviolet light?

26. Photons of ultraviolet radiation have greater energy than photons of microwave radiation. So why do we use microwave ovens and not ultraviolet ovens?

Multiple-Choice Questions

27. | A circular loop of wire has an area of 0.30 m². It is tilted by 45° with respect to a uniform 0.40 T magnetic field. What is the magnetic flux through the loop?
   A. 0.085 T · m²
   B. 0.12 T · m²
   C. 0.38 T · m²
   D. 0.75 T · m²
   E. 1.3 T · m²

28. ‖ In Figure Q25.28, a square loop is rotating in the plane of the page around an axis through its center. A uniform magnetic field is directed into the page. What is the direction of the induced current in the loop?
   A. Clockwise.
   B. Counterclockwise.
   C. There is no induced current.

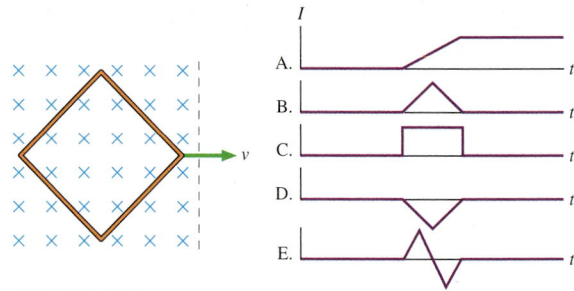

FIGURE Q25.28

29. ‖ A diamond-shaped loop of wire is pulled at a constant velocity through a region where the magnetic field is directed into the paper in the left half and is zero in the right half, as shown in Figure Q25.29. As the loop moves from left to right, which graph best represents the induced current in the loop as a function of time? Let a clockwise current be positive and a counterclockwise current be negative.

FIGURE Q25.29

30. | Figure Q25.30 shows a triangular loop of wire in a uniform magnetic field. If the field strength changes from 0.30 to 0.10 T in 50 ms, what is the induced emf in the loop?
   A. 0.08 V
   B. 0.12 V
   C. 0.16 V
   D. 0.24 V
   E. 0.36 V

FIGURE Q25.30

31. ‖ A device called a *flip coil* can be used to measure the earth's magnetic field. The coil has 100 turns and an area of 0.010 m². It is oriented with its plane perpendicular to the earth's magnetic field, then flipped 180° so the field goes through the coil in the opposite direction. The earth's magnetic field is 0.050 mT, and the coil flips over in 0.50 s. What is the average emf induced in the coil during the flip?
   A. 0.050 mV   B. 0.10 mV   C. 0.20 mV   D. 1.0 mV

32. | The electromagnetic waves that carry FM radio range in frequency from 87.9 MHz to 107.9 MHz. What is the range of wavelengths of these radio waves?
   A. 500–750 nm   B. 0.87–91.08 m
   C. 2.78–3.41 m   D. 278–341 m
   E. 234–410 km

33. | Two electromagnetic waves travel in a vacuum. Which wave travels faster?
   A. The wave with the higher frequency.
   B. The wave with the larger wavelength.
   C. The wave with the larger electric field.
   D. The wave with the larger magnetic field.
   E. Both waves travel at the same speed.

34. | A spacecraft in orbit around the moon measures its altitude by reflecting a pulsed 10 MHz radio signal from the surface. If the spacecraft is 10 km high, what is the time between the emission of the pulse and the detection of the echo?
    A. 33 ns   B. 67 ns   C. 33 μs   D. 67 μs

35. | A 6.0 mW vertically polarized laser beam passes through a polarizing filter whose axis is 75° from vertical. What is the laser-beam power after passing through the filter?
    A. 0.40 mW   B. 1.0 mW   C. 1.6 mW   D. 5.6 mW

36. | Communication with submerged submarines via radio waves is difficult because seawater is conductive and absorbs electromagnetic waves. Penetration into the ocean is greater at longer wavelengths, so the United States has radio installations that transmit at 76 Hz for submarine communications. What is the approximate wavelength of those extremely low-frequency waves?
    A. 500 km   B. 1000 km   C. 2000 km   D. 4000 km

37. | How many photons are emitted during 5.0 s of operation of a red laser pointer? The device outputs 2.8 mW at a 635 nm wavelength.
    A. 4.5×10^{10} B. 4.5×10^{11}
    C. 4.5×10^{15} D. 4.5×10^{16}

PROBLEMS

Section 25.1 Induced Currents

Section 25.2 Motional emf

1. | A potential difference of 0.050 V is developed across the 10-cm-long wire in Figure P25.1 as it moves through a magnetic field at 5.0 m/s. The magnetic field is perpendicular to the axis of the wire. What are the direction and strength of the magnetic field?

FIGURE P25.1

2. | A scalloped hammerhead shark swims at a steady speed of
BIO 1.5 m/s with its 85-cm-wide head perpendicular to the earth's 50 μT magnetic field. What is the magnitude of the emf induced between the two sides of the shark's head?

3. | A 10-cm-long wire is pulled along a U-shaped conducting rail in a perpendicular magnetic field. The total resistance of the wire and rail is 0.20 Ω. Pulling the wire with a force of 1.0 N causes 4.0 W of power to be dissipated in the circuit.
    a. What is the speed of the wire when pulled with 1.0 N?
    b. What is the strength of the magnetic field?

4. | Figure P25.4 shows a 15-cm-long metal rod pulled along two frictionless, conducting rails at a constant speed of 3.5 m/s. The rails have negligible resistance, but the rod has a resistance of 0.65 Ω.
    a. What is the current induced in the rod?
    b. What force is required to keep the rod moving at a constant speed?

FIGURE P25.4

5. || A 50 g horizontal metal bar, 12 cm long, is free to slide up and down between two tall, vertical metal rods. A 0.060 T magnetic field is directed perpendicular to the plane of the rods. The bar is raised to near the top of the rods, and a 1.0 Ω resistor is connected across the two rods at the top. Then the bar is dropped. What is the terminal speed at which the bar falls? Assume the bar remains horizontal and in contact with the rods at all times.

6. | A delivery truck with 2.8-m-high aluminum sides is driving west at 75 km/hr in a region where the earth's magnetic field is $\vec{B} = (5.0 \times 10^{-5}$ T, north).
    a. What is the potential difference between the top and the bottom of the truck's side panels?
    b. Will the tops of the panels be positive or negative relative to the bottoms?

Section 25.3 Magnetic Flux

7. | What is the magnetic flux through the loop shown in Figure P25.7?

FIGURE P25.7

8. |||| The 2.0-cm-diameter solenoid in Figure P25.8 passes through the center of a 6.0-cm-diameter loop. The magnetic field inside the solenoid is 0.20 T. What is the magnetic flux through the loop (a) when it is perpendicular to the solenoid and (b) when it is tilted at a 60° angle?

FIGURE P25.8

9. |||| An MRI solenoid produces a magnetic field of 1.6 T. What
BIO is the magnitude of the flux through the head of a patient inside the solenoid? We can approximate the cross section of the head as a circle with diameter 18 cm.

10. || At a typical location in the United States, the earth's magnetic field has a magnitude of 5.0×10^{-5} T and is at a 65° angle from the downward vertical. What is the flux through the 22 cm × 28 cm front cover of your textbook if it is flat on your desk?

11. | The metal equilateral triangle in Figure P25.11, 20 cm on each side, is halfway into a 0.10 T magnetic field.
    a. What is the magnetic flux through the triangle?
    b. If the magnetic field strength decreases, what is the direction of the induced current in the triangle?

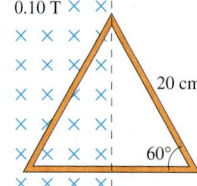

FIGURE P25.11

Section 25.4 Faraday's Law

12. ||| Figure P25.12 shows a 10-cm-diameter loop in three different magnetic fields. The loop's resistance is 0.10 Ω. For each case, determine the induced emf, the induced current, and the direction of the current.

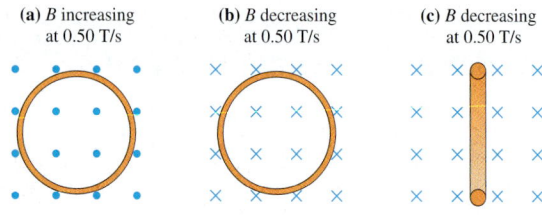

(a) B increasing at 0.50 T/s   (b) B decreasing at 0.50 T/s   (c) B decreasing at 0.50 T/s

FIGURE P25.12

13. || A loop of wire is perpendicular to a magnetic field. Rank, from greatest to least, the magnitudes of the loop's induced emf for the following situations:
    A. The magnetic field strength increases from 0 to 1 T in 6 s.
    B. The magnetic field strength increases from 1 T to 4 T in 2 s.
    C. The magnetic field strength remains at 4 T for 1 min.
    D. The magnetic field strength decreases from 4 T to 3 T in 4 s.
    E. The magnetic field strength decreases from 3 T to 0 T in 1 s.

14. | Five loops of wire are in a region where the same time-varying magnetic field passes through each of them. Rank, from greatest to least, the loop's resistances.
    A. Area 0.2 m², induced current 1 mA.
    B. Area 0.2 m², induced current 2 mA.
    C. Area 0.4 m², induced current 1 mA.
    D. Area 0.4 m², induced current 2 mA.

15. ||| A 1000-turn coil of wire 2.0 cm in diameter is in a magnetic field that drops from 0.10 T to 0 T in 10 ms. The axis of the coil is parallel to the field. What is the emf of the coil?

16. ||| The resistance of the loop in Figure P25.16 is 0.10 Ω. Is the magnetic field strength increasing or decreasing? At what rate (T/s)?

150 mA

8.0 cm

FIGURE P25.16   8.0 cm

17. ||| A square wire loop with 5.0-cm-long sides and resistance 20 Ω is perpendicular to a magnetic field. The field strength is 0.40 T at $t = 0$ s, and it is increasing at the rate 0.010 T/s.
    a. What is the magnetic flux through the loop at $t = 0$ s?
    b. What is the induced current in the loop at $t = 0$ s?
    c. What is the induced current in the loop at $t = 5.0$ s?

18. ||| The circuit shown in Figure P25.18 encloses an area of 25 cm². The magnetic field increases steadily from 0 T to 0.50 T in 10 ms. What is the current in the resistor during this time?

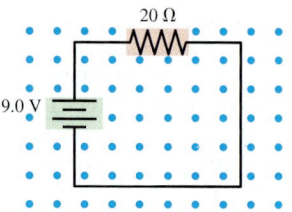

20 Ω

9.0 V

FIGURE P25.18

Section 25.5 Induced Fields and Electromagnetic Waves

Section 25.6 Properties of Electromagnetic Waves

19. | What is the magnetic field amplitude of an electromagnetic wave whose electric field amplitude is 10 V/m?

20. | What is the electric field amplitude of an electromagnetic wave whose magnetic field amplitude is 2.0 mT?

21. | The magnetic field of an electromagnetic wave in a vacuum is $B_z = (3.0\ \mu\text{T})\sin((1.0 \times 10^7)x - 2\pi ft)$, where x is in m and t is in s. What are the wave's (a) wavelength, (b) frequency, and (c) electric field amplitude?

22. || The electric field of an electromagnetic wave in a vacuum is $E_y = (20\ \text{V/m})\sin((6.28 \times 10^8)x - 2\pi ft)$, where x is in m and t is in s. What are the wave's (a) wavelength, (b) frequency, and (c) magnetic field amplitude?

23. | a. What is the magnetic field amplitude of an electromagnetic wave whose electric field amplitude is 100 V/m?
    b. What is the intensity of the wave?

24. || A radio receiver can detect signals with electric field amplitudes as small as 300 μV/m. What is the intensity of the smallest detectable signal?

25. || A 200 MW laser pulse is focused with a lens to a diameter of 2.0 μm.
    a. What is the laser beam's electric field amplitude at the focal point?
    b. What is the ratio of the laser beam's electric field to the electric field that keeps the electron bound to the proton of a hydrogen atom? The radius of the electron's orbit is 0.053 nm.

26. || A radio antenna broadcasts a 1.0 MHz radio wave with 25 kW of power. Assume that the radiation is emitted uniformly in all directions.
    a. What is the wave's intensity 30 km from the antenna?
    b. What is the electric field amplitude at this distance?

27. || At what distance from a 10 W point source of electromagnetic waves is the electric field amplitude (a) 100 V/m and (b) 0.010 V/m?

28. | The intensity of a polarized electromagnetic wave is 10 W/m². What will be the intensity after passing through a polarizing filter whose axis makes the following angles with the plane of polarization? (a) $\theta = 0°$ (b) $\theta = 30°$ (c) $\theta = 45°$ (d) $\theta = 60°$ (e) $\theta = 90°$.

29. || Only 25% of the intensity of a polarized light wave passes through a polarizing filter. What is the angle between the electric field and the axis of the filter?

30. || A 200 mW horizontally polarized laser beam passes through a polarizing filter whose axis is 25° from vertical. What is the power of the laser beam as it emerges from the filter?

31. ‖ Unpolarized light with intensity 350 W/m² passes first through a polarizing filter with its axis vertical, then through a polarizing filter with its axis 30° from vertical. What light intensity emerges from the second filter?

Section 25.7 The Photon Model of Electromagnetic Waves

32. | What is the energy (in eV) of a photon of visible light that has a wavelength of 500 nm?

33. | What is the energy (in eV) of an x-ray photon that has a wavelength of 1.0 nm?

34. | What is the wavelength of a photon whose energy is twice that of a photon with a 600 nm wavelength?

35. | One recent study has shown that x rays with a wavelength
BIO of 0.0050 nm can produce mutations in human cells.
   a. Calculate the energy in eV of a photon of radiation with this wavelength.
   b. Assuming that the bond energy holding together a water molecule is typical, use Table 25.1 to estimate how many molecular bonds could be broken with this energy.

36. | Rod cells in the retina of the eye detect light using a pho-
BIO topigment called rhodopsin. 1.8 eV is the lowest photon energy that can trigger a response in rhodopsin. What is the maximum wavelength of electromagnetic radiation that can cause a transition? In what part of the spectrum is this?

37. | What frequency and wavelength of electromagnetic radiation have photons with an energy of 2.0 eV? What part of the spectrum does this correspond to?

38. ‖ What is the energy of 1 mol of photons that have a wavelength of 1.0 μm?

39. ‖ The intensity of electromagnetic radiation from the sun reaching the earth's upper atmosphere is 1.37 kW/m². Assuming an average wavelength of 680 nm for this radiation, find the number of photons per second that strike a 1.00 m² solar panel directly facing the sun on an orbiting satellite.

40. ‖ The human eye can barely detect a star whose intensity at
BIO the earth's surface is 1.6×10^{-11} W/m². If the dark adapted eye has a pupil diameter of 7.0 mm, how many photons per second enter the eye from the star? Assume the starlight has a wavelength of 550 nm.

Section 25.8 The Electromagnetic Spectrum

41. | The spectrum of a glowing filament has its peak at a wavelength of 1200 nm. What is the temperature of the filament, in °C?

42. ‖ While using a dimmer switch to investigate a new type of incandescent light bulb, you notice that the light changes both its spectral characteristics and its brightness as the voltage is increased.
   a. If the wavelength of maximum intensity decreases from 1800 nm to 1600 nm as the bulb's voltage is increased, by how many °C does the filament temperature increase?
   b. By what factor does the total radiation from the filament increase due to this temperature change?

43. | The photon energies used in different types of medical
BIO x-ray imaging vary widely, depending upon the application. Single dental x rays use photons with energies of about 25 keV. The energies used for x-ray microtomography, a process that allows repeated imaging in single planes at vary-

ing depths within the sample, is 2.5 times greater. What are the wavelengths of the x rays used for these two purposes?

General Problems

44. ‖ A 10 cm × 10 cm square is bent at a 90° angle as shown in Figure P25.44. A uniform 0.050 T magnetic field points downward at a 45° angle. What is the magnetic flux through the loop?

FIGURE P25.44   **FIGURE P25.45**

45. | What is the magnetic flux through the loop shown in Figure P25.45?

46. ‖ a. A circular loop antenna has a diameter of 20 cm. If the plane of the loop is perpendicular to the earth's 50 μT magnetic field, what is the flux through the loop?
   b. What is the flux if the loop is rotated by 30°?

47. | An 1.1-m-diameter MRI solenoid with a length of 2.4 m has a magnetic field of 1.5 T along its axis. If the current is turned off in a time of 1.2 s, what is the emf in one turn of the solenoid's windings?

48. ‖ A magnet and a coil are oriented as shown in Figure P25.48. The magnet is moved rapidly into the coil, held stationary in the coil for a short time, and then rapidly pulled back out of the coil.

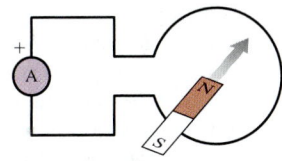

FIGURE P25.48

Sketch a graph showing the reading of the ammeter as a function of time. The ammeter registers a positive value when current goes into the "+" terminal.

49. | A wire loop with an area of 0.020 m² is perpendicular to a uniform magnetic field of 0.30 T. If the field drops to zero in 45 ms, what is the average induced emf in the loop?

50. ‖ A 100-turn, 2.0-cm-diameter coil is at rest in a horizontal plane. A uniform magnetic field 60° away from vertical increases from 0.50 T to 1.50 T in 0.60 s. What is the induced emf in the coil?

51. ‖‖ A 25-turn, 10.0-cm-diameter coil is oriented in a vertical plane with its axis aligned east-west. A magnetic field pointing to the northeast decreases from 0.80 T to 0.20 T in 2.0 s. What is the emf induced in the coil?

52. ‖‖ A 100-turn, 8.0-cm-diameter coil is made of 0.50-mm-
INT diameter copper wire. A magnetic field is perpendicular to the coil. At what rate must B increase to induce a 2.0 A current in the coil? The resistivity of copper is 1.7×10^{-8} Ω·m.

53. ‖‖‖ The loop in Figure P25.53 is being pushed into the 0.20 T magnetic field at 50 m/s. The resistance of the loop is 0.10 Ω. What are the direction and magnitude of the current in the loop?

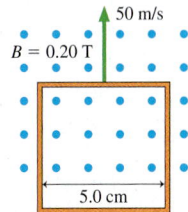

FIGURE P25.53

54. ‖ A spool of insulated copper wire falls out of a truck and rolls to a stop in a position such that the earth's magnetic field ($B = 5.0 \times 10^{-5}$ T) points along the axis of the spool. The spool's diameter is 8.0 cm, and it holds 10,000 turns of wire. Soon thereafter, another truck runs over the spool and crushes it flat in 0.063 s. Calculate the emf that was induced between the ends of the wire while it was being crushed.

55. ‖ A science museum exhibit illustrates induced currents by lighting a bulb when a 200-turn square coil of wire, 8.5 cm on a side, is pulled out of the 0.15 T magnetic field created by a large permanent magnet. The bulb, which sits on top of the coil, is connected to the two ends of the coil's wire and completes the circuit. The total resistance of the coil and the bulb is 4.5 Ω.
   a. When the coil is in the magnet, the magnetic field completely fills the area inside the coil and is zero outside. The coil is then pulled out of the field at a speed of 2.5 m/s. What is the emf induced in the coil as it is withdrawn?
   b. What is the current in the bulb as the coil is withdrawn from the field?

56. ‖ For your lab final, you've been given a 10-m-long wire and three plastic cylinders having radii of 4.0 mm, 1.0 cm, and 5.0 cm. Your task is to make a coil, by wrapping the wire around a cylinder, that will produce the largest induced emf when placed in a magnetic field that falls from 0.40 T to 0.10 T in 2.5 s. Because the wire must make an integer number of turns around the cylinder, you can trim off any excess wire that doesn't make a full turn. Which cylinder should you use?

57. ‖ A 20-cm-long, zero-resistance wire is pulled outward, on
INT zero-resistance rails, at a steady speed of 10 m/s in a 0.10 T magnetic field. (See Figure P25.57.) On the opposite side, a 1.0 Ω carbon resistor completes the circuit by connecting the two rails. The mass of the resistor is 50 mg.
   a. What is the induced current in the circuit?
   b. How much force is needed to pull the wire at this speed?
   c. How much does the temperature of the carbon increase if the wire is pulled for 10 s? The specific heat of carbon is 710 J/kg · K. Neglect thermal energy transfer out of the resistor.

FIGURE P25.57 **FIGURE P25.58**

58. ‖ The 10-cm-wide, zero-resistance wire shown in Figure P25.58 is pushed toward the 2.0 Ω resistor at a steady speed of 0.50 m/s. The magnetic field strength is 0.50 T.
   a. How big is the pushing force?
   b. How much power does the pushing force supply to the wire?
   c. What are the direction and magnitude of the induced current?
   d. How much power is dissipated in the resistor?

59. ‖‖‖ Experiments to study
BIO vision often need to track the movements of a subject's eye. One way of doing so is to have the subject sit in a magnetic field while wearing special contact lenses that have a coil of very fine wire circling the

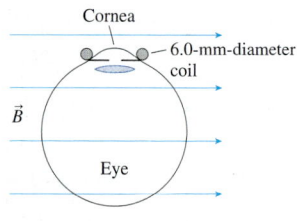

FIGURE P25.59

edge. A current is induced in the coil each time the subject rotates his eye. Consider an experiment in which a 20-turn, 6.0-mm-diameter coil of wire circles the subject's cornea while a 1.0 T magnetic field is directed as shown in Figure P25.59. The subject begins by looking straight ahead. What emf is induced in the coil if the subject shifts his gaze by 5.0° in 0.20 s?

60. ‖ Assume that a 100 W light bulb radiates all its energy as a single wavelength of visible light. Estimate the electric and magnetic field strength at the surface of the bulb.

61. ‖ A LASIK vision correction system uses a laser that emits
BIO 10-ns-long pulses of light, each with 2.5 mJ of energy. The
INT laser is focused to a 0.85-mm-diameter circle. (a) What is the average power of each laser pulse? (b) What is the electric field strength of the laser light at the focus point?

62. ‖‖‖‖ A digital cell phone emits 0.60 W of 1.9 GHz radio waves. What is the wave intensity at a distance of 10 cm? What is the amplitude of the electric field at this distance?

63. ‖ When the Voyager 2 spacecraft passed Neptune in 1989, it was 4.5×10^9 km from the earth. Its radio transmitter, with which it sent back data and images, broadcast with a mere 21 W of power. Assuming that the transmitter broadcast equally in all directions,
   a. What signal intensity was received on the earth?
   b. What electric field amplitude was detected?
   (The received signal was slightly stronger than your result because the spacecraft used a directional antenna.)

64. ‖‖‖‖ In reading the instruction manual that came with your garage-door opener, you see that the transmitter unit in your car produces a 250 mW signal and that the receiver unit is supposed to respond to a radio wave of the correct frequency if the electric field amplitude exceeds 0.10 V/m. You wonder if this is really true. To find out, you put fresh batteries in the transmitter and start walking away from your garage while opening and closing the door. Your garage door finally fails to respond when you're 42 m away. Are the manufacturer's claims true?

65. ‖ Unpolarized light of intensity I_0 is incident on three polarizing filters. The axis of the first is vertical, that of the second is 45° from vertical, and that of the third is horizontal. What light intensity emerges from the third filter?

66. ‖ a. What is the wavelength of a gamma-ray photon with energy 1.0×10^{-13} J?
   b. How many visible-light photons with a wavelength of 500 nm would you need to match the energy of this one gamma-ray photon?

67. ‖ Gamma rays with the very high energy of 2.0×10^{13} eV are occasionally observed from distant astrophysical sources. What are the wavelength and frequency corresponding to this photon energy?

68. ‖ A 1000 kHz AM radio station broadcasts with a power of 20 kW. How many photons does the transmitting antenna emit each second?

69. | Fireflies emit flashes of light to communicate, directly con-
BIO verting chemical energy into the energy of visible-light pho-
    tons. A firefly emits a 0.15-s-long pulse of light with an
    average wavelength of 590 nm. During the flash, the emitted
    power is 40 μW.
    a. What is the energy of one light photon having the average wavelength?
    b. Assume that all photons have the average wavelength. How many photons does the firefly emit during each flash?

Passage Problems

The Metal Detector

Metal detectors use induced currents to sense the presence of any metal—not just magnetic materials such as iron. A metal detector, shown in Figure P25.70, consists of two coils: a transmitter coil and a receiver coil. A high-frequency oscillating current in the transmitter coil generates an oscillating magnetic field along the axis and a changing flux through the receiver coil. Consequently, there is an oscillating induced current in the receiver coil.

FIGURE P25.70

If a piece of metal is placed between the transmitter and the receiver, the oscillating magnetic field in the metal induces eddy currents in a plane parallel to the transmitter and receiver coils. The receiver coil then responds to the superposition of the transmitter's

magnetic field and the magnetic field of the eddy currents. Because the eddy currents attempt to prevent the flux from changing, in accordance with Lenz's law, the net field at the receiver decreases when a piece of metal is inserted between the coils. Electronic circuits detect the current decrease in the receiver coil and set off an alarm.

70. | The metal detector will not detect insulators because
    A. Insulators block magnetic fields.
    B. No eddy current can be produced in an insulator.
    C. No emf can be produced in an insulator.
    D. An insulator will increase the field at the receiver.

71. | A metal detector can detect the presence of metal screws used to repair a broken bone inside the body. This tells us that
    A. The screws are made of magnetic materials.
    B. The tissues of the body are conducting.
    C. The magnetic fields of the device can penetrate the tissues of the body.
    D. The screws must be perfectly aligned with the axis of the device.

72. | Suppose the magnetic field from the transmitter coil in Figure P25.70 points toward the receiver coil and is increasing with time. As viewed along this axis, the induced currents are
    A. Clockwise in the metal, clockwise in the receiver coil.
    B. Clockwise in the metal, counterclockwise in the receiver coil.
    C. Counterclockwise in the metal, clockwise in the receiver coil.
    D. Counterclockwise in the metal, counterclockwise in the receiver coil.

73. | Which of the following changes would *not* produce a larger eddy current in the metal?
    A. Increasing the frequency of the oscillating current in the transmitter coil.
    B. Increasing the magnitude of the oscillating current in the transmitter coil.
    C. Increasing the resistivity of the metal.
    D. Decreasing the distance between the metal and the transmitter.

STOP TO THINK ANSWERS

Stop to Think 25.1: E. According to the right-hand rule, the magnetic force on a positive charge carrier is to the right.

Stop to Think 25.2: D. The field of the bar magnet emerges from the north pole and points upward. As the coil moves toward the pole, the flux through it is upward and increasing. To oppose the increase, the induced field must point downward. This requires a clockwise (negative) current. As the coil moves away from the pole, the upward flux is decreasing. To oppose the decrease, the induced field must point upward. This requires a counterclockwise (positive) current.

Stop to Think 25.3: E. The right-hand rule requires $\vec{B}$ to point into the page if the wave is to propagate upward.

Stop to Think 25.4: $I_D > I_A > I_B = I_C$. The intensity depends upon $\cos^2\theta$, where θ is the angle *between* the axes of the two filters. The filters in D have $\theta = 0°$, so all light is transmitted. The two filters in both B and C are crossed ($\theta = 90°$) and transmit no light at all.

Stop to Think 25.5: A. The photon energy is proportional to the frequency. The photons of the 90.5 MHz station each have lower energy, so more must be emitted per second.

Stop to Think 25.6: D. A hotter object emits more radiation across the *entire* spectrum than a cooler object. The 6000 K star has its maximum intensity in the blue region of the spectrum, but it still emits more red radiation than the somewhat cooler stars.

26

AC CIRCUITS

Transmission lines carry alternating current at voltages as high as 500,000 V. Why are such high voltages used? And why can birds perch safely on such high-voltage wires?

Looking Ahead ▶▶

The goal of Chapter 26 is to understand and apply basic ideas of AC circuits. In this chapter you will learn to:

▶ Calculate the AC current in resistors, capacitors, and inductors.
▶ Understand how electricity gets to and is used in the home.
▶ Understand the circumstances under which electricity can be dangerous.
▶ Understand inductors and inductance.
▶ Analyze *RLC* circuits and discuss their resonance.

Looking Back ◀◀

The material in this chapter depends on the fundamentals of circuits, on the properties of resistors and capacitors, and on electromagnetic induction. The mathematical representation of AC circuits is based on simple harmonic motion. Please review:

◀ Sections 14.1 and 14.5–14.6 Simple harmonic motion, damped oscillations, and resonance.
◀ Section 21.8 Capacitors.
◀ Sections 23.2 and 23.5 Fundamentals of circuit analysis and household electricity.
◀ Section 25.3–25.4 Lenz's law and Faraday's law.

Thomas Edison built the first large-scale electric generating station in 1882, in New York City. His entrepreneurial motive was to sell light bulbs, which he had invented a few years earlier. Edison's company, which later became General Electric, is still one of the largest manufacturers of electrical equipment.

Edison soon had competition from George Westinghouse, another name that probably looks familiar. Whereas Edison's system used *direct current* (DC), Westinghouse favored *alternating current* (AC). The technological debate between these two men and their companies lasted nearly 20 years, but eventually alternating current proved to be superior for the long-distance transmission of electric energy.

Today, more than a century later, a "grid" of AC electrical distribution systems spans the United States and other countries. Any device that plugs into an electric outlet uses an AC circuit. In this chapter, you will learn some of the basic techniques for analyzing AC circuits. You'll also learn when electricity can be dangerous and how safety features are built into your home wiring and the appliances you use.

26.1 Alternating Current

One of the examples of Faraday's law in Chapter 25 was an electric generator. A turbine, which might be powered by expanding steam or falling water, causes a coil of wire to rotate in a magnetic field. As the coil spins, the emf and the induced current oscillate sinusoidally. The emf is alternately positive and then negative, causing the charges to flow in one direction and then, a half cycle later, in the other. The oscillation frequency in North and South America is $f = 60$ Hz, whereas most of the rest of the world uses a 50 Hz oscillation.

The generator's emf—its voltage—is determined by the magnetic field strength and the number of turns in the generator coil, so it might seem logical to call a generator an *alternating voltage source*. Nonetheless, circuits powered by a sinusoidal emf are called **AC circuits,** where AC stands for *alternating current.* In contrast, the steady-current circuits you studied in Chapter 23 are called **DC circuits,** for *direct current.*

AC circuits are not limited to the use of 50 Hz or 60 Hz power-line voltages. Audio, radio, television, and telecommunication equipment all make extensive use of AC circuits, with frequencies ranging from approximately 10^2 Hz in audio circuits to approximately 10^9 Hz in cell phones. These devices use *electrical oscillators* rather than generators to produce a sinusoidal emf, but the basic principles of circuit analysis are the same.

You can think of an AC generator or oscillator as a battery whose output voltage undergoes sinusoidal oscillations. The instantaneous emf of an AC generator or oscillator, shown graphically in Figure 26.1, can be written

$$\mathcal{E} = \mathcal{E}_0\cos(2\pi f t) = \mathcal{E}_0\cos\left(\frac{2\pi t}{T}\right) \qquad (26.1)$$

SINUSOIDAL
p. 460

Emf of an AC voltage source

where $\mathcal{E}_0$ is the peak or maximum emf (recall that the units of emf are volts), T is the period of oscillation, and $f = 1/T$ is the oscillation frequency (cycles per second). As Figure 26.1 and Equation 26.1 suggest, the mathematics of AC circuit analysis are going to be very similar to the mathematics of simple harmonic motion.

Resistor Circuits

In Chapter 23, you learned to analyze a circuit in terms of the current I and potential difference ΔV. Now, because the current and voltage are oscillating, we will use a lowercase i to represent the *instantaneous* current through a circuit element, that is, the value of the current at a particular instant of time. Similarly, we will use a lowercase v for the circuit element's instantaneous voltage.

Figure 26.2 shows the instantaneous current i_R through a resistor R. The potential difference across the resistor, which we call the *resistor voltage* v_R, is given by Ohm's law:

$$v_R = i_R R \qquad (26.2)$$

The potential *decreases* in the direction of the current.

Figure 26.3 shows a resistor R connected across an AC emf $\mathcal{E}$. The circuit symbol for an AC generator is —$\bigcirc$— . We can analyze this circuit in exactly the same way we analyzed a DC resistor circuit. Kirchhoff's loop law says that the sum of all the potential differences around a closed path is zero:

$$\sum \Delta V = \Delta V_{\text{source}} + \Delta V_R = \mathcal{E} - v_R = 0 \qquad (26.3)$$

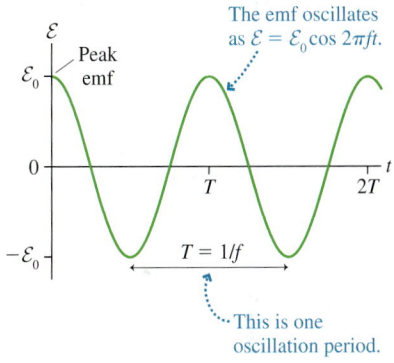

FIGURE 26.1 The emf of an AC voltage source.

FIGURE 26.2 The instantaneous current through a resistor.

FIGURE 26.3 An AC resistor circuit.

The minus sign appears, just as it did in the equation for a DC circuit, because the potential *decreases* when we travel through a resistor in the direction of the current. Thus we find from the loop law that $v_R = \mathcal{E} = \mathcal{E}_0 \cos 2\pi ft$. This isn't surprising, because the resistor is connected directly across the terminals of the emf.

Because the resistor voltage is a sinusoidal voltage at frequency f, it will be useful to write

$$v_R = V_R \cos 2\pi ft \tag{26.4}$$

where V_R is the peak or maximum voltage. You can see that $V_R = \mathcal{E}_0$ in the single-resistor circuit of Figure 26.3. Thus the current through the resistor is

$$i_R = \frac{v_R}{R} = \frac{V_R \cos 2\pi ft}{R} = I_R \cos 2\pi ft \tag{26.5}$$

where $I_R = V_R/R$ is the peak current.

> **NOTE** ▶ It is important to understand the distinction between instantaneous and peak quantities. The instantaneous current i_R, for example, is a quantity that is changing with time according to Equation 26.5. The peak current I_R is the maximum value that the instantaneous current reaches. Thus the instantaneous current oscillates between $+I_R$ and $-I_R$. ◀

The resistor's instantaneous current and voltage are *in phase,* both oscillating as $\cos 2\pi ft$. Figure 26.4 shows the voltage and the current simultaneously on a graph. The fact that the current peak current I_R is drawn as being less than V_R has no significance. Current and voltage are measured in different units, so in a graph like this you can't compare the value of one to the value of the other. Showing the two different quantities on a single graph—a tactic that can be misleading if you're not careful—is simply to show that they oscillate in phase: **The current is at its maximum value when the voltage is at its maximum, and the current is at its minimum value when the voltage is at its minimum.**

FIGURE 26.4 Graph of the current through and voltage across a resistor.

AC Power in Resistors

In Chapter 23, you also learned that the power consumed by a resistor is $P = I\,\Delta V_R = I^2 R$. In an AC circuit, the resistor current i_R and voltage v_R are constantly changing, as we saw in Figure 26.4, so the instantaneous power loss $p = i_R v_R = i_R^2 R$ (note the lowercase p) is constantly changing as well. We can use Equations 26.4 and 26.5 to write this instantaneous power as

$$p = i_R^2 R = (I_R \cos 2\pi ft)^2 R = I_R^2 R \cos^2 2\pi ft \tag{26.6}$$

Figure 26.5 shows the instantaneous power graphically. You can see that, because the cosine is squared, the power oscillates *twice* during every cycle of the emf: The energy dissipation peaks both when $i_R = I_R$ and when $i_R = -I_R$. This shows that energy dissipation doesn't depend on the current's direction through the resistor, a result that seems reasonable.

In many applications, we're not interested in the rapid fluctuations of the power shown in Figure 26.5. For instance, the current in a household light bulb reverses direction 120 times per second (twice per cycle), so the power has a maximum 120 times a second. In practice, we're more interested in the *average power* than in the instantaneous power. The **average power** P_R is the total energy dissipated by the resistor each second. You can see from Figure 26.5 that the average power is

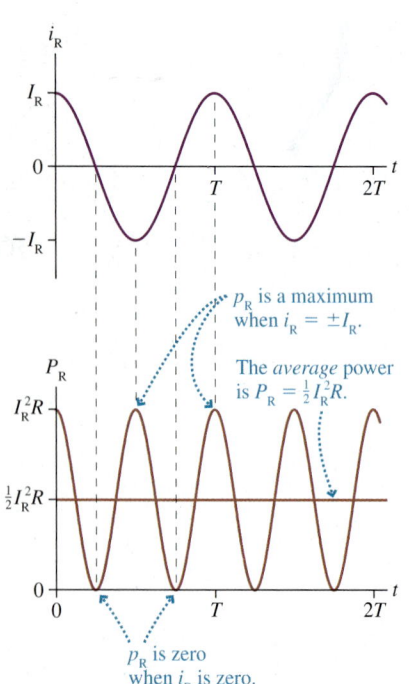

FIGURE 26.5 The instantaneous power loss in a resistor.

$$P_R = \frac{1}{2} I_R^2 R \tag{26.7}$$

Average power loss in a resistor with peak current I_R

INVERSE
p. 118

QUADRATIC
p. 50

In a DC circuit, the power is $P_R = I^2 R$. It's conventional to write Equation 26.7 in a form that resembles this DC expression. We do so by writing

$$P_R = \left(\frac{I_R}{\sqrt{2}}\right)^2 R = (I_{rms})^2 R \qquad (26.8)$$

where the quantity

$$I_{rms} = \frac{I_R}{\sqrt{2}} \qquad (26.9)$$

is called the **root-mean-square current,** or rms current, I_{rms}. Technically, an rms quantity is the square root of the average, or mean, of the quantity squared. We've seen this idea before in Chapter 11, where we introduced the rms speed of molecules in a gas.

The rms current allows us to compare Equation 26.8 directly to the energy dissipated by a resistor in a DC circuit: $P = I^2 R$. You can see that the average power loss of a resistor in an AC circuit with $I_{rms} = 1$ A is the same as in a DC circuit with $I = 1$ A. **As far as power is concerned, an rms current is equivalent to an equal DC current.**

Similarly, we can define the root-mean-square voltage:

$$V_{rms} = \frac{V_R}{\sqrt{2}} \qquad (26.10)$$

The resistor's average power loss can be written in terms of the rms quantities as

$$P_R = (I_{rms})^2 R = \frac{(V_{rms})^2}{R} = I_{rms} V_{rms} \qquad (26.11)$$

AC voltages and currents are usually given as rms values. For instance, household lamps and appliances in the United States operate at the power-line voltage of 120 V present at wall outlets. This voltage is the rms value $\mathcal{E}_{rms}$; the peak voltage is larger by a factor of $\sqrt{2}$, so $\mathcal{E}_0 = 170$ V.

EXAMPLE 26.1 The resistance and current of a light bulb
A lamp with a 100 W light bulb is plugged into a 120 V wall outlet.

a. What is the resistance of the bulb's filament?
b. What are the rms and peak currents through the lamp?

PREPARE The power rating of a light bulb is the average power dissipated by its filament when operated at 120 V. We model the filament as a resistance R, with an rms voltage $V_{rms} = 120$ V across it. We can then use Equation 26.11 to relate the average power to R and the rms values of the current and voltage.

SOLVE

a. From Equation 26.11 we have

$$R = \frac{(V_{rms})^2}{P_R} = \frac{(120 \text{ V})^2}{100 \text{ W}} = 144 \ \Omega$$

b. From Equation 26.8 we have $(I_{rms})^2 = P_R/R$, or

$$I_{rms} = \sqrt{\frac{P_R}{R}} = \sqrt{\frac{100 \text{ W}}{144 \ \Omega}} = 0.833 \text{ A}$$

From Equation 26.9, the peak current I_R is

$$I_R = \sqrt{2} I_{rms} = \sqrt{2}(0.833 \text{ A}) = 1.18 \text{ A}$$

ASSESS As long as we work with rms voltages and currents, all the expressions you learned for DC power carry over to AC power.

The "120 V" on this light bulb is its operating rms voltage. The 100 W stamp refers to its average power dissipation.

An AC current with a peak value of 1.0 A passes through a light bulb. A DC current of 1.0 A passes through a second, identical bulb. The power loss by the second bulb is

A. Two times that of the first bulb.   B. $\sqrt{2}$ times that of the first bulb.
C. Equal to that of the first bulb.   D. $1/\sqrt{2}$ times that of the first bulb.
E. Half that of the first bulb.

26.2 The Transmission and Use of Electricity

AC electricity won out over DC electricity because AC can be transmitted long distances with little energy loss. Only AC electricity can use *transformers* to boost the voltage to very high values for efficient long-distance transmission, then to decrease it to safe values for use in the home. A **transformer** is a device that takes an *alternating* voltage as an input, and produces either a higher or lower voltage as its output. The operation of a transformer depends on changing magnetic fields, and can be explained using Faraday's law.

Figure 26.6 shows a simplified version of a transformer, consisting of two coils of wire wrapped on an iron core. The left coil is called the **primary coil** (or simply the *primary*). It has N_1 turns and is driven by an AC voltage of amplitude V_1. The magnetic field of the primary follows the iron core and passes through the right coil, which has N_2 turns and is called the **secondary coil** (or the *secondary*). The alternating current through the primary coil causes an oscillating magnetic field in the iron core. The field lines are then guided by the core through the secondary, where their changing flux induces an emf in the secondary coil. The induced emf of the secondary coil is delivered, as an AC voltage of amplitude V_2, to a *load* with a resistance R.

How is the voltage V_1 at the primary related to the voltage V_2 at the secondary? Suppose at some instant of time the instantaneous current through the primary is i_1. This current creates a magnetic flux Φ through the primary coil. This flux is changing, because i_1 is changing, and the change induces an emf across the coil. According to Faraday's law (Equation 25.11), the instantaneous voltage v_1 across the N_1 turns of the primary coil is

$$v_1 = \mathcal{E}_1 = N_1 \frac{\Delta \Phi}{\Delta t} \tag{26.12}$$

Ideally, all the flux through the primary coil is guided by the iron core through the secondary coil. Consequently, the rate at which the flux changes through the secondary coil is also $\Delta\Phi/\Delta t$. This changing flux induces an emf across the secondary coil given by

$$v_2 = \mathcal{E}_2 = N_2 \frac{\Delta \Phi}{\Delta t} \tag{26.13}$$

Because $\Delta\Phi/\Delta t$ is the same in Equations 26.12 and 26.13, we can write

$$\frac{v_1}{v_2} = \frac{N_1}{N_2} \tag{26.14}$$

Equation 26.14 gives the ratio of the instantaneous voltages. But this equation also applies at the instant at which these voltages are at their peak values V_1 and V_2, so we can write

$$V_2 = \frac{N_2}{N_1} V_1 \qquad \text{or} \qquad (V_2)_{\text{rms}} = \frac{N_2}{N_1} (V_1)_{\text{rms}} \tag{26.15}$$

Transformer voltages for primary and secondary coils with N_1 and N_2 turns

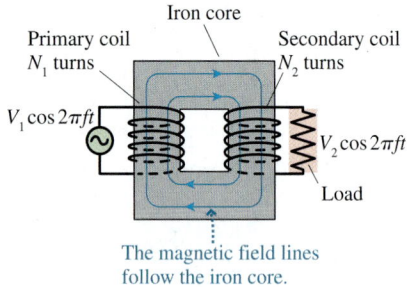

Iron core
Primary coil N_1 turns
Secondary coil N_2 turns
$V_1 \cos 2\pi ft$
$V_2 \cos 2\pi ft$
Load

The magnetic field lines follow the iron core.

FIGURE 26.6 A transformer.

▶ **Getting a charge** The batteries in some recent devices, such as the electric toothbrush shown, get charged from their base unit without having any metal contacts to carry current. The charging occurs via transformer action. Primary coils in the base create an AC magnetic field that induces an alternating emf in a secondary coil in the brush's handle. This emf, after conversion to DC, is used to charge the toothbrush's battery. With no metal contacts to corrode or get dirty, this method of charging is quite reliable.

Depending on the ratio N_2/N_1, the voltage V_2 across the load can be *transformed* to a higher or a lower voltage than V_1. A *step-up transformer,* with $N_2 > N_1$, increases the voltage while a *step-down transformer* $(N_2 < N_1)$ lowers the voltage.

Equation 26.15 relates the voltage at a transformer's secondary to the voltage at its primary. We can also relate the *currents* in the secondary and primary by using considerations of energy. Suppose a transformer is connected to a load, such as a resistor, that draws an rms current $(I_2)_{rms}$ from the secondary. The average *power* supplied to the load by the transformer, given by Equation 26.11, is

$$P_2 = (V_2)_{rms}(I_2)_{rms}$$

Now the primary draws a current $(I_1)_{rms}$ from the voltage source to which it's connected. This source provides power

$$P_1 = (V_1)_{rms}(I_1)_{rms}$$

to the transformer.

In an **ideal transformer** for which no electrical energy is lost, $P_1 = P_2$, or $(V_1)_{rms}(I_1)_{rms} = (V_2)_{rms}(I_2)_{rms}$. Thus we have

$$(I_2)_{rms} = \frac{(V_1)_{rms}}{(V_2)_{rms}}(I_1)_{rms} = \frac{N_1}{N_2}(I_1)_{rms} \qquad (26.16)$$

Here we used Equation 26.15 to relate the ratio of voltages to the ratio of turns. Comparing Equations 26.15 and 26.16, you can see that a step-up transformer *raises* voltage but *lowers* current. It must, to conserve energy. Similarly, a step-down transformer increases the current.

EXAMPLE 26.2 Analyzing a step-up transformer

A transformer has 100 turns on its primary and 500 turns on its secondary, which is connected to a 10 kΩ resistor. The resistor consumes an average power of 0.20 W.

a. What is the rms voltage at the transformer's primary?
b. How much current flows in the primary coil?

PREPARE Assume an ideal transformer. We can use Equation 26.11 to relate the average power loss to the rms voltage at the secondary. Then we can use Equation 26.15 to find the primary voltage.

SOLVE

a. The power consumed by the resistor is $P_R = (V_2)_{rms}^2/R$, so

$$(V_2)_{rms} = \sqrt{P_R R} = \sqrt{(0.20 \text{ W})(10^4 \ \Omega)} = 45 \text{ V}$$

Then from Equation 26.15 we have

$$(V_1)_{rms} = \left(\frac{N_1}{N_2}\right)(V_2)_{rms} = \left(\frac{100}{500}\right)(45 \text{ V}) = 9.0 \text{ V}$$

b. For an ideal transformer, the input power equals the output power. Thus we have

$$P_1 = (I_1)_{rms}(V_1)_{rms} = P_2$$

so that

$$(I_1)_{rms} = \frac{P_2}{(V_1)_{rms}} = \frac{0.20 \text{ W}}{9.0 \text{ V}} = 0.022 \text{ A}$$

ASSESS Because its secondary has more turns than its primary, this is a step-up transformer. Its secondary voltage is higher than the primary voltage but, to conserve energy, its secondary current is *lower* than the primary current.

Power Transmission

The transmission of electrical power over long distances is very inefficient at low voltages. The efficiency can be greatly increased by stepping up the voltage from the power station to very high voltages, such as 250 kV or even 500 kV (half a million volts!). To understand why, consider the simple model of a power plant, transmission line, and city shown in Figure 26.7. To provide power to the city, the power plant generates an emf $\mathcal{E}$ and delivers a current I_{rms} through the wires. If the power plant is 50 km from the city, a fairly typical distance, the 100 km of copper wire used in the transmission line has a resistance of about 7 Ω.

FIGURE 26.7 How electrical power is delivered from the power plant to a city.

A modest-sized city of 100,000 inhabitants uses approximately 120 MW of electric power. If this power were transmitted at the outlet voltage of 120 V, we can use Equation 26.10 to find that the current in the transmission line would have to be

$$I_{rms} = \frac{P_{city}}{V_{rms}} = \frac{120 \times 10^6 \text{ W}}{120 \text{ V}} = 10^6 \text{ A}$$

This is an extraordinarily large current. If a 10^6 A current passed through a transmission line with 7 Ω of resistance, the power loss in the wire would be

$$P_{wire} = (I_{rms})^2 R = (10^6 \text{ A})^2 (7 \text{ Ω}) = 7 \times 10^{12} \text{ W}$$

In other words, 60,000 times more power would be lost in the transmission lines than is used by the city! Of course, this much current would immediately melt any real power line.

But suppose we use a step-up transformer at the power plant to boost the voltage to 500,000 V. To transmit 120 MW of power at 500 kV requires a current of only

$$I_{rms} = \frac{P_{city}}{V_{rms}} = \frac{120 \times 10^6 \text{ W}}{500,000 \text{ V}} = 240 \text{ A}$$

This is still a large current, but one that can be handled by transmission line cables with a diameter of roughly 1 in. At this current, the power loss in the transmission line is $P_{wire} = (I_{rms})^2 R = 400$ kW. This is less than 1% of the power used by the city, an acceptable loss.

Because power loss in the transmission lines depends on the *square* of the current, higher voltages and thus lower currents provide a huge improvement in efficiency. Step-down transformers at the city then lower the voltage to 120 V. Only AC electricity can use transformers to change voltages and currents, making AC the clear preference for long-distance transmission of electric power.

Household Electricity

Once the 500 kV transmission line reaches your city, large transformers step the voltage down to 7200 V for distribution around the city. Near your house, smaller transformers step the voltage down further to 120 V (rms); you may have noticed these gray cylinders atop utility poles in your neighborhood.

Figure 26.8 shows how the transformer works. The secondary of the transformer produces an rms voltage of 240 V. However, the secondary coil has a

FIGURE 26.8 Converting distribution voltages to household voltages.

center tap that is connected to ground, at 0 V. With this center tap at 0 V, when the instantaneous voltage at one end of the secondary is v, the voltage at the other end is at $-v$. In other words, the two ends of the secondary windings are *out of phase*, as shown in the graph in Figure 26.8.

Thus one of the three wires exiting the transformer secondary—the *neutral* wire—is at ground, or 0 V. The other two wires, called the *hot* wires, each oscillate at 60 Hz with a 120 V rms potential difference with respect to ground. Because they're out of phase, the rms potential difference between the two hot wires is 240 V.

The path followed by these three wires as they enter your house is shown in Figure 26.9. After passing through an electric meter, which measures the total electric energy used, the wires enter the *load center*. The red and black wires are the two hot wires, and each is connected to a metal *bus*. Both bus A and bus B are at 120 V AC, but, as we've seen, these two voltages are out of phase with each other. The white wire is from the grounded center tap of the transformer; it is grounded again inside the house and so is at 0 V.

FIGURE 26.9 Wiring inside the home.

The wiring that is present throughout the home is organized into *circuits*. A circuit is a complete loop from one of the two 120 V buses, through an outlet or lighting fixture, and back to the grounded bar in the load center. A circuit typically serves several outlets that, as you learned in Chapter 23, must be connected in *parallel* so that each works when the others are not being used. To balance the load, about half the circuits in your house operate from the A bus and half from the B bus.

Appliances such as clothes dryers and stove cooktops require large amounts of electrical power. They would require a very large current if they operated at 120 V. Instead of being connected between one 120 V bus and neutral, the heating elements of such appliances are connected (using a different type of plug) between the two 120 V buses. As we've seen, the AC voltage between these two buses is 240 V, because they are out of phase. With twice the voltage, an appliance need only draw half the current to get the same power. (The motor and electronics of these appliances are powered by 120 V between one bus and neutral.)

FIGURE 26.10 High voltages are not necessarily dangerous.

26.3 Biological Effects and Electrical Safety

You can handle an ordinary 9 V battery without the slightest danger, but the 120 volts from an electrical outlet can lead to a nasty shock or even death. Yet the girl in Figure 26.10 is safely touching a Van de Graaff generator at a potential of 400,000 V. What makes electricity either safe or dangerous?

Although "voltage kills" is a common belief, in reality **it is currents passing through the body that are dangerous.** Current—the flow of charges through the body—is what produces physiological effects and damage because it mimics nerve impulses and causes muscles to involuntarily contract. Current also causes burns from the $I^2 R$ resistive heating as it passes through the body.

Higher voltages are generally more dangerous than lower voltages because they tend to produce larger currents, but the amount of current also depends on the body's resistance and on the ability of the voltage source to deliver current. The Van de Graaff generator in Figure 26.10 has a very high voltage, but it's charged very slowly and thus generates very little current. It may be a bit scary, making sparks and popping noises, but it's not dangerous. An innocent looking 120 V power line, although a much smaller voltage, can be lethal because it can deliver a much larger current.

Many electrocutions occur when the victim inadvertently grasps a wire or other metal object that is at a dangerous voltage. The current through the hand then causes involuntary contractions of the muscles of the fingers, forcing the hand to grasp the dangerous conductor even more tightly; the victim is unable to let go. The current at which this happens is called the "let-go" current. For DC, the let-go current is about 75 mA. The let-go current is less for 60 Hz AC, only around 15 mA, making alternating current more dangerous than an equal direct current.

Currents through the chest cavity are particularly dangerous because they can interfere with the proper rhythm of the heart. At an AC current of 60–100 mA, fibrillation of the heart can set in. The heart begins to beat in a rapid, uncontrolled fashion that it cannot stop on its own. As we saw in Chapter 21, a *defibrillator* is a medical device that sends a large DC pulse of current (several amps) through the heart, momentarily stopping the heart. The heart can better resume a normal beating pattern from a dead stop than when it is in an episode of fibrillation.

Table 26.1 summarizes the physiological effects of current.

We can model the body as several resistors in series. Because of its high saltwater content, the resistance of the interior of the body is low. This resistance depends somewhat on the current path through the body, but is about 500 Ω from a hand to the feet. But current must pass through the skin before getting inside the body, and the skin generally has a fairly high resistance. It is thus the body's primary defense against large currents. For dry skin grasping a wire, the skin's resistance might be >1 MΩ. When the skin is wet, this value can drop to <10 kΩ, and sweaty skin can further reduce this value.

Figure 26.11a on the next page shows a worker grasping a metal pipe that has accidentally touched a live 120 V wire. We can model this situation using the resistor circuit shown in Figure 26.11b. If his hands are sweaty, the resistance of his skin might be as low as 500 Ω. But he is wearing rubber-soled shoes and he is standing on a wood floor, both insulators. These add, say, 10 MΩ to the circuit. According to Ohm's law, the current through the worker is

$$I = \frac{\Delta V}{R_{\text{total}}} = \frac{\Delta V}{R_{\text{skin}} + R_{\text{body}} + R_{\text{shoes \& floor}}} \approx \frac{120 \text{ V}}{10 \text{ M}\Omega} \approx 12 \text{ }\mu\text{A}$$

TABLE 26.1 Physiological effects of currents passing through the body.

Physiological effect	AC current (mA)	DC current (mA)
Threshold of sensation	5	1
Let-go current	15	75
Paralysis of respiratory muscles	20	100
Heart fibrillation, likely fatal	>50	>300

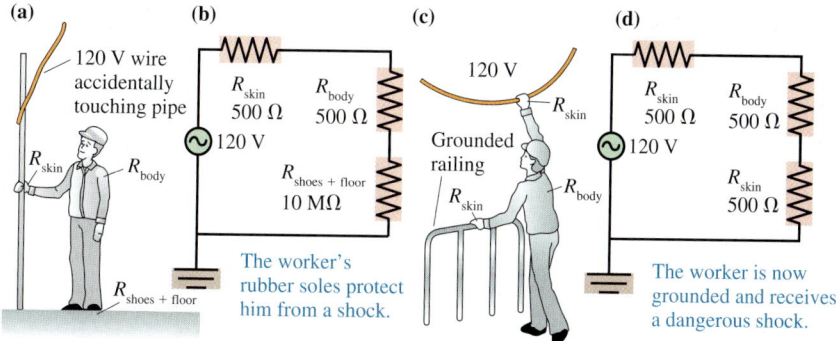

FIGURE 26.11 Depending on the situation, a live wire can be safe or dangerous.

The resistance is dominated by the shoes and the floor, and the worker would not receive a noticeable shock.

But suppose the worker had steadied himself by holding a grounded metal railing, shown in Figure 26.11c. The equivalent resistor circuit is shown in Figure 26.11d. Now the current is

$$I = \frac{\Delta V}{R_{total}} = \frac{\Delta V}{R_{skin} + R_{body} + R_{skin}} \approx \frac{120 \text{ V}}{1500 \text{ }\Omega} \approx 80 \text{ mA}$$

From Table 26.1 we see that this is a very dangerous, possibly fatal, current.

The lineman working on these high-voltage wires wears long rubber gloves. The gloves have a very high resistance, preventing any substantial current through him if a hot wire is accidentally touched.

CONCEPTUAL EXAMPLE 26.1 A bird on a high-voltage wire

Birds are often seen sitting on high-voltage electrical transmission lines. Why can they do this safely?

REASON A high-voltage line can be at 500,000 V, but this voltage is measured between *ground* and the wire. A bird's feet contact only the 500 kV line, so there is almost no potential difference between its feet.

To estimate this potential difference we can use the numbers given earlier for a transmission line. A 100-km-long transmission line has a resistance of 7 Ω and carries 200 A. Thus the voltage drop along the entire 100-km-long line is

$$\Delta V_{line} = IR = (200 \text{ A})(7 \text{ }\Omega) = 1400 \text{ V}$$

If the bird's feet are 2 cm apart, the voltage between its feet is only

$$\Delta V_{bird} = \left(\frac{2 \text{ cm}}{100 \text{ km}}\right)\Delta V_{line} = 2.8 \times 10^{-4} \text{ V}$$

This voltage is far too small to be dangerous.

ASSESS It is not potential that produces a current, but a potential *difference*. If an object were to be connected from the high-voltage line to ground, or between two different lines, then the large potential difference would cause an enormous current to flow. Indeed, birds are sometimes electrocuted if their wings contact two parallel wires.

▶ **The lightning crouch** BIO If you're caught outdoors in a lightning storm, and can't get to safety, experts recommend that you assume the crouched position shown in the photo. Many people are injured not by a direct strike, but by a nearby strike whose current travels through the ground, creating a large electric field E in the ground. If you are standing with your feet separated by a large distance d, a large potential difference $\Delta V = Ed$ will develop between your feet, causing a large current through your body. By placing your feet as close together as possible, as shown, and lifting your heels off the ground, you can minimize the potential difference and thus the current.

FIGURE 26.12 A standard plug will fit the outlet in only one orientation.

TRY IT YOURSELF

Testing GFI circuits You can test a GFI circuit by pressing the black "Test" button. This creates an electrical connection between the hot wire and the ground connection, so the current in the hot and neutral wires will not be equal. You should hear a click as the circuit disconnects. You can then reset the outlet by pressing the red button. If the outlet does not respond like this, the outlet should be replaced.

(a) The instantaneous current to and from the capacitor

The instantaneous capacitor voltage is $v_C = q/C$. The potential decreases from + to −.

(b)　$\mathcal{E} = \mathcal{E}_0 \cos 2\pi ft$

FIGURE 26.13 An AC capacitor circuit.

Household Electrical Safety

Household electricity is normally quite safe, but there are conditions under which it can be dangerous. To minimize these conditions, many safety devices are built into the electrical system in your home.

In Chapter 23, you learned about two important safety features built into every wall outlet. First, standard outlets are *polarized*. As Figure 26.12 shows, one slot (the neutral side) is a little wider than the other (the hot side), and the plug can fit only one way. This ensures, for example, that when you change a light bulb in a table lamp, the hot terminal is the little tab at the very bottom of the socket while the more easily contacted threaded part of the socket remains neutral.

Second, the metal case of many appliances is *grounded* through the third hole in the wall outlet. This contact is directly connected to ground, as shown by the copper-colored wires in Figure 26.9. If a hot wire inside the device should accidentally touch the case, a large current would flow to ground and trip the circuit breaker. Some household appliances and tools are now *double insulated* and don't require a grounding plug. These appliances have internal plastic insulation that makes it impossible for a broken internal wire to cause a shock.

GFI Circuits

If you are standing in good electrical contact with the ground, you are grounded. If you then touch a hot wire with your hand, the small resistance between your hand and ground could allow a dangerous current to pass through you. In kitchens and bathrooms, where grounding on damp floors is a good possibility, building codes require *ground fault interrupter* outlets, abbreviated GFI. Some devices, such as hair dryers, are constructed with a GFI in the power cord.

GFI outlets have a built-in sensing circuit that compares the current in the hot and neutral wires of the outlet. In normal operation, all the current coming in through the hot wire passes through the electrical device and then back out through the neutral. Thus the currents in the hot and neutral wires should always be equal. If the current in the hot wire does *not* equal the current in the neutral wire, some current from the hot wire is finding a path to ground—perhaps through a person. This is a *ground fault,* and the GFI disconnects the circuit. GFIs are set to trip at current differences of about 5 mA, which, as Table 26.1 shows, is less than the let-go current. GFIs also respond much more quickly—in only a few hundredths of a second—than a circuit breaker.

26.4 Capacitor Circuits

Figure 26.13a shows a current i_C charging a capacitor with capacitance C. The instantaneous capacitor voltage is $v_C = q/C$, where $\pm q$ is the charge on the two capacitor plates at this instant of time. It is useful to compare Figure 26.13a to Figure 26.2 for a resistor.

Figure 26.13b, where capacitance C is connected across an AC source of emf $\mathcal{E}$, is the most basic capacitor circuit. The capacitor is in parallel with the source, so the capacitor voltage equals the emf: $v_C = \mathcal{E} = \mathcal{E}_0 \cos 2\pi ft$. It will be useful to write

$$v_C = V_C \cos 2\pi ft \tag{26.17}$$

where V_C is the peak or maximum voltage across the capacitor. You can see that $V_C = \mathcal{E}_0$ in this single-capacitor circuit.

As Figure 26.13a shows, charge flows to and from the capacitor plates but not through the gap between the plates. Nonetheless, for circuit analysis it is convenient to think of the current as if it continues *through* the capacitor. We can find the current i_C through the capacitor by considering how the charge q on the capacitor varies with time. In Figure 26.14a we plot the oscillating voltage v_C across the capacitor. Because the charge and the capacitor voltage are directly proportional, with $q = Cv_C$, the graph of q, shown in Figure 26.14b, looks like the graph of v_C. We say that the charge is *in phase* with the voltage.

In Chapter 22, we defined current to be

$$i_C = \frac{\Delta q}{\Delta t}$$

where the subscript C arises because we're considering the current through the capacitor. The main factor determining i_C is Δq, the *change* of charge during a time interval Δt, not the amount of charge q. When charge is flowing rapidly onto the capacitor, the charge on the capacitor is *changing* rapidly and the current—the flow of charge—is large and positive. This is the situation at $t = 3T/4$ in Figure 26.14c. The current is large and negative at $t = T/4$ because that's when charge is flowing rapidly out of the capacitor as the voltage drops. Notice that the capacitor charge is *changing* most rapidly when $q \approx 0$.

When the capacitor is fully charged positive or negative, the total charge q may be large but there's very little *change* in q. No additional charges are flowing onto the capacitor, and the current is small. This is the situation at $t = 0$ and $t = T/2$ in Figure 26.14c.

A capacitor's voltage and current are *not* in phase, as they were for a resistor. You can see from Figure 26.14 that the current peaks one-quarter of a period *before* the voltage peaks. That is, the current peaks at $t = 3T/4$ whereas the voltage doesn't peak until $t = T$. Because sinusoidal oscillations are associated with motion around a circle, and one-quarter of a circle is 90°, we say that

The AC current through a capacitor *leads* the capacitor voltage by 90°.

Capacitive Reactance

The phase of the current is not the whole story. We would like to know its peak value, which we can also determine from $i_C = \Delta q/\Delta t$. Because $q = Cv_C$, this can also be written as

$$i_C = C\frac{\Delta v_C}{\Delta t} \tag{26.18}$$

Let's now reason by analogy. In Chapter 14, the position of a simple harmonic oscillator was given by $v = A\cos(2\pi ft)$. Further, the oscillator's velocity $v = \Delta x/\Delta t$ was found to be $x = -v_{max}\sin(2\pi ft)$, with a maximum or peak value $v_{max} = 2\pi fA$. Here we have an oscillating voltage $v_C = V_C\cos(2\pi ft)$, analogous to position, and we need to find the quantity $\Delta v_C/\Delta t$, analogous to velocity. Thus $\Delta v_C/\Delta t$ must have a maximum or peak value $(\Delta v_C/\Delta t)_{max} = 2\pi fV_C$.

The peak capacitor current I_C occurs when $\Delta v_C/\Delta t$ is maximum, so the peak current is

$$I_C = C(2\pi fV_C) = (2\pi fC)V_C \tag{26.19}$$

For a resistor, the peak current and voltage are related through Ohm's law:

$$I_R = \frac{V_R}{R}$$

We can write Equation 26.19 in a form similar to Ohm's law if we define **capacitive reactance** X_C to be

$$X_C = \frac{1}{2\pi fC} \tag{26.20}$$

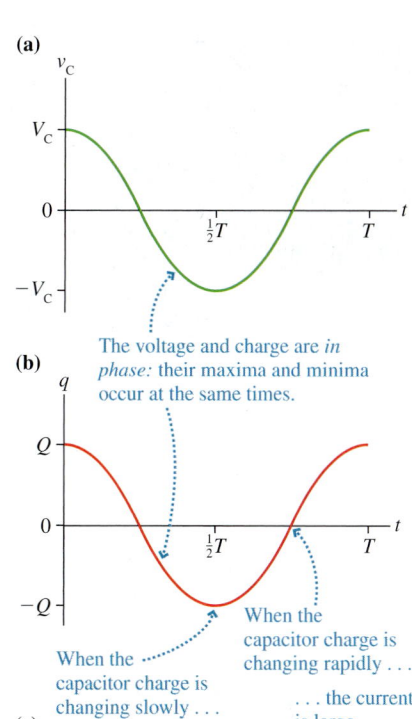

The voltage and charge are *in phase:* their maxima and minima occur at the same times.

When the capacitor charge is changing slowly . . .

. . . the current is small.

When the capacitor charge is changing rapidly . . .

. . . the current is large.

FIGURE 26.14 Voltage, charge, and current graphs for a capacitor in an AC circuit.

INVERSE
p. 118

◄**Replacing a mouse** Under the surface of the trackpad on your laptop computer is an array of thousands of tiny capacitors. An AC voltage source is applied to each, causing a current given by Equation 26.21. When your finger is brought near the surface, your finger's high dielectric constant (it's largely water, with $\kappa \approx 80$) changes the capacitance of those capacitors nearby, and hence the current in each. This tells the computer the location of your finger. (A plastic pen, with its low value of κ, usually won't affect a trackpad.)

With this definition of capacitive reactance, Equation 26.19 becomes

$$I_C = \frac{V_C}{X_C} \quad \text{or} \quad V_C = I_C X_C \qquad (26.21)$$

Peak current through or voltage across a capacitor

The units of reactance, like those of resistance, are ohms.

NOTE ▶ Reactance relates the *peak* voltage V_C and current I_C. It does *not* relate the *instantaneous* capacitor voltage and current because they are out of phase. That is, $v_C \neq i_C X_C$. ◄

A resistor's resistance R is independent of the emf frequency. In contrast, as Figure 26.15 shows, a capacitor's reactance X_C depends inversely on the frequency. The reactance becomes very large at low frequencies (i.e., the capacitor is a large impediment to current). This makes sense, because $f = 0$ would be a nonoscillating, DC circuit, and we know that a steady DC current cannot pass through a capacitor. The reactance decreases as the frequency increases until, at very high frequencies, $X_C \approx 0$ and the capacitor begins to act like an ideal wire. This result has important consequences for how capacitors are used in many circuits.

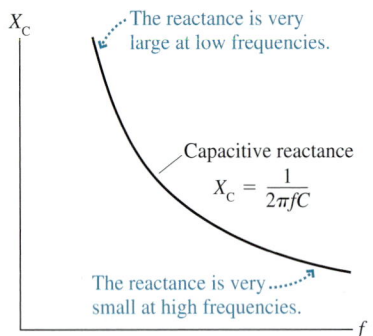

X_C
The reactance is very large at low frequencies.

Capacitive reactance
$X_C = \frac{1}{2\pi f C}$

The reactance is very small at high frequencies.

FIGURE 26.15 The capacitive reactance as a function of frequency.

EXAMPLE 26.3 Finding the capacitive reactance

What is the capacitive reactance of a 0.100 μF capacitor at a 100 Hz audio frequency and at a 100 MHz FM-radio frequency?

SOLVE At 100 Hz,

$$X_C(\text{at } 100 \text{ Hz}) = \frac{1}{2\pi f C} = \frac{1}{2\pi(100 \text{ Hz})(1.00 \times 10^{-7} \text{ F})}$$

$$= 15,900 \ \Omega$$

Increasing the frequency by a factor of 10^6 decreases X_C by a factor of 10^6, giving

$$X_C(\text{at } 100 \text{ MHz}) = 0.0159 \ \Omega$$

ASSESS A capacitor with a substantial reactance at audio frequencies has virtually no reactance at FM-radio frequencies.

EXAMPLE 26.4 Finding a capacitor's current

A 10 μF capacitor is connected to a 1000 Hz oscillator with a peak emf of 5.0 V. What is the peak current through the capacitor?

PREPARE Figure 26.13b showed the circuit diagram. It is a simple one-capacitor circuit.

SOLVE The capacitive reactance at $f = 1000$ Hz is

$$X_C = \frac{1}{2\pi f C} = \frac{1}{2\pi(1000 \text{ Hz})(10 \times 10^{-6} \text{ F})} = 16 \ \Omega$$

The peak voltage across the capacitor is $V_C = \mathcal{E}_0 = 5.0$ V, hence the peak current is

$$I_C = \frac{V_C}{X_C} = \frac{5.0 \text{ V}}{16 \ \Omega} = 0.31 \text{ A}$$

ASSESS Using reactance is just like using Ohm's law, but don't forget it applies only to the *peak* current and voltage, not the instantaneous values. Further, reactance, unlike resistance, depends on the frequency of the signal.

A capacitor is attached to an AC voltage source. Which change will result in a doubling of the current?

A. Halving the voltage and doubling the frequency.
B. Doubling the frequency.
C. Halving the frequency.
D. Doubling the voltage and halving the frequency.

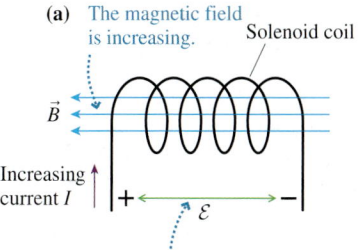

(a) The magnetic field is increasing.   Solenoid coil

$\vec{B}$

Increasing current I

$\mathcal{E}$

The increasing flux through the loop causes an emf to develop. By Lenz's law, the sign of the emf is such as to oppose further increases in I.

(b) The magnetic field is decreasing.

$\vec{B}$

Decreasing current

$\mathcal{E}$

The direction of the emf is opposite to the case where the current is increasing.

FIGURE 26.16 A changing current through a solenoid induces an emf across the solenoid.

26.5 Inductors and Inductor Circuits

Figure 26.16 shows a length of wire formed into a coil, making a solenoid. In Chapter 24 you learned that a current flowing in a solenoid creates a magnetic field inside the solenoid. If this current is *increasing,* as shown in Figure 26.16a, then the magnetic field—and thus the flux—inside the coil increases as well. According to Faraday's law, this changing flux causes an emf—a potential difference—to develop across the coil. The *direction* of the emf can be inferred from Lenz's law: Its direction must be such to *oppose* the increase in the flux; that is, it must oppose the increase in the current. The emf will have the opposite sign if the current through the coil is decreasing, as shown in Figure 26.16b.

Coils of this kind, called **inductors,** are widely used in AC circuits. The circuit symbol for an inductor is ⎓⎓⎓⎓. There are two primary things to remember about an inductor. First, **an inductor develops a potential difference across it if the current through it is *changing.*** Second, because the direction of this potential difference opposes the change in the current, **an inductor resists changes in the current through it.**

Inductance

By Faraday's law, the voltage developed across an inductor is proportional to the rate at which the flux through the coil changes. And, because the flux is proportional to the coil's current, the instantaneous inductor voltage v_L must be proportional to $\Delta i_L / \Delta t$, the rate at which the current through the inductor changes. Thus we can write

$$v_L = L \frac{\Delta i_L}{\Delta t} \qquad (26.22)$$

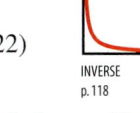

INVERSE
p. 118

The constant of proportionality L is called the **inductance** of the inductor. The inductance is determined by the shape and size of the coil. A coil with many turns, for instance, has a higher inductance than a similarly sized coil with fewer turns. Inductors often have an iron core inside their windings to increase their inductance. The magnetic field from the current magnetizes the iron core, which greatly increases the overall field through the windings. This gives a larger change in flux through the windings and hence a larger induced emf. Equation 26.22 shows that this implies a larger value of L.

From Equation 26.22 we see that inductance has units of V · s/A. It's convenient to define an SI unit of inductance called the **henry,** in honor of Joseph Henry, an early investigator of magnetism. We have

$$1 \text{ henry} = 1 \text{ H} = 1 \text{ V} \cdot \text{s/A} = 1 \ \Omega \cdot \text{s}$$

Is anybody up there? Detectors are often installed beneath the pavement to sense the presence of cars waiting at intersections. A slot is cut in the pavement in the familiar pattern shown in the photo. A wire is then sealed into the slot, forming an inductor consisting of a single loop. The steel in a car over the loop acts just like an iron core for an ordinary inductor, greatly increasing the inductance of the loop. This change in inductance signals that a car is present.

(a)

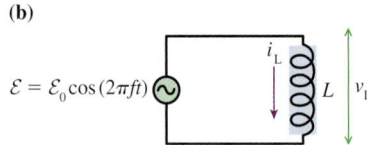

The instantaneous current through the inductor

i_L

L

$+$ $-$

The instantaneous inductor voltage is $v_L = L(\Delta i_L/\Delta t)$.

(b)

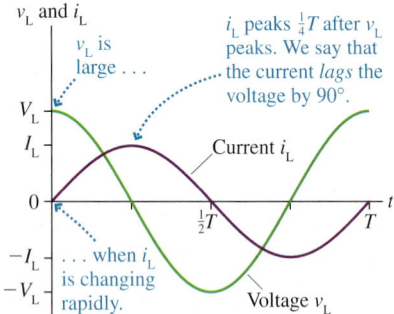

$\mathcal{E} = \mathcal{E}_0 \cos(2\pi ft)$

i_L

L v_L

FIGURE 26.17 Using an inductor in an AC circuit.

v_L and i_L

v_L is large . . .

i_L peaks $\frac{1}{4}T$ after v_L peaks. We say that the current *lags* the voltage by 90°.

V_L

I_L

Current i_L

0

$\frac{1}{2}T$ T t

$-I_L$

. . . when i_L is changing rapidly.

$-V_L$

Voltage v_L

FIGURE 26.18 Voltage and current graphs for an inductor in an AC circuit.

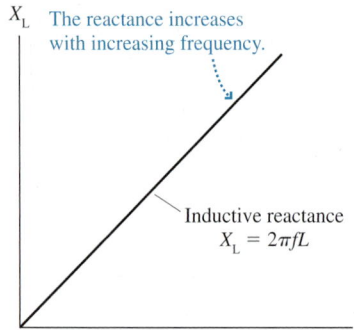

X_L The reactance increases with increasing frequency.

Inductive reactance $X_L = 2\pi fL$

f

FIGURE 26.19 The inductive reactance as a function of frequency.

This last definition of the henry, $1\ \text{H} = 1\ \Omega \cdot \text{s}$, is based on the definition of the ohm, $1\ \Omega = 1\ \text{V/A}$, and it will be useful later when we consider inductors in AC circuits. Practical inductances are usually in the range of millihenries (mH) or microhenries (μH).

Inductor Circuits

Figure 26.17a shows the instantaneous current i_L through an inductor. If the current is changing, the instantaneous inductor voltage is given by Equation 26.22.

Figure 26.17b, where inductance L is connected across an AC source of emf $\mathcal{E}$, is the simplest inductor circuit. The inductor is in parallel with the source, so the inductor voltage equals the emf: $v_L = \mathcal{E} = \mathcal{E}_0 \cos 2\pi ft$. We can write

$$v_L = V_L \cos 2\pi ft \qquad (26.23)$$

where V_L is the peak or maximum voltage across the inductor. You can see that $V_L = \mathcal{E}_0$ in this single-inductor circuit.

We can find the inductor current i_L by considering again Equation 26.22, which tells us that the inductor voltage is large when the current is changing rapidly (i.e., $\Delta i_L/\Delta t$ is large), and is small when the current is changing slowly (i.e., $\Delta i_L/\Delta t$ is small). From this, we can graphically find the current, as shown in Figure 26.18.

Just as for a capacitor, the current and voltage are not in phase. There is again a phase difference of 1/4 of a cycle, but for an inductor the current peaks one-quarter of a period *after* the voltage peaks. Thus we can say

The AC current through an inductor *lags* the inductor voltage by 90°.

Inductive Reactance

To find the relationship between the peak values of the inductor's current and voltage, we can use Equation 26.22 and, as we did for the capacitor, the analogy with simple harmonic motion. For the oscillating capacitor voltage, with peak value $(v_C)_{max} = V_C$, the maximum value of $\Delta v_C/\Delta t$ was $(\Delta v_C/\Delta t)_{max} = 2\pi f V_C$. If an oscillating inductor current has peak value $(i_L)_{max} = I_L$, then, by exactly the same reasoning, the maximum value of $\Delta i_L/\Delta t$ is $(\Delta i_L/\Delta t)_{max} = 2\pi f I_L$. If we use this result in Equation 26.22, we see that the maximum or peak value of the inductor voltage is

$$V_L = L(2\pi f I_L) = (2\pi fL)I_L \qquad (26.24)$$

We can write Equation 26.24 in a form reminiscent of Ohm's law as

$$I_L = \frac{V_L}{X_L} \quad \text{or} \quad V_L = I_L X_L \qquad (26.25)$$

where the **inductive reactance,** analogous to the capacitive reactance, is defined to be

$$X_L = 2\pi fL \qquad (26.26)$$

INVERSE
p. 118

Figure 26.19 shows that the inductive reactance increases linearly as the frequency increases. This makes sense. Faraday's law tells us that the induced voltage across a coil increases as the rate of change of $\vec{B}$ increases, and $\vec{B}$ is directly proportional to the inductor current. For a given peak current I_L, $\vec{B}$ changes more rapidly at higher frequencies than at lower frequencies, and thus V_L is larger at higher frequencies than at lower frequencies.

► **Cleaning up a power line** Computers generate high-frequency voltages, with frequencies in the range of a few MHz to a few GHz. These voltages can leak through the computer's power supply and propagate out through your household electricity, causing problems with other equipment that is sensitive to these high frequencies. To help prevent this, the power cord entering your computer often has an inductor attached in series; it's the gray cylinder seen in the photo. At high frequencies the inductive reactance of this inductor is very high and, according to Equation 26.25, the high-frequency current leaving the computer will be reduced accordingly.

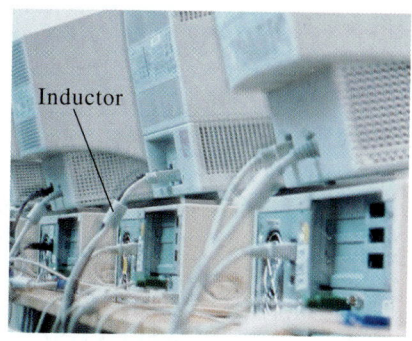

Inductor

EXAMPLE 26.5 Finding the current and voltage of a radio's inductor

A 0.25 μH inductor is used in an FM radio circuit that oscillates at 100 MHz. The current through the inductor reaches a peak value of 2.0 mA at $t = 5.0$ ns. What is the peak inductor voltage, and when, closest to $t = 5.0$ ns, does it occur?

PREPARE The inductor current lags the voltage by 90°, or, equivalently, the voltage reaches its peak value one-quarter period *before* the current. The circuit looks like Figure 26.17b.

SOLVE The inductive reactance at $f = 100$ MHz $= 1.0 \times 10^8$ Hz is

$$X_L = 2\pi f L = 2\pi (1.0 \times 10^8 \text{ Hz})(0.25 \times 10^{-6} \text{ H}) = 160 \ \Omega$$

Thus the peak voltage is $V_L = I_L X_L = (0.0020 \text{ A})(160 \ \Omega) = 0.32$ V. The voltage peak occurs one-quarter period before the current peaks, and we know that the current peaks at $t = 5.0$ ns. The period of a 100 MHz oscillation is 10 ns, so the voltage peaks at

$$t = 5.0 \text{ ns} - \frac{10 \text{ ns}}{4} = 2.5 \text{ ns}$$

STOP TO THINK 26.3 An inductor is attached to an AC voltage source. Which change will result in a halving of the current?

A. Halving the voltage and doubling the frequency.
B. Doubling the frequency.
C. Halving the frequency.
D. Doubling the voltage and halving the frequency.

26.6 Oscillation Circuits

Resistors, capacitors, inductors, and AC sources can be wired into circuits in an infinite variety of ways. Of particular importance are circuits in which the current and voltage oscillate. All telecommunication devices, from radios to television to cell phones, are based on these circuits.

A *Tesla coil* uses a resonant *LC* circuit to produce very large voltages, high enough to produce miniature lightning strikes.

LC Circuits

You learned in Chapter 23 that the voltage across a charged capacitor decays exponentially if the capacitor is connected to a resistor to form an *RC* circuit. Something very different occurs if the resistor is replaced with an *inductor*. Instead of decaying to zero, the capacitor voltage now undergoes sinusoidal *oscillations*.

To understand how this occurs, let's start with the capacitor and inductor shown in the **LC circuit** of Figure 26.20. Initially, the capacitor has charge Q_0 and there is no current in the inductor. Then, at $t = 0$, the switch is closed. How does the circuit respond?

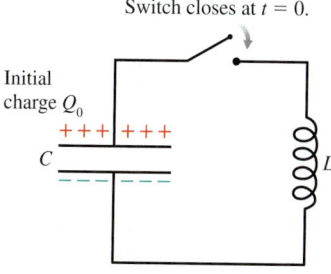

FIGURE 26.20 An *LC* circuit.

FIGURE 26.21 The capacitor charge oscillates much like a block attached to a spring.

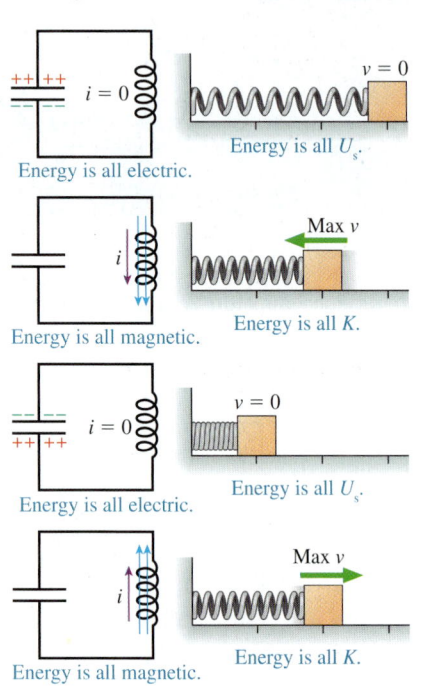

FIGURE 26.22 The energies of an *LC* circuit and a block on a spring.

As Figure 26.21 shows, the inductor provides a conducting path for discharging the capacitor. However, the discharge current has to pass through the inductor, and, as we've seen, an inductor resists changes in current. Consequently, the current doesn't stop when the capacitor charge reaches zero.

A block attached to a stretched spring is a useful mechanical analogy. Closing the switch to discharge the capacitor is like releasing the block. But the block doesn't stop when it reaches the origin. Its momentum keeps it going until the spring is fully compressed. Likewise, the current continues until it has recharged the capacitor with the opposite polarization. This process repeats over and over, charging the capacitor first one way, then the other. That is, the charge and current *oscillate*.

Let's further pursue this analogy between an *LC* circuit and a block on a spring. In Chapter 14 you learned that the energy of an oscillating block on a spring is constantly transformed back and forth between the kinetic energy *K* of the block and the potential energy U_s of the spring. As Figure 26.22 shows, the energy of an *LC* oscillator is also constantly transforming between the electric potential energy of the capacitor and the energy of the magnetic fields in the inductor. Recall also that the oscillation frequency of a mass *m* on a spring with spring constant *k* is

$$f = \frac{1}{2\pi}\sqrt{\frac{k}{m}}$$

The oscillation frequency thus depends only on the two basic parameters k and m of the system and not, for example, on the amplitude of the oscillation. Similarly, the frequency of an LC oscillator is determined solely by the values of its inductance and capacitance. We would expect larger values of L and C to cause an oscillation with a lower frequency, because a larger inductance means the current takes longer to change, and a larger capacitance takes longer to discharge. A detailed analysis shows that the frequency has a form quite similar to that for a mass and spring:

$$f = \frac{1}{2\pi\sqrt{LC}} \qquad (26.27)$$

Frequency of an LC oscillator

EXAMPLE 26.6 The frequency of an LC oscillator

An LC circuit consists of a 10 μH inductor and a 500 pF capacitor. What is the oscillator's frequency?

SOLVE From Equation 26.27 we have

$$f = \frac{1}{2\pi\sqrt{LC}} = \frac{1}{2\pi\sqrt{(10\times10^{-6}\,\text{H})(500\times10^{-12}\,\text{F})}} = 2.3\times10^{6}\,\text{Hz}$$

ASSESS The frequencies of LC oscillators are generally *much* higher than those of mechanical oscillators. Frequencies in the MHz or even GHz (10^{9} Hz) range are typical. All devices that send or receive electromagnetic waves use an LC circuit just like this to establish the correct frequency.

FIGURE 26.23 An RLC circuit.

The RLC Circuit

Once the oscillations of the ideal LC circuit of Figure 26.21 are started, they will continue forever. All real circuits, however, have some resistance, either from the small resistance of the wires that make up the circuit or from actual circuit resistors. These resistors dissipate energy, causing the amplitudes of the voltages and currents to decay.

We can model this situation by adding a series resistor R to our LC circuit, as shown in Figure 26.23. The circuit still oscillates, but the peak values of the voltage and current decrease with time as the current through the resistor transforms the electric and magnetic energy of the circuit into thermal energy. Because the power loss is proportional to R, a larger resistance causes the oscillations to decay more rapidly, as shown in Figure 26.24. This behavior is completely analogous to the damped harmonic oscillator discussed in Chapter 14.

To keep the circuit oscillating, we need to add an AC source to the circuit. This makes the **driven *RLC* circuit** of Figure 26.25. Because the reactances of the capacitor and inductor vary with the frequency of the AC source, the current in this circuit varies with frequency as well.

Recall that the reactance of a capacitor or inductor plays the same role for the peak quantities I and V as does the resistance of a resistor. When the reactance is large, the current through the capacitor or inductor is small. Because this is a series circuit, with the same current throughout, any circuit element with a large resistance or reactance can block the current.

If the AC source frequency f is very small, the capacitor's reactance $X_C = 1/(2\pi f C)$ is extremely large. If the source frequency f becomes very large, the inductor's reactance $X_L = 2\pi f L$ becomes extremely large. This has two consequences. First, the current in the driven RLC circuit will approach zero at very

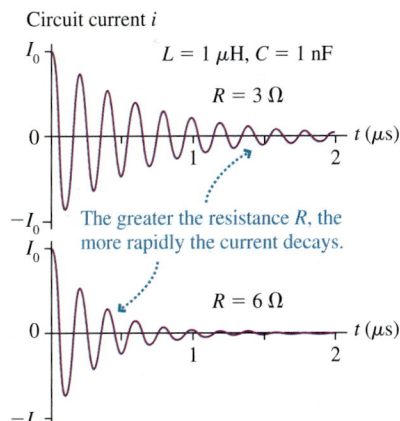

FIGURE 26.24 An RLC circuit exhibits damped oscillations.

FIGURE 26.25 A driven RLC circuit.

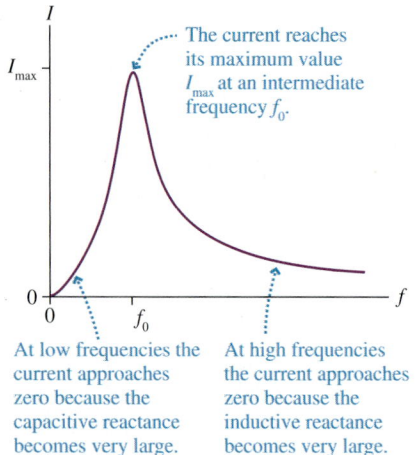

FIGURE 26.26 The current in a driven *RLC* circuit as a function of the source frequency *f*.

low and very high frequencies. Second, there must be some intermediate frequency, where neither X_C nor X_L is too large, at which the circuit current I is a maximum. This behavior of the current is shown in Figure 26.26.

Recall that a driven mechanical oscillator has a *resonance,* a large-amplitude response, at one particular frequency. You can see the same thing happening in Figure 26.26. The frequency f_0 at which the current is at its maximum value is called the **resonance frequency.**

Resonance occurs at the frequency at which the capacitive reactance equals the inductive reactance. If the reactances are equal, the capacitor voltage and the inductor voltage are the same. However, these two voltages are each 90° out of phase with the current, with the current *leading* the capacitor voltage by 90° and *lagging* the inductor voltage by 90°. Thus the capacitor and inductor voltages have equal magnitudes but are 180° out of phase with each other. When we add all the voltages around the loop in Kirchhoff's loop law, the capacitor and inductor voltages cancel and the current is then limited only by the resistance R. Thus the condition for resonance is $X_C = X_L$, or

$$\frac{1}{2\pi f_0 C} = 2\pi f_0 L$$

which gives

$$f_0 = \frac{1}{2\pi \sqrt{LC}} \tag{26.28}$$

Note that this is the same frequency at which the circuit would oscillate if it had no resistor; that is, if it were an ideal *LC* circuit.

At resonance the current is limited only by the resistor. Thus the maximum current is

$$I_{max} = \frac{\mathcal{E}_0}{R} \tag{26.29}$$

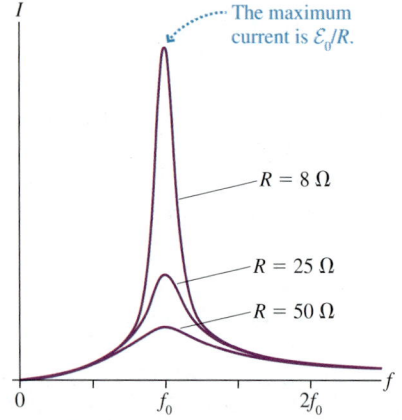

FIGURE 26.27 A graph of the current I versus emf frequency for a series *RLC* circuit.

The driven *RLC* circuit is directly analogous to the driven, damped oscillator that you studied in Chapter 14. A mechanical oscillator exhibits *resonance* by having a large-amplitude response when the driving frequency matches the system's natural frequency. Equation 26.28 is the natural frequency of the driven *RLC* circuit, the frequency at which the current would like to oscillate. Consequently, the circuit has a large current response when the oscillating emf matches this frequency.

Figure 26.27 shows the peak current I of a driven *RLC* circuit as the emf frequency f is varied. Notice how the current increases until reaching a maximum at frequency f_0, then decreases. This is the hallmark of a resonance.

As R decreases, causing the damping to decrease, the maximum current becomes larger and the curve in Figure 26.27 becomes narrower. You saw exactly the same behavior for a driven mechanical oscillator. The emf frequency must be very close to f_0 in order for a lightly damped system to respond, but the response at resonance is very large.

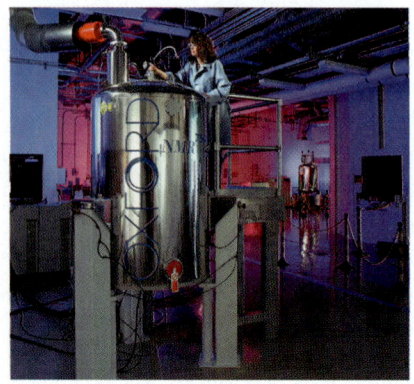

◄ **Nuclear magnetic resonance** BIO or *nmr*, is an important analytic technique in chemistry and biology. In a large magnetic field, the magnetic moments of atomic nuclei rotate at tens to hundreds of MHz. This motion generates a changing magnetic field that induces an AC emf in a coil placed around the sample. Each kind of nucleus rotates at a characteristic frequency, so by the addition of a capacitor to the coil an *RLC* resonant circuit is formed that responds mainly to one frequency—and hence to one kind of nucleus. The technique is thus extremely useful for understanding the elemental composition of a sample.

It is possible to derive an expression for the current shown in Figure 26.27. It is found that the peak current in the RLC circuit is

$$I = \frac{\mathcal{E}_0}{\sqrt{R^2 + (X_L - X_C)^2}} = \frac{\mathcal{E}_0}{\sqrt{R^2 + (2\pi f L - 1/2\pi f C)^2}} \quad (26.30)$$

Peak current in an RLC circuit

Note that the denominator is smallest, and hence the current the largest, when $X_L = X_C$; that is, at resonance. The three peak voltages, if you need them, are then found from $V_R = IR$, $V_L = IX_L$ and $V_C = IX_C$.

 14.2, 14.3

EXAMPLE 26.7 Designing a radio receiver

An AM radio antenna picks up a 1000 kHz signal with a peak voltage of 5.0 mV. The tuning circuit consists of a 60 μH inductor in series with a variable capacitor. The inductor coil has a resistance of 0.25 Ω, and the resistance of the rest of the circuit is negligible.

a. To what value should the capacitor be tuned to listen to this radio station?
b. What is the peak current through the circuit at resonance?
c. A stronger station at 1050 kHz produces a 10 mV antenna signal. What is the current at this frequency when the radio is tuned to 1000 MHz?

PREPARE The inductor's 0.25 Ω resistance can be modeled as a resistance in series with the inductance, hence we have a series RLC circuit. The antenna signal at $f = 1000$ kHz $= 10^6$ Hz is the emf. The circuit looks like Figure 26.25.

SOLVE

a. The capacitor needs to be tuned to where it and the inductor are resonant at $f_0 = 1000$ kHz. Because $f_0 = 1/2\pi\sqrt{LC}$, the appropriate capacitance is

$$C = \frac{1}{L(2\pi f_0)^2} = \frac{1}{(60 \times 10^{-6}\,\text{H})(2\pi \times 10^6\,\text{Hz})^2}$$
$$= 4.2 \times 10^{-10}\,\text{F} = 420\,\text{pF}$$

b. $X_L = X_C$ at resonance, so the maximum current is

$$I_{\text{max}} = \frac{\mathcal{E}_0}{R} = \frac{5.0 \times 10^{-3}\,\text{V}}{0.25\,\Omega} = 0.020\,\text{A} = 20\,\text{mA}$$

c. The 1050 kHz signal is "off resonance," so we need to compute $X_L = 2\pi f L = 396\,\Omega$ and $X_C = 1/2\pi f C = 361\,\Omega$ at $f = 1050$ kHz. The peak voltage of this signal is $\mathcal{E}_0 = 10$ mV. With these values, Equation 26.30 for the peak current is

$$I = \frac{\mathcal{E}_0}{\sqrt{R^2 + (X_L - X_C)^2}} = 0.29\,\text{mA}$$

ASSESS These are realistic values for the input stage of an AM radio. You can see that the signal from the 1050 kHz station is strongly suppressed when the radio is tuned to 1000 kHz.

STOP TO THINK 26.4   A series RLC circuit has $V_C = 5.0$ V, $V_R = 7.0$ V, and $V_L = 9.0$ V. Is the frequency above, below, or equal to the resonance frequency?

The goal of Chapter 26 has been to understand and apply basic ideas of AC circuits.

IMPORTANT CONCEPTS

AC circuits are driven by an emf that oscillates with frequency f.

The emf oscillates as $\mathcal{E} = \mathcal{E}_0 \cos 2\pi ft$.

- Peak values of voltages and currents are denoted by capital letters: I, V.
- Instantaneous values of voltages and currents are denoted by small letters: i, v.

Circuit elements used in AC circuits

	Resistor	Capacitor	Inductor
Symbol	—⋀⋀⋀—	—||—	—0000—
Reactance	Resistance R is constant	$X_C = 1/(2\pi fC)$	$X_L = 2\pi fL$
I and V	$V_R = I_R R$	$V_C = I_C X_C$	$V_L = I_L X_L$
Graph	v_R and i_R; i_R is in phase with v.	v_C and i_C; i_C leads v by 90°.	v_L and i_L; i_L lags v by 90°.

Power in AC resistor circuits

The average power dissipated by a resistor is

$$P_R = (I_{rms})^2 R = \frac{(V_{rms})^2}{R} = I_{rms} V_{rms}$$

where $I_{rms} = I_R/\sqrt{2}$ and $V_{rms} = V_R/\sqrt{2}$ are the root-mean-square (rms) voltage and current.

Electrical safety and biological effects

It is currents passing through the body that are dangerous. Currents greater than about 15 mA (AC) and 75 mA (DC) are potentially fatal.

The body acts like a resistor with a resistance as low as $\sim 150\ \Omega$ if the skin is wet. Then the current is given by Ohm's law as $I = V/R$.

High voltages are dangerous if the voltage source can provide a dangerous current. Household electricity at 120 V can do so. Static electricity that builds up on your body can be 10 kV or more, but can produce only miniscule currents.

APPLICATIONS

Transformers are used to increase or decrease an AC voltage. The rms voltage at the secondary is related to the rms voltage on the primary by

$$(V_2)_{rms} = \frac{N_2}{N_1}(V_1)_{rms}$$

and the currents are related by

$$(I_2)_{rms} = \frac{N_1}{N_2}(I_1)_{rms}$$

Primary coil, N_1 turns

Secondary coil, N_2 turns

The driven *RLC* circuit

If an AC source of amplitude $\mathcal{E}_0$ is placed in an *RLC* circuit, then the voltages and currents oscillate continuously.

$\mathcal{E}_0 \cos(2\pi ft)$

At the **resonant frequency** f_0 the current reaches its maximum value I_{max}.

At low frequencies, X_C is large and the current in the circuit is small.

At high frequencies, X_L is large and the current in the circuit is small.

- The resonant frequency is $f_0 = \dfrac{1}{2\pi\sqrt{LC}}$.
- The maximum value of the current is $I_{max} = \mathcal{E}_0/R$.
- The peak current at any frequency f is given by

$$I = \frac{\mathcal{E}_0}{\sqrt{R^2 + (X_L - X_C)^2}}$$

LC and *RLC* circuits

In an *LC* circuit, the current and voltages oscillate with frequency

$$f = \frac{1}{2\pi\sqrt{LC}}$$

In the *RLC* circuit, the oscillations decay as energy is dissipated in the resistor.

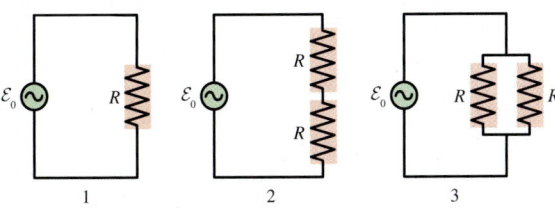

For instructor-assigned homework, go to
www.masteringphysics.com

Problem difficulty is labeled as | (straightforward) to |||| (challenging).

Problems labeled INT integrate significant material
from earlier chapters; BIO are of biological or medical
interest.

QUESTIONS

Conceptual Questions

1. Identical resistors are connected to separate 12 V AC sources. One source operates at 60 Hz, the other at 120 Hz. In which circuit, if either, does the resistor dissipate the greater average power?

2. Consider the three circuits in Figure Q26.2. Rank in order, from largest to smallest, the average total powers P_1 to P_3 dissipated by all the resistors in each circuit. Explain.

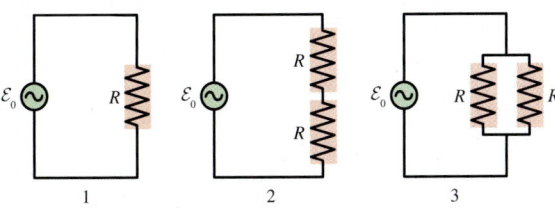

FIGURE Q26.2

3. Lightbulbs in AC circuits "flicker" in intensity, though the flickering is too fast for your eyes to sense it. If a light bulb is connected to a 120 V/60 Hz power supply, how many times a second does the bulb reach peak brightness?

4. A household wall outlet can provide no more than 15 A. But from this outlet, you can power a small welder that uses 150 A. How is this possible?

5. A 12 V DC power supply is connected to the primary coil of a transformer. The primary coil has 100 turns and the secondary coil has 200. What is the rms voltage across the secondary?

6. If you take Figure 26.6 literally, you see that it represents a "1 : 1 transformer," with the same number of turns on the primary coil as on the secondary coil. Suppose that the load resistance, in series with the secondary coil, is a lightbulb, and that an identical lightbulb is placed in series with the primary coil and the alternating voltage source.
   a. Which bulb is brighter? Or are they the same? Explain.
   b. If the bulb in series with the primary coil is unscrewed from its socket, does the other bulb glow more brightly, less brightly, go out, or is its brightness unchanged? Explain.
   c. If the bulb in series with the secondary coil is unscrewed from its socket, does the other bulb glow more brightly, less brightly, go out, or is it unchanged? Explain.

7. Figure Q26.7 shows three wires wrapped around an iron core. The number of turns and the directions of the windings in the figure should be carefully noted. At one particular instant, $V_A - V_B = 20$ V. At that same instant, what is $V_C - V_D$?

FIGURE Q26.7

8. BIO Can a person safely hang from a high-voltage electrical transmission line, with both hands grabbing the line and feet dangling in the air? Does it matter if the person's hands are dry or wet?

9. Your computer has a built-in "surge protector." A surge is a sudden increase in voltage on the power line. This causes a "current spike" that can destroy sensitive electronic components. After the AC electricity is converted to DC in your computer's power supply, the current going to the computer processor must pass through a coil of wire wrapped around an iron core. Explain how this coil works as a surge suppressor.

10. The peak current through a resistor is 2.0 A. What is the peak current if
    a. The resistance R is doubled?
    b. The peak emf $\mathcal{E}_0$ is doubled?
    c. The frequency f is doubled?

11. The average power dissipated by a resistor is 4.0 W. What is the average power if
    a. The resistance R is doubled?
    b. The peak emf $\mathcal{E}_0$ is doubled?
    c. Both are doubled simultaneously?

12. The peak current through a capacitor is 2.0 A. What is the peak current if
    a. The peak emf $\mathcal{E}_0$ is doubled?
    b. The capacitance C is doubled?
    c. The frequency f is doubled?

13. Consider the four circuits in Figure Q26.13. Rank in order, from largest to smallest, the capacitive reactances $(X_C)_1$ to $(X_C)_4$. Explain.

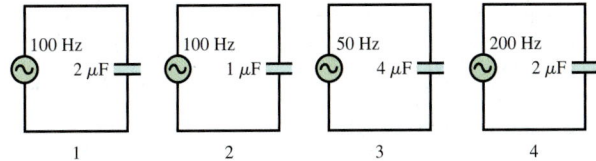

FIGURE Q26.13

14. An inductor is placed across a 120 V/60 Hz wall outlet in the U.S. Would the peak current be larger, smaller, or unchanged if this inductor were placed across a wall outlet in a country where the voltage is 120 V at 50 Hz? Explain.

15. Figure Q26.15 shows two inductors and the potential difference across them at time $t = 0$ s.
    a. Can you tell which of these inductors has the larger current flowing through it at $t = 0$ s? If so, which one? If not, why not?
    b. Can you tell through which inductor the current is changing more rapidly at $t = 0$ s? If so, which one? If not, why not?

FIGURE Q26.15

16. The peak current passing through an inductor is 2.0 A. What is the peak current if:
    a. The peak emf $\mathcal{E}_0$ is doubled?
    b. The inductance L is doubled?
    c. The frequency f is doubled?

17. Consider the four circuits in Figure Q26.17. Rank in order, from largest to smallest, the inductive reactances $(X_L)_1$ to $(X_L)_4$. Explain.

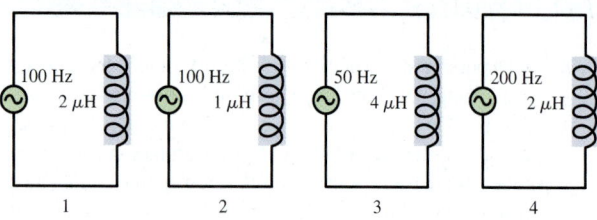

FIGURE Q26.17

18. The tuning circuit in a radio uses an *RLC* circuit. If you need to change the circuit's resonance frequency, will adjusting the resistance have any effect?

19. The resonance frequency of an *RLC* circuit is 1000 Hz. What is the resonance frequency if
    a. The resistance R is doubled?
    b. The inductance L is doubled?
    c. The capacitance C is doubled?
    d. The peak emf $\mathcal{E}_0$ is doubled?
    e. The emf frequency f is doubled?

20. Consider the four circuits in Figure Q26.20. They all have the same resonance frequency f_0. Rank in order, from largest to smallest, the maximum currents $(I_{max})_1$ to $(I_{max})_4$. Explain.

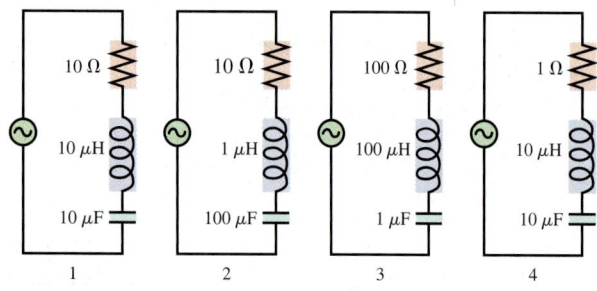

FIGURE Q26.20

Multiple-Choice Questions

21. | A resistor dissipates an average power of 10 W. What is the peak power dissipated by the resistor?
    A. 5.0 W
    B. 7.1 W
    C. 10 W
    D. 14 W
    E. 20 W

22. ‖ A 10 Ω resistor is connected to an AC source. In 15 min, the resistor dissipates 135 J of energy. What is the peak current in the resistor?
    A. 0.17 A   B. 0.35 A   C. 0.52 A
    D. 1.3 A   E. 3.8 A

23. ‖ A transformer has 1000 turns on the primary and 100 turns on the secondary. If the primary is connected to a 120 V outlet and draws 0.050 A, what are the voltage and current of the secondary?
    A. 1200 V, 0.0050 A
    B. 1200 V, 0.50 A
    C. 12 V, 0.0050 A
    D. 12 V, 0.50 A

24. | An inductor is connected to an AC generator. As the generator's frequency is increased, the current in the inductor
    A. Increases.
    B. Decreases.
    C. Does not change.

25. | A capacitor is connected to an AC generator. As the generator's frequency is increased, the current in the capacitor
    A. Increases.
    B. Decreases.
    C. Does not change.

26. | A capacitor and inductor are connected in series to an AC generator. As the generator's frequency is increased, the current in the capacitor
    A. Increases.
    B. Decreases.
    C. Does not change.
    D. Not enough information to tell.

27. ‖ In the circuit shown in Figure Q26.27, at what frequency f of the emf are the peak currents through the inductor and resistor the same?
    A. 3.2 mHz
    B. 0.26 Hz
    C. 2.0 Hz
    D. 8.0 Hz
    E. 13 Hz

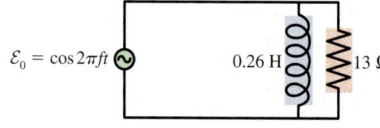

FIGURE Q26.27

28. | The circuit shown in Figure Q26.28 has a resonance frequency of 15 kHz. What is the value of L?
    A. 1.6 μH
    B. 2.4 μH
    C. 5.2 μH
    D. 18 μH
    E. 59 μH

FIGURE Q26.28

PROBLEMS

Section 26.1 Alternating Current

1. | In an AC circuit the rms current through a resistor is 0.80 A. What is the peak current?

2. | A resistor in an AC circuit has a peak voltage of 24 V across it. What is the rms voltage?

3. | A 200 Ω resistor is connected to an AC source with $\mathcal{E}_0 = 10$ V. What is the peak current through the resistor if the emf frequency is (a) 100 Hz? (b) 100 kHz?

4. ‖ Figure P26.4 shows voltage and current graphs for a resistor.
   a. What is the value of the resistance R?
   b. What is the emf frequency f?

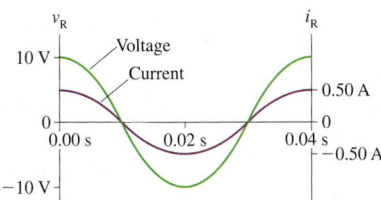

FIGURE P26.4

5. ‖ A resistor dissipates 2.00 W when the rms voltage of the emf is 10.0 V. At what rms voltage will the resistor dissipate 10.0 W?

6. | The heating element of a hair drier dissipates 1500 W when connected to a 120 V/60 Hz power line. What is its resistance?

7. | A toaster oven is rated at 1600 W for operation at 120 V/60 Hz.
   a. What is the resistance of the oven heater element?
   b. What is the peak current through it?
   c. What is the peak power dissipated by the oven?

8. | A 100 Ω resistor is connected to a 120 V/60 Hz power line. What is its average power loss?

9. | A generator produces 40 MW of power and sends it to town at an rms voltage of 75 kV. What is the rms current in the transmission lines?

Section 26.2 The Transmission and Use of Electricity

10. | The primary coil of a transformer is connected to a 120 V/60 Hz wall outlet. The secondary coil is connected to a lamp that dissipates 60 W. What is the rms current in the primary coil?

11. | The primary coil of a transformer is connected to a 120 V/60 Hz wall outlet. Its secondary coil has 300 turns. How many turns must the primary have so that the transformer's output voltage is 50 V rms?

12. ‖ A power pack charging a cell phone battery has an output of 0.40 A at 5.2 V (both rms). What is the rms current at the 120 V/60 Hz wall outlet where the power pack is plugged in?

13. ‖ An x-ray machine that plugs into a 120 V/60 Hz outlet uses 250 W of power. The voltage coming from the electrical substation to the transformer outside the hospital is 7200 V. By how much does the rms current in the primary coil of the transformer increase when the machine is turned on?

14. | The "power cube" transformer for a portable CD player has an output of 4.5 V and 600 mA (both rms) when plugged into a 120 V/60 Hz outlet.
   a. If the primary coil has 1250 turns of wire, how many turns are on the secondary coil?
   b. What is the peak current in the primary coil?

Section 26.3 Biological Effects and Electrical Safety

15. | Estimate the current that will flow through your body if
BIO your two hands touch the opposite terminals of a 9 V battery. Is this current likely to be dangerous?

16. | Estimate the minimum potential difference between a per-
BIO son's hands that could lead to a dangerous current through the person's body if his or her skin is (a) dry and (b) wet.

Section 26.4 Capacitor Circuits

17. ‖ A 0.30 μF capacitor is connected across an AC generator that produces a peak voltage of 10.0 V. What is the peak current through the capacitor if the emf frequency is (a) 100 Hz? (b) 100 kHz?

18. | A 20 μF capacitor is connected across an AC generator that produces a peak voltage of 6.0 V. The peak current is 0.20 A. What is the oscillation frequency in Hz?

19. | The peak current through a capacitor is 10.0 mA. What is the current if
   a. The emf frequency is doubled?
   b. The emf peak voltage is doubled (at the original frequency)?
   c. The frequency is halved and, at the same time, the emf is doubled?

20. ‖ A 20 nF capacitor is connected across an AC generator that produces a peak voltage of 5.0 V.
   a. At what frequency f is the peak current 50 mA?
   b. What is the instantaneous value of the emf at the instant when $i_C = I_C$?

21. ‖ A capacitor is connected to a 15 kHz oscillator that produces an rms voltage of 6.0 V. The peak current is 65 mA. What is the value of the capacitance C?

22. | The peak current through a capacitor is 8.0 mA when connected to an AC source with a peak voltage of 1.0 V. What is the capacitive reactance of the capacitor?

Section 26.5 Inductors and Inductor Circuits

23. | What is the potential difference across a 10 mH inductor if the current through the inductor drops from 150 mA to 50 mA in 10 μs?

24. | A 20 mH inductor is connected across an AC generator that produces a peak voltage of 10.0 V. What is the peak current through the inductor if the emf frequency is (a) 100 Hz? (b) 100 kHz?

25. | The peak current through an inductor is 10.0 mA. What is the current if
   a. The emf frequency is doubled?
   b. The emf peak voltage is doubled (at the original frequency)?
   c. The frequency is halved and, at the same time, the emf is doubled?

26. | A 500 μH inductor is connected across an AC generator that produces a peak voltage of 5.0 V.
    a. At what frequency f is the peak current 50 mA?
    b. What is the instantaneous value of the emf at the instant when $i_L = I_L$?

27. || An inductor is connected to a 15 kHz oscillator that produces an rms voltage of 6.0 V. The peak current is 65 mA. What is the value of the inductance L?

28. | The peak current through an inductor is 12.5 mA when connected to an AC source with a peak voltage of 1.0 V. What is the inductive reactance of the inductor?

29. ||| The superconducting magnet in a magnetic resonance
    BIO imaging system consists of a solenoid with 5.6 H of inductance. The solenoid is energized by connecting it to a 1.2 V DC power supply. Starting from zero, how long does it take for the solenoid to reach its operating current of 100 A?

Section 26.6 Oscillation Circuits

30. || In a nuclear magnetic resonance spectrometer (see Sec-
    BIO tion 26.6), a 15 mH coil is designed to detect an emf oscillating at 400 MHz. The coil is part of an *RLC* circuit. What value of the capacitance C should be used so that the coil circuit is resonates at 400 MHz?

31. || A 2.0 mH inductor is connected in parallel with a variable capacitor. The capacitor can be varied from 100 pF to 200 pF. What is the range of oscillation frequencies for this circuit?

32. || An FM radio station broadcasts at a frequency of 100 MHz. What inductance should be paired with a 10 pF capacitor to build a receiver circuit for this station?

33. | The inductor in the *RLC* tuning circuit of an AM radio has a value of 350 mH. What should be the value of the variable capacitor in the circuit to tune the radio to 740 kHz?

34. | At what frequency f do a 1.0 μF capacitor and a 1.0 μH inductor have the same reactance? What is the value of the reactance at this frequency?

35. || Two *RLC* circuits have identical capacitors but different inductors. Circuit 1 has a resonance frequency f_0 while circuit 2 has a resonance at $2f_0$. What is the ratio L_2/L_1 of the inductance in circuit 2 to the inductance in circuit 1?

36. ||| What capacitor in series with a 100 Ω resistor and a 20 mH inductor will give a resonance frequency of 1000 Hz?

37. || What inductor in series with a 100 Ω resistor and a 2.5 μF capacitor will give a resonance frequency of 1000 Hz?

38. || A series *RLC* circuit has a 200 kHz resonance frequency. What is the resonance frequency if
    a. The resistor value is doubled?
    b. The capacitor value is doubled?

39. | A series *RLC* circuit has a 200 kHz resonance frequency. What is the resonance frequency if
    a. The resistor value is doubled?
    b. The capacitor value is doubled and, at the same time, the inductor value is halved?

40. || A series *RLC* circuit consists of a 280 Ω resistor, a 25 μH inductor, and a 18 μF capacitor. What is the rms current if the emf is supplied by a standard 120 V/60 Hz wall outlet?

General Problems

41. ||| Lightbulbs labeled 40 W, 60 W, and 100 W are connected to
    INT a 120 V/60 Hz power line as shown in Figure P26.41. What is the rate at which energy is dissipated in each bulb?

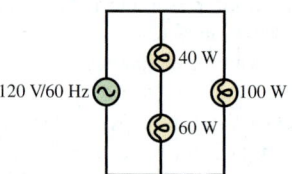

FIGURE P26.41

42. || A 40 Hz alternating current passes through a resistor. The current is zero at $t = 0$ s. Rank in order, from largest to smallest, the *magnitudes* of the potential differences across the resistor, $|\Delta V_A|$ to $|\Delta V_E|$, at the following times: $t_A = 10$ ms, $t_B = 20$ ms, $t_C = 30$ ms, $t_D = 40$ ms, and $t_E = 50$ ms. Indicate any ties.

43. ||| A 60 Hz power source with a peak voltage of 12.0 V is connected to a resistor. At a particular instant the voltage across the resistor is 10.0 V and decreasing. What is the voltage 1.5 ms later?

44. ||| The outer membranes of cells are quite irregular at the sub-
    BIO micron level, so the surface area of an apparently spherical cell differs from the value one would calculate from geometry. To determine the actual surface area of a cell membrane, electrophysiologists use intracellular electrodes to measure the membrane's capacitive reactance. They take advantage of the fact that the capacitance per unit area of all biological membranes is about 1.0 μF/cm^2.
    a. What is the surface area of a 20-μm-diameter sphere? Give your answer in square centimeters.
    b. The capacitive reactance of the outer membrane of a 20-μm-diameter cell is measured to be $6.4 \times 10^7\ \Omega$ at 1.0 kHz. What is the surface area of the membrane, in cm^2? By what factor is the area greater than your answer to part a?
    An important biological process is the release of neurotransmitters that are involved in chemical communication between nerve cells and muscles. This release occurs when intracellular vesicles containing the neurotransmitter chemical fuse with the outer cell membrane and release their contents. The vesicle's membrane becomes part of the cell's outer membrane, increasing the latter's area. As a result, the fusion of vesicles with the outer membrane can be detected by monitoring the outer membrane's capacitive reactance. Indeed, the fusion of a single vesicle can be resolved.
    c. What is the percentage increase in the capacitive reactance of the cell considered in part b when a single vesicle's membrane, a smooth sphere 0.10 μm in diameter, fuses with the outer membrane of the cell?

45. ||| A 13 μF capacitor is connected to a 250 Hz oscillator that produces a peak voltage of 5.5 V. What is the instantaneous capacitor current i_C when $\mathcal{E} = -5.5$ V?

46. ‖‖‖ When two capacitors are connected in parallel across a 10.0 V rms, 1.00 kHz oscillator, the oscillator supplies a total rms current of 545 mA. When the same two capacitors are connected to the oscillator in series, the oscillator supplies an rms current of 126 mA. What are the capacitances of the two capacitors?

    Hint: Just as for resistors, capacitors in parallel have the same *voltage* across them, while capacitors in series have the same *current* through them.

47. ‖‖‖ a. What is the peak current supplied by the emf in Figure P26.47?

    INT b. What is the peak voltage across the 3.0 μF capacitor?

FIGURE P26.47

48. ‖ The current through a 1000 Ω resistor is described by the equation $i_R = (60 \text{ mA})\cos(20t)$, where t is in seconds.

    a. What is the voltage across the resistor at $t = 0.20$ s?

    b. What is the average power dissipated by this resistor?

49. ‖ The voltage across a 60 μF capacitor is described by the equation $v_C = (18 \text{ V})\cos(200t)$, where t is in seconds.

    a. What is the voltage across the capacitor at $t = 0.010$ s?

    b. What is the capacitive reactance?

50. ‖ The voltage across a 75 μH inductor is described by the equation $v_L = (25 \text{ V})\cos(60t)$, where t is in seconds.

    a. What is the voltage across the inductor at $t = 0.10$ s?

    b. What is the inductive reactance?

51. ‖‖‖ An LC circuit is built with a 20 mH inductor and an 8.0 μF capacitor. The current has its maximum value of 0.50 A at $t = 0$ s. How long does it take until the capacitor is fully charged?

52. ‖ A coil has a resistance of 2.10 Ω. The potential difference across the coil is 11.6 V at an instant when the current through the coil is 5.00 A and is decreasing at a rate of 36.0 A/s. What is the coil's inductance?

53. ‖ An electronics hobbyist is building a radio set to receive the AM band, with frequencies from 520 to 1700 kHz. What range variable capacitor will she need to go with a 230 μH inductor?

54. ‖ FM radio frequencies extend from 88.0 MHz to 108 MHz. If a radio tuning circuit has a 1.80 μH inductor and a capacitor whose capacitance can be varied, what is the range of values that the capacitor needs to have for the circuit to cover the full FM band?

55. ‖ For the circuit of Figure P26.55,

    a. What is the resonance frequency?

    b. At resonance, what is the peak current through the circuit?

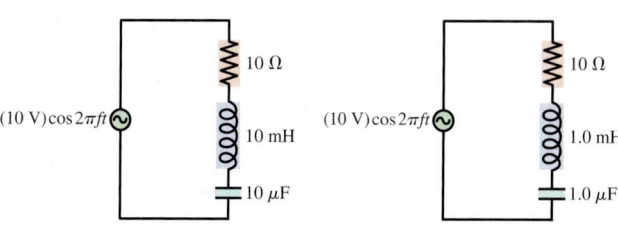

FIGURE P26.55   **FIGURE P26.56**

56. ‖ For the circuit of Figure P26.56,

    a. What is the resonance frequency?

    b. At resonance, what is the peak current through the circuit?

57. ‖‖‖ An RLC circuit consists of a 48 Ω resistor, a 200 μF capacitor, and an inductor. The current is 2.5 A rms when the circuit is connected to a 120 V/60 Hz outlet.

    a. What is the inductance?

    b. What would be the current if this circuit were used in France, where the outlets are 220 V/50 Hz?

Passage Problems

Halogen Bulbs

Halogen bulbs have some differences from standard light bulbs. They are generally smaller, the filament runs at a higher temperature, and they have a quartz (rather than glass) envelope. They may also operate at lower voltage. Consider a 12 V, 50 W halogen bulb for use in a desk lamp. The lamp plugs into a 120 V/60 Hz outlet, and it has a transformer in its base.

58. ‖ The 12 V rating of the bulb refers to the rms voltage. What is the peak voltage across the bulb?

    A. 8.5 V   B. 12 V   C. 17 V   D. 24 V

59. ‖ Suppose the transformer in the base of the lamp has 500 turns of wire on its primary coil. How many turns are on the secondary coil?

    A. 50   B. 160   C. 500   D. 5000

60. ‖ How much current is drawn by the lamp at the outlet? That is, what is the rms current in the primary?

    A. 0.42 A   B. 1.3 A   C. 4.2 A   D. 13 A

61. ‖ What will be the voltage across the bulb if the lamp's power cord is accidentally plugged into a 240 V/60 Hz outlet?

    A. 12 V   B. 24 V   C. 36 V   D. 48 V

STOP TO THINK ANSWERS

Stop to Think 26.1: A. The power in the AC circuit is proportional to the square of the rms current, or to $(I_{rms})^2 = (I_R/\sqrt{2})^2 = \frac{1}{2}I_R^2 = \frac{1}{2}(1 \text{ A})^2$. The power in the DC circuit is proportional to the square of the DC current, or to $I_R^2 = (1 \text{ A})^2$. Thus the power in the DC circuit is twice that in the AC circuit.

Stop to Think 26.2: B. The current is $I_C = V_C/X_C = 2\pi fCV_C$. Thus I_C is proportional to both f and V_C. Doubling the frequency while keeping V_C constant will double the current.

Stop to Think 26.3: B. The current is $I_L = V_L/X_L = V_L/2\pi fL$. Thus I_L is *inversely* proportional to f. Doubling the frequency while keeping V_L constant will halve the current.

Stop to Think 26.4: Above. $V_L > V_C$ tells us that $X_L > X_C$. This is the condition above resonance, where X_L is increasing with f while X_C is decreasing.

Electricity and Magnetism

Mass and charge are the two most fundamental properties of matter. The first four parts of this text were about properties and interactions of masses. Part VI has been a study of charge—what charge is and how charges interact.

Electric and magnetic fields were introduced to enable us to understand the long-range forces of electricity and magnetism. One charge—the source charge—alters the space around it by creating an electric field and, if the charge is moving, a magnetic field. Other charges experience forces exerted *by the fields*. Electric and magnetic fields are the agents by which charges interact.

In addition to the electric field, we often describe electric interactions in terms of the electric potential. This is a particularly fruitful concept for dealing with electric circuits in which charges flow through wires, resistors, etc.

Electric and magnetic fields are *real* and can exist independently of charges. The clearest evidence for their independent existence is electromagnetic waves—the quintessential electromagnetic phenomenon. All electromagnetic waves, including light, are similar in structure, but they span a wide range of wavelengths and frequencies. In Part VI we got our first hints that electromagnetic waves might have a particle-like nature, a concept we will explore further in Part VII.

Part VI has introduced many new phenomena, concepts, and laws. The knowledge structure below draws together the major ideas. These ideas build on what we've learned in the first five parts of this text. All the pieces are now in place to support the development of the ideas of modern physics in Part VII.

KNOWLEDGE STRUCTURE VI Electricity and Magnetism

BASIC GOALS How do charged particles interact? How do electric circuits work?
What are the properties and characteristics of electric and magnetic fields?

GENERAL PRINCIPLES

Forces between charges:
Coulomb's law

$$F_{1\,\text{on}\,2} = F_{2\,\text{on}\,1} = \frac{K|q_1||q_2|}{r^2}$$

The force is along the line connecting the charges. For like charges, the force is repulsive; for opposite charges, attractive.

Electric force on a charge:

$$\vec{F} = q\vec{E}$$

The force is in the direction of the field for a positive charge, opposite the field for a negative charge.

Magnetic force on moving charge:

$$F = |q|vB\sin\alpha$$

The force is perpendicular to the velocity and the field, with direction as specified by the right-hand rule for forces.

Induced emf:
Faraday's law

$$\mathcal{E} = \left|\frac{\Delta\Phi}{\Delta t}\right|$$

The induced current $I = \mathcal{E}/R$ is such that the induced magnetic field opposes the *change* in the magnetic flux. This is **Lenz's law.**

Electric and magnetic fields

Charges and changing magnetic fields create electric fields.

- Electric fields exert forces on charges and torques on dipoles.
- The electric field is perpendicular to equipotential surfaces and points in the direction of decreasing potential.
- The electric field causes charges to move in conductors but not insulators.

Currents and permanent magnets create magnetic fields.

- Magnetic fields exert forces on currents (and moving charged particles) and torques on magnetic dipoles.
- A compass needle or other magnetic dipole will line up with a magnetic field.

Electric potential

The interaction of charged particles can also be described in terms of an electric potential V.

- Only potential differences ΔV are important.
- If the potential of a particle of charge q changes by ΔV, its potential energy changes by $\Delta U = q\,\Delta V$.
- Two points separated by Δd parallel to a uniform electric field of strength E have a potential difference $\Delta V = E\,\Delta d$.

Current and circuits

Potential differences ΔV drive current in circuits. Though electrons are the charge carriers in metals, the **current** I is defined to be the motion of positive charges.

- Circuits obey Kirchhoff's loop law (conservation of energy) and Kirchhoff's junction law (conservation of charge).
- The current through a resistor is $I = \Delta V_R/R$. This is **Ohm's law.**

Electromagnetic waves

An electromagnetic wave is a self-sustaining oscillation of electric and magnetic fields.

- $\vec{E}$ and $\vec{B}$ are perpendicular to each other and to the direction of travel.
- All electromagnetic waves travel at the same speed, c
- The **electromagnetic spectrum** is the spread of wavelengths and frequencies of electromagnetic waves, from radio waves through visible light to gamma rays.

The Greenhouse Effect and Global Warming

Not only are electromagnetic waves real, we depend on them for our very existence; energy carried by electromagnetic waves from the sun provides the basis for all life on earth. Because of the sun's high surface temperature, it emits most of its thermal radiation in the visible portion of the electromagnetic spectrum. As the figure below shows, the earth's atmosphere is transparent to the visible and near-infrared radiation, so most of this energy travels through the atmosphere and warms the earth's surface.

Although seasons come and go, *on average* the earth's climate is very steady. To maintain this stability, the earth must radiate thermal energy—electromagnetic waves—back into space at exactly the same average rate that it receives energy from the sun. Because the earth is much cooler than the sun, its thermal radiation is long-wavelength infrared radiation that we cannot see. A straightforward calculation using Stefan's law finds that the average temperature of the earth should be −18°C, or 0°F, for the incoming and outgoing radiation to be in balance.

This result is clearly not correct; at this temperature, the entire earth would be covered in snow and ice. The measured global average temperature is actually a balmier 15°C, or 59°F. The straightforward calculation fails because it neglects to consider the earth's atmosphere. At visible wavelengths, as the figure shows, the atmosphere has a wide "window" of transparency, but this is not true at the infrared wavelengths of the earth's thermal radiation. The atmosphere lets in the visible radiation from the sun, but the outgoing infrared radiation from the earth sees a much smaller "window." Most of the infrared is absorbed in the atmosphere.

Thermal radiation curves for the sun and the earth. The white and gray bars show regions for which the atmosphere is transparent (white) or opaque (tan) to electromagnetic radiation.

Because it's easier for visible radiant energy to get in than for infrared to get out, the earth is warmer than it would be without the atmosphere. The additional warming of the earth's surface because of the atmosphere is called the **greenhouse effect.** The greenhouse effect is a natural part of the earth's physics; it has nothing to do with human activities, although it's doubtful any advanced life forms would have evolved without it.

The atmospheric gases most responsible for the greenhouse effect are carbon dioxide and water vapor, both strong absorbers of infrared radiation. These **greenhouse gases** are of concern today because humans, through the burning of fossil fuels (oil, coal, and natural gas) are rapidly increasing the amount of carbon dioxide in the atmosphere. Preserved air samples show that carbon dioxide made up 0.027% of the atmosphere before the industrial revolution. In the last 150 years, human activities have increased the carbon dioxide to 0.038%, a 40% increase. By 2050, the carbon dioxide concentration will likely increase to 0.054%, double the pre-industrial value, unless the use of fossil fuels is substantially reduced.

Carbon dioxide is a powerful absorber of infrared radiation. Adding more carbon dioxide makes it even harder for emitted thermal radiation to escape, increasing the average surface temperature of the earth. The net result is **global warming.**

There is strong evidence that the earth has warmed nearly 1°C in the last 100 years because of increased greenhouse gases. What happens next? Climate scientists, using sophisticated models of the earth's atmosphere and oceans, calculate that a doubling of the carbon dioxide concentration will likely increase the earth's average temperature by an additional 2°C ($\approx$3°F) to 6°C ($\approx$9°F). There is some uncertainty in these calculations; the earth is a large and complex system. Perhaps the earth will get cloudier as the temperature increases, moderating the increase. Or perhaps the arctic ice cap will melt, making the earth less reflective and leading to an even more dramatic temperature increase.

But the basic physics that leads to the greenhouse effect, and to global warming, is quite straightforward. Carbon dioxide in the atmosphere keeps the earth warm; more carbon dioxide will make it warmer. How much warmer? That's an important question, one that many scientists around the world are attempting to answer with ongoing research. But large or small, change *is* coming. Global warming is one of the most serious challenges facing scientists, engineers, and all citizens in the 21st century.

MODERN PHYSICS

New Ways of Looking at the World

Newton's mechanics and Maxwell's electromagnetism are remarkable theories that explain a wide range of physical phenomena, as we have seen in the past 26 chapters—but our story doesn't stop there. In the early 20th century, a series of discoveries profoundly altered our understanding of the universe at the most fundamental level, forcing scientists to reconsider the very nature of space and time and to develop new models of light and matter.

Relativity and Quantum Physics

The idea of measuring distance with a meter stick and time with a clock or stopwatch seems quite self-evident. But Albert Einstein, as a young, unknown scientist, realized that Maxwell's theory of electromagnetism could be consistent only if an additional rather odd assumption was made: that the speed of light is the same for all observers, no matter how they might be moving with respect to each other or to the source of the light. This assumption changes the way that we think about space and time. When we study Einstein's theory of *relativity,* you will see how different observers can disagree about lengths and time intervals. We need to go beyond stopwatches and meter sticks. Time can pass at different rates for different observers; time is, as you will see, relative. Our exploration will end with the most famous equation in physics, Einstein's $E = mc^2$. Matter can be converted to energy, and energy to matter.

We've seen that light, a wave, sometimes acts like a particle, a photon. We'll now find that particles such as electrons or atoms sometimes behave like waves. All of the characteristic of waves, such as diffraction and interference, will also apply to particles. This odd notion—that there's no clear distinction between particles and waves—is the core of a new model of light and matter called *quantum physics.* The wave nature of particles will lead to the *quantization* of energy. A particle confined in a box—or an electron in an atom—can only have certain energies. This idea will be a fruitful one for us, allowing us to understand the spectra of gases and phenomena such as bioluminescence.

Atoms, Nuclei, and Particles

We have frequently used the atomic model, explaining the properties of matter by considering the behavior of the atoms that comprise it. But when you get right down to it, what *is* an atom? And what's inside an atom? As you know, an atom has a tiny core called a *nucleus.* We'll look at what goes on inside the nucleus. One remarkable discovery will be that the nuclei of certain atoms spontaneously decay, turning the atom from one element to another. This phenomenon of *radioactivity* will give us a window into the nature of atoms and the particles that comprise them.

Once we know that the nucleus of an atom is composed of protons and neutrons, there's a natural next question to ask: What's inside a proton or neutron? There is an answer to this question, an answer we will learn in the final chapter of this book.

27

RELATIVITY

For a wildlife conservation study, this turtle has been fitted with a collar containing a global positioning system (GPS) receiver, allowing the turtle to be tracked over thousands of miles with an accuracy of ±20 m. How does the theory of relativity affect the accuracy of a GPS receiver?

Looking Ahead ▶▶

The goal of Chapter 27 is to understand how Einstein's theory of relativity changes our concepts of space and time. In this chapter you will learn to:

▶ Use the principle of relativity.

▶ Understand how time dilation and length contraction change our concepts of space and time.

▶ Use the Lorentz velocity transformation.

▶ Calculate relativistic momentum and energy.

▶ Understand how mass and energy are equivalent.

Looking Back ◀◀

Relativity extends dynamics to the realm of objects moving at very high speeds. It thus builds on ideas introduced in our study of mechanics. Please review:

◀ Section 3.5 Relative motion.

◀ Section 9.2 Momentum.

◀ Sections 10.3 and 10.5 Conservation of energy and kinetic energy.

S pace and time seem like straightforward ideas. You can measure lengths with a ruler or meter stick. You can time events with a stopwatch. Nothing could be simpler.

So it seemed to everyone until 1905, when an unknown young scientist had the nerve to suggest that this simple view of space and time was in conflict with other principles of physics. In the century since, Einstein's theory of relativity has radically altered our understanding of some of the most fundamental ideas in physics.

Relativity, despite its esoteric reputation, has very real implications for modern technology. Global positioning system (GPS) satellites depend on relativity, as do the navigation systems used by airliners. Nuclear reactors make tangible use of Einstein's famous equation $E = mc^2$ to generate 20% of the electricity used in the United States. The annihilation of matter in positron-emission tomography (PET scanners) has given neuroscientists an unprecedented ability to monitor activity within the brain.

The theory of relativity is fascinating, perplexing, and challenging. It is also vital to our contemporary understanding of the universe in which we live.

27.1 Relativity: What's It All About?

What do you think of when you hear the phrase "theory of relativity"? A white-haired Einstein? $E = mc^2$? Black holes? Time travel? Perhaps you've heard that the theory of relativity is so complicated and abstract that only a handful of people in the whole world really understand it.

There is, without doubt, a certain mystique associated with relativity, an aura of the strange and exotic. The good news is that understanding the ideas of relativity is well within your grasp. Einstein's *special theory of relativity,* the portion of relativity we'll study, is not mathematically difficult at all. The challenge is conceptual because relativity questions deeply held assumptions about the nature of space and time. In fact, that's what relativity is all about—space and time.

In one sense, relativity is not a new idea at all. Certain ideas about relativity are part of Newtonian mechanics: Newton's laws appear to hold just as well on a fast-moving train as they do in a stationary laboratory. Einstein, however, thought that relativity should apply to *all* the laws of physics, not just mechanics. The difficulty, as you'll see, is that some aspects of relativity appear to be incompatible with the laws of electromagnetism, particularly the laws governing the propagation of light waves.

Lesser scientists might have concluded that relativity simply doesn't apply to electromagnetism. Einstein's genius was to see that the incompatibility arises from *assumptions* about space and time, assumptions no one had ever questioned because they seem so obviously true. Rather than abandon the ideas of relativity, Einstein changed our understanding of space and time.

Fortunately, you need not be a genius to follow a path that someone else has blazed. However, we will have to exercise the utmost care with regard to logic and precision. We will need to state very precisely just how it is that we know things about the physical world, then ruthlessly follow the logical consequences. The challenge is to stay on this path, not to let our prior assumptions—assumptions that are deeply ingrained in all of us—lead us astray.

Albert Einstein (1879–1955) was one of the most influential thinkers in history.

What's Special About Special Relativity?

Einstein's first paper on relativity, in 1905, dealt exclusively with inertial reference frames, reference frames that move relative to each other with constant velocity. Ten years later, Einstein published a more encompassing theory of relativity that considers accelerated motion and its connection to gravity. The second theory, because it's more general in scope, is called *general relativity.* General relativity is the theory that describes black holes, curved spacetime, and the evolution of the universe. It is a fascinating theory but, alas, very mathematical and outside the scope of this textbook.

Motion at constant velocity is a "special case" of motion; namely, motion for which the acceleration is zero. Hence Einstein's first theory of relativity has come to be known as *special relativity.* It is special in the sense of being a restricted, special case of his more general theory, not special in the everyday sense of meaning distinctive or exceptional. Special relativity, with its conclusions about time dilation and length contraction, is what we will study.

27.2 Galilean Relativity

Galileo was the first to understand how the laws of physics depended on the relative motion between different observers. A firm grasp of *Galilean relativity* is necessary if we are to appreciate and understand what is new in Einstein's theory. Thus we begin with the ideas of relativity that are embodied in Newtonian mechanics.

You may have had the experience of sitting on a train and looking up to see another train moving slowly past. It can be hard to tell if the other train is moving past your stationary train, or if you're moving in the opposite direction past a stationary train. Only the *relative* velocity between the trains has meaning.

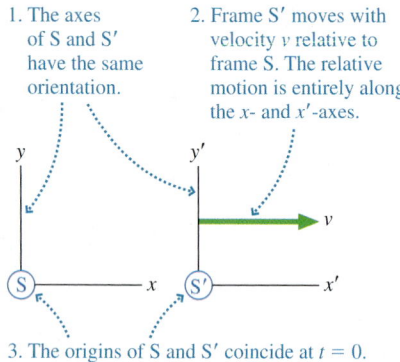

1. The axes of S and S′ have the same orientation.

2. Frame S′ moves with velocity *v* relative to frame S. The relative motion is entirely along the *x*- and *x*′-axes.

3. The origins of S and S′ coincide at *t* = 0. This is our definition of *t* = 0.

FIGURE 27.1 The standard reference frames S and S′.

(a)

The ball stays in place.

A ball with no horizontal forces stays at rest in an airplane cruising at constant velocity. The airplane is an inertial reference frame.

(b)

The ball rolls to the back.

The ball rolls to the back of the plane during takeoff. An accelerating plane is not an inertial reference frame.

FIGURE 27.2 Two reference frames.

Reference Frames

Suppose you're passing me as we both drive in the same direction along a freeway. My car's speedometer reads 55 mph while your speedometer shows 60 mph. Is 60 mph your "true" speed? That is certainly your speed relative to someone standing beside the road, but your speed relative to me is only 5 mph. Your speed is 120 mph relative to a driver approaching from the other direction at 60 mph.

An object does not have a "true" speed or velocity. The very definition of velocity, $v = \Delta x / \Delta t$, assumes the existence of a coordinate system in which, during some time interval Δt, the displacement Δx is measured. The best we can manage is to specify an object's velocity relative to, or with respect to, the coordinate system in which it is measured.

Let's define a **reference frame** to be a coordinate system in which experimenters equipped with meter sticks, stopwatches, and any other needed equipment make position and time measurements on moving objects. Three ideas are implicit in our definition of a reference frame:

- A reference frame extends infinitely far in all directions.
- The experimenters are at rest in the reference frame.
- The number of experimenters and the quality of their equipment are sufficient to measure positions and velocities to any level of accuracy needed.

The first two points are especially important. It is often convenient to say "the laboratory reference frame" or "the reference frame of the rocket." These are shorthand expressions for "a reference frame, infinite in all directions, in which the laboratory (or the rocket) and a set of experimenters happen to be at rest."

NOTE ▶ A reference frame is not the same thing as a "point of view." That is, each person or each experimenter does not have his or her own private reference frame. **All experimenters at rest relative to each other share the same reference frame. ◀**

Figure 27.1 shows how we represent two reference frames, S and S′, that are in relative motion. The coordinate axes in S are *x*, *y*, *z* and those in S′ are *x*′, *y*′, *z*′. Reference frame S′ moves with velocity *v* relative to S or, equivalently, S moves with velocity −*v* relative to S′. There's no implication that either reference frame is "at rest." Notice that the zero of time, when experimenters start their stopwatches, is the instant when the origins of S and S′ coincide.

Inertial Reference Frames

Certain reference frames are especially simple. Figure 27.2a shows a physics student cruising at constant velocity in an airplane. If the student places a ball on the floor, it stays there. There are no horizontal forces, and the ball remains at rest relative to the airplane. That is, $\vec{a} = \vec{0}$ in the airplane's coordinate system when $\vec{F}_{net} = \vec{0}$, so Newton's first law is satisfied. Similarly, if the student drops the ball, it falls straight down—relative to the student—with an acceleration of magnitude *g*, satisfying Newton's second law.

We define an **inertial reference frame** as one in which Newton's first law is valid. That is, an inertial reference frame is one in which an isolated particle, one on which there are no forces, either remains at rest or moves in a straight line at constant speed, as measured by experimenters at rest in that frame.

Not all reference frames are inertial. The physics student in Figure 27.2b conducts the same experiment during takeoff. He carefully places the ball on the floor just as the airplane starts to accelerate down the runway. You can imagine what happens. The ball rolls to the back of the plane as the passengers are being pressed back into their seats. If the student measures the ball's motion using a meter stick attached to the plane, he will find that the ball accelerates *in the plane's reference frame.* Yet he would be unable to identify any force on the ball that would act to accelerate it toward the back of the plane. This violates Newton's first law, so the plane is *not* an inertial reference frame during take-off. **In general, accelerating reference frames are not inertial reference frames.**

> **NOTE** ▶ An inertial reference frame is an idealization. A true inertial reference frame would need to be floating in deep space, far from any gravitational influence. In practice, an earthbound laboratory is a good approximation of an inertial reference frame because the accelerations associated with the earth's rotation and motion around the sun are too small to influence most experiments. ◀

These ideas are in accord with your everyday experience. If you're in a jet flying smoothly at 600 mph—an inertial reference frame—Newton's laws are valid: You can toss and catch a ball, or pour a cup of coffee, exactly as you would on the ground. But if the plane were diving, or shaking from turbulence, simple "experiments" like these would fail. A ball thrown straight up would land far from your hand, and the stream of coffee would bend and turn on its way to missing the cup. These apparently simple observations can be stated as the *Galilean principle of relativity:*

> **Galilean principle of relativity** Newton's laws of motion are valid in all inertial reference frames.

In our study of relativity, we will restrict our attention to inertial reference frames. This implies that the relative velocity v between two reference frames is constant. Any reference frame that moves at constant velocity with respect to an inertial reference frame is itself an inertial reference frame. Conversely, a reference frame that accelerates with respect to an inertial reference frame is not an inertial reference frame. Although special relativity can be used for accelerating reference frames, we will confine ourselves here to the simple case of inertial reference frames moving with respect to each other at constant velocity.

This flight attendant pours wine on a smoothly flying airplane moving at 600 mph just as easily as she does at the terminal. These ideas were first discussed by Galileo in 1632 in the context of pouring water while on a moving ship.

> **STOP TO THINK 27.1**   Which of these is an inertial reference frame (or a very good approximation)?
>
> A. Your bedroom.
> B. A car rolling down a steep hill.
> C. A train coasting along a level track.
> D. A rocket being launched.
> E. A roller coaster going over the top of a hill.
> F. A sky diver falling at terminal speed.

The Galilean Velocity Transformation

Special relativity is largely concerned with how physical quantities such as position and time are measured by experimenters in different reference frames. Let's begin by studying how, within Galilean relativity, the *velocity* of an object is measured in different reference frames. We have already touched on these ideas back in Section 3.5.

Suppose Sue is standing beside a highway as Jim drives by at 50 mph, as shown in Figure 27.3. Let S be Sue's reference frame—a reference frame attached to the ground—and S′ be the reference frame moving with Jim, attached to his car. We see that the velocity of reference frame S′ relative to S is $v = 50$ mph.

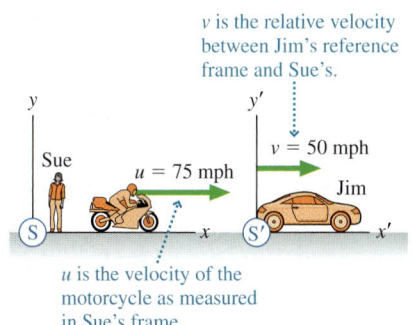

v is the relative velocity between Jim's reference frame and Sue's.

$v = 50$ mph

$u = 75$ mph

u is the velocity of the motorcycle as measured in Sue's frame.

FIGURE 27.3 A motorcycle's velocity as seen by Sue and by Jim.

Now suppose a motorcyclist blasts down the highway, traveling in the same direction as Jim. Sue measures the motorcycle's velocity to be $u = 75$ mph. What is the cycle's velocity u' measured relative to Jim? We can answer this on the basis of common sense. If you're driving at 50 mph, and someone passes you going 75 mph, then his speed *relative to you* is 25 mph. This is the *difference* between his speed relative to the ground and your speed relative to the ground. Thus Jim measures the motorcycle's velocity to be $u' = 25$ mph.

NOTE ▶ In this chapter, we will use v to represent the velocity of one reference frame relative to another. We will use u and u' to represent the velocities of objects with respect to reference frames S and S'. In addition, we will assume that all motion is parallel to the x-axis. ◀

We can state this idea as a general rule. If u is the velocity of an object as measured in reference frame S, and u' its velocity as measured in S', then the two velocities are related by

$$u' = u - v \quad \text{and} \quad u = u' + v \qquad (27.1)$$

Galilean transformation of velocity,
where v is the velocity of S' relative to S

Equations 27.1 are the *Galilean transformation of velocity*. If you know the velocity of a particle as measured by the experimenters in one inertial reference frame, you can use Equations 27.1 to find the velocity that would be measured by experimenters in any other inertial reference frame.

EXAMPLE 27.1 Finding the speed of sound

An airplane is flying at speed 200 m/s with respect to the ground. Sound wave 1 is approaching the plane from the front, while sound wave 2 is catching up from behind. Both waves travel at 340 m/s relative to the ground. What is the velocity of each wave relative to the plane?

PREPARE Assume that the earth (frame S) and the airplane (frame S') are inertial reference frames. Frame S', in which the airplane is at rest, moves with velocity $v = 200$ m/s relative to frame S. Figure 27.4 shows the airplane and the sound waves.

FIGURE 27.4 Experimenters in the plane measure different speeds for the sound waves than do experimenters on the ground.

SOLVE The speed of a mechanical wave, such as a sound wave or a wave on a string, is its speed *relative to its medium*. Thus the *speed of sound* is the speed of a sound wave through a reference frame in which the air is at rest. This is reference frame S, where wave 1 travels with velocity $u_1 = -340$ m/s and wave 2 travels with velocity $u_2 = +340$ m/s. Notice that the Galilean transformations use *velocities*, with appropriate signs, not just speeds.

The airplane travels to the right with reference frame S' at velocity v. We can use the Galilean transformations of velocity to find the velocities of the two sound waves in frame S':

$$u_1' = u_1 - v = -340 \text{ m/s} - 200 \text{ m/s} = -540 \text{ m/s}$$
$$u_2' = u_2 - v = 340 \text{ m/s} - 200 \text{ m/s} = 140 \text{ m/s}$$

Thus wave 1 approaches the plane with a *speed* of 540 m/s while wave 2 approaches with a speed of 140 m/s.

ASSESS This isn't surprising. If you're driving 50 mph, a car coming the other way at 55 mph is approaching you at 105 mph. A car coming up behind you at 55 mph seems to be gaining on you at the rate of only 5 mph. Wave speeds behave the same. Notice that a mechanical wave would appear to be stationary to a person moving at the wave speed. To a surfer, the crest of the ocean wave remains at rest under his or her feet.

STOP TO THINK 27.2   Ocean waves are approaching the beach at 10 m/s. A boat heading out to sea travels at 6 m/s. How fast are the waves moving in the boat's reference frame?

A. 16 m/s   B. 10 m/s   C. 6 m/s   D. 4 m/s

27.3 Einstein's Principle of Relativity

The 19th century was an era of optics and electromagnetism. Thomas Young demonstrated in 1801 that light is a wave, and by midcentury scientists had devised techniques for measuring the speed of light. Faraday discovered electromagnetic induction in 1831, setting in motion a train of events leading to Maxwell's conclusion, in 1864, that light is an electromagnetic wave.

If light is a wave, what is the medium in which it travels? This was perhaps the most important scientific question of the second half of the 19th century. The medium in which light waves were assumed to travel was called the *ether*. Experiments to measure the speed of light were assumed to be measuring its speed through the ether. But just what is the ether? What are its properties? Can we collect a jar full of ether to study? Despite the significance of these questions, experimental efforts to detect the ether or measure its properties kept coming up empty-handed.

Maxwell's theory of electromagnetism didn't help the situation. The crowning success of Maxwell's theory was his prediction that light waves travel with speed

$$c = \frac{1}{\sqrt{\epsilon_0 \mu_0}} = 3.00 \times 10^8 \text{ m/s}$$

This is a very specific prediction with no wiggle room. The difficulty with such a specific prediction was the implication that Maxwell's laws of electromagnetism are valid *only* in the reference frame of the ether. After all, as Figure 27.5 shows, the light speed should certainly be larger or smaller than c in a reference frame moving through the ether, just as the sound speed is different to someone moving through the air.

As the 19th century closed, it appeared that Maxwell's theory did not obey the classical principle of relativity. There was just one reference frame, the reference frame of the ether, in which the laws of electromagnetism seemed to be true. And to make matters worse, the fact that no one had been able to detect the ether meant that no one could identify the one reference frame in which Maxwell's equations "worked."

It was in this muddled state of affairs that a young Albert Einstein made his mark on the world. Even as a teenager, Einstein had wondered how a light wave would look to someone "surfing" the wave, traveling alongside the wave at the wave speed. You can do that with a water wave or a sound wave, but light waves seemed to present a logical difficulty. An electromagnetic wave sustains itself by virtue of the fact that a changing magnetic field induces an electric field and a changing electric field induces a magnetic field. But to someone moving with the wave, *the fields would not change.* How could there be an electromagnetic wave under these circumstances?

Several years of thinking about the connection between electromagnetism and reference frames led Einstein to the conclusion that *all* the laws of physics, not just the laws of mechanics, should obey the principle of relativity. In other words, the principle of relativity is a fundamental statement about the nature of the physical universe. The Galilean principle of relativity stated only that Newton's laws hold in any inertial reference frame. Einstein was able to state a much more general principle:

Principle of relativity All the laws of physics are the same in all inertial reference frames.

All of the results of Einstein's theory of relativity flow from this one simple statement.

FIGURE 27.5 It seems as if the speed of light should differ from c in a reference frame moving through the ether.

In certain rivers, tides send waves upriver that can be surfed for miles. From the reference frame of these surfers, the waves are standing still. If you could move along with a light wave, would the electric and magnetic fields appear motionless?

The Constancy of the Speed of Light

If Maxwell's equations of electromagnetism are laws of physics, and there's every reason to think they are, then, according to the principle of relativity, Maxwell's equations must be true in *every* inertial reference frame. On the surface this seems to be an innocuous statement, equivalent to saying that the law of conservation of momentum is true in every inertial reference frame. But follow the logic:

1. Maxwell's equations are true in all inertial reference frames.
2. Maxwell's equations predict that electromagnetic waves, including light, travel at speed $c = 3.00 \times 10^8$ m/s.
3. Therefore, **light travels at speed c in all inertial reference frames.**

Figure 27.6 shows the implications of this conclusion. *All* experimenters, regardless of how they move with respect to each other, find that *all* light waves, regardless of the source, travel in their reference frame with the *same* speed c. If Cathy's velocity toward Bill and away from Amy is $v = 0.9c$, Cathy finds, by making measurements in her reference frame, that the light from Bill approaches her at speed c, not at $c + v = 1.9c$. And the light from Amy, which left Amy at speed c, catches up from behind at speed c *relative to Cathy,* not the $c - v = 0.1c$ you would have expected.

Although this prediction goes against all shreds of common sense, the experimental evidence for it is strong. Laboratory experiments are difficult because even the highest laboratory speed is insignificant in comparison to c. In the 1930s, however, the physicists R. J. Kennedy and E. M. Thorndike realized that they could use the earth itself as a laboratory. The earth's speed as it circles the sun is about 30,000 m/s. The velocity of the earth in January differs by 60,000 m/s from its velocity in July, when the earth is moving in the opposite direction. Kennedy and Thorndike were able to use a very sensitive and stable interferometer to show that the numerical values of the speed of light in January and July differ by less than 2 m/s.

More recent experiments have used unstable elementary particles, called π mesons, that decay into high-energy photons, or particles, of light. The π mesons, created in a particle accelerator, move through the laboratory at 99.975% the speed of light, or $v = 0.99975c$, as they emit photons at speed c in the π meson's reference frame. As Figure 27.7 shows, you would expect the photons to travel through the laboratory with speed $c + v = 1.99975c$. Instead, the measured speed of the photons in the laboratory was, within experimental error, 3.00×10^8 m/s.

In summary, *every* experiment designed to compare the speed of light in different reference frames has found that light travels at 3.00×10^8 m/s in every inertial reference frame, regardless of how the reference frames are moving with respect to each other.

This light wave leaves Amy at speed c relative to Amy. It approaches Cathy at speed c relative to Cathy.

$v = 0.9c$

Amy   Cathy   Bill

This light wave leaves Bill at speed c relative to Bill. It approaches Cathy at speed c relative to Cathy.

FIGURE 27.6 Light travels at speed c in all inertial reference frames, regardless of how the reference frames are moving with respect to the light source.

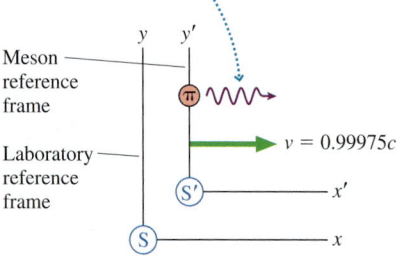

A photon is emitted at speed c relative to the π meson. Measurements find that the photon's speed in the laboratory reference frame is also c.

Meson reference frame

Laboratory reference frame

$v = 0.99975c$

FIGURE 27.7 Experiments find that the photons travel through the laboratory with speed c, not the speed $1.99975c$ that you might expect.

How Can This Be?

You're in good company if you find this impossible to believe. Suppose I shot a ball forward at 50 m/s while driving past you at 30 m/s. You would certainly see the ball traveling at 80 m/s relative to you and the ground. What we're saying with regard to light is equivalent to saying that the ball travels at 50 m/s relative to my car and *at the same time* travels at 50 m/s relative to the ground, even though the car is moving across the ground at 30 m/s. It seems logically impossible.

You might think that this is merely a matter of semantics. If we can just get our definitions and use of words straight, then the mystery and confusion will disappear. Or perhaps the difficulty is a confusion between what we "see" versus what "really happens." In other words, a better analysis, one that focuses on what really happens, would find that light "really" travels at different speeds in different reference frames.

Alas, what "really happens" is that light travels at 3.00×10^8 m/s in every inertial reference frame, regardless of how the reference frames are moving with respect to each other. It's not a trick. There remains only one way to escape the logical contradictions.

The definition of velocity is $u = \Delta x / \Delta t$, the ratio of a distance traveled to the time interval in which the travel occurs. Suppose you and I both make measurements on an object as it moves, but you happen to be moving relative to me. Perhaps I'm standing on the corner, you're jogging to the right, and we're both trying to measure the velocity of a bicycle moving to the right as it passes both of us. Further, suppose we have agreed in advance to measure the bicycle as it moves from the tree to the lamppost in Figure 27.8. Your $\Delta x'$ differs from my Δx because of your motion relative to me, causing you to calculate a bicycle velocity u' in your reference frame that differs from its velocity u in my reference frame. This is just the Galilean transformations showing up again.

Measurements made in frame S, in which the tree and lamppost are at rest. The bicycle's velocity is $u = \Delta x / \Delta t$.

Measurements made in frame S', which moves to the right relative to frame S. The bicycle's velocity is $u' = \Delta x' / \Delta t$.

FIGURE 27.8 Measuring the velocity of an object by appealing to the basic definition $u = \Delta x / \Delta t$.

Now let's repeat the measurements, but this time let's measure the velocity of a light wave as it travels from the tree to the lamppost. Once again, your $\Delta x'$ differs from my Δx, although the difference will be pretty small unless you happen to be jogging at a speed that is an appreciable fraction of the speed of light. The obvious conclusion is that your light speed u' differs from my light speed u. But it doesn't. The experiments show that, for a light wave, we'll get the *same* values: $u' = u$.

The only way this can be true is if your Δt is not the same as my Δt. If the time it takes the light to move from the tree to the lamppost in your reference frame, a time we'll now call $\Delta t'$, differs from the time Δt it takes the light to move from the tree to the lamppost in my reference frame, then we might find that $\Delta x' / \Delta t' = \Delta x / \Delta t$. That is, $u' = u$ even though you are moving with respect to me.

We've assumed, since the beginning of this textbook, that time is simply time. It flows along like a river, and all experimenters in all reference frames simply use it. For example, suppose the tree and the lamppost both have big clocks that we both can see. Shouldn't we be able to agree on the time interval Δt the light needs to move from the tree to the lamppost?

Perhaps not. It's demonstrably true that $\Delta x' \neq \Delta x$. It's experimentally verified that $u' = u$ for light waves. Something must be wrong with *assumptions* that we've made about the nature of time. The principle of relativity has painted us into a corner, and our only way out is to reexamine our understanding of time.

27.4 Events and Measurements

To question some of our most basic assumptions about space and time requires extreme care. We need to be certain that no assumptions slip into our analysis unnoticed. Our goal is to describe the motion of a particle in a clear and precise way, making the barest minimum of assumptions.

Events

The fundamental entity of relativity is called an **event.** An event is a physical activity that takes place at a definite point in space and at a definite instant of time. A firecracker exploding is an event. A collision between two particles is an event. A light wave hitting a detector is an event.

Events can be observed and measured by experimenters in different reference frames. An exploding firecracker is as clear to you as you drive by in your car as it is to me standing on the street corner. We can quantify where and when an event occurs with four numbers: the coordinates (x, y, z) and the instant of time t. These four numbers, illustrated in Figure 27.9, are called the **spacetime coordinates** of the event.

The spatial coordinates of an event measured in reference frames S and S′ may differ. But it now appears that the instant of time recorded in S and S′ may also differ. Thus the spacetime coordinates of an event measured by experimenters in frame S are (x, y, z, t) and the spacetime coordinates of the *same event* measured by experimenters in frame S′ are (x', y', z', t').

The motion of a particle can be described as a sequence of two or more events. We introduced this idea in the previous section when we agreed to measure the velocity of a bicycle and then of a light wave by comparing the object passing the tree (first event) to the object passing the lamppost (second event).

Measurements

Events are what "really happen," but how do we learn about an event? That is, how do the experimenters in a reference frame determine the spacetime coordinates of an event? This is a problem of *measurement.*

We defined a reference frame to be a coordinate system in which experimenters can make position and time measurements. That's a good start, but now we need to be more precise as to *how* the measurements are made. Imagine that a reference frame is filled with a cubic lattice of meter sticks, as shown in Figure 27.10. At every intersection is a clock, and all the clocks in a reference frame are *synchronized.* We'll return in a moment to consider how to synchronize the clocks, but assume for the moment it can be done.

Now, with our meter sticks and clocks in place, we can use a two-part measurement scheme:

- The (x, y, z) coordinates of an event are determined by the intersection of meter sticks closest to the event.
- The event's time t is the time displayed on the clock nearest the event.

You can imagine, if you wish, that each event is accompanied by a flash of light to illuminate the face of the nearest clock and make its reading known.

An event has spacetime coordinates (x, y, z, t) in frame S and different spacetime coordinates (x', y', z', t') in frame S′.

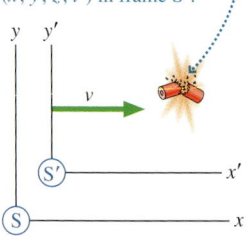

FIGURE 27.9 The location and time of an event are described by its spacetime coordinates.

The spacetime coordinates of this event are measured by the nearest meter stick intersection and the nearest clock.

Synchronized clocks

Meter sticks

Reference frame S

Reference frame S′ has its own meter sticks and its own clocks.

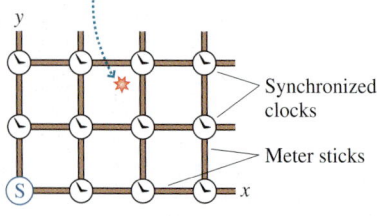

Reference frame S′

FIGURE 27.10 The spacetime coordinates of an event are measured by a lattice of meter sticks and clocks.

Several important issues need to be noted:

1. The clocks and meter sticks in each reference frame are imaginary, so they have no difficulty passing through each other.
2. Measurements of position and time made in one reference frame must use only the clocks and meter sticks in that reference frame.
3. There's nothing special about the sticks being 1 m long and the clocks 1 m apart. The lattice spacing can be altered to achieve whatever level of measurement accuracy is desired.
4. We'll assume that the experimenters in each reference frame have assistants sitting beside every clock to record the position and time of nearby events.
5. Perhaps most important, t is the time at which the event *actually happens,* not the time at which an experimenter sees the event or at which information about the event reaches an experimenter.
6. All experimenters in one reference frame agree on the spacetime coordinates of an event. In other words, **an event has a unique set of spacetime coordinates in each reference frame.**

STOP TO THINK 27.3   A carpenter is working on a house two blocks away. You notice a slight delay between seeing the carpenter's hammer hit the nail and hearing the blow. At what time does the event "hammer hits nail" occur?

A. At the instant you hear the blow.
B. At the instant you see the hammer hit.
C. Very slightly before you see the hammer hit.
D. Very slightly after you see the hammer hit.

Clock Synchronization

It's important that all the clocks in a reference frame be **synchronized,** meaning that all clocks in the reference frame have the same reading at any one instant of time. We would not be able to use a sequence of events to track the motion of a particle if the clocks differed in their readings. Thus we need a method of synchronization. One idea that comes to mind is to designate the clock at the origin as the *master clock.* We could then carry this clock around to every clock in the lattice, adjust that clock to match the master clock, and finally return the master clock to the origin.

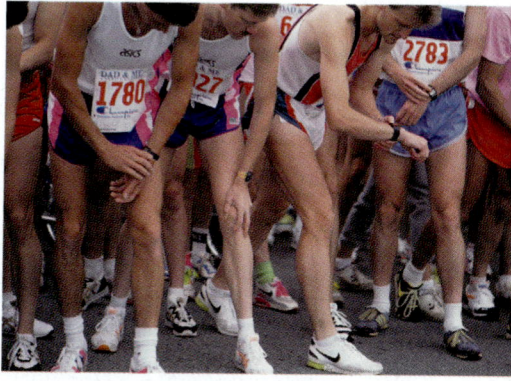

It's easy to synchronize clocks that are all in one place, but synchronizing distant clocks takes some care.

This would be a perfectly good method of clock synchronization in Newtonian mechanics, where time flows along smoothly, the same for everyone. But we've been driven to reexamine the nature of time by the possibility that time is different in reference frames moving relative to each other. Because the master clock would *move,* we cannot assume that the master clock keeps time in the same way as the stationary clocks.

We need a synchronization method that does not require moving the clocks. Fortunately, such a method is easy to devise. Each clock is resting at the intersection of meter sticks, so by looking at the meter sticks, the assistant knows, or can calculate, exactly how far each clock is from the origin. Once the distance is known, the assistant can calculate exactly how long a light wave will take to travel from the origin to each clock. For example, light will take $1.00 \ \mu s$ to travel to a clock 300 m from the origin.

NOTE ▶ It's handy for many relativity problems to know that the speed of light is $c = 300 \ \text{m}/\mu s$. ◀

To synchronize the clocks, the assistants begin by setting each clock to display the light travel time from the origin, but they don't start the clocks. Next, as

1. This clock is preset to 1.00 μs, the time it takes light to travel 300 m.

Clock at origin

300 m

2. A light flashes at the origin and the origin clock starts running at $t = 0$ s.

Wave front

3. The clock starts when the light wave reaches it. It is now synchronized with the origin clock.

FIGURE 27.11 Synchronizing the clocks.

Figure 27.11 shows, a light flashes at the origin and, simultaneously, the clock at the origin starts running from $t = 0$ s. The light wave spreads out in all directions at speed c. A photodetector on each clock recognizes the arrival of the light wave and, without delay, starts the clock. The clock had been preset with the light travel time, so each clock as it starts reads exactly the same as the clock at the origin. Thus all the clocks will be synchronized after the light wave has passed by.

Events and Observations

We noted above that t is the time the event *actually happens*. This is an important point, one that bears further discussion. Light waves take time to travel. Messages, whether they're transmitted by light pulses, telephone, or courier on horseback, take time to be delivered. An experimenter *observes* an event, such as an exploding firecracker, only *at a later time* when light waves reach his or her eyes. But our interest is in the event itself, not the experimenter's observation of the event. The time at which the experimenter sees the event or receives information about the event is not when the event actually occurred.

Suppose at $t = 0$ s a firecracker explodes at $x = 300$ m. The flash of light from the firecracker will reach an experimenter at the origin at $t_1 = 1.0$ μs. The sound of the explosion will reach the experimenter at $t_2 = 0.88$ s. Neither of these is the time t_{event} of the explosion, although the experimenter can work backward from these times, using known wave speeds, to determine t_{event}. In this example, the spacetime coordinates of the event—the explosion—are (300 m, 0 m, 0 m, 0 s).

EXAMPLE 27.2 Finding the time of an event

Experimenter A in reference frame S stands at the origin looking in the positive x-direction. Experimenter B stands at $x = 900$ m looking in the negative x-direction. A firecracker explodes somewhere between them. Experimenter B sees the light flash at $t = 3.00$ μs. Experimenter A sees the light flash at $t = 4.00$ μs. What are the spacetime coordinates of the explosion?

PREPARE Experimenters A and B are in the same reference frame and have synchronized clocks. Figure 27.12 shows the two experimenters and the explosion at unknown position x.

Wavefront reaches A at $t = 4.00$ μs.

Wavefront reaches B at $t = 3.00$ μs.

A

B

Explosion at position x, time t

0 m

900 m

x

SOLVE The two experimenters observe light flashes at two different instants, but there's only one event. Light travels 300 m/μs, so the additional 1.00 μs needed for the light to reach experimenter A implies that distance $(x - 0$ m$)$ from x to A is 300 m longer than distance $(900$ m $- x)$ from B to x. That is,

$$(x - 0 \text{ m}) = (900 \text{ m} - x) + 300 \text{ m}$$

This is easily solved to give $x = 600$ m as the position coordinate of the explosion. The light takes 1.00 μs to travel 300 m to experimenter B, 2.00 μs to travel 600 m to experimenter A. The light is received at 3.00 μs and 4.00 μs, respectively, hence it was emitted by the explosion at $t = 2.00$ μs. The spacetime coordinates of the explosion are (600 m, 0 m, 0 m, 2.00 μs).

ASSESS Although the experimenters *see* the explosion at different times, they agree that the explosion actually *happened* at $t = 2.00$ μs.

◄ **FIGURE 27.12** The light wave reaches the experimenters at different times. Neither of these is the time at which the event actually happened.

Simultaneity

Two events 1 and 2 that take place at different positions x_1 and x_2 but at the *same time $t_1 = t_2$*, as measured in some reference frame, are said to be **simultaneous** in that reference frame. Simultaneity is determined by when the events actually happen, not when they are seen or observed. In general, simultaneous events are not *seen* at the same time because of the difference in light travel times from the events to an experimenter.

CONCEPTUAL EXAMPLE 27.1 Are the explosions simultaneous?

An experimenter in reference frame S stands at the origin looking in the positive x-direction. At $t = 3.0$ μs she sees firecracker 1 explode at $x = 600$ m. A short time later, at $t = 5.0$ μs, she sees firecracker 2 explode at $x = 1200$ m. Are the two explosions simultaneous? If not, which firecracker exploded first?

PREPARE Light from both explosions travels toward the experimenter at 300 m/μs.

SOLVE The experimenter *sees* two different explosions, but perceptions of the events are not the events themselves. When did the explosions *actually* occur? Using the fact that light travels 300 m/μs, it's easy to see that firecracker 1 exploded at $t_1 = 1.0$ μs and firecracker 2 also exploded at $t_2 = 1.0$ μs. The events *are* simultaneous.

STOP TO THINK 27.4   A tree and a pole are 3000 m apart. Each is suddenly hit by a bolt of lightning. Mark, who is standing at rest midway between the two, sees the two lightning bolts at the same instant of time. Nancy is at rest under the tree. Define event 1 to be "lightning strikes tree" and event 2 to be "lightning strikes pole." For Nancy, does event 1 occur before, after, or at the same time as event 2?

27.5 The Relativity of Simultaneity

We've now established a means for measuring the time of an event in a reference frame, so let's begin to investigate the nature of time. The following "thought experiment" is very similar to one suggested by Einstein.

Figure 27.13 shows a long railroad car traveling to the right with a velocity v that may be an appreciable fraction of the speed of light. A firecracker is tied to each end of the car, right above the ground. Each firecracker is powerful enough that, when it explodes, it will make a burn mark on the ground at the position of the explosion.

Ryan is standing on the ground, watching the railroad car go by. Peggy is standing in the exact center of the car with a special box at her feet. This box has two light detectors, one facing each way, and a signal light on top. The box works as follows:

1. If a flash of light is received at the right detector before a flash is received at the left detector, then the light on top of the box will turn green.
2. If a flash of light is received at the left detector before a flash is received at the right detector, or if two flashes arrive simultaneously, the light on top will turn red.

The firecrackers explode as the railroad car passes Ryan, and he sees the two light flashes from the explosions simultaneously. He then measures the distances to the two burn marks and finds that he was standing exactly halfway between the marks. Because light travels equal distances in equal times, Ryan concludes that the two explosions were simultaneous in his reference frame, the reference frame of the ground. Further, because he was midway between the two ends of the car, he was directly opposite Peggy when the explosions occurred.

Figure 27.14a on the next page shows the sequence of events in Ryan's reference frame. Light travels at speed c in all inertial reference frames, so, although the firecrackers were moving, the light waves are spheres centered on the burn marks. Ryan determines that the light wave coming from the right reaches Peggy and the box before the light wave coming from the left. Thus, according to Ryan, the signal light on top of the box turns green.

The firecrackers will make burn marks on the ground at the positions where they explode.

FIGURE 27.13 A railroad car traveling to the right with velocity v.

(a) The events in Ryan's frame

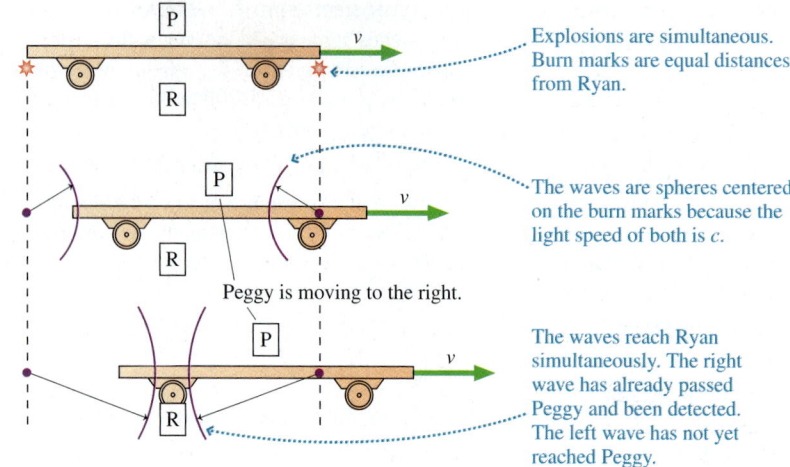

Explosions are simultaneous. Burn marks are equal distances from Ryan.

The waves are spheres centered on the burn marks because the light speed of both is *c*.

Peggy is moving to the right.

The waves reach Ryan simultaneously. The right wave has already passed Peggy and been detected. The left wave has not yet reached Peggy.

(b) The events in Peggy's frame

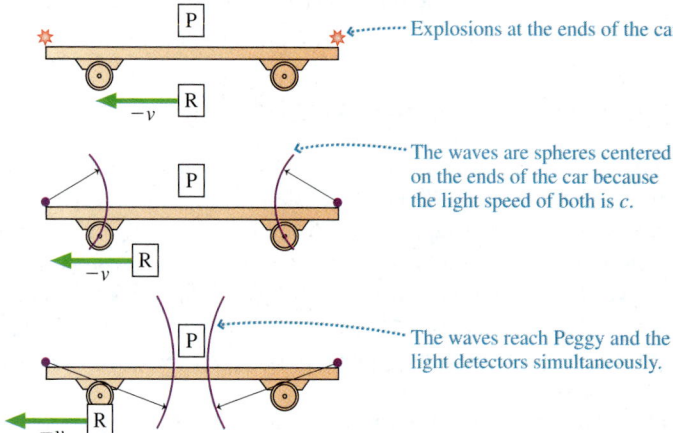

Explosions at the ends of the car

The waves are spheres centered on the ends of the car because the light speed of both is *c*.

The waves reach Peggy and the light detectors simultaneously.

FIGURE 27.14 Exploding firecrackers seen in two different reference frames.

How do things look in Peggy's reference frame, a reference frame moving to the right at velocity *v* relative to the ground? As Figure 27.14b shows, Peggy sees Ryan moving to the left with speed *v*. Light travels at speed *c* in all inertial reference frames, so the light waves are spheres centered on the ends of the car. If the explosions are simultaneous, as Ryan has determined, the two light waves reach her and the box simultaneously. Thus, according to Peggy, the signal light on top of the box turns red!

Now the light on top must be either green or red. *It can't be both!* Later, after the railroad car has stopped, Ryan and Peggy can place the box in front of them. Either it has a red light or a green light. Ryan can't see one color while Peggy sees the other. Hence we have a paradox. It's impossible for Peggy and Ryan both to be right. But who is wrong, and why?

What do we know with absolute certainty?

1. Ryan detected the flashes simultaneously.
2. Ryan was halfway between the firecrackers when they exploded.
3. The light from the two explosions traveled toward Ryan at equal speeds.

The conclusion that the explosions were simultaneous in Ryan's reference frame is unassailable. The light is green.

Peggy, however, made an assumption. It's a perfectly ordinary assumption, one that seems sufficiently obvious that you probably didn't notice, but an assumption nonetheless. Peggy assumed that the explosions were simultaneous.

Didn't Ryan find them to be simultaneous? Indeed, he did. Suppose we call Ryan's reference frame S, the explosion on the right event R, and the explosion on the left event L. Ryan found that $t_R = t_L$. But Peggy has to use a different set of clocks, the clocks in her reference frame S′, to measure the times t'_R and t'_L at which the explosions occurred. The fact that $t_R = t_L$ in frame S does *not* allow us to conclude that $t'_R = t'_L$ in frame S′.

In fact, the right firecracker must explode *before* the left firecracker in frame S′. Figure 27.14, with its assumption about simultaneity, was incorrect. Figure 27.15 shows the situation in Peggy's reference frame with the right firecracker exploding first. Now the wave from the right reaches Peggy and the box first, as Ryan had concluded, and the light on top turns green.

One of the most disconcerting conclusions of relativity is that **two events occurring simultaneously in reference frame S are** *not* **simultaneous in any reference frame S′ that is moving relative to S.** This is called the **relativity of simultaneity.**

The two firecrackers *really* explode at the same instant of time in Ryan's reference frame. And the right firecracker *really* explodes first in Peggy's reference frame. It's not a matter of when they see the flashes. Our conclusion refers to the times at which the explosions actually occur.

The paradox of Peggy and Ryan contains the essence of relativity, and it's worth careful thought. First, review the logic until you're certain that there *is* a paradox, a logical impossibility. Then convince yourself that the only way to resolve the paradox is to abandon the assumption that the explosions are simultaneous in Peggy's reference frame. If you understand the paradox and its resolution, you've made a big step toward understanding what relativity is all about.

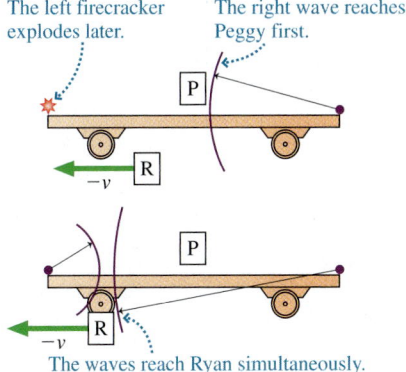

The right firecracker explodes first.

The left firecracker explodes later.

The right wave reaches Peggy first.

The waves reach Ryan simultaneously. The left wave has not reached Peggy.

FIGURE 27.15 The real sequence of events in Peggy's reference frame.

> **STOP TO THINK 27.5** A tree and a pole are 3000 m apart. Each is suddenly hit by a bolt of lightning. Mark, who is standing at rest midway between the two, sees the two lightning bolts at the same instant of time. Nancy is flying her rocket at $v = 0.5c$ in the direction from the tree toward the pole. The lightning hits the tree just as she passes by it. Define event 1 to be "lightning strikes tree" and event 2 to be "lightning strikes pole." For Nancy, does event 1 occur before, after, or at the same time as event 2?

27.6 Time Dilation

The principle of relativity has driven us to the logical conclusion that the time at which an event occurs may not be the same for two reference frames moving relative to each other. Our analysis thus far has been mostly qualitative. It's time to start developing some quantitative tools that will allow us to compare measurements in one reference frame to measurements in another reference frame.

Figure 27.16a shows a special clock called a **light clock.** The light clock is a box of height h with a light source at the bottom and a mirror at the top. The light source emits a very short pulse of light that travels to the mirror and reflects back to a light detector beside the source. The clock advances one "tick" each time the detector receives a light pulse, and it immediately, with no delay, causes the light source to emit the next light pulse.

Our goal is to compare two measurements of the interval between two ticks of the clock: one taken by an experimenter standing next to the clock and the other by an experimenter moving with respect to the clock. To be specific, Figure 27.16b shows the clock at rest in reference frame S′. We call this the **rest frame** of the clock. Reference frame S′ moves to the right with velocity v relative to reference frame S.

Relativity requires us to measure *events,* so let's define event 1 to be the emission of a light pulse and event 2 to be the detection of that light pulse.

(a) A light clock

Mirror

Time display

Light source   Light detector

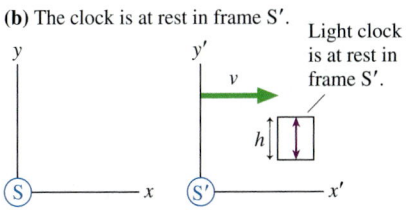

(b) The clock is at rest in frame S′.

Light clock is at rest in frame S′.

FIGURE 27.16 The ticking of a light clock can be measured by experimenters in two different reference frames.

Experimenters in both reference frames are able to measure where and when these events occur *in their frame.* In frame S, the time interval $\Delta t = t_2 - t_1$ is one tick of the clock. Similarly, one tick in frame S′ is $\Delta t' = t_2' - t_1'$.

It's simple to calculate the tick interval $\Delta t'$ observed in frame S′, the rest frame of the clock, because the light simply goes straight up and back down. The total distance traveled by the light is $2h$ so the time of one tick is

$$\Delta t' = \frac{2h}{c} \tag{27.2}$$

Figure 27.17 shows the light clock as seen in frame S. As seen in S, the clock is moving to the right at speed v. Thus the mirror has moved a distance $\frac{1}{2}v(\Delta t)$ during the time $\frac{1}{2}(\Delta t)$ in which the light pulse moves from the source to the mirror. To move from the source to the mirror, as seen from frame S the light must move along the *diagonal path* shown. That is, the light must travel *farther* from source to mirror than it did in the rest frame of the clock.

The length of this diagonal is easy to calculate because, according to special relativity, the speed of light is equal to c in all inertial frames. Thus the diagonal length is simply

$$\text{distance} = \text{speed} \times \text{time} = c\left(\frac{1}{2}\,\Delta t\right) = \frac{1}{2}c\,\Delta t$$

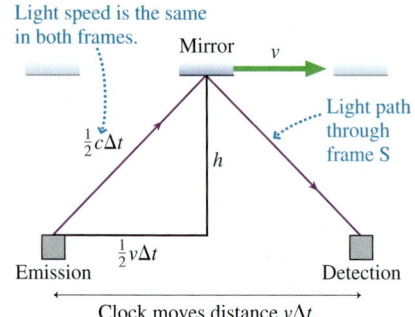

FIGURE 27.17 A light clock analysis in which the speed of light is the same in all reference frames.

We can then apply the Pythagorean theorem to the right triangle in Figure 27.17 to find that

$$h^2 + \left(\frac{1}{2}v\,\Delta t\right)^2 = \left(\frac{1}{2}c\,\Delta t\right)^2 \tag{27.3}$$

We can solve for Δt by first rewriting Equation 27.3 as

$$h^2 = \left(\frac{1}{2}c\,\Delta t\right)^2 - \left(\frac{1}{2}v\,\Delta t\right)^2 = \left[\left(\frac{1}{2}c\right)^2 - \left(\frac{1}{2}v\right)^2\right]\Delta t^2 = \left(\frac{1}{2}\right)^2(c^2 - v^2)\Delta t^2$$

so that

$$\Delta t^2 = \frac{h^2}{(\frac{1}{2})^2(c^2 - v^2)} = \frac{(2h)^2}{(c^2 - v^2)} = \frac{(2h/c)^2}{(1 - v^2/c^2)}$$

from which we find

$$\Delta t = \frac{2h/c}{\sqrt{1 - v^2/c^2}} = \frac{\Delta t'}{\sqrt{1 - v^2/c^2}} \tag{27.4}$$

where we have used Equation 27.2 to write $2h/c$ as $\Delta t'$. The time interval between two ticks in frame S is *not* the same as in frame S′.

It's useful to define $\beta = v/c$, the velocity as a fraction of the speed of light. For example, a reference frame moving with $v = 2.4 \times 10^8$ m/s has $\beta = 0.80$. In terms of β, Equation 27.4 is

$$\Delta t = \frac{\Delta t'}{\sqrt{1 - \beta^2}} \tag{27.5}$$

If reference frame S′ is at rest relative to frame S, then $\beta = 0$ and $\Delta t = \Delta t'$. In other words, experimenters in both reference frames measure time to be the same. But two experimenters moving relative to each other will measure *different* time intervals between the same two events. We're unaware of these differences in our everyday lives because our typical speeds are so small compared to c. But the differences are easily measured in a laboratory, and they do affect the precise timekeeping needed to make accurate location measurements with a GPS receiver.

Proper Time

Frame S' has one important distinction. It is the *one and only* inertial reference frame in which the clock is at rest. Consequently, it is the one and only inertial reference frame in which the times of both events—the emission of the light and the detection of the light—are measured at the *same* position. You can see that the light pulse in Figure 27.16, the rest frame of the clock, starts and ends at the same position, while in Figure 27.17 the emission and detection take place at *different* positions in frame S.

The time interval between two events that occur at the *same position* is called the **proper time** $\Delta\tau$. Only one inertial reference frame measures the proper time, and it can do so with a single clock that is present at both events. Experimenters in an inertial reference frame moving with velocity $v = \beta c$ relative to the proper-time frame must use two clocks to measure the time interval because the two events occur at different positions. We can rewrite Equation 27.5, where the time interval $\Delta t'$ is the proper time $\Delta\tau$, to find that the time interval in any other reference frame is

Activ
Physics
17.1

$$\Delta t = \frac{\Delta\tau}{\sqrt{1-\beta^2}} \geq \Delta\tau \tag{27.6}$$

Time dilation in terms of proper time $\Delta\tau$ (where $\beta = v/c$)

Because $\beta = v/c$, and v is always less than c, β is always less than one. This means that the factor $1/\sqrt{1-\beta^2}$ appearing in Equation 27.6 is always *equal to* (when $v = 0$) *or greater than* one. Thus $\Delta t \geq \Delta\tau$. Recalling that Δt is the time between clock ticks in a frame such as S in which the clock is moving, while the proper time $\Delta\tau$ is the time between ticks in a frame at which the clock is at rest, we can interpret Equation 27.6 as saying that **the time interval between two ticks is the shortest in the reference frame in which the clock is at rest.** The time interval between two ticks is longer when it is measured in any reference frame in which the clock is moving. Because a longer tick interval implies a clock that runs more slowly, we can also say that a **moving clock runs slowly compared to an identical clock at rest.** This "stretching out" of the time interval implied by Equation 27.6 is called **time dilation.**

> **NOTE** ▶ If $v > c$, so that $\beta > 1$, the factor $1 - \beta^2$ appearing in Equation 27.6 would be *negative*. In this case Δt would contain the square root of a negative number; that is, it would be an *imaginary* number. Time intervals certainly have to be real numbers, suggesting that $v > c$ is not physically possible. One of the predictions of relativity, as you've undoubtedly heard, is that nothing can travel faster than the speed of light. Now you can begin to see why. We'll examine this issue more closely later in the chapter. ◄

Equation 27.6 was derived using a light clock because the operation of a light clock is clear and easy to analyze. But the conclusion is really about time itself. Any clock, regardless of how it operates, behaves the same. For example, suppose you and a light clock are traveling in a very fast spaceship. The light clock happens to tick at the same rate as your heart beats—say 60 times a minute, or once a second. Because the light clock is at rest in your frame, it measures the proper time between two successive beats of your heart; that is, $\Delta\tau = 1$ s. But to an experimenter stationed on the ground, watching you pass by at an enormous speed, the time interval $\Delta t = \Delta\tau/\sqrt{1-\beta^2}$ between two ticks of the clock—and hence two beats of your heart—would be *longer*. If, for instance, $\Delta t = 2$ s, the ground-based experimenter would conclude that your heat is beating only 30 times per minute. To the experimenter, *all* processes on your spaceship, including all your biological processes, would appear to run slowly.

EXAMPLE 27.3 **Journey time from the sun to Saturn**

Saturn is 1.43×10^{12} m from the sun. A rocket travels along a line from the sun to Saturn at a constant speed of exactly $0.9c$ relative to the solar system. How long does the journey take as measured by an experimenter on earth? As measured by an astronaut on the rocket?

PREPARE Let the solar system be in reference frame S and the rocket be in reference frame S' that travels with velocity $v = 0.9c$ relative to S. Relativity problems must be stated in terms of *events*. Let event 1 be "the rocket and the sun coincide" (the experimenter on earth says that the rocket passes the sun; the astronaut on the rocket says that the sun passes the rocket) and event 2 be "the rocket and Saturn coincide."

Figure 27.18 shows the two events as seen from the two reference frames. Notice that the two events occur at the *same position* in S', the position of the rocket.

SOLVE The time interval measured in the solar system reference frame, which includes the earth, is simply

$$\Delta t = \frac{\Delta x}{v} = \frac{1.43 \times 10^{12} \text{ m}}{0.9 \times (3.00 \times 10^8 \text{ m/s})} = 5300 \text{ s}$$

Relativity hasn't abandoned the basic definition $v = \Delta x/\Delta t$, although we do have to be sure that Δx and Δt are measured in just one reference frame and refer to the same two events.

How are things in the rocket's reference frame? The two events occur at the *same position* in S'. Thus the time measured by the astronauts is the *proper time* $\Delta \tau$ between the two events. We can then use Equation 27.6 with $\beta = 0.9$ to find

$$\Delta \tau = \sqrt{1 - \beta^2}\, \Delta t = \sqrt{1 - 0.9^2}\,(5300 \text{ s}) = 2310 \text{ s}$$

ASSESS The time interval measured between these two events by the astronauts is less than half the time interval measured by experimenters on earth. The difference has nothing to do with when earthbound astronomers *see* the rocket pass the sun and Saturn. Δt is the time interval from when the rocket actually passes the sun, as measured by a clock at the sun, until it actually passes Saturn, as measured by a synchronized clock at Saturn. The interval between *seeing* the events from earth, which would have to allow for light travel times, would be something other than 5300 s. Δt and $\Delta \tau$ are different because *time is different* in two reference frames moving relative to each other.

Rocket journey in frame S

Event 1

The time between these two events is Δt.

Event 2

$\Delta x = v\Delta t$

Rocket journey in frame S'

Event 1

The time between these two events is $\Delta t'$.

Event 2

$\Delta x' = 0$

FIGURE 27.18 Visual overview of the trip as seen in frames S and S'.

STOP TO THINK 27.6 Molly flies her rocket past Nick at constant velocity v. Molly and Nick both measure the time it takes the rocket, from nose to tail, to pass Nick. Which of the following is true?

A. Both Molly and Nick measure the same amount of time.
B. Molly measures a shorter time interval than Nick.
C. Nick measures a shorter time interval than Molly.

Experimental Evidence

Is there any evidence for the crazy idea that clocks moving relative to each other tell time differently? Indeed, there's plenty. An experiment in 1971 sent an atomic clock around the world on a jet plane while an identical clock remained in the laboratory. This was a difficult experiment because the traveling clock's speed was so small compared to c, but measuring the small differences between the time intervals was just barely within the capabilities of atomic clocks. It was also a more complex experiment than we've analyzed because the clock accelerated as it moved around a circle. Nonetheless, the traveling clock, upon its return, was 200 ns behind the clock that stayed at home, which was exactly as predicted by relativity.

Very detailed studies have been done on unstable particles called *muons* that are created at the top of the atmosphere, at a height of about 60 km, when high-energy cosmic rays collide with air molecules. It is well known, from laboratory

The global positioning system (GPS) If you've ever used a GPS receiver, you know it can pinpoint your location anywhere in the world. The system uses a set of orbiting satellites whose positions are very accurately known. The satellites orbit at a speed of about 14,000 km/h, enough to make the moving satellite's clocks run slow by about 7 μs a day. This may not seem like much, but it would introduce an error of 2000 m into your position! To function properly, the clocks are carefully corrected for effects due to relativity (including effects due to general relativity).

studies, that stationary muons decay with a *half-life* of 1.5 μs. That is, half the muons decay within 1.5 μs, half of those remaining decay in the next 1.5 μs, and so on. The decays can be used as a clock.

The muons travel down through the atmosphere at very nearly the speed of light. The time needed to reach the ground, assuming $v \approx c$, is $\Delta t \approx$ (60,000 m)/(3 × 10⁸ m/s) = 200 μs. This is 133 half lives, so the fraction of muons reaching the ground should be $\approx (1/2)^{133} = 10^{-40}$. That is, only 1 out of every 10^{40} muons should reach the ground. In fact, experiments find that about 1 in 10 muons reach the ground, an experimental result that differs by a factor of 10^{39} from our prediction!

The discrepancy is due to time dilation. In Figure 27.19, the two events "muon is created" and "muon hits ground" take place at two different places in the earth's reference frame. However, these two events occur at the *same position* in the muon's reference frame. (The muon is like the rocket in Example 27.3.) Thus the muon's internal clock measures the proper time. The time-dilated interval $\Delta t = 200\,\mu$s in the earth's reference frame corresponds to a time of only $\Delta t' \approx 5\,\mu$s in the muon's reference frame. That is, in the muon's reference frame it takes only 5 μs from its creation at the top of the atmosphere until the ground runs into it. This is 3.3 half-lives, so the fraction of muons reaching the ground is $(1/2)^{3.3} = 0.1$, or 1 out of 10. We wouldn't detect muons at the ground at all if not for time dilation.

The details are beyond the scope of this textbook, but dozens of high-energy particle accelerators around the world that study quarks and other elementary particles have been designed and built on the basis of Einstein's theory of relativity. The fact that they work exactly as planned is strong testimony to the reality of time dilation.

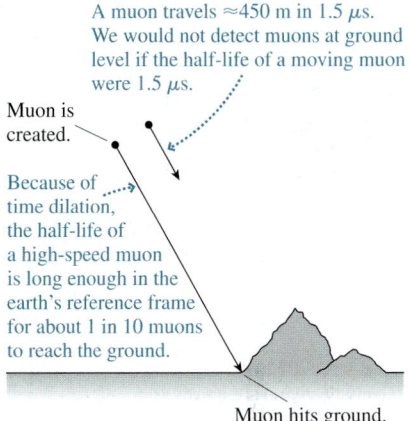

A muon travels ≈450 m in 1.5 μs. We would not detect muons at ground level if the half-life of a moving muon were 1.5 μs.

Muon is created.

Because of time dilation, the half-life of a high-speed muon is long enough in the earth's reference frame for about 1 in 10 muons to reach the ground.

Muon hits ground.

FIGURE 27.19 We wouldn't detect muons at the ground if not for time dilation.

The Twin Paradox

The most well-known relativity paradox is the twin paradox. George and Helen are twins. On their 25th birthday, Helen departs on a starship voyage to a distant star. Let's imagine, to be specific, that her starship accelerates almost instantly to a speed of 0.95c and that she travels to a star that is 9.5 light years (9.5 ly) from earth. Upon arriving, she discovers that the planets circling the star are inhabited by fierce aliens, so she immediately turns around and heads home at 0.95c.

A **light year,** abbreviated ly, is the distance that light travels in one year. A light year is vastly larger than the diameter of the solar system. The distance between two neighboring stars is typically a few light years. For our purpose, we can write the speed of light as $c = 1$ ly/year. That is, light travels 1 light year per year.

This value for c allows us to determine how long, according to George and his fellow earthlings, it takes Helen to travel out and back. Her total distance is 19 ly and, due to her rapid acceleration and rapid turn around, she travels essentially the entire distance at speed $v = 0.95c = 0.95$ ly/year. Thus the time she's away, as measured by George, is

$$\Delta t_G = \frac{19\ \text{ly}}{0.95\ \text{ly/year}} = 20\ \text{years} \qquad (27.7)$$

George will be 45 years old when his sister Helen returns with tales of adventure.

While she's away, George takes a physics class and studies Einstein's theory of relativity. He realizes that time dilation will make Helen's clocks run more slowly than his clocks, which are at rest relative to him. Her heart—a clock—will beat fewer times and the minute hand on her watch will go around fewer times. In other words, she's aging more slowly than he is. Although she is his twin, she will be younger than he is when she returns.

Alpha Centauri (arrow) is one of the closest stars to the sun, at a distance of 4.3 ly. If you traveled there and back at 0.99c, your earthbound friends would be 8.6 years older, while you would have aged only 1.2 years.

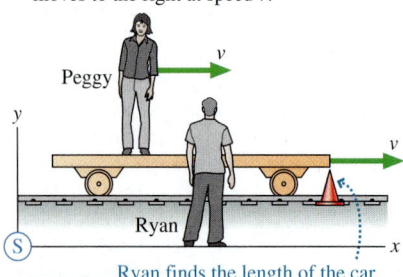

Helen is moving relative to me at 0.95c. Her clocks are running more slowly than mine, and when she returns she'll be younger than I am.

0.95c

9.5 ly

George is moving relative to me at 0.95c. His clocks are running more slowly than mine, and when I return he'll be younger than I am.

FIGURE 27.20 The twin paradox.

(a) In Ryan's frame S, Peggy moves to the right at speed v.

Peggy

v

y

v

Ryan

S

x

Ryan finds the length of the car by measuring the time Δt it takes to pass the cone. Then $L = v\Delta t$.

(b) In Peggy's frame S', Ryan moves to the left at speed v.

y'

S'

x'

$-v$

Peggy finds the length of the car by measuring the time $\Delta t'$ it takes the cone, moving at speed v, to pass her car. Then $L' = v\Delta t'$.

FIGURE 27.21 The length of a train car as measured by Ryan and by Peggy.

Calculating Helen's age is not hard. We simply have to identify Helen's clock, because it's always with Helen as she travels, as the clock that measures proper time $\Delta\tau$. From Equation 27.6,

$$\Delta t_H = \Delta\tau = \sqrt{1 - \beta^2}\,\Delta t_G = \sqrt{1 - 0.95^2}\,(20\text{ years}) = 6.25\text{ years} \quad (27.8)$$

George will have just celebrated his 45th birthday as he welcomes home his 31-year-and-3-month-old twin sister.

This may be unsettling, because it violates our commonsense notion of time, but it's not a paradox. There's no logical inconsistency in this outcome. So why is it called "the twin paradox"? Read on.

Helen, knowing that she had quite of bit of time to kill on her journey, brought along several physics books to read. As she learns about relativity, she begins to think about George and her friends back on earth. Relative to her, they are all moving away at 0.95c. Later they'll come rushing toward her at 0.95c. Time dilation will cause their clocks to run more slowly than her clocks, which are at rest relative to her. In other words, as Figure 27.20 shows, Helen concludes that people on earth are aging more slowly than she is. Alas, she will be much older than they when she returns.

Finally, the big day arrives. Helen lands back on earth and steps out of the starship. George is expecting Helen to be younger than he is. Helen is expecting George to be younger than she is.

Here's the paradox! It's logically impossible for each to be younger than the other at the time when they are reunited. Where, then, is the flaw in our reasoning? It seems to be a symmetrical situation—Helen moves relative to George and George moves relative to Helen—but symmetrical reasoning has led to a conundrum.

But are the situations really symmetrical? George goes about his business day after day without noticing anything unusual. Helen, on the other hand, experiences three distinct periods during which the starship engines fire, she's crushed into her seat, and free dust particles that had been floating inside the starship are no longer, in the starship's reference frame, at rest or traveling in a straight line at constant speed. In other words, George spends the entire time in an inertial reference frame, *but Helen does not*. The situation is *not* symmetrical.

The principle of relativity applies *only* to inertial reference frames. Our discussion of time dilation was for inertial reference frames. Thus George's analysis and calculations are correct. Helen's analysis and calculations are *not* correct because she was trying to apply an inertial reference frame result to a noninertial reference frame.

Helen is younger than George when she returns. This is strange, but not a paradox. It is a consequence of the fact that time flows differently in two reference frames moving relative to each other.

27.7 Length Contraction

We've seen that relativity requires us to rethink our idea of time. Now let's turn our attention to the concepts of space and distance. Consider again Peggy on her train car, which is reference frame S', moving past Ryan, who is at rest in frame S, at relative speed v. Ryan wants to measure the length L of Peggy's car as it moves past him. As shown in Figure 27.21a, he can do so by measuring the time Δt that it takes the car to pass the fixed cone; he then calculates that $L = v\Delta t$.

Figure 27.21b shows the situation in Peggy's reference frame S', where the car is at rest. Peggy wants to measure the length L' of her car; we'll soon see that L' need not be the same as L. Peggy can measure L' by finding the time $\Delta t'$ that the cone, moving at speed v, takes to move from one end of the car to the other. In this way, she finds that $L' = v\Delta t'$.

Speed v is the relative speed between S and S′ and is the same for both Ryan and Peggy. From Ryan's and Peggy's measurements of the car's length we can then write

$$v = \frac{L}{\Delta t} = \frac{L'}{\Delta t'} \qquad (27.9)$$

Activ Physics ONLINE   17.2

The time interval Δt measured in Ryan's frame S is the proper time $\Delta \tau$ because the two events that define the time intervals—the front end and back end of the car passing the cone—occur at the same position (the cone) in Ryan's frame. We can use the time-dilation result, Equation 27.6, to relate $\Delta \tau$ measured by Ryan to $\Delta t'$ measured by Peggy. Equation 27.9 then becomes

$$\frac{L}{\Delta \tau} = \frac{L'}{\Delta t'} = \frac{L'}{\Delta \tau / \sqrt{1 - \beta^2}} \qquad (27.10)$$

The $\Delta \tau$ cancels, so the car's length L in Ryan's frame is related to its length L' in Peggy's frame by

$$L = \sqrt{1 - \beta^2} L' \qquad (27.11)$$

Surprisingly, we find that **the length of the car in Ryan's frame is different from its length in Peggy's frame.**

Peggy's frame S′, in which the car's length is L', has one important distinction. It is the *one and only* inertial reference frame in which the car is at rest. Experimenters in frame S′ can take all the time they need to measure L' because the car isn't moving. The length of an object measured in the reference frame in which the object is at rest is called the **proper length** ℓ. When you measure the length of an everyday object, such as a curtain rod or tabletop, it is usually at rest in your reference frame, so everyday length measurements are of proper length.

We can use the proper length ℓ to write Equation 27.11 as

$$L = \sqrt{1 - \beta^2}\,\ell \le \ell \qquad (27.12)$$

Length contraction in terms of proper length ℓ

Because $\beta \ge 0$, the factor $\sqrt{1 - \beta^2}$ is less than or equal to one. This means that $L \le \ell$. Because ℓ, the proper length, is measured in a reference frame in which the object is at rest while L is measured in a frame in which the object is moving, we see that **the length of an object is greatest in the reference frame in which the object is at rest.** This "shrinking" of the length of an object or the distance between two objects, as measured by an experimenter moving with respect to the object(s), is called **length contraction.**

NOTE ▶ A moving object's length is contracted only in the direction in which it's moving (its length along the x-axis in Figure 27.21). The object's length in the y- and z-directions doesn't change. ◀

Two perspectives on a relativistic trip
The Stanford Linear Accelerator is a 3.2-km-long electron accelerator that accelerates electrons to a speed of $0.99999999995c$. From our perspective, the electrons take a time $\Delta t = (3200 \text{ m})/c = 11 \text{ μs}$ to make the trip. However, we see their "clocks" run slowly, ticking off only 110 ps during the trip. What do the electrons see? In their reference frame the end of the accelerator is coming toward them at $0.99999999995c$, but its length is contracted to only 3.3 cm. Thus it arrives in a time $(3.3 \text{ cm})/c = 110$ ps. The same result, but from a different perspective.

EXAMPLE 27.4 Length contraction of a ladder

Dan holds a 5.0-m-long ladder parallel to the ground. He then gets up to a good sprint, eventually reaching 98% of the speed of light. How long is the ladder according to Dan, once he is running, and according to Carmen, who is standing on the ground as Dan goes by?

PREPARE Let reference frame S′ be attached to Dan. The ladder is at rest in this reference frame, so Dan measures the proper length of the ladder: $\ell = 5.0$ m. Dan's frame S′ moves relative to Carmen's frame S with velocity $v = 0.98c$.

SOLVE We can find the length of the ladder in Carmen's frame from Equation 27.12. We have

$$L = \sqrt{1 - \beta^2}\,\ell = \sqrt{1 - 0.98^2}\,(5.0 \text{ m}) = 1.0 \text{ m}$$

ASSESS The length of the moving ladder as measured by Carmen is only one-fifth its length as measured by Dan. These lengths are different because *space is different* in two reference frames moving relative to each other.

The conclusion that space is different in reference frames moving relative to each other is a direct consequence of the fact that time is different. Experimenters in both reference frames agree on the relative velocity v, leading to Equation 27.9: $v = L/\Delta t = L'/\Delta t'$. Because of time dilation, Ryan (who measures proper time) finds that $\Delta t < \Delta t'$. Thus L *has* to be less than L'. That is the only way Ryan and Peggy can reconcile their measurements.

The Binomial Approximation

A useful mathematical tool is the **binomial approximation.** Suppose we need to evaluate the quantity $(1 + x)^n$. If x is much less than 1, it turns out that an excellent approximation is

$$(1 + x)^n \approx 1 + nx \qquad \text{if} \qquad x \ll 1 \qquad (27.13)$$

You can try this on your calculator. Suppose you need to calculate 1.01^2. Comparing this expression with Equation 27.13, we see that $x = 0.01$ and $n = 2$. Equation 27.13 then tells us that

$$1.01^2 \approx 1 + 2 \times 0.01 = 1.02$$

The exact result, using a calculator, is 1.0201. The approximate answer is good to about 99.99%! The smaller the value of x, the better the approximation.

The binomial approximation is very useful when we need to calculate a relativistic expression for a velocity much less than c, so that $v \ll c$. Because $\beta = v/c$, a reference frame moving with $v^2/c^2 \ll 1$ has $\beta^2 \ll 1$. In these cases, we can write

$$\sqrt{1 - \beta^2} = (1 - v^2/c^2)^{1/2} \approx 1 - \frac{1}{2}\frac{v^2}{c^2}$$

$$\frac{1}{\sqrt{1 - \beta^2}} = (1 - v^2/c^2)^{-1/2} \approx 1 + \frac{1}{2}\frac{v^2}{c^2} \qquad (27.14)$$

The following example illustrates the use of the binomial approximation.

EXAMPLE 27.5 The shrinking school bus

An 8.0-m-long school bus drives past at 30 m/s. By how much is its length contracted?

PREPARE The school bus is at rest in an inertial reference frame S' moving at velocity $v = 30$ m/s relative to the ground frame S. The given length, 8.0 m, is the proper length ℓ in frame S'.

SOLVE In frame S, the school bus is length contracted to

$$L = \sqrt{1 - \beta^2}\,\ell$$

The bus's velocity v is much less than c, so we can use the binomial approximation to write

$$L \approx \left(1 - \frac{1}{2}\frac{v^2}{c^2}\right)\ell = \ell - \frac{1}{2}\frac{v^2}{c^2}\ell$$

The *amount* of the length contraction is

$$\ell - L = \frac{1}{2}\frac{v^2}{c^2}\ell = \left(\frac{30\text{ m/s}}{3.0 \times 10^8\text{ m/s}}\right)^2 (4.0\text{ m})$$

$$= 4.0 \times 10^{-14}\text{ m} = 40\text{ fm}$$

where 1 fm = 1 femtometer = 10^{-15} m.

ASSESS The amount the bus "shrinks" is only slightly larger than the diameter of the nucleus of an atom. It's no wonder that we're not aware of length contraction in our everyday lives. If you had tried to calculate this number exactly, your calculator would have shown $\ell - L = 0$. The difficulty is that the difference between ℓ and L shows up only in the 14th decimal place. A scientific calculator determines numbers to 10 or 12 decimal places, but that isn't sufficient to show the difference. The binomial approximation provides an invaluable tool for finding the very tiny difference between two numbers that are nearly identical.

27.8 Velocities of Objects in Special Relativity

In Section 27.2 we discussed Galilean relativity, which is applicable to objects that are moving at speeds much less than the speed of light. We found that if the velocity of an object is u in reference frame S, then its velocity measured in frame S′, moving at velocity v relative to frame S, is $u' = u - v$.

But we soon learned that this expression is invalid for objects moving at an appreciable fraction of the speed of light. In particular, light itself moves at speed c as measured by *all* observers, independent of their relative velocities. The Galilean transformation of velocity needs to be modified for objects moving at relativistic speeds.

Although a proof is beyond the scope of this text, Einstein's relativity includes a velocity-addition expression valid for *any* velocities. If u is the velocity of an object as measured in reference frame S, and u' its velocity as measured in S′, moving at velocity v relative to frame S, then the two velocities are related by

$$u' = \frac{u - v}{1 - uv/c^2} \quad \text{or} \quad u = \frac{u' + v}{1 + u'v/c^2} \qquad (27.15)$$

Lorentz transformation of velocity

These equations were discovered by Dutch physicist H. A. Lorentz a few years before Einstein published his theory of relativity, but Lorentz didn't completely understand their implications for space and time. Notice that the denominator is ≈ 1 if either u or v is much less than c. In other words, these equations agree with the Galilean velocity transformation when velocities are nonrelativistic (i.e., $\ll c$), but they differ as velocities approach the speed of light.

NOTE ▶ It is important to distinguish carefully between v, which is the relative velocity of the reference frames in which measurements are carried out, and u and u', which are the velocities of an *object* as measured in two different reference frames. ◀

EXAMPLE 27.6 A speeding bullet

A rocket flies past the earth at precisely $0.9c$. As it goes by, the rocket fires a bullet in the forward direction at precisely $0.95c$ with respect to the rocket. What is the bullet's speed with respect to the earth?

PREPARE The rocket and the earth are inertial reference frames. Let the earth be frame S and the rocket be frame S′. The velocity of frame S′ relative to frame S is $v = 0.9c$. The bullet's velocity in frame S′ is $u' = 0.95c$.

SOLVE We can use the Lorentz velocity transformation to find

$$u = \frac{u' + v}{1 + u'v/c^2} = \frac{0.95c + 0.90c}{1 + (0.95c)(0.90c)/c^2} = 0.997c$$

The bullet's speed with respect to the earth is 99.7% of the speed of light.

NOTE ▶ Many relativistic calculations are much easier when velocities are specified as a fraction of c. ◀

ASSESS The Galilean transformation of velocity would give $u = 1.85c$. Now, despite the very high speed of the rocket and of the bullet with respect to the rocket, the bullet's speed with respect to the earth remains less than c. This is yet more evidence that objects cannot exceed the speed of light.

Suppose the rocket in Example 27.6 fired a laser beam in the forward direction as it traveled past the earth at velocity v. The laser beam would travel away from the rocket at speed $u' = c$ in the rocket's reference frame S'. What is the laser beam's speed in the earth's frame S? According to the Lorentz velocity transformation, it must be

$$u = \frac{u' + v}{1 + u'v/c^2} = \frac{c + v}{1 + cv/c^2} = \frac{c + v}{1 + v/c} = \frac{c + v}{(c + v)/c} = c$$

Light travels at speed c in both frame S and frame S'. This important consequence of the principle of relativity is "built into" the Lorentz velocity transformation.

27.9 Relativistic Momentum

In Newtonian mechanics, the total momentum of a system is a conserved quantity. Further, the law of conservation of momentum, $P_f = P_i$, is true in all inertial reference frames *if* the particle velocities in different reference frames are related by the Galilean velocity transformation.

The difficulty, of course, is that the Galilean transformation is not consistent with the principle of relativity. It is a reasonable approximation when all velocities are much less than c, but the Galilean transformation fails dramatically as velocities approach c. It's not hard to show that $P_f' \neq P_i'$ if the particle velocities in frame S' are related to the particle velocities in frame S by the Lorentz velocity transformation.

There are two possibilities:

1. The so-called law of conservation of momentum is not really a law of physics. It is approximately true at low velocities but fails as velocities approach the speed of light.
2. The law of conservation of momentum really is a law of physics, but the expression $p = mu$ is not the correct way to calculate momentum when the particle velocity u becomes a significant fraction of c.

In this photograph, the track (shown in red) of a high-energy proton enters from the lower right. The proton then collides with other protons, sending them in all directions, where further collisions occur. Momentum is conserved, but at these high speeds the relativistic expression for momentum must be used.

Momentum conservation is such a central and important feature of mechanics that it seems unlikely to fail in relativity. How else might the momentum of a particle be defined?

The classical momentum, for one-dimensional motion, is $p = mu = m(\Delta x/\Delta t)$. Δt is the time needed to move distance Δx. That seemed clear enough within a Newtonian framework, but now we've learned that experimenters in different reference frames disagree about the amount of time needed. So whose Δt should we use?

One possibility is to use the time measured *by the particle*. This is the proper time $\Delta\tau$ because the particle is at rest in its own reference frame. With this in mind, let's redefine the momentum of a particle of mass m moving with velocity $u = \Delta x/\Delta t$ to be

$$p = m\frac{\Delta x}{\Delta\tau} \tag{27.16}$$

We can relate this new expression for p to the familiar Newtonian expression by using the time-dilation result $\Delta\tau = (1 - u^2/c^2)^{1/2}\Delta t$ to relate the proper time interval measured by the particle to the more practical time interval Δt measured by experimenters in frame S. With this substitution, Equation 27.16 becomes

$$p = m\frac{\Delta x}{\Delta\tau} = m\frac{\Delta x}{\sqrt{1 - u^2/c^2}\,\Delta t} = \frac{mu}{\sqrt{1 - u^2/c^2}} \tag{27.17}$$

You can see that Equation 27.17 reduces to the classical expression $p = mu$ when the particle's speed $u \ll c$. That is an important requirement, but whether this is the "correct" expression for p depends on whether the total momentum P is conserved when the velocities of a system of particles are transformed with the Lorentz velocity transformation equations. The proof is rather long and tedious, so we will assert, without actual proof, that the momentum defined in Equation 27.17 is, indeed, conserved. **The law of conservation of momentum is still valid in all inertial reference frames *if* the momentum of each particle is calculated with Equation 27.17.**

To simplify our notation, let's define the quantity

$$\gamma = \frac{1}{\sqrt{1 - u^2/c^2}} \tag{27.18}$$

With this definition of γ, the momentum of a particle is

$$p = \gamma mu \tag{27.19}$$

Relativistic momentum for a particle with mass m and speed u

$v = \frac{d}{t}$

EXAMPLE 27.7 Momentum of a subatomic particle

Electrons in a particle accelerator reach a speed of $0.999c$ relative to the laboratory. One collision of an electron with a target produces a muon that moves forward with a speed of $0.950c$ relative to the laboratory. The muon mass is 1.90×10^{-28} kg. What is the muon's momentum in the laboratory frame and in the frame of the electron beam?

PREPARE Let the laboratory be reference frame S. The reference frame S' of the electron beam (i.e., a reference frame in which the electrons are at rest) moves in the direction of the electrons at $v = 0.999c$. The muon velocity in frame S is $u = 0.95c$.

SOLVE γ for the muon in the laboratory reference frame is

$$\gamma = \frac{1}{\sqrt{1 - u^2/c^2}} = \frac{1}{\sqrt{1 - 0.95^2}} = 3.20$$

Thus the muon's momentum in the laboratory is

$$p = \gamma mu$$

$$= (3.20)(1.90 \times 10^{-28}\text{ kg})(0.95 \times 3.00 \times 10^8\text{ m/s})$$

$$= 1.73 \times 10^{-19}\text{ kg} \cdot \text{m/s}$$

The momentum is a factor of 3.2 larger than the Newtonian momentum mu. To find the momentum in the electron-beam reference frame, we must first use the velocity transformation equation to find the muon's velocity in frame S':

$$u' = \frac{u - v}{1 - uv/c^2} = \frac{0.95c - 0.999c}{1 - (0.95c)(0.999c)/c^2} = -0.962c$$

In the laboratory frame, the faster electrons are overtaking the slower muon. Hence the muon's velocity in the electron-beam frame is negative. γ' for the muon in frame S' is

$$\gamma' = \frac{1}{\sqrt{1 - u'^2/c^2}} = \frac{1}{\sqrt{1 - 0.962^2}} = 3.66$$

The muon's momentum in the electron-beam reference frame is

$$p' = \gamma' mu'$$

$$= (3.66)(1.90 \times 10^{-28}\text{ kg})(-0.962 \times 3.00 \times 10^8\text{ m/s})$$

$$= -2.01 \times 10^{-19}\text{ kg} \cdot \text{m/s}$$

ASSESS From the laboratory perspective, the muon moves only slightly slower than the electron beam. But it turns out that the muon moves faster with respect to the electrons, although in the opposite direction, than it does with respect to the laboratory.

The Cosmic Speed Limit

Figure 27.22a on the next page is a graph of momentum versus velocity. For a Newtonian particle, with $p = mu$, the momentum is directly proportional to the velocity. The relativistic expression for momentum agrees with the Newtonian value if $u \ll c$, but p approaches ∞ as $u \to c$.

(a)

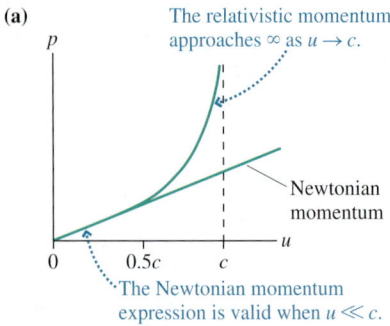

The relativistic momentum approaches ∞ as $u \rightarrow c$.

Newtonian momentum

The Newtonian momentum expression is valid when $u \ll c$.

(b)

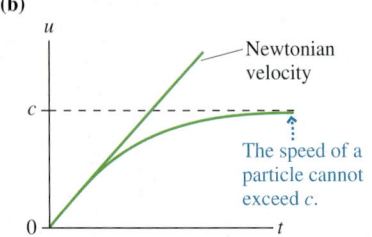

Newtonian velocity

The speed of a particle cannot exceed c.

FIGURE 27.22 The speed of a particle cannot reach the speed of light.

The implications of this graph become clear when we relate momentum to force. Consider a particle subjected to a constant force, such as a rocket that never runs out of fuel. From the impulse-momentum theorem we have $\Delta p = F\Delta t$, or $p = mu = Ft$ if the rocket starts from rest at $t = 0$. If Newtonian physics were correct, a particle would go faster and faster as its velocity $u = p/m = (F/m)t$ increased without limit. But the relativistic result, shown in Figure 27.22b, is that the particle's velocity approaches the speed of light ($u \rightarrow c$) as p approaches ∞. Relativity gives a very different outcome than Newtonian mechanics.

The speed c is a "cosmic speed limit" for material particles. A force cannot accelerate a particle to a speed higher than c because the particle's momentum becomes infinitely large as the speed approaches c. The amount of effort required for each additional increment of velocity becomes larger and larger until no amount of effort can raise the velocity any higher.

Actually, at a more fundamental level, c is a speed limit for *any* kind of **causal influence**. If I throw a rock and break a window, my throw is the *cause* of the breaking window and the rock is the *causal influence*. If I shoot a laser beam at a light detector that is wired to a firecracker, the light wave is the *causal influence* that leads to the explosion. A causal influence can be any kind of particle, wave, or information that travels from A to B and allows A to be the cause of B.

For two unrelated events—a firecracker explodes in Tokyo and a balloon bursts in Paris—the relativity of simultaneity tells us that they may be simultaneous in one reference frame but not in others. Or in one reference frame the firecracker may explode before the balloon bursts, but in some other reference frame the balloon may burst first. These possibilities violate our commonsense view of time, but they're not in conflict with the principle of relativity.

For two causally related events—A *causes* B—it would be nonsense for an experimenter in any reference frame to find that B occurs before A. No experimenter in any reference frame, no matter how it is moving, will find that you are born before your mother is born. If A causes B, then it must be the case that $t_A < t_B$ in *all* reference frames.

Suppose there exists some kind of causal influence that *can* travel at speed $u > c$. For example, Figure 27.23 shows Peggy traveling in the exact center of a 1000-m-long (as measured in the reference frame of the ground) train car, so that her position at $t = 0$ s is $x = 500$ m. The car is moving at $v = (2/3)c$, or 200 m/μs, so she moves 200 m to the right every microsecond.

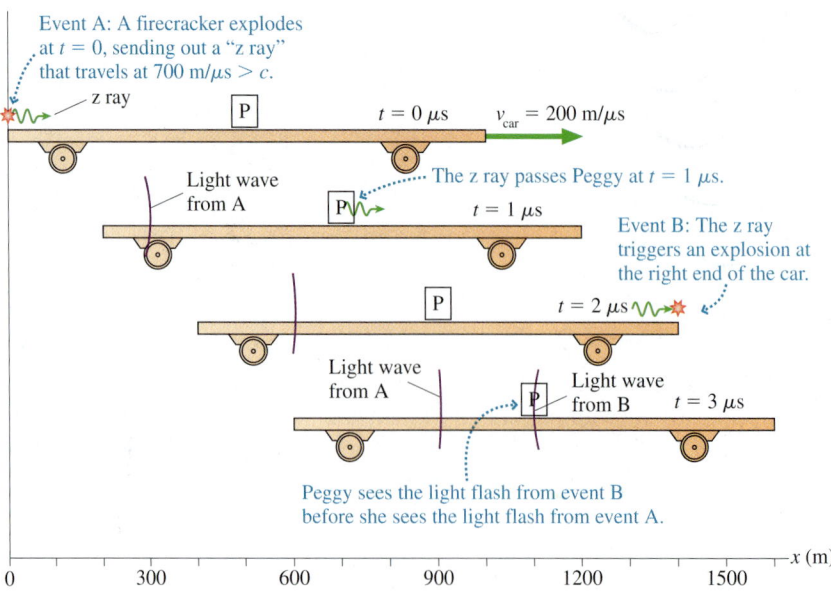

Event A: A firecracker explodes at $t = 0$, sending out a "z ray" that travels at 700 m/μs > c.

z ray

$t = 0$ μs $v_{car} = 200$ m/μs

Light wave from A   The z ray passes Peggy at $t = 1$ μs.

$t = 1$ μs

Event B: The z ray triggers an explosion at the right end of the car.

$t = 2$ μs

Light wave from A   Light wave from B $t = 3$ μs

Peggy sees the light flash from event B before she sees the light flash from event A.

FIGURE 27.23 Faster-than-light travel violates causality.

At $t = 0$ s, as the left end of the car passes $x = 0$ m, a firecracker explodes at this point. Call this explosion event A. The explosion emits a mysterious "z ray" that travels faster than light at $u = (7/3)c$, or 700 m/μs. At $t = 1$ μs, the z ray reaches Peggy who is now at $x = 700$ m; note that the light wave from the explosion is lagging far behind. At $t = 2$ μs, the z ray reaches the far end of the car, now at $x = 1400$ m, where it causes a second firecracker to explode. Call this event B. This explosion sends a light wave back toward Peggy. The light wave from event B reaches Peggy at $t = 3$ μs, and the light wave from event A, the original explosion, finally catches up with her at $t = 5$ μs (not shown).

How does Peggy interpret these events? We noted earlier that *seeing* an event is not the same as the event itself because light takes time to travel from the event to an experimenter. The order in which you see events is not necessarily the order in which they happened. But Peggy knows two important pieces of information that allow her to turn her *observations* of light flashes into *knowledge* about events A and B. First, she is in the center of the train car, so both explosions occurred at equal distances from her. Second, the speed of light is the same in all reference frames. The analysis of Figure 27.23 showed light moving at 300 m/μs in the ground's reference frame, but it also moves at 300 m/μs in Peggy's reference frame. Because she sees flash B before flash A, events A and B were equally distant, and the light from both traveled to her at the same speed, Peggy concludes—correctly—that event B happened before event A. But this is a logical impossibility. Event A *caused* event B, so it's not possible for B to occur before A.

This example shows that if a causal influence can travel faster than c, then there exist reference frames in which the effect happens before the cause. We know this can't happen, so our assumption $u > c$ must be wrong. **No causal influence of any kind—particle, wave, or yet-to-be-discovered z rays—can travel faster than c.**

The existence of a cosmic speed limit is one of the most interesting consequences of the theory of relativity. "Hyperdrive," in which a spaceship suddenly leaps to faster-than-light velocities, is simply incompatible with the theory of relativity. Rapid travel to the stars will remain in the realm of science fiction unless future scientific discoveries find flaws in Einstein's theory and open the doors to yet-undreamed-of theories. While we can't say with certainty that a scientific theory will never be overturned, there is currently not even a hint of evidence that disagrees with the special theory of relativity.

In the *Star Wars* movies, the heroes escaped the imperial forces by jumping to hyperspace, a process in which the spaceship evidently exceeds the speed of light. According to relativity, this would allow Luke to go back before he was born, and maybe convince his dad to be a better guy.

27.10 Relativistic Energy

Energy is our final topic in this chapter on relativity. Space, time, velocity, and momentum are changed by relativity, so it seems inevitable that we'll need a new view of energy. Indeed, one of the most profound results of relativity, and perhaps the one with the most far-reaching consequences, was Einstein's discovery of the fundamental relationship between energy and mass.

Consider an object of mass m moving at speed u. Einstein found that the **total energy** of such an object is

$$E = \frac{mc^2}{\sqrt{1 - u^2/c^2}} = \gamma mc^2 \qquad (27.20)$$

Total energy of an object of mass m moving at speed u

where $\gamma = 1/(1 - u^2/c^2)^{1/2}$ was defined in Equation 27.18.

This fuel rod for a nuclear power reactor contains about 5 kg of uranium. Its usable energy content, which comes from the conversion of a small fraction of the uranium's mass to energy, is equivalent to that of about 10 million kg of coal.

To understand this expression, let's start by examining its behavior for objects traveling at speeds much less than the speed of light. In this case we can use the binomial approximation you learned in Section 27.7 to write

$$\gamma = \frac{1}{\sqrt{1 - u^2/c^2}} \approx 1 + \frac{1}{2}\frac{u^2}{c^2}$$

For low speeds u, then, the object's total energy is

$$E \approx mc^2 + \frac{1}{2}mu^2 \qquad (27.21)$$

The second term in this expression is the familiar Newtonian kinetic energy $K = \frac{1}{2}mu^2$, written here in terms of velocity u rather than v. But there is an additional term in the total energy, the **rest energy** given by

$$E_0 = mc^2 \qquad (27.22)$$

When a particle is at rest, with $u = 0$, it still has energy E_0. Indeed, because c is so large, the rest energy can be enormous. Equation 27.22 is, of course, Einstein's famous $E = mc^2$, perhaps the most famous equation in all of physics. It tells us that there is a fundamental equivalence between mass and energy, an idea we'll explore later in this section.

EXAMPLE 27.8 The rest energy of an apple

What is the rest energy of a 200 g apple?

SOLVE From Equation 27.22 we have

$$E_0 = mc^2 = (0.20 \text{ kg})(3.0 \times 10^8 \text{ m/s})^2 = 1.8 \times 10^{16} \text{ J}$$

ASSESS This is an enormous energy, enough to power a medium-sized city for about a year.

Equation 27.21 suggests that the total energy of an object is the sum of a rest energy, which is a new idea, and the familiar kinetic energy. But Equation 27.21 is valid only when the object's speed is low compared to c. For higher speeds, we need to use the full energy expression, Equation 27.20. We can use Equation 27.20 to find a relativistic expression for the kinetic energy K by subtracting the rest energy E_0 from the total energy. Doing so gives

$$K = E - E_0 = \gamma mc^2 - mc^2 = (\gamma - 1)mc^2 = (\gamma - 1)E_0 \qquad (27.23)$$

Thus we can write the total energy of an object of mass m as

$$E = \underbrace{mc^2}_{\text{Rest energy } E_0} + \underbrace{(\gamma - 1)mc^2}_{\text{Kinetic energy } K} \qquad (27.24)$$

EXAMPLE 27.9 Comparing energies of a ball and an electron

Calculate the rest energy and the kinetic energy of (a) a 100 g ball moving with a speed of 100 m/s and (b) an electron with a speed of $0.999c$.

PREPARE The ball, with $u \ll c$, is a classical particle. We don't need to use the relativistic expression for its kinetic energy. The electron is highly relativistic.

SOLVE

a. For the ball, with $m = 0.100$ kg,

$$E_0 = mc^2 = 9.00 \times 10^{15} \text{ J}$$

$$K = \frac{1}{2}mu^2 = 500 \text{ J}$$

b. For the electron, we start by calculating

$$\gamma = \frac{1}{\sqrt{1 - u^2/c^2}} = 22.4$$

Then, using $m_e = 9.11 \times 10^{-31}$ kg,

$$E_0 = mc^2 = 8.20 \times 10^{-14} \text{ J}$$

$$K = (\gamma - 1)E_0 = 175 \times 10^{-14} \text{ J}$$

ASSESS The ball's kinetic energy is a typical kinetic energy. Its rest energy, by contrast, is a staggeringly large number. For a relativistic electron, on the other hand, the kinetic energy is more important than the rest energy.

STOP TO THINK 27.7   An electron moves through the lab at a speed such that $\gamma = 1.5$. The electron's kinetic energy is

  A. Greater than its rest energy.
  B. Equal to its rest energy.
  C. Less than its rest energy.

The Equivalence of Mass and Energy

Now we're ready to explore the significance of Einstein's famous equation $E = mc^2$. Figure 27.24 shows an experiment that has been done countless times in the last 50 years at particle accelerators around the world. An electron that has been accelerated to $u \approx c$ is aimed at a target material. When a high-energy electron collides with an atom in the target, it can easily knock one of the electrons out of the atom. Thus we would expect to see two electrons leaving the target: the incident electron and the ejected electron. Instead, *four* particles emerge from the target: three electrons and a positron. A *positron*, or positive electron, is the antimatter version of an electron, identical to an electron in all respects other than having charge $q = +e$. In particular, a positron has the same mass m_e as an electron.

The tracks of elementary particles in a bubble chamber show the creation of an electron-positron pair. The negative electron and positive positron spiral in opposite directions in the magnetic field.

FIGURE 27.24 An inelastic collision between electrons can create an electron-positron pair.

In chemical-reaction notation, the collision is

$$e^-(\text{fast}) + e^-(\text{at rest}) \rightarrow e^- + e^- + e^- + e^+$$

An electron and a positron have been created, apparently out of nothing. Mass $2m_e$ before the collision has become mass $4m_e$ after the collision. (Notice that charge has been conserved in this collision.)

Although the mass has increased, it wasn't created "out of nothing." If you measured the energies before and after the collision, you would find that the kinetic energy before the collision was *greater* than the kinetic energy after. In

FIGURE 27.25 The annihilation of an electron-positron pair.

fact, the decrease in kinetic energy is exactly equal to the rest energy of the two particles that have been created: $\Delta K = 2m_e c^2$. The new particles have been created *out of energy!*

Not only can particles be created from energy, particles can return to energy. Figure 27.25 shows an electron colliding with a positron, its antimatter partner. When a particle and its antiparticle meet, they *annihilate* each other. The mass disappears, and the energy equivalent of the mass is transformed into two high-energy photons. Photons have no mass, and represent pure energy. Positron-electron annihilation is also the basis of the medical procedure known as a positron-emission tomography, or PET scans. We'll study this important diagnostic tool in detail in Chapter 30.

Conservation of Energy

The creation and annihilation of particles with mass, processes strictly forbidden in Newtonian mechanics, are vivid proof that neither mass nor the Newtonian definition of energy is conserved. Even so, the *total* energy—the kinetic energy *and* the energy equivalent of mass—remains a conserved quantity.

> **Law of conservation of total energy** The energy $E = \sum E_i$ of an isolated system is conserved, where $E_i = \gamma_i m_i c^2$ is the total energy of particle i.

Mass and energy are not the same thing, but, as the last few examples have shown, they are *equivalent* in the sense that mass can be transformed into energy and energy can be transformed into mass as long as the total energy is conserved.

Probably the most well-known application of the conservation of total energy is nuclear fission. The uranium isotope ^{236}U, containing 236 protons and neutrons, does not exist in nature. It can be created when a ^{235}U nucleus absorbs a neutron, increasing its atomic mass from 235 to 236. The ^{236}U nucleus quickly fragments into two smaller nuclei and several extra neutrons, a process known as **nuclear fission.** The nucleus can fragment in several ways, but one is

$$n + {}^{235}\text{U} \rightarrow {}^{236}\text{U} \rightarrow {}^{144}\text{Ba} + {}^{89}\text{Kr} + 3n \qquad (27.25)$$

Ba and Kr are the atomic symbols for barium and krypton.

This reaction seems like an ordinary chemical reaction—until you check the masses. The masses of atomic isotopes are known with great precision from many decades of measurement in instruments called mass spectrometers. As shown in Table 27.1, if you add up the masses on both sides, you find that the mass of the products is 0.186 u smaller than the mass of the initial neutron and ^{235}U, where, you will recall, 1 u = 1.66×10^{-27} kg is the atomic mass unit. Converting to kilograms gives us the mass loss of 3.07×10^{-28} kg.

Mass has been lost, but the energy equivalent of the mass has not. As Figure 27.26 shows, the mass has been converted to kinetic energy, causing the two product nuclei and three neutrons to be ejected at very high speeds. The kinetic energy is easily calculated: $\Delta K = m_{\text{lost}}c^2 = 2.8 \times 10^{-11}$ J.

This is a very tiny amount of energy, but it is the energy released from *one* fission. The number of nuclei in a macroscopic sample of uranium is on the order of N_A, Avogadro's number. Hence the energy available if *all* the nuclei fission is enormous. This energy, of course, is the basis for both nuclear power reactors and nuclear weapons.

We started this chapter with an expectation that relativity would challenge our basic notions of space and time. We end by finding that relativity changes our understanding of mass and energy. Most remarkable of all is that each and every one of these new ideas flows from one simple statement: The laws of physics are the same in all inertial reference frames.

TABLE 27.1 Mass before and after fission of ^{235}U

Initial nucleus	Initial mass (u)	Final nucleus	Final mass (u)
^{235}U	235.0439	^{144}B	143.9229
n	1.0087	^{89}K	88.9176
		3n	3.0260
Total	236.0526		235.8665

The mass of the reactants is 0.186 u more than the mass of the products.

n •→ ^{235}U

^{236}U

^{144}Ba ^{89}Kr

0.186 u of mass has been converted into kinetic energy.

FIGURE 27.26 In nuclear fission, the energy equivalent of lost mass is converted into kinetic energy.

The goal of Chapter 27 has been to understand how Einstein's theory of relativity changes our concepts of space and time.

GENERAL PRINCIPLES

Principle of Relativity

All the laws of physics are the same in all inertial reference frames.

- The speed of light c is the same in all inertial reference frames.
- No particle or causal influence can travel at a speed greater than c.

IMPORTANT CONCEPTS

Time

Time measurements depend on the motion of the experimenter relative to the events.

Proper time $\Delta\tau$ is the time interval between two events measured in a reference frame in which the events occur at the same position. The time interval Δt between the same two events, in a frame moving with relative velocity v, is

$$\Delta t = \Delta\tau/\sqrt{1 - \beta^2} \geq \Delta\tau$$

where $\beta = v/c$. This is called **time dilation.**

Space

Spatial measurements depend on the motion of the experimenter relative to the events.

Proper length ℓ is the length of an object measured in a reference frame in which the object is at rest. The length L in a frame in which the object moves with velocity v is

$$L = \sqrt{1 - \beta^2}\,\ell \leq \ell$$

This is called **length contraction.**

Momentum

The law of conservation of momentum is valid in all inertial reference frames if the momentum of a particle with velocity u is $p = \gamma mu$, where

$$\gamma = 1/\sqrt{1 - u^2/c^2}$$

The momentum approaches ∞ as $u \to c$.

Energy

The **total energy** of a particle is $E = \gamma mc^2$. This can be written as

$$E = \underbrace{mc^2}_{\text{Rest energy } E_0} + \underbrace{(\gamma - 1)mc^2}_{\text{Kinetic energy } K}$$

K approaches ∞ as $u \to c$.

The total energy of an isolated system is conserved.

Simultaneity

Events that are simultaneous in reference frame S are not simultaneous in frame S′ moving relative to S.

Mass-energy equivalence

Mass m can be transformed into energy $E = mc^2$.

Energy can be transformed into mass $m = E/c^2$.

APPLICATIONS

An **event** happens at a specific place in space and time. Spacetime coordinates are (x, t) in frame S and (x', t') in frame S′.

A **reference frame** is a coordinate system with meter sticks and clocks for measuring events. Experimenters at rest relative to each other share the same reference frame.

If an object has velocity u in frame S and u' in frame S′, the two velocities are related by the **Lorentz velocity transformation:**

$$u' = \frac{u - v}{1 - uv/c^2} \qquad u = \frac{u' + v}{1 + u'v/c^2}$$

where v is the relative velocity between the two frames.

For instructor-assigned homework, go to www.masteringphysics.com

Problem difficulty is labeled as I (straightforward) to IIII (challenging).

Problems labeled INT integrate significant material from earlier chapters; BIO are of biological or medical interest.

QUESTIONS

Conceptual Questions

1. You are in an airplane cruising smoothly at 600 mph. What experiment, if any, could you do that would demonstrate that you are moving, while those on the ground are at rest?

2. Frame S' moves relative to frame S as shown in Figure Q27.2.
   a. A ball is at rest in frame S'. What are the speed and direction of the ball in frame S?
   b. A ball is at rest in frame S. What are the speed and direction of the ball in frame S'?

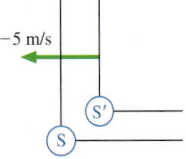

FIGURE Q27.2

3. a. Two balls move as shown in Figure Q27.3. What are the speed and direction of each ball in a reference frame that moves with ball 1?
   b. What are the speed and direction of each ball in a reference frame that moves with ball 2?

FIGURE Q27.3

4. A lighthouse beacon alerts ships to the danger of a rocky coastline.
   a. According to the lighthouse keeper, with what speed does the light leave the lighthouse?
   b. A boat is approaching the coastline at speed $0.5c$. According to the captain, with what speed is the light from the beacon approaching her boat?

5. As a rocket passes the earth at $0.75c$, it fires a laser perpendicular to its direction of travel as shown in Figure Q27.5.
   a. What is the speed of the laser beam relative to the rocket?
   b. What is the speed of the laser beam relative to the earth?

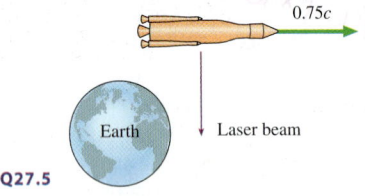

FIGURE Q27.5

6. At the instant that a clock standing next to you reads $t = 1.0 \ \mu s$, you look at a second clock, 300 m away, and see that it reads $t = 0 \ \mu s$. Are the two clocks synchronized? If not, which one is ahead?

7. Firecracker 1 is 300 m from you. Firecracker 2 is 600 m from you in the same direction. You see both explode at the same time. Define event 1 to be "firecracker 1 explodes" and event 2 to be "firecracker 2 explodes." Does event 1 occur before, after, or at the same time as event 2? Explain.

8. Firecrackers 1 and 2 are 600 m apart. You are standing exactly halfway between them. Your lab partner is 300 m on the other side of firecracker 1. You see two flashes of light, from the two explosions, at exactly the same instant of time. Define event 1 to be "firecracker 1 explodes" and event 2 to be "firecracker 2 explodes." According to your lab partner, based on measurements he or she makes, does event 1 occur before, after, or at the same time as event 2? Explain.

9. Your clocks and calendars are synchronized with the clocks and calendars in a star system exactly 10 ly from earth that is at rest relative to the earth. You receive a TV transmission from the star system that shows a date and time display. The date it shows is June 17, 2050. When you glance over at your own wall calendar, what date does it show?

10. Two trees are 600 m apart. You are standing exactly halfway between them and your lab partner is at the base of tree 1. Lightning strikes both trees.
    a. Your lab partner, based on measurements he makes, determines that the two lightning strikes were simultaneous. What did you see? Did you see the lightning hit tree 1 first, hit tree 2 first, or hit them both at the same instant of time? Explain.
    b. Lightning strikes again. This time your lab partner sees both flashes of light at the same instant of time. What did you see? Did you see the lightning hit tree 1 first, hit tree 2 first, or hit them both at the same instant of time? Explain.
    c. In the scenario of part b, were the lightning strikes simultaneous? Explain.

11. Figure Q27.11 shows Peggy standing at the center of her railroad car as it passes Ryan on the ground. Firecrackers attached to the ends of the car explode. A short time later, the flashes from the two explosions arrive at Peggy at the same time.
    a. Were the explosions simultaneous in Peggy's reference frame? If not, which exploded first? Explain.
    b. Were the explosions simultaneous in Ryan's reference frame. If not, which exploded first? Explain.

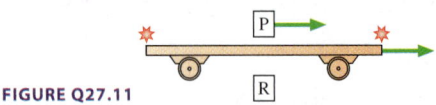

FIGURE Q27.11

12. In Figure Q27.12, clocks C_1 and C_2 in frame S are synchronized. Clock C' moves at speed v relative to frame S. Clocks C' and C_1 read exactly the same as C' goes past. As C' passes C_2, is the time shown on C' earlier, later, or the same as the time shown on C_2? Explain.

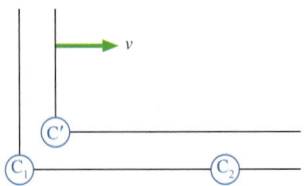

FIGURE Q27.12

13. A meter stick passes you at a speed of 0.5c. Explain clearly how you would measure the length of this fast-moving object.

14. You're passing a car on the highway. You want to know how much time is required to completely pass the car, from no overlap between the cars to no overlap between the cars. Call your car A, the car you are passing B.
    a. Specify two events that can be given spacetime coordinates. In describing the events, refer to cars A and B and to their front bumpers and rear bumpers.
    b. In either reference frame, is there *one* clock that is present at both events?
    c. Who, if anyone, measures the proper time between the events?

15. Your friend flies from Los Angeles to New York. He determines the distance using the tried-and-true $d = vt$. You and your assistants on the ground also measure the distance using meter sticks and surveying equipment.
    a. Who, if anyone, measures the proper length?
    b. Who, if anyone, measures the shorter distance?

16. A 100-m-long train is heading for an 80-m-long tunnel. If the train moves sufficiently fast, is it possible, according to experimenters on the ground, for the entire train to be inside the tunnel at one instant of time? Explain.

17. Dan runs at 0.99c with a 5-m-long ladder held parallel to the ground, but *perpendicular* to his direction of motion. Will he be able to pass through a barn door that is 1.0 m wide?

18. The rocket speeds shown in Figure Q27.18 are relative to the earth. Is the speed of A relative to B greater than, less than, or equal to 0.8c?

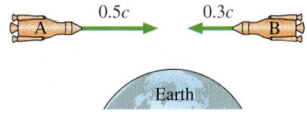

FIGURE Q27.18

19. Can a particle of mass m have total energy less than mc^2? Explain.

20. In your chemistry classes, you have probably learned that, in a chemical reaction, the mass of the products is equal to the mass of the reactants. That is, the mass of the substances produced in a chemical reaction is equal to the mass of the substances consumed in the reaction. Is this absolutely true, or is there actually a small difference? Explain.

Multiple-Choice Questions

21. ‖ Lee and Leigh are twins. At their first birthday party, Lee is placed on a spaceship that travels away from the earth and back at a steady 0.866c. The spaceship eventually returns, landing in the swimming pool at Leigh's eleventh birthday party. When Lee emerges from the ship, it is discovered that:
    A. He is still only 1 year old.   B. He is 6 years old.
    C. He is also 11 years old.   D. He is 21 years old.

22. ‖ A space cowboy wants to eject from his spacecraft 100,000 km after passing a space buoy, as seen by spectators at rest with respect to the buoy. To do this, the cowboy sets a timer on his craft that will start as he passes the buoy. He plans to cruise by the buoy at 0.300c. How much time should he allow between passing the buoy and ejecting?
    A. 1.01 s   B. 1.06 s   C. 1.11 s
    D. 1.33 s   E. 1.58 s

23. ‖‖ Event 1 occurs at (0 m, 0 s) in reference frame S. The following events occur, also measured in reference frame S. Which of them could either be causes of event 1 or be caused by event 1?
    A. (1 m, 2 ns)   B. (−1 m, −2 ns)   C. (−1 m, 2 ns)
    D. All of the above.
    E. None of the above.

24. ‖ Energy in the sun is produced by the fusion of four protons into a helium nucleus. The process involves several steps, but the net reaction is simply $4p \rightarrow {}^4\text{He}$ + energy. Given this, you can say that
    A. One helium atom has more mass than four hydrogen atoms.
    B. One helium atom has less mass than four hydrogen atoms.
    C. One helium atom has the same mass as four hydrogen atoms.

25. ‖ A particle moving at speed 0.40c has momentum p_0. The speed of the particle is increased to 0.80c. Its momentum is now
    A. Less than $2p_0$.   B. Exactly $2p_0$.   C. Greater than $2p_0$.

26. ‖ A particle moving at speed 0.40c has kinetic energy K_0. The speed of the particle is increased to 0.80c. The kinetic energy is now
    A. Less than $4K_0$.   B. Exactly $4K_0$.   C. Greater than $4K_0$.

PROBLEMS

Section 27.2 Galilean Relativity

1. ‖‖ A sprinter crosses the finish line of a race. The roar of the crowd in front approaches her at a speed of 360 m/s. The roar from the crowd behind her approaches at 330 m/s. What are the speed of sound and the speed of the sprinter?

2. ‖ A baseball pitcher can throw a ball with a speed of 40 m/s. He is in the back of a pickup truck that is driving away from you. He throws the ball in your direction, and it floats toward you at a lazy 10 m/s. What is the speed of the truck?

3. ‖ A boy on a skateboard coasts along at 5 m/s. He has a ball that he can throw at a speed of 10 m/s. What is the ball's speed relative to the ground if he throws the ball
    a. Forward?   b. Backward?

4. ‖‖ A boat takes 3.0 hours to travel 30 km down a river, then 5.0 hours to return. How fast is the river flowing?

5. ‖ When the moving sidewalk at the airport is broken, as it often seems to be, it takes you 50 s to walk from your gate to baggage claim. When it is working and you stand on the moving sidewalk the entire way, without walking, it takes 75 s to travel the same distance. How long will it take you to travel from the gate to baggage claim if you walk while riding on the moving sidewalk?

6. ‖ An assembly line has a staple gun that rolls to the left at 1.0 m/s while parts to be stapled roll past it to the right at 3.0 m/s. The staple gun fires 10 staples per second. How far apart are the staples in the finished part?

Section 27.3 Einstein's Principle of Relativity

7. ‖ An out-of-control alien spacecraft is diving into a star at a speed of 1.0×10^8 m/s. At what speed, relative to the spacecraft, is the starlight approaching?

8. | A starship blasts past the earth at 2.0×10^8 m/s. Just after passing the earth, the starship fires a laser beam out its back. With what speed does the laser beam approach the earth?

9. | A positron moving in the positive x-direction at 2.0×10^8 m/s collides with an electron at rest. The positron and electron annihilate, producing two gamma-ray photons. Photon 1 travels in the positive x-direction and photon 2 travels in the negative x-direction. What is the speed of each photon?

Section 27.4 Events and Measurements

Section 27.5 Relativity of Simultaneity

10. | Your job is to synchronize the clocks in a reference frame. You are going to do so by flashing a light at the origin at $t = 0$ s. To what time should the clock at $(x, y, z) = (30$ m, 40 m, 0 m$)$ be preset?

11. | Bjorn is standing at $x = 600$ m. Firecracker 1 explodes at the origin and firecracker 2 explodes at $x = 900$ m. The flashes from both explosions reach Bjorn's eye at $t = 3.0$ μs. At what time did each firecracker explode?

12. ‖ Bianca is standing at $x = 600$ m. Firecracker 1, at the origin, and firecracker 2, at $x = 900$ m, explode simultaneously. The flash from firecracker 1 reaches Bianca's eye at $t = 3.0$ μs. At what time does she see the flash from firecracker 2?

13. ‖ You are standing at $x = 9.0$ km. Lightning bolt 1 strikes at $x = 0$ km and lightning bolt 2 strikes at $x = 12.0$ km. Both flashes reach your eye at the same time. Your assistant is standing at $x = 3.0$ km. Does your assistant see the flashes at the same time? If not, which does she see first and what is the time difference between the two?

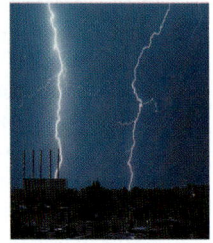

14. ‖ You are standing at $x = 9.00$ km and your assistant is standing at $x = 3.00$ km. Lightning bolt 1 strikes at $x = 0$ km and lightning bolt 2 strikes at $x = 12.00$ km. You see the flash from bolt 2 at $t = 10$ μs and the flash from bolt 1 at $t = 50$ μs. According to your assistant, were the lightning strikes simultaneous? If not, which occurred first and what was the time difference between the two?

15. ‖ Jose is looking to the east. Lightning bolt 1 strikes a tree 300 m from him. Lightning bolt 2 strikes a barn 900 m from him in the same direction. Jose sees the tree strike 1.0 μs before he sees the barn strike. According to Jose, were the lightning strikes simultaneous? If not, which occurred first and what was the time difference between the two?

16. ‖ You are flying your personal rocketcraft at $0.90c$ from Star A toward Star B. The distance between the stars, in the stars' reference frame, is 1.0 ly. Both stars happen to explode simultaneously in your reference frame at the instant you are exactly halfway between them. Do you see the flashes simultaneously? If not, which do you see first and what is the time difference between the two?

Section 27.6 Time Dilation

17. ‖ A cosmic ray travels 60 km through the earth's atmosphere in 400 μs, as measured by experimenters on the ground. How long does the journey take according to the cosmic ray?

18. | At what speed relative to a laboratory does a clock tick at half the rate of an identical clock at rest in the laboratory? Give your answer as a fraction of c.

19. ‖ An astronaut travels to a star system 4.5 ly away at a speed of $0.90c$. Assume that the time needed to accelerate and decelerate is negligible.
   a. How long does the journey take according to Mission Control on earth?
   b. How long does the journey take according to the astronaut?
   c. How much time elapses between the launch and the arrival of the first radio message from the astronaut saying that she has arrived?

20. ‖ A starship voyages to a distant planet 10 ly away. The explorers stay 1 yr, return at the same speed, and arrive back on earth 26 yr after they left. Assume that the time needed to accelerate and decelerate is negligible.
   a. What is the speed of the starship?
   b. How much time has elapsed on the astronauts' chronometers?

Section 27.7 Length Contraction

21. | At what speed, as a fraction of c, will a moving rod have a length 60% that of an identical rod at rest?

22. | Jill claims that her new rocket is 100 m long. As she flies past your house, you measure the rocket's length and find that it is only 80 m. Should Jill be cited for exceeding the $0.5c$ speed limit?

23. ‖ A muon travels 60 km through the atmosphere at a speed of $0.9997c$. According to the muon, how thick is the atmosphere?

24. ‖ The Stanford Linear Accelerator (SLAC) accelerates electrons to $v = 0.99999997c$ in a 3.2-km-long tube. If they travel the length of the tube at full speed (they don't, because they are accelerating), how long is the tube in the electrons' reference frame?

25. | Our Milky Way galaxy is 100,000 ly in diameter. A spaceship crossing the galaxy measures the galaxy's diameter to be a mere 1.0 ly.
   a. What is the speed of the spaceship relative to the galaxy?
   b. How long is the crossing time as measured in the galaxy's reference frame?

26. ‖ An optical interferometer can detect a displacement of about 50 nm. At what speed would a meter stick "shrink" by 50 nm?
   Hint: Use the binomial approximation.

Section 27.8 Velocities of Objects in Special Relativity

27. ‖ A rocket cruising past earth at $0.800c$ shoots a bullet out the back door, opposite the rocket's motion, at $0.900c$ relative to the rocket. What is the bullet's speed relative to the earth?

28. | A laboratory experiment shoots an electron to the left at $0.900c$. What is the electron's speed relative to a proton moving to the right at $0.900c$?

29. ‖‖ A solar flare blowing out from the sun at $0.90c$ is overtaking a rocket as it flies away from the sun at $0.80c$. According to the crew on board, with what speed is the flare gaining on the rocket?

Section 27.9 Relativistic Momentum

30. ‖ A proton is accelerated to $0.999c$.
    a. What is the proton's momentum?
    b. By what factor does the proton's momentum exceed its Newtonian momentum?
31. ‖‖ A 1.0 g particle has momentum 400,000 kg · m/s. What is the particle's speed?
32. ‖ At what speed is a particle's momentum twice its Newtonian value?
33. ‖‖ What is the speed of a particle whose momentum is mc?

Section 27.10 Relativistic Energy

34. ‖ What are the kinetic energy, the rest energy, and the total energy of a 1.0 g particle with a speed of $0.80c$?
35. ‖ A quarter-pound hamburger with all the fixings has a mass of 200 g. The food energy of the hamburger (480 food calories) is 2 MJ.
    a. What is the energy equivalent of the mass of the hamburger?
    b. By what factor does the energy equivalent exceed the food energy?
36. ‖ How fast must an electron move so that its total energy is 10% more than its rest mass energy?
37. ‖‖ At what speed is a particle's kinetic energy twice its rest energy?

General Problems

38. ‖‖‖ A firecracker explodes at $x = 0$, $t = 0$. A second explodes at $x = 300$ m, $t = 2$ μs. What is the proper time between these events?
39. ‖ You're standing on an asteroid when you see your best friend rocketing by in her new spaceship. As she goes by, you notice that the front and rear of her ship coincide exactly with the 400-m-diameter of another nearby asteroid that is stationary with respect to you. However, you happen to know that your friend's spaceship measured 500 m long in the showroom. What is your friend's speed relative to you?
40. ‖ A subatomic particle moves through the laboratory at $0.90c$. Laboratory experimenters measure its lifetime, from creation to annihilation, to be 2.3 ps (1 ps = 10^{-12} s). According to the particle, how long did it live?
41. ‖ The diameter of the solar system is 10 light hours. A spaceship crosses the solar system in 15 hours, as measured on earth. How many hours does the passage take according to passengers on the spaceship?
    Hint: $c = 1$ light hour per hour.

42. ‖‖ A 30-m-long rocket train car is traveling from Los Angeles to New York at $0.50c$ when a light at the center of the car flashes. When the light reaches the front of the car, it immediately rings a bell. Light reaching the back of the car immediately sounds a siren.
    a. Are the bell and siren simultaneous events for a passenger seated in the car? If not, which occurs first and by how much time?
    b. Are the bell and siren simultaneous events for a bicyclist waiting to cross the tracks? If not, which occurs first and by how much time?
43. ‖ Two events in reference frame S occur 10 μs apart at the same point in space. Frame S′ travels at speed $v = 0.90c$ relative to frame S.
    a. What is the time interval between the events in reference frame S′?
    b. What is the distance between the events in frame S′?
44. ‖ a. How fast must a rocket travel on a journey to and from a distant star so that the astronauts age 10 years while the Mission Control workers on earth age 120 years?
    b. As measured by Mission Control, how far away is the distant star?
45. ‖‖‖‖ You fly 5000 km across the United States on an airliner at 250 m/s. You return two days later at the same speed.
    a. Have you aged more or less than your friends at home?
    b. By how much?
    Hint: Use the binomial approximation.
46. ‖‖ Two clocks are synchronized. One is placed in a race car that drives around a 2.0-km-diameter track at 100 m/s for 24 hr. Afterward, by how much do the two clocks differ?
    Hint: Use the binomial approximation.
47. ‖‖ In Section 27.6 we explained that muons can reach the ground because of time dilation. But how do things appear in the muon's reference frame, where the muon's half-life is only 1.5 μs? How can a muon travel the 60 km to reach the earth's surface before decaying? Resolve this apparent paradox. Be as quantitative as you can in your answer.
48. ‖ A cube has a density of 2000 kg/m³ while at rest in the laboratory. What is the cube's density as measured by an experimenter in the laboratory as the cube moves through the laboratory at 90% of the speed of light in a direction perpendicular to one of its faces?
49. ‖ In an attempt to reduce the extraordinarily long travel times for voyaging to distant stars, some people have suggested traveling at close to the speed of light. Suppose you wish to visit the red giant star Betelgeuse, which is 430 ly away, and that you want your 20,000 kg rocket to move so fast that you age only 20 years during the round trip.
    a. How fast must the rocket travel relative to earth?
    b. How much energy is needed to accelerate the rocket to this speed?
    c. How many times larger is this energy than the total energy used by the United States in the year 2000, which was roughly 1.0×10^{20} J?
50. ‖ A rocket traveling at $0.500c$ sets out for the nearest star, Alpha Centauri, which is 4.25 ly away from earth. It will return to earth immediately after reaching Alpha Centauri. What distance will the rocket travel and how long will the journey last according to (a) stay-at-home earthlings and (b) the rocket crew? (c) Which answers are the correct ones, those in part a or those in part b?

51. ‖ A distant quasar is found to be moving away from the earth at 0.80c. A galaxy closer to the earth and along the same line of sight is moving away from us at 0.20c. What is the recessional speed of the quasar as measured by astronomers in the other galaxy?

52. | Two rockets approach each other. Each is traveling at 0.75c in the earth's reference frame. What is the speed of one rocket relative to the other?

53. | A military jet traveling at 1500 m/s has engine trouble and the pilot must bail out. Her ejection seat shoots her forward at 300 m/s relative to the jet. How fast is she traveling relative to the ground according to the relativistic velocity transformation? Would the pilot or anyone else be able to tell the difference between the relativistic and the nonrelativistic result?

54. ‖ Two rockets, A and B, approach the earth from opposite directions at speed 0.800c. The length of each rocket measured in its rest frame is 100 m. What is the length of rocket A as measured by the crew of rocket B?

55. ‖ Event A occurs at spacetime coordinates (300 m, 2.0 μs).
    a. Event B occurs at spacetime coordinates (1200 m, 6.0 μs). Could A possibly be the cause of B?
    b. Event C occurs at spacetime coordinates (2400 m, 8.0 μs). Could A possibly be the cause of C?

56. | What is the speed of an electron after being accelerated from rest through a 20×10^6 V potential difference?

57. ‖‖ What is the speed of a proton after being accelerated from rest through a 50×10^6 V potential difference?

58. ‖ The half-life of a muon at rest is 1.5 μs. Muons that have been accelerated to a very high speed and are then held in a circular storage ring have a half-life of 7.5 μs.
    a. What is the speed of the muons in the storage ring?
    b. What is the total energy of a muon in the storage ring? The mass of a muon is 207 times the mass of an electron.

59. | What is the momentum of a particle with speed 0.95c and total energy 2.0×10^{-10} J?

60. ‖‖ What is the momentum of a particle whose total energy is four times its rest energy? Give your answer as a multiple of mc.

61. ‖ What is the total energy, in MeV, of
    a. A proton traveling at 99.0% of the speed of light?
    b. An electron traveling at 99.0% of the speed of light?

62. ‖ What is the velocity, as a fraction of c, of
    a. A proton with 500 GeV total energy?
    b. An electron with 2.0 GeV total energy?

63. ‖‖‖ At what speed is the kinetic energy of a particle twice its Newtonian value?

64. | What is the speed of an electron whose total energy equals the rest energy of a proton?

65. ‖‖ The factor γ appears in many relativistic expressions. A value $\gamma = 1.01$ implies that relativity changes the Newtonian values by approximately 1% and that relativistic effects can no longer be ignored. At what kinetic energy, in MeV, is $\gamma = 1.01$ for (a) an electron, and (b) a proton?

66. | A typical nuclear power plant generates electricity at the rate of 1000 MW. The efficiency of transforming thermal energy into electrical energy is $\frac{1}{3}$ and the plant runs at full capacity for 80% of the year. (Nuclear power plants are down about 20% of the time for maintenance and refueling.)
    a. How much thermal energy does the plant generate in one year?
    b. What mass of uranium is transformed into energy in one year?

67. | The sun radiates energy at the rate 3.8×10^{26} W. The source of this energy is fusion, a nuclear reaction in which mass is transformed into energy. The mass of the sun is 2.0×10^{30} kg.
    a. How much mass does the sun lose each year?
    b. What percentage is this of the sun's total mass?
    c. Estimate the lifetime of the sun.

68. ‖ The radioactive element radium (Ra) decays by a process known as *alpha decay,* in which the nucleus emits a helium nucleus. (These high-speed helium nuclei were named alpha particles when radioactivity was first discovered, long before the identity of the particles was established.) The reaction is ^{226}Ra → ^{222}Rn + ^{4}He, where Rn is the element radon. The accurately measured atomic masses of the three atoms are 226.025, 222.017, and 4.003. How much energy is released in each decay? (The energy released in radioactive decay is what makes nuclear waste "hot.")

69. ‖ The nuclear reaction that powers the sun is the fusion of four protons into a helium nucleus. The process involves several steps, but the net reaction is simply 4p → ^{4}He + energy. The mass of a helium nucleus is known to be 6.64×10^{-27} kg.
    a. How much energy is released in each fusion?
    b. What fraction of the initial rest mass energy is this energy?

70. ‖‖ When antimatter (which we'll learn more about in Chapter 30) interacts with an equal mass of ordinary matter, both matter and antimatter are converted completely into energy in the form of photons. In an antimatter-fueled spaceship, a staple of science fiction, the newly created photons are shot from the back of the ship, propelling it forward. Suppose such a ship has a mass of 2.0×10^6 kg, and carries a mass of fuel equal to 1% of its mass, or 1.0×10^4 kg of matter and an equal mass of antimatter.
    a. What is the final speed of the ship, assuming it starts from rest, if all energy released in the matter-antimatter annihilation is transformed into the kinetic energy of the ship?
    b. Not only do photons have energy, as you learned in Chapter 25, they also have momentum. Explain why, when energy and momentum conservation are both considered, the final speed of the ship will be less than you calculated in part a.

Passage Problems

Pion Therapy BIO

Subatomic particles called *pions* are created when protons, accelerated to speeds very near c in a particle accelerator, smash into the nucleus of a target atom. Charged pions are unstable particles that decay into muons with a half-life of 1.8×10^{-8} s. Pions have been investigated for use in cancer treatment because they pass through tissue doing minimal damage until they decay, releasing significant energy at that point. The speed of the pions can be adjusted so that the most likely place for the decay is in a tumor.

Suppose pions are created in an accelerator, then directed into a medical bay 30 m away. The pions travel at the very high speed of $0.99995c$. Without time dilation, half of the pions would have decayed after traveling only 5.4 m, not far enough to make it to the medical bay. Time dilation allows them to survive long enough to reach the medical bay, enter tissue, slow down, and then decay where they are needed, in a tumor.

71. ‖ What is the half-life of a pion in the reference frame of the patient undergoing pion therapy?
    A. 1.8×10^{-10} s   B. 1.8×10^{-8} s
    C. 1.8×10^{-7} s   D. 1.8×10^{-6} s

72. ‖ According to the pion, what is the distance it travels from the accelerator to the medical bay?
    A. 0.30 m   B. 3.0 m   C. 30 m   D. 3000 m

73. ‖ The proton collision that creates the pion also creates a gamma-ray photon traveling in the same direction as the pion. The photon will get to the medical bay first because it is moving faster. What is the speed of the photon in the pion's reference frame?
    A. $0.00005c$ B. $0.5c$
    C. $0.99995c$ D. c

74. ‖ If the pion slows down to $0.99990c$, about what percentage of its kinetic energy is lost?
    A. 0.03%   B. 0.3%   C. 3%   D. 30%

STOP TO THINK ANSWERS

Stop to Think 27.1: A, C, and **F.** These move at constant velocity, or very nearly so. The others are accelerating.

Stop to Think 27.2: A. $u' = u - v = -10$ m/s $- 6$ m/s $= -16$ m/s. The *speed* is 16 m/s.

Stop to Think 27.3: C. Even the light has a slight travel time. The event is the hammer hitting the nail, not your seeing the hammer hit the nail.

Stop to Think 27.4: At the same time. Mark is halfway between the tree and the pole, so the fact that he *sees* the lightning bolts at the same time means they *happened* at the same time. It's true that Nancy *sees* event 1 before event 2, but the events actually occurred before she sees them. Mark and Nancy share a reference frame, because they are at rest relative to each other, and all experimenters in a reference frame, after correcting for any signal delays, *agree* on the spacetime coordinates of an event.

Stop to Think 27.5: After. This is the same as the case of Peggy and Ryan. In Mark's reference frame, as in Ryan's, the events are simultaneous. Nancy *sees* event 1 first, but the time when an event is seen is not when the event actually happens. Because all experimenters in a reference frame agree on the spacetime coordinates of an event, Nancy's position in her reference frame cannot affect the order of the events. If Nancy had been passing Mark at the instant the lightning strikes occur in Mark's frame, then Nancy would be equivalent to Peggy. Event 2, like the firecracker at the front of Peggy's railroad car, occurs first in Nancy's reference frame.

Stop to Think 27.6: C. Nick measures proper time because Nick's clock is present at both the "nose passes Nick" event and the "tail passes Nick" event. Proper time is the smallest measured time interval between two events.

Stop to Think 27.7: C. The kinetic energy is $(\gamma - 1)mc^2 = 0.5mc^2$, which is less than the rest energy mc^2.

28

QUANTUM PHYSICS

This false-color image showing individual rod cells (green) and cone cells (blue) on the human retina was made with an electron microscope. Such exquisite detail would not be possible in an image created with a light microscope. Why is greater resolution possible in an image made with a beam of electrons?

Looking Ahead ▶▶

The goal of Chapter 28 is to understand the quantization of energy for light and matter. In this chapter you will learn to:

▶ Understand how x-ray diffraction reveals the atomic structure of matter.

▶ Use the photon model of light to understand the photoelectric effect.

▶ Recognize the experimental evidence for the wave nature of matter.

▶ Understand how de Broglie's matter waves lead to the quantization of energy.

▶ Recognize the inherent limitations to our knowledge imposed by the uncertainty principle.

Looking Back ◀◀

The material in this chapter will make frequent reference to wave optics topics and will build on the wave and particle models of light. Please review:

◀ Sections 17.2–17.5 Diffraction and interference of light.

◀ Section 25.7 The photon model of electromagnetic waves.

◀ Section 25.8 X rays.

Except for relativity, everything we have studied until this point in the book was known by 1900. Newtonian mechanics, thermodynamics, and the theory of electromagnetism form what we call *classical physics*. It is an impressive body of knowledge with immense explanatory power and a vast number of applications. Many scientists of the late 1800s thought they could use these theories to explain just about anything, and some even felt there was nothing significant left to discover.

But a spate of discoveries right around 1900 showed that classical physics, though remarkable, was incomplete. Investigations into the nature of light and matter led to many astonishing discoveries that classical physics simply could not explain. New experiments called into question the very nature of waves and particles. Sometimes, as you will see, light refuses to act like a wave and seems more like a collection of particles. Other experiments found that electrons sometimes behave like waves.

These discoveries eventually led to a radical new theory of light and matter called *quantum physics*. Today, a century later, quantum physics is the most successful physical theory ever devised. This chapter will introduce you to this strange but wonderful quantum world. We will take a more historical approach than in previous chapters. As we introduce new ideas, we will describe in some detail the key experiments and the evolution of theories to explain them. This will be a chance for you to develop your understanding not just of quantum physics but of how a new scientific discipline takes shape.

28.1 X Rays and X-Ray Diffraction

In 1895, the German physicist Wilhelm Röntgen made a remarkable discovery. Röntgen was studying how electrons (called *cathode rays* at the time) traveled through a vacuum. He sealed an electron-producing cathode and a metal target electrode into a vacuum tube. A high voltage pulled electrons from the cathode and accelerated them to very high speed before they struck the target. One day, by chance, Röntgen left a sealed envelope containing film near the vacuum tube. He was later surprised to discover that the film had been exposed even though it had never been removed from the envelope. Some sort of penetrating radiation from the tube had exposed the film.

Röntgen had no idea what was coming from the tube, so he called them **x rays,** using the algebraic symbol *x* as meaning "unknown." X rays were unlike anything, particle or wave, ever discovered before. Röntgen was not successful at reflecting the rays or at focusing them with a lens. He showed that they travel in straight lines, like particles, but they also pass right through most solid materials with virtually no absorption, something no known particle could do.

The experiments of Röntgen and others led scientists to conclude that these mysterious rays were electromagnetic waves with very short wavelengths, as we learned in Chapter 25. These short-wavelength waves were produced in Röntgen's apparatus by the collision of fast electrons with a metallic target. X rays are still produced this way, as shown in the outline of the operation of a modern x-ray tube in Figure 28.1.

Electrons come off a hot wire . . .   . . . accelerate through a large potential difference . . .   . . . and produce x rays when they strike a metal target.

FIGURE 28.1 The operation of a modern x-ray tube.

X-Ray Images and CAT Scans

The penetration of x rays depends on the material. Substances with high atomic numbers, such as lead or the minerals in bone, are effective at stopping them; materials with low atomic numbers, such as the beryllium window of the x-ray tube in Figure 28.1 or the water and organic compounds of soft tissues in the body, diminish them only slightly.

Röntgen immediately realized that x rays could be used to create an image of the interior of the body. One of Röntgen's first images showed the bones in his wife's hand, dramatically demonstrating the medical potential of these newly discovered rays. As illustrated in Figure 28.2, an x-ray image is essentially a shadow of the bones and dense components of the body; where these tissues stop the x rays, the film is not exposed. The basic procedure for producing an x-ray image on film is little changed from Röntgen's day.

In the 1970s, as powerful computers became available, physicists had a new idea for creating more informative x-ray images. The process, illustrated in Figure 28.3a on the next page, is called *computerized axial tomography,* but it is better known by the acronym *CAT scan.* Combining penetration data from x rays at all angles in the rotating system can produce an image of incredible detail, as in Figure 28.3b.

The part of the body to be imaged is on top of a piece of film. Dense tissues pass few x rays; the film is not exposed below these tissues.

When the film is developed, the film is light where dense tissues or metal have blocked the x rays.

FIGURE 28.2 Creating an x-ray image.

X-Ray Diffraction

At about the same time scientists were first concluding that x rays were very-short-wavelength electromagnetic waves, researchers were also deducing that the size of an atom is ≈0.1 nm, and it was suggested that solids might consist of

(a) Producing a CAT scan

X rays penetrate a thin slice of tissue of the body.

X-ray tube

Detectors measure the penetration.

Monitor

Collimator

Detectors

Computer

The x-ray tube and the detectors rotate around the body.

Image processing creates a detailed recreation of the tissues in the imaged slice.

(b) A CAT scan of the chest

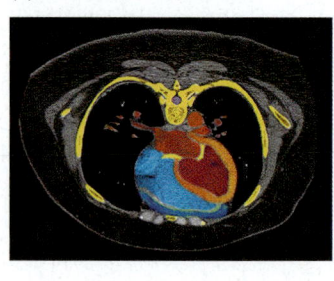

FIGURE 28.3 A CAT scan.

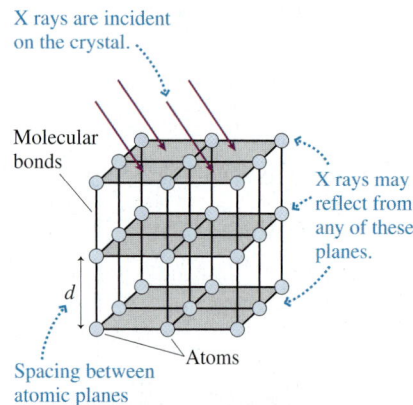

X rays are incident on the crystal.

Molecular bonds

X rays may reflect from any of these planes.

d

Atoms

Spacing between atomic planes

FIGURE 28.4 X rays incident on a simple cubic lattice crystal.

atoms arranged in a regular crystalline *lattice*. In 1912, the German scientist Max von Laue noted that x rays passing through a crystal ought to undergo diffraction from the "three-dimensional grating" of the crystal in much the same way that visible light diffracts from a diffraction grating. Such x-ray diffraction by crystals was soon confirmed experimentally, and measurements confirmed that x rays are indeed electromagnetic waves with wavelengths in the range 0.01 nm to 10 nm—a much shorter wavelength than visible light.

To understand x-ray diffraction, we begin by looking at the arrangement of atoms in a solid. Figure 28.4 shows x rays striking a crystal with a *simple cubic lattice*. This is a very straightforward arrangement, with the atoms in planes with spacing d between them.

Figure 28.5a on the next page shows a side view of the x rays striking the crystal, with the x rays incident at angle θ. Most of the x rays are transmitted through the plane, but a small fraction of the wave is reflected, much like the weak reflection of light from a sheet of glass. The reflected wave obeys the law of reflection—the angle of reflection equals the angle of incidence—and the figure has been drawn accordingly.

As we saw in Figure 28.4, a solid has not one single plane of atoms but many parallel planes. As x rays pass through a solid, a small fraction of the wave reflects from each of the parallel planes of atoms shown in Figure 28.5b. The *net* reflection from the solid is the *superposition* of the waves reflected by each atomic plane. For most angles of incidence, the reflected waves are out of phase and their superposition is very nearly zero. However, as in the thin-film interference we studied in Chapter 17, there are a few specific angles of incidence for which the reflected waves are in phase. For these angles of incidence, the reflected waves interfere constructively to produce a strong reflection. This strong x-ray reflection at a few specific angles of incidence is called **x-ray diffraction.**

NOTE ▶ The basic physical phenomenon that produces the reflection is the diffraction of the x rays by individual atoms. Though the mathematics are those of reflection, it is correct to call it diffraction. ◀

You can see from Figure 28.5b that the wave reflecting from any particular plane travels an extra distance $\Delta r = 2d\cos\theta$ before combining with the reflection from the plane immediately above it, where d is the spacing between the atomic planes. If $\Delta r = m\lambda$, these two waves will be in phase when they recombine. But if the reflections from two neighboring planes are in phase, then *all* the reflections from *all* the planes are in phase and will interfere constructively to produce a strong reflection. Consequently, x rays will reflect from the crystal when the angle of incidence θ_m satisfies the **Bragg condition:**

$$\Delta r = 2d\cos\theta_m = m\lambda \qquad m = 1, 2, 3, \ldots \qquad (28.1)$$

The Bragg condition for constructive
interference of x rays reflected from a solid

NOTE ▶ This formula is similar to that for constructive interference for light passed through a grating that we saw in Chapter 17. In both cases, we get constructive interference only at a few well-defined angles. ◀

EXAMPLE 28.1 Analyzing x-ray diffraction

X rays with a wavelength of 0.105 nm are diffracted by a crystal with a simple cubic lattice. Diffraction maxima are observed at angles 31.6° and 55.4° and at no angles between these two. What is the spacing between the atomic planes causing this diffraction?

PREPARE The angles must satisfy the Bragg condition. We don't know the values of m, but we know that they are two consecutive values. In Equation 28.1 θ_m *decreases* as m increases, so 31.6° corresponds to the larger value of m. We will assume that 55.4° corresponds to m and 31.6° to $m + 1$.

SOLVE The values of d and λ are the same for both diffractions, so we can use the Bragg condition to find

$$\frac{m + 1}{m} = \frac{\cos 31.6°}{\cos 55.4°} = 1.50 = \frac{3}{2}$$

Thus 55.4° is the second-order diffraction and 31.6° is the third-order diffraction. With this information we can use the Bragg condition again to find

$$d = \frac{2\lambda}{2\cos\theta_2} = \frac{0.105 \text{ nm}}{\cos 55.4°} = 0.185 \text{ nm}$$

ASSESS We learned above that the spacing of atoms in a solid is ≈ 0.1 nm, so this is a reasonable value for the atomic spacing in a crystal.

Example 28.1 shows that an x-ray diffraction pattern reveals details of the crystal that produced it. The structure of the crystal was quite simple, so the example was straightforward. More complex crystals produce correspondingly complex patterns that can help reveal the structure of the crystals that produced them. X-ray diffraction is widely used as a tool to decipher the three-dimensional structure of biological molecules such as proteins.

STOP TO THINK 28.1 The first-order diffraction of x rays from two crystals with simple cubic structure is measured. The first-order diffraction from crystal A occurs at an angle of 20°. The first-order diffraction of the same x rays from crystal B occurs at 30°. Which crystal has the larger atomic spacing?

28.2 The Photoelectric Effect

In Chapter 25, we introduced the idea that light can be thought of as *photons,* packets of energy of a particular size. This is an idea that you have likely heard before, but when it was first introduced, it was truly revolutionary. For such an odd idea to find broad acceptance, truly compelling experimental evidence was

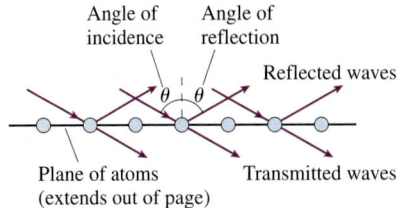

(a) X rays are transmitted and reflected at one plane of atoms.

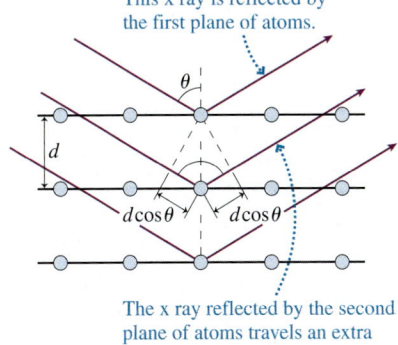

(b) The reflections from parallel planes interfere.

FIGURE 28.5 X-ray reflections from parallel atomic planes.

X marks the spot **BIO** Rosalind Franklin obtained this x-ray diffraction pattern for DNA in 1953. The cross of dark bands in the center of the diffraction pattern reveals something about the arrangement of atoms in the DNA molecule—that the molecule has the structure of a helix. This x-ray diffraction image was a key piece of information in the effort to unravel the structure of the DNA molecule.

 17.3

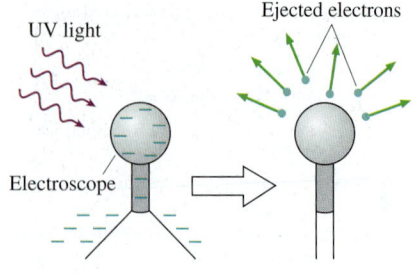

Ultraviolet light discharges a negatively charged electroscope by causing it to emit electrons.

FIGURE 28.6 Ultraviolet light discharges a negatively charged electroscope.

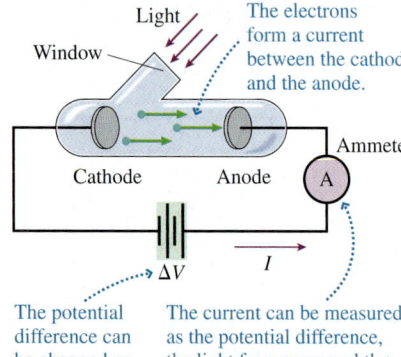

The potential difference can be changed or reversed.

The current can be measured as the potential difference, the light frequency, and the light intensity are varied.

FIGURE 28.7 An experimental device to study the photoelectric effect.

needed. This evidence was provided by studies of the *photoelectric effect,* which we will explore in detail in this section to recognize the rationale for and the impact of this startling new concept.

The first hints about the photon nature of light came in the late 1800s with the discovery that a negatively charged electroscope could be discharged by shining ultraviolet light on it. The English physicist J. J. Thomson found that the ultraviolet light was causing the electroscope to emit electrons, as illustrated in Figure 28.6. The emission of electrons from a substance due to light striking its surface came to be called the **photoelectric effect.** This seemingly minor discovery became a pivotal event that opened the door to the new ideas we discuss in this chapter.

Characteristics of the Photoelectric Effect

Figure 28.7 shows an evacuated glass tube with two facing electrodes and a window. When ultraviolet light shines on the cathode, a steady counterclockwise current (clockwise flow of electrons) passes through the ammeter. There are no junctions in this circuit, so the current must be the same all the way around the loop. The current in the space between the cathode and the anode consists of electrons moving freely through space (i.e., not inside a wire) at the *same rate* as the current in the wire. There is no current if the electrodes are in the dark, so electrons don't spontaneously leap off the cathode. Instead, the light causes electrons to be ejected from the cathode at a steady rate.

The battery in Figure 28.7 establishes an adjustable potential difference ΔV between the two electrodes. With it, we can study how the current I varies as the potential difference and the light's wavelength and intensity are changed. Doing so reveals the following characteristics of the photoelectric effect:

1. The current I is directly proportional to the light intensity. If the light intensity is doubled, the current also doubles.
2. The current appears without delay when the light is applied.
3. Electrons are emitted *only* if the light frequency f exceeds a **threshold frequency** f_0. This is shown in the graph of Figure 28.8a.
4. The value of the threshold frequency f_0 depends on the type of metal from which the cathode is made.
5. If the potential difference ΔV is positive (anode positive with respect to the cathode), the current changes very little as ΔV is increased. If ΔV is made negative (anode negative with respect to the cathode), by reversing the battery, the current decreases until, at some voltage $\Delta V = -V_{\text{stop}}$ the current reaches zero. The value of V_{stop} is called the **stopping potential.** This behavior is shown in Figure 28.8b.

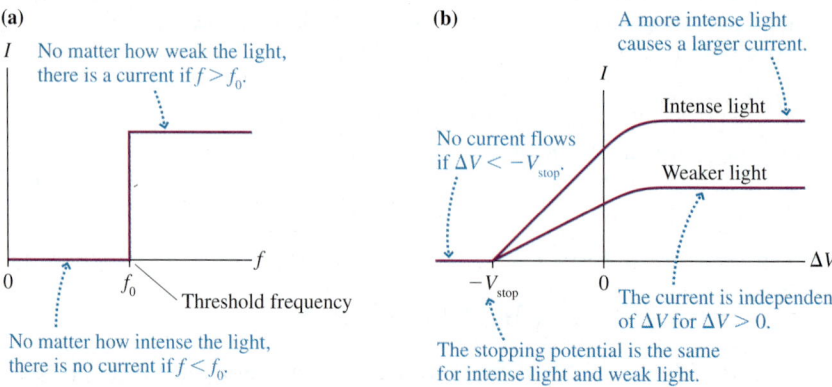

FIGURE 28.8 The photoelectric current dependence on the light frequency f and the battery potential difference ΔV.

6. The value of V_{stop} is the same for both weak light and intense light. A more intense light causes a larger current, but in both cases the current ceases when $\Delta V = -V_{stop}$.

NOTE ▶ We're defining V_{stop} to be a *positive* number. The potential difference that stops the electrons is $\Delta V = -V_{stop}$, with an explicit minus sign. ◀

Understanding the Photoelectric Effect

You learned in Chapter 22 that electrons are the charge carriers in a metal and move around freely inside like a sea of negatively charged particles. The electrons are bound inside the metal and do not spontaneously spill out of an electrode at room temperature. But a piece of metal heated to a sufficiently high temperature *does* emit electrons in a process called **thermal emission.** The electron gun in a television or computer display terminal starts with the thermal emission of electrons from a hot tungsten filament, as does the electron gun in an x-ray tube.

A useful analogy, shown in Figure 28.9, is the water in a swimming pool. Water molecules do not spontaneously leap out of the pool if the water is calm. To remove a water molecule, you must do *work* on it to lift it upward, against the force of gravity, to the edge of the pool. A minimum energy is needed to extract a water molecule, namely the energy needed to lift a molecule that is right at the surface. Removing a water molecule that is deeper requires more than the minimum energy. If sufficient energy is added, such as by people playing in the pool, a few of the water molecules may gain enough energy to splash over the edge and leave the pool.

Similarly, a *minimum* energy is needed to free an electron from a metal. To extract an electron, you would need to exert a force on it and pull it (i.e., do *work* on it) until its speed is large enough to escape. The minimum energy E_0 needed to free an electron is called the **work function** of the metal. Some electrons, like the deeper water molecules, may require more energy than E_0 to escape, but all will require *at least* E_0. Table 28.1 lists the work functions in eV of a few elements. (Recall that the conversion to joules is $1 \text{ eV} = 1.60 \times 10^{-19} \text{ J}$.)

Heating a metal, like splashing in the pool, increases the thermal energy of the electrons. At a sufficiently high temperature, the kinetic energy of a few of the electrons may exceed the work function. These electrons can "make it out of the pool" and leave the metal, and this is what happens in thermal emission.

Now, let's return to the photoelectric effect experiment of Figure 28.7. When ultraviolet light shines on the cathode, electrons leave with some kinetic energy. An electron with energy E_{elec} inside the metal loses energy ΔE as it escapes, so it emerges as an electron with kinetic energy $K = E_{elec} - \Delta E$. The work function energy E_0 is the *minimum* energy needed to remove an electron, so the *maximum* possible kinetic energy of an ejected electron is

$$K_{max} = E_{elec} - E_0$$

The electrons, after leaving the cathode, move out in all directions, as shown in Figure 28.10 on the next page. If the potential difference between the cathode and the anode is $\Delta V = 0$, there will be no electric field between the plates. Some electrons will reach the anode, creating a measurable current, but many do not. Additionally, the panels in the figure show that

- If the anode is positive, it attracts *all* of the electrons to the anode. A further increase in ΔV does not cause any more electrons to reach the anode and thus does not cause a further increase in the current I.
- If the anode is negative, it repels the electrons. However, an electron leaving the cathode with sufficient kinetic energy can still reach the anode, just as a ball hits

The cost of convenience Television sets and computer monitors may take some time to "warm up"; time is needed for the filament in the electron gun to get hot enough to emit electrons by thermal emission. Older televisions had many electron tubes, meaning a lengthy wait between turning the switch and seeing a picture. Modern devices generally keep a small current in the electron gun filament even when the power is off. The already-warm filament permits a faster start time. Your television thus probably uses power even when turned off—a "shadow load" that may add significantly to your electric bill.

The *minimum* energy to remove a drop of water from the pool is *mgh*.

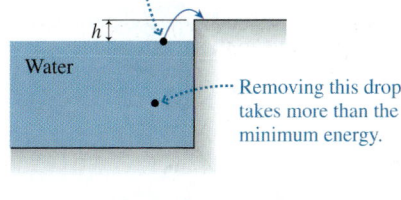

Removing this drop takes more than the minimum energy.

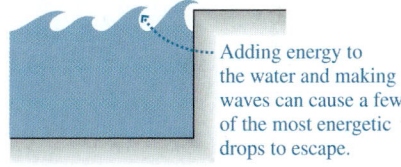

Adding energy to the water and making waves can cause a few of the most energetic drops to escape.

FIGURE 28.9 A swimming pool analogy of electrons in a metal.

TABLE 28.1 The work function for some metals

Element	E_0 (eV)
Potassium	2.30
Sodium	2.75
Aluminum	4.28
Tungsten	4.55
Copper	4.65
Iron	4.70
Gold	5.10

$\Delta V = 0$: The electrons leave the cathode in all directions. Only some reach the anode.

$\Delta V > 0$: Biasing the anode positive creates an electric field that pushes all the electrons to the anode.

$\Delta V < 0$: Biasing the anode negative repels the electrons. Only the very fastest make it to the anode.

FIGURE 28.10 The effect of different voltages between the anode and cathode.

the ceiling if you toss it upward with sufficient kinetic energy. A slightly negative anode voltage turns back only the slowest electrons. The current steadily decreases as the anode voltage becomes increasingly negative until, at the stopping potential, *all* electrons are turned back and the current ceases.

We can use conservation of energy to relate the maximum kinetic energy to the stopping potential. When ΔV is negative, as in the bottom panel of Figure 28.10, electrons are "going uphill," converting kinetic energy to potential energy as they slow down. That is, $\Delta U = -e\,\Delta V = -\Delta K$, where we've used $q = -e$ for electrons and ΔK is negative because the electrons are losing kinetic energy. When $\Delta V = -V_{stop}$, where the current ceases, the very fastest electrons, with K_{max}, are being turned back *just* as they reach the anode. They're converting 100% of their kinetic energy into potential energy, so $\Delta K = -K_{max}$. Thus $e\,\Delta V_{stop} = K_{max}$, or

$$V_{stop} = \frac{K_{max}}{e} \qquad (28.2)$$

In other words, **measuring the stopping potential tells us the maximum kinetic energy of the electrons.**

Limits of the Classical Explanation of the Photoelectric Effect

In the photoelectric effect, what makes the electron leave the metal? Perhaps the light source is simply heating the metal surface and causing it to emit electrons. The electrons then move through the tube with a certain kinetic energy, and a large enough voltage will stop them from reaching the anode.

This is a completely classical explanation based on the thermal emission of electrons from a metal, requiring no new physics. But it has three major shortcomings:

1. Nothing in the classical explanation suggests that there should be a threshold frequency. If a weak intensity at a frequency just slightly above f_0 can generate a current, then certainly a strong intensity at a frequency just slightly below f_0 should be able to do so—it should heat the metal even more. There is no reason that a slight change in frequency should matter. Yet the experimental evidence shows a sharp threshold at f_0.
2. If the electrons are due to thermal emission, it should take some length of time for the light to raise the temperature sufficiently high for electrons to escape. The experimental evidence is in sharp disagreement. If $f > f_0$, the current starts instantly for both weak light and intense light.
3. More intense light would be expected to heat the electrons to a higher temperature. Doing so should increase the maximum kinetic energy of the electrons and increase the stopping potential V_{stop}. Instead, the stopping potential is the same for strong light as it is for weak light.

Early investigators considered this and other explanations, but, ultimately, **classical physics is unable to explain the observed properties of the photoelectric effect.** Some new physics is needed.

Einstein's Explanation

In 1905, Albert Einstein wrote a paper in which he offered an exceedingly simple but amazingly bold idea to explain the photoelectric-effect data. The German physicist Max Planck had found that he could explain the spectrum of a glowing, incandescent object that we saw in Chapter 25 only if he assumed that the oscillating atoms inside the heated solid vibrated in a particular way. The energy of an atom vibrating with frequency f had to be one of the specific energies $E = 0, hf,$

2*hf*, 3*hf*, . . . , where *h* is a constant. That is, the vibration energies are **quantized.** The constant *h*, now called **Planck's constant,** is

$$h = 6.63 \times 10^{-34} \text{ J} \cdot \text{s} = 4.14 \times 10^{-15} \text{ eV} \cdot \text{s}$$

The first value, with SI units, is the proper one for most calculations, but you will find the second to be useful when energies are expressed in eV.

Einstein was the first to take Planck's idea seriously and to suggest that the quantization is real. Einstein went even further and suggested that **electromagnetic radiation itself is quantized!** That is, light is not really a continuous wave but, instead, arrives in small packets or bundles of energy. Einstein called each packet of energy a **light quantum,** and he postulated that the energy of one light quantum is directly proportional to the frequency of the light. That is, each quantum of light, which is now known as a **photon,** has energy

$$E = hf \qquad (28.3)$$

The energy of a photon, a quantum of light, of frequency *f*

where *h* is Planck's constant. Higher-frequency light will be composed of higher-energy photons—it will be composed of bundles of higher energy. This seemingly simple assumption allowed Einstein to explain the properties of the photoelectric effect. It can also explain many other observations about electromagnetic waves of different frequencies.

EXAMPLE 28.2 Finding the energy of ultraviolet photons

Ultraviolet light at 290 nm does 250 times as much cellular damage as an equal intensity of ultraviolet at 310 nm; there is a clear threshold for damage at about 300 nm. What is the energy, in eV, of photons with a wavelength of 300 nm?

PREPARE The energy of a photon is related to its frequency by $E = hf$.

SOLVE The frequency at a wavelength 300 nm is

$$f = \frac{c}{\lambda} = \frac{3.00 \times 10^8 \text{ m/s}}{300 \times 10^{-9} \text{ m}} = 1.00 \times 10^{15} \text{ Hz}$$

We can now use Equation 28.3 to calculate the energy, using the value of *h* in eV · s:

$$E = hf = (4.14 \times 10^{-15} \text{ eV} \cdot \text{s})(1.00 \times 10^{15} \text{ Hz}) = 4.14 \text{ eV}$$

ASSESS For shorter wavelengths, the energy of ultraviolet photons is quite damaging to the genetic material of cells; for longer wavelengths, the photon energies are too low to cause much damage.

▶ **Not all ultraviolet is created equal** **BIO** The sharp threshold for ultraviolet damage to tissue means that ultraviolet light sources with small differences in wavelength can have very different biological effects. Tanning beds emit nearly all of their energy at wavelengths greater than 315 nm. This light stimulates cells to produce melanin—resulting in a tan—but produces little short-term cell damage. Germicidal lamps use ultraviolet peaked at 254 nm, which will damage and even kill cells. Exposure to such a source will result in very painful sunburn.

Einstein's "Miracle Year" Albert Einstein was a little-known young man of 26 in 1905. This photograph from the time bears little resemblance to the familiar picture of a white-haired older Einstein. In 1905, within the span of a single year, Einstein published three papers on three different topics, all of which would revolutionize physics. One was his initial paper on the theory of relativity, a subject you learned about in Chapter 27. Though relativity is the subject with which Einstein is most associated in the public mind, this paper received less attention at the time than the other two. A second paper used statistical mechanics to explain a phenomenon called *Brownian motion,* the random motion of small particles suspended in water. It is Einstein's third paper of 1905, on the nature of light, in which we are most interested in this chapter.

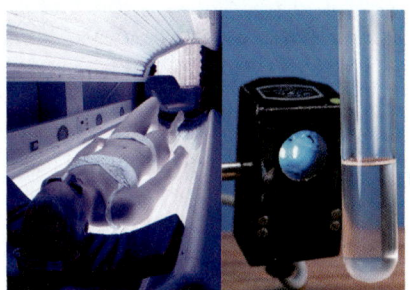

Einstein's Postulates

The idea that light is quantized is something that is now widely understood and accepted. But at the time of Einstein's paper, it was a truly revolutionary idea. Though we have used the photon model before, it is worthwhile to look at the theoretical underpinnings in more detail. In his 1905 paper, Einstein framed three postulates about light quanta and their interaction with matter:

1. Light of frequency f consists of discrete quanta, each of energy $E = hf$. Each photon travels at the speed of light c.
2. Light quanta are emitted or absorbed on an all-or-nothing basis. A substance can emit 1 or 2 or 3 quanta, but not 1.5. Similarly, an electron in a metal cannot absorb half a quantum but, instead, only an integer number.
3. A light quantum, when absorbed by a metal, delivers its entire energy to *one* electron.

NOTE ▶ These three postulates—that light comes in chunks, that the chunks cannot be divided, and that the energy of one chunk is delivered to one electron—are crucial for understanding the new ideas that will lead to quantum physics. They are completely at odds with the concepts of classical physics, where energy can be continuously divided and shared, so they deserve careful thought. ◀

Let's look at how Einstein's postulates apply to the photoelectric effect. We now think of the light shining on the metal as a torrent of photons, each of energy hf. Each photon is absorbed by *one* electron, giving that electron an energy $E_{elec} = hf$. This leads us to several interesting conclusions:

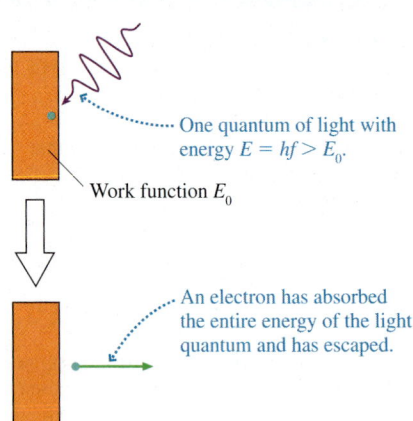

One quantum of light with energy $E = hf > E_0$.

Work function E_0

An electron has absorbed the entire energy of the light quantum and has escaped.

FIGURE 28.11 The ejection of an electron.

Seeing the world in a different light BIO
Many processes that are triggered by light have threshold frequencies. Plants use photosynthesis to convert the energy of light to chemical energy. Photons of visible light have sufficient energy to trigger the necessary molecular transitions, but photons of infrared do not. The leaves of trees absorb most of the visible light that falls on them, so trees will appear quite dark in a normal black and white landscape photo. But this photo wasn't made by visible light—the film was exposed by infrared. Infrared photons do not have enough energy to cause photosynthesis, so they are reflected, not absorbed, making the trees appear a ghostly white.

1. An electron that has just absorbed a quantum of light energy has $E_{elec} = hf$. Figure 28.11 shows that this electron can escape from the metal if its energy exceeds the work function E_0, or if

$$E_{elec} = hf \geq E_0 \tag{28.4}$$

In other words, there is a *threshold frequency*

$$f_0 = \frac{E_0}{h} \tag{28.5}$$

for the ejection of electrons. If f is less than f_0, even by just a small amount, none of the electrons will have sufficient energy to escape no matter how intense the light. But even very weak light with $f \geq f_0$ will give a few electrons sufficient energy to escape **because each photon delivers all of its energy to one electron.** This threshold behavior is exactly what the data show.
2. A more intense light delivers a larger number of photons to the surface. These eject a larger number of electrons and cause a larger current, exactly as observed.
3. There is a distribution of kinetic energies, because different electrons require different amounts of energy to escape, but the *maximum* kinetic energy is

$$K_{max} = E_{elec} - E_0 = hf - E_0 \tag{28.6}$$

As we noted in Equation 28.2, the stopping potential V_{stop} is a measure of K_{max}. Einstein's theory predicts that the stopping potential is related to the light frequency by

$$V_{stop} = \frac{K_{max}}{e} = \frac{hf - E_0}{e} \tag{28.7}$$

According to Equation 28.7, the stopping potential does *not* depend on the intensity of the light. Both weak light and intense light will have the same stopping potential. This agrees with the data.
4. If each photon transfers its energy hf to just one electron, that electron immediately has enough energy to escape. The current should begin instantly, with no delay, exactly as experiments had found.

Ultimately, Einstein's postulates are able to explain all of the observed features of the data for the photoelectric effect, though they require us to think of light in a very different way.

Let's use the swimming pool analogy again to help us visualize the photon model. Figure 28.12 shows a pebble being thrown into the pool. The pebble increases the energy of the water, but the increase is shared among all the molecules in the pool. The increase in the water's energy is barely enough to make ripples, not nearly enough to splash water out of the pool. But suppose *all* the pebble's energy could go to *one drop* of water that didn't have to share it. That one drop of water would easily have enough energy to leap out of the pool. Einstein's hypothesis that a light quantum transfers all its energy to one electron is equivalent to the pebble transferring all its energy to one drop of water.

Einstein was awarded the Nobel Prize in 1921 not for his theory of relativity, as many would suppose, but for his explanation of the photoelectric effect. Einstein showed convincingly that energy is quantized and that light, even though it exhibits wave-like interference, comes in the particle-like packets of energy we now call photons. Although not widely recognized at first, the age of classical physics had ended and a new era of quantum physics had begun.

Classically, the energy of the pebble is shared by all the water molecules. One pebble causes only very small waves.

If the pebble could give *all* its energy to one drop, that drop could easily splash out of the pool.

FIGURE 28.12 A pebble transfers energy to the water.

EXAMPLE 28.3 Finding the photoelectric threshold frequency

What are the threshold frequencies and wavelengths for electron emission from sodium and from aluminum?

PREPARE Table 28.1 gives the work function for sodium as $E_0 = 2.75$ eV, that for aluminum as $E_0 = 4.28$ eV.

SOLVE We can use Equation 28.5, with h in units of eV · s, to calculate

$$f_0 = \frac{E_0}{h} = \begin{cases} 6.64 \times 10^{14} \text{ Hz} & \text{sodium} \\ 10.34 \times 10^{14} \text{ Hz} & \text{aluminum} \end{cases}$$

These frequencies are converted to wavelengths with $\lambda = c/f$, giving

$$\lambda = \begin{cases} 452 \text{ nm} & \text{sodium} \\ 290 \text{ nm} & \text{aluminum} \end{cases}$$

ASSESS The photoelectric effect can be observed with sodium for $\lambda < 452$ nm. This includes blue and violet visible light but not red, orange, yellow, or green. Aluminum, with a larger work function, needs ultraviolet wavelengths $\lambda < 290$ nm.

EXAMPLE 28.4 Determining the maximum electron speed

What are the maximum electron speed and the stopping potential if sodium is illuminated with light of 300 nm?

SOLVE The light frequency is $f = c/\lambda = 1.00 \times 10^{15}$ Hz, so each light quantum has energy $hf = 4.14$ eV. The maximum kinetic energy of an electron is

$$K_{max} = hf - E_0 = 4.14 \text{ eV} - 2.75 \text{ eV} = 1.39 \text{ eV}$$

$$= 2.22 \times 10^{-19} \text{ J}$$

Because $K = \frac{1}{2}mv^2$, where m is the electron's mass, not the mass of the sodium atom, the maximum speed of an electron leaving the cathode is

$$v_{max} = \sqrt{\frac{2K_{max}}{m}} = 6.99 \times 10^5 \text{ m/s}$$

Note that we had to convert K_{max} to SI units of J before calculating a speed in m/s.

Now that we know the maximum kinetic energy of the electrons, we can use Equation 28.7 to calculate the stopping potential:

$$V_{stop} = \frac{K_{max}}{e} = 1.39 \text{ V}$$

An anode voltage of -1.39 V will be just sufficient to stop the very fastest electrons and thus reduce the current to zero.

ASSESS Dividing by e gives a potential difference with the *same numerical value* as K_{max} expressed in eV, which makes sense. Accelerating an electron through a potential difference of 1.39 V changes the kinetic energy by 1.39 eV. This is a good check on our results.

STOP TO THINK 28.2   The work functions of metals A, B, and C are 3.0 eV, 4.0 eV, and 5.0 eV, respectively. Ultraviolet light shines on all three metals, causing electrons to be emitted. Rank in order, from largest to smallest, the stopping voltages for A, B, and C.

28.3 Photons

We've now seen compelling evidence for the photon nature of light, but this leaves an important question: Just what *are* photons? To begin our explanation, let's return to the experiment that showed most dramatically the wave nature of light—Young's double-slit interference experiment. We will make a change, though: We will dramatically lower the light intensity by inserting filters between the light source and the slits. The fringes will be too dim to see with the naked eye, so we will use a detector that can build up an image over time. (This is the same sort of detector we imagined using for the extremely low-light photograph in Chapter 25.)

Figure 28.13 shows the outcome of such an experiment at three different times. At early times, very little light has reached the detector, and it does not show bands at all. Instead, it shows dots; the detector is registering the arrival of particle-like objects.

As the detector builds up the image for a longer time, we see that the dots are not entirely random. They are grouped into bands at *exactly* the positions where we expect to see bright constructive interference fringes. As the detector continues to gather light, the light and dark fringes become quite distinct.

The dots of light on the screen, which we'll attribute to the arrival of individual photons, are particle-like, but the overall picture clearly does not mesh with the classical idea of a particle. A classical particle, when faced with Young's double-slit apparatus, would go through one slit or the other. If light consisted of classical particles, we would see two bright areas on the screen, corresponding to light that has gone through one or the other slit. Instead, we see particle-like dots forming wave-like interference fringes.

We can perform an experiment like this with a light level so low that only one photon at a time is passing through the apparatus. If particle-like photons arrive at the detector in a banded pattern as a consequence of wave-like interference, as Figure 28.13 shows, but if only one photon at a time is passing through the experiment, what is it interfering with? The only possible answer is that the photon is somehow interfering *with itself.* Nothing else is present. But if each photon interferes with itself, rather than with other photons, then each photon, despite the fact that it is a particle-like object, must somehow go through *both* slits! This is something only a wave could do.

This all seems pretty crazy. But crazy or not, this is the way light behaves in real experiments. Sometimes it exhibits particle-like behavior and sometimes it exhibits wave-like behavior. The thing we call *light* is stranger and more complex than it first appeared, and there just is no way for us to reconcile these seemingly contradictory behaviors. We have to accept nature as it is, rather than hoping that nature will conform to our expectations. Furthermore, as we will see, this half-wave/half-particle behavior is not restricted to light.

(a) Image after a very short time

(b) Image is built up for a longer time

(c) The image after a considerably longer time

FIGURE 28.13 A simulation of a double-slit interference experiment.

◄ **Do we see light waves or photons?** BIO In this schematic view of the center of the retina of the eye, each dot represents a light-sensing cone cell. There are three types of cones with maximum sensitivities in different parts of the spectrum. The spacing between cones is close to what we would compute by treating light as a wave and calculating the maximum resolution of the eye given the pupil size and the wavelength of light. But the structure of the cones depends on light's particle nature. Cones detect individual photons by the transition of a single molecule. Different cones use molecules "tuned" to different photon energies, and thus different colors.

The Photon Rate

Light consists of a stream of photons. For monochromatic light of frequency f, each photon has energy $E = hf$. Suppose R is the *rate* at which photons arrive or, equivalently, the number of photons per second. Then the rate at which light energy is delivered, which is the *power* of the light (in joules per second, or watts) is

$$P = Rhf \qquad (28.8)$$

EXAMPLE 28.5 Finding the photon rate for a laser beam

The 1.0 mW light beam from a laser pointer ($\lambda = 670$ nm) shines on a screen. How many photons strike the screen each second?

SOLVE The light-beam power, or energy delivered per second, is $P = 1.0$ mW $= 0.0010$ J/s. This is a realistic value. The frequency of the light is $f = c/\lambda = 4.5 \times 10^{14}$ Hz. The number of

photons striking the screen per second, which is the *rate* of arrival of photons, is

$$R = \frac{P}{hf} = 3.4 \times 10^{15} \text{ photons per second}$$

ASSESS Each photon carries a small amount of energy, so there must be a huge number of photons per second to produce even this modest power.

Most light sources with which you are familiar emit such vast numbers of photons that you are only aware of their wave-like superposition, just as you only notice the roar of a heavy rain on your roof and not the individual raindrops. Only at extremely low intensities does the light begin to appear as a stream of individual photons, like the random patter of raindrops when it is barely sprinkling.

CONCEPTUAL EXAMPLE 28.1 Comparing photon rates

A red laser pointer and a green laser pointer have the same power. Which one emits a larger number of photons per second?

REASON Red light has a longer wavelength than green light, and so has a lower frequency. The photon rate is inversely proportional to the frequency, so the red laser emits more photons per second.

ASSESS This makes sense. Photons of green light have larger energy than photons of red light, so the green laser emits light in larger "chunks." The red laser must emit more chunks per second to have the same power.

Photodetectors

Early light detectors, which used the photoelectric effect directly, consisted of a polished metal plate in a vacuum tube. When light fell on the plate, an electron current was generated that could trigger an action, such as opening a door, or provide a measurement of the light intensity.

Modern photodetectors work on similar principles. Light fixtures that automatically turn on after dark generally use a *photoresistor,* a device whose resistance changes with light intensity. When light falls on the photoresistor, electrons are liberated, providing a higher density of charge carriers and thus a reduced resistance. Other devices use what is called a *photodiode.* An incoming photon gives its energy to an electron, lifting it into a higher energy state. The photon energy must exceed some threshold to cause this transition, so photodiodes have a threshold frequency, just as a device does that uses the photoelectric effect directly.

The *solar cell* is really a large-scale version of a photodiode. Recall the charge-escalator model of a battery in Chapter 22. The solar cell works much like a battery, but the energy input is from photons rather than chemical reactions, as in Figure 28.14. For a silicon-based solar cell, the most common type, the energy threshold is about 1.1 eV. This corresponds to a wavelength of about 1200 nm, in the near infrared.

A solar cell converts the energy of photons in sunlight into electrical energy.

Photons with energy greater than the threshold give their energy to charge carriers, increasing their potential energy and lifting them to the positive terminal of the solar cell.

Charge carriers move "downhill" through the circuit. Their energy can be used to run useful devices.

▶ **FIGURE 28.14** The operation of a solar cell.

EXAMPLE 28.6   Finding the current from a solar cell

1.0 W of monochromatic light of wavelength 550 nm illuminates a solar cell, driving a current in a circuit. What is the maximum possible current this light could produce?

PREPARE The photons have energy greater than the threshold energy, so the photons can cause charge carriers to flow. Each photon of the incident light can give its energy to a single charge carrier. The maximum number of charge carriers that can flow in each second is thus equal to the number of photons that arrive each second—the photon rate.

SOLVE The power of the light is $P = 1.0\ \text{W} = 1.0\ \text{J/s}$; the frequency is $f = c/\lambda = 5.5 \times 10^{14}\ \text{Hz}$. The number of photons arriving per second is thus

$$R = \frac{P}{hf} = 2.8 \times 10^{18}\ \text{photons per second}$$

Each photon can set at most one charge carrier into motion, so the maximum current is the photon rate times the charge:

$$I_{\text{max}} = eR = 0.44\ \text{A}$$

ASSESS The current will be less than this in a real solar cell because some photons will be reflected or otherwise "lost" and will not transfer their energy to charge carriers.

TRY IT YOURSELF

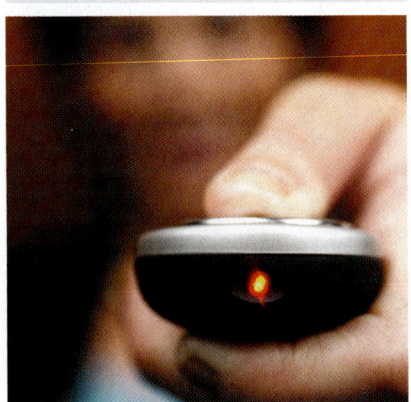

Photographing photons
Photodetectors based on silicon can be triggered by photons with energy as low as 1.1. eV, corresponding to a wavelength in the infrared. The light-sensing chip in your digital camera can detect the infrared signal given off by a remote control. Press a button on your remote control, aim it at your digital camera, and snap a picture. The picture will clearly show the infrared emitted by the remote, though this signal is invisible to your eye.

The *charge-coupled device* (CCD) or *complementary metal oxide semiconductor* (CMOS) detector in a digital camera consists of millions of *pixels,* each a microscopic silicon-based photodetector. Each photon hitting a pixel (if its frequency exceeds the threshold frequency) liberates one electron. These electrons are stored inside the pixel, and the total accumulated charge is directly proportional to the light intensity—the number of photons—hitting the pixel. After the exposure, the charge in each pixel is read and the value stored in memory, then the pixel is reset to be ready for the next picture.

STOP TO THINK 28.3   The intensity of a beam of light is increased but the light's frequency is unchanged. Which one (or perhaps more than one) of the following is true?

A. The photons travel faster.
B. Each photon has more energy.
C. The photons are larger.
D. There are more photons per second.

28.4 Matter Waves

Prince Louis-Victor de Broglie was a French graduate student in 1924. It had been 19 years since Einstein had shaken the world of physics by introducing photons and thus blurring the distinction between a particle and a wave. As de Broglie thought about these issues, it seemed that nature should have some kind of symmetry. If light waves could have a particle-like nature, why shouldn't material particles have some kind of wave-like nature? In other words, could **matter waves** exist?

With no experimental evidence to go on, de Broglie reasoned by analogy with Einstein's equation $E = hf$ for the photon and with some of the ideas of his theory of relativity. De Broglie determined that *if* a material particle of momentum $p = mv$ has a wave-like nature, its wavelength must be given by

$$\lambda = \frac{h}{p} = \frac{h}{mv} \tag{28.9}$$

De Broglie wavelength for a moving particle

LINEAR
p. 38

where h is Planck's constant. This wavelength is called the **de Broglie wavelength.**

EXAMPLE 28.7 Calculating the de Broglie wavelength of an electron

What is the de Broglie wavelength of an electron with a kinetic energy of 1.0 eV?

SOLVE An electron with kinetic energy $K = \frac{1}{2}mv^2 = 1.0\,\text{eV} = 1.6 \times 10^{-19}\,\text{J}$ has speed

$$v = \sqrt{\frac{2K}{m}} = 5.9 \times 10^5 \,\text{m/s}$$

Although fast by macroscopic standards, the electron gains this speed by accelerating through a potential difference of a mere 1 V. The de Broglie wavelength is

$$\lambda = \frac{h}{mv} = 1.2 \times 10^{-9}\,\text{m} = 1.2\,\text{nm}$$

ASSESS The electron's wavelength is small, but it is larger than the wavelengths of x rays and larger than the approximately 0.1 nm spacing of atoms in a crystal. We can observe x-ray diffraction, so if an electron has a wave nature, it should be easily observable.

What would it mean for matter—an electron or a proton or a baseball—to have a wavelength? Would it obey the principle of superposition? Would it exhibit diffraction and interference? Surprisingly, **matter exhibits all of the properties that we associate with waves.** For example, Figure 28.15 shows the intensity pattern recorded after 50 keV electrons passed through two narrow slits separated by 1.0 μm. The pattern is clearly a double-slit interference pattern, and the spacing of the fringes is exactly as the theory of Chapter 17 would predict for a wavelength given by de Broglie's formula. **The electrons are behaving like waves!**

If matter waves are real, why don't we see the interference of matter all around us? The following example answers this question.

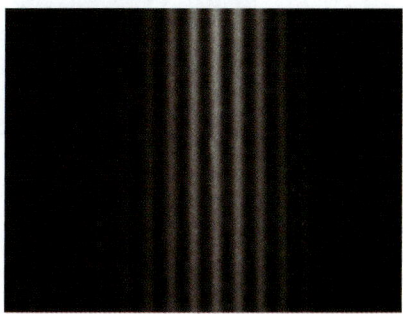

FIGURE 28.15 A double-slit interference pattern created with electrons.

EXAMPLE 28.8 Calculating the de Broglie wavelength of a smoke particle

One of the smallest macroscopic particles we could imagine using for an experiment would be a very small smoke or soot particle. These are $\approx 1\,\mu$m in diameter, too small to see with the naked eye and just barely at the limits of resolution of a microscope. A particle this size has mass $m \approx 10^{-18}$ kg. Estimate the de Broglie wavelength for a 1-μm-diameter particle moving at the very slow speed of 1 mm/s.

SOLVE The particle's momentum is $p = mv \approx 10^{-21}\,\text{kg}\cdot\text{m/s}$. The de Broglie wavelength of a particle with this momentum is

$$\lambda = \frac{h}{p} \approx 7 \times 10^{-13}\,\text{m}$$

ASSESS This wavelength is $\approx 1\%$ the size of an atom. We can't shoot a 1-μm-diameter particle though an atom-size hole, so we certainly don't expect to see any wave-like behavior in this case.

The previous example shows that a very small particle moving at a very slow speed has a wavelength that is too small to be of consequence. For larger objects moving at higher speeds, the wavelength is even smaller. A pitched baseball will have a wavelength of about 10^{-34} m, so a batter cannot use the wave nature of the ball as an excuse for not getting a hit. With such unimaginably small wavelengths, it is little wonder that we do not see macroscopic objects exhibiting wave-like behavior.

The Interference and Diffraction of Matter

17.5 **Activ Physics ONLINE**

Though de Broglie made his hypothesis in the absence of experimental data, experimental evidence was soon forthcoming. The English physicist G. P. Thomson performed a diffraction experiment with an electron beam transmitted through a crystal. Figures 28.16a and 28.16b show diffraction patterns produced by x rays and electrons passing through an aluminum-foil target. The primary observation to make from Figure 28.16 is that **electrons diffract and interfere exactly like x rays.**

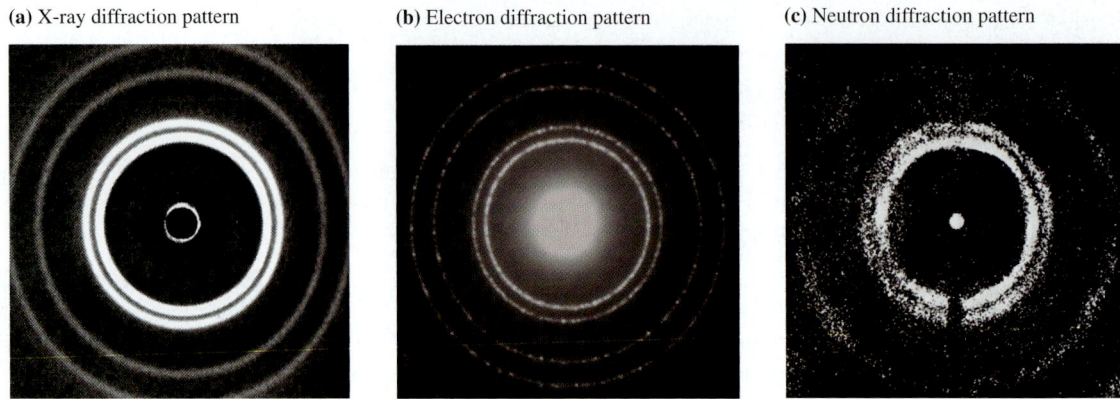

(a) X-ray diffraction pattern   **(b)** Electron diffraction pattern   **(c)** Neutron diffraction pattern

FIGURE 28.16 The diffraction patterns produced by x rays, electrons, and neutrons passing through an aluminum-foil target.

Later experiments demonstrated that de Broglie's hypothesis applies to other material particles as well. Neutrons have a much larger mass than electrons, which tends to decrease their de Broglie wavelength, but it is possible to generate very slow neutrons. The much smaller speed compensates for the heavier mass, so neutron wavelengths can be made comparable to electron wavelengths. Figure 28.16c shows a neutron diffraction pattern. It is similar to the x-ray and electron diffraction patterns, although of lower quality because neutrons are harder to detect. A neutron, too, is a matter wave. In recent years it has become possible to observe the interference and diffraction of atoms and even large molecules!

The Electron Microscope

Ray optics is based on the idea that light travels in straight lines—light rays—except when it crosses the boundary between two transparent media. Refraction at the boundary bends the rays, and we can use this idea to design lenses that bring parallel rays to a focus at a single point. If light really followed the ray model, carefully designed lenses would allow us to build a microscope with unlimited resolution and magnification. However, real microscopes are limited by the fact that light has wave-like properties. We learned in Chapter 19 that diffraction, a wave behavior, limits the resolving power of a microscope to, at best, about half the wavelength. For visible light, the smallest feature that can be resolved, even with perfect lenses, is about 200 or 250 nm. But the picture of the retina at the start of the chapter can show details much finer than this because it wasn't made with light—it was made with a beam of electrons.

The electron microscope, invented in the 1930s, works much like a light microscope. In the absence of electric or magnetic fields, electrons travel through a vacuum in straight lines much like light rays. Electron trajectories can be bent with electric or magnetic fields. A coil of wire carrying a current can produce a magnetic field that bends parallel electron trajectories so that they all cross at a single point; we call this an *electron lens*. An electron lens focuses electrons in the same way a glass lens bends and focuses light rays.

Figure 28.17 shows how a *transmission electron microscope (TEM)* works. This is purely classical physics; the electrons experience electric and magnetic forces,

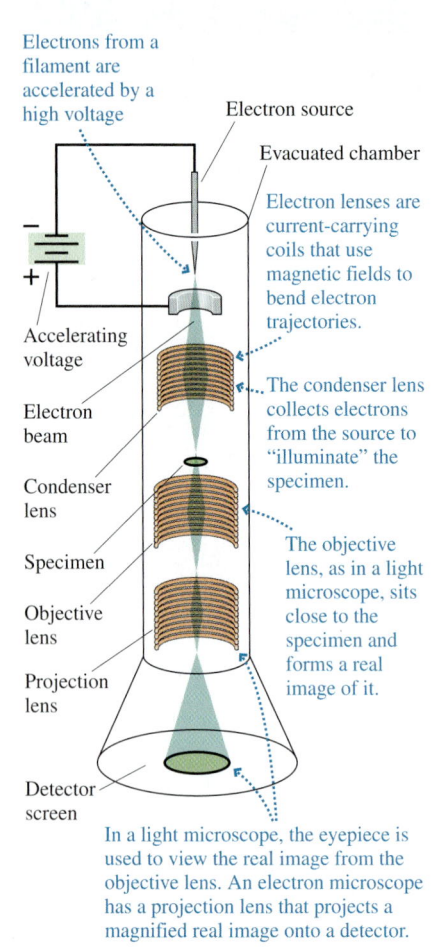

Electrons from a filament are accelerated by a high voltage

Electron source

Evacuated chamber

Electron lenses are current-carrying coils that use magnetic fields to bend electron trajectories.

Accelerating voltage

Electron beam

The condenser lens collects electrons from the source to "illuminate" the specimen.

Condenser lens

Specimen

The objective lens, as in a light microscope, sits close to the specimen and forms a real image of it.

Objective lens

Projection lens

Detector screen

In a light microscope, the eyepiece is used to view the real image from the objective lens. An electron microscope has a projection lens that projects a magnified real image onto a detector.

FIGURE 28.17 The electron microscope.

and they follow trajectories given by Newton's second law. Our ability to control electron trajectories allows electron microscopes to have magnifications far exceeding those of light microscopes. But just as in a light microscope, the resolution is ultimately limited by wave effects. Electrons are not classical point particles; they have wave-like properties and a de Broglie wavelength $\lambda = h/p$.

CONCEPTUAL EXAMPLE 28.2 Changing the wavelength of an electron

If an electron is accelerated to a higher speed, does its de Broglie wavelength increase or decrease?

REASON The wavelength is inversely proportional to the speed, so an increase in speed results in a decrease in wavelength.

ASSESS Creating an electron micrograph requires high-speed electrons. Increasing the speed decreases the wavelength, allowing for better resolution.

The reasoning used in Chapter 19 to determine maximum resolution applies equally well to electrons. Thus the resolving power is, at best, about half the electrons' de Broglie wavelength. For a 100 kV accelerating voltage, which is fairly typical, the de Broglie wavelength is $\lambda \approx 0.004$ nm (the electrons are moving fast enough that the momentum has to be calculated using relativity) and thus the theoretical resolving power of an electron microscope is about 0.002 nm.

In practice, the resolving powers of the best electron microscopes are limited by imperfections in the electron lenses to about 0.2 nm, not quite sufficient to resolve individual atoms with diameters of about 0.1 nm. This resolving power is about 1000 times smaller than can be achieved with light microscopes, as noted in Figure 28.18. Good light microscopes function at their theoretical limit, but there's still room to improve electron microscopes if a clever scientist or engineer can make a better electron lens.

The great resolving power of the microscope allows the imaging of incredibly fine detail, in this case the actual structure of a molecule.

→ 20 nm

The total image size is 1/10 the size of the smallest feature that can be resolved by a light microscope. A light microscope could not detect this individual molecule, let alone show its structure.

FIGURE 28.18 TEM image of a pigment molecule from a crustacean shell.

STOP TO THINK 28.4   A beam of electrons, a beam of protons, and a beam of oxygen atoms each pass at the same speed through a 1-μm-wide slit. Which will produce the widest central maximum on a detector behind the slit?

A. The beam of electrons.
B. The beam of protons.
C. The beam of oxygen atoms.
D. All three patterns will be the same.
E. None of the beams will produce a diffraction pattern.

28.5 Energy Is Quantized

De Broglie hypothesized that material particles have wave-like properties, and you've now seen experimental evidence that this must be true. Not only is this bizarre, the implications are profound.

You learned in Chapter 16 that the waves on a string fixed at both ends form standing waves. Wave reflections from both ends create waves traveling in both directions, and the superposition of two oppositely directed waves produces a standing wave.

Could we do something like this with particles? Is there such a thing as a "standing matter wave"? Let's consider a very simple physical system called a "particle in a box." For simplicity, we'll consider one-dimensional motion, a particle that moves back and forth along the x-axis. The "box" is defined by two fixed ends, and the particle bounces back and forth between these boundaries as in Figure 28.19 on the next page. We'll assume that collisions with the ends of the box are perfectly elastic, with no loss of kinetic energy.

(a) A classical particle of mass m bounces back and forth between two boundaries.

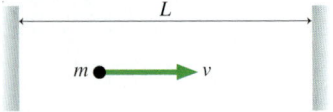

(b) Matter waves moving in opposite directions create standing waves.

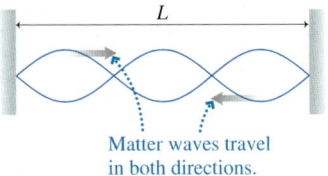

Matter waves travel in both directions.

FIGURE 28.19 A particle of mass m confined in a box of length L.

Figure 28.19a shows a classical particle, such as a ball or a dust particle, in the box. This particle simply bounces back and forth at constant speed. But if particles have wave-like properties, perhaps we should consider a *wave* reflecting back and forth from the ends of the box. The reflections will create the standing wave shown in Figure 28.19b. This standing wave is analogous to the standing wave on a string that is tied at both ends.

What can we say about the properties of this standing matter wave? A full treatment is beyond the scope of this text, but we can use what we know about matter waves and standing waves to make some deductions.

For waves on a string, we saw that there were only certain possible modes. The same will be true for the particle in a box; only certain states are possible. In Chapter 16, we found that the wavelength of a standing wave is related to the length L of the string by

$$\lambda_n = \frac{2L}{n} \qquad n = 1, 2, 3, 4, \ldots \qquad (28.10)$$

The wavelength of the particle in a box will follow the same formula, but the wave describing the particle must also satisfy the de Broglie condition $\lambda = h/p$. Equating these two expressions for the wavelength gives

$$\frac{h}{p} = \frac{2L}{n} \qquad (28.11)$$

Solving Equation 28.11 for the particle's momentum p, we find

$$p_n = n\left(\frac{h}{2L}\right) \qquad n = 1, 2, 3, 4, \ldots \qquad (28.12)$$

The particle's energy, which is entirely kinetic energy, is related to its momentum by

$$E = \frac{1}{2}mv^2 = \frac{p^2}{2m} \qquad (28.13)$$

If we use Equation 28.12 for the momentum, we find that the particle's energy is restricted to the discrete values

$$E_n = \frac{1}{2m}\left(\frac{hn}{2L}\right)^2 = \frac{h^2}{8mL^2}n^2 \qquad n = 1, 2, 3, 4, \ldots \qquad (28.14)$$

QUADRATIC
p. 50

Allowed energies of a particle in a box

This conclusion is one of the most profound discoveries of physics. Because of the wave nature of matter, **a confined particle can have only certain energies.** This result, that a confined particle can have only discrete values of energy, is called the **quantization** of energy. More informally, we say that energy is *quantized*. The number n is called the **quantum number,** and each value of n characterizes one **energy level** of the particle in the box.

Not only is the energy quantized, we see from Equation 28.14 that the energy of the particle in the box cannot be reduced below

$$E_1 = \frac{h^2}{8mL^2} \qquad (28.15)$$

E_1 is the *least* kinetic energy a particle can have. Because $E_1 > 0$, **the particle always has kinetic energy; the kinetic energy cannot be reduced to zero!** These properties of a wave-like particle in a box are in stark contrast to those of a classical Newtonian particle, for which the possible energies are continuous and the minimum kinetic energy is zero. In terms of E_1, the allowed energies are

$$E_n = n^2 E_1 \qquad (28.16)$$

$$L^2 = \frac{h^2}{8m\varepsilon_n} \times n^2$$

This result is analogous to our earlier finding that standing waves can exist for only the discrete frequencies $f_n = nf_1$.

EXAMPLE 28.9 Finding the allowed energies of a confined electron

An electron is confined to a region of space of length 0.19 nm—comparable in size to an atom. What are the first three allowed energies of the electron?

SOLVE The mass of an electron is $m = 9.11 \times 10^{-31}$ kg. Thus the first allowed energy is

$$E_1 = \frac{h^2}{8mL^2} = 1.6 \times 10^{-18} \text{ J} = 10 \text{ eV}$$

This is the lowest allowed energy. The next two allowed energies are

$$E_2 = 2^2 E_1 = 40 \text{ eV}$$

$$E_3 = 3^2 E_1 = 90 \text{ eV}$$

ASSESS These energies are rather significant; E_1 is larger than the work function of any metal in Table 28.1. As this electron is confined to a region about the size of an atom, we can see that the wave nature of matter is of vital importance for understanding what goes on at an atomic scale.

The energies allowed by Equation 28.14 are inversely proportional to both m and L^2. Both m and L have to be exceedingly small before energy quantization has any significance. Classical physics still works for baseballs! It is only at the atomic scale that quantization effects become important.

EXAMPLE 28.10 Determining the minimum energy of a smoke particle

What is the first allowed energy of the very small 1-μm-diameter particle of Example 28.8 if it is confined to a very small box 10 μm in length?

SOLVE In Example 28.8 we noted that such a particle has $m \approx 10^{-18}$ kg. The first allowed energy, $n = 1$, is

$$E_1 = \frac{h^2}{8mL^2} \approx 5 \times 10^{-40} \text{ J}$$

ASSESS This is an unimaginably small amount of energy. By comparison, the kinetic energy of a 1-μm-diameter particle moving at a barely perceptible speed of 1 mm/s is $K = 5 \times 10^{-25}$ J, a factor of 10^{15} larger. There is no way we could ever observe or measure discrete energies this small, so it is not surprising that we are unaware of energy quantization for macroscopic objects.

An atom is certainly more complicated than a simple one-dimensional box, but an electron is "confined" within an atom rather as a particle is in a box. Thus the electron orbits must, in some sense, be standing waves, and **the energy of the electrons in an atom must be quantized.** As we noted at the beginning of this section, the wave-like properties of matter have profound implications.

28.6 Energy Levels and Quantum Jumps

Einstein and de Broglie introduced revolutionary new ideas—a blurring of the distinction between waves and particles, and the quantization of energy—but the first to develop a full-blown theory of quantum physics, in 1925, was the Austrian physicist Erwin Schrödinger. Schrödinger's theory, now called *quantum mechanics,* provided answers to three fundamental questions:

- If particles have a wave nature, what exactly is waving?
- For a given system, whether it's a particle in a box or a hydrogen atom or a uranium nucleus, how do you calculate the allowed energies?
- How does a quantized system gain or lose energy?

Erwin Schrödinger, one of the early architects of quantum mechanics.

Quantum mechanics, as it has developed since 1925, is the most successful physical theory ever devised. It is the theory that describes everything from subatomic particles to atoms to semiconductors, often with exquisite precision. We will close this chapter with mostly qualitative descriptions of some of the major ideas that we will apply to the study of atoms and nuclei in the final two chapters.

What Is Waving?

This is perhaps the hardest question to answer. The first waves we studied were mechanical waves, such as sound and water, that travel through a medium via the oscillation of particles in the medium. Electromagnetic waves were trickier because they require no medium; they are a self-sustaining oscillation of the electromagnetic field itself. The wave in quantum physics is yet more subtle.

Figure 28.20 shows a double-slit experiment with electrons, like the one used to form the interference fringes seen in Figure 28.15. This is not at all how we expect electrons to behave, but it would be unsurprising behavior for a light wave. The dots in the center picture in Figure 28.20 show the arrival positions of individual electrons—each arriving as a particle—as they form the interference fringes. Below this is a graph of the wave intensity that would be seen on the detector screen if a wave with the de Broglie wavelength λ passed through the two slits. You saw graphs exactly like this in Chapter 17 to explain the interference fringes of light waves.

The arrival position of any particular electron is *unpredictable*. That is, nothing about how the experiment is set up or conducted allows us to predict exactly where the dot of an individual electron will appear on the detector. Yet there is clearly an overall pattern. There are some positions at which an electron is *more likely* to arrive, other positions where electrons are never found. Further, the positions where the electrons are most likely to be found are exactly the same as the positions where the wave intensity is maximum.

According to Schrödinger's theory, atomic particles, such as electrons or protons, are characterized by a *wave function*, represented as $\psi(x)$. The symbol ψ is the Greek letter psi. The square of the wave function, shown on the graph of Figure 28.20 as $|\psi(x)|^2$, gives the *probability* of finding the particle at position x. Thus electrons are detected at and near the peaks of $|\psi(x)|^2$, because that is where the probability of finding them is highest, and no electrons are detected where $|\psi(x)|^2 = 0$. **The probability of finding or detecting a *particle* is governed by the behavior of a *wave*.**

Thus the answer to "what is waving" is that no "thing" is waving. There is no disturbance of a medium. The wave function $\psi(x)$ is a *wave-like function* (i.e., it oscillates) that can be used to make probabilistic predictions about atomic particles.

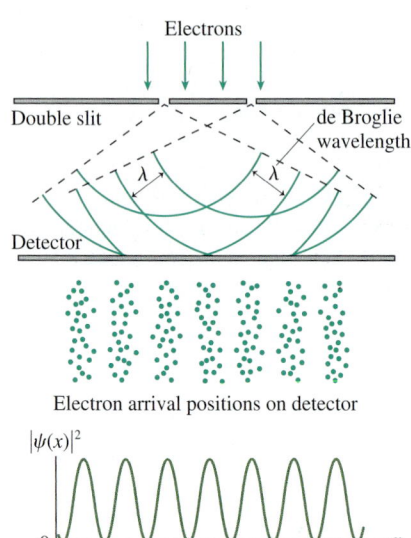

FIGURE 28.20 The wave-like interference fringes govern the probabilities of detecting electrons in a double-slit experiment.

There are 38 wells on the spinning roulette wheel, but the ball will only land in one. Where will it stop? Nobody knows—the final position is unpredictable. You can't predict the final result for one spin, but a record of the results of several spins would show a clear pattern, with the ball landing in each with equal probability. The wave nature of matter means that the behavior of atomic particles is similarly unpredictable, and is similarly governed by the laws of probability.

Energy-Level Diagrams

We used the idea of a standing de Broglie wave to find the allowed energies of a particle in a one-dimensional box. The full theory of quantum mechanics is needed to predict the allowed energy of more realistic physical systems, such as atoms or semiconductors, but the final results share a key property: The energy is quantized. Only certain energies are allowed while all other energies are forbidden.

An **energy-level diagram** is a useful visual representation of the quantized energies. As an example, Figure 28.21 on the next page is the energy-level diagram for an electron in a 0.19-nm-long box. We computed these energies in Example 28.9. An energy-level diagram is less a graph than it is a picture. The vertical axis represents energy, but the horizontal axis is not a scale. Think of this as a ladder in which the energies are the rungs of the ladder. The lowest rung, with energy E_1, is called the **ground state.** Higher rungs, called **excited states,** are labeled by their quantum numbers, $n = 2, 3, 4, \ldots$ Whether it is a particle in a box, an atom, or the nucleus of an atom, quantum physics requires the system to be on one of the rungs of the ladder.

If a quantum system changes from one state to another, there is a change in energy. One thing that has not changed in quantum physics is the conservation of energy—energy is still conserved in the quantum world. If a system drops from a higher energy level to a lower, the excess energy ΔE_{system} must go somewhere. In the systems we will consider, this energy generally ends up in the form of an emitted photon. A quantum system in energy level E_i that "jumps down" to energy level E_f loses an energy $\Delta E_{\text{system}} = \left| E_f - E_i \right|$. This jump must correspond to the emission of a photon of frequency

$$f_{\text{photon}} = \frac{\Delta E_{\text{system}}}{h} \qquad (28.17)$$

Conversely, if the system absorbs a photon, it can "jump up" to a higher energy level. In this case, the frequency of the absorbed photon must follow Equation 28.17 as well. Such jumps are called **transitions** or **quantum jumps.** Figure 28.22 shows some transitions for the particle in a box system of Figure 28.21.

Notice that Equation 28.17 links Schrödinger's quantum theory to Einstein's earlier idea about the quantization of light energy. According to Einstein, a photon of frequency f has energy $E_{\text{photon}} = hf$. If a particle jumps from an initial state with energy E_i to a final state with *lower* energy E_f, energy will be conserved if the system emits a photon with $E_{\text{photon}} = \Delta E_{\text{system}}$. The photon must have exactly the frequency given by Equation 28.17 if it is to carry away exactly the right amount of energy. As we'll see in the next chapter, these photons form the *emission spectrum* of the quantum system.

Similarly, a particle can conserve energy while jumping to a higher energy state, for which additional energy is needed, by absorbing a photon of frequency $f_{\text{photon}} = \Delta E_{\text{system}}/h$. The photon will not be absorbed unless it has exactly this frequency. The frequencies absorbed in these upward transitions form the system's *absorption spectrum.*

Let's summarize what quantum physics has to say about the properties of atomic-level systems:

1. **The energies are quantized.** Only certain energies are allowed, all others are forbidden. This is a consequence of the wave-like properties of matter.
2. **The ground state is stable.** Quantum systems seek the lowest possible energy state. A particle in an excited state, if left alone, will jump to lower and lower energy states until it reaches the ground state. Once in its ground state, there are no states of any lower energy to which a particle can jump.
3. **Quantum systems emit and absorb a *discrete spectrum* of light.** Only those photons whose frequencies match the energy *intervals* between the allowed energy levels can be emitted or absorbed. Photons of other frequencies cannot be emitted or absorbed without violating energy conservation.

We'll use these ideas in the next two chapters to understand the properties of atoms and nuclei.

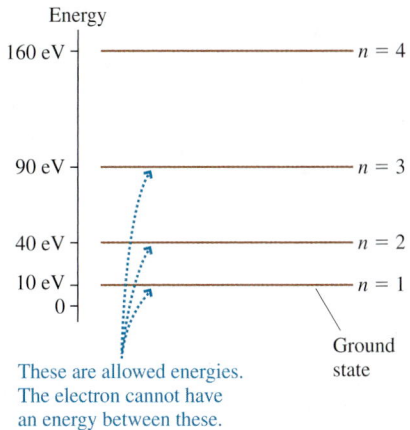

These are allowed energies. The electron cannot have an energy between these.

FIGURE 28.21 The energy-level diagram of an electron in a 0.19-nm-long box.

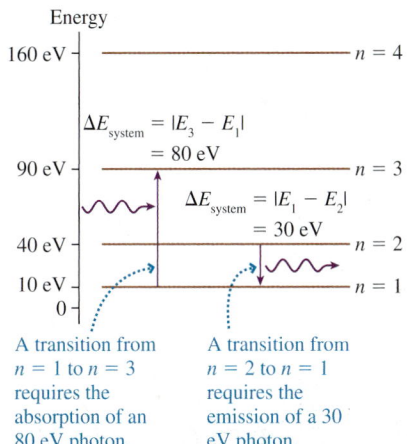

A transition from $n = 1$ to $n = 3$ requires the absorption of an 80 eV photon.

A transition from $n = 2$ to $n = 1$ requires the emission of a 30 eV photon.

FIGURE 28.22 A particle can jump between energy levels by emitting or absorbing a photon.

EXAMPLE 28.11 Determining an emission spectrum from quantum states

An electron in a quantum system has allowed energies $E_1 = 1.0$ eV, $E_2 = 4.0$ eV, and $E_3 = 6.0$ eV. What wavelengths are observed in the emission spectrum of this system?

PREPARE Figure 28.23 shows the energy-level diagram for this system. Photons are emitted when the system undergoes a quantum jump from a higher energy level to a lower energy level. There are three possible transitions.

SOLVE This system will emit photons on the $3 \rightarrow 1$, $2 \rightarrow 1$, and $3 \rightarrow 2$ transitions, with $\Delta E_{3 \rightarrow 1} = 5.0$ eV, $\Delta E_{2 \rightarrow 1} = 3.0$ eV, and $\Delta E_{3 \rightarrow 2} = 2.0$ eV. From $f_{photon} = \Delta E_{system}/h$ and $\lambda = c/f$, we find that the wavelengths in the emission spectrum are

$$3 \rightarrow 1 \quad f = 5.0 \text{ eV}/h = 1.2 \times 10^{15} \text{ Hz}$$
$$\lambda = 250 \text{ nm (ultraviolet)}$$
$$2 \rightarrow 1 \quad f = 3.0 \text{ eV}/h = 7.2 \times 10^{14} \text{ Hz}$$
$$\lambda = 420 \text{ nm (blue)}$$
$$3 \rightarrow 2 \quad f = 2.0 \text{ eV}/h = 4.8 \times 10^{14} \text{ Hz}$$
$$\lambda = 620 \text{ nm (orange)}$$

ASSESS Transitions with a small energy difference, like $3 \rightarrow 2$, correspond to lower photon energies and thus longer wavelengths than transitions with a large energy difference like $3 \rightarrow 1$, as we would expect.

FIGURE 28.23 The system's energy-level diagram and quantum jumps.

STOP TO THINK 28.5 A photon with a wavelength of 420 nm has energy $E_{photon} = 3.0$ eV. Do you expect to see a spectral line with $\lambda = 420$ nm in the emission spectrum of the system represented by this energy-level diagram?

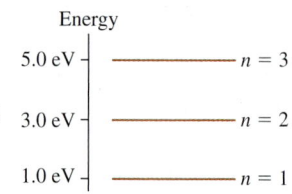

28.7 The Uncertainty Principle

One of the strangest aspects of the quantum view of the world is an inherent limitation on our knowledge: **For a particle such as an electron, if you know where it is, you can't know where it is going.** This very counterintuitive notion is a result of the wave nature of matter and is worth a bit of explanation.

Suppose we set up an experiment in which we try to pin down an electron's position by passing it through a narrow slit—if the electron goes through the slit, we have some idea of where it is. Figure 28.24 shows such an experiment in which electrons moving along the y-axis with velocity v_y and momentum $p_y = mv_y$ are forced to pass through a slit of width a. Just as electrons passing through a double slit make a double-slit interference pattern, the electrons in this experiment make a single-slit diffraction pattern characterized by the de Broglie wavelength $\lambda = h/p_y$.

A light wave spreads out after passing through a narrow slit, as you learned in Chapter 17, and the wave nature of electrons means they will do the same thing. The electrons, which had been moving along the y-axis with $v_x = 0$ before reaching the slit, now have a component of velocity along the x-axis. Because the electrons strike the screen over a range of positions, the value of v_x must vary from electron to electron. **By trying to pin down the electron's horizontal position, we've created an uncertainty in its horizontal velocity!**

As an electron passes through the slit, the uncertainty in its position is $\Delta x = a$.

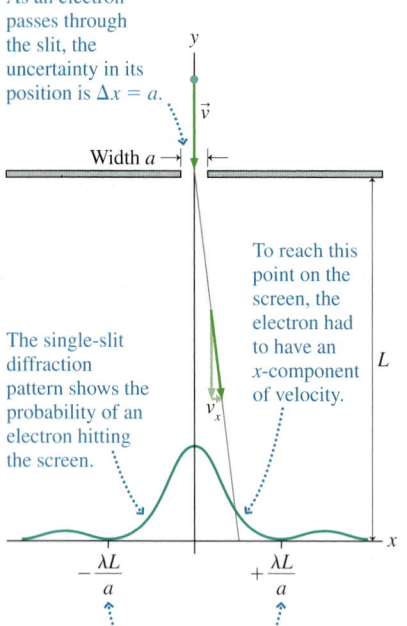

The single-slit diffraction pattern shows the probability of an electron hitting the screen.

To reach this point on the screen, the electron had to have an x-component of velocity.

We can't predict with certainty where an electron will hit, but most land between these two minima.

FIGURE 28.24 An experiment to illustrate the uncertainty principle.

CONCEPTUAL EXAMPLE 28.3 Changing the uncertainty

Suppose we narrow the slit in the above experiment, allowing us to determine the electron's horizontal position more precisely. How does this affect the diffraction pattern? How does this change in the diffraction pattern affect the uncertainty in the velocity?

REASON If the slit is narrower, the diffraction pattern will be wider, as we learned in Chapter 17. Because narrowing the slit makes the pattern wider, determining the position to greater precision means we know *less* about the velocity. The pattern is more spread out, so the uncertainty in the velocity is greater.

ASSESS Improving our knowledge of the position decreases our knowledge of the velocity! This is the hallmark of the *uncertainty principle*.

Let's analyze the situation of Figure 28.24 in a bit more detail. As the electron passes through the slit, its x-coordinate could be anywhere between $x = -a/2$ and $x = +a/2$. In other words, the electron's position in the x-direction is *uncertain,* with uncertainty $\Delta x = a$.

Now let's look at where the electrons hit the screen. Most of the intensity of a single-slit diffraction pattern lies within the central maximum, which we found in Chapter 17 to lie between $x = \pm\lambda L/a$. We cannot predict exactly where any particular electron will land on the screen, but nearly all will land between these two points, as noted in Figure 28.24.

The electrons travel to the screen in a time $\Delta t = L/v_y$. Consider the path of the particular electron shown in Figure 28.25. To have moved off the y-axis, the electron must have acquired an x-component of velocity v_x as it passed through the slit. The electron's x-velocity is

$$v_x = \frac{\text{distance}}{\text{time}} = \frac{x}{\Delta t} = \frac{\lambda L/a}{L/v_y} = \lambda\frac{v_y}{a}$$

If we substitute the de Broglie wavelength for the electron, we find

$$v_x = \lambda\frac{v_y}{a} = \frac{h}{mv_y}\frac{v_y}{a} = \frac{h}{ma} \qquad (28.18)$$

Equation 28.18 is specifically the velocity needed to reach the edge of the central maximum. Most electrons land somewhere between the two edges, so most electrons leave the slit with an x-velocity somewhere *between* $-h/ma$ and $+h/ma$. That is, the x-velocity is *uncertain* by roughly $\Delta v_x \approx 2h/ma$. If we multiply both sides by m and use our finding that the uncertainty in the position is $\Delta x = a$, we find

$$m\Delta v_x = \Delta p_x \approx \frac{2h}{a} = \frac{2h}{\Delta x}$$

which can be rewritten as

$$\Delta x\,\Delta p_x \approx 2h \qquad (28.19)$$

Our argument has been based on a specific experiment. In 1927, the German physicist Werner Heisenberg proved a more general consequence of quantum mechanics: For any particle, the product of the uncertainty Δx in its position and the uncertainty Δp_x in its y-momentum has a lower limit fixed by the following expression:

$$\Delta x\,\Delta p_x \geq \frac{h}{2\pi} \qquad (28.20)$$

Heisenberg uncertainty principle for position and momentum

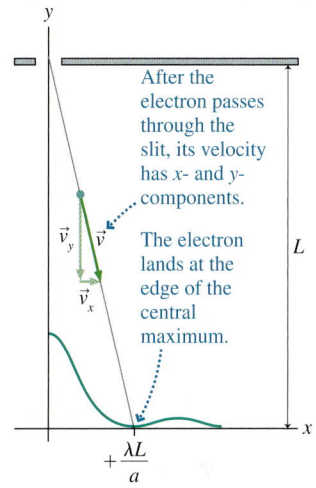

FIGURE 28.25 The path of an electron in the experiment of Figure 28.24.

After the electron passes through the slit, its velocity has x- and y-components.

The electron lands at the edge of the central maximum.

  17.6

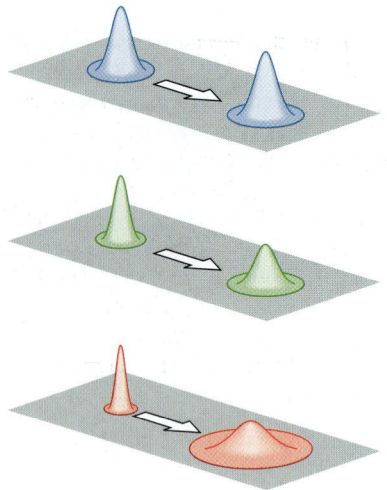

◀ **If I know where you are, I don't know where you're going** This series of pictures shows simulations of the evolution of three traveling wave functions. The top picture shows the wave function of a particle that is not precisely localized. The wave function is quite wide, meaning there is a significant uncertainty as to its position. The uncertainty in momentum (and thus velocity) is therefore small, and the wave function doesn't spread out much as it travels. The lower graphs show particles with more sharply peaked wave functions. As these particles have less uncertainty in their initial position, they will have larger uncertainties in velocity, so their wave functions spread out more quickly.

NOTE ▶ In statements of the uncertainty principle, the right side is sometimes $h/2\pi$, as we have it, but other formulations have $h/4\pi$ or $h/2$. The specific number on the right side depends on exactly how Δx and Δp are defined. Different conventions result in slightly different statements of the uncertainty principle. We needn't worry about these small differences in a treatment at this level. For us, the important idea is that the product of Δx and Δp_x for a particle cannot be significantly less than Planck's constant h. ◀

Uncertainties are associated with all experimental measurements, but better procedures and techniques can reduce those uncertainties. Classical physics places no limits on how small the uncertainties can be. A classical particle at any instant of time has an exact position x and an exact momentum p_x, and with sufficient care we can measure both x and p_x with such precision that we can make the product $\Delta x \, \Delta p_x$ as small as we like. There are no inherent limits to our knowledge about a classical particle.

In the quantum world, it's not so simple. No matter how clever you are, and no matter how good your experiment, you *cannot* measure both x and p_x simultaneously with arbitrarily good precision. Any measurements you make are limited by the condition that $\Delta x \, \Delta p_x \geq h/2\pi$. **The position and the momentum of a particle are *inherently* uncertain.**

Why? Because of the wave-like nature of matter! The "particle" is spread out in space, so there simply is not a precise value of its position x. Our belief that position and momentum have precise values is tied to our classical concept of a particle. As we revise our ideas of what atomic particles are like, we must also revise our ideas about position and momentum.

Let's revisit the case of particles in a one-dimensional "box," now looking at uncertainties.

EXAMPLE 28.12 Determining uncertainties

a. What range of velocities might an electron have if confined to a 0.10-nm-wide region, about the size of an atom?
b. A 1.0-μm-diameter dust particle ($m \approx 10^{-15}$ kg) is confined within a 10-μm-long box. Can we know with certainty if the particle is at rest? If not, within what range is its velocity likely to be found?

PREPARE Localizing a particle means specifying its position with some accuracy—so there must be an uncertainty in the velocity. We can estimate the uncertainty by using Heisenberg's uncertainty principle.

SOLVE a. The electron's position is located within an interval $\Delta x = 1.0 \times 10^{-10}$ m. The uncertainty principle thus specifies that the least possible uncertainty in the momentum is

$$\Delta p_x = \frac{h}{2\pi \, \Delta x}$$

The uncertainty in the velocity is thus approximately

$$\Delta v_x = \frac{\Delta p_x}{m} \approx \frac{h}{2\pi m \, \Delta x} \approx 1 \times 10^6 \text{ m/s}$$

Because the *average* velocity is zero, the best we can do is to say that the electron's velocity is somewhere in the interval -5×10^5 m/s $\leq v_x \leq 5 \times 10^5$ m/s. **It is simply not possible to specify the electron's velocity more precisely than this.**

b. If we know *for sure* that the dust particle is at rest, then $p_x = 0$ with no uncertainty. That is, $\Delta p_x = 0$. But then, according to the uncertainty principle, the uncertainty in our knowledge of the particle's position would have to be $\Delta x = \infty$. In other words, we would have no knowledge at all about the particle's position—it could be anywhere! But that is not the case. We know the particle is *somewhere* in the box, so the uncertainty in our knowledge of its position is at most $\Delta x = L = 10 \ \mu$m. With a finite Δx, the uncertainty Δp_x *cannot* be zero. **We cannot know with certainty if the particle is at rest inside the box.** No matter how hard we try to bring the particle to rest, the uncertainty in our knowledge of the particle's momentum will be approximately $\Delta p_x \approx h/(2\pi \Delta x) = h/(2\pi L)$. Consequently, the range of possible velocities is

$$\Delta v_x = \frac{\Delta p_x}{m} \approx \frac{h}{2\pi mL} \approx 1.0 \times 10^{-14} \text{ m/s}$$

This range of possible velocities will be centered on $v_x = 0$ m/s if we have done our best to have the particle be at rest. Therefore all we can know with certainty is that the particle's velocity is somewhere within the interval -5×10^{-15} m/s $\leq v_x \leq 5 \times 10^{-15}$ m/s.

ASSESS Our uncertainty about the electron's velocity is enormous—nearly 1% of the speed of light! For an electron confined to a region of this size, the best we can do is to state that its speed is less than one million miles per hour! The uncertainty principle clearly sets real, practical limits on our ability to describe electrons. The situation for the dust particle is different. We can't say for certain that the particle is absolutely at rest. But knowing that its speed is less than 5×10^{-15} m/s means that the particle is at rest for all practical purposes. At this speed, the dust particle would require nearly 6 hours to travel the width of one atom! Again we see that the quantum view has profound implications at the atomic scale but need not affect the way we think of macroscopic objects.

STOP TO THINK 28.6   The speed of an electron and a proton have been measured to the same uncertainty. Which one has a larger uncertainty in position?

A. The proton, because it's more massive.
B. The electron, because it's less massive.
C. The uncertainty in position is the same, as the uncertainty in velocity is the same.

28.8 Applications and Implications of Quantum Theory

Quantum theory seems rather bizarre to those of us living at a scale where the rather different rules of classical physics apply. In this section we consider some of the implications of quantum theory and some applications that confirm these unusual notions.

Tunneling and the Scanning Tunneling Microscope

The fact that particles have a wave nature allows for imaging at remarkably small scales—the scale of single atoms! The *scanning tunneling microscope* doesn't work like other microscopes you have seen, but instead builds an image of a solid surface by scanning a probe near the surface.

Figure 28.26 shows the tip of a very, very thin metal needle, called the *probe tip*, positioned above the surface of a solid sample. The space between the tip and the surface is about 0.5 nm, only a few atomic diameters. Electrons in the sample are attracted to the positive probe tip, but no current should flow, according to classical physics, because the electrons cannot cross the gap between the sample and the probe; it is an incomplete circuit. As we found with the photoelectric effect, the electrons are bound inside the sample and not free to leave.

However, electrons are not classical particles. The probability of finding an electron is governed by the electron's wave function, and waves don't have sharp edges. One startling prediction of Schrödinger's quantum mechanics is that the electrons' wave functions extend very slightly beyond the edge of the sample; we can't pin down exactly where an electron is. When the probe tip comes close enough to the surface, close enough to poke into an electron's wave function, an

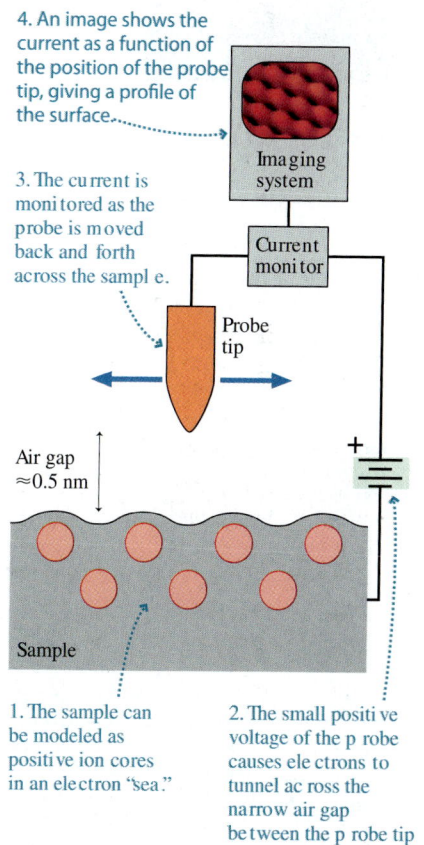

4. An image shows the current as a function of the position of the probe tip, giving a profile of the surface.

Imaging system

3. The current is monitored as the probe is moved back and forth across the sample.

Current monitor

Probe tip

Air gap ≈0.5 nm

Sample

+

1. The sample can be modeled as positive ion cores in an electron "sea."

2. The small positive voltage of the probe causes electrons to tunnel across the narrow air gap between the probe tip and the sample.

FIGURE 28.26 The scanning tunneling microscope.

The hexagonal arrangement of atoms is clearly visible.

(a) Surface of graphite

The light-colored peaks show the right-handed spiral of a DNA molecule.

(b) DNA molecule

FIGURE 28.27 STM images.

An electron and a positron attract each other. When they meet . . .

e⁻

e⁺

. . . the energy equivalent of their mass is converted into two gamma rays directed in opposite directions.

FIGURE 28.28 The annihilation of an electron and a positron.

electron that had been in the sample might suddenly find itself in the probe tip. In other words, a quantum electron *can* cross the gap between the sample and the probe tip, thus causing a current to flow in the circuit. This process is called **tunneling** because it is rather like tunneling through an uncrossable mountain barrier to get to the other side. Tunneling is completely forbidden by the laws of classical physics, so the fact that it occurs is a testament to the reality of quantum ideas.

The probability that an electron will tunnel across the gap is very sensitive to the size of the gap, which makes the **scanning tunneling microscope,** or STM, possible. When the probe tip passes over an atom or over an atomic-level bump on the surface, the gap narrows and the current increases as more electrons are able to tunnel across. Similarly, the tunneling probability decreases when the probe tip passes across an atomic-size valley, and the current falls. The current-versus-position data are used to construct an image of the surface.

The STM was the first technology that allowed imaging of individual atoms, and it was one of a handful of inventions in the 1980s that jump-started the current interest in nanotechnology. STM images offer a remarkable view of the world at an atomic scale. The STM image of Figure 28.27a clearly shows the hexagonal arrangement of the individual atoms on the surface of pyrolytic graphite. The image of a DNA molecule in Figure 28.27b shows the actual twists of the double-helix structure. Current research efforts aim to develop methods for sequencing DNA with scanning tunneling microscopes and other nanoprobes—to directly "read" a single strand of DNA!

Antimatter

In 1929, the English physicist Paul Dirac had developed a theory that unified the new disciplines of relativity and quantum theory. One of the remarkable predictions of the theory was the existence of positive electrons—particles that seem identical to the electron but that have an opposite charge.

Such positive electrons, now called **positrons,** were first found in **cosmic rays,** very energetic particles that hit the earth's upper atmosphere. In 1932, a cosmic-ray experiment detected particles that seemed identical to the electron but that curved the opposite direction in a magnetic field, the direction of positive charges. These particles were the positrons predicted by Dirac's synthesis of quantum mechanics and relativity.

The experimental discovery not only vindicated Dirac's theory, it was the first of many discoveries by which we now know that **every subatomic particle has an *antiparticle.*** A positron is an *antielectron* or, said another way, the *antimatter* version of an electron.

In what sense is a positron an *anti*electron? It was soon discovered that whenever a particle and its antiparticle meet, they annihilate each other in a subatomic fireball and turn the energy equivalent of their masses into the pure energy of two photons. Mass disappears and light appears in one of the most spectacular confirmations of Einstein's relativity.

Figure 28.28 shows the process. Suppose an electron and a positron are at rest, or nearly so. They have total momentum $p \approx 0$. We learned in Chapter 27 that a particle of mass m has rest energy given by Einstein's famous equation $E = mc^2$, so the two particles have total energy $E = 2mc^2$. The two particles attract each other, meet, and then completely annihilate—but energy and momentum are still conserved.

One consequence of the theory of relativity is that photons have momentum. Momentum can't be conserved if all of the energy goes to a single photon because that photon would leave the scene of the annihilation with momentum. Instead, the annihilation produces two photons, each with energy $E_{photon} = mc^2$, that are shot out back-to-back with no net momentum.

Electrons and positrons, each with $m = 9.11 \times 10^{-31}$ kg, have rest energy

$$E = mc^2 = 8.19 \times 10^{-14}\,\text{J} = 511\,\text{keV}$$

Consequently, the annihilation of an electron and positron produces two photons with $E_{photon} = 511$ keV. These high-energy photons are *gamma rays*.

In the following decades, physicists discovered antiprotons (with $q = -e$), antineutrons (also neutral, but not the same as a regular neutron), and antimatter versions of all the various subatomic particles that we will learn about in Chapter 30.

Wave-Particle Duality

One common theme that has run through this chapter is the idea that, in quantum theory, things we think of as being waves have a particle nature, while things we think of as being particles have a wave nature. What is the true nature of light, or an electron? Are they particles or waves?

The various objects of classical physics are *either* particles *or* waves. There's no middle ground. Planets and baseballs are particles or collections of particles, while sound and light are clearly waves. Particles follow trajectories given by Newton's laws; waves obey the principle of superposition and exhibit interference. This wave-particle dichotomy seemed obvious until physicists encountered irrefutable evidence that light sometimes acts like a particle and, even stranger, that matter sometimes acts like a wave.

You might at first think that light and matter are *both* a wave *and* a particle, but that idea doesn't quite work. The basic definitions of particleness and waviness are mutually exclusive. Two sound waves can pass through each other, and can overlap to produce a larger-amplitude sound wave; two baseballs can't. It is more profitable to conclude that light and matter are *neither* a wave *nor* a particle. At the microscopic scale of atoms and their constituents—a physical scale not directly accessible to our five senses—the classical concepts of particles and waves turn out to be simply too limited to explain the subtleties of nature.

Although matter and light have both wave-like aspects and particle-like aspects, they show us only one face at a time. If we arrange an experiment to measure a wave-like property, such as interference, we find photons and electrons acting like waves, not particles. An experiment to look for particles will find photons and electrons acting like particles, not waves. These two aspects of light and matter are *complementary* to each other, like a two-piece jigsaw puzzle. Neither the wave nor the particle model alone provides an adequate picture of light or matter, but taken together they provide us with a basis for understanding these elusive but most fundamental constituents of nature. This two-sided point of view is called *wave-particle duality*.

For over two hundred years, scientists and nonscientists alike felt that the clockwork universe of Newtonian physics was a fundamental description of reality. But wave-particle duality, along with Einstein's relativity, undermines the basic assumptions of the Newtonian worldview. The certainty and predictability of classical physics have given way to a new understanding of the universe in which chance and uncertainty play key roles—the universe of quantum physics.

A cold quantum condensate The wave nature of matter is responsible for one of the oddest forms of matter ever produced, a Bose-Einstein condensate. A sample of atoms held in a trap are cooled to extremely cold temperatures, only a few *nano*kelvin. The pictures above show the distribution of atoms in the trap. As the temperature is lowered, the wave functions of thousands of the atoms move into the ground state, producing the combined large-amplitude wave function that gives the peak on the right. As their wave functions overlap, the atoms in the trap lose their individual identities to produce a single standing matter wave consisting of thousands of atoms! The atoms are *overlapping in the same place at the same time,* something that is simply not possible in the classical view of matter.

The dual nature of a buckyball Treating atomic-level structures involves frequent shifts between particle and wave views. 60 carbon atoms can create the molecule diagrammed at left, known as C_{60}, or *buckminsterfullerene*. We rely on the wave nature of electrons to produce the scanning electron microscope image of a C_{60} molecule on the right. But the image is a particle-like view of the molecule with individual carbon atoms clearly visible. The C_{60} molecule, though we can make a picture of it—showing the atoms that make it up—also has a wave nature. A beam of C_{60} sent through a grating will produce a diffraction pattern!

SUMMARY

The goal of Chapter 28 has been to understand the quantization of energy for light and matter.

GENERAL PRINCIPLES

Light has particle-like properties

- The energy of a light wave comes in discrete packets (light quanta) we call **photons.**
- For light of frequency f, the energy of each photon is $E = hf$, where h is **Planck's constant.**
- For a light wave that delivers power P, photons arrive at rate R such that $P = Rhf$.

Matter has wave-like properties

- The **de Broglie wavelength** of a particle of mass m is $\lambda = h/mv$.
- The wave-like nature of matter is seen in the interference patterns of electrons, protons, and other particles.

Quantization of energy

When a particle is confined, it sets up a de Broglie standing wave.

The fact that standing waves can have only certain allowed wavelengths leads to the conclusion that a confined particle can have only certain allowed energies.

Wave-particle duality

- Experiments designed to measure wave properties will show the wave nature of light and matter.
- Experiments designed to measure particle properties will show the particle nature of light and matter.

Heisenberg uncertainty principle

A particle with wave-like characteristics does not have a precise value of position x or a precise value of momentum p_x. Both are uncertain. The position uncertainty Δx and momentum uncertainty Δp_x are related by

$$\Delta x\, \Delta p_x \geq \frac{h}{2\pi}$$

The more you pin down the value of one, the less precisely the other can be known.

IMPORTANT CONCEPTS

Photoelectric effect

Light can eject electrons from a metal only if $f \geq f_0 = E_0/h$, where E_0 is the metal's **work function.** Electrons will be ejected even if the intensity of the light is very small.

The **stopping potential** that stops even the fastest electrons is

$$V_{\text{stop}} = \frac{h}{e}(f - f_0)$$

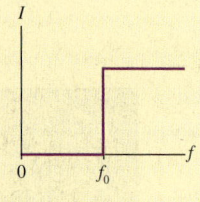

The details of the photoelectric effect could not be explained with classical physics. New models were needed.

X-ray diffraction

X rays with wavelength λ undergo strong reflections from atomic planes spaced by d when the angle of incidence satisfies the **Bragg condition:**

$$2d\cos\theta = m\lambda$$
$$m = 1, 2, 3, \ldots$$

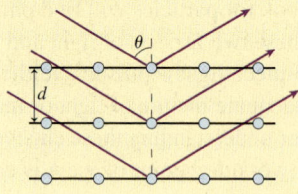

Energy levels and quantum jumps

The localization of electrons leads to quantized energy levels. An electron can exist only in certain energy states. An electron can jump to a higher level if a photon is absorbed, or to a lower level if a photon is emitted. The energy difference between the levels equals the photon energy.

Antiparticles

Every particle has an associated antiparticle. The two annihilate if they meet, creating a pair of photons.

APPLICATIONS

The wave nature of light limits the resolution of a light microscope. A more detailed image may be made with an **electron microscope** because of the very small de Broglie wavelength of fast electrons.

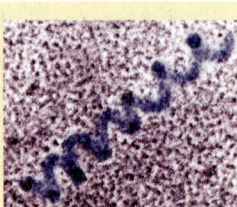

The wave nature of electrons allows them to **tunnel** across an insulating layer of air to the tip of a **scanning tunneling microscope,** revealing details of the atoms on a surface.

For instructor-assigned homework, go to www.masteringphysics.com

Problem difficulty is labeled as ▥ (straightforward) to ▮ (challenging).

Problems labeled **INT** integrate significant material from earlier chapters; **BIO** are of biological or medical interest.

QUESTIONS

Conceptual Questions

1. The first-order x-ray diffraction of monochromatic x rays from a crystal occurs at angle θ_1. The crystal is then compressed, causing a slight reduction in its volume. Does θ_1 increase, decrease, or stay the same? Explain.

2. Explain the reasoning by which we claim that the stopping potential V_{stop} measures the maximum kinetic energy of the electrons in a photoelectric-effect experiment.

3. Consider the following characteristics of the photoelectric effect:
   A. The ejection of electrons from a metal surface when light shines on it.
   B. The existence of a threshold frequency.
   C. The photoelectric current increases with increasing light intensity.
   D. The photoelectric current is independent of ΔV for $\Delta V > 0$.
   E. The photoelectric current decreases slowly as ΔV becomes more negative.
   F. The stopping potential is independent of the light intensity.
   G. Electrons begin to be ejected instantly when the light is turned on.

   Which of these *cannot* be explained by classical physics? Explain.

4. How would the graph of Figure 28.8a look *if* classical physics provided the correct description of the photoelectric effect? Draw the graph and explain your reasoning. Assume that the light intensity remains constant as its frequency and wavelength are varied.

5. Figure Q28.5 shows the typical photoelectric behavior of a metal as the anode-cathode potential difference ΔV is varied.
   a. Why do the curves become horizontal for $\Delta V > 0$ V? Shouldn't the current increase as the potential difference increases? Explain.
   b. Why doesn't the current immediately drop to zero for $\Delta V < 0$ V? Shouldn't $\Delta V < 0$ V prevent the electrons from reaching the anode? Explain.
   c. The current is zero for $\Delta V < -V_{stop}$. Where do the electrons go? Are no electrons emitted if $\Delta V < -V_{stop}$? Or if they are, why is there no current? Explain.

FIGURE Q28.5

6. In the photoelectric effect experiment, as illustrated by Figure Q28.6, a current is measured while light is shining on the cathode. But this does not appear to be a complete circuit, so how can there be a current? Explain.

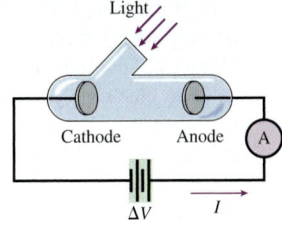

FIGURE Q28.6

7. Why are only electrons emitted in the photoelectric effect experiment? Why are positive ions never emitted?

8. Metal 1 has a larger work function than metal 2. Both are illuminated with the same short-wavelength ultraviolet light. Do electrons from metal 1 have a higher speed, a lower speed, or the same speed as electrons from metal 2? Explain.

9. A gold cathode (work function 5.1 eV) is illuminated with light of wavelength of 250 nm. It is found that the current is zero when $\Delta V = 1.0$ V. Would the current change if
   a. The light intensity is doubled?
   b. The anode-cathode potential difference is increased to $\Delta V = 5.5$ V?

10. Three laser beams have wavelengths $\lambda_1 = 400$ nm, $\lambda_2 = 600$ nm, and $\lambda_3 = 800$ nm. The power of each laser beam is 1 W.
    a. Rank in order, from largest to smallest, the photon energies E_1, E_2, and E_3 in these three laser beams. Explain.
    b. Rank in order, from largest to smallest, the number of photons per second N_1, N_2, and N_3 delivered by the three laser beams. Explain.

11. When we say that a photon is a "quantum of light," what does that mean? What is quantized?

12. The intensity of a beam of light is increased but the light's frequency is unchanged. As a result,
    A. The photons travel faster.
    B. Each photon has more energy.
    C. The photons are larger.
    D. There are more photons per second.
    Which of these (perhaps none, perhaps more than one) are true? Explain.

13. The frequency of a beam of light is increased but the light's intensity is unchanged. As a result,
    A. The photons travel faster.
    B. Each photon has more energy.
    C. There are fewer photons per second.
    D. There are more photons per second.
    Which of these (perhaps none, perhaps more than one) are true? Explain.

14. A neutron and an electron have the same de Broglie wavelength. Which is moving faster?

15. To have the best resolution, should an electron microscope use very fast electrons or very slow electrons? Explain.

16. Electron 1 is accelerated from rest through a potential difference of 100 V. Electron 2 is accelerated from rest through a potential difference of 200 V. Afterward, which electron has the larger de Broglie wavelength? Explain.

17. An electron and a proton are accelerated from rest through potential differences of the same magnitude. Afterward, which particle has the larger de Broglie wavelength? Explain.

18. A neutron is shot straight up with an initial speed of 100 m/s. As it rises, does its de Broglie wavelength increase, decrease, or not change? Explain.

19. Double-slit interference of electrons occurs because:
    A. The electrons passing through the two slits repel each other.
    B. Electrons collide with each other behind the slits.
    C. Electrons collide with the edges of the slits.
    D. Each electron goes through both slits.
    E. The energy of the electrons is quantized.
    F. Only certain wavelengths of the electrons fit through the slits.

    Which of these (perhaps none, perhaps more than one) are correct? Explain.

20. Can an electron with a de Broglie wavelength of 2 μm pass through a slit that is 1 μm wide? Explain.

21. a. For the allowed energies of a particle in a box to be large, should the box be very big or very small? Explain.

    b. Which is likely to have larger values for the allowed energies: An atom in a molecule, an electron in an atom, or a proton in a nucleus? Explain.

22. Figure Q28.22 shows the standing de Broglie wave of a particle in a box.

    a. What is the quantum number?

    b. Can you determine from this picture whether the "classical" particle is moving to the right or to the left? If so, which is it? If not, why not?

FIGURE Q28.22

23. A particle in a box of length L_a has $E_1 = 2$ eV. The same particle in a box of length L_b has $E_2 = 50$ eV. What is the ratio L_a/L_b?

24. Imagine that the horizontal box of Figure 28.19 is instead oriented vertically. Also imagine the box to be on a neutron star where the gravitational field is so strong that the particle in the box slows significantly, nearly stopping, before it hits the top of the box. Make a *qualitative* sketch of the $n = 3$ de Broglie standing wave of a particle in this box.

    Hint: The nodes are *not* uniformly spaced.

25. Figure Q28.25 shows a standing de Broglie wave.

    a. Does this standing wave represent a particle that travels back and forth between the boundaries with a constant speed or a changing speed? Explain.

    b. If the speed is changing, at which end is the particle moving faster and at which end is it moving slower?

FIGURE Q28.25

Multiple-Choice Questions

26. | A light sensor is based on a photodiode that requires a minimum photon energy of 1.7 eV to create mobile electrons. What is the longest wavelength of electromagnetic radiation that the sensor can detect?
    A. 500 nm   B. 730 nm
    C. 1200 nm   D. 2000 nm

27. | In a photoelectric effect experiment, the frequency of the light is increased. As a result,
    A. There are more electrons.
    B. The electrons are faster.
    C. Both A and B.
    D. Neither A nor B.

28. | In a photoelectric effect experiment, the intensity of the light is increased. As a result,
    A. There are more electrons.
    B. The electrons are faster.
    C. Both A and B.
    D. Neither A nor B.

29. | In the photoelectric effect, electrons are never emitted from a metal if the frequency of the incoming light is below a certain threshold value. This is because
    A. Photons of lower-frequency light don't have enough energy to eject an electron.
    B. The electric field of low-frequency light does not vibrate the electrons rapidly enough to eject them.
    C. The number of photons in low-frequency light is too small to eject electrons.
    D. Low-frequency light does not penetrate far enough into the metal to eject electrons.

30. | Visible light has a wavelength of about 500 nm. A typical radio wave has a wavelength of about 1.0 m. How many photons of the radio wave are needed to equal the energy of one photon of visible light?
    A. 2,000   B. 20,000
    C. 200,000   D. 2,000,000

31. | Two radio stations have the same power output from their antennas. One broadcasts AM at a frequency of 1000 kHz and one broadcasts FM at a frequency of 100 MHz. Which statement is true?
    A. The FM station emits more photons per second.
    B. The AM station emits more photons per second.
    C. The two stations emit the same number of photons per second.

32. | The following particles are traveling at the same speed. Which has the shortest de Broglie wavelength?
    A. Electron
    B. Proton
    C. Alpha particle
    D. Uranium atom

33. ‖ You shoot a beam of electrons through a double slit to make an interference pattern. After noting the properties of the pattern, you then double the speed of the electrons. What effect would this have?
    A. The fringes would get closer together.
    B. The fringes would get farther apart.
    C. The positions of the fringes would not change.

34. | Photon P in Figure Q28.34 moves an electron from energy level $n = 1$ to energy level $n = 3$. The electron jumps to $n = 2$, emitting photon Q, and then jumps to $n = 1$, emitting photon R. The spacing between energy levels is drawn to scale. What is the correct relationship among the wavelengths of the photons?

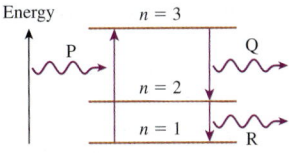

FIGURE Q28.34

A. $\lambda_P < \lambda_Q < \lambda_R$ B. $\lambda_R < \lambda_P < \lambda_Q$
C. $\lambda_Q < \lambda_P < \lambda_R$ D. $\lambda_R < \lambda_Q < \lambda_P$

35. | An electron, a proton, a helium atom, and a uranium atom are all moving at the same speed v and each has the same uncertainty in speed Δv. The uncertainty in position is largest for the
A. Electron
B. Proton
C. Helium atom
D. Uranium atom

PROBLEMS

Section 28.1 X Rays and X-Ray Diffraction

1. | X rays with a wavelength of 0.12 nm undergo first-order diffraction from a crystal at a 68° angle of incidence. What is the angle of second-order diffraction?

2. | X rays with a wavelength of 0.20 nm undergo first-order diffraction from a crystal at a 54° angle of incidence. At what angle does first-order diffraction occur for x rays with a wavelength of 0.15 nm?

3. | X rays diffract from a crystal in which the spacing between atomic planes is 0.175 nm. The second-order diffraction occurs at 45.0°. What is the angle of the first-order diffraction?

4. | The spacing between atomic planes in a crystal is 0.110 nm. If 12.0 keV x rays are diffracted by this crystal, what are the angles of (a) first-order and (b) second-order diffraction?

5. | X rays with a wavelength of 0.085 nm diffract from a crystal in which the spacing between atomic planes is 0.18 nm. How many diffraction orders are observed?

Section 28.2 The Photoelectric Effect

6. | Which metals in Table 28.1 exhibit the photoelectric effect for (a) light with $\lambda = 400$ nm and (b) light with $\lambda = 250$ nm?

7. | Electrons are emitted when a metal is illuminated by light with a wavelength less than 388 nm. What is the metal's work function?

8. ||| Electrons in a photoelectric-effect experiment emerge from a copper surface with a maximum kinetic energy of 1.10 eV. What is the wavelength of the light?

9. | You need to design a photodetector that can respond to the entire range of visible light. What is the maximum possible work function of the cathode?

10. || A photoelectric-effect experiment finds a stopping potential of 1.93 V when light of 200 nm wavelength is used to illuminate the cathode.
    a. From what metal is the cathode made?
    b. What is the stopping potential if the intensity of the light is doubled?

11. | Zinc has a work function of 4.3 eV.
    a. What is the longest wavelength of light that will release an electron from a zinc surface?
    b. A 4.7 eV photon strikes the surface and an electron is emitted. What is the maximum possible speed of the electron?

12. | Light with a wavelength of 550 nm shines on a metal surface, which emits electrons. The maximum kinetic energy of the electrons is 1.20 eV. What is the work function of this metal?

13. || Light with a wavelength of 350 nm shines on a metal surface, which emits electrons. The stopping potential is measured to be 1.25 V.
    a. What is the speed of electrons with 1.25 eV of kinetic energy?
    b. Calculate the work function and identify the metal.

Section 28.3 Photons

14. | When an ultraviolet photon is absorbed by a molecule of
BIO DNA, the photon's energy can be converted into vibrational energy of the molecular bonds. Excessive vibration damages the molecule by causing the bonds to break. Ultraviolet light of wavelength less than 290 nm causes significant damage to DNA; ultraviolet light of longer wavelength causes minimal damage. What is the threshold photon energy, in eV, for DNA damage?

15. | A firefly glows by the
BIO direct conversion of chemical energy to light. The light emitted by a firefly has peak intensity at a wavelength of 550 nm.
    a. What is the minimum chemical energy, in eV, required to generate each photon?
    b. One molecule of ATP provides 0.30 eV of energy when it is metabolized in a cell. What is the minimum number of ATP molecules that must be consumed in the reactions that lead to one photon of 550 nm light?

16. | The spacing between atoms in graphite is approximately 0.25 nm. What is the energy of an x-ray photon with this wavelength?

17. | What is the wavelength of a 5.0 keV x-ray photon?

18. | What is the wavelength, in nm, of a photon with energy (a) 0.30 eV, (b) 3.0 eV, and (c) 30 eV? For each, is this wavelength visible, ultraviolet, or infrared light?

19. | What is the ratio of the energy of a photon of light at the far red end of the visible spectrum (700 nm) to that of a photon at the far blue end of the visible spectrum (400 nm)?

20. || Station KAIM in Hawaii broadcasts on the AM dial at 870 kHz, with a maximum power of 50,000 W. At maximun power, how many photons does the transmitting antenna emit each second?

21. | What is the energy, in eV, of (a) a 100 MHz radio-frequency photon, (b) a visible-light photon with a wavelength of 500 nm, and (c) an x-ray photon with a wavelength of 0.10 nm?

22. | 550 nm is the average wavelength of visible light.
    a. What is the energy of a photon with a wavelength of 550 nm?
    b. A typical incandescent light bulb emits about 1 J of visible light energy every second. Estimate the number of visible photons emitted per second.

23. ||| For what wavelength of light does a 100 mW laser deliver 2.5×10^{17} photons per second?

Section 28.4 Matter Waves

24. | Estimate your de Broglie wavelength while walking at a speed of 1 m/s.

25. | a. What is the de Broglie wavelength of a 200 g baseball with a speed of 30 m/s?
    b. What is the speed of a 200 g baseball with a de Broglie wavelength of 0.20 nm?

26. | a. What is the speed of an electron with a de Broglie wavelength of 0.20 nm?
    b. What is the speed of a proton with a de Broglie wavelength of 0.20 nm?

27. || What is the kinetic energy, in eV, of an electron with a de Broglie wavelength of 1.0 nm?

28. || At what speed is an electron's de Broglie wavelength (a) 1.0 nm, (b) 1.0 μm, and (c) 1.0 mm?

29. || Through what potential difference must an electron be accelerated from rest to have a de Broglie wavelength of 500 nm?

30. || The diameter of the nucleus is about 10 fm (1 fm = 10^{-15} m). What is the kinetic energy, in MeV, of a proton with a de Broglie wavelength of 10 fm?

31. || A paramecium is covered with
BIO motile hairs called cilia that propel
INT it at a speed of 1 mm/s. If the paramecium has a volume of 2×10^{-13} m^3 and a density equal that of water, what is its de Broglie wavelength? What fraction of the paramecium's 150 μm length does this wavelength represent?

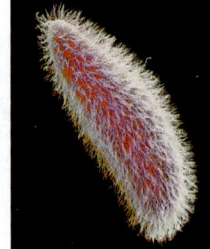

Section 28.5 Energy Is Quantized

32. || What is the length of a box in which the minimum energy of an electron is 1.5×10^{-18} J?

33. ||| What is the length of a one-dimensional box in which an electron in the $n = 1$ state has the same energy as a photon with a wavelength of 600 nm?

34. |||| An electron confined in a one-dimensional box is observed, at different times, to have energies of 12 eV, 27 eV, and 48 eV. What is the length of the box?

35. || The smallest allowed energy of a hydrogen atom (atomic mass number 1) in a box of length L is 1.0×10^{-20} J. What is the smallest allowed energy of a helium atom (atomic mass number 4) in a box of length $L/2$?

36. | The nucleus of an atom is 5.0 fm (1 fm = 10^{-15} m) in diameter. A very simple model of the nucleus is a one-dimensional box in which protons are confined. Estimate the energy of a proton in the nucleus by finding the first three allowed energies of a proton in a 5.0-fm-long box.

Section 28.6 Energy Levels and Quantum Jumps

37. || The allowed energies of a quantum system are 1.0 eV, 2.0 eV, 4.0 eV, and 7.0 eV. What wavelengths appear in the system's emission spectrum?

38. | Figure P28.38 is an energy-level diagram for a quantum system. What wavelengths appear in the system's emission spectrum?

FIGURE P28.38

39. || A quantum system's lowest three allowed energies are as shown in Figure P28.38. What possible energies of the $n = 4$ state would lead to an emission line at 420 nm?

40. | The allowed energies of a quantum system are 0.0 eV, 4.0 eV, and 6.0 eV.
    a. Draw the system's energy-level diagram. Label each level with the energy and the quantum number.
    b. What wavelengths appear in the system's emission spectrum?

41. || The allowed energies of a quantum system are 0.0 eV, 1.5 eV, 3.0 eV, and 6.0 eV. How many different wavelengths appear in the emission spectrum?

Section 28.7 The Uncertainty Principle

42. | The speed of an electron is known to be between 3.0×10^6 m/s and 3.2×10^6 m/s. Estimate the uncertainty in its position.

43. |||| A spherical virus has a diameter of 50 nm. It is contained
BIO inside a long, narrow cell of length 1×10^{-4} m. What uncer-
INT tainty does this imply for the velocity of the virus along the length of the cell? Assume the virus has a density equal to that of water.

44. | Andrea thinks she's sitting at rest in her 5.0-m-long dorm room as she does her physics homework. Can Andrea be sure she's at rest? If not, within what range is her velocity likely to be?

45. |||| What is the smallest box in which you can confine an electron if you want to know for certain that the electron's speed is no more than 10 m/s?

46. | A thin solid barrier in the xy-plane has a 10-μm-diameter circular hole. An electron traveling in the z-direction with $v_x = 0$ m/s passes through the hole. Afterward, is v_x still zero? If not, within what range is v_x likely to be?

47. ||| A proton is confined within an atomic nucleus of diameter 4 fm (1 fm $= 10^{-15}$ m). Estimate the smallest range of speeds you might find for a proton in the nucleus.

General Problems

48. | X-ray photons with energies of 1.50×10^{-15} J are incident on a crystal. The spacing between the atomic planes in the crystal is 0.210 nm. At what angles of incidence will the x rays diffract from the crystal?

49. || X rays with a wavelength of 0.0700 nm diffract from a crystal. Two adjacent angles of x-ray diffraction are 45.6° and 21.0°. What is the distance in nm between the atomic planes responsible for the diffraction?

50. || Potassium and gold cathodes are used in a photoelectric-effect experiment. For each cathode, find:
    a. The threshold frequency.
    b. The threshold wavelength.
    c. The maximum electron ejection speed if the light has a wavelength of 220 nm.
    d. The stopping potential if the wavelength is 220 nm.

51. ||| The maximum kinetic energy of electrons is 2.8 eV. When the wavelength of the light is increased by 50%, the maximum energy decreases to 1.1 eV. What are (a) the work function of the cathode and (b) the initial wavelength?

52. ||| In a photoelectric-effect experiment, the stopping potential at a wavelength of 400 nm is 25.7% of the stopping potential at a wavelength of 300 nm. Of what metal is the cathode made?

53. ||| Light of constant intensity but varying wavelength was used to illuminate the cathode in a photoelectric-effect experiment. The graph of Figure P28.53 shows how the stopping potential depended on the frequency of the light. What is the work function, in eV, of the cathode?

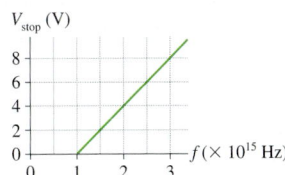

FIGURE P28.53

54. | What is the de Broglie wavelength of a red blood cell with a
BIO mass of 1.00×10^{-11} g that is moving with a speed of 0.400 cm/s? Do we need to be concerned with the wave nature of the blood cells when we describe the flow of blood in the body?

55. || Suppose you need to image the structure of a virus with a
BIO diameter of 50 nm. For a sharp image, the wavelength of the probing wave must be 5.0 nm or less. We have seen that, for imaging such small objects, this short wavelength is obtained by using an electron beam in an electron microscope. Why don't we simply use short-wavelength electromagnetic waves?

There's a problem with this approach: As the wavelength gets shorter, the energy of a photon of light gets greater and could damage or destroy the object being studied. Let's compare the energy of a photon and an electron that can provide the same resolution.
    a. For light of wavelength 5.0 nm, what is the energy (in eV) of a single photon?
    b. For an electron with a de Broglie wavelength of 5.0 nm, what is the kinetic energy (in eV).
    c. Explain why electrons can produce sharp images with less damage to the objects being imaged.

56. | Gamma rays are photons with very high energy.
    a. What is the wavelength of a gamma-ray photon with energy 625 keV?
    b. How many visible-light photons with a wavelength of 500 nm would you need to match the energy of this one gamma-ray photon?

57. || How long does it take a 670 nm laser with a 1.0 mW output power to emit 1.0 mol of photons?

58. | A helium-neon laser emits a light beam with a wavelength of 633 nm. The power of the laser beam is 1.00 mW.
    a. What is the energy of one photon of laser light?
    b. How many photons does the laser emit each second?

59. || A typical incandescent lightbulb emits approximately
BIO 3×10^{18} visible-light photons per second. Your eye, when it is fully dark adapted, can barely see the light from an incandescent lightbulb 10 km away. How many photons per second are incident at the image point on your retina? The diameter of a dark-adapted pupil is 6 mm.

60. ||| The intensity of sunlight hitting the surface of the earth on a
BIO cloudy day is about 0.50 kW/m². Assuming your pupil can close down to a diameter of 2.0 mm and that the average wavelength of visible light is 550 nm, how many photons per second of visible light enter your eye if you look up at the sky?

61. | A red laser with a wavelength of 650 nm and a blue laser with a wavelength of 450 nm emit laser beams with the same light power. How do their rates of photon emission compare? Answer this by computing R_{red}/R_{blue}.

62. || A ruby laser emits an intense pulse of light that lasts a mere 10 ns. The light has a wavelength of 690 nm, and each pulse has an energy of 500 mJ.
    a. How many photons are emitted in each pulse?
    b. What is the *rate* of photon emission, in photons per second, during the 10 ns that the laser is "on"?

63. || The wavelength of the radiation in a microwave oven is
INT 12 cm. How many photons are absorbed by 200 g of water as it's heated from 20°C to 90°C?

64. ||| Electrons with a speed of 2.0×10^6 m/s pass through a
INT double-slit apparatus. Interference fringes are detected with a fringe spacing of 1.5 mm.
    a. What will the fringe spacing be if the electrons are replaced by neutrons with the same speed?
    b. What speed must neutrons have to produce interference fringes with a fringe spacing of 1.5 mm?

65. || Electrons pass through a 1.0-μm-wide slit with a speed of
INT 1.5×10^6 m/s. How wide is the electron diffraction pattern on a detector 1.0 m behind the slit?

66. ‖ The electron interference pattern of Figure 28.15 was made
INT by shooting electrons with 50 keV of kinetic energy through two slits spaced 1.0 μm apart. The fringes were recorded on a detector 1.0 m behind the slits.
   a. What was the speed of the electrons? (The speed is large enough to justify using relativity, but for simplicity do this as a nonrelativistic calculation.)
   b. Figure 28.15 is greatly magnified. What was the actual spacing on the detector between adjacent bright fringes?

67. ‖‖ The electron beam in a cathode-ray tube is accelerated
INT through a potential difference of 250 V. The electrons then pass through a small circular hole and are viewed on the screen. The central bright spot you observe on the screen is the base of a cone that has its apex at the hole. The outer edge of the cone makes a $0.50°$ angle with the original direction of the electron beam. What is the diameter of the hole?

68. ‖ A neutron has a de Broglie wavelength of 7.0×10^{-12} m.
INT a. What is the neutron's speed?
   b. Suppose that speed is the rms speed for a neutron gas. What is the temperature of the gas?

69. ‖‖ It is stated in the text that special relativity must be used to
INT calculate the de Broglie wavelength of electrons in an electron microscope. Let us discover how much of an effect relativity has. Consider an electron accelerated through a potential difference of 1.00×10^5 V.
   a. Using the Newtonian (nonrelativistic) expressions for kinetic energy and momentum, what is the electron's de Broglie wavelength?
   b. The de Broglie wavelength is $\lambda = h/p$, but the momentum of a relativistic particle is not mv. Using the relativistic expressions for kinetic energy and momentum, what is the electron's de Broglie wavelength?

70. ‖ a. What are the energies of the first three energy levels of an electron confined in a one-dimensional box of length 0.70 nm?
   b. How much energy must the electron lose to jump from the $n = 2$ energy level to the $n = 1$ energy level?
   c. Suppose that an electron can jump from the $n = 2$ level to the $n = 1$ level by emitting a photon of light. If energy is conserved, what must the photon's wavelength be? Give your answer in nm.

71. ‖ a. What is the minimum energy of a 2.7 g Ping-Pong ball in a 10-cm-long box?
   b. At what speed does it have this energy?

72. ‖ What is the length of a box in which the difference between an electron's first and second allowed energies is 1.0×10^{-19} J?

73. ‖ Two adjacent allowed energies of an electron in a one-dimensional box are 2.0 eV and 4.5 eV. What is the length of the box?

74. ‖ An electron confined to a box has an energy of 1.28 eV. Another electron confined to an identical box has an energy of 2.88 eV. What is the smallest possible length for those boxes?

75. ‖‖ Consider a small virus having a diameter of 10 nm. The
BIO atoms of the intracelluar fluid are confined within this "box."
INT Suppose we model the virus as a one-dimensional box of length 10 nm. What is the ground state energy (in eV) of a sodium ion confined in such a box?

76. ‖ It can be shown that the allowed energies of a particle of mass m in a two-dimensional square box of side L are

$$E_{nm} = \frac{h^2}{8mL^2}(n^2 + l^2)$$

The energy depends on two quantum numbers, n and l, both of which must have an integer value 1, 2, 3, . . .
   a. What is the minimum energy for a particle in a two-dimensional square box of side L?
   b. What are the five lowest allowed energies? Give your values as multiples of E_{min}.

77. ‖‖ An electron confined in a one-dimensional box emits a 200 nm photon in a quantum jump from $n = 2$ to $n = 1$. What is the length of the box?

78. ‖‖ A proton confined in a one-dimensional box emits a 2.0 MeV gamma-ray photon in a quantum jump from $n = 2$ to $n = 1$. What is the length of the box?

79. ‖‖ As an electron in a one-dimensional box of length 0.600 nm jumps between two energy levels, a photon of energy 8.36 eV is emitted. What are the quantum numbers of the two levels?

80. ‖ What are (a) the energy, in eV, and (b) the frequency of the
INT two photons produced when a proton and an antiproton encounter one another and are annihilated? Assume the two particles are essentially at rest before colliding.

Passage Problems

Compton Scattering

Further support for the photon model of electromagnetic waves comes from *Compton scattering,* in which x rays scatter from electrons, changing direction and frequency in the process. Classical electromagnetic wave theory cannot explain the change in frequency of the x rays on scattering, but the photon model can.

Suppose an x-ray photon is moving to the right. It has a collision with a slow-moving electron, as in Figure P28.81. The photon transfers energy and momentum to the electron, which recoils at a high speed. The x-ray photon loses energy, and the photon energy formula $E = hf$ tells us that its frequency must decrease. The collision looks very much like the collision between two particles.

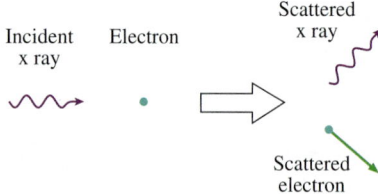

FIGURE P28.81

81. ‖ When the x-ray photon scatters from the electron,
   A. Its speed increases.
   B. Its speed decreases.
   C. Its speed stays the same.

82. ‖ When the x-ray photon scatters from the electron,
   A. Its wavelength increases.
   B. Its wavelength decreases.
   C. Its wavelength stays the same.

83. ‖ When the electron is struck by the x-ray photon,
    A. Its de Broglie wavelength increases.
    B. Its de Broglie wavelength decreases.
    C. Its de Broglie wavelength stays the same.
84. ‖ X-ray diffraction can also change the direction of a beam of x rays. Which statement offers the best comparison between Compton scattering and x-ray diffraction?

A. X-ray diffraction changes the wavelength of x rays; Compton scattering does not.
B. Compton scattering changes the speed of x rays; x-ray diffraction does not.
C. X-ray diffraction relies on the particle nature of the x rays; Compton scattering relies on the wave nature.
D. X-ray diffraction relies on the wave nature of the x rays; Compton scattering relies on the particle nature.

STOP TO THINK ANSWERS

Stop to Think 28.1: A. The Bragg condition $2d\sin\theta_1 = \lambda$ tells us that larger values of d go with smaller values of θ_1.

Stop to Think 28.2: $V_A > V_B > V_C$. For a given wavelength of light, electrons are ejected faster from metals with smaller work functions because it takes less energy to remove an electron. Faster electrons need a larger negative voltage to stop them.

Stop to Think 28.3: D. Photons always travel at c, and a photon's energy depends only on the light's frequency, not its intensity.

Stop to Think 28.4: A. The widest diffraction pattern occurs for the largest wavelength. The de Broglie wavelength is inversely proportional to the particle's mass, and so will be largest for the least massive particle.

Stop to Think 28.5: No. The energy of an emitted photon is the energy *difference* between two allowed energies. The three possible quantum jumps have energy differences of 2.0 eV, 2.0 eV, and 4.0 eV.

Stop to Think 28.6: B. Because $\Delta p_x = m\Delta v_x$, the uncertainty in position is $\Delta x = \dfrac{h}{\Delta p_x} = \dfrac{h}{m\Delta v_x}$. A more massive particle has a smaller position uncertainty.

29

ATOMS AND MOLECULES

This microscope image is a kidney cell from an African green monkey. The blue and green show, respectively, nuclear DNA and filamentous actin. What makes the different parts of this cell so brightly colored?

Looking Ahead ▶▶

The goal of Chapter 29 is to use quantum physics to understand the properties of atoms, molecules, and their spectra. In this chapter you will learn to:

▶ Recognize continuous and discrete spectra.

▶ Understand how we know about atoms and their structure.

▶ Understand the Bohr model of the atom.

▶ Use a quantum-mechanical model of the hydrogen atom.

▶ Understand multielectron atoms and their spectra.

▶ Apply quantum ideas to molecules and molecular fluorescence.

▶ Understand how lasers work.

Looking Back ◀◀

The material in this chapter depends on understanding the use of diffraction gratings, the emission of electromagnetic waves, and the basic ideas of quantum physics. Please review:

◀ Section 17.3 Diffraction gratings.

◀ Section 25.8 Thermal radiation.

◀ Sections 28.5–28.6 Quantization and energy levels.

We cannot see individual atoms or molecules, even with the very best microscope, yet all the light we see comes from atoms and molecules. Understanding how atoms absorb and emit light was the driving force behind the development of quantum physics a century ago, and today it remains the focus of much of modern technology.

X rays, lasers, and fluorescence microscopy, which produced the photo shown here, are important tools for biotechnology, industry, and fundamental research. To understand these tools, we have to understand the quantum physics of atoms and molecules. Some of the ideas in this chapter will be familiar to you from your chemistry classes. Our goal is to see where these ideas come from and to learn how atoms and molecules interact with light.

29.1 Spectroscopy

The interference and diffraction of light were well understood by the end of the 19th century, and the knowledge was used to design practical tools for measuring wavelengths with great accuracy. The primary instrument for measuring the wavelengths of light is a **spectrometer,** such as the one shown in Figure 29.1. The heart of a spectrometer is a diffraction grating that causes different wavelengths of light to diffract at different angles.

Each wavelength is focused to a different position on the photographic plate or, more likely today, a CCD detector like the one in your digital camera. The distinctive pattern of wavelengths emitted by a source of light and recorded on the detector is called the **spectrum** of the light. Spectroscopists discovered very early that there are two types of spectra, continuous spectra and discrete spectra:

- As you learned in Section 25.8, hot, self-luminous objects, such as the sun or an incandescent lightbulb, emit a **continuous spectrum** in which a rainbow is formed by light being emitted at every possible wavelength.
- In contrast, the light emitted by a gas discharge tube (such as those used to make neon signs) contains only certain discrete, individual wavelengths. Such a spectrum is called a **discrete spectrum.**

Figure 29.2 shows examples of spectra as they would appear on the photographic plate of a spectrometer. Each bright line in a discrete spectrum, called a **spectral line,** represents *one* specific wavelength present in the light emitted by the source. A discrete spectrum is sometimes called a *line spectrum* because of its appearance on the plate. You can see that a neon light has its familiar reddish-orange color because nearly all of the wavelengths emitted by neon atoms fall within the wavelength range 600–700 nm that we perceive as orange and red.

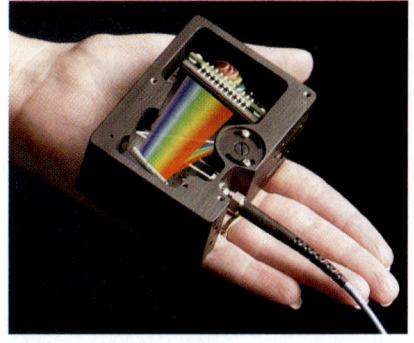

Some modern spectrometers are small enough to hold in your hand. (The rainbow has been added to show the paths that different colors take.)

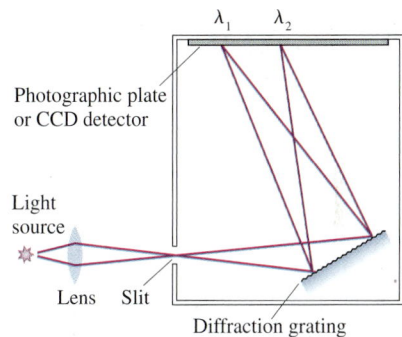

FIGURE 29.1 A diffraction spectrometer.

(a) Incandescent light bulb

400 nm   500 nm   600 nm   700 nm
Violet   Blue   Green   Yellow   Orange   Red

(b) Hydrogen

400 nm   500 nm   600 nm   700 nm

(c) Mercury

400 nm   500 nm   600 nm   700 nm

(d) Neon

400 nm   500 nm   600 nm   700 nm

FIGURE 29.2 Examples of spectra in the visible wavelength range 400–700 nm. (a) is a continuous spectrum, while (b)–(d) are discrete spectra.

Continuous Spectra and Blackbody Radiation

In Chapter 25 you learned that hot objects emit electromagnetic radiation. As physicists began to study this radiation, and especially its *spectrum,* they discovered a remarkable fact. Consider an ideal **blackbody,** an object that absorbs all light incident upon it, reflecting none. Because we usually see objects when light reflects from them, such a nonreflective object appears perfectly black. Coal and charcoal are nearly ideal blackbodies, as you might expect, but so are non-reflective metals like the tungsten wires used in light bulb filaments, whose spectrum you studied in Chapter 25.

Cool lava is nearly black, but it glows brightly when hot.

A hotter object has a much greater intensity, peaked at shorter wavelengths.

FIGURE 29.3 The spectrum of a blackbody at different temperatures.

A heated blackbody begins to glow, becoming visible by virtue of being *self-luminous* rather than by reflecting light. Physicists found that **the spectrum of a blackbody depends only on its temperature,** as shown in Figure 29.3. All blackbodies, regardless of their material, emit the same spectrum, called **blackbody radiation,** if heated to the same temperature. This is why we could plot the spectra in Figure 29.3 without specifying the material that the blackbody was made of.

The fact that blackbody radiation depends only on an object's temperature, and nothing else, suggests that we should be able to use the laws of thermodynamics and electromagnetism to explain the shape of the spectrum. However, all attempts to do so in the late 1800s failed. Finally, in the year 1900, the German physicist Max Planck successfully derived an equation giving the shape of the blackbody spectrum, but, as we saw in Chapter 28, to do so he had to make a most unusual assumption.

It was known that electromagnetic waves are emitted by "atomic oscillators." According to classical physics, an atomic oscillator, just like an oscillating mass on a spring, can have any energy; there are no restrictions. But Planck found that he could explain the shape of the blackbody spectrum only by assuming that an atomic oscillator with frequency f is restricted to having energy

$$E = nhf \qquad n = 1, 2, 3, \ldots$$

where h was a constant. That is, the energy of the atomic oscillators is *quantized,* restricted to being integer multiples of hf.

Planck thought this was a calculation trick, something that further research would explain. He never really accepted energy quantization as real. It was Einstein, five years later in 1905, who applied Planck's idea to the light itself, explaining the photoelectric effect with the claim that light comes in quanta—photons—of energy hf. The constant h, now called Planck's constant, has the modern value

$$h = 6.63 \times 10^{-34} \, \text{J} \cdot \text{s} = 4.14 \times 10^{-15} \, \text{eV} \cdot \text{s}$$

A simple effort to explain the spectrum of glowing coals was the spark that ignited the quantum revolution.

Discrete Spectra of the Elements

If a high voltage is applied to two electrodes sealed in a glass tube filled with a low-pressure gas, the gas begins to glow and emits light with a color that is unique to that gas. *Gas discharge tubes* were developed in the late 19th century, and scientists soon realized that the light was being emitted by the atoms in the gas. These *atomic spectra* were a new way to study the properties of the elements.

When the hydrogen discharge tube at the left is viewed through a diffraction grating, its unique discrete spectrum at the right is observed.

Two important conclusions had been established by the end of the 19th century:

1. The light emitted by atoms in a gas discharge tube has a discrete spectrum. Figure 29.2b–d showed some examples.
2. Every element in the periodic table has its own unique spectrum.

The fact that each element emits a unique spectrum means that atomic spectra can be used as "fingerprints" to identify elements. Consequently, atomic spectroscopy is the basis of many contemporary technologies for analyzing the composition of unknown materials, monitoring air pollutants, and studying the atmospheres of the earth and other planets.

Gases can also *absorb* light. In an absorption experiment, as shown in Figure 29.4a, a white-light source emits a continuous spectrum that, in the absence of a gas, exposes the film completely and uniformly. When a sample of gas is placed in the light's path, any wavelengths absorbed by the gas are missing and the film is dark at that wavelength.

NOTE ▶ A peculiarity of the English language is that the process of *absorbing* light is called *absorption,* not "absorbtion." ◀

Gases not only emit discrete wavelengths, they also absorb discrete wavelengths. But there is an important difference between the emission spectrum and the absorption spectrum of a gas: **Every wavelength that is absorbed by the gas is also emitted, but *not* every emitted wavelength is absorbed.** The wavelengths in the absorption spectrum appear as a subset of the wavelengths in the emission spectrum. As an example, Figure 29.4b shows both the emission and the absorption spectra of sodium atoms. All of the absorption wavelengths are prominent in the emission spectrum, but there are many emission lines for which no absorption occurs.

What causes atoms to emit or absorb light? Why a discrete spectrum? Why are some wavelengths emitted but not absorbed? Why is each element different? Nineteenth-century physicists struggled with these questions but could not answer them. Ultimately, classical physics was simply incapable of providing an understanding of atoms.

The only encouraging sign came from an unlikely source. While the spectra of other atoms have dozens or even hundreds of wavelengths, the visible spectrum of hydrogen, between 400 nm and 700 nm, consists of a mere four spectral lines (see Figure 29.2b and Table 29.1). If any spectrum could be understood, it should be that of the first element in the periodic table. The breakthrough came in 1885, not by an established and recognized scientist but by a Swiss school teacher, Johann Balmer. Balmer showed that the wavelengths in the hydrogen spectrum could be represented by the simple formula

$$\lambda = \frac{91.18 \text{ nm}}{\left(\frac{1}{2^2} - \frac{1}{n^2}\right)}, \qquad n = 3, 4, 5, \ldots \tag{29.1}$$

Later experimental evidence, as ultraviolet and infrared spectroscopy developed, showed that Balmer's result could be generalized to

$$\lambda = \frac{91.18 \text{ nm}}{\left(\frac{1}{m^2} - \frac{1}{n^2}\right)} \quad \begin{cases} m \text{ can be } 1, 2, 3, \ldots \\ \text{If } m = 1, n \text{ can be } 2, 3, 4, \ldots \\ \text{If } m = 2, n \text{ can be } 3, 4, 5, \ldots \\ \text{If } m = 3, n \text{ can be } 4, 5, 6, \ldots \\ \text{and so on.} \end{cases} \tag{29.2}$$

We now refer to Equation 29.2 as the **Balmer formula,** although Balmer himself only suggested the original version in which $m = 2$. Other than at the very highest levels of resolution, where new details appear that need not concern us in this text, the Balmer formula accurately describes *every* wavelength in the emission spectrum of hydrogen.

The Balmer formula is what we call *empirical knowledge.* It is an accurate mathematical representation found through experimental evidence, but it does not rest on any physical principles or physical laws. Balmer's formula was useful, but no one was able to *derive* Balmer's formula from Newtonian mechanics or the theory of electromagnetism. Yet the formula was so simple that it must, everyone agreed, have a simple explanation. It would take 30 years to find it.

STOP TO THINK 29.1 These spectra are due to the same element. Which one is an emission spectrum and which is an absorption spectrum?

(a) Measuring an absorption spectrum

$\lambda_1 \quad \lambda_2$

Photographic plate or CCD detector

Glass cell filled with sample gas

White light

(b) Absorption and emission spectra of sodium

Absorption

Emission

300 nm   400 nm   500 nm   600 nm   700 nm

Ultraviolet ┆ Visible

FIGURE 29.4 Measuring an absorption spectrum.

TABLE 29.1 Wavelengths of visible lines in the hydrogen spectrum

656.5 nm
486.3 nm
434.2 nm
410.3 nm

Astronomical colors The red color of this nebula is due to the emission of light from hydrogen atoms. The atoms are excited by intense ultraviolet light from the star in the center. They then emit red light, with $\lambda = 656$ nm, as predicted by the Balmer formula with $m = 2$ and $n = 3$. Spectroscopy of glowing gases from astronomical objects is an invaluable tool for astronomers, because it allows the gases present in objects many light years distant to be identified from earth.

29.2 Atoms

It was the ancient Greeks who first had the idea of atoms as indivisible units of matter, but experimental evidence for atoms didn't appear until Dalton, Avogadro, and others began to formulate the laws of chemistry in the early 19th century. The existence of atoms with diameters of approximately 10^{-10} m was widely accepted by 1890, but it was still unknown if atoms were indivisible little spheres or if they had some kind of internal structure.

The 1897 discovery of the electron by J. J. Thomson had two important implications:

1. Atoms are not indivisible; they are built of smaller pieces. The electron was the first *subatomic* particle to be discovered.
2. The constituents of the atom are *charged particles.* Hence the atom must be held together by electric forces.

Within a few years, measurements of the electron's charge

$$e = 1.60 \times 10^{-19} \text{ C}$$

and mass

$$m_e = 9.11 \times 10^{-31} \text{ kg}$$

revealed that the electron is much less massive than even the smallest atom.

Because the electrons are very small and light compared to the whole atom, it seemed reasonable to think that the positively charged part (protons were not yet known) had most of the mass and would take up most of the space. Thomson suggested that the atom consists of a spherical "cloud" of positive charge, roughly 10^{-10} m in diameter, in which the smaller negative electrons are embedded. The positive charge exactly balances the negative, so the atom as a whole has no net charge. This model of the atom has often been called the "plum-pudding model" or the "raisin-cake model" for reasons that should be clear from the picture of Figure 29.5. However, Thomson's model of the atom did not stand the tests of time.

Almost simultaneously with Thomson's discovery of the electron, and just one year after Röntgen's discovery of x rays, the French physicist Antoine Henri Becquerel announced his discovery that some new form of "rays" were emitted by crystals of uranium. These rays, like x rays, could expose film and pass through objects, but they were emitted continuously from the uranium without having to "do" anything to it. This was the discovery of **radioactivity,** a topic we'll study in Chapter 30.

One of Thomson's students, Ernest Rutherford, began a study of these new rays and quickly discovered that a uranium crystal actually emits two *different* rays. The first, which he called **alpha rays,** were easily absorbed by a piece of paper. The second, **beta rays,** could penetrate through at least 0.1 inch of metal and through much greater thicknesses of soft materials. The beta rays turned out to be high-speed electrons emitted by the uranium crystal. Rutherford soon showed that alpha rays are *positively* charged particles, and by 1906 he established that alpha rays (or alpha particles, as we now call them) consist of doubly ionized helium atoms (bare helium nuclei with mass $m = 6.64 \times 10^{-27}$ kg) emitted at high speed ($\approx 3 \times 10^7$ m/s) from the sample.

It had been quite a shock to discover that atoms are not indivisible, that they have an inner structure. Now, with the discovery of radioactivity, it appeared that some atoms were not even stable but could spit out various kinds of charged particles!

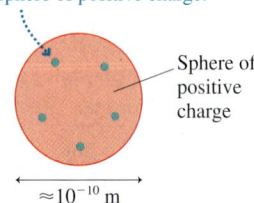

Thomson proposed that small, negative electrons are embedded in a sphere of positive charge.

Sphere of positive charge

$\approx 10^{-10}$ m

FIGURE 29.5 Thomson's raisin-cake model of the atom.

EXAMPLE 29.1 The speed of an alpha particle

Alpha particles are usually characterized by their kinetic energy in MeV. What is the speed of an 8.3 MeV alpha particle?

SOLVE Recall that 1 eV is the energy acquired by an electron accelerating through a 1 V potential difference, with the conversion $1.00 \text{ eV} = 1.60 \times 10^{-19}$ J. First, convert the energy to joules:

$$K = 8.3 \times 10^6 \text{ eV} \times \frac{1.60 \times 10^{-19} \text{ J}}{1.00 \text{ eV}} = 1.3 \times 10^{-12} \text{ J}$$

Now, using the alpha-particle mass $m = 6.64 \times 10^{-27}$ kg given above, we can find the speed:

$$K = \frac{1}{2}mv^2 = 1.3 \times 10^{-12} \text{ J}$$

$$v = \sqrt{\frac{2K}{m}} = 2.0 \times 10^7 \text{ m/s}$$

ASSESS This is quite fast, about 7% of the speed of light.

The First Nuclear Physics Experiment

Rutherford soon realized that he could use these high-speed alpha particles as projectiles to probe inside other atoms. In 1909, Rutherford and his students Hans Geiger and Ernest Marsden set up the experiment shown in Figure 29.6 to shoot alpha particles through very thin metal foils. The alpha particle is charged, and it experiences electric forces from the positive and negative charges of the atoms as it passes through the foil. According to Thomson's raisin-cake model of the atom, the forces exerted on the alpha particle by the positive atomic charges should roughly cancel the forces from the negative electrons, causing the alpha particles to experience only slight deflections. Indeed, this was the experimenters' initial observation.

At Rutherford's suggestion, Geiger and Marsden set up the apparatus to see if any alpha particles were deflected at *large* angles. It took only a few days to find the answer. Not only were alpha particles deflected at large angles, a very few were reflected almost straight backward toward the source!

How can we understand this result? Figure 29.7a shows that an alpha particle passing through a Thomson atom would experience only a small deflection. But if an atom has a small, positive core, such as the one in Figure 29.7b, a few of the alpha particles can come very close to the core. Because the electric force varies with the inverse square of the distance, the very large force of this very close approach can cause a large-angle scattering or even a backward deflection of the alpha particle. This is what Geiger and Marsden were observing.

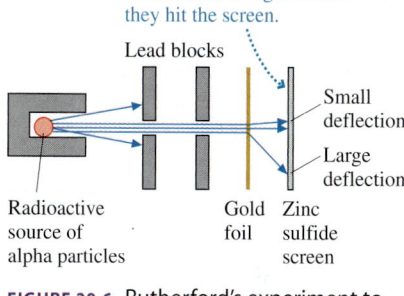

The alpha particles make little flashes of light where they hit the screen.

FIGURE 29.6 Rutherford's experiment to shoot high-speed alpha particles through a thin gold foil.

I remember two or three days later Geiger coming to me in great excitement and saying, "We have been able to get some of the alpha particles coming backward." It was quite the most incredible event that has ever happened to me in my life. It was almost as if you fired a 15-inch shell at a piece of tissue paper and it came back and hit you. . . . It was then that I had the idea of an atom with a minute massive center, carrying a charge.

Ernest Rutherford

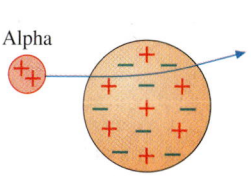

Alpha

The alpha particle is only slightly deflected by a Thomson atom because forces from the spread-out positive and negative charges nearly cancel.

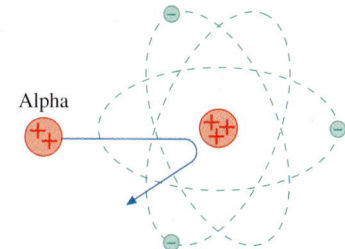

Alpha

If the atom has a concentrated positive nucleus, some alpha particles will be able to come very close to the nucleus and thus feel a very strong repulsive force.

FIGURE 29.7 Alpha particles interact differently with a concentrated positive nucleus than they would with the spread-out charge in Thomson's model.

The discovery of large-angle scattering of alpha particles quickly led Rutherford to envision an atom in which negative electrons orbit an unbelievably small, massive, positive **nucleus,** rather like a miniature solar system. This is the **nuclear model of the atom.** Further experiments showed that the diameter of the atomic nucleus is $\approx 1 \times 10^{-14}$ m = 10 fm (1 fm = 1 femtometer = 10^{-15} m), a mere 0.01% the diameter of the atom itself. Thus nearly all of the atom is merely empty space—the void!

Activ ONLINE Physics   19.1

EXAMPLE 29.2 **Going for the gold!**

An 8.3 MeV alpha particle is shot directly toward the nucleus of a gold atom (atomic number 79). What is the distance of closest approach of the alpha particle to the nucleus?

PREPARE Energy is conserved in electric interactions. Assume that the gold nucleus, which is much more massive than the alpha particle, does not move. Also recall that the electric field and potential of a sphere of charge can be found by treating the total charge as a point charge. Figure 29.8 is a before-and-after visual overview; the "before" situation is when the alpha is very far from the gold nucleus, and the "after" situation is when the alpha is at its distance of closest approach to the nucleus. The motion is in and out along a straight line.

FIGURE 29.8 A before-and-after visual overview of an alpha particle colliding with a nucleus.

SOLVE Electric potential energy decreases rapidly with increasing separation, so initially, when the alpha particle is very far away, the system has only the initial kinetic energy of the alpha particle. At the moment of closest approach the alpha particle has just come to rest and so the system has only potential energy. The conservation of energy statement $K_f + U_f = K_i + U_i$ is

$$0 + \frac{1}{4\pi\epsilon_0}\frac{q_\alpha q_{Au}}{r_{min}} = \frac{1}{2}mv_i^2 + 0$$

where q_α is the alpha-particle charge and we've treated the gold nucleus as a point charge q_{Au}. The solution for r_{min} is

$$r_{min} = \frac{1}{4\pi\epsilon_0}\frac{2q_\alpha q_{Au}}{mv_i^2}$$

The mass of the alpha particle is $m = 6.64 \times 10^{-27}$ kg and its charge is $q_\alpha = 2e = 3.20 \times 10^{-19}$ C. Gold has atomic number 79, so $q_{Au} = 79e = 1.26 \times 10^{-17}$ C. In Example 29.1 we found that an 8.3 MeV alpha particle has speed $v = 2.0 \times 10^7$ m/s. With this information, we can calculate

$$r_{min} = 2.7 \times 10^{-14}\text{ m}$$

This is only about 1/10,000 the size of the atom itself!

ASSESS We ignored the gold atom's electrons in this example. In fact, they make almost no contribution to the alpha particle's trajectory. The alpha particle is exceedingly massive compared to the electrons, and the electrons are spread out over a distance very large compared to the size of the nucleus. Hence the alpha particle easily pushes them aside without any noticeable change in its velocity.

It may seem surprising to you that the Rutherford model of the atom, with its solar system analogy, was not Thomson's original choice. However, scientists at the time could not imagine matter having the extraordinarily high density implied by a small nucleus. Neither could they understand what holds the nucleus together, why the positive charges do not push each other apart. Thomson's model, in which the positive charge was spread out and balanced by the negative electrons, actually made more sense. It would be several decades before the forces holding the nucleus together began to be understood, but Rutherford's evidence for a very small nucleus was indisputable.

Using the Nuclear Model

The nuclear model of the atom makes it easy to picture atoms and understand such processes as ionization. For example, the **atomic number** of an element, its position in the periodic table of the elements, is the number of orbiting electrons (of a neutral atom) and the number of units of positive charge in the nucleus. The atomic number is represented by Z. Hydrogen, with $Z = 1$, has one electron orbiting a nucleus with charge $+1e$. Helium, with $Z = 2$, has two orbiting electrons and a nucleus with charge $+2e$. Because the orbiting electrons are very light, an x-ray photon or a rapidly moving particle, such as another electron, can knock one of the electrons away, creating a positive *ion*. Removing one electron makes a singly charged ion, with $q_{ion} = +e$. Removing two electrons creates a doubly charged ion, with $q_{ion} = +2e$. This is shown for lithium ($Z = 3$) in Figure 29.9.

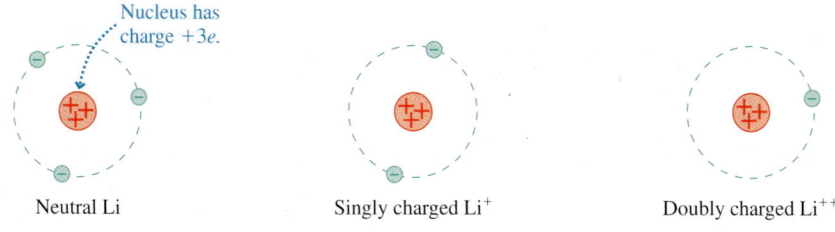

FIGURE 29.9 Different ionization stages of the lithium atom ($Z = 3$).

Experiments soon led to the recognition that the positive charge of the nucleus is associated with a positive subatomic particle called the **proton.** The proton's charge is $+e$, equal in magnitude but opposite in sign to the electron's charge. Further, because nearly all the atomic mass is associated with the nucleus, the proton is about 1800 times more massive than the electron: $m_p = 1.67 \times 10^{-27}$ kg. Atoms with atomic number Z have Z protons in the nucleus, giving the nucleus charge $+Ze$.

This nuclear model of the atom allows us to understand why, during chemical reactions and when an object is charged by rubbing, electrons are easily transferred but protons are not. The protons are tightly bound in the nucleus, shielded by all the electrons, but outer electrons are easily stripped away. Rutherford's nuclear model has explanatory power that was lacking in Thomson's model.

But there was a problem. Helium, with atomic number 2, has twice as many electrons and protons as hydrogen. Lithium, $Z = 3$, has three electrons and protons. If a nucleus contains Z protons to balance the Z orbiting electrons, and if nearly all the atomic mass is contained in the nucleus, then helium should be twice as massive as hydrogen and lithium three times as massive. But it was known from chemistry measurements that helium is *four times* as massive as hydrogen and lithium is *seven times* as massive.

This difficulty was not resolved until the discovery, in 1932, of a third subatomic particle. This particle has essentially the same mass as a proton but *no* electric charge. It is called the **neutron.** Neutrons reside in the nucleus, with the protons, where they contribute to the mass of the atom but not to its charge. As you'll see in Chapter 30, neutrons help provide the "glue" that holds the nucleus together.

We now know that a nucleus contains Z protons plus N neutrons, as shown in Figure 29.10, giving the atom a **mass number** $A = Z + N$. The mass number, which is a dimensionless integer, is *not* the same thing as the atomic mass m. But because the proton and neutron masses are both ≈ 1 u, where the *atomic mass unit* is

$$1 \text{ u} = 1 \text{ atomic mass unit} = 1.66 \times 10^{-27} \text{ kg}$$

the mass number A is *approximately* the mass in atomic mass units. For example, helium, with two protons and two neutrons ($A = 4$), has atomic mass $m = 6.64 \times 10^{-27}$ kg $= 4.03$ u ≈ 4 u.

There are a *range* of neutron numbers that happily form a nucleus with Z protons, creating a series of nuclei with the same Z-value (i.e., they are all the same chemical element) but different masses. Such a series of nuclei are called **isotopes.** The notation used to label isotopes is ^AZ, where the mass number A is given as a *leading* superscript. The proton number Z is not specified by an actual number but, equivalently, by the chemical symbol for that element. The most common isotope of neon has $Z = 10$ protons and $N = 10$ neutrons. Thus it has mass number $A = 20$ and is labeled ^{20}Ne. The neon isotope ^{22}Ne has $Z = 10$ protons (that's what makes it neon) and $N = 12$ neutrons. Helium has the two isotopes shown in Figure 29.11. The rare ^{3}He is only 0.0001% abundant, but it can be isolated and has important uses in scientific research.

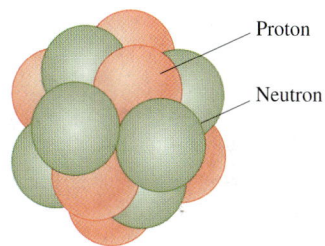

FIGURE 29.10 The nucleus of an atom contains protons and neutrons.

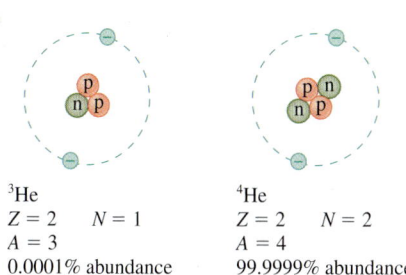

^{3}He
$Z = 2$ $N = 1$
$A = 3$
0.0001% abundance

^{4}He
$Z = 2$ $N = 2$
$A = 4$
99.9999% abundance

FIGURE 29.11 The two isotopes of helium. ^{3}He is only 0.0001% abundant.

Carbon is the sixth element in the periodic table. How many protons and how many neutrons are there in a nucleus of the isotope ^{14}C?

29.3 Bohr's Model of Atomic Quantization

18.1 Activ Physics

Rutherford's nuclear model was an important step toward understanding atoms, but it had two serious shortcomings. First, electrons orbiting the nucleus in a Rutherford atom are oscillating charged particles. According to Maxwell's theory of electricity and magnetism, these orbiting electrons should act as small antennas and radiate electromagnetic waves. That sounds encouraging, because we know that atoms can emit light, but it was easy to show that a Rutherford atom would radiate a *continuous* rainbow-like spectrum. Thus one failure of Rutherford's model was an inability to predict the discrete nature of emission and absorption spectra.

In addition, atoms would continuously lose energy as they radiated electromagnetic waves. As Figure 29.12 shows, this would cause the electrons to spiral into the nucleus! Calculations showed that a Rutherford atom can last no more than about a microsecond. In other words, classical Newtonian mechanics and electromagnetism predict that an atom with electrons orbiting a nucleus would be highly unstable and would immediately self-destruct. This clearly does not happen.

The experimental efforts of the late 19th century and early 20th had been impressive, and there could be no doubt about the existence of electrons, about the small positive nucleus, and about the unique discrete spectrum emitted by each atom. But the theoretical framework for understanding such observations had lagged behind. As the new century dawned, physicists could not explain the structure of atoms, could not explain the stability of matter, could not explain discrete spectra or why an element's absorption spectrum differs from its emission spectrum, and could not explain the origin of x rays or radioactivity.

A missing piece of the puzzle, although not recognized as such for a few years, was Einstein's 1905 introduction of light quanta. If light comes in discrete packets of energy, which we now call photons, and if atoms emit and absorb light, what does that imply about the structure of the atoms? This was the question posed by the Danish physicist Niels Bohr.

Bohr wanted to understand how a solar-system-like atom could be stable and not radiate away all its energy. He soon recognized that Einstein's light quanta had profound implications about the structure of atoms, and in 1913 Bohr proposed a radically new model of the atom in which he added quantization to Rutherford's nuclear atom. The basic assumptions of the **Bohr model of the atom** are as follows:

1. An atom consists of negative electrons orbiting a very small positive nucleus, as in the Rutherford model.
2. Atoms can exist only in certain **stationary states.** Each stationary state corresponds to a particular set of electron orbits around the nucleus. These states are distinct and can be numbered $n = 1, 2, 3, \ldots$ where n is the *quantum number.*
3. Each stationary state has a discrete, well-defined energy E_n. That is, atomic energies are *quantized.* The stationary states of an atom are numbered in order of increasing energy: $E_1 < E_2 < E_3 < E_4 < \cdots$
4. The *ground state* of the atom, with energy E_1, is stable and can persist indefinitely. Other stationary states with energies E_2, E_3, E_4, . . . are the *excited states* of the atom.

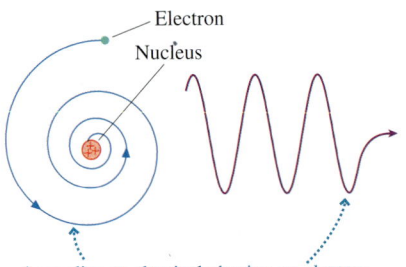

Electron
Nucleus

According to classical physics, an electron would spiral into the nucleus while radiating energy as an electromagnetic wave.

FIGURE 29.12 The fate of a Rutherford atom.

5. An atom can undergo a *transition* or a *quantum jump* from one stationary state with energy E_i to another with energy E_f by emitting or absorbing a photon of frequency

$$f_{\text{photon}} = \frac{\Delta E_{\text{atom}}}{h} \qquad (29.3)$$

where h is Planck's constant and $\Delta E_{\text{atom}} = |E_f - E_i|$. Figure 29.13a is a schematic view of the emission and absorption of photons in an atom with stationary states.

6. An atom also can move from a lower energy state to a higher energy state by absorbing energy $\Delta E_{\text{atom}} = E_f - E_i$ in an *inelastic collision* with an electron or another atom. This process, called **collisional excitation,** is shown in Figure 29.13b.

7. Atoms will seek the lowest energy state. An atom in an excited state, if left alone, will jump to lower and lower energy states, emitting a photon at each jump, until it reaches the ground state.

These ideas, many of which you'll recognize from the previous chapter, were incorporated 10 years later in Schrödinger's quantum mechanics, but they were first introduced by Bohr to understand atoms.

Assumption 5 of Bohr's model merges Einstein's idea of light quanta with the law of conservation of energy. Thus **an atom can jump from one allowed state to another by emitting or absorbing a photon of just the right frequency to conserve energy.** If an atom jumps from an initial state with energy E_i to a final state with *lower* energy E_f, energy will be conserved if the atom emits a photon with $E_{\text{photon}} = \Delta E_{\text{atom}}$. This photon must have exactly the frequency given by Equation 29.3 if it is to carry away exactly the right amount of energy. Similarly, an atom can jump to a higher energy state, for which additional energy is needed, by absorbing a photon of frequency $f_{\text{photon}} = \Delta E_{\text{atom}}/h$. The total energy of the atom-plus-light system is conserved.

The Bohr model also introduces *collisions* as another important mechanism by which a quantum system can gain energy. For example, a fluorescent lightbulb contains mercury vapor at a very low pressure. When you turn on the light, electrons flow back and forth inside the tube. As they do, they occasionally collide with the mercury atoms. These are inelastic collisions in which some of the electron's energy is transferred to the mercury atom, kicking the atom up to a higher energy level. The atom then jumps back to a lower energy level, emitting an ultraviolet photon. The ultraviolet light causes the white paint on the inside surface of the tube to *fluoresce,* a process we'll study later in the chapter.

The implications of Bohr's model are profound. In particular:

1. Matter is stable. Once an atom is in its ground state, there are no states of any lower energy to which it can jump. It can remain in the ground state forever.

2. Atoms emit and absorb a *discrete spectrum*. Only those photons whose frequencies match the energy *intervals* between the stationary states can be emitted or absorbed. Photons of other frequencies cannot be emitted or absorbed without violating energy conservation.

3. Emission spectra can be produced by collisions. In a gas discharge tube, the current-carrying electrons moving through the tube occasionally collide with the atoms. A collision transfers energy to an atom and can kick it to an excited state. Once the atom is in an excited state, it can emit photons of light—a discrete emission spectrum—as it jumps back down to lower-energy states.

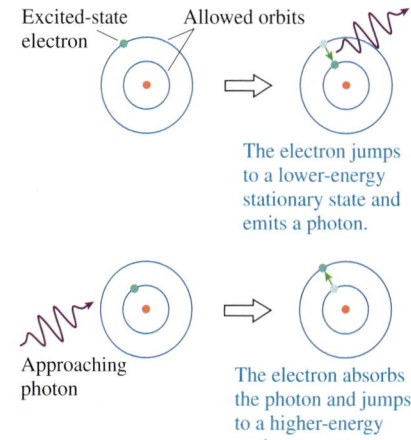

(a) Emission and absorption of light

Excited-state electron   Allowed orbits

The electron jumps to a lower-energy stationary state and emits a photon.

Approaching photon

The electron absorbs the photon and jumps to a higher-energy stationary state.

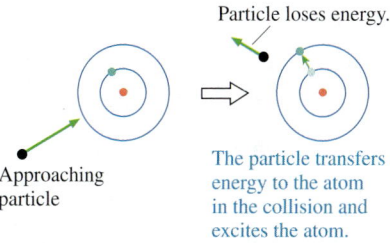

(b) Collisional excitation

Particle loses energy.

Approaching particle

The particle transfers energy to the atom in the collision and excites the atom.

FIGURE 29.13 An atom can change stationary states by emitting or absorbing a photon or by undergoing a collision.

4. **Absorption wavelengths are a subset of the wavelengths in the emission spectrum.** Recall that all the lines seen in an absorption spectrum are also seen in emission, but many emission lines are *not* seen in absorption. According to Bohr's model, **most atoms, most of the time, are in their lowest energy state,** the $n = 1$ ground state. Thus the absorption spectrum consists of *only* those transitions such as $1 \rightarrow 2$, $1 \rightarrow 3$, . . . in which the atom jumps from $n = 1$ to a higher value of n by absorbing a photon. Transitions such as $2 \rightarrow 3$ are not observed because there are essentially no atoms in $n = 2$ at any instant of time to do the absorbing. On the other hand, atoms that have been excited to the $n = 3$ state by collisions can emit photons corresponding to transitions $3 \rightarrow 1$ *and* $3 \rightarrow 2$. Thus the wavelength corresponding to $\Delta E_{atom} = E_3 - E_1$ is seen in both emission and absorption, but photons with $\Delta E_{atom} = E_3 - E_2$ occur in emission only.

5. **Each element in the periodic table has a unique spectrum.** The energies of the stationary states are just the energies of the orbiting electrons. The atom has no other form of energy. Different elements, with different numbers of electrons, have different stable orbits and thus different stationary states. States with different energies will emit and absorb photons of different wavelengths.

EXAMPLE 29.3 Wavelengths in emission and absorption spectra

An atom has stationary states $E_1 = 0.0$ eV, $E_2 = 2.0$ eV, and $E_3 = 5.0$ eV. What wavelengths are observed in the absorption spectrum and in the emission spectrum of this atom?

PREPARE Figure 29.14 shows an energy-level diagram for the atom. Photons are emitted when an atom undergoes a quantum jump from a higher energy level to a lower energy level. Photons are absorbed in a quantum jump from a lower energy level to a higher energy level. However, most of the atoms are in the $n = 1$ ground state, so the only quantum jumps seen in the absorption spectrum start from the $n = 1$ state.

$n = 3$ ———————— 5.0 eV

$n = 2$ ———————— 2.0 eV

$n = 1$ ———————— 0.0 eV

Absorption transitions must start from $n = 1$.

Emission transitions can start and end anywhere.

FIGURE 29.14 The atom's energy-level diagram.

SOLVE This atom absorbs photons on the $1 \rightarrow 2$ and $1 \rightarrow 3$ transitions, with $\Delta E_{1 \rightarrow 2} = 2.0$ eV and $\Delta E_{1 \rightarrow 3} = 5.0$ eV. From $f_{photon} = \Delta E_{atom}/h$ and $\lambda = c/f$, we find that the wavelengths in the absorption spectrum are

$$1 \rightarrow 3 \quad f_{photon} = 5.0 \text{ eV}/h = 1.2 \times 10^{15} \text{ Hz}$$
$$\lambda = 250 \text{ nm (ultraviolet)}$$

$$1 \rightarrow 2 \quad f_{photon} = 2.0 \text{ eV}/h = 4.8 \times 10^{14} \text{ Hz}$$
$$\lambda = 620 \text{ nm (orange)}$$

The emission spectrum also has the 620 nm and 250 nm wavelengths due to the $2 \rightarrow 1$ and $3 \rightarrow 1$ quantum jumps. In addition, the emission spectrum contains the $3 \rightarrow 2$ quantum jump with $\Delta E_{3 \rightarrow 2} = -3.0$ eV that is *not* seen in absorption because there are too few atoms in the $n = 2$ state to absorb. A similar calculation finds $f_{photon} = 7.2 \times 10^{14}$ Hz and $\lambda = c/f = 420$ nm. Thus the emission wavelengths are

$$2 \rightarrow 1 \quad \lambda = 620 \text{ nm (orange)}$$
$$3 \rightarrow 2 \quad \lambda = 420 \text{ nm (blue)}$$
$$3 \rightarrow 1 \quad \lambda = 250 \text{ nm (ultraviolet)}$$

STOP TO THINK 29.3 A photon with a wavelength of 420 nm has energy $E_{photon} = 3.0$ eV. Do you expect to see a spectral line with $\lambda = 420$ nm in the emission spectrum of the atom represented by this energy-level diagram? If so, what transition or transitions will emit it? Do you expect to see a spectral line with $\lambda = 420$ nm in the absorption spectrum? If so, what transition or transitions will absorb it?

$n = 4$ ———————— 6.0 eV
$n = 3$ ———————— 5.0 eV

$n = 2$ ———————— 2.0 eV

$n = 1$ ———————— 0.0 eV

29.4 The Bohr Hydrogen Atom

Bohr's hypothesis was a bold new idea, yet there was still one enormous stumbling block: What *are* the stationary states of an atom? Everything in Bohr's model hinges on the existence of these stationary states, of there being only certain electron orbits that are allowed. But nothing in classical physics provides any basis for such orbits. And Bohr's model describes only the *consequences* of having stationary states, not how to find them. If such states really exist, we will have to go beyond classical physics to find them.

To address this problem, Bohr did an explicit analysis of the hydrogen atom. The hydrogen atom, with only a single electron, was known to be the simplest atom. Furthermore, as we discussed in Section 29.1, Balmer had discovered a fairly simple formula that characterizes the wavelengths in the hydrogen emission spectrum. Anyone with a successful model of an atom was going to have to *derive* Balmer's formula for the hydrogen atom.

Bohr's paper followed a rather circuitous line of reasoning. That is not surprising, because he had little to go on at the time. But our goal is a clear explanation of the ideas, not a historical study of Bohr's methods, so we are going to follow a different analysis using de Broglie's matter waves. De Broglie did not propose matter waves until 1924, 11 years after Bohr, but with the clarity of hindsight we can see that treating the electron as a wave provides a more straightforward analysis of the hydrogen atom. Although our route will be different from Bohr's, we will arrive at the same point, and, in addition, we will be in a much better position to understand the work that came after Bohr.

NOTE ▶ Bohr's analysis of the hydrogen atom is sometimes called the *Bohr atom*. It's important not to confuse this analysis, which applies only to hydrogen, with the more general postulates of the *Bohr model of the atom*. Those postulates, which we looked at in the previous section, apply to any atom. To make the distinction clear, we'll call Bohr's analysis of hydrogen the *Bohr hydrogen atom*. ◀

The Stationary States of the Hydrogen Atom

Figure 29.15 shows a Rutherford hydrogen atom, with a single electron orbiting a proton. We will assume a circular orbit of radius r and speed v. We will also assume the proton remains stationary while the electron revolves around it, a reasonable assumption because the proton is roughly 1800 times as massive as the electron. With these assumptions, the atom's energy is the kinetic energy of the electron plus the potential energy of the electron-proton interaction. This is

$$E = K + U = \frac{1}{2}mv^2 + \frac{1}{4\pi\epsilon_0}\frac{q_{elec}\,q_{proton}}{r} = \frac{1}{2}mv^2 - \frac{e^2}{4\pi\epsilon_0 r} \quad (29.4)$$

where we used $q_{elec} = -e$ and $q_{proton} = +e$.

NOTE ▶ m is the mass of the electron, *not* the mass of the entire atom. ◀

Now the electron, as we are coming to understand it, has both particle-like and wave-like properties. First, let us treat the electron as a charged particle. The proton exerts a Coulomb electric force on the electron,

$$\vec{F}_{elec} = \left(\frac{1}{4\pi\epsilon_0}\frac{e^2}{r^2}, \text{ toward center}\right) \quad (29.5)$$

This force gives the electron an acceleration $\vec{a}_{elec} = \vec{F}_{elec}/m$ that also points to the center. This is a centripetal acceleration, causing the particle to move in its

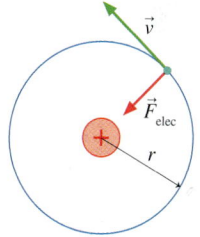

FIGURE 29.15 A Rutherford hydrogen atom. The size of the nucleus is greatly exaggerated.

circular orbit. The centripetal acceleration of a particle moving in a circle of radius r at speed v is v^2/r, so that

$$a_{elec} = \frac{F_{elec}}{m} = \frac{e^2}{4\pi\epsilon_0 mr^2} = \frac{v^2}{r}$$

Rearranging, we find

$$v^2 = \frac{e^2}{4\pi\epsilon_0 mr} \tag{29.6}$$

Equation 29.6 is a *constraint* on the motion. The speed v and radius r must obey Equation 29.6 if the electron is to move in a circular orbit. This constraint is not unique to atoms; we earlier found a similar relationship between v and r for orbiting satellites.

Now let's treat the electron as a de Broglie wave. In Section 28.5 we found that a particle confined to a one-dimensional box sets up a standing wave as it reflects back and forth. A standing wave, you will recall, consists of two traveling waves moving in opposite directions. When the round-trip distance in the box is equal to an integer number of wavelengths ($2L = n\lambda$), the two oppositely traveling waves interfere constructively to set up the standing wave.

Suppose that, instead of traveling back and forth along a line, our wave-like particle travels around the circumference of a circle. The particle will set up a standing wave, just like the particle in the box, if there are waves traveling in both directions and if the round-trip distance is an integer number of wavelengths. This is the idea we want to carry over from the particle in a box. As an example, Figure 29.16 shows a standing wave around a circle with $n = 10$ wavelengths.

The mathematical condition for a circular standing wave is found by replacing the round-trip distance $2L$ in a box with the round-trip distance $2\pi r$ on a circle. Thus a circular standing wave will occur when

$$2\pi r = n\lambda \qquad n = 1, 2, 3, \ldots \tag{29.7}$$

But the de Broglie wavelength for a particle *has* to be $\lambda = h/p = h/mv$. Thus the standing-wave condition for a de Broglie wave is

$$2\pi r = n\frac{h}{mv}$$

This condition is true only if the electron's speed is

$$v = \frac{nh}{2\pi mr} \qquad n = 1, 2, 3, \ldots \tag{29.8}$$

In other words, the electron cannot have just any speed, only the discrete values given by Equation 29.8.

The quantity $h/2\pi$ occurs so often in quantum physics that it is customary to give it a special name. We define the quantity $\hbar$, pronounced "h bar," as

$$\hbar = \frac{h}{2\pi} = 1.06 \times 10^{-34}\,\text{J} \cdot \text{s} = 6.58 \times 10^{-16}\,\text{eV} \cdot \text{s}$$

With this definition, we can write Equation 29.8 as

$$v = \frac{n\hbar}{mr} \qquad n = 1, 2, 3, \ldots \tag{29.9}$$

This, like Equation 29.6, is another relationship between v and r. This is the constraint that arises from treating the electron as a wave.

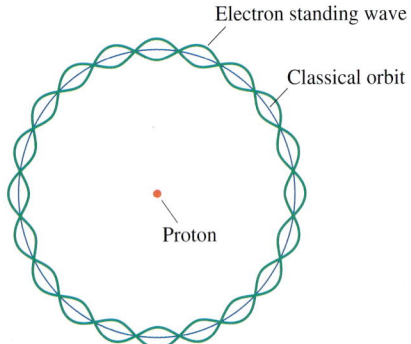

Electron standing wave

Classical orbit

Proton

FIGURE 29.16 An $n = 10$ electron standing wave around the orbit's circumference.

Now if the electron can act as both a particle *and* a wave, then both the Equation 29.6 *and* Equation 29.9 constraints have to be obeyed. That is, v^2 as given by the Equation 29.6 particle constraint has to equal v^2 of the Equation 29.9 wave constraint. Equating these gives

$$v^2 = \frac{e^2}{4\pi\epsilon_0 mr} = \frac{n^2\hbar^2}{m^2 r^2}$$

We can solve this equation to find that the radius r is

$$r_n = n^2 \frac{4\pi\epsilon_0\hbar^2}{me^2} \qquad n = 1, 2, 3, \ldots \qquad (29.10)$$

where we have added a subscript n to the radius r to indicate that it depends on the integer n.

The right-hand side of Equation 29.10, except for the n^2, is just a collection of constants. Let's group them all together and define the **Bohr radius** a_B to be

$$a_B = \text{Bohr radius} = \frac{4\pi\epsilon_0\hbar^2}{me^2} = 5.29 \times 10^{-11} \text{ m} = 0.0529 \text{ nm}$$

With this definition, Equation 29.10 for the radius of the electron's orbit becomes

$$r_n = n^2 a_B \qquad n = 1, 2, 3, \ldots \qquad (29.11)$$

Allowed radii of the Bohr hydrogen atom

QUADRATIC
p. 50

The first few allowed values of r are

$$r_n = \begin{cases} 0.053 \text{ nm} & n = 1 \\ 0.212 \text{ nm} & n = 2 \\ 0.476 \text{ nm} & n = 3 \\ \vdots & \vdots \end{cases}$$

We have discovered stationary states! That is, **a hydrogen atom can exist *only* if the radius of the electron's orbit is one of the values given by Equation 29.11.** Intermediate values of the radius, such as $r = 0.100$ nm, cannot exist because the electron cannot set up a standing wave around the circumference. The possible orbits are *quantized,* and integer n is the quantum number.

Hydrogen Atom Energy Levels

Now we can make progress quickly. Knowing the possible radii, we can return to Equation 29.9 and find the possible electron speeds to be

$$v_n = \frac{n\hbar}{mr_n} = \frac{1}{n}\frac{\hbar}{ma_B} = \frac{v_1}{n} \qquad n = 1, 2, 3, \ldots \qquad (29.12)$$

where $v_1 = \hbar/ma_B = 2.19 \times 10^6$ m/s is the electron's speed in the $n = 1$ orbit. The speed decreases as n increases.

Finally, we can determine the energies of the stationary states by using Equations 29.11 and 29.12 for r and v in Equation 29.4 for the energy. The algebra is rather messy, but the result simplifies to

$$E_n = \frac{1}{2}mv_n^2 - \frac{e^2}{4\pi\epsilon_0 r_n} = -\frac{1}{n^2}\left(\frac{1}{4\pi\epsilon_0}\frac{e^2}{2a_B}\right) \qquad (29.13)$$

Let's define

$$E_1 = \frac{1}{4\pi\epsilon_0} \frac{e^2}{2a_B} = 13.60 \text{ eV}$$

We can then write the energy levels of the stationary states of the hydrogen atom as

$$E_n = -\frac{E_1}{n^2} = -\frac{13.60 \text{ eV}}{n^2} \qquad n = 1, 2, 3, \ldots \qquad (29.14)$$

Allowed energies of the Bohr hydrogen atom

This has been a lot of math, so we need to see where we are and what we have learned. Table 29.2 shows values of r_n, v_n, and E_n for quantum numbers $n = 1$ to 4. We do indeed seem to have discovered stationary states of the hydrogen atom. Each state, characterized by its quantum number n, has a unique radius, speed, and energy. These are displayed graphically in Figure 29.17, in which the orbits are drawn to scale. Notice how the atom's diameter increases very rapidly as n increases. At the same time, the electron's speed decreases.

TABLE 29.2 Radii, speeds, and energies for the first four states of the Bohr hydrogen atom

n	r_n (nm)	v_n (m/s)	E_n (eV)
1	0.053	2.19×10^6	-13.60
2	0.212	1.09×10^6	-3.40
3	0.476	0.73×10^6	-1.51
4	0.846	0.55×10^6	-0.85

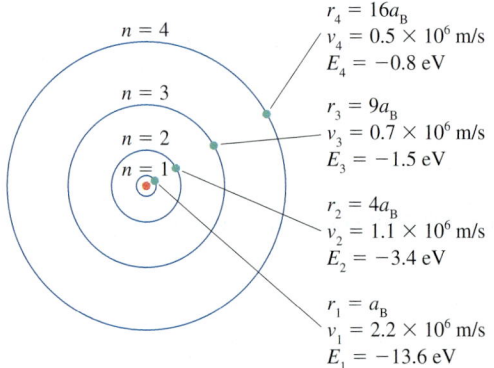

$r_4 = 16a_B$
$v_4 = 0.5 \times 10^6$ m/s
$E_4 = -0.8$ eV

$r_3 = 9a_B$
$v_3 = 0.7 \times 10^6$ m/s
$E_3 = -1.5$ eV

$r_2 = 4a_B$
$v_2 = 1.1 \times 10^6$ m/s
$E_2 = -3.4$ eV

$r_1 = a_B$
$v_1 = 2.2 \times 10^6$ m/s
$E_1 = -13.6$ eV

FIGURE 29.17 The first four stationary states, or allowed orbits, of the Bohr hydrogen atom drawn to scale.

EXAMPLE 29.4 Possible electron speeds in a hydrogen atom

Can an electron in a hydrogen atom have a speed of 3.60×10^5 m/s? If so, what are its energy and the radius of its orbit? What about a speed of 3.65×10^5 m/s?

PREPARE To be in a stationary state, the electron must have speed $v_n = v_1/n$, with $n = v_1/v$ an integer. Only if v_1/v is an integer is v an allowed electron speed.

SOLVE A speed of 3.60×10^5 m/s would require quantum number

$$n = \frac{v_1}{v} = \frac{2.19 \times 10^6 \text{ m/s}}{3.60 \times 10^5 \text{ m/s}} = 6.08$$

This is not an integer, so the electron can *not* have this speed. But if $v = 3.65 \times 10^5$ m/s, then

$$n = \frac{2.19 \times 10^6 \text{ m/s}}{3.65 \times 10^5 \text{ m/s}} = 6$$

This is the speed of an electron in the $n = 6$ excited state. An electron in this state has energy

$$E_6 = -\frac{13.60 \text{ eV}}{6^2} = -0.378 \text{ eV}$$

and the radius of its orbit is

$$r_6 = 6^2 a_B = 6^2(0.0529 \text{ nm}) = 1.90 \text{ nm}$$

It is important to understand why the energies of the stationary states are negative. An electron and a proton bound into an atom have *less* energy than they do when they're separated. We know this because we would have to do work (i.e., *add* energy) to pull the electron and proton apart.

When the electron and proton are completely separated ($r \to \infty$) and at rest ($v = 0$), their potential energy $U = q_1 q_2/4\pi\epsilon_0 r$ and kinetic energy $K = mv^2/2$ are zero. As the electron moves closer to the proton to form a hydrogen atom, its potential energy *decreases,* becoming negative, while the kinetic energy of the orbiting electron increases. Equation 29.13, however, shows that the potential energy decreases faster than the kinetic energy increases, leading to an overall negative energy for the atom. In general, negative energies are characteristic of *bound systems.*

Quantization of Angular Momentum

The angular momentum of a particle in circular motion, whether it is a planet or an electron, is

$$L = mvr$$

You will recall that angular momentum is conserved in orbital motion because a force directed toward a central point exerts no torque on the particle. Bohr used conservation of energy explicitly in his analysis of the hydrogen atom, but what role does conservation of angular momentum play?

The condition that a de Broglie wave for the electron set up a standing wave around the circumference was given, in Equation 29.7, as

$$2\pi r = n\lambda = n\frac{h}{mv}$$

We can rewrite this equation as

$$mvr = n\frac{h}{2\pi} = n\hbar \qquad (29.15)$$

But *mvr* is the angular momentum *L* for a particle in a circular orbit. It appears that the angular momentum of an orbiting electron cannot have just any value. Instead, it must satisfy

$$L = n\hbar \qquad n = 1, 2, 3, \ldots \qquad (29.16)$$

Quantization of angular momentum
for the Bohr hydrogen atom

Thus angular momentum is also quantized! The atom's angular momentum must be an integer multiple of Planck's constant $\hbar$.

The quantization of angular momentum is a direct consequence of the wave-like nature of the electron. We will find that the quantization of angular momentum plays a major role in the behavior of more complex atoms, leading to the idea of electron shells that you likely have studied in chemistry.

STOP TO THINK 29.4   What is the quantum number of this hydrogen atom?

The Hydrogen Spectrum

The most important experimental evidence that we have about the hydrogen atom is its spectrum, so the primary test of the Bohr hydrogen atom is whether it correctly predicts the spectrum. Figure 29.18 on the next page is an energy-level diagram for the hydrogen atom. As we noted in Chapter 28, the energies are like the

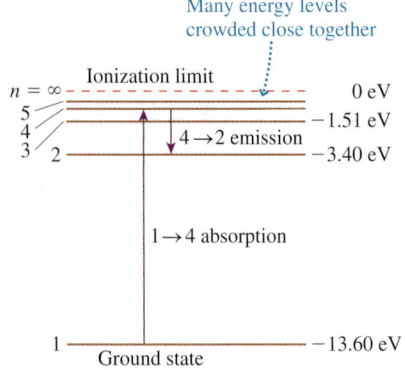

FIGURE 29.18 The energy-level diagram of the hydrogen atom.

rungs of a ladder. The lowest rung is the ground state, with $E_1 = -13.60$ eV. The top rung, with $E = 0$ eV, corresponds to a hydrogen ion in the limit $n \to \infty$. This top rung is called the **ionization limit.** In principle there are an infinite number of rungs, but only the lowest few are shown. The higher values of n are all crowded together just below the ionization limit at $n = \infty$.

The figure shows a $1 \to 4$ transition in which a photon is absorbed and a $4 \to 2$ transition in which a photon is emitted. For two quantum states m and n, where $n > m$ and E_n is the higher-energy state, an atom can *emit* a photon in an $n \to m$ transition or *absorb* a photon in an $m \to n$ transition.

According to the fifth assumption of Bohr's model of atomic quantization, the frequency of the photon emitted in an $n \to m$ transition is

$$f_{\text{photon}} = \frac{\Delta E_{\text{atom}}}{h} = \frac{E_n - E_m}{h} \tag{29.17}$$

We can use Equation 29.14 for the energies E_n and E_m, to predict that the emitted photon has frequency

$$f_{\text{photon}} = \frac{1}{h}\left(\frac{-13.60 \text{ eV}}{n^2} - \frac{-13.60 \text{ eV}}{m^2} \right) = \frac{13.60 \text{ eV}}{h}\left(\frac{1}{m^2} - \frac{1}{n^2} \right)$$

The frequency is a positive number because $m < n$ and thus $1/m^2 > 1/n^2$.

We are more interested in wavelength than frequency, because wavelengths are the quantity measured by experiment. The wavelength of the photon emitted in an $n \to m$ quantum jump is

$$\lambda_{n \to m} = \frac{c}{f_{\text{photon}}} = \frac{\lambda_0}{\left(\dfrac{1}{m^2} - \dfrac{1}{n^2} \right)} \qquad \begin{array}{l} m = 1, 2, 3, \ldots \\ n = m + 1, m + 2, \ldots \end{array} \tag{29.18}$$

with $\lambda_0 = 91.12$ nm. This should look familiar. It is the Balmer formula, Equation 29.2! However, there is one *slight* difference: Bohr's analysis of the hydrogen atom has predicted $\lambda_0 = 91.12$ nm whereas Balmer found, from experiment, that $\lambda_0 = 91.18$ nm. Could Bohr have come this close but then failed to predict the Balmer formula correctly?

The problem, it turns out, is in our assumption that the proton remains at rest while the electron orbits it. In fact, *both* particles rotate about their common center of mass, rather like a dumbbell with a big end and a small end. The center of mass is very close to the proton, which is far more massive than the electron, but the proton is not entirely motionless. The good news is that a more advanced analysis can account for the proton's motion. It changes the energies of the stationary states ever so slightly—about 1 part in 2000—but that is precisely what is needed to give a revised value:

$$\lambda_0 = 91.18 \text{ nm when corrected for the nuclear motion}$$

It works! Unlike previous atomic models, **the Bohr hydrogen atom correctly predicts the discrete spectrum of the hydrogen atom.** Figure 29.19 shows two series of transitions that give rise to wavelengths in the spectrum. The *Balmer series,* consisting of transitions ending on the $m = 2$ state, gives visible wavelengths, and this is the series that Balmer initially analyzed. The *Lyman series,* ending on the $m = 1$ ground state, is in the ultraviolet region of the spectrum and was not measured until later. These series, as well as others in the infrared, are observed in a discharge tube where collisions with electrons excite the atoms upward from the ground state to state n. They then decay downward by emitting photons. Only the Lyman series is observed in the absorption spectrum because, as noted previously, essentially all the atoms in a quiescent gas are in the ground state.

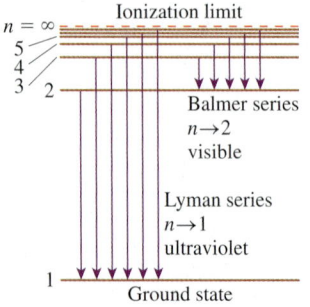

FIGURE 29.19 Transitions producing the Lyman series and the Balmer series of lines in the hydrogen spectrum.

EXAMPLE 29.5 **Wavelengths in galactic hydrogen absorption**

Whenever astronomers look at distant galaxies, they find that the light has been strongly absorbed at the wavelength of the $1 \rightarrow 2$ transition in the Lyman series of hydrogen. This absorption tells us that interstellar space is filled with vast clouds of hydrogen left over from the Big Bang. What is the wavelength of the $1 \rightarrow 2$ absorption in hydrogen?

SOLVE Equation 29.18 predicts the *absorption* spectrum of hydrogen if we let $m = 1$. The absorption seen by astronomers is from the ground state of hydrogen ($m = 1$) to its first excited state ($n = 2$). The wavelength is

$$\lambda_{1 \rightarrow 2} = \frac{91.18 \text{ nm}}{\left(\dfrac{1}{1^2} - \dfrac{1}{2^2} \right)} = 122 \text{ nm}$$

ASSESS This wavelength is far into the ultraviolet. Ground-based astronomy cannot observe this region of the spectrum because the wavelengths are strongly absorbed by the atmosphere, but with space-based telescopes, first widely used in the 1970s, astronomers see 122 nm absorption in nearly every direction they look.

Success and Failure

Bohr's analysis of the hydrogen atom seemed to be a resounding success. By introducing stationary states, together with Einstein's ideas about light quanta, Bohr was able to provide the first solid understanding of discrete spectra and, in particular, to predict the Balmer formula for the wavelengths in the hydrogen spectrum. And the Bohr hydrogen atom, unlike Rutherford's model, was stable. There was clearly some validity to the idea of stationary states.

But Bohr was completely unsuccessful at explaining the spectra of any other atom. His method did not work even for helium, the second element in the periodic table with a mere two electrons. Something inherent in Bohr's assumptions seemed to work correctly for a single electron but not in situations with two or more electrons.

It is important to make a distinction between the Bohr model of the atom, described in Section 29.3, and the Bohr hydrogen atom. The Bohr model assumes that stationary states exist, but it does not say how to find them. We found the stationary states of a hydrogen atom by requiring that an integer number of de Broglie waves fit around the circumference of the orbit, setting up standing waves. The difficulty with more complex atoms is not the Bohr model but the method of finding the stationary states. Bohr's model of the atomic quantization remains valid, and we will continue to use it, but the procedure of fitting standing waves to a circle is just too simple to find the stationary states of complex atoms.

Einstein, de Broglie, and Bohr carried physics into uncharted waters. Their successes made it clear that the microscopic realm of light and atoms is governed by quantization, discreteness, and a blurring of the distinction between particles and waves. Although Bohr was clearly on the right track, his inability to extend the Bohr hydrogen atom to more complex atoms made it equally clear that the complete and correct theory remained to be discovered. Bohr's theory was what we now call "semiclassical," a hybrid of classical Newtonian mechanics with the new ideas of quanta. Still missing was a complete theory of motion and dynamics in a quantized universe—a *quantum* mechanics. Schrödinger provided this theory 12 years later, and with it a means for calculating the stationary states of any atom.

29.5 The Quantum-Mechanical Hydrogen Atom

In 1925, Erwin Schrödinger introduced a general theory of quantum mechanics, a theory capable of calculating the allowed energy levels (i.e., the stationary states) of any system. The calculations are mathematically difficult, even for a system as simple as the hydrogen atom, and we will present results without proof.

The Bohr hydrogen atom was characterized by a single quantum number n. In contrast, the quantum-mechanical hydrogen atom is characterized by *four* quantum numbers and *four* quantized parameters. Schrödinger's analysis of the hydrogen atom found four conditions that must be met:

1. The atom's energy must be one of the values

$$E_n = -\frac{13.60 \text{ eV}}{n^2} \qquad n = 1, 2, 3, \ldots \qquad (29.19)$$

The integer n is called the **principal quantum number.** These energies are the same as those in the Bohr hydrogen atom.

2. The angular momentum L of the electron's orbit must be one of the values

$$L = \sqrt{l(l + 1)}\, \hbar \qquad l = 0, 1, 2, 3, \ldots, n - 1 \qquad (29.20)$$

The integer l is called the **orbital quantum number.**

3. The component of the orbital angular momentum along any defined direction, such as the z-axis, must be one of the values

$$L_z = m\hbar \qquad m = -l, -l + 1, \ldots, 0, \ldots, l - 1, l \qquad (29.21)$$

The integer m is called the **magnetic quantum number** because it becomes important when the atom is placed in a magnetic field.

4. The component of the electron's *spin* angular momentum along the z-axis must be one of the values

$$S_z = m_s\hbar \qquad m_s = -\frac{1}{2} \text{ or } +\frac{1}{2} \qquad (29.22)$$

The number m_s is called the **spin quantum number;** electron spin, discovered by Uhlenbeck and Goudsmit in 1925, is discussed below.

In other words, each stationary state of the hydrogen atom is identified by a quartet of quantum numbers (n, l, m, m_s), and each quantum number is associated with a physical property of the atom.

NOTE ▶ The energy of the stationary state depends only on the principal quantum number n, not on l, m, or m_s. ◀

EXAMPLE 29.6 Listing quantum numbers

List all possible states of a hydrogen atom that have energy $E = -3.40$ eV.

SOLVE Energy depends only on the principal quantum number n. From Equation 29.19, states with $E = -3.40$ eV have

$$n = \sqrt{\frac{-13.60 \text{ eV}}{-3.40 \text{ eV}}} = 2$$

An atom with principal quantum number $n = 2$ could have either $l = 0$ or $l = 1$, but $l \geq 2$ is ruled out. If $l = 0$, the only

possible value for the magnetic quantum number m is $m = 0$. If $l = 1$, then the atom could have $m = -1$, $m = 0$, or $m = +1$. For each of these, the spin quantum number could be $m_s = +\frac{1}{2}$ or $m_s = -\frac{1}{2}$. Thus the possible sets of quantum numbers are

n	l	m	m_s	n	l	m	m_s
2	0	0	$+\frac{1}{2}$	2	0	0	$-\frac{1}{2}$
2	1	1	$+\frac{1}{2}$	2	1	1	$-\frac{1}{2}$
2	1	0	$+\frac{1}{2}$	2	1	0	$-\frac{1}{2}$
2	1	−1	$+\frac{1}{2}$	2	1	−1	$-\frac{1}{2}$

These eight states all have the same energy.

Hydrogen turns out to be unique. For all other elements, the allowed energies depend on both *n and l* (but not *m* or m_s). Consequently, it is useful to label the stationary states by their values of *n* and *l*. The lowercase letters shown in Table 29.3 are customarily used to represent the various values of quantum number *l*. These symbols come from spectroscopic notation used in prequantum-mechanics days, when some spectral lines were classified as **s**harp, others as **p**rincipal, and so on.

Using these symbols, the ground state of the hydrogen atom, with *n* = 1 and *l* = 0, is called the 1*s* state. The 3*d* state has *n* = 3, *l* = 2. In Example 29.6, we found two 2*s* states (with *l* = 0) and six 2*p* states (with *l* = 1), all with the same energy.

TABLE 29.3 Symbols used to represent quantum number *l*

l	Symbol
0	*s*
1	*p*
2	*d*
3	*f*

Energy and Angular Momentum Are Quantized

The energy of the hydrogen atom is quantized. The allowed energies of hydrogen, given by Equation 29.19, depend only on the principal quantum number *n*, but for other atoms the energies will depend on both *n* and *l*. In anticipation of using both quantum numbers, Figure 29.20 is an energy-level diagram for the hydrogen atom in which the rows are labeled by *n* and the columns by *l*. The left column contains all of the *l* = 0 *s* states, the next column is the *l* = 1 *p* states, and so on.

Because the quantum condition of Equation 29.20 requires *n* > *l*, the *s* states begin with *n* = 1, the *p* states begin with *n* = 2, and the *d* states with *n* = 3. That is, the lowest-energy *d* state is 3*d* because states with *n* = 1 or *n* = 2 cannot have *l* = 2. For hydrogen, where the energy levels do not depend on *l*, the energy-level diagram shows that the 3*s*, 3*p*, and 3*d* states have equal energy. Figure 29.20 shows only the first few energy levels for each value of *l*, but there really are an infinite number of levels, as *n* → ∞, crowding together beneath *E* = 0. The dotted line at *E* = 0 is the atom's *ionization limit,* the energy of a hydrogen atom in which the electron has been moved infinitely far away to form an H^+ ion.

You should compare this energy-level diagram to Figure 29.18 for the Bohr atom. Because the energy levels of the quantum-mechanical hydrogen atom are exactly the same as for the Bohr atom, the quantum-mechanical solution also correctly predicts the discrete spectrum of hydrogen.

Classically, the angular momentum *L* of an orbiting electron can have any value. Not so in quantum mechanics. Equation 29.20 tells us that **the electron's orbital angular momentum is quantized.** The magnitude of the orbital angular momentum must be one of the discrete values

$$L = \sqrt{l(l+1)}\,\hbar = 0, \sqrt{2}\hbar, \sqrt{6}\hbar, \sqrt{12}\hbar, \ldots$$

where *l* is an integer. The Bohr atom also predicted quantized angular momentum, but the precise values of that prediction turned out to be in error. The quantum-mechanical prediction for *L* is more complex, but it agrees with experimental observations.

A particularly interesting prediction is that the ground state of hydrogen, with *l* = 0, has *no* angular momentum. A classical particle cannot orbit unless it has angular momentum, but apparently a quantum particle does not have this requirement.

FIGURE 29.20 Energy-level diagram for the hydrogen atom.

STOP TO THINK 29.5   What are the quantum numbers *n* and *l* for a hydrogen atom with $E = -(13.60/9)$ eV and $L = \sqrt{2}\hbar$?

The Electron Spin

You learned in Chapter 25 that a current loop creates a *magnetic dipole moment,* acting like a bar magnet with a north and south pole. In the 1920s, physicists studying the magnetic properties of atoms discovered that the electron has a magnetic dipole moment. The electron has a mass, which allows it to experience gravitational forces, and an electric charge, which allows it to experience electric forces, so perhaps it's not surprising that the electron comes with a magnetic dipole moment, allowing it to experience magnetic forces.

It was first thought that the electron was a very tiny ball of negative charge spinning on its axis. This would give the electron both a magnetic dipole moment and *spin angular momentum.* However, a spinning ball of charge would violate the laws of relativity and other physical laws. As far as we know today, the electron is truly a point particle that happens to have an intrinsic magnetic dipole moment and angular momentum *as if* it were spinning. This inherent angular momentum is called the *electron spin,* but it is a convenient figure of speech, not a factual statement. **The electron has a spin, but it is not a spinning electron.**

The two possible spin quantum numbers $m_s = \pm\frac{1}{2}$ mean that the component of the spin angular momentum along the z-axis is either $+\frac{1}{2}\hbar$ or $-\frac{1}{2}\hbar$. These two orientations of the spin angular momentum are called *spin up* and *spin down.* It is convenient to picture a little angular momentum vector that can be drawn ↑ for a spin-up state and ↓ for a spin-down state. We will use this notation in the next section.

Hydrogen Atom Probabilities

You learned in Chapter 28 that quantum mechanics and the uncertainty principle prevent us from knowing exactly where a particle is. Instead, quantum mechanics predicts the *probability* of finding the particle at a particular point in space. For the hydrogen atom, quantum mechanics predicts the probability of finding the electron at different distances from the proton. Portraying the probability of a three-dimensional wave function is a challenge, but one way to do so, shown in Figure 29.21, is to use denser shading to indicate regions of larger probability. These figures show the probabilities of locating electrons in the 1s, 2s, and 2p states of hydrogen. As you can see, the three-dimensional probability creates what is often called an **electron cloud** around the nucleus.

An electron in the 1s state is most likely to be found at the origin.

An electron in a 2s state is likely to be found either at the origin or in a surrounding shell.

The p electrons are more likely to be found in some directions than in others.

1s
m = 0

2s
m = 0

2p
m = 0

2p
m = ±1

FIGURE 29.21 The probabilities of locating the electron in the 1s, 2s, and 2p states of hydrogen.

These figures contain a lot of information. For example, notice how the p electrons have directional properties. These directional properties allow p electrons to "reach out" toward nearby atoms, forming molecular bonds. In contrast, the spherical symmetry of the s electrons suggests why these electrons, with $l = 0$, have no angular momentum.

29.6 Multielectron Atoms

The quantum-mechanical solution for the hydrogen atom matches the experimental evidence, but so did the Bohr hydrogen atom. The real test of Schrödinger's theory is how well it works for multielectron atoms.

A neutral multielectron atom with atomic number Z consists of Z electrons surrounding a nucleus with Z protons and charge $+Ze$. A multielectron atom might seem like the solar system, with many planets orbiting the sun, but the electron-electron interaction makes the atomic-structure problem more difficult than the solar-system problem, and it proved to be the downfall of the simple Bohr atom. The planets in the solar system do exert attractive gravitational forces on each other, but their masses are so much less than that of the sun that these planet-planet forces are insignificant for all but the most precise calculations. Not so in an atom. The electron charge is the same as the proton charge, so the electron-electron repulsion is just as important to atomic structure as is the electron-nucleus attraction.

One of the first big successes of Schrödinger's quantum mechanics was an ability to calculate the stationary states and energy levels of multielectron atoms. A major difference between multielectron atoms and the simple one-electron hydrogen is that the energy of an electron in a multielectron atom depends on both quantum numbers n *and* l. Whereas the $2s$ and $2p$ states in hydrogen have the same energy, their energies are different in a multielectron atom. The difference arises from the electron-electron interactions that do not exist in a single-electron hydrogen atom.

Figure 29.22 shows an energy-level diagram for the electrons in a multielectron atom. (Compare this to the hydrogen energy-level diagram in Figure 29.20.) For comparison, the hydrogen-atom energies are shown on the right edge of the figure. Two features of this diagram are of particular interest:

1. For each n, the energy increases as l increases until the maximum-l state has an energy very nearly that of the same n in hydrogen. States with small values of l are significantly lower in energy than the corresponding state in hydrogen.

2. As the energy increases, states with different n begin to alternate in energy. For example, the $3s$ and $3p$ states have lower energy than a $4s$ state, but the energy of an electron in a $3d$ state is slightly higher. This will have important implications for the structure of the periodic table of the elements.

FIGURE 29.22 An energy-level diagram for electrons in a multielectron atom.

The Pauli Exclusion Principle

By definition, the ground state of a quantum system is the state of lowest energy. What is the ground state of an atom having Z electrons and Z protons? Because the $1s$ state is the lowest energy state, it seems that the ground state should be one in which all Z electrons are in the $1s$ state. However, this idea is not consistent with the experimental evidence.

In 1925, the Austrian physicist Wolfgang Pauli hypothesized that no two electrons in a quantum system can be in the same quantum state. That is, **no two electrons can have exactly the same set of quantum numbers (n, l, m, m_s).** If one electron is present in a state, it *excludes* all others. This statement is called the **Pauli exclusion principle.** It turns out to be an extremely profound statement about the nature of matter.

The exclusion principle is not applicable to hydrogen, where there is only a single electron, but in helium, with $Z = 2$ electrons, we must make sure that the two electrons are in different quantum states. This is not difficult. For a $1s$ state, with $l = 0$, the only possible value of the magnetic quantum number is $m = 0$. But there are *two* possible values of m_s, namely $-\frac{1}{2}$ and $+\frac{1}{2}$. If a first electron is in

(a) He ground state

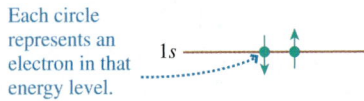

The horizontal lines are the allowed energies.

Each circle represents an electron in that energy level.

(b) He excited state

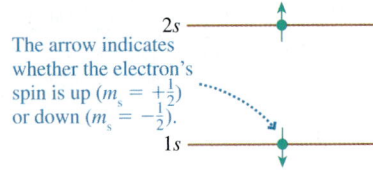

The arrow indicates whether the electron's spin is up ($m_s = +\frac{1}{2}$) or down ($m_s = -\frac{1}{2}$).

FIGURE 29.23 The ground state and first excited state of helium.

(a) Li ground state

(b) Li excited state

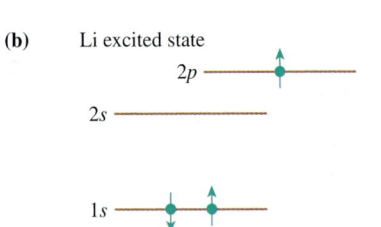

FIGURE 29.24 The ground state and first excited state of lithium.

(a) Li ground state $1s^2 2s$ **(b)** Li excited state $1s^2 2p$

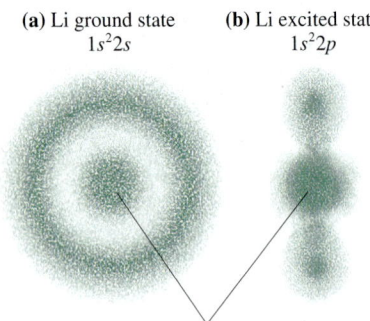

Inner core of two $1s$ electrons

FIGURE 29.25 Electron clouds for the lithium electron configurations $1s^2 2s$ and $1s^2 2p$.

the spin-up $1s$ state $\left(1, 0, 0, -\frac{1}{2}\right)$, a second $1s$ electron can still be added to the atom as long as it is in the spin-down state $\left(1, 0, 0, +\frac{1}{2}\right)$. This is shown schematically in Figure 29.23a, where the dots represent electrons on the rungs of the "energy ladder" and the arrows represent spin up or spin down.

The Pauli exclusion principle does not prevent both electrons of helium from being in the $1s$ state as long as they have opposite values of m_s, so we predict this to be the ground state. A list of an atom's occupied energy levels is called its **electron configuration.** The electron configuration of the helium ground state is written $1s^2$, where the superscript 2 indicates two electrons in the $1s$ energy level. An excited state of the helium atom might be the electron configuration $1s2s$. This state is shown in Figure 29.23b. Here, because the two electrons have different values of n, there is no restriction on their values of m_s.

The states $\left(1, 0, 0, -\frac{1}{2}\right)$ and $\left(1, 0, 0, +\frac{1}{2}\right)$ are the only two states with $n = 1$. The ground state of helium has one electron in each of these states, so all the possible $n = 1$ states are filled. Consequently, the electron configuration $1s^2$ is called a **closed shell.**

The next element, lithium, has $Z = 3$ electrons. The first two electrons can go into $1s$ states, with opposite values of m_s, but what about the third electron? The $1s^2$ shell is closed, and there are no additional quantum states having $n = 1$. The only option for the third electron is the next energy state, $n = 2$. The $2s$ and $2p$ states had equal energies in the hydrogen atom, but they do *not* in a multielectron atom. As Figure 29.22 showed, a lower-l state has lower energy than a higher-l state with the same n. The $2s$ state of lithium is lower in energy than $2p$, so lithium's third ground-state electron will be $2s$. This requires $l = 0$ and $m = 0$ for the third electron, but the value of m_s is not relevant because there is only a single electron in $2s$. Figure 29.24a shows the electron configuration with the $2s$ electron being spin-up, but it could equally well be spin-down. The electron configuration for the lithium ground state is written $1s^2 2s$. This indicates two $1s$ electrons and a single $2s$ electron.

Figure 29.25a shows the probability distribution of electrons in the $1s^2 2s$ ground state of lithium. You can see the $2s$ electron shell surrounding the inner $1s^2$ core. For comparison, Figure 29.25b shows the *first excited state* of lithium, in which the $2s$ electron has been excited to the $2p$ energy level. This forms the $1s^2 2p$ configuration, also shown in Figure 29.24b.

The Periodic Table of the Elements

The 19th century was a time when chemists were discovering new elements and studying their chemical properties. The century opened with the atomic model still not completely validated, and with no one having any idea how many elements there might be. But chemistry developed quickly, and by mid-century it was clear that there were dozens of elements, but not hundreds.

The Russian chemist Dmitri Mendeléev was the first to propose, in 1867, a *periodic* arrangement of the elements based on the regular recurrence of chemical properties. He did so by explicitly pointing out "gaps" where, according to his hypothesis, undiscovered elements should exist. He could then predict the expected properties of the missing elements. The subsequent discovery of these elements verified Mendeléev's organizational scheme, which came to be known as the *periodic table of the elements.*

Figure 29.26 on the next page shows a modern periodic table; a larger version is printed in Appendix B. The quantum mechanics of multielectron atoms explains why the periodic table looks this way. We need three basic ideas to see how this works:

FIGURE 29.26 The modern periodic table of the elements, showing the atomic number Z of each.

1. The energy levels of an atom are found by solving the quantum mechanics equation for multielectron atoms. Figure 29.22, a very important figure for understanding the periodic table, shows that the energy depends on quantum numbers n and l.

2. For each value l of the orbital quantum number, there are $2l + 1$ possible values of the magnetic quantum number m and, for each of these, two possible values of the spin quantum number m_s. Consequently, each energy *level* in Figure 29.22 is actually $2(2l + 1)$ different *states* that, taken together, are called a *subshell*. Table 29.4 lists the number of states in each subshell. Each of these states has the same energy.

3. The ground state of the atom is the lowest-energy electron configuration that is consistent with the Pauli exclusion principle.

TABLE 29.4 Number of states in each subshell of an atom

Subshell	l	Number of states
s	0	2
p	1	6
d	2	10
f	3	14

We used these ideas in the last section to look at the elements helium ($Z = 2$) and lithium ($Z = 3$). Four-electron beryllium ($Z = 4$) comes next. The first two electrons go into $1s$ states, forming a closed shell, and the third goes into $2s$. There is room in the $2s$ level for a second electron as long as its spin is opposite that of the first $2s$ electron. Thus the third and fourth electrons occupy states $(2, 0, 0, -\frac{1}{2})$ and $(2, 0, 0, +\frac{1}{2})$. These are the only two possible $2s$ states. All of the states with the same values and n and l are called a **subshell,** so the fourth electron closes the $2s$ subshell. (The outer two electrons are called a subshell, rather than a shell, because they complete only the $2s$ possibilities; there are still spaces for $2p$ electrons.) The ground state of beryllium, shown in Figure 29.27, is $1s^2 2s^2$.

Boron ($1s^2 2s^2 2p$) opens the $2p$ subshell. The remaining possible $2p$ states are filled as we continue across the second row of the periodic table. These elements are shown in Figure 29.28 on the next page. With neon ($1s^2 2s^2 2p^6$), which has six $2p$ electrons, the $n = 2$ shell is complete, and we have another closed shell. The second row of the periodic table is eight elements wide because of the two $2s$ electrons *plus* the six $2p$ electrons needed to fill the $n = 2$ shell.

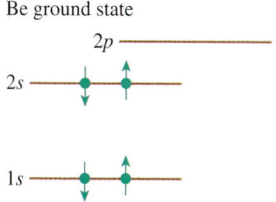

FIGURE 29.27 The ground state of beryllium ($Z = 4$).

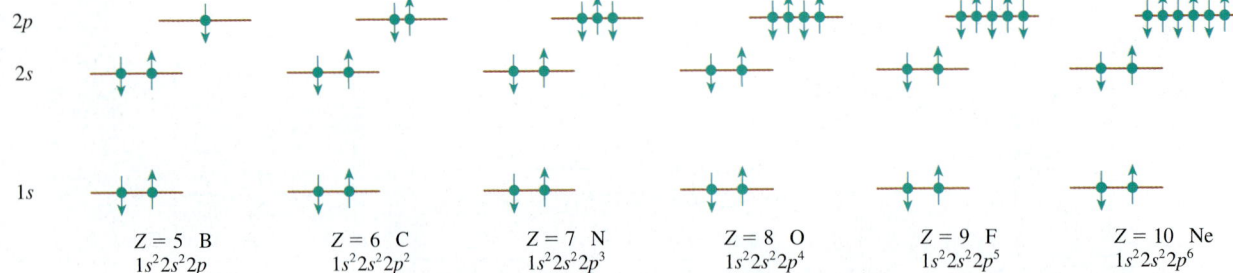

FIGURE 29.28 Filling the 2p subshell in the elements boron ($Z = 5$) through neon ($Z = 10$).

Elements with $Z > 10$

The third row of the periodic table is similar to the second. The two 3s states are filled in sodium and magnesium. The two columns on the left of the periodic table represent the two electrons that can go into an s subshell. Then the six 3p states are filled, one by one, in aluminum through argon. The six columns on the right represent the six electrons of the p subshell.

The fourth row is where the periodic table begins to get complicated. You might expect the closure of the 3p subshell in argon to be followed, starting with potassium ($Z = 19$), by filling the 3d subshell. But if you look back at Figure 29.22, where the energies of the different nl states are shown, you will see that the 3d state is slightly *higher* in energy than the 4s state. Because the ground state is the *lowest energy state* consistent with the Pauli exclusion principle, potassium finds it more favorable to fill a 4s state than to fill a 3d state. Thus the ground state configuration of potassium is $1s^2 2s^2 2p^6 3s^2 3p^6 4s$ rather than the expected $1s^2 2s^2 2p^6 3s^2 3p^6 3d$.

At this point, we begin to see a competition between increasing n and decreasing l. Because $E_{4s} < E_{3d} < E_{4p}$, the states across the fourth row are filled in the order 4s, then 3d, then finally 4p. The 3d subshell "splits open" the periodic table to form the 10-element-wide group of *transition elements*. Most commonly occurring metals are transition elements, and their metallic properties are determined by their partially filled d subshell. The 3d subshell closes with zinc, at $Z = 30$, then the next six elements fill the 4p subshell up to krypton, at $Z = 36$.

Things get even more complex starting in the sixth row, but the ideas are familiar. The $l = 3$ subshell (f electrons) becomes a possibility with $n = 4$, but it turns out that the 5s, 5p, and 6s states are all lower in energy than 4f. Not until barium ($Z = 56$) fills the 6s subshell (and lanthanum ($Z = 57$) adds a 5d electron) is it energetically favorable to add a 4f electron. Immediately after lanthanum you have to switch down to the *lanthanides* at the bottom of the table. The lanthanides fill in the 4f states.

The 4f subshell is complete with $Z = 71$ lutetium. Then $Z = 72$ hafnium through $Z = 80$ mercury complete the transition-element 5d subshell, followed by the 6p subshell in the six elements thallium through radon at the end of the sixth row. Radon, the last inert gas, has $Z = 86$ electrons and the ground-state configuration

radon ($Z = 86$): $1s^2 2s^2 2p^6 3s^2 3p^6 4s^2 3d^{10} 4p^6 5s^2 4d^{10} 5p^6 6s^2 4f^{14} 5d^{10} 6p^6$

This is frightening to behold, but we can now understand it!

EXAMPLE 29.7 The ground state of arsenic

Predict the ground-state electron configuration of arsenic.

SOLVE The periodic table shows that arsenic (As) has $Z = 33$, so we must identify the states of 33 electrons. Arsenic is in the fourth row, following the first group of transition elements. Argon ($Z = 18$) filled the $3p$ subshell, then calcium ($Z = 20$) filled the $4s$ subshell. The next 10 elements, through zinc ($Z = 30$), filled the $3d$ subshell. The $4p$ subshell starts filling with gallium ($Z = 31$), and arsenic is the third element in this group, so it will have three $4p$ electrons. Thus the ground-state configuration of arsenic is

$$1s^2 2s^2 2p^6 3s^2 3p^6 4s^2 3d^{10} 4p^3$$

The entire periodic table is well explained by quantum mechanics. Figure 29.29 summarizes the results, showing the subshells as they are filled. It is especially important to note the significance of the electron's spin. Although the introduction of the electron's spin may have seemed obscure and unnecessary, we now find that the spin quantum number m_s is absolutely essential for understanding the periodic table.

FIGURE 29.29 Summary of the order in which subshells are filled in the periodic table.

STOP TO THINK 29.6 Is the electron configuration $1s^2 2s^2 2p^4 3s$ a ground-state configuration or an excited-state configuration?

A. Ground-state.
B. Excited-state.
C. It's not possible to tell without knowing which element it is.

29.7 Excited States and Spectra

The periodic table organizes information about the *ground states* of the elements. These states are chemically most important because most atoms spend most of the time in their ground states. All the chemical ideas of valence, bonding, reactivity, and so on are consequences of these ground-state atomic structures. But the periodic table does not tell us anything about the excited states of atoms. It is the excited states that hold the key to understanding atomic spectra, and that is the topic to which we turn next.

 18.2

The dots of light are being emitted by two beryllium ions held in a device called an ion trap. Each ion, which is excited by an invisible ultraviolet laser, emits about 10^6 visible-light photons per second.

FIGURE 29.30 The 3s ground state of the sodium atom and some of the excited states.

Sodium ($Z = 11$) is a multielectron atom that we will use as a prototypical atom. The ground-state electron configuration of sodium is $1s^2 2s^2 2p^6 3s$. The first 10 electrons completely fill the $n = 1$ and $n = 2$ shells, creating a *neon core*, while the 3s electron is a valence electron. It is customary to represent this configuration as [Ne]3s or, more simply, as just 3s.

The excited states of sodium are produced by raising the valence electron to a higher energy level. The electrons in the neon core are unchanged. Thus the excited states can be labeled [Ne]nl or, more simply, as just nl. Figure 29.30 is an energy-level diagram showing the ground state and some of the excited states of sodium. The 1s, 2s, and 2p states of the neon core are not shown on the diagram. These states are filled and unchanging, so only the states available to the valence electron are shown. Notice that the zero of energy has been shifted to the ground state. As we have discovered before, the zero of energy can be located where it is most convenient. With this choice, the excited-state energies tell us how far each state is above the ground state. The ionization limit now occurs at the value of the atom's ionization energy, which is 5.14 eV for sodium.

The first energy level above 3s is 3p, so the *first excited state* of sodium is $1s^2 2s^2 2p^6 3p$, written as [Ne]3p or, more simply, 3p. The valence electron is excited, while the core electrons are unchanged. This state is followed, in order of increasing energy, by [Ne]4s, [Ne]3d, and [Ne]4p. Notice that the order of excited states is exactly the same order (3p-4s-3d-4p) that explained the fourth row of the periodic table.

Other atoms have energy-level diagrams similar to that of sodium, although things do get more complicated when there is more than one valence electron. The point to remember is that you can *utilize* the information shown on an energy-level diagram without having to understand precisely *why* each level is where it is.

Excitation by Absorption

Left to itself, an atom will be in its lowest-energy ground state. How does an atom get into an excited state? The process of getting it there is called **excitation**, and there are two basic mechanisms: absorption and collision. We'll begin by looking at excitation by absorption.

One of the postulates of the basic Bohr model is that an atom can jump from one stationary state, of energy E_1, to a higher-energy state E_2 by absorbing a photon of frequency $f_{photon} = \Delta E_{atom}/h$. Because we are interested in spectra, it is more useful to write this in terms of the wavelength:

$$\lambda = \frac{c}{f_{photon}} = \frac{hc}{\Delta E_{atom}} = \frac{1240 \text{ eV} \cdot \text{nm}}{\Delta E_{atom}} \tag{29.23}$$

The final expression, which uses the value $hc = 1240$ eV · nm, gives the wavelength in nanometers *if* ΔE_{atom} is in electron volts.

Bohr's idea of quantum jumps remains an integral part of our interpretation of the results of quantum mechanics. By absorbing a photon, an atom jumps from its ground state to one of its excited states. However, a quantum-mechanical analysis of how the electrons in an atom interact with a light wave shows that not every conceivable transition can occur. The allowed transitions must satisfy the following **selection rule:** A transition (either absorption or emission) from a state in which the valence electron has orbital quantum number l_1 to another with orbital quantum number l_2 is allowed only if

$$\Delta l = |l_2 - l_1| = 1 \tag{29.24}$$

Selection rule for emission and absorption

For example, an atom in an s state ($l = 0$) can absorb a photon and be excited to a p state ($l = 1$) but *not* to another s state or to a d state. An atom in a p state ($l = 1$) can emit a photon by dropping to a lower-energy s state ($l = 0$) *or* to a lower-energy d state ($l = 2$) but not to another p state.

EXAMPLE 29.8 Analyzing absorption in sodium

What are the two longest wavelengths in the absorption spectrum of sodium? What are the transitions?

SOLVE The sodium ground state is $[\mathrm{Ne}]3s$. Starting from an s state, the selection rule permits quantum jumps only to p states. The lowest excited state is the 3p state and $3s \rightarrow 3p$ is an allowed transition ($\Delta l = 1$), so this will be the longest wavelength. You can see from the data in Figure 29.30 that $\Delta E_{\mathrm{atom}} = 2.10 \,\mathrm{eV} - 0.00 \,\mathrm{eV} = 2.10 \,\mathrm{eV}$ for this transition.

The corresponding wavelength is

$$\lambda = \frac{1240 \,\mathrm{eV} \cdot \mathrm{nm}}{2.10 \,\mathrm{eV}} = 589 \,\mathrm{nm}$$

(This calculation actually rounds to 590 nm, but more precise values of hc and ΔE_{atom} give 589 nm, which is the experimentally observed wavelength.) The next excited state is 4s, but a

$3s \rightarrow 4s$ transition is not allowed by the selection rule. The next allowed transition is $3s \rightarrow 4p$, with $\Delta E_{\mathrm{atom}} = 3.76 \,\mathrm{eV}$. The wavelength of this transition is

$$\lambda = \frac{1240 \,\mathrm{eV} \cdot \mathrm{nm}}{3.76 \,\mathrm{eV}} = 330 \,\mathrm{nm}$$

ASSESS If you look at the sodium spectrum shown in Figure 29.31 (which is the same as that shown earlier in Figure 29.4), you will see that 589 nm and 330 nm are, indeed, the two longest wavelengths in the absorption spectrum.

300 nm   400 nm   500 nm   600 nm   700 nm

FIGURE 29.31 Absorption spectrum of sodium.

Collisional Excitation

An electron traveling with a speed of 1.0×10^6 m/s has a kinetic energy of 2.85 eV. If this electron collides with a ground-state sodium atom, a portion of its energy can be used to excite the atom to a higher-energy state, such as its 3p state. This process is called **collisional excitation** of the atom.

Collisional excitation differs from excitation by absorption in one very fundamental way. In absorption, the photon disappears. Consequently, *all* of the photon's energy must be transferred to the atom. Conservation of energy then requires $E_{\mathrm{photon}} = \Delta E_{\mathrm{atom}}$. In contrast, the electron is still present after collisional excitation and can still have some kinetic energy. That is, the electron does *not* have to transfer its entire energy to the atom. If the electron has an incident kinetic energy of 2.85 eV, it could transfer 2.10 eV to the sodium atom, thereby exciting it to the 3p state, and still depart the collision with a speed of 5.1×10^5 m/s and 0.75 eV of kinetic energy.

To excite the atom, the incident energy of the electron (or any other matter particle) merely has to *exceed* ΔE_{atom}. That is $E_{\mathrm{particle}} \geq \Delta E_{\mathrm{atom}}$. There's a threshold energy for exciting the atom, but no upper limit. It is all a matter of energy conservation. Figure 29.32 shows the idea graphically.

Collisional excitation by electrons is the predominant method of excitation in electrical discharges such as fluorescent lights, street lights, and neon signs. A gas is sealed in a tube at reduced pressure (≈ 1 mm of Hg), then a fairly high voltage (≈ 1000 V) between electrodes at the ends of the tube causes the gas to ionize, creating a current in which both ions and electrons are charge carriers. The electrons accelerate in the electric field, gaining several eV of kinetic energy, then transfer some of this energy to the gas atoms upon collision. The process does not work at atmospheric pressure because the average distance between collisions is too short for the electrons to gain enough kinetic energy to excite the atoms.

NOTE ▶ In contrast to photon absorption, there are no selection rules for collisional excitation. Any state can be excited if the colliding particle has sufficient energy. ◀

The photon disappears. Energy conservation requires $E_{\mathrm{photon}} = E_2 - E_1$.

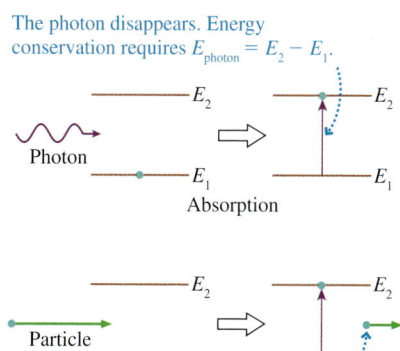

The particle carries away energy. Energy conservation requires $E_{\mathrm{particle}} \geq E_2 - E_1$.

FIGURE 29.32 Excitation by photon absorption and electron collision.

Can an electron with a kinetic energy of 11.4 eV cause a hydrogen atom to emit the prominent red spectral line (λ = 656 nm, E_{photon} = 1.89 eV) in the Balmer series?

REASON The electron must have sufficient energy to excite the upper state of the transition. The electron's energy of 11.4 eV is significantly larger than the 1.89 eV energy of a photon with wavelength 656 nm, but don't confuse the energy of the photon with the energy of the excitation. The red spectral line in the Balmer series is emitted in an $n = 3 \rightarrow 2$ quantum jump with ΔE_{atom} = 1.89 eV, but to cause this emission, the electron must

excite an atom from its *ground state*, with $n = 1$, up to the $n = 3$ level. From Figure 29.18, the necessary excitation energy is

$$\Delta E_{atom} = E_3 - E_1 = (-1.51 \text{ eV}) - (-13.60 \text{ eV})$$
$$= 12.09 \text{ eV}$$

The electron does *not* have sufficient energy to excite the atom to the state from which the emission would occur.

ASSESS As our discussion of absorption spectra showed, almost all excitations of atoms begin from the ground state. Quantum jumps down in energy, however, can begin and end at any two states allowed by selection rules.

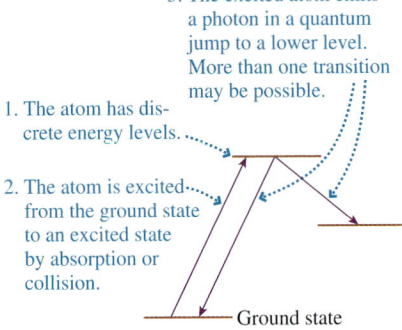

3. The excited atom emits a photon in a quantum jump to a lower level. More than one transition may be possible.

1. The atom has discrete energy levels.

2. The atom is excited from the ground state to an excited state by absorption or collision.

Ground state

FIGURE 29.33 Generation of an emission spectrum.

Emission Spectra

The absorption of light is an important process, but it is the emission of light that really gets our attention. The overwhelming bulk of sensory information that we perceive comes to us in the form of light. With the small exception of cosmic rays, all of our knowledge about the cosmos comes to us in the form of light and other electromagnetic waves emitted in various processes. And emission spectra are more than just scientific curiosities. Many of today's artificial light sources, from fluorescent lights to lasers, are applications of emission spectra.

Understanding emission hinges upon the three ideas shown in Figure 29.33. Once we have determined the energy levels of an atom, from quantum mechanics, we can immediately predict its emission spectrum. We might also ask *how long* an atom remains in an excited state before undergoing a quantum jump to a lower state and emitting a photon. Just as the uncertainty principle prevents us from knowing exactly where an electron is, we also can't determine exactly how long any particular atom spends in the excited state. However, we can determine, both theoretically and experimentally, the *average* time, and the average time an atom spends in the excited state before emitting a photon is called the **lifetime** of that state. A typical excited-state lifetime is just a few nanoseconds.

As an example, Figure 29.34a shows some of the transitions and wavelengths observed in the emission spectrum of sodium. This diagram makes the point that each wavelength represents a quantum jump between two well-defined energy levels. Notice that the selection rule $\Delta l = 1$ is being obeyed in the sodium spectrum. The 5p levels can undergo quantum jumps to 3s, 4s, or 3d but *not* to 3p or 4p.

Figure 29.34b shows the emission spectrum of sodium as it would be recorded in a spectrometer. (Many of the lines seen in this spectrum start from higher excited states that are not seen in the rather limited energy-level diagram of Figure 29.34a.) By comparing the spectrum to the energy-level diagram, you can recognize that the spectral lines at 589 nm, 330 nm, 286 nm, and 268 nm form a *series* of lines due to all the possible $np \rightarrow 3s$ transitions. They are the dominant features in the sodium spectrum.

The most obvious visual feature of sodium emission is its bright yellow color, produced by the 589 nm photons emitted in the $3p \rightarrow 3s$ transition. (The lifetime of the 3p state happens to be 17 ns.) This is the basis of the *flame test* used in chemistry to test for sodium: A sample is held in a Bunsen burner, and a bright yellow glow indicates the presence of sodium. The 589 nm emission is also prominent in the pinkish-yellow glow of the common sodium-vapor street lights. These operate by creating an electrical discharge in sodium vapor. Most sodium-vapor lights use high-pressure lamps to increase their light output. The high pressure, however, causes the formation of Na_2 molecules, and these molecules emit the pinkish portion of the light.

(a)

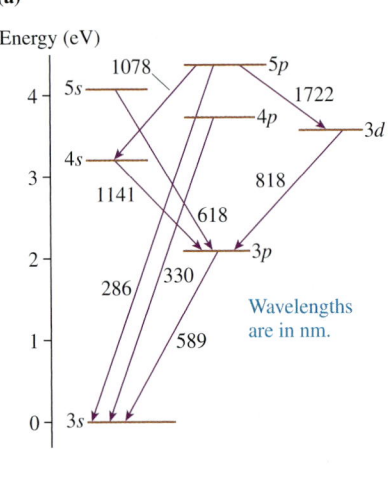

Energy (eV)

1078 5s 5p
4 — 1722
4p
3d
4s 818
3 — 1141
618
286 3p
330
2 — Wavelengths are in nm.
589
1 —

0 — 3s

(b)

268 nm 6p → 3s
286 nm 5p → 3s
330 nm 4p → 3s
589 nm 3p → 3s

300 nm   400 nm   500 nm   600 nm   700 nm

FIGURE 29.34 The emission spectrum of sodium.

◀ **Seeing the light** Some cities close to astronomical observatories use low-pressure sodium lights, and these emit the distinctively yellow 589 nm light of sodium. The glow of city lights is a severe problem for astronomers, but the very specific 589 nm emission from sodium is easily removed with a *sodium filter,* an interference filter that lets all colors pass except the 589 nm yellow sodium light. The photos show the sky near a sodium streetlight without (left) and with (right) a sodium filter. The constellation of Orion, nearly obscured in the left photo, is clearly visible in the right photo.

X Rays

Chapter 28 noted that x rays are produced by causing very-high-speed electrons, accelerated with potential differences of many thousands of volts, to crash into metal targets. Rather than exciting the atom's valence electrons, such as happens in a gas discharge tube, these high-speed projectiles are capable of knocking inner-shell electrons out of the target atoms, producing an *inner-shell vacancy.* As Figure 29.35 shows for copper atoms, this vacancy is filled when an electron from a higher shell undergoes a quantum jump into the vacancy, emitting a photon in the process.

In heavy elements, such as copper or iron, the energy difference between the inner and outer shells is very large—typically 10 keV. Consequently, the photon has energy $E_{photon} \approx 10$ keV and wavelength $\lambda \approx 0.1$ nm. These high-energy photons are the x rays discovered by Röntgen. X-ray photons are about 10,000 times more energetic than visible-light photons, and the wavelengths are about 10,000 times smaller. Even so, the underlying physics is the same: A photon is emitted when an electron in an atom undergoes a quantum jump.

FIGURE 29.35 The generation of x rays from copper atoms.

STOP TO THINK 29.7 In this hypothetical atom, what is the photon energy E_{photon} of the longest-wavelength photons emitted by atoms in the 5p state?

A. 1.0 eV
B. 2.0 eV
C. 3.0 eV
D. 4.0 eV

29.8 Molecules

Quantum mechanics applies to molecules just as it does to atoms, but molecules are more complex because they have internal modes of storing energy. In particular, molecules can *rotate* about their center of mass, and the atoms can *vibrate* back and forth as if the molecular bonds holding them together were little springs. For the most part, we'll overlook this internal motion and focus our attention on the electrons and the electron energy levels.

Molecular hydrogen H_2, the joining of two hydrogen atoms, is the simplest neutral molecule, with two protons and two electrons. Figure 29.36 shows the electron cloud of a hydrogen molecule. Just as for the atomic-hydrogen electron clouds of Figure 29.21, which you should compare, this image shows the probability of finding an electron. The important thing to notice is the large probability of finding the electrons *between* the protons. The electrons' electrostatic attraction to the protons is the "glue" of the molecular bond, preventing the two protons from pushing each other apart.

There is a high probability of finding an electron between the protons. There they form the "glue" that holds the protons together.

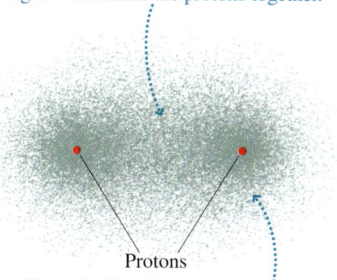

There shading density represents the probability of finding an electron.

FIGURE 29.36 The electron cloud of a hydrogen molecule.

Other molecules, whether they're simple inorganic molecules such as CO_2 or complex biological molecules, are bound together in much the same way, although the details become very complex in molecules with hundreds of atoms and thousands of electrons. A quantum-mechanical analysis of molecules finds the following general results:

- The energy is quantized. Electrons can exist only in certain allowed energy levels.
- The number of energy levels is so extraordinarily high, and the energy levels are jammed so close together, that for all practical purposes the energy levels group into *band*s of allowed energy.
- In thermal equilibrium, nearly all molecules are in the very lowest energy levels.

Figure 29.37 shows a generic molecular energy-level diagram for a medium-size molecule. Whereas an atom has a well-defined ground state, a molecule has a broad band of lower energy levels. Similarly, a single excited state, such as the 2*s* state of the hydrogen atom, has been replaced by a band of excited energy levels. Despite the vast number of allowed energy levels, nearly all molecules spend nearly all their time in the very lowest energy levels.

Figure 29.38 uses the energy-level diagram to explain two important phenomena of molecular spectroscopy: absorption and fluorescence. Whereas the absorption spectrum of an atom consists of discrete spectral lines, a molecule has a continuous *absorption band*. The absorption of light at a higher frequency (shorter wavelength) followed by the emission of light at a lower frequency (longer wavelength) is called **fluorescence.** Fluorescence occurs in molecules, but not atoms, because molecules can transform some of the absorbed energy into the vibrational energy of the atoms and thus increase the thermal energy of the molecules.

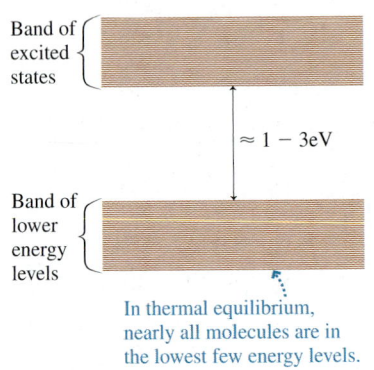

Band of excited states

$\approx 1 - 3\text{eV}$

Band of lower energy levels

In thermal equilibrium, nearly all molecules are in the lowest few energy levels.

FIGURE 29.37 The molecular energy-level diagram for a medium-size molecule.

The molecules can absorb light over a range of wavelengths. This makes an *absorption band*.

The molecules rapidly transform some of the absorbed energy into molecular vibrations, causing the molecules to fall to the bottom edge of the excited band.

Quantum jumps back to the lower band have less energy than the original jumps up. Thus the *emission band* is at longer wavelength than the absorption band. This is *fluoresence*.

Absorption band
Emission band

Relative intensity

300 400 500 600
Wavelength (nm)

FIGURE 29.38 Molecular absorption and fluorescence.

Fluorescence has many important applications. In biology, fluorescent dyes are used to stain tissue samples for microscopy. In the brilliantly colored photo that opened this chapter, the cells were stained with three different dyes, one of which (blue) was preferentially absorbed in the nucleus and another (green) by actin. The red dye was absorbed by other cell structures. When illuminated with ultraviolet light, which we can't see, each dye fluoresces with its own characteristic color.

Many biological molecules are fluorescent. A good example is chlorophyll, the green pigment in plants that allows photosynthesis to happen. When illuminated by blue or ultraviolet light, chlorophyll fluoresces an unexpected dark red color. This property of chlorophyll is used by marine biologists to measure the concentration of phytoplankton—microscopic plants—in sea water.

▶ **Whiter than white** Laundry detergents often contain fluorescent dyes that absorb invisible ultraviolet light, and then fluoresce in a broad band of visible wavelengths; that is, they give off white light. Sunlight contains lots of UV, so outdoors your white clothes not only reflect the white visible sunlight, but fluoresce even more white light—"whiter than white." In the photo, you can see how brightly the girl's shirt glows when illuminated with UV light (and, evidently, your hair and teeth fluoresce this way as well).

Quite recently, scientists isolated a fluorescent protein from a species of jelly-fish. This protein fluoresces green when illuminated with ultraviolet light, so it's been dubbed GFP, for *green fluorescent protein*. GFP has become an important tool in genetics because it can be used to identify when particular genes are being expressed in living cells. This is done by fusing the jellyfish GFP gene to the gene being studied. When the gene is active, the cell manufactures GFP in addition to the usual protein coded by the gene. If the cells are observed under ultraviolet light, a bright green glow from the GFP indicates that the gene is turned on. The cells are dark where the gene is not active. Figure 29.39 shows an example.

The color of GFP may remind you of *bioluminescence*—the summer flashes of fireflies or the green glow of various deep-sea fish—and, indeed, there is a connection. Bioluminescence is actually a form of *chemiluminescence,* the production of light in chemical reactions. Some chemical reactions create reactant molecules in an excited state. These molecules then emit light as they jump to lower energy levels, just as if they had first absorbed shorter-wavelength light. The emitted light has exactly the same spectrum as fluorescence, but the method by which the molecules are excited is different. Light sticks, which come in many colors, are an example of chemiluminescence.

Bioluminescence is just chemiluminescence in a biological organism. Some biochemical reaction—often a catalyzed reaction in an organism—produces a molecule in an excited state, and light is emitted as the molecule jumps to a lower state. In fireflies, the reaction involves an enzyme with the intriguing name *Luciferase.* In the jellyfish *Aequoria victoria,* the source of GFP, bioluminescent reactions actually create blue light. However, the blue light is absorbed by the green fluorescent protein—just as predicted by the absorption curve in Figure 29.38—and re-emitted as green fluorescence, giving the jellyfish an eerie green glow when seen in the dark ocean.

FIGURE 29.39 Green fluorescent protein shows the locations in mosquito larvae at which a particular gene is being expressed.

29.9 Stimulated Emission and Lasers

We have seen that an atom can jump from a lower-energy level E_1 to a higher-energy level E_2 by absorbing a photon. Figure 29.40a illustrates the basic absorption process, with a photon of frequency $f = \Delta E_{atom}/h$ disappearing as the atom jumps from level 1 to level 2. Once in level 2, as shown in Figure 29.40b, the atom can emit a photon of the same frequency as it jumps back to level 1. Because this transition occurs spontaneously, without the introduction of outside energy, it is called **spontaneous emission.**

In 1917, four years after Bohr's proposal of stationary states in atoms but still prior to de Broglie and Schrödinger, Einstein was puzzled by how quantized atoms reach thermodynamic equilibrium in the presence of electromagnetic radiation. Einstein found that the processes of absorption and spontaneous emission were not sufficient to allow a collection of atoms to reach thermodynamic equilibrium. To resolve this difficulty, Einstein proposed a third mechanism for the interaction of atoms with light.

The left half of Figure 29.40c shows a photon with frequency $f = \Delta E_{atom}/h$ approaching an *excited* atom. If a photon can induce the $1 \rightarrow 2$ transition of absorption, then Einstein proposed that it should also be able to induce a $2 \rightarrow 1$ transition. In a sense, this transition is a *reverse absorption.* But to undergo a reverse absorption, the atom must *emit* a photon of frequency $f = \Delta E_{atom}/h$. The

(a) Absorption

E_2 ——————

〜〜〜▶ ⇨

Photon

E_1 ——————

(b) Spontaneous emission

E_2 ——————

⇨   〜〜〜▶

Photon

E_1 ——————

(c) Stimulated emission

E_2 ——————

〜〜〜▶ ⇨   〜〜〜▶
〜〜〜▶

Photon   Two identical

E_1 ——————   photons

FIGURE 29.40 Three types of radiative transitions.

18.3 Activ ONLINE Physics

end result, as seen in the right half of Figure 29.40c, is an atom in level 1 plus *two* photons! Because the first photon induced the atom to emit the second photon, this process is called **stimulated emission.**

Stimulated emission occurs only if the first photon's frequency exactly matches the $E_2 - E_1$ energy difference of the atom. This is precisely the same condition that absorption has to satisfy. More interesting, the emitted photon is *identical* to the incident photon. This means that as the two photons leave the atom they have exactly the same frequency and wavelength, are traveling in exactly the same direction, and are exactly in phase with each other. In other words, **stimulated emission produces a second photon that is an exact clone of the first.**

Stimulated emission is of no importance in most practical situations. Atoms typically spend only a few nanoseconds in an excited state before undergoing spontaneous emission, so the atom would need to be in an extremely intense light wave for stimulated emission to occur prior to spontaneous emission. Ordinary light sources are not nearly intense enough for stimulated emission to be more than a minor effect, hence it was many years before Einstein's prediction was confirmed. No one had doubted Einstein, because he had clearly demonstrated that stimulated emission was necessary to make the energy equations balance, but it seemed no more important than would pennies to a millionaire balancing her checkbook. At least, that is, until 1960, when a revolutionary invention appeared that made explicit use of stimulated emission: the laser.

Lasers

The word **laser** is an acronym for the phrase **l**ight **a**mplification by the **s**timulated **e**mission of **r**adiation. The first laser, a ruby laser, was demonstrated in 1960, and several other kinds of lasers appeared within a few months. Today, lasers do everything from being the light source in fiber-optic communications to measuring the distance to the moon, and from playing your CD to performing delicate eye surgery.

But what is a laser? Basically it is a device that produces a beam of highly *coherent* and essentially monochromatic (single-color) light as a result of stimulated emission. **Coherent** light is light in which all the electromagnetic waves have the same phase, direction, and amplitude. It is the coherence of a laser beam that allows it to be very tightly focused or to be rapidly modulated for communications.

Let's take a brief look at how a laser works. Figure 29.41 represents a system of atoms that have a lower energy level E_1 and a higher energy level E_2. Suppose that there are N_1 atoms in level 1 and N_2 atoms in level 2. Left to themselves, all the atoms would soon end up in level 1 because of the spontaneous emission $2 \rightarrow 1$. To prevent this, we can imagine that some type of excitation mechanism, perhaps an electrical discharge, is continuing to produce new excited atoms in level 2.

Let a photon of frequency $f = (E_2 - E_1)/h$ be incident on this group of atoms. Because it has the correct frequency, it could be absorbed by one of the atoms in level 1. Another possibility is that it could cause stimulated emission from one of the level 2 atoms. Ordinarily $N_2 \ll N_1$, so absorption events far outnumber stimulated emission events. Even if a few photons were generated by stimulated emission, they would quickly be absorbed by the vastly larger group of atoms in level 1.

But what if we could somehow arrange to place *every* atom in level 2, making $N_1 = 0$? Then the incident photon, upon encountering its first atom, will cause stimulated emission. Where there was initially one photon of frequency f, now

A spectacular laser light show depends on three key properties of coherent laser light: It can be very intense, its color is extremely pure, and the laser beam is narrow with little divergence.

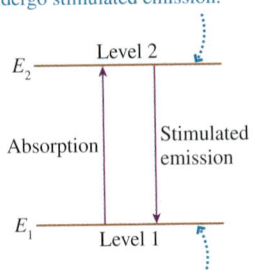

N_2 atoms in level 2. Photons of energy $E_{photon} = E_2 - E_1$ can cause these atoms to undergo stimulated emission.

N_1 atoms in level 1. These atoms can absorb photons of energy $E_{photon} = E_2 - E_1$.

FIGURE 29.41 Energy levels 1 and 2, with populations N_1 and N_2.

there are two. These will strike two additional excited-state atoms, again causing stimulated emission. Then there will be four photons. As Figure 29.42 shows, there will be a *chain reaction* of stimulated emission until all N_2 atoms emit a photon of frequency f.

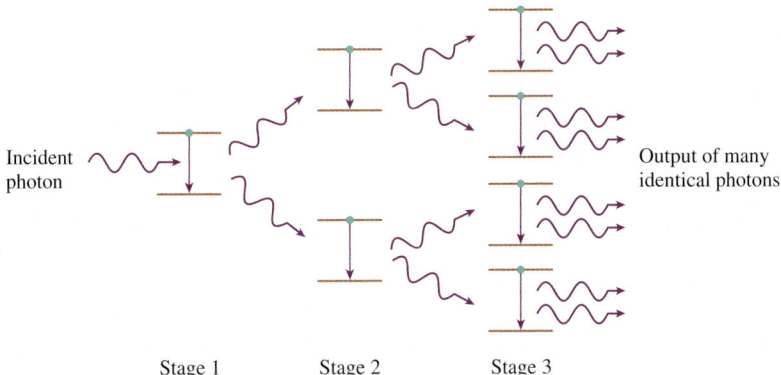

FIGURE 29.42 Stimulated emission creates a chain reaction of photon production in a population of excited atoms.

In stimulated emission, each emitted photon is *identical* to the incident photon. The chain reaction of Figure 29.42 will lead not just to N_2 photons of frequency f, but to N_2 *identical* photons, all traveling together in the same direction with the same phase. If N_2 is a large number, as would be the case in any practical device, the one initial photon will have been *amplified* into a gigantic, coherent pulse of light!

Although the chain reaction of Figure 29.42 illustrates the idea most clearly, it is not necessary for every atom to be in level 2 for amplification to occur. All that is needed is to have $N_2 > N_1$ so that stimulated emission exceeds absorption. Such a situation is called a **population inversion.** The stimulated emission is sustained by placing the *lasing medium*—the sample of atoms that emits the light—in an **optical cavity** consisting of two facing mirrors. As Figure 29.43 shows, the photons interact repeatedly with the atoms in the medium as they bounce back and forth. This repeated interaction is necessary for the light intensity to build up to a high level. If one of the mirrors is partially transmitting, some of the light emerges as the *laser beam*.

The Helium-Neon Laser

The familiar red laser used in lecture demonstrations, laboratories, and supermarket checkout scanners is the helium-neon laser, often called a HeNe laser, with a wavelength of 633 nm. The medium of a HeNe laser is a mixture of $\approx$90% helium and $\approx$10% neon gases. As Figure 29.44a shows, the gases are sealed in a glass tube, then an electrical discharge is established along the bore of the tube. Two mirrors are bonded to the ends of the discharge tube, one a total reflector and the other having $\approx$2% transmission so that the laser beam can be extracted.

The atoms that "lase" in a HeNe are the neon atoms, but the production of a population inversion involves the helium atoms. The electrons in the discharge easily excite the $1s2s$ excited state of helium. This state happens to have an abnormally long lifetime—a *metastable state*—so it is possible to build up a fairly large population (but not an inversion) of excited helium atoms in this state. The energy of the $1s2s$ state is 20.6 eV.

Interestingly, an excited state of neon, the $5s$ state, also has an energy of 20.6 eV. If a $1s2s$ excited helium atom collides with a ground state neon atom, as

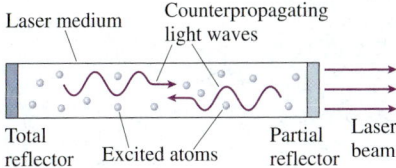

FIGURE 29.43 Lasing takes place in an optical cavity.

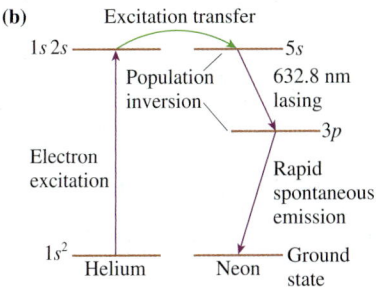

FIGURE 29.44 A HeNe laser.

frequently happens, the excitation energy can be transferred from one atom to the other! Written as a chemical reaction, the process is

$$He^* + Ne \rightarrow He + Ne^*$$

where the asterisk indicates the atom is in an excited state. This process, called **excitation transfer,** is very efficient for the $5s$ state because the process is *resonant*—a perfect energy match. Thus the two-step process of collisional excitation of helium, followed by excitation transfer between helium and neon, pumps the neon atoms into the excited $5s$ state. This is shown in Figure 29.44b.

The $5s$ energy level in neon is ≈ 1.95 eV above the $3p$ state. The $3p$ state is very nearly empty of population, both because it is not efficiently populated in the discharge and because it undergoes very rapid spontaneous emission to the $3s$ states. Thus the large number of atoms pumped into the $5s$ state creates a population inversion with respect to the lower $3p$ state. These are the necessary conditions for laser action.

Because the lower level of the laser transition is normally empty of population, or very nearly so, placing only a small fraction of the neon atoms in the $5s$ state creates a population inversion. Thus a fairly modest excitation is sufficient to create the inversion and start the laser. Furthermore, a HeNe laser can maintain a *continuous* inversion and thus sustain continuous lasing. The electrical discharge continuously creates $5s$ excited atoms in the upper level, and the rapid spontaneous decay of the $3p$ atoms from the lower level keeps its population low enough to sustain the inversion.

A typical helium-neon laser has a power output of 1 mW = 10^{-3} J/s at 633 nm in a 1-mm-diameter laser beam. This output corresponds to the emission of 3.2×10^{15} photons per second. Other continuous lasers operate by similar principles, but can produce much more power. The argon laser, which is widely used in scientific research, can produce up to 20 W of power at green and blue wavelengths. The carbon dioxide laser produces output power in excess of 1000 W at the infrared wavelength of 10.6 μm. It is used in industrial applications for cutting and welding.

Lasers in Medicine

The invention of lasers was followed almost immediately by medical applications. Even a small-power laser beam can produce a significant amount of very localized *heating* if focused with a lens. More powerful lasers can easily *cut* through tissue by literally vaporizing it, replacing a stainless steel scalpel with a beam of light. Not only can laser surgery be very precise, it generally has less blood loss than conventional surgery because the heat of the laser seals the blood vessels and capillaries.

One common medical use of lasers is to remove plaque from artery walls, thus reducing the risk of stroke or heart attack. In this procedure, an optical fiber is threaded through arteries to reach the site. A powerful laser beam is then fired through the fiber to carefully vaporize the plaque. Laser beams traveling through optical fibers are also used to kill cancer cells in *photodynamic therapy*. In this case, light-sensitive chemicals are injected into the bloodstream and are preferentially taken up by cancer cells. The optical fiber is positioned next to the tumor and illuminates it with just the right wavelength to activate the light-sensitive chemicals and kill the cells. These procedures are minimally invasive, and they can reach areas of the body not readily accessible by conventional surgery.

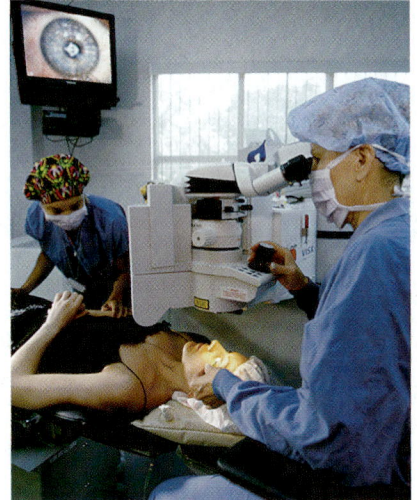

Laser vision BIO Laser-based LASIK surgery can correct for vision defects, such as near- or far-sightedness, that result from an incorrect refractive power of the eye. You learned in Chapter 19 that the majority of your eye's focusing power occurs at the surface of the cornea. In LASIK, a special knife first cuts a small, thin flap in the cornea, and this flap is folded out of the way. A computer-controlled ultraviolet laser very carefully vaporizes the underlying corneal tissue to give it the desired shape, then the flap is folded back into place. The procedure takes only a few minutes and requires only a few numbing drops in the eye.

The goal of Chapter 29 has been to use quantum physics to understand the properties of atoms, molecules, and their spectra.

IMPORTANT CONCEPTS

The Structure of an Atom

An atom consists of a very small, positively charged nucleus, surrounded by orbiting electrons.

- The number of protons is the atom's **atomic number** Z.
- The **atomic mass number** A is the number of protons + the number of neutrons.

The Bohr Atom

In Bohr's model,

- The atom can exist only in certain **stationary states.** These states correspond to different electron orbits. Each state is numbered by **quantum number** $n = 1, 2, 3 \ldots$
- Each state has a discrete, well-defined energy E_n.
- The atom can change its energy by undergoing a **quantum jump** between two states by emitting or absorbing a photon of energy $E_{photon} = \Delta E_{atom} = |E_f - E_i|$.

The Hydrogen Atom

In Bohr's model of the hydrogen atom the stationary states are found by requiring an integer number of de Broglie wavelengths to fit around the circumference of the electron's orbit: $2\pi r = n\lambda$. The integer n is the *principal quantum number.*

This leads to energy quantization with

$$r_n = n^2 a_B \qquad E_n = -\frac{13.60 \text{ eV}}{n^2}$$

where $a_B = 0.053$ nm is the **Bohr radius.**

The wavelengths of light in the hydrogen atom spectrum are given by the **Balmer formula:**

$$\lambda_{n \to m} = \frac{91.18 \text{ nm}}{\left(\frac{1}{m^2} - \frac{1}{n^2}\right)} \qquad \begin{array}{l} m = 1, 2, 3, \ldots \\ n = m + 1, m + 2, \ldots \end{array}$$

Beyond the Bohr model, quantum mechanics adds several quantized parameters, each with its own quantum number:

- The *orbital angular momentum,* quantum number l:

$$L = \sqrt{l(l+1)}\hbar \qquad l = 0, 1, 2, 3, \ldots, n - 1$$

- The *z-component of angular momentum,* quantum number m:

$$L_z = m\hbar \qquad m = -l, -l+1, \ldots, 0, \ldots, l-1, l$$

- The *z-component of the electron* **spin,** quantum number m_s:

$$S_z = m_s\hbar \qquad m_s = -\tfrac{1}{2} \text{ or } +\tfrac{1}{2}$$

The energy of a hydrogen atom depends only on n:

Multielectron atoms

Each electron is described by the same quantum numbers (n, l, m, m_s) used for the hydrogen atom, but the energy now depends on l as well as n.

The **Pauli exclusion principle** states that no more than one electron can occupy each quantum state.

Molecules

In molecules, the states are spaced very closely into **bands** of states. Because electrons can be excited to and from many states, the spectra of molecules are broad, not discrete.

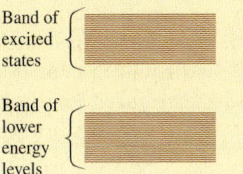

APPLICATIONS

Atomic emission spectra are generated by excitation followed by a photon-emitting quantum jump.

- **Excitation** occurs by absorption of a photon or by collision.
- A quantum jump can only occur if $\Delta l = \pm 1$.
- Quantized energies give rise to a **discrete spectrum.**

Excitation occurs from the lowest-energy, or *ground,* state.

Emission occurs back to the ground state or other states.

Lasers

A photon with energy $E_{photon} = E_2 - E_1$ can induce **stimulated emission** of a second photon identical to the first. These photons can then induce more atoms to emit photons. If more atoms are in state 2 than in state 1, this process can rapidly build up an intense beam of identical photons. This is the principle behind the laser.

For instructor-assigned homework, go to
www.masteringphysics.com

Problem difficulty is labeled as | (straightforward) to ||||| (challenging).

Problems labeled INT integrate significant material from earlier chapters; BIO are of biological or medical interest.

QUESTIONS

Conceptual Questions

1. A neon discharge emits a bright reddish-orange spectrum. But a glass tube filled with neon is completely transparent. Why doesn't the neon in the tube absorb orange and red wavelengths?

2. The two spectra shown in Figure Q29.2 belong to the same element, a fictional Element X. Explain why they are different.

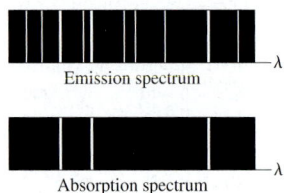

Emission spectrum

Absorption spectrum

FIGURE Q29.2

3. Is a spectral line with wavelength 656.5 nm seen in the absorption spectrum of hydrogen atoms? Why or why not?

4. J. J. Thomson studied the ionization of atoms in collisions with electrons. He accelerated electrons through a potential difference, shot them into a gas of atoms, then used a mass spectrometer to detect any ions produced in the collisions. By using different gases, he found that he could produce singly ionized atoms of all the elements that he tried. When he used higher accelerating voltages, he was able to produce doubly ionized atoms of all elements *except* hydrogen.

    a. Why did Thomson have to use higher accelerating voltages to detect doubly ionized atoms than to detect singly ionized atoms?

    b. What conclusion or conclusions about hydrogen atoms can you draw from these observations? Be specific as to how your conclusions are related to the observations.

5. Bohr did not include the gravitational force in his analysis of the hydrogen atom. Is this one of the reasons that his model of the hydrogen atom had only limited success? Explain.

6. If an electron is in a *stationary state* of an atom, is the electron at rest? If not, what does the term mean?

7. The $n = 3$ state of hydrogen has $E_3 = -1.51$ eV.

    a. Why is the energy negative?

    b. What is the physical significance of the specific number 1.51 eV?

8. For a hydrogen atom, list all possible states (n, l, m, m_s) that have $E = -1.51$ eV.

9. What are the n and l values of the following states of a hydrogen atom: (a) $4d$, (b) $5f$, (c) $6s$?

10. How would you label the hydrogen-atom states with the following (n, l, m) quantum numbers: (a) $(4, 3, 0)$, (b) $(3, 2, 1)$, (c) $(3, 2, -1)$?

11. Consider the two hydrogen-atom states $5d$ and $4f$. Which has the higher energy? Explain.

12. Does each diagram in Figure Q29.12 represent a possible electron configuration of an element? If so, (i) identify the element and (ii) determine if this is the ground state or an excited state. If not, why not?

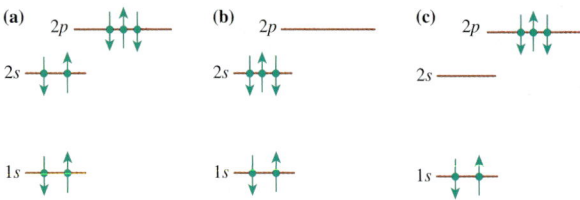

FIGURE Q29.12

13. Do the following electron configurations represent a possible state of an element? If so, (i) identify the element and (ii) determine if this is the ground state or an excited state. If not, why not?

    a. $1s^2 2s^2 2p^6 3s^2$ b. $1s^2 2s^2 2p^7 3s$ c. $1s^2 2s^2 2p^4 3s^2 3p^2$

14. Why is the section of the periodic table labeled as "transition elements" exactly 10 elements wide in all rows?

15. An electron is in an f state. Can it undergo a quantum jump to an s state? A p state? A d state? Explain.

16. Figure Q29.16 shows the energy level diagram of Element X.

    a. What is the ionization energy of Element X?

    b. An atom in the ground state absorbs a photon, then emits a photon with a wavelength of 1240 nm. What conclusion can you draw about the energy of the photon that was absorbed?

    c. An atom in the ground state has a collision with an electron, then emits a photon with a wavelength of 1240 nm. What conclusion can you draw about the initial kinetic energy of the electron?

FIGURE Q29.16

17. a. Which states of a hydrogen atom can be excited by a collision with an electron with kinetic energy $K = 12.5$ eV? Explain.

    b. After the collision the atom is not in its ground state. What happens to the electron? (i) It bounces off with $K > 12.5$ eV, (ii) It bounces off with $K = 12.5$ eV, (iii) It bounces off with $K < 12.5$ eV, (iv) It is absorbed by the atom. Explain your choice.

    c. After the collision, the atom emits a photon. List all the possible $n \rightarrow m$ transitions that might occur as a result of this collision.

18. What *is* an atom's ionization energy? In other words, if you know the ionization energy of an atom, what is it that you know about the atom?

19. Figure Q29.19 shows the energy levels of a hypothetical atom.

    a. What *minimum* kinetic energy (in eV) must an electron have to collisionally excite this atom and cause the emission of a 620 nm photon? Explain.

    b. Can an electron with $K = 6$ eV cause the emission of 620 nm light? If so, what is the final kinetic energy of the electron? If not, why not?

    c. Can a 6 eV photon cause the emission of 620 nm light from this atom? Why or why not?

    d. Can a 7 eV photon cause the emission of 620 nm light from this atom? Why or why not?

FIGURE Q29.19

20. Seven possible transitions are identified on the energy level diagram in Figure Q29.20. For each, is this an allowed transition? If allowed, is it an emission or an absorption transition, and is the photon infrared, visible, or ultraviolet? If not allowed, why not?

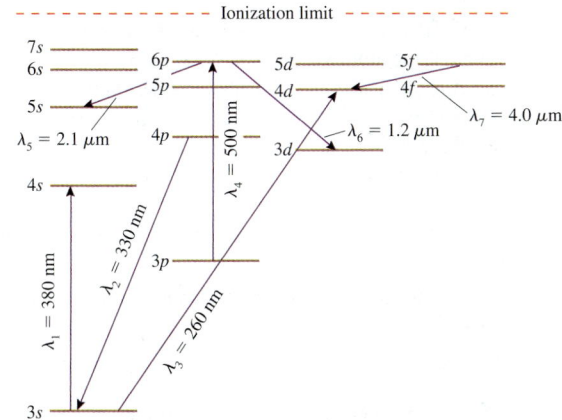

FIGURE Q29.20

21. A 2.0 eV photon is incident on an atom in the *p*-state, as shown in the energy level diagram in Figure Q29.21. Does the atom undergo an absorption transition, a stimulated emission transition, or neither? Explain.

FIGURE Q29.21

22. A glass tube contains 2×10^{11} atoms, some of which are in the ground state and some of which are excited. Figure Q29.22 shows the populations for the atoms' three energy levels. Is it possible for these atoms to be a laser? If so, on which transition would laser action occur? If not, why not?

$$N_3 = 8 \times 10^{10} \quad \text{Level 3}$$
$$s\text{-state}$$
$$\text{Level 2}$$
$$p\text{-state}$$
$$N_2 = 2 \times 10^{10}$$

$$N_1 = 10 \times 10^{10}$$
$$s\text{-state}$$
$$\text{Level 1}$$

FIGURE Q29.22

Multiple-Choice Questions

23. ‖ An electron collides with an atom in its ground state. The atom then emits a photon of energy E_{photon}. In this process the *change* ΔE_{elec} in the electron's energy is
    A. Greater than E_{photon}.
    B. Greater than or equal to E_{photon}.
    C. Equal to E_{photon}.
    D. Less than or equal to E_{photon}.
    E. Less than E_{photon}.

24. ‖ How many states are in the $l = 4$ subshell?
    A. 8   B. 9   C. 16   D. 18   E. 22

25. ‖ What is the ground-state electron configuration of calcium $(Z = 20)$?
    A. $1s^2 2s^2 2p^6 3s^2 3p^8$ B. $1s^2 2s^2 2p^6 3s^2 3p^6 4s^1 4p^1$
    C. $1s^2 2s^2 2p^6 3s^2 3p^6 4s^2$ D. $1s^2 2s^2 2p^6 3s^2 3p^6 4p^2$

26. ‖ An atom emits a photon with a wavelength of 275 nm. By how much does the atom's energy change?
    A. 0.72 eV   B. 1.06 eV   C. 2.29 eV
    D. 3.06 eV   E. 4.51 eV

27. ‖ The energy of a hydrogen atom is -3.40 eV. What is the electron's kinetic energy?
    A. 1.70 eV   B. 2.62 eV   C. 3.40 eV
    D. 5.73 eV   E. 6.80 eV

28. ‖ The angular momentum of an electron in a Bohr hydrogen atom is 3.18×10^{-34} kg · m²/s. What is the atom's energy?
    A. -13.60 eV   B. -6.73 eV   C. -3.40 eV
    D. -1.51 eV   E. -0.47 eV

29. ‖ A "soft x-ray" photon with an energy of 41.8 eV is absorbed by a hydrogen atom in its ground state, knocking the atom's electron out. What is the speed of the electron as it leaves the atom?
    A. 1.84×10^5 m/s   B. 3.08×10^5 m/s   C. 8.16×10^5 m/s
    D. 3.15×10^6 m/s   E. 3.83×10^6 m/s

PROBLEMS

Section 29.1 Spectroscopy

1. ‖ Figure 29.2b and Table 29.1 showed the wavelengths of the first four lines in the visible spectrum of hydrogen.
    a. Determine the Balmer formula n and m values for these wavelengths.
    b. Predict the wavelength of the fifth line in the spectrum.

2. │ The wavelengths in the hydrogen spectrum with $m = 1$ form a series of spectral lines called the Lyman series. Calculate the wavelengths of the first four members of the series.

3. │ The Paschen series is analogous to the Balmer series, but with $m = 3$. Calculate the wavelengths of the first three members in the Paschen series. What part(s) of the electromagnetic spectrum are these in?

Section 29.2 Atoms

4. | How many electrons, protons, and neutrons are contained in the following atoms or ions: (a) ^{6}Li, (b) ^{13}C$^+$, and (c) ^{18}O^{++}?

5. | How many electrons, protons, and neutrons are contained in the following atoms or ions: (a) ^{9}Be$^+$, (b) ^{12}C, and (c) ^{15}N^{+++}?

6. | Write the symbol for an atom or ion with:
   a. three electrons, three protons, and five neutrons.
   b. five electrons, six protons, and eight neutrons.

7. | Write the symbol for an atom or ion with:
   a. one electron, one proton, and one neutron.
   b. five electrons, seven protons, and seven neutrons.

Section 29.3 Bohr's Model of Atomic Quantization

8. | Figure P29.8 is an energy level diagram for a simple atom. What wavelengths appear in the atom's (a) emission spectrum and (b) absorption spectrum?

FIGURE P29.8

9. | An electron with 2.0 eV of kinetic energy collides with the atom shown in Figure P29.8.
   a. Is the electron able to kick the atom to an excited state? Why or why not?
   b. If your answer to part a was yes, what is the electron's kinetic energy after the collision?

10. ‖ The allowed energies of a simple atom are 0.0 eV, 4.0 eV, and 6.0 eV.
   a. Draw the atom's energy-level diagram. Label each level with the energy and the principal quantum number.
   b. What wavelengths appear in the atom's emission spectrum?
   c. What wavelengths appear in the atom's absorption spectrum?

11. ‖ The allowed energies of a simple atom are 0.0 eV, 4.0 eV, and 6.0 eV. An electron traveling with a speed of 1.3×10^6 m/s collides with the atom. Can the electron excite the atom to the $n = 2$ stationary state? The $n = 3$ stationary state? Explain.

Section 29.4 The Bohr Hydrogen Atom

12. | For relativity to have a significant effect, speeds must exceed 1% of the speed of light. Do we need to consider relativity when we calculate the orbits of an electron in a hydrogen atom?

13. | A hydrogen atom is in the $n = 3$ state. In the Bohr model, how many electron wavelengths fit around this orbit?

14. ‖ A hydrogen atom is in its $n = 1$ state. In the Bohr model, what is the ratio of its kinetic energy to its potential energy?

15. | Show, by actual calculation, that the Bohr radius is 0.0529 nm and that the ground-state energy of hydrogen is -13.60 eV.

16. | a. What quantum number of the hydrogen atom comes closest to giving a 500-nm-diameter electron orbit?
   b. What are the electron's speed and energy in this state?

17. | a. Calculate the de Broglie wavelength of the electron in the $n = 1, 2,$ and 3 states of the hydrogen atom. Use the information in Table 29.2.
   b. Show numerically that the circumference of the orbit for each of these stationary states is exactly equal to n de Broglie wavelengths.
   c. Sketch the de Broglie standing wave for the $n = 3$ orbit.

18. | Show, by calculation, that the first three states of the hydrogen atom have angular momenta $\hbar$, $2\hbar$, and $3\hbar$, respectively.

19. | Determine all possible wavelengths of photons that can be emitted from the $n = 4$ state of a hydrogen atom.

Section 29.5 The Quantum-Mechanical Hydrogen Atom

20. | List the quantum numbers of (a) all possible $3p$ states and (b) all possible $3d$ states.

21. ‖ When all quantum numbers are considered, how many different quantum states are there for a hydrogen atom with $n = 1$? With $n = 2$? With $n = 3$? List the quantum numbers of each state.

22. | What is the angular momentum of a hydrogen atom in (a) a $4p$ state and (b) a $5f$ state? Give your answers as a multiple of $\hbar$.

23. | A hydrogen atom has orbital angular momentum 3.65×10^{-34} J · s.
   a. What letter ($s, p, d,$ or f) describes the electron?
   b. What is the atom's minimum possible energy? Explain.

24. | What is the maximum possible angular momentum L (as a multiple of $\hbar$) of a hydrogen atom with energy -0.544 eV?

25. | What are E and L (as a multiple of $\hbar$) of a hydrogen atom in the $6f$ state?

Section 29.6 Multielectron Atoms

26. | Predict the ground-state electron configurations of Mg, Sr, and Ba.

27. ‖ Predict the ground-state electron configurations of Si, Ge, and Pb.

28. | Identify the element for each of these electron configurations. Then determine whether this configuration is the ground state or an excited state.
   a. $1s^2 2s^2 2p^5$
   b. $1s^2 2s^2 2p^6 3s^2 3p^6 3d^{10} 4s^2 4p$

29. ‖ Identify the element for each of these electron configurations. Then determine whether this configuration is the ground state or an excited state.
   a. $1s^2 2s^2 2p^4 3d$
   b. $1s^2 2s^2 2p^6 3s^2 3p^6 3d^8 4s^2$

30. ‖ Identify the element for each of these electron configurations. Then determine whether this configuration is the ground state or an excited state.
   a. $1s^2 2s^2 2p^6 3s^2 3p^6 4s^2 3d^9$
   b. $1s^2 2s^2 2p^6 3s^2 3p^6 4s^2 3d^{10} 4p^6 5s^2 4d^{10} 5p^6 6s^2 4f^{14} 5d^7$

Section 29.7 Excited States and Spectra

31. ‖ An electron with a speed of 5.00×10^6 m/s collides with an atom. The collision excites the atom from its ground state (0 eV) to a state with an energy of 3.80 eV. What is the speed of the electron after the collision?

32. ▮▮▮ Hydrogen gas absorbs light of wavelength 103 nm. Afterward, what wavelengths are seen in the emission spectrum?

33. ▮▮ What is the electron configuration of the second excited state of lithium?

34. ▮ Show that $hc = 1240 \text{ eV} \cdot \text{nm}$.

35. ▮ a. Is a $4p \rightarrow 4s$ transition allowed in sodium? If so, what is its wavelength? If not, why not?
    b. Is a $3d \rightarrow 4s$ transition allowed in sodium? If so, what is its wavelength? If not, why not?

Section 29.8 Molecules

36. ▮▮ Figure P29.36 shows a molecular energy-level diagram. What are the longest and shortest wavelengths in (a) the molecule's absorption spectrum and (b) the molecule's fluorescence spectrum?

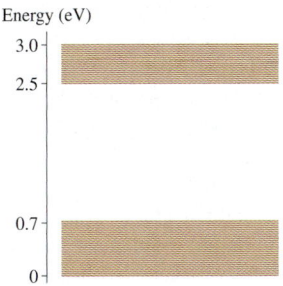

FIGURE P29.36

Section 29.9 Stimulated Emission and Lasers

37. ▮▮ A 1.00 mW helium neon-laser emits a visible laser beam with a wavelength of 633 nm. How many photons are emitted per second?

38. ▮ A 1000 W carbon dioxide laser emits an infrared laser beam with a wavelength of 10.6 μm. How many photons are emitted per second?

39. ▮▮ A ruby laser emits an intense pulse of light that lasts a mere 10 ns. The light has a wavelength of 690 nm, and each pulse has an energy of 500 mJ.
    a. How many photons are emitted in each pulse?
    b. What is the *rate* of photon emission, in photons per second, during the 10 ns that the laser is "on"?

General Problems

40. ▮▮ A 2.55 eV photon is emitted from a hydrogen atom. What are the Balmer formula n and m values corresponding to this emission?

41. ▮ Two of the wavelengths emitted by a hydrogen atom are 102.6 nm and 1876 nm.
    a. What are the Balmer formula n and m values for each of these wavelengths?
    b. For each of these wavelengths, is the light infrared, visible, or ultraviolet?

42. ▮▮▮ In Example 29.2 it was assumed that the initially stationary
    INT gold nucleus would remain motionless during a head-on collision with an 8.3 MeV alpha particle. What is the actual recoil speed of the gold nucleus after that elastic collision? Assume

that the mass of a gold nucleus is exactly 50 times the mass of an alpha particle.
    Hint: Review the discussion of perfectly elastic collisions in Chapter 10.

43. ▮ Consider the gold isotope ^{197}Au.
    INT a. How many electrons, protons, and neutrons are in a neutral ^{197}Au atom?
    b. The gold nucleus has a diameter of 14.0 fm. What is the density of matter in a gold nucleus?
    c. The density of lead is 11,400 kg/m^3. How many times the density of lead is your answer to part b?

44. ▮▮▮ Consider the lead isotope ^{207}Pb.
    INT a. How many electrons, protons, and neutrons are in a neutral ^{207}Pb atom?
    b. The lead nucleus has a diameter of 14.2 fm. What are the electric potential and the electric field strength at the surface of a lead nucleus?

45. ▮▮▮ The diameter of an atom is 1.2×10^{-10} m and the diameter of its nucleus is 1.0×10^{-14} m. What percent of the atom's volume is occupied by mass and what percent is empty space?

46. ▮ The charge-to-mass ratio of a nucleus, in units of e/u, is $q/m = Z/A$. For example, a hydrogen nucleus has $q/m = 1/1 = 1$.
    a. Make a graph of charge-to-mass ratio versus proton number Z for nuclei with $Z = 5, 10, 15, 20, \ldots, 90$. For A, use the average atomic mass shown on the periodic table of elements in Appendix B. Show each of these 18 nuclei as a dot, but don't connect the dots together as a curve.
    b. Describe any trend that you notice in your graph.
    c. What's happening in the nuclei that is responsible for this trend?

47. ▮▮▮ If the nucleus is a few fm in diameter, the distance between
    INT the centers of two protons must be ≈ 2 fm.
    a. Calculate the electric force between two protons that are 2.0 fm apart.
    b. Calculate the gravitational force between two protons that are 2.0 fm apart. Could gravity be the force that holds the nucleus together?

48. ▮ In a head-on collision, the closest approach of a 6.24 MeV alpha particle to the center of a nucleus is 6.00 fm. The nucleus is in an atom of what element? Assume that the nucleus is heavy enough to remain stationary during the collision.

49. ▮▮▮ Through what potential difference would you need to
    INT accelerate an alpha particle, starting from rest, so that it will just reach the surface of a 15-fm-diameter ^{238}U nucleus? Assume that the nucleus is heavy enough to remain stationary during the collision.

50. ▮▮▮ The oxygen nucleus ^{16}O has a radius of 3.0 fm.
    a. With what speed must a proton be fired toward an oxygen nucleus to have a turning point 1.0 fm from the surface? Assume that the nucleus is heavy enough to remain stationary during the collision.
    b. What is the proton's kinetic energy in MeV?

51. ▮▮ The absorption spectrum of an atom consists of the wavelengths 200 nm, 300 nm, and 500 nm.
    a. Draw the atom's energy-level diagram.
    b. What wavelengths are seen in the atom's emission spectrum?

52. ▮▮▮ The first three energy levels of the fictitious element X are shown in Figure P29.52.

    a. What wavelengths are observed in the absorption spectrum of element X? Give your answers in nm.

    b. State whether each of your wavelengths in part b corresponds to ultraviolet, visible, or infrared light.

    E (eV)

    ---- 0

    $n = 3$ ——— -2.0
    $n = 2$ ——— -3.0

    $n = 1$ ——— -6.5

    FIGURE P29.52

    c. An electron with a speed of 1.4×10^6 m/s collides with an atom of element X. Shortly afterward, the atom emits a 1240 nm photon. What was the electron's speed after the collision? Assume that, because the atom is so much more massive than the electron, the recoil of the atom is negligible.

    Hint: The energy of the photon is *not* the energy transferred to the atom in the collision.

53. ▮ Show that Planck's constant $\hbar$ has units of angular momentum.

54. ▮▮ What is the energy of a Bohr hydrogen atom with a 5.18 nm diameter?

55. ▮▮ What is the wavelength of the series limit (i.e., the shortest possible wavelength) of the Lyman series in hydrogen?

56. ▮▮▮ The He^+ ion is a hydrogen-like atom, a single electron orbiting a nucleus with $q_{nucleus} = +2e$. Following Bohr's analysis of the hydrogen atom, determine

    a. The radius of the two lowest-energy states of the ion.

    b. The speed of the electron associated with those two states.

    c. The energy difference between the ground state and the first excited state.

57. ▮▮▮ A hydrogen atom in the ground state absorbs a 12.75 eV photon. Immediately after the absorption, the atom undergoes a quantum jump to the next-lowest energy level. What is the wavelength of the photon emitted in this quantum jump?

58. ▮ a. Calculate the orbital radius and the speed of an electron
    INT in both the $n = 99$ and the $n = 100$ state of hydrogen.

    b. Determine the orbital frequency of the electron in each of these states.

    c. Calculate the frequency of a photon emitted in a $100 \rightarrow 99$ transition.

    d. Compare the photon frequency of part c to the *average* of your two orbital frequencies from part b. By what percent do they differ?

59. ▮▮▮ Two hydrogen atoms collide head on. The collision brings
    INT both atoms to a halt. Immediately after the collision, both atoms emit a 121.6 nm photon. What was the speed of each atom just before the collision?

60. ▮▮▮ A beam of electrons is incident on a gas of hydrogen atoms.
    INT a. What minimum speed must the electrons have to cause the emission of 656 nm light from the $3 \rightarrow 2$ transition of hydrogen?

    b. Through what potential difference must the electrons be accelerated to have this speed?

61. ▮▮ A hydrogen atom in its fourth excited state emits a photon with a wavelength of 1282 nm. What is the atom's maximum possible orbital angular momentum after the emission? Give your answer as a multiple of $\hbar$.

62. ▮▮ Draw a series of pictures, similar to Figure 29.28, for the ground states of K, Ti, Fe, Ge, and Br.

63. ▮▮ Draw a series of pictures, similar to Figure 29.28, for the ground states of Ca, Sc, Co, Zn, and Kr.

64. ▮▮ A particular emission line in the hydrogen spectrum has a wavelength of 656.5 nm. What are all possible transitions (e.g., $6d \rightarrow 2s$) that could give rise to this emission?

65. ▮▮ a. What downward transitions are possible for a sodium atom in the 6s state? (See Figure 29.30.)

    b. What are the wavelengths of the photons emitted in each of these transitions?

66. ▮▮ The $5d \rightarrow 3p$ transition in the emission spectrum of sodium (see Figure 29.30) has a wavelength of 499 nm. What is the energy of the 5d state?

67. ▮ A sodium atom (see Figure 29.30) emits a photon with wavelength 818 nm shortly after being struck by an electron. What minimum speed did the electron have before the collision?

68. ▮▮ Figure P29.68 shows a few energy levels of the mercury atom. One valence electron is always in the 6s state; the other electron changes states.

    a. Make a table showing all the allowed transitions in the emission spectrum. For each transition, indicate the photon wavelength, in nm.

    b. What minimum speed must an electron have to excite the 492-nm-wavelength blue emission line in the Hg spectrum?

FIGURE P29.68

69. ▮▮ Figure P29.69 shows the first few energy levels of the lithium atom. Make a table showing all the allowed transitions in the emission spectrum. For each transition, indicate

    a. The wavelength, in nm.

    b. Whether the transition is in the infrared, the visible, or the ultraviolet spectral region.

    c. Whether or not the transition would be observed in the lithium absorption spectrum.

    Energy (eV)

    $4s$ —— 4.34
    $3p$ —— 3.83 $3d$ —— 3.88
    $3s$ —— 3.37

    $2p$ —— 1.85

    Energies for each level are in eV.

    $2s$ —— 0.00

    FIGURE P29.69

70. | A laser emits 1.00×10^{19} photons per second from an excited state with energy $E_2 = 1.17$ eV. The lower energy level is $E_1 = 0$ eV.
    a. What is the wavelength of this laser?
    b. What is the power output of this laser?

71. ‖‖ Fluorescence microscopy, discussed in Section 29.8, is an
BIO important tool in modern cell biology. A variation on this technique depends on a phenomenon known as two-photon excitation. If two photons are absorbed simultaneously (i.e., within about 10^{-16} s), their energies can add. A molecule that is normally excited by a 350 nm photon can be excited by two photons each having half as much energy. For this process to be useful, photons must illuminate the sample at the very high rate of at least 10^{29} photons/m$^2 \cdot$ s. This is achieved by focusing a laser beam to a small spot and by concentrating the power of the laser into very short (10^{-13} s) pulses that are fired 10^8 times each second. Suppose a biologist wants to use two-photon excitation to excite a molecular species that would be excited by 500 nm light in normal one-photon fluorescence microscopy. What minimum intensity (W/m^2) must the laser beam have during each pulse?

Passage Problems

Light-Emitting Diodes

Light-emitting diodes, known by the acronym LED, produce the familiar green and red indicator lights used in a wide variety of consumer electronics. LEDs are semiconductor devices in which the electrons can exist only in certain energy levels. Much like molecules, the energy levels are packed together close enough to form what appears to be a continuous band. Energy supplied to an LED in a circuit excites electrons from a *valence band* into a *conduction band*. An electron can emit a photon by undergoing a quantum jump from a state in the conduction band into an open state in the valence band, as shown in Figure P29.72.

The size of the band gap ΔE_{band} determines the possible energies—and thus the wavelengths—of the emitted photons. Most LEDs emit a narrow range of wavelengths and thus have a distinct color. This makes them well-suited for traffic lights and other applications where a certain color is desired, but it makes them less desirable for general illumination. One way to make a "white"

LED is to combine a blue LED with a substance that fluoresces yellow when illuminated with the blue light. The combination of the two colors makes light that appears reasonably white.

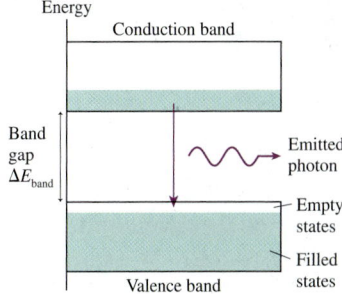

FIGURE P29.72 Energy-level diagram of an LED.

72. ‖ An LED emits green light. Increasing the size of the band gap could change the color of the emitted light to
    A. Red   B. Orange   C. Yellow   D. Blue

73. ‖ Suppose the LED band gap is 2.5 eV, which corresponds to a wavelength of 500 nm. Consider the possible electron transitions in Figure P29.72. 500 nm is the
    A. Maximum wavelength of the LED.
    B. Average wavelength of the LED.
    C. Minimum wavelength of the LED.

74. ‖ The same kind of semiconducting material used to make an LED can also be used to convert absorbed light into electrical energy, essentially operating as an LED in reverse. In this case, the absorption of a photon causes an electron transition from a filled state in the valence band to an unfilled state in the conduction band. If $\Delta E_{band} = 1.4$ eV, what is the minimum wavelength of electromagnetic radiation that could lead to electrical energy output?
    A. 140 nm   B. 890 nm   C. 1400 nm   D. 8900 nm

75. ‖ The efficiency of a light source is the percentage of its energy input that gets radiated as visible light. If some of the blue light in an LED is used to cause a fluorescent material to glow,
    A. The overall efficiency of the LED is increased.
    B. The overall efficiency of the LED does not change.
    C. The overall efficiency of the LED decreases.

STOP TO THINK ANSWERS

Stop to Think 29.1: A is emission, B is absorption. All wavelengths in the absorption spectrum are seen in the emission spectrum, but not all wavelengths in the emission spectrum are seen in the absorption spectrum.

Stop to Think 29.2: 6 protons and 8 neutrons. The number of protons is the atomic number, which is 6. That leaves $14 - 6 = 8$ neutrons.

Stop to Think 29.3: In emission from the $n = 3$ to $n = 2$ transition, but not in absorption. The photon energy has to match the energy *difference* between two energy levels. Absorption is from the ground state, at $E_1 = 0$ eV. There's no energy level at 3 eV to which the atom could jump.

Stop to Think 29.4: $n = 3$. Each antinode is half a wavelength, so this standing wave has three full wavelengths in one circumference.

Stop to Think 29.5: $n = 3$, $l = 1$, or a $3p$ state.

Stop to Think 29.6: B. The atom would have less energy if the $3s$ electron were in a $2p$ state.

Stop to Think 29.7: C. Emission is a quantum jump to a lower-energy state. The $5p \rightarrow 4p$ transition is not allowed because $\Delta l = 0$ violates the selection rule. The lowest-energy allowed transition is $5p \rightarrow 3d$, with $E_{photon} = \Delta E_{atom} = 3.0$ eV.

30

NUCLEAR PHYSICS

This is a *bone scan,* not an x ray. It was created using the radioactive decay of a particular type of nucleus inside the body. How can a process that occurs in the nucleus of an atom allow us to create an image of tissues of the human body?

Looking Ahead ▶▶

The goal of Chapter 30 is to understand the physics of the nucleus and some of the applications of nuclear physics. In this chapter you will learn to:

▶ Interpret the basic structure of the nucleus.

▶ Understand how the strong force holds the nucleus together and why some nuclei are unstable.

▶ Calculate the half-lives of radioactive decay.

▶ Understand the nature and properties of subatomic particles.

▶ Apply nuclear physics to biology and medicine.

Looking Back ◀◀

This chapter builds on models of the atom and the nucleus introduced in the previous chapter. Please review:

◀ Section 29.2 Models of the atom and the nucleus.

◀ Section 29.6 Multielectron atoms.

The nucleus of the atom is extremely remote from our everyday experience. Thus it comes as something of a surprise to notice the extent to which nuclear physics has become part of our modern lives. A nucleus that was created in a nuclear reactor is a key element of your smoke detector. The measurement of nuclear isotopes allows us to determine ancient water temperatures or the age of artifacts. The use of radioactive isotopes in medicine allows for carefully targeted imaging and treatment. The bone scan above doesn't just show the structure of the skeleton; this nuclear image reveals regions of active skeletal growth, an important aid to diagnosis, as we will see in the final section of this chapter.

The development of the model of the nucleus is a story that we started to tell in the last chapter. We will continue the tale here. Since the history is recent and well-documented, this is a wonderful opportunity to illustrate the development of a new scientific model. As you will see, there were surprises and missteps along the way, but the final model now rests on a solid foundation of careful experimental and theoretical justification.

30.1 Nuclear Structure

The 1890s was a decade of mysterious rays. Cathode rays were being studied in several laboratories, and, in 1895, Röntgen discovered x rays. In 1896, after hearing of Röntgen's discovery, the French scientist A. H. Becquerel wondered if mineral crystals that fluoresce after exposure to sunlight were emitting x rays. He put a piece of film in an opaque envelope, then placed a crystal on top and left it in the sun. To his delight, the film in the envelope was exposed.

Becquerel thought he had discovered x rays coming from crystals, but his joy was short lived. He soon found that the film could be exposed equally well simply by being stored in a closed drawer with the crystals. Further investigation showed that the crystal, which happened to be a mineral containing uranium, was spontaneously emitting some new kind of ray. Rather than finding x rays, as he had hoped, Becquerel had discovered what became known as *radioactivity*.

Ernest Rutherford soon took up the investigation and found not one but three distinct kinds of rays emitted from crystals containing uranium. Not knowing what they were, he named them for their ability to penetrate matter and ionize air. The first, which caused the most ionization and penetrated the least, he called *alpha rays*. The second, with intermediate penetration and ionization, were *beta rays,* and the third, with the least ionization but the largest penetration, became *gamma rays.*

Within a few years, Rutherford was able to show that alpha rays are helium nuclei emitted from the crystal at very high velocities. These became the projectiles that he used in 1909 to probe the structure of the atom. The outcome of that experiment, as you learned in Chapter 29, was Rutherford's discovery that atoms have a very small, dense nucleus at the center.

Rutherford's discovery of the nucleus may have settled the question of atomic structure, but it raised many new issues for scientific research. Foremost among them were

- What is nuclear matter? What are its properties?
- What holds the nucleus together? Why doesn't the repulsive electrostatic force blow it apart?
- What is the connection between the nucleus and radioactivity?

These were the beginnings of **nuclear physics,** the study of the properties of the atomic nucleus.

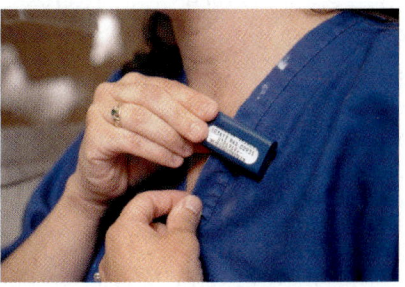

Picturing exposure BIO The fact that radiation can expose film provides a simple and effective means of determining a person's radiation exposure. This badge contains a square of very sensitive film. If the person wearing the badge is exposed to radiation, the film will be exposed; this will be revealed when the film is developed. Such film badges were the standard means of monitoring radiation exposure for many years.

Nucleons

The nucleus is a tiny speck in the center of a vastly larger atom. As Figure 30.1 shows, the nuclear diameter of roughly 10^{-14} m is only about 1/10,000 the diameter of the atom. What we call *matter* is overwhelmingly empty space!

Let's review some information about the atom and the nucleus from Section 29.2. The nucleus is composed of two types of particles: protons and neutrons. Together, these are referred to as **nucleons.** The role of the neutrons, which have nothing to do with keeping electrons in orbit, is an important issue that we'll address in this chapter. The number of protons Z is the element's atomic number. In fact, an element is identified by the number of protons in the nucleus, not by the number of orbiting electrons. Electrons are easily added and removed, forming negative and positive ions, but doing so doesn't change the element. The mass number A is defined to be $A = Z + N$, where N is the neutron number. The mass number is the total number of nucleons in a nucleus.

> **NOTE** ▶ The mass number, which is dimensionless, is *not* the same thing as the atomic mass m. We'll look at actual atomic masses below. ◀

This picture of an atom would need to be 10 m in diameter if it were drawn to the same scale as the dot representing the nucleus.

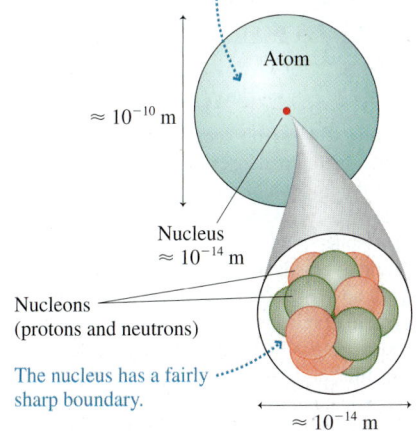

FIGURE 30.1 The nucleus is a tiny speck within an atom.

TABLE 30.1 Protons and neutrons

	Proton	**Neutron**
Number	Z	N
Charge q	$+e$	0
Spin	$\frac{1}{2}$	$\frac{1}{2}$
Mass, in u	1.00728	1.00866

Protons and neutrons are virtually identical other than the fact that the proton has one unit of the fundamental charge e whereas the neutron is electrically neutral. The neutron is slightly more massive than the proton, but the difference is very small, only about 0.1%. Notice that the proton and neutron, like the electron, have an *inherent angular momentum* and magnetic moment with spin quantum number $s = \frac{1}{2}$. As a consequence, protons and neutrons obey the Pauli exclusion principle. Table 30.1 summarizes the basic properties of protons and neutrons.

Isotopes

As we learned in Chapter 29, not all atoms of the same element (and thus the same Z) have the same mass. There is a *range* of neutron numbers that happily form a nucleus with Z protons, creating a series of nuclei having the same Z-value (i.e., they are all the same chemical element) but different A-values. Each A-value in a series of nuclei with the same Z-value is called an *isotope*. Isotopes for many of the elements are given in a table in Appendix C.

The notation used to label isotopes uses the mass number A as a *leading* superscript. Hence ordinary carbon, which has six protons and six neutrons in the nucleus (and thus has $A = 12$), is written ^{12}C and pronounced "carbon twelve." The radioactive form of carbon used in carbon dating is ^{14}C. It has six protons, making it carbon, and eight neutrons, for a total of 14 nucleons. The isotope ^{2}H is a hydrogen atom in which the nucleus is not simply a proton but a proton and a neutron. Although the isotope is a form of hydrogen, it is called **deuterium.** Sometimes, for clarity, we will find it useful to include the atomic number as a leading *subscript*. Ordinary carbon is then written as $^{12}_{6}$C; deuterium as $^2_1$H.

> **NOTE** ▶ Adding the leading subscript doesn't provide any additional information. If the element is carbon, we know that it has 6 protons. But when we "balance" nuclear equations, we will find that the subscript is a useful tool and is worth including. ◀

The chemical behavior of an atom is largely determined by the orbiting electrons. Different isotopes of the same element have very similar *chemical* properties. ^{14}C will form the same chemical compounds as ^{12}C, and will generally be treated the same by the body, a fact that permits the use of ^{14}C to determine the age of a sample. But the *nuclear* properties of these two isotopes are quite different, as we will see.

Most elements have multiple naturally occurring isotopes. For each element, the fraction of naturally occurring nuclei represented by one particular isotope is called the **natural abundance** of that isotope. For instance, oxygen has two primary isotopes, ^{16}O and ^{18}O. The data in Appendix C show that the natural abundance of ^{16}O is 99.76%, meaning that 9976 out of every 10,000 naturally occurring oxygen atoms are the isotope ^{16}O. Most of the remaining 0.24% of naturally occurring oxygen is the isotope ^{18}O, which has two extra neutrons.

The different masses of isotopes of an element can lead to some subtle differences in macroscopic behaviors. A water molecule made with the heavier ^{18}O isotope is slightly heavier as well, and will behave slightly differently than water made with the predominant ^{16}O isotope. Atmospheric water vapor is always deficient in ^{18}O compared to water in the ocean because the lighter molecules containing ^{16}O evaporate slightly more readily. Different atmospheric conditions lead to rain and snow (and thus lakes, rivers, and glaciers) with slightly different fractions of ^{18}O.

More than 3000 isotopes are known. The majority of these are **radioactive,** meaning that the nucleus is not stable but, after some period of time, will either fragment or emit some kind of subatomic particle in an effort to reach a more stable state. Many of these radioactive isotopes are created by nuclear reactions in the laboratory and have only a fleeting existence. Only 266 isotopes are **stable**

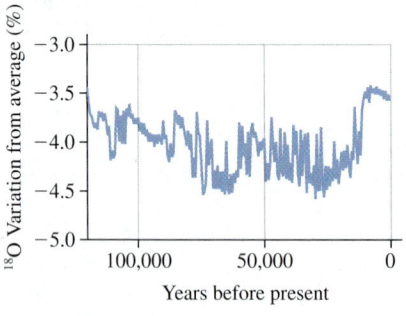

Taking the earth's temperature When water vapor condenses to make snow crystals, the fraction of molecules containing ^{18}O is greater for snow that forms at higher atmospheric temperatures. Snow accumulating over tens of thousands of years has built up a thick ice sheet in Greenland. A core sample of this ice gives a record of the isotopic composition of the snow that fell over this time period. The above graph shows the isotopic composition of a single long core taken from the ice sheet, averaged over 50-year intervals; higher numbers correspond to higher temperatures. Broad trends, such as the increase in temperature at the end of the last ice age, are clearly seen.

(i.e., nonradioactive) and occur in nature. In addition, there are a handful of radioactive isotopes with such long half-lives, measured in billions of years, that they also occur naturally.

Atomic Mass

You learned in Chapter 12 that atomic masses are specified in terms of the *atomic mass unit* u, defined such that the atomic mass of the isotope ^{12}C is exactly 12 u. The conversion to SI units is

$$1 \text{ u} = 1.6605 \times 10^{-27} \text{ kg}$$

Alternatively, we can use Einstein's $E_0 = mc^2$ to express masses in terms of their energy equivalent. The energy equivalent of 1 u of mass is

$$\begin{aligned} E_0 &= (1.6605 \times 10^{-27} \text{ kg})(2.9979 \times 10^8 \text{ m/s})^2 \\ &= 1.4924 \times 10^{-10} \text{ J} = 931.49 \text{ MeV} \end{aligned} \tag{30.1}$$

By noting that Einstein's formula implies $m = E_0/c^2$, we can write 1 u in the following form:

$$1 \text{ u} = \frac{E_0}{c^2} = 931.49 \left(\frac{\text{MeV}}{c^2} \right) \tag{30.2}$$

It may seem unusual, but the units MeV/c^2 are units of mass. This will be a useful unit for us when we need to compute energy equivalents. The energy equivalent of mass 1 MeV/c^2 is simply 1 MeV.

> **NOTE** ▶ In the above equations, we have included more significant figures than usual. Many nuclear calculations look for the small difference between two masses that are almost the same. Those two masses must be calculated or specified to four or five significant figures if their difference is to be meaningful. ◄

Table 30.2 shows some important atomic mass values. Notice that the mass of a hydrogen atom is equal to the sum of the masses of a proton and an electron. But a quick calculation shows that the mass of a helium atom (2 protons, 2 neutrons, and 2 electrons) is 0.03038 u *less* than the sum of the masses of its constituents. The difference is due to the *binding energy* of the nucleus, a topic we'll look at in Section 30.2.

The *chemical* atomic mass shown on the periodic table of the elements is the *weighted average* of the atomic masses of all naturally occurring isotopes. For example, chlorine has two stable isotopes: ^{35}Cl, with atomic mass $m = 34.97$ u, has an abundance of 75.8% and ^{37}Cl, at 36.97 u, has an abundance of 24.2%. The average, weighted by abundance, is 0.758×34.97 u + 0.242×36.97 u = 35.45 u. This is the value shown on the periodic table and is the correct value for most chemical calculations, but it is not the mass of any particular isotope of chlorine.

> **NOTE** ▶ The atomic masses of the proton and the neutron are both ≈ 1 u. Consequently, the value of the mass number A is *approximately* the atomic mass in u. The approximation $m \approx A$ u is sufficient in many contexts, such as when we're calculating the masses of atoms in the kinetic theory of gases. But in nuclear physics calculations we almost always need the more accurate mass values such as those in Table 30.2 or Appendix C. ◄

TABLE 30.2 Some atomic masses

Particle	Symbol	Mass (u)	Mass (MeV/c^2)
Electron	e	0.00055	0.51
Proton	p	1.00728	938.28
Neutron	n	1.00866	939.57
Hydrogen	^{1}H	1.00783	938.79
Helium	^{4}He	4.00260	3728.40

STOP TO THINK 30.1 Three electrons orbit a neutral ^{6}Li atom. How many electrons orbit a neutral ^{7}Li atom?

30.2 Nuclear Stability

We've already noted that less than 10% of the known nuclei are stable (i.e., not radioactive). Because nuclei are characterized by two independent numbers, N and Z, it is useful to show the known nuclei on a plot of neutron number N versus proton number Z. Figure 30.2 shows such a plot. Stable nuclei are represented by blue diamonds and unstable, radioactive nuclei by red dots.

FIGURE 30.2 Stable and unstable nuclei shown on a plot of neutron number N versus proton number Z.

We can make several observations from this graph:

- The stable nuclei cluster very close to the curve called the **line of stability.**
- There are no stable nuclei with $Z > 83$ (bismuth). Heavier elements (up to $Z = 92$ (uranium)) are found in nature, but they are radioactive.
- Unstable nuclei are in bands along both sides of the line of stability.
- The lightest elements, with $Z < 16$, are stable when $N \approx Z$. The familiar isotopes ^{4}He, ^{12}C, and ^{16}O all have equal numbers of protons and neutrons.
- As Z increases, the number of neutrons needed for stability grows increasingly larger than the number of protons. The N/Z ratio is ≈ 1.2 at $Z = 40$ but has grown to ≈ 1.5 at $Z = 80$.

These observations—especially the fact that $N \approx Z$ for small Z but $N > Z$ for large Z—cry out for an explanation. The quantum-mechanical model of the nucleus that we'll develop in Section 30.3 will provide the explanation we seek.

Binding Energy

A nucleus is a *bound system.* That is, you would need to supply energy to disperse the nucleons by breaking the nuclear bonds between them. Figure 30.3 shows this idea schematically.

You learned a similar idea in atomic physics. The energy levels of the hydrogen atom are negative numbers because the bound system has less energy than a free proton and electron. The energy you must supply to an atom to remove an electron is called the *ionization energy.*

In much the same way, the energy you would need to supply to a nucleus to disassemble it into individual protons and neutrons is called the **binding energy.** Whereas ionization energies of atoms are only a few eV, the binding energies of nuclei are tens or hundreds of MeV, energies large enough that their mass equivalent is not negligible.

The binding energy is the energy that would be needed to disassemble a nucleus into individual nucleons.

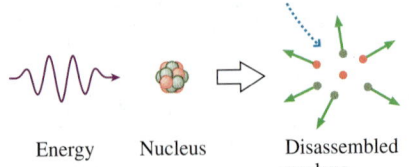

Energy   Nucleus   Disassembled nucleus

FIGURE 30.3 The nuclear binding energy.

We noted earlier that the mass of a helium atom is less than the mass of its constituents. Suppose we break a helium atom into two hydrogen atoms (taking account of the two protons and the two electrons) and two free neutrons as shown in Figure 30.4. The mass of the separated components is more than that of the helium atom. The difference in mass Δm arises from the energy that was put into the system to separate the tightly bound nucleons. The difference in masses is thus a measure of the binding energy:

$$\text{Binding energy of helium nucleus} = \Delta m \cdot c^2$$

We can evaluate this energy B by converting the mass to MeV/c^2:

$$B = \Delta m \cdot c^2$$

$$= \left((0.03038 \text{ u})\left(931.49 \, \frac{\text{MeV}/c^2}{\text{u}}\right)\right)c^2 = (0.03038 \text{ u})(931.49 \text{ MeV/u})$$

$$= 28.30 \text{ MeV}$$

Generally, the nuclear binding energy is computed by considering the mass difference between the atom and its separated components, Z hydrogen atoms and N neutrons:

Electron

Separate into components

Neutron   Proton

Helium atom   **2 hydrogen atoms, 2 neutrons**

Mass: Mass:
4.00260 u   2 H atoms: 2.01566 u
               + 2 neutrons: 2.01732 u
               Total mass: 4.03298 u

Difference in mass:
$\Delta m = 0.03038 \text{ u}$

FIGURE 30.4 The binding energy of the helium nucleus.

Activ Physics ONLINE   19.2

$$B = (Zm_{\text{H}} + Nm_{\text{n}} - m_{\text{atom}}) \times (931.49 \text{ MeV/u})$$

(30.3)

Nuclear binding energy for an atom of
mass m_{atom} with Z protons and N neutrons

EXAMPLE 30.1 Finding the binding energy of iron

What is the nuclear binding energy of ^{56}Fe?

PREPARE Appendix C gives the atomic mass of ^{56}Fe as 55.9349 u. Iron has atomic number 26, so an atom of ^{56}Fe could be separated into 26 hydrogen atoms and 30 neutrons. The mass of the separated components is more than that of the iron nucleus; the difference allows us to find the binding energy.

SOLVE We solve for the binding energy using Equation 30.3. The mass of the hydrogen atom and the neutron is given in Table 30.2. We find:

$$B = (26(1.0078 \text{ u}) + 30(1.0087 \text{ u}) - 55.9349 \text{ u})(931.49 \text{ MeV/u})$$

$$= (0.528 \text{ u})(931.49 \text{ MeV/u}) = 492 \text{ MeV}$$

ASSESS The difference in mass between the nucleus and its components is a small fraction of the mass of the nucleus, so we must use several significant figures in our mass values. The mass difference is small—about half that of a proton—but the energy equivalent, the binding energy, is enormous.

How much energy is 492 MeV? To make a comparison with another energy value we have seen, the binding energy of a single iron nucleus is equivalent to the energy released in the metabolism of nearly *2 billion* molecules of ATP! The energy scale of nuclear processes is clearly quite different from that of chemical processes. In nuclear reactors, a small amount of nuclear "fuel" can produce a tremendous amount of energy.

As A increases the nuclear binding energy increases, simply because there are more nuclear bonds. A more useful measure for comparing one nucleus to another is the quantity B/A, called the *binding energy per nucleon*. Iron, with $B = 492$ MeV and $A = 56$, has 8.79 MeV per nucleon. This is the amount of energy, on average, you would need to supply in order to remove *one* nucleon from the nucleus. Nuclei with larger values of B/A are more tightly held together than nuclei with smaller values of B/A.

A nuclear fusion weight-loss plan The sun's energy comes from reactions that combine four hydrogen atoms to create a single atom of helium—a process called **nuclear fusion.** Because energy is released, the mass of the helium atom is less than that of the four hydrogen atoms. As the fusion reactions continue, the mass of the sun decreases—by 130 trillion tons per year! That's a lot of mass, but given the sun's enormous size, this change will only amount to a few hundredths of a percent of the sun's mass over its ten-billion-year lifetime.

Activ Physics ONLINE   19.3

FIGURE 30.5 The curve of binding energy for the stable nuclei.

Figure 30.5 is a graph of the binding energy per nucleon versus mass number A. The dotted line connecting the points is often called the **curve of binding energy.**

For small values of A, adding more nucleons increases the binding energy per nucleon. If two light nuclei can be joined together to make a single, larger nucleus, the final nucleus will have a higher binding energy per nucleon. Because the final nucleus is more tightly bound, energy will be released in this *nuclear fusion* process. Nuclear fusion of hydrogen to helium is the basic reaction that powers the sun, as we have seen.

The curve of binding energy has a broad maximum at $A \approx 60$. For larger values of A, adding nucleons *decreases* the binding energy per nucleon. If you take a nucleus with $A \gg 60$ and break it into two smaller pieces, the two final nuclei will have a higher binding energy per nucleon. As the final nuclei are more tightly bound, energy will be released. Certain nuclei can be induced to split into two fragments through *nuclear fission,* which you first saw in Chapter 27. But, in fact, **all nuclei larger than a certain size will spontaneously fragment.** The structure of the curve of binding energy will be critical to our understanding of radioactivity.

◀ **A formidable chain reaction** Uranium and plutonium nuclei are so large that they have no stable isotopes; all isotopes of both elements undergo radioactive decay. Certain isotopes of uranium and plutonium will also support nuclear fission. A nucleus struck by a neutron can split into two more tightly bound fragments with the release of a great deal of energy. Extra neutrons are left over after the split, each of which can cause the fission of another nucleus—releasing more neutrons, which can produce further reactions. The net result is a nuclear fission **chain reaction.** A controlled reaction is the energy source of a nuclear power plant. An uncontrolled reaction is responsible for the terrible destructive power of a nuclear explosion, shown here in an above-ground test in Nevada in 1957.

30.3 Forces and Energy in the Nucleus

The nucleus of the atom is made of protons, which are positively charged, and neutrons, which have no charge. Why doesn't the repulsive force of the protons simply cause the nucleus to fly apart? This was a puzzle faced by early investigators. It soon became clear that a previously unknown force of nature operates within the nucleus to hold the nucleons together. This new force had to be stronger than the repulsive electrostatic force, hence it was named the **strong nuclear force,** or just the strong force.

The strong force has several important properties:

1. It is an *attractive* force between any two nucleons.
2. It does not act on electrons.
3. It is a *short-range* force, acting only over nuclear distances. We see no evidence for nuclear forces outside the nucleus.
4. Over the range where it acts, it is *stronger* than the electrostatic force that tries to push two protons apart.

Figure 30.6 summarizes the three interactions that take place within the nucleus. Many decades of research have shown that the strong force between two nucleons is independent of whether they are protons or neutrons. Charge is the basis for electromagnetic interactions, but it is of no relevance to the strong force. Protons and neutrons are identical as far as nuclear forces are concerned.

Protons throughout the nucleus exert repulsive electrostatic forces on each other, but, because of the short range of the strong force, a proton feels an attractive force only from the very few other protons with which it is in close contact. A nucleus with too many protons will be unstable, as the repulsive electrostatic forces will overcome the attractive strong forces. Because neutrons participate in the strong force but exert no repulsive forces, **the neutrons provide the extra**

The attractive strong force is the same for any two nucleons.

Two protons also experience a smaller electrostatic repulsive force.

FIGURE 30.6 The strong force is the same between any two nucleons.

"glue" that holds the nucleus together. In small nuclei, one neutron per proton is sufficient for stability. Hence small nuclei have $N \approx Z$. But as the nucleus grows, the repulsive force increases faster than the binding energy. More neutrons are needed for stability, causing heavy nuclei to have $N > Z$.

The Shell Model

Calculating a multielectron atom's energy levels is difficult because we must consider not only the electron-nucleus attraction but also the electron-electron repulsion. The problem of finding the allowed energy levels of the nucleus is even more complex. Nucleons attract other nucleons via the strong force while the protons exert repulsive electrostatic forces on each other. Even so, a **shell model** of the nucleus, using multielectron atoms as an analogy, was proposed in 1949 by Marie Goeppert-Mayer, one of the first prominent women in physics. The shell model considers each nucleon to move independently with an *average* potential energy due to the forces of all the other nucleons.

Figure 30.7 shows this average potential energy. For distances greater than the nuclear radius, the potential energy is zero. In other words, nuclear forces are *short range*, and a nucleon outside the nuclear radius is a free particle that does not interact with the nucleus. Inside the nucleus, the nucleon has an average potential energy of about -50 MeV, meaning the nucleon is a *bound particle*. You would have to supply 50 MeV of energy to remove it from the nucleus. Because the potential energy is constant while the total energy is conserved, the nucleon moves about inside the nucleus with constant kinetic energy. In essence, the nucleons in the nucleus are very similar to a three-dimensional version of the "particle in a box" that we considered in Chapter 28.

The task of quantum mechanics is to solve for the allowed energy levels of the nucleons. Once the energy levels are found, we build up the nuclear state, just as we did with atoms, by placing all the nucleons in the lowest energy levels consistent with the Pauli principle, which applies to nucleons just as it did electrons. Each energy level can hold only a certain number of spin-up particles and spin-down particles, depending on the quantum numbers. Additional nucleons must go into higher energy levels.

Marie Goeppert-Mayer shows her Nobel Prize.

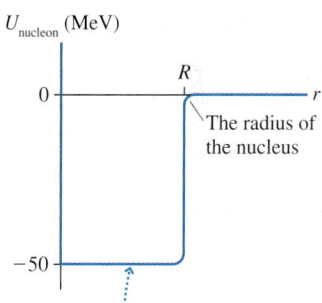

The potential energy is approximately constant up to the edge of the nucleus.

FIGURE 30.7 The average potential energy of a nucleon.

Low-Z Nuclei

As an example, we'll consider the energy levels of low-Z nuclei ($Z < 8$). Because these nuclei have so few protons, we can neglect the electrostatic potential energy due to proton-proton repulsion and consider only the much larger nuclear potential energy. In that case, the proton and neutron potential-energy wells and energy levels are the same.

Figure 30.8 shows the three lowest allowed energy levels for protons and neutrons and the maximum number of protons and neutrons the Pauli principle allows in each. Energy values vary from nucleus to nucleus, but the spacing between these levels is several MeV. (Contrast this with atoms, where the spacing between energy levels is usually several eV.)

Suppose we look at a series of nuclei, all with $A = 12$ but with different numbers of protons and neutrons: ^{12}B, ^{12}C, ^{12}N. All have 12 nucleons, but only ^{12}C is a stable isotope; the other two are not. Why do we see this difference in stability?

Figure 30.9 on the next page shows the energy-level diagrams of ^{12}B, ^{12}C, and ^{12}N. Look first at ^{12}C, a nucleus with six protons and six neutrons. You can see that exactly six protons are allowed in the $n = 1$ and $n = 2$ proton energy levels. The same is true for the six neutrons. No other arrangements of the nucleons would lower the total energy, so this nucleus is stable.

The proton potential energy is nearly identical to the neutron potential energy when Z is small.

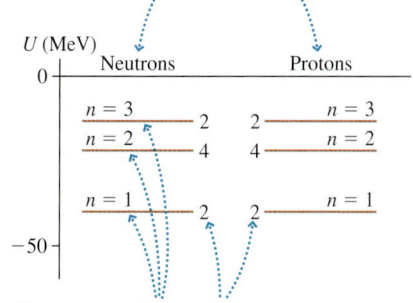

These are the first three allowed energy levels. They are spaced several MeV apart.

These are the maximum number of protons and neutrons allowed by the Pauli principle.

FIGURE 30.8 The three lowest energy levels of a low-Z nucleus. The neutron energy levels are on the left, the proton energy levels on the right.

A ^{12}C nucleus is in its lowest possible energy state.

Maximum numbers of neutrons and protons in each level.

A ^{12}B nucleus could lower its energy if a neutron could turn into a proton.

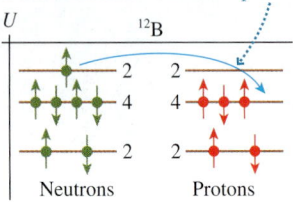

A ^{12}N nucleus could lower its energy if a proton could turn into a neutron.

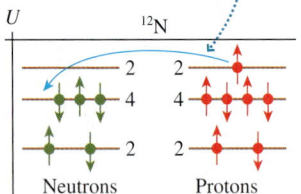

FIGURE 30.9 Nuclear energy-level diagrams of ^{12}B, ^{12}C, and ^{12}N.

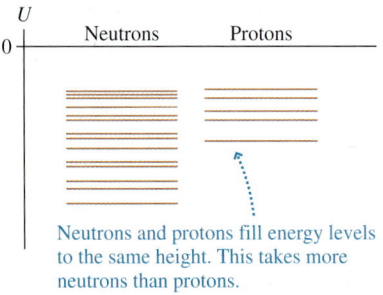

Neutrons and protons fill energy levels to the same height. This takes more neutrons than protons.

FIGURE 30.10 The proton energy levels are displaced upward in a high-Z nucleus.

TABLE 30.3 Three types of radiation

Radiation	Identification	Charge
Alpha, α	^{4}He nucleus	$+2e$
Beta, β	Electron	$-e$
Gamma, γ	High-energy photon	0

^{12}B has five protons and seven neutrons. The sixth neutron fills the $n = 2$ neutron shell, so the seventh neutron has to go into the $n = 3$ energy level. (This is just like the third electron in Li having to go to the $n = 2$ shell because the first two electrons have filled the $n = 1$ shell.) The $n = 2$ proton shell has one vacancy because there are only five protons.

In atoms, electrons in higher energy levels move to lower energy levels by emitting a photon as the electron undergoes a quantum jump. That can't happen here because the higher-energy nucleon in ^{12}B is a neutron whereas the vacant lower energy level is that of a proton. But an analogous process could occur *if a neutron could somehow turn into a proton*, allowing it to move to a lower energy level. ^{12}N is just the opposite, with the seventh proton in the $n = 3$ energy level. If a proton could somehow turn into a neutron, it could move to a lower energy level, lowering the energy of the nucleus.

You can see from the diagrams that the ^{12}B and ^{12}N nuclei have significantly more energy—by several MeV—than ^{12}C. If a neutron could turn into a proton, and vice-versa, these nuclei could move to a lower-energy state, that of ^{12}C. In fact, that's exactly what happens! These nuclei are not stable. We'll explore the details in Section 30.5, but both ^{12}B and ^{12}N decay into the more stable ^{12}C in the process known as *beta decay*.

High-Z Nuclei

We can use the shell model to give a qualitative explanation for one more observation. Figure 30.10 shows the neutron and proton energy levels of a high-Z nucleus. In a nucleus with many protons, the increasing electrostatic potential energy raises the proton energy levels but not the neutron energy levels. Protons and neutrons now have a different set of energy levels.

As a nucleus is "built," by the adding of protons and neutrons, the proton energy levels and the neutron energy levels must fill to just about the same height. If there were neutrons in energy levels above vacant proton levels, the nucleus would lower its energy by changing neutrons into protons. Similarly, a proton would turn into a neutron if there were a vacant neutron energy level beneath a filled proton level. **The net result is that the filled levels for protons and neutrons are at just about the same height.**

Because the neutron energy levels start at a lower energy, *more neutron states are available* than proton states. Consequently, a high-Z nucleus will have more neutrons than protons. This conclusion is consistent with our observation in Figure 30.2 that $N > Z$ for heavy nuclei.

> **STOP TO THINK 30.2**   Based on the model of nuclear shells and transitions you have seen, would you expect ^{13}C to be stable?

30.4 Radiation and Radioactivity

Becquerel's 1896 discovery of "rays" from crystals of uranium prompted a burst of activity. In England, J. J. Thomson and Ernest Rutherford worked to characterize the unknown rays. There were different rays with different penetrating abilities, as we noted. The different rays also experienced varying deflection by magnetic fields, as shown in Figure 30.11 on the next page, giving valuable clues as to their identities. Alpha and beta rays were determined to be *particles,* so we generally no longer use the word "ray" to describe them. Alpha particles are helium nuclei and beta particles are electrons. Gamma rays were found to be similar to x rays, high-energy electromagnetic waves best described as photons. Table 30.3 summarizes the properties of alpha and beta particles and gamma rays.

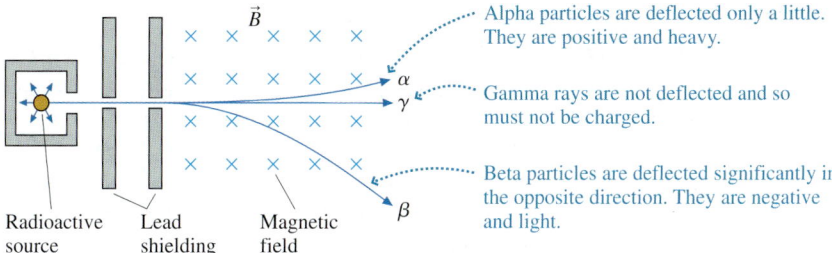

FIGURE 30.11 Identifying radiation by its deflection in a magnetic field.

Within a few years, as Rutherford and others deduced the basic structure of the atom, it also became clear that these emissions of radiation were coming from the atomic nucleus. We now define *radioactivity* or *radioactive decay* to be the spontaneous emission of particles or high-energy photons from unstable nuclei as they decay from higher-energy to lower-energy states. **Radioactivity has nothing to do with orbiting valence electrons.**

Nuclear Radiation Is a Form of Ionizing Radiation

The energies of the alpha and beta particles and the gamma-ray photons of nuclear decay are typically in the range 0.1–10 MeV. These energies are much larger than the ionization energies of atoms and molecules, which, as we saw in Chapter 25, are ≈10 eV. When the sun shines on your skin, it warms it; the low-energy photons are absorbed, their energy converted to thermal energy. The much higher energies of alpha and beta particles and gamma rays cause them to interact very differently with matter, *ionizing* atoms and *breaking* molecular bonds, leaving long trails of ionized atoms and molecules behind them before finally stopping. A particle with 1 MeV of kinetic energy can ionize ≈100,000 atoms or molecules. Alpha and beta particles and gamma rays are, like x rays, examples of ionizing radiation.

Ionization is the basis for the **Geiger counter,** the most well-known detector of nuclear radiation. Figure 30.12 shows how a Geiger counter detects the passage of a beta particle. A Geiger counter, like other nuclear radiation detectors, measures only *ionizing radiation.*

When ionizing radiation enters the body, it causes damage in two ways. First, the ions drive chemical reactions that wouldn't otherwise occur. These reactions may damage the machinery of cells. Very large doses of ionizing radiation upset the delicate ionic balance that drives cellular transport, and can rapidly lead to cell death. For this reason, large doses of penetrating gamma rays are sometimes used to sterilize medical equipment inside and out.

Second, ionizing radiation can damage DNA molecules by ionizing them and breaking bonds. If the damage is extensive, cellular repair mechanisms will not be able to cope, and the DNA will be permanently damaged, possibly creating a mutation or a tumor. Tissues with rapidly proliferating cells, such as bone marrow, are quite sensitive to ionizing radiation. Those with less-active cell reproduction, such as the nervous system, are much less sensitive.

NOTE ▶ Ionizing radiation causes damage to materials and tissues, but **objects irradiated with alpha, beta, or gamma radiation do not become radioactive.** Ionization drives chemical processes involving the electrons. An object could become radioactive only if its nuclei were somehow changed, and that does not happen. ◀

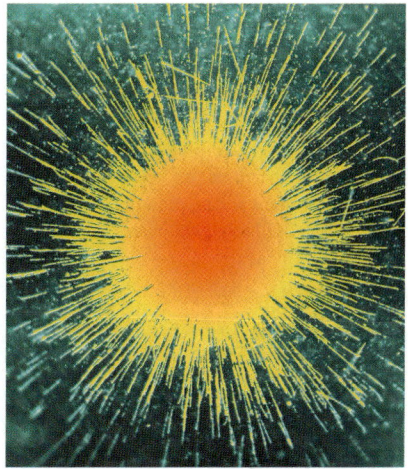

A speck of radium placed on a photographic plate emits alpha particles that leave clearly visible ionization trails.

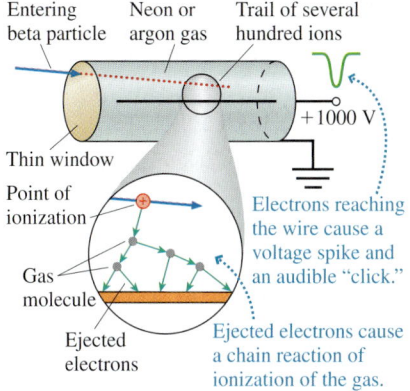

FIGURE 30.12 The operation of a Geiger counter.

Powered by decay The spontaneous radio-active decay of the nuclei in this plutonium pellet releases a lot of energy, warming the pellet enough that it glows. This pellet is intended for a thermoelectric generator, in which the heat of the pellet is used to produce electricity. The radioactive decay of short-lived isotopes of plutonium has provided power for spacecraft on voyages far from the sun. For many years, plutonium "batteries" were used to power heart pacemakers as well.

Nuclear Decay and Half-Lives

The decay of nuclei is different from other types of decay you are familiar with. A tree branch that falls to the forest floor decays. It darkens, becomes soft, and crumbles. You could tell just by looking at it about how long it had been decaying. Nuclear decay is different. The nucleus doesn't "age" in any sense. Instead, a nucleus has a certain probability that, within the next second, it will spontaneously turn into a different nucleus and, in the process, eject an alpha or beta particle or gamma ray. This spontaneous transformation of the nucleus is called *nuclear decay.*

If you toss a coin, it always has a 50% probability of showing tails, no matter what previous tosses might have been. You might "expect" a head if you've tossed 10 tails in a row, but the 11th toss still has a 50% chance of being a tail. Likewise with nuclei. If a nucleus doesn't decay in this second, it is no more or less likely to decay in the next. The nucleus remains just as it was, without any change, until the decay finally occurs. An "old" nucleus is identical to a "young" one.

Let's use the symbol r to represent the probability that one particular nucleus will decay in the next 1 s. For example, $r = 0.010$ s^{-1} means that a nucleus has a 1% chance of decay in the next second. Notice that r has units s^{-1}, making it a *rate;* we call r the **decay rate.**

The probability that a particular nucleus decays during the small interval of time Δt is

$$\text{Probability of decay(in } \Delta t) = r\Delta t \qquad (30.4)$$

If there are N independent nuclei, the number of nuclei expected to decay during Δt is

$$\text{number of decays} = N \times \text{Probability of decay} = rN\Delta t \qquad (30.5)$$

Each decay *decreases* the number of radioactive nuclei in the sample, hence the change in the number of radioactive nuclei during Δt is

$$\Delta N = -rN\Delta t \qquad (30.6)$$

The negative sign shows that N, the number of nuclei, decreases due to the decays. The *rate of change* in the number of radioactive nuclei is

$$\frac{\Delta N}{\Delta t} = -rN \qquad (30.7)$$

The rate of change depends both on the decay rate (a larger probability of decay means more decays per second) and on the number of radioactive nuclei present (more nuclei mean that more are available to decay). The rate of change, $\Delta N/\Delta t$, is negative because N is decreasing.

We can also characterize the decay by a *time constant τ* that is the inverse of the rate:

$$\tau = \frac{1}{r} \qquad (30.8)$$

Because r has units s^{-1}, the units of τ are s; τ thus represents an interval of time.

Suppose we have a sample of N_0 radioactive nuclei at a time $t = 0$. How many nuclei will still be present, undecayed, at a later time t? To answer this question we would need to "solve" Equation 30.7. The solution, beyond the scope of this book, is an *exponential decay* of the number of nuclei:

$$N = N_0 e^{-t/\tau} \qquad (30.9)$$

Exponential decay of nuclei in a radioactive sample

EXPONENTIAL
p.471

Notice that the exponential decay of the number of nuclei is analogous to the exponential decay of the capacitor voltage in an *RC* circuit or the amplitude of a damped harmonic oscillator.

Figure 30.13 shows the decrease of N with time. The number of radioactive nuclei decreases from N_0 at $t = 0$ to $N = N_0 e^{-1} = 0.37N_0$ at time $t = \tau$. In practical terms, the number decreases by roughly two-thirds during one time constant.

NOTE ▶ There is no natural "starting time" for an exponential decay; you can choose any instant you wish to be $t = 0$. The number of radioactive nuclei present at that instant is N_0. If at one instant you have 10,000 radioactive nuclei whose time constant is $\tau = 10$ min, you'll have roughly 3700 nuclei 10 min later. The fact that you may have had more than 10,000 nuclei at an earlier time isn't relevant. ◀

If you look up properties of radioactive nuclei, you won't find the decay rate or the time constant listed. Instead, you will see a much more convenient measure, the *half-life*. The **half-life,** symbol $t_{1/2}$, is the time interval in which half of a sample of radioactive atoms decays. This time is shown on the decay graph in Figure 30.13.

We can relate the half-life to the time constant τ because we know, by definition, that $N = \frac{1}{2}N_0$ at $t = t_{1/2}$. Thus, according to Equation 30.9

$$\frac{N_0}{2} = N_0 e^{-t_{1/2}/\tau}$$

The N_0 cancels, and we can then take the natural logarithm of both sides to find

$$\ln\left(\frac{1}{2}\right) = -\ln 2 = -\frac{t_{1/2}}{\tau}$$

With one final rearrangement we have

$$t_{1/2} = \tau \ln 2 = 0.693\tau \tag{30.10}$$

As the decay rate r is the inverse of the time constant, we find

$$r = \frac{1}{\tau} = \frac{0.693}{t_{1/2}} \tag{30.11}$$

A shorter half-life implies a larger decay rate, as we would expect.

We can use the definition of the half-life to formulate an alternative expression for Equation 30.9:

$$N = N_0\left(\frac{1}{2}\right)^{t/t_{1/2}} \tag{30.12}$$

Decay of nuclei in a radioactive sample in terms of half-life

Thus $N = N_0/2$ at $t = t_{1/2}$, $N = N_0/4$ at $t = 2t_{1/2}$, $N = N_0/8$ at $t = 3t_{1/2}$, and so on. **No matter how many nuclei there are, the number decays by half during the next half-life.** This is shown graphically in Figure 30.14. This figure also conveys two other important ideas:

1. Nuclei don't vanish when they decay. The decayed nuclei have merely become some other kind of nuclei.
2. The decay process is random. We can predict that half the nuclei will decay in one half-life, but we can't predict which ones.

NOTE ▶ Depending on the isotope, half-lives may be expressed in seconds, minutes, hours, days, or years. In Equation 30.12, the times appear as a ratio, so any of these units will do as long as all times are in the same units. ◀

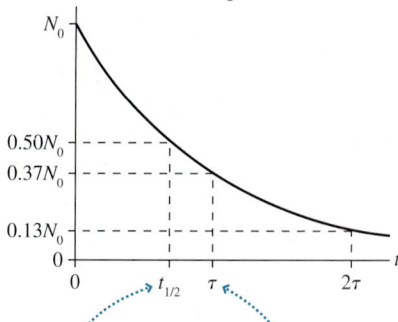

The half-life is the time in which half the nuclei decay.

The time constant is the time at which the number of nuclei is e^{-1}, or 37%, of the initial number.

FIGURE 30.13 The number of radioactive atoms decreases exponentially with time.

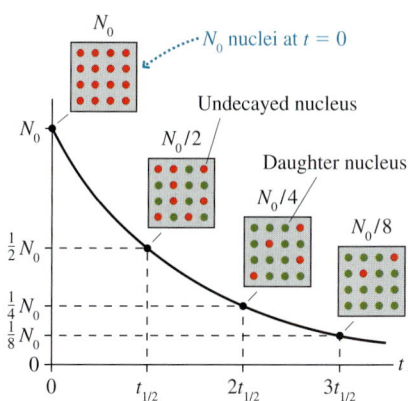

FIGURE 30.14 Half the nuclei decay during each half-life.

EXAMPLE 30.2 **Determining the decay of radioactive iodine**

Patients with Graves disease have an overactive thyroid gland. A common treatment uses radioactive iodine, which is taken up by the thyroid. The radiation emitted in its decay will destroy the tissues of the gland. A single pill is produced with 4.0×10^{14} atoms of the isotope ^{131}I, which has a half-life of 8.0 days.

a. How many atoms remain 24 hours after the pill's creation, when the pill is delivered to a hospital?
b. Although the iodine in the pill is constantly decaying, it is still usable as long as it contains at least 1.1×10^{14} ^{131}I atoms. What is the maximum delay before the pill is no longer usable?

PREPARE The atoms in the sample undergo exponential decay, decreasing steadily in number.

SOLVE a. The half-life is $t_{1/2} = 8.0$ days $= 192$ h. Using Equation 30.12, we can find the number of atoms remaining after 24 h have elapsed,

$$N = (4.0 \times 10^{14})\left(\frac{1}{2}\right)^{24/192} = 3.7 \times 10^{14} \text{ atoms}$$

b. The time after which 1.1×10^{14} atoms remain is given by

$$1.1 \times 10^{14} = (4.0 \times 10^{14})\left(\frac{1}{2}\right)^{t/192}$$

To solve for t, write this as

$$\frac{1.1 \times 10^{14}}{4.0 \times 10^{14}} = \left(\frac{1}{2}\right)^{t/192}$$

or

$$0.275 = \left(\frac{1}{2}\right)^{t/192}$$

Now, take the natural logarithm of both sides:

$$\ln(0.275) = \ln\left(\left(\frac{1}{2}\right)^{t/192}\right)$$

We can solve for t by using the fact that $\ln(a^x) = x\ln(a)$. This allows us to "pull out" the $t/192$ exponent to find

$$\ln(0.275) = \left(\frac{t}{192}\right)\ln\left(\frac{1}{2}\right)$$

$$t = 192\frac{\ln(0.275)}{\ln(1/2)} = 360 \text{ h} = 15 \text{ days}$$

ASSESS The weakest usable concentration of iodine is approximately 1/4 of the initial concentration. This means that the decay time should be approximately equal to 2 half-lives, which is what we found.

Activity

The **activity** R of a radioactive sample is the number of decays per second. Each decay corresponds to an alpha, beta, or gamma emission, so the activity is also a measure of how much radiation is being given off. Equation 30.7 gave the rate of decay of the nuclei in a sample; the number of decays is just the absolute value of this expression:

$$R = rN = \frac{0.693N}{t_{1/2}} \tag{30.13}$$

The activity is inversely proportional to the half-life. Isotopes with shorter half-lives have higher activities. We can combine Equation 30.13 with Equation 30.9 to obtain an expression for the variation of activity with time:

$$R = rN = rN_0 e^{-t/\tau} = R_0 e^{-t/\tau} = R_0\left(\frac{1}{2}\right)^{t/t_{1/2}} \tag{30.14}$$

where $R_0 = rN_0$ is the activity at $t = 0$. This equation has the same form as that for the decay of the sample. **The activity of a sample decreases exponentially** along with the number of remaining nuclei.

The SI unit of activity is the **becquerel,** defined as

$$1 \text{ becquerel} = 1 \text{ Bq} = 1 \text{ decay/s or } 1 \text{ s}^{-1}$$

An older unit of activity, but one that continues in widespread use, is the **curie.** The conversion factor is

$$1 \text{ curie} = 1 \text{ Ci} = 3.7 \times 10^{10} \text{ Bq}$$

1 Ci is a substantial amount of radiation. The radioactive samples used in laboratory experiments are typically $\approx 1\ \mu Ci$, or, equivalently, $\approx 40,000$ Bq. These samples can be handled with only minor precautions. Larger sources of activity require thick shielding and other special precautions to prevent exposure to high levels of radiation.

CONCEPTUAL EXAMPLE 30.1 Relative activities of isotopes in the body

^{40}K ($t_{1/2} = 1.3 \times 10^9$ yr) and ^{14}C ($t_{1/2} = 5.7 \times 10^3$ yr) are two radioactive isotopes found in measurable quantities in your body. Suppose you have one mole of each. Which is more radioactive—that is, which has a greater activity?

REASON Equation 30.13 shows that the activity of a sample is proportional to the number of atoms and inversely proportional to the half-life. Because both samples have the same number of atoms, the sample of ^{14}C, with its much shorter half-life, has a much greater activity.

EXAMPLE 30.3 Determining the decay of activity

A ^{60}Co (half-life 5.3 yr) source used to provide gamma rays to irradiate tumors has an activity of 0.43 Ci.

a. How many ^{60}Co atoms are in the source?
b. What will be the activity of the source after 10 yr?

PREPARE The activity of the source depends on the half-life and the number of atoms. If we know the activity and the half-life, we can compute the number of atoms. The number of atoms will exponentially decay, and so will the activity.

SOLVE a. Equation 30.13 relates the activity and the number of atoms. Rewriting the equation, we can relate the initial activity and the initial number of atoms as $N_0 = t_{1/2}R_0/0.693$. To use this equation, we need numbers in SI units. In Bq, the initial activity of the source is

$$R_0 = (0.43\ Ci)\left(\frac{3.7 \times 10^{10}\ Bq}{1\ Ci}\right) = 1.6 \times 10^{10}\ Bq$$

The half-life $t_{1/2}$ in s is

$$t_{1/2} = (5.3\ yr)\left(\frac{3.15 \times 10^7\ s}{1\ yr}\right) = 1.7 \times 10^8\ s$$

Thus the initial number of ^{60}Co atoms in the source is

$$N_0 = \frac{t_{1/2}R_0}{0.693} = \frac{(1.7 \times 10^8\ s)(1.6 \times 10^{10}\ Bq)}{0.693}$$

$$= 3.9 \times 10^{18}\ atoms$$

b. The variation of activity with time is given by Equation 30.14. After 10 yr, the activity is

$$R = R_0\left(\frac{1}{2}\right)^{t/t_{1/2}} = (1.6 \times 10^{10}\ Bq)\left(\frac{1}{2}\right)^{10/5.3} = 4.3 \times 10^9\ Bq$$

ASSESS Although N_0 is a very large number, it is a very small fraction of a mole. The sample contains only about 400 μg of ^{60}Co. In the first part of the question we needed to convert the half-life to s. The second part of the question used a ratio of two times, so we can use any units we like as long as the units of both times are the same.

Radioactive Dating

Many geological and archeological samples can be dated by measuring the decays of naturally occurring radioactive isotopes.

The most well-known dating technique uses the radioactive carbon isotope ^{14}C and is known as carbon dating or **radiocarbon dating.** ^{14}C has a half-life of 5730 years, so any ^{14}C present when the earth formed 4.5 billion years ago has long since decayed away. Nonetheless, ^{14}C is present in atmospheric carbon dioxide because high-energy cosmic rays collide with gas molecules high in the atmosphere to create ^{14}C nuclei from nuclear reactions with nitrogen and oxygen nuclei. The creation and decay of ^{14}C have reached a steady state in which the $^{14}C/^{12}C$ ratio is relatively stable at 1.3×10^{-12}.

All living organisms constantly exchange carbon dioxide with the atmosphere, so the $^{14}C/^{12}C$ ratio in living organisms is also 1.3×10^{-12}. As soon as an organism dies, the ^{14}C in its tissue begins to decay and no new ^{14}C is added. As time goes on, the ^{14}C decays at a well-known rate. Thus, a measurement of the activity of an ancient organic sample permits a determination of the age. "New" samples have a higher fraction of ^{14}C than "old" samples.

A researcher is extracting a small sample of an ancient bone. By measuring the carbon isotopes present in the sample she will determine the age of the bone.

The first step in radiocarbon dating is to extract and purify carbon from the sample. The carbon is then placed in a shielded chamber and its activity measured. This activity is then compared to the activity of an identical modern sample. Equation 30.14 relates the activity of a sample at a time t to its initial activity. If we assume that the original activity R_0 was the same as the activity of the modern sample, we can determine the time since the decay began—and thus the age of the sample.

EXAMPLE 30.4 Carbon dating a tooth

A rear molar from a mammoth skeleton is dated using a measurement of its ^{14}C content. Carbon from the tooth is chemically extracted and formed into benzene. The benzene sample is placed in a shielded chamber. Decays from the sample come at an average rate of 11.5 counts per minute. A modern benzene sample of the exact same size gives 54.9 counts per minute. What is the age of the skeleton?

PREPARE We can assume that, thousands of years ago, the sample had an initial activity of 54.9 counts per minute—the activity of a modern sample. The present activity is lower due to the decay of the ^{14}C since the death of the mammoth.

SOLVE Equation 30.14 gives the decrease of the activity as a function of time as $R = R_0(1/2)^{t/t_{1/2}}$. The current activity is $R = 11.5$ counts per minute, and we assume that the initial activity was $R_0 = 54.9$. t is the time since the mammoth stopped growing—the age of the skeleton. We solve for t by

rearranging terms and computing a natural logarithm, as in Example 30.2:

$$\frac{R}{R_0} = \left(\frac{1}{2}\right)^{t/t_{1/2}}$$

$$\ln\left(\frac{R}{R_0}\right) = \left(\frac{t}{t_{1/2}}\right)\ln(1/2)$$

We then solve for the time t:

$$t = \frac{t_{1/2}}{\ln(1/2)}\ln\left(\frac{R}{R_0}\right) = \frac{5730\ \text{yr}}{\ln(1/2)}\ln\left(\frac{11.5}{54.9}\right) = 12,900\ \text{yr}$$

ASSESS The final time is in years, the same unit as we used for the half-life. This is a realistic example of how such radiocarbon dating is done; the numbers and details used in this example come from an actual experimental measurement.

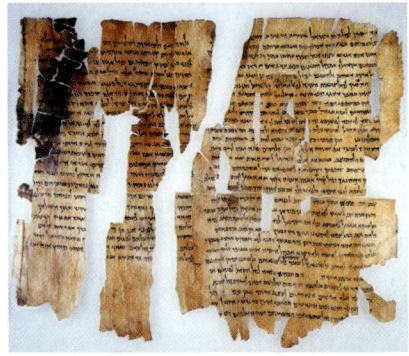

Responsible dating BIO Measuring the activity of the carbon in a sample to determine the fraction of ^{14}C can require a significant amount of organic material—perhaps 25 g. For an artifact of great historical importance, such as this parchment fragment from the Dead Sea scrolls, this would be unacceptable. Instead, dates are obtained by using a mass spectrometer to directly measure the ratio of carbon isotopes, from which the age can be determined. This can be done with excellent accuracy on as little as 0.1 g of material.

CONCEPTUAL EXAMPLE 30.2 Source contamination

One possible problem with carbon dating is contamination with modern carbon sources. Suppose an archaeologist has unearthed and carbon-dated a fragment of wood that has absorbed carbon of recent vintage from organic molecules in groundwater. Does he underestimate or overestimate the age of the wood?

REASON Because the wood has absorbed modern carbon, it will have more ^{14}C than it would had it decayed undisturbed. The present activity is higher than it would otherwise be. This will lead to an underestimate of the age of the wood.

Carbon dating can be used to date skeletons, wood, paper, fur, food material, and anything else made of organic matter. It is quite accurate for ages to about 15,000 years, about 3 half-lives. Beyond that, the difficulty of measuring the small remaining fraction of ^{14}C and some uncertainties about the cosmic ray flux in the past combine to decrease the accuracy. Even so, items are dated to about 50,000 years with a fair degree of reliability.

Isotopes with longer half-lives are used to date geological samples. Potassium-argon dating, using ^{40}K with a half-life of 1.25 billion years, is especially useful for dating rocks of volcanic origin.

STOP TO THINK 30.3 A sample starts with 1000 radioactive atoms. How many half-lives have elapsed when 750 atoms have decayed?

A. 0.25   B. 1.5   C. 2.0   D. 2.5

30.5 Nuclear Decay Mechanisms

Early investigators into the phenomenon of radioactivity discovered that a radioactive sample in an evacuated container was soon accompanied by helium gas. No helium had been present before, so something in the decay had produced it. The helium was discovered to be the accumulated alpha particles from many decays. The idea that atoms of one element could spawn atoms of another element was quite surprising, and was a very exciting development in the understanding of the nature of the atom.

Actlv
Physlcs   19.4

In this section, we will explore the mechanisms that result in the three different forms of radioactive decay, considering the changes in the nuclei that result from these decays.

Alpha Decay

An alpha particle, symbolized as α, is a ^{4}He nucleus, a strongly bound system of two protons and two neutrons. An unstable nucleus that ejects an alpha particle will lose two protons and two neutrons, so we can write the decay as

$$^A_Z X \rightarrow \, ^{A-4}_{Z-2} Y + \alpha + \text{energy} \qquad (30.15)$$

Alpha decay of a nucleus

Figure 30.15 shows the alpha-decay process. The original nucleus X is called the **parent nucleus** and the decay-product nucleus Y is the **daughter nucleus.**

Energy conservation tells us that an alpha decay can occur only when the mass of the parent nucleus is greater than the mass of the daughter nucleus plus the mass of the alpha particle. This requirement is met for heavy, high-Z nuclei beyond the maximum of the curve of binding energy of Figure 30.5. It is energetically favorable for these nuclei to eject an alpha particle because the daughter nucleus is more tightly bound than the parent nucleus.

The daughter nucleus, which is much more massive than an alpha particle, undergoes only a slight recoil, as we see in Figure 30.15. Consequently, **the energy released in an alpha decay ends up mostly as the kinetic energy of the alpha particle.** We can compute this energy by looking at the mass energy difference between the initial and final states:

$$K_\alpha \approx \Delta E = (m_X - m_Y - m_{He})c^2 \qquad (30.16)$$

Before:

$^A_Z X$ Parent nucleus

The alpha particle, a fast helium nucleus, carries away most of the energy released in the decay.

After:

$^{A-4}_{Z-2} Y$

The daughter nucleus has two fewer protons and four fewer nucleons. It has a small recoil.

FIGURE 30.15 Alpha decay.

EXAMPLE 30.5 Analyzing alpha decay in a smoke detector

Americium, atomic number 95, doesn't exist in nature; it is produced in nuclear reactors. An isotope of americium, ^{241}Am, is part of the sensing circuit in most smoke detectors. ^{241}Am decays with a half-life of 432 yr by emitting an alpha particle. What is the daughter nucleus?

SOLVE Equation 30.15 shows that an alpha decay causes the atomic number to decrease by 2 and the atomic weight by 4. Let's write an equation for the decay showing the alpha particle as a helium nucleus, including the atomic weight superscript and the atomic number subscript for each element. There is no change in the total number of neutrons or protons, so the subscripts and superscripts must "balance" in the reaction:

$$^{241}_{95}\text{Am} \rightarrow \, ^{237}_{93}? + \, ^4_2\text{He} + \text{energy}$$

A quick glance at the periodic table reveals the unknown element in this equation, the daughter nucleus, to be an isotope of neptunium, $^{237}_{93}$Np.

ASSESS Balancing the two sides of the above reaction is similar to balancing the equation for a chemical reaction. Because the reaction is spontaneous, the daughter nucleus will be more stable than the parent; ^{237}Np has a half-life of over 2 million years.

EXAMPLE 30.6 Finding the energy of an emitted alpha particle

The uranium isotope ^{238}U undergoes alpha decay to an isotope of thorium, ^{234}Th. What is the kinetic energy, in MeV, of the alpha particle?

PREPARE The energy of the alpha particle depends on the change in mass in the reaction. Appendix C gives the atomic mass of ^{238}U as 238.0508 u and that of ^{234}Th as 234.0436 u. The atomic mass of helium, from Table 30.2, is 4.0026 u.

SOLVE We can calculate the kinetic energy of the alpha particle using Equation 30.16:

$$K_\alpha = (238.0508 \text{ u} - 234.0436 \text{ u} - 4.0026 \text{ u})c^2$$

$$= (0.0046 \text{ u})c^2$$

If we convert from u to MeV/c^2, using the conversion factor 1 u = 931.49 MeV/c^2, the c^2 cancels, and we end up with

$$K_\alpha = \left(0.0046 \text{ u} \times \frac{931.49 \text{ MeV/}c^2}{1 \text{ u}}\right)c^2 = 4.3 \text{ MeV}$$

ASSESS This is a typical alpha-particle energy, corresponding to a speed of about 5% of the speed of light. Notice that with a careful use of conversion factors we never had to evaluate c^2.

Beta Decay

Beta decay was initially associated with the emission of an electron e$^-$, the beta particle. It was later discovered that some nuclei can undergo beta decay by emitting a positron e$^+$, the antiparticle of the electron, although this decay mode is not as common. As we saw in Chapter 28, a positron is identical to an electron except that it has a positive charge. To be precise, the emission of an electron is called *beta-minus decay* and the emission of a positron is *beta-plus decay*.

A typical example of beta-minus decay occurs in the carbon isotope ^{14}C, which undergoes the beta-decay process ^{14}C → ^{14}N + e$^-$. Carbon has $Z = 6$ and nitrogen has $Z = 7$. Because Z increases by 1 but A doesn't change, it appears that a neutron within the nucleus has changed itself into a proton and electron. That is, the basic beta-minus decay process appears to be

$$n \rightarrow p^+ + e^- \tag{30.17}$$

The electron is ejected from the nucleus but the proton is not. Thus the decay process, shown in Figure 30.16a, is

$$^A_Z X \rightarrow \,_{Z+1}^{\,A} Y + e^- + \text{energy} \tag{30.18}$$

Beta-minus decay of a nucleus

Do neutrons *really* turn into protons? It turns out that a free neutron—one not bound in a nucleus—is *not* a stable particle. It decays into a proton and an electron, with a half-life of approximately 10 minutes. This decay conserves energy, because $m_n > m_p + m_e$. Furthermore, it conserves charge.

Whether a neutron *within* a nucleus can decay depends not only on the masses of the neutron and proton but also on the masses of the parent and daughter nuclei, because energy has to be conserved for the entire nuclear system. **Beta decay occurs only if $m_X > m_Y$.** ^{14}C can undergo beta decay to ^{14}N because $m(^{14}\text{C}) > m(^{14}\text{N})$. But $m(^{12}\text{C}) < m(^{12}\text{N})$, so ^{12}C is stable and its neutrons will not decay.

Beta-plus decay is the conversion of a proton into a neutron and a positron:

$$p^+ \rightarrow n + e^+ \tag{30.19}$$

The full decay process, shown in Figure 30.16b, is

$$^A_Z X \rightarrow \,_{Z-1}^{\,A} Y + e^+ + \text{energy} \tag{30.20}$$

Beta-plus decay of a nucleus

(a) Beta-minus decay

Before: $^A_Z X$

A neutron changes into a proton and an electron. The electron is ejected from the nucleus.

After: $^A_{Z+1} Y$ e$^-$

(b) Beta-plus decay

Before: $^A_Z X$

A proton changes into a neutron and a positron. The positron is ejected from the nucleus.

After: $^A_{Z-1} Y$ e$^+$

FIGURE 30.16 Beta decay.

Beta-plus decay does *not* happen for a free proton because $m_p < m_n$. It *can* happen within a nucleus as long as energy is conserved for the entire nuclear system, but it is far less common than beta-minus decay.

In our earlier discussion of Section 30.3 we noted that the ^{12}B and ^{12}N nuclei could reach a lower energy state if a proton could change into a neutron and vice versa. Now we see that such changes can occur if the energy conditions are favorable. And, indeed, ^{12}B undergoes beta-minus decay to ^{12}C while ^{12}N undergoes beta-plus decay to ^{12}C.

In general, beta decay is a process of nuclei with too many neutrons or too many protons that moves them closer to the line of stability in Figure 30.2.

NOTE ▶ The electron emitted in beta decay has nothing to do with the atom's valence electrons. The beta particle is created in the nucleus and ejected directly from the nucleus when a neutron is transformed into a proton and an electron. ◀

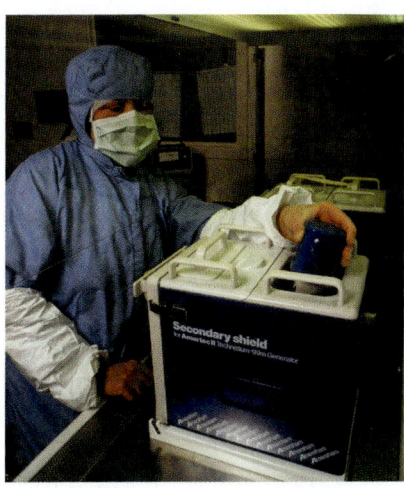

Isotopes on demand BIO Many hospitals keep a radioactive molybdenum ^{99}Mo source on hand. The ^{99}Mo undergoes beta decay, but it's not the energy or the electron released in the decay that the hospital wants—it's the technetium daughter nucleus, ^{99}Tc, which is used in medical imaging procedures. ^{99}Tc has a short 6.0 hr half-life, so hospitals that use it must produce it on site. This happens in a radioactive "cow," a column with chemically bound ^{99}Mo. The column is inside a shielded enclosure, which protects the technician from ionizing radiation as she extracts the ^{99}Tc produced in the decay of the ^{99}Mo source.

EXAMPLE 30.7 Analyzing beta decay in the human body

Your body contains several radioactive isotopes. Approximately 20% of the radiation dose you receive each year comes from the radioactive decay of these atoms. Most of this dose comes from one potassium isotope, ^{40}K, which decays by beta-minus emission. What is the daughter nucleus?

SOLVE Rewriting Equation 30.18 as

$$^{40}_{19}K \rightarrow ^{40}_{20}? + e^- + energy$$

we see that the daughter nucleus must be the calcium isotope $^{40}_{20}$Ca.

Gamma Decay

Gamma decay is similar to quantum processes you saw in previous chapters. In Chapter 28, you learned that an atomic system can emit a photon with $E_{photon} = \Delta E_{atom}$ when an electron undergoes a quantum jump from an excited energy level to a lower energy level. Nuclei are no different. A proton or a neutron in an excited nuclear state, such as the one shown in Figure 30.17, can undergo a quantum jump to a lower-energy state by emitting a high-energy photon. This is the gamma-decay process.

The spacing between atomic energy levels is only a few eV. Nuclear energy levels, by contrast, are on the order of 1 MeV apart, meaning gamma ray photons will have energies $E_{gamma} \sim 1$ MeV. Photons with this much energy are quite penetrating and deposit an extremely large amount of energy at the point where they are finally absorbed.

Nuclei left to themselves are usually in their ground states and thus cannot emit gamma-ray photons. However, alpha and beta decay often leave the daughter nucleus in an excited nuclear state, so gamma emission is often found to accompany alpha and beta emission.

Let's look at an example. One of the most important isotopes for medical imaging is ^{99}Tc, an isotope of the element technetium. An excited state of ^{99}Tc is produced in the beta decay of the molybdenum isotope ^{99}Mo:

$$^{99}Mo \rightarrow ^{99}Tc^* + e^- + energy \qquad (t_{1/2} = 66 \text{ hours})$$

The asterisk signifies that the technetium nucleus is in an excited state. The excited nucleus then makes a transition to a lower-energy state via the emission of a 140 keV gamma ray:

$$^{99}Tc^* \rightarrow ^{99}Tc + \gamma \qquad (t_{1/2} = 6.0 \text{ hours})$$

FIGURE 30.17 Gamma decay.

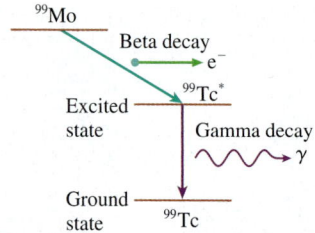

FIGURE 30.18 ^{99}Tc*, a gamma emitter, is produced in the beta decay of ^{99}Mo.

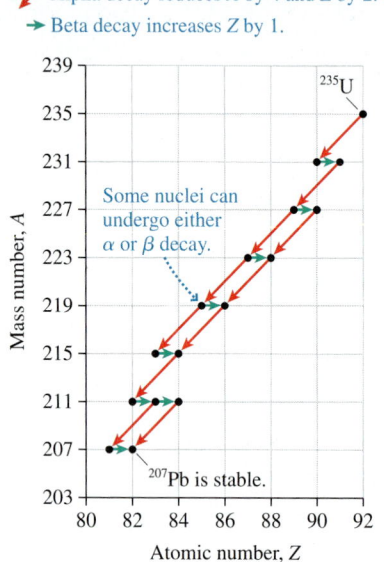

FIGURE 30.19 The nuclear decay series of ^{235}U.

19.5 Activ Physics ONLINE

The full decay process is shown in Figure 30.18. The final state of the technetium nucleus is much more stable than the excited state; it undergoes beta decay with a half-life of 211,000 years.

Decay Series

A radioactive nucleus decays into a daughter nucleus. In many cases, the daughter nucleus is also radioactive and decays to produce its own daughter nucleus. The process continues until reaching a daughter nucleus that is stable. The sequence of isotopes, starting with the original unstable isotope and ending with the stable isotope, is called a **decay series.**

The elements with $Z > 83$ present in the earth's crust are part of the decay series of a few long-lived isotopes of uranium and thorium. As an example, Figure 30.19 shows the decay series of ^{235}U, an isotope of uranium with a 700-million-year half-life. This is a very long time, but it is only about 15% the age of the earth, so most—but not all—of the ^{235}U nuclei present when the earth was formed have now decayed. There are many unstable nuclei in the decay series. Ultimately, all ^{235}U nuclei end as the ^{207}Pb isotope of lead, a stable nucleus.

Notice that some nuclei can decay by either alpha *or* beta decay. Thus there are a variety of paths that a decay can follow, but they all end at the same point.

STOP TO THINK 30.4 The cobalt isotope ^{60}Co ($Z = 27$) decays to the nickel isotope ^{60}Ni ($Z = 28$). The decay process is

A. Alpha decay. B. Beta-minus decay.
C. Beta-plus decay. D. Gamma decay.

30.6 Subatomic Particles

When we presented beta decay, we did so as if it were perfectly normal for one kind of matter to change spontaneously into a completely different kind of matter, simply because the conservation of energy allows it. But there must be more to the story! It would be energetically favorable for a large truck to spontaneously turn into a Cadillac and a VW Beetle, ejecting the Beetle at high speed. But this doesn't happen. Once you stop to think of it, the process n → p$^+$ + e$^-$ seems ludicrous, not because it violates mass-energy conservation but because we have no idea *how* a neutron could turn into a proton. We're suggesting that one of the basic building blocks of matter can somehow morph into a different basic building block.

But are protons and neutrons *really* basic building blocks? Molecules are made of atoms, atoms are made of a cloud of electrons surrounding a positively charged nucleus. The nucleus is composed of protons and neutrons. Where does this process end? Are electrons, protons, and neutrons the basic building blocks of matter, or are they made of still smaller subunits?

This question takes us into the domain of what is known as **particle physics**—the branch of physics that deals with the basic constituents of matter and the relationships among them. Particle physics starts with the constituents of the atom, the proton, neutron, and the electron, but there are many other particles below the scale of the atom. We call these particles **subatomic particles.**

Antiparticles

The positron, which you met in Chapter 28, is the *antiparticle* to the electron. You also learned that protons and neutrons also have associated antiparticles; in fact, so do all subatomic particles. A fundamental symmetry of nature is that **every** subatomic particle has an antiparticle that is opposite to it in every key way. The notation to represent an antiparticle is a bar over the top of the symbol. A proton is represented as p, an antiproton as p̄.

Antiparticles provide interesting opportunities for creating "exotic" subatomic particles. When a particle meets its associated antiparticle, the two annihilate, leaving nothing but their energy behind. This energy must go somewhere. Sometimes it is emitted as gamma-ray photons, but this energy can also be used to create other particles.

The major tool for creating and studying subatomic particles is the *particle accelerator.* An accelerator uses electric and magnetic fields to accelerate particles and their antiparticles, such as e and ē, or p and p̄, to speeds very close to the speed of light. These particles then collide head on. As they collide and annihilate, their mass-energy and kinetic energy combine to produce new, quite exotic particles that are not part of ordinary matter. These particles come in a dizzying variety—pions, kaons, lambda particles, sigma particles, and dozens of others—each with its own antiparticle. Most live no more than a trillionth of a second.

Subatomic crash tests The above picture is the record of a collision between an electron and a positron (paths represented by the green arrows) brought to high speeds in an accelerator. The particles collided in the center of a detector that measured the paths of the particles produced in the collision. In this case, the annihilation of the electron and positron created a particle known as a Z boson. The Z boson then quickly decayed into the two jets of particles seen coming out of the detector.

EXAMPLE 30.8 Determining a possible outcome of a proton-antiproton collision

When a proton and an antiproton annihilate, the resulting energy can be used to create new particles. One possibility is the creation of electrically neutral particles called *neutral pions.* A neutral pion has a rest mass of 135 MeV/c^2. How many neutral pions could be produced in the annihilation of a proton and an antiproton? Assume the proton and antiproton are moving very slowly as they collide.

PREPARE The mass of a proton is given in Table 30.2 as 938 MeV/c^2. The mass of an antiproton is the same. Because the proton and antiproton are moving slowly, with essentially no kinetic energy, the total energy available for creating new particles is the energy equivalent of the masses of the proton and the antiproton.

SOLVE The total energy from the annihilation of a proton and an antiproton is the energy equivalent of their masses:

$$E = (m_{proton} + m_{antiproton})c^2 = (938 \text{ MeV}/c^2 + 938 \text{ MeV}/c^2)c^2 = 1876 \text{ MeV}$$

It takes 135 MeV to create a neutral pion. The ratio

$$\frac{\text{Energy available}}{\text{Energy required to create a pion}} = \frac{1876 \text{ MeV}}{135 \text{ MeV}} = 13.9$$

tells us that we have enough energy to produce 13 neutral pions from this process, but not quite enough to produce 14.

ASSESS Because the mass of a pion is much less than that of a proton or an antiproton, the annihilation of a proton and antiproton can produce many more pions than the number of particles at the start. Though the production of 13 neutral pions is a possible outcome of a proton-antiproton interaction, it is not a likely one. In addition to the conservation of energy, there are many other physical laws that determine what types of particles, and in what quantities, are likely to be produced.

Neutrinos

The most abundant particle in the universe is not the electron, proton, or neutron; it is a particle you may have never heard of. Vast numbers of these particles pass through your body every day, only in extremely rare cases leaving any trace of their passage. This elusive particle is the *neutrino,* a neutral, nearly massless particle that interacts only weakly with matter.

Some of the first studies of beta decay, in the 1930s, found that neither energy nor momentum seemed to be conserved in the process. The Italian physicist Enrico Fermi correctly suggested that, in addition to the electron, the beta-decay

A big detector for a small particle The rubber raft in the photo is floating inside a particle detector designed to measure neutrinos. Neutrinos are so weakly interacting that a neutrino produced in a nuclear reaction in the center of the sun will likely pass through the entire mass of the sun and escape. Of course, the neutrino's weakly interacting nature also means that it is likely to pass right through a detector. The Super Kamiokande experiment in Japan monitors interactions in an enormous volume of water in order to spot a very small number of neutrino interactions.

process emits a particle that, at that time, had not been detected. This new particle had to be electrically neutral, in order to conserve charge, and it had to be much less massive than an electron. Fermi called it the **neutrino,** meaning "little neutral one."

The neutrino is represented by the symbol ν, a lowercase Greek nu. There are three types of neutrinos. The neutrino involved in beta decay is the *electron neutrino* ν_e; it shows up in processes involving electrons and positrons. The electron neutrino of course has an antiparticle, the antineutrino $\bar{\nu}_e$. The full descriptions of beta-minus and beta-plus decays, including the neutrinos, are as follows:

$$n \rightarrow p^+ + e^- + \bar{\nu}_e$$

$$p^+ \rightarrow n + e^+ + \nu_e$$

Figure 30.20 shows that the electron and antineutrino (or positron and neutrino) share the energy released in the decay. If we are concerned with the accurate balance of energy and momentum in a beta decay, we must include the antineutrino, but we can generally ignore the presence of this weakly interacting particle for the problems we solve.

It was initially thought that the neutrino, like the photon, was a *massless* particle. However, experiments in the last few years have shown that the neutrino mass, while tiny, is not zero. The best current evidence suggests a mass about one-millionth the mass of an electron. Experiments now under way will determine a more precise value. Because the neutrino is so abundant in the universe, this small mass is of great cosmological significance.

The antineutrino carries away a small amount of energy and momentum.

FIGURE 30.20 An accurate picture of beta decay must include the antineutrino.

Quarks

Modeling the nucleus as being made of protons and neutrons allows a description of all of the elements in the periodic table in terms of just three basic particles—protons, neutrons, and electrons. As far as chemistry and biology are concerned, these are the three fundamental units of matter. But beta decay, in which a neutron can change into a proton, and vice-versa, gives a hint that the neutron and the proton are *not* fundamental units but are made of smaller subunits.

There is another reason to imagine such subunits: the existence of dozens of subatomic particles—muons, pions, kaons, omega particles, and so on. Just as the periodic table explains the many different atomic elements in terms of three basic particles, perhaps it is possible to do something similar for this "subatomic zoo." In the 1960s, the American physicist Murray Gell-Mann and the Japanese physicist Kazuhiko Nishijima independently postulated the existence of particles that could be combined to make protons, neutrons, and other subatomic particles. Their ideas were soon confirmed in particle-accelerator experiments showing that neutrons and protons seem to have internal structure—they appear to be made of three distinct subparticles.

We now understand protons and neutrons to be composed of smaller charged particles whimsically named **quarks.** The quarks that form protons and neutrons are called **up quarks** and **down quarks,** symbolized as u and d, respectively. The nature of these quarks and the composition of the neutron and the proton are shown in Figure 30.21.

NOTE ▶ It seems surprising that the charges of quarks are *fractions* of e. Don't charges have to be integer multiples of e? It's true that atoms, molecules, and all macroscopic matter must have $q = Ne$ because these entities are constructed from electrons and protons. But no law of nature prevents other types of matter from having other amounts of charge. ◀

A proton is made of two up quarks and one down quark.

A neutron is made of one up quark and two down quarks.

An up quark has charge $+\frac{2}{3}e$.

A down quark has charge $-\frac{1}{3}e$.

FIGURE 30.21 The quark content of the proton and neutron.

A neutron and a proton differ by one quark. Beta decay can now be understood as a process in which a down quark changes to an up quark, or vice versa. Beta-minus decay of a neutron can be written as

$$d \rightarrow u + e^- + \bar{\nu}_e$$

The existence of quarks thus provides an explanation of how a neutron can turn into a proton.

CONCEPTUAL EXAMPLE 30.3   Quarks and beta-plus decay

What is the quark description of beta-plus decay?

REASON In beta-plus decay, a proton turns into a neutron, with the emission of a positron and an electron neutrino. To turn a proton into a neutron requires the conversion of an up quark into a down quark; the total reaction is thus

$$u \rightarrow d + e^+ + \nu_e$$

The quark model is very successful, but no one has ever seen a solitary quark. They don't exist alone, only inside other particles. In fact, the theory that describes quarks and their interactions specifies that we will *never* see a single quark, all by itself. The force between quarks increases as they move apart, so an infinite amount of energy would be required to separate two quarks. Quarks must always be bound with other quarks, a principle known as **quark confinement.**

Particles and Forces

Our current understanding of the truly fundamental particles—the ones that cannot be broken down into smaller subunits—is that they come in three basic types: **leptons** (particles like the electron), quarks (which combine to form particles like the proton and the neutron), and **bosons** (which are responsible for the forces between particles). The leptons and quarks are described in Table 30.4. A few points are worthy of note:

- Each particle has an associated antiparticle.
- There are three *families* of leptons. The first is the electron and its associated neutrino, and their antiparticles. The other families are based on the muon and the tau, heavier siblings to the electron. Only the electron and positron are stable particles.
- There are also three families of quarks. The first is the up-down family that makes all "normal" matter. The other families are pairs of heavier quarks that form more exotic particles

Current theory suggests that all of the forces of nature are due to the exchange of bosons. One type of boson is already quite familiar to you: The force of electromagnetism is produced by the exchange of photons, a type of boson. We have spoken about the strong nuclear force; there is another, weaker force that is important in certain nuclear reactions, called the **weak nuclear force.** The weak and strong forces are produced by the exchange of the bosons noted in Table 30.5. These particles have all been observed experimentally. We also presume that there is a boson responsible for the force of gravity, the **graviton,** though it has not been experimentally observed.

TABLE 30.4 Leptons and quarks

Leptons		Antileptons	
Electron	e^-	Positron	e^+
Electron neutrino	ν_e	Electron antineutrino	$\bar{\nu}_e$
Muon	μ^-	Antimuon	μ^+
Muon neutrino	ν_μ	Muon antineutrino	$\bar{\nu}_\mu$
Tau	τ^-	Antitau	τ^+
Tau neutrino	ν_τ	Tau antineutrino	$\bar{\nu}_\tau$

Quarks		Antiquarks	
Up	u	Antiup	$\bar{u}$
Down	d	Antidown	$\bar{d}$
Strange	s	Antistrange	$\bar{s}$
Charm	c	Anticharm	$\bar{c}$
Bottom	b	Antibottom	$\bar{b}$
Top	t	Antitop	$\bar{t}$

TABLE 30.5 Forces and exchange particles

Force	Associated boson
Electromagnetism	Photon
Weak nuclear force	W^+, W^-, Z bosons
Strong nuclear force	Gluon

30.7 Medical Applications of Nuclear Physics

Nuclear physics has brought both peril and promise to society. Radioactivity can cause tumors. At the same time, radiation can be used to diagnose and cure some cancers. This section is a brief survey of medical applications of nuclear physics.

A gamma source used for irradiating medical equipment to sterilize it. The blue glow is due to the ionization of the air around the source.

Radiation Dose

Nuclear radiation disrupts a cell's machinery by altering and damaging biological molecules, as we saw in Section 30.4. Beta and gamma radiation can penetrate the entire body and damage internal organs. Alpha radiation has less penetrating ability, but it deposits all its energy in a very small, localized volume. Internal organs are usually safe from alpha radiation unless the source of radiation is inhaled or ingested.

The biological effects of radiation depend on two factors. The first is the physical factor of how much energy is absorbed by the body. The second is the biological factor of how tissue reacts to different forms of radiation.

Suppose a beta particle travels through tissue, losing kinetic energy as it ionizes atoms it passes. The energy lost by the beta particle is a good measure of the number of ions produced and thus the amount of damage done. In a certain volume of tissue, more ionization means more damage. For this reason, we define the radiation **dose** as the energy from ionizing radiation absorbed by 1 kg of tissue. The SI unit for the dose is the **gray,** abbreviated Gy. The Gy is defined as

$$1 \text{ Gy} = 1.00 \text{ J/kg of absorbed energy}$$

The number of Gy depends only on the energy absorbed, not on the type of radiation or on what the absorbing material is. Another common unit for dose is the *rad;* 1 rad = 0.01 Gy.

A 1 Gy dose of gamma rays and a 1 Gy dose of alpha particles have different biological consequences. To account for such differences, the **relative biological effectiveness,** RBE, is defined as the biological effect of a given dose relative to the biological effect of an equal dose of x rays.

Table 30.6 shows the relative biological effectiveness of different forms of radiation. Larger values correspond to larger biological effects. Alpha and beta radiation have a range of values because the biological effect varies with the energy of the particle. Alpha radiation has the largest RBE because the energy is deposited in a localized region.

The radiation **dose equivalent** is the product of the energy dose in Gy and the relative biological effectiveness. Dose equivalent is measured in **sieverts,** abbreviated Sv. To be precise,

$$\text{Dose equivalent in Sv} = \text{dose in Gy} \times \text{RBE}$$

One Sv of radiation produces the same biological damage regardless of the type of radiation. Another common unit of dose equivalent (also called biologically equivalent dose) is the *rem;* 1 rem = 0.01 Sv.

NOTE ▶ In practice, the term "dose" is often used for both dose and dose equivalent. Use the units as a guide. If the unit is Sv or rem, it is a dose equivalent; if Gy or rad, a dose. ◀

TABLE 30.6 Relative biological effectiveness of radiation

Radiation type	RBE
X rays	1
Gamma rays	1
Beta particles	1–2
Alpha particles	10–20

EXAMPLE 30.9 Finding energy deposited in radiation exposure

A 75 kg patient is given a bone scan. A phosphorus compound containing the gamma-emitter ^{99}Tc is injected into the patient. It is taken up by the bones, and the emitted gamma rays are measured. The procedure exposes the patient to 3.6 mSv (360 mrem) of radiation. What is the total energy deposited in the patient's body, in J and in eV?

PREPARE The exposure is given in Sv, so it is a dose equivalent, a combination of deposited energy and biological effectiveness. The RBE for gamma rays is 1. Gamma rays are penetrating, and the source is distributed throughout the body, so this is a whole-body exposure. Each kg of the patient's body will receive approximately the same energy.

SOLVE The dose in Gy is the dose equivalent in Sv divided by the RBE. In this case, because the RBE = 1 the dose in Gy is numerically equal to the equivalent dose in Sv. The dose is thus 3.6 mGy = 3.6×10^{-3} J/kg. The radiation energy absorbed in the patient's body is

$$\text{Absorbed energy} = (3.6 \times 10^{-3}\ \text{J/kg})(75\ \text{kg}) = 0.27\ \text{J}$$

In eV, this is

$$\text{Absorbed energy} = (0.27\ \text{J})(1\ \text{eV}/1.6 \times 10^{-19}\ \text{J})$$
$$= 1.7 \times 10^{18}\ \text{eV}$$

ASSESS The total energy deposited, 0.27 J, is quite small; there will be negligible heating of tissue. But radiation produces its effects in other ways, as we have seen. Because it only takes ≈10 eV to ionize an atom, this dose is enough energy to ionize over 10^{17} atoms, meaning it can cause significant disruption to the cells of the body.

The question inevitably arises, "What is a safe dose?" Unfortunately, there is no simple or clear definition of a safe dose. A prudent policy is to avoid unnecessary exposure, and to weigh the significance of an exposure in relation to the *natural background.* We are all exposed to continual radiation from cosmic rays and from naturally occurring radioactive atoms in the ground, the atmosphere, and even the food we eat. This background averages about 3 mSv (300 mrem) per year, although there are wide regional variations depending on the soil type and the elevation. (Higher elevations have less atmospheric shielding, and thus have a larger exposure to cosmic rays. On the moon, with no protective atmosphere, the yearly dose would be 50 mSv.)

Table 30.7 lists the expected exposure from several different sources. A dental x ray subjects a person to approximately 1% of the yearly natural background that he or she would normally receive, and is likely not a cause for significant worry. Mammograms involve a much larger dose, concentrated in a small region of the body. A nuclear medicine procedure, like a PET scan (which is discussed below) may involve an exposure that is much larger than the typical yearly background dose. This is a significant dose, which must be weighed against the medical benefits of the procedure.

Nuclear Medicine

The tissues in the body most susceptible to radiation are those that are rapidly proliferating—including tumors. The goal in *radiation therapy* is to apply a large enough dose of radiation to destroy or shrink a tumor while producing minimal damage to surrounding healthy tissue.

In Figure 30.22a, a patient with a brain tumor is fitted with a metal *collimator* that absorbs gamma rays except for those traveling along desired paths. The collimator is fashioned so that gamma rays from an external source will be concentrated on the tumor, as shown in Figure 30.22b. Because the rapidly dividing cells of a tumor are much more sensitive to radiation than the tissues of the brain, and surgical options carry risks of significant complications, radiation is a common means of treating tumors of the brain.

Other tumors are treated by surgically implanting radioactive "seeds" within the tumor. One common type of seed contains ^{125}I, which undergoes a nuclear decay followed by emission of a 27.5 keV photon. The photon has a very short range so that it will only damage tissue close to the seed. Careful placement of the seeds permits a significant dose to a tumor with only minimal exposure to surrounding tissue.

TABLE 30.7 Radiation exposure

Radiation source	Typical exposure (mSv)
PET scan	7.0
Natural background (1 year)	3.0
Mammogram	0.70
Chest x ray	0.30
Transatlantic airplane flight	0.050
Dental x ray	0.030

(a)

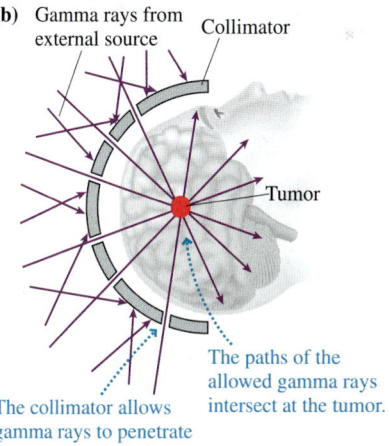
(b) Gamma rays from external source   Collimator

Tumor

The paths of the allowed gamma rays intersect at the tumor.

The collimator allows gamma rays to penetrate only along certain lines.

FIGURE 30.22 The use of gamma rays to treat a tumor in the brain.

Metal "seeds" containing radioactive ^{125}I in the prostate gland can shrink a tumor.

The detector is measuring gamma radiation emitted by isotopes taken up by tissues in the woman's head and neck.

Some tissues in the body will preferentially take up certain isotopes, allowing for treatment by isotope ingestion. A common treatment for hyperthyroidism, in which the thyroid gland is overactive, is to damage the gland with the isotope ^{131}I, a beta emitter with a half-life of 8.0 days. A patient is given a tablet containing ^{131}I. The iodine in the blood is taken up and retained by the thyroid gland, resulting in a reduction of the gland's activity with minimal disruption of surrounding tissue.

A promising new technique is the "tagging" of antibodies with short-half-life isotopes. The antibodies bind to proteins on tumor cells, placing the isotope where its decay directly damages the tumor.

Nuclear Imaging

X rays from an external source may be used to make an image of the body, as described in Chapter 28. *Nuclear imaging* uses an internal source—radiation from isotopes in the body—to produce an image of tissues in the body. The bone-scan image that opened the chapter is an example of nuclear imaging.

There is a key difference between x rays and nuclear imaging procedures. **An x ray is an image of anatomical structure;** it is excellent for identifying structural issues such as broken bones. **Nuclear imaging creates an image of the biological activity of tissues in the body.** For example, nuclear imaging can detect reduced metabolic activity of brain tissue after a stroke.

Let's look at an example that illustrates the difference between a conventional x-ray image or CAT scan and an image made with a nuclear imaging technique. Suppose a doctor suspects a patient has cancerous tissue in the bones. An x ray does not show anything out of the ordinary; the tumors may be too small or may appear similar to normal bone. The doctor then orders a scan with a **gamma camera,** a device that can measure and produce an image from gamma rays emitted within the body.

The patient is given a dose of a phosphorus compound labeled with the gamma emitter ^{99}Tc. This compound is taken up and retained in bone tissue where active growth is occurring. The ^{99}Tc will be concentrated in the bones where there has been recent injury or inflammation—or where a tumor is growing. The patient is then scanned with a gamma camera. Figure 30.23a shows how the gamma camera can pinpoint the location of the gamma emitting isotopes in the body and produce an image that reveals their location and intensity. A typical image is shown in Figure 30.23b. The bright spots show high concentrations of ^{99}Tc, revealing areas of tumor growth. The tumors may be too small to show up on an x ray, but their activity is easily detected with the gamma camera. With such early detection, the patient's chance of a cure is greatly improved.

(a) 2. The position of the radioisotope is determined by the position of the active sensor.

Sensors

Processing   Display

3. A record of the number and position of gamma rays is processed into an image.

Collimator

1. The lead plates of the collimator prevent gamma rays from reaching a sensor unless the source is directly below.

Radioisotope presence

(b)

The bright spots show areas of active tumor growth.

FIGURE 30.23 The operation of a gamma camera.

A conventional x ray can be improved by computer analysis of x rays transmitted through the body, resulting in the greater resolution of a CAT scan. Similar analysis is applied to a gamma scan in the technique known as *SPECT,* for *single photon emission computed tomography.* Even better resolution is possible from the two-photon technique described in the next section.

Positron-Electron Tomography

We have seen that a small number of radioactive isotopes decay by the emission of a positron. Such isotopes can be used for an imaging technique known as *positron-emission tomography,* or *PET.* PET is particularly important for imaging the brain.

Most PET scans use the fluorine isotope ^{18}F, which emits a positron as it undergoes beta-plus decay to ^{18}O with a half-life of 110 minutes. ^{18}F is used to create an analog of glucose called fluorine-18 fluoro-deoxy-glucose (F-18 FDG). This compound is taken up by tissues in the brain. Areas that are more active are using more glucose, so the F-18 FDG is concentrated in active brain regions. When a fluorine atom in the F-18 FDG decays, the emitted positron immediately collides with a regular electron. The two annihilate to produce two gamma rays that travel out of the brain in opposite directions, as shown in Figure 30.24a.

Figure 30.24b shows a patient's head surrounded by a ring of gamma-ray detectors. Because the gamma rays from the positron's annihilation are emitted back to back, simultaneous detection of two gamma rays on opposite sides of the subject indicates that the annihilation occurred somewhere along the line between those detectors. Recording many such pairs of gamma rays shows with great accuracy where the decays are occurring. A full scan will show more activity in regions of the brain where metabolic activity is enhanced, less activity in regions where metabolic activity is depressed. An analysis of these scans can provide a conclusive diagnosis of stroke, injury, or Alzheimer's disease.

Physics is a very practical discipline. The early chapters of this book, in which you learned about forces and motion, had very obvious applications to things in your daily life. But in these past few chapters we see that even modern discoveries—discoveries such as antimatter that may seem like science fiction—can be put to very practical use. As we come to the close of this book, we hope that you have gained an appreciation not only for what physics tells us about the world, but also for the wide range of problems it can be used to solve.

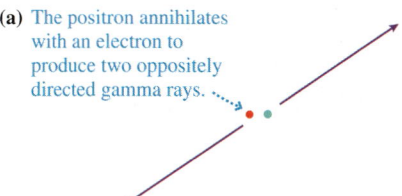

(a) The positron annihilates with an electron to produce two oppositely directed gamma rays.

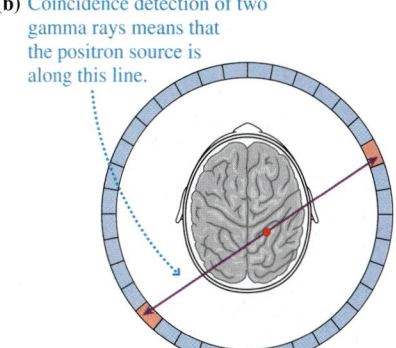

(b) Coincidence detection of two gamma rays means that the positron source is along this line.

FIGURE 30.24 Positron-emission tomography.

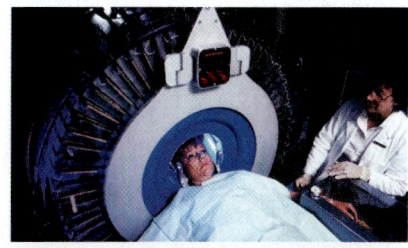

STOP TO THINK 30.5   A patient ingests a radioactive isotope to treat a tumor. The isotope provides a dose of 0.10 Gy. Which type of radiation will give the highest dose equivalent in Sv?

A. Alpha particles.   B. Beta particles.   C. Gamma rays.

▶ **This is your brain on PET** BIO The woman in the top photo is undergoing a PET scan not to diagnose disease but to probe the workings of the brain. While undergoing a PET scan, a subject is asked to perform different mental tasks. The lower panels show functional images from a PET scan superimposed on anatomical images from a CAT scan. The subject first listened to speech in an unfamiliar language; the active areas of the brain were those responsible for hearing. Next, she listened to speech in a familiar language, resulting in activity in the parts of the brain responsible for speech and comprehension.

The goal of Chapter 30 has been to understand the physics of the nucleus and some of the applications of nuclear physics.

GENERAL PRINCIPLES

The Nucleus

The nucleus is a small, dense, positive core at the center of an atom.

Z protons, charge $+e$, spin $\frac{1}{2}$
N neutrons, charge 0, spin $\frac{1}{2}$

The mass number is

$$A = Z + N$$

Isotopes of an element have the same value of Z but different values of N.

The strong force holds nuclei together. It:

- Acts between any two nucleons.
- Is short range.

Adding neutrons to a nucleus allows the strong force to overcome the repulsive Coulomb force between protons.

The binding energy B of a nucleus depends on the mass difference between an atom and its constituents:

$$B = (Zm_H + Nm_n - m_{atom}) \times (931.49 \text{ MeV/u})$$

Nuclear Stability

Most nuclei are not stable. Unstable nuclei undergo radioactive decay. Stable nuclei cluster along the **line of stability** in a plot of the isotopes.

Low-Z nuclei move closer to the line of stability by beta decay. Line of stability. Alpha decay is energetically favorable for high-Z nuclei.

Mechanisms by which unstable nuclei decay:

Decay	Particle	Penetration
alpha	^{4}He nucleus	low
beta-minus	e^-	medium
beta-plus	e^+	medium
gamma	photon	high

Alpha and beta decay change the nucleus; the daughter nucleus is a different element.

Alpha decay:

$$^A_Z X \rightarrow ^{A-4}_{Z-2} Y + \alpha + \text{energy}$$

Beta-minus decay:

$$^A_Z X \rightarrow ^A_{Z+1} Y + \beta + \text{energy}$$

IMPORTANT CONCEPTS

Shell model

Nucleons fill nuclear energy levels, similar to filling electron energy levels in atoms. Nucleons can move to other levels, and will do so if this can reduce total energy.

U ^{12}C   Neutrons   Protons

The quark model

Nucleons (and other particles) are made of quarks. Quarks, leptons, and bosons are the fundamental particles.

Proton   Up quark

Neutron   Down quark

APPLICATIONS

Radioactive decay

The number of undecayed nuclei decreases exponentially with time t.

$$N = N_0 e^{-t/\tau}$$

$$N = N_0 \left(\frac{1}{2}\right)^{t/t_{1/2}}$$

The **time constant** τ is $1/r$, where r is the **decay rate.** The **half-life**

$$t_{1/2} = \tau \ln 2 = 0.693\tau$$

is the time in which half of any sample decays.

Measuring radiation

The **activity** $R = rN$ of a radioactive sample, measured in becquerels or curies, is the number of decays per second.

The radiation **dose** is measured in grays, where

$$1 \text{ Gy} = 1.00 \text{ J/kg of absorbed energy}$$

The **relative biological effectiveness** RBE is the biological effect of a dose relative to the biological effects of x rays. The **dose equivalent** is measured in sieverts, where

$$\text{Dose equivalent in Sv} = \text{dose in Gy} \times \text{RBE}$$

One Sv of radiation produces the same biological effect regardless of the type of radiation.

For instructor-assigned homework, go to www.masteringphysics.com

Problem difficulty is labeled as I (straightforward) to IIII (challenging).

Problems labeled INT integrate significant material from earlier chapters; BIO are of biological or medical interest.

QUESTIONS

Conceptual Questions

1. Atom A has a larger atomic mass than atom B. Does this mean that atom A also has a larger atomic number? Explain.

2. Given that $m_H = 1.00783$ u, is the mass of a hydrogen atom ^{1}H greater than, less than, or equal to 1/12 the mass of a ^{12}C atom? Explain.

3. a. Is there a $^{30}_{3}$Li nucleus? If so, is it stable or radioactive? If not, why not?
   b. Is there a $^{184}_{92}$U nucleus? If so, is it stable or radioactive? If not, why not?

4. Rounding slightly, the nucleus ^{3}He has a binding energy of 2.5 MeV/nucleon and the nucleus ^{6}Li has a binding energy of 5 MeV/nucleon.
   a. What is the binding energy of ^{3}He?
   b. What is the binding energy of ^{6}Li?
   c. Is it energetically possible for two ^{3}He nuclei to join or fuse together into a ^{6}Li nucleus? Explain.
   d. Is it energetically possible for a ^{6}Li nucleus to split or fission into two ^{3}He nuclei? Explain.

5. How do we know the strong force exists?

6. For each nuclear energy-level diagram in Figure Q30.6, state whether it represents a nuclear ground state, an excited nuclear state, or an impossible nucleus.

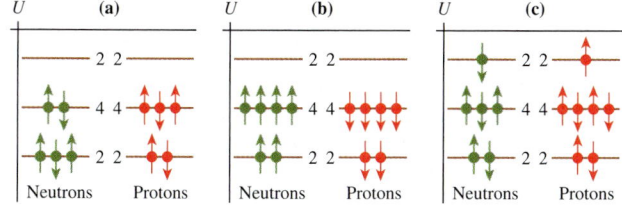

FIGURE Q30.6

7. Figure Q30.7 shows how the number of nuclei of one particular isotope depends on time. What is the half-life of the nucleus?

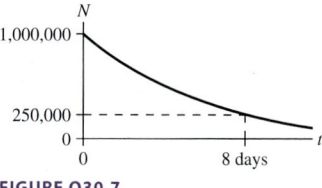

FIGURE Q30.7

8. A radioactive sample has a half-life of 10 s. 10,000 nuclei are present at $t = 20$ s.
   a. How many nuclei were there at $t = 0$ s?
   b. How many nuclei will there be at $t = 40$ s?

9. Nucleus A decays into the stable nucleus B with a half-life of 10 s. At $t = 0$ s there are 1000 A nuclei and no B nuclei. At what time will there be 750 B nuclei?

10. A radioactive sample's half-life is 1.0 min. That is, each nucleus has a 50% chance of undergoing a decay sometime between $t = 0$ and $t = 1$ min. One particular nucleus has not decayed at $t = 15$ min. What is the probability this nucleus will decay between $t = 15$ and $t = 16$ min?

11. Four samples each contain a single radioactive isotope. Sample A has 1 mol of matter and an activity of 100 Bq. Sample B has 10 mol and 100 Bq, sample C has 100 mol and 100 Bq, and sample D has 100 mol and 1000 Bq. Rank in order, from largest to smallest, the half-lives of these four isotopes. Explain.

12. Oil and coal generally contain no measurable ^{14}C. What does this tell us about how long they have been buried?

13. Radiocarbon dating assumes that the abundance of ^{14}C in the environment has been constant. Suppose ^{14}C was less abundant 10,000 years ago than it is today. Would this cause a lab using radiocarbon dating to overestimate or underestimate the age of a 10,000-year-old artifact? (In fact, the abundance of ^{14}C in the environment does vary slightly with time. But the issue has been well studied, and the ages of artifacts are adjusted to compensate for this variation.)

14. Identify the unknown X in the following decays:
   a. $^{222}_{86}$Rn $\rightarrow$ $^{218}_{84}$Po + X   b. $^{228}_{88}$Ra $\rightarrow$ $^{228}_{89}$Ac + X
   c. $^{140}_{54}$Xe $\rightarrow$ $^{140}_{55}$Cs + X   d. $^{64}_{29}$Cu $\rightarrow$ $^{64}_{28}$Ni + X

15. Are the following decays possible? If not, why not?
   a. $^{232}_{90}$Th $\rightarrow$ $^{236}_{92}$U + α b. $^{238}_{94}$Pu $\rightarrow$ $^{236}_{92}$U + α
   c. $^{33}_{15}$P $\rightarrow$ $^{32}_{16}$S + e$^-$

16. Is an alpha emitter or a beta emitter more likely to expose photographic film stored in a nearby plastic container?

17. Smoke detectors make use of the man-made element ^{241}Am, an alpha emitter. If you bring a Geiger counter near a smoke detector, you will not detect an increase in the reading. Why not?

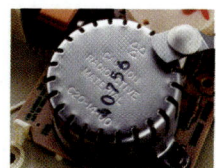

18. What kind of decay, if any, can occur for the nuclei with the energy level diagrams shown in Figure Q30.18?

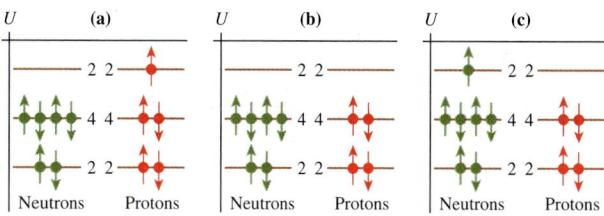

FIGURE Q30.18

19. The nuclei of ^{4}He and ^{16}O are very stable and are often referred to as "doubly magic" nuclei. Use the shell model to explain what is special about these particular nuclei.

20. BIO A and B are fresh apples. Apple A is strongly irradiated by nuclear radiation for 1 hour. Apple B is not irradiated. Afterward, in what ways are apples A and B different?

21. BIO A patient's tumor is irradiated with gamma rays from an external source. Afterward, is his body radioactive? Explain.

22. Four radiation doses are as follows: Dose A is 10 rad with an RBE of 1, dose B is 20 rad with an RBE of 1, dose C is 10 rad with an RBE of 2, and dose D is 20 rad with an RBE of 2.
    a. Rank in order, from largest to smallest, the amount of energy delivered by these four doses.
    b. Rank in order, from largest to smallest, the biological damage caused by these four doses.

23. Two different sources of radiation give the same dose equivalent in Sv. Does this mean that the radiation from each source has the same RBE? Explain.

24. Consider the following three decays: (i) ^{212}Po $\rightarrow$ ^{208}Pb + α, (ii) ^{137}Cs $\rightarrow$ ^{137}Ba + e$^-$ + γ, (iii) ^{90}Sr $\rightarrow$ ^{90}Y + e$^-$. Which of the three isotopes, ^{212}Po, ^{137}Cs, or ^{90}Sr, would be most useful for nuclear imaging? Why?

25. Some types of MRI can produce images of resolution and detail similar to PET. Though the images are similar, MRI is generally preferred over PET for studies of brain function involving healthy subjects. Why?

26. Sulfur colloid particles tagged with ^{99}Tc are taken up and retained by cells in the liver and spleen. A patient is suspected of having a liver tumor that would destroy these cells. Explain how a gamma camera scan could be used to confirm or rule out the existence of a tumor.

27. The first two letters in the acronym SPECT, which describes a nuclear imaging technique, stand for "single photon." Is a SPECT done with a gamma emitter or a positron emitter?

Multiple-Choice Questions

28. A significant fraction of the radiation dose you will receive during your life comes from radioactive materials in your body. The most important source of this radiation is the potassium isotope ^{40}K, which decays to the stable calcium isotope ^{40}Ca. What form of radiation is emitted in the decay?
    A. An alpha particle. B. A beta particle.
    C. A gamma photon. D. A calcium ion.

29. A watch's luminous glow is due to zinc sulfide paint that is energized by beta particles given off by *tritium,* the radioactive hydrogen isotope ^{3}H, which has a half-life of 12.3 years. This glow has about 1/10 of its initial brightness. How many years old is the watch?
    A. 20 yr B. 30 yr C. 40 yr D. 50 yr

30. The uranium in the earth's crust is 0.7% ^{235}U and 99.3% ^{238}U. Two billion years ago, ^{235}U comprised approximately 3% of the uranium in the earth's crust. This tells you something about the relative half-lives of the two isotopes. Suppose you have a sample of ^{235}U and a sample of ^{238}U, each with exactly the same number of atoms.
    A. The sample of ^{235}U has a higher activity.
    B. The sample of ^{238}U has a higher activity.
    C. The two samples have the same activity.

31. The quark composition of the proton and neutron are, respectively, uud and udd, where u is an up quark (charge $+\frac{2}{3}e$) and d is a down quark (charge $-\frac{1}{3}e$). There are also anti-up $\bar{u}$ (charge $-\frac{2}{3}e$) and anti-down $\bar{d}$ (charge $+\frac{1}{3}e$) quarks. The combination of a quark and an antiquark is called a *meson.* The mesons known as *pions* have the composition $\pi^+ = u\bar{d}$ and $\pi^- = \bar{u}d$. Suppose a proton collides with an antineutron. During such collisions, the various quarks and antiquarks annihilate whenever possible. When the remaining quarks combine to form a single particle, it is a
    A. Proton B. Neutron C. π^+ D. π^-

32. Suppose you have a 1 g sample of ^{226}Ra, half-life 1600 years. How long will it be until only 0.1 g of radium is left?
    A. 1600 yr B. 3200 yr C. 5300 yr D. 16,000 yr

33. A sample of ^{131}I, half-life 8.0 days, is registering 100 counts per second on a Geiger counter. How long will it be before the sample registers only 1 count per second?
    A. 8 days B. 53 days C. 80 days D. 800 days

34. Which of these patients would be the best candidate for using a nuclear-imaging technique to diagnose his or her disease instead of an x ray or CAT scan?
    A. A patient suspected of having a broken bone.
    B. A patient suspected of having an enlarged heart.
    C. A patient suspected of having decreased activity of cells in the kidneys.
    D. A patient suspected of having fluid in the lungs.

35. The decay of the radioactive hydrogen isotope *tritium* may be described as ^{3}H $\rightarrow$ ^{3}He + X + Y. The symbols X and Y represent
    A. $X = e^+, Y = \bar{\nu}_e$
    B. $X = e^-, Y = \nu_e$
    C. $X = e^+, Y = \nu_e$
    D. $X = e^-, Y = \bar{\nu}_e$

PROBLEMS

Section 30.1 Nuclear Structure

1. How many protons and how many neutrons are in (a) ^{3}H, (b) ^{40}Ar, (c) ^{40}Ca, and (d) ^{239}Pu?

2. How many protons and how many neutrons are in (a) ^{3}He, (b) ^{20}Ne, (c) ^{60}Co, and (d) ^{226}Ra?

3. Estimate the number of protons and the number of neutrons in 1 m^3 of air.

4. Estimate the number of protons and the number of neutrons in your body.

5. Use the data in Appendix C to calculate the chemical atomic mass of nitrogen.

6. Use the data in Appendix C to calculate the chemical atomic mass of oxygen.

Section 30.2 Nuclear Stability

7. Calculate (in MeV) the total binding energy and the binding energy per nucleon (a) for ^{3}H and (b) for ^{3}He.

8. Calculate (in MeV) the total binding energy and the binding energy per nucleon (a) for ^{40}Ar and (b) for ^{40}K.

9. Calculate (in MeV) the binding energy per nucleon for ^{3}He and ^{4}He. Which is more tightly bound?

10. Calculate (in MeV) the binding energy per nucleon for ^{12}C and ^{13}C. Which is more tightly bound?

11. Calculate (in MeV) the binding energy per nucleon for (a) ^{14}N, (b) ^{56}Fe, and (c) ^{207}Pb.

Section 30.3 Forces and Energy in the Nucleus

12. ‖ Draw energy-level diagrams, similar to Figure 30.9, for all $A = 10$ nuclei listed in Appendix C. Show all the occupied neutron and proton levels. Which of these nuclei do you expect to be stable?

13. ‖ Draw energy-level diagrams, similar to Figure 30.9, for all $A = 14$ nuclei listed in Appendix C. Show all the occupied neutron and proton levels. Which of these nuclei do you expect to be stable?

Section 30.4 Radiation and Radioactivity

14. ⎮ The barium isotope ^{133}Ba has a half-life of 10.5 years. A sample begins with 1.0×10^{10} ^{133}Ba atoms. How many are left after (a) 2 years, (b) 20 years, and (c) 200 years?

15. ⎮ The cadmium isotope ^{109}Cd has a half-life of 462 days. A sample begins with 1.0×10^{12} ^{109}Cd atoms. How many are left after (a) 50 days, (b) 500 days, and (c) 5000 days?

16. ‖ How many half-lives must elapse until (a) 90% and (b) 99% of a radioactive sample of atoms has decayed?

17. ⦀ What is the age in years of a bone in which the ^{14}C/^{12}C ratio is measured to be 1.65×10^{-13}?

18. ⦀ What is the half-life in days of a radioactive sample with 5.0×10^{15} atoms and an activity of 5.0×10^8 Bq?

19. ‖ Many medical PET scans use the isotope ^{18}F, which has a
BIO half-life of 1.8 hr. A sample prepared at 10:00 A.M. has an activity of 20 mCi. What is the activity at 1:00 P.M., when the patient is injected?

Section 30.5 Nuclear Decay Mechanisms

20. ⎮ The radioactive hydrogen isotope ^{3}H is called *tritium*. It decays by beta-minus decay with a half-life of 12.3 years.
    a. What is the daughter nucleus of tritium?
    b. What are the time constant and the decay rate of tritium?

21. ⎮ Technetium, element 43, has no stable isotopes. ^{99}Tc, the longest-lived isotope, decays by beta-minus decay with a half-life of 211,000 years. What is the daughter nucleus of the decay?

22. ⎮ ^{15}O is a positron-emitter used in positron emission tomography. What is the daughter nucleus of the decay?

23. ⎮ Spacecraft have been powered with energy from the alpha decay of ^{238}Pu. What is the daughter nucleus?

24. ⎮ Identify the unknown isotope X in the following decays.
    a. ^{234}U $\rightarrow X + \alpha$.   b. ^{32}P $\rightarrow X + e^-$
    c. $X \rightarrow {}^{30}$Si $+ e^+$ d. ^{24}Na $\rightarrow {}^{24}$Mg $+ e^- \rightarrow X + \gamma$

25. ⎮ Identify the unknown isotope X in the following decays.
    a. $X \rightarrow {}^{224}$Ra $+ \alpha$ b. $X \rightarrow {}^{207}$Pb $+ e^-$
    c. ^{7}Be $+ e^- \rightarrow X$ d. $X \rightarrow {}^{60}$Ni $+ \gamma$

26. ‖ What is the energy (in MeV) released in the alpha decay of ^{239}Pu?

27. ⎮ What is the energy (in MeV) released in the alpha decay of ^{228}Th?

28. ⦀ What is the total energy (in MeV) released in the beta decay of a neutron?

29. ⎮ Medical gamma imaging is generally done with the tech-
BIO netium isotope ^{99}Tc*, which decays by emitting a gamma-ray photon with energy 140 keV. What is the mass loss of the nucleus, in u, upon emission of this gamma ray?

Section 30.6 Subatomic Particles

30. ⦀ Positive and negative pions, denoted π^+ and π^-, are anti-particles of each other. Each has a rest mass of 140 MeV/c^2. Suppose a collision between an electron and positron, each with kinetic energy K, produces a π^+, π^- pair. What is the smallest possible value for K?

31. ⦀ In a particular beta-minus decay of a free neutron (that is, one not part of an atomic nucleus), the emitted electron has exactly the same kinetic energy as the emitted electron antineutrino. What is the value, in MeV, of that kinetic energy? Assume that the recoiling proton has negligible kinetic energy.

32. ‖ The masses of the neutrinos are still not precisely deter-
INT mined, but let us assume for the purpose of this problem that the mass of an electron neutrino is one millionth the mass of an electron. What is the kinetic energy, in eV, of an electron neutrino moving at $0.999c$?

Section 30.7 Medical Applications of Nuclear Physics

33. ‖ A 50 kg laboratory worker is exposed to 20 mJ of beta
BIO radiation with RBE = 1.5. What is the dose in mSv?

34. ‖ How many rad of gamma-ray photons cause the same bio-
BIO logical damage as 30 rad of alpha radiation with an RBE of 15?

35. ⎮ 150 rad of gamma radiation are directed into a 150 g tumor.
BIO How much energy does the tumor absorb?

36. ‖ During the 1950s, nuclear bombs were tested on islands in
BIO the South Pacific. In one test, personnel on a nearby island received 10 mGy per hour of beta and gamma radiation. At this rate, how long would it take to receive a potentially lethal dose equivalent of 4.5 Sv? Assume RBE = 1 for the beta particles.

37. ⎮ The doctors planning a radiation therapy treatment have
BIO determined that a 100 g tumor needs to receive 0.20 J of gamma radiation. What is the dose in Gy?

General Problems

38. ⎮ The chemical atomic mass of hydrogen, with the two stable isotopes ^{1}H and ^{2}H (deuterium), is 1.00798. Use this value to determine the natural abundance of these two isotopes.

39. ‖ You learned in Chapter 29 that the binding energy of the electron in a hydrogen atom is 13.6 eV.
    a. By how much does the mass decrease when a hydrogen atom is formed from a proton and an electron? Give your answer both in atomic mass units and as a percentage of the mass of the hydrogen atom.
    b. By how much does the mass decrease when a helium nucleus is formed from two protons and two neutrons? Give your answer both in atomic mass units and as a percentage of the mass of the helium nucleus.
    c. Compare your answers to parts a and b. Why do you hear it said that mass is "lost" in nuclear reactions but not in chemical reactions?

40. ⦀⎮ Use the graph of binding energy of Figure 30.5 to estimate the total energy released if a nucleus with mass number 240 fissions into two nuclei with mass number 120.

41. ‖ Use the graph of binding energy of Figure 30.5 to estimate the total energy released if three ^{4}He nuclei fuse together to form a ^{12}C nucleus.

42. ⎮ Could a ^{56}Fe nucleus fission into two ^{28}Al nuclei? Your answer, which should include some calculations, should be based on the curve of binding energy of Figure 30.5.

43. | a. What are the isotopic symbols of all isotopes in Appendix C with $A = 17$?
    b. Which of these are stable nuclei?
    c. For those that are not stable, identify both the decay mode and the daughter nucleus.

44. | Repeat Problem 43 with $A = 19$.

45. || What is the activity in Bq and in Ci of a 2.0 mg sample of ^{3}H?

46. | The activity of an ^{39}Ar sample is 1.25×10^9 Bq. What is the mass of the sample?

47. || The activity of a sample of the cesium isotope ^{137}Cs is 2.0×10^8 Bq. Many years later, after the sample has fully decayed, how many beta particles will have been emitted?

48. ||| BIO A 115 mCi sample of a radioactive isotope is made in a reactor. When delivered to a hospital 16 hours later, its activity is 95 mCi. The lowest usable activity level is 10 mCi.
    a. What is the isotope's half-life?
    b. For how long after delivery is the sample usable?

49. || BIO Radioactive tracers are used to study ion transport across cell membranes. The beta-minus emitting isotope ^{42}K ($t_{1/2} = 12.4$ h) has been particularly important for measuring the movement of potassium ions. A saline solution that contains the isotope bathes cells in a test tube, and the amount of ^{42}K that has accumulated inside the cells is measured after a short time. A sample of liquid extracted from the cells must have an activity of at least 100 Bq for an accurate measurement to be made. The activity of a sample just after extraction is 10,000 Bq. If a scientist wishes to confirm the result by measuring the sample a week later, will there be sufficient beta-minus emission?

50. ||| BIO You are assisting in an anthropology lab over the summer by carrying out ^{14}C dating. A graduate student found a bone he believes to be 20,000 years old. You extract the carbon from the bone and prepare an equal-mass sample of carbon from modern organic material. To determine the activity of a sample with the accuracy your supervisor demands, you need to measure the time it takes for 10,000 decays to occur.
    a. The activity of the modern sample is 1.06 Bq. How long does that measurement take?
    b. It turns out that the graduate student's estimate of the bone's age was accurate. How long does it take to measure the activity of the ancient carbon?

51. || BIO A sample of wood from an archaeological excavation is dated by using a mass spectrometer to measure the fraction of ^{14}C atoms. Suppose 100 atoms of ^{14}C are found for every 1.0×10^{15} atoms of ^{12}C in the sample. What is the wood's age?

52. ||| INT The radium isotope ^{223}Ra, an alpha emitter, has a half-life of 11.43 days. You happen to have a 1.0 g cube of ^{223}Ra, so you decide to use it to boil water for tea. You fill a well-insulated container with 100 mL of water at 18°C and drop in the cube of radium.
    a. How long will it take the water to boil?
    b. Will the water have been altered in any way by this method of boiling? If so, how?

53. | A sample of 1.0×10^{10} atoms that decay by alpha emission has a half-life of 100 min. How many alpha particles are emitted between $t = 50$ min and $t = 200$ min?

54. |||| A sample contains radioactive atoms of two types, A and B. Initially there are five times as many A atoms as there are B atoms. Two hours later, the numbers of the two atoms are equal. The half-life of A is 0.50 hours. What is the half-life of B?

55. ||| Suppose a 100 g sample contains ^{131}Ba, with a half-life of 12 days, and ^{47}Ca, with a half-life of 4.5 days. If there are initially twice as many calcium atoms as there are barium atoms, what will be the ratio of calcium atoms to barium atoms 2.5 weeks later?

56. |||| The technique known as potassium-argon dating is used to date old lava flows and thus any fossilized skeletons found in them, like this 1.8-million-year old hominid skull. The potassium isotope ^{40}K has a 1.28 billion year half-life and is naturally present at very low levels. ^{40}K decays by beta emission into the stable isotope 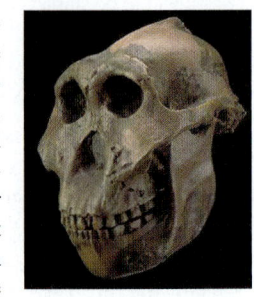 ^{40}Ar. Argon is a gas, and there is no argon in flowing lava because the gas escapes. Once the lava solidifies, any argon produced in the decay of ^{40}K is trapped inside and cannot escape. What is the age of a piece of solidified lava with a ^{40}Ar/^{40}K of 0.12?

57. || The half-life of the uranium isotope ^{235}U is 700 million years. The earth is approximately 4.5 billion years old. How much more ^{235}U was there when the earth formed than there is today? Give your answer as the then-to-now ratio.

58. |||| All the very heavy atoms found in the earth were created long ago by nuclear fusion reactions in a supernova, an exploding star. The debris spewed out by the supernova later coalesced to form the sun and the planets of our solar system. Nuclear physics suggests that the uranium isotopes ^{235}U ($t_{1/2} = 7.04 \times 10^8$ yr) and ^{238}U ($t_{1/2} = 4.47 \times 10^9$ yr) should have been created in roughly equal amounts. Today, 99.28% of uranium is ^{238}U and 0.72% is ^{235}U. How long ago did the supernova occur?

59. ||| BIO A 75 kg patient swallows a 30 μCi beta emitter with a half-life of 5.0 days. The average beta particle has energy 0.35 MeV and an RBE of 1.5. Ninety percent of the beta particles are absorbed within the patient's body and 10% escape. What dose equivalent does the patient receive?

60. || BIO INT Ground beef may irradiated with high-energy electrons from a linear accelerator to kill pathogens. In a standard treatment, 1.0 kg of beef receives 4.5 kGy of radiation in 40 s.
    a. How much energy is deposited in the beef?
    b. What is the average rate (in W) of energy deposition?
    c. Estimate the temperature increase of the beef due to this procedure. The specific heat of beef is approximately 3/4 of that of water.

61. |||| BIO A typical 70 kg human body contains 140 g of potassium. Potassium has a chemical atomic mass of 39.1 and has three naturally occurring isotopes. One of those isotopes, ^{40}K, is radioactive with a half-life of 1.3 billion years and a natural abundance of 0.012%. Each ^{40}K decay deposits, on average, 1.0 MeV of energy into the body. What yearly dose in Gy does the typical person receive from the decay of ^{40}K in the body?

62. |||| INT What dose in rads of gamma radiation must be absorbed by a block of ice at 0°C to transform the entire block to liquid water at 0°C?

63. || BIO A chest x ray uses 10 keV photons with an RBE of 0.85. A 60 kg person receives a 30 mrem dose from one x ray that exposes 25% of the patient's body. How many x-ray photons are absorbed in the patient's body?

64. ▌▌▌ The rate at which a radioactive isotope is lost from a
BIO patient's body is the rate at which the isotope decays *plus* the
rate at which the element is excreted from the body. Medical
experiments have shown that stable isotopes of a particular
element are excreted with a 6.0 day half-life. A radioactive
isotope of the same element has a half-life of 9.0 days. What
is the effective half-life of the isotope in a patient's body?
Hint: The total *rate* of loss is the rate due to excretion plus the
rate due to decay.

65. ▌ The plutonium isotope ^{239}Pu has a half-life of 24,000 years
BIO and decays by the emission of a 5.2 MeV alpha particle. Plutonium is not especially dangerous if handled because the
activity is low and the alpha radiation doesn't penetrate the
skin. But the tiniest speck of plutonium can cause problems if
it is inhaled and lodges deep in the lungs. Let's see why.
 a. Soot particles are roughly 1 μm in diameter, and it is
known that these particles can go deep into the lungs.
How many ^{239}Pu atoms are in a 1.0-μm diameter particle
of ^{239}Pu? The density of plutonium is 19,800 kg/m^3.
 b. What is the activity, in Bq, of this 1.0-μm-diameter particle?
 c. The activity of the particle is very small, but the penetrating power of alpha particles is also very small, so the damage is concentrated. The alpha particles deposit their
energy in a 50-μm-diameter sphere around the plutonium
particle. In one year, what is the dose equivalent in mSv to
this small sphere of tissue in the lungs? Use an RBE of 15
and assume that the tissue density is that of water.
 d. How does the exposure to this tissue compare to the natural background exposure?

66. ▌▌▌ Uranium is naturally present at low levels in many soils and
BIO rocks. The ^{238}U decay series includes the short-lived radon
isotope ^{222}Rn, with $t_{1/2} = 3.82$ days. Radon is a gas, and it can
seep into basements. The Environmental Protection Agency
recommends that homeowners take steps to remove radon if
the radon activity exceeds 4 pCi per liter of air. The daughter
nuclei from radon decay are of significant concern, but the
radon itself does provide some exposure.
 a. How many ^{222}Rn atoms are there in 1.0 m^3 of air if the
activity is 4.0 pCi/L?
 b. The range of alpha particles in air is 3 cm. Let's model a
person as a 180-cm-tall, 25-cm-diameter cylinder with a
mass of 65 kg. Only decays within 3 cm of the cylinder
cause exposure, and only 50% of the decays direct the
alpha particle toward the person. What is the dose equivalent (in mSv) for a person who spends an entire year in a
room where the activity is 4 pCi/L. Assume an RBE of 15.
 c. Does the EPA recommendation seem appropriate? Explain.

Passage Problems

Nuclear Fission

The uranium isotope ^{235}U can *fission*—break into two smaller-
mass components and free neutrons—if it is struck by a free neutron. A typical reaction is

$$^1_0n + {}^{235}_{92}U \rightarrow {}^{141}_{56}Ba + {}^{92}_{36}Kr + 3{}^1_0n$$

As you can see, the subscripts (the number of protons) and the
superscripts (the number of nucleons) "balance" before and after
the fission event; there is no change in the number of protons or
neutrons. Significant energy is released in this reaction. If a fission
event happens in a large chunk of ^{235}U, the neutrons released may
induce the fission of other ^{235}U atoms, resulting in a chain reaction.
This is how a nuclear reactor works.

The number of neutrons required to create a stable nucleus
increases with atomic number. When the heavy ^{235}U nucleus fissions, the lighter reaction products are thus neutron rich and are
likely unstable. Many of the short-lived radioactive nuclei used in
medicine are produced in fission reactions in nuclear reactors.

67. ▌▌ What statement can be made about the masses of atoms in
the above reaction?
 A. $m({}^{235}_{92}U) > m({}^{141}_{56}Ba) + m({}^{92}_{36}Kr) + 2m({}^1_0n)$
 B. $m({}^{235}_{92}U) < m({}^{141}_{56}Ba) + m({}^{92}_{36}Kr) + 2m({}^1_0n)$
 C. $m({}^{235}_{92}U) = m({}^{141}_{56}Ba) + m({}^{92}_{36}Kr) + 2m({}^1_0n)$
 D. $m({}^{235}_{92}U) = m({}^{141}_{56}Ba) + m({}^{92}_{36}Kr) + 3m({}^1_0n)$

68. ▌▌ Because the decay products in the above fission reaction
are neutron rich, they will likely decay by what process?
 A. Alpha decay.   B. Beta decay.   C. Gamma decay.

69. ▌▌ If the medically useful isotope $^{131}_{53}I$ is produced in the fission
reaction $^1_0n + {}^{235}_{92}U \rightarrow {}^{131}_{53}I + X + 3{}^1_0n$, what is the other fission product X?
 A. $^{86}_{37}Rb$ B. $^{102}_{37}Rb$ C. $^{89}_{39}Y$ D. $^{102}_{39}Y$

70. ▌▌ ^{235}U is radioactive, with a long half-life of 704 million
years. The decay products of a ^{235}U fission reaction typically
have half-lives of a few minutes. This means that the decay
products of a fission reaction have
 A. Much higher activity than the original uranium.
 B. Much lower activity than the original uranium.
 C. The same activity as the original uranium.

71. ▌▌ If a $^{238}_{92}U$ nucleus is struck by a neutron, it may absorb the
neutron. The resulting nucleus then rapidly undergoes beta-
minus decay. The daughter nucleus of that decay is
 A. $^{239}_{91}Pa$ B. $^{239}_{92}U$ C. $^{239}_{93}Np$ D. $^{239}_{94}Pu$

STOP TO THINK ANSWERS

Stop to Think 30.1: 3. Different isotopes of an element have different numbers of neutrons but the same number of protons. The number of electrons in a neutral atom matches the number of protons.

Stop to Think 30.2: Yes. ^{12}C has filled levels of protons and neutrons; the neutron we add to make ^{13}C will be in a higher energy
level, but there is no "hole" in a lower state for it to move to, so we
expect this nucleus to be stable.

Stop to Think 30.3: C. One-quarter of the atoms are left. This is
one-half of one-half, or $(1/2)^2$, so two half-lives have elapsed.

Stop to Think 30.4: B. An increase of Z with no change in A
occurs when a neutron changes to a proton and an electron, ejecting the electron.

Stop to Think 30.5: A. Dose equivalent is the product of dose in
Gy (the same for each) and RBE (highest for alpha particles).

Modern Physics

A common theme runs through the final chapters of this book: Nature is stranger than we thought it was. From the bizarre paradoxes of special relativity to the dizzying array of subatomic particles, the physical description of the world around us has taken on an almost science-fiction air in Part VII. But this material isn't science fiction; it's real, the product of decades of careful experiments.

Relativity requires us to stretch our notions of space and time. Time really does slow down for particles moving at high speeds, as decades of experiments have shown. Classical Newtonian physics has a comforting predictability, but this is left behind in quantum theory. We simply can't know an electron's position and velocity at the same time. Many decades of clever experiments have shown conclusively that *no* underlying laws can restore the predictability of classical physics at the atomic scale. And nuclear physics has shown that the alchemist's dream is true—you *can* turn one element into another!

This new physics is surprisingly practical. Relativistic corrections allow GPS systems to give extraordinarily accurate measurements of your position anywhere on the earth. Quantum mechanics is the theory underlying the development of computer chips and other modern electronics. And your smoke detector probably contains a small amount of an element not found in nature, an element created in a nuclear reactor.

As we conclude our journey, the knowledge structure for Part VII summarizes the key ideas of relativity, quantum physics, and nuclear physics. These are the theories behind the emerging technologies of the 21st century.

KNOWLEDGE STRUCTURE VII Modern Physics

BASIC GOALS	What are the properties and characteristics of space and time? What do we know about the nature of light and atoms? How are atomic and nuclear phenomena explained by energy levels, wave functions, and photons?
GENERAL PRINCIPLES	**Principle of relativity** All the laws of physics are the same in all inertial reference frames. **Quantization of energy** Particles of matter and photons of light only have certain allowed energies. **Uncertainty principle** $\Delta x \, \Delta p \geq h/2\pi$ **Pauli exclusion principle** No more than one electron can occupy the same quantum state.

Relativity

- The speed of light c is the same in all inertial reference frames.
- No particle or causal influence can travel faster than c.
- Length contraction: The length of an object in a reference frame in which the object moves with speed v is

$$L = \sqrt{1 - \beta^2}\,\ell \leq \ell$$

where ℓ is the proper length.

- Time dilation: The proper time interval $\Delta\tau$ between two events is measured in a reference frame in which the two events occur at the same position. The time interval Δt in a frame moving with relative speed v is

$$\Delta t = \Delta\tau/\sqrt{1 - \beta^2} \geq \Delta\tau$$

- Particles have energy even when at rest. Mass can be transformed into energy and vice versa: $E = mc^2$

Quantum physics

- Matter has wave-like properties. A particle has a de Broglie wavelength:

$$\lambda = \frac{h}{mv}$$

- Light has particle-like properties. A photon of light of frequency f has energy:

$$E_{\text{photon}} = hf = \frac{hc}{\lambda}$$

- The wave nature of matter leads to quantized energy levels in atoms and nuclei. A transition between quantized energy levels involves the emission or absorption of a photon.

Properties of atoms

- Quantized energy levels depend on quantum numbers n and l.
- An atom can jump from one state to another by emitting or absorbing a photon of energy $E_{\text{photon}} = \Delta E_{\text{atom}}$
- The ground-state electron configuration is the lowest-energy configuration consistent with the Pauli principle.

Properties of nuclei

- The nucleus is the small, dense, positive core at the center of an atom. The nucleus is held together by the strong force, an attractive short-range force between any two nucleons.

Proton

Neutron

- Unstable nuclei decay by alpha, beta, or gamma decay. The number of nuclei decreases exponentially with time.

$$N = N_0\left(\frac{1}{2}\right)^{t/t_{1/2}}$$

The Physics of Very Cold Atoms

Modern physics is a study of extremes. Relativity deals with the physics of objects traveling at near-light speeds. Quantum mechanics is about the physics of matter and energy at very small scales. Nuclear physics involves energies that dwarf anything dreamed of in previous centuries.

Some of the most remarkable discoveries of recent years are at another extreme—very low temperatures, mere billionths of a degree above absolute zero. Let's look at how such temperatures are achieved and some new physics that emerges.

You learned in Part III that the temperature of a gas depends on the speeds of the atoms in the gas. Very cold atoms move at very low speeds. Suppose we start with atoms at or above room temperature. How can we drastically reduce their speeds, bringing them nearly to a halt? The trick is to slow them, thus cooling them, using the interactions between light and atoms that we explored in Chapters 28 and 29.

Photons have momentum, and that momentum is transferred to an atom when a photon is absorbed. Part (a) of the figure shows an atom moving "upstream" against a laser beam tuned to an atomic transition. Photon absorptions transfer momentum, slowing the atom. Subsequent photon emissions give the atom a "kick," but in random directions, so on average the emissions won't speed up the atom in the same way that the absorptions slow it. A beam of atoms moving "upstream" against a correctly tuned laser beam is slowed down—the "hot" beam of atoms is cooled.

Once a laser cools the atoms, a different configuration of laser beams can trap them. Part (b) of the figure shows six overlapped laser beams, each tuned slightly *below* the frequency of an atomic absorption line. If an atom tries to leave the overlap region, it will be moving "upstream" against one of the laser beams. The atom will see that laser beam Doppler shifted to a higher frequency, matching the transition frequency. The atom will then absorb photons from this laser beam and the resulting kick will nudge it back into the overlap region. The atoms are trapped in what is known as *optical molasses* or, more generally, an *atom trap.* More effective traps can be made by adding magnetic fields. The final cooling of the atoms is by evaporation—letting the more energetic atoms leave the trap.

Ultimately, these techniques produce a diffuse gas of atoms moving only ≈1 mm/s. This corresponds to a nearly unbelievable temperature of just a few nanokelvin—billionths of a degree above absolute zero. This is colder than outer space. The coldest spot in the universe is inside an atom trap in a physics lab.

Once the atoms are cooled, some very remarkable things happen. As we saw in Part VII, all particles, including atoms, have a wave nature. As the atoms slow, their wavelengths increase. In a correctly prepared gas at a low enough temperature, the de Broglie wavelength of an individual atom is larger than the spacing between atoms, and the wave functions of multiple atoms overlap. When this happens, some atoms undergo *Bose-Einstein condensation,* coalescing into one "super atom," with thousands of atoms occupying the same quantum state. An example of the resulting Bose-Einstein condensate is shown in part (c) of the figure. In the condensate, the atoms—that is, their wave functions—are all in the same place at the same time! This truly bizarre state of matter is a remarkable example of the counterintuitive nature of the quantum world.

Are there applications for Bose-Einstein condensation? Current talk of atom lasers and other futuristic concepts aside, no one really knows, just as the early architects of quantum mechanics didn't know that their theory would be used to design the chips that power personal computers.

At the start of the 20th century, there was a worry that everything in physics had been discovered. There is no such worry at the start of the 21st century, which promises to be full of wonderful discoveries and remarkable applications. What do you imagine the final chapter of a physics textbook will look like 100 years from now?

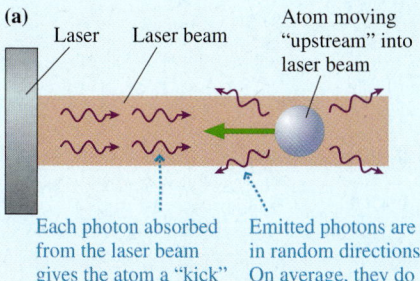

(a) Laser   Laser beam   Atom moving "upstream" into laser beam

Each photon absorbed from the laser beam gives the atom a "kick" opposite its direction of travel. Thus, each absorption slightly slows the atom.

Emitted photons are in random directions. On average, they do not affect the atom's momentum.

(b) Lasers

(c) The intersection of laser beams creates an atom trap. An atom moving out of the trap will be moving upstream into one of the laser beams, and will be pushed back.

The dense collection of atoms in the center of the trap is a Bose-Einstein condensate.

Laser cooling and trapping.

Mathematics Review

Algebra

Using exponents:

$$a^{-x} = \frac{1}{a^x} \qquad a^x a^y = a^{(x+y)} \qquad \frac{a^x}{a^y} = a^{(x-y)} \qquad (a^x)^y = a^{xy}$$

$$a^0 = 1 \qquad a^1 = a \qquad a^{1/n} = \sqrt[n]{a}$$

Fractions:

$$\left(\frac{a}{b}\right)\left(\frac{c}{d}\right) = \frac{ac}{bd} \qquad \frac{a/b}{c/d} = \frac{ad}{bc} \qquad \frac{1}{1/a} = a$$

Logarithms:

Natural (base e) logarithms: If $a = e^x$, then $\ln(a) = x \qquad \ln(e^x) = x \qquad e^{\ln(x)} = x$

Base 10 logarithms: If $a = 10^x$, then $\log_{10}(a) = x \qquad \log_{10}(10^x) = x \qquad 10^{\log_{10}(x)} = x$

The following rules hold for both natural and base 10 algorithms:

$$\ln(ab) = \ln(a) + \ln(b) \qquad \ln\left(\frac{a}{b}\right) = \ln(a) - \ln(b) \qquad \ln(a^n) = n\ln(a)$$

The expression $\ln(a + b)$ cannot be simplified.

Linear equations:

The graph of the equation $y = ax + b$ is a straight line.
a is the slope of the graph. b is the y-intercept.

Proportionality:

To say that y is proportional to x, written $y \propto x$, means that $y = ax$, where a is a constant. Proportionality is a special case of linearity. A graph of a proportional relationship is a straight line that passes through the origin. If $y \propto x$, then

$$\frac{y_1}{y_2} = \frac{x_1}{x_2}$$

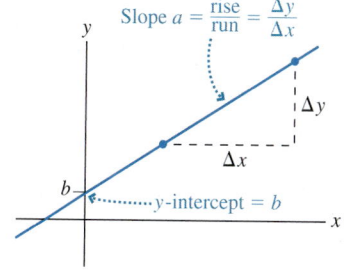

Slope $a = \dfrac{\text{rise}}{\text{run}} = \dfrac{\Delta y}{\Delta x}$

Δy

Δx

y-intercept $= b$

Quadratic equation: The quadratic equation $ax^2 + bx + c = 0$ has the two solutions $x = \dfrac{-b \pm \sqrt{b^2 - 4ac}}{2a}$.

Geometry and Trigonometry

Area and volume:

Rectangle
$A = ab$

Rectangular box
$V = abc$

Triangle
$A = \frac{1}{2}ab$

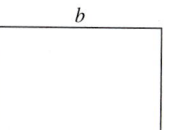

Right circular cylinder
$V = \pi r^2 l$

Circle
$C = 2\pi r$
$A = \pi r^2$

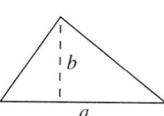

Sphere
$A = 4\pi r^2$
$V = \frac{4}{3}\pi r^3$

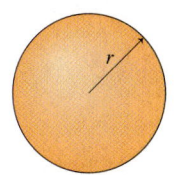

A APPENDIX

Arc length and angle: The angle θ in radians is defined as $\theta = s/r$.

The arc length that spans angle θ is $s = r\theta$.

2π rad $= 360°$

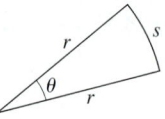

Right triangle: Pythagorean theorem $\quad c = \sqrt{a^2 + b^2}$ or $a^2 + b^2 = c^2$

$$\sin\theta = \frac{b}{c} = \frac{\text{far side}}{\text{hypotenuse}} \qquad \theta = \sin^{-1}\left(\frac{b}{c}\right)$$

$$\cos\theta = \frac{a}{c} = \frac{\text{adjacent side}}{\text{hypotenuse}} \qquad \theta = \cos^{-1}\left(\frac{a}{c}\right)$$

$$\tan\theta = \frac{b}{a} = \frac{\text{far side}}{\text{adjacent side}} \qquad \theta = \tan^{-1}\left(\frac{b}{a}\right)$$

In general, if it is known that sine of an angle θ is x, so $x = \sin\theta$, then we can find θ by taking the *inverse sine* of x, denoted $\sin^{-1} x$. Thus $\theta = \sin^{-1} x$. Similar relations apply for cosines and tangents.

General triangle: $\alpha + \beta + \gamma = 180° = \pi$ rad

Identities:

$$\tan\alpha = \frac{\sin\alpha}{\cos\alpha} \qquad\qquad \sin^2\alpha + \cos^2\alpha = 1$$

$$\sin(-\alpha) = -\sin\alpha \qquad\qquad \cos(-\alpha) = \cos\alpha$$

$$\sin(2\alpha) = 2\sin\alpha\cos\alpha \qquad\qquad \cos(2\alpha) = \cos^2\alpha - \sin^2\alpha$$

Expansions and Approximations

Binomial approximation: $(1 + x)^n \approx 1 + nx \quad$ if $\quad x \ll 1$

Small-angle approximation: If $\alpha \ll 1$ rad, then $\sin\alpha \approx \tan\alpha \approx \alpha$ and $\cos\alpha \approx 1$.

The small-angle approximation is excellent for $\alpha < 5°$ (≈ 0.1 rad) and generally acceptable up to $\alpha \approx 10°$.

Periodic Table of Elements

Key:
Atomic number — 27
Symbol — Co
Atomic mass — 58.9

Period	1	2	3	4	5	6	7	8	9	10	11	12	13	14	15	16	17	18
1	1 H 1.0																	2 He 4.0
2	3 Li 6.9	4 Be 9.0											5 B 10.8	6 C 12.0	7 N 14.0	8 O 16.0	9 F 19.0	10 Ne 20.2
3	11 Na 23.0	12 Mg 24.3											13 Al 27.0	14 Si 28.1	15 P 31.0	16 S 32.1	17 Cl 35.5	18 Ar 39.9
4	19 K 39.1	20 Ca 40.1	21 Sc 45.0	22 Ti 47.9	23 V 50.9	24 Cr 52.0	25 Mn 54.9	26 Fe 55.8	27 Co 58.9	28 Ni 58.7	29 Cu 63.5	30 Zn 65.4	31 Ga 69.7	32 Ge 72.6	33 As 74.9	34 Se 79.0	35 Br 79.9	36 Kr 83.8
5	37 Rb 85.5	38 Sr 87.6	39 Y 88.9	40 Zr 91.2	41 Nb 92.9	42 Mo 95.9	43 Tc 96.9	44 Ru 101.1	45 Rh 102.9	46 Pd 106.4	47 Ag 107.9	48 Cd 112.4	49 In 114.8	50 Sn 118.7	51 Sb 121.8	52 Te 127.6	53 I 126.9	54 Xe 131.3
6	55 Cs 132.9	56 Ba 137.3	57 La 138.9	72 Hf 178.5	73 Ta 180.9	74 W 183.9	75 Re 186.2	76 Os 190.2	77 Ir 192.2	78 Pt 195.1	79 Au 197.0	80 Hg 200.6	81 Tl 204.4	82 Pb 207.2	83 Bi 209.0	84 Po 209.0	85 At 210.0	86 Rn 222.0
7	87 Fr 223.0	88 Ra 226.0	89 Ac 227.0	104 Rf 261	105 Db 262	106 Sg 263	107 Bh 264	108 Hs 269	109 Mt 268	110 Ds 271	111 Rg 272	112 285						

Transition elements

Inner transition elements:

Lanthanides (6)

58 Ce 140.1	59 Pr 140.9	60 Nd 144.2	61 Pm 144.9	62 Sm 150.4	63 Eu 152.0	64 Gd 157.3	65 Tb 158.9	66 Dy 162.5	67 Ho 164.9	68 Er 167.3	69 Tm 168.9	70 Yb 173.0	71 Lu 175.0

Actinides (7)

90 Th 232.0	91 Pa 231.0	92 U 238.0	93 Np 237.0	94 Pu 239.1	95 Am 241.1	96 Cm 244.1	97 Bk 249.1	98 Cf 252.1	99 Es 257.1	100 Fm 257.1	101 Md 258.1	102 No 259.1	103 Lr 262.1

Atomic and Nuclear Data

Atomic Number (Z)	Element	Symbol	Mass Number (A)	Atomic Mass (u)	Percent Abundance	Decay Mode	Half-Life $t_{1/2}$
0	(Neutron)	n	1	1.008 665		β^-	10.4 min
1	Hydrogen	H	1	1.007 825	99.985	stable	
	Deuterium	D	2	2.014 102	0.015	stable	
	Tritium	T	3	3.016 049		β^-	12.33 yr
2	Helium	He	3	3.016 029	0.000 1	stable	
			4	4.002 602	99.999 9	stable	
			6	6.018 886		β^-	0.81 s
3	Lithium	Li	6	6.015 121	7.50	stable	
			7	7.016 003	92.50	stable	
			8	8.022 486		β^-	0.84 s
4	Beryllium	Be	9	9.012 174	100	stable	
			10	10.013 534		β^-	1.5×10^6 yr
5	Boron	B	10	10.012 936	19.90	stable	
			11	11.009 305	80.10	stable	
			12	12.014 352		β^-	0.020 2 s
6	Carbon	C	10	10.016 854		β^+	19.3 s
			11	11.011 433		β^+	20.4 min
			12	12.000 000	98.90	stable	
			13	13.003 355	1.10	stable	
			14	14.003 242		β^-	5 730 yr
			15	15.010 599		β^-	2.45 s
7	Nitrogen	N	12	12.018 613		β^+	0.011 0 s
			13	13.005 738		β^+	9.96 min
			14	14.003 074	99.63	stable	
			15	15.000 108	0.37	stable	
			16	16.006 100		β^-	7.13 s
			17	17.008 450		β^-	4.17 s
8	Oxygen	O	15	15.003 065		β^+	122 s
			16	15.994 915	99.76	stable	
			17	16.999 132	0.04	stable	
			18	17.999 160	0.20	stable	
			19	19.003 577		β^-	26.9 s
9	Fluorine	F	18	18.000 937		β^+	109.8 min
			19	18.998 404	100	stable	
			20	19.999 982		β^-	11.0 s
10	Neon	Ne	19	19.001 880		β^+	17.2 s
			20	19.992 435	90.48	stable	
			21	20.993 841	0.27	stable	
			22	21.991 383	9.25	stable	
17	Chlorine	Cl	35	34.968 853	75.77	stable	
			36	35.968 307		β^-	3.0×10^5 yr
			37	36.965 903	24.23	stable	

Atomic Number (Z)	Element	Symbol	Mass Number (A)	Atomic Mass (u)	Percent Abundance	Decay Mode	Half-Life $t_{1/2}$
18	Argon	Ar	36	35.967 547	0.34	stable	
			38	37.962 732	0.06	stable	
			39	38.964 314		β^-	269 yr
			40	39.962 384	99.60	stable	
			42	41.963 049		β^-	33 yr
19	Potassium	K	39	38.963 708	93.26	stable	
			40	39.964 000	0.01	β^+	1.28×10^9 yr
			41	40.961 827	6.73	stable	
26	Iron	Fe	54	54.939 613	5.9	stable	
			56	55.934 940	91.72	stable	
			57	56.935 396	2.1	stable	
			58	57.933 278	0.28	stable	
			60	59.934 072		β^-	1.5×10^6 yr
27	Cobalt	Co	59	58.933 198	100	stable	
			60	59.933 820		β^-	5.27 yr
53	Iodine	I	127	126.904 474	100	stable	
			129	128.904 984		β^-	1.6×10^7 yr
			131	130.906 124		β^-	8 days
54	Xenon	Xe	128	127.903 531	1.9	stable	
			129	128.904 779	26.4	stable	
			130	129.903 509	4.1	stable	
			131	130.905 069	21.2	stable	
			132	131.904 141	26.9	stable	
			133	132.905 906		β^-	5.4 days
			134	133.905 394	10.4	stable	
			136	135.907 215	8.9	stable	
55	Cesium	Cs	133	132.905 436	100	stable	
			137	136.907 078		β^-	30 yr
82	Lead	Pb	204	203.973 020	1.4	stable	
			206	205.974 440	24.1	stable	
			207	206.975 871	22.1	stable	
			208	207.976 627	52.4	stable	
			210	209.984 163		α, β^-	22.3 yr
			211	210.988 734		β^-	36.1 min
83	Bismuth	Bi	209	208.980 374	100	stable	
			211	210.987 254		α	2.14 min
			215	215.001 836		β^-	7.4 min
86	Radon	Rn	219	219.009 477		α	3.96 s
			220	220.011 369		α	55.6 s
			222	222.017 571		α, β^-	3.823 days
88	Radium	Ra	223	223.018 499		α	11.43 days
			224	224.020 187		α	3.66 days
			226	226.025 402		α	1 600 yr
			228	228.031 064		β^-	5.75 yr

Atomic Number (Z)	Element	Symbol	Mass Number (A)	Atomic Mass (u)	Percent Abundance	Decay Mode	Half-Life $t_{1/2}$
90	Thorium	Th	227	227.027 701		α	18.72 days
			228	228.028 716		α	1.913 yr
			229	229.031 757		α	7 300 yr
			230	230.033 127		α	75.000 yr
			231	231.036 299		α, β^-	25.52 hr
			232	232.038 051	100	α	1.40×10^{10} yr
			234	234.043 593		β^-	24.1 days
92	Uranium	U	233	233.039 630		α	1.59×10^5 yr
			234	234.040 946		α	2.45×10^5 yr
			235	235.043 924	0.72	α	7.04×10^8 yr
			236	236.045 562		α	2.34×10^7 yr
			238	238.050 784	99.28	α	4.47×10^9 yr
93	Neptunium	Np	237	237.048 168		α	2.14×10^6 yr
			238	238.050 946		β^-	2.12 days
			239	239.052 939		β^-	2.36 days
94	Plutonium	Pu	238	238.049 555		α	87.7 yr
			239	239.052 157		α	2.412×10^4 yr
			240	240.053 808		α	6 560 yr
			242	242.058 737		α	3.73×10^6 yr
			244	244.064 200		α	8.1×10^7 yr
95	Americium	Am	241	241.056 823		α	432.21 yr
			243	243.061 375		α	73.070 yr

Answers

Chapter 1

Answers to odd-numbered multiple-choice questions
21. C
23. A
25. B
27. D

Answers to odd-numbered problems

1.

Skid begins   Stops

3.
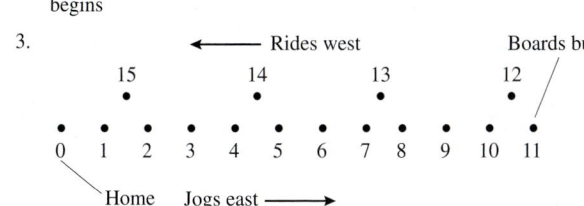

Rides west   Boards bus

Home   Jogs east →

5. -22 m
7. 5 mi
9. Bicycle, ball, cat, toy
11. -1 m/s
13. 15 s
15. a. 0.20 m   b. 10 m/s   c. 27 m/s
17. 1 km/ks $<$ 1 cm/ms $<$ 1 mm/μs
19. a. Three   b. Four   c. Five   d. Three
21. 3.81×10^2 m
23. 8.89 m/s, 9.01 m/s
25. 50 ft, 15 m
27. 1×10^{-9} m/s, 4 μm/h
29. 3.30 km
31. 10 m
33. (100 m, 61° north of east)
37. **Pictorial representation**   **Motion diagram**

y

y_i, v_i, t_i

Known

$y_i = 10$ m
$v_i = 0$
$t_i = 0$
$a = -9.8$ m/s^2
$y_f = 0$

Find

v_f

$\vec{v}$ ---•--- Starts

a

Watermelon

y_f, v_f, t_f

0

Hits sidewalk

41.
Known

$\theta = 20°$
$x_i = 0 \quad v_0 = 10$ m/s
$t_i = 0 \quad a < 0$
$v_f = 0$

Find

$h = x_f \sin \theta$

Pictorial representation

Motion diagram
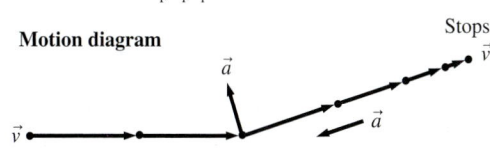

Stops

51. 12 in
53. 43 mph
55. a.
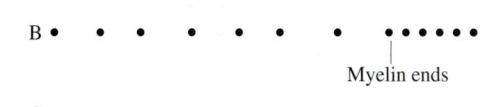

Myelin ends

b. C   c. A
57. a. AB and CD   b. All segments   c. AB and CD
59. 30 m
61. 12.0 min
63. B
65. A

Chapter 2

Answers to odd-numbered multiple-choice questions
17. C
19. D
21. B
23. A
25. D

Answers to odd-numbered problems

1. b.
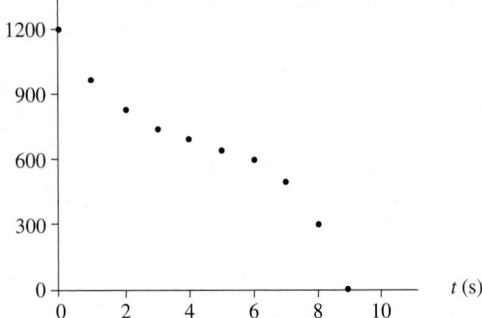
x (m)

5. a. 23 m   b. 40 m
7. a. -200 yd/min $+330$ yd/min
9. 8.3 min
11. 1.71 km
13. a.
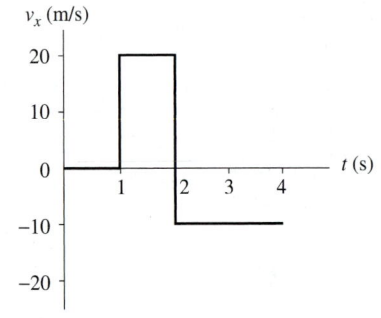
v_x (m/s)

b. There is one turning point at $t = 2$ s

15.
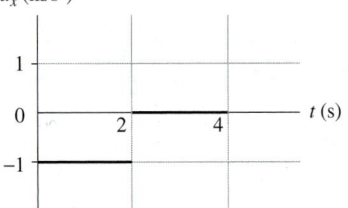
a_x (m/s^2)

17. a. 6 m   b. 4 m/s   c. 2 m/s^2
19. a. 6.1 m/s^2 b. 2.5 m/s^2 c. 1.5 m/s^2
21. a. 36 m/s^2 b. 0.22 m
23. a. 36.0 m   b. 324 m
25. -2.8 m/s^2
27. a. 5 m   b. 21.0 m/s
29. 10.0 s
31. 0.31 m
33. a. Accelerations are equal   b. Speeds are equal
35. a. 3.0 s   b. 15.2 m/s   c. 31 m/s   35 m/s

37.   **Position**   **Velocity**

A

B

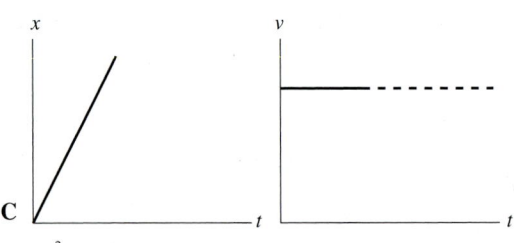
C

41. a. 83 m/s^2 b. $8.5g$
43. b.
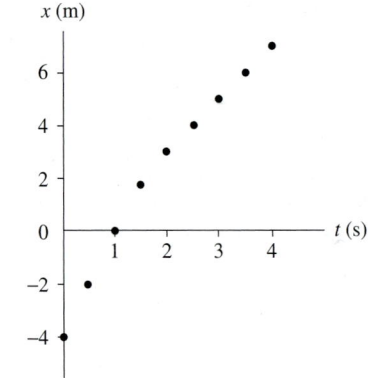
x (m)

c. 4 m   d. 4 m   e. 4 m/s   f. 2 m/s   g. -2 m/s^2
45. a. 4 s and 8 s
b.
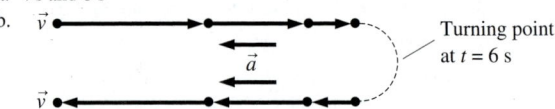
Turning point at $t = 6$ s

47. a. 180 mph   b. 23 m/s^2 c. 35 s   d. No
49. a. 350 d   b. 4.6×10^{15} m   c. 0.49 ly
51. a. 1.0 cm   b. 35 m/s^2 c. 0.84 m/s
53. a. 1000 m/s   b. 0.0010 s   c. 5.1 cm
55. 0.0 m/s^2, no deceleration
57. a. -24 m/s   b. 4.5 s
59. 12 m/s
61. a. 12.5 m   b. 45 s
63. -1.0 m/s^2
65. a. 4.1 s   b. Both same
67. a. 4.0 s   b. 16 m

69. a. 100 m b. x (m)

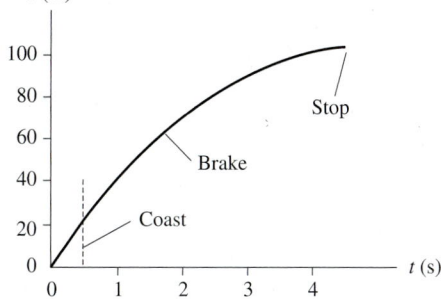

71. c. -8 m/s^2
73. c. 17 m/s
75. B

Chapter 3

Answers to odd-numbered multiple-choice questions

17. D
19. a. D b. A c. C d. A
21. a. C b. D
23. B
25. D

Answers to odd-numbered problems

3. b. Greater
5. The acceleration vector is toward the center in each case.

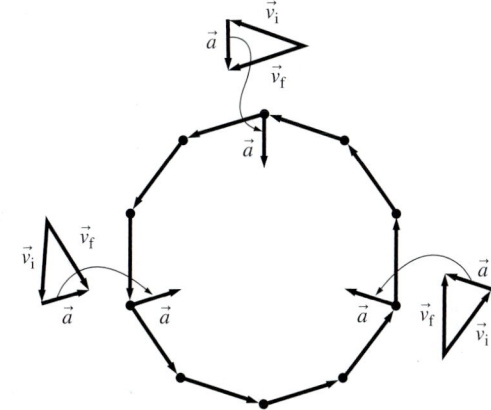

7. 12 m/s
9. 87 m/s
11. a. $d_x = 70.7$ m, $d_y = -70.7$ m b. $v_x = 282$ m/s, $v_y = 103$ m/s
    c. $a_x = 0.0$ m/s^2, $a_y = -5.0$ m/s^2
13. a. $v = 45$ mph, $\theta = 63°$ b. $a = 6.3$ m/s^2, $\theta = -72°$
15. 5.1 km
17. 530 m
19. 1.3 s
21. Ball 1: 5 m/s Ball 2: 15 m/s
23. 2.0 km/h
25. 0.4 m

27. a.

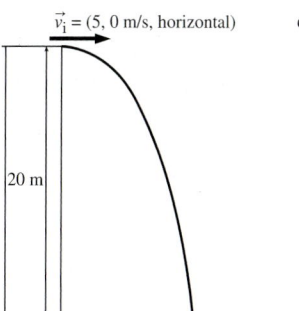

$\vec{v}_i = (5, 0$ m/s, horizontal)   d. 10 m

29. 240 m
31. a. 0.064 s b. 780 m/s
33. a. 15 s b. 330 m c. 280 m
35. 4.7 cm
37. a. 4.0 m/s^2 b. 32 m/s^2
39. Magnitude = 0.004 m/s^2, direction toward center of circle
41. a. $D_x = 8$ $D_y = 7$
    b.

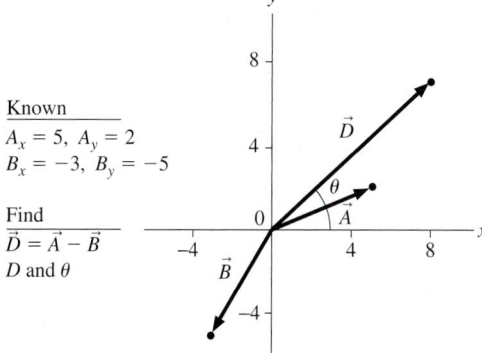

    c. $D = 11$ $\theta = 41°$ north of east
43. a. $B_x = -2$ $B_y = 2$ b. $B = 2.8$ $\theta = 45°$ north of west
45. a. 100 m, lower
    b. (500 m, east) + (5000 m, north) $-$ (100 m, vertical)
47. 7.5 m
49. 31° south of east for 20 min
51. a. 63.5 m b. 7.05 s
53. No
55. a. 39 mi b. 20 mph
57. 49° west of south
59. a. 19 mph, 27° south of west b. 8.5 mph, 3.3° east of north
61. a. $(v_x)_0 = 2.0$ m/s $(v_y)_0 = 4.0$ m/s
       $(v_x)_2 = 2.0$ m/s $(v_y)_2 = 0.0$ m/s
       $(v_x)_3 = 2.0$ m/s $(v_y)_3 = -2.0$ m/s
    b. 2.0 m/s c. 63.4° north of east
63. a. $v = L\sqrt{g/2h}$ b. 6.0 m/s, c. Yes, it is reasonable
65. 6.0 m
67. 15° and 75°
69. No
71. a. 6.1 m/s b. 1.8 m
73. b. 235 m
75. C
77. B
79. B

Chapter 4

Answers to odd-numbered multiple-choice questions
21. C
23. D
25. A
27. B
29. C

Answers to odd-numbered problems
1. First is rear-end; second is head-on
3. People in back can fly forward and injure people in front
5.

7. Weight, tension force by rope
9. Weight, normal force by ground, kinetic friction force by ground
11. Weight, normal force by ground, kinetic friction force by ground
13. $m_1 = 0.08$ kg, $m_3 = 0.50$ kg
15. a. 2.4 m/s^2 b. 0.6 m/s^2
17. a. 16 m/s^2 b. 4 m/s^2 c. 8 m/s^2 d. 32 m/s^2
19. 2.5 m/s^2
21. 0.025 kg
23. a. 25 N   b. 100 N   c. 70 N
25. 0.02 m/s^2
29.

Force identification	**Free-body diagram**

35.

Force identification	**Free-body diagram**

37. The pairs are identified on the diagram.

(i) Earth

(ii)

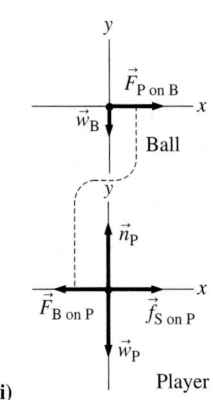

(iii)

39. **Motion diagram**   41. **Motion diagram**

43.   45.

47.

49.

53.

55.

59.

61.

63.

67. a.
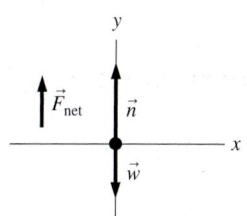

b. Greater
69. C
71. C

Chapter 5

Answers to odd-numbered multiple-choice questions

25. a. C   b. D
27. a. C   b. B
29. C
31. C
33. D

Answers to odd-numbered problems

1. $T_1 = 87$ N $T_2 = 50$ N
3. 110 N each
5. 510 N each
7.
9.
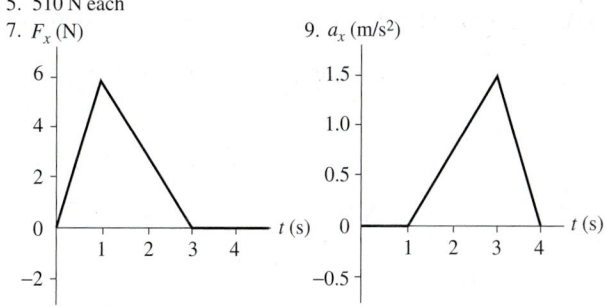

11. $a_x = 1.0$ m/s^2 $a_y = 0.0$ m/s^2
13. a. 0.0 N   b. 0.0 N   c. 250 N
15. 9800 N toward the rear
17. a. 540 N   b. 89 N
19. a. 780 N   b. 1600 N
21. a. 780 N   b. 1100 N
23. a. 5.9 N   b. 5.1 N
25. 0.25
27. 136 m
29. a.   b.
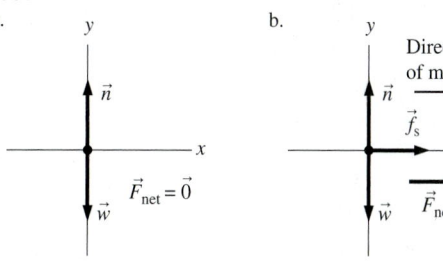

c. 4.9 m/s^2 d. 2.9 m/s^2

31. 170 m/s
33. a. 5.2×10^{-8} m   b. Much less
35. a. 6 N   b. 10 N
37. a. 20 N   b. 21 N
39. a. 530 N   b. 5300 N
41. At $t = 1$ s, $F_{net} = 8$ N; at $t = 4$ s, $F_{net} = 0$ N; at $t = 7$ s, $F_{net} = -12$ N
43. a. 490 N   b. 740 N
45. a. 20000 m/s²   b. 2800 N   c. 2800 N   d. Forehead no, cheek yes
47. a. 230 N   b. 0.20 m/s
49. a. 3.96 N   b. 2.32 N
51. a. 1.3 m/s²   b. 2.0 m/s²
53. a. 9.8 m/s²   b. -9.8 m/s²   c. 20 m/s²   d. 2.5 m/s   e. 0.31 m
55. a. 17 m/s   b. 150 m
57. Stay at rest
59. 1.9×10^5 N
61. a. $-5g$ b. $3g$
63. 9800 N
65. b. 0.36 s
67. c. 100 m
69. c. 2.8 m/s²
71. A
73. D

47. 22 m/s
49. a. 1.7 m/s²
    b.   c. 0.012 N, up   d. 0.0098 N

51. 54 N
53. a. 9.0 N
55. 2400 m
57. a. 3.0×10^{24} kg   b. 0.90 m/s²
59. a. 3.8 m/s²   b. 1.6 m/s
61. (12 cm, 0 cm)
63. 6.5×10^{23} kg
65. 0.48 m/s
67. b. 19.8 m/s
69. b. 1.00×10^8 m
71. C
73. A

Chapter 6

Answers to odd-numbered multiple-choice questions

23. C
25. B
27. A
29. D
31. C

Answers to odd-numbered problems

1. 1.7×10^{-3} rad/s
3. a. 1.3 rad   72°   b. 3.9×10^{-5} rad/s
5. 3.0 rad
7. 3.9 m/s
9. a. 20 s   b. 2.5 m/s
11. $v = 5.7$ m/s $a = 110$ m/s²
13. $T_3 > T_1 = T_4 > T_2$
15. 9400 N, toward center, static friction
17. a. 1.7×10^3 m/s²   b. 240 N
19. 12 m/s
21. 20 m/s
23. 1.58 m/s²
25. 1/2
27. 6.0×10^{-4}
29. a. 3.5×10^{22} N   b. 2.0×10^{20} N   c. 0.56%
31. 1600 earth days
33. 92 min   7700 m/s
35. 4.2 h
39. North pole, by 2.5 N
41. 5.5 m/s
43. a. Two; upward normal force and inward normal force   b. 0.24 N
45. a. 5.0 N   b. 30 rpm

Chapter 7

Answers to odd-numbered multiple-choice questions

17. C
19. A
21. A
23. C

Answers to odd-numbered problems

1. 11 m/s
3. a. 157 rad/s²   b. 50 rev
5. $\tau_1 < \tau_2 = \tau_3 < \tau_4$
7. -0.20 N·m
9. 5.7 N
11. a. 0.036 N·m   b. 100 N·m
13. 5.5 N·m, counterclockwise
15. a. 1.1 m   b. 1.1 m
17. 12 N·m
19. -98 N·m
21. a. 34 N·m   b. 24 N·m
23. a. 1.7 m   b. 833 N·m
25. 7.2×10^{-7} kg·m²
27. 120 kg·m²
29. 32
31. a. 6.9 kg·m²   b. 0.75 rad/s²
33. 1.6 kg·m²
35. 0.047 N·m
37. 17 rad/s²
39. 0.50 s
41. 2.2 m/s²
43. 16.8 m/s

45.

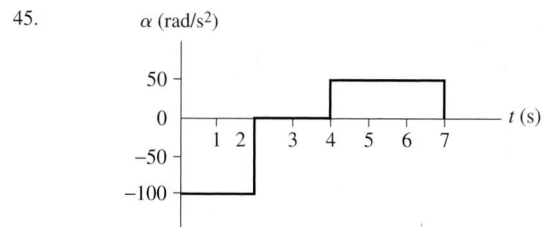

47. $-0.94 \text{ N} \cdot \text{m}$
49. 7.5 cm
51. a. (8.7 cm, 8.2 cm)   c. 43°
53. $0.018 \text{ kg} \cdot \text{m}^2$
55. $1.1 \text{ kg} \cdot \text{m}^2$
57. $6.0 \times 10^{-4} \text{ kg} \cdot \text{m}^2$
59. a. 0.039 N   b. 38 rpm
61. 3.4 m/s
63. $-0.28 \text{ N} \cdot \text{m}$
65. 1.1 s
67. B
69. B

Chapter 8

Answers to odd-numbered multiple-choice questions
15. B
17. B
19. C

Answers to odd-numbered problems
1. Right 470 N; left 160 N
3. 140 N
5. No, there is a net torque
7. 590 N
9. 1
11. 8.9°
13. 15 cm
15. 830 N/m
17. a. 13 cm   b. 8 cm
19. 18 cm
21. 1.5 m/s^2
23. a. 2.0 mm   b. 0.25 mm
25. 7900 N
27. a. 4000 N   b. 4.0 mm
29. 15 cm
31. 0.0078%
33. 1.4 mm
35. a. $F_1 = 750 \text{ N}$ b. 1000 N
37. 94 N
39. 350 N
41. $81 \text{ N} \cdot \text{m}$
43. 4.0 m
45. a. 49 N   b. 1500 N/m   c. 3.4 cm
47. 17.4 N
49. a. 1.0 cm   b. 4.5 N, to the left
51. 0.31 m
53. a. 8.2 m/s   b. 30 N
55. 2.7 cm
57. 25%
59. 30%
61. a. $4.8 \times 10^{-2} \text{ m}^2$ b. 70000 kg   c. 14%
63. D
65. A

Chapter 9

Answers to odd-numbered multiple-choice questions
19. D
21. D
23. B
25. B

Answers to odd-numbered problems
1. 75 m/s
3. 6.0 N
5. a. $1.5 \times 10^4 \text{ kg} \cdot \text{m/s}$ b. $8.0 \text{ kg} \cdot \text{m/s}$
7. a. 1.5 m/s, to the right   b. 0.5 m/s, to the right
9. 110 N, directed opposite the original motion
11. a. 19000 N, directed opposite the original motion,
    b. 280000 N, directed opposite the original motion
13. 0.205 m/s
15. 0.081 m/s, no
17. 0.143 m/s
19. 0.200 m/s
21. 4.8 m/s
23. 2.3 m/s
25. 300 m/s
27. $(p_{1x})_f = -2 \text{ kg} \cdot \text{m/s}$, $(p_{1y})_f = 4 \text{ kg} \cdot \text{m/s}$
29. (14 m/s, 45°)
31. $510 \text{ kg} \cdot \text{m}^2/\text{s}$
33. $(0.025 \text{ kg} \cdot \text{m}^2/\text{s}$, into page)
35. 1.3 rev/s
37. 0.20 s
39. 940 N
41. 3.8 m/s
43. a.

	p_x	p_y
After launch	$1.08 \text{ kg} \cdot \text{m/s}$	$0.625 \text{ kg} \cdot \text{m/s}$
At top	$1.08 \text{ kg} \cdot \text{m/s}$	$0.00 \text{ kg} \cdot \text{m/s}$
Before landing	$1.08 \text{ kg} \cdot \text{m/s}$	$-0.625 \text{ kg} \cdot \text{m/s}$

   b. p_x is constant because no forces act on the ball in the x-direction
45. $7.5 \times 10^{-10} \text{ kg} \cdot \text{m/s}$
47. a. 6.4 m/s   b. 360 N
49. 3.6 m/s
51. 30 m/s
53. 2.0 m/s
55. a. $6.7 \times 10^{-8} \text{ m/s}$ b. $2 \times 10^{-10}\%$
57. 13 s
59. 440 m/s
61. 20.0 m/s, downward
63. 200 m/s
65. $1.5 \times 10^7 \text{ m/s}$, to the right
67. 0.81 m/s, 73° below $+x$-axis
69. 4.5 rpm
71. a. 2.8 m/s   b. 2.2 m/s, 25 cm
73. 50 rpm
75. 18 rev/s, clockwise
77. B
79. C

Chapter 10

Answers to odd-numbered multiple-choice questions
29. C
31. C
33. C

Answers to odd-numbered problems

1. 0 J
3. 12,500 J by the weight, −7920 J by $\vec{T}_1$, −4580 J by $\vec{T}_2$
5. a. 0 J   b. 2600 J   c. −2600 J
7. The bullet
9. 2
11. 0 J
13. 1.8 kg · m²
15. a. 6.8×10^5 J   b. 46 m   c. No
17. a. 13 m/s   b. 14 m/s
19. 31 m/s
21. a. Yes   b. 14 m/s
23. 0.63 m
25. 3.0 m/s
27. 9.7 J
29. 17 m/s
31. a. Potential energy is transformed to kinetic and thermal energy.
    b. 550 J
33. 0.86 m/s and 2.9 m/s
35. Proton: 1.7×10^7 m/s to left; Carbon: 3.1×10^6 m/s to right
39. a. 180 J   b. 59 W
41. 45 kW
43. a. 30 N   b. 45 W
45. 2.0×10^4 W
47. a. 7.7 m/s   b. 6.6 m/s
49. 2.3 m/s
51. a. 0.20 kJ   b. 98 N   c. 2.0 m   d. 0.20 kJ
53. 15 m/s
55. 51 cm
57. a. 15 m/s   b. They will go hungry.
59. 3.8 m, not dependent on mass
61. a. $\sqrt{\dfrac{(m + M)kd^2}{m^2}}$ b. 200 m/s
    c. 0.9975 kinetic energy transformed to thermal energy
63. 7.9 m/s
65. a. $0.048v_i$ b. 95%
67. a. 100 N   b. 0.20 kW   c. 1.2 kW
69. 5.5×10^4 L
71. B
73. C
75. A
77. B
79. C

Chapter 11

Answers to odd-numbered multiple-choice questions

21. A
23. B
25. C

Answers to odd-numbered problems

1. 8.4×10^5 J
3. 3.3%
5. a. 4.2×10^6 J   b. 43 km
7. 230,000 J = 55,000 cal = 55 Cal
9. 1.4 km
11. 710 m
13. a. 200 J   b. 16,000 J/day   c. 0.0095 donuts/day
15. 1.0 K
17. 30 K
19. −700 J

21. −150 J
23. a. 15 kJ   b. 27%
25. 25%
27. a. 7°C   b. −4°C
29. 1.5
31. a. 200 J   b. 250 J
33. a. (b) only   b. (a) only
35. b. 0.125
37. 0.011 kg
39. a. 1.4×10^6 J   b. 1.4×10^6 J
41. a. 190 kJ   b. 760 kJ   c. 180 Cal   d. 140 Cal   e. 1200 W
43. 180 m/s, 1×10^{-22} J
45. 1.2×10^4 m
47. a. 3400 J   b. 3400 J
49. 320 K
51. 2/3
53. 30°C
55. 1/2
57. a. 110 m   b. 32%
59. 8.0%
61. C
63. A
65. C

Chapter 12

Answers to odd-numbered multiple-choice questions

27. D
29. A
31. E
33. D.

Answers to odd-numbered problems

1. Carbon
3. 3.54×10^{24}
5. 0.024 m³
7. 4.7 m
9. 0.83%
11. 49.7 psi
13. 1.1 N
15. a. −141°C   b. 2.7×10^{-21} J
17. a. 1.2 mol   b. 1.5 mol   c. 5.3 mol
19. 17 L
21. a. $T_2 = T_1$ b. $p_2 = p_1/2$
23. a. 98 cm³
    b.

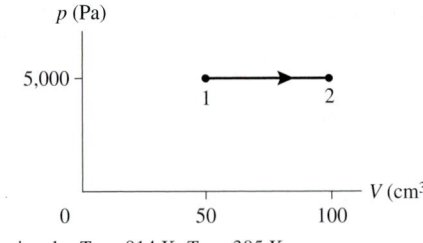

25. a. Isochoric   b. $T_1 = 914$ K, $T_2 = 305$ K
27. a. Isobaric   b. 391 K   c. 9.4 mol
29. 289 K = 16°C
31. 6900 J
33. 2300 W
35. 260000 J
37. 0.218 kg
39. Iron
41. a. 31 J   b. 60°C
43. 0.080°C or K

45. 41 J
47. 830 W
49. 24 W
51. 6.0 W
53. a. 0.050 m³   b. 1.3 atm
55. a. 1.6 mol   b. 9.4 × 10²³   c. 200 kPa
57. 6.8 × 10⁶ J
59. 17 m
61. 32 psi
63. No
65. 200 cm³
67. a. 600 cm³
    b.

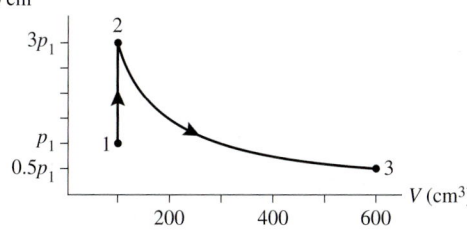

69. a. 0.51 atm   b. −112°C
    c.

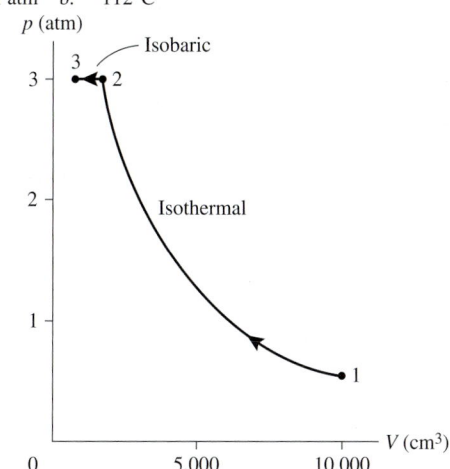

71. 2.4 × 10⁶ kg/min
73. 0.023 K, no
75. a. 83 J/(kg · K)   b. 2.0 × 10⁵ J/kg
77. 0.97 kg
79. 61 g
81. a. −3.1 atm   b. 9.7 L
83. a. 150 J added   b. 91 J removed
85. 0.026 W/(m · K)
87. 2
89. b. *p* (atm)   c. 6 atm

91. b. *p*   c. 420 cm³

93. A
95. C

Chapter 13

Answers to odd-numbered multiple-choice questions

31. A
33. D
35. B
37. B
39. A

Answers to odd-numbered problems

1. 1200 kg/m³
3. 11 kg/m³
5. a. 810 kg/m³   b. 840 kg/m³
7. 8.0%
9. a. 6.3 m³   b. 1.2 × 10⁵ Pa
11. 210 N
13. 3.2 km
15. a. 2.9 × 10⁴ N   b. 30
17. a. 110 kPa   b. 4400 Pa for both A–B and A–C
19. 3.7 mm
21. 89 mm Hg
23. 2 cm
25. 1.9 N
27. 46 kg
29. 1.0 m/s
31. 97 min
33. 110 kPa
35. 3.0 m/s
37. 260 Pa
39. 2.17 cm
41. 4.8 × 10²³
43. 27 cm
45. a. 5800 N   b. 6000 N
47. 1.8 cm
49. 750 kg/m³
51. 44 N
53. a. 0.38 N   b. 20 m/s
55. $\frac{4}{\pi}v_1 \approx 1.3v_1$
57. a. p_{atmos} b. 4.6 m
59. 4.4 cm
61. −28°C
63. 13 mm Hg
65. B
67. A

Chapter 14

Answers to odd-numbered multiple-choice questions

19. a. B   b. C   c. B   d. B   e. B
21. a. D   b. B   c. B
23. B

Answers to odd-numbered problems

1. 2.3 ms
3. 0.30 Hz
5. c. 12°   d. 0° to 10°
7. a. 0.13 m   b. 9.0 cm
9. a. 20 cm   b. 0.25 Hz
11.

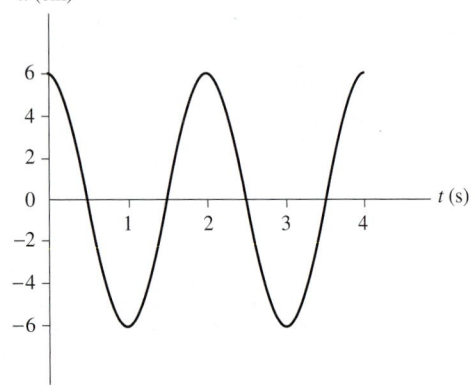

13. a. 0.2 m/s²   b. $\frac{1}{50}g$
15. a. $\frac{1}{4}$ is potential; $\frac{3}{4}$ is kinetic   b. $A/\sqrt{2}$
17. a. 2.83 s   b. 1.41 s   c. 2.00 s   d. 1.41 s
19. a. 2.0 cm   b. 0.63 s   c. 5.0 N/m   d. 20 cm/s   e. 1.0×10^{-3} J
    f. 15 cm/s
21. a. 1.00 s   b. 0.628 m/s   c. 0.100 J
23. a. 4.00 s   b. 5.66 s   c. 2.83 s   d. 4.00 s
25. a. 0.20 rad   b. 0.80 Hz   c. 0.39 m   d. −0.084 rad
27. 33.1 cm
29. 1.6 s
31. a. 0.124 rad   b. 2.00 s   c. 1.49 m
33. 10.0 s
37. 2.8 s
39. a. 20 N/m   b. 0.40 s
41. a. 24.5 N/m   b. 0.898 s   c. 0.700 m/s through equilibrium position
43. 3.5 Hz

45.

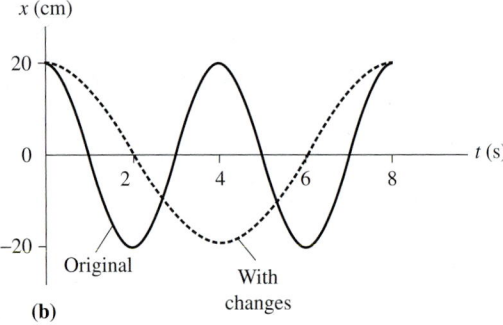

47. a. 0.25 Hz; 3.0 s   b. 6.0 s; 1.5 s   c. 9/4
49. a. 6.40 cm   28.3 cm/s
51. 0.08 N/m
53. 0.11 m; 1.7 s
55. 0.64 Hz
57. 8.1 N/m
59. a. 2.00709 s   b. 2.00721 s   c. 4.6 h
61. a.

	f (Hz)	T (s)
24.8 cm rod	1.00	1.00
38.8 cm rod	0.800	1.25

    b. 5.00 s   c. 0.200 Hz
63. 6.0 cm
65. 250 oscillations
67. 0.0453 J
69. C
71. B

Chapter 15

Answers to odd-numbered multiple-choice questions

19. D
21. D
23. D

Answers to odd-numbered problems

1. 280 m/s
3. 140 m/s
5. 6060 m/s

7.

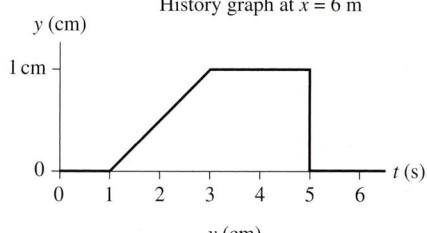

History graph at $x = 6$ m

9.

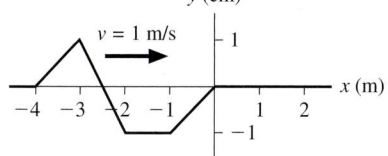

Snapshot graph at $t = 1.0$ s

11. 10 m/s
13. a. 8.6 mm   b. 15 ms
15. a. 11 Hz   b. 1.1 m   c. 13 m/s
17. $A = 4.0$ cm; $\lambda = 12$ m; $f = 2.0$ Hz
19. 1.5 cm
21. a. 1700 Hz   b. 1.5 GHz   c. 990 nm
23. a. 10 GHz   b. 0.17 ms
25. 5.5×10^{-15} J
27. 6.0×10^5 J
29. 14.2 W/m^2
31. 2.5×10^{-11} W/m^2; 14 dB
33. a. 1.1×10^{-3} W/m^2; 90 dB
35. a. 650 Hz   b. 560 Hz
37. Away at 16 m/s
39. 4.5 m/s
41. 86 m
43. 790 m
45. 880 g
47. $L_1 = 1.7$ m; $L_2 = 2.3$ m
49.

$y(x, t)$

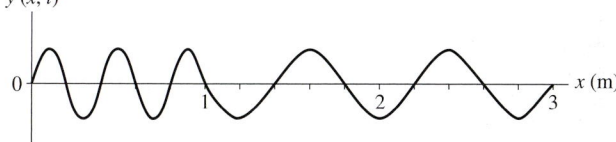

51. $f_{water}/f_{air} = 1.00$; $v_{water}/v_{air} = 4.31$; $\lambda_{water}/\lambda_{air} = 4.31$
53. 0.07°C
55. a. 3 s   b. 9 m/s
57. $y(x, t) = (5.0 \text{ cm}) \cos\left[2\pi\left(\dfrac{x}{50 \text{ cm}} + \dfrac{t}{0.125 \text{ s}}\right)\right]$
59. b. 33 m/s to the right
61. 0.068 m/s
63. 2.0×10^{-5} W/m^2
65. a. 5.0×10^7 W   b. 5.0 W   c. 6.4×10^{17} W/m^2 d. 5.8×10^{14}
67. 1.0 W/m^2
69. a. 1.9 W/m^2 b. 120 dB   c. 69 m
71. 620 Hz and 580 Hz
73. 110 m/s
75. B
77. C

Chapter 16

Answers to odd-numbered multiple-choice questions
19. A
21. D
23. A
25. D

Answers to odd-numbered problems
1.

5. 4 s

7.

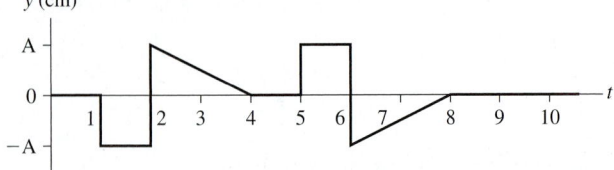

9. 60 Hz

11. 53 m/s

13. a. 4.80 m, 2.40 m, 1.60 m   b. 75 Hz

15. a. 0.807 m   b. 69.7 N

17. 4.8 cm

19. 10.60 μm; 2.830 × 10^{13} Hz

21. a. 16 Hz   b. 10 ft

23. a. 2.42 m, 1.21 m, 0.807 m   b. 4.84 m, 1.61 m, 0.968 m

25. 510 Hz

27. 32 Hz

29. 43 cm

31. a. 80 cm   b. 100 cm

33.

	r_1	r_2	Δr	C/D
P	3λ	4λ	λ	C
Q	$\frac{7}{2}\lambda$	2λ	$\frac{3}{2}\lambda$	D
R	$\frac{5}{2}\lambda$	$\frac{7}{2}\lambda$	λ	C

35. 203 Hz

37. 880 m/s

39. 1.2 m

41. 260 N

43. 12 kg

45. 8.2 m/s^2

47. 19 kN

49. a. 2.8 m   b. 1.4 m/s

51. 13.0 cm

53. 3800 Hz

55. 1210 Hz

57. 18 cm

59. 1400 Hz

61. a. 43.00 m   b. Strong   c. Weak

63. 20

65. a. No   b. 10 cm

67. a. 5 beats/s   b. 4.6 mm

69. 7.0 m/s

71. 0.40 m/s

73. B

75. B

Chapter 17

Answers to odd-numbered multiple-choice questions

17. B

19. C

21. B

23. D

Answers to odd-numbered problems

1. a. 1.5 × 10^{-11} s   b. 3.4 mm

3. 0.050 ns

5. 67 μs

7. 0.020 rad; 1.1°

9. 500 nm

11. 0.40 mm

13. $\theta_1 = 3.3°; \theta_2 = 6.3°$

15. 530

17. 224

19. a. 1.3 m   b. 7

21. 216 nm

23. 85.9 nm

25. 121 nm

27. 97.8 nm

29. 94 μm

31. 610 nm

33. 4.9 mm

35. 78 cm

37. 6.43 cm

39. 500 nm

41. 450 nm

43.

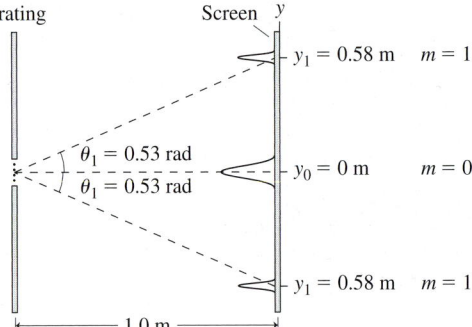

45. 19 cm

47. 43.2°

49. 43 cm

51. 500 nm

53. 25 cm

55. 1.8 μm

57. 500 nm

59. 667.8 nm

61. 690 nm and 410 nm; purplish

63. 2.9°

65. 7.6 m

67. 0.10 mm

69. a. Double   b. 0.15 mm

71. 0.40 mm

73. a. No; diffraction   b. 0.029°   c. 0.31 cm   d. 1.0 m

75. D

77. A

Chapter 18

Answers to odd-numbered multiple-choice questions

19. D

21. B

23. E

25. C

Answers to odd-numbered problems

1. 8.0 cm

3. 11 ft

5. a. 5.8 mm   b. The object must be getting closer.

7. B and C

9. 430 cm

11. a. 3   b. B = (1 m, −2 m), C = (−1 m, 2 m), D = (1 m, 2 m)

c.

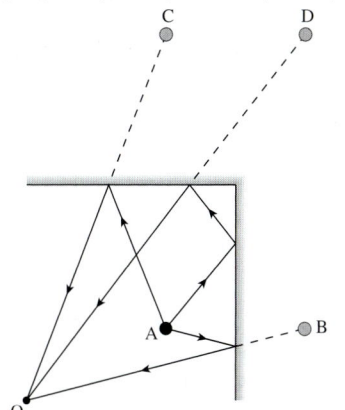

13. 1.37
15. 4.0 m
17. Yes, by filling the room you are in with water
19. 76.7°
21. Violet
23. 23 cm
25. a. Closer   b. Farther
27. $s' = 15$ cm, inverted and real
29. $s' = -6.7$ cm, upright and virtual
31. $s' = -40$ cm, upright and virtual
33. $s = 18$ cm
35. $s' = -1.3$ m, upright and virtual
37. a. 4.4 cm   b. Closer   c. Make the camera longer
39. 1.0 m
41. 42.0°
43. 20°
45. 60°
47. a. 32° from the normal   b. No
49. 42°
51. 23°
53. 35°
55. 83°
57. a. Yes   b. 60 cm
59. a. $\beta = \sin^{-1}\left(n\sin\left(\frac{1}{2}\right)\alpha\right)$ b. 1.58
61. 1.52
65. $2f$
67. $s' = 20$ cm, $h' = 1.0$ cm
69. $s' = -4.3$ cm, $h' = 0.57$ cm
71. $s' = -0.50$ m, $h' = 0.40$ m
73. C

Chapter 19

Answers to odd-numbered multiple-choice questions

15. B
17. A
19. B
21. C

Answers to odd-numbered problems

1. $s' = 40$ cm, $h' = 2.0$ cm
3. $s' = -60$ cm, $h' = 8.0$ cm
5. $s' = -8.6$ cm, $h' = 1.1$ cm

7. $s' = -9.4$ cm, $h' = 1.9$ cm
9. $s' = -38$ cm, $h' = 7.5$ cm.
11. $h'_A > h'_D > h'_B > h'_C$
13. $d_C > d_D > d_B > d_A$
15. 1/30 s
17. 1/250 s
19. 2.0 D
21. 3
23. a. Farsighted   b. ∞
25. 5.0 cm
27. a. 1.20 cm   b. 1.67 cm
29. 11 mm
31. a. 0.058 D   b. 17 mm
33. a. 14°   b. 6.4 mm
35. a. 380 nm   b. 270 nm
37. $f = 44$ cm, 67 cm
39. a. 5.9 cm   b. 6.0 cm
41. b. $s'' = -20$ cm, $h'' = 2.0$ cm
43. $s'' = -20$ cm, $h'' = 5.0$ cm, upright
45. $s'' = 7.5$ cm behind the second lens
47. a. 180 cm   b. 26 m   c. 4.7 cm   d. 1.5 m   e. 0.026 rad
    f. 0.058 rad
    g. The angular size of the image is smaller in the convex mirror than the angular size of the image in a flat mirror would be.
49. a. 1.7×10^{-4} rad   b. 4.1 μm
    c. Yes, the distance between cones is just over half of the distance between these two images.
51. a. Objective: +3.0 D; eyepiece: +4.5 D   b. −1.5   c. 0.56 m
53. 24 cm
55. 2.8×10^7 rad
57. 11 km
59. No
61. A
63. B

Chapter 20

Answers to odd-numbered multiple-choice questions

23. a. B   b. C   c. B   d. C
25. B
27. C
29. A

Answers to odd-numbered problems

1. a. Removed   b. 3.1×10^{10}
3. 1.5×10^6 C
5. a. Electrons from sphere to rod   b. 2.5×10^{10}
7. A has −160 nC, B has 0 C
9. a. 9.0×10^{-3} N   b. 9.0×10^{-3} m/s^2
11. 2.6 cm
13. 1.0×10^{-8} C
15. 8.0 nC
17. a. 1.7×10^{14} m/s^2, away from bead
    b. 3.2×10^{17} m/s^2, toward bead
19. 1.4×10^5 N/C, away from bead
21. a. 1.5×10^{-3} N/C, away from proton
    b. 1.4×10^{-3} N/C, toward electron
23. a. 3.6×10^4 N/C, 1.8×10^4 N/C, 1.8×10^4 N/C

b.

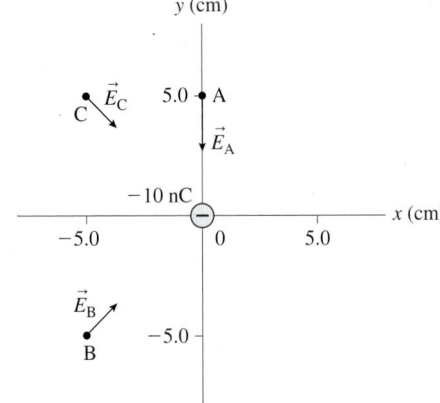

25. 2500 N/C, 90° below horizontal
27. a. 1.0×10^{-7} N/C, upward   b. 5.6×10^{-11} N/C, downward
29. a. 2   b. $\dfrac{1}{4}$ c. 1
31. 1.9 cm
35. 7.2×10^{-15} N
37. a. Zero   b. 5.6×10^{-27} N · m
39. 1.4×10^5 C
41. a. 500 N   b. 3.0×10^{29} m/s^2
43. a. 0.45 N   b. $q_A = 1.0 \times 10^{-6}$ C, $q_B = 0.50 \times 10^{-6}$ C
45. 4000 N/C, 9.3° above the horizontal
47. 3.1×10^{-4} N, upward
49. Zero
51. 1.1×10^{-5} N, upward
53. 0.68 nC
55. a.

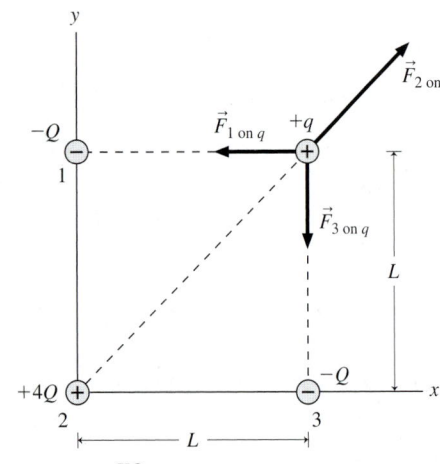

b. $F_{net} = (2 - \sqrt{2})\dfrac{KQq}{L^2}$

57. 6.6×10^{15} rev/s
59. (−6.0 cm, 3.0 cm)
61. 5.3×10^5 N/C
63. 33 nC
65. 180 nC
67. b. 1000
69. b. 5.0 cm
71. D
73. A

Chapter 21

Answers to odd-numbered multiple-choice questions

21. A
23. A
25. C
27. A
29. C

Answers to odd-numbered problems

1. $-3.0\ \mu$C
3. +530 V
5. a. 1000 eV   b. 1.6×10^{-16} J   c. 4.4×10^5 m/s
7. a. Lower   b. 0.71 V   c. 0.71 eV
11. a. $E = 2.0 \times 10^5$ V/m, $\Delta V_C = 200$ V   b. $E = 2.0 \times 10^5$ V/m, $\Delta V_C = 400$ V
13. a. 0.11 m, 0.075 m, 0.056 m
15. −5800 V
17. a. 1.5 V   b. 8.3×10^{-12} C
19. 200 V
21. 10000 V/m to the left
23. a. Positive   b. About zero   c. Positive
25. a. 7.1 pF   b. 0.71 nC
27. 4.8 cm
29. 24 V
31. 1.5×10^{-5} C
33. 13 nF
35. a. $Q = 5.0 \times 10^{-10}$ C $\Delta V_C = 9.0$ V $E = 90\,000$ V/m
    b. $Q = 2.5 \times 10^{-10}$ C $\Delta V_C = 9.0$ V $E = 90\,000$ V/m
37. 1400 V
39. 1/2
41. 1200 V/m
43. 3.2×10^{-5} J
45. -2.2×10^{-19} J
47. 1.4×10^{-3} N
49. ±12 cm
51. a. 103 V   b. 5.40×10^4 N/C along +x-axis
53. a. 27 V   b. -4.4×10^{-18} J
55. a. 3100 V   b. 5.0×10^{-16} J
57. 9600 V/m
59. 0.075 eV, increase
61. 1.4×10^5 m/s
63. 7.1×10^4 m/s
65. a. 1000 V   b. 1.4×10^{-9} C   c. 7.0×10^6 m/s
67. −0.200 V
69. 4.0×10^7 m/s
71. a. 2.1×10^{-10} C, 3000 V/m, 15 V   b. 1.1×10^{-10} C, 1500 V, 15 V
73. a. Smaller   b. Smaller   c. 5.5 V/cm, in the +x-direction
75. a. 0.80 mm   b. 1.6 mm
77. 22×10^{-6} F
79. b. 1.7×10^6 m/s
81. b. 0.022 mm
83. A
85. D

Chapter 22

Answers to odd-numbered multiple-choice questions

23. C
25. C
27. D

Answers to odd-numbered problems

1. 3000 C, 1.9×10^{22}
3. 1.9×10^{20}
5. 0.17 μA
7. 13 A
9. 120 C
11. $I_B = 5$ A, $I_C = -2$ A
13. 9.4×10^{18}
15. 12 V
17. 32
19. a. 1 b. 1/2
21. a. 0.087 Ω b. 0.0097 Ω
23. 4100 Ω
25. 13 mV
27. 3.2 mA
29. a. 2.0 Ω b.

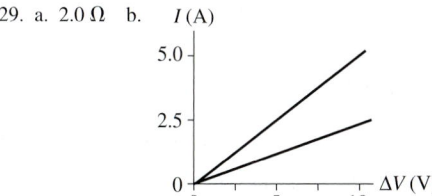

31. a. 200 V/m b. 100 V/m c. 0.50 A d. 480 Ω
33. More than 15 A
35. 15000 J
37. 23 mA
39. 94 s
41. 42 GΩ
43. 1800 mA · h
45. 1.4 Ω · m
47. 1.9 V
49. a. 3.1×10^{14} b. 9.1×10^5 N/C c. 0.23 W
51. 0.62 mm
53. Iron
55. 28 cm
57. 8.4×10^{-8} Ω
59. a. 14 C b. 27 J
61. a. 0.24 mA b. 120 mA c. 30 mW
63. D
65. A

Chapter 23

Answers to odd-numbered multiple-choice questions

29. A
31. B
33. B
35. C

Answers to odd-numbered problems

1.

3.

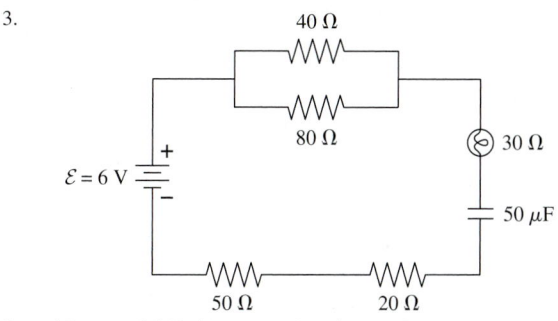

5. a. $\Delta V_{12} = -2.0$ V, $\Delta V_{23} = -1.0$ V, $\Delta V_{34} = 0$ V
   b. $\Delta V_{12} = 0$ V, $\Delta V_{23} = -3.0$ V, $\Delta V_{34} = 0$ V
7. a. 0.50 A clockwise
   b.

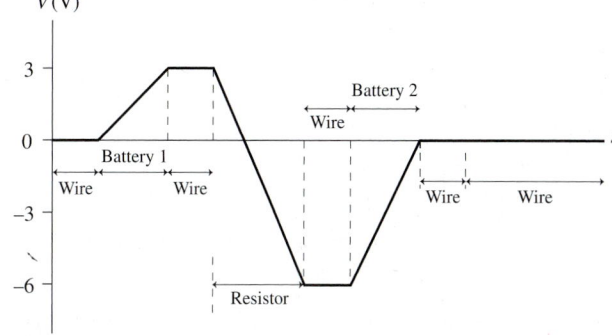

11. a. 1.0 Ω b. 1.0 Ω c. 0.5 Ω
13. Four in parallel
15. The three 6.0 Ω resistors in parallel with each other, and then that combination in series with the 3.0 Ω resistor.
17. 24 Ω
19. a. 2.0 A b. 5.0 A
21. 20 Ω, 60 V
23.

R	I (A)	ΔV (V)
R_1	2.0	12
R_2	0.8	12
R_3	1.2	7.2
R_4	1.2	4.8

25. a. 12 A b. 10 Ω
27. Yes
29. $0.15
31. 3.0 μF
33. 150 μF in series
35. 9.0 μF
37. 40 kΩ
39. 7.0 ms
41. 0.72 pF
43. 330 Hz
45. 19 W
47. a. 1.0 V b. 1/3
49. 1.2 A
51. 500 Ω
53. 180 Ω
55. 7.0 Ω
57. a. 0.505 Ω b. 0.50 Ω
59. a. 10 A b. 80 W
61.

R	I (A)	ΔV (V)
24 Ω	0.25	6.0
3.0 Ω	1.0	3.0
5.0 Ω	0.75	3.8
4.0 Ω	0.56	2.3
12 Ω	0.19	2.3

63. 1.3 pF
65. a. 6.9 ms   b. 3.5 ms
67. 18 μF
69. 0.69 ms
71. 6100
73. a. 4.2×10^{-11} C   b. 2.7×10^{8} sodium ions
75. C
77. B
79. A
81. B

Chapter 24

Answers to odd-numbered multiple-choice questions

31. a. C   b. B
33. C
35. D
37. C

Answers to odd-numbered problems

1. $I_{earthsurface} = 2.5$ A; $I_{refrigerator} = 250$ A;
   $I_{laboratory} = 5000$ A $- 50,000$ A; $I_{magnet} = 500,000$ A
3. a. 19.9 A   b. 1.59×10^{-3} m
5. Point 1: $(6.67 \times 10^{-5}$ T, out of page), point 2: $(2.0 \times 10^{-4}$ T, into page), point 3: $(6.67 \times 10^{-5}$ T, out of page)
7. a. 2.0 μT   b. 4.0%   c. 2.0 μT
   d. The two magnetic fields produced will be in opposite diresctions, so they will cancel each other out.
9. 0.50 cm   b. 8.0 cm
11. 1.6 kA
13. 3.8×10^{-7} A
15. 4.1×10^{-4} T
17. a. 2.7×10^{-5} T   b. Smaller
19. 3180
21. 12.5 T
23. a. $(8.0 \times 10^{-13}$ N, $+z$-direction)
    b. $(8.0 \times 10^{-13}$ N, $45°$ cw from $+z$-axis in yz plane)
25. a. 11.4 cm   b. 10.4 m
27. 0.086 T
29. He^{+}
31. 2.75 m
33. Wire 1: $(2.5 \times 10^{-4}$ N, up), wire 2: 0 N, wire 3: $(5.2 \times 10^{-4}$ N, down)
35. $62.7°$
37. 0.216 N
39. 7.5×10^{-4} N $\cdot$ m
41. 1.13×10^{10} A
43. 0.036 mm
45. a. 10×10^{-4} T
49. 3.0 Ω
51. (2.0 mT, into page)
53. a. $(2.4 \times 10^{10}$ m/s^2, down)   b. $(2.2 \times 10^{11}$ m/s^2, up)
55. $F_{wires} < F_{earth}$
57. 1.01 T
59. 0.00864 T
61. b. Attractive   c. To the left, toward the solenoid
    d. It will grow stronger
63. B
65. C

Chapter 25

Answers to odd-numbered multiple-choice questions

27. A
29. B
31. C
33. E
35. A
37. D

Answers to odd-numbered problems

1. 0.10 T out of the page
3. a. 4.0 m/s,   b. 2.2 T
5. 9500 m/s
7. 0.040 Wb
9. 0.041 Wb
11. a. 8.7×10^{-4} Wb   b. Clockwise
13. $|\mathcal{E}_E| > |\mathcal{E}_B| > |\mathcal{E}_D| > |\mathcal{E}_A| > |\mathcal{E}_C|$
15. 3.1 V
17. a. 1.0×10^{-3} Wb   b. 1.3 μA   c. 1.3 μA
19. 3.3×10^{-8} T
21. a. 630 nm   b. 4.8×10^{14} Hz   c. 900 V/m
23. a. 3.3×10^{-7} T   b. 13 W/m^2
25. a. 2.2×10^{11} V/M   b. Almost half
27. a. 0.25 m   b. 2500 m
29. $60°$
31. 130 W/m^2
33. 1.3 keV
35. a. 250 keV   b. 1 000000 bonds
37. 4.8×10^{14} Hz, 620 nm, visible
39. 4.68×10^{21} photons
41. $2100°$C
43. 25 keV photons: 50 pm; 62.5 keV photons: 20 pm
45. *Bab*
47. 1.4 Wb
49. 0.13 V
51. 42 mV
53. 5.0 A, clockwise
55. a. 6.4 V   b. 1.4 A
57. a. 0.20 A   b. 4.0×10^{-3} N   c. 11 K
59. 2.5×10^{-4} V
61. a. 250 kW   b. 1.8×10^{7} V/m
63. a. 8.3×10^{-26} W/m^2 b. 7.9×10^{-12} V/m
65. $\frac{1}{8}I_0$
67. 6.2×10^{-20} m, 4.8×10^{27} Hz
69. a. 3.4×10^{-19} J   b. 1.8×10^{13} photons/flash
71. C
73. C

Chapter 26

Answers to odd-numbered multiple-choice questions

21. E
23. D
25. A
27. D

Answers to odd-numbered problems

1. 1.1 A
3. a. 50 mA   b. 50 mA
5. 22.4 V
7. a. 9 Ω b. 19 A   c. 3200 W
9. 530 A

11. 720 turns
13. 0.035 A
15. 9 μA, no
17. a. 1.9 mA   b. 1.9 A
19. a. 20.0 mA   b. 20.0 mA   c. 10.0 mA
21. 81 nF
23. -100 V
25. a. 5.0 mA   b. 20.0 mA   c. 40.0 mA
27. 1.4 mH
29. 7.8 min
31. 2.5×10^5 Hz to 3.6×10^5 Hz
33. 0.13 pF
35. 1/4
37. 1.0×10^{-3} H
39. a. 200 kHz   b. 200 kHz
41. $P_{40} = 14$ W, $P_{60} = 9.6$ W, $P_{100} = 100$ W
43. 4.89 V
45. Zero
47. a. 25 mA   b. 6.7 V
49. a. -7.5 V   b. 83 Ω
51. 0.63 ms
53. 38 pF to 0.41 nF
55. a. 500 Hz   b. 0.71 A
57. a. 0.035 H   b. 4.6 A
59. A
61. B

Chapter 27

Answers to odd-numbered multiple-choice questions
21. B
23. E
25. C

Answers to odd-numbered problems
1. $v_{sound} = 345$ m/s, $v_{sprinter} = 15$ m/s
3. a. 15 m/s   b. -5 m/s
5. 30 s
7. 3.00×10^8 m/s (or c)
9. 3.00×10^8 m/s (or c)
11. $t_1 = 1\ \mu$s, $t_2 = 2\ \mu$s
13. She sees flash 2 40 μs after flash 1
15. Barn 1 μs before tree
17. 350 μs
19. a. 5.0 yr   b. 2.2 yr   c. 9.5 yr
21. $0.80c$
23. 1.5 km
25. a. $0.99999999995c$ b. 100 000 yr
27. $0.360c$
29. $0.36c$
31. $0.80c$
33. $0.71c$
35. a. 1.8×10^{16} J   b. 9.0×10^9
37. $0.94c$
39. $0.60c = 1.8 \times 10^8$ m/s
41. 11 h
43. a. 13 μs   b. 1.9×10^8 m/s
45. a. Less   b. 14 ns
47. In the muon's frame the atmosphere is only 1.5 km thick
49. a. $0.99973c$ b. 3.3×10^{24} J   c. 33 000 times more
51. $0.71c$
53. $(1800$ m/s$)(1 - 5.0 \times 10^{-12})$; no
55. a. Yes   b. No

57. $0.31c$
59. 6.3×10^{-19} kg $\cdot$ m/s
61. a. 6.7 GeV   b. 3.6 MeV
63. $0.79c$
65. a. 0.0051 MeV   b. 9.4 MeV
67. a. 1.3×10^{17} kg   b. 6.7×10^{-12}%   c. 1.5×10^{13} yr
69. a. 3.6×10^{-12} J   b. 0.60%
71. D
73. D

Chapter 28

Answers to odd-numbered multiple-choice questions
27. B
29. A
31. B
33. A
35. A

Answers to odd-numbered problems
1. $41°$
3. $69°$
5. Four
7. 3.20 eV
9. 1.8 eV
11. a. 290 nm   b. 3.7×10^5 m/s
13. a. 6.63×10^5 m/s   b. 2.30 eV, potassium
15. a. 2.3 eV   b. 8
17. 0.25 nm
19. $\frac{4}{7}$
21. a. 4.1×10^{-7} eV   b. 2.5 eV   c. 12 eV
23. 500 nm
25. a. 1.1×10^{-34} m   b. 1.7×10^{-23} m/s
27. 1.5 eV
29. 6.0×10^{-6} V
31. 3.3×10^{-21} m, 2.2×10^{-17}
33. 0.40 nm
35. 1.0×10^{-20} J
37. 210 nm, 250 nm, 410 nm, 620 nm, 1200 nm
39. 4.5 eV or 7.0 eV
41. Four
43. The uncertainty in speed is 1.6×10^{-11} m/s
45. $l = 6.0\ \mu$m
47. 0 to 8.0×10^6 m/s
49. 0.150 nm
51. a. 2.3 eV   b. 240 nm
53. 4.1 eV
55. a. 250 eV   b. 0.060 eV   c. Less energy
57. 1.8×10^7 s
59. 7×10^4
61. 1.4
63. 3.5×10^{28}
65. 0.97 mm
67. 1.1×10^{-8} m
69. a. 3.88×10^{-12} m   b. 3.71×10^{-12} m
71. a. 5.5×10^{-64} J   b. 3.3×10^{-31} m/s
73. 0.87 nm
75. 9.0×10^{-8} eV
77. 0.43 nm
79. $m = 3$ and $n = 1$
81. C
83. B

Chapter 29

Answers to odd-numbered multiple-choice questions

23. A
25. C
27. C
29. E

Answers to odd-numbered problems

1. a.

Wavelength (nm)	m	n
656.6	2	3
486.3	2	4
434.2	2	5
410.3	2	6

   b. 397.1 nm

3. Infrared

5.

	Electrons	Protons	Neutrons
a. $^9\text{Be}^+$	3	4	5
b. ^{12}C	6	6	6
c. $^{15}\text{N}^{+++}$	4	7	8

7. a. ^2H b. $^{14}\text{N}^{++}$

9. a. Yes b. 0.50 eV

11. Yes, no

13. 3

15. a. $a_B = \dfrac{4\pi\epsilon_0 \hbar^2}{me^2}$ b. $E_1 = \dfrac{-e^2}{4\pi\epsilon_0(2a_B)}$

17. a. $\lambda_1 = 0.332$ nm, $\lambda_2 = 0.665$ nm, $\lambda_3 = 0.997$ nm
    b. 0.0529 nm $\times 2\pi = 0.332$ nm $= \lambda_1$; similarly,
    0.212 nm $\times 2\pi = 1.332$ nm $= 2\lambda_2 = 2(0.665$ nm),
    and $3\lambda_3 = 2\pi r_3$

    c.

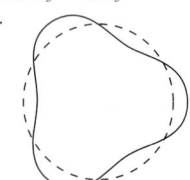

19.

Transition	Wavelength (nm)
$4 \rightarrow 1$	97.3
$4 \rightarrow 2$	486
$4 \rightarrow 3$	1876

21. $n = 1$: $(1, 0, 0, \pm\frac{1}{2})$ $n = 2$: $(2, 0, 0, \pm\frac{1}{2})$, $(2, 1, -1, \pm\frac{1}{2})$,
    $(2, 1, 0, \pm\frac{1}{2})$, $(2, 1, 1, \pm\frac{1}{2})$ $n = 3$: $(3, 0, 0, \pm\frac{1}{2})$, $(3, 1, -1, \pm\frac{1}{2})$,
    $(3, 1, 0, \pm\frac{1}{2})$, $(3, 1, 1, \pm\frac{1}{2})$, $(3, 2, -2, \pm\frac{1}{2})$, $(3, 2, -1, \pm\frac{1}{2})$,
    $(3, 2, 0, \pm\frac{1}{2})$, $(3, 2, 1, \pm\frac{1}{2})$, $(3, 2, 2, \pm\frac{1}{2})$

23. a. f b. -0.850 eV

25. $E_6 = -0.378$ eV, $L = \sqrt{12}\hbar$

27. Si: $1s^2 2s^2 2p^6 3s^2 3p^2$, Ge: $1s^2 2s^2 2p^6 3s^2 3p^6 4s^2 3d^{10} 4p^2$,
    Pb: $1s^2 2s^2 2p^6 3s^2 3p^6 4s^2 3d^{10} 4p^6 5s^2 5p^6 6s^2 4f^{14} 5d^{10} 6p^2$

29. a. F, excited state b. Ni, ground state

31. 4.86×10^6 m/s

33. $1s^2 3s^1$

35. a. Yes, 2.2 μm b. No, violates $\Delta l = 1$

37. 3.2×10^{15}

39. a. 1.7×10^{18} b. 1.7×10^{26} photons/s

41. a. 102.6 nm: $m = 1, n = 3$, 1876 nm: $m = 3, n = 4$
    b. Ultraviolet, infrared

43. a. 79 electrons, 79 protons, 118 neutrons b. 2.29×10^{17} kg/m^3
    c. 2.0×10^{13} times

45. 5.8×10^{-11}% occupied; 99.999999999942% empty

47. a. 58 N b. 4.7×10^{-35} N

49. 1.8×10^7 V

51. b. $\lambda_{21} = 500$ nm, $\lambda_{31} = 300$ nm, $\lambda_{32} = 753$ nm, $\lambda_{41} = 200$ nm,
    $\lambda_{42} = 334$ nm, $\lambda_{43} = 601$ nm

55. 91.2 nm

57. 1876 nm

59. 44 260 m/s

61. $\sqrt{6}\hbar$

63.

Ca

Sc

Co

Zn

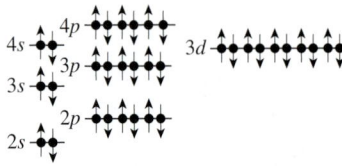

Kr

65. a. $6s \to 5p, 6s \to 4p, 6s \to 3p$ b. 7290 nm, 1630 nm, 515 nm
67. 1.13×10^6 m/s

69.
Transition	a. Wavelength	b. Type	c. Absorption
$2p \to 2s$	670 nm	VIS	Yes
$3s \to 2p$	816 nm	IR	No
$3p \to 2s$	324 nm	UV	Yes
$3p \to 3s$	2696 nm	IR	No
$3d \to 2p$	611 nm	VIS	No
$3d \to 3p$	25 μm	IR	No
$4s \to 2p$	498 nm	VIS	No
$4s \to 3p$	2430 nm	IR	No

71. 2.0×10^{23} W/m^2
73. A
75. C

Chapter 30

Answers to odd-numbered multiple-choice questions

29. C
31. C
33. B
35. D

Answers to odd-numbered problems

1. a. 1 proton, 2 neutrons   b. 18 protons, 22 neutrons
   c. 20 protons, 20 neutrons   d. 94 protons, 145 neutrons
3. 3.6×10^{26}, 3.6×10^{26}
5. 14 u
7. a. ^{3}H: total is 8.48 MeV, per nucleon is 2.83 MeV
   b. ^{3}He: total is 7.72 MeV, per nucleon is 2.57 MeV

9. ^{4}He
11. a. 7.5 MeV   b. 8.8 MeV   c. 7.9 MeV
13. ^{14}N is stable
15. a. 9.3×10^{11} b. 4.7×10^{11} c. 5.5×10^8
17. 17 100 yr
19. 6.3 mCi
21. ^{99}Ru
23. ^{234}U
25. a. ^{228}Th   b. ^{207}Tl   c. ^{7}Li   d. ^{60}Ni
27. 5.52 MeV
29. 1.5×10^{-4} u
31. 0.39 MeV
33. 0.60 mSv
35. 0.23 J
37. 2.0 Gy
39. a. 1.46×10^{-8} u, 1.45×10^{-6}%   b. 0.0304 u, 0.76%
41. 6.0 MeV
43. a. ^{17}N, ^{17}O, ^{17}F   b. ^{17}O
    c. ^{17}N decays by beta-minus decay to ^{17}O; ^{17}F decays by electron capture to ^{17}O
45. 7.2×10^{11} Bq = 19 Ci
47. 2.7×10^{17}
49. No
51. 21 000 yr
53. 4.6×10^9
55. 0.37
57. 86
59. 70 mrem
61. 3.2×10^{-4} Gy
63. 3.3×10^{12}
65. a. 2.6×10^{10} b. 0.024 Bq   c. 1.4×10^8 mSv
67. A
69. D
71. C

Credits

Index

Astronomical Data

Planetary body	Mean distance from sun (m)	Period (years)	Mass (kg)	Mean radius (m)
Sun	—	—	1.99×10^{30}	6.96×10^{8}
Moon	3.84×10^{8}*	27.3 days	7.36×10^{22}	1.74×10^{6}
Mercury	5.79×10^{10}	0.241	3.18×10^{23}	2.43×10^{6}
Venus	1.08×10^{11}	0.615	4.88×10^{24}	6.06×10^{6}
Earth	1.50×10^{11}	1.00	5.98×10^{24}	6.37×10^{6}
Mars	2.28×10^{11}	1.88	6.42×10^{23}	3.37×10^{6}
Jupiter	7.78×10^{11}	11.9	1.90×10^{27}	6.99×10^{7}
Saturn	1.43×10^{12}	29.5	5.68×10^{26}	5.85×10^{7}
Uranus	2.87×10^{12}	84.0	8.68×10^{25}	2.33×10^{7}
Neptune	4.50×10^{12}	165	1.03×10^{26}	2.21×10^{7}

*Distance from earth

Typical Coefficients of Friction

Material	Static μ_s	Kinetic μ_k	Rolling μ_r
Rubber on concrete	1.00	0.80	0.02
Steel on steel (dry)	0.80	0.60	0.002
Steel on steel (lubricated)	0.10	0.05	
Wood on wood	0.50	0.20	
Wood on snow	0.12	0.06	
Ice on ice	0.10	0.03	
Synovial joints in the body	0.01	0.003	

Melting/Boiling Temperatures, Heats of Transformation

Substance	T_m (°C)	L_f (J/kg)	T_b (°C)	L_v (J/kg)
Water	0	3.33×10^{5}	100	22.6×10^{5}
Nitrogen (N_2)	−210	0.26×10^{5}	−196	1.99×10^{5}
Ethyl alcohol	−114	1.09×10^{5}	78	8.79×10^{5}
Mercury	−39	0.11×10^{5}	357	2.96×10^{5}
Lead	328	0.25×10^{5}	1750	8.58×10^{5}

Properties of Materials

Substance	ρ (kg/m³)	c (J/kg·K)	v_{sound} (m/s)
Helium gas (1 atm, 0°C)	0.179		970
Air (1 atm, 0°C)	1.28		331
Air (1 atm, 20°C)	1.20		343
Ethyl alcohol	790	2400	1170
Gasoline	680		
Glycerin	1260		
Mercury	13,600	140	1450
Oil (typical)	900		
Water ice	920	2090	3500
Liquid water	1000	4190	1480
Seawater	1030		1500
Blood	1060		
Muscle	1040	3600	
Fat	920	3000	
Mammalian body	1005	3400	1540
Granite	2750	790	6000
Aluminum	2700	900	5100
Copper	8920	385	
Gold	19,300	129	
Iron	7870	449	
Lead	11,300	128	1200
Diamond	3520	510	12,000